List of Elements with Their Symbols and Ato...

Element	Symbol	Atomic Number	Atomic Mass	Element	Symbol	Atomic Number	Atomic Mass
Actinium	Ac	89	227.03[a]	Molybdenum	Mo	42	95.94
Aluminum	Al	13	26.98	Neodymium	Nd	60	144.24
Americium	Am	95	243.06[a]	Neon	Ne	10	20.18
Antimony	Sb	51	121.76	Neptunium	Np	93	237.05[a]
Argon	Ar	18	39.95	Nickel	Ni	28	58.69
Arsenic	As	33	74.92	Niobium	Nb	41	92.91
Astatine	At	85	209.99[a]	Nitrogen	N	7	14.01
Barium	Ba	56	137.33	Nobelium	No	102	259.10[a]
Berkelium	Bk	97	247.07[a]	Osmium	Os	76	190.23
Beryllium	Be	4	9.012	Oxygen	O	8	16.00
Bismuth	Bi	83	208.98	Palladium	Pd	46	106.42
Bohrium	Bh	107	264.12[a]	Phosphorus	P	15	30.97
Boron	B	5	10.81	Platinum	Pt	78	195.08
Bromine	Br	35	79.90	Plutonium	Pu	94	244.06[a]
Cadmium	Cd	48	112.41	Polonium	Po	84	208.98[a]
Calcium	Ca	20	40.08	Potassium	K	19	39.10
Californium	Cf	98	251.08[a]	Praseodymium	Pr	59	140.91
Carbon	C	6	12.01	Promethium	Pm	61	145[a]
Cerium	Ce	58	140.12	Protactinium	Pa	91	231.04
Cesium	Cs	55	132.91	Radium	Ra	88	226.03[a]
Chlorine	Cl	17	35.45	Radon	Rn	86	222.02[a]
Chromium	Cr	24	52.00	Rhenium	Re	75	186.21
Cobalt	Co	27	58.93	Rhodium	Rh	45	102.91
Copper	Cu	29	63.55	Roentgenium	Rg	111	272[a]
Curium	Cm	96	247.07[a]	Rubidium	Rb	37	85.47
Darmstadtium	Ds	110	271[a]	Ruthenium	Ru	44	101.07
Dubnium	Db	105	262.11[a]	Rutherfordium	Rf	104	261.11[a]
Dysprosium	Dy	66	162.50	Samarium	Sm	62	150.36
Einsteinium	Es	99	252.08[a]	Scandium	Sc	21	44.96
Erbium	Er	68	167.26	Seaborgium	Sg	106	266.12[a]
Europium	Eu	63	151.96	Selenium	Se	34	78.96
Fermium	Fm	100	257.10[a]	Silicon	Si	14	28.09
Fluorine	F	9	19.00	Silver	Ag	47	107.87
Francium	Fr	87	223.02[a]	Sodium	Na	11	22.99
Gadolinium	Gd	64	157.25	Strontium	Sr	38	87.62
Gallium	Ga	31	69.72	Sulfur	S	16	32.07
Germanium	Ge	32	72.64	Tantalum	Ta	73	180.95
Gold	Au	79	196.97	Technetium	Tc	43	98[a]
Hafnium	Hf	72	178.49	Tellurium	Te	52	127.60
Hassium	Hs	108	269.13[a]	Terbium	Tb	65	158.93
Helium	He	2	4.003	Thallium	Tl	81	204.38
Holmium	Ho	67	164.93	Thorium	Th	90	232.04
Hydrogen	H	1	1.008	Thulium	Tm	69	168.93
Indium	In	49	114.82	Tin	Sn	50	118.71
Iodine	I	53	126.90	Titanium	Ti	22	47.87
Iridium	Ir	77	192.22	Tungsten	W	74	183.84
Iron	Fe	26	55.85	Uranium	U	92	238.03
Krypton	Kr	36	83.80	Vanadium	V	23	50.94
Lanthanum	La	57	138.91	Xenon	Xe	54	131.293
Lawrencium	Lr	103	262.11[a]	Ytterbium	Yb	70	173.04
Lead	Pb	82	207.2	Yttrium	Y	39	88.91
Lithium	Li	3	6.941	Zinc	Zn	30	65.41
Lutetium	Lu	71	174.97	Zirconium	Zr	40	91.22
Magnesium	Mg	12	24.31	*[b]		112	277[a]
Manganese	Mn	25	54.94	*[b]		113	284[a]
Meitnerium	Mt	109	268.14[a]	*[b]		114	289[a]
Mendelevium	Md	101	258.10[a]	*[b]		115	288[a]
Mercury	Hg	80	200.59	*[b]		116	292[a]

[a]Mass of longest-lived or most important isotope.
[b]The names of these elements have not yet been decided.

CHEMISTRY

CHEMISTRY
A Molecular Approach

NIVALDO J. TRO

Westmont College

With special contributions by

Robert S. Boikess
Rutgers University

Joseph H. Bularzik
Purdue University, Calumet

William M. Cleaver
The University of Vermont

PEARSON

Prentice Hall

Upper Saddle River, NJ 07458

Library of Congress Cataloging-in-Publication Data
Tro, Nivaldo J.
      Chemistry : a molecular approach / Nivaldo J. Tro.
          p.   cm.
      Includes index.
      ISBN 0-13-100065-9
      1. Chemistry, Physical and theoretical.   I. Title.

QD453.2.T744 2008
540—dc22   2006039675

Editor in Chief: Nicole Folchetti
Senior Editor: Kent Porter Hamann
Development Editor: Dan Schiller
Editor in Chief, Development: Ray Mullaney
Executive Managing Editor: Kathleen Schiaparelli
Marketing Manager: Elizabeth Averbeck
Assistant Managing Editor, Science Supplements: Karen Bosch
Assistant Managing Editor, Media: Rich Barnes
Project Manager: Carol Dupont
Director of Creative Services: Paul Belfanti
Art Director and Cover Designer: Kenny Beck
Creative Director: Juan López
Art Production Editor: Connie Long
Editorial Assistant: James McBride
Interior Designer: Suzanne Behnke
Cover and Chapter Opening Illustrations: Quade Paul
Manufacturing Manager: Alexis Heydt-Long
Manufacturing Buyer: Alan Fischer
Director, Image Resource Center: Melinda Reo
Manager, Rights and Permissions: Zina Arabia
Manager, Visual Research: Beth Boyd-Brenzel
Manager, Cover Visual Research & Permissions: Karen Sanatar
Image Permission Coordinator: Nancy Seise
Art Studios: Artworks and Precision Graphics
   Production Manager: Ronda Whitson
   Manager, Production Technologies: Matthew Haas
   Art Development Editor: Jay McElroy
   Illustrators: Royce Copenheaver, Daniel Knopsnyder, Mark Landis
   Illustrator (EPMs): Professor Richard Johnson, University of New Hampshire
Senior Managing Editor, Art Production and Management: Patricia Burns
Managing Editor, Art Management: Abigail Bass
Production Supervision/Composition: Preparé, Inc.
   Production Editor: Rosaria Cassinese

© 2008 Pearson Education, Inc.
Pearson Prentice Hall
Pearson Education, Inc.
Upper Saddle River, New Jersey 07458

Printed in the United States of America

10 9 8 7

ISBN 0-13-100065-9

Pearson Education LTD., *London*
Pearson Education Australia PTY, Limited, *Sydney*
Pearson Education Singapore, Pte. Ltd
Pearson Education North Asia Ltd, *Hong Kong*
Pearson Education Canada, Ltd., *Toronto*
Pearson Educación de Mexico, S.A. de C.V.
Pearson Education—Japan, *Tokyo*
Pearson Education Malaysia, Pte. Ltd

To Michael, Ali, Kyle, and Kaden

About the Author

Nivaldo Tro is a Professor of Chemistry at Westmont College in Santa Barbara, California, where he has been a faculty member since 1990. He received his Ph.D. in chemistry from Stanford University for work on developing and using optical techniques to study the adsorption and desorption of molecules to and from surfaces in ultrahigh vacuum. He then went on to the University of California at Berkeley, where he did postdoctoral research on ultrafast reaction dynamics in solution. Since coming to Westmont, Professor Tro has been awarded grants from the American Chemical Society Petroleum Research Fund, from Research Corporation, and from the National Science Foundation to study the dynamics of various processes occurring in thin adlayer films adsorbed on dielectric surfaces. He has twice been honored as Westmont's outstanding teacher of the year and has also received the college's outstanding researcher of the year award. Professor Tro lives in Santa Barbara with his wife, Ann, and their four children, Michael, Ali, Kyle, and Kaden. In his leisure time, Professor Tro enjoys reading good literature to his children and being outdoors with his family.

Brief Contents

1 Matter, Measurement, and Problem Solving   2

2 Atoms and Elements   44

3 Molecules, Compounds, and Chemical Equations   82

4 Chemical Quantities and Aqueous Reactions   130

5 Gases   184

6 Thermochemistry   236

7 The Quantum-Mechanical Model of the Atom   280

8 Periodic Properties of the Elements   318

9 Chemical Bonding I: Lewis Theory   362

10 Chemical Bonding II: Molecular Shapes, Valence
   Bond Theory, and Molecular Orbital Theory   404

11 Liquids, Solids, and Intermolecular Forces   460

12 Solutions   518

13 Chemical Kinetics   568

14 Chemical Equilibrium   616

15 Acids and Bases   662

16 Aqueous Ionic Equilibrium   716

17 Free Energy and Thermodynamics   774

18 Electrochemistry   818

19 Radioactivity and Nuclear Chemistry   864

20 Organic Chemistry   902

21 Biochemistry   954

22 Chemistry of the Nonmetals   988

23 Metals and Metallurgy   1026

24 Transition Metals and Coordination Compounds   1050

   Appendix I: Common Mathematical Operations
   in Chemistry   A-1

   Appendix II: Useful Data   A-7

   Appendix III: Answers to Selected Exercises   A-16

   Appendix IV: Answers to In-Chapter Practice Problems   A-47

   Glossary   G-1

   Credits   C-1

   Index   I-1

Contents

Preface   xix

1

Matter, Measurement, and Problem Solving   2

| 1.1 | Atoms and Molecules | 3 |
| 1.2 | The Scientific Approach to Knowledge | 5 |

The Nature of Science: Thomas S. Kuhn and
Scientific Revolutions   7

| 1.3 | The Classification of Matter | 7 |

The States of Matter: Solid, Liquid, and Gas 7
Classifying Matter According to Its Composition:
Elements, Compounds, and Mixtures 9
Separating Mixtures 10

1.4	Physical and Chemical Changes and Physical and Chemical Properties	11
1.5	Energy: A Fundamental Part of Physical and Chemical Change	14
1.6	The Units of Measurement	15

The Standard Units 16   The Meter: A Measure
of Length 16   The Kilogram: A Measure of
Mass 16   The Second: A Measure of Time 16
The Kelvin: A Measure of Temperature 17
Prefix Multipliers 19   Derived Units: Volume
and Density 19   Calculating Density 20

Chemistry and Medicine: Bone Density   21

| 1.7 | The Reliability of a Measurement | 22 |

Counting Significant Figures 23   Exact
Numbers 24   Significant Figures in
Calculations 25   Precision and Accuracy 27

Chemistry in Your Day: Integrity in Data
Gathering   28

| 1.8 | Solving Chemical Problems | 28 |

Converting from One Unit to Another 28
General Problem-Solving Strategy 30   Units
Raised to a Power 31   Order of Magnitude
Estimations 33   Problems Involving an
Equation 34

Chapter in Review   35

Key Terms 35   Key Concepts 36   Key Equations
and Relationships 36   Key Skills 37

Exercises   37

Review Questions 37   Problems by Topic 38
Cumulative Problems 41   Challenge Problems 42
Conceptual Problems 43

2

Atoms and Elements   44

2.1	Imaging and Moving Individual Atoms	45
2.2	Early Ideas about the Building Blocks of Matter	47
2.3	Modern Atomic Theory and the Laws That Led to It	47

The Law of Conservation of Mass 47   The Law
of Definite Proportions 48   The Law of Multiple
Proportions 49   John Dalton and the Atomic
Theory 51

Chemistry in Your Day: Atoms and Humans   51

| 2.4 | The Discovery of the Electron | 51 |

Cathode Rays 51   Millikan's Oil Drop
Experiment: The Charge of the Electron 53

| 2.5 | The Structure of the Atom | 54 |
| 2.6 | Subatomic Particles: Protons, Neutrons, and Electrons in Atoms | 56 |

Elements: Defined by Their Numbers of Protons
57   Isotopes: When the Number of Neutrons
Varies 58   Ions: Losing and Gaining
Electrons 60

| 2.7 | Finding Patterns: The Periodic Law and the Periodic Table | 60 |

Ions and the Periodic Table 65

Chemistry and Medicine: The Elements of Life 67

| 2.8 | Atomic Mass: The Average Mass of an Element's Atoms | 67 |

Mass Spectrometry: Measuring the Mass of
Atoms and Molecules 68

2.9   Molar Mass: Counting Atoms by Weighing Them   69

The Mole: A Chemist's "Dozen" 70   Converting between Number of Moles and Number of Atoms 70   Converting between Mass and Amount (Number of Moles) 71

Chapter in Review   75

Key Terms 75   Key Concepts 75   Key Equations and Relationships 76   Key Skills 76

Exercises   77

Review Questions 77   Problems by Topic 77   Cumulative Problems 80   Challenge Problems 81   Conceptual Problems 81

3

Molecules, Compounds, and Chemical Equations   82

3.1   Hydrogen, Oxygen, and Water   83
3.2   Chemical Bonds   84

Ionic Bonds 85   Covalent Bonds 85

3.3   Representing Compounds: Chemical Formulas and Molecular Models   86

Types of Chemical Formulas 86   Molecular Models 87

3.4   An Atomic-Level View of Elements and Compounds   89
3.5   Ionic Compounds: Formulas and Names   92

Writing Formulas for Ionic Compounds 92   Naming Ionic Compounds 93   Naming Binary Ionic Compounds Containing a Metal That Forms Only One Type of Cation 94   Naming Binary Ionic Compounds Containing a Metal That Forms More than One Kind of Cation 94   Naming Ionic Compounds Containing Polyatomic Ions 95   Hydrated Ionic Compounds 96

3.6   Molecular Compounds: Formulas and Names   96

Naming Molecular Compounds 97   Naming Acids 98   Naming Binary Acids 99   Naming Oxyacids 99

Chemistry in the Environment: Acid Rain   100

3.7   Formula Mass and the Mole Concept for Compounds   101

Molar Mass of a Compound 101   Using Molar Mass to Count Molecules by Weighing 102

3.8   Composition of Compounds   103

Mass Percent Composition as a Conversion Factor 105   Conversion Factors from Chemical Formulas 106

Chemistry and Medicine: Methylmercury in Fish   107

3.9   Determining a Chemical Formula from Experimental Data   109

Calculating Molecular Formulas for Compounds 110   Combustion Analysis 112

3.10   Writing and Balancing Chemical Equations   114

How to Write Balanced Chemical Equations 115

3.11   Organic Compounds   117

Hydrocarbons 118   Functionalized Hydrocarbons 119

Chapter in Review   121

Key Terms 121 Key Concepts 121   Key Equations and Relationships 122   Key Skills 123

Exercises   124

Review Questions 124   Problems by Topic 124   Cumulative Problems 128   Challenge Problems 129   Conceptual Problems 129

4

Chemical Quantities and Aqueous Reactions   130

4.1   Global Warming and the Combustion of Fossil Fuels   131
4.2   Reaction Stoichiometry: How Much Carbon Dioxide?   133

Making Pizza: The Relationship among Ingredients 133   Making Molecules: Mole-to-Mole Conversions 133   Making Molecules: Mass-to-Mass Conversions 134

4.3   Limiting Reactant, Theoretical Yield, and Percent Yield   137

Limiting Reactant, Theoretical Yield, and Percent Yield from Initial Reactant Masses 139

Chemistry in the Environment: MTBE in Gasoline   143

4.4   Solution Concentration and Solution Stoichiometry   144

Solution Concentration 144   Using Molarity in Calculations 145   Solution Stoichiometry 149

4.5   Types of Aqueous Solutions and Solubility   150

Electrolyte and Nonelectrolyte Solutions 151   The Solubility of Ionic Compounds 154

4.6   Precipitation Reactions   155

4.7   Representing Aqueous Reactions: Molecular,
        Ionic, and Complete Ionic Equations   159

4.8   Acid–Base and Gas-Evolution Reactions   160
        Acid–Base Reactions 160   Gas-Evolution
        Reactions 165

4.9   Oxidation–Reduction Reactions   167
        Oxidation States 169   Identifying Redox
        Reactions 171

      Chemistry in Your Day: Bleached Blonde   173
        Combustion Reactions 173

      Chapter in Review   174
        Key Terms 174   Key Concepts 175   Key Equations
        and Relationships 175   Key Skills 176

      Exercises   176
        Review Questions 176 Problems by Topic 177
        Cumulative Problems 180   Challenge Problems 182
        Conceptual Problems 182

5

Gases   184

5.1   Water from Wells: Atmospheric Pressure at Work 185

5.2   Pressure: The Result of Molecular Collisions   187
        Pressure Units 188   The Manometer: A Way to
        Measure Pressure in the Laboratory 190

      Chemistry and Medicine: Blood Pressure   191

5.3   The Simple Gas Laws: Boyle's Law, Charles's Law,
        and Avogadro's Law   191
        Boyle's Law: Volume and Pressure 192

      Chemistry in Your Day: Extra-long Snorkels   194
        Charles's Law: Volume and Temperature 195
        Avogadro's Law: Volume and Amount (in Moles)
        197

5.4   The Ideal Gas Law   198

5.5   Applications of the Ideal Gas Law: Molar Volume,
        Density, and Molar Mass of a Gas   201
        Molar Volume at Standard Temperature and
        Pressure 201   Density of a Gas 201   Molar
        Mass of a Gas 203

5.6   Mixtures of Gases and Partial Pressures   204
        Deep-Sea Diving and Partial Pressure 206
        Collecting Gases over Water 209

5.7   Gases in Chemical Reactions: Stoichiometry
        Revisited   211
        Molar Volume and Stoichiometry 212

5.8   Kinetic Molecular Theory: A Model for Gases   214
        Kinetic Molecular Theory and the Ideal Gas Law
        215   Temperature and Molecular Velocities 217

5.9   Mean Free Path, Diffusion, and Effusion of Gases 220

5.10 Real Gases: The Effects of Size and Intermolecular
        Forces   221
        The Effect of the Finite Volume of Gas Particles 222
        The Effect of Intermolecular Forces 223
        Van der Waals Equation 224   Real Gases 224

5.11 Chemistry of the Atmosphere: Air Pollution and
        Ozone Depletion   225
        Air Pollution 225   Ozone Depletion 226

      Chapter in Review   228
        Key Terms 228   Key Concepts 228   Key Equations
        and Relationships 229   Key Skills 229

      Exercises   230
        Review Questions 230   Problems by Topic 230
        Cumulative Problems 233   Challenge Problems 235
        Conceptual Problems 235

6

Thermochemistry   236

6.1   Light the Furnace: The Nature of Energy and Its
        Transformations   237
        The Nature of Energy: Key Definitions 238
        Units of Energy 239

6.2   The First Law of Thermodynamics: There Is No Free
        Lunch   240

      Chemistry in Your Day: Redheffer's Perpetual
        Motion Machine   241
        Internal Energy 241

6.3   Quantifying Heat and Work   246
        Heat 246   Work: Pressure–Volume Work 249

6.4   Measuring ΔE for Chemical Reactions:
        Constant-Volume Calorimetry   251

6.5   Enthalpy: The Heat Evolved in a Chemical Reaction
        at Constant Pressure   253
        Exothermic and Endothermic Processes: A
        Molecular View 255   Stoichiometry Involving
        ΔH: Thermochemical Equations 255

6.6   Constant-Pressure Calorimetry: Measuring ΔH_{rxn} 257

6.7   Relationships Involving ΔH_{rxn} 259

6.8   Enthalpies of Reaction from Standard Heats
        of Formation   262
    Standard States and Standard Enthalpy Changes 262
    Calculating the Standard Enthalpy Change
    for a Reaction 264

6.9   Energy Use and the Environment   267
    Energy Consumption 267   Environmental
    Problems Associated with Fossil Fuel Use 268

    Chemistry in the Environment:
        Renewable Energy   270

    Chapter in Review   271
    Key Terms 271   Key Concepts 271   Key Equations
    and Relationships 272   Key Skills 273

    Exercises   273
    Review Questions 273   Problems by Topic 274
    Cumulative Problems 277   Challenge Problems 278
    Conceptual Problems 279

7

The Quantum-Mechanical Model of the Atom   280

7.1   Quantum Mechanics: A Theory That Explains
        the Behavior of the Absolutely Small   281

7.2   The Nature of Light   282
    The Wave Nature of Light 282   The
    Electromagnetic Spectrum 285

    Chemistry and Medicine: Radiation Treatment
        for Cancer   286

    Interference and Diffraction 287   The Particle
    Nature of Light 288

7.3   Atomic Spectroscopy and the Bohr Model   292

    Chemistry in Your Day: Atomic Spectroscopy,
        a Bar Code for Atoms   294

7.4   The Wave Nature of Matter: The de Broglie Wavelength,
        the Uncertainty Principle, and Indeterminacy   295
    The de Broglie Wavelength 297   The
    Uncertainty Principle 298   Indeterminacy and
    Probability Distribution Maps 299

7.5   Quantum Mechanics and the Atom   301
    Solutions to the Schrödinger Equation for the
    Hydrogen Atom 301   Atomic Spectroscopy
    Explained 303

7.6   The Shapes of Atomic Orbitals   306
    s Orbitals ($l = 0$) 307 p Orbitals ($l = 1$) 308
    d Orbitals ($l = 2$) 310 f Orbitals ($l = 3$) 310

    Chapter in Review   312
    Key Terms 312   Key Concepts 312   Key Equations
    and Relationships 313   Key Skills 313

    Exercises   314
    Review Questions 314   Problems by Topic 314
    Cumulative Problems 315   Challenge Problems 316
    Conceptual Problems 317

8

Periodic Properties of the Elements   318

8.1   Nerve Signal Transmission   319
8.2   The Development of the Periodic Table   320
8.3   Electron Configurations: How Electrons Occupy
        Orbitals   321
    Electron Spin and the Pauli Exclusion Principle
    322   Sublevel Energy Splitting in Multielectron
    Atoms 323   Electron Configurations for
    Multielectron Atoms 326

8.4   Electron Configurations, Valence Electrons, and the
        Periodic Table   328
    Orbital Blocks in the Periodic Table 330
    Writing an Electron Configuration for an
    Element from Its Position in the Periodic Table
    331   The Transition and Inner Transition
    Elements 332

8.5   The Explanatory Power of the Quantum-Mechanical
        Model   332

8.6   Periodic Trends in the Size of Atoms and Effective
        Nuclear Charge   334
    Effective Nuclear Charge 336   Atomic Radii and
    the Transition Elements 337

8.7   Ions: Electron Configurations, Magnetic Properties,
        Ionic Radii, and Ionization Energy   338
    Electron Configurations and Magnetic Properties
    of Ions 339   Ionic Radii 341   Ionization
    Energy 343   Trends in First Ionization Energy 344
    Exceptions to Trends in First Ionization Energy 346
    Trends in Second and Successive Ionization
    Energies 347

8.8   Electron Affinities and Metallic Character   348
    Electron Affinity 348   Metallic Character 349

8.9   Some Examples of Periodic Chemical Behavior:
        The Alkali Metals, the Halogens, and the
        Noble Gases   351

The Alkali Metals (Group 1A) 352
The Halogens (Group 7A) 353

Chemistry and Medicine: Potassium Iodide in
Radiation Emergencies   354

The Noble Gases (Group 8A) 355

Chapter in Review   356
Key Terms 356   Key Concepts 356   Key Equations
and Relationships 357   Key Skills 357

Exercises   358
Review Questions 358   Problems by Topic 359
Cumulative Problems 360   Challenge Problems 361
Conceptual Problems 361

9

Chemical Bonding I: Lewis Theory 362

9.1   Bonding Models and AIDS Drugs   363
9.2   Types of Chemical Bonds   364
9.3   Representing Valance Electrons with Dots   366
9.4   Ionic Bonding: Lewis Structures and
      Lattice Energies   367
      Ionic Bonding and Electron Transfer 367
      Lattice Energy: The Rest of the Story 368
      The Born–Haber Cycle 369   Trends in Lattice
      Energies: Ion Size 371   Trends in Lattice
      Energies: Ion Charge 371   Ionic Bonding:
      Models and Reality 372

      Chemistry and Medicine: Ionic Compounds
      in Medicine   374

9.5   Covalent Bonding: Lewis Structure   374
      Single Covalent Bonds 374   Double and Triple
      Covalent Bonds 375   Covalent Bonding: Models
      and Reality 375

9.6   Electronegativity and Bond Polarity   377
      Electronegativity 377   Bond Polarity, Dipole
      Moment, and Percent Ionic Character 379

9.7   Lewis Structures of Molecular Compounds and
      Polyatomic Ions   382
      Writing Lewis Structures for Molecular
      Compounds 382   Writing Lewis Structures for
      Polyatomic Ions 383

9.8   Resonance and Formal Charge   384
      Resonance 384   Formal Charge 386

9.9   Exceptions to the Octet Rule: Odd-Electron Species,
      Incomplete Octets, and Expanded Octets   387
      Odd-Electron Species 388

      Chemistry in the Environment: Free Radicals
      and the Atmospheric Vacuum Cleaner   388
      Incomplete Octets 389   Expanded Octets 390

9.10 Bond Energies and Bond Lengths   391
      Bond Energy 392   Using Average Bond Energies
      to Estimate Enthalpy Changes for Reactions 393
      Bond Lengths 395

      Chemistry in the Environment: The Lewis
      Structure of Ozone   396

9.11 Bonding in Metals: The Electron Sea Model   397

Chapter in Review   397
Key Terms 397   Key Concepts 397   Key Equations
and Relationships 398   Key Skills 399

Exercises   399
Review Questions 399   Problems by Topic 400
Cumulative Problems 401   Challenge Problems 403
Conceptual Problems 403

10

Chemical Bonding II: Molecular Shapes, Valence Bond Theory, and Molecular Orbital Theory 404

10.1 Artificial Sweeteners: Fooled by Molecular Shape   405
10.2 VSEPR Theory: The Five Basic Shapes   406
      Two Electron Groups: Linear Geometry 407
      Three Electron Groups: Trigonal Planar
      Geometry 407   Four Electron Groups:
      Tetrahedral Geometry 408   Five Electron
      Groups: Trigonal Bipyramidal Geometry 409
      Six Electron Groups: Octahedral Geometry 409

10.3 VSEPR Theory: The Effect of Lone Pairs   410
      Four Electron Groups with Lone Pairs 410
      Five Electron Groups with Lone Pairs 412
      Six Electron Groups with Lone Pairs 413

10.4 VSEPR Theory: Predicting Molecular Geometries 415
      Predicting the Shapes of Larger Molecules 417

10.5 Molecular Shape and Polarity   418

      Chemistry in Your Day: How Soap Works   422

10.6 Valence Bond Theory: Orbital Overlap as a
      Chemical Bond   423

10.7 Valence Bond Theory: Hybridization of Atomic
      Orbitals   425
      sp^3 Hybridization 426 sp^2 Hybridization and
      Double Bonds 428

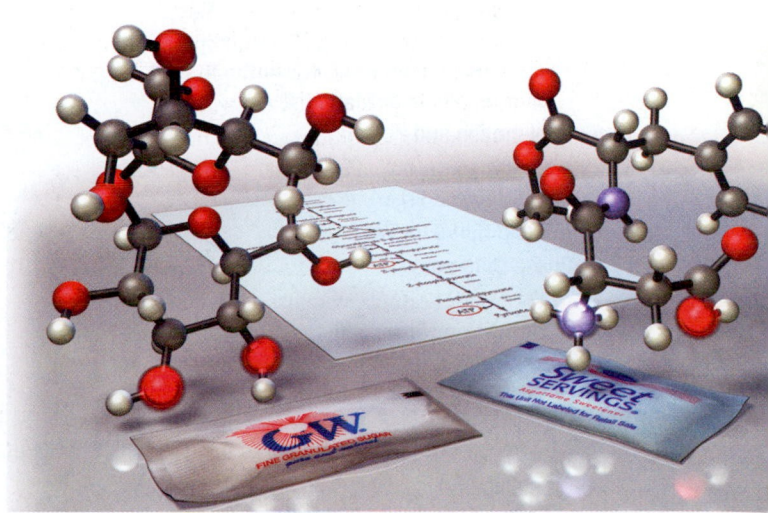

Chemistry in Your Day: The Chemistry of Vision   432

sp Hybridization and Triple Bonds 432 sp^3d and sp^3d^2 Hybridization 434   Writing Hybridization and Bonding Schemes 435

10.8   Molecular Orbital Theory: Electron Delocalization   438

Linear Combination of Atomic Orbitals (LCAO) 439   Period Two Homonuclear Diatomic Molecules 443   Second-Period Heteronuclear Diatomic Molecules 448   Polyatomic Molecules 450

Chapter in Review   451

Key Terms 451   Key Concepts 452   Key Equations and Relationships 452   Key Skills 452

Exercises   453

Review Questions 453   Problems by Topic 453   Cumulative Problems 456   Challenge Problems 458   Conceptual Problems 459

11

Liquids, Solids, and Intermolecular Forces   460

11.1   Climbing Geckos and Intermolecular Forces   461

11.2   Solids, Liquids, and Gases: A Molecular Comparison   462

Changes between Phases 464

11.3   Intermolecular Forces: The Forces That Hold Condensed Phases Together   465

Dispersion Force 466   Dipole–Dipole Force 468   Hydrogen Bonding 470   Ion–Dipole Force 472

Chemistry and Medicine: Hydrogen Bonding in DNA   473

11.4   Intermolecular Forces in Action: Surface Tension, Viscosity, and Capillary Action   474

Surface Tension 474   Viscosity 475

Chemistry in Your Day: Viscosity and Motor Oil   476

Capillary Action 476

11.5   Vaporization and Vapor Pressure   477

The Process of Vaporization 477   The Energetics of Vaporization 478   Vapor Pressure and Dynamic Equilibrium 480   The Critical Point: The Transition to an Unusual Phase of Matter 486

11.6   Sublimation and Fusion   487

Sublimation 487   Fusion 487   Energetics of Melting and Freezing 488

11.7   Heating Curve for Water   488

11.8   Phase Diagrams   491

The Major Features of a Phase Diagram 491   Navigation within a Phase Diagram 492   The Phase Diagrams of Other Substances 493

11.9   Water: An Extraordinary Substance   494

Chemistry in the Environment: Water Pollution   495

11.10   Crystalline Solids: Determining Their Structure by X-Ray Crystallography   495

11.11   Crystalline Solids: Unit Cells and Basic Structures   498

Closest-Packed Structures 502

11.12   Crystalline Solids: The Fundamental Types   504

Molecular Solids 505   Ionic Solids 505   Atomic Solids 506

11.13   Crystalline Solids: Band Theory   508

Doping: Controlling the Conductivity of Semiconductors 509

Chapter in Review   509

Key Terms 509   Key Concepts 510   Key Equations and Relationships 510   Key Skills 511

Exercises   511

Review Questions 511   Problems by Topic 512   Cumulative Problems 516   Challenge Problems 517   Conceptual Problems 517

12

Solutions   518

12.1   Thirsty Solutions: Why You Should Not Drink Seawater   519

12.2   Types of Solutions and Solubility   521

Nature's Tendency toward Mixing: Entropy 522   The Effect of Intermolecular Forces 522

12.3   Energetics of Solution Formation   526

Aqueous Solutions and Heats of Hydration 528

12.4   Solution Equilibrium and Factors Affecting Solubility   529

The Temperature Dependence of the Solubility of Solids 531   Factors Affecting the Solubility of Gases in Water 531

Chemistry in the Environment: Lake Nyos   534

12.5   Expressing Solution Concentration   535

Molarity 535   Molality 536   Parts by Mass and Parts by Volume 536

Chemistry in the Environment: The Dirty Dozen   538

Mole Fraction and Mole Percent 539

12.6 Vapor Pressure of Solutions 540

Ionic Solutes and Vapor Pressure 544   Ideal and
Nonideal Solutions 545

**12.7 Freezing Point Depression, Boiling Point Elevation,
and Osmosis** 548

Freezing Point Depression 549

Chemistry in Your Day: Antifreeze in Frogs 550

Boiling Point Elevation 551   Osmosis 552
Colligative Properties of Ionic Solutions 554
Colligative Properties and Medical Solutions 555

12.8 Colloids 557

Chapter in Review 559

Key Terms 559   Key Concepts 560   Key Equations
and Relationships 560   Key Skills 561

Exercises 561

Review Questions 561   Problems by Topic 562
Cumulative Problems 565   Challenge Problems 566
Conceptual Problems 566

13

Chemical Kinetics

568

13.1 Catching Lizards 569
13.2 Rate of a Chemical Reaction 570

Measuring Reaction Rates 574

**13.3 The Rate Law: The Effect of Concentration on
Reaction Rate** 575

Determining the Order of a Reaction 577
Reaction Order for Multiple Reactants 578

**13.4 The Integrated Rate Law: The Dependence of
Concentration on Time** 580

The Half-Life of a Reaction 584

13.5 The Effect of Temperature on Reaction Rate 587

Arrhenius Plots: Experimental Measurements of
the Frequency Factor and the Activation Energy 590
The Collision Model: A Closer Look at the Frequency
Factor 592

13.6 Reaction Mechanisms 594

Rate Laws for Elementary Steps 594
Rate-Determining Steps and Overall Reaction
Rate Laws 595   Mechanisms with a Fast Initial
Step 596

13.7 Catalysis 599

Homogeneous and Heterogeneous Catalysis 600

Chemistry and Medicine: Enzyme Catalysis
and the Role of Chymotrypsin in Digestion 604

Chapter in Review 605

Key Terms 605   Key Concepts 605   Key Equations
and Relationships 606   Key Skills 606

Exercises 606

Review Questions 606   Problems by Topic 607
Cumulative Problems 612   Challenge Problems 614
Conceptual Problems 615

14

Chemical Equilibrium

616

14.1 Fetal Hemoglobin and Equilibrium 617
14.2 The Concept of Dynamic Equilibrium 619
14.3 The Equilibrium Constant (K) 621

Expressing Equilibrium Constants for Chemical
Reactions 622   The Significance of the
Equilibrium Constant 623

Chemistry and Medicine: Life and Equilibrium 624

Relationships between the Equilibrium Constant
and the Chemical Equation 624

**14.4 Expressing the Equilibrium Constant in Terms
of Pressure** 626

Units of K 628

**14.5 Heterogeneous Equilibria: Reactions Involving
Solids and Liquids** 629

**14.6 Calculating the Equilibrium Constant from
Measured Equilibrium Concentrations** 630

**14.7 The Reaction Quotient: Predicting the Direction
of Change** 633

14.8 Finding Equilibrium Concentrations 636

Finding Equilibrium Concentrations When You
Are Given the Equilibrium Constant and All but
One Equilibrium Concentrations of the Reactants
or Products 636   Finding Equilibrium
Concentrations When You Are Given the
Equilibrium Constant and Initial Concentrations
or Pressures 637   Simplifying Approximations in
Working Equilibrium Problems 641

**14.9 Le Châtelier's Principle: How a System at
Equilibrium Responds to Disturbances** 645

The Effect of a Concentration Change on
Equilibrium 646   The Effect of a Volume (or
Pressure) Change on Equilibrium 649   The Effect of
a Temperature Change on Equilibrium 650

Chapter in Review 653

Key Terms 653   Key Concepts 653   Key Equations
and Relationships 653   Key Skills 654

Exercises   655
    Review Questions 655   Problems by Topic 655
    Cumulative Problems 659   Challenge Problems 660
    Conceptual Problems 661

15

Acids and Bases   662

15.1 Heartburn   663
15.2 The Nature of Acids and Bases   664
15.3 Definitions of Acids and Bases   666
    The Arrhenius Definition 666   The
    Brønsted–Lowry Definition 667
15.4 Acid Strength and the Acid Ionization
        Constant (K_a)   669
    Strong Acids 669   Weak Acids 670   The Acid
    Ionization Constant (K_a) 671
15.5 Autoionization of Water and pH   672
    The pH Scale: A Way to Quantify Acidity and
    Basicity 674   pOH and Other p Scales 676
    Chemistry and Medicine: Ulcers   677
15.6 Finding the [H_3O^+] and pH of Strong and Weak
        Acid Solutions   677
    Percent Ionization of a Weak Acid 683   Mixtures
    of Acids 684
15.7 Base Solutions   687
    Strong Bases 687   Weak Bases 687   Finding
    the [OH^-] and pH of Basic Solutions 688
15.8 The Acid–Base Properties of Ions and Salts   690
    Chemistry and Medicine: What's in My
        Antacid?   691
    Anions as Weak Bases 691   Cations as Weak
    Acids 695   Classifying Salt Solutions as Acidic,
    Basic, or Neutral 696
15.9 Polyprotic Acids   698
    Finding the pH of Polyprotic Acid Solutions 698
    Finding the Concentration of the Anions for a
    Weak Diprotic Acid Solution 701
15.10 Acid Strength and Molecular Structure   702
    Binary Acids 703   Oxyacids 703
15.11 Lewis Acids and Bases   704
    Molecules That Act as Lewis Acids 705   Cations
    That Act as Lewis Acids 706
15.12 Acid Rain   706
    Effects of Acid Rain 707   Acid Rain
    Legislation 707
    Chapter in Review   708
    Key Terms 708   Key Concepts 708   Key Equations
    and Relationships 709   Key Skills 709
    Exercises   710
    Review Questions 710   Problems by Topic 711
    Cumulative Problems 714   Challenge Problems 715
    Conceptual Problems 715

16

Aqueous Ionic Equilibrium   716

16.1 The Danger of Antifreeze   717
16.2 Buffers: Solutions That Resist pH Change   718
    Calculating the pH of a Buffer Solution 720
    The Henderson–Hasselbalch Equation 721
    Calculating pH Changes in a Buffer Solution 724
    Buffers Containing a Base and Its Conjugate
    Acid 728
16.3 Buffer Effectiveness: Buffer Range and
        Buffer Capacity   729
    Relative Amounts of Acid and Base 729
    Absolute Concentrations of the Acid and
    Conjugate Base 730   Buffer Range 731   Buffer
    Capacity 732
    Chemistry and Medicine: Buffer Effectiveness
        in Human Blood   732
16.4 Titrations and pH Curves   733
    The Titration of a Strong Acid with a Strong Base
    734   The Titration of a Weak Acid with a Strong
    Base 737   Titration of a Polyprotic Acid 743
    Indicators: pH-Dependent Colors 744
16.5 Solubility Equilibria and the Solubility Product
        Constant   747
    K_{sp} and Molar Solubility 748
    Chemistry in Your Day: Hard Water   749
    K_{sp} and Relative Solubility 750   The Effect of a
    Common Ion on Solubility 751   The Effect of
    pH on Solubility 752
16.6 Precipitation   753
    Selective Precipitation 755
16.7 Qualitative Chemical Analysis   756
    Group 1: Insoluble Chlorides 758   Group 2:
    Acid-Insoluble Sulfides 758   Group 3: Base-
    Insoluble Sulfides and Hydroxides 758   Group 4:
    Insoluble Phosphates 758   Group 5: Alkali
    Metals and NH_4^+ 758

16.8   Complex Ion Equilibria   759

The Effect of Complex Ion Equilibria on
Solubility 761   The Solubility of Amphoteric
Metal Hydroxides 761

Chapter in Review   763

Key Terms 763   Key Concepts 764   Key Equations
and Relationships 764   Key Skills 764

Exercises   765

Review Questions 765   Problems by Topic 766
Cumulative Problems 771   Challenge Problems 772
Conceptual Problems 773

17

Free Energy and Thermodynamics

                                                                   774

17.1   Nature's Heat Tax: You Can't Win and You Can't
        Break Even   775
17.2   Spontaneous and Nonspontaneous Processes   777
17.3   Entropy and the Second Law of Thermodynamics   779

Entropy 780   The Entropy Change Associated
with a Change in State 784

17.4   Heat Transfer and Changes in the Entropy of the
        Surroundings   785

The Temperature Dependence of ΔS_{surr} 786
Quantifying Entropy Changes in the
Surroundings 787

17.5   Gibbs Free Energy   788

The Effect of ΔH, ΔS, and T on Spontaneity 790

17.6   Entropy Changes in Chemical Reactions: Calculating
        ΔS°_{rxn} 793

Standard Molar Entropies (S°) and the Third Law
of Thermodynamics 793

17.7   Free Energy Changes in Chemical Reactions:
        Calculating ΔG°_{rxn} 797

Calculating Free Energy Changes using
$\Delta G^{\circ}_{rxn} = \Delta H^{\circ}_{rxn} - T\Delta S^{\circ}_{rxn}$ 798   Calculating
ΔG°_{rxn} using Tabulated Values of Free Energies of
Formation 799   Determining ΔG°_{rxn} for a
Stepwise Reaction from the Changes in Free
Energy for Each of the Steps 801

Chemistry in Your Day: Making a Nonspontaneous
Process Spontaneous   802

Why Free Energy Is "Free" 803

17.8   Free Energy Changes for Nonstandard States: The
        Relationship between ΔG°_{rxn} and ΔG_{rxn} 804

The Free Energy of Reaction under Nonstandard
Conditions 804

17.9   Free Energy and Equilibrium: Relating ΔG°_{rxn} to the
        Equilibrium Constant (K)   807

The Temperature Dependence of the Equilibrium
Constant 809

Chapter in Review   810

Key Terms 810   Key Concepts 810   Key Equations
and Relationships 811   Key Skills 811

Exercises   812

Review Questions 812   Problems by Topic 812
Cumulative Problems 815   Challenge Problems 816
Conceptual Problems 817

18

Electrochemistry

                                                                   818

18.1   Pulling the Plug on the Power Grid   819
18.2   Balancing Oxidation–Reduction Equations   820
18.3   Voltaic (or Galvanic) Cells: Generating Electricity
        from Spontaneous Chemical Reactions   823

Electrochemical Cell Notation 826

18.4   Standard Reduction Potentials   827

Predicting the Spontaneous Direction of an
Oxidation–Reduction Reaction 831   Predicting
whether a Metal Will Dissolve in Acid 833

18.5   Cell Potential, Free Energy, and the Equilibrium
        Constant   834

The Relationship between ΔG° and E°_{cell} 834
The Relationship between E°_{cell} and K 836

18.6   Cell Potential and Concentration   837

Concentration Cells 840

Chemistry and Medicine: Concentration Cells in
Human Nerve Cells   842

18.7   Batteries: Using Chemistry to Generate Electricity   842

Dry-Cell Batteries 842   Lead–Acid Storage
Batteries 843   Other Rechargeable Batteries 844
Fuel Cells 845

Chemistry in Your Day: The Fuel-Cell
Breathalyzer   846

18.8   Electrolysis: Driving Nonspontaneous Chemical
        Reactions with Electricity   846

Predicting the Products of Electrolysis 849
Stoichiometry of Electrolysis 852

18.9   Corrosion: Undesirable Redox Reactions   854

Preventing Corrosion 855

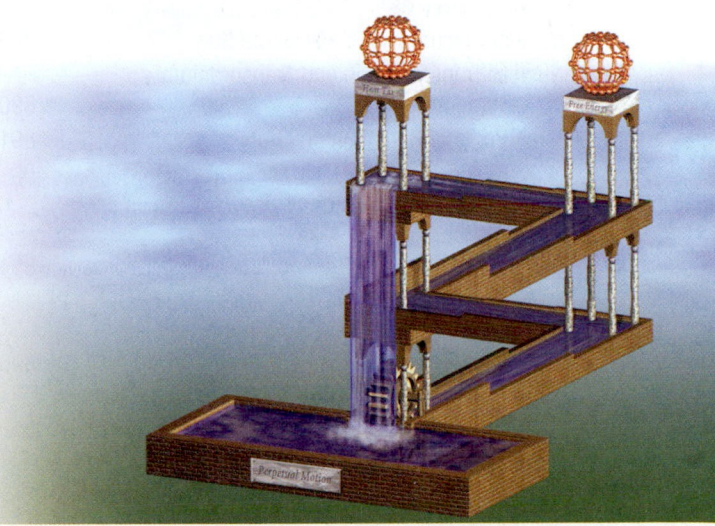

Chapter in Review   856
   Key Terms 856   Key Concepts 856   Key Equations
   and Relationships 857   Key Skills 857
Exercises   858
   Review Questions 858   Problems by Topic 858
   Cumulative Problems 862   Challenge Problems 863
   Conceptual Problems 863

19
Radioactivity and Nuclear Chemistry   864

19.1   Diagnosing Appendicitis   865
19.2   The Discovery of Radioactivity   866
19.3   Types of Radioactivity   867
       Alpha (α) Decay 868   Beta (β) Decay 869
       Gamma (γ) Ray Emission 870   Positron
       Emission 870   Electron Capture 870
19.4   The Valley of Stability: Predicting the Type of
       Radioactivity   872
       Magic Numbers 874   Radioactive Decay Series 874
19.5   Detecting Radioactivity   874
19.6   The Kinetics of Radioactive Decay and Radiometric
       Dating   875
       Chemistry in the Environment: Environmental
       Radon   877
       The Integrated Rate Law 877   Radiocarbon
       Dating: Using Radioactivity to Measure the Age of
       Fossils and Artifacts 879
       Chemistry in Your Day: Radiocarbon Dating
       and the Shroud of Turin   881
       Uranium/Lead Dating 881
19.7   The Discovery of Fission: The Atomic Bomb and
       Nuclear Power   883
       Nuclear Power: Using Fission to Generate Electricity 884
19.8   Converting Mass to Energy: Mass Defect and Nuclear
       Binding Energy   887
       Mass Defect 887
19.9   Nuclear Fusion: The Power of the Sun   889
19.10 Nuclear Transmutation and Transuranium
       Elements   890
19.11 The Effects of Radiation on Life   891
       Acute Radiation Damage 892   Increased Cancer
       Risk 892   Genetic Defects 892   Measuring
       Radiation Exposure 892
19.12 Radioactivity in Medicine and Other Applications 894
       Diagnosis in Medicine 894   Radiotherapy in
       Medicine 895   Other Applications 895
       Chapter in Review   896
          Key Terms 896   Key Concepts 896   Key Equations
          and Relationships 897   Key Skills 897
       Exercises   898
          Review Questions 898   Problems by Topic 898
          Cumulative Problems 900   Challenge Problems 901
          Conceptual Problems 901

20
Organic Chemistry   902

20.1   Fragrances and Odors   903
20.2   Carbon: Why It Is Unique   904
       Chemistry in Your Day: Vitalism and the
          Perceived Difference between Organic and
          Inorganic   905
20.3   Hydrocarbons: Compounds Containing Only
          Carbon and Hydrogen   906
       Drawing Hydrocarbon Structures 906
       Stereoisomerism and Optical Isomerism 909
20.4   Alkanes: Saturated Hydrocarbons   912
       Naming Alkanes 914
20.5   Alkenes and Alkynes   917
       Naming Alkenes and Alkynes 919   Geometric
       (Cis–Trans) Isomerism in Alkenes 920
20.6   Hydrocarbon Reactions   922
       Reactions of Alkanes 922   Reactions of Alkenes
       and Alkynes 923
20.7   Aromatic Hydrocarbons   924
       Naming Aromatic Hydrocarbons 925   Reactions
       of Aromatic Compounds 927
20.8   Functional Groups   927
20.9   Alcohols   928
       Naming Alcohols 929   About Alcohols 929
       Alcohol Reactions 929
20.10 Aldehydes and Ketones   931
       Naming Aldehydes and Ketones 932   About Aldehydes
       and Ketones 932   Aldehyde and Ketone Reactions 933
20.11 Carboxylic Acids and Esters   934
       Naming Carboxylic Acids and Esters 935   About
       Carboxylic Acids and Esters 935   Carboxylic
       Acid and Ester Reactions 936
20.12 Ethers   937
       Naming Ethers 937   About Ethers 937

20.13 Amines   938
    Amine Reactions 938
20.14 Polymers   938
    Chemistry in Your Day: Kevlar   941
    Chapter in Review   941
    Key Terms 941   Key Concepts 942   Key Equations
    and Relationships 943   Key Skills 944
    Exercises   944
    Review Questions 944   Problems by Topic 945
    Cumulative Problems 950   Challenge Problems 952
    Conceptual Problems 953

21

Biochemistry   954

21.1 Diabetes and the Synthesis of Human Insulin   955
21.2 Lipids   956
    Fatty Acids 956   Fats and Oils 958
    Chemistry and Medicine: Dietary Fat: The Good,
    the Bad, and the Ugly   959
    Other Lipids 960
21.3 Carbohydrates   961
    Simple Carbohydrates: Monosaccharides and
    Disaccharides 961   Complex Carbohydrates 964
21.4 Proteins and Amino Acids   966
    Amino Acids: The Building Blocks of Proteins 966
    Peptide Bonding between Amino Acids 969
    Chemistry and Medicine: The Essential Amino
    Acids   970
21.5 Protein Structure   970
    Primary Structure 972   Secondary Structure 972
    Tertiary Structure 973   Quaternary Structure 973
21.6 Nucleic Acids: Blueprints for Proteins   974
    The Basic Structure of Nucleic Acids 974   The
    Genetic Code 977
21.7 DNA Replication, the Double Helix, and Protein
    Synthesis   978
    DNA Replication and the Double Helix 978
    Protein Synthesis 979
    Chemistry and Medicine: The Human Genome
    Project   980
    Chapter in Review   981
    Key Terms 981   Key Concepts 981   Key Skills 982
    Exercises   982
    Review Questions 982   Problems by Topic 983
    Cumulative Problems 986   Challenge Problems 987
    Conceptual Problems 987

22

Chemistry of the Nonmetals   988

22.1 Insulated Nanowires   989
22.2 The Main-Group Elements: Bonding and
    Properties   990
    Atomic Size and Types of Bonds 991
22.3 Silicates: The Most Abundant Matter in
    Earth's Crust   992
    Quartz and Glass 992   Aluminosilicates 993
    Individual Silicate Units, Silicate Chains, and
    Silicate Sheets 993
22.4 Boron: An Intersting Group 3A Element and
    Its Remarkable Structures   996
    Elemental Boron 996   Boron Compounds:
    Trihalides 997   Boron–Oxygen Compounds 997
    Boron–Hydrogen Compounds: Boranes 998
22.5 Carbon, Carbides, and Carbonates   999
    Carbon 999   Carbides 1002   Carbon Oxides
    1003   Carbonates 1004
22.6 Nitrogen and Phosphorus: Essential Elements
    for Life   1004
    Elemental Nitrogen and Phosphorus 1005
    Nitrogen Compounds 1006   Phosphorus
    Compounds 1009
22.7 Oxygen   1011
    Elemental Oxygen 1011   Uses for Oxygen 1012
    Oxides 1012   Ozone 1012
22.8 Sulfur: A Dangerous but Useful Element   1013
    Elemental Sulfur 1013   Hydrogen Sulfide and
    Metal Sulfides 1015   Sulfur Dioxide 1016
    Sulfuric Acid 1016
22.9 Halogens: Reactive Elements with High
    Electronegativity   1017
    Elemental Fluorine and Hydrofluoric Acid 1018
    Elemental Chlorine 1019   Halogen Compounds
    1019
    Chapter in Review   1021
    Key Terms 1021   Key Concepts 1021
    Key Skills 1022
    Exercises   1022
    Review Questions 1022   Problems by Topic
    1023   Cumulative Problems 1024   Challenge
    Problems 1025   Conceptual Problems 1025

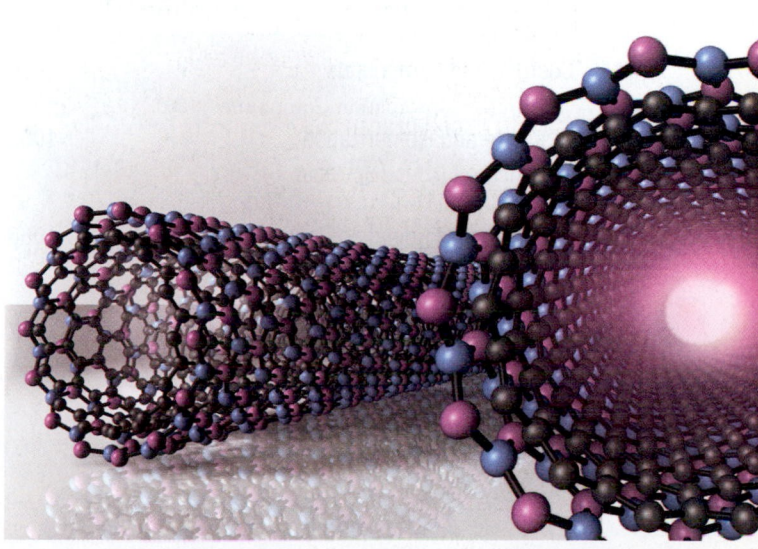

23

Metals and Metallurgy   1026

23.1	Vanadium: A Problem and an Opportunity	1027
23.2	The General Properties and Natural Distribution of Metals	1028
23.3	Metallurgical Processes	1029

Separation 1030   Pyrometallurgy 1030
Hydrometallurgy 1031   Electrometallurgy 1032
Powder Metallurgy 1033

23.4	Metal Structures and Alloys	1033

Alloys 1034   Substitutional Alloys: Miscible
Solid Solutions 1034   Alloys with Limited
Solubility 1035   Interstitial Alloys 1038

23.5	Sources, Properties, and Products of Some of the 3d Transition Metals	1039

Titanium 1039   Chromium 1040   Manganese
1042   Cobalt 1042   Copper 1043   Nickel 1044
Zinc 1044

Chapter in Review   1045

Key Terms 1045   Key Concepts 1045
Key Equations and Relationships 1046
Key Skills 1046

Exercises   1046

Review Questions 1046   Problems by Topic
1047   Cumulative Problems 1048   Challenge
Problems 1049   Conceptual Problems 1049

24

Transition Metals and Coordination Compounds   1050

24.1	The Colors of Rubies and Emeralds	1051
24.2	Properties of Transition Metals	1052

Electron Configurations 1052 Atomic Size 1054
Ionization Energy 1055   Electronegativity 1055
Oxidation States 1056

24.3	Coordination Compounds	1056

Naming Coordination Compounds 1060

24.4	Structure and Isomerization	1062

Structural Isomerism 1062
Stereoisomerism 1062

24.5	Bonding in Coordination Compounds	1066

Valance Bond Theory 1066   Crystal Field
Theory 1067   Octahedral Complexes 1067   The
Color of Complex Ions and Crystal Field Strength
1068   Magnetic Properties 1071   Tetrahedral
and Square Planar Complexes 1072

24.6	Applications of Coordination Compounds	1073

Chelating Agents 1073   Chemical Analysis 1073
Coloring Agents 1074   Biomolecules 1074

Chapter in Review   1076

Key Terms 1076   Key Concepts 1077   Key Equations
and Relationships 1077   Key Skills 1077

Exercises   1078

Review Questions 1078   Problems by Topic 1078
Cumulative Problems 1080   Challenge Problems 1080
Conceptual Problems 1081

Appendix I: Common Mathematical Operations in Chemistry   A-1

A	Scientific Notation	A-1
B	Logarithms	A-3
C	Quadratic Equations	A-5
D	Graphs	A-5

Appendix II: Useful Data   A-7

A	Atomic Colors	A-7
B	Standard Thermodynamic Quantities for Selected Substances at 25 °C	A-7
C	Aqueous Equilibrium Constants at 25 °C	A-12
D	Standard Reduction Half-Cell Potentials at 25 °C	A-15
E	Vapor Pressure of Water at various Temperatures	A-15

Appendix III: Answers to Selected Exercises   A-16

Appendix IV: Answers to In-Chapter Practice Problems   A-47

Glossary   G-1

Credits   C-1

Index   I-1

Preface

To the Student

As you begin this course, I invite you to think about your reasons for enrolling in it. Why are you taking general chemistry? More generally, why are you pursuing a college education? If you are like most college students taking general chemistry, part of your answer is probably that this course is required for your major and that you are pursuing a college education so you can get a good job some day. While these are good reasons, I would like to suggest a better one. I think the primary reason for your education is to prepare you to *live a good life*. You should understand chemistry—not for what it can *get* you— but for what it can *do* for you. Understanding chemistry, I believe, is an important source of happiness and fulfillment. Let me explain.

Understanding chemistry helps you to live life to its fullest for two basic reasons. The first is *intrinsic*: through an understanding of chemistry, you gain a powerful appreciation for just how rich and extraordinary the world really is. The second reason is *extrinsic:* understanding chemistry makes you a more informed citizen—it allows you to engage with many of the issues of our day. In other words, understanding chemistry makes *you* a deeper and richer person and makes your country and the world a better place to live. These reasons have been the foundation of education from the very beginnings of civilization.

How does chemistry help prepare you for a rich life and conscientious citizenship? Let me explain with two examples. My first one comes from the very first page of Chapter 1 of this book. There, I ask the following question: What is the most important idea in all of scientific knowledge? My answer to that question is this: **the behavior of matter is determined by the properties of molecules and atoms**. That simple statement is the reason I love chemistry. We humans have been able to study the substances that compose the world around us and explain their behavior by reference to particles so small that they can hardly be imagined. If you have never realized the remarkable sensitivity of the world we *can* see to the world we *cannot*, you have missed out on a fundamental truth about our universe. To have never encountered this truth is like never having read a play by Shakespeare or seen a sculpture by Michelangelo—or, for that matter, like never having discovered that the world is round. It robs you of an amazing and unforgettable experience of the world and the human ability to understand it.

My second example demonstrates how science literacy helps you to be a better citizen. Although I am largely sympathetic to the environmental movement, a lack of science literacy within some sectors of that movement, and the resulting antienvironmental backlash, creates confusion that impedes real progress and opens the door to what could be misinformed policies. For example, I have heard conservative pun-

dits say that volcanoes emit more carbon dioxide—the most significant greenhouse gas—than does petroleum combustion. I have also heard a liberal environmentalist say that we have to stop using hairspray because it is causing holes in the ozone layer that will lead to global warming. Well, the claim about volcanoes emitting more carbon dioxide than petroleum combustion can be refuted by the basic tools you will learn to use in Chapter 4 of this book. We can easily show that volcanoes emit only 1/50th as much carbon dioxide as petroleum combustion. As for hairspray depleting the ozone layer and thereby leading to global warming: the chlorofluorocarbons that deplete ozone have been banned from hairspray since 1978, and ozone depletion has nothing to do with global warming anyway. People with special interests or axes to grind can conveniently distort the truth before an ill-informed public, which is why we all need to be knowledgeable.

So this is why I think you should take this course. Not just to satisfy the requirement for your major, and not just to get a good job some day, but to help you to lead a fuller life and to make the world a little better for everyone. I wish you the best as you embark on the journey to understand the world around you at the molecular level. The rewards are well worth the effort.

To the Professor

Teaching general chemistry would be much easier if all of our students had exactly the same level of preparation and ability. But alas, that is not the case. Even though I teach at a relatively selective institution, my courses are populated with students with a range of backgrounds and abilities in chemistry. The challenge of successful teaching, in my opinion, is therefore figuring out how to instruct and challenge the best students while not losing those with lesser backgrounds and abilities. My strategy has always been to set the bar relatively high, while at the same time providing the motivation and support necessary to reach the high bar. That is exactly the philosophy of this book. We do not have to compromise away rigor in order to make chemistry accessible to our students. In this book, I have worked hard to combine rigor with accessibility—to create a book that does not dilute the content, yet can be used and understood by any student willing to put in the necessary effort.

Chemistry: A Molecular Approach is first and foremost a *student-oriented book*. My main goal is to motivate students and get them to achieve at the highest possible level. As we all know, many students take general chemistry because it is a requirement; they do not see the connection between chemistry and their lives or their intended careers. *Chemistry: A Molecular Approach* strives to make those connections consistently and effectively. Unlike other books, which often teach chemistry as something that happens only in the laboratory or in industry, this book teaches chemistry in the context of relevance. It shows students *why* chemistry is important to them, to their future careers, and to their world.

Chemistry: A Molecular Approach is secondly a *pedagogically driven book*. In seeking to develop problem-solving skills, a consistent approach (Sort, Strategize, Solve, and Check) is applied, usually in a two- or three-column format. In the two-column format, the left column shows the student how to analyze the problem and devise a solution strategy. It also lists the steps of the solution, explaining the rationale for each one, while the right column shows the implementation of each step. In the three-column format, the left column outlines a general procedure for solving an important category of problems that is then applied to two side-by-side examples. This strategy allows students to see both the general pattern and the slightly different ways in which the procedure may be applied in differing contexts. The aim is to help students understand both the *concept of the problem* (through the formulation of an explicit conceptual plan for each problem) and the *solution to the problem*.

Chemistry: A Molecular Approach is thirdly a *visual book*. Wherever possible, images are used to deepen the student's insight into chemistry. In developing chemical principles, multi-part images help to show the connection between everyday processes visible to the unaided eye and what atoms and molecules are actually doing. Many of these images have three parts: macroscopic, molecular, and symbolic. This combination helps students to see the relationships between the formulas they write down on paper (symbolic), the world they see around them (macroscopic), and the atoms and molecules that compose that world (molecular). In addition, most figures are designed to teach rather than just to illustrate. They are rich with annotations and labels intended to help the student grasp the most important processes and the principles that underlie them. The resulting images are rich with information, but also uncommonly clear and quickly understood.

Chemistry: A Molecular Approach is fourthly a *"big picture" book*. At the beginning of each chapter, a short paragraph helps students to see the key relationships between the different topics they are learning. Through focused and concise narrative, I strive to make the basic ideas of every chapter clear to the student. Interim summaries are provided at selected spots in the narrative, making it easier to grasp (and review) the main points of important discussions. And to make sure that students never lose sight of the forest for the trees, each chapter includes several *Conceptual Connections*, which ask them to think about concepts and solve problems without doing any math. I want students to learn the concepts, not just plug numbers into equations to churn out the right answer.

Chemistry: A Molecular Approach is lastly a book that delivers the depth of coverage faculty want. We do not have to cut corners and water down the material in order to get our students interested. We simply have to meet them where they are, challenge them to the highest level of achievement, and then support them with enough pedagogy to allow them to succeed.

The best new books, in my opinion, are evolutionary—they take what is already there and make it better. *Chemistry: A Molecular Approach* is designed to be such a book. The founda-

tions of the general chemistry curriculum have already been laid. This text presents those foundations in new and pedagogically innovative ways that make the subject clear, stimulating, and relevant to today's student.

I hope that this book supports you in your vocation of teaching students chemistry. I am increasingly convinced of the importance of our task. Please feel free to email me with any questions or comments about the book.

Nivaldo J. Tro
tro@westmont.edu

Supplements

For the Instructor and the Student

MasteringGeneralChemistry (http://www.masteringgeneralchemistry.com) For instructor-assigned homework, MasteringGeneralChemistry™ provides the first adaptive-learning online tutorial and assessment system. Based on extensive research of precise concepts students struggle with, the system is able to coach students with feedback specific to their needs and with simpler problems upon request. The result is targeted tutorial help to optimize study time and maximize learning for students.

Mastering General Chemistry Instructor Access Kit (0-13-615668-1)

Mastering General Chemistry Student Access Kit (0-13-615669-X)

For the Instructor

Instructor's Edition (0-13-615491-3)

Instructor's Resource Center on CD/DVD (0-13-615107-8) This fully searchable and integrated collection of resources, organized in one easy-to-access place, includes everything you need designed to help you make efficient and effective use of your lecture presentation.

Instructor's Resource Manual (0-13-615582-0) This useful guide describes all the different resources available to instructors and demonstrates how to integrate them into your course.

Test Item File (0-13-615109-4) This printed test bank, containing multiple-choice, short-answer, and matching test questions, has been fully reviewed for accuracy.

Transparency Pack (0-13-100178-7) This set of transparencies has been chosen specifically to focus on the illustrations that provide a visual perspective of the key principles; it is designed to save you time while preparing your lectures.

Blackboard® Test Item File (0-13-615117-5)

WebCT® Test Item File (0-13-615119-1)

Solutions Manual (0-13-61515110-8) This manual contains step-by-step solutions to all complete, end-of-chapter exercises. With instructor permission, this manual may be made available to students.

Virtual ChemLab, General Chemistry, v2.5

(0-13-228010-8) Virtual ChemLab provides students with an opportunity to apply the general chemistry concepts learned in the classroom in a safe and level-appropriate laboratory setting, lets students perform experiments before entering the lab, and allows them to perform experiments to which they would not normally have access. At the same time, it gives instructors a new tool for teaching and reinforcing important concepts.

COURSE MANAGEMENT

Blackboard® (0-13-100180-9)

WebCT® (0-13-100179-5)

For the Student

Selected Solutions Manual (0-13-615116-7) This manual for students contains complete, step-by-step solutions to selected odd-numbered end-of-chapter problems.

Virtual ChemLab, General Chemistry, v2.5

(0-13-228009-4) *Virtual ChemLab* is a set of realistic and sophisticated simulations covering general chemistry laboratories. In these laboratories, students are put into a virtual environment where they are free to make the choices and decisions that they would confront in an actual laboratory setting and, in turn, experience the resulting consequences.

Acknowledgments

The book you hold in your hands bears my name on the cover, but I am really only one member of a large team that carefully crafted the book over several years. Most importantly, I thank my editor, Kent Porter Hamann, who has believed in me and encouraged me from day one. She has been intimately and creatively involved with every aspect of the creation of this book—I am forever indebted to her. I also owe much to Dan Schiller, a development editor with incredible breadth and depth of knowledge, as well as keen layout and art development skills. I am grateful to Dan for his ideas, his persistence, and especially for the many hours he has poured into this project. This book would be much different without him. I would also like to express my appreciation for the manifold contributions of Ray Mullaney, whose wisdom and gentle guidance have steered this project throughout. Ray is a consummate professional, and I feel tremendously privileged to have worked with him. I am particularly grateful to Paul Corey. Paul is a man of incredible energy and vision—he has been an inspiration to me. Paul told me to dream big, and then he provided the resources I needed to make those dreams come true. Thanks, Paul.

New to the team is Liz Averbeck, and although we have worked together for only a short while, I am already impressed by her energy in marketing this book. A special word of thanks is owed to Glenn and Meg Turner of Burrston House, ideal collaborators whose contributions to this project were extremely important and much appreciated. I am deeply grateful to Kenny Beck, Juan Lopez, and Sue Behnke for their great patience, creativity, and hard work in crafting the design of this text. I owe an enormous debt to Rosaria Cassinese and her co-workers at Preparé, whose skill and diligence gave this book its physical existence; to Patty Burns, Connie Long, Jay McElroy, and the staff of Artworks; and to JC Morgan and his colleagues at Precision Graphics. All of these people labored tirelessly under great pressure to produce a book of uncommon elegance. I am also greatly indebted to my copy editor, Connie Parks, for her dedication and professionalism, and to Jerry Marshall, for his exemplary photo research. Heartfelt thanks are due to Abby Bass, Patty Gutierrez, Clara Bartunek, Karen Stephens, Karen Noferi, Joanne Del Ben, Julita Nazario, and Tom Benfatti for their invaluable and timely help; to Kathleen Schiaparelli, who kept all the plates spinning and the balls rolling; and to the rest of the Prentice Hall team. They are part of a first-class operation—this text has benefited immeasurably from their talents and hard work. I owe a special debt of gratitude to Quade and Emiko Paul, who made my ideas come alive in their art.

I would like to acknowledge the help of my colleagues Allan Nishimura and David Marten, who have supported me in my department while I worked on this book. I am also grateful to those who have supported me personally. First on that list is my wife, Ann. Her patience and love for me are beyond description, and without her, this book would never have been written. I am also indebted to my children, Michael, Ali, Kyle, and Kaden, whose smiling faces and love of life always inspire me. I come from a large Cuban family whose closeness and support most people would envy. Thanks to my parents, Nivaldo and Sara; my siblings, Sarita, Mary, and Jorge; my siblings-in-law, Jeff, Nachy, Karen, and John; my nephews and nieces, Germain, Danny, Lisette, Sara, and Kenny. These are the people with whom I celebrate life.

I would like to thank all of the general chemistry students who have been in my classes throughout my 17 years as a professor at Westmont College. You have taught me much about teaching that is now in this book. I am especially grateful to students Jon Rea and Audrey Farkas, who put in many hours proofreading my manuscript, working problems, and assembling data for tables, figures, and graphs. I would also like to express my appreciation to the following students who played various roles in the development of the manuscript: Julie Ray, Chris Osborne, Jessica Osborne, Callan Kaut, Roberto Valladares, Kevin Woodard, and Mathew Gunner.

Lastly, I am indebted to the many reviewers, listed on the following pages, whose ideas are imbedded throughout this book. They have corrected me, inspired me, and sharpened my thinking on how best to teach this subject we call chemistry. I deeply appreciate their commitment to this project. I am particularly grateful to Bob Boikess, Bill Cleaver, and Joseph Bularzik for their important contributions to the book, and to Norb Pienta in his role as media consultant. Thanks also to Frank Lambert for helping us all to think more clearly about entropy and for his review of the entropy sections of the book. Last but by no means least, I would like to record my gratitude to Margaret Asirvatham, Michelle Driessen, Louis Kirschenbaum, Tim Kreider, and Kathy Thrush Shaginaw, whose alertness, keen eyes, and scientific astuteness made this a much better book, and to Gabriele Backes, Melita Balch, Elizabeth Hairfield, Rafiq Ladhani, Philip Lukeman, Stephen Mezyk, George Pasles, Jerry Sarquis, Steven Socol, and Marcy Whitney, whose careful and painstaking scrutiny of the first printing enabled us to achieve a level of accuracy and freedom from errors not normally attainable in the first edition of a text.

Reviewers

Michael R. Adams, *Xavier University of Louisiana*
Patricia G. Amateis, *Virginia Tech*
Margaret R. Asirvatham, *University of Colorado*
Monica H. Baloga, *Florida Institute of Technology*
Mufeed M. Basti, *North Carolina Agricultural & Technological State University*
Amy E. Beilstein, *Centre College*
Kyle A. Beran, *University of Texas of the Permian Basin*
Christine V. Bilicki, *Pasadena City College*
Robert E. Blake, *Texas Tech University*
Angela E. Boerger, *Loyola University*
Robert S. Boikess, *Rutgers University*
Michelle M. Brooks, *College of Charleston*
Joseph H. Bularzik, *Purdue University, Calumet*
Cindy M. Burkhardt, *Radford University*
Andrew E. Burns, *Kent State University, Stark Campus*
Kim C. Calvo, *University of Akron*
Stephen C. Carlson, *Lansing Community College*
Eric G. Chesloff, *Villanova University*
William M. Cleaver, *University of Vermont*
Samuel R. Cron, *Arkansas State*
Darwin B. Dahl, *Western Kentucky University*
Robert F. Dias, *Old Dominion University*
Daniel S. Domin, *Tennessee State University*
Alan D. Earhart, *Southeast Community College*
Amina K. El-Ashmawy, *Collin County Community College*
Joseph P. Ellison, *United States Military Academy, West Point*
Joseph M. Eridon, *Albuquerque TVI*
Deborah B. Exton, *University of Oregon*
William A. Faber, *Grand Rapids Community College*
Maria C. Fermin-Ennis, *Gordon College*
Jan Florian, *Loyola University*
Candice E. Fulton, *Midwestern State*
Eric S. Goll, *Brookdale Community College*
Pierre Y. Goueth, *Santa Monica College*
Thomas J. Greenbowe, *Iowa State*
Jason A. Halfen, *University of Wisconsin, Eau Clair*
Lois Hansen-Polcar, *Cuyahoga Community College West*
Monte L. Helm, *Fort Lewis College*
David E. Henderson, *Trinity College*
Susan K. Henderson, *Quinnipiac University*
Peter M. Hierl, *University of Kansas*
Todd A. Hopkins, *Butler University*
Byron E. Howell, *Tyler Junior College*
Ralph Isovitsch, *Xavier University of Louisiana*
Kenneth C. Janda, *University of California, Irvine*
Jason A. Kautz, *University of Nebraska, Lincoln*
Catherine A. Keenan, *Chaffey College*
Steven W. Keller, *University of Missouri, Columbia*
Louis J. Kirschenbaum, *University of Rhode Island*
Tim Krieder
Bette Kreuz, *University of Michigan, Dearborn*
Sergiy Kryatov, *Tufts University*
Clifford B. Lemaster, *Boise State University*
Benjamin R. Martin, *Texas State*
Lydia J. Martinez-Rivera, *University of Texas, San Antonio*

Marcus T. McEllistrem, *University of Wisconsin, Eau Claire*
Danny G. McGuire, *Cameron University*
Charles W. McLaughlin, *University of Nebraska, Lincoln*
Curt L. McLendon, *Saddleback College*
Robert C. McWilliams, *United States Military Academy*
Ray Mohseni, *East Tennessee State University*
Elisabeth A. Morlino, *The University of the Sciences, Philadelphia*
James E. Murphy, *Santa Monica College*
Maria C. Nagan, *Truman State University*
Kenneth S. Overway, *Bates College*
Yasmin Patell, *Kansas State University*
Glenn A. Petrie, *Central Missouri State*
Norbert J. Pienta, *University of Iowa*
Louis H. Pignolet, *University of Minnesota*
Dana L. Richter-Egger, *University of Nebraska*
A. Timothy Royappa, *University of West Florida*
Stephen P. Ruis, *American River College*
Alan E. Sadurski, *Ohio Northern University*
Thomas W. Schleich, *University of California, Santa Cruz*
Tom Selegue, *Pima Community College, West*
Anju H. Sharma, *Stevens Institute of Technology*
Sherril A. Soman, *Grand Valley State University*
Michael S. Sommer, *University of Wyoming*
Jie S. Song, *University of Michigan, Flint*
Mary Kay Sorenson, *University of Wisconsin, Milwaukee*
Stacy E. Sparks, *University of Texas, Austin*
William H. Steel, *York College of Pennsylvania*
Tamar Y. Susskind, *Oakland Community College*
Jacquelyn Thomas, *Southwestern College*
Kathleen Thrush Shaginaw, *Villanova University*
Lydia Tien, *Monroe Community College*
Ramaiyer Venkatraman, *Jackson State University*
John B. Vincent, *University of Alabama, Tuscaloosa*
Kent S. Voelkner, *Lake Superior College*
Sheryl K. Wallace, *South Plains College*
Wayne E. Wesolowski, *University of Arizona*
Sarah E. West, *Notre Dame University*
Kurt J. Winkelmann, *Florida Institute of Technology*
Troy D. Wood, *University of Buffalo*
Servet M. Yatin, *Quincy College*
Kazushige Yokoyama, *SUNY Geneseo*

Focus Group Participants

We would like to thank the following professors for contributing their valuable time to meet with the author and the publishing team in order to provide a meaningful perspective on the most important challenges they face in teaching general chemistry and give us insight into creating a new general chemistry text that successfully responds to those challenges.

Focus Group 1

Michael R. Abraham, *University of Oklahoma*
Steven W. Keller, *University of Missouri, Columbia*
Roy A. Lacey, *State University of New York, Stonybrook*
Norbert J. Pienta, *University of Iowa*

Cathrine E. Reck, *Indiana University*
Reva A. Savkar, *Northern Virginia Community College*

Focus Group 2

Amina K. El-Ashmawy, *Collin County Community College*
Steven W. Keller, *University of Missouri, Columbia*
Joseph L. March, *University of Alabama, Birmingham*
Norbert J. Pienta, *University of Iowa*

Focus Group 3

James A. Armstrong, *City College of San Francisco*
Roberto A. Bogomolni, *University of California, Santa Cruz*
Kate Deline, *College of San Mateo*
Greg M. Jorgensen, *American River College*
Dianne Meador, *American River College*
Heino Nitsche, *University of California at Berkeley*
Thomas W. Schleich, *University of California, Santa Cruz*

Focus Group 4

Ramesh D. Arasasingham, *University of California, Irvine*
Raymond F. Glienna, *Glendale Community College*
Pierre Y. Goueth, *Santa Monica College*
Catherine A. Keenan, *Chaffey College*
Ellen Kime-Hunt, *Riverside Community College, Riverside Campus*
David P. Licata, *Coastline Community College*
Curtis L. McLendon, *Saddleback College*
John A. Milligan, *Los Angeses Valley College*

Focus Group 5

Eric S. Goll, *Brookdale Community College*
Kamal Ismail, *CUNY, Bronx Community College*
Sharon K. Kapica, *County College of Morris*
Richard Rosso, *St. John's University*
Steven Rowley, *Middlesex County College*
David M. Sarno, *CUNY, Queensborough Community College*
Donald L. Siegel, *Rutgers University, New Brunswick*
Servet M. Yatin, *Quincy College*

Focus Group 6

William Eck, *University of Wisconsin, Marshfield/Wood County*
Richard W. Frazee, *Rowan University*
Barbara A. Gage, *Prince George's Community College*
John A. W. Harkless, *Howard University*
Patrick M. Lloyd, *CUNY, Kingsborough Community College*
Boon H. Loo, *Towson University*
Elisabeth A. Morlino, *University of the Sciences, Philadelphia*
Benjamin E. Rusiloski, *Delaware Valley College*
Louise S. Sowers, *Richard Stockton College of New Jersey*
William H. Steel, *York College of Pennsylvania*
Galina G. Talanova, *Howard University*
Kathleen Thrush Shaginaw, *Villanova University*

Focus Group 7

Stephen C. Carlson, *Lansing Community College*
Darwin B. Dahl, *Western Kentucky University*
Robert J. Eierman, *University of Wisconsin, Eau Clair*
William A. Faber, *Grand Rapids Community College*
Jason A. Halfen, *University of Wisconsin, Eau Clair*
Todd A. Hopkins, *Butler University*
Michael E. Lipschutz, *Purdue University*
Jack F. McKenna, *St. Cloud State University*
Claire A. Tessier, *University of Akron*

Focus Group 8

Charles E. Carraher, *Florida Atlantic University*
Jerome E. Haky, *Florida Atlantic University*
Paul I. Higgs, *Barry University*
Moheb Ishak, *St. Petersburg College, St. Petersburg*
Peter J. Krieger, *Palm Beach Community College, Lake Worth*
Jeanette C. Madea, *Broward Community College, North*
Alice J. Monroe, *St. Petersburg College, Clearwater*
Mary L. Sohn, *Florida Institute of Technology*

Focus Group 9

Silas C. Blackstock, *University of Alabama*
Kenneth Capps, *Central Florida Community College*
Ralph C. Dougherty, *Florida State University*
W. Tandy Grubbs, *Stetson University*
Norris W. Hoffman, *University of South Alabama*
Tony Holland, *Wallace Community College*
Paul I. Higgs, *Barry University*
James L. Mack, *Fort Valley State University*
Karen Sanchez, *Florida Community College, Jacksonville*
Richard E. Sykora, *University of South Alabama*
Gary L. Wood, *Valdosta State University*

Student Focus Groups

We are very grateful to the students who gave part of their day to share with the chemistry team their experience in using textbooks and their ideas on how to make a general chemistry text a more valuable reference.

Bryan Aldea, *Brookdale Community College*
Corinthia Andres, *University of the Sciences of Philadelphia*
Hadara Biala, *Brookdale Community College*
Eric Bowes, *Villanova University*
Adrian Danemayer, *Drexel University*
Daniel Fritz, *Middlesex County College*
Olga Ginsburg, *Rutgers University*
Kira Gordin, *University of the Sciences of Philadelphia*
Geoffrey Haas, *Villanova University*
Hadi Dharma Halim, *Middlesex County College*
Heather Hartman, *Bucks County Community College*
Stephen A. Horvath, *Rutgers University*
Mark Howell, *Villanova University*

Gene Iucci, *Rutgers University*
Adrian Kochan, *Villanova University*
Jeffrey D. Laszczyk Jr., *University of the Sciences of Philadelphia*
Allison Lucci, *Drexel University*
Mallory B. McDonnell, *Villanova University*
Brian McLaughlin, *Brookdale Community College*
Michael McVann, *Villanova University*
Stacy L. Molnar, *Bucks County Community College*
Jenna Munnelly, *Villanova University*
Lauren Papa, *Rutgers University*
Ankur Patel, *Drexel University*
Janaka P. Peiris, *Middlesex County College*
Ann Mary Sage, *Brookdale Community College*
Salvatore Sansone, *Bucks County Community College*
Michael Scarneo, *Drexel University*
Puja Shahi, *Drexel University*
Rebeccah G. Steinberg, *Brookdale Community College*
Alyssa J. Urick, *University of the Sciences of Philadelphia*
Padma Vemuri, *Villanova University*
Joni Vitale, *Brookdale Community College*
Kyle Wright, *Rowan Uninversity*
Joseph L. Yobb, *Bucks County Community College*

Reviewer Conference Participants

We would also like to give great thanks to the participants in our Reviewer Conference, who unselfishly spent three days with the author and the chemistry team reviewing all 24 chapters of the late-draft manuscript and sharing their teaching methodologies, their feedback on the author's presentation, and their creative ideas.

Mufeed M. Basti, *North Carolina Agricultural & Technical State University*
Robert S. Boikess, *Rutgers University*
Jason A. Kautz, *University of Nebraska, Lincoln*
Curtis L. McLendon, *Saddleback College*
Norbert J. Pienta, *University of Iowa*
Alan E. Sadurski, *Ohio Northern University*
Jie S. Song, *University of Michigan, Flint*
John B. Vincent, *University of Alabama, Tuscaloosa*

First Printing Reviewers

Gabriele R. Backes, *Portland Community College*
Melita M. Balch, *University of Illinois, Chicago*
Elizabeth M. Hairfield, *Mary Baldwin College*
Rafiq Ladhani, *Editorial Services*
Philip S. Lukeman, *New York University*
Stephen Mezyk, *California State University*
George Pasles, *Editorial Services*
Jerry Sarquis, *Miami University, Ohio*
Steven M. Socol, *McHenry County College*
Marcy Whitney, *University of Alabama*

How This Text Was Developed, and Why It Should Make a Difference to You

Today's college textbooks, especially in the sciences, are designed to convey sophisticated concepts and teach critical skills with great efficiency. The books are used not only by students to learn the material, but also by professors to craft their courses. The material must be clearly presented, accurate, well organized, and compelling. The art program must be attractive, and it must convey the subject matter (often quantitative) in a pedagogically effective manner that complements the narrative. A single author (or team of authors) could never create such a book without the aid and support of many other educators and skilled publishing professionals.

Every good textbook must start with the author's vision. But that vision and its detailed implementation must be continuously tested and refined to ensure that the book meets its primary goal—to teach the material in new ways that result in better student learning. Ideas and features that work are enhanced to assist teaching and promote learning even more effectively; any elements that do not work are discarded. Prentice Hall is deeply committed to helping each of its authors realize their unique vision in a way that serves the needs of instructors and students as fully as possible.

The first step in the development of the book you are now holding was a small meeting of several professors who teach general chemistry and want to see it taught better. At the meeting, the participants discussed how that might be achieved. They bounced ideas off one another and brainstormed about how to draw college students into the discipline that each of them loved. They argued about the topics that deserved the most attention and the ones that deserved the least, and evaluated pedagogical features to determine which would be most effective for students. They discussed problem-solving techniques, common student misconceptions, and how best to convey the central themes and the "big picture" of chemistry. In short, this group of professors spent two intense days dreaming about what the ideal general chemistry book should look like.

At about the same time that this initial gathering was taking place, a development editor was assigned to the project. At Prentice Hall, the job of the development editor is, quite simply, to assist an author or authors—in every way possible—in the creation of a successful text. The development editor helps the author and the acquisitions editor to define the challenges of a particular course, evaluate the strengths and weaknesses of existing texts, formulate the essential characteristics of the new

text, and develop an appropriate pedagogical apparatus. He or she also plays a central role in analyzing the many reviews that are commissioned for each draft of the manuscript, helping the author refine the illustration program, and orchestrating the numerous complex details necessary to guide the book through production.

The most important task of the development editor, however, is *to serve as a surrogate for, and advocate for, the ultimate user of the text: the student.* As such, he or she must analyze and edit each chapter of the manuscript from the perspective of a student who will eventually use the text to ensure that it sets out, in the clearest possible way, the fundamental principles of the discipline. The development editor helps to keep the author's focus on the student, pointing out just how the current draft of the manuscript succeeds or falls short from the student's point of view.

Once an editor was assigned, the development of the text took the form of a series of interlocking feedback loops. Each chapter draft was subjected to an initial round of developmental editing, with emphasis on organization, clarity, and consistency of level. The chapters were then revised by the author. The new drafts, in turn, were exposed to intensive scrutiny by reviewers, including experts in the content of each chapter (whose main job was to evaluate the material for accuracy and currency) and instructors with extensive teaching experience (whose main job was to evaluate the material for teaching effectiveness). These reviewers were encouraged to focus on such issues as the breadth and depth of the coverage, the appropriateness of the worked examples in number, topic, and level, and the overall pedagogical effectiveness of the chapter. The reviews were analyzed by the author and development editor to determine what changes were still needed to improve each chapter and how they could best be implemented. Revisions were made, and the process was then repeated—at least once for all chapters, and several times for the most critical chapters.

As the chapters of this book began to assume a more polished form, additional focus groups and reviewer conferences were organized by Burrston House, an independent editorial and market development company whose extensive experience in college publishing made them an ideal partner for this phase of the project. Some of these conferences involved professors, who were asked not only to comment on the general issues that confront them in the classroom, but also to analyze in detail groups of chapters, and to discuss candidly their analysis with the author and editors. These sessions produced valuable insights that would be difficult to obtain in any other way, and were the inspiration for many significant improvements.

Other conferences were held with students. As the ultimate users of the book under development, students can provide a uniquely valuable perspective. Their responses represented one of the most crucial and stringent tests of the book's effectiveness.

To ensure that the important visual component of the book was fully integrated with the text, supporting and enhancing it at every point, development of the illustration

program proceeded hand in hand with development of the chapters. Every figure began as a simple sketch. Working collaboratively, the author, development editor, and artist decided how each sketch should be elaborated and refined so that the finished illustration was attractive, had a clear central teaching point, and made the best pedagogical use of color and other graphic elements. But this process, shown below for text Figure 5.3, did not take place in a vacuum. As the visual style, graphic vocabulary, and pedagogical devices of the art program took shape, they were subjected to evaluation and validation by the end users of the text, including students (in focus groups) and instructors (via WebEx conferences). Both groups evaluated

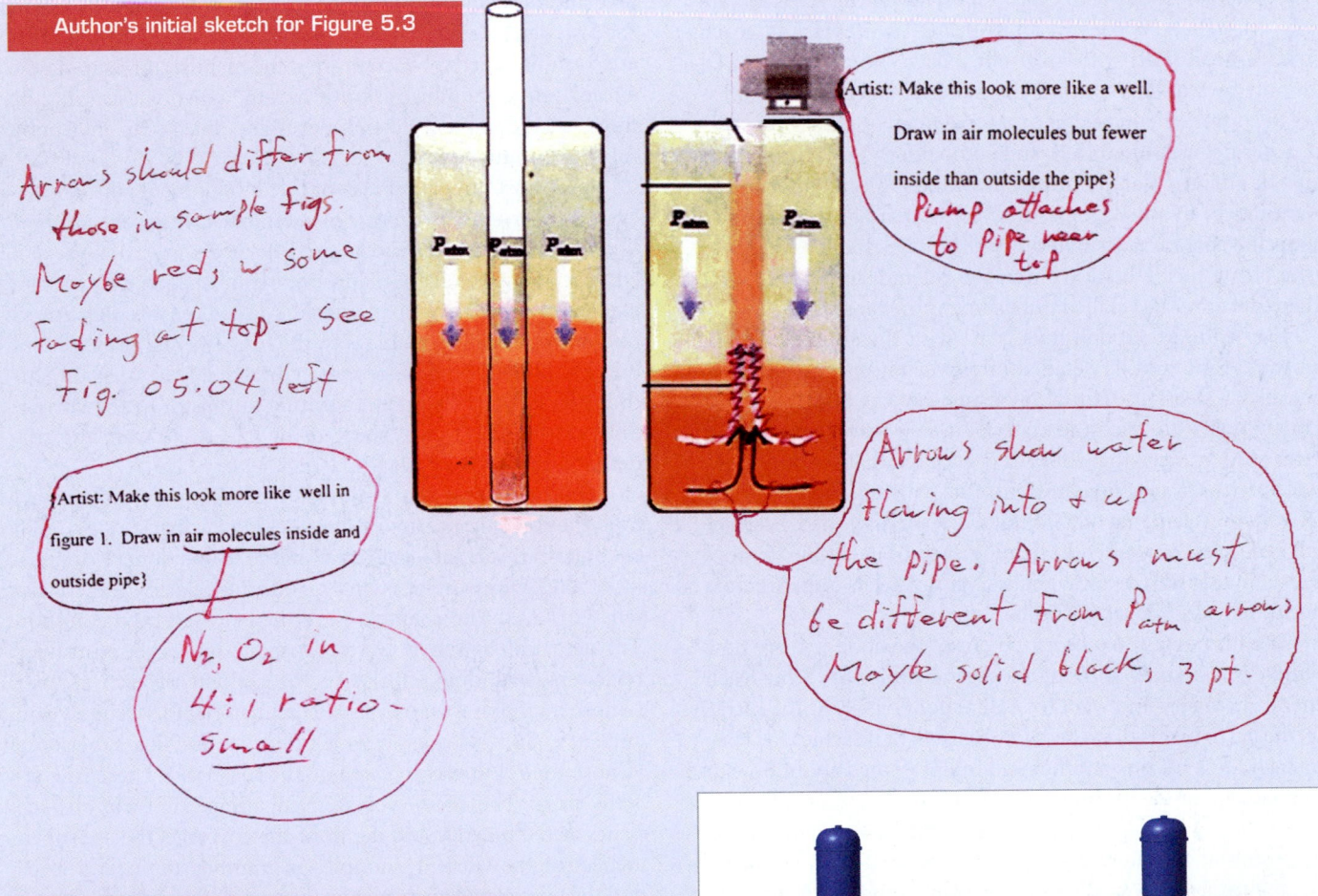

▶ **FIGURE 5.3 The Operation of Shallow Wells** When the pump is operating, the pressure within the pipe is lower than the pressure of the atmosphere on the water outside the pipe. This pressure difference causes water to flow up the pipe.

and critiqued representative figures at critical stages of their evolution.

Since the arrangement of textual and graphic elements on the page has an enormous impact on the pedagogical effectiveness of a book, each page of the text was carefully reviewed by the author and development editor. The guiding principle was that students should not have to turn pages to see important figures, nor have their concentration interrupted by uncertainty as to where to look next. No effort was spared to ensure that text, art, tables, and other elements worked together in a fully integrated way, making it easy to navigate from one page to the next. At the same time, each stage of page proof was closely scrutinized by accuracy checkers to catch any errors that might have crept in at some point in the production process.

The development of a text in this way is a lengthy, painstaking endeavor, typically taking several years and involving the work of many talented people. While other publishers claim that they provide developmental editing, few if any can match Prentice Hall's commitment of time, resources, and intellectual capital to the process. It is our conviction, however, that nothing less is sufficient to produce a book that is truly an outstanding pedagogical tool.

Author's and editor's revisions

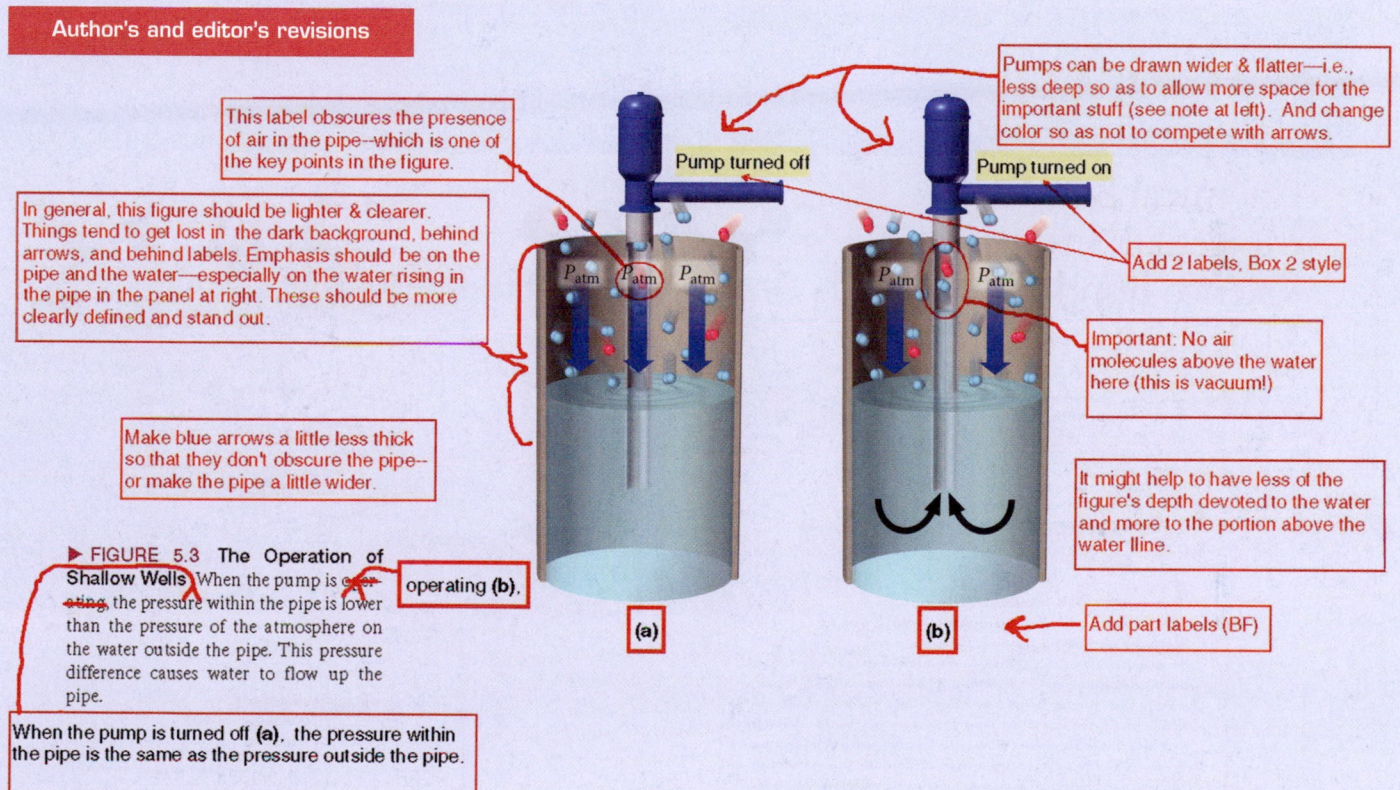

Pumps can be drawn wider & flatter—i.e., less deep so as to allow more space for the important stuff (see note at left). And change color so as not to compete with arrows.

This label obscures the presence of air in the pipe--which is one of the key points in the figure.

In general, this figure should be lighter & clearer. Things tend to get lost in the dark background, behind arrows, and behind labels. Emphasis should be on the pipe and the water—especially on the water rising in the pipe in the panel at right. These should be more clearly defined and stand out.

Add 2 labels, Box 2 style

Important: No air molecules above the water here (this is vacuum!)

Make blue arrows a little less thick so that they don't obscure the pipe-- or make the pipe a little wider.

It might help to have less of the figure's depth devoted to the water and more to the portion above the water lline.

▶ FIGURE 5.3 The Operation of Shallow Wells When the pump is ~~ope-rating~~, the pressure within the pipe is lower than the pressure of the atmosphere on the water outside the pipe. This pressure difference causes water to flow up the pipe.

operating (b).

Add part labels (BF)

When the pump is turned off (a), the pressure within the pipe is the same as the pressure outside the pipe.

The final rendering

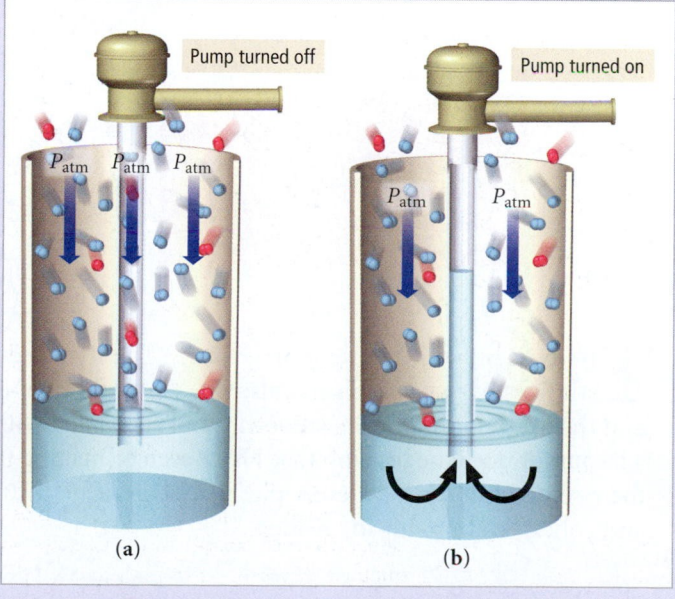

▶ FIGURE 5.3 The Operation of Shallow Wells When the pump is turned off (a), the pressure within the pipe is the same as the pressure outside the pipe. When the pump is operating (b), the pressure within the pipe is lower than the pressure of the atmosphere on the water outside the pipe. This pressure difference causes water to flow up the pipe.

A STUDENT'S GUIDE TO USING THIS TEXT

The following pages walk you through some of the main features of this text. Using the learning system in this book that was designed with you, the student, in mind will help you develop the essential knowledge and skills you need to succeed in and enjoy chemistry.

Chapter Outline
A list of the chapter's main sections provides a convenient overview of the topics to be covered.

Chapter Openers
The opening quotation, usually from a well-known scientist, is relevant to the chapter material and chosen to capture student interest.

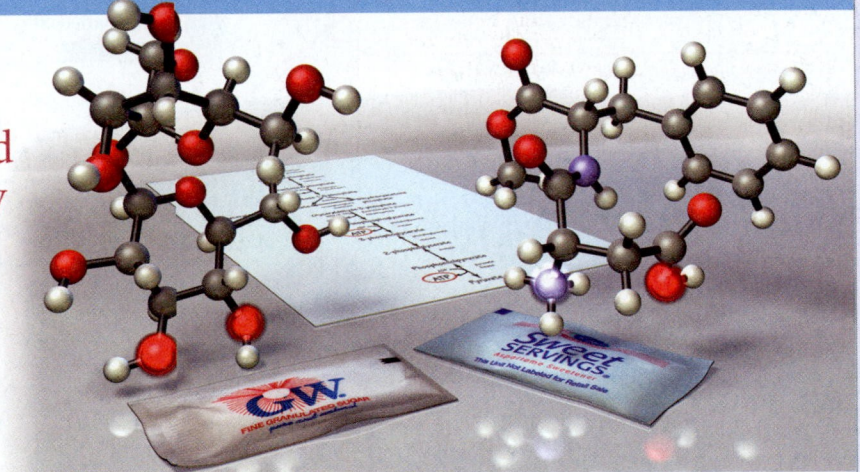

10

No theory ever solves all the puzzles with which it is confronted at a given time; nor are the solutions already achieved often perfect.
—Thomas Kuhn (1922–1996)

Chemical Bonding II: Molecular Shapes, Valence Bond Theory, and Molecular Orbital Theory

10.1 Artificial Sweeteners: Fooled by Molecular Shape
10.2 VSEPR Theory: The Five Basic Shapes
10.3 VSEPR Theory: The Effect of Lone Pairs
10.4 VSEPR Theory: Predicting Molecular Geometries
10.5 Molecular Shape and Polarity
10.6 Valence Bond Theory: Orbital Overlap as a Chemical Bond
10.7 Valence Bond Theory: Hybridization of Atomic Orbitals
10.8 Molecular Orbital Theory: Electron Delocalization

I N CHAPTER 9, WE LEARNED a simple model for chemical bonding called Lewis theory. We saw how this model helps us explain and predict the combinations of atoms that form stable molecules. When we combine Lewis theory with the idea that valence electron groups repel one another—the basis of an approach known as VSEPR theory—we can predict the general shape of a molecule from its Lewis structure. We address molecular shapes and their importance in the first part of this chapter. We then move on to learn two additional bonding theories—called valence bond theory and molecular orbital theory—that are progressively more sophisticated, but at the cost of being more complex, than Lewis theory. As you work through this chapter, our second on chemical bonding, keep in mind the importance of this topic. In our universe, elements join together to form compounds, and that makes many things possible, including our own existence.

404

Similarities in the shape of sugar and aspartame give both molecules the ability to stimulate a sweet taste sensation.

10.1 Artificial Sweeteners: Fooled by Molecular Shape

Artificial sweeteners, such as aspartame (Nutrasweet), taste sweet but have few or no calories. Why? *Because taste and caloric value are independent properties of foods.* The caloric value of a food depends on the amount of energy released when the food is metabolized. For example, sucrose (table sugar) is metabolized by oxidation to carbon dioxide and water:

$$C_{12}H_{22}O_{11} + 12\,O_2 \longrightarrow 12\,CO_2 + 11\,H_2O \qquad \Delta H^\circ_{rxn} = -5644\,kJ$$

When your body metabolizes a mole of sucrose, it obtains 5644 kJ of energy. Some artificial sweeteners, such as saccharin, for example, are not metabolized at all—they just pass

Chapter Overview
The chapter overview is designed to help students see "the big picture" of the chapter content.

The opening section introduces the chapter by means of a vivid example showing the relevance of the chapter material.

"Tro leads off with an attention getter…. The style and readability are the strongest virtues of this manuscript. The author has a knack for introducing and then developing topics smoothly and succinctly without sacrificing rigor. The author does the best job that I have seen of making the presentation of the essential features of these topics in General Chemistry into an interesting and cohesive 'story' for the readers."

CURT MCLENDON, SADDLEBACK COLLEGE

ANNOTATED MOLECULAR ART

Many of the important illustrations include three parts: a macroscopic image (what you can see with your eyes); a molecular image (what the molecules are doing); and a symbolic representation (the way that chemists represent the process with symbols and equations).

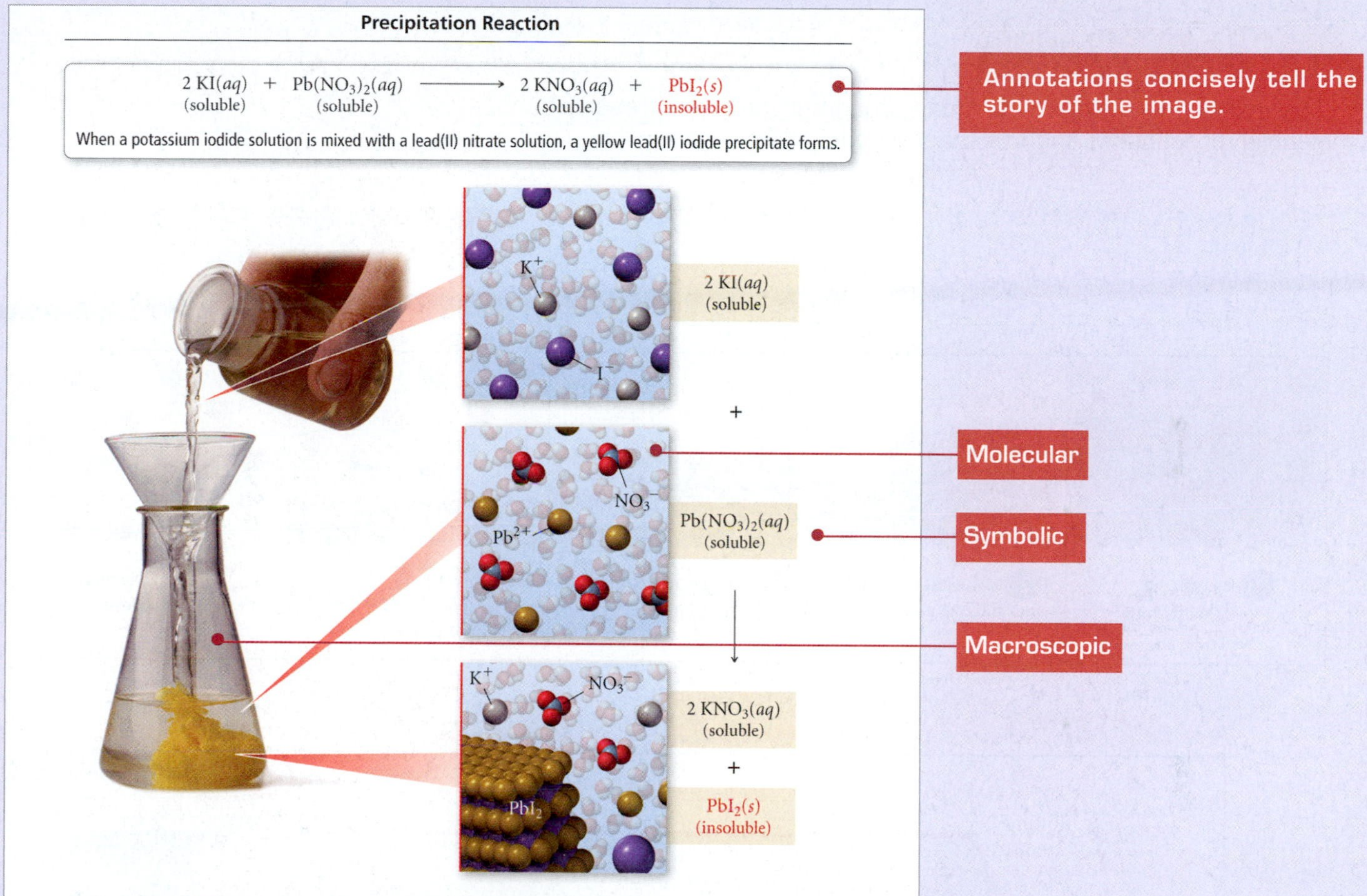

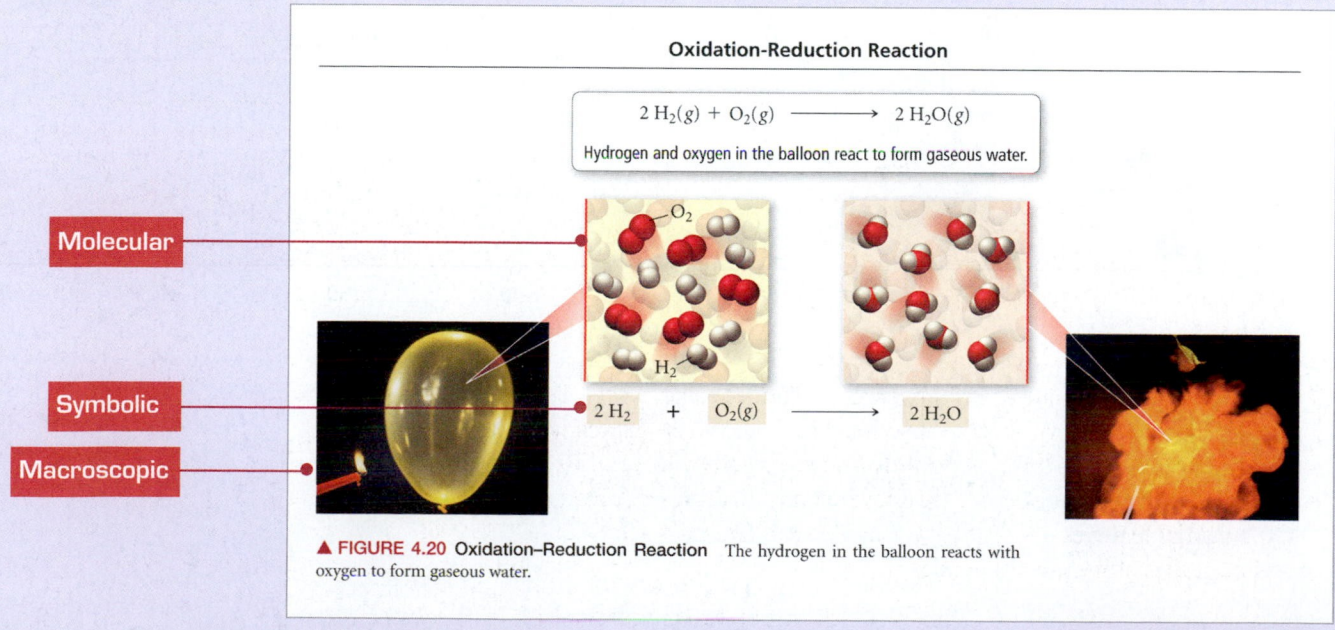

Precipitation Reaction

$$2 \text{ KI}(aq) + \text{Pb(NO}_3)_2(aq) \longrightarrow 2 \text{ KNO}_3(aq) + \text{PbI}_2(s)$$
(soluble)   (soluble)   (soluble)   (insoluble)

When a potassium iodide solution is mixed with a lead(II) nitrate solution, a yellow lead(II) iodide precipitate forms.

Annotations concisely tell the story of the image.

$2 \text{ KI}(aq)$
(soluble)

$\text{Pb(NO}_3)_2(aq)$
(soluble)

$2 \text{ KNO}_3(aq)$
(soluble)

$\text{PbI}_2(s)$
(insoluble)

Molecular

Symbolic

Macroscopic

▲ FIGURE 4.14 **Precipitation of Lead(II) Iodide**   When a potassium iodide solution is mixed with a lead(II) nitrate solution, a yellow lead(II) iodide precipitate forms.

Oxidation-Reduction Reaction

$$2 \text{ H}_2(g) + \text{O}_2(g) \longrightarrow 2 \text{ H}_2\text{O}(g)$$

Hydrogen and oxygen in the balloon react to form gaseous water.

Molecular

Symbolic

Macroscopic

$2 \text{ H}_2 + \text{O}_2(g) \longrightarrow 2 \text{ H}_2\text{O}$

▲ FIGURE 4.20 **Oxidation–Reduction Reaction**   The hydrogen in the balloon reacts with oxygen to form gaseous water.

"The visual presentations not only enhance the text but can almost stand alone as 'Teaching Models.' The use of macro and micro help students correlate their experiences/observations to the molecular level."

KYLE BERAN, UNIVERSITY OF TEXAS, PERMIAN BASIN

MULTIPART IMAGES

Multipart images make connections among graphical representations, molecular processes, and the macroscopic world.

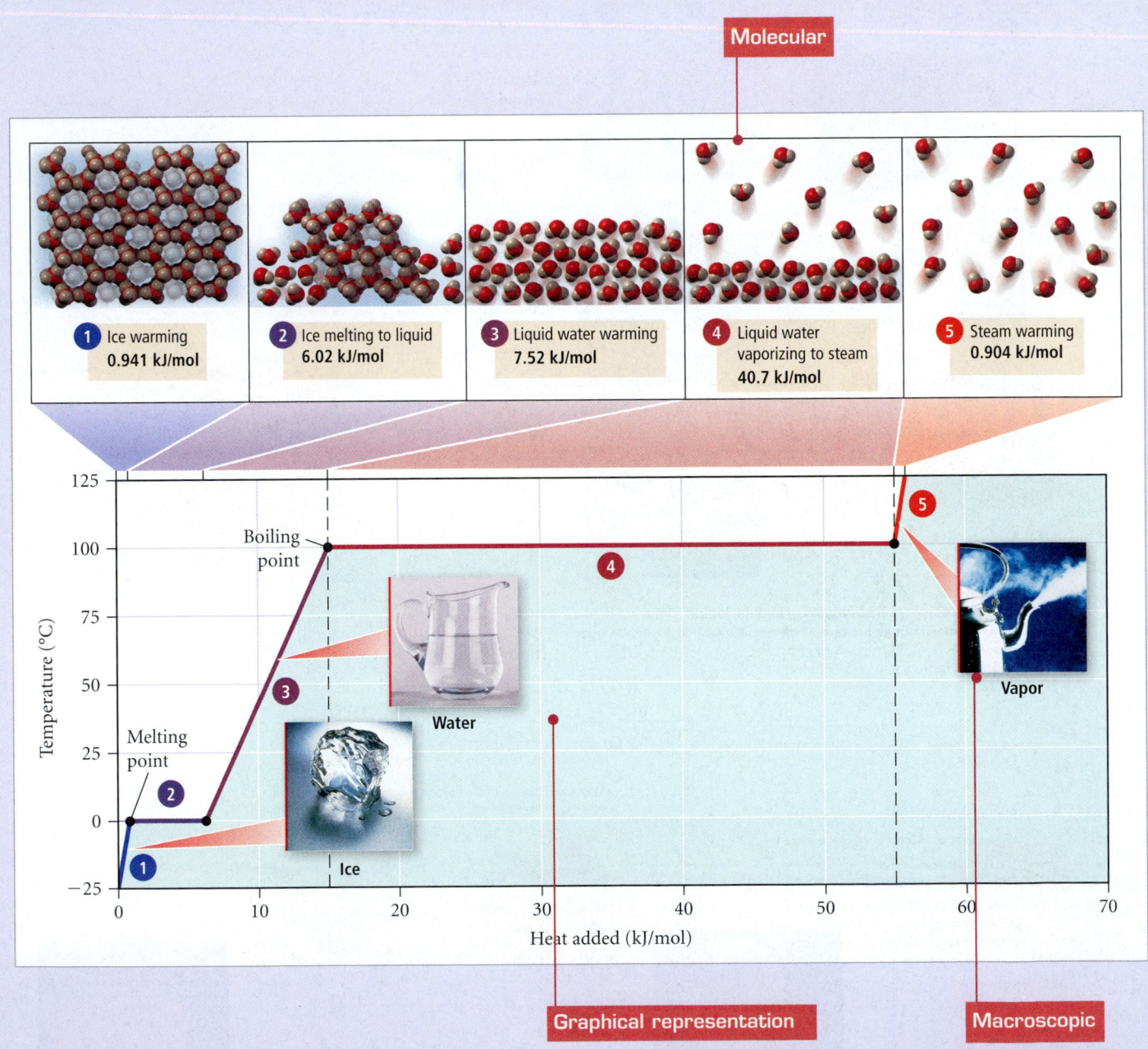

"I think students will find it helpful—but more importantly I think
it will help students understand how to solve problems rather than just
answer questions. If students take the time to thoroughly read the text
I believe they will find that Dr. Tro's problem-solving approach will
prepare them to solve a multitude of chemical problems."

DR. WILLIAM STEEL, YORK COLLEGE

PROBLEM-SOLVING STRATEGIES

A consistent approach to problem solving is used throughout the book.

TWO-COLUMN EXAMPLE

The left column explains how the problem is solved.

The right column shows the implementation of the steps explained in the left column.

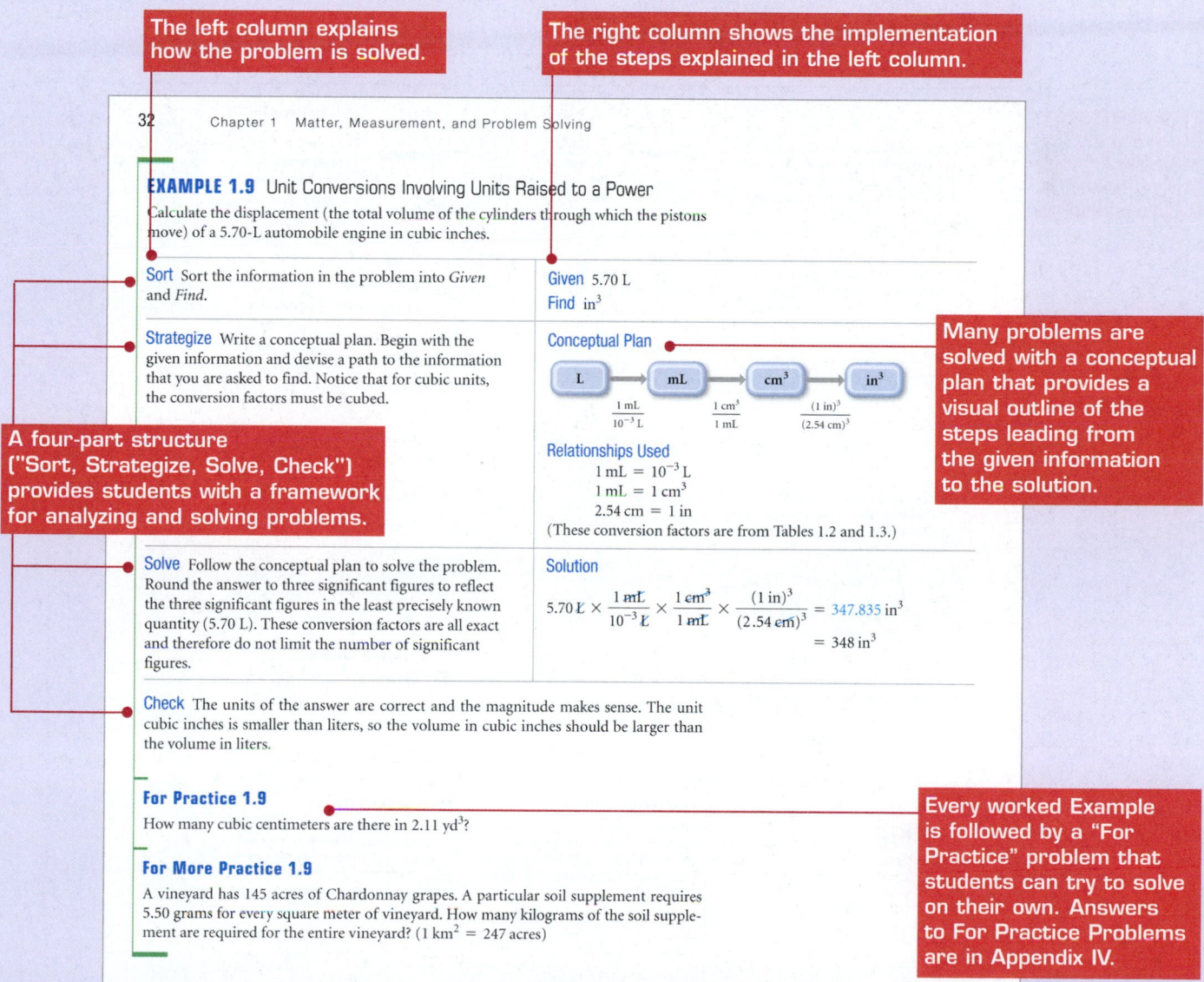

32   Chapter 1   Matter, Measurement, and Problem Solving

EXAMPLE 1.9 Unit Conversions Involving Units Raised to a Power

Calculate the displacement (the total volume of the cylinders through which the pistons move) of a 5.70-L automobile engine in cubic inches.

Sort Sort the information in the problem into *Given* and *Find*.

Given 5.70 L
Find in^3

Strategize Write a conceptual plan. Begin with the given information and devise a path to the information that you are asked to find. Notice that for cubic units, the conversion factors must be cubed.

Conceptual Plan

$$L \rightarrow mL \rightarrow cm^3 \rightarrow in^3$$

$$\frac{1\,mL}{10^{-3}\,L} \quad \frac{1\,cm^3}{1\,mL} \quad \frac{(1\,in)^3}{(2.54\,cm)^3}$$

Relationships Used
$1\,mL = 10^{-3}\,L$
$1\,mL = 1\,cm^3$
$2.54\,cm = 1\,in$
(These conversion factors are from Tables 1.2 and 1.3.)

A four-part structure ("Sort, Strategize, Solve, Check") provides students with a framework for analyzing and solving problems.

Many problems are solved with a conceptual plan that provides a visual outline of the steps leading from the given information to the solution.

Solve Follow the conceptual plan to solve the problem. Round the answer to three significant figures to reflect the three significant figures in the least precisely known quantity (5.70 L). These conversion factors are all exact and therefore do not limit the number of significant figures.

Solution

$$5.70\,\cancel{L} \times \frac{1\,\cancel{mL}}{10^{-3}\,\cancel{L}} \times \frac{1\,\cancel{cm^3}}{1\,\cancel{mL}} \times \frac{(1\,in)^3}{(2.54\,\cancel{cm})^3} = 347.835\,in^3$$
$$= 348\,in^3$$

Check The units of the answer are correct and the magnitude makes sense. The unit cubic inches is smaller than liters, so the volume in cubic inches should be larger than the volume in liters.

For Practice 1.9

How many cubic centimeters are there in 2.11 yd^3?

For More Practice 1.9

A vineyard has 145 acres of Chardonnay grapes. A particular soil supplement requires 5.50 grams for every square meter of vineyard. How many kilograms of the soil supplement are required for the entire vineyard? (1 km^2 = 247 acres)

Every worked Example is followed by a "For Practice" problem that students can try to solve on their own. Answers to For Practice Problems are in Appendix IV.

"I like the step-by-step discussion of how to address the problem.
Students seem to like to have the steps outlined in a straightforward,
simple, step-by-step method, even for something as intuitive as this."

JOHN VINCENT, UNIVERSITY OF ALABAMA AT TUSCALOOSA

THREE-COLUMN EXAMPLES

Problem-Solving Procedure boxes for important categories of problems allow students to see how the same reasoning is applied to different problems.

Two worked examples, side by side make it easy to see how differences are handled.

The general procedure is shown in the left column.

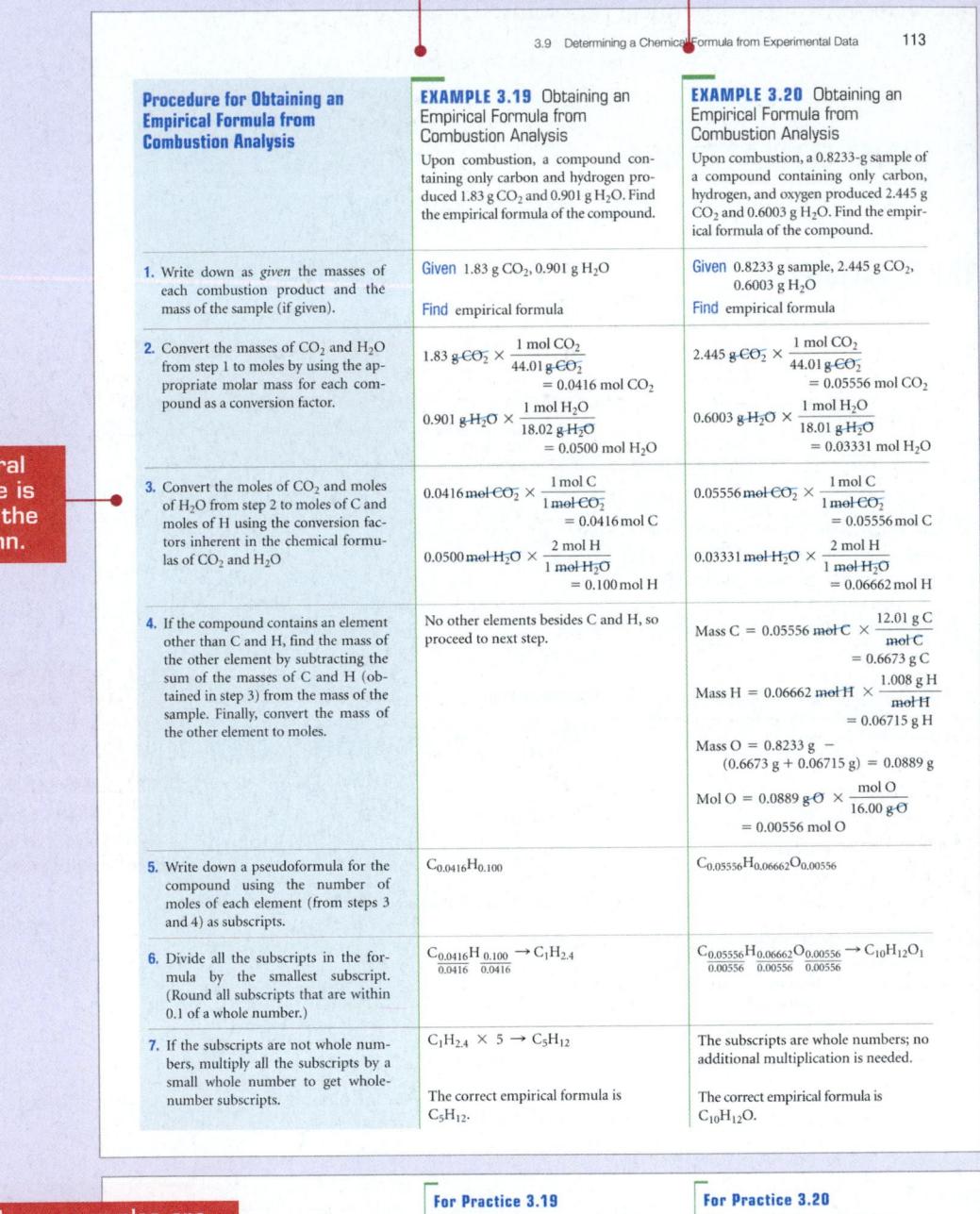

3.9   Determining a Chemical Formula from Experimental Data   113

Procedure for Obtaining an Empirical Formula from Combustion Analysis	EXAMPLE 3.19 Obtaining an Empirical Formula from Combustion Analysis	EXAMPLE 3.20 Obtaining an Empirical Formula from Combustion Analysis
	Upon combustion, a compound containing only carbon and hydrogen produced 1.83 g CO_2 and 0.901 g H_2O. Find the empirical formula of the compound.	Upon combustion, a 0.8233-g sample of a compound containing only carbon, hydrogen, and oxygen produced 2.445 g CO_2 and 0.6003 g H_2O. Find the empirical formula of the compound.
1. Write down as *given* the masses of each combustion product and the mass of the sample (if given).	Given 1.83 g CO_2, 0.901 g H_2O Find empirical formula	Given 0.8233 g sample, 2.445 g CO_2, 0.6003 g H_2O Find empirical formula
2. Convert the masses of CO_2 and H_2O from step 1 to moles by using the appropriate molar mass for each compound as a conversion factor.	$1.83 \text{ g } CO_2 \times \dfrac{1 \text{ mol } CO_2}{44.01 \text{ g } CO_2}$ $= 0.0416 \text{ mol } CO_2$ $0.901 \text{ g } H_2O \times \dfrac{1 \text{ mol } H_2O}{18.02 \text{ g } H_2O}$ $= 0.0500 \text{ mol } H_2O$	$2.445 \text{ g } CO_2 \times \dfrac{1 \text{ mol } CO_2}{44.01 \text{ g } CO_2}$ $= 0.05556 \text{ mol } CO_2$ $0.6003 \text{ g } H_2O \times \dfrac{1 \text{ mol } H_2O}{18.01 \text{ g } H_2O}$ $= 0.03331 \text{ mol } H_2O$
3. Convert the moles of CO_2 and moles of H_2O from step 2 to moles of C and moles of H using the conversion factors inherent in the chemical formulas of CO_2 and H_2O	$0.0416 \text{ mol } CO_2 \times \dfrac{1 \text{ mol C}}{1 \text{ mol } CO_2}$ $= 0.0416 \text{ mol C}$ $0.0500 \text{ mol } H_2O \times \dfrac{2 \text{ mol H}}{1 \text{ mol } H_2O}$ $= 0.100 \text{ mol H}$	$0.05556 \text{ mol } CO_2 \times \dfrac{1 \text{ mol C}}{1 \text{ mol } CO_2}$ $= 0.05556 \text{ mol C}$ $0.03331 \text{ mol } H_2O \times \dfrac{2 \text{ mol H}}{1 \text{ mol } H_2O}$ $= 0.06662 \text{ mol H}$
4. If the compound contains an element other than C and H, find the mass of the other element by subtracting the sum of the masses of C and H (obtained in step 3) from the mass of the sample. Finally, convert the mass of the other element to moles.	No other elements besides C and H, so proceed to next step.	Mass C $= 0.05556 \text{ mol C} \times \dfrac{12.01 \text{ g C}}{\text{mol C}}$ $= 0.6673 \text{ g C}$ Mass H $= 0.06662 \text{ mol H} \times \dfrac{1.008 \text{ g H}}{\text{mol H}}$ $= 0.06715 \text{ g H}$ Mass O $= 0.8233 \text{ g } -$ $(0.6673 \text{ g} + 0.06715 \text{ g}) = 0.0889 \text{ g}$ Mol O $= 0.0889 \text{ g O} \times \dfrac{\text{mol O}}{16.00 \text{ g O}}$ $= 0.00556 \text{ mol O}$
5. Write down a pseudoformula for the compound using the number of moles of each element (from steps 3 and 4) as subscripts.	$C_{0.0416}H_{0.100}$	$C_{0.05556}H_{0.06662}O_{0.00556}$
6. Divide all the subscripts in the formula by the smallest subscript. (Round all subscripts that are within 0.1 of a whole number.)	$C_{\frac{0.0416}{0.0416}}H_{\frac{0.100}{0.0416}} \rightarrow C_1H_{2.4}$	$C_{\frac{0.05556}{0.00556}}H_{\frac{0.06662}{0.00556}}O_{\frac{0.00556}{0.00556}} \rightarrow C_{10}H_{12}O_1$
7. If the subscripts are not whole numbers, multiply all the subscripts by a small whole number to get whole-number subscripts.	$C_1H_{2.4} \times 5 \rightarrow C_5H_{12}$ The correct empirical formula is C_5H_{12}.	The subscripts are whole numbers; no additional multiplication is needed. The correct empirical formula is $C_{10}H_{12}O$.

Three-column examples are followed by two For Practice problems for students to try on their own.

For Practice 3.19

Upon combustion, a compound containing only carbon and hydrogen produced 1.60 g CO_2 and 0.819 g H_2O. Find the empirical formula of the compound.

For Practice 3.20

Upon combustion, a 0.8009-g sample of a compound containing only carbon, hydrogen, and oxygen produced 1.6004 g CO_2 and 0.6551 g H_2O. Find the empirical formula of the compound

"I personally like the detailed explanation of pedagogically effective strategies that I will not need to spend time on in lecture; this is an important feature in textbook selection. The use of two examples and providing answers for practice problems is an excellent aspect of this textbook."

MARGARET ASIRVATHAM, UNIVERSITY OF COLORADO, BOULDER

"**I** am excited to see a textbook that encourages students to stop and think about WHY the concepts they're learning are relevant."

AIMEE MILLER, MILLERSVILLE UNIVERSITY

CONCEPTUAL CONNECTION QUESTIONS

Conceptual Connection questions pose problems that students can solve by reasoning, with little or no calculation, to test their grasp of key ideas.

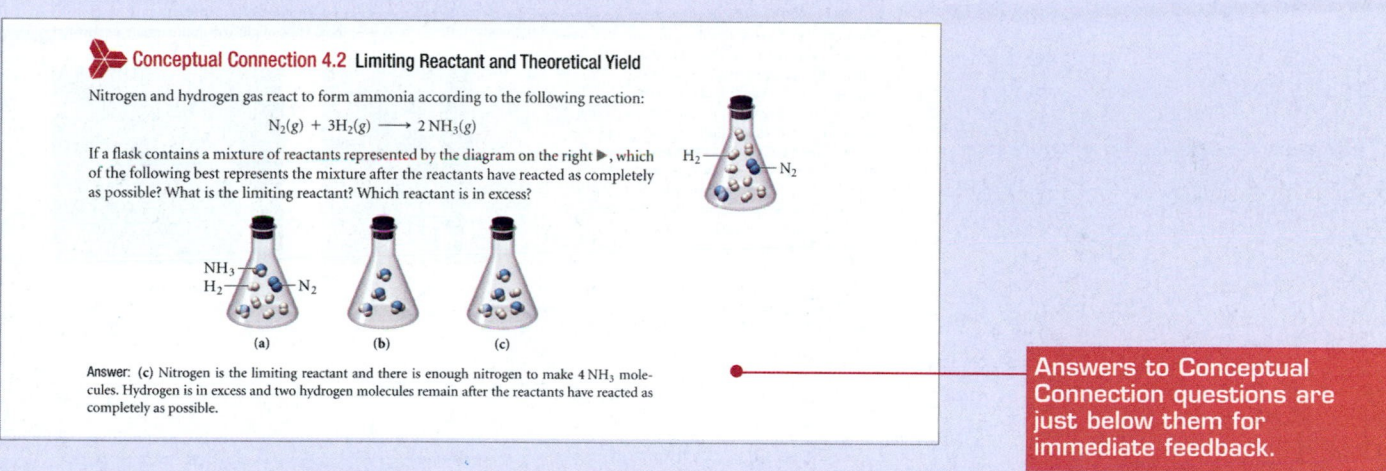

> **Conceptual Connection 4.2** Limiting Reactant and Theoretical Yield
>
> Nitrogen and hydrogen gas react to form ammonia according to the following reaction:
>
> $$N_2(g) + 3H_2(g) \longrightarrow 2NH_3(g)$$
>
> If a flask contains a mixture of reactants represented by the diagram on the right ▶, which of the following best represents the mixture after the reactants have reacted as completely as possible? What is the limiting reactant? Which reactant is in excess?
>
> (a)   (b)   (c)
>
> **Answer:** (c) Nitrogen is the limiting reactant and there is enough nitrogen to make $4\,NH_3$ molecules. Hydrogen is in excess and two hydrogen molecules remain after the reactants have reacted as completely as possible.

Answers to Conceptual Connection questions are just below them for immediate feedback.

INTERIM SUMMARIES

Interim summaries highlight the key points of important discussions.

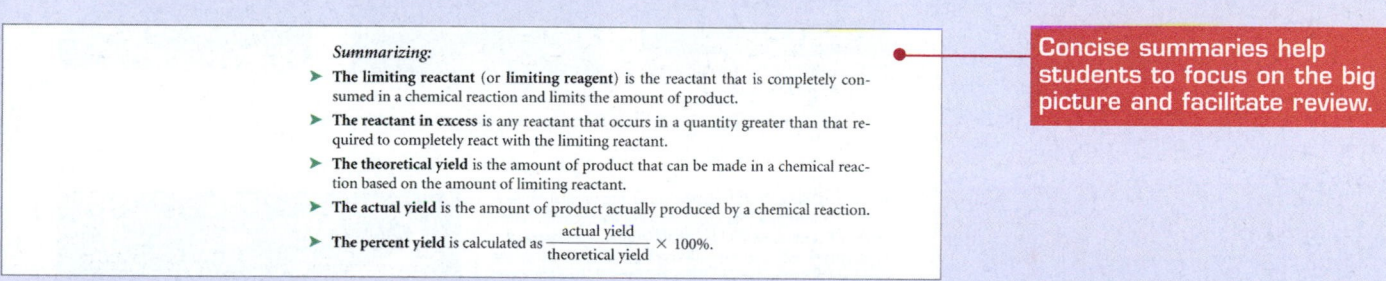

> *Summarizing:*
> ➤ **The limiting reactant** (or **limiting reagent**) is the reactant that is completely consumed in a chemical reaction and limits the amount of product.
> ➤ **The reactant in excess** is any reactant that occurs in a quantity greater than that required to completely react with the limiting reactant.
> ➤ **The theoretical yield** is the amount of product that can be made in a chemical reaction based on the amount of limiting reactant.
> ➤ **The actual yield** is the amount of product actually produced by a chemical reaction.
> ➤ **The percent yield** is calculated as $\dfrac{\text{actual yield}}{\text{theoretical yield}} \times 100\%$.

Concise summaries help students to focus on the big picture and facilitate review.

"**T**his feature does help reinforce the conceptual approach because it forces students to think about conceptual ideas instead of just plugging numbers into an equation. It also helps the instructor develop questions that are based on concepts."

SUSAN HENDERSON, QUINNIPIAC UNIERSITY

STUDENT INTEREST

Interesting descriptions of how chemistry appears in the modern world show students the relevance of chemistry.

Embedded questions connect the material in the box to the content of the chapter.

Chemistry and Medicine boxes feature applications relevant to biomedical and health-related topics.

Chemistry and Medicine
Bone Density

Osteoporosis—which means *porous bone*—is a condition in which bone density becomes too low. The healthy bones of a young adult have a density of about 1.0 g/cm³. Patients suffering from osteoporosis, however, can have bone densities as low as 0.22 g/cm³. These low densities mean the bones have deteriorated and weakened, resulting in increased susceptibility to fractures, especially hip fractures. Patients suffering from osteoporosis can also experience height loss and disfiguration such as dowager's hump, a condition in which the patient becomes hunched over because of compression of the vertebrae. Osteoporosis is most common in postmenopausal women, but it can also occur in people (including men) who have certain diseases, such as insulin-dependent diabetes, or who take certain medications, such as prednisone. Osteoporosis is usually tested by taking hip X-rays. Low-density bones absorb less of the X-rays than do high-density bones, producing characteristic differences in the X-ray image that allow diagnosis. Treatments for osteoporosis include additional calcium and vitamin D, drugs that prevent bone weakening, exercise and strength training, and, in extreme cases, hip-replacement surgery.

Question
Suppose you find a large animal bone in the woods, too large to fit in a beaker or flask. How might you approximate its density?

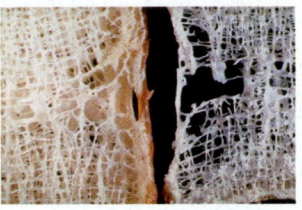

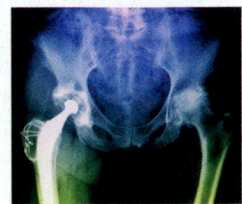

▲ Magnified views of the bone matrix in a normal vertebra (left) and one weakened by osteoporosis (right).

▲ Severe osteoporosis can necessitate surgery to implant an artificial hip joint, seen at left in this X-ray image.

Chemistry and the Environment boxes relate chapter topics to current environmental and societal issues.

Chemistry and the Environment
Renewable Energy

Parabolic troughs

Solar power tower

Dish/engine

▲ The sun's energy, concentrated by reflective surfaces in various arrangements, can produce enough heat to generate electricity.

Because of their limited supply and environmental impacts, fossil fuels will not be our major source of energy in the future. What will replace them? Although the answer is not clear, several alternative energy technologies are beginning to emerge. Unlike fossil fuels, these technologies are renewable, which means that they can be used indefinitely.

Our planet's greatest source of renewable energy is the sun. If we could capture and harness just a small fraction of the total sunlight falling on Earth, our energy needs would be met several times over. The main problem with solar energy, however, is its diffuseness—the sun's energy falls over an enormous area. How do we concentrate and store it? In California, some of the state's electricity is generated by parabolic troughs, solar power towers, and dish/engines (shown above).

These devices use reflective surfaces to focus the sun's energy to produce enough heat to generate electricity. Although the direct cost of generating electricity this way is higher than using fossil fuels, the benefits to the environment are obvious.

▲ A General Motors SUV prototype called the Sequel is powered by a fuel cell that runs on hydrogen gas and produces only water as exhaust.

range of 250 miles on one tank of fuel. In 2005, General Motors

"Real examples are used as a prelude to introduce new concepts and explain their importance. The constantly demonstrated connection of the chemical science with the real world and everyday activities is a very strong aspect of this text."

SERGIY KRYATOV, TUFTS UNIVERSITY

Chemistry in Your Day boxes demonstrate the importance of chemistry In everyday situations.

28   Chapter 1   Matter, Measurement, and Problem Solving

- The results of student B are precise (close to one another in value) but inaccurate. The inaccuracy is the result of **systematic error**, error that tends toward being either too high or too low. Systematic error does not average out with repeated trials. For example, if a balance is not properly calibrated, it may systematically read too high or too low.
- The results of student C display little systematic error or random error—they are both accurate and precise.

Chemistry in Your Day
Integrity in Data Gathering

Most scientists spend many hours collecting data in the laboratory. Often, the data do not turn out exactly as the scientist had expected (or hoped). A scientist may then be tempted to "fudge" his or her results. For example, suppose you are expecting a particular set of measurements to follow a certain pattern. After working hard over several days or weeks to make the measurements, you notice that a few of them do not quite fit the pattern that you anticipated. You might find yourself wishing that you could simply change or omit the "faulty" measurements to better fit your expectations. Altering data in this way is considered highly unethical in the scientific community and, when discovered, is usually punished severely.

In 2004, Dr. Hwang Woo Suk, a stem cell researcher at the Seoul National University in Korea, published a research paper in *Science* (a highly respected research journal) claiming that he and his colleagues had cloned human embryonic stem cells. As part of his evidence, he showed photographs of the cells. The paper was hailed as an incredible breakthrough, and Dr. Hwang traveled the world lecturing on his work. *Time* magazine even named him among their "people that matter" for 2004. Several months later, however, one of his co-workers revealed that the photographs were fraudulent. According to the co-worker, the photographs came from a computer data bank of stem cell photographs, not from a cloning experiment. A university panel investigated the results and confirmed that the photographs and other data had indeed been faked. Dr. Hwang was forced to resign his prestigious post at the university.

Although not common, incidents like this do occur from time to time. They are damaging to a community that is largely built on trust. Research papers are reviewed by peers (other researchers in similar fields), but usually reviewers are judging whether the data support the conclusion—they assume that the experimental measurements are authentic. The pressure to succeed sometimes leads researchers to betray that trust. However, over time, the tendency of scientists to reproduce and build upon one another's work results in the discovery of the fraudulent data. When that happens, the researchers at fault are usually banished from the community and their careers are ruined.

1.8 Solving Chemical Problems

Learning to solve problems is one of the most important skills you will acquire in this course. No one succeeds in chemistry—or in life, really—without the ability to solve problems. Although no simple formula applies to every problem, you can learn problem-solving strategies and begin to develop some chemical intuition. Many of the problems you will solve in this course can be thought of as *unit conversion problems*, where you are given one or more quantities and asked to convert them into different units. Other problems require the use of *specific equations* to get to the information you are trying to find. In the sections that follow, you will find strategies to help you solve both of these types of problems. Of course, many problems contain both conversions and equations, requiring the combination of these strategies.

Converting from One Unit to Another

In Section 1.6, we learned the SI unit system, the prefix multipliers, and a few other units. Knowing how to work with and manipulate these units in calculations is central to solving chemical problems. In calculations, units help to determine correctness. Using units as a guide to solving problems is often called **dimensional analysis**. Units should always be included in calculations; they are multiplied, divided, and canceled like any other algebraic quantity.

Consider converting 12.5 inches (in) to centimeters (cm). We know from Table 1.3 that 1 in = 2.54 cm (exact), so we can use this quantity in the calculation as follows:

$$12.5 \text{ in} \times \frac{2.54 \text{ cm}}{1 \text{ in}} = 31.8 \text{ cm}$$

"These are wonderful in helping students realize that chemistry is all around them. It is not just a subject to study in school. It helps to answer the question of why they are required to take this course"

KATHLEEN THRUSH SHAGINAW, VILLANOVA UNIVERSITY

COMPREHENSIVE END-OF-CHAPTER REVIEW SECTION

The end-of-chapter review section is designed to help students study the concepts and skills presented in the chapter in a systematic way that is ideal for test preparation.

---

228   Chapter 5   Gases

Chapter in Review

Key Terms

Section 5.1
pressure (185)

Section 5.2
millimeter of mercury (mmHg) (188)
barometer (188)
torr (188)
atmosphere (atm) (189)
pascal (Pa) (189)
manometer (190)

Section 5.3
Boyle's law (192)
Charles's law (195)
Avogadro's law (197)

Section 5.4
ideal gas law (198)
ideal gas constant (198)

Section 5.5
standard temperature and pressure (STP) (201)

molar volume (201)

Section 5.6
partial pressure (P_n) (204)
Dalton's law of partial pressures (204)
mole fraction (χ_a) (205)
hypoxia (206)
oxygen toxicity (207)
nitrogen narcosis (207)
vapor pressure (209)

Section 5.8
kinetic molecular theory (214)

Section 5.9
mean free path (220)
diffusion (220)
effusion (220)
Graham's law of effusion (220)

Section 5.10
van der Waals equation (224)

> The Key Terms section lists all of the chapter's bold-face terms, organized by section in order of their appearance, with page references. Definitions are found in the Glossary.

Key Concepts

Pressure (5.1, 5.2)

Gas pressure is the force per unit area that results from gas particles colliding with the surfaces around them. Pressure is measured in a number of units including mmHg, torr, Pa, psi, in Hg, and atm.

The Simple Gas Laws (5.3)

The simple gas laws express relationships between pairs of variables when the other variables are held constant. Boyle's law states that the volume of a gas is inversely proportional to its pressure. Charles's law states that the volume of a gas is directly proportional to its temperature. Avogadro's law states the volume of a gas is directly proportional to the amount (in moles).

The Ideal Gas Law and Its Applications (5.4, 5.5)

The ideal gas law, $PV = nRT$, gives the relationship among all four gas variables and contains the simple gas laws within it. The ideal gas law can be used to find one of the four variables given the other three. It can also be used to compute the molar volume of an ideal gas, which is 22.4 L at STP, and used to calculate the density and molar mass of a gas.

Mixtures of Gases and Partial Pressures (5.6)

In a mixture of gases, each gas acts independently of the others so that any overall property of the mixture is simply the sum of the properties of the individual components. The pressure of any individual component is called its partial pressure.

Gas Stoichiometry (5.7)

In reactions involving gaseous reactants and products, quantities are often reported in volumes at specified pressures and temperatures. These quantities can be converted to amounts (in moles) using the ideal gas law. Then the stoichiometric coefficients from the balanced equation can be used to determine the stoichiometric amounts of other reactants or products. The general form for these types of calculations is often as follows: volume A → amount A (in moles) → amount B (in moles) → quantity of B (in desired units). In cases where the reaction is carried out at STP, the molar volume at STP (22.4 L = 1 mol) can be used to convert between volume in liters and amount in moles.

Kinetic Molecular Theory and Its Applications (5.8, 5.9)

Kinetic molecular t[...]
has three main assu[...]
(2) the average kin[...]
temperature in kel[...]
another is complete[...]
gas laws all follow f[...]

The theory can [...]
mean square veloci[...]
tional to the molar [...]
ture—smaller gas p[...]
larger ones. The kin[...]
mean free path of a[...]
sions) and relative r[...]

Real Gases (5.[...]

Real gases differ fro[...]
assumptions of kir[...]
break down at high[...]
ed because the par[...]
the space between [...]
temperatures wher[...]
attraction between[...]
causes partially inel[...]
to predict gas prope[...]

The Atmosphe[...]

Our atmosphere is [...]
(21%). Common g[...]
dioxide, carbon mo[...]
tants affect exposee[...]
cardiovascular syste[...]
fluorocarbons tha[...]
ozone, which result[...]
crease in the risk of[...]

> The Key Concepts section summarizes the chapter's most important ideas.

Chapter in Review   229

Key Equations and Relationships

Relationship between Pressure (P), Force (F), and Area (A) (5.2)
$$P = \frac{F}{A}$$

Boyle's Law: Relationship between Pressure (P) and Volume (V) (5.3)
$$V \propto \frac{1}{P}$$
$$P_1V_1 = P_2V_2$$

Charles's Law: Relationship between Volume (V) and Temperature (T) (5.3)
$$V \propto T \quad \text{(in K)}$$
$$\frac{V_1}{T_1} = \frac{V_2}{T_2}$$

Avogadro's Law: Relationship between Volume (V) and Amount in Moles (n) (5.3)
$$V \propto n$$
$$\frac{V_1}{n_1} = \frac{V_2}{n_2}$$

Ideal Gas Law: Relationship between Volume (V), Pressure (P), Temperature (T), and Amount (n) (5.4)
$$PV = nRT$$

Dalton's Law: Relationship between Partial Pressures (P_n) in Mixture of Gases and Total Pressure (P_{total}) (5.6)
$$P_{total} = P_a + P_b + P_c + \cdots$$
$$P_a = \frac{n_aRT}{V} \quad P_b = \frac{n_bRT}{V} \quad P_c = \frac{n_cRT}{V}$$

Mole Fraction (χ_a) (5.6)
$$\chi_a = \frac{n_a}{n_{total}}$$
$$P_a = \chi_aP_{total}$$

Average Kinetic Energy (KE_{avg}) (5.8)
$$KE_{avg} = \frac{3}{2}RT$$

Relationship between Root Mean Square Velocity (u_{rms}) and Temperature (T) (5.8)
$$u_{rms} = \sqrt{\frac{3RT}{\mathcal{M}}}$$

Relationship of Effusion Rates of Two Different Gases (5.9)
$$\frac{rate\,A}{rate\,B} = \sqrt{\frac{\mathcal{M}_B}{\mathcal{M}_A}}$$

Van der Waals Equation: The Effects of Volume and Intermolecular Forces on Nonideal Gas Behavior (5.10)
$$[P + a(n/V)^2] \times (V - nb) = nRT$$

> The Key Equations and Relationships section lists each of the key equations and important quantitative relationships from the chapter.

Key Skills

Converting between Pressure Units (5.2)
• Example 5.1 • For Practice 5.1 • For More Practice 5.1 • Exercises 29–32

Relating Volume and Pressure: Boyle's Law (5.3)
• Example 5.2 • For Practice 5.2 • Exercises 35, 36

Relating Volume and Temperature: Charles's Law (5.3)
• Example 5.3 • For Practice 5.3 • Exercises 37, 38

Relating Volume and Moles: Avogadro's Law (5.3)
• Example 5.4 • For Practice 5.4 • Exercises 39, 40

Determining P, V, n, or T using the Ideal Gas Law (5.4)
• Examples 5.5, 5.6 • For Practice 5.5, 5.6 • For More Practice 5.6 • Exercises 41–48, 51, 52

Relating the Density of a Gas to Its Molar Mass (5.5)
• Example 5.7 • For Practice 5.7 • For More Practice 5.7 • Exercises 55, 56

Calculating the Molar Mass of a Gas with the Ideal Gas Law (5.5)
• Example 5.8 • For Practice 5.8 • Exercises 57–60

Calculating Total Pressure, Partial Pressures, and Mole Fractions of Gases in a Mixture (5.6)
• Examples 5.9, 5.10, 5.11 • For Practice 5.9, 5.10, 5.11 • Exercises 61, 62, 65, 67, 68, 70

Relating the Amounts of Reactants and Products in Gaseous Reactions: Stoichiometry (5.7)
• Examples 5.12, 5.13 • For Practice 5.12, 5.13 • For More Practice 5.12 • Exercises 71–77

Calculating the Root Mean Square Velocity of a Gas (5.8)
• Example 5.14 • For Practice 5.14 • Exercises 81, 82

Calculating the Effusion Rate or the Ratio of Effusion Rates of Two Gases (5.9)
• Example 5.15 • For Practice 5.15 • Exercises 83–86

> The Key Skills section lists the major types of problems that students should be able to solve, together with the chapter examples that demonstrate the needed techniques and the For Practice problems and end-of-chapter exercises that students can work to practice those skills.

END-OF-CHAPTER EXERCISES

End-of-Chapter Exercises provide a full range of assessment opportunities.

---

230   Chapter 5   Gases

Exercises

Review Questions

1. What is pressure? What causes pressure?
2. Explain how a shallow well works. What forces the water to the surface?
3. How deep can a shallow well be and still function? Why does this limit exist?
4. What are some common units of pressure? List these in order of smallest to largest unit.
5. What is a manometer? How does a manometer measure the pressure of a sample of gas?

16. Why do deep-sea divers breathe a mixture of helium and oxygen?
17. When a gas is collected over water, is the gas pure? Why or why not? How can the partial pressure of the desired gas be determined?
18. If a reaction occurs in the gas phase at STP, the mass of a product can be determined from the volumes of reactants. Explain.
19. What are the basic postulates of kinetic molecular theory? How does the concept of pressure follow from kinetic molecular theory?

> The Review Questions are designed for the student to use in reviewing the chapter content.

Problems by Topic

Converting between Pressure Units

29. The pressure in Denver, Colorado (elevation 5280 ft), averages about 24.9 in Hg. Convert this pressure to
   a. atm   b. mmHg   c. psi   d. Pa

30. The pressure on top of Mt. Everest averages about 235 mmHg. Convert this pressure to
   a. torr   b. psi   c. in Hg   d. atm

31. The North American record for highest recorded barometric pressure is 31.85 in Hg, set in 1989 in Northway, Alaska. Convert this pressure to
   a. mmHg   b. atm
   c. torr   d. kPa (kilopascals)

32. The world record for lowest pressure (at sea level) was 652.5 mmHg recorded inside Typhoon Tip on October 12, 1979, in the Western Pacific Ocean. Convert this pressure to
   a. torr   b. atm   c. in Hg   d. psi

33. Given a barometric pressure of 755.3 mmHg, calculate the pressure of each of the following gas samples as indicated by the manometer.

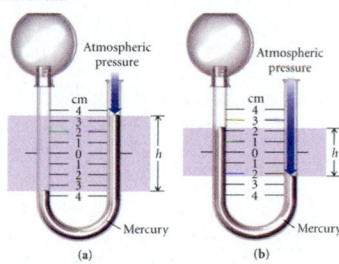

> The Problems by Topic are paired, with answers to the odd-numbered questions appearing in the appendix.

> Cumulative Problems combine material from different parts of the chapter, and often from previous chapters as well, allowing students to test their ability to integrate the course material.

> Challenge Problems are designed to challenge the best students.

Cumulative Problems

93. Modern pennies are composed of zinc coated with copper. A student determines the mass of a penny to be 2.482 g and then makes several scratches in the copper coating (to expose the underlying zinc). The student puts the scratched penny in hydrochloric acid, where the following reaction occurs between the zinc and the HCl (the copper remains undissolved):

$$Zn(s) + 2HCl(aq) \longrightarrow H_2(g) + ZnCl_2(aq)$$

The student collects the hydrogen produced over water at 25 °C. The collected gas occupies a volume of 0.899 L at a total pressure of 791 mmHg. Calculate the percent zinc in the penny. (Assume that all the Zn in the penny dissolves.)

94. A 2.85-g sample of an unknown chlorofluorocarbon is decomposed and produces 564 mL of chlorine gas at a pressure of 752 mmHg

and a temperature of 298 K. What is the percent chlorine (by mass) in the unknown chlorofluorocarbon?

95. The mass of an evacuated 255-mL flask is 143.187 g. The mass of the flask filled with 267 torr of an unknown gas at 25 °C is 143.289 g. Calculate the molar mass of the unknown gas.

96. A 118-mL flask is evacuated and found to have a mass of 97.129 g. When the flask is filled with 768 torr of helium gas at 35 °C, it is found to have a mass of 97.171 g. Was the helium gas pure?

97. A gaseous hydrogen and carbon containing compound is decomposed and found to contain 82.66% carbon and 17.34% hydrogen by mass. The mass of 158 mL of the gas, measured at 556 mmHg and 25 °C, was found to be 0.275 g. What is the molecular formula of the compound?

Challenge Problems

124. The world burns approximately 9.0×10^{12} kg of fossil fuel per year. Use the combustion of octane as the representative reaction and determine the mass of carbon dioxide (the most significant greenhouse gas) formed per year by this combustion. The current concentration of carbon dioxide in the atmosphere is approximately 387 ppm (by volume). By what percentage does the concentration increase in one year due to fossil fuel combustion? Approximate the average properties of the entire atmosphere by assuming that the atmosphere extends from sea level to 15 km and that it has an average pressure of 381 torr and average temperature of 275 K. Assume Earth is a perfect sphere with a radius of 6371 km.

125. The atmosphere slowly oxidizes hydrocarbons in a number of steps that eventually convert the hydrocarbon into carbon dioxide and water. The overall reactions of a number of such steps for methane gas is as follows:

$$CH_4(g) + 5O_2(g) + 5NO(g) \longrightarrow$$
$$CO_2(g) + H_2O(g) + 5NO_2(g) + 2OH(g)$$

Suppose that an atmospheric chemist combines 155 mL of methane at STP, 885 mL of oxygen at STP, and 55.5 mL of NO at

1654 psi and 298 K. [...] stand for several wee[...] completion (90.0% [...] are the partial pressures of each of the reactants and products in the flask at 275 K? What is the total pressure in the flask?

126. Two identical balloons are filled to the same volume, one with air and one with helium. The next day, the volume of the air-filled balloon has decreased by 5.0%. By what percent has the volume of the helium-filled balloon decreased? (Assume that the air is four-fifths nitrogen and one-fifth oxygen, and that the temperature did not change.)

127. A mixture of $CH_4(g)$ and $C_2H_6(g)$ has a total pressure of 0.53 atm. Just enough $O_2(g)$ is added to the mixture to bring about its complete combustion to $CO_2(g)$ and $H_2O(g)$. The total pressure of the two product gases is found to be 2.2 atm. Assuming constant volume and temperature, find the mole fraction of CH_4 in the mixture.

128. A sample of $C_2H_2(g)$ has a pressure of 7.8 kPa. After some time a portion of it reacts to form $C_6H_6(g)$. The total pressure of the mixture of gases is then 3.9 kPa. Assume the volume and the temperature do not change. Find the fraction of $C_2H_2(g)$ that has undergone reaction.

Conceptual Problems

129. When the driver of an automobile applies the brakes, the passengers are pushed toward the front of the car, but a helium balloon is pushed toward the back of the car. Upon forward acceleration, the passengers are pushed toward the back of the car, but the helium balloon is pushed toward the front of the car. Why?

130. Suppose that a liquid is 10 times denser than water. If you were to sip this liquid at sea level using a straw, what would be the maximum length of the straw?

131. The following reaction occurs in a closed container:

$$A(g) + 2B(g) \longrightarrow 2C(g)$$

A reaction mixture initially contains 1.5 L of A and 2.0 L of B. Assuming that the volume and temperature of the reaction mixture remain constant, what is the percent change in pressure if the reaction goes to completion?

132. One mole of nitrogen and one mole of neon are combined in a closed container at STP. How big is the container?

133. Exactly equal amounts (in moles) of gas A and gas B are combined in a 1-L container at room temperature. Gas B has a molar mass that is twice that of gas A. Which of the following is true for the mixture of gases and why?
   a. The molecules of gas B have greater kinetic energy than those of gas A.
   b. Gas B has a greater partial pressure than gas A.
   c. The molecules of gas B have a greater average velocity than those of gas A.
   d. Gas B makes a greater contribution to the average density of the mixture than gas A.

134. Which of the following gases would you expect to deviate most from ideal behavior under conditions of low temperature: F_2, Cl_2, or Br_2? Explain.

> Conceptual Problems let students test their grasp of key chapter concepts, often by means of reasoning that involves little or no math.

> "The questions are of excellent quality, and the quantity is sufficient. I like the separation of question types; having initial review questions lets the student organize the material in a systematic fashion while mentally getting an understanding of the material— not just crunching numbers immediately!"
>
> DARWIN DAHL, WESTERN KENTUCKY UNIVERSITY

CHEMISTRY

1

Matter, Measurement, and Problem Solving

1.1 Atoms and Molecules

1.2 The Scientific Approach to Knowledge

1.3 The Classification of Matter

1.4 Physical and Chemical Changes and Physical and Chemical Properties

1.5 Energy: A Fundamental Part of Physical and Chemical Change

1.6 The Units of Measurement

1.7 The Reliability of a Measurement

1.8 Solving Chemical Problems

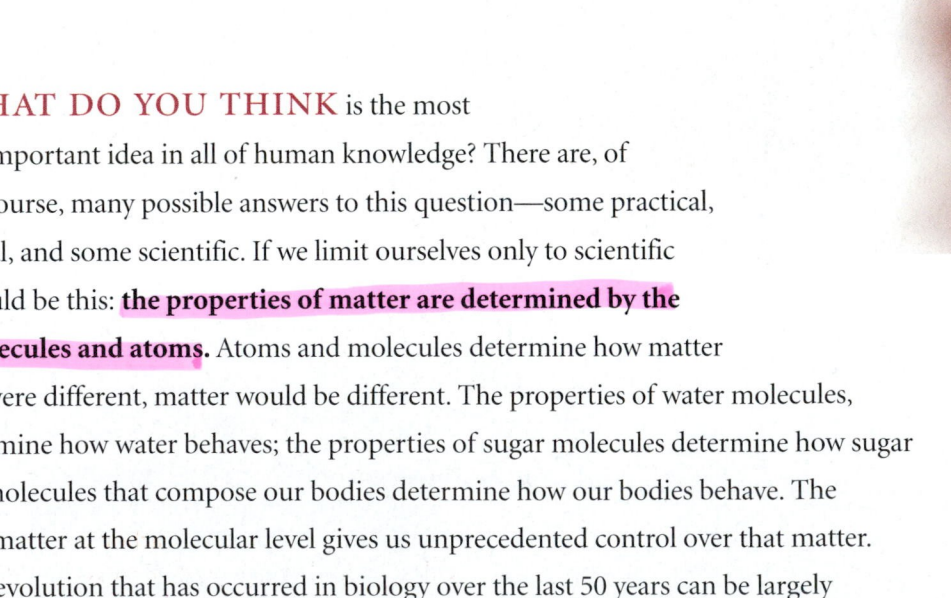

WHAT DO YOU THINK is the most important idea in all of human knowledge? There are, of course, many possible answers to this question—some practical, some philosophical, and some scientific. If we limit ourselves only to scientific answers, mine would be this: **the properties of matter are determined by the properties of molecules and atoms.** Atoms and molecules determine how matter behaves—if they were different, matter would be different. The properties of water molecules, for example, determine how water behaves; the properties of sugar molecules determine how sugar behaves; and the molecules that compose our bodies determine how our bodies behave. The understanding of matter at the molecular level gives us unprecedented control over that matter. For example, the revolution that has occurred in biology over the last 50 years can be largely attributed to understanding the details of the molecules that compose living organisms.

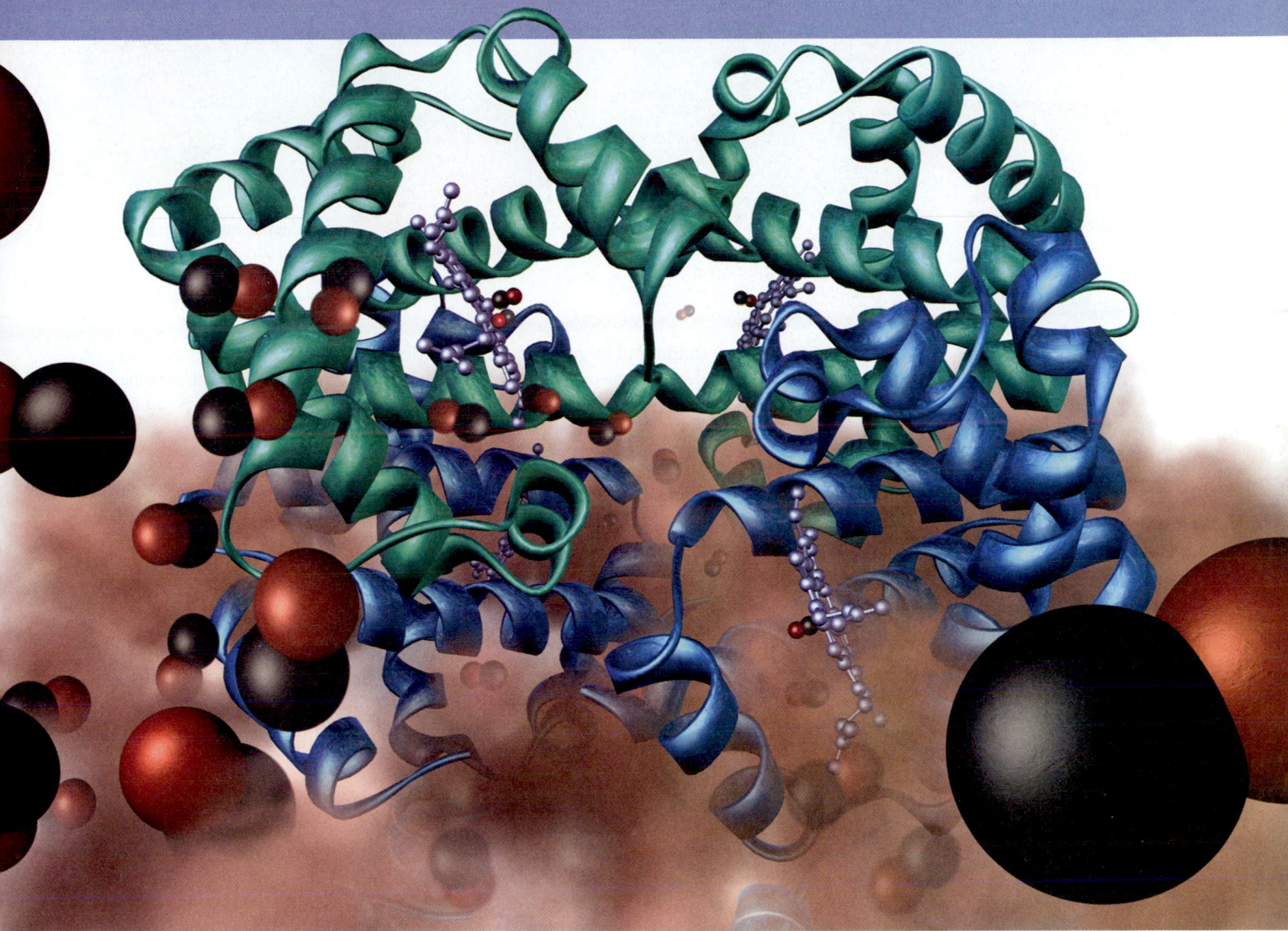

The most incomprehensible thing about the universe is that it is comprehensible.

—ALBERT EINSTEIN (1879–1955)

Hemoglobin, the oxygen-carrying protein in blood (depicted schematically above), can also bind carbon monoxide molecules (the linked red and black spheres).

1.1 Atoms and Molecules

The air over most U.S. cities, including my own, contains at least some pollution. A significant component of that pollution is carbon monoxide, a colorless gas emitted in the exhaust of cars and trucks. Carbon monoxide gas is composed of carbon monoxide molecules, each of which contains a carbon *atom* and an oxygen *atom* held together by a chemical bond. **Atoms** are the submicroscopic particles that constitute the fundamental building blocks of ordinary matter. They are most often found in **molecules**, two or more atoms joined in a specific geometrical arrangement.

The properties of the substances around us depend on the atoms and molecules that compose them, so the properties of carbon monoxide *gas* depend on the properties of carbon monoxide *molecules*. Carbon monoxide molecules happen to be just the right size and shape, and happen to have just the right chemical properties, to fit neatly into cavities within hemoglobin—the oxygen-carrying molecule in blood—that are

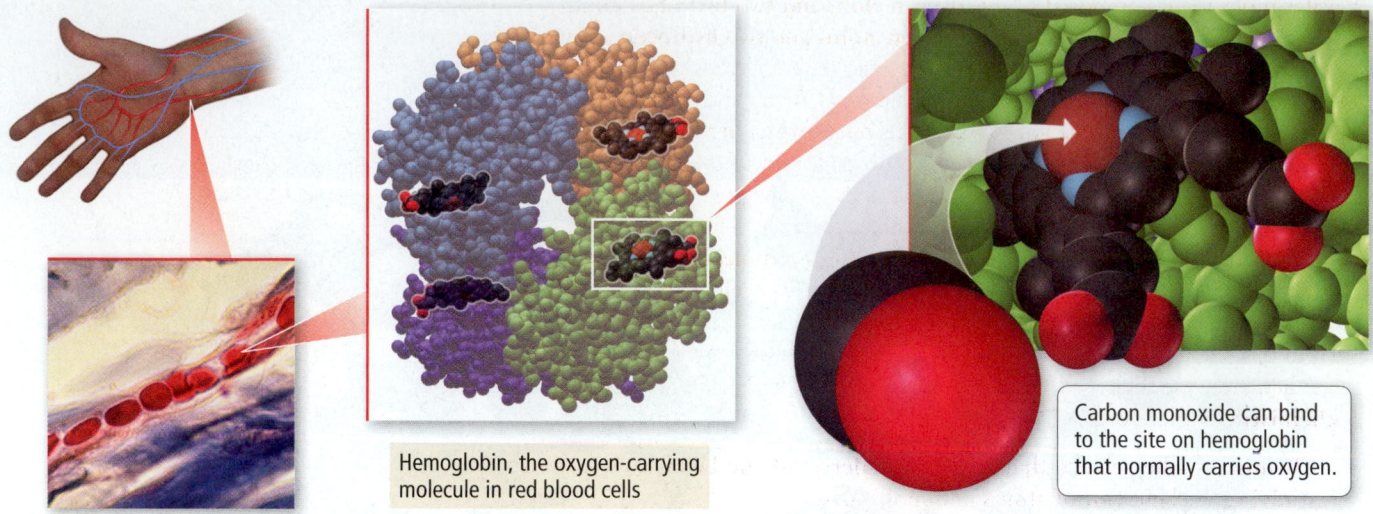

Hemoglobin, the oxygen-carrying molecule in red blood cells

Carbon monoxide can bind to the site on hemoglobin that normally carries oxygen.

▲ **FIGURE 1.1 Binding of Oxygen and Carbon Monoxide to Hemoglobin**   Hemoglobin, a large protein molecule, is the oxygen carrier in red blood cells. Each subunit of the hemoglobin molecule contains an iron atom to which oxygen binds. Carbon monoxide molecules can take the place of oxygen, thus reducing the amount of oxygen reaching the body's tissues.

Carbon monoxide molecule

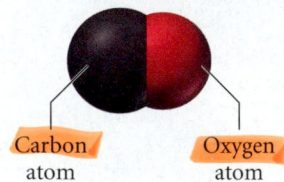

Carbon atom   Oxygen atom

Carbon dioxide molecule

Oxygen atom   Oxygen atom

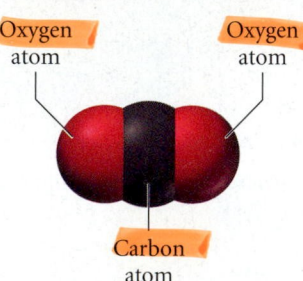

Carbon atom

In the study of chemistry, atoms are often portrayed as colored spheres, with each color representing a different kind of atom. For example, a black sphere represents a carbon atom, a red sphere represents an oxygen atom, and a white sphere represents a hydrogen atom. For a complete color code of atoms, see Appendix IIA.

normally reserved for oxygen molecules (Figure 1.1 ▲). Consequently, carbon monoxide diminishes the oxygen-carrying capacity of blood. Breathing air containing too much carbon monoxide (greater than 0.04% by volume) can lead to unconsciousness and even death because not enough oxygen reaches the brain. Carbon monoxide deaths have occurred, for example, as a result of running an automobile in a closed garage or using a propane burner in an enclosed space for too long. In smaller amounts, carbon monoxide causes the heart and lungs to work harder and can result in headache, dizziness, weakness, and confused thinking.

Cars and trucks emit another closely related molecule, called carbon dioxide, in far greater quantities than carbon monoxide. The only difference between carbon dioxide and carbon monoxide is that carbon dioxide molecules contain two oxygen atoms instead of just one. However, this extra oxygen atom dramatically affects the properties of the gas. We breathe much more carbon dioxide—which is naturally 0.03% of air, and a product of our own respiration as well—than carbon monoxide, yet it does not kill us. Why? Because the presence of the second oxygen atom prevents carbon dioxide from binding to the oxygen-carrying site in hemoglobin, making it far less toxic. Although high levels of carbon dioxide (greater than 10% of air) can be toxic for other reasons, lower levels can enter the bloodstream with no adverse effects. Such is the molecular world. Any changes in molecules—such as the addition of an oxygen atom to carbon monoxide—are likely to result in large changes in the properties of the substances they compose.

As another example, consider two other closely related molecules, water and hydrogen peroxide:

Water molecule

Oxygen atom

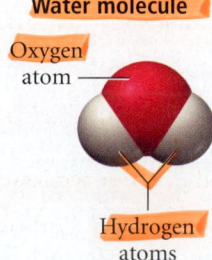

Hydrogen atoms

Hydrogen peroxide molecule

Oxygen atoms

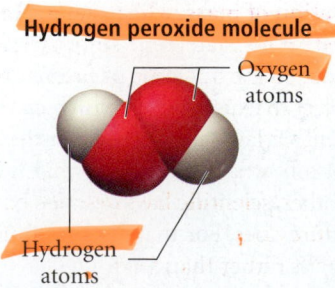

Hydrogen atoms

A water molecule is composed of *one* oxygen atom and two hydrogen atoms. A hydrogen peroxide molecule is composed of *two* oxygen atoms and two hydrogen atoms. This seemingly small molecular difference results in a huge difference between water and hydrogen peroxide. Water is the familiar and stable liquid we all drink and bathe in. Hydrogen peroxide, in contrast, is an unstable liquid that, in its pure form, burns the skin on contact and is used in rocket fuel. When you pour water onto your hair, your hair simply becomes wet. However, if you put hydrogen peroxide in your hair—which you may have done if you have bleached your hair—a chemical reaction occurs that turns your hair blonde.

The hydrogen peroxide used as an antiseptic or bleaching agent is considerably diluted.

The details of how specific atoms bond to form a molecule—in a straight line, at a particular angle, in a ring, or in some other pattern—as well as the type of atoms in the molecule, determine everything about the substance that the molecule composes. If we want to understand the substances around us, we must understand the atoms and molecules that compose them—this is the central goal of chemistry. A good simple definition of **chemistry** is, therefore,

Chemistry—the science that seeks to understand the behavior of matter by studying the behavior of atoms and molecules.

1.2 The Scientific Approach to Knowledge

Throughout history, humans have approached knowledge about the physical world in different ways. For example, the Greek philosopher Plato (427–347 B.C.) thought that the best way to learn about reality was not through the senses, but through reason. He believed that the physical world was an imperfect representation of a perfect and transcendent world (a world beyond space and time). For him, true knowledge came, not through observing the real physical world, but through reasoning and thinking about the ideal one.

Although some Greek philosophers, such as Aristotle, did use observation to attain knowledge, they did not emphasize experiment and measurement to the extent that modern science does.

The *scientific* approach to knowledge, however, is exactly the opposite of Plato's. Scientific knowledge is empirical—that is, it is based on *observation* and *experiment*. Scientists observe and perform experiments on the physical world to learn about it. Some observations and experiments are qualitative (noting or describing how a process happens), but many are quantitative (measuring or quantifying something about the process.) For example, Antoine Lavoisier (1743–1794), a French chemist who studied combustion, made careful measurements of the mass of objects before and after burning them in closed containers. He noticed that there was no change in the total mass of material within the container during combustion. Lavoisier made an important *observation* about the physical world.

Observations often lead scientists to formulate a **hypothesis**, a tentative interpretation or explanation of the observations. For example, Lavoisier explained his observations on combustion by hypothesizing that when a substance combusts, it combines with a component of air. A good hypothesis is *falsifiable*, which means that it makes predictions that can be confirmed or refuted by further observations. Hypotheses are tested by **experiments**, highly controlled procedures designed to generate such observations. The results of an experiment may support a hypothesis or prove it wrong—in which case the hypothesis must be modified or discarded.

In some cases, a series of similar observations can lead to the development of a **scientific law**, a brief statement that summarizes past observations and predicts future ones. For example, Lavoisier summarized his observations on combustion with the **law of conservation of mass**, which states, "In a chemical reaction, matter is neither created nor destroyed." This statement summarized Lavoisier's observations on chemical reactions and predicted the outcome of future observations on reactions. Laws, like hypotheses, are also subject to experiments, which can add support to them or prove them wrong.

Scientific laws are not *laws* in the same sense as civil or governmental laws. Nature does not follow laws in the way that we obey the laws against speeding or passing on the right. Rather, scientific laws *describe* how nature behaves—they are generalizations about what nature does. For that reason, some people find it more appropriate to refer to them as *principles* rather than *laws*.

▲ A painting of the French chemist Antoine Lavoisier with his wife, Marie, who helped him in his work by illustrating his experiments and translating scientific articles from English. Lavoisier, who also made significant contributions to agriculture, industry, education, and government administration, was executed during the French Revolution. (The Metropolitan Museum of Art)

One or more well-established hypotheses may form the basis for a scientific **theory**. A scientific theory is a model for the way nature is and tries to explain not merely what nature does but why. As such, well-established theories are the pinnacle of scientific knowledge, often predicting behavior far beyond the observations or laws from which they were developed. A good example of a theory is the **atomic theory** proposed by English chemist John Dalton (1766–1844). Dalton explained the law of conservation of mass, as well as other laws and observations of the time, by proposing that matter was composed of small, indestructible particles called atoms. Since these particles were merely rearranged in chemical changes (and not created or destroyed), the total amount of mass would remain the same. Dalton's theory is a model for the physical world—it gives us insight into how nature works, and therefore *explains* our laws and observations.

Finally, the scientific approach returns to observation to test theories. Theories are validated by experiments, though they can never be conclusively proved—there is always the possibility that a new observation or experiment will reveal a flaw. For example, the atomic theory can be tested by trying to isolate single atoms, or by trying to image them (both of which, by the way, have already been accomplished). Notice that the scientific approach to knowledge begins with observation and ends with observation, because an experiment is simply a highly controlled procedure for generating critical observations designed to test a theory or hypothesis. Each new set of observations allows refinement of the original model. This approach, often called the **scientific method**, is summarized in Figure 1.2 ▼. Scientific laws, hypotheses, and theories are all subject to continued experimentation. If a law, hypothesis, or theory is proved wrong by an experiment, it must be revised and tested with new experiments. Over time, poor theories and laws are eliminated or corrected and good theories and laws—those consistent with experimental results—remain.

Established theories with strong experimental support are the most powerful pieces of scientific knowledge. You may have heard the phrase, "That is just a theory," as if theories were easily dismissible. However, such a statement reveals a deep misunderstanding of the nature of a scientific theory. Well-established theories are as close to truth as we get in science. The idea that all matter is made of atoms is "just a theory," but it has over 200 years of experimental evidence to support it. It is a powerful piece of scientific knowledge on which many other scientific ideas have been built.

One last word about the scientific method: some people wrongly imagine science to be a strict set of rules and procedures that automatically lead to inarguable, objective facts. This is not the case. Even our diagram of the scientific method is only an idealization of real science, useful to help us see the key distinctions of science. Doing real science requires hard work, care, creativity, and even a bit of luck. Scientific theories do not just fall out of data—they are crafted by men and women of great genius and creativity. A great theory is not unlike a master painting and many see a similar kind of beauty in both. (For more on this aspect of science, see the box entitled *Thomas S. Kuhn and Scientific Revolutions*.)

In Dalton's time, atoms were thought to be indestructible. Today, because of nuclear reactions, we know that atoms can be broken apart into their smaller components.

Gallium arsenide   Cesium atoms

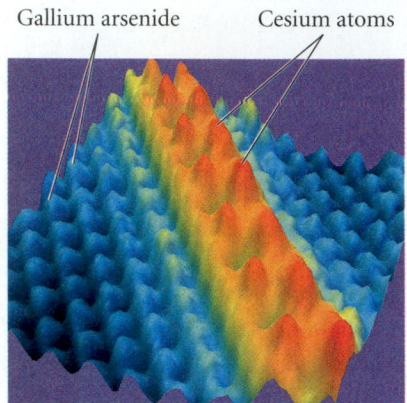

▲ Dalton's atomic theory has been validated in many ways, including the actual imaging of atoms by means of the scanning tunneling microscope (STM).

The Scientific Method

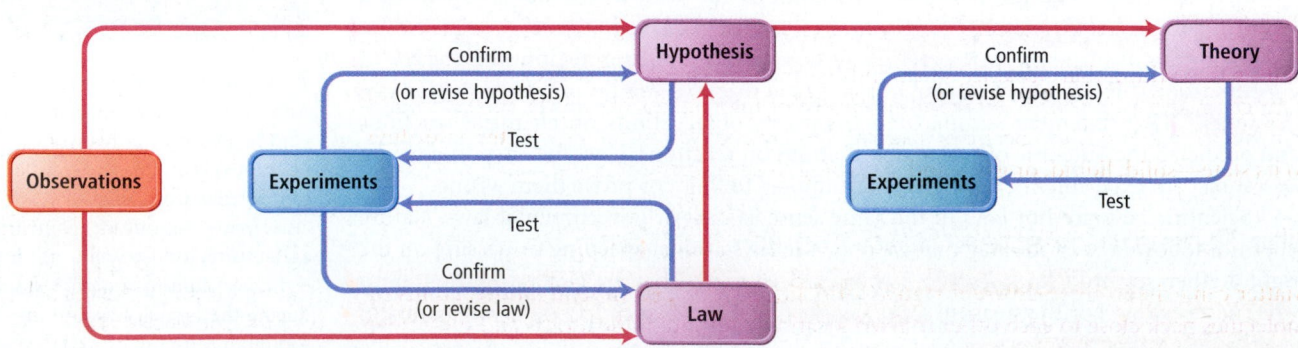

▲ **FIGURE 1.2** The Scientific Method

Conceptual Connection 1.1 Laws and Theories

Which of the following best explains the difference between a law and a theory?

(a) A law is truth whereas a theory is mere speculation.

(b) A law summarizes a series of related observations, while a theory gives the underlying reasons for them.

(c) A theory describes *what* nature does; a law describes *why* nature does it.

Answer: **(b)** A law simply summarizes a series of related observations, while a theory gives the underlying reasons for them.

The Nature of Science
Thomas S. Kuhn and Scientific Revolutions

When scientists talk about science, we often talk in ways that imply that our theories are "true." Further, we talk as if we arrive at theories in logical and unbiased ways. For example, a central theory to chemistry is John Dalton's atomic theory—the idea that all matter is composed of atoms. Is this theory "true"? Was it reached in logical, unbiased ways? Will this theory still be around in 200 years?

The answers to these questions depend on how you view science and its development. One way to view science—let's call it the *traditional view*—is as the continual accumulation of knowledge and the building of increasingly precise theories. In this view, a scientific theory is a model of the world that reflects what is *actually in* nature. New observations and experiments result in gradual adjustments to theories. Over time, theories get better, giving us a more accurate picture of the physical world.

In the twentieth century, however, a different view of scientific knowledge began to develop. In particular, a book by Thomas Kuhn, entitled *The Structure of Scientific Revolutions*, challenged the traditional view. Kuhn's ideas came from his study of the history of science, which, he argues, does not support the idea that science progresses in a smooth cumulative way. According to Kuhn, science goes through fairly quiet periods that he calls *normal science*. In these periods, scientists make their data fit the reigning theory, or paradigm. Small inconsistencies are swept aside during periods of normal science. However, when too many inconsistencies and anomalies develop, a crisis emerges. The crisis brings about a *revolution* and a new reigning theory. According to Kuhn, the new theory is usually quite different from the old one; it not only helps us to make sense of new or anomalous information, but also enables us to see accumulated data from the past in a dramatically new way.

Kuhn further contends that theories are held for reasons that are not always logical or unbiased, and that theories are not *true* models—in the sense of a one-to-one mapping—of the physical world. Because new theories are often so different from the ones they replace, he argues, and because old theories always make good sense to those holding them, they must not be "True" with a capital *T*, otherwise "truth" would be constantly changing.

Kuhn's ideas created a controversy among scientists and science historians that continues to this day. Some, especially postmodern philosophers of science, have taken Kuhn's ideas one step further. They argue that scientific knowledge is *completely* biased and lacks any objectivity. Most scientists, including Kuhn, would disagree. Although Kuhn points out that scientific knowledge has *arbitrary elements*, he also says, "*Observation . . . can and must drastically restrict the range of admissible scientific belief, else there would be no science.*" In other words, saying that science contains arbitrary elements is quite different from saying that science itself is arbitrary.

Question

In his book, Kuhn states, "*A new theory . . . is seldom or never just an increment to what is already known.*" Can you think of any examples of this from your knowledge of the history of science? In other words, can you think of instances in which a new theory or model was drastically different from the one it replaced?

1.3 The Classification of Matter

Matter is anything that occupies space and has mass. For example, this book, your desk, your chair, and even your body are all composed of matter. Less obviously, the air around you is also matter—it too occupies space and has mass. We can classify matter according to its state—solid, liquid, or gas—and according to its composition.

The States of Matter: Solid, Liquid, and Gas

Matter can exist in three different **states: solid, liquid,** and **gas.** In *solid matter*, atoms or molecules pack close to each other in fixed locations. Although the atoms and molecules in a solid vibrate, they do not move around or past each other. Consequently, a solid has a fixed volume and rigid shape. Ice, aluminum, and diamond are good examples of solids.

The state of matter changes from solid to liquid to gas with increasing temperature.

Solid matter

Liquid matter

Gaseous matter

▶ In a solid, the atoms or molecules are fixed in place and can only vibrate. In a liquid, although the atoms or molecules are closely packed, they can move past one another, allowing the liquid to flow and assume the shape of its container. In a gas, the atoms or molecules are widely spaced, making gases compressible as well as fluid.

Glasses and other amorphous solids can be thought of, from one point of view, as intermediate between solids and liquids—their atoms are fixed in position at room temperature, but they have no long-range structure and do not have sharp melting points.

Solid matter may be **crystalline**, in which case its atoms or molecules are arranged in patterns with long-range, repeating order (Figure 1.3a ▼), or it may be **amorphous**, in which case its atoms or molecules do not have any long-range order (Figure 1.3b ▼). Examples of *crystalline* solids include table salt and diamond; the well-ordered geometric shapes of salt and diamond crystals reflect the well-ordered geometric arrangement of their atoms. Examples of *amorphous* solids include glass, plastic, and charcoal.

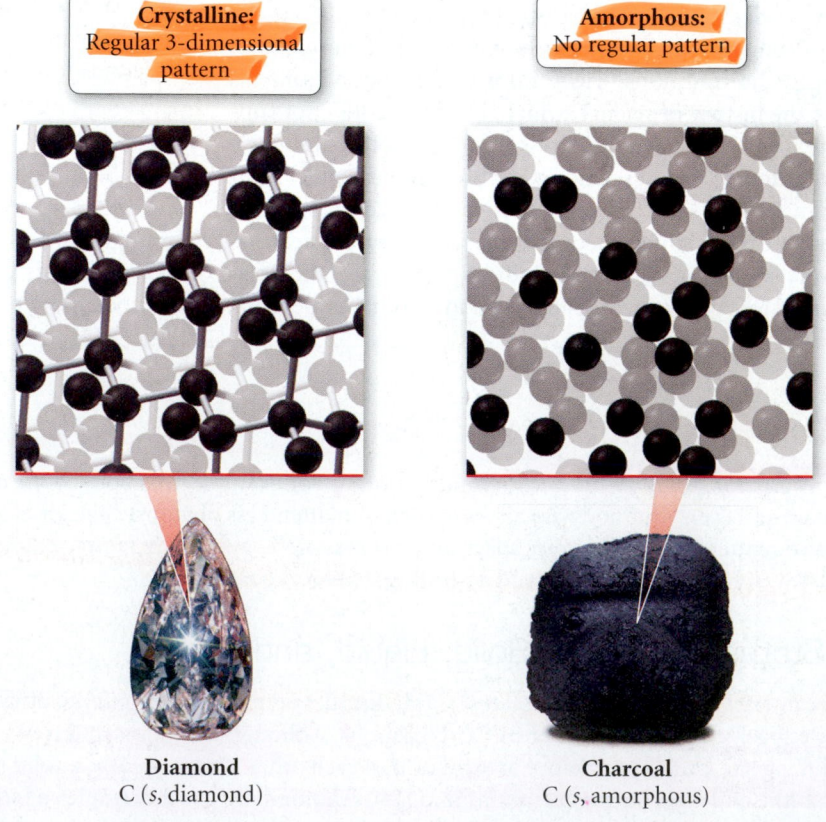

Crystalline:
Regular 3-dimensional pattern

Amorphous:
No regular pattern

Diamond
C (*s*, diamond)

Charcoal
C (*s*, amorphous)

▶ FIGURE 1.3 **Crystalline and Amorphous Solids** Diamond is a crystalline solid composed of carbon atoms arranged in a regular, repeating pattern. Charcoal is an amorphous solid composed of carbon atoms with no long-range order.

In *liquid matter*, atoms or molecules pack about as closely as they do in solid matter, but they are free to move relative to each other, giving liquids a fixed volume but not a fixed shape. Liquids assume the shape of their container. Water, alcohol, and gasoline are all good examples of substances that are liquids at room temperature.

In *gaseous matter*, atoms or molecules have a lot of space between them and are free to move relative to one another, making gases *compressible* (Figure 1.4 ▶). When you squeeze a balloon or sit down on an air mattress, you force the atoms and molecules into a smaller space, so that they are closer together. Gases always assume the shape *and* volume of their container. Examples of gases at room temperature include helium, nitrogen (the main component of air), and carbon dioxide.

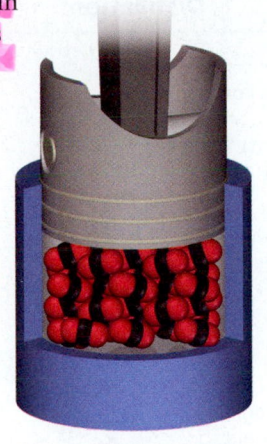 

Solid–not compressible   Gas–compressible

Conceptual Connection 1.2 The Mass of a Gas

A drop of water is put into a container and the container is sealed. The drop of water then vaporizes. Does the mass of the sealed container and its contents change upon vaporization?

Answer: No. The water vaporizes and becomes a gas, but the water molecules are still present within the flask and have the same mass.

▲ FIGURE 1.4 **The Compressibility of Gases** Gases can be compressed—squeezed into a smaller volume—because there is so much empty space between atoms or molecules in the gaseous sate.

Classifying Matter According to Its Composition: Elements, Compounds, and Mixtures

In addition to classifying matter according to its state, we can classify it according to its composition, as shown in the following chart:

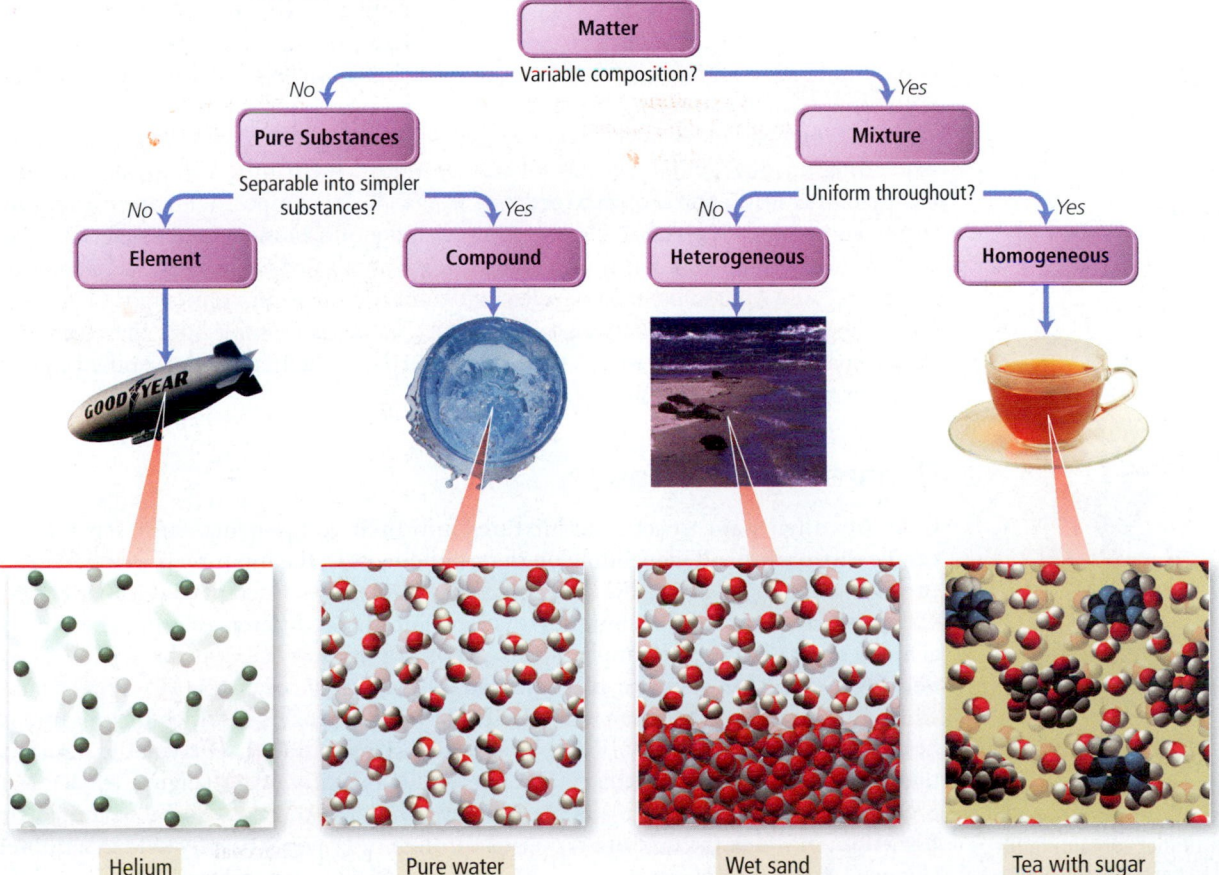

Distillation

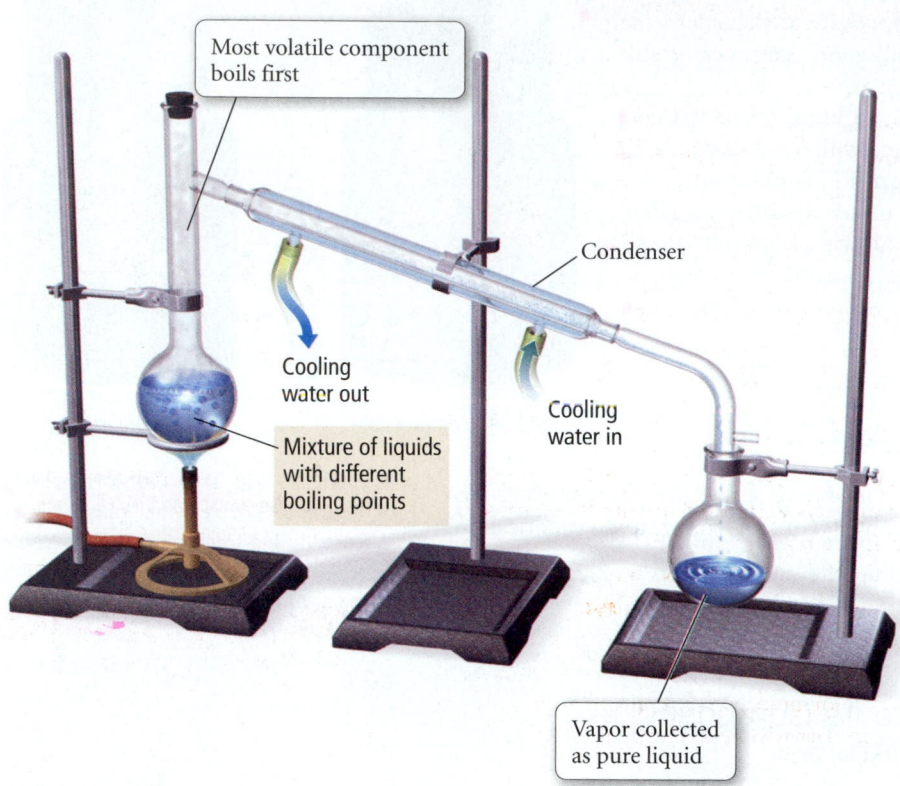

▲ **FIGURE 1.5 Separating Substances by Distillation**   When a liquid mixture is heated, the component with the lowest boiling point vaporizes first, leaving behind less volatile liquids or dissolved solids. The vapor is then cooled, condensing it back to a liquid, and collected.

Filtration

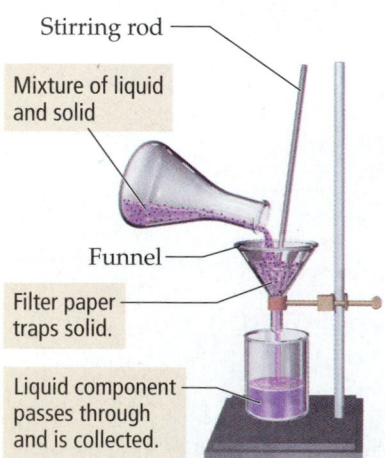

▲ **FIGURE 1.6 Separating Substances by Filtration**   A solid and liquid mixture can be separated by pouring the mixture through a funnel containing filter paper designed to pass only the liquid.

The first division in the classification of matter depends on whether or not the composition can vary from one sample to another. For example, the composition of distilled (or pure) water never varies—it is always 100% water and is therefore a **pure substance,** one composed of only a single type of atom or molecule. In contrast, the composition of sweetened tea can vary substantially from one sample to another, depending, for instance, on the strength of the tea or how much sugar has been added. Sweetened tea is an example of a **mixture,** a substance composed of two or more different types of atoms or molecules that can be combined in variable proportions.

Pure substances can be divided into two types—elements and compounds—depending on whether or not they can be broken down into simpler substances. The helium in a blimp or party balloon is a good example of an **element,** a substance that cannot be chemically broken down into simpler substances. Water is a good example of a **compound,** a substance composed of two or more elements (hydrogen and oxygen) in fixed, definite proportions. On Earth, compounds are more common than pure elements because most elements combine with other elements to form compounds.

Mixtures can be divided into two types—heterogeneous and homogeneous—depending on how uniformly the substances within them mix. Wet sand is a good example of a **heterogeneous mixture,** one in which the composition varies from one region to another. Sweetened tea is a good example of a **homogeneous mixture,** one with the same composition throughout. Homogeneous mixtures have uniform compositions because the atoms or molecules that compose them mix uniformly. Heterogeneous mixtures form distinct regions because the atoms or molecules that compose them separate. Here again we see that the properties of matter are determined by the atoms or molecules that compose it.

Separating Mixtures

Chemists often want to separate mixtures into their components. Such separations can be easy or difficult, depending on the components in the mixture. In general, mixtures are separable because the different components have different physical or chemical properties. Various techniques that exploit these differences can be used to achieve separation. For example, a mixture of sand and water can be separated by **decanting**—carefully pouring off—the water into another container. Mixtures of miscible liquids can usually be separated by **distillation,** a process in which the mixture is heated to boil off the more **volatile** (easily vaporizable) liquid. The volatile liquid is then recondensed in a condenser and collected in a separate flask (Figure 1.5 ▲). If a mixture is composed of an insoluble solid and a liquid, the two can be separated by **filtration,** in which the mixture is poured through filter paper usually held in a funnel (Figure 1.6 ◄).

1.4 Physical and Chemical Changes and Physical and Chemical Properties

Every day we witness changes in matter: ice melts, iron rusts, gasoline burns, fruit ripens, and water evaporates. What happens to the molecules that compose these samples of matter during such changes? The answer depends on the type of change. Changes that alter only state or appearance, but not composition, are called **physical changes**. The atoms or molecules that compose a substance *do not change* their identity during a physical change. For example, when water boils, it changes its state from a liquid to a gas, but the gas remains composed of water molecules, so this a physical change (Figure 1.7 ▼).

In contrast, changes that alter the composition of matter are called **chemical changes**. During a chemical change, atoms rearrange, transforming the original substances into different substances. For example, the rusting of iron is a chemical change. The atoms that compose iron (iron atoms) combine with oxygen molecules from air to form iron oxide, the orange substance we normally call rust (Figure 1.8 ▼). Some other examples of physical and chemical changes are shown in Figure 1.9 (page 12).

Physical and chemical changes are manifestations of physical and chemical properties. A **physical property** is one that a substance displays without changing its composition, whereas a **chemical property** is one that a substance displays only by changing its composition via a chemical change. For example, the smell of gasoline is a physical property—gasoline does not change its composition when it exhibits its odor. The flammability of gasoline, in contrast, is a chemical property—gasoline does change its composition when it burns, turning into completely new substances (primarily carbon dioxide and water). Physical properties include odor, taste, color, appearance, melting point, boiling point, and density. Chemical properties include corrosiveness, flammability, acidity, toxicity, and other such characteristics.

The differences between physical and chemical changes are not always apparent. Only chemical examination can confirm whether any particular change is physical or chemical. In many cases, however, we can identify chemical and physical changes based on what we know about the changes. Changes in the state of matter, such as melting or boiling, or changes in the physical condition of matter, such as those that result from cutting or crushing, are always physical changes. Changes involving chemical reactions—often evidenced by heat exchange or color changes—are chemical changes.

In Chapter 19 we will also learn about *nuclear changes*, which can involve atoms of one element changing into atoms of a different element.

A physical change results in a different form of the same substance, while a chemical change results in a completely different substance.

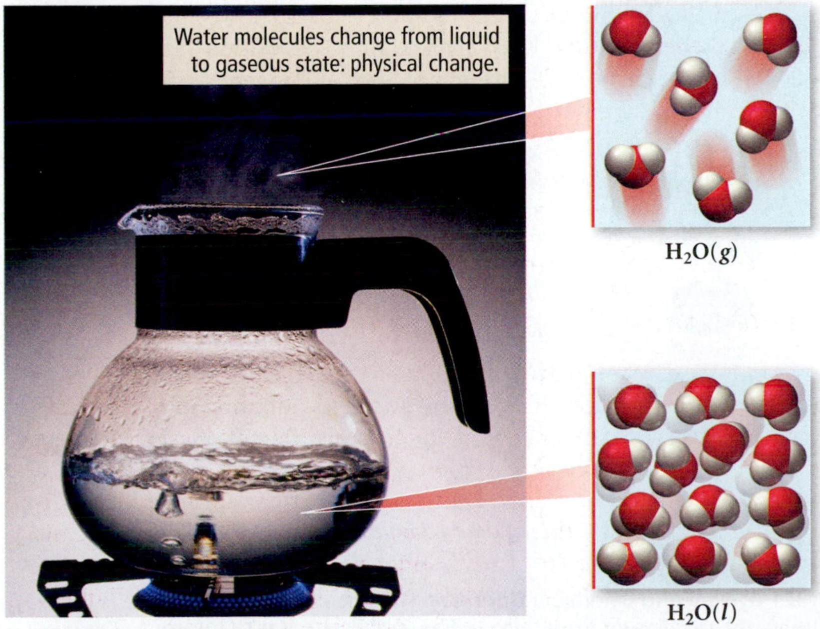

Water molecules change from liquid to gaseous state: physical change.

$H_2O(g)$

$H_2O(l)$

▲ FIGURE 1.7 **Boiling, a Physical Change** When water boils, it turns into a gas but does not alter its chemical identity—the water molecules are the same in both the liquid and gaseous states. Boiling is thus a physical change, and the boiling point of water is a physical property.

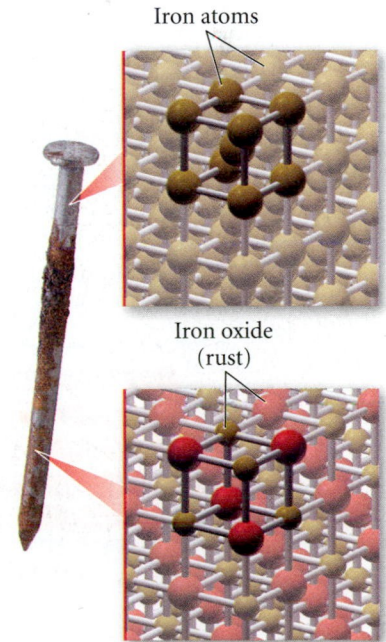

Iron atoms

Iron oxide (rust)

▲ FIGURE 1.8 **Rusting, a Chemical Change** When iron rusts, the iron atoms combine with oxygen atoms to form a different chemical substance, the compound iron oxide. Rusting is therefore a chemical change, and the tendency of iron to rust is a chemical property.

Physical Change and Chemical Change

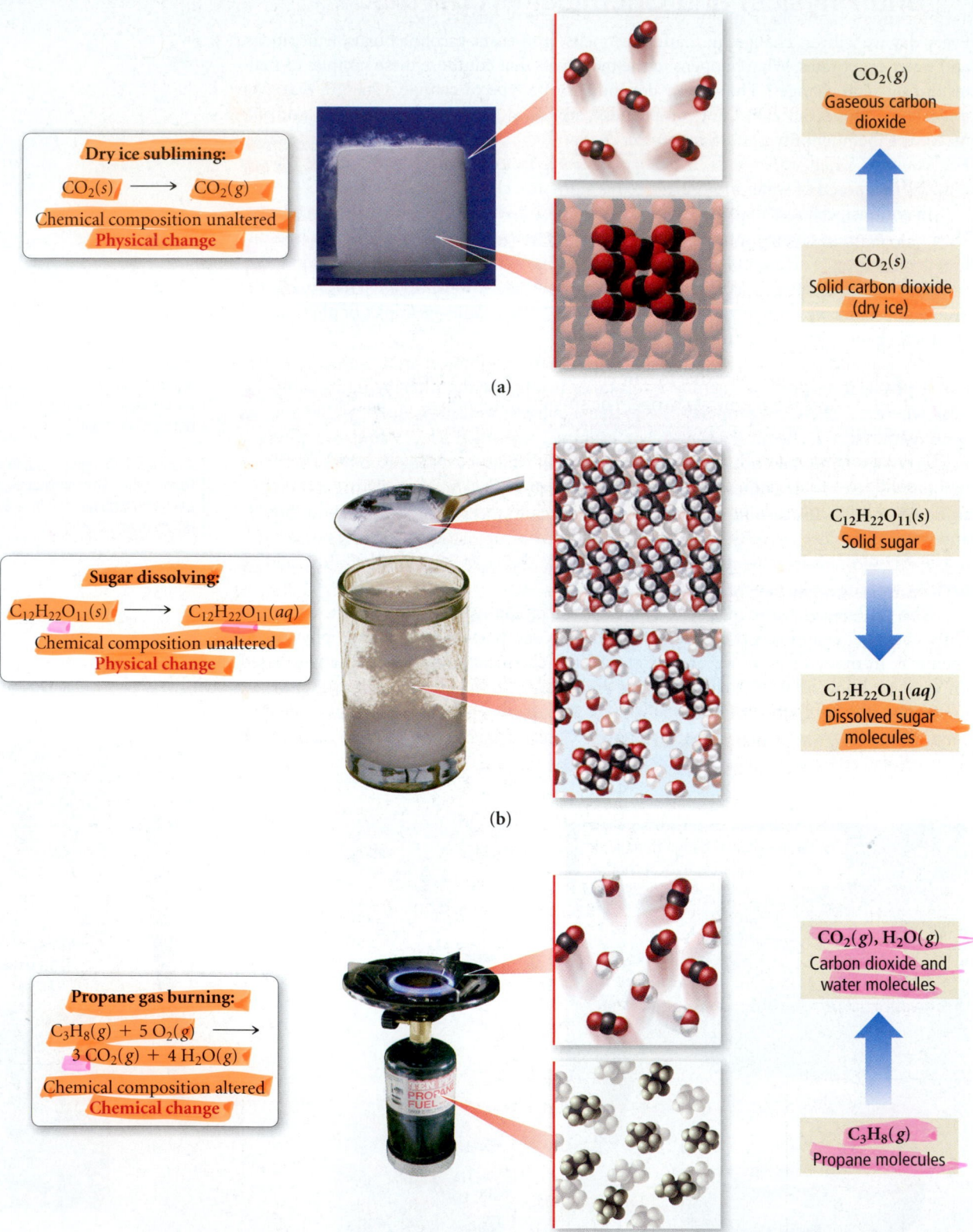

Dry ice subliming:

$$CO_2(s) \longrightarrow CO_2(g)$$

Chemical composition unaltered
Physical change

$CO_2(g)$
Gaseous carbon dioxide

$CO_2(s)$
Solid carbon dioxide (dry ice)

(a)

Sugar dissolving:

$$C_{12}H_{22}O_{11}(s) \longrightarrow C_{12}H_{22}O_{11}(aq)$$

Chemical composition unaltered
Physical change

$C_{12}H_{22}O_{11}(s)$
Solid sugar

$C_{12}H_{22}O_{11}(aq)$
Dissolved sugar molecules

(b)

Propane gas burning:

$$C_3H_8(g) + 5\,O_2(g) \longrightarrow$$
$$3\,CO_2(g) + 4\,H_2O(g)$$

Chemical composition altered
Chemical change

$CO_2(g),\ H_2O(g)$
Carbon dioxide and water molecules

$C_3H_8(g)$
Propane molecules

(c)

▲ **FIGURE 1.9 Physical and Chemical Changes**   (a) The sublimation of dry ice (solid CO_2) is a physical change. (b) The dissolution of sugar is a physical change. (c) The burning of propane is a chemical change.

EXAMPLE 1.1 Physical and Chemical Changes and Properties

Determine whether each of the following changes is physical or chemical.

(a) the evaporation of rubbing alcohol

(b) the burning of lamp oil

(c) the bleaching of hair with hydrogen peroxide

(d) the forming of frost on a cold night

Solution

(a) When rubbing alcohol evaporates, it changes from liquid to gas, but it remains alcohol—this is a physical change. The volatility (or ability to evaporate easily) of alcohol is a therefore a physical property.

(b) Lamp oil burns because it reacts with oxygen in air to form carbon dioxide and water—this is a chemical change. The flammability of lamp oil is therefore a chemical property.

(c) Applying hydrogen peroxide to hair changes pigment molecules in hair that give it color—this is a chemical change. The susceptibility of hair to bleaching is therefore a chemical property.

(d) Frost forms on a cold night because water vapor in air changes its state to form solid ice—this is a physical change. The temperature at which water freezes is therefore a physical property.

For Practice 1.1

Determine whether each of the following is a physical or chemical change. What kind of property (chemical or physical) is being demonstrated in each case?

(a) A copper wire is hammered flat.

(b) A nickel dissolves in acid to form a blue-green solution.

(c) Dry ice vaporizes without melting.

(d) A match ignites when struck on a flint.

Answers to For Practice and For More Practice problems can be found in Appendix IV.

 Conceptual Connection 1.3 Chemical and Physical Changes

The diagram to the right represents liquid water molecules in a pan.

Which of the following diagrams best represents the water molecules after they have been vaporized by the boiling of liquid water?

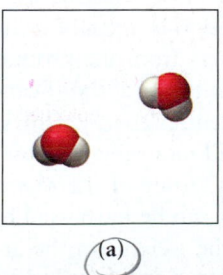

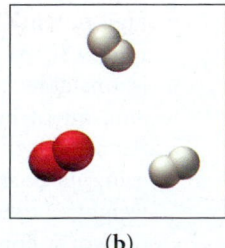

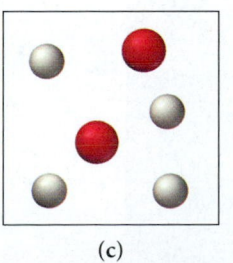

| (a) | (b) | (c) |

Answer: View **(a)** best represents the water after vaporization. Vaporization is a physical change, so the molecules must remain the same before and after the change.

1.5 Energy: A Fundamental Part of Physical and Chemical Change

The physical and chemical changes that we have just discussed are usually accompanied by energy changes. For example, when water evaporates from your skin (a physical change), the water molecules absorb energy, cooling your skin. When you burn natural gas on the stove (a chemical change), energy is released, heating the food you are cooking. Understanding the physical and chemical changes of matter—that is, understanding chemistry—requires that we also understand energy changes and energy flow.

The scientific definition of **energy** is *the capacity to do work*. **Work** is defined as the action of a force through a distance. For instance, when you push a box across the floor or when you pedal your bicycle down the street, you have done work.

The *total energy* of an object is a sum of its **kinetic energy**, the energy associated with its motion, and its **potential energy**, the energy associated with its position or composition. For example, a weight held at several meters from the ground has potential energy due to its position within Earth's gravitational field (Figure 1.10 ▼). If the weight is dropped, it accelerates, and the potential energy is converted to kinetic energy. When the weight hits the ground, its kinetic energy is converted primarily to **thermal energy**, the energy associated with the temperature of an object. Thermal energy is actually a type of kinetic energy because it arises from the motion of the individual atoms or molecules that make up an object. In other words, when the weight hits the ground its kinetic energy is essentially transferred to the atoms and molecules that compose the ground, raising the temperature of the ground ever so slightly.

The first principle to note about the way that energy changes as the weight falls to the ground is that *energy is neither created nor destroyed*. The potential energy of the weight becomes kinetic energy as the weight accelerates toward the ground. The kinetic energy then becomes thermal energy when the weight hits the ground. The total amount of thermal energy that is released through the process is exactly equal to the initial potential energy of the weight. The observation that energy is neither created nor destroyed is known as the **law of conservation of energy**. Although energy can change from one kind to another, and although it can flow from one object to another, the *total quantity* of energy does not change—it remains constant.

The second principle to note is *the tendency of systems with high potential energy to change in a way that lowers their potential energy*. For this reason, objects or systems with high potential energy tend to be *unstable*. The weight lifted several meters from the ground is unstable because it contains a significant amount of localized potential energy. Unless restrained, the weight will naturally fall, lowering its potential energy. Some of the raised weight's potential energy can be harnessed to do work. For example, the weight can be attached to a rope that turns a paddle wheel or spins a drill as the weight falls. After it falls to the ground, the weight contains less potential energy—it has become more *stable*.

Force acts through distance; work is done.

We will find in Chapter 19 that energy conservation is actually part of a more general law that allows for the interconvertibility of mass and energy.

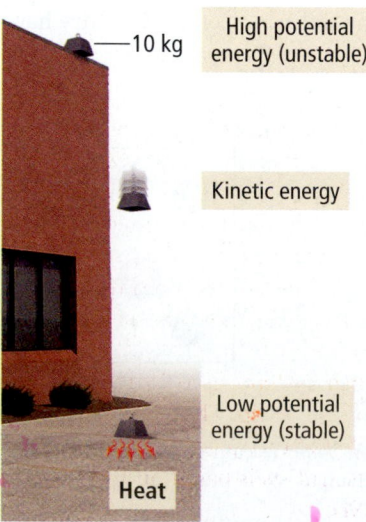

10 kg

High potential energy (unstable)

Kinetic energy

Low potential energy (stable)

Heat

▶ **FIGURE 1.10 Energy Conversions** Gravitational potential energy is converted into kinetic energy when the weight is released. The kinetic energy is converted mostly to thermal energy when the weight strikes the ground.

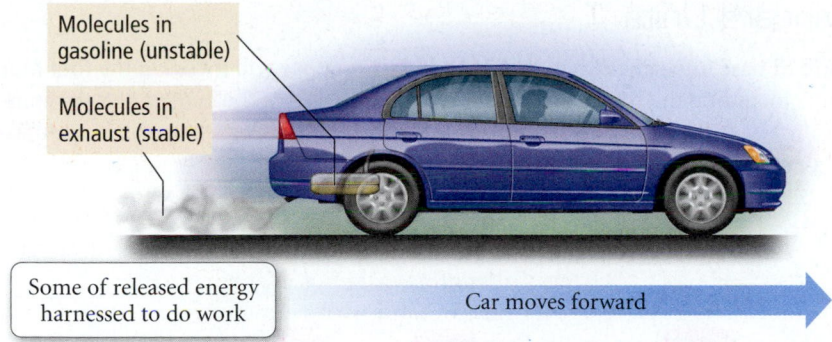

Molecules in gasoline (unstable)

Molecules in exhaust (stable)

Some of released energy harnessed to do work

Car moves forward

▲ FIGURE 1.11 **Using Chemical Energy to Do Work**   The compounds produced when gasoline burns have less chemical potential energy than the gasoline molecules.

Some chemical substances are like the raised weight just described. For example, the molecules that compose gasoline have a relatively high potential energy—energy is concentrated in them just as energy is concentrated in the raised weight. The molecules in the gasoline therefore tend to undergo chemical changes (specifically combustion) that will lower their potential energy. As the energy of the molecules is released, some of it can be harnessed to do work, such as moving a car down the street (Figure 1.11 ▲). The molecules that result from the chemical change have less potential energy than the original molecules in gasoline and are therefore more stable.

Chemical potential energy, such as that contained in the molecules that compose gasoline, arises primarily from electrostatic forces between the electrically charged particles (protons and electrons) that compose atoms and molecules. We will learn more about those particles, as well as the properties of electrical charge, in Chapter 2, but for now, know that molecules contain specific, sometimes complex, arrangements of these charged particles. Some of these arrangements—such as the one within the molecules that compose gasoline—have a much higher potential energy than others. When gasoline undergoes combustion the arrangement of these particles changes, creating molecules with much lower potential energy and transferring a great deal of energy (mostly in the form of heat) to the surroundings.

Summarizing:

➤ Energy is always conserved in a physical or chemical change; it is neither created nor destroyed.

➤ Systems with high potential energy tend to change in a direction of lower potential energy, releasing energy into the surroundings.

1.6 The Units of Measurement

In 1999, NASA lost the $125 million *Mars Climate Orbiter* (pictured here) because of confusion between English and metric units. The chairman of the commission that investigated the disaster concluded, "The root cause of the loss of the spacecraft was a failed translation of English units into metric units." As a result, the orbiter—which was supposed to monitor weather on Mars—descended too far into the Martian atmosphere and burned up. In chemistry as in space exploration, **units**—standard quantities used to specify measurements—are critical. If you get them wrong, the consequences can be enormous.

The two most common unit systems are the **English system**, used in the United States, and the **metric system**, used in most of the rest of the world. The English system consists of units such as inches, yards, and pounds, while the metric system uses centimeters, meters, and kilograms. The unit system used by scientists is based on the metric system and is called the **International System of Units (SI)**.

▲ The $125 million *Mars Climate Orbiter* was lost in the Martian atmosphere in 1999 because two groups of engineers failed to communicate to each other the units that they used in their calculations.

The abbreviation *SI* comes from the French, *Système International d'Unités.*

The Standard Units

The standard SI base units are shown in Table 1.1. For now, we will focus on the first four of these units including the *meter* as the standard unit of length, the *kilogram* as the standard unit of mass, the *second* as the standard unit of time, and the *kelvin* as the standard unit of temperature.

TABLE 1.1 SI Base Units

Quantity	Unit	Symbol
Length	Meter	m
Mass	Kilogram	kg
Time	Second	s
Temperature	Kelvin	K
Amount of substance	Mole	mol
Electric current	Ampere	A
Luminous intensity	Candela	cd

The Meter: A Measure of Length

The **meter (m)** is slightly longer than a yard (1 yard is 36 inches while 1 meter is 39.37 inches).

Yardstick

Meterstick

Thus, a 100-yard football field measures only 91.4 meters. The meter was originally defined as 1/10,000,000 of the distance from the equator to the north pole (through Paris). It is now defined more precisely as the distance light travels through a vacuum in a certain period of time, 1/299,792,458 second. A tall human is about 2 m tall and the Empire State Building stands 443 m tall (including its mast).

Scientists commonly deal with a wide range of lengths and distances. For example, the separation between the sun and the closest star (Proxima Centauri) is about 3.8×10^{16} m, while many chemical bonds measure about 1.5×10^{-10} m.

The Kilogram: A Measure of Mass

The **kilogram (kg)** is defined as the mass of a metal cylinder kept at the International Bureau of Weights and Measures at Sèvres, France. The kilogram is a measure of *mass*, a quantity different from *weight*. The **mass** of an object is a measure of the quantity of matter within it, while the weight of an object is a measure of the *gravitational pull* on the matter within it. If you weigh yourself on the moon, for example, its weaker gravity pulls on you with less force than does Earth's gravity, resulting in a lower weight. A 130-pound (lb) person on Earth weighs 21.5 lb on the moon. However, the person's mass—the quantity of matter in his or her body—remains the same. One kilogram of mass is the equivalent of 2.205 lb of weight on Earth, so if we express mass in kilograms, a 130-lb person has a mass of approximately 59 kg and this book has a mass of about 2.5 kg. A second common unit of mass is the gram (g). One gram is 1/1000 kg. A nickel (5¢) has a mass of about 5 g.

The Second: A Measure of Time

To those of us who live in the United States, the **second (s)** is perhaps the most familiar SI unit. The second was originally defined in terms of the day and the year, but it is now defined more precisely as the duration of 9,192,631,770 periods of the radiation emitted from a certain transition in a cesium-133 atom. Scientists measure time on a large range of scales. For example, the human heart beats about once every second; the age of the universe is estimated to be about 4.3×10^{17} s (13.7 billion years); and some molecular bonds break or form in time periods as short as 1×10^{-15} s.

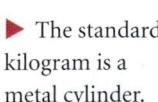

443 m

2 m

▲ The Empire State Building is 443 meters tall. A basketball player stands about 2 meters tall.

The velocity of light in a vacuum is 3.00×10^8 m/s.

▶ The standard kilogram is a metal cylinder.

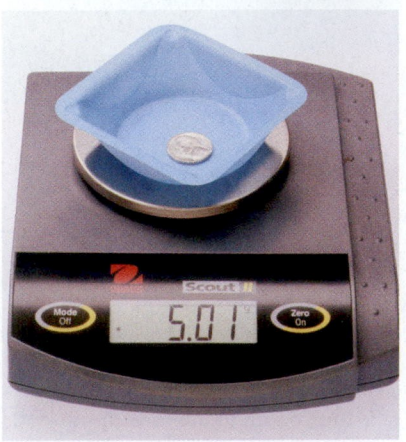

▲ A nickel (5 cents) weighs about 5 grams.

The Kelvin: A Measure of Temperature

The **kelvin (K)** is the SI unit of **temperature**. The temperature of a sample of matter is a measure of the amount of average kinetic energy—the energy due to motion—of the atoms or molecules that compose the matter. For example, the molecules in a *hot* glass of water are, on average, moving faster than the molecules in a *cold* glass of water. Temperature is a measure of this molecular motion.

Temperature also determines the direction of thermal energy transfer, or what we commonly call *heat*. Thermal energy transfers from hot objects to cold ones. For example, when you touch another person's warm hand (and yours is cold), thermal energy flows *from their hand to yours*, making your hand feel warmer. However, if you touch an ice cube, thermal energy flows *out of your hand* to the ice, cooling your hand (and possibly melting some of the ice cube).

The three common temperature scales are shown in Figure 1.12 ▼. The most familiar in the United States is the **Fahrenheit (°F) scale,** shown on the left. On the Fahrenheit scale, water freezes at 32 °F and boils at 212 °F at sea level. Room temperature is approximately 72 °F. The Fahrenheit scale was originally determined by assigning 0 °F to the freezing point of a concentrated saltwater solution and 96 °F to normal body temperature.

The scale most often used by scientists and by most countries other than the United States is the **Celsius (°C) scale,** shown in the middle. On this scale, pure water freezes at 0 °C and boils at 100 °C (at sea level). Room temperature is approximately 22 °C. The Fahrenheit scale and the Celsius scale differ both in the size of their respective degrees and the temperature each designates as "zero." Both the Fahrenheit and Celsius scales allow for negative temperatures.

The SI unit for temperature, as we have seen, is the kelvin, shown on the right in Figure 1.12. The **Kelvin scale** (sometimes also called the *absolute scale*) avoids negative temperatures by assigning 0 K to the coldest temperature possible, absolute zero. Absolute zero (−273 °C or −459 °F) is the temperature at which molecular motion virtually stops.

> Normal body temperature was later measured more accurately to be 98.6 °F

> Molecular motion does not *completely* stop at absolute zero because of the uncertainty principle in quantum mechanics, which we will discuss in Chapter 7.

Temperature Scales

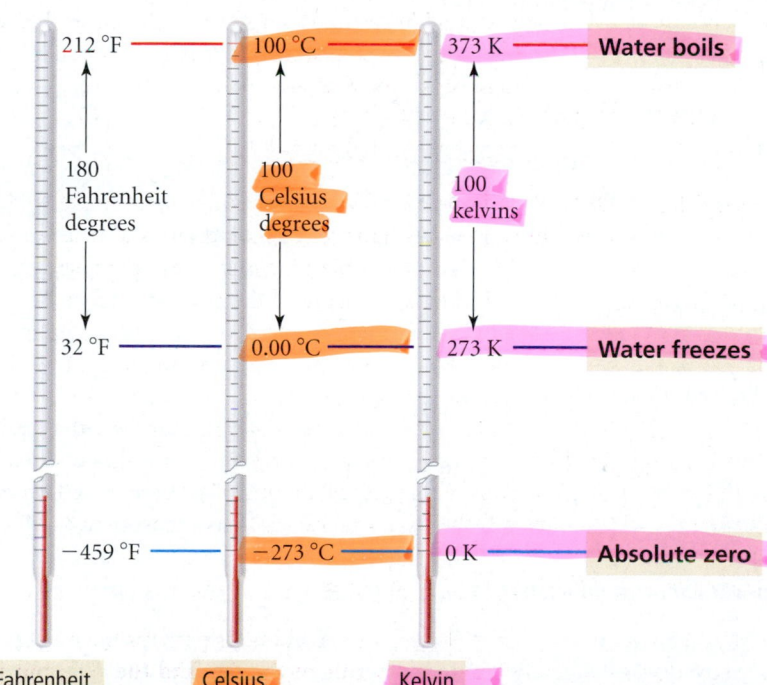

▲ FIGURE 1.12 Comparison of the Fahrenheit, Celsius, and Kelvin Temperature Scales
The Fahrenheit degree is five-ninths the size of the Celsius degree and the kelvin. The zero point of the Kelvin scale is absolute zero (the lowest possible temperature), whereas the zero point of the Celsius scale is the freezing point of water.

The Celsius Temperature Scale

0 °C – Water freezes

10 °C – Brisk fall day

22 °C – Room temperature

40 °C – Summer day in Death Valley

Note that we give Kelvin temperatures in kelvins (*not* "degrees Kelvin") or K (*not* °K).

Lower temperatures do not exist. The size of the kelvin is identical to that of the Celsius degree—the only difference is the temperature that each designates as zero. You can convert between the temperature scales with the following formulas:

$$°C = \frac{(°F - 32)}{1.8}$$

$$K = °C + 273.15$$

Throughout this book you will see examples worked out in formats that are designed to help you develop problem-solving skills. The most common format uses two columns to guide you through the worked example. The left column describes the thought processes and steps used in solving the problem while the right column shows the implementation. The first example in this two-column format follows.

EXAMPLE 1.2   Converting between Temperature Scales

A sick child has a temperature of 40.00 °C. What is the child's temperature in (a) K and (b) °F?

Solution

(a) Begin by finding the equation that relates the quantity that is given (°C) and the quantity you are trying to find (K).

$$K = °C + 273.15$$

Since this equation gives the temperature in K directly, simply substitute in the correct value for the temperature in °C and compute the answer.

$$K = °C + 273.15$$
$$K = 40.00 + 273.15 = 313.15 \text{ K}$$

(b) To convert from °C to °F, first find the equation that relates these two quantities.

$$°C = \frac{(°F - 32)}{1.8}$$

Since this equation expresses °C in terms of °F, you must solve the equation for °F.

$$°C = \frac{(°F - 32)}{1.8}$$
$$1.8(°C) = (°F - 32)$$
$$°F = 1.8(°C) + 32$$

Now substitute °C into the equation and compute the answer.

Note: The number of digits reported in this answer follows significant figure conventions, covered in Section 1.7.

$$°F = 1.8(°C) + 32$$
$$°F = 1.8(40.00 \text{ °C}) + 32 = 104.00 \text{ °F}$$

For Practice 1.2

Gallium is a solid metal at room temperature, but it will melt to a liquid in your hand. The melting point of gallium is 85.6 °F. What is this temperature on (a) the Celsius scale and (b) the Kelvin scale?

Prefix Multipliers

While scientific notation allows us to express very large or very small quantities in a compact manner, it requires us to use very large positive or negative exponents to do so. For example, the diameter of a hydrogen atom can be written as 1.06×10^{-10} m. The International System of Units uses the **prefix multipliers** shown in Table 1.2 with the standard units. These multipliers change the value of the unit by powers of 10. For example, the kilometer has the prefix "kilo" meaning 1000 or 10^3. Therefore,

$$1 \text{ kilometer} = 1000 \text{ meters} = 10^3 \text{ meters}$$

Similarly, the millimeter has the prefix "milli" meaning 0.001 or 10^{-3}.

$$1 \text{ millimeter} = 0.001 \text{ meters} = 10^{-3} \text{ meters}$$

When reporting a measurement, choose a prefix multiplier close to the size of the quantity being measured. For example, to state the diameter of a hydrogen atom, which is 1.06×10^{-10} m, use picometers (106 pm) or nanometers (0.106 nm) rather than micrometers or millimeters. Choose the prefix multiplier that is most convenient for a particular number.

TABLE 1.2 SI Prefix Multipliers

Prefix	Symbol	Multiplier	
exa	E	1,000,000,000,000,000,000	(10^{18})
peta	P	1,000,000,000,000,000	(10^{15})
tera	T	1,000,000,000,000	(10^{12})
giga	G	1,000,000,000	(10^{9})
mega	M	1,000,000	(10^{6})
kilo	k	1,000	(10^{3})
deci	d	0.1	(10^{-1})
centi	c	0.01	(10^{-2})
milli	m	0.001	(10^{-3})
micro	μ	0.000001	(10^{-6})
nano	n	0.000000001	(10^{-9})
pico	p	0.000000000001	(10^{-12})
femto	f	0.000000000000001	(10^{-15})
atto	a	0.000000000000000001	(10^{-18})

Derived Units: Volume and Density

A **derived unit** is a combination of other units. For example, the SI unit for speed is meters per second (m/s), a derived unit. Notice that this unit is formed from two other SI units— meters and seconds—put together. We are probably more familiar with speed in miles/hour or kilometers/hour—these are also examples of derived units. Two other common derived units are those for volume (SI base unit is m^3) and density (SI base unit is kg/m^3). We will look at each of these individually.

Volume **Volume** is a measure of space. Any unit of length, when cubed (raised to the third power), becomes a unit of volume. Thus, the cubic meter (m^3), cubic centimeter (cm^3), and cubic millimeter (mm^3) are all units of volume. The cubic nature of volume is not always intuitive, and studies have shown that our brains are not naturally wired to think abstractly, as required to think about volume. For example, consider the following question: How many small cubes measuring 1 cm on each side are required to construct a large cube measuring 10 cm (or 1 dm) on a side?

The answer to this question, as you can see by carefully examining the unit cube in Figure 1.13 ▶, is 1000 small cubes. When you go from a linear, one-dimensional distance to three-dimensional volume, you must raise both the linear dimension *and* its unit to the third power (not multiply by 3). Thus the volume of a cube is equal to the length of its edge cubed:

$$\text{volume of cube} = (\text{edge length})^3$$

A cube with a 10-cm edge length has a volume of $(10 \text{ cm})^3$ or 1000 cm^3, and a cube with a 100-cm edge length has a volume of $(100 \text{ cm})^3 = 1,000,000 \text{ cm}^3$.

Other common units of volume in chemistry are the **liter (L)** and the **milliliter (mL)**. One milliliter (10^{-3} L) is equal to 1 cm^3. A gallon of gasoline contains 3.785 L. Table 1.3 lists some common units—for volume and other quantities—and their equivalents.

Relationship between Length and Volume

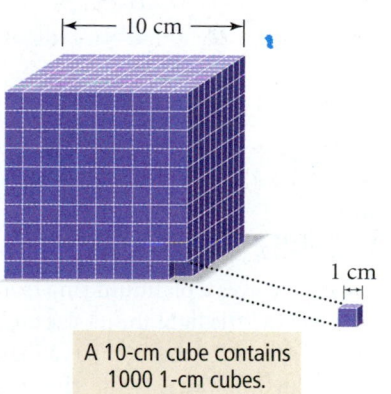

A 10-cm cube contains 1000 1-cm cubes.

▲ FIGURE 1.13 The Relationship between Length and Volume

TABLE 1.3 Some Common Units and Their Equivalents

Length

1 kilometer (km) = 0.6214 mile (mi)
1 meter (m) = 39.37 inches (in) = 1.094 yards (yd)
1 foot (ft) = 30.48 centimeters (cm)
1 inch (in) = 2.54 centimeters (cm) (exact)

Mass

1 kilogram (kg) = 2.205 pounds (lb)
1 pound (lb) = 453.59 grams (g)
1 ounce (oz) = 28.35 grams (g)

Volume

1 liter (L) = 1000 mL = 1000 cm^3
1 liter (L) = 1.057 quarts (qt)
1 U.S. gallon (gal) = 3.785 liters (L)

Density An old riddle asks, "Which weighs more, a ton of bricks or a ton of feathers?" The answer, of course, is neither—they both weigh the same (1 ton). If you answered bricks, you confused weight with density. The **density** (d) of a substance is the ratio of its mass (m) to its volume (V):

$$\text{Density} = \frac{\text{mass}}{\text{volume}} \quad \text{or} \quad d = \frac{m}{V}$$

Density is a characteristic physical property of materials and differs from one substance to another, as you can see in Table 1.4. The density of a substance also depends on its temperature. Density is an example of an **intensive property**, one that is *independent* of the amount of the substance. The density of aluminum, for example, is the same whether you have an ounce or a ton. Intensive properties are often used to identify substances because these properties depend only on the type of substance, not on the amount of it. For example, from Table 1.4 you can see that pure gold has a density of 19.3 g/cm³. One way to determine whether a substance is pure gold is to measure its density and compare it to 19.3 g/cm³. Mass, in contrast, is an **extensive property**, one that depends on the amount of the substance.

The units of density are those of mass divided by volume. Although the SI derived unit for density is kg/m³, the density of liquids and solids is most often expressed in g/cm³ or g/mL. (Remember that cm³ and mL are equivalent units: 1 cm³ = 1 mL.) Aluminum is among the least dense structural metals with a density of 2.7 g/cm³, while platinum is among the densest metals with a density of 21.4 g/cm³.

> Note that the *m* in this equation is in italic type, meaning that it stands for mass rather than for meters. In general, the symbols for units such as meters (m), seconds (s), or kelvins (K) appear in regular type while those for variables such as mass (*m*), volume (*V*), and time (*t*) appear in italics.

TABLE 1.4 The Density of Some Common Substances at 20 °C

Substance	Density (g/cm³)
Charcoal (from oak)	0.57
Ethanol	0.789
Ice	0.917 (at 0 °C)
Water	1.00 (at 4 °C)
Sugar (sucrose)	1.58
Table salt (sodium chloride)	2.16
Glass	2.6
Aluminum	2.70
Titanium	4.51
Iron	7.86
Copper	8.96
Lead	11.4
Mercury	13.55
Gold	19.3
Platinum	21.4

Calculating Density

The density of a substance is calculated by dividing the mass of a given amount of the substance by its volume. For example, suppose a small nugget suspected to be gold has a mass of 22.5 g and a volume of 2.38 cm³. To find its density, we divide the mass by the volume:

$$d = \frac{m}{V} = \frac{22.5 \text{ g}}{2.38 \text{ cm}^3} = 9.45 \text{ g/cm}^3$$

In this case, the density reveals that the nugget is not pure gold.

EXAMPLE 1.3 Calculating Density

A man receives a platinum ring from his fiancée. Before the wedding, he notices that the ring feels a little light for its size and decides to measure its density. He places the ring on a balance and finds that it has a mass of 3.15 grams. He then finds that the ring displaces 0.233 cm³ of water. Is the ring made of platinum? (Note: The volume of irregularly shaped objects is often measured by the displacement of water. To use this method, the object is placed in water and the change in volume of the water is measured. This increase in the total volume represents the volume of water *displaced* by the object, and is equal to the volume of the object.)

Set up the problem by writing the important information that is *given* as well as the information that you are asked to *find*. In this case, we are to find the density of the ring and compare it to that of platinum. *Note: This standard way of setting up problems is discussed in detail in Section 1.7.*	**Given** $m = 3.15$ g $V = 0.233$ cm **Find** Density in g/cm³
Next, write down the equation that defines density.	**Equation** $d = \dfrac{m}{V}$

Solve the problem by substituting the correct values of mass and volume into the expression for density.	Solution $d = \dfrac{m}{V} = \dfrac{3.15\,\text{g}}{0.233\,\text{cm}^3} = 13.5\,\text{g/cm}^3$

The density of the ring is much too low to be platinum (platinum density is 21.4 g/cm^3), and the ring is therefore a fake.

For Practice 1.3

The woman in the above example is shocked that the ring is fake and returns it. She buys a new ring that has a mass of 4.53 g and a volume of 0.212 cm^3. Is this ring genuine?

For More Practice 1.3

A metal cube has an edge length of 11.4 mm and a mass of 6.67 g. Calculate the density of the metal and use Table 1.4 to determine the likely identity of the metal.

Conceptual Connection 1.4 Density

The density of copper decreases with increasing temperature (as does the density of most substances). Which of the following will be true upon changing the temperature of a sample of copper from room temperature to 95 °C?

(a) the copper sample will become lighter
(b) the copper sample will become heavier
(c) the copper sample will expand
(d) the copper sample will contract

Answer: **(c)** The sample expands because its mass remains constant, but its density decreases.

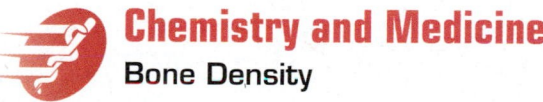
Chemistry and Medicine
Bone Density

Osteoporosis—which means *porous bone*—is a condition in which bone density becomes too low. The healthy bones of a young adult have a density of about 1.0 g/cm^3. Patients suffering from osteoporosis, however, can have bone densities as low as 0.22 g/cm^3. These low densities mean the bones have deteriorated and weakened, resulting in increased susceptibility to fractures, especially hip fractures. Patients suffering from osteoporosis can also experience height loss and disfiguration such as dowager's hump, a condition in which the patient becomes hunched over because of compression of the vertebrae. Osteoporosis is most common in postmenopausal women, but it can also occur in people (including men) who have certain diseases, such as insulin-dependent diabetes, or who take certain medications, such as prednisone. Osteoporosis is usually tested by taking hip X-rays. Low-density bones absorb less of the X-rays than do high-density bones, producing characteristic differences in the X-ray image that allow diagnosis. Treatments for osteoporosis include additional calcium and vitamin D, drugs that prevent bone weakening, exercise and strength training, and, in extreme cases, hip-replacement surgery.

Question

Suppose you find a large animal bone in the woods, too large to fit in a beaker or flask. How might you approximate its density?

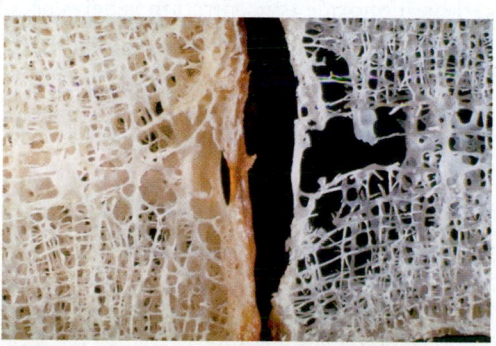
▲ Magnified views of the bone matrix in a normal vertebra (left) and one weakened by osteoporosis (right).

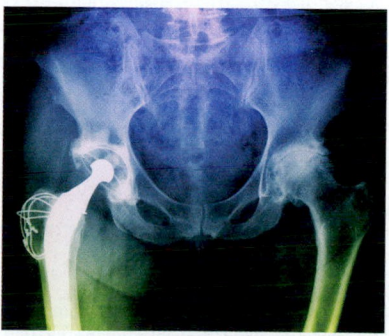
▲ Severe osteoporosis can necessitate surgery to implant an artificial hip joint, seen at left in this X-ray image.

1.7 The Reliability of a Measurement

Recall from our opening example (Section 1.1) that carbon monoxide is a colorless gas emitted by motor vehicles and found in polluted air. The table below shows carbon monoxide concentrations in Los Angeles County as reported by the U.S. Environmental Protection Agency (EPA) over the period 1997–2002:

Year	Carbon Monoxide Concentration (ppm)*
1997	15.0
1998	11.5
1999	11.1
2000	9.9
2001	7.2
2002	6.5

*Second maximum, 8 hour average; ppm = *parts per million*, defined as mL pollutant per million mL of air.

The first thing you should notice about these values is that they are decreasing over time. For this decrease, we can thank the Clean Air Act and its amendments, which have resulted in more efficient engines, in specially blended fuels, and consequently in cleaner air in all major U.S. cities over the last 30 years. The second thing you might notice is the number of digits to which the measurements are reported. The number of digits in a reported measurement indicates the certainty associated with that measurement. For example, a less certain measurement of carbon monoxide levels might be reported as follows:

Year	Carbon Monoxide Concentration (ppm)
1997	15
1998	12
1999	11
2000	10
2001	7
2002	7

Notice that the first set of data is reported to the nearest 0.1 ppm while the second set is reported to the nearest 1 ppm. Scientists agree to a standard way of reporting measured quantities in which the number of reported digits reflects the certainty in the measurement: more digits, more certainty; fewer digits, less certainty. Numbers are usually written so that the uncertainty is in the last reported digit. (That uncertainty is assumed to be ±1 in the last digit unless otherwise indicated.) For example, by reporting the 1997 carbon monoxide concentration as 15.0 ppm, the scientists mean 15.0±0.1 ppm. The carbon monoxide concentration is between 14.9 and 15.1 ppm—it might be 15.1 ppm, for example, but it could not be 16.0 ppm. In contrast, if the reported value was 15 ppm (without the .0), this would mean 15±1 ppm, or between 14 and 16 ppm. In general,

Scientific measurements are reported so that every digit is certain except the last, which is estimated.

For example, consider the following reported number:

$$5.213$$

certain   estimated

The first three digits are certain; the last digit is estimated. The number of digits reported in a measurement depends on the measuring device. For example, consider weighing a pistachio nut on two different balances (Figure 1.14 on page 23). The balance on the left has marks every 1 gram, while the balance on the right has marks every 0.1 gram. For the balance on the left, we mentally divide the space between the 1- and 2-gram marks into ten equal spaces and estimate that the pointer is at about 1.2 grams. We then write the measurement as 1.2 grams indicating that we are sure of the "1" but have estimated the ".2." The balance on the right, with marks every *tenth* of a gram,

In this situation, students will sometimes mentally divide the space between 1 and 2 into quarters or thirds, and make an estimate of 1.25 or 1.33. However, such an estimate is incorrect, because it really involves dividing the space into *hundredths* (25/100 or 33/100). Estimates should involve only the *first* decimal place beyond the last certain digit.

Estimation in Weighing

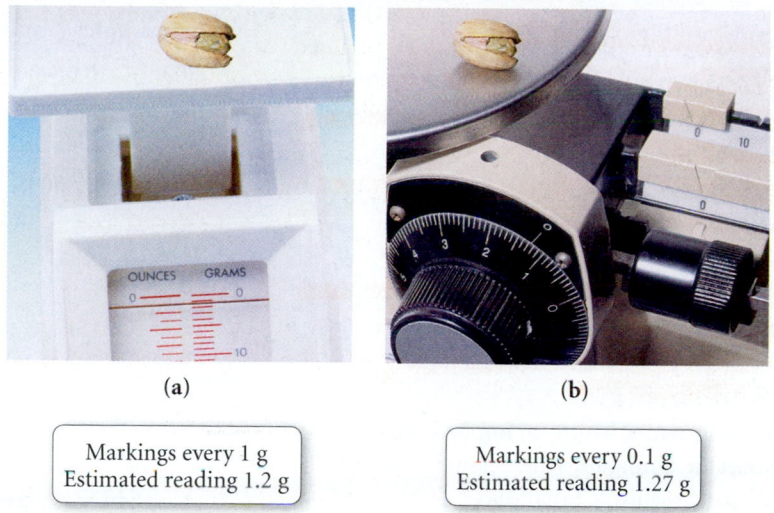

(a)

(b)

Markings every 1 g
Estimated reading 1.2 g

Markings every 0.1 g
Estimated reading 1.27 g

◀ **FIGURE 1.14 Estimation in Weighing**   (a) This scale has markings every 1 g, so we estimate to the tenths place by mentally dividing the space into ten equal spaces to estimate the last digit. This reading is 1.2 g. (b) Because this balance has markings every 0.1 g, we estimate to the hundredths place. This reading is 1.27 g.

requires us to write the result with more digits. The pointer is between the 1.2-gram mark and the 1.3-gram mark. We again divide the space between the two marks into ten equal spaces and estimate the third digit. For the figure shown, we report 1.27 g.

EXAMPLE 1.4 Reporting the Correct Number of Digits

The graduated cylinder shown at right has markings every 0.1 mL. Report the volume (which is read at the bottom of the meniscus) to the correct number of digits. (Note: The meniscus is the crescent-shaped surface at the top of a column of liquid.)

Solution

Since the bottom of the meniscus is between the 4.5 and 4.6 mL markings, mentally divide the space between the markings into ten equal spaces and estimate the next digit. In this case, you should report the result as 4.57 mL.

What if you estimated a little differently and wrote 4.56 mL? In general, one unit difference in the last digit is acceptable because the last digit is estimated and different people might estimate it slightly differently. However, if you wrote 4.63 mL, you would have misreported the measurement.

For Practice 1.4

Record the temperature on the thermometer shown at right to the correct number of significant digits.

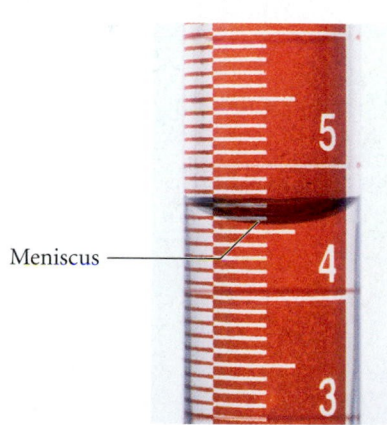

Meniscus

Counting Significant Figures

The precision of a measurement—which depends on the instrument used to make the measurement—must be preserved, not only when recording the measurement, but also when performing calculations that use the measurement. The preservation of this precision is conveniently accomplished by using *significant figures*. In any reported measurement, the non-place-holding digits—those that are not simply marking the decimal place—are called **significant figures** (or **significant digits**). *The greater the number of significant figures, the greater is the certainty of the measurement.* For example, the number 23.5 has three significant figures while the number 23.56 has four. To determine the number of significant figures in a number containing zeroes, we must distinguish between zeroes that are significant and those that simply mark the decimal place. For example, in the number 0.0008, the leading zeroes mark the decimal place but *do not* add to the certainty of the measurement and are therefore not significant; this number has only one significant figure. In contrast, the trailing zeroes in the number 0.000800 *do add* to the certainty of the measurement and are therefore counted as significant; this number has three significant figures.

To determine the number of significant figures in a number, follow these rules (with examples shown on the right).

Significant Figure Rules	Examples
1. All nonzero digits are significant.	28.03 0.0540
2. Interior zeroes (zeroes between two numbers) are significant.	408 7.0301
3. Leading zeroes (zeroes to the left of the first nonzero number) are not significant. They only serve to locate the decimal point.	0.0032 0.00006 not significant
4. Trailing zeroes (zeroes at the end of a number) are categorized as follows:	
• Trailing zeroes after a decimal point are always significant	45.000 3.5600
• Trailing zeroes before an implied decimal point are ambiguous and should be avoided by using scientific notation.	1200 ambiguous 1.2×10^3 2 significant figures 1.20×10^3 3 significant figures 1.200×10^3 4 significant figures
• Some textbooks put a decimal point after one or more trailing zeroes if the zeroes are to be considered significant. We avoid that practice in this book, but you should be aware of it.	1200. 4 significant figures (common in some textbooks)

Exact Numbers

Exact numbers have no uncertainty, and thus do not limit the number of significant figures in any calculation. In other words, we can regard an exact number as having an unlimited number of significant figures. Exact numbers originate from three sources:

- From the accurate counting of discrete objects. For example, 3 atoms means 3.00000 . . . atoms.
- From defined quantities, such as the number of centimeters in 1 m. Because 100 cm is defined as 1 m,

$$100 \text{ cm} = 1 \text{ m} \quad \text{means} \quad 100.00000 \ldots \text{cm} = 1.0000000 \ldots \text{m}$$

- From integral numbers that are part of an equation. For example, in the equation,

$$radius = \frac{diameter}{2},$$ the number 2 is exact and therefore has an unlimited number of significant figures.

EXAMPLE 1.5   Determining the Number of Significant Figures in a Number

How many significant figures are in each of the following?

(a) 0.04450 m   4
(b) 5.0003 km   5
(c) 10 dm = 1 m   exact
(d) 1.000×10^5 s   4
(e) 0.00002 mm   1
(f) 10,000 m   ambiguous

Solution

(a) 0.04450 m	*Four significant figures.* The two 4's and the 5 are significant (rule 1). The trailing zero is after a decimal point and is therefore significant (rule 4). The leading zeroes only mark the decimal place and are therefore not significant (rule 3).
(b) 5.0003 km	*Five significant figures.* The 5 and 3 are significant (rule 1) as are the three interior zeroes (rule 2).

(c) 10 dm = 1 m	*Unlimited significant figures.* Defined quantities have an unlimited number of significant figures.
(d) 1.000×10^5 s	*Four significant figures.* The 1 is significant (rule 1). The trailing zeroes are after a decimal point and therefore significant (rule 4).
(e) 0.00002 mm	*One significant figure.* The 2 is significant (rule 1). The leading zeroes only mark the decimal place and are therefore not significant (rule 3).
(f) 10,000 m	*Ambiguous.* The 1 is significant (rule 1) but the trailing zeroes occur before an implied decimal point and are therefore ambiguous (rule 4). Without more information, we would assume 1 significant figure. It is better to write this as 1×10^5 to indicate one significant figure or as 1.0000×10^5 to indicate five (rule 4).

For Practice 1.5

How many significant figures are in each of the following numbers?

(a) 554 km

(b) 7 pennies

(c) 1.01×10^5 m 3

(d) 0.00099 s 2

(e) 1.4500 km 3

(f) 21,000 m ambiguous

Significant Figures in Calculations

When you use measured quantities in calculations, the results of the calculation must reflect the precision of the measured quantities. You should not lose or gain precision during mathematical operations. Follow these rules when carrying significant figures through calculations.

Rules for Calculations

Examples

1. In multiplication or division, the result carries the same number of significant figures as the factor with the fewest significant figures.

$$1.052 \times 12.054 \times 0.53 = 6.7208 = 6.7$$
(4 sig. figures) (5 sig. figures) (2 sig. figures)   (2 sig. figures)

$$2.0035 \div 3.20 = 0.626094 = 0.626$$
(5 sig. figures) (3 sig. figures)   (3 sig. figures)

2. In addition or subtraction the result carries the same number of decimal places as the quantity with the fewest decimal places.

```
  2.345              5.9
  0.07             −0.221
+2.9975            5.679 = 5.7
 5.4125 = 5.41
```

In addition and subtraction, it is helpful to draw a line next to the number with the fewest decimal places. This line determines the number of decimal places in the answer.

3. When rounding to the correct number of significant figures, round down if the last (or leftmost) digit dropped is four or less; round up if the last (or leftmost) digit dropped is five or more.

To two significant figures:

5.37 rounds to 5.4

5.34 rounds to 5.3

5.35 rounds to 5.4

5.349 rounds to 5.3

Notice in the last example that only the *last (or leftmost) digit being dropped* determines in which direction to round—ignore all digits to the right of it.

A few books recommend a slightly different rounding procedure for cases where the last digit is 5. However, the procedure presented here is consistent with electronic calculators and will be used throughout this book.

4. To avoid rounding errors in multistep calculations round only the final answer—do not round intermediate steps. If you write down intermediate answers, keep track of significant figures by underlining the least significant digit.

$$6.78 \times 5.903 \times (5.489 - 5.01)$$
$$= 6.78 \times 5.903 \times 0.479$$
$$= 19.1707$$
$$= 19$$

underline least significant digit

Notice that for multiplication or division, the quantity with the fewest *significant figures* determines the number of *significant figures* in the answer, but for addition and subtraction, the quantity with the fewest *decimal places* determines the number of *decimal places* in the answer. In multiplication and division, we focus on significant figures, but in addition and subtraction we focus on decimal places. When a problem involves addition or subtraction, the answer may have a different number of significant figures than the initial quantities. Keep this in mind in problems that involve both addition or subtraction and multiplication or division. For example,

$$\frac{1.002 - 0.999}{3.754} = \frac{0.003}{3.754}$$

$$= 7.99 \times 10^{-4}$$

$$= 8 \times 10^{-4}$$

The answer has only one significant figure, even though the initial numbers had three or four.

EXAMPLE 1.6   Significant Figures in Calculations

Perform the following calculations to the correct number of significant figures.

(a) $1.10 \times 0.5120 \times 4.0015 \div 3.4555$

(b)   0.355
      $+105.1$
      -100.5820

(c) $4.562 \times 3.99870 \div (452.6755 - 452.33)$

(d) $(14.84 \times 0.55) - 8.02$

Solution

(a) Round the intermediate result (in blue) to three significant figures to reflect the three significant figures in the least precisely known quantity (1.10).	$1.10 \times 0.5120 \times 4.0015 \div 3.4555$ $= 0.65219$ $= 0.652$
(b) Round the intermediate answer (in blue) to one decimal place to reflect the quantity with the fewest decimal places (105.1). Notice that 105.1 is *not* the quantity with the fewest significant figures, but it has the fewest decimal places and therefore determines the number of decimal places in the answer.	0.3|55 $+105.1$ $-100.5$820 4.8|730 $= 4.9$
(c) Mark the intermediate result to two decimal places to reflect the number of decimal places in the quantity within the parentheses having the fewest number of decimal places (452.33). Round the final answer to two significant figures to reflect the two significant figures in the least precisely known quantity (0.3455).	$4.562 \times 3.99870 \div (452.6755 - 452.33)$ $= 4.562 \times 3.99870 \div 0.3455$ $= 52.79904$ $= 53$ 2 places of the decimal
(d) Mark the intermediate result to two significant figures to reflect the number of significant figures in the quantity within the parentheses having the fewest number of significant figures (0.55). Round the final answer to one decimal place to reflect the one decimal place in the least precisely known quantity (8.162).	$(14.84 \times 0.55) - 8.02$ $= 8.162 - 8.02$ $= 0.142$ $= 0.1$

For Practice 1.6

Perform the following calculations to the correct number of significant figures.

(a) $3.10007 \times 9.441 \times 0.0301 \div 2.31$

(b)
$$
\begin{array}{r}
0.881 \\
+132.1 \\
-12.02 \\
\hline
\end{array}
$$

(c) $2.5110 \times 21.20 \div (44.11 + 1.223)$

(d) $(12.01 \times 0.3) + 4.811$

Precision and Accuracy

Scientific measurements are often repeated several times to increase confidence in the result. We can distinguish between two different kinds of certainty—called accuracy and precision—associated with such measurements. **Accuracy** refers to how close the measured value is to the actual value. **Precision** refers to how close a series of measurements are to one another or how reproducible they are. A series of measurements can be precise (close to one another in value and reproducible) but not accurate (not close to the true value). For example, consider the results of three students who repeatedly weighed a lead block known to have a true mass of 10.00 g (indicated by the solid horizontal blue line on the graphs).

	Student A	Student B	Student C
Trial 1	10.49 g	9.78 g	10.03 g
Trial 2	9.79 g	9.82 g	9.99 g
Trial 3	9.92 g	9.75 g	10.03 g
Trial 4	10.31 g	9.80 g	9.98 g
Average	**10.13 g**	**9.79 g**	**10.01 g**

- The results of student A are both inaccurate (not close to the true value) and imprecise (not consistent with one another). The inconsistency is the result of **random error**, error that has equal probability of being too high or too low. Almost all measurements have some degree of random error. Random error can, with enough trials, average itself out.

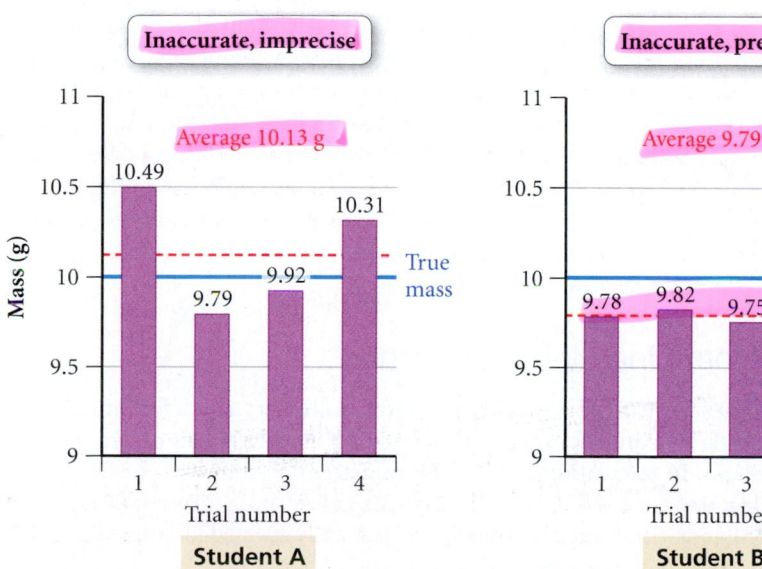

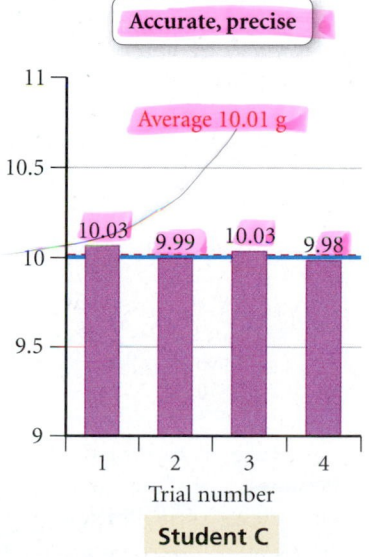

▲ Measurements are said to be precise if they are consistent with one another, but they are accurate only if they are close to the actual value.

- The results of student B are precise (close to one another in value) but inaccurate. The inaccuracy is the result of **systematic error**, error that tends toward being either too high or too low. Systematic error does not average out with repeated trials. For example, if a balance is not properly calibrated, it may systematically read too high or too low.
- The results of student C display little systematic error or random error—they are both accurate and precise.

Chemistry in Your Day
Integrity in Data Gathering

Most scientists spend many hours collecting data in the laboratory. Often, the data do not turn out exactly as the scientist had expected (or hoped). A scientist may then be tempted to "fudge" his or her results. For example, suppose you are expecting a particular set of measurements to follow a certain pattern. After working hard over several days or weeks to make the measurements, you notice that a few of them do not quite fit the pattern that you anticipated. You might find yourself wishing that you could simply change or omit the "faulty" measurements to better fit your expectations. Altering data in this way is considered highly unethical in the scientific community and, when discovered, is usually punished severely.

In 2004, Dr. Hwang Woo Suk, a stem cell researcher at the Seoul National University in Korea, published a research paper in *Science* (a highly respected research journal) claiming that he and his colleagues had cloned human embryonic stem cells. As part of his evidence, he showed photographs of the cells. The

paper was hailed as an incredible breakthrough, and Dr. Hwang traveled the world lecturing on his work. *Time* magazine even named him among their "people that matter" for 2004. Several months later, however, one of his co-workers revealed that the photographs were fraudulent. According to the co-worker, the photographs came from a computer data bank of stem cell photographs, not from a cloning experiment. A university panel investigated the results and confirmed that the photographs and other data had indeed been faked. Dr. Hwang was forced to resign his prestigious post at the university.

Although not common, incidents like this do occur from time to time. They are damaging to a community that is largely built on trust. Research papers are reviewed by peers (other researchers in similar fields), but usually reviewers are judging whether the data support the conclusion—they assume that the experimental measurements are authentic. The pressure to succeed sometimes leads researchers to betray that trust. However, over time, the tendency of scientists to reproduce and build upon one another's work results in the discovery of the fraudulent data. When that happens, the researchers at fault are usually banished from the community and their careers are ruined.

1.8 Solving Chemical Problems

Learning to solve problems is one of the most important skills you will acquire in this course. No one succeeds in chemistry—or in life, really—without the ability to solve problems. Although no simple formula applies to every problem, you can learn problem-solving strategies and begin to develop some chemical intuition. Many of the problems you will solve in this course can be thought of as *unit conversion problems*, where you are given one or more quantities and asked to convert them into different units. Other problems require the use of *specific equations* to get to the information you are trying to find. In the sections that follow, you will find strategies to help you solve both of these types of problems. Of course, many problems contain both conversions and equations, requiring the combination of these strategies.

Converting from One Unit to Another

In Section 1.6, we learned the SI unit system, the prefix multipliers, and a few other units. Knowing how to work with and manipulate these units in calculations is central to solving chemical problems. In calculations, units help to determine correctness. Using units as a guide to solving problems is often called **dimensional analysis**. Units should always be included in calculations; they are multiplied, divided, and canceled like any other algebraic quantity.

Consider converting 12.5 inches (in) to centimeters (cm). We know from Table 1.3 that 1 in = 2.54 cm (exact), so we can use this quantity in the calculation as follows:

$$12.5 \; \text{in} \times \frac{2.54 \; \text{cm}}{1 \; \text{in}} = 31.8 \; \text{cm}$$

The unit, in, cancels and we are left with cm as our final unit. The quantity $\frac{2.54\ cm}{1\ in}$ is a **conversion factor**—a fractional quantity with the units we are *converting from* on the bottom and the units we are *converting to* on the top. Conversion factors are constructed from any two equivalent quantities. In this example, 2.54 cm = 1 in, so we construct the conversion factor by dividing both sides of the equality by 1 in and canceling the units

$$2.54\ cm = 1\ in$$

$$\frac{2.54\ cm}{1\ in} = \frac{1\ \cancel{in}}{1\ \cancel{in}}$$

$$\frac{2.54\ cm}{1\ in} = 1$$

The quantity $\frac{2.54\ cm}{1\ in}$ is equivalent to 1, so multiplying by the conversion factor affects only the units, not the actual quantity. To convert the other way, from centimeters to inches, we must—using units as a guide—use a different form of the conversion factor. If you accidentally use the same form, you will get the wrong result, indicated by erroneous units. For example, suppose that you want to convert 31.8 cm to inches.

$$31.8\ cm \times \frac{2.54\ cm}{1\ in} = \frac{80.8\ cm^2}{in}$$

The units in the above answer (cm^2/in), as well as the value of the answer, are obviously wrong. When you solve a problem, always look at the final units. Are they the desired units? Always look at the magnitude of the numerical answer as well. Does it make sense? In this case, our mistake was the form of the conversion factor. It should have been inverted so that the units cancel as follows:

$$31.8\ \cancel{cm} \times \frac{1\ in}{2.54\ \cancel{cm}} = 12.5\ in$$

Conversion factors can be inverted because they are equal to 1 and the inverse of 1 is 1. Therefore,

$$\frac{2.54\ cm}{1\ in} = 1 = \frac{1\ in}{2.54\ cm}$$

Most unit conversion problems take the following form:

Information given × conversion factor(s) = information sought

$$\text{Given unit} \quad \times \quad \frac{\text{desired unit}}{\text{given unit}} \quad = \text{desired unit}$$

In this book, we diagram a problem solution using a *conceptual plan*. A conceptual plan is a visual outline that helps you to see the general flow of the problem. For unit conversions, the conceptual plan focuses on units and the conversion from one unit to another. The conceptual plan for converting in to cm is as follows:

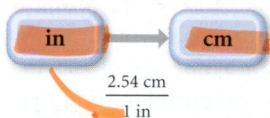

The conceptual plan for converting the other way, from cm to in, is just the reverse, with the reciprocal conversion factor:

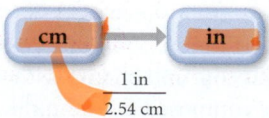

Each arrow in a conceptual plan for a unit conversion has an associated conversion factor with the units of the previous step in the denominator and the units of the following step in the numerator. In the following section, we incorporate the idea of a conceptual plan into an overall approach to solving numerical chemical problems.

General Problem-Solving Strategy

In this book, we use a standard problem-solving procedure that can be adapted to many of the problems encountered in general chemistry and beyond. Solving any problem essentially requires you to assess the information given in the problem and devise a way to get to the information asked for. In other words, you must

- Identify the starting point (the *given* information).
- Identify the end point (what you must *find*).
- Devise a way to get from the starting point to the end point using what is given as well as what you already know or can look up. (We call this guide the *conceptual plan*.)

In graphic form, we can represent this progression as

$$\text{Given} \longrightarrow \text{Conceptual Plan} \longrightarrow \text{Find}$$

One of the main difficulties beginning students have when trying to solve problems in general chemistry is simply not knowing where to start. While no problem-solving procedure is applicable to all problems, the following four-step procedure can be helpful in working through many of the numerical problems you will encounter in this book.

1. **Sort.** Begin by sorting the information in the problem. *Given* information is the basic data provided by the problem—often one or more numbers with their associated units. *Find* indicates what information you will need for your answer.

2. **Strategize.** This is usually the hardest part of solving a problem. In this process, you must develop a *conceptual plan*—a series of steps that will get you from the given information to the information you are trying to find. You have already seen conceptual plans for simple unit conversion problems. Each arrow in a conceptual plan represents a computational step. On the left side of the arrow is the quantity you had before the step; on the right side of the arrow is the quantity you will have after the step; and below the arrow is the information you need to get from one to the other—the relationship between the quantities.

Often such relationships will take the form of conversion factors or equations. These may be given in the problem, in which case you will have written them down under "Given" in step 1. Usually, however, you will need other information—which may include physical constants, formulas, or conversion factors—to help get you from what you are given to what you must find. You must recall this information from what you have learned or look it up in the chapter or in tables within the book.

In some cases, you may get stuck at the strategize step. If you cannot figure out how to get from the given information to the information you are asked to find, you might try working backwards. For example, you may want to look at the units of the quantity you are trying to find and try to find conversion factors to get to the units of the given quantity. You may even try a combination of strategies; work forward, backward, or some of both. If you persist, you will develop a strategy to solve the problem.

3. **Solve.** This is the easiest part of solving a problem. Once you set up the problem properly and devise a conceptual plan, you simply follow the plan to solve the problem. Carry out any mathematical operations (paying attention to the rules for significant figures in calculations) and cancel units as needed.

4. **Check.** This is the step most often overlooked by beginning students. Experienced problem solvers always go one step further and ask, does this answer make physical sense? Are the units correct? Is the number of significant figures correct? When solving multistep problems, errors easily creep into the solution. You can catch most of these errors by simply checking the answer. For example, suppose you are calculating the number of atoms in a gold coin and end up with an answer of 1.1×10^{-6} atoms. Could the gold coin really be composed of one-millionth of one atom?

Below we apply this problem-solving procedure to unit conversion problems. The procedure is summarized in the left column and two examples of applying the procedure are shown in the middle and right columns. This three-column format will be used in selected examples throughout the text. It allows you to see how a particular procedure can be applied

Most problems can be solved in more than one way. The solutions we derive in this book will tend to be the most straightforward but certainly not the only way to solve the problem.

to two different problems. Work through one problem first (from top to bottom) and then see how the same procedure is applied to the other problem. Being able to see the commonalities and differences between problems is a key part of developing problem-solving skills.

Procedure for Solving Unit Conversion Problems	**EXAMPLE 1.7** Unit Conversion Convert 1.76 yards to centimeters.	**EXAMPLE 1.8** Unit Conversion Convert 1.8 quarts to cubic centimeters.
Sort Begin by sorting the information in the problem into *given* and *find*.	**Given** 1.76 yd **Find** cm	**Given** 1.8 qt **Find** cm^3
Strategize Devise a *conceptual plan* for the problem. Begin with the *given* quantity and symbolize each conversion step with an arrow. Below each arrow, write the appropriate conversion factor for that step. Focus on the units. The conceptual plan should end at the *find* quantity and its units. In these examples, the other information needed consists of relationships between the various units as shown.	**Conceptual Plan** $$\boxed{yd} \rightarrow \boxed{m} \rightarrow \boxed{cm}$$ $\quad\quad \dfrac{1\,m}{1.094\,yd} \quad \dfrac{100\,cm}{1\,m}$ **Relationships Used** 1.094 yd = 1 m 1 m = 100 cm (These conversion factors are from Tables 1.2 and 1.3.)	**Conceptual Plan** $$\boxed{qt} \rightarrow \boxed{L} \rightarrow \boxed{mL} \rightarrow \boxed{cm^3}$$ $\quad \dfrac{1\,L}{1.057\,qt} \quad \dfrac{1000\,mL}{1\,L} \quad \dfrac{1\,cm^3}{1\,mL}$ **Relationships Used** 1.057 qt = 1 L 1 L = 1000 mL 1 ml = 1 cm^3 (These conversion factors are from Tables 1.2 and 1.3.)
Solve Follow the conceptual plan. Begin with the *given* quantity and its units. Multiply by the appropriate conversion factor(s), canceling units, to arrive at the *find* quantity. Round the answer to the correct number of significant figures by following the rules in Section 1.7. Remember that exact conversion factors do not limit significant figures.	**Solution** $1.76\,\cancel{yd} \times \dfrac{1\,\cancel{m}}{1.094\,\cancel{yd}} \times \dfrac{100\,cm}{1\,\cancel{m}}$ $= 160.8775\,cm$ $160.8775\,cm = 161\,cm$	**Solution** $1.8\,\cancel{qt} \times \dfrac{1\,\cancel{L}}{1.057\,\cancel{qt}} \times \dfrac{1000\,\cancel{mL}}{1\,\cancel{L}}$ $\times \dfrac{1\,cm^3}{1\,\cancel{mL}} = 1.70293 \times 10^3\,cm^3$ $1.70293 \times 10^3\,cm^3 = 1.7 \times 10^3\,cm^3$
Check Check your answer. Are the units correct? Does the answer make physical sense?	The units (cm) are correct. The magnitude of the answer (161) makes physical sense because a centimeter is a much smaller unit than a yard.	The units (cm^3) are correct. The magnitude of the answer (1700) makes physical sense because a cubic centimeter is a much smaller unit than a quart.
	For Practice 1.7 Convert 288 cm to yards.	**For Practice 1.8** Convert 9255 cm^3 to gallons.

Units Raised to a Power

When building conversion factors for units raised to a power, remember to raise both the number and the unit to the power. For example, to convert from in^2 to cm^2, we construct the conversion factor as follows:

$$2.54\,cm = 1\,in$$
$$(2.54\,cm)^2 = (1\,in)^2$$
$$(2.54)^2\,cm^2 = 1^2\,in^2$$
$$6.45\,cm^2 = 1\,in^2$$
$$\frac{6.45\,cm^2}{1\,in^2} = 1$$

The following example shows how to use conversion factors involving units raised to a power.

EXAMPLE 1.9 Unit Conversions Involving Units Raised to a Power

Calculate the displacement (the total volume of the cylinders through which the pistons move) of a 5.70-L automobile engine in cubic inches.

Sort Sort the information in the problem into *Given* and *Find*.	**Given** 5.70 L **Find** in^3
Strategize Write a conceptual plan. Begin with the given information and devise a path to the information that you are asked to find. Notice that for cubic units, the conversion factors must be cubed.	**Conceptual Plan** $$L \longrightarrow mL \longrightarrow cm^3 \longrightarrow in^3$$ $$\frac{1\ mL}{10^{-3}\ L} \qquad \frac{1\ cm^3}{1\ mL} \qquad \frac{(1\ in)^3}{(2.54\ cm)^3}$$ **Relationships Used** $$1\ mL = 10^{-3}\ L$$ $$1\ mL = 1\ cm^3$$ $$2.54\ cm = 1\ in$$ (These conversion factors are from Tables 1.2 and 1.3.)
Solve Follow the conceptual plan to solve the problem. Round the answer to three significant figures to reflect the three significant figures in the least precisely known quantity (5.70 L). These conversion factors are all exact and therefore do not limit the number of significant figures.	**Solution** $$5.70\ \cancel{L} \times \frac{1\ \cancel{mL}}{10^{-3}\ \cancel{L}} \times \frac{1\ \cancel{cm^3}}{1\ \cancel{mL}} \times \frac{(1\ in)^3}{(2.54\ \cancel{cm})^3} = 347.835\ in^3$$ $$= 348\ in^3$$

Check The units of the answer are correct and the magnitude makes sense. The unit cubic inches is smaller than liters, so the volume in cubic inches should be larger than the volume in liters.

For Practice 1.9

How many cubic centimeters are there in 2.11 yd^3?

For More Practice 1.9

A vineyard has 145 acres of Chardonnay grapes. A particular soil supplement requires 5.50 grams for every square meter of vineyard. How many kilograms of the soil supplement are required for the entire vineyard? ($1\ km^2 = 247$ acres)

EXAMPLE 1.10 Density as a Conversion Factor

The mass of fuel in a jet must be calculated before each flight to ensure that the jet is not too heavy to fly. A 747 is fueled with 173,231 L of jet fuel. If the density of the fuel is 0.768 g/cm^3, what is the mass of the fuel in kilograms?

Sort Begin by *sorting* the information in the problem into *Given* and *Find*.	**Given** fuel volume = 173,231 L density of fuel = 0.768 g/cm^3 **Find** mass in kg

Strategize Draw the conceptual plan by beginning with the given quantity, in this case the volume in liters (L). The overall goal of this problem is to find the mass. You can convert between volume and mass using density (g/cm^3). However, you must first convert the volume to cm^3. Once you have converted the volume to cm^3, use the density to convert to g. Finally convert g to kg.

Conceptual Plan

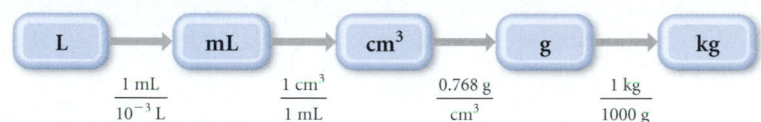

$$\frac{1\ mL}{10^{-3}\ L} \qquad \frac{1\ cm^3}{1\ mL} \qquad \frac{0.768\ g}{cm^3} \qquad \frac{1\ kg}{1000\ g}$$

Relationships Used

$1\ mL = 10^{-3}\ L$

$1\ mL = 1\ cm^3$

$d = 0.768\ g/cm^3$

$1000\ g = 1\ kg$

(These conversion factors are from Tables 1.2 and 1.3.)

Solve Follow the conceptual plan to solve the problem. Round the answer to three significant figures to reflect the three significant figures in the density.

Solution

$$173{,}231\ \cancel{L} \times \frac{1\ \cancel{mL}}{10^{-3}\ \cancel{L}} \times \frac{1\ \cancel{cm^3}}{1\ \cancel{mL}} \times \frac{0.768\ \cancel{g}}{1\ \cancel{cm^3}} \times \frac{1\ kg}{1000\ \cancel{g}} = 1.33 \times 10^5\ kg$$

Check The units of the answer (kg) are correct. The magnitude makes sense because the mass $(1.33 \times 10^5\ kg)$ is similar in magnitude to the given volume (173,231 L or $1.73231 \times 10^5\ L$), as expected for a density close to one $(0.768\ g/cm^3)$.

For Practice 1.10

Backpackers often use canisters of white gas to fuel a cooking stove's burner. If one canister contains 1.45 L of white gas, and the density of the gas is $0.710\ g/cm^3$, what is the mass of the fuel in kilograms?

For More Practice 1.10

A drop of gasoline has a mass of 22 mg and a density of $0.754\ g/cm^3$. What is its volume in cubic centimeters?

Order of Magnitude Estimations

Calculation plays a major role in chemical problem solving. But precise numerical calculation is not always necessary, or even possible. Sometimes data are only approximate, so there is no point in trying to determine an extremely precise answer. At other times, you simply don't need a high degree of precision—a rough estimate or a simplified "back of the envelope" calculation is enough. Scientists often use these kinds of calculations to get an initial feel for a problem, or as a quick check to see whether a proposed solution is "in the right ballpark."

One way to make such estimates is to simplify the numbers so that they can be manipulated easily. The technique known as *order-of-magnitude estimation* is based on focusing only on the exponential part of numbers written in scientific notation, according to the following guidelines:

- If the decimal part of the number is less than 5, just drop it. Thus, 4.36×10^5 becomes 10^5 and 2.7×10^{-3} becomes 10^{-3}.

- If the decimal part is 5 or more, round it up to 10 and rewrite the number as a power of 10. Thus, 5.982×10^7 becomes $10 \times 10^7 = 10^8$, and 6.1101×10^{-3} becomes $10 \times 10^{-3} = 10^{-2}$.

When you make these approximations, you are left with powers of 10, which are easily multiplied and divided—often in your head. It's important to remember, however, that your answer is only as reliable as the numbers used to get it, so never assume that the results of an order-of-magnitude calculation are accurate to more than an order of magnitude.

Suppose, for example, that you want to estimate the number of atoms an immortal being could have counted in the 14 billion (1.4×10^{10}) years that the universe has been in existence, assuming a counting rate of 10 atoms per second. Since a year has 3.2×10^7 seconds, you can approximate the number of atoms counted as follows:

$$10^{10}\ \cancel{\text{years}} \quad \times \quad 10^7 \frac{\cancel{\text{seconds}}}{\cancel{\text{year}}} \quad \times \quad 10^1 \frac{\text{atoms}}{\cancel{\text{second}}} \quad \approx \quad 10^{18}\ \text{atoms}$$

(number of years)   (number of seconds per year)   (number of atoms counted per second)

A million trillion atoms (10^{18}) may seem like a lot, but as you will see in Chapter 2, a million trillion atoms are nearly impossible to see without a microscope.

In our general problem-solving procedure, the last step is to check whether the results seem reasonable. Order-of-magnitude estimations can often help you catch the kinds of mistakes that may happen in a detailed calculation, such as entering an incorrect exponent or sign into your calculator, or multiplying when you should have divided.

Problems Involving an Equation

Problems involving equations can be solved in much the same way as problems involving conversions. Usually, in problems involving equations, you must find one of the variables in the equation, given the others. The *conceptual plan* concept outlined above can be used for problems involving equations. For example, suppose you are given the mass (m) and volume (V) of a sample and asked to calculate its density. The conceptual plan shows how the *equation* takes you from the *given* quantities to the *find* quantity.

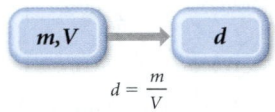

$$d = \frac{m}{V}$$

Here, instead of a conversion factor under the arrow, this conceptual plan has an equation. The equation shows the *relationship* between the quantities on the left of the arrow and the quantities on the right. Note that at this point, the equation need not be solved for the quantity on the right (although in this particular case it is). The procedure that follows, as well as the two examples, will guide you in developing a strategy to solve problems involving equations. We again use the three-column format here. Work through one problem from top to bottom and then see how the same general procedure is applied to the second problem.

Procedure for Solving Problems Involving Equations	EXAMPLE 1.11 Problems with Equations	EXAMPLE 1.12 Problems with Equations
	Find the radius (r), in centimeters, of a spherical water droplet with a volume (V) of 0.058 cm^3. For a sphere, $V = (4/3)\pi r^3$.	Find the density (in g/cm^3) of a metal cylinder with a mass of 8.3 g, a length (l) of 1.94 cm, and a radius (r) of 0.55 cm. For a cylinder, $V = \pi r^2 l$.
Sort Begin by sorting the information in the problem into *Given* and *Find*.	Given $V = 0.058$ cm^3 Find r in cm	Given $m = 8.3$ g $l = 1.94$ cm $r = 0.55$ cm Find d in g/cm^3

Strategize	Conceptual Plan	Conceptual Plan

Strategize

Write a *conceptual plan* for the problem. Focus on the equation(s). The conceptual plan shows how the equation takes you from the *given* quantity (or quantities) to the *find* quantity. The conceptual plan may have several parts, involving other equations or required conversions. In these examples, you must use the geometrical relationships given in the problem statements as well as the definition of density, $d = m/V$, which you learned in this chapter.

Conceptual Plan

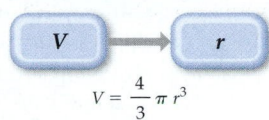

$$V = \frac{4}{3}\pi r^3$$

Relationships Used

$$V = \frac{4}{3}\pi r^3$$

Conceptual Plan

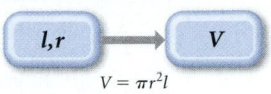

$$V = \pi r^2 l$$

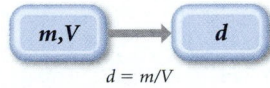

$$d = m/V$$

Relationships Used $V = \pi r^2 l$

$$d = \frac{m}{V}$$

Solve

Follow the conceptual plan. Solve the equation(s) for the *find* quantity (if it is not already). Gather each of the quantities that must go into the equation in the correct units. (Convert to the correct units if necessary.) Substitute the numerical values and their units into the equation(s) and compute the answer.

   Round the answer to the correct number of significant figures.

Solution

$$V = \frac{4}{3}\pi r^3$$

$$r^3 = \frac{3}{4\pi}V$$

$$r = \left(\frac{3}{4\pi}V\right)^{1/3}$$

$$= \left(\frac{3}{4\pi}0.058\ \text{cm}^3\right)^{1/3} = 0.24013\ \text{cm}$$

$$0.24013\ \text{cm} = 0.24\ \text{cm}$$

Solution

$$V = \pi r^2 l$$

$$= \pi(0.55\ \text{cm})^2(1.94\ \text{cm})$$

$$= 1.8436\ \text{cm}^3$$

$$d = \frac{m}{V}$$

$$= \frac{8.3\ \text{g}}{1.8436\ \text{cm}^3} = 4.50195\ \text{g/cm}^3$$

$$4.50195\ \text{g/cm}^3 = 4.5\ \text{g/cm}^3$$

Check

Check your answer. Are the units correct? Does the answer make physical sense?

The units (cm) are correct and the magnitude seems right.

The units (g/cm^3) are correct. The magnitude of the answer seems correct for one of the lighter metals (see Table 1.4).

For Practice 1.11

Find the radius (r) of an aluminum cylinder that is 2.00 cm long and has a mass of 12.4 g. For a cylinder, $V = \pi r^2 l$.

For Practice 1.12

Find the density, in g/cm^3, of a metal cube with a mass of 50.3 g and an edge length (l) of 2.65 cm. For a cube, $V = l^3$.

Chapter in Review

Key Terms

Section 1.1

atoms (3)
molecules (3)
chemistry (5)

Section 1.2

hypothesis (5)
experiment (5)
scientific law (5)

law of conservation
of mass (5)
theory (6)
atomic theory (6)
scientific method (6)

Section 1.3

matter (7)
state (7)

solid (7)
liquid (7)
gas (7)
crystalline (8)
amorphous (8)
pure substance (10)
mixture (10)
element (10)
compound (10)

heterogeneous mixture (10)
homogeneous mixture (10)
decanting (10)
distillation (10)
volatile (10)
filtration (10)

Section 1.4

physical change (11)

chemical change (11)
physical property (11)
chemical property (11)

Section 1.5

energy (14)
work (14)
kinetic energy (14)
potential energy (14)
thermal energy (14)
law of conservation
    of energy (14)

Section 1.6

units (15)
English system (15)
metric system (15)
International System of
    Units (SI) (15)
meter (m) (16)
kilogram (kg) (16)
mass (16)
second (s) (16)
kelvin (K) (17)
temperature (17)

Fahrenheit (°F)
    scale (17)
Celsius (°C) scale (17)
Kelvin scale (17)
prefix multipliers (19)
derived unit (19)
volume (19)
liter (L) (19)
milliliter (mL) (19)
density (d) (20)
intensive property (20)
extensive property (20)

Section 1.7

significant figures
    (significant digits) (23)
exact numbers (24)
accuracy (27)
precision (27)
random error (27)
systematic error (28)

Section 1.8

dimensional analysis (28)
conversion factor (29)

Key Concepts

Atoms and Molecules (1.1)

All matter is composed of atoms and molecules. Chemistry is the science that investigates the properties of matter by examining the atoms and molecules that compose it.

The Scientific Method (1.2)

Science begins with the observation of the physical world. A number of related observations can often be subsumed in a summary statement or generalization called a scientific law. Observations may suggest a hypothesis, a tentative interpretation or explanation of the observed phenomena. One or more well-established hypotheses may prompt the development of a scientific theory, a model for nature that explains the underlying reasons for observations and laws. Laws, hypotheses, and theories all give rise to predictions that can be tested by experiments, carefully controlled procedures designed to produce critical new observations. If the predictions are not confirmed, the law, hypothesis, or theory must be modified or replaced.

The Classification of Matter (1.3)

Matter can be classified according to its state (solid, liquid, or gas) or according to its composition (pure substance or mixture). A pure substance can either be an element, which is not decomposable into simpler substances, or a compound, which is composed of two or more elements in fixed proportions. A mixture can be either homogeneous, with the same composition throughout, or heterogeneous, with different compositions in different regions.

The Properties of Matter (1.4)

The properties of matter can be divided into two kinds: physical and chemical. Matter displays its physical properties without changing its composition. Matter displays its chemical properties only through changing its composition. Changes in matter in which its composition does not change are called physical changes. Changes in matter in which its composition does change are called chemical changes.

Energy (1.5)

In chemical and physical changes, matter often exchanges energy with its surroundings. In these exchanges, the total energy is always conserved; energy is neither created nor destroyed. Systems with high potential energy tend to change in the direction of lower potential energy, releasing energy into the surroundings.

The Units of Measurement and Significant Figures (1.6, 1.7)

Scientists use primarily SI units, which are based on the metric system. The SI base units include the meter (m) for length, the kilogram (kg) for mass, the second (s) for time, and the kelvin (K) for temperature. Derived units are those formed from a combination of other units. Common derived units include volume (cm^3 or m^3) and density (g/cm^3). Measured quantities are reported so that the number of digits reflects the uncertainty in the measurement. The non-place-holding digits in a reported number are called significant figures.

Key Equations and Relationships

Relationship between Kelvin (K) and Celsius (°C) Temperature Scales (1.6)

$$K = °C + 273.15$$

Relationship between Celsius (°C) and Fahrenheit (°F) Temperature Scales (1.6)

$$°C = \frac{(°F - 32)}{1.8}$$

Relationship between Density (d), Mass (m), and Volume (V) (1.6)

$$d = \frac{m}{V}$$

Key Skills

Determining Physical and Chemical Changes and Properties (1.4)

• Example 1.1   • For Practice 1.1   • Exercises 43–50

Converting between the Temperature Scales: Fahrenheit, Celsius, and Kelvin (1.6)

• Example 1.2   • For Practice 1.2   • Exercises 51–54

Calculating the Density of a Substance (1.6)

• Example 1.3   • For Practice 1.3   • For More Practice 1.3   • Exercises 61–64

Reporting Scientific Measurements to the Correct Digit of Uncertainty (1.7)

• Example 1.4   • For Practice 1.4   • Exercises 67, 68

Working with Significant Figures (1.7)

• Examples 1.5, 1.6   • For Practice 1.5, 1.6   • Exercises 71, 72, 74, 77–82

Using Conversion Factors (1.8)

• Examples 1.7, 1.8, 1.9, 1.10   • For Practice 1.7, 1.8, 1.9, 1.10   • For More Practice 1.9, 1.10
• Exercises 83, 84, 88–91, 93, 94

Solving Problems Involving Equations (1.8)

• Examples 1.11, 1.12   • For Practice 1.11, 1.12   • Exercises 109, 110

Exercises

Review Questions

1. Explain the following statement in your own words and give an example. *The properties of the substances around us depend on the atoms and molecules that compose them.*

2. Explain the main goal of chemistry.

3. Describe the scientific approach to knowledge. How does it differ from other approaches?

4. Explain the differences between a hypothesis, a law, and a theory.

5. What observations did Antoine Lavoisier make? What law did he formulate?

6. What theory did John Dalton formulate?

7. What is wrong with the expression, "That is just a theory," if by theory you mean a scientific theory?

8. What are two different ways to classify matter?

9. How do solids, liquids, and gases differ?

10. What is the difference between a crystalline solid and an amorphous solid?

11. Explain the difference between a pure substance and a mixture.

12. Explain the difference between an element and a compound.

13. Explain the difference between a homogeneous and a heterogeneous mixture.

14. What kind of mixtures can be separated by filtration?

15. Explain how distillation works to separate mixtures.

16. What is the difference between a physical property and a chemical property?

17. What is the difference between a physical change and a chemical change? Give some examples of each.

18. Explain the significance of the law of conservation of energy.

19. What kind of energy is chemical energy? In what way is a raised weight similar to a tank of gasoline?

20. What are the standard SI base units of length, mass, time, and temperature?

21. What are the three common temperature scales? Does the size of a degree differ among them?

22. What are prefix multipliers? Give some examples.

23. What is a derived unit? Give an example.

24. Explain the difference between density and mass.

25. Explain the difference between *intensive* and *extensive* properties.

26. What is the meaning of the number of digits reported in a measured quantity?

27. When multiplying or dividing measured quantities, what determines the number of significant figures in the result?

28. When adding or subtracting measured quantities, what determines the number of significant figures in the result?

29. What are the rules for rounding off the results of calculations?

30. Explain the difference between precision and accuracy.

31. Explain the difference between random error and systematic error.

32. What is dimensional analysis?

Problems by Topic

Note: Answers to all odd-numbered Problems, numbered in blue, can be found in Appendix III. Exercises in the Problems by Topic section are paired, with each odd-numbered problem followed by a similar even-numbered problem. Exercises in the Cumulative Problems section are also paired, but somewhat more loosely. (Challenge Problems and Conceptual Problems, because of their nature, are unpaired.)

The Scientific Approach to Knowledge

33. Classify each of the following as an observation, a law, or a theory.
  a. All matter is made of tiny, indestructible particles called atoms.
  b. When iron rusts in a closed container, the mass of the container and its contents does not change.
  c. In chemical reactions, matter is neither created nor destroyed.
  d. When a match burns, heat is evolved.

34. Classify each of the following as an observation, a law, or a theory.
  a. Chlorine is a highly reactive gas.
  b. If elements are listed in order of increasing mass of their atoms, their chemical reactivity follows a repeating pattern.
  c. Neon is an inert (or nonreactive) gas.
  d. The reactivity of elements depends on the arrangement of their electrons.

35. A chemist decomposes several samples of carbon monoxide into carbon and oxygen and weighs the resultant elements. The results are shown below:

Sample	Mass of Carbon (g)	Mass of Oxygen (g)
1	6	8
2	12	16
3	18	24

  a. Do you notice a pattern in these results?

Next, the chemist decomposes several samples of hydrogen peroxide into hydrogen and oxygen. The results are shown below:

Sample	Mass of Hydrogen (g)	Mass of Oxygen (g)
1	0.5	8
2	1	16
3	1.5	24

  b. Do you notice a similarity between these results and those for carbon monoxide in part a?
  c. Can you formulate a law from the observations in a and b?
  d. Can you formulate a hypothesis that might explain your law in c?

36. When astronomers observe distant galaxies, they can tell that most of them are moving away from one another. In addition, the more distant the galaxies, the more rapidly they are likely to be moving away from each other. Can you devise a hypothesis to explain these observations?

The Classification and Properties of Matter

37. Classify each of the following as a pure substance or a mixture. If it is a pure substance, classify it as an element or a compound. If it is a mixture, classify it as homogeneous or heterogeneous.
  a. sweat
  b. carbon dioxide
  c. aluminum
  d. vegetable soup

38. Classify each of the following as a pure substance or a mixture. If it is a pure substance, classify it as an element or a compound. If it is a mixture, classify it as homogeneous or heterogeneous.
  a. wine
  b. beef stew
  c. iron
  d. carbon monoxide

39. Complete the following table.

Substance	Pure or mixture	Type (element or compound)
aluminum	pure	element
apple juice		
hydrogen peroxide		
chicken soup		

40. Complete the following table.

Substance	Pure or mixture	Type (element or compound)
water	pure	compound
coffee		
ice		
carbon		

41. Classify each of the following molecular diagrams as a pure substance or a mixture. If it is a pure substance, classify it as an element or a compound. If it is a mixture, classify it as homogeneous or heterogeneous.

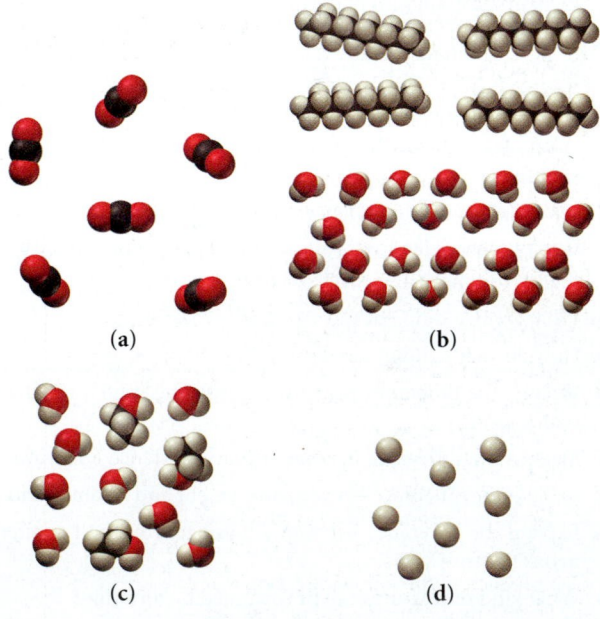

(a)   (b)

(c)   (d)

42. Classify each of the following molecular diagrams as a pure substance or a mixture. If it is a pure substance, classify it as an element or a compound. If it is a mixture, classify it as homogeneous or heterogeneous.

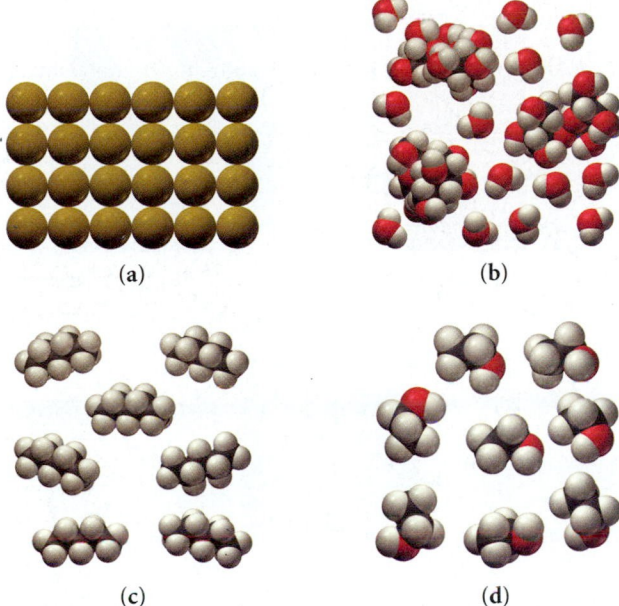

(a)   (b)

(c)   (d)

43. Several properties of isopropyl alcohol (also known as rubbing alcohol) are listed below. Classify each of the properties as physical or chemical.
   a. colorless
   b. flammable
   c. liquid at room temperature   **d.** density = 0.79 g/mL
   e. mixes with water

44. Several properties of ozone (a pollutant in the lower atmosphere, but part of a protective shield against UV light in the upper atmosphere) are listed below. Which are physical and which are chemical?
   a. bluish color
   b. pungent odor
   c. very reactive
   d. decomposes on exposure to ultraviolet light
   e. gas at room temperature

45. Classify each of the following properties as physical or chemical.
   a. the tendency of ethyl alcohol to burn
   b. the shine of silver
   c. the odor of paint thinner
   d. the flammability of propane gas

46. Classify each of the following properties as physical or chemical.
   a. the boiling point of ethyl alcohol
   b. the temperature at which dry ice evaporates
   c. the tendency of iron to rust
   d. the color of gold

47. Classify each of the following changes as physical or chemical.
   a. Natural gas burns in a stove.
   b. The liquid propane in a gas grill evaporates because the user left the valve open.
   c. The liquid propane in a gas grill burns in a flame.
   d. A bicycle frame rusts on repeated exposure to air and water.

48. Classify each of the following changes as physical or chemical.
   a. Sugar burns when heated on a skillet.
   b. Sugar dissolves in water.
   c. A platinum ring becomes dull because of continued abrasion.
   d. A silver surface becomes tarnished after exposure to air for a long period of time.

49. Based on the molecular diagram, classify each change as physical or chemical.

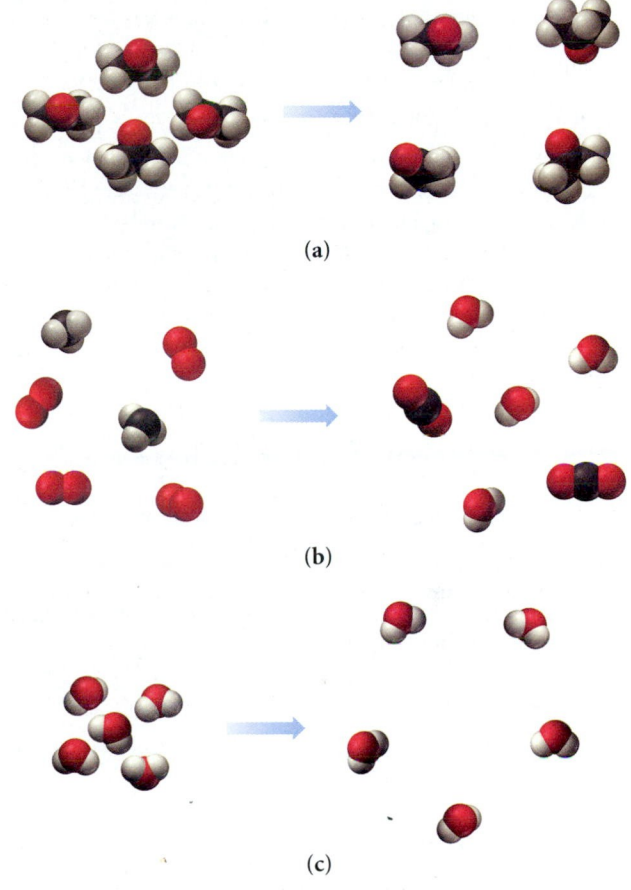

(a)

(b)

(c)

50. Based on the molecular diagram, classify each change as physical or chemical.

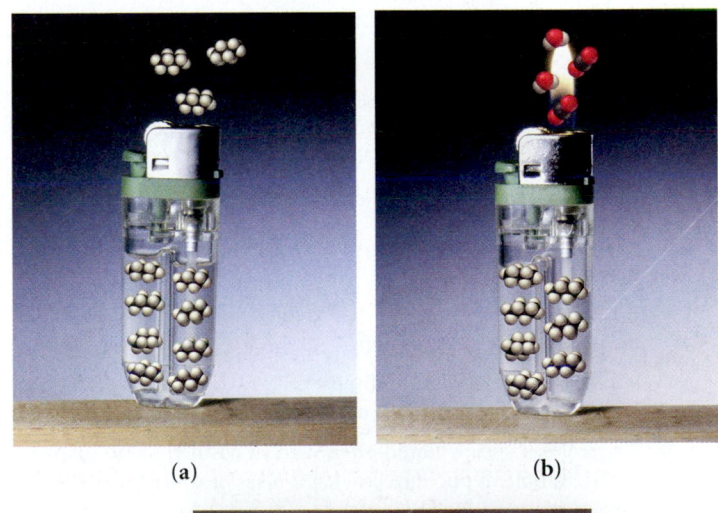

(a)   (b)

(c)

Units in Measurement

51. Perform each of the following temperature conversions.
  a. 32 °F to °C (temperature at which water freezes)
  b. 77 K to °F (temperature of liquid nitrogen)
  c. −109 °F to °C (temperature of dry ice)
  d. 98.6 °F to K (body temperature)

52. Perform each of the following temperature conversions.
  a. 212 °F to °C (temperature of boiling water at sea level)
  b. 22 °C to K (approximate room temperature)
  c. 0.00 K to °F (coldest temperature possible, also known as absolute zero)
  d. 2.735 K to °C (average temperature of the universe as measured from background black body radiation)

53. The coldest temperature ever measured in the United States is −80 °F on January 23, 1971, in Prospect Creek, Alaska. Convert that temperature to °C and K. (Assume that −80 °F is accurate to two significant figures.)

54. The warmest temperature ever measured in the United States is 134 °F on July 10, 1913, in Death Valley, California. Convert that temperature to °C and K.

55. Use the prefix multipliers to express each of the following measurements without any exponents.
  a. 1.2×10^{-9} m   **b.** 22×10^{-15} s
  c. 1.5×10^{9} g   **d.** 3.5×10^{6} L

56. Use scientific notation to express each of the following quantities with only the base units (no prefix multipliers).
  a. 4.5 ns   **b.** 18 fs   **c.** 128 pm   **d.** 35 μm
                                      1.28×10^{-12}

57. Complete the following table:
  a. 1245 kg 1.245×10^{6} g 1.245×10^{9} mg
  b. 515 km   _____ dm 5.15×10^{3} _____ cm
  c. 122.355 s   _____ ms   _____ ks
  d. 3.345 kJ   _____ J   _____ mJ

58. Express the quantity 254,998 m in each of the following:
  a. km   **b.** Mm   **c.** mm   **d.** cm

59. How many 1-cm squares would it take to construct a square that is 1 m on each side?

60. How many 1-cm cubes would it take to construct a cube that is 4 cm on edge?

Density

61. A new penny has a mass of 2.49 g and a volume of 0.349 cm³. Is the penny made of pure copper?

62. A titanium bicycle frame displaces 0.314 L of water and has a mass of 1.41 kg. What is the density of the titanium in g/cm³?

63. Glycerol is a syrupy liquid often used in cosmetics and soaps. A 3.25-L sample of pure glycerol has a mass of 4.10×10^{3} g. What is the density of glycerol in g/cm³?

64. A supposedly gold nugget is tested to determine its density. It is found to displace 19.3 mL of water and has a mass of 371 grams. Could the nugget be made of gold?

65. Ethylene glycol (antifreeze) has a density of 1.11 g/cm³.
  a. What is the mass in g of 417 mL of this liquid?
  b. What is the volume in L of 4.1 kg of this liquid?

66. Acetone (nail polish remover) has a density of 0.7857 g/cm³.
  a. What is the mass, in g, of 28.56 mL of acetone?
  b. What is the volume, in mL, of 6.54 g of acetone?

The Reliability of a Measurement and Significant Figures

67. Read each of the following to the correct number of significant figures. Laboratory glassware should always be read from the bottom of the meniscus.

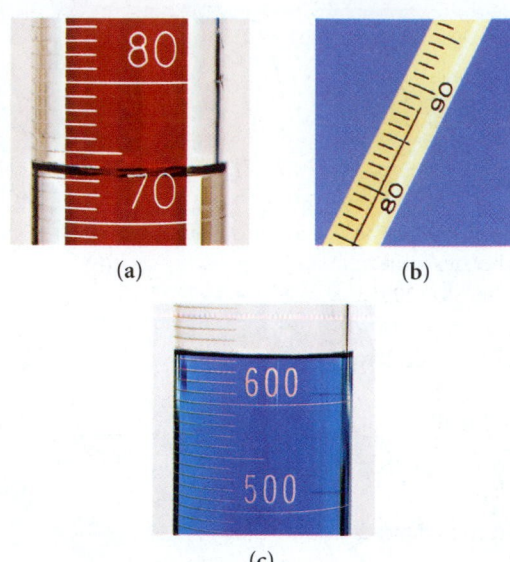

(a)   (b)

(c)

68. Read each of the following to the correct number of significant figures. Note: Laboratory glassware should always be read from the bottom of the meniscus. Digital balances normally display mass to the correct number of significant figures for that particular balance.

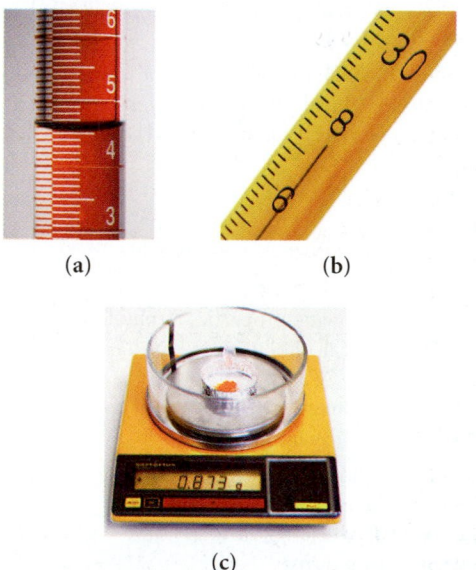

(a)   (b)

(c)

69. For each of the following measurements, underline the zeroes that are significant and draw an x through the zeroes that are not:
  a. 1,050,501 km   **b.** 0.0020 m
  c. 0.000000000000002 s   **d.** 0.001090 cm

70. For each of the following numbers, underline the zeroes that are significant and draw an x through the zeroes that are not:
  a. 180,701 mi   **b.** 0.001040 m
  c. 0.005710 km   **d.** 90,201 m

71. How many significant figures are in each of the following numbers?
  a. 0.000312 m   **b.** 312,000 s
  c. 3.12×10^{5} km   **d.** 13127 s
  e. 2000

72. How many significant figures are in each of the following numbers?

a. 0.1111 s **b.** 0.007 m

c. 108,700 km **d.** 1.563300×10^{11} m

e. 30,800

73. Which of the following numbers are exact numbers and therefore have an unlimited number of significant figures?

a. $\pi = 3.14$

b. 12 inches = 1 foot

c. EPA gas mileage rating of 26 miles per gallon

d. 1 gross = 144

74. Indicate the number of significant figures in each of the following numbers. If the number is an exact number, indicate an unlimited number of significant figures.

a. 284,796,887 (2001 U.S. population)

b. 2.54 cm = 1 in

c. 11.4 g/cm^3 (density of lead)

d. 12 = 1 dozen

75. Round each of the following numbers to four significant figures.

a. 156.852 **b.** 156.842

c. 156.849 **d.** 156.899

76. Round each to three significant figures.

a. 79,845.82 **b.** 1.548937×10^7

c. 2.3499999995 **d.** 0.000045389

Significant Figures in Calculations

77. Perform the following calculations to the correct number of significant figures.

a. $9.15 \div 4.970$

b. $1.54 \times 0.03060 \times 0.69$

c. $27.5 \times 1.82 \div 100.04$

d. $(2.290 \times 10^6) \div (6.7 \times 10^4)$

78. Perform the following calculations to the correct number of significant figures.

a. $89.3 \times 77.0 \times 0.08$

b. $(5.01 \times 10^5) \div (7.8 \times 10^2)$

c. $4.005 \times 74 \times 0.007$

d. $453 \div 2.031$

79. Perform the following calculations to the correct number of significant figures.

a. $43.7 - 2.341$

b. $17.6 + 2.838 + 2.3 + 110.77$

c. $19.6 + 58.33 - 4.974$

d. $5.99 - 5.572$

80. Perform the following calculations to the correct number of significant figures.

a. $0.004 + 0.09879$ **b.** $1239.3 + 9.73 + 3.42$

c. $2.4 - 1.777$ **d.** $532 + 7.3 - 48.523$

81. Perform the following calculations to the correct number of significant figures.

a. $(24.6681 \times 2.38) + 332.58$

b. $(85.3 - 21.489) \div 0.0059$

c. $(512 \div 986.7) + 5.44$

d. $[(28.7 \times 10^5) \div 48.533] + 144.99$

82. Perform the following calculations to the correct number of significant figures.

a. $[(1.7 \times 10^6) \div (2.63 \times 10^5)] + 7.33$

b. $(568.99 - 232.1) \div 5.3$

c. $(9443 + 45 - 9.9) \times 8.1 \times 10^6$

d. $(3.14 \times 2.4367) - 2.34$

Unit Conversions

83. Perform each of the following conversions:

a. 154 cm to in **b.** 3.14 kg to g

c. 3.5 L to qt **d.** 109 mm to in

84. Perform each of the following conversions:

a. 1.4 in to mm **b.** 116 ft to cm

c. 1845 kg to lb **d.** 815 yd to km

85. A runner wants to run 10.0 km. She knows that her running pace is 7.5 miles per hour. How many minutes must she run?

86. A cyclist rides at an average speed of 24 miles per hour. If she wants to bike 195 km, how long (in hours) must she ride?

87. A European automobile has a gas mileage of 14 km/L. What is the gas mileage in miles per gallon?

88. A gas can holds 5.0 gallons of gasoline. What is this quantity in cm^3?

89. A modest-sized house has an area of 195 m^2. What is its area in:

a. km^2 **b.** dm^2 **c.** cm^2

90. A bedroom has a volume of 115 m^3 What is its volume in:

a. km^3 **b.** dm^3 **c.** cm^3

91. The average U.S. farm occupies 435 acres. How many square miles is this? (1 acre = $43,560 \text{ ft}^2$, 1 mile = 5280 ft)

92. Total U.S. farmland occupies 954 million acres. How many square miles is this? (1 acre = $43,560 \text{ ft}^2$, 1 mile = 5280 ft). Total U.S. land area is 3.537 million square miles. What percentage of U.S. land is farmland?

93. An infant acetaminophen suspension contains 80 mg/0.80 mL suspension. The recommended dose is 15 mg/kg body weight. How many mL of this suspension should be given to an infant weighing 14 lb? (Assume two significant figures.)

94. An infant ibuprofen suspension contains 100 mg/5.0 mL suspension. The recommended dose is 10 mg/kg body weight. How many mL of this suspension should be given to an infant weighing 18 lb? (Assume two significant figures.)

Cumulative Problems

95. There are exactly 60 seconds in a minute, there are exactly 60 minutes in an hour, there are exactly 24 hours in a mean solar day, and there are 365.24 solar days in a solar year. Find the number of seconds in a solar year. Be sure to give your answer with the correct number of significant figures.

96. Use exponential notation to indicate the number of significant figures in the following statements:

a. Fifty million Frenchmen can't be wrong.

b. "For every ten jokes, thou hast got an hundred enemies" (Laurence Sterne, 1713–1768).

c. The diameter of a Ca atom is 1.8 one hundred millionths of a centimeter.

d. Sixty thousand dollars is a lot of money to pay for a car.

e. The density of platinum (Table 1.4)

97. Classify the following as intensive or extensive properties.
    a. volume
    b. boiling point
    c. temperature
    d. electrical conductivity
    e. energy

98. At what temperatures will the readings on the Fahrenheit and Celsius thermometers be the same?

99. Suppose you have designed a new thermometer called the X thermometer. On the X scale the boiling point of water is 130 °X and the freezing point of water is 10 °X. At what temperature will the readings on the Fahrenheit and X thermometers be the same?

100. On a new Jekyll temperature scale, water freezes at 17 °J and boils at 97 °J. On another new temperature scale, the Hyde scale, water freezes at 0 °H and boils at 120 °H. If methyl alcohol boils at 84 °H, what is its boiling point on the Jekyll scale?

101. Force is defined as mass times acceleration. Starting with SI base units, derive a unit for force. Using SI prefixes suggest a convenient unit for the force resulting from a collision with a 10-ton trailer truck moving at 55 miles per hour and for the force resulting from the collision of a molecule of mass around 10^{-20} kg moving almost at the speed of light (3×10^8 m/s) with the wall of its container. (Assume a 1 second deceleration time for both collisions.)

102. A temperature measurement of 25 °C has three significant figures, while a temperature measurement of −196 °C has only two significant figures. Explain.

103. Do each of the following calculations without using your calculator and give the answers to the correct number of significant figures.
    a. $1.76 \times 10^{-3}/8.0 \times 10^2$
    b. $1.87 \times 10^{-2} + 2 \times 10^{-4} - 3.0 \times 10^{-3}$
    c. $[(1.36 \times 10^5)(0.000322)/0.082](129.2)$

104. The value of the Euro is currently $1.25 U.S. and the price of 1 liter of gasoline in France is 0.97 Euro. What is the price of 1 gallon of gasoline in U.S. dollars in France?

105. A thief uses a can of sand to replace a solid gold cylinder that sits on a weight-sensitive, alarmed pedestal. The can of sand and the gold cylinder have exactly the same dimensions (length = 22 cm and radius = 3.8 cm).

    a. Calculate the mass of each cylinder (ignore the mass of the can itself). (density of gold = 19.3 g/cm³, density of sand = 3.00 g/cm³)
    b. Did the thief set off the alarm? Explain.

106. The proton has a radius of approximately 1.0×10^{-13} cm and a mass of 1.7×10^{-24} g. Determine the density of a proton. For a sphere $V = (4/3)\pi r^3$.

107. The density of titanium is 4.51 g/cm³. What is the volume (in cubic inches) of 3.5 lb of titanium?

108. The density of iron is 7.86 g/cm³. What is its density in pounds per cubic inch (lb/in³)?

109. A steel cylinder has a length of 2.16 in, a radius of 0.22 in, and a mass of 41 g. What is the density of the steel in g/cm³?

110. A solid aluminum sphere has a mass of 85 g. Use the density of aluminum to find the radius of the sphere in inches.

111. A backyard swimming pool holds 185 cubic yards (yd³) of water. What is the weight of the water in pounds?

112. An iceberg has a volume of 7655 cubic feet. What is the mass of the ice (in kg) composing the iceberg?

113. The Toyota Prius, a hybrid electric vehicle, has an EPA gas mileage rating of 52 mi/gal in the city. How many kilometers can the Prius travel on 15 liters of gasoline?

114. The Honda Insight, a hybrid electric vehicle, has an EPA gas mileage rating of 57 mi/gal in the city. How many kilometers can the Insight travel on the amount of gasoline that would fit in a soda pop can? The volume of a soda pop can is 355 mL.

115. The single proton that forms the nucleus of the hydrogen atom has a radius of approximately 1.0×10^{-13} cm. The hydrogen atom itself has a radius of approximately 52.9 pm. What fraction of the space within the atom is occupied by the nucleus?

116. A sample of gaseous neon atoms at atmospheric pressure and 0 °C contains 2.69×10^{22} atoms per liter. The atomic radius of neon is 69 pm. What fraction of the space is occupied by atoms themselves? What does this reveal about the separation between atoms in the gaseous phase?

Challenge Problems

117. In 1999, scientists discovered a new class of black holes with masses 100 to 10,000 times the mass of our sun, but occupying less space than our moon. Suppose that one of these black holes has a mass of 1×10^3 suns and a radius equal to one-half the radius of our moon. What is the density of the black hole in g/cm³? The radius of our sun is 7.0×10^5 km and it has an average density of 1.4×10^3 kg/m³. The diameter of the moon is 2.16×10^3 miles.

118. Section 1.7 showed that in 1997 Los Angeles County air had carbon monoxide (CO) levels of 15.0 ppm. An average human inhales about 0.50 L of air per breath and takes about 20 breaths per minute. How many milligrams of carbon monoxide does the average person inhale in an 8 hour period for this level of carbon monoxide pollution? Assume that the carbon monoxide has a density of 1.2 g/L. (Hint: 15.0 ppm CO means 15.0 L CO per 10^6 L air.)

119. Nanotechnology, the field of trying to build ultrasmall structures one atom at a time, has progressed in recent years. One potential application of nanotechnology is the construction of artificial cells. The simplest cells would probably mimic red blood cells, the body's oxygen transporters. For example, nanocontainers, perhaps constructed of carbon, could be pumped full of oxygen and injected into a person's bloodstream. If the person needed additional oxygen—due to a heart attack perhaps, or for the purpose of space travel—these containers could slowly release oxygen into the blood, allowing tissues that would otherwise die to remain alive. Suppose that the nanocontainers were cubic and had an edge length of 25 nanometers.

a. What is the volume of one nanocontainer? (Ignore the thickness of the nanocontainer's wall.)

b. Suppose that each nanocontainer could contain pure oxygen pressurized to a density of 85 g/L. How many grams of oxygen could be contained by each nanocontainer?

c. Normal air contains about 0.28 g of oxygen per liter. An average human inhales about 0.50 L of air per breath and takes about 20 breaths per minute. How many grams of oxygen does a human inhale per hour? (Assume two significant figures.)

d. What is the minimum number of nanocontainers that a person would need in their bloodstream to provide 1 hour's worth of oxygen?

e. What is the minimum volume occupied by the number of nanocontainers computed in part d? Is such a volume feasible, given that total blood volume in an adult is about 5 liters?

Conceptual Problems

120. A volatile liquid (one that easily evaporates) is put into a jar and the jar is then sealed. Does the mass of the sealed jar and its contents change upon the vaporization of the liquid?

121. The following diagram represents solid carbon dioxide, also known as dry ice.

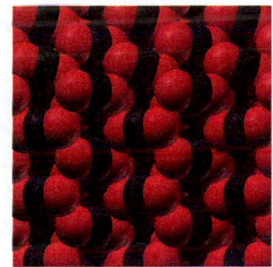

Which of the following diagrams best represents the dry ice after it has sublimed into a gas?

(a)

(b)

(c)

122. A cube has an edge length of 7 cm. If it is divided up into 1-cm cubes, how many 1-cm cubes would there be?

123. Substance A has a density of 1.7 g/cm³. Substance B has a density of 1.7 kg/m³. Without doing any calculations, determine which substance is most dense.

124. For each box below, examine the blocks attached to the balances. Based on their positions and sizes, determine which block is more dense (the dark block or the lighter-colored block), or if the relative densities cannot be determined. (Think carefully about the information being shown.)

(a)

(b)

(c)

125. Identify each of the following as being most like an observation, a law, or a theory.

a. All coastal areas experience two high tides and two low tides each day.

b. The tides in Earth's oceans are caused mainly by the gravitational attraction of the moon.

c. Yesterday, high tide in San Francisco Bay occurred at 2:43 A.M. and 3:07 P.M.

d. Tides are higher at the full moon and new moon than at other times of the month.

2

Atoms and Elements

2.1 Imaging and Moving Individual Atoms

2.2 Early Ideas about the Building Blocks of Matter

2.3 Modern Atomic Theory and the Laws That Led to It

2.4 The Discovery of the Electron

2.5 The Structure of the Atom

2.6 Subatomic Particles: Protons, Neutrons, and Electrons in Atoms

2.7 Finding Patterns: The Periodic Law and the Periodic Table

2.8 Atomic Mass: The Average Mass of an Element's Atoms

2.9 Molar Mass: Counting Atoms by Weighing Them

IF YOU CUT A PIECE OF GRAPHITE from the tip of a pencil into smaller and smaller pieces, how far could you go? Could you divide it forever? Would you eventually run into some basic particles that were no longer divisible, not because of their sheer smallness, but because of the nature of matter? This fundamental question about matter has been asked by thinkers for over two millennia. The answer given, however, has varied over time. On the scale of everyday objects, matter appears continuous, or infinitely divisible. Until about 200 years ago, many scientists thought that matter was indeed continuous—but they were wrong. Eventually, as you divide the graphite from your pencil tip into smaller and smaller pieces, you end up with carbon atoms. The word atom comes from the Greek *atomos*, meaning "indivisible." You cannot divide a carbon atom into smaller pieces and still have carbon. Atoms compose all ordinary matter—if you want to understand matter, you must begin by understanding atoms.

These observations have tacitly led to the conclusion which seems universally adopted, that all bodies of sensible magnitude . . . are constituted of a vast number of extremely small particles, or atoms of matter . . . .

—JOHN DALTON (1766–1844)

The tip of a scanning tunneling microscope (STM) moves across an atomic surface.

2.1 Imaging and Moving Individual Atoms

On March 16, 1981, Gerd Binnig and Heinrich Rohrer worked late into the night in their laboratory at IBM in Zurich, Switzerland. They were measuring how an electrical current—flowing between a sharp metal tip and a flat metal surface—varied with the separation between the tip and the surface. The results of that night's experiment and subsequent results over the next several months won Binnig and Rohrer a share of the 1986 Nobel Prize in Physics. They had discovered *scanning tunneling microscopy (STM)*, a technique that can image, and even move, individual atoms and molecules.

A scanning tunneling microscope works by moving an extremely sharp electrode (often tipped with a single atom) over a surface and measuring the resulting *tunneling current,* the electrical current that flows between the tip of the electrode and the surface even though the two are not in physical contact (Figure 2.1, page 46).

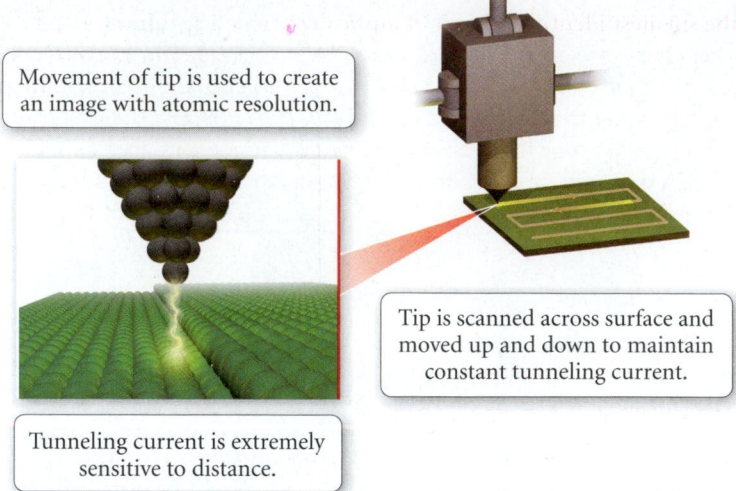

Movement of tip is used to create an image with atomic resolution.

Tip is scanned across surface and moved up and down to maintain constant tunneling current.

Tunneling current is extremely sensitive to distance.

▶ **FIGURE 2.1 Scanning Tunneling Microscopy** In this technique, an atomically sharp tip is scanned across a surface. The tip is kept at a fixed distance from the surface by moving it up and down so as to maintain a constant tunneling current. The motion of the tip is recorded to create an image of the surface with atomic resolution.

The tunneling current, as Binnig and Rohrer found that night in their laboratory at IBM, is extremely sensitive to distance, making it possible to maintain a precise separation of approximately two atomic diameters between the tip and the surface simply by moving the tip so as to keep the current constant. By measuring the up-and-down movement of the tip as it is scanned horizontally across a surface, an image of the surface—with atomic resolution—can be created, like the one shown in Figure 2.2(a) ▼.

In other words, Binnig and Rohrer had discovered a type of microscope that could "see" atoms. Later work by other scientists showed that the STM could also be used to pick up and move individual atoms or molecules, allowing structures and patterns to be made one atom at a time. Figure 2.2(b), for example, shows the Kanji characters for the word "atom" written with individual iron atoms on top of a copper surface. If all of the words in the books in the Library of Congress—29 million books on 530 miles of shelves—were written in letters the size of this Kanji character, they would fit in an area of about 5 square millimeters.

▶ **FIGURE 2.2 Imaging Atoms**
(**a**) A scanning tunneling microscope image of iodine atoms (green) on a platinum surface. (**b**) The Japanese characters for "atom" written with atoms.

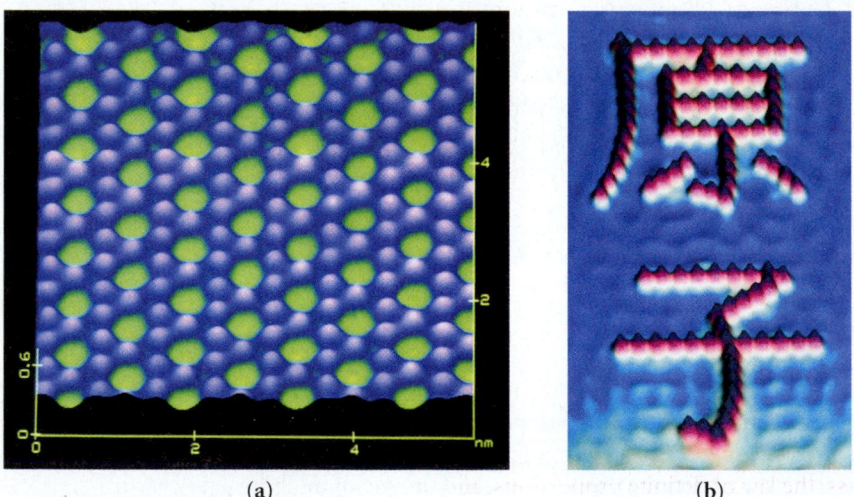

(a)   (b)

As we saw in Chapter 1, it was only 200 years ago that John Dalton proposed his atomic theory. Today we can not only image and move individual atoms, we are even beginning to build tiny machines out of just a few dozen atoms (an area of research called nanotechnology).

These atomic machines, and the atoms that compose them, are almost unimaginably small. To get an idea of the size of an atom, imagine picking up a grain of sand at your favorite beach. That grain contains more atoms than you could count in a lifetime. In fact, the number of atoms in one sand grain far exceeds the number of grains on the entire beach. If every atom within the sand grain were the size of the grain itself, the sand grain would be the size of a large mountain range.

Despite their size, atoms are the key to connecting the macroscopic and microscopic worlds. An *atom* is the smallest identifiable unit of an *element*. There are about 91 different naturally occurring elements. In addition, scientists have succeeded in making over 20 synthetic elements (not found in nature). In this chapter, we learn about atoms: what they are made of, how they differ from one another, and how they are structured. We also learn about the elements made up of these different kinds of atoms, and about some of the characteristics of those elements. We will find that the elements can be organized in a way that reveals patterns in their properties and helps us to understand what underlies those properties.

The exact number of naturally occurring elements is controversial, because some elements that were first discovered when they were synthesized are believed to be present in trace amounts in nature.

2.2 Early Ideas about the Building Blocks of Matter

The first people to propose that matter was composed of small, indestructible particles were Leucippus (fifth century B.C., exact dates unknown) and his student Democritus (460–370 B.C.). These Greek philosophers theorized that matter was ultimately composed of small, indivisible particles called *atomos*. Democritus wrote, "Nothing exists except atoms and empty space; everything else is opinion." Leucippus and Democritus proposed that many different kinds of atoms existed, each different in shape and size, and that they moved randomly through empty space. However, other influential Greek thinkers of the time, such as Plato and Aristotle, did not embrace the work of Leucippus and Democritus. Instead, they held that matter had no smallest parts and that different substances were composed of various proportions of fire, air, earth, and water. Since there was no experimental way to test the relative merits of the two ideas, Aristotle's view prevailed, partly because he was so influential. The idea that matter was composed of atoms took a back seat in intellectual thought for nearly 2000 years.

In the sixteenth century modern science began to emerge. A greater emphasis on observation led Nicolaus Copernicus (1473–1543) to publish *On the Revolution of the Heavenly Orbs* in 1543. That book—which proposed that the sun, not Earth, was at the center of the universe—marks the beginning of what we now call the *scientific revolution*. The next 200 years—with the works of scientists such as Francis Bacon (1561–1626), Johannes Kepler (1571–1630), Galileo Galilei (1564–1642), Robert Boyle (1627–1691), and Isaac Newton (1642–1727)—brought rapid advancement as the scientific method became the established way to learn about the physical world. By the early 1800s certain observations led the English chemist John Dalton (1766–1844) to offer convincing evidence that supported the early atomic ideas of Leucippus and Democritus.

2.3 Modern Atomic Theory and the Laws That Led to It

Recall the discussion of the scientific method from Chapter 1. The theory that all matter is composed of atoms grew out of observations and laws. The three most important laws that led to the development and acceptance of the atomic theory were the **law of conservation of mass**, the law of definite proportions, and the law of multiple proportions.

The Law of Conservation of Mass

In 1789, as we saw in Chapter 1, Antoine Lavoisier formulated the law of conservation of mass, which states the following:

> **In a chemical reaction, matter is neither created nor destroyed.**

In other words, when you carry out any chemical reaction, the total mass of the substances involved in the reaction does not change. For example, consider the reaction between sodium and chlorine to form sodium chloride.

We will see in Chapter 19 that this law is a slight oversimplification. However, the changes in mass in ordinary chemical processes are so minute that they can be ignored for all practical purposes.

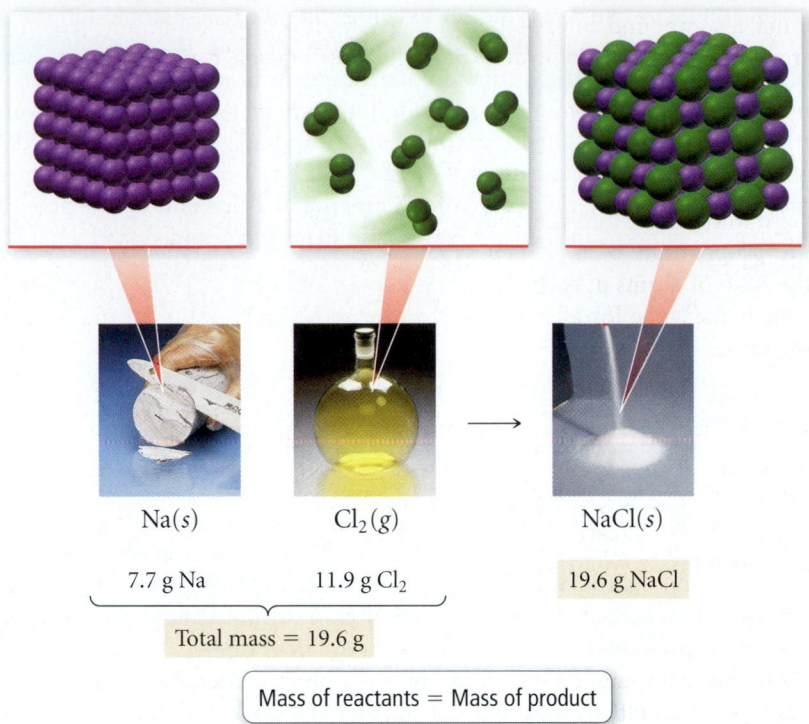

Na(s)   Cl₂(g)   NaCl(s)

7.7 g Na   11.9 g Cl₂   19.6 g NaCl

Total mass = 19.6 g

Mass of reactants = Mass of product

The combined mass of the sodium and chlorine that react exactly equals the mass of the sodium chloride that results from the reaction. This law is consistent with the idea that matter is composed of small, indestructible particles. The particles rearrange during a chemical reaction, but the amount of matter is conserved because the particles are indestructible (at least by chemical means).

 Conceptual Connection 2.1 The Law of Conservation of Mass

When a small log completely burns in a campfire, the mass of the ash is much less than the mass of the log. What happened to the matter that composed the log?

Answer: Most of the matter that composed the log underwent a chemical change by reacting with oxygen molecules in the air and was released as gases into the air, mostly carbon dioxide and water.

The Law of Definite Proportions

The law of definite proportions is sometimes called the law of constant composition.

In 1797, a French chemist named Joseph Proust (1754–1826) made observations on the composition of compounds. He found that the elements composing a given compound always occurred in fixed (or definite) proportions in all samples of the compound. In contrast, the components of a mixture could be present in any proportions whatsoever. He summarized his observations in the **law of definite proportions,** which states the following:

All samples of a given compound, regardless of their source or how they were prepared, have the same proportions of their constituent elements.

For example, the decomposition of 18.0 g of water results in 16.0 g of oxygen and 2.0 g of hydrogen, or an oxygen-to-hydrogen mass ratio of:

$$\text{Mass ratio} = \frac{16.0 \, \text{g O}}{2.0 \, \text{g H}} = 8.0 \quad \text{or} \quad 8:1$$

This ratio holds for any sample of pure water, regardless of its origin. The law of definite proportions applies not only to water, but also to every compound. Consider ammonia, a compound composed of nitrogen and hydrogen. Ammonia contains 14.0 g of nitrogen for every 3.0 g of hydrogen, resulting in a nitrogen-to-hydrogen mass ratio of:

$$\text{Mass ratio} = \frac{14.0\,\text{g N}}{3.0\,\text{g H}} = 4.7 \quad \text{or} \quad 4.7{:}1$$

Again, this ratio is the same for every sample of ammonia. The law of definite proportions also hints at the idea that matter might be composed of atoms. Compounds have definite proportions of their constituent elements because they are composed of a definite ratio of atoms of each element, each with its own specific mass. Since the ratio of atoms is the same for all samples of a particular compound, the ratio of masses is also the same.

EXAMPLE 2.1 Law of Definite Proportions

Two samples of carbon dioxide are decomposed into their constituent elements. One sample produces 25.6 g of oxygen and 9.60 g of carbon, and the other produces 21.6 g of oxygen and 8.10 g of carbon. Show that these results are consistent with the law of definite proportions.

Solution

To show this, compute the mass ratio of one element to the other for both samples by dividing the mass of one element by the mass of the other. It is usually more convenient to divide the larger mass by the smaller one.

For the first sample:

$$\frac{\text{Mass oxygen}}{\text{Mass carbon}} = \frac{25.6}{9.60} = 2.67 \quad \text{or} \quad 2.67{:}1$$

For the second sample:

$$\frac{\text{Mass oxygen}}{\text{Mass carbon}} = \frac{21.6}{8.10} = 2.67 \quad \text{or} \quad 2.67{:}1$$

The ratios are the same for the two samples, so these results are consistent with the law of definite proportions.

For Practice 2.1

Two samples of carbon monoxide were decomposed into their constituent elements. One sample produced 17.2 g of oxygen and 12.9 g of carbon, and the other sample produced 10.5 g of oxygen and 7.88 g of carbon. Show that these results are consistent with the law of definite proportion.

Answers to For Practice and For More Practice Problems can be found in Appendix IV.

The Law of Multiple Proportions

In 1804, John Dalton published his **law of multiple proportions,** which asserts the following principle:

> When two elements (call them A and B) form two different compounds, the masses of element B that combine with 1 g of element A can be expressed as a ratio of small whole numbers.

Dalton already suspected that matter was composed of atoms, so that when two elements A and B combined to form more than one compound, an atom of A combined with one, two, three, or more atoms of B (AB_1, AB_2, AB_3, etc.). Therefore the masses of B that reacted with a fixed mass of A would always be related to one another as small whole-number ratios. For example, consider the compounds carbon monoxide and carbon dioxide, which we discussed in the opening section of Chapter 1 as well as in Example 2.1 and its For Practice problem. Carbon monoxide and carbon dioxide are two compounds composed of the same two elements: carbon and oxygen. We saw in Example 2.1 that the mass ratio of oxygen to carbon in carbon dioxide is 2.67:1; therefore, 2.67 g of oxygen would react with 1 g of carbon. In carbon monoxide, however, the mass ratio of oxygen to carbon is 1.33:1, or 1.33 g of oxygen to every 1 g of carbon.

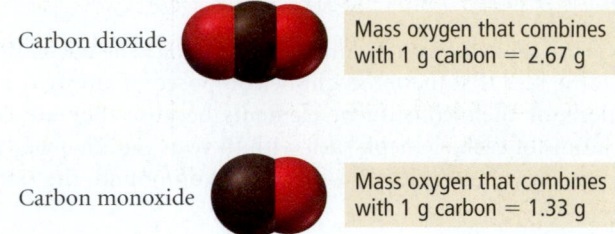

Carbon dioxide — Mass oxygen that combines with 1 g carbon = 2.67 g

Carbon monoxide — Mass oxygen that combines with 1 g carbon = 1.33 g

The ratio of these two masses is itself a small whole number.

$$\frac{\text{Mass oxygen to 1 g carbon in carbon dioxide}}{\text{Mass oxygen to 1 g carbon in carbon monoxide}} = \frac{2.67}{1.33} = 2$$

With the help of the molecular models, we can see why the ratio is 2:1—carbon dioxide contains two oxygen atoms to every carbon atom while carbon monoxide contains only one.

EXAMPLE 2.2 Law of Multiple Proportions

Nitrogen forms several compounds with oxygen, including nitrogen dioxide and dinitrogen monoxide. Nitrogen dioxide contains 2.28 g oxygen to every 1.00 g nitrogen while dinitrogen monoxide contains 0.570 g oxygen to every 1.00 g nitrogen. Show that these results are consistent with the law of multiple proportions.

Solution

To show this, simply compute the ratio of the mass of oxygen from one compound to the mass of oxygen in the other. Always divide the larger of the two masses by the smaller one.

$$\frac{\begin{array}{c}\text{Mass oxygen to 1 g}\\ \text{nitrogen in nitrogen dioxide}\end{array}}{\begin{array}{c}\text{Mass oxygen to 1 g}\\ \text{nitrogen in dinitrogen monoxide}\end{array}} = \frac{2.28}{0.570} = 4.00$$

The ratio is a small whole number (4); therefore, these results are consistent with the law of multiple proportions.

For Practice 2.2

Hydrogen and oxygen form both water and hydrogen peroxide. A sample of water is decomposed and forms 0.125 g hydrogen to every 1.00 g oxygen. A sample of hydrogen peroxide is decomposed and forms 0.250 g hydrogen to every 1.00 g oxygen. Show that these results are consistent with the law of multiple proportions.

Conceptual Connection 2.2 The Laws of Definite and Multiple Proportions

Explain the difference between the law of definite proportions and the law of multiple proportions.

Answer: The law of definite proportions applies to two or more samples of the *same compound* and states that the ratio of one element to the other in the samples will always be the same. The law of multiple proportions applies to two *different compounds* containing the same two elements (A and B) and states that the masses of B that combine with 1 g of A are related as a small whole-number ratio.

John Dalton and the Atomic Theory

In 1808, John Dalton explained the laws just discussed with his **atomic theory**, which included the following concepts:

1. Each element is composed of tiny, indestructible particles called atoms.
2. All atoms of a given element have the same mass and other properties that distinguish them from the atoms of other elements.
3. Atoms combine in simple, whole-number ratios to form compounds.
4. Atoms of one element cannot change into atoms of another element. In a chemical reaction, atoms change the way that they are bound together with other atoms to form a new substance.

Today, as we have seen in the opening section of this chapter, the evidence for the atomic theory is overwhelming. Matter is indeed composed of atoms.

Chemistry in Your Day
Atoms and Humans

We are all composed of atoms. We get those atoms from the food we eat over the years. Yesterday's cheeseburger contributes to today's skin, muscle, and hair. The carbon atoms in our own bodies have been used by other living organisms before we got them and will be used by still others when we are done with them. In fact, it is likely that at this moment, your body contains some carbon atoms (over one trillion*) that were at one time part of your chemistry professor.

The idea that humans are composed of atoms acting in accord with the laws of chemistry and physics has significant implications and raises important questions. If atoms compose our brains, for example, do they determine our thoughts and emotions? Are our feelings caused by atoms acting according to the laws of chemistry and physics?

Richard Feynman (1918–1988), a Nobel Prize–winning physicist, asserted "The most important hypothesis in all of biology is that everything that animals do, atoms do. In other words, there is nothing that living things do that cannot be understood from the point of view that they are made of atoms acting according to the laws of physics." Indeed, biology has undergone a revolution throughout the last 50 years, mostly through the investigation of the atomic and molecular basis for life. Some have seen the atomic view of life as a devaluation of human life. We have always wanted to distinguish ourselves from everything else, and the idea that we are made of the same basic particles as all other matter takes something away from that distinction . . . . Or does it?

* This calculation assumes that all of the carbon atoms metabolized by your professor over the last 40 years have been uniformly distributed into atmospheric carbon dioxide, and then incorporated into plants that you eat.

Questions

Do you find the idea that you are made of recycled atoms disturbing? Why or why not? *Reductionism* refers to the explanation of complex systems by reducing them to their parts. Is reductionism a good way to understand humans? Is it the only way?

2.4 The Discovery of the Electron

By the end of the nineteenth century, scientists were convinced that matter was composed of atoms, the permanent, supposedly indestructible building blocks that composed everything. However, further experiments revealed that the atom itself was composed of even smaller, more fundamental particles.

Cathode Rays

In the late 1800s an English physicist named J. J. Thomson (1856–1940), working at Cambridge University, performed experiments to probe the properties of **cathode rays**. Cathode rays are produced when a high electrical voltage is applied between two electrodes within a partially evacuated tube called a **cathode ray tube** shown in Figure 2.3 (page 52).

Cathode rays travel from the negatively charged electrode, called the cathode, to the positively charged electrode, called the anode. Thomson found that these rays were actually streams of particles with the following properties: they traveled in straight lines; they were independent of the composition of the material from which they originated; and

Cathode rays

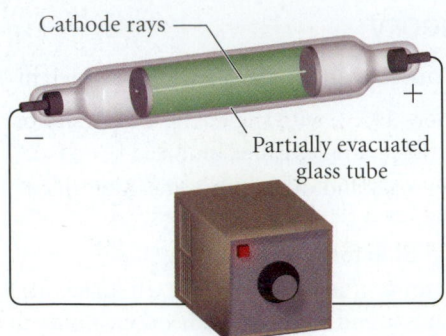
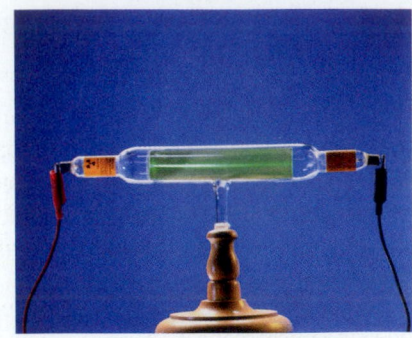

Partially evacuated
glass tube

High voltage

▶ FIGURE 2.3 Cathode Ray Tube

they carried a negative **electrical charge**. Electrical charge is a fundamental property of some of the particles that compose the atom and results in attractive and repulsive forces—called *electrostatic forces*—between them. The characteristics of electrical charge are summarized in the figure in the margin.

You have probably experienced excess electrical charge when brushing your hair on a dry day. The brushing action causes the accumulation of charged particles in your hair, which repel each other, causing your hair to stand on end.

J. J. Thomson measured the charge-to-mass ratio of the particles within cathode rays by deflecting them using electric and magnetic fields, as shown in Figure 2.4 ▼.

The value he measured, -1.76×10^8 coulombs (C) per gram, implied that the cathode ray particle was about 2000 times lighter than hydrogen, the lightest known atom. These results were incredible—the indestructible atom could apparently be chipped! J. J. Thomson had discovered the **electron**, a negatively charged, low mass particle present within all atoms. Thomson wrote, "We have in the cathode rays matter in a new state, a state in which the subdivision of matter is carried very much further . . . a state in which all matter . . . is of one and the same kind; this matter being the substance from which all the chemical elements are built up."

| The coulomb (C) is the SI unit for charge.

Properties of Electrical Charge

Positive (red) and negative (yellow) electrical charges attract one another.

Positive charges repel one another. Negative charges repel one another.

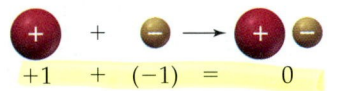

$+1 \quad + \quad (-1) \quad = \quad 0$

Positive and negative charges of exactly the same magnitude sum to zero when combined.

Charge-to-Mass Ratio of the Electron

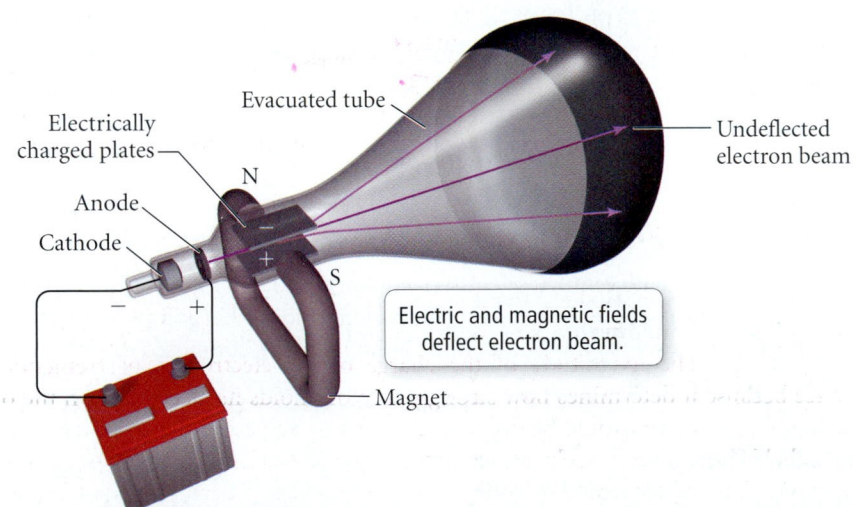

Electrically charged plates

Evacuated tube

Undeflected electron beam

Anode

Cathode

N

S

Electric and magnetic fields deflect electron beam.

Magnet

▲ FIGURE 2.4 Thomson's Measurement of the Charge-to-Mass Ratio of the Electron
J. J. Thomson used electric and magnetic fields to deflect the electron beam in a cathode ray tube. By measuring the field strengths at which the two effects canceled exactly, leaving the beam undeflected, he was able to calculate the charge-to-mass ratio of the electron.

Millikan's Oil Drop Experiment: The Charge of the Electron

In 1909, American physicist Robert Millikan (1868–1953), working at the University of Chicago, performed his now famous oil drop experiment in which he deduced the charge of a single electron. The apparatus for the oil drop experiment is shown in Figure 2.5 ▼.

Oil was sprayed into fine droplets using an atomizer. The droplets were allowed to fall under the influence of gravity through a small hole into the lower portion of the apparatus where they could be viewed with the help of a light source and a viewing microscope. During their fall, the drops would acquire electrons that had been produced by the interaction of high-energy radiation with air. These charged drops now interact- ed with two electrically charged plates within the apparatus. (Remember that like charges repel each other.) The negatively charged plate at the bottom of the appara- tus repelled the negatively charged drops. By varying the voltage on the plates, the fall of the charged drops could be slowed, stopped, or even reversed.

From the voltage required to halt the free fall of the drops, and from the masses of the drops themselves (determined from their radii and density), Millikan calculated the charge of each drop. He then reasoned that, since each drop must contain an integral (or whole) number of electrons, the charge of each drop must be a whole- number multiple of the electron's charge. Indeed, Mil- likan was correct; the measured charge on any drop was always a whole-number multiple of -1.60×10^{-19} C, the fundamental charge of a single electron. With this number in hand, and knowing Thomson's mass-to-charge ratio for electrons, we can deduce the mass of an electron as follows:

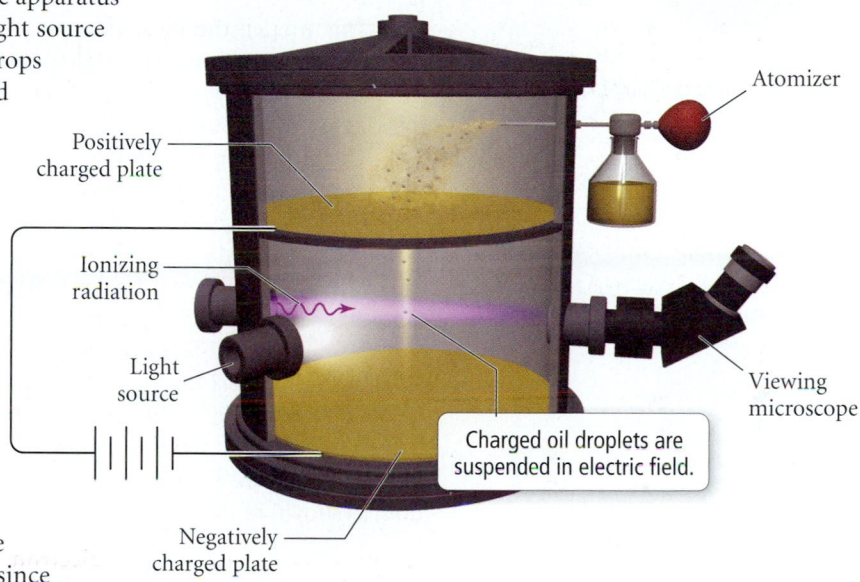

▲ **FIGURE 2.5 Millikan's Measurement of the Electron's Charge** Millikan calculated the charge on oil droplets falling in an electric field. He found that it was always a whole-number multiple of -1.60×10^{-19} C, the charge of a single electron.

$$\text{Charge} \times \frac{\text{mass}}{\text{charge}} = \text{mass}$$

$$-1.60 \times 10^{-19}\ \text{C} \times \frac{\text{g}}{-1.76 \times 10^{8}\ \text{C}} = 9.10 \times 10^{-28}\ \text{g}$$

As Thomson had correctly deduced, this mass is about 2000 times lighter than hydrogen, the lightest atom.

Why did scientists work so hard to measure the charge of the electron? Since the elec- tron is a fundamental building block of matter, scientists want to know its properties, in- cluding its charge. The magnitude of the charge of the electron is of tremendous importance because it determines how strongly an atom holds its electrons. On the one hand, imagine how matter would be different if electrons had a much smaller charge, so that atoms held them more loosely. Many atoms might not even be stable. On the other hand, imagine how matter would be different if electrons had a much greater charge, so that atoms held them more tightly. Since atoms form compounds by exchanging and sharing electrons (more on this in Chapter 3), the result could be fewer compounds or maybe even none. Without the abundant diversity of compounds, life would not be pos- sible. So, the magnitude of the charge of the electron—even though it may seem like an insignificantly small number—has great importance.

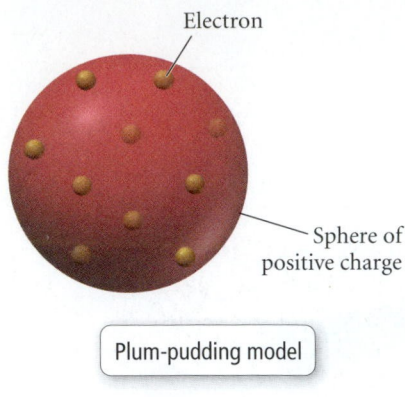

Electron

Sphere of
positive charge

Plum-pudding model

Alpha particles are about 7000 times more
massive than electrons.

2.5 The Structure of the Atom

The discovery of negatively charged particles within atoms raised a new question. Since atoms are charge-neutral, they must contain positive charge that neutralized the negative charge of the electrons—but how did the positive and negative charges within the atom fit together? Were atoms just a jumble of even more fundamental particles? Were they solid spheres? Did they have some internal structure? J. J. Thomson proposed that the negatively charged electrons were small particles held within a positively charged sphere, as shown here.

This model, the most popular of the time, became known as the plum-pudding model (plum pudding is an English dessert). The picture suggested by Thomson, to those of us not familiar with plum pudding, was more like a blueberry muffin, where the blueberries are the electrons and the muffin is the positively charged sphere.

The discovery of **radioactivity**—the emission of small energetic particles from the core of certain unstable atoms—by scientists Henri Becquerel (1852–1908) and Marie Curie (1867–1934) at the end of the nineteenth century allowed the structure of the atom to be experimentally probed. At the time, three different types of radioactivity had been identified: alpha (α) particles, beta (β) particles, and gamma (γ) rays. We will discuss these and other types of radioactivity in more detail in Chapter 19. For now, just know that α particles are positively charged and that they are by far the most massive of the three.

In 1909, Ernest Rutherford (1871–1937), who had worked under Thomson and subscribed to his plum-pudding model, performed an experiment in an attempt to confirm it. His experiment, which employed α particles, proved it wrong instead. In the experiment, Rutherford directed the positively charged α particles at an ultrathin sheet of gold foil, as shown in Figure 2.6 ▼.

Rutherford's Gold Foil Experiment

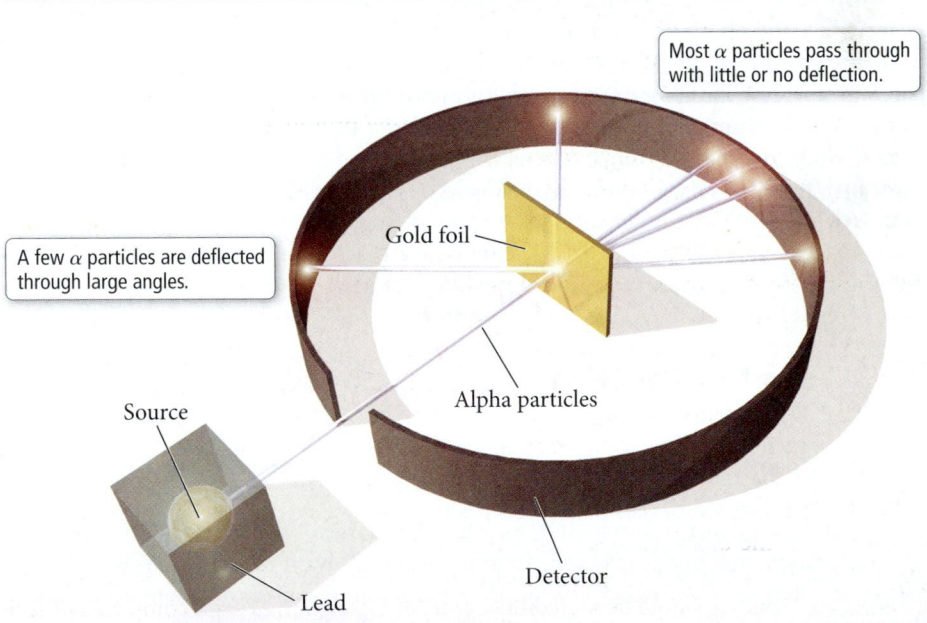

Most α particles pass through with little or no deflection.

A few α particles are deflected through large angles.

Gold foil

Alpha particles

Source

Detector

Lead

▶ FIGURE 2.6 Rutherford's Gold Foil Experiment Alpha particles were directed at a thin sheet of gold foil. Most of the particles passed through the foil, but a small fraction were deflected, and a few even bounced backward.

These particles were to act as probes of the gold atoms' structure. If the gold atoms were indeed like blueberry muffins or plum pudding—with their mass and charge spread throughout the entire volume of the atom—these speeding probes should pass right through the gold foil with minimum deflection.

Rutherford performed the experiment, but the results were not as he expected. A majority of the particles did pass directly through the foil, but some particles were deflected, and some (1 in 20,000) even bounced back. The results puzzled Rutherford, who wrote that they were "about as credible as if you had fired a 15-inch shell at a piece of tissue paper and it came back and hit you." What must the structure of the atom be in order to explain this odd behavior?

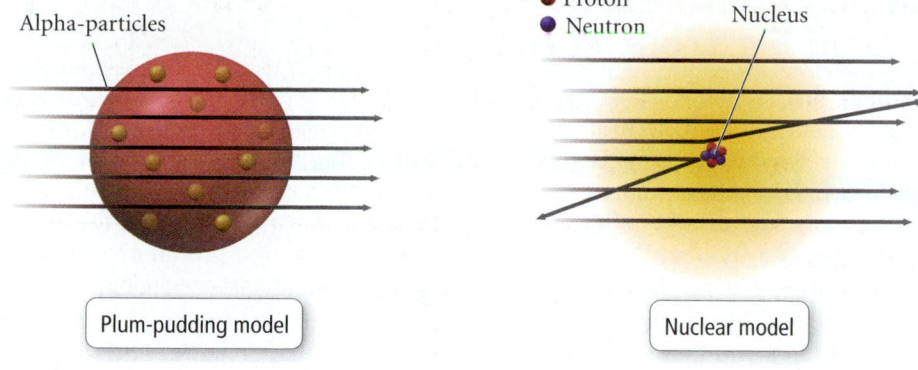

Alpha-particles

● Proton
● Neutron

Nucleus

Plum-pudding model

Nuclear model

◀ FIGURE 2.7 **The Nuclear Atom** Rutherford's results could not be explained by the plum-pudding model. Instead, they suggested that the atom must have a small, dense nucleus.

Rutherford created a new model—a modern version of which is shown in Figure 2.7 ▲ alongside the plum-pudding model—to explain his results.

He realized that to account for the deflections he observed, the mass and positive charge of an atom must all be concentrated in a space much smaller than the size of the atom itself. He concluded that, in contrast to the plum-pudding model, matter must not be as uniform as it appears. It must contain large regions of empty space dotted with small regions of very dense matter. Using this idea, he proposed the **nuclear theory** of the atom, with three basic parts:

1. Most of the atom's mass and all of its positive charge are contained in a small core called the **nucleus**.

2. Most of the volume of the atom is empty space, throughout which tiny, negatively charged electrons are dispersed.

3. There are as many negatively charged electrons outside the nucleus as there are positively charged particles (named **protons**) within the nucleus, so that the atom is electrically neutral.

Although Rutherford's model was highly successful, scientists realized that it was incomplete. For example, hydrogen atoms contain one proton and helium atoms contain two, yet the helium-to-hydrogen mass ratio is 4:1. The helium atom must contain some additional mass. Later work by Rutherford and one of his students, British scientist James Chadwick (1891–1974), demonstrated that the missing mass was composed of **neutrons**, neutral particles within the nucleus. The dense nucleus contains over 99.9% of the mass of the atom, but it occupies very little of its volume. For now, we can think of the electrons that surround the nucleus in analogy to the water droplets that make up a cloud—although their mass is almost negligibly small, they are dispersed over a very large volume. Consequently, the atom, like the cloud, is mostly empty space.

Rutherford's nuclear theory was a success and is still valid today. The revolutionary part of this theory is the idea that matter—at its core—is much less uniform than it appears. If the nucleus of the atom were the size of the period at the end of this sentence, the average electron would be about 10 meters away. Yet the period would contain almost the entire mass of the atom. Imagine what matter would be like if atomic structure broke down. What if matter were composed of atomic nuclei piled on top of each other like marbles? Such matter would be incredibly dense; a single grain of sand composed of solid atomic nuclei would have a mass of 5 million kilograms (or a weight of about 11 million pounds). Astronomers believe there are some objects in the universe composed of such matter—they are called neutron stars.

If matter really is mostly empty space, as Rutherford suggested, then why does it appear so solid? Why can we tap our knuckles on a table and feel a solid thump? Matter appears solid because the variation in its density is on such a small scale that our eyes cannot see it.

Imagine a scaffolding 100 stories high and the size of a football field as shown in the margin. It is mostly empty space. Yet if you viewed it from an airplane, it would appear as a solid mass. Matter is similar. When you tap your knuckle on the table, it is much like one giant scaffolding (your finger) crashing into another (the table). Even though they are both primarily empty space, one does not fall into the other.

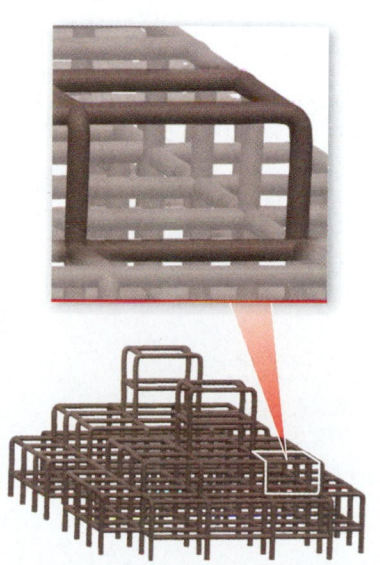

If a proton had the mass of a baseball, an electron would have the mass of a rice grain.

2.6 Subatomic Particles: Protons, Neutrons, and Electrons in Atoms

We have just seen that all atoms are composed of the same subatomic particles: protons, neutrons, and electrons. Protons and neutrons have nearly identical masses. In SI units, the mass of the proton is 1.67262×10^{-27} kg, and the mass of the neutron is 1.67493×10^{-27} kg. A more common unit to express these masses, however, is the **atomic mass unit (amu)**, defined as 1/12 the mass of a carbon atom containing six protons and six neutrons. Expressed in this unit, the mass of a proton or neutron is approximately 1 amu. Electrons, in contrast, have an almost negligible mass of 0.00091×10^{-27} kg or 0.00055 amu.

The proton and the electron both have electrical *charge*. We know from Millikan's oil drop experiment that the electron has a charge of -1.60×10^{-19} C. In atomic or relative units, the electron is assigned a charge of −1 and the proton is assigned a charge of +1. The charge of the proton and the electron are equal in magnitude but opposite in sign, so that when the two particles are paired, the charges sum to zero. The neutron has no charge.

Notice that matter is usually charge-neutral (it has no overall charge) because protons and electrons are normally present in equal numbers. When matter does acquire charge imbalances, these imbalances usually equalize quickly, often in dramatic ways. For example, the shock you receive when touching a doorknob during dry weather is the equalization of a charge imbalance that developed as you walked across the carpet. Lightning, as shown below, is simply an equalization of charge imbalances that develop during electrical storms.

▶ When the normal charge balance of matter is disturbed, as happens during an electrical storm, it quickly equalizes, often in dramatic ways.

Negative charge builds up on clouds.

Electrical discharge equalizes charge imbalance.

Positive charge builds up on ground.

If you had a sample of matter—even a tiny sample, such as a sand grain—that was composed only of protons or only of electrons, the repulsive forces inherent in that matter would be extraordinary, and the matter would be unstable. Luckily, that is not how matter is. The properties of protons, neutrons, and electrons are summarized in Table 2.1.

TABLE 2.1 Subatomic Particles

	Mass (kg)	Mass (amu)	Charge (relative)	Charge (C)
Proton	1.67262×10^{-27}	1.00727	+1	$+1.60218 \times 10^{-19}$
Neutron	1.67493×10^{-27}	1.00866	0	0
Electron	0.00091×10^{-27}	0.00055	−1	-1.60218×10^{-19}

Elements: Defined by Their Numbers of Protons

If all atoms are composed of the same subatomic particles, then what makes the atoms of one element different from those of another? The answer is the *number* of these particles. The number that is most important for the identity of an atom is the number of protons in its nucleus. In fact, the number of protons defines the element. For example, an atom with 2 protons in its nucleus is a helium atom; an atom with 6 protons in its nucleus is a carbon atom; and an atom with 92 protons in its nucleus is a uranium atom (Figure 2.8 ▶). The number of protons in an atom's nucleus is called the **atomic number** and is given the symbol Z. The atomic numbers of known elements range from 1 to 116 (although additional elements may still be discovered), as shown in the periodic table of the elements (Figure 2.9 ▼). In the periodic table, described in more detail in Section 2.7, the elements are arranged so that those with similar properties occur in the same column. An element, identified by a unique atomic number, is represented with a unique **chemical symbol**, a one- or two-letter abbreviation or the element that is listed directly below its atomic number on the periodic table. The chemical symbol for helium is He; for carbon,

The Number of Protons Defines the Element

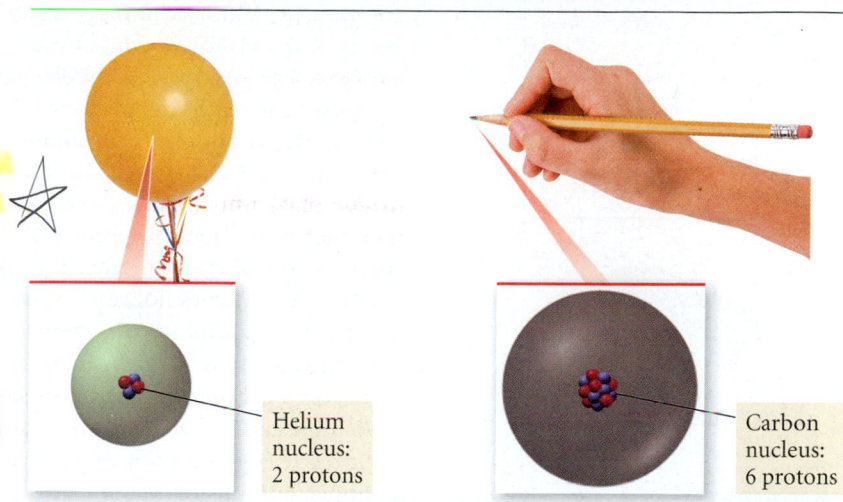

Helium nucleus: 2 protons

Carbon nucleus: 6 protons

▲ **FIGURE 2.8 How Elements Differ**   Each element is defined by a unique atomic number (Z), the number of protons in the nucleus of every atom of that element.

The Periodic Table

Atomic number (Z)

4

Be ── Chemical symbol

beryllium

Name

1 H hydrogen																	2 He helium
3 Li lithium	4 Be beryllium											5 B boron	6 C carbon	7 N nitrogen	8 O oxygen	9 F fluorine	10 Ne neon
11 Na sodium	12 Mg magnesium											13 Al aluminum	14 Si silicon	15 P phosphorus	16 S sulfur	17 Cl chlorine	18 Ar argon
19 K potassium	20 Ca calcium	21 Sc scandium	22 Ti titanium	23 V vanadium	24 Cr chromium	25 Mn manganese	26 Fe iron	27 Co cobalt	28 Ni nickel	29 Cu copper	30 Zn zinc	31 Ga gallium	32 Ge germanium	33 As arsenic	34 Se selenium	35 Br bromine	36 Kr krypton
37 Rb rubidium	38 Sr strontium	39 Y yttrium	40 Zr zirconium	41 Nb niobium	42 Mo molybdenum	43 Tc technetium	44 Ru ruthenium	45 Rh rhodium	46 Pd palladium	47 Ag silver	48 Cd cadmium	49 In indium	50 Sn tin	51 Sb antimony	52 Te tellurium	53 I iodine	54 Xe xenon
55 Cs cesium	56 Ba barium	57 La lanthanum	72 Hf hafnium	73 Ta tantalum	74 W tungsten	75 Re rhenium	76 Os osmium	77 Ir iridium	78 Pt platinum	79 Au gold	80 Hg mercury	81 Tl thallium	82 Pb lead	83 Bi bismuth	84 Po polonium	85 At astatine	86 Rn radon
87 Fr francium	88 Ra radium	89 Ac actinium	104 Rf rutherfordium	105 Db dubnium	106 Sg seaborgium	107 Bh bohrium	108 Hs hassium	109 Mt meitnerium	110 Ds darmstadtium	111 Rg roentgenium	112 **		114 **		116 **	

58 Ce cerium	59 Pr praseodymium	60 Nd neodymium	61 Pm promethium	62 Sm samarium	63 Eu europium	64 Gd gadolinium	65 Tb terbium	66 Dy dysprosium	67 Ho holmium	68 Er erbium	69 Tm thulium	70 Yb ytterbium	71 Lu lutetium
90 Th thorium	91 Pa protactinium	92 U uranium	93 Np neptunium	94 Pu plutonium	95 Am americium	96 Cm curium	97 Bk berkelium	98 Cf californium	99 Es einsteinium	100 Fm fermium	101 Md mendelevium	102 No nobelium	103 Lr lawrencium

▲ **FIGURE 2.9 The Periodic Table**   Each element is represented by its symbol and atomic number. Elements in the same column have similar properties.

96
Cm
Curium

▲ Element 96 is named curium, after Marie Curie, co-discoverer of radioactivity.

it is C; and for uranium, it is U. The chemical symbol and the atomic number always go together. If the atomic number is 2, the chemical symbol must be He. If the atomic number is 6, the chemical symbol must be C. This is just another way of saying that the number of protons defines the element.

Most chemical symbols are based on the English name of the element. For example, the symbol for sulfur is S; for oxygen, O; and for chlorine, Cl. Several of the oldest known elements, however, have symbols based on their original Latin names. Thus, the symbol for sodium is Na from the Latin *natrium*, and the symbol for tin is Sn from the Latin *stannum*. The names of elements were often given to describe their properties. For example, argon originates from the Greek word *argos* meaning inactive, referring to argon's chemical inertness (it does not react with other elements). Chlorine originates from the Greek word *chloros* meaning pale green, referring to chlorine's pale green color (see p. 48). Other elements, including helium, selenium, and mercury, were named after figures from Greek or Roman mythology or astronomical bodies. Still others (such as europium, polonium, and berkelium) were named for the places where they were discovered or where their discoverer was born. More recently, elements have been named after scientists; for example, curium for Marie Curie, einsteinium for Albert Einstein, and rutherfordium for Ernest Rutherford.

Isotopes: When the Number of Neutrons Varies

All atoms of a given element have the same number of protons; however, they do not necessarily have the same number of neutrons. Since neutrons have nearly the same mass as protons (1 amu), this means that—contrary to what John Dalton originally proposed in his atomic theory—all atoms of a given element *do not* have the same mass. For example, all neon atoms contain 10 protons, but they may have 10, 11, or 12 neutrons. All three types of neon atoms exist, and each has a slightly different mass. Atoms with the same number of protons but different numbers of neutrons are called **isotopes**. Some elements, such as beryllium (Be) and aluminum (Al), have only one naturally occurring isotope, while other elements, such as neon (Ne) and chlorine (Cl), have two or more.

Fortunately, the relative amount of each different isotope in a naturally occurring sample of a given element is usually the same. For example, in any natural sample of neon atoms, 90.48% of them are the isotope with 10 neutrons, 0.27% are the isotope with 11 neutrons, and 9.25% are the isotope with 12 neutrons. These percentages are called the **natural abundance** of the isotopes. Each element has its own characteristic natural abundance of isotopes.

The sum of the number of neutrons and protons in an atom is called the **mass number** and is given the symbol A:

$$A = \text{number of protons (p)} + \text{number of neutrons (n)}$$

For neon, with 10 protons, the mass numbers of the three different naturally occurring isotopes are 20, 21, and 22, corresponding to 10, 11, and 12 neutrons, respectively.

Isotopes are often symbolized in the following way:

where X is the chemical symbol, A is the mass number, and Z is the atomic number. Therefore, the symbols for the neon isotopes are

$$^{20}_{10}\text{Ne} \quad ^{21}_{10}\text{Ne} \quad ^{22}_{10}\text{Ne}$$

Notice that the chemical symbol, Ne, and the atomic number, 10, are redundant: if the atomic number is 10, the symbol must be Ne. The mass numbers, however, are different for different isotopes, reflecting the different number of neutrons in each one.

A second common notation for isotopes is the chemical symbol (or chemical name) followed by a dash and the mass number of the isotope.

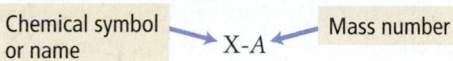

In this notation, the neon isotopes are

Ne-20	Ne-21	Ne-22
neon-20	neon-21	neon-22

We can summarize what we have learned about the neon isotopes in the following table:

Symbol	Number of Protons	Number of Neutrons	A (Mass Number)	Natural Abundance(%)
Ne-20 or $^{20}_{10}\text{Ne}$	10	10	20	90.48
Ne-21 or $^{21}_{10}\text{Ne}$	10	11	21	0.27
Ne-22 or $^{22}_{10}\text{Ne}$	10	12	22	9.25

Notice that all isotopes of a given element have the same number of protons (otherwise they would be different elements). Notice also that the mass number is the *sum* of the number of protons and the number of neutrons. The number of neutrons in an isotope is the difference between the mass number and the atomic number ($A - Z$). The different isotopes of an element generally exhibit the same chemical behavior—the three isotopes of neon, for example, all exhibit the same chemical inertness.

EXAMPLE 2.3 Atomic Numbers, Mass Numbers, and Isotope Symbols

(a) What are the atomic number (Z), mass number (A), and symbol of the chlorine isotope with 18 neutrons?

(b) How many protons, electrons, and neutrons are present in an atom of $^{52}_{24}\text{Cr}$?

Solution

(a) From the periodic table, we find that the atomic number (Z) of chlorine is 17, so chlorine atoms have 17 protons.	$Z = 17$
The mass number (A) for the isotope with 18 neutrons is the sum of the number of protons (17) and the number of neutrons (18).	$A = 17 + 18 = 35$
The symbol for the chlorine isotope is its two-letter abbreviation with the atomic number (Z) in the lower left corner and the mass number (A) in the upper left corner.	$^{35}_{17}\text{Cl}$
(b) For $^{52}_{24}\text{Cr}$, the number of protons is the lower left number. Since this is a neutral atom, there are an equal number of electrons.	Number of protons $= Z = 24$ Number of electrons $= 24$ (neutral atom)
The number of neutrons is equal to the upper left number minus the lower left number.	Number of neutrons $= 52 - 24 = 28$

For Practice 2.3

(a) What are the atomic number, mass number, and symbol for the carbon isotope with 7 neutrons?

(b) How many protons and neutrons are present in an atom of $^{39}_{19}\text{K}$?

Ions: Losing and Gaining Electrons

The number of electrons in a neutral atom is equal to the number of protons in its nucleus (given by the atomic number Z). During chemical changes, however, atoms often lose or gain electrons to form charged particles called **ions**. For example, neutral lithium (Li) atoms contain 3 protons and 3 electrons; however, in many chemical reactions lithium atoms lose one electron (e^-) to form Li^+ ions.

$$Li \rightarrow Li^+ + 1 e^-$$

The charge of an ion is indicated in the upper right corner of the symbol. The Li^+ *ion* contains 3 protons but only 2 electrons, resulting in a charge of 1+. The charge of an ion depends on how many electrons were gained or lost in forming the ion. Neutral fluorine (F) atoms contain 9 protons and 9 electrons; however, in many chemical reactions fluorine atoms gain one electron to form F^- ions.

$$F + 1e^- \rightarrow F^-$$

The F^- *ion* contains 9 protons and 10 electrons, resulting in a charge of 1−. For many elements, such as lithium and fluorine, the ion is much more common than the neutral atom. In fact, virtually all of the lithium and fluorine in nature are in the form of their ions.

Positively charged ions, such as Li^+, are called **cations** and negatively charged ions, such as F^-, are called **anions**. Ions act very differently than the atoms from which they are formed. Neutral sodium atoms, for example, are extremely unstable, reacting violently with most things they contact. Sodium cations (Na^+), in contrast, are relatively inert—we eat them all the time in sodium chloride (table salt). In ordinary matter, cations and anions always occur together so that matter is charge-neutral overall.

 Conceptual Connection 2.3 The Nuclear Atom, Isotopes, and Ions

In light of the nuclear model for the atom, which of the following statements is most likely to be true?

(a) The isotope of an atom with a greater number of neutrons is larger than one with a smaller number of neutrons.

(b) The size of an anion is greater than the size of the corresponding neutral atom.

(c) The size of a cation is greater than the size of the corresponding neutral atom.

Answer: **(b)** The number of neutrons in the nucleus of an atom does not affect the atom's size because the nucleus is miniscule compared to the atom itself. The number of electrons, however, does affect the size of the atom because most of the volume of the atom is occupied by electrons, and electrons repel one another. Therefore, an anion, with a greater number of electrons, is larger than the corresponding neutral atom.

2.7 Finding Patterns: The Periodic Law and the Periodic Table

The modern periodic table grew out of the work of Dmitri Mendeleev (1834–1907), a nineteenth-century Russian chemistry professor. In his time, about 65 different elements had been discovered. Through the work of a number of chemists, many of the properties of these elements—such as their relative masses, their chemical activity, and some of their physical properties—were known. However, there was no systematic way of organizing them.

▲ Dmitri Mendeleev, a Russian chemistry professor who proposed the periodic law and arranged early versions of the periodic table, was honored on a Soviet postage stamp.

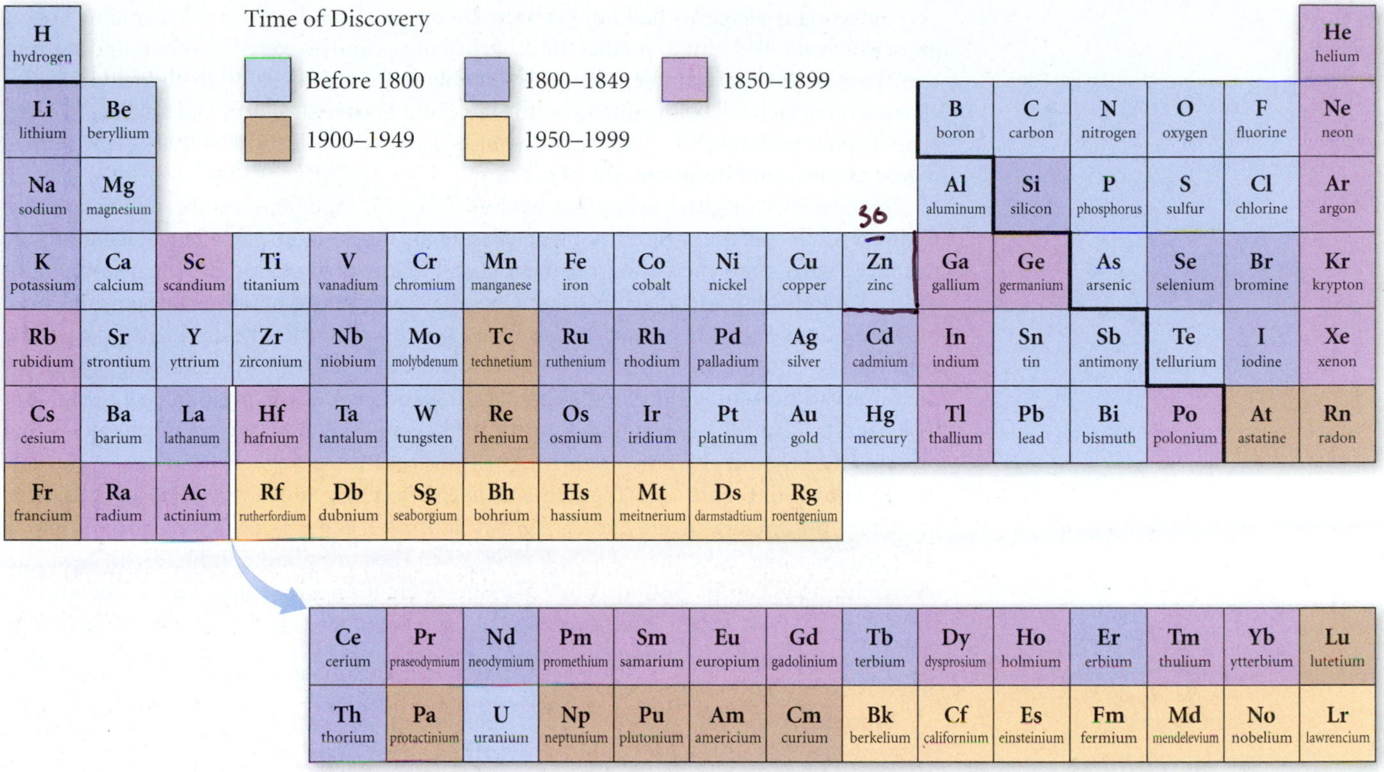

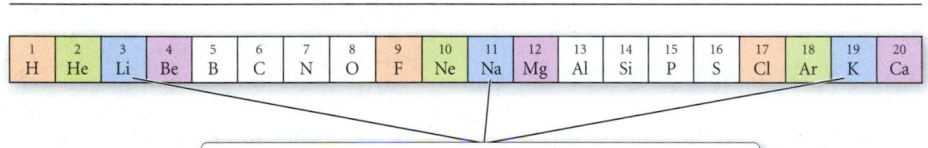

The first large periodic table with "Time of Discovery" legend:

Time of Discovery

- Before 1800
- 1800–1849
- 1850–1899
- 1900–1949
- 1950–1999

▲ Many of the elements that we know today were discovered during Mendeleev's lifetime.

In 1869, Mendeleev noticed that certain groups of elements had similar properties. Mendeleev found that when he listed elements in order of increasing mass, their properties recurred in a periodic pattern (Figure 2.10 ▼).

> Periodic means exhibiting a repeating pattern.

The Periodic Law

1	2	3	4	5	6	7	8	9	10	11	12	13	14	15	16	17	18	19	20
H	He	Li	Be	B	C	N	O	F	Ne	Na	Mg	Al	Si	P	S	Cl	Ar	K	Ca

Elements with similar properties recur in a regular pattern.

▲ **FIGURE 2.10 Recurring Properties** These elements are listed in order of increasing atomic number. Elements with similar properties are shown in the same color. Notice that the colors form a repeating pattern, much like musical notes, which form a repeating pattern on a piano keyboard.

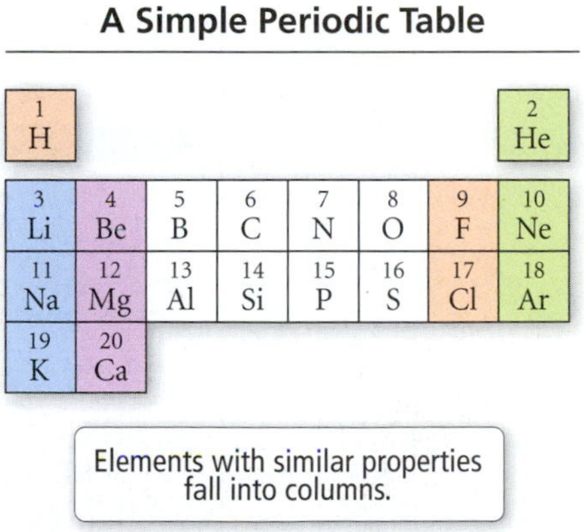

Mendeleev summarized these observations in the **periodic law**, which states the following:

When the elements are arranged in order of increasing mass, certain sets of properties recur periodically.

Mendeleev then organized all the known elements in a table consisting of a series of rows in which mass increased from left to right. The rows were arranged so that elements with similar properties aligned in the same vertical columns (Figure 2.11 ▶).

A Simple Periodic Table

1 H							2 He
3 Li	4 Be	5 B	6 C	7 N	8 O	9 F	10 Ne
11 Na	12 Mg	13 Al	14 Si	15 P	16 S	17 Cl	18 Ar
19 K	20 Ca						

Elements with similar properties fall into columns.

▲ **FIGURE 2.11 Making a Periodic Table** The elements in Figure 2.10 can be arranged in a table in which atomic number increases from left to right and elements with similar properties (as represented by the different colors) are aligned in columns.

Eka means the one beyond or the next one in a family of elements. So, eka-silicon means the element beyond silicon in the same family as silicon.

Since many elements had not yet been discovered, Mendeleev's table contained some gaps, which allowed him to predict the existence of yet undiscovered elements and some of their properties. For example, Mendeleev predicted the existence of an element he called eka-silicon, which fell below silicon on the table and between gallium and arsenic. In 1886, eka-silicon was discovered by German chemist Clemens Winkler (1838–1904), who named it germanium, after his home country.

Mendeleev's original listing has evolved into the modern periodic table shown in Figure 2.12 ▼. In the modern table, elements are listed in order of increasing atomic number rather than increasing relative mass. The modern periodic table also contains more elements than Mendeleev's original table because more have been discovered since his time. Mendeleev's periodic law was based on observation. Like all scientific laws, the periodic law summarized many observations but did not give the underlying reason for the observations—only theories do that. For now, we accept the periodic law as it is, but in Chapters 7 and 8 we examine a powerful theory—called quantum mechanics—that explains the law and gives the underlying reasons for it.

As shown in Figure 2.12, the elements in the periodic table can be broadly classified as metals, nonmetals, or metalloids. **Metals**, found on the lower left side and middle of the periodic table, have the following properties: they are good conductors of heat and electricity; they can be pounded into flat sheets (malleability); they can be drawn into wires (ductility); they are often shiny; and they tend to lose electrons when they undergo chemical changes. Good examples of metals include chromium, copper, strontium, and lead.

Nonmetals are found on the upper right side of the periodic table. The dividing line between metals and nonmetals is the zigzag diagonal line running from boron to astatine. Nonmetals have more varied properties—some are solids at room temperature,

Major Divisions of the Periodic Table

Metals | Metalloids | Nonmetals

	1A 1	2A 2	3B 3	4B 4	5B 5	6B 6	7B 7	8B 8	8B 9	8B 10	1B 11	2B 12	3A 13	4A 14	5A 15	6A 16	7A 17	8A 18
1	1 H																	2 He
2	3 Li	4 Be											5 B	6 C	7 N	8 O	9 F	10 Ne
3	11 Na	12 Mg											13 Al	14 Si	15 P	16 S	17 Cl	18 Ar
4	19 K	20 Ca	21 Sc	22 Ti	23 V	24 Cr	25 Mn	26 Fe	27 Co	28 Ni	29 Cu	30 Zn	31 Ga	32 Ge	33 As	34 Se	35 Br	36 Kr
5	37 Rb	38 Sr	39 Y	40 Zr	41 Nb	42 Mo	43 Tc	44 Ru	45 Rh	46 Pd	47 Ag	48 Cd	49 In	50 Sn	51 Sb	52 Te	53 I	54 Xe
6	55 Cs	56 Ba	57 La	72 Hf	73 Ta	74 W	75 Re	76 Os	77 Ir	78 Pt	79 Au	80 Hg	81 Tl	82 Pb	83 Bi	84 Po	85 At	86 Rn
7	87 Fr	88 Ra	89 Ac	104 Rf	105 Db	106 Sg	107 Bh	108 Hs	109 Mt	110 Ds	111 Rg	112		114		116		

Lanthanides	58 Ce	59 Pr	60 Nd	61 Pm	62 Sm	63 Eu	64 Gd	65 Tb	66 Dy	67 Ho	68 Er	69 Tm	70 Yb	71 Lu
Actinides	90 Th	91 Pa	92 U	93 Np	94 Pu	95 Am	96 Cm	97 Bk	98 Cf	99 Es	100 Fm	101 Md	102 No	103 Lr

▲ **FIGURE 2.12 Metals, Nonmetals, and Metalloids** The elements in the periodic table fall into these three broad classes.

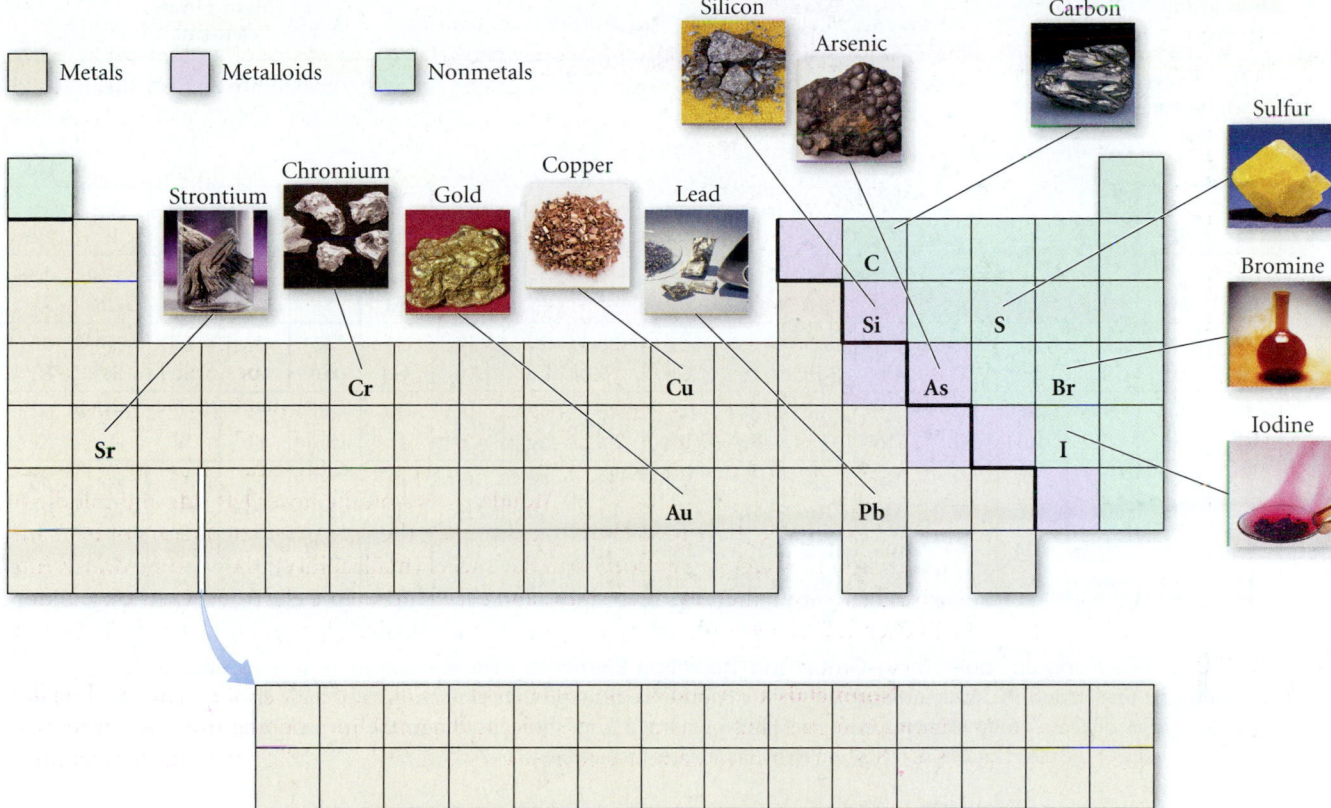

▲ Some representative metals, metalloids, and nonmetals.

others are liquids or gases—but as a whole they tend to be poor conductors of heat and electricity and they all tend to gain electrons when they undergo chemical changes. Good examples of nonmetals include oxygen, carbon, sulfur, bromine, and iodine.

Many of the elements that lie along the zigzag diagonal line that divides metals and nonmetals are called **metalloids** and show mixed properties. Several metalloids are also classified as **semiconductors** because of their intermediate (and highly temperature-dependent) electrical conductivity. The ability to change and control the conductivity of semiconductors makes them useful in the manufacture of the electronic chips and circuits central to computers, cellular telephones, and many other modern devices. Good examples of metalloids include silicon, arsenic, and antimony.

Metalloids are sometimes called semimetals.

The periodic table, as shown in Figure 2.13 on page 64, can also be broadly divided into **main-group elements**, whose properties tend to be largely predictable based on their position in the periodic table, and **transition elements** or **transition metals**, whose properties tend to be less predictable based simply on their position in the periodic table. Main-group elements are in columns labeled with a number and the letter A. Transition elements are in columns labeled with a number and the letter B. An alternative numbering system does not use letters, but only the numbers 1–18. Both numbering systems are shown in most of the periodic tables in this book. Each column within the main-group regions of the periodic table is called a **family** or **group** of elements.

The elements within a group usually have similar properties. For example, the group 8A elements, called the **noble gases**, are mostly unreactive. The most familiar noble gas is probably helium, used to fill buoyant balloons. Helium is chemically stable—it does not combine with other elements to form compounds—and is therefore safe to put into balloons. Other noble gases include neon (often used in electronic signs), argon (a small component of our atmosphere), krypton, and xenon.

Main-group elements		Transition elements										Main-group elements					

Periods

	1A	Group number															8A
1	1 H	2A										3A	4A	5A	6A	7A	2 He
2	3 Li	4 Be											5 B	6 C	7 N	8 O	9 F	10 Ne
3	11 Na	12 Mg	3B	4B	5B	6B	7B	⌐ 8B ⌐			1B	2B	13 Al	14 Si	15 P	16 S	17 Cl	18 Ar
4	19 K	20 Ca	21 Sc	22 Ti	23 V	24 Cr	25 Mn	26 Fe	27 Co	28 Ni	29 Cu	30 Zn	31 Ga	32 Ge	33 As	34 Se	35 Br	36 Kr
5	37 Rb	38 Sr	39 Y	40 Zr	41 Nb	42 Mo	43 Tc	44 Ru	45 Rh	46 Pd	47 Ag	48 Cd	49 In	50 Sn	51 Sb	52 Te	53 I	54 Xe
6	55 Cs	56 Ba	57 La	72 Hf	73 Ta	74 W	75 Re	76 Os	77 Ir	78 Pt	79 Au	80 Hg	81 Tl	82 Pb	83 Bi	84 Po	85 At	86 Rn
7	87 Fr	88 Ra	89 Ac	104 Rf	105 Db	106 Sg	107 Bh	108 Hs	109 Mt	110 Ds	111 Rg	112		114		116		

▲ **FIGURE 2.13 The Periodic Table: Main-Group and Transition Elements**   The elements in the periodic table fall into columns. The two columns at the left and the six columns at the right comprise the main-group elements. Each of these eight columns is a group or family. The properties of main-group elements can generally be predicted from their position in the periodic table. The properties of the elements in the middle of the table, known as transition elements, are less predictable.

The group 1A elements, called the **alkali metals**, are all reactive metals. A marble-sized piece of sodium explodes violently when dropped into water. Other alkali metals include lithium, potassium, and rubidium.

The group 2A elements, called the **alkaline earth metals**, are also fairly reactive, although not quite as reactive as the alkali metals. Calcium for example, reacts fairly vigorously when dropped into water but will not explode as dramatically as sodium. Other alkaline earth metals include magnesium (a common low-density structural metal), strontium, and barium.

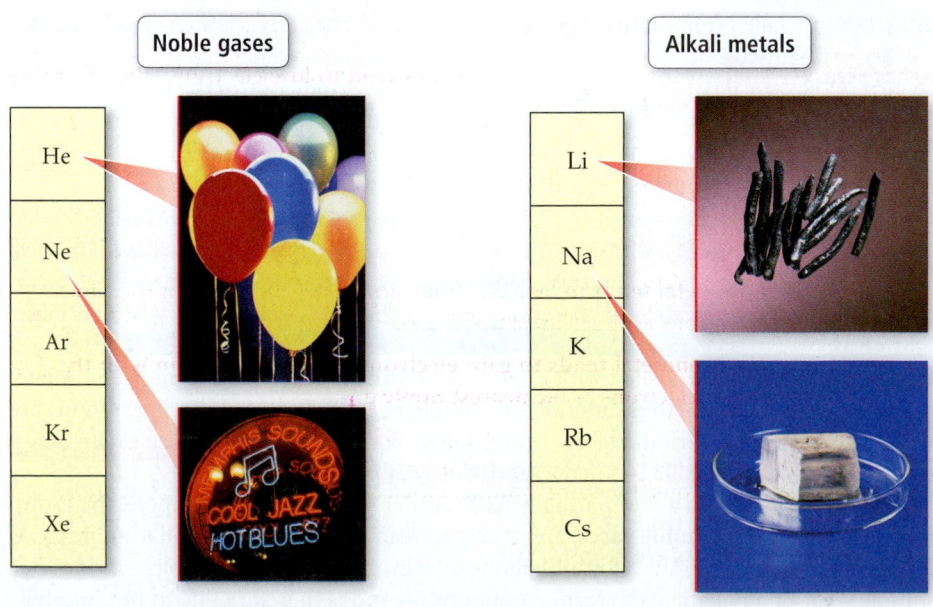

Noble gases

He
Ne
Ar
Kr
Xe

Alkali metals

Li
Na
K
Rb
Cs

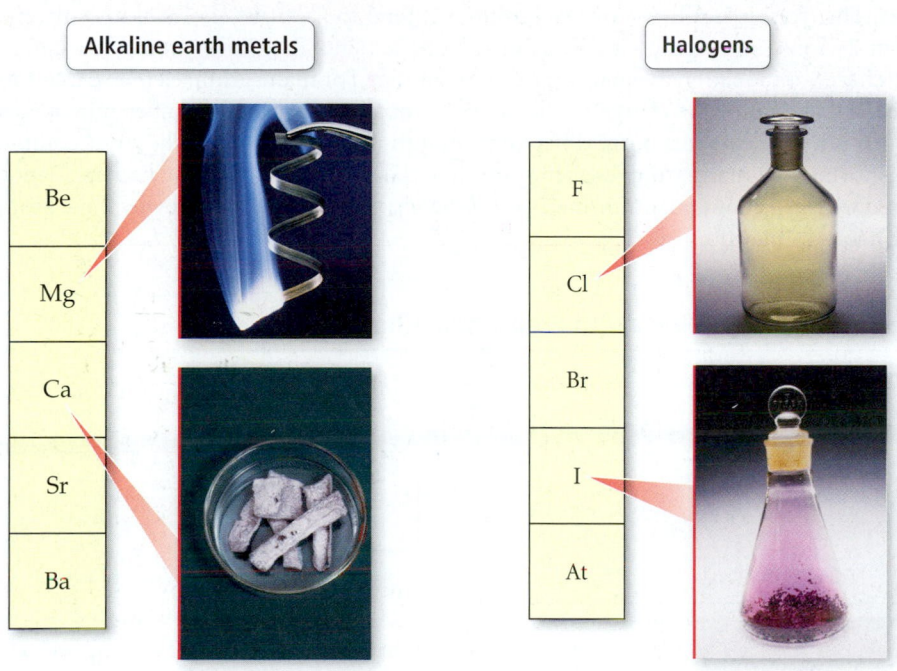

The group 7A elements, called the **halogens**, are very reactive nonmetals. The most familiar halogen is probably chlorine, a greenish-yellow gas with a pungent odor. Because of its reactivity, chlorine is often used as a sterilizing and disinfecting agent. Other halogens include bromine, a red-brown liquid that easily evaporates into a gas; iodine, a purple solid; and fluorine, a pale-yellow gas.

Ions and the Periodic Table

We have learned that, in chemical reactions, metals tend to lose electrons (thus forming cations) and nonmetals tend to gain them (thus forming anions). The number of electrons lost or gained, and therefore the charge of the resulting ion, is often predictable for a given element, especially main-group elements. Main-group elements tend to form ions that have the same number of electrons as the nearest noble gas (the noble gas that has the number of electrons closest to that of the element).

- A main-group metal tends to lose electrons, forming a cation with the same number of electrons as the nearest noble gas.
- A main-group nonmetal tends to gain electrons, forming an anion with the same number of electrons as the nearest noble gas.

For example, lithium, a metal with three electrons, tends to lose one electron to form a 1+ cation having two electrons, the same number of electrons as helium. Chlorine, a nonmetal with 17 electrons, tends to gain one electron to form a 1− anion having 18 electrons, the same number of electrons as argon.

In general, the alkali metals (group 1A) tend to lose one electron and therefore form 1+ ions. The alkaline earth metals (group 2A) tend to lose two electrons and therefore form 2+ ions. The halogens (group 7A) tend to gain one electron and therefore form 1– ions. The oxygen family nonmetals (group 6A) tend to gain two electrons and therefore form 2– ions. More generally, for main-group elements that form predictable cations, the charge of the cation is equal to the group number. For main-group elements that form predictable anions, the charge of the anion is equal to the group number minus eight. Transition elements may form different ions with different charges. The most common ions formed by main-group elements are shown in Figure 2.14 ▼. In Chapters 7 and 8, when we learn about quantum-mechanical theory, you will understand why these groups form ions as they do.

Elements That Form Ions with Predictable Charges

1A	2A	Transition metals	3A	4A	5A	6A	7A	8A
H^+							H^-	N o b l e G a s e s
Li^+					N^{3-}	O^{2-}	F^-	
Na^+	Mg^{2+}		Al^{3+}			S^{2-}	Cl^-	
K^+	Ca^{2+}					Se^{2-}	Br^-	
Rb^+	Sr^{2+}					Te^{2-}	I^-	
Cs^+	Ba^{2+}							

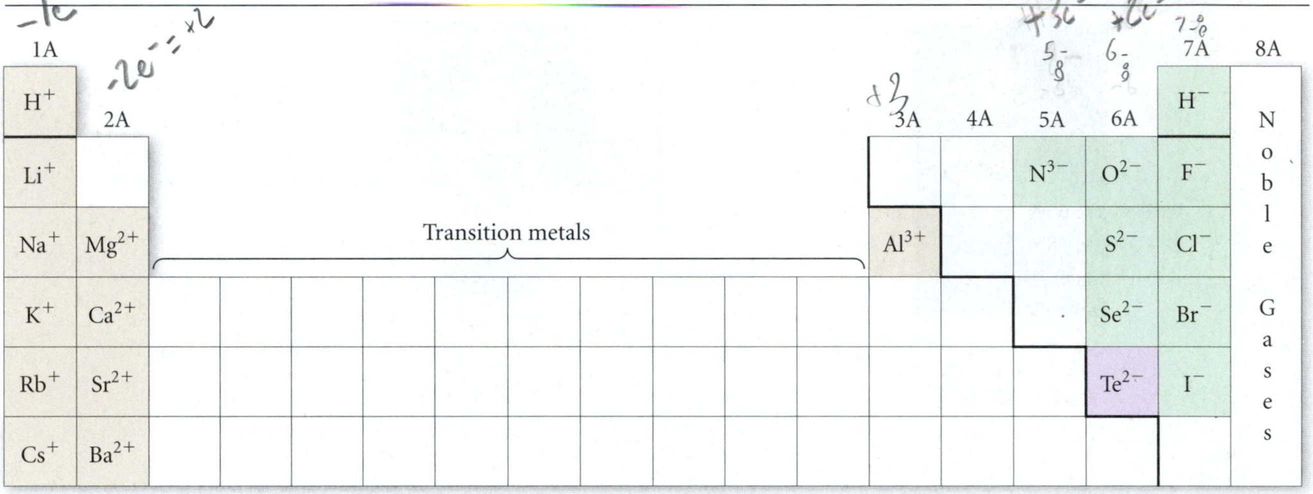

▲ FIGURE 2.14 Elements That Form Ions with Predictable Charges

---

EXAMPLE 2.4 Predicting the Charge of Ions

Predict the charges of the monoatomic ions formed by the following main-group elements.

(a) Al   (b) S

Solution

(a) Aluminum is a main-group metal and will therefore tend to lose electrons to form a cation with the same number of electrons as the nearest noble gas. Aluminum atoms have 13 electrons and the nearest noble gas is neon, which has 10 electrons. Therefore aluminum will tend to lose 3 electrons to form a cation with a +3 charge (Al^{3+}).

(b) Sulfur is a nonmetal and will therefore tend to gain electrons to form an anion with the same number of electrons as the nearest noble gas. Sulfur atoms have 16 electrons and the nearest noble gas is argon, which has 18 electrons. Therefore sulfur will tend to gain 2 electrons to form an anion with a −2 charge (S^{2-}).

For Practice 2.4

Predict the charges of the monoatomic ions formed by the following main-group elements.

(a) N   (b) Rb

Chemistry and Medicine
The Elements of Life

What are the atoms that compose living things? That compose humans? The chemistry of life revolves around the chemistry of carbon, an element that forms a disproportionately large number of compounds with a few other elements such as hydrogen, oxygen, and nitrogen. We will first encounter the chemistry of carbon—called organic chemistry—in Chapter 3 and then ex-

amine it in more detail in Chapter 20. By mass, our bodies are mostly oxygen atoms because of the large amount of water in our bodies. Carbon comes in second and hydrogen third, as shown in Figure 2.15 ▼ and Table 2.2.

Because the atoms of different elements have different masses (more on this in Sections 2.8 and 2.9), the elemental composition of humans looks different when expressed by number of atoms. When expressed this way, hydrogen comes in first (because hydrogen atoms are so light), with oxygen second and carbon third.

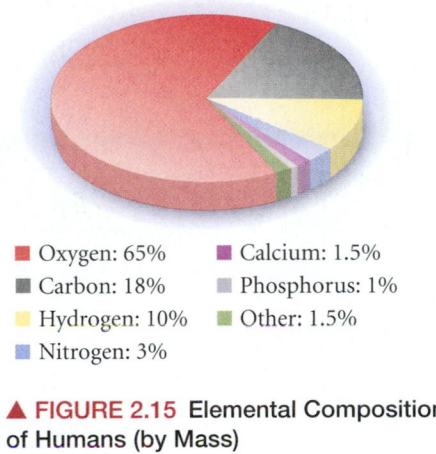

■ Oxygen: 65%   ■ Calcium: 1.5%
■ Carbon: 18%   ■ Phosphorus: 1%
■ Hydrogen: 10% ■ Other: 1.5%
■ Nitrogen: 3%

▲ **FIGURE 2.15** Elemental Composition of Humans (by Mass)

TABLE 2.2 Approximate Percent Elemental Composition of Humans

Element	% by Mass	% by Number of Atoms
Oxygen	65	26.4
Carbon	18	9.2
Hydrogen	10	62.3
Nitrogen	3	1.4
Calcium	1.5	0.2
Phosphorus	1	0.3
Other	1.5	0.2

2.8 Atomic Mass: The Average Mass of an Element's Atoms

An important part of Dalton's atomic theory was that all atoms of a given element have the same mass. However, in Section 2.6, we learned that because of isotopes, the atoms of a given element often have different masses, so Dalton was not completely correct. We can, however, calculate an average mass—called the **atomic mass**—for each element.

The atomic mass of each element is listed directly beneath the element's symbol in the periodic table and represents the average mass of the isotopes that compose that element, *weighted according to the natural abundance of each isotope.* For example, the periodic table lists the atomic mass of chlorine as 35.45 amu. Naturally occurring chlorine consists of 75.77% chlorine-35 atoms (mass 34.97 amu) and 24.23% chlorine-37 atoms (mass 36.97 amu). Its atomic mass is computed as follows:

> Atomic mass is sometimes called *atomic weight, average atomic mass,* or *average atomic weight.*

17
Cl
35.45
chlorine

$$\text{Atomic mass} = 0.7577(34.97\text{ amu}) + 0.2423(36.97\text{ amu}) = 35.45\text{ amu}$$

Notice that the atomic mass of chlorine is closer to 35 than 37. Naturally occurring chlorine contains more chlorine-35 atoms than chlorine-37 atoms, so the weighted average mass of chlorine is closer to 35 amu than to 37 amu.

When percentages are used in calculations, they are converted to their decimal value by dividing by 100.

In general, the atomic mass is calculated according to the following equation:

$$\text{Atomic mass} = \sum_{n} (\text{fraction of isotope } n) \times (\text{mass of isotope } n)$$

$$= (\text{fraction of isotope 1} \times \text{mass of isotope 1})$$
$$+ (\text{fraction of isotope 2} \times \text{mass of isotope 2})$$
$$+ (\text{fraction of isotope 3} \times \text{mass of isotope 3}) + \cdots$$

where the fractions of each isotope are the percent natural abundances converted to their decimal values. Atomic mass is useful because it allows us to assign a characteristic mass to each element, and as we will see shortly, it allows us to quantify the number of atoms in a sample of that element.

EXAMPLE 2.5 Atomic Mass

Copper has two naturally occurring isotopes: Cu-63 with mass 62.9396 amu and a natural abundance of 69.17%, and Cu-65 with mass 64.9278 amu and a natural abundance of 30.83%. Calculate the atomic mass of copper.

Solution

Convert the percent natural abundances into decimal form by dividing by 100.	$\text{Fraction Cu-63} = \dfrac{69.17}{100} = 0.6917$
	$\text{Fraction Cu-65} = \dfrac{30.83}{100} = 0.3083$
Compute the atomic mass using the equation given in the text.	$\text{Atomic mass} = 0.6917(62.9396 \text{ amu}) + 0.3083(64.9278 \text{ amu})$ $= 43.5353 \text{ amu} + 20.0172 \text{ amu} = 63.5525 = 63.55 \text{ amu}$

For Practice 2.5

Magnesium has three naturally occurring isotopes with masses of 23.99 amu, 24.99 amu, and 25.98 amu and natural abundances of 78.99%, 10.00%, and 11.01%, respectively. Calculate the atomic mass of magnesium.

For More Practice 2.5

Gallium has two naturally occurring isotopes: Ga-69 with a mass of 68.9256 amu and a natural abundance of 60.11%, and Ga-71. Use the atomic mass of gallium listed in the periodic table to find the mass of Ga-71.

Mass Spectrometry: Measuring the Mass of Atoms and Molecules

The masses of atoms and the percent abundances of isotopes of elements are measured using **mass spectrometry**. In a mass spectrometer, such as the one diagrammed in Figure 2.16 ▶, the sample (containing the atoms whose mass is to be measured) is injected into the instrument and vaporized. The vaporized atoms are then ionized by an electron beam. The electrons in the beam collide with the vaporized atoms, removing electrons from the atoms and creating positively charged ions. Charged plates with slits in them accelerate the positively charged ions into a magnetic field, which deflects them. The amount of deflection depends on the mass of the ions—lighter ions are deflected more than heavier ones.

In the diagram shown here, you can see three different paths, each corresponding to atoms of different mass. Finally, the ions strike a detector and produce an electrical signal

Mass Spectrometer

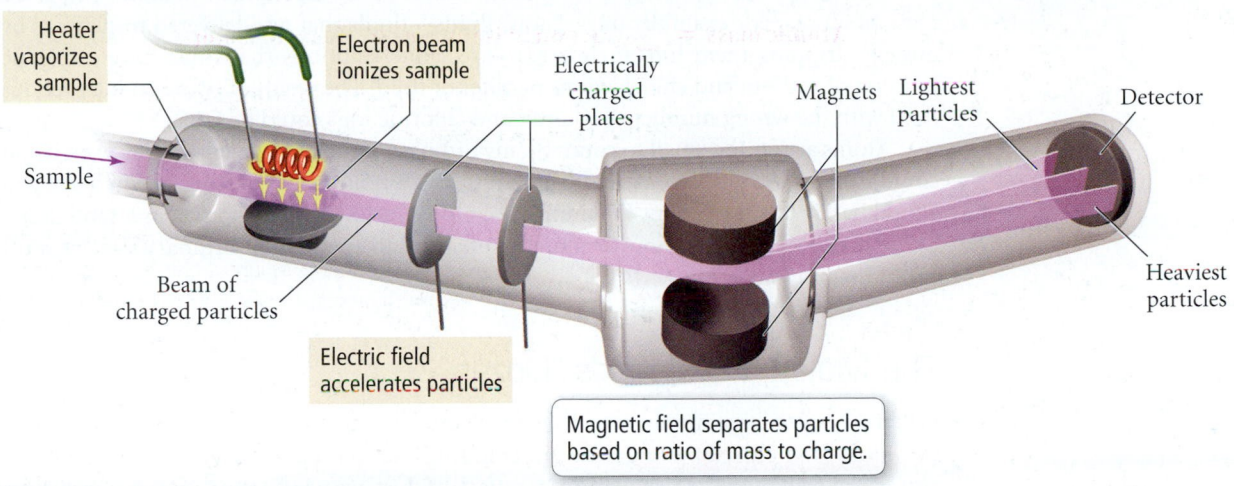

▲ **FIGURE 2.16 The Mass Spectrometer** Atoms are converted to positively charged ions, accelerated, and passed through a magnetic field that deflects their path. The heaviest ions undergo the least deflection.

that is recorded. The result is the separation of the atoms in the sample according to their mass, producing a mass spectrum such as the one shown in Figure 2.17 ▶. The *position of each peak on the x-axis gives the mass of the isotope* that was ionized, and the *intensity* (indicated by the height of the peak) gives the *relative abundance of that isotope.*

Mass spectrometry can also be used on molecules. Because molecules often fragment (break apart) during ionization, the mass spectrum of a molecule usually contains many peaks representing the masses of different parts of the molecule, as well as a peak representing the mass of the molecule as a whole. The fragments that form upon ionization, and therefore the corresponding peaks that appear in the mass spectrum, are specific to the molecule, so that a mass spectrum is like a molecular fingerprint. In other words, mass spectroscopy can be used to identify an unknown molecule and to determine how much of it is present in a particular sample. For example, mass spectrometry has been used to detect organic compounds present in meteorites, a puzzling observation which some scientists speculate may be evidence of life outside of our planet. Most scientists, however, think that the compounds probably formed in the same way as the first organic molecules on Earth, indicating that the formation of organic molecules may be common in the universe.

Since the early 1990s, mass spectrometry has also been successfully applied to biological molecules, including proteins (the workhorse molecules in cells) and nucleic acids (the molecules that carry genetic information). For a long time, these molecules could not be analyzed by mass spectrometry because they were difficult to vaporize and ionize without destroying them, but modern techniques have overcome this problem. A tumor, for example, can now be instantly analyzed to determine whether it contains specific proteins associated with cancer.

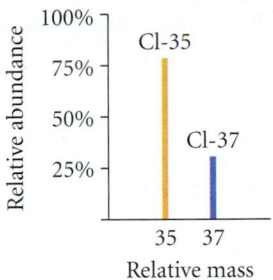

▲ **FIGURE 2.17 The Mass Spectrum of Chlorine** The position of each peak on the x-axis indicates the mass of the isotope. The intensity (or height) of the peak indicates the relative abundance of the isotope.

2.9 Molar Mass: Counting Atoms by Weighing Them

Have you ever bought shrimp and noticed a quantity called *count*? Shrimp is normally sold by count, which tells you the number of shrimp per pound. For example, 41–50 count shrimp means that there are between 41 and 50 shrimp per pound. The smaller the count, the larger the shrimp. The big tiger prawns have counts as low as 10–15, which means that each shrimp can weigh up to 1/10 of a pound. The nice thing about categorizing shrimp in this way is that you can count the shrimp by weighing them. For example, two pounds of 41–50 count shrimp contains between 82 and 100 shrimp.

A similar (but more precise) concept exists for atoms. Counting atoms is much more difficult than counting shrimp, yet we often need to know the number of atoms in a given mass of atoms. For example, intravenous fluids—fluids that are delivered to patients by directly dripping them into their veins—are saline solutions that must have a specific number of sodium and chloride ions per liter of fluid. The result of using an intravenous fluid with the wrong number of sodium and chloride ions could be fatal.

Atoms are far too small to count by any ordinary means. As we saw earlier, even if you could somehow count atoms, and counted them 24 hours a day as long as you lived, you would barely begin to count the number of atoms in something as small as a sand grain. Therefore, if we want to know the number of atoms in anything of ordinary size, we must count them by weighing.

The Mole: A Chemist's "Dozen"

When we count large numbers of objects, we often use units such as a dozen (12 objects) or a gross (144 objects) to organize our counting and to keep our numbers smaller. With atoms, quadrillions of which may be in a speck of dust, we need a much larger number for this purpose. The chemist's "dozen" is called the **mole** (abbreviated mol) and is defined as the *amount* of material containing 6.0221421×10^{23} particles.

$$1 \text{ mol} = 6.0221421 \times 10^{23} \text{ particles}$$

This number is also called **Avogadro's number**, named after Italian physicist Amedeo Avogadro (1776–1856), and is a convenient number to use when working with atoms, molecules, and ions. In this book, we will usually round Avogadro's number to four significant figures or 6.022×10^{23}. Notice that the definition of the mole is an *amount* of a substance. We will often refer to the number of moles of substance as the *amount* of the substance.

The first thing to understand about the mole is that it can specify Avogadro's number of anything. For example, 1 mol of marbles corresponds to 6.022×10^{23} marbles, and 1 mol of sand grains corresponds to 6.022×10^{23} sand grains. *One mole of anything is 6.022×10^{23} units of that thing.* One mole of atoms, ions, or molecules, however, makes up objects of everyday sizes. For example, 22 copper pennies contain approximately 1 mol of copper atoms and a tablespoon of water contains approximately 1 mol of water molecules.

The second, and more fundamental, thing to understand about the mole is how it gets its specific value.

The numerical value of the mole is defined as being equal to the number of atoms in exactly 12 grams of pure carbon-12 (12 g C = 1 mol C atoms = 6.022×10^{23} C atoms).

The definition of the mole gives us a relationship between mass (grams of carbon) and number of atoms (Avogadro's number). This relationship, as we will see shortly, allows us to count atoms by weighing them.

Converting between Number of Moles and Number of Atoms

Converting between number of moles and number of atoms is similar to converting between dozens of shrimp and number of shrimp. To convert between moles of atoms and number of atoms we simply use the conversion factors:

$$\frac{1 \text{ mol atoms}}{6.022 \times 10^{23} \text{ atoms}} \quad \text{or} \quad \frac{6.022 \times 10^{23} \text{ atoms}}{1 \text{ mol atoms}}$$

The following example shows how to use these conversion factors.

Twenty-two copper pennies contain approximately 1 mol of copper atoms.

Beginning in 1982, pennies became almost all zinc, with only a copper coating. Before 1982, however, pennies were mostly copper.

One tablespoon of water contains approximately one mole of water molecules.

One tablespoon is approximately 15 mL; one mole of water occupies 18 mL.

EXAMPLE 2.6   Converting between Number of Moles and Number of Atoms

Calculate the number of copper atoms in 2.45 mol of copper.

Sort You are given the amount of copper in moles and asked to find the number of copper atoms.	Given 2.45 mol Cu Find Cu atoms
Strategize Convert between number of moles and number of atoms by using Avogadro's number as a conversion factor.	Conceptual Plan mol Cu ⟶ Cu atoms $$\frac{6.022 \times 10^{23} \text{ Cu atoms}}{1 \text{ mol Cu}}$$ Relationships Used $6.022 \times 10^{23} = 1$ mol (Avogadro's number)
Solve Follow the conceptual plan to solve the problem. Begin with 2.45 mol Cu and multiply by Avogadro's number to get to Cu atoms.	Solution $$2.45 \text{ mol Cu} \times \frac{6.022 \times 10^{23} \text{ Cu atoms}}{1 \text{ mol Cu}} = 1.48 \times 10^{24} \text{ Cu atoms}$$

Check Since atoms are small, it makes sense that the answer is large. The number of moles of copper is almost 2.5, so the number of atoms is almost 2.5 times Avogadro's number.

For Practice 2.6

A pure silver ring contains 2.80×10^{22} silver atoms. How many moles of silver atoms does it contain?

Converting between Mass and Amount (Number of Moles)

To count atoms by weighing them, we need one other conversion factor—the mass of 1 mol of atoms. For the isotope carbon-12, we know that this mass is exactly 12 grams, which is numerically equivalent to carbon-12's atomic mass in atomic mass units. Since the masses of all other elements are defined relative to carbon-12, the same relationship holds for all elements.

The mass of 1 mol of atoms of an element is called the **molar mass.**

The value of an element's molar mass in grams per mole is numerically equal to the element's atomic mass in atomic mass units.

For example, copper has an atomic mass of 63.55 amu and a molar mass of 63.55 g/mol. One mole of copper atoms therefore has a mass of 63.55 g. Just as the count for shrimp depends on the size of the shrimp, so the mass of 1 mol of atoms depends on the element: 1 mol of aluminum atoms (which are lighter than copper atoms) has a mass of 26.98 g; 1 mol of carbon atoms (which are even lighter than aluminum atoms) has a mass of 12.01 g; and 1 mol of helium atoms (lighter yet) has a mass of 4.003 g.

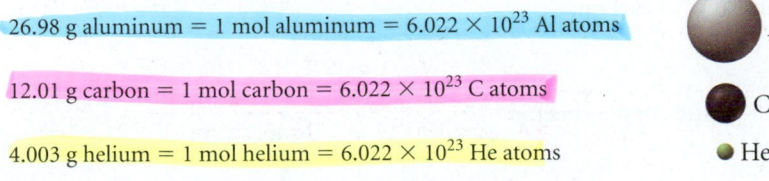

26.98 g aluminum = 1 mol aluminum = 6.022×10^{23} Al atoms   Al

12.01 g carbon = 1 mol carbon = 6.022×10^{23} C atoms   C

4.003 g helium = 1 mol helium = 6.022×10^{23} He atoms   He

The lighter the atom, the less mass it takes to make 1 mol.

1 dozen marbles

1 dozen peas

▶ The two pans contain the same number of objects (12), but the masses are different because peas are less massive than marbles. Similarly, a mole of light atoms will have less mass than a mole of heavier atoms.

Therefore, the molar mass of any element becomes a conversion factor between the mass (in grams) of that element and the amount (in moles) of that element. For carbon:

$$12.01\ g\ C = 1\ mol\ C \quad \text{or} \quad \frac{12.01\ g\ C}{mol\ C} \quad \text{or} \quad \frac{1\ mol\ C}{12.01\ g\ C}$$

The following example shows how to use these conversion factors.

EXAMPLE 2.7   Converting between Mass and Amount (Number of Moles)

Calculate the amount of carbon (in moles) contained in a 0.0265-g pencil "lead." (Assume that the pencil lead is made of pure graphite, a form of carbon.)

Sort You are given the mass of carbon and asked to find the amount of carbon in moles.	**Given** 0.0265 g C **Find** mol C
Strategize Convert between mass and amount (in moles) of an element by using the molar mass of the element.	**Conceptual Plan** $$\boxed{g\ C} \longrightarrow \boxed{mol\ C}$$ $$\frac{1\ mol}{12.01\ g}$$ **Relationships Used** $12.01\ g\ C = 1\ mol\ C$ (carbon molar mass)
Solve Follow the conceptual plan to solve the problem.	**Solution** $$0.0265\ g\ C \times \frac{1\ mol\ C}{12.01\ g\ C} = 2.21 \times 10^{-3}\ mol\ C$$

Check The given mass of carbon is much less than the molar mass of carbon. Therefore the answer (the amount in moles) is much less than 1 mol of carbon.

For Practice 2.7

Calculate the amount of copper (in moles) in a 35.8-g pure copper sheet.

For More Practice 2.7

Calculate the mass (in grams) of 0.473 mol of titanium.

We now have all the tools to count the number of atoms in a sample of an element by weighing it. First, obtain the mass of the sample. Then convert it to amount in moles using the element's molar mass. Finally, convert to number of atoms using Avogadro's number. The conceptual plan for these kinds of calculations takes the following form:

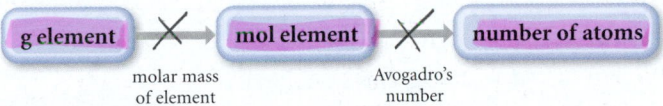

The examples that follow demonstrate these conversions.

EXAMPLE 2.8 The Mole Concept—Converting between Mass and Number of Atoms

How many copper atoms are in a copper penny with a mass of 3.10 g? (Assume that the penny is composed of pure copper.)

Sort You are given the mass of copper and asked to find the number of copper atoms.	**Given** 3.10 g Cu **Find** Cu atoms
Strategize Convert between the mass of an element in grams and the number of atoms of the element by first converting to moles (using the molar mass of the element) and then to number of atoms (using Avogadro's number).	**Conceptual Plan** g Cu → mol Cu → number of Cu atoms $\dfrac{1\ \text{mol Cu}}{63.55\ \text{g Cu}}$ $\dfrac{6.022 \times 10^{23}\ \text{Cu atoms}}{1\ \text{mol Cu}}$ **Relationships Used** 63.55 g Cu = 1 mol Cu (molar mass of copper) 6.022×10^{23} = 1 mol (Avogadro's number)
Solve Finally, follow the conceptual plan to solve the problem. Begin with 3.10 g Cu and multiply by the appropriate conversion factors to arrive at the number of Cu atoms.	**Solution** $3.10\ \cancel{\text{g Cu}} \times \dfrac{1\ \cancel{\text{mol Cu}}}{63.55\ \cancel{\text{g Cu}}} \times \dfrac{6.022 \times 10^{23}\ \text{Cu atoms}}{1\ \cancel{\text{mol Cu}}}$ $= 2.94 \times 10^{22}\ \text{Cu atoms}$

Check The answer (the number of copper atoms) is less than 6.022×10^{23} (one mole). This is consistent with the given mass of copper atoms, which is less than the molar mass of copper.

For Practice 2.8

How many carbon atoms are there in a 1.3-carat diamond? Diamonds are a form of pure carbon. (1 carat = 0.20 grams)

For More Practice 2.8

Calculate the mass of 2.25×10^{22} tungsten atoms.

Notice that numbers with large exponents, such as 6.022×10^{23}, are deceptively large. Twenty-two copper pennies contain 6.022×10^{23} or 1 mol of copper atoms, but 6.022×10^{23} pennies would cover Earth's entire surface to a depth of 300 m. Even objects small by everyday standards occupy a huge space when we have a mole of them. For example, a grain of sand has a mass of less than 1 mg and a diameter of less than 0.1 mm, yet 1 mol of sand grains would cover the state of Texas to a depth of several feet. For every increase of 1 in the exponent of a number, the number increases by a factor of 10, so 10^{23} is incredibly large. One mole has to be a large number, however, if it is to have practical value, because atoms are so small.

EXAMPLE 2.9 The Mole Concept

An aluminum sphere contains 8.55×10^{22} aluminum atoms. What is the radius of the sphere in centimeters? The density of aluminum is 2.70 g/cm^3.

Sort You are given the number of aluminum atoms in a sphere and the density of aluminum. You are asked to find the radius of the sphere.	**Given** 8.55×10^{22} Al atoms $d = 2.70 \text{ g/cm}^3$ **Find** radius (r) of sphere

Strategize The heart of this problem is density, which relates mass to volume, and though you aren't given the mass directly, you are given the number of atoms, which you can use to find mass.

(1) Convert from number of atoms to number of moles using Avogadro's number as a conversion factor.

(2) Convert from number of moles to mass using molar mass as a conversion factor.

(3) Convert from mass to volume (in cm^3) using density as a conversion factor.

(4) Once you compute the volume, find the radius from the volume using the formula for the volume of a sphere.

Conceptual Plan

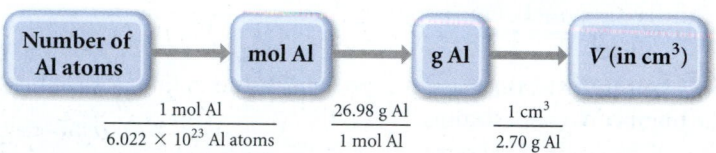

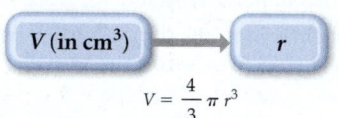

$$V = \frac{4}{3}\pi r^3$$

Relationships and Equations Used

$6.022 \times 10^{23} = 1 \text{ mol}$ (Avogadro's number)

$26.98 \text{ g Al} = 1 \text{ mol Al}$ (molar mass of aluminum)

2.70 g/cm^3 (density of aluminum)

$V = \frac{4}{3}\pi r^3$ (volume of a sphere)

Solve Finally, follow the conceptual plan to solve the problem. Begin with 8.55×10^{22} Al atoms and multiply by the appropriate conversion factors to arrive at volume in cm^3.

Then solve the equation for the volume of a sphere for r and substitute the volume to compute r.

Solution

$$8.55 \times 10^{22} \text{ Al atoms} \times \frac{1 \text{ mol Al}}{6.022 \times 10^{23} \text{ Al atoms}}$$

$$\times \frac{26.98 \text{ g Al}}{1 \text{ mol Al}} \times \frac{1 \text{ cm}^3}{2.70 \text{ g Al}} = 1.4187 \text{ cm}^3$$

$$V = \frac{4}{3}\pi r^3$$

$$r = \sqrt[3]{\frac{3V}{4\pi}} = \sqrt[3]{\frac{3(1.4187 \text{ cm}^3)}{4\pi}} = 0.697 \text{ cm}$$

Check The units of the answer (cm) are correct. The magnitude cannot be estimated accurately, but a radius of about one-half of a centimeter is reasonable for just over one-tenth of a mole of aluminum atoms.

For Practice 2.9

A titanium cube contains 2.86×10^{23} atoms. What is the edge length of the cube? The density of titanium is 4.50 g/cm^3.

For More Practice 2.9

Find the number of atoms in a copper rod with a length of 9.85 cm and a radius of 1.05 cm. The density of copper is 8.96 g/cm^3.

76   Chapter 2   At...

The Electron (2.4...

J. J. Thomson disc...
ments examini...
electrons w...
mass rati...
which...
of...

 Conceptual Connection 2.4 Avogadro's Number

Why is Avogadro's number defined as 6.022×10^{23} and not a simpler round number such as 1.00×10^{23}?

Answer: Remember that Avogadro's number is defined with respect to carbon-12—it is the number equal to the number of atoms in exactly 12 g of carbon-12. If Avogadro's number were defined as 1.00×10^{23} (a nice round number), it would correspond to 1.99 g of carbon-12 atoms (an inconvenient number). Avogadro's number is defined with respect to carbon-12 because, as you recall from Section 2.6, the amu (the basic mass unit used for all atoms) is defined relative to carbon-12. Therefore, the mass in grams of 1 mol of *any* element is equal to its atomic mass. As we have seen, these two definitions together make it possible to determine the number of atoms in a known mass of any element.

Conceptual Connection 2.5 The Mole

Without doing any calculations, determine which of the following contains the most atoms.

(a) a 1-g sample of copper

(b) a 1-g sample of carbon

(c) a 10-g sample of uranium

Answer: (b) The carbon sample contains more atoms than the copper sample because carbon has a lower molar mass than copper. Carbon atoms are lighter than copper atoms, so a 1-g sample of carbon contains more atoms than a 1-g sample of copper. The carbon sample also contains more atoms than the uranium sample because, even though the uranium sample has 10 times the mass of the carbon sample, a uranium atom is more than 10 times as massive (238 g/mol for U versus 12 g/mol for carbon).

Chapter in Review

Key Terms

Section 2.3
law of conservation of mass (47)
law of definite proportions (48)
law of multiple proportions (49)
atomic theory (51)

Section 2.4
cathode rays (51)
cathode ray tube (51)
electrical charge (52)
electron (52)

Section 2.5
radioactivity (54)
nuclear theory (55)
nucleus (55)
proton (55)
neutron (55)

Section 2.6
atomic mass unit (amu) (56)
atomic number (Z) (57)
chemical symbol (57)
isotope (58)
natural abundance (58)
mass number (A) (58)

ion (60)
cation (60)
anion (60)

Section 2.7
periodic law (61)
metal (62)
nonmetal (62)
metalloid (63)
semiconductor (63)
main-group elements (63)
transition elements (transition metals) (63)
family (group) (63)

noble gases (63)
alkali metals (64)
alkaline earth metals (64)
halogens (65)

Section 2.8
atomic mass (67)
mass spectrometry (68)

Section 2.9
mole (mol) (70)
Avogadro's number (70)
molar mass (71)

Key Concepts

Imaging and Moving Individual Atoms (2.1)

Although it was only 200 years ago that John Dalton proposed his atomic theory, technology has progressed to the level where individual atoms can be imaged and moved by techniques such as *scanning tunneling microscopy* (STM).

The Atomic Theory (2.2, 2.3)

The idea that all matter is composed of small, indestructible particles called atoms dates back to the fifth century B.C.; however, at the time the

atomic idea was rejected by most Greek thinkers. At about A.D. 1800 certain observations and laws including the law of conservation of mass, the law of constant composition, and the law of multiple proportions led John Dalton to reformulate the atomic theory with the following postulates: (1) each element is composed of indestructible particles called atoms; (2) all atoms of a given element have the same mass and other properties; (3) atoms combine in simple, whole-number ratios to form compounds; and (4) atoms of one element cannot change into atoms of another element. In a chemical reaction, atoms change the way that they are bound together with other atoms to form a new substance.

...vered the electron in the late 1800s through experi-
...ng the properties of cathode rays. He deduced that
...re negatively charged, and then measured their charge-to-
...o. Later, Robert Millikan measured the charge of the electron,
...—in conjunction with Thomson's results—led to the calculation
...he mass of an electron.

The Nuclear Atom (2.5)

In 1909, Ernest Rutherford probed the inner structure of the atom by working with a form of radioactivity called alpha radiation and thereby developed the nuclear theory of the atom. This theory states that the atom is mainly empty space, with most of its mass concentrated in a tiny region called the nucleus and most of its volume occupied by the relatively light electrons.

Subatomic Particles (2.6)

Atoms are composed of three fundamental particles: the proton (1 amu, +1 charge), the neutron (1 amu, 0 charge), and the electron (~0 amu, −1 charge). The number of protons in the nucleus of the atom is called the atomic number (Z) and defines the element. The sum of the number of protons and neutrons is called the mass number (A).

Atoms of an element that have different numbers of neutrons (and therefore different mass numbers) are called isotopes. Atoms that have lost or gained electrons become charged and are called ions. Cations are positively charged and anions are negatively charged.

The Periodic Table (2.7)

The periodic table tabulates all known elements in order of increasing atomic number. The periodic table is arranged so that similar elements are grouped together in columns. Elements on the left side and in the center of the periodic table are metals and tend to lose electrons in their chemical changes. Elements on the upper right side of the periodic table are nonmetals and tend to gain electrons in their chemical changes. Elements located on the boundary between these two classes are called metalloids.

Atomic Mass and the Mole (2.8, 2.9)

The atomic mass of an element, listed directly below its symbol in the periodic table, is a weighted average of the masses of the naturally occurring isotopes of the element.

One mole of an element is the amount of that element that contains Avogadro's number (6.022×10^{23}) of atoms. Any sample of an element with a mass (in grams) that equals its atomic mass contains one mole of the element. For example, the atomic mass of carbon is 12.011 amu, therefore 12.011 grams of carbon contains 1 mol of carbon atoms.

Key Equations and Relationships

Relationship between Mass Number (A), Number of Protons (p), and Number of Neutrons (n) (2.6)

$$A = \text{number of protons (p)} + \text{number of neutrons (n)}$$

Atomic Mass (2.8)

$$\text{Atomic mass} = \sum_{n} (\text{fraction of isotope } n) \times (\text{mass of isotope } n)$$

Avogadro's Number (2.9)

$$1 \text{ mol} = 6.0221421 \times 10^{23} \text{ particles}$$

Key Skills

Using the Law of Definite Proportions (2.3)
  • Example 2.1 • For Practice 2.1 • Exercises 33, 34

Using the Law of Multiple Proportions (2.3)
  • Example 2.2 • For Practice 2.2 • Exercises 37–40

Working with Atomic Numbers, Mass Numbers, and Isotope Symbols (2.6)
  • Example 2.3 • For Practice 2.3 • Exercises 53–60

Predicting the Charge of Ions (2.7)
  • Example 2.4 • For Practice 2.4 • Exercises 61–64

Calculating Atomic Mass (2.8)
  • Example 2.5 • For Practice 2.5 • For More Practice 2.5 • Exercises 73, 74, 76–78

Converting between Moles and Number of Atoms (2.9)
  • Example 2.6 • For Practice 2.6 • Exercises 79, 80

Converting between Mass and Amount (in Moles) (2.9)
  • Example 2.7 • For Practice 2.7 • For More Practice 2.7 • Exercises 81, 82

Using the Mole Concept (2.9)
  • Examples 2.8, 2.9 • For Practice 2.8, 2.9 • For More Practice 2.8, 2.9 • Exercises 83–90, 106, 107

Exercises

Review Questions

1. What is scanning tunneling microscopy? How does a scanning tunneling microscope work?

2. Why is it important to understand the nature of atoms?

3. Summarize the history of the atomic idea. How was Dalton able to convince others to accept an idea that had been controversial for 2000 years?

4. State the law of conservation of mass and explain what it means.

5. State the law of definite proportions and explain what it means.

6. State the law of multiple proportions and explain what it means.

7. Explain the difference between the law of definite proportions and the law of multiple proportions.

8. What are the main ideas in Dalton's atomic theory? How did they help explain the laws of conservation of mass, of constant composition, and of definite proportions?

9. How and by whom was the electron discovered? What basic properties of the electron were reported with its discovery?

10. Explain Millikan's oil drop experiment and how it led to the measurement of the electron's charge. Why is the magnitude of the charge of the electron so important?

11. Explain the plum-pudding model of the atom.

12. Describe Rutherford's gold foil experiment. How did the experiment show that the plum-pudding model of the atom was wrong?

13. Describe Rutherford's nuclear model of the atom. What was revolutionary about his model?

14. If matter is mostly empty space, as suggested by Rutherford, then why does it appear so solid?

15. List the three subatomic particles that compose atoms and give the basic properties (mass and charge) of each.

16. What defines an element?

17. Explain the difference between Z (the atomic number) and A (the mass number).

18. Where do elements get their names?

19. What are isotopes? What is percent natural abundance?

20. Describe the two different notations used to specify isotopes and give an example of each.

21. What is an ion? A cation? An anion?

22. What is the periodic law? How did it lead to the periodic table?

23. What are the characteristic properties of metals, nonmetals, and metalloids?

24. What are the characteristic properties of each of the following groups?

   a. noble gases   **b.** alkali metals
   c. alkaline earth metals   **d.** halogens

25. How do you predict the charges of ions formed by main-group elements?

26. What is atomic mass? How is it computed?

27. Explain how a mass spectrometer works.

28. What kind of information is contained in a mass spectrum?

29. What is a mole? How is it useful in chemical calculations?

30. Why is the mass corresponding to a mole of one element different from the mass corresponding to a mole of another element?

Problems by Topic

Note: Answers to all odd-numbered Problems, numbered in blue, can be found in Appendix III. Exercises in the Problems by Topic section are paired, with each odd-numbered problem followed by a similar even-numbered problem. Exercises in the Cumulative Problems section are also paired, but somewhat more loosely. (Challenge Problems and Conceptual Problems, because of their nature, are unpaired.)

The Laws of Conservation of Mass, Definite Proportions, and Multiple Proportions

31. A hydrogen-filled balloon was ignited and 1.50 g of hydrogen reacted with 12.0 g of oxygen. How many grams of water vapor were formed? (Assume that water vapor is the only product.)

32. An automobile gasoline tank holds 21 kg of gasoline. When the gasoline burns, 84 kg of oxygen is consumed and carbon dioxide and water are produced. What is the total combined mass of carbon dioxide and water that is produced?

33. Two samples of carbon tetrachloride were decomposed into their constituent elements. One sample produced 38.9 g of carbon and 448 g of chlorine, and the other sample produced 14.8 g of carbon and 134 g of chlorine. Are these results consistent with the law of definite proportions? Show why or why not.

34. Two samples of sodium chloride were decomposed into their constituent elements. One sample produced 6.98 g of sodium and 10.7 g of chlorine, and the other sample produced 11.2 g of sodium and 17.3 g of chlorine. Are these results consistent with the law of definite proportions?

35. The mass ratio of sodium to fluorine in sodium fluoride is 1.21:1. A sample of sodium fluoride produced 28.8 g of sodium upon decomposition. How much fluorine (in grams) was formed?

36. Upon decomposition, one sample of magnesium fluoride produced 1.65 kg of magnesium and 2.57 kg of fluorine. A second sample produced 1.32 kg of magnesium. How much fluorine (in grams) did the second sample produce?

37. Two different compounds containing osmium and oxygen have the following masses of oxygen per gram of osmium: 0.168 and 0.3369 g. Show that these amounts are consistent with the law of multiple proportions.

38. Palladium forms three different compounds with sulfur. The mass of sulfur per gram of palladium in each compound is listed below:

Compound	Grams S per Gram Pd
A	0.603
B	0.301
C	0.151

Show that these masses are consistent with the law of multiple proportions.

39. Sulfur and oxygen form both sulfur dioxide and sulfur trioxide. When samples of these were decomposed the sulfur dioxide produced 3.49 g oxygen and 3.50 g sulfur, while the sulfur trioxide produced 6.75 g oxygen and 4.50 g sulfur. Calculate the mass of oxygen per gram of sulfur for each sample and show that these results are consistent with the law of multiple proportions.

40. Sulfur and fluorine form several different compounds including sulfur hexafluoride and sulfur tetrafluoride. Decomposition of a sample of sulfur hexafluoride produced 4.45 g of fluorine and 1.25 g of sulfur, while decomposition of a sample of sulfur tetrafluoride produced 4.43 g of fluorine and 1.87 g of sulfur. Calculate the mass of fluorine per gram of sulfur for each sample and show that these results are consistent with the law of multiple proportions.

Atomic Theory, Nuclear Theory, and Subatomic Particles

41. Which of the following statements are *consistent* with Dalton's atomic theory as it was originally stated? Why?

   a. Sulfur and oxygen atoms have the same mass.
   b. All cobalt atoms are identical.
   c. Potassium and chlorine atoms combine in a 1:1 ratio to form potassium chloride.
   d. Lead atoms can be converted into gold.

42. Which of the following statements are *inconsistent* with Dalton's atomic theory as it was originally stated? Why?

   a. All carbon atoms are identical.
   b. An oxygen atom combines with 1.5 hydrogen atoms to form a water molecule.
   c. Two oxygen atoms combine with a carbon atom to form a carbon dioxide molecule.
   d. The formation of a compound often involves the destruction of one or more atoms.

43. Which of the following statements are *consistent* with Rutherford's nuclear theory as it was originally stated? Why?

   a. The volume of an atom is mostly empty space.
   b. The nucleus of an atom is small compared to the size of the atom.
   c. Neutral lithium atoms contain more neutrons than protons.
   d. Neutral lithium atoms contain more protons than electrons.

44. Which of the following statements are *inconsistent* with Rutherford's nuclear theory as it was originally stated? Why?

   a. Since electrons are smaller than protons, and since a hydrogen atom contains only one proton and one electron, it must follow that the volume of a hydrogen atom is mostly due to the proton.

   b. A nitrogen atom has seven protons in its nucleus and seven electrons outside of its nucleus.
   c. A phosphorus atom has 15 protons in its nucleus and 150 electrons outside of its nucleus.
   d. The majority of the mass of a fluorine atom is due to its nine electrons.

45. A chemist in an imaginary universe, where electrons have a different charge than they do in our universe, performs the Millikan oil drop experiment to measure the electron's charge. The charges of several drops are recorded below. What is the charge of the electron in this imaginary universe?

Drop #	Charge
A	-6.9×10^{-19} C
B	-9.2×10^{-19} C
C	-11.5×10^{-19} C
D	-4.6×10^{-19} C

46. Imagine a unit of charge called the zorg. A chemist performs the oil drop experiment and measures the charge of each drop in zorgs. Based on the results below, what is the charge of the electron in zorgs (z)? How many electrons are in each drop?

Drop #	Charge
A	-4.8×10^{-9} z
B	-9.6×10^{-9} z
C	-6.4×10^{-9} z
D	-12.8×10^{-9} z

47. On a dry day, your body can accumulate static charge from walking across a carpet or from brushing your hair. If your body develops a charge of $-15\ \mu$C (microcoulombs), how many excess electrons has it acquired? What is their collective mass?

48. How many electrons are necessary to produce a charge of -1.0 C? What is the mass of this many electrons?

49. Which of the following statements about subatomic particles are true?

   a. If an atom has an equal number of protons and electrons, it will be charge-neutral.
   b. Electrons are attracted to protons.
   c. Electrons are much lighter than neutrons.
   d. Protons have twice the mass of neutrons.

50. Which of the following statements about subatomic particles are false?

   a. Protons and electrons have charges of the same magnitude but opposite sign.
   b. Protons have about the same mass as neutrons.
   c. Some atoms don't have any protons.
   d. Protons and neutrons have charges of the same magnitude but opposite sign.

51. How many electrons would it take to the equal the mass of a proton?

52. A helium nucleus has two protons and two neutrons. How many electrons would it take to equal the mass of a helium nucleus?

Isotopes and Ions

53. Write isotopic symbols of the form $_Z^A X$ for each of the following isotopes.

   a. the sodium isotope with 12 neutrons
   b. the oxygen isotope with 8 neutrons
   c. the aluminum isotope with 14 neutrons
   d. the iodine isotope with 74 neutrons

54. Write isotopic symbols of the form X-*A* (e.g., C-13) for each of the following isotopes.

   a. the argon isotope with 22 neutrons
   b. the plutonium isotope with 145 neutrons
   c. the phosphorus isotope with 16 neutrons
   d. the fluorine isotope with 10 neutrons

55. Determine the number of protons and neutrons in each of the following isotopes.

   a. $_7^{14}N$ **b.** $_{11}^{23}Na$ **c.** $_{86}^{222}Rn$ **d.** $_{82}^{208}Pb$

56. Determine the number of protons and neutrons in each of the following isotopes.

   a. $_{19}^{40}K$ **b.** $_{88}^{226}Ra$ **c.** $_{43}^{99}Tc$ **d.** $_{15}^{33}P$

57. The amount of carbon-14 in artifacts and fossils is often used to establish their age. Determine the number of protons and neutrons in a carbon-14 isotope and write its symbol in the form $_Z^A X$.

58. Uranium-235 is used in nuclear fission. Determine the number of protons and neutrons in uranium-235 and write its symbol in the form $_Z^A X$.

59. Determine the number of protons and electrons in each of the following ions.

   a. Ni^{2+} **b.** S^{2-} **c.** Br^- **d.** Cr^{3+}

60. Determine the number of protons and electrons in each of the following.

   a. Al^{3+} **b.** Se^{2-} **c.** Ga^{3+} **d.** Sr^{2+}

61. Predict the charge of the ion formed by each of the following elements.

   a. O   **b.** K   **c.** Al   **d.** Rb

62. Predict the charge of the ion formed by each of the following elements.

   a. Mg   **b.** N   **c.** F   **d.** Na

63. Fill in the blanks to complete the following table.

Symbol	Ion Formed	Number of Electrons in Ion	Number of Protons in Ion
Ca	Ca^{2+}	_____	_____
_____	Be^{2+}	2	_____
Se	_____	_____	34
In	_____	_____	49

64. Fill in the blanks to complete the following table.

Symbol	Ion Formed	Number of Electrons in Ion	Number of Protons in Ion
Cl	_____	_____	17
Te	_____	54	_____
Br	Br^-	_____	_____
_____	Sr^{2+}	_____	38

The Periodic Table and Atomic Mass

65. Write the name of each of the following elements and classify it as a metal, nonmetal, or metalloid.

   a. Na   **b.** Mg   **c.** Br   **d.** N   **e.** As

66. Write the symbol for each of the following elements and classify it as a metal, nonmetal, or metalloid.

   a. lead   **b.** iodine   **c.** potassium
   d. silver   **e.** xenon

67. Which of the following elements are main-group elements?

   a. tellurium   **b.** potassium   **c.** vanadium   **d.** manganese

68. Which of the following elements are transition elements?

   a. Cr   **b.** Br   **c.** Mo   **d.** Cs

69. Classify each of the following elements as an alkali metal, alkaline earth metal, halogen, or noble gas.

   a. sodium   **b.** iodine   **c.** calcium
   d. barium   **e.** krypton

70. Classify each of the following elements as an alkali metal, alkaline earth metal, halogen, or noble gas.

   a. F   **b.** Sr   **c.** K   **d.** Ne   **e.** At

71. Which of the following pairs of elements do you expect to be most similar? Why?

   a. N and Ni   **b.** Mo and Sn   **c.** Na and Mg
   d. Cl and F   **e.** Si and P

72. Which of the following pairs of elements do you expect to be most similar? Why?

   a. nitrogen and oxygen   **b.** titanium and gallium
   c. lithium and sodium   **d.** germanium and arsenic
   e. argon and bromine

73. Rubidium has two naturally occurring isotopes with the following masses and natural abundances:

Isotope	Mass (amu)	Abundance (%)
Rb-85	84.9118	72.15
Rb-87	86.9092	27.85

Calculate the atomic mass of rubidium and sketch its mass spectrum.

74. Silicon has three naturally occurring isotopes with the following masses and natural abundances:

Isotope	Mass (amu)	Abundance (%)
Si-28	27.9769	92.2
Si-29	28.9765	4.67
Si-30	29.9737	3.10

Calculate the atomic mass of silicon and sketch its mass spectrum.

75. The atomic mass of fluorine is 18.998 amu and its mass spectrum shows a large peak at this mass. The atomic mass of chlorine is 35.45 amu, yet the mass spectrum of chlorine does not show a peak at this mass. Explain the difference.

76. The atomic mass of copper is 63.546 amu. Do any copper isotopes have a mass of 63.546 amu? Explain.

77. An element has two naturally occurring isotopes. Isotope 1 has a mass of 120.9038 amu and a relative abundance of 57.4%, and isotope 2 has a mass of 122.9042 amu. Find the atomic mass of this element and, by comparison to the periodic table, identify it.

78. Bromine has two naturally occurring isotopes (Br-79 and Br-81) and has an atomic mass of 79.904 amu. The mass of Br-81 is 80.9163 amu, and its natural abundance is 49.31%. Calculate the mass and natural abundance of Br-79.

The Mole Concept

79. How many sulfur atoms are there in 3.8 mol of sulfur?

80. How many moles of aluminum do 5.8×10^{24} aluminum atoms represent?

81. What is the amount, in moles, of each of the following?

    a. 11.8 g Ar   b. 3.55 g Zn   c. 26.1 g Ta   d. 0.211 g Li

82. What is the mass, in grams, of each of the following?

    a. 2.3×10^{-3} mol Sb   b. 0.0355 mol Ba
    c. 43.9 mol Xe   d. 1.3 mol W

83. How many silver atoms are there in 3.78 g of silver?

84. What is the mass of 4.91×10^{21} platinum atoms?

85. How many atoms are there in each of the following?

    a. 5.18 g P   b. 2.26 g Hg   c. 1.87 g Bi   d. 0.082 g Sr

86. Calculate the mass, in grams, of each of the following.

    a. 1.1×10^{23} gold atoms   b. 2.82×10^{22} helium atoms
    c. 1.8×10^{23} lead atoms   d. 7.9×10^{21} uranium atoms

87. How many carbon atoms are there in a diamond (pure carbon) with a mass of 52 mg?

88. How many helium atoms are there in a helium blimp containing 536 kg of helium?

89. Calculate the average mass, in grams, of a platinum atom.

90. Using scanning tunneling microscopy, scientists at IBM wrote the initials of their company with 35 individual xenon atoms (as shown below). Calculate the total mass of these letters in grams.

Cumulative Problems

91. A 7.83-g sample of HCN is found to contain 0.290 g of H and 4.06 g of N. Find the mass of carbon in a sample of HCN with a mass of 3.37 g.

92. The ratio of sulfur to oxygen by mass in SO_2 is 1.0:1.0.

    a. Find the ratio of sulfur to oxygen by mass in SO_3.
    b. Find the ratio of sulfur to oxygen by mass in S_2O.

93. The ratio of oxygen to carbon by mass in carbon monoxide is 1.33:1.00. Find the formula of an oxide of carbon in which the ratio by mass of oxygen to carbon is 2.00:1.00.

94. The ratio of the mass of a nitrogen atom to the mass of an atom of ^{12}C is 7:6 and the ratio of the mass of nitrogen to oxygen in N_2O is 7:4. Find the mass of 1 mol of oxygen atoms.

95. An α particle, $^4He^{2+}$, has a mass of 4.00151 amu. Find the value of its charge-to-mass ratio in C/kg..

96. Naturally occurring iodine has an atomic mass of 126.9045. A 12.3849-g sample of iodine is accidentally contaminated with 1.00070 g of ^{129}I, a synthetic radioisotope of iodine used in the treatment of certain diseases of the thyroid gland. The mass of ^{129}I is 128.9050 amu. Find the apparent "atomic mass" of the contaminated iodine.

97. Nuclei with the same number of *neutrons* but different mass numbers are called *isotones*. Write the symbols of four isotones of ^{236}Th.

98. Fill in the blanks to complete the following table.

Symbol	Z	A	Number of p	Number of e⁻	Number of n	Charge
Si	14	___	___	14	14	___
S²⁻	___	32	___	___	___	2−
Cu²⁺	___	___	___	___	24	2+
___	15	___	___	15	16	___

99. Fill in the blanks to complete the following table.

Symbol	Z	A	Number of p	Number of e⁻	Number of n	Charge
___	8	___	___	___	8	2−
Ca²⁺	20	___	___	___	20	___
Mg²⁺	___	25	___	___	13	2+
N³⁻	___	14	___	10	___	___

100. Neutron stars are believed to be composed of solid nuclear matter, primarily neutrons. Assume the radius of a neutron to be approximately 1.0×10^{-13} cm, and calculate the density of a neutron. [Hint: For a sphere $V = (4/3)\pi r^3$.] Assuming that a neutron star has the same density as a neutron, calculate the mass (in kg) of a small piece of a neutron star the size of a spherical pebble with a radius of 0.10 mm.

101. Carbon-12 contains 6 protons and 6 neutrons. The radius of the nucleus is approximately 2.7 fm (femtometers) and the radius of the atom is approximately 70 pm (picometers). Calculate the volume of the nucleus and the volume of the atom. What percentage of the carbon atom's volume is occupied by the nucleus? (Assume two significant figures.)

102. A penny has a thickness of approximately 1.0 mm. If you stacked Avogadro's number of pennies one on top of the other on Earth's surface, how far would the stack extend (in km)? [For comparison, the sun is about 150 million km from Earth and the nearest star (Proxima Centauri) is about 40 trillion km from Earth.]

103. Consider the stack of pennies in the previous problem. How much money (in dollars) would this represent? If this money were equally distributed among the world's population of 6.5 billion people, how much would each person receive? Would each person be a millionaire? Billionaire? Trillionaire?

104. The mass of an average blueberry is 0.75 g and the mass of an automobile is 2.0×10^3 kg. Find the number of automobiles whose total mass is the same as 1.0 mol blueberries.

105. Suppose that atomic masses were based on the assignment of a mass of 12.000 g to 1 mol of carbon, rather than 1 mol of ^{12}C. Find the atomic mass of oxygen.

106. A pure titanium cube has an edge length of 2.78 in. How many titanium atoms does it contain? Titanium has a density of 4.50 g/cm^3.

107. A pure copper sphere has a radius 0.935 in. How many copper atoms does it contain? [The volume of a sphere is $(4/3)\pi r^3$ and the density of copper is 8.96 g/cm^3.]

108. Boron has only two naturally occurring isotopes. The mass of boron-10 is 10.01294 amu and the mass of boron-11 is 11.00931 amu. Use the atomic mass of boron to calculate the relative abundances of the two isotopes.

109. Lithium has only two naturally occurring isotopes. The mass of lithium-6 is 6.01512 amu and the mass of lithium-7 is 7.01601 amu. Use the atomic mass of lithium to calculate the relative abundances of the two isotopes.

Challenge Problems

110. In Section 2.9, it was stated that 1 mol of sand grains would cover the state of Texas to several feet. Estimate how many feet by assuming that the sand grains are roughly cube-shaped, each one with an edge length of 0.10 mm. Texas has a land area of 268,601 square miles.

111. Use the concepts in this chapter to obtain an estimate for the number of atoms in the universe. Make the following assumptions: (a) Assume that all of the atoms in the universe are hydrogen atoms in stars. (This is not a ridiculous assumption because over three-fourths of the atoms in the universe are in fact hydrogen. Gas and dust between the stars represent only about 15% of the visible matter of our galaxy, and planets compose a far tinier fraction.) (b) Assume that the sun is a typical star composed of pure hydrogen with a density of 1.4 g/cm^3 and a radius of 7×10^8 m. (c) Assume that each of the roughly 100 billion stars in the Milky Way galaxy contains the same number of atoms as our sun. (d) Assume that each of the 10 billion galaxies in the visible universe contains the same number of atoms as our Milky Way galaxy.

112. Below is a representation of 50 atoms of a fictitious element called westmontium (Wt). The red spheres represent Wt-296, the blue spheres Wt-297, and the green spheres Wt-298.

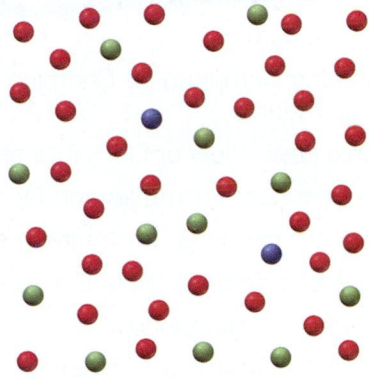

a. Assuming that the sample is statistically representative of a naturally occurring sample, calculate the percent natural abundance of each Wt isotope.

b. Draw the mass spectrum for a naturally occurring sample of Wt.

c. The mass of each Wt isotope is measured relative to C-12 and tabulated below. Use the mass of C-12 to convert each of the masses to amu and calculate the atomic mass of Wt.

Isotope	Mass
Wt-296	24.6630 × Mass(^{12}C)
Wt-297	24.7490 × Mass(^{12}C)
Wt-298	24.8312 × Mass(^{12}C)

Conceptual Problems

113. Which of the following is an example of the law of multiple proportions? Explain.

a. Two different samples of water are found to have the same ratio of hydrogen to oxygen.

b. When hydrogen and oxygen react to form water, the mass of water formed is exactly equal to the mass of hydrogen and oxygen that reacted.

c. The mass ratio of oxygen to hydrogen in water is 8:1. The mass ratio of oxygen to hydrogen in hydrogen peroxide (a compound that only contains hydrogen and oxygen) is 16:1.

114. The mole is defined as the amount of a substance containing the same number of particles as exactly 12 grams of C-12. The amu is defined as 1/12 of the mass of an atom of C-12. Why is it important that both of these definitions reference the same isotope? What would be the result, for example, of defining the mole with respect to C-12, but the amu with respect to Ne-20?

115. Without doing any calculations, determine which of the following samples contains the greatest amount of the element in moles. Which contains the greatest mass of the element?

a. 55.0 g Cr   b. 45.0 g Ti   c. 60.0 g Zn

116. The atomic radii of the isotopes of an element are identical to one another. However, the atomic radii of the ions of an element are significantly different from the atomic radii of the neutral atom of the element. Explain this behavior.

3

Molecules, Compounds, and Chemical Equations

3.1 Hydrogen, Oxygen, and Water

3.2 Chemical Bonds

3.3 Representing Compounds: Chemical Formulas and Molecular Models

3.4 An Atomic-Level View of Elements and Compounds

3.5 Ionic Compounds: Formulas and Names

3.6 Molecular Compounds: Formulas and Names

3.7 Formula Mass and the Mole Concept for Compounds

3.8 Composition of Compounds

3.9 Determining a Chemical Formula from Experimental Data

3.10 Writing and Balancing Chemical Equations

3.11 Organic Compounds

Periods

HOW MANY DIFFERENT substances exist? We learned in Chapter 2 that there are about 91 different elements in nature, so there are at least 91 different substances. However, the world would be dull—not to mention lifeless—with only 91 different substances. Fortunately, elements combine with each other to form compounds. Just as combinations of only 26 letters in our English alphabet allow for an almost limitless number of words, each with its own specific meaning, so combinations of the 91 naturally occurring elements allow for an almost limitless number of compounds, each with its own specific properties. The great diversity of substances found in nature is a direct result of the ability of elements to form compounds. Life, for example, could not exist with just 91 different elements. It takes compounds, in all of their diversity, to make life possible.

Almost all aspects of life are engineered at the molecular level, and without understanding molecules we can only have a very sketchy understanding of life itself.

—Francis Harry Compton Crick (1916–2004)

When a balloon filled with H_2 and O_2 is ignited, the two elements react violently to form H_2O.

3.1 Hydrogen, Oxygen, and Water

Hydrogen (H_2) is an explosive gas used as a fuel in the space shuttle. Oxygen (O_2), also a gas, is a natural component of air. Oxygen is not itself flammable, but it must be present for combustion to occur. Hydrogen and oxygen both have extremely low boiling points, as you can see from the table below. When hydrogen and oxygen combine to form the compound water (H_2O), however, a dramatically different substance results.

Selected Properties of Hydrogen	Selected Properties of Oxygen	Selected Properties of Water
Boiling point, −253 °C	Boiling point, −183 °C	Boiling point, 100 °C
Gas at room temperature	Gas at room temperature	Liquid at room temperature
Explosive	Supports combustion	Used to extinguish flame

First of all, water is a liquid rather than a gas at room temperature, and its boiling point is hundreds of degrees above the boiling points of hydrogen and oxygen. Second, instead of being flammable (like hydrogen gas) or supporting combustion (like oxygen gas), water actually smothers flames. Water is nothing like the hydrogen and oxygen from which it was formed.

The properties of compounds are generally very different from the properties of the elements that compose them. When two elements combine to form a compound, an entirely new substance results. Common table salt, for example, is a compound composed of sodium and chlorine. Sodium is a highly reactive, silvery metal that can explode on contact with water. Chlorine is a corrosive, greenish-yellow gas that can be fatal if inhaled. Yet the compound that results from the combination of these two elements is sodium chloride (or table salt), a flavor enhancer that we sprinkle on our food.

Although some of the substances that we encounter in everyday life are elements, most are not—they are compounds. Free atoms are rare on Earth. As we learned in Chapter 1, a compound is different from a mixture of elements. In a compound, elements combine in fixed, definite proportions; in a mixture, elements can mix in any proportions whatsoever. For example, consider the difference between a hydrogen–oxygen mixture and water. A hydrogen–oxygen mixture can have any proportions of hydrogen and oxygen gas. Water, by contrast, is composed of water molecules that always contain 2 hydrogen atoms to every 1 oxygen atom. Water has a definite proportion of hydrogen to oxygen.

In this chapter we will learn about compounds: how to represent them, how to name them, how to distinguish between their different types, and how to write chemical equations showing how they form and change. We will also learn how to quantify the composition of a compound according to its constituent elements. This is important, for example, to patients with high blood pressure who often have to reduce their sodium ion intake. Since the sodium ion is normally consumed as sodium chloride, a high blood pressure patient needs to know how much sodium is in a given amount of sodium chloride. Similarly, an iron-mining company needs to know how much iron they can recover from a given amount of iron ore. This chapter will give us the tools to understand and solve these kinds of problems.

3.2 Chemical Bonds

Compounds are composed of atoms held together by *chemical bonds*. Chemical bonds are the result of interactions between the charged particles—electrons and protons—that compose atoms. We can broadly classify most chemical bonds into two types: ionic and covalent. *Ionic bonds*—which occur between metals and nonmetals—involve the *transfer* of electrons from one atom to another. *Covalent bonds*—which occur between two or more nonmetals—involve the *sharing* of electrons between two atoms.

Mixtures and Compounds

Hydrogen and Oxygen Mixture	Water (A Compound)
Can have any ratio of hydrogen to oxygen.	Water molecules have a fixed ratio of hydrogen (2 atoms) to oxygen (1 atom).

  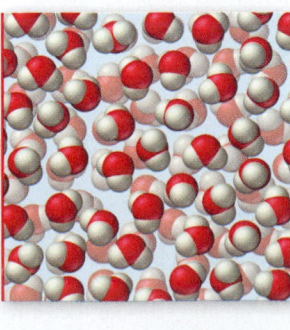

▶ The balloon in this illustration is filled with a mixture of hydrogen gas and oxygen gas. The proportions of hydrogen and oxygen are variable. The glass is filled with water, a compound of hydrogen and oxygen. The ratio of hydrogen to oxygen in water is fixed: Water molecules always have two hydrogen atoms for each oxygen atom.

Ionic Bonds

We learned in Chapter 2 that metals have a tendency to lose electrons and that nonmetals have a tendency to gain them. Therefore, when a metal interacts with a nonmetal, it can transfer one or more of its electrons to the nonmetal. The metal atom then becomes a *cation* (a positively charged ion) and the nonmetal atom becomes an *anion* (a negatively charged ion) as shown in Figure 3.1 ▼. These oppositely charged ions are then attracted to one another by electrostatic forces—they form an **ionic bond.** The result is an ionic compound which in the solid phase is composed of a lattice—a regular three-dimensional array—of alternating cations and anions.

Covalent Bonds

When a nonmetal bonds with another nonmetal, neither atom transfers its electron to the other. Instead some electrons are *shared* between the two bonding atoms. The shared electrons interact with the nuclei of both atoms, lowering their potential energy through electrostatic interactions with the nuclei. The resulting bond is called a **covalent bond.** We can understand the stability of a covalent bond by considering the most stable (lowest potential energy) arrangement of two protons and an electron. As you can see from

The Formation of an Ionic Compound

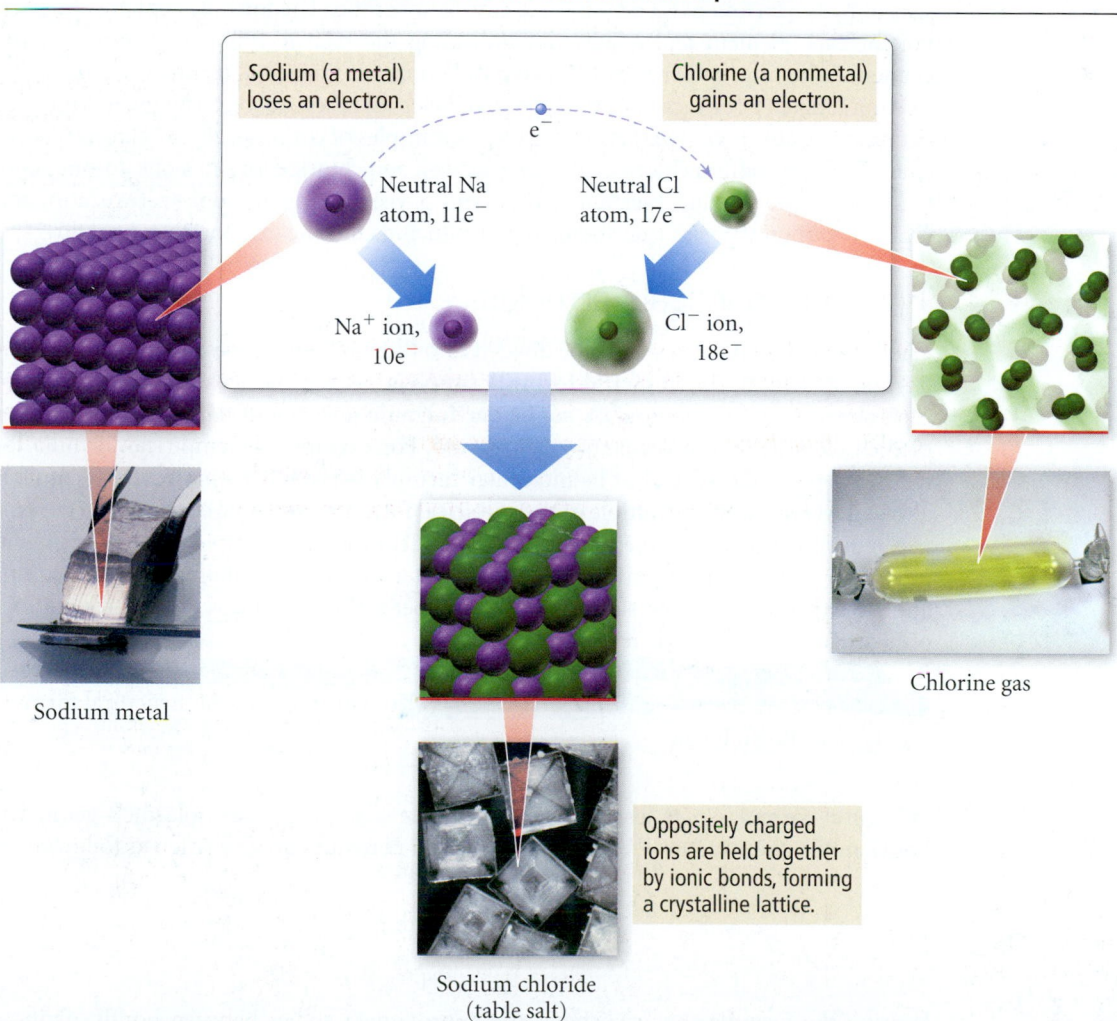

Sodium (a metal) loses an electron.

Chlorine (a nonmetal) gains an electron.

e^-

Neutral Na atom, $11e^-$

Neutral Cl atom, $17e^-$

Na^+ ion, $10e^-$

Cl^- ion, $18e^-$

Sodium metal

Chlorine gas

Oppositely charged ions are held together by ionic bonds, forming a crystalline lattice.

Sodium chloride (table salt)

▲ **FIGURE 3.1 The Formation of an Ionic Compound**   An atom of sodium (a metal) loses an electron to an atom of chlorine (a nonmetal), creating a pair of oppositely charged ions. The sodium cation is then attracted to the chloride anion and the two are held together as part of a crystalline lattice.

▶ **FIGURE 3.2 The Stability of a Covalent Bond** The potential energy of the electron is lowest when its position is between the two protons. The shared electron essentially holds the protons together.

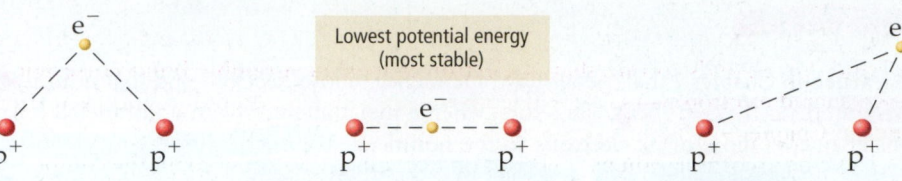

Lowest potential energy (most stable)

Figure 3.2 ▲, the arrangement in which the electron lies *between* the two protons has the lowest potential energy because the negatively charged electron can interact with *both protons*. In a sense, the electron holds the two protons together because its negative charge attracts the positive charges of both protons. Similarly, shared electrons in a covalent chemical bond hold the bonding atoms together by attracting the positively charged nuclei of the bonding atoms. The atoms within all molecular compounds are held together by covalent bonds.

3.3 Representing Compounds: Chemical Formulas and Molecular Models

The quickest and easiest way to represent a compound is with its **chemical formula,** which indicates the elements present in the compound and the relative number of atoms or ions of each. For example, H_2O is the chemical formula for water—it indicates that water consists of hydrogen and oxygen atoms in a two-to-one ratio. The formula contains the symbol for each element and a subscript indicating the relative number of atoms of the element. When the subscript is 1, it is typically omitted. Chemical formulas normally list the more metallic (or more positively charged) elements first, followed by the less metallic (or more negatively charged) elements. Other examples of common chemical formulas include NaCl for sodium chloride, meaning sodium and chloride ions in a one-to-one ratio; CO_2 for carbon dioxide, meaning carbon and oxygen atoms in a one-to-two ratio; and CCl_4 for carbon tetrachloride, indicating carbon and chlorine in a one-to-four ratio.

Types of Chemical Formulas

Chemical formulas can generally be divided into three different types: empirical, molecular, and structural. An **empirical formula** simply gives the *relative* number of atoms of each element in a compound. A **molecular formula** gives the *actual* number of atoms of each element in a molecule of a compound. For example, the empirical formula for hydrogen peroxide is HO, but its molecular formula is H_2O_2. The molecular formula is always a whole-number multiple of the empirical formula. For some compounds, the empirical formula and the molecular formula are identical. For example, the empirical and molecular formula for water is H_2O because water molecules contain 2 hydrogen atoms and 1 oxygen atom, and no simpler whole-number ratio can express the relative number of hydrogen atoms to oxygen atoms.

A **structural formula,** using lines to represent the covalent bonds, shows how atoms in a molecule are connected or bonded to each other. For example, the structural formula for H_2O_2 is shown below:

$$H—O—O—H$$

Structural formulas may also be written to give a sense of the molecule's geometry. For example, the structural formula for hydrogen peroxide can be written as follows:

$$\begin{array}{c} H \\ \backslash \\ O—O \\ \quad\backslash \\ \quad H \end{array}$$

Writing the formula this way shows the approximate angles between bonds, giving a sense of the molecule's shape. Structural formulas can also show different types of bonds that occur between molecules. For example, the structural formula for carbon dioxide is as follows:

$$O{=}C{=}O$$

The two lines between the carbon and oxygen atoms represent a double bond, which is generally stronger and shorter than a single bond (represented by a single line). A single bond corresponds to one shared electron pair while a double bond corresponds to two shared electron pairs. We will learn more about single, double, and even triple bonds in Chapter 9.

The type of formula you use depends on how much you know about the compound and how much you want to communicate. Notice that a structural formula communicates the most information, while an empirical formula communicates the least.

EXAMPLE 3.1 Molecular and Empirical Formulas

Write empirical formulas for the compounds represented by the following molecular formulas.

(a) C_4H_8 (b) B_2H_6 (c) CCl_4

Solution

To get the empirical formula from a molecular formula, divide the subscripts by the greatest common factor (the largest number that divides exactly into all of the subscripts).

(a) For C_4H_8, the greatest common factor is 4. The empirical formula is therefore CH_2.

(b) For B_2H_6, the greatest common factor is 2. The empirical formula is therefore BH_3.

(c) For CCl_4, the only common factor is 1, so the empirical formula and the molecular formula are identical.

For Practice 3.1

Write the empirical formula for the compounds represented by the following molecular formulas.

(a) C_5H_{12} (b) Hg_2Cl_2 (c) $C_2H_4O_2$

Answers to For Practice and For More Practice problems can be found in Appendix IV.

Molecular Models

The most accurate and complete way to specify compounds is with molecular models. **Ball-and-stick models** represent atoms as balls and chemical bonds as sticks; how the two connect reflects a molecule's shape. The balls are normally color-coded to specific elements. For example, carbon is customarily black, hydrogen is white, nitrogen is blue, and oxygen is red. (For a complete list of colors of elements in the molecular models used in this book see Appendix IIA.)

In **space-filling molecular models**, atoms fill the space between them to more closely represent our best estimates for how a molecule might appear if scaled to a visible size. For example, consider the following ways to represent a molecule of methane, the main component of natural gas:

 Hydrogen

 Carbon

 Nitrogen

 Oxygen

 Fluorine

 Phosphorus

 Sulfur

 Chlorine

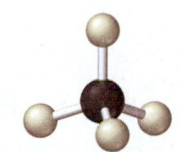

CH_4

Molecular formula | Structural formula | Ball-and-stick model | Space-filling model

The molecular formula of methane shows the number and type of each atom in the molecule: one carbon atom and four hydrogen atoms. The structural formula shows how the atoms are connected: the carbon atom is bonded to the four hydrogen atoms. The ball-and-stick model clearly shows the geometry of the molecule: the carbon atom sits in the center of a *tetrahedron* formed by the four hydrogen atoms. The space-filling model gives the best sense of the relative sizes of the atoms and how they merge together in bonding.

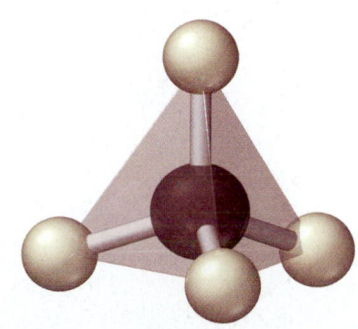

▲ A tetrahedron is a three-dimensional geometrical shape characterized by four equivalent triangular faces.

Throughout this book, you will see molecules represented in all of these ways. As you look at these representations, keep in mind what you learned in Chapter 1: the details about a molecule—the atoms that compose it, the lengths of the bonds between atoms, the angles of the bonds between atoms, and its overall shape—determine the properties of the substance that the molecule composes. Change any of these details and those properties change. Table 3.1 shows various compounds represented in the different ways we have just discussed.

TABLE 3.1   Benzene, Acetylene, Glucose, and Ammonia

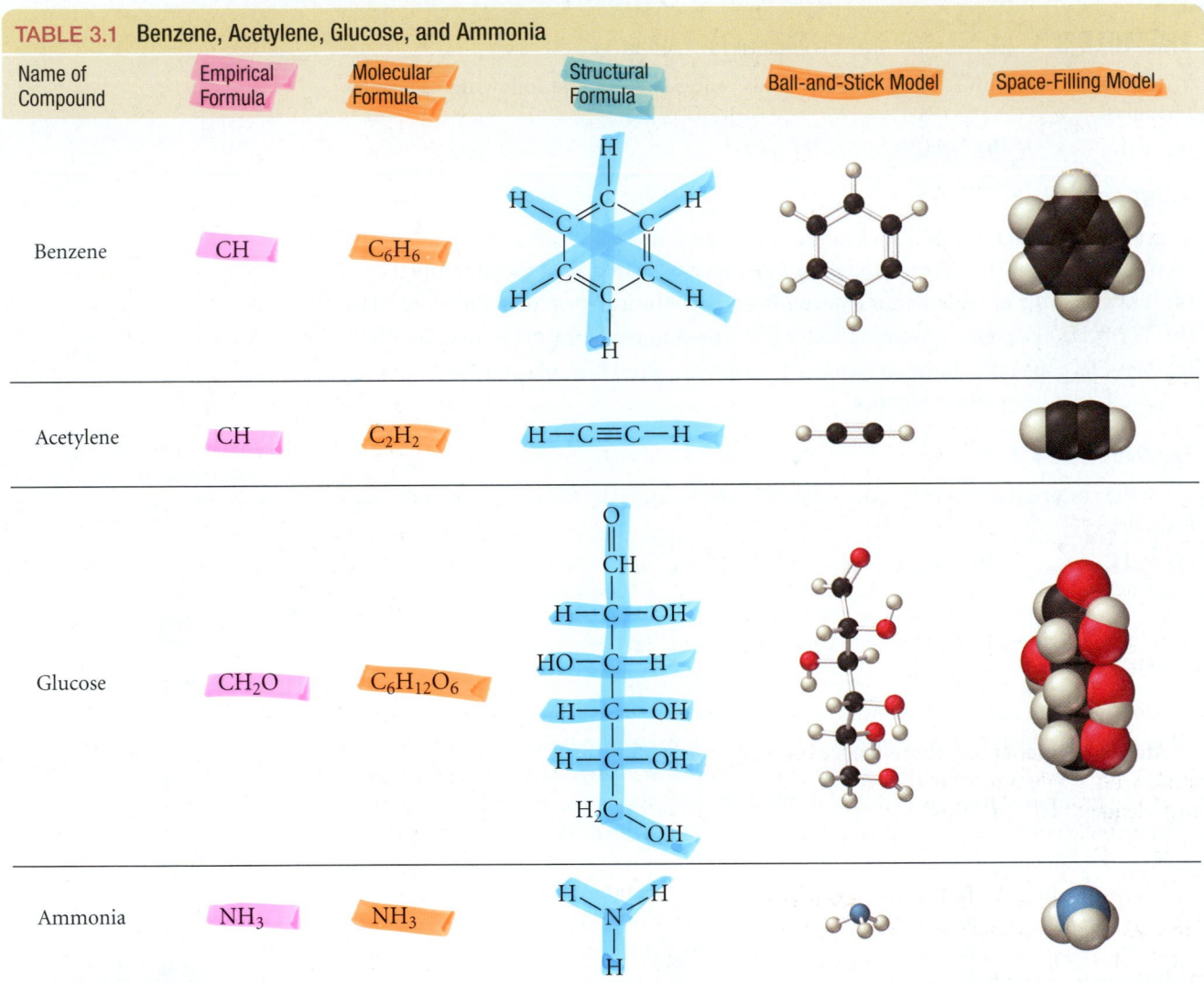

Name of Compound	Empirical Formula	Molecular Formula	Structural Formula	Ball-and-Stick Model	Space-Filling Model
Benzene	CH	C_6H_6			
Acetylene	CH	C_2H_2	$H{-}C{\equiv}C{-}H$		
Glucose	CH_2O	$C_6H_{12}O_6$			
Ammonia	NH_3	NH_3			

◆ Conceptual Connection 3.1 Representing Molecules

Based on what you learned in Chapter 2 about atoms, what part of the atom do you think the spheres in the above molecular models represent? If you were to superimpose a nucleus on one of these spheres, how big would you draw it?

Answer: The spheres represent the electron cloud of the atom. It would be nearly impossible to draw a nucleus to scale on any of the space-filling molecular models—on this scale, the nucleus would be too small to see.

3.4 An Atomic-Level View of Elements and Compounds

In Chapter 1, we learned that pure substances could be divided into elements and compounds. We can further subdivide elements and compounds according to the basic units that compose them, as shown in Figure 3.3 ▼. Elements may be either atomic or molecular. Compounds may be either molecular or ionic.

Classification of Elements and Compounds

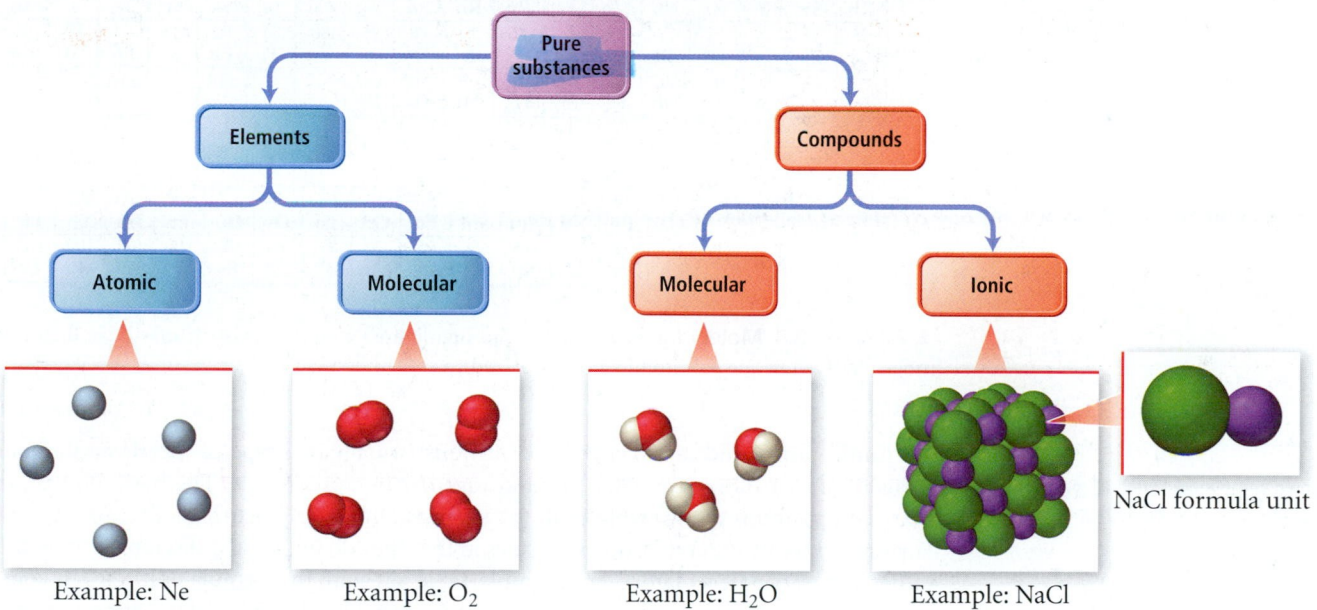

Example: Ne   Example: O_2 Example: H_2O Example: NaCl

▲ **FIGURE 3.3** A Molecular View of Elements and Compounds

Atomic elements are those that exist in nature with single atoms as their basic units. Most elements fall into this category. For example, helium is composed of helium atoms, aluminum is composed of aluminum atoms, and iron is composed of iron atoms. **Molecular elements** do not normally exist in nature with single atoms as their basic units. Instead, these elements exist as molecules, two or more atoms of the element bonded together. Most molecular elements exist as *diatomic* molecules. For example, hydrogen is composed of H_2 molecules, nitrogen is composed of N_2 molecules, and chlorine is composed of Cl_2 molecules. A few molecular elements exist as *polyatomic molecules.* Phosphorus, for example, exists as P_4 and sulfur exists as S_8. The elements that exist primarily as diatomic or polyatomic molecules are shown in Figure 3.4 on page 90.

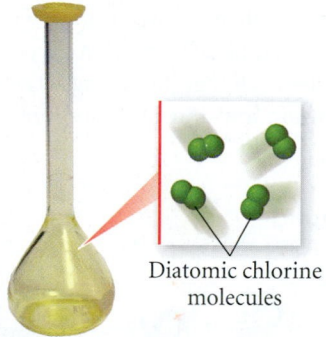

Diatomic chlorine molecules

▲ The basic units that compose chlorine gas are diatomic chlorine molecules.

Molecular compounds are usually composed of two or more covalently bonded nonmetals. The basic units of molecular compounds are molecules composed of the constituent atoms. For example, water is composed of H_2O molecules, dry ice is composed of CO_2 molecules, and propane (often used as a fuel for grills) is composed of C_3H_8 molecules as shown in Figure 3.5(a) on page 90.

Molecular Elements

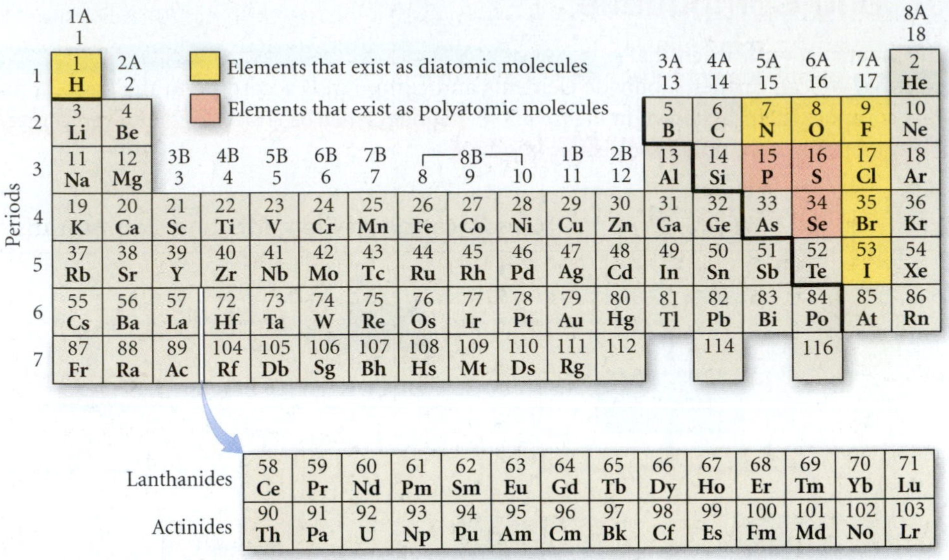

	1A 1													3A 13	4A 14	5A 15	6A 16	7A 17	8A 18

Periods

1	1 H	2A 2																	2 He
2	3 Li	4 Be												5 B	6 C	7 N	8 O	9 F	10 Ne
3	11 Na	12 Mg	3B 3	4B 4	5B 5	6B 6	7B 7	8	—8B— 9	10	1B 11	2B 12		13 Al	14 Si	15 P	16 S	17 Cl	18 Ar
4	19 K	20 Ca	21 Sc	22 Ti	23 V	24 Cr	25 Mn	26 Fe	27 Co	28 Ni	29 Cu	30 Zn		31 Ga	32 Ge	33 As	34 Se	35 Br	36 Kr
5	37 Rb	38 Sr	39 Y	40 Zr	41 Nb	42 Mo	43 Tc	44 Ru	45 Rh	46 Pd	47 Ag	48 Cd		49 In	50 Sn	51 Sb	52 Te	53 I	54 Xe
6	55 Cs	56 Ba	57 La	72 Hf	73 Ta	74 W	75 Re	76 Os	77 Ir	78 Pt	79 Au	80 Hg		81 Tl	82 Pb	83 Bi	84 Po	85 At	86 Rn
7	87 Fr	88 Ra	89 Ac	104 Rf	105 Db	106 Sg	107 Bh	108 Hs	109 Mt	110 Ds	111 Rg	112		114		116			

Elements that exist as diatomic molecules
Elements that exist as polyatomic molecules

Lanthanides	58 Ce	59 Pr	60 Nd	61 Pm	62 Sm	63 Eu	64 Gd	65 Tb	66 Dy	67 Ho	68 Er	69 Tm	70 Yb	71 Lu
Actinides	90 Th	91 Pa	92 U	93 Np	94 Pu	95 Am	96 Cm	97 Bk	98 Cf	99 Es	100 Fm	101 Md	102 No	103 Lr

▲ **FIGURE 3.4 Molecular Elements** The highlighted elements exist primarily as diatomic molecules (yellow) or polyatomic molecules (red).

Some ionic compounds, such as K_2NaPO_4, for example, contain more than one type of metal ion.

People occasionally refer to formula units as molecules, but this is *not* correct since ionic compounds do not contain distinct molecules.

Ionic compounds are composed of cations (usually one type of metal) and anions (usually one or more nonmetals) bound together by ionic bonds. The basic unit of an ionic compound is the **formula unit**, the smallest, electrically neutral collection of ions. Formula units are different from molecules in that they do not exist as discrete entities, but rather as part of a larger lattice. For example, table salt, with the formula unit NaCl, is composed of Na^+ and Cl^- ions in a one-to-one ratio. In table salt, Na^+ and Cl^- ions exist in a three-dimensional alternating array. However, because ionic bonds are not directional, no one Na^+ ion pairs with a specific Cl^- ion. Rather, as you can see from Figure 3.5(b) ▼, any one Na^+ cation is surrounded by Cl^- anions and vice versa.

Many common ionic compounds contain ions that are themselves composed of a group of covalently bonded atoms with an overall charge. For example, the active ingredient in household bleach is sodium hypochlorite, which acts to chemically alter color-causing molecules in clothes (bleaching action) and to kill bacteria (disinfection). Hypochlorite is a **polyatomic ion**—an ion composed of two or more atoms—with the formula ClO^-.

A Molecular Compound
An Ionic Compound

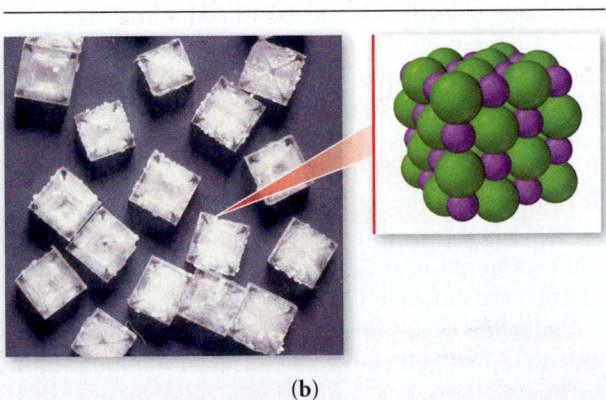

(a) (b)

▲ **FIGURE 3.5 Molecular and Ionic Compounds** (a) Propane is an example of a molecular compound. The basic units that compose propane gas are propane (C_3H_8) molecules. (b) Table salt (NaCl) is an ionic compound. Its formula unit is the simplest charge-neutral collection of ions: one Na^+ ion and one Cl^- ion.

(Note that the charge on the hypochlorite ion is a property of the whole ion, not just the oxygen atom. This is true for all polyatomic ions.) The hypochlorite ion is often found as a unit in other compounds as well [such as KClO and $Mg(ClO)_2$]. Other polyatomic ion–containing compounds found in everyday products include sodium bicarbonate ($NaHCO_3$), also known as baking soda, sodium nitrite ($NaNO_2$), an inhibitor of bacterial growth in packaged meats, and calcium carbonate ($CaCO_3$), the active ingredient in antacids such as Tums and Alka-Mints.

DANGER: CORROSIVE. HARMFUL IF SWALLOWED.
Ingredients: Sodium Hypochlorite, Sodium Hydroxide
May cause severe irritation or damage to eyes, skin, and mucous membranes. Avoid contact with eyes, skin and clothing. Do not ingest. For prolonged use, wear gloves.
FIRST AID: EYES–Rinse with plenty of water for 15 minutes. IF SWALLOWED–Do not induce vomiting. Drink a glassful of water. In either case, call a physician or poison control center immediately. SKIN–Remove contaminated clothing and wash skin thoroughly with water.
PHYSICAL AND CHEMICAL HAZARDS: Ultra Clorox® Fresh Wildflowers™ bleach contains a strong oxidizer. Always flush drains before and after use. **Do not use or mix with other household chemicals,** such as toilet bowl cleaners, rust removers, acids, or products containing ammonia. To do so will release hazardous gases. Prolonged contact with metal may cause pitting or discoloration. Not harmful to septic systems.
STORAGE: Store ultra Clorox® Fresh Wildflowers™ bleach upright in a cool, dry place. **Store away from children. Reclose cap tightly after each use.**
DISPOSAL: Offer empty container for recycling. If recycling is not available, discard in trash.
Clorox is a reg. trademark of The Clorox Co. Mfd. for & © 1999 The Clorox Company. 1221 Broadway Oakland, CA 94612. Made in U.S.A.

0 44600 02470 7

▲ Polyatomic ions are common in household products such as bleach, which contains sodium hypochlorite.

EXAMPLE 3.2 Classifying Substances as Atomic Elements, Molecular Elements, Molecular Compounds, or Ionic Compounds

Classify each of the following substances as an atomic element, molecular element, molecular compound, or ionic compound.

(a) xenon   (b) $NiCl_2$ (c) bromine   (d) NO_2 (e) $NaNO_3$

Solution

(a) Xenon is an element and it is not one of the elements that exist as diatomic molecules (Figure 3.4); therefore, it is an atomic element.

(b) $NiCl_2$ is a compound composed of a metal (left side of the periodic table) and nonmetal (right side of the periodic table); therefore, it is an ionic compound.

(c) Bromine is one of the elements that exist as diatomic molecules; therefore, it is a molecular element.

(d) NO_2 is a compound composed of a nonmetal and a nonmetal; therefore, it is a molecular compound.

(e) $NaNO_3$ is a compound composed of a metal and a polyatomic ion; therefore, it is an ionic compound.

For Practice 3.2

Classify each of the following substances as an atomic element, molecular element, molecular compound, or ionic compound.

(a) fluorine   (b) N_2O (c) silver   (d) K_2O (e) Fe_2O_3

✦ Conceptual Connection 3.2 Ionic and Molecular Compounds

Which of the following statements best captures the difference between ionic and molecular compounds?

(a) Molecular compounds contain highly directional covalent bonds, which results in the formation of molecules—discrete particles that do not covalently bond to each other. Ionic compounds contain nondirectional ionic bonds, which results (in the solid phase) in the formation of ionic lattices—extended networks of alternating cations and anions.

(b) Molecular compounds contain covalent bonds in which one of the atoms shares an electron with the other one, resulting in a new force that holds the atoms together in a covalent molecule. Ionic compounds contain ionic bonds in which one atom donates an electron to the other, resulting in a new force that holds the ions together in pairs (in the solid phase).

(c) The main difference between ionic and covalent compounds is the types of elements that compose them, not the way that the atoms bond together.

(d) A molecular compound is composed of covalently bonded molecules. An ionic compound is composed of ionically bonded molecules (in the solid phase).

Answer: Choice **(a)** best describes the difference between ionic and molecular compounds. The **(b)** answer is incorrect because there are no "new" forces in bonding (just rearrangements that result in lower potential energy), and because ions do not group together in pairs in the solid phase. The **(c)** answer is incorrect because the main difference between ionic and molecular compounds is the way that the atoms bond. The **(d)** answer is incorrect because ionic compounds do not contain molecules.

3.5 Ionic Compounds: Formulas and Names

Ionic compounds occur throughout Earth's crust as minerals. Examples include limestone ($CaCO_3$), a type of sedimentary rock, gibbsite [$Al(OH)_3$], an aluminum-containing mineral, and soda ash (Na_2CO_3), a natural deposit.

▲ Calcite (left) is the main component of limestone, marble, and other forms of calcium carbonate ($CaCO_3$) commonly found in Earth's crust. Trona (right) is a crystalline form of hydrated sodium carbonate ($Na_2CO_3 \cdot 2H_2O$).

Ionic compounds are also found in the foods that we eat. Examples include table salt ($NaCl$), the most common flavor enhancer, calcium carbonate ($CaCO_3$), a source of calcium necessary for bone health, and potassium chloride (KCl), a source of potassium necessary for fluid balance and muscle function. Ionic compounds are generally very stable because the attractions between cations and anions within ionic compounds are strong, and because each ion interacts with several oppositely charged ions in the crystalline lattice.

▲ Ionic compounds are common in food and consumer products such as light salt (a mixutre of NaCl and KCl) and Tums™ (CaCO₃).

Writing Formulas for Ionic Compounds

See Figure 2.14 to review the elements that form ions with a predictable charge.

Since ionic compounds are charge-neutral, and since many elements form only one type of ion with a predictable charge, the formulas for many ionic compounds can be deduced from their constituent elements. For example, the formula for the ionic compound composed of sodium and chlorine must be NaCl and not anything else because, in compounds, Na always forms 1+ cations and Cl always forms 1− anions. In order for the compound to be charge-neutral, it must contain one Na^+ cation to every one Cl^- anion. The formula for the ionic compound composed of calcium and chlorine must be CaCl₂ because Ca always forms 2+ cations and Cl always forms 1− anions. In order for this compound to be charge-neutral, it must contain one Ca^{2+} cation to every two Cl^- anions.

Summarizing:

➤ Ionic compounds always contain positive and negative ions.

➤ In a chemical formula, the sum of the charges of the positive ions (cations) must always equal the sum of the charges of the negative ions (anions).

➤ The formula reflects the smallest whole-number ratio of ions.

To write the formula for an ionic compound, follow the procedure in the left column below. Two examples of how to apply the procedure are provided in the center and right columns.

Procedure for Writing Formulas for Ionic Compounds	**EXAMPLE 3.3** Writing Formulas for Ionic Compounds Write a formula for the ionic compound that forms between aluminum and oxygen.	**EXAMPLE 3.4** Writing Formulas for Ionic Compounds Write a formula for the ionic compound that forms between calcium and oxygen.
1. Write the symbol for the metal cation and its charge followed by the symbol for the nonmetal anion and its charge. Obtain charges from the element's group number in the periodic table (refer to Figure 2.14).	Al^{3+} O^{2-}	Ca^{2+} O^{2-}
2. Adjust the subscript on each cation and anion to balance the overall charge.	Al^{3+} O^{2-} ↓ Al_2O_3	Ca^{2+} O^{2-} ↓ CaO
3. Check that the sum of the charges of the cations equals the sum of the charges of the anions.	cations: $2(3+) = 6+$ anions: $3(2-) = 6-$ The charges cancel.	cations: $2+$ anions: $2-$ The charges cancel.
	For Practice 3.3 Write a formula for the compound formed between potassium and sulfur.	**For Practice 3.4** Write a formula for the compound formed between aluminum and nitrogen.

Naming Ionic Compounds

Some ionic compounds—such as NaCl (table salt) and $NaHCO_3$ (baking soda)—have **common names**, which are a sort of nickname that can be learned only through familiarity. However, chemists have developed **systematic names** for different types of compounds including ionic ones. Systematic names can be determined simply by looking at the chemical formula of a compound. Conversely, the formula of a compound can be deduced from its systematic name.

The first step in naming an ionic compound is identifying it as one. Remember, *ionic compounds are usually formed between metals and nonmetals*; any time you see a metal and one or more nonmetals together in a chemical formula, you can assume you have an ionic compound. Ionic compounds can be divided into two types, depending on the metal in the compound. The first type contains a metal whose charge is invariant from one compound to another—the metal only forms one ion.

Since the charge of the metal is always the same, it need not be specified in the name of the compound. Sodium, for instance, has a 1+ charge in all of its compounds. Some examples of these types of metals are listed in Table 3.2 (page 94), and their charges can be inferred from their group number in the periodic table.

The second type of ionic compound contains a metal with a charge that can be different in different compounds—the metal can form more than one kind of cation and the charge must therefore be specified for a given compound. Iron, for instance, has a 2+ charge in some of its compounds and a 3+ charge in others. Metals of this type are often found in the section of the periodic table known as the *transition metals* (Figure 3.6 ▶).

However, some transition metals, such as Zn and Ag, have the same charge in all of their compounds (as shown in Table 3.2), and some main group metals, such as lead and tin, have charges that can vary from one compound to another.

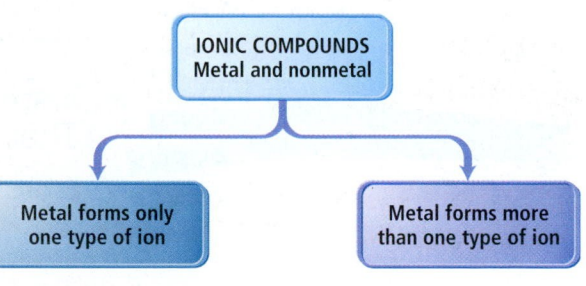

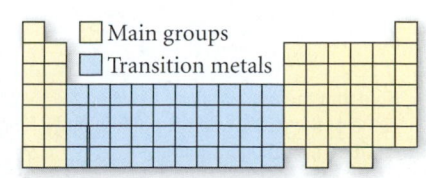

▲ **FIGURE 3.6 Metals That Form More than One Cation** Metals that can have different charges in different compounds are usually transition metals.

TABLE 3.2 Metals Whose Charge Is Invariant from One Compound to Another

Metal	Ion	Name	Group Number
Li	Li^+	Lithium	1A
Na	Na^+	Sodium	1A
K	K^+	Potassium	1A
Rb	Rb^+	Rubidium	1A
Cs	Cs^+	Cesium	1A
Be	Be^{2+}	Beryllium	2A
Mg	Mg^{2+}	Magnesium	2A
Ca	Ca^{2+}	Calcium	2A
Sr	Sr^{2+}	Strontium	2A
Ba	Ba^{2+}	Barium	2A
Al	Al^{3+}	Aluminum	3A
Zn	Zn^{2+}	Zinc	*
Sc	Sc^{3+}	Scandium	*
Ag**	Ag^+	Silver	*

* The charge of these metals cannot be inferred from their group number.

** Silver does sometimes form compounds with other charges, but these are rare.

TABLE 3.3 Some Common Anions

Nonmetal	Symbol for Ion	Base Name	Anion Name
Fluorine	F^-	fluor	Fluoride
Chlorine	Cl^-	chlor	Chloride
Bromine	Br^-	brom	Bromide
Iodine	I^-	iod	Iodide
Oxygen	O^{2-}	ox	Oxide
Sulfur	S^{2-}	sulf	Sulfide
Nitrogen	N^{3-}	nitr	Nitride
Phosphorus	P^{3-}	phosph	Phosphide

Note that there is no space between the name of the cation and the parenthetical number indicating its charge.

Naming Binary Ionic Compounds Containing a Metal That Forms Only One Type of Cation

Binary compounds are those containing only two different elements. The names for binary ionic compounds have the following form:

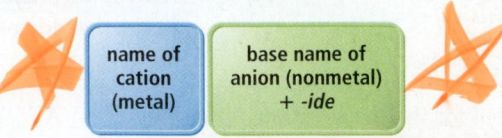

For example, the name for KCl consists of the name of the cation, *potassium*, followed by the base name of the anion, *chlor*, with the ending *-ide*. The full name is *potassium chloride.*

$$KCl \qquad potassium\ chloride$$

The name for CaO consists of the name of the cation, *calcium*, followed by the base name of the anion, *ox*, with the ending *-ide*. The full name is *calcium oxide.*

$$CaO \qquad calcium\ oxide$$

The base names for various nonmetals, and their most common charges in ionic compounds, are shown in Table 3.3.

EXAMPLE 3.5 Naming Ionic Compounds Containing a Metal That Forms Only One Type of Cation

Give the name for the compound $CaBr_2$.

Solution

The cation is *calcium*. The anion is from bromine, which becomes *bromide*. The correct name is *calcium bromide.*

For Practice 3.5

Give the name for the compound Ag_3N.

For More Practice 3.5

Write the formula for rubidium sulfide.

Naming Binary Ionic Compounds Containing a Metal That Forms More than One Kind of Cation

For these types of metals, the name of the cation is followed by a roman numeral (in parentheses) indicating its charge in that particular compound. For example, we distinguish between Fe^{2+} and Fe^{3+} as follows:

$$Fe^{2+} \qquad iron(II)$$
$$Fe^{3+} \qquad iron(III)$$

The full names therefore have the following form:

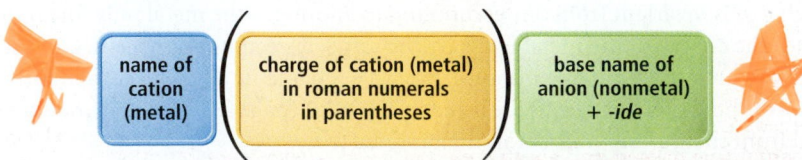

The charge of the metal cation is obtained by inference from the sum of the charges of the nonmetal anions—remember that the sum of all the charges must be zero. Table 3.4 shows some of the metals that form more than one cation and the values of their most common charges. For example, in $CrBr_3$, the charge of chromium must be 3+ in order for the compound to be charge-neutral with three Br^- anions. The cation is therefore named as follows:

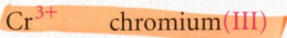

$$Cr^{3+} \qquad chromium(III)$$

The full name of the compound is

$CrBr_3$ chromium(III) bromide

Similarly, in CuO, the charge of copper must be 2+ in order for the compound to be charge-neutral with one O^{2-} anion. The cation is therefore named as follows:

Cu^{2+} copper(II)

The full name of the compound is

CuO   copper(II) oxide

EXAMPLE 3.6 Naming Ionic Compounds Containing a Metal That Forms More than One Kind of Cation

Give the name for the compound $PbCl_4$.

Solution

The charge on Pb must be 4+ for the compound to be charge-neutral with 4 Cl^- anions. The name for $PbCl_4$ consists of the name of the cation, *lead*, followed by the charge of the cation in parentheses *(IV)*, followed by the base name of the anion, *chlor*, with the ending *-ide*. The full name is *lead(IV) chloride*.

$PbCl_4$ lead(IV) chloride

For Practice 3.6

Give the name for the compound FeS.

For More Practice 3.6

Write the formula for ruthenium(IV) oxide.

TABLE 3.4 Some Metals That Form Cations with Different Charges

Metal	Ion	Name	Older Name[*]
Chromium	Cr^{2+}	Chromium(II)	Chromous
	Cr^{3+}	Chromium(III)	Chromic
Iron	Fe^{2+}	Iron(II)	Ferrous
	Fe^{3+}	Iron(III)	Ferric
Cobalt	Co^{2+}	Cobalt(II)	Cobaltous
	Co^{3+}	Cobalt(III)	Cobaltic
Copper	Cu^{+}	Copper(I)	Cuprous
	Cu^{2+}	Copper(II)	Cupric
Tin	Sn^{2+}	Tin(II)	Stannous
	Sn^{4+}	Tin(IV)	Stannic
Mercury	Hg_2^{2+}	Mercury(I)	Mercurous
	Hg^{2+}	Mercury(II)	Mercuric
Lead	Pb^{2+}	Lead(II)	Plumbous
	Pb^{4+}	Lead(IV)	Plumbic

[*]An older naming system substitutes the names found in this column for the name of the metal and its charge. Under this system, chromium(II) oxide is named chromous oxide. In this system, the suffix *-ous* indicates the ion with the lesser charge and *-ic* indicates the ion with the greater charge. We will *not* use the older system in this text.

Naming Ionic Compounds Containing Polyatomic Ions

Ionic compounds containing polyatomic ions are named in the same way as other ionic compounds, except that the name of the polyatomic ion is used whenever it occurs. Table 3.5 lists common polyatomic ions and their formulas. For example, $NaNO_2$ is named according to its cation, Na^+, *sodium*, and its polyatomic anion, NO_2^-, *nitrite*. The full name is *sodium nitrite*.

$NaNO_2$ sodium nitrite

$FeSO_4$ is named according to its cation, *iron*, its charge *(II)*, and its polyatomic ion *sulfate*. The full name is *iron(II) sulfate*.

$FeSO_4$ iron(II) sulfate

If the compound contains both a polyatomic cation and a polyatomic anion, simply use the names of both polyatomic ions. For example, NH_4NO_3 is named *ammonium nitrate*.

NH_4NO_3 ammonium nitrate

You must be able to recognize polyatomic ions in a chemical formula, so become familiar with Table 3.5. Most polyatomic ions are **oxyanions**, anions containing oxygen and another element. Notice that when a series of oxyanions contains different numbers of oxygen atoms, they are named systematically according to the number of oxygen atoms in the ion. If there are only two ions in the series, the one with more oxygen atoms is given the ending *-ate* and the one with fewer is given the ending *-ite*. For example, NO_3^- is called *nitrate* and NO_2^- is called *nitrite*.

NO_3^- nitrate
NO_2^- nitrite

TABLE 3.5 Some Common Polyatomic Ions

Name	Formula	Name	Formula
Acetate	$C_2H_3O_2^-$	Hypochlorite	ClO^-
Carbonate	CO_3^{2-}	Chlorite	ClO_2^-
Hydrogen carbonate (or bicarbonate)	HCO_3^-	Chlorate	ClO_3^-
Hydroxide	OH^-	Perchlorate	ClO_4^-
Nitrite	NO_2^-	Permanganate	MnO_4^-
Nitrate	NO_3^-	Sulfite	SO_3^{2-}
Chromate	CrO_4^{2-}	Hydrogen sulfite (or bisulfite)	HSO_3^-
Dichromate	$Cr_2O_7^{2-}$	Sulfate	SO_4^{2-}
Phosphate	PO_4^{3-}	Hydrogen sulfate (or bisulfate)	HSO_4^-
Hydrogen phosphate	HPO_4^{2-}	Cyanide	CN^-
Dihydrogen phosphate	$H_2PO_4^-$	Peroxide	O_2^{2-}
Ammonium	NH_4^+		

If there are more than two ions in the series then the prefixes *hypo-*, meaning *less than*, and *per-*, meaning *more than*, are used. So ClO^- is called hypochlorite meaning less oxygen than chlorite and ClO_4^- is called perchlorate meaning more oxygen than chlorate.

The other halides (halogen ions) form similar series with similar names. Thus, IO_3^- is called iodate and BrO_3^- is called bromate.

ClO^-	*hypo*chlor*ite*
ClO_2^-	chlor*ite*
ClO_3^-	chlor*ate*
ClO_4^-	*per*chlor*ate*

EXAMPLE 3.7   Naming Ionic Compounds That Contain a Polyatomic Ion

Give the name for the compound $Li_2Cr_2O_7$.

Solution

The name for $Li_2Cr_2O_7$ consists of the name of the cation, *lithium*, followed by the name of the polyatomic ion, *dichromate*. The full name is *lithium dichromate*.

$$Li_2Cr_2O_7 \qquad \text{lithium dichromate}$$

For Practice 3.7

Give the name for the compound $Sn(ClO_3)_2$.

For More Practice 3.7

Write a formula for cobalt(II) phosphate.

Hydrated Ionic Compounds

Some ionic compounds—called **hydrates**—contain a specific number of water molecules associated with each formula unit. For example, Epsom salts has the formula $MgSO_4 \cdot 7H_2O$ and the systematic name magnesium sulfate heptahydrate. The seven H_2O molecules associated with the formula unit are called *waters of hydration*. Waters of hydration can usually be removed by heating the compound. Figure 3.7 ◀, for example, shows a sample of cobalt(II) chloride hexahydrate ($CoCl_2 \cdot 6H_2O$) before and after heating. The hydrate is pink and the anhydrous salt (the salt without any associated water molecules) is blue. Hydrates are named just as other ionic compounds, but they are given the additional name "*prefix*hydrate," where the prefix indicates the number of water molecules associated with each formula unit.

Some other common examples of hydrated ionic compounds and their names are as follows:

$CaSO_4 \cdot \frac{1}{2}H_2O$	calcium sulfate hemihydrate
$BaCl_2 \cdot 6H_2O$	barium chloride hexahydrate
$CuSO_4 \cdot 5H_2O$	copper(II) sulfate pentahydrate

Hydrate	Anhydrous
$CoCl_2 \cdot 6H_2O$	$CoCl_2$

▲ **FIGURE   3.7 Hydrates**   Cobalt(II) chloride hexahydrate is pink, but heating the compound removes the waters of hydration, leaving the blue anhydrous cobalt(II) chloride.

Common hydrate prefixes
hemi = 1/2
mono = 1
di = 2
tri = 3
tetra = 4
penta = 5
hexa = 6
hepta = 7
octa = 8

3.6 Molecular Compounds: Formulas and Names

In contrast to ionic compounds, the formula for a molecular compound *cannot* easily be determined based on its constituent elements because the same elements may form many different molecular compounds, each with a different formula. For example, we learned in Chapter 1 that carbon and oxygen form both CO and CO_2, and that hydrogen and oxygen form both H_2O and H_2O_2. Nitrogen and oxygen form all of the following unique molecular compounds: NO, NO_2, N_2O, N_2O_3, N_2O_4, and N_2O_5. In Chapter 9, we will learn how to understand the stability of these various combinations of the same elements. For now, we focus on naming a molecular compound based on its formula or writing its formula based on its name.

Naming Molecular Compounds

Like ionic compounds, many molecular compounds have common names. For example, H_2O and NH_3 have the common names *water* and *ammonia*, which are routinely used. However, the sheer number of existing molecular compounds—numbering in the millions—requires a systematic approach to naming them.

The first step in naming a molecular compound is identifying it as one. Remember, *molecular compounds form between two or more nonmetals.* In this section, we learn how to name binary (two-element) molecular compounds. Their names have the following form:

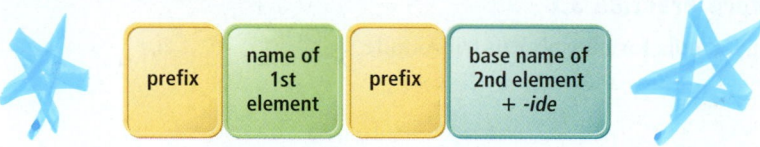

| prefix | name of 1st element | prefix | base name of 2nd element + *-ide* |

When writing the name of a molecular compound, as when writing the formula, the first element is the more metal-like one (toward the left and bottom of the periodic table). Always write the name of the element with the smallest group number first. If the two elements lie in the same group, then write the element with the greatest row number first. The prefixes given to each element indicate the number of atoms present:

mono = 1   hexa = 6

di = 2   hepta = 7

tri = 3   octa = 8

tetra = 4   nona = 9

penta = 5   deca = 10

These prefixes are the same as those used in naming hydrates.

If there is only one atom of the *first element* in the formula, the prefix *mono-* is normally omitted. For example, NO_2 is named according to the first element, *nitrogen*, with no prefix because *mono-* is omitted for the first element, followed by the prefix *di*, to indicate two oxygen atoms, followed by the base name of the second element, *ox*, with the ending *-ide*. The full name is *nitrogen dioxide.*

$$NO_2 \qquad \text{nitrogen dioxide}$$

The compound N_2O, sometimes called laughing gas, is named similarly except that we use the prefix *di-* before nitrogen to indicate two nitrogen atoms and the prefix *mono-* before oxide to indicate one oxygen atom. The entire name is *dinitrogen monoxide.*

When a prefix ends with "o" and the base name begins "o," the first "o" is often dropped. So mono-oxide becomes *monoxide.*

$$N_2O \qquad \text{dinitrogen monoxide}$$

EXAMPLE 3.8   Naming Molecular Compounds

Name each of the following.

(a) NI_3 **(b)** PCl_5 **(c)** P_4S_{10}

Solution

(a) The name of the compound is the name of the first element, *nitrogen*, followed by the base name of the second element, *iod,* prefixed by *tri-* to indicate three and given the suffix *-ide.*

$$NI_3 \qquad \text{nitrogen triiodide}$$

(b) The name of the compound is the name of the first element, *phosphorus*, followed by the base name of the second element, *chlor,* prefixed by *penta-* to indicate five and given the suffix *-ide.*

$$PCl_5 \qquad \text{phosphorus pentachloride}$$

(c) The name of the compound is the name of the first element, *phosphorus*, prefixed by *tetra-* to indicate four, followed by the base name of the second element, *sulf,* prefixed by *deca* to indicate ten and given the suffix *-ide*.

$$P_4S_{10} \qquad \text{tetra}\text{phosphorus }\text{deca}\text{sulfide}$$

For Practice 3.8

Name the compound N_2O_5.

For More Practice 3.8

Write a formula for phosphorus tribromide.

Naming Acids

Acids are molecular compounds that release hydrogen ions (H^+) when dissolved in water. They are composed of hydrogen, usually written first in their formula, and one or more nonmetals, written second. For example, HCl is a molecular compound that, when dissolved in water, forms $H^+(aq)$ and $Cl^-(aq)$ ions, where *aqueous* (*aq*) simply means *dissolved in water*. Therefore, HCl is an acid when dissolved in water. To distinguish between gaseous HCl (which is named hydrogen chloride because it is a molecular compound) and HCl in solution (which is named as an acid), we write the former as HCl(g) and the latter as HCl(aq).

Acids are characterized by their sour taste and their ability to dissolve many metals. Like all acids, HCl(aq) has a characteristically sour taste. Since HCl(aq) is present in stomach fluids, its sour taste becomes painfully obvious during vomiting. Hydrochloric acid also dissolves some metals. For example, if you put a strip of zinc into a test tube of HCl(aq), it slowly dissolves as the $H^+(aq)$ ions convert the zinc metal into $Zn^{2+}(aq)$ cations (Figure 3.8 ▼).

Acids Dissolve Many Metals

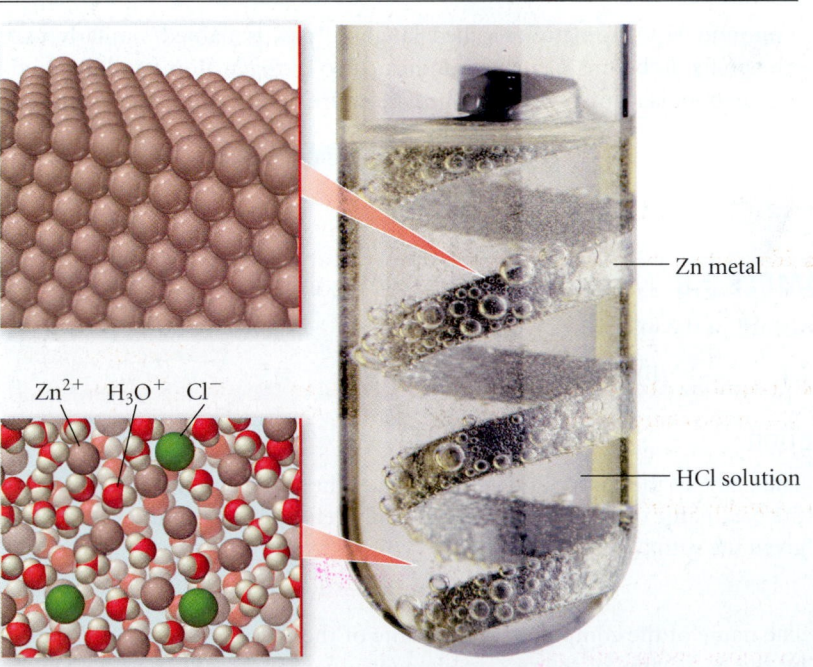

Zn metal

HCl solution

Zn^{2+} H_3O^+ Cl^-

▲ **FIGURE 3.8 Hydrochloric Acid Dissolving Zinc Metal**   The zinc atoms are ionized to zinc ions, which dissolve in the water. The HCl forms H_2 gas, as you can see from the bubbles.

Acids are present in many foods such as lemons and limes and are used in household products such as toilet bowl cleaner and Lime-Away. In this section, we learn how to name them; in Chapter 15 we will learn more about their properties. Acids can be divided into two categories, binary acids and oxyacids.

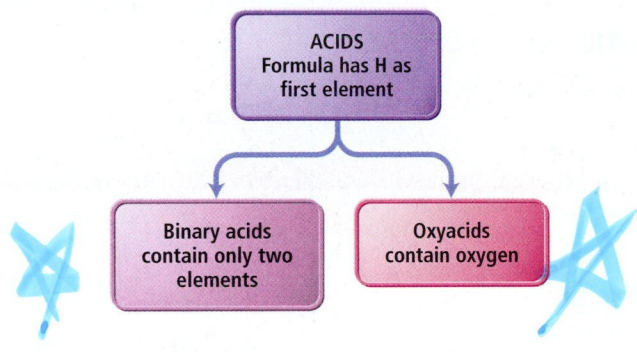

▲ Many fruits are acidic and have the characteristically sour taste of acids.

Naming Binary Acids

Binary acids are composed of hydrogen and a nonmetal. The names for binary acids have the following form:

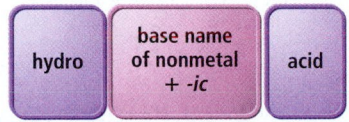

For example, HCl(*aq*) is named hydro*chlor*ic acid and HBr(*aq*) is named hydro*brom*ic acid.

| HCl(*aq*) | hydrochloric acid | HBr(*aq*) | hydrobromic acid |

EXAMPLE 3.9 Naming Binary Acids

Give the name of HI(*aq*).

Solution

The base name of I is *iod* so the name is hydroiodic acid.

HI(*aq*)   hydroiodic acid

For Practice 3.9

Give the name of HF(*aq*).

Naming Oxyacids

Oxyacids contain hydrogen and an oxyanion (an anion containing a nonmetal and oxygen). The common oxyanions are listed in the table of polyatomic ions (Table 3.5). For example, $HNO_3(aq)$ contains the nitrate (NO_3^-) ion, $H_2SO_3(aq)$ contains the sulfite (SO_3^{2-}) ion, and $H_2SO_4(aq)$ contains the sulfate (SO_4^{2-}) ion. Notice that these acids are simply a combination of one or more H^+ ions with an oxyanion. The number of H^+ ions depends on the charge of the oxyanion so that the formula is always charge-neutral. The names of oxyacids depend on the ending of the oxyanion and have the following forms:

oxyanions ending with *-ate*

+ic

oxyanions ending with *-ite*

+ ous

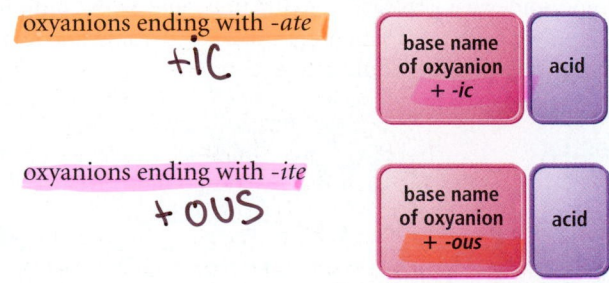

So $HNO_3(aq)$ is named nitric acid (oxyanion is nitrate), and $H_2SO_3(aq)$ is named sulfurous acid (oxyanion is sulfite).

$HNO_3(aq)$ nitric acid $H_2SO_3(aq)$ sulfurous acid

EXAMPLE 3.10 Naming Oxyacids

Give the name of $HC_2H_3O_2(aq)$.

Solution

The oxyanion is acetate, which ends in -*ate*; therefore, the name of the acid is *acetic acid*.

$HC_2H_3O_2(aq)$ acetic acid

For Practice 3.10

Give the name of $HNO_2(aq)$.

For More Practice 3.10

Write the formula for perchloric acid.

Chemistry in the Environment
Acid Rain

Certain pollutants—such as NO, NO_2, and SO_2—form acids when mixed with water. NO and NO_2, primarily from vehicular emissions, combine with atmospheric oxygen and water to form nitric acid, $HNO_3(aq)$. SO_2, primarily from coal-powered electricity generation, combines with atmospheric oxygen and water to form sulfuric acid, $H_2SO_4(aq)$. Both $HNO_3(aq)$ and $H_2SO_4(aq)$ cause rainwater to become acidic. The problem is greatest in the northeastern United States where pollutants from midwestern electrical power plants combine with rainwater to produce rain with acid levels that are up to 10 times normal.

When acid rain falls or flows into lakes and streams, it makes them more acidic. Some species of aquatic animals—such as trout, bass, snails, salamanders, and clams—cannot tolerate the increased acidity and die. This then disturbs the ecosystem of the lake, resulting in imbalances that may lead to the death of other aquatic species. Acid rain also weakens trees by dissolving nutrients in the soil (and washing them away) and by damaging their leaves. Appalachian red spruce trees have been the hardest hit, with many forests showing significant acid rain damage.

Acid rain also degrades building materials. Acids dissolve $CaCO_3$ (limestone), a main component of marble and concrete, and iron, the main component of steel. Consequently, many statues, buildings, and bridges in the northeastern United States have been harmed by acid rain. For example, some historical gravestones, made of limestone, are barely legible due to acid rain damage.

Although acid rain has been a problem for many years, recent legislation has offered hope for change. In 1990, Congress passed several amendments to the Clean Air Act that included provisions requiring electrical utilities to lower SO_2 emissions. Since then, SO_2 emissions have decreased and rain in the northeastern United States has become somewhat less acidic. With time, and with continued enforcement of the acid rain program, lakes, streams, and forests damaged by acid rain should recover.

▲ Acid rain damages building materials such as the limestone that composes many statues.

▶ A forest damaged by acid rain.

Question

Provide the names for each of the compounds given here as formulas: NO, NO_2, SO_2, H_2SO_4, HNO_3, $CaCO_3$.

3.7 Formula Mass and the Mole Concept for Compounds

In Chapter 2, we defined the average mass of an atom of an element as the *atomic mass* for that element. Similarly, we now define the average mass of a molecule (or a formula unit) of a compound as the **formula mass** for that compound. Also in common use are the terms *molecular mass* or *molecular weight*, which have the same meaning as formula mass. For any compound, the formula mass is simply the sum of the atomic masses of all the atoms in its chemical formula.

$$\text{Formula mass} = \left(\begin{array}{c} \text{Number of atoms} \\ \text{of 1st element in} \\ \text{chemical formula} \end{array} \times \begin{array}{c} \text{Atomic mass} \\ \text{of} \\ \text{1st element} \end{array} \right) + \left(\begin{array}{c} \text{Number of atoms} \\ \text{of 2nd element in} \\ \text{chemical formula} \end{array} \times \begin{array}{c} \text{Atomic mass} \\ \text{of} \\ \text{2nd element} \end{array} \right) + \dots$$

For example, the formula mass of carbon dioxide, CO_2, is

$$\text{Formula mass} = 12.01 \text{ amu} + 2(16.00 \text{ amu})$$
$$= 44.01 \text{ amu}$$

and that of sodium oxide, Na_2O, is

$$\text{Formula mass} = 2(22.99 \text{ amu}) + 16.00 \text{ amu}$$
$$= 61.98 \text{ amu}$$

EXAMPLE 3.11 Calculating Formula Mass

Calculate the formula mass of glucose, $C_6H_{12}O_6$.

Solution

To find the formula mass, we sum the atomic masses of each atom in the chemical formula:

$$\text{Formula mass} = 6 \times (\text{atomic mass C}) + 12 \times (\text{atomic mass H}) + 6 \times (\text{atomic mass O})$$
$$= 6(12.01 \text{ amu}) + 12(1.008 \text{ amu}) + 6(16.00 \text{ amu})$$
$$= 180.16 \text{ amu}$$

For Practice 3.11

Calculate the formula mass of calcium nitrate.

Molar Mass of a Compound

In Chapter 2 (Section 2.9), we learned that an element's molar mass—the mass in grams of one mole of its atoms—is numerically equivalent to its atomic mass. We then used the molar mass in combination with Avogadro's number to determine the number of atoms in a given mass of the element. The same concept applies to compounds. The *molar mass of a compound*—the mass in grams of 1 mol of its molecules or formula units—is numerically equivalent to its formula mass. For example, we just calculated the formula mass of CO_2 to be 44.01 amu. The molar mass is, therefore,

$$CO_2 \text{ molar mass} = 44.01 \text{ g/mol}$$

Remember, ionic compounds do not contain individual molecules. In casual language, the smallest electrically neutral collection of ions is sometimes called a molecule but is more correctly called a formula unit.

Using Molar Mass to Count Molecules by Weighing

The molar mass of CO_2 provides us with a conversion factor between mass (in grams) and amount (in moles) of CO_2. Suppose we want to find the number of CO_2 molecules in a sample of dry ice (solid CO_2) with a mass of 10.8 g. This calculation is analogous to Example 2.8, where we found the number of atoms in a sample of copper of a given mass. We begin with the mass of 10.8 g and use the molar mass to convert to the amount in moles. Then we use Avogadro's number to convert to number of molecules. The conceptual plan is as follows:

Conceptual Plan

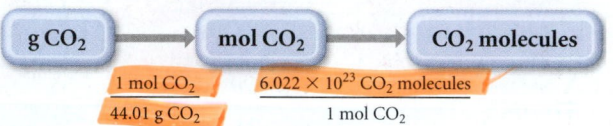

To solve the problem, we follow the conceptual plan, beginning with 10.8 g CO_2, converting to moles, and then to molecules.

Solution

$$10.8 \text{ g } CO_2 \times \frac{1 \text{ mol } CO_2}{44.01 \text{ g } CO_2} \times \frac{6.022 \times 10^{23} CO_2 \text{ molecules}}{1 \text{ mol } CO_2}$$

$$= 1.48 \times 10^{23} CO_2 \text{ molecules}$$

EXAMPLE 3.12 The Mole Concept—Converting between Mass and Number of Molecules

An aspirin tablet contains 325 mg of acetylsalicylic acid ($C_9H_8O_4$). How many acetylsalicylic acid molecules does it contain?

Sort You are given the mass of acetylsalicylic acid and asked to find the number of molecules.	**Given** 325 mg $C_9H_8O_4$ **Find** number of $C_9H_8O_4$ molecules
Strategize Convert between mass and the number of molecules of a compound by first converting to moles (using the molar mass of the compound) and then to the number of molecules (using Avogadro's number). You will need both the molar mass of acetylsalicylic acid and Avogadro's number as conversion factors. You will also need the conversion factor between g and mg.	**Conceptual Plan** mg $C_9H_8O_4$ → g $C_9H_8O_4$ $\dfrac{10^{-3} \text{ g}}{1 \text{ mg}}$ $\dfrac{1 \text{ mol } C_9H_8O_4}{180.15 \text{ g } C_9H_8O_4}$ mol $C_9H_8O_4$ → number of $C_9H_8O_4$ molecules $\dfrac{6.022 \times 10^{23} C_9H_8O_4 \text{ molecules}}{1 \text{ mol } C_9H_8O_4}$ **Relationships Used** $1 \text{ mg} = 10^{-3} \text{ g}$ $C_9H_8O_4$ molar mass $= 9(12.01) + 8(1.008) + 4(16.00)$ $= 180.15 \text{ g/mol}$ $6.022 \times 10^{23} = 1 \text{ mol}$
Solve Follow the conceptual plan to solve the problem.	**Solution** $325 \text{ mg } C_9H_8O_4 \times \dfrac{10^{-3} \text{ g}}{1 \text{ mg}} \times \dfrac{1 \text{ mol } C_9H_8O_4}{180.15 \text{ g } C_9H_8O_4} \times$ $\dfrac{6.022 \times 10^{23} C_9H_8O_4 \text{ molecules}}{1 \text{ mol } C_9H_8O_4} = 1.09 \times 10^{21} C_9H_8O_4 \text{ molecules}$

Check The units of the answer, $C_9H_8O_4$ molecules, are correct. The magnitude seems appropriate because it is smaller than Avogadro's number, as expected, since we have less than one molar mass of acetylsalicylic acid.

For Practice 3.12

Find the number of ibuprofen molecules in a tablet containing 200.0 mg of ibuprofen $(C_{13}H_{18}O_2)$.

For More Practice 3.12

What is the mass of a drop of water containing 3.55×10^{22} H_2O molecules?

 Conceptual Connection 3.3 Molecular Models and the Size of Molecules

Throughout this book, we use space-filling molecular models to represent molecules. Which of the following is a good estimate for the scaling factor used in these models? For example, by approximately what number would you have to multiply the radius of an actual oxygen atom to get the radius of the sphere used to represent the oxygen atom in the water molecule shown here?

   (a) 10   **(b)** 10^4 **(c)** 10^8 **(d)** 10^{16}

Answer: (c) Atomic radii range in the hundreds of picometers while the spheres in these models have radii of less than a centimeter. The scaling factor is therefore about 10^8.

3.8 Composition of Compounds

A chemical formula, in combination with the molar masses of its constituent elements, gives the relative amount of each element in a compound, which is extremely useful information. For example, about 25 years ago, scientists began to suspect that synthetic compounds known as chlorofluorocarbons (or CFCs) were destroying ozone (O_3) in Earth's upper atmosphere.

Upper atmospheric ozone is important because it acts as a shield to protect life on Earth from the sun's harmful ultraviolet light. CFCs are chemically inert compounds that were used primarily as refrigerants and industrial solvents. Over time, however, CFCs began to accumulate in the atmosphere. In the upper atmosphere sunlight breaks bonds within CFCs, resulting in the release of chlorine atoms. The chlorine atoms then react with ozone, converting it into O_2. So the harmful part of CFCs is the chlorine atoms that they carry. How do you determine the amount of chlorine in a given amount of a CFC?

One way to express how much of an element is in a given compound is to use the element's mass percent composition for that compound. The **mass percent composition** or simply **mass percent** of an element is that element's percentage of the compound's total mass. The mass percent of element X in a compound can be computed from the chemical formula as follows:

$$\text{mass percent of element X} = \frac{\text{mass of element X in 1 mol of compound}}{\text{mass of 1 mol of the compound}} \times 100\%$$

Suppose, for example, that we want to calculate the mass percent composition of Cl in the chlorofluorocarbon CCl_2F_2. The mass percent Cl is given by

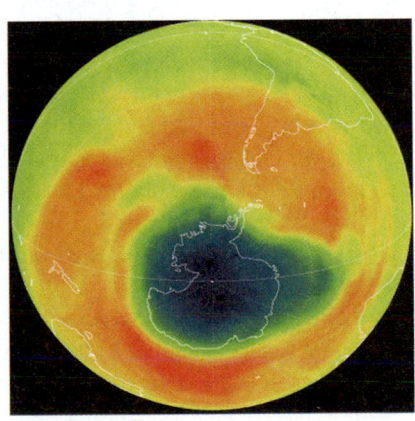

$$\text{Mass percent Cl} = \frac{2 \times \text{Molar mass Cl}}{\text{Molar mass } CCl_2F_2} \times 100\%$$

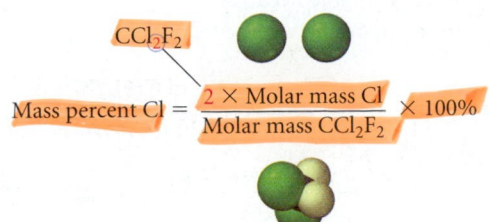

▲ The ozone hole over Antarctica is caused by the chlorine in chlorofluorocarbons. The dark blue color indicates depressed ozone levels.

The molar mass of Cl must be multiplied by two because the chemical formula has a subscript of 2 for Cl, meaning that 1 mol of CCl_2F_2 contains 2 mol of Cl atoms. The molar mass of CCl_2F_2 is computed as follows:

$$\text{Molar mass} = 12.01 \text{ g/mol} + 2(35.45 \text{ g/mol}) + 2(19.00 \text{ g/mol})$$

$$= 120.91 \text{ g/mol}$$

So the mass percent of Cl in CCl_2F_2 is

$$\text{Mass percent Cl} = \frac{2 \times \text{molar mass Cl}}{\text{molar mass } CCl_2F_2} \times 100\%$$

$$= \frac{2 \times 35.45 \text{ g/mol}}{120.91 \text{ g/mol}} \times 100\%$$

$$= 58.64\%$$

EXAMPLE 3.13 Mass Percent Composition

Calculate the mass percent of Cl in Freon-112 ($C_2Cl_4F_2$), a CFC refrigerant.

Sort You are given the molecular formula of Freon-112 and asked to find the mass percent of Cl.	**Given** $C_2Cl_4F_2$ **Find** mass percent Cl
Strategize The molecular formula tells you that there are 4 mol of Cl in each mole of Freon-112. Find the mass percent composition from the chemical formula by using the equation that defines mass percent. The conceptual plan shows how the mass of Cl in 1 mol of $C_2Cl_4F_2$ and the molar mass of $C_2Cl_4F_2$ are used to find the mass percent of Cl.	**Conceptual Plan** $\text{Mass \% Cl} = \dfrac{4 \times \text{molar mass Cl}}{\text{molar mass } C_2Cl_4F_2} \times 100\%$ **Relationships Used** $\text{Mass percent of element X} = \dfrac{\text{mass of element X in 1 mol of compound}}{\text{mass of 1 mol of compound}} \times 100\%$
Solve Calculate the necessary parts of the equation and substitute the values into the equation to find mass percent Cl.	**Solution** $4 \times \text{molar mass Cl} = 4(35.45 \text{ g/mol}) = 141.8 \text{ g/mol}$ $\text{Molar mass } C_2C_4F_2 = 2(12.01 \text{ g/mol}) + 4(35.45 \text{ g/mol}) + 2(19.00 \text{ g/mol})$ $\qquad = 24.02 \text{ g/mol} + 141.8 \text{ g/mol} + 38.00 \text{ g/mol} = 203.8 \text{ g/mol}$ $\text{Mass \% Cl} = \dfrac{4 \times \text{molar mass Cl}}{\text{molar mass } C_2Cl_4F_2} \times 100\%$ $\qquad = \dfrac{141.8 \text{ g/mol}}{203.8 \text{ g/mol}} \times 100\%$ $\qquad = 69.58\%$

Check The units of the answer (%) are correct and the magnitude is reasonable because (a) it is between 0 and 100% and (b) chlorine is the heaviest atom in the molecule and there are four of them.

For Practice 3.13

Acetic acid ($C_2H_4O_2$) is the active ingredient in vinegar. Calculate the mass percent composition of oxygen in acetic acid.

For More Practice 3.13

Calculate the mass percent composition of sodium in sodium oxide.

 Conceptual Connection 3.4 Mass Percent Composition

In For Practice 3.13 you calculated the mass percent of oxygen in acetic acid ($C_2H_4O_2$). Without doing any calculations, predict whether the mass percent of carbon in acetic acid would be greater or smaller. Explain.

Answer: The mass percent of carbon in acetic acid will be smaller than the mass percent of oxygen because even though the formula ($C_2H_4O_2$) contains the same molar amounts of the two elements, carbon is lighter (it has a lower molar mass) than oxygen.

Mass Percent Composition as a Conversion Factor

The mass percent composition of an element in a compound is a conversion factor between mass of the element and mass of the compound. For example, we saw that the mass percent composition of Cl in CCl_2F_2 is 58.64%. Since percent means *per hundred*, we know that there are 58.64 g Cl *per hundred* grams CCl_2F_2, which can be expressed as the following ratio:

$$58.64 \text{ g Cl} : 100 \text{ g } CCl_2F_2$$

or, in fractional form:

$$\frac{58.64 \text{ g Cl}}{100 \text{ g } CCl_2F_2} \quad \text{or} \quad \frac{100 \text{ g } CCl_2F_2}{58.64 \text{ g Cl}}$$

These ratios can function as conversion factors between grams of Cl and grams of CCl_2F_2. For example, to calculate the mass of Cl in 1.00 kg CCl_2F_2, we use the following conceptual plan

Conceptual Plan

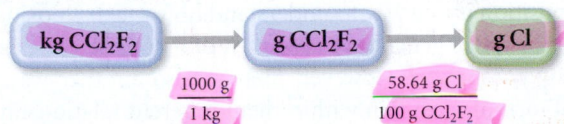

Notice that the mass percent composition acts as a conversion factor between grams of the compound and grams of the constituent element. To compute grams Cl, simply follow the conceptual plan.

Solution

$$1.00 \text{ kg } CCl_2F_2 \times \frac{1000 \text{ g}}{1 \text{ kg}} \times \frac{58.64 \text{ g Cl}}{100 \text{ g } CCl_2F_2} = 5.86 \times 10^2 \text{ g Cl}$$

EXAMPLE 3.14 Using Mass Percent Composition as a Conversion Factor

The U.S. Food and Drug Administration (FDA) recommends that you consume less than 2.4 g of sodium per day. What mass of sodium chloride (in grams) can you consume and still be within the FDA guidelines? Sodium chloride is 39% sodium by mass.

Sort You are given a mass of sodium and the mass percent of sodium in sodium chloride. You are asked to find the mass of NaCl that contains the given mass of sodium.	**Given** 2.4 g Na Mass % Na in NaCl = 39% **Find** g NaCl
Strategize Convert between mass of a constituent element and mass of a compound by using mass percent composition as a conversion factor.	**Conceptual Plan** 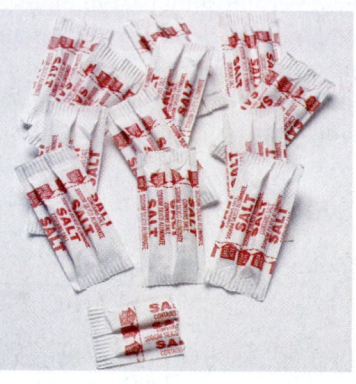 **Relationships Used** 39 g Na : 100 g NaCl
Solve Follow the conceptual plan to solve the problem.	**Solution** $2.4 \text{ g Na} \times \dfrac{100 \text{ g NaCl}}{39 \text{ g Na}} = 6.2 \text{ g NaCl}$ You can consume 6.2 g NaCl and still be within the FDA guidelines.

Check The units of the answer are correct. The magnitude seems reasonable because it is bigger than the amount of sodium, as expected because sodium is only one of the elements in NaCl.

For Practice 3.14

What mass (in grams) of iron(III) oxide contains 58.7 grams of iron? Iron(III) oxide is 69.94% iron by mass.

For More Practice 3.14

If someone consumes 22 g of sodium chloride per day, what mass (in grams) of sodium does that person consume? Sodium chloride is 39% sodium by mass.

▲ 12.5 packets of salt contain 6.2 g of NaCl.

Conversion Factors from Chemical Formulas

Mass percent composition is one way to understand how much chlorine is in a particular chlorofluorocarbon or, more generally, how much of a constituent element is present in a given mass of any compound. However, there is also another way to approach this problem. Chemical formulas contain within them inherent relationships between atoms (or moles of atoms) and molecules (or moles of molecules). For example, the formula for CCl_2F_2 tells us that 1 mol of CCl_2F_2 contains 2 mol of Cl atoms. We write the ratio as follows:

$$1 \text{ mol } CCl_2F_2 : 2 \text{ mol Cl}$$

With ratios such as these—that come from the chemical formula—we can directly determine the amounts of the constituent elements present in a given amount of a compound without having to compute mass percent composition. For example, we compute the number of moles of Cl in 38.5 mol of CCl_2F_2 as follows:

Conceptual Plan

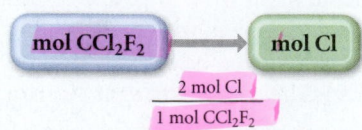

Solution

$$38.5 \ \text{mol CCl}_2\text{F}_2 \times \frac{2 \ \text{mol Cl}}{1 \ \text{mol CCl}_2\text{F}_2} = 77.0 \ \text{mol Cl}$$

We often want to know, however, not the *amount in moles* of an element in a certain number of moles of compound, but the *mass in grams* (or other units) of a constituent element in a given *mass* of the compound. For example, suppose we want to know the mass (in grams) of Cl contained in 25.0 g CCl_2F_2. *The relationship inherent in the chemical formula (2 mol Cl : 1 mol CCl_2F_2) applies to amount in moles, not to mass.* Therefore, we must first convert the mass of CCl_2F_2 to moles CCl_2F_2. Then we use the conversion factor from the chemical formula to convert to moles Cl. Finally, we use the molar mass of Cl to convert to grams Cl. The calculation proceeds as follows:

Conceptual Plan

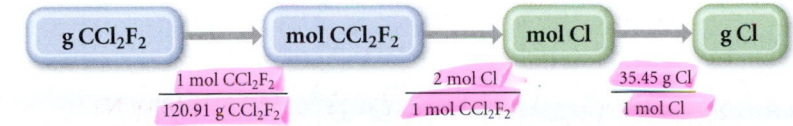

Solution

$$25.0 \ \text{g CCl}_2\text{F}_2 \times \frac{1 \ \text{mol CCl}_2\text{F}_2}{120.91 \ \text{g CCl}_2\text{F}_2} \times \frac{2 \ \text{mol Cl}}{1 \ \text{mol CCl}_2\text{F}_2} \times \frac{35.45 \ \text{g Cl}}{1 \ \text{mol Cl}} = 14.7 \ \text{g Cl}$$

Notice that we must convert from g CCl_2F_2 to mol CCl_2F_2 *before* we can use the chemical formula as a conversion factor. *The chemical formula gives us a relationship between the amounts (in moles) of substances, not between the masses (in grams) of them.*

The general form for solving problems where you are asked to find the mass of an element present in a given mass of a compound is

Mass compound → moles compound → moles element → mass element

The conversions between mass and moles are accomplished using the atomic or molar mass and the conversion between moles and moles is accomplished using the relationships inherent in the chemical formula.

Chemistry and Medicine
Methylmercury in Fish

In the last few years, the U.S. Environmental Protection Agency (EPA) has grown increasingly concerned about mercury levels in fish. Mercury—which is present in fish as methylmercury—is toxic to the central nervous system, especially in children and developing fetuses. In a developing fetus, excessive mercury exposure can result in slowed mental development and even retardation. Some lakes now have warnings about eating too much fish caught in the lake.

Recent regulations have forced many fish vendors to alert customers about the dangers of eating too much of certain kinds of commercial fish, including shark, tuna, and mackerel. These fish tend to harbor high levels of methylmercury and therefore should be eaten in moderation, especially by children and pregnant women. The U.S. Food and Drug Administration (FDA) action level—the level below which the FDA claims the food has no adverse health effects—for methylmercury in fish is 1.0 ppm or 1.0 g of methylmercury per million grams of fish. However, a number of environmental advocates, including the U.S. EPA, have suggested that, while this level may be safe for adults, it is too high for children and pregnant women.

▶ Lakes containing mercury—either from natural sources or from pollution—often post limits to the number of fish from the lake that can be eaten.

Consequently, the FDA suggests that pregnant women limit their intake of fish to 12 ounces per week.

Question

The levels of methylmercury in fish are normally tested by laboratory techniques that measure only the mercury (Hg). Suppose a lab analyzes a 14.5-g, sample of fish and finds that it contains 1.03×10^{-5} g of mercury. What is the level of methylmercury ($HgCH_3Cl$) in the fish in parts per million (ppm)? Is this above the FDA action level?

EXAMPLE 3.15 Chemical Formulas as Conversion Factors

Hydrogen is a potential future fuel that may one day replace gasoline. Most major automobile companies, therefore, are developing vehicles that run on hydrogen. These cars are environmentally friendly because their only emission is water vapor. One way to obtain hydrogen for fuel is to use an environmentally benign energy source such as wind power to split hydrogen from water. What mass of hydrogen (in grams) is contained in 1.00 gallon of water? (The density of water is 1.00 g/mL.)

Sort You are given a volume of water and asked to find the mass of hydrogen it contains. You are also given the density of water.	**Given** 1.00 gal H_2O $d_{H_2O} = 1.00$ g/mL **Find** g H

Strategize The first part of the conceptual plan shows how you can convert the units of volume from gallons to liters and then to mL. It also shows how you can then use the density to convert mL to g.

The second part of the conceptual plan is the basic sequence of mass → moles → moles → mass. Convert between moles and mass using the appropriate molar masses, and convert from mol H_2O to mol H using the conversion factor derived from the molecular formula.

Conceptual Plan

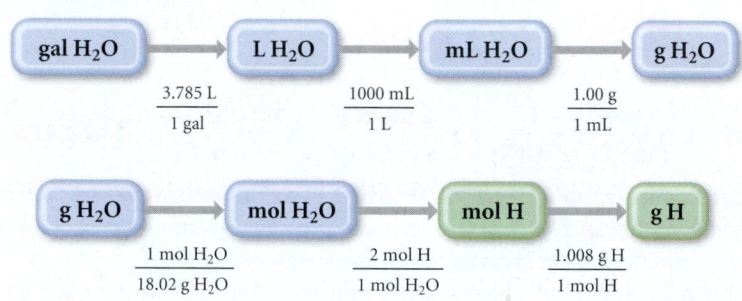

Relationships Used

3.785 L = 1 gal (Table 1.3)

1000 mL = 1 L

1.00 g H_2O = 1 mL H_2O (density of H_2O)

Molar mass H_2O = 2(1.008) + 16.00 = 18.02 g/mol

2 mol H : 1 mol H_2O

1.008 g H = 1 mol H

Solve Follow the conceptual plan to solve the problem.

Solution

$$1.00 \text{ gal } H_2O \times \frac{3.785 \text{ L}}{1 \text{ gal}} \times \frac{1000 \text{ mL}}{1 \text{ L}} \times \frac{1.0 \text{ g}}{\text{mL}} = 3.785 \times 10^3 \text{g } H_2O$$

$$3.785 \times 10^3 \text{g } H_2O \times \frac{1 \text{ mol } H_2O}{18.02 \text{ g } H_2O} \times \frac{2 \text{ mol H}}{1 \text{ mol } H_2O} \times \frac{1.008 \text{ g H}}{1 \text{ mol H}} = 4.23 \times 10^2 \text{ g H}$$

Check The units of the answer (g H) are correct. Since a gallon of water is about 3.8 L, its mass is about 3.8 kg. H is a light atom, so its mass should be significantly less than 3.8 kg, as it is in the answer.

For Practice 3.15

Determine the mass of oxygen in a 7.2-g sample of $Al_2(SO_4)_3$.

For More Practice 3.15

Butane (C_4H_{10}) is used as a liquid fuel in lighters. How many grams of carbon are present within a lighter containing 7.25 mL of butane? (The density of liquid butane is 0.601 g/mL.)

Conceptual Connection 3.5 Chemical Formulas and Elemental Composition

The molecular formula for water is H_2O. Which of the following ratios can be correctly derived from this formula? Explain.

(a) $2\,g\,H : 1\,g\,H_2O$

(b) $2\,mL\,H : 1\,mL\,H_2O$

(c) $2\,mol\,H : 1\,mol\,H_2O$

Answer: **(c)** The chemical formula for a compound gives relationships between atoms or moles of atoms. The chemical formula for water states that water molecules contain 2 H atoms to every 1 O atom or 2 mol H to every 1 mol H_2O. This does not imply a 2-to-1 relationship between *masses* of hydrogen and oxygen because these atoms have different masses. It also does not imply a 2-to-1 relationship between volumes.

3.9 Determining a Chemical Formula from Experimental Data

In Section 3.8, we learned how to calculate mass percent composition from a chemical formula. But can we also do the reverse? Can we calculate a chemical formula from mass percent composition? This question is important because laboratory analyses of compounds do not often give chemical formulas directly, but only the relative masses of each element present in a compound. For example, if we decompose water into hydrogen and oxygen in the laboratory, we can measure the masses of hydrogen and oxygen produced. Can we get a chemical formula from this kind of data? The answer is a qualified yes. We can get a chemical formula, but it is an empirical formula (not a molecular formula). To get a molecular formula, we need additional information, such as the molar mass of the compound.

Suppose we decompose a sample of water in the laboratory and find that it produces 0.857 g of hydrogen and 6.86 g of oxygen. How do we get an empirical formula from these data? We know that an empirical formula represents a ratio of atoms or a ratio of moles of atoms, but not a ratio of masses. So the first thing we must do is convert our data from mass (in grams) to amount (in moles). How many moles of each element are present in the sample? To convert to moles, simply divide each mass by the molar mass of that element:

$$\text{Moles H} = 0.857\,g\,H \times \frac{1\,mol\,H}{1.01\,g\,H} = 0.849\,mol\,H$$

$$\text{Moles O} = 6.86\,g\,O \times \frac{1\,mol\,O}{16.00\,g\,O} = 0.429\,mol\,O$$

From these data, we know there are 0.849 mol H for every 0.429 mol O. We can now write a pseudoformula for water:

$$H_{0.849}O_{0.429}$$

To get the smallest whole-number subscripts in our formula, we simply divide all the subscripts by the smallest one, in this case 0.429:

$$H_{\frac{0.849}{0.429}}O_{\frac{0.429}{0.429}} = H_{1.98}O = H_2O$$

Our empirical formula for water, which also happens to be the molecular formula, is H_2O. The following procedure can be used to obtain the empirical formula of any compound from experimental data giving the relative masses of the constituent elements. The left column outlines the procedure, and the center and right columns show two examples of how to apply the procedure.

Procedure for Obtaining an Empirical Formula from Experimental Data

EXAMPLE 3.16 Obtaining an Empirical Formula from Experimental Data

A compound containing nitrogen and oxygen is decomposed in the laboratory and produces 24.5 g nitrogen and 70.0 g oxygen. Calculate the empirical formula of the compound.

EXAMPLE 3.17 Obtaining an Empirical Formula from Experimental Data

A laboratory analysis of aspirin determined the following mass percent composition:

$$C \quad 60.00\%$$
$$H \quad 4.48\%$$
$$O \quad 35.52\%$$

Find the empirical formula.

---

1. Write down (or compute) as *given* the masses of each element present in a sample of the compound. If you are given mass percent composition, assume a 100-g sample and compute the masses of each element from the given percentages.

Given 24.5 g N, 70.0 g O

Find empirical formula

Given In a 100 g sample: 60.00 g C, 4.48 g H, 35.52 g O

Find empirical formula

---

2. Convert each of the masses in step 1 to moles by using the appropriate molar mass for each element as a conversion factor.

$$24.5 \text{ g N} \times \frac{1 \text{ mol N}}{14.01 \text{ g N}} = 1.75 \text{ mol N}$$

$$70.0 \text{ g O} \times \frac{1 \text{ mol O}}{16.00 \text{ g O}} = 4.38 \text{ mol O}$$

$$60.00 \text{ g C} \times \frac{1 \text{ mol C}}{12.01 \text{ g C}} = 4.996 \text{ mol C}$$

$$4.48 \text{ g H} \times \frac{1 \text{ mol H}}{1.008 \text{ g H}} = 4.44 \text{ mol H}$$

$$35.52 \text{ g O} \times \frac{1 \text{ mol O}}{16.00 \text{ g O}} = 2.220 \text{ mol O}$$

---

3. Write down a pseudoformula for the compound using the number of moles of each element (from step 2) as subscripts.

$$N_{1.75}O_{4.38}$$

$$C_{4.996}H_{4.44}O_{2.220}$$

---

4. Divide all the subscripts in the formula by the smallest subscript.

$$N_{\frac{1.75}{1.75}}O_{\frac{4.38}{1.75}} \rightarrow N_1 O_{2.5}$$

$$C_{\frac{4.996}{2.220}}H_{\frac{4.44}{2.220}}O_{\frac{2.220}{2.220}} \rightarrow C_{2.25}H_2O_1$$

---

5. If the subscripts are not whole numbers, multiply all the subscripts by a small whole number (see table) to get whole-number subscripts.

$$N_1 O_{2.5} \times 2 \rightarrow N_2O_5$$

The correct empirical formula is N_2O_5.

$$C_{2.25}H_2O_1 \times 4 \rightarrow C_9H_8O_4$$

The correct empirical formula is $C_9H_8O_4$.

Fractional Subscript	Multiply by This
0.20	5
0.25	4
0.33	3
0.40	5
0.50	2
0.66	3
0.75	4
0.80	5

For Practice 3.16

A sample of a compound is decomposed in the laboratory and produces 165 g carbon, 27.8 g hydrogen, and 220.2 g oxygen. Calculate the empirical formula of the compound.

For Practice 3.17

Ibuprofen, an aspirin substitute, has the following mass percent composition: C 75.69%, H 8.80%, O 15.51%. What is the empirical formula of ibuprofen?

Calculating Molecular Formulas for Compounds

You can find the molecular formula of a compound from the empirical formula if you also know the molar mass of the compound. Recall from Section 3.3 that the molecular formula is always a whole-number multiple of the empirical formula:

$$\text{Molecular formula} = \text{empirical formula} \times n, \quad \text{where } n = 1, 2, 3,\ldots$$

Suppose we want to find the molecular formula for fructose (a sugar found in fruit) from its empirical formula, CH_2O, and its molar mass, 180.2 g/mol. We know that the molecular formula is a whole-number multiple of CH_2O:

$$\text{Molecular formula} = (CH_2O) \times n$$
$$= C_nH_{2n}O_n$$

We also know that the molar mass is a whole-number multiple of the **empirical formula molar mass**, the sum of the masses of all the atoms in the empirical formula.

$$\text{Molar mass} = \text{empirical formula molar mass} \times n$$

For a particular compound, the value of n in both cases is the same. Therefore, we can find n by computing the ratio of the molar mass to the empirical formula molar mass:

$$n = \frac{\text{molar mass}}{\text{empirical formula molar mass}}$$

For fructose, the empirical formula molar mass is

empirical formula molar mass

$$= 12.01\text{ g/mol} + 2(1.01\text{ g/mol}) + 16.00\text{ g/mol} = 30.03\text{ g/mol}$$

Therefore, n is

$$n = \frac{180.2\text{ g/mol}}{30.03\text{ g/mol}} = 6$$

We can then use this value of n to find the molecular formula:

$$\text{Molecular formula} = (CH_2O) \times 6 = C_6H_{12}O_6$$

EXAMPLE 3.18 Calculating a Molecular Formula from an Empirical Formula and Molar Mass

Butanedione—a main component in the smell and taste of butter and cheese—contains the elements carbon, hydrogen, and oxygen. The empirical formula of butanedione is C_2H_3O and its molar mass is 86.09 g/mol. Find its molecular formula.

Sort You are given the empirical formula and molar mass of butanedione and asked to find the molecular formula.	**Given** Empirical formula $= C_2H_3O$ molar mass $= 86.09$ g/mol **Find** molecular formula
Strategize A molecular formula is always a whole-number multiple of the empirical formula. Divide the molar mass by the empirical formula mass to get the whole number.	Molecular formula $=$ empirical formula $\times n$ $n = \dfrac{\text{molar mass}}{\text{empirical formula mass}}$
Solve Compute the empirical formula mass.	Empirical formula molar mass $= 2(12.01\text{ g/mol}) + 3(1.008\text{ g/mol}) + 16.00\text{ g/mol} = 43.04\text{ g/mol}$
Divide the molar mass by the empirical formula mass to find n.	$n = \dfrac{\text{molar mass}}{\text{empirical formula mass}} = \dfrac{86.09\text{ g/mol}}{43.04\text{ g/mol}} = 2$
Multiply the empirical formula by n to obtain the molecular formula.	Molecular formula $= C_2H_3O \times 2$ $= C_4H_6O_2$

Check Check the answer by computing the molar mass of the computed formula as follows:

$$4(12.01\text{ g/mol}) + 6(1.008\text{ g/mol}) + 2(16.00\text{ g/mol}) = 86.09\text{ g/mol}$$

The computed molar mass is in agreement with the given molar mass. The answer is correct.

For Practice 3.18

A compound has the empirical formula CH and a molar mass of 78.11 g/mol. Find its molecular formula.

For More Practice 3.18

A compound with the percent composition shown below has a molar mass of 60.10 g/mol. Find its molecular formula.

C, 39.97%

H, 13.41%

N, 46.62%

 Conceptual Connection 3.6 Chemical Formula and Mass Percent Composition

Without doing any calculations, order the elements in the following compound in order of decreasing mass percent composition.

$$C_6H_6O$$

Answer: C > O > H. Since carbon and oxygen differ in atomic mass by only 4 amu, and since there are 6 carbon atoms in the formula, we can conclude that carbon must constitute the greatest fraction of the mass. Oxygen is next because its mass is 16 times that of hydrogen and there only 6 hydrogen atoms to every 1 oxygen atom.

Combustion Analysis

In the previous section, we learned how to compute the empirical formula of a compound from the relative masses of its constituent elements. Another common (and related) way of obtaining empirical formulas for unknown compounds, especially those containing carbon and hydrogen, is through **combustion analysis**. In combustion analysis, the unknown compound undergoes combustion (or burning) in the presence of pure oxygen, as shown in Figure 3.9 ▼. All of the carbon in the sample is converted to CO_2, and all of the hydrogen is converted to H_2O. The CO_2 and H_2O produced are weighed and the numerical relationships between moles inherent in the formulas for CO_2 and H_2O (1 mol CO_2 : 1 mol C and 1 mol H_2O : 2 mol H) are used to determine the amounts of C and H in the original sample. Any other elemental constituents, such as O, Cl, or N, can be determined by subtracting the original mass of the sample from the sum of the masses of C and H. The examples below show how to perform these calculations for a sample containing only C and H and for a sample containing C, H, and O.

▼ FIGURE 3.9 **Combustion Analysis Apparatus** The sample to be analyzed is placed in a furnace and burned in oxygen. The water and carbon dioxide produced are absorbed into separate containers and weighed.

Combustion Analysis

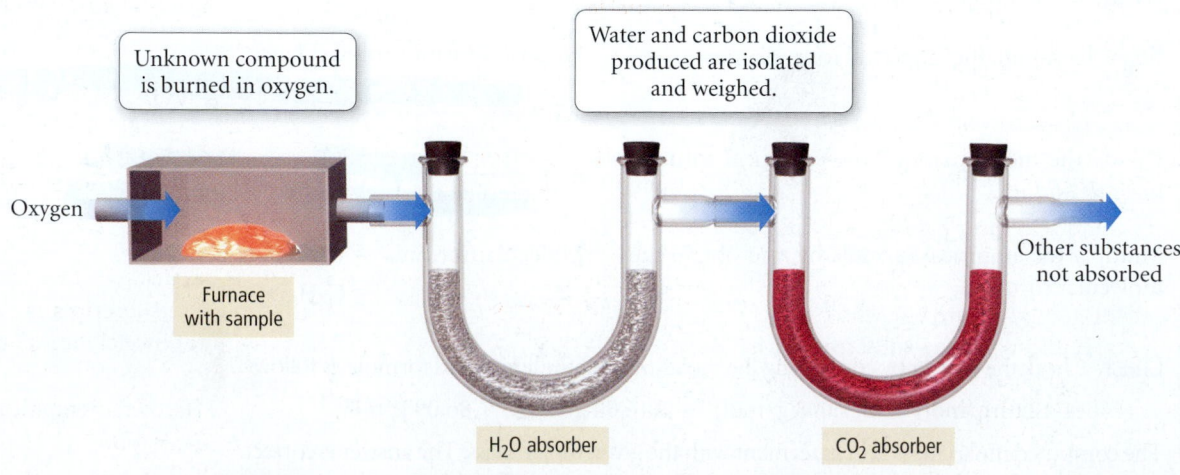

Procedure for Obtaining an Empirical Formula from Combustion Analysis	**EXAMPLE 3.19** Obtaining an Empirical Formula from Combustion Analysis Upon combustion, a compound containing only carbon and hydrogen produced 1.83 g CO_2 and 0.901 g H_2O. Find the empirical formula of the compound.	**EXAMPLE 3.20** Obtaining an Empirical Formula from Combustion Analysis Upon combustion, a 0.8233-g sample of a compound containing only carbon, hydrogen, and oxygen produced 2.445 g CO_2 and 0.6003 g H_2O. Find the empirical formula of the compound.
1. Write down as *given* the masses of each combustion product and the mass of the sample (if given).	**Given** 1.83 g CO_2, 0.901 g H_2O **Find** empirical formula	**Given** 0.8233 g sample, 2.445 g CO_2, 0.6003 g H_2O **Find** empirical formula
2. Convert the masses of CO_2 and H_2O from step 1 to moles by using the appropriate molar mass for each compound as a conversion factor.	$1.83 \text{ g } CO_2 \times \dfrac{1 \text{ mol } CO_2}{44.01 \text{ g } CO_2}$ $= 0.0416 \text{ mol } CO_2$ $0.901 \text{ g } H_2O \times \dfrac{1 \text{ mol } H_2O}{18.02 \text{ g } H_2O}$ $= 0.0500 \text{ mol } H_2O$	$2.445 \text{ g } CO_2 \times \dfrac{1 \text{ mol } CO_2}{44.01 \text{ g } CO_2}$ $= 0.05556 \text{ mol } CO_2$ $0.6003 \text{ g } H_2O \times \dfrac{1 \text{ mol } H_2O}{18.01 \text{ g } H_2O}$ $= 0.03331 \text{ mol } H_2O$
3. Convert the moles of CO_2 and moles of H_2O from step 2 to moles of C and moles of H using the conversion factors inherent in the chemical formulas of CO_2 and H_2O	$0.0416 \text{ mol } CO_2 \times \dfrac{1 \text{ mol C}}{1 \text{ mol } CO_2}$ $= 0.0416 \text{ mol C}$ $0.0500 \text{ mol } H_2O \times \dfrac{2 \text{ mol H}}{1 \text{ mol } H_2O}$ $= 0.100 \text{ mol H}$	$0.05556 \text{ mol } CO_2 \times \dfrac{1 \text{ mol C}}{1 \text{ mol } CO_2}$ $= 0.05556 \text{ mol C}$ $0.03331 \text{ mol } H_2O \times \dfrac{2 \text{ mol H}}{1 \text{ mol } H_2O}$ $= 0.06662 \text{ mol H}$
4. If the compound contains an element other than C and H, find the mass of the other element by subtracting the sum of the masses of C and H (obtained in step 3) from the mass of the sample. Finally, convert the mass of the other element to moles.	No other elements besides C and H, so proceed to next step.	$\text{Mass C} = 0.05556 \text{ mol C} \times \dfrac{12.01 \text{ g C}}{\text{mol C}}$ $= 0.6673 \text{ g C}$ $\text{Mass H} = 0.06662 \text{ mol H} \times \dfrac{1.008 \text{ g H}}{\text{mol H}}$ $= 0.06715 \text{ g H}$ $\text{Mass O} = 0.8233 \text{ g } -$ $(0.6673 \text{ g} + 0.06715 \text{ g}) = 0.0889 \text{ g}$ $\text{Mol O} = 0.0889 \text{ g O} \times \dfrac{\text{mol O}}{16.00 \text{ g O}}$ $= 0.00556 \text{ mol O}$
5. Write down a pseudoformula for the compound using the number of moles of each element (from steps 3 and 4) as subscripts.	$C_{0.0416}H_{0.100}$	$C_{0.05556}H_{0.06662}O_{0.00556}$
6. Divide all the subscripts in the formula by the smallest subscript. (Round all subscripts that are within 0.1 of a whole number.)	$C_{\frac{0.0416}{0.0416}}H_{\frac{0.100}{0.0416}} \rightarrow C_1H_{2.4}$	$C_{\frac{0.05556}{0.00556}}H_{\frac{0.06662}{0.00556}}O_{\frac{0.00556}{0.00556}} \rightarrow C_{10}H_{12}O_1$
7. If the subscripts are not whole numbers, multiply all the subscripts by a small whole number to get whole-number subscripts.	$C_1H_{2.4} \times 5 \rightarrow C_5H_{12}$ The correct empirical formula is C_5H_{12}.	The subscripts are whole numbers; no additional multiplication is needed. The correct empirical formula is $C_{10}H_{12}O$.

For Practice 3.19

Upon combustion, a compound containing only carbon and hydrogen produced 1.60 g CO_2 and 0.819 g H_2O. Find the empirical formula of the compound.

For Practice 3.20

Upon combustion, a 0.8009-g sample of a compound containing only carbon, hydrogen, and oxygen produced 1.6004 g CO_2 and 0.6551 g H_2O. Find the empirical formula of the compound

3.10 Writing and Balancing Chemical Equations

The method of combustion analysis (just examined) employs a **chemical reaction**, a process in which one or more substances are converted into one or more different ones. Compounds form and change through chemical reactions. As we have seen, water can be made by the reaction of hydrogen with oxygen. A **combustion reaction** is a particular type of chemical reaction in which a substance combines with oxygen to form one or more oxygen-containing compounds. Combustion reactions also emit heat, which makes them useful as a way to provide the energy that our society needs. The heat produced in the combustion of gasoline, for example, helps expand the gaseous combustion products in an engine's cylinders, which push the pistons and propel the car. The heat released by the combustion of *natural gas* is used to cook food or to heat our homes.

A chemical reaction is represented by a **chemical equation**. The combustion of natural gas is represented by the following equation:

$$CH_4 + O_2 \rightarrow CO_2 + H_2O$$

$$\underbrace{}_{\text{reactants}} \qquad \underbrace{}_{\text{products}}$$

The substances on the left side of the equation are called the **reactants** and the substances on the right side are called the **products.** We often specify the states of each reactant or product in parentheses next to the formula as follows:

$$CH_4(g) + O_2(g) \rightarrow CO_2(g) + H_2O(g)$$

The (g) indicates that these substances are gases in the reaction. The common states of reactants and products and their symbols used in chemical equations are summarized in Table 3.6.

If we look more closely at our equation for the combustion of natural gas, we should immediately notice a problem.

$$CH_4(g) + O_2(g) \longrightarrow CO_2(g) + H_2O(g)$$

| 2 O atoms | 2 O atoms + 1 O atom = |
| | 3 O atoms |

The left side of the equation has two oxygen atoms while the right side has three. The reaction as written violates the law of conservation of mass because an oxygen atom formed out of nothing. Notice also that there are four hydrogen atoms on the left and only two on the right.

$$CH_4(g) + O_2(g) \longrightarrow CO_2(g) + H_2O(g)$$

| 4 H atoms | 2 H atoms |

Two hydrogen atoms have vanished, again violating mass conservation. To correct these problems—that is, to write an equation that more closely represents *what actually happens*—we must **balance** the equation. We must change the coefficients—*not the subscripts*—to ensure that the number of each type of atom on the left side of the equation is equal to the number on the right side. New atoms do not form during a reaction, nor do atoms vanish—matter is always conserved.

TABLE 3.6 States of Reactants and Products in Chemical Equations

Abbreviation	State
(g)	Gas
(l)	Liquid
(s)	Solid
(aq)	Aqueous (water solution)

The reason that you cannot change the subscripts when balancing a chemical equation is that changing the subscripts changes the substance itself, while changing the coefficients simply changes the number of molecules of the substance. For example, 2 H_2O is simply two water molecules, but H_2O_2 is hydrogen peroxide, a drastically different compound.

When we add coefficients to the reactants and products to balance an equation, we change the number of molecules in the equation but not the *kind of* molecules. To balance the equation for the combustion of methane, we put the coefficient 2 before O_2 in the reactants, and the coefficient 2 before H_2O in the products.

$$CH_4(g) + 2\,O_2(g) \longrightarrow CO_2(g) + 2\,H_2O(g)$$

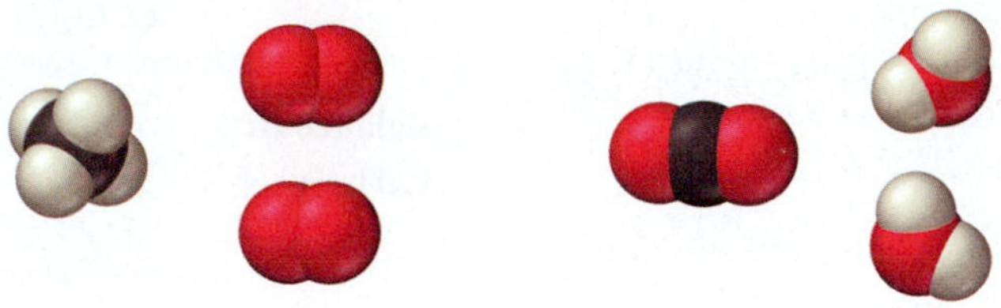

The equation is now balanced because the numbers of each type of atom on either side of the equation are equal. The balanced equation tells us that one CH_4 molecule reacts with 2 O_2 molecules to form 1 CO_2 molecule and 2 H_2O molecules. We verify that the equation is balanced by summing the number of each type of atom on each side of the equation.

$CH_4(g) + 2\,O_2(g)$	$\rightarrow$	$CO_2(g) + 2\,H_2O(g)$
Reactants		**Products**
1 C atom (1 × CH_4)		1 C atom (1 × CO_2)
4 H atoms (1 × CH_4)		4 H atoms (2 × H_2O)
4 O atoms (2 × O_2)		4 O atoms (1 × CO_2 + 2 × H_2O)

The number of each type of atom on both sides of the equation is now equal—the equation is balanced.

How to Write Balanced Chemical Equations

We balance many chemical equations simply by trial and error. However, some guidelines can be useful. For example, balancing the atoms in the most complex substances first and the atoms in the simplest substances (such as pure elements) last often makes the process shorter. The following illustrations of how to balance chemical equations are presented in a three-column format. The general guidelines are shown on the left, with two examples of how to apply them on the right. This procedure is meant only as a flexible guide, not a rigid set of steps.

Procedure For Balancing Chemical Equations	**EXAMPLE 3.21** Balancing Chemical Equations	**EXAMPLE 3.22** Balancing Chemical Equations
	Write a balanced equation for the reaction between solid cobalt(III) oxide and solid carbon to produce solid cobalt and carbon dioxide gas.	Write a balanced equation for the combustion of gaseous butane (C_4H_{10}), a fuel used in portable stoves and grills, in which it combines with gaseous oxygen to form gaseous carbon dioxide and gaseous water.
1. Write a skeletal equation by writing chemical formulas for each of the reactants and products. Review Sections 3.5 and 3.6 for nomenclature rules. (If a skeletal equation is provided, go to step 2.)	$Co_2O_3(s) + C(s) \rightarrow$ $Co(s) + CO_2(g)$	$C_4H_{10}(g) + O_2(g) \rightarrow$ $CO_2(g) + H_2O(g)$

2. Balance atoms that occur in more complex substances first. Always balance atoms in compounds before atoms in pure elements.

Begin with O:

$$Co_2O_3(s) + C(s) \rightarrow$$
$$Co(s) + CO_2(g)$$
3 O atoms → 2 O atoms

To balance O, put a 2 before $Co_2O_3(s)$ and a 3 before $CO_2(g)$.

$$\textbf{2}\,Co_2O_3(s) + C(s) \rightarrow$$
$$Co(s) + \textbf{3}\,CO_2(g)$$
6 O atoms → 6 O atoms

Begin with C:

$$C_4H_{10}(g) + O_2(g) \rightarrow$$
$$CO_2(g) + H_2O(g)$$
4 C atoms → 1 C atom

To balance C, put a 4 before $CO_2(g)$.

$$C_4H_{10}(g) + O_2(g) \rightarrow$$
$$\textbf{4}\,CO_2(g) + H_2O(g)$$
4 C atoms → 4 C atoms

Balance H:

$$C_4H_{10}(g) + O_2(g) \rightarrow$$
$$4\,CO_2(g) + H_2O(g)$$
10 H atoms → 2 H atoms

To balance H, put a 5 before $H_2O(g)$:

$$C_4H_{10}(g) + O_2(g) \rightarrow$$
$$4\,CO_2(g) + \textbf{5}H_2O(g)$$
10 H atoms → 10 H atoms

---

3. Balance atoms that occur as free elements on either side of the equation last. Always balance free elements by adjusting the coefficient on the free element.

Balance Co:

$$2\,Co_2O_3(s) + C(s) \rightarrow$$
$$Co(s) + 3\,CO_2(g)$$
4 Co atoms → 1 Co atom

To balance Co, put a 4 before $Co(s)$.

$$2\,Co_2O_3(s) + C(s) \rightarrow$$
$$\textbf{4}\,Co(s) + 3\,CO_2(g)$$
4 Co atoms → 4 Co atoms

Balance C:

$$2\,Co_2O_3(s) + C(s) \rightarrow$$
$$4\,Co(s) + 3\,CO_2(g)$$
1 C atoms → 3 C atoms

To balance C, put a 3 before $C(s)$.

$$2\,Co_2O_3(s) + \textbf{3}\,C(s) \rightarrow$$
$$4\,Co(s) + 3\,CO_2(g)$$

Balance O:

$$C_4H_{10}(g) + O_2(g) \rightarrow$$
$$4\,CO_2(g) + 5\,H_2O(g)$$
2 O atoms → 8 O + 5 O = 13 O atoms

To balance O, put a 13/2 before $O_2(g)$:

$$C_4H_{10}(g) + \textbf{13/2}\,O_2(g) \rightarrow$$
$$4\,CO_2(g) + 5\,H_2O(g)$$
13 O atoms → 13 O atoms

---

4. If the balanced equation contains coefficient fractions, clear these by multiplying the entire equation by the denominator of the fraction.

This step is not necessary in this example. Proceed to step 5.

$$[C_4H_{10}(g) + 13/2\,O_2(g) \rightarrow$$
$$4\,CO_2(g) + 5\,H_2O(g)] \times 2$$

$$2\,C_4H_{10}(g) + 13\,O_2(g) \rightarrow$$
$$8\,CO_2(g) + 10\,H_2O(g)$$

---

5. Check to make certain the equation is balanced by summing the total number of each type of atom on both sides of the equation.

$$2\,Co_2O_3(s) + 3\,C(s) \rightarrow$$
$$4\,Co(s) + 3\,CO_2(g)$$

Left	Right
4 Co atoms	4 Co atoms
6 O atoms	6 O atoms
3 C atoms	3 C atoms

The equation is balanced.

$$2\,C_4H_{10}(g) + 13\,O_2(g) \rightarrow$$
$$8\,CO_2(g) + 10\,H_2O(g)$$

Left	Right
8 C atoms	8 C atoms
20 H atoms	20 H atoms
26 O atoms	26 O atoms

The equation is balanced

For Practice 3.21

Write a balanced equation for the reaction between solid silicon dioxide and solid carbon to produce solid silicon carbide and carbon monoxide gas.

For Practice 3.22

Write a balanced equation for the combustion of gaseous ethane (C_2H_6), a minority component of natural gas, in which it combines with gaseous oxygen to form gaseous carbon dioxide and gaseous water.

Conceptual Connection 3.7 **Balanced Chemical Equations**

Which of the following must always be the same on both sides of a chemical equation?

(a) the number of atoms of each type
(b) the number of molecules of each type
(c) the number of moles of each type of molecule
(d) the sum of the masses of all substances involved

Answer: Both **(a)** and **(d)** are correct. When the number of atoms of each type is balanced, the sum of the masses of the substances involved will be the same on both sides of the equation. Since molecules change during a chemical reaction, their number is not the same on both sides, nor is the number of moles necessarily the same.

3.11 Organic Compounds

Early chemists divided compounds into two types: organic and inorganic. Organic compounds came from living things. Sugar—obtained from sugarcane or the sugar beet—is a common example of an organic compound. Inorganic compounds, on the other hand, came from the earth. Salt—mined from the ground or from the ocean—is a common example of an inorganic compound.

Not only were organic and inorganic compounds different in their origin, they were also different in their properties. Organic compounds were easily decomposed. Sugar, for instance, easily decomposes into carbon and water when heated. Inorganic compounds, however, were typically more difficult to decompose. Salt must be heated to very high temperatures before it decomposes. Moreover, eighteenth-century chemists could synthesize inorganic compounds in the laboratory, but not organic compounds, so a great division existed between the two different types of compounds. Today, chemists can synthesize both organic and inorganic compounds, and even though organic chemistry is a subfield of chemistry, the differences between organic and inorganic compounds are primarily organizational (not fundamental).

Organic compounds are common in everyday substances. For example, many smells—such as those in perfumes, spices, and foods—are caused by organic compounds. When you sprinkle cinnamon onto your French toast, some cinnamaldehyde—an organic compound present in cinnamon—evaporates into the air. You inhale some of the cinnamaldehyde molecules and experience the unique smell of cinnamon. Organic compounds are also the main components of most of our fuels, such as gasoline, oil, and natural gas, and they are the main ingredients in most pharmaceuticals, such as aspirin and ibuprofen. Organic compounds are also the major components of living organisms.

Organic compounds are composed of carbon and hydrogen and a few other elements, including nitrogen, oxygen, and sulfur. The key element to organic chemistry, however, is carbon. In its compounds, carbon always forms four bonds. For example, the simplest organic compound is methane or CH_4.

▲ Sugar is an organic compound that is easily decomposed by heating.

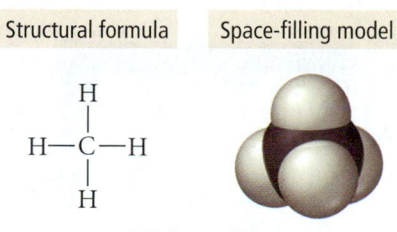
Structural formula   Space-filling model

Methane, CH_4

◄ The organic compound cinnamaldehyde is largely responsible for the taste and smell of cinnamon.

The chemistry of carbon is unique and complex because carbon frequently bonds to itself to form chain, branched, and ring structures:

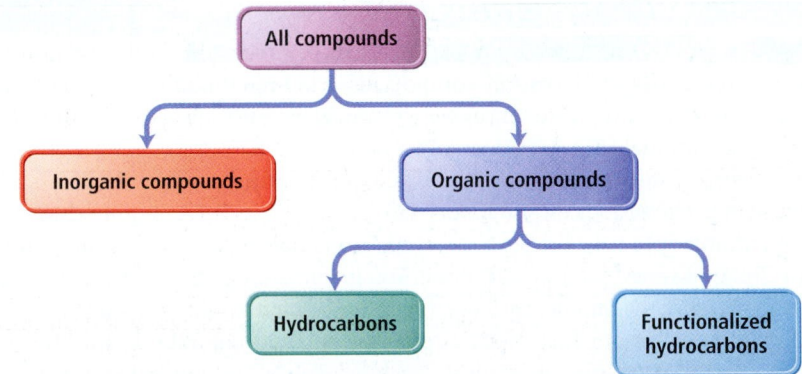

Propane (C_3H_8)   Isobutane (C_4H_{10})   Cyclohexane (C_6H_{12})

Carbon can also form double bonds and triple bonds with itself and with other elements.

Ethene (C_2H_4)   Ethyne (C_2H_2)   Acetic acid (CH_3COOH)

This versatility allows carbon to be the backbone of millions of different chemical compounds, which is why even a survey of organic chemistry requires a yearlong course.

Hydrocarbons

We can begin to scratch the surface of organic chemistry by dividing organic compounds into types: hydrocarbons and functionalized hydrocarbons.

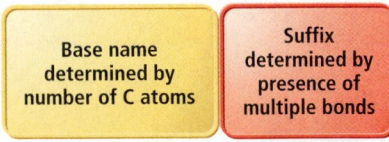

Hydrocarbons are organic compounds that contain only carbon and hydrogen. Hydrocarbons compose common fuels such as oil, gasoline, liquid propane gas, and natural gas. Hydrocarbons containing only single bonds are called **alkanes**, while those containing double or triple bonds are called **alkenes** and **alkynes**, respectively. The names of simple, straight-chain hydrocarbons consist of a base name, which is determined by the number of carbon atoms in the chain, and a suffix, determined by whether the hydrocarbon is an alkane (*-ane*), alkene (*-ene*), or alkyne (*-yne*).

Base name determined by number of C atoms	Suffix determined by presence of multiple bonds

The base names for a number of hydrocarbons are given below:

1	meth	2	eth
3	prop	4	but
5	pent	6	hex
7	hept	8	oct
9	non	10	dec

▲ Gasoline is composed mostly of hydrocarbons.

Table 3.7 lists some common hydrocarbons, their names, and their uses.

TABLE 3.7   Common Hydrocarbons

Name	Molecular Formula	Structural Formula	Space-filling Model	Common Uses
Methane	CH_4	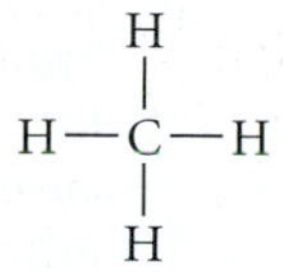 H—C—H (with H above and below)		Primary component of natural gas
Propane	C_3H_8	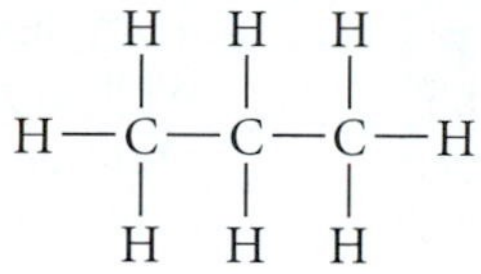 H—C—C—C—H (with H above and below each C)	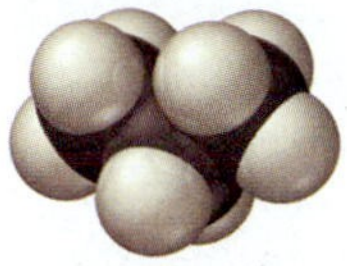	LP gas for grills and outdoor stoves
n-Butane*	C_4H_{10}	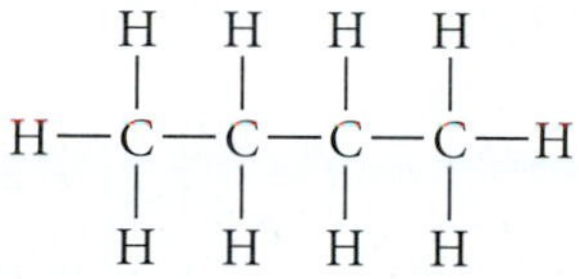 H—C—C—C—C—H (with H above and below each C)		Common fuel for lighters
n-Pentane*	C_5H_{12}	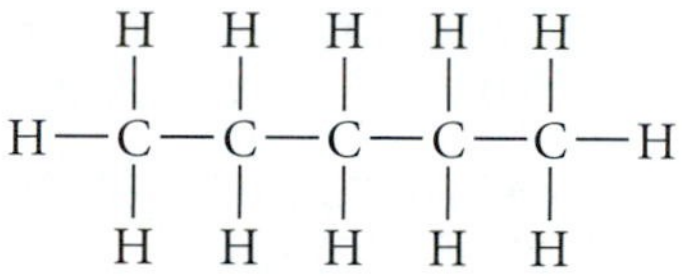 H—C—C—C—C—C—H (with H above and below each C)	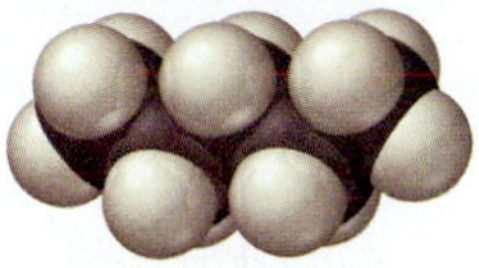	Component of gasoline
Ethene	C_2H_4	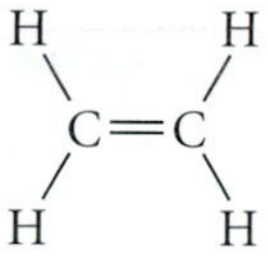 C=C (with H's)	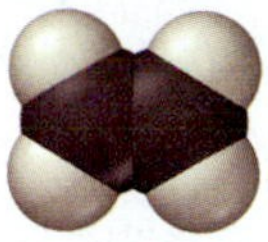	Ripening agent in fruit
Ethyne	C_2H_2	H—C≡C—H		Fuel for welding torches

*The "n" in the names of these hydrocarbons stands for normal, which means straight chain.

Functionalized Hydrocarbons

Functionalized hydrocarbons can be thought of as hydrocarbons in which a **functional group**—a characteristic atom or group of atoms—has been incorporated into the hydrocarbon. For example, families of organic compounds called **alcohols** have an –OH functional group. If we designate the hydrocarbon portion of the molecule as "R," the general formula for an alcohol can be written R—OH. Some specific examples of alcohols include methanol (also known as methyl alcohol or wood alcohol) and isopropanol (also known as isopropyl alcohol or rubbing alcohol):

The term *functional group* derives from the functionality or chemical character that a specific atom or group of atoms imparts to an organic compound. Therefore even a carbon–carbon double bond can justifiably be called a "functional group."

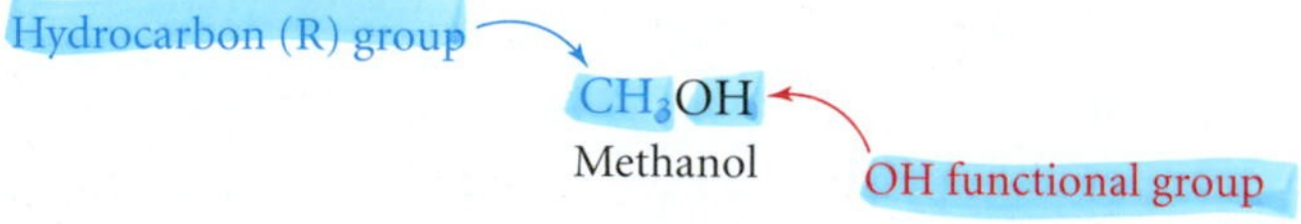

Hydrocarbon (R) group → CH_3OH ← OH functional group
Methanol

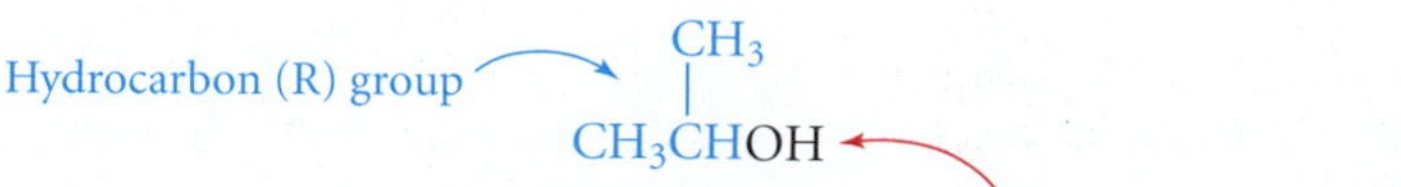

Hydrocarbon (R) group → CH_3 / CH_3CHOH ← OH functional group
Isopropanol
(2-proponal)

▲ Rubbing alcohol is isopropyl alcohol.

A group of organic compounds with the same functional group forms a **family.** Methanol and isopropyl alcohol are both members of the family of alcohols.

The addition of a functional group to a hydrocarbon usually alters the properties of the compound significantly. For example, *methanol*—which can be thought of as methane with an –OH group substituted for one of the hydrogen atoms—is a liquid at room temperature while *methane* is a gas. While each member of a family is unique, their common functional group gives them some chemical similarities. The names of functional groups have suffixes or endings that are unique to that functional group. For example, alcohols always have names that end in *-ol*. Table 3.8 lists some common functional groups, their general formulas, their characteristic suffixes or endings, and examples of each.

TABLE 3.8 Families of Organic Compunds

Family	Name Ending	General Formula	Example	Name	Occurrence/Use
Alcohols	-ol	R—OH	CH_3CH_2—OH	Ethanol (ethyl alcohol)	Alcohol in fermented beverages
Ethers	ether	R—O—R′	CH_3H_2C—O—CH_2CH_3	Diethyl ether	Anesthetic; laboratory solvent
Aldehydes	-al	$R-\overset{\overset{\displaystyle O}{\|\|}}{C}-H$	$H_3C-\overset{\overset{\displaystyle O}{\|\|}}{C}-H$	Ethanal (acetaldehyde)	Perfumes; flavors
Ketones	-one	$R-\overset{\overset{\displaystyle O}{\|\|}}{C}-R'$	$H_3C-\overset{\overset{\displaystyle O}{\|\|}}{C}-CH_3$	Propanone (acetone)	Fingernail polish remover
Carboxylic acids	acid	$R-\overset{\overset{\displaystyle O}{\|\|}}{C}-OH$	$H_3C-\overset{\overset{\displaystyle O}{\|\|}}{C}-OH$	Acetic acid	Vinegar
Esters	-ate	$R-\overset{\overset{\displaystyle O}{\|\|}}{C}-OR'$	$H_3C-\overset{\overset{\displaystyle O}{\|\|}}{C}-OCH_3$	Methyl acetate	Laboratory solvent
Amines	amine	RNH_2	$CH_3H_2C-\overset{\overset{\displaystyle H}{\|}}{N}-H$	Ethyl amine	Smell of rotten fish

Chapter in Review

Key Terms

Section 3.2
ionic bond (85)
covalent bond (85)

Section 3.3
chemical formula (86)
empirical formula (86)
molecular formula (86)
structural formula (86)
ball-and-stick model (87)
space-filling molecular model (87)

Section 3.4
atomic element (89)
molecular element (89)

molecular compound (89)
ionic compound (90)
formula unit (90)
polyatomic ion (91)

Section 3.5
common name (93)
systematic name (93)
binary compound (94)
oxyanion (95)
hydrate (96)

Section 3.6
acid (98)
binary acid (99)
oxyacid (99)

Section 3.7
formula mass (101)

Section 3.8
mass percent composition (mass percent) (103)

Section 3.9
empirical formula molar mass (111)
combustion analysis (112)

Section 3.10
chemical reaction (114)
combustion reaction (114)
chemical equation (114)

reactants (114)
products (114)
balanced chemical equation (114)

Section 3.11
hydrocarbon (118)
alkane (118)
alkene (118)
alkyne (118)
functional group (119)
alcohol (119)
family (120)

Key Concepts

Chemical Bonds (3.2)

Chemical bonds, the forces that hold atoms together in compounds, arise from the interactions between nuclei and electrons in atoms. In an ionic bond, one or more electrons are *transferred* from one atom to another, forming a cation (positively charged) and an anion (negatively charged). The two ions are then drawn together by the attraction between the opposite charges. In a covalent bond, one or more electrons are *shared* between two atoms. The atoms are held together by the attraction between their nuclei and the shared electrons.

Representing Molecules and Compounds (3.3, 3.4)

A compound is represented with a chemical formula, which indicates the elements present and the number of atoms of each. An empirical formula gives only the *relative* number of atoms, while a molecular formula gives the *actual* number present in the molecule. Structural formulas show how the atoms are bonded together, while molecular models show the geometry of the molecule.

Compounds can be divided into two types: molecular compounds, formed between two or more covalently bonded nonmetals; and ionic compounds, usually formed between a metal ionically bonded to one or more nonmetals. The smallest identifiable unit of a molecular compound is a molecule, and the smallest identifiable unit of an ionic compound is a formula unit: the smallest electrically neutral collection of ions. Elements can also be divided into two types: molecular elements, which occur as (mostly diatomic) molecules; and atomic elements, which occur as individual atoms.

Naming Inorganic Ionic and Molecular Compounds and Acids (3.5, 3.6)

A flowchart for naming simple inorganic compounds is shown at the end of this section. Use this chart to name inorganic compounds.

Formula Mass and Mole Concept for Compounds (3.7)

The formula mass of a compound is the sum of the atomic masses of all the atoms in the chemical formula. Like the atomic masses of elements, the formula mass characterizes the average mass of a molecule (or a formula unit). The mass of one mole of a compound is called the molar mass and equals its formula mass (in grams).

Chemical Composition (3.8, 3.9)

The mass percent composition of a compound is each element's percentage of the total compound's mass. The mass percent composition can be obtained from the compound's chemical formula and the molar masses of its elements. The chemical formula of a compound provides the relative number of atoms (or moles) of each element in a compound, and can therefore be used to determine numerical relationships between moles of the compound and moles of its constituent elements. This relationship can be extended to mass by using the molar masses of the compound and its constituent elements. The calculation can also go the other way—if the mass percent composition and molar mass of a compound are known, its empirical and molecular formulas can be determined.

Writing and Balancing Chemical Equations (3.10)

In chemistry, we represent chemical reactions with chemical equations. The substances on the left hand side of a chemical equation are called the reactants and the substances on the right hand side are called the products. Chemical equations are balanced when the number of each type of atom on the left side of the equation is equal to the number on the right side.

Organic Compounds (3.11)

Organic compounds—originally derived only from living organisms but now readily synthesized in the laboratory—are composed of carbon, hydrogen, and a few other elements such as nitrogen, oxygen, and sulfur. The simplest organic compounds are hydrocarbons, compounds composed of only carbon and hydrogen. Hydrocarbons can be divided into three types based on the bonds they contain: alkanes contain single bonds, alkenes contain double bonds, and alkynes contain triple bonds. All other organic compounds can be thought of as hydrocarbons with one or more functional groups, characteristic atoms or groups of atoms. Common functionalized hydrocarbons include alcohols, ethers, aldehydes, ketones, carboxylic acids, esters, and amines.

Key Equations and Relationships

Formula Mass (3.7)

$$\left(\begin{array}{c}\text{\# atoms of 1st element} \\ \text{in chemical formula}\end{array} \times \begin{array}{c}\text{atomic mass} \\ \text{of 1st element}\end{array}\right) + \left(\begin{array}{c}\text{\# atoms of 2nd element} \\ \text{in chemical formula}\end{array} \times \begin{array}{c}\text{atomic mass} \\ \text{of 2nd element}\end{array} + \cdots\right)$$

Mass Percent Composition (3.8)

$$\text{Mass \% of element X} = \frac{\text{mass of X in 1 mol compound}}{\text{mass of 1 mol compound}} \times 100\%$$

Empirical Formula Molar Mass (3.9)

$$\text{Molecular formula} = n \times (\text{empirical formula})$$

$$n = \frac{\text{molar mass}}{\text{empirical formula molar mass}}$$

Inorganic Nomenclature Summary Chart

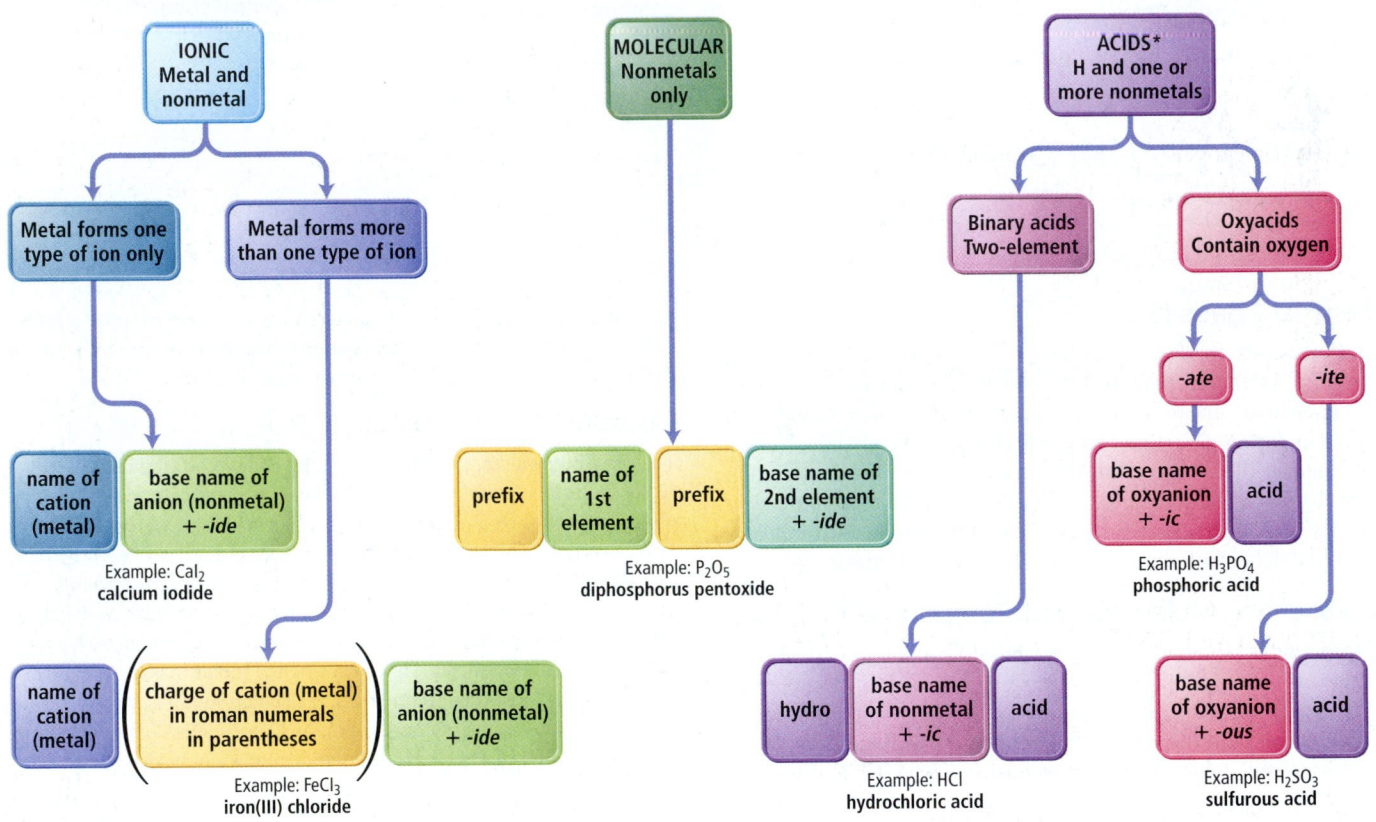

Using the Flowchart

The examples below show how to name compounds using the flowchart. The path through the flowchart is shown below each compound followed by the correct name for the compound.

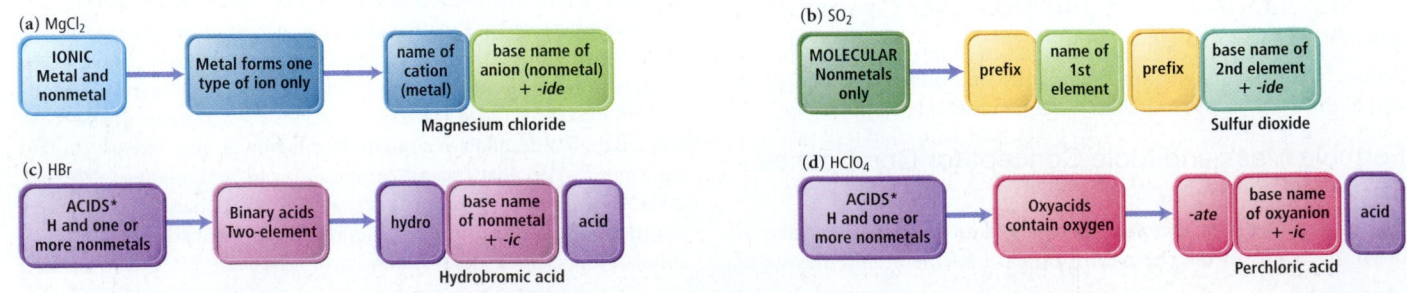

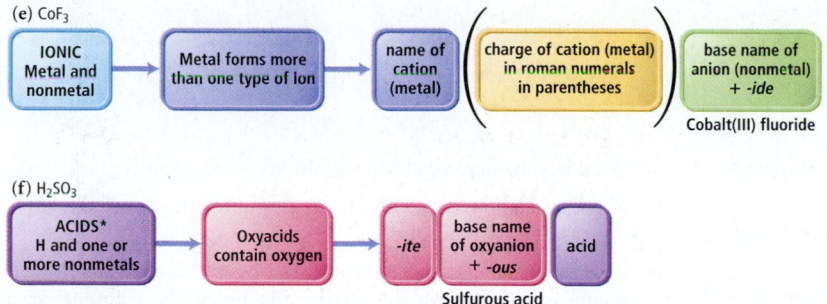

(e) CoF₃

Cobalt(III) fluoride

(f) H₂SO₃

Sulfurous acid

Key Skills

Writing Molecular and Empirical Formulas (3.3)
• Example 3.1   • For Practice 3.1   • Exercises 4, 23–26

Classifying Substances as Atomic Elements, Molecular Elements, Molecular Compounds, or Ionic Compounds (3.4)
• Example 3.2   • For Practice 3.2   • Exercises 27–32

Writing Formulas for Ionic Compounds (3.5)
• Examples 3.3, 3.4   • For Practice 3.3, 3.4   • Exercises 33–36, 45, 46

Naming Ionic Compounds (3.5)
• Examples 3.5, 3.6   • For Practice 3.5, 3.6   • For More Practice 3.5, 3.6   • Exercises 37–42

Naming Ionic Compounds Containing Polyatomic Ions (3.5)
• Example 3.7   • For Practice 3.7   • For More Practice 3.7   • Exercises 43–46

Naming Molecular Compounds (3.6)
• Example 3.8   • For Practice 3.8   • For More Practice 3.8   • Exercises 49–52

Naming Acids (3.6)
• Examples 3.9, 3.10   • For Practice 3.9, 3.10   • For More Practice 3.10   • Exercises 53–56

Calculating Formula Mass (3.7)
• Example 3.11   • For Practice 3.11   • Exercises 57, 58

Using Formula Mass to Count Molecules by Weighing (3.7)
• Example 3.12   • For Practice 3.12   • For More Practice 3.12   • Exercises 59–64

Calculating Mass Percent Composition (3.8)
• Example 3.13   • For Practice 3.13   • For More Practice 3.13   • Exercises 65–70

Using Mass Percent Composition as a Conversion Factor (3.8)
• Example 3.14   • For Practice 3.14   • For More Practice 3.14   • Exercises 71, 72

Using Chemical Formulas as Conversion Factors (3.8)
• Example 3.15   • For Practice 3.15   • For More Practice 3.15   • Exercises 77, 78

Obtaining an Empirical Formula from Experimental Data (3.9)
• Examples 3.16, 3.17   • For Practice 3.16, 3.17   • Exercises 79–84

Calculating a Molecular Formula from an Empirical Formula and Molar Mass (3.9)
• Example 3.18   • For Practice 3.18   • For More Practice 3.18   • Exercises 85–86

Obtaining an Empirical Formula from Combustion Analysis (3.9)
• Examples 3.19, 3.20   • For Practice 3.19, 3.20   • Exercises 87–90

Balancing Chemical Equations (3.10)
• Examples 3.21, 3.22   • For Practice 3.21, 3.22   • Exercises 91–100

Exercises

Review Questions

1. How do the properties of compounds compare to the properties of the elements from which they are composed?

2. What is a chemical bond? Explain the difference between an ionic bond and a covalent bond.

3. Explain the different ways to represent compounds. Why are there so many?

4. What is the difference between an empirical formula and a molecular formula?

5. Define and provide an example for each of the following: atomic element, molecular element, ionic compound, molecular compound.

6. Explain how to write a formula for an ionic compound given the names of the metal and nonmetal (or polyatomic ion) in the compound.

7. Explain how to name binary ionic compounds. How do you name the compound if it contains a polyatomic ion?

8. Why do the names of some ionic compounds include the charge of the metal ion while others do not?

9. Explain how to name molecular inorganic compounds.

10. How many atoms does each of the following prefixes specify? mono, di, tri, tetra, penta, hexa.

11. Explain how to name binary and oxy acids.

12. What is the formula mass for a compound? Why is it useful?

13. Explain how the information in a chemical formula can be used to determine how much of a particular element is present in a given amount of a compound. Give some examples demonstrating why this might be important.

14. What is mass percent composition? Why is it useful?

15. What kinds of conversion factors are inherent in chemical formulas? Give an example.

16. What kind of chemical formula can be obtained from experimental data showing the relative masses of the elements in a compound?

17. How can a molecular formula be obtained from an empirical formula? What additional information is required?

18. What is combustion analysis? What is it used for?

19. Explain the difference between organic and inorganic compounds according to early chemists. What elements are normally present in organic compounds?

20. Explain the difference between an alkane, an alkene, and an alkyne.

21. What are functionalized hydrocarbons? Give an example of a functionalized hydrocarbon.

22. Write a generic formula for each of the following families of organic compounds.
   a. alcohols
   b. ethers
   c. aldehydes
   d. ketones
   e. carboxylic acids
   f. esters
   g. amines

Problems by Topic

Note: Answers to all odd-numbered Problems, numbered in blue, can be found in Appendix III. Exercises in the Problems by Topic section are paired, with each odd-numbered problem followed by a similar even-numbered problem. Exercises in the Cumulative Problems section are also paired, but somewhat more loosely. (Challenge Problems and Conceptual Problems, because of their nature, are unpaired.)

Chemical Formulas and Molecular View of Elements and Compounds

23. Determine the number of each type of atom in each of the following formulas:
   a. $Ca_3(PO_4)_2$ b. $SrCl_2$
   c. KNO_3 d. $Mg(NO_2)_2$

24. Determine the number of each type of atom in each of the following formulas:
   a. $Ba(OH)_2$ b. NH_4Cl
   c. $NaCN$ d. $Ba(HCO_3)_2$

25. Write a chemical formula for each of the following molecular models. (See Appendix II.A for color codes.)

(a)   (b)   (c)

26. Write a chemical formula for each of the following molecular models. (See Appendix II.A for color codes.)

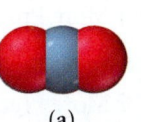   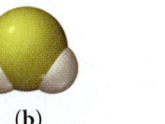   

(a)   (b)   (c)

27. Classify each of the following elements as atomic or molecular.
   a. neon   b. fluorine
   c. potassium   d. nitrogen

28. Which of the following elements have molecules as their basic units?
   a. hydrogen   b. iodine
   c. lead   d. oxygen

29. Classify each of the following compounds as ionic or molecular.
   a. CO_2 b. $NiCl_2$ c. NaI d. PCl_3

30. Classify each of the following compounds as ionic or molecular.
   a. CF_2Cl_2 b. CCl_4 c. PtO_2 d. SO_3

31. Based on the following molecular views, classify each substance as an atomic element, a molecular element, an ionic compound, or a molecular compound.

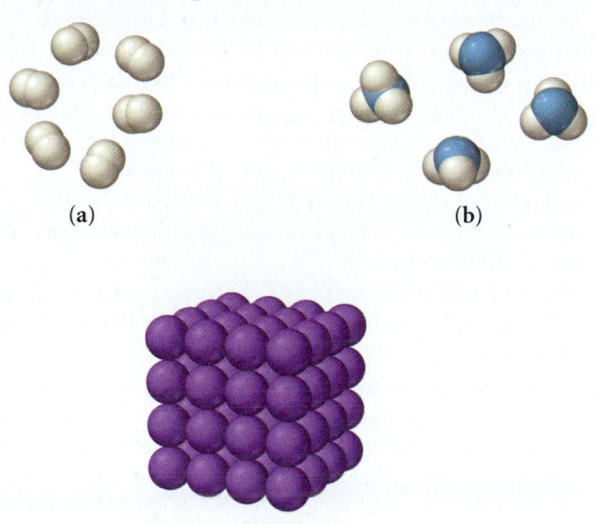

(a)   (b)

(c)

32. Based on the following molecular views, classify each substance as an atomic element, a molecular element, an ionic compound, or a molecular compound.

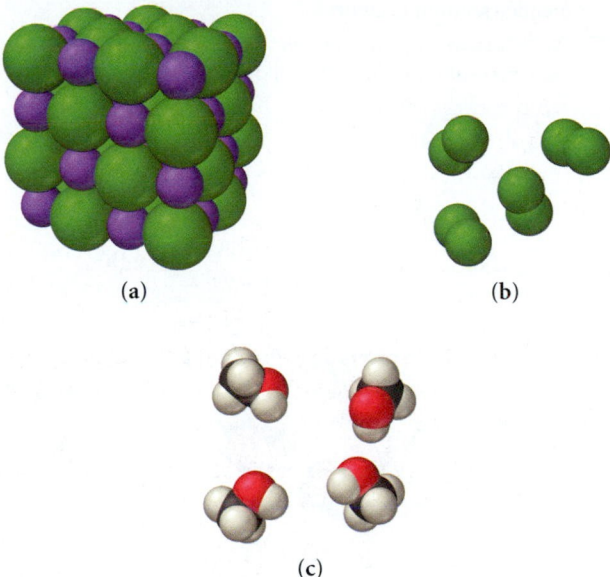

(a)   (b)

(c)

Formulas and Names for Ionic Compounds

33. Write a formula for the ionic compound that forms between each of the following pairs of elements.
a. magnesium and sulfur
b. barium and oxygen
c. strontium and bromine
d. beryllium and chlorine

34. Write a formula for the ionic compound that forms between each of the following pairs of elements.
a. aluminum and sulfur
b. aluminum and oxygen
c. sodium and oxygen
d. strontium and iodine

35. Write a formula for the compound that forms between barium and each of the following polyatomic ions:
a. hydroxide   b. chromate
c. phosphate   d. cyanide

36. Write a formula for the compound that forms between sodium and each of the following polyatomic ions:
a. carbonate   b. phosphate
c. hydrogen phosphate   d. acetate

37. Name each of the following ionic compounds.
a. Mg_3N_2 b. KF   c. Na_2O d. Li_2S

38. Name each of the following ionic compounds.
a. CsF   b. KI   c. $SrCl_2$ d. $BaCl_2$

39. Name each of the following ionic compounds.
a. $SnCl_4$ b. PbI_2 c. Fe_2O_3 d. CuI_2

40. Name each of the following ionic compounds.
a. SnO_2 b. $HgBr_2$ c. $CrCl_2$ d. $CrCl_3$

41. Give each ionic compound an appropriate name.
a. SnO   b. Cr_2S_3 c. RbI   d. $BaBr_2$

42. Give each ionic compound an appropriate name.
a. BaS   b. $FeCl_3$ c. PbI_4 d. $SrBr_2$

43. Name each of the following ionic compounds containing a polyatomic ion.
a. $CuNO_2$ b. $Mg(C_2H_3O_2)_2$
c. $Ba(NO_3)_2$ d. $Pb(C_2H_3O_2)_2$
e. $KClO_3$ f. $PbSO_4$

44. Name each of the following ionic compounds containing a polyatomic ion.
a. $Ba(OH)_2$ b. NH_4I
c. $NaBrO_4$ d. $Fe(OH)_3$
e. $CoSO_4$ f. $KClO$

45. Write a formula for each of the following ionic compounds:
a. sodium hydrogen sulfite
b. lithium permanganate
c. silver nitrate
d. potassium sulfate
e. rubidium hydrogen sulfate
f. potassium hydrogen carbonate

46. Write a formula for each of the following ionic compounds:
a. copper(II) chloride
b. copper(I) iodate
c. lead(II) chromate
d. calcium fluoride
e. potassium hydroxide
f. iron(II) phosphate

47. Give the name from the formula or the formula from the name for each of the following hydrated ionic compounds:
a. $CoSO_4 \cdot 7H_2O$
b. iridium(III) bromide tetrahydrate
c. $Mg(BrO_3)_2 \cdot 6H_2O$
d. potassium carbonate dihydrate

48. Give the name from the formula or the formula from the name for each of the following hydrated ionic compounds:
a. cobalt(II) phosphate octahydrate
b. $BeCl_2 \cdot 2H_2O$
c. chromium(III) phosphate trihydrate
d. $LiNO_2 \cdot H_2O$

Formulas and Names for Molecular Compounds and Acids

49. Name each of the following molecular compounds.
    a. CO   b. NI_3 c. $SiCl_4$
    d. N_4Se_4 e. I_2O_5

50. Name each of the following molecular compounds.
    a. SO_3 b. SO_2 c. BrF_5
    d. NO   e. XeO_3

51. Write a formula for each of the following molecular compounds.
    a. phosphorus trichloride
    b. chlorine monoxide
    c. disulfur tetrafluoride
    d. phosphorus pentafluoride
    e. diphosphorus pentasulfide

52. Write a formula for each of the following molecular compounds.
    a. boron tribromide   b. dichlorine monoxide
    c. xenon tetrafluoride   d. carbon tetrabromide
    e. diboron tetrachloride

53. Name each of the following acids.
    a. HI   b. HNO_3
    c. H_2CO_3 d. $HC_2H_3O_2$

54. Name each of the following acids.
    a. HCl   b. $HClO_2$
    c. H_2SO_4 d. HNO_2

55. Write formulas for each of the following acids.
    a. hydrofluoric acid   b. hydrobromic acid
    c. sulfurous acid

56. Write formulas for each of the following acids.
    a. phosphoric acid   b. hydrocyanic acid
    c. chlorous acid

Formula Mass and the Mole Concept for Compounds

57. Calculate the formula mass for each of the following compounds.
    a. NO_2 b. C_4H_{10}
    c. $C_6H_{12}O_6$ d. $Cr(NO_3)_3$

58. Calculate the formula mass for each of the following compounds.
    a. $MgBr_2$ b. HNO_2
    c. CBr_4 d. $Ca(NO_3)_2$

59. How many molecules are in each of the following?
    a. 6.5 g H_2O b. 389 g CBr_4
    c. 22.1 g O_2 d. 19.3 g C_8H_{10}

60. Calculate the mass (in g) of each of the following.
    a. 5.94×10^{20} SO_3 molecules
    b. 2.8×10^{22} H_2O molecules
    c. 4.5×10^{25} O_3 molecules
    d. 9.85×10^{19} CCl_2F_2 molecules

61. Calculate the mass (in g) of a single water molecule.

62. Calculate the mass (in g) of a single glucose molecule ($C_6H_{12}O_6$).

63. A sugar crystal contains approximately 1.8×10^{17} sucrose ($C_{12}H_{22}O_{11}$) molecules. What is its mass in mg?

64. A salt crystal has a mass of 0.12 mg. How many NaCl formula units does it contain?

Composition of Compounds

65. Calculate the mass percent composition of carbon in each the following carbon compounds:
    a. CH_4 b. C_2H_6 c. C_2H_2 d. C_2H_5Cl

66. Calculate the mass percent composition of nitrogen in each of the following nitrogen compounds:
    a. N_2O b. NO   c. NO_2 d. HNO_3

67. Most fertilizers consist of nitrogen-containing compounds such as NH_3, $CO(NH_2)_2$, NH_4NO_3, and $(NH_4)_2SO_4$. The nitrogen content in these compounds is needed for protein synthesis in plants. Calculate the mass percent composition of nitrogen in each of the fertilizers named above. Which fertilizer has the highest nitrogen content?

68. Iron is mined from the earth as iron ore. Common ores include Fe_2O_3 (hematite), Fe_3O_4 (magnetite), and $FeCO_3$ (siderite). Calculate the mass percent composition of iron for each of these iron ores. Which ore has the highest iron content?

69. Copper(II) fluoride contains 37.42% F by mass. Use this percentage to calculate the mass of fluorine (in g) contained in 55.5 g of copper(II) fluoride.

70. Silver chloride, often used in silver plating, contains 75.27% Ag. Calculate the mass of silver chloride required to plate 155 mg of pure silver.

71. The iodide ion is a dietary mineral essential to good nutrition. In countries where potassium iodide is added to salt, iodine deficiency or goiter has been almost completely eliminated. The recommended daily allowance (RDA) for iodine is 150 μg/day. How much potassium iodide (76.45% I) should be consumed to meet the RDA?

72. The American Dental Association recommends that an adult female should consume 3.0 mg of fluoride (F^-) per day to prevent tooth decay. If the fluoride is consumed as sodium fluoride (45.24% F), what amount of sodium fluoride contains the recommended amount of fluoride?

73. Write a ratio showing the relationship between the amounts of each element for each of the following:

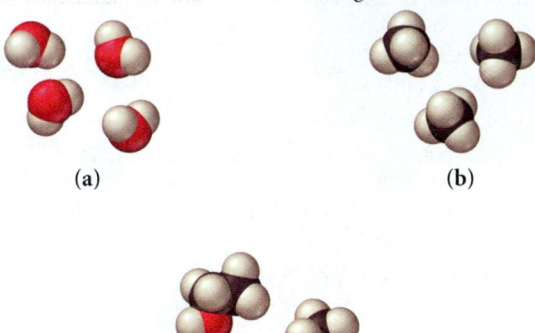

(a)   (b)

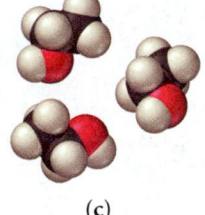

(c)

74. Write a ratio showing the relationship between the amounts of each element for each of the following:

(a)   (b)

(c)

75. Determine the number of moles of hydrogen atoms in each of the following:
   a. 0.0885 mol C_4H_{10} **b.** 1.3 mol CH_4
   c. 2.4 mol C_6H_{12} **d.** 1.87 mol C_8H_{18}

76. Determine the number of moles of oxygen atoms in each of the following:
   a. 4.88 mol H_2O_2 **b.** 2.15 mol N_2O
   c. 0.0237 mol H_2CO_3 **d.** 24.1 mol CO_2

77. Calculate the number of grams of sodium in 8.5 g of each of the following sodium-containing food additives:
   a. NaCl (table salt)
   b. Na_3PO_4 (sodium phosphate)
   c. $NaC_7H_5O_2$ (sodium benzoate)
   d. $Na_2C_6H_6O_7$ (sodium hydrogen citrate)

78. How many kilograms of chlorine are in 25 kg of each of the following chlorofluorocarbons (CFCs)?
   a. CF_2Cl_2 **b.** $CFCl_3$ **c.** $C_2F_3Cl_3$ **d.** CF_3Cl

Chemical Formulas from Experimental Data

79. Samples of several compounds are decomposed and the masses of their constituent elements are shown below. Calculate the empirical formula for each compound.
   a. 1.651 g Ag, 0.1224 g O
   b. 0.672 g Co, 0.569 g As, 0.486 g O
   c. 1.443 g Se, 5.841 g Br

80. Samples of several compounds are decomposed and the masses of their constituent elements are shown below. Calculate the empirical formula for each compound.
   a. 1.245 g Ni, 5.381 g I
   b. 2.677 g Ba, 3.115 g Br
   c. 2.128 g Be, 7.557 g S, 15.107 g O

81. Calculate the empirical formula for each of the following stimulants based on their elemental mass percent composition:
   a. nicotine (found in tobacco leaves):
     C 74.03%, H 8.70%, N 17.27%
   b. caffeine (found in coffee beans):
     C 49.48%, H 5.19%, N 28.85%, O 16.48%

82. Calculate the empirical formula for each of the following natural flavors based on their elemental mass percent composition:
   a. methyl butyrate (component of apple taste and smell):
     C 58.80% H 9.87% O 31.33%
   b. vanillin (responsible for the taste and smell of vanilla):
     C 63.15% H 5.30% O 31.55%.

83. A 0.77-mg sample of nitrogen reacts with chlorine to form 6.61 mg of the chloride. What is the empirical formula of the nitrogen chloride?

84. A 45.2-mg sample of phosphorus reacts with selenium to form 131.6 mg of the selenide. What is the empirical formula of the phosphorus selenide?

85. The empirical formula and molar mass of several compounds are listed below. Find the molecular formula of each compound.
   a. C_6H_7N, 186.24 g/mol   **b.** C_2HCl, 181.44 g/mol
   c. $C_5H_{10}NS_2$, 296.54 g/mol

86. The molar mass and empirical formula of several compounds are listed below. Find the molecular formula of each compound.
   a. C_4H_9, 114.22 g/mol   **b.** CCl, 284.77 g/mol
   c. C_3H_2N, 312.29 g/mol

87. Combustion analysis of a hydrocarbon produced 33.01 g CO_2 and 13.51 g H_2O. Calculate the empirical formula of the hydrocarbon.

88. Combustion analysis of naphthalene, a hydrocarbon used in mothballs, produced 8.80 g CO_2 and 1.44 g H_2O. Calculate the empirical formula for naphthalene.

89. The foul odor of rancid butter is due largely to butyric acid, a compound containing carbon, hydrogen, and oxygen. Combustion analysis of a 4.30-g sample of butyric acid produced 8.59 g CO_2 and 3.52 g H_2O. Find the empirical formula for butyric acid.

90. Tartaric acid is the white, powdery substance that coats sour candies such as Sour Patch Kids. Combustion analysis of a 12.01-g sample of tartaric acid—which contains only carbon, hydrogen, and oxygen—produced 14.08 g CO_2 and 4.32 g H_2O. Find the empirical formula for tartaric acid.

Writing and Balancing Chemical Equations

91. Sulfuric acid is a component of acid rain formed when gaseous sulfur dioxide pollutant reacts with gaseous oxygen and liquid water to form aqueous sulfuric acid. Write a balanced chemical equation for this reaction.

92. Nitric acid is a component of acid rain that forms when gaseous nitrogen dioxide pollutant reacts with gaseous oxygen and liquid water to form aqueous nitric acid. Write a balanced chemical equation for this reaction.

93. In a popular classroom demonstration, solid sodium is added to liquid water and reacts to produce hydrogen gas and aqueous sodium hydroxide. Write a balanced chemical equation for this reaction.

94. When iron rusts, solid iron reacts with gaseous oxygen to form solid iron(III) oxide. Write a balanced chemical equation for this reaction.

95. Write a balanced chemical equation for the fermentation of sucrose ($C_{12}H_{22}O_{11}$) by yeasts in which the aqueous sugar reacts with water to form aqueous ethyl alcohol (C_2H_5OH) and carbon dioxide gas.

96. Write a balanced equation for the photosynthesis reaction in which gaseous carbon dioxide and liquid water react in the presence of chlorophyll to produce aqueous glucose ($C_6H_{12}O_6$) and oxygen gas.

97. Write a balanced chemical equation for each of the following:
   a. Solid lead(II) sulfide reacts with aqueous hydrobromic acid to form solid lead(II) bromide and dihydrogen monosulfide gas.
   b. Gaseous carbon monoxide reacts with hydrogen gas to form gaseous methane (CH_4) and liquid water.
   c. Aqueous hydrochloric acid reacts with solid manganese(IV) oxide to form aqueous manganese(II) chloride, liquid water, and chlorine gas.
   d. Liquid pentane (C_5H_{12}) reacts with gaseous oxygen to form carbon dioxide and liquid water.

98. Write a balanced chemical equation for each of the following:
   a. Solid copper reacts with solid sulfur to form solid copper(I) sulfide.
   b. Solid iron(III) oxide reacts with hydrogen gas to form solid iron and liquid water.
   c. Sulfur dioxide gas reacts with oxygen gas to form sulfur trioxide gas.
   d. Gaseous ammonia (NH_3) reacts with gaseous oxygen to form gaseous nitrogen monoxide and gaseous water.

99. Balance each of the following chemical equations:
   a. $CO_2(g) + CaSiO_3(s) + H_2O(l) \rightarrow SiO_2(s) + Ca(HCO_3)_2(aq)$
   b. $Co(NO_3)_3(aq) + (NH_4)_2S(aq) \rightarrow Co_2S_3(s) + NH_4NO_3(aq)$
   c. $Cu_2O(s) + C(s) \rightarrow Cu(s) + CO(g)$
   d. $H_2(g) + Cl_2(g) \rightarrow HCl(g)$

100. Balance each of the following chemical equations:

a. $Na_2S(aq) + Cu(NO_3)_2(aq) \rightarrow NaNO_3(aq) + CuS(s)$

b. $N_2H_4(l) \rightarrow NH_3(g) + N_2(g)$

c. $HCl(aq) + O_2(g) \rightarrow H_2O(l) + Cl_2(g)$

d. $FeS(s) + HCl(aq) \rightarrow FeCl_2(aq) + H_2S(g)$

Organic Compounds

101. Classify each of the following compounds as organic or inorganic:

a. $CaCO_3$ b. C_4H_8 c. $C_4H_6O_6$ d. LiF

102. Classify each of the following compounds as organic or inorganic:

a. C_8H_{18} b. CH_3NH_2 c. CaO d. $FeCO_3$

103. Classify each of the following hydrocarbons as an alkane, alkene, or alkyne:

a. $H_2C{=}CH{-}CH_3$ b. $H_3C{-}CH_2{-}CH_3$

c. $HC{\equiv}C{-}CH_3$ d. $H_3C{-}CH_2{-}CH_2{-}CH_3$

104. Classify each of the following hydrocarbons as an alkane, alkene, or alkyne:

a. $HC{\equiv}CH$

b. $H_3C{-}CH{=}C{-}CH_3$

c. $H_3C{-}\overset{\overset{\displaystyle CH_3}{|}}{CH}{-}CH_3$

d. $H_3C{-}C{\equiv}C{-}CH_3$

105. Write a formula based on the name, or a name based on the formula, for each of the following hydrocarbons:

a. propane   b. $CH_3CH_2CH_3$

c. octane   d. $CH_3CH_2CH_2CH_2CH_3$

106. Write a formula based on the name, or a name based on the formula, for each of the following hydrocarbons:

a. CH_3CH_3 b. pentane

c. $CH_3CH_2CH_2CH_2CH_2CH_3$ d. heptane

107. Classify each of the following as a hydrocarbon or a functionalized hydrocarbon. For functionalized hydrocarbons, identify the family to which the compound belongs.

a. $H_3C{-}CH_2OH$

b. $H_3C{-}CH_3$

c. $H_3C{-}\overset{\overset{\displaystyle O}{||}}{C}{-}CH_2{-}CH_3$

d. $H_3C{-}NH_2$

108. Classify each of the following as a hydrocarbon or a functionalized hydrocarbon. For functionalized hydrocarbons, identify the family to which the compound belongs.

a. $H_3C{-}CH_2{-}\overset{\overset{\displaystyle O}{||}}{C}{-}OH$

b. $H_3C{-}\overset{\overset{\displaystyle O}{||}}{CH}$

c. $H_3C{-}\overset{\overset{\displaystyle CH_3}{|}}{\underset{\underset{\displaystyle CH_3}{|}}{C}}{-}CH_3$

d. $H_3C{-}CH_2{-}O{-}CH_3$

Cumulative Problems

109. How many molecules of ethanol (C_2H_5OH) (the alcohol in alcoholic beverages) are present in 145 mL of ethanol? The density of ethanol is $0.789 \, g/cm^3$.

110. A drop of water has a volume of approximately 0.05 mL. How many water molecules does it contain? The density of water is $1.0 \, g/cm^3$.

111. Determine the chemical formula of each of the following compounds and then use it to calculate the mass percent composition of each constituent element:

a. potassium chromate   b. lead(II) phosphate

c. sulfurous acid   d. cobalt(II) bromide

112. Determine the chemical formula of each of the following compounds and then use it to calculate the mass percent composition of each constituent element:

a. perchloric acid   b. phosphorus pentachloride

c. nitrogen triiodide   d. carbon dioxide

113. A Freon leak in the air conditioning system of an older car releases 25 g of CF_2Cl_2 per month. What mass of chlorine is emitted into the atmosphere each year by this car?

114. A Freon leak in the air-conditioning system of a large building releases 12 kg of CHF_2Cl per month. If the leak were allowed to continue, how many kilograms of Cl would be emitted into the atmosphere each year?

115. A metal (M) forms a compound with the formula MCl_3. If the compound contains 65.57% Cl by mass, what is the identity of the metal?

116. A metal (M) forms an oxide with the formula M_2O. If the oxide contains 16.99% O by mass, what is the identity of the metal?

117. Estradiol is a female sexual hormone that causes maturation and maintenance of the female reproductive system. Elemental analysis of estradiol gave the following mass percent composition: C 79.37%, H 8.88%, O 11.75%. The molar mass of estradiol is 272.37 g/mol. Find the molecular formula of estradiol.

118. Fructose is a common sugar found in fruit. Elemental analysis of fructose gave the following mass percent composition: C 40.00%, H 6.72%, O 53.28%. The molar mass of fructose is 180.16 g/mol. Find the molecular formula of fructose.

119. Combustion analysis of a 13.42-g sample of equilin (which contains only carbon, hydrogen, and oxygen) produced 39.61 g CO_2 and 9.01 g H_2O. The molar mass of equilin is 268.34 g/mol. Find the molecular formula for equilin.

120. Estrone, which contains only carbon, hydrogen, and oxygen, is a female sexual hormone that occurs in the urine of pregnant women. Combustion analysis of a 1.893-g sample of estrone produced 5.545 g of CO_2 and 1.388 g H_2O. The molar mass of estrone is 270.36 g/mol. Find the molecular formula for estrone.

121. Epsom salts is a hydrated ionic compound with the following formula: $MgSO_4 \cdot xH_2O$ A sample of Epsom salts with a mass of 4.93 was heated to drive off the water of hydration. The mass of the sample after complete dehydration was 2.41 g. Find the number of waters of hydration (x) in Epsom salts.

122. A hydrate of copper(II) chloride has the following formula: $CuCl_2 \cdot xH_2O$. The water in a 3.41-g sample of the hydrate was driven off by heating. The remaining sample had a mass of 2.69 g. Find the number of waters of hydration (x) in the hydrate.

123. A compound of molar mass 177 g/mol contains only carbon, hydrogen, bromine, and oxygen. Analysis reveals that the compound contains 8 times as much carbon as hydrogen by mass. Find the molecular formula.

124. The following data were obtained from experiments to find the molecular formula of benzocaine, a local anesthetic, which contains only carbon, hydrogen, nitrogen, and oxygen. Complete combustion of a 3.54-g sample of benzocaine with excess O_2 formed 8.49 g of CO_2 and 2.14 g H_2O. Another sample of mass 2.35 g was found to contain 0.199 g of N. The molar mass of benzocaine was found to be 165. Find the molar formula of benzocaine.

125. Find the total number of atoms in a sample of cocaine hydrochloride, $C_{17}H_{22}ClNO_4$, of mass 23.5 mg.

126. Vanadium forms four different oxides in which the percent by mass of vanadium is respectively 76%, 68%, 61%, and 56%. Find the formula and give the name of each one of these oxides.

127. The chloride of an unknown metal is believed to have the formula MCl_3. A 2.395-g sample of the compound is found to contain 3.606×10^{-2} mol Cl. Find the atomic mass of M.

128. Write the structural formulas of three different compounds that have the molecular formula C_5H_{12}.

Challenge Problems

129. A mixture of NaCl and NaBr has a mass of 2.00 g and is found to contain 0.75 g of Na. What is the mass of NaBr in the mixture?

130. Three pure compounds are formed when 1.00-g samples of element X combine with, respectively, 0.472 g, 0.630 g, and 0.789 g of element Z. The first compound has the formula X_2Z_3. Find the empirical formulas of the other two compounds.

131. A mixture of $CaCO_3$ and $(NH_4)_2CO_3$ is 61.9% CO_3 by mass. Find the mass percent of $CaCO_3$ in the mixture.

132. A mixture of 50.0 g of S and 1.00×10^2 g of Cl_2 reacts completely to form S_2Cl_2 and SCl_2. Find the mass of S_2Cl_2 formed.

133. Because of increasing evidence of damage to the ozone layer, chlorofluorocarbon (CFC) production was banned in 1996. However, there are about 100 million auto air conditioners that still use CFC-12 (CF_2Cl_2). These air conditioners are recharged from stockpiled supplies of CFC-12. If each of the 100 million automobiles contains 1.1 kg of CFC-12 and leaks 25% of its

CFC-12 into the atmosphere per year, how much chlorine, in kg, is added to the atmosphere each year due to auto air conditioners? (Assume two significant figures in your calculations.)

134. A particular coal contains 2.55% sulfur by mass. When the coal is burned, it produces SO_2 emissions which combine with rainwater to produce sulfuric acid. Use the formula of sulfuric acid to calculate the mass percent of S in sulfuric acid. Then determine how much sulfuric acid (in metric tons) is produced by the combustion of 1.0 metric ton of this coal. (A metric ton is 1000 kg.)

135. Lead is found in Earth's crust as several different lead ores. Suppose a certain rock is composed of 38.0% PbS (galena), 25.0% $PbCO_3$ (cerussite), and 17.4% $PbSO_4$ (anglesite). The remainder of the rock is composed of substances containing no lead. How much of this rock (in kg) must be processed to obtain 5.0 metric tons of lead? (A metric ton is 1000 kg.)

Conceptual Problems

136. When molecules are represented by molecular models, what does each sphere represent? How big is the nucleus of an atom in comparison to the sphere used to represent an atom in a molecular model?

137. Without doing any calculations, determine which element in each of the following compounds will have the highest mass percent composition.

  a. CO   **b.** N_2O **c.** $C_6H_{12}O_6$ **d.** NH_3

138. Explain the problem with the following statement and correct it. "The chemical formula for ammonia (NH_3) indicates that ammonia contains three grams of hydrogen to each gram of nitrogen."

139. Explain the problem with the following statement and correct it. "When a chemical equation is balanced, the number of molecules of each type on both sides of the equation will be equal."

140. Without doing any calculations, arrange the elements in H_2SO_4 in order of decreasing mass percent composition.

4

Chemical Quantities and Aqueous Reactions

4.1 Global Warming and the Combustion of Fossil Fuels

4.2 Reaction Stoichiometry: How Much Carbon Dioxide?

4.3 Limiting Reactant, Theoretical Yield, and Percent Yield

4.4 Solution Concentration and Solution Stoichiometry

4.5 Types of Aqueous Solutions and Solubility

4.6 Precipitation Reactions

4.7 Representing Aqueous Reactions: Molecular, Ionic, and Complete Ionic Equations

4.8 Acid–Base and Gas-Evolution Reactions

4.9 Oxidation–Reduction Reactions

WHAT ARE THE RELATIONSHIPS between the amounts of reactants in a chemical reaction and the amounts of products that are formed? How do we best describe and understand these relationships? The first half of this chapter focuses on chemical stoichiometry—the numerical relationships between the amounts of reactants and products in chemical reactions. In Chapter 3, we learned how to write balanced chemical equations for chemical reactions. Now we examine more closely the meaning of those balanced equations. In the second half of this chapter, we turn to describing chemical reactions that occur in water. You have probably witnessed many of these types of reactions in your daily life because they are so common. For example, have you ever mixed baking soda with vinegar and observed the subsequent bubbling? Or have you ever seen hard water deposits form on your plumbing fixtures? These reactions—and many others, including those that occur within the watery environment of living cells—are aqueous chemical reactions, the subject of the second half of this chapter.

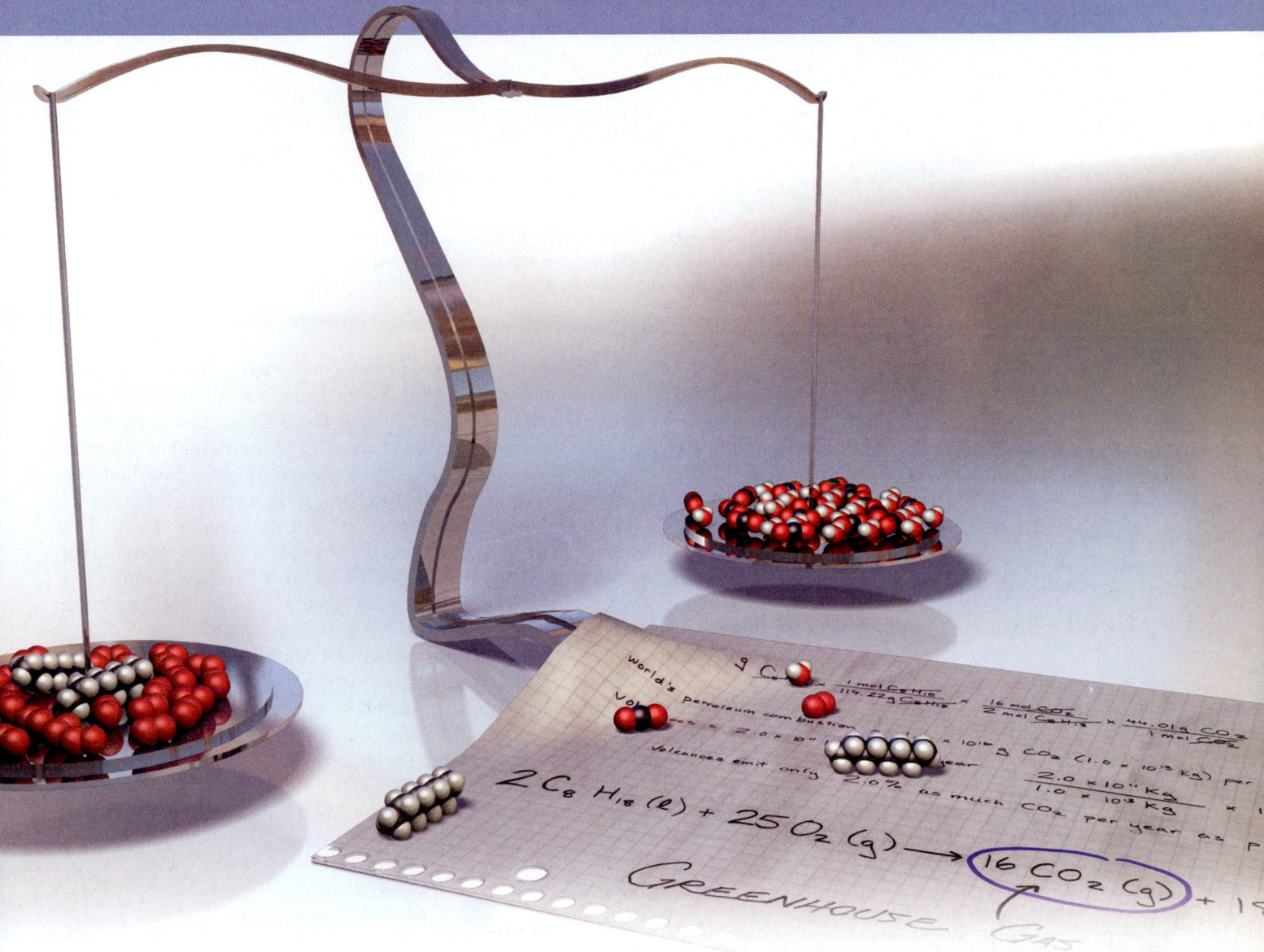

The molecular models on this balance represent the reactants and products in the combustion of
octane, a component of petroleum. One of the products, carbon dioxide, is the main greenhouse
gas implicated in global warming.

4.1 Global Warming and the Combustion of Fossil Fuels

The temperature outside my office today is a cool 48 °F, lower than normal for this time of
year on the West Coast. However, today's "chill" pales in comparison with how cold it
would be without the presence of *greenhouse gases* in the atmosphere. These gases act like
the glass of a greenhouse, allowing sunlight to enter the atmosphere and warm Earth's
surface, but preventing some of the heat generated by the sunlight from escaping, as
shown in Figure 4.1 (page 132). The balance between incoming and outgoing energy
from the sun then determines Earth's average temperature.

The Greenhouse Effect

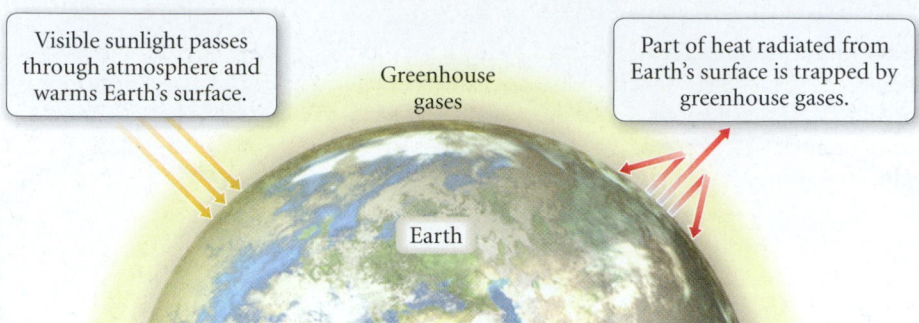

Visible sunlight passes through atmosphere and warms Earth's surface.

Greenhouse gases

Part of heat radiated from Earth's surface is trapped by greenhouse gases.

Earth

▲ **FIGURE 4.1 The Greenhouse Effect**   Greenhouse gases in the atmosphere act as a one-way filter. They allow visible light to pass through and warm Earth's surface, but they prevent heat energy from being radiated back out into space.

The extremely cold temperatures of Mars are as much caused by its lack of atmosphere as by its being farther from the sun than Earth. Conversely, Venus is an inferno partly because its thick atmosphere is rich in greenhouse gases.

Without greenhouse gases in the atmosphere, more heat energy would escape, and Earth's average temperature would be about 60 °F colder than it is now. The temperature outside of my office today would be below 0 °F (-17.8 °C), and even the sunniest U.S. cities would most likely be covered with snow. However, if the concentration of greenhouse gases in the atmosphere were to increase, Earth's average temperature would rise.

In recent years scientists have become concerned because the amount of atmospheric carbon dioxide (CO_2)—Earth's most significant greenhouse gas in terms of its contribution to climate—is rising. More CO_2 enhances the atmosphere's ability to hold heat and may therefore lead to *global warming*, an increase in Earth's average temperature. Since 1860, atmospheric CO_2 levels have risen by 35% (Figure 4.2 ▼), and Earth's average temperature has risen by 0.6 °C (about 1.1 °F), as shown in Figure 4.3 ▼.

Most scientists now believe that the primary cause of rising atmospheric CO_2 concentration is the burning of fossil fuels (natural gas, petroleum, and coal), which provide 90% of our society's energy. Some people, however, have suggested that fossil fuel combustion does not significantly contribute to global warming. They argue, for example, that the amount of carbon dioxide emitted into the atmosphere by volcanic eruptions far exceeds that from fossil fuel combustion. We could judge the validity of this argument if we could compute the amount of carbon dioxide emitted by fossil fuel combustion and compare it to that released by volcanic eruptions.

Atmospheric Carbon Dioxide

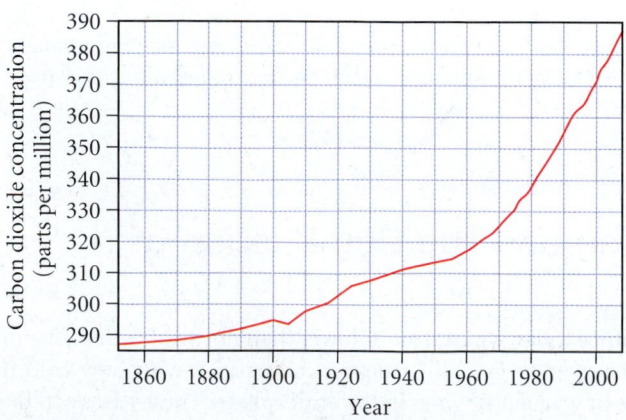

▲ **FIGURE 4.2 Carbon Dioxide Concentrations in the Atmosphere**   The rise in carbon dioxide levels shown in this graph is due largely to fossil fuel combustion.

Global Temperature

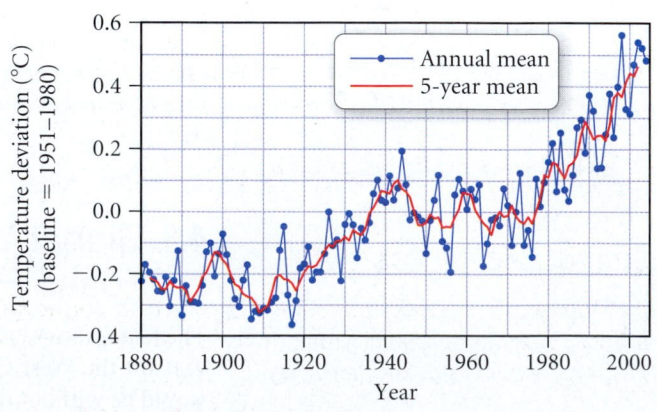

▲ **FIGURE 4.3 Global Temperature**   Average temperatures worldwide have risen by about 0.6 °C since 1880.

4.2 Reaction Stoichiometry: How Much Carbon Dioxide?

The amount of carbon dioxide emitted by fossil fuel combustion is related to the amount of fossil fuel that is burned—the balanced chemical equations for the combustion reactions give the exact relationships between these amounts. In this discussion, we use octane (a component of gasoline) as a representative fossil fuel. The balanced equation for the combustion of octane is as follows:

$$2\,C_8H_{18}(l) + 25\,O_2(g) \longrightarrow 16\,CO_2(g) + 18\,H_2O(g)$$

The balanced chemical equation shows that 16 CO_2 molecules are produced for every 2 molecules of octane burned. This numerical relationship between molecules can be extended to the amounts in moles as follows:

The coefficients in a chemical reaction specify the relative amounts in moles of each of the substances involved in the reaction.

In other words, from the equation, we know that 16 moles of CO_2 are produced for every 2 moles of octane burned. The numerical relationships between chemical amounts in a balanced chemical equation are called reaction **stoichiometry**. Stoichiometry allows us to predict the amounts of products that will form in a chemical reaction based on the amounts of reactants that undergo the reaction. Stoichiometry also allows us to determine the amount of reactants necessary to form a given amount of product. These calculations are central to chemistry, allowing chemists to plan and carry out chemical reactions to obtain products in the desired quantities.

Stoichiometry is pronounced stoy-kee-AHM-e-tree.

Making Pizza: The Relationships among Ingredients

The concepts of stoichiometry are similar to those in a cooking recipe. Calculating the amount of carbon dioxide produced by the combustion of a given amount of a fossil fuel is analogous to calculating the number of pizzas that can be made from a given amount of cheese. For example, suppose we use the following pizza recipe:

$$1\text{ crust } + 5\text{ ounces tomato sauce } + 2\text{ cups cheese } \longrightarrow 1\text{ pizza}$$

The recipe shows the numerical relationships between the pizza ingredients. It says that if we have 2 cups of cheese—and enough of everything else—we can make 1 pizza. We can write this relationship as a ratio between the cheese and the pizza:

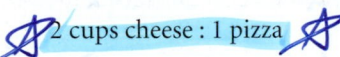

$$2\text{ cups cheese} : 1\text{ pizza}$$

What if we have 6 cups of cheese? Assuming that we have enough of everything else, we can use the above ratio as a conversion factor to calculate the number of pizzas:

$$6\text{ cups cheese} \times \frac{1\text{ pizza}}{2\text{ cups cheese}} = 3\text{ pizzas}$$

Six cups of cheese are sufficient to make 3 pizzas. The pizza recipe contains numerical ratios between other ingredients as well, including the following:

$$1\text{ crust} : 1\text{ pizza}$$
$$5\text{ ounces tomato sauce} : 1\text{ pizza}$$

Making Molecules: Mole-to-Mole Conversions

In a balanced chemical equation, we have a "recipe" for how reactants combine to form products. From our balanced equation for the combustion of octane, for example, we can write the following stoichiometric ratio:

$$2\text{ mol } C_8H_{18}(l) : 16\text{ mol } CO_2(g)$$

We can use this ratio to determine how many moles of CO_2 are produced for a given number of moles of C_8H_{18} burned. Suppose that we burn 22.0 moles of C_8H_{18}; how many moles of CO_2 are produced? We can use the ratio from the balanced chemical equation in the same way that we used the ratio from the pizza recipe. This ratio acts as a conversion factor allowing us to convert from the amount in moles of the reactant (C_8H_{18}) to the amount in moles of the product (CO_2):

$$22.0 \text{ mol } C_8H_{18} \times \frac{16 \text{ mol } CO_2}{2 \text{ mol } C_8H_{18}} = 176 \text{ mol } CO_2$$

The combustion of 22 moles of C_8H_{18} adds 176 moles of CO_2 to the atmosphere.

Making Molecules: Mass-to-Mass Conversions

According to the U.S. Department of Energy, the world burned 3.0×10^{10} barrels of petroleum in 2004, the equivalent of approximately 3.4×10^{15} g of gasoline. Let's estimate the mass of CO_2 emitted into the atmosphere from burning this much gasoline by using the combustion of 3.4×10^{15} g octane as the representative reaction. This calculation is similar to the one we just did, except that we are given the *mass* of octane instead of the *amount* of octane in moles. Consequently, we must first convert the mass (in grams) to the amount (in moles). The general conceptual plan for calculations in which you are given the mass of a reactant or product in a chemical reaction and asked to find the mass of a different reactant or product is as follows:

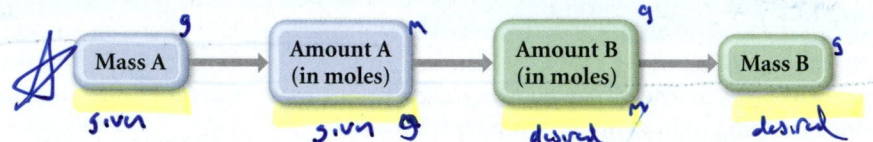

where A and B are two different substances involved in the reaction. We use the molar mass of A to convert from the mass of A to the amount of A (in moles). We use the appropriate ratio from the balanced chemical equation to convert from the amount of A (in moles) to the amount of B (in moles). And finally, we use the molar mass of B to convert from the amount of B (in moles) to the mass of B. To calculate the mass of CO_2 emitted upon the combustion of 3.4×10^{15} g of octane, therefore, we use the following conceptual plan:

Conceptual Plan

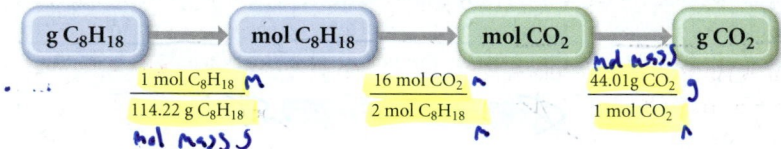

Mass-to-Mass Conversion

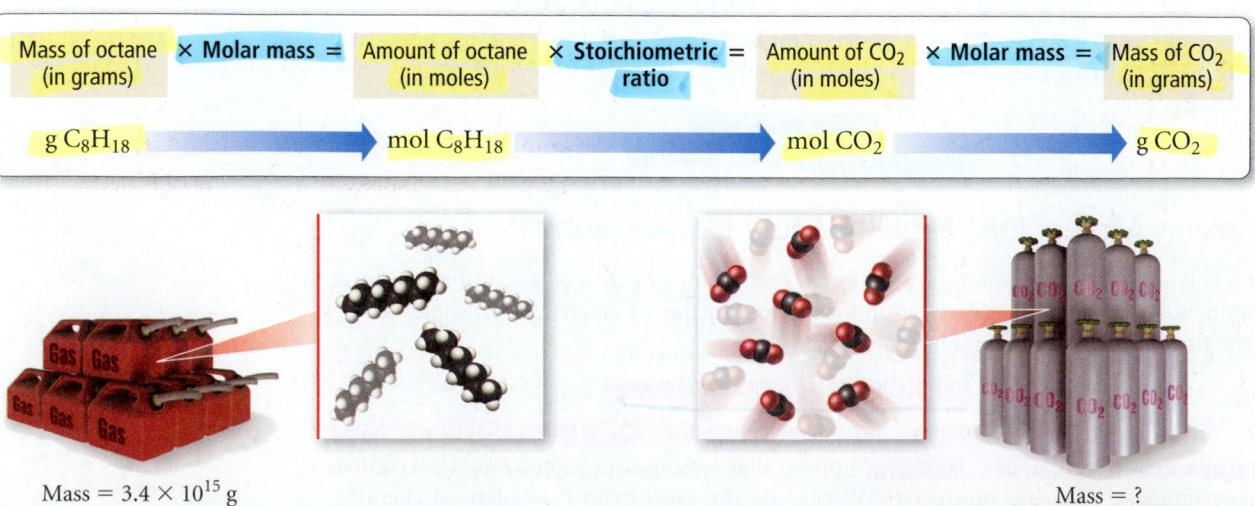

| Mass of octane (in grams) | × Molar mass = | Amount of octane (in moles) | × Stoichiometric ratio = | Amount of CO_2 (in moles) | × Molar mass = | Mass of CO_2 (in grams) |

g C_8H_{18} → mol C_8H_{18} → mol CO_2 → g CO_2

Mass = 3.4×10^{15} g

Mass = ?

Relationships Used

2 mol C_8H_{18} : 16 mol CO_2 (from the chemical equation)

molar mass C_8H_{18} = 114.22 g/mol

molar mass CO_2 = 44.01 g/mol

Solution

We then follow the conceptual plan to solve the problem, beginning with g C_8H_{18} and canceling units to arrive at g CO_2:

$$3.4 \times 10^{15} \text{ g } C_8H_{18} \times \frac{1 \text{ mol } C_8H_{18}}{114.22 \text{ g } C_8H_{18}} \times \frac{16 \text{ mol } CO_2}{2 \text{ mol } C_8H_{18}} \times \frac{44.01 \text{ g } CO_2}{1 \text{ mol } CO_2} = 1.0 \times 10^{16} \text{ g } CO_2$$

The world's petroleum combustion produces 1.0×10^{16} g CO_2 (1.0×10^{13} kg) per year. In comparison, volcanoes produce about 2×10^{11} kg CO_2 per year.* In other words,

volcanoes emit only $\dfrac{2.0 \times 10^{11} \text{ kg}}{1.0 \times 10^{13} \text{ kg}} \times 100\% = 2.0\%$ as much CO_2 per year as petroleum

combustion. The argument that volcanoes emit more carbon dioxide than fossil fuel combustion is blatantly mistaken. Additional examples of stoichiometric calculations follow.

> The percentage of CO_2 emitted by volcanoes relative to all fossil fuels is even less than 2% because CO_2 is also emitted by the combustion of coal and natural gas.

EXAMPLE 4.1 Stoichiometry

In photosynthesis, plants convert carbon dioxide and water into glucose ($C_6H_{12}O_6$) according to the following reaction:

$$6\, CO_2(g) + 6\, H_2O(l) \xrightarrow{\text{sunlight}} 6\, O_2(g) + C_6H_{12}O_6(aq)$$

Suppose you determine that a particular plant consumes 37.8 g CO_2 in one week. Assuming that there is more than enough water present to react with all of the CO_2, what mass of glucose (in grams) can the plant synthesize from the CO_2?

Sort The problem gives the mass of carbon dioxide and asks you to find the mass of glucose that can be produced.

Given 37.8 g CO_2

Find g $C_6H_{12}O_6$

Strategize The conceptual plan follows the general pattern of mass A → amount A (in moles) → amount B (in moles) → mass B. From the chemical equation, you can deduce the relationship between moles of carbon dioxide and moles of glucose. Use the molar masses to convert between grams and moles.

Conceptual Plan

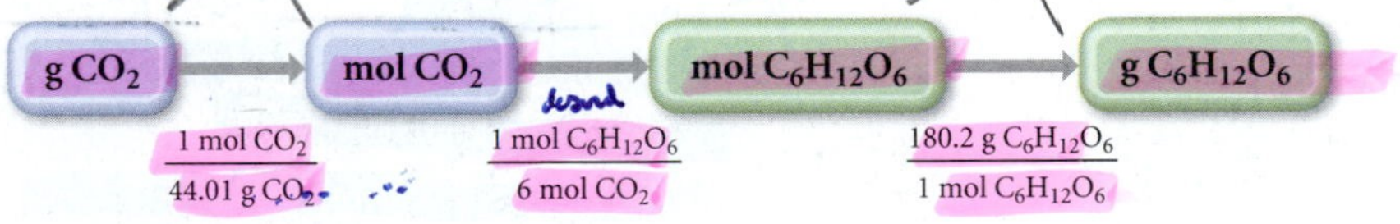

g CO_2	mol CO_2	mol $C_6H_{12}O_6$	g $C_6H_{12}O_6$
	$\dfrac{1 \text{ mol } CO_2}{44.01 \text{ g } CO_2}$	$\dfrac{1 \text{ mol } C_6H_{12}O_6}{6 \text{ mol } CO_2}$	$\dfrac{180.2 \text{ g } C_6H_{12}O_6}{1 \text{ mol } C_6H_{12}O_6}$

Relationships Used

molar mass CO_2 = 44.01 g/mol

6 mol CO_2 : 1 mol $C_6H_{12}O_6$

molar mass $C_6H_{12}O_6$ = 180.2 g/mol

Solve Follow the conceptual plan to solve the problem. Begin with g CO_2 and use the conversion factors to arrive at g $C_6H_{12}O_6$.

Solution

$$37.8 \text{ g } CO_2 \times \frac{1 \text{ mol } CO_2}{44.01 \text{ g } CO_2} \times \frac{1 \text{ mol } C_6H_{12}O_6}{6 \text{ mol } CO_2} \times \frac{180.2 \text{ g } C_6H_{12}O_6}{1 \text{ mol } C_6H_{12}O_6} = 25.8 \text{ g } C_6H_{12}O_6$$

Check The units of the answer are correct. The magnitude of the answer (25.8 g) is less than the initial mass of CO_2 (37.8 g). This is reasonable because each carbon in CO_2 has two oxygen atoms associated with it, while in $C_6H_{12}O_6$ each carbon has only one oxygen atom associated with it and two hydrogen atoms, which are much lighter than oxygen. Therefore the mass of glucose produced should be less than the mass of carbon dioxide for this reaction.

For Practice 4.1

Magnesium hydroxide, the active ingredient in milk of magnesia, neutralizes stomach acid, primarily HCl, according to the following reaction:

$$Mg(OH)_2(aq) + 2\,HCl(aq) \longrightarrow 2\,H_2O(l) + MgCl_2(aq)$$

What mass of HCl, in grams, can be neutralized by a dose of milk of magnesia containing 3.26 g $Mg(OH)_2$?

EXAMPLE 4.2 Stoichiometry

Sulfuric acid (H_2SO_4) is a component of acid rain that forms when SO_2, a pollutant, reacts with oxygen and water according to the following simplified reaction:

$$2\,SO_2(g) + O_2(g) + 2\,H_2O(l) \longrightarrow 2\,H_2SO_4(aq)$$

The generation of the electricity needed to power a medium-sized home produces about 25 kg of SO_2 per year. Assuming that there is more than enough O_2 and H_2O, what mass of H_2SO_4, in kg, can form from this much SO_2?

Sort The problem gives the mass of sulfur dioxide and asks you to find the mass of sulfuric acid.	**Given** 25 kg SO_2 **Find** kg H_2SO_4

Strategize The conceptual plan follows the standard format of mass $\rightarrow$ amount (in moles) $\rightarrow$ amount (in moles) $\rightarrow$ mass. Since the original quantity of SO_2 is given in kg, you must first convert to grams. You can deduce the relationship between moles of sulfur dioxide and moles of sulfuric acid from the chemical equation. Since the final quantity is requested in kg, convert to kg at the end.	**Conceptual Plan** 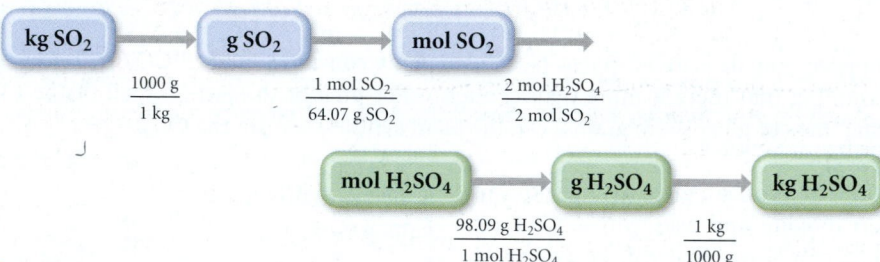 **Relationships Used** 1 kg = 1000 g molar mass SO_2 = 64.07 g/mol 2 mol SO_2 : 2 mol H_2SO_4 molar mass H_2SO_4 = 98.09 g/mol

Solve Follow the conceptual plan to solve the problem. Begin with the given amount of SO_2 in kilograms and use the conversion factors to arrive at kg H_2SO_4.	**Solution** $25\ \text{kg SO}_2 \times \dfrac{1000\ \text{g}}{1\ \text{kg}} \times \dfrac{1\ \text{mol SO}_2}{64.07\ \text{g SO}_2} \times \dfrac{2\ \text{mol H}_2\text{SO}_4}{2\ \text{mol SO}_2}$ $\times \dfrac{98.09\ \text{g H}_2\text{SO}_4}{1\ \text{mol H}_2\text{SO}_4} \times \dfrac{1\ \text{kg}}{1000\ \text{g}} = 38\ \text{kg H}_2\text{SO}_4$

Check The units of the final answer are correct. The magnitude of the final answer (38 kg H_2SO_4) is larger than the amount of SO_2 given (25 kg). This is reasonable because in the reaction each SO_2 molecule "gains weight" by reacting with O_2 and H_2O.

For Practice 4.2

Another component of acid rain is nitric acid, which forms when NO_2, also a pollutant, reacts with oxygen and water according to the following simplified equation:

$$4\,NO_2(g) + O_2(g) + 2\,H_2O(l) \longrightarrow 4\,HNO_3(aq)$$

The generation of the electricity needed to power a medium-sized home produces about 16 kg of NO_2 per year. Assuming that there is plenty of O_2 and H_2O, what mass of HNO_3, in kg, can form from this amount of NO_2 pollutant?

Conceptual Connection 4.1 **Stoichiometry**

Sodium reacts with oxygen to form sodium oxide according to the following reaction:

$$4\,Na(s) + O_2(g) \longrightarrow 2\,Na_2O(s)$$

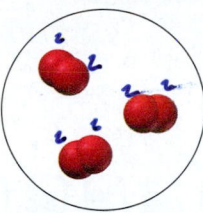

A flask contains the amount of oxygen represented by the diagram on the right ▶.

Which of the following best represents the amount of sodium required to completely react with all of the oxygen in the flask?

$2Na \overset{?}{=} 1O_2$

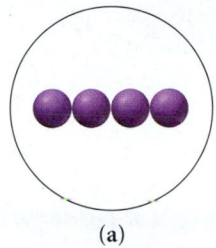   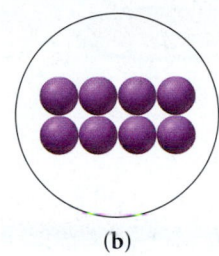   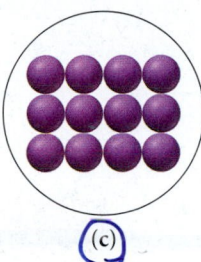   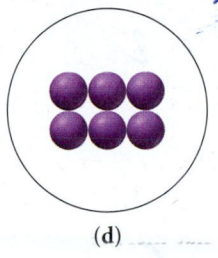

(a)   (b)   (c)   (d)

Answer: **(c)** Since each O_2 molecule reacts with 4 Na atoms, 12 Na atoms are required to react with 3 O_2 molecules.

4.3 Limiting Reactant, Theoretical Yield, and Percent Yield

Let's return to our pizza analogy to understand three more concepts important in reaction stoichiometry: *limiting reactant, theoretical yield,* and *percent yield*. Recall our pizza recipe from Section 4.2:

$$1\text{ crust} + 5\text{ ounces tomato sauce} + 2\text{ cups cheese} \rightarrow 1\text{ pizza}$$

Suppose that we have 4 crusts, 10 cups of cheese, and 15 ounces of tomato sauce. How many pizzas can we make?

We have enough crusts to make:

$$4\text{ crusts} \times \frac{1\text{ pizza}}{1\text{ crust}} = 4\text{ pizzas}$$

We have enough cheese to make:

$$10\text{ cups cheese} \times \frac{1\text{ pizza}}{2\text{ cups cheese}} = 5\text{ pizzas}$$

We have enough tomato sauce to make:

$$15\text{ ounces tomato sauce} \times \frac{1\text{ pizza}}{5\text{ ounces tomato sauce}} = 3\text{ pizzas}$$

Limiting reactant   Smallest number of pizzas

We have enough crusts for 4 pizzas, enough cheese for 5 pizzas, but enough tomato sauce for only 3 pizzas. Consequently, unless we get more ingredients, we can make only 3 pizzas. The tomato sauce *limits* how many pizzas we can make. If this were a chemical reaction, the tomato sauce would be the **limiting reactant,** the reactant that limits the amount of product in a chemical reaction. Notice that the limiting reactant is simply the reactant that makes *the least amount of product*. The reactants that *do not* limit the amount of product—such as the crusts and the cheese in this example—are said to be *in excess*. If this were a chemical reaction, 3 pizzas would be the **theoretical yield,** the amount of product that can be made in a chemical reaction based on the amount of limiting reactant.

The term *limiting reagent* is sometimes used in place of *limiting reactant*.

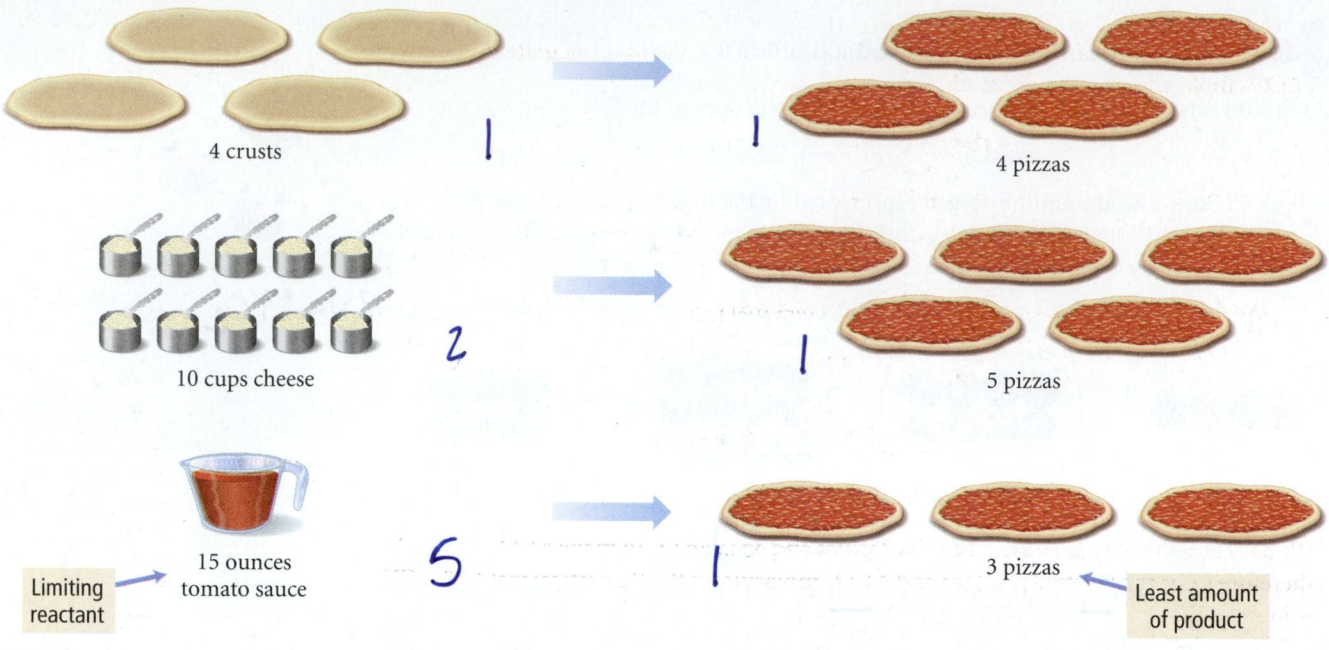

4 crusts

I

4 pizzas

10 cups cheese

2

I

5 pizzas

Limiting
reactant

15 ounces
tomato sauce

5   *I*

3 pizzas

Least amount
of product

▲ The ingredient that makes the least amount of pizza will determine how many pizzas you can make.

Let us carry this analogy one step further. Suppose we go on to cook our pizzas and accidentally burn one of them. So even though we theoretically have enough ingredients for 3 pizzas, we end up with only 2. If this were a chemical reaction, the 2 pizzas would be our **actual yield**, the amount of product actually produced by a chemical reaction. (The actual yield is always equal to or less than the theoretical yield because a small amount of product is usually lost to other reactions or does not form during a reaction.) Finally, our **percent yield**, the percentage of the theoretical yield that was actually attained, is calculated as follows:

$$\% \text{ yield} = \frac{2 \text{ pizzas } (actual)}{3 \text{ pizzas } (theoretical)} \times 100\% = 67\%$$

Since one of our pizzas burned, we obtained only 67% of our theoretical yield.

Summarizing:

➤ **The limiting reactant** (or **limiting reagent**) is the reactant that is completely consumed in a chemical reaction and limits the amount of product.

➤ **The reactant in excess** is any reactant that occurs in a quantity greater than that required to completely react with the limiting reactant.

➤ **The theoretical yield** is the amount of product that can be made in a chemical reaction based on the amount of limiting reactant.

➤ **The actual yield** is the amount of product actually produced by a chemical reaction.

➤ **The percent yield** is calculated as $\dfrac{\text{actual yield}}{\text{theoretical yield}} \times 100\%$.

Now let's apply these concepts to a chemical reaction. Recall from Section 3.10 our balanced equation for the combustion of methane:

$$\overset{I}{CH_4(g)} + \overset{2}{2\,O_2(g)} \longrightarrow \overset{I}{CO_2(g)} + 2\,H_2O(l)$$

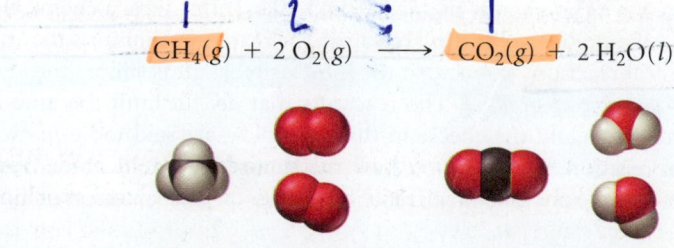

If we start out with 5 CH$_4$ molecules and 8 O$_2$ molecules, what is our limiting reactant? What is our theoretical yield of carbon dioxide molecules? We first calculate the number of CO$_2$ molecules that can be made from 5 CH$_4$ molecules:

$$5 \text{ CH}_4 \times \frac{1 \text{ CO}_2}{1 \text{ CH}_4} = 5 \text{ CO}_2$$

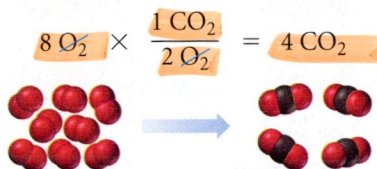

We then calculate the number of CO$_2$ molecules that can be made from 8 O$_2$ molecules:

$$8 \text{ O}_2 \times \frac{1 \text{ CO}_2}{2 \text{ O}_2} = 4 \text{ CO}_2$$

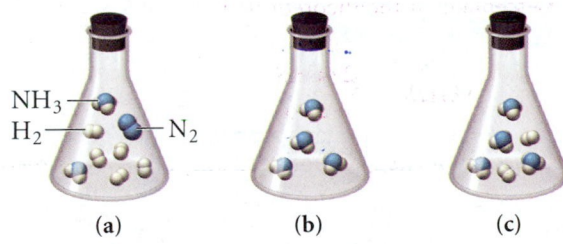

We have enough CH$_4$ to make 5 CO$_2$ molecules and enough O$_2$ to make 4 CO$_2$ molecules; therefore O$_2$ is the limiting reactant and 4 CO$_2$ molecules is the theoretical yield. The CH$_4$ is in excess.

 ## Conceptual Connection 4.2 Limiting Reactant and Theoretical Yield

Nitrogen and hydrogen gas react to form ammonia according to the following reaction:

$$\overset{1}{\text{N}_2(g)} + \overset{3}{3\text{H}_2(g)} \overset{:}{\longrightarrow} \overset{2}{2\text{NH}_3(g)}$$

If a flask contains a mixture of reactants represented by the diagram on the right ▶, which of the following best represents the mixture after the reactants have reacted as completely as possible? What is the limiting reactant? Which reactant is in excess?

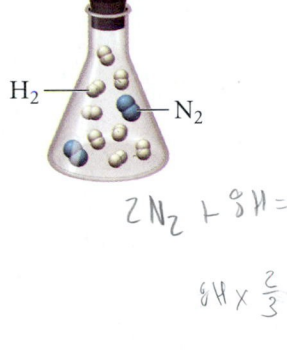

$$2 N_2 + 8 H =$$

$$8 H \times \frac{2}{3}$$

(a)   (b)   (c)

Answer: (c) Nitrogen is the limiting reactant and there is enough nitrogen to make 4 NH$_3$ molecules. Hydrogen is in excess and two hydrogen molecules remain after the reactants have reacted as completely as possible.

Limiting Reactant, Theoretical Yield, and Percent Yield from Initial Reactant Masses

When working in the laboratory, we normally measure the initial quantities of reactants in grams, not in number of molecules. To find the limiting reactant and theoretical yield from initial masses, we must first convert the masses to amounts in moles. Consider, for example, the following reaction:

$$\overset{2}{2\text{Mg}(s)} + \overset{1}{\text{O}_2(g)} \overset{:}{\longrightarrow} \overset{2}{2\text{MgO}(s)}$$

If we have 42.5 g Mg and 33.8 g O$_2$, what is the limiting reactant and theoretical yield?

To solve this problem, we must determine which of the reactants makes the least amount of product.

Conceptual Plan

We find the limiting reactant by calculating how much product can be made from each reactant. However, since we are given the initial quantities in grams, and stoichiometric relationships are between moles, we must first convert to moles. We then convert from moles of the reactant to moles of product. The reactant that makes the *least amount of product* is the limiting reactant. The conceptual plan is as follows:

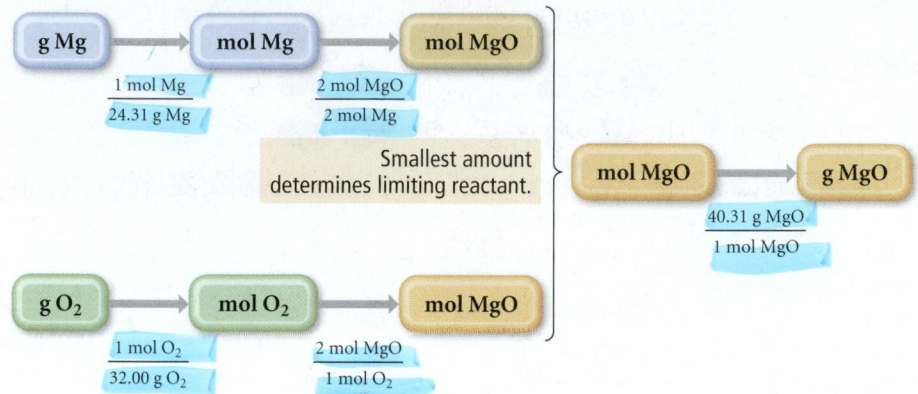

In the above plan, we compare the number of moles of MgO made by each reactant, and convert only the smaller amount to grams. (Alternatively, you can convert both quantities to grams and determine the limiting reactant based on the mass of the product.)

Relationships Used

molar mass Mg $= 24.31$ g Mg

molar mass $O_2 = 32.00$ g O_2

2 mol Mg : 2 mol MgO

1 mol O_2 : 2 mol MgO

molar mass MgO $= 40.31$ g MgO

Solution

Beginning with the masses of each reactant, we follow the conceptual plan to calculate how much product can be made from each:

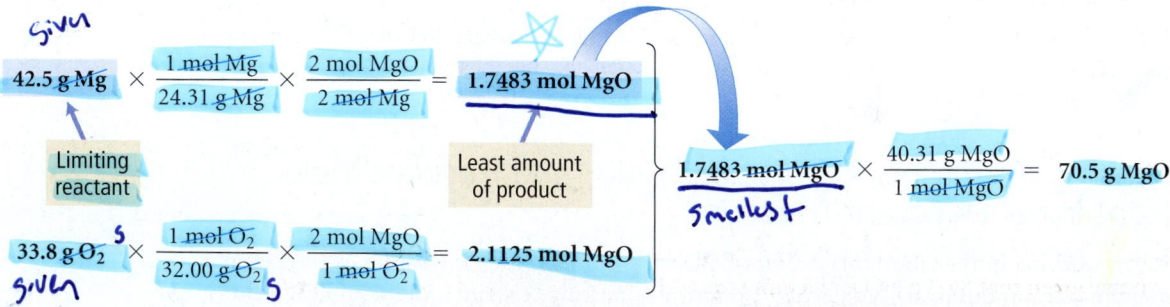

Since Mg makes the least amount of product, it is the limiting reactant and O_2 is in excess. Notice that the limiting reactant is not necessarily the reactant with the least mass. In this case, the mass of O_2 is less than the mass of Mg, yet Mg is the limiting reactant because it makes the least amount of MgO. The theoretical yield is therefore 70.5 g of MgO, the mass of product possible based on the limiting reactant.

Now suppose that when the synthesis is carried out, the actual yield of MgO is 55.9 g. What is the percent yield? The percent yield is computed as follows:

$$\% \text{ yield} = \frac{\text{actual yield}}{\text{theoretical yield}} \times 100\% = \frac{55.9 \text{ g}}{70.5 \text{ g}} \times 100\% = 79.3\%$$

EXAMPLE 4.3 Limiting Reactant and Theoretical Yield

Ammonia, NH_3, can be synthesized by the following reaction:

$$2\,NO(g) + 5\,H_2(g) \longrightarrow 2\,NH_3(g) + 2\,H_2O(g)$$

Starting with 86.3 g NO and 25.6 g H_2, find the theoretical yield of ammonia in grams.

Sort You are given the mass of each reactant in grams and asked to find the theoretical yield of a product.

Given 86.3 g NO, 25.6 g H_2

Find theoretical yield of NH_3 (g)

Strategize Determine which reactant makes the least amount of product by converting from grams of each reactant to moles of the reactant to moles of the product. Use molar masses to convert between grams and moles and use the stoichiometric relationships (deduced from the chemical equation) to convert between moles of reactant and moles of product. The reactant that makes *the least amount of product* is the limiting reactant. Convert the number of moles of product obtained using the limiting reactant to grams of product.

Conceptual Plan

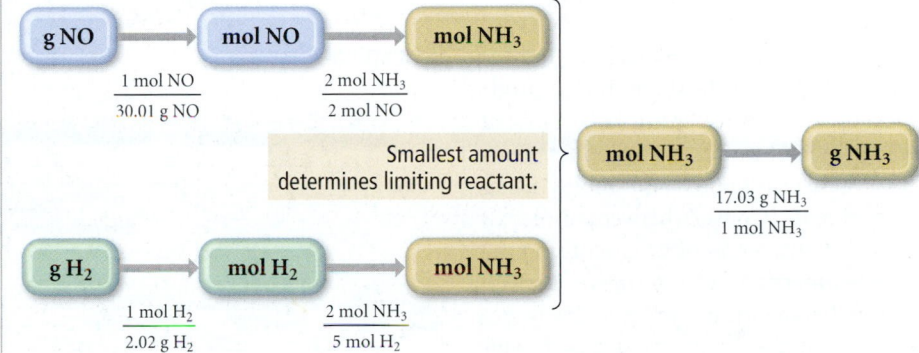

Relationships Used

molar mass NO = 30.01 g/mol

molar mass H_2 = 2.02 g/mol

2 mol NO : 2 mol NH_3 (from chemical equation)

5 mol H_2 : 2 mol NH_3 (from chemical equation)

molar mass NH_3 = 17.03 g/mol

Solve Beginning with the given mass of each reactant, calculate the amount of product that can be made in moles. Convert the amount of product made by the limiting reactant to grams—this is the theoretical yield.

Solution

limiting

$$86.3\,\text{g NO} \times \frac{1\,\text{mol NO}}{30.01\,\text{g NO}} \times \frac{2\,\text{mol NH}_3}{2\,\text{mol NO}} = 2.8757\,\text{mol NH}_3$$

Limiting reactant   Least amount of product

$$2.8757\,\text{mol NH}_3 \times \frac{17.03\,\text{g NH}_3}{\text{mol NH}_3} = 49.0\,\text{g NH}_3$$

Theoretical

$$25.6\,\text{g H}_2 \times \frac{1\,\text{mol H}_2}{2.02\,\text{g H}_2} \times \frac{2\,\text{mol NH}_3}{5\,\text{mol H}_2} = 5.0693\,\text{mol NH}_3$$

Since NO makes the least amount of product, it is limiting reactant, and the theoretical yield of ammonia is 49.0 g.

Check The units of the answer (g NH_3) are correct. The magnitude (49.0 g) seems reasonable given that 86.3 g NO is the limiting reactant. NO contains one oxygen atom per nitrogen atom and NH_3 contains three hydrogen atoms per nitrogen atom. Since three hydrogen atoms have less mass than one oxygen atom, it is reasonable that the mass of NH_3 obtained is less than the mass of NO.

For Practice 4.3

Ammonia can also be synthesized by the following reaction:

$$3\,H_2(g) + N_2(g) \longrightarrow 2\,NH_3(g)$$

What is the theoretical yield of ammonia, in kg, that can be synthesized from 5.22 kg of H_2 and 31.5 kg of N_2?

EXAMPLE 4.4 Limiting Reactant and Theoretical Yield

Titanium metal can be obtained from its oxide according to the following balanced equation:

$$TiO_2(s) + 2\,C(s) \longrightarrow Ti(s) + 2\,CO(g)$$

When 28.6 kg of C is allowed to react with 88.2 kg of TiO_2, 42.8 kg of Ti is produced. Find the limiting reactant, theoretical yield (in kg), and percent yield.

Sort You are given the mass of each reactant and the mass of product formed. You are asked to find the limiting reactant, theoretical yield and percent yield.	**Given** 28.6 kg C, 88.2 kg TiO_2, 42.8 kg Ti produced **Find** limiting reactant, theoretical yield, % yield

Strategize Determine which of the reactants makes the least amount of product by converting from kilograms of each reactant to moles of product. Convert between grams and moles using molar mass. Convert between moles of reactant and moles of product using the stoichiometric relationships derived from the chemical equation. The reactant that makes the *least amount of product* is the limiting reactant.

Determine the theoretical yield (in kg) by converting the number of moles of product obtained with the limiting reactant to kilograms of product.

Conceptual Plan

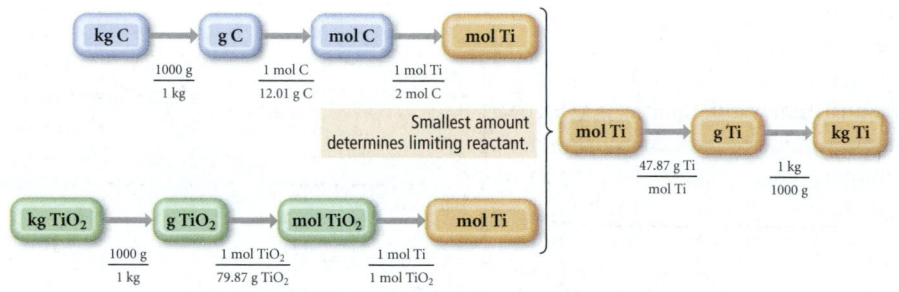

Relationships Used

1000 g = 1 kg

molar mass of C = 12.01 g/mol

molar mass of TiO_2 = 79.87 g/mol

1 mol TiO_2 : 1 mol Ti

2 mol C : 1 mol Ti

molar mass of Ti = 47.87 g/mol

Solve Beginning with the actual amount of each reactant, calculate the amount of product that can be made in moles. Convert the amount of product made by the limiting reactant to kilograms—this is the theoretical yield.

Solution

$$28.6\ \text{kg C} \times \frac{1000\ \text{g}}{1\ \text{kg}} \times \frac{1\ \text{mol C}}{12.01\ \text{g C}} \times \frac{1\ \text{mol Ti}}{2\ \text{mol C}} = 1.1907 \times 10^3\ \text{mol Ti}$$

Limiting reactant

Least amount of product

$$88.2\ \text{kg TiO}_2 \times \frac{1000\ \text{g}}{1\ \text{kg}} \times \frac{1\ \text{mol TiO}_2}{79.87\ \text{g TiO}_2} \times \frac{1\ \text{mol Ti}}{1\ \text{mol TiO}_2} = 1.1043 \times 10^3\ \text{mol Ti}$$

$$1.1043 \times 10^3\ \text{mol Ti} \times \frac{47.87\ \text{g Ti}}{1\ \text{mol Ti}} \times \frac{1\ \text{kg}}{1000\ \text{g}} = 52.9\ \text{kg Ti}$$

Since TiO_2 makes the least amount of product, it is the limiting reactant and 52.9 kg Ti is the theoretical yield.

Calculate the percent yield by dividing the actual yield (42.8 kg Ti) by the theoretical yield.	% yield = $\dfrac{\text{actual yield}}{\text{theoretical yield}} \times 100\%$ $= \dfrac{42.8\ \text{g}}{52.9\ \text{g}} \times 100\% = 80.9\%$

Check The theoretical yield has the correct units (kg Ti) and has a reasonable magnitude compared to the mass of TiO_2. Since Ti has a lower molar mass than TiO_2, the amount of Ti made from TiO_2 should have a lower mass. The percent yield is reasonable (under 100% as it should be).

For Practice 4.4

The following reaction is used to obtain iron from iron ore:

$$Fe_2O_3(s) + 3\,CO(g) \longrightarrow 2\,Fe(s) + 3\,CO_2(g)$$

The reaction of 167 g Fe_2O_3 with 85.8 g CO produces 72.3 g Fe. Find the limiting reactant, theoretical yield, and percent yield. *actual*

Chemistry in the Environment
MTBE in Gasoline

We have seen that the balanced chemical equation for the combustion of octane, a component of gasoline, is as follows:

$$2\,C_8H_{18}(l) + 25\,O_2(g) \longrightarrow 16\,CO_2(g) + 18\,H_2O(g)$$

The above equation shows that 25 moles of O_2 are required to completely react with 2 moles of C_8H_{18}. What if there is not enough O_2 in an automobile cylinder to fully react with the amount of octane flowing into it? For many reactions, a shortage of one reactant simply means that less product forms—in other words, oxygen becomes the limiting reactant. However, for some reactions, a shortage of one reactant causes side reactions to occur along with the desired reaction. In the case of octane and the other major components of gasoline, those side reactions result in pollutants such as carbon monoxide (CO) and unburned hydrocarbon fragments that lead to the formation of ozone (O_3).

In 1990, the U. S. Congress, in efforts to lower air pollution, passed amendments to the Clean Air Act requiring oil companies to add substances to gasoline that prevent these side reactions. Because these additives have the effect of increasing the amount of oxygen during combustion, the resulting gasoline is called an oxygenated fuel. The additive of choice among oil companies was a compound called MTBE (methyl tertiary butyl ether).

◀ MTBE, a gasoline additive that promotes complete combustion.

The immediate results of adding MTBE to gasoline were positive. Carbon monoxide and ozone levels in many major cities decreased significantly.

Over time, however, MTBE—a compound that does not readily biodegrade—began to appear in drinking water supplies across the nation. MTBE got into drinking water through gasoline spills at gas stations, from boat motors, and from leaking underground storage tanks. The consequences have been significant. MTBE, even at low levels, imparts a turpentinelike odor and foul taste to drinking water. It is also a suspected carcinogen.

Public response has been swift and dramatic. Several class action lawsuits have been filed against the manufacturers of MTBE, against gas stations suspected of leaking it, and against the oil companies that put MTBE into gasoline. Many states have banned MTBE from gasoline and the U.S. Congress is in the process of imposing a federal ban on MTBE. This does raise a problem, however. MTBE was added to gasoline as a way to meet the requirements of the 1990 Clean Air Act amendments. Should the federal government temporarily remove these requirements until a substitute for MTBE is found? Or should the government completely remove the requirements, weakening the Clean Air Act? The answers are not simple. However, ethanol, made from the fermentation of grains, is a ready substitute for MTBE that has many of the same pollution-reducing effects without the associated health hazards.

Question

How many kilograms of oxygen (O_2) is required to completely react with 48 kg of octane (approximate capacity of a 15-gallon automobile gasoline tank)?

▲ Oxygenated fuel contains compounds that increase the amount of oxygen available for combustion, reducing the formation of by-products such as carbon monoxide and hydrocarbon fragments.

4.4 Solution Concentration and Solution Stoichiometry

Chemical reactions in which the reactants are dissolved in water are among the most common and important. For example, the reactions that occur in lakes, streams, and oceans, as well as the reactions that occur in every cell of our bodies, take place in water. A homogeneous mixture of two substances—such as salt and water—is called a **solution.** The majority component of the mixture is called the **solvent,** and the minority component is called the **solute.** An **aqueous solution** is a solution in which water acts as the solvent. In this section, we first examine how to quantify the concentration of a solution (the amount of solute relative to solvent) and then turn to applying the principles of stoichiometry, which we learned in the previous section, to reactions occurring in solution.

Solution Concentration

Concentrated and Dilute Solutions

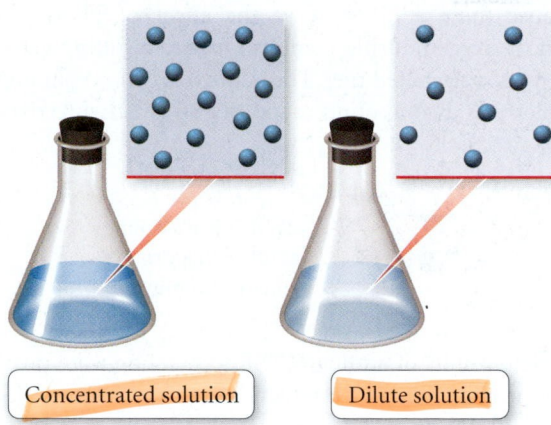

Concentrated solution   Dilute solution

▲ **FIGURE 4.4 Concentrated and Dilute Solutions** A concentrated solution contains a relatively large amount of solute relative to solvent. A dilute solution contains a relatively small amount of solute relative to solvent.

The amount of solute in a solution is variable. For example, you can add just a little salt to water to make a **dilute solution,** one that contains a small amount of solute relative to the solvent, or you can add a lot of salt to water to make a **concentrated solution,** one that contains a large amount of solute relative to the solvent (Figure 4.4 ◄). A common way to express solution concentration is **molarity (M),** the amount of solute (in moles) divided by the volume of solution (in liters).

$$\text{Molarity (M)} = \frac{\text{amount of solute (in mol)}}{\text{volume of solution (in L)}}$$

Note that molarity is a ratio of the amount of solute per liter of *solution,* not per liter of solvent. To make an aqueous solution of a specified molarity, you usually put the solute into a flask and then add water to the desired volume of solution. For example, to make 1 L of a 1 M NaCl solution, you add 1 mol of NaCl to a flask and then add water to make 1 L of solution (Figure 4.5 ▼). You *do not* combine 1 mol of NaCl with 1 L of water because the resulting solution would have a total volume exceeding 1 L and therefore a molarity of less than 1 M. To calculate molarity, simply divide the amount of the solute in moles by the volume of the solution (solute *and* solvent) in liters, as shown in the following example.

Preparing a Solution of Specified Concentration

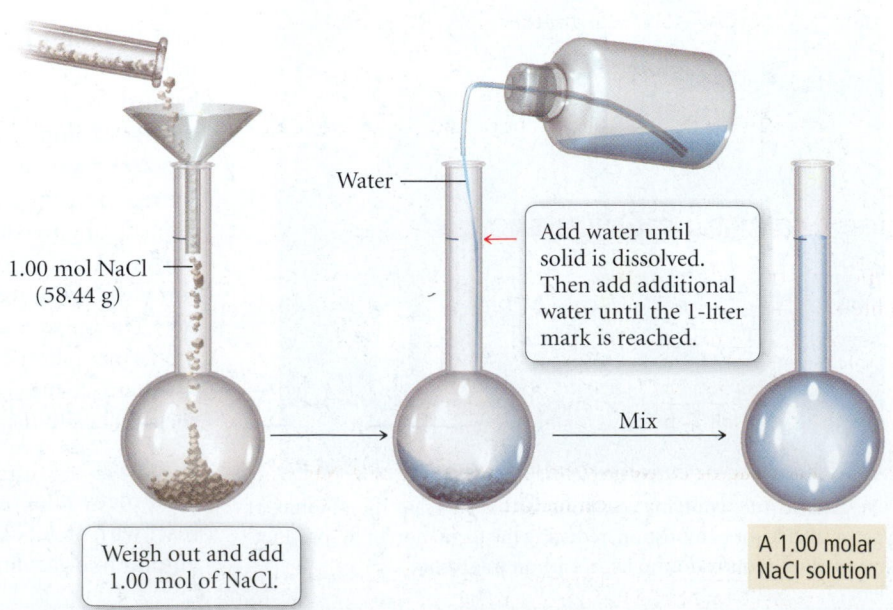

Water

1.00 mol NaCl (58.44 g)

Add water until solid is dissolved. Then add additional water until the 1-liter mark is reached.

Mix

Weigh out and add 1.00 mol of NaCl.

A 1.00 molar NaCl solution

▶ **FIGURE 4.5 Preparing a 1 Molar NaCl Solution**

EXAMPLE 4.5 Calculating Solution Concentration

If 25.5 g KBr is dissolved in enough water to make 1.75 L of solution, what is the molarity of the solution?

Sort You are given the mass of KBr and the volume of a solution and asked to find its molarity.	**Given** 25.5 g KBr, 1.75 L of solution **Find** molarity (M)

Strategize When formulating the conceptual plan, think about the definition of molarity, the amount of solute *in moles* per liter of solution.

You are given the mass of KBr, so first use the molar mass of KBr to convert from g KBr to mol KBr.

Then use the number of moles of KBr and liters of solution to find the molarity.

Conceptual Plan

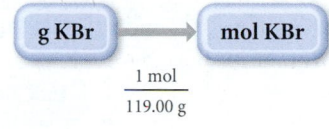

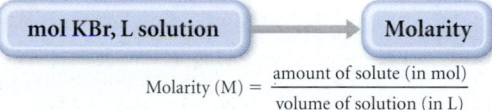

$$\text{Molarity (M)} = \frac{\text{amount of solute (in mol)}}{\text{volume of solution (in L)}}$$

Relationships Used molar mass of KBr = 119.00 g/mol

Solve Follow the conceptual plan. Begin with g KBr and convert to mol KBr, then use mol KBr and L solution to compute molarity.

Solution

$$25.5 \ \text{g KBr} \times \frac{1 \ \text{mol KBr}}{119.00 \ \text{g KBr}} = 0.21429 \ \text{mol KBr}$$

$$\text{molarity (M)} = \frac{\text{amount of solute (in mol)}}{\text{volume of solution (in L)}}$$

$$= \frac{0.21429 \ \text{mol KBr}}{1.75 \ \text{L solution}}$$

$$= 0.122 \ \text{M}$$

Check

The units of the answer (M) are correct. The magnitude is reasonable. Common solutions range in concentration from 0 to about 18 M. Concentrations significantly above 18 M are suspect and should be double-checked.

For Practice 4.5

Calculate the molarity of a solution made by adding 45.4 g of $NaNO_3$ to a flask and dissolving with water to a total volume of 2.50 L.

For More Practice 4.5

What mass of KBr (in grams) should be used to make 250.0 mL of a 1.50 M KBr solution?

Using Molarity in Calculations

The molarity of a solution can be used as a conversion factor between moles of the solute and liters of the solution. For example, a 0.500 M NaCl solution contains 0.500 mol NaCl for every liter of solution:

This conversion factor converts from L solution to mol NaCl. If you want to go the other way, simply invert the conversion factor:

The following example shows how to use molarity in this way.

EXAMPLE 4.6 Using Molarity in Calculations

How many liters of a 0.125 M NaOH solution contains 0.255 mol of NaOH?

Sort You are given the concentration of a NaOH solution. You are asked to find the volume of the solution that contains a given amount (in moles) of NaOH.	**Given** 0.125 M NaOH solution, 0.255 mol NaOH **Find** volume of NaOH solution (in L)
Strategize The conceptual plan begins with mol NaOH and shows the conversion to L of solution using the molarity as a conversion factor.	**Conceptual Plan** $$\boxed{\text{mol NaOH}} \longrightarrow \boxed{\text{L solution}}$$ $$\dfrac{1\ \text{L solution}}{0.125\ \text{mol NaOH}}$$ **Relationships Used** $0.125\ \text{M NaOH} = \dfrac{0.125\ \text{mol NaOH}}{1\ \text{L solution}}$
Solve Follow the conceptual plan. Begin with mol NaOH and convert to L solution.	**Solution** $$0.255\ \text{mol NaOH} \times \dfrac{1\ \text{L solution}}{0.125\ \text{mol NaOH}} = 2.04\ \text{L solution}$$

Check The units of the answer (L) are correct. The magnitude seems reasonable because the solution contains 0.125 mol per liter. Therefore, roughly 2 L contains the given amount of moles (0.255 mol).

For Practice 4.6

How many grams of sucrose ($C_{12}H_{22}O_{11}$) are contained in 1.55 L of 0.758 M sucrose solution?

For More Practice 4.6

How many mL of a 0.155 M KCl solution contains 2.55 g KCl?

 Conceptual Connection 4.3 Solutions

If 25 grams of salt is dissolved in 251 grams of water, what is the mass of the resulting solution?

(a) 251 g   (b) 276 g   (c) 226 g

Answer: (b) The mass of a solution is equal to the mass of the solute plus the mass of the solvent. Although the solute seems to disappear, it really does not, and its mass becomes part of the mass of the solution, in accordance with the law of mass conservation.

Solution Dilution To save space in laboratory storerooms, solutions are often stored in concentrated forms called **stock solutions**. For example, hydrochloric acid is often stored as a 12 M stock solution. However, many lab procedures call for much less concentrated hydrochloric acid solutions, so chemists must dilute the stock solution to the required concentration. How do we know how much of the stock solution to use? The easiest way to solve dilution problems is to use the following dilution equation:

When diluting acids, always add the concentrated acid to the water. Never add water to concentrated acid solutions, as the heat generated may cause the concentrated acid to splatter and burn your skin.

$$M_1 V_1 = M_2 V_2 \tag{4.1}$$

where M_1 and V_1 are the molarity and volume of the initial concentrated solution and M_2 and V_2 are the molarity and volume of the final diluted solution. This equation works because the molarity multiplied by the volume gives the number of moles of solute, which is the same in both solutions.

$$M_1 V_1 = M_2 V_2$$
$$mol_1 = mol_2$$

In other words, the number of moles of solute does not change when we dilute a solution.

For example, suppose a laboratory procedure calls for 3.00 L of a 0.500 M $CaCl_2$ solution. How should we prepare this solution from a 10.0 M stock solution? We can solve Equation 4.1 for V_1, the volume of the stock solution required for the dilution, and then substitute in the correct values to compute it.

$$M_1 V_1 = M_2 V_2$$

$$V_1 = \frac{M_2 V_2}{M_1}$$

$$= \frac{0.500 \ mol/L \times 3.00 \ L}{10.0 \ mol/L}$$

$$= 0.150 \ L$$

Consequently, we make the solution by diluting 0.150 L of the stock solution to a total volume of 3.00 L (V_2). The resulting solution will be 0.500 M in $CaCl_2$ (Figure 4.6 ▼).

Diluting a Solution

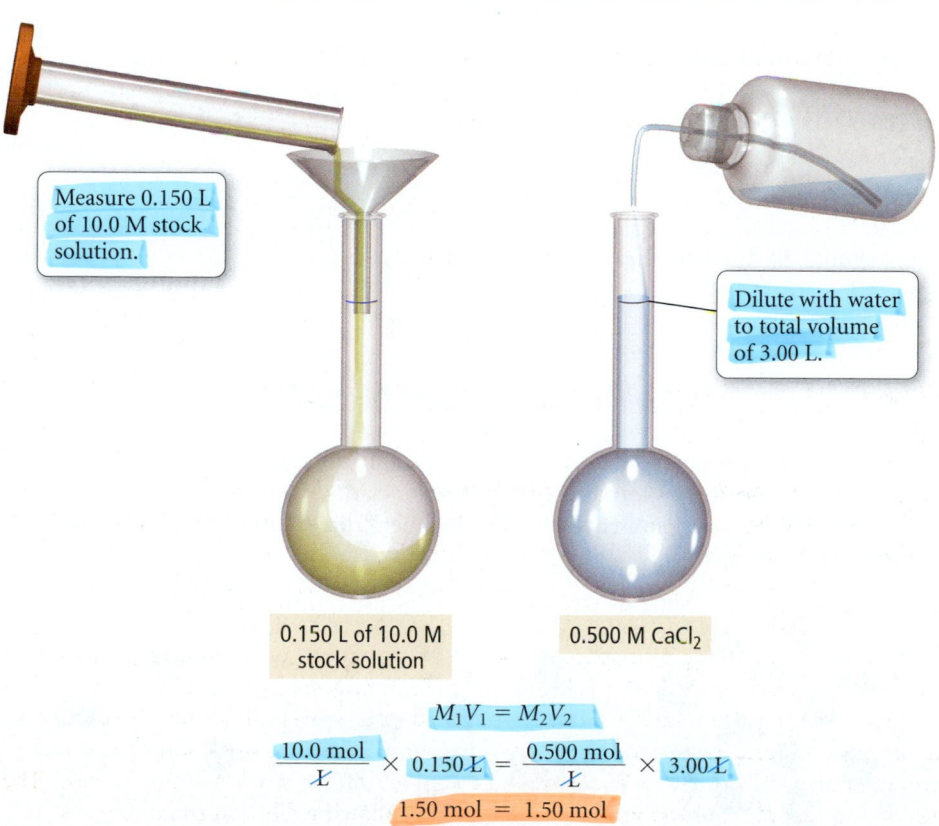

Measure 0.150 L of 10.0 M stock solution.

Dilute with water to total volume of 3.00 L.

0.150 L of 10.0 M stock solution

0.500 M $CaCl_2$

$$M_1 V_1 = M_2 V_2$$

$$\frac{10.0 \ mol}{L} \times 0.150 \ L = \frac{0.500 \ mol}{L} \times 3.00 \ L$$

$$1.50 \ mol = 1.50 \ mol$$

▲ **FIGURE 4.6 Preparing 3.00 L of 0.500 M $CaCl_2$ from a 10.0 M Stock Solution**

EXAMPLE 4.7 Solution Dilution

To what volume should you dilute 0.200 L of a 15.0 M NaOH solution to obtain a 3.00 M NaOH solution?

Sort You are given the initial volume, initial concentration, and final concentration of a solution, and you need to find the final volume.	**Given** $V_1 = 0.200$ L $M_1 = 15.0$ M $M_2 = 3.00$ M **Find** V_2
Strategize Equation 4.1 relates the initial and final volumes and concentrations for solution dilution problems. You are asked to find V_2. The other quantities (V_1, M_1, and M_2) are all given in the problem.	**Conceptual Plan** $\boxed{V_1, M_1, M_2} \longrightarrow \boxed{V_2}$ $\quad\quad\quad M_1V_1 = M_2V_2$ **Relationships Used** $M_1V_1 = M_2V_2$
Solve Begin with the solution dilution equation and solve it for V_2. Substitute in the required quantities and compute V_2. Make the solution by diluting 0.200 L of the stock solution to a total volume of 1.00 L (V_2). The resulting solution will have a concentration of 3.00 M.	**Solution** $M_1V_1 = M_2V_2$ $V_2 = \dfrac{M_1V_1}{M_2}$ $\quad = \dfrac{15.0 \text{ mol/L} \times 0.200 \text{ L}}{3.00 \text{ mol/L}}$ $\quad = 1.00$ L

Check The final units (L) are correct. The magnitude of the answer is reasonable because the solution is diluted from 15.0 M to 3.00 M, a factor of five. Therefore the volume should increase by a factor of five.

For Practice 4.7

To what volume (in mL) should you dilute 100.0 mL of a 5.00 M $CaCl_2$ solution to obtain a 0.750 M $CaCl_2$ solution?

For More Practice 4.7

What volume of a 6.00 M $NaNO_3$ solution should be used to make 0.525 L of a 1.20 M $NaNO_3$ solution?

✦ Conceptual Connection 4.4 Solution Dilution

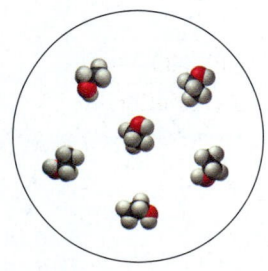

The figure at left represents a small volume within 500 mL of aqueous ethanol (CH_3CH_2OH) solution. (The water molecules have been omitted for clarity.)

Which picture best represents the same volume of the solution after adding an additional 500 mL of water?

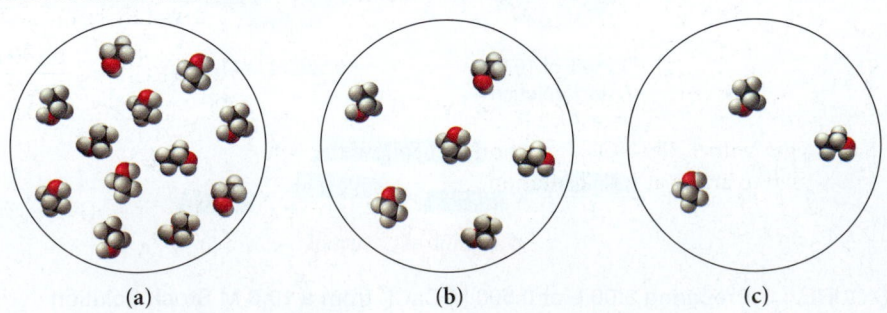

(a)   (b)   (c)

Answer: (c) Since the volume has doubled, the concentration is halved, so the same volume should contain half as many solute molecules.

Solution Stoichiometry

In Section 4.2 we learned how the coefficients in chemical equations are used as conversion factors between the amounts of reactants (in moles) and the amounts of products (in moles). In reactions involving aqueous reactants and products, it is often convenient to specify their quantities in terms of volumes and concentrations. We can then use these quantities to calculate the amounts in moles of reactants or products and use the stoichiometric coefficients to convert these to amounts of other reactants or products. The general conceptual plan for these kinds of calculations is as follows:

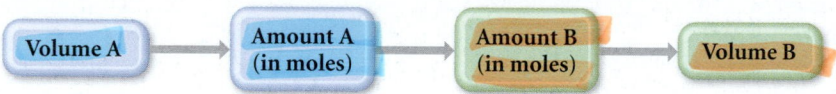

The conversions between solution volumes and amounts of solute in moles are made using the molarities of the solutions. The conversions between amounts in moles of A and B are made using the stoichiometric coefficients from the balanced chemical equation. The following example demonstrates solution stoichiometry.

EXAMPLE 4.8 Solution Stoichiometry

What volume (in L) of 0.150 M KCl solution is required to completely react with 0.150 L of a 0.175 M $Pb(NO_3)_2$ solution according to the following balanced chemical equation?

$$2 \, KCl(aq) + Pb(NO_3)_2(aq) \longrightarrow PbCl_2(s) + 2 \, KNO_3(aq)$$

Sort You are given the volume and concentration of a $Pb(NO_3)_2$ solution. You are asked to find the volume of KCl solution (of a given concentration) required to react with it.	**Given** 0.150 L of $Pb(NO_3)_2$ solution, 0.175 M $Pb(NO_3)_2$ solution, 0.150 M KCl solution **Find** volume KCl solution (in L)
Strategize The conceptual plan has the following form: volume A → amount A (in moles) → amount B (in moles) → volume B. The molar concentrations of the KCl and $Pb(NO_3)_2$ solutions can be used as conversion factors between the number of moles of reactants in these solutions and their volumes. The stoichiometric coefficients from the balanced equation are used to convert between number of moles of $Pb(NO_3)_2$ and number of moles of KCl.	**Conceptual Plan** 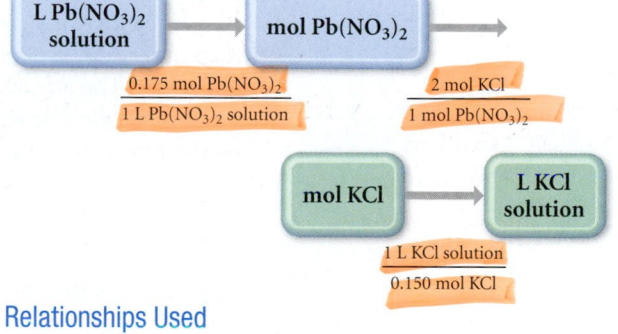 **Relationships Used** $$M\,[Pb(NO_3)_2] = \frac{0.175 \text{ mol } Pb(NO_3)_2}{1 \text{ L } Pb(NO_3)_2 \text{ solution}}$$ 2 mol KCl : 1 mol $Pb(NO_3)_2$ $$M\,(KCl) = \frac{0.150 \text{ mol KCl}}{1 \text{ L KCl solution}}$$
Solve Begin with L $Pb(NO_3)_2$ solution and follow the conceptual plan to arrive at L KCl solution.	**Solution** $$0.150 \text{ L } \cancel{Pb(NO_3)_2 \text{ solution}} \times \frac{0.175 \text{ mol } \cancel{Pb(NO_3)_2}}{1 \text{ L } \cancel{Pb(NO_3)_2 \text{ solution}}}$$ $$\times \frac{2 \text{ mol } \cancel{KCl}}{1 \text{ mol } \cancel{Pb(NO_3)_2}} \times \frac{1 \text{ L KCl solution}}{0.150 \text{ mol } \cancel{KCl}}$$ $$= 0.350 \text{ L KCl solution}$$

Check The final units (L KCl solution) are correct. The magnitude (0.350 L) seems reasonable because the reaction stoichiometry requires 2 mol of KCl per mole of $Pb(NO_3)_2$. Since the concentrations of the two solutions are not very different (0.150 M compared to 0.175 M), the volume of KCl required should be roughly two times the 0.150 L of $Pb(NO_3)_2$ given in the problem.

For Practice 4.8

What volume (in mL) of a 0.150 M HNO_3 solution is required to completely react with 35.7 mL of a 0.108 M Na_2CO_3 solution according to the following balanced chemical equation?

$$Na_2CO_3(aq) + 2\,HNO_3(aq) \longrightarrow 2\,NaNO_3(aq) + CO_2(g) + H_2O(l)$$

For More Practice 4.8

In the reaction above, what mass (in grams) of carbon dioxide is formed?

4.5 Types of Aqueous Solutions and Solubility

Consider two familiar aqueous solutions: salt water and sugar water. Salt water is a homogeneous mixture of NaCl and H_2O, and sugar water is a homogeneous mixture of $C_{12}H_{22}O_{11}$ and H_2O. You may have made these solutions yourself by adding solid table salt or solid sugar to water. As you stir either of these two substances into the water, it seems to disappear. However, you know that the original substance is still present because you can taste saltiness or sweetness in the water. How do solids such as salt and sugar dissolve in water?

When a solid is put into a liquid solvent, the attractive forces that hold the solid together (the solute–solute interactions) come into competition with the attractive forces between the solvent molecules and the particles that compose the solid (the solvent–solute interactions), as shown in Figure 4.7 ▼. For example, when sodium chloride is put into water, there is a competition between the attraction of Na^+ cations and Cl^- anions to each other (due to their opposite charges) and the attraction of Na^+ and Cl^- to water molecules. The attraction of Na^+ and Cl^- to water is based on the *polar nature* of the water molecule. For reasons we discuss later in this book (Section 9.6), the oxygen atom in water is electron-rich, giving it a partial negative charge (δ^-), as shown in Figure 4.8 ▼. The hydrogen atoms, in contrast, are electron-poor, giving them a partial positive charge (δ^+). As a result, the positively charged sodium ions are strongly attracted to the oxygen side of the water molecule (which has a partial negative charge), and the negatively charged chloride ions are attracted to the hydrogen side of the water molecule (which has a partial positive charge), as shown in

Solute and Solvent Interactions

Solvent-solute interactions

Solute-solute interactions

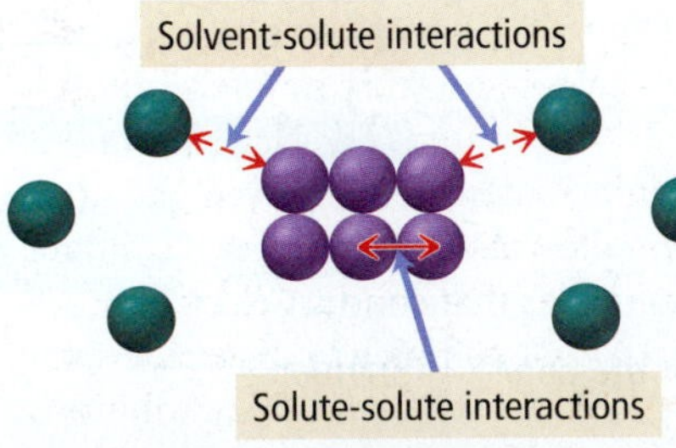

▲ **FIGURE 4.7 Solute and Solvent Interactions** When a solid is put into a solvent, the interactions between solvent and solute particles compete with the interactions among the solute particles themselves.

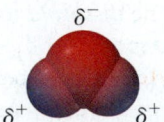

▲ **FIGURE 4.8 Charge Distribution in a Water Molecule** An uneven distribution of electrons within the water molecule causes the oxygen side of the molecule to have a partial negative charge and the hydrogen side to have a partial positive charge.

Interactions in a Sodium Chloride Solution

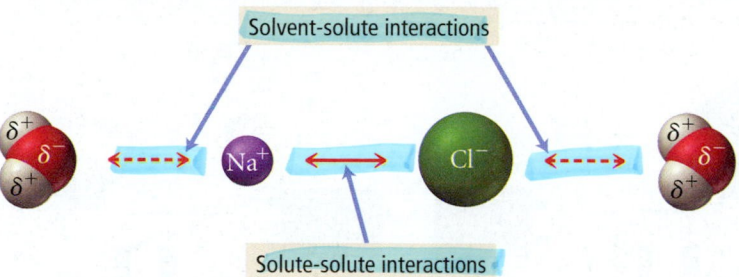

Solvent-solute interactions

Solute-solute interactions

◀ **FIGURE 4.9 Solute and Solvent Interactions in a Sodium Chloride Solution** When sodium chloride is put into water, the attraction of Na^+ and Cl^- ions to water molecules competes with the attraction among the oppositely charged ions themselves.

Figure 4.9 ▲. In the case of NaCl, the attraction between the separated ions and the water molecules overcomes the attraction of sodium and chloride ions to each other, and the sodium chloride dissolves in the water (Figure 4.10 ▼).

Dissolution of an Ionic Compound

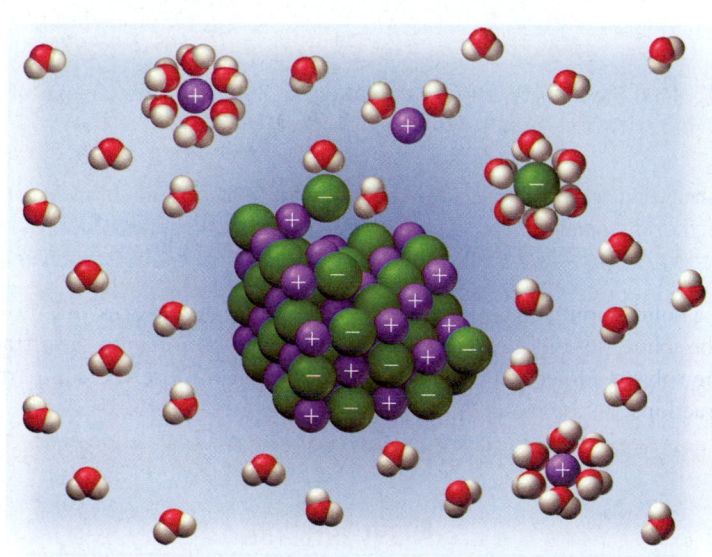

◀ **FIGURE 4.10 Sodium Chloride Dissolving in Water** The attraction between water molecules and the ions of sodium chloride causes NaCl to dissolve in the water.

Electrolyte and Nonelectrolyte Solutions

The difference in the way that salt (an ionic compound) and sugar (a molecular compound) dissolve in water illustrates a fundamental difference between types of solutions. As Figure 4.11 (page 152) shows, a salt solution conducts electricity while a sugar solution does not. As we have just seen exemplified with sodium chloride, ionic compounds dissociate into their component ions when they dissolve in water. An NaCl solution, represented as NaCl(aq), does not contain any NaCl units, but rather dissolved Na^+ ions and Cl^- ions. The dissolved ions act as charge carriers, allowing the solution to conduct electricity. Substances that dissolve in water to form solutions that conduct electricity are called **electrolytes.** Substances such as sodium chloride that completely dissociate into ions when they dissolve in water are called **strong electrolytes,** and the resulting solutions are called strong electrolyte solutions.

   In contrast to sodium chloride, sugar is a molecular compound. Most molecular compounds—with the important exception of acids, which we discuss shortly—dissolve in water as intact molecules. Sugar dissolves because the attraction between sugar molecules and water molecules— both of which contain a distribution of electrons that results in partial positive and partial negative charges —overcomes the attraction of sugar molecules

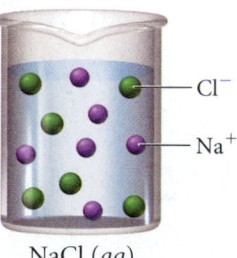

NaCl (aq)

Electrolyte and Nonelectrolyte Solutions

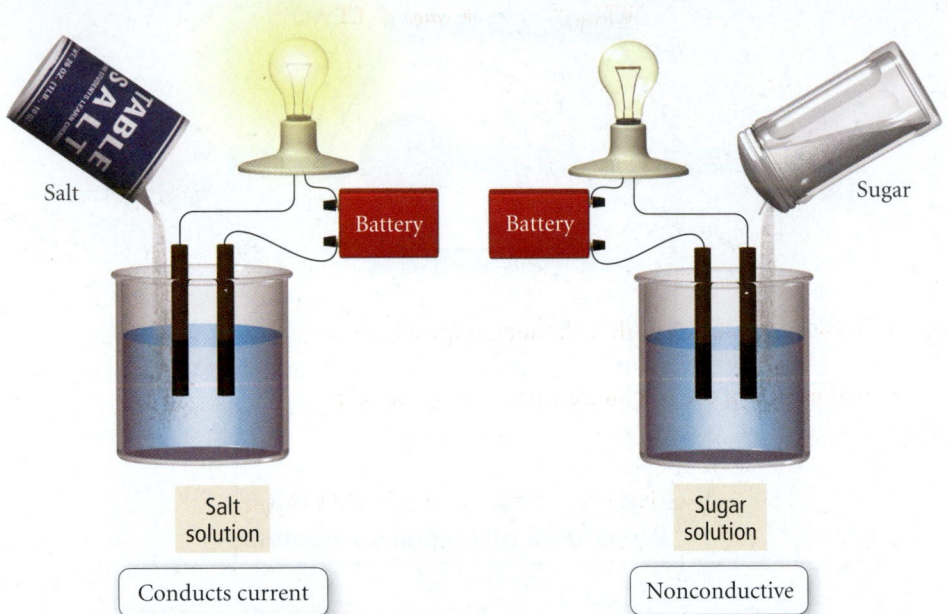

▲ FIGURE 4.11 Electrolyte and Nonelectrolyte Solutions   A solution of salt (an electrolyte) conducts electrical current. A solution of sugar (a nonelectrolyte) does not.

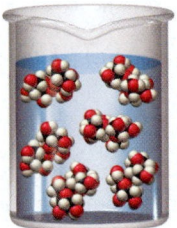

$C_{12}H_{22}O_{11}$ (*aq*)

to each other (Figure 4.12 ▼). However, in contrast to a sodium chloride solution (which is composed of dissociated ions), a sugar solution is composed of intact $C_{12}H_{22}O_{11}$ molecules homogeneously mixed with the water molecules. Compounds such as sugar that do not dissociate into ions when dissolved in water are called **nonelectrolytes,** and the resulting solutions—called *nonelectrolyte solutions*—do not conduct electricity.

Acids, first encountered in Section 3.6, are molecular compounds that ionize—form ions—when they dissolve in water. For example, hydrochloric acid (HCl) is a molecular compound that ionizes into H^+ and Cl^- when it dissolves in water. HCl is an example of a **strong acid,** one that completely ionizes in solution. Since they completely ionize in

Sugar Solution

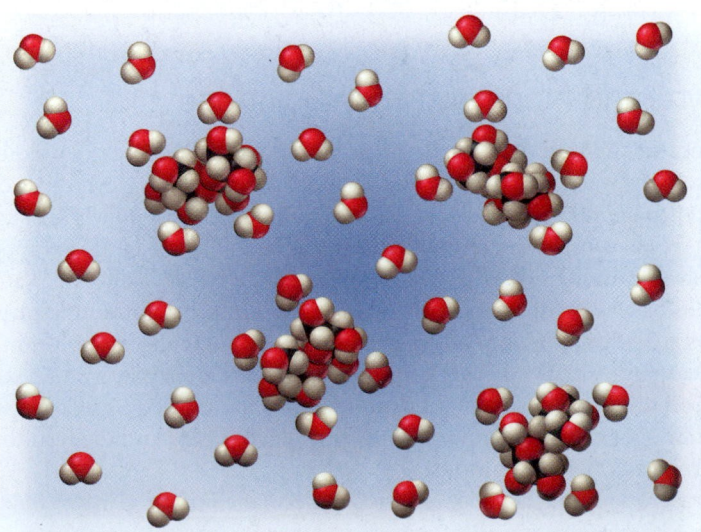

▲ FIGURE 4.12   A Sugar Solution   Sugar dissolves because the attractions between sugar molecules and water molecules, which both contain a distribution of electrons that results in partial positive and partial negative charges, overcomes the attractions between sugar molecules to each other.

solution, strong acids are also strong electrolytes. We represent the complete ionization of a strong acid with a single reaction arrow between the acid and its ionized form:

$$HCl(aq) \longrightarrow H^+(aq) + Cl^-(aq)$$

Unlike soluble ionic compounds, which contain ions and therefore *dissociate* in water, acids are molecular compounds that *ionize* in water.

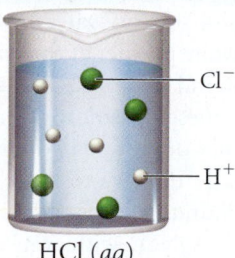

HCl (*aq*)

Many acids are **weak acids**; they do not completely ionize in water. For example, acetic acid ($HC_2H_3O_2$), the acid present in vinegar, is a weak acid. A solution of a weak acid is composed mostly of the nonionized acid—only a small percentage of the acid molecules ionize. We represent the partial ionization of a weak acid with opposing half arrows between the reactants and products:

$$HC_2H_3O_2(aq) \rightleftharpoons H^+(aq) + C_2H_3O_2^-(aq)$$

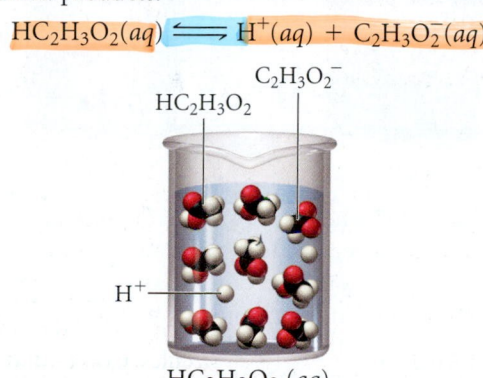

$HC_2H_3O_2$ $C_2H_3O_2^-$

H^+

$HC_2H_3O_2$ (*aq*)

Weak acids are classified as **weak electrolytes** and the resulting solutions—called *weak electrolyte solutions*—conduct electricity only weakly. Figure 4.13 ▼ summarizes the electrolytic properties of solutions.

▼ **FIGURE 4.13 Electrolytic Properties of Solutions**

Electrolytic Properties of Solutions

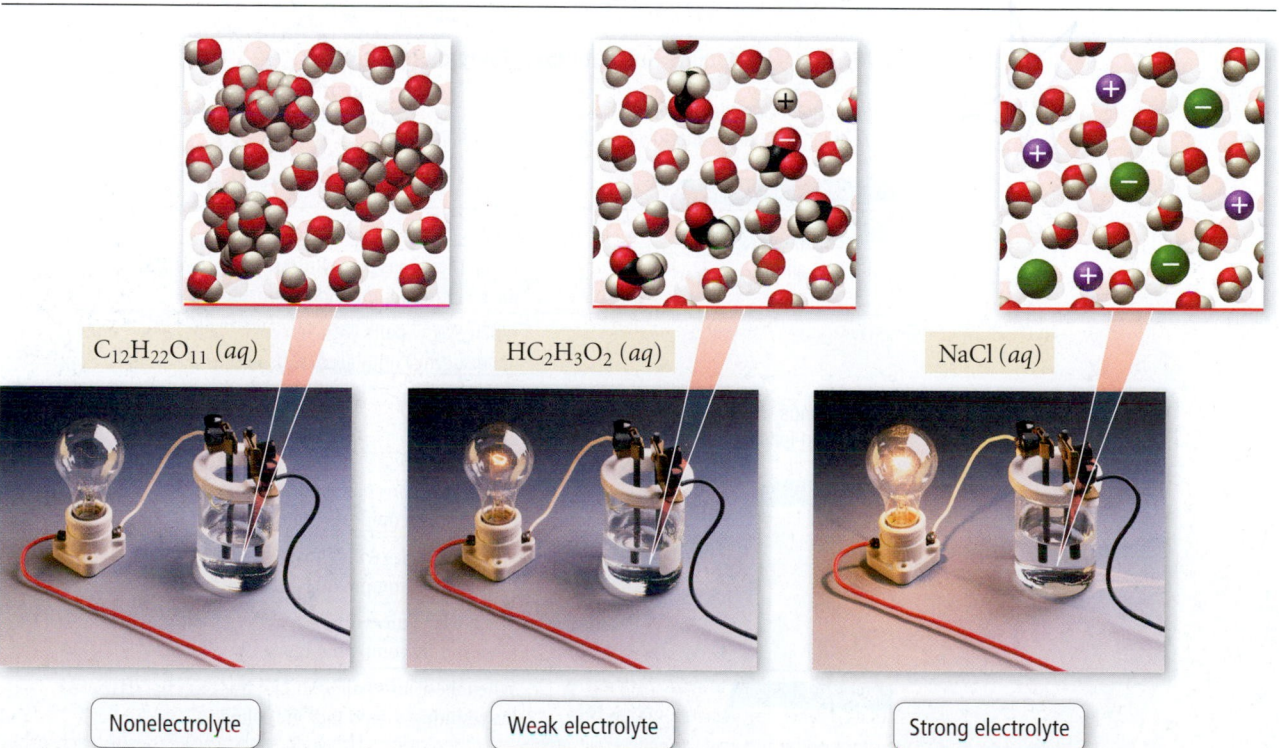

$C_{12}H_{22}O_{11}$ (*aq*) $HC_2H_3O_2$ (*aq*)   NaCl (*aq*)

Nonelectrolyte   Weak electrolyte   Strong electrolyte

▲ AgCl does not dissolve in water, but remains as a white powder at the bottom of the beaker.

The Solubility of Ionic Compounds

We have just seen that, when an ionic compound dissolves in water, the resulting solution contains, not the intact ionic compound itself, but its component ions dissolved in water. However, not all ionic compounds dissolve in water. If we add AgCl to water, for example, it remains solid and appears as a white powder at the bottom of the water.

In a general sense, a compound is termed **soluble** if it dissolves in water and **insoluble** if it does not. However, these classifications are a bit of an oversimplification. In reality, solubility is a continuum. Compounds exhibit a very wide range of solubilities, and even "insoluble" compounds do dissolve to some extent, though usually orders of magnitude less than soluble compounds. By this classification scheme, for example, silver nitrate is soluble. If we mix solid $AgNO_3$ with water, it dissolves and forms a strong electrolyte solution. Silver chloride, on the other hand, is almost completely insoluble. If we mix solid AgCl with water, virtually all of it remains as a solid within the liquid water.

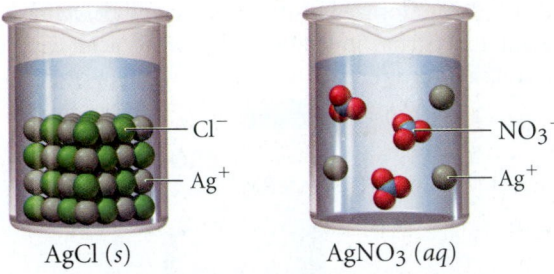

AgCl (s) $AgNO_3$ (aq)

There is no easy way to tell whether a particular compound is soluble or insoluble just by looking at its formula. In Section 12.3, we examine more closely the energetics of solution formation. For now, however, we can follow a set of empirical rules that have been inferred from observations on many ionic compounds. These are called *solubility rules* and are summarized in Table 4.1.

For example, the solubility rules state that compounds containing the sodium ion are soluble. That means that compounds such as NaBr, $NaNO_3$, Na_2SO_4, NaOH, and Na_2CO_3 all dissolve in water to form strong electrolyte solutions. Similarly, the solubility rules state that compounds containing the NO_3^- ion are soluble. That means that compounds such as $AgNO_3$, $Pb(NO_3)_2$, $NaNO_3$, $Ca(NO_3)_2$, and $Sr(NO_3)_2$ all dissolve in water to form strong electrolyte solutions.

TABLE 4.1   Solubility Rules for Ionic Compounds in Water

Compounds Containing the Following Ions Are Generally Soluble	Exceptions
Li^+, Na^+, K^+, and NH_4^+	None
NO_3^- and $C_2H_3O_2^-$	None
Cl^-, Br^-, and I^-	When these ions pair with Ag^+, Hg_2^{2+} or Pb^{2+}, the resulting compounds are insoluble.
SO_4^{2-}	When SO_4^{2-} pairs with Sr^{2+}, Ba^{2+}, Pb^{2+}, Ag^+, or Ca^{2+}, the resulting compound is insoluble.

Compounds Containing the Following Ions Are Generally Insoluble	Exceptions
OH^- and S^{2-}	When these ions pair with Li^+, Na^+, K^+, or NH_4^+, the resulting compounds are soluble.
	When S^{2-} pairs with Ca^{2+}, Sr^{2+}, or Ba^{2+}, the resulting compound is soluble.
	When OH^- pairs with Ca^{2+}, Sr^{2+}, or Ba^{2+}, the resulting compound is slightly soluble.
CO_3^{2-} and PO_4^{3-}	When these ions pair with Li^+, Na^+, K^+, or NH_4^+, the resulting compounds are soluble.

Notice that when compounds containing polyatomic ions such as NO_3^- dissolve, the polyatomic ions dissolve as intact units.

The solubility rules also state that, with some exceptions, compounds containing the CO_3^{2-} ion are insoluble. Therefore, compounds such as $CuCO_3$, $CaCO_3$, $SrCO_3$, and $FeCO_3$ do not dissolve in water. Note that the solubility rules contain many exceptions. For example, compounds containing CO_3^{2-} *are* soluble when paired with Li^+, Na^+, K^+, or NH_4^+. Thus Li_2CO_3, Na_2CO_3, K_2CO_3, and $(NH_4)_2CO_3$ are all soluble.

EXAMPLE 4.9 Predicting whether an Ionic Compound Is Soluble

Predict whether each of the following compounds is soluble or insoluble.

(a) $PbCl_2$ (b) $CuCl_2$ (c) $Ca(NO_3)_2$ (d) $BaSO_4$

Solution

(a) Insoluble. Compounds containing Cl^- are normally soluble, but Pb^{2+} is an exception.

(b) Soluble. Compounds containing Cl^- are normally soluble and Cu^{2+} is not an exception.

(c) Soluble. Compounds containing NO_3^- are always soluble.

(d) Insoluble. Compounds containing SO_4^{2-} are normally soluble, but Ba^{2+} is an exception.

For Practice 4.9

Predict whether each of the following compounds is soluble or insoluble.

(a) NiS (b) $Mg_3(PO_4)_2$ (c) Li_2CO_3 (d) NH_4Cl

4.6 Precipitation Reactions

Have you ever taken a bath in hard water? Hard water contains dissolved ions such as Ca^{2+} and Mg^{2+} that diminish the effectiveness of soap. These ions react with soap to form a gray curd that often appears as "bathtub ring" when you drain the tub. Hard water is particularly troublesome for washing clothes. Imagine how your white shirt would look covered with the gray curd that is normally in your bathtub. Consequently, most laundry detergents include substances designed to remove Ca^{2+} and Mg^{2+} from the laundry mixture. The most common substance used for this purpose is sodium carbonate, which dissolves in water to form sodium cations (Na^+) and carbonate (CO_3^{2-}) anions:

$$Na_2CO_3(aq) \longrightarrow 2\,Na^+(aq) + CO_3^{2-}(aq)$$

Sodium carbonate is soluble, but calcium carbonate and magnesium carbonate are not (see the solubility rules in Table 4.1). Consequently, the carbonate anions react with dissolved Mg^{2+} and Ca^{2+} ions in hard water to form solids that *precipitate* from (or come out of) solution:

$$Mg^{2+}(aq) + CO_3^{2-}(aq) \longrightarrow MgCO_3(s)$$
$$Ca^{2+}(aq) + CO_3^{2-}(aq) \longrightarrow CaCO_3(s)$$

The precipitation of these ions prevents their reaction with the soap, eliminating curd and keeping shirts white instead of gray.

The reactions between CO_3^{2-} and Mg^{2+} and Ca^{2+} are examples of **precipitation reactions, ones in which a solid or precipitate forms upon mixing two solutions.** Precipitation reactions are common in chemistry. As another example, consider potassium iodide and lead(II) nitrate, which both form colorless, strong electrolyte solutions when dissolved in water. When the two solutions are combined, however, a brilliant yellow precipitate forms (Figure 4.14, page 156). This precipitation reaction can be described with the following chemical equation:

$$2\,KI(aq) + Pb(NO_3)_2(aq) \longrightarrow 2\,KNO_3(aq) + PbI_2(s)$$

▲ The reaction of ions in hard water with soap produces a gray curd that can often be seen after draining the bathwater.

Precipitation Reaction

$$2\,KI(aq) + Pb(NO_3)_2(aq) \longrightarrow 2\,KNO_3(aq) + PbI_2(s)$$
(soluble)   (soluble)   (soluble)   (insoluble)

When a potassium iodide solution is mixed with a lead(II) nitrate solution, a yellow lead(II) iodide precipitate forms.

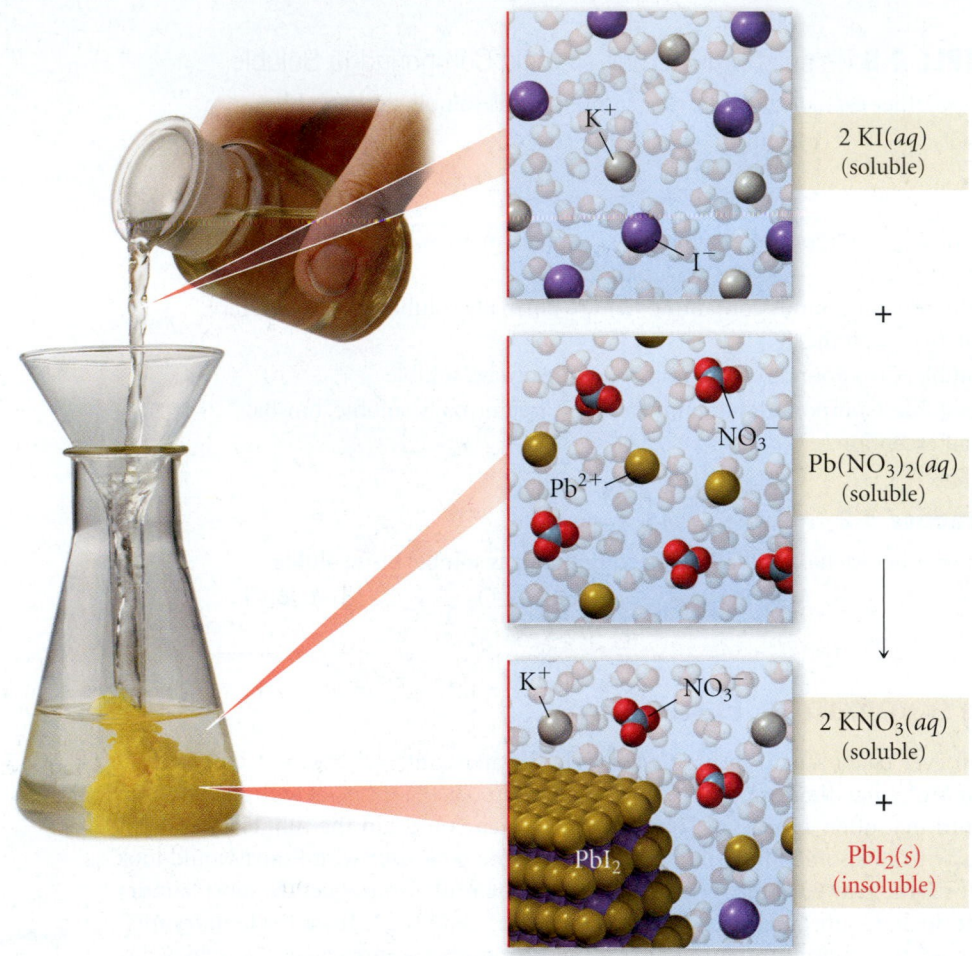

$2\,KI(aq)$
(soluble)

+

$Pb(NO_3)_2(aq)$
(soluble)

$2\,KNO_3(aq)$
(soluble)

+

$PbI_2(s)$
(insoluble)

▲ **FIGURE 4.14 Precipitation of Lead(II) Iodide**   When a potassium iodide solution is mixed with a lead(II) nitrate solution, a yellow lead(II) iodide precipitate forms.

Precipitation reactions do not always occur when two aqueous solutions are mixed. For example, if solutions of $KI(aq)$ and $NaCl(aq)$ are combined, nothing happens (Figure 4.15, page 157):

$$KI(aq) + NaCl(aq) \longrightarrow NO\ REACTION$$

The key to predicting precipitation reactions is to understand that only *insoluble* compounds form precipitates. In a precipitation reaction, two solutions containing soluble compounds combine and an insoluble compound precipitates. For example, consider the precipitation reaction described previously:

$$2\,KI(aq) + Pb(NO_3)_2(aq) \longrightarrow PbI_2(s) + 2\,KNO_3(aq)$$
soluble   soluble   insoluble   soluble

No Reaction

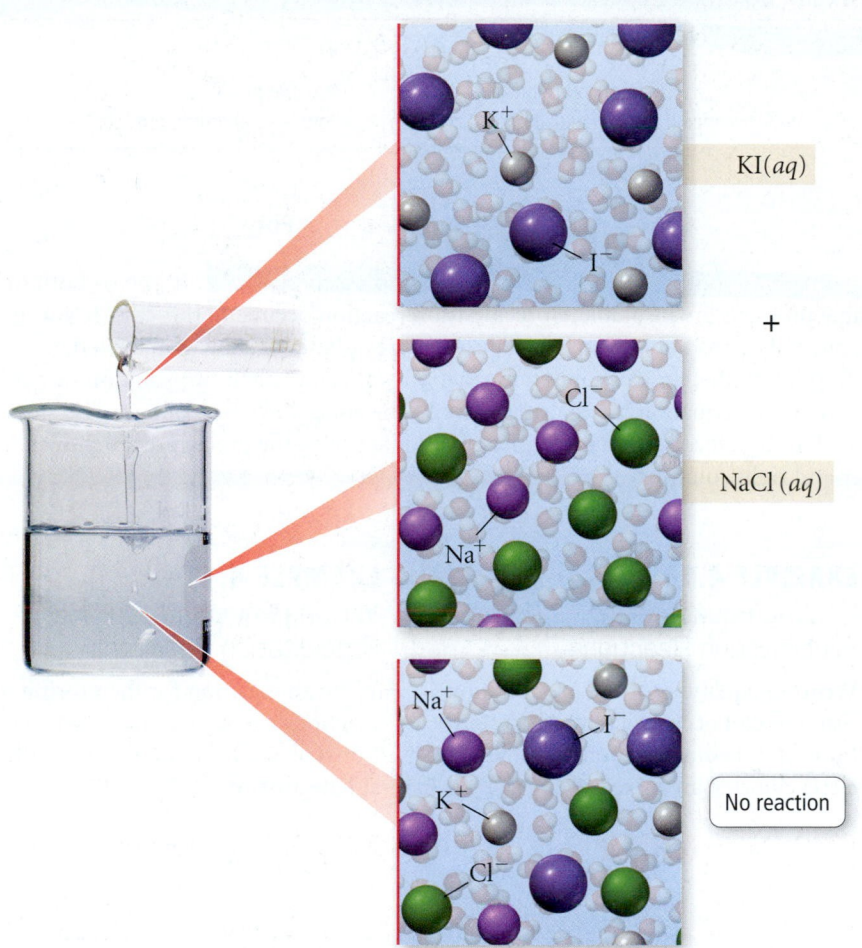

KI(*aq*)

+

NaCl (*aq*)

No reaction

◀ FIGURE 4.15 **No Precipitation** When a potassium iodide solution is mixed with a sodium chloride solution, no reaction occurs.

KI and $Pb(NO_3)_2$ are both soluble, but the precipitate, PbI_2, is insoluble. Before mixing, KI(*aq*) and $Pb(NO_3)_2(aq)$ are both dissociated in their respective solutions:

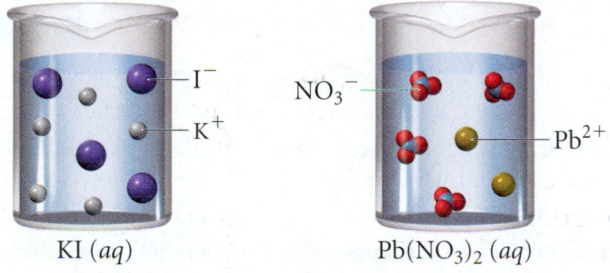

KI (*aq*) $Pb(NO_3)_2$ (*aq*)

The instant that the solutions are mixed, all four ions are present:

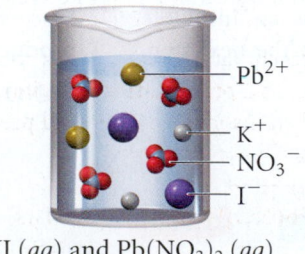

KI (*aq*) and $Pb(NO_3)_2$ (*aq*)

However, new compounds—one or both of which might be insoluble—are now possible. Specifically, the cation from each compound can pair with the anion from the other compound to form possibly insoluble products:

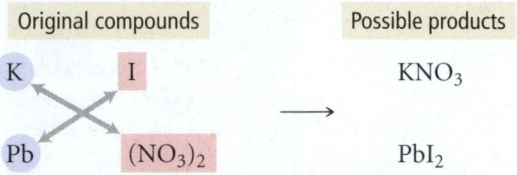

If the possible products are both soluble, then no reaction occurs. If one or both of the possible products are insoluble, a precipitation reaction occurs. In this case, KNO_3 is soluble, but PbI_2 is insoluble. Consequently, PbI_2 precipitates as shown on the left.

To predict whether a precipitation reaction will occur when two solutions are mixed and to write an equation for the reaction, use the procedure that follows. The steps are outlined in the left column, and two examples of applying the procedure are shown in the center and right columns.

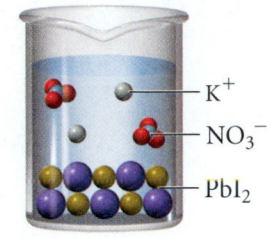

PbI_2 (s) and KNO_3 (aq)

Procedure for Writing Equations for Precipitation Reactions	EXAMPLE 4.10 Writing Equations for Precipitation Reactions	EXAMPLE 4.11 Writing Equations for Precipitation Reactions
	Write an equation for the precipitation reaction that occurs (if any) when solutions of potassium carbonate and nickel(II) chloride are mixed.	Write an equation for the precipitation reaction that occurs (if any) when solutions of sodium nitrate and lithium sulfate are mixed.
1. Write the formulas of the two compounds being mixed as reactants in a chemical equation.	$K_2CO_3(aq) + NiCl_2(aq) \longrightarrow$	$NaNO_3(aq) + Li_2SO_4(aq) \longrightarrow$
2. Below the equation, write the formulas of the products that could form from the reactants. Obtain these by combining the cation from each reactant with the anion from the other. Make sure to write correct formulas for these ionic compounds, as described in Section 3.5.	$K_2CO_3(aq) + NiCl_2(aq) \longrightarrow$ Possible products KCl $NiCO_3$	$NaNO_3(aq) + Li_2SO_4(aq) \longrightarrow$ Possible products $LiNO_3$ Na_2SO_4
3. Use the solubility rules to determine whether any of the possible products are insoluble.	KCl is soluble. (Compounds containing Cl^- are usually soluble and K^+ is not an exception.) $NiCO_3$ is insoluble. (Compounds containing CO_3^{2-} are usually insoluble and Ni^{2+} is not an exception.)	$LiNO_3$ is soluble. (Compounds containing NO_3^- are soluble and Li^+ is not an exception.) Na_2SO_4 is soluble. (Compounds containing SO_4^{2-} are generally soluble and Na^+ is not an exception.)
4. If all of the possible products are soluble, there will be no precipitate. Write NO REACTION after the arrow.	Since this example has an insoluble product, we proceed to the next step.	Since this example has no insoluble product, there is no reaction. $NaNO_3(aq) + Li_2SO_4(aq) \longrightarrow$ NO REACTION
5. If any of the possible products are insoluble, write their formulas as the products of the reaction using (s) to indicate solid. Write any soluble products with (aq) to indicate aqueous.	$K_2CO_3(aq) + NiCl_2(aq) \longrightarrow$ $NiCO_3(s) + KCl(aq)$	

6. Balance the equation. Remember to adjust only coefficients here, not subscripts.

$$K_2CO_3(aq) + NiCl_2(aq) \longrightarrow$$
$$NiCO_3(s) + 2\,KCl(aq)$$

For Practice 4.10

Write an equation for the precipitation reaction that occurs (if any) when solutions of ammonium chloride and iron(III) nitrate are mixed.

For Practice 4.11

Write an equation for the precipitation reaction that occurs (if any) when solutions of sodium hydroxide and copper(II) bromide are mixed.

4.7 Representing Aqueous Reactions: Molecular, Ionic, and Complete Ionic Equations

Consider the following equation for a precipitation reaction:

$$Pb(NO_3)_2(aq) + 2\,KCl(aq) \longrightarrow PbCl_2(s) + 2\,KNO_3(aq)$$

This equation is a **molecular equation,** an equation showing the complete neutral formulas for each compound in the reaction as if they existed as molecules. However, in actual solutions of soluble ionic compounds, dissociated substances are present as ions. Therefore, equations for reactions occurring in aqueous solution can be written to better show the dissociated nature of dissolved ionic compounds. For example, the above equation can be rewritten as follows:

$$Pb^{2+}(aq) + 2\,NO_3^-(aq) + 2\,K^+(aq) + 2Cl^-(aq) \longrightarrow PbCl_2(s) + 2\,K^+(aq) + 2\,NO_3^-(aq)$$

Equations such as this, which list individually all of the ions present as either reactants or products in a chemical reaction, are called **complete ionic equations.**

Notice that in the complete ionic equation some of the ions in solution appear unchanged on both sides of the equation. These ions are called **spectator ions** because they do not participate in the reaction.

$$Pb^{2+}(aq) + 2\,NO_3^-(aq) + 2\,K^+(aq) + 2\,Cl^-(aq) \longrightarrow$$
$$PbCl_2(s) + 2\,K^+(aq) + 2\,NO_3^-(aq)$$

Spectator ions

To simplify the equation, and to show more clearly what is happening, spectator ions can be omitted:

$$Pb^{2+}(aq) + 2\,Cl^-(aq) \longrightarrow PbCl_2(s)$$

Equations such as this one, which show only the species that actually change during the reaction, are called **net ionic equations.**

As another example, consider the following reaction between $HCl(aq)$ and $KOH(aq)$:

$$HCl(aq) + KOH(aq) \longrightarrow H_2O(l) + KCl(aq)$$

Since HCl, KOH, and KCl all exist in solution primarily as independent ions, the complete ionic equation is as follows:

$$H^+(aq) + Cl^-(aq) + K^+(aq) + OH^-(aq) \longrightarrow H_2O(l) + K^+(aq) + Cl^-(aq)$$

To write the net ionic equation, we remove the spectator ions, those that are unchanged on both sides of the equation:

$$H^+(aq) + Cl^-(aq) + K^+(aq) + OH^-(aq) \longrightarrow H_2O(l) + K^+(aq) + Cl^-(aq)$$

Spectator ions

The net ionic equation is $H^+(aq) + OH^-(aq) \longrightarrow H_2O(l)$.

Summarizing:

➤ A **molecular equation** is a chemical equation showing the complete, neutral formulas for every compound in a reaction.

➤ A **complete ionic equation** is a chemical equation showing all of the species as they are actually present in solution.

➤ A **net ionic equation** is an equation showing only the species that actually change during the reaction.

EXAMPLE 4.12 Writing Complete Ionic and Net Ionic Equations

Consider the following precipitation reaction occurring in aqueous solution:

$$3\, SrCl_2(aq) + 2\, Li_3PO_4(aq) \longrightarrow Sr_3(PO_4)_2(s) + 6\, LiCl(aq)$$

Write the complete ionic equation and net ionic equation for this reaction.

Solution Write the complete ionic equation by separating aqueous ionic compounds into their constituent ions. The $Sr_3(PO_4)_2(s)$, since it precipitates as a solid, remains as one unit.	**Complete ionic equation:** $$3\, Sr^{2+}(aq) + 6\, Cl^-(aq) + 6\, Li^+(aq) + 2\, PO_4^{3-}(aq) \longrightarrow$$ $$Sr_3(PO_4)_2(s) + 6\, Li^+(aq) + 6\, Cl^-(aq)$$
Write the net ionic equation by eliminating the spectator ions, those that do not change from one side of the reaction to the other.	**Net ionic equation:** $$3\, Sr^{2+}(aq) + 2\, PO_4^{3-}(aq) \longrightarrow Sr_3(PO_4)_2(s)$$

For Practice 4.12

Consider the following reaction occurring in aqueous solution:

$$2\, HI(aq) + Ba(OH)_2(aq) \longrightarrow 2\, H_2O(l) + BaI_2(aq)$$

Write the complete ionic equation and net ionic equation for this reaction.

For More Practice 4.12

Write complete ionic and net ionic equations for the following reaction occurring in aqueous solution:

$$2\, AgNO_3(aq) + MgCl_2(aq) \longrightarrow 2\, AgCl(s) + Mg(NO_3)_2(aq)$$

▲ In a gas-evolution reaction, such as the reaction of hydrochloric acid with limestone ($CaCO_3$) to produce CO_2, bubbling typically occurs as the gas is released.

4.8 Acid–Base and Gas-Evolution Reactions

Two other important classes of reactions that occur in aqueous solution are acid–base reactions and gas-evolution reactions. In an **acid–base reaction** (also called a **neutralization reaction**), an acid reacts with a base and the two neutralize each other, producing water (or in some cases a weak electrolyte). In a **gas-evolution reaction**, a gas forms, resulting in bubbling. In both cases, as in precipitation reactions, the reactions occur when the anion from one reactant combines with the cation of the other. In addition, many gas-evolution reactions are also acid–base reactions.

Acid–Base Reactions

Our stomachs contain hydrochloric acid, which acts in the digestion of food. Certain foods and stress, however, can increase the stomach's acidity to uncomfortable levels, causing acid stomach or heartburn. Antacids are over-the-counter medicines that work by reacting with and neutralizing stomach acid. Antacids employ different *bases*—substances that produce hydroxide (OH^-) ions in water—as neutralizing agents. Milk of magnesia, for example, contains $Mg(OH)_2$ and Mylanta contains $Al(OH)_3$. All antacids, however, have the same effect of neutralizing stomach acid through *acid-base reactions* and relieving heartburn.

We learned in Chapter 3 that an acid forms H^+ ions in solution, and we learned above that a base is a substance that produces OH^- ions in solution:

➤ **Acid**   **Substance that produces H^+ ions in aqueous solution**
➤ **Base**   **Substance that produces OH^- ions in aqueous solution**

These definitions of acids and bases are called the **Arrhenius definitions**, after Swedish chemist Svante Arrhenius (1859–1927). In Chapter 15, we will learn more general definitions of acid–base behavior, but these are sufficient to describe neutralization reactions.

Under the Arrhenius definition, HCl is an acid because it produces H^+ ions in solution:

$$HCl(aq) \longrightarrow H^+(aq) + Cl^-(aq)$$

An H^+ ion is a bare proton. Protons associate with water molecules in solution to form **hydronium ions** (Figure 4.16 ▼):

$$H^+(aq) + H_2O \longrightarrow H_3O^+(aq)$$

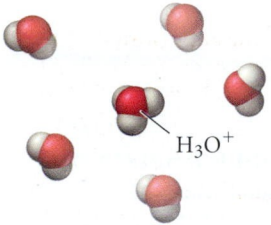

◀ **FIGURE 4.16 The Hydronium Ion**   Protons normally associate with water molecules in solution to form H_3O^+ ions, which in turn interact with other water molecules.

H_3O^+

In water, H^+ ions *always* associate with H_2O molecules to form hydronium ions. Chemists often use $H^+(aq)$ and $H_3O^+(aq)$ interchangeably, however, to mean the same thing—a hydronium ion. The ionization of HCl and other acids is often written to show the association of the proton with a water molecule to form the hydronium ion:

$$HCl(aq) + H_2O \longrightarrow H_3O^+(aq) + Cl^-(aq)$$

Some acids—called **polyprotic acids**—contain more than one ionizable proton and release them sequentially. For example, sulfuric acid, H_2SO_4, is a **diprotic acid.** It is strong in its first ionizable proton, but weak in its second:

$$H_2SO_4(aq) \longrightarrow H^+(aq) + HSO_4^-(aq)$$
$$HSO_4^-(aq) \rightleftharpoons H^+(aq) + SO_4^{2-}(aq)$$

Under the Arrhenius definition, NaOH is a base because it produces OH^- ions in solution:

$$NaOH(aq) \longrightarrow Na^+(aq) + OH^-(aq)$$

In analogy to diprotic acids, some bases, such as $Sr(OH)_2$, for example, produce two moles of OH^- per mole of the base.

$$Sr(OH)_2(aq) \longrightarrow Sr^{2+}(aq) + 2\,OH^-(aq)$$

Common acids and bases are listed in Table 4.2. Acids and bases are found in many everyday substances. Foods such as citrus fruits, and vinegar contain acids. Soap, baking soda, and milk of magnesia all contain bases.

H^+ and H_3O^+ are used interchangeably because, even though H^+ associates with water to form hydronium, it is the H^+ part that reacts with other substances such as bases.

▲ Acids are found in lemons, limes, and vinegar. Vitamin C and aspirin are also acids.

TABLE 4.2 Some Common Acids and Bases

Name of Acid	Formula	Name of Base	Formula
Hydrochloric acid	HCl	Sodium hydroxide	NaOH
Hydrobromic acid	HBr	Lithium hydroxide	LiOH
Hydroiodic acid	HI	Potassium hydroxide	KOH
Nitric acid	HNO_3	Calcium hydroxide	$Ca(OH)_2$
Sulfuric acid	H_2SO_4	Barium hydroxide	$Ba(OH)_2$
Perchloric acid	$HClO_4$	Ammonia*	NH_3 (weak base)
Acetic acid	$HC_2H_3O_2$ (weak acid)		
Hydrofluoric acid	HF (weak acid)		

*Ammonia does not contain OH^-, but it produces OH^- in a reaction with water that occurs only to a small extent: $NH_3(aq) + H_2O(l) \rightleftharpoons NH_4^+(aq) + OH^-(aq)$.

▲ Many common household products are bases.

When an acid and base are mixed, the $H^+(aq)$ from the acid—whether it is weak or strong—combines with the $OH^-(aq)$ from the base to form $H_2O(l)$ (Figure 4.17 ▼). For example, consider the reaction between hydrochloric acid and sodium hydroxide:

$$HCl(aq) + NaOH(aq) \longrightarrow H_2O(l) + NaCl(aq)$$

Acid   Base   Water   Salt

> The word *salt* in this sense applies to any ionic compound and is therefore more general than the common usage, which refers only to table salt (NaCl).

Acid–base reactions generally form water and an ionic compound—called a **salt**—that usually remains dissolved in the solution. The net ionic equation for many acid–base reactions is

$$H^+(aq) + OH^-(aq) \longrightarrow H_2O(l)$$

Another example of an acid-base reaction is that between sulfuric acid and potassium hydroxide:

$$H_2SO_4(aq) + 2\,KOH(aq) \longrightarrow 2\,H_2O(l) + K_2SO_4(aq)$$

acid   base   water   salt

Acid-Base Reaction

$$HCl(aq) + NaOH(aq) \longrightarrow H_2O(l) + NaCl(aq)$$

The reaction between hydrochloric acid and sodium hydroxide forms water and a salt, sodium chloride, which remains dissolved in the solution.

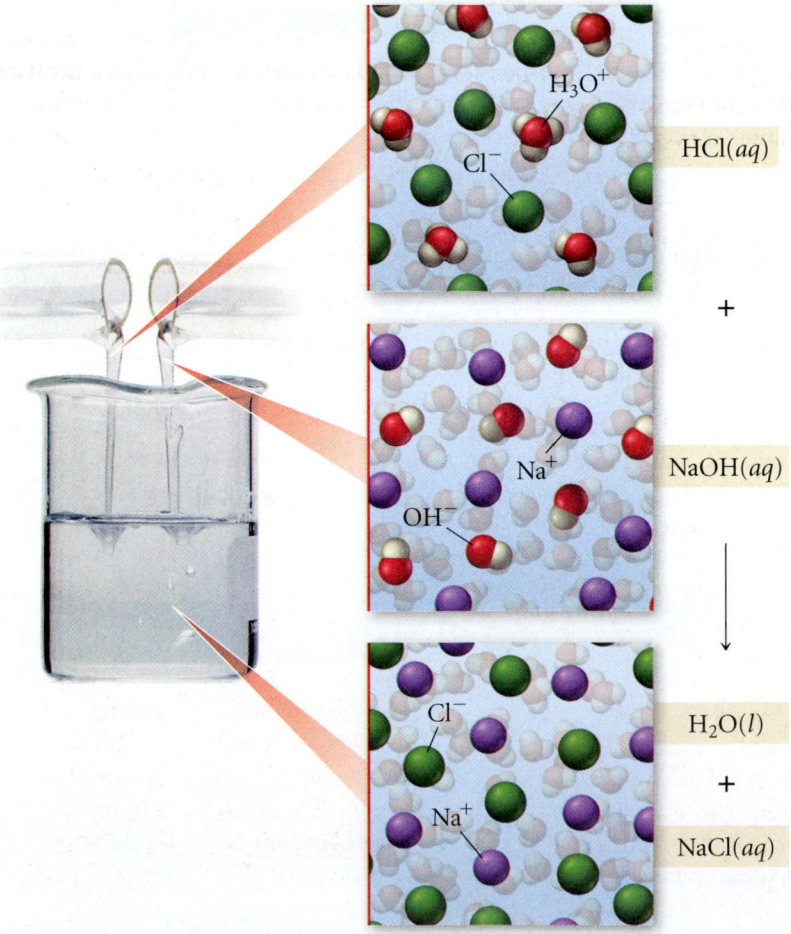

▲ **FIGURE 4.17 Acid–Base Reaction**   The reaction between hydrochloric acid and sodium hydroxide forms water and a salt, sodium chloride, which remains dissolved in the solution.

Again, notice the pattern of acid and base reacting to form water and a salt.

$$\text{Acid} + \text{Base} \longrightarrow \text{Water} + \text{Salt} \qquad \text{(acid–base reactions)}$$

When writing equations for acid–base reactions, write the formula of the salt using the procedure for writing formulas of ionic compounds given in Section 3.5.

EXAMPLE 4.13 Writing Equations for Acid–Base Reactions

Write a molecular and net ionic equation for the reaction between aqueous HI and aqueous $Ba(OH)_2$.

Solution You must first recognize these substances as an acid and a base. Begin by writing the skeletal reaction in which the acid and the base combine to form water and a salt.	$\underset{\text{acid}}{HI(aq)} + \underset{\text{base}}{Ba(OH)_2(aq)} \longrightarrow \underset{\text{water}}{H_2O(l)} + \underset{\text{salt}}{BaI_2(aq)}$
Next, balance the equation; this is the molecular equation.	$2\,HI(aq) + Ba(OH)_2(aq) \longrightarrow 2\,H_2O(l) + BaI_2(aq)$
Write the net ionic equation by removing the spectator ions.	$2\,H^+(aq) + 2\,OH^-(aq) \longrightarrow 2\,H_2O(l)$ or simply $H^+(aq) + OH^-(aq) \longrightarrow H_2O(l)$

For Practice 4.13

Write a molecular and a net ionic equation for the reaction that occurs between aqueous H_2SO_4 and aqueous LiOH.

Acid–Base Titrations The principles of acid-base neutralization and stoichiometry can be applied to a common laboratory procedure called a *titration*. In a **titration,** a substance in a solution of known concentration is reacted with another substance in a solution of unknown concentration. For example, consider the following acid–base reaction:

$$HCl(aq) + NaOH(aq) \rightarrow H_2O(l) + NaCl(aq)$$

The net ionic equation for this reaction is as follows:

$$H^+(aq) + OH^-(aq) \rightarrow H_2O$$

Suppose you have an HCl solution represented by the following molecular diagram (the Cl^- ions and the H_2O molecules not involved in the reaction have been omitted from this representation for clarity):

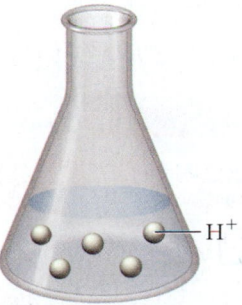

In titrating this sample, we slowly add a solution of known OH^- concentration, as represented by the molecular diagrams to the right.

As the OH^- is added, it reacts with and neutralizes the H^+, forming water. At the **equivalence point**—the point in the titration when the number of moles of

Acid-Base Titration

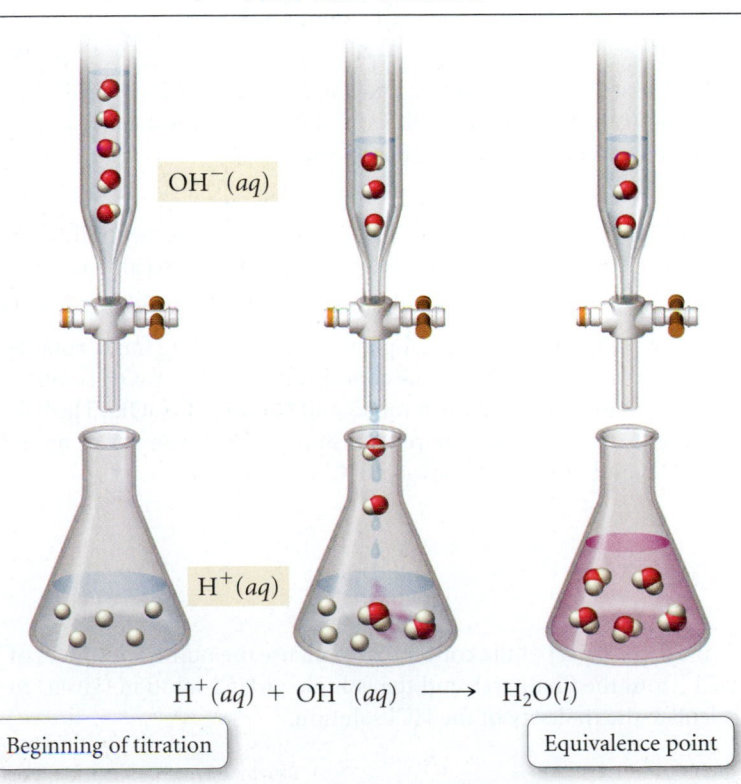

$$H^+(aq) + OH^-(aq) \longrightarrow H_2O(l)$$

Beginning of titration

Equivalence point

Indicator in Titration

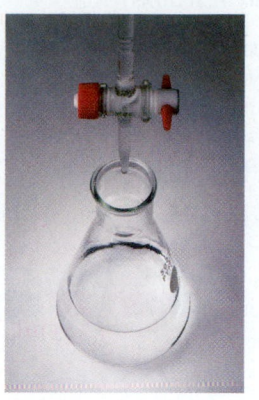

▲ FIGURE 4.18 **Titration**   In this titration, NaOH is added to a dilute HCl solution. When the NaOH and HCl reach stoichiometric proportions (the equivalence point), the phenolphthalein indicator changes color to pink.

At the equivalence point, neither reactant is present in excess, and both are limiting. The number of moles of the reactants are related by the reaction stoichiometry.

OH^- equals the number of moles of H^+ in solution—the titration is complete. The equivalence point is usually signaled by an **indicator,** a dye whose color depends on the acidity or basicity of the solution (Figure 4.18 ▲).

Acid–base titrations and indicators are covered in more detail in Chapter 16. In most laboratory titrations, the concentration of one of the reactant solutions is unknown and the concentration of the other is precisely known. By carefully measuring the volume of each solution required to reach the equivalence point, the concentration of the unknown solution can be determined, as demonstrated in the following example.

EXAMPLE 4.14   Acid–Base Titration

The titration of a 10.00-mL sample of an HCl solution of unknown concentration requires 12.54 mL of a 0.100 M NaOH solution to reach the equivalence point. What is the concentration of the unknown HCl solution in M?

Sort   You are given the volume and concentration of NaOH solution required to titrate a given volume of HCl solution. You are asked to find the concentration of the HCl solution.

Given   12.54 mL of NaOH solution, 0.100 M NaOH solution, 10.00 mL of HCl solution

Find   concentration of HCl solution

Strategize   Since this problem implies an acid–base neutralization reaction between HCl and NaOH, you must start by writing the balanced equation, using the techniques covered earlier in this section.

The first part of the conceptual plan has the following form: volume A → moles A → moles B. The concentration of the NaOH solution is a conversion factor between moles and volume of NaOH. The balanced equation provides the relationship between number of moles of NaOH and number of moles of HCl.

$$HCl(aq) + NaOH(aq) \longrightarrow H_2O(l) + NaCl(aq)$$

Conceptual Plan

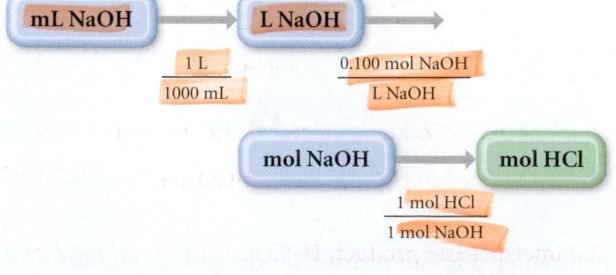

In the second part of the conceptual plan, use the number of moles of HCl (from the first part), and the volume of HCl solution (given) to calculate the molarity of the HCl solution.

Relationships Used

$1 \text{ L} = 1000 \text{ mL}$

$$M \text{ (NaOH)} = \frac{0.100 \text{ mol NaOH}}{\text{L NaOH}}$$

$1 \text{ mol HCl} : 1 \text{ mol NaOH}$

$$\text{Molarity (M)} = \frac{\text{moles of solute (mol)}}{\text{volume of solution (L)}}$$

Solve The first part of the solution gives you the number of moles of HCl in the unknown solution.

Solution

$$12.54 \text{ mL NaOH} \times \frac{1 \text{ L}}{1000 \text{ mL}} \times \frac{0.100 \text{ mol NaOH}}{\text{L NaOH}}$$

$$\times \frac{1 \text{ mol HCl}}{1 \text{ mol NaOH}} = 1.25 \times 10^{-3} \text{ mol HCl}$$

In the second part of the solution, divide the number of moles of HCl by the volume of the HCl solution in L. 10.0 mL is equivalent to 0.010 L.

$$\text{Molarity} = \frac{1.25 \times 10^{-3} \text{ mol HCl}}{0.01000 \text{ L}} = 0.125 \text{ M HCl}$$

Check The units of the answer (M HCl) are correct. The magnitude of the answer (0.125 M) seems reasonable because it is similar to the molarity of the NaOH, solution, as expected from the reaction stoichiometry (1 mol HCl reacts with 1 mol NaOH) and the similar volumes of NaOH and HCl.

For Practice 4.14

The titration of a 20.0-mL sample of an H_2SO_4 solution of unknown concentration requires 22.87 mL of a 0.158 M KOH solution to reach the equivalence point. What is the concentration of the unknown H_2SO_4 solution?

For More Practice 4.14

What volume (in mL) of 0.200 M NaOH is required to titrate 35.00 mL of 0.140 M HBr to the equivalence point?

Gas-Evolution Reactions

Aqueous reactions that form a gas upon mixing two solutions are called *gas-evolution reactions*. Some gas-evolution reactions form a gaseous product directly when the cation of one reactant combines with the anion of the other. For example, when sulfuric acid reacts with lithium sulfide, dihydrogen sulfide gas is formed:

$$H_2SO_4(aq) + Li_2S(aq) \longrightarrow \underset{\text{gas}}{H_2S(g)} + Li_2SO_4(aq)$$

Other gas-evolution reactions often form an intermediate product that then decomposes into a gas. For example, when aqueous hydrochloric acid is mixed with aqueous sodium bicarbonate the following reaction occurs (Figure 4.19, page 166):

$$HCl(aq) + NaHCO_3(aq) \longrightarrow H_2CO_3(aq) + NaCl(aq) \longrightarrow H_2O(l) + \underset{\text{gas}}{CO_2(g)} + NaCl(aq)$$

The intermediate product, H_2CO_3, is not stable and decomposes into H_2O and gaseous CO_2. Other important gas-evolution reactions form either H_2SO_3 or NH_4OH as intermediate products:

$$HCl(aq) + NaHSO_3(aq) \longrightarrow H_2SO_3(aq) + NaCl(aq) \longrightarrow H_2O(l) + SO_2(g) + NaCl(aq)$$

$$NH_4Cl(aq) + NaOH(aq) \longrightarrow NH_4OH(aq) + NaCl(aq) \longrightarrow H_2O(l) + NH_3(g) + NaCl(aq)$$

Many gas-evolution reactions such as this one are also acid–base reactions. In Chapter 15 we will learn how ions such as CO_3^{2-} act as bases in aqueous solution.

The intermediate product NH_4OH provides a convenient way to think about this reaction, but the extent to which it actually forms is debatable.

Gas-Evolution Reaction

$$\text{NaHCO}_3(aq) + \text{HCl}(aq) \longrightarrow \text{H}_2\text{O}(l) + \text{NaCl}(aq) + \text{CO}_2(g)$$

When aqueous sodium bicarbonate is mixed with aqueous hydrochloric acid gaseous CO_2 bubbles are the result of the reaction.

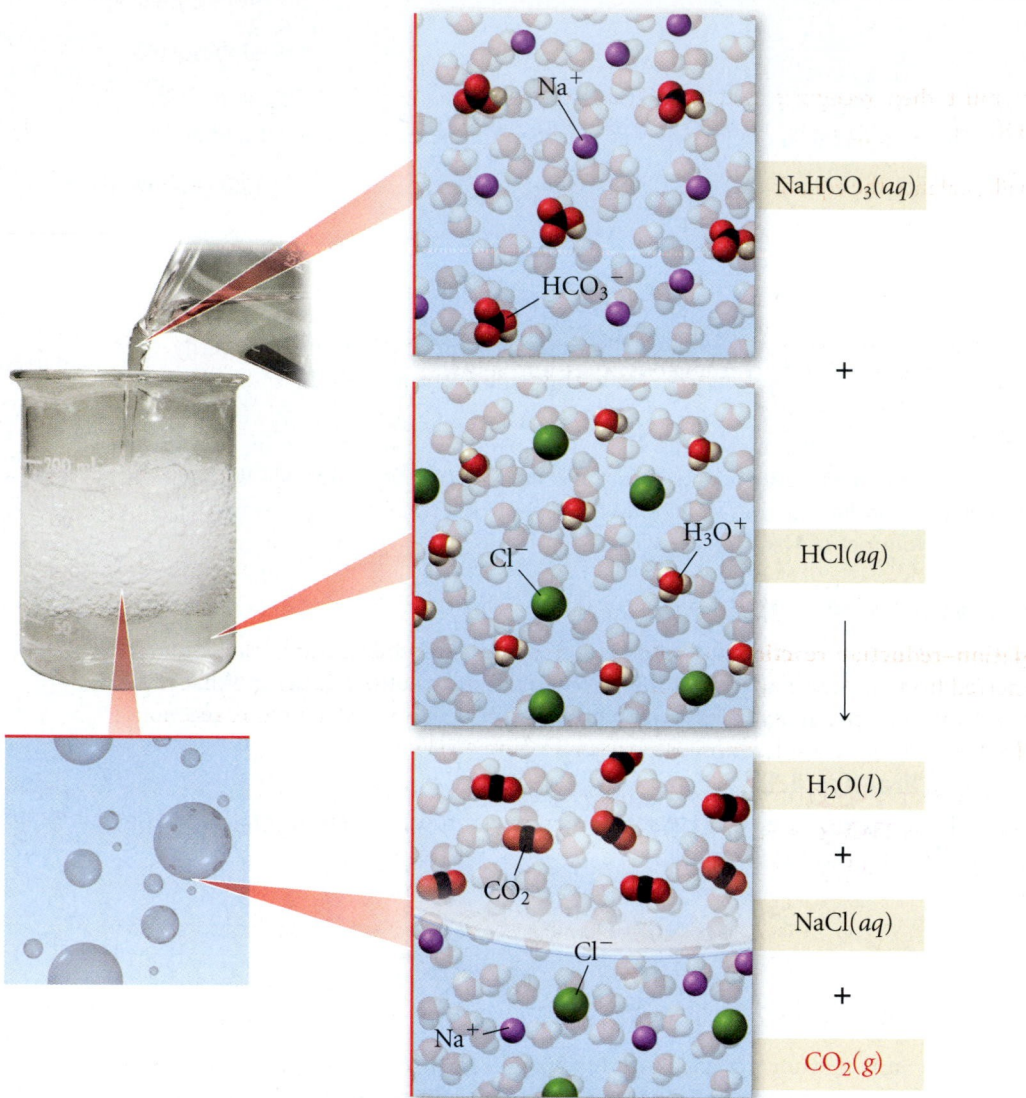

$\text{NaHCO}_3(aq)$

$+$

$\text{HCl}(aq)$

$\text{H}_2\text{O}(l)$

$+$

$\text{NaCl}(aq)$

$+$

$\text{CO}_2(g)$

▲ FIGURE 4.19 **Gas-Evolution Reaction** When aqueous hydrochloric acid is mixed with aqueous sodium bicarbonate, gaseous CO_2 bubbles out of the reaction mixture.

The main types of compounds that form gases in aqueous reactions, as well as the gases formed, are listed in Table 4.3.

Reactant Type	Intermediate Product	Gas Evolved	Example
Sulfides	None	H_2S	$2\,\text{HCl}(aq) + \text{K}_2\text{S}(aq) \rightarrow \text{H}_2\text{S}(g) + 2\,\text{KCl}(aq)$
Carbonates and bicarbonates	H_2CO_3	CO_2	$2\,\text{HCl}(aq) + \text{K}_2\text{CO}_3(aq) \rightarrow \text{H}_2\text{O}(l) + \text{CO}_2(g) + 2\,\text{KCl}(aq)$
Sulfites and bisulfites	H_2SO_3	SO_2	$2\,\text{HCl}(aq) + \text{K}_2\text{SO}_3(aq) \rightarrow \text{H}_2\text{O}(l) + \text{SO}_2(g) + 2\,\text{KCl}(aq)$
Ammonium	NH_4OH	NH_3	$\text{NH}_4\text{Cl}(aq) + \text{KOH}(aq) \rightarrow \text{H}_2\text{O}(l) + \text{NH}_3(g) + \text{KCl}(aq)$

TABLE 4.3 Types of Compounds That Undergo Gas-Evolution Reactions

EXAMPLE 4.15 Writing Equations for Gas-Evolution Reactions

Write a molecular equation for the gas-evolution reaction that occurs when you mix aqueous nitric acid and aqueous sodium carbonate.

Begin by writing a skeletal equation in which the cation of each reactant combines with the anion of the other.	$HNO_3(aq) + Na_2CO_3(aq) \longrightarrow$ $H_2CO_3(aq) + NaNO_3(aq)$
You must then recognize that $H_2CO_3(aq)$ decomposes into $H_2O(l)$ and $CO_2(g)$ and write these products into the equation.	$HNO_3(aq) + Na_2CO_3(aq) \longrightarrow$ $H_2O(l) + CO_2(g) + NaNO_3(aq)$
Finally, balance the equation.	$2\,HNO_3(aq) + Na_2CO_3(aq) \longrightarrow$ $H_2O(l) + CO_2(g) + 2\,NaNO_3(aq)$

For Practice 4.15

Write a molecular equation for the gas-evolution reaction that occurs when you mix aqueous hydrobromic acid and aqueous potassium sulfite.

For More Practice 4.15

Write a net ionic equation for the reaction that occurs when you mix hydroiodic acid with calcium sulfide.

4.9 Oxidation–Reduction Reactions

Oxidation–reduction reactions or **redox reactions** are reactions in which electrons are transferred from one reactant to the other. The rusting of iron, the bleaching of hair, and the production of electricity in batteries involve redox reactions. Many redox reactions involve the reaction of a substance with oxygen (Figure 4.20 ▼):

$$4\,Fe(s) + 3\,O_2(g) \longrightarrow 2\,Fe_2O_3(s) \qquad \text{(rusting of iron)}$$
$$2\,C_8H_{18}(l) + 25\,O_2(g) \longrightarrow 16\,CO_2(g) + 18\,H_2O(g) \qquad \text{(combustion of octane)}$$
$$2\,H_2(g) + O_2(g) \longrightarrow 2\,H_2O(g) \qquad \text{(combustion of hydrogen)}$$

Oxidation–reduction reactions are covered in more detail in Chapter 18.

Oxidation-Reduction Reaction

$$2\,H_2(g) + O_2(g) \longrightarrow 2\,H_2O(g)$$

Hydrogen and oxygen in the balloon react to form gaseous water.

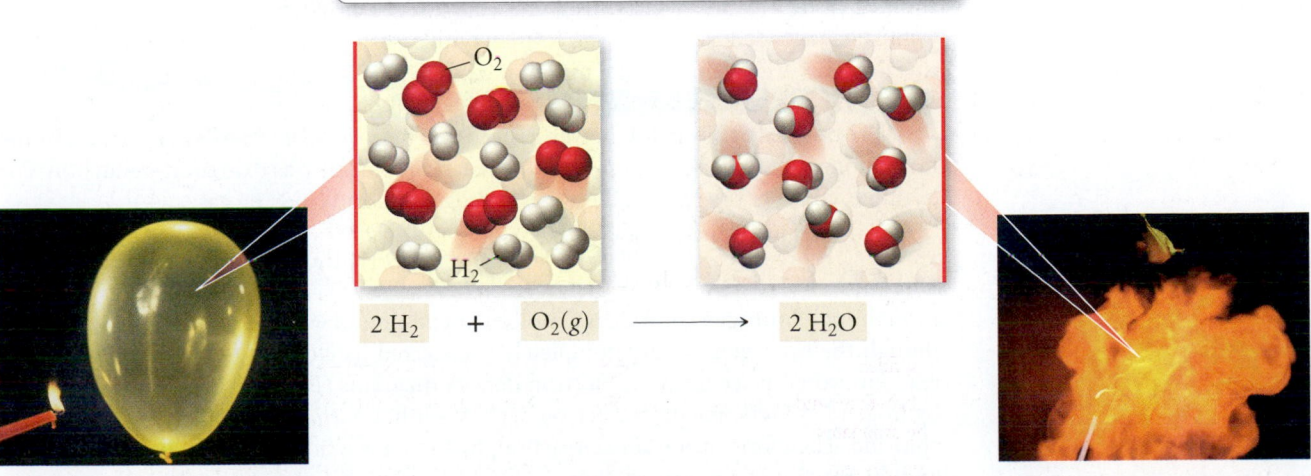

$$2\,H_2 \quad + \quad O_2(g) \longrightarrow 2\,H_2O$$

▲ **FIGURE 4.20 Oxidation–Reduction Reaction** The hydrogen in the balloon reacts with oxygen to form gaseous water.

Oxidation-Reduction Reaction without Oxygen

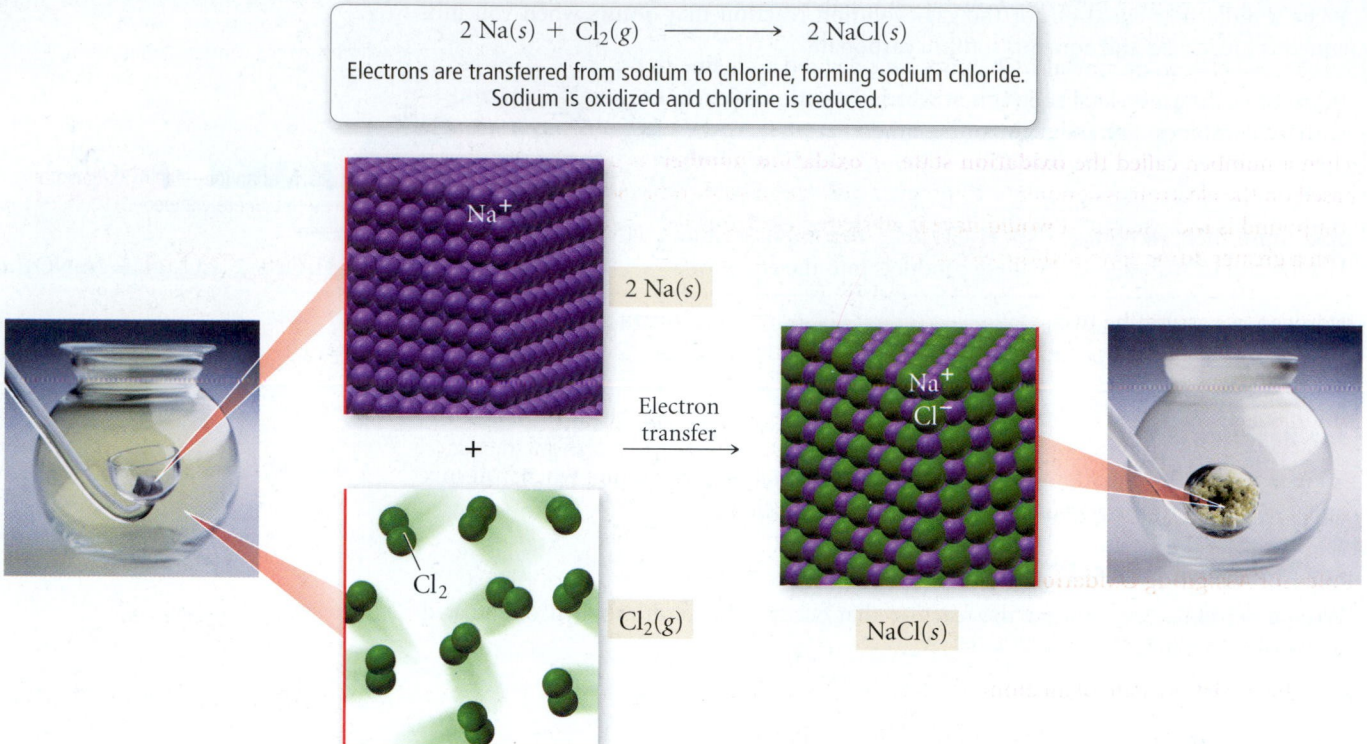

$$2\,\text{Na}(s) + \text{Cl}_2(g) \longrightarrow 2\,\text{NaCl}(s)$$

Electrons are transferred from sodium to chlorine, forming sodium chloride. Sodium is oxidized and chlorine is reduced.

Na$^+$

2 Na(s)

Electron transfer

+

Cl$_2$

Cl$_2$(g)

Na$^+$
Cl$^-$

NaCl(s)

▲ **FIGURE 4.21 Oxidation–Reduction without Oxygen** When sodium reacts with chlorine, electrons are transferred from the sodium to the chlorine, resulting in the formation of sodium chloride. In this redox reaction, sodium is oxidized and chlorine is reduced.

However, redox reactions need not involve oxygen. Consider, for example, the reaction between sodium and chlorine to form sodium chloride (NaCl), depicted in Figure 4.21 ▲:

$$2\,\text{Na}(s) + \text{Cl}_2(g) \longrightarrow 2\,\text{NaCl}(s)$$

This reaction is similar to the reaction between sodium and oxygen to form sodium oxide:

$$4\,\text{Na}(s) + \text{O}_2(g) \longrightarrow 2\,\text{Na}_2\text{O}(s)$$

Helpful Mnemonics O I L R I G— Oxidation Is Loss; Reduction Is Gain.

In both cases, a metal (which has a tendency to lose electrons) reacts with a nonmetal (which has a tendency to gain electrons). In both cases, metal atoms lose electrons to nonmetal atoms. A fundamental definition of **oxidation** is the loss of electrons, and a fundamental definition of **reduction** is the gain of electrons.

The transfer of electrons, however, need not be a *complete* transfer (as occurs in the formation of an ionic compound) for the reaction to qualify as oxidation–reduction. For example, consider the reaction between hydrogen gas and chlorine gas:

$$\text{H}_2(g) + \text{Cl}_2(g) \longrightarrow 2\,\text{HCl}(g)$$

Even though hydrogen chloride is a molecular compound with a covalent bond, and even though the hydrogen has not completely transferred its electron to chlorine during the reaction, you can see from the electron density diagrams (Figure 4.22, on page 169) that hydrogen has lost some of its electron density—it has *partially* transferred its electron to chlorine. Therefore, in the above reaction, hydrogen is oxidized and chlorine is reduced and the reaction is a redox reaction.

Oxidation States

Identifying a reaction between a metal and a nonmetal as a redox reaction is fairly straightforward because of ion formation. However, how do we identify redox reactions that occur between nonmetals? Chemists have devised a scheme to track electrons before and after a chemical reaction. In this scheme—which is like bookkeeping for electrons—all shared electrons are assigned to the atom that attracts the electrons most strongly. Then a number, called the **oxidation state** or **oxidation number**, is given to each atom based on the electron assignments. In other words, the oxidation number of an atom in a compound is the "charge" it would have if all shared electrons were assigned to the atom with a greater attraction for those electrons.

For example, consider HCl. Since chlorine attracts electrons more strongly than hydrogen, we assign the two shared electrons in the bond to chlorine; then H (which has lost an electron in our assignment) has an oxidation state of $+1$ and Cl (which has gained one electron in our assignment) has an oxidation state of -1. Notice that, in contrast to ionic charges, which are usually written with the sign of the charge after the magnitude ($1+$ and $1-$, for example), oxidation states are written with the sign of the charge before the magnitude ($+1$ and -1, for example). The following rules can be used to assign oxidation states to atoms in elements and compounds.

The ability of an element to attract electrons in a chemical bond is called electronegativity. Electronegativity is covered in more detail in Section 8.6.

Do not confuse oxidation state with ionic charge. Unlike ionic charge—which is a real property of an ion—the oxidation state of an atom is merely a theoretical (but useful) construct.

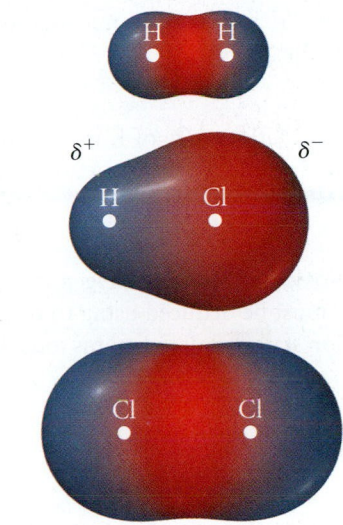

▲ FIGURE 4.22 **Redox with Partial Electron Transfer** When hydrogen bonds to chlorine, the electrons are unevenly shared, resulting in an increase of electron density (reduction) for chlorine and a decrease in electron density (oxidation) for hydrogen.

Rules for Assigning Oxidation States	**Examples**

(These rules are hierarchical. If any two rules conflict, follow the rule that is higher on the list.)

1. The oxidation state of an atom in a free element is 0.

    Cu 0 ox state Cl_2 0 ox state

2. The oxidation state of a monoatomic ion is equal to its charge.

    Ca^{2+} +2 ox state Cl^- −1 ox state

3. The sum of the oxidation states of all atoms in:

    • A neutral molecule or formula unit is 0.

    H_2O 2(H ox state) + 1(O ox state) = 0

    • An ion is equal to the charge of the ion.

    NO_3^- 1(N ox state) + 3(O ox state) = −1

4. In their compounds, metals have positive oxidation states.

    • Group 1A metals *always* have an oxidation state of $+1$.

    $NaCl$ +1 ox state

    • Group 2A metals *always* have an oxidation state of $+2$.

    CaF_2 +2 ox state

5. In their compounds, nonmetals are assigned oxidation states according to the table at right. Entries at the top of the table take precedence over entries at the bottom of the table.

When assigning oxidation states, keep these points in mind:

• The oxidation state of any given element generally depends on what other elements are present in the compound. (The exceptions are the group 1A and 2A metals, which are *always* $+1$ and $+2$, respectively.)

• Rule 3 must always be followed. Therefore, when following the hierarchy shown in rule 5, give priority to the element(s) highest on the list and then assign the oxidation state of the element lowest on the list using rule 3.

• When assigning oxidation states to elements that are not covered by rules 4 and 5 (such as carbon) use rule 3 to deduce their oxidation state once all other oxidation states have been assigned.

Nonmetal	Oxidation State	Example
Fluorine	−1	MgF_2 −1 ox state
Hydrogen	+1	H_2O +1 ox state
Oxygen	−2	CO_2 −2 ox state
Group 7A	−1	CCl_4 −1 ox state
Group 6A	−2	H_2S −2 ox state
Group 5A	−3	NH_3 −3 ox state

EXAMPLE 4.16 Assigning Oxidation States

Assign an oxidation state to each atom in each of the following compounds.

(a) Cl_2 **(b)** Na^+ **(c)** KF   **(d)** CO_2 **(e)** SO_4^{2-} **(f)** K_2O_2

Solution

Since Cl_2 is a free element, the oxidation state of both Cl atoms is 0 (rule 1)	**(a)** Cl_2 ClCl 0 0
Since Na^+ is a monoatomic ion, so the oxidation state of the Na^+ ion is +1 (rule 2).	**(b)** Na^+ Na^+ +1
The oxidation state of K is +1 (rule 4). The oxidation state of F is −1 (rule 5). Since this is a neutral compound, the sum of the oxidation states is 0.	**(c)** KF KF +1 −1 sum: +1 −1 = 0
The oxidation state of oxygen is −2 (rule 5). The oxidation state of carbon must be deduced using rule 3, which states that the sum of the oxidation states of all the atoms must be 0.	**(d)** CO_2 (C ox state) + 2(O ox state) = 0 (C ox state) + 2(−2) = 0 C ox state = +4 CO_2 +4 −2 sum: +4 + 2(−2) = 0
The oxidation state of oxygen is −2 (rule 5). We would ordinarily expect the oxidation state of S to be −2 (rule 5). However, if that were the case, the sum of the oxidation states would not equal the charge of the ion. Since O is higher on the list than S, it takes priority and the oxidation state of sulfur is computed by setting the *sum* of all of the oxidation states equal to −2 (the charge of the ion).	**(e)** SO_4^{2-} (S ox state) + 4(O ox state) = −2 (S ox state) + 4(−2) = −2 S ox state = +6 SO_4^{2-} +6 −2 sum: +6 + 4(−2) = −2
The oxidation state of potassium is +1 (rule 4). We would ordinarily expect the oxidation state of O to be −2 (rule 5), but rule 4 takes priority, and we deduce the oxidation state of O by setting the sum of all of the oxidation states equal to 0.	**(f)** K_2O_2 2(K ox state) + 2(O ox state) = 0 2(+1) + 2(O ox state) = 0 O ox state = −1 K_2O_2 +1 −1 sum: 2(+1) + 2(−1) = 0

For Practice 4.16

Assign an oxidation state to each atom in the following species.

(a) Cr   **(b)** Cr^{3+} **(c)** CCl_4 **(d)** $SrBr_2$ **(e)** SO_3 **(f)** NO_3^-

In most cases, oxidation states are positive or negative integers; however, on occasion an atom within a compound can have a fractional oxidation state. For example, consider KO_2 and Fe_3O_4. The oxidation states are assigned as follows:

$$KO_2 \qquad\qquad Fe_3O_4$$
$$+1 \; -\tfrac{1}{2} \qquad\qquad +\tfrac{8}{3} \; -2$$
$$\text{sum: } +1 + 2(-\tfrac{1}{2}) = 0 \qquad \text{sum: } 3(+\tfrac{8}{3}) + 4(-2) = 0$$

In KO_2, oxygen has a $-\tfrac{1}{2}$ oxidation state and in Fe_3O_4, iron has $+\tfrac{8}{3}$ oxidation state. Although these seem unusual, they are acceptable because oxidation states are merely an imposed electron bookkeeping scheme, not an actual physical quantity.

Identifying Redox Reactions

Oxidation states can be used to identify redox reactions, even between nonmetals. For example, is the following reaction between carbon and sulfur a redox reaction?

$$C + 2S \longrightarrow CS_2$$

If so, what element is oxidized? What element is reduced? We can use the oxidation state rules to assign oxidation states to all elements on both sides of the equation.

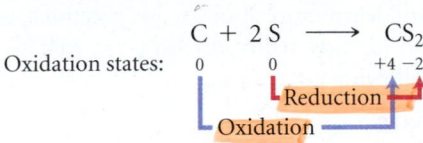

Carbon changed from an oxidation state of 0 to an oxidation state of $+4$. In terms of our electron bookkeeping scheme (the assigned oxidation state), carbon *lost electrons* and was *oxidized*. Sulfur changed from an oxidation state of 0 to an oxidation state of -2. In terms of our electron bookkeeping scheme, sulfur *gained electrons* and was *reduced*. In terms of oxidation states, oxidation and reduction are defined as follows.

➤ **Oxidation   An increase in oxidation state**

➤ **Reduction   A decrease in oxidation state**

Remember that a reduction is a *reduction* in oxidation state.

EXAMPLE 4.17 Using Oxidation States to Identify Oxidation and Reduction

Use oxidation states to identify the element that is being oxidized and the element that is being reduced in the following redox reaction.

$$Mg(s) + 2H_2O(l) \longrightarrow Mg(OH)_2(aq) + H_2(g)$$

Solution

Begin by assigning oxidation states to each atom in the reaction.

Oxidation states:

$$Mg(s) + 2H_2O(l) \longrightarrow Mg(OH)_2(aq) + H_2(g)$$
$$\quad 0 \qquad\quad +1\ -2 \qquad\quad +2\ -2\ +1 \qquad\quad 0$$

Reduction
Oxidation

Since Mg increased in oxidation state, it was oxidized. Since H decreased in oxidation state, it was reduced.

For Practice 4.17

Use oxidation states to identify the element that is being oxidized and the element that is being reduced in the following redox reaction.

$$Sn(s) + 4HNO_3(aq) \longrightarrow SnO_2(s) + 4NO_2(g) + 2H_2O(g)$$

For More Practice 4.17

Which of the following is a redox reaction? If the reaction is a redox reaction, identify which element is oxidized and which is reduced.

(a) $Hg_2(NO_3)_2(aq) + 2KBr(aq) \longrightarrow Hg_2Br_2(s) + 2KNO_3(aq)$

(b) $4Al(s) + 3O_2(g) \longrightarrow 2Al_2O_3(s)$

(c) $CaO(s) + CO_2(g) \longrightarrow CaCO_3(s)$

Notice that *oxidation and reduction must occur together.* If one substance loses electrons (oxidation) then another substance must gain electrons (reduction). A substance that causes the oxidation of another substance is called an **oxidizing agent.** Oxygen, for example, is an excellent oxidizing agent because it causes the oxidation of many other substances. In a redox reaction, *the oxidizing agent is always reduced.* A substance that causes the reduction of another substance is called a **reducing agent.** Hydrogen, for example, as well as the group 1A and group 2A metals (because of their tendency to lose electrons) are excellent reducing agents. In a redox reaction, *the reducing agent is always oxidized.*

In Section 18.2 you will learn more about redox reactions, including how to balance them. For now, be able to identify redox reactions, as well as oxidizing and reducing agents, according to the following guidelines.

Redox reactions include:
- Any reaction in which there is a change in the oxidation states of atoms in going from the reactants to the products.

In a redox reaction:
- The oxidizing agent oxidizes another substance (and is itself reduced).
- The reducing agent reduces another substance (and is itself oxidized).

EXAMPLE 4.18 Identifying Redox Reactions, Oxidizing Agents, and Reducing Agents

Determine whether each of the following reactions is an oxidation–reduction reaction. If the reaction is an oxidation–reduction, identify the oxidizing agent and the reducing agent.

(a) $2\,Mg(s) + O_2(g) \longrightarrow 2\,MgO(s)$

(b) $2\,HBr(aq) + Ca(OH)_2(aq) \longrightarrow 2\,H_2O(l) + CaBr_2(aq)$

(c) $Zn(s) + Fe^{2+}(aq) \longrightarrow Zn^{2+}(aq) + Fe(s)$

Solution

This is a redox reaction because magnesium increases in oxidation number (oxidation) and oxygen decreases in oxidation number (reduction).	(a) $2\,Mg(s) + O_2(g) \longrightarrow 2\,MgO(s)$ Oxidizing agent: O_2 Reducing agent: Mg
This is not a redox reaction because none of the atoms undergoes a change in oxidation number	(b) $2\,HBr(aq) + Ca(OH)_2(aq) \rightarrow 2\,H_2O(l) + CaBr_2(aq)$
This is a redox reaction because zinc increases in oxidation number (oxidation) and iron decreases in oxidation number (reduction).	(c) $Zn(s) + Fe^{2+}(aq) \longrightarrow Zn^{2+}(aq) + Fe(s)$ Oxidizing agent: Fe^{2+} Reducing agent: Zn

For Practice 4.18

Which of the following is a redox reaction? For all redox reactions, identify the oxidizing agent and the reducing agent.

(a) $2\,Li(s) + Cl_2(g) \longrightarrow 2\,LiCl(s)$

(b) $2\,Al(s) + 3\,Sn^{2+}(aq) \longrightarrow 2\,Al^{3+}(aq) + 3\,Sn(s)$

(c) $Pb(NO_3)_2(aq) + 2\,LiCl(aq) \longrightarrow PbCl_2(s) + 2\,LiNO_3(aq)$

(d) $C(s) + O_2(g) \longrightarrow CO_2(g)$

Conceptual Connection 4.5 Oxidation and Reduction

Which of the following statements is true regarding redox reactions?

(a) A redox reaction involves *either* the transfer of an electron *or* a change in the oxidation state of an element.

(b) If any of the reactants or products in a reaction contains oxygen, the reaction is a redox reaction.

(c) In a reaction, oxidation can occur independently of reduction.

(d) In a redox reaction, any increase in the oxidation state of a reactant must be accompanied by a decrease in the oxidation state of a reactant.

Answer: **(d)** Since oxidation and reduction must occur together, an increase in the oxidation state of a reactant will always be accompanied by a decrease in the oxidation state of a reactant.

Chemistry in Your Day
Bleached Blonde

Have you ever bleached your hair? Most home kits for hair bleaching contain hydrogen peroxide (H_2O_2), an excellent oxidizing agent. When applied to hair, hydrogen peroxide oxidizes melanin, the dark pigment that gives hair color. Once melanin is oxidized, it no longer imparts a dark color to hair, leaving the hair with the familiar bleached look. Hydrogen peroxide also oxidizes other components of hair. For example, protein molecules in hair contain —SH groups called thiols. Hydrogen peroxide oxidizes these thiol groups to sulfonic acid groups, —SO_3H. The oxidation of thiol groups to sulfonic acid groups causes changes in the proteins that compose hair, making the hair more brittle and easier to tangle. Consequently, people with heavily bleached hair must use conditioners, which contain compounds that form thin, lubricating coatings on individual hair shafts. These coatings prevent tangling and make hair softer and more manageable.

Question

The following is a reaction of hydrogen peroxide with an alkene:

$$H_2O_2 + C_2H_4 \longrightarrow C_2H_4O + H_2O$$

Can you see why this reaction is a redox reaction? Can you identify the oxidizing and reducing agents?

▶ The bleaching of hair involves a redox reaction in which melanin—the main pigment in hair—is oxidized.

Combustion Reactions

We encountered combustion reactions, a type of redox reaction, in the opening section of this chapter. Combustion reactions are important because most of our society's energy is derived from them (Figure 4.23 ▼).

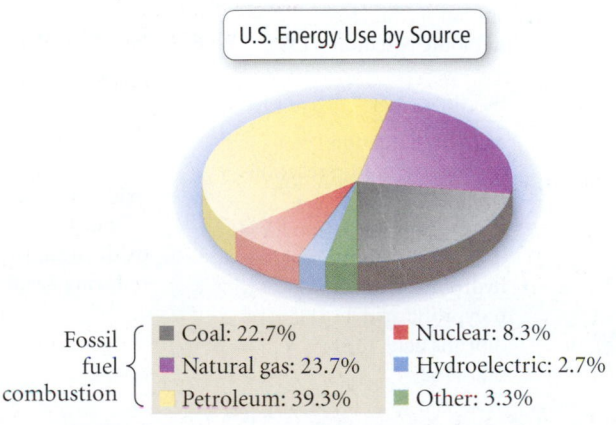

U.S. Energy Use by Source

Fossil fuel combustion {
■ Coal: 22.7%
■ Natural gas: 23.7%
■ Petroleum: 39.3%

■ Nuclear: 8.3%
■ Hydroelectric: 2.7%
■ Other: 3.3%

◀ **FIGURE 4.23 U. S. Energy Consumption**
Over 85% of the energy used in the United States is produced by combustion reactions.

As we learned in Section 4.1, *combustion reactions* are characterized by the reaction of a substance with O_2 to form one or more oxygen-containing compounds, often including water. Combustion reactions also emit heat. For example, as we saw earlier in this chapter, natural gas (CH_4) reacts with oxygen to form carbon dioxide and water:

$$CH_4(g) + 2\,O_2(g) \longrightarrow CO_2(g) + 2\,H_2O(g)$$

Oxidation state: $-4\ +1$ 0 $+4\ -2$ $+1\ -2$

In this reaction, carbon is oxidized and oxygen is reduced. Ethanol, the alcohol in alcoholic beverages, also reacts with oxygen in a combustion reaction to form carbon dioxide and water:

$$C_2H_5OH(l) + 3\,O_2(g) \longrightarrow 2\,CO_2(g) + 3\,H_2O(g)$$

Compounds containing carbon and hydrogen—or carbon, hydrogen, and oxygen—always form carbon dioxide and water upon complete combustion. Other combustion reactions include the reaction of carbon with oxygen to form carbon dioxide:

$$C(s) + O_2(g) \longrightarrow CO_2(g)$$

and the reaction of hydrogen with oxygen to form water:

$$2\,H_2(g) + O_2(g) \longrightarrow 2\,H_2O(g)$$

EXAMPLE 4.19 Writing Equations for Combustion Reactions

Write a balanced equation for the combustion of liquid methyl alcohol (CH_3OH).

Solution

Begin by writing a skeletal equation showing the reaction of CH_3OH with O_2 to form CO_2 and H_2O.	$CH_3OH(l) + O_2(g) \longrightarrow CO_2(g) + H_2O(g)$
Balance the skeletal equation using the guidelines in Section 3.10.	$2\,CH_3OH(l) + 3\,O_2(g) \longrightarrow 2\,CO_2(g) + 4\,H_2O(g)$

For Practice 4.19

Write a balanced equation for the complete combustion of liquid C_2H_5SH.

Chapter in Review

Key Terms

Section 4.2

stoichiometry (133)

Section 4.3

limiting reactant (137)
theoretical yield (137)
actual yield (138)
percent yield (138)

Section 4.4

solution (144)
solvent (144)
solute (144)
aqueous solution (144)
dilute solution (144)
concentrated solution (144)

molarity (M) (144)
stock solution (146)

Section 4.5

electrolyte (151)
strong electrolyte (151)
nonelectrolyte (152)
strong acid (152)
weak acid (153)
weak electrolyte (153)
soluble (154)
insoluble (154)

Section 4.6

precipitation reaction (155)
precipitate (155)

Section 4.7

molecular equation (159)
complete ionic equation (159)
spectator ion (159)
net ionic equation (159)

Section 4.8

acid–base reaction
(neutralization reaction)
(160)
gas-evolution reaction (160)
Arrhenius definitions (161)
hydronium ion (161)
polyprotic acid (161)
diprotic acid (161)
salt (162)

titration (163)
equivalence point (163)
indicator (164)

Section 4.9

oxidation–reduction (redox)
reaction (167)
oxidation (167)
reduction (167)
oxidation state (oxidation
number) (169)
oxidizing agent (172)
reducing agent (172)

Key Concepts

Global Warming and the Combustion of Fossil Fuels (4.1)

Greenhouse gases are not in themselves harmful; they warm Earth by trapping some of the sunlight that penetrates Earth's atmosphere. However, global warming, caused by rising atmospheric carbon dioxide levels, is potentially harmful. The largest carbon dioxide source is, arguably, the burning of fossil fuels. This can be verified by reaction stoichiometry.

Reaction Stoichiometry (4.2)

Reaction stoichiometry refers to the numerical relationships between the reactants and products in a balanced chemical equation. Reaction stoichiometry allows us to predict, for example, the amount of product that can be formed for a given amount of reactant, or how much of one reactant is required to react with a given amount of another.

Limiting Reactant, Theoretical Yield, and Percent Yield (4.3)

When a chemical reaction actually occurs, the reactants are usually not in the exact stoichiometric ratios specified by the balanced chemical equation. The limiting reactant is the one that is present in the smallest stoichiometric quantity—it will be completely consumed in the reaction and it limits the amount of product that can be made. Any reactant that does not limit the amount of product is said to be in excess. The amount of product that can be made from the limiting reactant is the theoretical yield. The actual yield—always equal to or less than the theoretical yield—is the amount of product that is actually made when the reaction is carried out. The percentage of the theoretical yield that is actually produced is the percent yield.

Solution Concentration and Stoichiometry (4.4)

The concentration of a solution is often expressed in molarity, the number of moles of solute per liter of solution. The molarities and volumes of reactant solutions can be used to predict the amount of product that will form in an aqueous reaction.

Aqueous Solutions and Precipitation Reactions (4.5, 4.6)

An aqueous solution is a homogeneous mixture of water, the solvent, with another substance, called the solute. Solutes that completely dissociate (or completely ionize in the case of the acids) to ions in solution are called strong electrolytes and are good conductors of electricity. Solutes that only partially dissociate (or partially ionize) are called weak electrolytes, and solutes that do not dissociate (or ionize) at all are called nonelectrolytes. A substance that dissolves in water to form a solution is said to be soluble.

In a precipitation reaction, two aqueous solutions are mixed and a solid—or precipitate—forms. The solubility rules are an empirical set of guidelines that help predict the solubilities of ionic compounds; these rules are especially useful when determining whether or not a precipitate will form.

Equations for Aqueous Reactions (4.7)

An aqueous reaction can be represented with a molecular equation, which shows the complete neutral formula for each compound in the reaction. Alternatively, it can be represented with a complete ionic equation, which shows the dissociated nature of the aqueous ionic compounds. A third representation is the net ionic equation, in which the spectator ions—those that do not change in the course of the reaction—are left out of the equation.

Acid–Base and Gas-Evolution Reactions (4.8)

In an acid–base reaction, an acid, a substance which produces H^+ in solution, reacts with a base, a substance which produces OH^- in solution, and the two neutralize each other, producing water (or in some cases a weak electrolyte). An acid–base titration is a laboratory procedure in which a reaction is carried to its equivalence point—the point at which the reactants are in exact stoichiometric proportions; titrations are useful in determining the concentrations of unknown solutions. In gas-evolution reactions, two aqueous solutions are combined and a gas is produced.

Oxidation–Reduction Reactions (4.9)

In oxidation–reduction reactions, one substance transfers electrons to another substance. The substance that loses electrons is oxidized and the one that gains them is reduced. An oxidation state is a fictitious charge given to each atom in a redox reaction by assigning all shared electrons to the atom with the greater attraction for those electrons. The oxidation state of an atom increases upon oxidation and decreases upon reduction. A combustion reaction is a specific type of oxidation-reduction reaction in which a substance reacts with oxygen— emitting heat and forming one or more oxygen containing products.

Key Equations and Relationships

Mass-to-Mass Conversion: Stoichiometry (4.2)

mass A $\rightarrow$ amount A (in moles) $\rightarrow$ amount B (in moles) $\rightarrow$ mass B

Percent Yield (4.3)

$$\% \text{ yield} = \frac{\text{actual yield}}{\text{theoretical yield}} \times 100\%$$

Molarity (M): Solution Concentration (4.4)

$$M = \frac{\text{amount of solute (in mol)}}{\text{volume of solution (in L)}}$$

Solution Dilution (4.4)

$$M_1 V_1 = M_2 V_2$$

Solution Stoichiometry (4.4)

volume A $\rightarrow$ amount A (in moles) $\rightarrow$

amount B (in moles) $\rightarrow$ volume B

Key Skills

Calculations Involving the Stoichiometry of a Reaction (4.2)
- Examples 4.1, 4.2 • For Practice 4.1, 4.2 • Exercises 31–36

Determining the Limiting Reactant and Calculating Theoretical and Percent Yield (4.3)
- Examples 4.3, 4.4 • For Practice 4.3, 4.4 • Exercises 43–47

Calculating and Using Molarity as a Conversion Factor (4.4)
- Examples 4.5, 4.6 • For Practice 4.5, 4.6 • For More Practice 4.5, 4.6 • Exercises 49–54

Determining Solution Dilutions (4.4)
- Example 4.7 • For Practice 4.7 • For More Practice 4.7 • Exercises 57, 58

Using Solution Stoichiometry to Find Volumes and Amounts (4.4)
- Example 4.8 • For Practice 4.8 • For More Practice 4.8 • Exercises 59–61

Predicting whether a Compound Is Soluble (4.5)
- Example 4.9 • For Practice 4.9 • Exercises 65, 66

Writing Equations for Precipitation Reactions (4.6)
- Examples 4.10, 4.11 • For Practice 4.10, 4.11 • Exercises 67–70

Writing Complete Ionic and Net Ionic Equations (4.7)
- Example 4.12 • For Practice 4.12 • For More Practice 4.12 • Exercises 71, 72

Writing Equations for Acid–Base Reactions (4.8)
- Example 4.13 • For Practice 4.13 • Exercises 75, 76

Calculations Involving Acid–Base Titrations (4.8)
- Example 4.14 • For Practice 4.14 • For More Practice 4.14 • Exercises 79, 80

Writing Equations for Gas-Evolution Reactions (4.8)
- Example 4.15 • For Practice 4.15 • For More Practice 4.15 • Exercises 81, 82

Assigning Oxidation States (4.9)
- Example 4.16 • For Practice 4.16 • Exercises 83–86

Identifying Redox Reactions, Oxidizing Agents, and Reducing Agents Using Oxidation States (4.9)
- Examples 4.17, 4.18 • For Practice 4.17, 4.18 • For More Practice 4.17 • Exercises 87, 88

Writing Equations for Combustion Reactions (4.9)
- Example 4.19 • For Practice 4.19 • Exercises 89, 90

Exercises

Review Questions

1. What is reaction stoichiometry? What is the significance of the coefficients in a balanced chemical equation?
2. In a chemical reaction, what is the limiting reactant? The theoretical yield? The percent yield? What is meant by a reactant being in excess?
3. The percent yield is normally calculated using the actual yield and theoretical yield in units of mass (g or kg). Would the percent yield be different if the actual yield and theoretical yield were in units of amount (moles)?
4. What is an aqueous solution? What is the difference between the solute and the solvent?
5. What is molarity? How is it useful?
6. Explain how a strong electrolyte, a weak electrolyte, and a non-electrolyte differ.
7. Explain the difference between a strong acid and a weak acid.
8. What does it mean for a compound to be soluble? Insoluble?
9. What are the solubility rules? How are they useful?
10. What are the cations and anions whose compounds are usually soluble? What are the exceptions? What are the anions whose compounds are mostly insoluble? What are the exceptions?

11. What is a precipitation reaction? Give an example.
12. How can you predict whether a precipitation reaction will occur upon mixing two aqueous solutions?
13. Explain how a molecular equation, a complete ionic equation, and a net ionic equation differ.
14. What are the Arrhenius definitions of an acid and a base?
15. What is an acid–base reaction? Give an example.
16. Explain the principles behind an acid–base titration. What is an indicator?
17. What is a gas-evolution reaction? Give an example.
18. What reactant types give rise to gas-evolution reactions?
19. What is an oxidation–reduction reaction? Give an example.
20. What are oxidation states?
21. How can oxidation states be used to identify redox reactions?
22. What happens to a substance when it becomes oxidized? Reduced?
23. In a redox reaction, which reactant is the oxidizing agent? The reducing agent?
24. What is a combustion reaction? Why are they important? Give an example.

Problems by Topic

Reaction Stoichiometry

25. Consider the following unbalanced equation for the combustion of hexane:

$$C_6H_{14}(g) + O_2(g) \rightarrow CO_2(g) + H_2O(g)$$

Balance the equation and determine how many moles of O_2 are required to react completely with 4.9 moles C_6H_{14}.

26. Consider the following unbalanced equation for the neutralization of acetic acid:

$$HC_2H_3O_2(aq) + Ba(OH)_2(aq) \rightarrow H_2O(l) + Ba(C_2H_3O_2)_2(aq)$$

Balance the equation and determine how many moles of $Ba(OH)_2$ are required to completely neutralize 0.107 mole of $HC_2H_3O_2$.

27. For the reaction shown, calculate how many moles of NO_2 form when each of the following completely reacts.

$$2\,N_2O_5(g) \rightarrow 4\,NO_2(g) + O_2(g)$$

a. 1.3 mol N_2O_5 **b.** 5.8 mol N_2O_5
c. 10.5 g N_2O_5 **d.** 1.55 kg N_2O_5

28. For the reaction shown, calculate how many moles of NH_3 form when each of the following completely reacts.

$$3\,N_2H_4(l) \rightarrow 4\,NH_3(g) + N_2(g)$$

a. 5.3 mol N_2H_4 **b.** 2.28 mol N_2H_4
c. 32.5 g N_2H_4 **d.** 14.7 kg N_2H_4

29. Consider the following balanced equation:

$$SiO_2(s) + 3\,C(s) \rightarrow SiC(s) + 2\,CO(g)$$

Complete the following table showing the appropriate number of moles of reactants and products. If the number of moles of a *reactant* is provided, fill in the required amount of the other reactant, as well as the moles of each product formed. If the number of moles of a *product* is provided, fill in the required amount of each reactant to make that amount of product, as well as the amount of the other product that is made.

Mol SiO₂	Mol C	Mol SiC	Mol CO
3			
	6		
			10
2.8			
	1.55		

30. Consider the following balanced equation:

$$2\,N_2H_4(g) + N_2O_4(g) \rightarrow 3\,N_2(g) + 4\,H_2O(g)$$

Complete the following table showing the appropriate number of moles of reactants and products. If the number of moles of a reactant is provided, fill in the required amount of the other reactant, as well as the moles of each product formed. If the number of moles of a product is provided, fill in the required amount of each reactant to make that amount of product, as well as the amount of the other product that is made.

Mol N₂H₄	Mol N₂O₄	Mol N₂	Mol H₂O
2			
	5		
			10
2.5			
	4.2		
		11.8	

31. Hydrobromic acid dissolves solid iron according to the following reaction:

$$Fe(s) + 2\,HBr(aq) \rightarrow FeBr_2(aq) + H_2(g)$$

What mass of HBr (in g) would you need to dissolve a 3.2-g pure iron bar on a padlock? What mass of H_2 would be produced by the complete reaction of the iron bar?

32. Sulfuric acid dissolves aluminum metal according to the following reaction:

$$2\,Al(s) + 3\,H_2SO_4(aq) \rightarrow Al_2(SO_4)_3(aq) + 3\,H_2(g)$$

Suppose you wanted to dissolve an aluminum block with a mass of 15.2 g. What minimum mass of H_2SO_4 (in g) would you need? What mass of H_2 gas (in g) would be produced by the complete reaction of the aluminum block?

33. For each of the reactions shown, calculate the mass (in grams) of the product formed when 2.5 g of the underlined reactant completely reacts. Assume that there is more than enough of the other reactant.
a. $\underline{Ba(s)} + Cl_2(g) \rightarrow BaCl_2(s)$
b. $\underline{CaO(s)} + CO_2(g) \rightarrow CaCO_3(s)$
c. $2\,\underline{Mg(s)} + O_2(g) \rightarrow 2\,MgO(s)$
d. $4\,\underline{Al(s)} + 3\,O_2(g) \rightarrow 2\,Al_2O_3(s)$

34. For each of the reactions shown, calculate the mass (in grams) of the product formed when 10.4 g of the underlined reactant completely reacts. Assume that there is more than enough of the other reactant.
a. $2\,K(s) + \underline{Cl_2(g)} \rightarrow 2\,KCl(s)$
b. $2\,K(s) + \underline{Br_2(l)} \rightarrow 2\,KBr(s)$
c. $4\,Cr(s) + 3\,\underline{O_2(g)} \rightarrow 2\,Cr_2O_3(s)$
d. $2\,\underline{Sr(s)} + O_2(g) \rightarrow 2\,SrO(s)$

35. For each of the following acid–base reactions, calculate the mass (in grams) of each acid necessary to completely react with and neutralize 4.85 g of the base.
a. $HCl(aq) + NaOH(aq) \rightarrow H_2O(l) + NaCl(aq)$
b. $2\,HNO_3(aq) + Ca(OH)_2(aq) \rightarrow 2\,H_2O(l) + Ca(NO_3)_2(aq)$
c. $H_2SO_4(aq) + 2\,KOH(aq) \rightarrow 2\,H_2O(l) + K_2SO_4(aq)$

36. For each of the following precipitation reactions, calculate how many grams of the first reactant are necessary to completely react with 55.8 g of the second reactant.
a. $2\,KI(aq) + Pb(NO_3)_2(aq) \rightarrow PbI_2(s) + 2\,KNO_3(aq)$
b. $Na_2CO_3(aq) + CuCl_2(aq) \rightarrow CuCO_3(s) + 2\,NaCl(aq)$
c. $K_2SO_4(aq) + Sr(NO_3)_2(aq) \rightarrow SrSO_4(s) + 2\,KNO_3(aq)$

Limiting Reactant, Theoretical Yield, and Percent Yield

37. For the reaction shown, find the limiting reactant for each of the following initial amounts of reactants.

$$2\,Na(s) + Br_2(g) \rightarrow 2\,NaBr(s)$$

a. 2 mol Na, 2 mol Br_2 b. 1.8 mol Na, 1.4 mol Br_2
c. 2.5 mol Na, 1 mol Br_2 d. 12.6 mol Na, 6.9 mol Br_2

38. For the reaction shown, find the limiting reactant for each of the following initial amounts of reactants.

$$4\,Al(s) + 3\,O_2(g) \rightarrow 2\,Al_2O_3(s)$$

a. 1 mol Al, 1 mol O_2 b. 4 mol Al, 2.6 mol O_2
c. 16 mol Al, 13 mol O_2 d. 7.4 mol Al, 6.5 mol O_2

39. Consider the following reaction:

$$4\,HCl(g) + O_2(g) \rightarrow 2\,H_2O(g) + 2\,Cl_2(g)$$

Each of the following molecular diagrams represents an initial mixture of the reactants. How many molecules of Cl_2 would be formed from the reaction mixture that produces the greatest amount of products?

(a)   (b)   (c)

40. Consider the following reaction:

$$2\,CH_3OH(g) + 3\,O_2(g) \rightarrow 2\,CO_2(g) + 4\,H_2O(g)$$

Each of the following molecular diagrams represents an initial mixture of the reactants. How many CO_2 molecules would be formed from the reaction mixture that produces the greatest amount of products?

(a)   (b)   (c)

41. For the reaction shown, compute the theoretical yield of the product (in moles) for each of the following initial amounts of reactants.

$$Ti(s) + 2\,Cl_2(g) \rightarrow TiCl_4(s)$$

a. 4 mol Ti, 4 mol Cl_2
b. 7 mol Ti, 17 mol Cl_2
c. 12.4 mol Ti, 18.8 mol Cl_2

42. For the reaction shown, compute the theoretical yield of product (in moles) for each of the following initial amounts of reactants.

$$2\,Mn(s) + 2\,O_2(g) \rightarrow 2\,MnO_2(s)$$

a. 3 mol Mn, 3 mol O_2
b. 4 mol Mn, 7 mol O_2
c. 27.5 mol Mn, 43.8 mol O_2

43. For the reaction shown, compute the theoretical yield of product (in grams) for each of the following initial amounts of reactants.

$$2\,Al(s) + 3\,Cl_2(g) \rightarrow 2\,AlCl_3(s)$$

a. 2.0 g Al, 2.0 g Cl_2
b. 7.5 g Al, 24.8 g Cl_2
c. 0.235 g Al, 1.15 g Cl_2

44. For the reaction shown, compute the theoretical yield of the product (in grams) for each of the following initial amounts of reactants.

$$Ti(s) + 2\,F_2(g) \rightarrow TiF_4(s)$$

a. 5.0 g Ti, 5.0 g F_2
b. 2.4 g Ti, 1.6 g F_2
c. 0.233 g Ti, 0.288 g F_2

45. Lead ions can be precipitated from solution with KCl according to the following reaction:

$$Pb^{2+}(aq) + 2\,KCl(aq) \rightarrow PbCl_2(s) + 2\,K^+(aq)$$

When 28.5 g KCl is added to a solution containing 25.7 g Pb^{2+}, a $PbCl_2$ precipitate forms. The precipitate is filtered and dried and found to have a mass of 29.4 g. Determine the limiting reactant, theoretical yield of $PbCl_2$, and percent yield for the reaction.

46. Magnesium oxide can be made by heating magnesium metal in the presence of the oxygen. The balanced equation for the reaction is

$$2\,Mg(s) + O_2(g) \rightarrow 2\,MgO(s)$$

When 10.1 g of Mg is allowed to react with 10.5 g O_2, 11.9 g MgO is collected. Determine the limiting reactant, theoretical yield, and percent yield for the reaction.

47. Urea (CH_4N_2O) is a common fertilizer that can be synthesized by the reaction of ammonia (NH_3) with carbon dioxide as follows:

$$2\,NH_3(aq) + CO_2(aq) \rightarrow CH_4N_2O(aq) + H_2O(l)$$

In an industrial synthesis of urea, a chemist combines 136.4 kg of ammonia with 211.4 kg of carbon dioxide and obtains 168.4 kg of urea. Determine the limiting reactant, theoretical yield of urea, and percent yield for the reaction.

48. Many computer chips are manufactured from silicon, which occurs in nature as SiO_2. When SiO_2 is heated to melting, it reacts with solid carbon to form liquid silicon and carbon monoxide gas. In an industrial preparation of silicon, 155.8 kg of SiO_2 is allowed to react with 78.3 kg of carbon to produce 66.1 kg of silicon. Determine the limiting reactant, theoretical yield, and percent yield for the reaction.

Solution Concentration and Solution Stoichiometry

49. Calculate the molarity of each of the following solutions.
a. 4.3 mol of LiCl in 2.8 L solution
b. 22.6 g $C_6H_{12}O_6$ in 1.08 L of solution
c. 45.5 mg NaCl in 154.4 mL of solution

50. Calculate the molarity of each of the following solutions.
a. 0.11 mol of $LiNO_3$ in 5.2 L of solution
b. 61.3 g C_2H_6O in 2.44 L of solution
c. 15.2 mg KI in 102 mL of solution

51. How many moles of KCl are contained in each of the following?
a. 0.556 L of a 2.3 M KCl solution
b. 1.8 L of a 0.85 M KCl solution
c. 114 mL of a 1.85 M KCl solution

52. What volume of 0.200 M ethanol solution contains each of the following number of moles of ethanol?
a. 0.45 mol ethanol
b. 1.22 mol ethanol
c. 1.2×10^{-2} mol ethanol

53. A laboratory procedure calls for making 400.0 mL of a 1.1 M $NaNO_3$ solution. What mass of $NaNO_3$ (in g) is needed?

54. A chemist wants to make 5.5 L of a 0.300 M $CaCl_2$ solution. What mass of $CaCl_2$ (in g) should the chemist use?

55. If 123 mL of a 1.1 M glucose solution is diluted to 500.0 mL, what is the molarity of the diluted solution?

56. If 3.5 L of a 4.8 M $SrCl_2$ solution is diluted to 45 L, what is the molarity of the diluted solution?

57. To what volume should you dilute 50.0 mL of a 12 M stock HNO_3 solution to obtain a 0.100 M HNO_3 solution?

58. To what volume should you dilute 25 mL of a 10.0 M H_2SO_4 solution to obtain a 0.150 M H_2SO_4 solution?

59. Consider the following precipitation reaction:

$$2\,Na_3PO_4(aq) + 3\,CuCl_2(aq) \rightarrow Cu_3(PO_4)_2(s) + 6\,NaCl(aq)$$

What volume of 0.175 M Na_3PO_4 solution is necessary to completely react with 95.4 mL of 0.102 M $CuCl_2$?

60. Consider the following reaction:

$$Li_2S(aq) + Co(NO_3)_2(aq) \rightarrow 2\,LiNO_3(aq) + CoS(s)$$

What volume of 0.150 M Li_2S solution is required to completely react with 125 mL of 0.150 M $Co(NO_3)_2$?

61. What is the minimum amount of 6.0 M H_2SO_4 necessary to produce 25.0 g of H_2 (g) according to the following reaction?

$$2\,Al(s) + 3\,H_2SO_4(aq) \rightarrow Al_2(SO_4)_3(aq) + 3\,H_2(g)$$

62. What is the molarity of $ZnCl_2$ that forms when 25.0 g of zinc completely reacts with $CuCl_2$ according to the following reaction? Assume a final volume of 275 mL.

$$Zn(s) + CuCl_2(aq) \rightarrow ZnCl_2(aq) + Cu(s)$$

Types of Aqueous Solutions and Solubility

63. Each of the following compounds is soluble in water. For each compound, do you expect the resulting aqueous solution to conduct electrical current?
   a. CsCl   **b.** CH_3OH **c.** $Ca(NO_2)_2$ **d.** $C_6H_{12}O_6$

64. Classify each of the following as a strong electrolyte or nonelectrolyte.
   a. $MgBr_2$ **b.** $C_{12}H_{22}O_{11}$ **c.** Na_2CO_3 **d.** KOH

65. Determine whether each of the following compounds is soluble or insoluble. If the compound is soluble, write the ions present in solution.
   a. $AgNO_3$ **b.** $Pb(C_2H_3O_2)_2$ **c.** KNO_3 **d.** $(NH_4)_2S$

66. Determine whether each of the following compounds is soluble or insoluble. For the soluble compounds, write the ions present in solution.
   a. AgI   **b.** $Cu_3(PO_4)_2$ **c.** $CoCO_3$ **d.** K_3PO_4

Precipitation Reactions

67. Complete and balance each of the following equations. If no reaction occurs, write NO REACTION.
   a. $LiI(aq) + BaS(aq) \rightarrow$
   b. $KCl(aq) + CaS(aq) \rightarrow$
   c. $CrBr_2(aq) + Na_2CO_3(aq) \rightarrow$
   d. $NaOH(aq) + FeCl_3(aq) \rightarrow$

68. Complete and balance each of the following equations. If no reaction occurs, write NO REACTION.
   a. $NaNO_3(aq) + KCl(aq) \rightarrow$
   b. $NaCl(aq) + Hg_2(C_2H_3O_2)_2(aq) \rightarrow$
   c. $(NH_4)_2SO_4(aq) + SrCl_2(aq) \rightarrow$
   d. $NH_4Cl(aq) + AgNO_3(aq) \rightarrow$

69. Write a molecular equation for the precipitation reaction that occurs (if any) when the following solutions are mixed. If no reaction occurs, write NO REACTION.
   a. potassium carbonate and lead(II) nitrate
   b. lithium sulfate and lead(II) acetate
   c. copper(II) nitrate and magnesium sulfide
   d. strontium nitrate and potassium iodide

70. Write a molecular equation for the precipitation reaction that occurs (if any) when the following solutions are mixed. If no reaction occurs, write NO REACTION.
   a. sodium chloride and lead(II) acetate
   b. potassium sulfate and strontium iodide
   c. cesium chloride and calcium sulfide
   d. chromium(III) nitrate and sodium phosphate

Ionic and Net Ionic Equations

71. Write balanced complete ionic and net ionic equations for each of the following reactions.
   a. $HCl(aq) + LiOH(aq) \rightarrow H_2O(l) + LiCl(aq)$
   b. $MgS(aq) + CuCl_2(aq) \rightarrow CuS(s) + MgCl_2(aq)$
   c. $NaOH(aq) + HNO_3(aq) \rightarrow H_2O(l) + NaNO_3(aq)$
   d. $Na_3PO_4(aq) + NiCl_2(aq) \rightarrow Ni_3(PO_4)_2(s) + NaCl(aq)$

72. Write balanced complete ionic and net ionic equations for each of the following reactions.
   a. $K_2SO_4(aq) + CaI_2(aq) \rightarrow CaSO_4(s) + KI(aq)$
   b. $NH_4Cl(aq) + NaOH(aq) \rightarrow H_2O(l) + NH_3(g) + NaCl(aq)$
   c. $AgNO_3(aq) + NaCl(aq) \rightarrow AgCl(s) + NaNO_3(aq)$
   d. $HC_2H_3O_2(aq) + K_2CO_3(aq) \rightarrow$
   $$H_2O(l) + CO_2(g) + KC_2H_3O_2(aq)$$

73. Mercury ions (Hg_2^{2+}) can be removed from solution by precipitation with Cl^-. Suppose that a solution contains aqueous $Hg_2(NO_3)_2$. Write complete ionic and net ionic equations to show the reaction of aqueous $Hg_2(NO_3)_2$ with aqueous sodium chloride to form solid Hg_2Cl_2 and aqueous sodium nitrate.

74. Lead ions can be removed from solution by precipitation with sulfate ions. Suppose that a solution contains lead(II) nitrate. Write complete ionic and net ionic equations to show the reaction of aqueous lead(II) nitrate with aqueous potassium sulfate to form solid lead(II) sulfate and aqueous potassium nitrate.

Acid–Base and Gas-Evolution Reactions

75. Write balanced molecular and net ionic equations for the reaction between hydrobromic acid and potassium hydroxide.

76. Write balanced molecular and net ionic equations for the reaction between nitric acid and calcium hydroxide.

77. Complete and balance each of the following equations for acid–base reactions:
   a. $H_2SO_4(aq) + Ca(OH)_2(aq) \rightarrow$
   b. $HClO_4(aq) + KOH(aq) \rightarrow$
   c. $H_2SO_4(aq) + NaOH(aq) \rightarrow$

78. Complete and balance each of the following equations for acid–base reactions:
   a. $HI(aq) + LiOH(aq) \rightarrow$
   b. $HC_2H_3O_2(aq) + Ca(OH)_2(aq) \rightarrow$
   c. $HCl(aq) + Ba(OH)_2(aq) \rightarrow$

79. A 15.0-mL sample of an unknown $HClO_4$ solution requires titration with 25.3 mL of 0.200 M NaOH to reach the equivalence point. What is the concentration of the unknown $HClO_4$ solution? The neutralization reaction is as follows:

$$HClO_4(aq) + NaOH(aq) \rightarrow H_2O(l) + NaClO_4(aq)$$

80. A 20.0-mL sample of an unknown H_3PO_4 solution is titrated with a 0.100 M NaOH solution. The equivalence point is reached when 18.45 mL of NaOH solution is added. What is the concentration of the unknown H_3PO_4 solution? The neutralization reaction is as follows:

$$H_3PO_4(aq) + 3\,NaOH(aq) \rightarrow 3\,H_2O(l) + Na_3PO_4(aq)$$

81. Complete and balance each of the following equations for gas-evolution reactions:
    a. $HBr(aq) + NiS(s) \rightarrow$
    b. $NH_4I(aq) + NaOH(aq) \rightarrow$
    c. $HBr(aq) + Na_2S(aq) \rightarrow$
    d. $HClO_4(aq) + Li_2CO_3(aq) \rightarrow$

82. Complete and balance each of the following equations for gas-evolution reactions:
    a. $HNO_3(aq) + Na_2SO_3(aq) \rightarrow$
    b. $HCl(aq) + KHCO_3(aq) \rightarrow$
    c. $HC_2H_3O_2(aq) + NaHSO_3(aq) \rightarrow$
    d. $(NH_4)_2SO_4(aq) + Ca(OH)_2(aq) \rightarrow$

Oxidation–Reduction and Combustion

83. Assign oxidation states to each atom in each of the following species.
    a. Ag b. Ag^+ c. CaF_2
    d. H_2S e. CO_3^{2-} f. CrO_4^{2-}

84. Assign oxidation states to each atom in each of the following species.
    a. Cl_2 b. Fe^{3+} c. $CuCl_2$
    d. CH_4 e. $Cr_2O_7^{2-}$ f. HSO_4^-

85. What is the oxidation state of Cr in each of the following compounds?
    a. CrO b. CrO_3 c. Cr_2O_3

86. What is the oxidation state of Cl in each of the following ions?
    a. ClO^- b. ClO_2^- c. ClO_3^- d. ClO_4^-

87. Which of the following reactions are redox reactions? For each redox reaction, identify the oxidizing agent and the reducing agent.
    a. $4 Li(s) + O_2(g) \rightarrow 2 Li_2O(s)$
    b. $Mg(s) + Fe^{2+}(aq) \rightarrow Mg^{2+}(aq) + Fe(s)$
    c. $Pb(NO_3)_2(aq) + Na_2SO_4(aq) \rightarrow PbSO_4(s) + 2 NaNO_3(aq)$
    d. $HBr(aq) + KOH(aq) \rightarrow H_2O(l) + KBr(aq)$

88. Which of the following reactions are redox reactions? For each redox reaction, identify the oxidizing agent and the reducing agent.
    a. $Al(s) + 3 Ag^+(aq) \rightarrow Al^{3+}(aq) + 3 Ag(s)$
    b. $SO_3(g) + H_2O(l) \rightarrow H_2SO_4(aq)$
    c. $Ba(s) + Cl_2(g) \rightarrow BaCl_2(s)$
    d. $Mg(s) + Br_2(l) \rightarrow MgBr_2(s)$

89. Complete and balance each of the following equations for combustion reactions:
    a. $S(s) + O_2(g) \rightarrow$ b. $C_3H_6(g) + O_2(g) \rightarrow$
    c. $Ca(s) + O_2(g) \rightarrow$ d. $C_5H_{12}S(l) + O_2(g) \rightarrow$

90. Complete and balance each of the following equations for combustion reactions:
    a. $C_4H_6(g) + O_2(g) \rightarrow$ b. $C(s) + O_2(g) \rightarrow$
    c. $CS_2(s) + O_2(g) \rightarrow$ d. $C_3H_8O(l) + O_2(g) \rightarrow$

Cumulative Problems

91. The density of a 20.0% by mass ethylene glycol ($C_2H_6O_2$) solution in water is 1.03 g/mL. Find the molarity of the solution.

92. Find the percent by mass of sodium chloride in a 1.35 M NaCl solution. The density of the solution is 1.05 g/mL.

93. Sodium bicarbonate is often used as an antacid to neutralize excess hydrochloric acid in an upset stomach. What mass of hydrochloric acid (in grams) can be neutralized by 2.5 g of sodium bicarbonate? (Hint: Begin by writing a balanced equation for the reaction between aqueous sodium bicarbonate and aqueous hydrochloric acid.)

94. Toilet bowl cleaners often contain hydrochloric acid to dissolve the calcium carbonate deposits that accumulate within a toilet bowl. What mass of calcium carbonate (in grams) can be dissolved by 3.8 g of HCl? (Hint: Begin by writing a balanced equation for the reaction between hydrochloric acid and calcium carbonate.)

95. The combustion of gasoline produces carbon dioxide and water. Assume gasoline to be pure octane (C_8H_{18}) and calculate the mass (in kg) of carbon dioxide that is added to the atmosphere per 1.0 kg of octane burned. (Hint: Begin by writing a balanced equation for the combustion reaction.)

96. Many home barbeques are fueled with propane gas (C_3H_8). What mass of carbon dioxide (in kg) is produced upon the complete combustion of 18.9 L of propane (approximate contents of one 5-gallon tank)? Assume that the density of the liquid propane in the tank is 0.621 g/mL. (Hint: Begin by writing a balanced equation for the combustion reaction.)

97. Aspirin can be made in the laboratory by reacting acetic anhydride ($C_4H_6O_3$) with salicylic acid ($C_7H_6O_3$) to form aspirin ($C_9H_8O_4$) and acetic acid ($C_2H_4O_2$). The balanced equation is

    $$C_4H_6O_3 + C_7H_6O_3 \rightarrow C_9H_8O_4 + C_2H_4O_2$$

    In a laboratory synthesis, a student begins with 3.00 mL of acetic anhydride (density = 1.08 g/mL) and 1.25 g of salicylic acid. Once the reaction is complete, the student collects 1.22 g of aspirin. Determine the limiting reactant, theoretical yield of aspirin, and percent yield for the reaction.

98. The combustion of liquid ethanol (C_2H_5OH) produces carbon dioxide and water. After 4.62 mL of ethanol (density = 0.789 g/mL) was allowed to burn in the presence of 15.55 g of oxygen gas, 3.72 mL of water (density = 1.00 g/mL) was collected. Determine the limiting reactant, theoretical yield of H_2O, and percent yield for the reaction. (Hint: Write a balanced equation for the combustion of ethanol.)

99. A loud classroom demonstration involves igniting a hydrogen-filled balloon. The hydrogen within the balloon reacts explosively with oxygen in the air to form water. If the balloon is filled with a mixture of hydrogen and oxygen, the explosion is even louder than if the balloon is filled only with hydrogen, and the intensity of the explosion depends on the relative amounts of oxygen and hydrogen within the balloon. Look at the following molecular

views representing different amounts of hydrogen and oxygen in four different balloons. Based on the balanced chemical equation, which balloon will make the loudest explosion?

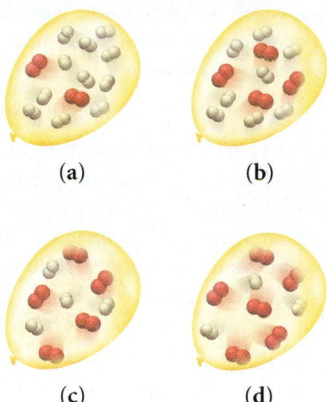

(a)   **(b)**

(c)   **(d)**

100. A hydrochloric acid solution will neutralize a sodium hydroxide solution. Look at the following molecular views showing one beaker of HCl and four beakers of NaOH. Which NaOH beaker will just neutralize the HCl beaker? Begin by writing a balanced chemical equation for the neutralization reaction.

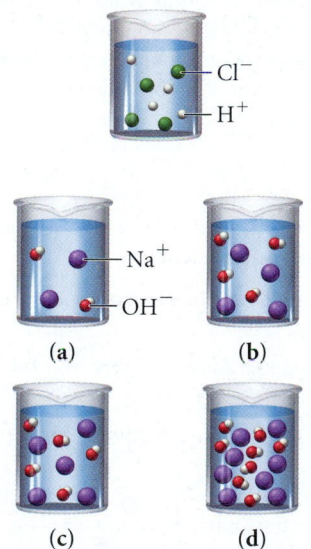

(a)   **(b)**

(c)   **(d)**

101. Predict the products of each of these reactions and write balanced molecular equations for each. If no reaction occurs, write NO RE-ACTION.
   a. $HCl(aq) + Hg_2(NO_3)_2(aq) \rightarrow$
   b. $KHSO_3(aq) + HNO_3(aq) \rightarrow$
   c. aqueous ammonium chloride and aqueous lead(II) nitrate
   d. aqueous ammonium chloride and aqueous calcium hydroxide

102. Predict the products of each of these reactions and write balanced molecular equations for each. If no reaction occurs, write NO RE-ACTION.
   a. $H_2SO_4(aq) + HNO_3(aq) \rightarrow$
   b. $Cr(NO_3)_3(aq) + LiOH(aq) \rightarrow$
   c. liquid pentanol ($C_5H_{12}O$) and gaseous oxygen
   d. aqueous strontium sulfide and aqueous copper(II) sulfate

103. Hard water often contains dissolved Ca^{2+} and Mg^{2+} ions. One way to soften water is to add phosphates. The phosphate ion forms insoluble precipitates with calcium and magnesium ions, removing them from solution. Suppose that a solution is 0.050 M in calcium chloride and 0.085 M in magnesium nitrate. What mass of sodium phosphate would have to be added to 1.5 L of this solution to completely eliminate the hard water ions? Assume complete reaction.

104. An acid solution is 0.100 M in HCl and 0.200 M in H_2SO_4. What volume of a 0.150 M KOH solution would have to be added to 500.0 mL of the acidic solution to neutralize completely all of the acid?

105. Find the mass of barium metal (in grams) that must react with O_2 to produce enough barium oxide to prepare 1.0 L of a 0.10 M solution of OH^-.

106. A solution contains Cr^{3+} ion and Mg^{2+} ion. The addition of 1.00 L of 1.51 M NaF solution is required to cause the complete precipitation of these ions as $CrF_3(s)$ and $MgF_2(s)$. The total mass of the precipitate is 49.6 g. Find the mass of Cr^{3+} in the original solution.

107. The nitrogen in sodium nitrate and in ammonium sulfate is available to plants as fertilizer. Which is the more economical source of nitrogen, a fertilizer containing 30.0% sodium nitrate by weight and costing $9.00 per 100 lb or one containing 20.0% ammonium sulfate by weight and costing $8.10 per 100 lb.

108. Find the volume of 0.110 M hydrochloric acid necessary to react completely with 1.52 g $Al(OH)_3$.

109. Treatment of gold metal with BrF_3 and KF produces Br_2 and $KAuF_4$, a salt of gold. Identify the oxidizing agent and the reducing agent in this reaction. Find the mass of the gold salt that forms when a 73.5-g mixture of equal masses of all three reactants is prepared.

110. A solution is prepared by mixing 0.10 L of 0.12 M sodium chloride with 0.23 L of a 0.18 M $MgCl_2$ solution. What volume of a 0.20 M silver nitrate solution is required to precipitate all the Cl^- ion in the solution as AgCl?

111. A solution contains one or more of the following ions: Ag^+, Ca^{2+}, and Cu^{2+}. When sodium chloride is added to the solution, no precipitate forms. When sodium sulfate is added to the solution, a white precipitate forms. The precipitate is filtered off and sodium carbonate is added to the remaining solution, producing a precipitate. Which ions were present in the original solution? Write net ionic equations for the formation of each of the precipitates observed.

112. A solution contains one or more of the following ions: Hg_2^{2+}, Ba^{2+}, and Fe^{2+}. When potassium chloride is added to the solution, a precipitate forms. The precipitate is filtered off and potassium sulfate is added to the remaining solution, producing no precipitate. When potassium carbonate is added to the remaining solution, a precipitate forms. Which ions were present in the original solution? Write net ionic equations for the formation of each of the precipitates observed.

Challenge Problems

113. Lakes that have been acidified by acid rain (HNO_3 and H_2SO_4) can be neutralized by a process called liming, in which limestone ($CaCO_3$) is added to the acidified water. What mass of limestone (in kg) would be required to completely neutralize a 15.2 billion-liter lake that is 1.8×10^{-5} M in H_2SO_4 and 8.7×10^{-6} M in HNO_3?

114. We learned in Section 4.6 that sodium carbonate is often added to laundry detergents to soften hard water and make the detergent more effective. Suppose that a particular detergent mixture is designed to soften hard water that is 3.5×10^{-3} M in Ca^{2+} and 1.1×10^{-3} M in Mg^{2+} and that the average capacity of a washing machine is 19.5 gallons of water. If the detergent requires using 0.65 kg detergent per load of laundry, determine what percentage (by mass) of the detergent should be sodium carbonate in order to completely precipitate all of the calcium and magnesium ions in an average load of laundry water.

115. Lead poisoning is a serious condition resulting from the ingestion of lead in food, water, or other environmental sources. It affects the central nervous system, leading to a variety of symptoms such as distractibility, lethargy, and loss of motor coordination. Lead poisoning is treated with chelating agents, substances that bind to metal ions, allowing it to be eliminated in the urine. A modern chelating agent used for this purpose is succimer ($C_4H_6O_4S_2$). Suppose you are trying to determine the appropriate dose for succimer treatment of lead poisoning. What minimum mass of succimer (in mg) is needed to bind all of the lead in a patient's bloodstream? Assume that patient blood lead levels are 45 µg/dL, that total blood volume is 5.0 L, and that one mole of succimer binds one mole of lead.

116. A particular kind of emergency breathing apparatus—often placed in mines, caves, or other places where oxygen might become depleted or where the air might become poisoned—works via the following chemical reaction:

$$4 KO_2(s) + 2 CO_2(g) \longrightarrow 2 K_2CO_3(s) + 3 O_2(g)$$

Notice that the reaction produces O_2, which can be breathed, and absorbs CO_2, a product of respiration. Suppose you work for a company interested in producing a self-rescue breathing apparatus (based on the above reaction) which would allow the user to survive for 10 minutes in an emergency situation. What are the important chemical considerations in designing such a unit? Estimate how much KO_2 would be required for the apparatus. (Find any necessary additional information—such as human breathing rates—from appropriate sources. Assume that normal air is 20% oxygen.)

117. Metallic aluminum reacts with MnO_2 at elevated temperatures to form manganese metal and aluminum oxide. A mixture of the two reactants is 67.2% mole percent Al. Find the theoretical yield (in grams) of manganese from the reaction of 250 g of this mixture.

118. Hydrolysis of the compound B_5H_9 forms boric acid, H_3BO_3. Fusion of boric acid with sodium oxide forms a borate salt, $Na_2B_4O_7$. Without writing complete equations, find the mass (in grams) of B_5H_9 required to form 151 g of the borate salt by this reaction sequence.

Conceptual Problems

119. Consider the following reaction:

$$4 K(s) + O_2(g) \rightarrow 2 K_2O(s)$$

The molar mass of K is 39.09 g/mol and that of O_2 is 32.00 g/mol. Without doing any calculations, pick the conditions under which potassium is the limiting reactant and explain your reasoning.

a. 170 g K, 31 g O_2 **b.** 16 g K, 2.5 g O_2
c. 165 kg K, 28 kg O_2 **d.** 1.5 g K, 0.38 g O_2

120. Consider the following reaction:

$$2 NO(g) + 5 H_2(g) \rightarrow 2 NH_3(g) + 2 H_2O(g)$$

A reaction mixture initially contains 5 moles of NO and 10 moles of H_2. Without doing any calculations, determine which of the following best represents the mixture after the reactants have reacted as completely as possible. Explain your reasoning.

a. 1 mol NO, 0 mol H_2, 4 mol NH_3, 4 mol H_2O
b. 0 mol NO, 1 mol H_2, 5 mol NH_3, 5 mol H_2O
c. 3 mol NO, 5 mol H_2, 2 mol NH_3, 2 mol H_2O
d. 0 mol NO, 0 mol H_2, 4 mol NH_3, 4 mol H_2O

121. The circle below represents 1.0 liter of a solution with a solute concentration of 1 M:

Explain what you would add (the amount of solute or volume of solvent) to the solution above so as to obtain a solution represented by each of the following:

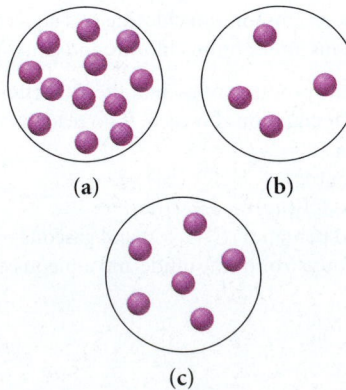

(a)   (b)

(c)

122. Consider the following reaction:

$$2\,N_2H_4(g) + N_2O_4(g) \longrightarrow 3\,N_2(g) + 4\,H_2O(g)$$

Consider the following representation of an initial mixture of N_2H_4 and N_2O_4:

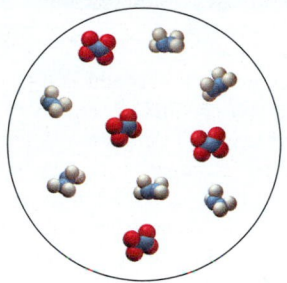

Which of the following best represents the reaction mixture after the reactants have reacted as completely as possible?

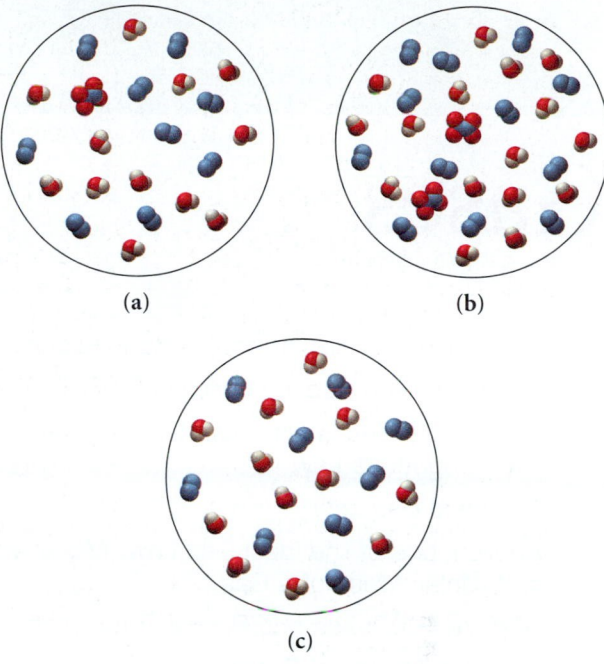

(a)   (b)

(c)

5

Gases

5.1 Water from Wells: Atmospheric Pressure at Work

5.2 Pressure: The Result of Molecular Collisions

5.3 The Simple Gas Laws: Boyle's Law, Charles's Law, and Avogadro's Law

5.4 The Ideal Gas Law

5.5 Applications of the Ideal Gas Law: Molar Volume, Density, and Molar Mass of a Gas

5.6 Mixtures of Gases and Partial Pressures

5.7 Gases in Chemical Reactions: Stoichiometry Revisited

5.8 Kinetic Molecular Theory: A Model for Gases

5.9 Mean Free Path, Diffusion, and Effusion of Gases

5.10 Real Gases: The Effects of Size and Intermolecular Forces

5.11 Chemistry of the Atmosphere: Air Pollution and Ozone Depletion

WE CAN SURVIVE FOR WEEKS without food, days without water, but only minutes without air. Fortunately, we live at the bottom of a vast ocean of air, held to Earth by gravity. We inhale a lungful of this air every few seconds, keep some of the molecules for our own needs, add some molecules we no longer need, and exhale the mixture back into the surrounding air. The air around us is matter in the gaseous state. What are the fundamental properties of these gases? What laws describe their behavior? What kind of theory can explain these properties and laws? The gaseous state is the simplest and best-understood state of matter. In this chapter, we examine that state.

So many of the properties of matter, especially when in the gaseous form, can be deduced from the hypothesis that their minute parts are in rapid motion, the velocity increasing with the temperature, that the precise nature of this motion becomes a subject of rational curiosity.

—JAMES CLERK MAXWELL (1831–1879)

The cork in this champagne bottle is expelled by the buildup of pressure, which results from the constant collisions of gas molecules with the surfaces around them.

5.1 Water from Wells: Atmospheric Pressure at Work

If you live in a city, you may not know where your water comes from. If you live *outside* of a city, however, you probably get your water from a well. Exactly how you pump water from the well depends on the depth of the well. Wells are usually separated into two main categories: shallow wells (less than 30 ft deep) and deep wells (greater than 30 ft deep).

In a shallow well, a pipe runs from a pump at the surface of the ground to the water below (Figure 5.1, page 186). The pump creates a partial vacuum in the pipe, much as your mouth creates a partial vacuum in a straw when you sip a beverage. The result is a *pressure difference* between the inside and the outside of the pipe. **Pressure** is the force exerted per unit area by gas molecules as they strike the surfaces around them (Figure 5.2, page 186). Just as a ball exerts a force when it bounces against a wall, so a molecule exerts a force when it collides with a surface. The sum of all these molecular collisions is pressure—a constant

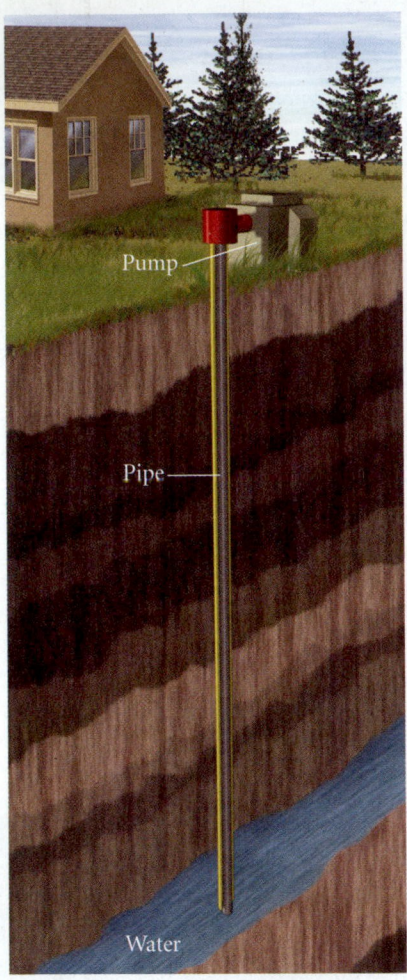

▲ FIGURE 5.1 **A Shallow Well**   In a shallow well, a pump at the surface creates a partial vacuum, allowing atmospheric pressure to drive water up the pipe.

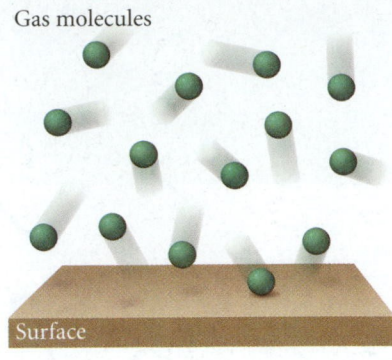

Gas molecules

Surface

◄ FIGURE 5.2 **Gas Pressure**   Pressure is the force per unit area exerted by gas molecules colliding with the surfaces around them.

force on the surfaces exposed to any gas. The total pressure exerted by a gas depends on several factors, including the concentration of gas molecules in the sample. On Earth at sea level, the gas molecules in our atmosphere exert an average pressure of 101,325 newtons per square meter (The newton is the SI unit of force–see Section 1.6.) In English units, this is 14.7 pounds per square inch. Above sea level, pressure decreases with increasing altitude as air becomes thinner (there are fewer molecules per unit volume).

When the pipe for a shallow well is lowered into the water below ground for the first time, the pressure is the same inside and outside the pipe. Therefore, the water does not rise within the pipe (Figure 5.3a ▼). When the pump is attached to the pipe and turned on, however, it removes some of the air molecules from inside the pipe, lowering the pressure there (Figure 5.3b). The pressure pushing down on the surface of the water *outside* the pipe, however, remains the same. The result is a pressure differential—the pressure outside the pipe is now greater than the pressure inside the pipe. This greater external pressure pushes the water up the pipe to ground level and eventually to the tap where the water is used.

In the shallow well just described, atmospheric pressure does the work of pushing the water up the pipe. The pump acts to reduce the pressure inside the pipe, allowing atmospheric pressure to drive the water to the surface, against the force of gravity. The maximum depth of the well, therefore, depends on the total atmospheric pressure. Even if the pump could create a perfect vacuum (zero pressure) within the pipe, normal atmospheric pressure can only push the water to a total height of about 10.3 meters (Figure 5.4 ▶) because a 10.3-meter column of water exerts the same pressure—101,325 newtons per square meter or 14.7 pounds per square inch—as the gas molecules in our atmosphere.

▶ FIGURE 5.3 **The Operation of Shallow Wells**   When the pump is turned off (**a**), the pressure within the pipe is the same as the pressure outside the pipe. When the pump is operating (**b**), the pressure within the pipe is lower than the pressure of the atmosphere on the water outside the pipe. This pressure difference causes water to flow up the pipe.

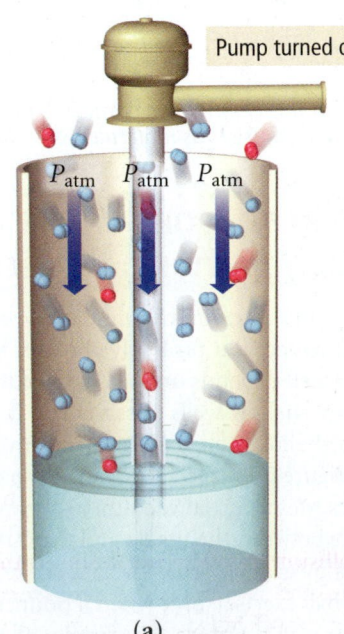

Pump turned off

P_{atm} P_{atm} P_{atm}

(a)

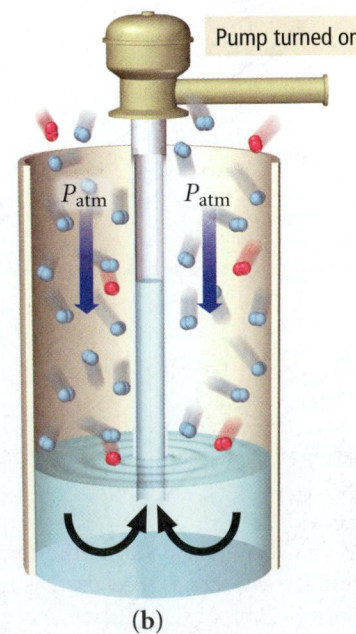

Pump turned on

P_{atm} P_{atm}

(b)

Atmospheric Pressure

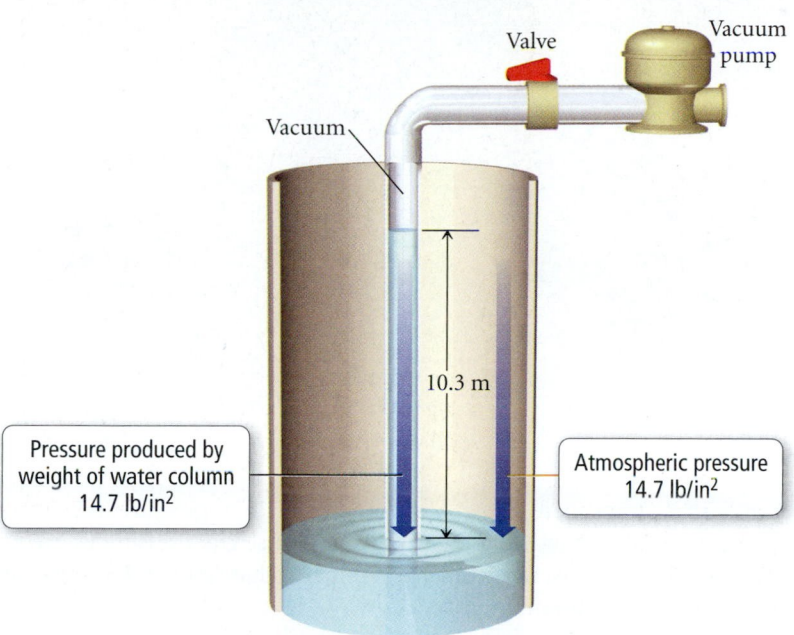

◀ **FIGURE 5.4 Atmospheric Pressure** Atmospheric pressure is equivalent to the pressure exerted by a 10.3-m column of water.

In other words, the water rises up the pipe until the pressure exerted by its weight exactly equals the pressure exerted by the atmosphere. On a normal day, with the pump creating a perfect vacuum, that height is 10.3 m (or 33.8 ft). Since atmospheric pressure can vary by a few percent due to weather changes, it is conceivable that a shallow well at the border of the depth limit would work on some days (when atmospheric pressure is high) and not on others (when atmospheric pressure is low).

If your well is in an area where the water table is deeper than 10.3 m, or if shallow water is polluted, you must install a deep-well pumping system. In this case, you cannot simply create a vacuum in the pipe and expect atmospheric pressure to push the water to the surface. Instead, you must put the pump underground, in the water. The pump—by creating pressure exceeding atmospheric pressure—pushes the water up the pipe and to your faucet.

Some deep water wells also use compressed air to push the water up.

 Conceptual Connection 5.1 Shallow Wells at Altitude

How would the maximum depth of a shallow well be different at different altitudes? For example, if you lived in Denver (at an altitude of approximately 1 mile above sea level), would the maximum depth of a shallow well be greater than, less than, or the same as at sea level? Explain.

Answer: The maximum depth of a shallow well is slightly less at higher altitude. Since a shallow well uses atmospheric pressure to drive water up a pipe, and since atmospheric pressure decreases with increasing altitude, the maximum depth of a shallow well decreases with increasing altitude.

5.2 Pressure: The Result of Molecular Collisions

Air can hold up a jumbo jet or knock down a building. How? Because, as we just discussed, air contains gas molecules in constant motion that collide with each other and with the surfaces around them. Each collision exerts only a small force, but when these forces are summed over the many molecules in air they can add up to a substantial force. As we have just seen, the result of the constant collisions between the atoms or molecules in a gas and the surfaces around them is called pressure. Because of pressure, we can drink from straws, inflate basketballs, and move air into and out of our lungs. Variation in

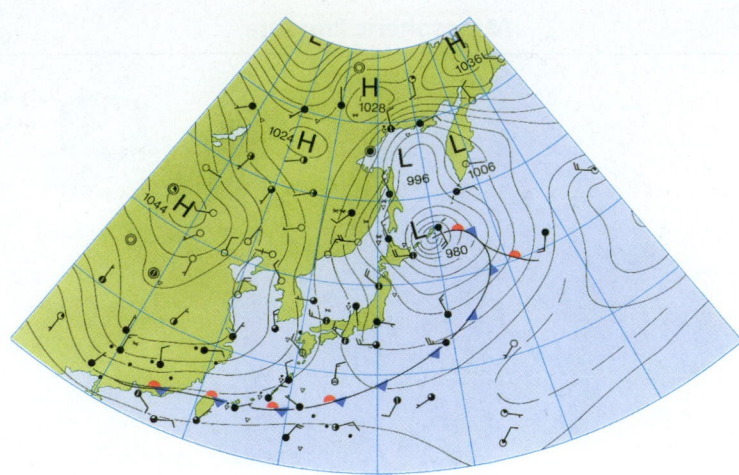

▶ Pressure variations in Earth's atmosphere create wind and weather. The H's in this map indicate regions of high pressure, usually associated with clear weather. The L's indicate regions of low pressure, usually associated with unstable weather. The map shows a typhoon off the northeast coast of Japan. The isobars, or lines of constant pressure, are labeled in hectopascals (100 Pa).

pressure in Earth's atmosphere creates wind, and changes in pressure help us to predict weather. Pressure is all around us and even inside of us. The pressure exerted by a gas sample, as defined previously, is the force per unit area that results from the collisions of gas particles with the surrounding surfaces:

$$\text{Pressure} = \frac{\text{force}}{\text{area}} = \frac{F}{A} \qquad [5.1]$$

Pressure and Density

Lower pressure   Higher pressure

▲ FIGURE 5.5 Pressure and Particle Density   A low density of gas particles results in low pressure. A high density of gas particles results in high pressure.

The pressure exerted by a gas depends on several factors, including the number of gas particles in a given volume—the fewer the gas particles, the lower the pressure (Figure 5.5 ◀). Pressure decreases with increasing altitude because there are fewer molecules per unit volume of air. Above 30,000 ft, for example, where most commercial airplanes fly, the pressure is so low that you could pass out for lack of oxygen. For this reason, most airplane cabins are artificially pressurized.

You can often feel the effect of a drop in pressure as a brief pain in your ears. This pain is caused by air-containing cavities within your ear (Figure 5.6 ▶). When you ascend a mountain, the external pressure (the pressure that surrounds you) drops, while the pressure within your ear cavities (the internal pressure) remains the same. This creates an imbalance—the greater internal pressure forces your eardrum to bulge outward, causing pain. With time, and with the help of a yawn or two, the excess air within your ear's cavities escapes, equalizing the internal and external pressure and relieving the pain.

Pressure Units

Pressure can be measured in a number of different units. A common unit of pressure, the **millimeter of mercury (mmHg)**, originates from how pressure is measured with a **barometer** (Figure 5.7 ▶). A barometer is an evacuated glass tube whose tip is submerged in a pool of mercury. As discussed in Section 5.1, a liquid in an evacuated tube is forced upward by atmospheric gas pressure on the liquid's surface. Water is pushed up to a height of 10.3 m by the average pressure at sea level. Because the density of mercury is 13.5 times that of water, however, atmospheric pressure can support a column of Hg only 1/13.5 times as high as a column of water. This shorter length—0.760 m, or 760 mm (about 30 in)—makes a column of mercury a convenient way to measure pressure. In a barometer, the mercury column rises with increasing atmospheric pressure and falls with decreasing atmospheric pressure. The unit millimeter of mercury is often called a **torr**, after the Italian physicist Evangelista Torricelli (1608–1647) who invented the barometer.

$$1 \text{ mmHg} = 1 \text{ torr}$$

Pressure Imbalance

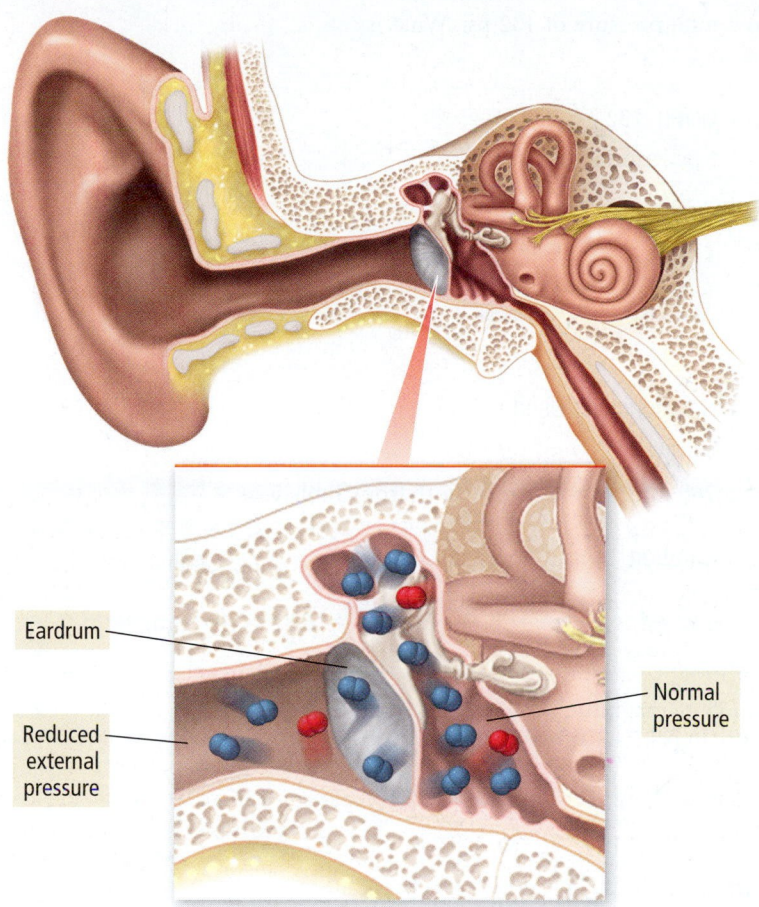

▲ FIGURE 5.6 Pressure Imbalance The pain you feel in your ears upon ascending a mountain is caused by a pressure imbalance between the cavities in your ears and the outside air.

The Mercury Barometer

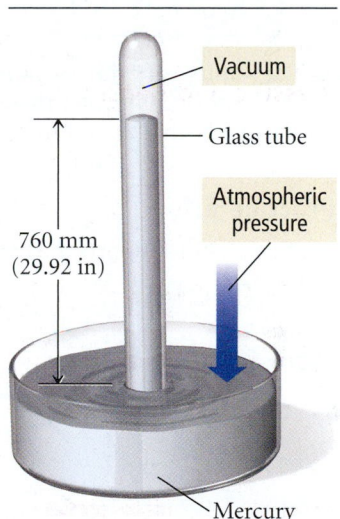

▲ FIGURE 5.7 The Mercury Barometer Average atmospheric pressure at sea level can support a column of mercury 760 mm in height.

A second unit of pressure is the **atmosphere (atm)**, the average pressure at sea level. Since one atmosphere of pressure pushes a column of mercury to a height of 760 mm, 1 atm and 760 mmHg are equal:

$$1 \text{ atm} = 760 \text{ mmHg}$$

In this unit, a fully inflated mountain bike tire has a pressure of about 6 atm, and the pressure at the top of Mt. Everest is about 0.31 atm.

The SI unit of pressure is the **pascal (Pa)**, defined as 1 newton (N) per square meter.

$$1 \text{ Pa} = 1 \text{ N/m}^2$$

The pascal is a much smaller unit of pressure than the atmosphere:

$$1 \text{ atm} = 101,325 \text{ Pa}$$

Other common units of pressure include inches of mercury (in Hg) and pounds per square inch (psi).

$$1 \text{ atm} = 29.92 \text{ in Hg} \qquad 1 \text{ atm} = 14.7 \text{ psi}$$

These units are summarized in Table 5.1.

TABLE 5.1 Common Units of Pressure

Unit	Abbreviation	Average Air Pressure at Sea Level
Pascal (1 N/m^2)	Pa	101,325 Pa
Pounds per square inch	psi	14.7 psi
Torr (1 mmHg)	torr	760 torr (exact)
Inches of mercury	in Hg	29.92 in Hg
Atmosphere	atm	1 atm

EXAMPLE 5.1 Converting between Pressure Units

A high-performance road bicycle tire is inflated to a total pressure of 132 psi. What is this pressure in mmHg?

Sort The problem gives a pressure in psi and asks you to convert the units to mmHg.	**Given** 132 psi **Find** mmHg
Strategize Since Table 5.1 does not have a direct conversion factor between psi and mmHg, but does provide relationships between both of these units and atmospheres, you can convert to atm as an intermediate step.	**Conceptual Plan** psi → atm → mmHg $$\frac{1\ atm}{14.7\ psi} \qquad \frac{760\ mmHg}{1\ atm}$$ **Relationships Used** 1 atm = 14.7 psi 760 mmHg = 1 atm (both from Table 5.1)
Solve Follow the conceptual plan to solve the problem. Begin with 132 psi and use the conversion factors to arrive at the pressure in mmHg.	**Solution** $$132\ \cancel{psi} \times \frac{1\ \cancel{atm}}{14.7\ \cancel{psi}} \times \frac{760\ mmHg}{1\ \cancel{atm}} = 6.82 \times 10^3\ mmHg$$

Check The units of the answer are correct. The magnitude of the answer (6.82×10^3 mmHg) is greater than the given pressure in psi. This is reasonable because mmHg is a much smaller unit than psi.

For Practice 5.1

The weather channel reports the barometric pressure as 30.44 in Hg. Convert this pressure to psi.

For More Practice 5.1

Convert a pressure of 23.8 in Hg to kPa.

The Manometer: A Way to Measure Pressure in the Laboratory

The pressure of a gas sample in the laboratory is often measured with a **manometer**. A manometer is a U-shaped tube containing a dense liquid, usually mercury, as shown in Figure 5.8 ◄.

In this manometer, one end of the tube is open to atmospheric pressure and the other is attached to a flask containing the gas sample. If the pressure of the gas sample is exactly equal to atmospheric pressure, then the mercury levels on both sides of the tube are the same. If the pressure of the sample is *greater than* atmospheric pressure, the mercury level on the left side of the tube is *higher than* on the right. If the pressure of the sample is *less than* atmospheric pressure, the mercury level on the left side is *lower than* on the right. This type of manometer always measures the pressure of the gas sample relative to atmospheric pressure. The difference in height between the two levels is equal to the pressure difference from atmospheric pressure. To accurately calculate the absolute pressure, you also need a barometer to measure atmospheric pressure (which can change from day to day).

The Manometer

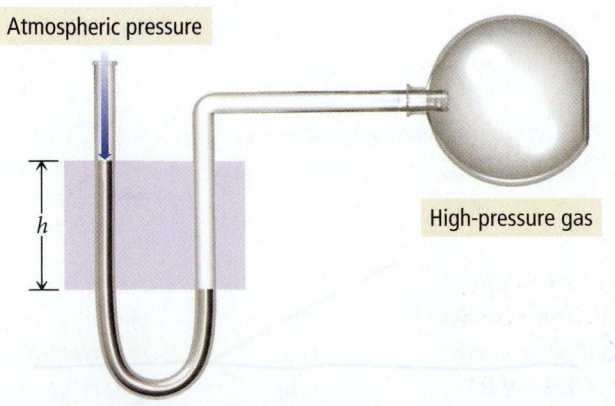

▲ **FIGURE 5.8 The Manometer** A manometer can be used to measure the pressure exerted by a sample of gas.

Chemistry and Medicine
Blood Pressure

Blood pressure is the force within arteries that drives the circulation of blood throughout the body. Blood pressure in the body is analogous to water pressure in a plumbing system. Just as water pressure pushes water through the pipes to faucets and fixtures throughout a house, so blood pressure pushes blood to muscles and other tissues throughout the body. However, unlike the water pressure in a plumbing system—which is normally nearly constant—blood pressure varies with each heartbeat. When the heart muscle contracts, blood pressure increases; between contractions it decreases. Systolic blood pressure refers to the peak pressure during a contraction, and diastolic blood pressure refers to the lowest pressure between contractions. Just as too high a water pressure in a plumbing system can damage pipes, so too a high blood pressure in a circulatory system can damage the heart and arteries, resulting in increased risk of stroke and heart attack.

Blood pressure is usually measured with an instrument called a sphygmomanometer—an inflatable cuff equipped with a pressure gauge—and a stethoscope. The cuff is wrapped around the patient's arm and inflated with air. As air is pumped into the cuff, the pressure in the cuff increases. The cuff tightens around the arm and compresses the artery, momentarily stopping blood flow. The person measuring the blood pressure then listens to the artery through the stethoscope while slowly releasing the pressure in the cuff. When the pressure in the cuff equals the systolic blood pressure, a pulse is heard through the stethoscope. The pulse is the sound of blood getting through the compressed artery during a contraction of the heart. The pressure reading at that exact moment is the systolic blood pressure. As the pressure in the cuff continues to decrease, the blood can flow through the compressed artery even between contractions, so the pulsing sound disappears. The pressure reading when the pulsing sound just stops is the diastolic blood pressure.

The result of a blood pressure measurement is usually expressed as the value of the two pressures, in mmHg, separated by a slash. For example, a result of 122/84 means that the systolic blood pressure is 122 mmHg and the diastolic blood pressure is 84 mmHg. Although the value of blood pressure can vary throughout the day, a healthy (or normal) value is usually considered to be below 120 mmHg for systolic and below 80 mmHg for diastolic (Table 5.2). High blood pressure, also called hypertension, entails the health risks mentioned previously.

Risk factors for hypertension include obesity, high salt (sodium) intake, high alcohol intake, lack of exercise, stress, a family history of high blood pressure, and age (blood pressure tends to increase as we get older). Mild hypertension can be managed with diet and exercise. Moderate to severe cases require doctor-prescribed medication.

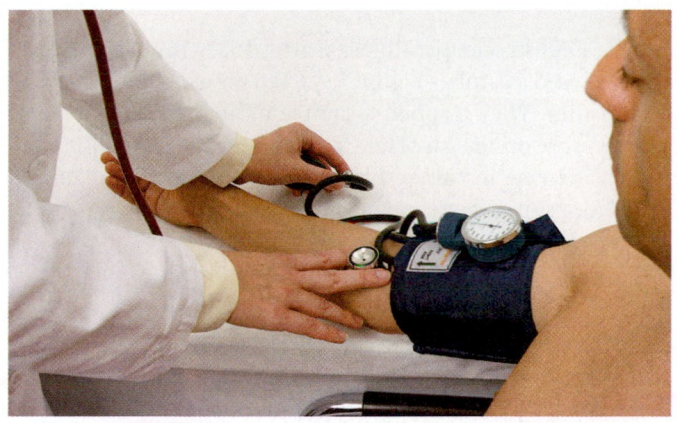

▲ Blood pressure is measured with an inflatable cuff that compresses the main artery in the arm. A stethoscope is used to listen for blood flowing through the artery with each heartbeat.

TABLE 5.2 Blood Pressure Ranges

Blood Pressure	Systolic (mmHg)	Diastolic (mmHg)
Hypotension	<100	<60
Normal	100–119	60–79
Prehypertension	120–139	80–89
Hypertension Stage 1	140–159	90–99
Hypertension Stage 2	>160	>100

5.3 The Simple Gas Laws: Boyle's Law, Charles's Law, and Avogadro's Law

A sample of gas has four basic physical properties: its pressure (P), volume (V), temperature (T), and amount in moles (n). These properties are interrelated—when you change one, it affects the others. The simple gas laws describe the relationships between pairs of these properties. For example, how does *volume* vary with *pressure* at constant temperature and amount of gas, or with *temperature* at constant pressure and amount of gas? Such questions can be answered by experiments in which two of the four basic properties are held constant in order to elucidate the relationship between the other two.

The J-Tube

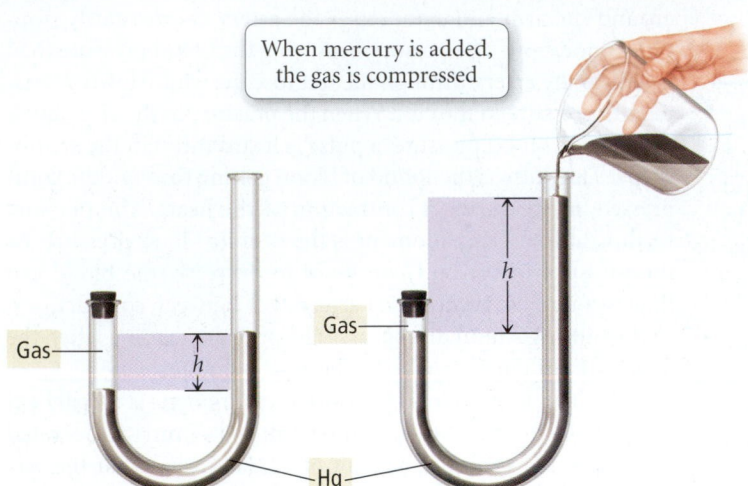

When mercury is added, the gas is compressed

Gas

Gas

Hg

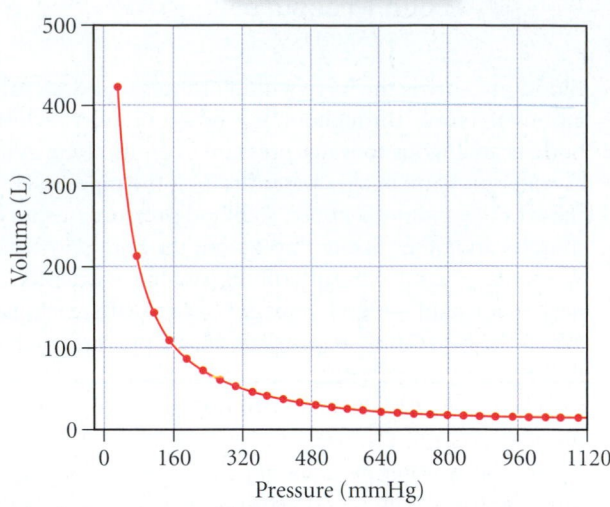

Boyle's Law
As pressure increases, volume decreases.

▲ **FIGURE 5.9 The J-Tube** In a J-tube, a sample of gas is trapped by a column of mercury. The pressure on the gas can be increased by increasing the amount of mercury in the column.

▲ **FIGURE 5.10 Volume versus Pressure** A plot of the volume of a gas sample—as measured in a J-tube—versus pressure. The plot shows that volume and pressure are inversely related.

Boyle's Law: Volume and Pressure

In the early 1660s, the pioneering English scientist Robert Boyle (1627–1691) and his assistant Robert Hooke (1635–1703) used a J-tube (Figure 5.9 ▲) to measure the volume of a sample of gas at different pressures. They trapped a sample of air in the J-tube and added mercury to increase the pressure on the gas. They found an *inverse relationship* between volume and pressure—an increase in one results in a decrease in the other—as shown in Figure 5.10 ▲. This relationship is now known as **Boyle's law.**

Boyle's law assumes constant temperature and constant amount of gas.

$$\text{Boyle's law: } V \propto \frac{1}{P} \quad \text{(constant } T \text{ and } n\text{)}$$

Boyle's law follows from the idea that pressure results from the collisions of the gas particles with the walls of their container. If the volume of a gas sample is decreased, the same number of gas particles is crowded into a smaller volume, resulting in more collisions with the walls and therefore an increase in the pressure (Figure 5.11 ▼).

Volume versus Pressure: A Molecular View

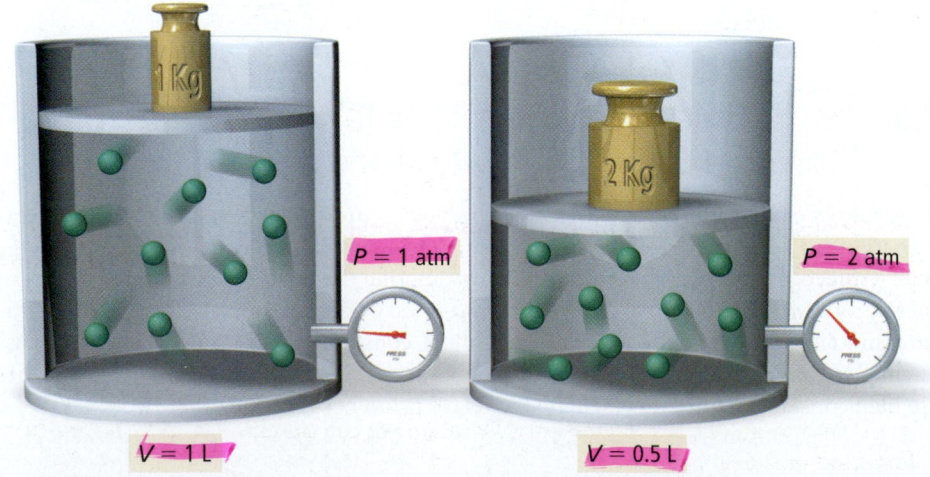

▶ **FIGURE 5.11 Molecular Interpretation of Boyle's Law** As the volume of a gas sample is decreased, gas molecules collide with surrounding surfaces more frequently, resulting in greater pressure.

1 Kg

2 Kg

$P = 1$ atm

$P = 2$ atm

$V = 1$ L

$V = 0.5$ L

Depth = 0 m
P = 1 atm

Depth = 20 m
P = 3 atm

◄ **FIGURE 5.12 Increase in Pressure with Depth** For every 10 m of depth, a diver experiences approximately one additional atmosphere of pressure due to the weight of the surrounding water. At 20 m, for example, the diver experiences approximately 3 atm of pressure (1 atm of normal atmospheric pressure plus an additional 2 atm due to the weight of the water).

Scuba divers learn about Boyle's law during certification because it explains why they should not ascend toward the surface without continuous breathing. For every 10 m of depth that a diver descends in water, she experiences an additional 1 atm of pressure due to the weight of the water above her (Figure 5.12 ▲). The pressure regulator used in scuba diving delivers air at a pressure that matches the external pressure; otherwise the diver could not inhale the air (see *Chemistry in Your Day: Extra-long Snorkels* on page 194). For example, when a diver is at a depth of 20 m below the surface, the regulator delivers air at a pressure of 3 atm to match the 3 atm of pressure around the diver (1 atm due to normal atmospheric pressure and 2 additional atmospheres due to the weight of the water at 20 m). Suppose that a diver inhaled a lungful of air at a pressure of 3 atm and swam quickly to the surface (where the pressure drops to 1 atm) while holding her breath. What would happen to the volume of air in her lungs? Since the pressure decreases by a factor of 3, the volume of the air in her lungs would increase by a factor of 3, severely damaging her lungs and possibly killing her. Of course, the volume increase in the diver's lungs would be so great that the diver would not be able to hold her breath all the way to the surface—the air would force itself out of her mouth. Nonetheless, the most important rule in diving is *never hold your breath*. Divers must ascend slowly and breathe continuously, allowing the regulator to bring the air pressure in their lungs back to 1 atm by the time they reach the surface.

Boyle's law can be used to compute the volume of a gas following a pressure change or the pressure of a gas following a volume change *as long as the temperature and the amount of gas remain constant*. For these calculations, we must write Boyle's law in a slightly different way.

$$\text{Since } \quad V \propto \frac{1}{P}, \quad \text{then } \quad V = (\text{constant}) \times \frac{1}{P} \quad \text{or} \quad V = \frac{(\text{constant})}{P}$$

If two quantities are proportional, then one is equal to the other multiplied by a constant.

If we multiply both sides by P, we get

$$PV = \text{constant}$$

This relationship shows that if the pressure increases, the volume decreases, but the product $P \times V$ is always equal to the same constant. For two different sets of conditions, we can say that

$$P_1 V_1 = \text{constant} = P_2 V_2$$

or

$$P_1 V_1 = P_2 V_2 \qquad\qquad [5.2]$$

where P_1 and V_1 are the initial pressure and volume of the gas and P_2 and V_2 are the final volume and pressure.

EXAMPLE 5.2 Boyle's Law

A cylinder equipped with a movable piston has a volume of 7.25 L under an applied pressure of 4.52 atm. What is the volume of the cylinder if the applied pressure is decreased to 1.21 atm?

To solve the problem, first solve Boyle's law (Equation 5.2) for V_2 and then substitute the given quantities to compute V_2.	**Solution** $P_1 V_1 = P_2 V_2$ $V_2 = \dfrac{P_1}{P_2} V_1$ $\quad = \dfrac{4.52 \ \text{atm}}{1.21 \ \text{atm}} 7.25 \ \text{L}$ $\quad = 27.1 \ \text{L}$

For Practice 5.2

A snorkeler takes a syringe filled with 16 mL of air from the surface, where the pressure is 1.0 atm, to an unknown depth. The volume of the air in the syringe at this depth is 7.5 mL. What is the pressure at this depth? If the pressure increases by an additional 1 atm for every 10 m of depth, how deep is the snorkeler?

Chemistry in Your Day

Extra-long Snorkels

Several episodes of *The Flintstones* cartoon featured Fred Flintstone and Barney Rubble snorkeling. Their snorkels, however, were not the modern kind, but long reeds that stretched from the surface of the water down to many meters of depth. Fred and Barney swam around in deep water while breathing air provided to them by these extra-long snorkels. Would this work? Why do people bother with scuba diving equipment if they could simply use 10-m snorkels as Fred and Barney did?

When we breathe, we expand the volume of our chest cavity, reducing the pressure on the outer surface of the lungs to less than 1 atm (Boyle's law). Because of this pressure differen-

tial, the lungs expand, the pressure in them falls, and air from outside our lungs then flows into them. Extra-long snorkels do not work because of the pressure exerted by water at depth. A diver at 10 m experiences an external pressure of 2 atm. This is more than the muscles of the chest cavity can overcome—the chest cavity and lungs are compressed, resulting in an air pressure within them of more than 1 atm. If the diver had a snorkel that went to the surface—where the air pressure is 1 atm—air would flow out of his lungs, not into them. It would be impossible to breathe.

Question

Suppose a diver takes a balloon with a volume of 2.5 L from the surface, where the pressure is 1.0 atm, to a depth of 20 m, where the pressure is 3.0 atm. What would happen to the volume of the balloon? What if the end of the balloon were on a long pipe that went to the surface and was attached to another balloon? Which way would air flow as the diver descended?

◄ In the popular cartoon *The Flintstones*, Fred and Barney used long reeds to breathe surface air while swimming at depth. This would not work because the increased pressure at depth would force air out of their lungs; the pressure would not allow them to inhale.

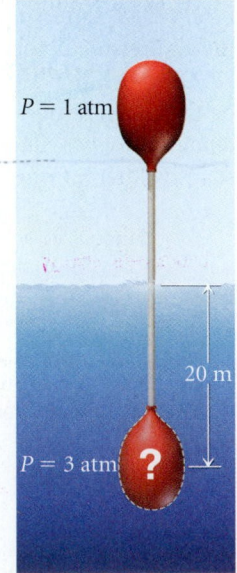

$P = 1$ atm

20 m

$P = 3$ atm   ?

► If two balloons were joined by a long tube and one end was submerged in water, what would happen to the volumes of the two balloons?

Charles's Law: Volume and Temperature

Suppose you keep the pressure of a gas sample constant and measure its volume at a number of different temperatures. The results of several such measurements are shown in Figure 5.13 ▶. From the plot we can see the relationship between volume and temperature: the volume of a gas increases with increasing temperature. Looking at the plot more closely, however, reveals more—volume and temperature are linearly related. If two variables are linearly related, then plotting one against the other produces a straight line.

Another interesting feature emerges if we extend or *extrapolate* the line in the plot backwards from the lowest measured temperature. The extrapolated line shows that the gas should have a zero volume at −273.15 °C. Recall from Chapter 1 that −273.15 °C corresponds to 0 K (zero on the Kelvin scale), the coldest possible temperature. The extrapolated line shows that below −273.15 °C, the gas would have a negative volume, which is physically impossible. For this reason, we refer to 0 K as *absolute zero*—colder temperatures do not exist.

The first person to carefully quantify the relationship between the volume of a gas and its temperature was J. A. C. Charles (1746–1823), a French mathematician and physicist. Charles was interested in gases and was among the first people to ascend in a hydrogen-filled balloon. The direct proportionality between volume and temperature is named **Charles's law** after him.

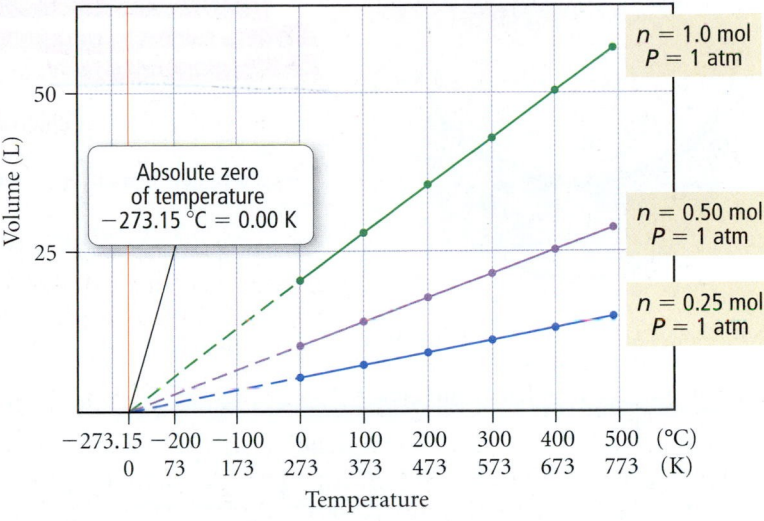

Charles's Law
As temperature increases, volume increases.

▲ **FIGURE 5.13 Volume versus Temperature** The volume of a fixed amount of gas at a constant pressure increases linearly with increasing temperature in kelvins. (The extrapolated lines could not be measured experimentally because all gases would condense into liquids before −273.15 °C is reached.)

Charles's law: $V \propto T$ (constant P and n)

When the temperature of a gas sample is increased, the gas particles move faster; collisions with the walls are more frequent, and the force exerted with each collision is greater. The only way for the pressure (the force per unit area) to remain constant is for the gas to occupy a larger volume, so that collisions become less frequent and occur over a larger area (Figure 5.14 ▼).

Charles's law explains why the second floor of a house is usually a bit warmer than the ground floor. According to Charles's law, when air is heated, its volume increases, resulting in a lower density. The warm, less dense air tends to rise in a room filled with colder, denser air. Similarly, Charles's law explains why a hot-air balloon can take flight. The gas that fills a hot air balloon is warmed with a burner, increasing its volume and lowering its density, and causing it to float in the colder, denser surrounding air.

Charles's law assumes constant pressure and constant amount of gas.

▲ A hot-air balloon floats because the hot air is less dense than the surrounding cold air.

Volume versus Temperature: A Molecular View

Low kinetic energy

High kinetic energy

Ice water

Boiling water

▲ **FIGURE 5.14 Molecular Interpretation of Charles's Law**   If a balloon is moved from an ice water bath to a boiling water bath, its volume will expand as the gas particles within the balloon move faster (due to the increased temperature) and collectively occupy more space.

▲ If a balloon is placed into liquid nitrogen (77 K), it shrivels up as the air within it cools and occupies less volume at the same external pressure.

You can experience Charles's law directly by holding a partially inflated balloon over a warm toaster. As the air in the balloon warms, you can feel the balloon expanding. Alternatively, you can a put an inflated balloon into liquid nitrogen and see that it becomes smaller as it cools.

Charles's law can be used to compute the volume of a gas following a temperature change or the temperature of a gas following a volume change *as long as the pressure and the amount of gas are constant.* For these calculations, we rearrange Charles's law as follows:

$$\text{Since } V \propto T, \quad \text{then } V = \text{constant} \times T$$

If we divide both sides by T, we get

$$V/T = \text{constant}$$

If the temperature increases, the volume increases in direct proportion so that the quotient, V/T, is always equal to the same constant. So, for two different measurements, we can say that

$$V_1/T_1 = \text{constant} = V_2/T_2,$$

or

$$\frac{V_1}{T_1} = \frac{V_2}{T_2}. \qquad [5.3]$$

where V_1 and T_1 are the initial volume and temperature of the gas and V_2 and T_2 are the final volume and temperature. *The temperatures must always be expressed in kelvins (K)*, because, as you can see from Figure 5.13 (page 195), the volume of a gas is directly proportional to its absolute temperature, not its temperature in °C. For example, doubling the temperature of a gas sample from 1 °C to 2 °C does not double its volume, but doubling the temperature from 200 K to 400 K does.

EXAMPLE 5.3 Charles's Law

A sample of gas has a volume of 2.80 L at an unknown temperature. When the sample is submerged in ice water at $T = 0.00$ °C, its volume decreases to 2.57 L. What was its initial temperature (in K and in °C)?

To solve the problem, first solve Charles's law for T_1.	**Solution** $$\frac{V_1}{T_1} = \frac{V_2}{T_2}$$ $$T_1 = \frac{V_1}{V_2} T_2$$
Before you substitute the numerical values to compute T_1, you must convert the temperature to kelvins (K). *Remember, gas law problems must always be worked with Kelvin temperatures.*	$$T_2(K) = 0.00 + 273.15 = 273.15 \text{ K}$$
Substitute T_2 and the other given quantities to compute T_1.	$$T_1 = \frac{V_1}{V_2} T_2$$ $$= \frac{2.80 \text{ L}}{2.57 \text{ L}} 273.15 \text{ K}$$ $$= 297.6 \text{ K}$$
Compute T_1 in °C by subtracting 273 from the value in kelvins.	$$T_1(°C) = 297.6 - 273.15 = 24 \text{ °C}$$

For Practice 5.3

A gas in a cylinder with a moveable piston has an initial volume of 88.2 mL. If the gas is heated from 35 °C to 155 °C, what is its final volume (in mL)?

Avogadro's Law: Volume and Amount (in Moles)

So far, we have learned the relationships between volume and pressure, and volume and temperature, but we have considered only a constant amount of a gas. What happens when the amount of gas changes? The volume of a gas sample (at constant temperature and pressure) as a function of the amount of gas (in moles) in the sample is shown in Figure 5.15 ▶. We can see that the relationship between volume and amount is linear. As we might expect, extrapolation to zero moles shows zero volume. This relationship, first stated formally by Amadeo Avogadro, is called **Avogadro's law:**

Avogadro's law: $V \propto n$ (constant T and P)

When the amount of gas in a sample is increased at constant temperature and pressure, its volume increases in direct proportion because the greater number of gas particles fill more space.

You experience Avogadro's law when you inflate a balloon, for example. With each exhaled breath, you add more gas particles to the inside of the balloon, increasing its volume. Avogadro's law can be used to compute the volume of a gas following a change in the amount of the gas *as long as the pressure and temperature of the gas are constant.* For these calculations, Avogadro's law is expressed as

$$\frac{V_1}{n_1} = \frac{V_2}{n_2} \qquad [5.4]$$

where V_1 and n_1 are the initial volume and number of moles of the gas and V_2 and n_2 are the final volume and number of moles. In calculations, Avogadro's law is used in a manner similar to the other gas laws, as shown in the following example.

> Avogadro's law assumes constant temperature and pressure and is independent of the nature of the gas.

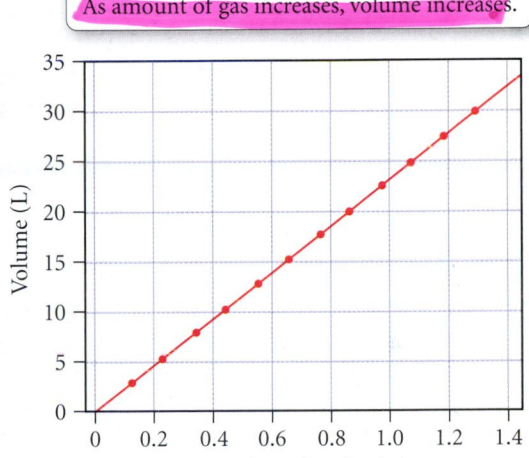

Avogadro's Law

As amount of gas increases, volume increases.

▲ **FIGURE 5.15 Volume versus Number of Moles** The volume of a gas sample increases linearly with the number of moles of gas in the sample.

EXAMPLE 5.4 Avogadro's Law

A 4.65-L sample of helium gas contains 0.225 mol of helium. How many additional moles of helium gas must be added to the sample to obtain a volume of 6.48 L? Assume constant temperature and pressure.

To solve the problem, first solve Avogadro's law for n_2. Then substitute the given quantities to compute n_2.	**Solution** $$\frac{V_1}{n_1} = \frac{V_2}{n_2}$$ $$n_2 = \frac{V_2}{V_1} n_1$$
Since the balloon already contains 0.225 mol of gas, compute the amount of gas to add by subtracting 0.225 mol from the value you calculated for n_2. (In Chapter 1, we introduced the practice of underlining the least (rightmost) significant digit of intermediate answers, but not rounding the final answer until the very end of the calculation. We continue that practice in this chapter. However, in order to avoid unneccessary notation, we will not carry additional digits in cases, such as this one, where doing so would not affect the final answer.)	$$= \frac{6.48 \text{ L}}{4.65 \text{ L}} 0.225 \text{ mol}$$ $$= 0.314 \text{ mol}$$ moles to add $= 0.314 \text{ mol} - 0.225 \text{ mol}$ $$= 0.089 \text{ mol}$$

For Practice 5.4

A chemical reaction occurring in a cylinder equipped with a moveable piston produces 0.621 mol of a gaseous product. If the cylinder contained 0.120 mol of gas before the reaction and had an initial volume of 2.18 L, what was its volume after the reaction? (Assume constant pressure and temperature and that the initial amount of gas completely reacts.)

5.4 The Ideal Gas Law

The relationships that we have learned so far can be combined into a single law that encompasses all of them. So far, we know that

$$V \propto \frac{1}{P} \quad \text{(Boyle's law)}$$

$$V \propto T \quad \text{(Charles's law)}$$

$$V \propto n \quad \text{(Avogadro's law)}$$

Combining these three expressions, we get

$$V \propto \frac{nT}{P}$$

The volume of a gas is directly proportional to the number of moles of gas and to the temperature of the gas, but is inversely proportional to the pressure of the gas. We can replace the proportionality sign with an equals sign by incorporating R, a proportionality constant called the *ideal gas constant*:

$$V = \frac{RnT}{P}$$

Rearranging, we get

$$PV = nRT \tag{5.5}$$

This equation is called the **ideal gas law**. The value of R, the **ideal gas constant**, is the same for all gases and has the following value:

$$R = 0.08206 \frac{L \cdot atm}{mol \cdot K}$$

L = liters
atm = atmospheres
mol = moles
K = kelvins

The ideal gas law contains within it the simple gas laws that we have learned. For example, recall that Boyle's law states that $V \propto 1/P$ when the amount of gas (n) and the temperature of the gas (T) are kept constant. We can rearrange the ideal gas law as follows:

$$PV = nRT$$

First, divide both sides by P:

$$V = \frac{nRT}{P}$$

Then put the variables that are constant, along with R, in parentheses:

$$V = (nRT)\frac{1}{P}$$

Since n and T are constant in this case, and since R is always a constant, we can write

$$V \propto (\text{constant}) \times \frac{1}{P}$$

which simply means that $V \propto 1/P$.

The ideal gas law also shows how other pairs of variables are related. For example, from Charles's law we know that $V \propto T$ at constant pressure and constant number of moles. But what if we heat a sample of gas at constant *volume* and constant number of moles? This question applies to the warning labels on aerosol cans such as hair spray or deodorants. These labels warn against excessive heating or incineration of the can, even after the contents are used up. Why? An "empty" aerosol can is not really empty but contains a fixed amount of gas trapped in a fixed volume. What would happen if you heated

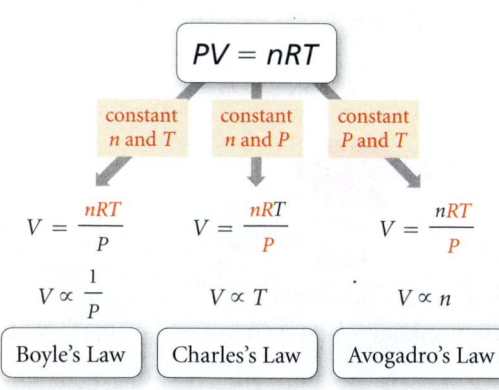

Ideal Gas Law

$$PV = nRT$$

constant *n* and *T*	constant *n* and *P*	constant *P* and *T*

$$V = \frac{nRT}{P} \qquad V = \frac{nRT}{P} \qquad V = \frac{nRT}{P}$$

$$V \propto \frac{1}{P} \qquad V \propto T \qquad V \propto n$$

Boyle's Law   Charles's Law   Avogadro's Law

▲ The ideal gas law contains the simple gas laws within it.

the can? Let's rearrange the ideal gas law to clearly see the relationship between pressure and temperature at constant volume and constant number of moles:

$$PV = nRT$$

$$P = \frac{nRT}{V} = \left(\frac{nR}{V}\right)T$$

| Divide both sides by *V*.

Since *n* and *V* are constant and since *R* is always a constant:

$$P = (\text{constant}) \times T$$

This relationship between pressure and temperature is also known as *Gay-Lussac's law*. As the temperature of a fixed amount of gas in a fixed volume increases, the pressure increases. In an aerosol can, this pressure increase can blow the can apart, which is why aerosol cans should not be heated or incinerated. They might explode.

The ideal gas law can also be used to determine the value of any one of the four variables (*P*, *V*, *n*, or *T*) given the other three. However, each of the quantities in the ideal gas law *must be expressed* in the units within *R*:

- pressure (*P*) in atm
- volume (*V*) in L
- moles (*n*) in mol
- temperature (*T*) in K

▲ The labels on most aerosol cans warn against incineration. Since the volume of the can is constant, the increase in temperature causes an increase in pressure, possibly causing the can to explode.

EXAMPLE 5.5 Ideal Gas Law I

Calculate the volume occupied by 0.845 mol of nitrogen gas at a pressure of 1.37 atm and a temperature of 315 K.

Sort The problem gives you the number of moles of nitrogen gas, the pressure, and the temperature. You are asked to find the volume.

Given $n = 0.845$ mol, $P = 1.37$ atm, $T = 315$ K
Find *V*

Strategize You are given three of the four variables (*P*, *T*, and *n*) in the ideal gas law and asked to find the fourth (*V*). The conceptual plan shows how the ideal gas law provides the relationship between the given quantities and the quantity to be found.

Conceptual Plan

n, *P*, *T* → *V*

$PV = nRT$

Relationships Used $PV = nRT$ (ideal gas law)

Solve To solve the problem, first solve the ideal gas law for *V*.

Solution

$$PV = nRT$$

$$\boxed{V = \frac{nRT}{P}}$$

Then substitute the given quantities to compute *V*.

$$V = \frac{0.845 \text{ mol} \times 0.08206 \frac{\text{L} \cdot \text{atm}}{\text{mol} \cdot \text{K}} \times 315 \text{ K}}{1.37 \text{ atm}}$$

$$= 15.9 \text{ L}$$

Check The units of the answer are correct. The magnitude of the answer (15.9 L) makes sense because, as you will see in the next section, one mole of an ideal gas under standard conditions (273 K and 1 atm) occupies 22.4 L. Although these are not standard conditions, they are close enough for a ballpark check of the answer. Since this gas sample contains 0.845 mol, a volume of 15.9 L is reasonable.

For Practice 5.5

An 8.50-L tire is filled with 0.552 mol of gas at a temperature of 305 K. What is the pressure (in atm and psi) of the gas in the tire?

EXAMPLE 5.6 Ideal Gas Law II

Calculate the number of moles of gas in a 3.24-L basketball inflated to a *total pressure* of 24.3 psi at 25 °C. (Note: The *total pressure* is not the same as the pressure read on a pressure gauge such as we use for checking a car or bicycle tire. That pressure, called the *gauge pressure*, is the *difference* between the total pressure and atmospheric pressure. In this case, if atmospheric pressure is 14.7 psi, the gauge pressure would be 9.6 psi. However, for calculations involving the ideal gas law, you must use the *total pressure* of 24.3 psi.)

Sort The problem gives you the pressure, the volume, and the temperature. You are asked to find the number of moles of gas.	**Given** $P = 24.3$ psi, $V = 3.24$ L, $T(°C) = 25°C$ **Find** n
Strategize The conceptual plan shows how the ideal gas law provides the relationship between the given quantities and the quantity to be found.	**Conceptual Plan** $\boxed{P, V, T} \longrightarrow \boxed{n}$ $PV = nRT$ **Relationship Used** $PV = nRT$ (ideal gas law)
Solve To solve the problem, first solve the ideal gas law for n. Before substituting into the equation, convert P and T into the correct units. Finally, substitute into the equation and compute n.	**Solution** $PV = nRT$ $n = \dfrac{PV}{RT}$ $P = 24.3 \text{ psi} \times \dfrac{1 \text{ atm}}{14.7 \text{ psi}} = 1.6531 \text{ atm}$ (Since rounding the intermediate answer would result in a slightly different final answer, we mark the least significant digit in the intermediate answer, but don't round until the end.) $T(K) = 25 + 273 = 298 \text{ K}$ $n = \dfrac{1.6531 \text{ atm} \times 3.24 \text{ L}}{0.08206 \dfrac{\text{L} \cdot \text{atm}}{\text{mol} \cdot \text{K}} \times 298 \text{ K}}$ $= 0.219 \text{ mol}$

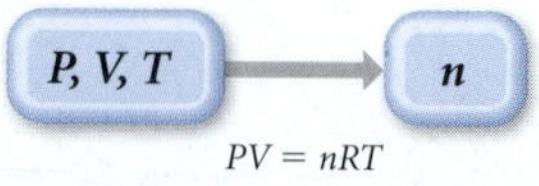

Check The units of the answer are correct. The magnitude of the answer (0.219 mol) makes sense because, as you will see in the next section, one mole of an ideal gas under standard conditions (273 K and 1 atm) occupies 22.4 L. At a pressure that is 65% higher than standard conditions, the volume of 1 mol of gas would be proportionally lower. Since this gas sample occupies 3.24 L, the answer of 0.219 mol is reasonable.

For Practice 5.6

What volume does 0.556 mol of gas occupy at a pressure of 715 mmHg and a temperature of 58 °C?

For More Practice 5.6

Find the pressure in mmHg of a 0.133-g sample of helium gas in a 648-mL container at a temperature of 32 °C.

5.5 Applications of the Ideal Gas Law: Molar Volume, Density, and Molar Mass of a Gas

We just examined how the ideal gas law can be used to calculate one of the variables (P, V, T, or n) given the other three. We now turn to three other applications of the ideal gas law: molar volume, density, and molar mass.

Molar Volume at Standard Temperature and Pressure

The volume occupied by one mole of gas at $T = 0\,°C$ (273 K) and $P = 1.00$ atm can be easily calculated using the ideal gas law. These conditions are called **standard temperature and pressure (STP)**, or simply *standard conditions*, and the volume occupied by one mole of gas under these conditions is called the **molar volume** of an ideal gas. Using the ideal gas law, molar volume is

> The molar volume of 22.4 L only applies at STP.

$$V = \frac{nRT}{P}$$

$$= \frac{1.00\ \text{mol} \times 0.08206\ \dfrac{\text{L} \cdot \text{atm}}{\text{mol} \cdot \text{K}} \times 273\ \text{K}}{1.00\ \text{atm}}$$

$$= 22.4\ \text{L}$$

The molar volume is useful not only because it gives the volume of an ideal gas under standard conditions (which we will use later in this chapter), but also because—as we saw in Examples 5.5 and 5.6—it gives us a way to approximate the volume of an ideal gas under conditions close to standard conditions.

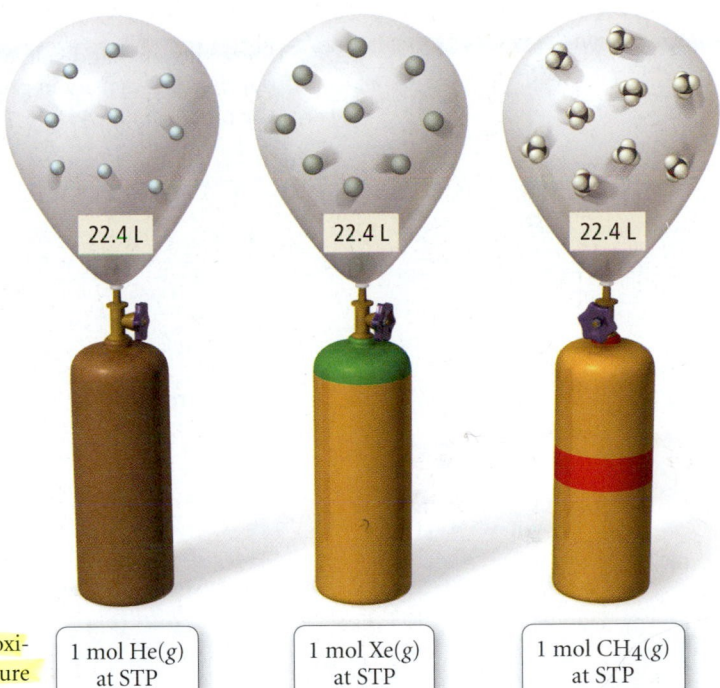

22.4 L   22.4 L   22.4 L

1 mol He(g) at STP   1 mol Xe(g) at STP   1 mol CH$_4$(g) at STP

▶ One mole of any gas occupies approximately 22.4 L at standard temperature (273 K) and pressure (1.0 atm).

❖ Conceptual Connection 5.2 Molar Volume

Assuming ideal behavior, which of the following gas samples will have the greatest volume at STP?

(a) 1 g of H$_2$ **(b)** 1 g of O$_2$ **(c)** 1 g of Ar

Answer: (a) Since 1 g of H$_2$ contains the greatest number of particles (due to H$_2$ having the lowest molar mass of the set), it will occupy the greatest volume.

Density of a Gas

Since one mole of an ideal gas occupies 22.4 L under standard conditions, the density of an ideal gas can be easily calculated under standard conditions. Since density is simply mass/volume, and since the mass of one mole of a gas is simply its molar mass, the *density of a gas under standard conditions* is given by the following relationship:

$$\text{Density} = \frac{\text{molar mass}}{\text{molar volume}}$$

For example, the densities of helium and nitrogen gas at STP are computed as follows:

$$d_{He} = \frac{4.00 \text{ g/mol}}{22.4 \text{ L/mol}} = 0.179 \text{ g/L} \qquad d_{N_2} = \frac{28.02 \text{ g/mol}}{22.4 \text{ L/mol}} = 1.25 \text{ g/L}$$

Notice that *the density of a gas is directly proportional to its molar mass.* The greater the molar mass of a gas, the more dense the gas. For this reason, a gas with a molar mass lower than that of air tends to rise in air. For example, both helium and hydrogen gas (molar masses of 4.00 and 2.01 g/mol, respectively) have molar masses that are lower than the average molar mass of air (approximately 28.8 g/mol). Therefore a balloon filled with either helium or hydrogen gas will float in air.

We can calculate the density of a gas more generally (under any conditions) by using the ideal gas law. For example, we can arrange the ideal gas law as follows:

$$PV = nRT$$

$$\frac{n}{V} = \frac{P}{RT}$$

Since the left-hand side of this equation has units of moles/liter, it represents the molar density. The density in grams/liter can be obtained from molar density by multiplying by the molar mass ($\mathcal{M}$):

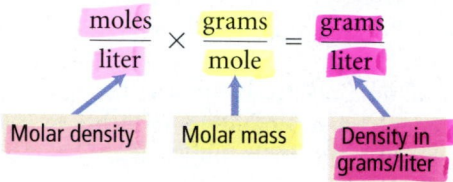

$\dfrac{\text{moles}}{\text{liter}}$	$\times$	$\dfrac{\text{grams}}{\text{mole}}$	$=$	$\dfrac{\text{grams}}{\text{liter}}$
Molar density		Molar mass		Density in grams/liter

> The detailed composition of air is covered in Section 5.6. The main components of air are nitrogen (about four-fifths) and oxygen (about one-fifth).

Density

$$\underset{\substack{\uparrow \\ \text{Molar density}}}{\frac{n}{V}} \, \underset{\substack{\uparrow \\ \text{Molar mass}}}{\mathcal{M}} = \frac{P\mathcal{M}}{RT}$$

$$d = \frac{P\mathcal{M}}{RT}$$

Therefore;

$$d = \frac{P\mathcal{M}}{RT} \qquad [5.6]$$

Notice that, as expected, density increases with increasing molar mass. Notice also that as we learned in Section 5.3, density decreases with increasing temperature.

EXAMPLE 5.7 Density

Calculate the density of nitrogen gas at 125 °C and a pressure of 755 mmHg.

Sort The problem gives you the temperature and pressure of a gas and asks you to find its density. The problem also states that the gas is nitrogen.	Given $T(°C) = 125$ °C, $P = 755$ mmHg Find d
Strategize Equation 5.6 provides the relationship between the density of a gas and its temperature, pressure, and molar mass. The temperature and pressure are given and you can compute the molar mass from the formula of the gas, which we know is N_2.	Conceptual Plan $$\boxed{P, T, \mathcal{M}} \longrightarrow \boxed{d}$$ $$d = \frac{P\mathcal{M}}{RT}$$ Relationships Used $$d = \frac{P\mathcal{M}}{RT} \quad \text{(density of a gas)}$$ Molar mass $N_2 = 28.02$ g/mol

Solve To solve the problem, you must gather each of the required quantities in the correct units. Convert the temperature to kelvins and the pressure to atmospheres.	**Solution** $T(K) = 125 + 273 = 398 \text{ K}$ $P = 755 \text{ mmHg} \times \dfrac{1 \text{ atm}}{760 \text{ mmHg}} = 0.99342 \text{ atm}$
Now simply substitute the quantities into the equation to compute density.	$d = \dfrac{P\mathcal{M}}{RT}$ $= \dfrac{0.99342 \text{ atm} \left(28.02 \dfrac{\text{g}}{\text{mol}}\right)}{0.08206 \dfrac{\text{L} \cdot \text{atm}}{\text{mol} \cdot \text{K}} (398 \text{ K})}$ $= 0.852 \text{ g/L}$

Check The units of the answer are correct. The magnitude of the answer (0.852 g/L) makes sense because earlier we calculated the density of nitrogen gas at STP as 1.25 g/L. Since the temperature is higher than standard conditions, the density should be lower.

For Practice 5.7

Calculate the density of xenon gas at a pressure of 742 mmHg and a temperature of 45 °C.

For More Practice 5.7

A gas has a density of 1.43 g/L at a temperature of 23 °C and a pressure of 0.789 atm. Calculate the molar mass of the gas.

Molar Mass of a Gas

The ideal gas law can be used in combination with mass measurements to calculate the molar mass of an unknown gas. Usually, the mass and volume of an unknown gas are measured under conditions of known pressure and temperature. Then, the amount of the gas in moles is determined from the ideal gas law. Finally, the molar mass is computed by dividing the mass (in grams) by the amount (in moles) as shown in the following example.

EXAMPLE 5.8   Molar Mass of a Gas

A sample of gas has a mass of 0.311 g. Its volume is 0.225 L at a temperature of 55 °C and a pressure of 886 mmHg. Find its molar mass.

Sort The problem gives you the mass of a gas sample, along with its volume, temperature, and pressure. You are asked to find the molar mass.	**Given** $m = 0.311 \text{ g}, V = 0.225 \text{ L}, T(°C) = 55 °C, P = 886 \text{ mmHg}$ **Find** molar mass (g/mol)
Strategize The conceptual plan has two parts. In the first part, use the ideal gas law to find the number of moles of gas. In the second part, use the definition of molar mass to find the molar mass.	**Conceptual Plan** 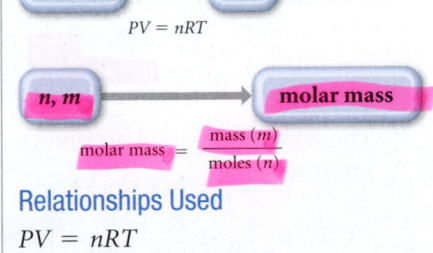 **Relationships Used** $PV = nRT$ $\text{Molar mass} = \dfrac{\text{mass } (m)}{\text{moles } (n)}$

Solve To find the number of moles, first solve the ideal gas law for n.

Solution

$$PV = nRT$$

$$n = \frac{PV}{RT}$$

Before substituting into the equation for n, convert the pressure to atm and the temperature to K.

$$P = 886 \text{ mmHg} \times \frac{1 \text{ atm}}{760 \text{ mmHg}} = 1.1658 \text{ atm}$$

$$T(\text{K}) = 55 + 273 = 328 \text{ K}$$

Now, substitute into the equation and compute n, the number of moles.

$$n = \frac{1.1658 \text{ atm} \times 0.225 \text{ L}}{0.08206 \dfrac{\text{L} \cdot \text{atm}}{\text{mol} \cdot \text{K}} \times 328 \text{ K}}$$

$$= 9.7454 \times 10^{-3} \text{ mol}$$

Finally, use the number of moles (n) and the given mass (m) to find the molar mass.

$$\text{molar mass} = \frac{\text{mass } (m)}{\text{moles } (n)}$$

$$= \frac{0.311 \text{ g}}{9.7454 \times 10^{-3} \text{ mol}}$$

$$= 31.9 \text{ g/mol}$$

Check The units of the answer are correct. The magnitude of the answer (31.9 g/mol) is a reasonable number for a molar mass. If you calculate some very small number (such as any number smaller than 1) or a very large number, you probably made some mistake. Most gases have molar masses between one and several hundred grams per mole.

For Practice 5.8

A sample of gas has a mass of 827 mg. Its volume is 0.270 L at a temperature of 88 °C and a pressure of 975 mmHg. Find its molar mass.

5.6 Mixtures of Gases and Partial Pressures

Many gas samples are not pure, but are mixtures of gases. Dry air, for example, is a mixture containing nitrogen, oxygen, argon, carbon dioxide, and a few other gases in smaller amounts (Table 5.3).

Because the molecules in an ideal gas do not interact, each of the components in an ideal gas mixture acts independently of the others. For example, the nitrogen molecules in air exert a certain pressure—78% of the total pressure—that is independent of the presence of the other gases in the mixture. Likewise, the oxygen molecules in air exert a certain pressure—21% of the total pressure—that is also independent of the presence of the other gases in the mixture. The pressure due to any individual component in a gas mixture is called the **partial pressure** (P_n) of that component and can be calculated from the ideal gas law by assuming that each gas component acts independently. For a multicomponent gas mixture, the partial pressure of each component can be computed from the ideal gas law and the number of moles of that component (n_n) as follows:

$$P_a = n_a\frac{RT}{V}; \quad P_b = n_b\frac{RT}{V}; \quad P_c = n_c\frac{RT}{V}; \quad \dots \qquad [5.7]$$

The sum of the partial pressures of the components in a gas mixture must equal the total pressure:

$$P_{\text{total}} = P_a + P_b + P_c + \dots \qquad [5.8]$$

where P_{total} is the total pressure and $P_a, P_b, P_c, \dots$, are the partial pressures of the components. This is known as **Dalton's law of partial pressures**.

TABLE 5.3 Composition of Dry Air

Gas	Percent by Volume (%)
Nitrogen (N_2)	78
Oxygen (O_2)	21
Argon (Ar)	0.9
Carbon dioxide (CO_2)	0.04

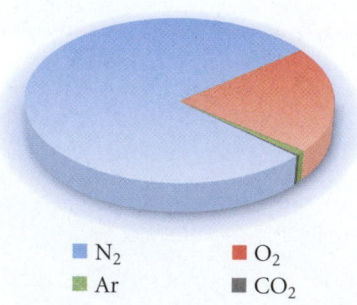

■ N_2 ■ O_2
■ Ar   ■ CO_2

Combining Equations 5.7 and 5.8, we get

$$P_{total} = P_a + P_b + P_c + \dots$$

$$= n_a\frac{RT}{V} + n_b\frac{RT}{V} + n_c\frac{RT}{V} + \dots$$

$$= (n_a + n_b + n_c + \dots)\frac{RT}{V}$$

$$= (n_{total})\frac{RT}{V} \qquad [5.9]$$

The total number of moles in the mixture, when substituted into the ideal gas law, gives the total pressure of the sample.

If we divide Equation 5.7 by Equation 5.9, we get the following result:

$$\frac{P_a}{P_{total}} = \frac{n_a(RT/V)}{n_{total}(RT/V)} = \frac{n_a}{n_{total}} \qquad [5.10]$$

The quantity n_a/n_{total}, the number of moles of a component in a mixture divided by the total number of moles in the mixture, is called the **mole fraction** (χ_a):

$$\chi_a = \frac{n_a}{n_{total}} \qquad [5.11]$$

Rearranging Equation 5.10 and substituting the definition of mole fraction gives the following:

$$\frac{P_a}{P_{total}} = \frac{n_a}{n_{total}}$$

$$P_a = \frac{n_a}{n_{total}}P_{total} = \chi_a P_{total}$$

or simply

$$P_a = \chi_a P_{total} \qquad [5.12]$$

The partial pressure of a component in a gaseous mixture is its mole fraction multiplied by the total pressure. For gases, the mole fraction of a component is equivalent to its percent by volume divided by 100%. Therefore, based on Table 5.3, we compute the partial pressure of nitrogen (P_{N_2}) in air at 1.00 atm as follows:

$$P_{N_2} = 0.78 \times 1.00 \text{ atm}$$

$$= 0.78 \text{ atm}$$

Similarly, the partial pressure of oxygen in air at 1.00 atm is 0.21 atm and the partial pressure of Ar in air is 0.01 atm. Applying Dalton's law of partial pressures to air at 1.00 atm:

$$P_{total} = P_{N_2} + P_{O_2} + P_{Ar}$$

$$P_{total} = 0.78 \text{ atm} + 0.21 \text{ atm} + 0.01 \text{ atm}$$

$$= 1.00 \text{ atm}$$

We can ignore the contribution of the CO_2 and other trace gases because they are so small.

EXAMPLE 5.9 Total Pressure and Partial Pressures

A 1.00-L mixture of helium, neon, and argon has a total pressure of 662 mmHg at 298 K. If the partial pressure of helium is 341 mmHg and the partial pressure of neon is 112 mmHg, what mass of argon is present in the mixture?

Sort The problem gives you the partial pressures of two of the three components in a gas mixture, along with the total pressure, the volume, and the temperature, and asks you find the mass of the third component.	Given $P_{He} = 341$ mmHg, $P_{Ne} = 112$ mmHg, $P_{total} = 662$ mmHg, $V = 1.00$ L, $T = 298$ K Find m_{Ar}

Strategize You can find the mass of argon from the number of moles of argon, which you can calculate from the partial pressure of argon and the ideal gas law.

Begin by finding the partial pressure of argon from Dalton's law of partial pressures.

Then use the partial pressure of argon together with the volume of the sample and the temperature to find the number of moles of argon.

Finally, use the molar mass of argon to compute the mass of argon from the number of moles of argon.

Conceptual Plan

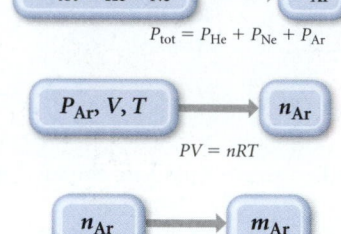

$$P_{tot} = P_{He} + P_{Ne} + P_{Ar}$$

$$PV = nRT$$

$$\frac{1 \text{ mol Ar}}{39.95 \text{ g Ar}}$$

Relationships Used

$P_{total} = P_{He} + P_{Ne} + P_{Ar}$ (Dalton's law)

$PV = nRT$ (ideal gas law)

molar mass Ar $= 39.95 \text{ g/mol}$

Solve Follow the conceptual plan. To find the partial pressure of argon, solve the equation for P_{Ar} and substitute the values of the other partial pressures to compute P_{Ar}.

Convert the partial pressure from mmHg to atm and use it in the ideal gas law to compute the amount of argon in moles.

Use the molar mass of argon to convert from amount of argon in moles to mass of argon.

Solution

$$P_{total} = P_{He} + P_{Ne} + P_{Ar}$$

$$P_{Ar} = P_{total} - P_{He} - P_{Ne}$$

$$= 662 \text{ mmHg} - 341 \text{ mmHg} - 112 \text{ mmHg}$$

$$= 209 \text{ mmHg}$$

$$209 \text{ mmHg} \times \frac{1 \text{ atm}}{760 \text{ mmHg}} = 0.275 \text{ atm}$$

$$n = \frac{PV}{RT} = \frac{0.275 \text{ atm} \,(1.00 \text{ L})}{0.08206 \dfrac{\text{L} \cdot \text{atm}}{\text{mol} \cdot \text{K}}(298 \text{ K})} = 1.125 \times 10^{-2} \text{ mol Ar}$$

$$1.125 \times 10^{-2} \text{ mol Ar} \times \frac{39.95 \text{ g Ar}}{1 \text{ mol Ar}} = 0.449 \text{ g Ar}$$

Check The units of the answer are correct. The magnitude of the answer makes sense because the volume is 1.0 L, which at STP would contain about 1/22 mol. Since the partial pressure of argon in the mixture is about 1/3 of the total pressure, we roughly estimate about 1/66 of one molar mass of argon, which is fairly close to the answer we got.

For Practice 5.9

A sample of hydrogen gas is mixed with water vapor. The mixture has a total pressure of 755 torr and the water vapor has a partial pressure of 24 torr. What amount (in moles) of hydrogen gas is contained in 1.55 L of this mixture at 298 K?

Deep-Sea Diving and Partial Pressures

Our lungs have evolved to breathe oxygen at a partial pressure of $P_{O_2} = 0.21 \text{ atm}$. If the total pressure decreases—when climbing a mountain, for example—the partial pressure of oxygen also decreases. On top of Mt. Everest, where the total pressure is only 0.311 atm, the partial pressure of oxygen (P_{O_2}) is only 0.065 atm. Low oxygen levels produce a physiological condition called **hypoxia** or oxygen starvation (Figure 5.16 ▶). Mild hypoxia causes dizziness, headache, and shortness of breath. Severe hypoxia, which occurs when P_{O_2} drops below 0.1 atm, may result in unconsciousness or even death. For this reason, climbers hoping to make the summit of Mt. Everest usually carry oxygen to breathe.

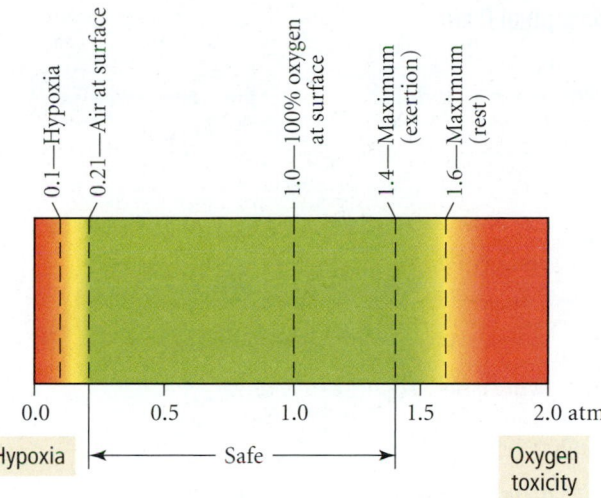

Hypoxia ← Safe → Oxygen toxicity

◀ **FIGURE 5.16 Oxygen Partial Pressure Limits** The partial pressure of oxygen in air at sea level is 0.21 atm. Partial pressures of oxygen below 0.1 atm and above 1.4 atm are dangerous to humans.

While not as dangerous as a lack of oxygen, too much oxygen can also cause physiological problems. Scuba divers, as we have learned, breathe pressurized air. At 30 m, a scuba diver breathes air at a total pressure of 4.0 atm, making P_{O_2} about 0.84 atm. This elevated partial pressure of oxygen raises the density of oxygen molecules in the lungs, resulting in a higher concentration of oxygen in body tissues. When P_{O_2} increases beyond 1.4 atm, the increased oxygen concentration in body tissues causes a condition called **oxygen toxicity**, which results in muscle twitching, tunnel vision, and convulsions. Divers who venture too deep without proper precautions have drowned because of oxygen toxicity.

A second problem associated with breathing pressurized air is the increase of nitrogen in the lungs. At 30 m, a scuba diver breathes nitrogen at $P_{N_2} = 3.12$ atm, which increases the nitrogen concentration in body tissues and fluids. When P_{N_2} increases beyond about 4 atm, a condition called **nitrogen narcosis** or *rapture of the deep* results. Divers describe this condition as feeling inebriated or drunk. A diver breathing compressed air at 60 m feels as if he has consumed too much wine.

To avoid oxygen toxicity and nitrogen narcosis, deep-sea divers—those venturing beyond 50 m—breathe specialized mixtures of gases. One common mixture is called heliox, a mixture of helium and oxygen. These mixtures usually contain a smaller percentage of oxygen than would be found in air, thereby lowering the risk of oxygen toxicity. Heliox also contains helium instead of nitrogen, eliminating the risk of nitrogen narcosis.

Surface	
P_{tot}	= 1 atm
P_{N_2}	= 0.78 atm
P_{O_2}	= 0.21 atm

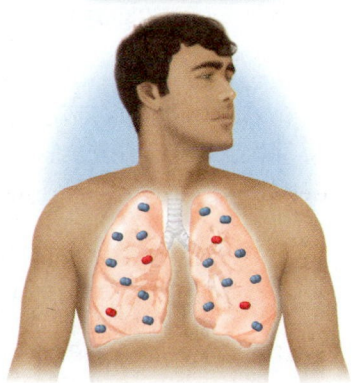

30 m	
P_{tot}	= 4 atm
P_{N_2}	= 3.12 atm
P_{O_2}	= 0.84 atm

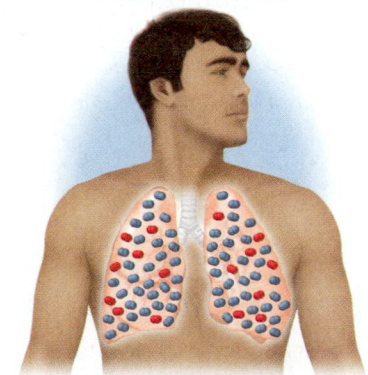

▲ When a diver breathes compressed air, the abnormally high partial pressure of oxygen in the lungs produces an elevated concentration of oxygen in body tissues.

EXAMPLE 5.10 Partial Pressures and Mole Fractions

A 12.5-L scuba diving tank is filled with a heliox mixture containing 24.2 g of He and 4.32 g of O_2 at 298 K. Calculate the mole fraction and partial pressure of each component in the mixture and calculate the total pressure.

Sort The problem gives the masses of two gases in a mixture and the volume and temperature of the mixture. You are asked to find the mole fraction and partial pressure of each component, as well as the total pressure.	**Given** $m_{He} = 24.2$ g, $m_{O_2} = 4.32$ g, $V = 12.5$ L, $T = 298$ K **Find** $\chi_{He}, \chi_{O_2}, P_{He}, P_{O_2}, P_{total}$

Strategize The conceptual plan has several parts. To calculate the mole fraction of each component, you must first find the number of moles of each component. Therefore, in the first part of the conceptual plan, convert the masses to moles using the molar masses.

In the second part, compute the mole fraction of each component using the mole fraction definition.

To calculate *partial pressures* you must calculate the *total pressure* and then use the mole fractions from the previous part to calculate the partial pressures. Calculate the total pressure from the sum of the moles of both components. (Alternatively, you could calculate the partial pressures of the components individually, using the number of moles of each component. Then you could sum them to obtain the total pressure.)

Last, use the mole fractions of each component and the total pressure to calculate the partial pressure of each component.

Conceptual Plan

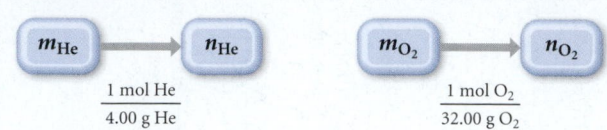

$$X_{He} = \frac{n_{He}}{n_{He} + n_{O_2}}; \quad X_{O_2} = \frac{n_{O_2}}{n_{He} + n_{O_2}}$$

$$P_{total} = \frac{(n_{He} + n_{O_2})RT}{V}$$

$$P_{He} = X_{He}P_{total}; \quad P_{O_2} = X_{O_2}P_{total}$$

Relationships Used

$X_a = n_a/n_{total}$ (mole fraction definition)

$P_{total}V = n_{total}RT$ (ideal gas law)

$P_a = X_a P_{total}$

Solve Follow the plan to solve the problem. Begin by converting each of the masses to amounts in moles.

Solution

$$24.2 \text{ g He} \times \frac{1 \text{ mol He}}{4.00 \text{ g He}} = 6.05 \text{ mol He}$$

$$4.32 \text{ g O}_2 \times \frac{1 \text{ mol O}_2}{32.00 \text{ g O}_2} = 0.135 \text{ mol O}_2$$

Compute each of the mole fractions.

$$X_{He} = \frac{n_{He}}{n_{He} + n_{O_2}} = \frac{6.05}{6.05 + 0.135} = 0.97817$$

$$X_{O_2} = \frac{n_{O_2}}{n_{He} + n_{O_2}} = \frac{0.135}{6.05 + 0.135} = 0.021827$$

Compute the total pressure.

$$P_{total} = \frac{(n_{He} + n_{O_2})RT}{V}$$

$$= \frac{(6.05 \text{ mol} + 0.135 \text{ mol})\left(0.08206 \dfrac{\text{L} \cdot \text{atm}}{\text{mol} \cdot \text{K}}\right)(298 \text{ K})}{12.5 \text{ L}}$$

$$= 12.099 \text{ atm}$$

Finally, compute the partial pressure of each component.

$$P_{He} = X_{He}P_{total} = 0.97817 \times 12.099 \text{ atm}$$
$$= 11.8 \text{ atm}$$

$$P_{O_2} = X_{O_2}P_{total} = 0.021827 \times 12.099 \text{ atm}$$
$$= 0.264 \text{ atm}$$

Check The units of the answers are correct and the magnitudes are reasonable.

For Practice 5.10

A diver breathes a heliox mixture with an oxygen mole fraction of 0.050. What must the total pressure be for the partial pressure of oxygen to be 0.21 atm?

Collecting a Gas Over Water

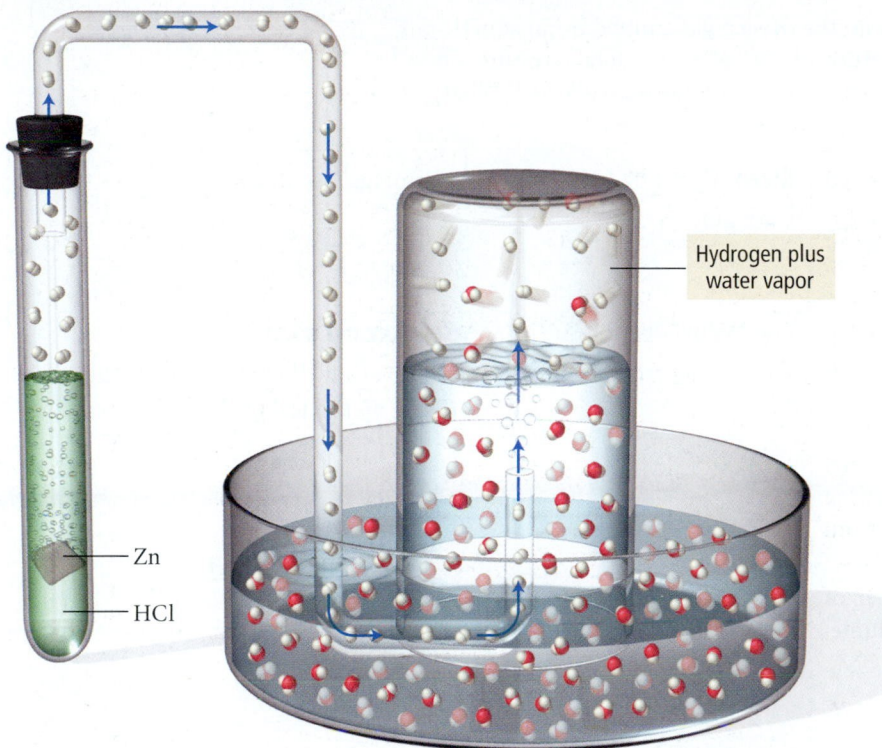

Hydrogen plus water vapor

Zn

HCl

◀ **FIGURE 5.17 Collecting a Gas over Water** When the gaseous product of a chemical reaction is collected over water, the product molecules become mixed with water molecules. The pressure of water in the final mixture is equal to the vapor pressure of water at the temperature at which the gas is collected.

Collecting Gases over Water

When the product of a chemical reaction is gaseous, it is often collected by the displacement of water. For example, suppose the following reaction is used as a source of hydrogen gas:

$$Zn(s) + 2\,HCl(aq) \longrightarrow ZnCl_2(aq) + H_2(g)$$

As the hydrogen gas forms, it bubbles through the water and gathers in the collection flask. The hydrogen gas collected in this way is not pure, however, but mixed with water vapor because some water molecules evaporate and become mixed with the hydrogen molecules (Figure 5.17 ▲).

The partial pressure of water in the mixture, called its **vapor pressure**, depends on the temperature (Table 5.4). Vapor pressure increases with increasing temperature because the higher temperatures cause more water molecules to evaporate.

Suppose we collect the hydrogen gas over water at a total pressure of 758.2 mmHg and a temperature of 25 °C. What is the partial pressure of the hydrogen gas? We know that the total pressure is 758.2 mmHg and that the partial pressure of water is 23.78 mmHg (its vapor pressure at 25 °C):

$$P_{total} = P_{H_2} + P_{H_2O}$$
$$758.2 \text{ mmHg} = P_{H_2} + 23.78 \text{ mmHg}$$

Therefore,

$$P_{H_2} = 758.2 \text{ mmHg} - 23.78 \text{ mmHg}$$
$$= 734.4 \text{ mmHg}$$

The partial pressure of the hydrogen in the mixture will be 734.4 mmHg.

TABLE 5.4 Vapor Pressure of Water versus Temperature

Temperature (°C)	Pressure (mmHg)	Temperature (°C)	Pressure (mmHg)
0	4.58	55	118.2
5	6.54	60	149.6
10	9.21	65	187.5
15	12.79	70	233.7
20	17.55	75	289.1
25	23.78	80	355.1
30	31.86	85	433.6
35	42.23	90	525.8
40	55.40	95	633.9
45	71.97	100	760.0
50	92.6		

Vapor pressure is covered in detail in Chapter 11.

EXAMPLE 5.11   Collecting Gases over Water

In order to determine the rate of photosynthesis, the oxygen gas emitted by an aquatic plant was collected over water at a temperature of 293 K and a total pressure of 755.2 mmHg. Over a specific time period, a total of 1.02 L of gas was collected. What mass of oxygen gas (in grams) was formed?

Sort The problem gives the volume of gas collected over water as well as the temperature and the pressure. You are asked to find the mass in grams of oxygen formed.	**Given** $V = 1.02$ L, $P_{total} = 755.2$ mmHg, $T = 293$ K **Find** gO_2

Strategize You can determine the mass of oxygen from the amount of oxygen in moles, which you can calculate from the ideal gas law if you know the partial pressure of oxygen. Since the oxygen is mixed with water vapor, you can find the partial pressure of oxygen in the mixture by subtracting the partial pressure of water at 293 K (20 °C) from the total pressure.

Conceptual Plan

$P_{O_2} = P_{total} - P_{H_2O}(20\ °C)$

Relationship Used

$P_{total} = P_a + P_b + P_c + \ldots$ (Dalton's law)
$PV = nRT$ (ideal gas law)

Next, use the ideal gas law to find the number of moles of oxygen from its partial pressure, volume, and temperature.

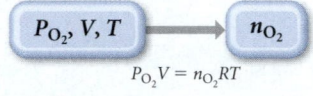

$$P_{O_2}V = n_{O_2}RT$$

Finally, use the molar mass of oxygen to convert the number of moles to grams.

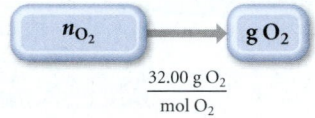

$$\frac{32.00\text{ g }O_2}{\text{mol }O_2}$$

Solve Follow the conceptual plan to solve the problem. Begin by calculating the partial pressure of oxygen in the oxygen/water mixture. You can find the partial pressure of water at 20 °C in Table 5.4.

Solution

$P_{O_2} = P_{total} - P_{H_2O}(20\ °C)$
$\quad = 755.2\text{ mmHg} - 17.55\text{ mmHg}$
$\quad = 737.65\text{ mmHg}$

Next, solve the ideal gas law for number of moles.

$n_{O_2} = \dfrac{P_{O_2}V}{RT}$

Before substituting into the ideal gas law, you must convert the partial pressure of oxygen from mmHg to atm.

$737.65\ \cancel{\text{mmHg}} \times \dfrac{1\text{ atm}}{760\ \cancel{\text{mmHg}}} = 0.97059\text{ atm}$

Next, substitute into the ideal gas law to find the number of moles of oxygen.

$n_{O_2} = \dfrac{P_{O_2}V}{RT} = \dfrac{0.97059\ \cancel{\text{atm}}\ (1.02\ \cancel{L})}{0.08206\ \dfrac{\cancel{L}\cdot\cancel{\text{atm}}}{\text{mol}\cdot\cancel{K}}(293\ \cancel{K})}$

$\quad = 4.1175 \times 10^{-2}\text{ mol}$

Finally, use the molar mass of oxygen to convert to grams of oxygen.

$4.1175 \times 10^{-2}\ \cancel{\text{mol }O_2} \times \dfrac{32.00\text{ g }O_2}{1\ \cancel{\text{mol }O_2}} = 1.32\text{ g }O_2$

Check The answer is in the correct units. We can quickly check the magnitude of the answer by using molar volume. Under STP one liter is about 1/22 of one mole. Therefore the answer should be about 1/22 the molar mass of oxygen (1/22 × 32 = 1.45). The magnitude of our answer seems reasonable.

For Practice 5.11

A common way to make hydrogen gas in the laboratory is to place a metal such as zinc in hydrochloric acid. The hydrochloric acid reacts with the metal to produce hydrogen gas, which is then collected over water. Suppose a student carries out this reaction and collects a total of 154.4 mL of gas at a pressure of 742 mmHg and a temperature of 25 °C. What mass of hydrogen gas (in mg) did the student collect?

5.7 Gases in Chemical Reactions: Stoichiometry Revisited

In Chapter 4, we learned how the coefficients in chemical equations can be used as conversion factors between number of moles of reactants and number of moles of products in a chemical reaction. These conversion factors can be used to determine, for example, the mass of product obtained in a chemical reaction based on a given mass of reactant, or the mass of one reactant needed to react completely with a given mass of another reactant. The general conceptual plan for these kinds of calculations is

where A and B are two different substances involved in the reaction and the conversion factor between amounts (in moles) of each comes from the stoichiometric coefficients in the balanced chemical equation.

In reactions involving gaseous reactant or products, it is often convenient to specify the quantity of a gas in terms of its volume at a given temperature and pressure. As we have seen, stoichiometric relationships always express relative amounts in moles. However, we can use the ideal gas law to find the amounts in moles from the volumes, or find the volumes from the amounts in moles.

$$n = \frac{PV}{RT} \qquad V = \frac{nRT}{P}$$

The general conceptual plan for these kinds of calculations is

> The pressures here could also be partial pressures.

The following examples show this kind of calculation.

EXAMPLE 5.12 Gases in Chemical Reactions

Methanol (CH_3OH) can be synthesized by the following reaction:

$$CO(g) + 2H_2(g) \longrightarrow CH_3OH(g)$$

What volume (in liters) of hydrogen gas, measured at a temperature of 355 K and a pressure of 738 mmHg, is required to synthesize 35.7 g of methanol?

Sort You are given the mass of methanol, the product of a chemical reaction. You are asked to find the required volume of one of the reactants (hydrogen gas) at a specified temperature and pressure.	**Given** 35.7 g CH_3OH, $T = 355$ K, $P = 738$ mmHg **Find** V_{H_2}
Strategize The required volume of hydrogen gas can be calculated from the number of moles of hydrogen gas, which can be obtained from the number of moles of methanol via the stoichiometry of the reaction. First, find the number of moles of methanol from its mass by using the molar mass. Then use the stoichiometric relationship from the balanced chemical equation to find the number of moles of hydrogen needed to form that quantity of methanol. Finally, substitute the number of moles of hydrogen together with the pressure and temperature into the ideal gas law to find the volume of hydrogen.	**Conceptual Plan** 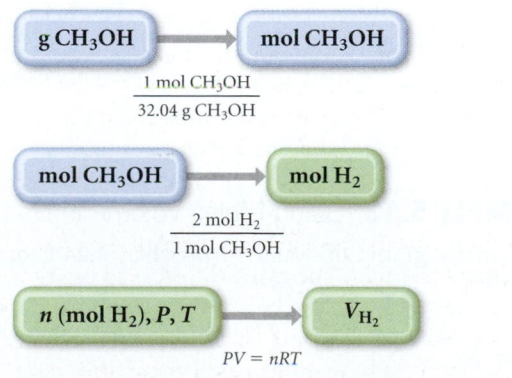

Relationships Used

$PV = nRT$ (ideal gas law)

$2 \text{ mol } H_2 : 1 \text{ mol } CH_3OH$
(from balanced chemical equation)

molar mass $CH_3OH = 32.04 \text{ g/mol}$

Solve Follow the conceptual plan to solve the problem. Begin by using the mass of methanol to get the number of moles of methanol.

Next, convert the number of moles of methanol to moles of hydrogen.

Finally, use the ideal gas law to find the volume of hydrogen. Before substituting into the equation, you must convert the pressure to atmospheres.

Solution

$$35.7 \text{ g CH}_3\text{OH} \times \frac{1 \text{ mol CH}_3\text{OH}}{32.04 \text{ g CH}_3\text{OH}} = 1.1\underline{1}42 \text{ mol CH}_3\text{OH}$$

$$1.1142 \text{ mol CH}_3\text{OH} \times \frac{2 \text{ mol H}_2}{1 \text{ mol CH}_3\text{OH}} = 2.2\underline{2}84 \text{ mol H}_2$$

$$V_{H_2} = \frac{n_{H_2}RT}{P}$$

$$P = 738 \text{ mmHg} \times \frac{1 \text{ atm}}{760 \text{ mmHg}} = 0.97\underline{1}05 \text{ atm}$$

$$V_{H_2} = \frac{(2.2\underline{2}84 \text{ mol})\left(0.08206\dfrac{\text{L} \cdot \text{atm}}{\text{mol} \cdot \text{K}}\right)(355 \text{ K})}{0.97105 \text{ atm}}$$

$$= 66.9 \text{ L}$$

Check The units of the answer are correct. The magnitude of the answer (66.9 L) seems reasonable for the following reason. You are given slightly more than one molar mass of methanol, which is therefore slightly more than one mole of methanol. From the equation you can see that 2 mol hydrogen is required to make 1 mol methanol. Therefore, the answer must be slightly greater than 2 mol hydrogen. Under standard conditions, slightly more than two mol hydrogen occupies slightly more than $2 \times 22.4 \text{ L} = 44.8 \text{ L}$. At a temperature greater than standard temperature, the volume would be even greater; therefore, the magnitude of the answer is reasonable.

For Practice 5.12

In the following reaction, 4.58 L of O_2 was formed at $P = 745 \text{ mmHg}$ and $T = 308 \text{ K}$. How many grams of Ag_2O must have decomposed?

$$2\,Ag_2O(s) \longrightarrow 4\,Ag(s) + O_2(g)$$

For More Practice 5.12

In the above reaction, what mass of $Ag_2O(s)$ (in grams) is required to form 388 mL of oxygen gas at $P = 734 \text{ mmHg}$ and 25.0 °C?

Molar Volume and Stoichiometry

In Section 5.5, we saw that, under standard conditions, 1 mol of an ideal gas occupies 22.4 L. Consequently, if a reaction is occurring at or near standard conditions, we can use 1 mol = 22.4 L as a conversion factor in stoichiometric calculations, as shown in the following example.

EXAMPLE 5.13 Using Molar Volume in Gas Stoichiometric Calculations

How many grams of water form when 1.24 L of H_2 gas at STP completely reacts with O_2?

$$2\,H_2(g) + O_2(g) \longrightarrow 2\,H_2O(g)$$

Sort You are given the volume of hydrogen gas (a reactant) at STP and asked to determine the mass of water that forms upon complete reaction.

Given 1.24 L H_2

Find g H_2O

Strategize Since the reaction occurs under standard conditions, you can convert directly from the volume (in L) of hydrogen gas to the amount in moles. Then use the stoichiometric relationship from the balanced equation to find the number of moles of water formed. Finally, use the molar mass of water to obtain the mass of water formed.

Conceputal Plan

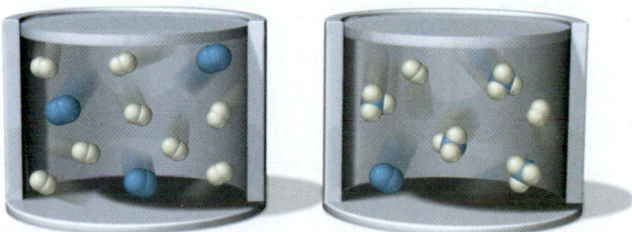

Relationships Used

1 mol = 22.4 L (at STP)

2 mol H_2 : 2 mol H_2O (from balanced equation)

molar mass H_2O = 18.02 g/mol

Solve Follow the conceptual plan to solve the problem.

Solution

$$1.24 \; \cancel{L \; H_2} \times \frac{1 \; \cancel{mol \; H_2}}{22.4 \; \cancel{L \; H_2}} \times \frac{2 \; \cancel{mol \; H_2O}}{2 \; \cancel{mol \; H_2}} \times \frac{18.02 \; g \; H_2O}{1 \; \cancel{mol \; H_2O}} = 0.998 \; g \; H_2O$$

Check The units of the answer are correct. The magnitude of the answer (0.998 g) is about 1/18 of the molar mass of water, roughly equivalent to the approximately 1/22 of a mole of hydrogen gas given, as expected for a stoichiometric relationship between number of moles of hydrogen and number of moles of water that is 1:1.

For Practice 5.13

How many liters of oxygen (at STP) is required to form 10.5 g of H_2O?

$$2 \, H_2(g) + O_2(g) \longrightarrow 2 \, H_2O(g)$$

 Conceptual Connection 5.3 Pressure and Number of Moles

Nitrogen and hydrogen react to form ammonia according the following equation:

$$N_2(g) + 3 \, H_2(g) \rightleftharpoons 2 \, NH_3(g)$$

Consider the following representations of the initial mixture of reactants and the resulting mixture after the reaction has been allowed to react for some time:

If the volume is kept constant, and nothing is added to the reaction mixture, what happens to the total pressure during the course of the reaction?

(a) the pressure increases   **(b)** the pressure decreases   **(c)** the pressure does not change

Answer: (b) Since the total number of gas molecules decreases (from 11 on the left side to 7 on the right), the total pressure—the sum of all the partial pressures—must also decrease.

5.8 Kinetic Molecular Theory: A Model for Gases

Kinetic Molecular Theory

▲ FIGURE 5.18 **A Model for Gas Behavior** In the kinetic molecular theory of gases, a gas sample is modeled as a collection of particles in constant straight-line motion. The size of the particles is negligibly small and their collisions are elastic.

In Chapter 1, we learned how the scientific method proceeds from observations to laws and eventually to theories. Remember that laws summarize behavior—for example, Charles's law summarizes *how* the volume of a gas depends on temperature—while theories give the underlying reasons for the behavior. A theory of gas behavior explains, for example, *why* the volume of a gas increases with increasing temperature.

The simplest model for the behavior of gases is provided by the **kinetic molecular theory**. In this theory, a gas is modeled as a collection of particles (either molecules or atoms, depending on the gas) in constant motion (Figure 5.18 ◄). A single particle moves in a straight line until it collides with another particle (or with the wall of the container). The basic postulates of kinetic molecular theory are as follows:

1. **The size of a particle is negligibly small.** Kinetic molecular theory assumes that the particles themselves occupy no volume, even though they have mass. This postulate is justified because, under normal pressures, the space between atoms or molecules in a gas is very large compared to the size of an atom or molecule itself. For example, in a sample of argon gas under STP conditions, only about 0.01% of the volume is occupied by atoms and the average distance from one argon atom to another is 3.3 nm. In comparison, the atomic radius of argon is 97 pm. If an argon atom were the size of a golf ball, its nearest neighbor would be, on average, just over 4 ft away at STP.

2. **The average kinetic energy of a particle is proportional to the temperature in kelvins.** The motion of atoms or molecules in a gas is due to thermal energy, which distributes itself among the particles in the gas. At any given moment, some particles are moving faster than others—there is a distribution of velocities—but the higher the temperature, the faster is the overall motion, and the greater is the average kinetic energy. Notice that *kinetic energy* ($\frac{1}{2}mv^2$)—not *velocity*—is proportional to temperature. The atoms in a sample of helium and a sample of argon at the same temperature have the same average kinetic energy, but not the same average velocity. Since the helium atoms are lighter, they must move faster to have the same kinetic energy as argon atoms.

3. **The collision of one particle with another (or with the walls) is completely elastic.** This means that when two particles collide, they may *exchange energy*, but there is no overall *loss of energy*. Any kinetic energy lost by one particle is completely gained by the other. This is the case because the particles have no "stickiness," and they are not deformed by the collision. In other words, an encounter between two particles in kinetic molecular theory is more like the collision between two billiard balls than between two lumps of clay (Figure 5.19 ▼). Between collisions, the particles do not exert any forces on one another.

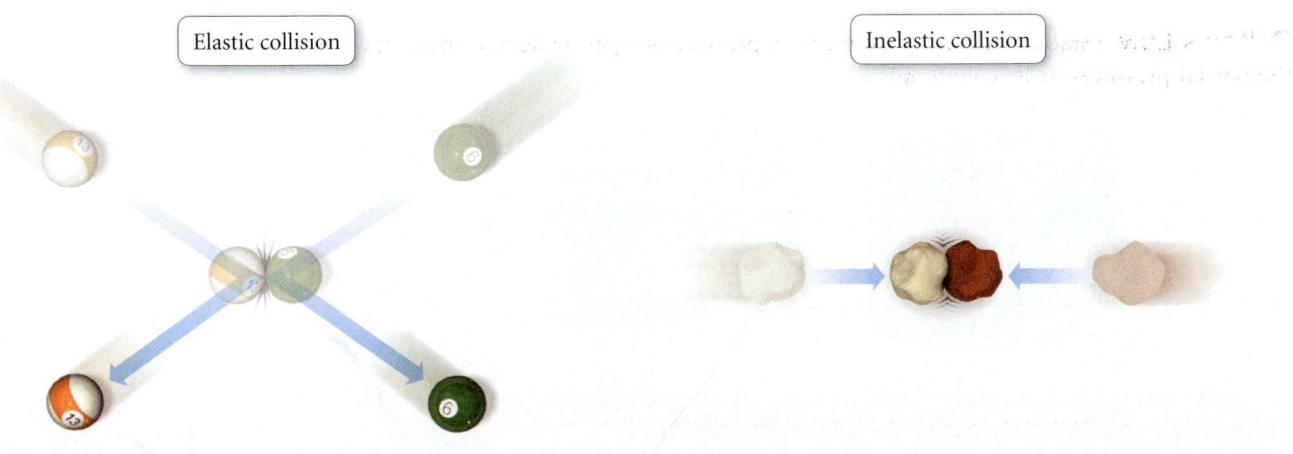

Elastic collision   Inelastic collision

▲ FIGURE 5.19 **Elastic versus Inelastic Collisions** When two billiard balls collide, the collision is elastic—the total kinetic energy of the colliding bodies is the same before and after the collision. When two lumps of clay collide, the collision is inelastic—the kinetic energy of the colliding bodies is dissipated as heat during the collision.

If you start with the postulates of kinetic molecular theory, you can mathematically derive the ideal gas law (as we show later). In other words, the ideal gas law follows directly from kinetic molecular theory, which gives us confidence that the assumptions of the theory are valid, at least under conditions where the ideal gas law works. Let's see how the concept of pressure and each of the gas laws we have examined follow conceptually from kinetic molecular theory.

The Nature of Pressure In Section 5.2, we defined pressure as force divided by area:

$$P = \frac{F}{A}$$

According to kinetic molecular theory, a gas is a collection of particles in constant motion. The motion results in collisions between the particles and the surfaces around them. As each particle collides with a surface, it exerts a force upon that surface. The result of many particles in a gas sample exerting forces on the surfaces around them is a constant pressure.

The force (F) associated with an individual collision is given by $F = ma$, where m is the mass of the particle and a is its acceleration as it changes its direction of travel due to the collision.

Boyle's Law Boyle's law states that, for a constant number of particles at constant temperature, the volume of a gas is inversely proportional to its pressure. If you decrease the volume of a gas, you force the gas particles to occupy a smaller space. It follows from kinetic molecular theory that, as long the temperature remains the same, the result is a greater number of collisions with the surrounding surfaces and therefore a greater pressure.

Charles's Law Charles's law states that, for a constant number of particles at constant pressure, the volume of a gas is proportional to its temperature. According to kinetic molecular theory, when you increase the temperature of a gas, the average speed, and thus the average kinetic energy, of the particles increases. Since this greater kinetic energy results in more frequent collisions and more force per collision, the pressure of the gas would increase if its volume were held constant (Gay-Lussac's law). The only way for the pressure to remain constant is for the volume to increase. The greater volume spreads the collisions out over a greater area, so that the pressure (defined as force per unit area) is unchanged.

Avogadro's Law Avogadro's law states that, at constant temperature and pressure, the volume of a gas is proportional to the number of particles. According to kinetic molecular theory, when you increase the number of particles in a gas sample, the number of collisions with the surrounding surfaces increases. Since the greater number of collisions would result in a greater overall force on surrounding surfaces, the only way for the pressure to remain constant is for the volume to increase so that the number of particles per unit volume (and thus the number of collisions) remains constant.

Dalton's Law Dalton's law states that the total pressure of a gas mixture is the sum of the partial pressures of its components. In other words, according to Dalton's law, the components in a gas mixture act identically to, and independently of, one another. According to kinetic molecular theory, the particles have negligible size and they do not interact. Consequently, the only property that would distinguish one type of particle from another is its mass. However, even particles of different masses have the same average kinetic energy at a given temperature, so they exert the same force upon collision with a surface. Consequently, adding components to a gas mixture—even different *kinds* of gases—has the same effect as simply adding more particles. The partial pressures of all the components sum to the overall pressure.

Kinetic Molecular Theory and the Ideal Gas Law

We have just seen how each of the gas laws conceptually follows from kinetic molecular theory. We can also *derive* the ideal gas law from the postulates of kinetic molecular theory. In other words, the kinetic molecular theory is a quantitative model that implies $PV = nRT$. We now explore the general form of this derivation.

Calculating Gas Pressure: A Molecular View

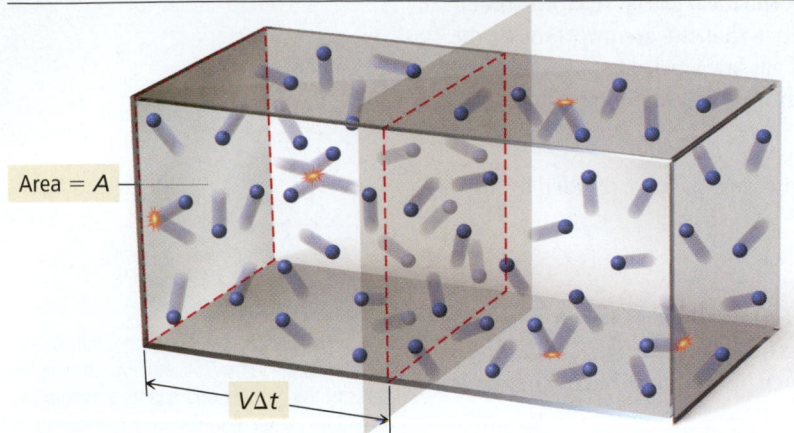

▲ FIGURE 5.20 The Pressure on the Wall of a Container The pressure on the wall of a container can be calculated by determining the total force due to collisions of the particles with the wall.

The pressure on a wall of a container (Figure 5.20 ◀) occupied by particles in constant motion is the total force on the wall (due to the collisions) divided by the area of the wall.

$$P = \frac{F_{total}}{A} \qquad [5.13]$$

From Newton's second law, the force (F) associated with an individual collision is given by $F = ma$, where m is the mass of the particle and a is its acceleration as it changes its direction of travel due to the collision. Since the acceleration for each collision is the change in velocity (Δv) divided by the time interval (Δt), the force imparted for each collision is

$$F_{collision} = m\frac{\Delta v}{\Delta t} \qquad [5.14]$$

If a particle collides elastically with the wall, it bounces off the wall with no loss of energy. For a straight-line collision, the change in velocity is $2v$ (the particle's velocity was v before the collision and $-v$ after the collision; therefore, the change is $2v$). So the force per collision is given by the following:

$$F_{collision} = m\frac{2v}{\Delta t} \qquad [5.15]$$

The total number of collisions in the time interval Δt on a wall of surface area A is proportional to the number of particles that can reach the wall in this time interval—in other words, all those within a distance of $v\Delta t$ of the wall. These particles occupy a volume given by $v\Delta t \times A$, and their total number is equal to this volume multiplied by the density of particles in the container (n/V):

Number of collisions $\propto$ number of particles within $v\,\Delta t$

$$\propto \underbrace{v\,\Delta t \times A}_{\text{Volume}} \times \underbrace{\frac{n}{V}}_{\substack{\text{Density of} \\ \text{particles}}} \qquad [5.16]$$

The *total force* on the wall is equal to the force per collision multiplied by the number of collisions:

$$F_{total} = F_{collision} \times \text{number of collisions}$$

$$\propto m\frac{2v}{\Delta t} \times v\Delta t \times A \times \frac{n}{V}$$

$$\propto mv^2 \times A \times \frac{n}{V} \qquad [5.17]$$

The pressure on the wall is equal to the total force divided by the surface area of the wall:

$$P = \frac{F_{total}}{A}$$

$$\propto \frac{mv^2 \times A \times \dfrac{n}{V}}{A}$$

$$P \propto mv^2 \times \frac{n}{V} \qquad [5.18]$$

Notice that Equation 5.18 contains within it Boyle's law ($P \propto 1/V$) and Avogadro's law ($V \propto n$). We can get the complete ideal gas law from postulate 2 of the kinetic molecular theory, which states that the average kinetic energy ($\frac{1}{2}mv^2$) is proportional to the temperature in kelvins (T):

$$mv^2 \propto T \qquad [5.19]$$

Therefore, by combining Equations 5.18 and 5.19, we get the following:

$$P \propto \frac{T \times n}{V}$$

$$PV \propto nT \qquad [5.20]$$

The proportionality can be replaced by an equals sign by providing the correct constant, R:

$$PV = nRT \qquad [5.21]$$

In other words, the kinetic molecular theory (a model for how gases behave) predicts behavior that is consistent with our observations and measurements of gases—the theory agrees with the experiment. Recall from Chapter 1 that a scientific theory is the most powerful kind of scientific knowledge. In the kinetic molecular theory we have a model for what a gas is like. Although the model is not perfect—indeed, it breaks down under certain conditions, as we shall see later in this chapter—it predicts a great deal about the behavior of gases. Therefore, the model is a good approximation of what a gas is actually like. A careful examination of the conditions under which the model breaks down gives us even more insight into the behavior of gases.

Temperature and Molecular Velocities

According to kinetic molecular theory, particles of different masses have the same average kinetic energy at a given temperature. The kinetic energy of a particle depends on its mass and velocity according to the following equation:

$$KE = \tfrac{1}{2}mv^2$$

The only way for particles of different masses to have the same kinetic energy is if they are traveling at different velocities.

  In a gas at a given temperature, lighter particles travel faster (on average) than heavier ones.

In kinetic molecular theory, we define the root mean square velocity (u_{rms}) of a particle as follows:

$$u_{rms} = \sqrt{\overline{u^2}} \qquad [5.22]$$

where $\overline{u^2}$ is the average of the squares of the particle velocities. Even though the root mean square velocity of a collection of particles is not identical to the average velocity, the two are close in value and conceptually similar. Root mean square velocity is simply a special type of average. The average kinetic energy of one mole of gas particles is then given by

$$KE_{avg} = \tfrac{1}{2}N_A m \overline{u^2} \qquad [5.23]$$

where N_A is Avogadro's number.

Postulate 2 of the kinetic molecular theory states that the average kinetic energy is proportional to the temperature in kelvins. The constant of proportionality in this relationship is $(3/2)\,R$:

$$KE_{avg} = (3/2)\,RT \qquad [5.24]$$

where R is the gas constant, but in different units ($R = 8.314$ J/mol·K). If we combine Equations 5.23 and 5.24, and solve for $\overline{u^2}$, we get the following:

$$(1/2)\,N_A m \overline{u^2} = (3/2)\,RT$$

$$\overline{u^2} = \frac{(3/2)\,RT}{(1/2)\,N_A m} = \frac{3RT}{N_A m}$$

The $(3/2)\,R$ proportionality constant comes from a derivation that is beyond our current scope.

The joule (J) is a unit of energy that is covered in more detail in Section 6.1.

$$\left(1\text{J} = 1\text{kg}\,\frac{\text{m}^2}{\text{s}^2}\right)$$

Variation of Velocity Distribution with Molar Mass

▶ FIGURE 5.21 Velocity Distribution for Several Gases at 25 °C At a given temperature, there is a distribution of velocities among the particles in a sample of gas. The exact shape and peak of the distribution varies with the molar mass of the gas.

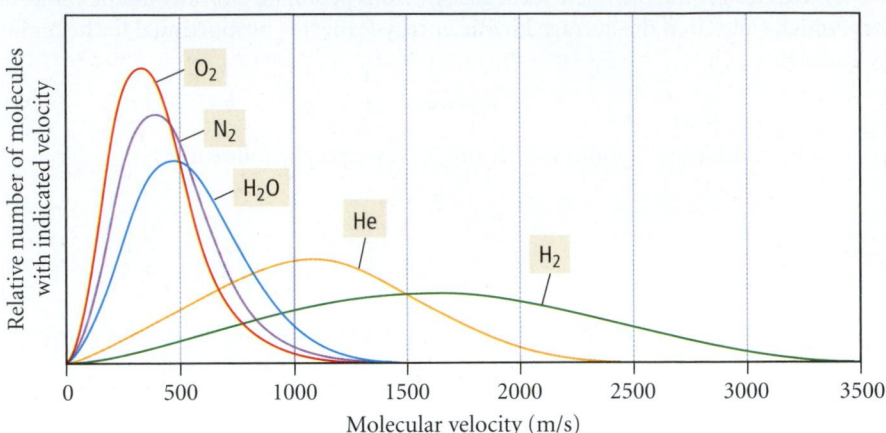

Variation of Velocity Distribution with Temperature

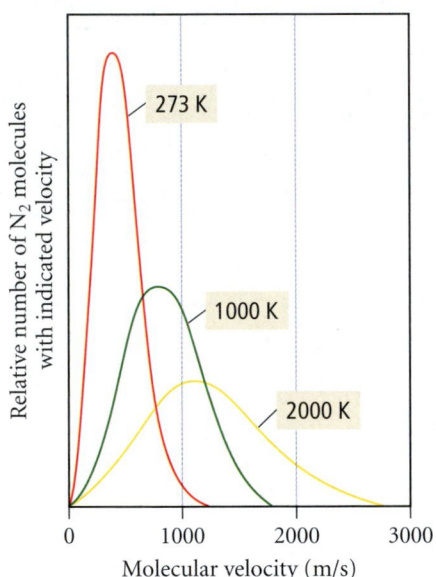

▲ FIGURE 5.22 Velocity Distribution for Nitrogen at Several Temperatures As the temperature of a gas sample is increased, the velocity distribution of the molecules shifts toward higher temperature and becomes less sharply peaked.

Taking the square root of both sides we get

$$\sqrt{\overline{u^2}} = u_{rms} = \sqrt{\frac{3RT}{N_A m}} \qquad [5.25]$$

In Equation 5.25, m is the mass of a particle in kg and N_A is Avogadro's number. The product $N_A m$, then, is simply the molar mass in kg/mol. If we call this quantity $\mathcal{M}$, then the expression for mean square velocity as a function of temperature becomes the following important result:

$$u_{rms} = \sqrt{\frac{3RT}{\mathcal{M}}} \qquad [5.26]$$

The root mean square velocity of a collection of gas particles is proportional to the square root of the temperature in kelvins and inversely proportional to the square root of the molar mass of the particles in kilograms per mole. The root mean square velocity of nitrogen molecules at 25 °C, for example, is 515 m/s (1152 mi/hr). The root mean square velocity of hydrogen molecules at room temperature is 1920 m/s (4295 mi/hr). Notice that the lighter molecules move much faster at a given temperature.

The root mean square velocity, as we have seen, is a kind of average velocity. Some particles are moving faster and some are moving slower than this average. The velocities of all the particles in a gas sample form a distribution like those shown in Figure 5.21 ▲. We can see from these distributions that some particles are indeed traveling at the root mean square velocity. However, many particles are traveling faster and many slower than the root mean square velocity. For lighter molecules, the velocity distribution is shifted toward higher velocities and the curve becomes broader, indicating a wider range of velocities. The velocity distribution for nitrogen at different temperatures is shown in Figure 5.22 ◀. As the temperature increases, the root mean square velocity increases and the distribution becomes broader.

EXAMPLE 5.14 Root Mean Square Velocity

Calculate the root mean square velocity of oxygen molecules at 25 °C.

Sort You are given the kind of molecule and the temperature and asked to find the root mean square velocity.	Given O_2, $t = 25\,°C$ Find u_{rms}

Strategize The conceptual plan for this problem simply shows how the molar mass of oxygen and the temperature (in kelvins) can be used with the equation that defines the root mean square velocity to compute root mean square velocity.

Conceptual Plan

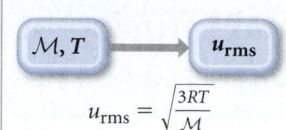

$$u_{rms} = \sqrt{\frac{3RT}{M}}$$

Relationship Used

$$u_{rms} = \sqrt{\frac{3RT}{M}} \text{ (Equation 5.26)}$$

Solve First gather the required quantities in the correct units. Note that molar mass must be in kg/mol.

Solution

$$T = 25 + 273 = 298\,K$$

$$M = \frac{32.00\text{ g O}_2}{1\text{ mol O}_2} \times \frac{1\text{ kg}}{1000\text{ g}} = \frac{32.00 \times 10^{-3}\text{ kg O}_2}{1\text{ mol O}_2}$$

Substitute the quantities into the equation to compute root mean square velocity. Note that $1\,J = 1\,kg \cdot m^2/s^2$.

$$u_{rms} = \sqrt{\frac{3RT}{M}}$$

$$= \sqrt{\frac{3\left(8.314\dfrac{J}{mol \cdot K}\right)(298\,K)}{\dfrac{32.00 \times 10^{-3}\text{ kg O}_2}{1\text{ mol O}_2}}}$$

$$= \sqrt{2.32 \times 10^5 \frac{J}{kg}}$$

$$= \sqrt{2.32 \times 10^5 \frac{\dfrac{kg \cdot m^2}{s^2}}{kg}}$$

$$= 482\text{ m/s}$$

Check The units of the answer (m/s) are correct. The magnitude of the answer seems reasonable because oxygen is slightly heavier than nitrogen and should therefore have a slightly lower root mean square velocity at the same temperature. Recall that earlier we stated the root mean square velocity of nitrogen to be 515 m/s at 25 °C.

For Practice 5.14

Calculate the root mean square velocity of gaseous xenon atoms at 25 °C.

 Conceptual Connection 5.4 Kinetic Molecular Theory

Which of the following samples of an ideal gas, all at the same temperature, will have the greatest pressure? Assume that the mass of the particle is proportional to its size.

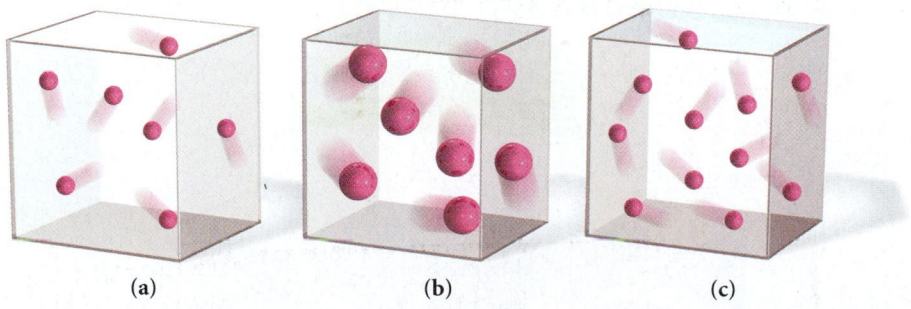

(a)   (b)   (c)

Answer: (c) Since the temperature and the volume are both constant, the ideal gas law tells us that the pressure will depend solely on the number of particles. Sample **(c)** has the greatest number of particles per unit volume, and so it will have the greatest pressure. The pressures of **(a)** and **(b)** at a given temperature will be identical. Even though the particles in **(b)** are more massive than those in **(a)**, they have the same average kinetic energy at a given temperature. The particles in **(b)** therefore move more slowly, and so exert the same pressure as the particles in **(a)**.

5.9 Mean Free Path, Diffusion, and Effusion of Gases

We have just learned that the root mean square velocity of gas molecules at room temperature is measured in hundreds of meters per second. However, suppose that your roommate just put on too much perfume in the bathroom only 2 m away. Why does it take a minute or two before you can smell the fragrance? Although most molecules in a perfume bottle have higher molar masses than nitrogen, their velocities are still hundreds of meters per second, so why the delay? The answer is that, even though gaseous particles travel at tremendous speeds, they also travel in very haphazard paths (Figure 5.23 ◄). To a perfume molecule, the path from the perfume bottle in the bathroom to your nose 2 m away is much like the path through a busy shopping mall during a clearance sale. The molecule travels only a short distance before it collides with another molecule, changes direction, only to collide again, and so on. In fact, at room temperature and atmospheric pressure, a molecule in the air experiences several billion collisions per second. The average distance that a molecule travels between collisions is called its mean free path. At room temperature and atmospheric pressure, the mean free path of a nitrogen molecule with a molecular diameter of 300 pm (four times the covalent radius) is 93 nm, or about 310 molecular diameters. If the nitrogen molecule were the size of a golf ball, it would travel about 40 ft between collisions. Mean free path increases with *decreasing* pressure. Under conditions of ultrahigh vacuum (10^{-10} torr), the mean free path of a nitrogen molecule is hundreds of kilometers.

The process by which gas molecules spread out in response to a concentration gradient is called **diffusion,** and even though the particles do not travel in a straight line, the root mean square velocity still influences the rate of diffusion. Heavier molecules diffuse more slowly than lighter ones, so the first molecules you would smell from a perfume mixture (in a room with no air currents) are the lighter ones.

A process related to diffusion is **effusion,** the process by which a gas escapes from a container into a vacuum through a small hole (Figure 5.24 ▼). The rate of effusion is also related to root mean square velocity—heavier molecules effuse more slowly than lighter ones. The rate of effusion—the amount of gas that diffuses in a given time—is inversely proportional to the square root of the molar mass of the gas as follows:

$$\text{rate} \propto \frac{1}{\sqrt{\mathcal{M}}}$$

The ratio of effusion rates of two different gases is given by **Graham's law of effusion**, named after Thomas Graham (1805–1869):

$$\frac{\text{rate}_A}{\text{rate}_B} = \sqrt{\frac{\mathcal{M}_B}{\mathcal{M}_A}}$$ [5.27]

The transport of molecules within a ventilated room is also enhanced by air currents.

Typical Gas Molecule Path

The average distance between collisions is the mean free path.

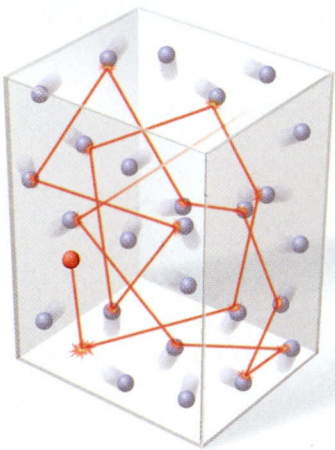

▲ FIGURE 5.23 **Mean Free Path** A molecule in a volume of gas follows a haphazard path, involving many collisions with other molecules.

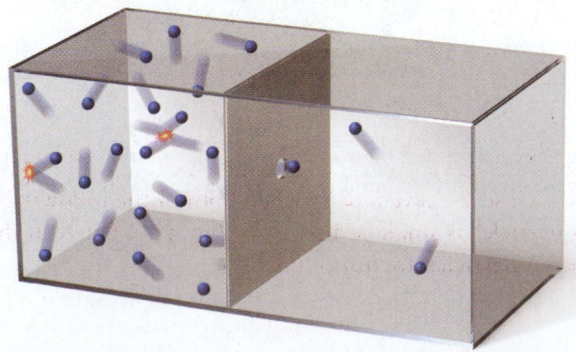

▶ FIGURE 5.24 **Effusion**   Effusion is the escape of a gas from a container into a vacuum through a small hole.

In this expression, rate_A and rate_B are the effusion rates of gases A and B and $\mathcal{M}_A$ and $\mathcal{M}_B$ are their molar masses.

EXAMPLE 5.15   Graham's law of Effusion

An unknown gas effuses at a rate that is 0.462 times that of nitrogen gas (at the same temperature). Calculate the molar mass of the unknown gas in g/mol.

Sort You are given the ratio of effusion rates for the unknown gas and nitrogen and asked to find the molar mass of the unknown gas.	**Given** $\dfrac{\text{Rate}_{unk}}{\text{Rate}_{N_2}} = 0.462$ **Find** $\mathcal{M}_{unk}$
Strategize The conceptual plan uses Graham's law of effusion. You are given the ratio of rates and you know the molar mass of the nitrogen. You can use Graham's law to find the molar mass of the unknown gas.	**Conceptual Plan** 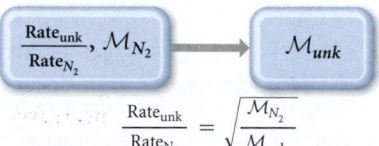 $\dfrac{\text{Rate}_{unk}}{\text{Rate}_{N_2}} = \sqrt{\dfrac{\mathcal{M}_{N_2}}{\mathcal{M}_{unk}}}$ **Relationship Used** $\dfrac{\text{rate}_A}{\text{rate}_B} = \sqrt{\dfrac{\mathcal{M}_B}{\mathcal{M}_A}}$ (Graham's law)
Solve Solve the equation for $\mathcal{M}_{unk}$ and substitute the correct values to compute it.	**Solution** $\dfrac{\text{rate}_{unk}}{\text{rate}_{N_2}} = \sqrt{\dfrac{\mathcal{M}_{N_2}}{\mathcal{M}_{unk}}}$ $\mathcal{M}_{unk} = \dfrac{\mathcal{M}_{N_2}}{\left(\dfrac{\text{rate}_{unk}}{\text{rate}_{N_2}}\right)^2}$ $= \dfrac{28.02 \text{ g/mol}}{(0.462)^2}$ $= 131 \text{ g/mol}$

Check The units of the answer are correct. The magnitude of the answer seems reasonable for the molar mass of a gas. In fact, we can even conclude that the gas is probably xenon, which has a molar mass of 131.29 g/mol.

For Practice 5.15

Find the ratio of effusion rates of hydrogen gas and krypton gas.

5.10 Real Gases: The Effects of Size and Intermolecular Forces

One mole of an ideal gas has a volume of 22.41 L at STP. Figure 5.25 on page 222 shows the molar volume of several real gases at STP. As you can see, most of these gases have a volume that is very close to 22.41 L, meaning that they are acting very nearly as ideal gases. Gases behave ideally when both of the following are true: (a) the volume of the gas particles is small compared to the space between them; and (b) the forces between the gas particles are not significant. At STP, these assumptions are valid for most common gases. However, these assumptions break down at higher pressures or lower temperatures.

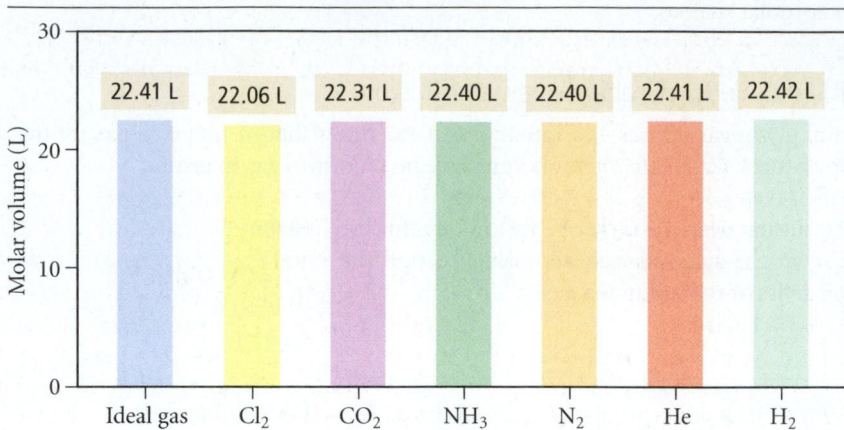

Molar Volume

▶ **FIGURE 5.25 Molar Volumes of Real Gases** The molar volumes of several gases at STP are all close to 22.414 L, indicating that their departures from ideal behavior are small.

The Effect of the Finite Volume of Gas Particles

The finite volume of gas particles becomes important at high pressure, when the particles themselves occupy a significant portion of the total gas volume (Figure 5.26 ◀). We can see the effect of particle volume by comparing the molar volume of argon to the molar volume of an ideal gas as a function of pressure at 500 K. At low pressures, the molar volume of argon is nearly identical to that of an ideal gas. But as the pressure increases, the molar volume of argon becomes *greater than* that of an ideal gas. At the higher pressures, the argon atoms themselves occupy a significant portion of the gas volume, making the actual volume greater than that predicted by the ideal gas law.

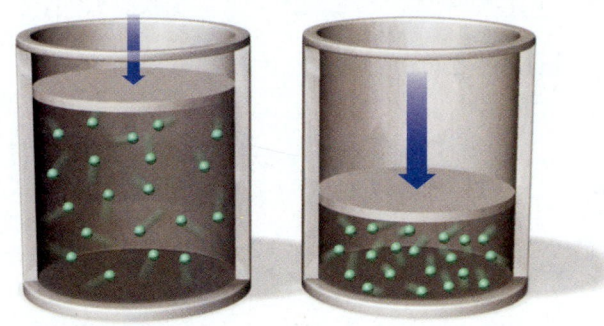

▲ **FIGURE 5.26 Particle Volume and Ideal Behavior** As a gas is compressed the gas particles themselves begin to occupy a significant portion of the total gas volume, leading to deviations from ideal behavior.

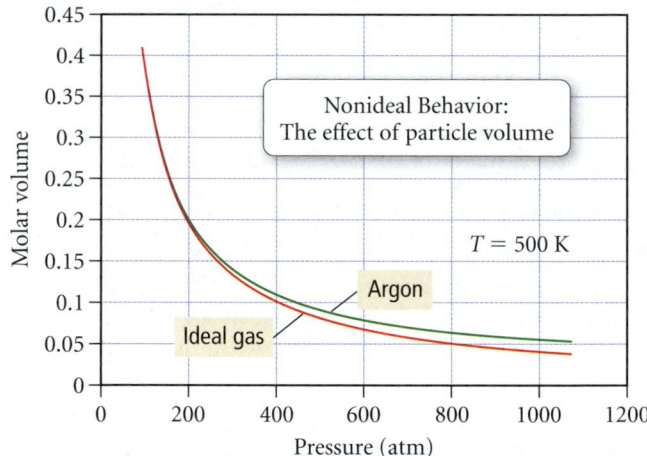

▶ At high pressures, 1 mol of argon occupies a larger volume than would 1 mol of an ideal gas because of the volume of the argon atoms themselves. (This example was chosen to minimize the effects of intermolecular forces, which are very small in argon at 500 K, thereby isolating the effect of particle volume.)

In 1873, Johannes van der Waals (1837–1923) modified the ideal gas equation to fit the behavior of real gases. From the graph for argon, we can see that the ideal gas law predicts a volume that is too small. Van der Waals suggested a small correction factor that accounts for the volume of the gas particles themselves:

Ideal behavior
$$V = \frac{nRT}{P}$$

Corrected for volume of gas particles
$$V = \frac{nRT}{P} + nb \qquad [5.28]$$

The correction adds the quantity nb to the volume, where n is the number of moles and b is a constant that depends on the gas (see Table 5.5, p. 223). We can rearrange the corrected equation as follows:

$$(V - nb) = \frac{nRT}{P} \qquad [5.29]$$

The Effect of Intermolecular Forces

Intermolecular forces are attractions between the atoms or molecules in a gas. These attractions are small and therefore do not matter much at low pressure, when the molecules are too far apart to "feel" the attractions. They also do not matter too much at high temperatures because the molecules have a lot of kinetic energy. When two particles with high kinetic energies collide, a weak attraction between them does not affect the collision very much. At lower temperatures, however, the collisions occur with less kinetic energy, and weak attractions do affect the collisions. The effect of these attractions between particles is a lower number of collisions with the surfaces of the container, thereby lowering the pressure compared to that of an ideal gas. We can see the effect of intermolecular forces by comparing the pressure of 1.0 mol of xenon gas to the pressure of 1.0 mol of an ideal gas as a function of temperature and at a fixed volume of 1.0 L, as shown in the graph below. At high temperature, the pressure of the xenon gas is nearly identical to that of an ideal gas. But at lower temperatures, the pressure of xenon is *less than* that of an ideal gas. At the lower temperatures, the xenon atoms spend more time interacting with each other and less time colliding with the walls, making the actual pressure less than that predicted by the ideal gas law.

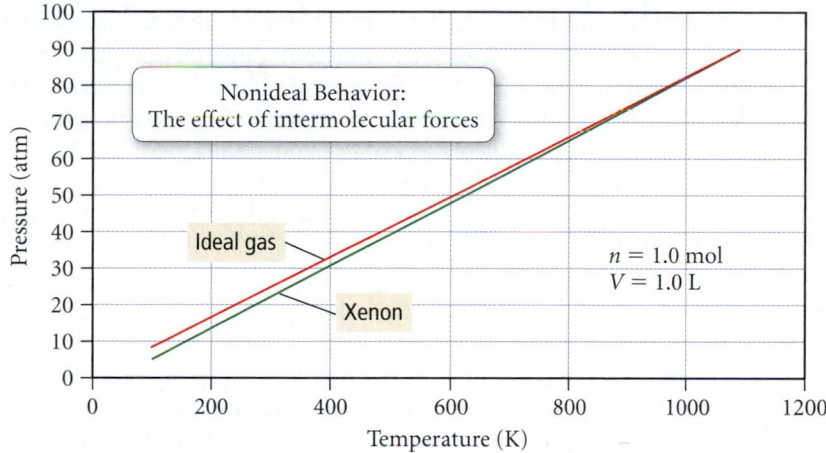

▲ At low temperatures, the pressure of xenon is less than an ideal gas would exert because interactions among xenon molecules reduce the number of collisions with the walls of the container.

From the above graph for xenon, we can see that the ideal gas law predicts a pressure that is too large at low temperatures. Van der Waals suggested a small correction factor that accounts for the intermolecular forces between gas particles:

Ideal behavior
$$P = \frac{nRT}{V}$$

Corrected for intermolecular forces
$$P = \frac{nRT}{V} - a\left(\frac{n}{V}\right)^2 \qquad [5.30]$$

The correction subtracts the quantity $a(n/V)^2$ from the pressure, where n is the number of moles, V is the volume, and a is a constant that depends on the gas (see Table 5.5). Notice that the correction factor increases as n/V (the number of moles of particles per unit volume) increases because a greater concentration of particles makes it more likely that they will interact with one another. We can rearrange the corrected equation as follows:

$$P + a\left(\frac{n}{V}\right)^2 = \frac{nRT}{V} \qquad [5.31]$$

TABLE 5.5 Van der Waals Constants for Common Gases

Gas	a (L$^2 \cdot$ atm/mol^2)	b (L/mol)
He	0.0342	0.02370
Ne	0.211	0.0171
Ar	1.35	0.0322
Kr	2.32	0.0398
Xe	4.19	0.0511
H_2	0.244	0.0266
N_2	1.39	0.0391
O_2	1.36	0.0318
Cl_2	6.49	0.0562
H_2O	5.46	0.0305
CH_4	2.25	0.0428
CO_2	3.59	0.0427
CCl_4	20.4	0.1383

Van der Waals Equation

We can now combine the effects of particle volume (Equation 5.29) and particle inter-molecular forces (Equation 5.31) into one equation that describes nonideal gas behavior:

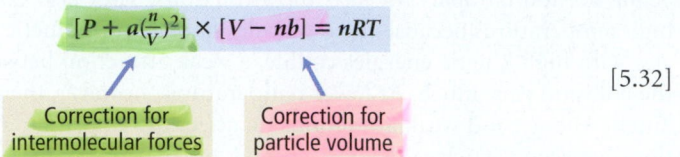

$$[P + a(\tfrac{n}{V})^2] \times [V - nb] = nRT$$

Correction for intermolecular forces   Correction for particle volume

[5.32]

The above equation is called the **van der Waals equation** and can be used to calculate the properties of a gas under nonideal conditions.

The Behavior of Real Gases

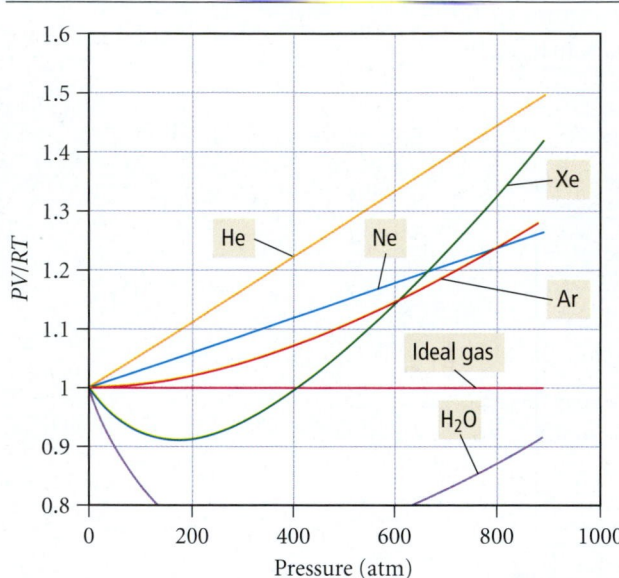

▲ **FIGURE 5.27 Real versus Ideal Behavior** For 1 mol of an ideal gas, PV/RT would be equal to 1. The combined effects of the volume of gas particles and the interactions among them cause each real gas to deviate from ideal behavior in a slightly different way. These curves were computed at a temperature of 273 K.

Real Gases

We can see the combined effects of particle volume and intermolecular forces by examining a plot of PV/RT versus P for 1 mol of a number of real gases (Figure 5.27 ◀). For an ideal gas, $PV/RT = n$, the number of moles of gas. Therefore, for 1 mol of an ideal gas, PV/RT is simply equal to 1, as shown in the plot. However, for real gases, PV/RT deviates from 1. Notice, however, that the deviations are not uniform. For example, water displays a large negative deviation from PV/RT. This is because, for water, the effect of intermolecular forces on lowering the pressure (relative to an ideal gas) is far greater than the effect of particle size on increasing the volume. Notice from Table 5.5 that water has a high value of a, the constant that corrects for intermolecular forces, but a moderate value of b, the constant that corrects for particle size. Therefore PV/RT is lower than predicted from the ideal gas law.

By contrast, consider the behavior of helium, which displays only a positive deviation from the ideal behavior. This is because helium has very weak intermolecular forces and their effect on lowering the pressure (relative to ideal gas) is small compared to the effect of parti-cle size on increasing the volume. Therefore PV/RT is greater than predicted from the ideal gas law.

✦ Conceptual Connection 5.5 Real Gases

The graph below shows PV/RT for water at three different temperatures. Rank the curves in order of increasing temperature.

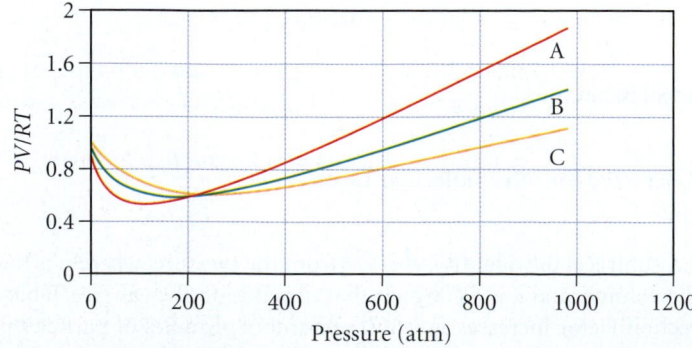

Answer: A < B < C. Curve A is the lowest temperature curve because it deviates the most from ide-ality. The tendency for the intermolecular forces in water to lower the pressure (relative to that of an ideal gas) is greatest at low temperature (because the molecules are moving more slowly and are therefore less able to overcome the intermolecular forces). As a result, the curve that dips the lowest must correspond to the lowest temperature.

5.11 Chemistry of the Atmosphere: Air Pollution and Ozone Depletion

In Section 5.6, we learned that our atmosphere is composed primarily of nitrogen and oxygen, with a few other minor components. However, human activities have polluted the air with other substances. In this section, we will take a brief look at two major environmental problems associated with the atmosphere: air pollution and ozone depletion. Air pollution is a problem that occurs primarily in the troposphere, the part of our atmosphere that is closest to Earth and ranges from ground level to about 10 km. The troposphere is the main recipient of gases emitted into the atmosphere from Earth. Ozone depletion occurs primarily in the stratosphere, the part of our atmosphere above the troposphere, ranging from 10 km to 50 km (Figure 5.28 ▼).

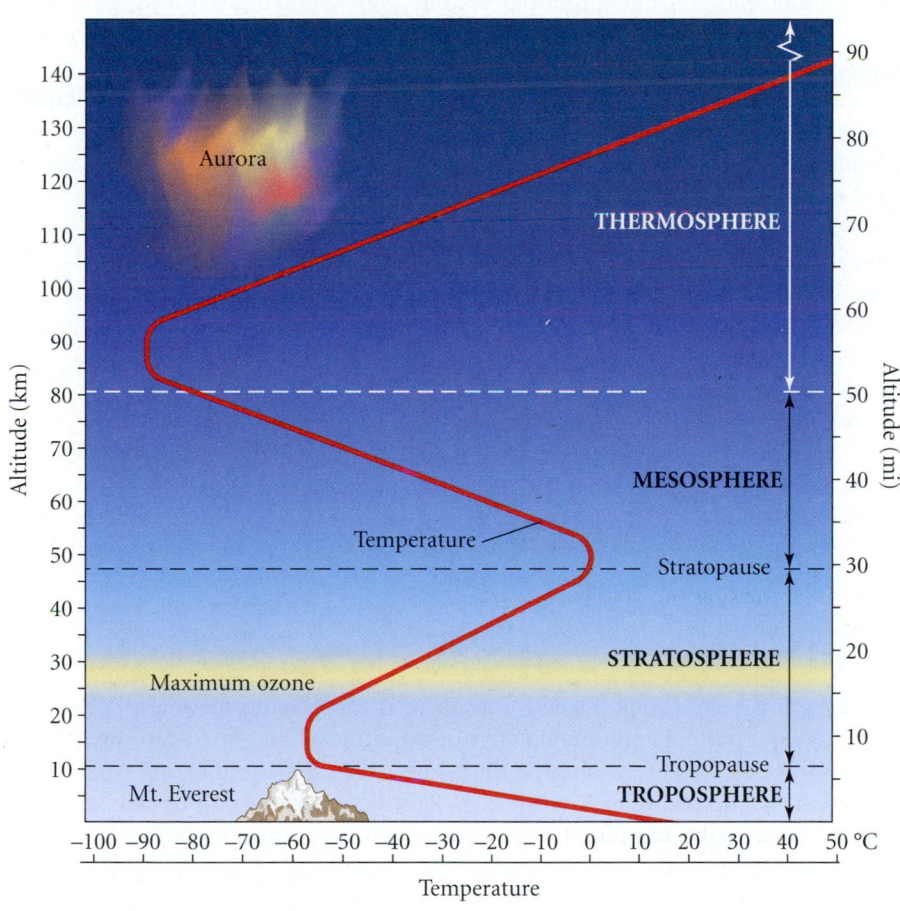

Structure of the Atmosphere

▲ FIGURE 5.28 **The Structure of the Atmosphere**   Earth's atmosphere can be divided by altitude into several different regions, including the troposphere and stratosphere.

Air Pollution

All major cities in the world have polluted air. This pollution comes from a number of sources including electricity generation, motor vehicles, and industrial waste. While there are many different kinds of air pollutants, the *major gaseous air pollutants* include the following.

Sulfur Oxides (SO_x)   Sulfur oxides include SO_2 and SO_3, which are produced chiefly during coal-fired electricity generation and industrial metal refining. They are lung and eye irritants that affect the respiratory system. Sulfur oxides are also the main precursors of acid rain (see Section 3.6).

Carbon Monoxide (CO) Carbon monoxide is formed by the incomplete combustion of fossil fuels (petroleum, natural gas, and coal). It is emitted mainly by motor vehicles. Carbon monoxide displaces oxygen in the blood, forcing the heart and lungs to work harder. At high levels, CO can cause sensory impairment, decreased thinking ability, unconsciousness, and even death.

Nitrogen Oxides (NO_x) Nitrogen oxides include NO and NO_2, which are emitted by motor vehicles, by fossil fuel-based electricity generation plants, and by any high-temperature combustion process that occurs in air. Nitrogen dioxide is an orange-brown gas that causes the dark haze often seen over polluted cities. Nitrogen oxides are eye and lung irritants and precursors of acid rain.

Ozone (O_3) Ozone is produced when some of the products of fossil fuel combustion, especially nitrogen oxides and unburned volatile organic compounds (VOCs), react in the presence of sunlight. The products of this reaction, which include ozone, are called *photochemical smog*. Ozone produced in this way—sometimes called ground-level ozone—should not be confused with upper atmospheric or *stratospheric* ozone. Although ozone is always the same molecule (O_3), stratospheric ozone is a natural part of our environment that protects Earth from harmful ultraviolet light. Stratospheric ozone does not harm us because we are not directly exposed to it. Ground-level ozone, on the other hand, is a pollutant to which we are directly exposed; it is an eye and lung irritant, and prolonged exposure has been shown to cause permanent lung damage.

In the United States, the U.S. Environmental Protection Agency (EPA) has set standards for these pollutants. Beginning in the 1970s, the U.S. Congress passed the Clean Air Act and its amendments, requiring U.S. cities to reduce their pollution and maintain levels below the standards set by the EPA. As a result of this legislation, pollutant levels in U.S. cities have decreased significantly over the last 25 years, even as the number of vehicles has increased. For example, according to the EPA's 2004 National Air Quality and Emissions Trends Report, the emission levels of the major pollutants in U.S. cities have decreased during the period 1980–2004. The amounts of these decreases are shown in Table 5.6.

Although the levels of pollutants (especially ozone) in some cities are still above what the EPA considers safe, much progress has been made. These trends demonstrate that good legislation can clean up our environment.

TABLE 5.6 Changes in Pollutant Emission Levels, 1980–2004

Pollutant	Change (%) in Emission Level
SO_x	−41
CO	−51
NO_x	−31
VOCs*	−50

*VOCs are volatile organic compounds, which are precursors to ground-level ozone formation.

Ozone Depletion

As we have just seen, ozone in the troposphere is a pollutant. However, ozone in the stratosphere is a natural and essential part of our atmosphere. The stratospheric ozone layer is important because it absorbs ultraviolet (UV) light originating from the sun (see Section 7.2). Excessive exposure to UV light increases the risk of skin cancer and cataracts, weakens the immune system, and causes premature wrinkling of the skin. Consequently, stratospheric ozone plays an important role in preventing these conditions. Stratospheric ozone absorbs UV light through the following reaction:

$$O_3 + UV \text{ light} \longrightarrow O_2 + O$$

Under normal conditions, O_2 and O recombine in the stratosphere to re-form O_3.

$$O_2 + O \longrightarrow O_3 + \text{heat}$$

Through these two reactions, ozone transforms UV light into heat and is regenerated to go through the cycle again.

In the 1970s, scientists began to worry that a class of compounds known as chlorofluorocarbons (CFCs) might be destroying stratospheric ozone. These compounds had become important as aerosols, foam-blowing agents, refrigerants (Freons), and industrial solvents. One of the advantages of CFCs for these applications was their relative chemical inertness—they did not react with most substances. Consequently, they were seen as stable, trouble-free compounds. But this seeming advantage turned out to be a major problem.

Unlike more reactive pollutants, which are decomposed by chemical attack in the troposphere, CFCs linger in the atmosphere—the Freon that leaks out of your refrigerator can stay in the atmosphere for over 20 years. Consequently, CFCs diffuse into the stratosphere and ozone layer. Once they reach this part of the atmosphere, UV light (which is less abundant below the ozone layer because the ozone absorbs it) breaks a carbon–chlorine bond in the CFC. For example, consider the reaction of CF_2Cl_2 with UV light:

$$CF_2Cl_2 + UV\ light \longrightarrow CF_2Cl + Cl$$

The newly released chlorine atom then reacts with ozone in the following cycle:

$$Cl + O_3 \longrightarrow ClO + O_2$$
$$O_3 + UV\ light \longrightarrow O_2 + O$$
$$ClO + O \longrightarrow Cl + O_2$$

Notice what happens here. In the first reaction, a chlorine atom destroys an ozone molecule. In the third reaction, the product of the first reaction reacts with an oxygen atom that would normally react with O_2 to re-form ozone. Consequently, a single chlorine atom destroys two ozone molecules and regenerates itself to repeat the process. In this way, a single chlorine atom can destroy hundreds of ozone molecules. Since the chlorine is not consumed in the process, it acts as a *catalyst*—a substance that enhances the rate of a chemical reaction without being consumed itself by it. The overall process is called the *catalytic destruction of ozone.*

The first evidence that ozone depletion was indeed occurring came from the south pole in the mid-1980s. At that time, scientists discovered a transitory "hole" in the ozone layer directly over the pole. The amount of ozone depletion within the hole, which appears during the Antarctic spring and generally peaks during the months of September and October, was a startling 50%. Analysis of previous years' data revealed that the hole had been forming every spring since 1977, and getting larger each year (Figure 5.29 ▶).

For a number of years, scientists wondered why the ozone hole existed only over the south pole and only in the spring. Over time, they discovered that special conditions present at the south pole exacerbated ozone depletion there. Specifically, the south pole has a cold, dark winter, which allows the formation of unique clouds called polar stratospheric clouds (PSCs). Unlike normal clouds, which form in the troposphere and are composed of water droplets, PSCs form in the stratosphere and are composed of ice crystals. During the Antarctic winter, the ice crystals act as catalysts that release molecular chlorine (Cl_2) from atmospheric chemical reservoirs (such as $ClONO_2$). When the sun rises in the Antarctic spring, sunlight breaks the relatively weak Cl—Cl bond, releasing chlorine atoms into the stratosphere. The chlorine atoms then deplete ozone through the catalytic cycle discussed previously. In other words, the degree of ozone depletion observed over the south pole does not happen all over the world because much of the chlorine that enters our atmosphere from chlorofluorocarbons is neutralized in atmospheric chemical reservoirs. Conditions at the south pole in the spring months happen to be just right for releasing that chlorine from its reservoirs. Once the sun melts the PSCs, however, the chlorine returns to its reservoirs and the ozone hole recovers . . . until the next spring.

Ozone depletion has also occurred over the entire globe, although to a lesser extent than over Antarctica. The amount of depletion depends primarily on latitude. Mid-northern latitudes have experienced decreases of about 6% since 1979. The clear scientific data demonstrating the ozone-depleting capability of chlorofluorocarbons, coupled with the immense depletion of ozone over Antarctica, forced lawmakers to take action. In 1992, President George H. W. Bush called for a complete ban on CFC production beginning in 1996. Many nations followed suit. Because of this legislation, atmospheric chlorine levels are expected to level out and then drop in coming years. The ozone hole is expected to recover by the year 2050.

The Antarctic Ozone Hole

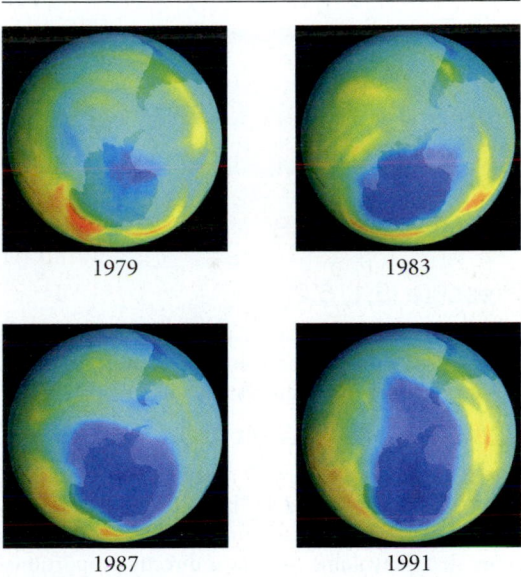

1979

1983

1987

1991

▲ **FIGURE 5.29 Spring in the Antarctic** Maps of the ozone layer in the spring of four different years reveal a startling decline in Antarctic ozone levels. (The darkest blue color represents the greatest ozone depletion.)

Chapter in Review

Key Terms

Section 5.1
pressure (185)

Section 5.2
millimeter of mercury (mmHg) (188)
barometer (188)
torr (188)
atmosphere (atm) (189)
pascal (Pa) (189)
manometer (190)

Section 5.3
Boyle's law (192)
Charles's law (195)
Avogadro's law (197)

Section 5.4
ideal gas law (198)
ideal gas constant (198)

Section 5.5
standard temperature and pressure (STP) (201)

molar volume (201)

Section 5.6
partial pressure (P_n) (204)
Dalton's law of partial pressures (204)
mole fraction (χ_a) (205)
hypoxia (206)
oxygen toxicity (207)
nitrogen narcosis (207)
vapor pressure (209)

Section 5.8
kinetic molecular theory (214)

Section 5.9
mean free path (220)
diffusion (220)
effusion (220)
Graham's law of effusion (220)

Section 5.10
van der Waals equation (224)

Key Concepts

Pressure (5.1, 5.2)

Gas pressure is the force per unit area that results from gas particles colliding with the surfaces around them. Pressure is measured in a number of units including mmHg, torr, Pa, psi, in Hg, and atm.

The Simple Gas Laws (5.3)

The simple gas laws express relationships between pairs of variables when the other variables are held constant. Boyle's law states that the volume of a gas is inversely proportional to its pressure. Charles's law states that the volume of a gas is directly proportional to its temperature. Avogadro's law states the volume of a gas is directly proportional to the amount (in moles).

The Ideal Gas Law and Its Applications (5.4, 5.5)

The ideal gas law, $PV = nRT$, gives the relationship among all four gas variables and contains the simple gas laws within it. The ideal gas law can be used to find one of the four variables given the other three. It can also be used to compute the molar volume of an ideal gas, which is 22.4 L at STP, and used to calculate the density and molar mass of a gas.

Mixtures of Gases and Partial Pressures (5.6)

In a mixture of gases, each gas acts independently of the others so that any overall property of the mixture is simply the sum of the properties of the individual components. The pressure of any individual component is called its partial pressure.

Gas Stoichiometry (5.7)

In reactions involving gaseous reactants and products, quantities are often reported in volumes at specified pressures and temperatures. These quantities can be converted to amounts (in moles) using the ideal gas law. Then the stoichiometric coefficients from the balanced equation can be used to determine the stoichiometric amounts of other reactants or products. The general form for these types of calculations is often as follows: volume A → amount A (in moles) → amount B (in moles) → quantity of B (in desired units). In cases where the reaction is carried out at STP, the molar volume at STP (22.4 L = 1 mol) can be used to convert between volume in liters and amount in moles.

Kinetic Molecular Theory and Its Applications (5.8, 5.9)

Kinetic molecular theory is a quantitative model for gases. The theory has three main assumptions: (1) the gas particles are negligibly small; (2) the average kinetic energy of a gas particle is proportional to the temperature in kelvins; and (3) the collision of one gas particle with another is completely elastic (the particles do not stick together). The gas laws all follow from the kinetic molecular theory.

The theory can also be used to derive the expression for the root mean square velocity of gas particles. This velocity is inversely proportional to the molar mass of the gas, and therefore—at a given temperature—smaller gas particles are (on average) moving more quickly than larger ones. The kinetic molecular theory also allows us to predict the mean free path of a gas particle (the distance it travels between collisions) and relative rates of diffusion or effusion.

Real Gases (5.10)

Real gases differ from ideal gases to the extent that they do not fit the assumptions of kinetic molecular theory. These assumptions tend to break down at high pressures, where the volume is higher than predicted because the particles are no longer negligibly small compared to the space between them. The assumptions also break down at low temperatures where the pressure is lower than predicted because the attraction between molecules combined with low kinetic energies causes partially inelastic collisions. Van der Waals equation can be used to predict gas properties under nonideal conditions.

The Atmosphere (5.11)

Our atmosphere is composed primarily of nitrogen (78%) and oxygen (21%). Common gaseous pollutants in our atmosphere include sulfur dioxide, carbon monoxide, ozone, and nitrogen dioxide. These pollutants affect exposed organs, such as the eyes and lungs, and force the cardiovascular system to work harder. Other pollutants include chlorofluorocarbons that contribute to the depletion of upper atmospheric ozone, which results in higher levels of ultraviolet radiation and an increase in the risk of skin cancer and cataracts.

Key Equations and Relationships

Relationship between Pressure (P), Force (F), and Area (A) (5.2)

$$P = \frac{F}{A}$$

Boyle's Law: Relationship between Pressure (P) and Volume (V) (5.3)

$$V \propto \frac{1}{P}$$

$$P_1V_1 = P_2V_2$$

Charles's Law: Relationship between Volume (V) and Temperature (T) (5.3)

$$V \propto T \quad (\text{in K})$$

$$\frac{V_1}{T_1} = \frac{V_2}{T_2}$$

Avogadro's Law: Relationship between Volume (V) and Amount in Moles (n) (5.3)

$$V \propto n$$

$$\frac{V_1}{n_1} = \frac{V_2}{n_2}$$

Ideal Gas Law: Relationship between Volume (V), Pressure (P), Temperature (T), and Amount (n) (5.4)

$$PV = nRT$$

Dalton's Law: Relationship between Partial Pressures (P_n) in Mixture of Gases and Total Pressure (P_{total}) (5.6)

$$P_{total} = P_a + P_b + P_c + \cdots$$

$$P_a = \frac{n_aRT}{V} \qquad P_b = \frac{n_bRT}{V} \qquad P_c = \frac{n_cRT}{V}$$

Mole Fraction (χ_a) (5.6)

$$\chi_a = \frac{n_a}{n_{total}}$$

$$P_a = \chi_a P_{total}$$

Average Kinetic Energy (KE_{avg}) (5.8)

$$KE_{avg} = \frac{3}{2}RT$$

Relationship between Root Mean Square Velocity (u_{rms}) and Temperature (T) (5.8)

$$u_{rms} = \sqrt{\frac{3RT}{M}}$$

Relationship of Effusion Rates of Two Different Gases (5.9)

$$\frac{rate\,A}{rate\,B} = \sqrt{\frac{M_B}{M_A}}$$

Van der Waals Equation: The Effects of Volume and Intermolecular Forces on Nonideal Gas Behavior (5.10)

$$[P + a(n/V)^2] \times (V - nb) = nRT$$

 Internal force

 Volume partial

Key Skills

Converting between Pressure Units (5.2)
- Example 5.1 • For Practice 5.1 • For More Practice 5.1 • Exercises 29–32

Relating Volume and Pressure: Boyle's Law (5.3)
- Example 5.2 • For Practice 5.2 • Exercises 35, 36

Relating Volume and Temperature: Charles's Law (5.3)
- Example 5.3 • For Practice 5.3 • Exercises 37, 38

Relating Volume and Moles: Avogadro's Law (5.3)
- Example 5.4 • For Practice 5.4 • Exercises 39, 40

Determining P, V, n, or T using the Ideal Gas Law (5.4)
- Examples 5.5, 5.6 • For Practice 5.5, 5.6 • For More Practice 5.6 • Exercises 41–48, 51, 52

Relating the Density of a Gas to Its Molar Mass (5.5)
- Example 5.7 • For Practice 5.7 • For More Practice 5.7 • Exercises 55, 56

Calculating the Molar Mass of a Gas with the Ideal Gas Law (5.5)
- Example 5.8 • For Practice 5.8 • Exercises 57–60

Calculating Total Pressure, Partial Pressures, and Mole Fractions of Gases in a Mixture (5.6)
- Examples 5.9, 5.10, 5.11 • For Practice 5.9, 5.10, 5.11 • Exercises 61, 62, 65, 67, 68, 70

Relating the Amounts of Reactants and Products in Gaseous Reactions: Stoichiometry (5.7)
- Examples 5.12, 5.13 • For Practice 5.12, 5.13 • For More Practice 5.12 • Exercises 71–77

Calculating the Root Mean Square Velocity of a Gas (5.8)
- Example 5.14 • For Practice 5.14 • Exercises 81, 82

Calculating the Effusion Rate or the Ratio of Effusion Rates of Two Gases (5.9)
- Example 5.15 • For Practice 5.15 • Exercises 83–86

Exercises

Review Questions

1. What is pressure? What causes pressure?

2. Explain how a shallow well works. What forces the water to the surface?

3. How deep can a shallow well be and still function? Why does this limit exist?

4. What are some common units of pressure? List these in order of smallest to largest unit.

5. What is a manometer? How does a manometer measure the pressure of a sample of gas?

6. Summarize each of the simple gas laws (Boyle's law, Charles's law, and Avogadro's law). For each law, explain the relationship between the two variables and also state which variables must be kept constant.

7. Explain the source of ear pain that is often experienced by a rapid change in altitude.

8. Explain why scuba divers should never hold their breath as they ascend to the surface.

9. Why would it be impossible to breathe air through an extralong snorkel (greater than a couple of meters) while swimming under water?

10. Explain why hot air balloons float above the ground and why the second story of a two-story home is often warmer than the ground story.

11. What is the ideal gas law? Why is it useful?

12. Explain how the ideal gas law contains within it the simple gas laws (you may want to show an example).

13. Define molar volume and give its value at STP.

14. How does the density of a gas depend on temperature? Pressure? How does it depend on the molar mass of the gas?

15. What is partial pressure? What is the relationship between the partial pressures of each gas in a sample and the total pressure of gas in the sample?

16. Why do deep-sea divers breathe a mixture of helium and oxygen?

17. When a gas is collected over water, is the gas pure? Why or why not? How can the partial pressure of the desired gas be determined?

18. If a reaction occurs in the gas phase at STP, the mass of a product can be determined from the volumes of reactants. Explain.

19. What are the basic postulates of kinetic molecular theory? How does the concept of pressure follow from kinetic molecular theory?

20. Explain how Boyle's law, Charles's law, Avogadro's law, and Dalton's law all follow from kinetic molecular theory.

21. How is the kinetic energy of a gas related to temperature? How is the root mean square velocity of a gas related to its molar mass?

22. Describe how the molecules in a perfume bottle travel from the bottle to your nose. What is mean free path?

23. Explain the difference between diffusion and effusion. How is the effusion rate of a gas related to its molar mass?

24. Deviations from the ideal gas law are often observed at high pressure and low temperature. Explain why in light of kinetic molecular theory.

25. What are the main atmospheric pollutants and their sources? What are the trends in the levels of these pollutants over U.S. cities?

26. Explain why ozone is a pollutant in the lower atmosphere but a necessary component of our upper atmosphere.

27. What compounds are blamed with depletion of stratospheric ozone? How do these compounds deplete ozone, and what is being done to prevent further depletion?

28. Why does an ozone hole form over the south pole every October?

Problems by Topic

Converting between Pressure Units

29. The pressure in Denver, Colorado (elevation 5280 ft), averages about 24.9 in Hg. Convert this pressure to
    a. atm   b. mmHg   c. psi   d. Pa

30. The pressure on top of Mt. Everest averages about 235 mmHg. Convert this pressure to
    a. torr   b. psi   c. in Hg   d. atm

31. The North American record for highest recorded barometric pressure is 31.85 in Hg, set in 1989 in Northway, Alaska. Convert this pressure to
    a. mmHg   b. atm
    c. torr   d. kPa (kilopascals)

32. The world record for lowest pressure (at sea level) was 652.5 mmHg recorded inside Typhoon Tip on October 12, 1979, in the Western Pacific Ocean. Convert this pressure to
    a. torr   b. atm   c. in Hg   d. psi

33. Given a barometric pressure of 755.3 mmHg, calculate the pressure of each of the following gas samples as indicated by the manometer.

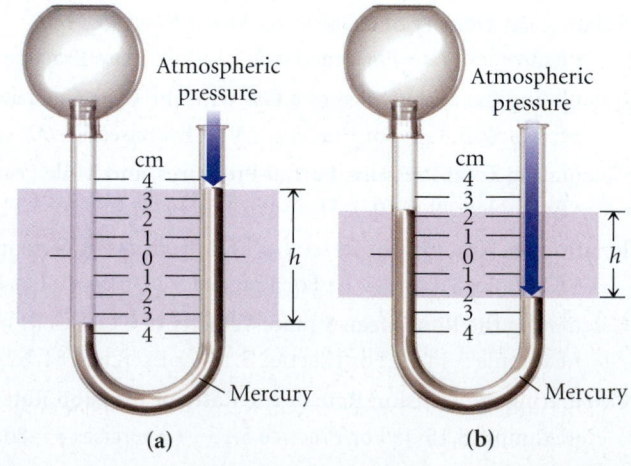

(a)   (b)

34. Given a barometric pressure of 748.2 mmHg, calculate the pressure of each of the following gas samples as indicated by the manometer.

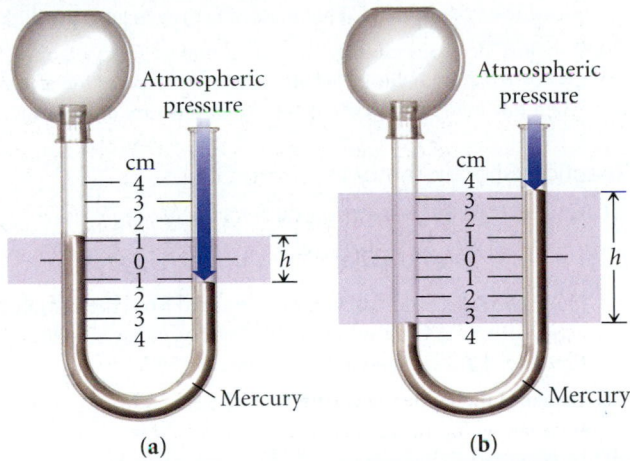

(a)   (b)

Simple Gas Laws

35. A sample of gas has an initial volume of 2.8 L at a pressure of 755 mmHg. If the volume of the gas is increased to 3.7 L, what will the pressure be?

36. A sample of gas has an initial volume of 32.6 L at a pressure of 1.3 atm. If the sample is compressed to a volume of 13.8 L, what will its pressure be?

37. A 48.3-mL sample of gas in a cylinder is warmed from 22 °C to 87 °C. What is its volume at the final temperature?

38. A syringe containing 1.55 mL of oxygen gas is cooled from 95.3 °C to 0.0 °C. What is the final volume of oxygen gas?

39. A balloon contains 0.128 mol of gas and has a volume of 2.76 L. If an additional 0.073 mol of gas is added to the balloon (at the same temperature and pressure), what will its final volume be?

40. A cylinder with a moveable piston contains 0.87 mol of gas and has a volume of 334 mL. What will its volume be if an additional 0.22 mol of gas is added to the cylinder? (Assume constant temperature and pressure.)

Ideal Gas Law

41. What is the volume occupied by 0.118 mol of helium gas at a pressure of 0.97 atm and a temperature of 305 K?

42. What is the pressure in a 10.0-L cylinder filled with 0.448 mol of nitrogen gas at a temperature of 315 K?

43. A cylinder contains 28.5 L of oxygen gas at a pressure of 1.8 atm and a temperature of 298 K. How much gas (in moles) is in the cylinder?

44. What is the temperature of 0.52 mol of gas at a pressure of 1.3 atm and a volume of 11.8 L?

45. An automobile tire has a maximum rating of 38.0 psi (gauge pressure). The tire is inflated (while cold) to a volume of 11.8 L and a gauge pressure of 36.0 psi at a temperature of 12.0 °C. While driving on a hot day, the tire warms to 65.0 °C and its volume expands to 12.2 L. Does the pressure in the tire exceed its maximum rating? (Note: The *gauge pressure* is the *difference* between the total pressure and atmospheric pressure. In this case, assume that atmospheric pressure is 14.7 psi.)

46. A weather balloon is inflated to a volume of 28.5 L at a pressure of 748 mmHg and a temperature of 28.0 °C. The balloon rises in the atmosphere to an altitude of approximately 25,000 feet, where the pressure is 385 mmHg and the temperature is −15.0 °C. Assuming the balloon can freely expand, calculate the volume of the balloon at this altitude.

47. A piece of dry ice (solid carbon dioxide) with a mass of 28.8 g is allowed to sublime (convert from solid to gas) into a large balloon. Assuming that all of the carbon dioxide ends up in the balloon, what will be the volume of the balloon at a temperature of 22 °C and a pressure of 742 mmHg?

48. A 1.0-L container of liquid nitrogen is kept in a closet measuring 1.0 m by 1.0 m by 2.0 m. Assuming that the container is completely full, that the temperature is 25.0 °C, and that the atmospheric pressure is 1.0 atm, calculate the percent (by volume) of air that would be displaced if all of the liquid nitrogen evaporated. (Liquid nitrogen has a density of 0.807 g/mL.)

49. Which of the following gas samples, all at the same temperature, will have the greatest pressure? Explain.

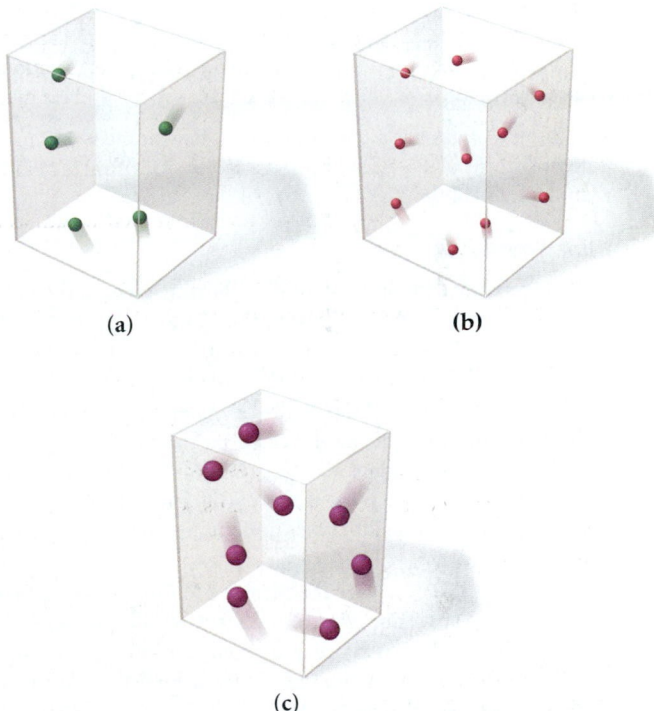

(a)   (b)

(c)

50. The following picture represents a sample of gas at a pressure of 1 atm, a volume of 1 L, and a temperature of 25 °C. Draw a similar picture showing what would happen if the volume were reduced to 0.5 L and the temperature increased to 250 °C. What would happen to the pressure?

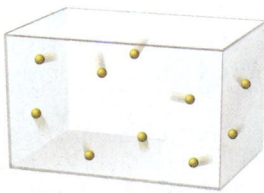

51. Aerosol cans carry clear warnings against incineration because of the high pressures that can develop upon heating. Suppose that a can contains a residual amount of gas at a pressure of 755 mmHg and a temperature of 25 °C. What would the pressure be if the can were heated to 1155 °C?

52. A sample of nitrogen gas in a 1.75-L container exerts a pressure of 1.35 atm at 25 °C. What is the pressure if the volume of the container is maintained constant and the temperature is raised to 355 °C?

Molar Volume, Density, and Molar Mass of a Gas

53. Use the molar volume of a gas at STP to determine the volume (in L) occupied by 10.0 g of neon at STP.

54. Use the molar volume of a gas at STP to calculate the density (in g/L) of carbon dioxide gas at STP.

55. What is the density (in g/L) of hydrogen gas at 20.0 °C and a pressure of 1655 psi?

56. A sample of N_2O gas has a density of 2.85 g/L at 298 K. What must be the pressure of the gas (in mmHg)?

57. An experiment shows that a 248-mL gas sample has a mass of 0.433 g at a pressure of 745 mmHg and a temperature of 28 °C. What is the molar mass of the gas?

58. An experiment shows that a 113-mL gas sample has a mass of 0.171 g at a pressure of 721 mmHg and a temperature of 32 °C. What is the molar mass of the gas?

59. A sample of gas has a mass of 38.8 mg. Its volume is 224 mL at a temperature of 55 °C and a pressure of 886 torr. Find the molar mass of the gas.

60. A sample of gas has a mass of 0.555 g. Its volume is 117 mL at a temperature of 85 °C and a pressure of 753 mmHg. Find the molar mass of the gas.

Partial Pressure

61. A gas mixture contains each of the following gases at the indicated partial pressures: N_2, 325 torr; O_2, 124 torr; and He, 209 torr. What is the total pressure of the mixture? What mass of each gas is present in a 1.05-L sample of this mixture at 25.0 °C?

62. A gas mixture with a total pressure of 755 mmHg contains each of the following gases at the indicated partial pressures: CO_2, 255 mmHg; Ar, 124 mmHg; and O_2, 167 mmHg. The mixture also contains helium gas. What is the partial pressure of the helium gas? What mass of helium gas is present in a 10.0-L sample of this mixture at 273 K?

63. A 1.20-g sample of dry ice is added to a 755-mL flask containing nitrogen gas at a temperature of 25.0 °C and a pressure of 725 mmHg. The dry ice is allowed to sublime (convert from solid to gas) and the mixture is allowed to return to 25.0 °C. What is the total pressure in the flask?

64. A 275-mL flask contains pure helium at a pressure of 752 torr. A second flask with a volume of 475 mL contains pure argon at a pressure of 722 torr. If the two flasks are connected through a stopcock and the stopcock is opened, what are the partial pressures of each gas and the total pressure?

65. A gas mixture contains 1.25 g N_2 and 0.85 g O_2 in a 1.55-L container at 18 °C. Calculate the mole fraction and partial pressure of each component in the gas mixture.

66. What is the mole fraction of oxygen gas in air (see Table 5.3)? What volume of air contains 10.0 g of oxygen gas at 273 K and 1.00 atm?

67. The hydrogen gas formed in a chemical reaction is collected over water at 30.0 °C at a total pressure of 732 mmHg. What is the partial pressure of the hydrogen gas collected in this way? If the total volume of gas collected is 722 mL, what mass of hydrogen gas is collected?

68. The air in a bicycle tire is bubbled through water and collected at 25 °C. If the total volume of gas collected is 5.45 L at a temperature of 25 °C and a pressure of 745 torr, how many moles of gas was in the bicycle tire?

69. The zinc within a copper-plated penny will dissolve in hydrochloric acid if the copper coating is filed down in several spots (so that the hydrochloric acid can get to the zinc). The reaction between

the acid and the zinc is as follows: $2\,H^+(aq) + Zn(s) \rightarrow H_2(g) + Zn^{2+}(aq)$. When the zinc in a certain penny dissolves, the total volume of gas collected over water at 25 °C was 0.951 L at a total pressure of 748 mmHg. What mass of hydrogen gas was collected?

70. A heliox deep-sea diving mixture contains 2.0 g of oxygen to every 98.0 g of helium. What is the partial pressure of oxygen when this mixture is delivered at a total pressure of 8.5 atm?

Reaction Stiochiometry Involving Gases

71. Consider the following chemical reaction:

$$C(s) + H_2O(g) \longrightarrow CO(g) + H_2(g)$$

How many liters of hydrogen gas is formed from the complete reaction of 15.7 g C? Assume that the hydrogen gas is collected at a pressure of 1.0 atm and a temperature of 355 K.

72. Consider the following chemical reaction.

$$2\,H_2O(l) \longrightarrow 2\,H_2(g) + O_2(g)$$

What mass of H_2O is required to form 1.4 L of O_2 at a temperature of 315 K and a pressure of 0.957 atm?

73. CH_3OH can be synthesized by the following reaction.

$$CO(g) + 2\,H_2(g) \longrightarrow CH_3OH(g)$$

What volume of H_2 gas (in L), measured at 748 mmHg and 86 °C, is required to synthesize 25.8 g CH_3OH? How many liters of CO gas, measured under the same conditions, is required?

74. Oxygen gas reacts with powdered aluminum according to the following reaction:

$$4\,Al(s) + 3\,O_2(g) \longrightarrow 2\,Al_2O_3(s)$$

What volume of O_2 gas (in L), measured at 782 mmHg and 25 °C, is required to completely react with 53.2 g Al?

75. Automobile air bags inflate following a serious impact. The impact triggers the following chemical reaction.

$$2\,NaN_3(s) \longrightarrow 2\,Na(s) + 3\,N_2(g)$$

If an automobile air bag has a volume of 11.8 L, what mass of NaN_3 (in g) is required to fully inflate the air bag upon impact? Assume STP conditions.

76. Lithium reacts with nitrogen gas according to the following reaction:

$$6\,Li(s) + N_2(g) \longrightarrow 2\,Li_3N(s)$$

What mass of lithium (in g) is required to react completely with 58.5 mL of N_2 gas at STP?

77. Hydrogen gas (a potential future fuel) can be formed by the reaction of methane with water according to the following equation:

$$CH_4(g) + H_2O(g) \longrightarrow CO(g) + 3\,H_2(g)$$

In a particular reaction, 25.5 L of methane gas (measured at a pressure of 732 torr and a temperature of 25 °C) is mixed with 22.8 L of water vapor (measured at a pressure of 702 torr and a temperature of 125 °C). The reaction produces 26.2 L of hydrogen gas measured at STP. What is the percent yield of the reaction?

78. Ozone is depleted in the stratosphere by chlorine from CF_3Cl according to the following set of equations:

$$CF_3Cl + UV\ light \longrightarrow CF_3 + Cl$$
$$Cl + O_3 \longrightarrow ClO + O_2$$
$$O_3 + UV\ light \longrightarrow O_2 + O$$
$$ClO + O \longrightarrow Cl + O_2$$

What total volume of ozone measured at a pressure of 25.0 mmHg and a temperature of 225 K can be destroyed when all of the chlorine from 15.0 g of CF_3Cl goes through ten cycles of the above reactions?

Kinetic Molecular Theory

79. Consider a 1.0-L sample of helium gas and a 1.0-L sample of argon gas, both at room temperature and atmospheric pressure.
    a. Do the atoms in the helium sample have the same *average kinetic energy* as the atoms in the argon sample?
    b. Do the atoms in the helium sample have the same *average velocity* as the atoms in the argon sample?
    c. Do the argon atoms, since they are more massive, exert a greater pressure on the walls of the container? Explain.
    d. Which gas sample would have the fastest rate of effusion?

80. A flask at room temperature contains exactly equal amounts (in moles) of nitrogen and xenon.
    a. Which of the two gases exerts the greater partial pressure?
    b. The molecules or atoms of which gas have the greater average velocity?
    c. The molecules of which gas have the greater average kinetic energy?
    d. If a small hole were opened in the flask, which gas would effuse more quickly?

81. Calculate the root mean square velocity and kinetic energy of F_2, Cl_2, and Br_2 at 298 K. Rank the three halogens with respect to their rate of effusion.

82. Calculate the root mean square velocity and kinetic energy of CO, CO_2, and SO_3 at 298 K. Which gas has the greatest velocity? The greatest kinetic energy? The greatest effusion rate?

83. Uranium-235 can be separated from U-238 by fluorinating the uranium to form UF_6 (which is a gas) and then taking advantage of the different rates of effusion and diffusion for compounds containing the two isotopes. Calculate the ratio of effusion rates for $^{238}UF_6$ and $^{235}UF_6$. The atomic mass of U-235 is 235.054 amu and that of U-238 is 238.051 amu.

84. Calculate the ratio of effusion rates for Ar and Kr.

85. A sample of neon effuses from a container in 76 seconds. The same amount of an unknown noble gas requires 155 seconds. Identify the gas.

86. A sample of N_2O effuses from a container in 42 seconds. How long would it take the same amount of gaseous I_2 to effuse from the same container under identical conditions?

87. The following graph shows the distribution of molecular velocities for two different molecules (A and B) at the same temperature. Which molecule has the higher molar mass? Which molecule would have the higher rate of effusion?

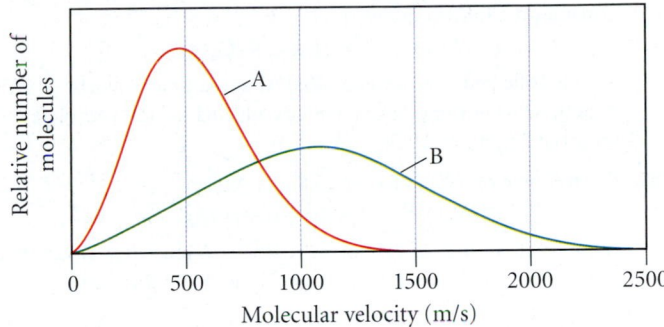

88. The following graph shows the distribution of molecular velocities for the same molecule at two different temperatures (T_1 and T_2). Which temperature is greater? Explain.

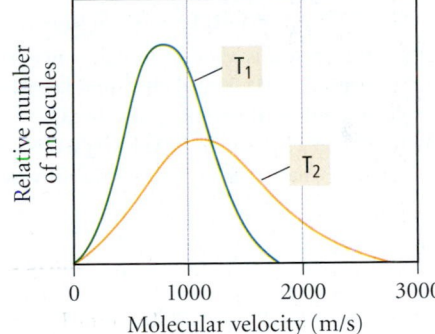

Real Gases

89. Which postulate of the kinetic molecular theory breaks down under conditions of high pressure? Explain.

90. Which postulate of the kinetic molecular theory breaks down under conditions of low temperature? Explain.

91. Use the van der Waals equation and the ideal gas equation to calculate the volume of 1.000 mol of neon at a pressure of 500.0 atm and a temperature of 355.0 K. Explain why the two values are different.

92. Use the van der Waals equation and the ideal gas equation to calculate the pressure exerted by 1.000 mol of Cl_2 in a volume of 5.000 L at a temperature of 273.0 K. Explain why the two values are different.

Cumulative Problems

93. Modern pennies are composed of zinc coated with copper. A student determines the mass of a penny to be 2.482 g and then makes several scratches in the copper coating (to expose the underlying zinc). The student puts the scratched penny in hydrochloric acid, where the following reaction occurs between the zinc and the HCl (the copper remains undissolved):

    $$Zn(s) + 2HCl(aq) \longrightarrow H_2(g) + ZnCl_2(aq)$$

    The student collects the hydrogen produced over water at 25 °C. The collected gas occupies a volume of 0.899 L at a total pressure of 791 mmHg. Calculate the percent zinc in the penny. (Assume that all the Zn in the penny dissolves.)

94. A 2.85-g sample of an unknown chlorofluorocarbon is decomposed and produces 564 mL of chlorine gas at a pressure of 752 mmHg and a temperature of 298 K. What is the percent chlorine (by mass) in the unknown chlorofluorocarbon?

95. The mass of an evacuated 255-mL flask is 143.187 g. The mass of the flask filled with 267 torr of an unknown gas at 25 °C is 143.289 g. Calculate the molar mass of the unknown gas.

96. A 118-mL flask is evacuated and found to have a mass of 97.129 g. When the flask is filled with 768 torr of helium gas at 35 °C, it is found to have a mass of 97.171 g. Was the helium gas pure?

97. A gaseous hydrogen and carbon containing compound is decomposed and found to contain 82.66% carbon and 17.34% hydrogen by mass. The mass of 158 mL of the gas, measured at 556 mmHg and 25 °C, was found to be 0.275 g. What is the molecular formula of the compound?

98. A gaseous hydrogen and carbon containing compound is decomposed and found to contain 85.63% C and 14.37% H by mass. The mass of 258 mL of the gas, measured at STP, was found to be 0.646 g. What is the molecular formula of the compound?

99. Consider the following reaction:

$$2\,NiO(s) \longrightarrow 2\,Ni(s) + O_2(g)$$

If O_2 is collected over water at 40.0 °C and a total pressure of 745 mmHg, what volume of gas will be collected for the complete reaction of 24.78 g of NiO?

100. The following reaction forms 15.8 g of $Ag(s)$:

$$2\,Ag_2O(s) \longrightarrow 4\,Ag(s) + O_2(g)$$

What total volume of gas forms if it is collected over water at a temperature of 25 °C and a total pressure of 752 mmHg?

101. When hydrochloric acid is poured over potassium sulfide, 42.9 mL of hydrogen sulfide gas is produced at a pressure of 752 torr and 25.8 °C. Write an equation for the gas-evolution reaction and determine how much potassium sulfide (in grams) reacted.

102. Consider the following reaction:

$$2\,SO_2(g) + O_2(g) \longrightarrow 2\,SO_3(g)$$

   a. If 285.5 mL of SO_2 is allowed to react with 158.9 mL of O_2 (both measured at 315 K and 50.0 mmHg), what is the limiting reactant and the theoretical yield of SO_3?

   b. If 187.2 mL of SO_3 is collected (measured at 315 K and 50.0 mmHg), what is the percent yield for the reaction?

103. Ammonium carbonate decomposes upon heating according to the following balanced equation:

$$(NH_4)_2CO_3(s) \longrightarrow 2\,NH_3(g) + CO_2(g) + H_2O(g)$$

Calculate the total volume of gas produced at 22 °C and 1.02 atm by the complete decomposition of 11.83 g of ammonium carbonate.

104. Ammonium nitrate decomposes explosively upon heating according to the following balanced equation:

$$2\,NH_4NO_3(s) \longrightarrow 2\,N_2(g) + O_2(g) + 4\,H_2O(g)$$

Calculate the total volume of gas (at 125 °C and 748 mmHg) produced by the complete decomposition of 1.55 kg of ammonium nitrate.

105. Olympic cyclists fill their tires with helium to make them lighter. Calculate the mass of air in an air-filled tire and the mass of helium in a helium-filled tire. What is the mass difference between the two? Assume that the volume of the tire is 855 mL, that it is filled to a total pressure of 125 psi, and that the temperature is 25 °C. Also, assume an average molar mass for air of 28.8 g/mol.

106. In a common classroom demonstration, a balloon is filled with air and submerged in liquid nitrogen. The balloon contracts as the gases within the balloon cool. Suppose the balloon initially contains 2.95 L of air at a temperature of 25.0 °C and a pressure of 0.998 atm. Calculate the expected volume of the balloon upon cooling to −196 °C (the boiling point of liquid nitrogen). When the demonstration is carried out, the actual volume of the balloon decreases to 0.61 L. How well does the observed volume of the balloon compare to your calculated value? Can you explain the difference?

107. Gaseous ammonia can be injected into the exhaust stream of a coal-burning power plant to reduce the pollutant NO to N_2 according to the following reaction:

$$4\,NH_3(g) + 4\,NO(g) + O_2(g) \longrightarrow 4\,N_2(g) + 6\,H_2O(g)$$

Suppose that the exhaust stream of a power plant has a flow rate of 335 L/s at a temperature of 955 K, and that the exhaust contains a partial pressure of NO of 22.4 torr. What should be the flow rate of ammonia delivered at 755 torr and 298 K into the stream to react completely with the NO if the ammonia is 65.2% pure (by volume)?

108. The emission of NO_2 by fossil fuel combustion can be prevented by injecting gaseous urea into the combustion mixture. The urea reduces NO (which oxidizes in air to form NO_2) according to the following reaction:

$$2\,CO(NH_2)_2(g) + 4\,NO(g) + O_2(g) \longrightarrow$$
$$4\,N_2(g) + 2\,CO_2(g) + 4\,H_2O(g)$$

Suppose that the exhaust stream of an automobile has a flow rate of 2.55 L/s at 655 K and contains a partial pressure of NO of 12.4 torr. What total mass of urea is necessary to react completely with the NO formed during 8.0 hours of driving?

109. An ordinary gasoline can measuring 30.0 cm by 20.0 cm by 15.0 cm is evacuated with a vacuum pump. Assuming that virtually all of the air can be removed from inside the can, and that atmospheric pressure is 14.7 psi, what is the total force (in pounds) on the surface of the can? Do you think that the can could withstand the force?

110. Twenty-five milliliters of liquid nitrogen (density = 0.807 g/mL) is poured into a cylindrical container with a radius of 10.0 cm and a length of 20.0 cm. The container initially contains only air at a pressure of 760.0 mmHg (atmospheric pressure) and a temperature of 298 K. If the liquid nitrogen completely vaporizes, what is the total force (in lb) on the interior of the container at 298 K?

111. A 160.0-L helium tank contains pure helium at a pressure of 1855 psi and a temperature of 298 K. How many 3.5-L helium balloons can be filled from the helium in the tank? (Assume an atmospheric pressure of 1.0 atm and a temperature of 298 K.)

112. A 11.5-mL sample of liquid butane (density = 0.573 g/mL) is evaporated in an otherwise empty container at a temperature of 28.5 °C. The pressure in the container following evaporation is 892 torr. What is the volume of the container?

113. A scuba diver creates a spherical bubble with a radius of 2.5 cm at a depth of 30.0 m where the total pressure (including atmospheric pressure) is 4.00 atm. What is the radius of the bubble when it reaches the surface of the water? (Assume atmospheric pressure to be 1.00 atm and the temperature to be 298 K.)

114. A particular balloon can be stretched to a maximum surface area of 1257 cm². The balloon is filled with 3.0 L of helium gas at a pressure of 755 torr and a temperature of 298 K. The balloon is then allowed to rise in the atmosphere. Assume an atmospheric temperature of 273 K and determine at what pressure the balloon will burst. (Assume the balloon to be in the shape of a sphere.)

115. A catalytic converter in an automobile uses a palladium or platinum catalyst to convert carbon monoxide gas to carbon dioxide according to the following reaction:

$$2\,CO(g) + O_2(g) \longrightarrow 2\,CO_2(g)$$

A chemist researching the effectiveness of a new catalyst combines a 2.0 : 1.0 mole ratio mixture of carbon monoxide and oxygen gas (respectively) over the catalyst in a 2.45-L flask at a total pressure of 745 torr and a temperature of 552 °C. When the reaction is complete, the pressure in the flask has dropped to 552 torr. What percentage of the carbon monoxide was converted to carbon dioxide?

116. A quantity of N_2 occupies a volume of 1.0 L at 300 K and 1.0 atm. The gas expands to a volume of 3.0 L as the result of a change in both temperature and pressure. Find the density of the gas at these new conditions.

117. A mixture of $CO(g)$ and $O_2(g)$ in a 1.0-L container at 1.0×10^3 K has a total pressure of 2.2 atm. After some time the total pressure falls to 1.9 atm as the result of the formation of CO_2. Find the amount of CO_2 that forms.

118. The radius of a xenon atom is 1.3×10^{-8} cm. A 100-mL flask is filled with Xe at a pressure of 1.0 atm and a temperature of 273 K. Calculate the fraction of the volume that is occupied by Xe atoms. (Hint: The atoms are spheres.)

119. A natural gas storage tank is a cylinder with a moveable top whose volume can change only as its height changes. Its radius remains fixed. The height of the cylinder is 22.6 m on a day when the temperature is 22 °C. The next day the height of the cylinder increas-es to 23.8 m as the gas expands because of a heat wave. Find the temperature, assuming that the pressure and amount of gas in the storage tank have not changed.

120. A mixture of 8.0 g CH_4 and 8.0 g Xe is placed in a container and the total pressure is found to be 0.44 atm. Find the partial pressure of CH_4.

121. A steel container of volume 0.35 L can withstand pressures up to 88 atm before exploding. Find the mass of helium that can be stored in this container at 299 K.

122. Binary compounds of alkali metals and hydrogen react with water to liberate $H_2(g)$. The H_2 from the reaction of a sample of NaH with an excess of water fills a volume of 0.490 L above the water. The temperature of the gas is 35 °C and the total pressure is 758 mmHg. Find the mass of H_2 liberated and the mass of NaH that reacted.

123. In a given diffusion apparatus, 15.0 mL of HBr gas diffused in 1.0 min. In the same apparatus and under the same conditions, 20.3 mL of an unknown gas diffused in 1.0 min. The unknown gas is a hydrocarbon. Find its molecular formula.

Challenge Problems

124. The world burns approximately 9.0×10^{12} kg of fossil fuel per year. Use the combustion of octane as the representative reaction and determine the mass of carbon dioxide (the most significant greenhouse gas) formed per year by this combustion. The current concentration of carbon dioxide in the atmosphere is approximately 387 ppm (by volume). By what percentage does the concentration increase in one year due to fossil fuel combustion? Approximate the average properties of the entire atmosphere by assuming that the atmosphere extends from sea level to 15 km and that it has an average pressure of 381 torr and average temperature of 275 K. Assume Earth is a perfect sphere with a radius of 6371 km.

125. The atmosphere slowly oxidizes hydrocarbons in a number of steps that eventually convert the hydrocarbon into carbon dioxide and water. The overall reactions of a number of such steps for methane gas is as follows:

$$CH_4(g) + 5\,O_2(g) + 5\,NO(g) \longrightarrow$$
$$CO_2(g) + H_2O(g) + 5\,NO_2(g) + 2\,OH(g)$$

Suppose that an atmospheric chemist combines 155 mL of methane at STP, 885 mL of oxygen at STP, and 55.5 mL of NO at STP in a 2.0-L flask. The reaction is allowed to stand for several weeks at 275 K. If the reaction reaches 90.0% of completion (90.0% of the limiting reactant is consumed), what are the partial pressures of each of the reactants and products in the flask at 275 K? What is the total pressure in the flask?

126. Two identical balloons are filled to the same volume, one with air and one with helium. The next day, the volume of the air-filled balloon has decreased by 5.0%. By what percent has the volume of the helium-filled balloon decreased? (Assume that the air is four-fifths nitrogen and one-fifth oxygen, and that the temperature did not change.)

127. A mixture of $CH_4(g)$ and $C_2H_6(g)$ has a total pressure of 0.53 atm. Just enough $O_2(g)$ is added to the mixture to bring about its complete combustion to $CO_2(g)$ and $H_2O(g)$. The total pressure of the two product gases is found to be 2.2 atm. Assuming constant volume and temperature, find the mole fraction of CH_4 in the mixture.

128. A sample of $C_2H_2(g)$ has a pressure of 7.8 kPa. After some time a portion of it reacts to form $C_6H_6(g)$. The total pressure of the mixture of gases is then 3.9 kPa. Assume the volume and the temperature do not change. Find the fraction of $C_2H_2(g)$ that has undergone reaction.

Conceptual Problems

129. When the driver of an automobile applies the brakes, the passengers are pushed toward the front of the car, but a helium balloon is pushed toward the back of the car. Upon forward acceleration, the passengers are pushed toward the back of the car, but the helium balloon is pushed toward the front of the car. Why?

130. Suppose that a liquid is 10 times denser than water. If you were to sip this liquid at sea level using a straw, what would be the maximum length of the straw?

131. The following reaction occurs in a closed container:

$$A(g) + 2\,B(g) \longrightarrow 2\,C(g)$$

A reaction mixture initially contains 1.5 L of A and 2.0 L of B. Assuming that the volume and temperature of the reaction mixture remain constant, what is the percent change in pressure if the reaction goes to completion?

132. One mole of nitrogen and one mole of neon are combined in a closed container at STP. How big is the container?

133. Exactly equal amounts (in moles) of gas A and gas B are combined in a 1-L container at room temperature. Gas B has a molar mass that is twice that of gas A. Which of the following is true for the mixture of gases and why?
   a. The molecules of gas B have greater kinetic energy than those of gas A.
   b. Gas B has a greater partial pressure than gas A.
   c. The molecules of gas B have a greater average velocity than those of gas A.
   d. Gas B makes a greater contribution to the average density of the mixture than gas A.

134. Which of the following gases would you expect to deviate most from ideal behavior under conditions of low temperature: F_2, Cl_2, or Br_2? Explain.

6

Thermochemistry

6.1 Light the Furnace: The Nature of Energy and Its Transformations

6.2 The First Law of Thermodynamics: There Is No Free Lunch

6.3 Quantifying Heat and Work

6.4 Measuring ΔE for Chemical Reactions: Constant-Volume Calorimetry

6.5 Enthalpy: The Heat Evolved in a Chemical Reaction at Constant Pressure

6.6 Constant-Pressure Calorimetry: Measuring ΔH_{rxn}

6.7 Relationships Involving ΔH_{rxn}

6.8 Enthalpies of Reaction from Standard Heats of Formation

6.9 Energy Use and the Environment

WE HAVE SPENT THE FIRST FEW CHAPTERS of this book examining one of the two major components of our universe—matter. We now turn our attention to the other major component—energy. As far as we know, matter and energy—which can be interchanged but not destroyed—are what make up the physical universe. Unlike matter, energy is not something we can touch with our fingers or hold in our hand, but we experience it in many ways. The warmth of sunlight, the feel of wind on our faces, and the force that presses us back when a car accelerates are all manifestations of energy and its interconversions. And of course energy and its uses are critical to society and to the world. The standard of living around the world is strongly correlated with access to and use of energy resources. Most of those resources, as we shall see, are chemical ones, and their advantages as well as their drawbacks can be understood in terms of chemistry.

There is a fact, or if you wish, a law, governing all natural phenomena that are known to date. There is no exception to this law—it is exact as far as we know. The law is called the conservation of energy. It states that there is a certain quantity, which we call energy, that does not change in the manifold changes which nature undergoes.
—RICHARD P. FEYNMAN (1918–1988)

Heating a house with a natural gas furnace involves many of the principles of thermochemistry.

6.1 Light the Furnace: The Nature of Energy and Its Transformations

The month may vary, depending on whether you live in Maine or Texas, but for most people, the annual fall ritual is pretty much the same. The days get shorter, the temperature drops, and soon it is time to light the household furnace. The furnace at my house, like many, is fueled by natural gas. Heating a home with natural gas involves many of the principles of **thermochemistry**, the study of the relationships between chemistry and energy. The combustion of natural gas gives off heat. Although some of the heat is lost through open doors, cracks in windows, or even directly through the walls (especially if the house is poorly insulated), most of it is transferred to the air in the house, resulting in a temperature increase. The magnitude of the temperature increase depends on how big the house is (and therefore how much air is in it) and how much natural gas is burned. We can begin to understand this process in more detail by examining the nature of energy.

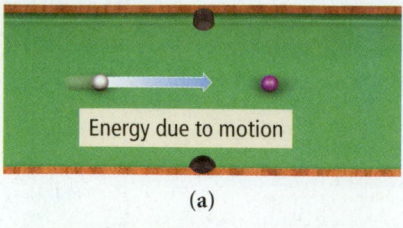

Energy due to motion

(a)

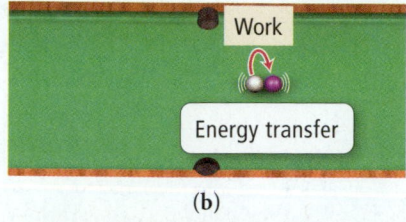

Work

Energy transfer

(b)

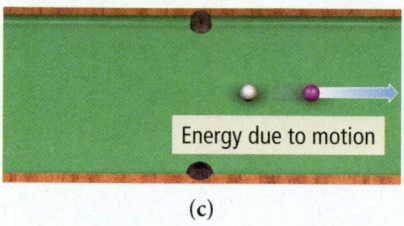

Energy due to motion

(c)

▲ (a) A rolling billiard ball has energy due to its motion. (b) When the ball collides with a second ball it does work, transferring energy to the second ball. (c) The second ball now has energy as it rolls away from the collision.

Einstein showed that it is mass–energy that is conserved—one can be converted into the other. This equivalence becomes important in nuclear reactions, discussed in Chapter 19. In ordinary chemical reactions, however, the interconversion of mass and energy is not a significant factor, and we can regard mass and energy as independently conserved.

The Nature of Energy: Key Definitions

We briefly examined energy in Section 1.5. Recall that we defined **energy** as the capacity to do work and defined **work** as the result of a force acting through a distance. For example, when you push a box across the floor you have done work. Consider as another example a billiard ball rolling across a billiard table and colliding straight on with a second, stationary billiard ball. The rolling ball has *energy* due to its motion. When it collides with another ball it does *work*, resulting in the *transfer* of energy from one ball to the other. The second billiard ball absorbs the energy and begins to roll across the table.

Energy can also be transferred through **heat**, the flow of energy caused by a temperature difference. For example, if you hold a cup of coffee in your hand, energy is transferred, in the form of heat, from the hot coffee to your cooler hand. Think of *energy* as something that an object or set of objects possesses. Think of *heat* and *work* as ways that objects or sets of objects *exchange* energy.

The energy contained in a rolling billiard ball is an example of **kinetic energy**, which is energy associated with the motion of an object. The energy contained in a hot cup of coffee is **thermal energy**, the energy associated with the temperature of an object. Thermal energy is actually a type of kinetic energy because it arises from the motions of atoms or molecules within a substance. If you raise a billiard ball off the table, you increase its **potential energy**, which is energy associated with the position or composition of an object. The potential energy of the billiard ball, for example, is a result of its position in Earth's gravitational field. Raising the ball off the table, against Earth's gravitational pull, gives it more potential energy. Another example of potential energy is the energy contained in a compressed spring. When you compress a spring, you push against the forces that tend to maintain the spring's shape, storing energy as potential energy. **Chemical energy**, the energy associated with the relative positions of electrons and nuclei in atoms and molecules, is also a form of potential energy. Some chemical compounds, such as the methane in natural gas, are like a compressed spring—they contain potential energy that can be released by a chemical reaction.

The **law of conservation of energy** states that energy can be neither created nor destroyed. However, energy can be transferred from one object to another, and it can assume different forms. For example, if you drop a raised billiard ball, some of its potential energy becomes kinetic energy as the ball falls toward the table, as shown in Figure 6.1 ▼. If you release a compressed spring, the potential energy becomes kinetic

▼ **FIGURE 6.1 Energy Transformation: Potential and Kinetic Energy I** (a) A billiard ball held above the ground has gravitational potential energy. (b) When the ball is released, the potential energy is transformed into kinetic energy, the energy of motion.

Energy Transformation I

Gravitational potential energy

Kinetic energy

(a)   (b)

Energy Transformation II	**Energy Transfer**

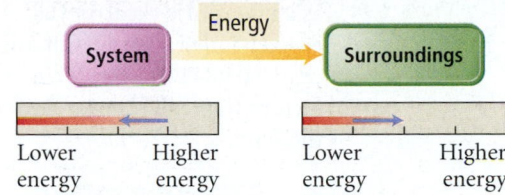

▲ FIGURE 6.3 Energy Transfer   When a system transfers energy to its surroundings, the energy of the system decreases while the energy of the surroundings increases. The total amount of energy is conserved.

▲ FIGURE 6.2 Energy Transformation: Potential and Kinetic Energy II   (a) A compressed spring has potential energy. (b) When the spring is released, the potential energy is transformed into kinetic energy.

energy as the spring expands outward, as shown in Figure 6.2 ▲. When you burn natural gas in a furnace, the chemical energy of the natural gas molecules becomes thermal energy that increases the temperature of the air.

A good way to understand and track energy changes is to define the **system** under investigation. For example, the system may be a beaker of chemicals in the lab, or it may be natural gas burning in a furnace. The system's **surroundings** are then everything else. For the beaker of chemicals in the lab, the surroundings may include the water that the chemicals are dissolved in (for aqueous solutions), the beaker itself, the lab bench on which the beaker sits, the air in the room, and so on. For the furnace, the surroundings include the air in the house, the furnishings in the house, and even the structure of the house itself.

In an energy exchange, energy is transferred between the system and the surroundings, as shown in Figure 6.3 ▲. If the system loses energy, the surroundings gain energy, and vice versa. When you burn natural gas in a home furnace, the system (the reactants and products) loses energy to the surroundings (the air in the house and the house itself), producing a temperature increase.

Units of Energy

The units of energy can be deduced from the definition of kinetic energy. An object of mass m, moving at velocity v, has a kinetic energy KE given by

$$KE = \frac{1}{2}mv^2 \qquad [6.1]$$

Because the SI unit of mass is the kg and the unit of velocity is m/s, the SI unit of energy is $kg \cdot m^2/s^2$, defined as the **joule (J)**, named after the English scientist James Joule (1818–1889).

$$1 \text{ kg} \frac{m^2}{s^2} = 1 \text{ J}$$

One joule is a relatively small amount of energy—for example, it takes 3.6×10^5 J to light a 100-watt lightbulb for 1 hour. Therefore, we often use kilojoules (kJ) in our energy discussions and calculations (1 kJ = 1000 J). A second unit of energy in common use is the **calorie (cal)**, originally defined as the amount of energy required to raise the temperature of 1 g of water by 1 °C. The current definition is 1 cal = 4.184 J (exact);

3.6×10^5 J or 0.10 kWh used in 1 hour

▲ A watt (W) is 1 J/s, so a 100-W lightbulb uses 100 J every second or 3.6×10^5 J every hour.

The "calorie" referred to on all nutritional labels (regardless of the capitalization) is always the capital *C* Calorie.

so, a calorie is a larger unit than a joule. A related energy unit is the nutritional, or capital *C*, **Calorie (Cal)**, equivalent to 1000 little *c* calories. The Calorie is therefore the same as a kilocalorie (kcal): 1 Cal = 1 kcal = 1000 cal. Electricity bills usually come in another, even larger, energy unit, the **kilowatt-hour (kWh)**: $1 \text{ kWh} = 3.60 \times 10^6$ J. Electricity costs \$0.08–\$0.15 per kWh. Table 6.1 shows various energy units and their conversion factors. Table 6.2 shows the amount of energy required for various processes in each of these units.

TABLE 6.1 Energy Conversion Factors*

1 calorie (cal)	= 4.184 joules (J)
1 Calorie (Cal) or kilocalorie (kcal)	= 1000 cal = 4184 J
1 kilowatt-hour (kWh)	= 3.60×10^6 J

* All conversion factors in this table are exact.

TABLE 6.2 Energy Uses in Various Units

Unit	Amount Required to Raise Temperature of 1 g of Water by 1 °C	Amount Required to Light 100-W Bulb for 1 Hour	Amount Used by Human Body in Running 1 Mile (Approximate)	Amount Used by Average U.S. Citizen in 1 Day
joule (J)	4.18	3.60×10^5	4.2×10^5	9.0×10^8
calorie (cal)	1.00	8.60×10^4	1.0×10^5	2.2×10^8
Calorie (Cal)	0.00100	86.0	100.	2.2×10^5
kilowatt-hour (KWh)	1.16×10^{-6}	0.100	0.12	2.5×10^2

6.2 The First Law of Thermodynamics: There Is No Free Lunch

The general study of energy and its interconversions is called **thermodynamics**. The laws of thermodynamics are among the most fundamental in all of science, governing virtually every process that involves change. The **first law of thermodynamics** is the law of energy conservation, stated as follows:

The total energy of the universe is constant.

In other words, because energy is neither created nor destroyed, and the universe does not exchange energy with anything else, its energy content does not change. The first law has many implications, the most important of which is that, with energy, you do not get something for nothing. The best we can do with energy is break even—there is no free lunch. According to the first law, a device that would continually produce energy with no energy input, sometimes known as a *perpetual motion machine*, cannot exist. Occasionally, the media report or speculate on the discovery of a perpetual motion machine. For example, you may have heard someone propose an electric car that recharges itself while driving, or a motor that creates usable electricity as well as electricity to power itself. Although some new hybrid cars (electric and gasoline powered) can capture energy from braking and use that energy to recharge their batteries, they could never run indefinitely without adding fuel. As for the motor that powers an external load as well as itself—no such thing exists. Our society has a continual need for energy, and as our current energy resources dwindle, new energy sources will be required. However, those sources must also follow the first law of thermodynamics—energy must be conserved.

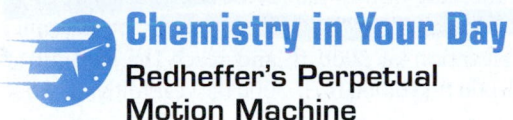

Chemistry in Your Day
Redheffer's Perpetual Motion Machine

In 1812, a man named Charles Redheffer appeared in Philadelphia with a machine that he claimed could run forever without any energy input—a perpetual motion machine. He set up the machine on the edge of town and charged admission to view it. He also appealed to the city for money to build a larger version of the machine. When city commissioners came out to inspect the machine, Redheffer did his best to keep them from viewing it too closely. Nonetheless, one of the commissioners noticed something suspicious: the gears that were supposed to run an external driveshaft were cut in the wrong direction. The driveshaft that the machine was supposedly powering was instead powering the machine. The city commissioners hired a local engineer and clockmaker named Isaiah Lukens to make a similar machine in order to show how Redheffer's deception was accomplished. Lukens's machine was even more ingenious than Redheffer's, and Redheffer left Philadelphia exposed as a fraud.

Redheffer was persistent, however, and took his machine to New York. In 1813, during a public display of the machine, the famous mechanical engineer Robert Fulton—who 6 years earlier had demonstrated the first successful steamboat—noticed a rhythm to the machine's motion. It seemed to speed up and slow down at regular intervals. Fulton recognized that such a motion would happen if the machine were powered by a manual crank. Fulton knocked away some boards in a wall next to the machine and discovered a long belt that led to an enclosed room where, indeed, an old man sat turning a crank. Redheffer's machine—like many other perpetual motion machines throughout history—was clearly a hoax.

Question
Can you think of any recent claims of perpetual motion or limitless free energy?

Internal Energy

The **internal energy (E)** of a system is the sum of the kinetic and potential energies of all of the particles that compose the system. Internal energy is a **state function**, which means that its value depends *only on the state of the system*, not on how the system arrived at that state. The state of a chemical system is specified by parameters such as temperature, pressure, concentration, and phase (solid, liquid, or gas). We can understand state functions with a mountain-climbing analogy. The elevation at any point during a climb is a state function. For example, when you reach 10,000 ft, your elevation is 10,000 ft, no matter how you got there. The distance you traveled to get there, by contrast, is not a state function. You could have climbed the mountain by any number of routes, each requiring you to cover a different distance.

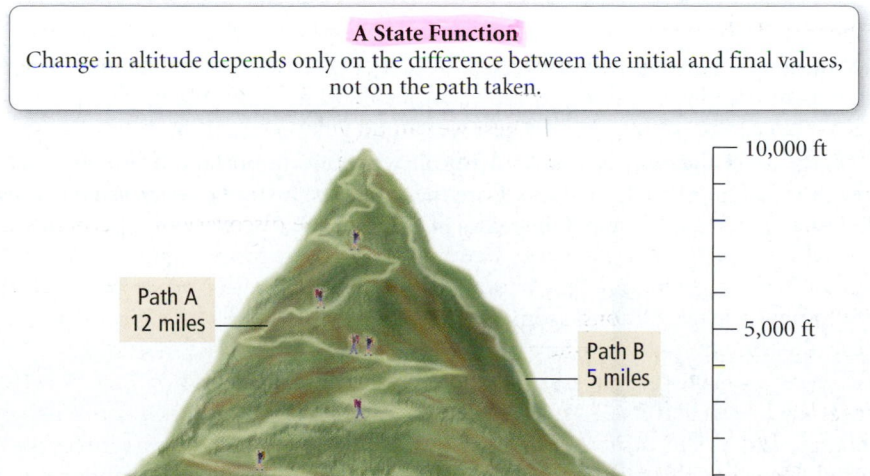

A State Function
Change in altitude depends only on the difference between the initial and final values, not on the path taken.

Path A
12 miles

Path B
5 miles

10,000 ft

5,000 ft

0 ft

◄ Altitude is a state function. The change in altitude during a climb depends only on the difference between the final and initial altitudes.

Since state functions depend only on the state of the system, the value of a change in a state function is always the difference between its final and initial values. For example, if you start climbing a mountain at an elevation of 3000 ft, and reach the summit at 10,000 feet, then your elevation change is 7000 ft (10,000 ft − 3000 ft), regardless of what path you took.

Like an altitude change, an internal energy change (ΔE) is given by the difference in internal energy between the final and initial states:

$$\Delta E = E_{\text{final}} - E_{\text{initial}}$$

In a chemical system, the reactants constitute the initial state and the products constitute the final state. So ΔE is simply the difference in internal energy between the products and the reactants.

$$\Delta E = E_{\text{products}} - E_{\text{reactants}} \qquad [6.2]$$

For example, consider the reaction between carbon and oxygen to form carbon dioxide:

$$C(s) + O_2(g) \longrightarrow CO_2(g)$$

The energy diagram showing the internal energies of the reactants and products for this reaction is as follows:

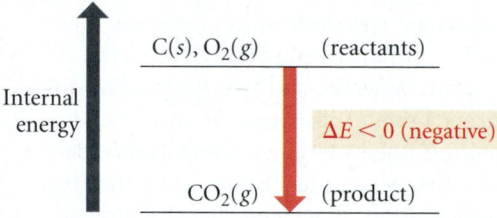

The reactants have a higher internal energy than the product. Therefore, when the reaction occurs in the forward direction, ΔE (that is, $E_{\text{products}} - E_{\text{reactants}}$) is *negative*.

Where does the energy lost by the reactants go? If we define the thermodynamic *system* as the reactants and products of the reaction, then energy must flow *out of the system* and *into the surroundings*.

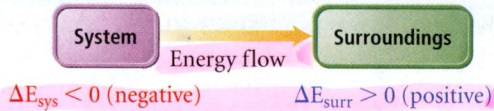

According to the first law, energy must be conserved. Therefore, the amount of energy lost by the system must exactly equal the amount gained by the surroundings.

$$\Delta E_{\text{sys}} = -\Delta E_{\text{surr}} \qquad [6.3]$$

Now, suppose the reaction is reversed:

$$CO_2(g) \longrightarrow C(s) + O_2(g)$$

The energy level diagram is nearly identical, with one important difference: $CO_2(g)$ is now the reactant and $C(s)$ and $O_2(g)$ are the products. Instead of decreasing in energy as the reaction occurs, the system increases in energy:

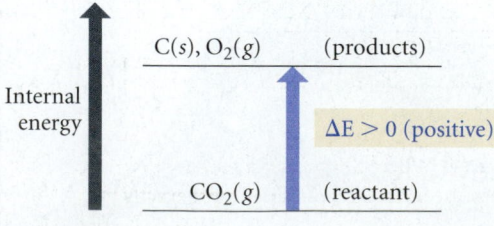

The difference ΔE is *positive* and energy flows *into the system* and *out of the surroundings*.

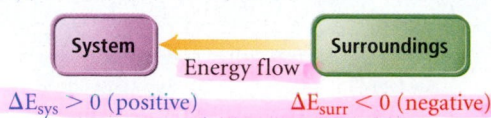

$\Delta E_{\text{sys}} > 0$ (positive) $\Delta E_{\text{surr}} < 0$ (negative)

Summarizing:

➤ If the reactants have a higher internal energy than the products, ΔE_{sys} is negative and energy flows out of the system into the surroundings.

➤ If the reactants have a lower internal energy than the products, ΔE_{sys} is positive and energy flows into the system from the surroundings.

You can think of the internal energy of the system in the same way you think about the balance in a checking account. Energy flowing *out of* the system is like a withdrawal and therefore carries a negative sign. Energy flowing *into* the system is like a deposit and carries a positive sign.

 ## Conceptual Connection 6.1   System and Surroundings

The following are fictitious internal energy gauges for a chemical system and its surroundings:

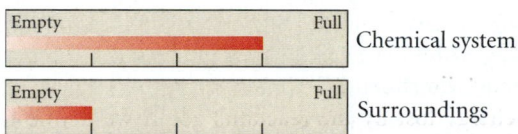

Which of the following best represents the energy gauges for the same system and surroundings following an energy exchange in which ΔE_{sys} is negative?

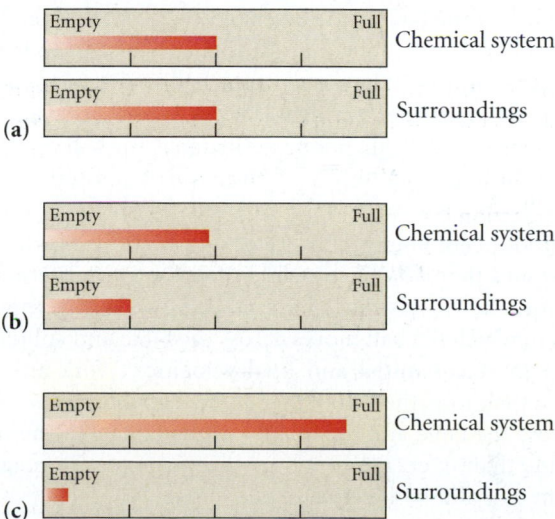

Answer: The correct answer is (**a**). When ΔE_{sys} is negative, energy flows out of the system and into the surroundings. The energy increase in the surroundings must exactly match the decrease in the system.

As we saw in Section 6.1, a system can exchange energy with its surroundings through *heat* and *work*:

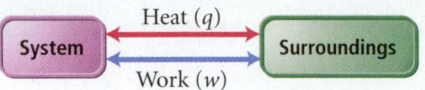

According to the first law of thermodynamics, the change in the internal energy of the system (ΔE) must be the sum of the heat transferred (q) and the work done (w):

$$\Delta E = q + w \qquad\qquad [6.4]$$

In the above equation, and from this point forward, we follow the standard convention that ΔE (with no subscript) refers to the internal energy change of the *system*. As shown in Table 6.3, energy entering the system through heat or work carries a positive sign, and energy leaving the system through heat or work carries a negative sign.

TABLE 6.3 Sign Conventions for q, w, and ΔE

q (heat)	+ system *gains* thermal energy	− system *loses* thermal energy
w (work)	+ work done *on* the system	− work done *by* the system
ΔE (change in internal energy)	+ energy flows *into* the system	− energy flows *out* of the system

Suppose we define our system as the previously discussed billiard ball rolling across a pool table. The rolling ball has a certain initial amount of kinetic energy. At the other end of the table, the rolling ball collides straight-on with a second ball. Let's assume that the first ball transfers all of its kinetic energy to the second ball through work, so that the first ball remains completely still at the point of collision with no kinetic energy. The total change in internal energy (ΔE) for the first ball is simply the difference between its initial kinetic energy and its final kinetic energy (which is zero). However, the amount of work that it does on the second ball depends on the quality of the billiard table, because as the first ball rolls across the table, it can lose some of its initial kinetic energy through collisions with minute bumps on the table surface. These collisions, which we call friction, slow the ball down by converting kinetic energy to heat. On a smooth, high-quality billiard table, the amount of energy lost in this way is relatively small, as shown in Figure 6.4(a) ▶. The speed of the ball is not greatly reduced, leaving a great deal of its original kinetic energy available to perform work when it collides with the second ball. In contrast, on a rough, poor-quality table, the ball may lose much of its initial kinetic energy as heat, leaving only a relatively small amount available for work, as shown in Figure 6.4(b) ▶.

Notice that the amounts of energy converted to heat and work depend on the details of the pool table and the path taken, while the change in internal energy of the rolling ball does not. In other words, since internal energy is a state function, the value of ΔE for the process in which the ball moves across the table and collides with another ball depends only on the ball's initial and final velocities. Work and heat, however, are *not* state functions; therefore, the values of q and w depend on the details of exactly how the ball rolls across the table and the quality of the table. On the smooth table, w is greater in magnitude than q; on the rough table, q is greater in magnitude than w. However, ΔE (the sum of q and w) is constant.

(a) Smooth table

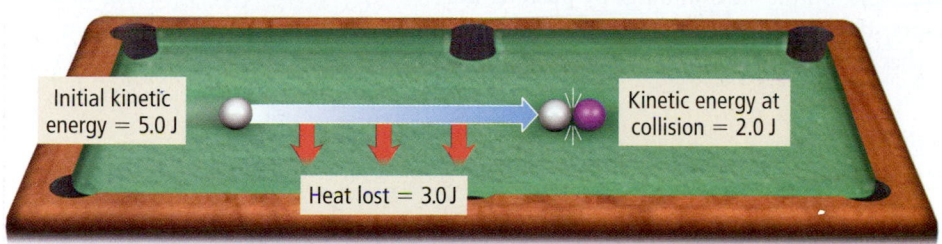

(b) Rough table

◀ **FIGURE 6.4 Energy, Work, and Heat** (a) On a smooth table, most of the first billiard ball's initial kinetic energy is transferred to the second ball as work. Only a small amount is lost to heat. (b) On a rough table, most of the first billiard ball's initial kinetic energy is lost to heat. Only a small amount is left to do work on the second billiard ball.

 Conceptual Connection 6.2 Heat and Work

Identify each of the following energy exchanges as heat or work and determine whether the sign of heat or work (relative to the system) is positive or negative.

(a) An ice cube melts and cools the surrounding beverage. (Ice cube is the system.)

(b) A metal cylinder is rolled up a ramp. (The metal cylinder is the system; assume no friction.)

(c) Steam condenses on skin, causing a burn. (The condensing steam is the system.)

Answer: (a) heat, sign is positive **(b)** work, sign is positive **(c)** heat, sign is negative

EXAMPLE 6.1   Internal Energy, Heat, and Work

The firing of a potato cannon provides a good example of the heat and work associated with a chemical reaction. In a potato cannon, a potato is stuffed into a long cylinder that is capped on one end and open at the other. Some kind of fuel is introduced under the potato at the capped end—usually through a small hole—and ignited. The potato then shoots out of the cannon, sometimes flying hundreds of feet, and heat is given off to the surroundings. If the burning of the fuel performs 855 J of work on the potato and produces 1422 J of heat, what is ΔE for the burning of the fuel? (Note: A potato cannon can be dangerous and should not be constructed without proper training and experience.)

Solution To solve the problem, simply substitute the values of q and w into the equation for ΔE. Since work is done by the system on the surroundings, w is negative. Similarly, since heat is released by the system to the surroundings, q is also negative.

$$\Delta E = q + w$$
$$= -1422\,\text{J} - 855\,\text{J}$$
$$= -2277\,\text{J}$$

For Practice 6.1

A cylinder and piston assembly (defined as the system) is warmed by an external flame. The contents of the cylinder expand, doing work on the surroundings by pushing the piston outward against the external pressure. If the system absorbs 559 J of heat and does 488 J of work during the expansion, what is the value of ΔE?

6.3 Quantifying Heat and Work

In the previous section, we calculated ΔE based on *given values of q and w*. We now turn to *calculating q (heat) and w (work)* based on changes in temperature and volume.

Heat

As we saw in Section 6.1, *heat* is the exchange of thermal energy between a system and its surroundings caused by a temperature difference. Notice the distinction between heat and temperature. Temperature is a *measure* of the thermal energy of a sample of matter. Heat is the *transfer* of thermal energy. Thermal energy always flows from matter at higher temperatures to matter at lower temperatures. For example, a hot cup of coffee transfers thermal energy—as heat—to the lower temperature surroundings as it cools down. Imagine a world where the cooler surroundings actually got colder as they transferred thermal energy to the hot coffee, which got hotter. Such a world exists only in our imaginations (or in the minds of science fiction writers), because the transfer of heat from a hotter object to a colder one is a fundamental principle of our universe—no exception has ever been observed. So the thermal energy in the molecules within the hot coffee distributes itself to the molecules in the surroundings. The heat transfer from the coffee to the surroundings stops when the two reach the same temperature, a condition called **thermal equilibrium.** At thermal equilibrium, there is no additional net transfer of heat.

The reason for this one-way transfer is related to the second law of thermodynamics, which states that energy tends to distribute itself among the greatest number of particles possible. We cover the second law of thermodynamics in more detail in Chapter 17.

Temperature Changes and Heat Capacity When a system absorbs heat (q) its temperature changes by ΔT:

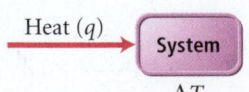

Heat (q) → System

ΔT

Experimental measurements demonstrate that the heat absorbed by a system and its corresponding temperature change are directly proportional: $q \propto \Delta T$. The constant of proportionality between q and ΔT is called the *heat capacity* (C), a measure of the system's ability to absorb thermal energy without undergoing a large change in temperature.

$$q = C \times \Delta T$$

Heat capacity

[6.5]

Notice that the higher the heat capacity of a system, the smaller the change in temperature for a given amount of absorbed heat. The **heat capacity (C)** of a system is usually defined as the quantity of heat required to change its temperature by 1 °C. As we can see by solving Equation 6.5 for heat capacity, the units of heat capacity are those of heat (usually J) divided by those of temperature (usually °C).

$$C = \frac{q}{\Delta T} = \frac{J}{°C}$$

In order to understand two important features of heat capacity, consider putting a steel saucepan on a kitchen flame. The saucepan's temperature rises rapidly as it absorbs heat from the flame. However, if you add some water to the saucepan, the temperature rises more slowly. Why? The first reason is that, when you add the water, the same amount of heat must now warm more matter, so the temperature rises more slowly. In other words, heat capacity is an extensive property—it depends on the amount of matter being heated (see Section 1.6). The second (and more fundamental) reason is that water is more resistant to temperature change than steel—water has an intrinsically higher capacity to absorb heat without undergoing a large temperature change. The measure of the *intrinsic capacity* of a substance to absorb heat is called its **specific heat capacity (Cs)**, the amount of heat required to raise the temperature of *1 gram* of the substance by 1 °C. The units of specific heat capacity (also called *specific heat*) are J/g · °C. Table 6.4 shows the values of the specific heat capacity for several substances. Heat capacity is sometimes reported as **molar heat capacity**, the amount of heat required to raise the temperature of *1 mole* of a substance by 1 °C. The units of molar heat capacity are J/mol · °C. You can see from these definitions that *specific* heat capacity and *molar* heat capacity are intensive properties—they depend on the *kind* of substance being heated, not on the amount.

Notice that water has the highest specific heat capacity of all the substances in Table 6.4—changing its temperature requires a lot of heat. If you have ever experienced the drop in temperature that occurs when traveling from an inland region to the coast during the summer, you have experienced the effects of water's high specific heat capacity. On a summer's day in California, for example, the temperature difference between Sacramento (an inland city) and San Francisco (a coastal city) can be 30 °F—San Francisco enjoys a cool 68 °F, while Sacramento bakes at nearly 100 °F. Yet the intensity of sunlight falling on these two cities is the same. Why the large temperature difference? San Francisco sits on a peninsula, surrounded by the water of the Pacific Ocean. Water, with its high heat capacity, absorbs much of the sun's heat without undergoing a large increase in temperature, keeping San Francisco cool. Sacramento, by contrast, is about 100 mi inland. The land surrounding Sacramento, with its low heat capacity, undergoes a large increase in temperature as it absorbs a similar amount of heat.

Similarly, only two U.S. states have never recorded a temperature above 100 °F. One of them is obvious: Alaska. It is too far north to get that hot. The other one, however, may come as a surprise. It is Hawaii. The high heat capacity of the water that surrounds the only island state moderates the temperature, keeping Hawaii from ever getting too hot.

The specific heat capacity of a substance can be used to quantify the relationship between the amount of heat added to a given amount of the substance and the corresponding temperature increase. The equation that relates these quantities is

$$\text{heat} = \text{mass} \times \text{specific heat capacity} \times \text{temperature change}$$

$$q = m \times C_s \times \Delta T$$

[6.6]

where q is the amount of heat in J, m is the mass of the substance in g, C_s is the specific heat capacity in J/g · °C, and ΔT is the temperature change in °C. The following example demonstrates the use of this equation.

TABLE 6.4 Specific Heat Capacities of Some Common Substances

Substance	Specific Heat Capacity, C_s (J/g · °C)*
Elements	
Lead	0.128
Gold	0.128
Silver	0.235
Copper	0.385
Iron	0.449
Aluminum	0.903
Compounds	
Ethanol	2.42
Water	4.18
Materials	
Glass (Pyrex)	0.75
Granite	0.79
Sand	0.84

*At 298 K.

▲ The high heat capacity of the water surrounding San Francisco results in relatively cool summer temperatures.

ΔT in °C is equal to ΔT in K, but not equal to ΔT in °F (Section 1.6).

EXAMPLE 6.2 Temperature Changes and Heat Capacity

Suppose you find a copper penny (pre-1982, when pennies were almost entirely copper) in the snow. How much heat is absorbed by the penny as it warms from the temperature of the snow, which is $-8.0\,°C$, to the temperature of your body, $37.0\,°C$? Assume the penny is pure copper and has a mass of $3.10\,g$.

Sort You are given the mass of copper as well as its initial and final temperature. You are asked to find the heat required for the given temperature change.	**Given** $m = 3.10\,g$ copper $T_i = -8.0\,°C$ $T_f = 37.0\,°C$ **Find** q
Strategize The equation $q = m \times C_s \times \Delta T$ gives the relationship between the amount of heat (q) and the temperature change (ΔT).	**Conceptual Plan** 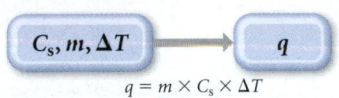 $$q = m \times C_s \times \Delta T$$ **Relationships Used** $q = m \times C_s \times \Delta T$ (Equation 6.6) $C_s = 0.385\,J/g \cdot °C$ (Table 6.4)
Solve Gather the necessary quantities for the equation in the correct units and substitute these into the equation to compute q.	**Solution** $\Delta T = T_f - T_i$ $\quad = 37.0\,°C - (-8.0\,°C) = 45.0\,°C$ $q = m \times C_s \times \Delta T$ $\quad = 3.10\,g \times 0.385\,\dfrac{J}{g \cdot °C} \times 45.0\,°C = 53.7\,J$

Check The units (J) are correct for heat. The sign of q is *positive*, as it should be since the penny *absorbed* heat from the surroundings.

For Practice 6.2

To determine whether a shiny gold-colored rock is actually gold, a chemistry student decides to measure its heat capacity. She first weighs the rock and finds it has a mass of $4.7\,g$. She then finds that upon absorption of $57.2\,J$ of heat, the temperature of the rock rises from $25\,°C$ to $57\,°C$. Find the specific heat capacity of the substance composing the rock and determine whether the value is consistent with the rock being pure gold.

For More Practice 6.2

A 55.0-g aluminum block initially at $27.5\,°C$ absorbs $725\,J$ of heat. What is the final temperature of the aluminum?

 Conceptual Connection 6.3 **The Heat Capacity of Water**

Suppose you are cold-weather camping and decide to heat some objects to bring into your sleeping bag for added warmth. You place a large water jug and a rock of equal mass near the fire. Over time, both the rock and the water jug warm to about $38\,°C$ ($100\,°F$). If you could bring only one into your sleeping bag, which one should you choose to keep you the warmest? Why?

Answer: Bring the water because it has the higher heat capacity and will therefore release more heat as it cools.

Work: Pressure–Volume Work

We have learned that energy transfer between a system and its surroundings occurs via heat (q) and work (w). We just saw how to calculate the *heat* associated with an observed *temperature* change. We now turn to calculating the *work* associated with an observed *volume* change. Although there are several types of work that a chemical reaction can do, for now we will limit ourselves to what is called **pressure–volume work**. We have already defined work as a force acting through a distance. Pressure–volume work occurs when the force is the result of a volume change against an external pressure. Pressure–volume work occurs, for example, in the cylinder of an automobile engine. The combustion of gasoline causes gases within the cylinders to expand, pushing the piston outward and ultimately moving the wheels of the car.

We can derive an equation for the value of pressure–volume work from the definition of work as a force (F) acting through a distance (D)

$$w = F \times D \qquad [6.7]$$

When the volume of a cylinder increases (Figure 6.5 ▼), the external force against which it pushes is caused by the external pressure (P), which is defined as force (F) divided by area (A):

$$P = \frac{F}{A} \quad \text{or} \quad F = P \times A$$

If we substitute this expression for force into the definition of work given by Equation 6.7, we get the following:

$$w = F \times D$$
$$= P \times A \times D$$

The distance through which the force acts is simply the change in the height of the piston as it moves during the expansion (Δh). Substituting Δh for D, we get

$$w = P \times A \times \Delta h$$

However, since the volume of a cylinder is the area of its base times its height, then $A \times \Delta h$ is actually the change in volume (ΔV) that occurs during the expansion. Thus, the expression for work becomes the following:

$$w = P \Delta V$$

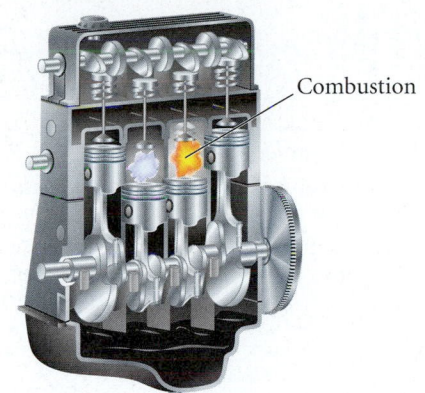

▲ The combustion of gasoline within an engine's cylinders does pressure–volume work that ultimately results in the car's motion.

The force in this equation must be a constant force.

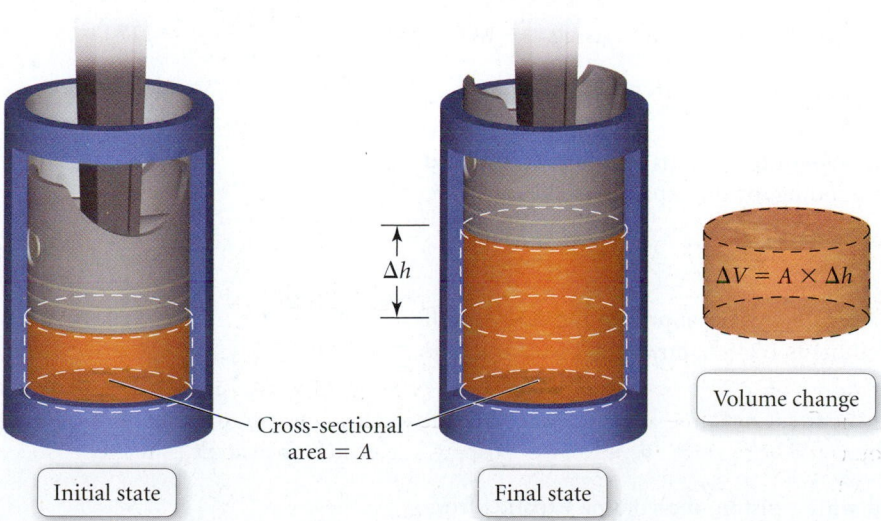

▲ **FIGURE 6.5** **Piston Moving within a Cylinder against an External Pressure**

Still missing from the equation is the sign of the work done by the expanding gases. As the volume of the cylinder increases, work is done *on* the surroundings *by* the system, so w should be negative. However, upon expansion, V_2 (the final volume) is greater than V_1 (the initial volume) so ΔV is positive. In order for w to be negative for a positive expansion, we must add a negative sign to our equation. In other words, w and ΔV must be opposite in sign.

$$w = -P\Delta V \qquad\qquad [6.8]$$

So the work caused by an expansion of volume is simply the negative of the pressure that the volume expands against multiplied by the change in volume that occurs during the expansion. The units of the work obtained by using this equation will be those of pressure (usually atm) multiplied by those of volume (usually L). To convert between L · atm and J, use the conversion factor 101.3 J = 1 L · atm.

EXAMPLE 6.3 Pressure–Volume Work

Inflating a balloon requires the inflator to do pressure–volume work on the surroundings. If a balloon is inflated from a volume of 0.100 L to 1.85 L against an external pressure of 1.00 atm, how much work is done (in joules)?

Sort You are given the initial and final volumes of the balloon and the pressure against which it expands. The balloon and its contents are the system.	**Given** $V_1 = 0.100$ L, $V_2 = 1.85$ L, $P = 1.00$ atm **Find** w
Strategize The equation $w = -P\Delta V$ specifies the amount of work done during a volume change against an external pressure.	**Conceptual Plan** 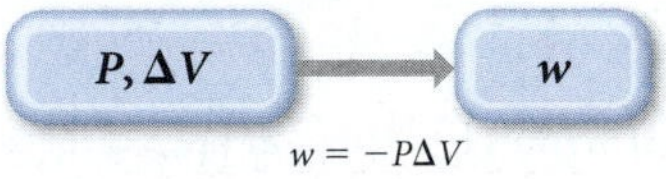 $\boxed{P, \Delta V} \longrightarrow \boxed{w}$ $w = -P\Delta V$
Solve To solve the problem, compute the value of ΔV and substitute it, together with P, into the equation.	**Solution** $\Delta V = V_2 - V_1$ $\quad = 1.85\ \text{L} - 0.100\ \text{L}$ $\quad = 1.75\ \text{L}$ $w = -P\Delta V$ $\quad = -1.00\ \text{atm} \times 1.75\ \text{L}$ $\quad = -1.75\ \text{L} \cdot \text{atm}$
The units of the answer (L · atm) can be converted to joules using 101.3 J = 1 L · atm.	$-1.75\ \cancel{\text{L} \cdot \text{atm}} \times \dfrac{101.3\ \text{J}}{1\ \cancel{\text{L} \cdot \text{atm}}} = -177\ \text{J}$

Check The units (J) are correct for work. The sign of the work is negative, as it should be for an expansion: work is done on the surroundings by the expanding balloon.

For Practice 6.3

A cylinder equipped with a piston expands against an external pressure of 1.58 atm. If the initial volume is 0.485 L and the final volume is 1.245 L, how much work (in J) is done?

For More Practice 6.3

When fuel is burned in a cylinder equipped with a piston, the volume expands from 0.255 L to 1.45 L against an external pressure of 1.02 atm. In addition, 875 J is emitted as heat. What is ΔE for the burning of the fuel?

6.4 Measuring ΔE for Chemical Reactions: Constant-Volume Calorimetry

We now have a complete picture of how a system exchanges energy with its surroundings via heat and pressure–volume work:

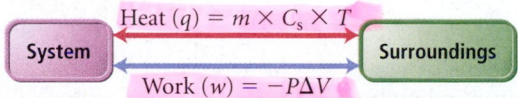

From Section 6.2, we know that the change in internal energy that occurs during a chemical reaction (ΔE) is a measure of *all of the energy* (heat and work) exchanged with the surroundings ($\Delta E = q + w$). Therefore, we can measure the changes in temperature (to calculate heat) and the changes in volume (to calculate work) that occur during a chemical reaction, and then sum them together to compute ΔE. However, an easier way to obtain the value of ΔE for a chemical reaction is to force all of the energy change associated with a reaction to appear as heat rather than work. We then simply measure the temperature change caused by the heat flow.

Recall that $\Delta E = q + w$ and that $w = -P\Delta V$. If a reaction is carried out at constant volume, then $\Delta V = 0$ and $w = 0$. The heat evolved, called the *heat at constant volume (q_v)*, is then equivalent to ΔE_{rxn}.

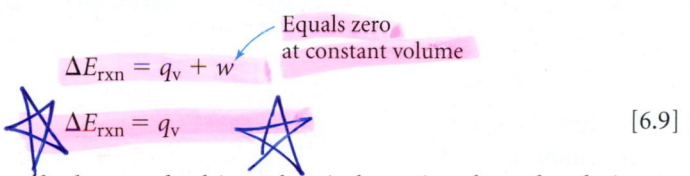

$$\Delta E_{rxn} = q_v + w$$

Equals zero at constant volume

$$\Delta E_{rxn} = q_v \qquad [6.9]$$

We can measure the heat evolved in a chemical reaction through *calorimetry*. In **calorimetry**, the thermal energy exchanged between the reaction (defined as the system) and the surroundings is measured by observing the change in temperature of the surroundings.

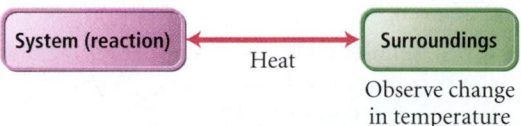

The magnitude of the temperature change in the surroundings depends on the magnitude of ΔE for the reaction and on the heat capacity of the surroundings.

Figure 6.6 ▶ shows a **bomb calorimeter**, a piece of equipment designed to measure ΔE for combustion reactions. In a bomb calorimeter, a tight-fitting, sealed lid forces the reaction to occur at constant volume. The sample to be burned (of known mass) is placed into a cup equipped with an ignition wire. The cup is sealed into a stainless steel container, called a *bomb*, filled with oxygen gas. The bomb is then placed in a water-filled, insulated container equipped with a stirrer and a thermometer. The sample is ignited using a wire coil, and the temperature is monitored with the thermometer. The temperature change (ΔT) is then related to the heat absorbed by the entire calorimeter assembly (q_{cal}) by the following equation:

$$q_{cal} = C_{cal} \times \Delta T \qquad [6.10]$$

where C_{cal} is the heat capacity of the entire calorimeter assembly (which is usually determined in a separate measurement involving the burning of a substance that gives off a known amount of heat). If no heat escapes from the calorimeter, the amount of heat *gained by* the calorimeter must exactly equal that *released by* the reaction (the two are equal in magnitude but opposite in sign):

$$q_{cal} = -q_{rxn} \qquad [6.11]$$

The Bomb Calorimeter

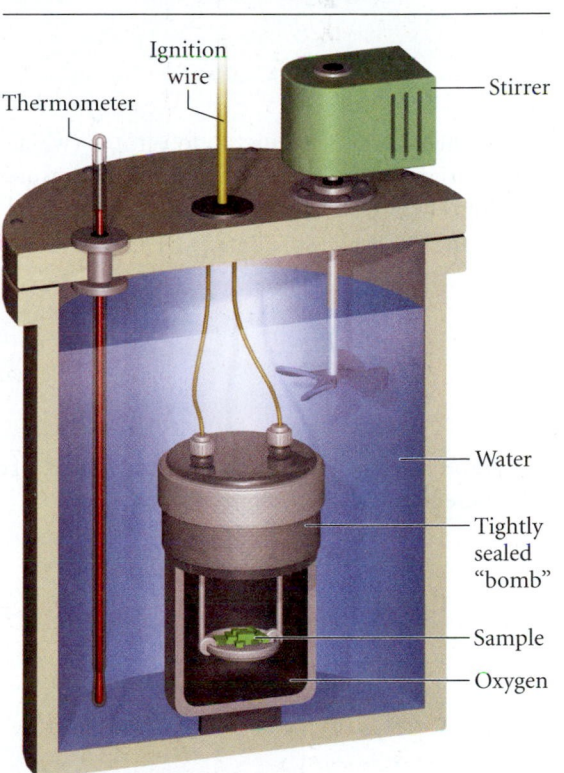

▲ **FIGURE 6.6 The Bomb Calorimeter** A bomb calorimeter is used to measure changes in internal energy for combustion reactions.

The heat capacity of the calorimeter, C_{cal}, has units of energy over temperature; its value accounts for all of the heat absorbed by all of the components within the calorimeter (including the water).

Since the reaction is occurring under conditions of constant volume, $q_{rxn} = q_v = \Delta E_{rxn}$. This measured quantity is the change in the internal energy of the reaction for the specific amount of reactant burned. To get ΔE_{rxn} per mole of a particular reactant—a more general quantity—you simply divide by the number of moles that actually reacted, as shown in Example 6.4.

EXAMPLE 6.4 Measuring ΔE_{rxn} in a Bomb Calorimeter

When 1.010 g of sucrose ($C_{12}H_{22}O_{11}$) undergoes combustion in a bomb calorimeter, the temperature rises from 24.92 °C to 28.33 °C. Find ΔE_{rxn} for the combustion of sucrose in kJ/mol sucrose. The heat capacity of the bomb calorimeter, determined in a separate experiment, is 4.90 kJ/°C. (You can ignore the heat capacity of the small sample of sucrose because it is negligible compared to the heat capacity of the calorimeter.)

Sort You are given the mass of sucrose, the heat capacity of the calorimeter, and the initial and final temperatures. You are asked to find the change in internal energy for the reaction.	**Given** 1.010 g $C_{12}H_{22}O_{11}$, T_i = 24.92 °C, T_f = 28.33 °C, C_{cal} = 4.90 kJ/°C **Find** ΔE_{rxn}

Strategize The conceptual plan has three parts. In the first part, use the temperature change and the heat capacity of the calorimeter to find q_{cal}.

Conceptual Plan

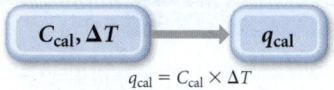

$$q_{cal} = C_{cal} \times \Delta T$$

In the second part, use q_{cal} to get q_{rxn} (which simply involves changing the sign). Since the bomb calorimeter ensures constant volume, q_{rxn} is equivalent to ΔE_{rxn} for the amount of sucrose burned.

In the third part, divide q_{rxn} by the number of moles of sucrose to get ΔE_{rxn} per mole of sucrose.

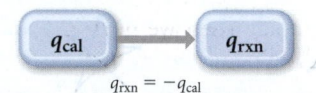

$$q_{rxn} = -q_{cal}$$

$$\Delta E_{rxn} = \frac{q_{rxn}}{\text{mol } C_{12}H_{22}O_{11}}$$

Relationships Used

$$q_{cal} = C_{cal} \times \Delta T = -q_{rxn}$$

molar mass $C_{12}H_{22}O_{11}$ = 342.3 g/mol

Solve Gather the necessary quantities in the correct units and substitute these into the equation to compute q_{cal}.

Solution

$$\Delta T = T_f - T_i$$
$$= 28.33 \,°C - 24.92 \,°C$$
$$= 3.41 \,°C$$

$$q_{cal} = C_{cal} \times \Delta T$$

$$q_{cal} = 4.90 \frac{kJ}{°C} \times 3.41 \,°C$$

$$= 16.7 \text{ kJ}$$

Find q_{rxn} by simply taking the negative of q_{cal}.

$$q_{rxn} = -q_{cal} = -16.7 \text{ kJ}$$

Find ΔE_{rxn} per mole of sucrose by dividing q_{rxn} by the number of moles of sucrose (calculated from the given mass of sucrose and its molar mass).

$$\Delta E_{rxn} = \frac{q_{rxn}}{\text{mol } C_{12}H_{22}O_{11}}$$

$$= \frac{-16.7 \text{ kJ}}{1.010 \text{ g } C_{12}H_{22}O_{11} \times \dfrac{1 \text{ mol } C_{12}H_{22}O_{11}}{342.3 \text{ g } C_{12}H_{22}O_{11}}}$$

$$= -5.66 \times 10^3 \text{ kJ/mol } C_{12}H_{22}O_{11}$$

Check The units of the answer (kJ) are correct for a change in internal energy. The sign of ΔE_{rxn} is negative, as it should be for a combustion reaction that gives off energy.

For Practice 6.4

When 1.550 g of liquid hexane (C_6H_{14}) undergoes combustion in a bomb calorimeter, the temperature rises from 25.87 °C to 38.13 °C. Find ΔE_{rxn} for the reaction in kJ/mol hexane. The heat capacity of the bomb calorimeter, determined in a separate experiment, is 5.73 kJ/°C.

For More Practice 6.4

The combustion of toluene has a ΔE_{rxn} of -3.91×10^3 kJ/mol. When 1.55 g of toluene (C_7H_8) undergoes combustion in a bomb calorimeter, the temperature rises from 23.12 °C to 37.57 °C. Find the heat capacity of the bomb calorimeter.

6.5 Enthalpy: The Heat Evolved in a Chemical Reaction at Constant Pressure

We have just seen that when a chemical reaction occurs in a sealed container under conditions of constant volume, the energy evolved in the reaction appears only as heat. However, when a chemical reaction occurs open to the atmosphere under conditions of constant pressure—for example, a reaction occurring in an open beaker or the burning of natural gas in a furnace—the energy evolved may appear as both heat and work. As we have also seen, ΔE_{rxn} is a measure of the *total energy change* (both heat and work) that occurs during the reaction. However, in many cases, we are interested only in the heat exchanged, not the work done. For example, when we burn natural gas in the furnace to heat our homes, we do not really care how much work the combustion reaction does on the atmosphere by expanding against it—we just want to know how much heat is given off to warm the home. Under conditions of constant pressure, a thermodynamic quantity called *enthalpy* provides exactly this.

The **enthalpy (H)** of a system is defined as the sum of its internal energy and the product of its pressure and volume:

$$H = E + PV \qquad [6.12]$$

Since internal energy, pressure, and volume are all state functions, enthalpy is also a state function. The *change in enthalpy* (ΔH) for any process occurring under constant pressure is therefore given by the following expression:

$$\Delta H = \Delta E + P\Delta V \qquad [6.13]$$

To better understand this expression, we can interpret the two terms on the right with the help of relationships already familiar to us. We saw previously that $\Delta E = q + w$. Thus, if we represent the heat at constant pressure as q_p, then the change in internal energy at constant pressure is simply $\Delta E = q_p + w$. In addition, from our definition of pressure–volume work, we know that $P\Delta V = -w$. Substituting these expressions into the expression for ΔH gives us

$$\Delta H = \Delta E + P\Delta V$$
$$= (q_p + w) + P\Delta V$$
$$= q_p + w - w$$
$$\Delta H = q_p \qquad [6.14]$$

So we can see that ΔH is equal to q_p, the heat at constant pressure.

Conceptually (and often numerically), ΔH and ΔE are similar: they both represent changes in a state function for the system. However, ΔE is a measure of *all of the energy* (heat and work) exchanged with the surroundings, while ΔH is a measure of only the heat exchanged under conditions of constant pressure. For chemical reactions that do not exchange much work with the surroundings—that is, those that do not cause a large change in the reaction volume as they occur—ΔH and ΔE are nearly identical in value.

For chemical reactions that produce or consume large amounts of gas, and therefore produce large volume changes, ΔH and ΔE can be slightly different in value.

 Conceptual Connection 6.4 The Difference between ΔH and ΔE

Lighters are usually fueled by butane (C_4H_{10}). When 1 mole of butane burns at constant pressure, it produces 2658 kJ of heat and does 3 kJ of work. What are the values of ΔH and ΔE for the combustion of one mole of butane?

Answer: ΔH represents only the heat exchanged; therefore $\Delta H = -2658$ kJ. ΔE represents the heat *and work* exchanged; therefore $\Delta E = -2661$ kJ. The signs of both ΔH and ΔE are negative because heat and work are flowing out of the system and into the surroundings. Notice that the values of ΔH and ΔE are similar in magnitude, as is the case for many chemical reactions.

The signs of ΔH and ΔE also follow the same conventions. A positive ΔH means that heat flows into the system as the reaction occurs. A chemical reaction with a positive ΔH, called an **endothermic reaction**, absorbs heat from its surroundings. A chemical cold pack provides a good example of an endothermic reaction. When a barrier separating the reactants in a chemical cold pack is broken, the substances mix, react, and absorb heat from the surroundings. The surroundings—possibly including your bruised wrist—get *colder*. A chemical reaction with a negative ΔH, called an **exothermic reaction**, gives off heat to its surroundings. The burning of natural gas is a good example of an exothermic reaction. As the gas burns, it gives off heat, raising the temperature of its surroundings.

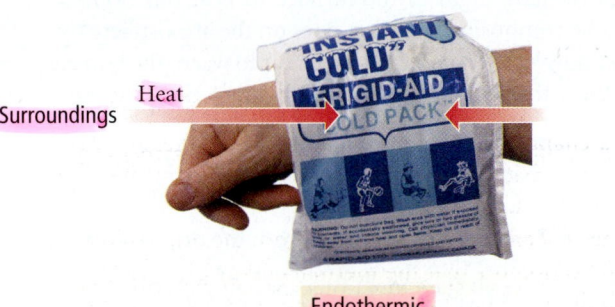
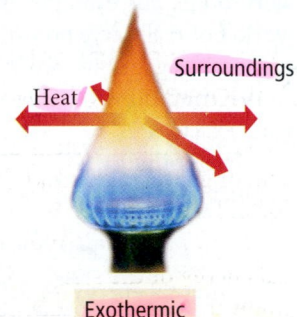

Endothermic

Exothermic

▲ The reaction that occurs in a chemical cold pack is endothermic—it absorbs energy from the surroundings. The combustion of natural gas is an exothermic reaction—it gives off energy to the surroundings.

Summarizing:

➤ The value of ΔH for a chemical reaction is the amount of heat absorbed or evolved in the reaction under conditions of constant pressure.

➤ An endothermic reaction has a *positive* ΔH and absorbs heat from the surroundings. An endothermic reaction feels cold to the touch.

➤ An exothermic reaction has a *negative* ΔH and gives off heat to the surroundings. An exothermic reaction feels warm to the touch.

 Conceptual Connection 6.5 Exothermic and Endothermic Reactions

If an endothermic reaction absorbs heat, then why does it feel cold to the touch?

Answer: An endothermic reaction feels cold to the touch because the reaction (acting here as the system) absorbs heat from the surroundings. When you touch the vessel in which the reaction occurs, you, being part of the surroundings, lose heat to the system (the reaction), which results in the feeling of cold. The heat absorbed by the reaction does not go to increase its temperature, but rather becomes potential energy stored in chemical bonds.

EXAMPLE 6.5 Exothermic and Endothermic Processes

Identify each of the following processes as endothermic or exothermic and indicate the sign of ΔH.

(a) sweat evaporating from skin
(b) water freezing in a freezer
(c) wood burning in a fire

Solution

(a) Sweat evaporating from skin cools the skin and is therefore endothermic, with a positive ΔH. The skin must supply heat to the water in order for it to continue to evaporate.
(b) Water freezing in a freezer releases heat and is therefore exothermic, with a negative ΔH. The refrigeration system in the freezer must remove this heat for the water to continue to freeze.
(c) Wood burning in a fire releases heat and is therefore exothermic, with a negative ΔH.

For Practice 6.5

Identify each of the following processes as endothermic or exothermic and indicate the sign of ΔH.

(a) an ice cube melting
(b) nail polish remover quickly evaporating after it is accidentally spilled on the skin
(c) a chemical hand warmer emitting heat after the mixing of substances within a small handheld package

Exothermic and Endothermic Processes: A Molecular View

When a chemical system undergoes a change in enthalpy, where does the energy come from or go to? For example, we just learned that an exothermic chemical reaction gives off *thermal energy*—what is the source of that energy?

First, we know that the emitted thermal energy *does not* come from the original thermal energy of the system. Recall from Section 6.1 that the thermal energy of a system is simply the composite kinetic energy of the atoms and molecules that compose the system. This kinetic energy *cannot* be the source of the energy given off in an exothermic reaction because, if the atoms and molecules that compose the system were to lose kinetic energy, their temperature would necessarily fall—the system would get colder. Yet, we know that in exothermic reactions, the temperature of the system and the surroundings rises. So there must be some other source of energy.

Recall also from Section 6.1 that the internal energy of a chemical system is the sum of its kinetic energy and its *potential energy*. It is this potential energy that is the energy source in an exothermic chemical reaction. Under normal circumstances, chemical potential energy (or simply chemical energy) arises primarily from the electrostatic forces between the protons and electrons that compose the atoms and molecules within the system. In an exothermic reaction, some bonds break and new ones form, and the protons and electrons go from an arrangement of higher potential energy to one of lower potential energy. As the molecules rearrange, their potential energy is converted into kinetic energy, the heat emitted in the reaction. In an endothermic reaction, the opposite happens: as some bonds break and others form, the protons and electrons go from an arrangement of lower potential energy to one of higher potential energy, absorbing thermal energy in the process.

Stoichiometry Involving ΔH: Thermochemical Equations

The enthalpy change for a chemical reaction, abbreviated ΔH_{rxn}, is also called the **enthalpy of reaction** or **heat of reaction** and is an extensive property, one that depends on the amount of material. In other words, the amount of heat generated or absorbed when a chemical reaction occurs depends on the *amounts* of reactants that actually react. We usually specify ΔH_{rxn} in combination with the balanced chemical equation for the reaction. The magnitude of ΔH_{rxn} is for the stoichiometric amounts of reactants and products for the reaction *as written*. For example, the balanced equation and ΔH_{rxn} for the combustion of propane (the main component of LP gas) are as follows:

$$C_3H_8(g) + 5\,O_2(g) \longrightarrow 3\,CO_2(g) + 4\,H_2O(g) \qquad \Delta H_{rxn} = -2044\ kJ$$

This means that when 1 mol of C_3H_8 reacts with 5 mol of O_2 to form 3 moles of CO_2 and 4 mol of H_2O, 2044 kJ of heat is emitted. We can write these relationships in the same way that we expressed stoichiometric relationships in Chapter 4, as ratios between two amounts. For example, for the reactants, we write:

$$1\ mol\ C_3H_8: -2044\ kJ \qquad or \qquad 5\ mol\ O_2: -2044\ kJ$$

The ratios mean that 2044 kJ of heat is evolved when 1 mol of C_3H_8 and 5 mol of O_2 completely react. These ratios can then be used to construct conversion factors between amounts of reactants or products and the quantity of heat emitted (for exothermic reactions) or absorbed (for endothermic reactions). To find out how much heat is emitted upon the combustion of a certain mass in grams of C_3H_8, we would use the following conceptual plan:

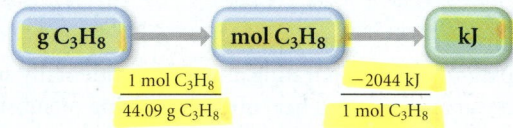

The molar mass is used to convert between grams and moles, and the stoichiometric relationship between moles of C_3H_8 and the heat of reaction is used to convert between moles and kilojoules, as shown in the following example.

EXAMPLE 6.6   Stoichiometry Involving ΔH

An LP gas tank in a home barbeque contains 13.2 kg of propane, C_3H_8. Calculate the heat (in kJ) associated with the complete combustion of all of the propane in the tank.

$$C_3H_8(g) + 5\,O_2(g) \longrightarrow 3\,CO_2(g) + 4\,H_2O(g) \qquad \Delta H_{rxn} = -2044\ kJ$$

Sort You are given the mass of propane and asked to find the heat evolved in its combustion.	**Given** 13.2 kg C_3H_8 **Find** q
Strategize Starting with kg C_3H_8, convert to g C_3H_8 and then use the molar mass of C_3H_8 to find the number of moles. Next, use the stoichiometric relationship between mol C_3H_8 and kJ to find the heat evolved.	**Conceptual Plan** 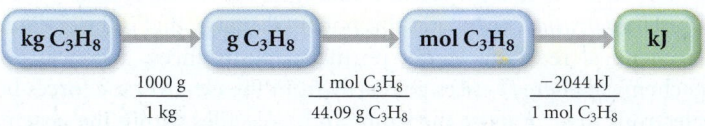 **Relationships Used** 1000 g = 1 kg molar mass C_3H_8 = 44.09 g/mol 1 mol C_3H_8: −2044 kJ (from balanced equation)
Solve Follow the conceptual plan to solve the problem. Begin with 13.2 kg C_3H_8 and multiply by the appropriate conversion factors to arrive at kJ.	**Solution** $13.2\ \cancel{kg}\ \cancel{C_3H_8} \times \dfrac{1000\ \cancel{g}}{1\ \cancel{kg}} \times \dfrac{1\ \cancel{mol\ C_3H_8}}{44.09\ \cancel{g}\ \cancel{C_3H_8}} \times \dfrac{-2044\ kJ}{1\ \cancel{mol\ C_3H_8}} = -6.12 \times 10^5\ kJ$

Check The units of the answer (kJ) are correct for energy. The answer is negative, as it should be for heat evolved by the reaction.

For Practice 6.6

Ammonia reacts with oxygen according to the following equation:

$$4\,NH_3(g) + 5\,O_2(g) \longrightarrow 4\,NO(g) + 6\,H_2O(g) \qquad \Delta H_{rxn} = -906\,kJ$$

Calculate the heat (in kJ) associated with the complete reaction of 155 g of NH_3.

For More Practice 6.6

What mass of butane in grams is necessary to produce 1.5×10^3 kJ of heat? What mass of CO_2 is produced?

$$C_4H_{10}(g) + 13/2\,O_2(g) \longrightarrow 4\,CO_2(g) + 5\,H_2O(g) \qquad \Delta H_{rxn} = -2658\,kJ$$

6.6 Constant-Pressure Calorimetry: Measuring ΔH_{rxn}

For many aqueous reactions, ΔH_{rxn} can be measured fairly simply using a **coffee-cup calorimeter** (Figure 6.7 ▼). The calorimeter consists of two Styrofoam coffee cups, one inserted into the other, to provide insulation from the laboratory environment. The calorimeter is equipped with a thermometer and a stirrer. The reaction is then carried out in a specifically measured quantity of solution within the calorimeter, so that the mass of the solution is known. During the reaction, the heat evolved (or absorbed) causes a temperature change in the solution, which is measured by the thermometer. If you know the specific heat capacity of the solution, normally assumed to be that of water, you can calculate q_{soln}, the heat absorbed by or lost from the solution (which is acting as the surroundings) using the following equation:

$$q_{soln} = m_{soln} \times C_{s,soln} \times \Delta T$$

The Coffee-Cup Calorimeter

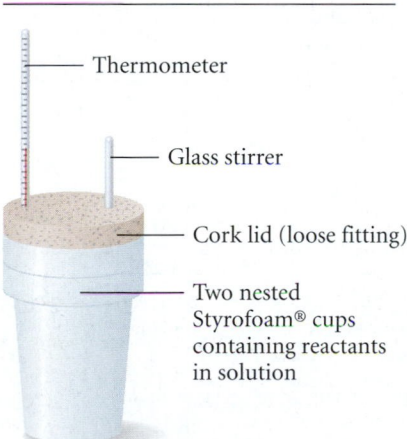

— Thermometer

— Glass stirrer

— Cork lid (loose fitting)

— Two nested Styrofoam® cups containing reactants in solution

◀ **FIGURE 6.7 The Coffee-Cup Calorimeter**   A coffee-cup calorimeter is used to measure enthalpy changes for chemical reactions in solution.

Since the insulated calorimeter prevents heat from escaping, you can assume that the heat gained by the solution equals that lost by the reaction (or vice versa):

$$q_{rxn} = -q_{soln}$$

Since the reaction happens under conditions of constant pressure (open to the atmosphere), $q_{rxn} = q_p = \Delta H_{rxn}$. This measured quantity is the heat of reaction for the specific amount (measured ahead of time) of reactants that reacted. To get ΔH_{rxn} per mole of a particular reactant—a more general quantity—you simply divide by the number of moles that actually reacted, as shown in the following example.

This equation assumes that no heat is lost to the calorimeter itself. If heat absorbed by the calorimeter is accounted for, the equation becomes $q_{rxn} = -(q_{soln} + q_{cal})$.

EXAMPLE 6.7   Measuring ΔH_{rxn} in a Coffee-Cup Calorimeter

Magnesium metal reacts with hydrochloric acid according to the following balanced equation:

$$Mg(s) + 2\,HCl(aq) \longrightarrow MgCl_2(aq) + H_2(g)$$

In an experiment to determine the enthalpy change for this reaction, 0.158 g of Mg metal is combined with enough HCl to make 100.0 mL of solution in a coffee-cup calorimeter. The HCl is sufficiently concentrated so that the Mg completely reacts. The temperature of the solution rises from 25.6 °C to 32.8 °C as a result of the reaction. Find ΔH_{rxn} for the reaction as written. Use 1.00 g/mL as the density of the solution and $C_{s,soln} = 4.18\ J/g \cdot °C$ as the specific heat capacity of the solution.

Sort You are given the mass of magnesium, the volume of solution, the initial and final temperatures, the density of the solution, and the heat capacity of the solution. You are asked to find the change in enthalpy for the reaction.

Given 0.158 g Mg
100.0 mL soln
$T_i = 25.6\ °C$
$T_f = 32.8\ °C$
$d = 1.00\ g/mL,$
$C_{s,soln} = 4.18\ J/g \cdot °C$

Find ΔH_{rxn}

Strategize The conceptual plan has three parts. In the first part, use the temperature change and the other given quantities, together with the equation $q = m \times C_s \times \Delta T$, to find q_{soln}.

In the second part, use q_{soln} to get q_{rxn} (which simply involves changing the sign). Because the pressure is constant, q_{rxn} is equivalent to ΔH_{rxn} for the amount of magnesium that reacted.

In the third part, divide q_{rxn} by the number of moles of magnesium to get ΔH_{rxn} per mole of magnesium.

Conceptual Plan

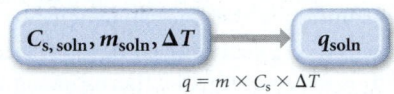

$$q = m \times C_s \times \Delta T$$

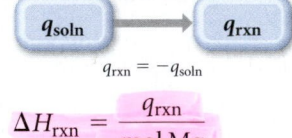

$$q_{rxn} = -q_{soln}$$

$$\Delta H_{rxn} = \frac{q_{rxn}}{mol\ Mg}$$

Relationships Used

$$q = m \times C_s \times \Delta T$$

$$q_{rxn} = -q_{soln}$$

Solve Gather the necessary quantities in the correct units for the equation $q = m \times C_s \times \Delta T$ and substitute these into the equation to compute q_{soln}. Notice that the sign of q_{soln} is *positive*, meaning that the solution *absorbed heat* from the reaction.

Solution

$C_{s,soln} = 4.18\ J/g \cdot °C$

$$m_{soln} = 100.0\ \text{mL soln} \times \frac{1.00\ g}{1\ \text{mL soln}} = 1.00 \times 10^2\ g$$

$$\Delta T = T_f - T_i = 32.8\ °C - 25.6\ °C = 7.2\ °C$$

$$q_{soln} = m_{soln} \times C_{s,soln} \times \Delta T$$

$$= 1.00 \times 10^2\ g \times 4.18\frac{J}{g \cdot °C} \times 7.2\ °C = 3.0 \times 10^3\ J$$

Find q_{rxn} by simply taking the negative of q_{soln}. Notice that q_{rxn} is negative, as expected for an *exothermic* reaction.

$$q_{rxn} = -q_{soln} = -3.0 \times 10^3\ J$$

Finally, find ΔH_{rxn} per mole of magnesium by dividing q_{rxn} by the number of moles of magnesium that reacted. Find the number of moles of magnesium from the given mass of magnesium and its molar mass.

$$\Delta H_{rxn} = \frac{q_{rxn}}{mol\ Mg}$$

$$= \frac{-3.0 \times 10^3\ J}{0.158\ \text{g Mg} \times \frac{1\ mol\ Mg}{24.31\ \text{g Mg}}}$$

Since the stoichiometric coefficient for magnesium in the balanced chemical equation is 1, the computed value represents ΔH_{rxn} for the reaction as written.

$$= -4.6 \times 10^5\ J/mol\ Mg$$

$$Mg(s) + 2\,HCl(aq) \longrightarrow MgCl_2(aq) + H_2(g) \quad \Delta H_{rxn} = -4.6 \times 10^5\ J$$

Check The units of the answer (J) are correct for the change in enthalpy of a reaction. The sign is negative, as expected for an exothermic reaction.

For Practice 6.7

The addition of hydrochloric acid to a silver nitrate solution precipitates silver chloride according to the following reaction:

$$AgNO_3(aq) + HCl(aq) \longrightarrow AgCl(s) + HNO_3(aq)$$

When 50.0 mL of 0.100 M $AgNO_3$ is combined with 50.0 mL of 0.100 M HCl in a coffee-cup calorimeter, the temperature changes from 23.40 °C to 24.21 °C. Calculate ΔH_{rxn} for the reaction as written. Use 1.00 g/mL as the density of the solution and $C = 4.18$ J/g·°C as the specific heat capacity.

 Conceptual Connection 6.6 **Constant-Pressure versus Constant-Volume Calorimetry**

The same reaction, with exactly the same amount of reactant, is conducted in a bomb calorimeter and in a coffee-cup calorimeter. In one of the measurements, $q_{rxn} = -12.5$ kJ and in the other $q_{rxn} = -11.8$ kJ. Which value was obtained in the bomb calorimeter? (Assume that the reaction has a positive ΔV in the coffee-cup calorimeter.)

Answer: The value of q_{rxn} with the greater magnitude (-12.5 kJ) must have come from the bomb calorimeter. Recall that $\Delta E_{rxn} = q_{rxn} + w_{rxn}$. In a bomb calorimeter, the energy change that occurs in the course of the reaction all takes the form of heat (q). In a coffee-cup calorimeter, the amount of energy released as heat may be smaller because some of the energy may be used to do work (w).

6.7 Relationships Involving ΔH_{rxn}

We now turn our attention to three quantitative relationships between a chemical equation and ΔH_{rxn}.

> 1. **If a chemical equation is multiplied by some factor, then ΔH_{rxn} is also multiplied by the same factor.**

We learned in Section 6.5 that ΔH_{rxn} is an extensive property; therefore, it depends on the quantity of reactants undergoing reaction. We also learned that ΔH_{rxn} is usually reported for a reaction involving stoichiometric amounts of reactants. For example, for a reaction A + 2 B $\longrightarrow$ C, ΔH_{rxn} is usually reported as the amount of heat emitted or absorbed when 1 mol A reacts with 2 mol B to form 1 mol C. Therefore, if a chemical equation is multiplied by a factor, then ΔH_{rxn} is also multiplied by the same factor. For example,

$$A + 2B \longrightarrow C \qquad \Delta H_1$$
$$2A + 4B \longrightarrow 2C \qquad \Delta H_2 = 2 \times \Delta H_1$$

> 2. **If a chemical equation is reversed, then ΔH_{rxn} changes sign.**

We learned in Section 6.5 that ΔH_{rxn} is a state function, which means that its value depends only on the initial and final states of the system.

$$\Delta H = H_{final} - H_{initial}$$

When a reaction is reversed, the final state becomes the initial state and vice versa. Consequently, ΔH_{rxn} changes sign, as exemplified by the following:

$$A + 2B \longrightarrow C \qquad \Delta H_1$$
$$C \longrightarrow A + 2B \qquad \Delta H_2 = -\Delta H_1$$

3. **If a chemical equation can be expressed as the sum of a series of steps, then ΔH_{rxn} for the overall equation is the sum of the heats of reactions for each step.**

This last relationship, known as **Hess's law** also follows from the enthalpy of reaction being a state function. Since ΔH_{rxn} is dependent only on the initial and final states, and not on the pathway the reaction follows, then ΔH obtained from summing the individual steps that lead to an overall reaction must be the same as ΔH for that overall reaction. For example,

$$A + 2B \longrightarrow C \qquad \Delta H_1$$
$$C \longrightarrow 2D \qquad \Delta H_2$$
$$\overline{A + 2B \longrightarrow 2D \qquad \Delta H_3 = \Delta H_1 + \Delta H_2}$$

We illustrate Hess's law with the energy level diagram shown in Figure 6.8 ▼.

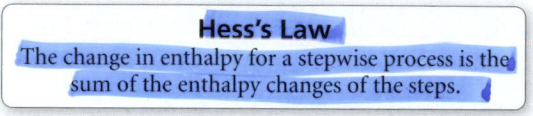

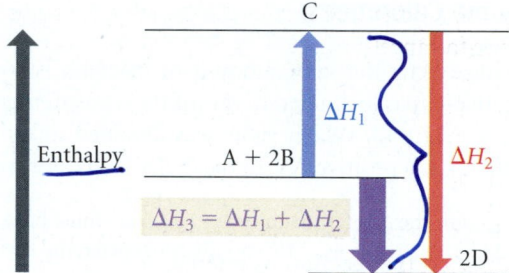

▶ **FIGURE 6.8 Hess's Law**
The change in enthalpy for a stepwise process is the sum of the enthalpy changes of the steps.

These three quantitative relationships make it possible to determine ΔH for a reaction without directly measuring it in the laboratory. (For some reactions, direct measurement can be difficult.) If you can find related reactions (with known ΔH's) that sum to the reaction of interest, you can find ΔH for the reaction of interest. For example, the following reaction between $C(s)$ and $H_2O(g)$ is an industrially important way to generate hydrogen gas:

$$C(s) + H_2O(g) \longrightarrow CO(g) + H_2(g) \qquad \Delta H_{rxn} = ?$$

We can find ΔH_{rxn} from the following reactions with known ΔH's:

$$C(s) + O_2(g) \longrightarrow CO_2(g) \qquad \Delta H = -393.5 \text{ kJ}$$
$$2CO(g) + O_2(g) \longrightarrow 2CO_2(g) \qquad \Delta H = -566.0 \text{ kJ}$$
$$2H_2(g) + O_2(g) \longrightarrow 2H_2O(g) \qquad \Delta H = -483.6 \text{ kJ}$$

We just have to determine how to sum these reactions to get the overall reaction of interest. We do this by manipulating the reactions with known ΔH's in such a way as to get the reactants of interest on the left, the products of interest on the right, and other species to cancel.

Since the first reaction has $C(s)$ as a reactant, and the reaction of interest also has $C(s)$ as a reactant, we simply write the first reaction unchanged.

$$C(s) + O_2(g) \longrightarrow CO_2(g) \qquad \Delta H = -393.5 \text{ kJ}$$

The second reaction has 2 mol of $CO(g)$ as a reactant. However, the reaction of interest has 1 mol of $CO(g)$ as a product. Therefore, we reverse the second reaction, change the sign of ΔH, and multiply the reaction and ΔH by $\frac{1}{2}$.

$$\frac{1}{2} \times [2CO_2(g) \longrightarrow 2CO(g) + O_2(g)] \qquad \Delta H = \frac{1}{2} \times (+566.0 \text{ kJ})$$

The third reaction has $H_2(g)$ as a reactant. In the reaction of interest, however, $H_2(g)$ is a product. Therefore, we reverse the equation and change the sign of ΔH. In addition, to obtain coefficients that match the reaction of interest, and to cancel O_2, we must multiply the reaction and ΔH by $\frac{1}{2}$.

$$\frac{1}{2} \times [2H_2O(g) \longrightarrow 2H_2(g) + O_2(g)] \qquad \Delta H = \frac{1}{2} \times (+483.6 \text{ kJ})$$

Lastly, we rewrite the three reactions after multiplying through by the indicated factors and show how they sum to the reaction of interest. ΔH for the reaction of interest is then just the sum of the ΔH's for the steps.

$$C(s) + O_2(g) \longrightarrow CO_2(g) \qquad \Delta H = -393.5 \text{ kJ}$$
$$CO_2(g) \longrightarrow CO(g) + \tfrac{1}{2}O_2(g) \qquad \Delta H = +283.0 \text{ kJ}$$
$$H_2O(g) \longrightarrow H_2(g) + \tfrac{1}{2}O_2(g) \qquad \Delta H = +241.8 \text{ kJ}$$

$$\overline{C(s) + H_2O(g) \longrightarrow CO(g) + H_2(g) \qquad \Delta H_{rxn} = +131.3 \text{ kJ}}$$

EXAMPLE 6.8 Hess's Law

Find ΔH_{rxn} for the following reaction:

$$3\,C(s) + 4\,H_2(g) \longrightarrow C_3H_8(g)$$

Use the following reactions with known ΔH's:

$$C_3H_8(g) + 5\,O_2(g) \longrightarrow 3\,CO_2(g) + 4\,H_2O(g) \qquad \Delta H = -2043 \text{ kJ}$$
$$C(s) + O_2(g) \longrightarrow CO_2(g) \qquad \Delta H = -393.5 \text{ kJ}$$
$$2\,H_2(g) + O_2(g) \longrightarrow 2\,H_2O(g) \qquad \Delta H = -483.6 \text{ kJ}$$

Solution

To work this and other Hess's law problems, manipulate the reactions with known ΔH's in such a way as to get the reactants of interest on the left, the products of interest on the right, and other species to cancel.

Since the first reaction has C_3H_8 as a reactant, and the reaction of interest has C_3H_8 as a product, reverse the first reaction and change the sign of ΔH.	$3\,CO_2(g) + 4\,H_2O(g) \longrightarrow C_3H_8(g) + 5\,O_2(g) \qquad \Delta H = +2043 \text{ kJ}$
The second reaction has C as a reactant and CO_2 as a product, just as required in the reaction of interest. However, the coefficient for C is 1, and in the reaction of interest, the coefficient for C is 3. Therefore, multiply this equation and its ΔH by 3.	$3 \times [C(s) + O_2(g) \longrightarrow CO_2(g)] \qquad \Delta H = 3 \times (-393.5 \text{ kJ})$
The third reaction has $H_2(g)$ as a reactant, as required. However, the coefficient for H_2 is 2, and in the reaction of interest, the coefficient for H_2 is 4. Therefore multiply this reaction and its ΔH by 2.	$2 \times [2\,H_2(g) + O_2(g) \longrightarrow 2\,H_2O(g)] \qquad \Delta H = 2 \times (-483.6 \text{ kJ})$
Lastly, rewrite the three reactions after multiplying through by the indicated factors and show how they sum to the reaction of interest. Then ΔH for the reaction of interest is just the sum of the ΔH's for the steps.	$3\,CO_2(g) + 4\,H_2O(g) \longrightarrow C_3H_8(g) + 5\,O_2(g) \qquad \Delta H = +2043 \text{ kJ}$ $3\,C(s) + 3\,O_2(g) \longrightarrow 3\,CO_2(g) \qquad \Delta H = -1181 \text{ kJ}$ $4\,H_2(g) + 2\,O_2(g) \longrightarrow 4\,H_2O(g) \qquad \Delta H = -967.2 \text{ kJ}$ $\overline{3\,C(s) + 4\,H_2(g) \longrightarrow C_3H_8(g) \qquad \Delta H_{rxn} = -105 \text{ kJ}}$

For Practice 6.8

Find ΔH_{rxn} for the following reaction:

$$N_2O(g) + NO_2(g) \longrightarrow 3\,NO(g)$$

Use the following reactions with known ΔH's:

$$2\,NO(g) + O_2(g) \longrightarrow 2\,NO_2(g) \qquad \Delta H = -113.1 \text{ kJ}$$
$$N_2(g) + O_2(g) \longrightarrow 2\,NO(g) \qquad \Delta H = +182.6 \text{ kJ}$$
$$2\,N_2O(g) \longrightarrow 2\,N_2(g) + O_2(g) \qquad \Delta H = -163.2 \text{ kJ}$$

For More Practice 6.8

Find ΔH_{rxn} for the following reaction:

$$3\,H_2(g) + O_3(g) \longrightarrow 3\,H_2O(g)$$

Use the following reactions with known ΔH's:

$$2\,H_2(g) + O_2(g) \longrightarrow 2\,H_2O(g) \qquad \Delta H = -483.6 \text{ kJ}$$
$$3\,O_2(g) \longrightarrow 2\,O_3(g) \qquad \Delta H = +285.4 \text{ kJ}$$

6.8 Enthalpies of Reaction from Standard Heats of Formation

We have seen two ways to determine ΔH for a chemical reaction: experimentally through calorimetry and inferentially through Hess's law. We now turn to a third and more convenient way to determine ΔH for a large number of chemical reactions: from tabulated *standard enthalpies of formation.*

Standard States and Standard Enthalpy Changes

Recall that ΔH is the *change* in enthalpy for a chemical reaction—the difference in enthalpy between the products and the reactants. Since we are interested in changes in enthalpy (and not in absolute values of enthalpy itself), we are free to define the *zero* of enthalpy as conveniently as possible. Returning to our mountain-climbing analogy, a change in altitude (like a change in enthalpy) is an absolute quantity. The altitude itself (like enthalpy itself), however, is a relative quantity, defined relative to some standard such as sea level. We must define a similar, albeit slightly more complex, standard for enthalpy. This standard has three parts: the **standard state**, the **standard enthalpy change** ($\Delta H°$), and the **standard enthalpy of formation** ($\Delta H_f°$).

1. **Standard State**

   - *For a Gas:* The standard state for a gas is the pure gas at a pressure of exactly 1 atmosphere.

   - *For a Liquid or Solid:* The standard state for a liquid or solid is the pure substance in its most stable form at a pressure of 1 atm and at the temperature of interest (often taken to be 25 °C).

   - *For a Substance in Solution:* The standard state for a substance in solution is a concentration of exactly 1 M.

2. **Standard Enthalpy Change ($\Delta H°$)**

   - The change in enthalpy for a process when all reactants and products are in their standard states. The degree sign indicates standard states.

3. **Standard Enthalpy of Formation ($\Delta H_f°$)**

   - *For a Pure Compound:* The change in enthalpy when 1 mole of the compound forms from its constituent elements in their standard states.

   - *For a Pure Element in its Standard State:* $\Delta H_f° = 0$.

> The standard state was changed in 1997 to a pressure of 1 bar, which is very close to 1 atm (1 atm = 1.013 bar). Both standards are now in common use.

Assigning the value of zero to the standard enthalpy of formation for an element in its standard state is the equivalent of assigning an altitude of zero to sea level—we can then measure all subsequent changes in altitude relative to sea level. Similarly, we can measure all changes in enthalpy relative to those of pure elements in their standard states. For example, consider the standard enthalpy of formation of methane gas at 25 °C:

$$C(s, \text{graphite}) + 2\,H_2(g) \longrightarrow CH_4(g) \qquad \Delta H_f° = -74.6\ \text{kJ/mol}$$

> The carbon in this equation must be graphite (the most stable form of carbon at 1 atm and 25 °C).

For methane, as with most compounds, $\Delta H_f°$ is negative. So if we think of pure elements in their standard states as *sea level*, then most compounds lie *below sea level*. The chemical equation for the enthalpy of formation of a compound is always written to form 1 mole of the compound, so $\Delta H_f°$ has the units of kJ/mol. Table 6.5 shows $\Delta H_f°$ values for some selected compounds. A more complete list can be found in Appendix IIB.

TABLE 6.5 Standard Enthalpies of Formation, ΔH_f°, at 298 K

Formula	ΔH_f° (kJ/mol)	Formula	ΔH_f° (kJ/mol)	Formula	ΔH_f° (kJ/mol)
Bromine		$C_3H_8O(l, \text{isopropanol})$	−318.1	*Oxygen*	
$Br(g)$	111.9	$C_6H_6(l)$	49.1	$O_2(g)$	0
$Br_2(l)$	0	$C_6H_{12}O_6(s, \text{glucose})$	−1273.3	$O_3(g)$	142.7
$HBr(g)$	−36.3	$C_{12}H_{22}O_{11}(s, \text{sucrose})$	−2226.1	$H_2O(g)$	−241.8
Calcium		*Chlorine*		$H_2O(l)$	−285.8
$Ca(s)$	0	$Cl(g)$	121.3	*Silver*	
$CaO(s)$	−634.9	$Cl_2(g)$	0	$Ag(s)$	0
$CaCO_3(s)$	−1207.6	$HCl(g)$	−92.3	$AgCl(s)$	−127.0
Carbon		*Fluorine*		*Sodium*	
$C(s, \text{graphite})$	0	$F(g)$	79.38	$Na(s)$	0
$C(s, \text{diamond})$	1.88	$F_2(g)$	0	$Na(g)$	107.5
$CO(g)$	−110.5	$HF(g)$	−273.3	$NaCl(s)$	−411.2
$CO_2(g)$	−393.5	*Hydrogen*		$Na_2CO_3(s)$	−1130.7
$CH_4(g)$	−74.6	$H(g)$	218.0	$NaHCO_3(s)$	−950.8
$CH_3OH(l)$	−238.6	$H_2(g)$	0	*Sulfur*	
$C_2H_2(g)$	227.4	*Nitrogen*		$S_8(s, \text{rhombic})$	0
$C_2H_4(g)$	52.4	$N_2(g)$	0	$S_8(s, \text{monoclinic})$	0.3
$C_2H_6(g)$	−84.68	$NH_3(g)$	−45.9	$SO_2(g)$	−296.8
$C_2H_5OH(l)$	−277.6	$NH_4NO_3(s)$	−365.6	$SO_3(g)$	−395.7
$C_3H_8(g)$	−103.85	$NO(g)$	91.3	$H_2SO_4(l)$	−814.0
$C_3H_6O(l, \text{acetone})$	−248.4	$N_2O(g)$	81.6		

EXAMPLE 6.9   Standard Enthalpies of Formation

Write an equation for the formation of (a) $MgCO_3(s)$ and (b) $C_6H_{12}O_6(s)$ from their elements in their standard states. Include the value of ΔH_f° for each equation.

Solution

(a) $MgCO_3(s)$

Write the equation with the elements in $MgCO_3$ in their standard states as the reactants and 1 mol of $MgCO_3$ as the product.

$$Mg(s) + C(s, \text{graphite}) + O_2(g) \longrightarrow MgCO_3(s)$$

Balance the equation and look up ΔH_f° in Appendix IIB. (Use fractional coefficients so that the product of the reaction is 1 mol of $MgCO_3$.)

$$Mg(s) + C(s, \text{graphite}) + \frac{3}{2}O_2(g) \longrightarrow MgCO_3(s)$$
$$\Delta H_f^{\circ} = -1095.8 \text{ kJ/mol}$$

(b) $C_6H_{12}O_6(s)$

Write the equation with the elements in $C_6H_{12}O_6$ in their standard states as the reactants and 1 mol of $C_6H_{12}O_6$ as the product.

$$C(s, \text{graphite}) + H_2(g) + O_2(g) \longrightarrow C_6H_{12}O_6(s)$$

Balance the equation and look up ΔH_f° in Appendix IIB.

$$6\,C(s, \text{graphite}) + 6\,H_2(g) + 3\,O_2(g) \longrightarrow C_6H_{12}O_6(s)$$
$$\Delta H_f^{\circ} = -1273.3 \text{ kJ/mol}$$

For Practice 6.9

Write an equation for the formation of (a) $NaCl(s)$ and (b) $Pb(NO_3)_2(s)$ from their elements in their standard states. Include the value of ΔH_f° for each equation.

Calculating the Standard Enthalpy Change for a Reaction

We have just seen that the standard heat of formation corresponds to the *formation* of a compound from its constituent elements in their standard states:

$$\text{elements} \longrightarrow \text{compound} \qquad \Delta H^\circ_f$$

Therefore, the *negative* of the standard heat of formation corresponds to the *decomposition* of a compound into its constituent elements in their standard states.

$$\text{compound} \longrightarrow \text{elements} \qquad -\Delta H^\circ_f$$

We can use these two concepts—the decomposing of a compound into its elements and the forming of a compound from its elements—to calculate the enthalpy change of any reaction by mentally taking the reactants through two steps. In the first step we *decompose the reactants* into their constituent elements in their standard states; in the second step we *form the products* from the constituent elements in their standard states.

$$\begin{aligned}\text{reactants} &\longrightarrow \text{elements} & \Delta H_1 &= -\Sigma\,\Delta H^\circ_f\,(\text{reactants})\\ \text{elements} &\longrightarrow \text{products} & \Delta H_2 &= +\Sigma\,\Delta H^\circ_f\,(\text{products})\\ \hline \text{reactants} &\longrightarrow \text{products} & \Delta H^\circ_{rxn} &= \Delta H_1 + \Delta H_2\end{aligned}$$

In these equations, Σ means "the sum of" so that ΔH_1 is the sum of the negatives of the heats of formation of the reactants and ΔH_2 is the sum of the heats of formation of the products.

We can demonstrate this by calculating the standard enthalpy change for a reaction, also called the standard heat of reaction (ΔH°_{rxn}), for the combustion of methane:

$$CH_4(g) + 2\,O_2(g) \longrightarrow CO_2(g) + 2\,H_2O(g) \qquad \Delta H^\circ_{rxn} = ?$$

The energy changes associated with the decomposition of the reactants and the formation of the products are shown in Figure 6.9 ▼. The first step (1) is the decomposition of 1 mol of methane into its constituent elements in their standard states. The change in enthalpy for this step is obtained by reversing the enthalpy of formation equation for methane and changing the sign of ΔH°_f:

$$(1)\; CH_4(g) \longrightarrow C(s, \text{graphite}) + 2\,H_2(g) \qquad -\Delta H^\circ_f = +74.6\,\text{kJ/mol}$$

The second step, the formation of the products from their constituent elements, has two parts: (a) the formation of 1 mol CO_2 and (b) and the formation of 2 mol H_2O. Since part (b) forms 2 mol H_2O, the ΔH°_f for that step must be multiplied by 2.

$$(2a)\; C(s, \text{graphite}) + O_2(g) \longrightarrow CO_2(g) \qquad \Delta H^\circ_f = -393.5\,\text{kJ/mol}$$
$$(2b)\; 2 \times [H_2(g) + \tfrac{1}{2}\,O_2(g) \longrightarrow H_2O(g)] \qquad 2 \times \Delta H^\circ_f = 2 \times (-241.8\,\text{kJ/mol})$$

As we know from Hess's law, the enthalpy of reaction for the overall reaction is the sum of the enthalpies of reaction of the individual steps:

$$\begin{aligned}(1) &\quad CH_4(g) \longrightarrow \cancel{C(s, \text{graphite})} + \cancel{2\,H_2(g)} & -\Delta H^\circ_f &= +74.6\,\text{kJ/mol}\\ (2a) &\quad \cancel{C(s, \text{graphite})} + O_2(g) \longrightarrow CO_2(g) & \Delta H^\circ_f &= -393.5\,\text{kJ/mol}\\ (2b) &\quad \cancel{2\,H_2(g)} + O_2(g) \longrightarrow 2\,H_2O(g) & 2 \times \Delta H^\circ_f &= -483.6\,\text{kJ/mol}\\ \hline &\quad CH_4(g) + 2\,O_2(g) \longrightarrow CO_2(g) + 2\,H_2O(g) & \Delta H^\circ_{rxn} &= -802.5\,\text{kJ}\end{aligned}$$

Calculating the Enthalpy Change for the Combustion of Methane

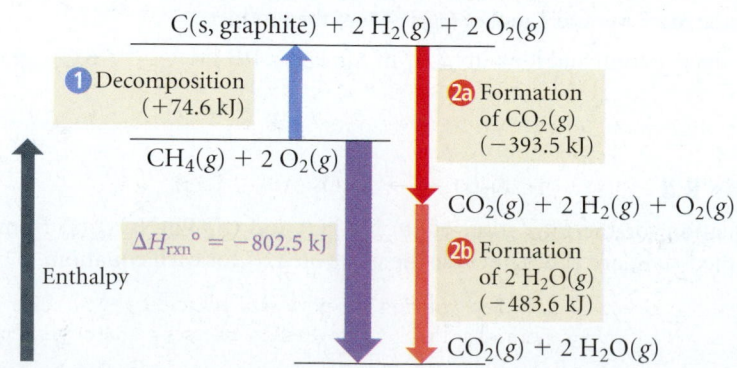

▶ FIGURE 6.9 Calculating the Enthalpy Change for the Combustion of Methane

We can streamline and generalize this process as follows:

To calculate ΔH°_{rxn}, subtract the heats of formations of the reactants multiplied by their stoichiometric coefficients from the heats of formation of the products multiplied by their stoichiometric coefficients.

In the form of an equation,

$$\Delta H^\circ_{rxn} = \Sigma\, n_p\, \Delta H^\circ_f\,(\text{products}) - \Sigma\, n_r\, \Delta H^\circ_f\,(\text{reactants}) \qquad [6.15]$$

In this equation, n_p represents the stoichiometric coefficients of the products, n_r represents the stoichiometric coefficients of the reactants and ΔH°_f represents the standard enthalpies of formation. Keep in mind when using this equation that elements in their standard states have $\Delta H^\circ_f = 0$. The following examples demonstrate this process.

EXAMPLE 6.10 ΔH°_{rxn} and Standard Enthalpies of Formation

Use the standard enthalpies of formation to determine ΔH°_{rxn} for the following reaction:

$$4\,NH_3(g) + 5\,O_2(g) \longrightarrow 4\,NO(g) + 6\,H_2O(g)$$

Sort You are given the balanced equation and asked to find the enthalpy of reaction.

Given $4\,NH_3(g) + 5\,O_2(g) \longrightarrow 4\,NO(g) + 6\,H_2O(g)$

Find ΔH°_{rxn}

Strategize To calculate ΔH°_{rxn} from standard enthalpies of formation, subtract the heats of formations of the reactants multiplied by their stoichiometric coefficients from the heats of formation of the products multiplied by their stoichiometric coefficients.

Conceptual Plan

$\Delta H^\circ_{rxn} = \Sigma\, n_p\, \Delta H^\circ_f\,(\text{products}) - \Sigma\, n_r\, \Delta H^\circ_f\,(\text{reactants})$

Solve Begin by looking up (in Appendix IIB) the standard enthalpy of formation for each reactant and product. Remember that the standard enthalpy of formation of pure elements in their standard state is zero. Compute ΔH°_{rxn} by substituting into the equation.

Solution

Reactant or product	ΔH°_f (kJ/mol, from Appendix IIB)
$NH_3(g)$	−45.9
$O_2(g)$	0.0
$NO(g)$	+91.3
$H_2O(g)$	−241.8

$$
\begin{aligned}
\Delta H^\circ_{rxn} &= \Sigma\, n_p\, \Delta H^\circ_f\,(\text{products}) - \Sigma\, n_r\, \Delta H^\circ_f\,(\text{reactants}) \\
&= [4(\Delta H^\circ_{f,\,NO(g)}) + 6(\Delta H^\circ_{f,\,H_2O(g)})] - [4(\Delta H^\circ_{f,\,NH_3(g)}) + 5(\Delta H^\circ_{f,\,O_2(g)})] \\
&= [4(+91.3\text{ kJ}) + 6(-241.8\text{ kJ})] - [4(-45.9\text{ kJ}) + 5(0.0\text{ kJ})] \\
&= -1085.6\text{ kJ} - (-183.6\text{ kJ}) \\
&= -902.0\text{ kJ}
\end{aligned}
$$

Check The units of the answer (kJ) are correct. The answer is negative, which means that the reaction is exothermic.

For Practice 6.10

The thermite reaction, in which powdered aluminum reacts with iron oxide, is highly exothermic.

$$2\,Al(s) + Fe_2O_3(s) \longrightarrow Al_2O_3(s) + 2\,Fe(s)$$

Use standard enthalpies of formation to find ΔH°_{rxn} for the thermite reaction.

▶ The reaction of powdered aluminum with iron oxide, known as the thermite reaction, releases a large amount of heat.

EXAMPLE 6.11 $\Delta H^{\circ}_{\text{rxn}}$ and Standard Enthalpies of Formation

A city of 100,000 people uses approximately 1.0×10^{11} kJ of energy per day. Suppose all of that energy comes from the combustion of liquid octane (C_8H_{18}) to form gaseous water and gaseous carbon dioxide. Use standard enthalpies of formation to calculate $\Delta H^{\circ}_{\text{rxn}}$ for the combustion of octane and then determine how many kilograms of octane would be necessary to provide this amount of energy.

Sort You are given the amount of energy used and asked to find the mass of octane required to produce the energy.	**Given** 1.0×10^{11} kJ **Find** kg C_8H_{18}

Strategize The conceptual plan has three parts. In the first part, write a balanced equation for the combustion of octane. In the second part, calculate $\Delta H^{\circ}_{\text{rxn}}$ from the $\Delta H^{\circ}_{\text{f}}$'s of the reactants and products. In the third part, convert from kilojoules of energy to moles of octane using the conversion factor found in step 2, and then convert from moles of octane to mass of octane using the molar mass.	**Conceptual Plan** (1) Write balanced equation. (2) 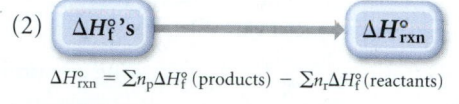 $\Delta H^{\circ}_{\text{rxn}} = \Sigma n_{\text{p}}\Delta H^{\circ}_{\text{f}} \text{(products)} - \Sigma n_{\text{r}}\Delta H^{\circ}_{\text{f}} \text{(reactants)}$ (3) 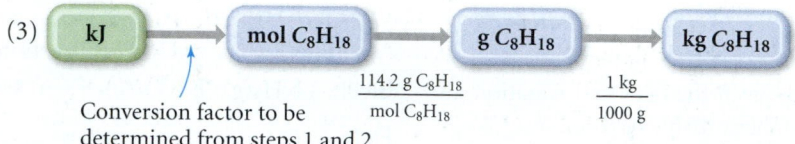 Conversion factor to be determined from steps 1 and 2 **Relationships Used** molar mass C_8H_{18} = 114.22 g/mol 1 kg = 1000 g

Solve Begin by writing the balanced equation for the combustion of octane. For convenience, do not clear the 25/2 fraction in order to keep the coefficient on octane as 1.	**Solution Step 1** $C_8H_{18}(l) + 25/2\, O_2(g) \longrightarrow 8\, CO_2(g) + 9\, H_2O(g)$

Look up (in Appendix IIB) the standard enthalpy of formation for each reactant and product and then compute $\Delta H^{\circ}_{\text{rxn}}$.	**Solution Step 2**

Reactant or product	$\Delta H^{\circ}_{\text{f}}$ (kJ/mol from Appendix IIB)
$C_8H_{18}(l)$	−250.1
$O_2(g)$	0.0
$CO_2(g)$	−393.5
$H_2O(g)$	−241.8

$$\Delta H^{\circ}_{\text{rxn}} = \Sigma\, n_{\text{p}}\, \Delta H^{\circ}_{\text{f}}\,(\text{products}) - \Sigma\, n_{\text{r}}\, \Delta H^{\circ}_{\text{f}}\,(\text{reactants})$$

$$= [8(\Delta H^{\circ}_{\text{f},CO_2(g)}) + 9(\Delta H^{\circ}_{\text{f},H_2O(g)})] - [1(\Delta H^{\circ}_{\text{f},C_8H_{18}(l)}) + \frac{25}{2}(\Delta H^{\circ}_{\text{f},O_2(g)})]$$

$$= [8(-393.5\ \text{kJ}) + 9(-241.8\ \text{kJ})] - [1(-250.1\ \text{kJ}) + \frac{25}{2}(0.0\ \text{kJ})]$$

$$= -5324.2\ \text{kJ} - (-250.1\ \text{kJ})$$

$$= -5074.1\ \text{kJ}$$

From steps 1 and 2 build a conversion factor between mol C_8H_{18} and kJ. Follow step 3 of the conceptual plan. Begin with -1.0×10^{11} kJ (since the city uses this much energy, the reaction must emit it, and therefore the sign is negative) and follow the steps to arrive at kg octane.	**Solution Step 3** 1 mol C_8H_{18} : −5074.1 kJ $-1.0 \times 10^{11}\ \text{kJ} \times \dfrac{1\ \text{mol}\ C_8H_{18}}{-5074.1\ \text{kJ}} \times \dfrac{114.22\ \text{g}\ C_8H_{18}}{1\ \text{mol}\ C_8H_{18}} \times \dfrac{1\ \text{kg}}{1000\ \text{g}} = 2.3 \times 10^6\ \text{kg}\ C_8H_{18}$

Check The units of the answer (kg C_8H_{18}) are correct. The answer is positive, as it should be for mass. The magnitude is fairly large, but it is expected to be so because this amount of octane is supposed to provide the energy for an entire city.

For Practice 6.11

Dry chemical hand warmers are small flexible pouches that produce heat when they are removed from their airtight plastic wrappers. They can be slipped into a mitten or glove and will keep your hands warm for up to 10 hours. They utilize the oxidation of iron to form iron oxide according to the following reaction: $4 Fe(s) + 3 O_2(g) \longrightarrow 2 Fe_2O_3(s)$. Calculate $\Delta H°_{rxn}$ for this reaction and compute how much heat is produced from a hand warmer containing 15.0 g of iron powder.

6.9 Energy Use and the Environment

We opened this chapter by considering how the combustion of natural gas heats a home. We now know how to calculate not only the heat of reaction for natural gas combustion, but also how much heat is released by the combustion of any given amount of natural gas, and even the rise in temperature of the air within a house for a given amount of natural gas burned (see Exercise 105). We have not, however, talked about the environmental impacts of burning natural gas or other fossil fuels. We now turn our attention to this topic.

Energy Consumption

According to the U.S. Department of Energy, the United States currently consumes close to 100 quads (1 quad = 1 quadrillion British thermal units = 1.06×10^{18} J) of energy per year in the categories shown in this chart.

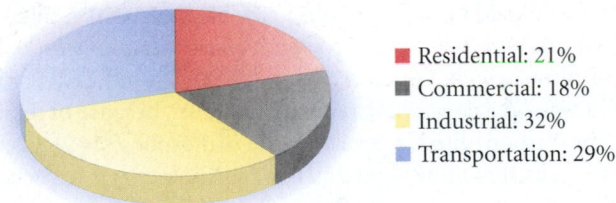

- ■ Residential: 21%
- ■ Commercial: 18%
- ■ Industrial: 32%
- ■ Transportation: 29%

Source: U.S. Energy Information Administration, Monthly Energy Review, February 2006.

This corresponds to over 100,000 kWh of energy use per person per year. If you used physical laborers to do the equivalent amount of work, you would need about 120 people. In other words, the average American enjoys the work output of 120 people, day and night, all year long! For this reason, Americans enjoy one of the highest standards of living in the world. However, as we learned earlier in the chapter, when it comes to energy, there is no free lunch. The consumption of energy carries significant environmental consequences.

Most U.S. energy comes from the combustion of fossil fuels, as shown here.

Source: U.S. Energy Information Administration, Annual Energy Review, 2004.

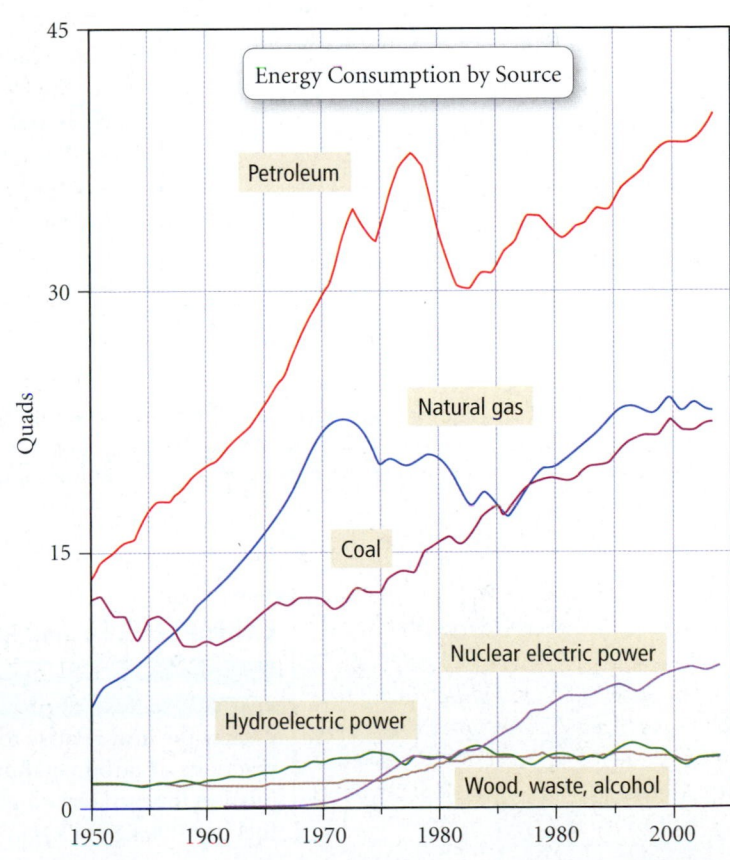

Energy Consumption by Source

Fossil fuels include petroleum, natural gas, and coal, all of which are convenient fuels because they are relatively abundant in Earth's crust (and therefore relatively inexpensive), they are easily transportable, and their combustion is highly exothermic. The reactions for the combustion of the main or representative components of several fossil fuels, and the associated enthalpies of reaction, are as follows:

Coal:	$C(s) + O_2(g) \longrightarrow CO_2(g)$	$\Delta H^\circ_{rxn} = -393.5 \text{ kJ}$
Natural gas:	$CH_4(g) + 2\,O_2(g) \longrightarrow CO_2(g) + 2\,H_2O(g)$	$\Delta H^\circ_{rxn} = -802.3 \text{ kJ}$
Petroleum:	$C_8H_{18}(l) + 25/2\,O_2(g) \longrightarrow 8\,CO_2(g) + 9\,H_2O(g)$	$\Delta H^\circ_{rxn} = -5074.1 \text{ kJ}$

Environmental Problems Associated with Fossil Fuel Use

One of the main problems associated with the burning of fossil fuels is that, even though they are abundant in Earth's crust, they are also finite. Fossil fuels originate from ancient plant and animal life and are a nonrenewable energy source—once they are all burned, they cannot be replenished. At current rates of consumption, oil and natural gas supplies will be depleted in 50 to 100 years. While there is enough coal to last much longer, it is a dirtier fuel and, because it is a solid, is less convenient (more difficult to transport and use) than petroleum and natural gas.

The other major problems associated with fossil fuel use are related to the products of combustion. The chemical equations for fossil fuel combustion all show that the products are carbon dioxide and water. However, these equations represent the reactions under ideal conditions and do not account for impurities in the fuel, side reactions, and incomplete combustion. When these are taken into account, we can identify three major environmental problems associated with the emissions of fossil fuel combustion: air pollution, acid rain, and global warming. We have already discussed air pollution in the previous chapter (see Section 5.11), and acid rain in Chapter 3 (see *Chemistry in the Environment: Acid Rain* in Section 3.6). Here we will address global warming, which we first touched on in Section 4.1.

One of the main products of fossil fuel combustion is carbon dioxide (CO_2). Carbon dioxide is a greenhouse gas, meaning that it allows visible light from the sun to enter Earth's atmosphere but prevents heat (in the form of infrared light) from escaping. The result is that carbon dioxide acts as a blanket, keeping Earth warm. However, because of fossil fuel combustion, carbon dioxide levels in the atmosphere have been steadily increasing, as shown in Figure 6.10 ◀.This increase is expected to raise Earth's average temperature. Current observations suggest that Earth has already warmed by about 0.6 °C in the last century, due to an increase of about 25% in atmospheric carbon dioxide. Computer models suggest that the warming could worsen if carbon dioxide emissions are not curbed. The possible effects of this warming include heightened storm severity, increasing numbers of floods and droughts, major shifts in agricultural zones, rising sea levels and coastal flooding, and profound changes in habitats that could result in the extinction of some plant and animal species.

Atmospheric Carbon Dioxide

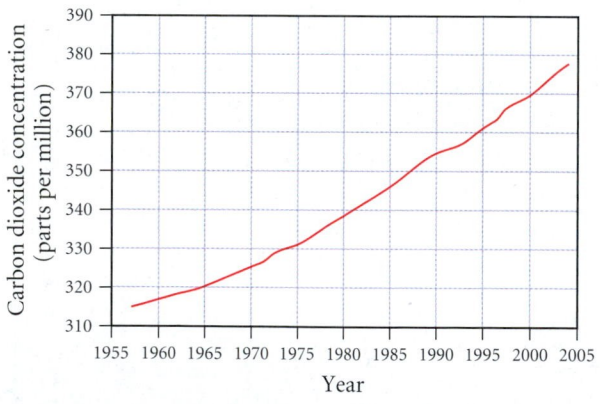

▲ **FIGURE 6.10 The Rise in Atmospheric Carbon Dioxide** Atmospheric carbon dioxide levels have been steadily increasing as a result of fossil fuel combustion.

EXAMPLE 6.12 Fossil Fuels and Global Warming

One way to evaluate fuels with respect to global warming is to determine how much heat they release during combustion relative to how much CO_2 they produce. The greater the heat relative to the amount of CO_2, the better the fuel. Use the combustion reactions of carbon, natural gas, and octane, in combination with the enthalpy of combustion for each reaction (all given earlier), to calculate the heat (in kJ) released by each fuel per 1.00 kg of CO_2 produced.

Sort You are given the mass of CO_2 emitted and asked to find the energy output for three different fuels.

Given 1.00 kg CO_2
Find kJ

Strategize You must first write the thermochemical equations for the combustion of each fuel given earlier in the chapter.

$$C(s) + O_2(g) \longrightarrow CO_2(g) \qquad \Delta H^\circ_{rxn} = -393.5 \text{ kJ}$$
$$CH_4(g) + 2\,O_2(g) \longrightarrow CO_2(g) + 2\,H_2O(g) \qquad \Delta H^\circ_{rxn} = -802.3 \text{ kJ}$$
$$C_8H_{18}(l) + 25/2\,O_2(g) \longrightarrow 8\,CO_2(g) + 9\,H_2O(g) \qquad \Delta H^\circ_{rxn} = -5074.1 \text{ kJ}$$

The conceptual plan has two parts. In the first part, use the molar mass of CO_2 to convert from mass of CO_2 to moles of CO_2. This part is the same for each fuel.

Conceptual Plan

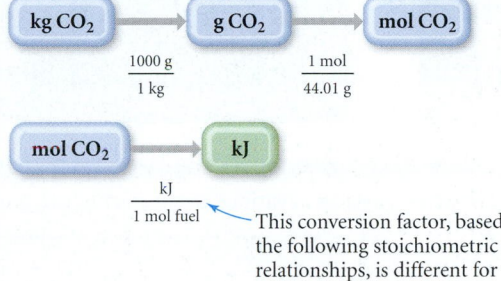

In the second part, use the stoichiometric relationship between moles of CO_2 produced and kilojoules of energy released to calculate the energy output. Repeat the second part for each fuel using the appropriate stoichiometric relationship from the balanced equations.

This conversion factor, based on the following stoichiometric relationships, is different for each fuel.

Stoichiometric Relationships

For C: 1 mol CO_2 : -393.5 kJ
For CH_4:   1 mol CO_2 : -802.3 kJ
For C_8H_{18}: 8 mol CO_2 : -5074.1 kJ

Other Relationships Used

1 kg = 1000 g
molar mass CO_2 = 44.01 g/mol

Solve Begin by converting kg CO_2 to mol CO_2.

Solution

$$1.00 \text{ kg } CO_2 \times \frac{1000 \text{ g}}{1 \text{ kg}} \times \frac{1 \text{ mol } CO_2}{44.01 \text{ g } CO_2} = 22.72 \text{ mol } CO_2$$

Then, for each fuel, convert mol CO_2 to kJ.

For C:

$$22.72 \text{ mol } CO_2 \times \frac{-393.5 \text{ kJ}}{1 \text{ mol } CO_2} = -8.94 \times 10^3 \text{ kJ}$$

For CH_4:

$$22.72 \text{ mol } CO_2 \times \frac{-802.3 \text{ kJ}}{1 \text{ mol } CO_2} = -1.82 \times 10^4 \text{ kJ}$$

For C_8H_{18}:

$$22.72 \text{ mol } CO_2 \times \frac{-5074.1 \text{ kJ}}{8 \text{ mol } CO_2} = -1.44 \times 10^4 \text{ kJ}$$

As you can see from the heat released in the production of 1 kg CO_2, CH_4 provides the most energy per kg CO_2; therefore, it is the best fuel with respect to global warming.

Check Each answer is in kJ, as it should be for heat produced. Each answer is negative, as expected for exothermic combustion reactions.

For Practice 6.12

What mass of CO_2 (in kg) is released into the atmosphere upon combustion of a 15-gallon tank of gasoline? Assume the gasoline is pure octane (C_8H_{18}) and that it has a density of 0.70 g/mL.

Chemistry and the Environment
Renewable Energy

Parabolic troughs

Solar power tower

Dish/engine

▲ The sun's energy, concentrated by reflective surfaces in various arrangements, can produce enough heat to generate electricity.

Because of their limited supply and environmental impacts, fossil fuels will not be our major source of energy in the future. What will replace them? Although the answer is not clear, several alternative energy technologies are beginning to emerge. Unlike fossil fuels, these technologies are renewable, which means that they can be used indefinitely.

Our planet's greatest source of renewable energy is the sun. If we could capture and harness just a small fraction of the total sunlight falling on Earth, our energy needs would be met several times over. The main problem with solar energy, however, is its diffuseness—the sun's energy falls over an enormous area. How do we concentrate and store it? In California, some of the state's electricity is generated by parabolic troughs, solar power towers, and dish/engines (shown above).

These devices use reflective surfaces to focus the sun's energy to produce enough heat to generate electricity. Although the direct cost of generating electricity this way is higher than using fossil fuels, the benefits to the environment are obvious. In addition, with time, the costs are expected to fall.

Another way to capture the sun's energy is in chemical bonds. For example, solar energy could be used to drive the decomposition of water:

$$H_2O(l) \longrightarrow H_2(g) + \tfrac{1}{2}O_2(g) \qquad \Delta H^\circ_{rxn} = +285.8 \text{ kJ}$$

The hydrogen gas could be stored until needed to provide energy by reforming water in the reverse reaction:

$$H_2(g) + \tfrac{1}{2}O_2(g) \longrightarrow H_2O(l) \qquad \Delta H^\circ_{rxn} = -285.8 \text{ kJ}$$

This reaction can be carried out in an electrochemical device called a fuel cell. In a fuel cell, hydrogen and oxygen gas combine to form water and produce electricity. In 2003, General Motors demonstrated a fuel cell–powered car called the Hydro-Gen3 to members of the U.S. Congress. The vehicle seats five passengers, has a top speed of 100 miles per hour, and has a

▲ A General Motors SUV prototype called the Sequel is powered by a fuel cell that runs on hydrogen gas and produces only water as exhaust.

range of 250 miles on one tank of fuel. In 2005, General Motors demonstrated the Sequel: a fuel cell SUV with a 300-mile range and quick acceleration (0–60 mph in 10 seconds). According to the company, the Sequel is quicker, easier to handle, easier to build, and safer than gasoline-powered vehicles, and its only emission is water vapor. Other automakers have similar prototype models in development.

Other renewable energy sources include wind power and hydroelectric power. Hydroelectric power plants—which generate approximately 8% of U.S. electricity—harness the gravitational potential energy of water held behind a dam. Water is released at a controlled rate. As it falls, it acquires kinetic energy that is used to spin a turbine, generating electricity. Wind power plants—which produce about 1% of California's electricity—consist of hundreds of turbines that are spun by the wind to generate electricity. Both of these technologies are cost competitive with fossil fuels, have no emissions, and are completely renewable.

Our energy future will probably see a combination of these technologies and some new ones, combined with greater efficiency and conservation. One thing, however, is clear—the future of fossil fuels is limited.

◀ Wind turbines such as these generate about 1% of California's electricity.

Chapter in Review

Key Terms

Section 6.1
thermochemistry (237)
energy (238)
work (238)
heat (238)
kinetic energy (238)
thermal energy (238)
potential energy (238)
chemical energy (238)
law of conservation of energy (238)
system (239)
surroundings (239)
joule (J) (239)

calorie (cal) (239)
Calorie (Cal) (240)
kilowatt-hour (kWh) (240)

Section 6.2
thermodynamics (240)
first law of thermodynamics (240)
internal energy (E) (241)
state function (241)

Section 6.3
thermal equilibrium (246)
heat capacity (C) (247)

specific heat capacity (C_s) (247)
molar heat capacity (247)
pressure–volume work (249)

Section 6.4
calorimetry (251)
bomb calorimeter (251)

Section 6.5
enthalpy (H) (253)
endothermic reaction (254)
exothermic reaction (254)
enthalpy (heat) of reaction (ΔH_{rxn}) (255)

Section 6.6
coffee-cup calorimeter (257)

Section 6.7
Hess's law (260)

Section 6.8
standard state (262)
standard enthalpy change ($\Delta H°$) (262)
standard enthalpy of formation (ΔH_f°) (262)

Key Concepts

The Nature of Energy and Thermodynamics (6.1, 6.2)

Energy, which has the SI unit of joules (J), is the capacity to do work. Work is the result of a force acting through a distance. Many different kinds of energy exist, including kinetic energy, thermal energy, potential energy, and chemical energy, a type of potential energy associated with the relative positions of electrons and nuclei in atoms and molecules. According to the first law of thermodynamics, energy can be converted from one form to another, but the total amount of energy is always conserved.

The internal energy (E) of a system is the sum of all of its kinetic and potential energy. Internal energy is a state function, which means that it depends only on the state of the system and not on the pathway by which it got to that state. A chemical system exchanges energy with its surroundings through heat (the transfer of thermal energy caused by a temperature difference) or work. The total change in internal energy is simply the sum of these two quantities.

Heat and Work (6.3)

Heat can be quantified using the equation $q = m \times C_s \times \Delta T$. In this expression, C_s is the specific heat capacity, the amount of heat required to change the temperature of 1 g of the substance by 1 °C. Compared to most substances, water has a very high heat capacity, meaning that it takes a lot of heat to change its temperature.

The type of work most characteristic of chemical reactions is pressure–volume work, which occurs when a gas expands against an external pressure. Pressure–volume work can be quantified with the equation $w = -P\Delta V$. The change in internal energy (ΔE) that occurs during a chemical reaction is the sum of the heat (q) exchanged and the work (w) done: $\Delta E = q + w$.

Enthalpy (6.5)

The heat evolved in a chemical reaction occurring at constant pressure is called the change in enthalpy (ΔH) for the reaction. Like internal energy, enthalpy is a state function. An endothermic reaction has a positive enthalpy of reaction, whereas an exothermic reaction has a negative enthalpy of reaction. The enthalpy of reaction can be used to determine stoichiometrically the heat evolved when a specific amount of reactant reacts.

Calorimetry (6.4, 6.6)

Calorimetry is a method of measuring ΔE or ΔH for a reaction. In bomb calorimetry, the reaction is carried out under conditions of constant volume, so $\Delta E = q_v$. The temperature change of the calorimeter can therefore be used to calculate ΔE for the reaction. When a reaction takes place at constant pressure, energy may be released both as heat and as work. In coffee-cup calorimetry, a reaction is carried out under atmospheric pressure in a solution, so $\Delta E = \Delta H$. The temperature change of the solution is then used to calculate ΔH for the reaction.

Calculating ΔH_{rxn} (6.7, 6.8)

The enthalpy of reaction (ΔH_{rxn}) can be calculated from known thermochemical data in two ways. The first way is by using the following relationships: (a) when a reaction is multiplied by a factor, ΔH_{rxn} is multiplied by the same factor; (b) when a reaction is reversed, ΔH_{rxn} changes sign; and (c) if a chemical reaction can be expressed as a sum of two or more steps, ΔH_{rxn} is the sum of the ΔH's for the individual steps (Hess's law). Together, these relationships can be used to determine the enthalpy change of an unknown reaction from reactions with known enthalpy changes. The second way to calculate ΔH_{rxn} from known thermochemical data is by using tabulated standard enthalpies of formation for the reactants and products of the reaction. These are usually tabulated for substances in their standard states, and the enthalpy of reaction is called the standard enthalpy of reaction (ΔH°_{rxn}). For any reaction, ΔH°_{rxn} is obtained by subtracting the sum of the enthalpies of formation of the reactants multiplied by their stoichiometric coefficients from the sum of the enthalpies of formation of the products multiplied by their stoichiometric coefficients.

Environmental Problems Associated with Fossil Fuel Use (6.9)

Fossil fuels are nonrenewable fuels; once they are consumed, they cannot be replaced. At current rates of consumption, natural gas and petroleum reserves will be depleted in 50–100 years. In addition to their limited supply, the products of the combustion of fossil fuels—directly or indirectly formed—create several environmental problems including air pollution, acid rain, and global warming, which is an increase in Earth's average temperature caused by CO_2.

Key Equations and Relationships

Kinetic Energy (6.1)

$$KE = \frac{1}{2} mv^2$$

Change in Internal Energy (ΔE) of a Chemical System (6.2)

$$\Delta E = E_{products} - E_{reactants}$$

Energy Flow between System and Surroundings (6.2)

$$\Delta E_{system} = -\Delta E_{surroundings}$$

Relationship between Internal Energy (ΔE), Heat (q), and Work (w) (6.2)

$$\Delta E = q + w$$

Relationship between Heat (q), Temperature (T), and Heat Capacity (C) (6.3)

$$q = C \times \Delta T$$

Relationship between Heat (q), Mass (m), Temperature (T), and Specific Heat Capacity of a Substance (C_s) (6.3)

$$q = m \times C_s \times \Delta T$$

Relationship between Work (w), Force (F), and Distance (D) (6.3)

$$w = F \times D$$

Relationship between Work (w), Pressure (P), and Change in Volume (ΔV) (6.3)

$$w = -P \Delta V$$

Change in Internal Energy (ΔE) of System at Constant Volume (6.4)

$$\Delta E = q_v$$

Heat of a Bomb Calorimeter (q_{cal}) (6.4)

$$q_{cal} = C_{cal} \times \Delta T$$

Heat Exchange between a Calorimeter and a Reaction (6.4)

$$q_{cal} = -q_{rxn}$$

Relationship between Enthalpy (ΔH), Internal Energy (ΔE), Pressure (P), and Volume (V) (6.5)

$$\Delta H = \Delta E + P \Delta V$$

Relationship between Enthalpy of a Reaction (ΔH°_{rxn}) and the Heats of Formation (ΔH°_f) (6.8)

$$\Delta H^\circ_{rxn} = \Sigma n_p \, \Delta H^\circ_f(\text{products}) - \Sigma n_r \, \Delta H^\circ_f(\text{reactants})$$

Key Skills

Calculating Internal Energy from Heat and Work (6.2)
- Example 6.1 • For Practice 6.1 • Exercises 39–42

Finding Heat from Temperature Changes (6.3)
- Example 6.2 • For Practice 6.2 • For More Practice 6.2 • Exercises 45–48

Finding Work from Volume Changes (6.3)
- Example 6.3 • For Practice 6.3 • For More Practice 6.3 • Exercises 49, 50

Using Bomb Calorimetry to Calculate ΔE_{rxn} (6.4)
- Example 6.4 • For Practice 6.4 • For More Practice 6.4 • Exercises 51, 52

Predicting Endothermic and Exothermic Processes (6.5)
- Example 6.5 • For Practice 6.5 • Exercises 55, 56

Determining heat from ΔH and Stoichiometry (6.5)
- Examples 6.6, 6.12 • For Practice 6.6, 6.12 • For More Practice 6.6 • Exercises 57–60

Finding ΔH_{rxn} Using Calorimetry (6.6)
- Example 6.7 • For Practice 6.7 • Exercises 65, 66

Finding ΔH_{rxn} Using Hess's Law (6.7)
- Example 6.8 • For Practice 6.8 • For More Practice 6.8 • Exercises 69–72

Finding ΔH°_{rxn} Using Standard Enthalpies of Formation (6.8)
- Examples 6.9, 6.10, 6.11 • For Practice 6.9, 6.10, 6.11 • Exercises 73–82

Exercises

Review Questions

1. What is thermochemistry? Why is it important?

2. What is energy? What is work? Give some examples of each.

3. What is kinetic energy? What is potential energy? Give some examples of each.

4. Explain the law of conservation of energy. How does it relate to energy exchanges between a thermodynamic system and its surroundings?

5. What is the SI unit of energy? List some other common units of energy.

6. What is the first law of thermodynamics? What are its implications?

7. Explain why any machine that supposedly produces energy without the need for energy input should be viewed with suspicion.

8. What is a state function? Give some examples of state functions.

9. How is internal energy defined?

10. If energy flows out of a chemical system and into the surroundings, what is the sign of ΔE_{system}?

11. If the internal energy of the products of a reaction is higher than the internal energy of the reactants, what is the sign of ΔE for the reaction? In which direction does energy flow?

12. What is heat? Explain the difference between heat and temperature.

13. How is the change in internal energy of a system related to heat and work?

14. Explain how the sum of heat and work can be a state function, even though heat and work are themselves not state functions.

15. What is heat capacity? Explain how the high heat capacity of water may affect the weather in coastal versus inland regions.

16. What is pressure–volume work? How is it computed?

17. What is calorimetry? Explain the difference between a coffee-cup calorimeter and a bomb calorimeter. What is each designed to measure?

18. What is the change in enthalpy (ΔH) for a chemical reaction? How is ΔH different from ΔE?

19. Explain the difference between an exothermic and an endothermic reaction. Give the sign of ΔH for each one.

20. From a molecular viewpoint, where does the energy emitted in an exothermic chemical reaction come from? Why do the reactants undergo an increase in temperature even though energy is emitted?

21. From a molecular viewpoint, where does the energy absorbed in an endothermic chemical reaction go? Why do the reactants undergo a decrease in temperature even though energy is absorbed?

22. Is the change in enthalpy for a reaction an extensive property? Explain the relationship between the ΔH for a reaction and the amounts of reactants and products that undergo reaction.

23. Explain how the value of ΔH for a reaction changes upon:
    a. multiplying the reaction by a factor
    b. reversing the reaction
    Why do these relationships hold?

24. What is Hess's law? Why is it useful?

25. What is a standard state? What is the standard enthalpy change for a reaction?

26. What is the standard enthalpy of formation for a compound? For a pure element in its standard state?

27. How can you calculate $\Delta H^\circ_{\text{rxn}}$ from tabulated standard enthalpies of formation?

28. What are the main sources of the energy consumed in the United States?

29. What are the main environmental problems associated with fossil fuel use?

30. Explain global warming. What causes global warming? Is there any evidence that global warming is occurring?

Problems by Topic

Energy Units

31. Perform each of the following conversions between energy units:
    a. 3.55×10^4 J to cal
    b. 1025 Cal to J
    c. 355 kJ to cal
    d. 125 kWh to J

32. Perform each of the following conversions between energy units:
    a. 1.58×10^3 kJ to kcal
    b. 865 cal to kJ
    c. 1.93×10^4 J to Cal
    d. 1.8×10^4 kJ to kWh

33. Suppose that a person eats a diet of 2155 Calories per day. Convert this energy into each of the following units:
    a. J
    b. kJ
    c. kWh

34. A frost-free refrigerator uses about 655 kWh of electrical energy per year. Express this amount of energy in each of the following units:
    a. J
    b. kJ
    c. Cal

Internal Energy, Heat, and Work

35. Which of the following is true of the internal energy of a system and its surroundings during an energy exchange with a negative ΔE_{sys}?
    a. The internal energy of the system increases and the internal energy of the surroundings decreases.
    b. The internal energy of both the system and the surroundings increases.
    c. The internal energy of both the system and the surroundings decreases.
    d. The internal energy of the system decreases and the internal energy of the surroundings increases.

36. During an energy exchange, a chemical system absorbs energy from its surroundings. What is the sign of ΔE_{sys} for this process? Explain.

37. Identify each of the following energy exchanges as primarily heat or work and determine whether the sign of ΔE is positive or negative for the system.
    a. Sweat evaporates from skin, cooling the skin. (The evaporating sweat is the system.)
    b. A balloon expands against an external pressure. (The contents of the balloon is the system.)
    c. An aqueous chemical reaction mixture is warmed with an external flame. (The reaction mixture is the system.)

38. Identify each of the following energy exchanges as heat or work and determine whether the sign of ΔE is positive or negative for the system.
    a. A rolling billiard ball collides with another billiard ball. The first billiard ball (defined as the system) stops rolling after the collision.

    b. A book is dropped to the floor (the book is the system).
    c. A father pushes his daughter on a swing (the daughter and the swing are the system).

39. A system releases 415 kJ of heat and does 125 kJ of work on the surroundings. What is the change in internal energy of the system?

40. A system absorbs 214 kJ of heat and the surroundings do 110 kJ of work on the system. What is the change in internal energy of the system?

41. The gas in a piston (defined as the system) is warmed and absorbs 655 J of heat. The expansion performs 344 J of work on the surroundings. What is the change in internal energy for the system?

42. The air in an inflated balloon (defined as the system) is warmed over a toaster and absorbs 115 J of heat. As it expands, it does 77 kJ of work. What is the change in internal energy for the system?

Heat, Heat Capacity, and Work

43. Two identical coolers are packed for a picnic. Each cooler is packed with twenty-four 12-ounce soft drinks and 5 pounds of ice. However, the drinks that went into cooler A were refrigerated for several hours before they were packed in the cooler, while the drinks that went into cooler B were packed at room temperature. When the two coolers are opened 3 hours later, most of the ice in cooler A is still ice, while nearly all of the ice in cooler B has melted. Explain this difference.

44. A kilogram of aluminum metal and a kilogram of water are each warmed to 75 °C and placed in two identical insulated containers. One hour later, the two containers are opened and the temperature of each substance is measured. The aluminum has cooled to 35 °C while the water has cooled only to 66 °C. Explain this difference.

45. How much heat is required to warm 1.50 L of water from 25.0 °C to 100.0 °C? (Assume a density of 1.0 g/mL for the water.)

46. How much heat is required to warm 1.50 kg of sand from 25.0 °C to 100.0 °C?

47. Suppose that 25 g of each of the following substances is initially at 27.0 °C. What is the final temperature of each substance upon absorbing 2.35 kJ of heat?
    a. gold
    b. silver
    c. aluminum
    d. water

48. An unknown mass of each of the following substances, initially at 23.0 °C, absorbs 1.95×10^3 J of heat. The final temperature is recorded as indicated. Find the mass of each substance.
    a. Pyrex glass ($T_f = 55.4$ °C)
    b. sand ($T_f = 62.1$ °C)
    c. ethanol ($T_f = 44.2$ °C)
    d. water ($T_f = 32.4$ °C)

49. How much work (in J) is required to expand the volume of a pump from 0.0 L to 2.5 L against an external pressure of 1.1 atm?

50. During a breath, the average human lung expands by about 0.50 L. If this expansion occurs against an external pressure of 1.0 atm, how much work (in J) is done during the expansion?

51. The air within a piston equipped with a cylinder absorbs 565 J of heat and expands from an initial volume of 0.10 L to a final volume of 0.85 L against an external pressure of 1.0 atm. What is the change in internal energy of the air within the piston?

52. A gas is compressed from an initial volume of 5.55 L to a final volume of 1.22 L by an external pressure of 1.00 atm. During the compression the gas releases 124 J of heat. What is the change in internal energy of the gas?

Enthalpy and Thermochemical Stoichiometry

53. When 1 mol of a fuel is burned at constant pressure, it produces 3452 kJ of heat and does 11 kJ of work. What are the values of ΔE and ΔH for the combustion of the fuel?

54. The change in internal energy for the combustion of 1.0 mol of octane at a pressure of 1.0 atm is 5084.3 kJ. If the change in enthalpy is 5074.1 kJ, how much work is done during the combustion?

55. Determine whether each of the following is exothermic or endothermic and indicate the sign of ΔH.
  a. natural gas burning on a stove
  b. isopropyl alcohol evaporating from skin
  c. water condensing from steam

56. Determine whether each of the following is exothermic or endothermic and indicate the sign of ΔH.
  a. dry ice evaporating
  b. a sparkler burning
  c. the reaction that occurs in a chemical cold pack often used to ice athletic injuries

57. Consider the following thermochemical equation for the combustion of acetone (C_3H_6O), the main ingredient in nail polish remover.

$$C_3H_6O(l) + 4\,O_2(g) \longrightarrow 3\,CO_2(g) + 3\,H_2O(g)$$
$$\Delta H^\circ_{rxn} = -1790\ kJ$$

If a bottle of nail polish remover contains 177 mL of acetone, how much heat would be released by its complete combustion? The density of acetone is 0.788 g/mL.

58. What mass of natural gas (CH_4) must you burn to emit 267 kJ of heat?

$$CH_4(g) + 2\,O_2(g) \longrightarrow CO_2(g) + 2\,H_2O(g)$$
$$\Delta H^\circ_{rxn} = -802.3\ kJ$$

59. The propane fuel (C_3H_8) used in gas barbeques burns according to the following thermochemical equation:

$$C_3H_8(g) + 5\,O_2(g) \longrightarrow 3\,CO_2(g) + 4\,H_2O(g)$$
$$\Delta H^\circ_{rxn} = -2217\ kJ$$

If a pork roast must absorb 1.6×10^3 kJ to fully cook, and if only 10% of the heat produced by the barbeque is actually absorbed by the roast, what mass of CO_2 is emitted into the atmosphere during the grilling of the pork roast?

60. Charcoal is primarily carbon. Determine the mass of CO_2 produced by burning enough carbon (in the form of charcoal) to produce 5.00×10^2 kJ of heat.

$$C(s) + O_2(g) \longrightarrow CO_2(g) \qquad \Delta H^\circ_{rxn} = -393.5\ kJ$$

Calorimetry

61. Exactly 1.5 g of a fuel is burned under conditions of constant pressure and then again under conditions of constant volume. In measurement A the reaction produces 25.9 kJ of heat, and in measurement B the reaction produces 23.3 kJ of heat. Which measurement (A or B) corresponds to conditions of constant pressure? Which one corresponds to conditions of constant volume? Explain.

62. In order to obtain the largest possible amount of heat from a chemical reaction in which there is a large increase in the number of moles of gas, should you carry out the reaction under conditions of constant volume or constant pressure? Explain.

63. When 0.514 g of biphenyl ($C_{12}H_{10}$) undergoes combustion in a bomb calorimeter, the temperature rises from 25.8 °C to 29.4 °C. Find ΔE_{rxn} for the combustion of biphenyl in kJ/mol biphenyl. The heat capacity of the bomb calorimeter, determined in a separate experiment, is 5.86 kJ/°C.

64. Mothballs are composed primarily of the hydrocarbon naphthalene ($C_{10}H_8$). When 1.025 g of naphthalene is burned in a bomb calorimeter, the temperature rises from 24.25 °C to 32.33 °C. Find ΔE_{rxn} for the combustion of naphthalene. The heat capacity of the calorimeter, determined in separate experiment, is 5.11 kJ/°C.

65. Zinc metal reacts with hydrochloric acid according to the following balanced equation.

$$Zn(s) + 2\,HCl(aq) \longrightarrow ZnCl_2(aq) + H_2(g)$$

When 0.103 g of $Zn(s)$ is combined with enough HCl to make 50.0 mL of solution in a coffee-cup calorimeter, all of the zinc reacts, raising the temperature of the solution from 22.5 °C to 23.7 °C. Find ΔH_{rxn} for this reaction as written. (Use 1.0 g/mL for the density of the solution and 4.18 J/g·°C as the specific heat capacity.)

66. Instant cold packs, often used to ice athletic injuries on the field, contain ammonium nitrate and water separated by a thin plastic divider. When the divider is broken, the ammonium nitrate dissolves according to the following endothermic reaction:

$$NH_4NO_3(s) \longrightarrow NH_4^+(aq) + NO_3^-(aq)$$

In order to measure the enthalpy change for this reaction, 1.25 g of NH_4NO_3 is dissolved in enough water to make 25.0 mL of solution. The initial temperature is 25.8 °C and the final temperature (after the solid dissolves) is 21.9 °C. Calculate the change in enthalpy for the reaction in kJ. (Use 1.0 g/mL as the density of the solution and 4.18 J/g·C as the specific heat capacity.)

Quantitative Relationships Involving ΔH and Hess's Law

67. For each of the following, determine the value of ΔH_2 in terms of ΔH_1.

  a.
  $$A + B \longrightarrow 2\,C \qquad \Delta H_1$$
  $$2C \longrightarrow A + B \qquad \Delta H_2 = ?$$

  b.
  $$A + \tfrac{1}{2}B \longrightarrow C \qquad \Delta H_1$$
  $$2\,A + B \longrightarrow 2C \qquad \Delta H_2 = ?$$

  c.
  $$A \longrightarrow B + 2\,C \qquad \Delta H_1$$
  $$\tfrac{1}{2}B + C \longrightarrow \tfrac{1}{2}A \qquad \Delta H_2 = ?$$

68. Consider the following generic reaction:

$$A + 2B \longrightarrow C + 3D \qquad \Delta H = 155 \text{ kJ}$$

Determine the value of ΔH for each of the following related reactions:

a. $3A + 6B \longrightarrow 3C + 9D$

b. $C + 3D \longrightarrow A + 2B$

c. $\frac{1}{2}C + 3/2 D \longrightarrow \frac{1}{2}A + B$

69. Calculate ΔH_{rxn} for the following reaction:

$$Fe_2O_3(s) + 3CO(g) \longrightarrow 2Fe(s) + 3CO_2(g)$$

Use the following reactions and given ΔH's.

$$2Fe(s) + 3/2 O_2(g) \longrightarrow Fe_2O_3(s) \qquad \Delta H = -824.2 \text{ kJ}$$
$$CO(g) + \frac{1}{2}O_2(g) \longrightarrow CO_2(g) \qquad \Delta H = -282.7 \text{ kJ}$$

70. Calculate ΔH_{rxn} for the following reaction:

$$CaO(s) + CO_2(g) \longrightarrow CaCO_3(s)$$

Use the following reactions and given ΔH's.

$$Ca(s) + CO_2(g) + \frac{1}{2}O_2(g) \longrightarrow CaCO_3(s)$$
$$\Delta H = -812.8 \text{ kJ}$$
$$2Ca(s) + O_2(g) \longrightarrow 2CaO(s) \qquad \Delta H = -1269.8 \text{ kJ}$$

71. Calculate ΔH_{rxn} for the following reaction:

$$5C(s) + 6H_2(g) \longrightarrow C_5H_{12}(l)$$

Use the following reactions and given ΔH's.

$$C_5H_{12}(l) + 8O_2(g) \longrightarrow 5CO_2(g) + 6H_2O(g)$$
$$\Delta H = -3505.8 \text{ kJ}$$
$$C(s) + O_2(g) \longrightarrow CO_2(g) \qquad \Delta H = -393.5 \text{ kJ}$$
$$2H_2(g) + O_2(g) \longrightarrow 2H_2O(g) \qquad \Delta H = -483.5 \text{ kJ}$$

72. Calculate ΔH_{rxn} for the following reaction:

$$CH_4(g) + 4Cl_2(g) \longrightarrow CCl_4(g) + 4HCl(g)$$

Use the following reactions and given ΔH's.

$$C(s) + 2H_2(g) \longrightarrow CH_4(g) \qquad \Delta H = -74.6 \text{ kJ}$$
$$C(s) + 2Cl_2(g) \longrightarrow CCl_4(g) \qquad \Delta H = -95.7 \text{ kJ}$$
$$H_2(g) + Cl_2(g) \longrightarrow 2HCl(g) \qquad \Delta H = -92.3 \text{ kJ}$$

Enthalpies of Formation and ΔH

73. Write an equation for the formation of each of the following compounds from their elements in their standard states, and find ΔH_f° for each from Appendix IIB.

a. $NH_3(g)$ **b.** $CO_2(g)$

c. $Fe_2O_3(s)$ **d.** $CH_4(g)$

74. Write an equation for the formation of each of the following compounds from their elements in their standard states, and find ΔH_f° for each from Appendix IIB.

a. $NO_2(g)$ **b.** $MgCO_3(s)$

c. $C_2H_4(g)$ **d.** $CH_3OH(l)$

75. Hydrazine (N_2H_4) is a fuel used by some spacecraft. It is normally oxidized by N_2O_4 according to the following equation:

$$N_2H_4(l) + N_2O_4(g) \longrightarrow 2N_2O(g) + 2H_2O(g)$$

Calculate ΔH_{rxn}° for this reaction using standard enthalpies of formation.

76. Pentane (C_5H_{12}) is a component of gasoline that burns according to the following balanced equation:

$$C_5H_{12}(l) + 8O_2(g) \longrightarrow 5CO_2(g) + 6H_2O(g)$$

Calculate ΔH_{rxn}° for this reaction using standard enthalpies of formation. (The standard enthalpy of formation of liquid pentane is -146.8 kJ/mol.)

77. Use standard enthalpies of formation to calculate ΔH_{rxn}° for each of the following reactions:

a. $C_2H_4(g) + H_2(g) \longrightarrow C_2H_6(g)$

b. $CO(g) + H_2O(g) \longrightarrow H_2(g) + CO_2(g)$

c. $3NO_2(g) + H_2O(l) \longrightarrow 2HNO_3(aq) + NO(g)$

d. $Cr_2O_3(s) + 3CO(g) \longrightarrow 2Cr(s) + 3CO_2(g)$

78. Use standard enthalpies of formation to calculate ΔH_{rxn}° for each of the following reactions:

a. $2H_2S(g) + 3O_2(g) \longrightarrow 2H_2O(l) + 2SO_2(g)$

b. $SO_2(g) + \frac{1}{2}O_2(g) \longrightarrow SO_3(g)$

c. $C(s) + H_2O(g) \longrightarrow CO(g) + H_2(g)$

d. $N_2O_4(g) + 4H_2(g) \longrightarrow N_2(g) + 4H_2O(g)$

79. During photosynthesis, plants use energy from sunlight to form glucose ($C_6H_{12}O_6$) and oxygen from carbon dioxide and water. Write a balanced equation for photosynthesis and calculate ΔH_{rxn}°.

80. Ethanol can be made from the fermentation of crops and has been used as a fuel additive to gasoline. Write a balanced equation for the combustion of ethanol and calculate ΔH_{rxn}°.

81. Top fuel dragsters and funny cars burn nitromethane as fuel according to the following balanced combustion equation:

$$2CH_3NO_2(l) + 3/2 O_2(g) \longrightarrow 2CO_2(g) + 3H_2O(g) + N_2(g)$$

The standard enthalpy of combustion for nitromethane is -709.2 kJ/mol. Calculate the standard enthalpy of formation (ΔH_f°) for nitromethane.

82. The explosive nitroglycerin ($C_3H_5N_3O_9$) decomposes rapidly upon ignition or sudden impact according to the following balanced equation:

$$4C_3H_5N_3O_9(l) \longrightarrow 12CO_2(g) + 10H_2O(g) + 6N_2(g) + O_2(g)$$
$$\Delta H_{rxn}^\circ = -5678 \text{ kJ}$$

Calculate the standard enthalpy of formation (ΔH_f°) for nitroglycerin.

Energy Use and The Environment

83. Determine the mass of CO_2 produced by burning enough of each of the following fuels to produce $1.00 \times 10^2 \text{ kJ}$ of heat. Which fuel contributes least to global warming per kJ of heat produced?

a. $CH_4(g) + 2O_2(g) \longrightarrow CO_2(g) + 2H_2O(g)$
$$\Delta H_{rxn}^\circ = -802.3 \text{ kJ}$$

b. $C_3H_8(g) + 5O_2(g) \longrightarrow 3CO_2(g) + 4H_2O(g)$
$$\Delta H_{rxn}^\circ = -2217 \text{ kJ}$$

c. $C_8H_{18}(l) + 25/2 O_2(g) \longrightarrow 8CO_2(g) + 9H_2O(g)$
$$\Delta H_{rxn}^\circ = -5074.1 \text{ kJ}$$

84. Methanol (CH_3OH) has been suggested as a fuel to replace gasoline. Write a balanced equation for the combustion of methanol, find ΔH_{rxn}°, and determine the mass of carbon dioxide emitted per kJ of heat produced. Use the information from the previous exercise to calculate the same quantity for octane, C_8H_{18}. How does methanol compare to octane with respect to global warming?

85. The world burns the fossil fuel equivalent of 7×10^{12} kg of petroleum per year. Assume that all of this petroleum is in the form of octane (C_8H_{18}) and calculate how much CO_2 (in kg) is produced by world fossil fuel combustion per year. (Hint: Begin by writing a balanced equation for the combustion of octane.) If the atmosphere currently contains approximately 3×10^{15} kg of CO_2, how long will it take for the world's fossil fuel combustion to double the amount of atmospheric carbon dioxide?

86. In a sunny location, sunlight has a power density of about 1 kW/m^2. Photovoltaic solar cells can convert this power into electricity with 15% efficiency. If a typical home uses 385 kWh of electricity per month, how many square meters of solar cells would be required to meet its energy requirements? Assume that electricity can be generated from the sunlight for 8 hours per day.

▲ What area of solar cells is needed to power a home?

Cumulative Problems

87. The kinetic energy of a rolling billiard ball is given by $KE = \frac{1}{2}mv^2$ Suppose a 0.17-kg billiard ball is rolling down a pool table with an initial speed of 4.5 m/s. As it travels, it loses some of its energy as heat. The ball slows down to 3.8 m/s and then collides straight-on with a second billiard ball of equal mass. The first billiard ball completely stops and the second one rolls away with a velocity of 3.8 m/s. Assume the first billiard ball is the system and calculate w, q, and ΔE for the process.

88. A 100-W lightbulb is placed in a cylinder equipped with a moveable piston. The lightbulb is turned on for 0.015 hour, and the assembly expands from an initial volume of 0.85 L to a final volume of 5.88 L against an external pressure of 1.0 atm. Use the wattage of the lightbulb and the time it is on to calculate ΔE in joules (assume that the cylinder and lightbulb assembly is the system and assume two significant figures). Calculate w and q.

89. Evaporating sweat cools the body because evaporation is an endothermic process:

$$H_2O(l) \longrightarrow H_2O(g) \qquad \Delta H^\circ_{rxn} = +44.01 \text{ kJ}$$

Estimate the mass of water that must evaporate from the skin to cool the body by 0.50 °C. Assume a body mass of 95 kg and assume that the specific heat capacity of the body is 4.0 J/g·°C.

90. LP gas burns according to the following exothermic reaction:

$$C_3H_8(g) + 5\,O_2(g) \longrightarrow 3\,CO_2(g) + 4\,H_2O(g)$$
$$\Delta H^\circ_{rxn} = -2044 \text{ kJ}$$

What mass of LP gas is necessary to heat 1.5 L of water from room temperature (25.0 °C) to boiling (100.0 °C)? Assume that during heating, 15% of the heat emitted by the LP gas combustion goes to heat the water. The rest is lost as heat to the surroundings.

91. Use standard enthalpies of formation to calculate the standard change in enthalpy for the melting of ice. (The ΔH°_f for $H_2O(s)$ is −291.8 kJ/mol.) Use this value to calculate the mass of ice required to cool 355 mL of a beverage from room temperature (25.0 °C) to 0.0 °C. Assume that the specific heat capacity and density of the beverage are the same as those of water.

92. Dry ice is solid carbon dioxide. Instead of melting, solid carbon dioxide sublimes according to the following equation:

$$CO_2(s) \longrightarrow CO_2(g)$$

When dry ice is added to warm water, heat from the water causes the dry ice to sublime more quickly. The evaporating carbon dioxide produces a dense fog often used to create special effects. In a simple dry ice fog machine, dry ice is added to warm water in a Styrofoam cooler. The dry ice produces fog until it evaporates away, or until the water gets too cold to sublime the dry ice quickly enough. Suppose that a small Styrofoam cooler holds 15.0 liters of water heated to 85 °C. Use standard enthalpies of formation to calculate the change in enthalpy for dry ice sublimation, and calculate the mass of dry ice that should be added to the water so that the dry ice completely sublimes away when the water reaches 25 °C. Assume no heat loss to the surroundings (The ΔH°_f for $CO_2(s)$ is −427.4 kJ/mol.

▲ When carbon dioxide sublimes, the gaseous CO_2 is cold enough to cause water vapor in the air to condense, forming fog.

93. A 25.5-g aluminum block is warmed to 65.4 °C and plunged into an insulated beaker containing 55.2 g water initially at 22.2 °C. The aluminum and the water are allowed to come to thermal equilibrium. Assuming that no heat is lost, what is the final temperature of the water and aluminum?

94. If 50.0 mL of ethanol (density = 0.789 g/mL) initially at 7.0 °C is mixed with 50.0 mL of water (density = 1.0 g/mL) initially at 28.4 °C in an insulated beaker, and assuming that no heat is lost, what is the final temperature of the mixture?

95. Palmitic acid ($C_{16}H_{32}O_2$) is a dietary fat found in beef and butter. The caloric content of palmitic acid is typical of fats in general. Write a balanced equation for the complete combustion of palmitic acid and calculate the standard enthalpy of combustion. What is the caloric content of palmitic acid in Cal/g? Do the same calculation for table sugar (sucrose, $C_{12}H_{22}O_{11}$). Which dietary substance (sugar or fat) contains more Calories per gram? The standard enthalpy of formation of palmitic acid is −208 kJ/mol and that of sucrose is −2226.1 kJ/mol. Use $H_2O(l)$ in the balanced chemical equations because the metabolism of these compounds produces liquid water.)

96. Hydrogen and methanol have both been proposed as alternatives to hydrocarbon fuels. Write balanced reactions for the complete combustion of hydrogen and methanol and use standard enthalpies of formation to calculate the amount of heat released per kilogram of the fuel. Which fuel contains the most energy in the least mass? How does the energy of these fuels compare to that of octane (C_8H_{18})?

97. Derive a relationship between ΔH and ΔE for a process in which the temperature of a fixed amount of an ideal gas changes.

98. Under certain nonstandard conditions, oxidation by $O_2(g)$ of 1 mol of $SO_2(g)$ to $SO_3(g)$ absorbs 89.5 kJ. The heat of formation of $SO_3(g)$ is −204.2 kJ under these conditions. Find the heat of formation of $SO_2(g)$.

99. One tablespoon of peanut butter has a mass of 16 g. It is combusted in a calorimeter whose heat capacity is 120.0 kJ/°C. The temperature of the calorimeter rises from 22.2 °C to 25.4 °C. Find the food caloric content of peanut butter.

100. A mixture of 2.0 mol of $H_2(g)$ and 1.0 mol of $O_2(g)$ is placed in a sealed evacuated container made of a perfect insulating material at 25 °C. The mixture is ignited with a spark and it reacts to form liquid water. Find the temperature of the water.

101. A 20.0-L volume of an ideal gas in a cylinder with a piston is at a pressure of 3.0 atm. Enough weight is suddenly removed from the piston to lower the external pressure to 1.5 atm. The gas then expands at constant temperature until its pressure is 1.5 atm. Find ΔE, ΔH, q, and w for this change in state.

102. When 10.00 g of phosphorus is burned in $O_2(g)$ to form $P_4O_{10}(s)$, enough heat is generated to raise the temperature of 2950 g of water from 18.0 °C to 38.0 °C. Calculate the heat of formation of $P_4O_{10}(s)$ under these conditions.

Challenge Problems

103. A typical frostless refrigerator uses 655 kWh of energy per year in the form of electricity. Suppose that all of this electricity is generated at a power plant that burns coal containing 3.2% sulfur by mass and that all of the sulfur is emitted as SO_2 when the coal is burned. If all of the SO_2 goes on to react with rainwater to form H_2SO_4, what mass of H_2SO_4 is produced by the annual operation of the refrigerator? (Hint: Assume that the remaining percentage of the coal is carbon and begin by calculating $\Delta H°_{rxn}$ for the combustion of carbon.)

104. A large sport utility vehicle has a mass of 2.5×10^3 kg. Calculate the mass of CO_2 emitted into the atmosphere upon accelerating the SUV from 0.0 mph to 65.0 mph. Assume that the required energy comes from the combustion of octane with 30% efficiency. (Hint: Use KE = $\frac{1}{2} mv^2$ to calculate the kinetic energy required for the acceleration.)

105. Combustion of natural gas (primarily methane) occurs in most household heaters. The heat given off in this reaction is used to raise the temperature of the air in the house. Assuming that all the energy given off in the reaction goes to heating up only the air in the house, determine the mass of methane required to heat the air in a house by 10.0 °C. Assume each of the following: house dimensions are 30.0 m × 30.0 m × 3.0 m; specific heat capacity of air is 30 J/K · mol; 1.00 mol of air occupies 22.4 L for all temperatures concerned.

106. When backpacking in the wilderness, hikers often boil water to sterilize it for drinking. Suppose that you are planning a backpacking trip and will need to boil 35 L of water for your group. What volume of fuel should you bring? Assume each of the following: the fuel has an average formula of C_7H_{16}; 15% of the heat generated from combustion goes to heat the water (the rest is lost to the surroundings); the density of the fuel is 0.78 g/mL; the initial temperature of the water is 25.0 °C; and the standard enthalpy of formation of C_7H_{16} is −224.4 kJ/mol.

107. An ice cube of mass 9.0 g is added to a cup of coffee, whose temperature is 90.0 °C and which contains 120.0 g of liquid. Assume the specific heat capacity of the coffee is the same as that of water. The heat of fusion of ice (the heat associated with ice melting) is 6.0 kJ/mol. Find the temperature of the coffee after the ice melts.

108. Find ΔH, ΔE, q, and w for the freezing of water at −10.0 °C. The specific heat capacity of ice is 2.04 J/g · °C and its heat of fusion is −332 J/g.

109. Starting from the relationship between temperature and kinetic energy for an ideal gas, find the value of the molar heat capacity of an ideal gas when its temperature is changed at constant volume. Find its molar heat capacity when its temperature is changed at constant pressure.

110. An amount of an ideal gas expands from 12.0 L to 24.0 L at a constant pressure of 1.0 atm. Then the gas is cooled at a constant volume of 24.0 L back to its original temperature. Then it contracts back to its original volume. Find the total heat flow for the entire process.

Conceptual Problems

111. Which of the following is true of the internal energy of the system and its surroundings following a process in which $\Delta E_{sys} = +65$ kJ. Explain.
     a. The system and the surroundings both lose 65 kJ of energy.
     b. The system and the surroundings both gain 65 kJ of energy.
     c. The system loses 65 kJ of energy and the surroundings gain 65 kJ of energy.
     d. The system gains 65 kJ of energy and the surroundings lose 65 kJ of energy.

112. The internal energy of an ideal gas depends only on its temperature. Which of the following is true of an isothermal (constant-temperature) expansion of an ideal gas against a constant external pressure? Explain.
     a. ΔE is positive
     b. w is positive
     c. q is positive
     d. ΔE is negative

113. Which of the following expressions describes the heat evolved in a chemical reaction when the reaction is carried out at constant pressure? Explain.
     a. $\Delta E - w$
     b. ΔE
     c. $\Delta E - q$

114. Two identical refrigerators are plugged in for the first time. Refrigerator A is empty (except for air) and refrigerator B is filled with jugs of water. The compressors of both refrigerators immediately turn on and begin cooling the interiors of the refrigerators. After two hours, the compressor of refrigerator A turns off while the compressor of refrigerator B continues to run. The next day, the compressor of refrigerator A can be heard turning on and off every few minutes, while the compressor of refrigerator B turns off and on every hour or so (and stays on longer each time.) Explain these observations.

115. A 1-kg cylinder of aluminum and 1-kg jug of water, both at room temperature, are put into a refrigerator. After one hour, the temperature of each object is measured. One of the objects is much cooler than the other. Which one and why?

116. When 1 mol of a gas burns at constant pressure, it produces 2418 J of heat and does 5 kJ of work. Identify ΔE, ΔH, q, and w for the process.

117. In an exothermic reaction, the reactants lose energy and the reaction feels hot to the touch. Explain why the reaction feels hot even though the reactants are losing energy. Where does the energy come from?

118. Which of the following is true of a reaction in which ΔV is positive? Explain.
     a. $\Delta H = \Delta E$
     b. $\Delta H > \Delta E$
     c. $\Delta H < \Delta E$

7

The Quantum-Mechanical Model of the Atom

7.1 Quantum Mechanics: A Theory That Explains the Behavior of the Absolutely Small

7.2 The Nature of Light

7.3 Atomic Spectroscopy and the Bohr Model

7.4 The Wave Nature of Matter: The de Broglie Wavelength, the Uncertainty Principle, and Indeterminacy

7.5 Quantum Mechanics and the Atom

7.6 The Shapes of Atomic Orbitals

THE EARLY PART OF THE TWENTIETH century brought changes that revolutionized how we think about physical reality, especially in the atomic realm. Before that time, all descriptions of the behavior of matter had been deterministic—the present set of conditions completely determined the future. Quantum mechanics changed that. This new theory suggested that for subatomic particles—electrons, neutrons, and protons—the present does NOT completely determine the future. For example, if you shoot one electron down a path and measure where it lands, a second electron shot down the same path under the same conditions will most likely land in a different place! Quantum-mechanical theory was developed by several unusually gifted scientists including Albert Einstein, Neils Bohr, Louis de Broglie, Max Planck, Werner Heisenberg, P. A. M. Dirac, and Erwin Schrödinger. These scientists did not necessarily feel comfortable with their own theory. Bohr said, "Anyone who is not shocked by quantum mechanics has not understood it." Schrödinger wrote, "I don't like it, and I'm sorry I ever had anything to do with it." Albert Einstein disbelieved the very theory he helped create, stating, " God does not play dice with the universe." In fact, Einstein attempted to disprove quantum mechanics—without success—until he died. However, quantum mechanics was able to account for fundamental observations, including the very stability of atoms, which could not be understood within the framework of classical physics. Today, quantum mechanics forms the foundation of chemistry—explaining, for example, the periodic table and the behavior of the elements in chemical bonding—as well as providing the practical basis for lasers, computers, and countless other applications.

Anyone who is not shocked by quantum mechanics has not understood it.

—NEILS BOHR (1885–1962)

Our universe contains objects that span an almost unimaginable range of sizes. This chapter focuses on the behavior of electrons, one of the smallest particles in existence.

7.1 Quantum Mechanics: A Theory That Explains the Behavior of the Absolutely Small

In everyday language, small is a relative term: something is small relative to something else. A car is smaller than a house, and a person is smaller than a car. But there are limits to how small something can be. For example, a house cannot be smaller than the bricks from which it is made.

Atoms and the particles that compose them are unimaginably small. As we have learned, electrons have a mass of less than a trillionth of a trillionth of a gram, and a size so small that it is immeasurable. A single speck of dust contains more electrons than the number of people that have existed on Earth over all the centuries of time. Electrons are *small* in the absolute sense of the word—they are among the smallest particles that make up matter. And yet, an atom's electrons determine many of its chemical and physical properties. If we are to understand these properties, we must understand how electrons exist within atoms.

The absolute smallness of electrons presents a problem in trying to understand them through observation. Consider the difference between observing a baseball, for example, and observing an electron. You can measure the position of a baseball by observing the light that strikes the ball, bounces off it, and enters your eye. The baseball is so large in comparison to the disturbance caused by the light that it is virtually unaffected by your observation. By contrast, imagine observing the position of an electron. If you attempt to measure its position using light, the light itself disturbs the electron. The interaction of the light with the electron actually changes its position, the very thing you are trying to measure.

The inability to observe electrons without disturbing them has significant implications. It means that when you observe an electron, it behaves differently than when you do not observe it—the act of observation changes what the electron does. It means that limits exist as to what we can know about electron behavior. It means that the *absolutely small* world of the electron is different from the *large* world that we are used to. We need to think about subatomic particles in a different way than we think about the macroscopic world.

In this chapter, we examine the **quantum-mechanical model** of the atom, a model that explains how electrons exist in atoms and how those electrons determine the chemical and physical properties of elements. We have already learned much about those properties. We know, for example, that some elements are metals and that others are nonmetals. We know that the noble gases are chemically inert and that the alkali metals are chemically reactive. We know that sodium tends to form 1+ ions and that fluorine tends to form 1− ions. But we do not know *why*. The quantum-mechanical model explains why. In doing so, it explains the modern periodic table and provides the basis for our understanding of chemical bonding.

7.2 The Nature of Light

Before we explore electrons and their behavior within the atom, we must understand a few things about light. Prior to the development of quantum mechanics, the nature of light was viewed as being very different from that of subatomic particles such as electrons. However, as quantum mechanics developed, light was found to have many characteristics in common with electrons. Chief among these is the *wave–particle duality* of light. Certain properties of light are best described by thinking of it as a wave, while other properties are best described by thinking of it as a particle. We will first explore the wave behavior of light, and then its particle behavior. We then turn to electrons to see how they also display the same wave–particle duality.

The Wave Nature of Light

Light is **electromagnetic radiation**, a type of energy embodied in oscillating electric and magnetic fields. An *electric field* is a region of space where an electrically charged particle experiences a force. A *magnetic field* is a region of space where a magnetic particle experiences a force. Electromagnetic radiation can be described as a wave composed of oscillating, mutually perpendicular electric and magnetic fields propagating through space, as shown in Figure 7.1 ▼. In a vacuum, these waves move at a constant speed of 3.00×10^8 m/s

Electromagnetic Radiation

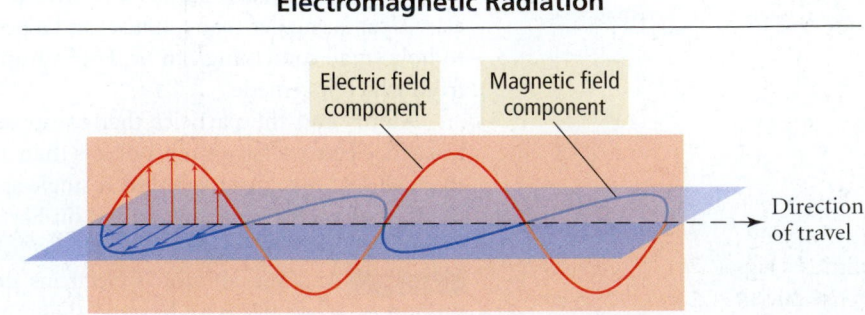

▶ **FIGURE 7.1 Electromagnetic Radiation** Electromagnetic radiation can be described as a wave composed of oscillating electric and magnetic fields. The fields oscillate in perpendicular planes.

Electric field component

Magnetic field component

Direction of travel

(186,000 mi/s)—fast enough to circle the Earth in one-seventh of a second. This great speed explains the delay between the moment when you see a firework in the sky and the moment when you hear the sound of its explosion. The light from the exploding firework reaches your eye almost instantaneously. The sound, traveling much more slowly (340 m/s), takes longer. The same thing happens in a thunder storm—you see the flash immediately, but the sound takes a few seconds to reach you.

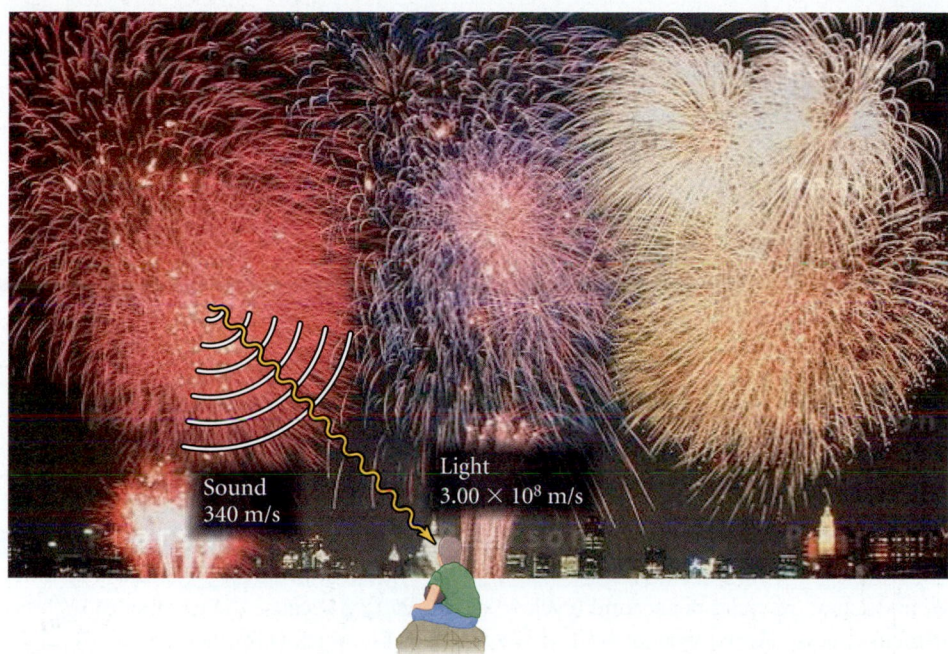

Sound
340 m/s

Light
3.00×10^8 m/s

▲ Because light travels nearly a million times faster than sound, light from an exploding firework reaches your eyes before the sound reaches your ears.

An electromagnetic wave, like all waves, can be characterized by its *amplitude* and its *wavelength*. In the graphical representation shown here, the **amplitude** of the wave is the vertical height of a crest (or depth of a trough). The amplitude of the electric and magnetic field waves in light determines the *intensity* or brightness of the light—the greater the amplitude, the greater the intensity. The **wavelength** (λ) of the wave is the distance in space between adjacent crests (or any two analogous points) and is measured in units of distance such as the meter, micrometer, or nanometer.

The symbol λ is the Greek letter lambda, pronounced "lamb-duh."

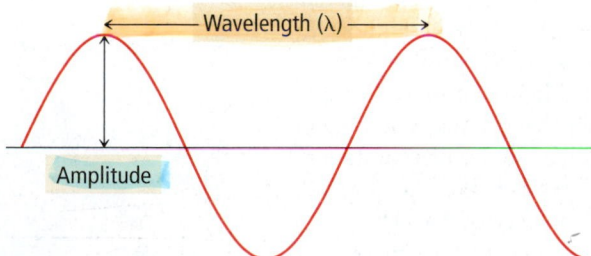

Wavelength (λ)

Amplitude

Wavelength and amplitude are both related to the amount of energy carried by a wave. Imagine trying to swim out from a shore that is being pounded by waves. Either greater amplitude (higher waves) or shorter wavelength (more closely spaced, and thus steeper, waves) will make the swim more difficult. Notice also that amplitude and wavelength can

Different wavelengths, different colors

Different amplitudes, different brightness

λ_A

λ_B

λ_C

▶ **FIGURE 7.2 Wavelength and Amplitude** Wavelength and amplitude are independent properties. The wavelength of light determines its color. The amplitude, or intensity, determines its brightness.

vary independently of one another, as shown in Figure 7.2 ▲. A wave can have a large amplitude and a long wavelength, or a small amplitude and a short wavelength. The most energetic waves have large amplitudes and short wavelengths.

Like all waves, light is also characterized by its **frequency (ν)**, the number of cycles (or wave crests) that pass through a stationary point in a given period of time. The units of frequency are cycles per second (cycle/s) or simply s^{-1} (because the number of cycles is dimensionless). An equivalent unit of frequency is the hertz (Hz), defined as 1 cycle/s. The frequency of a wave is directly proportional to the speed at which the wave is traveling—the faster the wave, the more crests will pass a fixed location per unit time. Frequency is also *inversely* proportional to the wavelength (λ)—the farther apart the crests, the fewer will pass a fixed location per unit time. For light, therefore, we can write

The symbol *ν* is the Greek letter nu, pronounced "noo."

$$\nu = \frac{c}{\lambda} \qquad [7.1]$$

where the speed of light, *c*, and the wavelength, λ, are expressed using the same unit of distance. Therefore, wavelength and frequency simply represent different ways of specifying the same information—if we know one, we can easily calculate the other.

For *visible light*—light that can be seen by the human eye—wavelength (or, alternatively, frequency) determines color. White light, as produced by the sun or by a lightbulb, contains a spectrum of wavelengths and therefore a spectrum of colors. We can see these colors—red, orange, yellow, green, blue, indigo, and violet—in a rainbow or when white light is passed through a prism (Figure 7.3 ◀). Red light, with a wavelength of 750 nanometers (nm), has the longest wavelength of visible light; violet light, with a wavelength of 400 nm, has the shortest. The presence of a variety of wavelengths in white light is responsible for the way we perceive colors in objects. When a substance absorbs some colors while reflecting others, it appears colored. For example, a red shirt appears red because it reflects predominantly red light while absorbing most other colors (Figure 7.4 ▶). Our eyes see only the reflected light, making the shirt appear red.

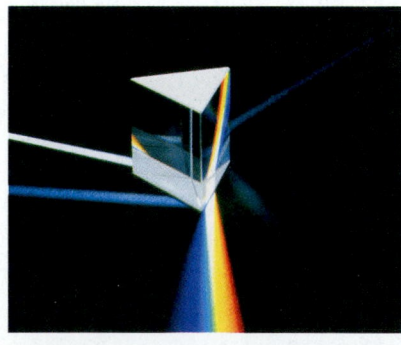

▶ **FIGURE 7.3 Components of White Light** White light can be decomposed into its constituent colors, each with a different wavelength, by passing it through a prism. The array of colors makes up the spectrum of visible light.

nano = 10^{-9}

▲ **FIGURE 7.4 The Color of an Object** A red shirt is red is because it reflects predominantly red light while absorbing most other colors.

EXAMPLE 7.1 Wavelength and Frequency

Calculate the wavelength (in nm) of the red light emitted by a barcode scanner that has a frequency of $4.62 \times 10^{14} \, s^{-1}$.

Solution

You are given the frequency of the light and asked to find its wavelength. Use Equation 7.1, which relates frequency to wavelength. You can convert the wavelength from meters to nanometers by using the conversion factor between the two (1 nm = 10^{-9} m)

$$\nu = \frac{c}{\lambda}$$

$$\lambda = \frac{c}{\nu} = \frac{3.00 \times 10^8 \, \text{m/s}}{4.62 \times 10^{14}/\text{s}}$$

$$= 6.49 \times 10^{-7} \, \text{m}$$

$$6.49 \times 10^{-7} \, \text{m} \times \frac{1 \, \text{nm}}{10^{-9} \, \text{m}} = 649 \, \text{nm}$$

For Practice 7.1

A laser used to the dazzle the audience in a rock concert emits green light with a wavelength of 515 nm. Calculate the frequency of the light.

The Electromagnetic Spectrum

Visible light makes up only a tiny portion of the entire **electromagnetic spectrum**, which includes all wavelengths of electromagnetic radiation. Figure 7.5 ▼ shows the main regions of the electromagnetic spectrum, ranging in wavelength from 10^{-15} m (gamma rays) to 10^5 m (radio waves). In Figure 7.5, short-wavelength, high-frequency radiation is on the right and long-wavelength, low-frequency radiation on the left. As you can see, visible light constitutes only a sliver in the middle.

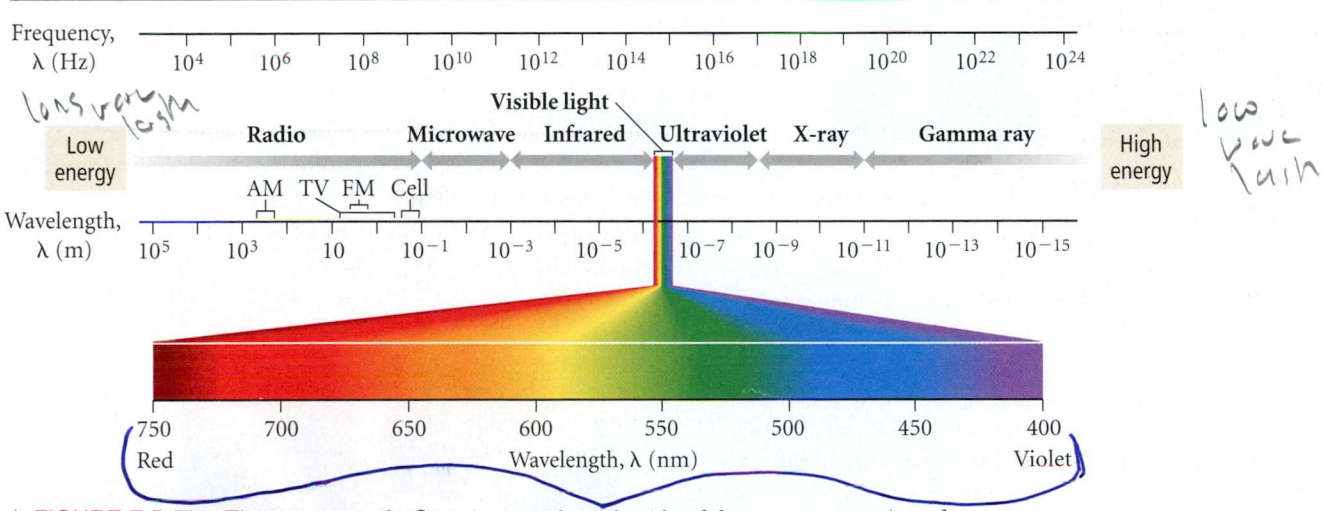

The Electromagnetic Spectrum

▲ FIGURE 7.5 The Electromagnetic Spectrum The right side of the spectrum consists of high-energy, high-frequency, short-wavelength radiation. The left side consists of low-energy, low-frequency, long-wavelength radiation. Visible light constitutes a small segment in the middle.

As we noted previously, short-wavelength light inherently has greater energy than long-wavelength light. Therefore, the most energetic forms of electromagnetic radiation have the shortest wavelengths. The form of electromagnetic radiation with the shortest wavelength is the **gamma (γ) ray**. Gamma rays are produced by the sun, other stars, and certain unstable atomic nuclei on Earth. Human exposure to gamma rays is dangerous because the high energy of gamma rays can damage biological molecules.

Gamma rays are discussed in more detail in Chapter 19.

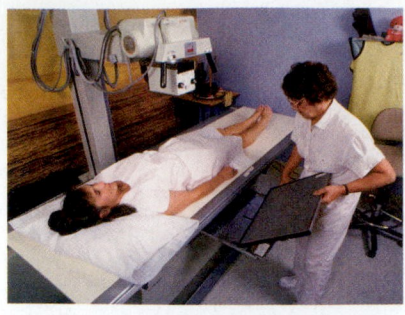

▲ To produce a medical X-ray, the patient is exposed to short-wavelength electromagnetic radiation that can pass through the skin to create an image of bones and internal organs.

High-intensity visible light, such as that emitted by a laser, can damage biological tissue by burning it.

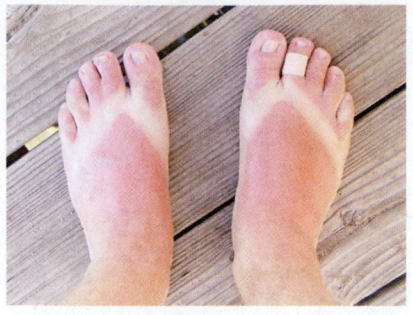

▲ Suntans and sunburns are produced by ultraviolet light from the sun.

Next on the electromagnetic spectrum, with longer wavelengths than gamma rays, are **X-rays**, familiar to us from their medical use. X-rays pass through many substances that block visible light and are therefore used to image bones and internal organs. Like gamma rays, X-rays are sufficiently energetic to damage biological molecules. While several yearly exposures to X-rays are relatively harmless, excessive exposure to X-rays increases cancer risk.

Sandwiched between X-rays and visible light in the electromagnetic spectrum is **ultraviolet (UV) radiation**, most familiar to us as the component of sunlight that produces a sunburn or suntan. While not as energetic as gamma rays or X-rays, ultraviolet light still carries enough energy to damage biological molecules. Excessive exposure to ultraviolet light increases the risk of skin cancer and cataracts and causes premature wrinkling of the skin.

Next on the spectrum is **visible light**, ranging from violet (shorter wavelength, higher energy) to red (longer wavelength, lower energy). Visible light—as long as the intensity is not too high—does not carry enough energy to damage biological molecules. It does,

Chemistry and Medicine
Radiation Treatment for Cancer

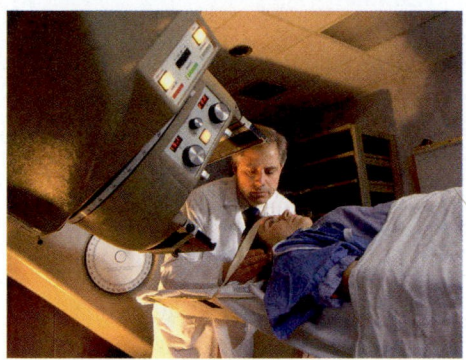

▲ In radiation therapy, highly energetic gamma rays are aimed at cancerous tumors.

X-rays and gamma rays are sometimes called *ionizing radiation* because their short wavelengths correspond to high energies that can ionize atoms and molecules. When ionizing radiation interacts with biological molecules, it can permanently change or even destroy them. Consequently, we normally try to limit our exposure to ionizing radiation. However, doctors use ionizing radiation to destroy molecules within unwanted cells such as cancer cells.

In radiation therapy (or radiotherapy) doctors aim X-ray or gamma-ray beams at cancerous tumors. The ionizing radiation damages the molecules within the tumor's cells that carry genetic information—information necessary for the cell to grow and divide. Consequently, the cell dies or stops dividing. Ionizing radiation also damages molecules in healthy cells, but cancerous cells divide more quickly than normal cells, making them more susceptible to genetic damage. Nonetheless, harm to healthy tissues during treatments can result in side effects such as fatigue,

skin lesions, hair loss, and organ damage. Doctors try to reduce such effects by appropriate shielding and by targeting the tumor from multiple directions, minimizing the exposure of healthy cells while maximizing the exposure of cancerous cells.

Another side effect of exposing healthy cells to radiation is that they too may become cancerous. If a treatment for cancer may cause cancer, why do we continue to use it? In radiation therapy, as in most other disease therapies, there is an associated risk. We take risks all the time, many for lesser reasons. For example, every time we fly in an airliner or drive in a car, we risk injury or even death. Why? Because we perceive the benefit—the convenience of being able to travel a significant distance in a short time—to be worth the relatively small risk. The situation is similar in cancer therapy, or any other therapy for that matter. The benefit of cancer therapy (possibly curing a cancer that might otherwise kill you) is worth the risk (a slight increase in the chance of developing a future cancer).

Question
Why would visible light not work to destroy cancerous tumors?

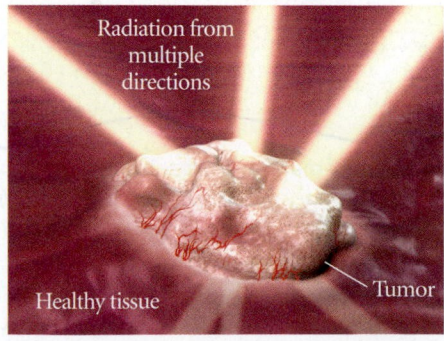

▲ During radiation therapy, a tumor is targeted from multiple directions in order to minimize the exposure of healthy cells while maximizing the exposure of cancerous cells.

however, cause certain molecules in our eyes to change their shape, sending a signal to our brains that results in vision.

Beyond visible light lies **infrared (IR) radiation**. The heat you feel when you place your hand near a hot object is infrared radiation. All warm objects, including human bodies, emit infrared light. Although infrared light is invisible to our eyes, infrared sensors can detect it and are often used in night vision technology to "see" in the dark.

Beyond infrared light, at longer wavelengths still, are **microwaves**, used for radar and in microwave ovens. Although microwave radiation has longer wavelengths and therefore lower energies than visible or infrared light, it is efficiently absorbed by water and can therefore heat substances that contain water. The longest wavelengths are those of **radio waves**, which are used to transmit the signals responsible for AM and FM radio, cellular telephones, television, and other forms of communication.

Interference and Diffraction

Waves, including electromagnetic waves, interact with each other in a characteristic way called **interference**: they can cancel each other out or build each other up, depending on their alignment upon interaction. For example, if waves of equal amplitude from two sources are *in phase* when they interact—that is, they align with overlapping crests—a wave with twice the amplitude results. This is called **constructive interference**.

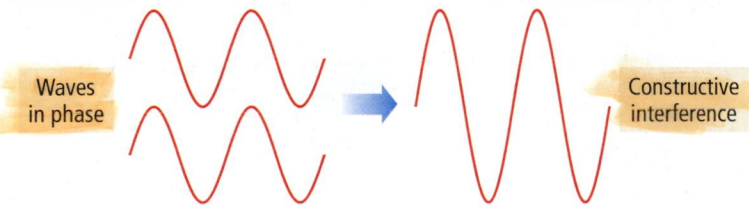

Waves in phase → Constructive interference

On the other hand, if the waves are completely *out of phase*—that is, they align so that the crest from one source overlaps the trough from the other source—the waves cancel by **destructive interference.**

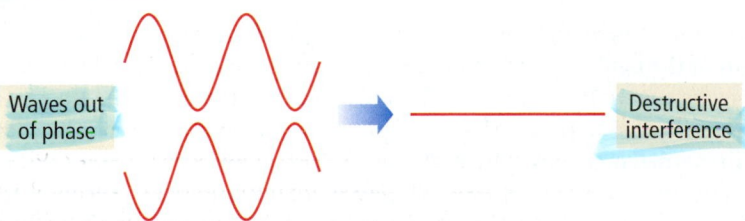

Waves out of phase → Destructive interference

When a wave encounters an obstacle or a slit that is comparable in size to its wavelength, it bends around it—a phenomenon called **diffraction** (Figure 7.6 ▼).

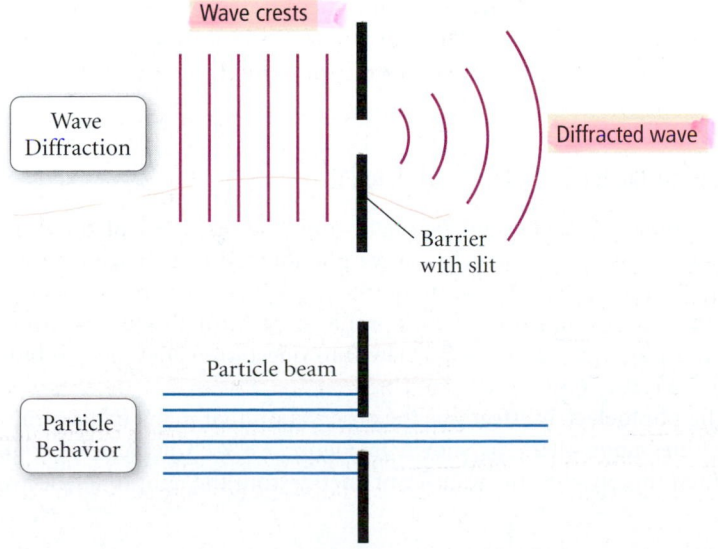

Wave Diffraction — Wave crests — Diffracted wave — Barrier with slit

Particle Behavior — Particle beam

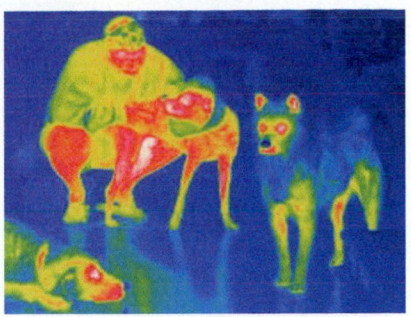

▲ Warm objects emit infrared light, which is invisible to the eye but can be captured on film or by detectors to produce an infrared photograph.

Understanding interference in waves is critical to understanding the wave nature of the electron, as we will soon see.

▲ When a reflected wave meets an incoming wave near the shore, the two waves interfere constructively for an instant, producing a large amplitude spike.

◄ **FIGURE 7.6 Diffraction** Waves are bent, or diffracted, when they encounter an obstacle or slit with a size comparable to their wavelength. Thus, when a wave passes through a small opening, it spreads out. Particles, by contrast, do not diffract; they simply pass through the opening.

Interference From Two Slits

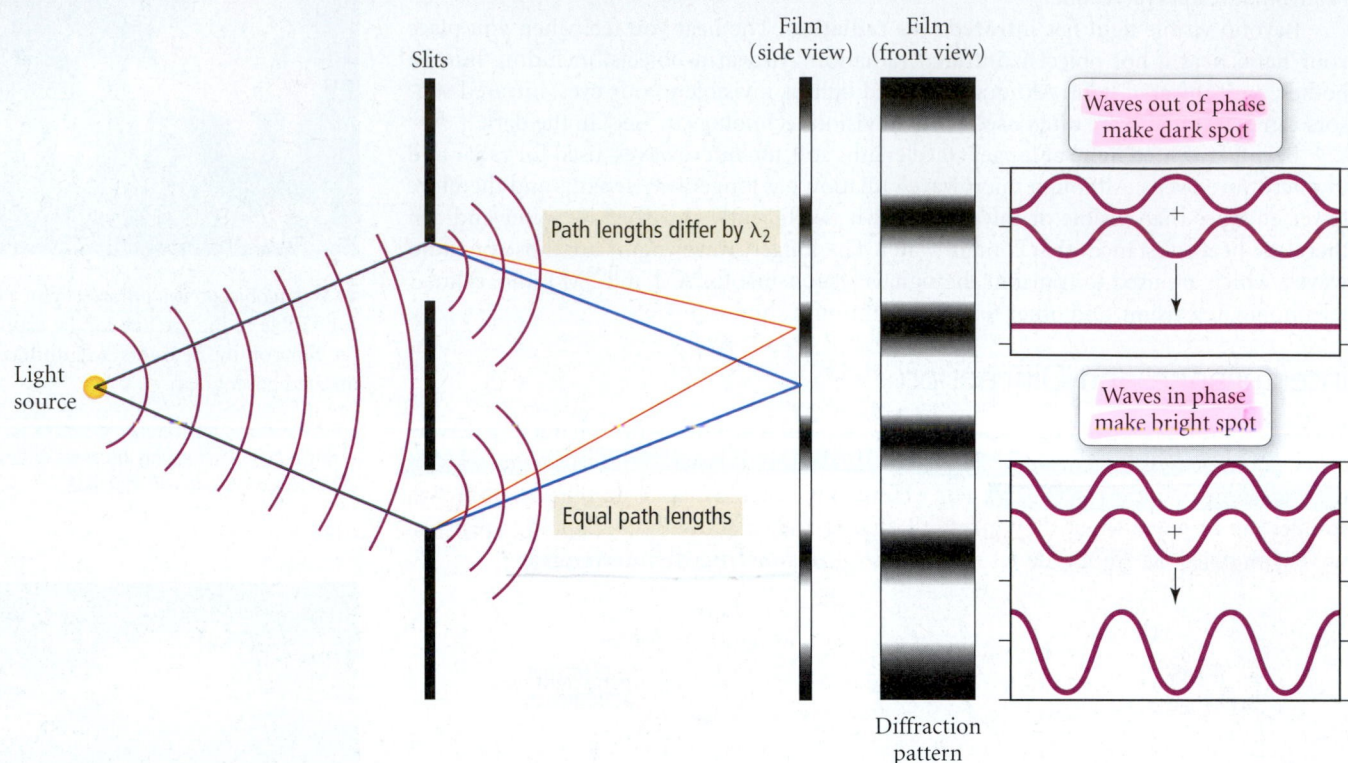

▲ **FIGURE 7.7 Interference from Two Slits**   When a beam of light passes through two small slits, the two resulting waves interfere with each other. Whether the interference is constructive or destructive at any given point depends on the difference in the path lengths traveled by the waves. The resulting interference pattern can be viewed as a series of bright and dark lines on a screen.

The diffraction of light through two slits separated by a distance comparable to the wavelength of the light results in an *interference pattern*, as shown in Figure 7.7 ▲. Each slit acts as a new wave source, and the two new waves interfere with each other. The resulting pattern consists of a series of bright and dark lines that can be viewed on a screen (or recorded on a film) placed at a short distance behind the slits. At the center of the screen, the two waves travel equal distances and interfere constructively to produce a bright line. However, a small distance away from the center in either direction, the two waves travel slightly different distances, so that they are out of phase. At the point where the difference in distance is one-half of a wavelength, the interference is destructive and a dark line appears on the screen. Moving a bit further away from the center produces constructive interference again because the difference between the paths is one whole wavelength. The end result is the interference pattern shown. Notice that interference is a result of the ability of a wave to diffract through the two slits—this is an inherent property of waves.

The Particle Nature of Light

The term *classical*, as in classical electromagnetic theory or classical mechanics, refers to descriptions of matter and energy before the advent of quantum mechanics.

Prior to the early 1900s, and especially after the discovery of the diffraction of light, light was thought to be purely a wave phenomenon. Its behavior was described adequately by classical electromagnetic theory, which treated the electric and magnetic fields that constitute light as waves propagating through space. However, a number of discoveries brought the classical view into question. Chief among those for light was the *photoelectric effect*.

The **photoelectric effect** was the observation that many metals emit electrons when light shines upon them, as shown in Figure 7.8 ▶. Classical electromagnetic theory attributed this effect to the transfer of energy from the light to an electron in the metal,

The Photoelectric Effect

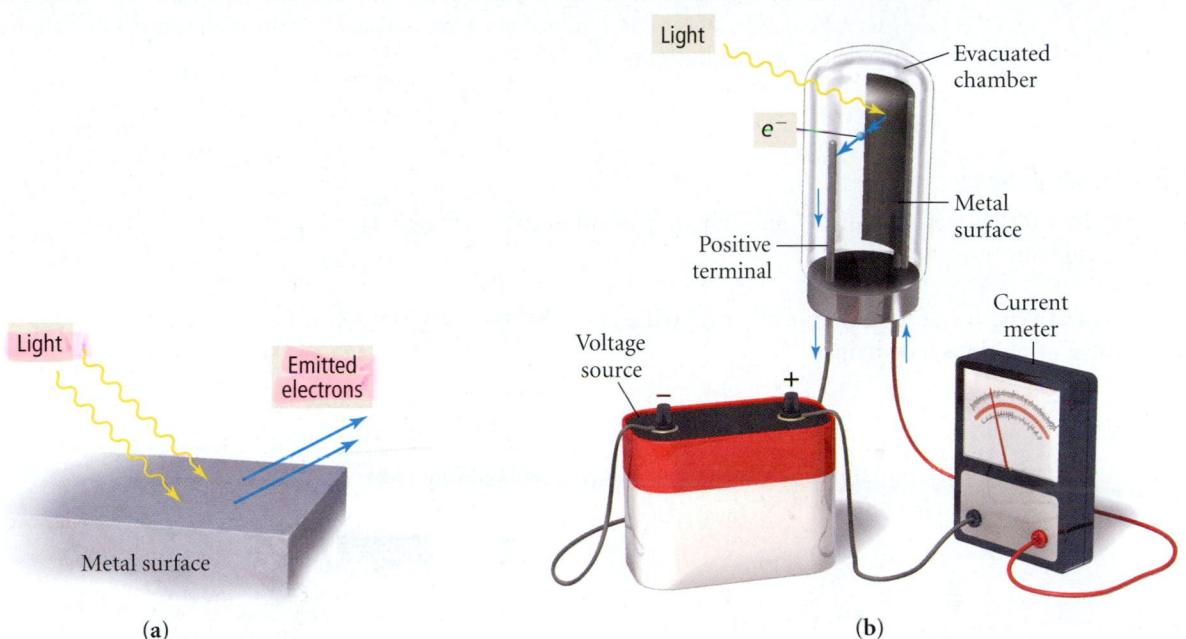

(a)   **(b)**

▲ **FIGURE 7.8 The Photoelectric Effect** **(a)** When sufficiently energetic light is shined on a metal surface, electrons are emitted. **(b)** The emitted electrons can be measured as an electrical current.

dislodging the electron. In this description, changing either the wavelength (color) or the amplitude (intensity) of the light should affect the emission of electrons. In other words, according to the classical description, the rate at which electrons were emitted from a metal due to the photoelectric effect could be increased by using either light of shorter wavelength or light of higher intensity (brighter light). If a dim light were used, the classical description predicted that there would be a *lag time* between the initial shining of the light and the subsequent emission of an electron. The lag time was the minimum of amount of time required for the dim light to transfer sufficient energy to the electron to dislodge it.

However, experiments showed that the light used to dislodge electrons in the photoelectric effect had a *threshold frequency*, below which no electrons were emitted from the metal, no matter how long the light shone on the metal. In other words, low-frequency (long-wavelength) light would not eject electrons from a metal regardless of its intensity or its duration. But high-frequency (short-wavelength) light would eject electrons, even if its intensity were low. Furthermore, the high-frequency, low-intensity light produced electrons without the predicted lag time. What could explain this odd behavior?

In 1905, Albert Einstein proposed a bold explanation: *light energy must come in packets.* According to Einstein, the amount of energy (E) in a light packet depends on its frequency (ν) according to the following equation:

$$E = h\nu \qquad [7.2]$$

where h, called *Planck's constant*, has the value $h = 6.626 \times 10^{-34}$ J·s. A *packet* of light is called a **photon** or a **quantum** of light. Since $\nu = c/\lambda$, the energy of a photon can also be expressed in terms of wavelength as follows:

$$E = \frac{hc}{\lambda} \qquad [7.3]$$

Einstein was not the first to suggest that energy was quantized. Planck used the idea in 1900 to account for certain characteristics of radiation from hot bodies. However, he did not suggest that light actually traveled in discrete packets.

The energy of a photon is directly proportional to its frequency.

The energy of a photon is inversely proportional to its wavelength.

Unlike classical electromagnetic theory, in which light was viewed purely as a wave whose intensity was *continuously variable*, Einstein suggested that light was *lumpy*. From this perspective, a beam of light is *not* a wave propagating through space, but a shower of particles, each with energy $h\nu$.

EXAMPLE 7.2 Photon Energy

A nitrogen gas laser pulse with a wavelength of 337 nm contains 3.83 mJ of energy. How many photons does it contain?

Sort You are given the wavelength and total energy of a light pulse and asked to find the number of photons it contains.	**Given** $E_{pulse} = 3.83$ mJ $\lambda = 337$ nm **Find** number of photons

Strategize In the first part of the conceptual plan, calculate the energy of an individual photon from its wavelength.

Conceptual Plan

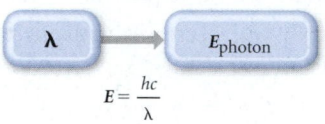

$$E = \frac{hc}{\lambda}$$

In the second part, divide the total energy of the pulse by the energy of a photon to get the number of photons in the pulse.

$$\frac{E_{pulse}}{E_{photon}} = \text{number of photons}$$

Relationships Used $E = hc/\lambda$ (Equation 7.3)

Solve To execute the first part of the conceptual plan, convert the wavelength to meters and substitute it into the equation to compute the energy of a 337-nm photon.

Solution

$$\lambda = 337 \text{ nm} \times \frac{10^{-9} \text{ m}}{1 \text{ nm}} = 3.37 \times 10^{-7} \text{ m}$$

$$E_{photon} = \frac{hc}{\lambda} = \frac{(6.626 \times 10^{-34} \text{ J} \cdot \text{s})(3.00 \times 10^8 \frac{\text{m}}{\text{s}})}{3.37 \times 10^{-7} \text{ m}}$$

$$= 5.8985 \times 10^{-19} \text{ J}$$

To execute the second part of the conceptual plan, convert the energy of the pulse from mJ to J. Then divide the energy of the pulse by the energy of a photon to obtain the number of photons.

$$3.83 \text{ mJ} \times \frac{10^{-3} \text{ J}}{1 \text{ mJ}} = 3.83 \times 10^{-3} \text{ J}$$

$$\text{number of photons} = \frac{E_{pulse}}{E_{photon}} = \frac{3.83 \times 10^{-3} \text{ J}}{5.8985 \times 10^{-19} \text{ J}}$$

$$= 6.49 \times 10^{15} \text{ photons}$$

For Practice 7.2

A 100-watt lightbulb radiates energy at a rate of 100 J/s. (The watt, a unit of power, or energy over time, is defined as 1 J/s.) If all of the light emitted has a wavelength of 525 nm, how many photons are emitted per second? (Assume three significant figures in this calculation.)

For More Practice 7.2

The energy required to dislodge electrons from sodium metal via the photoelectric effect is 275 kJ/mol. What wavelength in nm of light has sufficient energy per photon to dislodge an electron from the surface of sodium?

EXAMPLE 7.3 Wavelength, Energy, and Frequency

Arrange the following three types of electromagnetic radiation—visible light, X-rays, and microwaves—in order of increasing:

(a) wavelength

(b) frequency

(c) energy per photon

Solution

Examine Figure 7.5 and note that that X-rays have the shortest wavelength, followed by visible light and then microwaves	**(a)** wavelength X-rays < visible < microwaves
Since frequency and wavelength are inversely proportional—the longer the wavelength the shorter the frequency—the ordering with respect to frequency is the reverse order with respect to wavelength.	**(b)** frequency microwaves < visible < X-rays
Energy per photon decreases with increasing wavelength, but increases with increasing frequency; therefore the ordering with respect to energy per photon is the same as for frequency.	**(c)** energy per photon microwaves < visible < X-rays

For Practice 7.3

Arrange the following colors of visible light—green, red, and blue—in order of increasing:

(a) wavelength

(b) frequency

(c) energy per photon

Einstein's idea that light was quantized elegantly explains the photoelectric effect. The emission of electrons depends on whether or not a single photon has sufficient energy (as given by $h\nu$) to dislodge a single electron. For an electron bound to the metal with binding energy ϕ, the threshold frequency is reached when the energy of the photon is equal to ϕ.

The symbol ϕ is the Greek letter phi, pronounced "fi."

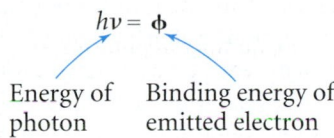

Threshold frequency condition

$$h\nu = \phi$$

Energy of photon   Binding energy of emitted electron

Low-frequency light will not eject electrons because no single photon has the minimum energy necessary to dislodge the electron. Increasing the *intensity* of low-frequency light simply increases the number of low-energy photons, but does not produce any single photon with greater energy. In contrast, increasing the *frequency* of the light, even at low intensity, increases the energy of each photon, allowing the photons to dislodge electrons with no lag time.

As the frequency of the light is increased past the threshold frequency, the excess energy of the photon (beyond what is needed to dislodge the electron) is transferred to the electron in the form of kinetic energy. The kinetic energy (KE) of the ejected electron, therefore, is simply the difference between the energy of the photon ($h\nu$) and the binding energy of the electron, as given by the equation

$$KE = h\nu - \phi$$

Although the quantization of light explained the photoelectric effect, the wave explanation of light continued to have explanatory power as well, depending on the circumstances of the particular observation. So what slowly emerged (albeit with some measure of resistance) is what we now call the *wave–particle duality of light*. Sometimes light appears to behave like a wave, at other times like a particle. Which behavior you observe depends on the particular experiment.

Conceptual Connection 7.1 The Photoelectric Effect

Light of three different wavelengths—325 nm, 455 nm, and 632 nm—was shined on a metal surface. The observations for each wavelength, labeled A, B, and C, were as follows:

Observation A: No photoelectrons were observed.

Observation B: Photoelectrons with a kinetic energy of 155 kJ/mol were observed.

Observation C: Photoelectrons with a kinetic energy of 51 kJ/mol were observed.

Which observation corresponds to which wavelength of light?

Answer: Observation A corresponds to 632 nm; observation B corresponds to 325 nm; and observation C corresponds to 455 nm. The shortest wavelength of light (highest energy per photon) must correspond to the photoelectrons with the greatest kinetic energy. The longest wavelength of light (lowest energy per photon) must correspond to the observation where no photoelectrons were observed.

7.3 Atomic Spectroscopy and the Bohr Model

The discovery of the particle nature of light began to break down the division that existed in nineteenth century physics between electromagnetic radiation, which was thought of as a wave phenomenon, and the small particles (protons, neutrons, and electrons) that compose atoms, which were thought to follow Newton's laws of motion. Just as the photoelectric effect suggested the particle nature of light, so certain observations of atoms began to suggest a wave nature for particles. The most important of these came from *atomic spectroscopy*, the study of the electromagnetic radiation absorbed and emitted by atoms. When an atom absorbs energy—in the form of heat, light, or electricity—it often reemits that energy as light. For example, a neon sign is composed of one or more glass tubes filled with neon gas. When an electric current is passed through the tube, the neon atoms absorb some of the electrical energy and reemit it as the familiar red light of a neon sign. If the atoms in the tube are different, the emitted light is a different color. In other words, atoms of each element emit light of a characteristic color. Mercury atoms, for example, emit light that appears blue, helium atoms emit light that appears violet, and hydrogen atoms emit light that appears reddish (Figure 7.9 ◀).

Closer inspection of the light emitted by various atoms reveals that each contains several distinct wavelengths. Just as the white light from a lightbulb can be separated into its constituent wavelengths by passing it through a prism, so can the light emitted by an element when it is heated, as shown in Figure 7.10 ▶. The result is a series of bright lines called an **emission spectrum**. The emission spectrum of a particular element is always the same and can be used to identify the element. For example, light arriving from a distant star contains the emission spectra of the elements that compose it. Analysis of the light allows us to identify the elements present in the star.

Notice the differences between a white light spectrum and the emission spectra of hydrogen, helium, and barium. The white light spectrum is *continuous*, meaning that there are no sudden interruptions in the intensity of the light as a function of wavelength—it consists of light of all wavelengths. The emission spectra of hydrogen, helium, and barium, however, are not continuous—they consist of bright lines at specific wavelengths, with complete darkness in between. That is, only certain discrete wavelengths of light are present. Classical physics could not explain why these spectra consisted of discrete lines. In fact, according to classical physics, an atom composed of an electron orbiting a nucleus should emit a continuous white light spectrum. Even more problematic, the electron should lose energy as it emitted the light and spiral into the nucleus.

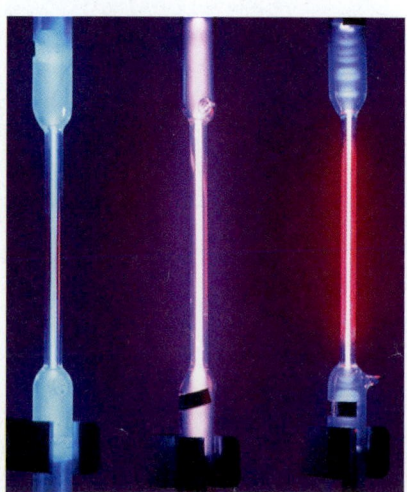

▲ The familiar red light from a neon sign is emitted by neon atoms that have absorbed electrical energy, which they reemit as visible radiation.

Remember that the color of visible light is determined by its wavelength.

▲ **FIGURE 7.9 Mercury, Helium, and Hydrogen** Each element emits a characteristic color.

Emission Spectra

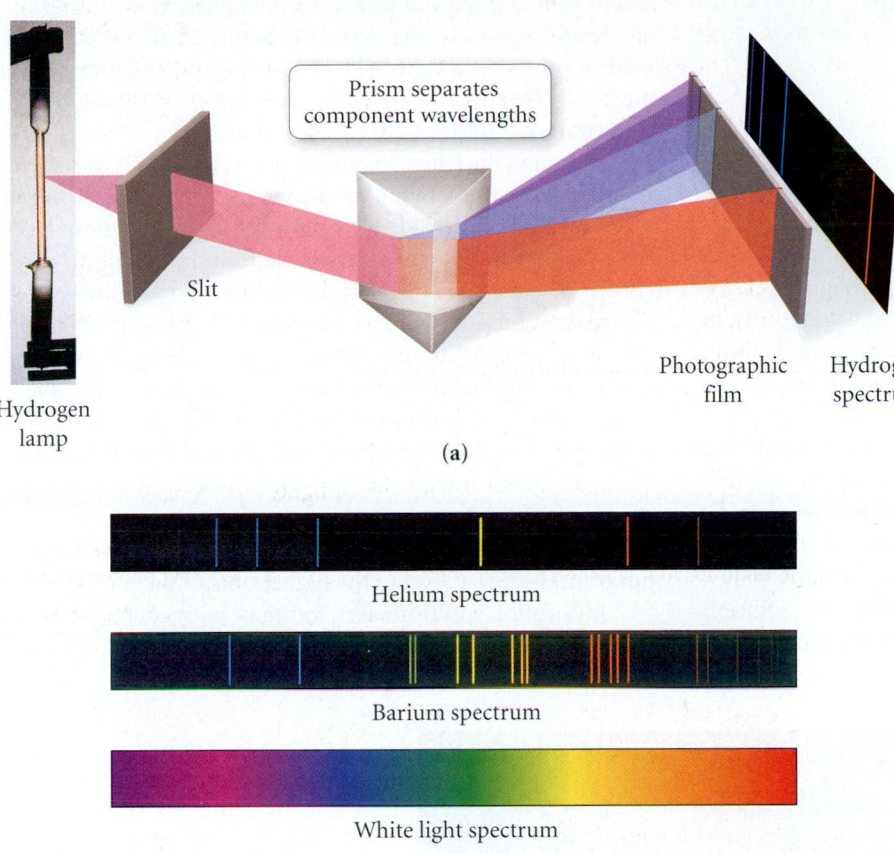

Prism separates
component wavelengths

Slit

Hydrogen
lamp

Photographic
film

Hydrogen
spectrum

(a)

Helium spectrum

Barium spectrum

White light spectrum

(b)

◀ **FIGURE 7.10 Emission Spectra**
(a) The light emitted from a hydrogen, helium, or barium lamp consists of specific wavelengths, which can be separated by passing the light through a prism. (b) The resulting bright lines constitute an emission spectrum characteristic of the element that produced it.

Johannes Rydberg, a Swedish mathematician, analyzed many atomic spectra and developed a simple equation (shown in the margin) that predicted the wavelengths of the hydrogen emission spectrum. However, his equation gave little insight into *why* atomic spectra were discrete, *why* atoms were stable, or *why* his equation worked.

The Danish physicist Neils Bohr (1885–1962) attempted to develop a model for the atom that explained atomic spectra. In his model, electrons travel around the nucleus in circular orbits (similar to those of the planets around the sun). However, in contrast to planetary orbits—which can theoretically exist at any distance from the sun—Bohr's orbits could exist only at specific, fixed distances from the nucleus. The energy of each Bohr orbit was also fixed, or *quantized*. Bohr called these orbits *stationary states* and suggested that, although they obeyed the laws of classical mechanics, they also possessed "a peculiar, mechanically unexplainable, stability." We now know that the stationary states were really manifestations of the wave nature of the electron, which we expand upon shortly. Bohr further proposed that, in contradiction to classical electromagnetic theory, no radiation was emitted by an electron orbiting the nucleus in a stationary state. It was only when an electron jumped, or made a *transition*, from one stationary state to another that radiation was emitted or absorbed (Figure 7.11 ▶).

The Rydberg equation is $1/\lambda = R(1/m^2 - 1/n^2)$, where R is the Rydberg constant ($1.097 \times 10^7 \text{m}^{-1}$), and m and n are integers.

The Bohr Model and Emission Spectra

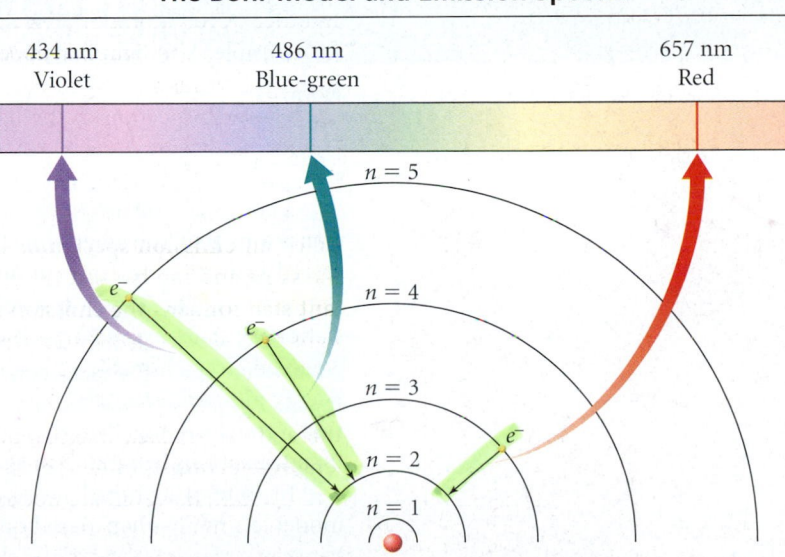

434 nm
Violet

486 nm
Blue-green

657 nm
Red

$n = 5$

$n = 4$

$n = 3$

$n = 2$

$n = 1$

▲ **FIGURE 7.11 The Bohr Model and Emission Spectra** In the Bohr model, each spectral line is produced when an electron falls from one stable orbit, or stationary state, to another of lower energy.

Chemistry in Your Day

Atomic Spectroscopy, a Bar Code for Atoms

When you check out of the grocery store, a laser scanner reads the bar code on the items that you have selected for purchase. Each item has a unique code that identifies it and its price. Similarly, each element in the periodic table has a spectrum unlike that of any other element. For example, Figure 7.12 ▼ shows the emission spectra of oxygen and neon. (Previously, in Figure 7.10, we saw the emission spectra of hydrogen, helium, and barium.) Notice that each spectrum is unique and can be used to identify the substance.

The presence of intense lines in the spectra of a number of metals is the basis for *flame tests*, simple tests used to identify elements in ionic compounds even without a precise spectral analysis. For example, the emission spectrum of sodium in the visible range features two closely spaced, bright yellow lines. When a crystal of a sodium salt (or a drop of a solution containing a sodium salt) is put into a flame, the flame glows

bright yellow (Figure 7.13 ▶). As the figure shows, other metals exhibit similarly characteristic colors in flame tests. Each color represents an especially bright spectral emission line (or a combination of two or more such lines). Similar emissions form the basis of the colors seen in fireworks.

Although the *emission* of light from elements is easier to detect, the *absorption* of light by elements is even more commonly used for purposes of identification. Whereas emission spectra consist of bright lines on a dark background, absorption spectra consist of dark lines on a bright background (Figure 7.14 ▶). An absorption spectrum is measured by passing white light through a sample and observing what wavelengths are *missing* due to absorption by the sample. Notice that, in the spectra shown here, the absorption lines are at the same wavelengths as the emission lines. This is because the processes that produce them are mirror images. In emission, an electron makes a transition from a higher energy level to a lower energy one. In absorption, the transition is between the same two energy levels, but from the lower level to the higher one.

Absorption spectrometers, found in most chemistry laboratories, typically produce a plot of the intensity of absorption as a

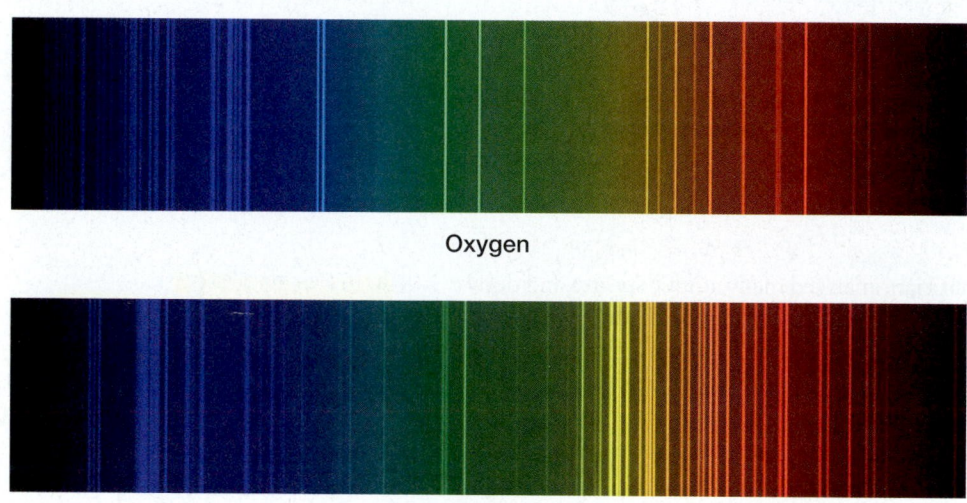

Oxygen

Neon

◀ **FIGURE 7.12 Emission Spectra of Oxygen and Neon**   The emission spectrum of each element is unique and can be used to identify the element.

The transitions between stationary states in a hydrogen atom are quite unlike any transitions that you might imagine in the macroscopic world. The electron is *never* observed *between states*, only in one state or the next—the transition between states is instantaneous. The emission spectrum of an atom consisted of discrete lines because the stationary states existed only at specific, fixed energies. The energy of the photon created when an electron made a transition from one stationary state to another was simply the energy difference between the two stationary states. Transitions between stationary states that were closer together, therefore, produced light of lower energy (longer wavelength) than transitions between stationary states that were farther apart.

In spite of its initial success in explaining the line spectrum of hydrogen, the Bohr model left many unanswered questions. It did, however, serve as an intermediate model between a classical view of the electron and a fully quantum-mechanical view, and therefore has great historical and conceptual importance. Nonetheless, it was ultimately replaced by a more complete quantum-mechanical theory that fully incorporated the wave nature of the electron.

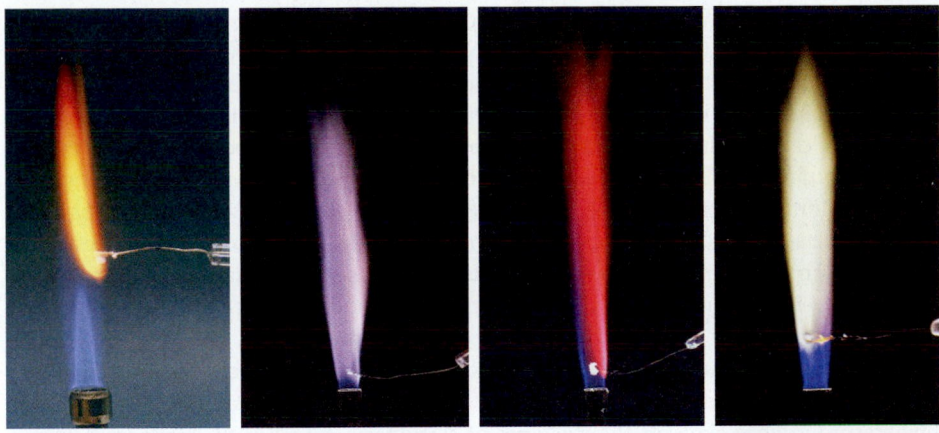

▲ **FIGURE 7.13 Flame Tests for Sodium, Potassium, Lithium, and Barium** Elements can often be identified by the characteristic color of the light they produce when heated. The colors derive from especially bright lines in their emission spectra.

◄ Fireworks typically contain salts of such metals as sodium, calcium, strontium, barium, and copper. Emissions from these elements produce the brilliant colors of pyrotechnic displays.

function of wavelength. Such plots are useful both for identifying substances (qualitative analysis) and for determining the concentration of substances (quantitative analysis). Quantitative analysis is possible because the amount of light absorbed by a sample depends on the concentration of the absorbing substance within the sample. For example, the concentration of Ca^{2+} in a hard water sample can be determined by measuring the quantity of light absorbed by the calcium ion at its characteristic wavelength.

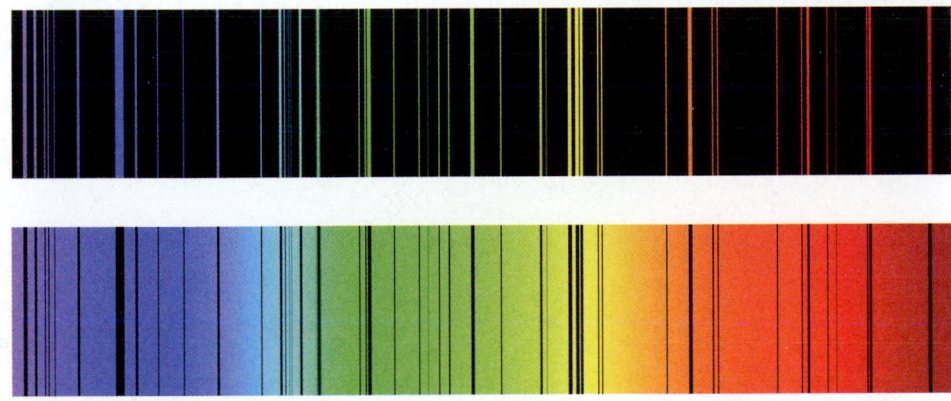

◄ **FIGURE 7.14 Emission and Absorption Spectrum of Mercury** Elements absorb light of the same wavelengths that they radiate when heated. When these wavelengths are subtracted from a beam of white light, the result is a pattern of dark lines called an absorption spectrum.

7.4 The Wave Nature of Matter: The de Broglie Wavelength, the Uncertainty Principle, and Indeterminacy

The heart of the quantum-mechanical theory that replaced Bohr's model is the wave nature of the electron, first proposed by Louis de Broglie (1892-1987) in 1924 and later confirmed by experiments in 1927. It seemed incredible at the time, but electrons—which were thought of as particles and known to have mass—also had a wave nature. The wave nature of the electron is seen most clearly in its diffraction. If an electron beam is aimed at two closely spaced slits, and a series (or array) of detectors is arranged to detect the electrons after they pass through the slits, an interference pattern similar to that observed

The first evidence of electron wave properties was provided by the Davisson-Germer experiment of 1927, in which electrons were observed to undergo diffraction by a metal crystal.

For interference to occur, the spacing of the slits has to be on the order of atomic dimensions.

for light is recorded behind the slits (Figure 7.15a ▼). The detectors at the center of the array (midway between the two slits) detect a large number of electrons—exactly the opposite of what you would expect for particles (Figure 7.15b ▼). Moving outward from this center spot, the detectors alternately detect small numbers of electrons and then large numbers again and so on, forming an interference pattern characteristic of waves.

It is critical to understand that the interference pattern described here is *not caused by pairs of electrons interfering with each other, but rather by single electrons interfering with themselves.* If the electron source is turned down to a very low level, so that electrons come out only one at a time, *the interference pattern remains.* In other words, we can design an experiment in which electrons come out of the source singly. We can then record where each electron strikes the detector after it has passed through the slits. If we record the positions of thousands of electrons over a long period of time, we find the same interference pattern shown in Figure 7.15(a). This leads us to an important conclusion: *The wave nature of the electron is an inherent property of individual electrons.* As it turns out, this wave nature is what explains the existence of stationary states (in the Bohr model) and prevents the electrons in an atom from crashing into the nucleus as predicted by classical physics. We now turn to three important manifestations of the electron's wave nature: the de Broglie wavelength, the uncertainty principle, and indeterminacy.

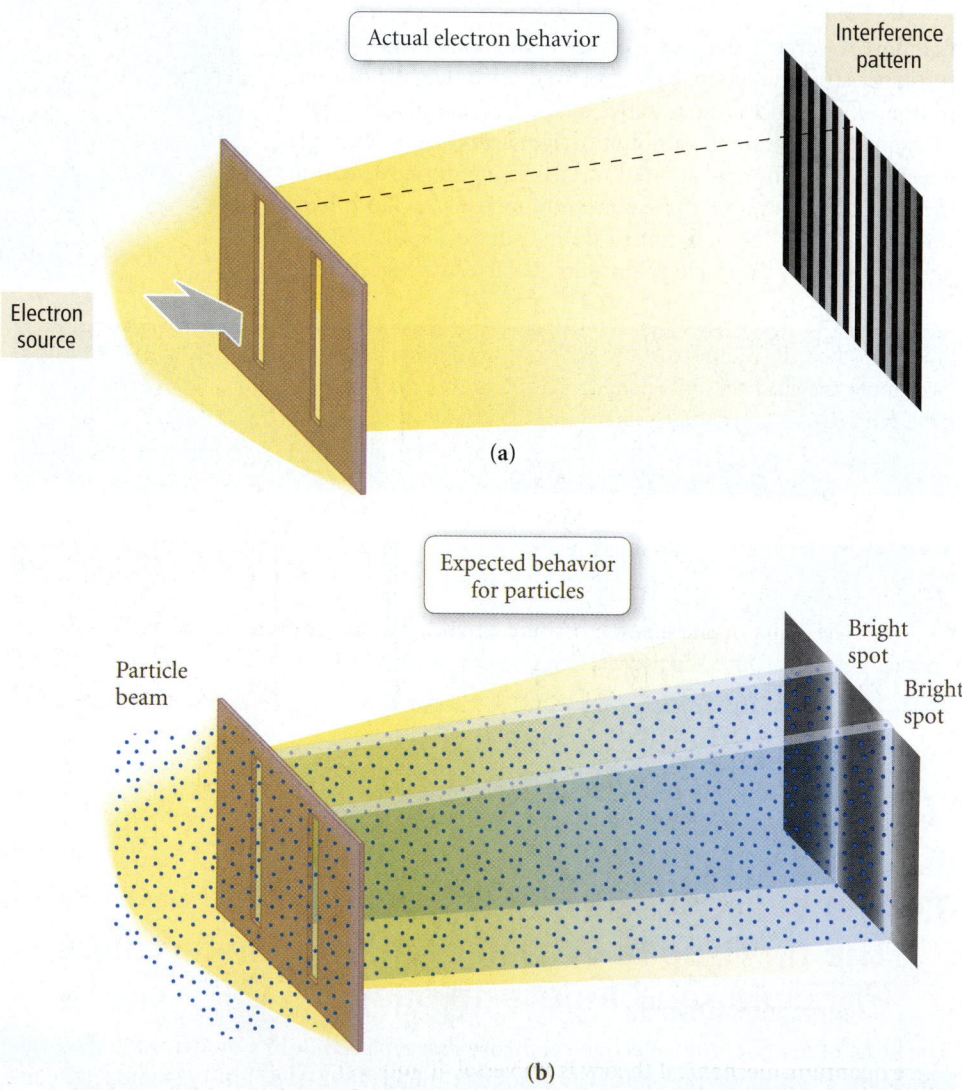

▲ **FIGURE 7.15 Electron Diffraction**   When a beam of electrons goes through two closely spaced slits (**a**), an interference pattern is created, as if the electrons were waves. By contrast, a beam of particles passing through two slits (**b**) should simply produce two smaller beams of particles. Notice that for particle beams, there is a dark line directly behind the center of the two slits, in contrast to wave behavior, which produces a bright line.

The de Broglie Wavelength

As we have seen, a single electron traveling through space has a wave nature; its wavelength is related to its kinetic energy (the energy associated with its motion). The faster the electron is moving, the higher its kinetic energy and the shorter its wavelength. The wavelength (λ) of an electron of mass m moving at velocity v is given by the **de Broglie relation**:

$$\lambda = \frac{h}{mv} \qquad \text{de Broglie relation} \qquad [7.4]$$

where h is Planck's constant. Notice that the velocity of a moving electron is related to its wavelength—knowing one is equivalent to knowing the other.

> The mass of an object (m) times its velocity (v) is its momentum. Therefore, the wavelength of an electron is inversely proportional to its momentum.

EXAMPLE 7.4 De Broglie Wavelength

Calculate the wavelength of an electron traveling with a speed of 2.65×10^6 m/s

Sort You are given the speed of an electron and asked to calculate its wavelength.	**Given** $v = 2.65 \times 10^6$ m/s **Find** λ
Strategize The conceptual plan shows how the de Broglie relation relates the wavelength of an electron to its mass and velocity.	**Conceptual Plan** $v \longrightarrow \lambda$ $\lambda = \dfrac{h}{mv}$ **Relationships Used** $\lambda = h/mv$ (de Broglie relation, Equation 7.4)
Solve Substitute the velocity, Planck's constant, and the mass of an electron to compute the electron's wavelength. To correctly cancel the units, break down the J in Planck's constant into its SI base units ($1\,\text{J} = 1\,\text{kg} \cdot \text{m}^2/\text{s}^2$)	**Solution** $\lambda = \dfrac{h}{mv} = \dfrac{6.626 \times 10^{-34} \dfrac{\text{kg} \cdot \text{m}^2}{\text{s}^2} \cdot \text{s}}{(9.11 \times 10^{-31}\,\text{kg})\left(2.65 \times 10^6\,\dfrac{\text{m}}{\text{s}}\right)}$ $= 2.74 \times 10^{-10}\,\text{m}$

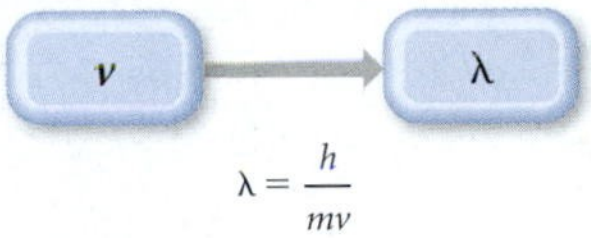

Check The units of the answer (m) are correct. The magnitude of the answer is very small, as expected for the wavelength of an electron.

For Practice 7.4

What is the velocity of an electron having a de Broglie wavelength that is approximately the length of a chemical bond? Assume this length to be 1.2×10^{-10} m.

✦ Conceptual Connection 7.2 The de Broglie Wavelength of Macroscopic Objects

Since quantum-mechanical theory is universal, it applies to all objects, regardless of size. Therefore, according to the de Broglie relation, a thrown baseball should also exhibit wave properties. Why do we not observe such properties at the ballpark?

Answer: Because of the baseball's large mass, its de Broglie wavelength is minuscule. (For a 150-g baseball, λ is on the order of 10^{-34} m.) This minuscule wavelength is insignificant compared to the size of the baseball itself, and therefore its effects are not measurable.

The Uncertainty Principle

The wave nature of the electron is difficult to reconcile with its particle nature. How can a single entity behave as both a wave and a particle? We can begin to answer this question by returning to the single-electron diffraction experiment. Specifically, we can ask the following question: how does a single electron aimed at a double slit produce an interference pattern? A possible hypothesis is that the electron splits into two, travels through both slits, and interferes with itself. This hypothesis seems testable. We simply have to observe the single electron as it travels through the slits. If it travels through both slits at the same time, we know our hypothesis is correct.

However, any experiment designed to observe the electron as it travels through the slits results in the detection of an electron "particle" traveling through a single slit and no interference pattern. For example, the following electron diffraction experiment is designed to "watch" which slit the electron travels through by using a laser beam placed directly behind the slits.

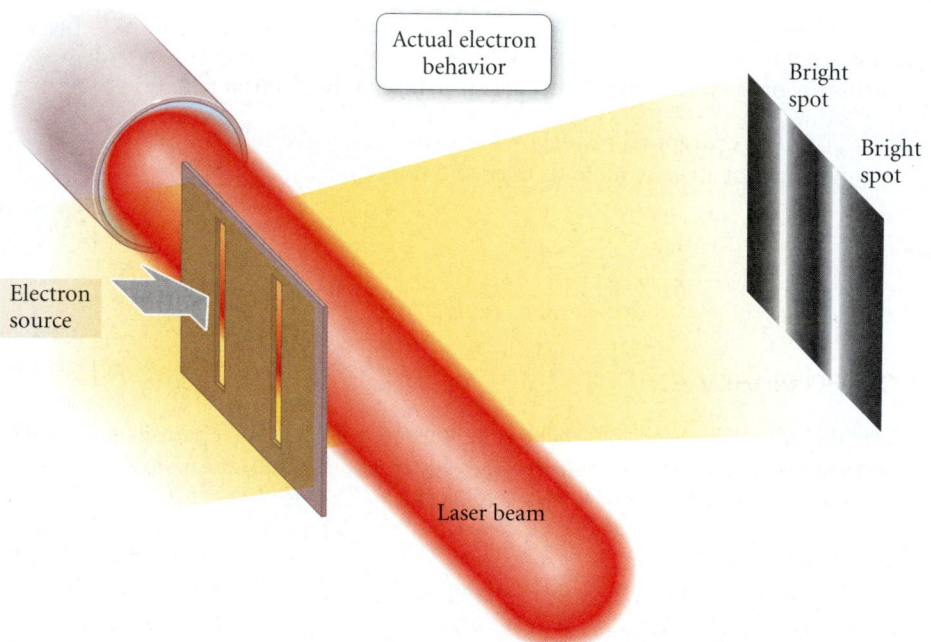

An electron that crosses a light beam produces a tiny "flash"—a single photon is scattered at the point of crossing. A flash behind a particular slit indicates an electron passing through that slit. However, when the experiment is performed, the flash always originates either from one slit *or* the other, but *never* from both at once. Futhermore, the interference pattern, which was present without the laser, is now absent. With the laser on, the electrons hit positions directly behind each slit, as if they were ordinary particles.

As it turns out, no matter how hard we try, or whatever method we set up, *we can never see the interference pattern and simultaneously determine which hole the electron goes through*. It has never been done, and most scientists agree that it never will. In the words of P. A. M. Dirac,

> There is a limit to the fineness of our powers of observation and the smallness of the accompanying disturbance—a limit which is inherent in the nature of things and can never be surpassed by improved technique or increased skill on the part of the observer.

We have encountered the absolutely small and have no way of determining what it is doing without disturbing it.

The single electron diffraction experiment demonstrates that you cannot simultaneously observe both the wave nature and the particle nature of the electron. When you try to observe which hole the electron goes through (associated with the particle nature of the electron) you lose the interference pattern (associated with the wave nature of the electron). When you try to observe the interference pattern, you cannot determine which hole the electron goes through. The wave nature and particle nature of the electron are said to be **complementary properties**. Complementary properties exclude one another—the more you know about one, the less you know about the other. Which of two complementary properties you observe depends on the experiment you perform—remember that in quantum mechanics, the observation of an event affects its outcome.

As we just saw in the de Broglie relation, the *velocity* of an electron is related to its *wave nature*. The *position* of an electron, however, is related to its *particle nature*. (Particles have well-defined position, but waves do not.) Consequently, our inability to observe the electron simultaneously as both a particle and a wave means that we cannot simultaneously measure its position and its velocity. Werner Heisenberg formalized this idea with the following equation:

$$\Delta x \times m\,\Delta v \geq \frac{h}{4\pi} \qquad \text{Heisenberg's uncertainty principle} \qquad [7.5]$$

where Δx is the uncertainty in the position, Δv is the uncertainty in the velocity, m is the mass of the particle, and h is Planck's constant. **Heisenberg's uncertainty principle** states that the product of Δx and $m\,\Delta v$ must be greater than or equal to a finite number ($h/4\pi$). In other words, the more accurately you know the position of an electron (the smaller Δx) the less accurately you can know its velocity (the bigger Δv) and vice versa. The complementarity of the wave nature and particle nature of the electron results in the complementarity of velocity and position.

Although Heisenberg's uncertainty principle may seem puzzling, it actually solves a great puzzle. Without the uncertainty principle, we are left with the following question: how can something be *both* a particle and a wave? Saying that an object is both a particle and a wave is like saying that an object is both a circle and a square, a contradiction. Heisenberg solved the contradiction by introducing complementarity—an electron is observed as *either* a particle or a wave, but never both at once.

▲ Werner Heisenberg (1901–1976)

Indeterminacy and Probability Distribution Maps

According to classical physics, and in particular Newton's laws of motion, particles move in a *trajectory* (or path) that is determined by the particle's velocity (the speed and direction of travel), its position, and the forces acting on it. Even if you are not familiar with Newton's laws, you probably have an intuitive sense of them. For example, when you chase a baseball in the outfield, you visually predict where the ball will land by observing its path. You do this by noting its initial position and velocity, watching how these are affected by the forces acting on it (gravity, air resistance, wind), and then inferring its trajectory, as shown in Figure 7.16 ▼. Notice that if you knew only the ball's velocity, or only its position (imagine a still photo of the baseball in the air), you could not predict its landing spot. In classical mechanics, both position and velocity are required to predict a trajectory.

Newton's laws of motion are **deterministic**—the present *determines* the future. This means that if two baseballs are hit consecutively with the same velocity from the same

Remember that velocity includes speed as well as direction of travel.

The Classical Concept of Trajectory

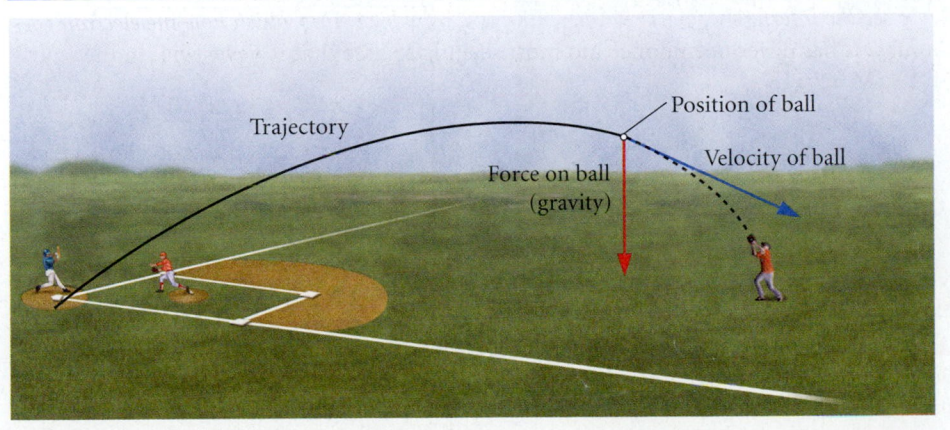

Trajectory

Position of ball

Velocity of ball

Force on ball
(gravity)

◀ **FIGURE 7.16 The Concept of Trajectory** In classical mechanics, the position and velocity of a particle determine its future trajectory, or path. Thus, an outfielder can catch a baseball by observing its position and velocity, allowing for the effects of forces acting on it, such as gravity, and estimating its trajectory. (For simplicity, air resistance and wind are not shown.)

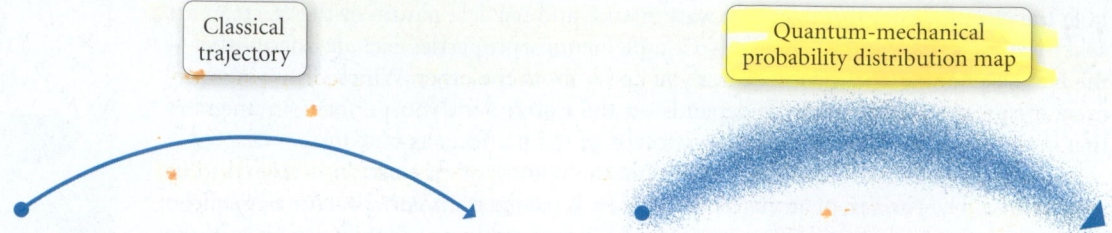

Classical trajectory

Quantum-mechanical probability distribution map

▲ **FIGURE 7.17 Trajectory versus Probability**   In quantum mechanics, we cannot calculate deterministic trajectories. Instead, it is necessary to think in terms of probability maps: statistical pictures of where a quantum-mechanical particle, such as an electron, is most likely to be found. In this hypothetical map, darker shading indicates greater probability.

▲ **FIGURE 7.18 Trajectory of a Macroscopic Object**   A baseball follows a well-defined trajectory from the hand of the pitcher to the mitt of the catcher.

position under identical conditions, they will land in exactly the same place. The same is not true of electrons. We have just seen that we cannot simultaneously know the position and velocity of an electron; therefore, we cannot know its trajectory. In quantum mechanics, trajectories are replaced with *probability distribution maps*, as shown in Figure 7.17 ▲. A probability distribution map is a statistical map that shows where an electron is likely to be found under a given set of conditions.

To understand the concept of a probability distribution map, let us return to baseball. Imagine a baseball thrown from the pitcher's mound to a catcher behind home plate (Figure 7.18 ◄.) The catcher can watch the baseball's path, predict exactly where it will cross home plate, and place his mitt in the correct place to catch it. As we have seen, this would be impossible for an electron. If an electron were thrown from the pitcher's mound to home plate, it would land in a different place every time, even if it were thrown in exactly the same way. This behavior is called **indeterminacy**. Unlike a baseball, whose future path is *determined* by its position and velocity when it leaves the pitcher's hand, the future path of an electron is indeterminate, and can only be described statistically.

In the quantum-mechanical world of the electron, the catcher could not know exactly where the electron will cross the plate for any given throw. However, if he kept track of hundreds of identical electron throws, the catcher could observe a reproducible, *statistical pattern* of where the electron crosses the plate. He could even draw a map of the strike zone showing the probability of an electron crossing a certain area, as shown in Figure 7.19 ▼. This would be a probability distribution map. In the sections that follow, we discuss quantum-mechanical electron *orbitals*, which are essentially probability distribution maps for electrons as they exist within atoms.

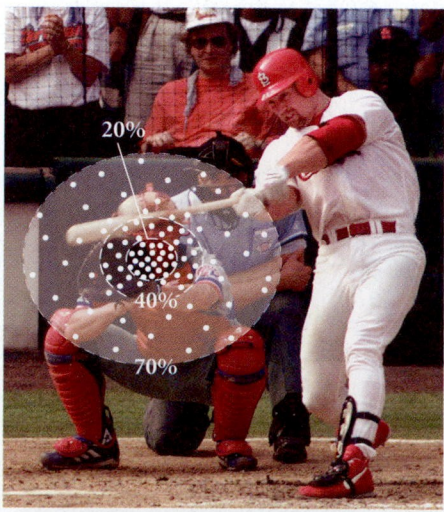

20%

40%

70%

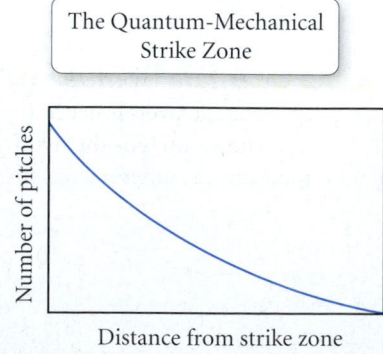

The Quantum-Mechanical Strike Zone

Number of pitches

Distance from strike zone

▲ **FIGURE 7.19 The Quantum-Mechanical Strike Zone**   An electron does not have a well-defined trajectory. However, we can construct a probability distribution map to show the relative probability of it crossing home plate at different points.

7.5 Quantum Mechanics and the Atom

As we have seen, the position and velocity of the electron are complementary properties—if we know one accurately, the other becomes indeterminate. Since velocity is directly related to energy (we have seen that kinetic energy equals $\frac{1}{2}mv^2$), position and *energy* are also complementary properties—the more you know about one, the less you know about the other. Many of the properties of an element, however, depend on the energies of its electrons. For example, whether an electron is transferred from one atom to another to form an ionic bond depends in part on the relative energies of the electron in the two atoms. In the following paragraphs, we describe the probability distribution maps for electron states in which the electron has well-defined energy, but not well-defined position. In other words, for each state, we can specify the *energy* of the electron precisely, but not its location at a given instant. Instead, the electron's position is described in terms of an **orbital**, a probability distribution map showing where the electron is likely to be found. Since chemical bonding often involves the sharing of electrons between atoms to form covalent bonds, the spatial distribution of atomic electrons is also important to bonding.

The mathematical derivation of energies and orbitals for electrons in atoms comes from solving the *Schrödinger* equation for the atom of interest. The general form of the Schrödinger equation is as follows:

$$\mathcal{H}\psi = E\psi \qquad [7.6]$$

The symbol $\mathcal{H}$ stands for the Hamiltonian operator, a set of mathematical operations that represent the total energy (kinetic and potential) of the electron within the atom. The symbol E is the actual energy of the electron. The symbol ψ is the **wave function**, a mathematical function that describes the wavelike nature of the electron. A plot of the wave function squared (ψ^2) represents an orbital, a position probability distribution map of the electron.

> These states are known as energy *eigenstates*.

> An operator is different from a normal algebraic entity. In general, an operator transforms a mathematical function into another mathematical function. For example, d/dx is an operator that means "take the deriviative of." When d/dx operates on a function (such as x^2) it returns another function ($2x$).

> The symbol ψ is the Greek letter psi, pronounced "sigh."

Solutions to the Schrödinger Equation for the Hydrogen Atom

When the Schrödinger equation is solved, it yields many solutions—many possible wave functions. The wave functions themselves are fairly complicated mathematical functions, and we will not examine them in detail in this book. Instead, we will introduce graphical representations (or plots) of the orbitals that correspond to the wave functions. Each orbital is specified by three interrelated **quantum numbers**: n, the **principal quantum number**; l, the **angular momentum quantum number** (sometimes called the *azimuthal quantum number*); and m_l the **magnetic quantum number**. These quantum numbers all have integer values, as had been hinted at by both the Rydberg equation and Bohr's model. We examine each of these quantum numbers individually.

The Principal Quantum Number (*n*)   The principal quantum number is an integer that determines the overall size and energy of an orbital. Its possible values are $n = 1, 2, 3, \ldots$ and so on. For the hydrogen atom, the energy of an electron in an orbital with quantum number n is given by

$$E_n = -2.18 \times 10^{-18}\,\text{J}\left(\frac{1}{n^2}\right) \qquad (n = 1, 2, 3, \ldots) \qquad [7.7]$$

The energy is negative because the energy of the electron in the atom is less than the energy of the electron when it is very far away from the atom (which is taken to be zero). Notice that orbitals with higher values of n have greater (less negative) energies, as shown in the energy level diagram on the right.

Notice also that, as n increases, the spacing between the energy levels becomes smaller.

Energy

$n = 4$ ——— $E_4 = -1.36 \times 10^{-19}$ J
$n = 3$ ——— $E_3 = -2.42 \times 10^{-19}$ J

$n = 2$ ——— $E_2 = -5.45 \times 10^{-19}$ J

$n = 1$ ——— $E_1 = -2.18 \times 10^{-18}$ J

The Angular Momentum Quantum Number (*l*) The angular momentum quantum number is an integer that determines the shape of the orbital. We will consider these shapes in Section 7.6. The possible values of l are $0, 1, 2, \ldots, (n - 1)$. In other words, for a given value of n, l can be any integer (including 0) up to $n - 1$. For example, if $n = 1$, then the only possible value of l is 0; if $n = 2$, the possible values of l are 0 and 1. In order to avoid confusion between n and l, values of l are often assigned letters as follows:

> The values of l beyond 3 are designated with letters in alphabetical order so that $l = 4$ is designated g, $l = 5$ is designated h, and so on.

Value of l	Letter Designation
$l = 0$	s
$l = 1$	p
$l = 2$	d
$l = 3$	f

The Magnetic Quantum Number (*m_l*) The magnetic quantum number is an integer that specifies the orientation of the orbital. We will consider these orientations in Section 7.6. The possible values of m_l are the integer values (including zero) ranging from $-l$ to $+l$. For example, if $l = 0$, then the only possible value of m_l is 0; if $l = 1$, the possible values of m_l are -1, 0, and $+1$; if $l = 2$, the possible values of m_l are $-2, -1, 0, +1$, and $+2$, and so on.

Each specific combination of n, l, and m_l specifies one atomic orbital. For example, the orbital with $n = 1$, $l = 0$, and $m_l = 0$ is known as the 1s orbital. The 1 in 1s is the value of n and the s specifies that $l = 0$. There is only one 1s orbital in an atom, and its m_l value is zero. Orbitals with the same value of n are said to be in the same **principal level** (or **principal shell**). Orbitals with the same value of n and l are said to be in the same **sublevel** (or **subshell**). The following diagram shows all of the orbitals in the first three principal levels.

Principal level (specified by n)

Sublevel (specified by n and l)

Orbital (specified by n, l, and m_l)

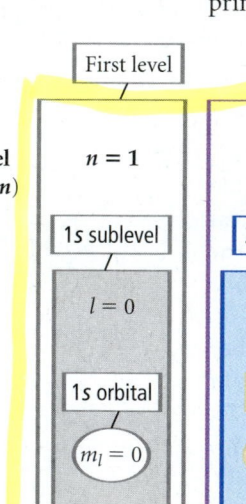
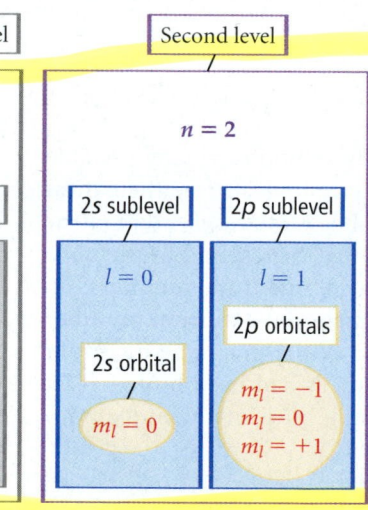
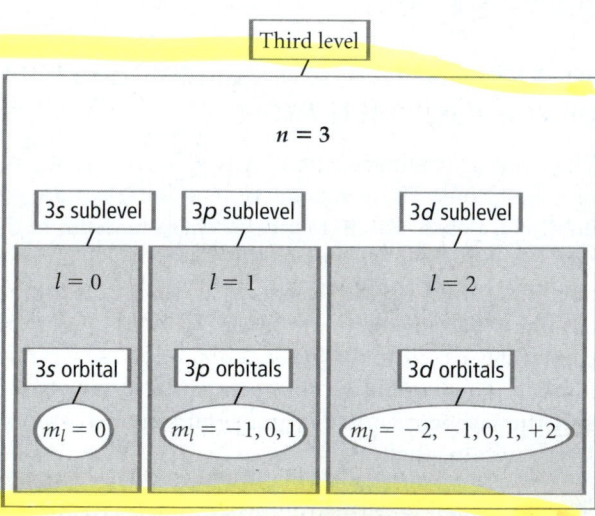

For example, the $n = 2$ level contains the $l = 0$ and $l = 1$ sublevels. Within the $n = 2$ level, the $l = 0$ sublevel—called the 2s sublevel—contains only one orbital (the 2s orbital), with $m_l = 0$. The $l = 1$ sublevel—called the 2p sublevel—contains three 2p orbitals, with $m_l = -1, 0, +1$.

In general, notice the following:

- The number of sublevels in any level is equal to n, the principal quantum number. Therefore, the $n = 1$ level has one sublevel, the $n = 2$ level has two sublevels, etc.
- The number of orbitals in any sublevel is equal to $2l + 1$. Therefore, the s sublevel ($l = 0$) has one orbital, the p sublevel ($l = 1$) has three orbitals, the d sublevel ($n = 2$) has five orbitals, etc.
- The number of orbitals in a level is equal to n^2. Therefore, the $n = 1$ level has one orbital, the $n = 2$ level has four orbitals, the $n = 3$ level has nine orbitals, etc.

EXAMPLE 7.5 Quantum Numbers I

What are the quantum numbers and names (for example, $2s$, $2p$) of the orbitals in the $n = 4$ principal level? How many $n = 4$ orbitals exist?

Solution

| We first determine the possible values of l (from the given value of n). We then determine the possible values of m_l for each possible value of l. For a given value of n, the possible values of l are $0, 1, 2, \ldots, (n - 1)$. | $n = 4$; therefore $l = 0, 1, 2,$ and 3 |

For a given value of l, the possible values of m_l are the integer values including zero ranging from $-l$ to $+l$. The name of an orbital is its principal quantum number (n) followed by the letter corresponding to the value l. The total number of orbitals is given by n^2.

l	possible m_l Values	Orbital name(s)
0	0	$4s$ (1 orbital)
1	$-1, 0, +1$	$4p$ (3 orbitals)
2	$-2, -1, 0, +1, +2$	$4d$ (5 orbitals)
3	$-3, -2, -1, 0, +1, +2, +3$	$4f$ (7 orbitals)

Total number of orbitals $= 4^2 = 16$

For Practice 7.5

List the quantum numbers associated with all of the $5d$ orbitals. How many $5d$ orbitals exist?

EXAMPLE 7.6 Quantum Numbers II

The following sets of quantum numbers are each supposed to specify an orbital. One set, however, is erroneous. Which one and why?

(a) $n = 3; l = 0; m_l = 0$ **(c)** $n = 1; l = 0; m_l = 0$

(b) $n = 2; l = 1; m_l = -1$ **(d)** $n = 4; l = 1; m_l = -2$

Solution

Choice **(d)** is erroneous because, for $l = 1$, the possible values of m_l are only $-1, 0,$ and $+1$.

For Practice 7.6

Each of the following sets of quantum numbers is supposed to specify an orbital. However, each set contains one quantum number that is not allowed. Replace the quantum number that is not allowed with one that is allowed.

(a) $n = 3; l = 3; m_l = +2$ **(b)** $n = 2; l = 1; m_l = -2$ **(c)** $n = 1; l = 1; m_l = 0$

Atomic Spectroscopy Explained

Quantum theory explains the atomic spectra of atoms discussed earlier. Each wavelength in the emission spectrum of an atom corresponds to an electron *transition* between quantum-mechanical orbitals. When an atom absorbs energy, an electron in a lower energy level is *excited* or promoted to a higher energy level, as shown in Figure 7.20 ▼.

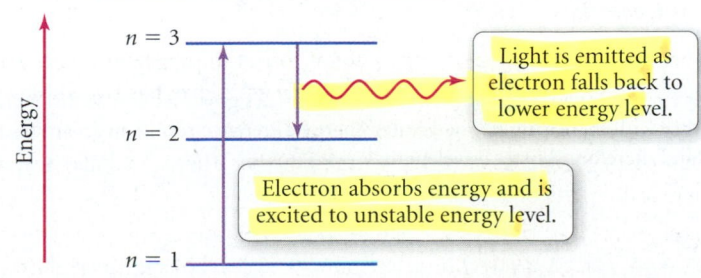

Excitation and Radiation

Light is emitted as electron falls back to lower energy level.

Electron absorbs energy and is excited to unstable energy level.

◄ **FIGURE 7.20 Excitation and Radiation** When an atom absorbs energy, an electron can be excited from an orbital in a lower energy level to an orbital in a higher energy level. The electron in this "excited state" is unstable, however, and relaxes to a lower energy level, releasing energy in the form of electromagnetic radiation.

In this new configuration, however, the atom is unstable, and the electron quickly falls back or *relaxes* to a lower energy orbital. As it does so, it releases a photon of light containing an amount of energy precisely equal to the energy difference between the two energy levels.

For example, suppose that an electron in a hydrogen atom relaxes from an orbital in the $n = 3$ level to an orbital in the $n = 2$ level. Recall that the energy of an orbital in the hydrogen atom depends only on n and is given by $E_n = -2.18 \times 10^{-18} \, \text{J}(1/n^2)$, where $n = 1, 2, 3, \ldots$. Therefore, ΔE, the energy difference corresponding to the transition from $n = 3$ to $n = 2$, is determined as follows:

$\mid \Delta E = E_{\text{final}} - E_{\text{initial}}$

$$\Delta E_{\text{atom}} = E_2 - E_3$$

$$= -2.18 \times 10^{-18} \, \text{J}\left(\frac{1}{2^2}\right) - \left[-2.18 \times 10^{-18} \, \text{J}\left(\frac{1}{3^2}\right)\right]$$

$$= -2.18 \times 10^{-18} \, \text{J}\left(\frac{1}{2^2} - \frac{1}{3^2}\right)$$

$$= -3.03 \times 10^{-19} \, \text{J}$$

The energy carries a negative sign because the atom *emits* the energy as it relaxes from $n = 3$ to $n = 2$. Since energy must be conserved, the exact amount of energy emitted by the atom is carried away by the photon:

$$\Delta E_{\text{atom}} = -E_{\text{photon}}$$

This energy then determines the frequency and wavelength of the photon. Since the wavelength of the photon is related to its energy as $E = hc/\lambda$, we calculate the wavelength of the photon as follows:

$$\lambda = \frac{hc}{E}$$

$$= \frac{(6.626 \times 10^{-34} \, \text{J} \cdot \text{s})(3.00 \times 10^8 \, \text{m/s})}{3.03 \times 10^{-19} \, \text{J}}$$

$$= 6.56 \times 10^{-7} \, \text{m} \quad \text{or} \quad 656 \, \text{nm}$$

Consequently, the light emitted by an excited hydrogen atom as it relaxes from an orbital in the $n = 3$ level to an orbital in the $n = 2$ level has a wavelength of 656 nm (red). The light emitted due to a transition from $n = 4$ to $n = 2$ can be calculated in a similar fashion to be 486 nm (green). Notice that transitions between orbitals that are further apart in energy produce light that is higher in energy, and therefore shorter in wavelength, than transitions between orbitals that are closer together. Figure 7.21 ▶ shows several of the transitions in the hydrogen atom and their corresponding wavelengths.

The Rydberg equation,
$1/\lambda = R(1/m^2 - 1/n^2)$, can be derived from the relationships just covered. We leave this derivation to an exercise (see Problem 7.90).

 ## Conceptual Connection 7.3 Emission Spectra

Which of the following transitions will result in emitted light with the shortest wavelength?

(a) $n = 5 \longrightarrow n = 4$

(b) $n = 4 \longrightarrow n = 3$

(c) $n = 3 \longrightarrow n = 2$

Answer: (c) The energy difference between $n = 3$ and $n = 2$ is greatest because the energy spacings get closer together with increasing n. The greater energy difference results in an emitted photon of greater energy and therefore shorter wavelength.

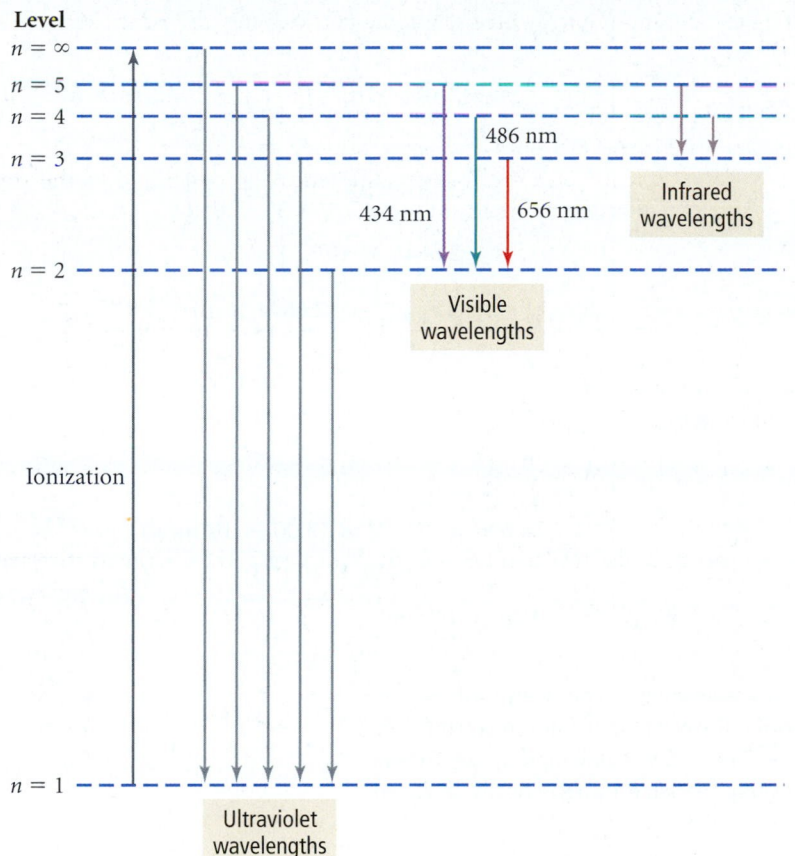

Hydrogen Energy Transitions and Radiation

Level

$n = \infty$

$n = 5$

$n = 4$

$n = 3$ — 486 nm

434 nm   656 nm

Infrared wavelengths

$n = 2$

Visible wavelengths

Ionization

$n = 1$

Ultraviolet wavelengths

◄ **FIGURE 7.21 Hydrogen Energy Transitions and Radiation** An atomic energy level diagram for hydrogen, showing some possible electron transitions between levels and the corresponding wavelengths of emitted light.

EXAMPLE 7.7 Wavelength of Light for a Transition in the Hydrogen Atom

Determine the wavelength of light emitted when an electron in a hydrogen atom makes a transition from an orbital in $n = 6$ to an orbital in $n = 5$.

Sort You are given the energy levels of an atomic transition and asked to find the wavelength of emitted light.	**Given** $n = 6 \longrightarrow n = 5$ **Find** λ

Strategize In the first part of the conceptual plan, calculate the energy of the electron in the $n = 6$ and $n = 5$ orbitals using Equation 7.7 and subtract to find ΔE_{atom}.

Conceptual Plan

$$n = 5, n = 6 \longrightarrow \Delta E_{atom}$$
$$\Delta E = E_5 - E_6$$

In the second part, find E_{photon} by taking the negative of ΔE_{atom}, and then calculate the wavelength corresponding to a photon of this energy using Equation 7.3. (The difference in sign between E_{photon} and ΔE_{atom} applies only to emission. *The energy of a photon must always be positive.*)

$$\Delta E_{atom} \longrightarrow E_{photon} \longrightarrow \lambda$$
$$\Delta E_{atom} = -E_{photon} \qquad E = \frac{hc}{\lambda}$$

Relationships Used

$$E_n = -2.18 \times 10^{-18} \, J(1/n^2)$$
$$E = hc/\lambda$$

Solve Follow the conceptual plan. Begin by computing ΔE_{atom}.

Solution

$$\Delta E_{atom} = E_5 - E_6$$

$$= -2.18 \times 10^{-18}\,\text{J}\left(\frac{1}{5^2}\right) - \left[-2.18 \times 10^{-18}\,\text{J}\left(\frac{1}{6^2}\right)\right]$$

$$= -2.18 \times 10^{-18}\text{J}\left(\frac{1}{5^2} - \frac{1}{6^2}\right)$$

$$= -2.6644 \times 10^{-20}\,\text{J}$$

Compute E_{photon} by changing the sign of ΔE_{atom}.

$$E_{photon} = -\Delta E_{atom} = +2.6644 \times 10^{-20}\,\text{J}$$

Solve the equation relating the energy of a photon to its wavelength for λ. Substitute the energy of the photon and compute λ.

$$E = \frac{hc}{\lambda}$$

$$\lambda = \frac{hc}{E}$$

$$= \frac{(6.626 \times 10^{-34}\,\text{J}\cdot\text{s})(3.00 \times 10^8\text{m/s})}{2.6644 \times 10^{-20}\,\text{J}}$$

$$= 7.46 \times 10^{-6}\,\text{m}$$

Check The units of the answer (m) are correct for wavelength. The magnitude seems reasonable because 10^{-6} m is in the infrared region of the electromagnetic spectrum. We know that transitions from $n = 3$ or $n = 4$ to $n = 2$ lie in the visible region, so it makes sense that a transition between levels of higher n value (which are energetically closer to one another) would result in light of longer wavelength.

For Practice 7.7

Determine the wavelength of the light absorbed when an electron in a hydrogen atom makes a transition from an orbital in $n = 2$ to an orbital in $n = 7$.

For More Practice 7.7

An electron in the $n = 6$ level of the hydrogen atom relaxes to a lower energy level, emitting light of $\lambda = 93.8$ nm. Find the principal level to which the electron relaxed.

7.6 The Shapes of Atomic Orbitals

As we noted previously, the shapes of atomic orbitals are important because covalent chemical bonds depend on the sharing of the electrons that occupy these orbitals. In one model of chemical bonding, for example, a bond consists of the overlap of atomic orbitals on adjacent atoms. Therefore the shapes of the overlapping orbitals determine the shape of the molecule. Although we limit ourselves in this chapter to the orbitals of the hydrogen atom, we will see in Chapter 8 that the orbitals of all atoms can be approximated as being hydrogen-like and therefore have very similar shapes to those of hydrogen.

The shape of an atomic orbital is determined primarily by l, the angular momentum quantum number. As we have seen, each value of l is assigned a letter that therefore corresponds to particular orbitals. For example, the orbitals with $l = 0$ are called s orbitals; those with $l = 1$, p orbitals; those with $l = 2$, d orbitals, etc. We now examine the shape of each of these orbitals.

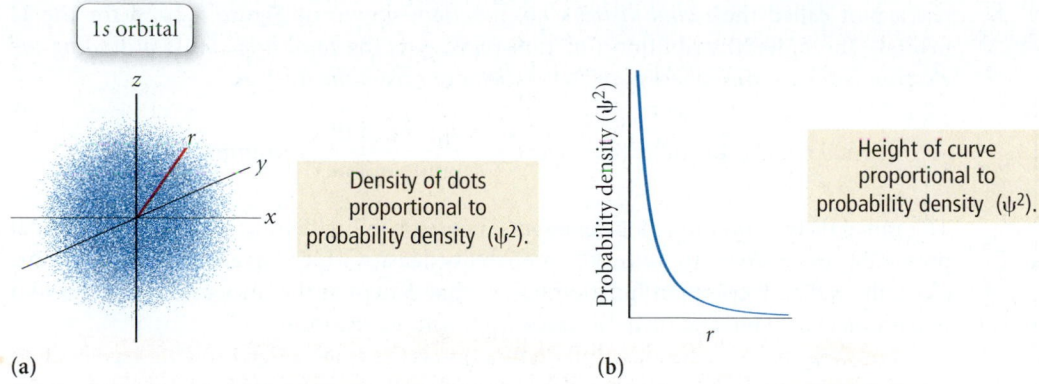

1s orbital

Density of dots proportional to probability density (ψ^2).

Height of curve proportional to probability density (ψ^2).

(a)

(b)

▲ **FIGURE 7.22 The 1s Orbital: Two Representations**   In **(a)** the dot density is proportional to the electron probability density. In **(b)**, the height of the curve is proportional to the electron probability density. The *x*-axis is *r*, the distance from the nucleus.

s Orbitals (*l* = 0)

The lowest energy orbital is the spherically symmetrical 1s orbital shown in Figure 7.22(a) ▲. This image is actually a three-dimensional plot of the wave function squared (ψ^2), which represents **probability density**, the probability (per unit volume) of finding the electron at a point in space.

$$\psi^2 = \text{probability density} = \frac{\text{probability}}{\text{unit volume}}$$

The magnitude of ψ^2 in this plot is proportional to the density of the dots shown in the image. The high dot density near the nucleus indicates a higher probability density for the electron there. As you move away from the nucleus, the probability density decreases. Figure 7.22(b) shows a plot of probability density (ψ^2) versus *r*, the distance from the nucleus. This is essentially a slice through the three-dimensional plot of ψ^2 and shows how the probability density decreases as *r* increases.

We can understand probability density with the help of a thought experiment. Imagine an electron in the 1s orbital located within the volume surrounding the nucleus. Imagine also taking a photograph of the electron every second for 10 or 15 minutes. In one photograph, the electron is very close to the nucleus, in another it is farther away, and so on. Each photo has a dot showing the electron's position relative to the nucleus when the photo was taken. Remember that you can never predict where the electron will be for any one photo. However, if you took hundreds of photos and superimposed all of them, you would have a plot similar to Figure 7.22(a)—a statistical representation of how likely the electron is to be found at each point.

An atomic orbital can also be represented by a geometrical shape that encompasses the volume where the electron is likely to be found most frequently—typically, 90% of the time. For example, the 1s orbital can be represented as the three-dimensional sphere shown in Figure 7.23 ▶. If we were to superimpose the dot-density representation of the 1s orbital on the shape representation, 90% of the dots would be within the sphere, meaning that when the electron is in the 1s orbital it has a 90% chance of being found within the sphere.

The plots we have just seen represent probability *density*. However, they are a bit misleading because they seem to imply that the electron is most likely to be found *at the nucleus*. To get a better idea of where the electron is most likely to be found, we can

When an orbital is represented as shown below, the surface shown is one of constant probability. The probability of finding the electron at any point on the surface is the same.

1s orbital surface

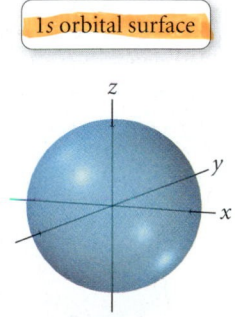

▲ **FIGURE 7.23 The 1s Orbital Surface**   In this representation, the surface of the sphere encompasses the volume where the electron is found 90% of the time.

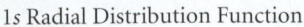

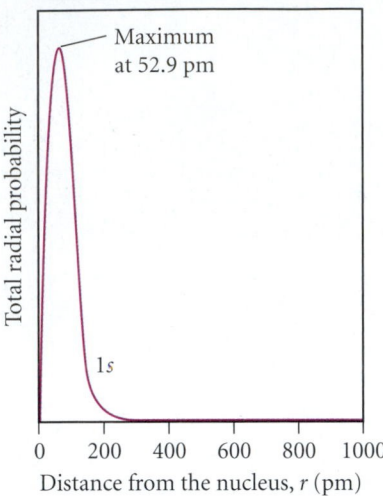

▲ **FIGURE 7.24 The Radial Distribution Function for the 1s Orbital** The curve shows the total probability of finding the electron within a thin shell at a distance r from the nucleus.

use a plot called the *radial distribution function*, shown in Figure 7.24 ◀ for the 1s orbital. The radial distribution function represents the *total probability of finding the electron within a thin spherical shell at a distance r from the nucleus.*

$$\text{Total radial probability (at a given } r) = \frac{\text{probability}}{\text{unit volume}} \times \text{volume of shell at } r$$

The radial distribution function represents, not probability density *at a point r*, but total probability *at a radius r*. In contrast to probability density, which has a maximum at the nucleus, the radial distribution function has a value of *zero* at the nucleus. It increases to a maximum at 52.9 pm and then decreases again with increasing r. The shape of the radial distribution function is the result of multiplying together two functions with opposite trends in r: (1) the probability density function (ψ^2), which is the probability per unit volume and decreases with increasing r, and (2) the volume of the thin shell, which increases with increasing r. At the nucleus ($r = 0$), for example, the probability *density* is at a maximum; however, the volume of a thin spherical shell is zero, so the radial distribution function is zero. As r increases, the volume of the thin spherical shell increases. We can see this by analogy to an onion. A spherical shell at a distance r from the nucleus is like a layer in an onion at a distance r from its center. If the layers of the onion are all the same thickness, then the volume of any one layer—think of this as the total amount of onion in the layer—is greater as r increases. Similarly, the volume of any one spherical shell in the radial distribution function increases with increasing distance from the nucleus, resulting in a greater total probability of finding the electron within that shell. Close to the nucleus, this increase in volume with increasing r outpaces the decrease in probability density, producing a maximum at 52.9 pm. Farther out, however, the density falls off faster than the volume increases.

The maximum in the radial distribution function, 52.9 pm, turns out to be the very same radius that Bohr had predicted for the innermost orbit of the hydrogen atom. However, there is a significant conceptual difference between the two radii. In the Bohr model, every time you probe the atom (in its lowest energy state), you would find the electron at a radius of 52.9 pm. In the quantum-mechanical model, you would generally find the electron at various radii, with 52.9 pm having the greatest probability.

The probability densities and radial distribution functions for the 2s and 3s orbitals are shown in Figure 7.25 ▶. Like the 1s orbital, these orbitals are spherically symmetric. Unlike the 1s orbital, however, these orbitals are larger in size, and they contain *nodes*. A **node** is a point where the wave function (ψ), and therefore the probability density (ψ^2) and radial distribution function, all go through zero. A node in a wave function is much like a node in a standing wave on a vibrating string. We can see nodes in an orbital most clearly by actually looking at a slice through the orbital. Plots of probability density and the radial distribution function as a function of r both reveal the presence of nodes. The probability of finding the electron at a node is zero.

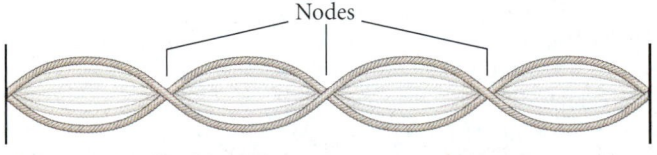

▲ The nodes in quantum-mechanical atomic orbitals are three-dimensional analogs of the nodes we find on a vibrating string.

p Orbitals ($l = 1$)

Each principal level with $n = 2$ or greater contains three p orbitals ($m_l = -1, 0, +1$). The three 2p orbitals and their radial distribution functions are shown in Figure 7.26 (page 310). The p orbitals are not spherically symmetric like the s orbitals, but have two *lobes* of electron density on either side of the nucleus and a node located at the nucleus. The three p orbitals differ only in their orientation and are orthogonal (mutually perpendicular) to one another. It is convenient to define an x, y, and z axis system and then label each p orbital as p_x, p_y, and p_z. The 3p, 4p, 5p, and higher p orbitals are all similar in shape to the 2p orbitals, but they contain additional nodes (like the higher s orbitals) and are progressively larger in size.

The 2s and 3s Orbitals

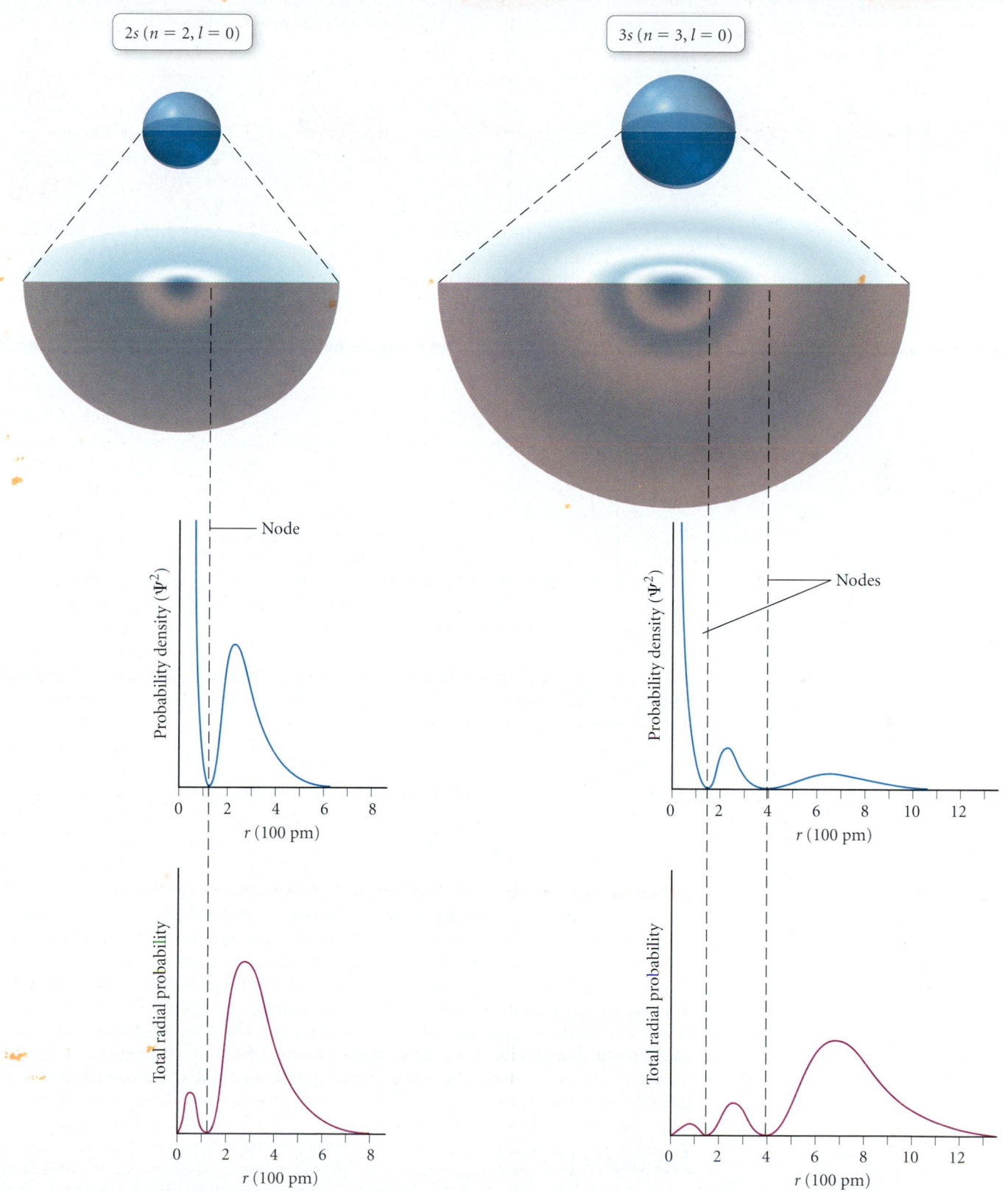

▲ FIGURE 7.25 Probability Densities and Radial Distribution Functions for the 2s and 3s Orbitals

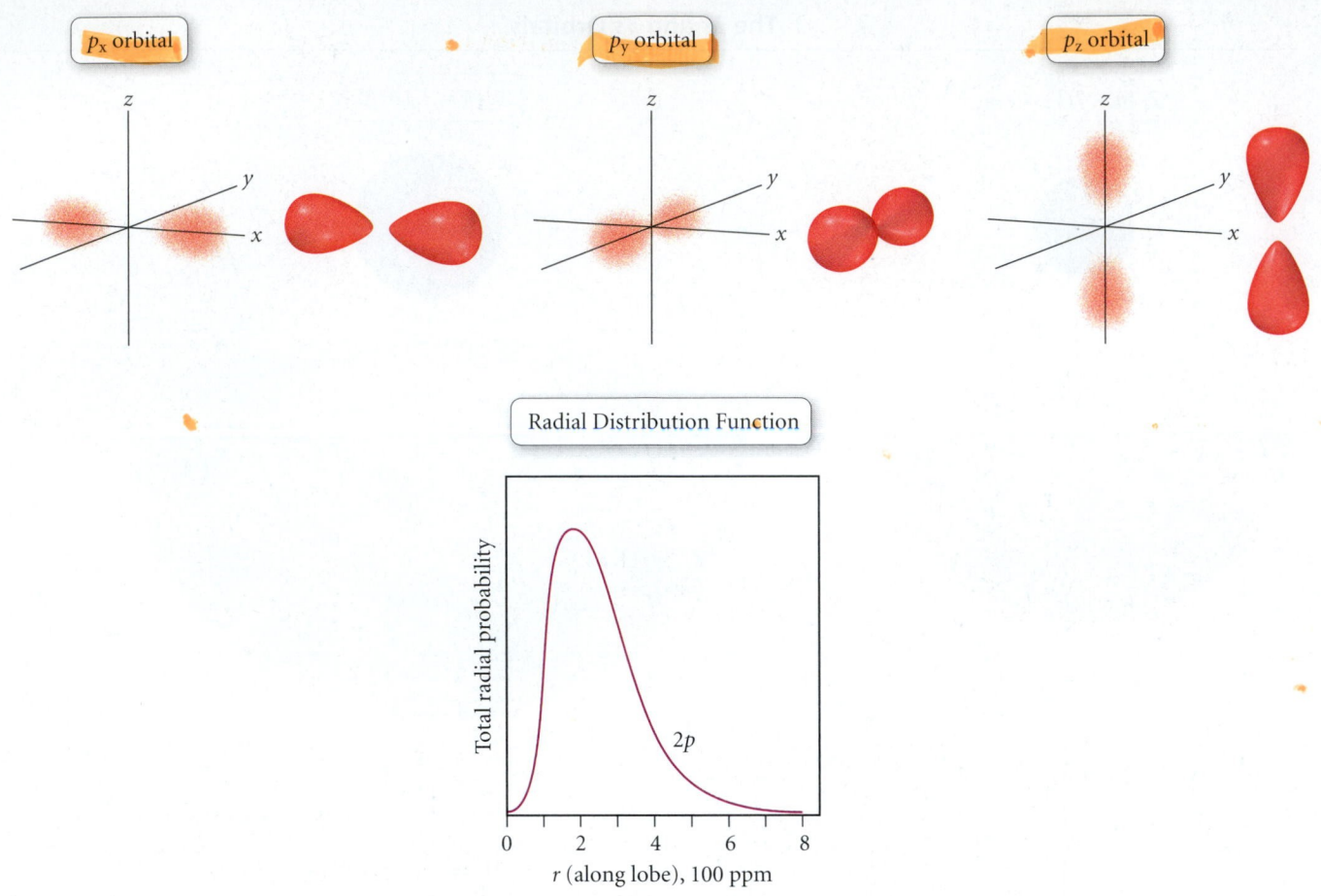

▲ **FIGURE 7.26 The 2p Orbitals and Their Radial Distribution Function**   The radial distribution function is the same for all three $2p$ orbitals when the x-axis of the graph is taken as the axis containing the lobes of the orbital.

d Orbitals ($l = 2$)

> A nodal plane is simply a plane where the electron probability density is zero. For example, in the d_{xy} orbitals, the nodal planes lie in the xz and yz planes.

Each principal level with $n = 3$ or greater contains five d orbitals ($m_l = -2, -1, 0, +1, +2$). The five $3d$ orbitals are shown in Figure 7.27 ▶. Four of these orbitals have a cloverleaf shape, with four lobes of electron density around the nucleus and two perpendicular nodal planes. The d_{xy}, d_{xz}, and d_{yz} orbitals are oriented along the xy, xz, and yz planes, respectively, and their lobes are oriented *between* the corresponding axes. The four lobes of the $d_{x^2-y^2}$ orbital are oriented along the x- and y-axes. The d_{z^2} orbital is different in shape from the other four, having two lobes oriented along the z-axis and a donut-shaped ring along the xy plane. The $4d$, $5d$, $6d$, etc., orbitals are all similar in shape to the $3d$ orbitals, but they contain additional nodes and are progressively larger in size.

f Orbitals ($l = 3$)

Each principal level with $n = 4$ or greater contains seven f orbitals ($m_l = -3, -2, -1, 0, +1, +2, +3$), as shown in Figure 7.28 ▶. Notice that as l increases, the number of lobes (and the number of nodes) increases.

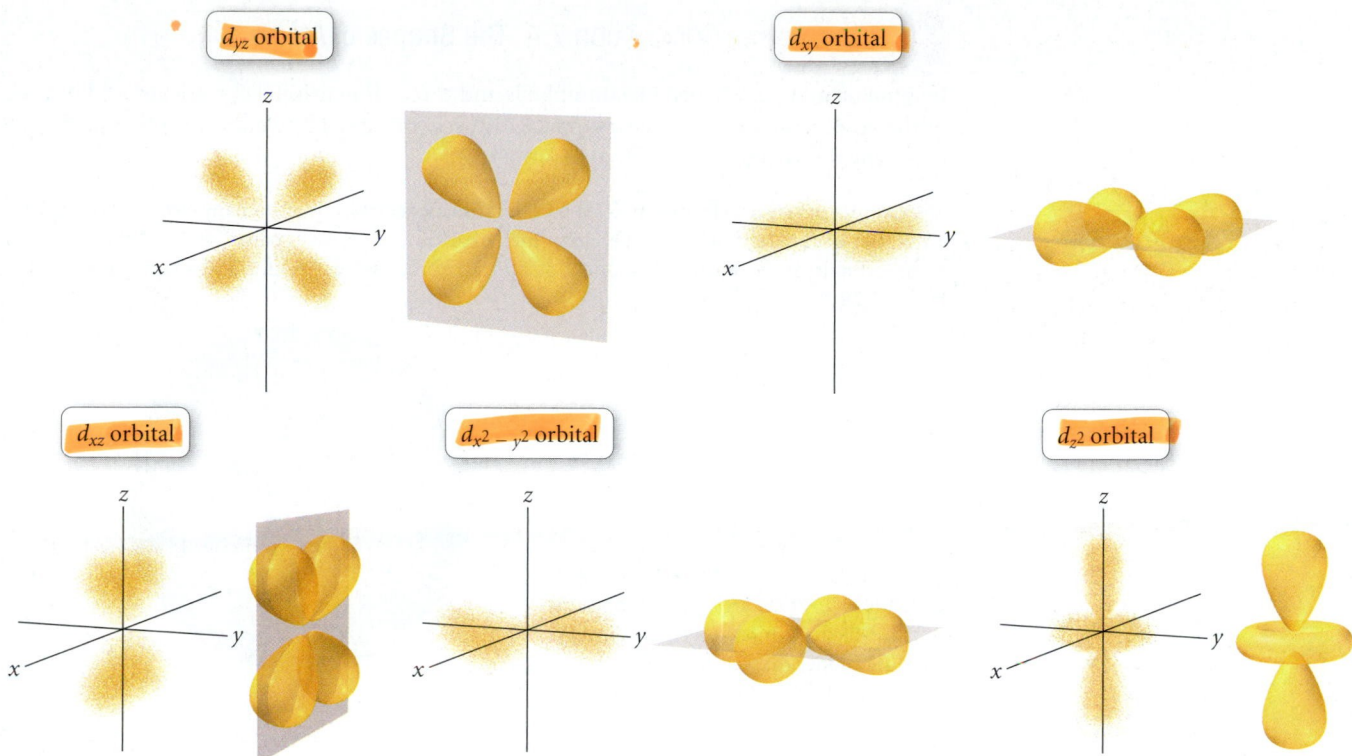

▲ FIGURE 7.27 The 3*d* Orbitals

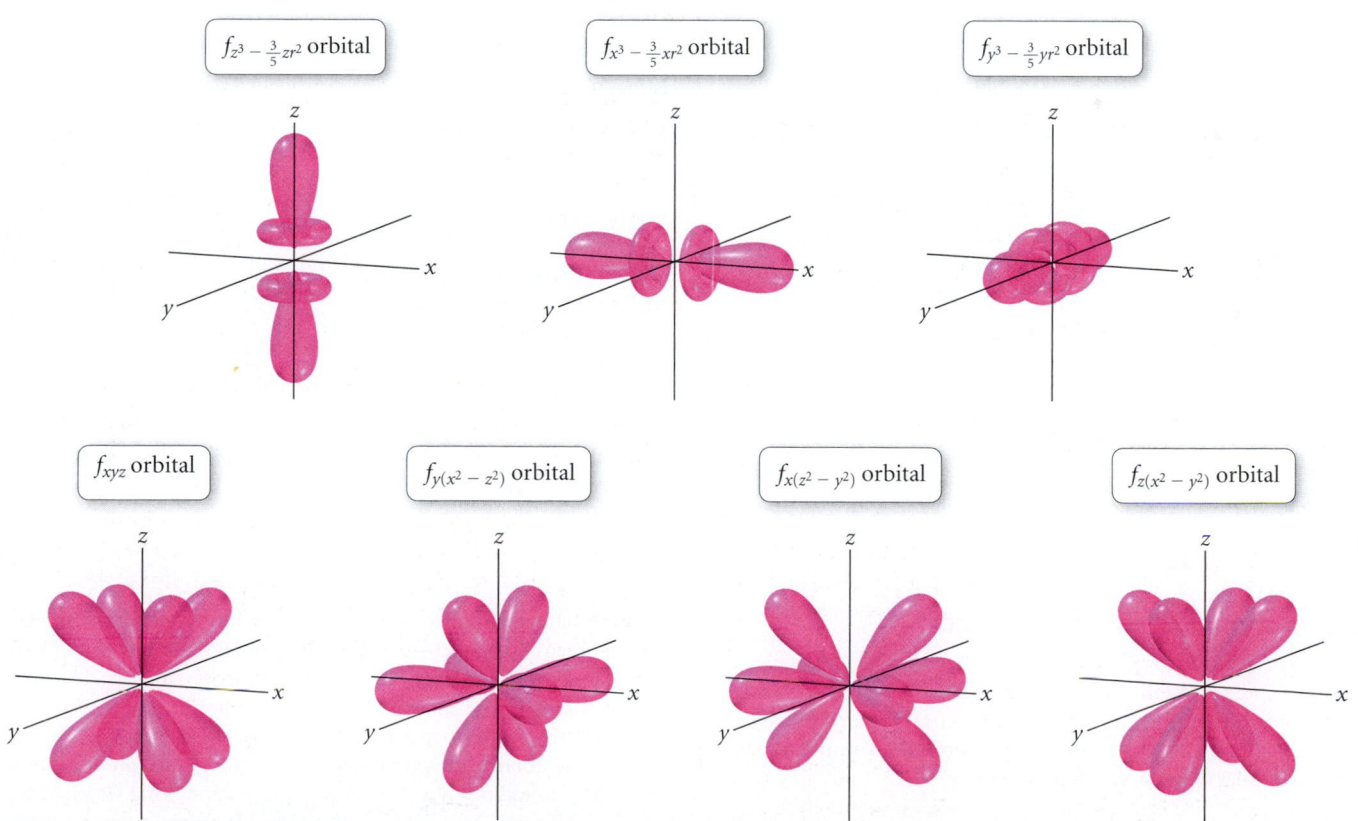

▲ FIGURE 7.28 The 4*f* Orbitals

Conceptual Connection 7.4 The Shapes of Atoms

If some orbitals are shaped like dumbbells and three-dimensional cloverleafs, and if most of the volume of an atom is empty space diffusely occupied by electrons in these orbitals, then why do we often depict atoms as spheres?

Answer: Atoms are usually drawn as spheres because most atoms contain many electrons occupying a number of different orbitals. Therefore, the shape of an atom is obtained by superimposing all of its orbitals. If we superimpose the *s*, *p*, and *d* orbitals we get a spherical shape, as shown in Figure 7.29 ▼.

▶ **FIGURE 7.29 Why Atoms Are Spherical** Atoms are roughly spherical because all the orbitals together make up a roughly spherical shape.

Chapter in Review

Key Terms

Section 7.1

quantum-mechanical model (282)

Section 7.2

electromagnetic radiation (282)
amplitude (283)
wavelength (λ) (283)
frequency (ν) (284)
electromagnetic spectrum (285)
gamma rays (285)
X-rays (286)
ultraviolet (UV) radiation (286)

visible light (286)
infrared (IR) radiation (287)
microwaves (287)
radio waves (287)
interference (287)
constructive interference (287)
destructive interference (287)
diffraction (287)
photoelectric effect (288)
photon (quantum) (289)

Section 7.3

emission spectrum (292)

Section 7.4

de Broglie relation (297)
complementary properties (298)
Heisenberg's uncertainty principle (299)
deterministic (299)
indeterminacy (300)

Section 7.5

orbital (301)
wave function (301)
quantum number (301)

principal quantum number (n) (301)
angular momentum quantum number (l) (301)
magnetic quantum number (m_l) (301)
principal level (shell) (302)
sublevel (subshell) (302)

Section 7.6

probability density (307)
node (308)

Key Concepts

The Realm of Quantum Mechanics (7.1)

The theory of quantum mechanics explains the behavior of particles in the atomic and subatomic realms. These particles include photons (particles of light) and electrons. Since the electrons of an atom determine many of its chemical and physical properties, quantum mechanics is foundational to understanding chemistry.

The Nature of Light (7.2)

Light is a type of electromagnetic radiation—a form of energy embodied in oscillating electric and magnetic fields that travels though space at 3.00×10^8 m/s. Light has both a wave nature and a particle nature.

The wave nature of light is characterized by its wavelength—the distance between wave crests—and the ability of light to experience interference (constructive or destructive) and diffraction. Its particle nature is characterized by the energy carried in each photon.

The electromagnetic spectrum includes all wavelengths of electromagnetic radiation from gamma rays (high energy per photon, short wavelength) to radio waves (low energy per photon, long wavelength). Visible light is a tiny sliver in the middle of the electromagnetic spectrum.

Atomic Spectroscopy (7.3)

Atomic spectroscopy is the study of the light absorbed and emitted by atoms when an electron makes a transition from one energy level to another. The wavelengths absorbed or emitted depend on the energy differences between the levels involved in the transition; large energy differences result in short wavelengths and small energy differences result in long wavelengths.

The Wave Nature of Matter (7.4)

Electrons have a wave nature with an associated wavelength, as quantified by the de Broglie relation. The wave nature and particle nature of matter are complementary, which means that the more you know of one, the less you know of the other. The wave–particle duality of electrons is quantified in Heisenberg's uncertainty principle, which states that there is a limit to how well we can know both the position of an electron (associated with the electron's particle nature) and the velocity times the mass of an electron (associated with the electron's wave nature)—the more accurately one is measured, the greater the uncertainty in the other. The inability to simultaneously know both the position and the velocity of an electron results in indeterminacy, the inability to predict a trajectory for an electron. Consequently electron behavior is described differently than the behavior of everyday-sized particles. The trajectory we normally associate with macroscopic objects is replaced, for electrons, with statistical descriptions that show, not the electron's path, but the region where it is most likely to be found.

The Quantum-Mechanical Model of the Atom (7.5, 7.6)

The most common way to describe electrons in atoms according to quantum mechanics is to solve the Schrödinger equation for the energy states of the electrons within the atom. When the electron is in these states, its energy is well-defined but its position is not. The position of an electron is described by a probability distribution map called an orbital.

The solutions to the Schrödinger equation (including the energies and orbitals) are characterized by three quantum numbers: n, l, and m_l. The principal quantum number (n) determines the energy of the electron and the size of the orbital; the angular momentum quantum number (l) determines the shape of the orbital; and the magnetic quantum number (m_l) determines the orientation of the orbital.

Key Equations and Relationships

Relationship between Frequency (ν), Wavelength (λ), and the Speed of Light (c) (7.2)

$$\nu = \frac{c}{\lambda}$$

Relationship between Energy (E), Frequency (ν), Wavelength (λ), and Planck's Constant (h) (7.2)

$$E = h\nu$$

$$E = \frac{hc}{\lambda}$$

De Broglie Relation: Relationship Between Wavelength (λ), Mass (m), and Velocity (v) of a Particle (7.4)

$$\lambda = \frac{h}{mv}$$

Heisenberg's Uncertainty Principle: Relationship between a Particle's Uncertainty in Position (Δx) and Uncertainty in Velocity (Δv) (7.4)

$$\Delta x \times m\,\Delta v \geq \frac{h}{4\pi}$$

Energy of an Electron in an Orbital with Quantum Number n in a Hydrogen Atom (7.5)

$$E_n = -2.18 \times 10^{-18}\,\text{J}\left(\frac{1}{n^2}\right) \quad (n = 1, 2, 3, \dots)$$

Key Skills

Calculating the Wavelength and Frequency of Light (7.2)

• Example 7.1   • For Practice 7.1   • Exercises 41, 42

Calculating the Energy of a Photon (7.2)

•Example 7.2   • For Practice 7.2   • For More Practice 7.2   • Exercises 43–48

Relating Wavelength, Energy, and Frequency to the Electromagnetic Spectrum (7.2)

•Example 7.3   • For Practice 7.3   • Exercises 39, 40

Using the de Broglie Relation to Calculate Wavelength (7.4)

•Example 7.4   • For Practice 7.4   • Exercises 51–54

Relating Quantum Numbers to One Another and to Their Corresponding Orbitals (7.5)

•Examples 7.5, 7.6   • For Practice 7.5, 7.6   • Exercises 57–60

Relating the Wavelength of Light to Transitions in the Hydrogen Atom (7.5)

•Example 7.7   • For Practice 7.7   • For More Practice 7.7   • Exercises 67–70

Exercises

Review Questions

1. What does it mean for a particle to be *absolutely* small? What particles fit this description?

2. Explain the difference between observing an object such as a baseball and observing a particle that is absolutely small.

3. Why is the quantum-mechanical model of the atom important for understanding chemistry?

4. What is light? How fast does it travel in a vacuum?

5. Define the wavelength and amplitude of a wave. How are these related to the energy carried by the wave?

6. Define the frequency of electromagnetic radiation. How is frequency related to wavelength?

7. What determines the color of light? For example, what is the difference between red light and blue light?

8. What determines the color of a colored object? For example, explain why grass appears green.

9. Give an approximate range of wavelengths for each of the following types of electromagnetic radiation and summarize the characteristics and/or the uses of each.

   a. gamma rays   b. X-rays
   c. ultraviolet radiation d. visible light
   e. infrared radiation   f. microwave radiation
   g. radio waves

10. Explain the wave behavior known as interference. Explain the difference between constructive and destructive interference.

11. Explain the wave behavior known as diffraction. Draw the diffraction pattern that occurs when light travels through two slits comparable in size and separation to the light's wavelength.

12. Describe the photoelectric effect. How did the experimental observations differ from the predictions of classical electromagnetic theory?

13. How did the photoelectric effect lead Einstein to propose that light is quantized?

14. What is a photon? How is the energy of a photon related to its wavelength? Its frequency?

15. What is an emission spectrum? How does an emission spectrum differ from a white light spectrum?

16. Describe the Bohr model for the atom. How did the Bohr model account for the emission spectra of atoms?

17. Explain electron diffraction.

18. What is the de Broglie wavelength of an electron? What determines the value of the de Broglie wavelength for an electron?

19. What are complementary properties? How does electron diffraction demonstrate the complementarity of the wave nature and particle nature of the electron?

20. Explain Heisenberg's uncertainty principle. What paradox is at least partially solved by the uncertainty principle?

21. What is a trajectory? What kind of information do you need to predict the trajectory of a particle?

22. Why does the uncertainty principle make it impossible to predict a trajectory for the electron?

23. Newton's laws of motion are said to be *deterministic*. What does this mean?

24. An electron behaves in ways that are at least partially indeterminate. What does this mean?

25. What is a probability distribution map?

26. For each solution to the Schrödinger equation, what can be precisely specified: the electron's energy or its position? Explain.

27. What is a quantum-mechanical orbital?

28. What is the Schrödinger equation? What is a wave function? How is a wave function related to an orbital?

29. What are the possible values of the principal quantum number? What does the principal quantum number determine?

30. What are the possible values of the angular momentum quantum number? What does the angular momentum quantum number determine?

31. What are the possible values of the magnetic quantum number? What does the magnetic quantum number determine?

32. List all the orbitals in each of the following principal levels. Specify the three quantum numbers for each orbital.

   a. $n = 1$ b. $n = 2$
   c. $n = 3$ d. $n = 4$

33. Explain the difference between a plot showing the probability density for an orbital and one showing the radial distribution function.

34. Make sketches of the general shapes of the s, p, and d orbitals.

35. List the four different sublevels and, given that only a maximum of two electrons can occupy an orbital, determine the maximum number of electrons that can exist in each sublevel.

36. Why are atoms usually portrayed as spheres when most orbitals are not spherically shaped?

Problems by Topic

Electromagnetic Radiation

37. The distance from the sun to Earth is 1.496×10^8 km. How long does it take light to travel from the sun to Earth?

38. The nearest star to our sun is Proxima Centauri, at a distance of 4.3 light-years from the sun. A light-year is the distance that light travels in one year (365 days). How far away, in km, is Proxima Centauri from the sun?

39. List the following types of electromagnetic radiation in order of (i) increasing wavelength and (ii) increasing energy per photon:
   a. radio waves   b. microwaves
   c. infrared radiation   d. ultraviolet radiation

40. List the following types of electromagnetic radiation in order of (i) increasing frequency and (ii) decreasing energy per photon:
   a. gamma rays   b. radio waves
   c. microwaves   d. visible light

41. Calculate the frequency of each of the following wave lengths of electromagnetic radiation:
    a. 632.8 nm (wavelength of red light from helium–neon laser)
    b. 503 nm (wavelength of maximum solar radiation)
    c. 0.052 nm (a wavelength contained in medical X-rays)

42. Calculate the wavelength of each of the following frequencies of electromagnetic radiation:
    a. 100.2 MHz (typical frequency for FM radio broadcasting)
    b. 1070 kHz (typical frequency for AM radio broadcasting) (assume four significant figures)
    c. 835.6 MHz (common frequency used for cell phone communication)

43. Calculate the energy of a photon of electromagnetic radiation at each of the wavelengths indicated in Problem 41.

44. Calculate the energy of a photon of electromagnetic radiation at each of the frequencies indicated in Problem 42.

45. A laser pulse with wavelength 532 nm contains 4.67 mJ of energy. How many photons are in the laser pulse?

46. A heat lamp produces 25.5 watts of power at a wavelength of 6.5 μm. How many photons are emitted per second? (1 watt = 1 J/s)

47. Determine the energy of 1 mol of photons for each of the following kinds of light. (Assume three significant figures.)
    a. infrared radiation (1500 nm)
    b. visible light (500 nm)
    c. ultraviolet radiation (150 nm)

48. How much energy is contained in 1 mol of each of the following?
    a. X-ray photons with a wavelength of 0.155 nm
    b. Γ-ray photons with a wavelength of 2.55×10^{-5} nm

The Wave Nature of Matter and the Uncertainty Principle

49. Make a sketch of the interference pattern that results from the diffraction of electrons passing through two closely spaced slits.

50. What happens to the interference pattern described in Problem 49 if the rate of electrons going through the slits is decreased to one electron per hour? What happens to the pattern if we try to determine which hole the electron goes through by using a laser placed directly behind the slits?

51. Calculate the wavelength of an electron traveling at 1.15×10^5 m/s.

52. An electron has a de Broglie wavelength of 225 nm. What is the speed of the electron?

53. Calculate the de Broglie wavelength of a 143-g baseball traveling at 95 mph. Why is the wave nature of matter not important for a baseball?

54. A 0.22-caliber handgun fires a 27-g bullet at a velocity of 765 m/s. Calculate the de Broglie wavelength of the bullet. Is the wave nature of matter significant for bullets?

Orbitals and Quantum Numbers

55. Which electron is, on average, closer to the nucleus: an electron in a 2s orbital or an electron in a 3s orbital?

56. Which electron is, on average, further from the nucleus: an electron in a 3p orbital or an electron in a 4p orbital?

57. What are the possible values of l for each of the following values of n?
    a. 1   b. 2   c. 3   d. 4

58. What are the possible values of m_l for each of the following values of l?
    a. 0   b. 1   c. 2   d. 3

59. Which set of quantum numbers *cannot* occur together to specify an orbital?
    a. $n = 2, l = 1, m_l = -1$ b. $n = 3, l = 2, m_l = 0$
    c. $n = 3, l = 3, m_l = 2$ d. $n = 4, l = 3, m_l = 0$

60. Which of the following combinations of n and l represent real orbitals and which are impossible?
    a. 1s   b. 2p   c. 4s   d. 2d

61. Make a sketch of the 1s and 2p orbitals. How would the 2s and 3p orbitals differ from the 1s and 2p orbitals?

62. Make a sketch of the 3d orbitals. How would the 4d orbitals differ from the 3d orbitals?

Atomic Spectroscopy

63. An electron in a hydrogen atom is excited with electrical energy to an excited state with $n = 2$. The atom then emits a photon. What is the value of n for the electron following the emission?

64. Determine whether each of the following transitions in the hydrogen atom corresponds to absorption or emission of energy.
    a. $n = 3 \longrightarrow n = 1$ b. $n = 2 \longrightarrow n = 4$
    c. $n = 4 \longrightarrow n = 3$

65. According to the quantum-mechanical model for the hydrogen atom, which of the following electron transitions would produce light with the longer wavelength: $2p \longrightarrow 1s$ or $3p \longrightarrow 1s$?

66. According to the quantum-mechanical model for the hydrogen atom, which of the following transitions would produce light with the longer wavelength: $3p \longrightarrow 2s$ or $4p \longrightarrow 3p$?

67. Calculate the wavelength of the light emitted when an electron in a hydrogen atom makes each of the following transitions and indicate the region of the electromagnetic spectrum (infrared, visible, ultraviolet, etc.) where the light is found.
    a. $n = 2 \longrightarrow n = 1$ b. $n = 3 \longrightarrow n = 1$
    c. $n = 4 \longrightarrow n = 2$ d. $n = 5 \longrightarrow n = 2$

68. Calculate the frequency of the light emitted when an electron in a hydrogen atom makes each of the following transitions:
    a. $n = 4 \longrightarrow n = 3$ b. $n = 5 \longrightarrow n = 1$
    c. $n = 5 \longrightarrow n = 4$ d. $n = 6 \longrightarrow n = 5$

69. An electron in the $n = 7$ level of the hydrogen atom relaxes to a lower energy level, emitting light of 397 nm. What is the value of n for the level to which the electron relaxed?

70. An electron in a hydrogen atom relaxes to the $n = 4$ level, emitting light of 114 THz. What is the value of n for the level in which the electron originated?

Cumulative Problems

71. Ultraviolet radiation and radiation of shorter wavelengths can damage biological molecules because they carry enough energy to break bonds within the molecules. A carbon–carbon bond requires 348 kJ/mol to break. What is the longest wavelength of radiation with enough energy to break carbon–carbon bonds?

72. The human eye contains a molecule called 11-*cis*-retinal that changes conformation when struck with light of sufficient energy. The change in conformation triggers a series of events that results in an electrical signal being sent to the brain. The minimum energy required to change the conformation of 11-*cis*-retinal within the eye is about 164 kJ/mol. Calculate the longest wavelength visible to the human eye.

73. An argon ion laser puts out 5.0 W of continuous power at a wavelength of 532 nm. The diameter of the laser beam is 5.5 mm. If the laser is pointed toward a pinhole with a diameter of 1.2 mm, how many photons will travel through the pinhole per second? Assume that the light intensity is equally distributed throughout the entire cross-sectional area of the beam. (1 W = 1 J/s)

74. A green leaf has a surface area of 2.50 cm^2. If solar radiation is 1000 W/m^2, how many photons strike the leaf every second? Assume three significant figures and an average wavelength of 504 nm for solar radiation.

75. In a technique used for surface analysis called Auger electron spectroscopy (AES), electrons are accelerated toward a metal surface. These electrons cause the emissions of secondary electrons—called auger electrons—from the metal surface. The kinetic energy of the auger electrons depends on the composition of the surface. The presence of oxygen atoms on the surface results in auger electrons with a kinetic energy of approximately 506 eV. What is the de Broglie wavelength of this electron?

[KE = $\frac{1}{2} mv^2$; 1 electron volt (eV) = 1.602 × 10^{-19} J]

76. An X-ray photon of wavelength 0.989 nm strikes a surface. The emitted electron has a kinetic energy of 969 eV. What is the binding energy of the electron in kJ/mol?

[KE = $\frac{1}{2} mv^2$; 1 electron volt (eV) = 1.602 × 10^{-19} J]

77. Ionization involves completely removing an electron from an atom. How much energy is required to ionize a hydrogen atom in its ground (or lowest energy) state? What wavelength of light contains enough energy in a single photon to ionize a hydrogen atom?

78. The energy required to ionize sodium is 496 kJ/mol. What minimum frequency of light is required to ionize sodium?

79. Suppose that in an alternate universe, the possible values of l were the integer values from 0 to n (instead of 0 to $n - 1$). Assuming no other differences from this universe, how many orbitals would exist in each of the following levels?
    a. $n = 1$ b. $n = 2$ c. $n = 3$

80. Suppose that, in an alternate universe, the possible values of m_l were the integer values including 0 ranging from $-l - 1$ to $l + 1$ (instead of simply $-l$ to $+l$). How many orbitals would exist in each of the following sublevels?
    a. s sublevel   b. p sublevel   c. d sublevel

81. An atomic emission spectrum of hydrogen shows the following three wavelengths: 1875 nm, 1282 nm, and 1093 nm. Assign these wavelengths to transitions in the hydrogen atom.

82. An atomic emission spectrum of hydrogen shows the following three wavelengths: 121.5 nm, 102.6 nm, and 97.23 nm. Assign these wavelengths to transitions in the hydrogen atom.

83. The binding energy of electrons in a metal is 193 kJ/mol. Find the threshold frequency of the metal.

84. In order for a thermonuclear fusion reaction of two deuterons ($^2_1H^+$) to take place, the deuterons must collide each with a velocity of about 1×10^6 m/s. Find the wavelength of such a deuteron.

85. The speed of sound in air is 344 m/s at room temperature. The lowest frequency of a large organ pipe is 30 s^{-1} and the highest frequency of a piccolo is 1.5×10^4 s^{-1}. Find the difference in wavelength between these two sounds.

86. The distance from Earth to the sun is 1.5×10^8 km. Find the number of crests in a light wave of frequency 1.0×10^{14} s^{-1} traveling from the sun to the Earth.

Challenge Problems

87. An electron confined to a one-dimensional box has energy levels given by the equation

$$E_n = n^2 h^2 / 8 mL^2$$

where n is a quantum number with possible values of 1, 2, 3, ..., m is the mass of the particle, and L is the length of the box.
    a. Calculate the energies of the $n = 1$, $n = 2$, and $n = 3$ levels for an electron in a box with a length of 155 pm.
    b. Calculate the wavelength of light required to make a transition from $n = 1 \longrightarrow n = 2$ and from $n = 2 \longrightarrow n = 3$. In what region of the electromagnetic spectrum do these wavelengths lie?

88. The energy of a vibrating molecule is quantized much like the energy of an electron in the hydrogen atom. The energy levels of a vibrating molecule are given by the equation

$$E_n = (n + \frac{1}{2}) h\nu$$

where n is a quantum number with possible values of 1, 2, ..., and ν is the frequency of vibration. The vibration frequency of HCl is approximately 8.85×10^{13} s^{-1}. What minimum energy is required to excite a vibration in HCl? What wavelength of light is required to excite this vibration?

89. The wave functions for the 1s and 2s orbitals are as follows:

1s $\psi = (1/\pi)^{1/2}(1/a_0^{3/2}) \exp(-r/a_0)$

2s $\psi = (1/32\pi)^{1/2}(1/a_0^{3/2})(2 - r/a_0) \exp(-r/a_0)$

where a_0 is a constant ($a_0 = 53$ pm) and r is the distance from the nucleus. Use Microsoft Excel to make a plot of each of these wave functions for values of r ranging from 0 pm to 200 pm. Describe the differences in the plots and identify the node in the 2s wave function.

90. Before quantum mechanics was developed, Johannes Rydberg developed the following equation that predicted the wavelengths (λ) in the atomic spectrum of hydrogen:

$$1/\lambda = R(1/m^2 - 1/n^2)$$

In this equation R is a constant and m and n are integers. Use the quantum-mechanical model for the hydrogen atom to derive the Rydberg equation.

91. Find the velocity of an electron emitted by a metal whose threshold frequency is 2.25×10^{14} s^{-1} when it is exposed to visible light of wavelength 5.00×10^{-7} m.

92. Water is exposed to infrared radiation of wavelength 2.8×10^{-4} cm. Assume that all the radiation is absorbed and converted to heat. How many photons will be required to raise the temperature of 2.0 g of water by 2.0 K?

Conceptual Problems

93. Explain the difference between the Bohr model for the hydrogen atom and the quantum-mechanical model. Is the Bohr model consistent with Heisenberg's uncertainty principle?

94. The light emitted from one of the following electronic transitions ($n = 4 \longrightarrow n = 3$ or $n = 3 \longrightarrow n = 2$) in the hydrogen atom caused the photoelectric effect in a particular metal while light from the other transition did not. Which transition was able to cause the photoelectric effect and why?

95. Determine whether an interference pattern is observed on the other side of the slits in each of the following experiments.
   a. An electron beam is aimed at two closely spaced slits. The beam is attenuated to produce only 1 electron per minute.
   b. An electron beam is aimed at two closely spaced slits. A light beam is placed at each slit to determine when an electron goes through the slit.
   c. A high-intensity light beam is aimed at two closely spaced slits.
   d. A gun is fired at a solid wall containing two closely spaced slits. (Will the bullets that pass through the slits form an interference pattern on the other side of the solid wall?)

8

Periodic Properties of the Elements

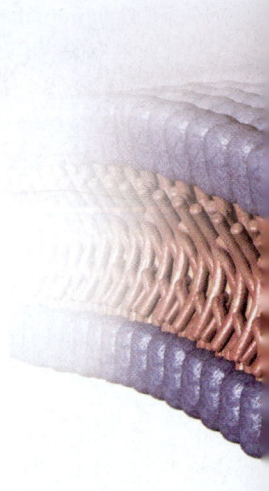

8.1 Nerve Signal Transmission

8.2 The Development of the Periodic Table

8.3 Electron Configurations: How Electrons Occupy Orbitals

8.4 Electron Configurations, Valence Electrons, and the Periodic Table

8.5 The Explanatory Power of the Quantum-Mechanical Model

8.6 Periodic Trends in the Size of Atoms and Effective Nuclear Charge

8.7 Ions: Electron Configurations, Magnetic Properties, Ionic Radii, and Ionization Energy

8.8 Electron Affinities and Metallic Character

8.9 Some Examples of Periodic Chemical Behavior: The Alkali Metals, the Halogens, and the Noble Gases

GREAT ADVANCES IN SCIENCE come not only when a scientist sees something new, but also when a scientist sees what everyone else has seen in a new way. In other words, great scientists often see patterns where others have seen only disjointed facts. Such was the case in 1869 when Dmitri Mendeleev, a Russian chemistry professor, saw a pattern in the properties of elements. Mendeleev's insight led to the periodic table, arguably the single most important tool for the chemist. Recall that science proceeds by devising theories that explain the underlying reasons for observations. If we think of Mendeleev's periodic table as a compact way to summarize a large number of observations, then quantum mechanics (covered in Chapter 7) is the theory that explains the underlying reasons for the periodic table. The concepts of quantum mechanics explain the arrangement of elements in the periodic table by reference to the electrons within the atoms that compose the elements. In this chapter, we see a continuation of the theme we have been developing since page one of this book—the properties of macroscopic substances (in this case, the elements in the periodic table) are explained by the properties of the particles that compose them (in this case, atoms and their electrons).

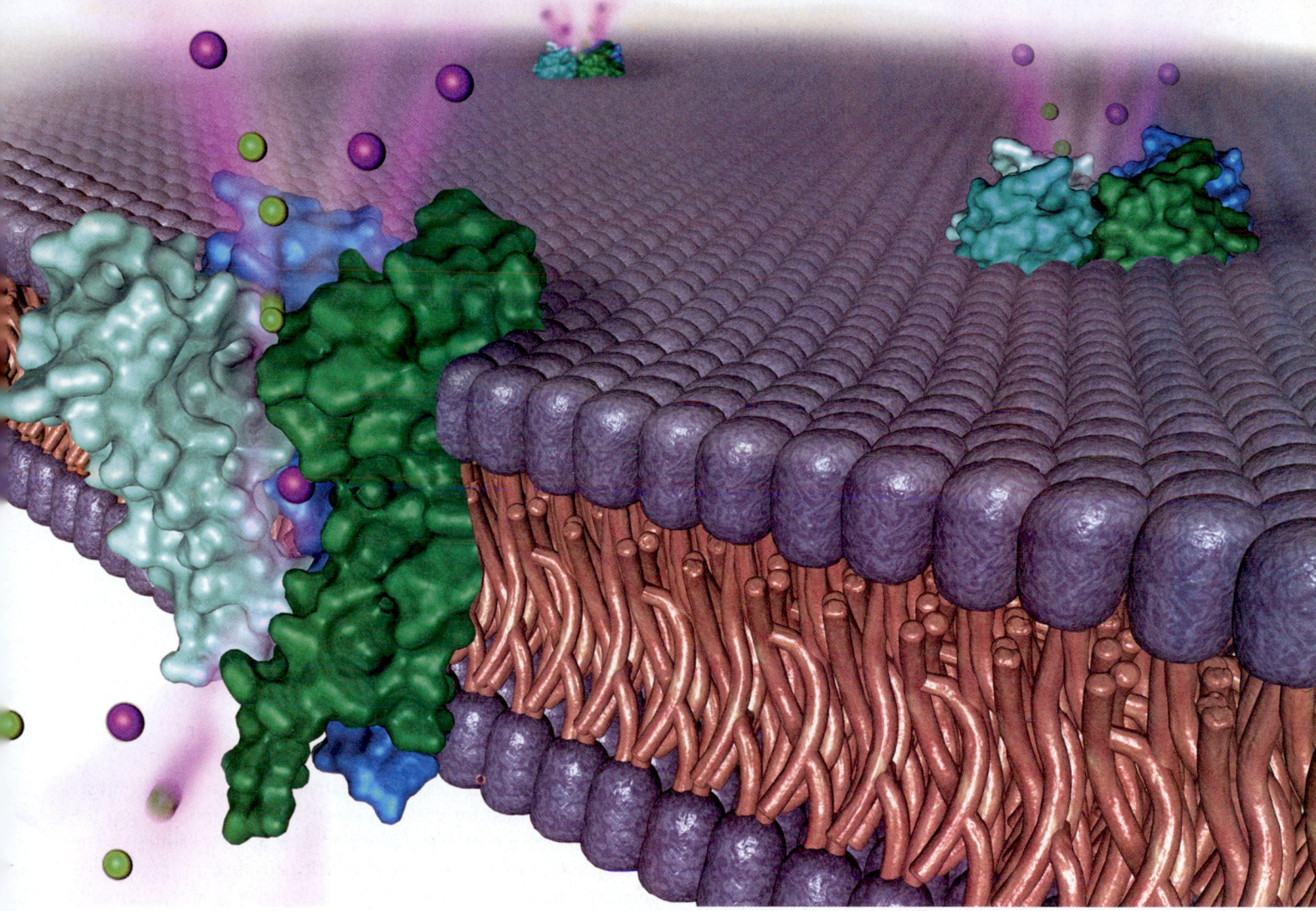

In order for a nerve cell to transmit a signal, sodium and potassium ions must flow in opposite directions through specific ion channels in the cell membrane.

8.1 Nerve Signal Transmission

As you sit reading this book, tiny pumps in the membranes of your cells are working hard to transport ions—especially sodium (Na^+) and potassium (K^+)—through those membranes. Amazingly, the ions are pumped in opposite directions. Sodium ions are pumped *out of cells,* while potassium ions are pumped *into cells.* The result is a *chemical gradient* for each ion: the concentration of sodium is higher outside the cell than within, while just the opposite is true for potassium. These ion pumps are analogous to water pumps in a high-rise building that pump water against the force of gravity to a tank on the roof. Other structures within the membrane, called ion channels, are like the building's faucets. When these open, sodium and potassium ions flow back down their gradients—sodium flowing

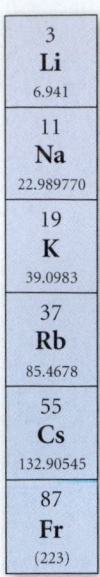

3
Li
6.941
11
Na
22.989770
19
K
39.0983
37
Rb
85.4678
55
Cs
132.90545
87
Fr
(223)

▲ The group 1A metals. Potassium lies directly beneath sodium in the periodic table.

The relationship between the size of the ion and its ability to move through an ion channel is complex, involving a number of factors that are beyond our current scope.

in and potassium flowing out. This movement of ions is the basis for the transmission of nerve signals in the brain and throughout the body. Every move you make, every thought you have, and every sensation you experience is mediated by these ion movements.

How do the pumps and channels differentiate between sodium and potassium ions so as to selectively move one out of the cell and the other into the cell? To answer this question, we must examine the ions more closely. Both are cations of group 1A metals. All group 1A metals tend to lose one electron to form cations with a 1+ charge, so that cannot be the decisive factor. However, potassium (atomic number 19) lies directly below sodium in the periodic table (atomic number 11), indicating that it has more protons, neutrons, and electrons than sodium. How do these additional subatomic particles affect the properties of potassium (relative to sodium)? As we will see in this chapter, although a higher atomic number does not always result in a larger ion (or atom), it does for sodium and potassium. The potassium ion has a radius of 133 pm while the sodium ion has a radius of 95 pm. (Recall that 1 pm = 10^{-12} m.) The pumps and channels within cell membranes are so sensitive that they can distinguish between the sizes of these two ions and selectively allow only one or the other to pass.

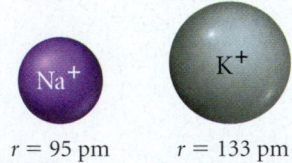

$r = 95$ pm $r = 133$ pm

The size of sodium and potassium ions is an example of a **periodic property**: one that is predictable based on an element's position within the periodic table. In this chapter, we examine several periodic properties of elements, including atomic radius, ionization energy, and electron affinity. We will see that these properties, as well as the overall arrangement of the periodic table, are explained by quantum-mechanical theory, which we examined in Chapter 7. The arrangement of the periodic table—originally based on similarities in the properties of the elements—reflects how electrons fill quantum-mechanical orbitals.

8.2 The Development of the Periodic Table

Prior to the 1700s, the number of known elements was relatively small, consisting mostly of the metals that had long been used for coinage, jewelry, and weapons. From the early 1700s to the mid-1800s, however, chemists discovered over 50 new elements. The first attempt to organize these elements according to similarities in their properties was made by the German chemist Johann Döbereiner (1780–1849), who grouped elements into *triads*: three elements with similar properties. For example, Döbereiner formed a triad out of barium, calcium, and strontium, three fairly reactive metals. A more complex approach was attempted about 50 years later, when the English chemist John Newlands (1837–1898) organized elements into *octaves*, in analogy to musical notes. When arranged this way, the properties of every eighth element were similar, much as every eighth note in the musical scale is similar. Newlands endured some ridicule for drawing an analogy between chemistry and music, including the derisive comments of one colleague who asked Newlands if he had ever tried ordering the elements according to the first letters of their names.

The modern periodic table is credited primarily to the Russian chemist Dmitri Mendeleev (1834–1907), even though a similar organization had been suggested by the German chemist Julius Lothar Meyer (1830–1895). As we saw in Chapter 2, Mendeleev's table is based on the periodic law, which states that when elements are arranged in order of increasing mass, their properties recur periodically. Mendeleev arranged the elements in a table in which mass increased from left to right and elements with similar properties fell in the same columns.

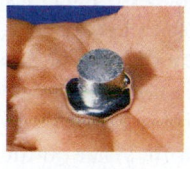

Gallium (eka-aluminum)

	Mendeleev's predicted properties	Actual properties
Atomic mass	About 68 amu	69.72 amu
Melting point	Low	29.8 °C
Density	5.9 g/cm³	5.90 g/cm³
Formula of oxide	X₂O₃	Ga₂O₃
Formula of chloride	XCl₃	GaCl₃

Germanium (eka-silicon)

	Mendeleev's predicted properties	Actual properties
Atomic mass	About 72 amu	72.64 amu
Density	5.5 g/cm³	5.35 g/cm³
Formula of oxide	XO₂	GeO₂
Formula of chloride	XCl₄	GeCl₄

▲ **FIGURE 8.1 Eka-aluminum and Eka-silicon**   Mendeleev's arrangement of elements in the periodic table allowed him to predict the existence of these elements, now known as gallium and germanium, and anticipate their properties.

Mendeleev's arrangement was a huge success, allowing him to predict the existence and properties of yet undiscovered elements such as eka-aluminum (later discovered and named gallium) and eka-silicon (later discovered and named germanium), whose properties are summarized in Figure 8.1 ▲. (As noted in Chapter 2, *eka* means "the one beyond" or "the next one" in a family of elements.) However, Mendeleev did have some difficulties. For example, according to accepted values of atomic masses, tellurium (with higher mass) should come *after* iodine. But based on their properties, Mendeleev placed tellurium *before* iodine and suggested that the mass of tellurium was erroneous. The mass was correct; later work by the English physicist Henry Moseley (1887–1915) showed that listing elements according to *atomic number*, rather than atomic mass, resolved this problem and resulted in even better correlation with elemental properties.

Notice the scientific development in the history of the periodic table. A number of related observations led to a scientific law—the periodic law. Mendeleev's table, which is really just an expression of the periodic law, had predictive power, as laws usually do. However, it did not explain *why* the properties of elements recurred, or *why* certain elements had similar properties. Recall from Chapter 1 that laws *summarize* behavior while theories *explain* behavior. The theory that explains the existence of the periodic law is quantum-mechanical theory, which we learned about in Chapter 7. We now turn to exploring the connection between the periodic table and quantum-mechanical theory.

▲ Dmitri Mendeleev is credited with the arrangement of the periodic table.

8.3 Electron Configurations: How Electrons Occupy Orbitals

Quantum-mechanical theory describes the behavior of electrons in atoms. Since chemical bonding involves the transfer or sharing of electrons, quantum-mechanical theory helps us understand and describe chemical behavior. As we saw in Chapter 7, electrons in atoms exist within orbitals. An **electron configuration** for an atom shows the particular orbitals that are occupied for that atom. For example, the **ground state**—or lowest energy state—electron configuration for a hydrogen atom is shown below.

H $1s^1$ ← Number of electrons in orbital

Orbital

The electron configuration tells us that hydrogen's one electron is in the $1s$ orbital. Electrons generally occupy the lowest energy orbitals available. Since the $1s$ orbital is the lowest energy orbital in hydrogen (see Section 7.5), the electron occupies that orbital. If we could write electron configurations for all the elements, we could see how the arrangements of the electrons within their atoms correlate with the element's chemical properties. However, the solutions to the Schrödinger equation (the atomic orbitals and their energies) that we described in Chapter 7 were for the hydrogen atom. What do the atomic orbitals of other atoms look like? What are their relative energies?

Unfortunately, the Schrödinger equation for multielectron atoms is so complicated—because of new terms introduced into the equation by the interactions of the electrons with one another—that it cannot be solved exactly. However, approximate solutions indicate that the orbitals in multielectron atoms are hydrogen-like—they are similar to the s, p, d, and f orbitals that we examined in Chapter 7. In order to see how the electrons in multielectron atoms occupy these hydrogen-like orbitals, we must examine two additional concepts: *electron spin*, a fundamental property of all electrons that affects the number of electrons that are allowed in one orbital; and *sublevel energy splitting*, which applies to multielectron atoms and determines the order of orbital filling within a level.

Electron Spin and the Pauli Exclusion Principle

The electron configuration of hydrogen ($1s^1$) can be represented in a slightly different way by an **orbital diagram,** which gives similar information, but symbolizes the electron as an arrow in a box that represents the orbital. The orbital diagram for a hydrogen atom is

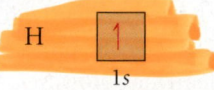

H   ↑
    $1s$

In orbital diagrams, the direction of the arrow (pointing up or pointing down) represents **electron spin,** a fundamental property of electrons. Electron spin was experimentally demonstrated in 1922 by the Stern–Gerlach experiment, shown in Figure 8.2 ▼. In this experiment a beam of silver atoms is split into two separate trajectories by a magnet. The spin of the electrons within the atoms creates a tiny magnetic field that interacts with the external field. One spin orientation causes the deflection of the beam in one direction, while the other orientation causes a deflection in the opposite direction. This experiment and others that followed it demonstrate two fundamental aspects of electron spin:

1. Spin, like negative electric charge, is a basic property of all electrons. One electron does not have more or less spin than another—all electrons have the same amount of spin.
2. The orientation of the electron's spin is quantized, with only two possibilities that we can call spin up and spin down.

> The idea of a "spinning" electron is something of a metaphor. A more correct way to express the same idea is to say that an electron has inherent angular momentum.

Stern–Gerlach Experiment

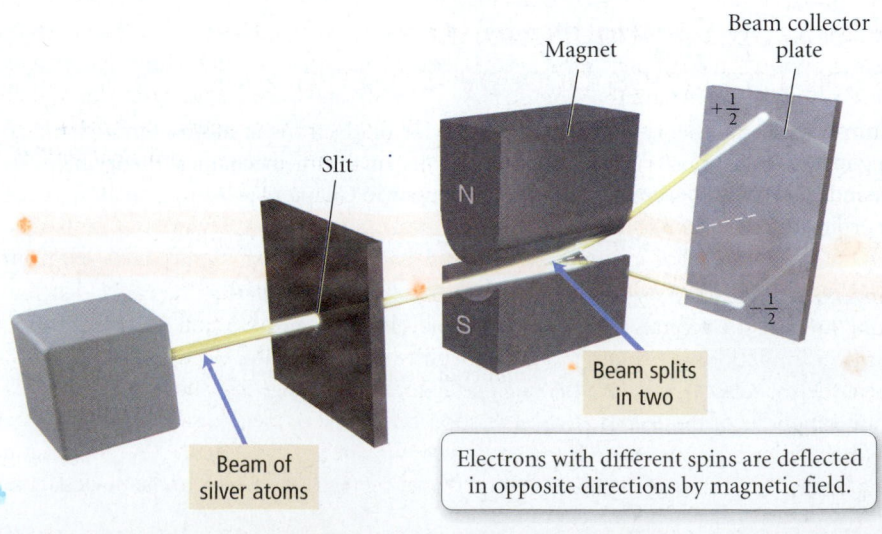

Magnet

Beam collector plate

Slit

N

$+\frac{1}{2}$

S

$-\frac{1}{2}$

Beam splits in two

Beam of silver atoms

Electrons with different spins are deflected in opposite directions by magnetic field.

▶ **FIGURE 8.2 The Stern–Gerlach Experiment**

The spin of an electron is specified by a fourth quantum number called the **spin quantum number (m_s)**. The possible values of m_s are $+\frac{1}{2}$ (spin up) and $-\frac{1}{2}$ (spin down). In an orbital diagram, $m_s = +\frac{1}{2}$ is represented with a half-arrow pointing up (↑), and $m_s = -\frac{1}{2}$ is represented with a half-arrow pointing down (↓). In a collection of hydrogen atoms, the electrons in about half of the atoms are spin up and the other half are spin down. Since no other electrons are present within the atom, we conventionally represent the hydrogen atom electron configuration with its one electron as spin up.

Helium is the first element on the periodic table that contains two electrons. The two electrons occupy the 1s orbital.

$$\text{He} \qquad 1s^2$$

How do the spins of the two electrons in helium align relative to each other? The answer to this question is described by the **Pauli exclusion principle,** formulated by Wolfgang Pauli in 1925:

> **Pauli exclusion principle: No two electrons in an atom can have the same four quantum numbers.**

Since two electrons occupying the same orbital have three identical quantum numbers (n, l, and m_l), they must have different spin quantum numbers. Since there are only two possible spin quantum numbers ($+\frac{1}{2}$ and $-\frac{1}{2}$), the Pauli exclusion principle implies that *each orbital can have a maximum of only two electrons, with opposing spins.* By applying the exclusion principle, we can write an electron configuration and orbital diagram for helium as follows:

Electron configuration	Orbital diagram
He $\quad 1s^2$	↑↓

The table below shows the four quantum numbers for each of the two electrons in helium.

n	l	m_l	m_s
1	0	0	$+\dfrac{1}{2}$
1	0	0	$-\dfrac{1}{2}$

The two electrons have three quantum numbers in common (because they are in the same orbital) but have different spin quantum numbers (as indicated by the opposing half-arrows in the orbital diagram).

Sublevel Energy Splitting in Multielectron Atoms

A major difference in the (approximate) solutions to the Schrödinger equation for multielectron atoms compared to the solutions for the hydrogen atom is the energy ordering of the orbitals. In the hydrogen atom, the energy of an orbital depends only on n, the principal quantum number. For example, the 3s, 3p, and 3d orbitals (which are empty for hydrogen in its lowest energy state) all have the same energy—they are said to be **degenerate.** The orbitals within a principal level of a *multielectron atom*, in contrast, are not degenerate—their energy depends on the value of l. In general, the lower the value of l within a principal level, the lower the energy of the corresponding orbital. Thus, for a given value of n:

$$E\,(s \text{ orbital}) < E\,(p \text{ orbital}) < E\,(d \text{ orbital}) < E\,(f \text{ orbital})$$

We can understand the energy splitting of the sublevels by considering the radial distribution functions for the orbitals within a principal level. Recall from Section 7.6 that the radial distribution function shows the total probability of finding the electron within a thin spherical shell at a distance r from the nucleus. Figure 8.3 on page 324 shows the radial distribution functions of the 2s and 2p orbitals superimposed on one another (with the radial distribution function of the 1s orbital also shown for comparison). Notice that, in general, an electron in a 2p orbital has a greater probability of being found closer to the nucleus than

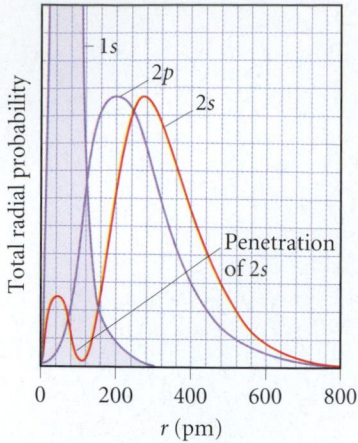

▲ FIGURE 8.3 **Radial Distribution Functions for the 1s, 2s, and 2p Orbitals**

an electron in a 2s orbital. We might initially expect, therefore, that the 2p orbital would be lower in energy. However, exactly the opposite is the case—the 2s orbital is actually lower in energy, *but only when the 1s orbital is occupied.* (When the 1s orbital is empty, the 2s and 2p orbitals are degenerate.) In other words, the splitting of the sublevels is caused by electron–electron repulsions.

For multielectron atoms, any one electron experiences both the positive charge of the nucleus (which is attractive) and the negative charges of the other electrons (which are repulsive). We can think of the repulsion of one electron by other electrons as equivalent to a *screening* or **shielding** of that electron from the full effects of the nuclear charge. For example, consider a lithium ion (Li^+). Since the lithium ion contains two electrons, its electron configuration is identical to that of helium:

$$Li^+ \qquad 1s^2$$

Now imagine bringing a third electron toward the lithium ion. When the third electron is far from the nucleus, it experiences the 3+ charge of the nucleus through the *screen* or *shield* of the 2− charge of the two 1s electrons, as shown in Figure 8.4(a) ▼. In effect, we

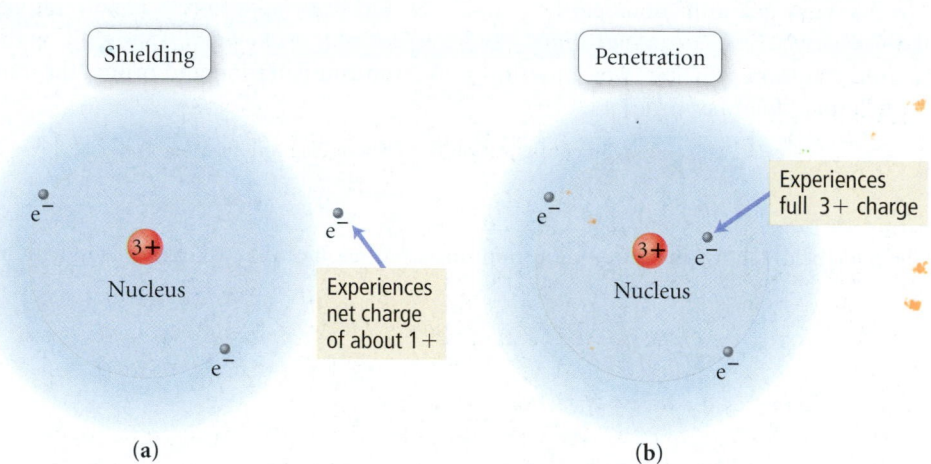

▶ FIGURE 8.4 **Shielding and Penetration** (a) An electron far from the nucleus is partly shielded by the electrons in the 1s orbital, reducing the effective net nuclear charge that it experiences. (b) An electron that penetrates the electron cloud of the 1s orbital experiences more of the nuclear charge.

Shielding

e^-

3+

Nucleus

Experiences net charge of about 1+

e^-

Penetration

e^-

3+ e^-

Experiences full 3+ charge

Nucleus

e^-

(a)   (b)

can think of the third electron as experiencing an **effective nuclear charge** (Z_{eff}) of approximately 1+ (3+ from the nucleus and 2− from the electrons, for a net of 1+). Imagine allowing this third electron to come close to the nucleus. As the electron *penetrates* the electron cloud of the 1s electrons it begins to experience the 3+ charge of the nucleus more fully because it is less shielded by the intervening electrons. If the electron could somehow get closer to the nucleus than the 1s electrons, it would experience the full 3+ charge, as shown in Figure 8.4(b) ▲.

Now look again at the radial distribution function for the 2s and 2p orbitals in Figure 8.3. Notice the bump near $r = 0$ (near the nucleus) for the 2s orbital. This bump represents a significant probability of the electron being found very close to the nucleus. Even more importantly, this area of the probability penetrates into the 1s orbital—it gets into the region where shielding by the 1s electrons is less effective. In contrast, most of the probability in the radial distribution function of the 2p orbital lies outside the radial distribution function of the 1s orbital. Consequently, almost all of the 2p orbital is shielded from nuclear charge by the 1s orbital. The end result is that the 2s orbital—since it experiences more of the nuclear charge due to its greater **penetration**—is lower in energy than the 2p orbital. The results are similar when we compare the 3s, 3p, and 3d orbitals. The s orbitals penetrate more fully than the p orbitals, which in turn penetrate more fully than the d orbitals, as shown in Figure 8.5 ◀.

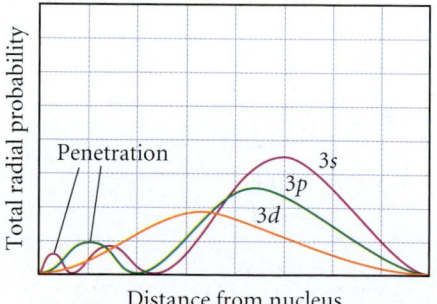

▲ FIGURE 8.5 **Radial Distribution Functions for the 3s, 3p, and 3d Orbitals** The 3s electrons penetrate most deeply into the inner orbitals, are least shielded, and experience the greatest effective nuclear charge. The 3d electrons penetrate least. This accounts for the energy ordering of the sublevels: $d > p > s$.

The following diagram shows the energy ordering of a number of orbitals in multi-electron atoms:

General Energy Ordering of Orbitals for Multielectron Atoms

Notice each of the following points about the diagram:

- Because of penetration, the sublevels of each principal level are *not* degenerate for multielectron atoms.
- In the fourth and fifth principal levels, the effects of penetration become so important that the 4*s* orbital lies lower in energy than the 3*d* orbitals and the 5*s* orbital lies lower in energy than the 4*d* orbitals.
- The energy separations between one set of orbitals and the next become smaller for 4*s* orbitals and beyond, and the relative energy ordering of these orbitals can actually vary among elements. These variations result in anomalies in the electron configurations of the transition metals and their ions (as we shall see later).

Conceptual Connection 8.1 Penetration and Shielding

Which of the following statements is true?

(a) An orbital that penetrates into the region occupied by core electrons is more shielded from nuclear charge than an orbital that does not penetrate and will therefore have a higher energy.

(b) An orbital that penetrates into the region occupied by core electrons is less shielded from nuclear charge than an orbital that does not penetrate and will therefore have a higher energy.

(c) An orbital that penetrates into the region occupied by core electrons is less shielded from nuclear charge than an orbital that does not penetrate and will therefore have a lower energy.

(d) An orbital that penetrates into the region occupied by core electrons is more shielded from nuclear charge than an orbital that does not penetrate and will therefore have a lower energy.

Answer: **(c)** Penetration results in less shielding from nuclear charge and therefore lower energy.

Electron Configurations for Multielectron Atoms

Unless otherwise specified, we will use the term "electron configuration" to mean the ground state (or lowest energy) configuration.

Now that we know the energy ordering of orbitals in multielectron atoms, we can build ground state electron configurations for the rest of the elements. Since we know that electrons occupy the lowest energy orbitals available when the atom is in its ground state, and that only two electrons (with opposing spins) are allowed in each orbital, we can systematically build up the electron configurations for the elements. The pattern of orbital filling that reflects what we have just learned is known as the **aufbau principle** (the German word *aufbau* means "build up"). For lithium, with three electrons, the electron configuration and orbital diagram are

Remember that the number of electrons in an atom is equal to its atomic number.

Electron configuration   Orbital diagram

Li $1s^2 2s^1$ $\boxed{\uparrow\downarrow}$ $\boxed{\uparrow}$

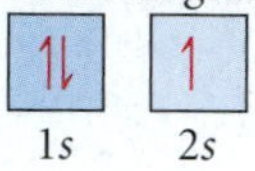

                                 1s   2s

For carbon, with atomic number 6 and therefore 6 electrons, the electron configuration and orbital diagram are

Electron configuration   Orbital diagram

C $1s^2 2s^2 2p^2$ $\boxed{\uparrow\downarrow}$ $\boxed{\uparrow\downarrow}$ $\boxed{\uparrow}\boxed{\uparrow}\boxed{}$

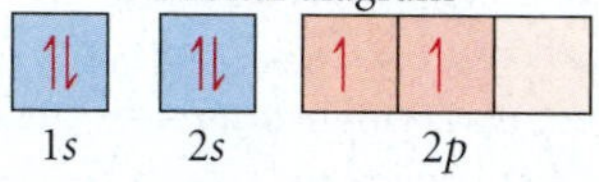

                               1s   2s   2p

Notice that the 2p electrons occupy the p orbitals (of equal energy) singly, rather than pairing in one orbital. This way of filling orbitals is in accord with **Hund's rule,** which states that *when filling degenerate orbitals, electrons fill them singly first, with parallel spins.* Hund's rule is a result of an atom's tendency to find the lowest energy state possible. When two electrons occupy separate orbitals of equal energy, the repulsive interaction between them is lower than when they occupy the same orbital because the electrons are spread out over a larger region of space.

Summarizing:

➤ Electrons occupy orbitals so as to minimize the energy of the atom; therefore, lower energy orbitals fill before higher energy orbitals. Orbitals fill in the following order: 1s 2s 2p 3s 3p 4s 3d 4p 5s 4d 5p 6s.

➤ Orbitals can hold no more than two electrons each. When two electrons occupy the same orbital, their spins are opposite. This is another way of expressing the Pauli exclusion principle (no two electrons in one atom can have the same four quantum numbers).

➤ When orbitals of identical energy are available, electrons first occupy these orbitals singly with parallel spins rather than in pairs. Once the orbitals of equal energy are half-full, the electrons start to pair (Hund's rule).

Consider the electron configurations and orbital diagrams for elements with atomic numbers 3–10.

Symbol	Number of electrons	Electron configuration	Orbital diagram
Li	3	$1s^2 2s^1$	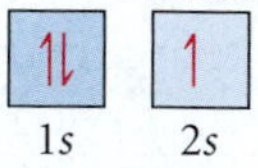 $\boxed{\uparrow\downarrow}$ $\boxed{\uparrow}$ 1s 2s
Be	4	$1s^2 2s^2$	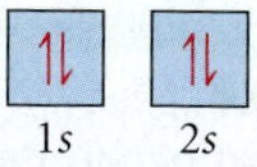 $\boxed{\uparrow\downarrow}$ $\boxed{\uparrow\downarrow}$ 1s 2s
B	5	$1s^2 2s^2 2p^1$	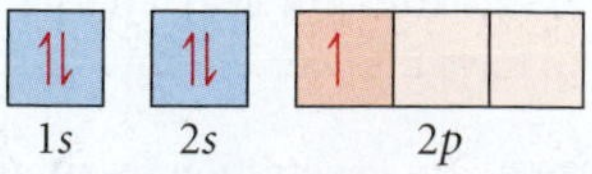 $\boxed{\uparrow\downarrow}$ $\boxed{\uparrow\downarrow}$ $\boxed{\uparrow}\boxed{}\boxed{}$ 1s 2s 2p
C	6	$1s^2 2s^2 2p^2$	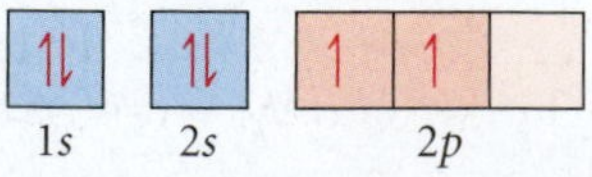 $\boxed{\uparrow\downarrow}$ $\boxed{\uparrow\downarrow}$ $\boxed{\uparrow}\boxed{\uparrow}\boxed{}$ 1s 2s 2p

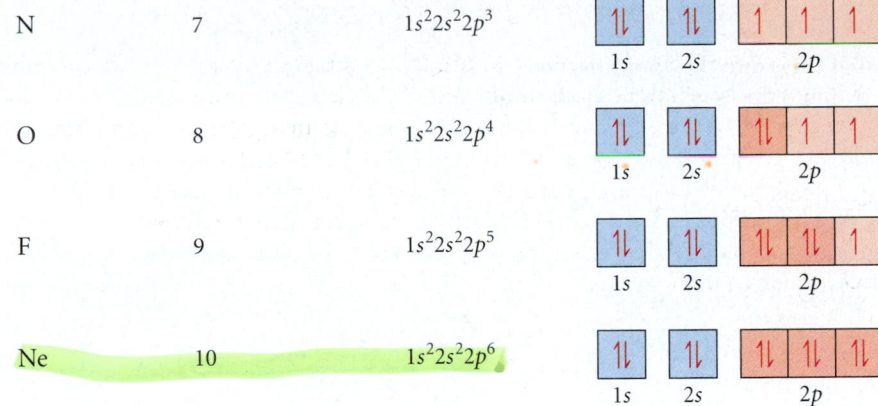

N	7	$1s^2 2s^2 2p^3$
O	8	$1s^2 2s^2 2p^4$
F	9	$1s^2 2s^2 2p^5$
Ne	10	$1s^2 2s^2 2p^6$

Notice that, as a result of Hund's rule, the p orbitals fill with single electrons before the electrons pair.

The electron configuration of neon represents the complete filling of the $n = 2$ principal level. When writing electron configurations for elements beyond neon, or beyond any other noble gas, the electron configuration of the previous noble gas—sometimes called the *inner electron configuration*—is often abbreviated by the symbol for the noble gas in square brackets. For example, the electron configuration of sodium is

$$\text{Na} \quad 1s^2 2s^2 2p^6 3s^1$$

This can also be written as:

$$\text{Na} \quad [\text{Ne}]\, 3s^1$$

where [Ne] represents $1s^2 2s^2 2p^6$, the electron configuration for neon.

To write an electron configuration for an element, first find its atomic number from the periodic table—this number equals the number of electrons. Then use the order of filling to distribute the electrons in the appropriate orbitals. Remember that each orbital can hold a maximum of 2 electrons. Consequently,

- The s sublevel has only one orbital and can therefore hold only 2 electrons.
- The p sublevel has three orbitals and can therefore hold 6 electrons.
- The d sublevel has five orbitals and can therefore hold 10 electrons.
- The f sublevel has seven orbitals and can therefore hold 14 electrons.

EXAMPLE 8.1 Electron Configurations

Write electron configurations for each of the following elements:

(a) Mg   (b) P   (c) Br   (d) Al

Solution

(a) Mg Magnesium has 12 electrons. Distribute two of these into the $1s$ orbital, two into the $2s$ orbital, six into the $2p$ orbitals, and two into the $3s$ orbital.	Mg $1s^2 2s^2 2p^6 3s^2$ or $[\text{Ne}]\, 3s^2$
(b) P Phosphorus has 15 electrons. Distribute two of these into the $1s$ orbital, two into the $2s$ orbital, six into the $2p$ orbitals, two into the $3s$ orbital, and three into the $3p$ orbitals.	P $1s^2 2s^2 2p^6 3s^2 3p^3$ or $[\text{Ne}]\, 3s^2 3p^3$
(c) Br Bromine has 35 electrons. Distribute two of these into the $1s$ orbital, two into the $2s$ orbital, six into the $2p$ orbitals, two into the $3s$ orbital, six into the $3p$ orbitals, two into the $4s$ orbital, ten into the $3d$ orbitals, and five into the $4p$ orbitals.	Br $1s^2 2s^2 2p^6 3s^2 3p^6 4s^2 3d^{10} 4p^5$ or $[\text{Ar}]\, 4s^2 3d^{10} 4p^5$

(d) Al Aluminum has 13 electrons. Distribute two of these into the $1s$ orbital, two into the $2s$ orbital, six into the $2p$ orbitals, two into the $3s$ orbital, and one into the $3p$ orbital.	Al $1s^2 2s^2 2p^6 3s^2 3p^1$ or $[Ne] 3s^2 3p^1$

For Practice 8.1

Write electron configurations for each of the following elements:

(a) Cl   **(b)** Si   **(c)** Sr   **(d)** O

EXAMPLE 8.2 Writing Orbital Diagrams

Write an orbital diagram for sulfur and determine the number of unpaired electrons.

Solution

Since sulfur is atomic number 16, it has 16 electrons and the electron configuration $1s^2 2s^2 2p^6 3s^2 3p^4$. Draw a box for each orbital putting the lowest energy orbital ($1s$) on the far left and proceeding to orbitals of higher energy to the right.	 $1s$ $2s$ $2p$ $3s$ $3p$
Distribute the 16 electrons into the boxes representing the orbitals, allowing a maximum of two electrons per orbital and remembering Hund's rule. You can see from the diagram that sulfur has two unpaired electrons.	 $1s$ $2s$ $2p$ $3s$ $3p$ Two unpaired electrons

For Practice 8.2

Write an orbital diagram for Ar and determine the number of unpaired electrons.

Conceptual Connection 8.2 Electron Configurations and Quantum Numbers

What are the four quantum numbers for each of the two electrons in a $4s$ orbital?

Answer: $n = 4, l = 0, m_l = 0, m_s = +\frac{1}{2}; n = 4, l = 0, m_l = 0, m_s = -\frac{1}{2}$

8.4 Electron Configurations, Valence Electrons, and the Periodic Table

We have seen that Mendeleev arranged the periodic table so that elements with similar chemical properties lie in the same column. We can begin to make the connection between an element's properties and its electron configuration by superimposing the electron configurations of the first 18 elements onto a partial periodic table, as shown in Figure 8.6 ▶. As you move to the right across a row, the orbitals are simply filling in the correct order. With each subsequent row, the highest principal quantum number increases by one. Notice that as you move down a column, *the number of electrons in the outermost principal energy level (highest n value) remains the same.* The key connection between the macroscopic world (an element's chemical properties) and the microscopic world (an atom's electronic structure) lies in these outermost electrons.

Outer Electron Configurations of Elements 1–18

1A							8A
1 **H** $1s^1$	2A	3A	4A	5A	6A	7A	2 **He** $1s^2$
3 **Li** $2s^1$	4 **Be** $2s^2$	5 **B** $2s^22p^1$	6 **C** $2s^22p^2$	7 **N** $2s^22p^3$	8 **O** $2s^22p^4$	9 **F** $2s^22p^5$	10 **Ne** $2s^22p^6$
11 **Na** $3s^1$	12 **Mg** $3s^2$	13 **Al** $3s^23p^1$	14 **Si** $3s^23p^2$	15 **P** $3s^23p^3$	16 **S** $3s^23p^4$	17 **Cl** $3s^23p^5$	18 **Ar** $3s^23p^6$

◀ **FIGURE 8.6 Outer Electron Configurations of the First 18 Elements in the Periodic Table**

We define an atom's **valence electrons** as those that are important in chemical bonding. *For main-group elements, the valence electrons are those in the outermost principal energy level* (Figure 8.6 ▲). For transition elements, we also count the outermost *d* electrons among the valence electrons (even though they are not in an outermost principal energy level). The chemical properties of an element depend on its valence electrons, which are important in bonding because they are held most loosely (and are therefore the easiest to lose or share). We can now see *why* the elements in a column of the periodic table have similar chemical properties: *they have the same number of valence electrons.*

Valence electrons are distinguished from all the other electrons in an atom, which are called **core electrons**. The core electrons are those in *complete* principal energy levels and those in *complete d and f* sublevels. For example, silicon, with the electron configuration $1s^22s^22p^63s^23p^2$ has 4 valence electrons (those in the $n = 3$ principal level) and 10 core electrons.

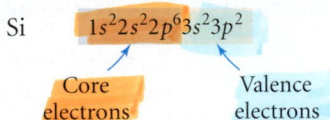

Si $\quad 1s^22s^22p^63s^23p^2$

Core electrons → | Valence electrons →

EXAMPLE 8.3 Valence Electrons and Core Electrons

Write an electron configuration for Ge and identify the valence electrons and the core electrons.

Solution

Write the electron configuration for Ge by determining the total number of electrons from germanium's atomic number (32) and then distributing them into the appropriate orbitals as described previously.	Ge $\quad 1s^22s^22p^63s^23p^64s^23d^{10}4p^2$
Since germanium is a main-group element, its valence electrons are those in the outermost principal energy level. For germanium, the $n = 1, 2,$ and 3 principal levels are complete (or full) and the $n = 4$ principal level is outermost. Consequently, the $n = 4$ electrons are valence electrons and the rest are core electrons.	4 valence electrons Ge $\quad 1s^22s^22p^63s^23p^64s^23d^{10}4p^2$ 28 core electrons

Note: In this book, electron configurations are always written with the orbitals in the order *of filling. However, writing electron configurations in* order *of increasing principal quantum number is also common. For example, the electron configuration of germanium written in order of increasing principal quantum number is*

Ge $\quad 1s^22s^22p^63s^23p^63d^{10}4s^24p^2$

For Practice 8.3

Write an electron configuration for phosphorus and identify the valence electrons and core electrons.

Orbital Blocks of the Periodic Table

FIGURE 8.7 The *s*, *p*, *d*, and *f* Blocks of the Periodic Table

Legend: s-block elements, p-block elements, d-block elements, f-block elements

Group	1 / 1A	2 / 2A	3 / 3B	4 / 4B	5 / 5B	6 / 6B	7 / 7B	8 / 8B	9 / 8B	10 / 8B	11 / 1B	12 / 2B	13 / 3A	14 / 4A	15 / 5A	16 / 6A	17 / 7A	18 / 8A
1	1 H $1s^1$																	2 He $1s^2$
2	3 Li $2s^1$	4 Be $2s^2$											5 B $2s^2 2p^1$	6 C $2s^2 2p^2$	7 N $2s^2 2p^3$	8 O $2s^2 2p^4$	9 F $2s^2 2p^5$	10 Ne $2s^2 2p^6$
3	11 Na $3s^1$	12 Mg $3s^2$											13 Al $3s^2 3p^1$	14 Si $3s^2 3p^2$	15 P $3s^2 3p^3$	16 S $3s^2 3p^4$	17 Cl $3s^2 3p^5$	18 Ar $3s^2 3p^6$
4	19 K $4s^1$	20 Ca $4s^2$	21 Sc $4s^2 3d^1$	22 Ti $4s^2 3d^2$	23 V $4s^2 3d^3$	24 Cr $4s^1 3d^5$	25 Mn $4s^2 3d^5$	26 Fe $4s^2 3d^6$	27 Co $4s^2 3d^7$	28 Ni $4s^2 3d^8$	29 Cu $4s^1 3d^{10}$	30 Zn $4s^2 3d^{10}$	31 Ga $4s^2 4p^1$	32 Ge $4s^2 4p^2$	33 As $4s^2 4p^3$	34 Se $4s^2 4p^4$	35 Br $4s^2 4p^5$	36 Kr $4s^2 4p^6$
5	37 Rb $5s^1$	38 Sr $5s^2$	39 Y $5s^2 4d^1$	40 Zr $5s^2 4d^2$	41 Nb $5s^1 4d^4$	42 Mo $5s^1 4d^5$	43 Tc $5s^2 4d^5$	44 Ru $5s^1 4d^7$	45 Rh $5s^1 4d^8$	46 Pd $4d^{10}$	47 Ag $5s^1 4d^{10}$	48 Cd $5s^2 4d^{10}$	49 In $5s^2 5p^1$	50 Sn $5s^2 5p^2$	51 Sb $5s^2 5p^3$	52 Te $5s^2 5p^4$	53 I $5s^2 5p^5$	54 Xe $5s^2 5p^6$
6	55 Cs $6s^1$	56 Ba $6s^2$	57 La $6s^2 5d^1$	72 Hf $6s^2 5d^2$	73 Ta $6s^2 5d^3$	74 W $6s^2 5d^4$	75 Re $6s^2 5d^5$	76 Os $6s^2 5d^6$	77 Ir $6s^2 5d^7$	78 Pt $6s^1 5d^9$	79 Au $6s^1 5d^{10}$	80 Hg $6s^2 5d^{10}$	81 Tl $6s^2 6p^1$	82 Pb $6s^2 6p^2$	83 Bi $6s^2 6p^3$	84 Po $6s^2 6p^4$	85 At $6s^2 6p^5$	86 Rn $6s^2 6p^6$
7	87 Fr $7s^1$	88 Ra $7s^2$	89 Ac $7s^2 6d^1$	104 Rf $7s^2 6d^2$	105 Db $7s^2 6d^3$	106 Sg $7s^2 6d^4$	107 Bh	108 Hs	109 Mt	110 Ds	111 Rg	112	114		116			

Lanthanides

58 Ce $6s^2 4f^1 5d^1$	59 Pr $6s^2 4f^3$	60 Nd $6s^2 4f^4$	61 Pm $6s^2 4f^5$	62 Sm $6s^2 4f^6$	63 Eu $6s^2 4f^7$	64 Gd $6s^2 4f^7 5d^1$	65 Tb $6s^2 4f^9$	66 Dy $6s^2 4f^{10}$	67 Ho $6s^2 4f^{11}$	68 Er $6s^2 4f^{12}$	69 Tm $6s^2 4f^{13}$	70 Yb $6s^2 4f^{14}$	71 Lu $6s^2 4f^{14} 6d^1$

Actinides

90 Th $7s^2 6d^2$	91 Pa $7s^2 5f^2 6d^1$	92 U $7s^2 5f^3 6d^1$	93 Np $7s^2 5f^4 6d^1$	94 Pu $7s^2 5f^6$	95 Am $7s^2 5f^7$	96 Cm $7s^2 5f^7 6d^1$	97 Bk $7s^2 5f^9$	98 Cf $7s^2 5f^{10}$	99 Es $7s^2 5f^{11}$	100 Fm $7s^2 5f^{12}$	101 Md $7s^2 5f^{13}$	102 No $7s^2 5f^{14}$	103 Lr $7s^2 5f^{14} 6d^1$

Orbital Blocks in the Periodic Table

A pattern similar to what we just saw for the first 18 elements exists for the entire periodic table, as shown in Figure 8.7 ▲. Note that, because of the filling order of orbitals, the periodic table can be divided into blocks representing the filling of particular sublevels. The first two columns on the left side of the periodic table comprise the *s* block, with outer electron configurations of ns^1 (the alkali metals) and ns^2 (the alkaline earth metals). The six columns on the right side of the periodic table comprise the *p* block, with outer electron configurations of $ns^2 np^1$, $ns^2 np^2$, $ns^2 np^3$, $ns^2 np^4$, $ns^2 np^5$ (halogens), and $ns^2 np^6$ (noble gases). The transition elements comprise the *d* block, and the lanthanides and actinides (also called the inner transition elements) comprise the *f* block. (For compactness, the *f* block is normally printed below the *d* block instead of being imbedded within it.)

You can see that *the number of columns in a block corresponds to the maximum number of electrons that can occupy the particular sublevel of that block.* The *s* block has 2 columns (corresponding to one *s* orbital holding a maximum of two electrons); the *p* block has 6 columns (corresponding to three *p* orbitals with two electrons each); the *d* block has 10 columns (corresponding to five *d* orbitals with two electrons each); and the *f* block has 14 columns (corresponding to seven *f* orbitals with two electrons each).

Notice also that, except for helium, *the number of valence electrons for any main-group element is equal to its lettered group number.* For example, we can tell that chlorine has 7 valence electrons because it is in group number 7A.

Lastly, note that, for main-group elements, *the row number in the periodic table is equal to the number (or n value) of the highest principal level.* For example, because chlorine is in row 3, its highest principal level is the $n = 3$ level.

Summarizing:

➤ The periodic table is divisible into four blocks corresponding to the filling of the four quantum sublevels (*s*, *p*, *d*, and *f*).

Helium is an exception. Even though it lies in the column with an outer electron configuration of $ns^2 np^6$, its electron configuration is simply $1s^2$.

Recall from Chapter 2 that main-group elements are those in the two far left columns (groups 1A, 2A) and the six far right columns (groups 3A–8A) of the periodic table.

▶ The group number of a main-group element is equal to the number of valence electrons for that element.

▶ The row number of a main-group element is equal to the highest principal quantum number of that element.

Writing an Electron Configuration for an Element from Its Position in the Periodic Table

The organization of the periodic table allows us to write the electron configuration for any element simply based on its position in the periodic table. For example, suppose we want to write an electron configuration for Cl. The *inner electron configuration* of Cl is simply that of the noble gas that precedes it in the periodic table, Ne. So we can represent the inner electron configuration with [Ne]. The *outer electron configuration*—the configuration of the electrons beyond the previous noble gas—is obtained by tracing the elements between Ne and Cl and assigning electrons to the appropriate orbitals, as shown here. Remember that the highest n value is given by the row number (3 for chlorine).

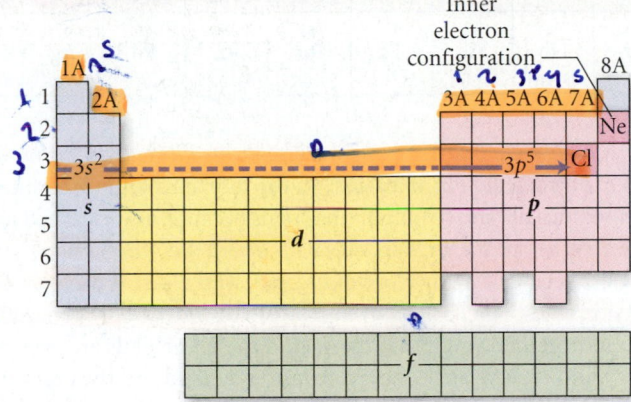

So, we begin with [Ne], then add in the two 3s electrons as we trace across the s block, followed by five 3p electrons as we trace across the p block to Cl, which is in the fifth column of the p block. The electron configuration is

$$Cl \qquad [Ne]\, 3s^2 3p^5$$

Notice that Cl is in column 7A and therefore has 7 valence electrons and an outer electron configuration of $ns^2 np^5$.

EXAMPLE 8.4 Writing Electron Configurations from the Periodic Table

Use the periodic table to write an electron configuration for selenium (Se).

Solution

The atomic number of Se is 34. The noble gas that precedes Se in the periodic table is argon, so the inner electron configuration is [Ar]. Obtain the outer electron configuration by tracing the elements between Ar and Se and assigning electrons to the appropriate orbitals. Begin with [Ar]. Because Se is in row 4, add two 4s electrons as you trace across the s block (n = row number). Next, add ten 3d electrons as you trace across the d block (n = row number − 1). Lastly, add four 4p electrons as you trace across the p block to Se, which is in the fourth column of the p block (n = row number).

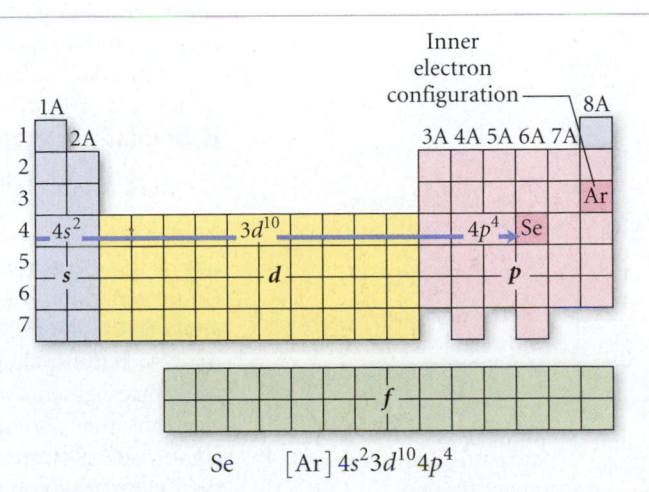

$$Se \qquad [Ar]\, 4s^2 3d^{10} 4p^4$$

For Practice 8.4

Use the periodic table to determine the electron configuration for bismuth (Bi).

For More Practice 8.4

Use the periodic table to write an electron configuration for iodine (I).

The Transition and Inner Transition Elements

The electron configurations of the transition elements (d block) and inner transition elements (f block) exhibit trends that differ somewhat from those of the main-group elements. As you move to the right across a row in the d block, the d orbitals fill as shown here:

21	22	23	24	25	26	27	28	29	30
Sc	**Ti**	**V**	**Cr**	**Mn**	**Fe**	**Co**	**Ni**	**Cu**	**Zn**
$4s^23d^1$	$4s^23d^2$	$4s^23d^3$	$4s^13d^5$	$4s^23d^5$	$4s^23d^6$	$4s^23d^7$	$4s^23d^8$	$4s^13d^{10}$	$4s^23d^{10}$
39	40	41	42	43	44	45	46	47	48
Y	**Zr**	**Nb**	**Mo**	**Tc**	**Ru**	**Rh**	**Pd**	**Ag**	**Cd**
$5s^24d^1$	$5s^24d^2$	$5s^14d^4$	$5s^14d^5$	$5s^24d^5$	$5s^14d^7$	$5s^14d^8$	$4d^{10}$	$5s^14d^{10}$	$5s^24d^{10}$

However, *the principal quantum number of the d orbital being filled across each row in the transition series is equal to the row number minus one.* In the fourth row, the $3d$ orbitals fill, and in the fifth row, the $4d$ orbitals fill, and so on. This happens because, as we learned in Section 8.3, the $4s$ orbital is generally lower in energy than the $3d$ orbital (because it more efficiently penetrates into the region occupied by the core electrons). The result is that the $4s$ orbital fills before the $3d$ orbital, even though its principal quantum number ($n = 4$) is higher.

We must keep in mind, however, that the $4s$ and the $3d$ orbitals are extremely close to each other in energy and their relative energy ordering depends on the exact species under consideration and the exact electron configuration (first discussed in Section 8.3); this causes some anomalous behavior in the transition metals. For example, notice that, in the first transition series of the d block, the outer configuration is $4s^23d^x$ with two exceptions: Cr is $4s^13d^5$ and Cu is $4s^13d^{10}$. This behavior is related to the closely spaced $3d$ and $4s$ energy levels and the stability associated with a half-filled or completely filled sublevel. Actual electron configurations are always determined experimentally (through spectroscopy) and do not always conform to simple patterns. Nonetheless, the patterns we have described allow us to predict electron configurations for most of the elements in the periodic table.

As you move across the f block, the f orbitals fill. However, the principal quantum number of the f orbital being filled across each row in the inner transition series is the row number *minus two.* (In the sixth row, the $4f$ orbitals fill, and in the seventh row, the $5f$ orbitals fill.) In addition, within the inner transition series, the close energy spacing of the $5d$ and $4f$ orbitals sometimes causes an electron to enter a $5d$ orbital instead of the expected $4f$ orbital. For example, the electron configuration of gadolinium is $[\text{Xe}]\,6s^24f^75d^1$ (instead of the expected $[\text{Xe}]\,6s^24f^8$).

8.5 The Explanatory Power of the Quantum-Mechanical Model

We can now see how the quantum-mechanical model accounts for the chemical properties of the elements, such as the inertness of helium or the reactivity of hydrogen, and (more generally) how it accounts for the periodic law. *The chemical properties of elements are largely determined by the number of valence electrons they contain.* Their properties are periodic because the number of valence electrons is periodic.

Since elements within a column in the periodic table have the same number of valence electrons, they also have similar chemical properties. The noble gases, for example, all have eight valence electrons, except for helium, which has two. Although we do not cover the quantitative (or numerical) aspects of the quantum-mechanical model in this book,

8A

2
He
$1s^2$
10
Ne
$2s^22p^6$
18
Ar
$3s^23p^6$
36
Kr
$4s^24p^6$
54
Xe
$5s^25p^6$
86
Rn
$6s^26p^6$

Noble gases

◀ The noble gases all have eight valence electrons except for helium, which has two. They have full outer energy levels and are particularly stable and unreactive.

calculations of the overall energy of atoms with eight valence electrons (or two for helium) show that they are particularly stable. In other words, when a quantum level is completely full, the overall energy of the electrons that occupy that level is particularly low. This means that those electrons *cannot* lower their energy by reacting with other atoms or molecules, so the corresponding atom is unreactive or inert. Consequently, the noble gases are the most chemically stable or relatively unreactive family in the periodic table.

Elements with electron configurations *close* to those of the noble gases are the most reactive because they can attain noble gas electron configurations by losing or gaining a small number of electrons. For example, alkali metals (group 1A) are among the most reactive metals because their outer electron configuration (ns^1) is one electron beyond a noble gas configuration. They react to lose the ns^1 electron, obtaining a noble gas configuration. This explains why—as we saw in Chapter 2—the group 1A metals tend to form 1+ cations. Similarly, alkaline earth metals, with an outer electron configuration of ns^2, also tend to be reactive metals, losing their ns^2 electrons to form 2+ cations. This does not mean that forming an ion with a noble gas configuration is in itself energetically favorable. In fact, forming cations always *requires* energy. But when the cation formed has a noble gas configuration, the energy cost of forming the cation is often less than the energy payback that occurs when that cation forms ionic bonds with anions, as we shall see in Chapter 9.

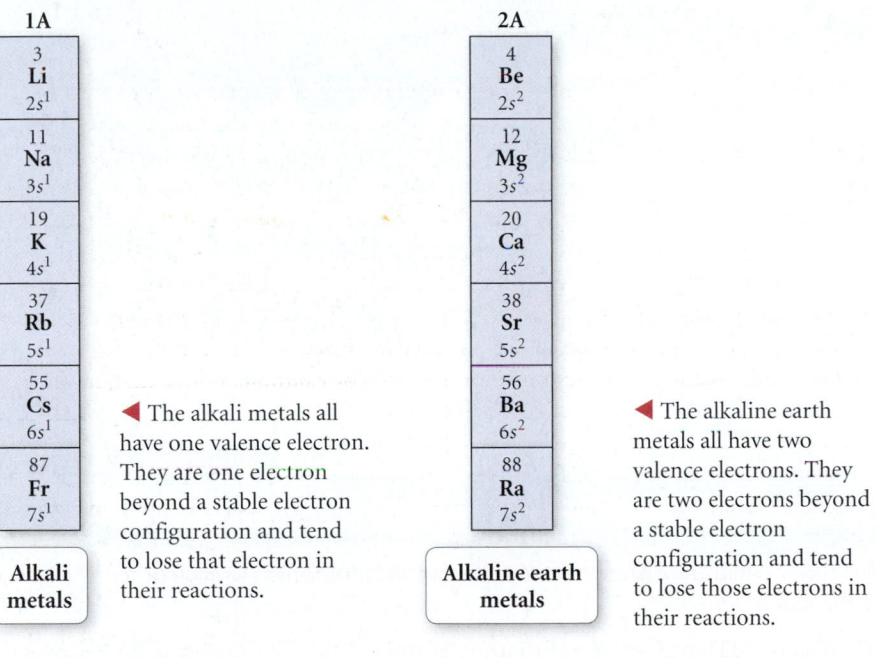

1A
3 **Li** $2s^1$
11 **Na** $3s^1$
19 **K** $4s^1$
37 **Rb** $5s^1$
55 **Cs** $6s^1$
87 **Fr** $7s^1$
Alkali metals

◄ The alkali metals all have one valence electron. They are one electron beyond a stable electron configuration and tend to lose that electron in their reactions.

2A
4 **Be** $2s^2$
12 **Mg** $3s^2$
20 **Ca** $4s^2$
38 **Sr** $5s^2$
56 **Ba** $6s^2$
88 **Ra** $7s^2$
Alkaline earth metals

◄ The alkaline earth metals all have two valence electrons. They are two electrons beyond a stable electron configuration and tend to lose those electrons in their reactions.

7A
9 **F** $2s^2 2p^5$
17 **Cl** $3s^2 3p^5$
35 **Br** $4s^2 4p^5$
53 **I** $5s^2 5p^5$
85 **At** $6s^2 6p^5$
Halogens

◄ The halogens all have seven valence electrons. They are one electron short of a stable electron configuration and tend to gain one electron in their reactions.

On the right side of the periodic table, halogens are among the most reactive nonmetals because of their $ns^2 np^5$ electron configurations. They are only one electron short of a noble gas configuration and tend to react to gain that one electron, forming 1− ions. Figure 8.8 ▼, first introduced in Chapter 2, shows the elements that form predictable ions. Notice how the charges of these ions reflect their electron configurations—in their reactions, these elements form ions with noble gas electron configurations.

Elements That Form Ions with Predictable Charges

	1A	2A										3A	4A	5A	6A	7A	8A
1	Li^+													N^{3-}	O^{2-}	F^-	
2	Na^+	Mg^{2+}	3B	4B	5B	6B	7B	⎯ 8B ⎯	1B	2B	Al^{3+}			S^{2-}	Cl^-	
3	K^+	Ca^{2+}												Se^{2-}	Br^-	
4	Rb^+	Sr^{2+}												Te^{2-}	I^-	
5	Cs^+	Ba^{2+}														

◄ **FIGURE 8.8 Elements That Form Ions with Predictable Charges** Notice that each ion has a noble gas electron configuration.

8.6 Periodic Trends in the Size of Atoms and Effective Nuclear Charge

In previous chapters, we saw that the volume of an atom is taken up primarily by its electrons (Chapter 2) occupying quantum-mechanical orbitals (Chapter 7). We also saw that these orbitals do not have a definite boundary, but represent only a statistical probability distribution for where the electron is found. So how do we define the size of an atom? One way to define atomic radii is to consider the distance between *nonbonding* atoms in molecules or atoms that are in direct contact. For example, krypton can be frozen into a solid in which the krypton atoms are touching each other but are not bonded together. The distance between the centers of adjacent krypton atoms—which can be determined from the solid's density (see Exercise 107)—is then twice the radius of a krypton atom. An atomic radius determined in this way is called the **nonbonding atomic radius** or the **van der Waals radius**. The van der Waals radius represents the radius of an atom when it is not bonded to another atom.

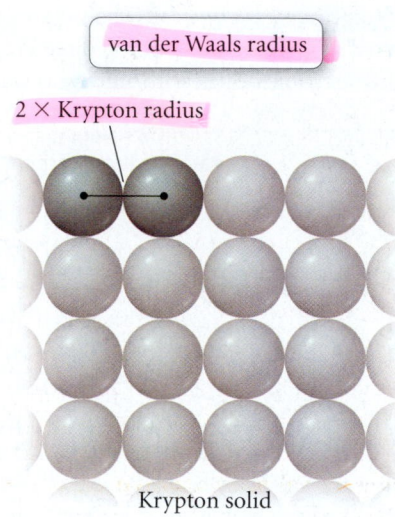

van der Waals radius

2 × Krypton radius

▶ The van der Waals radius of an atom is one-half the distance between adjacent nuclei in the atomic solid.

Krypton solid

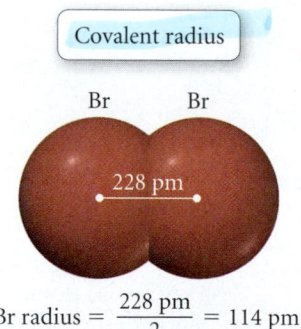

Covalent radius

Br   Br

228 pm

Br radius = $\dfrac{228 \text{ pm}}{2}$ = 114 pm

▲ The covalent radius of bromine is one-half the distance between two bonded bromine atoms.

The bonding radii of some elements, such as helium and neon, must be approximated since they do not form either chemical bonds or metallic crystals.

Another way to define the size of an atom, called the **bonding atomic radius** or **covalent radius**, is defined differently for nonmetals and metals, as follows:

Nonmetals: one-half the distance between two of the atoms bonded together

Metals: one-half the distance between two of the atoms next to each other in a crystal of the metal

For example, the distance between Br atoms in Br_2 is 228 pm; therefore, the Br covalent radius is assigned to be one-half of 228 pm or 114 pm.

Similar radii can be assigned to all elements in the periodic table that form chemical bonds or form metallic crystals. A more general term, the **atomic radius,** refers to a set of average bonding radii determined from measurements on a large number of elements and compounds. The atomic radius represents the radius of an atom when it is bonded to another atom and is always smaller than the van der Waals radius. The approximate bond length of any two covalently bonded atoms is simply the sum of their atomic radii. For example, the approximate bond length for ICl is iodine's atomic radius (133 pm) plus chlorine's atomic radius (99 pm), for a bond length of 232 pm. (The actual experimentally measured bond length in ICl is 232.07 pm.)

Figure 8.9 ▶ shows the atomic radius plotted as a function of atomic number for the first 57 elements in the periodic table. Notice the periodic trend in the radii. Atomic radii peak with each alkali metal. Figure 8.10 ▶ is a relief map of atomic radii for most of the elements in the periodic table. The general trends in the atomic radii of main-group elements, which are the same as trends observed in van der Waals radii, are stated below.

1. As you move down a column (or family) in the periodic table, atomic radius increases.

2. As you move to the right across a period (or row) in the periodic table, atomic radius decreases.

Atomic Radii

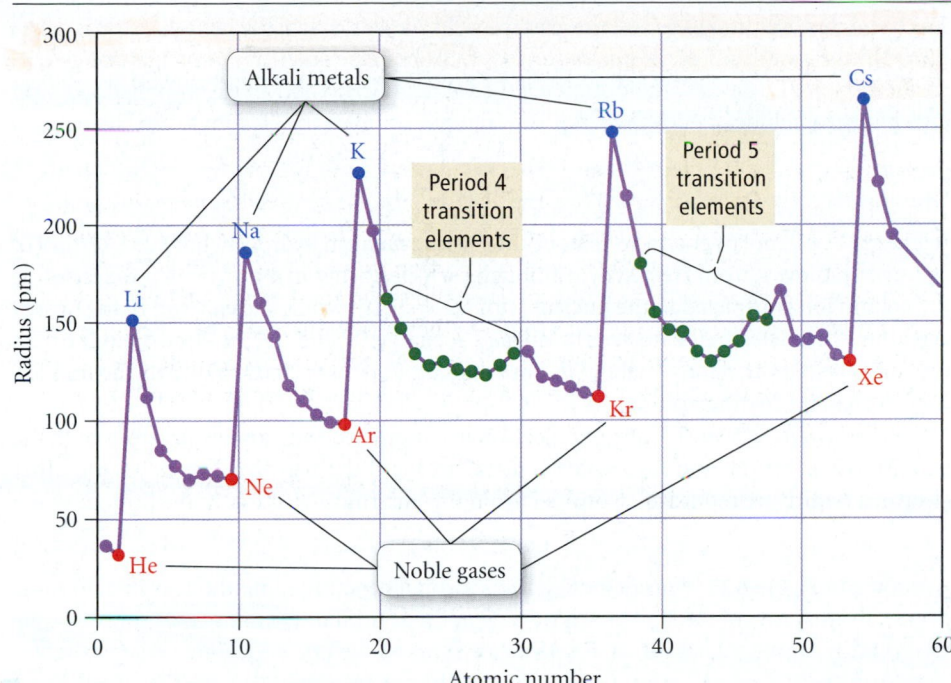

◀ **FIGURE 8.9 Atomic Radius versus Atomic Number** Notice the periodic trend in atomic radius, starting at a peak with each alkali metal and falling to a minimum with each noble gas.

The observed trend in radius as you move down a column can be understood in light of the trends in the sizes of atomic orbitals. The atomic radius is largely determined by the valence electrons, the electrons farthest from the nucleus. As you move down a column in the periodic table, the highest principal quantum number (n) of the valence electrons increases. Consequently, the valence electrons occupy larger orbitals, resulting in larger atoms.

The observed trend in atomic radius as you move to the right across a row, however, is bit more complex. To understand this trend, we revisit some concepts from Section 8.3, including effective nuclear charge and shielding.

Trends in Atomic Radius

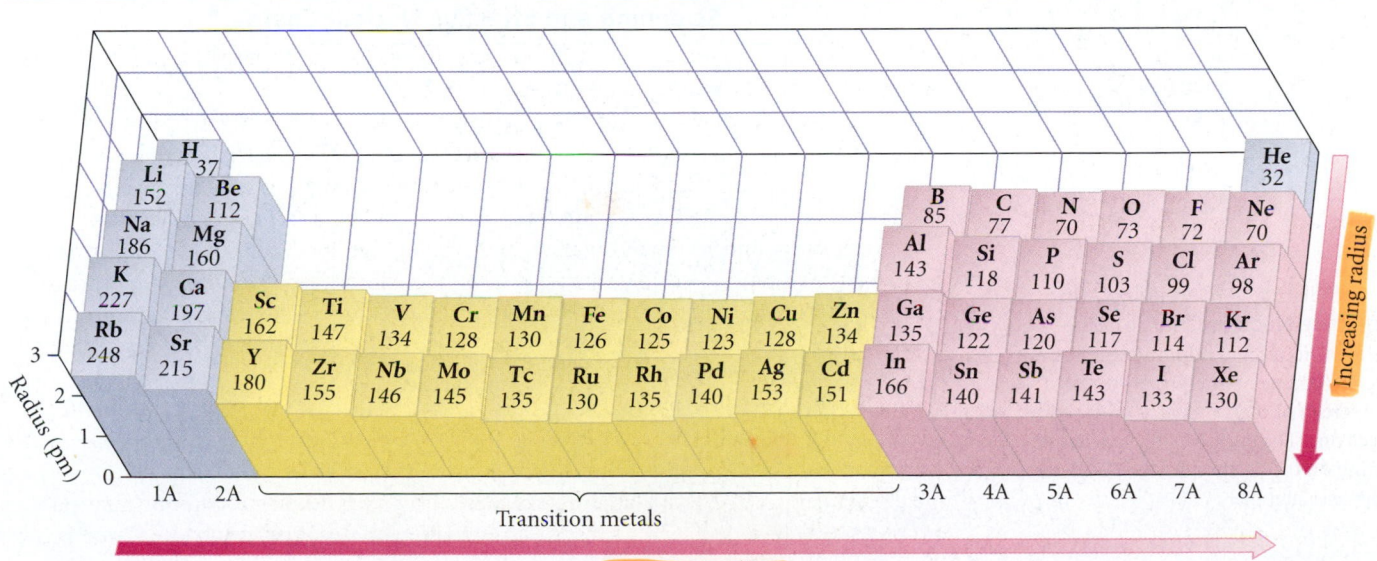

▲ **FIGURE 8.10 Trends in Atomic Radius** In general, atomic radii increase as you move down a column and decrease as you move to the right across a period in the periodic table.

Effective Nuclear Charge

The trend in atomic radius as you move to the right across a row in the periodic table is determined by the inward pull of the nucleus on the electrons in the outermost principal energy level (highest n value). The coulombic attraction between a nucleus and an electron increases with increasing magnitude of nuclear charge. For example, compare the H atom to the He^+ ion.

$$
\begin{array}{ll}
H & 1s^1 \\
He^+ & 1s^1
\end{array}
$$

It takes 1312 kJ/mol of energy to remove the $1s$ electron from hydrogen, but 5251 kJ/mol of energy to remove it from He^+. Why? Although each electron is in a $1s$ orbital, the electron in the helium ion is attracted to the nucleus with a 2+ charge, while the electron in the hydrogen atom is attracted to the nucleus by only a 1+ charge. Therefore, the electron in the helium ion is held more tightly, making it more difficult to remove and making the helium ion smaller than the hydrogen atom.

As we saw in Section 8.3, any one electron in a multielectron atom experiences both the positive charge of the nucleus (which is attractive) and the negative charges of the other electrons (which are repulsive). Consider again the outermost electron in the lithium atom:

$$
Li \qquad 1s^2 2s^1
$$

As shown in Figure 8.11 ▼, even though the $2s$ orbital penetrates into the $1s$ orbital to some degree, the majority of the $2s$ orbital is outside of the $1s$ orbital. Therefore the electron in the $2s$ orbital is partially *screened* or *shielded* from the 3+ charge of the nucleus by the 2− charge of the $1s$ (or core) electrons, reducing the net charge experienced by the $2s$ electron.

As we have seen, we can define the average or net charge experienced by an electron as the *effective nuclear charge*. The effective nuclear charge experienced by a particular electron in an atom is simply the *actual nuclear charge (Z)* minus *the charge shielded by other electrons (S)*:

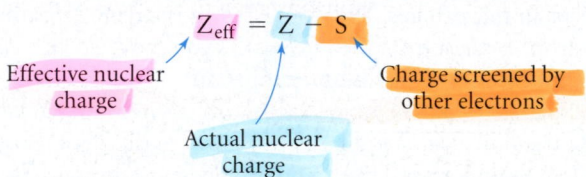

$$
Z_{eff} = Z - S
$$

Effective nuclear charge   Charge screened by other electrons

Actual nuclear charge

For lithium, we can estimate that the two core electrons shield the valence electron from the nuclear charge with high efficiency (S is nearly 2). The effective nuclear charge experienced by lithium's valence electron is therefore slightly greater than 1+.

Screening and Effective Nuclear Charge

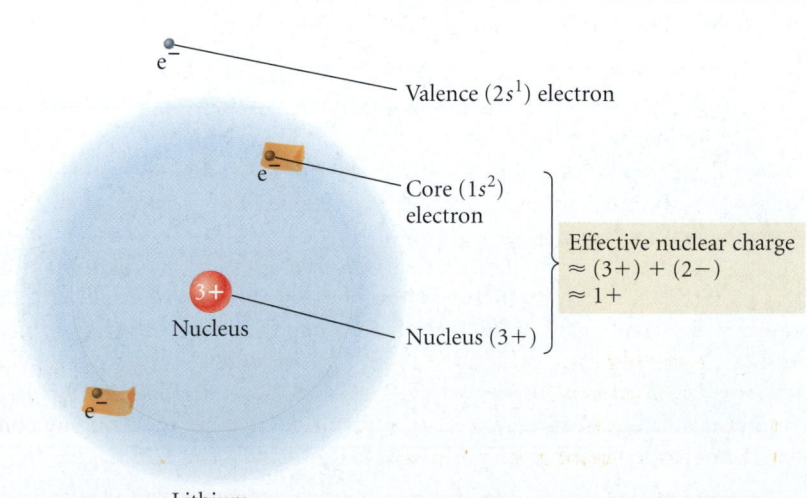

e^-
Valence ($2s^1$) electron

e^-
Core ($1s^2$) electron

3+
Nucleus

Nucleus (3+)

Effective nuclear charge
$\approx (3+) + (2-)$
$\approx 1+$

e^-

Lithium

▶ **FIGURE 8.11 Screening and Effective Nuclear Charge** The valence electron in lithium experiences the 3+ charge of the nucleus through the screen of the 2− charge of the core electrons. The effective nuclear charge acting on the valence electron is therefore approximately 1+.

Now consider the valence electrons in beryllium (Be), with atomic number 4. Its electron configuration is

$$\text{Be} \qquad 1s^2 2s^2$$

To estimate the effective nuclear charge experienced by the $2s$ electrons, we must distinguish between two different types of shielding: (1) the shielding of the outermost electrons by the core electrons and (2) the shielding of the outermost electrons by *each other*. The key to understanding the trend in atomic radius is the difference between these two types of shielding. In general;

> Core electrons efficiently shield electrons in the outermost principal energy level from nuclear charge, but outermost electrons do not efficiently shield one another from nuclear charge.

In other words, the two outermost electrons in beryllium experience the 4+ charge of the nucleus through the shield of the two $1s$ core electrons without shielding each other from that charge very much. We can therefore estimate that the shielding (S) experienced by any one of the outermost electrons due to the core electrons is nearly 2, but that the shielding due to the other outermost electron is nearly zero. The effective nuclear charge experienced by beryllium's outermost electrons is therefore slightly greater than 2+.

Notice that the effective nuclear charge experienced by beryllium's outermost electrons is greater than that experienced by lithium's outermost electron. Consequently, beryllium's outermost electrons are held more tightly than lithium's, resulting in a smaller atomic radius for beryllium. The effective nuclear charge experienced by an atom's outermost electrons continues to become more positive as you move to the right across the rest of the second row in the periodic table, resulting in successively smaller atomic radii. The same trend is generally observed in all main-group elements.

Summarizing, for main-group elements:

➤ As you move down a column in the periodic table, the principal quantum number (n) of the electrons in the outermost principal energy level increases, resulting in larger orbitals and therefore larger atomic radii.

➤ As you move to the right across a row in the periodic table, the effective nuclear charge (Z_{eff}) experienced by the electrons in the outermost principal energy level increases, resulting in a stronger attraction between the outermost electrons and the nucleus and therefore smaller atomic radii.

Atomic Radii and the Transition Elements

From Figure 8.10, you can see that as you go down the first two rows of a column within the transition metals, the elements follow the same general trend in atomic radii as the main-group elements (the radii get larger). However, with the exception of the first couple of elements in each transition series, the atomic radii of the transition elements *do not* follow the same trend as the main-group elements as you move to the right across a row. Instead of decreasing in size, *the radii of transition elements stay roughly constant across each row*. Why? The difference is that, across a row of transition elements, the number of electrons in the outermost principal energy level (highest n value) is nearly constant (recall from Section 8.3, for example, that the $4s$ orbital fills before the $3d$). As another proton is added to the nucleus with each successive element, another electron is added as well, but the electron goes into an $n_{highest} - 1$ orbital. The number of outermost electrons stays constant and they experience a roughly constant effective nuclear charge, keeping the radius approximately constant.

EXAMPLE 8.5 Atomic Size

On the basis of periodic trends, choose the larger atom from each of the following pairs (if possible). Explain your choices.

(a) N or F   (b) C or Ge   (c) N or Al   (d) Al or Ge

Solution

(a) N atoms are larger than F atoms because, as you trace the path between N and F on the periodic table, you move to the right within the same period. As you move to the right across a period, the effective nuclear charge experienced by the outermost electrons increases, resulting in a smaller radius.

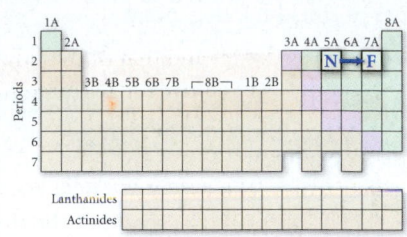

(b) Ge atoms are larger than C atoms because, as you trace the path between C and Ge on the periodic table, you move down a column. Atomic size increases as you move down a column because the outermost electrons occupy orbitals with a higher principal quantum number that are therefore larger, resulting in a larger atom.

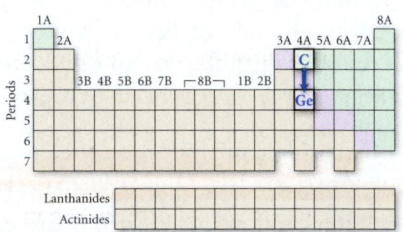

(c) Al atoms are larger than N atoms because, as you trace the path between N and Al on the periodic table, you move down a column (atomic size increases) and then to the left across a period (atomic size increases). These effects add together for an overall increase.

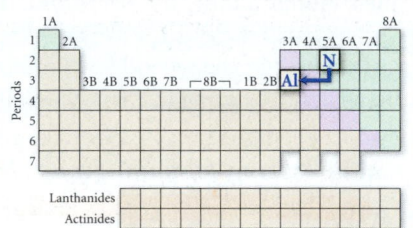

(d) Based on periodic trends alone, you cannot tell which atom is larger, because as you trace the path between Al and Ge you go to the right across a period (atomic size decreases) and then down a column (atomic size increases). These effects tend to oppose each other, and it is not easy to tell which will predominate.

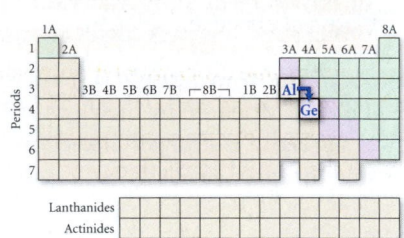

For Practice 8.5

On the basis of periodic trends, choose the larger atom from each of the following pairs (if possible):

(a) Sn or I   (b) Ge or Po   (c) Cr or W   (d) F or Se

For More Practice 8.5

Arrange the following elements in order of decreasing radius: S, Ca, F, Rb, Si.

8.7 Ions: Electron Configurations, Magnetic Properties, Ionic Radii, and Ionization Energy

As we have seen, ions are simply atoms (or groups of atoms) that have lost or gained electrons. In this section, we examine periodic trends in ionic electron configurations, magnetic properties, ionic radii, and ionization energies.

Electron Configurations and Magnetic Properties of Ions

The electron configuration of a main-group monoatomic ion can be deduced from the electron configuration of the neutral atom and the charge of the ion. For anions, we simply *add* the number of electrons required by the magnitude of the charge of the anion. For example, the electron configuration of fluorine (F) is $1s^2 2s^2 2p^5$ and that of the fluoride ion (F^-) is $1s^2 2s^2 2p^6$.

The electron configuration of cations is obtained by *subtracting* the number of electrons required by the magnitude of the charge. For example, the electron configuration of lithium (Li) is $1s^2 2s^1$ and that of the lithium ion (Li^+) is $1s^2 2s^0$ (or simply $1s^2$). For main-group cations, we remove the required number of electrons in the reverse order of filling. However, an important exception occurs for transition metal cations. When writing the electron configuration of a transition metal cation, *remove the electrons in the highest n-value orbitals first, even if this does not correspond to the reverse order of filling.* For example, the electron configuration of vanadium is as follows:

$$V \quad [Ar]\, 4s^2 3d^3$$

The V^{2+} ion, however, has the following electron configuration:

$$V^{2+} \quad [Ar]\, 4s^0 3d^3$$

In other words, for transition metal cations, the order in which electrons are removed upon ionization is *not* the reverse of the filling order. During filling, we normally fill the $4s$ orbital before the $3d$ orbital. When a fourth period transition metal ionizes, however, it normally loses its $4s$ electrons before its $3d$ electrons. Why this odd behavior? The full answer to this question is beyond our scope, but the following two factors contribute to this behavior.

- As discussed previously, the ns and $(n-1)d$ orbitals are extremely close in energy and, depending on the exact configuration, can vary in relative energy ordering.
- As the $(n-1)d$ orbitals begin to fill in the first transition series, the increasing nuclear charge stabilizes the $(n-1)d$ orbitals relative to the ns orbitals. This happens because the $(n-1)d$ orbitals are not outermost (or highest n) orbitals and are therefore not effectively shielded from the increasing nuclear charge by the ns orbitals.

The bottom-line experimental observation is that an $ns^0(n-1)d^x$ configuration is lower in energy than an $ns^2(n-1)d^{x-2}$ configuration for transition metal ions. Therefore, when writing electron configurations for transition metals, remove the ns electrons before the $(n-1)d$ electrons.

The magnetic properties of transition metal ions support these assignments. Recall from Section 8.3 that an unpaired electron generates a magnetic field due to its spin. Consequently, if an atom or ion contains unpaired electrons, it is attracted by an external magnetic field, and we say that the atom or ion is **paramagnetic**. For example, the magnetic properties of silver—which resulted in the splitting of a beam of silver atoms in the Stern–Gerlach experiment discussed in Section 8.3—is caused by silver's unpaired $5s$ electron.

$$Ag \quad [Kr]\, 5s^1 4d^{10}$$

An atom or ion in which all electrons are paired is not attracted to an external magnetic field—it is in fact slightly repelled—and we say that the atom or ion is **diamagnetic**. The zinc atom, for example, is diamagnetic.

$$Zn \quad [Ar]\, 4s^2 3d^{10}$$

The magnetic properties of the zinc ion provide confirmation that the $4s$ electrons are indeed lost before $3d$ electrons in the ionization of zinc. If zinc lost two $3d$ electrons upon ionization, then the Zn^{2+} would become paramagnetic (because the two electrons would come out of two different filled *d* orbitals, leaving each of them with one unpaired electron). However, the zinc ion, like the zinc atom, is diamagnetic because the 4*s* electrons are lost instead.

$$Zn^{2+} \quad [Ar] \quad 4s^0 3d^{10}$$

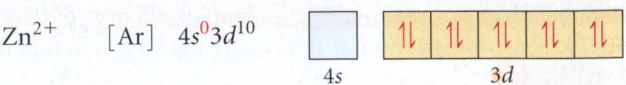

4*s*   3*d*

Similar observations in other transition metals confirm that the *ns* electrons are lost before the $(n-1)d$ electrons upon ionization.

EXAMPLE 8.6 Electron Configurations and Magnetic Properties for Ions

Write the electron configuration and orbital diagram for each of the following ions and determine whether the ion is diamagnetic or paramagnetic.

(a) Al^{3+} **(b)** S^{2-} **(d)** Fe^{3+}

Solution

(a) Al^{3+}

Begin by writing the electron configuration of the neutral atom.

Since this ion has a 3+ charge, remove three electrons to write the electron configuration of the ion. Write the orbital diagram by drawing half-arrows to represent each electron in boxes representing the orbitals. Because there are no unpaired electrons, Al^{3+} is diamagnetic.

Al	$[Ne] \, 3s^2 3p^1$
Al^{3+}	$[Ne]$ or $[He] \, 2s^2 2p^6$
Al^{3+}	$[He]$ ⊞ (2*s*) ⊞⊞⊞ (2*p*)

Diamagnetic

(b) S^{2-}

Begin by writing the electron configuration of the neutral atom.

Since this ion has a 2− charge, add two electrons to write the electron configuration of the ion. Write the orbital diagram by drawing half-arrows to represent each electron in boxes representing the orbitals. Because there are no unpaired electrons, S^{2-} is diamagnetic.

S	$[Ne] \, 3s^2 3p^4$
S^{2-}	$[Ne] \, 3s^2 3p^6$
S^{2-}	$[Ne]$ ⊞ (3*s*) ⊞⊞⊞ (3*p*)

Diamagnetic

(c) Fe^{3+}

Begin by writing the electron configuration of the neutral atom.

Since this ion has a 3+ charge, remove three electrons to write the electron configuration of the ion. Since it is a transition metal, remove the electrons from the 4*s* orbital before removing electrons from the 3*d* orbitals. Write the orbital diagram by drawing half-arrows to represent each electron in boxes representing the orbitals. Because there are unpaired electrons, Fe^{3+} is paramagnetic.

Fe	$[Ar] \, 4s^2 3d^6$
Fe^{3+}	$[Ar] \, 4s^0 3d^5$
Fe^{3+}	$[Ar]$ ⬜ (4*s*) ⬆⬆⬆⬆⬆ (3*d*)

Paramagnetic

For Practice 8.6

Write the electron configuration and orbital diagram for each of the following ions and predict whether the ion will be paramagnetic or diamagnetic.

(a) Co^{2+} **(b)** N^{3-} **(c)** Ca^{2+}

Ionic Radii

What happens to the radius of an atom when it becomes a cation? An anion? Consider, for example, the difference between the Na atom and the Na^+ ion. Their electron configurations are as follows:

$$Na \qquad [Ne] 3s^1$$
$$Na^+ \qquad [Ne]$$

The sodium atom has an outer $3s$ electron and a neon core. Since the $3s$ electron is the outermost electron, and since it is shielded from the nuclear charge by the core electrons, it contributes greatly to the size of the sodium atom. The sodium cation, having lost the outermost $3s$ electron, has only the neon core and carries a charge of $1+$. Without the $3s$ electron, the sodium cation (ionic radius = 95 pm) becomes much smaller than the sodium atom (covalent radius = 186 pm). The trend is the same with all cations and their atoms, as shown in Figure 8.12 ▼. In general,

Cations are much smaller than their corresponding atoms.

Radii of Atoms and Their Cations (pm)

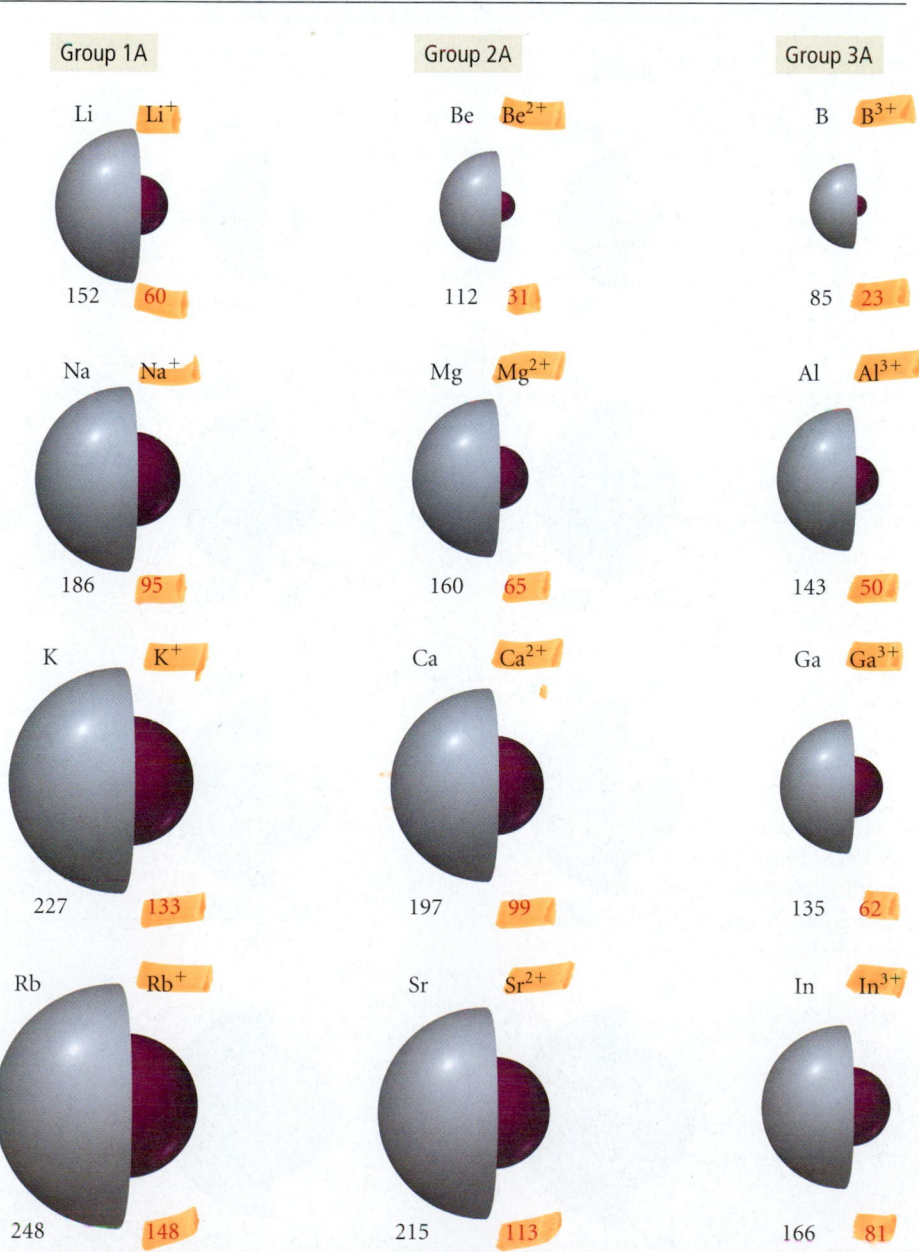

Group 1A

| Li | Li⁺ |
| 152 | 60 |

| Na | Na⁺ |
| 186 | 95 |

| K | K⁺ |
| 227 | 133 |

| Rb | Rb⁺ |
| 248 | 148 |

Group 2A

| Be | Be²⁺ |
| 112 | 31 |

| Mg | Mg²⁺ |
| 160 | 65 |

| Ca | Ca²⁺ |
| 197 | 99 |

| Sr | Sr²⁺ |
| 215 | 113 |

Group 3A

| B | B³⁺ |
| 85 | 23 |

| Al | Al³⁺ |
| 143 | 50 |

| Ga | Ga³⁺ |
| 135 | 62 |

| In | In³⁺ |
| 166 | 81 |

◄ **FIGURE 8.12 Sizes of Atoms and Their Cations** Atomic and ionic radii (pm) for the first three columns of main-group elements.

What about anions? Consider, for example, the difference between Cl and Cl⁻. Their electron configurations are as follows:

$$Cl \qquad [Ne]\,3s^2 3p^5$$
$$Cl^- \qquad [Ne]\,3s^2 3p^6$$

The chlorine anion has one additional outermost electron, but no additional proton to increase the nuclear charge. The extra electron increases the repulsions among the outermost electrons, resulting in a chloride anion that is larger than the chlorine atom. The trend is the same with all anions and their atoms, as shown in Figure 8.13 ▼. In general,

Anions are much larger than their corresponding atoms.

We can observe an interesting trend in ionic size by examining the radii of an *isoelectronic* series of ions—ions with the same number of electrons. For example, consider the following ions and their radii:

S^{2-} (184 pm)	Cl^- (181 pm)	K^+ (133 pm)	Ca^{2+} (99 pm)
18 electrons	18 electrons	18 electrons	18 electrons
16 protons	17 protons	19 protons	20 protons

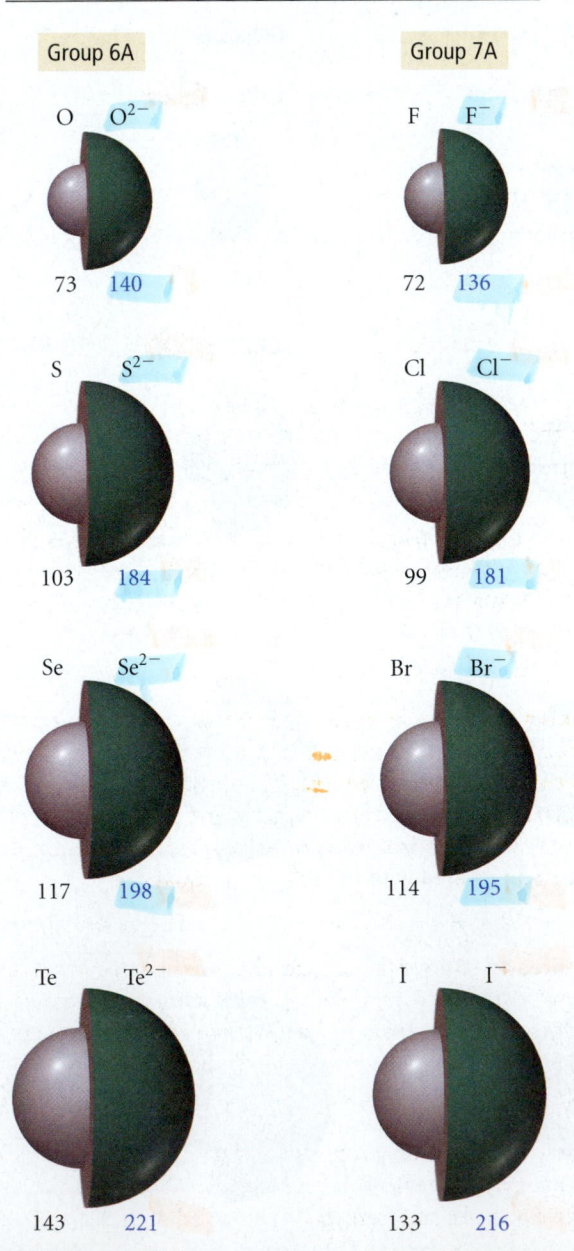

Radii of Atoms and Their Anions (pm)

Group 6A

| O | O^{2-} |
| 73 | 140 |

| S | S^{2-} |
| 103 | 184 |

| Se | Se^{2-} |
| 117 | 198 |

| Te | Te^{2-} |
| 143 | 221 |

Group 7A

| F | F^- |
| 72 | 136 |

| Cl | Cl^- |
| 99 | 181 |

| Br | Br^- |
| 114 | 195 |

| I | I^- |
| 133 | 216 |

▶ **FIGURE 8.13 Sizes of Atoms and Their Anions**   Atomic and ionic radii for groups 6A and 7A in the periodic table.

Each of these ions has 18 electrons in exactly the same orbitals, but the radius of the ions gets successively smaller. Why? The reason is the progressively greater number of protons. The S^{2-} ion has 16 protons, and therefore a charge of 16+ pulling on 18 electrons. The Ca^{2+} ion, however, has 20 protons, and therefore a charge of 20+ pulling on the same 18 electrons. The result is a much smaller radius. For a given number of electrons, a greater nuclear charge results in a smaller atom or ion.

EXAMPLE 8.7   Ion Size

Choose the larger atom or ion from each of the following pairs:

(a) S or S^{2-} (b) Ca or Ca^{2+} (c) Br^- or Kr

Solution

(a) The S^{2-} ion is larger than an S atom because anions are larger than the atoms from which they are formed.

(b) A Ca atom is larger than Ca^{2+} because cations are smaller than the atoms from which they are formed.

(c) A Br^- ion is larger than a Kr atom because, although they are isoelectronic, Br^- has one fewer proton than Kr, resulting in a lesser pull on the electrons and therefore a larger radius.

For Practice 8.7

Choose the larger atom or ion from each of the following pairs:

(a) K or K^+ (b) F or F^- (c) Ca^{2+} or Cl^-

For More Practice 8.7

Arrange the following in order of decreasing radius: Ca^{2+}, Ar, Cl^-.

Conceptual Connection 8.3   Ions, Isotopes, and Atomic Size

In the previous sections, we have seen how the number of electrons and the number of protons affects the size of an atom or ion. However, we have not considered how the number of neutrons affects the size of an atom. Why not? Would you expect isotopes— for example, C-12 and C-13—to have different atomic radii?

Answer: The isotopes of an element all have the same radii for two reasons: (1) neutrons are negligibly small compared to the size of an atom and therefore extra neutrons do not increase atomic size; (2) neutrons have no charge and therefore do not attract electrons in the way that protons do.

Ionization Energy

The **ionization energy (IE)** of an atom or ion is the energy required to remove an electron from the atom or ion in the gaseous state. The ionization energy is always positive because removing an electron always takes energy. (The process is similar to an endothermic reaction, which absorbs heat and therefore has a positive ΔH.) The energy required to remove the first electron is called the *first ionization energy* (IE_1). For example, the first ionization of sodium can be represented with the following equation:

$$Na(g) \longrightarrow Na^+(g) + 1\,e^- \qquad IE_1 = 496 \text{ kJ/mol}$$

The energy required to remove the second electron is called the *second ionization energy* (IE_2), the energy required to remove the third electron is called the *third ionization energy* (IE_3), and so on. For example, the second ionization energy of sodium can be represented as follows:

$$Na^+(g) \longrightarrow Na^{2+}(g) + 1\,e^- \qquad IE_2 = 4560 \text{ kJ/mol}$$

Notice that the second ionization energy is not the energy required to remove *two* electrons from sodium (that quantity would be the sum of IE_1 and IE_2), but rather the energy required to remove one electron from Na^+. We look at trends in IE_1 and IE_2 separately.

First Ionization Energies

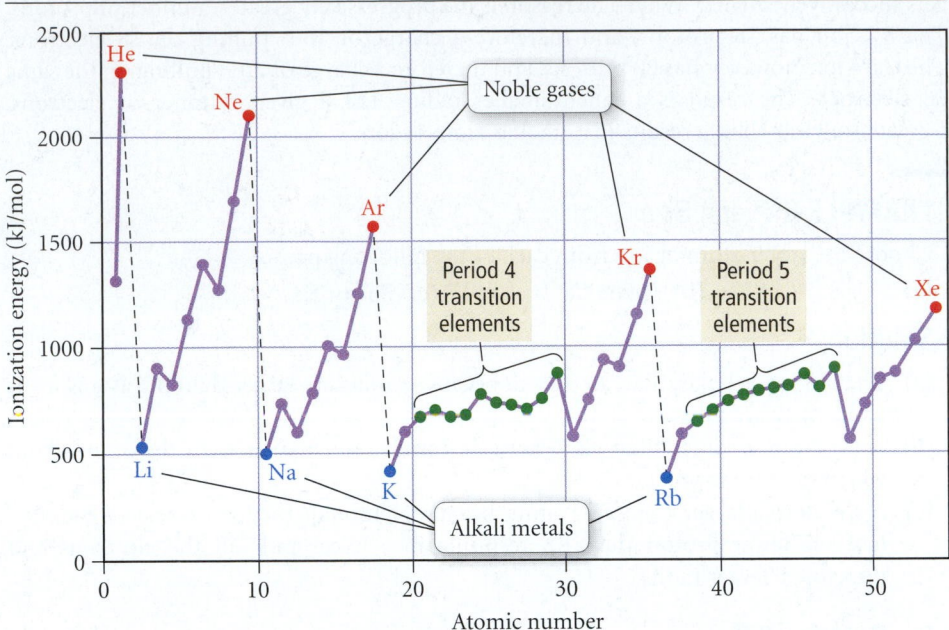

▲ **FIGURE 8.14 First Ionization Energy versus Atomic Number for the Elements through Xenon**   Ionization starts at a minimum with each alkali metal and rises to a peak with each noble gas. (Compare with Figure 8.9.)

Trends in First Ionization Energy

The first ionization energies of the elements through Xe are shown in Figure 8.14 ▲. Notice again the periodic trend in ionization energy, peaking at each noble gas. Based on what we have learned about electron configurations and effective nuclear charge, how can we account for the observed trend? As we have seen, the principal quantum number, n, increases as we move down a column. Within a given sublevel, orbitals with higher principal quantum numbers are larger than orbitals with smaller principal quantum numbers. Consequently, electrons in the outermost principal level are farther away from the positively charged nucleus—and are therefore held less tightly—as you move down a column. This results in a lower ionization energy as you move down a column, as shown in Figure 8.15 ▶.

What about the trend as we move to the right across a row? For example, would it take more energy to remove an electron from Na or from Cl, two elements on either end of the third row in the periodic table? We know that Na has an outer electron configuration of $3s^1$ and Cl has an outer electron configuration of $3s^2 3p^5$. As discussed previously, the outermost electrons in chlorine experience a higher effective nuclear charge than the outermost electrons in sodium (which is why chlorine has a smaller atomic radius than sodium). Consequently, we would expect chlorine to have a higher ionization energy than sodium, which is in fact the case. A similar argument can be made for other main-group elements so that ionization energy generally increases as you move to the right across a row in the periodic table, as shown in Figure 8.15 .

Summarizing, for main-group elements:

➤ Ionization energy generally *decreases* as you move down a column (or family) in the periodic table because electrons in the outermost principal level become farther away from the positively charged nucleus and are therefore held less tightly.

➤ Ionization energy generally *increases* as you move to the right across a period (or row) in the periodic table because electrons in the outermost principal energy level generally experience a greater effective nuclear charge (Z_{eff}).

Trends in First Ionization Energy

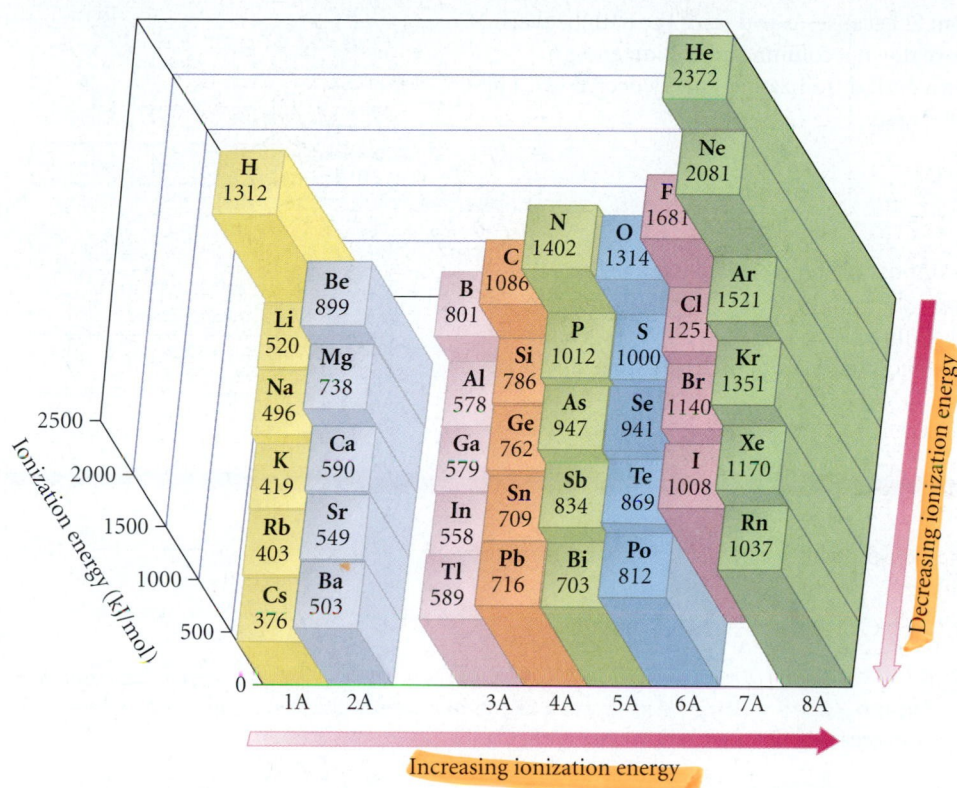

▲ **FIGURE 8.15 Trends in Ionization Energy**   Ionization energy increases as you move to the right across a period and decreases as you move down a column in the periodic table.

EXAMPLE 8.8   Ionization Energy

On the basis of periodic trends, choose the element with the higher first ionization energy from each of the following pairs (if possible):

(a) Al or S   **(b)** As or Sb   **(c)** N or Si   **(d)** O or Cl

Solution

(a) Al or S

S has a higher ionization energy than Al because, as you trace the path between Al and S on the periodic table, you move to the right within the same period. Ionization energy increases as you go to the right because of increasing effective nuclear charge.

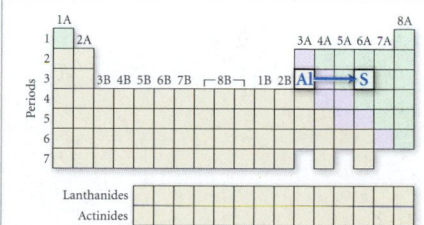

(b) As or Sb

As has a higher ionization energy than Sb because, as you trace the path between As and Sb on the periodic table, you move down a column. Ionization energy decreases as you go down a column because of the increasing size of orbitals with increasing n.

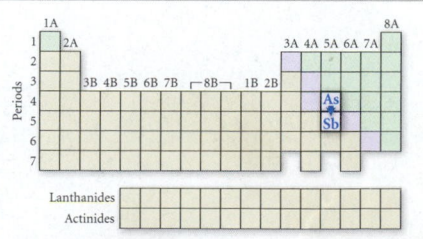

(c) N or Si

N has a higher ionization energy than Si because, as you trace the path between N and Si on the periodic table, you move down a column (ionization energy decreases) and then to the left across a period (ionization energy decreases). These effects sum together for an overall decrease.

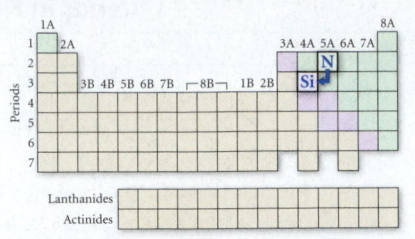

(d) O or Cl

Based on periodic trends alone, it is impossible to tell which has a higher ionization energy because, as you trace the path between O and Cl you go to the right across a period (ionization energy increases) and then down a column (ionization energy decreases). These effects tend to oppose each other, and it is not obvious which will dominate.

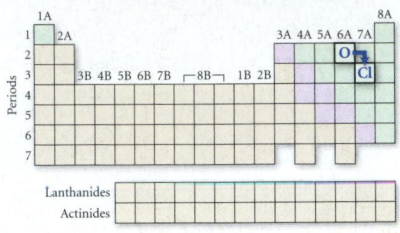

For Practice 8.8

On the basis of periodic trends, choose the element with the higher first ionization energy from each of the following pairs (if possible):

(a) Sn or I   **(b)** Ca or Sr   **(c)** C or P   **(d)** F or S

For More Practice 8.8

Arrange the following elements in order of decreasing first ionization energy: S, Ca, F, Rb, Si.

Exceptions to Trends in First Ionization Energy

By carefully examining Figure 8.15, we can see some exceptions to the trends in first ionization energies. For example, boron has a smaller ionization energy than beryllium, even though it lies to the right of beryllium in the same row. This exception is caused by the change in going from the s block to the p block. Recall from Section 7.6 that the $2s$ orbital penetrates into the nuclear region more than the $2p$ orbital. The result is that the electrons in the $2s$ orbital shield the electron in the $2p$ orbital from nuclear charge, making the electron in the $2p$ orbital easier to remove. Similar exceptions occur for aluminum and gallium, both directly below boron in group 3A.

Another exception occurs between nitrogen and oxygen: although oxygen is to the right of nitrogen in the same row, it has a lower ionization energy. This exception is caused by the repulsion between electrons when they must occupy the same orbital. Examine the electron configurations and orbital diagrams of nitrogen and oxygen:

N $1s^2 2s^2 2p^3$ ⇅ | ⇅ | ↑ ↑ ↑
                              $1s$ $2s$ $2p$

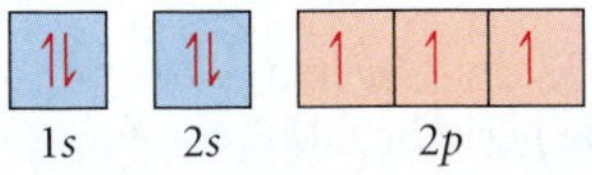

O $1s^2 2s^2 2p^4$ ⇅ | ⇅ | ⇅ ↑ ↑
                              $1s$ $2s$ $2p$

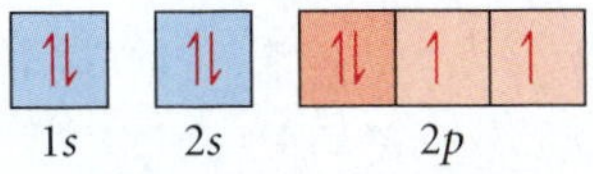

Nitrogen has three electrons in three p orbitals, while oxygen has four. In nitrogen, the $2p$ orbitals are half-filled (which makes the configuration particularly stable). Oxygen's fourth electron must pair with another electron, making it easier to remove. Similar exceptions occur for S and Se, directly below oxygen in group 6A.

Trends in Second and Successive Ionization Energies

Notice the trends in the first, second, and third ionization energies of sodium (group 1A) and magnesium (group 2A), as shown at right.

For sodium, there is a huge jump between the first and second ionization energies. For magnesium, the ionization energy roughly doubles from the first to the second, but then a huge jump occurs between the second and third ionization energies. What is the reason for these jumps?

We can understand these trends by examining the electron configurations of sodium and magnesium, which are as follows:

$$Na \quad [Ne]\,3s^1 \quad -1e = Ne$$
$$Mg \quad [Ne]\,3s^2$$

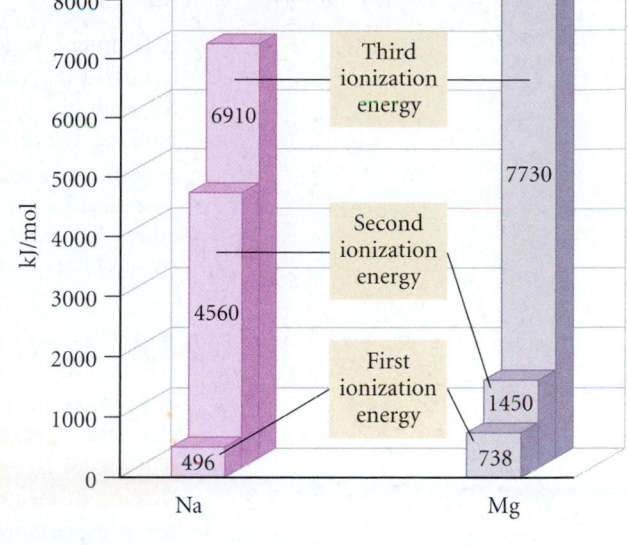

The first ionization of sodium involves removing the valence electron in the $3s$ orbital. Recall that these valence electrons are held more loosely than the core electrons, and that the resulting ion has a noble gas configuration, which is particularly stable. Consequently, the first ionization energy is fairly low. The second ionization of sodium, however, involves removing a core electron from an ion with a noble gas configuration. This requires a tremendous amount of energy, making the value of IE_2 very high.

As with sodium, the first ionization of magnesium involves removing a valence electron in the $3s$ orbital. This requires a bit more energy than the corresponding ionization of sodium because of the trends in Z_{eff} that we discussed earlier (Z_{eff} increases as you move to the right across a row). The second ionization of magnesium also involves removing an outer electron in the $3s$ orbital, but this time from an ion with a 1+ charge (instead of from a neutral atom). This requires roughly twice the energy as removing the electron from the neutral atom. The third ionization of magnesium is analogous to the second ionization of sodium—it requires removing a core electron from an ion with a noble gas configuration. This requires a tremendous amount of energy, making the value of IE_3 very high.

As shown in Table 8.1, similar trends exist for the successive ionization energies of many elements. The ionization energy increases fairly uniformly with each successive removal of an outermost electron, but then it takes a large jump with the removal of the first core electron.

TABLE 8.1 Successive Values of Ionization Energies for the Elements Sodium through Argon (kJ/mol)

Element	IE_1	IE_2	IE_3	IE_4	IE_5	IE_6	IE_7
Na	496	4560					
Mg	738	1450	7730		Core electrons		
Al	578	1820	2750	11,600			
Si	786	1580	3230	4360	16,100		
P	1012	1900	2910	4960	6270	22,200	
S	1000	2250	3360	4560	7010	8500	27,100
Cl	1251	2300	3820	5160	6540	9460	11,000
Ar	1521	2670	3930	5770	7240	8780	12,000

Conceptual Connection 8.4 Ionization Energies and Chemical Bonding

In Chapter 3, we learned that chemical bonding involves the transfer of electrons (to form ionic bonds) or the sharing of electrons (to form covalent bonds). Earlier in this chapter, we saw that for main-group elements, the outermost or valence electrons are most important in chemical bonding. Use the trend in successive ionization energies to explain why this is so.

Answer: As we can see from the successive ionization energies of any element, valence electrons are held most loosely and can therefore be transferred or shared most easily. Core electrons, on the other hand, are held tightly and are not easily transferred or shared. Consequently, valence electrons are most important to chemical bonding.

8.8 Electron Affinities and Metallic Character

Two other properties that exhibit periodic trends are electron affinity and metallic character. Electron affinity is important because it is a measure of how easily an atom will accept an additional electron, and since chemical bonding involves the transfer or sharing of electrons, electron affinity is crucial to chemical bonding. Metallic character is important because of the high proportion of metals in the periodic table and the enormous role they play in our lives. Of the roughly 110 elements, 87 are metals. We examine each of these periodic properties individually.

Electron Affinity

The **electron affinity (EA)** of an atom or ion is the energy change associated with the gaining of an electron by the atom in the gaseous state. The electron affinity is usually—though not always—negative because an atom or ion usually releases energy when it gains an electron. (The process is analogous to an exothermic reaction, which releases heat and therefore has a negative ΔH.) In other words, the coulombic attraction between the nucleus of an atom and the incoming electron usually results in the release of energy as the electron is gained. For example, the electron affinity of chlorine can be represented with the following equation:

$$\text{Cl}(g) + 1\,e^- \longrightarrow \text{Cl}^-(g) \qquad EA = -349\,\text{kJ/mol}$$

Electron Affinities (kJ/mol)

1A							8A
H −73	2A	3A	4A	5A	6A	7A	**He** >0
Li −60	**Be** >0	**B** −27	**C** −122	**N** >0	**O** −141	**F** −328	**Ne** >0
Na −53	**Mg** >0	**Al** −43	**Si** −134	**P** −72	**S** −200	**Cl** −349	**Ar** >0
K −48	**Ca** −2	**Ga** −30	**Ge** −119	**As** −78	**Se** −195	**Br** −325	**Kr** >0
Rb −47	**Sr** −5	**In** −30	**Sn** −107	**Sb** −103	**Te** −190	**I** −295	**Xe** >0

▲ **FIGURE 8.16** Electron Affinities of Selected Main-Group Elements

Electron affinities for a number of main-group elements are shown in Figure 8.16 ◀. As you can see from this figure, the trends in electron affinity are not as regular as trends in other properties we have examined. For example, we might expect electron affinities to become relatively more positive (so that the addition of an electron is less exothermic) as we move down a column because the electron is entering orbitals with successively higher principal quantum numbers, and will therefore be farther from the nucleus. This trend applies to the group 1A metals but does not hold for the other columns in the periodic table.

There is a more regular trend in electron affinity as you move to the right across a row, however. Based on the periodic properties we have learned so far, would you expect more energy to be released when an electron is gained by Na or Cl? We know that Na has an outer electron configuration of $3s^1$ and Cl has an outer electron configuration of $3s^2 3p^5$. Since adding an electron to chlorine gives it a noble gas configuration and adding an electron to sodium does not, and since the outermost electrons in chlorine experience a higher Z_{eff} than the outermost electrons in sodium, we would expect chlorine to have a more negative electron affinity—the process should be more exothermic for chlorine. This is in fact the case. For main-group elements, electron affinity generally becomes more negative (more exothermic) as you move to the right across a row in the periodic table. The halogens (group 7A) therefore have the most negative electron affinities. However, exceptions do occur. For example, notice that nitrogen and the other group 5A elements do not follow the general trend. These elements have $ns^2 np^3$ outer electron configurations. When an electron is added to this configuration, it must pair with another electron in an already occupied p orbital. The repulsion between two electrons occupying the same orbital causes the electron affinity to be more positive than for elements in the previous column.

Summarizing, for main-group elements:

➤ Most groups of the periodic table do not exhibit any definite trend in electron affinity. Among the group 1A metals, however, electron affinity becomes more positive as you move down the column (adding an electron becomes less exothermic).

➤ Electron affinity generally becomes more negative (adding an electron becomes more exothermic) as you move to the right across a period (or row) in the periodic table.

Metallic Character

As we learned in Chapter 2, metals are good conductors of heat and electricity; they can be pounded into flat sheets (malleability); they can be drawn into wires (ductility); they are often shiny; and they tend to lose electrons in chemical reactions. Nonmetals, in contrast, have more varied physical properties; some are solids at room temperature, others are gases, but in general they tend to be poor conductors of heat and electricity, and they all tend to gain electrons in chemical reactions. As you move to the right across a period in the periodic table, ionization energy increases and electron affinity becomes more negative, which means that elements on the left side of the periodic table are more likely to lose electrons than elements on the right side of the periodic table, which are more likely to gain them. The other properties associated with metals follow the same general trend (even though we do not quantify them here). Consequently, as shown in Figure 8.17 ▼,

> **As you move to the right across a period (or row) in the periodic table, metallic character decreases.**

As you move down a column in the periodic table, ionization energy decreases, making electrons more likely to be lost in chemical reactions. Consequently,

> **As you move down a column (or family) in the periodic table, metallic character increases.**

These trends, based on the quantum-mechanical model, explain the distribution of metals and nonmetals that we learned in Chapter 2. Metals are found on the left side and toward the center of the periodic table and nonmetals on the upper right side. The change in chemical behavior from metallic to nonmetallic can be seen most clearly as you proceed to the right across period 3, or down along group 5A, of the periodic table, as can be seen in Figure 8.18 (page 350).

Trends in Metallic Character I

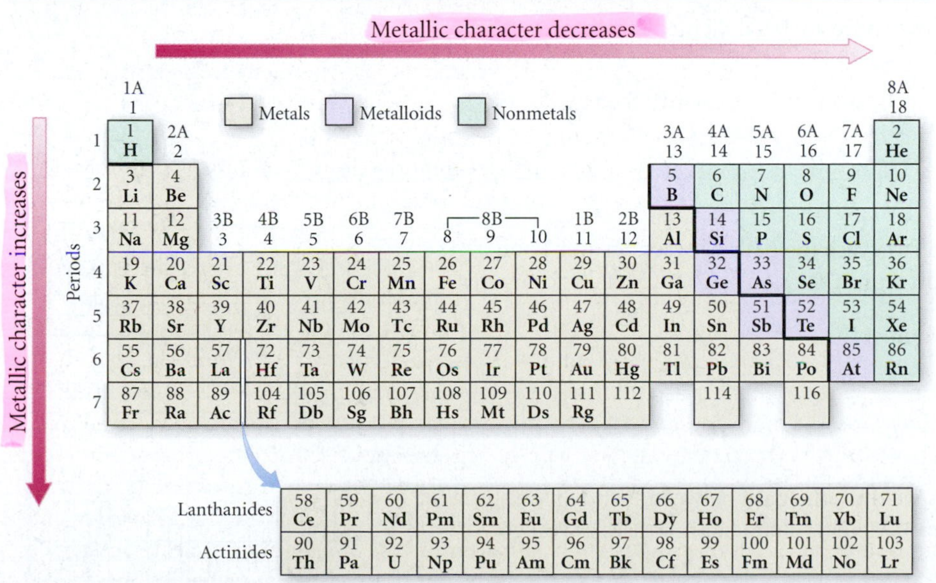

▲ **FIGURE 8.17 Trends in Metallic Character I** Metallic character decreases as you move to the right across a period and increases as you move down a column in the periodic table.

Trends in Metallic Character II

Period 3

Group 5A

Metallic character increases

Metallic character decreases

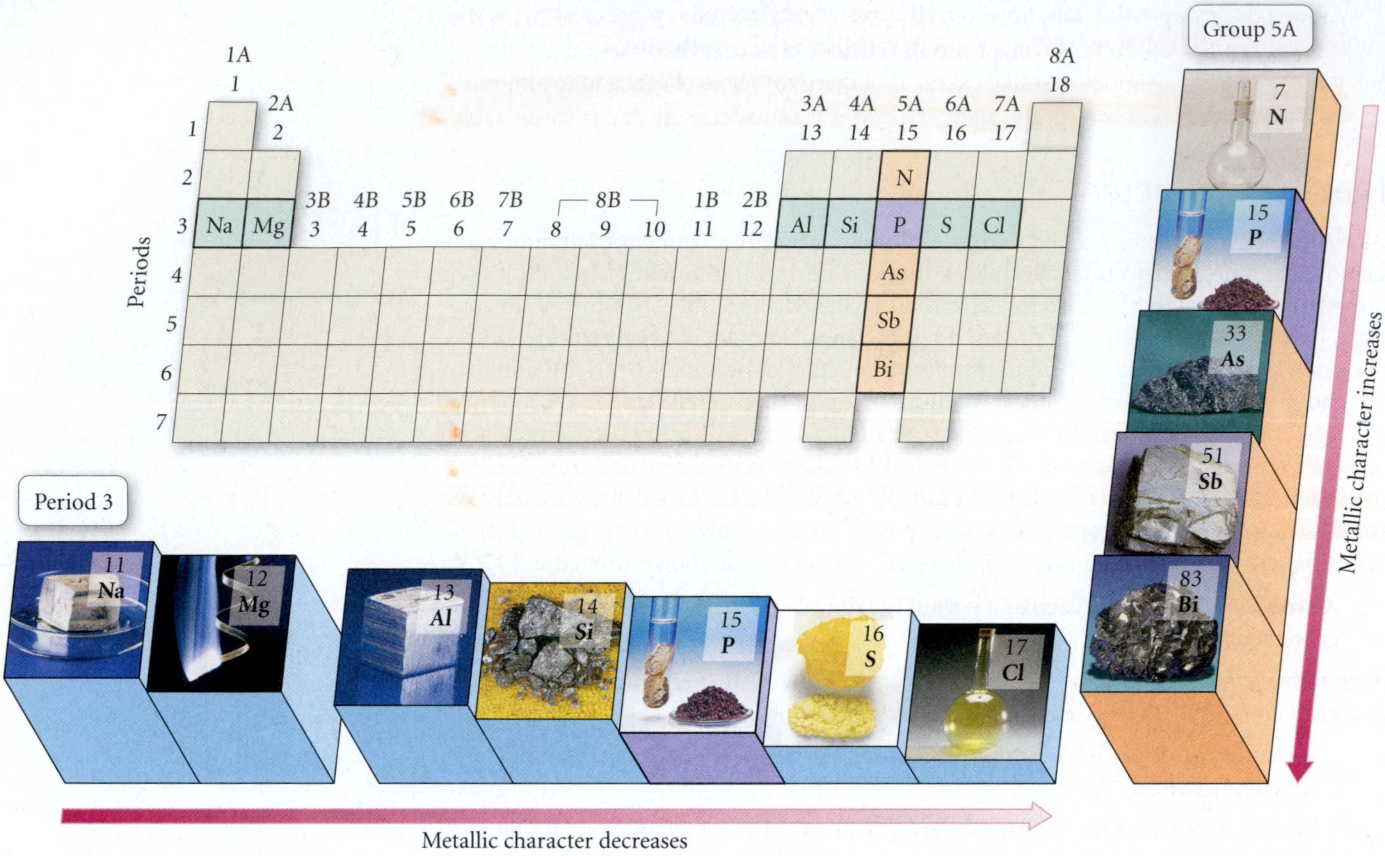

▲ **FIGURE 8.18 Trends in Metallic Character II**   As you move down group 5A in the periodic table, metallic character increases. As you move across period 3 in the periodic table, metallic character decreases.

EXAMPLE 8.9 Metallic Character

On the basis of periodic trends, choose the more metallic element from each of the following pairs (if possible):

(a) Sn or Te   **(b)** P or Sb   **(c)** Ge or In   **(d)** S or Br

Solution

(a) Sn or Te

Sn is more metallic than Te because, as we trace the path between Sn and Te on the periodic table, we move to the right within the same period. Metallic character decreases as you go to the right.

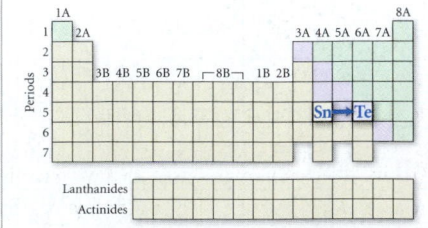

(b) P or Sb

Sb is more metallic than P because, as we trace the path between P and Sb on the periodic table, we move down a column. Metallic character increases as you go down a column.

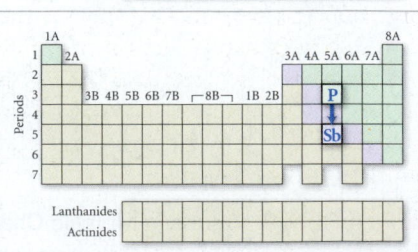

(c)   Ge or In

In is more metallic than Ge because, as we trace the path between Ge and In on the periodic table, we move down a column (metallic character increases) and then to the left across a period (metallic character increases). These effects add together for an overall increase.

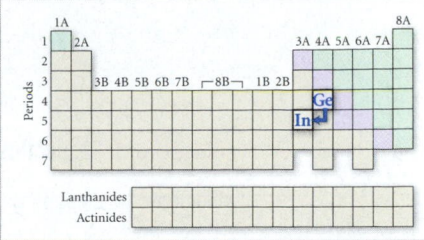

(d)   S or Br

Based on periodic trends alone, we cannot tell which is more metallic because as we trace the path between S and Br, we go to the right across a period (metallic character decreases) and then down a column (metallic character increases). These effects tend to oppose each other, and it is not easy to tell which will predominate.

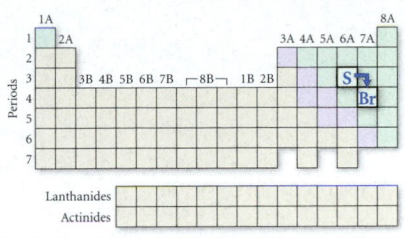

For Practice 8.9

On the basis of periodic trends, choose the more metallic element from each of the following pairs (if possible):

(a) Ge or Sn   **(b)** Ga or Sn   **(c)** P or Bi   **(d)** B or N

For More Practice 8.9

Arrange the following elements in order of increasing metallic character: Si, Cl, Na, Rb.

 Conceptual Connection 8.5 **Periodic Trends**

Use the trends in ionization energy and electron affinity to explain why sodium chloride has the formula NaCl and not Na_2Cl or $NaCl_2$.

Answer: The $3s$ electron in sodium has a relatively low ionization energy (496 kJ/mol) because it is a valence electron. The energetic cost for sodium to lose a second electron is extraordinarily high (4560 kJ/mol) because the next electron to be lost is a core electron ($2p$). Similarly, the electron affinity of chlorine to gain one electron (-349 kJ/mol) is highly exothermic because the added electron completes chlorine's valence shell. The gain of a second electron by the negatively charged chlorine anion would not be so favorable. Therefore, we would expect sodium and chlorine to combine in a 1:1 ratio.

8.9 Some Examples of Periodic Chemical Behavior: The Alkali Metals, the Halogens, and the Noble Gases

In this section, we explore some of the properties and chemical reactions of three families in the periodic table: the alkali metals, the halogens, and the noble gases. These families exemplify the connection between chemical behavior and electron configuration. The alkali metals (group 1A) have ns^1 outer electron configurations. The single valence electron that keeps these metals from having noble gas configurations is easily removed (the metals have low ionization energies), making these elements the most active metals in the periodic table. The halogens (group 7A) have ns^2np^5 outer electron configurations. The one electron needed to attain noble gas configurations is easily acquired (the halogens have highly negative electron affinities), making these elements among the most active nonmetals in the periodic table. The noble gases (group 8A) have electron configurations with full outer principal quantum levels (ns^2np^6), and so are the most chemically inert family in the periodic table. We examine the properties of each of these groups separately. (Even though hydrogen is often listed in group 1A, it behaves like a nonmetal because of its high ionization energy: 1312 kJ/mol. We therefore do not include hydrogen in our discussion of the group 1A metals.)

The Alkali Metals (Group 1A)

Some properties of the alkali metals are listed in Table 8.2.

TABLE 8.2 Properties of the Alkali Metals*

Element	Electron Configuration	Atomic Radius (pm)	IE_1 (kJ/mol)	Density at 25 °C (g/cm^3)	Melting Point (°C)
Li	$[He]\, 2s^1$	152	520	0.535	181
Na	$[Ne]\, 3s^1$	186	496	0.968	102
K	$[Ar]\, 4s^1$	227	419	0.856	98
Rb	$[Kr]\, 5s^1$	248	403	1.532	39
Cs	$[Xe]\, 6s^1$	265	376	1.879	29

*Francium is omitted because it has no stable isotopes.

Notice that, in general, the properties of the alkali metals vary fairly smoothly as you proceed down the column. As expected from periodic trends, the atomic radius increases steadily while the first ionization energy decreases steadily.

With the exception of potassium, density increases as you move down the column. This is a general trend that occurs in other columns within the periodic table. As you move down a column, the increase in mass (due to the additional protons and neutrons) outpaces the increase in volume caused by greater atomic radius. The result is successively greater densities. The melting points of the alkali metals as a group are anomalously low for metals, and they show a steady decrease as you move down the column. (This is not a general trend for the rest of the periodic table, which shows more irregular patterns in melting points.)

Because of their generally low ionization energies, the alkali metals are excellent reducing agents—they are readily oxidized, losing electrons to other substances. Consequently, the alkali metals are normally found in nature in their oxidized state, either in compounds or as dissolved ions in seawater. Since ionization energy *decreases* as you go down the column, the relative reactivities of the alkali metals tend to *increase* as you move down the column. In other words, the lower the ionization energy of an alkali metal, the greater tendency it will have to lose its electron and the more reactive it will be.

The reactions of the alkali metals with nonmetals are vigorous. For example, the alkali metals (M) react with halogens (X) according to the following reaction:

$$2\,M + X_2 \longrightarrow 2\,MX$$

The reaction of sodium and chlorine to form sodium chloride is typical:

$$2\,Na(s) + Cl_2(g) \longrightarrow 2\,NaCl(s)$$

This reaction emits heat and sparks as it occurs (Figure 8.19 ◀). Each successive alkali metal reacts even more vigorously with chlorine. The alkali metals also react with water to form the dissolved alkali metal ion, the hydroxide ion, and hydrogen gas:

$$2\,M(s) + 2\,H_2O(l) \longrightarrow 2\,M^+(aq) + 2\,OH^-(aq) + H_2(g)$$

The reaction is highly exothermic and can be explosive because the heat from the reaction can ignite the hydrogen gas. The reaction becomes more explosive as you move down the column from one metal to the next, as shown in Figure 8.20 ▶.*

▲ FIGURE 8.19 Reaction of Sodium and Chlorine to Form Sodium Chloride

*The rate of the alkali metal reaction with water, and therefore its vigor, is enhanced by the successively lower melting points of the alkali metals as you move down the column. The low melting points of the heavier metals allow the emitted heat to actually melt the metal, increasing the reaction rate.

Reactions of the Alkali Metals with Water

Lithium

Sodium

Potassium

◀ **FIGURE 8.20 Reactions of the Alkali Metals with Water** The reaction becomes progressively more vigorous as you move down the group.

The Halogens (Group 7A)

Some properties of the first four halogens are listed in Table 8.3.

TABLE 8.3 Properties of the Halogens*

Element	Electron Configuration	Atomic Radius (pm)	EA (kJ/mol)	Melting Point (°C)	Boiling Point (°C)	Density of Liquid (g/cm^3)
F	$[\text{He}]\,2s^2 2p^5$	72	−328	−219	−188	1.51
Cl	$[\text{Ne}]\,3s^2 3p^5$	99	−349	−101	−34	2.03
Br	$[\text{Ar}]\,4s^2 4p^5$	114	−325	−7	59	3.19
I	$[\text{Kr}]\,5s^2 4p^5$	133	−295	114	184	3.96

*At is omitted because it is rare and radioactive.

Notice that the properties of the halogens, like those of the alkali metals, vary fairly smoothly as you proceed down the column. As expected from periodic trends, the atomic radius and the density increase for each successive halogen. You can see from the melting and boiling points that fluorine and chlorine are both gases at room temperature, bromine is a liquid, and iodine is a solid.

All of the halogens are powerful oxidizing agents—they are readily reduced, gaining electrons from other substances in their reactions. Fluorine is the most powerful oxidizing agent of the group—reacting with almost everything including the heavier noble gases—and iodine is the least. The halogens all react with metals to form *metal halides* according to the following equation:

$$2\,\text{M} + n\,\text{X}_2 \longrightarrow 2\,\text{MX}_n$$

where M is the metal, X is the halogen, and MX_n is the metal halide. For example, chlorine reacts with iron according to the following equation:

$$2\,\text{Fe}(s) + 3\,\text{Cl}_2(g) \longrightarrow 2\,\text{FeCl}_3(s)$$

Since metals tend to lose electrons and the halogens tend to gain them, the metal halides—like all compounds that form between metals and nonmetals—contain ionic bonds.

The halogens also react with hydrogen to form *hydrogen halides* according to the following equation:

$$\text{H}_2(g) + \text{X}_2 \longrightarrow 2\,\text{HX}(g)$$

Chlorine   Bromine   Iodine

The hydrogen halides—like all compounds that form between two nonmetals—contain covalent bonds. As we saw in Chapter 3, all of the hydrogen halides form acidic solutions when combined with water.

The halogens also react with each other to form *interhalogen compounds*. For example, bromine reacts with fluorine according to the following equation:

$$Br_2(l) + F_2(g) \longrightarrow 2\,BrF(g)$$

Again, like all compounds that form between two nonmetals, the interhalogen compounds contain covalent bonds.

EXAMPLE 8.10 Alkali Metal and Halogen Reactions

Write a balanced chemical equation for each of the following:

(a) The reaction between potassium metal and bromine gas.

(b) The reaction between rubidium metal and liquid water.

(c) The reaction between gaseous chlorine and solid iodine.

Solution

(a) Alkali metals react with halogens to form metal halides. Write the formulas for the reactants and the metal halide product (making sure to write the correct ionic chemical formula for the metal halide, as outlined in Section 3.5), and then balance the equation.	$2\,K(s) + Br_2(g) \longrightarrow 2\,KBr(s)$
(b) Alkali metals react with water to form the dissolved metal ion, the hydroxide ion, and hydrogen gas. Write the skeletal equation including each of these and then balance it.	$2\,Rb(s) + 2\,H_2O(l) \longrightarrow 2\,Rb^+(aq) + 2\,OH^-(aq) + H_2(g)$
(c) Halogens react with each other to form interhalogen compounds. Write the skeletal equation with each of the halogens as the reactants and the interhalogen compound as the product and balance the equation.	$Cl_2(g) + I_2(s) \longrightarrow 2\,ICl(g)$

For Practice 8.10

Write a balanced chemical equation for each of the following:

(a) the reaction between aluminum metal and chlorine gas

(b) the reaction between lithium metal and liquid water

(c) the reaction between gaseous hydrogen and liquid bromine

Chemistry and Medicine
Potassium Iodide in Radiation Emergencies

Since the attack on the World Trade Center on September 11, 2001, the United States has lived under the threat of additional terrorist strikes, including the possibility of nuclear attack. One danger of such an attack is radiation from the decay of radioiso-topes released by a nuclear device—especially a so-called dirty bomb. This radiation can produce elevated rates of many cancers for years following exposure. The risk of developing thyroid cancer after ingesting radioactive isotopes of iodine, for example, is particularly high, especially in children. The number of thyroid cancers among children and adolescents in Belarus and Ukraine (areas affected by the radioactive plume from the 1986 nuclear accident at Chernobyl in the former Soviet Union) has been 30–100 times higher than in the normal population.

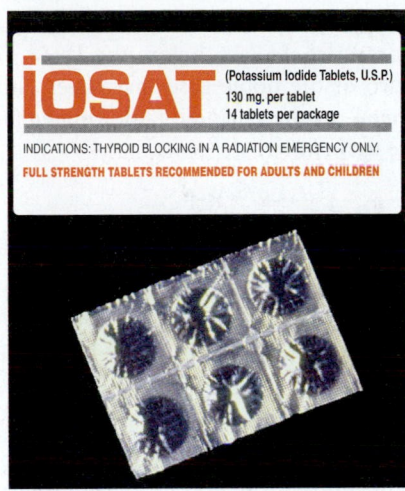

▲ The U.S. Food and Drug Administration recommends taking potassium iodide pills in the event of a nuclear emergency.

▲ Each molecule of thyroxine, a thyroid hormone that plays a key role in metabolism, contains four iodine atoms.

The U.S. Food and Drug Administration (FDA), in cooperation with other federal agencies, recommends the administration of potassium iodide (KI) to citizens in the event of a nuclear radiation emergency. Although KI does not prevent exposure to radiation from a nuclear event, it does decrease the risk of thyroid cancer that follows the intake of radioactive isotopes, particularly I-131. The chief function of the thyroid gland is to synthesize and release the hormone thyroxine, which regulates many aspects of human metabolism. Because thyroxine contains iodine, the thyroid normally accumulates iodine in concentrations far greater than those found elsewhere in the body.

When potassium iodide is taken in the recommended doses, it floods the thyroid with nonradioactive iodine, preventing the thyroid from absorbing the cancer-causing radioactive iodine, which is then excreted in the urine.

In the United States, the FDA prioritizes KI treatment in the event of a nuclear emergency based on age. Infants, children, and pregnant females are at highest risk and are therefore treated at the lowest threshold exposure levels. Adults aged 18 to 40 are treated at slightly higher exposure levels, and those over 40 are only treated if the exposure level is actually high enough to destroy the thyroid.

The federal government has purchased stockpiles of potassium iodide for all states with nuclear reactors. Potassium iodide, also available over the counter, works best if taken 3–4 hours after exposure. Because of the increased threat of terrorist attacks after 9/11, potassium iodide pills have been distributed to residents, schools, and businesses within a 10-mile radius of a nuclear reactor. If a terrorist attack occurs on a nuclear reactor, residents are advised to take the potassium iodide pill and evacuate the area as soon as possible.

The Noble Gases (Group 8A)

Some properties of the noble gases are listed in Table 8.4.

TABLE 8.4   Properties of the Noble Gases*

Element	Electron Configuration	Atomic Radius (pm)**	IE_1 (kJ/mol)	Boiling Point (K)	Density of Gas (g/L at STP)
He	$1s^2$	32	2372	4.2	0.18
Ne	$[He]\,2s^2 2p^6$	70	2081	27.1	0.90
Ar	$[Ne]\,3s^2 3p^6$	98	1521	87.3	1.78
Kr	$[Ar]\,4s^2 4p^6$	112	1351	119.9	3.74
Xe	$[Kr]\,5s^2 5p^6$	130	1170	165.1	5.86

*Radon is omitted because it is radioactive.
**Since only the heavier noble gases form compounds, covalent radii for the smaller noble gases are estimated.

Notice that the properties of the noble gases, like those of the alkali metals and halogens, vary fairly smoothly as you proceed down the column. As expected from periodic trends, the atomic radius and the density increase for each successive noble gas, and the ionization energy decreases. As you can see from the boiling points, all of the noble gases are gases at room temperature and must be cooled to extremely low temperatures before they liquefy. For this reason some noble gases are used as cryogenic liquids—liquids used to cool other substances to low temperatures. For example, researchers often submerse samples of interest in boiling liquid helium to cool them down to 4.2 K and study their properties at this extremely low temperature.

Xenon can also be forced to react with oxygen to form XeO_3 and XeO_4.

▲ Liquid helium is used as a cryogenic liquid to cool substances to temperatures as low as 1.2 K.

The high ionization energies of the noble gases and their completely full outer quantum levels make them exceptionally unreactive. In fact, before the 1960s, no noble gas compounds were known. Since then, two of the noble gases have been shown to react with fluorine (the most reactive nonmetal on the periodic table) under fairly extreme conditions. Krypton reacts with fluorine to form KrF_2:

$$Kr \ + \ F_2 \ \longrightarrow \ KrF_2$$

Similarly, Xe reacts with fluorine to form three different xenon fluorides:

$$Xe \ + \ F_2 \ \longrightarrow \ XeF_2$$
$$Xe \ + \ 2\,F_2 \ \longrightarrow \ XeF_4$$
$$Xe \ + \ 3\,F_2 \ \longrightarrow \ XeF_6$$

The inertness of the noble gases has led to their use in situations where reactions would be undesirable. For example, argon is used in lightbulbs to prevent the hot tungsten filament from oxidizing, and helium is part of the mixture breathed by deep-sea divers to prevent the toxicity caused by too much oxygen and nitrogen under high pressures.

Chapter in Review

Key Terms

Section 8.1
periodic property (320)

Section 8.3
electron configuration (321)
groud state (321)
orbital diagram (322)
electron spin (322)
spin quantum number (m_s) (323)

Pauli exclusion principle (323)
degenerate (323)
shielding (323)
effective nuclear charge (Z_{eff}) (324)
penetration (324)
aufbau principle (326)
Hund's rule (326)

Section 8.4
valence electrons (329)
core electrons (329)

Section 8.6
van der Waals radius (nonbonding atomic radius) (334)
covalent radius (bonding atomic radius) (334)

atomic radius (334)

Section 8.7
paramagnetic (339)
diamagnetic (339)
ionization energy (IE) (343)

Section 8.8
electron affinity (EA) (348)

Key Concepts

Periodic Properties and the Development of the Periodic Table (8.1, 8.2)

The periodic table was primarily developed by Dmitri Mendeleev in the nineteenth century. Mendeleev arranged the elements in a table so that atomic mass increased from left to right in a row and elements with similar properties fell in the same columns. Periodic properties are those that are predictable based on an element's position within the periodic table. Periodic properties include atomic radius, ionization energy, electron affinity, density, and metallic character.

Electron Configurations (8.3)

An electron configuration for an atom simply shows which quantum-mechanical orbitals are occupied by the atom's electrons. For example, the electron configuration of helium ($1s^2$) shows that helium's two electrons exist within the $1s$ orbital. The order of filling quantum-mechanical orbitals in multielectron atoms is as follows: $1s\ 2s\ 2p\ 3s\ 3p\ 4s\ 3d\ 4p\ 5s\ 4d\ 5p\ 6s$. According to the Pauli exclusion principle, each

orbital can hold a maximum of two electrons with opposing spins. According to Hund's rule, orbitals of the same energy first fill singly with electrons with parallel spins, before pairing.

Electron Configurations and the Periodic Table (8.4, 8.5)

Because quantum-mechanical orbitals fill sequentially with increasing atomic number, the electron configuration of an element can be inferred from its position in the periodic table. Quantum-mechanical calculations of the relative energies of electron configurations show that the most stable configurations are those with completely full principal energy levels. Therefore, the most stable and unreactive elements—those with the lowest energy electron configurations—are the noble gases. Elements with one or two valence electrons are among the most active metals, readily losing their valence electrons to attain noble gas configurations. Elements with six or seven valence electrons are among the most active nonmetals, readily gaining enough electrons to attain a noble gas configuration.

Effective Nuclear Charge and Periodic Trends in Atomic Size (8.6)

The size of an atom is largely determined by its outermost electrons. As you move down a column in the periodic table, the principal quantum number (n) of the outermost electrons increases, resulting in successively larger orbitals and therefore larger atomic radii. As you move across a row in the periodic table, atomic radii decrease because the effective nuclear charge—the net or average charge experienced by the atom's outermost electrons—increases. The atomic radii of the transition elements stay roughly constant across each row because, as you move across a row, electrons are added to the $n_{highest} - 1$ orbitals while the number of highest n electrons stays roughly constant.

Ion Properties (8.7)

The electron configuration of an ion can be determined by adding or subtracting the corresponding number of electrons to the electron configuration of the neutral atom. For main-group ions, the order of removing electrons is the same as the order in which they are added in building up the electron configuration. For transition metal atoms, the ns electrons are removed before the $(n - 1)d$ electrons. The radius of a cation is much *smaller* than that of the corresponding atom, and the radius of an anion is much *larger* than that of the corresponding atom. The ionization energy—the energy required to remove an electron from an atom in the gaseous state—generally decreases as you move down a column in the periodic table and increases when moving to the right across a row. Successive ionization energies for valence electrons increase smoothly from one to the next, but the ionization energy increases dramatically for the first core electron.

Electron Affinities and Metallic Character (8.8)

Electron affinity—the energy associated with an element in its gaseous state gaining an electron—does not show a general trend as you move down a column in the periodic table, but it generally becomes more negative (more exothermic) to the right across a row. Metallic character—the tendency to lose electrons in a chemical reaction—generally increases down a column in the periodic table and decreases to the right across a row.

The Alkali Metals, Halogens, and Noble Gases (8.9)

The most active metals are the alkali metals (group 1A) and the most active nonmetals are the halogens (group 7A). The alkali metals are powerful reducing agents, reacting with many nonmetals—including the halogens and water— to form ionic compounds. The halogens are powerful oxidizing agents, reacting with many metals to form ionic compounds. The halogens also react with many nonmetals to form covalent compounds. The noble gases are relatively unreactive; only krypton and xenon form compounds and mostly with fluorine, the most reactive element in the periodic table.

Key Equations and Relationships

Order of Filling Quantum-Mechanical Orbitals (8.3)

$$1s\ 2s\ 2p\ 3s\ 3p\ 4s\ 3d\ 4p\ 5s\ 4d\ 5p\ 6s$$

Key Skills

Writing Electron Configurations (8.3)
• Example 8.1   • For Practice 8.1   • Exercises 41, 42

Writing Orbital Diagrams (8.3)
• Example 8.2   • For Practice 8.2   • Exercises 43, 44

Valence Electrons and Core Electrons (8.4)
• Example 8.3   • For Practice 8.3   • Exercises 51, 52

Electron Configurations from the Periodic Table (8.4)
• Example 8.4   • For Practice 8.4   • For More Practice 8.4   • Exercises 45, 46

Using Periodic Trends to Predict Atomic Size (8.6)
• Example 8.5   • For Practice 8.5   • For More Practice 8.5   • Exercises 59–62

Writing Electron Configurations for Ions (8.7)
• Example 8.6   • For Practice 8.6   • Exercises 63, 64

Using Periodic Trends to Predict Ion Size (8.7)
• Example 8.7   • For Practice 8.7   • For More Practice 8.7   • Exercises 67–70

Using Periodic Trends to Predict Relative Ionization Energies (8.7)
• Example 8.8   • For Practice 8.8   • For More Practice 8.8   • Exercises 71–74

Predicting Metallic Character Based on Periodic Trends (8.8)
• Example 8.9   • For Practice 8.9   • For More Practice 8.9   • Exercises 79–82

Writing Reactions for Alkali Metal and Halogen Reactions (8.9)
• Example 8.10   • For Practice 8.10   • Exercises 83–88

Exercises

Review Questions

1. What are periodic properties?

2. Which periodic property is particularly important to nerve signal transmission? Why?

3. What were the contributions of Johann Döbereiner and John Newlands to the organization of elements according to their properties?

4. Who is credited with arranging the periodic table? How were elements arranged in this table?

5. Explain the contributions of Meyer and Moseley to the periodic table.

6. The periodic table is a result of the periodic law. What observations led to the periodic law? What theory explains the underlying reasons for the periodic law?

7. What is electron spin? Explain the difference between an electron with $m_s = +\frac{1}{2}$ and $m_s = -\frac{1}{2}$.

8. Describe the Stern–Gerlach experiment. How did the experiment demonstrate that the orientation of electron spin was quantized?

9. What is an electron configuration? Give an example.

10. What is meant by the terms shielding and penetration?

11. Why are the sublevels within a principal level split into different energies for multielectron atoms but not for the hydrogen atom?

12. What is an orbital diagram? Give an example.

13. Why is electron spin important when writing electron configurations? Explain in terms of the Pauli exclusion principle.

14. What are degenerate orbitals? According to Hund's rule, how are degenerate orbitals occupied?

15. List all orbitals from 1s through 5s according to increasing energy for multielectron atoms.

16. What are valence electrons? Why are they important?

17. Copy the following blank periodic table onto a sheet of paper and label each of the blocks within the table: s block, p block, d block, and f block.

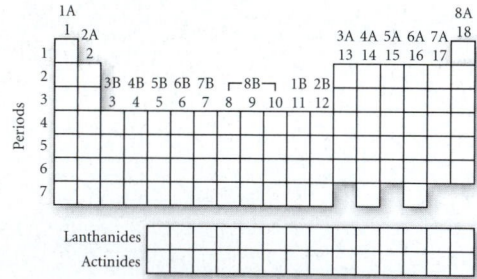

18. Explain why the s block in the periodic table has only two columns while the p block has six.

19. Why do the rows in the periodic table get progressively longer as you move down the table? For example, the first row contains two elements, the second and third rows each contain eight elements, and the fourth and fifth rows each contain eighteen elements. Explain.

20. Explain the relationship between a main-group element's lettered group number (the number of the element's column) and its valence electrons.

21. Explain the relationship between an element's row number in the periodic table and the highest principal quantum number in the element's electron configuration. How is this relationship different for main-group elements, transition elements, and inner transition elements?

22. Which of the transition elements in the first transition series have anomalous electron configurations?

23. Explain how to write an electron configuration for an element from its position in the periodic table.

24. Explain the relationship between the properties of an element and the number of valence electrons that it contains.

25. Give the number of valence electrons for each of the following families in the periodic table, and explain the relationship between the number of valence electrons and the resulting chemistry of the elements in the family.
    a. alkali metals   b. alkaline earth metals
    c. halogens   d. oxygen family

26. How is atomic radius defined? For main-group elements, give the observed trends in atomic radius as you:
    a. move across a period in the periodic table
    b. move down a column in the periodic table

27. What is effective nuclear charge? What is shielding?

28. Use the concepts of effective nuclear charge, shielding, and n value of the valence orbital to explain the trend in atomic radius as you move across a period in the periodic table.

29. For transition elements, give the trends in atomic radius as you:
    a. move across a period in the periodic table
    b. move down a column in the periodic table
    Explain the reasons for the trends in a and b.

30. How is the electron configuration of an anion different from that of the corresponding neutral atom? How is the electron configuration of a cation different?

31. Explain how to write an electron configuration for a transition metal cation. Is the order of electron removal upon ionization simply the reverse of electron addition upon filling? Why or why not?

32. What is the relationship between the following?
    a. the radius of a cation and that of the atom from which it is formed
    b. the radius of an anion and that of the atom from which it is formed

33. What is ionization energy? What is the difference between first ionization energy and second ionization energy?

34. What is the general trend in ionization energy as you move down a column in the periodic table? As you move across a row?

35. What are the exceptions to the periodic trends in ionization energy? Why do they occur?

36. Examination of the first few successive ionization energies for a given element usually reveals a large jump between two ionization energies. For example, the successive ionization energies of magnesium show a large jump between IE_2 and IE_3. The successive ionization energies of aluminum show a large jump between IE_3 and IE_4. Explain why these jumps occur and how you might predict them.

37. What is electron affinity? What are the observed periodic trends in electron affinity?

38. What is metallic character? What are the observed periodic trends in metallic character?

39. Write a general equation for the reaction of an alkali metal with:
    a. a halogen   b. water

40. Write a general equation for the reaction of a halogen with:
    a. a metal   b. hydrogen
    c. another halogen

Problems by Topic

Electron Configurations

41. Write full electron configurations for each of the following elements:
   a. P   **b.** C   **c.** Na   **d.** Ar

42. Write full electron configurations for each of the following elements:
   a. O   **b.** Si   **c.** Ne   **d.** K

43. Write full orbital diagrams for each of the following elements:
   a. N   **b.** F   **c.** Mg   **d.** Al

44. Write full orbital diagrams for each of the following elements:
   a. S   **b.** Ca   **c.** Ne   **d.** He

45. Use the periodic table to write electron configurations for each of the following elements. Represent core electrons with the symbol of the previous noble gas in brackets.
   a. P   **b.** Ge   **c.** Zr   **d.** I

46. Use the periodic table to determine the element corresponding to each of the following electron configurations.
   a. $[Ar] 4s^2 3d^{10} 4p^6$ **b.** $[Ar] 4s^2 3d^2$
   c. $[Kr] 5s^2 4d^{10} 5p^2$ **d.** $[Kr] 5s^2$

47. Use the periodic table to determine each of the following:
   a. The number of $2s$ electrons in Li
   b. The number of $3d$ electrons in Cu
   c. The number of $4p$ electrons in Br
   d. The number of $4d$ electrons in Zr

48. Use the periodic table to determine each of the following:
   a. The number of $3s$ electrons in Mg
   b. The number of $3d$ electrons in Cr
   c. The number of $4d$ electrons in Y
   d. The number of $6p$ electrons in Pb

49. Name an element in the fourth period (row) of the periodic table with:
   a. five valence electrons   **b.** four $4p$ electrons
   c. three $3d$ electrons   **d.** a complete outer shell

50. Name an element in the third period (row) of the periodic table with:
   a. three valence electrons
   b. four $3p$ electrons
   c. six $3p$ electrons
   d. two $3s$ electrons and no $3p$ electrons

Valence Electrons and Simple Chemical Behavior from the Periodic Table

51. Determine the number of valence electrons in each of the following elements.
   a. Ba   **b.** Cs   **c.** Ni   **d.** S

52. Determine the number of valence electrons for each of the following elements. Which elements do you expect to lose electrons in their chemical reactions? Which do you expect to gain electrons?
   a. Al   **b.** Sn   **c.** Br   **d.** Se

53. Which of the following outer electron configurations would you expect to belong to a reactive metal? To a reactive nonmetal?
   a. ns^2 **b.** $ns^2 np^6$ **c.** $ns^2 np^5$ **d.** $ns^2 np^2$

54. Which of the following outer electron configurations would you expect to belong to a noble gas? To a metalloid?
   a. ns^2 **b.** $ns^2 np^6$ **c.** $ns^2 np^5$ **d.** $ns^2 np^2$

Effective Nuclear Charge and Atomic Radius

55. Which electrons experience a greater effective nuclear charge, the valence electrons in beryllium, or the valence electrons in nitrogen? Why?

56. Arrange the following atoms according to decreasing effective nuclear charge experienced by their valence electrons: S, Mg, Al, Si.

57. If core electrons completely shielded valence electrons from nuclear charge (i.e., if each core electron reduced nuclear charge by 1 unit) and if valence electrons did not shield one another from nuclear charge at all, what would be the effective nuclear charge experienced by the valence electrons of the following atoms?
   a. K   **b.** Ca   **c.** O   **d.** C

58. In Section 8.6, we estimated the effective nuclear charge on beryllium's valence electrons to be slightly greater than 2+. What would a similar treatment predict for the effective nuclear charge on boron's valence electrons? Would you expect the effective nuclear charge to be different for boron's $2s$ electrons compared to its $2p$ electron? How so? (Hint: Consider the shape of the $2p$ orbital compared to that of the $2s$ orbital.)

59. Choose the larger atom from each of the following pairs:
   a. Al or In   **b.** Si or N   **c.** P or Pb   **d.** C or F

60. Choose the larger atom from each of the following pairs:
   a. Sn or Si   **b.** Br or Ga   **c.** Sn or Bi   **d.** Se or Sn

61. Arrange the following elements in order of increasing atomic radius: Ca, Rb, S, Si, Ge, F.

62. Arrange the following elements in order of decreasing atomic radius: Cs, Sb, S, Pb, Se.

Ionic Electron Configurations, Ionic Radii, Magnetic Properties, and Ionization Energy

63. Write electron configurations for each of the following ions:
   a. O^{2-} **b.** Br^- **c.** Sr^{2+} **d.** Co^{3+}
   e. Cu^{2+}

64. Write electron configurations for each of the following ions:
   a. Cl^- **b.** P^{3-} **c.** K^+ **d.** Mo^{3+}
   e. V^{3+}

65. Write orbital diagrams for each of these ions and determine if the ion is diamagnetic or paramagnetic.
   a. V^{5+} **b.** Cr^{3+} **c.** Ni^{2+} **d.** Fe^{3+}

66. Write orbital diagrams for each of these ions and determine if the ion is diamagnetic or paramagnetic.
   a. Cd^{2+} **b.** Au^+ **c.** Mo^{3+} **d.** Zr^{2+}

67. Pick the larger species from each of the following pairs:
   a. Li or Li^+ **b.** I^- or Cs^+
   c. Cr or Cr^{3+} **d.** O or O^{2-}

68. Pick the larger species from each of the following pairs:
   a. Sr or Sr^{2+} **b.** N or N^{3-}
   c. Ni or Ni^{2+} **d.** S^{2-} or Ca^{2+}

69. Arrange the following isoelectronic series in order of decreasing radius: F^-, Ne, O^{2-}, Mg^{2+}, Na^+.

70. Arrange the following isoelectronic series in order of increasing atomic radius: Se^{2-}, Kr, Sr^{2+}, Rb^+, Br^-.

71. Choose the element with the highest first ionization energy from each of the following pairs:
   a. Br or Bi   **b.** Na or Rb
   c. As or At   **d.** P or Sn

72. Choose the element with the highest first ionization energy from each of the following pairs:
   a. P or I   **b.** Si or Cl
   c. P or Sb   **d.** Ga or Ge

73. Arrange the following elements in order of increasing first ionization energy: Si, F, In, N.

74. Arrange the following elements in order of decreasing first ionization energy: Cl, S, Sn, Pb.

75. For each of the following elements, predict where the "jump" occurs for successive ionization energies. (For example, does the jump occur between the first and second ionization energies, the second and third, or the third and fourth?)

a. Be   b. N   c. O   d. Li

76. Consider the following set of successive ionization energies:

$$IE_1 = 578 \text{ kJ/mol}$$
$$IE_2 = 1,820 \text{ kJ/mol}$$
$$IE_3 = 2,750 \text{ kJ/mol}$$
$$IE_4 = 11,600 \text{ kJ/mol}$$

To which third period element do these ionization values belong?

Electron Affinities and Metallic Character

77. Choose the element with the more negative (more exothermic) electron affinity from each of the following pairs:

a. Na or Rb   b. B or S
c. C or N   d. Li or F

78. Choose the element with the more negative (more exothermic) electron affinity from each of the following pairs:

a. Mg or S   b. K or Cs
c. Si or P   d. Ga or Br

79. Choose the more metallic element from each of the following pairs:

a. Sr or Sb   b. As or Bi
c. Cl or O   d. S or As

80. Choose the more metallic element from each of the following pairs:

a. Sb or Pb   b. K or Ge
c. Ge or Sb   d. As or Sn

81. Arrange the following elements in order of increasing metallic character: Fr, Sb, In, S, Ba, Se.

82. Arrange the following elements in order of decreasing metallic character: Sr, N, Si, P, Ga, Al.

Chemical Behavior of the Alkali Metals and the Halogens

83. Write a balanced chemical equation for the reaction of solid strontium with iodine gas.

84. Based on the ionization energies of the alkali metals, which alkali metal would you expect to undergo the most exothermic reaction with chlorine gas? Write a balanced chemical equation for the reaction.

85. Write a balanced chemical equation for the reaction of solid lithium with liquid water.

86. Write a balanced chemical equation for the reaction of solid potassium with liquid water.

87. Write a balanced equation for the reaction of hydrogen gas with bromine gas.

88. Write a balanced equation for the reaction of chlorine gas with fluorine gas.

Cumulative Problems

89. Bromine is a highly reactive liquid while krypton is an inert gas. Explain the difference based on their electron configurations.

90. Potassium is a highly reactive metal while argon is an inert gas. Explain the difference based on their electron configurations.

91. Both vanadium and its 3+ ion are paramagnetic. Use electron configurations to explain why this is so.

92. Use electron configurations to explain why copper is paramagnetic while its 1+ ion is not.

93. Suppose you were trying to find a substitute for K^+ in nerve signal transmission. Where would you begin your search? What ions would be most like K^+? For each ion you propose, explain the ways in which it would be similar to K^+ and the ways it would be different. Use periodic trends in your discussion.

94. Suppose you were trying to find a substitute for Na^+ in nerve signal transmission. Where would you begin your search? What ions would be most like Na^+? For each ion you propose, explain the ways in which it would be similar to Na^+ and the ways it would be different. Use periodic trends in your discussion.

95. Life on Earth evolved around the element carbon. Based on periodic properties, what two or three elements would you expect to be most like carbon?

96. Which of the following pairs of elements would you expect to have the most similar atomic radii, and why?

a. Si and Ga   b. Si and Ge   c. Si and As

97. Consider the following elements: N, Mg, O, F, Al.

a. Write an electron configuration for each element.
b. Arrange the elements in order of decreasing atomic radius.
c. Arrange the elements in order of increasing ionization energy.
d. Use the electron configurations in part a to explain the differences between your answers to parts b and c.

98. Consider the following elements: P, Ca, Si, S, Ga.

a. Write an electron configuration for each element.
b. Arrange the elements in order of decreasing atomic radius.

c. Arrange the elements in order of increasing ionization energy.
d. Use the electron configurations in part a to explain the differences between your answers to parts b and c.

99. Explain why atomic radius decreases as you move to the right across a period for main-group elements but not for transition elements.

100. Explain why vanadium (radius = 134 pm) and copper (radius = 128 pm) have nearly identical atomic radii, even though the atomic number of copper is about 25% higher than that of vanadium. What would you predict about the relative densities of these two metals? Look up the densities in a reference book, periodic table, or on the Web. Are your predictions correct?

101. The lightest noble gases, such as helium and neon, are completely inert—they do not form any chemical compounds whatsoever. The heavier noble gases, in contrast, do form a limited number of compounds. Explain this difference in terms of trends in fundamental periodic properties.

102. The lightest halogen is also the most chemically reactive, and reactivity generally decreases as you move down the column of halogens in the periodic table. Explain this trend in terms of periodic properties.

103. Write general outer electron configurations ($ns^x np^y$) for groups 6A and 7A in the periodic table. The electron affinity of each group 7A element is more negative than that of each corresponding group 6A element. Use the electron configurations to explain why this is so.

104. The electron affinity of each group 5A element is more positive than that of each corresponding group 4A element. Use the outer electron configurations for these columns to suggest a reason for this behavior.

105. Elements 35 and 53 have similar chemical properties. Based on their electronic configurations predict the atomic number of a heavier element that also should have these chemical properties.

106. Write the electronic configurations of the six cations that form from sulfur by the loss of one to six electrons. For those cations that have unpaired electrons, write orbital diagrams.

Challenge Problems

107. Consider the densities and atomic radii of the following noble gases at 25 °C:

Element	Atomic Radius (pm)	Density (g/L)
He	32	0.18
Ne	70	0.90
Ar	98	—
Kr	112	3.75
Xe	130	—
Rn	—	9.73

  a. Estimate the densities of argon and xenon by interpolation from the data.

  b. Provide an estimate of the density of the yet undiscovered element with atomic number 118 by extrapolation from the data.

  c. Use the molar mass of neon to estimate the mass of a neon atom. Then use the atomic radius of neon to calculate the average density of a neon atom. How does this density compare to the density of neon gas? What does this comparison suggest about the nature of neon gas?

  d. Use the densities and molar masses of krypton and neon to calculate the number of atoms of each found in a volume of 1.0 L. Use these values to estimate the number of atoms that occur in 1.0 L of Ar. Now use the molar mass of argon to estimate the density of Ar. How does this estimate compare to that in part a?

108. As we have seen, the periodic table is a result of empirical observation (i.e., the periodic law), but quantum-mechanical theory explains *why* the table is so arranged. Suppose that, in another universe, quantum theory was such that there were one *s* orbital but only two *p* orbitals (instead of three) and only three *d* orbitals (instead of five). Draw out the first four periods of the periodic table in this alternative universe. Which elements would be the equivalent of the noble gases? Halogens? Alkali metals?

109. Consider the metals in the first transition series. Use periodic trends to predict a trend in density as you move to the right across the series.

110. Imagine a universe in which the value of m_s can be $+\frac{1}{2}$, 0, and $-\frac{1}{2}$. Assuming that all the other quantum numbers can take only the values possible in our world and that the Pauli exclusion principle applies, give the following:

  a. the new electronic configuration of neon

  b. the atomic number of the element with a completed $n = 2$ shell

  c. the number of unpaired electrons in fluorine

111. A carbon atom can absorb radiation of various wavelengths with resulting changes in its electronic configuration. Write orbital diagrams for the electronic configuration of carbon that would result from absorption of the three longest wavelengths of radiation it can absorb.

112. Only trace amounts of the synthetic element darmstadtium, atomic number 110, have been obtained. The element is so highly unstable that no observations of its properties have been possible. Based on its position in the periodic table, propose three different reasonable valence electron configurations for this element.

113. What is the atomic number of the as yet undiscovered element in which the 8*s* and 8*p* electron energy levels fill? Predict the chemical behavior of this element.

114. The trend in second ionization energy for the elements from lithium to fluorine is not a smooth one. Predict which of these elements has the highest second ionization energy and which has the lowest and explain. Of the elements N, O, and F, O has the highest and N the lowest second ionization energy. Explain.

115. Unlike the elements in groups 1A and 2A, those in group 3A do not show a smooth decrease in first ionization energy in going down the column. Explain the irregularities.

116. Using the data in Figures 8.15 and 8.16, calculate ΔE for the reaction

$$Na(g) + Cl(g) \longrightarrow Na^+(g) + Cl^-(g)$$

117. Despite the fact that adding two electrons to O or S forms an ion with a noble gas electron configuration, the second electron affinity of both of these elements is positive. Explain.

118. In Section 2.7 we discussed the metalloids, which form a diagonal band separating the metals from the nonmetals. There are other instances in which elements such as lithium and magnesium that are diagonal to each other have comparable metallic character. Suggest an explanation for this observation.

Conceptual Problems

119. Imagine that in another universe, atoms and elements are identical to ours, except that atoms with six valence electrons have particular stability (in contrast to our universe where atoms with eight valence electrons have particular stability). Give an example of an element in the alternative universe that corresponds to each of the following:

  a. a noble gas   **b.** a reactive nonmetal

  c. a reactive metal

120. Determine whether each of the following is true or false regarding penetration and shielding. (Assume that all lower energy orbitals are fully occupied.)

  a. An electron in a 3*s* orbital is more shielded than an electron in a 2*s* orbital.

  b. An electron in a 3*s* orbital penetrates into the region occupied by core electrons more than electrons in a 3*p* orbital.

  c. An electron in an orbital that penetrates closer to the nucleus will always experience more shielding than an electron in an orbital that does not penetrate as far.

  d. An electron in an orbital that penetrates close to the nucleus will tend to experience a higher effective nuclear charge than one that does not.

121. Give a combination of four quantum numbers that could be assigned to an electron occupying a 5*p* orbital. Do the same for an electron occupying a 6*d* orbital.

122. Use the trends in ionization energy and electron affinity to explain why calcium fluoride has the formula CaF_2 and not Ca_2F or CaF.

9

Chemical Bonding I: Lewis Theory

9.1 Bonding Models and AIDS Drugs

9.2 Types of Chemical Bonds

9.3 Representing Valence Electrons with Dots

9.4 Ionic Bonding: Lewis Structures and Lattice Energies

9.5 Covalent Bonding: Lewis Structures

9.6 Electronegativity and Bond Polarity

9.7 Lewis Structures of Molecular Compounds and Polyatomic Ions

9.8 Resonance and Formal Charge

9.9 Exceptions to the Octet Rule: Odd-Electron Species, Incomplete Octets, and Expanded Octets

9.10 Bond Energies and Bond Lengths

9.11 Bonding in Metals: The Electron Sea Model

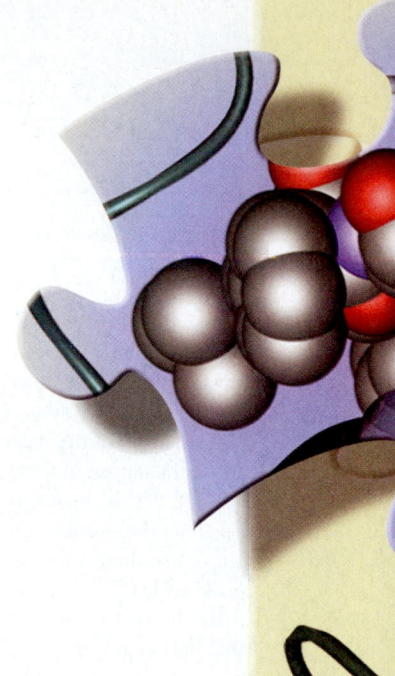

CHEMICAL BONDING IS AT THE HEART of chemistry. The bonding theories that we examine in the next two chapters are—as Karl Popper eloquently states on the facing page—nets cast to understand the world. We examine three theories, with successively finer "meshes." The first is Lewis theory, which can be done on the back of an envelope. With just a few dots, dashes, and chemical symbols, we can understand and predict a myriad of chemical observations. The second is valence bond theory, which treats electrons in a more quantum-mechanical manner, but stops short of viewing them as belonging to the entire molecule. The third is molecular orbital theory, essentially a full quantum-mechanical treatment of the molecule and its electrons as a whole. Molecular orbital theory has great predictive power, but at the expense of great complexity and intensive computational requirements. Which theory is right? Remember that theories are models that help us understand and predict behavior. All three of these theories are extremely useful, depending on exactly what aspect of chemical bonding we want to predict or understand.

The AIDS drug Indinavir—shown here as the missing piece in a puzzle depicting the protein HIV-protease—was developed with the help of chemical bonding theories.

9.1 Bonding Models and AIDS Drugs

In 1989, researchers used X-ray crystallography—a technique in which X-rays are scattered from crystals of the molecule of interest—to determine the structure of a molecule called HIV-protease. HIV-protease is a protein (a class of large biological molecules) synthesized by the human immunodeficiency virus (HIV). This particular protein is crucial to the virus's ability to multiply and cause acquired immune deficiency syndrome, or AIDS. Without HIV-protease, HIV cannot spread in the human body because the virus cannot replicate. In other words, without HIV-protease, AIDS can't develop.

With knowledge of the HIV-protease structure, drug companies set out to create a molecule that would disable HIV-protease by sticking to the working part of the molecule, called the active site. To design such a molecule, researchers used *bonding theories*—models that predict how atoms bond together to form molecules—to simulate the shape

Proteins are discussed in more detail in Chapter 21.

of potential drug molecules and how they would interact with the protease molecule. By the early 1990s, these companies had developed several drug molecules that seemed to work. Since these molecules inhibit the action of HIV-protease, they are called *protease inhibitors*. In human trials, protease inhibitors, when given in combination with other drugs, have decreased the viral count in HIV-infected individuals to undetectable levels. Although protease inhibitors are not a cure for AIDS, many AIDS patients are still alive today because of these drugs.

Bonding theories are central to chemistry because they explain how atoms bond together to form molecules. They explain why some combinations of atoms are stable and others are not. For example, bonding theories explain why table salt is NaCl and not $NaCl_2$ and why water is H_2O and not H_3O. Bonding theories also predict the shapes of molecules—a topic of our next chapter—which in turn determine many of their physical and chemical properties. The bonding theory you will learn in this chapter is called **Lewis theory**, named after the American chemist who developed it, G. N. Lewis (1875–1946). In Lewis theory, we represent valence electrons as dots and draw what are called **Lewis electron-dot structures** (or simply **Lewis structures**) to represent molecules. These structures, which are fairly simple to draw, have tremendous predictive power. In just a few minutes, you can use Lewis theory to predict whether a particular set of atoms will form a stable molecule and what that molecule might look like. Although we will also learn about more advanced theories in our next chapter, Lewis theory remains the simplest method for making quick, everyday predictions about most molecules.

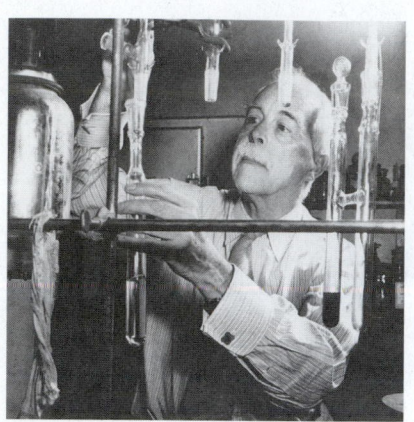

▲ G.N. Lewis

9.2 Types of Chemical Bonds

We begin our discussion of chemical bonding by asking why bonds form in the first place. This seemingly simple question is vitally important. Imagine a universe without chemical bonding. There would be just 91 different kinds of substances (the 91 naturally occurring elements). With such a poor diversity of substances, life would be impossible, and we would not be around to wonder why. The answer to this question, however, is not simple and involves not only quantum mechanics but also some thermodynamics that we do not introduce until Chapter 17. Nonetheless, an important *part* of the answer is that *chemical bonds form because they lower the potential energy between the charged particles that compose atoms.*

The potential energy (E) of two charged particles with charges q_1 and q_2 separated by a distance r is given by the following equation, a form of Coulomb's law:

$$E = \frac{1}{4\pi\epsilon_0} \frac{q_1 q_2}{r}$$ [9.1]

In this equation, ϵ_0 is a constant ($\epsilon_0 = 8.85 \times 10^{-12} \, C^2/J \cdot m$). Notice that potential energy is positive for charges of the same sign (plus × plus or minus × minus) and negative for charges of opposite sign (plus × minus or minus × plus). Notice also that, because r is in the denominator, the *magnitude* of the potential energy depends inversely on the separation between the charged particles. As the particles get closer, the magnitude of the potential energy increases. From this we can draw two important conclusions:

- For like charges, the potential energy (E) is positive and decreases as the particles get *farther apart* (as r increases). Since systems tend toward lower potential energy, like charges repel each other (in much the same way that like poles of two magnets repel each other), as shown in Figure 9.1 ▶.

- For opposite charges, the potential energy is negative and becomes more negative as the particles get *closer together* (as r decreases). Therefore opposite charges (like opposite poles on a magnet) *attract each other*, as shown in Figure 9.2 ▶.

As we already know, atoms are composed of particles with positive charges (the protons in the nucleus) and negative charges (the electrons). When two atoms approach each other, the electrons of one atom are attracted to the nucleus of the other and vice versa. However, at the same time, the electrons of each atom repel the electrons of the other, and the nucleus of each atom repels the nucleus of the other. The result is a complex set of interactions among a potentially large number of charged particles.

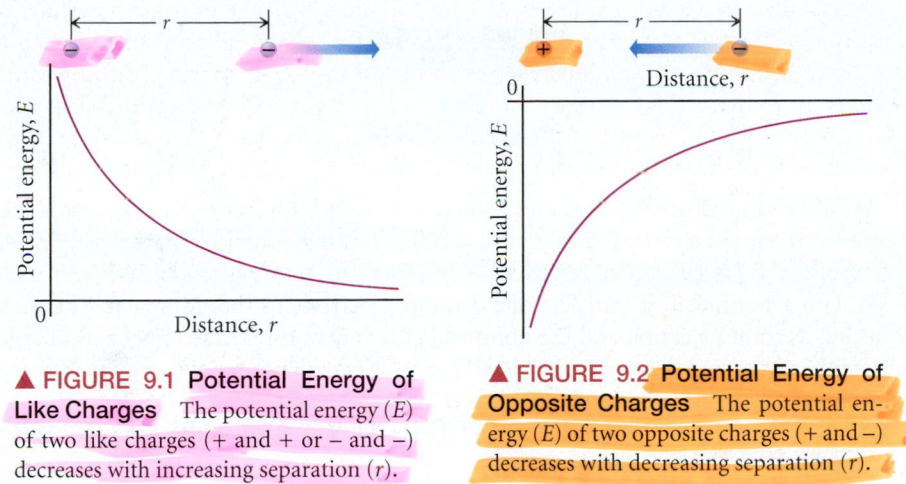

▲ FIGURE 9.1 **Potential Energy of Like Charges**   The potential energy (E) of two like charges (+ and + or − and −) decreases with increasing separation (r).

▲ FIGURE 9.2 **Potential Energy of Opposite Charges**   The potential energy (E) of two opposite charges (+ and −) decreases with decreasing separation (r).

If these interactions lead to an overall net reduction of energy between the charged particles, a chemical bond forms. If not, then a chemical bond does not form. Bonding theories help us to predict the circumstances under which bonds form and also the properties of the resultant molecules.

We can broadly classify chemical bonds into three types depending on the kind of atoms involved in the bonding (Figure 9.3 ▼).

Types of Atoms	Type of Bond	Characteristic of Bond
Metal and nonmetal	Ionic	Electrons transferred
Nonmetal and nonmetal	Covalent	Electrons shared
Metal and metal	Metallic	Electrons pooled

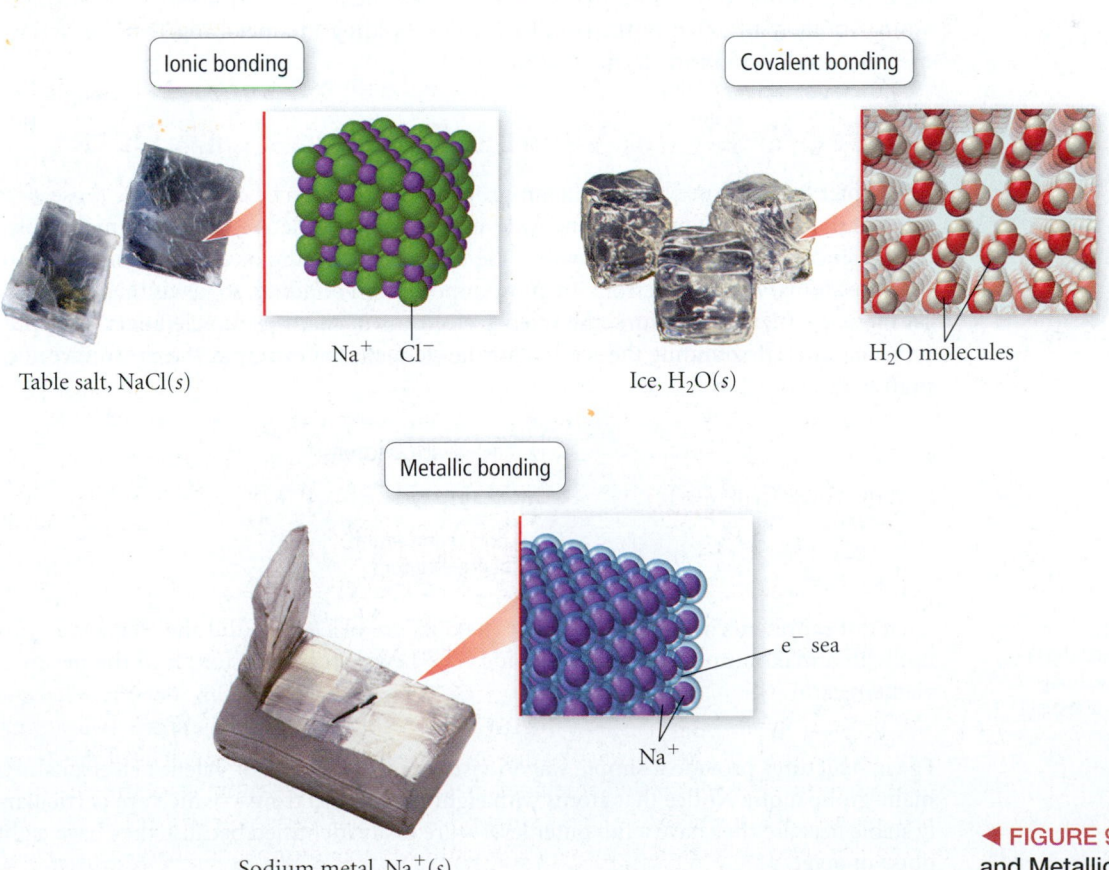

Ionic bonding

Table salt, $NaCl(s)$

Na^+ Cl^-

Covalent bonding

Ice, $H_2O(s)$

H_2O molecules

Metallic bonding

Sodium metal, $Na^+(s)$

e^- sea

Na^+

◀ FIGURE 9.3 **Ionic, Covalent, and Metallic Bonding**

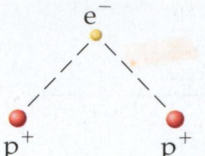
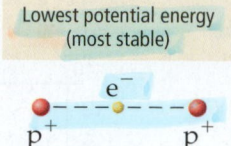
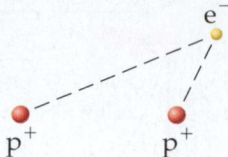

Lowest potential energy (most stable)

▶ FIGURE 9.4 **Some Possible Configurations of One Electron and Two Protons**

We learned in Chapter 8 that metals tend to have low ionization energies (it is relatively easy to remove electrons from them) and that nonmetals tend to have negative electron affinities (it is energetically favorable for them to gain electrons). When a metal bonds with a nonmetal, it transfers one or more electrons to the nonmetal. The metal atom thus becomes a cation and the nonmetal atom an anion. These oppositely charged ions are then attracted to one another, lowering their overall potential energy as indicated by Coulomb's law. The resulting bond is called an **ionic bond**.

We also learned in Chapter 8 that nonmetals tend to have high ionization energies (it is difficult to remove electrons from them). Therefore when a nonmetal bonds with another nonmetal, neither atom transfers electrons to the other. Instead, some electrons are *shared* between the two bonding atoms. The shared electrons interact with the nuclei of both of the bonding atoms, lowering their potential energy in accordance with Coulomb's law. The resulting bond is called a **covalent bond**. Recall from Section 3.2 that we can understand the stability of a covalent bond by considering the most stable arrangement (the one with the lowest potential energy) of two protons and an electron. As you can see from Figure 9.4 ▲, the arrangement in which the electron lies *between* the two protons has the lowest potential energy because the negatively charged electron interacts most strongly with *both protons*. In a sense, the electron holds the two protons together because its negative charge attracts the positive charges of the protons. Similarly, shared electrons in a covalent chemical bond *hold* the bonding atoms together by attracting the positive charges of the bonding atoms' nuclei.

A third type of bonding, called **metallic bonding**, occurs in metals. Since metals have low ionization energies, they tend to lose electrons easily. In the simplest model for metallic bonding—called the *electron sea* model—all of the atoms in a metal lattice pool their valence electrons to form an *electron sea*. These pooled electrons are no longer localized on a single atom, but delocalized over the entire metal. The positively charged metal atoms are then attracted to the sea of electrons, holding the metal together. We discuss metallic bonding in more detail in Section 9.11.

9.3 Representing Valence Electrons with Dots

In Chapter 8, we learned that, for main-group elements, valence electrons are those electrons in the outermost principal energy level. Since valence electrons are held most loosely, and since chemical bonding involves the transfer or sharing of electrons between two or more atoms, valence electrons are most important in bonding, so Lewis theory focuses on these. In a Lewis structure, the valence electrons of main-group elements are represented as dots surrounding the symbol for the element. For example, the electron configuration of O is

$$1s^2 2s^2 2p^4$$

6 valence electrons

And the Lewis structure is

·Ö:

6 dots representing valence electrons

Remember, the number of valence electrons for any main group is equal to the group number of the element (except for helium, which is in group 8A but has only two valence electrons).

Each dot represents a valence electron. The dots are placed around the element's symbol with a maximum of two dots per side. The Lewis structures for all of the period 2 elements are

Li·   Be: Ḃ: ·Ċ: ·N̈:   ·Ö: :F̈:   :N̈e:

While the exact location of dots is not critical, in this book we will place the first two dots on the right side of the atomic symbol and then fill in the rest of the dots singly first before pairing.

Lewis structures provide a simple way to visualize the number of valence electrons in a main-group atom. Notice that atoms with eight valence electrons—which are particularly stable because they have a full outer level—are easily identified because they have eight dots, an **octet**.

Helium is somewhat of an exception. Its electron configuration and Lewis structure are

$$1s^2 \qquad \text{He:}$$

The Lewis structure of helium contains only two dots (a **duet**). For helium, a duet represents a stable electron configuration because the $n = 1$ quantum level fills with only two electrons.

In Lewis theory, a **chemical bond** is the sharing or transfer of electrons to attain stable electron configurations for the bonding atoms. If electrons are transferred, as occurs between a metal and a nonmetal, the bond is an *ionic bond*. If the electrons are shared, as occurs between two nonmetals, the bond is a *covalent bond*. In either case, the bonding atoms obtain stable electron configurations; since the stable configuration is usually eight electrons in the outermost shell, this is known as the **octet rule**. Notice that Lewis theory does not attempt to take account of attractions and repulsions between electrons and nuclei on neighboring atoms. The energy changes that occur because of these interactions are central to chemical bonding (as we saw previously), yet Lewis theory ignores them because calculating these energy changes is extremely complicated. Instead Lewis theory uses the simple octet rule, a practical approach which, as we will see, predicts what we see in nature for a large number of compounds—hence the success and longevity of Lewis theory.

9.4 Ionic Bonding: Lewis Structures and Lattice Energies

Although Lewis theory's strength is in modeling covalent bonding, it can also be applied to ionic bonding. In Lewis theory, we represent ionic bonding by moving electron dots from the metal to the nonmetal and then allowing the resultant ions to form a crystalline lattice composed of alternating cations and anions.

Ionic Bonding and Electron Transfer

To see how ionic bonding is formulated in terms of Lewis theory, consider potassium and chlorine, which have the following Lewis structures:

$$\text{K·} \qquad \text{:Cl:}$$

When potassium and chlorine bond, potassium transfers its valence electron to chlorine:

$$\text{K·} + \text{:Cl:} \longrightarrow \text{K}^+ \left[\text{:Cl:}\right]^-$$

The transfer of the electron gives chlorine an octet (shown as eight dots around chlorine) and leaves potassium without any valence electrons but with an octet in the *previous* principal energy level (which is now the outermost level).

$$\text{K} \qquad 1s^2 2s^2 2p^6 3s^2 3p^6 4s^1$$

$$\text{K}^+ \qquad 1s^2 2s^2 2p^6 \underbrace{3s^2 3p^6}_{\text{Octet in previous level}} 4s^0$$

The potassium, because it has lost an electron, becomes positively charged, while the chlorine, which has gained an electron, becomes negatively charged. The Lewis structure of an anion is usually written within brackets with the charge in the upper right-hand corner, outside the brackets. The positive and negative charges attract one another, resulting in the compound KCl.

Lewis theory predicts the correct chemical formulas for ionic compounds. For the compound that forms between K and Cl, for example, Lewis theory predicts one potassium cation to every one chlorine anion, KCl. In nature, when we examine the compound formed between potassium and chlorine, we indeed find one potassium ion to every chloride ion. As another example, consider the ionic compound formed between sodium and sulfur. The Lewis structures for sodium and sulfur are

$$\text{Na·} \qquad \text{·S:}$$

Recall that solid ionic compounds do not contain distinct molecules, but rather are composed of alternating positive and negative ions in a three-dimensional crystalline array.

Notice that sodium must lose its one valence electron in order to have an octet (in the previous principal shell), while sulfur must gain two electrons to get an octet. Consequently, the compound that forms between sodium and sulfur requires two sodium atoms to every one sulfur atom. The Lewis structure is

$$2\,Na^+ \left[:\ddot{\underset{..}{S}}:\right]^{2-}$$

The two sodium atoms each lose their one valence electron while the sulfur atom gains two electrons and gets an octet. Lewis theory predicts that the correct chemical formula is Na_2S. When we look at the compound formed between sodium and sulfur in nature, the formula predicted by Lewis theory is exactly what we see.

EXAMPLE 9.1 Using Lewis Theory to Predict the Chemical Formula of an Ionic Compound.

Use Lewis theory to predict the formula for the compound that forms between calcium and chlorine.

Solution

Draw Lewis structures for calcium and chlorine based on their number of valence electrons, obtained from their group number in the periodic table.	Ca: :$\ddot{\underset{..}{Cl}}$:
Calcium must lose its two valence electrons (to be left with an octet in its previous principal shell), while chlorine only needs to gain one electron to get an octet. Thus, draw two chlorine anions, each with an octet and a 1− charge, and one calcium cation with a 2+ charge. Draw the chlorine anions in brackets and indicate the charges on each ion.	$Ca^{2+}\ 2\left[:\ddot{\underset{..}{Cl}}:\right]^-$
Finally, write the formula with subscripts to indicate the number of atoms.	$CaCl_2$

For Practice 9.1

Use Lewis theory to predict the formula for the compound that forms between magnesium and nitrogen.

Lattice Energy: The Rest of the Story

The formation of an ionic compound from its constituent elements is usually quite exothermic. For example, when sodium chloride (table salt) forms from elemental sodium and chlorine, 411 kJ of heat is evolved in the following violent reaction:

$$Na(s) + {}^1\!/_2\,Cl_2(g) \longrightarrow NaCl(s) \qquad \Delta H_f^\circ = -411\ kJ/mol$$

Where does this energy come from? We may think that it comes solely from the tendency of metals to lose electrons and nonmetals to gain electrons—but it does not. In fact, the transfer of an electron from sodium to chlorine—by itself—actually absorbs energy. The first ionization energy of sodium is +496 kJ/mol, and the electron affinity of Cl is only −349 kJ/mol. Thus, based only on these energies, the reaction should be *endothermic* by +147 kJ/mol. So why is the reaction so exothermic?

The answer lies in the **lattice energy**—the energy associated with forming a crystalline lattice of alternating sodium cations and chlorine anions from the gaseous ions. Since the sodium ions are positively charged and the chlorine ions negatively charged, there is a lowering of potential energy—as prescribed by Coulomb's law—when these ions come together to form a lattice. That energy is emitted as heat when the lattice forms, as shown in Figure 9.5 ▶. The exact value of the lattice energy, however, is not simple to determine because it involves a large number of interactions among many charged particles in a lattice. The easiest way to calculate lattice energy is with the *Born–Haber cycle*.

Lattice Energy of an Ionic Compound

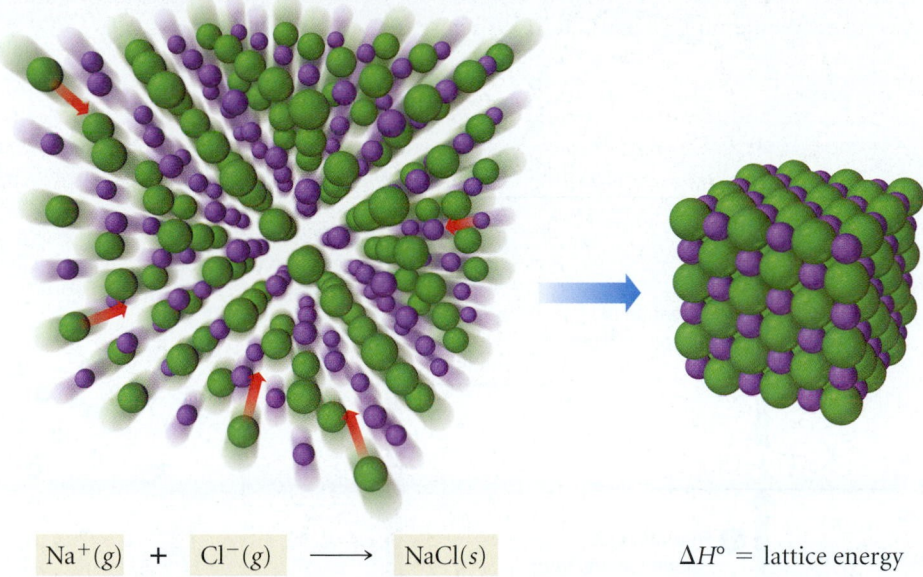

$$Na^+(g) \ + \ Cl^-(g) \ \longrightarrow \ NaCl(s) \qquad\qquad \Delta H° = \text{lattice energy}$$

▲ **FIGURE 9.5 Lattice Energy**   The lattice energy of an ionic compound is the energy associated with forming a crystalline lattice of the compound from the gaseous ions.

The Born–Haber Cycle

The **Born–Haber cycle** is a hypothetical series of steps that represents the formation of an ionic compound from its constituent elements. The steps are chosen so that the change in enthalpy of each step is known except for the last one, which is the lattice energy. The change in enthalpy for the overall process is also known. Using Hess's law, we can therefore determine the enthalpy change for the unknown last step, the lattice energy.

Consider the formation of NaCl from its constituent elements in their standard states. The enthalpy change for the overall reaction is just the standard enthalpy of formation of NaCl(s):

$$Na(s) \ + \ {}^1\!/_2\,Cl_2(g) \ \longrightarrow \ NaCl(s) \qquad \Delta H_f° = -411 \text{ kJ/mol}$$

Now consider the following set of steps—the Born–Haber cycle—from which NaCl(s) can also be made from Na(s) and $Cl_2(g)$.

- The first step is the formation of gaseous sodium from solid sodium.

$$Na(s) \ \longrightarrow \ Na(g) \qquad \Delta H°_{step\,1} \text{ (sublimation energy of Na)} = +108 \text{ kJ}$$

- The second step is the formation of a chlorine atom from a chlorine molecule.

$${}^1\!/_2\,Cl_2(g) \ \longrightarrow \ Cl(g) \qquad \Delta H°_{step\,2} \text{ (bond energy of } Cl_2 \times {}^1\!/_2) = +122 \text{ kJ}$$

- The third step is the ionization of gaseous sodium. The enthalpy change for this step is simply the ionization energy of sodium.

$$Na(g) \ \longrightarrow \ Na^+(g) \ + \ e^- \qquad \Delta H°_{step\,3} \text{ (ionization energy of Na)} = +496 \text{ kJ}$$

- The fourth step is the addition of an electron to gaseous chlorine. The enthalpy change for this step is the electron affinity of chlorine.

$$Cl(g) \ + \ e^- \ \longrightarrow \ Cl^-(g) \qquad \Delta H°_{step\,4} \text{ (electron affinity of Cl)} = -349 \text{ kJ}$$

- The fifth and final step is the formation of the crystalline solid from the gaseous ions. The enthalpy change for this step is the lattice energy, the unknown quantity.

$$Na^+(g) \ + \ Cl^-(g) \ \longrightarrow \ NaCl(s) \qquad \Delta H°_{step\,5} = \Delta H°_{lattice} = ?$$

Born-Haber Cycle for Production of NaCl from Na(s) and Cl₂(g)

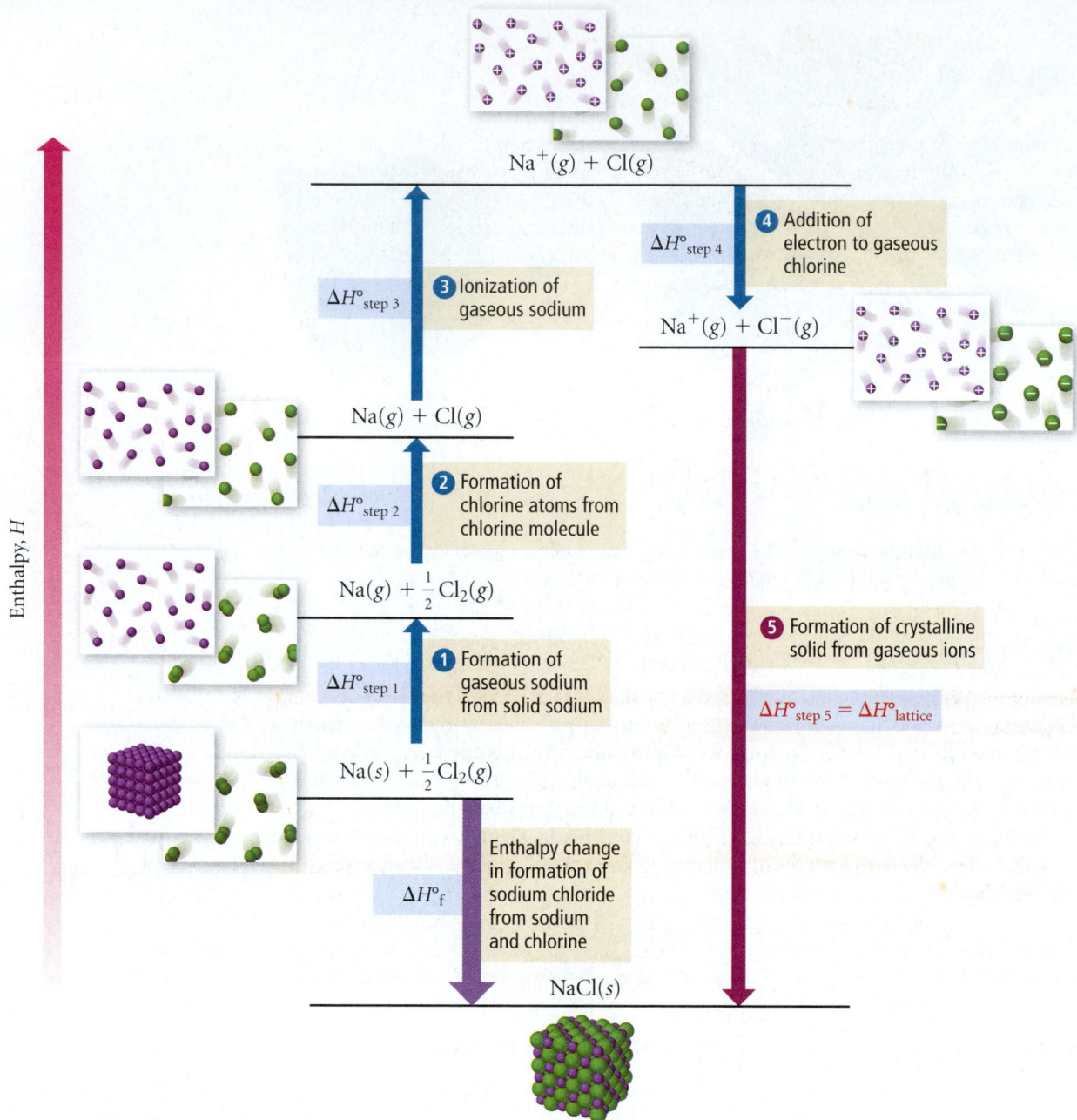

▲ FIGURE 9.6 **Born–Haber Cycle for Sodium Chloride**   The sum of the steps is the formation of NaCl from elemental Na and Cl. The enthalpy change of the last step is the lattice energy.

The entire Born–Haber cycle for NaCl is shown in Figure 9.6 ▲.

Since the overall reaction obtained by summing the steps in the Born–Haber cycle is equivalent to the formation of NaCl from its constituent elements, we can use Hess's law to set the overall enthalpy of formation for NaCl(s) equal to the sum of the steps in the Born–Haber cycle:

Lattice energy

$$\Delta H_f^\circ = \Delta H_{\text{step 1}}^\circ + \Delta H_{\text{step 2}}^\circ + \Delta H_{\text{step 3}}^\circ + \Delta H_{\text{step 4}}^\circ + \Delta H_{\text{step 5}}^\circ$$

We then solve this equation for $\Delta H^\circ_{step\,5}$, which is $\Delta H^\circ_{lattice}$, and substitute the appropriate values to calculate the lattice energy.

$$\Delta H^\circ_{lattice} = \Delta H^\circ_{step\,5} = \Delta H^\circ_f - (\Delta H^\circ_{step\,1} + \Delta H^\circ_{step\,2} + \Delta H^\circ_{step\,3} + \Delta H^\circ_{step\,4})$$
$$= -411\,kJ - (+108\,kJ + 122\,kJ + 496\,kJ - 349\,kJ)$$
$$= -788\,kJ$$

Notice that the value of the lattice energy is a large negative number. The formation of the crystalline NaCl lattice from sodium cations and chloride anions is highly exothermic and more than compensates for the endothermicity of the electron transfer process. In other words, the formation of ionic compounds is not exothermic because sodium "wants" to lose electrons and chlorine "wants" to gain them; rather, it is exothermic because of the large amount of heat released when sodium and chlorine ions coalesce to form a crystalline lattice.

Trends in Lattice Energies: Ion Size

Consider the lattice energies of the following alkali metal chlorides:

Metal Chloride	Lattice Energy (kJ/mol)
LiCl	−834
NaCl	−787
KCl	−701
CsCl	−657

Why do you suppose that the magnitude of the lattice energy decreases as you move down the column? We know from the periodic trends discussed in Chapter 8 that ionic radius increases as we move down a column in the periodic table (see Section 8.7). We also know, from our discussion of Coulomb's law in Section 9.2, that the potential energy of oppositely charged ions becomes less negative (or more positive) as the distance between the ions increases. As the size of the alkali metal ions increases down the column, so necessarily does the distance between the metal cations and the chloride anions. Therefore, the magnitude of the lattice energy of the chlorides decreases accordingly, making the formation of the chlorides less exothermic and the compounds less stable. In other words, as the ionic radii increase as you move down the column, the ions cannot get as close to each other and therefore do not release as much energy when the lattice forms.

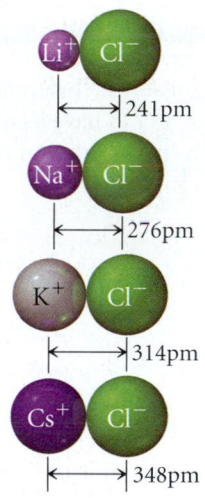

241pm

276pm

314pm

348pm

▲ Bond lengths of the group 1A metal chlorides.

Trends in Lattice Energies: Ion Charge

Consider the lattice energies of the following two compounds:

Compound	Lattice Energy (kJ/mol)
NaF	−910
CaO	−3414

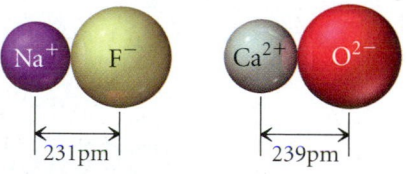

231pm   239pm

Why is the magnitude of the lattice energy of CaO so much greater than the lattice energy of NaF? Na^+ has a radius of 95 pm and F^- has a radius of 136 pm, resulting in a distance between ions of 231 pm. Ca^{2+} has a radius of 99 pm and O^{2-} has a radius of 140 pm, resulting in a distance between ions of 239 pm. Even though the separation between the calcium and oxygen is slightly greater (which would tend to lower the lattice energy), the lattice energy is almost four times greater. The explanation lies in the charges of the ions. Recall from the coulombic equation that the magnitude of the potential energy of two interacting charges depends not only on the distance between the charges, but also on the product of the charges:

$$E = \frac{1}{4\pi\epsilon_0}\frac{q_1 q_2}{r}$$

For NaF, E is proportional to $(1+)(1-) = 1-$, while for CaO, E is proportional to $(2+)(2-) = 4-$, so the relative stabilization for CaO relative to NaF should be roughly four times greater, as observed in the lattice energy.

Summarizing trends in lattice energies:

➤ *Lattice energies become less exothermic (less negative) with increasing ionic radius.*

➤ *Lattice energies become more exothermic (more negative) with increasing magnitude of ionic charge.*

EXAMPLE 9.2 Predicting Relative Lattice Energies

Arrange the following ionic compounds in order of increasing *magnitude* of lattice energy: CaO, KBr, KCl, SrO.

Solution

KBr and KCl should have lattice energies of smaller magnitude than CaO and SrO because of their lower ionic charges (1+, 1− compared to 2+, 2−). Between KBr and KCl, we expect KBr to have a lattice energy of lower magnitude due to the larger ionic radius of the bromide ion relative to the chloride ion. Between CaO and SrO, we expect SrO to have a lattice energy of lower magnitude due to the larger ionic radius of the strontium ion relative to the calcium ion.

Order of increasing *magnitude* of lattice energy:

$$KBr < KCl < SrO < CaO$$

Actual lattice energy values:

Compound	Lattice Energy (kJ/mol)
KBr	−671
KCl	−701
SrO	−3217
CaO	−3414

For Practice 9.2

Arrange the following in order of increasing magnitude of lattice energy: LiBr, KI, and CaO.

For More Practice 9.2

Which compound has the lattice energy of greatest magnitude, NaCl or $MgCl_2$?

Ionic Bonding: Models and Reality

In this section, we developed a model for ionic bonding. The value of a model is in how well it accounts for what we see in experiments. The ionic bonding model explains the properties of ionic compounds, including their high melting and boiling points, their tendency *not to conduct* electricity as solids, and their tendency *to conduct* electricity when dissolved in water.

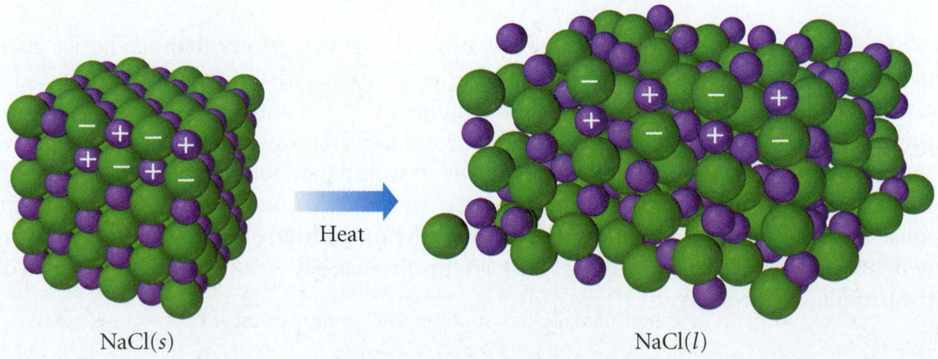

➤ The melting of solid ionic compounds such as sodium chloride requires enough heat to overcome the electrical forces holding the anions and cations together in a lattice. Thus, the melting points of ionic compounds are relatively high.

NaCl(*s*)

Heat

NaCl(*l*)

We modeled ionic solids as a lattice of individual ions held together by coulombic forces which are equal in all directions. To melt the solid, these forces must be overcome, which requires a significant amount of heat. Therefore, our model accounts for the high melting points of ionic solids. In our model, electrons are transferred from the metal to the nonmetal, but the transferred electrons remain localized on one atom. In other words, our model does not include any free electrons that might conduct electricity, and the ions themselves are fixed in place; therefore, our model accounts for the nonconductivity of ionic solids. When our idealized ionic solid dissolves in water, however, the cations and anions dissociate, forming free ions in solution. These ions can move in response to electrical forces, creating an electrical current. Thus, our model predicts that solutions of ionic compounds conduct electricity.

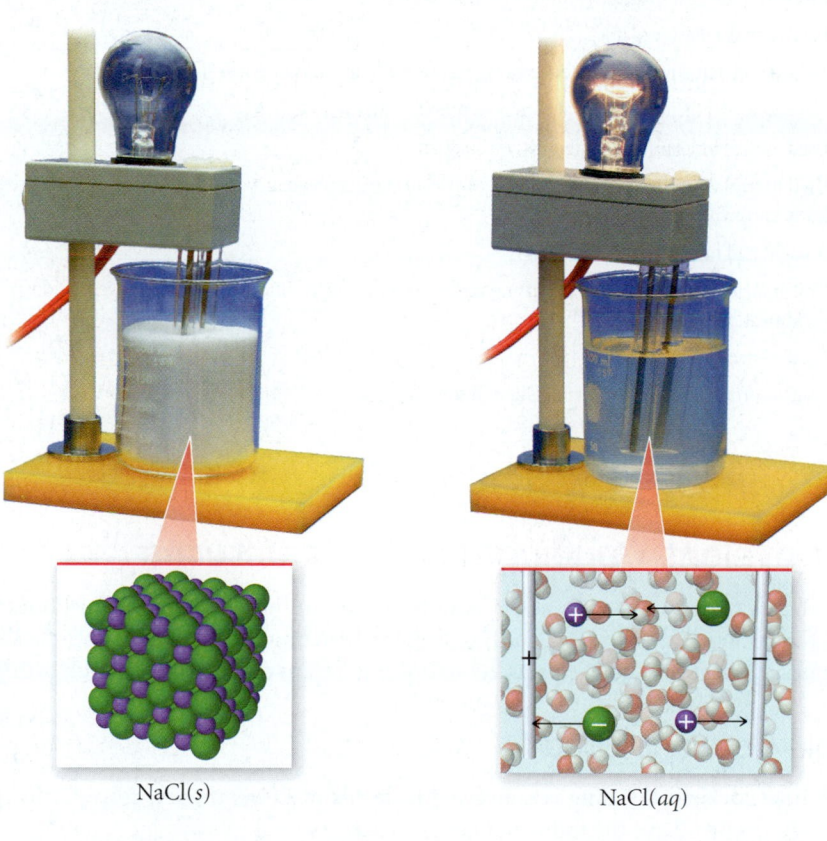

NaCl(s)

NaCl(aq)

▲ Solid sodium chloride does not conduct electricity.

▲ When sodium chloride dissolves in water, the resulting solution contains mobile ions that can create an electric current.

✦ Conceptual Connection 9.1 Melting Points of Ionic Solids

Use the ionic bonding model to determine which has the higher melting point, NaCl or MgO. Explain the relative ordering.

Answer: We would expect MgO to have the higher melting point because, in our bonding model, the magnesium and oxygen ions are held together in a crystalline lattice by charges of 2+ for magnesium and 2− for oxygen. In contrast, the NaCl lattice is held together by charges of 1+ for sodium and 1− for chlorine. The experimentally measured melting points of these compounds are 801 °C for NaCl and 2852 °C for MgO, in accordance with our model.

Chemistry and Medicine
Ionic Compounds in Medicine

Although most drugs are molecular compounds, a number of ionic compounds have medical uses. Consider the following partial list of ionic compounds used in medicine. Notice that many of these compounds contain polyatomic ions. The bonding between a metal and a polyatomic ion is ionic. However, the bonding within a polyatomic ion is covalent, the topic of our next section.

Formula	Name	Medical Use
$AgNO_3$	Silver nitrate	Topical anti-infective agent; in solution, silver nitrate is used to treat and prevent eye infection, especially in newborn infants
$BaSO_4$	Barium sulfate	Given as a contrast medium— or image enhancer—in X-rays
$CaSO_4$	Calcium sulfate	Used to make plaster casts
$KMnO_4$	Potassium permanganate	Topical anti-infective agent; often used to treat fungal infections on the feet
KI	Potassium iodide	Antiseptic and disinfectant; given orally to prevent radiation sickness
Li_2CO_3	Lithium carbonate	Used to treat bipolar (manic-depressive) disorders
$MgSO_4$	Magnesium sulfate	Used to treat eclampsia (a condition that can occur during pregnancy in which elevated blood pressure leads to convulsions)
$Mg(OH)_2$	Magnesium hydroxide	Antacid and mild laxative
$NaHCO_3$	Sodium bicarbonate	Oral antacid used to treat heartburn and acid stomach; injected into blood to treat severe acidosis (acidification of the blood)
NaF	Sodium fluoride	Used to strengthen teeth
ZnO	Zinc oxide	Used as protection from ultraviolet light in sun

9.5 Covalent Bonding: Lewis Structures

Lewis theory provides us with a very simple and useful model for covalent bonding. In Lewis theory, we represent covalent bonding by depicting neighboring atoms as sharing some (or all) of their valence electrons in order to attain octets (or duets for hydrogen).

Single Covalent Bonds

To see how covalent bonding is conceived in terms of Lewis theory, consider hydrogen and oxygen, which have the following Lewis structures:

$$H\cdot \qquad \cdot \ddot{O}:$$

In water, hydrogen and oxygen share their unpaired valence electrons so that each hydrogen atom gets a duet and the oxygen atom gets an octet:

$$H:\ddot{O}:H$$

The shared electrons—those that appear in the space between the two atoms—count towards the octets (or duets) of *both of the atoms*.

$$H\,\ddot{O}\,H$$
Duet   Octet   Duet

A pair of electrons that is shared between two atoms is called a **bonding pair**, while a pair that is associated with only one atom—and therefore not involved in bonding—is called a **lone pair**.

Sometimes lone pair electrons are also called nonbonding electrons.

Bonding pair
$$H:\ddot{O}:H$$
Lone pair

Bonding pair electrons are often represented by dashes to emphasize that they constitute a chemical bond.

Keep in mind that *one* dash always stands for *two* electrons (a bonding pair).

$$H—\ddot{O}—H$$

Lewis theory also explains why the halogens form diatomic molecules. Consider the Lewis structure of chlorine:

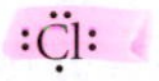

$$:\ddot{\underset{..}{C}l}:$$

If two Cl atoms pair together, they can each get an octet:

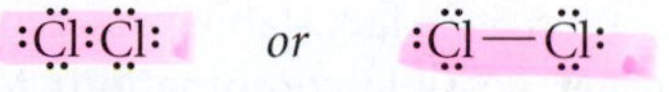

$$:\ddot{\underset{..}{C}l}:\ddot{\underset{..}{C}l}: \quad or \quad :\ddot{\underset{..}{C}l}—\ddot{\underset{..}{C}l}:$$

When we examine elemental chlorine, it indeed exists as a diatomic molecule, just as Lewis theory predicts. The same is true for the other halogens.

Similarly, Lewis theory predicts that hydrogen, which has the Lewis structure

$$H·$$

should exist as H_2. When two hydrogen atoms share their valence electrons, each gets a duet, a stable configuration for hydrogen.

$$H:H \quad or \quad H—H$$

Again, Lewis theory is correct. In nature, elemental hydrogen exists as H_2 molecules.

Double and Triple Covalent Bonds

In Lewis theory, two atoms may share more than one electron pair to get octets. For example, if we pair two oxygen atoms together, they must share two electron pairs in order for each oxygen atom to have an octet. Each oxygen atom now has an octet because the additional bonding pair counts toward the octet of both oxygen atoms.

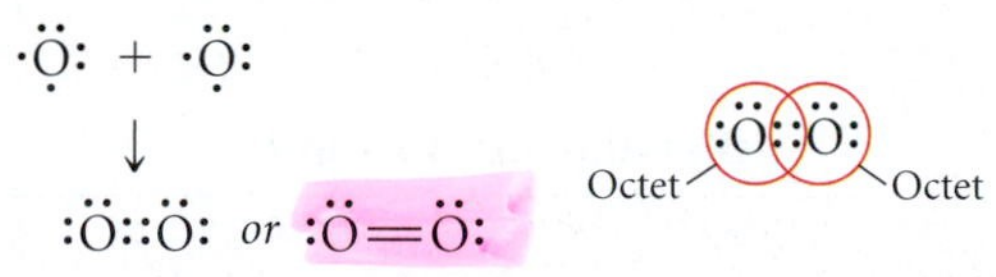

$$·\ddot{O}: + ·\ddot{O}:$$
$$\downarrow$$
$$:\ddot{O}::\ddot{O}: \quad or \quad :\ddot{O}=\ddot{O}:$$

Octet ⟨:Ö::Ö:⟩ Octet

When two electron pairs are shared between two atoms, the resulting bond is a **double bond**. In general, double bonds are shorter and stronger than single bonds. Atoms can also share three electron pairs. Consider the Lewis structure of N_2. Since each N atom has five valence electrons, the Lewis structure for N_2 has 10 electrons. Both nitrogen atoms can attain octets only by sharing three electron pairs:

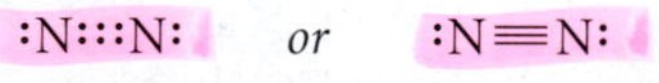

$$:N:::N: \quad or \quad :N\equiv N:$$

We will explore the characteristics of multiple bonds more fully in Section 9.10.

The bond is called a **triple bond**. Triple bonds are even shorter and stronger than double bonds. When we examine nitrogen in nature, we find that it indeed exists as a diatomic molecule with a very strong bond between the two nitrogen atoms. The bond is so strong, in fact, that it is difficult to break, making N_2 a relatively unreactive molecule.

Covalent Bonding: Models and Reality

Lewis theory predicts the properties of molecular compounds in many ways. First, it shows us why particular combinations of atoms form molecules and others do not. For example, why is water H_2O and not H_3O ? We can write a good Lewis structure for H_2O, but not for H_3O.

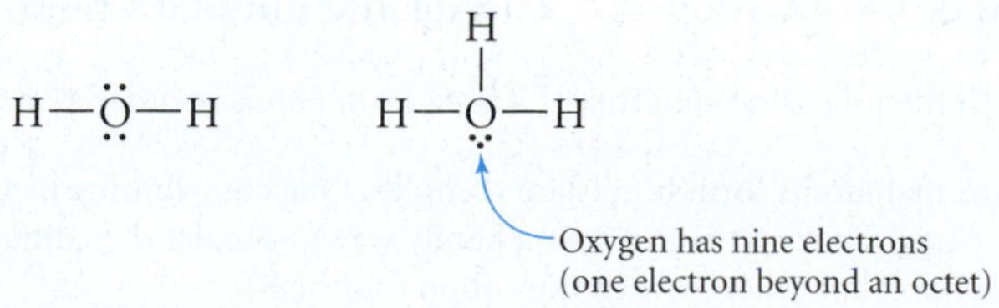

$$H—\ddot{O}—H \qquad H—\underset{|}{\overset{H}{O}}—H$$

Oxygen has nine electrons
(one electron beyond an octet)

So Lewis theory predicts that H_2O should be stable, while H_3O should not be, and that is in fact the case. However, if we remove an electron from H_3O, we get H_3O^+, which should be stable (according to Lewis theory) because, by removing the extra electron, oxygen gets an octet.

$$\left[\begin{array}{c} H \\ | \\ H-\overset{..}{O}-H \end{array} \right]^+$$

This ion, called the hydronium ion, is in fact stable in aqueous solutions (see Section 4.8). Lewis theory also predicts other possible combinations for hydrogen and oxygen. For example, we can write a Lewis structure for H_2O_2 as follows:

$$H-\overset{..}{\underset{..}{O}}-\overset{..}{\underset{..}{O}}-H$$

Indeed, H_2O_2, or hydrogen peroxide, exists and is often used as a disinfectant and a bleach.

Lewis theory also shows that covalent bonds are highly *directional*. The attraction between two covalently bonded atoms is due to the sharing of one or more electron pairs in the space between them. Thus, each bond links just one specific pair of atoms—*in contrast to ionic bonds, which are nondirectional and hold together the entire array of ions*. As a result, the fundamental units of covalently bonded compounds are individual molecules. These molecules can interact with one another in a number of different ways that we cover in Chapter 11. However, the interactions *between* molecules (intermolecular forces) are generally much weaker than the bonding interactions within a molecule (intramolecular forces), as shown in Figure 9.7 ▼. When a molecular compound melts or boils, the molecules themselves remain intact—only the relatively weak interactions between molecules must be overcome. Consequently, molecular compounds tend to have lower melting and boiling points than ionic compounds.

Molecular Compound

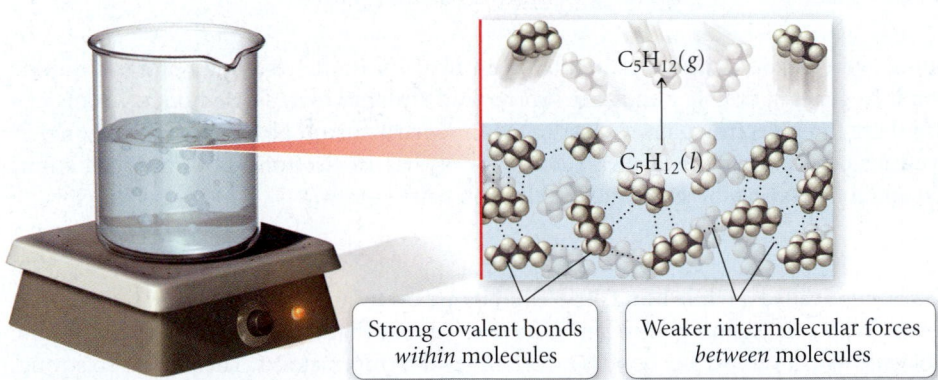

Strong covalent bonds *within* molecules

Weaker intermolecular forces *between* molecules

▲ **FIGURE 9.7 Intermolecular and Intramolecular Forces** The covalent bonds between atoms of a molecule are much stronger than the interactions between molecules. To boil a molecular substance, you simply have to overcome the relatively weak intermolecular forces, so molecular compounds generally have low boiling points.

◆ Conceptual Connection 9.2 Energy and the Octet Rule

What is wrong with the following statement? *Atoms form bonds in order to satisfy the octet rule.*

Answer: The reasons that atoms form bonds are complex. One contributing factor is the lowering of their potential energy. The octet rule is just a handy way to predict the combinations of atoms that will have a lower potential energy when they bond together.

9.6 Electronegativity and Bond Polarity

We know from Chapter 7 that representing electrons with dots, as we do in Lewis theory, is a drastic oversimplification. As we have already discussed, this does not invalidate Lewis theory—which is an extremely useful theory—but we must recognize and compensate for its inherent limitations. One limitation of representing electrons as dots, and covalent bonds as two dots shared between two atoms, is that the shared electrons always appear to be *equally* shared. Such is not the case. For example, consider the Lewis structure of hydrogen fluoride.

$$H : \ddot{\ddot{F}} :$$

The two shared electron dots sitting between the H and the F atoms appear to be equally shared between hydrogen and fluorine. However, based on laboratory measurements, we know they are not. When HF is put in an electric field, the molecules orient as shown in Figure 9.8 ▼. From this observation, we know that the hydrogen side of the molecule must have a slight positive charge and the fluorine side of the molecule must have a slight negative charge. We represent this as follows:

$$\overset{+\longrightarrow}{H-F} \quad or \quad \overset{\delta^+ \quad \delta^-}{H-F}$$

The arrow on the left, with a positive sign on the tail, shows that the left side of the molecule has a partial positive charge and that the right side of the molecule (the side the arrow is pointing *toward*) has a partial negative charge. Similarly, the $\delta+$ (delta plus) represents a partial positive charge and the $\delta-$ (delta minus) represents a partial negative charge. Does this make the bond ionic? No. In an ionic bond, the electron is essentially *transferred* from one atom to another. In HF, it is simply *unequally shared*. In other words, even though the Lewis structure of HF portrays the bonding electrons as residing *between* the two atoms, in reality the electron density is greater on the fluorine atom than on the hydrogen atom (Figure 9.9 ▼). The bond is said to be *polar*—having a positive pole and a negative pole. A **polar covalent bond** is intermediate in nature between a pure covalent bond and an ionic bond. In fact, the categories of pure covalent and ionic are really two extremes within a broad continuum. Most covalent bonds between dissimilar atoms are actually *polar covalent*, somewhere between the two extremes.

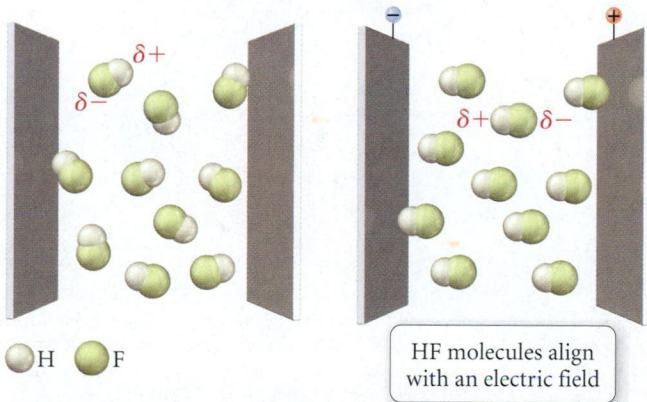

H ⬤ F

HF molecules align with an electric field

▲ **FIGURE 9.8 Orientation of Gaseous Hydrogen Fluoride in an Electric Field** Because one side of the HF molecule has a slight positive charge and the other side a slight negative charge, the molecules will align themselves with an external electric field.

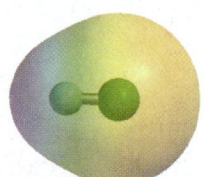

▲ **FIGURE 9.9 Electron Density Plot for the HF Molecule** The F end of the molecule, with its partial negative charge, is pink; the H end, with its partial positive charge, is blue.

Electronegativity

The ability of an atom to attract electrons to itself in a chemical bond (which results in polar bonds) is called **electronegativity**. We say that fluorine is more *electronegative* than hydrogen because it takes a greater share of the electron density in HF.

Electronegativity was quantified by the American chemist Linus Pauling in his classic book, *The Nature of the Chemical Bond*. Pauling compared the bond energy—the energy required to break a bond—of a heteronuclear diatomic molecule such as HF with the bond

Notice the difference between *electronegativity* (the ability of an atom to attract electrons to itself in a covalent bond) and *electron affinity* (the energy associated with the addition of an electron to a gas phase atom). The two terms are related but not identical.

We cover the concept of bond energy in more detail in Section 9.10.

Pauling's "average" bond energy was actually calculated a little bit differently than the normal average shown here. He took the square root of the product of the bond energies of the homologs as the "average."

energies of its homonuclear counterparts, in this case H_2 and F_2. The bond energies of H_2 and F_2 are 436 kJ/mol and 155 kJ/mol, respectively. Pauling reasoned that if the HF bond were purely covalent—that is, if the electrons were shared exactly equally—the bond energy of HF should simply be an average of the bond energies of H_2 and F_2, which would then be 296 kJ/mol. However, the bond energy of HF is experimentally measured to be 565 kJ/mol. Pauling suggested that the additional bond energy was due to the *ionic character* of the bond. Based on many such comparisons of bond energies, and by arbitrarily assigning an electronegativity of 4.0 to fluorine (the most electronegative element on the periodic table), Pauling developed the electronegativity values shown in Figure 9.10 ▼.

For main-group elements, notice the following periodic trends in electronegativity from Figure 9.10:

- Electronegativity generally increases across a period in the periodic table.
- Electronegativity generally decreases down a column in the periodic table.
- Fluorine is the most electronegative element.
- Francium is the least electronegative element (sometimes called the most *electropositive*).

The periodic trends in electronegativity are consistent with other periodic trends we have seen. In general, electronegativity is inversely related to atomic size—the larger the atom, the less ability it has to attract electrons to itself in a chemical bond.

Trends in Electronegativity

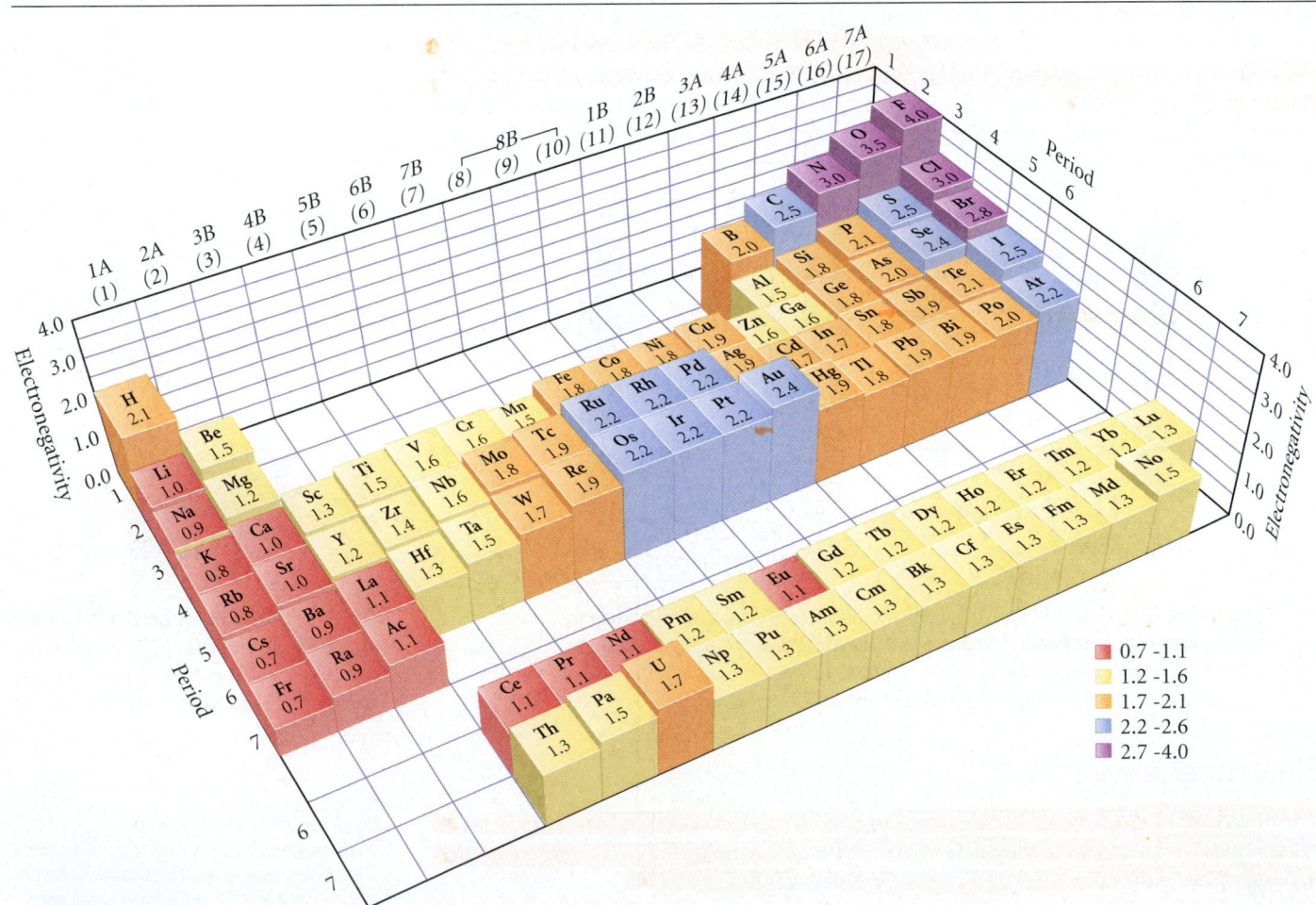

▲ **FIGURE 9.10 Electronegativities of the Elements**   Electronegativity generally increases as you move across a row in the periodic table and decreases as you move down a column.

Bond Polarity, Dipole Moment, and Percent Ionic Character

The degree of polarity in a chemical bond depends on the electronegativity difference (sometimes abbreviated ΔEN) between the two bonding elements. The greater the electronegativity difference, the more polar the bond. If two elements with identical electronegativities form a covalent bond, they share the electrons equally, and the bond is purely covalent or *nonpolar*. For example, the chlorine molecule, composed of two chlorine atoms (which necessarily have identical electronegativities), has a covalent bond in which electrons are evenly shared.

If there is a large electronegativity difference between the two elements in a bond, such as normally occurs between a metal and a nonmetal, the electron from the metal is almost completely transferred to the nonmetal, and the bond is ionic. For example, sodium and chlorine form an ionic bond.

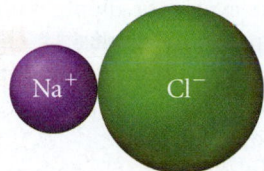

If there is an intermediate electronegativity difference between the two elements, such as between two different nonmetals, then the bond is polar covalent. For example, HCl has a polar covalent bond.

While all attempts to divide the bond polarity continuum into specific regions are necessarily arbitrary, it is helpful to classify bonds as covalent, polar covalent, and ionic, based on the electronegativity difference between the bonding atoms as shown in Table 9.1 and Figure 9.11 ▼.

TABLE 9.1   The Effect of Electronegativity Difference on Bond Type

Electronegativity Difference (ΔEN)	Bond Type	Example
Small (0–0.4)	Covalent	Cl_2
Intermediate (0.4–2.0)	Polar covalent	HCl
Large (2.0+)	Ionic	NaCl

The Continuum of Bond Types

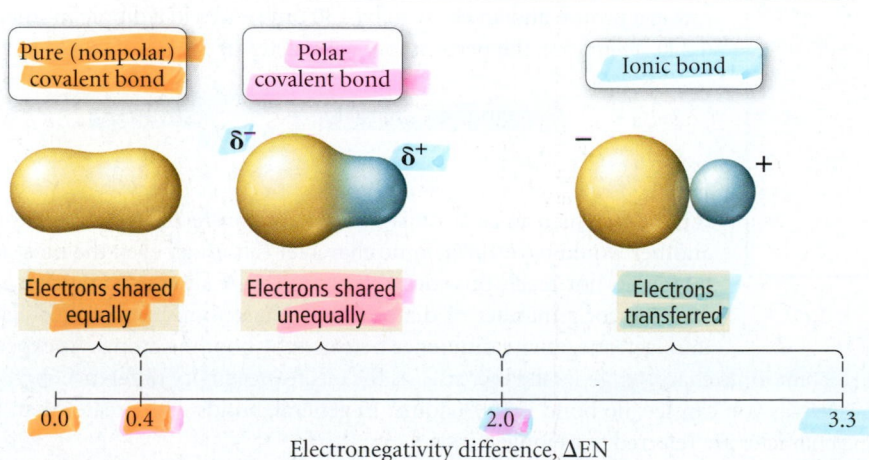

We can quantify the polarity of a bond by the size of its dipole moment. A **dipole moment (μ)** occurs anytime there is a separation of positive and negative charge. The magnitude of a dipole moment created by separating two particles of equal but opposite charges of magnitude q by a distance r is given by the following equation:

$$\mu = qr \qquad\qquad\qquad [9.2]$$

We can get a sense for the dipole moment of a completely ionic bond by calculating the dipole moment that results from separating a proton and an electron ($q = 1.6 \times 10^{-19}$ C) by a distance of $r = 130$ pm (the approximate length of a short chemical bond)

$$
\begin{aligned}
\mu &= qr \\
&= (1.6 \times 10^{-19} \text{ C})(130 \times 10^{-12} \text{ m}) \\
&= 2.1 \times 10^{-29} \text{ C} \cdot \text{m} \\
&= 6.2 \text{ D}
\end{aligned}
$$

The debye (D) is a common unit used for reporting dipole moments (1 D = 3.34 × 10^{-30} C · m). We would therefore expect the dipole moment of completely ionic bonds with bond lengths close to 130 pm to be about 6 D. The smaller the magnitude of the charge separation, and the smaller the distance the charges are separated by, the smaller the dipole moment. Table 9.2 shows the dipole moments of several molecules along with the electronegativity differences of their atoms.

TABLE 9.2 Dipole Moments of Several Molecules in the Gas Phase

Molecule	ΔEN	Dipole Moment (D)
Cl_2	0	0
ClF	1.0	0.88
HF	1.9	1.82
LiF	3.0	6.33

By comparing the *actual* dipole moment of a bond to what the dipole moment would be if the electron were completely transferred from one atom to the other, we can get a sense of the degree to which the electron is transferred (or the degree to which the bond is ionic). A quantity called the **percent ionic character** is defined as the ratio of a bond's actual dipole moment to the dipole moment it would have if the electron were completely transferred from one atom to the other, multiplied by 100%:

$$\text{Percent ionic character} = \frac{\text{measured dipole moment of bond}}{\text{dipole moment if electron were completely transferred}} \times 100\%$$

For example, suppose a diatomic molecule with a bond length of 130 pm has a dipole moment of 3.5 D. We previously calculated that separating a proton and an electron by 130 pm results in a dipole moment of 6.2 D. Therefore, the percent ionic character of the bond would be:

$$
\begin{aligned}
\text{Percent ionic character} &= \frac{3.5 \text{ D}}{6.2 \text{ D}} \times 100\% \\
&= 56\%
\end{aligned}
$$

A bond in which an electron is completely transferred from one atom to another would have 100% ionic character (although even the most ionic bonds do not reach this ideal). Figure 9.12 ◄ shows the percent ionic character of a number of diatomic gas-phase molecules plotted against the electronegativity difference between the bonding atoms. As expected, the percent ionic character generally increases as the electronegativity difference increases. However, as you can see, no bond is 100% ionic. In general, bonds with greater than 50% ionic character are referred to as ionic bonds.

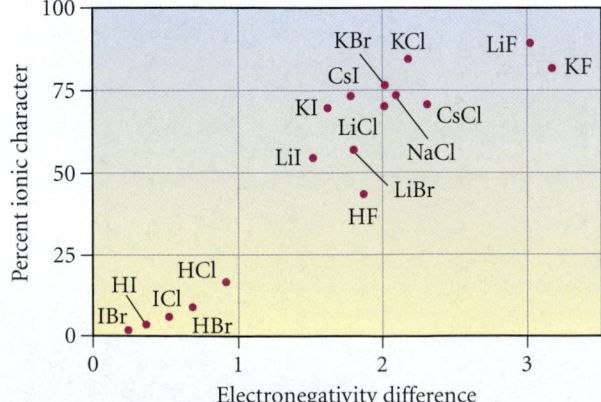

▲ FIGURE 9.12 Percent Ionic Character versus Electronegativity Difference for Some Ionic Compounds

EXAMPLE 9.3 Classifying Bonds as Pure Covalent, Polar Covalent, or Ionic

Determine whether the bond formed between each of the following pairs of atoms is covalent, polar covalent, or ionic.

(a) Sr and F   (b) N and Cl   (c) N and O

Solution

(a) From Figure 9.10, we find the electronegativity of Sr (1.0) and of F (4.0).
The electronegativity difference (ΔEN) is ΔEN = 4.0 − 1.0 = 3.0.
Using Table 9.1, we classify this bond as ionic.

(b) From Figure 9.10, we find the electronegativity of N (3.0) and of Cl (3.0).
The electronegativity difference (ΔEN) is ΔEN = 3.0 − 3.0 = 0.
Using Table 9.1, we classify this bond as covalent.

(c) From Figure 9.10, we find the electronegativity of N (3.0) and of O (3.5).
The electronegativity difference (ΔEN) is ΔEN = 3.5 − 3.0 = 0.5.
Using Table 9.1, we classify this bond as polar covalent.

For Practice 9.3

Determine whether the bond formed between each of the following pairs of atoms is pure covalent, polar covalent, or ionic.

(a) I and I   (b) Cs and Br   (c) P and O

We discuss the difference in properties between polar and nonpolar molecules in more detail in Chapter 11. Here we point out that a polar molecule, because of its dipole moment, interacts with an electric field. We saw previously that HF molecules will align with an external electric field. Another interaction can be seen by flowing a thin stream of water past a glass rod that has been rubbed to develop static charge, as shown in Figure 9.13(a) ▼. The stream of water is "bent" by the electric field around the charged rod. By contrast, a stream of a nonpolar liquid, such as hexane, is not bent by the charged rod, as shown in Figure 9.13(b).

The exact reason for the deflection of the polar liquid stream is beyond our scope. (Under certain circumstances, even nonpolar liquids can be deflected in this experiment.)

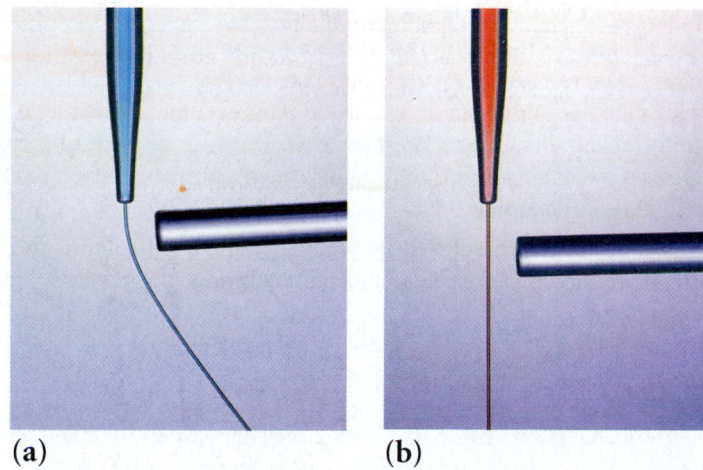

(a)   (b)

◀ **FIGURE 9.13 Polar and Nonpolar Molecules** (a) A glass rod charged with static electricity deflects a stream of water because H_2O is a highly polar molecule. (b) A stream of hexane (C_6H_{14}), which is nonpolar, is not deflected. (The water and hexane were colored blue and red, respectively, to make them more visible)

✸ Conceptual Connection 9.3 Percent Ionic Character

The HCl(g) molecule has a bond length of 127 pm and a dipole moment of 1.08 D. Without doing detailed calculations, determine which one of the following is the best estimate for its percent ionic character.

(a) 5%   (b) 15%   (c) 50%   (d) 80%

Answer: (b) We are given that the dipole moment of the HCl bond is about 1 D and that the bond length is 127 pm. Previously we calculated the dipole moment for a 130-pm bond that is 100% ionic to be about 6.2 D. We can therefore estimate the bond's ionic character as 1/6 × 100%, which is closest to 15%.

9.7 Lewis Structures of Molecular Compounds and Polyatomic Ions

We have now seen the basic ideas in Lewis theory and how they work to explain and predict chemical bonding in nature. We now turn to the basic sequence of steps involved in actually writing Lewis structures for given combinations of atoms.

Writing Lewis Structures for Molecular Compounds

To write a Lewis structure for a molecular compound, follow these steps:

Often, chemical formulas are written in a way that provides clues to how the atoms are bonded together. For example, CH_3OH indicates that three hydrogen atoms and the oxygen atom are bonded to the carbon atom, but the fourth hydrogen atom is bonded to the oxygen atom.

There are a few exceptions to this rule, such as diborane (B_2H_6), which contains *bridging hydrogens*, but these are rare and cannot be treated by simple Lewis theory.

1. **Write the correct skeletal structure for the molecule.** The Lewis structure of a molecule must have the atoms in the correct positions. For example, you could not write a Lewis structure for water if you started with the hydrogen atoms next to each other and the oxygen atom at the end (H H O). In nature, oxygen is the central atom and the hydrogen atoms are *terminal* (at the ends). The correct skeletal structure is H O H. The only way to determine the skeletal structure of a molecule with absolute certainty is to examine its structure experimentally. However, we can write likely skeletal structures by remembering two guidelines. First, *hydrogen atoms will always be terminal*. Hydrogen does not ordinarily occur as a central atom because central atoms must form at least two bonds, and hydrogen, which has only a single valence electron to share and requires only a duet, can form just one. Second, *put the more electronegative elements in terminal positions* and the less electronegative elements (other than hydrogen) in the central position. Later in this section, you will learn how to distinguish between competing skeletal structures based on the concept of formal charge.

2. **Calculate the total number of electrons for the Lewis structure by summing the valence electrons of each atom in the molecule.** Remember that the number of valence electrons for any main-group element is equal to its group number in the periodic table. *If you are writing a Lewis structure for a polyatomic ion, the charge of the ion must be considered when calculating the total number of electrons.* Add one electron for each negative charge and subtract one electron for each positive charge. Don't worry about which electron comes from which atom—only the total number is important.

3. **Distribute the electrons among the atoms, giving octets (or duets for hydrogen) to as many atoms as possible.** Begin by placing two electrons between every two atoms. These represent the minimum number of bonding electrons. Then distribute the remaining electrons as lone pairs, first to terminal atoms, and then to the central atom, giving octets (or duets for hydrogen) to as many atoms as possible.

4. **If any atoms lack an octet, form double or triple bonds as necessary to give them octets.** Do this by moving lone electron pairs from terminal atoms into the bonding region with the central atom.

Sometimes distributing all the remaining electrons to the central atom results in more than an octet. This is called an expanded octet and is covered in Section 9.9.

A brief version of the above procedure is shown below in the left column and two examples of applying it are shown in the center and right columns.

Procedure for Writing Lewis Structures for Covalent Compounds	**EXAMPLE 9.4** Writing Lewis Structures	**EXAMPLE 9.5** Writing Lewis Structures
	Write a Lewis Structure for CO_2.	Write a Lewis Structure for NH_3.
	Solution	**Solution**
1. Write the correct skeletal structure for the molecule.	Because carbon is the less electronegative atom, we put it in the central position. O C O	Since hydrogen is always terminal, we put nitrogen in the central position. H N H H
2. Calculate the total number of electrons for the Lewis structure by summing the valence electrons of each atom in the molecule.	Total number of electrons for Lewis structure = $$\begin{pmatrix} \text{number of} \\ \text{valence} \\ \text{e}^- \text{ for C} \end{pmatrix} + 2 \begin{pmatrix} \text{number of} \\ \text{valence} \\ \text{e}^- \text{ for O} \end{pmatrix}$$ $= 4 + 2(6) = 16$	Total number of electrons for Lewis structure = $$\begin{pmatrix} \text{number of} \\ \text{valence} \\ \text{e}^- \text{ for N} \end{pmatrix} + 3 \begin{pmatrix} \text{number of} \\ \text{valence} \\ \text{e}^- \text{ for H} \end{pmatrix}$$ $= 5 + 3(1) = 8$

3. Distribute the electrons among the atoms, giving octets (or duets for hydrogen) to as many atoms as possible. Begin with the bonding electrons, and then proceed to lone pairs on terminal atoms, and finally to lone pairs on the central atom.	Bonding electrons are first. O:C:O (4 of 16 electrons used) Lone pairs on terminal atoms are next. :Ö:C:Ö: (16 of 16 electrons used)	Bonding electrons are first. H:N:H :H (6 of 8 electrons used) Lone pairs on terminal atoms are next, but none are needed on hydrogen. Lone pairs on central atom are last. H—N̈—H 	 H (8 of 8 electrons used)
4. If any atom lacks an octet, form double or triple bonds as necessary to give them octets.	Move lone pairs from the oxygen atoms to bonding regions to form double bond 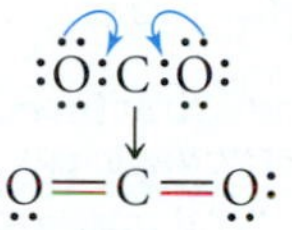 :Ö:C:Ö: ↓ :O=C=O: **For Practice 9.4** Write a Lewis structure for CO.	Since all of the atoms have octets (or duets for hydrogen), the Lewis structure for NH_3 is complete as shown above. **For Practice 9.5** Write a Lewis structure for H_2CO.

Writing Lewis Structures for Polyatomic Ions

We write Lewis structures for polyatomic ions by following the same procedure, but we pay special attention to the charge of the ion when calculating the number of electrons for the Lewis structure. We add one electron for each negative charge and subtract one electron for each positive charge. The Lewis structure for a polyatomic ion is usually written within brackets with the charge of the ion in the upper right-hand corner, outside the bracket.

EXAMPLE 9.6   Writing Lewis Structures for Polyatomic Ions

Write the Lewis structure for the NH_4^+ ion.

Solution

Begin by writing the skeletal structure. Since hydrogen is always terminal, put the nitrogen atom in the central position.	H H N H H
Calculate the total number of electrons for the Lewis structure by summing the number of valence electrons for each atom and subtracting 1 for the 1+ charge.	Total number of electrons for Lewis structure = (number of valence e^- in N) + (number of valence e^- in H) − 1 = 5 + 4(1) − 1 = 8 Subtract 1 e^- to account for 1^+ charge of ion.
Place two bonding electrons between every two atoms. Since all of the atoms have complete octets, no double bonds are necessary.	H H:N:H :H (8 of 8 electrons used)

Lastly, write the Lewis structure in brackets with the charge of the ion in the upper right-hand corner.

$$\left[\begin{array}{c} \text{H} \\ | \\ \text{H}-\text{N}-\text{H} \\ | \\ \text{H} \end{array}\right]^{+}$$

For Practice 9.6

Write a Lewis structure for the hypochlorite ion, ClO^-.

9.8 Resonance and Formal Charge

We need two additional concepts to write the best possible Lewis structures for a large number of compounds. The concepts are *resonance,* used when two or more valid Lewis structures can be drawn for the same compound, and *formal charge,* an electron book-keeping system that allows us to discriminate between alternative Lewis structures.

Resonance

When writing Lewis structures, we may find that, for some molecules, we can write more than one valid Lewis structure. For example, consider writing a Lewis structure for O_3. The following two Lewis structures, with the double bond on alternate sides, are equally correct:

$$:\ddot{O}=\ddot{O}-\ddot{O}: \qquad :\ddot{O}-\ddot{O}=\ddot{O}:$$

In cases such as this—where we can write two or more Lewis structures for the same molecule—we find that, in nature, the molecule exists as an *average* of the two Lewis structures. Any *one* of the two Lewis structures for O_3 would predict that O_3 should contain two different kinds of bonds (one double bond and one single bond). However, when we experimentally examine the structure of O_3, we find that both bonds are equivalent and intermediate in strength and length between a double bond and single bond. We account for this in Lewis theory by representing the molecule with both structures, called **resonance structures**, with a double-headed arrow between them:

$$:\ddot{O}=\ddot{O}-\ddot{O}: \longleftrightarrow :\ddot{O}-\ddot{O}=\ddot{O}:$$

The actual structure of the molecule is intermediate between the two resonance structures and is called a **resonance hybrid**. The term *hybrid* comes from breeding and means the offspring of two animals or plants of different varieties or breeds. If you breed a Labrador retriever with a German shepherd, you get a *hybrid* that is intermediate between the two breeds (Figure 9.14(a) ▼). Similarly, the structure of a resonance hybrid is intermediate between the two resonance structures (Figure 9.14(b)). However, the only structure that actually exists is the hybrid structure—the individual resonance structures do not exist and are merely a convenient way to describe the real structure.

◀ **FIGURE 9.14 Hybridization** Just as the offspring of two different dog breeds is a hybrid that is intermediate between the two breeds (**a**), the structure of a resonance hybrid is intermediate between that of the contributing resonance structures (**b**).

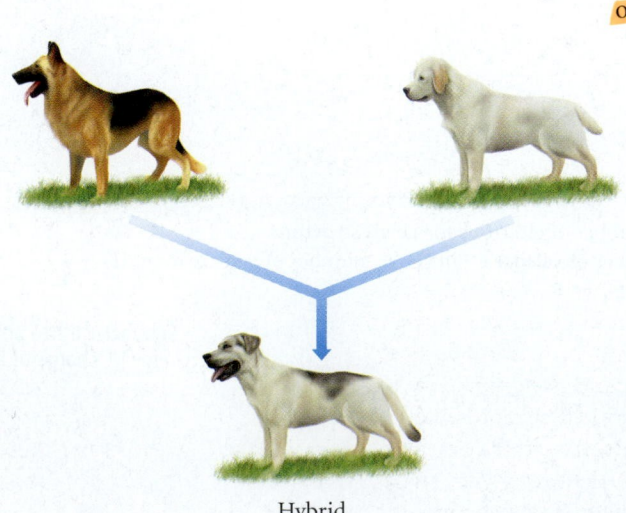

$$:\ddot{O}=\ddot{O}-\ddot{O}: \qquad\qquad :\ddot{O}-\ddot{O}=\ddot{O}:$$

Hybrid
(a)

$$O=\!\!=\!\!=O=\!\!=\!\!=O$$

Resonance hybrid structure
(b)

EXAMPLE 9.7 Writing Resonance Structures

Write a Lewis structure for the NO_3^- ion. Include resonance structures.

Solution

Begin by writing the skeletal structure. Since nitrogen is the least electronegative atom, put it in the central position.	 O O N O
Calculate the total number of electrons for the Lewis structure by summing the number of valence electrons for each atom and adding 1 for the 1− charge.	Total number of electrons for Lewis structure = (number of valence e⁻ in N) + 3 (number of valence e⁻ in O) + 1 = 5 + 3(6) + 1 = 24 Add 1 e⁻ to account for 1− charge of ion.
Place two bonding electrons between each pair of atoms.	 O O:N:O (6 of 24 electrons used)
Distribute the remaining electrons, first to terminal atoms. There are not enough electrons to complete the octet on the central atom.	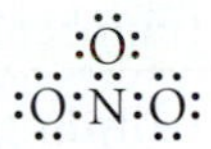 :Ö: :Ö:N:Ö: (24 of 24 electrons used)
Form a double bond by moving a lone pair from one of the oxygen atoms into the bonding region with nitrogen. Enclose the structure in brackets and include the charge.	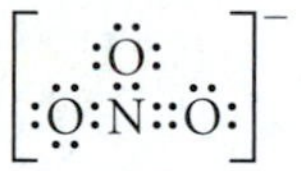 [:Ö: :Ö:N::Ö:]⁻ or [:Ö: :Ö—N=Ö:]⁻ 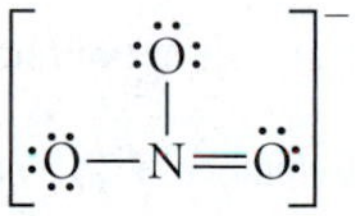
Since the double bond can equally well form with any of the three oxygen atoms, write all three structures as resonance structures.	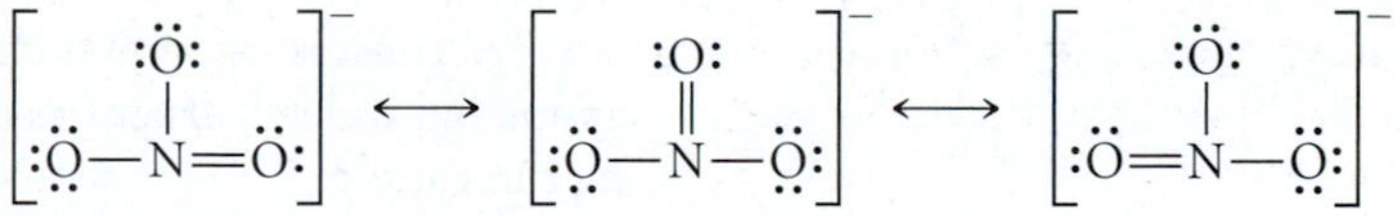 [:Ö: :Ö—N=Ö:]⁻ ⟷ [:O: :Ö—N—Ö:]⁻ ⟷ [:Ö: :Ö=N—Ö:]⁻

For Practice 9.7

Write a Lewis structure for the NO_2^- ion. Include resonance structures.

In the examples of resonance hybrids that we have examined so far, the contributing structures have been equivalent (or equally valid) Lewis structures. In these cases, the true structure is an equally weighted average of the resonance structures. In some cases, however, we can write resonance structures that are not equivalent. For reasons we cover below—such as formal charge, for example—one possible resonance structure may be somewhat better than another. In such cases, the true structure may still be represented as an average of the resonance structures, but with the better resonance structure contributing more to the true structure. In other words, multiple nonequivalent resonance structures may be weighted differently in their contributions to the true overall structure of a molecule (see Example 9.8).

Formal Charge

Formal charge is a fictitious charge assigned to each atom in a Lewis structure that helps us to distinguish among competing Lewis structures. The **formal charge** of an atom in a Lewis structure is *the charge it would have if all bonding electrons were shared equally between the bonded atoms*. In other words, formal charge is the computed charge for an atom if you completely ignore the effects of electronegativity. For example, we know that because fluorine is more electronegative than hydrogen, HF has a dipole moment—the hydrogen atom has a slight positive charge and the fluorine atom has a slight negative charge. However, the *formal charges* of hydrogen and fluorine in HF (the charges computed if we ignore their differences in electronegativity) are both zero.

$$H\!:\!\ddot{\underset{\cdot\cdot}{F}}\!:$$

Formal charge = 0   Formal charge = 0

Formal charge can be calculated simply by taking the number of valence electrons in the atom and subtracting the number of electrons that it "owns" in a Lewis structure. An atom in a Lewis structure can be thought of as "owning" all of its lone pair electrons and one-half of its bonding electrons.

Formal charge = number of valence electrons −
(number of lone pair electrons + $\frac{1}{2}$ number of bonding electrons)

So the formal charge of hydrogen in HF is computed as follows:

$$= 1 - \left[0 + \tfrac{1}{2}(2)\right] = 0$$

Number of valence   Number of electrons that H
electrons for H   "owns" in the Lewis structure

Similarly, the formal charge of fluorine in HF is computed as follows:

$$\text{Formal charge} = 7 - \left[6 + \tfrac{1}{2}(2)\right] = 0$$

Number of valence electrons for F   Number of electrons that F
"owns" in the Lewis structure

The concept of formal charge is useful because it can help us distinguish between competing skeletal structures or competing resonance structures. In general, the following rules apply:

1. The sum of all formal charges in a neutral molecule must be zero.
2. The sum of all formal charges in an ion must equal the charge of the ion.
3. Small (or zero) formal charges on individual atoms are better than large ones.
4. When formal charge cannot be avoided, negative formal charge should reside on the most electronegative atom.

Let's use formal charge to decide between the competing skeletal structures for hydrogen cyanide shown below. Notice that both skeletal structures equally satisfy the octet rule. The formal charge of each atom in the structure is computed below it.

	Structure A			Structure B		
	H —	C ≡	N:	H —	N ≡	C:
number of valence e⁻	1	4	5	1	5	4
− number of lone pair e⁻	−0	−0	−2	−0	−0	−2
− ½(number of bonding e⁻)	−½(2)	−½(8)	−½(6)	−½(2)	−½(8)	−½(6)
Formal charge	0	0	0	0	+1	−1

As required, the sum of the formal charges for each of these structures is zero (as it should be for neutral molecules). However, structure B has formal charges on both the N atom and the C atom, while structure A has no formal charges on any atom. Furthermore, in structure B, the negative formal charge is not on the most electronegative element (nitrogen is more electronegative than carbon). Consequently, structure A is the better Lewis structure. Since atoms in the middle of a molecule tend to have more bonding electrons and fewer lone pairs, they will also tend to have more positive formal charges. Consequently, the best skeletal structure will usually have the least electronegative atom in the central position, as we learned in step 1 of our procedure for writing Lewis structures.

Both HCN and HNC exist, but—as predicted by formal charge—HCN is more stable than HNC.

EXAMPLE 9.8 Assigning Formal Charges

Assign formal charges to each atom in the following resonance forms of the cyanate ion (OCN^-). Which resonance form is likely to contribute most to the correct structure of OCN^-?

A $[:\ddot{O}-C≡N:]^-$ B $[:\ddot{O}=C=\ddot{N}:]^-$ C $[:O≡C-\ddot{\ddot{N}}:]^-$

Solution

Calculate the formal charge on each atom by finding the number of valence electrons and subtracting the number of lone pair electrons and one-half the number of bonding electrons.

	A $[:\ddot{O}-C≡N:]^-$			B $[:\ddot{O}=C=\ddot{N}:]^-$			C $[:O≡C-\ddot{\ddot{N}}:]^-$		
Number of valence e$^-$	6	4	5	6	4	5	6	4	5
− number of lone pair e$^-$	−6	−0	−2	−4	−0	−4	−2	−0	−6
−$\frac{1}{2}$(number of bond e$^-$)	−1	−4	−3	−2	−4	−2	−3	−4	−1
Formal charge	−1	0	0	0	0	−1	+1	0	−2

The sum of all formal charges for each structure is −1, as it should be for a 1− ion. Structures A and B have the least amount of formal charge and are therefore to be preferred over structure C. Structure A is preferable to B because it has the negative formal charge on the more electronegative atom. We therefore expect structure A to make the biggest contribution to the resonance forms of the cyanate ion.

For Practice 9.8

Assign formal charges to each atom in the following resonance forms of N_2O. Which resonance form is likely to contribute most to the correct structure of N_2O?

A $:\ddot{N}=N=\ddot{O}:$ B $:N≡N-\ddot{O}:$ C $:\ddot{\ddot{N}}-N≡O:$

For More Practice 9.8

Assign formal charges to each of the atoms in the nitrate ion (NO_3^-). The Lewis structure for the nitrate ion is shown in Example 9.7.

9.9 Exceptions to the Octet Rule: Odd-Electron Species, Incomplete Octets, and Expanded Octets

The octet rule in Lewis theory has some exceptions that must be accommodated. They include (1) *odd-electron species*, molecules or ions with an odd number of electrons; (2) *incomplete octets*, molecules or ions with *fewer than eight electrons* around an atom; and (3) *expanded octets*, molecules or ions with *more than eight electrons* around an atom. We examine each of these exceptions individually.

Odd-Electron Species

The unpaired electron in nitrogen monoxide is put on the nitrogen rather than the oxygen in order to minimize formal charges.

Molecules and ions with an odd number of electrons in their Lewis structures are called **free radicals** (or simply *radicals*). For example, nitrogen monoxide—a pollutant found in motor vehicle exhaust—has 11 electrons. If we try to write a Lewis structure for nitrogen monoxide the best we can do is as follows:

$$:\dot{N}::\ddot{O}: \quad\quad \text{or} \quad\quad :\dot{N}=\ddot{O}:$$

The nitrogen atom does not have an octet, so this Lewis structure does not satisfy the octet rule. Yet, nitrogen monoxide exists, especially in polluted air. Why? As with any simple theory, Lewis theory is not sophisticated enough to model every single case. It is impossible to write good Lewis structures for free radicals, yet some of these molecules exist in nature. Perhaps it is a testament to Lewis theory, however, that *relatively few* such molecules exist and that, in general, they tend to be somewhat unstable and reactive. NO, for example, reacts with oxygen in the air to form NO_2, another odd-electron molecule represented with the following 17-electron resonance structures:

$$:\ddot{O}=\dot{N}-\ddot{O}: \longleftrightarrow :\ddot{O}-\dot{N}=\ddot{O}:$$

In turn, NO_2 reacts with water to form nitric acid (a component of acid rain) and also reacts with other atmospheric pollutants to form peroxyacetylnitrate (PAN), an active component of photochemical smog. For free radicals, such as NO and NO_2, we simply write the best Lewis structure that we can.

Chemistry and the Environment

Free Radicals and the Atmospheric Vacuum Cleaner

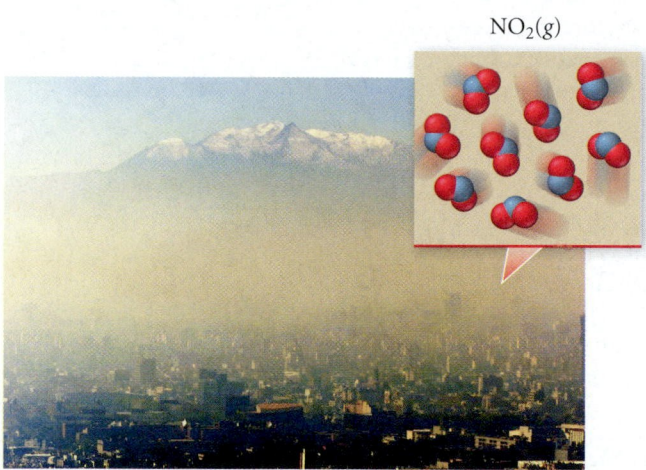

$NO_2(g)$

▲ $NO_2(g)$ is a pollutant found in urban air.

Free radicals play important roles in much of the chemistry of the atmosphere. The free radical that is most important to atmospheric reactions is the hydroxyl radical:

$$:\ddot{O}-H$$

Many free radical structures are abbreviated by writing a single dot with the formula. For example, the hydroxyl radical is often abbreviated as follows:

$$\cdot OH$$

In the atmosphere, the hydroxyl radical forms when excited oxygen atoms—formed from the photodecomposition of ozone—react with water vapor.

$$O_3 \xrightarrow{\text{UV light}} O_2 + O^\star$$
$$O^\star + H_2O \longrightarrow 2 \cdot OH$$

The * next to the O above simply means that the oxygen atom has excess energy.

The resulting hydroxyl radical is reactive toward a wide variety of molecules that are present in the atmosphere, either from natural sources or from air pollution. For example, the hydroxyl radical reacts with carbon monoxide, an atmospheric pollutant that we first encountered in Chapter 1 of this book, in the following two step-process:

$$CO + \cdot OH \longrightarrow HOCO\cdot$$
$$HOCO\cdot + O_2 \longrightarrow CO_2 + HOO\cdot$$

You can see from this reaction that the hydroxyl radical converts toxic CO into relatively nontoxic CO_2. The HOO· free radical generated by the second reaction is converted back into the hydroxyl radical by reactions with other atmospheric substances and the process repeats itself. Therefore, a single hydroxyl radical can convert a lot of CO into CO_2.

Do you ever wonder what happens to the hydrocarbons you accidentally spill when filling your car's gas tank or to the natural gas that is released into the atmosphere as you light your kitchen stove? Hydrocarbons released into the atmosphere are converted to CO_2 and H_2O in a series of steps initiated by the hydroxyl free radical. Consider the following representative reaction of methane, the main hydrocarbon in natural gas:

$$CH_4 + 5 O_2 + NO\cdot + 2 \cdot OH \xrightarrow{\text{UV light}}$$
$$CO_2 + H_2O + NO_2\cdot + 4 HOO\cdot$$

$C_8H_{18}(g)$

◀ Hydrocarbons such as octane evaporate into the atmosphere when a motor vehicle is fueled. What happens to them?

Notice the similarity between this reaction and the direct combustion (or burning) of methane:

$$CH_4 + 2\,O_2 \longrightarrow CO_2 + 2\,H_2O$$

As you can see, the free radical reaction initiates a slow "burning" of CH_4 in a series of steps that produce carbon dioxide and water and some additional free radicals. The hydroxyl radical initiates similar reactions with other pollutants as well as undesirable naturally occurring atmospheric gases. Without the hydroxyl free radical—sometimes called *the atmospheric vacuum cleaner*—our atmosphere would be a much dirtier place.

Question

Draw the best possible Lewis structures for the following free radicals important in atmospheric chemistry: NO, NO_2, HOO, OH, CH_3.

Incomplete Octets

Another significant exception to the octet rule involves those elements that tend to form *incomplete octets.* The most important such element is boron, which forms compounds with only six electrons around B, rather than eight. For example, BF_3 and BH_3 lack an octet for B.

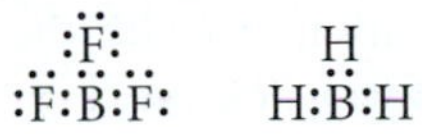

> Beryllium compounds, such as BeH_2, also have incomplete octets.

You might be wondering why we don't simply form double bonds to increase the number of electrons around B. For BH_3, of course, we can't, because there are no additional electrons to move into the bonding region. For BF_3, however, we could attempt to give B an octet by moving a lone pair from an F atom into the bonding region with B.

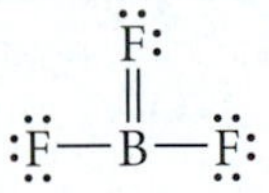

This Lewis structure has octets for all atoms, including boron. However, when we assign formal charges to this structure, we get the following:

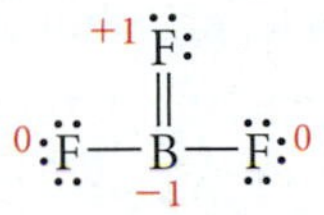

In this Lewis structure, fluorine—the most electronegative element in the periodic table—has a positive formal charge, making this an unfavorable structure. This leaves us with the following question: do we complete the octet on B at the expense of giving fluorine a positive formal charge? Or do we leave B without an octet in order to avoid the positive formal charge on fluorine? The answers to these kinds of questions are not always clear because we are pushing the limits of Lewis theory. In the case of boron, we usually accept the incomplete octet as the better Lewis structure. However, doing so does not rule out the possibility that the doubly bonded Lewis structure might be a minor contributing resonance structure. The ultimate answer to these kinds of issues must come from experiment. Experimental measurements of the B—F bond length in BF_3 suggest that the bond may be slightly shorter than expected for a single B—F bond, indicating that it may indeed have a small amount of double-bond character.

BF$_3$ can complete its octet in another way—via a chemical reaction. Lewis theory predicts that BF$_3$ might react in ways to complete its octet, and indeed it does. BF$_3$ reacts with NH$_3$ as follows:

When nitrogen bonds to boron, the nitrogen atom provides both of the electrons. This kind of bond is called a *coordinate covalent bond*, which we discuss further in Chapter 24.

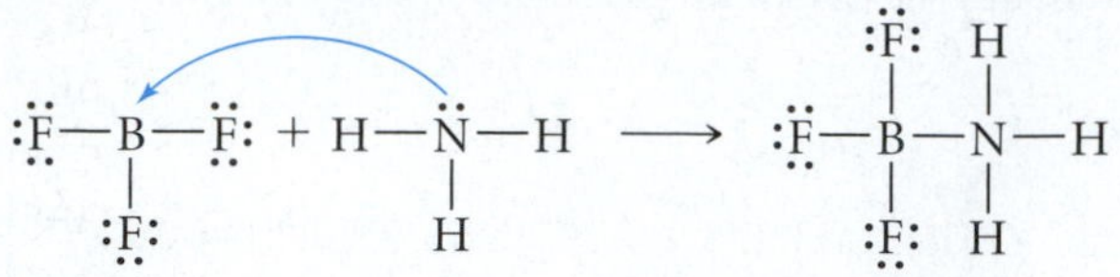

The product has complete octets for all atoms in the structure.

Expanded Octets

Elements in the third row of the periodic table and beyond often exhibit *expanded* octets of up to 12 (and occasionally 14) electrons. For example, consider the Lewis structures of arsenic pentafluoride and sulfur hexafluoride.

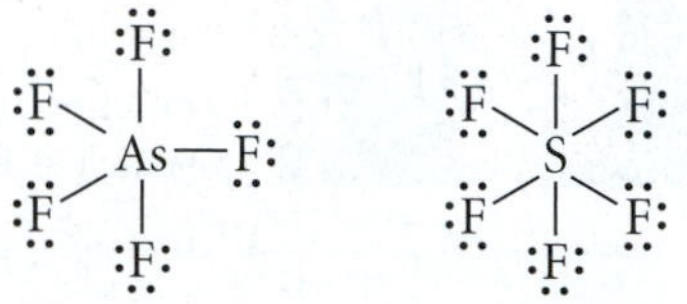

In AsF$_5$ arsenic has an expanded octet of 10 electrons, and in SF$_6$ sulfur has an expanded octet of 12 electrons. Both of these compounds exist and are stable. Ten- and twelve-electron expanded octets are common in third-period elements and beyond because the *d* orbitals in these elements are energetically accessible (they are not much higher in energy than the orbitals occupied by the valence electrons) and can accommodate the extra electrons (see Section 8.3). Expanded octets *never* occur in second-period elements.

In some Lewis structures, we are faced with the decision as to whether or not to expand an octet in order to lower formal charge. For example, consider the Lewis structure of H$_2$SO$_4$.

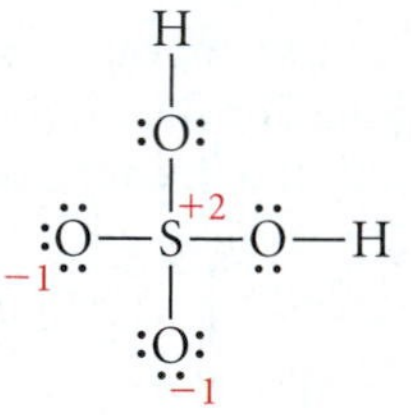

Notice that both of the oxygen atoms have a -1 formal charge and that sulfur has a $+2$ formal charge. While this amount of formal charge is acceptable, especially since the negative formal charge resides on the more electronegative atom, it is possible to eliminate the formal charge by expanding the octet on sulfur as follows:

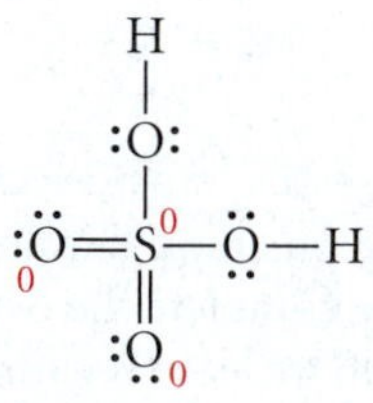

Which of these two Lewis structures for H$_2$SO$_4$ is better? Again, the answer is not straightforward. Experiments show that the sulfur–oxygen bond lengths in the two sulfur–oxygen bonds without the hydrogen atoms are shorter than expected for sulfur–oxygen single bonds, indicating that the double-bonded Lewis structure plays an important role in describing the bonding in H_2SO_4. In general, it is acceptable to expand octets in third-row (or beyond) elements in order to lower formal charge. However, you must *never* expand the octets of second-row elements. Second-row elements do not have energetically accessible *d* orbitals and never exhibit expanded octets.

EXAMPLE 9.9   Writing Lewis Structures for Compounds Having Expanded Octets

Write the Lewis structure for XeF_2.

Solution

Begin by writing the skeletal structure. Since xenon is the less electronegative atom, put it in the central position.	F Xe F
Calculate the total number of electrons for the Lewis structure by summing the number of valence electrons for each atom.	Total number of electrons for Lewis structure = (number of valence e⁻ in Xe) + 2(number of valence e⁻ in F) = 8 + 2(7) = 22
Place two bonding electrons between the atoms of each pair of atoms.	F:Xe:F (4 of 22 electrons used)
Distribute the remaining electrons to give octets to as many atoms as possible, beginning with terminal atoms and finishing with the central atom. Arrange additional electrons (beyond an octet) around the central atom, giving it an expanded octet of up to 12 electrons.	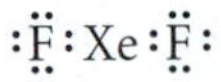 :F̈:Xe:F̈: (16 of 22 electrons used) 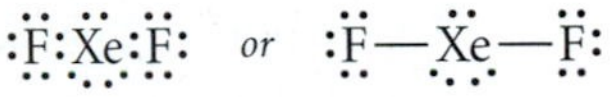 :F̈:Xe:F̈: *or* :F̈—Ẍe—F̈: (22 of 22 electrons used)

For Practice 9.9

Write a Lewis structure for XeF_4.

For More Practice 9.9

Write a Lewis structure for H_3PO_4. If necessary, expand the octet on any appropriate atoms to lower formal charge.

9.10 Bond Energies and Bond Lengths

In Chapter 6, we learned how to calculate the standard enthalpy change for a chemical reaction ($\Delta H°_{rxn}$) from tabulated standard enthalpies of formation. However, at times it may not be possible to find standard enthalpies of formation for all of the reactants and products of a reaction. In such cases, we can use individual *bond energies* to estimate enthalpy changes of reaction. In this section, we introduce the concept of bond energy and then show how bond energies can be used to calculate enthalpy changes of reaction. We also look at average bond lengths for a number of commonly encountered bonds.

Bond Energy

Bond energy is also called bond enthalpy or bond dissociation energy.

The **bond energy** of a chemical bond is the energy required to break 1 mole of the bond in the gas phase. For example, the bond energy of the Cl—Cl bond in Cl_2 is 243 kJ/mol.

$$Cl_2(g) \longrightarrow 2Cl(g) \qquad \Delta H = 243 \text{ kJ}$$

The bond energy of HCl is 431 kJ/mol.

$$HCl(g) \longrightarrow H(g) + Cl(g) \qquad \Delta H = 431 \text{ kJ}$$

Bond energies are always positive, because it always takes energy to break a bond. We say that the HCl bond is *stronger* than the Cl_2 bond because it requires more energy to break it. In general, compounds with stronger bonds tend to be more chemically stable, and therefore less chemically reactive, than compounds with weaker bonds. The triple bond in N_2 has a bond energy of 946 kJ/mol.

$$N_2(g) \longrightarrow N(g) + N(g) \qquad \Delta H = 946 \text{ kJ}$$

It is a very strong and stable bond, which explains nitrogen's relative inertness.

The bond energy of a particular bond in a polyatomic molecule is a little more difficult to determine because a particular type of bond can have different bond energies in different molecules. For example, consider the C—H bond. In CH_4, the energy required to break one C—H bond is 438 kJ/mol.

$$H_3C{-}H(g) \longrightarrow H_3C(g) + H(g) \qquad \Delta H = 438 \text{ kJ}$$

However, the energy required to break a C—H bond in other molecules varies slightly, as shown here.

$$F_3C{-}H(g) \longrightarrow F_3C(g) + H(g) \qquad \Delta H = 446 \text{ kJ}$$
$$Br_3C{-}H(g) \longrightarrow Br_3C(g) + H(g) \qquad \Delta H = 402 \text{ kJ}$$
$$Cl_3C{-}H(g) \longrightarrow Cl_3C(g) + H(g) \qquad \Delta H = 401 \text{ kJ}$$

It is useful to calculate an *average bond energy* for a chemical bond, which is an average of the bond energies for that bond in large number of compounds. For example, for our limited number of compounds listed above, we calculate an average C—H bond energy of 422 kJ/mol. Table 9.3 lists average bond energies for a number of common chemical bonds averaged over a large number of compounds. Notice that the C—H bond energy

TABLE 9.3 Average Bond Energies

Bond	Bond Energy (kJ/mol)	Bond	Bond Energy (kJ/mol)	Bond	Bond Energy (kJ/mol)
H—H	436	N—N	163	Br—F	237
H—C	414	N=N	418	Br—Cl	218
H—N	389	N≡N	946	Br—Br	193
H—O	464	N—O	222	I—Cl	208
H—S	368	N=O	590	I—Br	175
H—F	565	N—F	272	I—I	151
H—Cl	431	N—Cl	200	Si—H	323
H—Br	364	N—Br	243	Si—Si	226
H—I	297	N—I	159	Si—C	301
C—C	347	O—O	142	S—O	265
C=C	611	O=O	498	Si=O	368
C≡C	837	O—F	190	S—O	523
C—N	305	O—Cl	203	Si—Cl	464
C=N	615	O—I	234	S—S	418
C≡N	891	F—F	159	S—F	327
C—O	360	Cl—F	253	S—Cl	253
C=O	736*	Cl—Cl	243	S—Br	218
C≡O	1072			S—S	266
C—Cl	339				

*799 in CO_2

is listed as 414 kJ/mol, which is not too different from the value we calculated from a limited number of compounds. Notice also that bond energies depend not only on the kind of atoms involved in the bond, but also on the type of bond: single, double, or triple. In general, for a particular pair of atoms, triple bonds are stronger than double bonds, which are in turn stronger than single bonds. For example, consider the bond energies of carbon–carbon triple, double, and single bonds listed at right.

Bond	Bond Energy (kJ/mol)
C≡C	837 kJ/mol
C=C	611 kJ/mol
C—C	347 kJ/mol

Using Average Bond Energies to Estimate Enthalpy Changes for Reactions

Average bond energies are useful in *estimating* the enthalpy change of a reaction. For example, consider the following reaction:

$$H_3C—H(g) + Cl—Cl(g) \longrightarrow H_3C—Cl(g) + H—Cl(g)$$

We can imagine this reaction occurring by the breaking of a C—H bond and a Cl—Cl bond and the forming of a C—Cl bond and an H—Cl bond. Since breaking bonds is endothermic (positive bond energy) and forming bonds is exothermic (negative bond energy) we can calculate the overall enthalpy change as a sum of the enthalpy changes associated with breaking the required bonds in the reactants and forming the required bonds in the products, as shown in Figure 9.15 ▼.

$$H_3C—H(g) + Cl—Cl(g) \longrightarrow H_3C—Cl(g) + H—Cl(g)$$

Bonds Broken		**Bonds Formed**	
C—H break	+414 kJ	C—Cl form	−339 kJ
Cl—Cl break	+243 kJ	H—Cl form	−431 kJ

Sum (Σ) ΔH's bonds broken: +657 kJ   *Sum (Σ) ΔH's bonds formed:* −770 kJ

$$\Delta H_{rxn} = \Sigma(\Delta H's\ bonds\ broken) + \Sigma(\Delta H's\ bonds\ formed)$$
$$= +657\ kJ - 770\ kJ$$
$$= -113\ kJ$$

Estimating the Enthalpy Change of a Reaction from Bond Energies

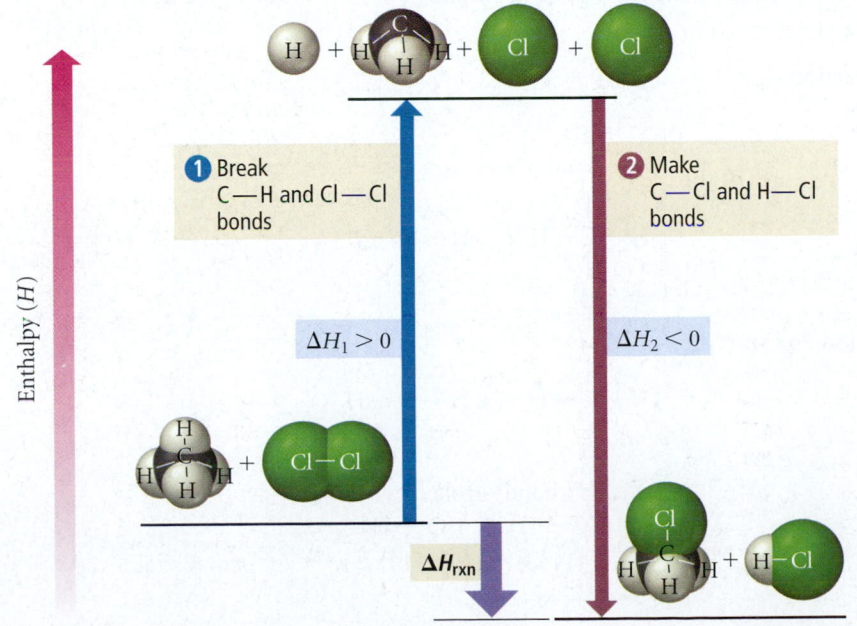

① Break C—H and Cl—Cl bonds

$\Delta H_1 > 0$

② Make C—Cl and H—Cl bonds

$\Delta H_2 < 0$

Enthalpy (H)

ΔH_{rxn}

◀ **FIGURE 9.15 Estimating** ΔH_{rxn} **from Bond Energies** The enthalpy change of a reaction can be approximated by summing up the enthalpy changes involved in breaking old bonds and those involved in forming new ones.

We find that $\Delta H_{rxn} = -113\,kJ$. Calculating $\Delta H°_{rxn}$ from tabulated enthalpies of formation—as we learned in Chapter 6—gives $\Delta H°_{rxn} = -101\,kJ$, fairly close to the value we obtained from average bond energies. In general, calculate ΔH_{rxn} from average bond energies by summing the changes in enthalpy for all of the bonds that are broken and adding the sum of the enthalpy changes for all of the bonds that are formed. Remember that ΔH is positive for breaking bonds and negative for forming them:

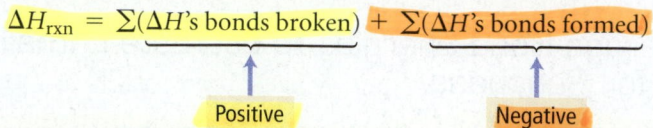

$$\Delta H_{rxn} = \underbrace{\Sigma(\Delta H\text{'s bonds broken})}_{\text{Positive}} + \underbrace{\Sigma(\Delta H\text{'s bonds formed})}_{\text{Negative}}$$

As you can see from the above equation:

• A reaction is *exothermic* when weak bonds break and strong bonds form.
• A reaction is *endothermic* when strong bonds break and weak bonds form.

Scientists often say that "energy is stored in chemical bonds or in a chemical compound," which may sound as if breaking the bonds in the compound releases energy. For example, we often hear in biology that energy is stored in glucose or in ATP. However, *breaking a chemical bond always requires energy.* When we say that energy is stored in a compound, or that a compound is energy rich, we mean that the compound can undergo a reaction in which weak bonds break and strong bonds form, thereby releasing energy. *It is always the forming of chemical bonds that releases energy.*

Conceptual Connection 9.4 Bond Energies and ΔH_{rxn}

The reaction between hydrogen and oxygen to form water is highly exothermic. Which of the following is true of the energies of the bonds that break and form during the reaction?

(a) The energy needed to break the required bonds is greater than the energy released when the new bonds form.

(b) The energy needed to break the required bonds is less than the energy released when the new bonds form.

(c) The energy needed to break the required bonds is about the same as the energy released when the new bonds form.

Answer: **(b)** In a highly exothermic reaction, the energy needed to break bonds is less than the energy released when the new bonds form, resulting a net release of energy.

EXAMPLE 9.10 Calculating ΔH_{rxn} from Bond Energies

Hydrogen gas, a potential fuel, can be made by the reaction of methane gas and steam.

$$CH_4(g) + 2\,H_2O(g) \longrightarrow 4\,H_2(g) + CO_2(g)$$

Use bond energies to calculate ΔH_{rxn} for this reaction.

Solution

Begin by rewriting the reaction using the Lewis structures of the molecules involved.	$$\begin{matrix} & H & \\ & | & \\ H\!-\!&C&\!-\!H + 2\,H\!-\!\ddot{O}\!-\!H \longrightarrow 4\,H\!-\!H + \ddot{O}\!=\!C\!=\!\ddot{O} \\ & | & \\ & H & \end{matrix}$$
Determine which bonds are broken in the reaction and sum the bond energies of these.	$$\begin{matrix} & H & \\ & | & \\ H\!-\!&C&\!-\!H + 2\,H\!-\!\ddot{O}\!-\!H \\ & | & \\ & H & \end{matrix}$$ $\Sigma(\Delta H\text{'s bonds broken})$ $= 4(C\!-\!H) + 4(O\!-\!H)$ $= 4(414\,kJ) + 4(464\,kJ)$ $= 3512\,kJ$

Determine which bonds are formed in the reaction and sum the negatives of the bond energies of these.	$4\,H\!-\!H + \ddot{O}\!=\!C\!=\!\ddot{O}$ $\Sigma(\Delta H\text{'s bonds formed})$ $= -4(H\!-\!H) - 2(C\!=\!O)$ $= -4(436\ \text{kJ}) - 2(799\ \text{kJ})$ $= -3342\ \text{kJ}$
Find ΔH_{rxn} by summing the results of the previous two steps.	$\Delta H_{rxn} = \Sigma(\Delta H\text{'s bonds broken}) + \Sigma(\Delta H\text{'s bonds formed})$ $= 3512 - 3342$ $= 1.70 \times 10^2\ \text{kJ}$

For Practice 9.10

Another potential future fuel is methanol (CH_3OH). Write a balanced equation for the combustion of gaseous methanol and use bond energies to calculate the enthalpy of combustion of methanol in kJ/mol.

For More Practice 9.10

Use bond energies to calculate ΔH_{rxn} for the following reaction:
$N_2(g) + 3\,H_2(g) \longrightarrow 2\,NH_3(g)$.

Bond Lengths

Just as we can tabulate average bond energies, which represent the average energy of a bond between two particular atoms in a large number of compounds, so we can tabulate average **bond lengths** (Table 9.4), which represent the average length of a bond between two particular atoms in a large number of compounds. Like bond energies, bond lengths depend not only on the kind of atoms involved in the bond, but also on the type of bond: single, double, or triple. In general, for a particular pair of atoms, ==triple bonds are shorter than double bonds, which are in turn shorter than single bonds.== For example, consider the bond lengths (along with bond strengths, repeated from earlier in this section) of carbon–carbon triple, double, and single bonds.

Bond	Bond Length (pm)	Bond Strength (kJ/mol)
C≡C	120 pm	837 kJ/mol
C=C	134 pm	611 kJ/mol
C—C	154 pm	347 kJ/mol

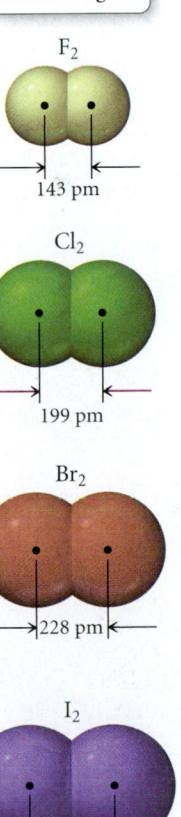

Bond Lengths

F_2
143 pm

Cl_2
199 pm

Br_2
228 pm

I_2
266 pm

▲ Bond lengths in the diatomic halogen molecules.

TABLE 9.4 Average Bond Lengths

Bond	Bond Length (pm)	Bond	Bond Length (pm)	Bond	Bond Length (pm)
H—H	74	C—C	154	N—N	145
H—C	110	C=C	134	N=N	123
H—N	100	C≡C	120	N≡N	110
H—O	97	C—N	147	N—O	136
H—S	132	C=N	128	N=O	120
H—F	92	C≡N	116	O—O	145
H—Cl	127	C—O	143	O=O	121
H—Br	141	C=O	120	F—F	143
H—I	161	C—Cl	178	Cl—Cl	199
				Br—Br	228
				I—I	266

Notice that, as the bond gets longer, it also becomes weaker. This relationship between the length of a bond and the strength of a bond does not necessarily hold for all bonds. For example, consider the following series of nitrogen–halogen single bonds:

Bond	Bond Length (pm)	Bond Strength (kJ/mol)
N — F	139	272
N — Cl	191	200
N — Br	214	243
N — I	222	159

Although the bonds generally get weaker as they get longer, the trend is not a smooth one.

Chemistry in the Environment
The Lewis Structure of Ozone

Ozone is a form of oxygen in which three oxygen atoms bond together. Its Lewis structure consists of the following two resonance structures:

$$:\ddot{O}{=}\ddot{O}{-}\ddot{O}{:} \quad \longleftrightarrow \quad :\ddot{O}{-}\ddot{O}{=}\ddot{O}{:}$$

Compare the Lewis structure of ozone to the Lewis structure of O_2:

$$:\ddot{O}{=}\ddot{O}{:}$$

Since double bonds are stronger and shorter than single bonds, O_2 must have a stronger bond because it is a double bond. O_3, on the other hand, has bonds that are intermediate between single and double, so O_3 has weaker bonds. The effects of this are significant. As we learned in Section 5.11, O_3 absorbs harmful

ultraviolet light entering Earth's atmosphere. Ozone is ideally suited to do this because photons at wavelengths of 280–320 nm (the most dangerous components of sunlight to humans) are just strong enough to break the bonds in the O_3 molecule:

$$:\ddot{O}{-}\ddot{O}{=}\ddot{O}{:} + \text{UV light} \longrightarrow :\ddot{O}{=}\ddot{O}{:} + \cdot\ddot{O}{:}$$

In this process, the photon is absorbed. O_2 and O then recombine to re-form O_3, which can in turn absorb more UV light. The same wavelengths of UV light, however, do not have sufficient energy to break the stronger double bond of O_2. No other molecules in our atmosphere can do the job that ozone does. Consequently, it is important that we continue, and even strengthen, the ban on ozone-depleting compounds.

Question
Calculate the average bond energy of one O_3 bond. What wavelength of light has just the right amount of energy to break this bond?

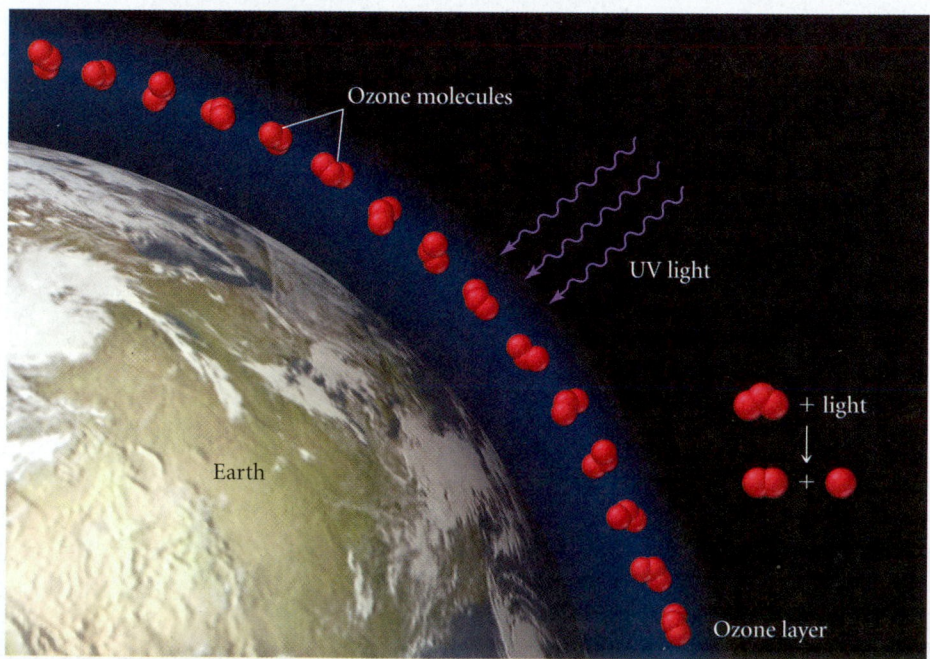

Ozone molecules

UV light

+ light

Earth

Ozone layer

► Ozone protects life on Earth from harmful ultraviolet light.

9.11 Bonding in Metals: The Electron Sea Model

So far, we have developed simple models for bonding between a metal and a nonmetal (ionic bonding) and for bonding between two nonmetals (covalent bonding). We have seen how these models account for and predict the properties of ionic and molecular compounds. The last type of bonding that we examine in this chapter is metallic bonding, which occurs between metals (this topic is covered in more detail in Chapter 24).

As we know, metals have a tendency to lose electrons, which means that they have relatively low ionization energies. When metal atoms bond together to form a solid, each metal atom donates one or more electrons to an *electron sea*. For example, we can think of sodium metal as an array of positively charged Na^+ ions immersed in a sea of negatively charged electrons (e^-), as shown in Figure 9.16 ▶. Each sodium atom donates its one valence electron to the "sea" and becomes a sodium ion. The sodium cations are then held together by their attraction to the sea of electrons.

Although this model is simple, it accounts for many of the properties of metals. For example, metals conduct electricity because—in contrast to ionic solids where electrons are localized on an ion—the electrons in a metal are free to move. The movement or flow of electrons in response to an electric potential (or voltage) is an electric current. Metals are also excellent conductors of heat, again because of the highly mobile electrons, which help to disperse thermal energy throughout the metal.

The *malleability* of metals (ability to be pounded into sheets) and the *ductility* of metals (ability to be drawn into wires) are also accounted for by this model. Since there are no localized or specific "bonds" in a metal, it can be deformed relatively easily by forcing the metal ions to slide past one another. The electron sea can easily accommodate these deformations by flowing into the new shape.

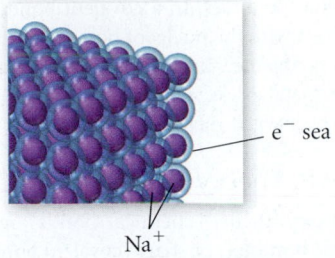

▲ **FIGURE 9.16 The Electron Sea Model for Sodium** In this model of metallic bonding, Na^+ ions are immersed in a "sea" of electrons.

▲ Copper can easily be drawn into fine strands like those used in household electrical cords.

Chapter in Review

Key Terms

Section 9.1
Lewis theory (364)
Lewis electron-dot structures (Lewis structures) (364)

Section 9.2
ionic bond (366)
covalent bond (366)
metallic bonding (366)

Section 9.3
octet (366)
duet (367)
chemical bond (367)
octet rule (367)

Section 9.4
lattice energy (368)
Born–Haber cycle (369)

Section 9.5
bonding pair (374)
lone pair (374)
double bond (375)
triple bond (375)

Section 9.6
polar covalent bond (377)
electronegativity (377)
dipole moment (μ) (380)
percent ionic character (380)

Section 9.8
resonance structures (384)
resonance hybrid (384)
formal charge (386)

Section 9.9
free radical (388)

Section 9.10
bond energy (392)
bond length (395)

Key Concepts

Bonding Models and AIDS Drugs (9.1)

Theories that predict how and why atoms bond together are central to chemistry, because they explain compound stability and molecule shape. Bonding theories have been useful in combating HIV because they help in the design of molecules that can bind to the active site of a protein crucial for the development of AIDS.

Types of Chemical Bonds (9.2)

Chemical bonds can be divided into three general types: ionic bonds, which occur between a metal and a nonmetal; covalent bonds, which occur between two nonmetals; and metallic bonds, which occur within metals. In an ionic bond, an electron is transferred from the metal to the nonmetal and the resultant ions are attracted to each other by coulombic forces. In a covalent bond, nonmetals share electrons that interact with the nuclei of both atoms via coulombic forces, holding the atoms together. In a metallic bond, the atoms form a lattice in which each metal loses electrons to an "electron sea." The attraction of the positively charged metal ions to the electron sea holds the metal together.

Lewis Theory and Electron Dots (9.3)

In Lewis theory, chemical bonds are formed when atoms transfer (ionic bonding) or share (covalent bonding) valence electrons to attain noble gas electron configurations. Lewis theory represents valence electrons as dots surrounding the symbol for an element. When two or more elements bond together, the dots are transferred or shared so that every atom gets eight dots, an octet (or two dots, a duet, in the case of hydrogen).

Ionic Lewis Structures and Lattice Energy (9.4)

In an ionic Lewis structure involving main-group metals, the metal transfers its valence electrons (dots) to the nonmetal. The formation of most ionic compounds is exothermic because of lattice energy, the energy released when metal cations and nonmetal anions coalesce to form the solid; the smaller the radius of the ions and the greater their charge, the more exothermic the lattice energy.

Covalent Lewis Structures, Electronegativity, and Polarity (9.5. 9.6, 9.7)

In a covalent Lewis structure, neighboring atoms share valence electrons to attain octets (or duets). A single shared electron pair constitutes a single bond, while two or three shared pairs constitute double or triple bonds, respectively. The shared electrons in a covalent bond are not always *equally* shared; when two dissimilar nonmetals form a covalent bond, the electron density is greater on the more electronegative element. The result is a polar bond, with one element carrying a partial positive charge and the other a partial negative charge. Electronegativity—the ability of an atom to attract electrons to itself in a chemical bond—increases as you move to the right across a period in the periodic table and decreases as you move down a column. Elements with very different electronegativities form ionic bonds; those with very similar electronegativities form nonpolar covalent bonds; and those with intermediate electronegativity differences form polar covalent bonds.

Resonance and Formal Charge (9.8)

Some molecules are best represented not by a single Lewis structure, but by two or more resonance structures. The actual structure of the molecule is then a resonance hybrid: a combination or average of the contributing structures. The formal charge of an atom in a Lewis structure is the charge the atom would have if all bonding electrons were shared equally between bonding atoms. In general, the best Lewis structures will have the fewest atoms with formal charge and any negative formal charge will be on the most electronegative atom.

Exceptions to the Octet Rule (9.9)

Although the octet rule is normally used in drawing Lewis structures, some exceptions occur. These exceptions include odd-electron species, which necessarily have Lewis structures with only 7 electrons around an atom. Such molecules, called free radicals, tend to be unstable and chemically reactive. Other exceptions to the octet rule include incomplete octets—usually totaling 6 electrons (especially important in compounds containing boron)—and expanded octets—usually 10 or 12 electrons (important in compounds containing elements from the third row of the periodic table and below). Expanded octets never occur in second-period elements.

Bond Energies and Bond Lengths (9.10)

The bond energy of a chemical bond is the energy required to break 1 mole of the bond in the gas phase. Average bond energies for a number of different bonds are tabulated and can be used to calculate enthalpies of reaction. Average bond lengths are also tabulated. In general, triple bonds are shorter and stronger than double bonds, which are in turn shorter and stronger than single bonds.

Bonding in Metals (9.11)

When metal atoms bond together to form a solid, each metal atom donates one or more electrons to an *electron sea*. The metal cations are then held together by their attraction to the sea of electrons. This simple model accounts for the electrical conductivity, thermal conductivity, malleability, and ductility of metals.

Key Equations and Relationships

Coulomb's Law: Potential Energy (E) of Two Charged Particles with Charges q_1 and q_2 Separated by a Distance r (9.2)

$$E = \frac{1}{4\pi\epsilon_0}\frac{q_1 q_2}{r} \qquad \epsilon_0 = 8.85 \times 10^{-12}\, \text{C}^2/\text{J} \cdot \text{m}$$

Dipole Moment (μ): Separation of Two Particles of Equal but Opposite Charges of Magnitude q by a Distance r (9.6)

$$\mu = qr$$

Percent Ionic Character (9.6)

$$\text{Percent ionic character} = \frac{\text{measured dipole moment of bond}}{\text{dipole moment if electron were completely transferred}} \times 100\%$$

Formal Charge (9.8)

Formal charge = number of valance electrons −
(number of lone pair electrons + $\frac{1}{2}$ number of shared electrons)

Enthalpy Change of a Reaction (ΔH_{rxn}): Relationship of Bond Energies (9.10)

$$\Delta H_{rxn} = \Sigma(\Delta H\text{'s bonds broken}) + \Sigma(\Delta H\text{'s bonds formed})$$

Key Skills

Predicting Chemical Formulas of an Ionic Compound (9.4)

•Example 9.1 • For Practice 9.1 • Exercises 43, 44

Predicting Relative Lattice Energies (9.4)

•Example 9.2 • For Practice 9.2 • For More Practice 9.2 • Exercise 48

Classifying Bonds: Pure Covalent, Polar Covalent, or Ionic (9.6)

•Example 9.3 • For Practice 9.3 • Exercises 55, 56

Writing Lewis Structures for Covalent Compounds (9.7)

•Examples 9.4, 9.5 • For Practice 9.4, 9.5 • Exercises 53, 54

Writing Lewis Structures for Polyatomic Ions (9.7)

•Example 9.6 • For Practice 9.6 • Exercises 59–62

Writing Resonance Lewis Structures (9.8)

•Example 9.7 • For Practice 9.7 • Exercises 61, 62

Assigning Formal Charges to Assess Competing Resonance Structures (9.8)

•Example 9.8 • For Practice 9.8 • For More Practice 9.8 • Exercises 63, 64

Writing Lewis Structures for Compounds Having Expanded Octets (9.9)

•Example 9.9 • For Practice 9.9 • For More Practice 9.9 • Exercises 71, 72

Calculating ΔH_{rxn} from Bond Energies (9.10)

•Example 9.10 • For Practice 9.10 • For More Practice 9.10 • Exercises 75–77

Exercises

Review Questions

1. Why are bonding theories important?

2. Give some examples of what bonding theories can predict.

3. Why do chemical bonds form? What basic forces are involved in bonding?

4. How does the potential energy of two charged particles depend on each of the following:
   a. magnitude of the charges
   b. sign of the charges
   c. separation between the charges

5. What are the three basic types of chemical bonds? What happens to electrons in the bonding atoms in each case?

6. How do you determine how many dots to put around the Lewis symbol of an element?

7. Describe the octet rule in Lewis theory.

8. According to Lewis theory, what is a chemical bond?

9. How do you draw an ionic Lewis structure?

10. How can Lewis theory be used to determine the formula of ionic compounds? Give an example.

11. What is lattice energy?

12. Why is the formation of solid sodium chloride from solid sodium and gaseous chlorine exothermic, even though it takes more energy to form the Na^+ ion than the amount of energy released upon formation of Cl^-?

13. What is the Born–Haber Cycle? List each of the steps in the cycle and show how the cycle is used to calculate lattice energy.

14. How does lattice energy depend on ionic radii? On ion charge?

15. How does the ionic bonding model explain the relatively high melting points of ionic compounds?

16. How does the ionic bonding model explain the nonconductivity of ionic solids, and at the same time the conductivity of ionic solutions?

17. Within a covalent Lewis structure, what is the difference between lone pair and bonding pair electrons?

18. In what ways are double and triple covalent bonds different from single covalent bonds?

19. How does the Lewis model for covalent bonding explain why certain combinations of atoms are stable while others are not?

20. How does the Lewis model for covalent bonding explain the relatively low melting and boiling points of molecular compounds (compared to ionic compounds)?

21. What is electronegativity? What are the periodic trends in electronegativity?

22. Explain the difference between a pure covalent bond, a polar covalent bond, and an ionic bond.

23. Explain what is meant by the percent ionic character of a bond. Do any bonds have 100% ionic character?

24. What is a dipole moment?

25. What is the magnitude of the dipole moment formed by separating a proton and an electron by 100 pm? 200 pm?

26. What is the basic procedure for writing a covalent Lewis structure?

27. How do you determine the number of electrons that go into the Lewis structure of a molecule? A polyatomic ion?

28. What are resonance structures? What is a resonance hybrid?

29. Do all resonance structures always contribute equally to the overall structure of a molecule? Explain.

30. What is formal charge? How is formal charge calculated? How is it helpful?

31. Why does the octet rule have exceptions? Give the three major categories of exceptions and an example of each.

32. What elements can have expanded octets? What elements should never have expanded octets?

33. What is bond energy? How can average bond energies be used to calculate enthalpies of reaction?

34. Explain the difference between endothermic reactions and exothermic reactions with respect to the bond energies of the bonds broken and formed.

35. What is the electron sea model for bonding in metals?

36. How does the electron sea model explain the conductivity of metals? The malleability and ductility of metals?

Problems by Topic

Valence Electrons and Dot Structures

37. Write an electron configuration for N. Then write a Lewis structure for N and show which electrons from the electron configuration are included in the Lewis structure.

38. Write an electron configuration for Ne. Then write a Lewis structure for Ne and show which electrons from the electron configuration are included in the Lewis structure.

39. Write a Lewis structure for each of the following:
    a. Al   b. Na^+ c. Cl   d. Cl^-

40. Write a Lewis structure for each of the following:
    a. S^{2-} b. Mg   c. Mg^{2+} d. P

Ionic Lewis Structures and Lattice Energy

41. Write a Lewis structure for each of the following ionic compounds.
    a. NaF   b. CaO   c. $SrBr_2$ d. K_2O

42. Write a Lewis structure for each of the following ionic compounds.
    a. SrO   b. Li_2S c. CaI_2 d. RbF

43. Use Lewis theory to determine the formula for the compound that forms between:
    a. Sr and Se   b. Ba and Cl
    c. Na and S   d. Al and O

44. Use Lewis theory to determine the formula for the compound that forms between:
    a. Ca and N   b. Mg and I
    c. Ca and S   d. Cs and F

45. Consider the following trend in the lattice energies of the alkaline earth metal oxides:

Metal Oxide	Lattice Energy (kJ/mol)
MgO	−3795
CaO	−3414
SrO	−3217
BaO	−3029

Explain this trend.

46. Rubidium iodide has a lattice energy of −617 kJ/mol, while potassium bromide has a lattice energy of −671 kJ/mol. Why is the lattice energy of potassium bromide more exothermic than the lattice energy of rubidium iodide?

47. The lattice energy of CsF is −744 kJ/mol, whereas that of BaO is −3029 kJ/mol. Explain this large difference in lattice energy.

48. Arrange the following substances in order of increasing magnitude of lattice energy: KCl, SrO, RbBr, CaO.

49. Use the Born–Haber cycle data from Appendix IIB and Chapters 8 and 9 to calculate the lattice energy of KCl.

50. Use the Born–Haber cycle and data from Appendix IIB and Table 9.3 to calculate the lattice energy of CaO.

Simple Covalent Lewis Structures, Electronegativity, and Bond Polarity

51. Use covalent Lewis structures to explain why each of the following elements (or families of elements) occur as diatomic molecules:
    a. hydrogen   b. the halogens
    c. oxygen   d. nitrogen

52. Use covalent Lewis structures to explain why the compound that forms between nitrogen and hydrogen has the formula NH_3. Show why NH_2 and NH_4 are not stable.

53. Write a Lewis structure for each of the following molecules:
    a. PH_3 b. SCl_2 c. HI   d. CH_4

54. Write a Lewis structure for each of the following molecules:
    a. NF_3 b. HBr   c. SBr_2 d. CCl_4

55. Determine whether a bond between each of the following pairs of atoms would be pure covalent, polar covalent, or ionic.
    a. Br and Br   b. C and Cl
    c. C and S   d. Sr and O

56. Determine whether a bond between each of the following pairs of atoms would be pure covalent, polar covalent, or ionic.
    a. C and N   b. N and S
    c. K and F   d. N and N

57. Draw a Lewis structure for CO with an arrow representing the dipole moment. Use Figure 9.12 to estimate the percent ionic character of the CO bond.

58. Draw a Lewis structure for BrF with an arrow representing the dipole moment. Use Figure 9.12 to estimate the percent ionic character of the BrF bond.

Covalent Lewis Structures, Resonance, and Formal Charge

59. Write a Lewis structure for each of the following molecules or ions:
    a. CI_4 b. N_2O c. SiH_4 d. Cl_2CO
    e. H_3COH f. OH^- g. BrO^-

60. Write a Lewis structure for each of the following molecules or ions:
    a. N_2H_2 b. N_2H_4 c. C_2H_2 d. C_2H_4
    e. H_3COCH_3 f. CN^- g. NO_2^-

61. Write a Lewis structure that obeys the octet rule for each of the following molecules or ions. Include resonance structures if necessary and assign formal charges to each atom.

   a. SeO_2 **b.** CO_3^{2-}

   c. ClO^- **d.** NO_2^-

62. Write a Lewis structure that obeys the octet rule for each of the following ions. Include resonance structures if necessary and assign formal charges to each atom.

   a. ClO_3^- **b.** ClO_4^-

   c. NO_3^- **d.** NH_4^+

63. Use formal charge to determine which of the following two Lewis structures is better:

$$H-\overset{\overset{\textstyle H}{|}}{C}=\ddot{\underset{..}{S}} \qquad H-\overset{\overset{\textstyle H}{|}}{S}=\ddot{C}$$

64. Use formal charge to determine which of the following two Lewis structures is better:

$$H-\overset{\overset{\textstyle H}{|}}{\underset{\underset{\textstyle H}{|}}{S}}-\ddot{C}-H \qquad H-\overset{\overset{\textstyle H}{|}}{\underset{\underset{\textstyle H}{|}}{C}}-\ddot{S}-H$$

65. How important is the following resonance structure to the overall structure of carbon dioxide? Explain.

$$:O\equiv C-\ddot{\underset{..}{O}}:$$

66. In N_2O, nitrogen is the central atom and the oxygen atom is terminal. In OF_2, however, oxygen is the central atom. Use formal charges to explain why this is so.

Odd-Electron Species, Incomplete Octets, and Expanded Octets

67. Write a Lewis structure for each of the following molecules that are exceptions to the octet rule.

   a. BCl_3 **b.** NO_2 **c.** BH_3

68. Write a Lewis structure for each of the following molecules that are exceptions to the octet rule.

   a. BBr_3 **b.** NO **c.** ClO_2

69. Write a Lewis structure for each of the following ions. Include resonance structures if necessary and assign formal charges to all atoms. If necessary, expand the octet on the central atom to lower formal charge.

   a. PO_4^{3-} **b.** CN^- **c.** SO_3^{2-} **d.** ClO_2^-

70. Write Lewis structures for each of the following molecules or ions. Include resonance structures if necessary and assign formal charges to all atoms. If necessary, expand the octet on the central atom to lower formal charge.

   a. SO_4^{2-} **b.** HSO_4^- **c.** SO_3 **d.** BrO_2^-

71. Write Lewis structures for each of the following species. Use expanded octets as necessary.

   a. PF_5 **b.** I_3^- **c.** SF_4 **d.** GeF_4

72. Write Lewis structures for each of the following species. Use expanded octets as necessary.

   a. ClF_5 **b.** AsF_6^- **c.** Cl_3PO **d.** IF_5

Bond Energies and Bond Lengths

73. Consider the following three compounds: HCCH, H_2CCH_2, H_3CCH_3. Order these compounds in order of increasing carbon–carbon bond strength and in order of decreasing carbon–carbon bond length.

74. Which of the following two compounds has the strongest nitrogen–nitrogen bond? The shortest nitrogen–nitrogen bond?

$$H_2NNH_2, \ HNNH$$

75. Hydrogenation reactions are used to add hydrogen across double bonds in hydrocarbons and other organic compounds. Use average bond energies to calculate ΔH_{rxn} for the following hydrogenation reaction:

$$H_2C=CH_2(g) + H_2(g) \longrightarrow H_3C-CH_3(g)$$

76. Ethanol is a possible fuel. Use average bond energies to calculate ΔH_{rxn} for the combustion of ethanol.

$$CH_3CH_2OH(g) + 5/2 \ O_2(g) \longrightarrow 2 \ CO_2(g) + 2 \ H_2O(g)$$

77. Hydrogen, a potential future fuel, can be produced from carbon (from coal) and steam by the following reaction:

$$C(s) + 2 \ H_2O(g) \longrightarrow 2 \ H_2(g) + CO_2(g)$$

Use average bond energies to calculate ΔH_{rxn} for this reaction.

78. In the *Chemistry and the Environment* box on free radicals in this chapter, we discussed the importance of the hydroxyl radical in reacting with and eliminating many atmospheric pollutants. However, the hydroxyl radical does not clean up everything. For example, chlorofluorocarbons—which destroy stratospheric ozone—are not attacked by the hydroxyl radical. Consider the following hypothetical reaction by which the hydroxyl radical might react with a chlorfluorocarbon:

$$OH(g) + CF_2Cl_2(g) \longrightarrow HOF(g) + CFCl_2(g)$$

Use bond energies to explain why this reaction is unlikely.

Cumulative Problems

79. Write an appropriate Lewis structure for each of the following compounds. Make certain to distinguish between ionic and covalent compounds.

   a. BI_3 **b.** K_2S **c.** $HCFO$ **d.** PBr_3

80. Write an appropriate Lewis structure for each of the following compounds. Make certain to distinguish between ionic and covalent compounds.

   a. Al_2O_3 **b.** ClF_5 **c.** MgI_2 **d.** XeO_4

81. Each of the following compounds contains both ionic and covalent bonds. Write ionic Lewis structures for each of them, including the covalent structure for the ion in brackets. Write resonance structures if necessary.

   a. $BaCO_3$ **b.** $Ca(OH)_2$ **c.** KNO_3 **d.** $LiIO$

82. Each of the following compounds contains both ionic and covalent bonds. Write ionic Lewis structures for each of them, including the covalent structure for the ion in brackets. Write resonance structures if necessary.

   a. $RbIO_2$ **b.** NH_4Cl **c.** KOH **d.** $Sr(CN)_2$

83. Carbon ring structures are common in organic chemistry. Draw a Lewis structure for each of the following carbon ring structures, including any necessary resonance structures.

   a. C_4H_8 **b.** C_4H_4 **c.** C_6H_{12} **d.** C_6H_6

84. Amino acids are the building blocks of proteins. The simplest amino acid is glycine (H_2NCH_2COOH). Draw a Lewis structure for glycine. (Hint: The central atoms in the skeletal structure are nitrogen bonded to carbon which is bonded to another carbon. The two oxygen atoms are bonded directly to the right-most carbon atom.)

85. Formic acid is responsible for the sting in biting ants. By mass, formic acid is 26.10% C, 4.38% H, and 69.52% O. The molar mass of formic acid is 46.02 g/mol. Find the molecular formula of formic acid and draw its Lewis structure.

86. Diazomethane is a highly poisonous, explosive compound. The reason it is explosive is that it readily evolves N_2. Diazomethane has the following composition by mass: 28.57% C; 4.80% H; and 66.64% N. The molar mass of diazomethane is 42.04 g/mol. Find the molecular formula of diazomethane, draw its Lewis structure, and assign formal charges to each atom. Why is diazomethane not very stable? Explain.

87. The reaction of $Fe_2O_3(s)$ with $Al(s)$ to form $Al_2O_3(s)$ and $Fe(s)$ is called the thermite reaction and is highly exothermic. What role does lattice energy play in the exothermicity of the reaction?

88. NaCl has a lattice energy -787 kJ/mol. Consider a hypothetical salt XY. X^{3+} has the same radius of Na^+ and Y^{3-} has the same radius as Cl^-. Estimate the lattice energy of XY.

89. Draw a Lewis structure for nitric acid (the hydrogen atom is attached to one of the oxygen atoms). Include all three resonance structures by alternating the double bond among the three oxygen atoms. Use formal charge to determine which of the resonance structures is most important to the structure of nitric acid.

90. Phosgene (Cl_2CO) is a poisonous gas that was used as a chemical weapon during World War I and is a potential agent for chemical terrorism. Draw the Lewis structure of phosgene. Include all three resonance forms by alternating the double bond among the three terminal atoms. Which resonance structure is the best?

91. The cyanate ion (OCN^-) and the fulminate ion (CNO^-) share the same three atoms, but have vastly different properties. The cyanate ion is stable, while the fulminate ion is unstable and forms explosive compounds. The resonance structures of the cyanate ion were explored in Example 9.8. Draw Lewis structures for the fulminate ion—including possible resonance forms—and use formal charge to explain why the fulminate ion is less stable (and therefore more reactive) than the cyanate ion.

92. Use Lewis structures to explain why Br_3^- and I_3^- are stable, while F_3^- is not.

93. Draw a Lewis structure for $HCSNH_2$. (The carbon and nitrogen atoms are bonded together and the sulfur atom is bonded to the carbon atom.) Label each bond in the molecule as polar or nonpolar.

94. Draw a Lewis structure for urea, H_2NCONH_2, the compound primarily responsible for the smell of urine. (The central carbon atom is bonded to both nitrogen atoms and to the oxygen atom.) Does urea contain polar bonds? Which bond in urea is most polar?

95. Some theories of aging suggest that free radicals cause certain diseases and perhaps aging in general. As you know from Lewis theory, such molecules are not chemically stable and will quickly react with other molecules. Free radicals may attack molecules within the cell, such as DNA, changing them and causing cancer or other diseases. Free radicals may also attack molecules on the surfaces of cells, making them appear foreign to the body's immune system. The immune system then attacks the cells and destroys them, weakening the body. Draw Lewis structures for each of the following free radicals implicated in this theory of aging.

  a. O_2^- b. O^- c. OH
  d. CH_3OO (unpaired electron on terminal oxygen)

96. Free radicals are important in many environmentally significant reactions (see the *Chemistry in the Environment box* on free radicals in this chapter). For example, photochemical smog—smog that forms as a result of the action of sunlight on air pollutants—is formed in part by the following two steps:

$$NO_2 \xrightarrow{\text{UV light}} NO + O$$

$$O + O_2 \longrightarrow O_3$$

The product of this reaction, ozone, is a pollutant in the lower atmosphere. (Upper atmospheric ozone is a natural part of the atmosphere that protects life on Earth from ultraviolet light.) Ozone is an eye and lung irritant and also accelerates the weathering of rubber products. Rewrite the above reactions using the Lewis structure of each reactant and product. Identify the free radicals.

97. If hydrogen were used as a fuel, it could be burned according to the following reaction:

$$H_2(g) + \tfrac{1}{2}O_2(g) \longrightarrow H_2O(g)$$

Use average bond energies to calculate ΔH_{rxn} for this reaction and also for the combustion of methane (CH_4). Which fuel yields more energy per mole? Per gram?

98. Calculate ΔH_{rxn} for the combustion of octane (C_8H_{18}), a component of gasoline, by using average bond energies and then calculate it again using enthalpies of formation from Appendix IIB. What is the percent difference between the two results? Which result would you expect to be more accurate?

99. Draw Lewis structures for the following compounds
  a. Cl_2O_7 (no Cl—Cl bond)
  b. H_3PO_3 (two OH bonds)
  c. H_3AsO_4

100. The azide ion, N_3^-, is a symmetrical ion, all of whose contributing structures have formal charges. Draw three important contributing structures for this ion.

101. List the following gas phase ion pairs in order of the quantity of energy released when they form from separated gas-phase ions. Start with the pair that releases the least energy. Na^+F^-, $Mg^{2+}F^-$, Na^+O^{2-}, $Mg^{2+}O^{2-}$, $Al^{3+}O^{2-}$.

102. Calculate $\Delta H°$ for the reaction $H_2(g) + Br_2(g) \longrightarrow 2\,HBr(g)$ using the bond energy values. The $\Delta H_f°$ of HBr(g) is not equal to one-half of the value calculated. Account for the difference.

Challenge Problems

103. The main component of acid rain (H_2SO_4) forms from SO_2 pollutant in the atmosphere via the following series of steps:

$$SO_2 + OH\cdot \longrightarrow HSO_3\cdot$$
$$HSO_3\cdot + O_2 \longrightarrow SO_3 + HOO\cdot$$
$$SO_3 + H_2O \longrightarrow H_2SO_4$$

Draw a Lewis structure for each of the species in these steps and use bond energies and Hess's law to estimate ΔH_{rxn} for the overall process. (Use 265 kJ/mol for the S—O single bond energy.)

104. A 0.167-g sample of an unknown acid requires 27.8 mL of 0.100 M NaOH to titrate to the equivalence point. Elemental analysis of the acid gives the following percentages by mass: 40.00% C; 6.71% H; 53.29% O. Determine the molecular formula, molar mass, and Lewis structure of the unknown acid.

105. Use the dipole moments of HF and HCl (given below) together with the percent ionic character of each bond (Figure 9.12) to estimate the bond length in each molecule. How well does your estimated bond length agree with the bond length given in Table 9.4?

$$HCl \quad \mu = 1.08\ D$$
$$HF \quad \mu = 1.82\ D$$

106. Use average bond energies together with the standard enthalpy of formation of $C(g)$ (718.4 kJ/mol) to estimate the standard enthalpy of formation of gaseous benzene, $C_6H_6(g)$. (Remember that average bond energies apply to the gas phase only.) Compare the value you obtain using average bond energies to the actual standard enthalpy of formation of gaseous benzene, 82.9 kJ/mol. What does the difference between these two values tell you about the stability of benzene?

107. The standard state of phosphorus at 25 °C is P_4. This molecule has four equivalent P atoms, no double or triple bonds, and no expanded octets. Draw its Lewis structure.

108. The standard heat of formation of $CaBr_2$ is −675 kJ/mol. The first ionization energy of Ca is 590 kJ/mol and its second ionization energy is 1145 kJ/mol. The heat of sublimation of Ca [$Ca(s) \longrightarrow Ca(g)$] is 178 kJ/mol. The bond energy of Br_2 is 193 kJ/mol, the heat of vaporization of $Br_2(l)$ is 31 kJ/mol, and the electron affinity of Br is −325 kJ/mol. Calculate the lattice energy of $CaBr_2$.

109. The standard heat of formation of $PI_3(s)$ is −24.7 kJ/mol and the PI bond energy in this molecule is 184 kJ/mol. The standard heat of formation of $P(g)$ is 334 kJ/mol and that of $I_2(g)$ is 62 kJ/mol. The I_2 bond energy is 151 kJ/mol. Calculate the heat of sublimation of $PI_3[PI_3(s) \longrightarrow PI_3(g)]$.

Conceptual Problems

110. Which of the following is true of an endothermic reaction?
  a. Strong bonds break and weak bonds form
  b. Weak bonds break and strong bonds form.
  c. The bonds that break and those that form are of approximately the same strength.

111. When a firecracker explodes, energy is obviously released. The compounds in the firecracker can be viewed as being "energy rich." What does this mean? Explain the source of the energy in terms of chemical bonds.

112. A fundamental difference between compounds containing ionic bonds and those containing covalent bonds is the existence of molecules. Explain why molecules exist in solid covalent compounds but do not exist in solid ionic compounds.

113. In the very first chapter of this book, we described the scientific method and put a special emphasis on scientific models or theories. In this chapter, we looked carefully at a model for chemical bonding (Lewis theory). Why is this theory successful? Can you name some of the limitations of the theory?

10

Chemical Bonding II: Molecular Shapes, Valence Bond Theory, and Molecular Orbital Theory

10.1 Artificial Sweeteners: Fooled by Molecular Shape

10.2 VSEPR Theory: The Five Basic Shapes

10.3 VSEPR Theory: The Effect of Lone Pairs

10.4 VSEPR Theory: Predicting Molecular Geometries

10.5 Molecular Shape and Polarity

10.6 Valence Bond Theory: Orbital Overlap as a Chemical Bond

10.7 Valence Bond Theory: Hybridization of Atomic Orbitals

10.8 Molecular Orbital Theory: Electron Delocalization

I N CHAPTER 9, WE LEARNED a simple model for chemical bonding called Lewis theory. We saw how this model helps us explain and predict the combinations of atoms that form stable molecules. When we combine Lewis theory with the idea that valence electron groups repel one another—the basis of an approach known as VSEPR theory—we can predict the general shape of a molecule from its Lewis structure. We address molecular shapes and their importance in the first part of this chapter. We then move on to learn two additional bonding theories—called valence bond theory and molecular orbital theory—that are progressively more sophisticated, but at the cost of being more complex, than Lewis theory. As you work through this chapter, our second on chemical bonding, keep in mind the importance of this topic. In our universe, elements join together to form compounds, and that makes many things possible, including our own existence.

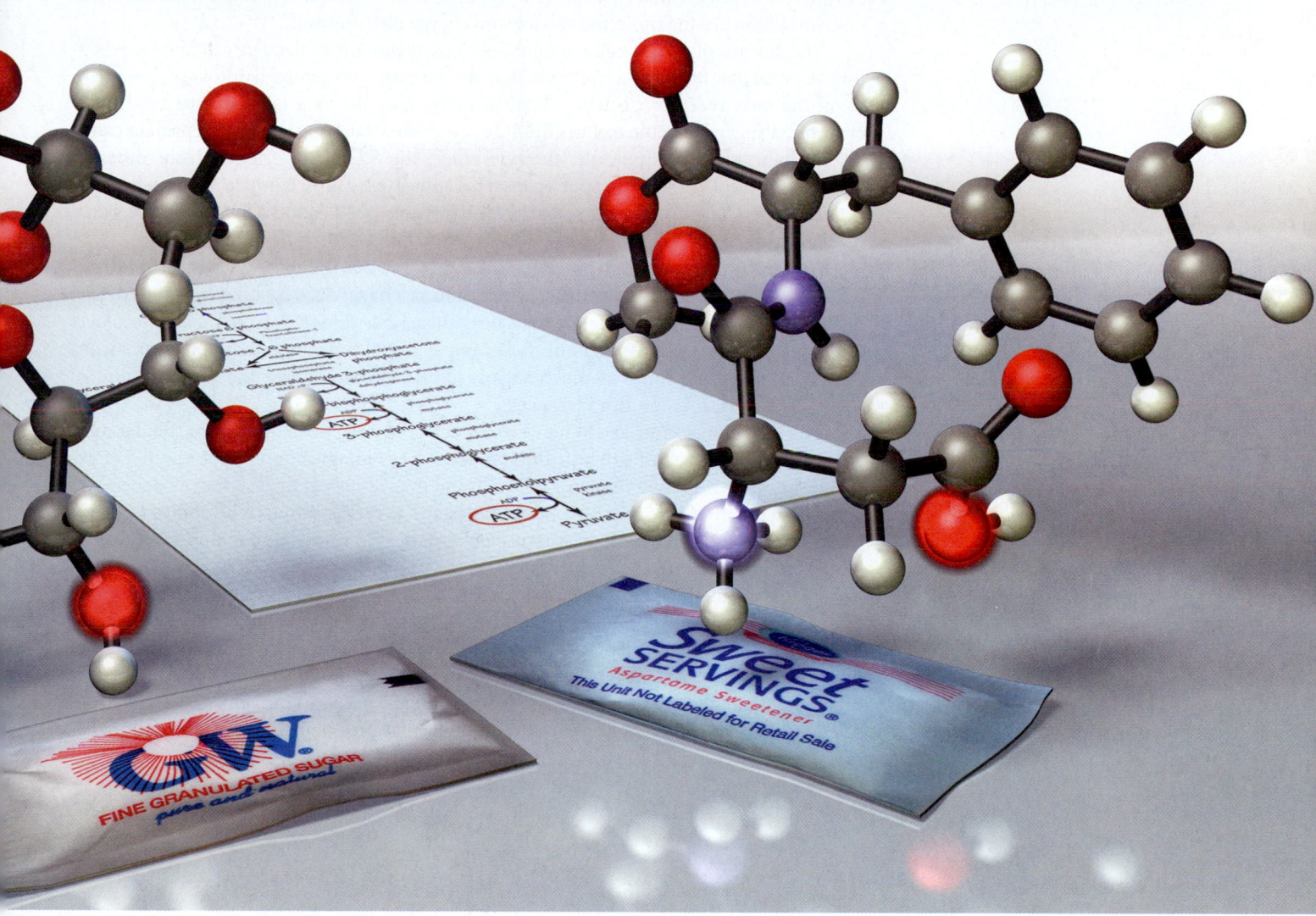

No theory ever solves all the puzzles with which it is confronted at a given time; nor are the solutions already achieved often perfect.
—Thomas Kuhn (1922–1996)

Similarities in the shape of sugar and aspartame give both molecules the ability to stimulate a sweet taste sensation.

10.1 Artificial Sweeteners: Fooled by Molecular Shape

Artificial sweeteners, such as aspartame (Nutrasweet), taste sweet but have few or no calories. Why? *Because taste and caloric value are independent properties of foods.* The caloric value of a food depends on the amount of energy released when the food is metabolized. For example, sucrose (table sugar) is metabolized by oxidation to carbon dioxide and water:

$$C_{12}H_{22}O_{11} + 12\,O_2 \longrightarrow 12\,CO_2 + 11\,H_2O \qquad \Delta H^\circ_{rxn} = -5644\ \text{kJ}$$

When your body metabolizes a mole of sucrose, it obtains 5644 kJ of energy. Some cial sweeteners, such as saccharin, for example, are not metabolized at all—the through the body unchanged—and therefore have no caloric value. Other artificial sweeteners, such as aspartame, are metabolized but have a much lower caloric content (for a given amount of sweetness) than sucrose.

The *taste* of a food, however, is independent of its metabolism. The sensation of taste originates in the tongue, where specialized cells called taste cells act as highly sensitive and specific molecular detectors. These cells can detect sugar molecules out of the thousands of different types of molecules present in a mouthful of food. The main factors for this discrimination are the molecule's shape and charge distribution.

The surface of a taste cell contains specialized protein molecules called taste receptors. A particular *tastant*—a molecule that we can taste—fits snuggly into a special pocket on the taste receptor protein called the *active site*, just as a key fits into a lock. For example, a sugar molecule just fits into the active site of the sugar receptor protein called T1r3. When the sugar molecule (the key) enters the active site (the lock), the different subunits of the T1r3 protein split apart. This split causes ion channels in the cell membrane to open, resulting in nerve signal transmission (see Section 8.1). The nerve signal reaches the brain and registers a sweet taste.

Artificial sweeteners taste sweet because they fit into the receptor pocket that normally binds sucrose. In fact, both aspartame and saccharin actually bind to the active site in the T1r3 protein more strongly than does sugar! For this reason, artificial sweeteners are "sweeter than sugar." Aspartame, for example, is 200 times sweeter than sugar, meaning that it takes 200 times as much sugar as aspartame to trigger the same amount of nerve signal transmission from taste cells.

The type of lock-and-key fit between the active site of a protein and a particular molecule is important not only to taste but to many other biological functions as well. For example, immune response, the sense of smell, and many types of drug action all depend on shape-specific interactions between molecules and proteins. In fact, the ability to determine the shapes of key biological molecules is largely responsible for the revolution in biology that has occurred over the last 50 years.

In this chapter, we look at ways to predict and account for the shapes of molecules. The molecules we examine are much smaller than the protein molecules we just discussed, but the same principles apply to both. The simple model we develop to account for molecular shape is called *valence shell electron pair repulsion* (VSEPR) theory, and it is used in conjunction with Lewis theory. We will then proceed to introduce two additional bonding theories: valence bond theory and molecular orbital theory. These bonding theories are more complex, but also more powerful, than Lewis theory. They also predict and account for molecular shape as well as other properties of molecules.

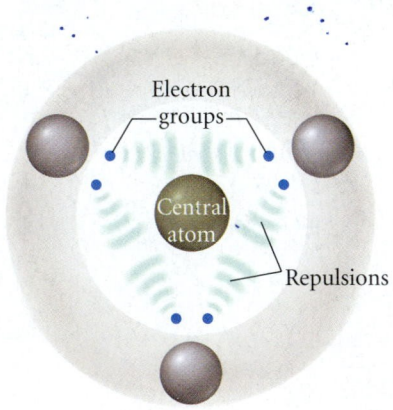

Electron groups

Central atom

Repulsions

▲ **FIGURE 10.1 Repulsion between Electron Groups** The basic idea of
s that repulsions between
s determine molecular

10.2 VSEPR Theory: The Five Basic Shapes

The first theory that we shall consider, **valence shell electron pair repulsion (VSEPR) theory**, is based on the simple idea that **electron groups**—which we define as lone pairs, single bonds, multiple bonds, and even single electrons—repel one another through coulombic forces. The repulsion between electron groups on interior atoms of a molecule, therefore, determines the geometry of the molecule (Figure 10.1 ◀). The preferred geometry is the one in which the electron groups have the maximum separation (and therefore the minimum energy) possible. Consequently, for molecules having just one interior atom—the central atom—molecular geometry depends on (a) the number of electron groups around the central atom and (b) how many of those electron groups are bonding groups and how many are lone pairs. We first look at the molecular geometries associated with two to six electron groups around the central atom when all of those groups are bonding groups (single or multiple bonds). The resulting geometries constitute the five basic shapes of molecules. We will then see how these basic shapes are modified if one or more of the electron groups are lone pairs.

Two Electron Groups: Linear Geometry

Consider the Lewis structure of $BeCl_2$, which has two electron groups (two single bonds) about the central atom:

$$:\ddot{Cl}:Be:\ddot{Cl}:$$

According to VSEPR theory, the geometry of $BeCl_2$ is determined by the repulsion between these two electron groups, which can maximize their separation by assuming a 180° bond angle or a **linear geometry**. Experimental measurements of the geometry of $BeCl_2$ indicate that the molecule is indeed linear, as predicted by the theory.

Molecules that form only two single bonds, with no lone pairs, are rare because they do not follow the octet rule. However, the same geometry is observed in all molecules that have two electron groups (and no lone pairs). For example, consider the Lewis structure of CO_2, which has two electron groups (the double bonds) around the central carbon atom:

$$:\ddot{O}=C=\ddot{O}:$$

According to VSEPR theory, the two double bonds repel each other (just as the two single bonds in $BeCl_2$ repel each other), resulting in a linear geometry for CO_2. Experimental observations confirm that CO_2 is indeed a linear molecule, as predicted by this simple theory.

Three Electron Groups: Trigonal Planar Geometry

Consider the Lewis structure of BF_3 (another molecule with an incomplete octet), which has three electron groups around the central atom:

These three electron groups can maximize their separation by assuming 120° bond angles in a plane—a **trigonal planar geometry**. Experimental observations of the structure of BF_3 are again in agreement with the predictions of VSEPR theory.

Consider also another molecule with three electron groups, formaldehyde, which has one double bond and two single bonds around the central atom:

$$\begin{array}{c} :O: \\ \| \\ H-C-H \end{array}$$

Since formaldehyde has three electron groups around the central atom, we initially predict that the bond angles should also be 120°. However, experimental observations show that the HCO bond angles are 121.9° and that the HCH bond angle is 116.2°. These bond angles are close to the idealized 120° that we originally predicted. However, the HCO bond angles are slightly greater than the HCH bond angle because the double bond contains more electron density than the single bond and therefore exerts a slightly greater repulsion on the single bonds. In general, *different types of electron groups exert slightly different repulsions*—the resulting bond angles reflect these differences.

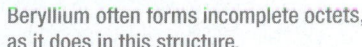

Beryllium often forms incomplete octets, as it does in this structure.

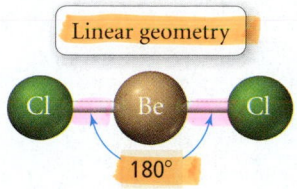

Linear geometry

A double bond counts as one electron group.

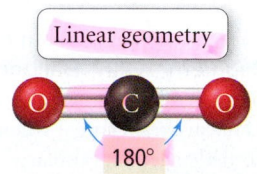

Linear geometry

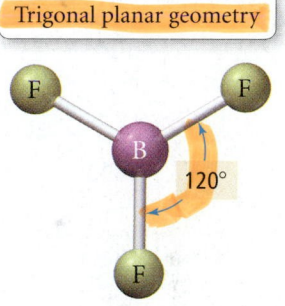

Trigonal planar geometry

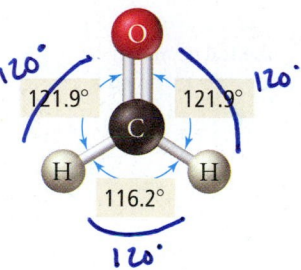

 Conceptual Connection 10.1 Electron Groups and Molecular Geometry

In determining electron geometry, why do we consider only the electron groups on the central atom? In other words, why don't we consider electron groups on terminal atoms?

Answer: The geometry of a molecule is determined by how the terminal atoms are arranged around the central atom, which is in turn determined by how the electron groups are arranged around the central atom. The electron groups on the terminal atoms do not affect this arrangement.

Four Electron Groups: Tetrahedral Geometry

The VSEPR geometries of molecules with two or three electron groups around the central atom are two-dimensional and can therefore easily be visualized and represented on paper. For molecules with four or more electron groups around the central atom, the geometries are three-dimensional and are therefore more difficult to imagine and to draw. One common way to help visualize these basic shapes is by analogy to balloons tied together. In this analogy, each electron group around a central atom is like a balloon tied to a central point.

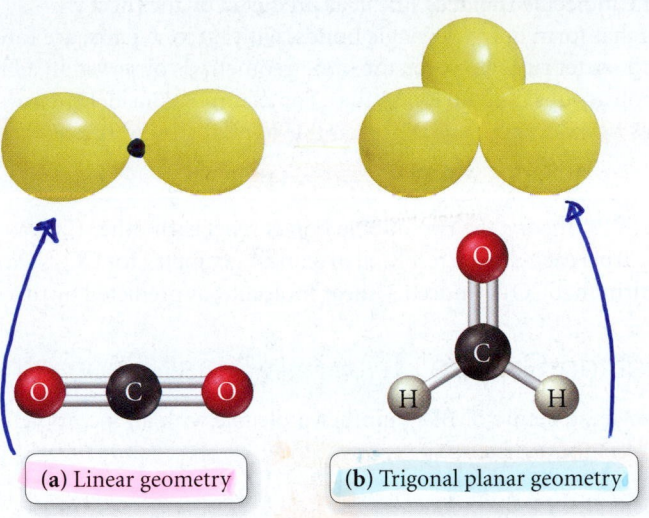

▶ **FIGURE 10.2 Representing Electron Geometry with Balloons (a)** The bulkiness of balloons causes them to assume a linear arrangement when two of them are tied together. Similarly, the repulsion between two electron groups produces a linear geometry. **(b)** Like three balloons tied together, three electron groups adopt a trigonal planar geometry.

(a) Linear geometry

(b) Trigonal planar geometry

The bulkiness of the balloons causes them to spread out as much as possible, much as the repulsion between electron groups causes them to position themselves as far apart as possible. For example, if you tie two balloons together, they assume a roughly linear arrangement, as shown in Figure 10.2(a) ▲, analogous to the linear geometry of $BeCl_2$ that we just examined. Notice that the balloons do not represent atoms, but *electron groups*. Similarly, if you tie three balloons together—in analogy to three electron groups—they assume a trigonal planar geometry, as shown in Figure 10.2(b), much like our BF_3 molecule. If you tie *four* balloons together, however, they assume a three-dimensional **tetrahedral geometry** with 109.5° angles between the balloons. That is, the balloons point toward the vertices of a *tetrahedron*—a geometrical shape with four identical faces, each an equilateral triangle, as shown at left.

109.5°

Tetrahedral geometry

Tetrahedron

As an example of a molecule with four electron groups around the central atom, consider methane:

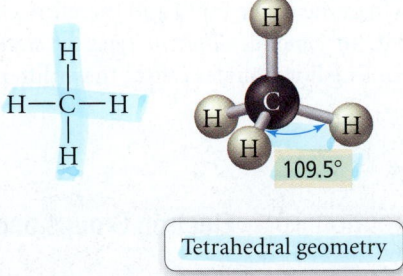

109.5°

Tetrahedral geometry

For four electron groups, the tetrahedron is the three-dimensional shape that allows the maximum separation among the groups. The repulsions among the four electron groups in the C—H bonds thus cause the molecule to assume the tetrahedral shape. When we write the Lewis structure of CH_4 on paper, it may seem that the molecule should be square planar, with bond angles of 90°. However, in three dimensions, the electron groups can get farther away from each other by forming the tetrahedral geometry, as shown by our balloon analogy.

Five Electron Groups: Trigonal Bipyramidal Geometry

Five electron groups around a central atom assume a **trigonal bipyramidal geometry**, like that of five balloons tied together. In this structure, three of the groups lie in a single plane, as in the trigonal planar configuration, while the other two are positioned above and below this plane. The angles in the trigonal bipyramidal structure are not all the same. The angles between the *equatorial positions* (the three bonds in the trigonal plane) are 120°, while the angle between the *axial positions* (the two bonds on either side of the trigonal plane) and the trigonal plane is 90°. As an example of a molecule with five electron groups around the central atom, consider PCl_5:

Trigonal bipyramidal geometry

Trigonal bipyramid

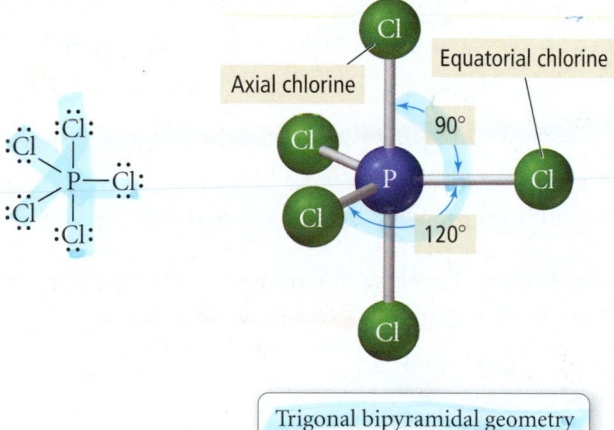

Trigonal bipyramidal geometry

The three equatorial chlorine atoms are separated by 120° bond angles and the two axial chlorine atoms are separated from the equatorial atoms by 90° bond angles.

Six Electron Groups: Octahedral Geometry

Six electron groups around a central atom assume an **octahedral geometry**, like that of six balloons tied together. In this structure—named after the eight-sided geometrical shape called the octahedron—four of the groups lie in a single plane, with one group above the plane and another below it. The angles in this geometry are all 90°. As an example of a molecule with six electron groups around the central atom, consider SF_6:

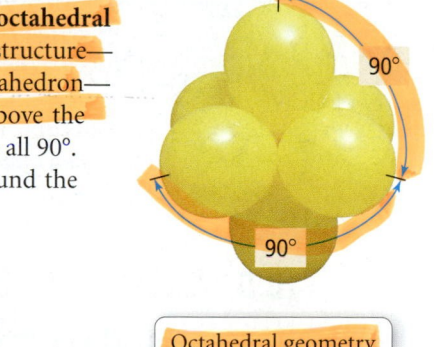

Octahedral geometry

Octahedron

Octahedral geometry

You can see that the structure of this molecule is highly symmetrical. All six bonds and therefore all six atoms are equivalent.

EXAMPLE 10.1 VSEPR Theory and the Basic Shapes

Determine the molecular geometry of NO_3^-.

Solution

The molecular geometry of NO_3^- is determined by the number of electron groups around the central atom (N). Begin by drawing a Lewis structure of NO_3^-.	NO_3^- has $5 + 3(6) + 1 = 24$ valence electrons. The Lewis structure is as follows: $$\left[:\ddot{O}-N-\ddot{O}: \atop \underset{:\ddot{O}:}{\|} \right]^- \longleftrightarrow \left[\ddot{O}=N-\ddot{O}: \atop \underset{:\ddot{O}:}{\|} \right]^- \longleftrightarrow \left[:\ddot{O}-N=\ddot{O} \atop \underset{:\ddot{O}:}{\|} \right]^-$$ The hybrid structure is intermediate between these three and has three equivalent bonds.
Use the Lewis structure, or any one of the resonance structures, to determine the number of electron groups around the central atom.	$$\left[:\ddot{O}-N-\ddot{O}: \atop \underset{:\ddot{O}:}{\|} \right]^-$$ The nitrogen atom has three electron groups.
Based on the number of electron groups, determine the geometry that minimizes the repulsions between the groups.	The electron geometry that minimizes the repulsions between three electron groups is trigonal planar. 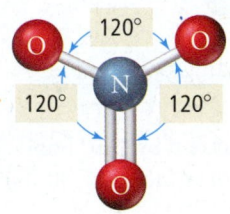 Since the three bonds are equivalent, they each exert the same repulsion on the other two and the molecule has three equal bond angles of 120°.

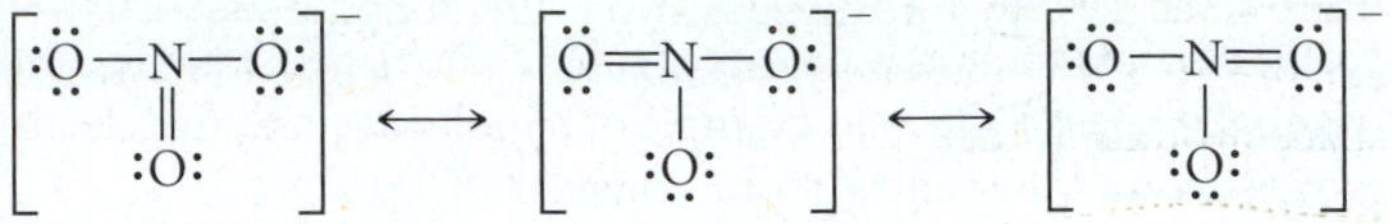

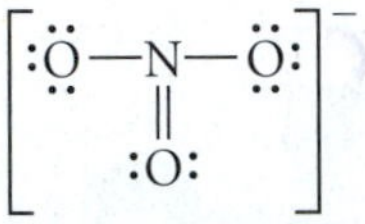

For Practice 10.1

Determine the molecular geometry of CCl_4.

10.3 VSEPR Theory: The Effect of Lone Pairs

Each of the examples we have just examined has only bonding electron groups around the central atom. What happens in molecules that also have lone pairs around the central atom? These lone pairs also repel other electron groups, as we see in the examples that follow.

Four Electron Groups with Lone Pairs

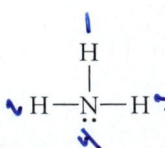

Consider the Lewis structure of ammonia, shown at left. The central nitrogen atom has four electron groups (one lone pair and three bonding pairs) that repel one another. If we do not distinguish between bonding electron groups and lone pairs, we find that the **electron geometry**—the geometrical arrangement of the *electron groups*—is still tetrahedral, as we expect for four electron groups. However, the **molecular geometry**—the geometrical arrangement of the atoms—is **trigonal pyramidal**, as shown on the next page.

Notice that although the electron geometry and the molecular geometry are different, *the electron geometry is relevant to the molecular geometry.*

In other words, the lone pair exerts its influence on the bonding pairs. As we saw previously, different kinds of electron groups generally result in different amounts of repulsion. Lone pair electrons generally exert slightly greater repulsions than bonding electrons. If all four electron groups in NH_3 exerted equal repulsions on one another, the bond angles in the molecule would all be the ideal tetrahedral angle, 109.5°. However, the actual angle between N—H bonds in ammonia is slightly smaller, 107°. A lone electron pair is more spread out in space than a bonding electron pair because the lone pair is attracted to only one nucleus while the bonding pair is attracted to two (Figure 10.3 ▼). In other words, the lone pair occupies more of the angular space around a nucleus, exerting a greater repulsive force on neighboring electrons and compressing the N—H bond angles.

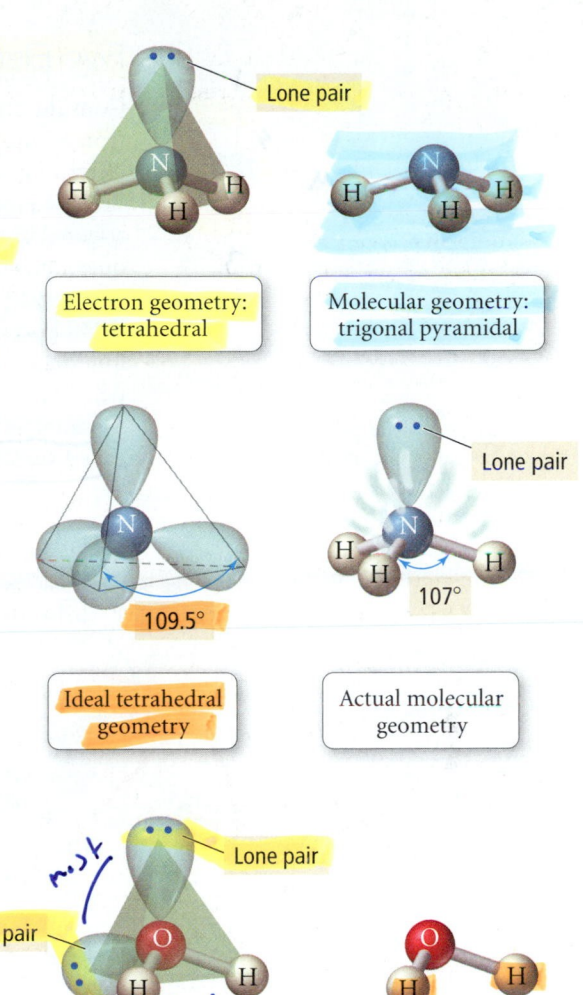

Electron geometry: tetrahedral

Molecular geometry: trigonal pyramidal

Ideal tetrahedral geometry — 109.5°

Actual molecular geometry — 107°

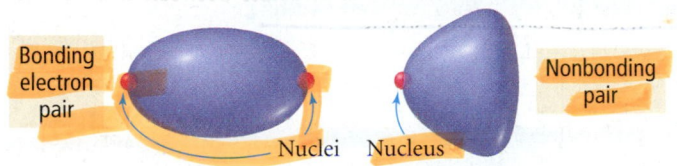

▲ FIGURE 10.3 Nonbonding versus Bonding Electron Pairs A nonbonding electron pair occupies more space than a bonding pair.

Consider also H_2O. Its Lewis structure is

$$ H-\ddot{O}-H $$

Since it has four electron groups (two bonding pairs and two lone pairs), its *electron geometry* is also tetrahedral, but its *molecular geometry* is **bent**, as shown at right. As in NH_3, the bond angles in H_2O are smaller (104.5°) than the ideal tetrahedral bond angles because of the greater repulsion exerted by the lone pair electrons. Notice that the bond angle in H_2O is even smaller than in NH_3 because H_2O has *two* lone pairs of electrons on the central oxygen atom. These lone pairs compress the H_2O bond angle to an even greater extent than in NH_3. In general, electron group repulsions vary as follows:

Lone pair–lone pair > Lone pair–bonding pair > Bonding pair–bonding pair
Most repulsive   Least repulsive

We see the effects of this ordering in the progressively smaller bond angles of CH_4, NH_3, and H_2O, as shown in Figure 10.4 ▼. The relative ordering of repulsions also plays a role in determining the geometry of molecules with five and six electron groups when one or more of those groups are lone pairs, as we shall now see.

Electron geometry: tetrahedral

Molecular geometry: bent

Ideal tetrahedral geometry — 109.5°

Actual molecular geometry — 104.5°

Effect of Lone Pairs on Molecular Geometry

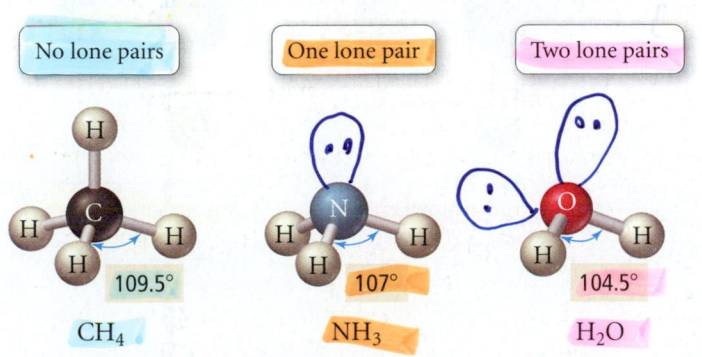

No lone pairs — CH_4 — 109.5°

One lone pair — NH_3 — 107°

Two lone pairs — H_2O — 104.5°

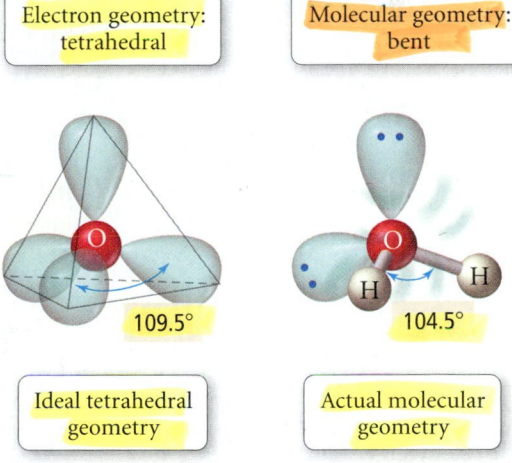

◄ FIGURE 10.4 The Effect of Lone Pairs on Molecular Geometry The bond angles get progressively smaller as the number of lone pairs on the central atom increases from zero in CH_4 to one in NH_3 to two in H_2O.

Five Electron Groups with Lone Pairs

Consider the Lewis structure of SF_4, shown at left. The central sulfur atom has five electron groups (one lone pair and four bonding pairs). The *electron geometry*, due to the five electron groups, is trigonal bipyramidal. In determining the molecular geometry, notice that the lone pair can occupy either an equatorial position or an axial position within the trigonal bipyramidal electron geometry. Which position is most favorable? To answer this question, we must consider that, as we have just seen, lone pair–bonding pair repulsions are greater than bonding pair–bonding pair repulsions. Consequently, the lone pair should occupy the position that minimizes its interaction with the bonding pairs. If the lone pair were in an axial position, it would have three 90° interactions with bonding pairs. In an equatorial position, however, it has only two 90° interactions. Consequently, the lone pair occupies an equatorial position and the resulting molecular geometry is called **seesaw**, because it resembles a seesaw (or teeter-totter).

The seesaw molecular geometry is sometimes called an *irregular tetrahedron*.

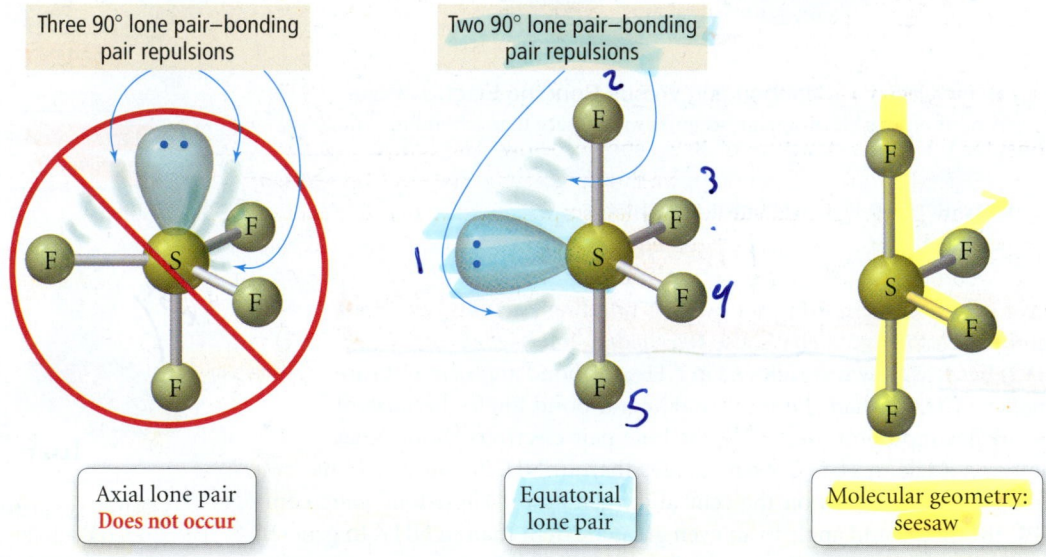

| Axial lone pair Does not occur | Equatorial lone pair | Molecular geometry: seesaw |

Three 90° lone pair–bonding pair repulsions

Two 90° lone pair–bonding pair repulsions

When two of the five electron groups around the central atom are lone pairs, as in BrF_3, the lone pairs occupy two of the three equatorial positions—again minimizing 90° interactions with bonding pairs and also avoiding a lone pair–lone pair 90° repulsion. The resulting molecular geometry is **T-shaped**.

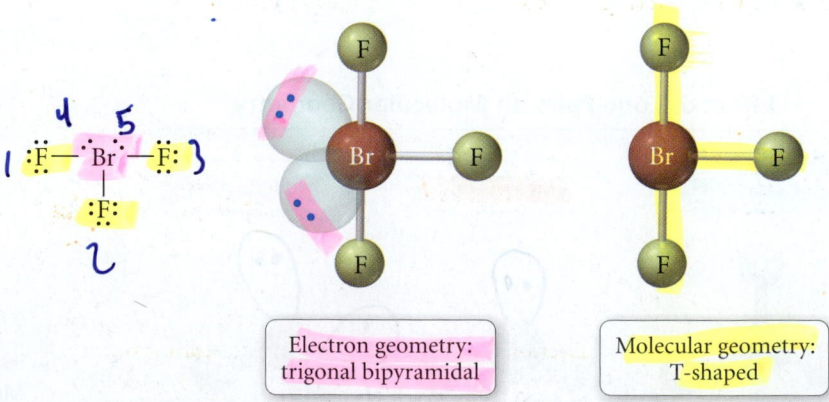

| Electron geometry: trigonal bipyramidal | Molecular geometry: T-shaped |

When three of the five electron groups around the central atom are lone pairs, as in XeF_2, the lone pairs occupy all three of the equatorial positions and the resulting molecular geometry is linear.

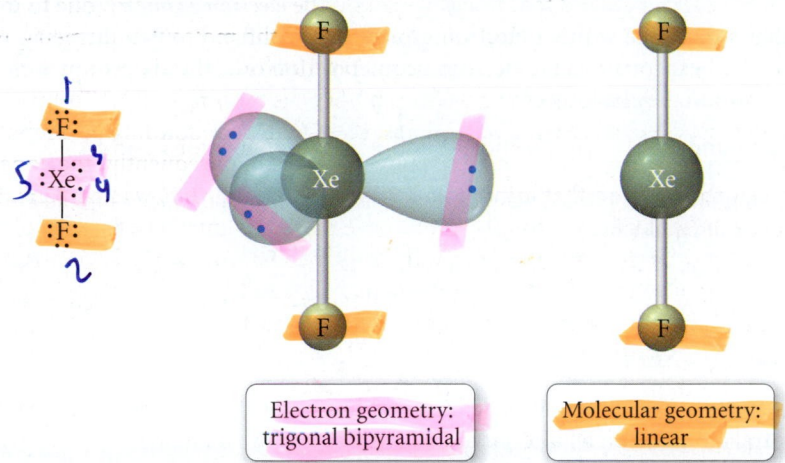

Electron geometry:
trigonal bipyramidal

Molecular geometry:
linear

Six Electron Groups with Lone Pairs

Consider the Lewis structure of BrF_5, shown below. The central bromine atom has six electron groups (one lone pair and five bonding pairs). The electron geometry, due to the six electron groups, is octahedral. Since all six positions in the octahedral geometry are equivalent, the lone pair can be situated in any one of these positions. The resulting molecular geometry is called **square pyramidal**.

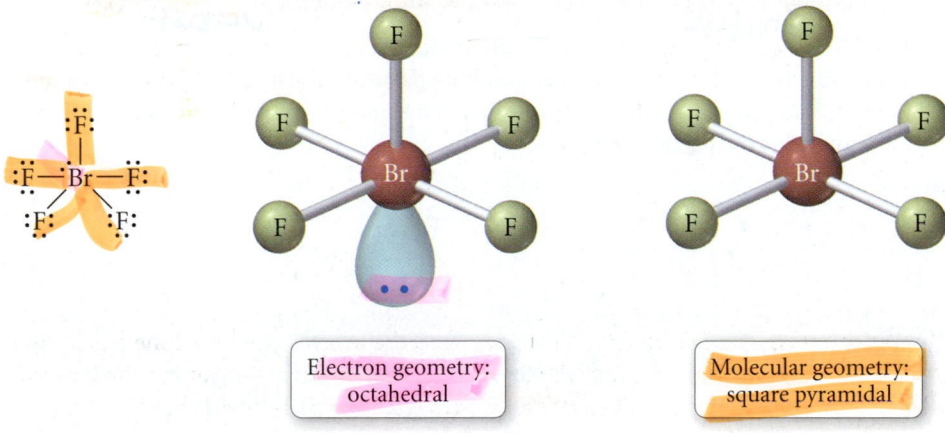

Electron geometry:
octahedral

Molecular geometry:
square pyramidal

When two of the six electron groups around the central atom are lone pairs, as in XeF_4, the lone pairs occupy positions across from one another (to minimize lone pair–lone pair repulsions), and the resulting molecular geometry is **square planar**.

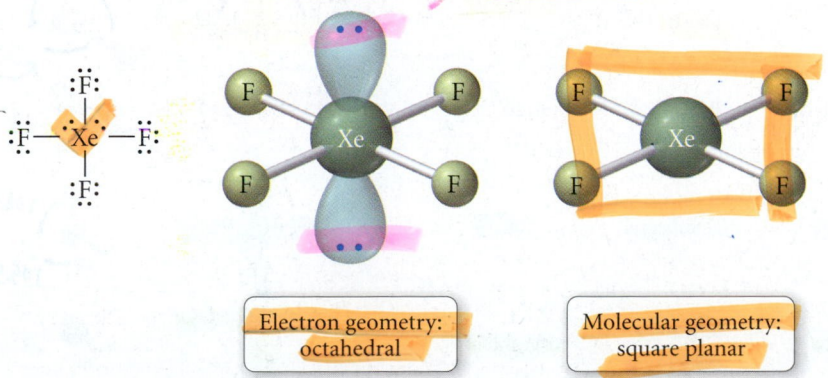

Electron geometry:
octahedral

Molecular geometry:
square planar

Conceptual Connection 10.2 Lone Pair Electrons and Molecular Geometry

Suppose that a molecule with six electron groups were confined to two dimensions and therefore had a hexagonal planar electron geometry. If two of the six groups were lone pairs, where would they be located?

(a) positions 1 and 2   **(b)** positions 1 and 3   **(c)** positions 1 and 4

Answer: **(c)** Positions 1 and 4 would put the greatest distance between the lone pairs and minimize lone pair–lone pair repulsions.

Summary of VSEPR Theory

➤ The geometry of a molecule is determined by the number of electron groups on the central atom (or on all interior atoms, if there is more than one).

➤ The number of electron groups can be determined from the Lewis structure of the molecule. If the Lewis structure contains resonance structures, use any one of the resonance structures to determine the number of electron groups.

➤ Each of the following counts as a single electron group: a lone pair, a single bond, a double bond, a triple bond, or a single electron.

➤ The geometry of the electron groups is determined by minimizing their repulsions as summarized in Table 10.1. In general, electron group repulsions vary as follows:

Lone pair–lone pair > lone pair–bonding pair > bonding pair–bonding pair

➤ Bond angles can vary from the idealized angles because double and triple bonds occupy more space than single bonds, and lone pairs occupy more space than bonding groups. The presence of lone pairs will usually make bond angles smaller than the ideal angle for the particular geometry.

TABLE 10.1   Electron and Molecular Geometries

Electron Groups*	Bonding Groups	Lone Pairs	Electron Geometry	Molecular Geometry	Approximate Bond Angles	Example
2	2	0	Linear	Linear	180°	$:\ddot{O}=C=\ddot{O}:$
3	3	0	Trigonal planar	Trigonal planar	120°	$:\ddot{F}-B-\ddot{F}:$ with $:F:$
3	2	1	Trigonal planar	Bent	<120°	$:\ddot{O}=\ddot{S}-\ddot{O}:$
4	4	0	Tetrahedral	Tetrahedral	109.5°	$H-C-H$ with H top and bottom
4	3	1	Tetrahedral	Trigonal pyramidal	<109.5°	$H-\ddot{N}-H$ with H below
4	2	2	Tetrahedral	Bent	<109.5°	$H-\ddot{O}-H$

90° 180°

5	5	0	Trigonal bipyramidal	Trigonal bipyramidal	120° (equatorial) 90° (axial)	
5	4	1	Trigonal bipyramidal	Seesaw	<120° (equatorial) <90° (axial)	
5	3	2	Trigonal bipyramidal	T-shaped	<90°	
5	2	3	Trigonal bipyramidal	Linear	180°	

90°

6	6	0	Octahedral	Octahedral	90°	
6	5	1	Octahedral	Square pyramidal	<90°	
6	4	2	Octahedral	Square planar	90°	

*Count only electron groups around the central atom. Each of the following is considered one electron group: a lone pair, a single bond, a double bond, a triple bond, or a single electron.

Conceptual Connection 10.3 Molecular Geometry and Electron Groups

Which of the following statements is always true according to VSEPR theory?

(a) The shape of a molecule is determined by repulsions among bonding electron groups.

(b) The shape of a molecule is determined by repulsions among nonbonding electron groups.

(c) The shape of a molecule is determined by the polarity of its bonds.

(d) The shape of a molecule is determined by repulsions among all electron groups on the central atom.

Answer: (d) All electron groups on the central atom determine the shape of a molecule according to VSEPR theory.

10.4 VSEPR Theory: Predicting Molecular Geometries

To determine the geometry of a molecule, follow the procedure below. As usual, we give the steps in the left column and provide two examples of applying the steps in the center and right columns.

Procedure for Predicting Molecular Geometries

1. Draw a Lewis structure for the molecule.

2. Determine the total number of electron groups around the central atom. Lone pairs, single bonds, double bonds, triple bonds, and single electrons each count as one group.

3. Determine the number of bonding groups and the number of lone pairs around the central atom. These should sum to the result from step 2. Bonding groups include single bonds, double bonds, and triple bonds.

4. Use Table 10.1 to determine the electron geometry and molecular geometry. If no lone pairs are present around the central atom, the bond angles will be that of the ideal geometry. If lone pairs are present, the bond angles may be smaller than the ideal geometry.

EXAMPLE 10.2
Predicting Molecular Geometries

Predict the geometry and bond angles of PCl_3.

PCl_3 has 26 valence electrons.

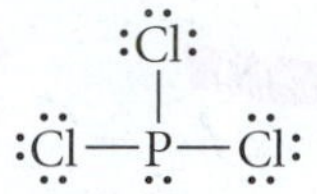

The central atom (P) has four electron groups.

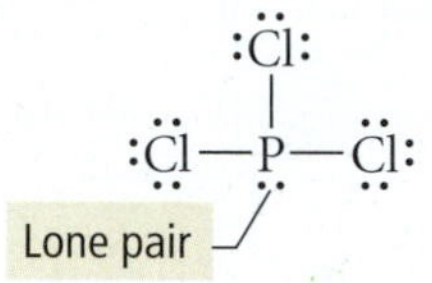

Three of the four electron groups around P are bonding groups and one is a lone pair.

The electron geometry is tetrahedral (four electron groups) and the molecular geometry—the shape of the molecule—is *trigonal pyramidal* (three bonding groups and one lone pair). Because of the presence of a lone pair, the bond angles are less than 109.5°.

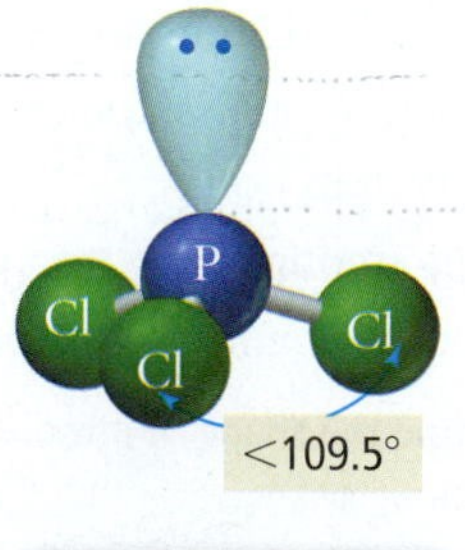

<109.5°

Trigonal pyramidal

For Practice 10.2

Predict the molecular geometry and bond angle of ClNO.

EXAMPLE 10.3
Predicting Molecular Geometries

Predict the geometry and bond angles of ICl_4^-

ICl_4^- has 36 valence electrons.

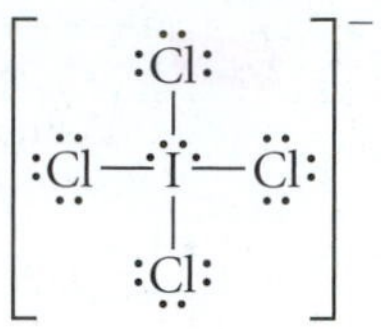

The central atom (I) has six electron groups.

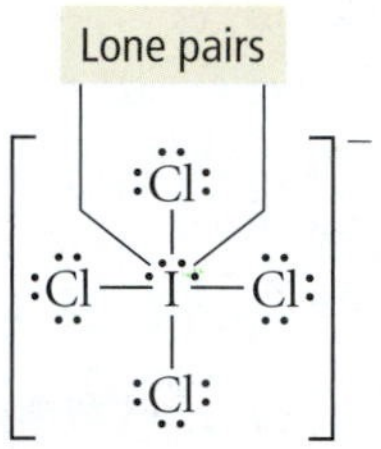

Four of the six electron groups around I are bonding groups and two are lone pairs.

The electron geometry is octahedral (six electron groups) and the molecular geometry—the shape of the molecule—is *square planar* (four bonding groups and two lone pairs). Even though lone pairs are present, the bond angles are 90° because the lone pairs are symmetrically arranged and do not compress the I—Cl bond angles.

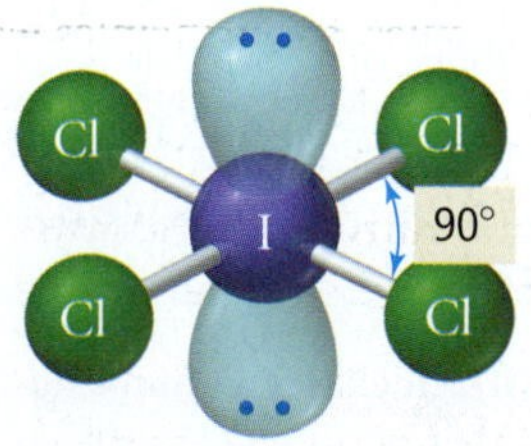

90°

Square planar

For Practice 10.3

Predict the molecular geometry of I_3^-.

Representing Molecular Geometries on Paper

Since molecular geometries are three-dimensional, they are often difficult to represent on two-dimensional paper. Many chemists use the following notation for bonds to indicate three-dimensional structures on two-dimensional paper.

Straight line Bond in plane of paper	*Hatched wedge* Bond going into the page	*Solid wedge* Bond coming out of the page

Some examples of the molecular geometries used in this book are shown below using this notation.

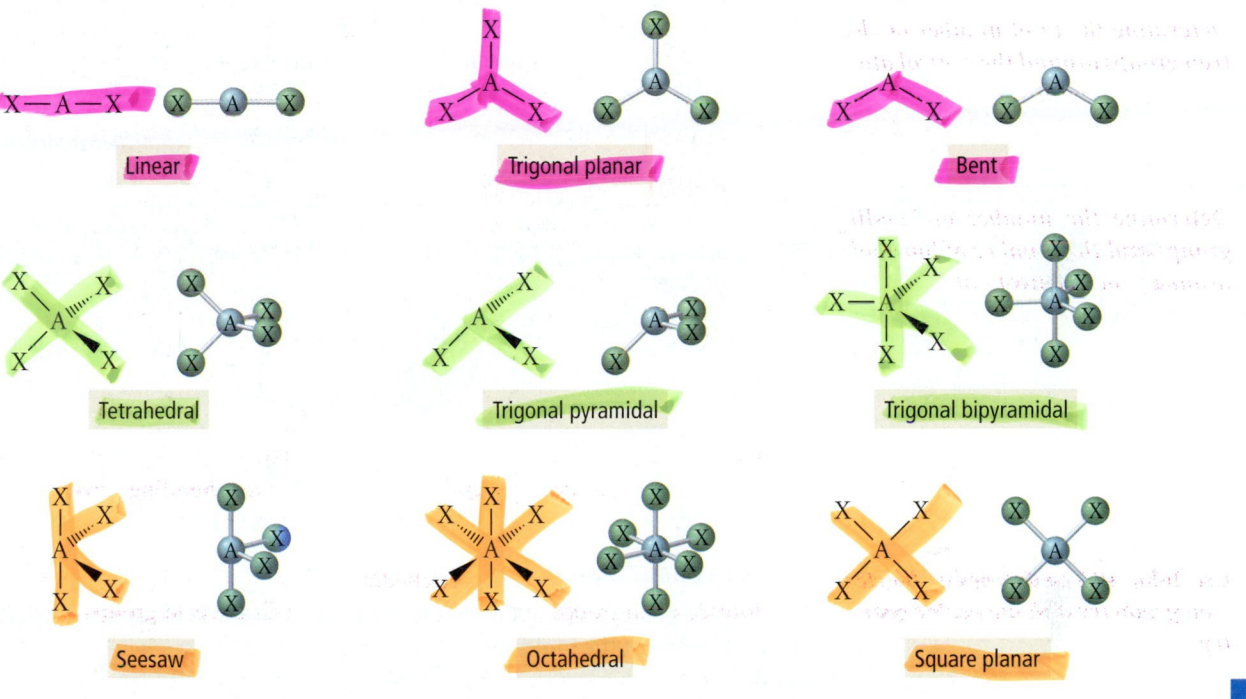

Linear   Trigonal planar   Bent

Tetrahedral   Trigonal pyramidal   Trigonal bipyramidal

Seesaw   Octahedral   Square planar

Predicting the Shapes of Larger Molecules

Larger molecules may have two or more *interior* atoms. When predicting the shapes of these molecules, the principles we just covered must be applied to each interior atom. For example, consider glycine, an amino acid found in many proteins such as those involved in taste that we discussed in Section 10.1. Glycine, shown at right, contains four interior atoms: one nitrogen atom, two carbon atoms, and an oxygen atom. To determine the shape of glycine, we must determine the geometry about each interior atom as follows:

Four interior atoms

Atom	Number of Electron Groups	Number of Lone Pairs	Molecular Geometry
Nitrogen	4	1	Trigonal pyramidal
Leftmost carbon	4	0	Tetrahedral
Rightmost carbon	3	0	Trigonal planar
Oxygen	4	2	Bent

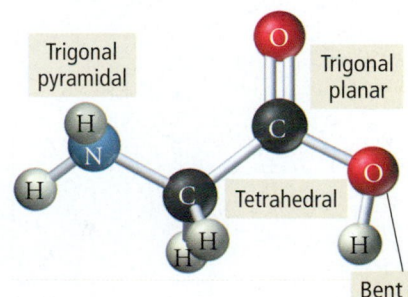

Using the geometries of each of these, we can determine the entire three-dimensional shape of the molecule as shown here.

EXAMPLE 10.4 Predicting the Shape of Larger Molecules

Predict the geometry about each interior atom in methanol (CH_3OH) and make a sketch of the molecule.

Solution

Begin by drawing the Lewis structure of CH_3OH. CH_3OH contains two interior atoms: one carbon atom and one oxygen atom. To determine the shape of methanol, determine the geometry about each interior atom as follows:

$$H-\overset{\overset{\textstyle H}{|}}{\underset{\underset{\textstyle H}{|}}{C}}-\overset{\cdot\cdot}{\underset{\cdot\cdot}{O}}-H$$

Atom	Number of Electron Groups	Number of Lone Pairs	Molecular Geometry
Carbon	4	0	Tetrahedral
Oxygen	4	2	Bent

Using the geometries of each of these, draw a three-dimensional sketch of the molecule as shown here.

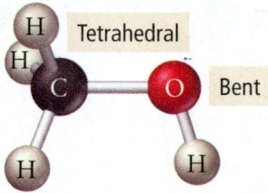

For Practice 10.4

Predict the geometry about each interior atom in acetic acid ($H_3C-\overset{\overset{\textstyle O}{||}}{C}-OH$) and make a sketch of the molecule.

10.5 Molecular Shape and Polarity

In Chapter 9, we discussed polar bonds. Entire molecules can also be polar, depending on their shape and the nature of their bonds. For example, if a diatomic molecule has a polar bond, the molecule as a whole will be polar.

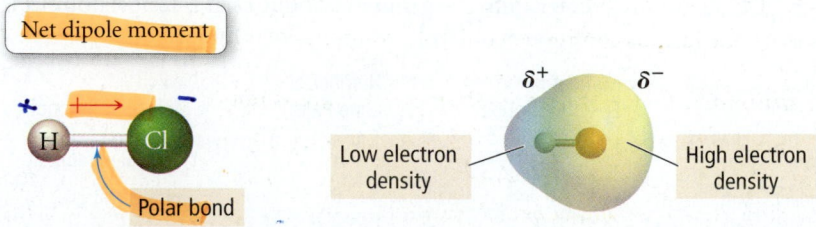

The figure at the right is an electron density model of HCl. Yellow indicates moderately high electron density, red indicates very high electron density, and blue indicates low electron density. Notice that the electron density is greater around the more electronegative atom (chlorine). Thus the molecule itself is polar. If the bond in a diatomic molecule is nonpolar, the molecule as a whole will be nonpolar.

In polyatomic molecules, the presence of polar bonds may or may not result in a polar molecule, depending on the molecular geometry. If the molecular geometry is such that the dipole moments of individual polar bonds sum together to a net dipole moment, then the molecule will be polar. However, if the molecular geometry is such that the dipole moments of the individual polar bonds cancel each other (that is, sum to zero), then the molecule will be nonpolar. It all depends on the geometry of the molecule. For example, consider carbon dioxide:

$$:\ddot{O}=C=\ddot{O}:$$

Each C=O bond in CO_2 is polar because oxygen and carbon have significantly different electronegativities (3.5 and 2.5, respectively). However, since CO_2 is a linear molecule, the polar bonds directly oppose one another and the dipole moment of one bond exactly opposes the dipole moment of the other—the two dipole moments sum to zero and the molecule is nonpolar. Dipole moments can cancel each other because they are *vector quantities*; they have both a magnitude and a direction. Think of each polar bond as a vector, pointing in the direction of the more electronegative atom. The length of the vector is proportional to the electronegativity difference between the bonds. In CO_2, we have two identical vectors pointing in exactly opposite directions—the vectors sum to zero, much as $+1$ and -1 sum to zero:

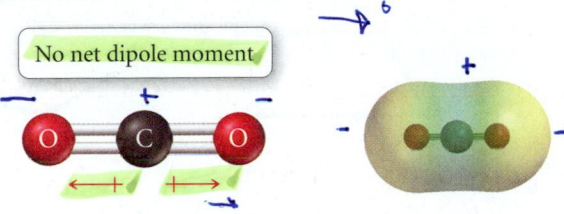

Notice that the electron density model shows regions of moderately high electron density (yellow) positioned symmetrically on either end of the molecule with a region of low electron density (blue) located in the middle.

In contrast, consider water:

$$H—\ddot{O}—H$$

The O—H bonds in water are also polar; oxygen and hydrogen have electronegativities of 3.5 and 2.1, respectively. However, the water molecule is not linear but bent, so the two dipole moments do not sum to zero. If we imagine each bond as a vector pointing toward oxygen (the more electronegative atom) we see that, because of the angle between the vectors, they do not cancel, but sum to an overall vector or a net dipole moment.

See the box on p. 420 for an explanation of how to add vectors.

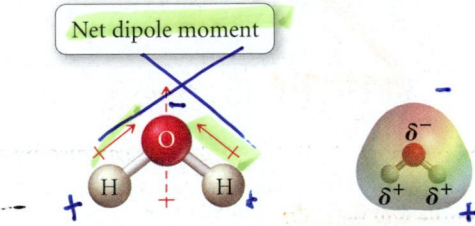

The electron density model shows a region of very high electron density at the oxygen end of the molecule. Consequently, water is a polar molecule. Table 10.2 (p. 421) summarizes whether or not various common geometries result in polar molecules.

In summary, to determine whether a molecule is polar:

▶ *Draw a Lewis structure for the molecule and determine the molecular geometry.*

▶ *Determine whether the molecule contains polar bonds.* A bond is polar if the two bonding atoms have sufficiently different electronegativities (see Figure 9.10). If the molecule contains polar bonds, superimpose a vector, pointing toward the more electronegative atom, on each bond. Make the length of the vector proportional to the electronegativity difference between the bonding atoms.

▶ *Determine whether the polar bonds add together to form a net dipole moment.* Sum the vectors corresponding to the polar bonds together. If the vectors sum to zero, the molecule is nonpolar. If the vectors sum to a net vector, the molecule is polar.

Vector Addition

As discussed previously, we can determine whether a molecule is polar by summing the vectors associated with the dipole moments of all the polar bonds in the molecule. If the vectors sum to zero, the molecule will be nonpolar. If they sum to a net vector, the molecule will be polar. In this box, we show how to add vectors together in one dimension and in two or more dimensions.

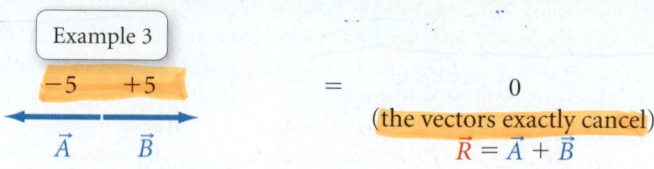

Example 3

-5 $+5$ $=$ 0

(the vectors exactly cancel)

$\vec{A}$ $\vec{B}$ $\vec{R} = \vec{A} + \vec{B}$

One Dimension

To add two vectors that lie on the same line, assign one direction as positive. Vectors pointing in that direction have positive magnitudes. Vectors pointing in the opposite direction are considered to have negative magnitudes. Then simply sum the vectors (always remembering to include their signs), as shown in the following examples.

Two or More Dimensions

To add two vectors, draw a parallelogram in which the two vectors form two of the four sides of the parallelogram. Draw the other two sides of the parallelogram parallel to and the same length as the two original vectors. Draw the resultant vector beginning at the origin and extending to the far corner of the parallelogram.

Example 1

$+5$ $+5$ $=$ $+10$

$\vec{A}$ $\vec{B}$ $\vec{R} = \vec{A} + \vec{B}$

Example 2

-5 $+10$ $=$ $+5$

$\vec{A}$ $\vec{B}$ $\vec{R} = \vec{A} + \vec{B}$

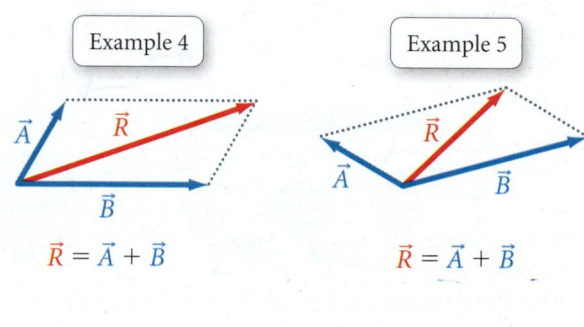

Example 4

$\vec{A}$ $\vec{R}$ $\vec{B}$

$\vec{R} = \vec{A} + \vec{B}$

Example 5

$\vec{R}$ $\vec{A}$ $\vec{B}$

$\vec{R} = \vec{A} + \vec{B}$

To add three or more vectors, add two of them together first, and then add the third vector to the result.

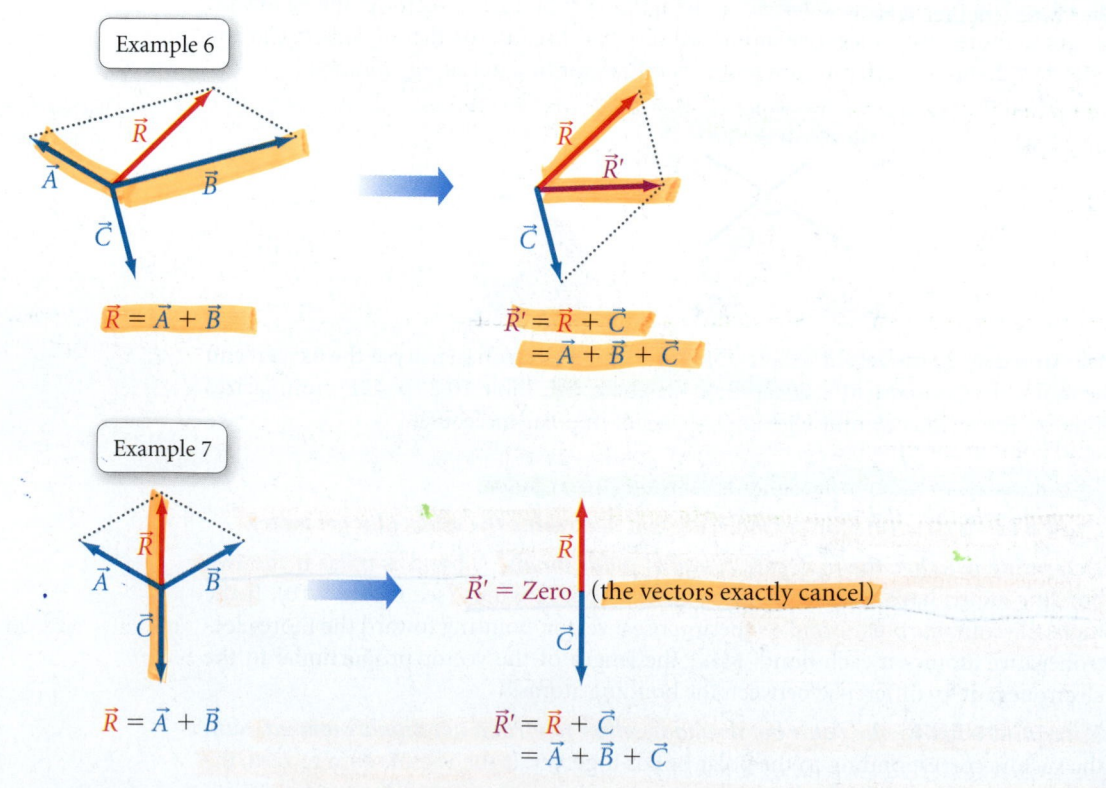

Example 6

$\vec{R}$ $\vec{A}$ $\vec{B}$ $\vec{C}$

$\vec{R} = \vec{A} + \vec{B}$

$\vec{R}$ $\vec{R}'$ $\vec{C}$

$\vec{R}' = \vec{R} + \vec{C}$
$= \vec{A} + \vec{B} + \vec{C}$

Example 7

$\vec{R}$ $\vec{A}$ $\vec{B}$ $\vec{C}$

$\vec{R} = \vec{A} + \vec{B}$

$\vec{R}$ $\vec{R}' = $ Zero   (the vectors exactly cancel) $\vec{C}$

$\vec{R}' = \vec{R} + \vec{C}$
$= \vec{A} + \vec{B} + \vec{C}$

TABLE 10.2 Common Cases of Adding Dipole Moments to Determine whether a Molecule Is Polar

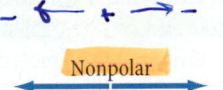

Nonpolar

The dipole moments of two identical polar bonds pointing in opposite directions will cancel. The molecule is nonpolar.

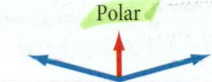

Polar

The dipole moments of two polar bonds with an angle of less than 180° between them will not cancel. The resultant dipole moment vector is shown in red. The molecule is polar.

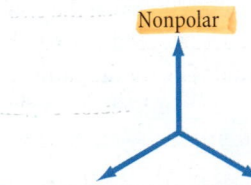

Nonpolar

The dipole moments of three identical polar bonds at 120° from each other will cancel. The molecule is nonpolar.

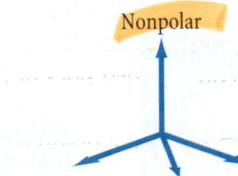

Nonpolar

The dipole moments of four identical polar bonds in a tetrahedral arrangement (109.5° from each other) will cancel. The molecule is nonpolar.

Polar

The dipole moments of three polar bonds in a trigonal pyramidal arrangement (109.5° from each other) will not cancel. The resultant dipole moment vector is shown in red. The molecule is polar.

Note: In all cases where the dipoles of two or more polar bonds cancel, the bonds are assumed to be identical. If one or more of the bonds are different from the other(s), the dipoles will not cancel and the molecule will be polar.

EXAMPLE 10.5 Determining whether a Molecule Is Polar

Determine whether NH_3 is polar.

Solution

Draw a Lewis structure for the molecule and determine the molecular geometry.	H | H—N—H .. The Lewis structure has three bonding groups and one lone pair about the central atom. Therefore the molecular geometry is trigonal pyramidal.
Determine whether the molecule contains polar bonds. Sketch the molecule and superimpose a vector for each polar bond. The relative length of each vector should be proportional to the electronegativity difference between the atoms forming each bond. The vector should point in the direction of the more electronegative atom.	The electronegativities of nitrogen and hydrogen are 3.0 and 2.1, respectively. Therefore the bonds are polar. 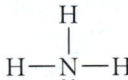
Determine whether the polar bonds add together to form a net dipole moment. Examine the symmetry of the vectors (representing dipole moments) and determine whether they cancel each other or sum to a net dipole moment.	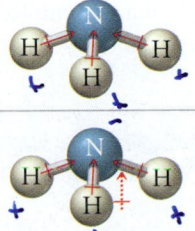 The three dipole moments sum to a net dipole moment. The molecule is polar.

For Practice 10.5

Determine whether CF_4 is polar.

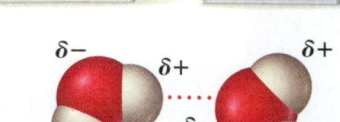

Opposite magnetic poles attract one another.

Opposite partial charges on molecules attract one another.

▲ FIGURE 10.5 Interaction of Polar Molecules   The north pole of one magnet attracts the south pole of another magnet. In an analogous way, the positively charged end of one molecule attracts the negatively charged end of another (although the forces involved are different). As a result of this electrical attraction, polar molecules interact strongly with one another.

Whether or not a molecule is polar is important because polar and nonpolar molecules have different properties. Water and oil do not mix, for example, because water molecules are polar and the molecules that compose oil are generally nonpolar. Polar molecules interact strongly with other polar molecules because the positive end of one molecule is attracted to the negative end of another, just as the south pole of a magnet is attracted to the north pole of another magnet (Figure 10.5 ◄). A mixture of polar and nonpolar molecules is similar to a mixture of small magnetic particles and nonmagnetic ones. The magnetic particles (which are like polar molecules) clump together, expelling the nonmagnetic particles (which are like nonpolar molecules) and separating into distinct regions.

Oil is nonpolar.

Water is polar.

▲ Oil and water do not mix because water molecules are polar and the molecules that compose oil are nonpolar.

▲ A mixture of polar and nonpolar molecules is analogous to a mixture of magnetic marbles (colored) and nonmagnetic marbles. As with the magnetic marbles, mutual attraction causes polar molecules to clump together, excluding the nonpolar molecules.

Chemistry in Your Day
How Soap Works

Imagine eating a greasy cheeseburger with both hands and no napkins. By the end of the meal, your hands are coated with grease and oil. If you try to wash them with only water, they remain greasy. However, if you add a little soap, the grease washes away. Why? As we just learned, water molecules are polar and the molecules that compose grease and oil are nonpolar. As a result, water and grease do not mix.

The molecules that compose soap, however, have a special structure that allows them to interact strongly with both water and grease. One end of a soap molecule is polar while the other end is nonpolar:

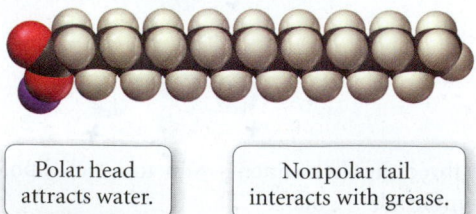

Polar head attracts water.

Nonpolar tail interacts with grease.

The nonpolar end is a long hydrocarbon chain. Hydrocarbons are always nonpolar because the electronegativity difference between carbon and hydrogen is small, and because the tetrahedral arrangement about each carbon atom tends to cancel any small dipole moments of individual bonds. The polar head of a soap molecule—usually, though not always, ionic—strongly attracts water molecules, while the nonpolar tail interacts more strongly with grease and oil molecules (the nature of these interactions is discussed in Chapter 11). Thus, soap is a sort of molecular liaison, one end interacting with water and the other end interacting with grease. Soap allows water and grease to mix, removing the grease from your hands and washing it down the drain.

Question

Consider the detergent molecule below. Which end do you think is polar? Which end is nonpolar?

$CH_3(CH_2)_{11}OCH_2CH_2OH$

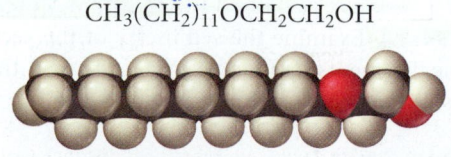

10.6 Valence Bond Theory: Orbital Overlap as a Chemical Bond

In Lewis theory, we use "dots" to represent electrons as they are transferred or shared between bonding atoms. We know from quantum-mechanical theory, however, that such a treatment is oversimplified. More advanced bonding theories treat electrons in a quantum-mechanical manner. In fact, these more advanced theories are actually extensions of quantum mechanics, but applied to molecules. Although a detailed quantitative treatment of these theories is beyond the scope of this book, we introduce them in a *qualitative* manner in the sections that follow. Keep in mind, however, that modern *quantitative* approaches to chemical bonding using these theories can accurately predict many of the properties of molecules—such as bond lengths, bond strengths, molecular geometries, and dipole moments—that we have been discussing in this book.

The simpler of the two more advanced bonding theories is called **valence bond theory**. In valence bond theory, electrons reside in quantum-mechanical orbitals localized on individual atoms. In many cases, these orbitals are simply the standard s, p, d, and f atomic orbitals that we learned about in Chapter 7. In other cases, these orbitals are *hybridized atomic orbitals*, which are a kind of blend or combination of two or more standard atomic orbitals.

When two atoms approach each other, the electrons and nucleus of one atom interact with the electrons and nucleus of the other atom. In valence bond theory, we calculate the effect of these interactions on the energies of the electrons in the atomic orbitals. If the energy of the system is lowered because of the interactions, then a chemical bond forms. If the energy of the system is raised by the interactions, then a chemical bond does not form.

The interaction energy is usually calculated as a function of the internuclear distance between the two bonding atoms. For example, Figure 10.6 ▼ shows the calculated interaction energy between two hydrogen atoms as a function of the distance between them.

> Valence bond theory is an application of a more general quantum-mechanical approximation method called *perturbation theory*. In perturbation theory, a simpler system (such as an atom) is viewed as being slightly altered (or perturbed) by some additional force or interaction.

Interaction Energy of Two Hydrogen Atoms

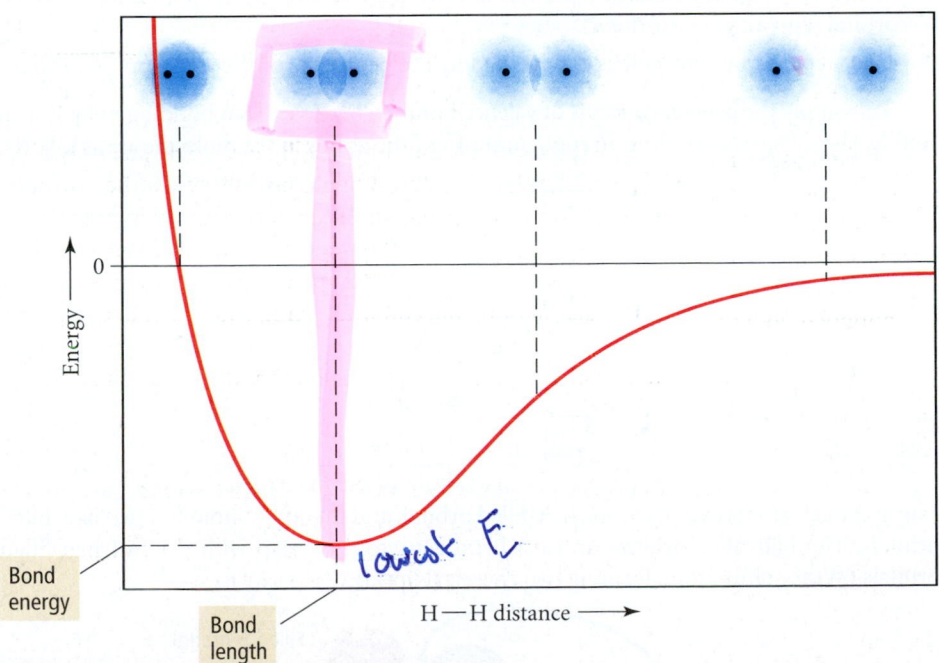

lowest E

Bond energy

Bond length

H—H distance ⟶

▲ **FIGURE 10.6 Interaction Energy Diagram for H₂**   The potential energy of two hydrogen atoms is lowest when they are separated by a distance that allows their 1s orbitals a substantial degree of overlap without too much repulsion between their nuclei. This distance, at which the system is most stable, is the bond length of the H_2 molecule.

The *y*-axis of the graph is the potential energy of the interaction between the electron and nucleus of one hydrogen atom and the electron and nucleus of the other. The *x*-axis is the separation (or internuclear distance) between the two atoms. As you can see from the graph, when the atoms are far apart (right side of the graph), the interaction energy is nearly zero because the two atoms do not interact to any significant extent. As the atoms get closer, the interaction energy becomes negative. This is a net stabilization that attracts one hydrogen atom to the other. If the atoms get too close, however, the interaction energy begins to rise, primarily because of the mutual repulsion of the two positively charged nuclei. The most stable point on the curve occurs at the minimum of the interaction energy—this is the equilibrium bond length. At this distance, the two atomic 1*s* orbitals have a significant amount of overlap and the electrons spend time in the internuclear region where they can interact with both nuclei. The value of the interaction energy at the equilibrium bond distance is the bond energy.

When valence bond theory is applied to a number of atoms and their corresponding molecules, we can make the following general observation: *the interaction energy is usually negative (or stabilizing) when the interacting atomic orbitals contain a total of two electrons that can spin-pair.* Most commonly, the two electrons come from the two half-filled orbitals, but in some cases, the two electrons can come from one filled orbital overlapping with a completely empty orbital (this is called a coordinate covalent bond and is covered in more detail in Chapter 24). In other words, when two atoms with half-filled orbitals approach each other, the half-filled orbitals *overlap*—parts of the orbitals occupy the same space—and the electrons occupying them align with opposite spins. This results in a net energy stabilization that constitutes a covalent chemical bond. The resulting geometry of the molecule emerges from the geometry of the overlapping orbitals.

> When *completely filled* orbitals overlap, the interaction energy is positive (or destabilizing) and no bond forms.

Summarizing the main concepts of valence bond theory:

➤ The valence electrons of the atoms in a molecule reside in quantum-mechanical atomic orbitals. The orbitals can be the standard *s*, *p*, *d*, and *f* orbitals or they may be hybrid combinations of these.

➤ A chemical bond results from the overlap of two half-filled orbitals with spin-pairing of the two valence electrons (or less commonly the overlap of a completely filled orbital with an empty orbital).

➤ The shape of the molecule is determined by the geometry of the overlapping orbitals.

Let's apply the general concepts of valence bond theory to explain bonding in hydrogen sulfide, H_2S. The valence electron configurations of the atoms in the molecule are as follows:

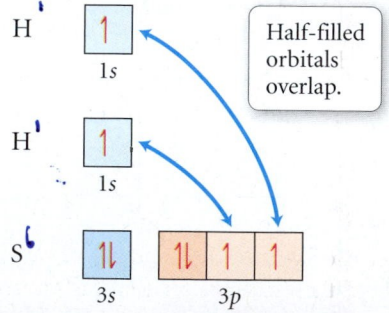

The hydrogen atoms each have one half-filled orbital, and the sulfur atom has two half-filled orbitals. The half-filled orbitals on each hydrogen atom overlap with the two half-filled orbitals on the sulfur atom, forming two chemical bonds:

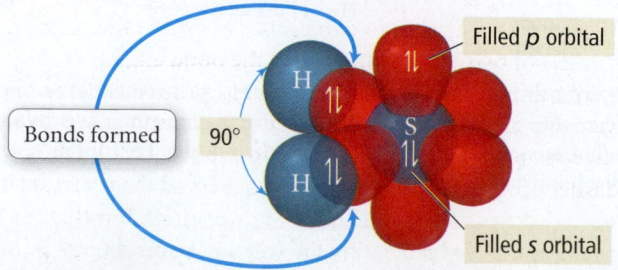

To show the spin-pairing of the electrons in the overlapping orbitals, we superimpose a half-arrow for each electron in each half-filled orbital and show that, within a bond, the electrons are spin-paired (one half-arrow pointing up and the other pointing down). We also super-impose paired half-arrows in the filled sulfur s and p orbitals to represent the lone pair electrons in those orbitals. (Since those orbitals are full, they are not involved in bonding.)

A quantitative calculation of H_2S using valence bond theory yields bond energies, bond lengths, and bond angles. In our more qualitative treatment, we simply show how orbital overlap leads to bonding and make a rough sketch of the molecule based on the overlapping orbitals. Notice that, because the overlapping orbitals on the central atom (sulfur) are p orbitals, and because p orbitals are oriented at 90° to one another, the predicted bond angle is 90°. The actual bond angle in H_2S is 92°. In the case of H_2S, a simple valence bond treatment matches well with the experimentally measured bond angle (in contrast to VSEPR theory, which simply predicts a bond angle of less than 109.5°).

 Conceptual Connection 10.4 What Is a Chemical Bond, Part I?

How you answer the question, *what is a chemical bond*, depends on the bonding model. Answer the following three questions:

(a) What is a covalent chemical bond according to Lewis theory?
(b) What is a covalent chemical bond according to valence bond theory?
(c) Why are the answers different?

Answer: (a) In Lewis theory, a covalent chemical bond is the sharing of electrons (represented by dots). **(b)** In valence bond theory, a covalent chemical bond is the overlap of half-filled atomic orbitals. **(c)** The answers are different because Lewis theory and valence bond theory are different *models* for chemical bonding. They both make useful and often similar predictions, but the assumptions of each model are different, and so are their respective descriptions of a chemical bond.

10.7 Valence Bond Theory: Hybridization of Atomic Orbitals

Although the overlap of half-filled *standard* atomic orbitals adequately explains the bonding in H_2S, it cannot adequately explain the bonding in many other molecules. For example, suppose we try to explain the bonding between hydrogen and carbon using the same approach. The valence electron configurations of H and C are as follows:

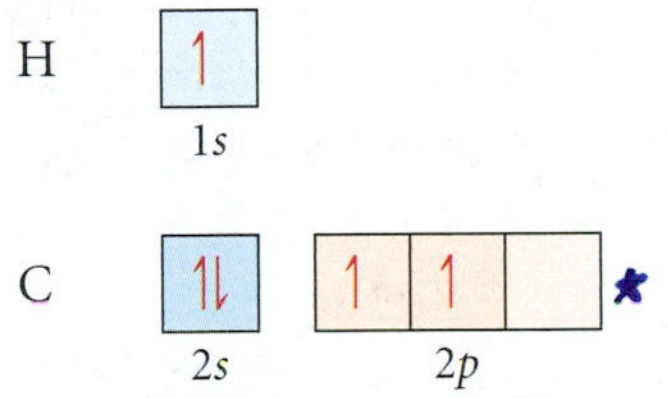

Carbon has only two half-filled orbitals and should therefore form only two bonds with two hydrogen atoms. We would therefore predict that carbon and hydrogen should form a molecule with the formula CH_2 and with a bond angle of 90° (corresponding to the angle between any two p orbitals).

However, from experiment, we know that the stable compound formed from carbon and hydrogen is CH_4 (methane), with bond angles of 109.5°. The experimental reality is different from our simple prediction in two ways. The first is that carbon forms bonds to four hydrogen atoms, not two. The second is that the bond angles are much larger than the angle between two p orbitals. Valence bond theory can account for the bonding in CH_4 and many other polyatomic molecules by incorporating a concept called *orbital hybridization*.

So far, we have assumed that the overlapping orbitals that form chemical bonds are simply the standard s, p, or d atomic orbitals. Valence bond theory treats the electrons in a molecule as if they occupied these standard atomic orbitals, but this is a major oversimplification. The concept of hybridization in valence bond theory is essentially a step

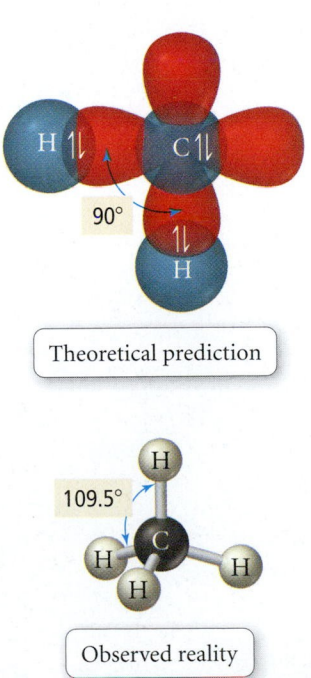

Theoretical prediction

Observed reality

In Section 10.8, we examine *molecular orbital theory*, which treats electrons in a molecule as occupying orbitals that belong to the molecule as a whole.

toward recognizing that *the orbitals in a molecule are not necessarily the same as the orbitals in an atom*. **Hybridization** is a mathematical procedure in which the standard atomic orbitals are combined to form new atomic orbitals called **hybrid orbitals** that correspond more closely to the actual distribution of electrons in chemically bonded atoms. Hybrid orbitals are still localized on individual atoms, but they have different shapes and energies from those of standard atomic orbitals.

Why do we hypothesize that electrons in some molecules occupy hybrid orbitals? In valence bond theory, a chemical bond is the overlap of two orbitals that together contain two electrons. The greater the overlap, the stronger the bond and the lower the energy. In hybrid orbitals, the electron probability density is more concentrated in a single directional lobe, allowing greater overlap with the orbitals of other atoms. In other words, hybrid orbitals *minimize* the energy of the molecule by *maximizing* the orbital overlap in a bond.

As we saw in Chapter 9, the word *hybrid* comes from breeding. A *hybrid* is an offspring of two animals or plants of different standard races or breeds. Similarly, a hybrid orbital is a product of mixing two or more standard atomic orbitals.

Hybridization, however, is not a free lunch—in most cases it actually costs some energy. So hybridization occurs only to the degree that the energy payback through bond formation is large. In general, therefore, the more bonds that an atom forms, the greater the tendency of its orbitals to hybridize. Central or interior atoms, which form the most bonds, have the greatest tendency to hybridize. Terminal atoms, which form the fewest bonds, have the least tendency to hybridize. *In this book, we will focus on the hybridization of interior atoms and assume that all terminal atoms—those bonding to only one other atom—are unhybridized.* Hybridization is particularly important in carbon, which tends to form four bonds in its compounds and therefore always hybridizes.

Although we cannot show the procedure for obtaining hybrid orbitals in its mathematical detail, we can make the following general statements regarding hybridization:

In a more detailed treatment, hybridization is not an all-or-nothing process—it can occur to varying degrees that are not always easy to predict. We saw earlier, for example, that sulfur does not hybridize very much in forming H_2S.

- The *number of standard atomic orbitals* added together always equals the *number of hybrid orbitals* formed. The total number of orbitals is conserved.

- The *particular combinations* of standard atomic orbitals added together determines the *shapes and energies* of the hybrid orbitals formed.

- The *particular type of hybridization that occurs* is the one that yields the *lowest overall energy for the molecule*. Since actual energy calculations are beyond the scope of this book, we will use electron geometries as determined by VSEPR theory to predict the type of hybridization.

*sp*³ hybridization

We can account for the tetrahedral geometry in CH_4 by the hybridization of the one *2s* orbital and the three *2p* orbitals on the carbon atom. The four new orbitals that result, called *sp*³ hybrids, are shown in the following energy diagram.

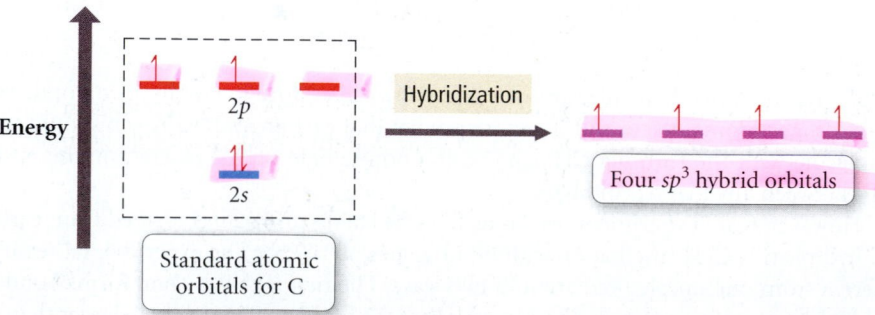

The notation "*sp*³" indicates that the hybrid orbitals are mixtures of one s orbital and three *p* orbitals. Notice that the hybrid orbitals all have the same energy—they are degenerate. The shapes of the *sp*³ hybrid orbitals are shown in Figure 10.7 ▶. Notice also that the four hybrid orbitals are arranged in a tetrahedral geometry with 109.5° angles between them.

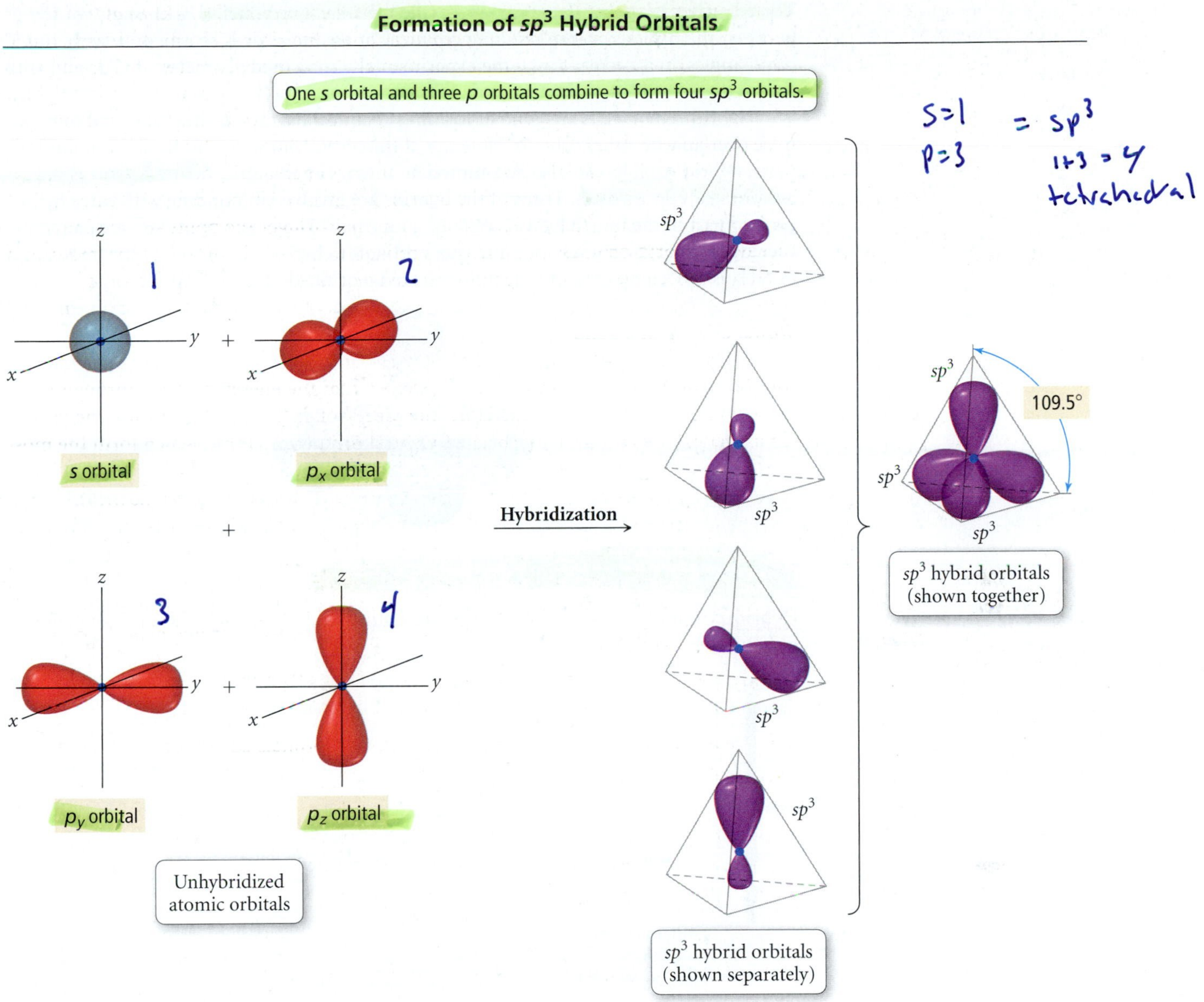

Formation of *sp³* Hybrid Orbitals

One *s* orbital and three *p* orbitals combine to form four *sp³* orbitals.

$S = 1$ $= sp^3$
$P = 3$ $1 + 3 = 4$
          tetrahedral

1
s orbital

2
p$_x$ orbital

3
p$_y$ orbital

4
p$_z$ orbital

Unhybridized atomic orbitals

Hybridization →

sp³

sp³

sp³

sp³

sp³ hybrid orbitals (shown separately)

sp³   109.5°
sp³
sp³
sp³ hybrid orbitals (shown together)

▲ **FIGURE 10.7 *sp³* Hybridization**   One *s* orbital and three *p* orbitals combine to form four *sp³* hybrid orbitals.

The orbital diagram for carbon using these hybrid orbitals is as follows:

C | ↑ | ↑ | ↑ | ↑ |
        sp³

Carbon's four valence electrons occupy these orbitals singly with parallel spins as dictated by Hund's rule. With this electron configuration, carbon has four half-filled orbitals and can form four bonds with four hydrogen atoms as follows:

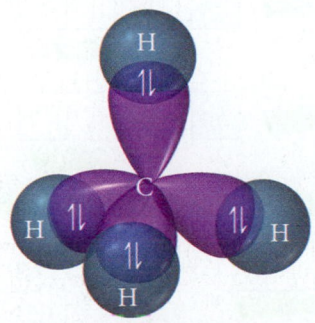

The geometry of the *overlapping orbitals* (the hybrids) is tetrahedral, with angles of 109.5° between the orbitals, so the *resulting geometry of the molecule* is tetrahedral, with 109.5° bond angles, in agreement with the experimentally measured geometry of CH$_4$ and with the predicted VSEPR geometry.

Hybridized orbitals are good at forming chemical bonds because they tend to maximize overlap with other orbitals. However, if the central atom of a molecule contains lone pairs, hybrid orbitals can also accommodate them. For example, the nitrogen orbitals in ammonia are sp^3 hybrids. Three of the hybrids are involved in bonding with three hydrogen atoms, but the fourth hybrid contains a lone pair. The presence of the lone pair, however, does lower the tendency of nitrogen's orbitals to hybridize. Therefore the bond angle in NH$_3$ is 107°, a bit closer to the unhybridized p orbital bond angle of 90°.

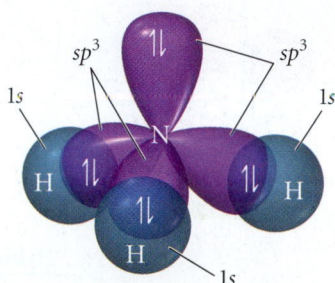

sp^2 Hybridization and Double Bonds

In valence bond theory, the particular hybridization scheme to follow (sp^2 versus sp^3 for example) for a given molecule is determined computationally, which is beyond our scope. In this book, we will determine the particular hybridization scheme from the VSEPR geometry of the molecule, as shown later in this section.

Hybridization of one s and two p orbitals results in three sp^2 hybrids and one leftover unhybridized p orbital.

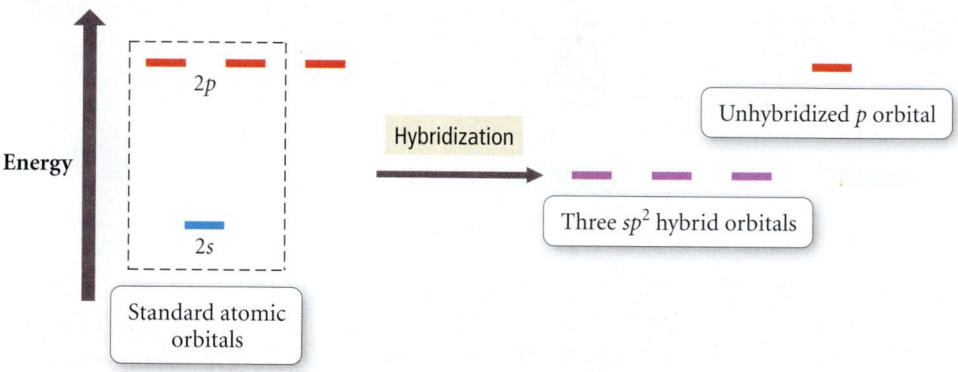

The notation "sp^2" indicates that the hybrids are mixtures of one s orbital and two p orbitals. The shapes of the sp^2 hybrid orbitals are shown in Figure 10.8 ▶. Notice that the three hybrid orbitals are arranged in a trigonal planar geometry with 120° angles between them. The unhybridized p orbital is oriented perpendicular to the three hybridized orbitals.

As an example of a molecule with sp^2 hybrid orbitals, consider H$_2$CO. The unhybridized valence electron configurations of each of the atoms are as follows:

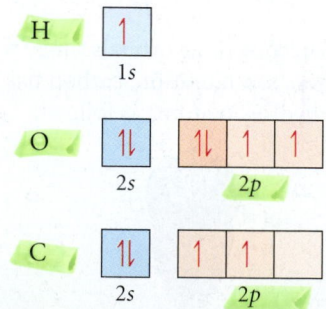

Carbon is the central atom and the hybridization of its orbitals is sp^2:

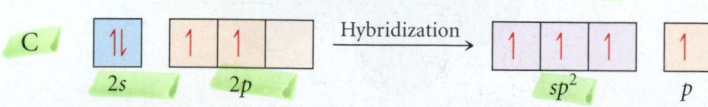

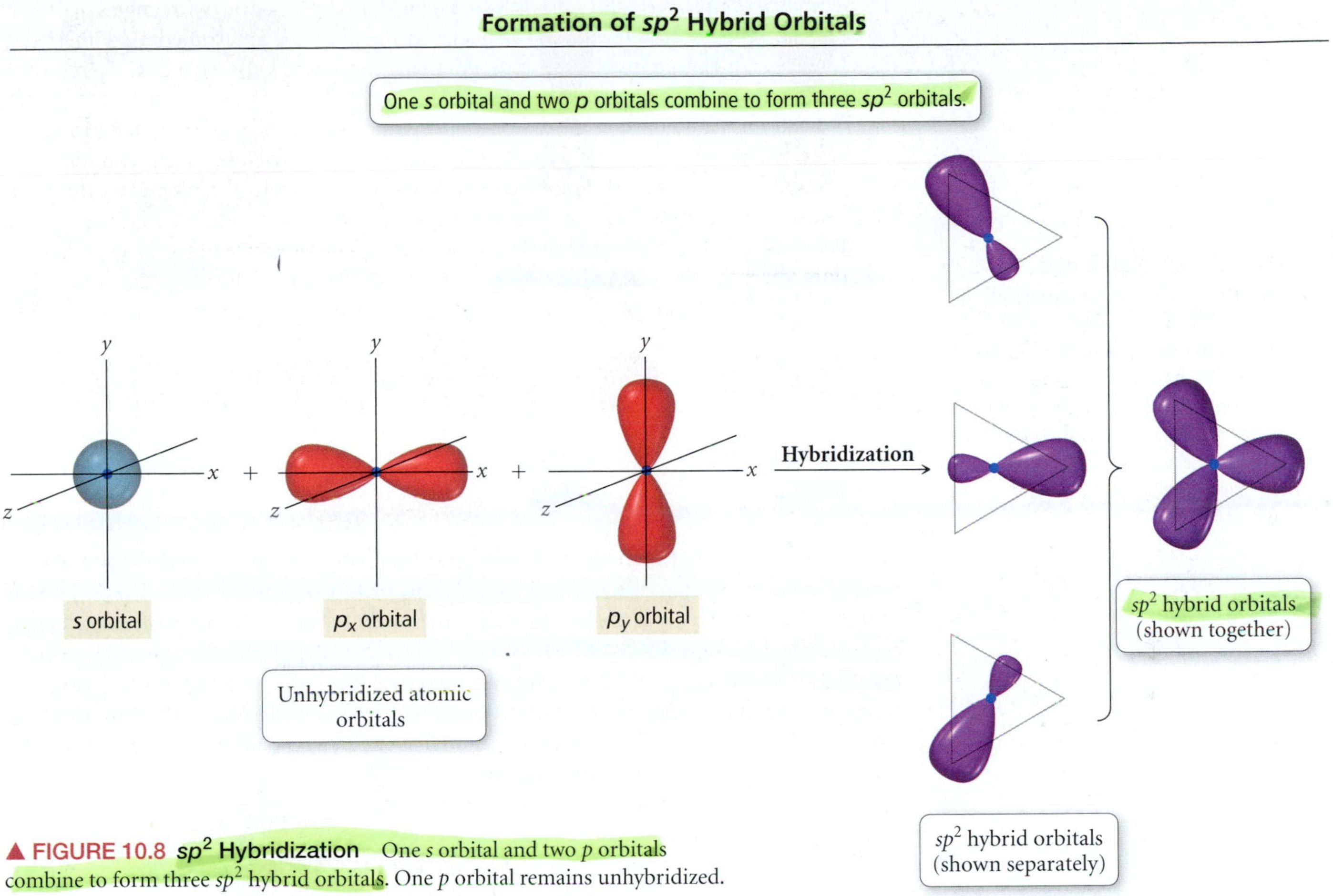

Formation of sp^2 Hybrid Orbitals

One s orbital and two p orbitals combine to form three sp^2 orbitals.

s orbital $\quad+\quad$ p_x orbital $\quad+\quad$ p_y orbital $\quad$ **Hybridization** $\rightarrow$

Unhybridized atomic orbitals

sp^2 hybrid orbitals (shown together)

sp^2 hybrid orbitals (shown separately)

▲ **FIGURE 10.8** sp^2 **Hybridization** One s orbital and two p orbitals combine to form three sp^2 hybrid orbitals. One p orbital remains unhybridized.

Each of the sp^2 orbitals is half-filled. The remaining electron occupies the leftover p orbital, even though it is slightly higher in energy. We can now see that the carbon atom has four half-filled orbitals and can therefore form four bonds: two with two hydrogen atoms and two (a double bond) with the oxygen atom. We draw the molecule and the overlapping orbitals as follows:

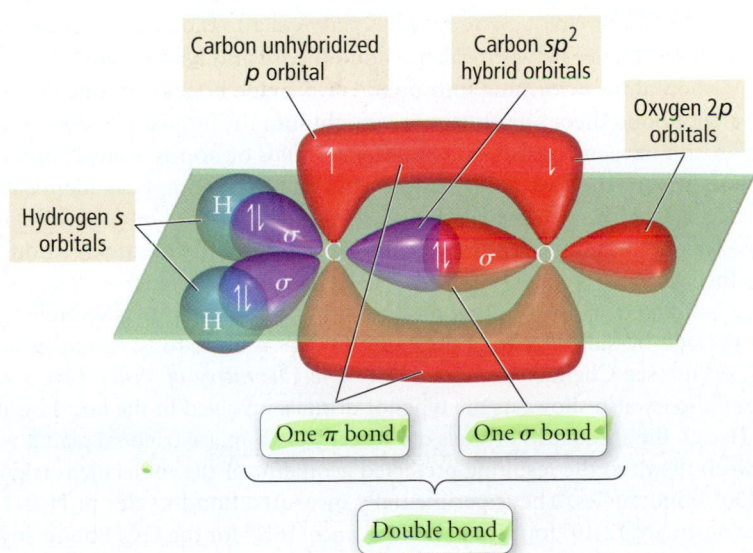

Carbon unhybridized p orbital

Carbon sp^2 hybrid orbitals

Oxygen $2p$ orbitals

Hydrogen s orbitals

One π bond

One σ bond

Double bond

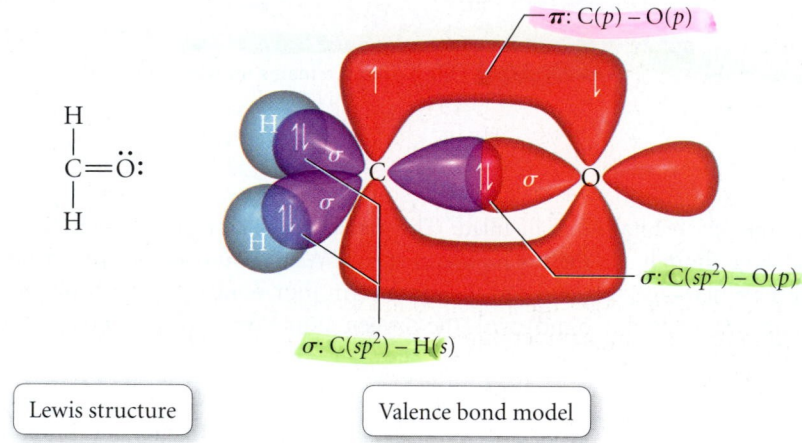

► FIGURE 10.9 Sigma and Pi Bonding When orbitals overlap end to end, they form a sigma bond. When orbitals overlap side by side, the result is a pi bond. Two atoms can form only one sigma bond. A single bond is a sigma bond; a double bond consists of a sigma bond and a pi bond; a triple bond consists of a sigma bond and two pi bonds.

Notice the overlap between the half-filled *p* orbitals on the carbon and oxygen atoms. When *p* orbitals overlap this way (side by side) the resulting bond is called a **pi (π) bond**, and the electron density is above and below the internuclear axis. When orbitals overlap end to end, as in all of the rest of the bonds in the molecule, the resulting bond is called a **sigma (σ) bond** (Figure 10.9 ▲). We can therefore label all the bonds in the molecule using a notation that specifies the type of bond (σ or π) as well as the type of overlapping orbitals. We have included this notation, as well as the Lewis structure of H_2CO for comparison, in the following bonding diagram for H_2CO:

Notice the correspondence between valence bond theory and Lewis theory. In both cases, the central carbon atom is forming four bonds: two single bonds and one double bond. However, valence bond theory gives us more insight into the bonds. The double bond between carbon and oxygen consists of two different *kinds* of bonds—one σ and one π—while in Lewis theory the two bonds within the double bond appear identical. *Double bonds in Lewis theory always correspond to one σ and one π bond in valence bond theory.* In this sense, valence bond theory gives us more insight into the nature of a double bond than Lewis theory.

One—and only one—σ bond forms between any two atoms. Additional bonds must be π bonds.

Valence bond theory shows us that rotation about a double bond is severely restricted. Because of the side-by-side overlap of the *p* orbitals, the π bond must essentially break for rotation to occur (see Chemistry in Your Day: *The Chemistry of Vision* box on p. 432.) Valence bond theory also shows us the types of orbitals involved in the bonding and their shapes. In H_2CO, the sp^2 hybrid orbitals on the central atom are trigonal planar with 120° angles between them, so the resulting predicted geometry of the molecule is trigonal planar with 120° bond angles. The experimentally measured bond angles in H_2CO, as discussed previously, are 121.9° for the HCO bond and 116.2° for the HCH bond angle, close to the predicted values.

Although rotation about a double bond is highly restricted, rotation about a single bond is relatively unrestricted. Consider, for example, the structures of two chlorinated hydrocarbons, 1,2-dichloroethane and 1,2-dichloroethene.

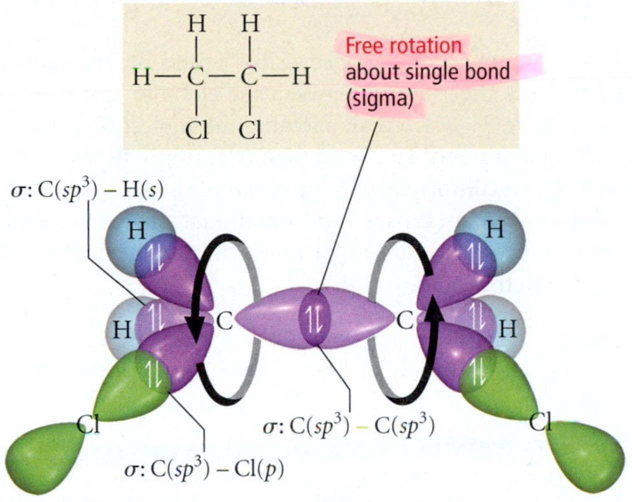

σ: C(sp^3) – H(s)

Free rotation about single bond (sigma)

σ: C(sp^3) – C(sp^3)

σ: C(sp^3) – Cl(p)

1,2-Dichloroethane

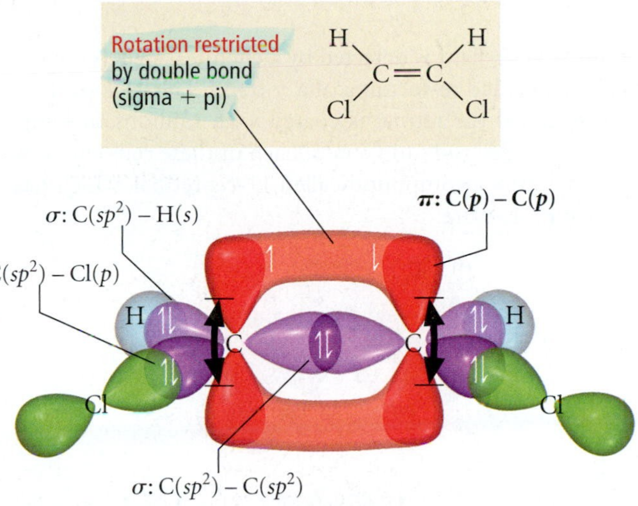

Rotation restricted by double bond (sigma + pi)

π: C(p) – C(p)

σ: C(sp^2) – H(s)

σ: C(sp^2) – Cl(p)

σ: C(sp^2) – C(sp^2)

1,2-Dichloroethene

The hybridization of the carbon atoms in 1,2-dichloroethane is sp^3, resulting in relatively free rotation about the sigma single bond. Consequently, there is no difference between the following two structures at room temperature because they quickly interconvert:

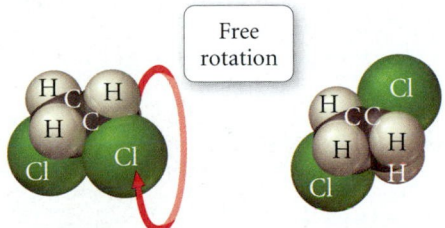

Free rotation

In contrast, rotation about the double bond (sigma + pi) in 1,2-dichloroethene is restricted, so that, at room temperature, 1,2-dichloroethene can exist in two forms:

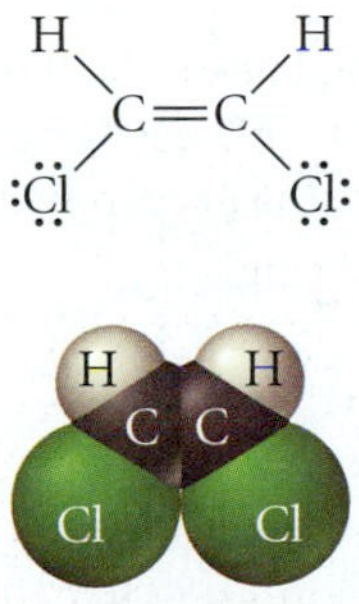

cis-1,2-Dichloroethene 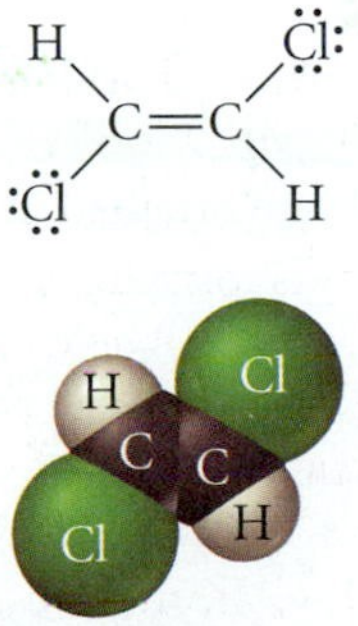   trans-1,2-Dichloroethene

These two forms of 1,2-dichloroethene are indeed different compounds with different properties. We distinguish between them with the designations *cis* (meaning "same side") and *trans* (meaning "opposite sides"). Compounds such as these, with the same molecular formula but different structures or different spatial arrangement of atoms, are called *isomers*. In other words, nature can—and does—make different compounds out of the same atoms by arranging the atoms in different ways. Isomerism is common throughout chemistry and especially important in organic chemistry, as we shall see in Chapter 20.

Chemistry in Your Day
The Chemistry of Vision

In the human eye, light is detected by a chemical switch involving the breaking and re-forming of a π bond. The back portion of the eye, called the retina, is coated with millions of light-sensitive cells called rods and cones. Each of these cells contains proteins that bind a compound called 11-*cis*-retinal, which has the following structure:

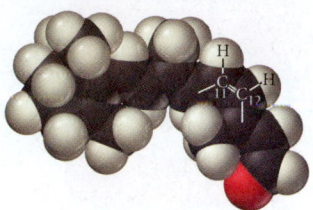

11-*cis*-Retinal

When a photon of sufficient energy strikes a rod or cone, it causes the isomerization of 11-*cis*-retinal to all-*trans*-retinal:

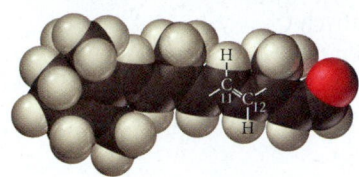

all-*trans*-Retinal

The isomerization occurs because visible light contains enough energy to break the π bond between the eleventh and twelfth carbon atom in 11-*cis*-retinal, allowing the molecule to freely rotate about that bond. The bond then re-forms with the molecule in the *trans* conformation. The different shape of the resultant all-*trans*-retinal causes conformational changes in the protein to which it is bound that results in an electrical signal transmitted to the brain.

Question

What is the hybridization of the eleventh and twelfth carbon atoms in retinal?

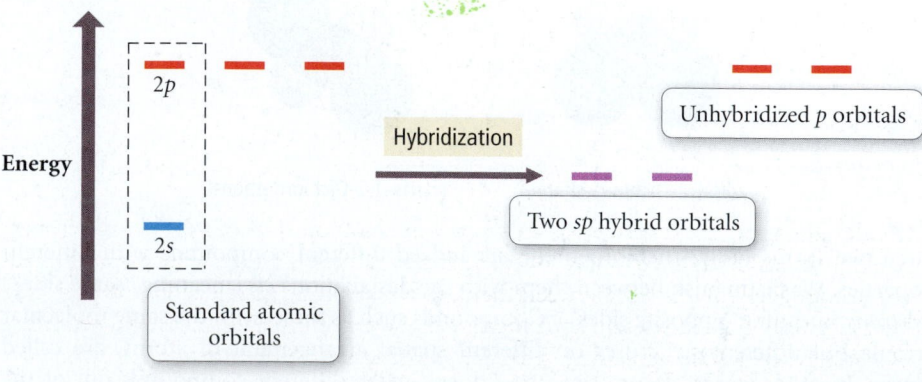

sp Hybridization and Triple Bonds

Hybridization of one *s* and one *p* orbital results in two *sp* hybrid orbitals and two leftover unhybridized *p* orbitals.

The shapes of the *sp* hybrid orbitals are shown in Figure 10.10 ▶. Notice that the two *sp* hybrid orbitals are arranged in a linear geometry with a 180° angle between them. The unhybridized *p* orbitals are oriented in the plane that is perpendicular to the hybridized *sp* orbitals.

Energy

2p

2s

Standard atomic orbitals

Hybridization

Unhybridized *p* orbitals

Two *sp* hybrid orbitals

Formation of *sp* Hybrid Orbitals

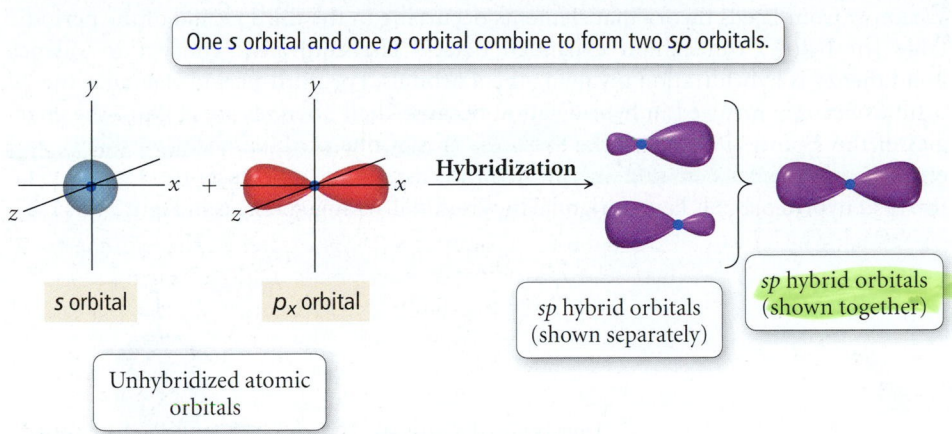

One *s* orbital and one *p* orbital combine to form two *sp* orbitals.

s orbital   *p_x* orbital

Hybridization

sp hybrid orbitals
(shown separately)

sp hybrid orbitals
(shown together)

Unhybridized atomic
orbitals

▲ **FIGURE 10.10 *sp* Hybridization**   One *s* orbital and one *p* orbital combine to form two *sp* hybrid orbitals. Two *p* orbitals (not shown) remain unhybridized.

As an example of a molecule with *sp* hybrid orbitals, consider acetylene, HC≡CH. The valence electron configurations (showing hybridization) of the atoms are as follows:

H   1
     1*s*

C   2*s*   2*p*   →(Hybridization)→   *sp*   2*p*

The two interior carbon atoms have *sp* hybridized orbitals, leaving two unhybridized 2*p* orbitals on each carbon atom. Each carbon atom then has four half-filled orbitals and can form four bonds: one with a hydrogen atom and three (a triple bond) with the other carbon atom. We draw the molecule and the overlapping orbitals as follows:

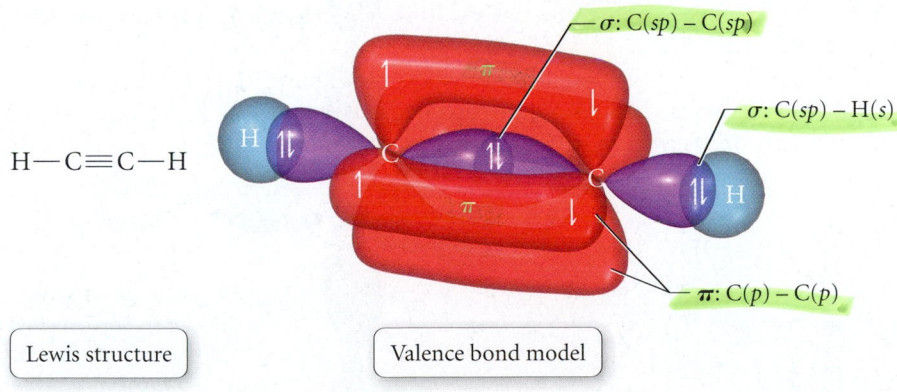

H—C≡C—H

σ: C(*sp*) – C(*sp*)

σ: C(*sp*) – H(*s*)

π: C(*p*) – C(*p*)

Lewis structure

Valence bond model

Notice that the triple bond between the two carbon atoms consists of two π bonds (overlapping *p* orbitals) and one σ bond (overlapping *sp* orbitals). The *sp* orbitals on the carbon atoms are linear with 180° between them, so the resulting geometry of the molecule is linear with 180° bond angles, in agreement with the experimentally measured geometry of HC≡CH, and also in agreement with the prediction of VSEPR theory.

sp^3d and sp^3d^2 Hybridization

We know from Lewis theory that elements occurring in the third period of the periodic table (or below) can exhibit expanded octets. The equivalent concept in valence bond theory is hybridization involving the d orbitals. For third-period elements, the $3d$ orbitals become involved in hybridization because their energies are close to the energies of the $3s$ and $3p$ orbitals. The hybridization of one s orbital, three p orbitals, and one d orbital results in sp^3d hybrid orbitals, as shown in Figure 10.11(a) ▼. The five sp^3d hybrid orbitals have a trigonal bipyramidal arrangement, as in Figure 10.11(b).

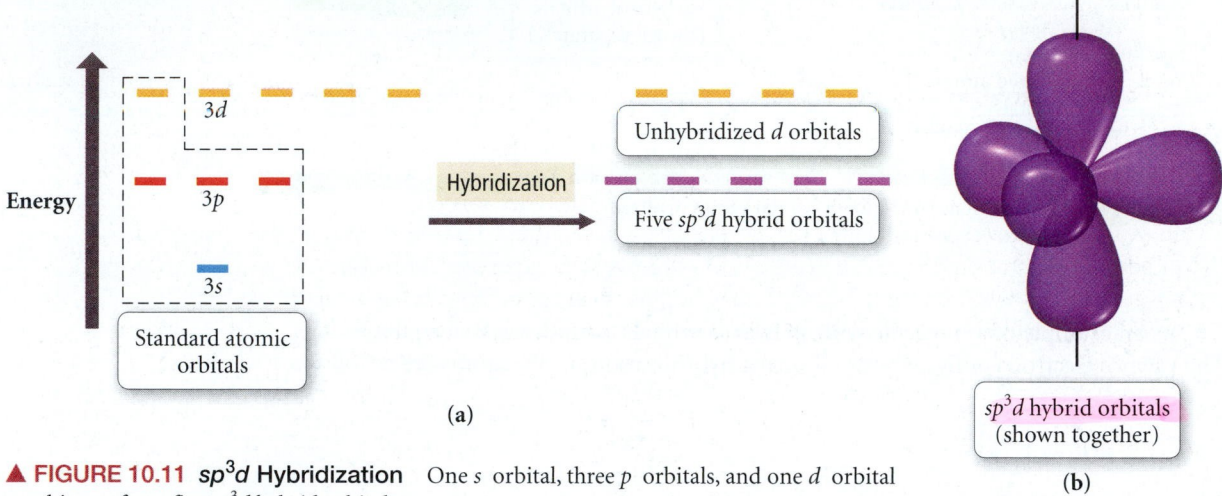

(a)

sp^3d hybrid orbitals (shown together)

(b)

▲ FIGURE 10.11 sp^3d Hybridization   One s orbital, three p orbitals, and one d orbital combine to form five sp^3d hybrid orbitals.

As an example of sp^3d hybridization, consider arsenic pentafluoride, AsF_5 The arsenic atom bonds to five fluorine atoms by overlap between the sp^3d hybrid orbitals on arsenic and p orbitals on the fluorine atoms, as shown here:

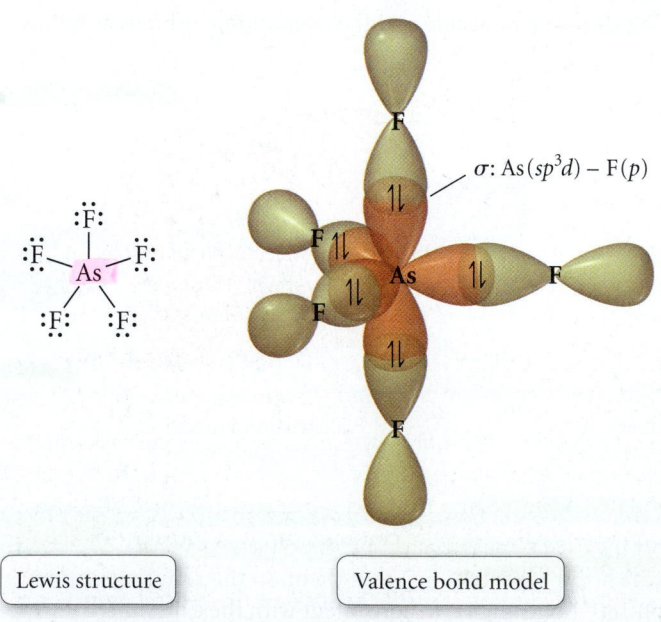

Lewis structure

Valence bond model

The sp^3d orbitals on the arsenic atom are trigonal bipyramidal, so the molecular geometry is trigonal bipyramidal.

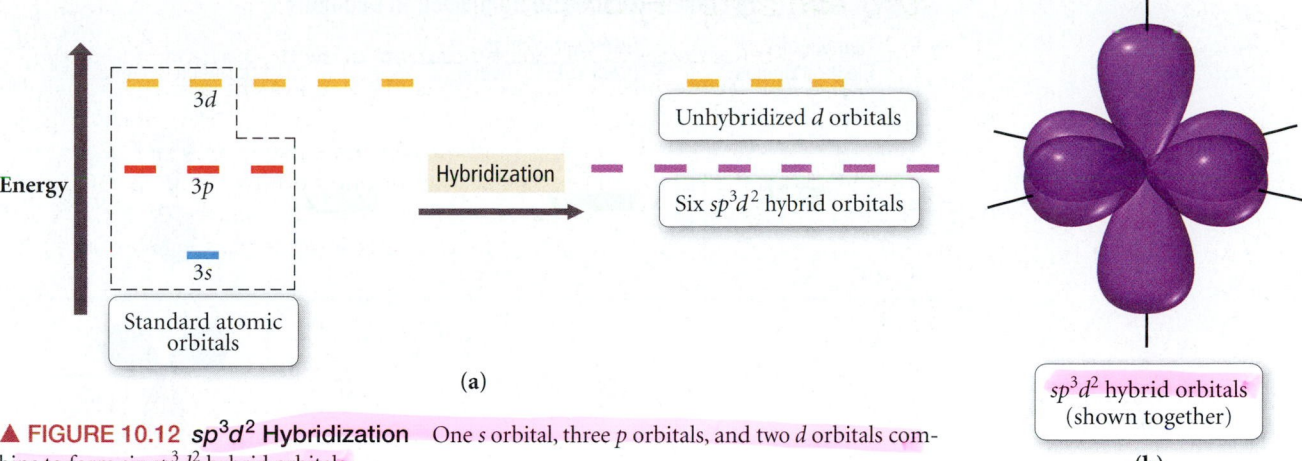

▲ **FIGURE 10.12** sp^3d^2 **Hybridization**   One s orbital, three p orbitals, and two d orbitals combine to form six sp^3d^2 hybrid orbitals.

The hybridization of one s orbital, three p orbitals, and *two d* orbitals results in sp^3d^2 hybrid orbitals, as shown in Figure 10.12(a) ▲. The six sp^3d^2 hybrid orbitals have an octahedral geometry, as in Figure 10.12(b). As an example of sp^3d^2 hybridization, consider sulfur hexafluoride, SF_6. The sulfur atom bonds to six fluorine atoms by overlap between the sp^3d^2 hybrid orbitals on sulfur and p orbitals on the fluorine atoms, as shown here:

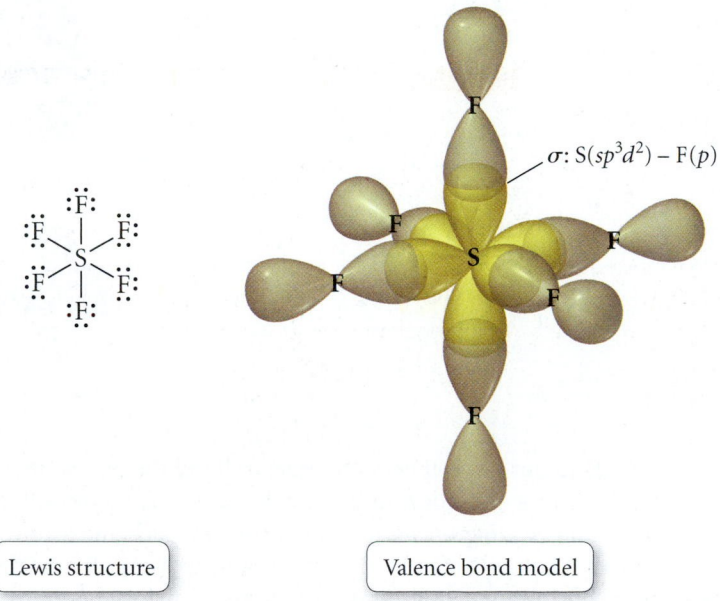

Lewis structure

Valence bond model

The sp^3d^2 orbitals on the sulfur atom are octahedral, so the molecular geometry is octahedral, again in agreement with VSEPR theory and with the experimentally observed geometry.

Writing Hybridization and Bonding Schemes

We have now seen examples of the five main types of atomic orbital hybridization. *But how do we know which hybridization scheme best describes the orbitals of a specific atom in a specific molecule?* In computational valence bond theory, the energy of the molecule is actually calculated using a computer, and the degree of hybridization as well as the type of hybridization are varied to find the combination that gives the molecule the lowest overall energy. For our purposes, we will assign a hybridization scheme from the electron geometry—determined using VSEPR theory—of the central atom (or interior atoms) of the molecule. The five VSEPR electron geometries and the corresponding hybridization schemes are shown in Table 10.3. For example, if the electron geometry of the central atom is tetrahedral, then the hybridization is sp^3; if the electron geometry is octahedral, then the hybridization is sp^3d^2, and so on.

TABLE 10.3   Hybridization Scheme from Electron Geometry

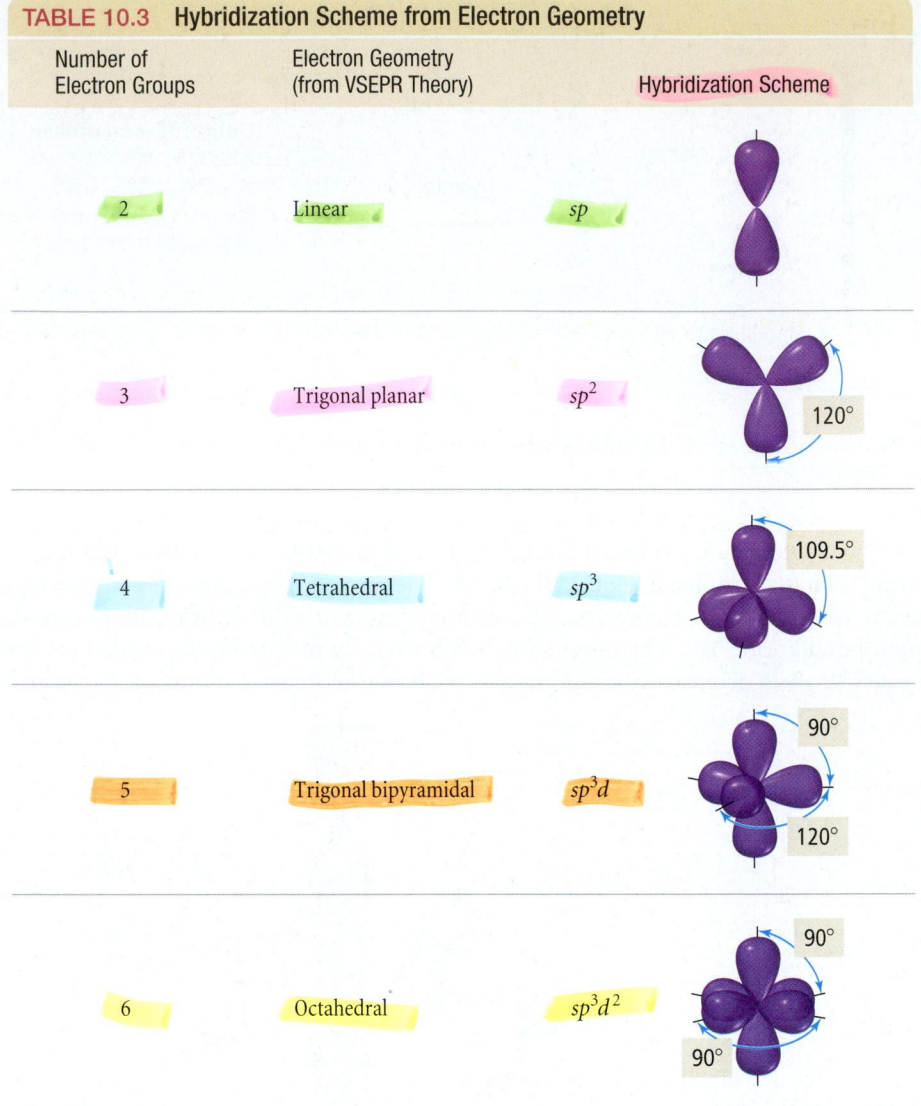

Number of Electron Groups	Electron Geometry (from VSEPR Theory)	Hybridization Scheme	
2	Linear	sp	
3	Trigonal planar	sp^2	120°
4	Tetrahedral	sp^3	109.5°
5	Trigonal bipyramidal	sp^3d	90° / 120°
6	Octahedral	sp^3d^2	90° / 90°

We are now ready to put Lewis theory and valence bond theory together to describe bonding in molecules. In the procedure and examples that follow, you will learn how to write a *hybridization and bonding scheme* for a molecule. This scheme involves drawing a Lewis structure for the molecule, determining its geometry using VSEPR theory, determining the correct hybridization of the interior atoms, drawing the molecule with its overlapping orbitals, and labeling each bond with the σ and π notation followed by the type of overlapping orbitals. As you can see, this procedure involves virtually everything you have learned about bonding in this chapter and the previous one. The procedure for writing a hybridization and bonding scheme is shown in the left column below, with two examples of how to apply the procedure in the columns to the right.

Hybridization and Bonding Scheme Procedure

EXAMPLE 10.6
Hybridization and Bonding Scheme

Write a hybridization and bonding scheme for bromine trifluoride, BrF_3.

EXAMPLE 10.7
Hybridization and Bonding Scheme

Write a hybridization and bonding scheme for acetaldehyde,

$$\begin{matrix} & & O & \\ & & \| & \\ H_3C & - & C & - H \end{matrix}$$

1. Write a Lewis structure for the molecule.

Solution

BrF_3 has 28 valence electrons and the following Lewis structure:

$$:\ddot{F}—\ddot{Br}—\ddot{F}:$$
$$|$$
$$:\ddot{F}:$$

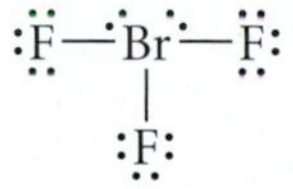

Solution

Acetaldehyde has 18 valence electrons and the following Lewis structure:

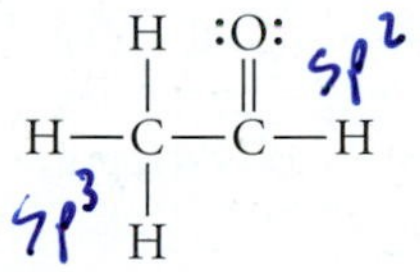

2. Use VSEPR theory to predict the electron geometry about the central atom (or interior atoms).

The bromine atom has five electron groups and therefore has a trigonal bipyramidal electron geometry.

The leftmost carbon atom has four electron groups and a tetrahedral electron geometry. The rightmost carbon atom has three electron groups and trigonal planar geometry.

3. Select the correct hybridization for the central atom (or interior atoms) based on the electron geometry.

A trigonal bipyramidal electron geometry corresponds to sp^3d hybridization.

The leftmost carbon atom is sp^3 hybridized, and the rightmost carbon atom is sp^2 hybridized.

4. Sketch the molecule, beginning with the central atom and its orbitals. Show overlap with the appropriate orbitals on the terminal atoms.

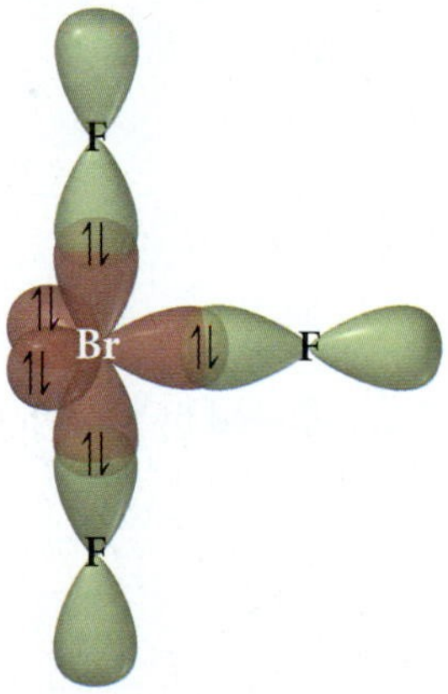

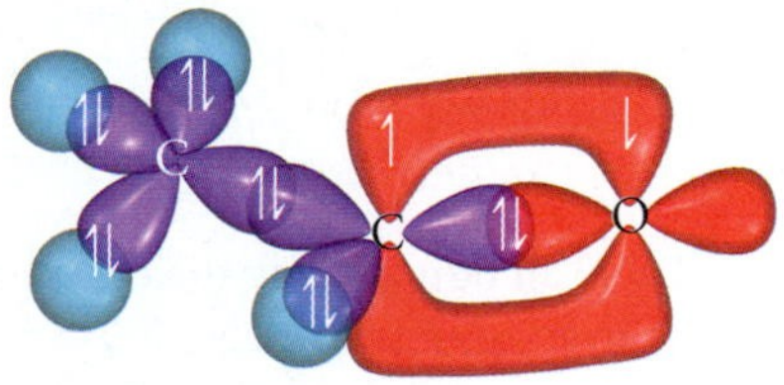

5. Label all bonds using the σ and π notation followed by the type of overlapping orbitals.

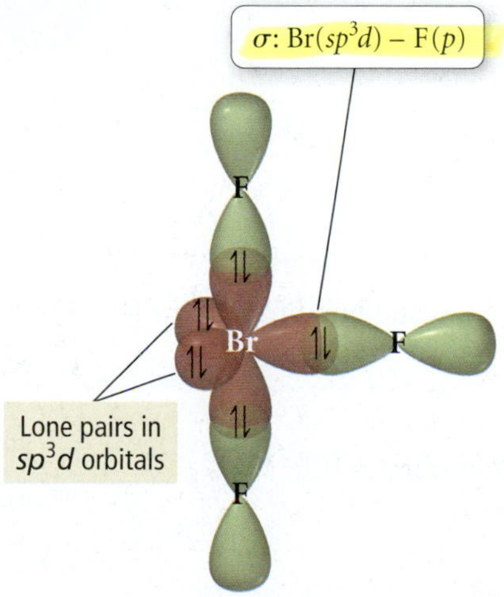

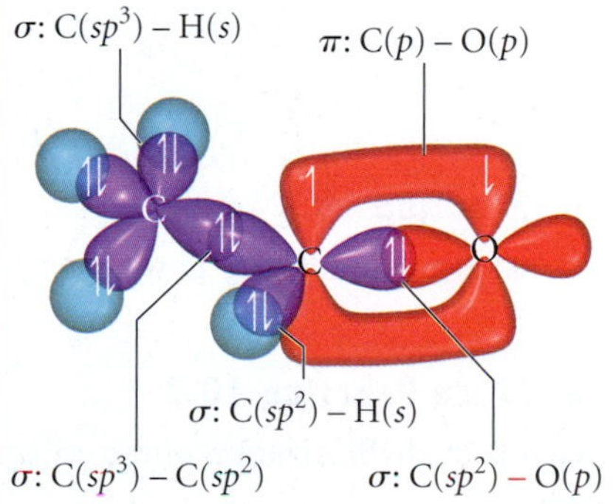

σ: $Br(sp^3d) - F(p)$

Lone pairs in sp^3d orbitals

σ: $C(sp^3) - H(s)$ π: $C(p) - O(p)$

σ: $C(sp^2) - H(s)$

σ: $C(sp^3) - C(sp^2)$ σ: $C(sp^2) - O(p)$

For Practice 10.6

Write a hybridization and bonding scheme for XeF_4.

For Practice 10.7

Write a hybridization and bonding scheme for HCN.

EXAMPLE 10.8 Hybridization and Bonding Scheme

Use valence bond theory to write a hybridization and bonding scheme for ethene, $H_2C{=}CH_2$.

Solution

1. Write a Lewis structure for the molecule.	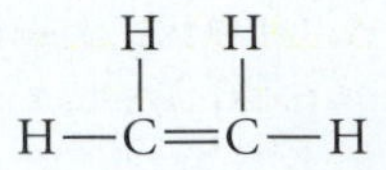 $\begin{array}{cc} H & H \\ \| & \| \\ H-C{=}C-H \end{array}$
2. Use VSEPR theory to predict the electron geometry about the central atom (or interior atoms).	The molecule has two interior atoms. Since each atom has three electron groups (one double bond and two single bonds), the electron geometry about each atom is trigonal planar.
3. Use Table 10.3 to select the correct hybridization for the central atom (or interior atoms) based on the electron geometry.	A trigonal planar geometry corresponds to sp^2 hybridization.
4. Sketch the molecule, beginning with the central atom and its orbitals. Show overlap with the appropriate orbitals on the terminal atoms.	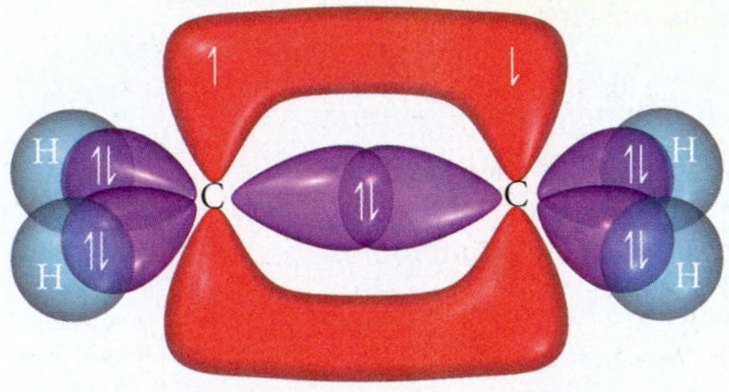
5. Label all bonds using the σ and π notation followed by the type of overlapping orbitals.	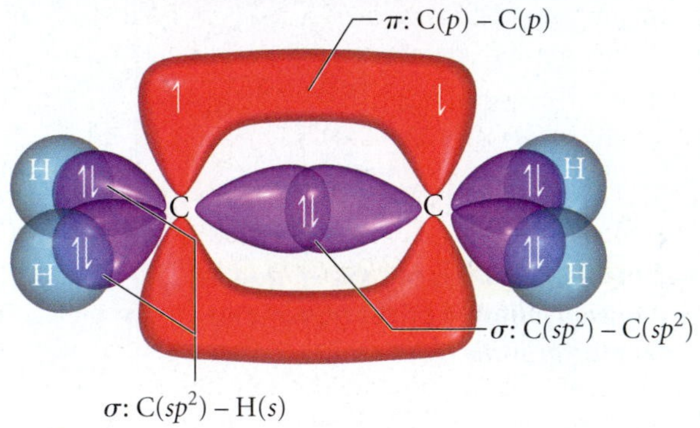 $\pi: C(p) - C(p)$ $\sigma: C(sp^2) - C(sp^2)$ $\sigma: C(sp^2) - H(s)$

For Practice 10.8

Use valence bond theory to write a hybridization and bonding scheme for CO_2.

For More Practice 10.8

What is the hybridization of the central iodine atom in I_3^-?

10.8 Molecular Orbital Theory: Electron Delocalization

Although we have seen how valence bond theory can explain a number of aspects of chemical bonding—such as the rigidity of a double bond—it also has limitations. Recall that in valence bond theory, we treat electrons as if they reside in the quantum-mechanical orbitals that we calculated *for atoms*. This is a significant oversimplification that we partially compensate for by hybridization. Nevertheless, we can do better.

In Chapter 7, we learned that the mathematical derivation of energies and orbitals for electrons *in atoms* comes from solving the Schrödinger equation for the atom of interest. For a molecule, you can theoretically do the same thing. The resulting orbitals would be the actual *molecular* orbitals of the molecule as a whole (in contrast to valence bond theory, in which the orbitals are those of individual atoms). As it turns out, however, solving the Schrödinger equation exactly for even the simplest molecules is impossible without making some approximations.

In **molecular orbital (MO) theory**, you do not actually solve the Schrödinger equation for a molecule directly. Instead, you use a trial function, an "educated guess" as to what the solution might be. In other words, instead of mathematically solving the Schrödinger equation, which would give you a mathematical function describing an orbital, you start with a trial mathematical function for the orbital. You then test the trial function to see how well it "works."

We can understand this process by analogy to solving an algebraic equation. Suppose you want to know x in the equation $4x+5 = 70$ without actually solving the equation. For an easy equation like this one, you might first estimate that x = 16. You can then determine how well your estimate works by substituting x = 16 into the equation. If the estimate did not work, you could try again until you found the right value of x. (In this case, you can quickly see that x must be a little more than 16.)

In molecular orbital theory, the estimating procedure is analogous. However, we need to add one more important concept to get at the heart of molecular orbital theory. In order to determine how well a trial function for an orbital "works" in molecular orbital theory, you calculate its energy. No matter how good your trial function, *you will never do better than nature at minimizing the energy of the orbital*. In other words, devise any trial function that you like for an orbital and calculate its energy. The energy you calculate for the devised orbital will always be greater than or (at best) equal to the energy of the actual orbital.

How does this help us? The best possible orbital will therefore be the one with the minimum energy. In modern molecular orbital theory, computer programs are designed to try many different variations of a guessed orbital and compare the energies of each one. The variation with the lowest energy is the best approximation for the actual molecular orbital.

> Molecular orbital theory is a specific application of a more general quantum-mechanical approximation technique called the variational method. In this method, the energy of a trial function within the Schrödinger equation is minimized.

> You calculate the energy of an estimated orbital by substituting it into the Schrödinger equation and solving for the energy.

Linear Combination of Atomic Orbitals (LCAO)

The simplest guesses that work reasonably well in molecular orbital theory turn out to be linear combinations of atomic orbitals, or LCAOs. An LCAO molecular orbital is simply a *weighted linear sum—analogous to a weighted average—of the valence atomic orbitals* of the atoms in the molecule. At first glance, this concept might seem very similar to that of hybridization in valence bond theory. However, in valence bond theory, hybrid orbitals are weighted linear sums of the valence atomic orbitals of a *particular atom*, and the hybrid orbitals remain *localized* on that atom. In molecular orbital theory, the molecular orbitals are weighted linear sums of the valence atomic orbitals of *all the atoms* in a molecule, and many of the molecular orbitals are *delocalized* over the entire molecule.

As an example, consider the H_2 molecule. One of the molecular orbitals for H_2 is simply an equally weighted sum of the 1s orbital from one atom and the 1s orbital from the other. We can represent this pictorially and energetically as follows:

> When molecular orbitals are computed mathematically, it is actually the *wave functions* corresponding to the orbitals that are combined.

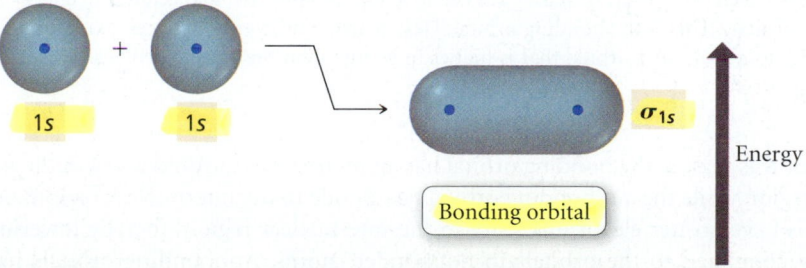

Bonding orbital

The name of this molecular orbital is σ_{1s}. The σ comes from the shape of the orbital, which looks like a σ bond in valence bond theory, and the 1s comes from its formation by a linear sum of 1s orbitals. The σ_{1s} orbital is lower in energy than either of the two 1s atomic orbitals from which it was formed. For this reason, this orbital is called a **bonding orbital**. When electrons occupy bonding molecular orbitals, the energy of the electrons is lower than it would be if they were occupying atomic orbitals.

You can think of a molecular orbital for a molecule in much the same way that you think about an atomic orbital in an atom. Electrons will seek the lowest energy molecular orbital available, but just as an atom has more than one atomic orbital (and some may be empty), so a molecule has more than one molecular orbital (and some may be empty). The next molecular orbital of H_2 is approximated by summing the $1s$ orbital on one hydrogen atom with the *negative* of the $1s$ orbital on the other hydrogen atom.

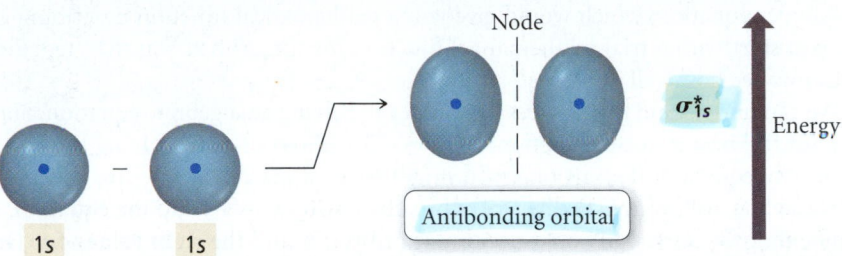

The name of this molecular orbital is σ_{1s}^*. The star indicates that this orbital is an **antibonding orbital**. Electrons in antibonding orbitals have higher energies than they did in their respective atomic orbitals and therefore tend to raise the energy of the system (relative to the unbonded atoms).

In general, when two atomic orbitals are added together to form molecular orbitals, one of the resultant molecular orbitals will be lower in energy (the bonding orbital) than the atomic orbitals and the other will be higher in energy (the antibonding orbital). Remember that electrons in orbitals behave like waves. The bonding molecular orbital arises out of constructive interference between the atomic orbitals, while the antibonding orbital arises out of destructive interference between the atomic orbitals (Figure 10.13).

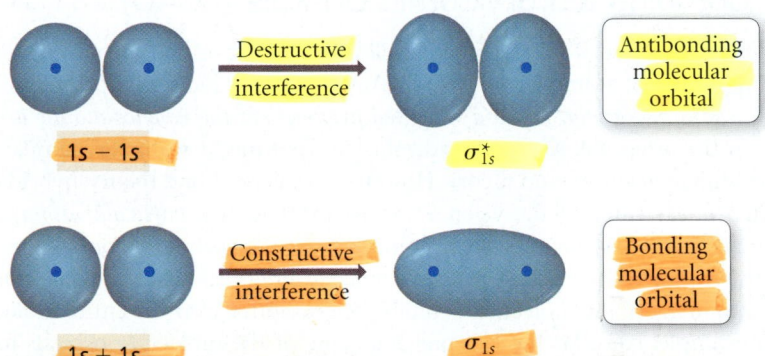

▲ **FIGURE 10.13 Formation of Bonding and Antibonding Orbitals**   Constructive interference between two atomic orbitals gives rise to a molecular orbital that is lower in energy than the atomic orbitals. This is the bonding orbital. Destructive interference between two atomic orbitals gives rise to a molecular orbital that is higher in energy than the atomic orbitals. This is the antibonding orbital.

For this reason, the bonding orbital has an increased electron density in the internuclear region while the antibonding orbital has a node in the internuclear region. Bonding orbitals have greater electron density in the internuclear region, thereby lowering their energy compared to the orbitals in nonbonded atoms. Antibonding orbitals have less electron density in the internuclear region, and their energies are generally higher than in the orbitals of nonbonded atoms.

We put all of this together in the molecular orbital energy diagram for H_2 as follows:

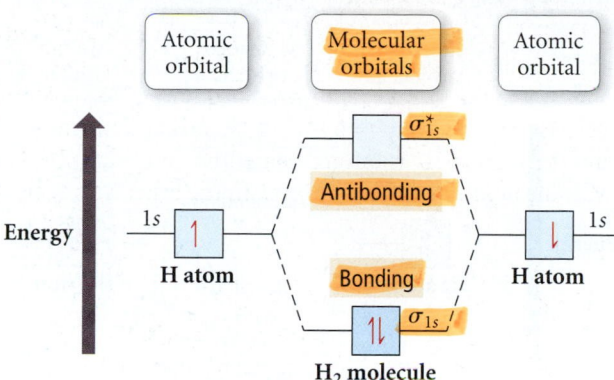

The molecular orbital (MO) diagram shows that two hydrogen atoms can lower their overall energy by forming H_2 because the electrons can move from higher energy atomic orbitals into the lower energy σ_{1s} bonding molecular orbital. In molecular orbital theory, we define the **bond order** of a diatomic molecule such as H_2 as follows:

$$\text{Bond order} = \frac{\text{(number of electrons in bonding MOs)} - \text{(number of electrons in antibonding MOs)}}{2}$$

For H_2, the bond order is

$$H_2 \text{ bond order} = \frac{2 - 0}{2} = 1$$

A positive bond order means that there are more electrons in bonding molecular orbitals than in antibonding molecular orbitals. The electrons will therefore have lower energy than they did in the orbitals of the isolated atoms and a chemical bond will form. In general, the higher the bond order, the stronger the bond. A negative or zero bond order indicates that a bond will *not* form between the atoms. For example, consider the MO diagram for He_2:

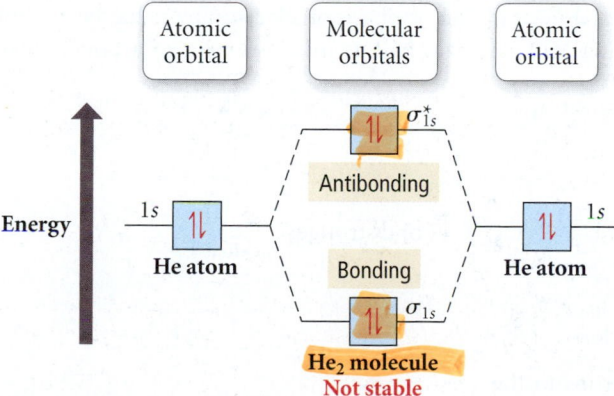

Notice that the two additional electrons must go into the higher energy antibonding orbital. There is no net stabilization by joining two helium atoms to form a helium molecule, as indicated by the bond order:

$$He_2 \text{ bond order} = \frac{2 - 2}{2} = 0$$

So according to MO theory, He_2 should not exist as a stable molecule, and indeed it does not. An interesting case is the helium–helium ion, He_2^+, with the following MO diagram:

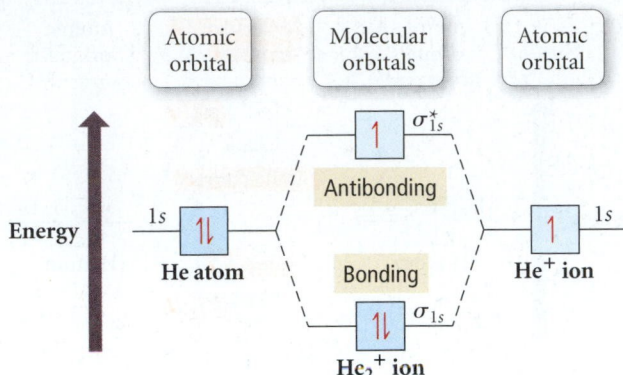

The bond order is $\frac{1}{2}$, indicating that He_2^+ should exist, and indeed it does.

Before we move on to applying MO theory to diatomic molecules from the second row of the periodic table, we formalize the main ideas in MO theory as follows.

Main Ideas in Applying LCAO–MO Theory

➤ Molecular orbitals (MOs) can be approximated by a linear combination of atomic orbitals (AOs). The total number of MOs formed from a particular set of AOs will always equal the number of AOs in the set.

➤ When two AOs combine to form two MOs, one MO will be lower in energy (the bonding MO) and the other will be higher energy (the antibonding MO).

➤ When assigning the electrons of a molecule to MOs, fill the lowest energy MOs first with a maximum of two spin-paired electrons per orbital.

➤ When assigning electrons to two MOs of the same energy, follow Hund's rule—fill the orbitals singly first, with parallel spins, before pairing.

➤ The bond order in a diatomic molecule is simply the number of electrons in bonding MOs minus the number in antibonding MOs divided by two. Stable bonds require a positive bond order (more electrons in bonding MOs than in antibonding MOs).

Notice the power of the molecular orbital approach. Every electron that enters a bonding MO stabilizes the molecule or polyatomic ion and every electron that enters an antibonding MO destabilizes it. The emphasis on electron pairs has been removed. One electron in a bonding MO stabilizes half as much as two, so a bond order of one-half is nothing mysterious.

EXAMPLE 10.9 Bond Order

Use molecular orbital theory to predict the bond order in H_2^-. Is the H_2^- bond stronger or weaker bond than the H_2 bond?

Solution

The H_2^- ion has three electrons. Assign the three electrons to the molecular orbitals, filling lower energy orbitals first and proceeding to higher energy orbitals.

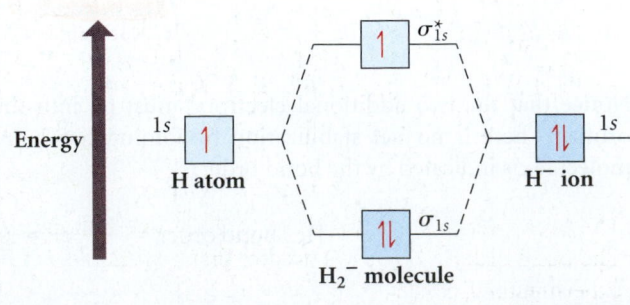

Calculate the bond order by subtracting the number of electrons in antibonding orbitals from the number in bonding orbitals and dividing the result by two.	H_2^- bond order $= \dfrac{2-1}{2} = +\dfrac{1}{2}$

Since the bond order is positive, H_2^- should be stable. However, the bond order of H_2^- is lower than the bond order of H_2 (which is 1); therefore, the bond in H_2^- is weaker than in H_2.

For Practice 10.9

Use molecular orbital theory to predict the bond order in H_2^+. Is the H_2^+ bond stronger or weaker bond than the H_2 bond?

Period Two Homonuclear Diatomic Molecules

The homonuclear diatomic molecules (molecules made up of two atoms of the same kind) formed from second-period elements have between 2 and 16 valence electrons. To explain bonding in these molecules, we must consider the next set of higher energy molecular orbitals, which can be approximated by linear combinations of the valence atomic orbitals of the period 2 elements.

The core electrons can be ignored because, as with other models for bonding, these electrons do not contribute significantly to chemical bonding.

We begin with Li_2. Even though lithium is normally a metal, we can use MO theory to predict whether or not the Li_2 molecule should exist in the gas phase. The molecular orbitals in Li_2 are approximated as linear combinations of the $2s$ atomic orbitals and look much like the H_2 molecular orbitals formed from linear combinations of the $1s$ orbitals. The MO diagram for Li_2 is therefore as follows:

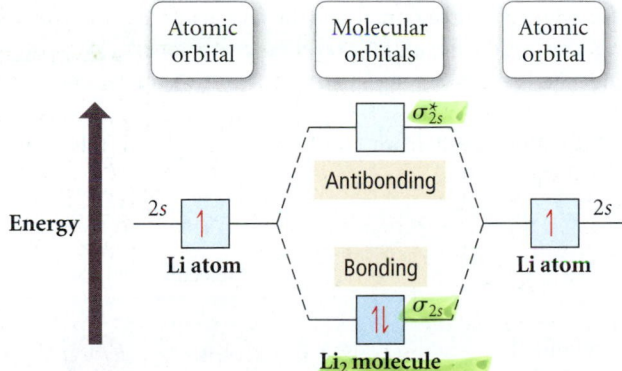

The two valence electrons of Li_2 occupy a bonding molecular orbital. We would predict that the Li_2 molecule is stable with a bond order of 1. Experiments confirm this prediction. In contrast, consider the MO diagram for Be_2:

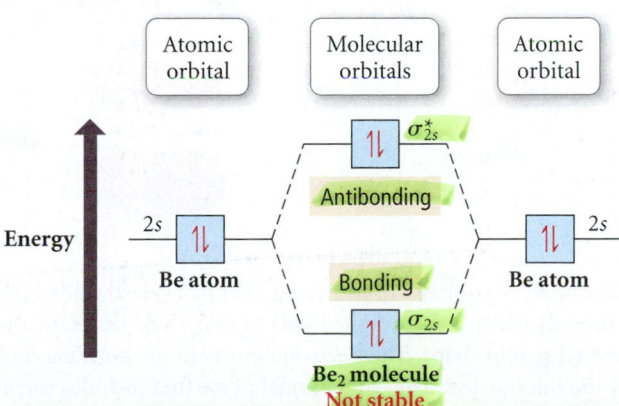

The four valence electrons of Be_2 occupy one bonding MO and one antibonding MO. The bond order is 0 and we predict that Be_2 should not be stable, again consistent with experimental findings.

The next homonuclear molecule composed of second row elements is B_2, which has six total valence electrons to accommodate. We can approximate the next higher energy molecular orbitals for B_2 and the rest of the period 2 diatomic molecules as linear combinations of the $2p$ orbitals taken pairwise. Since the three $2p$ orbitals orient along three orthogonal axes, we must assign similar axes to the molecule. In this book, we assign the internuclear axis to be the x direction. Then the LCAO–MOs that result from combining the $2p_x$ orbitals—the ones that lie along the internuclear axis—from each atom are represented pictorially as follows:

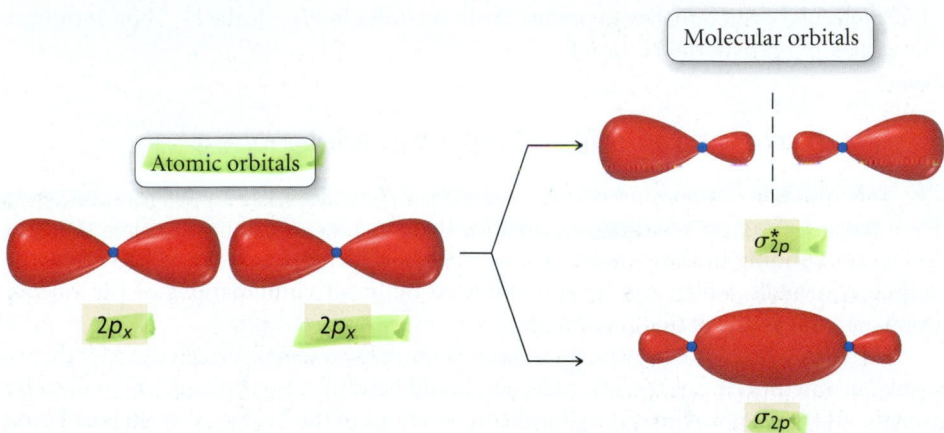

The bonding MO in this pair looks something like a candy wrapper, with increased electron density in the internuclear region. It has the characteristic σ shape (it is cylindrically symmetrical about the bond axis) and is therefore called the σ_{2p} bonding orbital. The antibonding orbital, called σ^*_{2p}, has a node between the two nuclei and is higher in energy than either of the $2p_x$ orbitals.

The LCAO–MOs that result from combining the $2p_y$ orbitals from each atom are represented pictorially as follows:

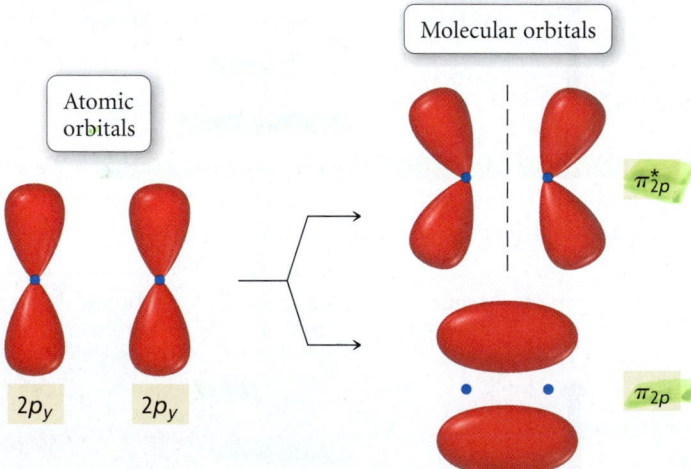

Notice that in this case the p orbitals are added together in a side-by-side orientation (in contrast to the $2p_x$ orbitals which were oriented end to end). The resultant molecular orbitals consequently have a different shape. The electron density in the bonding molecular orbital is above and below the internuclear axis with a nodal plane that includes the internuclear axis. This orbital resembles the electron density distribution of a π bond in valence bond theory. We call this orbital the π_{2p} orbital. The corresponding antibonding orbital has an additional node *between* the nuclei (perpendicular to the internuclear axis) and is called the π^*_{2p} orbital.

The LCAO–MOs that result from combining the $2p_z$ orbitals from each atom are represented pictorially as follows:

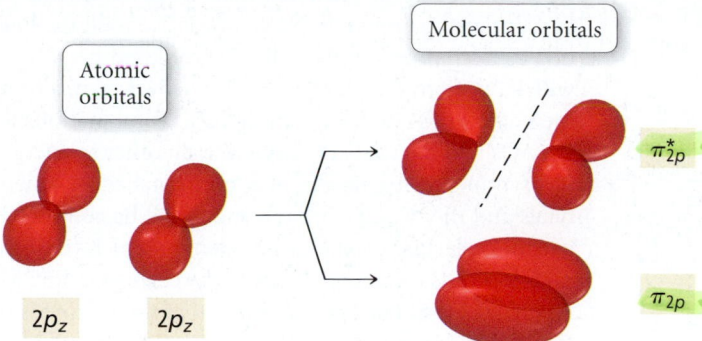

The only difference between the $2p_z$ and the $2p_y$ atomic orbitals is a 90° rotation about the internuclear axis. Consequently, the only difference between the resulting MOs is a 90° rotation about the internuclear axis. The energies and the names of the bonding and antibonding MOs obtained from the combination of the $2p_z$ AOs are identical to those obtained from the combination of the $2p_y$ AOs.

Before we can draw MO diagrams for B_2 and the other second-period diatomic molecules, we must determine the relative energy ordering of the MOs obtained from the $2p$ AO combinations. This is not a simple issue. The relative ordering of MOs obtained from LCAO–MO theory is usually determined computationally. There is no single order that will work for all molecules. For second-period diatomic molecules, computations reveal that the energy ordering for B_2, C_2, and N_2 is slightly different than that for O_2, F_2, and Ne_2 as follows:

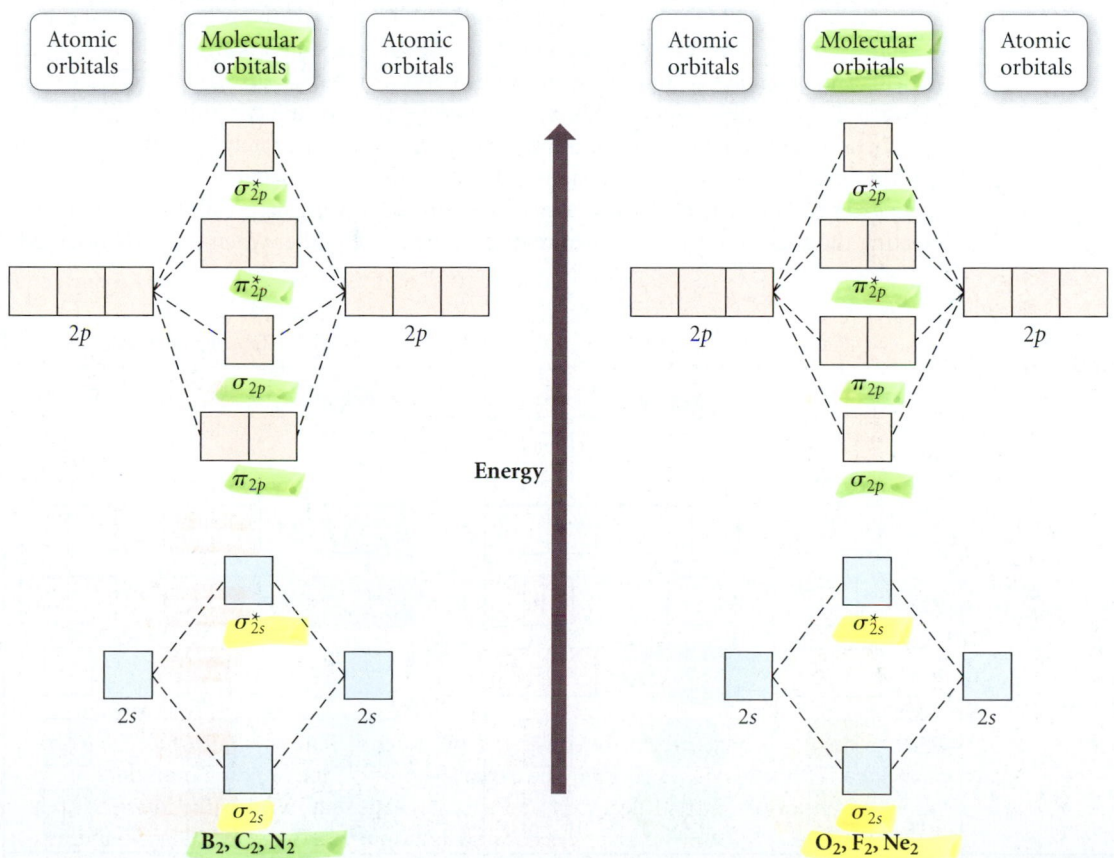

▲ Molecular orbital energy diagrams for second-period diatomic molecules show that the energy ordering of the π_{2p} and σ_{2p} molecular orbitals can vary.

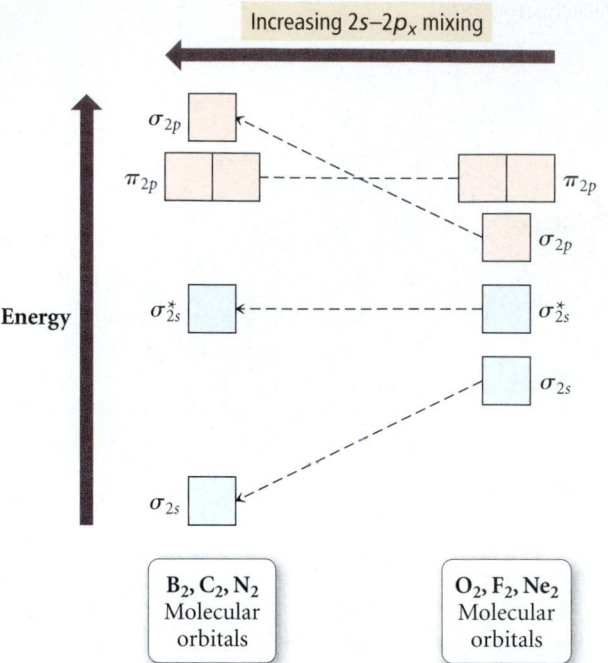

Increasing 2s–2p$_x$ mixing

Energy

σ_{2p}
π_{2p}
σ_{2p}
σ_{2s}^*
σ_{2s}^*
σ_{2s}
σ_{2s}
σ_{2s}

B$_2$, C$_2$, N$_2$
Molecular
orbitals

O$_2$, F$_2$, Ne$_2$
Molecular
orbitals

▲ **FIGURE 10.14 The Effects of 2s—2p Mixing** The degree of mixing between two orbitals decreases with increasing energy difference between them. Mixing of the 2s and 2p$_x$ orbitals is therefore greater in B$_2$, C$_2$, and N$_2$ than in O$_2$, F$_2$, and Ne$_2$ because in B, C, and N the energy levels of the atomic orbitals are more closely spaced than in O, F, and Ne. This mixing produces a change in energy ordering for the π_{2p} and σ_{2p} molecular orbitals.

The reason for the difference in energy ordering can only be explained by going back to our LCAO–MO model. In our simplified treatment, we assumed that the MOs that result from the second-period AOs could be calculated pairwise. In other words, we took the linear combination of a 2s from one atom with the 2s from another, a 2p$_x$ from one atom with a 2p$_x$ from the other and so on. However, in a more detailed treatment, the MOs are formed from linear combinations that include all of the AOs that are relatively close to each other in energy and of the correct symmetry. Specifically, in a more detailed treatment, the two 2s orbitals and the two 2p$_x$ orbitals should all be combined to form a total of four molecular orbitals. The extent to which you include this type of mixing affects the energy levels of the corresponding MOs, as shown in Figure 10.14 ◄. The bottom line is that s–p mixing is significant in B$_2$, C$_2$, and N$_2$ but not in O$_2$, F$_2$, and Ne$_2$. The result is a different energy ordering, depending on the specific molecule.

The MO energy diagrams for all of the second-period homonuclear diatomic molecules, as well as their bond orders, bond energies, and bond lengths, are shown in Figure 10.15 ▼. Notice that as bond order increases, the bond gets stronger (greater bond energy) and shorter (smaller bond length). For B$_2$, with six electrons, the bond order is 1. For C$_2$, the bond order is 2, and for N$_2$, the bond order reaches a maximum with a value of 3. Recall that the Lewis structure of N$_2$ has a triple bond, so both Lewis theory and molecular orbital theory predict a strong bond for N$_2$, which is experimentally observed.

In O$_2$, the two additional electrons occupy antibonding orbitals and the bond order is 2. Notice that these two electrons are unpaired—they occupy the π_{2p}^* orbitals *singly with parallel spins*, as indicated by Hund's rule. The presence of unpaired electrons in the molecular orbital diagram of oxygen is significant because oxygen is known from experiment to be **paramagnetic**, which means that it is attracted to a magnetic field. The paramagnetism of oxygen can be demonstrated by suspending liquid oxygen between the poles of a magnet. This magnetic property is the direct result of *unpaired electrons*, whose spin and movement around the nucleus (more accurately known as orbital angular momentum) generate tiny magnetic fields. When a paramagnetic substance is placed in an external magnetic field, the magnetic fields of each atom align with the external field, creating the attraction (much as two magnets attract each other when properly oriented).

▶ **FIGURE 10.15 Molecular Orbital Energy Diagrams for Second-Row Homonuclear Diatomic Molecules**

	Large 2s–2p$_x$ interaction				Small 2s–2p$_x$ interaction		
	B$_2$	**C$_2$**	**N$_2$**		**O$_2$**	**F$_2$**	**Ne$_2$**
σ_{2p}^*				σ_{2p}^*			⇅
π_{2p}^*				π_{2p}^*	↑ ↑	⇅ ⇅	⇅ ⇅
σ_{2p}			⇅	π_{2p}	⇅ ⇅	⇅ ⇅	⇅ ⇅
π_{2p}	↑ ↑	⇅ ⇅	⇅ ⇅	σ_{2p}	⇅	⇅	⇅
σ_{2s}^*	⇅	⇅	⇅	σ_{2s}^*	⇅	⇅	⇅
σ_{2s}	⇅	⇅	⇅	σ_{2s}	⇅	⇅	⇅
Bond order	1	2	3		2	1	0
Bond energy (kJ/mol)	290	620	941		495	155	—
Bond length (pm)	159	131	110		121	143	—

In contrast, when the electrons in an atom are all *paired*, the magnetic fields caused by electron spin and orbital angular momentum tend to cancel each other, resulting in diamagnetism. A **diamagnetic** substance is not attracted to a magnetic field (and is, in fact, slightly repelled).

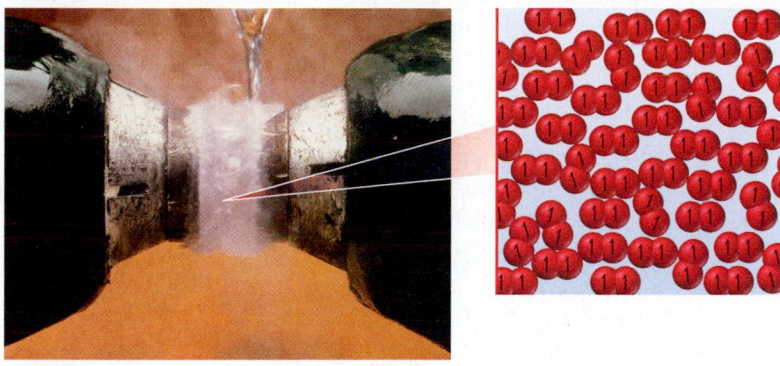

◀ Liquid oxygen can be suspended between the poles of a magnet because it is paramagnetic. It contains unpaired electrons that generate tiny magnetic fields, which align with and interact with the external field.

In the Lewis structure of O_2, as well as in the valence bond model of O_2, all of its electrons seem to be paired:

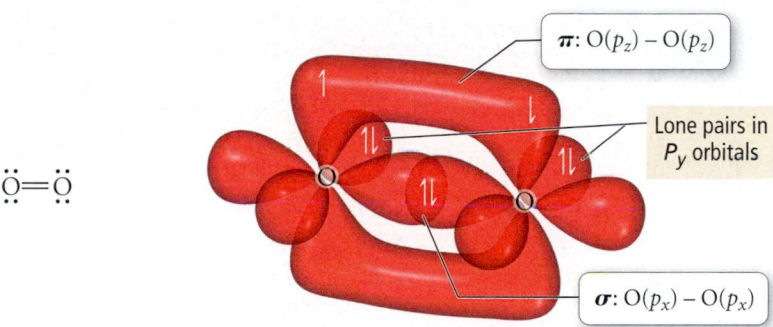

In the MO diagram for O_2, however, we can see the unpaired electrons. Molecular orbital theory is more powerful in that it can account for the paramagnetism of O_2—it gives us a picture of bonding that more closely corresponds to what we see in experiment. Continuing along the second-row homonuclear diatomic molecules we see that F_2 has a bond order of 1 and Ne_2 has a bond order of 0, again consistent with experiment since F_2 exists and Ne_2 does not.

EXAMPLE 10.10 Molecular Orbital Theory

Draw an MO energy diagram and determine the bond order for the N_2^- ion. Do you expect the bond to be stronger or weaker than in the N_2 molecule? Is N_2^- diamagnetic or paramagnetic?

Solution

Write an energy level diagram for the molecular orbitals in N_2^-. Use the energy ordering for N_2.

σ_{2p}^*

π_{2p}^*

σ_{2p}

π_{2p}

σ_{2s}^*

σ_{2s}

The N_2^- ion has 11 valence electrons (5 for each nitrogen atom plus 1 for the negative charge). Assign the electrons to the molecular orbitals beginning with the lowest energy orbitals and following Hund's rule.

Calculate the bond order by subtracting the number of electrons in antibonding orbitals from the number in bonding orbitals and dividing the result by two.

$$N_2^- \text{ bond order} = \frac{8 - 3}{2} = +2.5$$

The bond order is 2.5, which is a lower bond order than in the N_2 molecule (bond order = 3); therefore, the bond is weaker. The MO diagram shows that the N_2^- ion has one unpaired electron and is therefore paramagnetic.

For Practice 10.10

Draw an MO energy diagram and determine the bond order for the N_2^+ ion. Do you expect the bond to be stronger or weaker than in the N_2 molecule? Is N_2^+ diamagnetic or paramagnetic?

For More Practice 10.10

Use molecular orbital theory to determine the bond order of Ne_2.

Second-Period Heteronuclear Diatomic Molecules

Molecular orbital theory can also be applied to heteronuclear diatomic molecules (two different atoms). For example, we can draw an MO diagram for NO as follows:

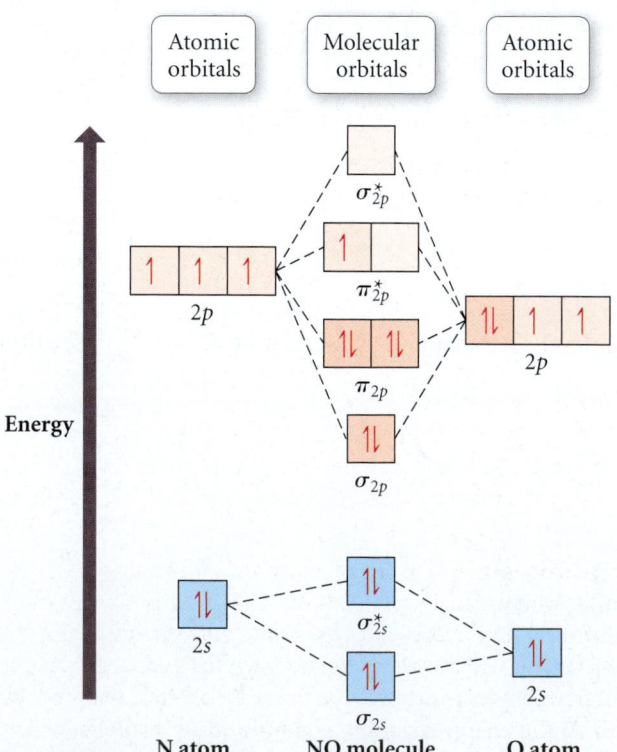

Since oxygen is more electronegative than nitrogen, its atomic orbitals are lower in energy than nitrogen's atomic orbitals. When two atomic orbitals are identical and of equal energy, then the weighting of each orbital in forming a molecular orbital is identical. However, when two atomic orbitals are different, the weighting of each orbital in forming a molecular orbital may be different. More specifically, when a molecular orbital is approximated as a linear combination of atomic orbitals of different energies, the lower energy atomic orbital makes a greater contribution to the bonding molecular orbital and the higher energy atomic orbital makes a greater contribution to the antibonding molecular orbital. For example, notice that the σ_{2s} bonding orbital is closer in energy to the oxygen 2s orbital than to the nitrogen 2s orbital. We can also see this unequal weighting in the shape of the resultant molecular orbital, in which the electron density is concentrated on the oxygen atom, as shown in Figure 10.16 ▼.

A given orbital will have lower energy in a more electronegative atom. For this reason, electronegative atoms have the ability to attract electrons to themselves.

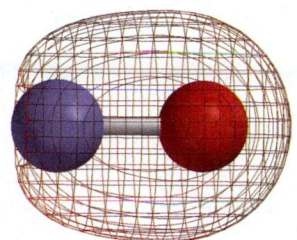

▲ FIGURE 10.16 Shape of σ_{2s} bonding orbital in NO   The molecular orbital shows more electron density at the oxygen end of the molecule because the atomic orbitals of oxygen, the more electronegative element, are lower in energy than those of nitrogen. They therefore contribute more to the bonding molecular orbital.

As another example of a heteronuclear diatomic molecule, consider the MO diagram for HF:

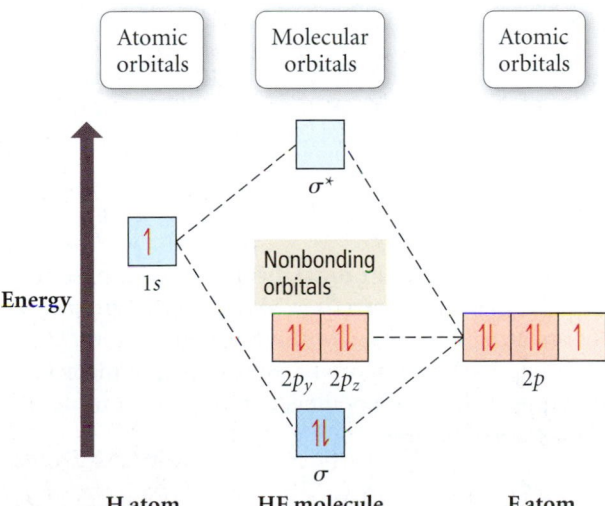

Fluorine is so electronegative that all of its atomic orbitals are lower in energy than hydrogen's atomic orbitals. In fact, fluorine's 2s orbital is so low in energy compared to hydrogen's 1s orbital that it does not contribute appreciably to the molecular orbitals. The molecular orbitals in HF are approximated by the linear combination of the fluorine $2p_x$ orbital and the hydrogen 1s orbital. The other 2p orbitals remain localized on the fluorine and appear in the energy diagram as **nonbonding orbitals**, which means that the electrons in those orbitals remain localized on the fluorine atom.

EXAMPLE 10.11 Molecular Orbital Theory for Heteronuclear Diatomic Molecules and Ions

Use molecular orbital theory to determine the bond order of the CN^- ion. Is the ion paramagnetic or diamagnetic?

Solution

Determine the number of valence electrons in the molecule or ion.	Number of valence electrons $= 4\,(\text{from C}) + 5\,(\text{from N}) +$ $1\,(\text{from negative charge})$ $= 10$
Write an energy level diagram using Figure 10.15 as a guide. Fill the orbitals beginning with the lowest energy orbital and progressing upward until all electrons have been assigned to an orbital. Remember to allow no more than two electrons (with paired spins) per orbital and to fill degenerate orbitals with single electrons (with parallel spins) before pairing.	$\square$ σ^*_{2p} $\square\,\square$ π^*_{2p} $\boxed{\uparrow\downarrow}$ σ_{2p} $\boxed{\uparrow\downarrow}\,\boxed{\uparrow\downarrow}$ π_{2p} $\boxed{\uparrow\downarrow}$ σ^*_{2s} $\boxed{\uparrow\downarrow}$ σ_{2s}
Compute the bond order using the following formula: Bond order = $\dfrac{(\text{number of } e^- \text{ in bonding MOs}) - (\text{number of } e^- \text{ in antibonding MOs})}{2}$	$CN^- \text{ bond order} = \dfrac{8-2}{2} = +3$
If the MO diagram has unpaired electrons, the molecule or ion is paramagnetic. If the electrons are all paired, the molecule or ion is diamagnetic.	Since the MO diagram has no unpaired electrons, the ion is diamagnetic.

For Practice 10.11

Use molecular orbital theory to determine the bond order of NO. (Use the energy ordering of N_2.) Is the molecule paramagnetic or diamagnetic?

Polyatomic Molecules

With the aid of computers, molecular orbital theory can be applied to polyatomic molecules and ions, yielding results that correlate very well with experimental measurements. These applications are beyond the scope of this text. However, the delocalizaton of electrons over an entire molecule is an important contribution of molecular orbital theory to our basic understanding of chemical bonding. For example, consider the Lewis structure and valence bond diagram of ozone:

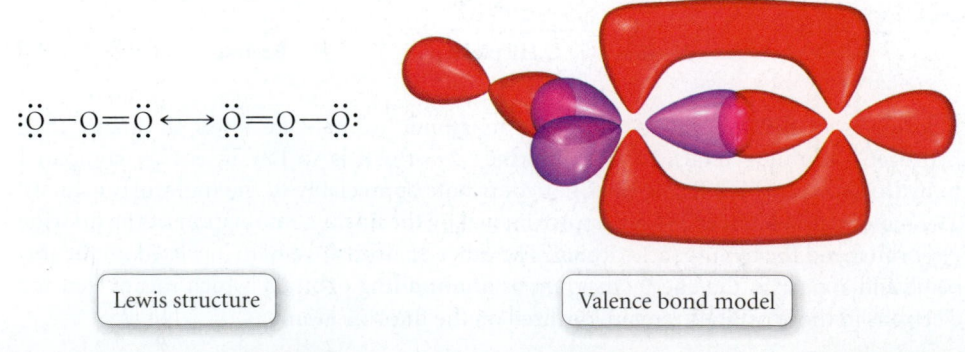

:Ö—O̤=Ö: ⟷ Ö=O̤—Ö:

Lewis structure   Valence bond model

In Lewis theory, we use resonance forms to represent the two equivalent bonds. In valence bond theory, it appears that the two oxygen–oxygen bonds should be different. In molecular orbital theory, however, the π molecular orbitals in ozone are formed from a linear combination of the three oxygen $2p$ orbitals and are delocalized over the entire molecule. The lowest energy π bonding molecular orbital is shown here.

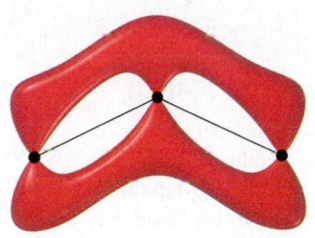

When we examine ozone in nature, we indeed find two equivalent bonds. A similar situation occurs with benzene (C_6H_6). In Lewis theory, we represent the structure with two resonance forms:

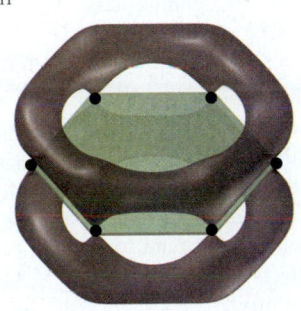

In molecular orbital theory, the π molecular orbitals in benzene are formed from a linear combination of the six carbon $2p$ orbitals and are delocalized over the entire molecule. The lowest energy π bonding molecular orbital is shown here.

Benzene is in fact a highly symmetric molecule with six identical carbon–carbon bonds. The best picture of the π electrons in benzene is one in which the electrons occupy roughly circular shaped orbitals above and below the plane of the molecule, as depicted in the MO approach.

 Conceptual Connection 10.5 **What Is a Chemical Bond, Part II?**

We have learned that Lewis theory portrays a chemical bond as the transfer or sharing of electrons represented as dots. Valence bond theory portrays a chemical bond as the overlap of two half-filled atomic orbitals. What is a chemical bond according to molecular orbital theory?

Answer: In MO theory, atoms will join together (or bond) when the electrons in the atoms can lower their energy by occupying the molecular orbitals of the resultant molecule. Unlike Lewis theory or valence bond theory, the chemical "bonds" in MO theory are not localized between atoms, but spread throughout the entire molecule.

Chapter in Review

Key Terms

Section 10.2
valence shell electron pair repulsion (VSEPR) theory (406)
electron groups (406)
linear geometry (407)
trigonal planar geometry (407)
tetrahedral geometry (408)
trigonal bipyramidal geometry (409)
octahedral geometry (409)

Section 10.3
electron geometry (410)
molecular geometry (410)
trigonal pyramidal geometry (410)
bent geometry (411)
seesaw geometry (412)
T-shaped geometry (412)
square pyramidal geometry (413)
square planar geometry (413)

Section 10.6
valence bond theory (423)

Section 10.7
hybridization (426)
hybrid orbitals (426)
pi (π) bond (430)
sigma (σ) bond (430)

Section 10.8
molecular orbital theory (439)
bonding orbital (440)
antibonding orbital (440)
bond order (441)
paramagnetic (446)
diamagnetic (447)
nonbonding orbitals (449)

Key Concepts

Molecular Shape and VSEPR Theory (10.1–10.4)

The properties of molecules are directly related to their shapes. In VSEPR theory, molecular geometries are determined by the repulsions between electron groups on the central atom. An electron group can be a single bond, double bond, triple bond, lone pair, or even a single electron. The five basic shapes are linear (two electron groups), trigonal planar (three electron groups), tetrahedral (four electron groups), trigonal bipyramidal (five electron groups), and octahedral (six electron groups). When lone pairs are present on the central atom, the *electron* geometry is still one of the five basic shapes, but one or more positions are occupied by lone pairs. The *molecular* geometry is therefore different from the electron geometry. Lone pairs are positioned so as to minimize repulsions with other lone pairs and with bonding pairs.

Polarity (10.5)

The polarity of a polyatomic molecule containing polar bonds depends on its geometry. If the dipole moments of the polar bonds are aligned in such a way that they cancel one another, the molecule will not be polar. If they are aligned in such a way as to add together, the molecule will be polar. Highly symmetric molecules tend to be nonpolar, while asymmetric molecules containing polar bonds tend to be polar. The polarity of a molecule dramatically affects its properties. For example, water (polar) and oil (nonpolar) do not mix.

Valence Bond Theory (10.6, 10.7)

In contrast to Lewis theory, in which a covalent chemical bond is the sharing of electrons represented by dots, a chemical bond in valence bond theory is the overlap of half-filled atomic orbitals (or in some cases the overlap between a completely filled orbital and an empty one). The overlapping orbitals may be the standard atomic orbitals, such as $1s$ or $2p$, or they may be hybridized atomic orbitals, which are mathematical combinations of the standard orbitals on a single atom. The basic hybridized orbitals are sp, sp^2, sp^3, sp^3d, and sp^3d^2. The geometry of the molecule is determined by the geometry of the overlapping orbitals. In our treatment of valence bond theory, we use the molecular geometry determined by VSEPR theory to determine the correct hybridization scheme. In valence bond theory, we distinguish between two types of bonds, σ (sigma) and π (pi). In a σ bond, the orbital overlap occurs in the region that lies directly between the two bonding atoms. In a π bond, formed from the side-by-side overlap of p orbitals, the overlap occurs above and below the region that lies directly between the two bonding atoms. Rotation about a σ bond is relatively free, while rotation about a π bond is restricted.

Molecular Orbital Theory (10.8)

In molecular orbital theory, we approximate solutions to the Schrödinger equation for the molecule *as a whole* by guessing the mathematical form of the orbitals. We differentiate between guesses by calculating the energies of guessed orbitals—the best guesses will have the lowest energy. Molecular orbitals obtained in this way are properties of the molecule and are often delocalized over the entire molecule. The simplest guesses that work well are linear combinations of atomic orbitals (LCAOs), weighted averages of the atomic orbitals of the different atoms in the molecule. When two atomic orbitals are combined to form molecular orbitals, they will form one molecular orbital of lower energy (the bonding orbital) and one of higher energy (the antibonding orbitals). A set of molecular orbitals are filled in much the same way as atomic orbitals. The stability of the molecule and the strength of the bond depend on the number of electrons in bonding orbitals compared to the number in antibonding orbitals.

Key Equations and Relationships

Bond Order of a Diatomic Molecule (10.8)

$$\text{Bond order} = \frac{(\text{number of electrons in bonding MOs}) - (\text{number of electrons in antibonding MOs})}{2}$$

Key Skills

Using VSEPR Theory to Predict the Basic Shapes of Molecules (10.2)
  • Example 10.1   • For Practice 10.1   • Exercises 29, 30

Predicting Molecular Geometries Using VSEPR Theory and the Effects of Lone Pairs (10.4)
  • Examples 10.2, 10.3   • For Practice 10.2, 10.3   • Exercises 33, 34

Predicting the Shapes of Larger Molecules (10.4)
  • Example 10.4   • For Practice 10.4   • Exercises 39, 40, 43, 44

Using Molecular Shape to Determine Polarity of a Molecule (10.5)
  • Example 10.5   • For Practice 10.5   • Exercises 47–50

Writing Hybridization and Bonding Schemes Using Valence Bond Theory (10.7)
  • Examples 10.6, 10.7, 10.8   • For Practice 10.6, 10.7, 10.8   • For More Practice 10.8   • Exercises 59–64

Drawing Molecular Orbital Diagrams to Predict Bond Order and Magnetism of a Diatomic Molecule (10.8)
  • Examples 10.9, 10.10, 10.11   • For Practice 10.9, 10.10, 10.11   • For More Practice 10.10   • Exercises 69, 70, 73–76, 79–80

Exercises

Review Questions

1. Why is molecular geometry important? Give some examples.

2. According to VSEPR theory, what determines the geometry of a molecule?

3. Give the name and make a sketch of the five basic electron geometries, and state the number of electron groups corresponding to each. What constitutes an *electron group*?

4. Explain the difference between electron geometry and molecular geometry and explain the circumstances under which they will not be the same.

5. Give the correct electron and molecular geometry that corresponds to each of the following sets of electron groups around the central atom of a molecule.

   a. four electron groups overall; three bonding groups and one lone pair
   b. four electron groups overall; two bonding groups and two lone pairs
   c. five electron groups overall; four bonding groups and one lone pair
   d. five electron groups overall; three bonding groups and two lone pairs
   e. five electron groups overall; two bonding groups and three lone pairs
   f. six electron groups overall; five bonding groups and one lone pair
   g. six electron groups overall; four bonding groups and two lone pairs

6. How do you use VSEPR theory to predict the shape of a molecule with more than one interior atom?

7. How do you determine whether a molecule is polar? Why is polarity important?

8. What is a chemical bond according to valence bond theory?

9. In valence bond theory, what determines the geometry of a molecule?

10. In valence bond theory, the interaction energy between the electrons and nucleus of one atom with the electrons and nucleus of another atom is usually negative (stabilizing) when _____.

11. What is hybridization? Why is hybridization necessary in valence bond theory?

12. How does hybridization of the atomic orbitals in the central atom of a molecule help lower the overall energy of the molecule?

13. How is the *number* of hybrid orbitals related to the number of standard atomic orbitals that are hybridized?

14. Make sketches of each of the following hybrid orbitals:

   a. sp b. sp^2 c. sp^3 d. sp^3d e. sp^3d^2

15. In Lewis theory, the two bonds in a double bond look identical. However, valence bond theory shows that they are not. Describe a double bond according to valence bond theory. Explain why rotation is restricted about a double bond, but not about a single bond.

16. Give the hybridization scheme that corresponds to each of the following electron geometries:

   a. linear   b. trigonal planar
   c. tetrahedral   d. trigonal bipyramidal
   e. octahedral

17. What is a chemical bond according to molecular orbital theory?

18. Explain the difference between hybrid atomic orbitals in valence bond theory and LCAO molecular orbitals in molecular orbital theory.

19. Explain the difference between bonding and antibonding molecular orbitals.

20. In molecular orbital theory, what is bond order? Why is it important?

21. How is the number of molecular orbitals approximated by a linear combination of atomic orbitals related to the number of atomic orbitals used in the approximation?

22. Make a sketch of each of the following molecular orbitals.

   a. σ_{2s} b. σ_{2s}^* c. σ_{2p} d. σ_{2p}^* e. π_{2p} f. π_{2p}^*

23. Draw an energy diagram for the molecular orbitals of period 2 diatomic molecules. Show the difference in ordering for B_2, C_2, and N_2 compared to O_2, F_2, and Ne_2.

24. Why does the energy ordering of the molecular orbitals of the period 2 diatomic molecules change in going from N_2 to O_2?

25. Explain the difference between a paramagnetic species and a diamagnetic one.

26. When applying molecular orbital theory to heteronuclear diatomic molecules, the atomic orbitals used may be of different energies. If two atomic orbitals of different energies are used to make two molecular orbitals, how are the energies of the molecular orbitals related to the energies of the atomic orbitals? How is the shape of the resultant molecular orbitals related to the shapes of the atomic orbitals?

27. In molecular orbital theory, what is a nonbonding orbital?

28. Write a short paragraph describing chemical bonding according to Lewis theory, valence bond theory, and molecular orbital theory. Indicate how the theories differ in their description of a chemical bond and indicate the strengths and weaknesses of each theory. Which theory is correct?

Problems by Topic

VSEPR Theory and Molecular Geometry

29. A molecule with the formula AB_3 has a trigonal pyramidal geometry. How many electron groups are on the central atom (A)?

30. A molecule with the formula AB_3 has a trigonal planar geometry. How many electron groups are on the central atom?

31. The following figures show several molecular geometries. For each geometry, give the number of total electron groups, the number of bonding groups, and the number of lone pairs on the central atom.

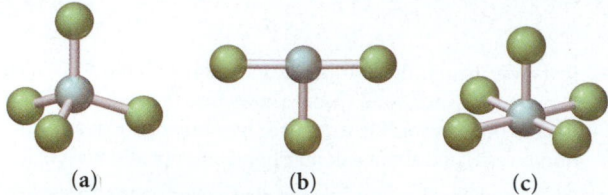

(a)   (b)   (c)

32. The following figures show several molecular geometries. For each geometry, give the number of total electron groups, the number of bonding groups, and the number of lone pairs on the central atom.

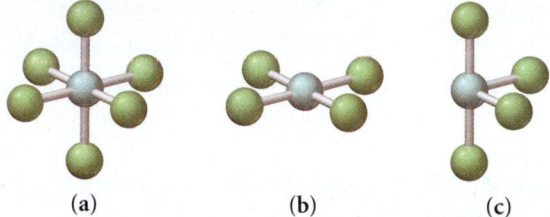

(a)   (b)   (c)

33. Determine the electron geometry, molecular geometry, and idealized bond angles for each of the following molecules. In which cases do you expect deviations from the idealized bond angle?
   a. PF_3 b. SBr_2 c. $CHCl_3$ d. CS_2

34. Determine the electron geometry, molecular geometry, and idealized bond angles for each of the following molecules. In which cases do you expect deviations from the idealized bond angle?
   a. CF_4 b. NF_3 c. OF_2 d. H_2S

35. Which species has the smaller bond angle, H_3O^+ or H_2O? Explain.

36. Which species has the smaller bond angle, ClO_4^- or ClO_3^-? Explain.

37. Determine the molecular geometry and make a sketch of each of the following molecules or ions, using the bond conventions shown in the Box in Section 10.4.
   a. SF_4 b. ClF_3 c. IF_2^- d. IBr_4^-

38. Determine the molecular geometry and make a sketch of each of the following molecules or ions, using the bond conventions shown in the Box in Section 10.4.
   a. BrF_5 b. SCl_6 c. PF_5 d. IF_4^+

39. Determine the molecular geometry about each interior atom and make a sketch of each of the following molecules.
   a. C_2H_2 (skeletal structure HCCH)
   b. C_2H_4 (skeletal structure H_2CCH_2)
   c. C_2H_6 (skeletal structure H_3CCH_3)

40. Determine the molecular geometry about each interior atom and make a sketch of each of the following molecules.
   a. N_2 b. N_2H_2 (skeletal structure HNNH)
   c. N_2H_4 (skeletal structure H_2NNH_2)

41. Each of the following ball-and-stick models shows the electron and molecular geometry of a generic molecule. Explain what is wrong with each molecular geometry and provide the correct molecular geometry, given the number of lone pairs and bonding groups on the central atom.

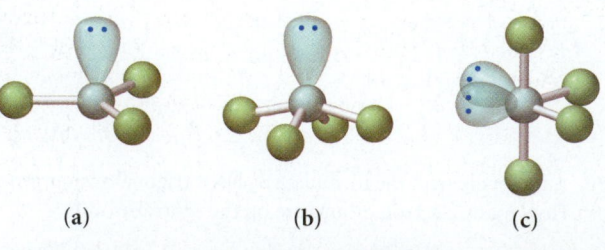

(a)   (b)   (c)

42. Each of the following ball-and-stick models shows the electron and molecular geometry of a generic molecule. Explain what is wrong with each molecular geometry and provide the correct molecular geometry, given the number of lone pairs and bonding groups on the central atom.

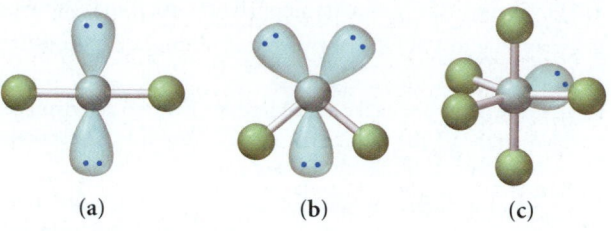

(a)   (b)   (c)

43. Determine the geometry about each interior atom in the following molecules and sketch the molecule. (Skeletal structure is indicated in parentheses.)
   a. CH_3OH (H_3COH)   b. CH_3OCH_3 (H_3COCH_3)
   c. H_2O_2 (HOOH)

44. Determine the geometry about each interior atom in the following molecules and sketch the molecule. (Skeletal structure is indicated in parentheses.)
   a. CH_3NH_2 (H_3CNH_2)
   b. $CH_3CO_2CH_3$ ($H_3CCOOCH_3$ both O atoms attached to second C)
   c. NH_2CO_2H (H_2NCOOH both O atoms attached to C)

Molecular Shape and Polarity

45. Explain why CO_2 and CCl_4 are both nonpolar even though they contain polar bonds.

46. CH_3F is a polar molecule, even though the tetrahedral geometry often leads to nonpolar molecules. Explain.

47. Determine whether each molecule in Exercise 33 is polar or nonpolar.

48. Determine whether each molecule in Exercise 34 is polar or nonpolar.

49. Determine whether each of the following molecules is polar or nonpolar.
   a. ClO_3^- b. SCl_2 c. SCl_4 d. $BrCl_5$

50. Determine whether each of the following molecules is polar or nonpolar.
   a. $SiCl_4$ b. CF_2Cl_2 c. SeF_6 d. IF_5

Valence Bond Theory

51. The valence electron configurations of several atoms are shown below. How many bonds can each atom make without hybridization?
   a. Be $2s^2$ b. P $3s^2 3p^3$ c. F $2s^2 2p^5$

52. The valence electron configurations of several atoms are shown below. How many bonds can each atom make without hybridization?
   a. B $2s^2 2p^1$ b. N $2s^2 2p^3$ c. O $2s^2 2p^4$

53. Write orbital diagrams (boxes with arrows in them) to represent the electron configurations—without hybridization—for all the atoms in PH_3. Circle the electrons involved in bonding. Draw a three-dimensional sketch of the molecule and show orbital overlap. What bond angle do you expect from the unhybridized orbitals? How well does valence bond theory agree with the experimentally measured bond angle of 93.3°?

54. Write orbital diagrams (boxes with arrows in them) to represent the electron configurations—without hybridization—for all the atoms in SF_2. Circle the electrons involved in bonding. Draw a three-dimensional sketch of the molecule and show orbital overlap. What bond angle do you expect from the unhybridized orbitals? How well does valence bond theory agree with the experimentally measured bond angle of 98.2°?

55. Write orbital diagrams (boxes with arrows in them) to represent the electron configuration of carbon before and after sp^3 hybridization.

56. Write orbital diagrams (boxes with arrows in them) to represent the electron configurations of carbon before and after sp hybridization.

57. Which of the following hybridization schemes allows the formation of at least one π bond?

$$sp^3, sp^2, sp^3d^2$$

58. Which of the following hybridization schemes allows the central atom to form more than four bonds?

$$sp^3, sp^3d, sp^2$$

59. Write a hybridization and bonding scheme for each of the following molecules. Sketch the molecule, including overlapping orbitals, and label all bonds using the notation shown in Examples 10.6 and 10.7.

   a. CCl_4 **b.** NH_3 **c.** OF_2 **d.** CO_2

60. Write a hybridization and bonding scheme for each of the following molecules. Sketch the molecule, including overlapping orbitals, and label all bonds using the notation shown in Examples 10.6 and 10.7.

   a. CH_2Br_2 **b.** SO_2 **c.** NF_3 **d.** BF_3

61. Write a hybridization and bonding scheme for each of the following molecules or ions. Sketch the structure, including overlapping orbitals, and label all bonds using the notation shown in Examples 10.6 and 10.7.

   a. $COCl_2$ (carbon is the central atom)
   b. BrF_5 **c.** XeF_2 **d.** I_3^-

62. Write a hybridization and bonding scheme for each of the following molecules or ions. Sketch the structure, including overlapping orbitals, and label all bonds using the notation shown in Examples 10.6 and 10.7.

   a. SO_3^{2-} **b.** PF_6^- **c.** BrF_3 **d.** HCN

63. Write a hybridization and bonding scheme for each of the following molecules containing more than one interior atom. Indicate the hybridization about each interior atom. Sketch the structure, including overlapping orbitals, and label all bonds using the notation shown in Examples 10.6 and 10.7.

   a. N_2H_2 (skeletal structure HNNH)
   b. N_2H_4 (skeletal structure H_2NNH_2)
   c. CH_3NH_2 (skeletal structure H_3CNH_2)

64. Write a hybridization and bonding scheme for each of the following molecules containing more than one interior atom. Indicate the hybridization about each interior atom. Sketch the structure, including overlapping orbitals, and label all bonds using the notation shown in Examples 10.6 and 10.7.

   a. C_2H_2 (skeletal structure HCCH)
   b. C_2H_4 (skeletal structure H_2CCH_2)
   c. C_2H_6 (skeletal structure H_3CCH_3)

65. Consider the structure of the amino acid alanine. Indicate the hybridization about each interior atom.

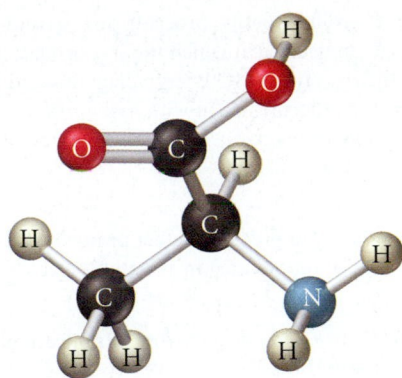

66. Consider the structure of the amino acid aspartic acid. Indicate the hybridization about each interior atom.

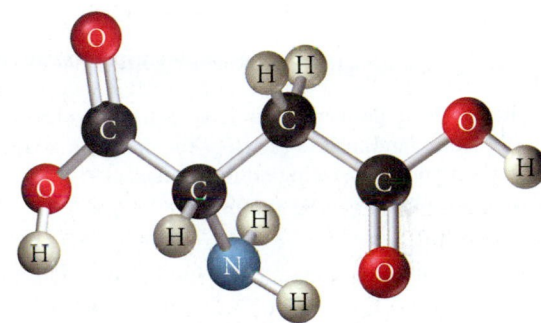

Molecular Orbital Theory

67. Sketch the bonding molecular orbital that results from the linear combination of two 1s orbitals. Indicate the region where interference occurs and state the kind of interference (constructive or destructive).

68. Sketch the antibonding molecular orbital that results from the linear combination of two 1s orbitals. Indicate the region where interference occurs and state the kind of interference (constructive or destructive).

69. Draw an MO energy diagram and predict the bond order of Be_2^+ and Be_2^-. Do you expect these molecules to exist in the gas phase?

70. Draw an MO energy diagram and predict the bond order of Li_2^+ and Li_2^-. Do you expect these molecules to exist in the gas phase?

71. Sketch the bonding and antibonding molecular orbitals that result from linear combinations of the $2p_x$ atomic orbitals in a homonuclear diatomic molecule. (The $2p_x$ orbitals are those whose lobes are oriented along the bonding axis.)

72. Sketch the bonding and antibonding molecular orbitals that result from linear combinations of the $2p_z$ atomic orbitals in a homonuclear diatomic molecule. (The $2p_z$ orbitals are those whose lobes are oriented perpendicular to the bonding axis.) How do these molecular orbitals differ from those obtained from linear combinations of the $2p_y$ atomic orbitals? (The $2p_y$ orbitals are also oriented perpendicular to the bonding axis, but also perpendicular to the $2p_z$ orbitals.)

73. Using the molecular orbital energy ordering for second-row homonuclear diatomic molecules in which the π_{2p} orbitals lie at *lower* energy than the σ_{2p}, draw MO energy diagrams and predict the bond order in a molecule or ion with each of the following numbers of total valence electrons. Will the molecule or ion be diamagnetic or paramagnetic?

   a. 4   **b.** 6   **c.** 8   **d.** 9

74. Using the molecular orbital energy ordering for second-row homonuclear diatomic molecules in which the π_{2p} orbitals lie at *higher* energy than the σ_{2p}, draw MO energy diagrams and predict the bond order in a molecule or ion with each of the following numbers of total valence electrons. Will the molecule or ion be diamagnetic or paramagnetic?

   a. 10   b. 12   c. 13   d. 14

75. Use molecular orbital theory to predict whether or not each of the following molecules or ions should exist in a relatively stable form.

   a. H_2^{2-} b. Ne_2 c. He_2^{2+} d. F_2^{2-}

76. Use molecular orbital theory to predict whether or not each of the following molecules or ions should exist in a relatively stable form.

   a. C_2^{2+} b. Li_2 c. Be_2^{2+} d. Li_2^{2-}

77. According to MO theory, which of the following has the highest bond order? Highest bond energy? Shortest bond length?

$$C_2, C_2^{+}, C_2^{-}$$

78. According to MO theory, which of the following ions has the highest bond order? Highest bond energy? Shortest bond length?

$$O_2, O_2^{-}, O_2^{2-}$$

79. Draw an MO energy diagram for CO. (Use the energy ordering of O_2.) Predict the bond order and make a sketch of the lowest energy bonding molecular orbital.

80. Draw an energy diagram for HCl. Predict the bond order and make a sketch of the lowest energy bonding molecular orbital.

Cumulative Problems

81. For each of the following compounds, draw an appropriate Lewis structure, determine the geometry using VSEPR theory, determine whether the molecule is polar, identify the hybridization of all interior atoms, and make a sketch of the molecule, according to valence bond theory, showing orbital overlap.

   a. COF_2 (carbon is the central atom)
   b. S_2Cl_2 (ClSSCl)   c. SF_4

82. For each of the following compounds, draw an appropriate Lewis structure, determine the geometry using VSEPR theory, determine whether the molecule is polar, identify the hybridization of all interior atoms, and make a sketch of the molecule, according to valence bond theory, showing orbital overlap.

   a. IF_5 b. CH_2CHCH_3 c. CH_3SH

83. Amino acids are biological compounds that link together to form proteins, the workhorse molecules in living organisms. The skeletal structures of several simple amino acids are shown below. For each skeletal structure, complete the Lewis structure, determine the geometry and hybridization about each interior atom, and make a sketch of the molecule, using the bond conventions of Section 10.4.

(a) serine

(b) asparagine

(c) cysteine

84. The genetic code is based on four different bases with the structures shown below. Assign a geometry and hybridization to each interior atom in these four bases.

   a. cytosine   b. adenine
   c. thymine   d. guanine

(a)   (b)

(c)   (d)

85. The structure of caffeine, present in coffee and many soft drinks, is shown below. How many pi bonds are present in caffeine? How many sigma bonds? Insert the lone pairs in the molecule. What kinds of orbitals do the lone pairs occupy?

86. The structure of acetylsalicylic acid (aspirin) is shown below. How many pi bonds are present in acetylsalicylic acid? How many sigma bonds? What parts of the molecule are free to rotate? What parts are rigid?

87. Most vitamins can be classified either as fat soluble, which tend to accumulate in the body (so that taking too much can be harmful), or water soluble, which tend to be quickly eliminated from the body in urine. Examine the following structural formulas and space-filling models of several vitamins and determine whether they are fat soluble (mostly nonpolar) or water soluble (mostly polar).

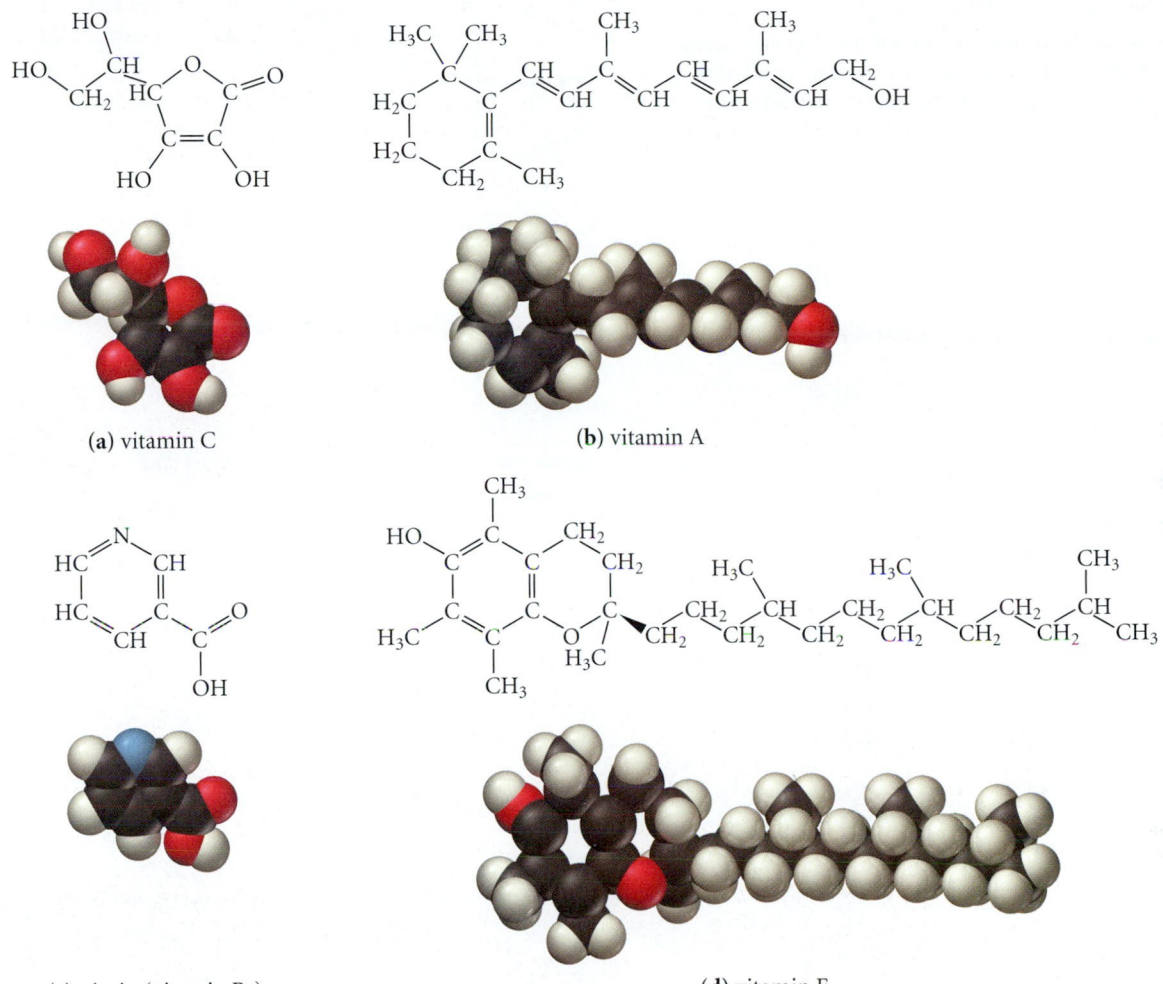

(**a**) vitamin C

(**b**) vitamin A

(**c**) niacin (vitamin B_3)

(**d**) vitamin E

88. Water does not easily remove grease from dishes or hands because grease is nonpolar and water is polar. The addition of soap to water, however, allows the grease to dissolve. Look at the structure of sodium stearate (a soap) below and suggest how it works.

$$CH_3(CH_2)_{16}\overset{\displaystyle O}{\overset{\|}{C}}-O^-Na^+$$

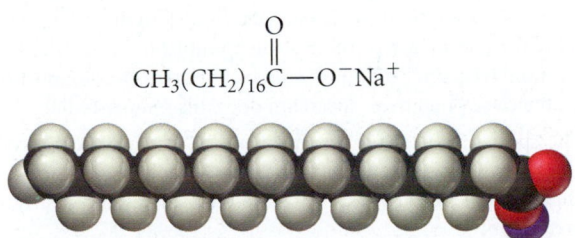

89. Draw a molecular orbital energy diagram for ClF. (Assume that the σ_p orbitals are lower in energy than the π orbitals.) What is the bond order in ClF?

90. Draw Lewis structures and MO diagrams for CN^+, CN, and CN^-. According to Lewis theory, which species is most stable? According to MO theory, which species is most stable? Do the two theories agree?

91. Bromine can form compounds or ions with any number of fluorine atoms from one to five. Write the formulas of all five of these species, assign a hybridization, and describe their electron and molecular geometry.

92. The compound C_3H_4 has two double bonds. Describe its bonding and geometry, using a valence bond approach.

Challenge Problems

93. In VSEPR theory, which uses Lewis theory to determine molecular geometry, the trend of decreasing bond angle in CH_4, NH_3, and H_2O is accounted for by the greater repulsion of lone pair electrons compared to bonding pair electrons. How would this trend be accounted for in valence bond theory?

94. The results of a molecular orbital calculation for H_2O are shown below. Examine each of the orbitals and classify them as bonding, antibonding, or nonbonding. Assign the correct number of electrons to the energy diagram. According to this energy diagram, is H_2O stable? Explain.

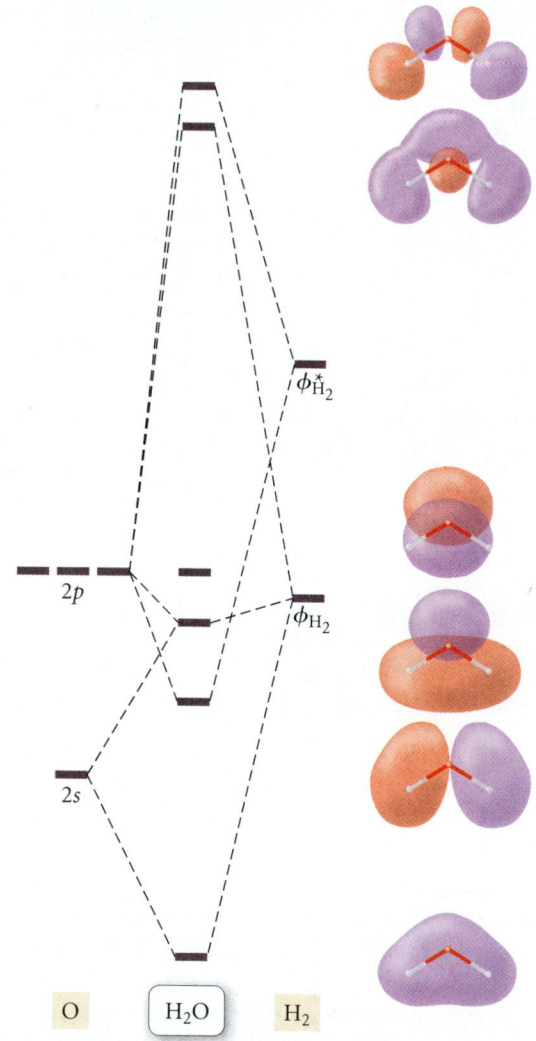

95. The results of a molecular orbital calculation for NH_3 are shown below. Examine each of the orbitals and classify them as bonding, antibonding, or nonbonding. Assign the correct number of electrons to the energy diagram. According to this energy diagram, is NH_3 stable? Explain.

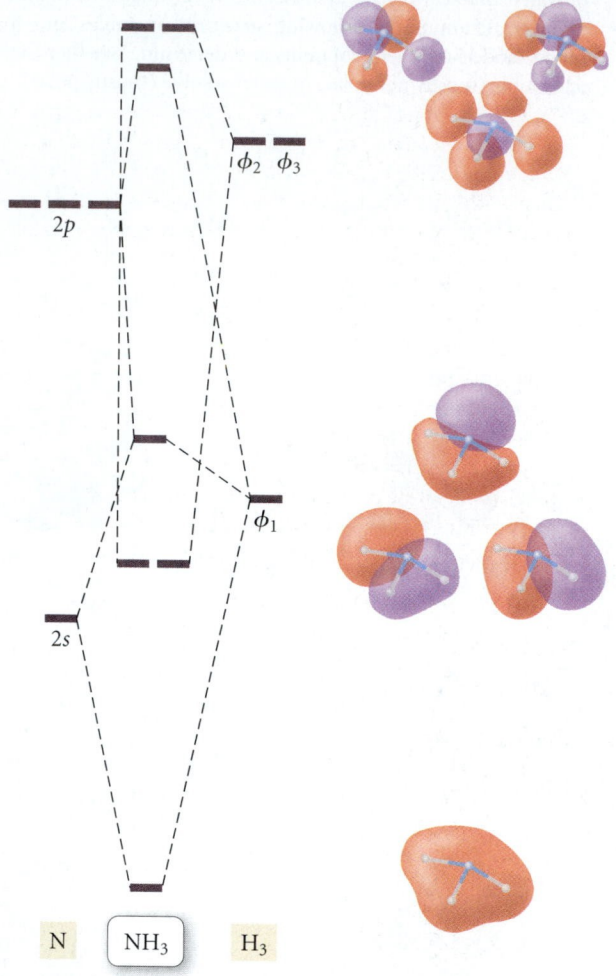

96. *cis*-2-Butene isomerizes to *trans*-2-butene via the following reaction:

$$H-\overset{\overset{\displaystyle H}{|}}{C}\quad \overset{\overset{\displaystyle H}{|}}{\underset{\underset{\displaystyle H}{|}}{C}}=\overset{\overset{\displaystyle H}{|}}{\underset{\underset{\displaystyle H}{|}}{C}}\quad \overset{\overset{\displaystyle H}{|}}{C}-H \longrightarrow$$

a. If isomerization requires breaking the pi bond, what minimum energy is required for isomerization in J/mol? In J/molecule?

b. If the energy for isomerization came from light, what minimum frequency of light would be required? In what portion of the electromagnetic spectrum does this frequency lie?

97. The species NO_2, NO_2^+, and NO_2^-, in which N is the central atom, have very different bond angles. Predict what these bond angles might be with respect to the ideal angles and justify your prediction.

98. The bond angles increase steadily in the series PF_3, PCl_3, PBr_3, and PI_3. After consulting the data on atomic radii in Chapter 8, provide an explanation for this observation.

Conceptual Problems

99. Pick the statement below that best captures the fundamental idea behind VSEPR theory. Explain what is wrong with each of the other statements.

   a. The angle between two or more bonds is determined primarily by the repulsions between the electrons within those bonds and other (lone pair) electrons on the central atom of a molecule. Each of these electron groups (bonding electrons or lone pair electrons) will lower its potential energy by maximizing its separation from other electron groups, thus determining the geometry of the molecule.

   b. The angle between two or more bonds is determined primarily by the repulsions between the electrons within those bonds. Each of these bonding electrons will lower its potential energy by maximizing its separation from other electron groups, thus determining the geometry of the molecule.

   c. The geometry of a molecule is determined by the shapes of the overlapping orbitals that form the chemical bonds. Therefore, to determine the geometry of a molecule, you must determine the shapes of the orbitals involved in bonding.

100. Suppose that a molecule has four bonding groups and one lone pair on the central atom. Suppose further that the molecule is confined to two dimensions (this is a purely hypothetical assumption for the sake of understanding the principles behind VSEPR theory). Make a sketch of the molecule and estimate the bond angles.

101. How does each of the three major bonding theories (Lewis theory, valence bond theory, and molecular orbital theory) define a single chemical bond? A double bond? A triple bond? How are these definitions similar? How are they different?

102. The most stable forms of the nonmetals in groups 4A, 5A, and 6A of the first period are molecules with multiple bonds. Beginning with the second-period, the most stable forms of the nonmetals of these groups are molecules without multiple bonds. Propose an explanation based on valence bond theory for this observation.

11

Liquids, Solids, and Intermolecular Forces

11.1 Climbing Geckos and Intermolecular Forces

11.2 Solids, Liquids, and Gases: A Molecular Comparison

11.3 Intermolecular Forces: The Forces That Hold Condensed Phases Together

11.4 Intermolecular Forces in Action: Surface Tension, Viscosity, and Capillary Action

11.5 Vaporization and Vapor Pressure

11.6 Sublimation and Fusion

11.7 Heating Curve for Water

11.8 Phase Diagrams

11.9 Water: An Extraordinary Substance

11.10 Crystalline Solids: Determining Their Structure by X-Ray Crystallography

11.11 Crystalline Solids: Unit Cells and Basic Structures

11.12 Crystalline Solids: The Fundamental Types

11.13 Crystalline Solids: Band Theory

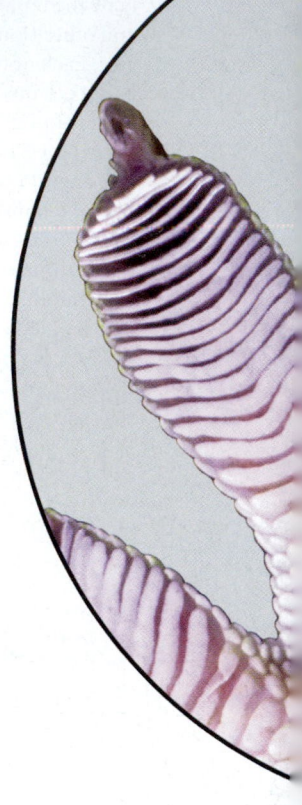

YOU LEARNED IN Chapter 1 that matter exists primarily in three phases: solid, liquid, and gas. In Chapter 5, we examined the gas phase. In this chapter we turn to the solid and liquid phases, known collectively as the condensed phases. The solid and liquid phases are more similar to each other than they are to the gas phase. In the gas phase, the constituent particles—atoms or molecules—are separated by large distances and do not interact with each other very much. In the condensed phases, the constituent particles are close together and exert moderate to strong attractive forces on one another. Unlike the gas phase, for which we have a good, simple quantitative model (kinetic molecular theory) to describe and predict behavior, we have no such model for the condensed phases. In fact, modeling the condensed phases is an active area of research today. In this chapter, we focus primarily on describing the condensed phases and their properties and on providing some qualitative guidelines to help us understand those properties.

> *It's a wild dance floor there at the molecular level.*
> —ROALD HOFFMANN (1937–)

Intermolecular forces

Recent studies suggest that the gecko's remarkable ability to climb walls and adhere to surfaces depends on intermolecular forces.

▲ Each of the millions of microhairs on a gecko's feet branches out to end in flattened tips called spatulae.

11.1 Climbing Geckos and Intermolecular Forces

The gecko shown above can run up a polished glass window in seconds or even walk across a ceiling. It can support its entire weight by a single toe in contact with a surface. How? Recent work by several scientists points to *intermolecular forces*—attractive forces that exist *between* all molecules and atoms—as the reason that the gecko can perform its gravity-defying feats. Intermolecular forces are the forces that hold many liquids and solids—such as water and ice, for example—together.

The key to the gecko's sticky feet lies in the millions of microhairs, called *setae*, that line its toes. Each seta is between 30 and 130 μm long and branches out to end in several hundred flattened tips called *spatulae*, as you can see in the photo. This unique structure allows the gecko's toes to have unusually close contact with the surfaces it climbs. The close contact allows intermolecular forces—which are significant only at short distances—to hold the gecko to the wall.

All living organisms depend on intermolecular forces, not for adhesion to walls, but for many physiological processes. For example, in Chapter 21 we describe how intermolecular forces help determine the shapes of protein molecules (the workhorse molecules in living organisms). Later in this chapter you will see how intermolecular forces are central to the structure of DNA, the inheritable molecule that carries the blueprints for life. (See the *Chemistry and Medicine* box in Section 11.3.)

Even more generally, intermolecular forces are responsible for the very existence of the condensed phases. The state of a sample of matter—solid, liquid, or gas—depends on the magnitude of intermolecular forces between the constituent particles relative to the amount of thermal energy in the sample. Recall from Chapter 6 that the molecules and atoms composing matter are in constant random motion that increases with increasing temperature. The energy associated with this motion is called *thermal energy*. When thermal energy is high relative to intermolecular forces, matter tends to be gaseous. When thermal energy is low relative to intermolecular forces, matter tends to be liquid or solid.

11.2 Solids, Liquids, and Gases: A Molecular Comparison

We are all familiar with solids and liquids. Water, gasoline, rubbing alcohol, and nail polish remover are all common liquids that you have probably encountered. Ice, dry ice, and diamond are familiar solids. To begin to understand the differences between the three common phases of matter, examine Table 11.1, which shows the density and molar volume of water in its three different phases, along with molecular representations of each phase. Notice that the densities of the solid and liquid phases are much greater than the density of the gas phase. Notice also that the solid and liquid phases are more similar in density and molar volume to one another than they are to the gas phase. The molecular representations show the reason for these differences. The molecules in liquid water and ice are in close contact with one another—essentially touching—while those in gaseous water are separated by large distances. The molecular representation of gaseous water in Table 11.1 is out of proportion—the water molecules in the figure should be much farther apart for their size. (Only a fraction of a molecule could be included in the figure if it were drawn to scale.) From the molar volumes, we know that 18.0 mL of liquid water (slightly more than a tablespoon) would occupy 30.5 L when converted to gas at 100 °C. The low density of gaseous water results from this large separation between molecules.

TABLE 11.1   The Three Phases of Water

Phase	Temperature (°C)	Density (g/cm³, at 1 atm)	Molar Volume	Molecular View
Gas (steam)	100	5.90×10^{-4}	30.5 L	
Liquid (water)	20	0.998	18.0 mL	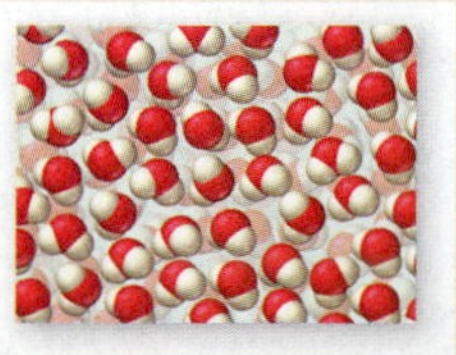
Solid (ice)	0	0.917	19.6 mL	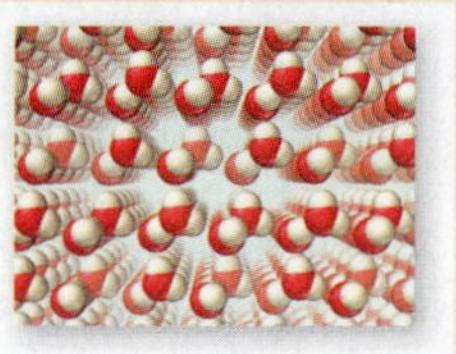

Notice also that, for water, the solid is slightly less dense than the liquid. This is *atypical* behavior. Most solids are slightly denser than their corresponding liquids because the molecules move closer together upon freezing. As we will see in Section 11.9, ice is less dense than liquid water because the unique crystal structure of ice results in water molecules moving slightly further apart upon freezing.

From the molecular perspective, one major difference between liquids and solids is the freedom of movement of the constituent molecules or atoms. Even though the atoms or molecules in a liquid are in close contact, thermal energy can partially overcome the attractions between them, allowing them to move around one another. Not so in solids, where the atoms or molecules are virtually locked in their positions, only vibrating back and forth about a fixed point. The molecular properties of solids and liquids result in the following macroscopic properties.

Properties of Liquids

- Liquids have high densities in comparison to gases.
- Liquids have an indefinite shape and assume the shape of their container.
- Liquids have a definite volume; they are not easily compressed.

Properties of Solids

- Solids have high densities in comparison to gases.
- Solids have a definite shape; they *do not* assume the shape of their container.
- Solids have a definite volume; they are not easily compressed.
- Solids may be crystalline (ordered) or amorphous (disordered).

These properties, as well as the properties of gases for comparison, are summarized in Table 11.2.

▲ FIGURE 11.1 **Liquids Assume the Shapes of Their Containers** When you pour water into a flask, it assumes the shape of the flask because water molecules are free to flow.

TABLE 11.2 Properties of the Phases of Matter

Phase	Density	Shape	Volume	Strength of Intermolecular Forces*
Gas	Low	Indefinite	Indefinite	Weak
Liquid	High	Indefinite	Definite	Moderate
Solid	High	Definite	Definite	Strong

*Relative to thermal energy.

Liquids assume the shape of their containers because the atoms or molecules that compose them are free to flow (or move around one another). When you pour water into a beaker, the water flows and assumes the shape of the beaker (Figure 11.1 ▶). Liquids are not easily compressed because the molecules or atoms that compose them are in close contact—they cannot be pushed much closer together. The molecules in a gas, by contrast, have a great deal of space between them and are easily forced into a smaller volume by an increase in external pressure (Figure 11.2 ▶).

Solids have a definite shape because, in contrast to liquids and gases, the molecules or atoms that compose

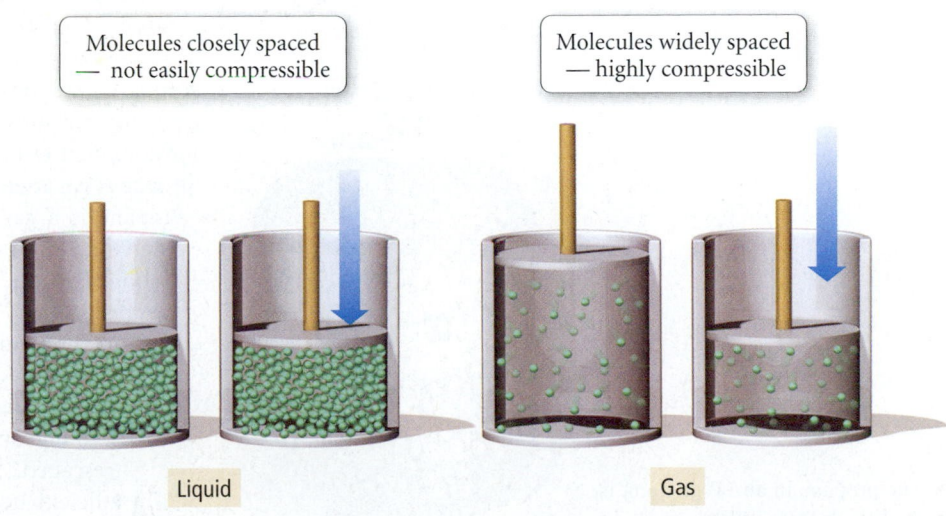

Molecules closely spaced — not easily compressible

Molecules widely spaced — highly compressible

Liquid

Gas

▲ FIGURE 11.2 **Gases Are Compressible** Molecules in a liquid are closely spaced and are not easily compressed. Molecules in a gas have a great deal of space between them, making gases compressible.

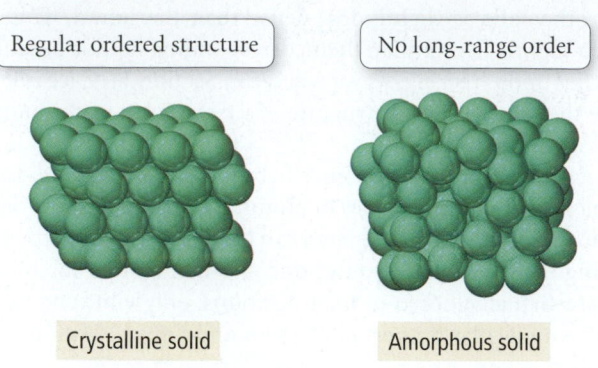

Regular ordered structure   No long-range order

Crystalline solid   Amorphous solid

▶ **FIGURE 11.3 Crystalline and Amorphous Solids** In a crystalline solid, the arrangement of the particles displays long-range order. In an amorphous solid, the arrangement of the particles has no long-range order.

By some definitions, an amorphous solid is considered a unique phase, different from the normal solid phase because it lacks any long-range order.

solids are fixed in place—each molecule or atom merely vibrates about a fixed point. Like liquids, solids have a definite volume and generally cannot be compressed because the molecules or atoms composing them are already in close contact. Solids may be **crystalline**, in which case the atoms or molecules that compose them are arranged in a well-ordered three-dimensional array, or they may be **amorphous**, in which case the atoms or molecules that compose them have no long-range order (Figure 11.3 ▲).

Changes between Phases

One phase of matter can be transformed to another by changing the temperature, pressure, or both. For example, solid ice can be converted to liquid water by heating, and liquid water can be converted to solid ice by cooling. The following diagram shows the three states of matter and the changes in conditions that commonly induce transitions between them.

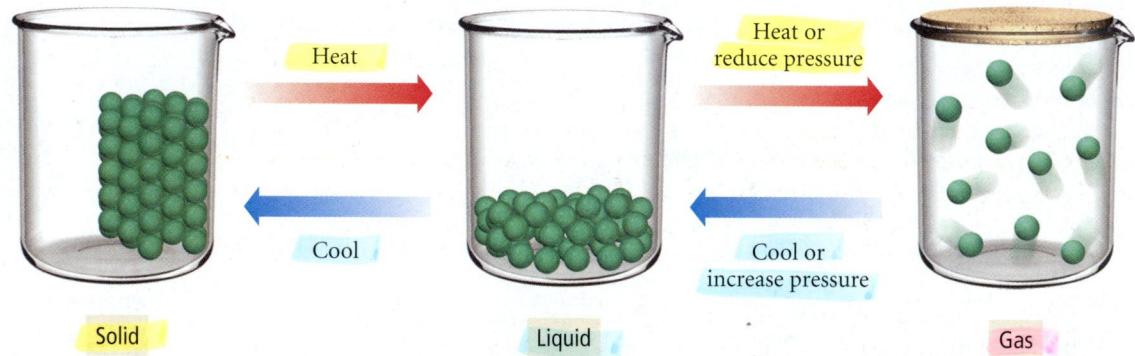

Heat

Heat or reduce pressure

Cool

Cool or increase pressure

Solid   Liquid   Gas

$C_3H_8 (g)$

Notice that transitions between the liquid and gas phase can be achieved not only by heating and cooling, but also through changes in pressure. In general, increases in pressure favor the denser phase, so increasing the pressure of a gas sample results in a transition to the liquid phase. The most familiar example of this phenomenon is the LP (liquid propane) gas used as a fuel for outdoor grills and lanterns. Propane is a gas at room temperature and atmospheric pressure. However, it liquefies at pressures exceeding about 2.7 atm. The propane you buy in a tank is under pressure and therefore in the liquid form. When you open the tank, some of the propane escapes as a gas, lowering the pressure in the tank for a brief moment. Immediately, however, some of the liquid propane evaporates, replacing the gas that escaped. Storing gases like propane as liquids is efficient because, in their liquid form, they occupy much less space.

▶ The propane in an LP gas tank is in the liquid phase. When you open the tank, some propane vaporizes and escapes as a gas.

$C_3H_8 (l)$

Conceptual Connection 11.1 Phase Changes

The molecular diagram at right shows a sample of liquid water.

Which of the following diagrams best depicts the vapor emitted from a pot of boiling of water?

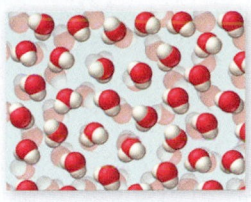

(a)  **(b)**  **(c)**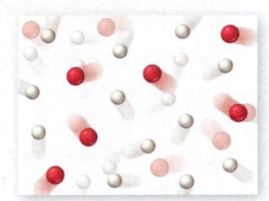

Answer: **(a)** When water boils, it simply changes phase from liquid to gas. Water molecules do not decompose during boiling.

11.3 Intermolecular Forces: The Forces That Hold Condensed Phases Together

The strength of the intermolecular forces between the molecules or atoms that compose a substance determines the phase—solid, liquid, or gas—of the substance at a given temperature. At room temperature, moderate to strong intermolecular forces tend to result in liquids and solids (high melting and boiling points) and weak intermolecular forces tend to result in gases (low melting and boiling points).

Intermolecular forces originate from the interactions between charges, partial charges, and temporary charges on molecules (or atoms and ions), much as bonding forces originate from interactions between charged particles in atoms. Recall from Section 9.2 that the potential energy (E) of two oppositely charged particles (with charges q_1 and q_2) decreases (becomes more negative) with increasing magnitude of charge and with decreasing separation (r) between them according to Coulomb's law:

$$E = \frac{1}{4\pi\epsilon_0}\frac{q_1q_2}{r}$$ (When q_1 and q_2 are opposite in sign, E is negative.)

Therefore, as we have seen, protons and electrons are attracted to each other because their potential energy decreases as they get closer together. Similarly, molecules with partial or temporary charges are attracted to each other because *their* potential energy decreases as they get closer together. However, intermolecular forces, even the strongest ones, are generally *much weaker* than bonding forces.

The reason for the relative weakness of intermolecular forces compared to bonding forces is also related to Coulomb's law. Bonding forces are the result of large charges (the charges on protons and electrons) interacting at very close distances. Intermolecular forces are the result of smaller charges (as we shall see in the following discussion) interacting at greater distances. For example, consider the interaction between two water molecules in liquid water:

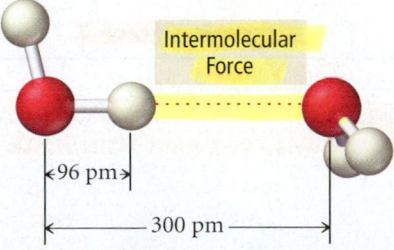

The length of an O—H bond in liquid water is 96 pm; however, the average distance between water molecules in liquid water is about 300 pm. The larger distances between molecules, as well as the smaller charges involved (partial charges on the hydrogen and oxygen atoms), result in weaker forces. To break the O—H bonds in water, for example, you have to heat the water to thousands of degrees Celsius. However, to completely overcome the intermolecular forces *between* water molecules, you have to heat water only to its boiling point, 100 °C. We will examine several different types of intermolecular forces, including dispersion forces, dipole–dipole forces, hydrogen bonding, and ion–dipole forces. The first three of these can potentially occur in all substances; the last one is found only in mixtures.

Dispersion Force

The nature of dispersion forces was first recognized by Fritz W. London (1900-1954), a German-American physicist.

The one intermolecular force present in all molecules and atoms is the **dispersion force** (also called the London force). Dispersion forces are the result of fluctuations in the electron distribution within molecules or atoms. Since all atoms and molecules have electrons, they all exhibit dispersion forces. The electrons in an atom or molecule may, *at any one instant*, be unevenly distributed. For example, imagine a frame-by-frame movie of a helium atom in which each "frame" captures the position of the helium atom's two electrons.

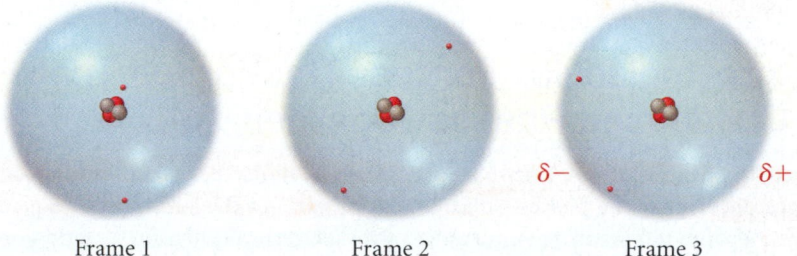

Frame 1   Frame 2   Frame 3

In any one frame, the electrons are not symmetrically arranged around the nucleus. In frame 3, for example, helium's two electrons are on the left side of the helium atom. At that instant, the left side will have a slightly negative charge ($\delta-$). The right side of the atom, which temporarily has no electrons, will be slightly positive ($\delta+$) because of the charge of the nucleus. This fleeting charge separation is called an *instantaneous dipole* or a *temporary dipole*. As shown in Figure 11.4 ▼, an instantaneous dipole on one helium atom induces an instantaneous dipole on its neighboring atoms because the positive end of the instantaneous dipole attracts electrons in the neighboring atoms. The neighboring atoms then attract one another—the positive end of one instantaneous dipole attracting the negative end of another. This attraction is the dispersion force.

To polarize means to form a dipole moment (see Section 9.6).

The *magnitude* of the dispersion force depends on how easily the electrons in the atom or molecule can move or *polarize* in response to an instantaneous dipole, which in turn depends on the size (or volume) of the electron cloud. A larger electron cloud results in a greater dispersion force because the electrons are held less tightly by the nucleus and can therefore polarize more easily. If all other variables are constant, the dispersion force increases with increasing molar mass because molecules or atoms of higher molar mass generally have more electrons dispersed over a greater volume. For example, consider the boiling points of the noble gases displayed in Table 11.3. As the molar masses and electron cloud volumes of the noble gases increase, the greater dispersion forces result in increasing boiling points.

Dispersion Force

An instantaneous dipole on any one helium atom induces instantaneous dipoles on neighboring atoms, which then attract one another.

▶ **FIGURE 11.4 Dispersion Interactions** The temporary dipole in one helium atom induces a temporary dipole in its neighbor. The resulting attraction between the positive and negative charges creates the dispersion force.

Molar mass alone, however, does not determine the magnitude of the dispersion force. For example, compare the molar masses and boiling points of *n*-pentane and neopentane:

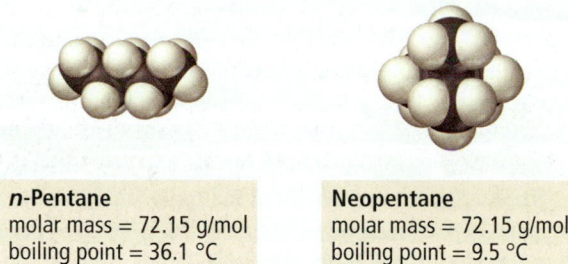

n-Pentane	Neopentane
molar mass = 72.15 g/mol	molar mass = 72.15 g/mol
boiling point = 36.1 °C	boiling point = 9.5 °C

These molecules have identical molar masses, but *n*-pentane has a higher boiling point than neopentane. The main reason for this difference is the different shapes of the molecules. The *n*-pentane molecules are long and can interact with one another along their entire length, as shown in Figure 11.5(a) ▼. In contrast, the bulky, round shape of neopentane molecules results in a smaller area of interaction between neighboring molecules, as shown in Figure 11.5(b). The result is a lower boiling point for neopentane.

TABLE 11.3   Boiling Points of the Noble Gases

Noble Gas		Molar Mass (g/mol)	Boiling Point (K)
He		4.00	4.2
Ne		20.18	27
Ar		39.95	87
Kr		83.80	120
Xe		131.30	165

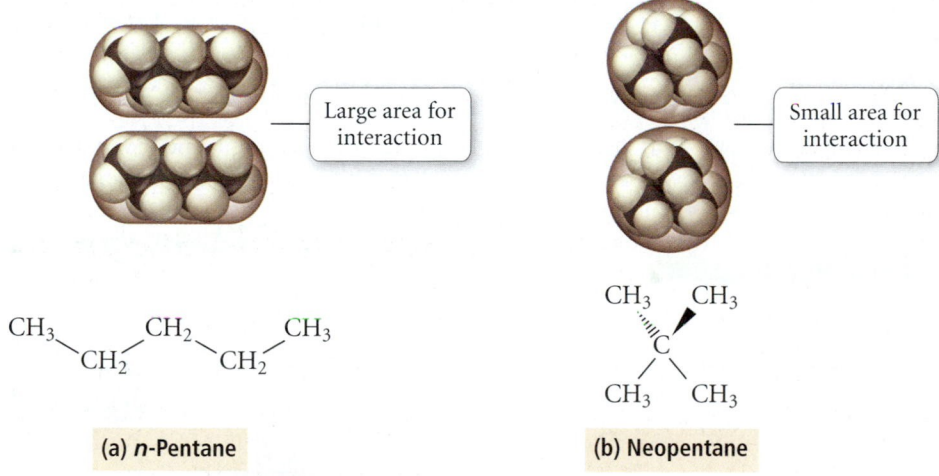

Large area for interaction

Small area for interaction

CH₃ CH₂ CH₂ CH₃

CH₃ CH₃
C
CH₃ CH₃

(a) *n*-Pentane

(b) Neopentane

◀ **FIGURE 11.5 Dispersion Force and Molecular Shape** (a) The straight shape of *n*-pentane molecules allows them to interact with one another along the entire length of the molecule. (b) The nearly spherical shape of neopentane molecules allows for only a small area of interaction. Thus, dispersion forces are weaker in neopentane than in *n*-pentane, resulting in a lower boiling point.

Although molecular shape and other factors must always be considered in determining the magnitude of dispersion forces, molar mass can act as a guide when comparing dispersion forces within a family of similar elements or compounds, as shown in Figure 11.6 ▼ for some selected *n*-alkanes.

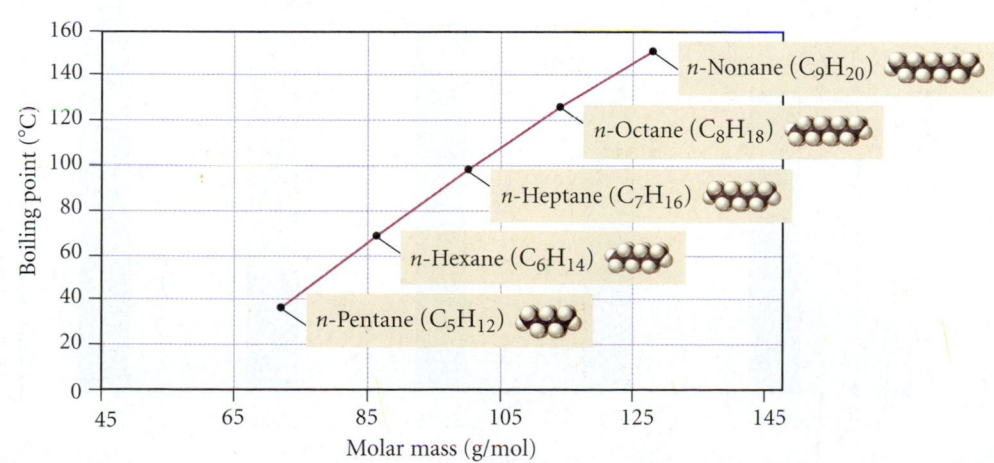

n-Nonane (C_9H_{20})

n-Octane (C_8H_{18})

n-Heptane (C_7H_{16})

n-Hexane (C_6H_{14})

n-Pentane (C_5H_{12})

◀ **FIGURE 11.6 Boiling Points of the *n*-Alkanes** The boiling points of the *n*-alkanes rise with increasing molar mass and the consequent stronger dispersion forces.

Dipole–Dipole Interaction

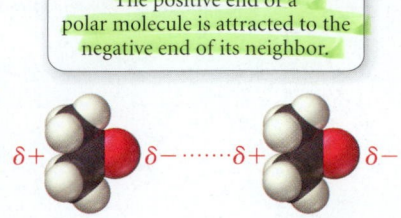

> The positive end of a polar molecule is attracted to the negative end of its neighbor.

$\delta+$ $\delta-$ ⋯⋯ $\delta+$ $\delta-$

▶ **FIGURE 11.7**
Dipole–Dipole Interaction
Molecules with permanent dipoles, such as acetone, are attracted to one another via dipole–dipole interactions.

| See Section 9.6 to review how to determine whether a molecule is polar.

Dipole–Dipole Force

The **dipole–dipole force** exists in all molecules that are polar. Polar molecules have **permanent dipoles** that interact with the permanent dipoles of neighboring molecules, as shown in Figure 11.7 ◀. The positive end of one permanent dipole is attracted to the negative end of another; this attraction is the dipole–dipole force. Polar molecules, therefore, have higher melting and boiling points than nonpolar molecules of similar molar mass. Remember that all molecules (including polar ones) have dispersion forces. Polar molecules have, *in addition*, dipole–dipole forces. This additional attractive force raises their melting and boiling points relative to nonpolar molecules of similar molar mass. For example, consider the following two compounds:

Name	Formula	Molar Mass (amu)	Structure	bp (°C)	mp (°C)
Formaldehyde	CH_2O	30.03	$\begin{array}{c} O \\ \| \\ H-C-H \end{array}$	−19.5	−92
Ethane	C_2H_6	30.07	$\begin{array}{c} H\ \ H \\ \|\ \ \| \\ H-C-C-H \\ \|\ \ \| \\ H\ \ H \end{array}$	−88	−172

Formaldehyde is polar, and therefore has a higher melting point and boiling point than nonpolar ethane even though the two compounds have the same molar mass. Figure 11.8 ▼ shows the boiling points of a series of molecules with similar molar mass but progressively greater dipole moments. Notice that the boiling points increase with increasing dipole moment.

▶ **FIGURE 11.8 Dipole Moment and Boiling Point** The molecules shown here all have similar molar masses but different dipole moments. The boiling points increase with increasing dipole moment.

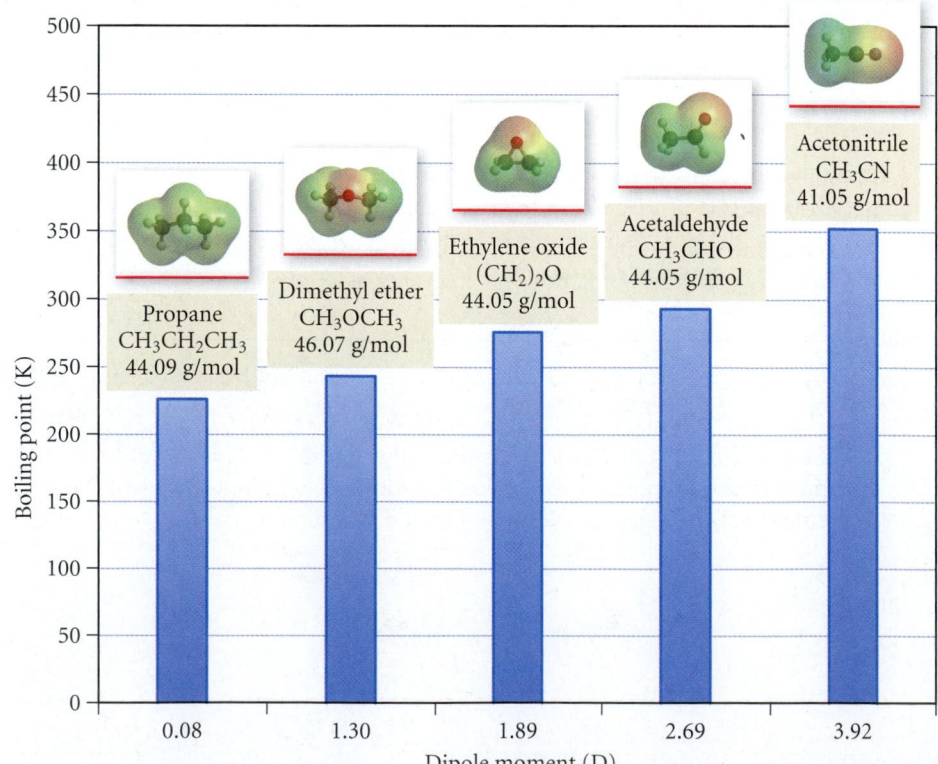

Propane
$CH_3CH_2CH_3$
44.09 g/mol

Dimethyl ether
CH_3OCH_3
46.07 g/mol

Ethylene oxide
$(CH_2)_2O$
44.05 g/mol

Acetaldehyde
CH_3CHO
44.05 g/mol

Acetonitrile
CH_3CN
41.05 g/mol

Boiling point (K)

Dipole moment (D)

The polarity of molecules composing liquids is also important in determining the **miscibility**—the ability to mix without separating into two phases—of liquids. In general, polar liquids are miscible with other polar liquids but are not miscible with nonpolar liquids. For example, water, a polar liquid, is not miscible with pentane (C_5H_{12}), a nonpolar liquid (Figure 11.9 ▼). Similarly, water and oil (also nonpolar) do not mix. Consequently, oily hands or oily stains on clothes cannot be washed with plain water. The water will not mix with the oil (see *Chemistry in Your Day: How Soap Works* in Section 10.5).

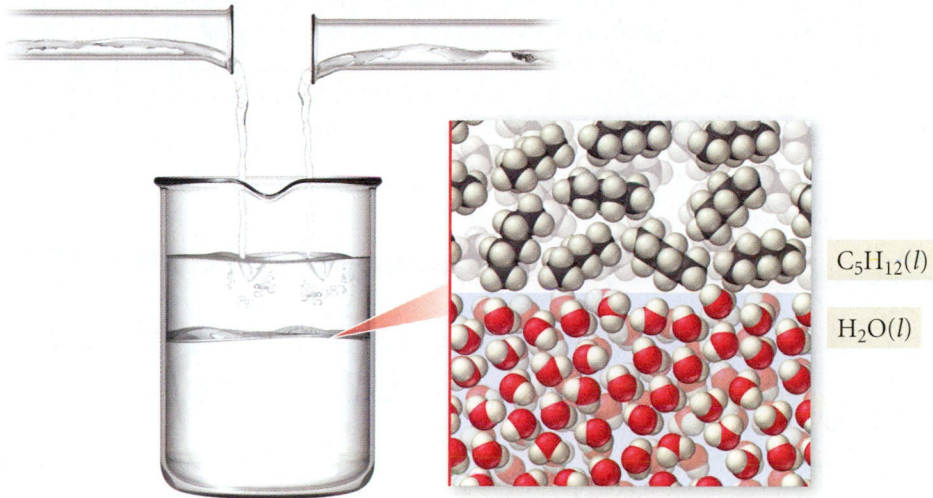

$C_5H_{12}(l)$

$H_2O(l)$

◀ FIGURE 11.9 **Polar and Nonpolar Compounds** Water and pentane do not mix because water molecules are polar and pentane molecules are nonpolar.

EXAMPLE 11.1 Dipole–Dipole Forces

Which of the following molecules have dipole–dipole forces?

(a) CO_2 **(b)** CH_2Cl_2 **(c)** CH_4

Solution

A molecule will have dipole–dipole forces if it is polar. To determine whether a molecule is polar, (1) *determine whether the molecule contains polar bonds* and (2) *determine whether the polar bonds add together to form a net dipole moment* (Section 9.6).

(a) CO_2 (1) Since the electronegativity of carbon is 2.5 and that of oxygen is 3.5 (Figure 9.10), CO_2 has polar bonds. (2) The geometry of CO_2 is linear. Consequently, the dipoles of the polar bonds cancel, so the molecule is *not polar* and does not have dipole–dipole forces.	(a) O=C=O 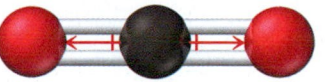 No dipole forces present
(b) CH_2Cl_2 (1) The electronegativity of C is 2.5, that of H is 2.1, and that of Cl is 3.5. Consequently, CH_2Cl_2 has two polar bonds (C—Cl) and two bonds that are nearly nonpolar (C—H). (2) The geometry of CH_2Cl_2 is tetrahedral. Since the C—Cl bonds and the C—H bonds are different, their dipoles do not cancel but sum to a net dipole moment. Therefore the molecule is polar and has dipole–dipole forces.	CH_2Cl_2 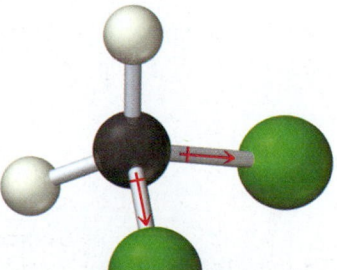 Dipole forces present

(c) CH$_4$

    (1) Since the electronegativity of C is 2.5 and that of hydrogen is 2.1 the C—H bonds are nearly nonpolar.

    (2) In addition, since the geometry of the molecule is tetrahedral, any slight polarities that the bonds might have will cancel. CH$_4$ is therefore nonpolar and does not have dipole–dipole forces.

CH$_4$

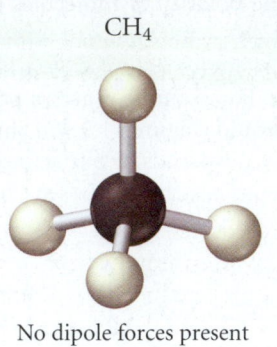

No dipole forces present

For Practice 11.1

Which of the following molecules have dipole–dipole forces?

(a) CI$_4$ **(b)** CH$_3$Cl   **(c)** HCl

Hydrogen Bonding

Polar molecules containing hydrogen atoms bonded directly to small electronegative atoms—most importantly fluorine, oxygen, or nitrogen—exhibit an intermolecular force called **hydrogen bonding**. HF, NH$_3$, and H$_2$O, for example, all undergo hydrogen bonding. The hydrogen bond is a sort of *super* dipole–dipole force. The large electronegativity difference between hydrogen and these electronegative elements means that the H atoms will have fairly large partial positive charges ($\delta+$), while the F, O, or N atoms will have fairly large partial negative charges ($\delta-$). In addition, since these atoms are all are quite small, they can approach one another very closely. The result is a strong attraction between the hydrogen in each of these molecules and the F, O, or N *on its neighbors*, an attraction called a **hydrogen bond**. For example, in HF, the hydrogen is strongly attracted to the fluorine on neighboring molecules (Figure 11.10 ◀).

    Hydrogen bonds should not be confused with chemical bonds. Chemical bonds occur *between individual atoms within a molecule*, whereas hydrogen bonds—like dispersion forces and dipole–dipole forces—are intermolecular forces that occur *between molecules*. A typical hydrogen bond is only 2–5% as strong as a typical covalent chemical bond. Hydrogen bonds are, however, the strongest of the three *intermolecular* forces we have discussed so far. Substances composed of molecules that form hydrogen bonds have higher melting and boiling points than substances composed of molecules that do not form hydrogen bonds. For example, consider the following two compounds:

Hydrogen Bonding

> When H bonds directly to F, O, or N, the bonding atoms acquire relatively large partial charges, giving rise to strong dipole-dipole attractions between neighboring molecules.

▲ **FIGURE 11.10 Hydrogen Bonding in HF** The hydrogen of one HF molecule, with its partial positive charge, is attracted to the fluorine of its neighbor, with its partial negative charge. This dipole–dipole interaction is an example of a hydrogen bond.

Name	Formula		Molar Mass (amu)	Structure	bp (°C)	mp (°C)
Ethanol	C$_2$H$_6$O		46.07	CH$_3$CH$_2$OH	78.3	−114.1
Dimethyl Ether	C$_2$H$_6$O		46.07	CH$_3$OCH$_3$	−22.0	−138.5

Since ethanol contains hydrogen bonded directly to oxygen, ethanol molecules form hydrogen bonds with each other. The hydrogen that is directly bonded to oxygen is strongly attracted to the oxygen on neighboring molecules, as shown in Figure 11.11 ▶. This strong attraction makes the boiling point of ethanol 78.3 °C. Consequently, ethanol is a liquid at room temperature. Dimethyl ether, in contrast, has an identical molar mass but does not exhibit hydrogen bonding because the oxygen atom is not bonded directly to hydrogen, resulting in lower boiling and melting points. Dimethyl ether is a gas at room temperature.

Water is another good example of a molecule with hydrogen bonding (Figure 11.12 ▶). Figure 11.13 ▼ shows the boiling points of the simple hydrogen compounds of the group 4A and group 6A elements. Notice that, in general, boiling points increase with increasing molar mass, as expected based on increasing dispersion forces. However, because of hydrogen bonding, the boiling point of water (100 °C) is much higher than expected based on its molar mass (18.0 g/mol). Without hydrogen bonding, all the water on our planet would be gaseous.

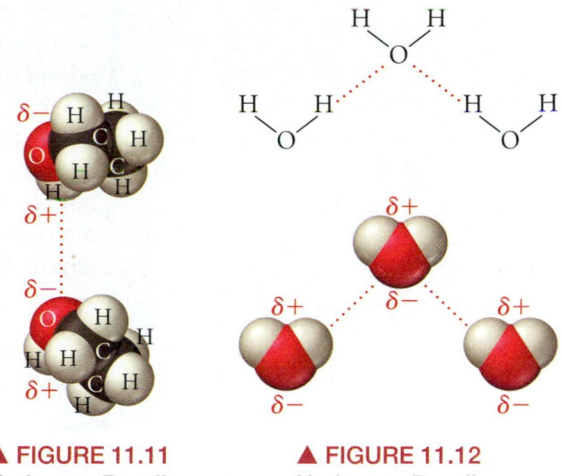

▲ FIGURE 11.11
Hydrogen Bonding in Ethanol

▲ FIGURE 11.12
Hydrogen Bonding in Water

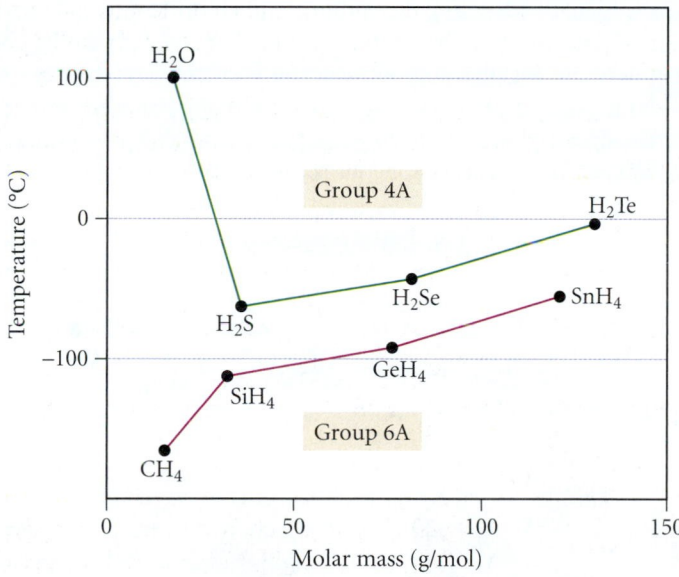

▲ FIGURE 11.13 **Boiling Points of Group 4A and 6A Compounds** Because of hydrogen bonding, the boiling point of water is anomalous compared to the boiling points of other hydrogen-containing compounds.

EXAMPLE 11.2 Hydrogen Bonding

One of the following compounds is a liquid at room temperature. Which one and why?

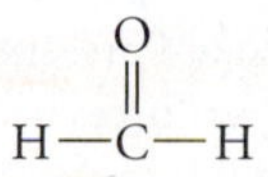

Formaldehyde 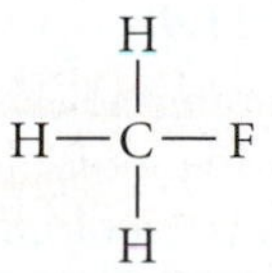   Fluoromethane 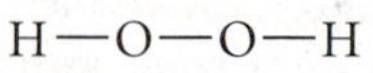   Hydrogen peroxide

Solution

The three compounds have similar molar masses:

Formaldehyde	30.03 g/mol
Fluoromethane	34.03 g/mol
Hydrogen peroxide	34.02 g/mol

So the strengths of their dispersion forces are similar. All three compounds are also polar, so they have dipole–dipole forces. Hydrogen peroxide, however, is the only compound that also contains H bonded directly to F, O, or N. Therefore it also has hydrogen bonding and is likely to have the highest boiling point of the three. Since the example stated that only one of the compounds was a liquid, we can safely assume that hydrogen peroxide is the liquid. Note that, although fluoromethane *contains* both H and F, H is not *directly bonded* to F, so fluoromethane does not have hydrogen bonding as an intermolecular force. Similarly, although formaldehyde *contains* both H and O, H is not *directly bonded* to O, so formaldehyde does not have hydrogen bonding either.

For Practice 11.2
Which has the higher boiling point, HF or HCl? Why?

Ion–Dipole Force

The **ion–dipole force** occurs when an ionic compound is mixed with a polar compound and is especially important in aqueous solutions of ionic compounds. For example, when sodium chloride is mixed with water, the sodium and chloride ions interact with water molecules via ion–dipole forces, as shown in Figure 11.14 ▼. Notice that the positive sodium ions interact with the negative poles of water molecules, while the negative chloride ions interact with the positive poles. Ion–dipole forces are the strongest of the types of intermolecular forces discussed here and are responsible for the ability of ionic substances to form solutions with water. We discuss aqueous solutions more thoroughly in Chapter 12.

Ion-Dipole Forces

The positively charged end of a polar molecule such as H_2O is attracted to negative ions and the negatively charged end of the molecule is attracted to positive ions.

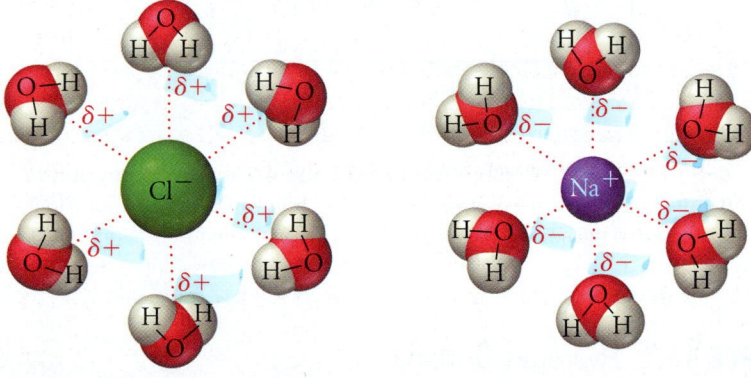

▶ **FIGURE 11.14 Ion–Dipole Forces**
Ion–dipole forces exist between Na^+ and the negative ends of H_2O molecules and between Cl^- and the positive ends of H_2O molecules.

Summarizing, as shown in Table 11.4:

➤ Dispersion forces, the weakest of the intermolecular forces, are present in all molecules and atoms and increase with increasing molar mass. These forces are always weak in small molecules but can be significant in molecules with high molar masses.

➤ Dipole–dipole forces are present in polar molecules.

➤ Hydrogen bonds, the strongest of the intermolecular forces that can occur in pure substances (second only to ion–dipole forces), are present in molecules containing hydrogen bonded directly to fluorine, oxygen, or nitrogen.

➤ Ion–dipole forces are present in mixtures of ionic compounds and polar compounds. These are very strong and are especially important in aqueous solutions of ionic compounds.

TABLE 11.4   Types of Intermolecular Forces

Type	Present in	Molecular perspective	Strength
Dispersion	All molecules and atoms		
Dipole-dipole	Polar molecules		
Hydrogen bonding	Molecules containing H bonded to F, O, or N		
Ion-dipole	Mixtures of ionic compounds and polar compounds		

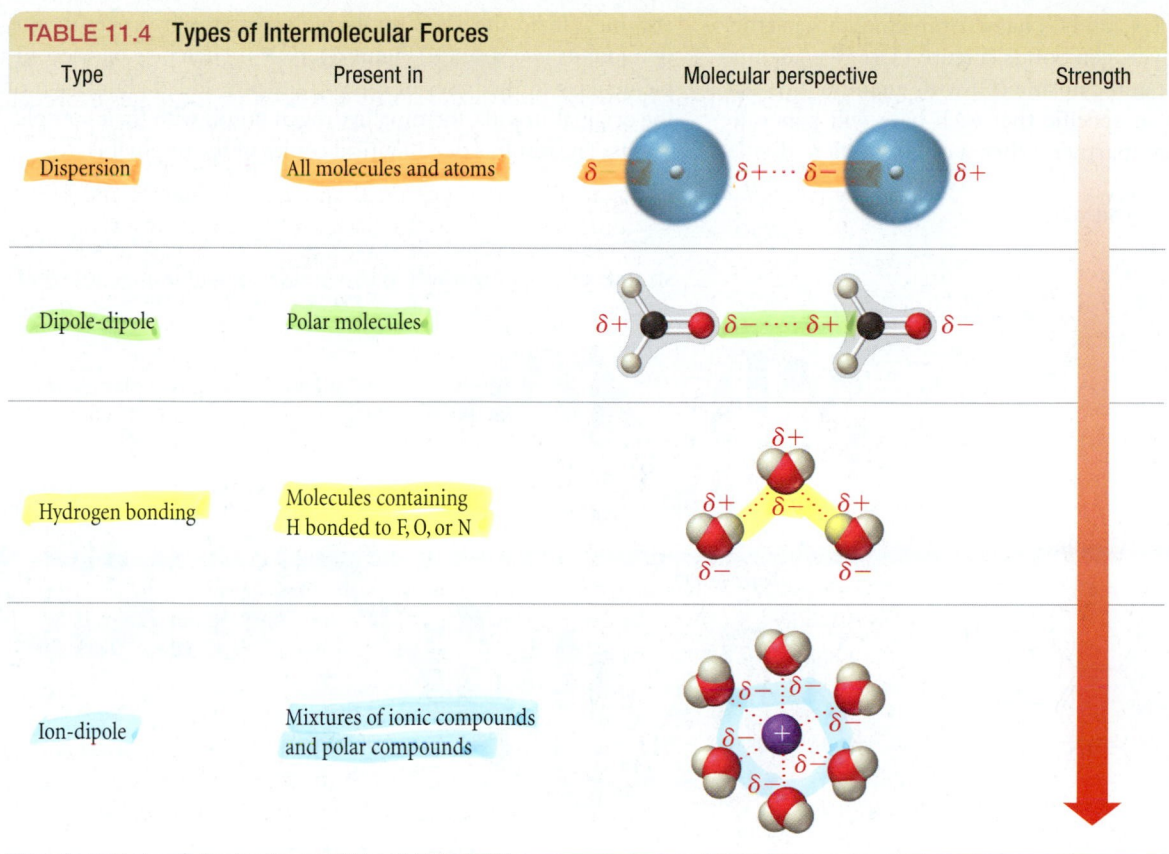

Chemistry and Medicine
Hydrogen Bonding in DNA

DNA is a long, chainlike molecule that acts as a blueprint for living organisms. Copies of DNA are passed from parent to off-spring, which is how we inherit traits from our parents. A DNA molecule is composed of thousands of repeating units called *nucleotides* (Figure 11.15 ▼). Each nucleotide contains one of four different *organic bases*: adenine, thymine, cytosine, and guanine (abbreviated A, T, C, and G). The order of these bases along DNA encodes the information that determines the nature of the proteins that are made in the body (proteins are the molecules that do most of the work in living organisms). Our proteins in turn determine many of our characteristics, including how we look, what diseases we are at risk of developing, and even how we behave. Consequently, human DNA is a blueprint for how humans are made.

The replicating mechanism of DNA is related to its structure, which was discovered in 1953 by James Watson and Francis Crick. DNA consists of two *complementary* strands, wrapped around each other in the now famous double helix and linked by hydrogen bonds between the bases on each

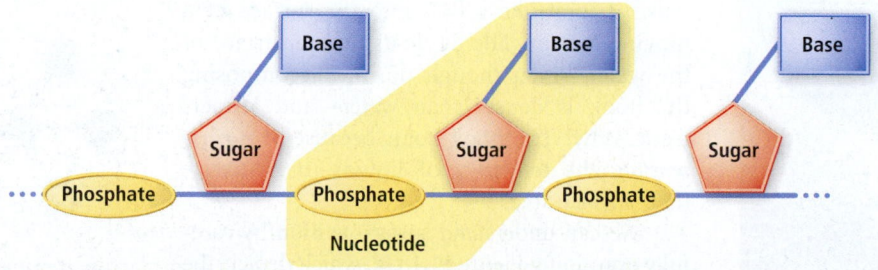

▲ **FIGURE 11.15 Nucleotides**   The individual units in a DNA polymer are called nucleotides. Each nucleotide contains one of four bases: adenine, thymine, cytosine, and guanine (abbreviated A, T, C, and G).

(continued on the next page)

strand. Each base (A, T, C, and G) has a complementary partner with which it forms hydrogen bonds (Figure 11.16 ▼): adenine (A) with thymine (T) and cytosine (C) with guanine (G). The hydrogen bonding is so specific that each base will pair only with its complementary partner. When a cell is going to divide,

enzymes unzip the DNA molecule across the hydrogen bonds that join its two strands (Figure 11.17 ▼). Then new bases, complementary to the bases in each strand, are added along each of the original strands, forming hydrogen bonds with their complements. The result is two identical copies of the original DNA.

Question

Why would dispersion forces not work as a way to hold the two strands of DNA together? Why would covalent bonds not work?

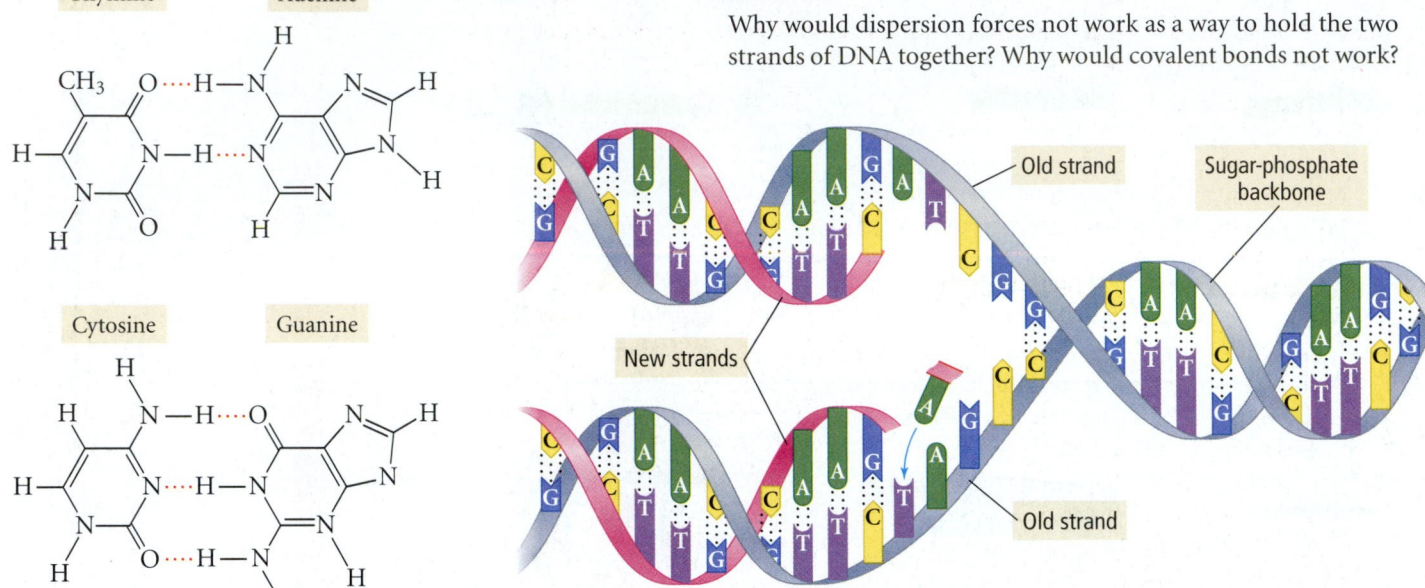

▲ **FIGURE 11.16 Complementary Base Pairing via Hydrogen Bonds**   The individual bases in DNA interact with one another via specific hydrogen bonds that form between A and T and between C and G.

▲ **FIGURE 11.17 Copying DNA**   The two strands of the DNA molecule can "unzip" by breaking the hydrogen bonds that join the base pairs. New bases complementary to the bases of each strand are assembled and joined together. The result is two molecules, each identical to the original one.

11.4 Intermolecular Forces in Action: Surface Tension, Viscosity, and Capillary Action

The most important manifestation of intermolecular forces is the very existence of liquids and solids. In liquids, we also observe several other manifestations of intermolecular forces including surface tension, viscosity, and capillary action.

Surface Tension

A fly fisherman delicately casts a small metal hook (with a few feathers and strings attached to make it look like a fly) onto the surface of a moving stream. The fly floats on the surface of the water—even though the metal composing the hook is denser than water—and attracts trout. Why? The hook floats because of *surface tension*, the tendency of liquids to minimize their surface area.

We can understand surface tension by carefully examining Figure 11.18 ▶, which depicts the intermolecular forces experienced by a molecule at the surface of the liquid compared to those experienced by a molecule in the interior. Notice that a molecule at the surface has relatively fewer neighbors with which to interact, because there are no

▲ A trout fly can float on water because of surface tension.

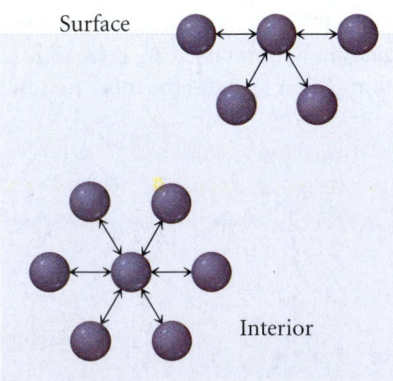

▲ **FIGURE 11.18 The Origin of Surface Tension**   Molecules at the liquid surface have a higher potential energy than those in the interior. As a result, liquids tend to minimize their surface area, and the surface behaves like a membrane or "skin."

molecules above it. Consequently, molecules at the surface are inherently less stable—they have higher potential energy—than those in the interior. In order to increase the surface area of the liquid, some molecules from the interior have to be moved to the surface, a process requiring energy. The **surface tension** of a liquid is the energy required to increase the surface area by a unit amount. For example, at room temperature, water has a surface tension of $72.8 \ mJ/m^2$—it takes 72.8 mJ to increase the surface area of water by one square meter.

Why does surface tension allow the fly fisherman's hook to float on water? Since molecules at the surface have a higher potential energy than those in the interior, liquids tend to minimize their surface area. This tendency creates a kind of skin at the surface that resists penetration. For the fisherman's hook to sink into the water, the water's surface area must increase slightly—an increase that is resisted by the surface tension. You can observe surface tension by carefully placing a paper clip on the surface of water (Figure 11.19 ▶). The paper clip, even though it is denser than water, will float on the surface of the water. A slight tap on the clip will overcome the surface tension and cause the clip to sink.

Surface tension decreases with decreasing intermolecular forces. You can't float a paper clip on benzene, for example, because the dispersion forces among the molecules composing benzene are significantly weaker than the hydrogen bonds among water molecules. The surface tension of benzene, for example, is only $28 \ mJ/m^2$—just 40% that of water.

Surface tension is also the reason that small water droplets (those not large enough to be distorted by gravity) form nearly perfect spheres. On the Space Shuttle, the complete absence of gravity allows even large samples of water to form nearly perfect spheres (Figure 11.20 ▶). Why? Just as gravity pulls the matter of a planet or star inward to form a sphere, so intermolecular forces among collections of water molecules pull the water into a sphere. The sphere is the geometrical shape with the smallest ratio of surface area to volume; therefore, the formation of a sphere minimizes the number of molecules at the surface, thus minimizing the potential energy of the system.

Viscosity

Another manifestation of intermolecular forces is **viscosity**, the resistance of a liquid to flow. Motor oil, for example, is more viscous than gasoline, and maple syrup is more viscous than water. Viscosity is measured in a unit called the poise (P), defined as $1 \ g/cm \cdot s$. The centipoise (cP) is a convenient unit because the viscosity of water at room temperature is approximately one centipoise. Viscosity is greater in substances with stronger intermolecular forces because molecules are more strongly attracted to each other, preventing them from flowing around each other as freely. Viscosity also depends on molecular shape, increasing in longer molecules that can interact over a greater area and possibly become entangled. Table 11.5 lists the viscosity of several hydrocarbons. Notice the increase in viscosity with increasing molar mass (and therefore increasing magnitude of dispersion forces) and with increasing length (and therefore increasing potential for molecular entanglement).

Viscosity also depends on temperature because thermal energy partially overcomes the intermolecular forces, allowing molecules to flow past each other more easily. Table 11.6 lists the viscosity of water as a function of temperature. Notice that the viscosity decreases with increasing temperature. Nearly all liquids become less viscous as temperature increases.

Recall from Section 11.3 that the interactions between molecules lower their potential energy in much the same way that the interaction between protons and electrons lowers their potential energy, in accordance with Coulomb's law.

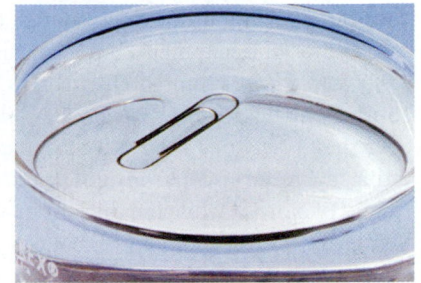

▲ **FIGURE 11.19 Surface Tension in Action**   A paper clip floats on water due to surface tension.

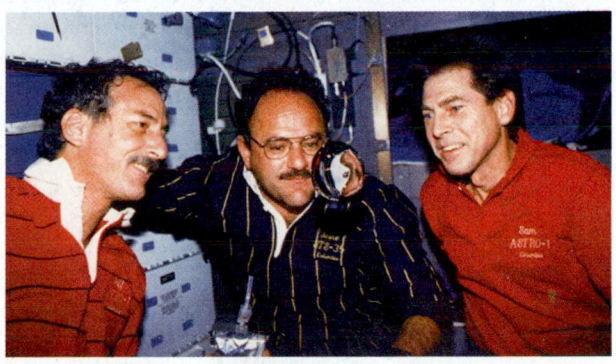

▲ **FIGURE 11.20 Spherical Water Droplets**   On the Space Shuttle in orbit, under weightless conditions, water coalesces into nearly perfect spheres held together by intermolecular forces between water molecules.

TABLE 11.5 Viscosity of Several Hydrocarbons at 20 °C

Hydrocarbon	Molar Mass (g/mol)	Formula	Viscosity (cP)
N-Pentane	72.15	$CH_3CH_2CH_2CH_2CH_3$	0.240
N-Hexane	86.17	$CH_3CH_2CH_2CH_2CH_2CH_3$	0.326
N-Heptane	100.2	$CH_3CH_2CH_2CH_2CH_2CH_2CH_3$	0.409
N-Octane	114.2	$CH_3CH_2CH_2CH_2CH_2CH_2CH_2CH_3$	0.542
N-Nonane	128.3	$CH_3CH_2CH_2CH_2CH_2CH_2CH_2CH_2CH_3$	0.711

TABLE 11.6 Viscosity of Liquid Water at Several Temperatures

Temperature (°C)	Viscosity (cP)
20	1.002
40	0.653
60	0.467
80	0.355
100	0.282

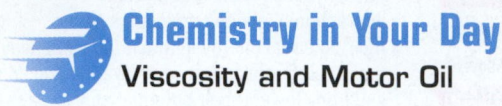

Chemistry in Your Day
Viscosity and Motor Oil

Viscosity is an important property of the motor oil you put into your car. The oil must be thick enough to adequately coat engine surfaces to lubricate them, but also thin enough to be pumped easily into all the required engine compartments. Motor oil viscosity is usually reported on a scale called the SAE scale (named after the Society of Automotive Engineers). The higher the SAE rating, the more viscous the oil. The thinnest motor oils have SAE ratings of 5 or 10, while the thickest have SAE ratings of up to 50. Before the 1950s, most automobile owners changed the oil in their engine to accommodate seasonal changes in weather—a higher SAE rating was required in the summer months and a lower rating in the winter. Today, the advent of multigrade oils allows car owners in many climates to keep the same oil all year long. Multigrade oils, such as the 10W-40 oil shown here, contain polymers (long molecules made up of repeating structural units) that coil at low temperatures but unwind at high temperatures.

At low temperatures, the coiled polymers—because of their compact shape—do not contribute very much to the viscosity of the oil. As the temperature increases, however, the molecules unwind and their long shape results in intermolecular forces and molecular entanglements that prevent the viscosity from decreasing as much as it would normally. The result is an oil whose viscosity is less temperature dependent than it would be otherwise, allowing the same oil to be used over a wider range of temperatures. The 10W-40 designation indicates that the oil has an SAE rating of 10 at low temperatures and 40 at high temperatures.

Capillary Action

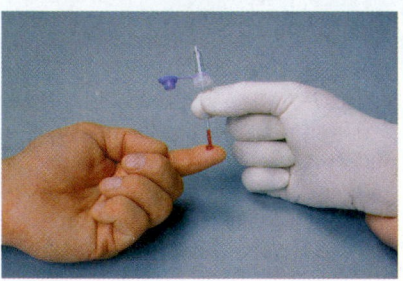

▲ Blood is drawn into a capillary tube by capillary action.

Medical technicians often take advantage of **capillary action**—the ability of a liquid to flow against gravity up a narrow tube—when taking a blood sample. The technician pokes the patient's finger with a pin, squeezes some blood out of the puncture, and then collects the blood with a thin tube. When the tube's tip comes into contact with the blood, the blood is drawn into the tube by capillary action. The same force plays a role when trees and plants draw water from the soil.

Capillary action results from a combination of two forces: the attraction between molecules in a liquid, called *cohesive forces*, and the attraction between these molecules and the surface of the tube, called *adhesive forces*. The adhesive forces cause the liquid to spread out over the surface of the tube, while the cohesive forces cause the liquid to stay together. If the adhesive forces are greater than the cohesive forces (as is the case for water in a glass tube), the attraction to the surface draws the liquid up the tube while the cohesive forces pull along those molecules not in direct contact with the tube walls (Figure 11.21 ◄). The water rises up the tube until the force of gravity balances the capillary action—the thinner the tube, the higher the rise. If the adhesive forces are smaller than the cohesive forces (as is the case for liquid mercury), the liquid does not rise up the tube at all (and in fact will drop to a level below the level of the surrounding liquid).

The result of the differences in the relative magnitudes of cohesive and adhesive forces can be seen by comparing the meniscus of water to the meniscus of mercury (Figure 11.22 ◄). (The meniscus is the curved shape of a liquid surface within a tube.) The meniscus of water is concave because the *adhesive forces* are greater than the cohesive forces, causing the edges of the

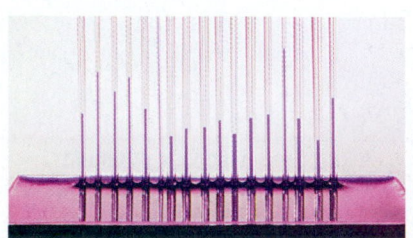

▲ **FIGURE 11.21 Capillary Action** The attraction of water molecules to the glass surface draws the liquid around the edge of the tube up the walls. The rest of the column is pulled along by the attraction of water molecules to one another. As can be seen above, the narrower the tube, the higher the liquid will rise.

▲ **FIGURE 11.22 Meniscuses of Water and Mercury** The meniscus of water (dyed red for visibility at left) is concave because water molecules are more strongly attracted to the glass wall than to one another. The meniscus of mercury is convex because mercury atoms are more strongly attracted to one another than to the glass walls.

water to creep up the sides of the tube a bit, forming the familiar cupped shape. The meniscus of mercury is convex because the *cohesive forces*—due to metallic bonding between the atoms—are greater than the adhesive forces. The mercury atoms crowd toward the interior of the liquid to maximize their interactions with each other, resulting in the upward bulge at the center of the surface.

11.5 Vaporization and Vapor Pressure

We now turn our attention to vaporization, the process by which thermal energy can overcome intermolecular forces and produce a phase change from liquid to gas. We will first discuss the process of vaporization itself, then the energetics of vaporization, and finally the concepts of vapor pressure, dynamic equilibrium, and critical point. As you will see, vaporization is a common occurrence that you experience every day and even depend on to maintain proper body temperature.

The Process of Vaporization

Imagine the water molecules within a beaker of water sitting on a table at room temperature and open to the atmosphere (Figure 11.23 ▶). The molecules are in constant motion due to thermal energy. If you could actually see the molecules at the surface, you would witness Roald Hoffmann's "wild dance floor" (see the chapter-opening quote) because of all the vibrating, jostling, and molecular movement. *The higher the temperature, the greater the average energy of the collection of molecules*. However, at any one time, some molecules will have more thermal energy than the average and some will have less.

The distributions of thermal energies for a sample of water at two different temperatures are shown in Figure 11.24 ▼. The molecules at the high end of the distribution curve have enough energy to break free from the surface—where molecules are held less tightly than in the interior due to fewer neighbor–neighbor interactions—and into the gas phase. This process is called **vaporization**, the phase transition from liquid to gas. Some of the water molecules in the gas phase, at the low end of the energy distribution curve for the gaseous molecules, can plunge back into the water and be captured by intermolecular forces. This process—the opposite of vaporization—is called **condensation**, the phase transition from gas to liquid.

▲ **FIGURE 11.23 Vaporization of Water**
Some molecules in an open beaker have enough kinetic energy to vaporize from the surface of the liquid.

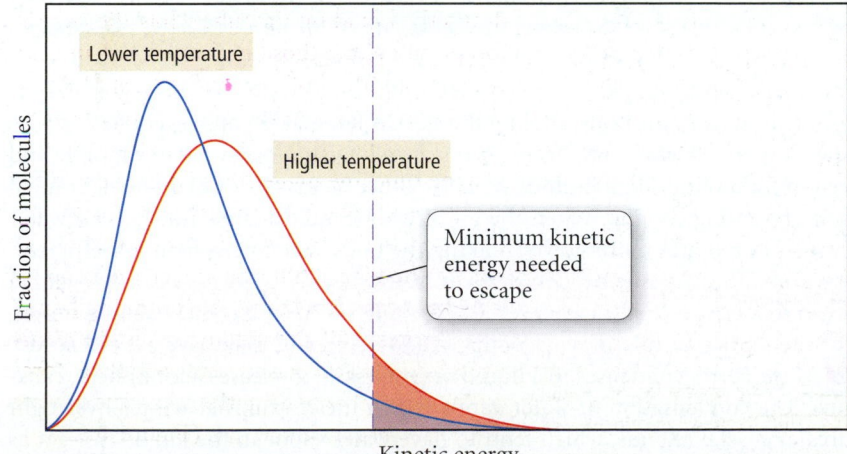

▲ **FIGURE 11.24 Distribution of Thermal Energy** The thermal energies of the molecules in a liquid are distributed over a range. The peak energy increases with increasing temperature.

Although both evaporation and condensation occur in a beaker open to the atmosphere, under normal conditions evaporation takes place at a greater rate because most of the newly evaporated molecules escape into the surrounding atmosphere and never come back. The result is a noticeable decrease in the water level within the beaker over time (usually several days).

What happens if we increase the temperature of the water within the beaker? Because of the shift in the energy distribution to higher energies (see Figure 11.24 on page 477), more molecules now have enough energy to break free and evaporate, so vaporization occurs more quickly. What happens if we spill the water on the table or floor? The same amount of water is now spread over a wider area, resulting in more molecules at the surface of the liquid. Since molecules at the surface have the greatest tendency to evaporate—because they are held less tightly—the vaporization also occurs more quickly. You probably know from experience that water in a beaker or glass may take many days to evaporate completely, while the same amount of water spilled on a table or floor probably evaporates within a few hours (depending on the exact conditions).

What happens if the liquid in the beaker is not water, but some other substance with weaker intermolecular forces, such as acetone? The weaker intermolecular forces allow more molecules to evaporate at a given temperature, again increasing the rate of vaporization. Liquids that vaporize easily are termed **volatile**, while those that do not vaporize easily are termed **nonvolatile**. Acetone, for example, is more volatile than water. Motor oil is virtually nonvolatile at room temperature.

Summarizing:
➤ The rate of vaporization increases with increasing temperature.
➤ The rate of vaporization increases with increasing surface area.
➤ The rate of vaporization increases with decreasing strength of intermolecular forces.

The Energetics of Vaporization

To understand the energetics of vaporization, consider again a beaker of water from the molecular point of view, except now let the beaker be thermally insulated so that heat from the surroundings cannot enter the beaker. What happens to the temperature of the water left in the beaker as molecules evaporate? To answer this question, think about the energy distribution curve again (see Figure 11.24). The molecules that leave the beaker are the ones at the high end of the energy curve—the most energetic. If no additional heat enters the beaker, the average energy of the entire collection of molecules goes down—much as the class average on an exam goes down if you eliminate the highest-scoring students. So vaporization is an *endothermic* process; it takes energy to vaporize the molecules in a liquid. Another way to understand the endothermicity of vaporization is to remember that vaporization requires overcoming the intermolecular forces that hold liquids together. Since energy must be absorbed to pull the molecules apart, the process is endothermic.

Our bodies use the endothermic nature of vaporization for cooling. When you overheat, you sweat, causing your skin to be covered with liquid water. As this water evaporates, it absorbs heat from the body, cooling the skin. A fan makes you feel cooler because it blows newly vaporized water away from your skin, allowing more sweat to vaporize and causing even more cooling. High humidity, on the other hand, slows down the net rate of evaporation, preventing cooling. When the air already contains large amounts of water vapor, the sweat evaporates more slowly, making the body's cooling system less efficient.

Condensation, the opposite of vaporization, is exothermic—heat is released when a gas condenses to a liquid. If you have ever accidentally put your hand above a steaming kettle, or opened a bag of microwaved popcorn too soon, you may have experienced a *steam burn*. As the steam condenses to a liquid on your skin, it releases a lot of heat, causing the burn. The condensation of water vapor is also the reason that winter overnight temperatures in coastal regions, which tend to have water vapor in the air, do not get as low as in deserts, which tend to have dry air. As the air temperature in a coastal area drops, water condenses out of the air, releasing heat and preventing the temperature from dropping further. In deserts, there is little moisture in the air to condense, so the temperature drop is greater.

See Chapter 6 to review endothermic and exothermic processes.

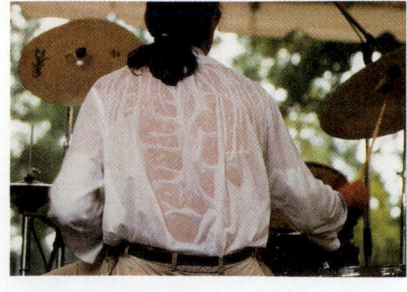

▲ When you sweat, water evaporates from the skin. Since evaporation is endothermic, the result is a cooling effect.

Heat of Vaporization The amount of heat required to vaporize one mole of a liquid to gas is called the **heat of vaporization** (ΔH_{vap}). The heat of vaporization of water at its normal boiling point of 100 °C is +40.7 kJ/mol:

$$H_2O(l) \longrightarrow H_2O(g) \qquad \Delta H_{vap} = +40.7 \text{ kJ/mol}$$

The heat of vaporization is always positive because the process is endothermic—energy must be absorbed to vaporize a substance. The heat of vaporization is somewhat temperature dependent. For example, at 25 °C the heat of vaporization of water is +44.0 kJ/mol, slightly more than at 100 °C because the water contains less thermal energy. Table 11.7 lists the heats of vaporization of several liquids at their boiling points and at 25 °C.

The sign conventions of ΔH were introduced in Chapter 6.

TABLE 11.7 Heats of Vaporization of Several Liquids at Their Boiling Points and at 25 °C

Liquid	Chemical Formula	Normal Boiling Point (°C)	ΔH_{vap} (kJ/mol) at Boiling Point	ΔH_{vap} (kJ/mol) at 25 °C
Water	H_2O	100	40.7	44.0
Rubbing alcohol (isopropyl alcohol)	C_3H_8O	82.3	39.9	45.4
Acetone	C_3H_6O	56.1	29.1	31.0
Diethyl ether	$C_4H_{10}O$	34.6	26.5	27.1

When a substance condenses from a gas to a liquid, the same amount of heat is involved, but the heat is emitted rather than absorbed.

$$H_2O(g) \longrightarrow H_2O(l) \qquad \Delta H = -\Delta H_{vap} = -40.7 \text{ kJ} \qquad \text{(at 100 °C)}$$

When one mole of water condenses, it releases 40.7 kJ of heat. The sign of ΔH in this case is negative because the process is exothermic.

The heat of vaporization of a liquid can be used to calculate the amount of heat energy required to vaporize a given mass of the liquid (or the amount of heat given off by the condensation of a given mass of liquid), using concepts similar to those covered in Section 6.5 (stoichiometry of ΔH). You can use the heat of vaporization as a conversion factor between number of moles of a liquid and the amount of heat required to vaporize it (or the amount of heat emitted when it condenses), as shown in the following example.

EXAMPLE 11.3 Using the Heat of Vaporization in Calculations

Calculate the mass of water (in g) that can be vaporized at its boiling point with 155 kJ of heat.

Sort You are given a certain amount of heat in kilojoules and asked to find the mass of water that can be vaporized.	**Given** 155 kJ **Find** g H_2O

Strategize The heat of vaporization gives the relationship between heat absorbed and moles of water vaporized. Begin with the given amount of heat (in kJ) and convert to moles of water that can be vaporized. Then use the molar mass as a conversion factor to convert from moles of water to mass of water.

Conceptual Plan

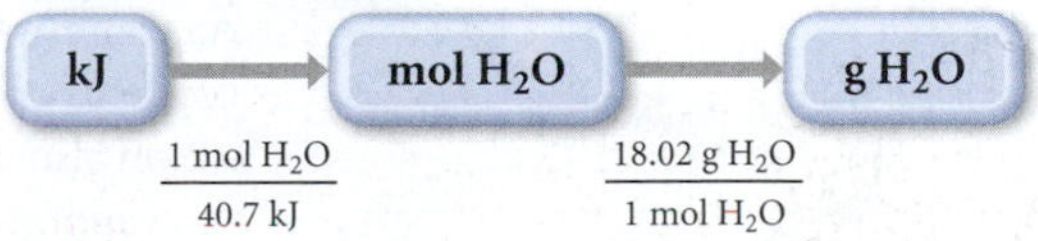

kJ → mol H_2O → g H_2O

$$\frac{1 \text{ mol } H_2O}{40.7 \text{ kJ}} \qquad \frac{18.02 \text{ g } H_2O}{1 \text{ mol } H_2O}$$

Relationships Used

$\Delta H_{vap} = 40.7$ kJ/mol (at 100 °C)

18.02 g H_2O = 1 mol H_2O

Solve Follow the conceptual plan to solve the problem.

Solution

$$155 \text{ kJ} \times \frac{1 \text{ mol } H_2O}{40.7 \text{ kJ}} \times \frac{18.02 \text{ g } H_2O}{1 \text{ mol } H_2O} = 68.6 \text{ g } H_2O$$

For Practice 11.3

Calculate the amount of heat (in kJ) required to vaporize 2.58 kg of water at its boiling point.

For More Practice 11.3

Suppose that 0.48 g of water at 25 °C condenses on the surface of a 55-g block of aluminum that is initially at 25 °C. If the heat released during condensation goes only toward heating the metal, what is the final temperature (in °C) of the metal block? (The specific heat capacity of aluminum is 0.903 J/g · °C.)

Vapor Pressure and Dynamic Equilibrium

We have already seen that if a container of water is left uncovered at room temperature, the water slowly evaporates away. But what happens if the container is sealed? Imagine a sealed evacuated flask—one from which the air has been removed—containing liquid water, as shown in Figure 11.25 ▼. Initially, the water molecules evaporate, as they did in the open beaker.

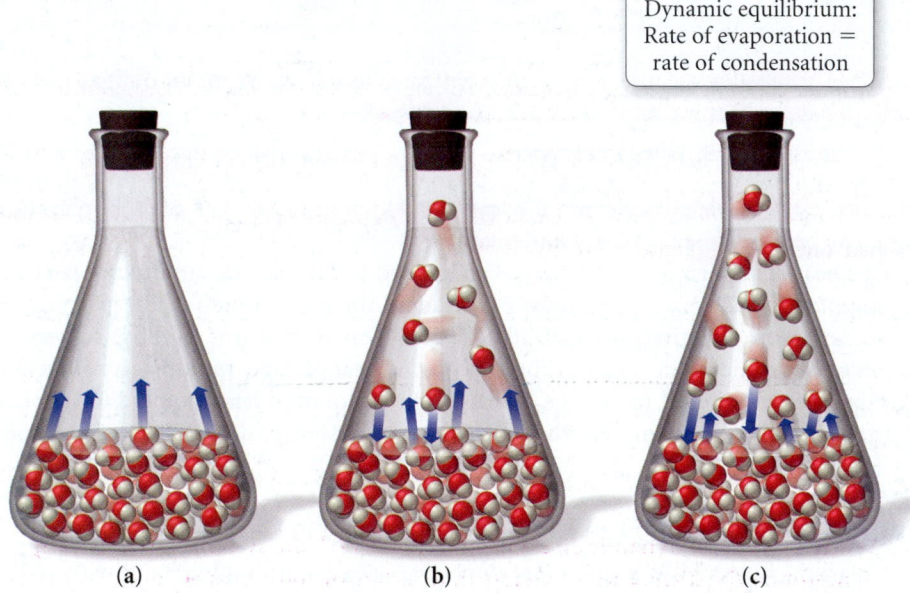

Dynamic equilibrium:
Rate of evaporation =
rate of condensation

(a)   (b)   (c)

▲ **FIGURE 11.25 Vaporization in a Sealed Flask** (a) When water is placed into a sealed container, water molecules begin to vaporize. (b) As water molecules build up in the gas phase, they begin to recondense into the liquid. (c) When the rate of evaporation equals the rate of condensation, dynamic equilibrium is reached.

However, because of the seal, the evaporated molecules cannot escape into the atmosphere. As water molecules enter the gas phase, some start condensing back into the liquid. As the concentration (or partial pressure) of gaseous water molecules increases, the rate of condensation also increases. However, as long as the water remains at a constant temperature, the rate of evaporation remains constant. Eventually the rate of condensation and the rate of vaporization become equal—**dynamic equilibrium** has been reached (Figure 11.26 ◄). Although condensation and vaporization continue, at equal rates, the concentration of water vapor above the liquid is constant.

The pressure of a gas in dynamic equilibrium with its liquid is called its **vapor pressure**. The vapor pressure of a particular liquid depends on the intermolecular forces present in the liquid and the temperature. Weak intermolecular forces result in volatile substances with high vapor pressures, because the intermolecular forces are easily overcome by thermal energy. Strong intermolecular forces result in nonvolatile substances with low vapor pressures.

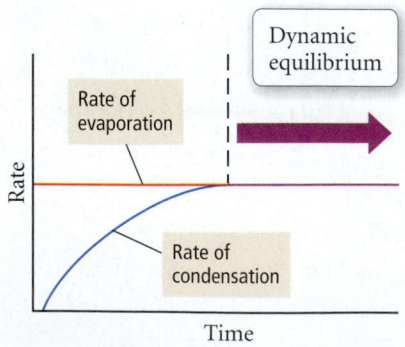

▲ **FIGURE 11.26 Dynamic Equilibrium** Dynamic equilibrium occurs when the rate of condensation is equal to the rate of evaporation.

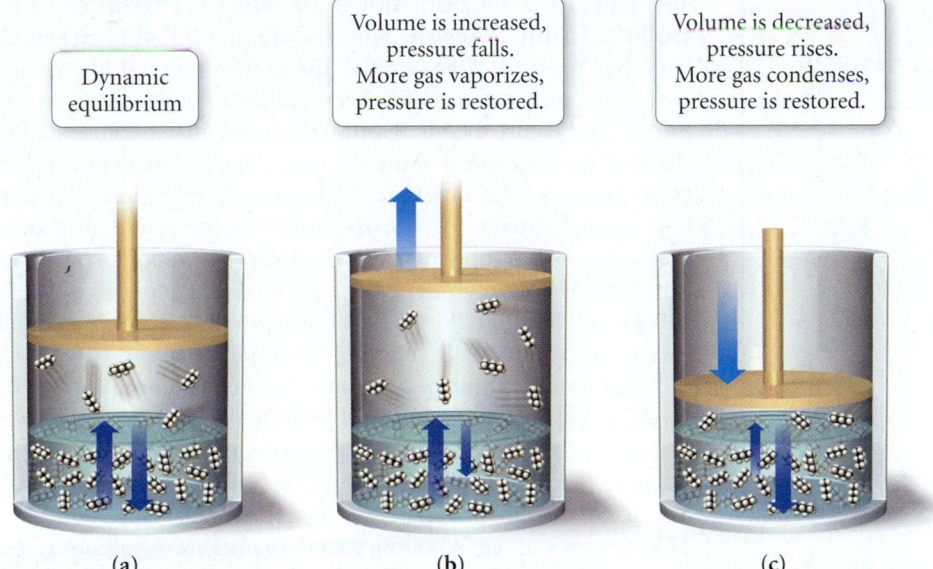

Dynamic equilibrium

Volume is increased, pressure falls. More gas vaporizes, pressure is restored.

Volume is decreased, pressure rises. More gas condenses, pressure is restored.

(a)   (b)   (c)

◀ **FIGURE 11.27 Dynamic Equilibrium in *n*-Pentane** **(a)** Liquid *n*-pentane is in dynamic equilibrium with its vapor. **(b)** When the volume is increased, the pressure drops and some liquid is converted to gas to bring the pressure back up. **(c)** When the volume is decreased, the pressure increases and some gas is converted to liquid to bring the pressure back down.

A liquid in dynamic equilibrium with its vapor is a balanced system that tends to return to equilibrium if disturbed. For example, consider a sample of *n*-pentane (a component of gasoline) at 25 °C in a cylinder equipped with a moveable piston (Figure 11.27a ▲). The cylinder contains no other gases except *n*-pentane vapor in dynamic equilibrium with the liquid. Since the vapor pressure of *n*-pentane at 25 °C is 510 mmHg, the pressure in the cylinder is 510 mmHg. Now, what happens when the piston is moved upward to expand the volume within the cylinder? Initially, the pressure in the cylinder drops below 510 mmHg, in accordance with Boyle's law. Then, however, more liquid vaporizes until equilibrium is reached once again (Figure 11.27b). If the volume of the cylinder is expanded again, the same thing happens—the pressure initially drops and more *n*-pentane vaporizes to bring the system back into equilibrium. Further expansion will cause the same result *as long as some liquid n-pentane remains in the cylinder*.

Conversely, what happens if the piston is lowered, decreasing the volume in the cylinder? Initially, the pressure in the cylinder rises above 510 mmHg, but then some of the gas condenses into liquid until equilibrium is reached again (Figure 11.27(c)).

We can describe the tendency of a system in dynamic equilibrium to return to equilibrium with the following general statement:

> When a system in dynamic equilibrium is disturbed, the system responds so as to minimize the disturbance and return to a state of equilibrium.

So, if the pressure above the system is decreased, the pressure increases (some of the liquid evaporates) so that the system returns to equilibrium. If the pressure above a liquid–vapor system in equilibrium is increased, the pressure of the system drops (some of the gas condenses) so that the system returns to equilibrium. This basic principle—often called Le Châtelier's principle—is applicable to any chemical system in equilibrium, as we will see in Chapter 14.

| Boyle's law is discussed in Section 5.3.

Conceptual Connection 11.2 Vapor Pressure

What happens to the vapor pressure of a substance when its surface area is increased at constant temperature?

(a) The vapor pressure increases.

(b) The vapor pressure remains the same.

(c) The vapor pressure decreases.

Answer: **(b)** Although the *rate of vaporization* increases with increasing surface area, the *vapor pressure* of a liquid is independent of surface area. An increase in surface increases both the rate of vaporization and the rate of condensation—the effects of surface area exactly cancel and the vapor pressure does not change.

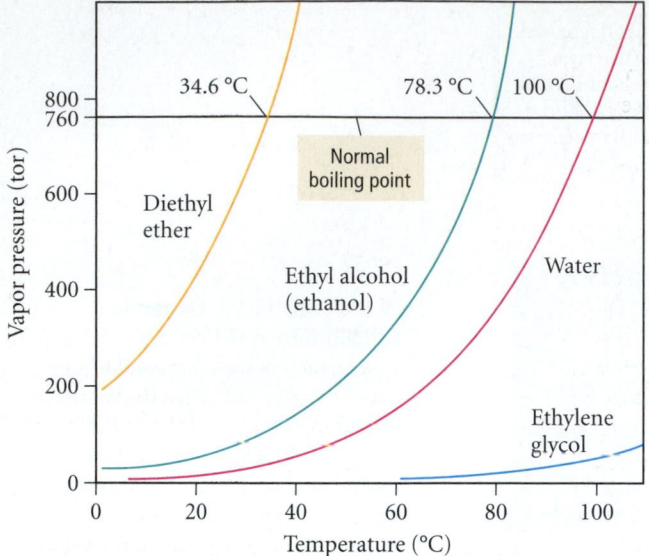

▲ **FIGURE 11.28 Vapor Pressure of Several Liquids at Different Temperatures** At higher temperatures, more molecules have enough thermal energy to escape into the gas phase, so vapor pressure increases with increasing temperature.

Sometimes you see bubbles begin to form in hot water below 100 °C. These bubbles are dissolved air—not gaseous water—leaving the liquid. Dissolved air comes out of water as you heat it because the solubility of a gas in a liquid decreases with increasing temperature (as we will see in Chapter 12).

▶ **FIGURE 11.29 Boiling** A liquid boils when thermal energy is high enough to cause molecules in the interior of the liquid to become gaseous, forming bubbles that rise to the surface.

Temperature Dependence of Vapor Pressure and Boiling Point

When the temperature of a liquid is increased, its vapor pressure rises because the higher thermal energy increases the number of molecules that have enough energy to vaporize (see Figure 11.24). Notice, however, that because of the shape of the energy distribution curve, a small change in temperature makes a large difference in the number of molecules that have enough energy to vaporize and therefore results in a large increase in vapor pressure. For example, the vapor pressure of water at 25 °C is 23.3 torr, while at 60 °C the vapor pressure is 149.4 torr. Figure 11.28 ◀ shows the vapor pressure of water and several other liquids as a function of temperature.

The **boiling point** of a liquid is *the temperature at which its vapor pressure equals the external pressure*. When a liquid reaches its boiling point, the thermal energy is enough for molecules in the interior of the liquid (not just those at the surface) to break free of their neighbors and enter the gas phase (Figure 11.29 ▼). The bubbles you see in boiling water are pockets of gaseous water that have formed within the liquid water. The bubbles float to the surface and leave as gaseous water or steam.

The **normal boiling point** of a liquid is *the temperature at which its vapor pressure equals 1 atm*. The normal boiling point of pure water is 100 °C. However, at a lower pressure, water boils at a lower temperature. In Denver, Colorado, for example, the average atmospheric pressure is about 83% of what it is at sea level, and water boils at approximately 94 °C. For this reason, it takes slightly longer to cook food in boiling water in Denver than in a city at sea level. Table 11.8 shows the boiling point of water at several locations of varied altitudes.

TABLE 11.8 Boiling Points of Water at Several Locations of Varied Altitudes

Location	Elevation (ft)	Approximate Pressure (atm)*	Approximate Boiling Point of Water (°C)
Mt. Everest, Tibet (highest mountain peak on Earth)	29,035	0.32	78
Mt. McKinley (Denali), Alaska (highest mountain peak in North America.)	20,320	0.46	83
Mt. Whitney, California (highest mountain peak in 48 contiguous U.S. states)	14,495	0.60	87
Denver, Colorado (mile high city)	5,280	0.83	94
Boston, Massachusetts (sea level)	20	1.0	100

*The atmospheric pressure in each of these locations is subject to weather conditions and can vary significantly from the values stated here.

Once the boiling point of a liquid is reached, additional heating only causes more rapid boiling; it does not raise the temperature of the liquid above its boiling point, as shown in the *heating curve* in Figure 11.30 ▶. Therefore, boiling water at 1 atm will always have a temperature of 100 °C. After all the water has been converted to steam, the temperature of the steam can continue to rise beyond 100 °C. *However, as long as liquid water is present, its temperature cannot rise above its boiling point.*

The Clausius–Clapeyron Equation

Now, let's return our attention to Figure 11.28. As you can see from the graph, the vapor pressure of a liquid increases with increasing temperature. However, *the relationship is not linear.* In other words, doubling the temperature results in more than a doubling of the vapor pressure. The relationship between vapor pressure and temperature is exponential, and can be expressed as follows:

$$P_{vap} = \beta \exp\left(\frac{-\Delta H_{vap}}{RT}\right) \qquad [11.1]$$

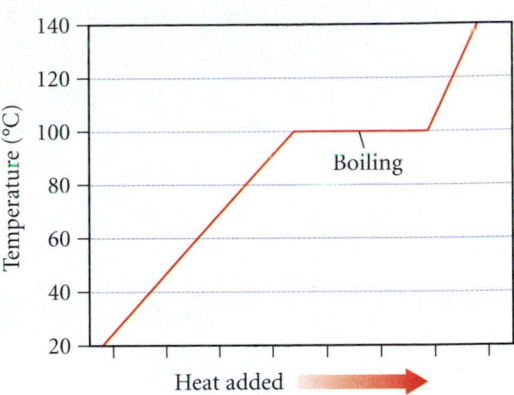

▲ **FIGURE 11.30 Temperature during Boiling** The temperature of water during boiling remains at 100 °C.

In this expression P_{vap} is the vapor pressure, β is a constant that depends on the gas, ΔH_{vap} is the heat of vaporization, R is the gas constant (8.314 J/mol·K), and T is the temperature in kelvins. Equation 11.1 can be rearranged by taking the natural logarithm of both sides as follows:

$$\ln P_{vap} = \ln\left[\beta \exp\left(\frac{-\Delta H_{vap}}{RT}\right)\right] \qquad [11.2]$$

Since $\ln AB = \ln A + \ln B$, we can rearrange the right side of Equation 11.2 as follows:

$$\ln P_{vap} = \ln \beta + \ln\left[\exp\left(\frac{-\Delta H_{vap}}{RT}\right)\right] \qquad [11.3]$$

Since $\ln e^x = x$ (see Appendix IB), we can simplify Equation 11.3 as follows:

$$\ln P_{vap} = \ln \beta + \frac{-\Delta H_{vap}}{RT} \qquad [11.4]$$

A slight additional rearrangement gives us the following important result:

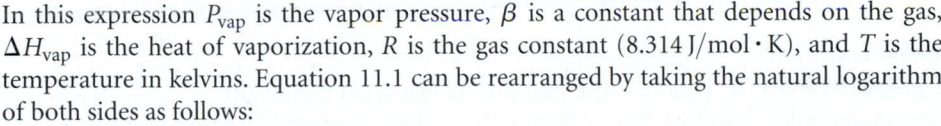

$$\ln P_{vap} = \frac{-\Delta H_{vap}}{R}\left(\frac{1}{T}\right) + \ln \beta \qquad \text{Clausius–Clapeyron equation}$$

$$y = \quad m \quad (x) \quad + b \qquad \text{(equation for a line)}$$

Notice the parallel relationship between the **Clausius–Clapeyron equation** and the equation for a straight line. Just as a plot of *y* versus *x* yields a straight line with slope *m* and intercept *b*, so a plot of $\ln P_{vap}$ (equivalent to *y*) versus $1/T$ (equivalent to *x*) gives a straight line with slope $-\Delta H_{vap}/R$ (equivalent to *m*) and y-intercept $\ln \beta$ (equivalent to *b*), as shown in Figure 11.31 on page 484. The Clausius–Clapeyron equation gives a linear relationship—not between the vapor pressure and the temperature (which have an exponential relationship)—but between the *natural log* of the vapor pressure and the *inverse* of temperature. This is a common technique in the analysis of chemical data. If two variables are not linearly related, it is often convenient to find ways to graph *functions of those variables* that are linearly related.

The Clausius–Clapeyron equation leads to a convenient way to measure the heat of vaporization in the laboratory. We simply measure the vapor pressure of a liquid as a function of temperature and create a plot of the natural log of the vapor pressure versus the inverse of the temperature. We can then determine the slope of the line to find the heat of vaporization, as shown in the following example.

Using the Clausius-Clapeyron equation in this way ignores the relatively small temperature dependence of ΔH_{vap}

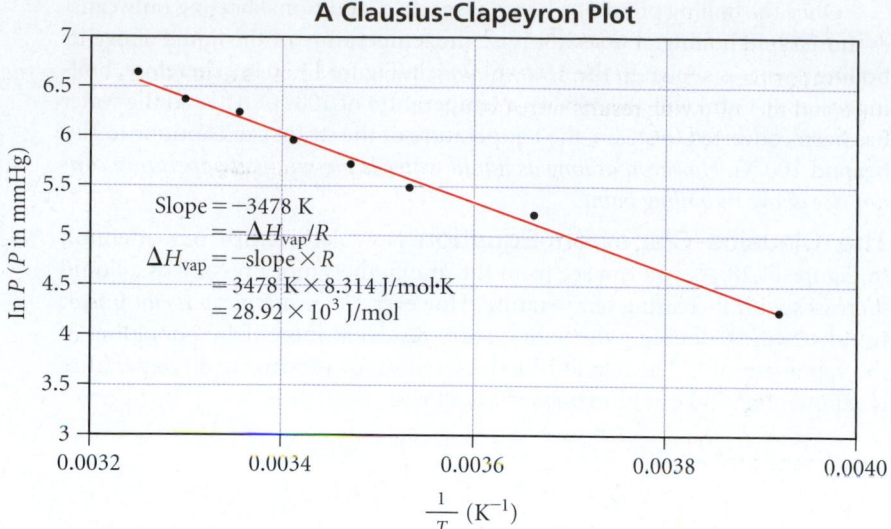

A Clausius–Clapeyron Plot

Slope $= -3478$ K
$= -\Delta H_{vap}/R$
$\Delta H_{vap} = -\text{slope} \times R$
$= 3478$ K $\times 8.314$ J/mol·K
$= 28.92 \times 10^3$ J/mol

▶ **FIGURE 11.31 A Clausius–Clapeyron Plot for Diethyl Ether ($CH_3CH_2OCH_2CH_3$)** A plot of the natural log of the vapor pressure versus the inverse of the temperature in K yields a straight line with slope $-\Delta H_{vap}/R$.

EXAMPLE 11.4 Using the Clausius–Clapeyron Equation to Determine Heat of Vaporization from Experimental Measurements of Vapor Pressure

The vapor pressure of dichloromethane was measured as a function of temperature, and the following results were obtained:

Temperature (K)	Vapor Pressure (torr)
200	0.8
220	4.5
240	21
260	71
280	197
300	391

Determine the heat of vaporization of dichloromethane.

Solution

To find the heat of vaporization, use an Excel spreadsheet or a graphing calculator to make a plot of the natural log of vapor pressure ($\ln P$) as a function of the inverse of the temperature in kelvins ($1/T$). Then fit the points to a line and determine the slope of the line. The slope of the best fitting line is -3805 K. Since the slope equals $-\Delta H_{vap}/R$, we find the heat of vaporization as follows:

$$\text{slope} = -\Delta H_{vap}/R$$

$$\Delta H_{vap} = -\text{slope} \times R$$
$$= -(-3805 \text{ K})(8.314 \text{ J/mol} \cdot \text{K})$$
$$= 3.16 \times 10^4 \text{ J/mol}$$
$$= 31.6 \text{ kJ/mol}$$

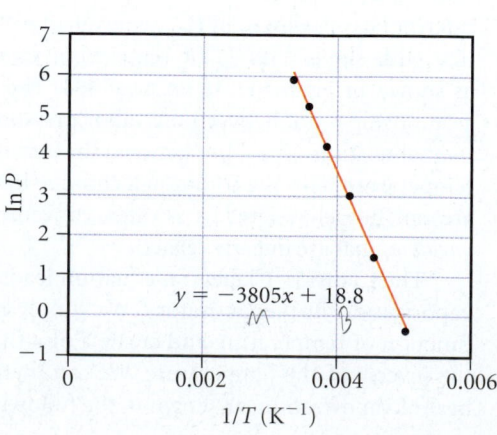

For Practice 11.4

The vapor pressure of carbon tetrachloride was measured as a function of the temperature and the following results were obtained:

Temperature (K)	Vapor Pressure (torr)
255	11.3
265	21.0
275	36.8
285	61.5
295	99.0
300	123.8

Determine the heat of vaporization of carbon tetrachloride.

The Clausius–Clapeyron equation can also be expressed in a two-point form that can
be used with just two measurements of vapor pressure and temperature to determine the
heat of vaporization.

$$\ln\frac{P_2}{P_1} = \frac{-\Delta H_{vap}}{R}\left(\frac{1}{T_2} - \frac{1}{T_1}\right)$$

Clausius–Clapeyron equation
(two-point form)   (Temp in K)

This form of the equation can also be used to predict the vapor pressure of a liquid at
any temperature if you know the enthalpy of vaporization and the normal boiling point
(or the vapor pressure at some other temperature), as shown in the following example.

> The two-point method is generally inferior to plotting multiple points because fewer data points result in greater possible error.

EXAMPLE 11.5 Using the Two-Point Form of the Clausius–Clapeyron Equation to Predict the Vapor Pressure at a Given Temperature

Methanol has a normal boiling point of 64.6 °C and a heat of vaporization (ΔH_{vap}) of
35.2 kJ/mol. What is the vapor pressure of methanol at 12.0 °C?

Sort The problem gives you the normal boiling point of methanol (the temperature at which the vapor pressure is 760 mmHg) and the heat of vaporization. You are asked to find the vapor pressure at a specified temperature which is also given.

Given $T_1(°C) = 64.6 °C$
$P_1 = 760$ torr
$\Delta H_{vap} = 35.2$ kJ/mol
$T_2(°C) = 12.0 °C$

Find P_2

Strategize The conceptual plan is essentially the Clausius–Clapeyron equation, which relates the given and find quantities.

Conceptual Plan

$$\ln\frac{P_2}{P_1} = \frac{-\Delta H_{vap}}{R}\left(\frac{1}{T_2} - \frac{1}{T_1}\right)$$

(Clausius–Clapeyron equation, two-point form)

Solve First, convert T_1 and T_2 from °C to K.

Solution
$T_1(K) = T_1(°C) + 273.15$
$\quad\quad = 64.6 + 273.15$
$\quad\quad = 337.8$ K
$T_2(K) = T_2(°C) + 273.15$
$\quad\quad = 12.0 + 273.15$
$\quad\quad = 285.2$ K

Then, substitute the required values into the Clausius–Clapeyron equation and solve for P_2.

$$\ln \frac{P_2}{P_1} = \frac{-\Delta H_{vap}}{R}\left(\frac{1}{T_2} - \frac{1}{T_1}\right)$$

$$\ln \frac{P_2}{P_1} = \frac{-35.2 \times 10^3 \frac{J}{mol}}{8.314\frac{J}{mol \cdot K}}\left(\frac{1}{285.2\ K} - \frac{1}{337.8\ K}\right)$$

$$= -2.31$$

$$\frac{P_2}{P_1} = e^{-2.31}$$

$$P_2 = P_1(e^{-2.31})$$

$$= 760\ torr(0.0993)$$

$$= 75.4\ torr$$

Check The units of the answer are correct. The magnitude of the answer makes sense because vapor pressure should be significantly lower at the lower temperature.

For Practice 11.5

Propane has a normal boiling point of $-42.0\ °C$ and a heat of vaporization (ΔH_{vap}) of $19.04\ kJ/mol$. What is the vapor pressure of propane at $25.0\ °C$?

The Critical Point: The Transition to an Unusual Phase of Matter

We have considered the vaporization of a liquid in a container open to the atmosphere with and without heating, and the vaporization of a liquid in a *sealed* container without heating. We now examine the vaporization of a liquid in a *sealed* container *while heating*. Consider liquid *n*-pentane in equilibrium with its vapor in a sealed container initially at 25 °C. At this temperature, the vapor pressure of *n*-pentane is 0.67 atm. What happens if the liquid is heated? As the temperature rises, more *n*-pentane vaporizes and the pressure within the container increases. At 100 °C, the pressure is 5.5 atm, and at 190 °C the pressure is 29 atm. As more and more gaseous *n*-pentane is forced into the same amount of space, the density of the *gas* becomes higher and higher. At the same time, the increasing temperature causes the density of the *liquid* to become lower and lower. At 197 °C, the meniscus between the liquid and gaseous *n*-pentane disappears and the gas and liquid phases commingle to form a *supercritical fluid* (Figure 11.32 ▼). For any substance, the *temperature* at which this transition occurs is called the **critical temperature**—it represents the temperature above which the liquid cannot exist (regardless of pressure). The *pressure* at which this transition occurs is called the **critical pressure** (P_c)—it represents the pressure required to bring about a transition to a liquid at the critical temperature.

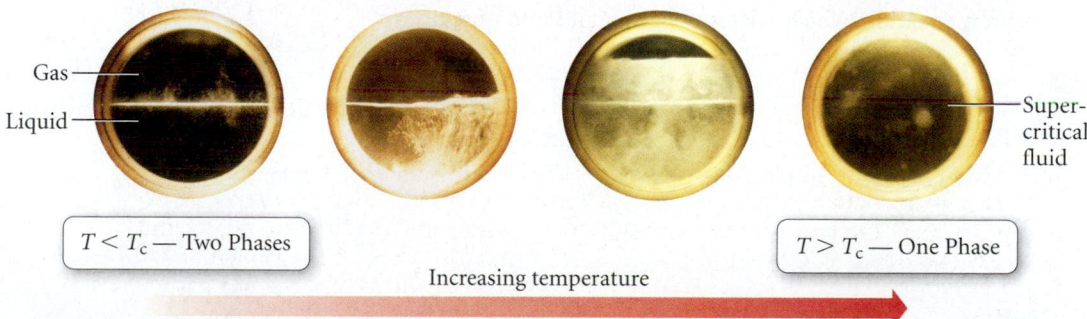

Gas
Liquid
$T < T_c$ — Two Phases
Increasing temperature
Super-critical fluid
$T > T_c$ — One Phase

▲ **FIGURE 11.32 Critical Point Transition** As *n*-pentane is heated in a sealed container, it undergoes a transition to a supercritical fluid. At the critical point, the meniscus separating the liquid and gas disappears, and the fluid becomes supercritical—neither a liquid nor a gas.

Supercritical fluids have drawn the interest of researchers because of their unique properties. A supercritical fluid has properties of both liquids and gases—it is in some sense intermediate between the two. Researchers have found that supercritical fluids can act as good solvents, selectively dissolving a number of compounds. For example, supercritical carbon dioxide is used as a solvent to extract caffeine from coffee beans. The caffeine dissolves in the supercritical carbon dioxide, but other substances—such as those responsible for the flavor of coffee—do not. Consequently, the caffeine can be removed without substantially altering the coffee's flavor. The supercritical carbon dioxide is easily removed from the mixture by simply lowering the pressure—below the critical pressure, the carbon dioxide just evaporates away, leaving no residue.

11.6 Sublimation and Fusion

In Section 11.5, we examined a beaker of liquid water at room temperature from the molecular viewpoint. Now, let's examine a block of ice at $-10\,^{\circ}C$ from the same molecular perspective, paying close attention to two common processes: sublimation and fusion.

Sublimation

Even though a block of ice is solid, the water molecules have thermal energy which causes each one to vibrate about a fixed point. The motion is much less than in a liquid, but significant nonetheless. As in liquids, at any one time some molecules will have more thermal energy than the average and some will have less. The molecules with high enough thermal energy can break free from the ice surface—where, as in liquids, molecules are held less tightly than in the interior due to fewer neighbor–neighbor interactions—and go directly into the gas phase (Figure 11.33 ▶). This process is called **sublimation**, the phase transition from solid to gas. Some of the water molecules in the gas phase (at the low end of the energy distribution curve for the gaseous molecules) can collide with the surface of the ice and be captured by the intermolecular forces with other molecules. This process—the opposite of sublimation—is called **deposition**, the phase transition from gas to solid. In analogy to a liquid, the pressure of a gas in dynamic equilibrium with its solid is the vapor pressure of the solid.

Although both sublimation and deposition are happening on the surface of an ice block open to the atmosphere at $-10\,^{\circ}C$, sublimation is usually happening at a greater rate because most of the newly sublimed molecules escape into the surrounding atmosphere and never come back. The result is a noticeable decrease in the size of the ice block over time (even though the temperature is below the melting point).

If you live in a cold climate, you may have noticed the disappearance of ice and snow from the ground even though the temperature remains below 0 °C. Similarly, ice cubes left in the freezer for a long time slowly shrink, even though the freezer is always below 0 °C. In both cases, the ice is subliming, turning directly into water vapor. Ice also sublimes out of frozen foods. You may have noticed, for example, the gradual growth of ice crystals on the *inside* of airtight plastic food-storage bags in your freezer. The ice crystals are composed of water that has sublimed out of the food and redeposited on the surface of the bag or on the surface of the food. For this reason, food that remains frozen for too long becomes dried out. Such dehydration can be avoided to some degree by freezing foods to colder temperatures, a process called deep-freezing. The colder temperature lowers the vapor pressure of ice and preserves the food longer. Freezer burn on meats is another common manifestation of sublimation. When meat is improperly stored (that is, when its container is not airtight) sublimation continues unabated. The result is the dehydration of the surface of the meat, which becomes discolored and loses flavor and texture.

A substance commonly associated with sublimation is solid carbon dioxide or dry ice, which does not melt under atmospheric pressure no matter what the temperature. However, at $-78\,^{\circ}C$ the CO_2 molecules have enough energy to leave the surface of the dry ice and become gaseous through sublimation.

Fusion

Now let's return to our ice block and examine what happens at the molecular level as we increase its temperature. The increasing thermal energy causes the water molecules to vibrate faster and faster. At the **melting point** (0 °C for water), the molecules have enough

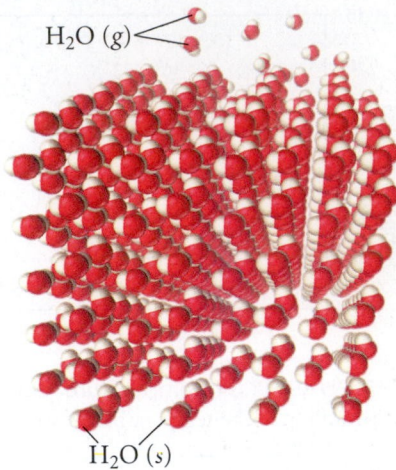

▲ **FIGURE 11.33 The Sublimation of Ice** The water molecules at the surface of an ice cube can sublime directly into the gas phase.

▲ The ice crystals that form on frozen food are due to sublimation of water from the food and redeposition on its surface.

▲ Dry ice (solid CO_2) sublimes but does not melt at atmospheric pressure.

The term fusion is used for melting because if you heat several crystals of a solid, they will *fuse* into a continuous liquid upon melting.

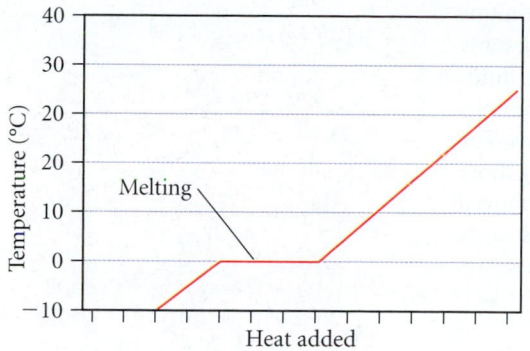

▲ **FIGURE 11.34 Temperature during Melting** The temperature of water during melting remains at 0.0 °C as long as both solid and liquid water remain.

thermal energy to overcome the intermolecular forces that hold them at their stationary points, and the solid turns into a liquid. This process is called **melting** or **fusion**, the phase transition from solid to liquid. The opposite of melting is **freezing**, the phase transition from liquid to solid. Once the melting point of a solid is reached, additional heating only causes more rapid melting; it does not raise the temperature of the solid above its melting point (Figure 11.34 ◄). Only after all of the ice has melted will additional heating raise the temperature of the liquid water past 0 °C. A mixture of water *and* ice will always have a temperature of 0 °C (at 1 atm pressure).

Energetics of Melting and Freezing

The most common way to cool a beverage quickly is to drop several ice cubes into it. As the ice melts, the drink cools because melting is endothermic—the melting ice absorbs heat from the liquid. The amount of heat required to melt 1 mol of a solid is called the **heat of fusion** (ΔH_{fus}). The heat of fusion for water is 6.02 kJ/mol:

$$H_2O(s) \longrightarrow H_2O(l) \qquad \Delta H_{fus} = 6.02 \text{ kJ/mol}$$

The heat of fusion is positive because melting is endothermic.

Freezing, the opposite of melting, is exothermic—heat is released when a liquid freezes into a solid. For example, as water in your freezer turns into ice, it releases heat, which must be removed by the refrigeration system of the freezer. If the refrigeration system did not remove the heat, the water would not completely freeze into ice. The heat released as the water began to freeze would warm the freezer, preventing further freezing. The change in enthalpy for freezing has the same magnitude as the heat of fusion but the opposite sign.

$$H_2O(l) \longrightarrow H_2O(s) \qquad \Delta H = -\Delta H_{fus} = -6.02 \text{ kJ/mol}$$

Different substances have different heats of fusion a shown in Table 11.9.

TABLE 11.9 Heats of Fusion of Several Substances

Liquid	Chemical Formula	Melting Point (°C)	ΔH_{fus}(kJ/mol)
Water	H_2O	0.00	6.02
Rubbing alcohol (isopropyl alcohol)	C_3H_8O	−89.5	5.37
Acetone	C_3H_6O	−94.8	5.69
Diethyl ether	$C_4H_{10}O$	−116.3	7.27

▼ **FIGURE 11.35 Heat of Fusion and Heat of Vaporization**

Typical heats of fusion are significantly less than heats of vaporization.

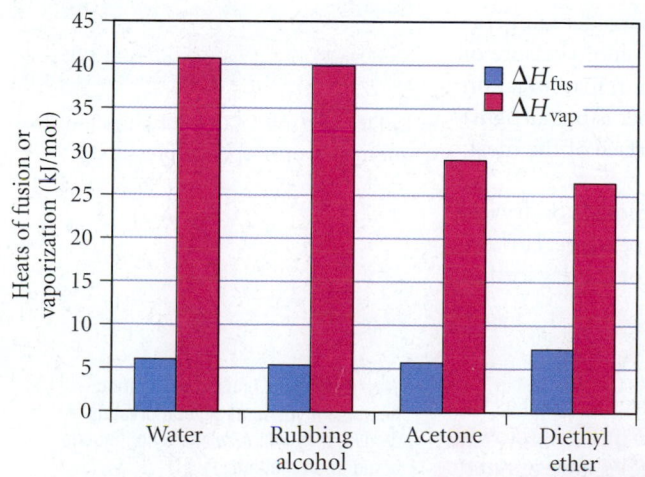

In general, the heat of fusion is significantly less than the heat of vaporization, as demonstrated by Figure 11.35 ◄. We have already mentioned that the solid and liquid phases are closer to each other in many ways than they are to the gas phase. It takes less energy to melt 1 mol of ice into liquid than it does to vaporize 1 mol of liquid water into gas because vaporization requires complete separation of molecules from one another, so the intermolecular forces must be completely overcome. Melting, however, requires that intermolecular forces be only partially overcome, allowing molecules to move around one another while still remaining in contact.

11.7 Heating Curve for Water

We can combine and build on the concepts from the previous two sections by examining the *heating curve* for 1.00 mol of water at 1.00 atm pressure shown in Figure 11.36 ►. The *y*-axis of the heating curve represents the temperature of the water sample.

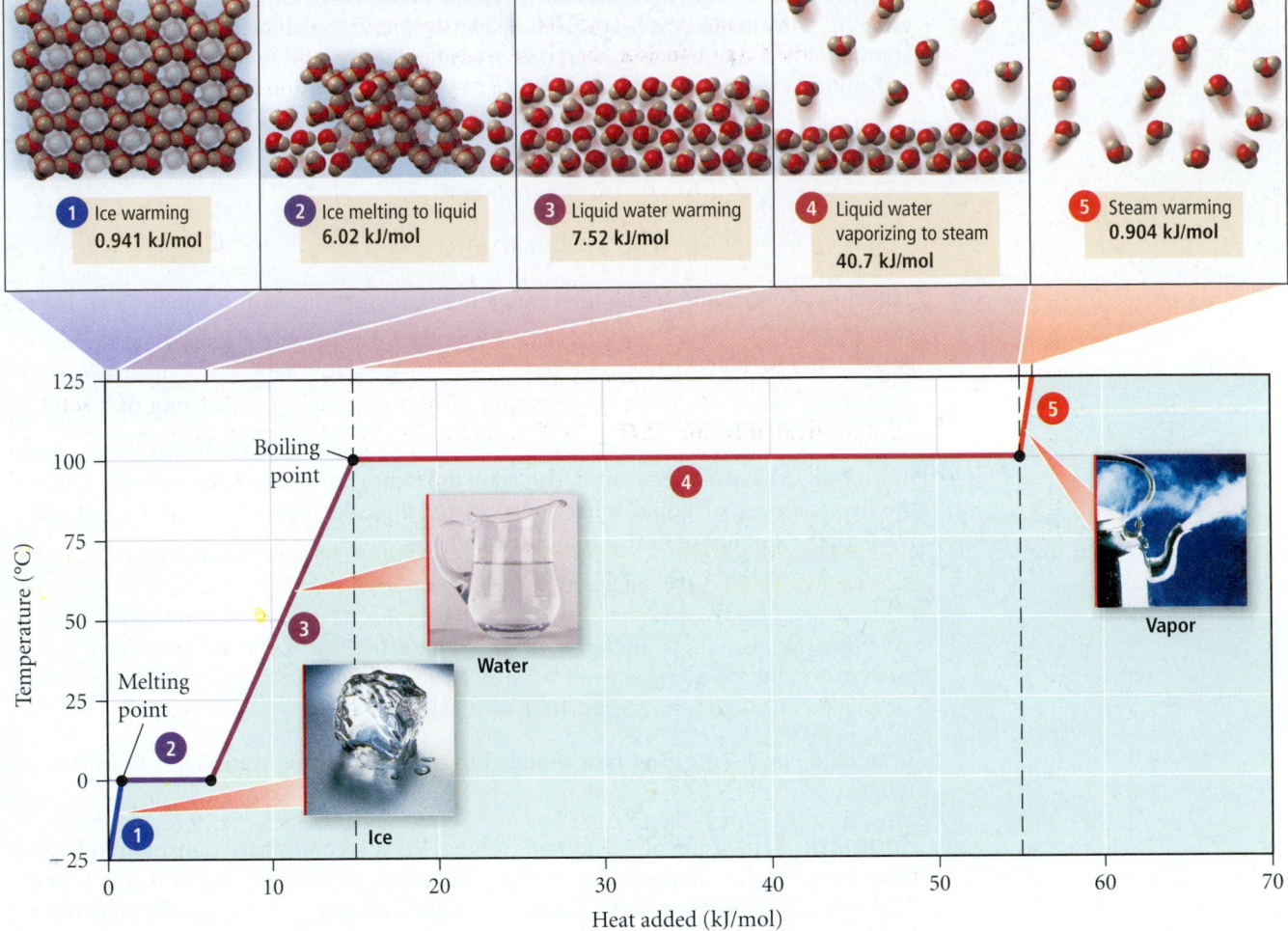

| 1 Ice warming 0.941 kJ/mol | 2 Ice melting to liquid 6.02 kJ/mol | 3 Liquid water warming 7.52 kJ/mol | 4 Liquid water vaporizing to steam 40.7 kJ/mol | 5 Steam warming 0.904 kJ/mol |

▲ **FIGURE 11.36 Heating Curve for Water**

The *x*-axis represents the amount of heat added (in kilojoules) during heating. As you can see from the diagram, the process can be divided into five segments: (1) ice warming; (2) ice melting into liquid water; (3) liquid water warming; (4) liquid water vaporizing into steam; and (5) steam warming. In two of these segments (ice melting and water vaporizing) the temperature is constant as heat is added. These segments represent phase transitions and the added heat goes into producing the phase transition, not increasing the temperature. In the other three segments, temperature increases linearly. These segments represent the heating of a single phase in which the deposited heat raises the temperature according to the heat capacity of the phase (which is represented by the slope of each line). We examine each of these segments individually.

Segment 1 In segment 1, solid ice is warmed from $-25\,°C$ to $0\,°C$. Since no phase transition occurs here, the amount of heat required to heat the solid ice is given by $q = mC_s\,\Delta T$ (see Section 6.3), where C_s is the specific heat capacity of ice ($C_{s,\,ice} = 2.09\ \text{J/g}\cdot°C$). For 1.00 mol of water (18.0 g), the amount of heat is computed as follows:

$$q = mC_{s,\,ice}\,\Delta T$$

$$= 18.0\ \text{g}\left(2.09\frac{\text{J}}{\text{g}\cdot°C}\right)[0.0\,°C - (-25.0\,°C)]$$

$$= 941\ \text{J} = 0.941\ \text{kJ}$$

So in segment 1, 0.941 kJ of heat is added to the ice, warming it from $-25\,°C$ to $0\,°C$.

Segment 2 In segment 2, the added heat does not change the temperature of the ice and water mixture because the heat is absorbed by the phase transition from solid to liquid. The amount of heat required to convert the ice to liquid water is given by $q = n\,\Delta H_{fus}$ where n is the number of moles of water and ΔH_{fus} is the heat of fusion (see Section 11.6).

$$\begin{aligned} q &= n\,\Delta H_{fus} \\ &= 1.00\ \cancel{\text{mol}}\left(\frac{6.02\ \text{kJ}}{\cancel{\text{mol}}}\right) \\ &= 6.02\ \text{kJ} \end{aligned}$$

Therefore, in segment 2, 6.02 kJ is added to the ice, melting it into liquid water. Notice that the temperature does not change during melting. The liquid and solid coexist at 0 °C as the melting occurs.

Segment 3 In segment 3, the liquid water is warmed from 0 °C to 100 °C. Since no phase transition occurs here, the amount of heat required to heat the liquid water is given by $q = mC_s\,\Delta T$, as in segment 1. The main difference, however, is that we must now use the heat capacity of liquid water (not ice) for the calculation. For 1.00 mol of water (18.0 g), the amount of heat is computed as follows:

$$\begin{aligned} q &= mC_{s,\,liq}\,\Delta T \\ &= 18.0\ \cancel{\text{g}}\left(4.18\frac{\text{J}}{\cancel{\text{g}}\cdot\cancel{°\text{C}}}\right)(100.0\ \cancel{°\text{C}} - 0.0\ \cancel{°\text{C}}) \\ &= 7.52 \times 10^{3}\ \text{J} = 7.52\ \text{kJ} \end{aligned}$$

So in segment 3, 7.52 kJ of heat is added to the liquid water, warming it from 0 °C to 100 °C.

Segment 4 In segment 4, the water undergoes a second phase transition, this time from liquid to gas. The amount of heat required to convert the liquid to gas is given by $q = n\,\Delta H_{vap}$, where n is the number of moles and ΔH_{vap} is the heat of vaporization (see Section 11.5).

$$\begin{aligned} q &= n\,\Delta H_{vap} \\ &= 1.00\ \cancel{\text{mol}}\left(\frac{40.7\ \text{kJ}}{\cancel{\text{mol}}}\right) \\ &= 40.7\ \text{kJ} \end{aligned}$$

Therefore, in segment 4, 40.7 kJ is added to the water, vaporizing it into steam. Notice that the temperature does not change during boiling. The liquid and gas coexist at 100 °C as the boiling occurs.

Segment 5 In segment 5, the steam is warmed from 100 °C to 125 °C. Since no phase transition occurs here, the amount of heat required to heat the steam is given by $q = mC_s\,\Delta T$ (as in segments 1 and 3) except that we must use the heat capacity of steam (2.01 J/g · °C).

$$\begin{aligned} q &= mC_{s,\,steam}\,\Delta T \\ &= 18.0\ \cancel{\text{g}}\left(2.01\frac{\text{J}}{\cancel{\text{g}}\cdot\cancel{°\text{C}}}\right)(125.0\ \cancel{°\text{C}} - 100.0\ \cancel{°\text{C}}) \\ &= 904 = 0.904\ \text{kJ} \end{aligned}$$

So in segment 5, 0.904 kJ of heat is added to the steam, warming it from 100 °C to 125 °C.

Some General Observations:

- When heat is added to a substance over a temperature range in which a phase transition does not occur (as in segments 1, 3, and 5), the kinetic energy of the molecules in the sample increases, and the temperature rises in accordance with the substance's heat capacity ($q = mC_s\,\Delta T$).

- When heat is added to a substance over a temperature range in which a phase transition occurs (as in segments 2 and 4), the temperature does *not* change during the phase transition. The two phases are in equilibrium during the phase change and the temperature remains constant. The amount of heat required to achieve the phase change is given by $q = n \, \Delta H$.

 Conceputal Connection 11.3 **Cooling of Water with Ice**

We just saw that the heat capacity of ice is $C_{s, ice} = 2.09 \text{ J/g} \cdot °C$ and that the heat of fusion of ice is 6.02 kJ/mol. When a small ice cube at $-10 °C$ is put into a cup of water at room temperature, which of the following plays a larger role in cooling the liquid water: the warming of the ice from $-10 °C$ to $0 °C$, or the melting of the ice?

Answer: The melting of the ice. The warming of the ice from $-10 °C$ to $0 °C$ absorbs only 20.9 J/g of ice. The melting of the ice, however, absorbs about 334 J/g of ice. (This value is obtained by dividing the heat of fusion by the molar mass of water.) Therefore, the melting of the ice produces a larger temperature decrease in the water than does the warming of the ice.

11.8 Phase Diagrams

Throughout most of this chapter, we have examined how the phase of a substance changes with temperature and pressure. We can combine both the temperature dependence and pressure dependence of the phase of a particular substance in a graph called a *phase diagram*. A **phase diagram** is simply a map of the phase of a substance as a function of pressure (on the *y*-axis) and temperature (on the *x*-axis). We first examine the major features of a phase diagram, then turn to navigating within a phase diagram, and finally examine and compare the phase diagrams of selected substances.

The Major Features of a Phase Diagram

Consider the phase diagram of water as an example to examine the major features within a phase diagram (Figure 11.37 ▼). The *y*-axis displays the pressure in torr and the *x*-axis shows the temperature in degrees Celsius. The main features of the phase diagram can be categorized as regions, lines, and points. We examine each of these individually.

Phase Diagram for Water

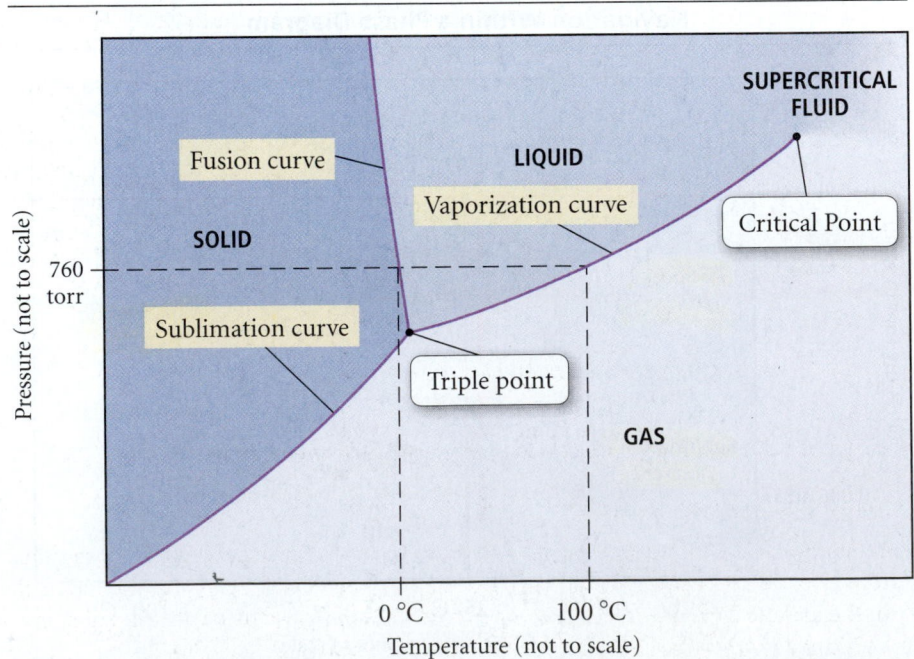

◀ **FIGURE 11.37**
Phase Diagram for Water

Regions *Any of the three main regions—solid, liquid, and gas—in the phase diagram represent conditions where that particular phase is stable.* For example, under any of the temperatures and pressures within the liquid region in the phase diagram of water, the liquid is the stable phase. Notice that the point 25 °C and 760 torr falls within the liquid region, as we know from everyday experience. In general, low temperature and high pressure favor the solid phase; high temperature and low pressure favor the gas phase; and intermediate conditions favor the liquid phase. A sample of matter that is not in the phase indicated by its phase diagram for a given set of conditions will convert to that phase. For example, steam that is cooled to room temperature at 1 atm will condense to liquid.

Lines *Each of the lines (or curves) in the phase diagram represents a set of temperatures and pressures at which the substance is in equilibrium between the two phases on either side of the line.* For example, in the phase diagram for water, consider the curved line beginning just beyond 0 °C separating the liquid from the gas. This line is simply the vapor pressure curve (also called the vaporization curve) for water that we examined in Section 11.5. At any of the temperatures and pressures that fall along this line, the liquid and gas phases of water are equally stable and in equilibrium. For example, at 100 °C and 760 torr pressure, water and its vapor are in equilibrium—they are equally stable and will coexist. The other two major lines in a phase diagram are the sublimation curve (separating the solid and the gas) and the fusion curve (separating the solid and the liquid).

The Triple Point *The point in the phase diagram called the **triple point** represents the unique set of conditions at which three phases are equally stable and in equilibrium.* For example, in the phase diagram for water, the triple point occurs at 0.0098 °C and 4.58 torr. Under these unique conditions (and only under these conditions), the solid, liquid, and gas phases of water are equally stable and will coexist in equilibrium.

The Critical Point *The point in the phase diagram called the **critical point** represents the temperature and pressure above which a supercritical fluid exists.* As we learned in Section 11.5, at the critical temperature and pressure, the liquid and gas phases coalesce into a *supercritical fluid.*

Navigation within a Phase Diagram

Changes in the temperature or pressure of a sample of water can be represented by movement within the phase diagram. For example, suppose you heat a block of ice initially at 1.0 atm and −25 °C. The change in temperature at constant pressure can be represented as movement along the line marked *A* in Figure 11.38 ▼. As the temperature rises, we

The triple point of a substance such as water can be reproduced anywhere to calibrate a thermometer or pressure gauge with a known temperature and pressure.

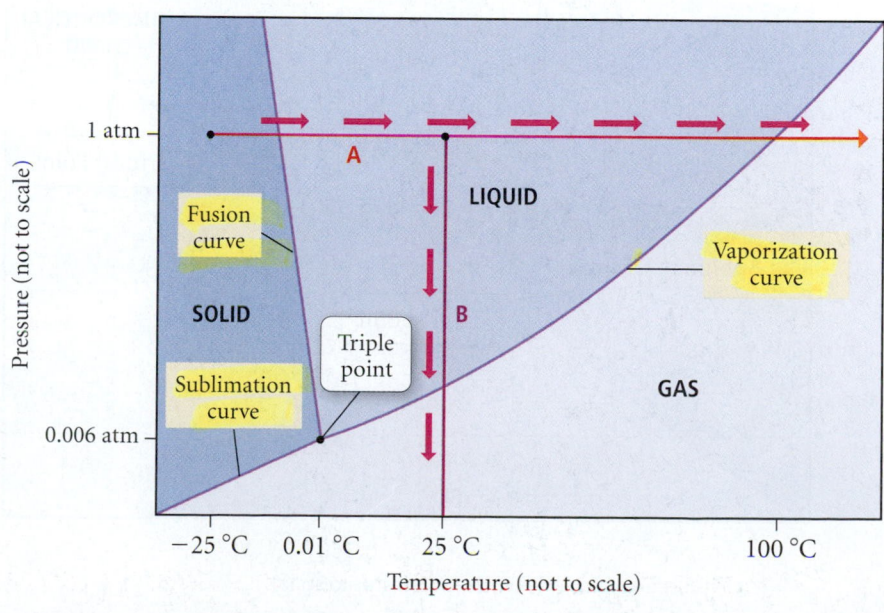

Navigation Within a Phase Diagram

▶ **FIGURE 11.38**
Navigation on the Phase Diagram for Water

move to the right along the line. At the fusion curve, the temperature stops rising and melting occurs until the solid ice is completely converted to liquid water. Notice that crossing the fusion curve requires the complete transition from solid to liquid. Once the ice has completely melted, the temperature of the liquid water can begin to rise until the vaporization curve is reached. The temperature again stops rising and boiling occurs until all the liquid is converted to gas.

A change in pressure can be represented by a vertical line on the phase diagram. For example, suppose you lower the pressure above a sample of water initially at 1.0 atm and 25 °C. The change in pressure at constant temperature can be represented as movement along the line marked *B* in Figure 11.38. As the pressure drops, we move down the line and approach the vaporization curve. At the vaporization curve, the pressure stops dropping and vaporization occurs until the liquid is completely converted to vapor. Notice that crossing the vaporization curve requires the complete transition from liquid to gas. Only after the liquid has all vaporized can the pressure continue to drop.

The Phase Diagrams of Other Substances

Examine the phase diagrams of iodine and carbon dioxide, shown in Figures 11.39(a) ▼ and 11.39(b), respectively. The phase diagrams are similar to that of water in most of their general features. However, a couple of significant differences are notable.

The fusion curves for both carbon dioxide and iodine have a positive slope—as the temperature increases the pressure also increases—in contrast to the fusion curve for water, which has a negative slope. The behavior of water is atypical. The fusion curve within the phase diagrams for most substances has a positive slope because increasing pressure favors the denser phase, which for most substances is the solid phase. For example, suppose the pressure on a sample of iodine is increased from 1 atm to 100 atm at 184 °C, as shown by line A in Figure 11.39(a). Notice that this change crosses the fusion curve, converting the liquid into a solid. In contrast, a pressure increase from 1 atm to 100 atm at −0.1 °C in water causes a phase transition from solid to liquid. Unlike most substances, the liquid phase of water is actually denser than the solid phase, resulting in the atypical behavior.

Both water and iodine have stable solid, liquid, and gaseous phases at a pressure of 1 atm. However, notice that carbon dioxide has no stable liquid phase at a pressure of 1 atm. If we increase the temperature of a block of solid carbon dioxide (dry ice) at 1 atm, as indicated by line B in Figure 11.39(b), we cross the sublimation curve at−78.5 °C. At this temperature, the solid sublimes to a gas, as you have probably observed with dry ice. Carbon dioxide will form a liquid only above pressures of 5.1 atm.

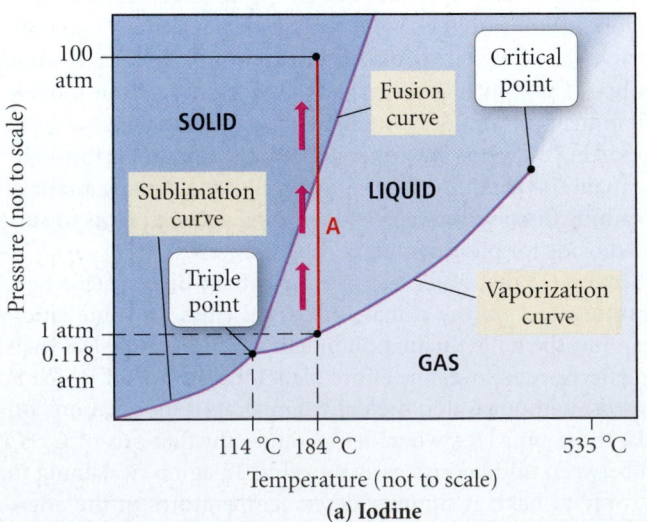

(a) Iodine

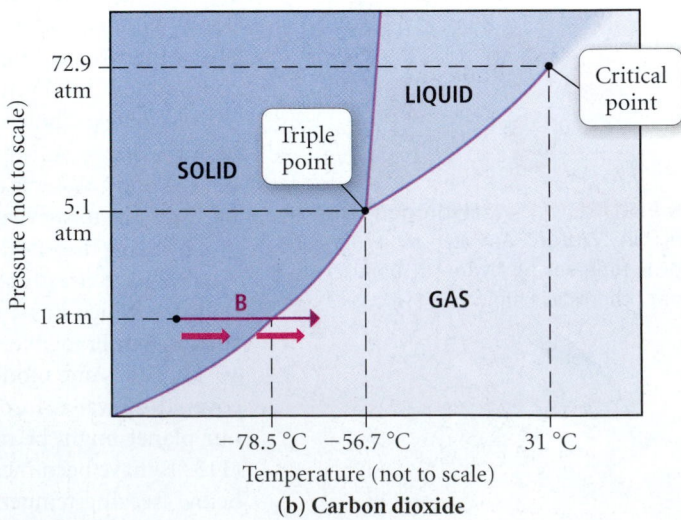

(b) Carbon dioxide

▲ FIGURE 11.39 **Phase Diagrams for Other Substances**   (a) Iodine, (b) Carbon dioxide.

11.9 Water: An Extraordinary Substance

▲ The *Mars Rover* found evidence suggesting that water could have existed on Mars in the past.

Water is easily the most common and important liquid on Earth. It fills our oceans, lakes, and streams. In its solid form, it caps our mountains, and in its gaseous form, it humidifies our air. We drink water, we sweat water, and we excrete bodily wastes dissolved in water. Indeed, the majority of our body mass *is* water. Life is impossible without water, and in most places on Earth where liquid water exists, life exists. Recent evidence for water on Mars in the past has fueled hopes of finding life or evidence of past life there. And though it may not be obvious to us (because we take water for granted), this familiar substance turns out to have many remarkable properties.

Among liquids, water is unique. It has a low molar mass (18.02 g/mol), yet it is a liquid at room temperature. Other main-group hydrides have higher molar masses but lower boiling points, as shown in Figure 11.40 ▼. No other substance of similar molar mass (except for HF) comes close to being a liquid at room temperature. Water's high boiling point for its molar mass can be understood by examining the structure of the water molecule. The bent geometry of the water molecule and the highly polar nature of the O—H bonds result in a molecule with a significant dipole moment. Water's two O—H bonds

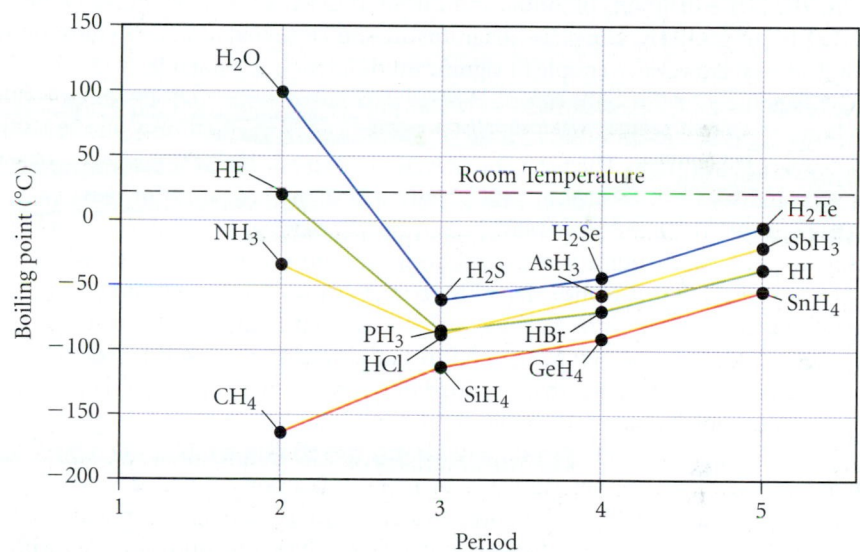

▲ FIGURE 11.40 **Boiling Points of Main Group Hydrides**   Water is the only common main-group hydride that is a liquid at room temperature.

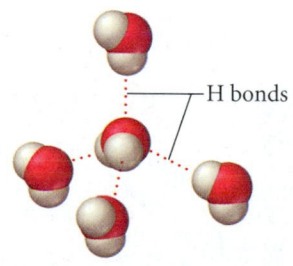

▲ FIGURE 11.41 **Hydrogen Bonding in Water**   A water molecule can form four strong hydrogen bonds with four other water molecules

(hydrogen directly bonded to oxygen) allow a water molecule to form strong hydrogen bonds with four other water molecules (Figure 11.41 ◀), resulting in a relatively high boiling point. Water's high polarity also allows it to dissolve many other polar and ionic compounds, and even a number of nonpolar gases such as oxygen and carbon dioxide (by inducing a dipole moment in their molecules). Consequently, water is the main solvent within living organisms, transporting nutrients and other important compounds throughout the body. Water is also the main solvent of the environment, allowing aquatic animals, for example, to survive by breathing dissolved oxygen and allowing aquatic plants to survive by using dissolved carbon dioxide for photosynthesis.

We have already seen in Section 6.3 that water has an exceptionally high specific heat capacity, which has a moderating effect on the climate of coastal cities. In some cities, such as San Francisco, for example, the daily fluctuation in temperature can be less than 10 °C. This same moderating effect occurs over the entire planet, two-thirds of which is covered by water. In other words, without water, the daily temperature fluctuations on our planet might be more like those on Mars, where temperature fluctuations of 63 °C (113 °F) have been measured between midday and early morning. Imagine awakening to below freezing temperatures, only to bake at summer desert temperatures in the afternoon! The presence of water on Earth, however, and its uniquely high specific heat capacity is largely responsible for our much smaller daily fluctuations.

Chemistry in the Environment
Water Pollution

Water quality is critical to human health. Many human diseases—especially in developing nations—are caused by poor water quality. Several kinds of pollutants, including biological and chemical contaminants, can get into water supplies.

▲ Fresh, sanitary water supplies are critical to human health.

Biological contaminants are microorganisms that cause diseases such as hepatitis, cholera, dysentery, and typhoid. They get into drinking water primarily from the dumping of human or animal waste into bodies of water. Drinking water in developed nations is usually chemically treated to kill microorganisms.

Water containing biological contaminants poses an immediate danger to human health and should not be consumed. Most biological contaminants can be eliminated from untreated water by boiling.

Chemical contaminants get into drinking water from sources such as industrial dumping, pesticide and fertilizer use, and household dumping. These contaminants include organic compounds, such as carbon tetrachloride and dioxin, and inorganic elements and compounds, such as mercury, lead, and nitrates. Since many chemical contaminants are neither volatile nor alive (like biological contaminants), they are usually *not* eliminated through boiling.

The U.S. Environmental Protection Agency (EPA), under the Safe Drinking Water Act of 1974 and its amendments, sets standards that specify the maximum contamination level (MCL) for nearly 100 biological and chemical contaminants in water. Water providers that serve more than 25 people must periodically test the water they deliver to their consumers for these contaminants. If levels exceed the standards set by the EPA, the water provider must notify the consumer and take appropriate measures to remove the contaminant from the water. According to the EPA, if water comes from a provider that serves more than 25 people, it should be safe to consume over a lifetime. If it is not safe to drink for a short period of time, you will be notified.

Question

Suppose a sample of water is contaminated by a nonvolatile contaminant such as lead. Why does boiling not eliminate the contaminant?

As we have seen, the way water freezes is also unique. Unlike other substances, which contract upon freezing, water expands upon freezing. Consequently, ice is less dense than liquid water, and it floats. This seemingly trivial property has significant consequences. The frozen layer of ice at the surface of a winter lake insulates the water in the lake from further freezing. If this ice layer sank, it would kill bottom-dwelling aquatic life and possibly allow the lake to freeze solid, eliminating virtually all life in the lake.

The expansion of water upon freezing, however, is one reason that most organisms do not survive freezing. When the water within a cell freezes, it expands and often ruptures the cell, just as water freezing within a pipe bursts the pipe. Many foods, especially those with high water content, do not survive freezing very well either. Have you ever tried, for example, to freeze your own vegetables? If you put lettuce or spinach in the freezer, it will be limp and damaged when you defrost it. The frozen-food industry gets around this problem by *flash freezing* vegetables and other foods. In this process, foods are frozen nearly instantaneously, not allowing water molecules to settle into their preferred crystalline structure. Consequently, the water does not expand very much and the food remains largely undamaged

▲ When lettuce freezes, the water within its cells expands, rupturing them.

11.10 Crystalline Solids: Determining Their Structure by X-Ray Crystallography

We have seen that crystalline solids are composed of atoms or molecules arranged in structures with long-range order (see Section 11.2). If you have ever visited the mineral section of a natural history museum and seen crystals with smooth faces and well-defined angles between them, or if you have carefully observed the hexagonal shapes of snowflakes, you have witnessed some of the effects of the underlying order in crystalline solids. The often beautiful geometric shapes that you see on the macroscopic scale are the result of specific

▲ The well-defined angles and smooth faces of crystalline solids reflect the underlying order of the atoms composing them.

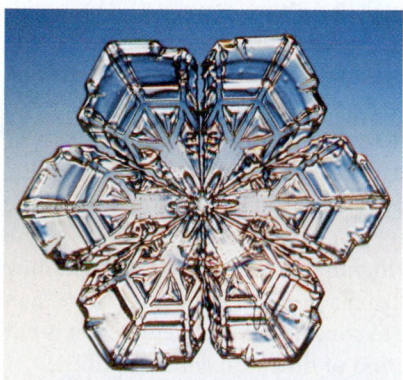

▲ The hexagonal shape of a snowflake derives from the hexagonal arrangement of water molecules in crystalline ice.

structural patterns on the molecular and atomic scales. But how do we study these patterns? How do we look into the atomic and molecular world to determine the arrangement of the atoms and measure the distances between them? In this section, we examine **X-ray diffraction**, a powerful laboratory technique that enables us to do exactly that.

Recall from Section 7.2 that electromagnetic waves (or light) interact with each other in a characteristic way called *interference*: they can cancel each other out or reinforce each other, depending on the alignment of their crests and troughs. *Constructive interference* occurs when two waves of identical wavelength interact with crests and troughs in alignment. *Destructive interference* occurs when two waves of identical wavelength interact in such a way that the crests of one align with the troughs of the other. Recall also from Section 7.2

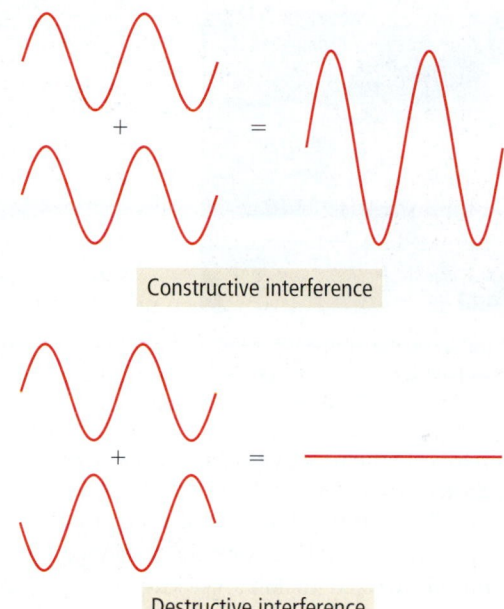

Constructive interference

Destructive interference

that when light encounters two slits separated by a distance comparable to the wavelength of the light, constructive and destructive interference between the resulting beams produces a characteristic *interference pattern*, consisting of alternating bright and dark lines.

Since atoms within crystal structures have spacings between them on the order of 10^2 pm, light of similar wavelength (which happens to fall in the X-ray region of the electromagnetic spectrum) forms interference patterns or *diffraction patterns* when it interacts with those atoms. The exact pattern of diffraction reveals the spacings between planes of atoms. Consider two planes of atoms within a crystalline lattice separated by a distance d, as shown in Figure 11.42 ◄. If two rays of light with wavelength λ that are initially in phase (that is, the crests of one wave are aligned with the crests of the other) diffract from the two layers, the diffracted rays may interfere with each other con-

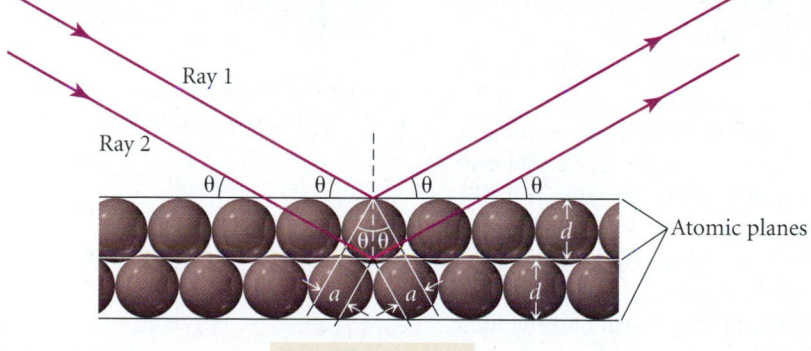

Path difference = 2a

▲ **FIGURE 11.42 Diffraction from a Crystal** When X-rays strike parallel planes of atoms in a crystal, constructive interference occurs if the difference in path length between beams reflected from adjacent planes is an integral number of wavelengths.

structively or destructively, depending on the difference between the path lengths traveled by each ray. If the difference between the two path lengths ($2a$) is an integral number (n) of wavelengths, then the interference will be constructive.

$$n\lambda = 2a \qquad \text{(criterion for constructive interference)} \qquad [11.5]$$

Using trigonometry, we can see that the angle of reflection (θ) is related to the distance a and the separation between layers (d) by the following relation:

$$\sin \theta = \frac{a}{d} \qquad [11.6]$$

Rearranging, we get:

$$a = d \sin \theta \qquad [11.7]$$

By substituting Equation 11.7 into Equation 11.5, we get the following important relationship:

$$n\lambda = 2d \sin \theta \qquad \text{Bragg's Law}$$

This equation is known as *Bragg's law*. For a given wavelength of light incident on atoms arranged in layers, we can then measure the angle that produces constructive interference (which appears as a bright spot on the X-ray diffraction pattern) and then compute *d*, the distance between the atomic layers.

$$d = \frac{n\lambda}{2 \sin \theta} \qquad [11.8]$$

In a modern X-ray diffractometer (Figure 11.43 ▶), the diffraction pattern from a crystal is collected and analyzed by a computer. By rotating the crystal and collecting the resulting diffraction patterns at different angles, the distances between various crystalline planes can be measured, eventually yielding the entire crystalline structure. This process is called X-ray crystallography. X-ray crystallography is used to determine not only the structures of simple atomic lattices, but also the structures of proteins, DNA, and other biologically important molecules. For example, the famous X-ray diffraction photograph shown below, obtained by Rosalind Franklin and Maurice Wilkins, helped Watson and Crick determine the double-helical structure of DNA. As we learned in Section 9.1, X-ray diffraction was also used to determine the structure of HIV protease, a protein critical to the reproduction of HIV and the development of AIDS. That structure was then used to design drug molecules that would inhibit the action of HIV protease, thus halting the advance of the disease.

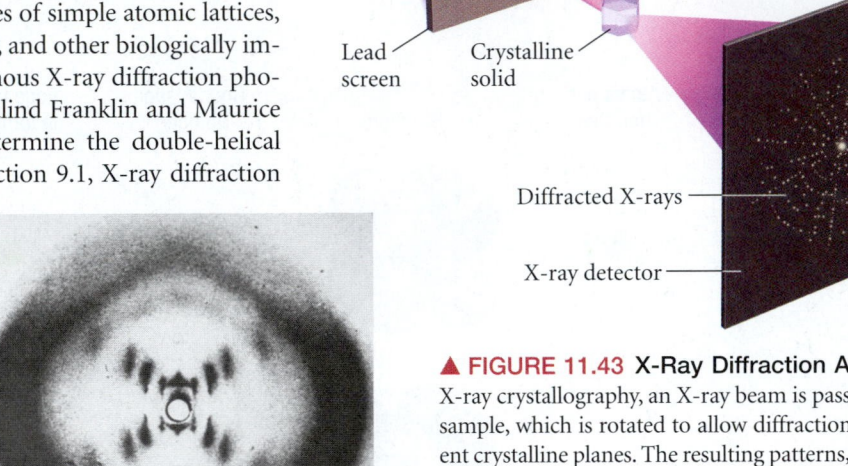

▲ **FIGURE 11.43 X-Ray Diffraction Analysis** In X-ray crystallography, an X-ray beam is passed through a sample, which is rotated to allow diffraction from different crystalline planes. The resulting patterns, representing constructive interference from various planes, are then analyzed to determine the crystalline structure.

EXAMPLE 11.6   Using Bragg's Law

When an X-ray beam of $\lambda = 154$ pm was incident on the surface of an iron crystal, it produced a maximum reflection at an angle of $\theta = 32.6°$. Assuming $n = 1$, calculate the separation between layers of iron atoms in the crystal.

Solution

To solve this problem, simply use Bragg's law in the form given by Equation 11.8. The distance, *d*, is the separation between layers in the crystal.

$$d = \frac{n\lambda}{2 \sin \theta}$$
$$= \frac{154 \text{ pm}}{2 \sin(32.6°)}$$
$$= 143 \text{ pm}$$

For Practice 11.6

The spacing between layers of molybdenum atoms is 157 pm. Compute the angle at which 154-pm X-rays would produce a maximum reflection for $n = 1$.

11.11 Crystalline Solids: Unit Cells and Basic Structures

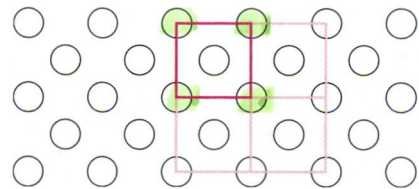

X-Ray crystallography allows us to determine the regular arrangements of atoms within a crystalline solid. This arrangement is called the **crystalline lattice**. The crystalline lattice of any solid is nature's solution to the problem of aggregating the particles that compose the lattice in a way that minimizes their energy. The crystalline lattice can be represented by a small collection of atoms, ions, or molecules called the **unit cell**. When the unit cell is repeated over and over—like the tiles of a floor or the pattern in a wallpaper design, but in three dimensions—the entire lattice can be reproduced. For example, consider the two-dimensional crystalline lattice shown at left. The unit cell for this lattice is the dark-colored square. Each circle represents a *lattice point,* a point in space occupied by an atom, ion, or molecule. Notice that repeating and moving the square throughout the two-dimensional space can generate the entire lattice.

Unit cells are often classified by their symmetry, and many different unit cells exist. In this book, we will focus primarily on *cubic unit cells* (although we will look at one hexagonal unit cell). Cubic unit cells are characterized by equal edge lengths and 90° angles at their corners. The three cubic unit cells—simple cubic, body-centered cubic, and face-centered cubic—along with some of their basic characteristics, are shown in Figure 11.44 ▼.

Cubic Cell Name	Atoms per Unit Cell	Structure	Coordination Number	Edge Length in terms of r	Packing Efficiency (fraction of volume occupied)
Simple Cubic	1		6	$2r$	52%
Body-centered Cubic	2		8	$\dfrac{4r}{\sqrt{3}}$	68%
Face-centered Cubic	4		12	$2\sqrt{2}r$	74%

▲ **FIGURE 11.44 The Cubic Crystalline Lattices**

The **simple cubic** unit cell (Figure 11.45 on page 499) consists of a cube with one atom at each corner. The atoms touch along each edge of the cube, so the edge length is simply twice the radius of the atoms ($l = 2r$). Note also that even though the unit cell may seem as though it contains eight atoms, it actually contains only one. Each corner atom is shared by eight other unit cells. In other words, any one unit cell actually contains only one-eighth of each of the eight atoms at its corners, for a total of only one atom per unit cell.

Simple Cubic Unit Cell

Coordination number = 6

Atoms per unit cell =
$$\frac{1}{8} \times 8 = 1$$

$\frac{1}{8}$ atom at each of 8 corners

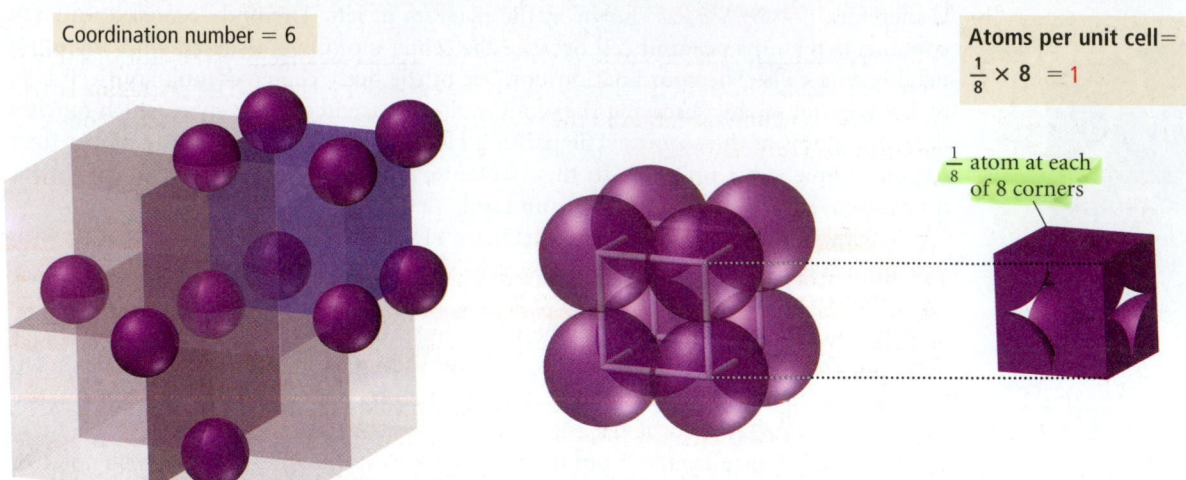

▲ **FIGURE 11.45 Simple Cubic Crystal Structure**

An important feature of any unit cell is the **coordination number**, the number of atoms with which each atom is in *direct contact*. The coordination number represents the number of atoms with which a particular atom can have a strong interaction. The coordination number for the simple cubic unit cell is 6, because any one atom touches only six others, as you can see in Figure 11.45. A quantity closely related to the coordination number is the **packing efficiency**, the percentage of the volume of the unit cell occupied by the spheres. The higher the coordination number, the greater the packing efficiency. The simple cubic unit cell has a packing efficiency of 52%—there is a lot of empty space in the simple cubic unit cell.

The **body-centered cubic** unit cell (Figure 11.46 ▼) consists of a cube with one atom at each corner and one atom of the same kind in the very center of the cube. Note that in the body-centered unit cell, the atoms *do not* touch along each edge of the cube,

Simple cubic

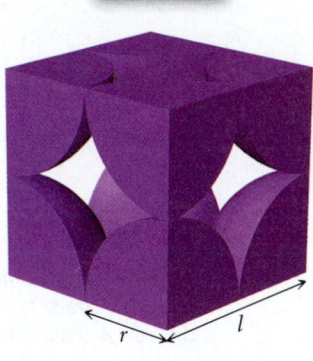

$l = 2r$

▲ In the simple cubic lattice, the atoms touch along each edge so that the edge length is 2r.

Body-Centered Cubic Unit Cell

Coordination number = 8

Atoms per unit cell =
$$\left(\frac{1}{8} \times 8\right) + 1 = 2$$

$\frac{1}{8}$ atom at each of 8 corners

1 atom at center

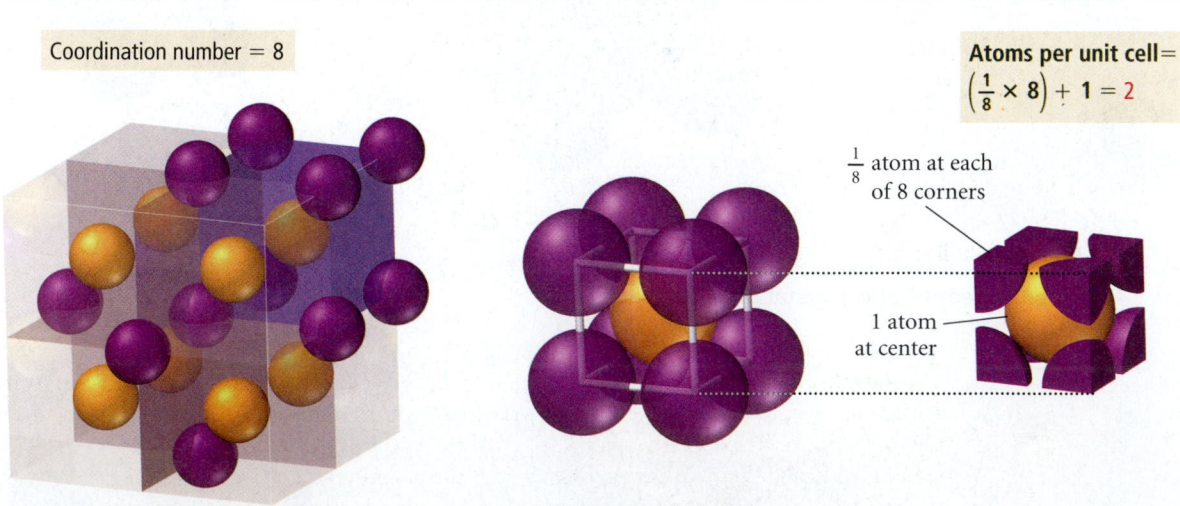

▲ **FIGURE 11.46 Body-Centered Cubic Crystal Structure**

Body-centered cubic

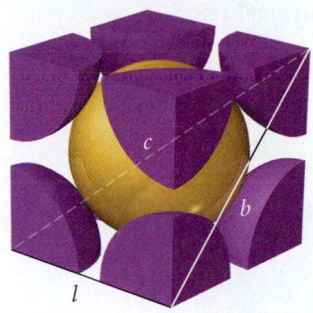

$$c^2 = b^2 + l^2 \qquad b^2 = l^2 + l^2$$
$$c = 4r \qquad\qquad b^2 = 2l^2$$
$$(4r)^2 = 2l^2 + l^2$$
$$(4r)^2 = 3l^2$$
$$l^2 = \frac{(4r)^2}{3}$$
$$l = \frac{4r}{\sqrt{3}}$$

▲ In the body-centered cubic lattice, the atoms touch only along the cube diagonal. The edge length is $4r/\sqrt{3}$.

but rather touch along the diagonal line that runs from one corner, through the middle of the cube, to the opposite corner. The edge length in terms of the atomic radius is therefore $l = 4r/\sqrt{3}$, as shown in the diagram at left. The body-centered unit cell contains two atoms per unit cell because the center atom is not shared with any other neighboring cells. The coordination number of the body-centered cubic unit cell is 8, which you can see by observing the atom in the very center of the cube, which touches the eight atoms at the corners. The packing efficiency is 68%, significantly higher than for the simple cubic unit cell. In this structure, any one atom strongly interacts with more atoms than in the simple cubic unit cell.

The **face-centered cubic** unit cell (Figure 11.47 ▼) is characterized by a cube with one atom at each corner and one atom of the same kind in the center of each cube face. Note that in the face-centered unit cell (like the body-centered unit cell), the atoms *do not* touch along each edge of the cube. Instead, the atoms touch *along the face diagonal*. The edge length in terms of the atomic radius is therefore $l = 2\sqrt{2}r$, as shown in the figure below. The face-centered unit cell contains four atoms per unit cell because the center atoms on each of the six faces are shared between two unit cells. So there are $1/2 \times 6 = 3$ face-centered atoms plus $1/8 \times 8 = 1$ corner atoms, for a total of four atoms per unit cell. The coordination number of the face-centered cubic unit cell is 12 and its packing efficiency is 74%. In this structure, any one atom strongly interacts with more atoms than in either the simple cubic unit cell or the body-centered cubic unit cell.

Face-Centered Cubic Unit Cell

Face-centered cubic:
extended structure
Coordination number = 12

Face-centered cubic: unit cell
Atoms/unit $= \left(\frac{1}{8} \times 8\right) + \left(\frac{1}{2} \times 6\right) = $ 4

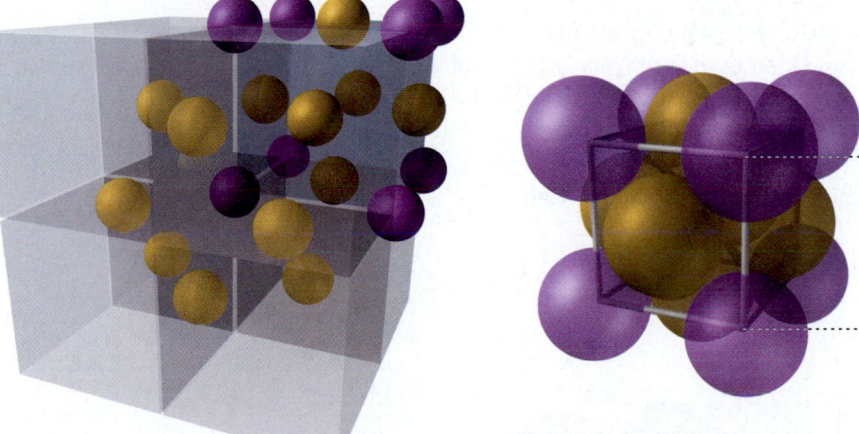

$\frac{1}{8}$ atom at 8 corners

$\frac{1}{2}$ atom at 6 faces

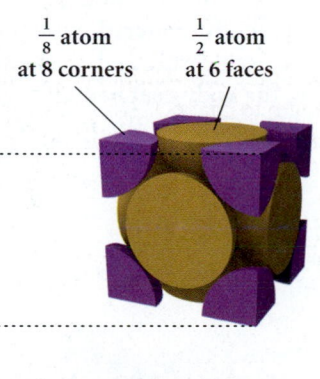

▲ FIGURE 11.47 Face-Centered Cubic Crystal Structure

▶ In the face-centered cubic lattice, the atoms touch along a face diagonal. The edge length is $2\sqrt{2}r$.

$$b^2 = l^2 + l^2 = 2l^2$$
$$b = 4r$$
$$(4r)^2 = 2l^2$$
$$l^2 = \frac{(4r)^2}{2}$$
$$l = \frac{4r}{\sqrt{2}}$$
$$= 2\sqrt{2}r$$

Face-centered cubic

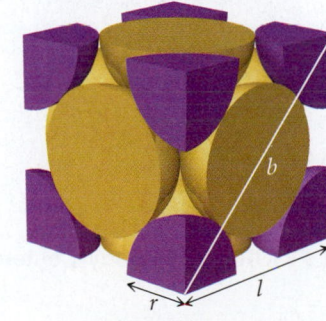

EXAMPLE 11.7   Relating Density to Crystal Structure

Aluminum crystallizes with a face-centered cubic unit cell. The radius of an aluminum atom is 143 pm. Calculate the density of solid crystalline aluminum in g/cm^3.

Sort You are given the radius of an aluminum atom and its crystal structure. You are asked to find the density of solid aluminum.	**Given:** $r = 143$ pm, face-centered cubic **Find:** d

Strategize The conceptual plan is based on the definition of density.
 	
Since the unit cell has the physical properties of the entire crystal, find the mass and volume of the unit cell and use these to calculate its density.
	Conceptual Plan
	$d = m/V$
	
	m = mass of unit cell
	  = number of atoms in unit cell $\times$ mass of each atom
	
	V = volume of unit cell
	  = (edge length)^3

Solve Begin by finding the mass of the unit cell. Obtain the mass of an aluminum atom from its molar mass. Since the face-centered cubic unit cell contains four atoms per unit cell, multiply the mass of aluminum by 4 to get the mass of a unit cell.	**Solution** $m(\text{Al atom}) = 26.98 \dfrac{g}{mol} \times \dfrac{1\,mol}{6.022 \times 10^{23}\ \text{atoms}}$ $\qquad\qquad = 4.481 \times 10^{-23}\ g$ $m(\text{unit cell}) = 4(4.481 \times 10^{-23}\ g)$ $\qquad\qquad = 1.792 \times 10^{-22}\ g$
Next, compute the edge length (l) of the unit cell (in m) from the atomic radius of aluminum. For the face-centered cubic structure, $l = 2\sqrt{2}r$.	$l = 2\sqrt{2}r$ $\quad = 2\sqrt{2}(143\ \text{pm})$ $\quad = 2\sqrt{2}\,(143 \times 10^{-12}\ m)$ $\quad = 4.045 \times 10^{-10}\ m$
Compute the volume of the unit cell (in cm) by converting the edge length to cm and cubing the edge length. (We use centimeters because we want to report the density in units of g/cm^3.)	$V = l^3$ $\quad = \left(4.045 \times 10^{-10}\,m \times \dfrac{1\ cm}{10^{-2}\ m}\right)^3$ $\quad = 6.618 \times 10^{-23}\ cm^3$
Finally, compute the density by dividing the mass of the unit cell by the volume of the unit cell.	$d = \dfrac{m}{V} = \dfrac{1.792 \times 10^{-22}\ g}{6.618 \times 10^{-23}\ cm^3}$ $\quad = 2.71\ g/cm^3$

Check The units of the answer are correct. The magnitude of the answer is reasonable because the density is greater than 1 g/cm^3 (as we would expect for metals), but still not too high (because aluminum is a low-density metal).

For Practice 11.7

Chromium crystallizes with a body-centered cubic unit cell. The radius of a chromium atom is 125 pm. Calculate the density of solid crystalline chromium in g/cm^3.

Closest-Packed Structures

Another way to envision crystal structures, especially useful in metals where bonds are not usually directional, is to think of the atoms as stacking in layers, much as fruit is stacked at the grocery store. For example, the simple cubic structure can be envisioned as one layer of atoms arranged in a square pattern with the next layer stacking directly over the first, so that the atoms in one layer align exactly on top of the atoms in the layer beneath it, as shown here.

As we saw previously, this crystal structure has a great deal of empty space—only 52% of the volume is occupied by the spheres, and the coordination number is 6.

More space-efficient packing can be achieved by aligning neighboring rows of atoms within a layer not in a square pattern, but in a pattern with one row offset from the next by one-half a sphere, as shown here.

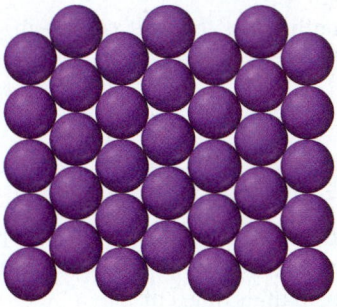

In this way, the atoms pack more closely to each other in any one layer. We can further increase the packing efficiency by placing the next layer *not directly on top of the first*, but again offset so that any one atom actually sits in the indentation formed by three atoms in the layer beneath it, as shown here.

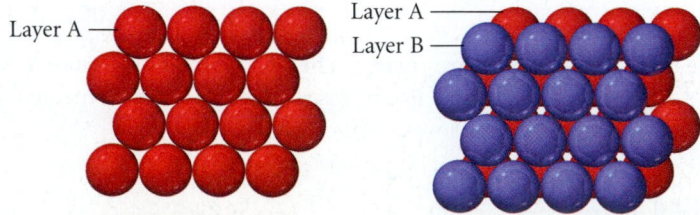

This kind of packing leads to two different crystal structures called *closest-packed structures,* both of which have packing efficiencies of 74% and coordination numbers of 12. In the first of these two closest-packed structures—called **hexagonal closest packing**—the third layer of atoms aligns exactly on top of the first, as shown as follow.

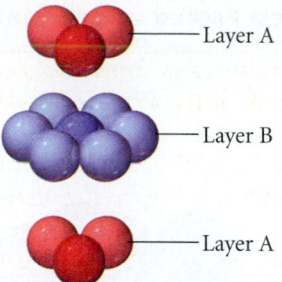

The pattern from one layer to the next is ABAB... with alternating layers aligning exactly on top of one another. Notice that the central atom in layer B of this structure is touching 6 atoms in its own layer, 3 atoms in the layer above it, and 3 atoms in the layer below, for a coordination number of 12. The unit cell for this crystal structure is not a cubic unit cell, but a hexagonal one, as shown in Figure 11.48 ▼.

Hexagonal Closest Packing

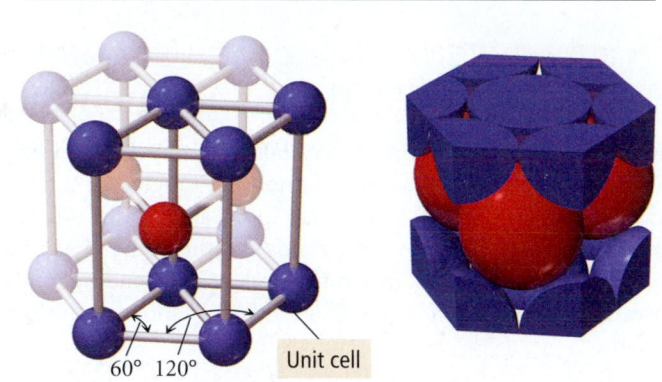

▲ **FIGURE 11.48 Hexagonal Closest-Packing Crystal Structure**
The unit cell is outlined in bold.

In the second of the two closest-packed structures—called **cubic closest packing**—the third layer of atoms is offset from the first, as shown here.

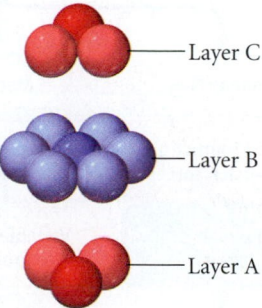

The pattern from one layer to the next is ABCABC... with every fourth layer aligning with the first. Although not simple to visualize, the unit cell for cubic closest packing is the face-centered cubic unit cell, as shown in Figure 11.49 on page 504. Therefore the cubic closest-packed structure is identical to the face-centered cubic unit cell structure.

Cubic Closest Packed = Face-Centered Cubic

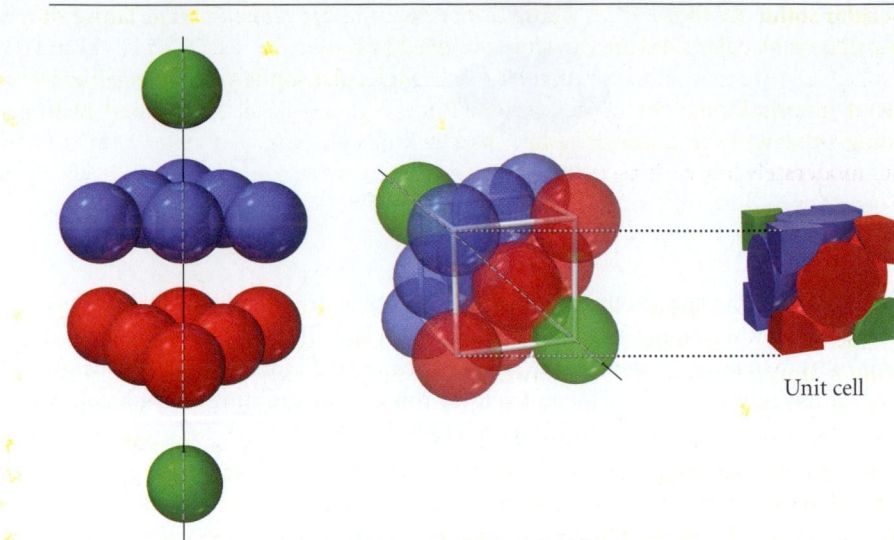

▶ **FIGURE 11.49 Cubic Closest-Packing Crystal Structure** The unit cell of the cubic closest-packed structure is face-centered cubic.

Unit cell

11.12 Crystalline Solids: The Fundamental Types

As we learned in Section 11.2, solids may be crystalline (comprising a well-ordered array of atoms or molecules) or amorphous (having no long-range order). Crystalline solids can be divided into three categories—molecular, ionic, and atomic—based on the individual units that compose the solid. Atomic solids can themselves be divided into three categories—nonbonded, metallic, and network covalent—depending on the types of interactions between atoms within the solid. Figure 11.50 ▼ shows the different categories of crystalline solids.

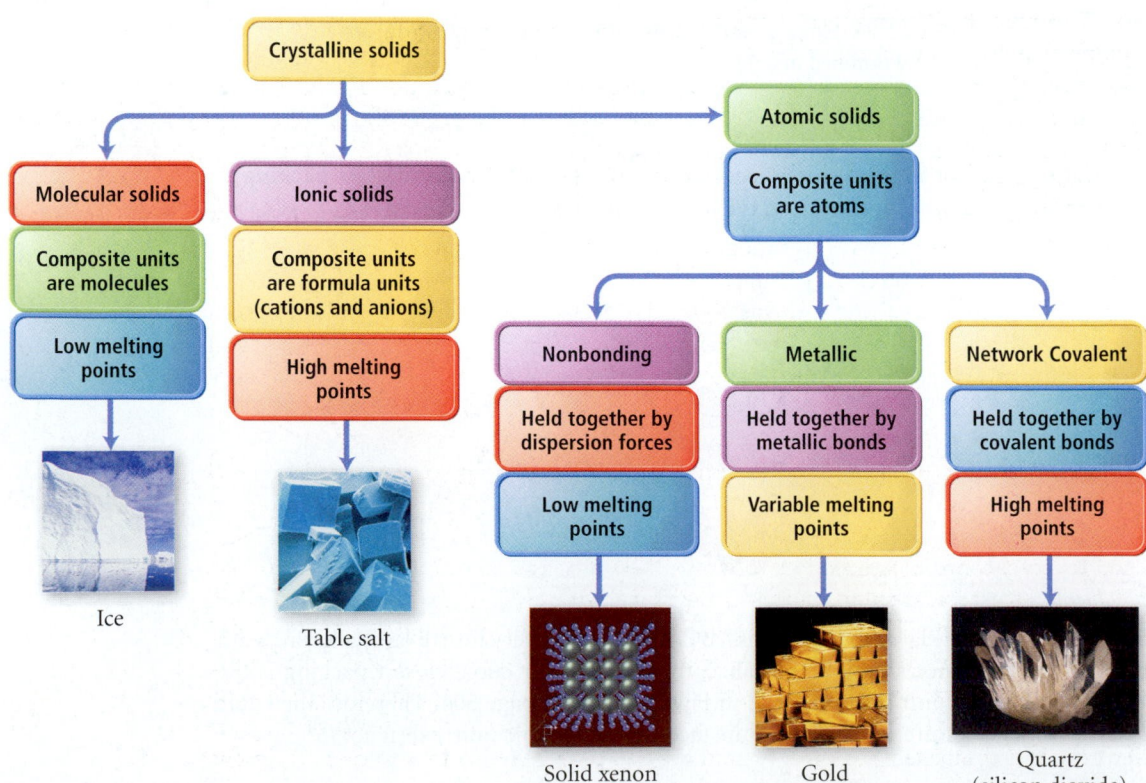

▲ **FIGURE 11.50 Types of Crystalline Solids**

Molecular Solids

Molecular solids are those solids whose composite units are *molecules*. The lattice sites in a crystalline molecular solid are therefore occupied by molecules. Ice (solid H_2O) and dry ice (solid CO_2) are examples of molecular solids. Molecular solids are held together by the kinds of intermolecular forces—dispersion forces, dipole–dipole forces, and hydrogen bonding—that we have discussed in this chapter. Molecular solids as a whole tend to have low to moderately low melting points. However, strong intermolecular forces (such as the hydrogen bonds in water) can increase the melting points of particular compounds.

Ionic Solids

Ionic solids are those solids whose composite units are ions. Table salt (NaCl) and calcium fluoride (CaF_2) are good examples of ionic solids. Ionic solids are held together by the coulombic interactions that occur between the cations and anions occupying the lattice sites in the crystal. The coordination number of the unit cell for an ionic compound, therefore, represents the number of close cation–anion interactions. Since these interactions lower potential energy, the crystal structure of a particular ionic compound will be the one that maximizes the coordination number, while accommodating both charge neutrality (each unit cell must be charge neutral) and the different sizes of the cations and anions that compose the particular compound. In general, the more similar the radii of the cation and the anion, the higher the possible coordination number.

Cesium chloride (CsCl) is a good example of an ionic compound containing cations and anions of similar size (Cs^+ radius = 167 pm; Cl^- radius = 181 pm). In the cesium chloride structure, the chloride ions occupy the lattice sites of a simple cubic cell and one cesium ion lies in the very center of the cell, as shown in Figure 11.51 ▶. The coordination number is 8, meaning that each cesium ion is in direct contact with eight chloride ions (and vice versa). Notice that the cesium chloride unit cell contains one chloride anion ($8 \times \frac{1}{8} = 1$) and one cesium cation (the cesium ion in the middle belongs entirely to the unit cell) for a ratio of Cs to Cl of 1:1, just as in the formula for the compound. Calcium sulfide (CaS) adopts the same structure as cesium chloride.

The crystal structure of sodium chloride must accommodate the more disproportionate sizes of Na^+ (radius = 97 pm) and Cl^- (radius = 181 pm). The larger chloride anion could theoretically fit many of the smaller sodium cations around it, but charge neutrality requires that each sodium cation be surrounded by an equal number of chloride anions. Therefore, the coordination number is limited by the number of chloride anions that can fit around the relatively small sodium cation. The structure that minimizes the energy is shown in Figure 11.52 ▶ and has a coordination number of 6 (each chloride anion is surrounded by six sodium cations and vice versa). You can visualize this structure, called the *rock salt* structure, as chloride anions occupying the lattice sites of a face-centered cubic structure with the smaller sodium cations occupying the holes between the anions. (Alternatively, you can visualize this structure as the *sodium cations* occupying the lattice sites of a face-centered cubic structure with the *larger chloride anions* occupying the spaces between the cations.) Each unit cell contains four chloride anions $[(8 \times \frac{1}{8}) + (6 \times \frac{1}{2}) = 4]$ and four sodium cations $[(12 \times \frac{1}{4}) + 1 = 4]$, resulting in a ratio of 1:1, just as in the formula of the compound. Other compounds exhibiting the sodium chloride structure include LiF, KCl, KBr, AgCl, MgO, and CaO.

An even greater disproportion between the sizes of the cations and anions in a compound makes a coordination number of even 6 physically impossible. For example, in ZnS (Zn^{2+} radius = 74 pm; S^{2-} radius = 184 pm) the crystal structure, shown in Figure 11.53 ▶, has a coordination number of only 4. You can visualize this structure, called the *zinc blende* structure, as sulfide anions occupying the lattice sites of a face-centered cubic structure with the smaller zinc cations occupying four of the eight tetrahedral holes located directly beneath each corner atom. A tetrahedral hole is the empty space that lies in the center of a tetrahedral arrangement of four atoms, as shown at right. Each unit cell contains four sulfide anions $[(8 \times \frac{1}{8}) + (6 \times \frac{1}{2} = 4)]$ and four zinc cations (each of the four zinc cations is completely contained within the unit cell), resulting in a ratio of 1:1, just as in the formula of the compound. Other compounds exhibiting the zinc blende structure include CuCl, AgI, and CdS.

Cesium chloride (CsCl)

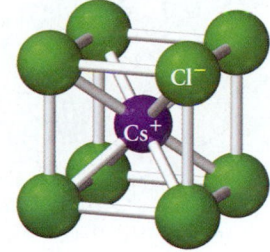

▲ **FIGURE 11.51**
Cesium Chloride Unit Cell

Sodium chloride (NaCl)

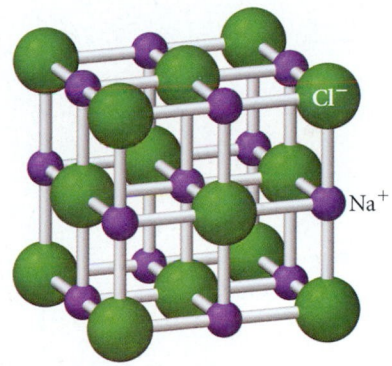

▲ **FIGURE 11.52**
Sodium Chloride Unit Cell

Zinc blende (ZnS)

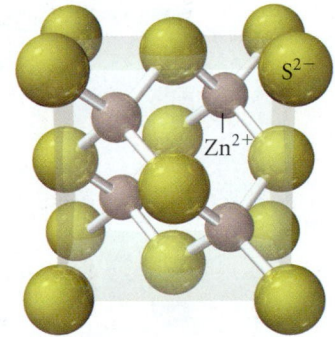

▲ **FIGURE 11.53 Zinc Sulfide (Zinc Blende) Unit Cell**

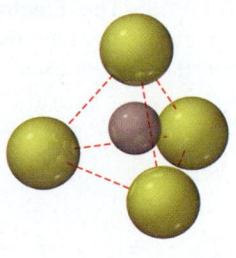

▲ A tetrahedral hole

Calcium fluoride (CaF₂)

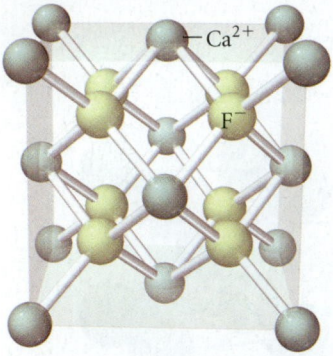

▲ FIGURE 11.54
Calcium Fluoride Unit Cell

When the ratio of cations to anions is not 1:1, the crystal structure must accommodate the unequal number of cations and anions. Many compounds that contain a cation to anion ratio of 1:2 adopt the *fluorite (CaF₂) structure* shown in Figure 11.54 ◄. You can visualize this structure as calcium cations occupying the lattice sites of a face-centered cubic structure with the larger fluoride anions occupying all eight of the tetrahedral holes located directly beneath each corner atom. Each unit cell contains four calcium cations $[(8 \times \frac{1}{8}) + (6 \times \frac{1}{2}) = 4]$ and eight fluoride anions (each of the eight fluoride anions is completely contained within the unit cell), resulting in a cation to anion ratio of 1:2, just as in the formula of the compound. Other compounds exhibiting the fluorite structure include PbF_2, SrF_2, and $BaCl_2$. Compounds with a cation to anion ratio of 2:1 often exhibit the *antifluorite structure*, in which the anions occupy the lattice sites of a face-centered cubic structure and the cations occupy the tetrahedral holes beneath each corner atom.

Since the forces holding ionic solids together are strong coulombic forces (or ionic bonds), and since these forces are much stronger than the intermolecular forces discussed previously, ionic solids tend to have much higher melting points than molecular solids. For example, sodium chloride melts at 801 °C, while carbon disulfide (CS_2)—a molecular solid with a higher molar mass—melts at −110 °C.

Atomic Solids

Those solids whose composite units are individual atoms are called **atomic solids**. Solid xenon (Xe), iron (Fe), and silicon dioxide (SiO_2) are good examples of atomic solids. Atomic solids can themselves be divided into three categories—*nonbonding atomic solids, metallic atomic solids, and network covalent atomic solids*—each held together by a different kind of force.

Nonbonding atomic solids, which include only the noble gases in their solid form, are held together by relatively weak dispersion forces. In order to maximize these interactions, nonbonding atomic solids form closest-packed structures, maximizing their coordination numbers and minimizing the distance between them. Nonbonding atomic solids have very low melting points which increase uniformly with molar mass. Argon, for example, has a melting point of −189 °C and xenon has a melting point of −112 °C.

Metallic atomic solids, such as iron or gold, are held together by *metallic bonds,* which in the simplest model are represented by the interaction of metal cations with a sea of electrons that surround them, as described in Section 9.11 (Figure 11.55 ◄).

Since metallic bonds are not directional, metals also tend to form closest-packed crystal structures. For example, nickel crystallizes in the cubic closest-packed structure and zinc crystallizes in the hexagonal closest-packed structure (Figure 11.56 ▶). Metallic bonds are of varying strengths. Some metals, such as mercury, have melting points below room temperature, whereas other metals, such as iron, have relatively high melting points (iron melts at 1809 °C).

Network covalent atomic solids, such as diamond, graphite, and silicon dioxide, are held together by covalent bonds. The crystal structures of these solids are more restricted by the geometrical constraints of the covalent

| We examine a more sophisticated model for bonding in metals in Section 11.13.

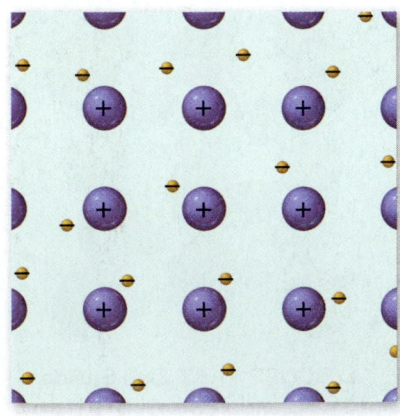

▲ FIGURE 11.55 The Electron Sea Model In the electron sea model for metals, the metal cations exist in a "sea" of electrons.

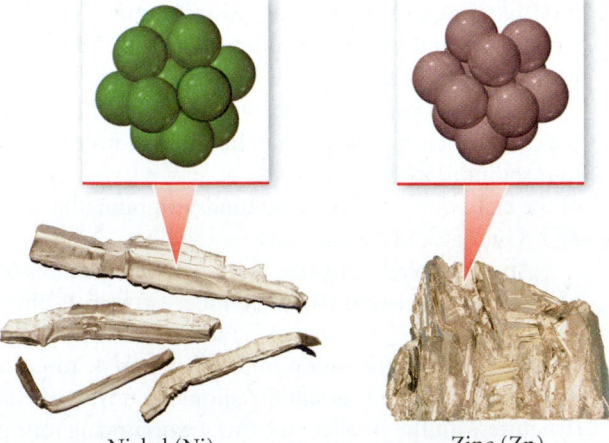

Nickel (Ni)   Zinc (Zn)

▲ FIGURE 11.56 Closest-Packed Crystal Structures in Metals Nickel crystallizes in the cubic closest-packed structure. Zinc crystallizes in the hexagonal closest-packed structure.

bonds (which tend to be more directional than intermolecular forces, ionic bonds, or metallic bonds) so they *do not* tend to form closest-packed structures.

In diamond (Figure 11.57a ▶), each carbon atom forms four covalent bonds to four other carbon atoms in a tetrahedral geometry. This structure extends throughout the entire crystal, so that a diamond crystal can be thought of as a giant molecule, held together by these covalent bonds. Since covalent bonds are very strong, covalent atomic solids have high melting points. Diamond is estimated to melt at about 3800 °C. The electrons in diamond are confined to the covalent bonds and are not free to flow. Therefore diamond does not conduct electricity.

In graphite (Figure 11.57b), carbon atoms are arranged in sheets. Within each sheet, carbon atoms are covalently bonded to each other by a network of sigma and pi bonds, similar to those in benzene. Just as the electrons within the pi bonds in benzene are delocalized over the entire molecule, so the pi bonds in graphite are delocalized over the entire sheet, making graphite a good electrical conductor along the sheets. The bond length between carbon atoms *within a sheet* is 142 pm. However, the bonding *between* sheets is much different. The separation between sheets is 341 pm. There are no covalent bonds between sheets, only relatively weak dispersion forces. Consequently, the sheets slide past each other relatively easily, which explains the slippery feel of graphite and its extensive use as a lubricant.

(a) Diamond   (b) Graphite

▲ FIGURE 11.57 **Network Covalent Atomic Solids** (a) In diamond, each carbon atom forms four covalent bonds to four other carbon atoms in a tetrahedral geometry. (b) In graphite, carbon atoms are arranged in sheets. Within each sheet, the atoms are covalently bonded to one another by a network of sigma and pi bonds. Neighboring sheets are held together by dispersion forces.

Sigma and pi bonds were discussed in Section 10.7.

The silicates (extended arrays of silicon and oxygen) are the most common network covalent atomic solids. Geologists estimate that 90% of Earth's crust is composed of silicates, and we cover these in more detail in Chapter 22. The basic silicon oxygen compound is silica (SiO_2), which in its most common crystalline form is called quartz. The structure of quartz consists of an array of SiO_4 tetrahedra with shared oxygen atoms, as shown in Figure 11.58(a) ▼. The strong silicon–oxygen covalent bonds that hold quartz together result in its high melting point of about 1600 °C. Common glass is also composed of SiO_2, but in its amorphous form (Figure 11.58b).

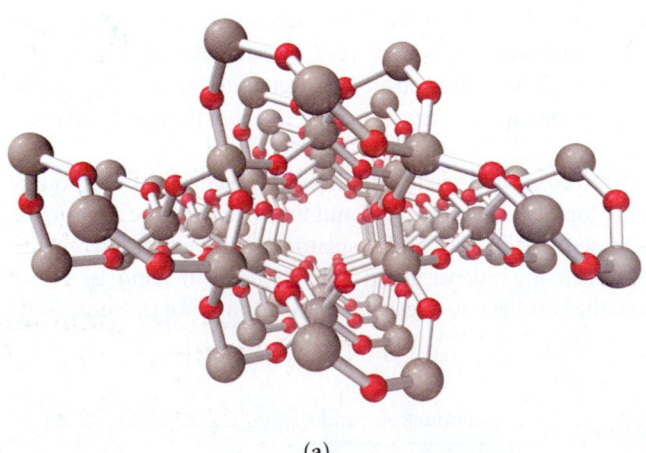

(a)

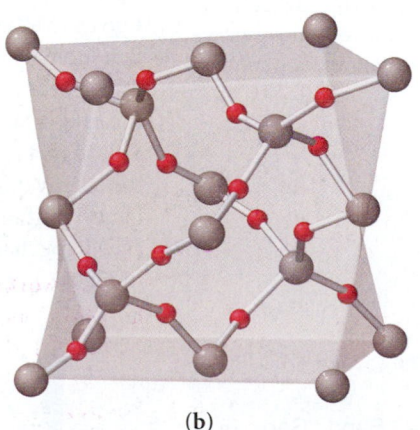

(b)

▲ FIGURE 11.58 **The Structure of Quartz** (a) Quartz consists of an array of SiO_4 tetrahedra with shared oxygen atoms. (b) Glass is amorphous SiO_2.

11.13 Crystalline Solids: Band Theory

In Section 9.11, we explored a model for bonding in metals called the *electron sea model*. We now turn to a model for bonding in solids that is both more sophisticated and more broadly applicable—it applies to both metallic solids and covalent solids. The model is called **band theory** and it grows out of molecular orbital theory, first covered in Section 10.8.

Recall that in molecular orbital theory, we combined the atomic orbitals of the atoms within a molecule to form molecular orbitals. These molecular orbitals were not localized on individual atoms, but *delocalized over the entire molecule*. Similarly, in band theory, we combine the atomic orbitals of the atoms within a solid crystal to form orbitals that are not localized on individual atoms, but delocalized over the entire *crystal*. In some sense then, the crystal is like a very large molecule and its valence electrons occupy the molecular orbitals formed from the atomic orbitals of each atom in the crystal.

We begin our discussion of band theory by considering a series of molecules constructed from individual lithium atoms. The energy levels of the atomic orbitals and resulting molecular orbitals for Li, Li_2, Li_3, Li_4, and Li_N (where N is a large number on the order of 10^{23}) are shown in Figure 11.59 ▼. The lithium atom has a single electron in a single $2s$ atomic orbital. The Li_2 molecule contains two electrons and two molecular orbitals. The electrons occupy the lower energy bonding orbital—the higher energy, or antibonding, molecular orbital is empty. The Li_4 molecule contains four electrons and four molecular orbitals. The electrons occupy the two bonding molecular orbitals—the two antibonding orbitals are completely empty.

The Li_N molecule contains N electrons and N molecular orbitals. However, because there are so many molecular orbitals, the energy spacings between them are infinitesimally small; they are no longer discrete energy levels, but rather form a *band* of energy levels. One half of the orbitals in the band ($N/2$) are bonding molecular orbitals and (at 0 K) contain the N valence electrons. The other $N/2$ molecular orbitals are antibonding and (at 0 K) are completely empty. If the atoms composing a solid have p orbitals available, then the same process leads to another band of orbitals at higher energies.

In band theory, electrons become mobile when they make a transition from the highest occupied molecular orbital into higher energy empty molecular orbitals. For this reason, the occupied molecular orbitals are often called the *valence band* and the unoccupied orbitals are called the *conduction band.* In lithium metal, the highest occupied molecular orbital lies in the middle of a band of orbitals, and the energy difference between it and the next higher energy orbital is infinitesimally small. Therefore, above 0 K, electrons can easily make the transition from the valence band to the conduction band. Since electrons in the conduction band are mobile, lithium, like all metals, is a good electrical conductor. Mobile electrons in the conduction band are also responsible for the thermal conductivity of metals. When a metal is heated, electrons are excited to higher energy molecular orbitals. These electrons can then quickly transport the thermal energy throughout the crystal lattice.

In metals, the valence band and conduction band are always energetically continuous—the energy difference between the top of the valence band and the bottom of the conduction band is infinitesimally small. In semiconductors and insulators, however, an energy gap, called the **band gap**, exists between the valence band and conduction band as shown in Figure 11.60 ▼. In insulators, the band gap is large, and electrons are not promoted into

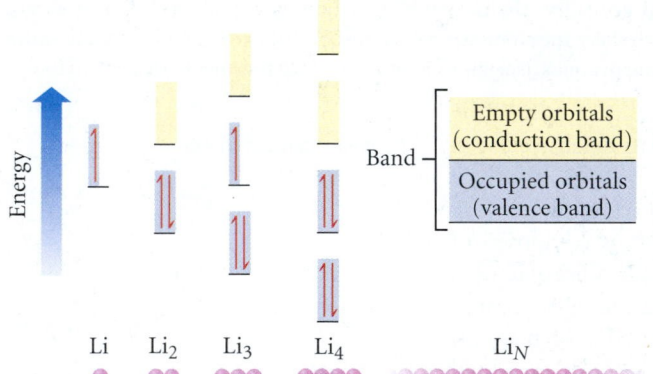

▲ FIGURE 11.59 Energy Levels of Molecular Orbitals in Lithium Molecules   When many Li atoms are present, the energy levels of the molecular orbitals are so closely spaced that they fuse to form a band. Half of the orbitals are bonding orbitals and contain valence electrons; the other half are antibonding orbitals and are empty.

▶ FIGURE 11.60 Band Gap In a conductor, there is no energy gap between the valence band and the conduction band. In semiconductors there is a small energy gap, and in insulators there is a large energy gap.

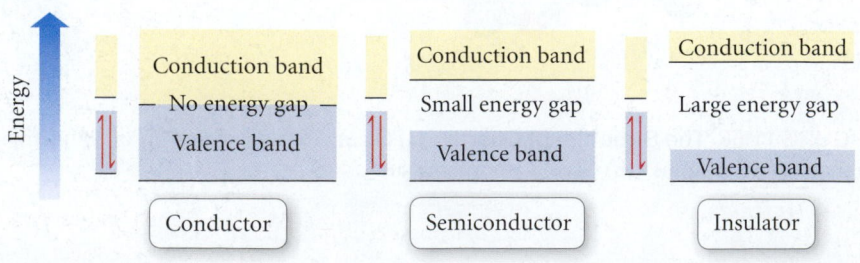

the conduction band at ordinary temperatures, resulting in no electrical conductivity. In semiconductors, the band gap is small, allowing some electrons to be promoted at ordinary temperatures and resulting in limited conductivity. However, the conductivity of semiconductors can be controlled by adding minute amounts of other substances, called *dopants*, to the semiconductor.

Doping: Controlling the Conductivity of Semiconductors

Doped semiconductors contain minute amounts of impurities that result in additional electrons in the conduction band or electron "holes" in the valence band. For example, silicon is a group 4A semiconductor. Its valence electrons just fill its valence band. The band gap in silicon is large enough that few electrons are promoted into the conduction band at room temperature; therefore silicon is a poor electrical conductor. However, silicon can be doped with phosphorus, a group 5A element with five valence electrons, to increase its conductivity. The phosphorus atoms are incorporated into the silicon crystal structure, but each phosphorus atom brings with it one additional electron. Since the valence band is completely full, the additional electrons must go into the conduction band. These electrons are then mobile and can conduct electrical current. This type of semiconductor is called an **n-type semiconductor** because the charge carriers are negatively charged electrons in the conduction band.

Silicon can also be doped with a group 3A element, such as gallium, which has only three valence electrons. When gallium is incorporated into the silicon crystal structure, it results in electron "holes," or empty molecular orbitals, in the valence band. The presence of holes also allows for the movement of electrical current because electrons in the valence band can move between holes. In this way, the holes move in the opposite direction as the electrons. This type of semiconductor is called a **p-type semiconductor** because the hole acts as a positive charge.

The heart of most modern electronic devices are silicon chips containing millions of **p–n junctions**, tiny spots that are p-type on one side and n-type on the other. These junctions can serve a number of functions including acting as **diodes**, elements that allow the flow of electrical current in only one direction, and acting as amplifiers, elements that amplify a small electrical current into a larger one.

Chapter in Review

Key Terms

Section 11.2
crystalline (464)
amorphous (464)

Section 11.3
dispersion force (466)
dipole–dipole force (468)
permanent dipole (468)
miscibility (469)
hydrogen bonding (470)
hydrogen bond (470)
ion–dipole force (472)

Section 11.4
surface tension (475)
viscosity (475)
capillary action (476)

Section 11.5
vaporization (477)
condensation (477)

volatile (478)
nonvolatile (478)
heat of vaporization (ΔH_{vap}) (479)
dynamic equilibrium (480)
vapor pressure (480)
boiling point (482)
normal boiling point (482)
Clausius–Clapeyron equation (483)
critical temperature (T_c) (486)
critical pressure(P_c) (486)

Section 11.6
sublimation (487)
deposition (487)
melting point (487)
melting (fusion) (487)
freezing (488)
heat of fusion (ΔH_{fus}) (488)

Section 11.8
phase diagram (491)
triple point (492)
critical point (492)

Section 11.10
X-ray diffraction (496)

Section 11.11
crystalline lattice (498)
unit cell (498)
simple cubic (498)
coordination number (499)
packing efficiency (499)
body-centered cubic (499)
face-centered cubic (500)
hexagonal closest packing (503)
cubic closest packing (503)

Section 11.12
molecular solids (505)
ionic solids (505)
atomic solids (506)
nonbonding atomic solids (506)
metallic atomic solids (506)
network covalent atomic solids (506)

Section 11.13
band theory (508)
band gap (508)
n-type semiconductor (509)
p-type semiconductor (509)
p–n junctions (509)
diodes (509)

Key Concepts

Solids, Liquids, and Intermolecular Forces (11.1, 11.2, 11.3)

The forces that hold molecules or atoms together in a liquid or solid are called intermolecular forces. The strength of the intermolecular forces in a substance determines its phase. Dispersion forces are always present because they result from the fluctuations in electron distribution within atoms and molecules. These are the weakest intermolecular forces, but they are significant in molecules with high molar masses. Dipole-dipole forces, generally stronger than dispersion forces, are present in all polar molecules. Hydrogen bonding occurs in polar molecules that contain hydrogen atoms bonded directly to fluorine, oxygen, or nitrogen. These are the strongest intermolecular forces. Ion–dipole forces occur in ionic compounds mixed with polar compounds, and they are especially important in aqueous solutions.

Surface Tension, Viscosity, and Capillary Action (11.4)

Surface tension results from the tendency of liquids to minimize their surface area in order to maximize the interactions between their constituent particles, thus lowering their potential energy. Surface tension causes water droplets to form spheres and allows insects and paper clips to "float" on the surface of water. Viscosity is the resistance of a liquid to flow. Viscosity increases with increasing strength of intermolecular forces and decreases with increasing temperature. Capillary action is the ability of a liquid to flow against gravity up a narrow tube. It is the result of adhesive forces, the attraction between the molecules and the surface of the tube, and cohesive forces, the attraction between the molecules in the liquid.

Vaporization and Vapor Pressure (11.5, 11.7)

Vaporization, the phase transition from liquid to gas, occurs when thermal energy overcomes the intermolecular forces present in a liquid. The opposite process is condensation. Vaporization is endothermic and condensation is exothermic. The rate of vaporization increases with increasing temperature, increasing surface area, and decreasing strength of intermolecular forces. The heat of vaporization (ΔH_{vap}) is the heat required to vaporize one mole of a liquid. In a sealed container, a solution and its vapor will come into dynamic equilibrium, at which point the rate of vaporization equals the rate of condensation. The pressure of a gas that is in dynamic equilibrium with its liquid is called its vapor pressure. The vapor pressure of a substance increases with increasing temperature and with decreasing strength of its intermolecular forces. The boiling point of a liquid is the temperature at which its vapor pressure equals the external pressure. The Clausius–Clapeyron equation expresses the relationship between the vapor pressure of a substance and its temperature and can be used to calculate the heat of vaporization

from experimental measurements. When a liquid is heated in a sealed container it eventually forms a supercritical fluid, which has properties intermediate between a liquid and a gas. This occurs at the critical temperature and critical pressure.

Fusion and Sublimation (11.6, 11.7)

Sublimation is the phase transition from solid to gas. The opposite process is deposition. Fusion, or melting, is the phase transition from solid to liquid. The opposite process is freezing. The heat of fusion (ΔH_{fus}) is the amount of heat required to melt one mole of a solid. Fusion is endothermic. The heat of fusion is generally less than the heat of vaporization because intermolecular forces do not have to be completely overcome for melting to occur.

Phase Diagrams (11.8)

A phase diagram is a map of the phases of a substance as a function of its pressure (y-axis) and temperature (x-axis). The regions in a phase diagram represent conditions under which a single stable phase (solid, liquid, gas) exists. The lines represent conditions under which two phases are in equilibrium. The triple point represents the conditions under which all three phases coexist. The critical point is the temperature and pressure above which a supercritical fluid exists.

The Uniqueness of Water (11.9)

Water is a liquid at room temperature despite its low molar mass. Water forms strong hydrogen bonds, resulting in its high boiling point. Its high polarity also enables it to dissolve many polar and ionic compounds, and even nonpolar gases. Water expands upon freezing, so that ice is less dense than liquid water. Water is critical both to the existence of life and to human health.

Crystalline Structures (11.10–11.13)

In X-ray crystallography, the diffraction pattern of X-rays is used to determine the crystal structure of solids. The crystal lattice is represented by a unit cell, a structure that can reproduce the entire lattice when repeated in all three dimensions. Three basic cubic unit cells are the simple cubic, the body-centered cubic, and the face-centered cubic. Some crystal lattices can also be depicted as closest-packed structures, including the hexagonal closest-packing structure (not cubic) and the cubic closest-packing structure (which has a face-centered cubic unit cell). The basic types of crystal solids are molecular, ionic, and atomic solids. Atomic solids can themselves be divided into three different types: nonbonded, metallic, and covalent. Band theory is a model for bonding in solids in which the atomic orbitals of the atoms are combined and delocalized over the entire crystal solid.

Key Equations and Relationships

Clausius–Clapeyron Equation: Relationship between Vapor Pressure (P_{vap}), the Heat of Vaporization (H_{vap}), and Temperature (T) (11.5)

$$\ln P_{vap} = \frac{-\Delta H_{vap}}{RT} + \ln \beta \quad (\beta \text{ is a constant})$$

$$\ln \frac{P_2}{P_1} = \frac{-\Delta H_{vap}}{R}\left[\frac{1}{T_2} - \frac{1}{T_1}\right]$$

Bragg's Law: Relationship between Light Wavelength (λ), Angle of Reflection (θ), and Distance (d) between the Atomic Layers (11.10)

$$n\lambda = 2d \sin \theta \quad (n = \text{integer})$$

Key Skills

Determining Whether a Molecule Has Dipole-Dipole Forces (11.3)

• Example 11.1   • For Practice 11.1   • Exercises 49-58

Determining Whether a Molecule Displays Hydrogen Bonding (11.3)

• Example 11.2   • For Practice 11.2   • Exercises 49–58

Using the Heat of Vaporization in Calculations (11.5)

• Example 11.3   • For Practice 11.3   • For More Practice 11.3   • Exercises 69–72

Using the Clausius–Clapeyron Equation (11.5)

• Examples 11.4, 11.5   • For Practice 11.4, 11.5   • Exercises 73–76

Using Bragg's Law in X-Ray Diffraction Calculations (11.10)

• Example 11.6   • For Practice 11.6   • Exercises 93, 94

Relating Density to Crystal Structure (11.11)

• Example 11.7   • For Practice 11.7   • Exercises 97–100

Exercises

Review Questions

1. Explain how a gecko can walk on a polished glass surface.

2. Why are intermolecular forces important?

3. What are the main properties of liquids (in contrast to gases and solids)?

4. What are the main properties of solids (in contrast to liquids and gases)?

5. What is the fundamental difference between an amorphous solid and a crystalline solid?

6. What factors cause changes between the solid and liquid phase? The liquid and gas phase?

7. Describe the relationship between the phase of a substance, its temperature, and the strength of its intermolecular forces.

8. From what kinds of interactions do intermolecular forces originate?

9. Why are intermolecular forces generally much weaker than bonding forces?

10. What is the dispersion force? What does the magnitude of the dispersion force depend on? How can you predict the magnitude of the dispersion force for closely related elements or compounds?

11. What is the dipole–dipole force? How can you predict the presence of dipole-dipole forces in a compound?

12. How is the miscibility of two liquids related to their polarity?

13. What is hydrogen bonding? How can you predict the presence of hydrogen bonding in a compound?

14. What is the ion–dipole force? Why is it important?

15. What is surface tension? How does surface tension arise out of intermolecular forces? How is it related to the strength of intermolecular forces?

16. What is viscosity? How does viscosity depend on intermolecular forces? What other factors affect viscosity?

17. What is capillary action? How does it depend on the relative strengths of adhesive and cohesive forces?

18. Explain what happens in the processes of vaporization and condensation. Why does the rate of vaporization increase with increasing temperature and surface area?

19. Why is vaporization endothermic? Why is condensation exothermic?

20. How is the volatility of a substance related to the intermolecular forces present within the substance?

21. What is the heat of vaporization for a liquid and why is it useful?

22. Explain the process of dynamic equilibrium. How is dynamic equilibrium related to vapor pressure?

23. What happens to a system in dynamic equilibrium when it is disturbed in some way?

24. How is vapor pressure related to temperature? What happens to the vapor pressure of a substance when the temperature is increased? Decreased?

25. Define the terms *boiling point* and *normal boiling point*.

26. What is the Clausius–Clapeyron equation and why is it important?

27. Explain what happens to a substance when it is heated in a closed container to its critical temperature.

28. What is sublimation? Give a common example of sublimation.

29. What is fusion? Is fusion exothermic or endothermic? Why?

30. What is the heat of fusion and why is it important?

31. Examine the heating curve for water in Section 11.7. Explain why the curve has two segments in which heat is added to the water but the temperature does not rise.

32. Examine the heating curve for water in Section 11.7. Explain the significance of the slopes of each of the three rising segments. Why are the slopes different?

33. What is a phase diagram? Draw a generic phase diagram and label each of the important features.

34. What is the significance of crossing a line in a phase diagram?

35. How do the properties of water differ from those of most other substances?

36. Explain the basic principles involved in X-ray crystallography. Include Bragg's law in your explanation.

37. What is a crystalline lattice? How is the lattice represented with the unit cell?

38. Make a drawing of each of the following unit cells: simple cubic, body-centered cubic, and face-centered cubic.

39. For each of the cubic cells in the previous problem, give the coordination number, edge length in terms of r, and number of atoms per unit cell.

40. What is the difference between hexagonal closest packing and cubic closest packing? What are the unit cells for each of these structures?

41. What are the three basic types of solids and the composite units of each? What types of forces hold each type of solid together?

42. In an ionic compound, how are the relative sizes of the cation and anion related to the coordination number of the crystal structure?

43. Show how the cesium chloride, sodium chloride, and zinc blende unit cells each contain a cation-to-anion ratio of 1:1.

44. Show how the fluorite structure contains a cation-to-anion ratio of 1:2.

45. What are the three basic subtypes of atomic solids? What kinds of forces hold each of these subtypes together?

46. In band theory of bonding for solids, what is a *band*? What is the difference between the *valence band* and the *conduction band*?

47. What is a band gap? How is the band gap different in metals, semiconductors, and insulators?

48. Explain how doping can increase the conductivity of a semiconductor. What is the difference between an n-type semiconductor and a p-type semiconductor?

Problems by Topic

Intermolecular Forces

49. Determine the kinds of intermolecular forces that are present in each of the following elements or compounds:
    a. Kr   b. NCl_3 c. SiH_4 d. HF
    e. N_2 f. NH_3 g. CO   h. CCl_4

50. Determine the kinds of intermolecular forces that are present in each of the following elements or compounds:
    a. HCl   b. H_2O c. Br_2 d. He
    e. PH_3 f. HBr   g. CH_3OH h. I_2

51. Arrange the following in order of increasing boiling point. Explain your reasoning.
    a. CH_4 b. CH_3CH_3
    c. CH_3CH_2Cl d. CH_3CH_2OH

52. Arrange the following in order of increasing boiling point. Explain your reasoning.
    a. H_2S b. H_2Se c. H_2O

53. For each pair of compounds, pick the one with the higher boiling point. Explain your reasoning.
    a. CH_3OH or CH_3SH b. CH_3OCH_3 or CH_3CH_2OH
    c. CH_4 or CH_3CH_3

54. For each pair of compounds, pick the one with the highest boiling point. Explain your reasoning.
    a. NH_3 or CH_4 b. CS_2 or CO_2
    c. CO_2 or NO_2

55. For each pair of compounds, pick the one with the higher vapor pressure at a given temperature. Explain your reasoning.
    a. Br_2 or I_2 b. H_2S or H_2O
    c. NH_3 or PH_3

56. For each pair of compounds, pick the one with the higher vapor pressure at a given temperature. Explain your reasoning.
    a. CH_4 or CH_3Cl
    b. $CH_3CH_2CH_2OH$ or CH_3OH
    c. CH_3OH or H_2CO

57. Which of the following pairs of substances would you expect to form homogeneous solutions when combined? For those that form homogeneous solutions, indicate the type of forces that are involved.
    a. CCl_4 and H_2O b. KCl and H_2O
    c. Br_2 and CCl_4 d. CH_3CH_2OH and H_2O

58. Which of the following pairs of compounds would you expect to form homogeneous solutions when combined? For those that form homogeneous solutions, indicate the type of forces that are involved.
    a. $CH_3CH_2CH_2CH_2CH_3$ and $CH_3CH_2CH_2CH_2CH_2CH_3$
    b. CBr_4 and H_2O
    c. $LiNO_3$ and H_2O
    d. CH_3OH and $CH_3CH_2CH_2CH_2CH_3$

Surface Tension, Viscosity, and Capillary Action

59. Which compound would you expect to have greater surface tension, acetone $[(CH_3)_2CO]$ or water (H_2O)? Explain.

60. Water (a) "wets" some surfaces and beads up on others. Mercury (b), in contrast, beads up on almost all surfaces. Explain this difference.

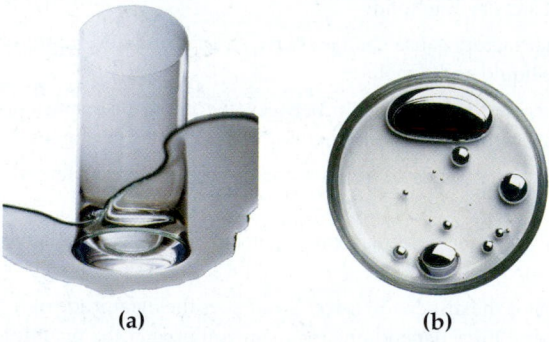

(a)   (b)

61. The structures of two isomers of heptane are shown below. Which of these two compounds would you expect to have the greater viscosity?

Compound A

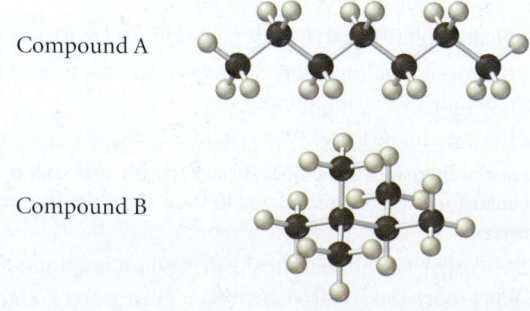

Compound B

62. Explain why the viscosity of multigrade motor oils is less temperature dependent than that of single-grade motor oils.

63. Water in a glass tube that contains grease or oil residue displays a flat meniscus (left); whereas water in a clean glass tube displays a concave meniscus (right). Explain this difference.

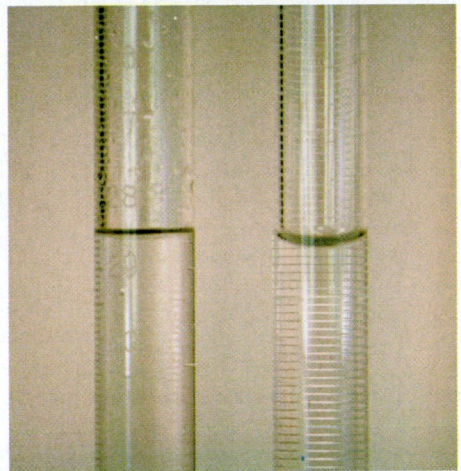

64. When a thin glass tube is put into water, the water rises 1.4 cm. When the same tube is put into hexane, the hexane rises only 0.4 cm. Explain the difference.

Vaporization and Vapor Pressure

65. Which will evaporate more quickly: 55 mL of water in a beaker with a diameter of 4.5 cm, or 55 mL of water in a dish with a diameter of 12 cm? Will the vapor pressure of the water be different in the two containers? Explain.

66. Which will evaporate more quickly: 55 mL of water (H_2O) in a beaker or 55 mL of acetone $[(CH_3)_2CO]$ in an identical beaker under identical conditions? Is the vapor pressure of the two substances different? Explain.

67. Spilling room-temperature water over your skin on a hot day will cool you down. Spilling vegetable oil (of the same temperature) over your skin on a hot day will not. Explain the difference.

68. Why is the heat of vaporization of water greater at room temperature than it is at its boiling point?

69. The human body obtains 955 kJ of energy from a candy bar. If this energy were used to vaporize water at 100.0 °C, how much water (in liters) could be vaporized? (Assume the density of water is 1.00 g/mL.)

70. A 55.0 mL sample of water is heated to its boiling point. How much heat (in kJ) is required to vaporize it? (Assume a density of 1.00 g/mL.)

71. Suppose that 0.88 g of water condenses on a 75.0-g block of iron that is initially at 22 °C. If the heat released during condensation goes only to warming the iron block, what is the final temperature (in °C) of the iron block? (Assume a constant enthalpy of vaporization for water of 44.0 kJ/mol.)

72. Suppose that 1.02 g of rubbing alcohol (C_3H_8O) evaporates from a 55.0-g aluminum block. If the aluminum block is initially at 25 °C, what is the final temperature of the block after the evaporation of the alcohol? Assume that the heat required for the vaporization of the alcohol comes only from the aluminum block and that the alcohol vaporizes at 25 °C.

73. The vapor pressure of ammonia at several different temperatures is shown in the following table. Use the data to determine the heat of vaporization and normal boiling point of ammonia.

Temperature (K)	Pressure (torr)
200	65.3
210	134.3
220	255.7
230	456.0
235	597.0

74. The vapor pressure of nitrogen at several different temperatures is shown below. Use the data to determine the heat of vaporization and normal boiling point of nitrogen.

Temperature (K)	Pressure (torr)
65	130.5
70	289.5
75	570.8
80	1028
85	1718

75. Ethanol has a heat of vaporization of 38.56 kJ/mol and a normal boiling point of 78.4 °C. What is the vapor pressure of ethanol at 15 °C?

76. Benzene has a heat of vaporization of 30.72 kJ/mol and a normal boiling point of 80.1 °C. At what temperature does benzene boil when the external pressure is 445 torr?

Sublimation and Fusion

77. How much energy is released when 47.5 g of water freezes?

78. Calculate the amount of heat required to completely sublime 25.0 g of solid dry ice (CO_2) at its sublimation temperature. The heat of sublimation for carbon dioxide is 32.3 kJ/mol.

79. An 8.5-g ice cube is placed into 255 g of water. Calculate the temperature change in the water upon the complete melting of the ice. Assume that all of the energy required to melt the ice comes from the water.

80. How much ice (in grams) would have to melt to lower the temperature of 352 mL of water from 25 °C to 5 °C? (Assume the density of water is 1.0 g/mL.)

81. How much heat (in kJ) is required to warm 10.0 g of ice, initially at −10.0 °C, to steam at 110.0 °C? The heat capacity of ice is 2.09 J/g · °C and that of steam is 2.01 J/g · °C.

82. How much heat (in kJ) is evolved in converting 1.00 mol of steam at 145.0 °C to ice at −50.0 °C? The heat capacity of steam is 2.01 J/g · °C and of ice is 2.09 J/g · °C.

Phase Diagrams

83. Consider the phase diagram shown below. Identify the phases present at points a through g

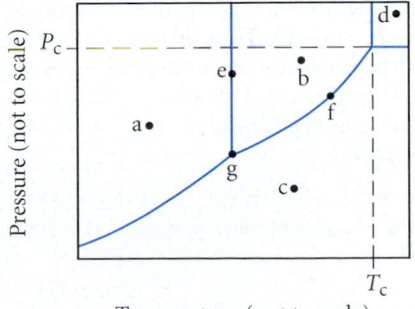

84. Consider the phase diagram for iodine shown below and answer each of the following questions.

   a. What is the normal boiling point for iodine?
   b. What is the melting point for iodine at 1 atm?
   c. What phase is present at room temperature and normal atmospheric pressure?
   d. What phase is present at 186 °C and 1.0 atm?

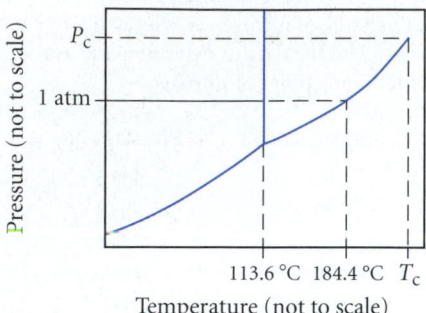

Temperature (not to scale)

85. Nitrogen has a normal boiling point of 77.3 K and a melting point (at 1 atm) of 63.1 K. Its critical temperature is 126.2 K and critical pressure is 2.55×10^4 torr. It has a triple point at 63.1 K and 94.0 torr. Sketch the phase diagram for nitrogen. Does nitrogen have a stable liquid phase at 1 atm?

86. Argon has a normal boiling point of 87.2 K and a melting point (at 1 atm) of 84.1 K. Its critical temperature is 150.8 K and critical pressure is 48.3 atm. It has a triple point at 83.7 K and 0.68 atm. Sketch the phase diagram for argon. Which has the greater density, solid argon or liquid argon?

87. The phase diagram for sulfur is shown below. The rhombic and monoclinic phases are two solid phases with different structures

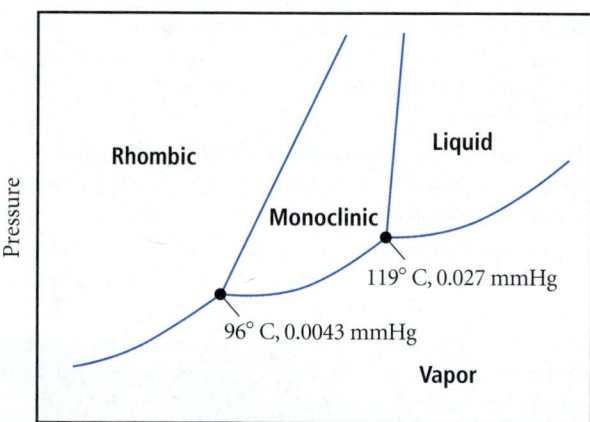

Temperature

   a. Below what pressure will solid sulfur sublime?
   b. Which of the two solid phases of sulfur is most dense?

88. The high-pressure phase diagram of ice is shown at the top of the next column. Notice that, under high pressure, ice can exist in several different solid forms. What three forms of ice are present at the triple point marked O? What is the density of ice II compared to ice I (the familiar form of ice.) Would ice III sink or float in liquid water?

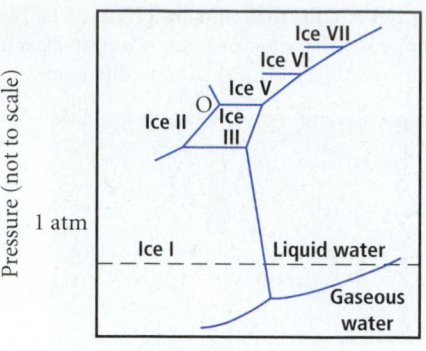

Temperature (not to scale)

The Uniqueness of Water

89. Water has a high boiling point for its relatively low molar mass. Why?

90. Water is a good solvent for many substances. What is the molecular basis for this property and why is it significant?

91. Explain the role of water in moderating Earth's climate.

92. How is the density of solid water compared to that of liquid water atypical among substances? Why is this significant?

Types of Solids and Their Structures

93. An X-ray beam with $\lambda = 154$ pm incident on the surface of a crystal produced a maximum reflection at an angle of $\theta = 28.3°$. Assuming $n = 1$, calculate the separation between layers of atoms in the crystal.

94. An X-ray beam of unknown wavelength is diffracted from a NaCl surface. If the interplanar distance in the crystal is 286 pm, and the angle of maximum reflection is found to be 7.23°, what is the wavelength of the X-ray beam? (Assume $n = 1$.)

95. Determine the number of atoms per unit cell for each of the following metals.

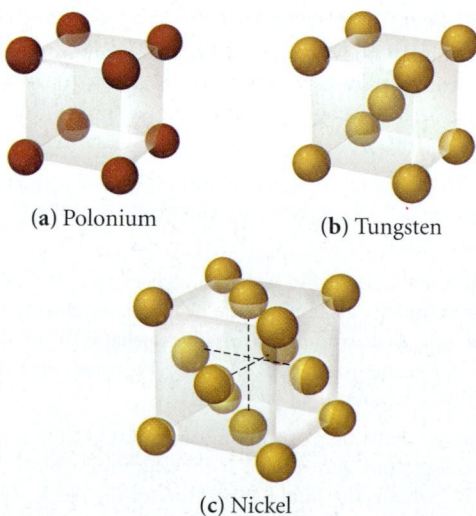

(a) Polonium   **(b)** Tungsten

(c) Nickel

96. Determine the coordination number for each of the following structures.

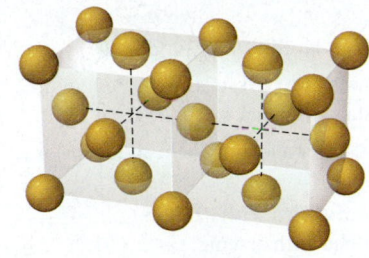

(a) Gold

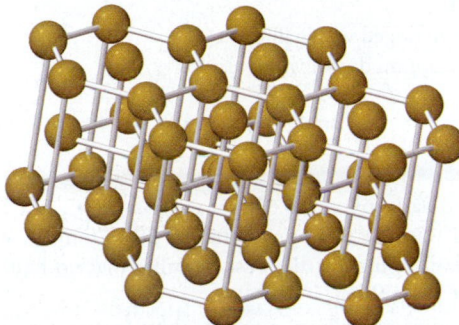

(b) Ruthenium

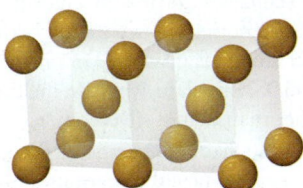

(c) Chromium

97. Platinum crystallizes with the face-centered cubic unit cell. The radius of a platinum atom is 139 pm. Calculate the edge length of the unit cell and the density of platinum in g/cm^3.

98. Molybdenum crystallizes with the body-centered unit cell. The radius of a molybdenum atom is 136 pm. Calculate the edge length of the unit cell and the density of molybdenum.

99. Rhodium has density of $12.41 \, g/cm^3$ and crystallizes with the face-centered cubic unit cell. Calculate the radius of a rhodium atom.

100. Barium has a density of $3.59 \, g/cm^3$ and crystallizes with the body-centered cubic unit cell. Calculate the radius of a barium atom.

101. Polonium crystallizes with a simple cubic structure. It has a density of $9.3 \, g/cm^3$, a radius of 167 pm, and a molar mass of $209 \, g/mol$. Use these data to estimate Avogadro's number (the number of atoms in one mole).

102. Palladium crystallizes with a face-centered cubic structure. It has a density of $12.0 \, g/cm^3$, a radius of 138 pm, and a molar mass of $106.42 \, g/mol$. Use these data to estimate Avogadro's number.

103. Identify each of the following solids as molecular, ionic, or atomic.
    a. $Ar(s)$ **b.** $H_2O(s)$ **c.** $K_2O(s)$ **d.** $Fe(s)$

104. Identify each of the following solids as molecular, ionic, or atomic.
    a. $CaCl_2(s)$ **b.** $CO_2(s)$ **c.** $Ni(s)$ **d.** $I_2(s)$

105. Which of the following solids has the highest melting point? Why?

$$Ar(s), CCl_4(s), LiCl(s), CH_3OH(s)$$

106. Which of the following solids has the highest melting point? Why?

$$C(s, \text{diamond}), Kr(s), NaCl(s), H_2O(s)$$

107. Of each pair of solids, which one has the higher melting point and why?
    a. $TiO_2(s)$ or $HOOH(s)$ **b.** $CCl_4(s)$ or $SiCl_4(s)$
    c. $Kr(s)$ or $Xe(s)$ **d.** $NaCl(s)$ or $CaO(s)$

108. Of each pair of solids, which one has the higher melting point and why?
    a. $Fe(s)$ or $CCl_4(s)$ **b.** $KCl(s)$ or $HCl(s)$
    c. $Ti(s)$ or $Ne(s)$ **d.** $H_2O(s)$ or $H_2S(s)$

109. An oxide of titanium crystallizes with the following unit cell (titanium = gray; oxygen = red). What is the formula of the oxide?

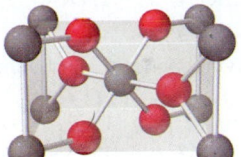

110. An oxide of rhenium crystallizes with the following unit cell (rhenium = gray; oxygen = red). What is the formula of the oxide?

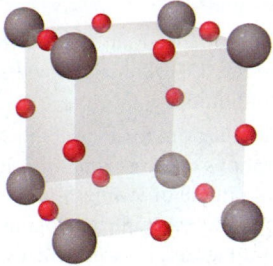

111. The unit cells for cesium chloride and barium(II) chloride are shown below. Show that the ratio of cations to anions in each unit cell corresponds to the ratio of cations to anions in the formula of each compound.

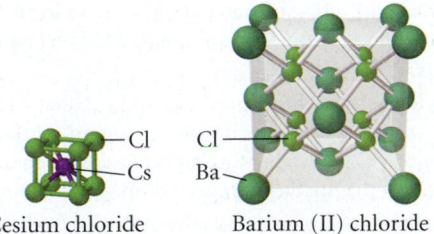

Cesium chloride   Barium (II) chloride

112. The unit cells for lithium oxide and silver iodide are shown below. Show that the ratio of cations to anions in each unit cell corresponds to the ratio of cations to anions in the formula of each compound

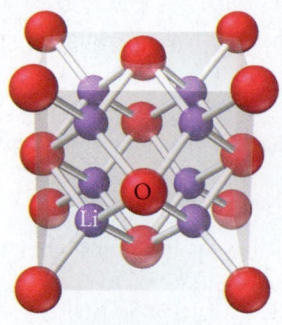
Lithium oxide

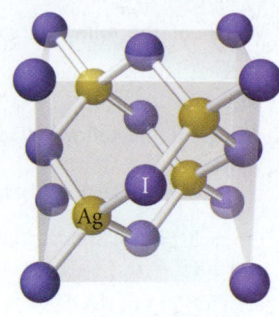
Silver iodide

Band Theory

113. Which of the following would you expect to have little or no band gap?

    a. $Zn(s)$ b. $Si(s)$ c. $As(s)$

114. How many molecular orbitals are present in the valence band of a sodium crystal with a mass of 5.45 g?

115. Would each of the following form an n-type or a p-type semiconductor?

    a. germanium doped with gallium
    b. silicon doped with arsenic

116. Would each of the following form an n-type or a p-type semiconductor?

    a. silicon doped with gallium
    b. germanium doped with antimony

Cumulative Problems

117. Explain the observed trend in the melting points of the alkyl halides.

HI	−50.8 °C
HBr	−88.5 °C
HCl	−114.8 °C
HF	−83.1 °C

118. Explain the following trend in the boiling points of the compounds listed.

H_2Te	−2 °C
H_2Se	−41.5 °C
H_2S	−60.7 °C
H_2O	+100 °C

119. The vapor pressure of water at 25 °C is 23.76 torr. If 1.25 g of water is enclosed in a 1.5-L container, will any liquid be present? If so, what mass of liquid?

120. The vapor pressure of CCl_3F at 300 K is 856 torr. If 11.5 g of CCl_3F is enclosed in a 1.0-L container, will any liquid be present? If so, what mass of liquid?

121. Examine the phase diagram for iodine shown in Figure 11.39(a). What phase transitions occur as you uniformly increase the pressure on a gaseous sample of iodine from 0.010 atm at 185 °C to 100 atm at 185 °C? Make a graph, analogous to the heating curve for water shown in Figure 11.36, in which you plot pressure versus time during the pressure increase.

122. Carbon tetrachloride displays a triple point at 249.0 K and a melting point (at 1 atm) of 250.3 K. Which phase of carbon tetrachloride is more dense, the solid or the liquid? Explain.

123. Four ice cubes at exactly 0 °C having a total mass of 53.5 g are combined with 115 g of water at 75 °C in an insulated container. If no heat is lost to the surroundings, what will be the final temperature of the mixture?

124. A sample of steam with a mass of 0.552 g and at a temperature of 100 °C condenses into an insulated container holding 4.25 g of water at 5.0 °C. Assuming that no heat is lost to the surroundings, what will be the final temperature of the mixture?

125. Air conditioners not only cool air, but dry it as well. Suppose that a room in a home measures 6.0 m × 10.0 m × 2.2 m. If the outdoor temperature is 30 °C and the vapor pressure of water in the air is 85% of the vapor pressure of water at this temperature, what mass of water must be removed from the air each time the volume of air in the room is cycled through the air conditioner? The vapor pressure for water at 30 °C is 31.8 torr.

126. A sealed flask contains 0.55 g of water at 28 °C. The vapor pressure of water at this temperature is 28.36 mmHg. What is the minimum volume of the flask in order that no liquid water be present in the flask?

127. Silver iodide crystallizes in the zinc blende structure. The separation between nearest neighbor cations and anions is approximately 325 pm and the melting point is 558 °C. Cesium chloride, by contrast, crystallizes in the cesium chloride structure shown in Figure 11.51. Even though the separation between nearest neighbor cations and anions is greater (348 pm), the melting point is still higher (645 °C). Explain why the melting point of cesium chloride is higher than that of silver iodide.

128. Copper iodide crystallizes in the zinc blende structure. The separation between nearest neighbor cations and anions is approximately 311 pm and the melting point is 606 °C. Potassium chloride, by contrast, crystallizes in the rock salt structure. Even though the separation between nearest neighbor cations and anions is greater (319 pm), the melting point is still higher (776 °C). Explain why the melting point of potassium chloride is higher than that of copper iodide.

129. Consider the face-centered cubic structure shown below:

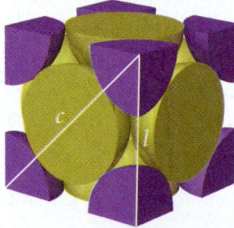

    a. What is the length of the line (labeled c) that runs diagonally across one of the faces of the cube in terms of r (the atomic radius)?
    b. Use the answer to part a and the Pythagorean theorem to derive the expression for the edge length (l) in terms of r.

130. Consider the body-centered cubic structure shown below:

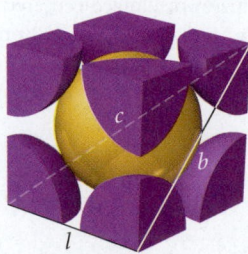

a. What is the length of the line (labeled c) that runs from one corner of the cube diagonally through the center of the cube to the other corner in terms of r (the atomic radius)?

b. Use the Pythagorean theorem to derive an expression for the length of the line (labeled b) that runs diagonally across one of the faces of the cube in terms the edge length (l).

c. Use the answer to part (a) and (b) along with the Pythagorean theorem to derive the expression for the edge length (l) in terms of r.

131. The unit cell in a crystal of diamond belongs to a crystal system different from any we have discussed. The volume of a unit cell of diamond is 0.0454 nm^3 and the density of diamond is 3.52 g/cm^3. Find the number of carbon atoms in a unit cell of diamond.

132. The density of an unknown metal is 12.3 g/cm^3 and its atomic radius is 0.134 nm. It has a face-centered cubic lattice. Find the atomic weight of this metal.

133. Based on the phase diagram of CO_2 shown in Figure 11.39(b), describe the phase changes that occur when the temperature of CO_2 is increased from 190 K to 350 K at a constant pressure of (a) 1 atm, (b) 5.1 atm, (c) 10 atm, (d) 100 atm.

134. Consider a planet where the pressure of the atmosphere at sea level is 2500 mmHg. Will water behave in a way that can sustain life on this planet?

Challenge Problems

135. Potassium chloride crystallizes in the rock salt structure. Estimate the density of potassium chloride using the ionic radii given in Chapter 8.

136. Butane (C_4H_{10}) has a heat of vaporization of 22.44 kJ/mol and a normal boiling point of $-0.4\,°C$. A 250 mL sealed flask contains 0.55 g of butane at $-22\,°C$. How much butane is present as a liquid? If the butane is warmed to 25 °C, how much is present as a liquid?

137. Liquid nitrogen can be used as a cryogenic substance to obtain low temperatures. Under atmospheric pressure, liquid nitrogen boils at 77 K, allowing low temperatures to be reached. However, if the nitrogen is placed in a sealed, insulated container connected to a vacuum pump, even lower temperatures can be reached. Why? If the vacuum pump has sufficient capacity, and is left on for an extended period of time, the liquid nitrogen will start to freeze. Explain.

138. Calculate the fraction of empty space in cubic closest packing to five significant figures.

139. A tetrahedral site in a close-packed lattice is formed by four spheres at the corners of a regular tetrahedron. This is equivalent to placing the spheres at alternate corners of a cube. In such a close-packed arrangement the spheres are in contact and if the spheres have a radius r, the diagonal of the face of the cube is $2r$. The tetrahedral hole is inside the middle of the cube. Find the length of the body diagonal of this cube and then find the radius of the tetrahedral hole.

140. Given that the heat of fusion of water is -6.02 kJ/mol, that the heat capacity of $H_2O(l)$ is 75.2 J/mol · K and that the heat capacity of $H_2O(s)$ is 37.7 J/mol · K, calculate the heat of fusion of water at $-10\,°C$.

Conceptual Problems

141. One prediction of global warming is the melting of global ice, which may result in coastal flooding. A criticism of this prediction is that the melting of icebergs does not increase ocean levels any more than the melting of ice in a glass of water increases the level of liquid in the glass. Is this a valid criticism? Does the melting of an ice cube in a cup of water raise the level of the liquid in the cup? Why or why not? In response to this criticism, scientists have asserted that they are not worried about melting icebergs, but rather the melting of ice sheets that sit on the continent of Antarctica. Would the melting of this ice increase ocean levels? Why or why not?

142. The rate of vaporization depends on the surface area of the liquid. However, the vapor pressure of a liquid does not depend on the surface area. Explain.

143. Substance A has a smaller heat of vaporization than substance B. Which of the two substances will undergo a larger change in vapor pressure for a given change in temperature?

144. The density of a substance is greater in its solid phase than in its liquid phase. If the triple point in the phase diagram of the substance is below 1.0 atm, then which will necessarily be at a lower temperature, the triple point or the normal melting point?

145. A substance has a heat of vaporization of ΔH_{vap} and heat of fusion of ΔH_{fus}. Express the heat of sublimation in terms of ΔH_{vap} and ΔH_{fus}.

146. Examine the heating curve for water in Section 11.7. If heat is added to the water at a constant rate, which of the three segments in which temperature is rising will have the least steep slope? Why?

147. A root cellar is an underground chamber used to store fruits, vegetables, and even meats. In extreme cold, farmers put large vats of water into the root cellar to prevent the fruits and vegetables from freezing. Explain why this works.

12

Solutions

12.1   Thirsty Solutions: Why You Should Not Drink Seawater

12.2   Types of Solutions and Solubility

12.3   Energetics of Solution Formation

12.4   Solution Equilibrium and Factors Affecting Solubility

12.5   Expressing Solution Concentration

12.6   Vapor Pressure of Solutions

12.7   Freezing Point Depression, Boiling Point Elevation, and Osmosis

12.8   Colloids

WE LEARNED IN Chapter 1 that most of the matter we encounter is in the form of mixtures. In this chapter, we focus on homogeneous mixtures, known as solutions. Solutions are mixtures in which atoms and molecules intermingle on the molecular and atomic scale. Some common examples of solutions include the salt water in the world's oceans, the gasoline we put into our cars, and the air we breathe. Why do solutions form? How are their properties different from the properties of the pure substances that compose them? As you read this chapter, keep in mind the great number of solutions that surround you at every moment, including those that exist within your own body.

One molecule of nonsaline substance (held in the solvent) dissolved in 100 molecules of any volatile liquid decreases the vapor pressure of this liquid by a nearly constant fraction, nearly 0.0105.

—FRANÇOIS-MARIE RAOULT (1830–1901)

Drinking seawater causes dehydration because seawater draws water out of body tissues.

12.1 Thirsty Solutions: Why You Should Not Drink Seawater

In a popular novel, *Life of Pi* by Yann Martel, the main character (whose name is Pi) is stranded on a lifeboat with a Bengal tiger in the middle of the Pacific Ocean for 227 days. He survives in part by distilling seawater for drinking using a solar still. In the first three days of his predicament, however, he becomes severely dehydrated from lack of water. Although he is surrounded by seawater, drinking that water would only have made his condition worse. Why? Seawater actually draws water *out of the body* as it passes through the stomach and intestines, resulting in diarrhea and further dehydration. We can think of seawater as a *thirsty solution*—it draws more water to itself. Consequently, seawater should never be consumed as drinking water.

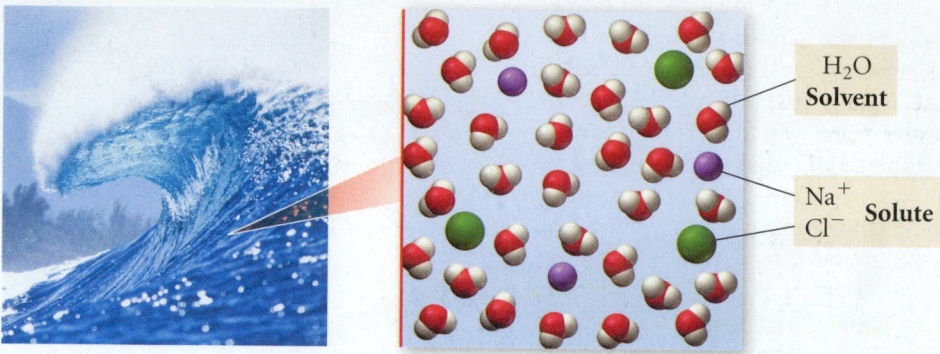

▲ FIGURE 12.1 A Typical Solution In seawater, sodium chloride is the primary solute. Water is the solvent.

In some cases, the concepts of solute and solvent are not useful. For example, a homogeneous mixture of water and ethanol can contain equal amounts of both components and neither component can then be identified as the solvent.

Seawater (Figure 12.1 ▲) is an example of a **solution**, a homogeneous mixture of two or more substances. A solution has at least two components. The majority component is usually called the **solvent** and the minority component is usually called the **solute**. In seawater, water is the solvent and sodium chloride is the primary solute.

The reason that seawater draws water to itself is related to nature's tendency toward spontaneous mixing, which we discuss in more detail later in this chapter and in Chapter 17. For now, we can point to the observation that, unless it is highly unfavorable energetically, substances tend to combine into uniform mixtures, not separate into pure substances. For example, suppose that pure water and a sodium chloride solution are enclosed in separate compartments with a removable barrier between them, as shown in Figure 12.2(a) ▼. If the barrier is removed, the two liquids will spontaneously mix together, eventually forming a more dilute salt solution of uniform concentration, as shown in Figure 12.2(b). The tendency toward mixing, therefore, results in a uniform concentration of the final solution.

Spontaneous Mixing

When the barrier is removed, spontaneous mixing occurs, producing a solution of uniform concentration.

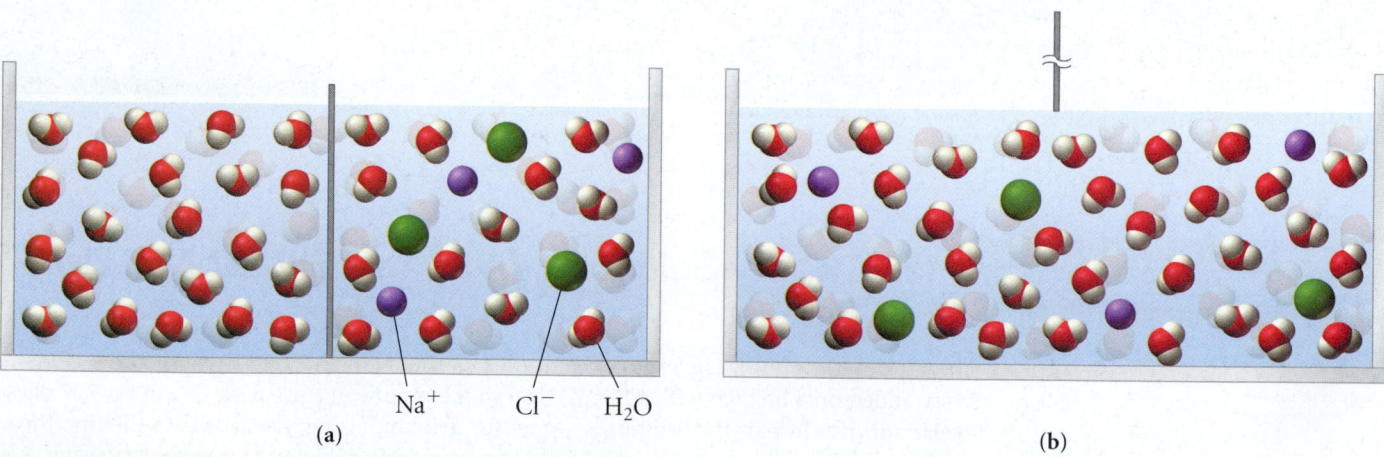

(a)   **(b)**

▲ FIGURE 12.2 The Tendency to Mix **(a)** Pure water and a sodium chloride solution are separated by a barrier. **(b)** When the barrier is removed, the two liquids spontaneously mix, producing a single solution of uniform concentration.

Seawater is a *thirsty* solution because of this tendency toward mixing.
As seawater moves through the intestine, it flows past cells that line the digestive tract, which consist of largely
fluid interiors surrounded by membranes. Although cellular fluids themselves contain dissolved ions, including
sodium and chloride, the fluids are
more dilute than seawater. Nature's
tendency toward mixing (which tends
to produce solutions of uniform concentration), together with the selective
permeability of the cell membranes
(which allow water to flow in and out,
but restrict the flow of dissolved solids), causes a *flow of solvent out of the body's cells and
into the seawater*. In this way, the two solutions become more similar in concentration
(as though they had mixed)—the solution in the intestine becomes somewhat more
dilute than it was and the solution in the cells becomes somewhat more concentrated.
The accumulation of extra fluid in the intestines causes diarrhea, and the decreased fluid
in the cells causes dehydration. If Pi had drunk the seawater instead of constructing the
solar still, neither he nor the Bengal tiger would have survived their ordeal.

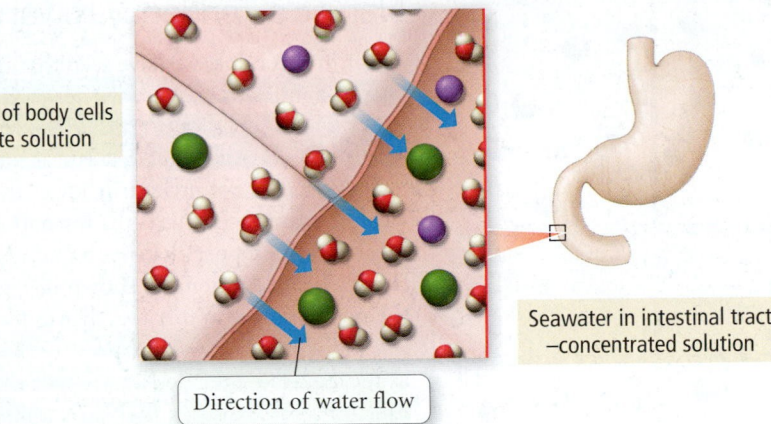

Interior of body cells
–dilute solution

Seawater in intestinal tract
–concentrated solution

Direction of water flow

▲ Seawater is a more concentrated solution than the fluids in body cells. As a
result, when seawater flows through the
digestive tract, it draws water out of the
surrounding tissues.

12.2 Types of Solutions and Solubility

A solution may be composed of a solid and a liquid (such as the salt and water that are the
primary components of seawater), but it may also be composed of a gas and a liquid,
two different liquids, or other combinations (see Table 12.1). **Aqueous solutions** contain
water as the solvent, and a solid, liquid, or gas as the solute. For example, solid sugar or salt
readily dissolves in water. Similarly, ethyl alcohol—the alcohol in alcoholic beverages—
readily mixes with water to form a solution, and carbon dioxide dissolves in water to form
the carbonated water that we know as club soda.

   You probably know from experience that a particular solvent, such as water, does not
dissolve all possible solutes. For example, if your hands are covered with automobile
grease, you cannot remove that grease with plain water. However, another solvent, such as
paint thinner, for example, easily dissolves the grease. We say that the grease is *insoluble* in
water but *soluble* in the paint thinner. The **solubility** of a substance is the amount of the
substance that will dissolve in a given amount of solvent. For example, the solubility
of sodium chloride in water at 25 °C is 36 g NaCl per 100 g water, while the solubility of
grease in water is nearly zero. The solubility of one substance in another depends both
on nature's tendency toward mixing that we discussed in Section 12.1 and on the types of
intermolecular forces that we discussed in Chapter 11.

CO_2 H_2O

▲ Club soda is a solution of carbon dioxide and water.

The general solubilities of a number of ionic
compounds are described by the solubility
rules in Section 4.5.

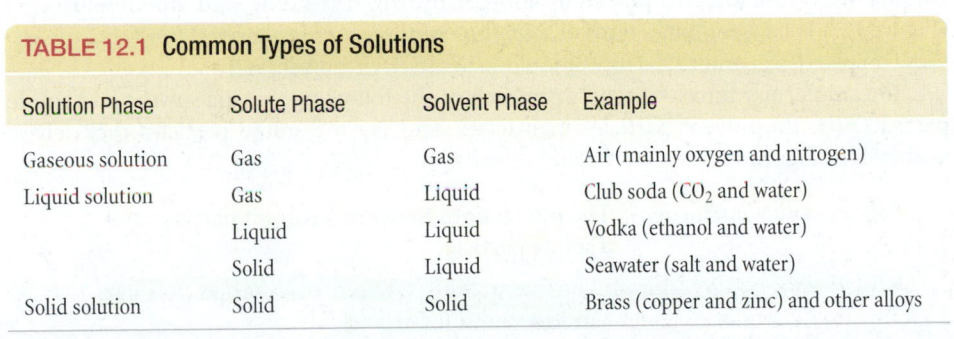

TABLE 12.1 Common Types of Solutions			
Solution Phase	Solute Phase	Solvent Phase	Example
Gaseous solution	Gas	Gas	Air (mainly oxygen and nitrogen)
Liquid solution	Gas	Liquid	Club soda (CO_2 and water)
	Liquid	Liquid	Vodka (ethanol and water)
	Solid	Liquid	Seawater (salt and water)
Solid solution	Solid	Solid	Brass (copper and zinc) and other alloys

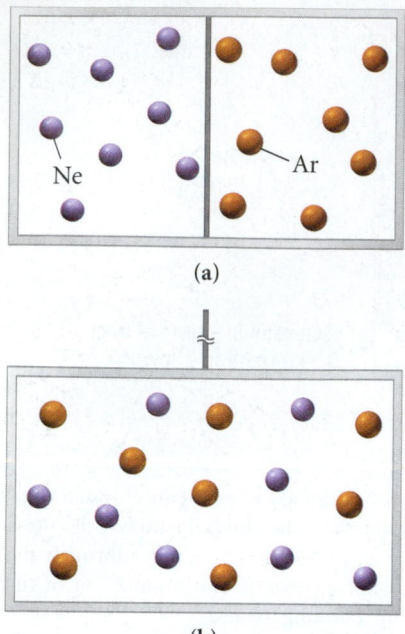

(a)

(b)

▲ FIGURE 12.3 Spontaneous Mixing of Two Ideal Gases (a) Neon and argon are separated by a barrier. (b) When the barrier is removed, the two gases spontaneously mix to form a uniform solution.

Nature's Tendency toward Mixing: Entropy

So far in this book, we have seen that many chemical systems tend toward lower *potential energy*. For example, two particles with opposite charges (such as a proton and an electron or a cation and an anion) move toward each other because their potential energy is lowered as their separation decreases according to Coulomb's law. However, the formation of a solution *does not necessarily* lower the potential energy of its constituent particles. The clearest example of this is the formation of a homogeneous mixture, or a *solution*, of two ideal gases. Suppose that neon and argon are enclosed in a container with a removable barrier between them, as shown in Figure 12.3(a) ◄. As soon as the barrier is removed, the neon and argon mix together to form a solution, as shown in Figure 12.3(b). *Why?*

Recall that at low pressures and moderate temperatures both neon and argon behave as ideal gases—they do not interact with each other in any way (that is, there are no significant forces between their constituent particles). When the barrier is removed, the two gases mix, but their potential energy remains unchanged. In other words, *we cannot think of the mixing of two ideal gases as lowering their potential energy*. The tendency to mix is related, rather, to a concept called *entropy*.

Entropy is a measure of energy randomization or energy dispersal in a system. Recall that a gas at any temperature above 0 K has kinetic energy due to the motion of its atoms. When neon and argon are confined to their individual compartments, their kinetic energies are also confined to those compartments. However, when the barrier between the compartments is removed, each gas—along with its kinetic energy—becomes *spread out* or *dispersed* over a larger volume. Therefore, the mixture of the two gases has greater energy dispersal, or greater *entropy*, than the separated components. *The pervasive tendency for all kinds of energy to spread out, or disperse, whenever it is not restrained from doing so* is the reason that two ideal gases mix. Another common example of the tendency toward energy dispersal is the transfer of thermal energy from hot to cold. If you heat one end of an iron rod, the thermal energy deposited at the end of the rod will spontaneously spread along the entire length of the rod. In contrast to the mixing of two ideal gases—where the kinetic energy of the particles becomes dispersed over a larger volume because the particles themselves become dispersed—the thermal energy in the rod, initially concentrated in relatively fewer particles, becomes dispersed by being distributed over a larger number of particles. The tendency for energy to disperse is the reason that thermal energy flows from the hot end of the rod to the cold one, and not the other way around. Imagine a metal rod that became spontaneously hotter on one end and ice cold on the other—it simply does not happen because energy does not spontaneously concentrate itself. In Chapter 17, we will see that the dispersal of energy is actually the fundamental criterion that ultimately determines the spontaneity of any process.

The Effect of Intermolecular Forces

We have just seen that, in the absence of intermolecular forces, two substances will spontaneously mix to form a homogeneous solution. We know from Chapter 11, however, that solids and liquids exhibit a number of different types of intermolecular forces including dispersion forces, dipole–dipole forces, hydrogen bonding, and ion–dipole forces (Figure 12.4 ▶). Depending on the exact nature of the forces in the solute and the solvent, these forces may promote the formation of a solution or prevent it.

Intermolecular forces exist between each of the following: (a) the solvent and solute particles, (b) the solvent particles themselves, and (c) the solute particles themselves, as shown in Figure 12.5 ▶

Solvent–solute interactions: The interactions between a solvent particle and a solute particle.

Solvent–solvent interactions: The interactions between a solvent particle and another solvent particle.

Solute–solute interactions: The interactions between a solute particle and another solute particle.

Intermolecular Forces

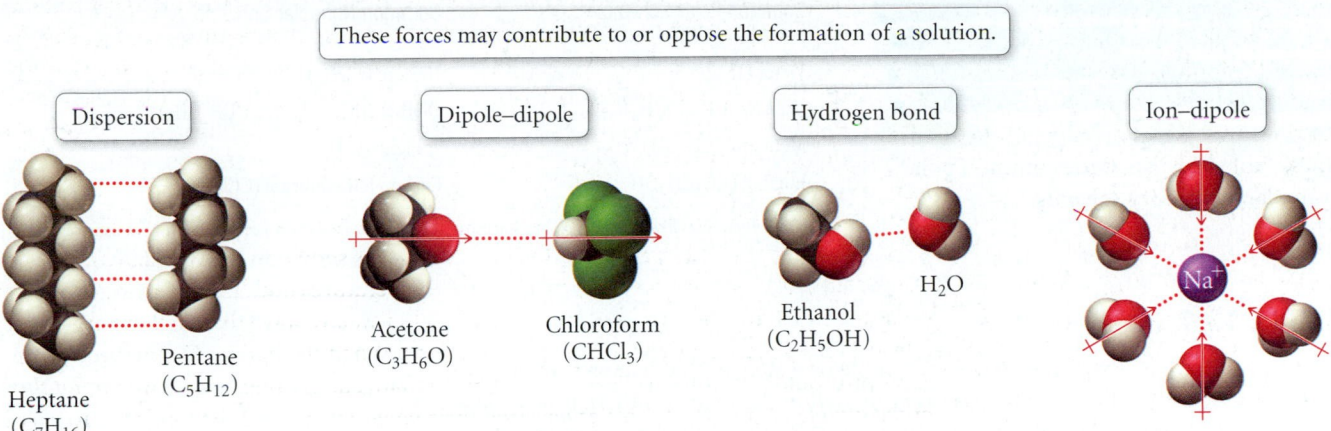

These forces may contribute to or oppose the formation of a solution.

Dispersion — Heptane (C_7H_{16}), Pentane (C_5H_{12})

Dipole–dipole — Acetone (C_3H_6O), Chloroform ($CHCl_3$)

Hydrogen bond — Ethanol (C_2H_5OH), H_2O

Ion–dipole — Na^+

▲ FIGURE 12.4 Intermolecular Forces Involved in Solutions

As summarized in Table 12.2, a solution always forms if the solvent–solute interactions are comparable to, or stronger than, the solvent–solvent interactions and the solute–solute interactions. For example, consider mixing the hydrocarbons pentane (C_5H_{12}) and heptane (C_7H_{16}.) The intermolecular forces present within both pentane and heptane are dispersion forces. Similarly, the intermolecular forces present *between* heptane and pentane are also dispersion forces. All three interactions are of similar magnitude and the two substances are soluble in each other in all proportions—they are said to be **miscible**. The formation of the solution is driven by the tendency toward mixing, or toward greater entropy, that we just discussed.

If the solvent–solute interactions are weaker than the solvent–solvent and solute–solute interactions—in other words, if the solvent molecules and the solute molecules each interact more strongly with molecules of their own kind than with molecules of the other kind—then a solution may still form, depending on the relative disparities between the interactions. If the disparity is small, the tendency to mix still results in the formation of a solution even though the process is energetically uphill. If the disparity is large, however, a solution will not form. For example, consider mixing hexane and water. The water molecules have strong hydrogen-bonding attractions to each other but cannot hydrogen bond to hexane. The energy required to pull water molecules away from one another is too great, and too little energy is returned when the water molecules interact with hexane molecules. As a result, a solution does not form when hexane and water are mixed. Although the tendency to mix is strong, it cannot overcome the large energy disparity between the powerful solvent–solvent interactions and the weak solvent–solute interactions.

Solution Interactions

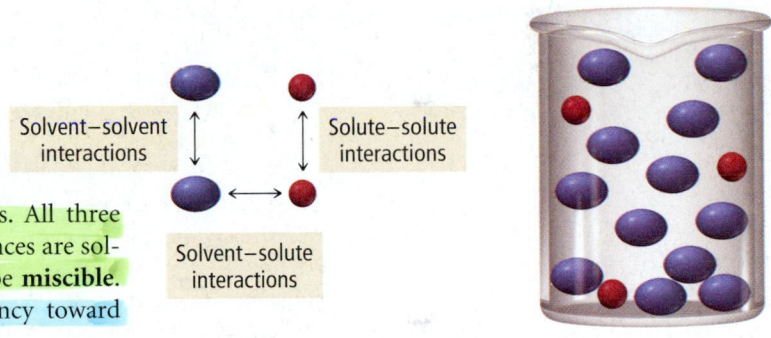

Solvent–solvent interactions

Solute–solute interactions

Solvent–solute interactions

Solution

▲ FIGURE 12.5 Forces in a Solution The relative strengths of these three interactions determine whether a solution will form.

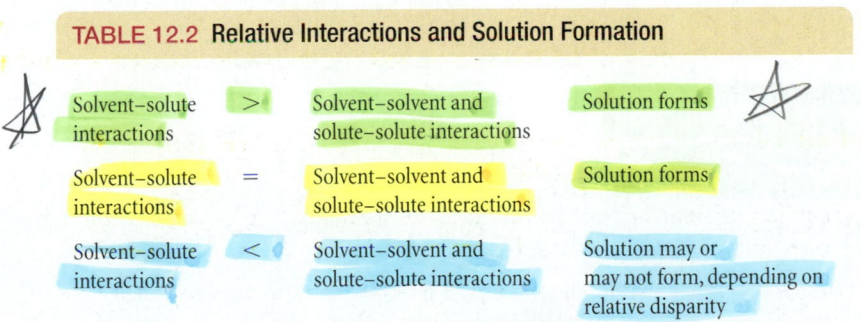

TABLE 12.2 Relative Interactions and Solution Formation

Solvent–solute interactions	>	Solvent–solvent and solute–solute interactions	Solution forms
Solvent–solute interactions	=	Solvent–solvent and solute–solute interactions	Solution forms
Solvent–solute interactions	<	Solvent–solvent and solute–solute interactions	Solution may or may not form, depending on relative disparity

In general, we can use the rule of thumb that *like dissolves like* when predicting the formation of solutions. Polar solvents, such as water, tend to dissolve many polar or ionic solutes, and nonpolar solvents, such as hexane, tend to dissolve many nonpolar solutes. Similar kinds of solvents dissolve similar kinds of solutes. Table 12.3 lists some common polar and nonpolar laboratory solvents.

TABLE 12.3 Common Laboratory Solvents

Common Polar Solvents	Common Nonpolar Solvents
Water (H_2O)	Hexane (C_6H_{14})
Acetone (CH_3COCH_3)	Diethyl ether ($CH_3CH_2OCH_2CH_3$)*
Methanol (CH_3OH)	Toluene (C_7H_8)
Ethanol (CH_3CH_2OH)	Carbon tetrachloride (CCl_4)

*Diethyl ether does have a small dipole moment and can be considered to be intermediate between polar and nonpolar.

EXAMPLE 12.1 Solubility

Vitamins are often categorized as either fat soluble or water soluble. Water-soluble vitamins dissolve in body fluids and are easily eliminated in the urine, so there is little danger of overconsumption. Fat-soluble vitamins, on the other hand, can accumulate in the body's fatty deposits. Overconsumption of a fat-soluble vitamin can be detrimental to health. Examine the structures of each of the following vitamins and classify them as either fat soluble or water soluble.

(a) Vitamin C

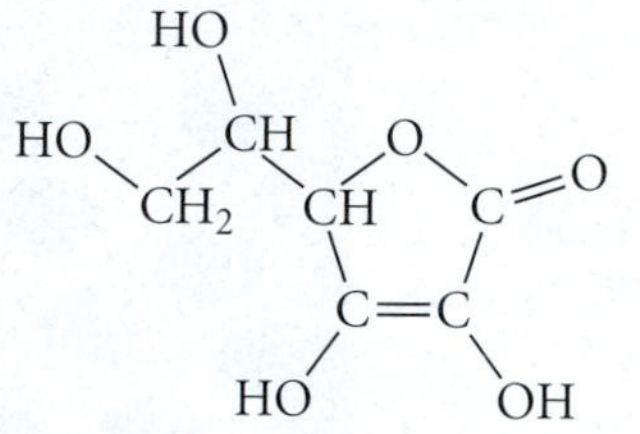

(b) Vitamin K_3

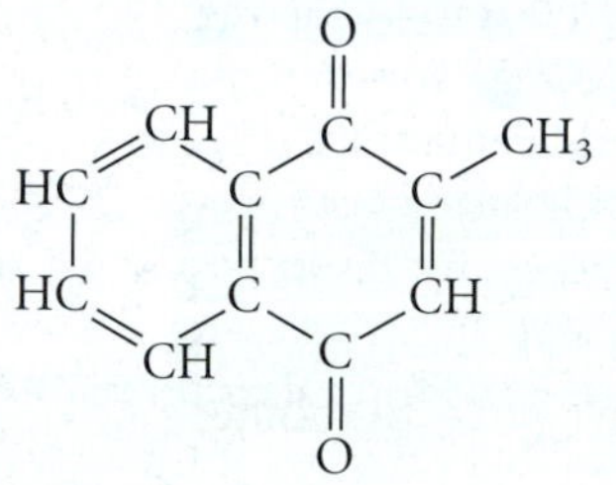

(c) Vitamin A

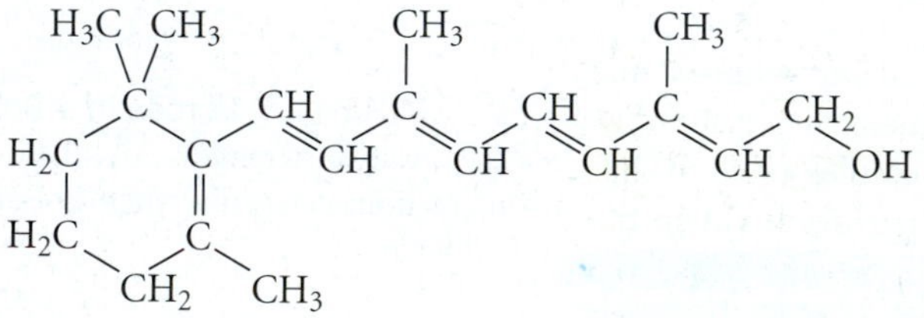

(d) Vitamin B_5

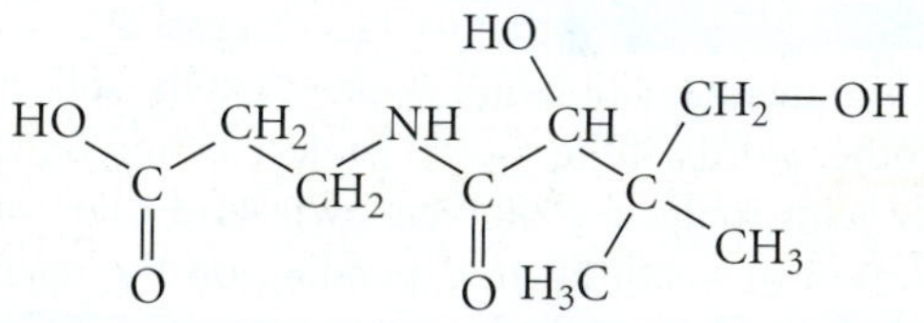

Solution

(a) The four —OH bonds in vitamin C make it highly polar and gives it the ability to hydrogen bond with water. Vitamin C is water soluble.

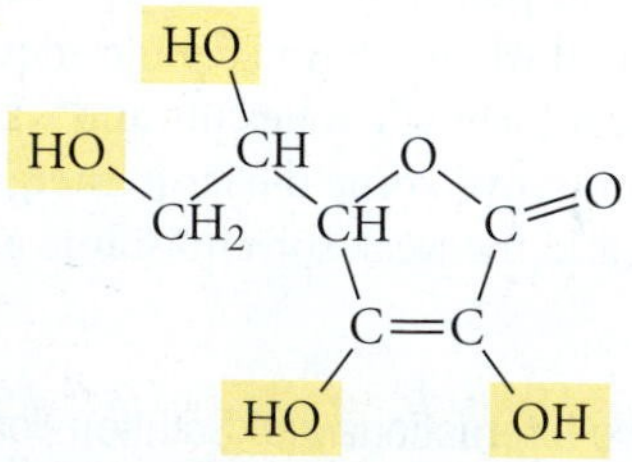

(b) The C—C bonds in vitamin K_3 are nonpolar and the C—H bonds nearly so. The C=O bonds are polar, but the bond dipoles oppose each other and therefore largely cancel, so the molecule is dominated by the nonpolar bonds. Vitamin K_3 is fat soluble.

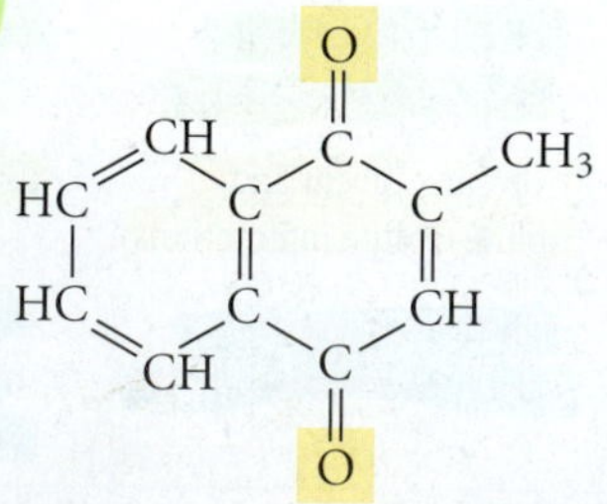

(c) The C—C bonds in vitamin A are nonpolar and the C—H bonds nearly so. The one polar —OH bond may increase the water solubility slightly, but overall vitamin A is nonpolar and therefore fat soluble.	
(d) The three —OH bonds and one —NH bond in vitamin B$_5$ make it highly polar and give it the ability to hydrogen bond with water. Vitamin B$_5$ is water soluble.	

For Practice 12.1

Determine whether each of the following compounds is soluble in hexane.

(a) water (H$_2$O)
(b) propane (CH$_3$CH$_2$CH$_3$)
(c) ammonia (NH$_3$)
(d) hydrogen chloride (HCl)

 Conceptual Connection 12.1 Solubility

Consider the following table showing the solubilities of several alcohols in water and in hexane. Explain the observed trend in terms of intermolecular forces.

Alcohol	Space-Filling Model	Solubility in H$_2$O (mol alcohol/100 g H$_2$O)	Solubility in Hexane (C$_6$H$_{14}$) (mol alcohol/100 g C$_6$H$_{14}$)
Methanol (CH$_3$OH)		Miscible	0.12
Ethanol (CH$_3$CH$_2$OH)		Miscible	Miscible
Propanol (CH$_3$CH$_2$CH$_2$OH)		Miscible	Miscible
Butanol (CH$_3$CH$_2$CH$_2$CH$_2$OH)		0.11	Miscible
Pentanol (CH$_3$CH$_2$CH$_2$CH$_2$CH$_2$OH)		0.030	Miscible

Answer: The first alcohol on the list is methanol, which is highly polar and forms hydrogen bonds with water. It is miscible in water and has only limited solubility in hexane, which is nonpolar. However, as the carbon chain gets longer in the series of alcohols, the OH group becomes less important relative to the growing nonpolar carbon chain. Therefore the alcohols become progressively less soluble in water and more soluble in hexane. This table demonstrates the rule of thumb, *like dissolves like*. Methanol is like water and therefore dissolves in water. It is unlike hexane and therefore has limited solubility in hexane. As you move down the list, the alcohols become increasingly like hexane and increasingly unlike water and therefore become increasingly soluble in hexane and increasingly insoluble in water.

12.3 Energetics of Solution Formation

In Chapter 6, we examined the energy changes associated with chemical reactions. Similar energy changes can occur upon the formation of a solution, depending on the relative interactions of the solute and solvent particles. For example, when sodium hydroxide is dissolved in water, heat is evolved—the solution process is *exothermic*. In contrast, when ammonium nitrate (NH_4NO_3) is dissolved in water, heat is absorbed—this solution process is *endothermic*. Other solutions, such as sodium chloride in water, barely absorb or evolve any heat upon formation. What causes these different behaviors?

We can understand the energy changes associated with solution formation by envisioning the process as occurring in the following three steps, each with an associated change in enthalpy:

1. Separating the solute into its constituent particles.

$\Delta H_{solute} > 0$

This step is always endothermic (positive ΔH) because energy is required to overcome the forces that hold the solute together.

2. Separating the solvent particles from each other to make room for the solute particles.

$\Delta H_{solvent} > 0$

This step is also endothermic because energy is required to overcome the intermolecular forces among the solvent particles.

3. Mixing the solute particles with the solvent particles.

$\Delta H_{mix} < 0$

This step is exothermic, because energy is released as the solute particles interact with the solvent particles through the various types of intermolecular forces.

According to Hess's law, the overall enthalpy change upon solution formation, called the **enthalpy of solution (ΔH_{soln})**, is then the sum of the changes in enthalpy for each step:

$$\Delta H_{soln} = \underset{\text{endothermic (+)}}{\Delta H_{solute}} + \underset{\text{endothermic (+)}}{\Delta H_{solvent}} + \underset{\text{exothermic (−)}}{\Delta H_{mix}}$$

Since the first two terms are endothermic (positive ΔH) and the third term is exothermic (negative ΔH), the overall sign of ΔH_{soln} depends on the magnitudes of the individual terms, as shown in Figure 12.6 ▼.

Energetics of Solution Formation

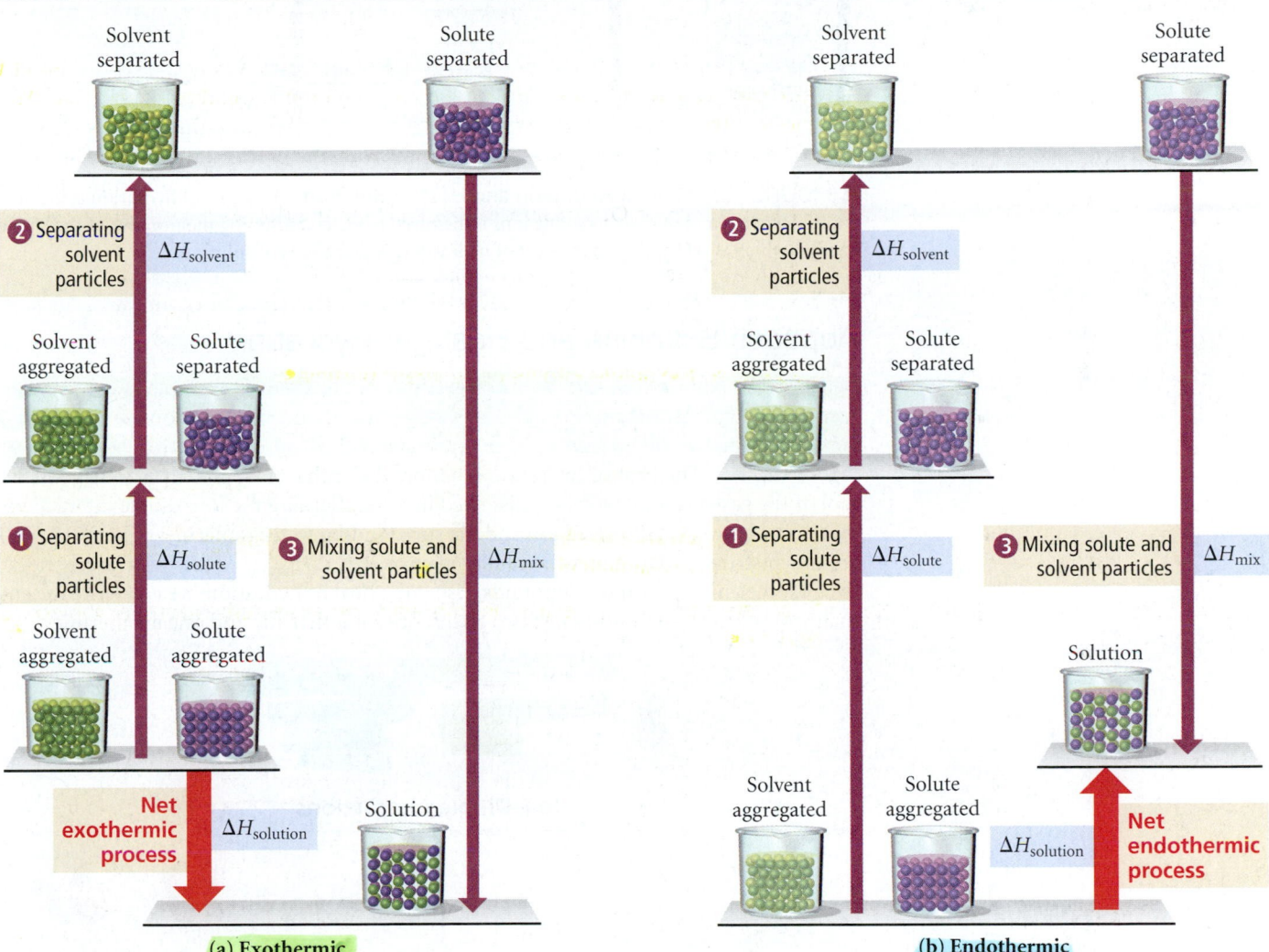

(a) Exothermic   **(b) Endothermic**

▲ **FIGURE 12.6 Energetics of the Solution Process** **(a)** When ΔH_{mix} is greater in magnitude than the sum of ΔH_{solute} and $\Delta H_{solvent}$, the heat of solution is negative (exothermic). **(b)** When ΔH_{mix} is smaller in magnitude than the sum of ΔH_{solute} and $\Delta H_{solvent}$, the heat of solution is positive (endothermic).

1. *If the sum of the endothermic terms is about equal in magnitude to the exothermic term, then ΔH_{soln} is about zero.* The increasing entropy upon mixing drives the solution process while the overall energy of the system remains nearly constant.

2. *If the sum of the endothermic terms is smaller in magnitude than the exothermic term, then ΔH_{soln} is negative and the solution process is exothermic.* In this case, both the tendency toward lower energy and the tendency toward greater entropy drive the formation of a solution.

3. *If the sum of the endothermic terms is greater in magnitude than the exothermic term, then ΔH_{soln} is positive and the solution process is endothermic.* In this case, as long as ΔH_{soln} is not too large, the tendency toward greater entropy will still drive the formation of a solution. If, on the other hand, ΔH_{soln} is too large, a solution will not form.

Heat of Hydration

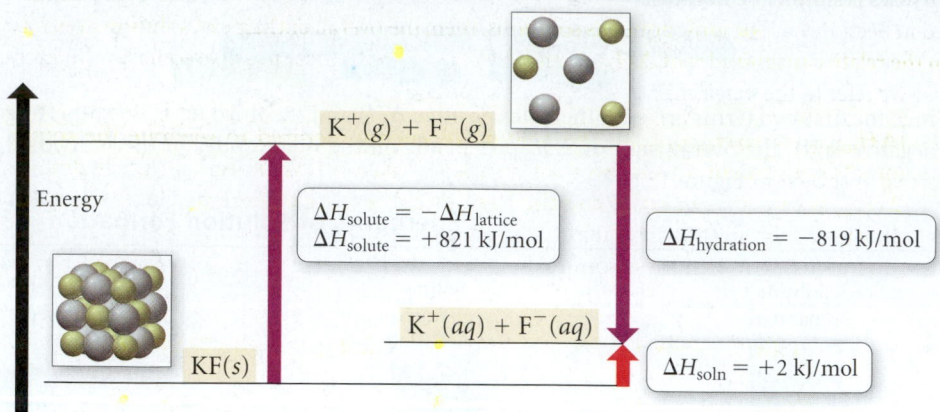

▲ **FIGURE 12.7 Heat of Hydration and Heat of Solution**   The heat of hydration is the heat emitted when 1 mol of gaseous solute ions is dissolved in water. Summing the negative of the lattice energy (which is ΔH_{solute}) and the heat of hydration gives the heat of solution.

Aqueous Solutions and Heats of Hydration

Many common solutions, such as the seawater from the opening example of this chapter, contain an ionic compound dissolved in water. In these aqueous solutions, $\Delta H_{solvent}$ and ΔH_{mix} can be combined into a single term called the **heat of hydration** ($\Delta H_{hydration}$) Figure 12.7 ▲). The heat of hydration is simply the enthalpy change that occurs when 1 mol of the gaseous solute ions are dissolved in water. Because the ion–dipole interactions that occur between a dissolved ion and the surrounding water molecules (Figure 12.8 ▼) are much stronger than the hydrogen bonds in water, $\Delta H_{hydration}$ is always largely negative (exothermic) for ionic compounds. Using the heat of hydration, we can write the enthalpy of solution as a sum of just two terms, one endothermic and one exothermic:

$$\Delta H_{soln} = \Delta H_{solute} + \underbrace{\Delta H_{solvent} + \Delta H_{mix}}$$

$$\Delta H_{soln} = \underset{\substack{\text{endothermic} \\ \text{(positive)}}}{\Delta H_{solute}} + \underset{\substack{\text{exothermic} \\ \text{(negative)}}}{\Delta H_{hydration}}$$

Ion–Dipole Interactions

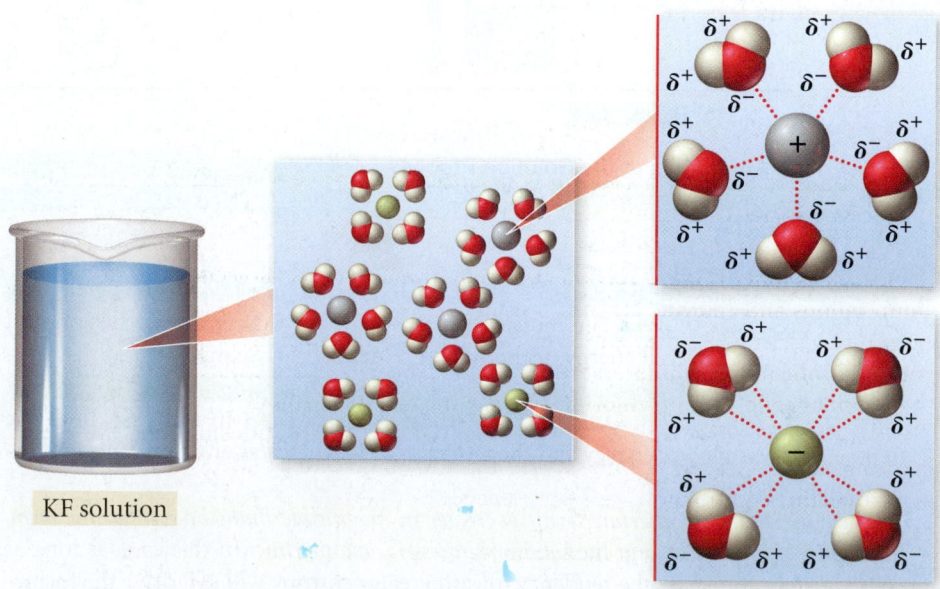

▶ **FIGURE 12.8 Ion–Dipole Interactions**   Ion–dipole interactions such as those between potassium ions, fluoride ions, and water molecules cause the heat of hydration to be largely negative (exothermic).

KF solution

For ionic compounds, ΔH_{solute}, the energy required to separate the solute into its constituent particles is simply the negative of the solute's lattice energy ($\Delta H_{solute} = -\Delta H_{lattice}$), discussed in Section 9.4. For ionic aqueous solutions, then, the overall enthalpy of solution depends on the relative magnitudes of ΔH_{solute} and $\Delta H_{hydration}$, with three possible scenarios (in each case we refer to the *magnitude* or *absolute value* of ΔH):

1. $|\Delta H_{solute}| < |\Delta H_{hydration}|$. The amount of energy required to separate the solute into its constituent ions is less than the energy given off when the ions are hydrated. ΔH_{soln} is therefore negative and the solution process is exothermic. Good examples of solutes with negative enthalpies of solution include lithium bromide and potassium hydroxide. When these solutes dissolve in water, the resulting solutions feel warm to the touch.

$$\text{LiBr}(s) \xrightarrow{\text{H}_2\text{O}} \text{Li}^+(aq) + \text{Br}^-(aq) \qquad \Delta H_{soln} = -48.78 \text{ kJ/mol}$$

$$\text{KOH}(s) \xrightarrow{\text{H}_2\text{O}} \text{K}^+(aq) + \text{OH}^-(aq) \qquad \Delta H_{soln} = -57.56 \text{ kJ/mol}$$

2. $|\Delta H_{solute}| > |\Delta H_{hydration}|$. The amount of energy required to separate the solute into its constituent ions is greater than the energy given off when the ions are hydrated. ΔH_{soln} is therefore positive and the solution process is endothermic (if a solution forms at all). Good examples of solutes that form aqueous solutions with positive enthalpies of solution include ammonium nitrate and silver nitrate. When these solutes dissolve in water, the resulting solutions feel cool to the touch.

$$\text{NH}_4\text{NO}_3(s) \xrightarrow{\text{H}_2\text{O}} \text{NH}_4^+(aq) + \text{NO}_3^-(aq) \qquad \Delta H_{soln} = +25.67 \text{ kJ/mol}$$

$$\text{AgNO}_3(s) \xrightarrow{\text{H}_2\text{O}} \text{Ag}^+(aq) + \text{NO}_3^-(aq) \qquad \Delta H_{soln} = +36.91 \text{ kJ/mol}$$

3. $|\Delta H_{solute}| \approx |\Delta H_{hydration}|$. The amount of energy required to separate the solute into its constituent ions is about equal to the energy given off when the ions are hydrated. ΔH_{soln} is therefore approximately zero and the solution process is neither appreciably exothermic nor appreciably endothermic. Good examples of solutes with enthalpies of solution near zero include sodium chloride and sodium fluoride. When these solutes dissolve in water, the resulting solutions do not undergo a noticeable change in temperature.

$$\text{NaCl}(s) \xrightarrow{\text{H}_2\text{O}} \text{Na}^+(aq) + \text{Cl}^-(aq) \qquad \Delta H_{soln} = +3.88 \text{ kJ/mol}$$

$$\text{NaF}(s) \xrightarrow{\text{H}_2\text{O}} \text{Na}^+(aq) + \text{F}^-(aq) \qquad \Delta H_{soln} = +0.91 \text{ kJ/mol}$$

12.4 Solution Equilibrium and Factors Affecting Solubility

The dissolution of a solute in a solvent is an equilibrium process similar to the equilibrium process associated with a phase change (discussed in Chapter 11). Imagine, from a molecular viewpoint, the dissolving of a solid solute such as sodium chloride in a liquid solvent such as water (Figure 12.9 on page 530). Initially, water molecules rapidly solvate sodium cations and chloride anions, resulting in a noticeable decrease in the amount of solid sodium chloride in the water. Over time, however, the concentration of dissolved sodium chloride in the solution increases. This dissolved sodium chloride can then begin to redeposit as solid sodium chloride. Initially the rate of dissolution far exceeds the rate of deposition. But as the concentration of dissolved sodium chloride increases, the rate of deposition also increases. Eventually the rates of dissolution and deposition become equal—**dynamic equilibrium** has been reached.

$$\text{NaCl}(s) \underset{\text{H}_2\text{O}}{\rightleftharpoons} \text{Na}^+(aq) + \text{Cl}^-(aq)$$

Solution Equilibrium

NaCl(s)

When sodium chloride is first added to water, sodium and chloride ions begin to dissolve into the water.

$$NaCl(s) \longrightarrow Na^+(aq) + Cl^-(aq)$$

As the solution becomes more concentrated, some of the sodium and chloride ions can begin to redeposit as solid sodium chloride.

$$NaCl(s) \rightleftharpoons Na^+(aq) + Cl^-(aq)$$

When the rate of dissolution equals the rate of deposition, dynamic equilibrium has been reached.

(a) Initial

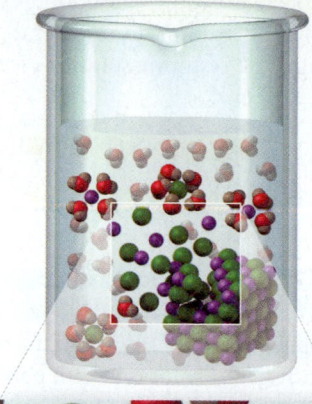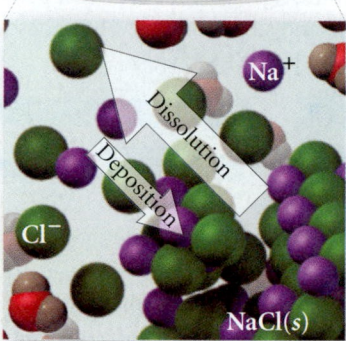

Rate of dissolution > Rate of deposition

(b) Dissolving

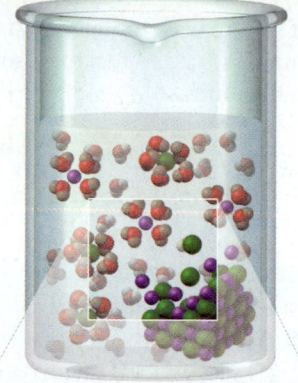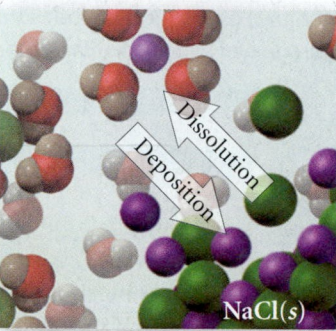

Rate of dissolution = Rate of deposition

(c) Dynamic equilibrium

▲ FIGURE 12.9 Dissolution of NaCl (a) When sodium chloride is first added to water, sodium and chloride ions begin to dissolve into the water. (b) As the solution becomes more concentrated, some of the sodium and chloride ions can begin to redeposit as solid sodium chloride. (c) When the rate of dissolution equals the rate of condensation, dynamic equilibrium has been reached.

A solution in which the dissolved solute is in dynamic equilibrium with the solid (or undissolved) solute is called a **saturated solution**. *If you add additional solute to a saturated solution, it will not dissolve.* A solution containing less than the equilibrium amount of solute is called an **unsaturated solution**. *If you add additional solute to an unsaturated solution, it will dissolve.*

Under certain circumstances, a **supersaturated solution**—one containing more than the equilibrium amount of solute—may form. Such solutions are unstable and the excess solute normally precipitates out of the solution. However, in some cases, if left undisturbed, a supersaturated solution can exist for an extended period of time. For example, a common classroom demonstration involves forming a supersaturated solution of sodium acetate. The addition of a tiny piece of solid sodium acetate to the solution triggers the precipitation of the solute, which crystallizes out of solution in an often beautiful and dramatic way (Figure 12.10 ◄).

▲ FIGURE 12.10 Precipitation from a Supersaturated Solution When a small piece of solid sodium acetate is added to a supersaturated sodium acetate solution, the excess solid precipitates out of the solution.

The Temperature Dependence of the Solubility of Solids

The solubility of solids in water can be highly dependent on temperature. Have you ever noticed how much more sugar you can dissolve in hot tea than in cold tea? Although there are several exceptions, *the solubility of most solids in water increases with increasing temperature,* as shown in Figure 12.11 ▼. For example, the solubility of potassium nitrate (KNO_3) at room temperature is about 37 g KNO_3 per 100 g of water. At 50 °C, however, the solubility rises to 88 g KNO_3 per 100 g of water.

In the case of sugar dissolving in water, the higher temperature increases both *how fast* the sugar dissolves and *how much* sugar dissolves.

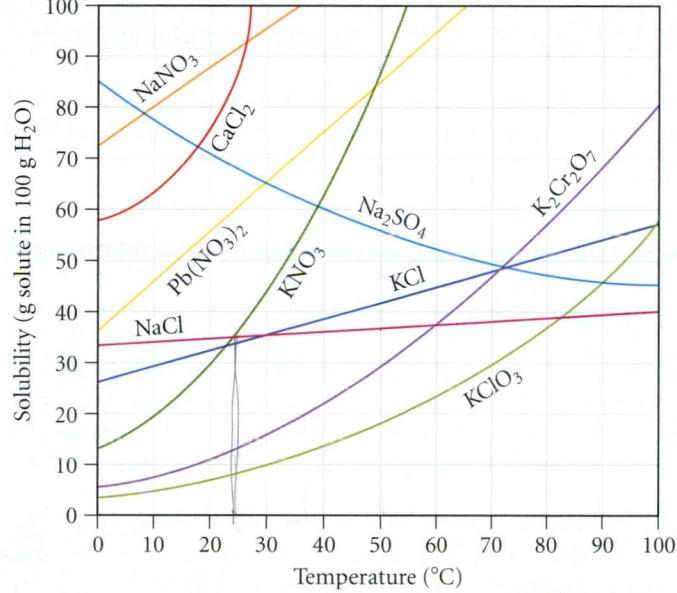

◀ **FIGURE 12.11 Solubility and Temperature**   The solubility of most solids increases with increasing temperature.

A common way to purify a solid is a technique called **recrystallization**. In this technique, the solid is put into water (or some other solvent) at an elevated temperature. Enough solid is added to the solvent to create a saturated solution at the elevated temperature. As the solution cools, it becomes supersaturated and the excess solid begins to come out of solution. If the solution cools slowly, the solid forms crystals as it comes out of solution. The crystalline structure tends to reject impurities, resulting in a purer solid.

The temperature dependence of the solubility of solids is also used in making rock candy. To make rock candy, a saturated sucrose (table sugar) solution is prepared at an elevated temperature. A string or stick is left to dangle in the solution, and the solution is allowed to cool and stand for several days. As the solution cools and solvent evaporates, the solution becomes supersaturated and sugar crystals grow on the string or stick. After several days, beautiful and delicious crystals or "rocks" of sugar cover the string, ready to be admired and eaten.

▲ Rock candy is formed by the recrystallization of sugar.

Factors Affecting the Solubility of Gases in Water

Solutions of gases dissolved in water are common. Club soda, for example, is a solution of carbon dioxide and water, and most liquids exposed to air contain some dissolved gases. Fish depend on the oxygen dissolved in lake or sea water for life, and our blood contains dissolved nitrogen, oxygen, and carbon dioxide. Even tap water contains dissolved gases from air. The solubility of a gas in a liquid is affected by both temperature and pressure.

The Effect of Temperature   The effect of temperature on the solubility of a gas in water can be observed by heating ordinary tap water on a stove. Before the water reaches its boiling point, you will see small bubbles develop in the water. These bubbles are the dissolved air (mostly nitrogen and oxygen) coming out of solution. (Once the water boils, the bubbling becomes more vigorous—these larger bubbles are composed of water vapor.) The dissolved air comes out of solution because—unlike solids, whose solubility generally increases with increasing temperature—*the solubility of gases in liquids decreases with increasing temperature.*

Cold soda
pop

Warm soda
pop

▲ Warm soda pop bubbles more than cold soda pop because carbon dioxide is less soluble in the warm solution.

The inverse relationship between gas solubility and temperature is the reason that warm soda pop bubbles more than cold soda pop when you open it and also the reason that warm beer goes flat faster than cold beer. More carbon dioxide comes out of solution at room temperature than at a lower temperature because the gas is less soluble at room temperature. The decreasing solubility of gases with increasing temperature is also the reason that fish don't bite much if the lake you are fishing in is too warm. The warm temperature results in a lower oxygen concentration. With lower oxygen levels, the fish become lethargic and do not strike at any lure or bait you might cast their way.

◆ Conceptual Connection 12.2 Solubility and Temperature

A solution is saturated in both nitrogen gas and potassium bromide at 75 °C. When the solution is cooled to room temperature, which of the following is most likely to occur?

(a) Some nitrogen gas bubbles out of solution.
(b) Some potassium bromide precipitates out of solution.
(c) Some nitrogen gas bubbles out of solution *and* some potassium bromide precipitates out of solution.
(d) Nothing happens.

Answer: (b) Some potassium bromide precipitates out of solution. The solubility of most solids decreases with decreasing temperature. However, the solubility of gases increases with decreasing temperature. Therefore, the nitrogen becomes more soluble and will not bubble out of solution.

The Effect of Pressure The solubility of gases also depends on pressure. The higher the pressure of a gas above a liquid, the more soluble the gas is in the liquid. In a sealed can of soda pop, for example, the carbon dioxide is maintained in solution by a high pressure of carbon dioxide within the can. When the can is opened, this pressure is released and the solubility of carbon dioxide decreases, resulting in bubbling (Figure 12.12 ▶).

The increased solubility of a gas in a liquid can be understood by considering the following cylinders containing water and carbon dioxide gas:

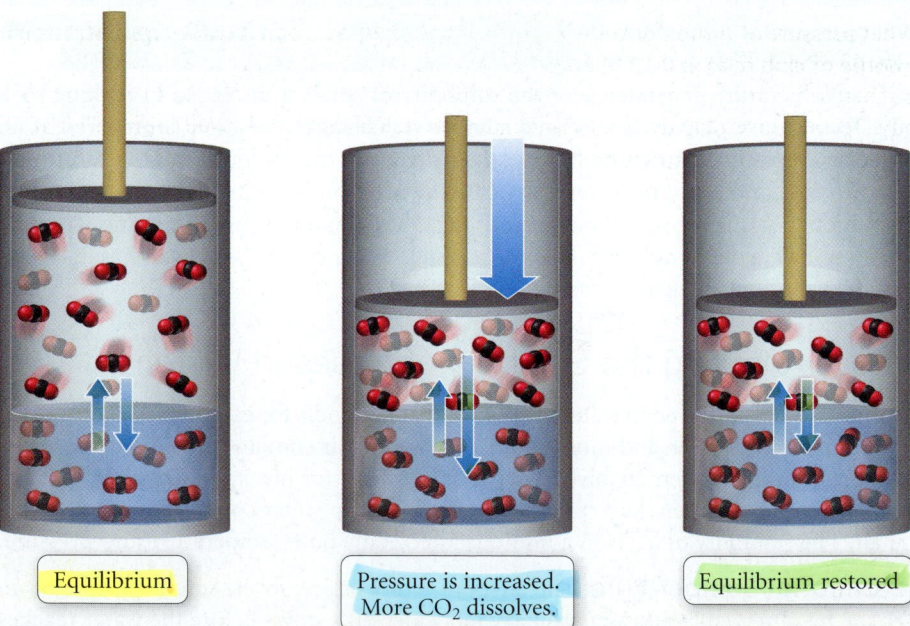

Equilibrium

Pressure is increased.
More CO₂ dissolves.

Equilibrium restored

The first cylinder represents an equilibrium between gaseous and dissolved carbon dioxide—the rate of carbon dioxide molecules entering solution exactly equals the rate of molecules leaving the solution. Now imagine decreasing the volume, as shown in the second cylinder. The pressure of carbon dioxide now increases, causing the rate of molecules entering the solution to rise. The number of molecules in solution increases until equilibrium is established again, as shown in the third cylinder. However, the amount of carbon dioxide in solution is now greater.

The solubility of gases with increasing pressure can be quantified with **Henry's law** as follows:

$$S_{gas} = k_H P_{gas}$$

where S_{gas} is the solubility of the gas (usually in M), k_H is a constant of proportionality (called the *Henry's law constant*) that depends on the specific solute and solvent and also on temperature, and P_{gas} is the partial pressure of the gas (usually in atm). The equation simply shows that the solubility of a gas in a liquid is directly proportional to the pressure of the gas above the liquid. Table 12.4 lists the Henry's law constants for several common gases.

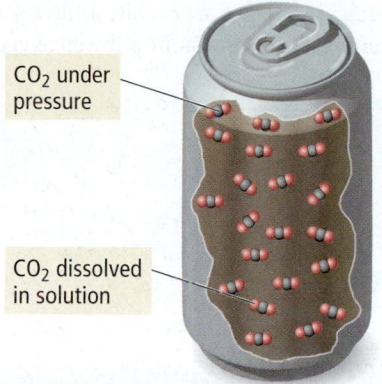
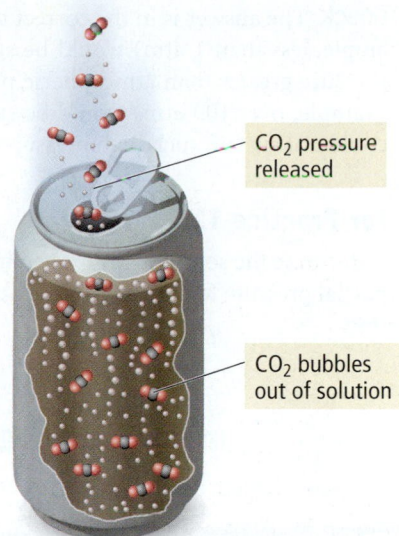

CO₂ under pressure

CO₂ dissolved in solution

CO₂ pressure released

CO₂ bubbles out of solution

▲ **FIGURE 12.12 Soda Fizz**   The bubbling that occurs when a can of soda is opened results from the reduced pressure of carbon dioxide over the liquid. At lower pressure, the carbon dioxide is less soluble and bubbles out of solution.

 Conceptual Connection 12.3 Henry's Law

Examine the Henry's law constants in Table 12.4. Why do you suppose that the constant for ammonia is bigger than the others?

Answer: Ammonia is the only compound on the list that is polar, so we would expect its solubility in water to be greater than those of the other gases (which are all nonpolar).

TABLE 12.4 Henry's Law Constants for Several Gases in Water at 25 °C	
Gas	k_H (M/atm)
O_2	1.3×10^{-3}
N_2	6.1×10^{-4}
CO_2	3.4×10^{-2}
NH_3	5.8×10^{1}
He	3.7×10^{-4}

EXAMPLE 12.2 Henry's Law

What pressure of carbon dioxide is required to keep the carbon dioxide concentration in a bottle of club soda at 0.12 M at 25 °C?

Sort You are given the desired solubility of carbon dioxide and asked to find the pressure required to achieve this solubility.

Given $S_{CO_2} = 0.12$ M
Find P_{CO_2}

Strategize Use Henry's law to find the required pressure from the solubility. You will need the Henry's law constant for carbon dioxide.

Conceptual Plan

$$S_{CO_2} \longrightarrow P_{CO_2}$$

$$S_{CO_2} = k_{H,CO_2} P_{CO_2}$$

Relationships Used
$S_{gas} = k_H P_{gas}$ (Henry's law)
$k_{H,\,CO_2} = 3.4 \times 10^{-2}$ M/atm (from Table 12.4)

Solve Solve the Henry's law equation for P_{CO_2} and substitute the other quantities to compute it.

Solution

$$S_{CO_2} = k_{H,\,CO_2}\, P_{CO_2}$$

$$P_{CO_2} = \frac{S_{CO_2}}{k_{H,\,CO_2}}$$

$$= \frac{0.12\ M}{3.4 \times 10^{-2}\ \dfrac{M}{atm}}$$

$$= 3.5\ atm$$

Check The answer is in the correct units and seems reasonable. A small answer (for example, less than 1 atm) would be suspect because you know that the soda is under a pressure greater than atmospheric pressure when you open it. A very large answer (for example, over 100 atm) would be suspect because an ordinary can or bottle probably could not sustain such high pressures without bursting.

For Practice 12.2

Determine the solubility of oxygen in water at 25 °C exposed to air at 1.0 atm. Assume a partial pressure for oxygen of 0.21 atm.

Chemistry in the Environment
Lake Nyos

Most people living near Lake Nyos in Cameroon, West Africa, began August 22, 1986, like any other day. Unfortunately, the day ended in tragedy. On that evening, a large cloud of carbon dioxide gas, burped up from the depths of Lake Nyos, killed over 1700 people and about 3000 cattle. Two years before that, a similar tragedy had occurred in Lake Monoun, just 60 miles away, killing 37 people. Today, scientists are taking steps to prevent these lakes, both of which are in danger of burping again, from accumulating the carbon dioxide that caused the disaster.

▲ Lake Nyos, in Cameroon, has a deceptively peaceful appearance, but in August 1986, more than 1700 people died around its shores.

Lake Nyos is a water-filled volcanic crater. Some 50 miles beneath the surface of the lake, molten volcanic rock (magma) produces carbon dioxide gas that seeps into the lake through the volcano's plumbing system. The carbon dioxide forms a solution with the lake water. The high pressure at the bottom of the deep lake allows the solution to become highly concentrated in carbon dioxide. Over time—either because of the high concentration itself or because of some

▲ In efforts to prevent another tragedy, scientists have built a plumbing system to slowly vent carbon dioxide from Lake Nyos.

other natural trigger such as a landslide or small earthquake—some gaseous carbon dioxide escaped. The rising bubbles disrupted the stratified layers of lake water, causing water at the bottom of the lake to rise to a region of lower pressure. The drop in pressure decreased the solubility of the carbon dioxide, so more carbon dioxide bubbles formed. This in turn caused more churning and still more carbon dioxide release. The result was a massive cloud of carbon dioxide gas that escaped from the lake. Since carbon dioxide is heavier than air, it traveled down the sides of the volcano and into the nearby valley, displacing air and asphyxiating many of the local residents.

In an effort to keep the events of August 1986 from recurring, scientists are building a piping system that slowly vents carbon dioxide from the lake bottom, preventing the buildup that led to the tragedy.

Question

Suppose that the water pressure at the bottom of Lake Nyos was 25 atm. What would the solubility of carbon dioxide be at that depth?

12.5 Expressing Solution Concentration

As we have seen, the amount of solute in a solution is an important property of the solution. For example, the amount of sodium chloride in a solution determines whether or not the solution will cause dehydration if consumed. A **dilute solution** is one containing small quantities of solute relative to the amount of solvent. Drinking a dilute sodium chloride solution will not cause dehydration. A **concentrated solution** is one containing large quantities of solute relative to the amount of solvent. Drinking a concentrated sodium chloride solution will cause dehydration. The common ways of reporting solution concentration include molarity, molality, parts by mass, parts by volume, mole fraction, and mole percent, as summarized in Table 12.5. We have seen two of these units before, molarity in Section 4.4 and mole fraction in Section 5.6. In the following section, we review the terms we have already covered and introduce the new ones.

TABLE 12.5 Solution Concentration Terms

Unit	Definition	Units
Molarity (M)	$\dfrac{\text{amount solute (in mol)}}{\text{volume solution (in L)}}$	$\dfrac{\text{mol}}{\text{L}}$
Molality (m)	$\dfrac{\text{amount solute (in mol)}}{\text{mass solvent (in kg)}}$	$\dfrac{\text{mol}}{\text{kg}}$
Mole fraction (χ)	$\dfrac{\text{amount solute (in mol)}}{\text{total amount of solute and solvent (in mol)}}$	None
Mole percent (mol %)	$\dfrac{\text{amount solute (in mol)}}{\text{total amount of solute and solvent (in mol)}} \times 100\%$	None
Parts by mass	$\dfrac{\text{mass solute}}{\text{mass solution}} \times$ multiplication factor	
Percent by mass (%)	Multiplication factor = 100	%
Parts per million by mass (ppm)	Multiplication factor = 10^6	ppm
Parts per billion by mass (ppb)	Multiplication factor = 10^9	ppb
Parts by volume (%, ppm, ppb)	$\dfrac{\text{volume solute}}{\text{volume solution}} \times$ multiplication factor*	

*Multiplication factors for parts by volume are identical to those for parts by mass.

Molarity

The **molarity (M)** of a solution is the amount of solute (in moles) divided by the volume of solution (in liters).

$$\text{Molarity (M)} = \frac{\text{amount solute (in mol)}}{\text{volume solution (in L)}}$$

Note that molarity is moles of solute per liter of *solution*, not liter of solvent. To make a solution of a specified molarity, you usually put the solute into a flask and then add water (or another solvent) to the desired volume of solution, as shown in Figure 12.13 ▼. Molarity is a convenient unit to use when making, diluting, and transferring solutions because it specifies the amount of solute per unit of solution transferred.

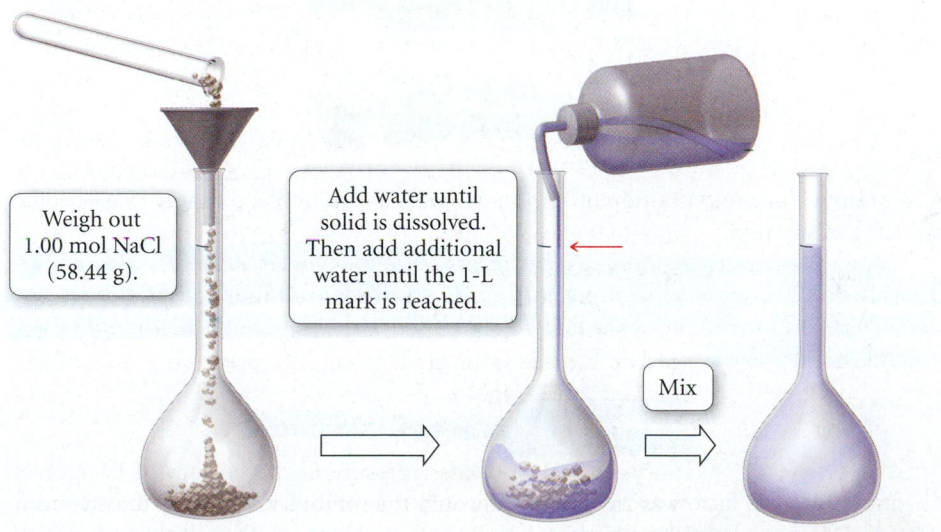

Weigh out 1.00 mol NaCl (58.44 g).

Add water until solid is dissolved. Then add additional water until the 1-L mark is reached.

Mix

A 1.00 molar NaCl solution

◀ FIGURE 12.13 Preparing a Solution of Known Concentration To make a 1 M NaCl solution, add 1 mol of the solid to a flask and dilute with water to make 1 L of solution.

Molality

Since molarity depends on volume, and since volume varies with temperature, molarity also varies with temperature. For example, a 1 M aqueous solution at room temperature will be slightly less than 1 M at an elevated temperature because the volume of the solution is greater at the elevated temperature. A concentration unit that is independent of temperature is **molality (m)**, the amount of solute (in moles) divided by the mass of solvent (in kilograms).

> Note that molality is abbreviated with a lowercase italic m while molarity is abbreviated with a capital M.

$$\text{Molality } (m) = \frac{\text{amount solute (in mol)}}{\text{mass solvent (in kg)}}$$

Notice that molality is defined with respect to kilograms *solvent*, not kilograms solution. Molality is particularly useful when concentrations must be compared over a range of different temperatures.

Parts by Mass and Parts by Volume

It is often convenient to report a concentration as a ratio of masses. A **parts by mass** concentration is the ratio of the mass of the solute to the mass of the solution, all multiplied by a multiplication factor:

$$\frac{\text{Mass solute}}{\text{Mass solution}} \times \text{multiplication factor}$$

The particular unit used, which determines the size of the multiplication factor, depends on the concentration of the solution. For example, to report a concentration as **percent by mass**, the multiplication factor is 100.

$$\text{Percent by mass} = \frac{\text{Mass solute}}{\text{Mass solution}} \times 100\%$$

Percent means *per hundred*, so a solution with a concentration of 14% by mass contains 14 g of solute per 100 g of solution.

For more dilute solutions, the concentration might be reported as **parts per million (ppm)**, which requires a multiplication factor of 10^6, or **parts per billion (ppb)**, which requires a multiplication factor of 10^9.

$$\text{ppm} = \frac{\text{mass solute}}{\text{mass solution}} \times 10^6$$

$$\text{ppb} = \frac{\text{mass solute}}{\text{mass solution}} \times 10^9$$

A solution with a concentration of 15 ppm by mass, for example, contains 15 g of solute per 10^6 g of solution.

Sometimes concentrations are reported as a ratio of volumes, especially for solutions in which both the solute and solvent are liquids. A **parts by volume** concentration is usually the ratio of the volume of the solute to the volume of the solution, all multiplied by a multiplication factor.

$$\frac{\text{Volume solute}}{\text{Volume solution}} \times \text{multiplication factor}$$

> For dilute aqueous solutions near room temperature, the units of ppm are equivalent to milligrams solute/per liter of solution. This is because the density of a dilute solution near room temperature is 1.0 g/mL, so that 1 L has a mass of 1000 g.

The multiplication factors are identical to those just described for parts by mass concentrations. For example, a 22% ethanol solution contains 22 mL of ethanol for every 100 mL of solution.

Using Parts by Mass (or Parts by Volume) in Calculations

The parts by mass (or parts by volume) concentration of a solution can be used as a conversion factor between mass (or volume) of the solute and mass (or volume) of the solution. For example, for a solution containing 3.5% sodium chloride by mass, we would write the following conversion factor:

$$\frac{3.5 \text{ g NaCl}}{100 \text{ g solution}} \quad \text{converts} \quad \boxed{\text{g solution}} \longrightarrow \boxed{\text{g NaCl}}$$

This conversion factor converts from grams solution to grams NaCl. To convert the other way, simply invert the conversion factor:

$$\frac{100 \text{ g solution}}{3.5 \text{ g NaCl}} \quad \text{converts} \quad \boxed{\text{g NaCl}} \longrightarrow \boxed{\text{g solution}}$$

EXAMPLE 12.3 Using Parts by Mass in Calculations

What volume (in mL) of a soft drink that is 10.5% sucrose ($C_{12}H_{22}O_{11}$) by mass contains 78.5 g of sucrose? (The density of the solution is 1.04 g/mL.)

Sort You are given a mass of sucrose and the concentration and density of a sucrose solution, and you are asked to find the volume of solution containing that mass.	**Given** 78.5 g $C_{12}H_{22}O_{11}$ 10.5% $C_{12}H_{22}O_{11}$ by mass density = 1.04 g/mL **Find** mL
Strategize Begin with the mass of sucrose in grams. Use the mass percent concentration of the solution (written as a ratio, as shown under relationships used) to find the number of grams of solution containing this quantity of sucrose. Then use the density of the solution to convert grams to milliliters of solution.	**Conceptual Plan** 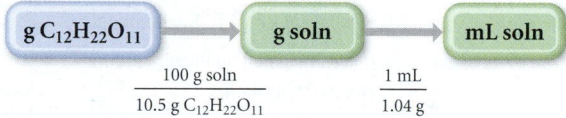 $\boxed{\text{g } C_{12}H_{22}O_{11}} \longrightarrow \boxed{\text{g soln}} \longrightarrow \boxed{\text{mL soln}}$ $\qquad \dfrac{100 \text{ g soln}}{10.5 \text{ g } C_{12}H_{22}O_{11}} \qquad \dfrac{1 \text{ mL}}{1.04 \text{ g}}$ **Relationships Used** $\dfrac{10.5 \text{ g } C_{12}H_{22}O_{11}}{100 \text{ g soln}}$ (percent by mass written as ratio) $\dfrac{1 \text{ mL}}{1.04 \text{ g}}$ (given density of the solution)
Solve Begin with 78.5 g $C_{12}H_{22}O_{11}$ and multiply by the conversion factors to arrive at the volume of solution.	**Solution** $78.5 \text{ g } C_{12}H_{22}O_{11} \times \dfrac{100 \text{ g soln}}{10.5 \text{ g } C_{12}H_{22}O_{11}} \times \dfrac{1 \text{ mL}}{1.04 \text{ g}} = 719 \text{ mL soln}$

Check The units of the answer are correct. The magnitude seems correct because the solution is approximately 10% sucrose by mass. Since the density of the solution is approximately 1 g/mL, the volume containing 78.5 g sucrose should be roughly 10 times larger, as calculated ($719 \approx 10 \times 78.5$).

For Practice 12.3

How much sucrose ($C_{12}H_{22}O_{11}$), in g, is contained in 355 mL (12 ounces) of a soft drink that is 11.5% sucrose by mass? (Assume a density of 1.04 g/mL.)

For More Practice 12.3

A water sample is found to contain the pollutant chlorobenzene with a concentration of 15 ppb (by mass). What volume of this water contains 5.00×10^2 mg of chlorobenzene? (Assume a density of 1.00 g/mL.)

Chemistry in the Environment
The Dirty Dozen

A number of potentially harmful chemicals—such as DDT, dioxin, and polychlorinated biphenyls (PCBs)—can make their way into our water sources from industrial dumping, atmospheric emissions, agricultural use, and household dumping. Since crops, livestock, and fish all rely on water, they too can accumulate these chemicals from water. Human consumption of food or water contaminated with harmful chemicals can lead to a number of diseases and adverse health effects such as increased cancer risk, liver damage, and central nervous system damage. Governments around the world have joined forces to ban a number of these kinds of chemicals—called persistent organic pollutants or POPs—from production. The original treaty targeted 12 such substances called the dirty dozen (Table 12.6).

TABLE 12.6 The Dirty Dozen

1. Aldrin—insecticide	7. Furan—industrial by-product
2. Chlordane—insecticide	8. Heptachlor—insecticide
3. DDT—insecticide	9. Hexachlorobenzene—fungicide, industrial by-product
4. Dieldrin—insecticide	10. Mirex—insecticide, fire retardant
5. Dioxin—industrial by-product	11. Polychlorinated biphenyls (PCBs)—electrical insulators
6. Eldrin—insecticide	12. Toxaphene—insecticide

▲ Potentially harmful chemicals can make their way into water sources by many routes.

One difficult problem with all of these chemicals is their persistence. These compounds are fairly stable and are not broken down under normal environmental conditions. Once they get into the environment, they stay there for a long time. A second problem is a process called *bioamplification*. Because these chemicals are nonpolar, they are stored and concentrated in the fatty tissues of the organisms that consume them. As larger organisms eat smaller ones they consume more of the stored chemicals. The result is an increase in the concentrations of these chemicals as they move up the food chain.

Under the treaty, nearly all intentional production of these chemicals has been banned. In the United States, the presence of these contaminants in water supplies is monitored under supervision of the Environmental Protection Agency (EPA). The EPA has set limits, called maximum contaminant levels (MCLs), for each of these in food and drinking water. Some MCLs for selected compounds in water supplies are listed in Table 12.7 at right. Notice the units that the EPA uses to express the concentration of the contaminants, mg/L. This unit is a conversion factor between liters of water consumed and the mass (in mg) of the pollutant. According to the EPA, as long as the contaminant concentrations are below these levels, the water is safe to drink.

TABLE 12.7 EPA Maximum Contaminant Level (MCL) for Several "Dirty Dozen" Chemicals

Chlordane	0.002 mg/L
Dioxin	0.00000003 mg/L
Heptachlor	0.0004 mg/L
Hexachlorobenzene	0.001 mg/L

Question

Using what you know about conversion factors, calculate how much of each of the chemicals in Table 12.7 at the MCL would be present in 715 L of water, the approximate amount of water consumed by an adult in one year.

Mole Fraction and Mole Percent

For some applications, especially those in which the ratio of solute to solvent can vary widely, the most useful way to express concentration is the amount of solute (in moles) divided by the total amount of solute and solvent (in moles). This ratio is called the **mole fraction** (X_{solute}):

The mole fraction can also be defined for the solvent:

$$X_{solvent} = \frac{n_{solvent}}{n_{solute} + n_{solvent}}$$

$$X_{solute} = \frac{\text{amount solute (in mol)}}{\text{total amount of solute and solvent (in mol)}}$$

$$= \frac{n_{solute}}{n_{solute} + n_{solvent}}$$

Also in common use is the **mole percent (mol %)**, which is simply the mole fraction × 100 percent.

$$\text{mol \%} = X_{solute} \times 100\%$$

EXAMPLE 12.4   Calculating Concentrations

A solution is prepared by dissolving 17.2 g of ethylene glycol ($C_2H_6O_2$) in 0.500 kg of water. The final volume of the solution is 515 mL. For this solution, calculate each of the following:

(a) molarity   (b) molality   (c) percent by mass   (d) mole fraction   (e) mole percent

Solution

(a) To calculate molarity, first find the amount of ethylene glycol in moles from the mass and molar mass.

$$\text{mol } C_2H_6O_2 = 17.2 \text{ g } C_2H_6O_2 \times \frac{1 \text{ mol } C_2H_6O_2}{62.07 \text{ g } C_2H_6O_2} = 0.2771 \text{ mol } C_2H_6O_2$$

Then divide the amount in moles by the volume of the solution in liters.

$$\text{Molarity (M)} = \frac{\text{amount solute (in mol)}}{\text{volume solution (in L)}}$$

$$= \frac{0.2771 \text{ mol } C_2H_6O}{0.515 \text{ L solution}}$$

$$= 0.538 \text{ M}$$

(b) To calculate molality, use the amount of ethylene glycol in moles from part a, and divide by the mass of the water in kilograms.

$$\text{Molality (m)} = \frac{\text{amount solute (in mol)}}{\text{mass solvent (in kg)}}$$

$$= \frac{0.2771 \text{ mol } C_2H_6O}{0.500 \text{ kg } H_2O}$$

$$= 0.554 \text{ m}$$

(c) To calculate percent by mass, divide the mass of the solute by the sum of the masses of the solute and solvent and multiply the ratio by 100%.

$$\text{Percent by mass} = \frac{\text{mass solute}}{\text{mass solution}} \times 100\%$$

$$= \frac{17.2 \text{ g}}{17.2 \text{ g} + 5.00 \times 10^2 \text{ g}} \times 100\%$$

$$= 3.33\%$$

(d) To calculate mole fraction, first determine the amount of water in moles from the mass of water and its molar mass.

$$\text{mol } H_2O = 5.00 \times 10^2 \text{ g } H_2O \times \frac{1 \text{ mol } H_2O}{18.02 \text{ g } H_2O} = 27.75 \text{ mol } H_2O$$

Then divide the amount of ethylene glycol in moles (from part a) by the total number of moles.

$$X_{solute} = \frac{n_{solute}}{n_{solute} + n_{solvent}}$$

$$= \frac{0.2771 \text{ mol}}{0.2771 \text{ mol} + 27.75 \text{ mol}}$$

$$= 9.89 \times 10^{-3}$$

(e) To calculate mole percent, simply multiply the mole fraction by 100%.

$$\text{mol \%} = X_{solute} \times 100\%$$

$$= 0.989\%$$

For Practice 12.4

A solution is prepared by dissolving 50.4 g sucrose ($C_{12}H_{22}O_{11}$) in 0.332 kg of water. The final volume of the solution is 355 mL. Calculate each of the following for this solution:
(**a**) molarity   (**b**) molality   (**c**) percent by mass   (**d**) mole fraction   (**e**) mole percent

The interrelationships between these different units are summarized in Figure 12.14 ▼.

Units of Concentration and Solution Quantities

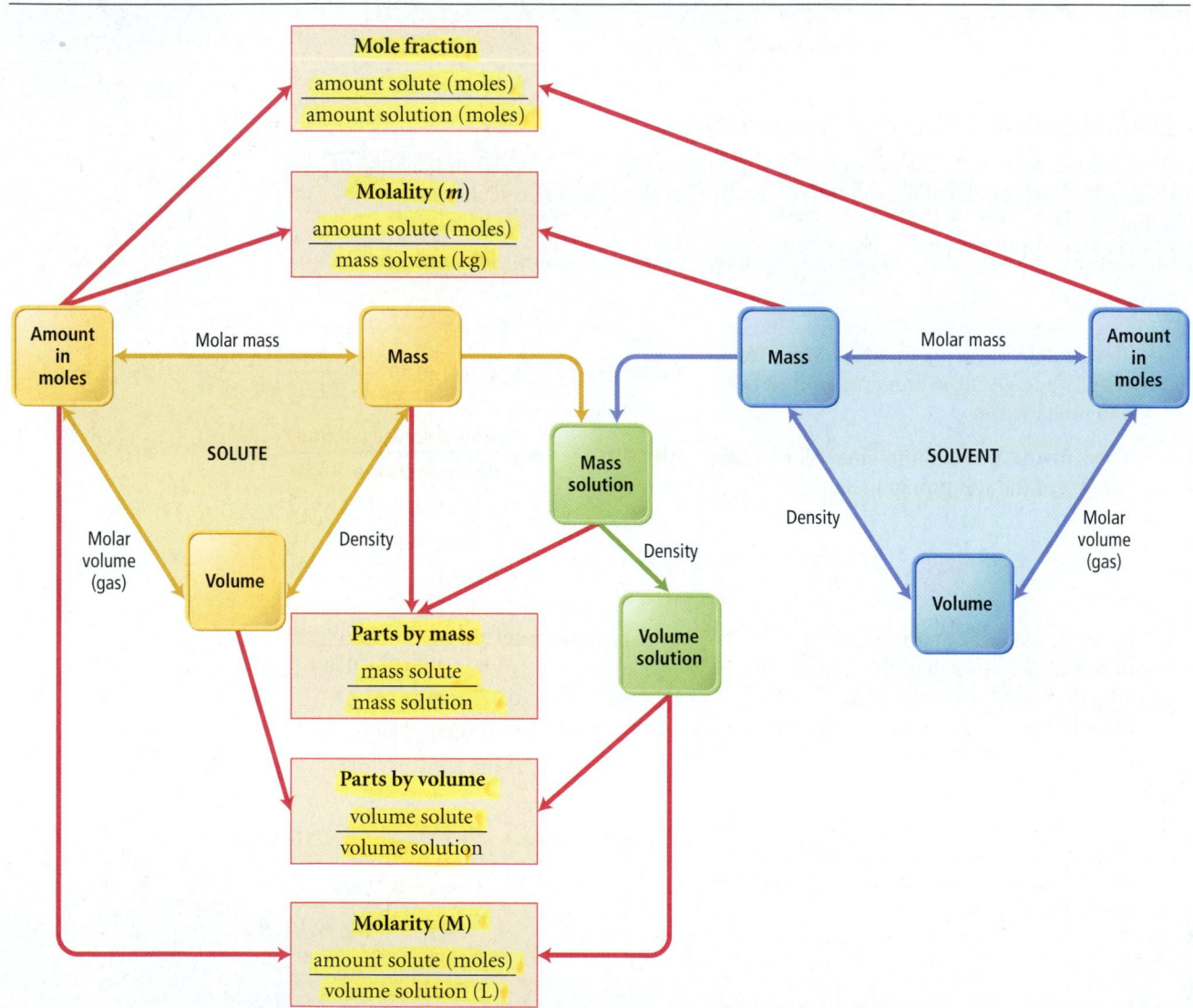

▲ FIGURE 12.14 Relationships among Solution Quantities and Concentration Units

12.6 Vapor Pressure of Solutions

Recall from Section 11.5 that the vapor pressure of a liquid is the pressure of the gas above the liquid when the two are in dynamic equilibrium (that is, when the rate of vaporization equals the rate of condensation). What is the effect of a nonvolatile solute on the vapor pressure of the liquid into which it dissolves? The basic answer to this question is that *the vapor pressure of the solution is lower than the vapor pressure of the pure solvent.* We can understand why this happens in two different ways.

The simplest explanation for the lowering of the vapor pressure of a solution relative to that of the pure solvent is related to the concept of dynamic equilibrium itself. Consider the following representation of a liquid in dynamic equilibrium with its vapor.

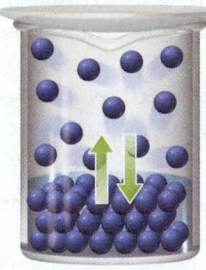

Dynamic
equilibrium

The rate of vaporization is equal to the rate of condensation. When a nonvolatile solute is added, however, the solute particles interfere with the ability of the solvent particles to vaporize, simply because they occupy some of the surface area formerly occupied by the solvent. The rate of vaporization is therefore diminished compared to that of the pure solvent.

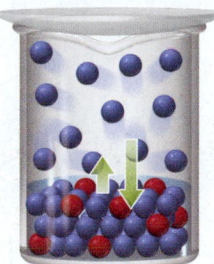

Rate of vaporization
reduced by solute

The change in the rate of vaporization creates an imbalance in the rates; the rate of condensation is now *greater* than the rate of vaporization. The net effect is that some of the molecules that were in the gas phase condense into the liquid. As they condense, the reduced number of molecules in the gas phase causes the rate of condensation to decrease. Eventually the two rates become equal again, but only after the concentration of molecules in the gas phase has decreased.

Equilibrium reestablished
but with fewer molecules
in gas phase

The result is a lower vapor pressure for the solution compared to the pure solvent.

A more fundamental explanation of why the vapor pressure of a solution is lower than that of the pure solvent is related to the tendency toward mixing (or toward greater entropy) that we discussed in Sections 12.1 and 12.2. Recall from Section 12.1 that a concentrated solution is a *thirsty* solution—it has the ability to draw solvent to itself. A dramatic demonstration of this tendency can be seen by placing both a concentrated solution of a nonvolatile solute and a beaker of the pure solvent in a sealed container, as shown below. Over time, the level of the pure solvent will drop and the level of the solution will rise as molecules vaporize out of the pure solvent and condense in the solution. Notice the similarity between this observation and the dehydration caused by drinking seawater. In both cases, a concentrated solution has the ability to draw solvent to itself. The reason is nature's tendency to mix. If a pure solvent and concentrated solution are combined in a beaker, they naturally form a mixture in which the solution is less concentrated than it was initially. Similarly, if a pure solvent and concentrated solution are combined in a sealed container—even though they are in separate beakers—the two mix to form a more dilute solution.

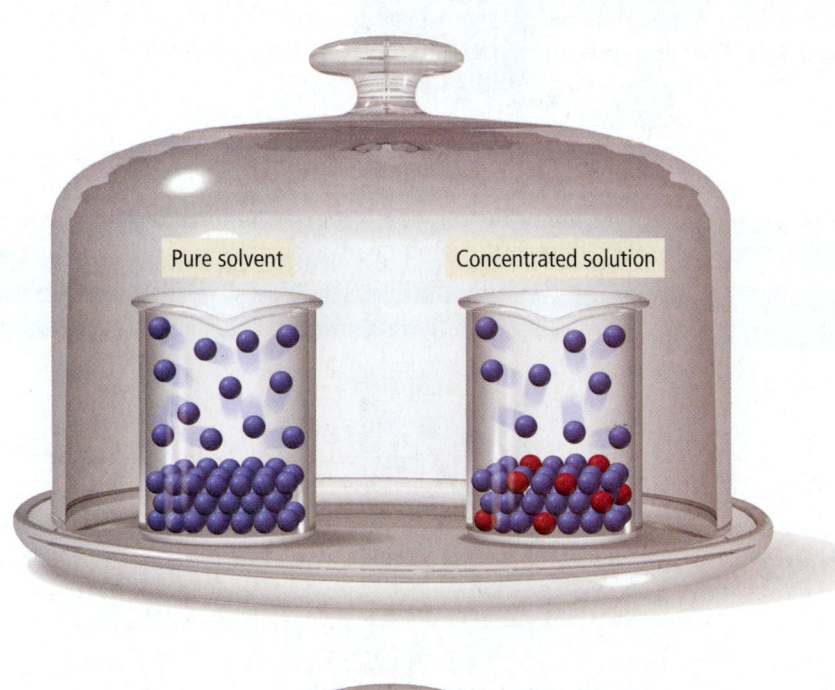

Pure solvent   Concentrated solution

The net transfer of solvent from the beaker containing pure solvent to the one containing the solution demonstrates that the vapor pressure of the solution is lower than that of the pure solvent. As solvent molecules vaporize, the vapor pressure in the container rises. Before dynamic equilibrium can be attained, however, the pressure exceeds the vapor pressure of the solution, causing molecules to condense into the solution. There-fore, molecules constantly vaporize from the pure solvent, but the solvent's vapor pressure is never reached because molecules are constantly entering the solution. The result is a continuous transfer of solvent molecules from the pure solvent to the solution.

We can quantify the vapor pressure of a solution with **Raoult's law:**

$$P_{solution} = \chi_{solvent}P^{\circ}_{solvent}$$

In this equation, $P_{solution}$ is the vapor pressure of the solution, $\chi_{solvent}$ is the mole fraction of the solvent, and $P^{\circ}_{solvent}$ is the vapor pressure of the pure solvent at the same temper-ature. For example, suppose a water sample at 25 °C contains 0.90 mol of water and 0.10 mol of a nonvolatile solute such as sucrose. The pure water would have a vapor pres-sure of 23.8 torr. The vapor pressure of the solution is calculated as follows:

$$\begin{aligned} P_{solution} &= \chi_{H_2O}P^{\circ}_{H_2O} \\ &= 0.90(23.8 \text{ torr}) \\ &= 21.4 \text{ torr} \end{aligned}$$

As you can see, the vapor pressure of the solution is lowered in direct proportion to the mole composition of the solvent—since the solvent particles compose 90% of all of the par-ticles in the solution, the vapor pressure is 90% of the vapor pressure of the pure solvent.

EXAMPLE 12.5 Calculating the Vapor Pressure of a Solution Containing a Nonionic and Nonvolatile Solute

Calculate the vapor pressure at 25 °C of a solution containing 99.5 g sucrose $(C_{12}H_{22}O_{11})$ and 300.0 mL water. The vapor pressure of pure water at 25 °C is 23.8 torr. Assume the density of water to be 1.00 g/mL.

Sort You are given the mass of sucrose and volume of water in a solution. You are also given the vapor pres-sure of pure water and asked to find the vapor pressure of the solution. The density of the pure water is also provided.	**Given** 99.5 g $C_{12}H_{22}O_{11}$ 300.0 mL H_2O $P^{\circ}_{H_2O} = 23.8$ torr at 25 °C $d_{H_2O} = 1.00$ g/mL **Find** $P_{solution}$
Strategize Raoult's law relates the vapor pressure of a solution to the mole fraction of the solvent and the vapor pressure of the pure solvent. Begin by calculat-ing the amount in moles of sucrose and water. Calculate the mole fraction of the solvent from the calculated amounts of solute and solvent. Then use Raoult's law to calculate the vapor pressure of the solution.	**Conceptual Plan** 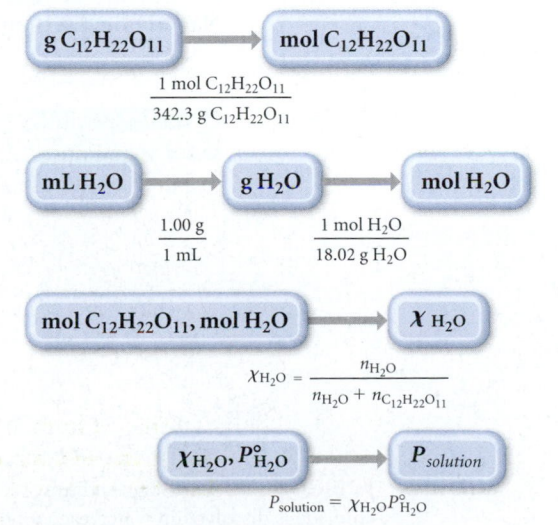

Solve Calculate the number of moles of each solution component.

Solution

$$99.5 \; g \; C_{12}H_{22}O_{11} \times \frac{1 \; mol \; C_{12}H_{22}O_{11}}{342.30 \; g \; C_{12}H_{22}O_{11}} = 0.2907 \; mol \; C_{12}H_{22}O_{11}$$

$$300.0 \; mL \; H_2O \times \frac{1.00 \; g}{1 \; mL} \times \frac{1 \; mol \; H_2O}{18.02 \; g \; H_2O} = 16.65 \; mol \; H_2O$$

Use the number of moles of each component to compute the mole fraction of the solvent (H_2O).

$$\chi_{H_2O} = \frac{n_{H_2O}}{n_{C_{12}H_{22}O_{11}} + n_{H_2O}}$$

$$\frac{16.65 \; mol}{0.2907 \; mol + 16.65 \; mol}$$

$$= 0.9828$$

Use the mole fraction of water and the vapor pressure of pure water to compute the vapor pressure of the solution.

$$P_{solution} = \chi_{H_2O} P^{\circ}_{H_2O}$$

$$= 0.9828(23.8 \; torr)$$

$$= 23.4 \; torr$$

Check The units of the answer are correct. The magnitude of the answer is about right because the calculated vapor pressure of the solution is just below that of the pure liquid, as expected for a solution with a large mole fraction of solvent.

For Practice 12.5

Calculate the vapor pressure at 25 °C of a solution containing 55.3 g ethylene glycol ($HOCH_2CH_2OH$) and 285.2 g water. The vapor pressure of pure water at 25 °C is 23.8 torr.

For More Practice 12.5

A solution containing ethylene glycol and water has a vapor pressure of 7.88 torr at 10 °C. Pure water has a vapor pressure of 9.21 torr at 10 °C. What is the mole fraction of ethylene glycol in the solution?

Ionic Solutes and Vapor Pressure

When the solute is ionic, it dissociates into its component ions upon formation of a solution. Consequently, the concentration of particles (for the same number of moles of solute) is greater for an ionic solute than for a nonionic solute. For example, a solution containing 0.10 mol of NaCl actually contains 0.20 mol of dissolved particles (0.10 mol of Na^+ ions and 0.10 mol of Cl^- ions) because of the following dissociation:

$$NaCl(s) \longrightarrow Na^+(aq) + Cl^-(aq)$$

Consequently the vapor pressure for a sodium chloride solution is lowered twice as much as for a nonionic solution of the same concentration. The following example shows how to calculate the vapor pressure of an ionic solution.

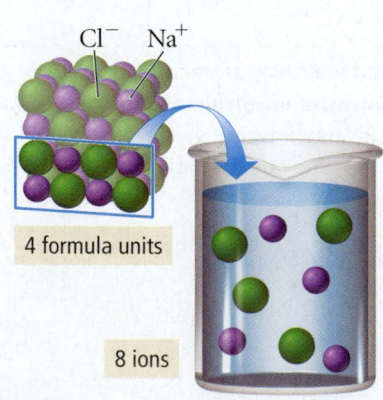

▶ **FIGURE 12.15** When sodium chloride is dissolved in water, each mole of NaCl produces 2 mol of particles: 1 mol of Na^+ cations and 1 mol of Cl^- anions.

EXAMPLE 12.6 Calculating the Vapor Pressure of a Solution Containing an Ionic Solute

A solution contains 0.102 mol $Ca(NO_3)_2$ and 0.927 mol H_2O. Calculate the vapor pressure of the solution at 55 °C. The vapor pressure of pure water at 55 °C is 118.1 torr.

Sort You are given the number of moles of each component of a solution and asked to find the vapor pressure of the solution. You are also given the vapor pressure of pure water at the desired temperature.	**Given** 0.102 mol $Ca(NO_3)_2$ 0.927 mol H_2O $P^\circ_{H_2O} = 118.1$ torr (at 55 °C) **Find** $P_{solution}$
Strategize To solve this problem, use Raoult's law as in Example 12.5. Calculate $\chi_{solvent}$ from the given amounts of solute and solvent.	**Conceptual Plan** $P_{solution} = \chi_{H_2O}P^\circ_{H_2O}$
Solve The key to this problem is to understand the dissociation of calcium nitrate. Write an equation showing the dissociation. Since 1 mol of calcium nitrate dissociates into 3 mol of dissolved particles, the number of moles of calcium nitrate must be multiplied by 3 when computing the mole fraction. Use the mole fraction of water and the vapor pressure of pure water to compute the vapor pressure of the solution.	**Solution** $Ca(NO_3)_2(s) \longrightarrow Ca^{2+}(aq) + 2\,NO_3^-(aq)$ $\chi_{H_2O} = \dfrac{n_{H_2O}}{3 \times n_{Ca(NO_3)_2} + n_{H_2O}}$ $\phantom{\chi_{H_2O}} = \dfrac{0.927\ \text{mol}}{3(0.102)\ \text{mol} + 0.927\ \text{mol}}$ $\phantom{\chi_{H_2O}} = 0.75\underline{1}8$ $P_{solution} = \chi_{H_2O}P^\circ_{H_2O}$ $\phantom{P_{solution}} = 0.75\underline{1}8(118.1\ \text{torr})$ $\phantom{P_{solution}} = 88.8\ \text{torr}$

Check The units of the answer are correct. The magnitude also seems right because the computed vapor pressure of the solution is significantly less than that of the pure solvent, as expected for a solution with a significant amount of solute.

For Practice 12.6

A solution contains 0.115 mol H_2O and an unknown number of moles of sodium chloride. The vapor pressure of the solution at 30 °C is 25.7 torr. The vapor pressure of pure water at 30 °C is 31.8 torr. Calculate the number of moles of sodium chloride in the solution.

Ideal and Nonideal Solutions

A solution that follows Raoult's law at all concentrations for both the solute and the solvent is called an **ideal solution** and is similar in concept to an ideal gas. Just as an ideal gas follows the ideal gas law exactly, so an ideal solution follows Raoult's law exactly. In an ideal solution, the solute–solvent interactions are similar in magnitude to the solute–solute and solvent–solvent interactions. In this type of solution, the solute simply dilutes the solvent and ideal behavior is observed. The vapor pressure of each of the solution components is given by Raoult's law throughout the entire composition range of the solution. For a two-component solution containing liquids A and B, we can write:

> Over a complete range of composition of a solution, it no longer makes any sense to designate a solvent and solute, so we simply label the two components A and B.

$$P_A = \chi_A P^\circ_A$$
$$P_B = \chi_B P^\circ_B$$

The total pressure above such a solution is simply the sum of the partial pressures of the components:

$$P_{tot} = P_A + P_B$$

Figure 12.16(a) ▼ shows a plot of vapor pressure versus solution composition for an ideal two-component solution.

If the solute–solvent interactions are particularly strong (stronger than solvent–solvent interactions), then the solute tends to prevent the solvent from vaporizing as easily as it would otherwise. If such a solution is sufficiently dilute, then the effect will be small and Raoult's law still works as an approximation. However, if the solution is not dilute, the effect will be significant and the vapor pressure of the solution will be *less than* that predicted by Raoult's law, as shown in Figure 12.16(b).

If, on the other hand, the solute–solvent interactions are particularly weak (weaker than solvent–solvent interactions), then the solute tends to allow more vaporization than would occur with just the solvent. If the solution is not dilute, the effect will be significant and the vapor pressure of the solution will be *greater than* predicted by Raoult's law, as shown in Figure 12.16(c).

Deviations from Raoult's Law

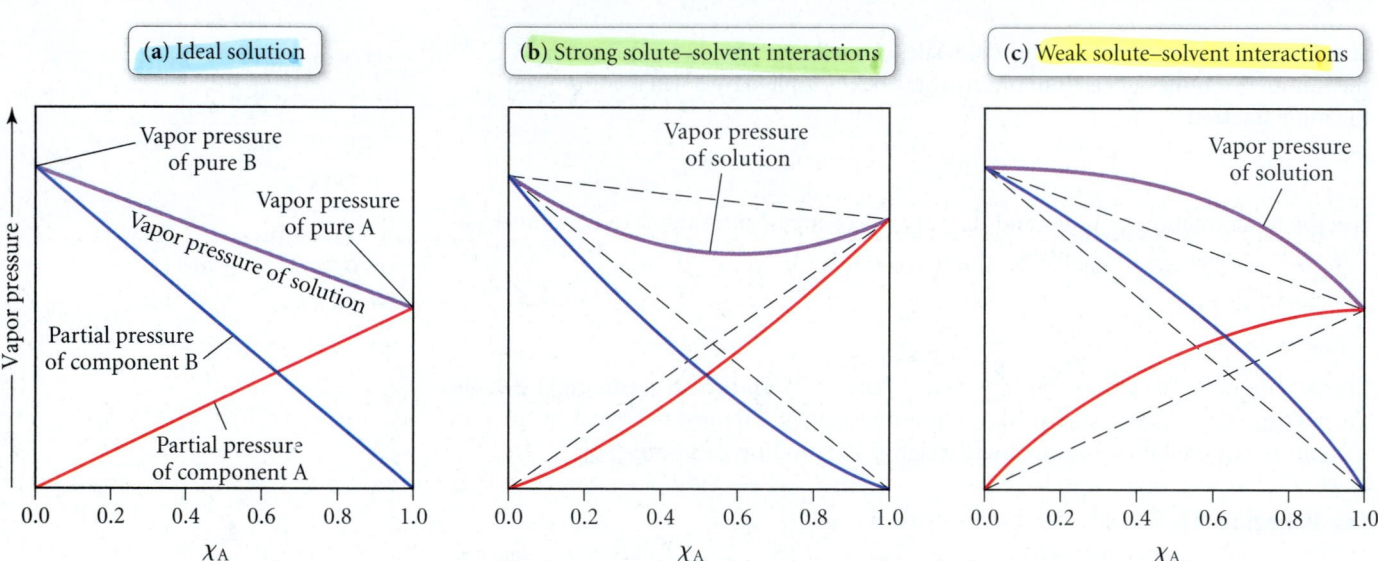

▲ **FIGURE 12.16 Behavior of Nonideal Solutions**   **(a)** An ideal solution follows Raoult's law for both components. **(b)** A solution with particularly strong solute–solvent interactions displays negative deviations from Raoult's law. **(c)** A solution with particularly weak solute–solvent interactions displays positive deviations from Raoult's law. (The dashed lines in parts b and c represent ideal behavior.)

EXAMPLE 12.7 Calculating the Vapor Pressure of a Two-Component Solution

A solution contains 3.95 g of carbon disulfide (CS_2) and 2.43 g of acetone (CH_3COCH_3) The vapor pressures at 35 °C of pure carbon disulfide and pure acetone are 515 torr and 332 torr, respectively. Assuming ideal behavior, calculate the vapor pressures of each of the components and the total vapor pressure above the solution. The experimentally measured total vapor pressure of the solution at 35 °C was 645 torr. Is the solution ideal? If not, what can you say about the relative strength of carbon disulfide–acetone interactions compared to the acetone–acetone and carbon disulfide–carbon disulfide interactions?

Sort You are given the masses and vapor pressures of carbon disulfide and acetone and are asked to find the vapor pressures of each component in the mixture and the total pressure assuming ideal behavior.

Given   3.95 g CS_2
2.43 g CH_3COCH_3
$P^\circ_{CS_2}$ = 515 torr (at 35 °C)
$P^\circ_{CH_3COCH_3}$ = 332 torr (at 35 °C)
P_{tot}(exp) = 645 torr (at 35 °C)

Find P_{CS_2}, $P_{CH_3COCH_3}$, P_{tot}(ideal)

Strategize This problem requires the use of Raoult's law to calculate the partial pressures of each component. In order to use Raoult's law, you must first compute the mole fractions of the two components. Convert the masses of each component to moles and then use the definition of mole fraction to compute the mole fraction of carbon disulfide. The mole fraction of acetone can easily be found because the mole fractions of the two components add up to 1.

Conceptual Plan

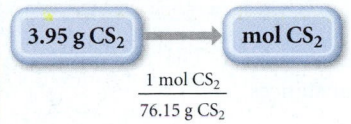

$$\frac{1 \text{ mol } CS_2}{76.15 \text{ g } CS_2}$$

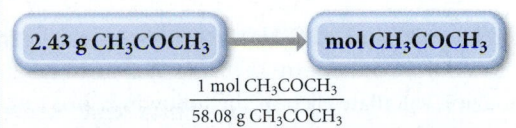

$$\frac{1 \text{ mol } CH_3COCH_3}{58.08 \text{ g } CH_3COCH_3}$$

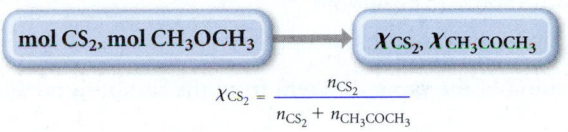

$$\chi_{CS_2} = \frac{n_{CS_2}}{n_{CS_2} + n_{CH_3COCH_3}}$$

Use the mole fraction of each component along with Raoult's law to compute the partial pressure of each component. The total pressure is simply the sum of the partial pressures.

$P_{CS_2} = \chi_{CS_2} P^\circ_{CS_2}$
$P_{CH_3COCH_3} = \chi_{CH_3COCH_3} P^\circ_{CH_3COCH_3}$
$P_{tot} = P_{CS_2} + P_{CH_3COCH_3}$

Relationships Used

$$\chi_A = \frac{n_A}{n_A + n_B} \text{ (mole fraction definition)}$$

$P_A = \chi_A P^\circ_A$ (Raoult's law)

Solve Begin by converting the masses of each component to the amounts in moles.

Solution

$$3.95 \text{ g } CS_2 \times \frac{1 \text{ mol } CS_2}{76.15 \text{ g } CS_2} = 0.05187 \text{ mol } CS_2$$

$$2.43 \text{ g } CH_3COCH_3 \times \frac{1 \text{ mol } CH_3COCH_3}{58.08 \text{ g } CH_3COCH_3} = 0.04184 \text{ mol } CH_3COCH_3$$

Then compute the mole fraction of carbon disulfide.

$$\chi_{CS_2} = \frac{n_{CS_2}}{n_{CS_2} + n_{CH_3COCH_3}}$$

$$= \frac{0.05187 \text{ mol}}{0.05187 \text{ mol} + 0.04184 \text{ mol}}$$

$$= 0.5535$$

Compute the mole fraction of acetone by subtracting the mole fraction of carbon disulfide from one.

$$\chi_{CH_3COCH_3} = 1 - 0.5535$$
$$= 0.4465$$

Compute the partial pressures of carbon disulfide and acetone by using Raoult's law and the given values of the vapor pressures of the pure substances.

$$P_{CS_2} = \chi_{CS_2} P^\circ_{CS_2}$$
$$= 0.5535(515 \text{ torr})$$
$$= 285 \text{ torr}$$

$$P_{CH_3COCH_3} = \chi_{CH_3COCH_3} P^\circ_{CH_3COCH_3}$$
$$= 0.4465(332 \text{ torr})$$
$$= 148 \text{ torr}$$

Compute the total pressure by summing the partial pressures.	$P_{tot}(\text{ideal}) = 285\text{ torr} + 148\text{ torr}$ $= 433\text{ torr}$
Lastly, compare the calculated total pressure for the ideal case to the experimentally measured total pressure. Since the experimentally measured pressure is greater than the calculated pressure, we can conclude that the interactions between the two components must be less than the interactions between the components themselves.	$P_{tot}(\text{exp}) = 645\text{ torr}$ $P_{tot}(\text{exp}) > P_{tot}(\text{ideal})$ The solution is not ideal and shows positive deviations from Raoult's law. Therefore, carbon disulfide–acetone interactions must be weaker than acetone–acetone and carbon disulfide–carbon disulfide interactions.

Check The units of the answer (torr) are correct. The magnitude seems reasonable given the partial pressures of the pure substances.

For Practice 12.7

A solution of benzene (C_6H_6) and toluene (C_7H_8) is 25.0% benzene by mass. The vapor pressures of pure benzene and pure toluene at 25 °C are 94.2 torr and 28.4 torr, respectively. Assuming ideal behavior, calculate each of the following:

(a) The vapor pressure of each of the solution components in the mixture.

(b) The total pressure above the solution.

(c) The composition of the vapor in mass percent.

Why is the composition of the vapor different from the composition of the solution?

Conceptual Connection 12.4 Raoult's Law

A solution contains equal amounts (in moles) of liquid components A and B. The vapor pressure of pure A is 100 mmHg and that of pure B is 200 mmHg. The experimentally measured vapor pressure of the solution is 120 mmHg. What can you say about the relative strengths of the solute–solute, solute–solvent, and solvent–solvent interactions in this solution?

Answer: The solute–solvent interactions must be stronger than the solute–solute and solvent–solvent interactions. The stronger interactions lower the vapor pressure from the expected ideal value of 150 mmHg.

12.7 Freezing Point Depression, Boiling Point Elevation, and Osmosis

▲ In winter, salt is often added to roads to lower the melting point of ice.

Have you ever wondered why salt is added to ice in an ice-cream maker? Or in cold climates, why salt is often scattered on icy roads? Salt actually lowers the melting point of ice. A salt and water solution will remain liquid even below 0 °C. By adding salt to ice in the ice-cream maker, you form an ice/water/salt mixture that can reach a temperature of about −10 °C, causing the cream to freeze. On the road, the salt allows the ice to melt, even if the ambient temperature is below freezing.

In order to understand this behavior, we return to a concept from Section 12.6, vapor pressure lowering. Recall that a nonvolatile solute lowers the vapor pressure of a solution relative to that of the pure solvent. The vapor pressure lowering occurs at all temperatures. We can see the effect of vapor pressure lowering over a range of temperatures by comparing the phase diagrams for pure water and for an aqueous solution containing a nonvolatile solute:

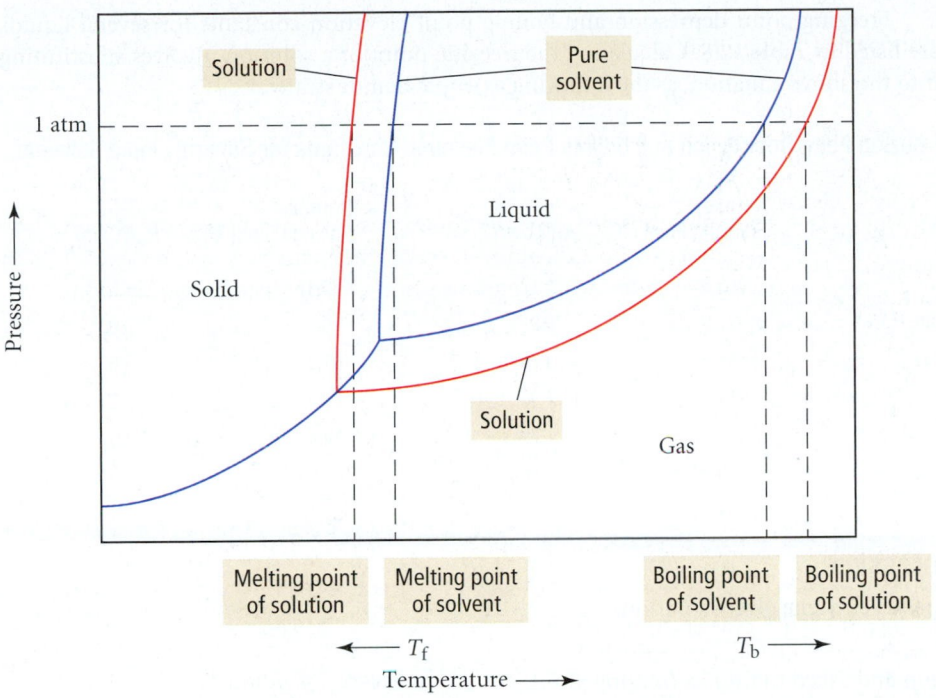

A nonvolatile solute lowers the vapor pressure of a solution, resulting in a lower freezing point and an elevated boiling point.

Notice that the vapor pressure for the solution is shifted downward compared to that of the pure solvent. Consequently, the vapor pressure curve intersects the solid–gas curve at a lower temperature. The net effect is that the solution has a *lower melting point* and a *higher boiling point* than the pure solvent. These effects are called **freezing point depression** and **boiling point elevation**. Freezing point depression and boiling point elevation depend only on the number of solute particles in solution, not on the type of solute particles. Properties that depend on the amount of solute and not the type of solute are called **colligative properties**.

Freezing Point Depression

The freezing point of a solution containing a nonvolatile solute is lower than the freezing point of the pure solvent. For example, antifreeze, used to prevent the freezing of engine blocks in cold climates, is an aqueous solution of ethylene glycol ($C_2H_6O_2$). The more concentrated the solution, the lower the freezing point becomes.

The amount that the freezing point is lowered for solutions is given by the equation

$$\Delta T_f = m \times K_f$$

where

- ΔT_f is the change in temperature of the freezing point in Celsius degrees (relative to the freezing point of the pure solvent), usually reported as a positive number;
- m is the molality of the solution in moles solute per kilogram solvent;
- K_f is the freezing point depression constant for the solvent.

For water,

$$K_f = 1.86\,°C/m$$

When an aqueous solution containing a dissolved solid solute freezes slowly, the solid that forms does not normally contain much of the solute. For example, when ice forms in ocean water, the ice is not salt water, but freshwater. As the ice forms, the crystal structure of the ice tends to exclude the solute particles. You can verify this yourself by partially freezing a salt water solution in the freezer. Take out the newly formed ice, rinse it several times, and taste it. Compare the taste of the ice to the taste of the original solution. The ice will be much less salty.

▲ Antifreeze is an aqueous solution of ethylene glycol. The solution has a lower freezing point and higher boiling point than pure water.

Freezing point depression and boiling point elevation constants for several liquids are listed in Table 12.8. Calculating the freezing point of a solution involves substituting into the above equation, as the following example demonstrates.

TABLE 12.8 Freezing Point Depression and Boiling Point Elevation Constants for Several Liquid Solvents

Solvent	Normal Freezing Point (°C)	$K_f (°C/m)$	Normal Boiling Point (°C)	$K_b (°C/m)$
Benzene (C_6H_6)	5.5	5.12	80.1	2.53
Carbon tetrachloride (CCl_4)	−22.9	29.9	76.7	5.03
Chloroform ($CHCl_3$)	−63.5	4.70	61.2	3.63
Ethanol (C_2H_5OH)	−114.1	1.99	78.3	1.22
Diethyl ether ($C_4H_{10}O$)	−116.3	1.79	34.6	2.02
Water (H_2O)	0.00	1.86	100.0	0.512

EXAMPLE 12.8 Freezing Point Depression

Calculate the freezing point of a 1.7 m aqueous ethylene glycol solution.

Sort You are given the molality of a solution and asked to find its freezing point.

Given 1.7 m solution

Find freezing point (from ΔT_f)

Strategize To solve this problem, use the freezing point depression equation.

Conceptual Plan

$$m \longrightarrow \Delta T_f$$

$$\Delta T_f = m \times K_f$$

Solve Simply substitute into the equation to compute ΔT_f.
The actual freezing point is the freezing point of pure water (0.00 °C) − ΔT_f.

Solution

$$\Delta T_f = m \times K_f$$
$$= 1.7\ m \times 1.86\ °C/m$$
$$= 3.2\ °C$$

$$\text{Freezing point} = 0.00\ °C − 3.2\ °C$$
$$= −3.2\ °C$$

Check The units of the answer are correct. The magnitude seems about right. The expected range for freezing points of an aqueous solution is anywhere from −10 °C to just below 0 °C. Any answers out of this range would be suspect.

For Practice 12.8

Calculate the freezing point of a 2.6 m aqueous sucrose solution.

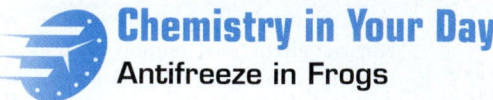

Chemistry in Your Day
Antifreeze in Frogs

On the outside, wood frogs (*Rana sylvatica*) look like most other frogs. They are only a few inches long and have characteristic greenish-brown skin. However, wood frogs survive cold winters in a remarkable way—they partially freeze. In this state, the frog has no heartbeat, no blood circulation, no breathing,

and no brain activity. Within 1–2 hours of thawing, however, these vital functions return and the frog hops off to find food. How does the wood frog do this?

Most cold-blooded animals cannot survive freezing temperatures because the water within their cells freezes. As we learned in Section 11.9, when water freezes, it expands, irreversibly damaging cells. When the wood frog hibernates for the winter, however, it produces large amounts of glucose that is secreted into the bloodstream and fills the interior of its cells.

When the temperature drops below freezing, extracellular body fluids, such as those in the abdominal cavity, freeze solid. Fluids within cells, however, remain liquid because the high glucose concentration lowers their freezing point. In other words, the concentrated glucose solution within the frog's cells acts as antifreeze, preventing the water within the cells from freezing and allowing the frog to survive.

Question

The wood frog can survive at body temperatures as low as $-8.0\,^{\circ}C$. Calculate the molality of a glucose solution ($C_6H_{12}O_6$) required to lower the freezing point of water to $-8.0\,^{\circ}C$.

◀ The wood frog survives winter by partially freezing. It protects its cells by flooding them with glucose, which acts as an antifreeze.

Boiling Point Elevation

The boiling point of a solution containing a nonvolatile solute is higher than the boiling point of the pure solvent. In automobiles, antifreeze not only prevents the freezing of water within engine blocks in cold climates, it also prevents the boiling of water within engine blocks in hot climates. The amount that the boiling point is raised for solutions is given by the equation

$$\Delta T_b = m \times K_b$$

where

- ΔT_b is the change in temperature of the boiling point in Celsius degrees (relative to the boiling point of the pure solvent);
- m is the molality of the solution in moles solute per kilogram solvent;
- K_b is the boiling point elevation constant for the solvent.

For water,

$$K_b = 0.512\,^{\circ}C/m$$

The boiling point of a solution is calculated by simply substituting into the above equation, as the following example demonstrates.

EXAMPLE 12.9   Boiling Point Elevation

How much ethylene glycol ($C_2H_6O_2$), in grams, must be added to 1.0 kg of water to produce a solution that boils at 105.0 °C?

Sort You are given the desired boiling point of an ethylene glycol solution containing 1.0 kg of water and asked to find the mass of ethylene glycol required to achieve the boiling point.	Given $\Delta T_b = 5.0\,^{\circ}C$, 1.0 kg H_2O Find g $C_2H_6O_2$

Strategize To solve this problem, use the boiling-point elevation equation to find the desired molality of the solution from ΔT_b.

Conceptual Plan

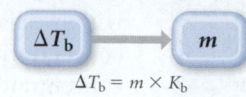

$$\Delta T_b = m \times K_b$$

Then use the molality you just found to determine how many moles of ethylene glycol are needed per kilogram of water. Finally, calculate the molar mass of ethylene glycol and use it to convert from moles of ethylene glycol to mass of ethylene glycol.

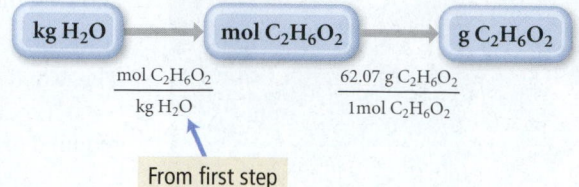

From first step

Relationships Used $C_2H_6O_2$ molar mass $= 62.07 \text{ g/mol}$

$\Delta T_b = m \times K_b$ (boiling point elevation)

Solve Begin by solving the boiling point elevation equation for molality and substituting the required quantities to compute m.

Solution

$$\Delta T_b = m \times K_b$$

$$m = \frac{\Delta T_b}{K_b} = \frac{5.0\ ^\circ C}{0.512 \dfrac{^\circ C}{m}}$$

$$= 9.\underline{7}7\ m$$

$$1.0\ \text{kg H}_2\text{O} \times \frac{9.\underline{7}7\ \text{mol } C_2H_6O_2}{\text{kg H}_2\text{O}} \times \frac{62.07\ \text{g } C_2H_6O_2}{1\ \text{mol } C_2H_6O_2} = 6.1 \times 10^2\ \text{g } C_2H_6O_2$$

Check The units of the answer are correct. The magnitude might seem a little high initially, but the boiling point elevation constant is so small that a lot of solute is required to raise the boiling point by a small amount.

For Practice 12.9

Calculate the boiling point of a 3.60 m aqueous sucrose solution.

Osmosis

The process discussed in the opening section of this chapter by which seawater causes dehydration is called *osmosis*. **Osmosis** can be formally defined as the flow of solvent from a solution of lower solute concentration to one of higher solute concentration. Concentrated solutions draw solvent from more dilute solutions because of nature's tendency to mix discussed previously.

Figure 12.17 ▶ shows an osmosis cell. The left side of the cell contains a concentrated saltwater solution and the right side of the cell contains pure water. A **semipermeable membrane**—a membrane that selectively allows some substances to pass through but not others—separates the two halves of the cell. Water flows by osmosis from the pure-water side of the cell through the semipermeable membrane and into the saltwater side. Over time, the water level on the left side of the cell rises, while the water level on the right side of the cell falls. If external pressure is applied to the water in the left cell, this process can be opposed and even stopped. The pressure required to stop the osmotic flow, called the **osmotic pressure**, is given by the following equation:

$$\Pi = MRT$$

where M is the molarity of the solution, T is the temperature (in kelvins), and R is the ideal gas constant (0.08206 L · atm/mol · K).

Osmosis and Osmotic Pressure

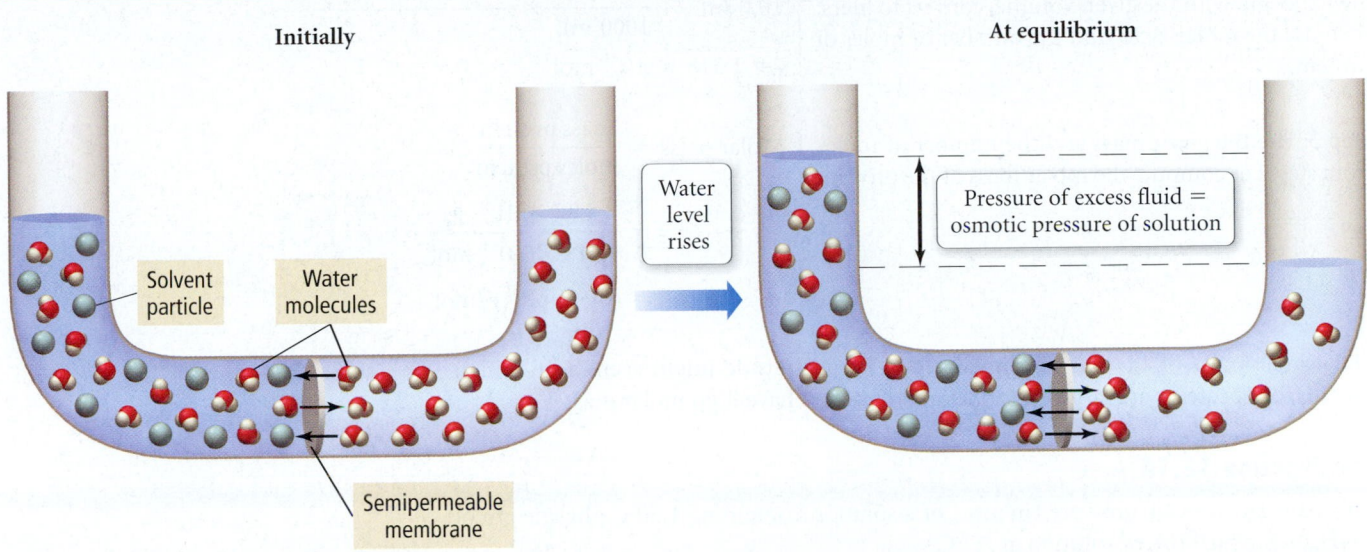

Initially

Solvent particle

Water molecules

Semipermeable membrane

Water level rises

At equilibrium

Pressure of excess fluid = osmotic pressure of solution

▲ **FIGURE 12.17 An Osmosis Cell**   In an osmosis cell, water flows from the pure-water side of the cell through the semipermeable membrane to the salt water side.

EXAMPLE 12.10   Osmotic Pressure

The osmotic pressure of a solution containing 5.87 mg of an unknown protein per 10.0 mL of solution was 2.45 torr at 25 °C. Find the molar mass of the unknown protein.

Sort You are given that a solution of an unknown protein contains 5.87 mg of the protein per 10.0 mL of solution. You are also given the osmotic pressure of the solution at a particular temperature and asked to find the molar mass of the unknown protein.

Given 5.87 mg protein
10.0 mL solution
$\Pi = 2.45$ torr
$T = 25\,°C$

Find molar mass of protein (g/mol)

Strategize Step 1: Use the given osmotic pressure and temperature to find the molarity of the protein solution.

Conceptual Plan

$$\boxed{\Pi, T} \longrightarrow \boxed{M}$$

$$\Pi = MRT$$

Step 2: Use the molarity calculated in step 1 to find the number of moles of protein in 10 mL of solution.

$$\boxed{\text{mL solution}} \longrightarrow \boxed{\text{L solution}} \longrightarrow \boxed{\text{mol protein}}$$

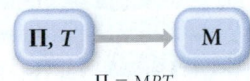

$$\frac{1\ L}{1000\ mL} \qquad \frac{\text{mol protein}}{\text{L solution}}$$

From first step

Step 3: Finally, use number of moles of the protein calculated in step 2 and the given mass of the protein in 10.0 mL of solution to find the molar mass.

$$\text{Molar mass} = \frac{\text{mass protein}}{\text{moles protein}}$$

Relationships Used $\Pi = MRT$ (osmotic pressure equation)

Solve

Step 1: Begin by solving the osmotic pressure equation for molarity and substituting in the required quantities in the correct units to compute M.

Solution

$\Pi = MRT$

$$M = \frac{\Pi}{RT} = \frac{2.45\ \text{torr} \times \dfrac{1\ \text{atm}}{760\ \text{torr}}}{0.08206\dfrac{L \cdot \text{atm}}{\text{mol} \cdot K}(298\ K)}$$

$$= 1.318 \times 10^{-4}\ M$$

Step 2: Begin with the given volume, convert to liters, then use the molarity to find the number of moles of protein.

$$10.0 \; \cancel{mL} \times \frac{1 \; \cancel{L}}{1000 \; \cancel{mL}} \times \frac{1.318 \times 10^{-4} \; \cancel{mol}}{\cancel{L}}$$

$$= 1.3\underline{1}8 \times 10^{-6} \; mol$$

Step 3: Use the given mass and the number of moles from step 2 to compute the molar mass of the protein.

$$Molar \; mass = \frac{mass \; protein}{moles \; protein}$$

$$= \frac{5.87 \times 10^{-3} \; g}{1.3\underline{1}8 \times 10^{-6} \; mol}$$

$$= 4.45 \times 10^{3} \; g/mol$$

Check The units of the answer are correct. The magnitude might seem a little high initially, but proteins are large molecules and therefore have high molar masses.

For Practice 12.10

Calculate the osmotic pressure (in atm) of a solution containing 1.50 g ethylene glycol ($C_2H_6O_2$) in 50.0 mL of solution at 25 °C.

Colligative Properties of Ionic Solutions

TABLE 12.9	Van't Hoff Factors at 0.05 *m* Concentration in Aqueous Solution

Solute	*i* Expected	*i* Measured
Nonelectrolyte	1	1
NaCl	2	1.9
$MgSO_4$	2	1.3
$MgCl_2$	3	2.7
K_2SO_4	3	2.6
$FeCl_3$	4	3.4

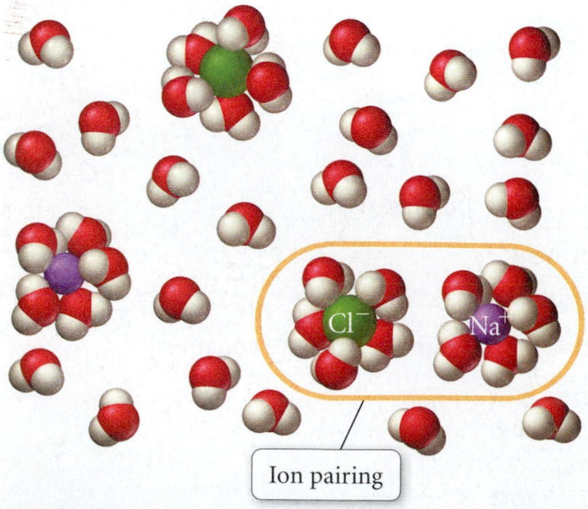

Recall from Section 12.6 that the vapor pressure of a solution containing an ionic solute is lower than the vapor pressure of a solution containing the same number of moles of a nonionic solute. The reason for this behavior is that vapor pressure lowering (like freezing point depression, boiling point elevation, and osmotic pressure) is a colligative property—it depends only on the number of solute particles, not the kind of solute particles. For the same reason, the magnitudes of freezing point depression, boiling point elevation, and osmotic pressure are all greater for ionic solutes than for nonionic solutes. For example, the freezing point depression of a 0.10 *m* sucrose solution is $\Delta T_f = 0.186$ °C. However, the freezing point depression of a 0.10 *m* sodium chloride solution is nearly twice this large. Why? Because 1 mol of sodium chloride dissociates into *nearly* 2 mol of ions in solution. The ratio of moles of particles in solution to moles of formula units dissolved is called the **van't Hoff factor** (*i*):

$$i = \frac{moles \; of \; particles \; in \; solution}{moles \; of \; fomula \; units \; dissolved}$$

Since 1 mol of NaCl produces 2 mol of particles in solution, we expect the van't Hoff factor to be exactly 2. In fact, this expected factor only occurs in very dilute solutions. For example, the van't Hoff factor for a 0.10 *m* NaCl solution is 1.87 and that for a 0.010 *m* NaCl in solution is 1.94. The van't Hoff factor approaches the expected value at infinite dilution. Table 12.9 lists the actual and expected van't Hoff factors for a number of solutes.

The reason that the van't Hoff factors are not exactly equal to the expected values is that some ions effectively pair in solution. In other words, ideally we expect the dissociation of an ionic compound to be complete in solution. In reality, however, the dissociation is not complete—at any moment, some cations are effectively combined with the corresponding anions (Figure 12.18 ◀), slightly reducing the number of particles in solution.

To calculate colligative properties of ionic solutions use the van't Hoff factor in each equation as follows:

$$\Delta T_f = im \times K_f \quad \text{(freezing point depression)}$$
$$\Delta T_b = im \times K_b \quad \text{(boiling point elevation)}$$
$$\Pi = iMRT \quad \text{(osmotic pressure)}$$

Ion pairing

▲ **FIGURE 12.18 Ion Pairing**
Hydrated anions and cations may get close enough together to effectively pair, lowering the concentration of particles below what would be expected ideally.

Conceptual Connection 12.5 Colligative Properties

Which of the following solutions will have the highest boiling point?

(a) 0.50 M $C_{12}H_{22}O_{11}$

(b) 0.50 M NaCl

(c) 0.50 M $MgCl_2$

Answer: (c) The 0.50 M $MgCl_2$ solution will have the highest boiling point because it has the highest concentration of particles. We expect 1 mol of $MgCl_2$ to form 3 mol of particles in solution (although it effectively forms slightly fewer).

EXAMPLE 12.11 Van't Hoff Factor and Freezing Point Depression

The freezing point of an aqueous 0.050 m $CaCl_2$ solution is −0.27 °C. What is the van't Hoff factor (i) for $CaCl_2$ at this concentration? How does it compare to the predicted value of i?

Sort You are given the molality of a solution and its freezing point. You are asked to find the value of i, the van't Hoff factor, and compare it to the predicted value.	**Given** 0.050 m $CaCl_2$ solution, $$\Delta T_f = 0.27 \degree C$$ **Find** i
Strategize To solve this problem, use the freezing point depression equation including the van't Hoff factor.	**Conceptual Plan** $$\Delta T_f = im \times K_f$$
Solve Solve the freezing point depression equation for i and substitute in the given quantities to compute its value. The expected value of i for $CaCl_2$ is 3 because calcium chloride forms 3 mol of ions for each mole of calcium chloride that dissolves. The experimental value is slightly less than 3, probably because of ion pairing.	**Solution** $$\Delta T_f = im \times K_f$$ $$i = \frac{\Delta T_f}{m \times K_f}$$ $$= \frac{0.27 \; \degree C}{0.050 \; m \times \dfrac{1.86 \; \degree C}{m}}$$ $$= 2.9$$

Check The answer has no units, as expected since i is a ratio. The magnitude is about right since it is close to the value you would expect upon complete dissociation of $CaCl_2$.

For Practice 12.11

Compute the freezing point of an aqueous 0.10 m $FeCl_3$ solution using a van't Hoff factor of 3.2.

Colligative Properties and Medical Solutions

Doctors and others working in health fields often administer solutions to patients. The osmotic pressure of these solutions is controlled for the desired effect on the patient. Solutions having osmotic pressures greater than those of body fluids are called *hyperosmotic*. These solutions tend to take water out of cells and tissues. When a human cell is placed in a hyperosmotic solution, it tends to shrivel as it loses water to the surrounding solution (Figure 12.19b on page 556). Solutions having osmotic pressures less than those of body fluids are called *hyposmotic*. These solutions tend to pump water into cells. When a human cell is placed in a hyposmotic solution—such as pure water, for example—water enters the cell, sometimes causing it to burst (Figure 12.19c).

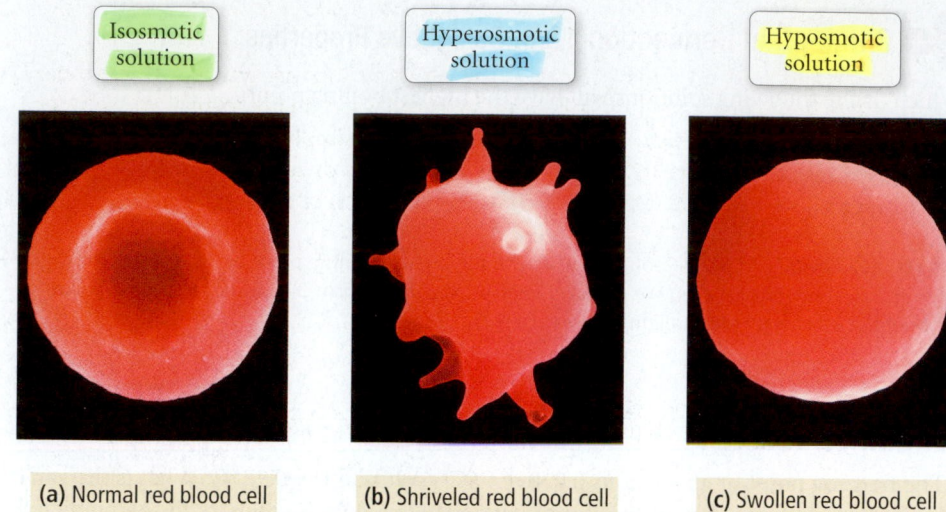

| Isosmotic solution | Hyperosmotic solution | Hyposmotic solution |

(a) Normal red blood cell   **(b)** Shriveled red blood cell   **(c)** Swollen red blood cell

▲ FIGURE 12.19 **Red Blood Cells and Osmosis**   (**a**) In an isosmotic solution, red blood cells have the normal shape shown here. In a hyperosmotic solution (**b**), they lose water and shrivel. In a hyposmotic solution (**c**), they swell up and may burst as water flows into the cell.

Intravenous solutions—those that are administered directly into a patient's veins—must have osmotic pressures equal to those of body fluids. These solutions are called *isosmotic* (or *isotonic*). When a patient is given an IV in a hospital, the majority of the fluid is usually an isosmotic saline solution—a solution containing 0.9 g NaCl per 100 mL of solution. In medicine and in other health-related fields, solution concentrations are often reported in units that indicate the mass of the solute per given volume of solution. Also common is *percent mass to volume*—which is simply the mass of the solute in grams divided by the volume of the solution in milliliters times 100%. In these units, the concentration of an isotonic saline solution is 0.9% mass/volume.

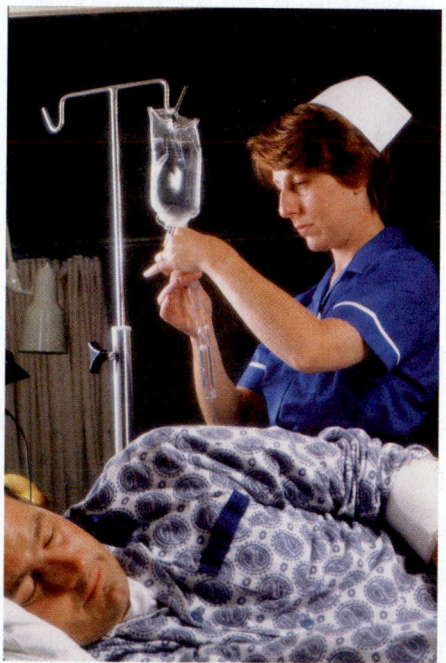

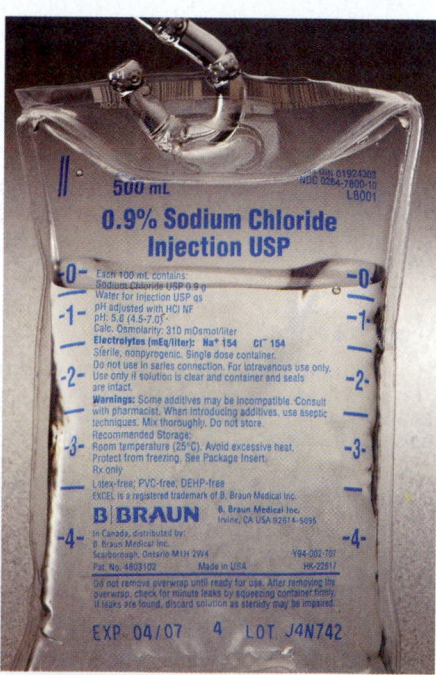

▲ Fluids used for intravenous transfusion must be isosmotic with bodily fluids—that is, they must have the same osmotic pressure.

12.8 Colloids

When water and soap are mixed together, the resulting mixture has a distinctive haze (Figure 12.20 ▶). Soapy water is hazy because soap and water do not form a true solution, but rather a *colloidal dispersion*. A **colloidal dispersion**, or simply a **colloid**, is a mixture in which a dispersed substance (which is solute-like) is finely divided in a dispersing medium (which is solvent-like). Other examples of colloids include fog, smoke, whipped cream, and milk, as shown in Table 12.10.

TABLE 12.10	Types of Colloidal Dispersions			
Classification	Dispersing Substance (Solute-like)	Dispersing Medium (Solvent-like)		Example
Aerosol	Liquid	Gas		Fog (water droplets in air)
Solid aerosol	Solid	Gas		Smoke (ash in air)
Foam	Gas	Liquid		Whipped cream (air bubbles in butterfat)
Emulsion	Liquid	Liquid		Milk (milk fat globules in water)
Solid emulsion	Liquid	Solid		Opal (water in silica glass)

▲ FIGURE 12.20 **A Colloid** A soapy water solution is an example of a colloidal dispersion. The haze is due to the scattering of light by the colloidal particles.

The easiest way to define a colloid is based on the size of the solute-like particles. If the particles are small (for example, individual small molecules), then the mixture is a solution. If the particles have a diameter greater than 1 μm (for example, grains of sand), then the mixture is a heterogeneous mixture. Sand stirred into water will slowly settle out of the water. *If the particles are between 1 nm and 1000 nm in size, the mixture is a colloid.* Colloidal particles are small enough to be kept dispersed throughout the dispersing medium by collisions with other molecules or atoms. When you view a colloidal particle dispersed in a liquid under a microscope, you can witness its jittery motion, which proceeds along a random path, as shown in Figure 12.21 ▶. This motion, called Brownian motion, is caused by collisions with molecules in the liquid. In beginning of the twentieth century, Brownian motion was a decisive factor in confirming the molecular and atomic nature of matter.

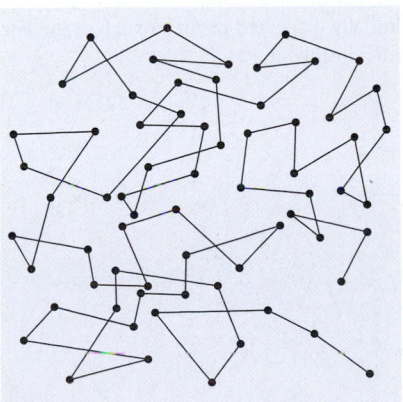

▲ FIGURE 12.21 **Brownian Motion** A colloidal particle exhibits Brownian motion, moving in a jerky, haphazard path as it undergoes collisions with other molecules.

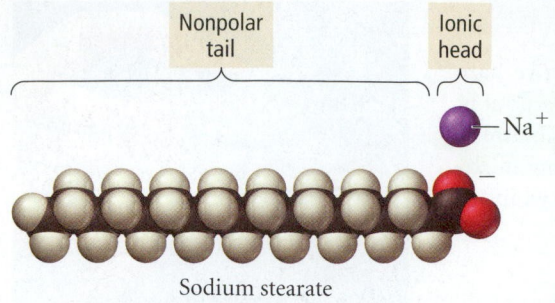

Nonpolar tail   Ionic head

—Na⁺

Sodium stearate

▲ **FIGURE 12.22** **Structure of a Soap**   A soap molecule has a charged ionic head and a long hydrocarbon tail that is nonpolar.

Soap does not dissolve in water because of its unique structure, shown in Figure 12.22 ◄. One end of the molecule is ionic and therefore interacts strongly with water molecules via ion–dipole interactions. However, the other end of the soap molecule is a long, nonpolar, hydrocarbon tail. When enough soap is added to water, the soap molecules aggregate in structures called *micelles* (Figure 12.23 ►). In a micelle, the nonpolar hydrocarbon tails crowd into the center of a sphere to maximize their interactions with one another. The ionic heads orient toward the surface of the sphere where they can interact with water molecules. The micelle structures are responsible for the haze seen in soapy water—they are too small to be seen by the naked eye, but they scatter light. This scattering of light by a colloidal dispersion is known as the **Tyndall effect** (Figure 12.24 ▼). The Tyndall effect can be observed in other colloids such as fog (water droplets dispersed in air) or dusty air. In fact, the Tyndall effect is often used as a test to determine whether a mixture is a solution or a colloid, since solutions contain completely dissolved solute molecules that are too small to scatter light.

The particles in a colloid need not be clusters of molecules. Some colloids, such as many protein solutions, contain dispersed macromolecules. For example, blood contains dispersed hemoglobin. The hemoglobin molecule is so large that it scatters light; therefore, blood is considered a colloid.

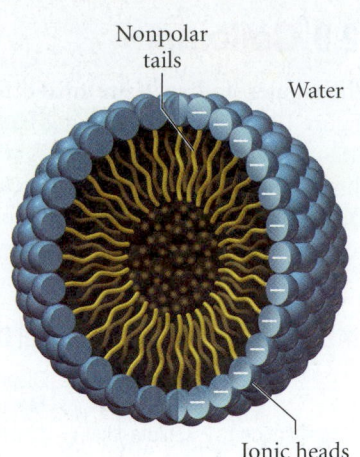

Nonpolar tails   Water

Ionic heads

▲ **FIGURE 12.23** **Micelle Structure** In a micelle, the nonpolar tails of soap molecules are oriented inward (where they can interact with one another) and the ionic heads are oriented outward (where they can interact with the polar water molecules).

▲ Light beams would be invisible if light were not sometimes scattered by colloidally dispersed particles such as dust or mist in the air.

▲ **FIGURE 12.24** **The Tyndall Effect**   When a light beam passes through a colloidal suspension (left), it is visible because the colloid particles scatter some of the light. The beam is not visible in pure water (right).

▲ FIGURE 12.25 **Micelle Repulsions**   Micelles do not coalesce because the charged surface of one micelle repels the charged surface of another.

Colloidal suspensions are kept stable by electrostatic repulsions that occur at their surfaces. For example, in a micelle, the ionic heads of many soap molecules compose the surface of the spherical particle (Figure 12.25 ▲). These ionic heads interact strongly with water molecules, but repel other colloid particles. However, heating a colloid can destroy it because collisions occur with more force that can overcome the electrostatic repulsions and allow the colloid particles to coalesce. Similarly, adding an electrolyte to a colloidal suspension can also disrupt the electrostatic repulsions that occur and destroy the colloid. For this reason, it is very difficult to use soap in a salt water solution.

Chapter in Review

Key Terms

Section 12.1

solution (520)
solvent (520)
solute (520)

Section 12.2

aqueous solution (521)
solubility (521)
entropy (522)
miscible (523)

Section 12.3

enthalpy of solution (ΔH_{soln}) (527)
heat of hydration ($\Delta H_{hydration}$) (528)

Section 12.4

dynamic equilibrium (529)
saturated solution (530)
unsaturated solution (530)
supersaturated solution (530)
recrystallization (531)
Henry's law (533)

Section 12.5

dilute solution (535)
concentrated solution (535)
molarity (M) (536)
molality (m) (536)
parts by mass (537)
percent by mass (537)
parts per million (ppm) (537)

parts per billion (ppb) (537)
parts by volume (537)
mole fraction (χ_{solute}) (539)
mole percent (mol %) (539)

Section 12.6

Raoult's law (543)
ideal solution (545)

Section 12.7

freezing point depression (549)
boiling point elevation (549)
colligative property (549)
osmosis (552)
semipermeable membrane (552)

osmotic pressure (552)
van't Hoff factor (i) (554)

Section 12.8

colloidal dispersion (colloid) (557)
Tyndall effect (558)

Key Concepts

Solutions (12.1, 12.2)

A solution is a homogeneous mixture of two or more substances. In a solution, the majority component is the solvent and the minority component is the solute. The tendency toward greater entropy (or greater energy dispersal) is the driving force for solution formation. Aqueous solutions contain water as a solvent and a solid, liquid, or gas as the solute.

Solubility and Energetics of Solution Formation (12.2, 12.3)

The solubility of a substance is the amount of the substance that will dissolve in a given amount of solvent. The solubility of one substance in another depends on the types of intermolecular forces that exist *between* the substances as well as *within* each substance. The overall enthalpy change upon solution formation can be determined by adding the enthalpy changes for the three steps of solution formation: (1) separation of the solute particles, (2) separation of the solvent particles, and (3) mixing of the solute and solvent particles. The first two steps are both endothermic, while the last is exothermic. In aqueous solutions of an ionic compound, the change in enthalpy for steps 2 and 3 can be combined as the heat of hydration ($\Delta H_{hydration}$), which is always negative.

Solution Equilibrium (12.4)

Dynamic equilibrium is the state in which the rates of dissolution and deposition in a solution are equal. A solution in this state is said to be saturated. Solutions containing less than or more than the equilibrium amount of solute are unsaturated or supersaturated, respectively. The solubility of most solids in water increases with increasing temperature. The solubility of gases in liquids generally decreases with increasing temperature, but increases with increasing pressure.

Concentration Units (12.5)

Common units used to express solution concentration include molarity (M), molality (m), mole fraction (χ), mole percent (mol%), percent (%) by mass or volume, parts per million (ppm) by mass or volume, and parts per billion (ppb) by mass or volume. These units are summarized in Table 12.5.

Vapor Pressure of Solutions (12.6)

The presence of a nonvolatile solute in a liquid results in a lower vapor pressure of the solution relative to the vapor pressure of the pure liquid. This lower vapor pressure is predicted by Raoult's law for an ideal solution. If the solute–solvent interactions are particularly strong, the actual vapor pressure is lower than that predicted by Raoult's law. If the solute–solvent interactions are particularly weak, the actual vapor pressure is higher than that predicted by Raoult's law.

Freezing Point Depression, Boiling Point Elevation, and Osmosis (12.7)

The addition of a nonvolatile solute to a liquid will result in a solution with a lower freezing point and a higher boiling point than those of the pure solvent. The flow of solvent from a solution of lower concentration to a solution of higher concentration is called osmosis. These phenomena depend only on the number of solute particles added, not the type of solute particles.

Colloids (12.8)

A colloid is a mixture in which a dispersed substance is finely divided in a dispersing medium. Colloidal mixtures occur when the dispersed substance ranges in size from 1 nm to 1000 nm. One way to identify colloidal mixtures is by their tendency to scatter light, known as the Tyndall effect.

Key Equations and Relationships

Henry's Law: Solubility of Gases with Increasing Pressure (12.4)

$$S_{gas} = k_H P_{gas} \qquad (k_H \text{ is Henry's law constant})$$

Molarity (M) of a Solution (12.5)

$$M = \frac{\text{amount solute (in mol)}}{\text{volume solution (in L)}}$$

Molality (m) of a Solution (12.5)

$$m = \frac{\text{amount solute (in mol)}}{\text{mass solvent (in kg)}}$$

Concentration of a Solution in Parts by Mass and Parts by Volume (12.5)

$$\text{Percent by mass} = \frac{\text{mass solute} \times 100\%}{\text{mass solution}}$$

$$\text{Parts per million (ppm)} = \frac{\text{mass solute} \times 10^6}{\text{mass solution}}$$

$$\text{Parts per billion (ppb)} = \frac{\text{mass solute} \times 10^9}{\text{mass solution}}$$

$$\text{Parts by volume} = \frac{\text{volume solute} \times \text{multiplication factor}}{\text{volume solution}}$$

Concentration of a Solution in Mole Fraction (χ) and Mole Percent (12.5)

$$\chi_{solute} = \frac{n_{solute}}{n_{solute} + n_{solvent}}$$

$$\text{Mol \%} = \chi \times 100\%$$

Raoult's Law: Relationship between the Vapor Pressure of a Solution ($P_{solution}$), the Mole Fraction of the Solvent ($\chi_{solvent}$), and the Vapor Pressure of the Pure Solvent ($P^\circ_{solvent}$) (12.6)

$$P_{solution} = \chi_{solvent} P^\circ_{solvent}$$

The Vapor Pressure of a Solution Containing Two Volatile Components (12.6)

$$P_A = \chi_A P^\circ_A$$
$$P_B = \chi_B P^\circ_B$$
$$P_{tot} = P_A + P_B$$

Relationship between Freezing Point Depression (ΔT_f), molality (m), and Freezing Point Depression Constant (K_f) (12.7)

$$\Delta T_f = m \times K_f$$

Relationship between Boiling Point Elevation (ΔT_b), Molality (*m*), and Boiling Point Elevation Constant (K_b) (12.7)

$$\Delta T_b = m \times K_b$$

Relationship between Osmotic Pressure (Π), Molarity (M), the Ideal Gas Constant (*R*), and Temperature (*T*, in K) (12.7)

$$\Pi = MRT \quad (R = 0.08206 \text{ L} \cdot \text{atm/mol} \cdot \text{K})$$

van't Hoff Factor (*i*): Ratio of Moles of Particles in Solution to Moles of Formula Units Dissolved (12.7)

$$i = \frac{\text{moles of particles in solution}}{\text{moles of formula units dissolved}}$$

Key Skills

Determining Whether a Solute Is Soluble in a Solvent (12.2)
•Example 12.1 • For Practice 12.1 • Exercises 31–34

Using Henry's Law to Predict the Solubility of Gases with Increasing Pressure (12.4)
•Example 12.2 • For Practice 12.2 • Exercises 49, 50

Calculating Concentrations of Solutions (12.5)
•Examples 12.3, 12.4 • For Practice 12.3, 12.4 • For More Practice 12.3 • Exercises 51–56, 63, 64

Determining the Vapor Pressure of a Solution Containing a Nonionic Solute (12.6)
•Example 12.5 • For Practice 12.5 • For More Practice 12.5 • Exercises 67, 68

Determining the Vapor Pressure of a Solution Containing an Ionic Solute (12.6)
•Example 12.6 • For Practice 12.6 • Exercises 69, 70

Determining the Vapor Pressure of a Two-Component Solution (12.6)
•Example 12.7 • For Practice 12.7 • Exercises 71–74

Calculating Freezing Point Depression (12.7)
•Example 12.8 • For Practice 12.8 • Exercises 75–78, 83, 84

Calculating Boiling Point Elevation (12.7)
•Example 12.9 • For Practice 12.9 • Exercises 75, 76, 83, 84

Determining the Osmotic Pressure (12.7)
•Example 12.10 • For Practice 12.10 • Exercises 79–82

Determining and Using the van't Hoff Factor (12.7)
•Example 12.11 • For Practice 12.11 • Exercises 85–88

Exercises

Review Questions

1. Explain why drinking seawater results in dehydration.

2. What is a solution? What are the solute and solvent?

3. What does it mean to say that a substance is soluble in another substance? What kinds of units are used in reporting solubility?

4. Why do two ideal gases thoroughly mix when combined? What drives the mixing?

5. What is entropy? Why is entropy important in discussing the formation of solutions?

6. What kinds of intermolecular forces are involved in solution formation?

7. Explain how the relative strengths of solute–solute interactions, solvent–solvent interactions, and solvent–solute interactions affect solution formation.

8. What does the statement, *like dissolves like*, mean with respect to solution formation?

9. What are three steps involved in evaluating the enthalpy changes associated with solution formation?

10. What is the heat of hydration ($\Delta H_{\text{hydration}}$)? How does the enthalpy of solution depend on the relative magnitudes of ΔH_{solute} and $\Delta H_{\text{hydration}}$?

11. Explain dynamic equilibrium with respect to solution formation. What is a saturated solution? An unsaturated solution? A supersaturated solution?

12. How does the solubility of a solid in a liquid depend on temperature? How is this temperature dependence exploited to purify solids through recrystallization?

13. How does the solubility of a gas in a liquid depend on temperature? How does this temperature dependence affect the amount of oxygen available for fish and other aquatic animals?

14. How does the solubility of a gas in a liquid depend on pressure? How does this pressure dependence account for the bubbling that occurs upon opening a can of soda?

15. What is Henry's law? For what kinds of calculations is Henry's law useful?

16. What are the common units for expressing solution concentration?

17. How are parts by mass and parts by volume used in calculations?

18. What is the effect of a nonvolatile solute on the vapor pressure of a liquid? Why is the vapor pressure of a solution different from the vapor pressure of the pure liquid solvent?

19. What is Raoult's law? For what kind of calculations is Raoult's law useful?

20. When calculating the vapor pressure of a solution containing an ionic solute, what additional factor must be considered besides those needed for calculating the vapor pressure of a solution containing a nonionic solute?

21. Explain the difference between an ideal and a nonideal solution. What is the effect on vapor pressure of a solution with particularly *strong* solute–solvent interactions? With particularly *weak* solute–solvent interactions?

22. Explain why the lower vapor pressure for a solution containing a nonvolatile solute also results in a higher boiling point and lower melting point compared to the pure solvent.

23. What are colligative properties?

24. What is osmosis? What is osmotic pressure?

25. Explain the role and meaning of the van't Hoff factor in determining the colligative properties of solutions containing ionic solutes.

26. Describe a colloidal dispersion. What is the difference between a colloidal dispersion and a true solution?

27. What is the Tyndall effect and how can it be used to help identify colloidal dispersions?

28. What keeps the particles in a colloidal dispersion from coalescing?

Problems by Topic

Solubility

29. Pick an appropriate solvent from Table 12.3 to dissolve each of the following. State the kind of intermolecular forces that would occur between the solute and solvent in each case.
    a. motor oil (nonpolar)
    b. ethanol (polar, contains an OH group)
    c. lard (nonpolar)
    d. potassium chloride (ionic)

30. Pick an appropriate solvent from Table 12.3 to dissolve:
    a. isopropyl alcohol (polar, contains an OH group)
    b. sodium chloride (ionic)
    c. vegetable oil (nonpolar)
    d. sodium nitrate (ionic)

31. Which of the following molecules would you expect to be more soluble in water, $CH_3CH_2CH_2OH$ or $HOCH_2CH_2CH_2OH$?

32. Which molecule would you expect to be more soluble in water, CCl_4 or CH_2Cl_2?

33. For each of the following molecules, would you expect greater solubility in water or in hexane? For each case, indicate the kinds of intermolecular forces that occur between the solute and the solvent in which the molecule is most soluble.

    a. glucose

34. For each of the following molecules, would you expect greater solubility in water or in hexane? For each case, indicate the kinds of intermolecular forces that would occur between the solute and the solvent in which the molecule is most soluble.

    b. naphthalene

    c. dimethyl ether

    d. alanine
       (an amino acid)

    a. toluene

b. sucrose
(table sugar)

c. isobutene

d. ethylene glycol

Energetics of Solution Formation

35. When ammonium chloride (NH_4Cl) is dissolved in water, the solution becomes colder.
   a. Is the dissolution of ammonium chloride endothermic or exothermic?
   b. What can you say about the relative magnitudes of the lattice energy of ammonium chloride and its heat of hydration?
   c. Sketch a qualitative energy diagram similar to Figure 12.7 for the dissolution of NH_4Cl.
   d. Why does the solution form? What drives the process?

36. When lithium iodide (LiI) is dissolved in water, the solution becomes hotter.
   a. Is the dissolution of lithium iodide endothermic or exothermic?
   b. What can you say about the relative magnitudes of the lattice energy of lithium iodide and its heat of hydration?
   c. Sketch a qualitative energy diagram similar to Figure 12.7 for the dissolution of LiI.
   d. Why does the solution form? What drives the process?

37. Silver nitrate has a lattice energy of $-820.$ kJ/mol and a heat of solution of $+22.6$ kJ/mol. Calculate the heat of hydration for silver nitrate.

38. Use the data below to calculate the heats of hydration of lithium chloride and sodium chloride. Which of the two cations, lithium or sodium, has stronger ion–dipole interactions with water? Why?

Compound	Lattice Energy (kJ/mol)	ΔH_{soln} (kJ/mol)
LiCl	-834	-37.0
NaCl	-769	$+3.88$

39. Lithium iodide has a lattice energy of -7.3×10^2 kJ/mol and a heat of hydration of -793 kJ/mol. Find the heat of solution for lithium iodide and determine how much heat is evolved or absorbed when 15.0 g of lithium iodide completely dissolves in water.

40. Potassium nitrate has a lattice energy of -163.8 kcal/mol and a heat of hydration of -155.5 kcal/mol. How much potassium nitrate has to dissolve in water to absorb 1.00×10^2 kJ of heat?

Solution Equilibrium and Factors Affecting Solubility

41. A solution contains 25 g of NaCl per 100.0 g of water at 25 °C. Is the solution unsaturated, saturated, or supersaturated? (Use Figure 12.11.)

42. A solution contains 32 g of KNO_3 per 100.0 g of water at 25 °C. Is the solution unsaturated, saturated, or supersaturated? (Use Figure 12.11.)

43. A KNO_3 solution containing 45 g of KNO_3 per 100.0 g of water is cooled from 40 °C to 0 °C. What will happen during cooling? (Use Figure 12.11.)

44. A KCl solution containing 42 g of KCl per 100.0 g of water is cooled from 60 °C to 0 °C. What will happen during cooling? (Use Figure 12.11.)

45. Some laboratory procedures involving oxygen-sensitive reactants or products call for using preboiled (and then cooled) water. Explain why this is so.

46. A person preparing a fish tank uses preboiled (and then cooled) water to fill it. When the fish is put into the tank, it dies. Explain.

47. Scuba divers breathing air at increased pressure can suffer from nitrogen narcosis—a condition resembling drunkenness—when the partial pressure of nitrogen exceeds about 4 atm. What property of gas/water solutions causes this to happen? How could the diver reverse this effect?

48. Scuba divers breathing air at increased pressure can suffer from oxygen toxicity—too much oxygen in their bloodstream—when the partial pressure of oxygen exceeds about 1.4 atm. What happens to the amount of oxygen in a diver's bloodstream when he or she breathes oxygen at elevated pressures? How can this be reversed?

49. Calculate the mass of nitrogen dissolved at room temperature in an 80.0 L home aquarium. Assume a total pressure of 1.0 atm and a mole fraction for nitrogen of 0.78.

50. Use Henry's law to determine the molar solubility of helium at a pressure of 1.0 atm and 25 °C.

Concentrations of Solutions

51. An aqueous NaCl solution is made using 133 g of NaCl diluted to a total solution volume of 1.00 L. Calculate the molarity, molality, and mass percent of the solution. (Assume a density of 1.08 g/mL for the solution.)

52. An aqueous KNO_3 solution is made using 88.4 g of KNO_3 diluted to a total solution volume of 1.50 L. Calculate the molarity, molality, and mass percent of the solution. (Assume a density of 1.05 g/mL for the solution.)

53. To what final volume should you dilute 50.0 mL of a 5.00 M KI solution so that 25.0 mL of the diluted solution contains 3.25 g of KI?

54. To what volume should you dilute 125 mL of an 8.00 M $CuCl_2$ solution so that 50.0 mL of the diluted solution contains 5.9 g $CuCl_2$?

55. Silver nitrate solutions are often used to plate silver onto other metals. What is the maximum amount of silver (in grams) that can be plated out of 4.8 L of an $AgNO_3$ solution containing 3.4% Ag by mass? Assume that the density of the solution is 1.01 g/mL.

56. A dioxin-contaminated water source contains 0.085% dioxin by mass. How much dioxin is present in 2.5 L of this water? Assume a density of 1.00 g/mL.

57. A hard water sample contains 0.0085% Ca by mass (in the form of Ca^{2+} ions). How much water (in grams) contains 1.2 g of Ca? (1.2 g of Ca is the recommended daily allowance of calcium for those between 19 and 24 years old.)

58. Lead is a toxic metal that affects the central nervous system. A Pb-contaminated water sample contains 0.0011% Pb by mass. How much of the water (in mL) contains 150 mg of Pb? (Assume a density of 1.0 g/mL.)

59. Nitric acid is usually purchased in a concentrated form that is 70.3% HNO_3 by mass and has a density of 1.41 g/mL Describe exactly how you would prepare 1.15 L of 0.100 M HNO_3 from the concentrated solution.

60. Hydrochloric acid is usually purchased in a concentrated form that is 37.0% HCl by mass and has a density of 1.20 g/mL. Describe exactly how you would prepare 2.85 L of 0.500 M HCl from the concentrated solution.

61. Describe how you would prepare each of the following solutions from the dry solute and the solvent.
    a. 1.00×10^2 mL of 0.500 M KCl
    b. 1.00×10^2 g of 0.500 m KCl
    c. 1.00×10^2 g of 5.0% KCl solution by mass

62. Describe how you would prepare each of the following solutions from the dry solute and the solvent.
    a. 125 mL of 0.100 M $NaNO_3$
    b. 125 g of 0.100 m $NaNO_3$
    c. 125 g of 1.0% $NaNO_3$ solution by mass

63. A solution is prepared by dissolving 28.4 g of glucose ($C_6H_{12}O_6$) in 355 g of water. The final volume of the solution is 378 mL. For this solution, calculate each of the following:
    a. molarity
    b. molality
    c. percent by mass
    d. mole fraction
    e. mole percent

64. A solution is prepared by dissolving 20.2 mL of methanol (CH_3OH) in 100.0 mL of water at 25 °C. The final volume of the solution is 118 mL. The densities of methanol and water at this temperature are 0.782 g/mL and 1.00 g/mL, respectively. For this solution, calculate each of the following:
    a. molarity
    b. molality
    c. percent by mass
    d. mole fraction
    e. mole percent

Vapor Pressure of Solutions

65. A beaker contains 100.0 mL of pure water. A second beaker contains 100.0 mL of seawater. The two beakers are left side by side on a lab bench for one week. At the end of the week, the liquid level in both beakers has decreased. However, the level has decreased more in one of the beakers than in the other. Which one and why?

66. Which one of the following solutions has the highest vapor pressure?
    a. 20.0 g of glucose ($C_6H_{12}O_6$) in 100.0 mL of water
    b. 20.0 g of sucrose ($C_{12}H_{22}O_{11}$) in 100.0 mL of water
    c. 10.0 g of potassium acetate $KC_2H_3O_2$ in 100.0 mL of water

67. Calculate the vapor pressure of a solution containing 28.5 g of glycerin ($C_3H_8O_3$) in 125 mL of water at 30.0 °C. The vapor pressure of pure water at this temperature is 31.8 torr. Assume that glycerin is not volatile and dissolves molecularly (i.e., it is not ionic) and use a density of 1.00 g/mL for the water.

68. A solution contains naphthalene ($C_{10}H_8$) dissolved in hexane (C_6H_{14}) at a concentration of 10.85% naphthalene by mass. Calculate the vapor pressure at 25 °C of hexane above the solution. The vapor pressure of pure hexane at 25 °C is 151 torr.

69. Calculate the vapor pressure at 25 °C of an aqueous solution that is 5.50% NaCl by mass.

70. An aqueous $CaCl_2$ solution has a vapor pressure of 81.6 mmHg at 50 °C. The vapor pressure of pure water at this temperature is 92.6 mmHg. What is the concentration of $CaCl_2$ in mass percent?

71. A solution contains 50.0 g of heptane (C_7H_{16}) and 50.0 g of octane (C_8H_{18}) at 25 °C. The vapor pressures of pure heptane and pure octane at 25 °C are 45.8 torr and 10.9 torr, respectively. Assuming ideal behavior, calculate each of the following:
    a. the vapor pressure of each of the solution components in the mixture
    b. the total pressure above the solution
    c. the composition of the vapor in mass percent
    d. Why is the composition of the vapor different from the composition of the solution?

72. A solution contains a mixture of pentane and hexane at room temperature. The solution has a vapor pressure of 258 torr. Pure pentane and hexane have vapor pressures of 425 torr and 151 torr, respectively, at room temperature. What is the mole fraction composition of the mixture? (Assume ideal behavior.)

73. A solution contains 4.08 g of chloroform ($CHCl_3$) and 9.29 g of acetone (CH_3COCH_3). The vapor pressures at 35 °C of pure chloroform and pure acetone are 295 torr and 332 torr, respectively. Assuming ideal behavior, calculate the vapor pressures of each of the components and the total vapor pressure above the solution. The experimentally measured total vapor pressure of the solution at 35 °C was 312 torr. Is the solution ideal? If not, what can you say about the relative strength of chloroform–acetone interactions compared to the acetone–acetone and chloroform–chloroform interactions?

74. A solution of methanol and water has a mole fraction of water of 0.312 and a total vapor pressure of 211 torr at 39.9 °C. The vapor pressures of pure methanol and pure water at this temperature are 256 torr and 55.3 torr, respectively. Is the solution ideal? If not, what can you say about the relative strengths of the solute–solvent interactions compared to the solute–solute and solvent–solvent interactions?

Freezing Point Depression, Boiling Point Elevation, and Osmosis

75. A glucose solution contains 55.8 g of glucose ($C_6H_{12}O_6$) in 455 g of water. Compute the freezing point and boiling point of the solution.

76. An ethylene glycol solution contains 21.2 g of ethylene glycol ($C_2H_6O_2$) in 85.4 mL of water. Compute the freezing point and boiling point of the solution. (Assume a density of 1.00 g/mL for water.)

77. An aqueous solution containing 17.5 g of an unknown molecular (nonelectrolyte) compound in 100.0 g of water was found to have a freezing point of −1.8 °C. Calculate the molar mass of the unknown compound.

78. An aqueous solution containing 35.9 g of an unknown molecular (nonelectrolyte) compound in 150.0 g of water was found to have a freezing point of −1.3 °C. Calculate the molar mass of the unknown compound.

79. Calculate the osmotic pressure of a solution containing 24.6 g of glycerin ($C_3H_8O_3$) in 250.0 mL of solution at 298 K.

80. What mass of sucrose ($C_{12}H_{22}O_{11}$) should be combined with 5.00×10^2 g of water to make a solution with an osmotic pressure of 8.55 atm at 298 K? (Assume a density of 1.0 g/mL for the solution.)

81. A solution containing 27.55 mg of an unknown protein per 25.0 mL solution was found to have an osmotic pressure of 3.22 torr at 25 °C. What is the molar mass of the protein?

82. Calculate the osmotic pressure of a solution containing 18.75 mg of hemoglobin in 15.0 mL of solution at 25 °C. The molar mass of hemoglobin is 6.5×10^4 g/mol.

83. Calculate the freezing point and boiling point of the following aqueous solutions, assuming complete dissociation.
    a. 0.100 m K_2S
    b. 21.5 g of $CuCl_2$ in 4.50×10^2 g water
    c. 5.5% $NaNO_3$ by mass (in water)

84. Calculate the freezing point and boiling point of the following solutions, assuming complete dissociation.
    a. 10.5 g $FeCl_3$ in 1.50×10^2 g water
    b. 3.5% KCl by mass (in water)
    c. 0.150 m MgF_2

85. Use the van't Hoff factors in Table 12.9 to compute each of the following:
    a. the melting point of a 0.100 m iron(III) chloride solution
    b. the osmotic pressure of a 0.085 M potassium sulfate solution at 298 K
    c. the boiling point of a 1.22% by mass magnesium chloride solution

86. Assuming the van't Hoff factors in Table 12.9, calculate the mass of each solute required to produce each of the following aqueous solutions:
    a. a sodium chloride solution containing 1.50×10^2 g of water that has a melting point of $-1.0\,°C$.
    b. 2.50×10^2 mL of a magnesium sulfate solution that has an osmotic pressure of 3.82 atm at 298 K
    c. an iron(III) chloride solution containing 2.50×10^2 g of water that has a boiling point of 102 °C

87. A 0.100 M ionic solution has an osmotic pressure of 8.3 atm at 25 °C. Calculate the van't Hoff factor (i) for this solution.

88. A solution contains 8.92 g of KBr in 500.0 mL of solution and has an osmotic pressure of 6.97 atm at 25 °C. Calculate the van't Hoff factor (i) for KBr at this concentration.

Cumulative Problems

89. The solubility of carbon tetrachloride (CCl_4) in water at 25 °C is 1.2 g/L. The solubility of chloroform ($CHCl_3$) at the same temperature is 10.1 g/L. Why is chloroform almost 10 times more soluble in water than is carbon tetrachloride?

90. The solubility of phenol in water at 25 °C is 8.7 g/L. The solubility of naphthol at the same temperature is only 0.074 g/L. Examine the structures of phenol and napthol shown below and explain why phenol is so much more soluble than naphthol.

Phenol   Naphthol

91. Potassium perchlorate ($KClO_4$) has a lattice energy of -599 kJ/mol and a heat of hydration of -548 kJ/mol. Find the heat of solution for potassium perchlorate and determine the temperature change that occurs when 10.0 g of potassium perchlorate is dissolved with enough water to make 100.0 mL of solution. (Assume a heat capacity of 4.05 J/g · °C for the solution and a density of 1.05 g/mL.)

92. Sodium hydroxide (NaOH) has a lattice energy of -887 kJ/mol and a heat of hydration of -932 kJ/mol How much solution could be heated to boiling by the heat evolved by the dissolution of 25.0 g of NaOH? (For the solution, assume a heat capacity of 4.0 J/g·°C, an initial temperature of 25.0 °C,, a boiling point of 100.0 °C, and a density of 1.05 g/mL.)

93. A saturated solution was formed when 0.0537 L of argon, at a pressure of 1.0 atm and temperature of 25 °C, was dissolved in 1.0 L of water. Calculate the Henry's law constant for argon.

94. A gas has a Henry's law constant of 0.112 M/atm. How much water would be needed to completely dissolve 1.65 L of the gas at a pressure of 725 torr and a temperature of 25 °C?

95. The Safe Drinking Water Act (SDWA) sets a limit for mercury—a toxin to the central nervous system—at 0.0020 ppm by mass. Water suppliers must periodically test their water to ensure that mercury levels do not exceed this limit. Suppose water becomes contaminated with mercury at twice the legal limit (0.0040 ppm). How much of this water would have to be consumed for someone to ingest 50.0 mg of mercury?

96. Water softeners often replace calcium ions in hard water with sodium ions. Since sodium compounds are soluble, the presence of sodium ions in water does not cause the white, scaly residues caused by calcium ions. However, calcium is more beneficial to human health than sodium because calcium is a necessary part of the human diet, while high levels of sodium intake are linked to increases in blood pressure. The U.S. Food and Drug Administration (FDA) recommends that adults ingest less than 2.4 g of sodium per day. How many liters of softened water, containing a sodium concentration of 0.050% sodium by mass, have to be consumed to exceed the FDA recommendation? (Assume a water density of 1.0 g/mL.)

97. An aqueous solution contains 12.5% NaCl by mass. What mass of water (in grams) is contained in 2.5 L of the vapor above this solution at 55 °C? The vapor pressure of pure water at 55 °C is 118 torr. (Assume complete dissociation of NaCl.)

98. The vapor above an aqueous solution contains 19.5 mg water per liter at 25 °C. Assuming ideal behavior, what is the concentration of the solute within the solution in mole percent?

99. What is the freezing point of an aqueous solution that boils at 106.5 °C?

100. What is the boiling point of an aqueous solution that has a vapor pressure of 20.5 torr at 25 °C? (Assume a nonvolatile solute.)

101. An isotonic solution contains 0.90% NaCl by mass per volume. Calculate the percent mass per volume for isotonic solutions containing each of the following solutes at 25 °C. Assume a van't Hoff factor of 1.9 for all *ionic* solutes.
    a. KCl   b. NaBr   c. Glucose ($C_6H_{12}O_6$)

102. Magnesium citrate, $Mg_3(C_6H_5O_7)_2$, belongs to a class of laxatives called *hyperosmotics*, which are used for rapid emptying of the bowel. When a concentrated solution of magnesium citrate is consumed, it passes through the intestines, drawing water and promoting diarrhea, usually within 6 hours. Calculate the osmotic pressure of a magnesium citrate laxative solution containing 28.5 g of magnesium citrate in 235 mL of solution at 37 °C (approximate body temperature). Assume complete dissociation of the ionic compound.

103. A solution is prepared from 4.5701 g of magnesium chloride and 43.238 g of water. The vapor pressure of water above this solution is found to be 0.3624 atm at 348.0 K. The vapor pressure of pure water at this temperature is 0.3804 atm. Find the value of the van't Hoff factor i for magnesium chloride in this solution.

104. When HNO_2 is dissolved in water it partially dissociates according to the equation $HNO_2 \rightleftharpoons H^+ + NO_2^-$. A solution is prepared that contains 7.050 g of HNO_2 in 1.000 kg of water. Its freezing point is found to be $-0.2929\ °C$. Calculate the fraction of HNO_2 that has dissociated.

105. A solution of a nonvolatile solute in water has a boiling point of 375.3 K. Calculate the vapor pressure of water above this solution at 338 K. The vapor pressure of pure water at this temperature is 0.2467 atm.

106. The density of a 0.438 M solution of potassium chromate (K_2CrO_4) at 298 K is 1.063 g/mL. Calculate the vapor pressure of water above the solution. The vapor pressure of pure water at this temperature is 0.0313 atm. Assume complete dissociation.

107. The vapor pressure of carbon tetrachloride, CCl_4, is 0.354 atm and the vapor pressure of chloroform, $CHCl_3$, is 0.526 atm at 316 K. A solution is prepared from equal masses of these two compounds at this temperature. Calculate the mole fraction of the chloroform in the vapor above the solution. If the vapor above the original solution is condensed and isolated into a separate flask, what would the vapor pressure of chloroform be above this new solution?.

108. Distillation is a method of purification based on successive separations and recondensations of vapor above a solution. Use the result of the previous problem to calculate the mole fraction of chloroform in the vapor above a solution obtained by three successive separations and condensations of the vapors above the original solution of carbon tetrachloride and chloroform. Show how this result explains the use of distillation as a separation method.

Challenge Problems

109. The small bubbles that form on the bottom of a water pot that is being heated (before boiling) are due to dissolved air coming out of solution. Use Henry's law and the solubilities given below to calculate the total volume of nitrogen and oxygen gas that should bubble out of 1.5 L of water upon warming from 25 °C to 50 °C. Assume that the water is initially saturated with nitrogen and oxygen gas at 25 °C and a total pressure of 1.0 atm. Assume that the gas bubbles out at a temperature of 50 °C. The solubility of oxygen gas at 50 °C is 27.8 mg/L. at an oxygen pressure of 1.00 atm. The solubility of nitrogen gas at 50 °C is 14.6 mg/L at a nitrogen pressure of 1.00 atm. Assume that the air above the water contains an oxygen partial pressure of 0.21 atm and a nitrogen partial pressure of 0.78 atm.

110. The vapor above a mixture of pentane and hexane at room temperature contains 35.5% pentane by mass. What is the mass percent composition of the solution? Pure pentane and hexane have vapor pressures of 425 torr and 151 torr, respectively, at room temperature.

111. A 1.10-g sample contains only glucose ($C_6H_{12}O_6$) and sucrose ($C_{12}H_{22}O_{11}$). When the sample is dissolved in water to a total solution volume of 25.0 mL, the osmotic pressure of the solution is 3.78 atm at 298 K. What is the percent composition of glucose and sucrose in the sample?

112. A solution is prepared by mixing 631 mL of methanol with 501 mL of water. The molarity of methanol in the resulting solution is 14.29 M. The density of methanol at this temperature is 0.792 g/mL. Calculate the difference in volume between this solution and the total volume of water and methanol that were mixed to prepare the solution.

113. Two alcohols, isopropyl alcohol and propyl alcohol, have the same molecular formula, C_3H_8O. A solution of the two that is two-thirds by mass isopropyl alcohol has a vapor pressure of 0.110 atm at 313 K. A solution that is one-third by mass isopropyl alcohol has a vapor pressure of 0.089 atm at 313 K. Calculate the vapor pressure of each pure alcohol at this temperature. Explain the difference given that the formula of propyl alcohol is $CH_3CH_2CH_2OH$ and that of isopropyl alcohol is $(CH_3)_2CHOH$.

114. A metal, M, of atomic weight 96 reacts with fluorine to form a salt that can be represented as MF_x. In order to determine x and therefore the formula of the salt, a boiling point elevation experiment is performed. A 9.18-g sample of the salt is dissolved in 100.0 g of water and the boiling point of the solution is found to be 374.38 K. Find the formula of the salt. Assume complete dissociation of the salt in solution.

Conceptual Problems

115. Substance A is a nonpolar liquid and has only dispersion forces among its constituent particles. Substance B is also a nonpolar liquid and has about the same magnitude of dispersion forces among its constituent particles. When substance A and B are combined, they spontaneously mix.
  a. Why do the two substances mix?
  b. Predict the sign and magnitude of ΔH_{soln}.
  c. Give the signs and relative magnitudes of ΔH_{solute}, $\Delta H_{solvent}$, and ΔH_{mix}.

116. A power plant built on a river uses river water as a coolant. The water is warmed as it is used in heat exchangers within the plant. Should the warm water be immediately cycled back into the river? Why or why not?

117. The vapor pressure of a 1 M ionic solution is different from the vapor pressure of a 1 M nonionic solution. In both cases, the solute is nonvolatile. Which of the following sets of diagrams best represents the differences between the two solutions and their vapors?

Ionic solute   Nonionic solute

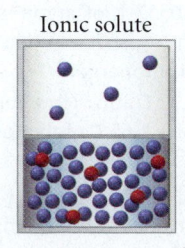
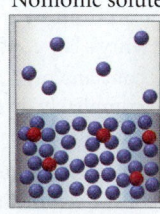

● Solvent particles

● Solute particles

(a)

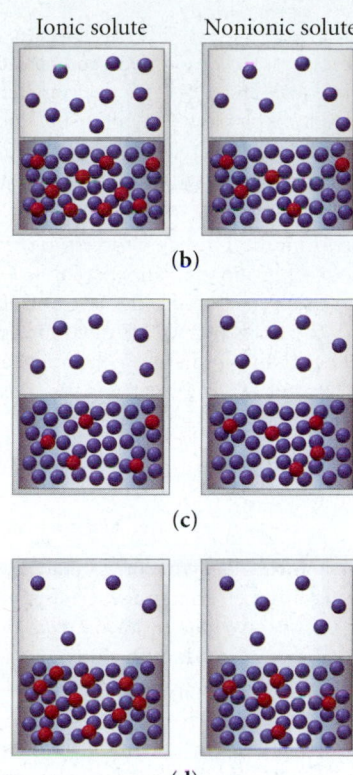

Ionic solute   Nonionic solute

(b)

(c)

(d)

118. If all of the following substances cost the same amount per kilogram, which would be most cost-effective as a way to lower the freezing point of water? (Assume complete dissociation for all ionic compounds.) Explain.

a. $HOCH_2CH_2OH$ **b.** NaCl
c. KCl   **d.** $MgCl_2$
e. $SrCl_2$

119. A helium balloon inflated on one day will fall to the ground by the next day. The volume of the balloon decreases somewhat overnight, but not by enough to explain why it no longer floats. (If you inflate a new balloon with helium to the same size as the balloon that fell to the ground, the newly inflated balloon floats.) Explain why the helium balloon falls to the ground overnight.

13

Chemical Kinetics

13.1 Catching Lizards

13.2 The Rate of a Chemical Reaction

13.3 The Rate Law: The Effect of Concentration on Reaction Rate

13.4 The Integrated Rate Law: The Dependence of Concentration on Time

13.5 The Effect of Temperature on Reaction Rate

13.6 Reaction Mechanisms

13.7 Catalysis

IN THE PASSAGE QUOTED on the facing page, Oxford chemistry professor Sir Cyril Hinshelwood calls attention to an aspect of chemistry often overlooked by the casual observer—the mystery of change with time. Since the opening chapter of this book, you have learned that the goal of chemistry is to understand the macroscopic world by examining the molecular one. In this chapter, we focus on understanding how this molecular world changes with time, a topic called chemical kinetics. The molecular world is anything but static. Thermal energy produces constant molecular motion, causing molecules to repeatedly collide with one another. In a tiny fraction of these collisions, something unique happens—the electrons on one molecule or atom are attracted to the nuclei of another. Some bonds weaken and new bonds form—a chemical reaction occurs. Chemical kinetics is the study of how these kinds of changes occur in time.

> *Nobody, I suppose, could devote many years to the study of chemical kinetics without being deeply conscious of the fascination of time and change: this is something that goes outside science into poetry. . . .*
>
> —Sir Cyril N. Hinshelwood (1897–1967)

Pouring ice water on a lizard slows it down, making it easier to catch.

13.1 Catching Lizards

The kids who live in my neighborhood (including my own) have a unique way of catching lizards. Armed with cups of ice water, they chase the cold-blooded reptiles into a corner, and then take aim and fire—or pour, actually. They pour the cold water directly onto the lizard's body. The lizard's body temperature drops and it becomes virtually immobilized—easy prey for little hands. The kids scoop up the lizard and place it in a tub filled with sand and leaves. They then watch as the lizard warms back up and becomes active again. They usually release the lizard back into the yard within hours. I guess you would call them catch-and-release lizard hunters.

Unlike mammals, which actively regulate their body temperature through metabolic activity, lizards are *ectotherms*—their body temperature depends on their surroundings. When splashed with cold water, a lizard's body simply gets colder. The drop in body temperature immobilizes the lizard because its movement depends on chemical reactions that occur within its muscles, and the rates of those reactions—how fast they occur—are highly sensitive to temperature. In other words, when the temperature drops, the reactions that produce movement occur more slowly; therefore, the movement itself slows down. Cold reptiles are lethargic, unable to move very quickly. For this reason, reptiles try to maintain their body temperature in a narrow range by moving between sun and shade.

The rates of chemical reactions, and especially the ability to *control* those rates, are important not just in reptile movement but in many other phenomena as well. For example, the launching of a rocket depends on controlling the rate at which fuel burns—too quickly and the rocket can explode, too slowly and it will not leave the ground. Chemists must always consider reaction rates when synthesizing compounds. No matter how stable a compound might be, its synthesis is impossible if the rate at which it forms is too slow. As we have seen with reptiles, reaction rates are important to life. In fact, the human body's ability to switch a specific reaction on or off at a specific time is achieved largely by controlling the rate of that reaction through the use of enzymes.

The first person to measure the rate of a chemical reaction carefully was Ludwig Wilhelmy, who in 1850 measured how fast sucrose, upon treatment with acid, hydrolyzed into glucose and fructose. This reaction occurred over several hours, and Wilhelmy was able to show how the rate depended on the initial amount of sugar present—the greater the initial amount, the faster the initial rate. Today we can measure the rates of reactions that occur in times as short as several femtoseconds (femto = 10^{-15}). The knowledge of reaction rates is not only practically important—giving us the ability to control how fast a reaction occurs—but also theoretically important. As you will see in Section 13.6, knowledge of the rate of a reaction can tell us much about how the reaction occurs on the molecular scale.

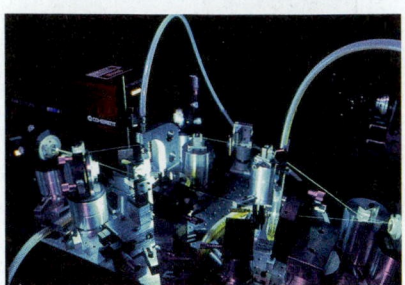

▲ This laser produces light pulses that are several femtoseconds long, allowing chemists to observe reactions that occur in less than one picosecond.

13.2 The Rate of a Chemical Reaction

The rate of a chemical reaction is a measure of how fast the reaction occurs. If a chemical reaction has a slow rate, only a relatively small fraction of molecules react to form products in a given period of time, as shown in Figure 13.1 ▶. If a chemical reaction has a fast rate, a large fraction of molecules react to form products in a given period of time.

When we measure how fast something occurs, or more specifically the *rate* at which it occurs, we usually express the measurement as a change in some quantity per unit of time. For example, we measure the speed of a car—the rate at which it travels—in *miles per hour*, and we measure how fast we might be losing weight in *pounds per week*. Notice that both of these rates are reported in units that represent the change in what we are measuring (distance or weight) divided by the change in time.

$$\text{Speed} = \frac{\text{change in distance}}{\text{change in time}} = \frac{\Delta x}{\Delta t} \qquad \text{Weight loss} = \frac{\text{change in weight}}{\text{change in time}} = \frac{\Delta \text{ weight}}{\Delta t}$$

Similarly, the rate of a chemical reaction is measured as a change in the amounts of reactants or products (usually in terms of concentration) divided by the change in time. For example, consider the following gas-phase reaction between $H_2(g)$ and $I_2(g)$:

$$H_2(g) + I_2(g) \longrightarrow 2\,HI(g)$$

We can define the rate of this reaction in the time interval t_1 to t_2 as follows:

| Recall that [A] means the concentration of A in M (mol/L).

$$\text{Rate} = -\frac{\Delta [H_2]}{\Delta t} = -\frac{[H_2]_{t_2} - [H_2]_{t_1}}{t_2 - t_1} \qquad\qquad [13.1]$$

In this expression, $[H_2]_{t_2}$ is the hydrogen concentration at time t_2 and $[H_2]_{t_1}$ is the hydrogen concentration at time t_1. Notice that the reaction rate is defined as *the negative* of the change in concentration of a reactant divided by the change in time. The negative sign is usually part of the definition when the reaction rate is defined with respect to a reactant because reactant concentrations decrease as a reaction proceeds; therefore *the change in the concentration of a reactant is negative*. The negative sign in the definition thus makes the overall *rate* positive. In other words, the negative sign is the result of the convention that reaction rates are usually reported as positive quantities.

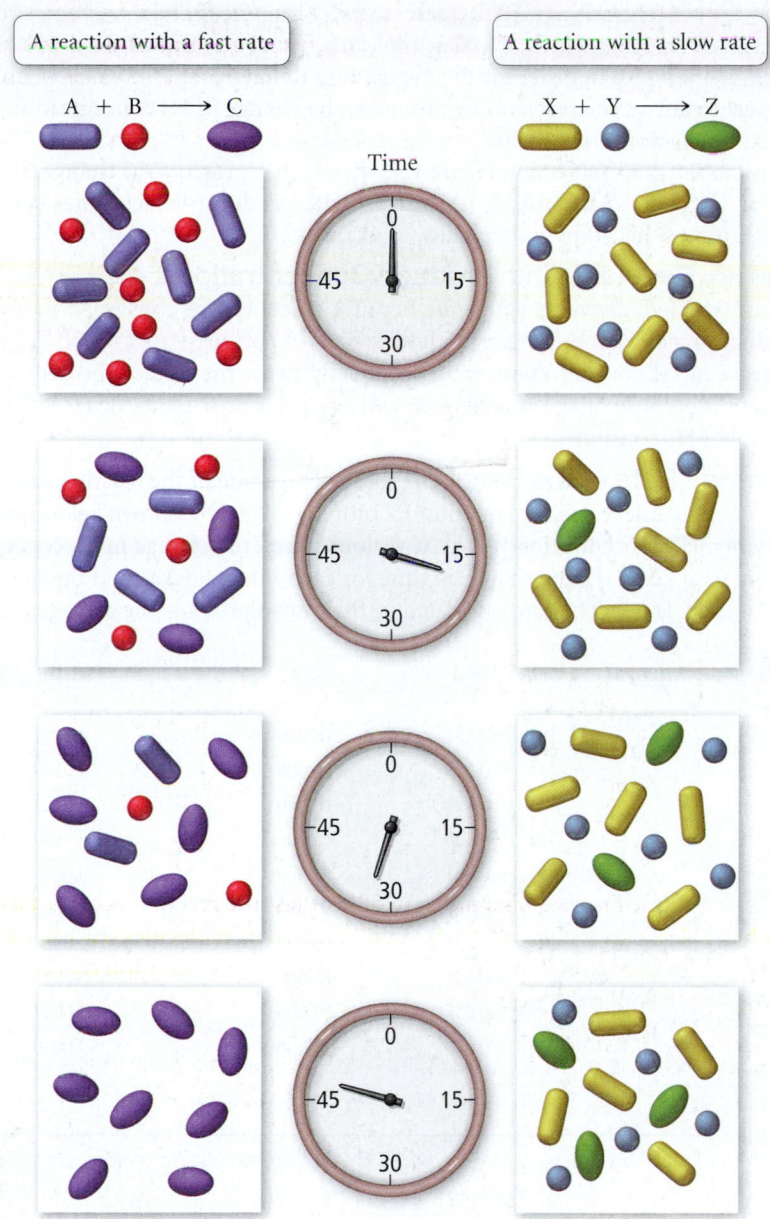

◀ FIGURE 13.1 The Rate of a Chemical Reaction

The reaction rate can also be defined with respect to the other reactant as follows:

$$\text{Rate} = -\frac{\Delta[I_2]}{\Delta t} \qquad [13.2]$$

Since 1 mol of H_2 reacts with 1 mol of I_2, the rates are defined in the same way. The rate can also be defined with respect to the *product* of the reaction as follows:

$$\text{Rate} = +\frac{1}{2}\frac{\Delta[HI]}{\Delta t} \qquad [13.3]$$

Notice that, because product concentrations *increase* as the reaction proceeds, the change in concentration of a product is positive. Therefore, when the rate is defined with respect to a product, we do not include a negative sign in the definition—the rate is naturally positive. Notice also the factor of $\frac{1}{2}$ in this definition. This factor is related to the stoichiometry of the reaction. In order to have a single rate for the entire reaction, the definition of the rate with respect to each reactant and product must reflect the stoichiometric coefficients of the reaction. For this particular reaction, 2 mol of HI is produced from 1 mol of H_2 and 1 mol of I_2.

Therefore the concentration of HI increases at twice the rate that the concentration of H_2 or I_2 decreases. In other words, if 100 I_2 molecules react per second, then 200 HI molecules form per second. In order for the overall rate to have the same value when defined with respect to any of the reactants or products, the change in HI concentration must be multiplied by a factor of one-half.

Consider the graph shown in Figure 13.2 ▼, which represents the changes in concentration for H_2 (one of the reactants) and HI (the product) versus time. We examine several features of this graph individually.

Change in Reactant and Product Concentrations

The reactant concentration, as expected, *decreases* with time because reactants are consumed in a reaction. The product concentration *increases* with time because products are formed in a reaction. The increase in HI concentration occurs at exactly twice the rate of the decrease in H_2 concentration because of the stoichiometry of the reaction—2 mol of HI is formed for every 1 mol of H_2 consumed.

The Average Rate of the Reaction

The average rate of the reaction can be calculated for any time interval using Equation 13.1 for H_2. The table shown below lists each of the following: H_2 concentration ($[H_2]$) at various times, the change in H_2 concentration for each interval ($\Delta[H_2]$), the change in time for each interval (Δt), and the rate for each interval ($-\Delta[H_2]/\Delta t$). The rate calculated in this way represents the average rate within

Time (s)	$[H_2]$ (M)	$\Delta[H_2]$	Δt	Rate $= -\Delta[H_2]/\Delta t$ (M/s)
0.000	1.000			
		−0.181	10.000	0.0181
10.000	0.819			
		−0.149	10.000	0.0149
20.000	0.670			
		−0.121	10.000	0.0121
30.000	0.549			
		−0.100	10.000	0.0100
40.000	0.449			
		−0.081	10.000	0.0081
50.000	0.368			
		−0.067	10.000	0.0067
60.000	0.301			
		−0.054	10.000	0.0054
70.000	0.247			
		−0.045	10.000	0.0045
80.000	0.202			
		−0.037	10.000	0.0037
90.000	0.165			
		−0.030	10.000	0.0030
100.000	0.135			

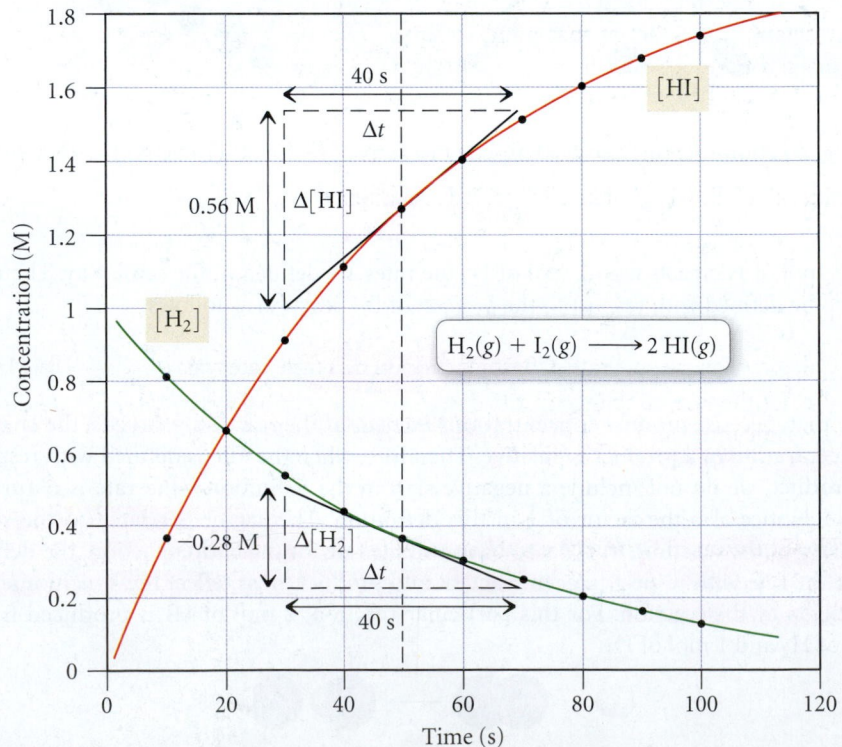

▶ **FIGURE 13.2 Reactant and Product Concentrations as a Function of Time**

the given time interval. For example, the average rate of the reaction in the time interval between 10 and 20 seconds is 0.0149 M/s, while the average rate in the time interval between 20 and 30 seconds is 0.0121 M/s. Notice that the average rate *decreases* as the reaction progresses. In other words, the reaction slows down as it proceeds. We discuss this further in the next section where we will see that, for most reactions, the rate depends on the concentrations of the reactants. As the reactants are consumed, their concentrations decrease, and the reaction slows down.

The Instantaneous Rate of the Reaction   The instantaneous rate of the reaction is the rate at any one point in time, represented by the instantaneous slope of the curve at that point. We can obtain the instantaneous rate from the slope of the tangent to the curve at the point of interest. In our graph, we have drawn the tangent lines for both [H_2] and [HI] at 50 seconds. We calculate the instantaneous rate at 50 seconds as follows:

Using [H_2]

$$\text{Instantaneous rate (at 50 s)} = -\frac{\Delta[H_2]}{\Delta t} = -\frac{-0.28 \text{ M}}{40 \text{ s}} = 0.0070 \text{ M/s}$$

Using [HI]

$$\text{Instantaneous rate (at 50 s)} = +\frac{1}{2}\frac{\Delta[HI]}{\Delta t} = +\frac{1}{2}\frac{0.56 \text{ M}}{40 \text{ s}} = 0.0070 \text{ M/s}$$

Notice that, as expected, the rate is the same whether we use one of the reactants or the product for the calculation. Notice also that the instantaneous rate at 50 seconds (0.0070 M/s) lies between the average rates calculated for the 10-second intervals just before and just after 50 seconds.

We can generalize our definition of reaction rates by using the following generic reaction:

$$aA + bB \longrightarrow cC + dD \qquad [13.4]$$

where A and B are reactants, C and D are products, and *a*, *b*, *c*, and *d* are the stoichiometric coefficients. The rate of the reaction is then defined as follows:

$$\text{Rate} = -\frac{1}{a}\frac{\Delta[A]}{\Delta t} = -\frac{1}{b}\frac{\Delta[B]}{\Delta t} = +\frac{1}{c}\frac{\Delta[C]}{\Delta t} = +\frac{1}{d}\frac{\Delta[D]}{\Delta t} \qquad [13.5]$$

Notice that knowing the rate of change in the concentration of any one reactant or product at a point in time allows you to determine the rate of change in the concentration of any other reactant or product at that point in time (from the balanced equation). *However, predicting the rate at some future time is not possible from just the balanced equation.*

EXAMPLE 13.1 Expressing Reaction Rates

Consider the following balanced chemical equation:

$$H_2O_2(aq) + 3\,I^-(aq) + 2\,H^+(aq) \longrightarrow I_3^-(aq) + 2\,H_2O(l)$$

In the first 10.0 seconds of the reaction, the concentration of I^- dropped from 1.000 M to 0.868 M.

(a) Calculate the average rate of this reaction in this time interval.

(b) Predict the rate of change in the concentration of H^+ (that is, $\Delta[H^+]/\Delta t$) during this time interval.

Solution

(a) Use Equation 13.5 to calculate the average rate of the reaction.

$$\text{Rate} = -\frac{1}{3}\frac{\Delta[I^-]}{\Delta t}$$

$$= -\frac{1}{3}\frac{(0.868 \text{ M} - 1.000 \text{ M})}{10.0 \text{ s}}$$

$$= 4.40 \times 10^{-3} \text{ M/s}$$

(b) Use Equation 13.5 again for the relationship between the rate of the reaction and $\Delta[H^+]/\Delta t$. After solving for $\Delta[H^+]/\Delta t$, substitute the computed rate from part (a) and compute $\Delta[H^+]/\Delta t$.

$$\text{Rate} = -\frac{1}{2}\frac{\Delta[H^+]}{\Delta t}$$

$$\frac{\Delta[H^+]}{\Delta t} = -2\,(\text{rate})$$

$$= -2(4.40 \times 10^{-3}\,\text{M/s})$$

$$= -8.80 \times 10^{-3}\,\text{M/s}$$

For Practice 13.1

For the above reaction, predict the rate of change in concentration of H_2O_2 ($\Delta[H_2O_2]/\Delta t$) and I_3^- ($\Delta[I_3^-]/\Delta t$) during this time interval.

Measuring Reaction Rates

In order to study the kinetics of a reaction, you must have an experimental way to measure the concentration of at least one of the reactants or products as a function of time. For example, Ludwig Wilhelmy, whose experiment on the rate of the conversion of sucrose to glucose and fructose was discussed briefly in Section 13.1, took advantage of sucrose's ability to rotate polarized light. When a beam of polarized light is passed through a sucrose solution, the polarization of the light is rotated clockwise. In contrast, the products of the reaction rotate polarized light to the left. By measuring the degree of polarization of light passing through a reacting solution—a technique known as polarimetry—Wilhelmy was able to determine the relative concentrations of the reactants and products as a function of time.

Perhaps the most common way to study the kinetics of a reaction is through spectroscopy. For example, the reaction of H_2 and I_2 to form HI can be followed spectroscopically because I_2 is violet and H_2 and HI are colorless. As I_2 reacts with H_2 to form HI, the violet color fades. The fading color can be monitored with a spectrometer, a device that passes light through a sample and measures how strongly the light is absorbed (Figure 13.3 ▼). If the sample contains the reacting mixture, the intensity of the light absorption will decrease as the reaction proceeds, providing a direct measure of the concentration of I_2 as a function of time. Because light travels so fast and current experimental techniques can produce very short pulses of light, spectroscopy can be used to measure reactions that happen on time scales as small as several femtoseconds.

Reactions in which the number of moles of gaseous reactants and products changes as the reaction proceeds can be readily monitored by measuring changes in pressure. For example, consider the reaction in which dinitrogen monoxide reacts to form nitrogen and oxygen gas:

$$2\,N_2O(g) \longrightarrow 2\,N_2(g) + O_2(g)$$

▼ **FIGURE 13.3 The Spectrometer** In a spectrometer, light of a specific wavelength is passed through the sample and the intensity of the transmitted light—which depends on how much light is absorbed by the sample—is measured and recorded.

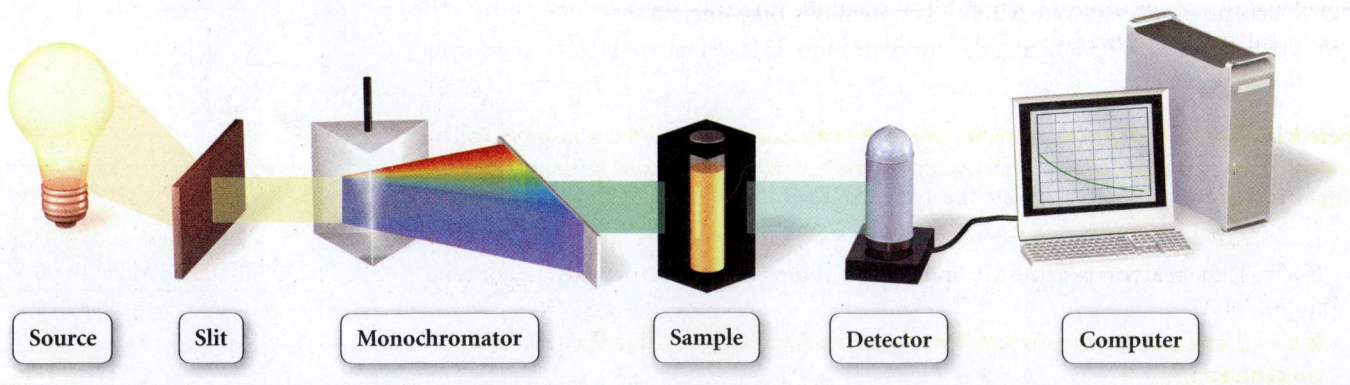

| Source | Slit | Monochromator | Sample | Detector | Computer |

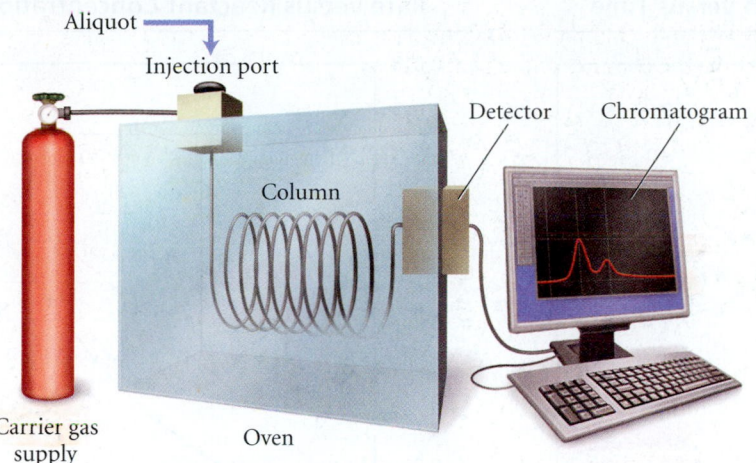

▲ FIGURE 13.4 The Gas Chromatograph   In a gas chromatograph (GC), a sample of the reaction mixture, or aliquot, is injected into a specially constructed column. Because of their characteristic physical and chemical properties, different components of the mixture pass through the column at different rates and thus exit at different times. As each component leaves the column, it is identified electronically and a chromatogram is recorded. The area under each peak in the chromatogram is proportional to the amount of one particular component in the sample mixture.

For every 2 mol of N_2O that reacts, the reaction vessel will contain one additional mole of gas. As the reaction proceeds, therefore, the pressure steadily rises. The rise in pressure can then be used to determine the relative concentrations of reactants and products as a function of time.

The three techniques mentioned above—polarimetry, spectroscopy, and pressure measurement—can all be used to monitor the reaction as it occurs in the reaction vessel. Some reactions, however, occur slowly enough that samples, or *aliquots,* can be periodically withdrawn from the reaction vessel and analyzed to determine the progress of the reaction. Instrumental techniques such as gas chromatography (Figure 13.4 ▲) or mass spectrometry, as well as wet chemical techniques such as titration, can be used to measure the relative amounts of reactants or products in the aliquot. By taking aliquots at regular time intervals, the relative amounts of reactants and products can be determined as a function of time.

13.3 The Rate Law: The Effect of Concentration on Reaction Rate

The rate of a reaction often depends on the concentration of one or more of the reactants. Wilhelmy noticed this effect in 1850 for the hydrolysis of sucrose. For simplicity, consider a reaction in which a single reactant, A, decomposes into products:

$$A \longrightarrow products$$

As long as the rate of the reverse reaction (in which the products return to reactants) is negligibly slow, we can write a relationship—called the **rate law**—between the rate of the reaction and the concentration of the reactant as follows:

$$Rate = k[A]^n \qquad [13.6]$$

where k is a constant of proportionality called the **rate constant** and n is a number called the **reaction order**. The value of n determines how the rate depends on the concentration of the reactant.

- If $n = 0$, the reaction is *zero order* and the rate is independent of the concentration of A.
- If $n = 1$, the reaction is *first order* and the rate is directly proportional to the concentration of A.
- If $n = 2$, the reaction is *second order* and the rate is proportional to the square of the concentration of A.

By definition, $[A]^0 = 1$, so the rate is equal to k regardless of $[A]$.

Reactant Concentration versus Time

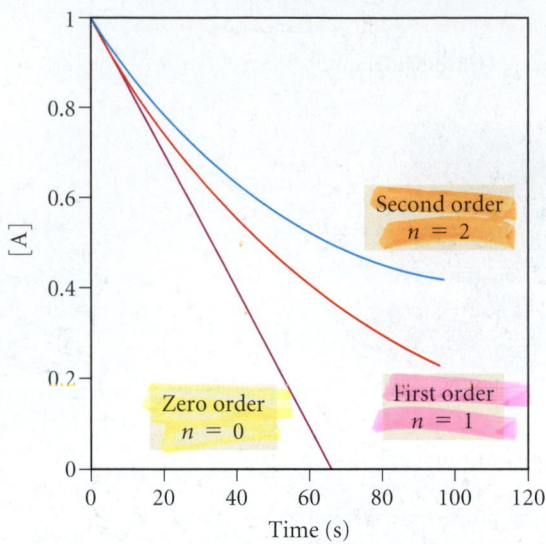

Rate versus Reactant Concentration

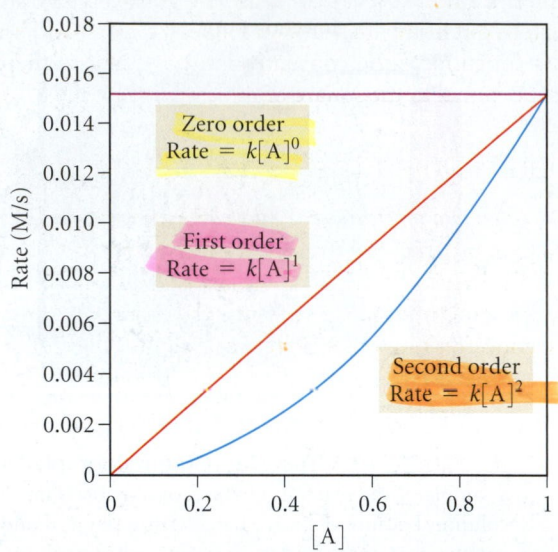

▲ FIGURE 13.5 **Reactant Concentration as a Function of Time for Different Reaction Orders.**

▲ FIGURE 13.6 **Reaction Rate as a Function of Reactant Concentration for Different Reaction Orders**

Although other orders are possible, including noninteger (or fractional) orders, these three are the most common.

Figure 13.5 ▲ shows three plots illustrating how the *concentration of A changes with time* for the three common reaction orders with identical values for the rate constant (k). Figure 13.6 ▲ has three plots showing the *rate of the reaction* (the slope of the lines in Figure 13.5) *as a function of the reactant concentration* for each reaction order.

Zero-Order Reaction   In a zero-order reaction, the rate of the reaction is independent of the concentration of the reactant:

$$\text{Rate} = k[A]^0 = k \qquad [13.7]$$

Consequently, for a zero-order reaction, the concentration of the reactant decreases linearly with time, as shown in Figure 13.5. Notice the constant slope in the plot—a constant slope means a constant rate. The rate is constant because the reaction does not slow down as the concentration of A decreases. The graph in Figure 13.6 shows that the rate of a zero-order reaction is the same at any concentration of A. Zero-order reactions occur under conditions where the amount of reactant actually *available for reaction* is unaffected by changes in the *overall quantity of reactant*. For example, sublimation is normally zero order because only molecules at the surface can sublime, and their concentration does not change when the amount of subliming substance decreases (Figure 13.7 ◄).

First-order reaction   In a first-order reaction, the rate of the reaction is directly proportional to the concentration of the reactant.

$$\text{Rate} = k[A]^1 \qquad [13.8]$$

Consequently, for a first-order reaction the rate slows down as the reaction proceeds because the concentration of the reactant decreases. You can see this in Figure 13.5 because the slope of the curve (the rate) becomes less steep (slower) with time. Figure 13.6 shows the rate as function of the concentration of A. Notice the linear relationship—the rate is directly proportional to the concentration.

Second-Order Reaction   In a second-order reaction, the rate of the reaction is proportional to the square of the concentration of the reactant.

$$\text{Rate} = k[A]^2 \qquad [13.9]$$

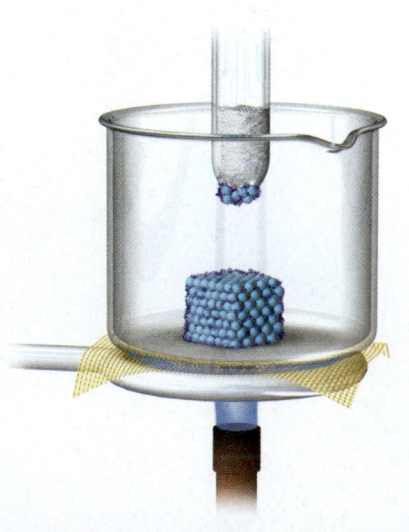

▲ FIGURE 13.7 **Sublimation**   When a layer of particles sublimes, another identical layer is just below it. Consequently, the number of particles available to sublime at any one time does not change with the total number of particles in the sample, and the process is zero order.

Consequently, for a second-order reaction, the rate is even more sensitive to the reactant concentration. You can see this in Figure 13.5 because the slope of the curve (the rate) flattens out more quickly than it does for a first-order reaction. Figure 13.6 shows the rate as a function of the concentration of A. Notice the quadratic relationship—the rate is proportional to the square of the concentration.

Determining the Order of a Reaction

The order of a reaction can be determined only by experiment. A common way to determine reaction order is by the *method of initial rates*. In this method, the initial rate—the rate for a short period of time at the beginning of the reaction—is measured by running the reaction several times with different initial reactant concentrations to determine the effect of the concentration on the rate. For example, let's return to our simple reaction in which a single reactant, A, decomposes into products:

$$A \longrightarrow products$$

In an experiment, the initial rate was measured at several different initial concentrations with the following results:

[A] (M)	Initial Rate (M/s)
0.10	0.015
0.20	0.030
0.40	0.060

Notice that, in this data set, when the concentration of A doubles, the rate doubles—the initial rate is directly proportional to the initial concentration. The reaction is therefore first order in A and the rate law is as follows:

$$Rate = k[A]^1$$

The value of the rate constant, k, can be determined by solving the rate law for k and substituting the concentration and the initial rate from any one of the three measurements. Using the first measurement, we get the following:

$$Rate = k[A]^1$$

$$k = \frac{rate}{[A]} = \frac{0.015 \text{ M/s}}{0.10 \text{ M}} = 0.15 \text{ s}^{-1}$$

Notice that the rate constant for a first-order reaction has units of s^{-1}.

The following two data sets show how measured initial rates would be different for zero-order and for second-order reactions:

Zero Order ($n = 0$)		Second Order ($n = 2$)	
[A] (M)	Initial Rate (M/s)	[A] (M)	Initial Rate (M/s)
0.10	0.015	0.10	0.015
0.20	0.015	0.20	0.060
0.40	0.015	0.40	0.240

For a zero-order reaction, the initial rate is independent of the reactant concentration—the rate is the same at all measured initial concentrations. For a second-order reaction, the initial rate quadruples for a doubling of the reactant concentration—the relationship between concentration and rate is quadratic. If you are unsure about how the initial rate is changing with the initial reactant concentration, or if the numbers are not as obvious as in these examples, you can substitute any two initial concentrations and the corresponding initial rates into a ratio of the rate laws to determine the order (n):

$$\frac{rate\ 2}{rate\ 1} = \frac{k[A]_2^n}{k[A]_1^n}$$

For example, we can substitute the last two measurements in the above data set for the second-order reaction as follows:

$$\frac{0.240 \ \cancel{M/s}}{0.060 \ \cancel{M/s}} = \frac{k(0.40 \ \cancel{M})^n}{k(0.20 \ \cancel{M})^n}$$

$$4.0 = \left(\frac{0.40}{0.20}\right)^n = 2^n$$

$$\log 4.0 = \log(2^n)$$

$$= n \log 2$$

$$n = \frac{\log 4}{\log 2}$$

$$= 2$$

| Remember that $\log x^n = n \log x$.

The rate constants for zero- and second-order reactions have different units than for first-order reactions. The rate constant for a zero-order reaction has units of $M \cdot s^{-1}$ and that for a second-order reaction has units of $M^{-1} \cdot s^{-1}$.

Reaction Order for Multiple Reactants

So far, we have considered a simple reaction with only one reactant. How is the rate law defined for reactions with more than one reactant? Consider the following reaction:

$$aA + bB \longrightarrow cC + dD$$

As long as the reverse reaction is negligibly slow, the rate law is defined as follows:

$$\text{Rate} = k[A]^m[B]^n \qquad [13.10]$$

where m is the reaction order with respect to A and n is the reaction order with respect to B. The **overall order** is simply the sum of the exponents $(m + n)$. For example, the reaction between hydrogen and iodine has been experimentally determined to be first order with respect to hydrogen, first order with respect to iodine, and thus second order overall.

$$H_2(g) + I_2(g) \longrightarrow 2\,HI(g) \qquad \text{Rate} = k[H_2]^1[I_2]^1$$

Similarly, the reaction between hydrogen and nitrogen monoxide has been experimentally determined to be first order with respect to hydrogen, second order with respect to nitrogen monoxide, and thus third order overall.

$$2\,H_2(g) + 2\,NO(g) \longrightarrow N_2(g) + 2\,H_2O(g) \qquad \text{Rate} = k[H_2]^1[NO]^2$$

The rate law for any reaction must always be determined by experiment, often by the method of initial rates described previously. There is no simple way merely to look at a chemical equation and determine the rate law for the reaction. When there are two or more reactants, the concentration of each reactant is usually varied independently of the others to determine the dependence of the rate on the concentration of that reactant. The following example uses the method of initial rates for determining the order of a reaction with multiple reactants.

EXAMPLE 13.2 Determining the Order and Rate Constant of a Reaction

Consider the following reaction between nitrogen dioxide and carbon monoxide:

$$NO_2(g) + CO(g) \longrightarrow NO(g) + CO_2(g)$$

The initial rate of the reaction was measured at several different concentrations of the reactants with the following results:

$[NO_2]$ (M)	$[CO]$ (M)	Initial Rate (M/s)
0.10	0.10	0.0021
0.20	0.10	0.0082
0.20	0.20	0.0083
0.40	0.10	0.033

From the data, determine each of the following:

(a) the rate law for the reaction   (b) the rate constant (k) for the reaction

Solution

(a) Begin by examining how the rate changes for each change in concentration. Between the first two experiments, the concentration of NO_2 doubled, the concentration of CO stayed constant, and the rate quadrupled, suggesting that the reaction is second order in NO_2.

Between the second and third experiments, the concentration of NO_2 stayed constant, the concentration of CO doubled, and the rate remained constant (the small change in the least significant figure is simply experimental error), suggesting that the reaction is zero order in CO. Between the third and fourth experiments, the concentration of NO_2 again doubled, the concentration of CO halved, yet the rate quadrupled again, confirming that the reaction is second order in NO_2 and zero order in CO.

$[NO_2]$	$[CO]$	Initial Rate (M/s)
0.10 M	0.10 M	0.0021
$\downarrow \times 2$	$\downarrow$ constant	$\downarrow \times 4$
0.20 M	0.10 M	0.0082 M
$\downarrow$ constant	$\downarrow \times 2$	$\downarrow \times 1$
0.20 M	0.20 M	0.0083 M
$\downarrow \times 2$	$\downarrow \times \frac{1}{2}$	$\downarrow \times 4$
0.40 M	0.10 M	0.033 M

Write the overall rate expression.

$$Rate = k[NO_2]^2[CO]^0 = k[NO_2]^2$$

(b) To determine the rate constant for the reaction, solve the rate law for k and substitute the concentration and the initial rate from any one of the four measurements. In this case, we use the first measurement.

$$Rate = k[NO_2]^2$$

$$k = \frac{rate}{[NO_2]^2} = \frac{0.0021 \text{ M/s}}{(0.10 \text{ M})^2} = 0.21 \text{ M}^{-1} \cdot \text{s}^{-1}$$

For Practice 13.2

Consider the following reaction:

$$CHCl_3(g) + Cl_2(g) \longrightarrow CCl_4(g) + HCl(g)$$

The initial rate of the reaction was measured at several different concentrations of the reactants with the following results:

$[CHCl_3]$ (M)	$[Cl_2]$ (M)	Initial Rate (M/s)
0.010	0.010	0.0035
0.020	0.010	0.0069
0.020	0.020	0.0098
0.040	0.040	0.027

From the data, determine each of the following:

(a) the rate law for the reaction   (b) the rate constant (k) for the reaction

 Conceptual Connection 13.1 Rate and Concentration

The following reaction was experimentally determined to be first order with respect to O_2 and second order with respect to NO.

$$O_2(g) + 2 NO(g) \longrightarrow 2 NO_2(g)$$

The following diagrams represent reaction mixtures in which the number of each type of molecule represents its relative initial concentration. Which mixture will have the fastest initial rate?

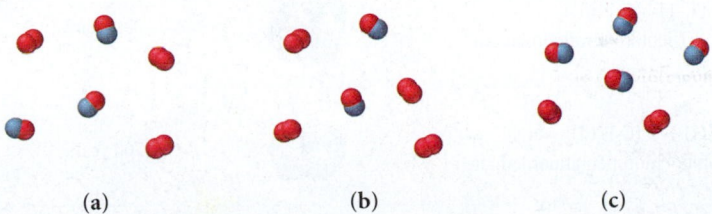

(a)   (b)   (c)

Answer: All three mixtures have the same total number of molecules, but mixture (c) has the greatest number of NO molecules. Since the reaction is second order in NO and only first order in O_2, mixture (c) will have the fastest initial rate.

13.4 The Integrated Rate Law: The Dependence of Concentration on Time

The rate laws we have examined so far show the relationship between *the rate of a reaction and the concentration of a reactant.* However, we often want to know the relationship between *the concentration of a reactant and time.* For example, the presence of chlorofluorocarbons (CFCs) in the atmosphere threatens the ozone layer. One of the reasons that CFCs pose such a significant threat is that the reactions that consume them are so slow (see Table 13.1). Even if we completely stopped adding CFCs to the atmosphere, their concentration in the atmosphere would decrease only very slowly. Nonetheless, we would like to be able to predict how their concentration changes with time. How much will be left in 20 years? In 50 years?

The **integrated rate law** for a chemical reaction is a relationship between the concentrations of the reactants and time. For simplicity, we return to a single reactant decomposing into products:

$$A \longrightarrow products$$

The integrated rate law for this reaction depends on the order of the reaction; we will examine each of the common reaction orders individually.

First-Order Integrated Rate Law   If our simple reaction is first order, the rate law is defined as follows:

$$Rate = k[A]$$

Since $Rate = -\Delta[A]/\Delta t$, we can write

$$-\frac{\Delta[A]}{\Delta t} = k[A] \qquad\qquad [13.11]$$

In this form, the rate law is also known as the *differential rate law.*

TABLE 13.1	Atmospheric Lifetimes of Several CFC's	
CFC Name	Structure	Atmospheric Lifetime*
CFC-11 (CCl_3F) Trichlorofluoromethane		45 years
CFC-12 (CCl_2F_2) Dichlorodifluoromethane		100 years
CFC-113 ($C_2F_3Cl_3$) 1,1,2-Trichloro-1,2,2-trifluoroethane		85 years
CFC-114 ($C_2F_4Cl_2$) 1,2-Dichlorotetrafluoroethane		300 years
CFC-115 (C_2F_5Cl) Monochloropentafluoroethane		1700 years

*Data taken from EPA site (under section 602 of Clean Air Act).

Although we do not show the steps here, we can use calculus to integrate the differential rate law to obtain the first-order *integrated rate law* shown here:

$$\ln[A]_t = -kt + \ln[A]_0 \qquad [13.12]$$

or

$$\ln\frac{[A]_t}{[A]_0} = -kt \qquad [13.13]$$

where $[A]_t$ is the concentration of A at any time t, k is the rate constant, and $[A]_0$ is the initial concentration of A. The two forms of the equation shown here are equivalent, as shown in the margin.

Notice that the integrated rate law shown in Equation 13.12 has the form of an equation for a straight line.

$$\ln[A]_t \;=\; -kt \;+\; \ln[A]_0$$
$$y \;=\; mx \;+\; b$$

Therefore, for a first-order reaction, a plot of the natural log of the reactant concentration as a function of time yields a straight line with a slope of $-k$ and a y-intercept of $\ln[A]_0$, as shown in Figure 13.8 ▶ (Note that the slope is negative but that the rate constant is always positive.).

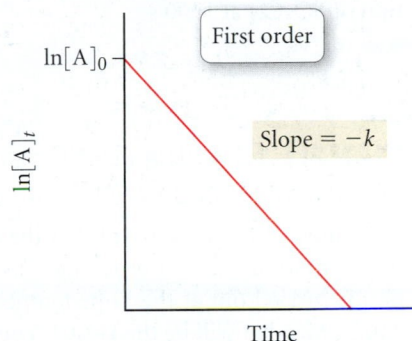

$$\ln[A]_t = -kt + \ln[A]_0$$
$$\ln[A]_t - \ln[A]_0 = -kt$$
$$\ln\frac{[A]_t}{[A]_0} = -kt$$

Remember that $\ln A - \ln B = \ln(A/B)$.

▲ **FIGURE 13.8 First-Order Integrated Rate Law** For a first-order reaction, a plot of the natural log of the reactant concentration as a function of time yields a straight line. The slope of the line is equal to $-k$ and the y-intercept is $\ln[A]_0$.

EXAMPLE 13.3 The First-Order Integrated Rate Law: Using Graphical Analysis of Reaction Data

Consider the following equation for the decomposition of SO_2Cl_2.

$$SO_2Cl_2(g) \longrightarrow SO_2(g) + Cl_2(g)$$

The concentration of SO_2Cl_2 was monitored at a fixed temperature as a function of time during the decomposition reaction and the following data were tabulated:

Time (s)	$[SO_2Cl_2]$ (M)	Time (s)	$[SO_2Cl_2]$ (M)
0	0.100	800	0.0793
100	0.0971	900	0.0770
200	0.0944	1000	0.0748
300	0.0917	1100	0.0727
400	0.0890	1200	0.0706
500	0.0865	1300	0.0686
600	0.0840	1400	0.0666
700	0.0816	1500	0.0647

Show that the reaction is first order and determine the rate constant for the reaction.

Solution

In order to show that the reaction is first order, prepare a graph of $\ln[SO_2Cl_2]$ versus time as shown below.

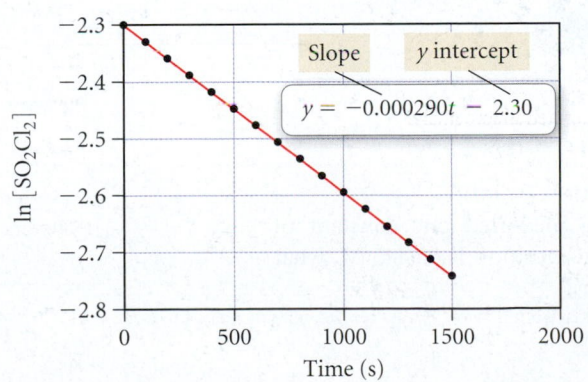

The plot is linear, confirming that the reaction is indeed first order. To obtain the rate constant, fit the data to a line. The slope of the line will be equal to $-k$. Since the slope of the best fitting line (which is most easily determined on a graphing calculator or with spreadsheet software such as Microsoft Excel) is $-2.90 \times 10^{-4}\,s^{-1}$, the rate constant is therefore $+2.90 \times 10^{-4}\,s^{-1}$.

For Practice 13.3

Use the graph and the best fitting line in the previous example to predict the concentration of SO_2Cl_2 at 1900 s.

EXAMPLE 13.4 The First-Order Integrated Rate Law: Determining the Concentration of a Reactant at a Given Time

In Example 13.3, we learned that the decomposition of SO_2Cl_2 (under the given reaction conditions) is first order and has a rate constant of $+2.90 \times 10^{-4}\,s^{-1}$. If the reaction is carried out at the same temperature, and the initial concentration of SO_2Cl_2 is 0.0225 M, what will be the SO_2Cl_2 concentration after 865 s?

Sort You are given the rate constant of a first-order reaction and the initial concentration of the reactant. You are asked to find the concentration at 865 seconds.	Given $k = +2.90 \times 10^{-4}\,s^{-1}$ $[SO_2Cl_2]_0 = 0.0225$ M Find $[SO_2Cl_2]$ at $t = 865$ s
Strategize Use the first-order integrated rate law to get from the given information to the information you are asked to find.	Equation $\ln[A]_t = -kt + \ln[A]_0$
Solve Substitute the rate constant, the initial concentration, and the time into the integrated rate law. Solve the integrated rate law for the concentration of $[SOCl_2]_t$.	Solution $\ln[SO_2Cl_2]_t = -kt + \ln[SO_2Cl_2]_0$ $\ln[SO_2Cl_2]_t = -(2.90 \times 10^{-4}\,s^{-1})865\,s + \ln(0.0225)$ $\ln[SO_2Cl_2]_t = -0.251 - 3.79$ $[SO_2Cl_2]_t = e^{-4.04}$ $\qquad = 0.0175$ M

Check The concentration is smaller than the original concentration as expected. If the concentration were larger than the initial concentration, this would indicate a mistake in the signs of one of the quantities on the right-hand side of the equation.

For Practice 13.4

Cyclopropane rearranges to form propene in the gas phase according to the following reaction:

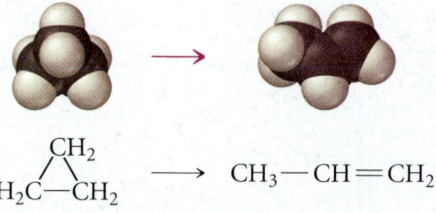

$$\begin{array}{c} CH_2 \\ / \quad \backslash \\ H_2C{-}CH_2 \end{array} \longrightarrow CH_3{-}CH{=}CH_2$$

The reaction is first order in cyclopropane and has a measured rate constant of $3.36 \times 10^{-5}\,s^{-1}$ at 720 K. If the initial cyclopropane concentration is 0.0445 M, what will be the cyclopropane concentration after 235.0 minutes?

Second-Order Integrated Rate Law If our simple reaction (A $\longrightarrow$ products) is second order, the rate law is defined as follows:

$$\text{Rate} = k[A]^2$$

Since Rate $= -\Delta[A]/\Delta t$, we can write:

$$-\frac{\Delta[A]}{\Delta t} = k[A]^2 \qquad [13.14]$$

Again, although we do not show the steps here, this differential rate law can be integrated to obtain the *second-order integrated rate law* shown here:

$$\frac{1}{[A]_t} = kt + \frac{1}{[A]_0} \qquad [13.15]$$

The second-order integrated rate law is also in the form of an equation for a straight line.

$$\begin{aligned}\frac{1}{[A]_t} &= kt &+& \frac{1}{[A]_0} \\ y &= mx &+& b\end{aligned}$$

Notice, however, that you must now plot the inverse of the concentration of the reactant as a function of time. The plot yields a straight line with a slope of k and an intercept of $1/[A]_0$, as shown in Figure 13.9 ▶.

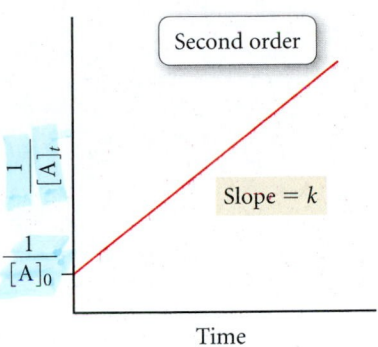

▲ **FIGURE 13.9 Second-Order Integrated Rate Law** For a second-order reaction, a plot of the inverse of the reactant concentration as a function of time yields a straight line. The slope of the line is equal to k and the y-intercept is $1/[A]_0$.

EXAMPLE 13.5 The Second-Order Integrated Rate Law: Using Graphical Analysis of Reaction Data

Consider the following equation for the decomposition of NO_2.

$$NO_2(g) \longrightarrow NO(g) + O(g)$$

The concentration of NO_2 was monitored at a fixed temperature as a function of time during the decomposition reaction and the data tabulated at right. Show by graphical analysis that the reaction is not first order and that it is second order. Determine the rate constant for the reaction.

Solution

In order to show that the reaction is *not* first order, prepare a graph of $\ln[NO_2]$ versus time as shown below.

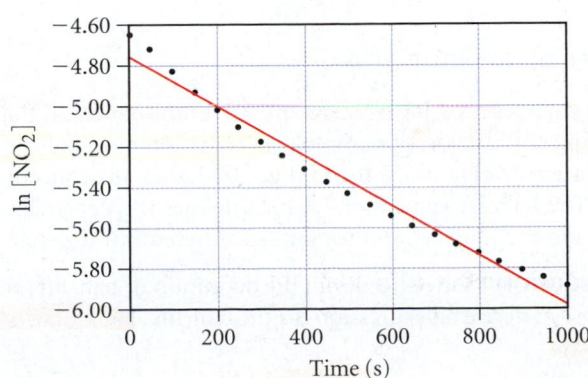

Time (s)	[NO₂] (M)
0	0.01000
50	0.00887
100	0.00797
150	0.00723
200	0.00662
250	0.00611
300	0.00567
350	0.00528
400	0.00495
450	0.00466
500	0.00440
550	0.00416
600	0.00395
650	0.00376
700	0.00359
750	0.00343
800	0.00329
850	0.00316
900	0.00303
950	0.00292
1000	0.00282

The plot is *not* linear (the straight line does not fit the data points), confirming that the reaction is not first order. In order to show that the reaction is second order, prepare a graph of $1/[NO_2]$ versus time as shown below.

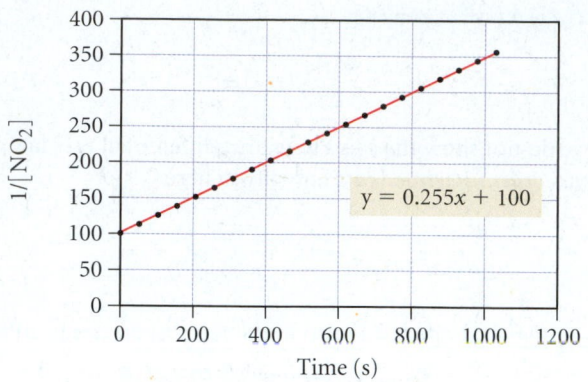

This graph is linear (the data points fit well to a straight line), confirming that the reaction is indeed second order. To obtain the rate constant, determine the slope of the best fitting line. The slope is $0.255 \ M^{-1} \cdot s^{-1}$; therefore, the rate constant is $0.255 \ M^{-1} \cdot s^{-1}$.

For Practice 13.5

Use the graph and the best fitting line in Example 13.5 to predict the concentration of NO_2 at 2000 s.

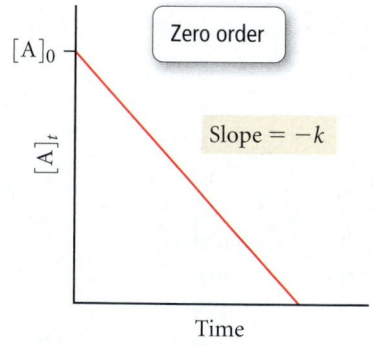

▲ FIGURE 13.10 Zero-Order Integrated Rate Law For a zero-order reaction, a plot of the reactant concentration as a function of time yields a straight line. The slope of the line is equal to $-k$ and the y-intercept is $[A]_0$.

Zero-Order Integrated Rate Law

If our simple reaction is zero order, the rate law is defined as follows:

$$Rate = k[A]^0$$

Since $Rate = -\Delta[A]/\Delta t$, we can write:

$$-\frac{\Delta[A]}{\Delta t} = k \qquad [13.16]$$

This differential rate law can be integrated to obtain the *zero-order integrated rate law* shown here:

$$[A]_t = -kt + [A]_0 \qquad [13.17]$$

The zero-order integrated rate law shown in Equation 13.17 is also in the form of an equation for a straight line. A plot of the concentration of the reactant as a function of time yields a straight line with a slope of $-k$ and an intercept of $[A]_0$, as shown in Figure 13.10 ◄.

The Half-Life of a Reaction

The **half-life** ($t_{1/2}$) of a reaction is the time required for the concentration of a reactant to fall to one-half of its initial value. For example, if a reaction has a half-life of 100 seconds, and if the initial concentration of the reactant is 1.0 M, then the concentration will fall to 0.50 M in 100 s. The half-life expression—which defines the dependence of half-life on the rate constant and the initial concentration—is different for different reaction orders.

First-Order Reaction Half-Life From the definition of half-life, and from the integrated rate law, we can derive an expression for the half-life. For a first-order reaction, the integrated rate law is

$$\ln\frac{[A]_t}{[A]_0} = -kt$$

At a time equal to the half-life ($t = t_{1/2}$), the concentration is exactly half of the initial concentration: $\left([A]_t = \frac{1}{2}[A]_0\right)$. Therefore, when $t = t_{1/2}$ we can write the following expression:

$$\ln \frac{\frac{1}{2}[A]_0}{[A]_0} = \ln \frac{1}{2} = -kt_{1/2} \qquad [13.18]$$

Solving for $t_{1/2}$, and substituting -0.693 for $\ln \frac{1}{2}$, we arrive at the following expression for the half-life of a first-order reaction:

$$t_{1/2} = \frac{0.693}{k} \qquad [13.19]$$

Notice that, for a first-order reaction, $t_{1/2}$ is independent of the initial concentration. For example, if $t_{1/2}$ is 100 s, and if the initial concentration is 1.0 M, the concentration falls to 0.50 M in 100 s, then to 0.25 M in another 100 s, then to 0.125 M in another 100 s, and so on (Figure 13.11 ▶). Even though the concentration is changing as the reaction proceeds, the half-life (how long it takes for the concentration to halve) is constant. A constant half-life is unique to first-order reactions, making the concept of half-life particularly useful for first-order reactions.

Half-Life for a First-Order Reaction

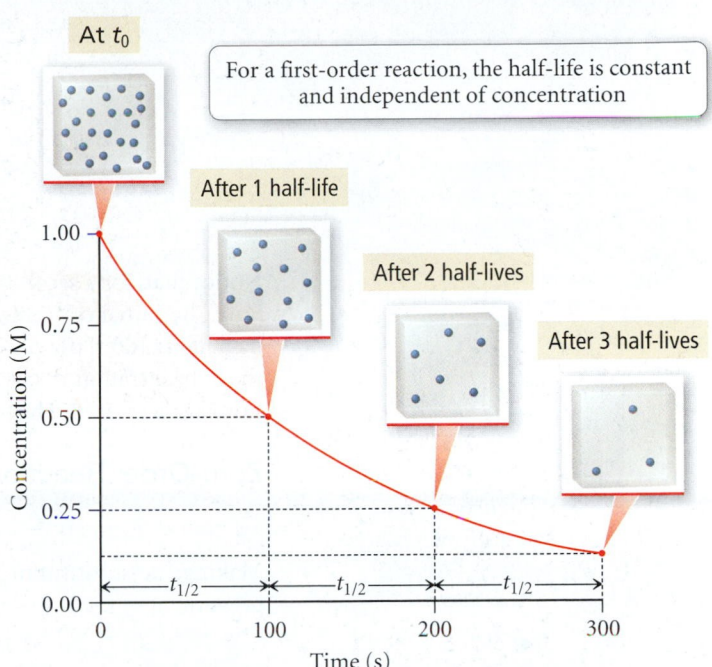

For a first-order reaction, the half-life is constant and independent of concentration

▲ FIGURE 13.11 Half-Life: Concentration versus Time for a First-Order Reaction For this reaction, the concentration falls by one-half every 100 seconds ($t_{1/2} = 100$ s). The blue spheres represent reactant molecules (the products are omitted for clarity).

EXAMPLE 13.6 Half-Life

Molecular iodine dissociates at 625 K with a first-order rate constant of 0.271 s^{-1}. What is the half-life of this reaction?

Solution

Since the reaction is first order, the half-life is given by Equation 13.19. Simply substitute the value of k into the expression and compute $t_{1/2}$.

$$t_{1/2} = \frac{0.693}{k}$$

$$= \frac{0.693}{0.271/s} = 2.56 \text{ s}$$

For Practice 13.6

A first-order reaction has a half-life of 26.4 seconds. How long will it take for the concentration of the reactant in the reaction to fall to one-eighth of its initial value?

Second-Order Reaction Half-Life For a second-order reaction, the integrated rate law is

$$\frac{1}{[A]_t} = kt + \frac{1}{[A]_0}$$

At a time equal to the half-life ($t = t_{1/2}$), the concentration is exactly one-half of the initial concentration ($[A]_t = \frac{1}{2}[A]_0$). We can therefore write the following expression at $t = t_{1/2}$:

$$\frac{1}{\frac{1}{2}[A]_0} = kt_{1/2} + \frac{1}{[A]_0} \qquad [13.20]$$

We can then solve for $t_{1/2}$:

$$kt_{1/2} = \frac{1}{\frac{1}{2}[A]_0} - \frac{1}{[A]_0}$$

$$kt_{1/2} = \frac{2}{[A]_0} - \frac{1}{[A]_0}$$

$$t_{1/2} = \frac{1}{k[A]_0} \qquad [13.21]$$

Notice that, for a second-order reaction, the half-life depends on the initial concentration. So if the initial concentration of a reactant in a second-order reaction is 1.0 M, and the half-life is 100 s, the concentration falls to 0.50 M in 100 s. However, the time it takes for the concentration to fall to 0.25 M is now *longer than 100 s*, because the initial concentration has decreased. The half-life continues to get longer as the concentration decreases.

Zero-Order Reaction Half-Life   For a zero-order reaction, the integrated rate law is

$$[A]_t = -kt + [A]_0$$

Making the substitutions ($t = t_{1/2}$; $[A]_t = \frac{1}{2}[A]_0$). We can then write the following expression at $t = t_{1/2}$:

$$\frac{1}{2}[A]_0 = -kt_{1/2} + [A]_0 \qquad [13.22]$$

We can then solve for $t_{1/2}$:

$$t_{1/2} = \frac{[A]_0}{2k} \qquad [13.23]$$

Notice that, for a zero-order reaction, the half-life also depends on the initial concentration.

Summarizing (see Table 13.2):

➤ The reaction order and rate law must be determined experimentally.

➤ The differential rate law relates the *rate* of the reaction to the *concentration* of the reactant(s).

➤ The integrated rate law (which is mathematically derived from the differential rate law) relates the *concentration* of the reactant(s) to *time*.

➤ The half-life is the time it takes for the concentration of a reactant to fall to one-half of its initial value.

➤ The half-life of a first-order reaction is independent of the initial concentration.

➤ The half-lives of zero-order and second-order reactions depend on the initial concentration.

◆ Conceptual Connections 13.2 Kinetics Summary

A decomposition reaction, whose rate is observed to slow down as the reaction proceeds, is found to have a half-life that depends on the initial concentration of the reactant. Which of the following is most likely to be true of this reaction?

(a) A plot of the natural log of the concentration of the reactant as a function of time will be linear.

(b) The half-life of the reaction increases as the initial concentration increases.

(c) A doubling of the initial concentration of the reactant results in a quadrupling of the rate.

Answer: (c) The reaction is most likely second order because its rate depends on the concentration (therefore it cannot be zero order), and its half-life depends on the initial concentration (therefore it cannot be first order). For a second-order reaction, a doubling of the initial concentration results in the quadrupling of the rate.

TABLE 13.2   Rate Law Summary Table

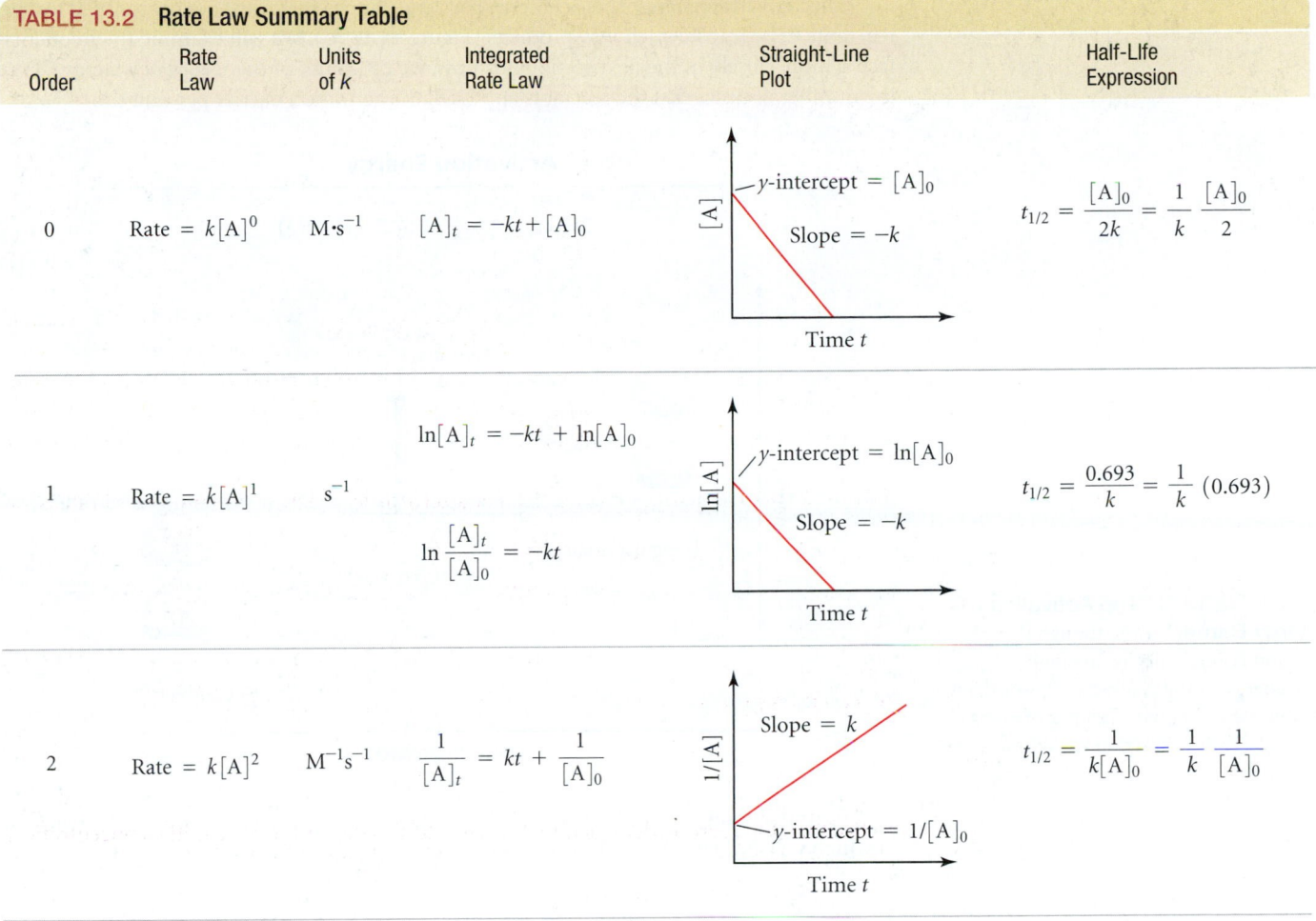

Order	Rate Law	Units of k	Integrated Rate Law	Straight-Line Plot	Half-Life Expression
0	Rate $= k[A]^0$	$M \cdot s^{-1}$	$[A]_t = -kt + [A]_0$	y-intercept $= [A]_0$ Slope $= -k$	$t_{1/2} = \dfrac{[A]_0}{2k} = \dfrac{1}{k}\dfrac{[A]_0}{2}$
1	Rate $= k[A]^1$	s^{-1}	$\ln[A]_t = -kt + \ln[A]_0$ $\ln\dfrac{[A]_t}{[A]_0} = -kt$	y-intercept $= \ln[A]_0$ Slope $= -k$	$t_{1/2} = \dfrac{0.693}{k} = \dfrac{1}{k}(0.693)$
2	Rate $= k[A]^2$	$M^{-1}s^{-1}$	$\dfrac{1}{[A]_t} = kt + \dfrac{1}{[A]_0}$	Slope $= k$ y-intercept $= 1/[A]_0$	$t_{1/2} = \dfrac{1}{k[A]_0} = \dfrac{1}{k}\dfrac{1}{[A]_0}$

13.5 The Effect of Temperature on Reaction Rate

In the opening section of this chapter, we saw that lizards become lethargic when their body temperature drops because the chemical reactions that control muscle movement slow down at lower temperatures. The rates of chemical reactions are, in general, highly sensitive to temperature. At around room temperature, a 10 °C increase in temperature increases the rate of a typical reaction by two or three times. How do we explain this highly sensitive temperature dependence?

We saw previously that the rate law for a reaction is Rate = $k[A]^n$. *The temperature dependence of the reaction rate is contained in the rate constant, k* (which is actually a constant only when the temperature remains constant). An increase in temperature generally results in an increase in k, which results in a faster rate. In 1889, Swedish chemist Svante Arrhenius wrote a paper quantifying the temperature dependence of the rate constant. The modern form of the **Arrhenius equation**, which relates the rate constant (k) and the temperature in kelvins (T), is as follows:

$$k = Ae^{\frac{-E_a}{RT}} \qquad\qquad [13.24]$$

Activation energy

Frequency factor

Exponential factor

where R is the gas constant (8.314 J/mol · K), A is a constant called the *frequency factor* (or the *pre-exponential factor*), and E_a is called the *activation energy* (or *activation barrier*).

The **activation energy** E_a is an energy barrier or hump that must be surmounted for the reactants to be transformed into products (Figure 13.12 ▼). We will examine the frequency factor more closely in the next section; for now, we can think of the **frequency factor (A)** as the number of times that the reactants approach the activation barrier per unit time.

Activation Energy

$$2H_2(g) + O_2(g) \rightleftharpoons 2H_2O(g)$$

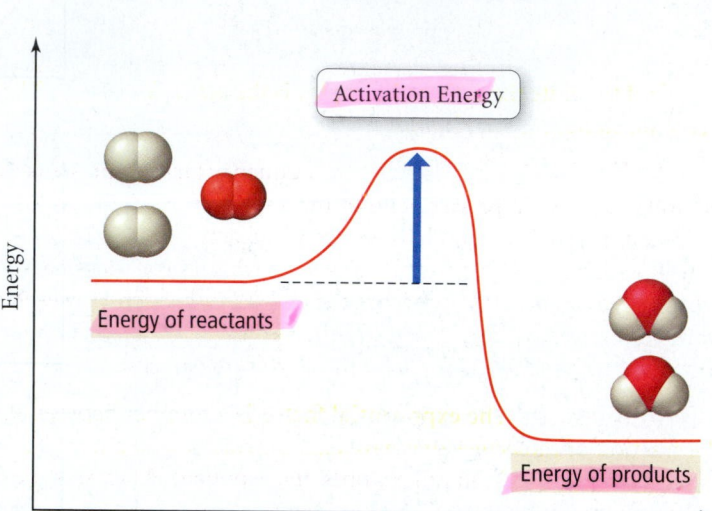

▶ **FIGURE 13.12 The Activation Energy Barrier** Even though the reaction is energetically favorable (the energy of the products is lower than that of the reactants), an input of energy is needed for the reaction to take place.

To understand each of these quantities better, consider the following simple reaction in which CH_3NC (methyl isonitrile) rearranges to form CH_3CN (acetonitrile):

$$CH_3 - N \equiv C \longrightarrow CH_3 - C \equiv N$$

Let's examine the physical meaning of the activation energy, frequency factor, and exponential factor for this reaction.

The Activation Energy Figure 13.13 ▼ shows the energy of the molecule as the reaction proceeds. The x-axis represents the progress of the reaction from left (reactant) to right (product). Notice that, to get from the reactant to the product, the molecule must go through a high-energy intermediate state called the **activated complex**, or **transition state**.

▶ **FIGURE 13.13 The Activated Complex** The reaction pathway includes a transitional state—the activated complex—that has a higher energy than either the reactant or the product.

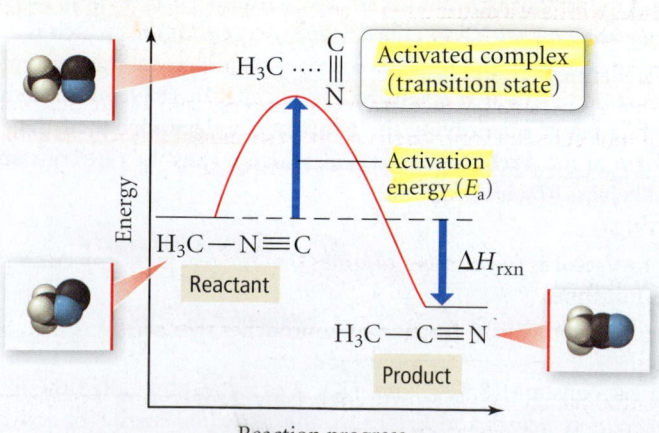

Even though the overall reaction is energetically downhill (exothermic), it must first go up-hill to reach the activated complex because energy is required to initially weaken the H_3C-N bond and allow the NC group to begin to rotate:

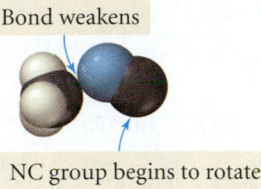

Bond weakens

NC group begins to rotate

The energy required to create the activated complex is the *activation energy. The higher the activation energy, the slower the reaction rate (at a given temperature).*

The Frequency Factor We just saw that the frequency factor represents the number of approaches to the activation barrier per unit time. Any time that the NC group begins to rotate, it approaches the activation barrier. For this reaction, the frequency factor represents the rate at which the NC part of the molecule vibrates (or rocks back and forth). With each vibration, the reactant approaches the activation barrier. However, approaching the activation barrier is not equivalent to surmounting it. Most of the approaches do not have enough total energy to make it over the activation barrier.

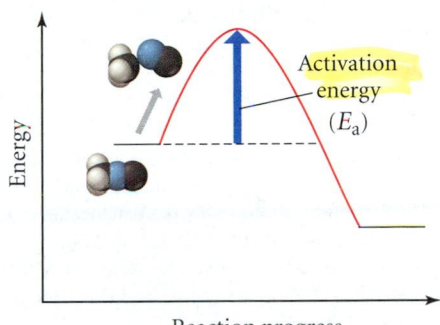

Reaction progress

The Exponential Factor The **exponential factor** is a number between 0 and 1 that represents the fraction of molecules that have enough energy to make it over the activation barrier on a given approach. In other words, the exponential factor is the fraction of approaches that are actually successful and result in the product. For example, if the frequency factor is 10^9/s and the exponential factor is 10^{-7} at a certain temperature, then the overall rate constant at that temperature is 10^9/s $\times$ 10^{-7} = 10^2/s. In this case, the CN group is "wagging" at a rate of 10^9/s. With each wag, the activation barrier is approached. However, only 1 in 10^7 molecules have sufficient energy to actually make it over the activation barrier for a given wag.

The exponential factor depends on both the temperature (T) and the activation energy (E_a) of the reaction.

$$\text{Exponential factor} = e^{-E_a/RT}$$

A low activation energy and a high temperature make the negative exponent small, so that the exponential factor approaches one. For example, if the activation energy is zero, then the exponent is zero, and the exponential factor is exactly one ($e^{-0} = 1$)—every approach to the activation barrier is successful. By contrast, a large activation energy and a low temperature make the exponent a very large negative number, so that the exponential factor becomes very small. For example, as the temperature approaches 0 K, the exponent approaches an infinitely large number, and the exponential factor approaches zero ($e^{-\infty} = 0$).

As the temperature increases, the number of molecules having enough thermal energy to surmount the activation barrier increases. At any given temperature, a sample of molecules will have a distribution of energies, as shown in Figure 13.14 ▶. Under common circumstances, only a small fraction of the molecules have enough energy to make it over the activation barrier. Because of the shape of the energy distribution curve, however, a small change in temperature results in a large difference in the number of molecules having enough energy to surmount the activation barrier. This explains the sensitivity of reaction rates to temperature.

Thermal Energy Distribution

As temperature increases, the fraction of molecules with enough energy to surmount the activation energy barrier also increases.

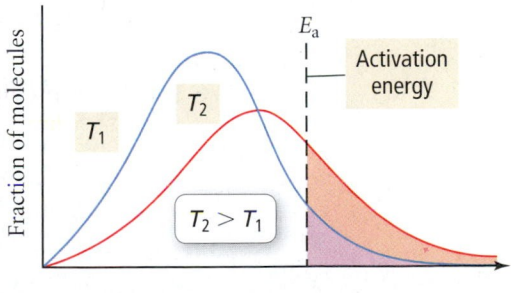

▲ **FIGURE 13.14 Thermal Energy Distribution** At any given temperature, the atoms or molecules in a gas sample will have a range of energies. The higher the temperature, the wider the energy distribution and the greater the average energy. The fraction of molecules with enough energy to surmount the activation energy barrier and react increases sharply as the temperature rises.

★ *Summarizing:* ★

➤ The frequency factor is the number of times that the reactants approach the activation barrier per unit time.

➤ The exponential factor is the fraction of approaches that are successful in surmounting the activation barrier and forming products.

➤ The exponential factor increases with increasing temperature, but decreases with an increasing value for the activation energy.

Arrhenius Plots: Experimental Measurements of the Frequency Factor and the Activation Energy

The frequency factor and activation energy are important quantities in understanding the kinetics of any reaction. To see how we can measure these factors in the laboratory, consider again Equation 13.24: $k = Ae^{-E_a/RT}$. Taking the natural log of both sides of this equation, we get the following:

$$\ln k = \ln(Ae^{-E_a/RT}) \qquad [13.25]$$

$$\ln k = \ln A + \ln e^{-E_a/RT}$$

$$\ln k = \ln A - \frac{E_a}{RT}$$

$$\ln k = -\frac{E_a}{R}\left(\frac{1}{T}\right) + \ln A \qquad [13.26]$$

$$y \quad = \quad mx \quad + \quad b$$

| Remember that $\ln(AB) = \ln A + \ln B$

| Remember that $\ln e^x = x$.

In an Arrhenius analysis, the pre-exponential factor (A) is assumed to be independent of temperature. Although the pre-exponential factor does depend on temperature to some degree, its temperature dependence is much less than that of the exponential factor and is often ignored.

Equation 13.26 is in the form of a straight line. *A plot of the natural log of the rate constant (ln k) versus the inverse of the temperature in kelvins (1/T) yields a straight line with a slope of* $-E_a/R$ *and a y-intercept of ln A.* Such a plot is called an **Arrhenius plot** and is commonly used in the analysis of kinetic data, as shown in the following example.

EXAMPLE 13.7 Using an Arrhenius Plot to Determine Kinetic Parameters

The decomposition of ozone is important to many atmospheric reactions:

$$O_3(g) \longrightarrow O_2(g) + O(g)$$

A study of the kinetics of the reaction resulted in the following data:

Temperature (K)	Rate Constant ($M^{-1} \cdot s^{-1}$)	Temperature (K)	Rate Constant ($M^{-1} \cdot s^{-1}$)
600	3.37×10^3	1300	7.83×10^7
700	4.85×10^4	1400	1.45×10^8
800	3.58×10^5	1500	2.46×10^8
900	1.70×10^6	1600	3.93×10^8
1000	5.90×10^6	1700	5.93×10^8
1100	1.63×10^7	1800	8.55×10^8
1200	3.81×10^7	1900	1.19×10^9

Determine the value of the frequency factor and activation energy for the reaction.

Solution

To find the frequency factor and activation energy, prepare a graph of the natural log of the rate constant (ln k) versus the inverse of the temperature (1/T), as shown

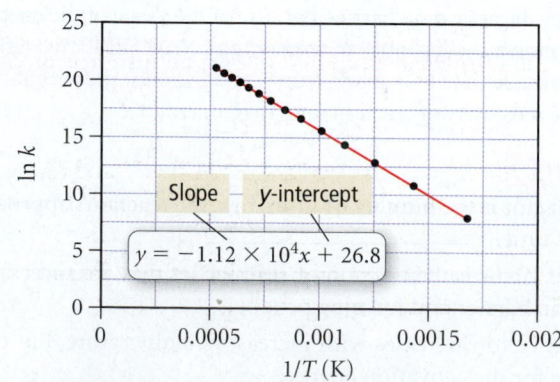

$y = -1.12 \times 10^4 x + 26.8$

The plot is linear, as expected for Arrhenius behavior. The best fitting line has a slope of -1.12×10^4 K and a y-intercept of 26.8. Calculate the activation energy from the slope by setting the slope equal to $-E_a/R$ and solving for E_a as follows:

$$-1.12 \times 10^4 \, K = \frac{-E_a}{R}$$

$$E_a = 1.12 \times 10^4 \, K \left(8.314 \frac{J}{mol \cdot K} \right)$$

$$= 9.31 \times 10^4 \, J/mol$$

$$= 93.1 \, kJ/mol$$

Calculate the frequency factor (A) by setting the intercept equal to $\ln A$.

$$26.8 = \ln A$$

$$A = e^{26.8}$$

$$= 4.36 \times 10^{11}$$

Since the rate constants were measured in units of $M^{-1} \cdot s^{-1}$, the frequency factor is in the same units. Consequently, we can conclude that the reaction has an activation energy of 93.1 kJ/mol and a frequency factor of $4.36 \times 10^{11} \, M^{-1} \cdot s^{-1}$.

For Practice 13.7

For the decomposition of ozone reaction in Example 13.7, use the results of the Arrhenius analysis to predict the rate constant at 298 K.

In some cases, where either data are limited or plotting capabilities are absent, the activation energy can be calculated from knowing the rate constant at just two different temperatures. The Arrhenius expression in Equation 13.25 can be applied to the two different temperatures as follows:

$$\ln k_2 = -\frac{E_a}{R} \left(\frac{1}{T_2} \right) + \ln A \qquad \ln k_1 = -\frac{E_a}{R} \left(\frac{1}{T_1} \right) + \ln A$$

We can then subtract $\ln k_1$ from $\ln k_2$ as follows:

$$\ln k_2 - \ln k_1 = \left[-\frac{E_a}{R} \left(\frac{1}{T_2} \right) + \ln A \right] - \left[-\frac{E_a}{R} \left(\frac{1}{T_1} \right) + \ln A \right]$$

Rearranging, we get the following two-point form of the Arrhenius equation:

$$\ln \frac{k_2}{k_1} = \frac{E_a}{R} \left(\frac{1}{T_1} - \frac{1}{T_2} \right) \qquad\qquad [13.27]$$

The following example shows how to use this equation to calculate the activation energy from experimental measurements of the rate constant at two different temperatures.

EXAMPLE 13.8   Using the Two-Point Form of the Arrhenius Equation

The reaction between nitrogen dioxide and carbon monoxide is given by the following equation:

$$NO_2(g) + CO(g) \longrightarrow NO(g) + CO_2(g)$$

The rate constant at 701 K was measured as $2.57 \, M^{-1} \cdot s^{-1}$ and that at 895 K was measured as $567 \, M^{-1} \cdot s^{-1}$. Find the activation energy for the reaction in kJ/mol.

Sort You are given the rate constant of a reaction at two different temperatures. You are asked to find the activation energy.	**Given** $T_1 = 701$ K, $k_1 = 2.57 \, M^{-1} \cdot s^{-1}$ $T_2 = 895$ K, $k_2 = 567 \, M^{-1} \cdot s^{-1}$ **Find** E_a

Strategize Use the two-point form of the Arrhenius equation, which relates the activation energy to the given information and R (a constant).	Equation $\ln \dfrac{k_2}{k_1} = \dfrac{E_a}{R}\left(\dfrac{1}{T_1} - \dfrac{1}{T_2}\right)$
Solve Substitute the two rate constants and the two temperatures into the equation.	Solution $$\ln \frac{567 \ M^{-1} \cdot s^{-1}}{2.57 \ M^{-1} \cdot s^{-1}} = \frac{E_a}{R}\left(\frac{1}{701 \text{ K}} - \frac{1}{895 \text{ K}}\right)$$ $$5.40 = \frac{E_a}{R}\left(\frac{3.09 \times 10^{-4}}{\text{K}}\right)$$
Solve the equation for E_a, the activation energy, and convert to kJ/mol.	$$E_a = 5.40\left(\frac{\text{K}}{3.09 \times 10^{-4}}\right)R$$ $$= 5.40\left(\frac{\text{K}}{3.09 \times 10^{-4}}\right)8.314\frac{\text{J}}{\text{mol} \cdot \text{K}}$$ $$= 1.45 \times 10^5 \text{ J/mol}$$ $$= 145 \text{ kJ/mol}$$

Check The magnitude of the answer is reasonable. Activation energies for most reactions range from tens to hundreds of kilojoules per mole.

For Practice 13.8

Use the results from the Example 13.8 and the given rate constant of the reaction at either of the two temperatures to predict the rate constant at 525 K.

The Collision Model: A Closer Look at the Frequency Factor

We suggested above that the frequency factor in the Arrhenius equation represents the number of approaches to the activation barrier per unit time. We now refine that idea for a reaction involving two gas-phase reactants:

$$A(g) + B(g) \longrightarrow \text{products}$$

In the **collision model**, a chemical reaction occurs after a sufficiently energetic collision between two reactant molecules (Figure 13.15 ◄). In collision theory, therefore, each approach to the activation barrier is a collision between the reactant molecules. Consequently, the value of the frequency factor should simply be the number of collisions that occur per second. However, the frequency factors of most (though not all) gas-phase chemical reactions tend to be smaller than the number of collisions that occur per second. Why?

In the collision model, the frequency factor can be separated into two separate parts, as shown in the following equations:

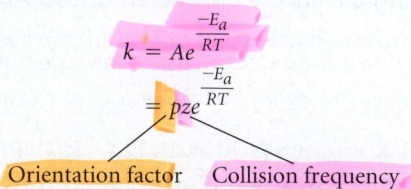

$$k = Ae^{\frac{-E_a}{RT}}$$
$$= pze^{\frac{-E_a}{RT}}$$

Orientation factor   Collision frequency

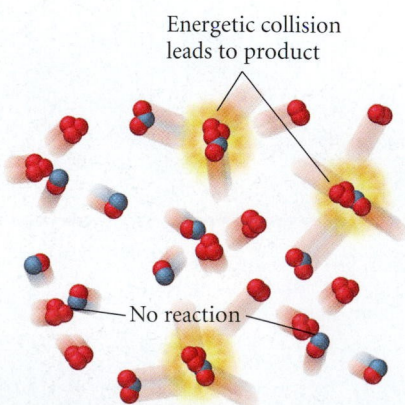

Energetic collision leads to product

No reaction

▲ FIGURE 13.15 The Collision Model In the collision model, two molecules react after a sufficiently energetic collision with the correct orientation to bring the reacting groups together.

where p is called the *orientation factor* and z is the *collision frequency*. The collision frequency is simply the number of collisions that occur per unit time, which can be calculated for a gas-phase reaction from the pressure of the gases and the temperature of the reaction mixture. Under typical conditions, a single molecule undergoes on the order of 10^9 collisions every second.

We can understand the orientation factor by considering the following reaction:

$$NOCl(g) + NOCl(g) \longrightarrow 2 NO(g) + Cl_2(g)$$

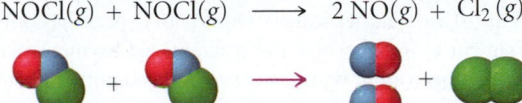

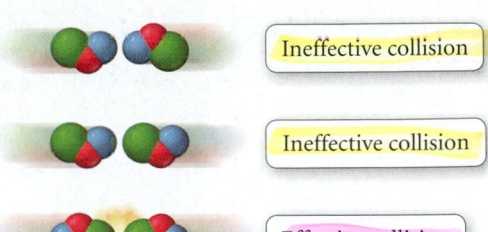

Ineffective collision

Ineffective collision

Effective collision

In order for the reaction to occur, two NOCl molecules must collide with sufficient energy. However, not all collisions with sufficient energy will lead to products because the reactant molecules must be properly oriented. For example, consider each of the possible orientations of the reactant molecules (shown at right) during a collision. The first two collisions shown, even if they occur with sufficient energy, will not result in a reaction because the reactant molecules are not oriented in a way that allows the chlorine atoms to bond. In other words, if two molecules are to react with each other, they must collide in such a way that allows the necessary bonds to break and form. For the reaction of $NOCl(g)$, the orientation factor is $p = 0.16$. This means that only 16 out of 100 sufficiently energetic collisions are actually successful in forming the products.

Some reactions have orientation factors that are much smaller than one. For example, consider the following reaction between hydrogen and ethene:

$$H_2(g) + CH_2{=}CH_2(g) \longrightarrow CH_3{-}CH_3(g)$$

The orientation factor for this reaction is 1.7×10^{-6}, which means that fewer than two out of each million sufficiently energetic collisions actually form products. The small orientation factor indicates that the orientational requirements for this reaction are very stringent—the molecules must be aligned in a *very specific way* for the reaction to occur.

Reactions between *individual atoms* usually have orientation factors of approximately 1, because atoms are spherically symmetric and any orientation can therefore lead to the formation of products. A few reactions have orientation factors greater than one. For example, consider the reaction between potassium and bromine.

$$K(g) + Br_2(g) \longrightarrow KBr(g) + Br(g)$$

This reaction has an orientation factor of $p = 4.8$. In other words, there are more reactions than collisions—the reactants do not even have to collide to react! Apparently, through a process dubbed *the harpoon mechanism,* a potassium atom can actually transfer an electron to a bromine molecule without a collision. The resulting positive charge on the potassium and the negative charge on the bromine cause the two species to attract each other and form a bond. In other words, the potassium atom *harpoons* a passing bromine molecule with an electron, and then *reels it in* through the electrostatic attraction between unlike charges.

We can picture a sample of reactive gases as a frenzy of collisions between the reacting atoms or molecules. At normal temperatures, the vast majority of these collisions do not have sufficient energy to overcome the activation barrier and the atoms or molecules simply bounce off one another. Of the collisions having sufficient energy to overcome the activation barrier, most do not have the proper orientation for the reaction to occur (for the majority of common reactions). When two molecules with sufficient energy *and* the correct orientation collide, something unique happens. The electrons on one of the atoms or molecules are attracted to the nuclei of the other; some bonds begin to weaken while other bonds begin to form and, if all goes well, the reactants go through the transition state and are transformed into the products. This is how a chemical reaction occurs.

▶ Conceptual Connections 13.3 Collision Theory

Which of the following reactions would you expect to have the smallest orientation factor?

(a) $H(g) + I(g) \longrightarrow HI(g)$

(b) $H_2(g) + I_2(g) \longrightarrow 2 HI(g)$

(c) $HCl(g) + HCl(g) \longrightarrow H_2(g) + Cl_2(g)$

Answer: **(c)** Since the reactants in part **(a)** are atoms, the orientation factor should be about one. The reactants in parts **(b)** and **(c)** are both molecules, so we expect orientation factors of less than one. Since the reactants in **(b)** are symmetrical, we would not expect the collision to have as specific an orientation requirement as in **(c)**, where the reactants are asymmetrical and must therefore collide in such way that a hydrogen atom is in close proximity to another hydrogen atom. Therefore, we expect **(c)** to have the smallest orientation factor.

13.6 Reaction Mechanisms

Most chemical reactions occur not in a single step, but through several steps. When we write a chemical equation to represent a chemical reaction, *we usually represent the overall reaction, not the series of individual steps by which the reaction occurs.* For example, consider the following reaction in which hydrogen gas reacts with iodine monochloride:

$$H_2(g) + 2\,ICl(g) \longrightarrow 2\,HCl(g) + I_2(g)$$

The overall equation simply shows the substances present at the beginning of the reaction and the substances formed by the reaction—it does not show the intermediate steps that may be involved. A **reaction mechanism** is a series of individual chemical steps by which an overall chemical reaction occurs. For example, the reaction between hydrogen and iodine monochloride occurs through the following proposed mechanism.

Step 1	$H_2(g) + ICl(g) \longrightarrow HI(g) + HCl(g)$
Step 2	$HI(g) + ICl(g) \longrightarrow HCl(g) + I_2(g)$

In the first step, an H_2 molecule collides with an ICl molecule and forms an HI molecule and HCl molecule. In the second step, the HI molecule formed in the first step collides with a second ICl molecule to form another HCl molecule and an I_2 molecule. Each step in a reaction mechanism is called an **elementary** step. Elementary steps cannot be broken down into simpler steps—they occur as they are written.

> An elementary step represents an actual interaction between the reactant molecules in the step. An overall reaction equation shows only the starting substances and the ending substances, not the path between them.

One of the requirements for a valid reaction mechanism is that the individual steps in the mechanism must add to the overall reaction. For example, the above mechanism sums to the overall reaction as shown here:

$$
\begin{array}{c}
H_2(g) + ICl(g) \longrightarrow HI(g) + HCl(g) \\
HI(g) + ICl(g) \longrightarrow HCl(g) + I_2(g) \\
\hline
H_2(g) + 2\,ICl(g) \longrightarrow 2\,HCl(g) + I_2(g)
\end{array}
$$

Species—such as HI—that are formed in one step of a mechanism and consumed in another are called **reaction intermediates**. An intermediate is not found in the balanced equation for the overall reaction but plays a key role in the mechanism. A reaction mechanism is a complete, detailed description of the reaction at the molecular level—it specifies the individual collisions and reactions that result in the overall reaction. As such, reaction mechanisms are highly sought-after pieces of chemical knowledge.

How do we determine reaction mechanisms? As mentioned in the opening section of this chapter, chemical kinetics are not only practically important (allowing us to control the rate of a particular reaction), but also theoretically important because they can help us determine the mechanism of the reaction. In other words, we can piece together a reaction mechanism by measuring the kinetics of the overall reaction and working backward to write a mechanism consistent with the measured kinetics.

Rate Laws for Elementary Steps

Elementary steps are characterized by their **molecularity**, the number of reactant particles involved in the step. The molecularity of the three most common types of elementary steps are as follows:

$A \longrightarrow$ products	**Unimolecular**
$A + A \longrightarrow$ products	**Bimolecular**
$A + B \longrightarrow$ products	**Bimolecular**

Elementary steps in which three reactant particles collide, called **termolecular** steps, are very rare because the probability of three particles simultaneously colliding is small.

Although the rate law for an overall chemical reaction cannot be deduced from the balanced chemical equation, the rate law for an elementary step can be. Since we know that an elementary step occurs through the collision of the reactant particles, the rate law is proportional to the product of the concentrations of those particles. For example, the rate law for the bimolecular elementary step in which A reacts with B is as follows:

$$A + B \longrightarrow products \quad Rate = k[A][B]$$

Similarly, the rate law for the bimolecular step in which A reacts with A is as follows:

$$A + A \longrightarrow products \quad Rate = k[A]^2$$

The rate laws for the common elementary steps, as well as those for the rare termolecular step, are summarized in Table 13.3. Notice that the molecularity of the elementary step is equal to the overall order of the step.

TABLE 13.3 Rate Laws for Elementary Step		
Elementary Step	Molecularity	Rate Law
A $\longrightarrow$ products	1	Rate = $k[A]$
A + A $\longrightarrow$ products	2	Rate = $k[A]^2$
A + B $\longrightarrow$ products	2	Rate = $k[A][B]$
A + A + A $\longrightarrow$ products	3 (rare)	Rate = $k[A]^3$
A + A + B $\longrightarrow$ products	3 (rare)	Rate = $k[A]^2[B]$
A + B + C $\longrightarrow$ products	3 (rare)	Rate = $k[A][B][C]$

Rate-Determining Steps and Overall Reaction Rate Laws

As we have noted, most chemical reactions occur through a series of elementary steps. In most cases, one of those steps—called the **rate-determining step**—is much slower than the others. The rate-determining step in a chemical reaction is analogous to the narrowest section on a freeway. If a freeway narrows from four lanes to two lanes, the rate at which cars travel along the freeway is limited by the rate at which they can travel through the narrow section (even though the rate could be much faster along the four-lane section). Similarly, the rate-determining step in a reaction mechanism limits the overall rate of the reaction (even though the other steps occur much faster) and therefore determines *the rate law for the overall reaction*.

As an example, consider the following reaction between nitrogen dioxide gas and carbon monoxide gas:

$$NO_2(g) + CO(g) \longrightarrow NO(g) + CO_2(g)$$

The experimentally determined rate law for this reaction is Rate = $k[NO_2]^2$. We can see from this rate law that the reaction must not be a single-step reaction—otherwise the rate law would be Rate = $k[NO_2][CO]$. A possible mechanism for this reaction is as follows:

$$NO_2(g) + NO_2(g) \longrightarrow NO_3(g) + NO(g) \quad Slow$$
$$NO_3(g) + CO(g) \longrightarrow NO_2(g) + CO_2(g) \quad Fast$$

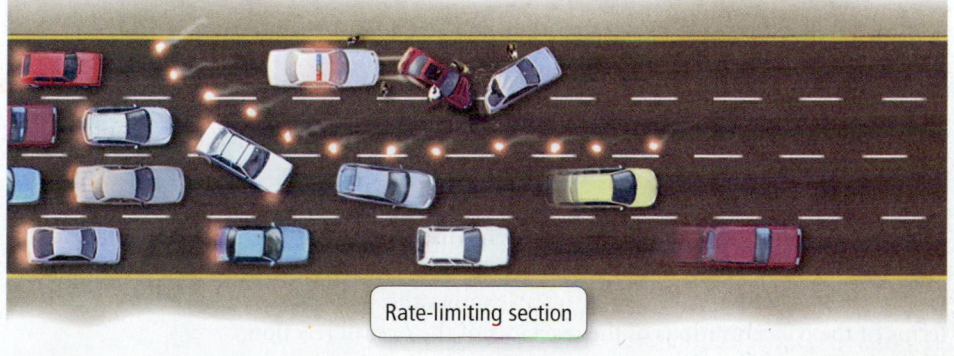

Rate-limiting section

◀ Just as the narrowest section of a highway limits the rate at which traffic can pass, so the rate-limiting step in a reaction mechanism limits the overall rate of the reaction.

Energy Diagram for a Two-Step Mechanism

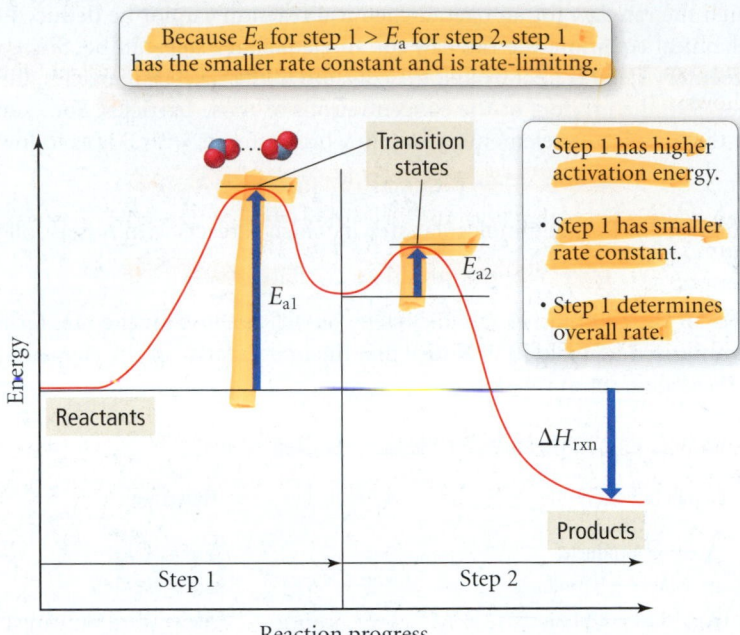

Because E_a for step 1 > E_a for step 2, step 1 has the smaller rate constant and is rate-limiting.

- Step 1 has higher activation energy.
- Step 1 has smaller rate constant.
- Step 1 determines overall rate.

Transition states

E_{a1}

E_{a2}

Reactants

ΔH_{rxn}

Products

Step 1 | Step 2

Reaction progress

▶ **FIGURE 13.16** Energy Diagram for a Two-Step Mechanism

The energy diagram accompanying this mechanism is shown in Figure 13.16 ▲. The first step has a much larger activation energy than the second step. The greater activation energy results in a much smaller rate constant for the first step compared to the second step. The first step determines the overall rate of the reaction, and the predicted rate law is therefore Rate = $k[NO_2]^2$, which is consistent with the observed experimental rate law.

For a proposed reaction mechanism such as the one above to be valid—mechanisms can only be validated, not proven—two conditions must be met:

1. **The elementary steps in the mechanism must sum to the overall reaction.**
2. **The rate law predicted by the mechanism must be consistent with the experimentally observed rate law.**

We have already seen that the rate law predicted by the above mechanism is consistent with the experimentally observed rate law. We can check whether the elementary steps sum to the overall reaction by simply adding them together:

$$NO_2(g) + NO_2(g) \longrightarrow NO_3(g) + NO(g) \qquad \text{Slow}$$
$$NO_3(g) + CO(g) \longrightarrow NO_2(g) + CO_2(g) \qquad \text{Fast}$$
$$NO_2(g) + CO(g) \longrightarrow NO(g) + CO_2(g) \qquad \text{Overall}$$

The mechanism fulfills both of the requirements and is therefore valid. A valid mechanism is not a *proven* mechanism. We can only say that a given mechanism is consistent with kinetic observations of the reaction and therefore possible. Other types of data—such as the experimental evidence for a proposed intermediate—can further strengthen the validity of a proposed mechanism.

Mechanisms with a Fast Initial Step

When the proposed mechanism for a reaction has a slow initial step—like the one shown above for the reaction between NO_2 and CO—the rate law predicted by the mechanism normally contains only reactants involved in the overall reaction. However, when a proposed mechanism contains a fast initial step, then some other step in the mechanism will be rate limiting. In these cases, the rate law specified by the rate-limiting step may contain reaction intermediates. Since reaction intermediates do not appear in the overall reaction equation, a rate law containing intermediates cannot correspond to the experimental rate law. Fortunately, however, we can often express the concentration of intermediates in terms of the concentrations of the reactants of the overall reaction.

In a multistep mechanism where the first step is fast, the products of the first step can build up, because the rate at which they are consumed is limited by some slower step further down the line. As those products build up, they can begin to react with one another to re-form the reactants. As long as the first step is fast enough compared to the rate-limiting step, the first-step reaction will reach equilibrium. We indicate the equilibrium as follows:

$$\text{Reactants} \underset{k_{-1}}{\overset{k_1}{\rightleftharpoons}} \text{products}$$

The double arrows indicate that both the forward reaction and the reverse reaction occur. If equilibrium is reached, then the rate of the forward reaction equals the rate of the reverse reaction.

As an example, consider the following reaction by which hydrogen reacts with nitrogen monoxide to form water and nitrogen gas:

$$2\,H_2(g) + 2\,NO(g) \longrightarrow 2\,H_2O(g) + N_2(g)$$

The experimentally observed rate law is Rate $= k[H_2][NO]^2$. The reaction is first order in hydrogen and second order in nitrogen monoxide. The proposed mechanism is as follows:

$$2\,NO(g) \underset{k_{-1}}{\overset{k_1}{\rightleftharpoons}} N_2O_2(g) \qquad \text{Fast}$$

$$H_2(g) + N_2O_2(g) \xrightarrow{k_2} H_2O(g) + N_2O(g) \qquad \text{Slow (rate limiting)}$$

$$N_2O(g) + H_2(g) \xrightarrow{k_3} N_2(g) + H_2O(g) \qquad \text{Fast}$$

$$\overline{2\,H_2(g) + 2\,NO(g) \longrightarrow 2\,H_2O(g) + N_2(g)} \qquad \text{Overall}$$

To determine whether the mechanism is valid, we must determine whether the two conditions described previously are met. The first condition is that the steps sum to the overall reaction. As you can see above, the steps do indeed sum to the overall reaction, so the first condition is met.

The second condition is that the rate law predicted by the mechanism must be consistent with the experimentally observed rate law. Since the second step is rate limiting, we write the following expression for the rate law:

$$\text{Rate} = k_2[H_2][N_2O_2] \qquad [13.28]$$

However, this rate law contains an intermediate (N_2O_2) and can therefore not be consistent with the experimentally observed rate law (which does not contain intermediates). Because of the equilibrium in the first step, however, *we can express the concentration of the intermediate in terms of the reactants of the overall equation.* Since the first step reaches equilibrium, the rate of the forward reaction in the first step equals the rate of the reverse reaction:

$$\text{Rate (forward)} = \text{rate (backward)}$$

The rate of the forward reaction is given by

$$\text{Rate} = k_1[NO]^2$$

The rate of the reverse reaction is given by

$$\text{Rate} = k_{-1}[N_2O_2]$$

Since these two rates are equal at equilibrium, we can write the following expression:

$$k_1[NO]^2 = k_{-1}[N_2O_2]$$

Rearranging, we get

$$[N_2O_2] = \frac{k_1}{k_{-1}}[NO]^2$$

We can now substitute this expression into Equation 13.28, the rate law obtained from the slow step:

$$\text{Rate} = k_2[H_2][N_2O_2]$$

$$= k_2[H_2]\frac{k_1}{k_{-1}}[NO]^2$$

$$= \frac{k_2k_1}{k_{-1}}[H_2][NO]^2$$

If we combine the individual rate constants into one overall rate constant, we get the following predicted rate law:

$$\text{Rate} = k[H_2][NO]^2 \qquad\qquad [13.29]$$

Since this rate law is consistent with the experimentally observed rate law, condition two is met and the proposed mechanism is valid.

EXAMPLE 13.9 Reaction Mechanisms

Ozone naturally decomposes to oxygen by the following reaction:

$$2\,O_3(g) \longrightarrow 3\,O_2(g)$$

The experimentally observed rate law for this reaction is as follows:

$$\text{Rate} = k[O_3]^2[O_2]^{-1}$$

Show that the following proposed mechanism is consistent with the experimentally observed rate law.

$$O_3(g) \underset{k_{-1}}{\overset{k_1}{\rightleftharpoons}} O_2(g) + O(g) \qquad \text{Fast}$$

$$O_3(g) + O(g) \underset{k_2}{\longrightarrow} 2\,O_2(g) \qquad \text{Slow}$$

Solution

To determine whether the mechanism is valid, you must first determine whether the steps sum to the overall reaction. Since the steps do indeed sum to the overall reaction, the first condition is met.	$$O_3(g) \underset{k_{-1}}{\overset{k_1}{\rightleftharpoons}} O_2(g) + \cancel{O}(g)$$ $$O_3(g) + \cancel{O}(g) \underset{k_2}{\longrightarrow} 2\,O_2(g)$$ $$\overline{2\,O_3(g) \longrightarrow 3\,O_2(g)}$$
The second condition is that the rate law predicted by the mechanism must be consistent with the experimentally observed rate law. Since the second step is rate limiting, write the rate law based on the second step.	$$\text{Rate} = k_2[O_3][O]$$
Since the rate law contains an intermediate (O), you must express the concentration of the intermediate in terms of the concentrations of the reactants of the overall reaction. To do this, set the rates of the forward reaction and the reverse reaction of the first step equal to each other. Solve the expression from the previous step for [O], the concentration of the intermediate.	$$\text{Rate (forward)} = \text{rate (backward)}$$ $$k_1[O_3] = k_{-1}[O_2][O]$$ $$[O] = \frac{k_1[O_3]}{k_{-1}[O_2]}$$
Finally, substitute [O] into the rate law predicted by the slow step.	$$\text{Rate} = k_2[O_3][O]$$ $$= k_2[O_3]\frac{k_1[O_3]}{k_{-1}[O_2]}$$ $$= k_2\frac{k_1}{k_{-1}}\frac{[O_3]^2}{[O_2]}$$ $$= k[O_3]^2[O_2]^{-1}$$

Check Since the two steps in the proposed mechanism sum to the overall reaction, and since the rate law obtained from the proposed mechanism is consistent with the experimentally observed rate law, the proposed mechanism is valid. The −1 reaction order with respect to [O_2] indicates that the rate slows down as the concentration of oxygen increases—oxygen inhibits, or slows down, the reaction.

For Practice 13.9

Predict the overall reaction and rate law that would result from the following two-step mechanism.

$$2 A \longrightarrow A_2 \quad \text{Slow}$$
$$A_2 + B \longrightarrow A_2B \quad \text{Fast}$$

13.7 Catalysis

Throughout this chapter, we have learned of ways to control the rates of chemical reactions. For example, we can speed up the rate of a reaction by increasing the concentration of the reactants or by increasing the temperature. However, these ways are not always feasible. There are limits to how concentrated we can make a reaction mixture, and increases in temperature may allow unwanted reactions—such as the decomposition of a reactant—to occur.

Alternatively, reaction rates can be increased by using a **catalyst**, a substance that increases the rate of a chemical reaction but is not consumed by the reaction. A catalyst works by providing an alternative mechanism for the reaction—one in which the rate-determining step has a lower activation energy. For example, consider the noncatalytic destruction of ozone in the upper atmosphere, discussed previously in Section 5.11:

$$O_3(g) + O(g) \longrightarrow 2 O_2(g)$$

In this reaction, an ozone molecule collides with an oxygen atom to form two oxygen molecules in a single elementary step. The reason that we have a protective ozone layer in the upper atmosphere is that the activation energy for this reaction is fairly high and the reaction, therefore, proceeds at a fairly slow rate. The ozone layer does not rapidly decompose into O_2. However, the addition of Cl atoms (which come from the photodissociation of man-made chlorofluorocarbons) to the upper atmosphere has made available another pathway by which O_3 can be destroyed. The first step in this pathway—called the catalytic destruction of ozone—is the reaction of Cl with O_3 to form ClO and O_2.

$$Cl + O_3 \longrightarrow ClO + O_2$$

This is followed by a second step in which ClO reacts with O, regenerating Cl.

$$ClO + O \longrightarrow Cl + O_2$$

Notice that if we add the two reactions, the overall reaction is identical to the noncatalytic reaction.

$$\cancel{Cl} + O_3 \longrightarrow \cancel{ClO} + O_2$$
$$\cancel{ClO} + O \longrightarrow \cancel{Cl} + O_2$$
$$\overline{O_3 + O \longrightarrow 2 O_2}$$

However, the activation energy for the rate-limiting step in this pathway is much smaller than for the first, uncatalyzed pathway (as shown in Figure 13.17 ▶), and therefore the reaction occurs at a much faster rate. Note that the Cl is not consumed in the overall reaction—this is characteristic of a catalyst.

In the case of the catalytic destruction of ozone, the catalyst speeds up a reaction that we *do not* want to happen. Most of the time, however, catalysts are used to speed up reactions that we *do* want to happen. For example, your car most likely has a catalytic converter in its exhaust system. The catalytic converter contains solid catalysts, such as

Energy Diagram for Catalyzed and Uncatalyzed Pathways

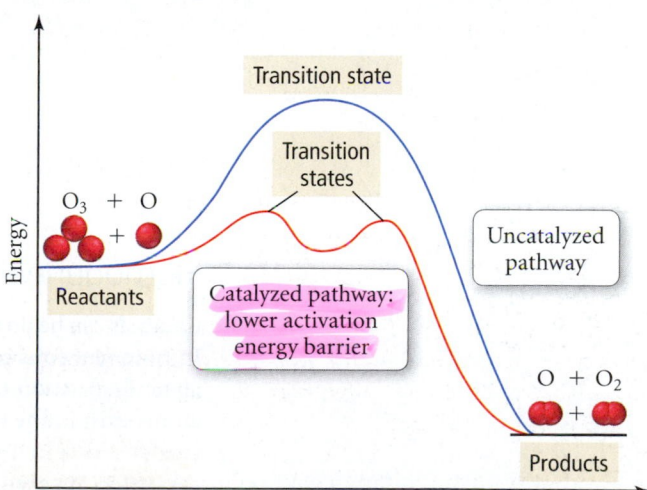

▲ **FIGURE 13.17 Catalyzed and Uncatalyzed Decomposition of Ozone** In the catalytic destruction of ozone, the activation barrier for the rate-limiting step is much lower than in the uncatalyzed process.

▶ The catalytic converter in the exhaust system of a car helps eliminate pollutants from the exhaust.

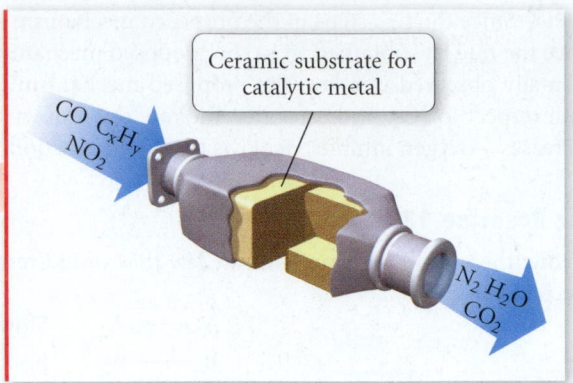

Ceramic substrate for catalytic metal

platinum, rhodium, or palladium, dispersed on an underlying high surface area ceramic structure. The catalysts convert exhaust pollutants such as nitrogen monoxide and carbon monoxide into less harmful substances:

$$2\,NO(g) + 2\,CO(g) \xrightarrow{\text{catalyst}} N_2(g) + 2\,CO_2(g)$$

The catalytic converter also promotes the complete combustion of any fuel fragments present in the exhaust:

$$\underset{\text{Fuel fragment}}{CH_3CH_2CH_3(g)} + 5\,O_2(g) \xrightarrow{\text{catalyst}} 3\,CO_2(g) + 4\,H_2O(g)$$

Fuel fragments in exhaust are harmful because they lead to the formation of ozone. We saw in Section 5.11 that although ozone is a natural part of our *upper* atmosphere that protects us from excess exposure to ultraviolet light, it is a pollutant in the *lower* atmosphere because it interferes with cardiovascular function and acts as an eye and lung irritant. The use of catalytic converters in motor vehicles has resulted in lower levels of these pollutants over most U.S. cities in the last 30 years even though the number of cars on the roadways has dramatically increased (see Table 13.4).

TABLE 13.4 Change in Pollutant Levels

Pollutant	Change 1983–1992	Change 1993–2002
NO$_2$	−21%	−11%
O$_3$	−22%	−2%
CO	−65%	−42 %

Source: *U.S. Environmental Protection Agency Air Trends Report.*

Homogeneous and Heterogeneous Catalysis

Catalysis can be divided into two types: homogeneous and heterogeneous (Figure 13.18 ▶). In **homogeneous catalysis,** the catalyst exists in the same phase as the reactants. The catalytic destruction of ozone by Cl is an example of homogeneous catalysis—the chlorine atoms exist in the gas phase with the gas-phase reactants. In **heterogeneous catalysis,** the catalyst exists in a phase different from the reactants. The solid catalysts used in catalytic converters are examples of heterogeneous catalysts—they exist as solids while the reactants are gases. The use of solid catalysts with gas-phase or solution-phase reactants is the most common type of heterogeneous catalysis.

Recent studies have shown that heterogeneous catalysis is most likely responsible for the ozone hole over Antarctica. After the discovery of the Antarctic ozone hole in 1985, scientists wondered why there was such a dramatic drop in ozone over Antarctica, but not over the rest of the planet. After all, the chlorine from chlorofluorocarbons that catalyzes ozone destruction is evenly distributed throughout the entire atmosphere.

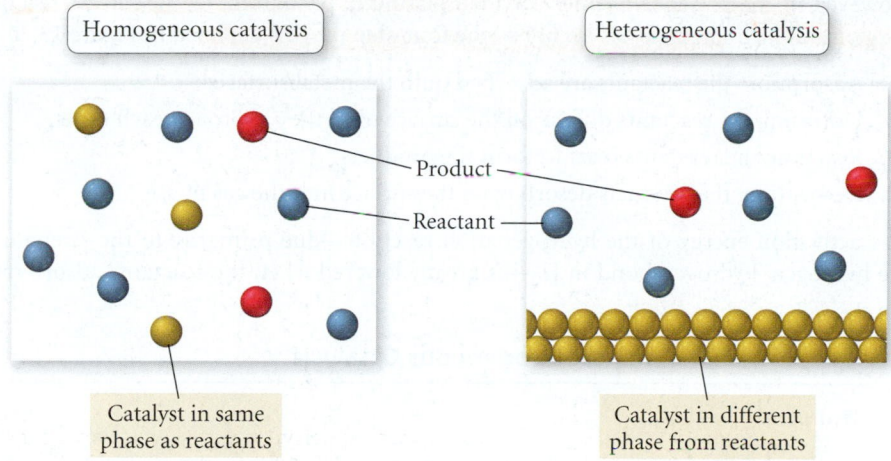

| Homogeneous catalysis | Heterogeneous catalysis |

Product
Reactant

Catalyst in same
phase as reactants

Catalyst in different
phase from reactants

▲ **FIGURE 13.18 Homogeneous and Heterogeneous Catalysis** A homogeneous catalyst exists in the same phase as the reactants. A heterogeneous catalyst exists in a different phase than the reactants. Often it provides a solid surface on which the reaction can take place.

As it turns out, most of the chlorine that enters the atmosphere from chlorofluorocarbons gets bound up in chlorine reservoirs, substances such as $ClONO_2$ that hold chlorine and prevent it from catalyzing ozone destruction. The unique conditions over Antarctica—especially the cold isolated air mass that exists during the long dark winter—result in clouds that contain solid ice particles. These unique clouds are called polar stratospheric clouds (or PSCs), and the surfaces of the ice particles within these clouds appear to catalyze the release of chlorine from their reservoirs:

$$ClONO_2 + HCl \xrightarrow[\text{PSCs}]{} Cl_2 + HNO_3$$

When the sun rises in the Antarctic spring, the sunlight dissociates the chlorine molecules into chlorine atoms:

$$Cl_2 \xrightarrow[\text{light}]{} 2\,Cl$$

The chlorine atoms then catalyze the destruction of ozone by the mechanism discussed previously. This continues until the sun melts the stratospheric clouds, allowing chlorine atoms to be reincorporated into their reservoirs. The result is an ozone hole that forms every spring and lasts about 6–8 weeks (Figure 13.19 ▼).

A second example of heterogeneous catalysis involves the **hydrogenation of double bonds within alkenes**. For example, the reaction between ethene and hydrogen is relatively slow at normal temperatures.

$$H_2C{=}CH_2(g) + H_2(g) \longrightarrow H_3C{-}CH_3(g) \qquad \text{Slow at room temperature}$$

▲ Polar stratospheric clouds contain ice particles that catalyze reactions by which chlorine is released from its atmospheric chemical reservoirs.

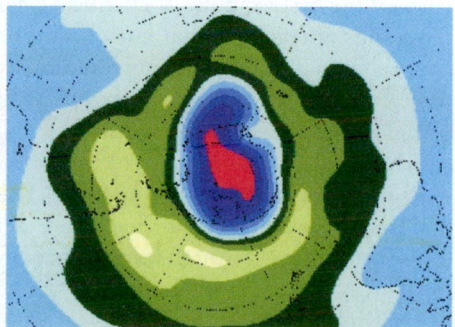

▲ **FIGURE 13.19 Ozone Depletion in the Antarctic Spring** The concentration of ozone over Antarctica shows a sharp drop during the month of October due to the catalyzed destruction of ozone by chlorine. The image at left shows the ozone levels in May 2004 while the image at right was produced in October of the same year. (The lowest ozone levels are represented by red and deep blue.)

However, in the presence of finely divided platinum, palladium, or nickel, the reaction happens rapidly. The catalysis occurs by the four-step process depicted in Figure 13.20 ▼.

1. Adsorption: the reactants are adsorbed onto the metal surface.
2. Diffusion: the reactants diffuse on the surface until they approach each other.
3. Reaction: the reactants react to form the products
4. Desorption: the products desorb from the surface into the gas phase.

The activation energy of the hydrogenation reaction—due primarily to the strength of the hydrogen–hydrogen bond in H_2—is greatly lowered when the reactants adsorb onto the surface.

Heterogeneous Catalysis

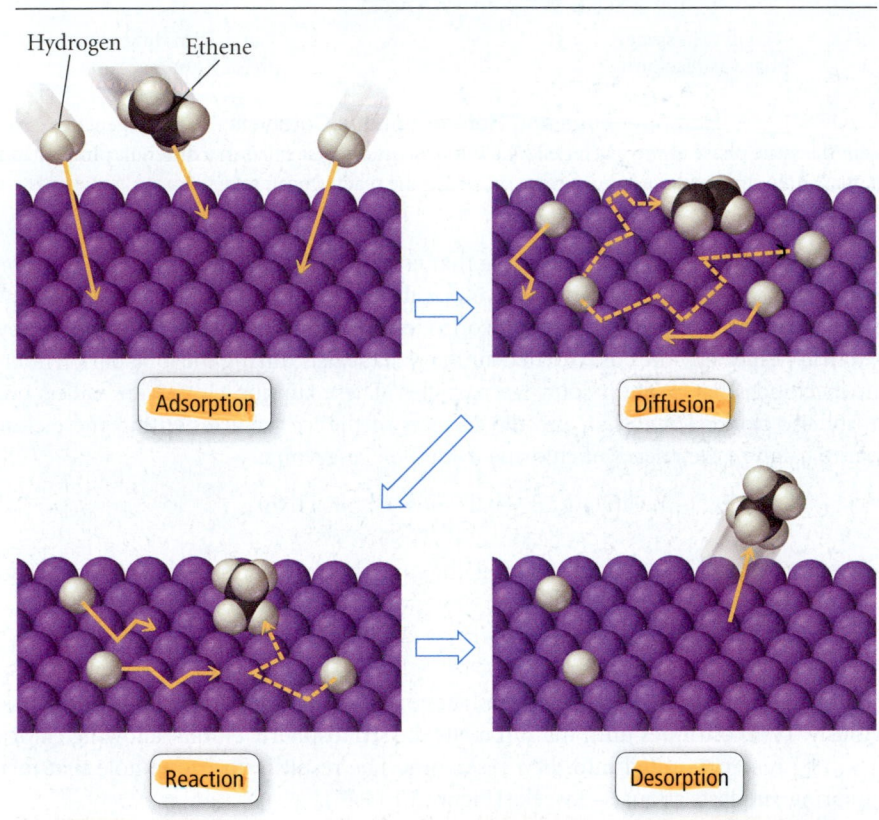

▲ FIGURE 13.20 Catalytic Hydrogenation of Ethene

Enzymes: Biological Catalysts

The strategies used to speed up chemical reactions in the laboratory—high temperatures, high pressures, strongly acidic or alkaline conditions—are not available to living organisms, since they would be fatal to cells.

Perhaps the best example of chemical catalysis is found in living organisms. Most of the thousands of reactions that must occur for an organism to survive would be too slow at normal temperatures. So living organisms use **enzymes**, biological catalysts that increase the rates of biochemical reactions. Enzymes are usually large protein molecules with complex three-dimensional structures. Within that structure is a specific area called the **active site**. The properties and shape of the active site are just right to bind the reactant molecule, usually called the **substrate**. The substrate fits into the active site in a manner that is analogous to a key fitting into a lock (Figure 13.21 ▶). When the substrate binds to the active site—through intermolecular forces such as hydrogen bonding and dispersion forces, or even covalent bonds—the activation energy of the reaction to be catalyzed is greatly lowered, allowing the reaction to occur at a much faster rate. The general mechanism by which an enzyme (E) binds a substrate (S) and then reacts to form the products (P) is as follows:

$$E + S \rightleftharpoons ES \qquad \text{Fast}$$
$$ES \longrightarrow E + P \qquad \text{Slow, rate limiting}$$

Enzyme-Substrate Binding

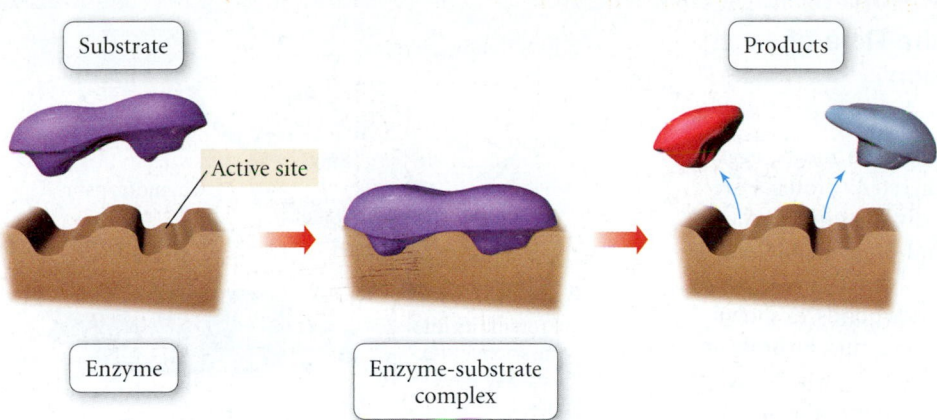

| Substrate | | Products |

Active site

| Enzyme | Enzyme-substrate complex |

▲ **FIGURE 13.21 Enzyme–Substrate Binding**   A substrate (or reactant) fits into the active site of an enzyme much as a key fits into a lock. It is held in place by intermolecular forces, forming an enzyme–substrate complex. (Sometimes temporary covalent bonding may also be involved.) After the reaction occurs, the products are released from the active site.

As an example of enzyme action, consider sucrase, an enzyme that catalyzes the breaking up of sucrose (table sugar) into glucose and fructose within the body. At body temperature, sucrose does not break into glucose and fructose because the activation

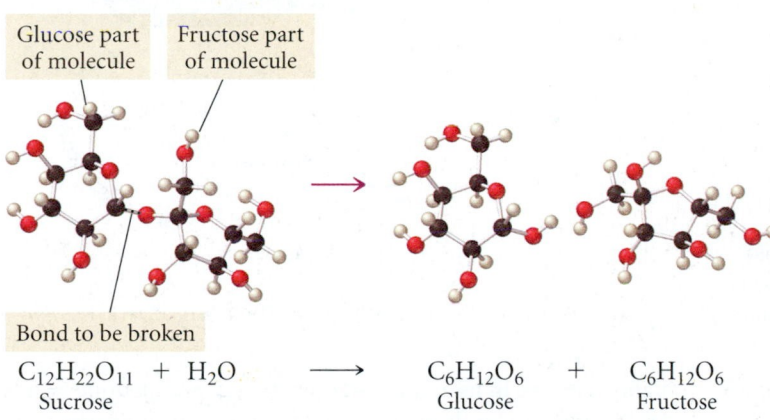

Glucose part of molecule   Fructose part of molecule

◀ Sucrose must break up into glucose and fructose during digestion.

Bond to be broken

$$C_{12}H_{22}O_{11} + H_2O \longrightarrow C_6H_{12}O_6 + C_6H_{12}O_6$$

Sucrose   Glucose   Fructose

energy is high, resulting in a slow reaction rate. However, when a sucrose molecule binds to the active site within sucrase, the bond between the glucose and fructose units weakens because glucose is forced into a geometry that stresses the bond. (Figure 13.22 ▶) Weakening of this bond lowers the activation energy for the reaction, increasing the reaction rate. The reaction can then proceed toward equilibrium—which favors the products—at a much lower temperature.

By allowing otherwise slow reactions to occur at a reasonable rate, enzymes give living organisms tremendous control over which reactions occur, and when they occur. Enzymes are extremely specific (each enzyme catalyzes only a single reaction) and efficient, speeding up reaction rates by factors of as much as a billion. If a living organism wants to turn a particular reaction on, it simply produces or activates the correct enzyme to catalyze that reaction. For this reason, many substances that inhibit the action of enzymes are highly toxic. Locking up a single enzyme molecule can stop the reaction of billions of substrates, much as one motorist stalled at a tollbooth can paralyze an entire highway full of cars. (For another example of enzyme action, see the *Chemistry and Medicine* box on the role of chymotrypsin in digestion.)

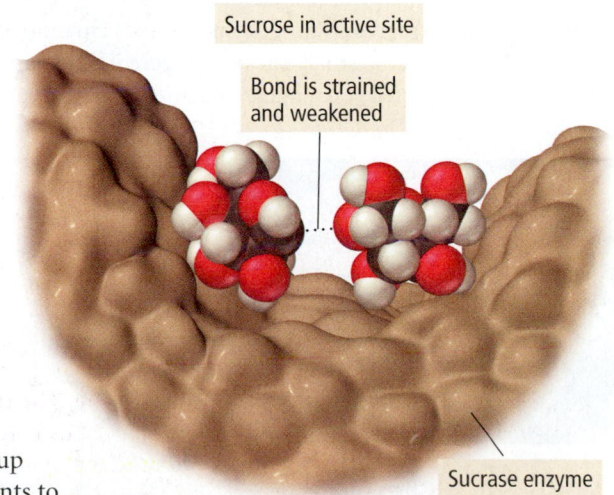

Sucrose in active site

Bond is strained and weakened

Sucrase enzyme

▲ **FIGURE 13.22 An Enzyme-Catalyzed Reaction**   Sucrase catalyzes the conversion of sucrose into glucose and fructose by weakening the bond that joins the two rings.

Chemistry and Medicine
Enzyme Catalysis and the Role of Chymotrypsin in Digestion

When we eat foods containing proteins—such as meats, eggs, beans, and nuts—the proteins must be digested. Proteins are large biological molecules composed of individual units called amino acids. (The structure of proteins and other biologically important molecules are discussed more fully in Chapter 21.) The amino acids are linked together via peptide bonds, as shown in Figure 13.23 ▼. During digestion, the protein must be broken up into individual amino acids (Figure 13.24 ▼), which then pass through the walls of the small intestine and into the bloodstream. However, the peptide bonds that link amino acids together are relatively stable, and under ordinary conditions the reaction is slow.

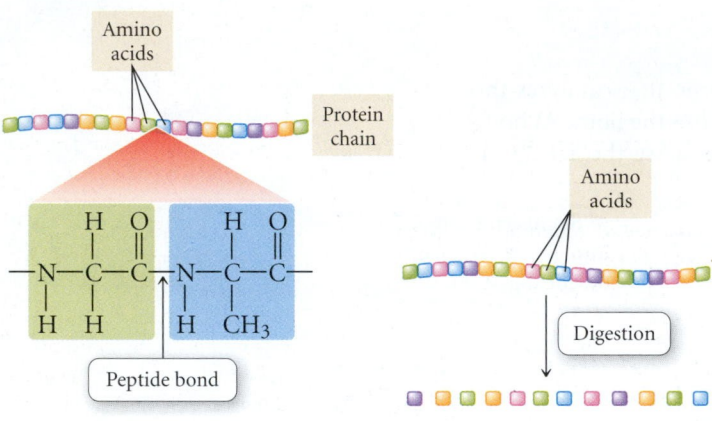

▲ FIGURE 13.23 The Structure of a Protein   Proteins are chains of amino acids linked together by peptide bonds.

▲ FIGURE 13.24 Protein Digestion   During digestion, a protein is broken down into its component amino acids.

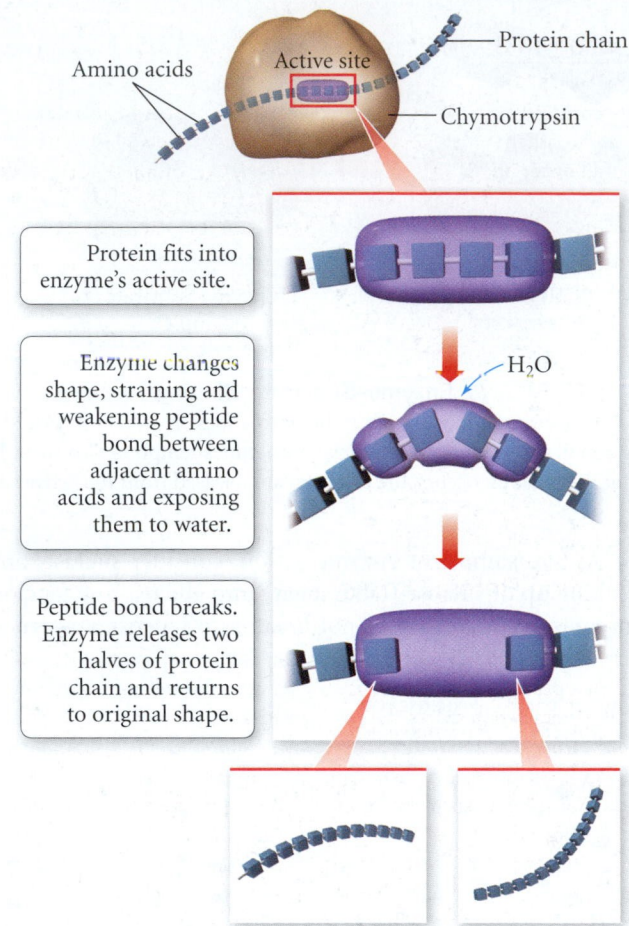

▲ FIGURE 13.26 The Action of Chymotrypsin

- Protein fits into enzyme's active site.
- Enzyme changes shape, straining and weakening peptide bond between adjacent amino acids and exposing them to water.
- Peptide bond breaks. Enzyme releases two halves of protein chain and returns to original shape.

The pancreas secretes an enzyme called chymotrypsin (Figure 13.25 ▼) into the small intestine, where it binds protein molecules to be digested. Like many enzymes, chymotrypsin is highly selective in its action—it operates only on peptide bonds between certain kinds of amino acids. When a protein molecule containing such a pair of amino acids is attached to the active site of chymotrypsin, the peptide bond between them is weakened as the chymotrypsin forms a covalent bond with the carbon in the peptide bond. A water molecule can then come in and cleave the bond, with an —OH from the water binding to the carbon atom and the remaining —H bonding to the nitrogen (Figure 13.26 ▲).

The result is that the amino acid chain is clipped at the peptide bond. The products of the reaction then leave the active site, another protein binds, and the process is repeated. Other digestive enzymes cleave protein chains between different pairs of amino acids. Together, these enzymes eventually reduce the entire protein to its constituent amino acids.

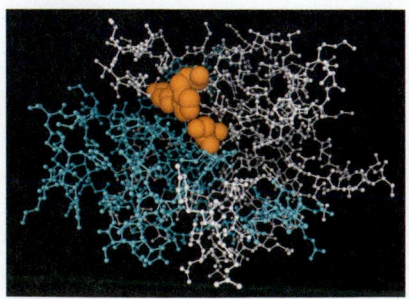

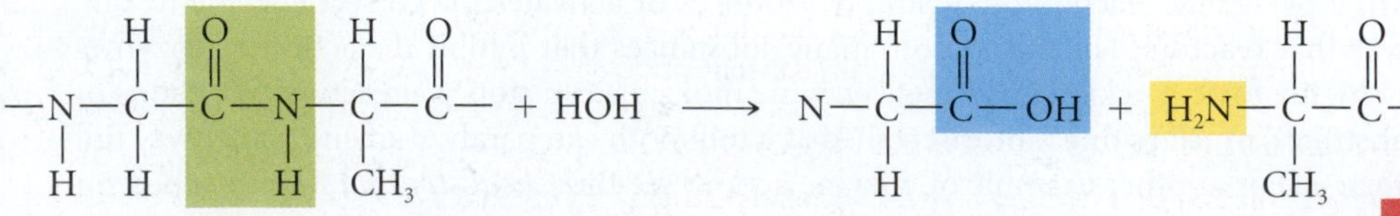

▲ FIGURE 13.25 Chymotrypsin, a Digestive Enzyme   This model of chymotrypsin shows a section of a protein substrate in the active site.

Chapter in Review

Key Terms

Section 13.3
rate law (575)
rate constant (k) (575)
reaction order (n) (575)
overall order (578)

Section 13.4
integrated rate law (580)
half-life ($t_{1/2}$) (585)

Section 13.5
Arrhenius equation (587)
activation energy (E_a) (588)
frequency factor (A) (588)
activated complex (transition
    state) (588)
exponential factor (589)
Arrhenius plot (590)
collision model (592)

Section 13.6
reaction mechanism (594)
elementary step (594)
reaction intermediates (594)
molecularity (594)
unimolecular (594)
bimolecular (594)
termolecular (594)
rate-determining step (595)

Section 13.7
catalyst (599)
homogeneous catalysis (600)
heterogeneous catalysis (600)
hydrogenation (601)
enzyme (602)
active site (602)
substrate (602)

Key Concepts

Reaction Rates, Orders, and Rate Laws (13.1–13.3)

The rate of a chemical reaction is a measure of how fast a reaction occurs. The rate reflects the change in the concentration of a reactant or
product per unit time, and is usually reported in units of M/s. Reaction
rates generally depend on the concentration of the reactants. The rate of
a first-order reaction is directly proportional to the concentration of the
reactant; the rate of a second-order reaction is proportional to the
square of the concentration of the reactant; and the rate of a zero-order
reaction is independent of the concentration of the reactant. For a reaction with more than one reactant, the order with respect to each reactant
is, in general, independent of the order with respect to other reactants.
The order of a reaction with respect to a particular reactant is usually determined by measuring the rate of the reaction at several different initial
concentrations of that reactant while holding the concentrations of
other reactants (if there are any) constant. The rate law shows the relationship between the rate and the concentrations of each reactant.

Integrated Rate Laws and Half-Life (13.4)

The rate law for a reaction gives the relationship between the rate of
the reaction and the concentrations of the reactants. The integrated
rate law, by contrast, gives the relationship between the concentration
of a reactant and time. The integrated rate law is different for each reaction order. The rate law for a zero-order reaction shows that the concentration of the reactant varies linearly with time. For a first-order
reaction, the *natural log* of the concentration of the reactant varies linearly with time, and for a second-order reaction, the *inverse* of the concentration of the reactant varies linearly with time. Therefore, the
order of a reaction can also be determined by plotting measurements
of concentration as a function of time in these three different ways.
The plot that is linear indicates the order of the reaction. The half-life
of a reaction can be derived from the integrated rate law and represents
the time required for the concentration of a reactant to fall to one-half
of its initial value. The half-life of a first-order reaction is *independent*
of initial concentration of the reactant. The half-life of a zero-order or
second-order reaction *depends* on the initial concentration of reactant.

The Effect of Temperature on Reaction Rate (13.5)

The rate constant of a reaction generally depends on temperature and
can be expressed by the Arrhenius equation, which consists of a frequency

factor and an exponential factor. The exponential factor depends on
both the temperature and the activation energy, a barrier that the reactants must overcome to become products. The frequency factor represents the number of times that the reactants approach the activation
barrier per unit time. The exponential factor is the fraction of approaches that are successful in surmounting the activation barrier and forming
products. The exponential factor increases with increasing temperature,
but decreases with an increasing value of the activation energy. The frequency factor and activation energy for a reaction can by determined by
measuring the rate constant at different temperatures and constructing
an Arrhenius plot. For reactions in the gas phase, Arrhenius behavior
can be modeled with the collision model. In this model, reactions occur
as a result of sufficiently energetic collisions. The colliding molecules
must be oriented in such a way that the reaction can occur. The frequency factor then contains two terms: p, which represents the fraction of collisions that have the proper orientation, and z, which represents the
number of collisions per unit time.

Reaction Mechanisms (13.6)

Most chemical reactions occur not in a single step, but through several
steps. The series of individual steps by which a reaction occurs is called
the reaction mechanism. In order for a mechanism to be valid, it must
fulfill two conditions: (a) the steps must sum to the overall reaction;
and (b) the mechanism must predict the experimentally observed rate
law. For mechanisms with a slow initial step, the predicted rate law is
derived from the slow step. For mechanisms with a fast initial step, the
predicted rate law is first written based on the slow step, and then equi-
libration of the fast steps is assumed in order to write concentrations of
intermediates in terms of the reactants.

Catalysis (13.7)

A catalyst is a substance that increases the rate of a chemical reaction
by providing an alternative mechanism that has a lower activation
energy for the rate-determining step. Catalysts can be divided into
two types: homogeneous and heterogeneous. A homogeneous catalyst exists in the same phase as the reactants and forms a homogeneous mixture with them. A heterogeneous catalyst generally exists
in a different phase than the reactants. Enzymes are biological catalysts capable of increasing the rate of specific biochemical reactions
by many orders of magnitude.

Key Equations and Relationships

The Rate of Reaction (13.2)

For a reaction, $aA + bB \longrightarrow cC + dD$, the rate is defined as

$$\text{Rate} = -\frac{1}{a}\frac{\Delta[A]}{\Delta t} = -\frac{1}{b}\frac{\Delta[B]}{\Delta t} = +\frac{1}{c}\frac{\Delta[C]}{\Delta t} = +\frac{1}{d}\frac{\Delta[D]}{\Delta t}$$

The Rate Law (13.3)

$$\text{Rate} = k[A]^n \quad \text{(single reactant)}$$
$$\text{Rate} = k[A]^m[B]^n \quad \text{(multiple reactants)}$$

Integrated Rate Laws and Half-Life (13.4)

Reaction Order	Integrated Rate Law	Units of k	Half-Life Expression
0	$[A]_t = -kt + [A]_0$	$M \cdot s^{-1}$	$t_{1/2} = \dfrac{[A]_0}{2k}$
1	$\ln[A]_t = -kt + \ln[A]_0$	s^{-1}	$t_{1/2} = \dfrac{0.693}{k}$
2	$\dfrac{1}{[A]_t} = kt + \dfrac{1}{[A]_0}$	$M^{-1} \cdot s^{-1}$	$t_{1/2} = \dfrac{1}{k[A]_0}$

Arrhenius Equation (13.5)

$$k = Ae^{-E_a/RT}$$

$$\ln k = -\frac{E_a}{R}\left(\frac{1}{T}\right) + \ln A \quad \text{(linearized form)}$$

$$\ln \frac{K_2}{K_1} = \frac{E_a}{R}\left(\frac{1}{T_1} - \frac{1}{T_2}\right) \quad \text{(two-point form)}$$

$$k = pz\,e^{-E_a/RT} \quad \text{(collision theory)}$$

Rate Laws for Elementary Steps (13.6)

Elementary Step	Molecularity	Rate Law
$A \longrightarrow$ products	1	$\text{Rate} = k[A]$
$A + A \longrightarrow$ products	2	$\text{Rate} = k[A]^2$
$A + B \longrightarrow$ products	2	$\text{Rate} = k[A][B]$
$A + A + A \longrightarrow$ products	3 (rare)	$\text{Rate} = k[A]^3$
$A + A + B \longrightarrow$ products	3 (rare)	$\text{Rate} = k[A]^2[B]$
$A + B + C \longrightarrow$ products	3 (rare)	$\text{Rate} = k[A][B][C]$

Key Skills

Expressing Reaction Rates (13.2)
- Example 13.1 • For Practice 13.1 • Exercises 25–32

Determining the Order, Rate Law, and Rate Constant of a Reaction (13.3)
- Example 13.2 • For Practice 13.2 • Exercises 39–42

Using Graphical Analysis of Reaction Data to Determine Reaction Order and Rate Constants (13.4)
- Examples 13.3, 13.5 • For Practice 13.3, 13.5 • Exercises 45–50

Determining the Concentration of a Reactant at a Given Time (13.4)
- Example 13.4 • For Practice 13.4 • Exercises 49–52

Working with the Half-Life of a Reaction (13.4)
- Example 13.6 • For Practice 13.6 • Exercises 51–54

Using the Arrhenius Equation to Determine Kinetic Parameters (13.5)
- Examples 13.7, 13.8 • For Practice 13.7, 13.8 • Exercises 57–66

Determining whether a Reaction Mechanism Is Valid (13.6)
- Example 13.9 • For Practice 13.9 • Exercises 69–72

Exercises

Review Questions

1. Explain why lizards become sluggish in cold weather. How is this phenomenon related to chemistry?

2. Why is knowledge of reaction rates important (both practically and theoretically)?

3. What units are typically used in expressing the rate of a reaction?

4. Why is the reaction rate for reactants defined as the *negative* of the change in reactant concentration with respect to time, whereas for products it is defined simply as the change in reactant concentration with respect to time (with a positive sign)?

5. Explain the difference between the average rate of reaction and the instantaneous rate of reaction.

6. Consider a simple reaction in which a reactant A forms products:

$$A \longrightarrow \text{products}$$

What is the rate law if the reaction is zero order with respect to A? First order? Second order? For each case, explain how a doubling of the concentration of A would affect the rate of reaction.

7. How is the order of a reaction generally determined?

8. For a reaction with multiple reactants, how is the overall order of the reaction defined?

9. Explain the difference between the rate law for a reaction and the integrated rate law for a reaction. What relationship does each kind of rate law express?

10. Write integrated rate laws for zero-order, first-order, and second-order reactions of the form A $\longrightarrow$ products.

11. What does the term *half-life* mean? Write the expressions for the half-lives of zero-order, first-order, and second-order reactions.

12. How do reaction rates typically depend on temperature? What part of the rate law is temperature dependent?

13. Explain the meaning of each of the following terms within the Arrhenius equation: activation energy, frequency factor, and exponential factor. Use these terms and the Arrhenius equation to explain why small changes in temperature can result in large changes in the reaction rate.

14. What is an Arrhenius plot? Explain the meaning of the slope and intercept of an Arrhenius plot.

15. Explain how a chemical reaction occurs according to the collision model. Explain the meaning of the orientation factor within this model.

16. Explain the difference between a normal chemical equation for a chemical reaction and the mechanism of that reaction.

17. In a reaction mechanism, what is an elementary step? Write down the three most common elementary steps and the corresponding rate law for each one.

18. What are the two requirements for a proposed mechanism to be valid for a given reaction?

19. What is an intermediate within a reaction mechanism?

20. What is a catalyst? How does a catalyst increase the rate of a chemical reaction?

21. Explain the difference between homogeneous catalysis and heterogeneous catalysis.

22. What are the four basic steps involved in heterogeneous catalysis?

23. What are enzymes? What is the active site of an enzyme? What does the term substrate mean?

24. What is the general two-step mechanism by which most enzymes work?

Problems by Topic

Reaction Rates

25. Consider the following reaction:

$$2\,HBr(g) \longrightarrow H_2(g) + Br_2(g)$$

a. Express the rate of the reaction with respect to each of the reactants and products.
b. In the first 15.0 s of this reaction, the concentration of HBr dropped from 0.500 M to 0.455 M. Calculate the average rate of the reaction in this time interval.
c. If the volume of the reaction vessel in part b was 0.500 L, what amount of Br_2 (in moles) was formed during the first 15.0 s of the reaction?

26. Consider the following reaction:

$$2\,N_2O(g) \longrightarrow 2\,N_2(g) + O_2(g)$$

a. Express the rate of the reaction with respect to each of the reactants and products.
b. In the first 10.0 s of the reaction, 0.018 mol of O_2 is produced in a reaction vessel with a volume of 0.250 L. What is the average rate of the reaction over this time interval?
c. Predict the rate of change in the concentration of N_2O over this time interval. In other words, what is $\Delta[N_2O]/\Delta t$?

27. For the reaction $2\,A(g) + B(g) \longrightarrow 3\,C(g)$,
a. Determine the expression for the rate of the reaction with respect to each of the reactants and products.
b. When A is decreasing at a rate of 0.100 M/s, how fast is B decreasing? How fast is C increasing?

28. For the reaction $A(g) + \frac{1}{2}\,B(g) \longrightarrow 2\,C(g)$,
a. Determine the expression for the rate of the reaction with respect to each of the reactants and products.
b. When C is increasing at a rate of 0.025 M/s, how fast is B decreasing? How fast is A decreasing?

29. Consider the following reaction:

$$C_4H_8(g) \longrightarrow 2\,C_2H_4(g)$$

The following data were collected for the concentration of C_4H_8 as a function of time:

Time (s)	[C_4H_8] (M)
0	1.000
10	0.913
20	0.835
30	0.763
40	0.697
50	0.637

a. What is the average rate of the reaction between 0 and 10 s? Between 40 and 50 s?
b. What is the rate of formation of C_2H_4 between 20 and 30 s?

30. Consider the following reaction:

$$NO_2(g) \longrightarrow NO(g) + \tfrac{1}{2}O_2(g)$$

The following data were collected for the concentration of NO_2 as a function of time:

Time (s)	[NO_2] (M)
0	1.000
10	0.951
20	0.904
30	0.860
40	0.818
50	0.778
60	0.740
70	0.704
80	0.670
90	0.637
100	0.606

a. What is the average rate of the reaction between 10 and 20 s? Between 50 and 60 s?
b. What is the rate of formation of O_2 between 50 and 60 s?

31. Consider the following reaction:

$$H_2(g) + Br_2(g) \longrightarrow 2\,HBr(g)$$

The graph below shows the concentration of Br_2 as a function of time.

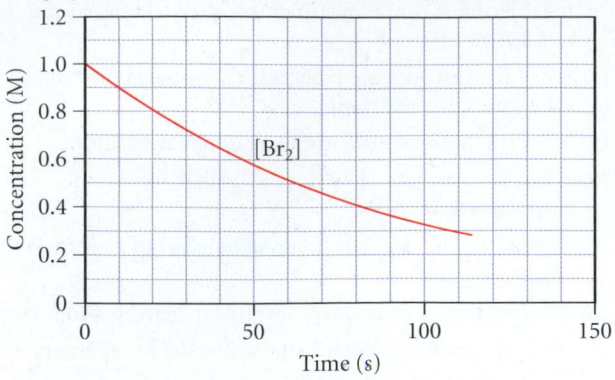

a. Use the graph to calculate the following:
  (i) The average rate of the reaction between 0 and 25 s.
  (ii) The instantaneous rate of the reaction at 25 s.
  (iii) The instantaneous rate of formation of HBr at 50 s.
b. Make a rough sketch of a curve representing the concentration of HBr as a function of time. Assume that the initial concentration of HBr is zero.

32. Consider the following reaction:

$$2\,H_2O_2(aq) \longrightarrow 2\,H_2O(l) + O_2(g)$$

The graph below shows the concentration of H_2O_2 as a function of time.

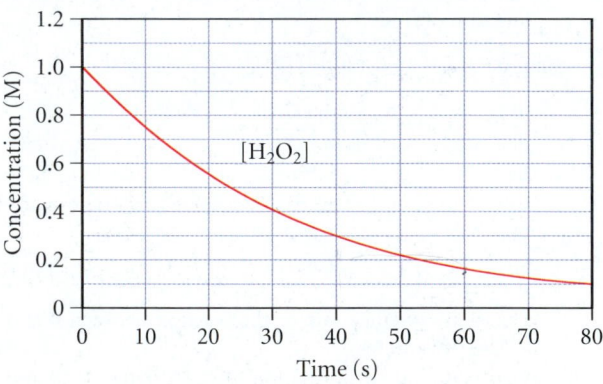

Use the graph to calculate the following:
a. The average rate of the reaction between 10 and 20 s.
b. The instantaneous rate of the reaction at 30 s.
c. The instantaneous rate of formation of O_2 at 50 s.
d. If the initial volume of the H_2O_2 is 1.5 L, what total amount of O_2 (in moles) is formed in the first 50 s of reaction?

The Rate Law and Reaction Orders

33. The graph below shows a plot of the rate of a reaction versus the concentration of the reactant A for the reaction A $\longrightarrow$ products.

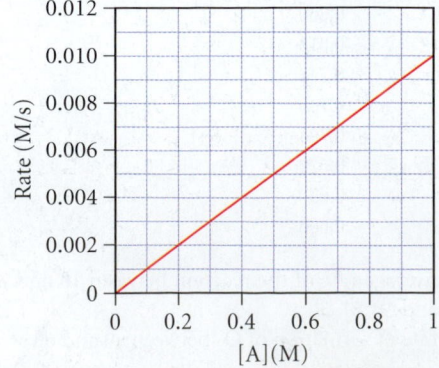

a. What is the order of the reaction with respect to A?
b. Make a rough sketch of how a plot of [A] versus *time* would appear.
c. Write a rate law for the reaction including an estimate for the value of k.

34. The graph below shows a plot of the rate of a reaction versus the concentration of the reactant.

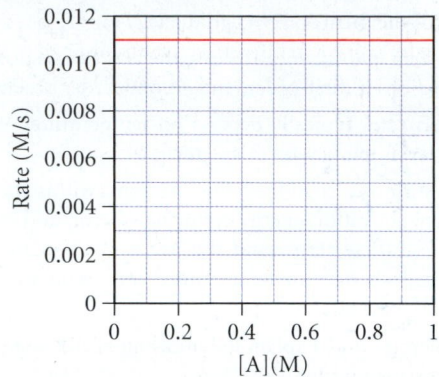

a. What is the order of the reaction with respect to A?
b. Make a rough sketch of how a plot of [A] versus *time* would appear.
c. Write a rate law for the reaction including the value of k.

35. What are the units of k for each of the following?
a. first-order reaction
b. second-order reaction
c. zero-order reaction

36. The following reaction is first order in N_2O_5:

$$N_2O_5(g) \longrightarrow NO_3(g) + NO_2(g)$$

The rate constant for the reaction at a certain temperature is 0.053/s.
a. Calculate the rate of the reaction when $[N_2O_5] = 0.055\,M$.
b. What would the rate of the reaction be at the same concentration as in part a if the reaction were second order? Zero order? (Assume the same *numerical* value for the rate constant with the appropriate units.)

37. A reaction in which A, B, and C react to form products is first order in A, second order in B, and zero order in C.
a. Write a rate law for the reaction.
b. What is the overall order of the reaction?
c. By what factor does the reaction rate change if [A] is doubled (and the other reactant concentrations are held constant)?
d. By what factor does the reaction rate change if [B] is doubled (and the other reactant concentrations are held constant)?
e. By what factor does the reaction rate change if [C] is doubled (and the other reactant concentrations are held constant)?
f. By what factor does the reaction rate change if the concentrations of all three reactants are doubled?

38. A reaction in which A, B, and C react to form products is zero order in A, one-half order in B, and second order in C.
a. Write a rate law for the reaction.
b. What is the overall order of the reaction?
c. By what factor does the reaction rate change if [A] is doubled (and the other reactant concentrations are held constant)?
d. By what factor does the reaction rate change if [B] is doubled (and the other reactant concentrations are held constant)?
e. By what factor does the reaction rate change if [C] is doubled (and the other reactant concentrations are held constant)?
f. By what factor does the reaction rate change if the concentrations of all three reactants are doubled?

39. Consider the following data showing the initial rate of a reaction (A $\longrightarrow$ products) at several different concentrations of A. What is the order of the reaction? Write a rate law for the reaction including the value of the rate constant, k.

[A] (M)	Initial Rate (M/s)
0.100	0.053
0.200	0.210
0.300	0.473

40. Consider the following data showing the initial rate of a reaction (A $\longrightarrow$ products) at several different concentrations of A. What is the order of the reaction? Write a rate law for the reaction including the value of the rate constant, k.

[A] (M)	Initial Rate (M/s)
0.15	0.008
0.30	0.016
0.60	0.032

41. The data below were collected for the following reaction:

$$2\,NO_2(g) + F_2(g) \longrightarrow 2\,NO_2F(g)$$

[NO₂] (M)	[F₂] (M)	Initial Rate (M/s)
0.100	0.100	0.026
0.200	0.100	0.051
0.200	0.200	0.103
0.400	0.400	0.411

Write an expression for the reaction rate law and calculate the value of the rate constant, k. What is the overall order of the reaction?

42. The data below were collected for the following reaction:

$$CH_3Cl(g) + 3\,Cl_2(g) \longrightarrow CCl_4(g) + 3\,HCl(g)$$

[CH₃Cl] (M)	[Cl₂] (M)	Initial Rate (M/s)
0.050	0.050	0.014
0.100	0.050	0.029
0.100	0.100	0.041
0.200	0.200	0.115

Write an expression for the reaction rate law and calculate the value of the rate constant, k. What is the overall order of the reaction?

The Integrated Rate Law and Half-Life

43. Indicate the order of reaction for each of the following observations.
  a. A plot of the concentration of the reactant versus time yields a straight line.
  b. The reaction has a half-life that is independent of initial concentration.
  c. A plot of the inverse of the concentration versus time yields a straight line.

44. Indicate the order of reaction for each of the following observations.
  a. The half-life of the reaction gets shorter as the initial concentration is increased.
  b. A plot of the natural log of the concentration of the reactant versus time yields a straight line.
  c. The half-life of the reaction gets longer as the initial concentration is increased.

45. The data below show the concentration of AB versus time for the following reaction:

$$AB(g) \longrightarrow A(g) + B(g)$$

Time (s)	[AB] (M)
0	0.950
50	0.459
100	0.302
150	0.225
200	0.180
250	0.149
300	0.128
350	0.112
400	0.0994
450	0.0894
500	0.0812

Determine the order of the reaction and the value of the rate constant. Predict the concentration of AB at 25 s.

46. The data below show the concentration of N_2O_5 versus time for the following reaction:

$$N_2O_5(g) \longrightarrow NO_3(g) + NO_2(g)$$

Time (s)	[N₂O₅] (M)
0	1.000
25	0.822
50	0.677
75	0.557
100	0.458
125	0.377
150	0.310
175	0.255
200	0.210

Determine the order of the reaction and the value of the rate constant. Predict the concentration of N_2O_5 at 250 s.

47. The data below show the concentration of cyclobutane (C_4H_8) versus time for the following reaction:

$$C_4H_8 \longrightarrow 2\,C_2H_4$$

Time (s)	[C₄H₈] (M)
0	1.000
10	0.894
20	0.799
30	0.714
40	0.638
50	0.571
60	0.510
70	0.456
80	0.408
90	0.364
100	0.326

Determine the order of the reaction and the value of the rate constant. What is the rate of reaction when $[C_4H_8] = 0.25$ M?

48. A reaction in which A $\longrightarrow$ products was monitored as a function of time and the results are shown below.

Time (s)	[A] (M)
0	1.000
25	0.914
50	0.829
75	0.744
100	0.659
125	0.573
150	0.488
175	0.403
200	0.318

Determine the order of the reaction and the value of the rate constant. What is the rate of reaction when $[A] = 0.10\ M$?

49. The following reaction was monitored as a function of time:

$$A \longrightarrow B + C$$

A plot of $\ln[A]$ versus time yields a straight line with slope $-0.0045/s$.
a. What is the value of the rate constant (k) for this reaction at this temperature?
b. Write the rate law for the reaction.
c. What is the half-life?
d. If the initial concentration of A is 0.250 M, what is the concentration after 225 s?

50. The following reaction was monitored as a function of time:

$$AB \longrightarrow A + B$$

A plot of $1/[AB]$ versus time yields a straight line with slope $0.055/M \cdot s$.
a. What is the value of the rate constant (k) for this reaction at this temperature?
b. Write the rate law for the reaction.
c. What is the half-life when the initial concentration is 0.55 M?
d. If the initial concentration of AB is 0.250 M, and the reaction mixture initially contains no products, what are the concentrations of A and B after 75 s?

51. The decomposition of SO_2Cl_2 is first order in SO_2Cl_2 and has a rate constant of $1.42 \times 10^{-4}\ s^{-1}$ at a certain temperature.
a. What is the half-life for this reaction?
b. How long will it take for the concentration of SO_2Cl_2 to decrease to 25% of its initial concentration?
c. If the initial concentration of SO_2Cl_2 is 1.00 M, how long will it take for the concentration to decrease to 0.78 M?
d. If the initial concentration of SO_2Cl_2 is 0.150 M, what is the concentration of SO_2Cl_2 after 2.00×10^2 s? After 5.00×10^2 s?

52. The decomposition of XY is second order in XY and has a rate constant of $7.02 \times 10^{-3}\ M^{-1} \cdot s^{-1}$ at a certain temperature.
a. What is the half-life for this reaction at an initial concentration of 0.100 M?
b. How long will it take for the concentration of XY to decrease to 12.5% of its initial concentration when the initial concentration is 0.100 M? When the initial concentration is 0.200 M?
c. If the initial concentration of XY is 0.150 M, how long will it take for the concentration to decrease to 0.062 M?
d. If the initial concentration of XY is 0.050 M, what is the concentration of XY after 5.0×10^1 s? After 5.50×10^2 s?

53. The half-life for the radioactive decay of U-238 is 4.5 billion years and is independent of initial concentration. How long will it take for 10% of the U-238 atoms in a sample of U-238 to decay? If a sample of U-238 initially contained 1.5×10^{18} atoms when the universe was formed 13.8 billion years ago, how many U-238 atoms will it contain today?

54. The half-life for the radioactive decay of C-14 is 5730 years. How long will it take for 25% of the C-14 atoms in a sample of C-14 to decay? If a sample of C-14 initially contains 1.5 mmol of C-14, how many millimoles will be left after 2255 years?

The Effect of Temperature and the Collision Model

55. The following diagram shows the energy of a reaction as the reaction progresses. Label each of the following in the diagram:

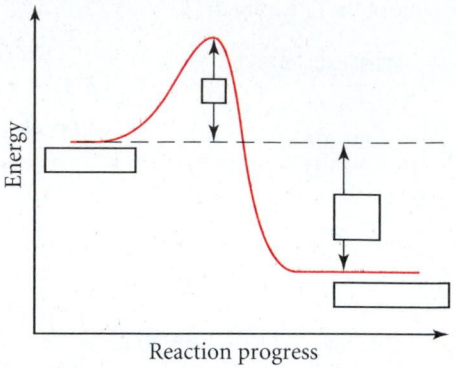

a. reactants
b. products
c. activation energy (E_a)
d. enthalpy of reaction (ΔH_{rxn})

56. A chemical reaction is endothermic and has an activation energy that is twice the value of the enthalpy of the reaction. Draw a diagram depicting the energy of the reaction as it progresses. Label the position of the reactants and products and indicate the activation energy and enthalpy of reaction.

57. The activation energy of a reaction is 56.8 kJ/mol and the frequency factor is $1.5 \times 10^{11}/s$. Calculate the rate constant of the reaction at 25 °C.

58. The rate constant of a reaction at 32 °C was measured to be $0.055/s$. If the frequency factor is $1.2 \times 10^{13}/s$, what is the activation barrier?

59. The data shown below were collected for the following first-order reaction:

$$N_2O(g) \longrightarrow N_2(g) + O(g)$$

Use an Arrhenius plot to determine the activation barrier and frequency factor for the reaction.

Temperature (K)	Rate Constant (1/s)
800	3.24×10^{-5}
900	0.00214
1000	0.0614
1100	0.955

60. The following data show the rate constant of a reaction measured at several different temperatures. Use an Arrhenius plot to determine the activation barrier and frequency factor for the reaction.

Temperature (K)	Rate Constant (1/s)
300	0.0134
310	0.0407
320	0.114
330	0.303
340	0.757

61. The data shown below were collected for the following second-order reaction:

$$Cl(g) + H_2(g) \longrightarrow HCl(g) + H(g)$$

Use an Arrhenius plot to determine the activation barrier and frequency factor for the reaction.

Temperature (K)	Rate Constant (L/mol · s)
90	0.00357
100	0.0773
110	0.956
120	7.781

62. The following data show the rate constant of a reaction measured at several different temperatures. Use an Arrhenius plot to determine the activation barrier and frequency factor for the reaction.

Temperature (K)	Rate Constant (1/s)
310	0.00434
320	0.0140
330	0.0421
340	0.118
350	0.316

63. A reaction has a rate constant of 0.0117/s at 400. K and 0.689/s at 450. K.
   a. Determine the activation barrier for the reaction.
   b. What is the value of the rate constant at 425 K?

64. A reaction has a rate constant of 0.000122/s at 27 °C and 0.228/s at 77 °C.
   a. Determine the activation barrier for the reaction.
   b. What is the value of the rate constant at 17 °C?

65. If a temperature increase from 10.0 °C to 20.0 °C doubles the rate constant for a reaction, what is the value of the activation barrier for the reaction?

66. If a temperature increase from 20.0 °C to 35.0 °C triples the rate constant for a reaction, what is the value of the activation barrier for the reaction?

67. Consider the following two gas-phase reactions:
   a. $AA(g) + BB(g) \longrightarrow 2\,AB(g)$
   b. $AB(g) + CD(g) \longrightarrow AC(g) + BD(g)$

   If the two reactions have identical activation barriers and are carried out under the same conditions, which one would you expect to have the faster rate?

68. Which of the following two reactions would you expect to have the smaller orientation factor? Explain.
   a. $O(g) + N_2(g) \longrightarrow NO(g) + N(g)$
   b. $NO(g) + Cl_2(g) \longrightarrow NOCl(g) + Cl(g)$

Reaction Mechanisms

69. Consider the following overall reaction which is experimentally observed to be second order in AB and zero order in C:

$$AB + C \longrightarrow A + BC$$

Determine whether the mechanism below is valid for this reaction.

$$AB + AB \xrightarrow[k_1]{} AB_2 + A \quad \text{Slow}$$

$$AB_2 + C \xrightarrow[k_2]{} AB + BC \quad \text{Fast}$$

70. Consider the following overall reaction which is experimentally observed to be second order in X and first order in Y:

$$X + Y \longrightarrow XY$$

   a. Does the reaction occur in a single step in which X and Y collide?
   b. Is the following two-step mechanism valid?

$$2\,X \underset{k_2}{\overset{k_1}{\rightleftharpoons}} X_2 \quad\quad \text{Fast}$$

$$X_2 + Y \xrightarrow[k_3]{} XY + X \quad \text{Slow}$$

71. Consider the following three-step mechanism for a reaction:

$$Cl_2(g) \underset{k_2}{\overset{k_1}{\rightleftharpoons}} 2\,Cl(g) \quad\quad \text{Fast}$$

$$Cl(g) + CHCl_3(g) \xrightarrow[k_3]{} HCl(g) + CCl_3(g) \quad \text{Slow}$$

$$Cl(g) + CCl_3(g) \xrightarrow[k_4]{} CCl_4(g) \quad\quad \text{Fast}$$

   a. What is the overall reaction?
   b. Identify the intermediates in the mechanism.
   c. What is the predicted rate law?

72. Consider the following two-step mechanism for a reaction:

$$NO_2(g) + Cl_2(g) \xrightarrow[k_1]{} ClNO_2(g) + Cl(g) \quad \text{Slow}$$

$$NO_2(g) + Cl(g) \xrightarrow[k_2]{} ClNO_2(g) \quad\quad \text{Fast}$$

   a. What is the overall reaction?
   b. Identify the intermediates in the mechanism.
   c. What is the predicted rate law?

Catalysis

73. Many heterogeneous catalysts are deposited on high surface-area supports. Why is a large surface area important in heterogeneous catalysis?

74. Suppose that the reaction $A \longrightarrow$ products is exothermic and has an activation barrier of 75 kJ/mol. Sketch an energy diagram showing the energy of the reaction as a function of the progress of the reaction. Draw a second energy curve showing the effect of a catalyst.

75. Suppose that a catalyst lowers the activation barrier of a reaction from 125 kJ/mol to 55 kJ/mol. By what factor would you expect the reaction rate to increase at 25 °C? (Assume that the frequency factors for the catalyzed and uncatalyzed reactions are identical.)

76. The activation barrier for the hydrolysis of sucrose into glucose and fructose is 108 kJ/mol. If an enzyme increases the rate of the hydrolysis reaction by a factor of 1 million, how much lower does the activation barrier have to be when sucrose is in the active site of the enzyme? (Assume that the frequency factors for the catalyzed and uncatalyzed reactions are identical and a termperature of 25 °C.)

Cumulative Problems

77. The data below were collected for the following reaction at 500 °C:

$$CH_3CN(g) \longrightarrow CH_3NC(g)$$

Time (h)	[CH$_3$CN] (M)
0.0	1.000
5.0	0.794
10.0	0.631
15.0	0.501
20.0	0.398
25.0	0.316

  a. Determine the order of the reaction and the value of the rate constant at this temperature.

  b. What is the half-life for this reaction (at the initial concentration)?

  c. How long will it take for 90% of the CH$_3$NC to convert to CH$_3$CN?

78. The data below were collected for the following reaction at a certain temperature:

$$X_2Y \longrightarrow 2X + Y$$

Time (h)	[X$_2$Y] (M)
0.0	0.100
1.0	0.0856
2.0	0.0748
3.0	0.0664
4.0	0.0598
5.0	0.0543

  a. Determine the order of the reaction and the value of the rate constant at this temperature.

  b. What is the half-life for this reaction (at the initial concentration)?

  c. What is the concentration of X after 10.0. hours?

79. Consider the following reaction:

$$A + B + C \longrightarrow D$$

The rate law for this reaction is as follows:

$$\text{Rate} = k\frac{[A][C]^2}{[B]^{1/2}}$$

Suppose the rate of the reaction at certain initial concentrations of A, B, and C is 0.0115 M/s. What is the rate of the reaction if the concentrations of A and C are doubled and the concentration of B is tripled?

80. Consider the following reaction:

$$2 O_3(g) \longrightarrow 3 O_2(g)$$

The rate law for this reaction is as follows:

$$\text{Rate} = k\frac{[O_3]^2}{[O_2]}$$

Suppose that a 1.0-L reaction vessel initially contains 1.0 mol of O_3 and 1.0 mol of O_2. What fraction of the O_3 will have reacted when the rate falls to one-half of its initial value?

81. Dinitrogen pentoxide decomposes in the gas phase to form nitrogen dioxide and oxygen gas. The reaction is first order in dinitrogen pentoxide and has a half-life of 2.81 h at 25 °C. If a 1.5-L reaction vessel initially contains 745 torr of N_2O_5 at 25 °C, what partial pressure of O_2 will be present in the vessel after 215 minutes?

82. Cyclopropane (C_3H_6) reacts to form propene (C_3H_6) in the gas phase. The reaction is first order in cyclopropane and has a rate constant of 5.87×10^{-4}/s at 485 °C. If a 2.5-L reaction vessel initially contains 722 torr of cyclopropane at 485 °C, how long will it take for the partial pressure of cylclopropane to drop to below 100. torr?

83. Iodine atoms will combine to form I_2 in liquid hexane solvent with a rate constant of 1.5×10^{10} L/mol·s. The reaction is second order in I. Since the reaction occurs so quickly, the only way to study the reaction is to create iodine atoms almost instantaneously, usually by photochemical decomposition of I_2. Suppose a flash of light creates an initial [I] concentration of 0.0100 M. How long will it take for 95% of the newly created iodine atoms to recombine to form I_2?

84. The hydrolysis of sucrose ($C_{12}H_{22}O_{11}$) into glucose and fructose in acidic water has a rate constant of $1.8 \times 10^{-4}\,s^{-1}$ at 25 °C. Assuming the reaction is first order in sucrose, determine the mass of sucrose that is hydrolyzed when 2.55 L of a 0.150 M sucrose solution is allowed to react for 195 minutes.

85. Consider the following energy diagram showing the energy of a reaction as it progresses:

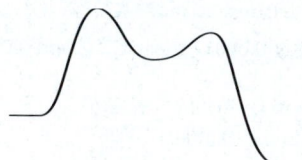

  a. How many elementary steps are involved in this reaction?

  b. Label the reactants, products, and intermediates.

  c. Which step is rate limiting?

  d. Is the overall reaction endothermic or exothermic?

86. Consider the following reaction in which HCl adds across the double bond of ethane:

$$HCl + H_2C{=}CH_2 \longrightarrow H_3C{-}CHCl$$

The following mechanism, with the energy diagram shown below, has been suggested for this reaction:

Step 1 $HCl + H_2C{=}CH_2 \longrightarrow H_3C{=}CH_2^+ + Cl^-$

Step 2 $H_3C{=}CH_2^+ + Cl^- \longrightarrow H_3C{-}CH_2Cl$

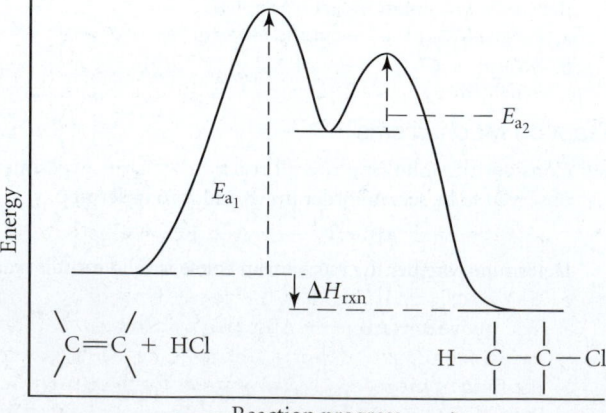

a. Based on the energy diagram, which step is rate limting?
b. What is the expected order of the reaction based on the proposed mechanism?
c. Is the overall reaction exothermic or endothermic?

87. The desorption of a single molecular layer of *n*-butane from a single crystal of aluminum oxide was found to be first order with a rate constant of 0.128/s at 150 K.
   a. What is the half-life of the desorption reaction?
   b. If the surface is initially completely covered with *n*-butane at 150 K, how long will it take for 25% of the molecules to desorb? For 50% to desorb?
   c. If the surface is initially completely covered, what fraction will remain covered after 10 s? After 20 s?

88. The evaporation of a 120-nm film of *n*-pentane from a single crystal of aluminum oxide was found to be zero order with a rate constant of 1.92×10^{13} molecules/cm$^2 \cdot$ s at 120 K.
   a. If the initial surface coverage is 8.9×10^{16} molecules/cm^2, how long will it take for one-half of the film to evaporate?
   b. What fraction of the film will be left after 10 s? Assume the same initial coverage as in part a.

89. The kinetics of the following reaction were studied as a function of temperature. (The reaction is first order in each reactant and second order overall.)

$$C_2H_5Br(aq) + OH^-(aq) \longrightarrow C_2H_5OH(l) + Br^-(aq)$$

Temperature (°C)	k (L/mol $\cdot$ s)
25	8.81×10^{-5}
35	0.000285
45	0.000854
55	0.00239
65	0.00633

   a. Determine the activation energy and frequency factor for the reaction.
   b. Determine the rate constant at 15 °C.
   c. If a reaction mixture is 0.155 M in C$_2$H$_5$Br, and 0.250 M in OH$^-$, what is the initial rate of the reaction at 75 °C?

90. The reaction $2 N_2O_5 \longrightarrow 2 N_2O_4 + O_2$ takes place at around room temperature in solvents such as CCl$_4$. The rate constant at 293 K is found to be 2.35×10^{-4} s^{-1} and at 303 K the rate constant is found to be 9.15×10^{-4} s^{-1}. Calculate the frequency factor for the reaction.

91. The following reaction has an activation energy of zero in the gas phase.

$$CH_3 + CH_3 \longrightarrow C_2H_6$$

   a. Would you expect the rate of this reaction to change very much with temperature?
   b. Can you think of a reason for why the activation energy is zero?
   c. What other types of reactions would you expect to have little or no activation energy?

92. Consider the following two reactions:

$$O + N_2 \longrightarrow NO + N \qquad E_a = 315\,kJ/mol$$
$$Cl + H_2 \longrightarrow HCl + H \qquad E_a = 23\,kJ/mol$$

   a. Can you suggest why the activation barrier for the first reaction is so much higher than that for the second?
   b. The frequency factors for these two reactions are very close to each other in value. Assuming that they are the same, compute the ratio of the reaction rate constants for these two reactions at 25 °C.

93. Anthropologists can estimate the age of a bone or other organic matter by its carbon-14 content. The carbon-14 in a living organism is constant until the organism dies, after which carbon-14 decays with first-order kinetics and a half-life of 5730 years. Suppose a bone from an ancient human contains 19.5% of the C-14 found in living organisms. How old is the bone?

94. Geologists can estimate the age of rocks by their uranium-238 content. The uranium is incorporated in the rock as it hardens and then decays with first-order kinetics and a half-life of 4.5 billion years. A rock is found to contain 83.2% of the amount of uranium-238 that it contained when it was formed. (The amount that the rock contained when it was formed can be deduced from the presence of the decay products of U-238.) How old is the rock?

95. Consider the following gas-phase reaction:

$$H_2(g) + I_2(g) \longrightarrow 2\,HI(g)$$

The reaction was experimentally determined to be first order in H$_2$ and first order in I$_2$. Consider the following proposed mechanisms.

Proposed mechanism I:

$$H_2(g) + I_2(g) \longrightarrow 2\,HI(g) \qquad \text{Single step}$$

Proposed mechanism II:

$$I_2(g) \underset{k_2}{\overset{k_1}{\rightleftharpoons}} 2\,I(g) \qquad \text{Fast}$$

$$H_2(g) + 2\,I(g) \underset{k_3}{\longrightarrow} 2\,HI(g) \qquad \text{Slow}$$

Proposed mechanism III:

$$I_2(g) \underset{k_2}{\overset{k_1}{\rightleftharpoons}} 2\,I(g) \qquad \text{Fast}$$

$$H_2(g) + I(g) \underset{k_4}{\overset{k_3}{\rightleftharpoons}} H_2I(g) \qquad \text{Fast}$$

$$H_2I(g) + I(g) \underset{k_5}{\longrightarrow} 2\,HI(g) \qquad \text{Slow}$$

   a. Show that all three of the proposed mechanisms are valid.
   b. What kind of experimental evidence might lead you to favor mechanisms II or III over mechanism I?

96. Consider the following reaction:

$$2\,NH_3(aq) + OCl^-(aq) \longrightarrow N_2H_4(aq) + H_2O(l) + Cl^-(aq)$$

The following three-step mechanism is proposed:

$$NH_3(aq) + OCl^-(aq) \underset{k_2}{\overset{k_1}{\rightleftharpoons}} NH_2Cl(aq) + OH^-(aq) \quad \text{Fast}$$

$$NH_2Cl(aq) + NH_3(aq) \underset{k_3}{\longrightarrow} N_2H_5^+(aq) + Cl^-(aq) \quad \text{Slow}$$

$$N_2H_5^+(aq) + OH^-(aq) \underset{k_4}{\longrightarrow} N_2H_4(aq) + H_2O(l) \quad \text{Fast}$$

   a. Show that the mechanism sums to the overall reaction.
   b. What is the rate law predicted by this mechanism?

97. A certain substance X decomposes. It is found that 50% of X remains after 100 minutes. How much X remains after 200 minutes if the reaction order with respect to X is (a) zero order, (b) first order, (c) second order?

98. The half-life for radioactive decay (a first-order process) of plutonium-239 is 24,000 years. How many years would it take for one mole of this radioactive material to decay so that just one atom remains?

99. The energy of activation for the decomposition of 2 mol of HI to H$_2$ and I$_2$ in the gas phase is 185 kJ. The heat of formation of HI(g) from H$_2$(g) and I$_2$(g) is -5.65 kJ/mol. Find the energy of activation for the reaction of 1 mol of H$_2$ and 1 mol of I$_2$ in the gas phase.

100. Ethyl chloride vapor decomposes by the first-order reaction

$$C_2H_5Cl \longrightarrow C_2H_4 + HCl$$

The activation energy is 249 kJ/mol and the frequency factor is $1.6 \times 10^{-14}\,s^{-1}$. Find the value of the specific rate constant at 710 K. Find the fraction of the ethyl chloride that decomposes in 15 minutes at this temperature. Find the temperature at which the rate of the reaction would be twice as fast.

Challenge Problems

101. In this chapter we have seen a number of reactions in which a single reactant forms products. For example, consider the following first-order reaction:

$$CH_3NC(g) \longrightarrow CH_3CN(g)$$

However, we also learned that gas-phase reactions occur through collisions.

a. One possible explanation is that two molecules of CH_3NC collide with each other and form two molecules of the product in a single elementary step. If that were the case, what reaction order would you expect?

b. Another possibility is that the reaction occurs through more than one step. For example, a possible mechanism involves one step in which the two CH_3NC molecules collide, resulting in the "activation" of one of them. In a second step, the activated molecule goes on to form the product. Write down this mechanism and determine which step must be rate determining in order for the kinetics of the reaction to be first order. Show explicitly how the mechanism predicts first-order kinetics.

102. The first-order *integrated* rate law for a reaction $A \longrightarrow$ products is derived from the rate law using calculus as follows:

$$\text{Rate} = k[A] \quad \text{(first-order rate law)}$$

$$\text{Rate} = -\frac{d[A]}{dt}$$

$$\frac{d[A]}{dt} = -k[A]$$

The above equation is a first-order, separable differential equation that can be solved by separating the variables and integrating as follows:

$$\frac{d[A]}{[A]} = -k\,dt$$

$$\int_{[A]_0}^{[A]} \frac{d[A]}{[A]} = -\int_0^t k\,dt$$

In the above integral, $[A]_0$ is simply the initial concentration of A. We then evaluate the integral as follows:

$$[\ln[A]]\big|_{[A]_0}^{[A]} = -k\,[t]_0^t$$

$$\ln[A] - \ln[A]_0 = -kt$$

$$\ln[A] = -kt + \ln[A]_0 \quad \text{(integrated rate law)}$$

a. Use a procedure similar to the one above to derive an integrated rate law for a reaction $A \longrightarrow$ products which is one-half order in the concentration of A (that is, Rate $= k[A]^{1/2}$).

b. Use the result from part a to derive an expression for the half-life of a one-half-order reaction.

103. The rate constant for the first-order decomposition of $N_2O_5(g)$ to $NO_2(g)$ and $O_2(g)$ is $7.48 \times 10^{-3}\,s^{-1}$ at a given temperature.

a. Find the length of time required for the total pressure in a system containing N_2O_5 at an initial pressure of 0.100 atm to rise to 0.145 atm.

b. To 0.200 atm.

c. Find the total pressure after 100 s of reaction.

104. Phosgene (Cl_2CO), a poison gas used in World War I, is formed by the reaction of Cl_2 and CO. The proposed mechanism for the reaction is

$$Cl_2 \rightleftharpoons 2\,Cl \quad \text{(fast, equilibrium)}$$
$$Cl + CO \rightleftharpoons ClCO \quad \text{(fast, equilibrium)}$$
$$ClCO + Cl_2 \longrightarrow Cl_2CO + Cl \quad \text{(slow)}$$

What rate law is consistent with this mechanism?

105. The rate of decomposition of $N_2O_3(g)$ to $NO_2(g)$ and NO(g) is followed by measuring $[NO_2]$ at different times. The following data are obtained.

$[NO_2]$ (mol/L)	0	0.193	0.316	0.427	0.784
t (s)	0	884	1610	2460	50,000

The reaction follows a first-order rate law. Calculate the rate constant. Assume that after 50,000 s all the $N_2O_3(g)$ had decomposed.

Conceptual Problems

106. Consider the following reaction:

$$CHCl_3(g) + Cl_2(g) \longrightarrow CCl_4(g) + HCl(g)$$

The reaction is first order in $CHCl_3$ and one-half order in Cl_2. Which of the following reaction mixtures would you expect to have the fastest initial rate?

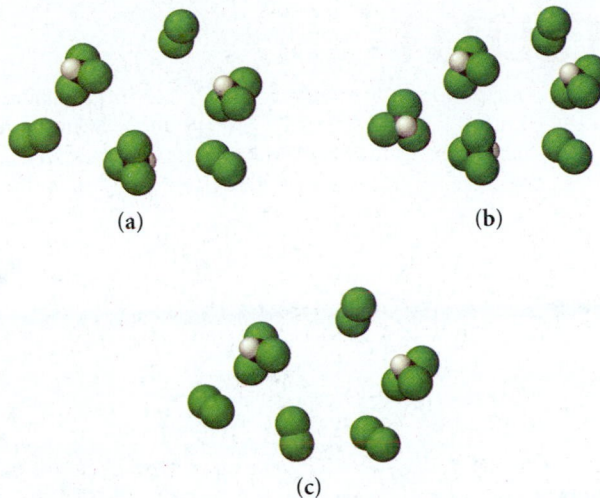

(a)   (b)

(c)

107. The graph below shows the concentration of a reactant as a function of time for two different reactions. One of the reactions is first order and the other is second order. Which of the two reactions is first order? Second order? How would you change each plot to make it linear?

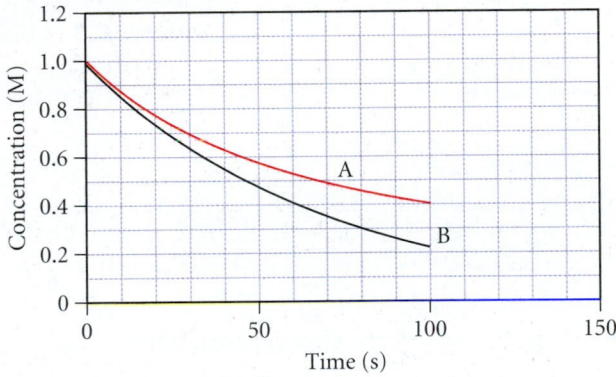

108. A particular reaction, A $\longrightarrow$ products, has a rate that slows down as the reaction proceeds. The half-life of the reaction is found to depend on the initial concentration of A. Determine whether each of the following is likely to be true or false for this reaction.

a. A doubling of the concentration of A doubles the rate of the reaction.

b. A plot of $1/[A]$ versus time is linear.

c. The half-life of the reaction gets longer as the initial concentration of A increases.

d. A plot of the concentration of A versus time has a constant slope.

14

Chemical Equilibrium

14.1 Fetal Hemoglobin and Equilibrium

14.2 The Concept of Dynamic Equilibrium

14.3 The Equilibrium Constant (K)

14.4 Expressing the Equilibrium Constant in Terms of Pressure

14.5 Heterogeneous Equilibria: Reactions Involving Solids and Liquids

14.6 Calculating the Equilibrium Constant from Measured Equilibrium Concentrations

14.7 The Reaction Quotient: Predicting the Direction of Change

14.8 Finding Equilibrium Concentrations

14.9 Le Châtelier's Principle: How a System at Equilibrium Responds to Disturbances

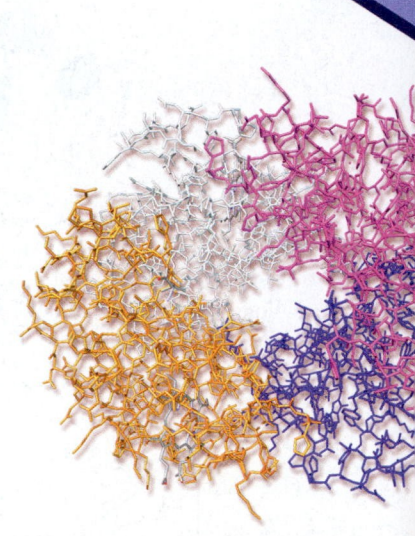

IN THE LAST CHAPTER, we examined *how fast* a chemical reaction occurs. In this chapter we examine *how far* a chemical reaction goes. The *speed* of a chemical reaction is determined by kinetics. The *extent* of a chemical reaction is determined by thermodynamics. In this chapter, we focus on describing and quantifying how far a chemical reaction goes based on an experimentally measurable quantity called the equilibrium constant. A reaction with a large equilibrium constant proceeds nearly to completion. A reaction with a small equilibrium constant barely proceeds at all (the reactants remain as reactants, forming hardly any product). In Chapter 17, we examine the underlying thermodynamics that determines the value of the equilibrium constant. In other words, for now we simply accept the equilibrium constant as an experimentally measurable quantity and learn how to use it to predict and quantify the extent of a reaction. In Chapter 17, we will examine the reasons underlying the magnitude of equilibrium constants.

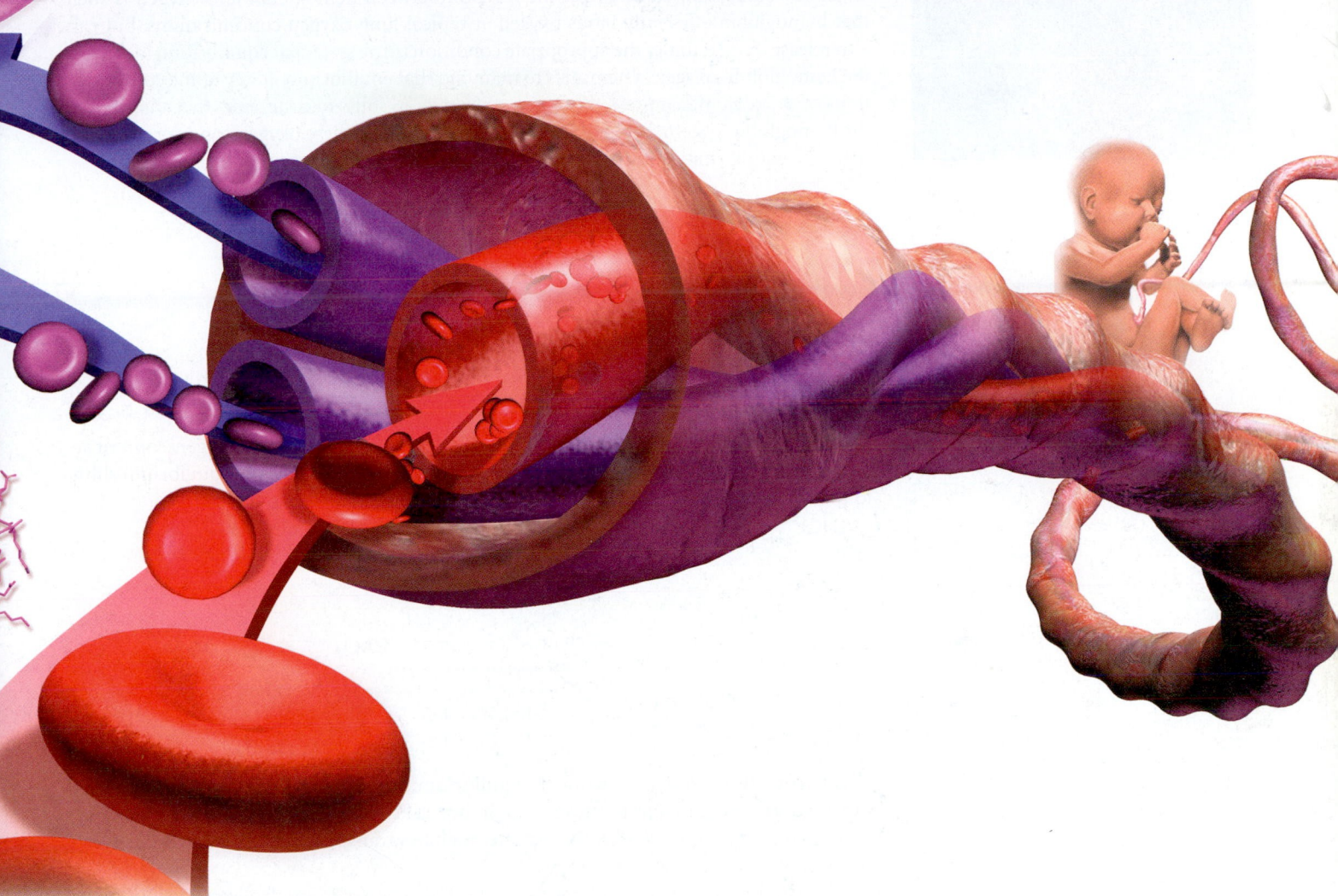

> *Every system in chemical equilibrium, under the influence of a change of any one of the factors of equilibrium, undergoes a transformation ... [that produces a change] ... in the opposite direction of the factor in question.*
>
> —HENRI LE CHÂTELIER (1850–1936)

A developing fetus gets oxygen from the mother's blood because the reaction between oxygen and fetal hemoglobin has a larger equilibrium constant than the reaction between oxygen and maternal hemoglobin.

14.1 Fetal Hemoglobin and Equilibrium

Have you ever wondered how a baby in the womb gets oxygen? Unlike you and me, a fetus does not breathe air. Yet, like you and me, a fetus needs oxygen. Where does that oxygen come from? In adults, air is inhaled into the lungs and diffuses into capillaries, where it comes into contact with the blood. Within red blood cells, a protein called hemoglobin (Hb) reacts with oxygen according to the following chemical equation:

$$Hb + O_2 \rightleftharpoons HbO_2$$

The double arrows in this equation indicate that the reaction can occur in both the forward and reverse directions and can reach chemical *equilibrium*. We have encountered this term in Chapters 11 and 12, and we define it more carefully in the next section. For now, understand that the concentrations of the reactants and products in a reaction at equilibrium are

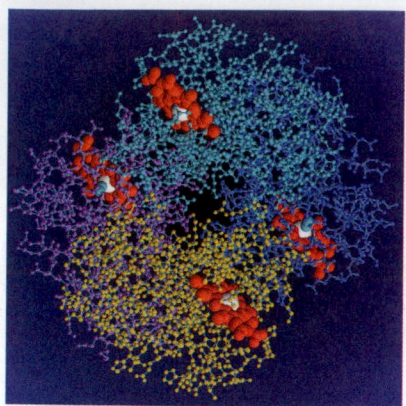

▲ Hemoglobin is the oxygen-carrying protein in red blood cells.

described by the *equilibrium constant, K*. A large value of K means that the reaction lies far to the right at equilibrium—a high concentration of products and a low concentration of reactants. A small value of K means that the reaction lies far to the left at equilibrium—a high concentration of reactants and a low concentration of products. In other words, the value of K is a measure of how far a reaction proceeds—the larger the value of K, the more the reaction proceeds toward the products.

The equilibrium constant for the reaction between hemoglobin and oxygen is such that hemoglobin efficiently binds oxygen at typical lung oxygen concentrations, but can also release oxygen under the appropriate conditions. Any system at equilibrium, including the hemoglobin–oxygen system, acts to maintain that equilibrium. If any of the concentrations of the reactants or products change, the reaction shifts to counteract that change. For the hemoglobin system, as blood flows through the lungs where oxygen concentrations are high, the equilibrium shifts to the right—hemoglobin binds oxygen:

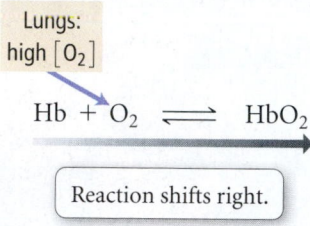

As blood flows out of the lungs and into muscles and organs where oxygen concentrations have been depleted (because muscles and organs use oxygen), the equilibrium shifts to the left—hemoglobin releases oxygen:

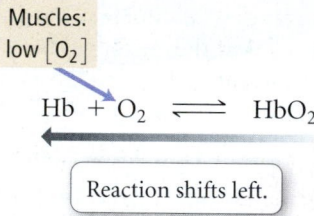

In other words, in order to maintain equilibrium, hemoglobin binds oxygen when the surrounding oxygen concentration is high, but releases oxygen when the surrounding oxygen concentration is low. In this way, hemoglobin transports oxygen from the lungs to all parts of the body that use oxygen.

A fetus has its own circulatory system. The mother's blood never flows into the fetus's body and the fetus cannot get any air in the womb. How, then, does the fetus get oxygen? The answer lies in the properties of fetal hemoglobin (HbF), which is slightly different from adult hemoglobin. Like adult hemoglobin, fetal hemoglobin is in equilibrium with oxygen:

$$HbF + O_2 \rightleftharpoons HbFO_2$$

However, the equilibrium constant for fetal hemoglobin is larger than the equilibrium constant for adult hemoglobin, meaning that the reaction tends to go farther in the direction of the product. Consequently, fetal hemoglobin loads oxygen at a lower oxygen concentration than does adult hemoglobin. In the placenta, fetal blood flows in close proximity to maternal blood, without the two ever mixing. Because of the different equilibrium constants, the maternal hemoglobin unloads oxygen which the fetal hemoglobin then binds and carries into its own circulatory system (Figure 14.1 ▶). Nature has thus evolved a chemical system through which the mother's hemoglobin can in effect *hand off* oxygen to the hemoglobin of the fetus.

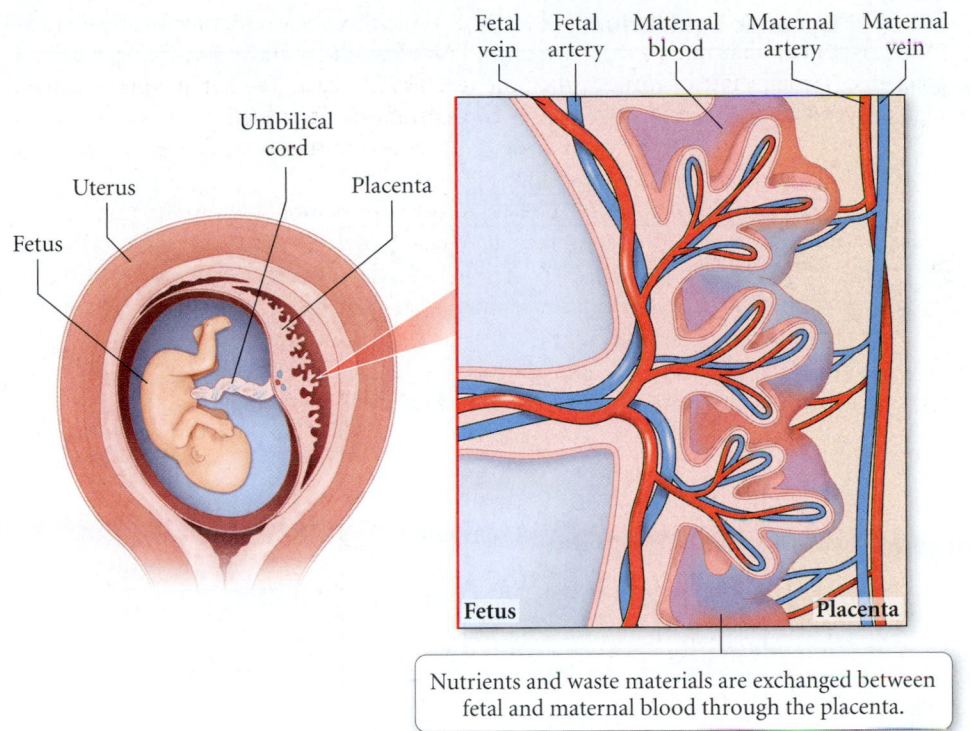

Fetal vein   Fetal artery   Maternal blood   Maternal artery   Maternal vein

Uterus   Umbilical cord   Placenta

Fetus

Fetus   Placenta

Nutrients and waste materials are exchanged between fetal and maternal blood through the placenta.

◀ **FIGURE 14.1 Oxygen Exchange between the Maternal and Fetal Circulation** In the placenta, the blood of the fetus comes into close proximity with that of the mother without the two ever mixing. Because the reaction of fetal hemoglobin with oxygen has a larger equilibrium constant than the reaction of maternal hemoglobin with oxygen, the fetus receives oxygen from the mother's blood.

14.2 The Concept of Dynamic Equilibrium

Recall from the previous chapter that reaction rates generally increase with increasing concentration of the reactants (unless the reaction is zero order) and decrease with decreasing concentration of the reactants. With this in mind, consider the following reaction between hydrogen and iodine:

$$H_2(g) + I_2(g) \rightleftharpoons 2\,HI(g)$$

In this reaction, H_2 and I_2 react to form 2 HI molecules, but the 2 HI molecules can also react to re-form H_2 and I_2. A reaction such as this one—that can proceed in both the forward and reverse directions—is said to be **reversible**. Suppose we begin with only H_2 and I_2 in a container (Figure 14.2a on page 620). What happens initially? H_2 and I_2 begin to react to form HI (Figure 14.2b). However, as H_2 and I_2 react their concentrations decrease, which in turn *decreases the rate of the forward reaction*. At the same time, HI begins to form. As the concentration of HI increases, the reverse reaction begins to occur at a faster and faster rate. Eventually the rate of the reverse reaction (which has been increasing) equals the rate of the forward reaction (which has been decreasing). At that point, **dynamic equilibrium** is reached (Figure 14.2c, d).

Dynamic equilibrium: For a chemical reaction, the condition in which the rate of the forward reaction equals the rate of the reverse reaction.

Dynamic equilibrium is called "dynamic" because the forward and reverse reactions are still occurring; however, they are occurring at the same rate. When dynamic equilibrium is reached, the concentrations of H_2, I_2, and HI no longer change. They remain the same because the reactants and products are formed at the same rate that they are depleted. However, just because the concentrations of reactants and products no longer change at equilibrium *does not imply that the concentrations of reactants and products are equal to one another* at equilibrium. Some reactions reach equilibrium only after most of the reactants have formed products. Others reach equilibrium when only a small fraction of the reactants have formed products. It depends on the reaction.

Nearly all chemical reactions are at least theoretically reversible. In many cases, however, the reversibility is so small that it can be ignored.

Dynamic Equilibrium

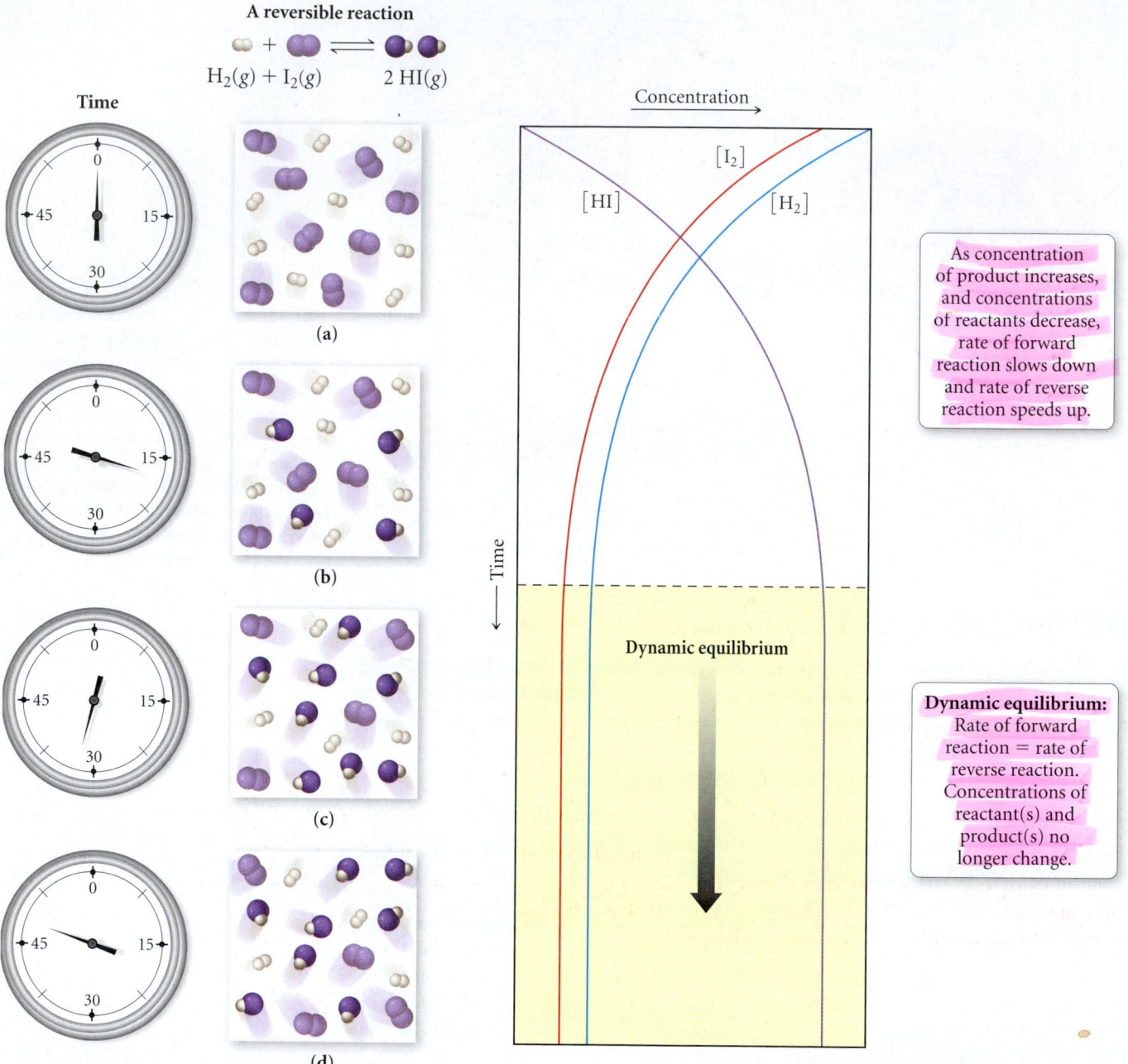

A reversible reaction

$$H_2(g) + I_2(g) \rightleftharpoons 2\,HI(g)$$

Time

(a)

(b)

(c)

(d)

Concentration

$[I_2]$

$[HI]$ $[H_2]$

Time

As concentration of product increases, and concentrations of reactants decrease, rate of forward reaction slows down and rate of reverse reaction speeds up.

Dynamic equilibrium

Dynamic equilibrium: Rate of forward reaction = rate of reverse reaction. Concentrations of reactant(s) and product(s) no longer change.

▲ **FIGURE 14.2 Dynamic Equilibrium**   Equilibrium is reached in a chemical reaction when the concentrations of the reactants and products no longer change. The molecular images on the left depict the progress of the reaction $H_2(g) + I_2(g) \rightleftharpoons 2\,HI(g)$. The graph on the right shows the concentrations of H_2, I_2, and HI as a function of time. When equilibrium is reached, both the forward and reverse reactions continue, but at equal rates, so the concentrations of the reactants and products remain constant.

We can better understand dynamic equilibrium with a simple analogy. Imagine two neighboring countries (A and B) with a closed border between them (Figure 14.3 ▶). Imagine further that Country A is overpopulated and that Country B is underpopulated. One day, the border between the two countries opens, and people immediately begin to leave Country A for Country B.

Country A ⟶ Country B

The population of Country A goes down as the population of Country B goes up. As people leave Country A, however, the *rate* at which they leave slows down, because as Country A becomes less crowded, the pool of potential emigrants gets smaller. On the

Dynamic Equilibrium: An Analogy

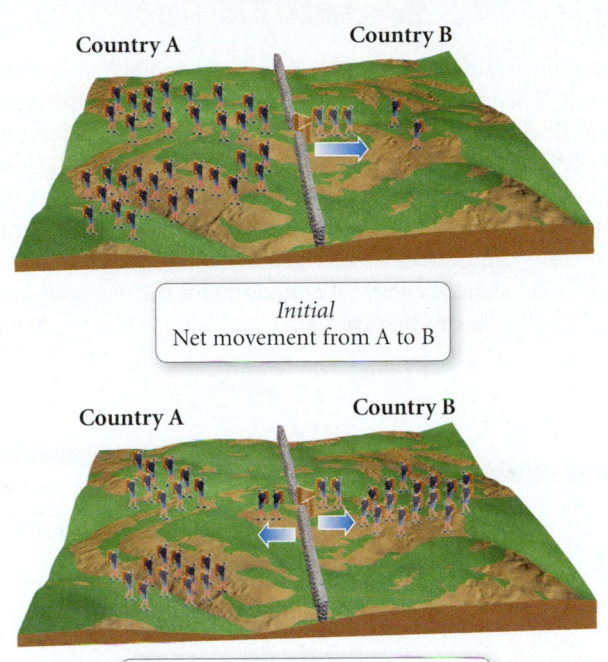

Country A   Country B

Initial
Net movement from A to B

Country A   Country B

Equilibrium
Equal movement in both directions

◀ **FIGURE 14.3 A Population Analogy for Chemical Equilibrium**   Because Country A is initially overpopulated, people migrate from Country A to Country B. As the population of Country A falls and that of Country B rises, the rate of migration from Country A to Country B decreases and the rate of migration from Country B to Country A increases. Eventually the two rates become equal. Equilibrium has been reached.

other hand, as people move into Country B, some decide it was not for them and begin to move back.

$$\text{Country A} \longleftarrow \text{Country B}$$

As Country B gets more crowded, the rate of people moving out of Country B gets faster because the pool of potential emigrants is increasing. Eventually the *rate* of people moving out of Country A (which has been slowing down as people leave) equals the *rate* of people moving out of Country B (which has been increasing as Country B gets more crowded). Dynamic equilibrium has been reached.

$$\text{Country A} \rightleftharpoons \text{Country B}$$

Notice that when the two countries reach dynamic equilibrium, their populations no longer change because the number of people moving out equals the number of people moving in. However, one country—because of its charm or the availability of good jobs or lower taxes, or for whatever other reason—may have a higher population than the other country, even when dynamic equilibrium is reached.

   Similarly, when a chemical reaction reaches dynamic equilibrium, the rate of the forward reaction equals the rate of the reverse reaction, and the relative concentrations of reactants and products become constant. But the concentrations of reactants and products will not necessarily be equal at equilibrium, just as the populations of the two countries are not necessarily equal at equilibrium.

14.3 The Equilibrium Constant (K)

We have just learned that the *concentrations of reactants and products* are not equal at equilibrium—rather, it is the *rates of the forward and reverse reactions* that are equal. But what about the concentrations? What can we know about them? The *equilibrium constant* is a way to quantify the concentrations of the reactants and products at equilibrium.

Consider the following general chemical equation,

$$aA + bB \rightleftharpoons cC + dD$$

where A and B are reactants, C and D are products, and a, b, c, and d are the respective stoichiometric coefficients in the chemical equation. The **equilibrium constant (K)** for the reaction is defined as the ratio—*at equilibrium*—of the concentrations of the products raised to their stoichiometric coefficients divided by the concentrations of the reactants raised to their stoichiometric coefficients.

In this notation, [A] means the molar concentration of A. The equilibrium constant quantifies the relative concentrations of reactants and products *at equilibrium*. The relationship between the balanced chemical equation and the expression of the equilibrium constant is known as the **law of mass action**.

> We distinguish between the equilibrium constant (K) and the kelvin unit of temperature (K) by italicizing the equilibrium constant.

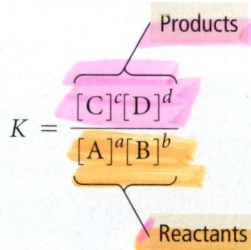

Products

$$K = \frac{[C]^c[D]^d}{[A]^a[B]^b}$$

Reactants

Expressing Equilibrium Constants for Chemical Reactions

To express an equilibrium constant for a chemical reaction, simply examine the balanced chemical equation and apply the law of mass action. For example, suppose we want to express the equilibrium constant for the following reaction:

$$2\,N_2O_5(g) \rightleftharpoons 4\,NO_2(g) + O_2(g)$$

The equilibrium constant is $[NO_2]$ raised to the fourth power multiplied by $[O_2]$ raised to the first power divided by $[N_2O_5]$ raised to the second power:

$$K = \frac{[NO_2]^4[O_2]}{[N_2O_5]^2}$$

Notice that the *coefficients* in the chemical equation become the *exponents* in the expression of the equilibrium constant.

EXAMPLE 14.1 Expressing Equilibrium Constants for Chemical Equations

Express the equilibrium constant for the following chemical equation.

$$CH_3OH(g) \rightleftharpoons CO(g) + 2\,H_2(g)$$

Solution

The equilibrium constant is defined as the concentrations of the products raised to their stoichiometric coefficients divided by the concentrations of the reactants raised to their stoichiometric coefficients.	$K = \dfrac{[CO][H_2]^2}{[CH_3OH]}$

For Practice 14.1

Express the equilibrium constant for the combustion of propane as shown by the following balanced chemical equation.

$$C_3H_8(g) + 5\,O_2(g) \rightleftharpoons 3\,CO_2(g) + 4\,H_2O(g)$$

The Significance of the Equilibrium Constant

Given the definition of an equilibrium constant, what does it mean? What, for example, does a large equilibrium constant ($K \gg 1$) imply about a reaction? It implies that the numerator of the equilibrium constant (which specifies the amount of products at equilibrium) is larger than the denominator (which specifies the amount of reactants at equilibrium). Therefore the forward reaction is favored. For example, consider the following reaction:

$$H_2(g) + Br_2(g) \rightleftharpoons 2\,HBr(g) \qquad K = 1.9 \times 10^{19} \text{ (at 25 °C)}$$

The equilibrium constant is large, meaning that the equilibrium point for the reaction lies far to the right—high concentrations of products, low concentrations of reactants (Figure 14.4 ▼). Remember that the equilibrium constant says nothing about *how fast* a reaction will reach equilibrium, only *how far* the reaction has proceeded once equilibrium is reached. A reaction with a large equilibrium constant may be kinetically very slow, meaning that it will take a long time to reach equilibrium.

$$H_2(g) + Br_2(g) \rightleftharpoons 2\,HBr(g)$$

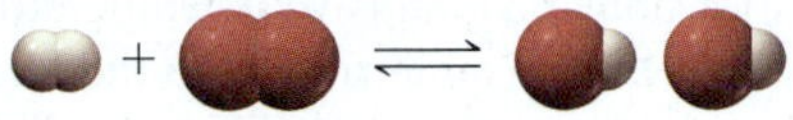

$$K = \frac{[HBr]^2}{[H_2][Br_2]} = \text{large number}$$

◀ **FIGURE 14.4 The Meaning of a Large Equilibrium Constant** If the equilibrium constant for a reaction is large, the equilibrium point of the reaction lies far to the right—the concentration of products is large and the concentration of reactants is small.

Conversely, what does a *small* equilibrium constant ($K \ll 1$) mean? It means that the reverse reaction is favored and that there will be more reactants than products when equilibrium is reached. For example, consider the following reaction:

$$N_2(g) + O_2(g) \rightleftharpoons 2\,NO(g) \qquad K = 4.1 \times 10^{-31} \text{ (at 25 °C)}$$

The equilibrium constant is very small, meaning that the equilibrium point for the reaction lies far to the left—high concentrations of reactants, low concentrations of products (Figure 14.5 ▼). This is fortunate because N_2 and O_2 are the main components of air. If this equilibrium constant were large, much of the N_2 and O_2 in air would react to form NO, a toxic gas.

$$N_2(g) + O_2(g) \rightleftharpoons 2\,NO(g)$$

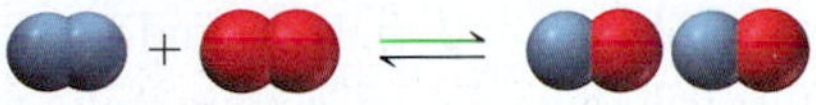

$$K = \frac{[NO]^2}{[N_2][O_2]} = \text{small number}$$

◀ **FIGURE 14.5 The Meaning of a Small Equilibrium Constant** If the equilibrium constant for a reaction is small, the equilibrium point of the reaction lies far to the left—the concentration of products is small and the concentration of reactants is large.

Chemistry and Medicine
Life and Equilibrium

Have you ever tried to define life? If you have, you know that a definition is elusive. How are living things different from nonliving things? You may try to define living things as those things that can move. But of course many living things do not move (most plants, for example), and some nonliving things, such as glaciers and Earth itself, do move. So motion is neither unique to nor definitive of life. You may try to define living things as those things that can reproduce. But again, many living things, such as mules or sterile humans, cannot reproduce; yet they are alive. In addition, some nonliving things, such as crystals, for example, reproduce (in some sense). So what is unique about living things?

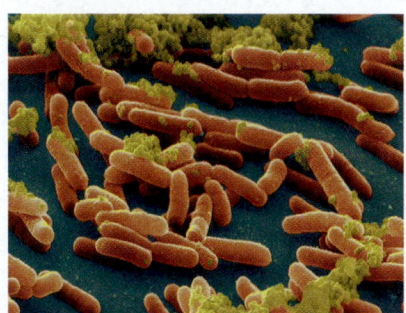

▲ What makes these cells alive?

One definition of life uses the concept of equilibrium—living things *are not* in equilibrium with their surroundings. Our body temperature, for example, is not the same as the temperature of our surroundings. If we jump into a swimming pool, the acidity of our blood does not become the same as the acidity of the surrounding water. Living things, even the simplest ones, maintain some measure of *disequilibrium* with their environment.

We must add one more concept, however, to complete our definition of life with respect to equilibrium. A cup of hot water is in disequilibrium with its environment with respect to temperature, yet it is not alive. The cup of hot water has no control over its disequilibrium, however, and will slowly come to equilibrium with its environment. In contrast, living things—as long as they are alive—maintain and *control* their disequilibrium. Your body temperature, for example, is not *only* in disequilibrium with your surroundings—it is in *controlled disequilibrium*. Your body maintains your temperature within a specific range that is not in equilibrium with the surrounding temperature.

So, one criterion for life is that living things are in *controlled disequilibrium* with their environment. Maintaining that disequilibrium is a main activity of living organisms, requiring energy obtained from their environment. Plants derive that energy from sunlight; animals eat plants (or other animals that have eaten plants), and thus they too ultimately derive their energy from the sun. A living thing comes into equilibrium with its surroundings only after it dies.

To summarize:

➤ $K \ll 1$ Reverse reaction is favored; forward reaction does not proceed very far.

➤ $K \approx 1$ Neither direction is favored; forward reaction proceeds about halfway.

➤ $K \gg 1$ Forward reaction is favored; forward reaction proceeds essentially to completion.

Conceptual Connection 14.1 Equilibrium Constants

The equilibrium constant for the reaction $A(g) \rightleftharpoons B(g)$ is 10. A reaction mixture initially contains 11 mol of A and 0 mol of B in a fixed volume of 1 L. When equilibrium is reached, which of the following is true?

(a) The reaction mixture will contain 10 mol of A and 1 mol of B.

(b) The reaction mixture will contain 1 mol of A and 10 mol of B.

(c) The reaction mixture will contain equal amounts of A and B.

Answer: (b) The reaction mixture will contain 1 mol of A and 10 mol of B so that $[B]/[A] = 10$.

Relationships between the Equilibrium Constant and the Chemical Equation

If a chemical equation is modified in some way, then the equilibrium constant for the equation must be changed to reflect the modification. The following modifications are common.

1. **If you reverse the equation, invert the equilibrium constant.** Consider the following chemical equation:

$$A + 2B \rightleftharpoons 3C$$

The expression for the equilibrium constant of this reaction is given by

$$K_{forward} = \frac{[C]^3}{[A][B]^2}$$

If we reverse the equation to the following:

$$3C \rightleftharpoons A + 2B$$

then, according to the law of mass action, the expression for the equilibrium constant becomes

$$K_{reverse} = \frac{[A][B]^2}{[C]^3} = \frac{1}{K_{forward}}$$

2. **If you multiply the coefficients in the equation by a factor, raise the equilibrium constant to the same factor.** Consider again the following chemical equation and corresponding expression for the equilibrium constant:

$$A + 2B \rightleftharpoons 3C \qquad K = \frac{[C]^3}{[A][B]^2}$$

If we multiply the equation by n, we get the following:

$$nA + 2nB \rightleftharpoons 3nC$$

Applying the law of mass action, the expression for the equilibrium constant becomes

$$K' = \frac{[C]^{3n}}{[A]^n[B]^{2n}} = \left(\frac{[C]^3}{[A][B]^2} \right)^n = K^n$$

If n is a fractional quantity, raise K to the same fractional quantity.

3. **If you add two or more individual chemical equations to obtain an overall equation, multiply the corresponding equilibrium constants by each other to obtain the overall equilibrium constant.** For example, consider the following two chemical equations and their corresponding equilibrium constant expressions:

$$A \rightleftharpoons 2B \qquad K_1 = \frac{[B]^2}{[A]}$$

$$2B \rightleftharpoons 3C \qquad K_2 = \frac{[C]^3}{[B]^2}$$

The two equations sum as follows:

$$A \rightleftharpoons \cancel{2B}$$
$$\cancel{2B} \rightleftharpoons 3C$$
$$A \rightleftharpoons 3C$$

According to the law of mass action, the equilibrium constant for this overall equation is then:

$$K_{overall} = \frac{[C]^3}{[A]}$$

Notice that $K_{overall}$ is simply the product of K_1 and K_2:

$$K_{overall} = K_1 \times K_2$$
$$= \frac{\cancel{[B]^2}}{[A]} \times \frac{[C]^3}{\cancel{[B]^2}}$$
$$= \frac{[C]^3}{[A]}$$

EXAMPLE 14.2 Manipulating the Equilibrium Constant to Reflect Changes in the Chemical Equation

Consider the following chemical equation and equilibrium constant for the synthesis of ammonia at 25 °C:

$$N_2(g) + 3H_2(g) \rightleftharpoons 2NH_3(g) \qquad K = 3.7 \times 10^8$$

Compute the equilibrium constant for the following reaction at 25 °C:

$$NH_3(g) \rightleftharpoons \tfrac{1}{2}N_2(g) + \tfrac{3}{2}H_2(g) \qquad K' = ?$$

Solution

We want to manipulate the given reaction and value of K to obtain the desired reaction and value of K. We can see that the given reaction is the reverse of the desired reaction, and its coefficients are twice those of the desired reaction.

Begin by reversing the given reaction and taking the inverse of the value of K.	$N_2(g) + 3H_2(g) \rightleftharpoons 2NH_3(g)$ $K = 3.7 \times 10^8$ $2NH_3(g) \rightleftharpoons N_2(g) + 3H_2(g)$ $K_{rev} = \dfrac{1}{3.7 \times 10^8}$
Next, multiply the reaction by $\tfrac{1}{2}$ and raise the equilibrium constant to the $\tfrac{1}{2}$ power.	$NH_3(g) \rightleftharpoons \tfrac{1}{2}N_2(g) + \tfrac{3}{2}H_2(g)$ $K' = K_{reverse}^{1/2} = \left(\dfrac{1}{3.7 \times 10^8}\right)^{1/2}$
Compute the value of K'.	$K' = 5.2 \times 10^{-5}$

For Practice 14.2

Consider the following chemical equation and equilibrium constant at 25°C:

$$2COF_2(g) \rightleftharpoons CO_2(g) + CF_4(g) \qquad K = 2.2 \times 10^6$$

Compute the equilibrium constant for the following reaction at 25 °C:

$$2CO_2(g) + 2CF_4(g) \rightleftharpoons 4COF_2(g) \qquad K' = ?$$

For More Practice 14.2

Predict the equilibrium constant for the first reaction given the equilibrium constants for the second and third reactions:

$$CO_2(g) + 3H_2(g) \rightleftharpoons CH_3OH(g) + H_2O(g) \qquad K_1 = ?$$
$$CO(g) + H_2O(g) \rightleftharpoons CO_2(g) + H_2(g) \qquad K_2 = 1.0 \times 10^5$$
$$CO(g) + 2H_2(g) \rightleftharpoons CH_3OH(g) \qquad K_3 = 1.4 \times 10^7$$

14.4 Expressing the Equilibrium Constant in Terms of Pressure

So far, we have expressed the equilibrium constant only in terms of the *concentrations* of the reactants and products. However, for gaseous reactions, the partial pressure of a particular gas is proportional to its concentration. Therefore, we can also express the equilibrium constant in terms of the *partial pressures* of the reactants and products. For example, consider the following gaseous reaction:

$$2SO_3(g) \rightleftharpoons 2SO_2(g) + O_2(g)$$

From this point on, we designate K_c as the equilibrium constant with respect to concentration in molarity. For the above reaction, K_c is easily expressed using the law of mass action:

$$K_c = \frac{[SO_2]^2[O_2]}{[SO_3]^2}$$

We now designate K_p as the equilibrium constant with respect to partial pressures in atmospheres. *The expression for K_p takes the form of the expression for K_c, except that we use the partial pressure of each gas in place of its concentration.* For the SO_3 reaction above, we write K_p as follows:

$$K_p = \frac{(P_{SO_2})^2 P_{O_2}}{(P_{SO_3})^2}$$

where P_A is simply the partial pressure of gas A in units of atmospheres.

Since the partial pressure of a gas in atmospheres is not the same as its concentration in molarity, the value of K_p for a reaction is not necessarily equal to the value of K_c. However, as long as the gases are behaving ideally, we can derive a relationship between the two constants. The concentration of an ideal gas A is simply the number of moles of A (n_A) divided by its volume (V) in liters:

$$[A] = \frac{n_A}{V} \text{ Moles}$$

From the ideal gas law, we can relate the quantity n_A/V to the partial pressure of A as follows:

$$P_A V = n_A RT$$

$$P_A = \frac{n_A}{V} RT$$

Since $[A] = n_A/V$, we can write

$$P_A = [A]RT \quad \text{or} \quad [A] = \frac{P_A}{RT} \qquad [14.1]$$

Now consider the following general equilibrium chemical equation:

$$aA + bB \rightleftharpoons cC + dD$$

According to the law of mass action, we write K_c as follows:

$$K_c = \frac{[C]^c[D]^d}{[A]^a[B]^b}$$

Substituting $[X] = P_X/RT$ for each concentration term, we get the following:

$$K_c = \frac{\left(\dfrac{P_C}{RT}\right)^c\left(\dfrac{P_D}{RT}\right)^d}{\left(\dfrac{P_A}{RT}\right)^a\left(\dfrac{P_B}{RT}\right)^b} = \frac{P_C^c P_D^d\left(\dfrac{1}{RT}\right)^{c+d}}{P_A^a P_B^b\left(\dfrac{1}{RT}\right)^{a+b}} = \frac{P_C^c P_D^d}{P_A^a P_B^b}\left(\dfrac{1}{RT}\right)^{c+d-(a+b)}$$

$$= K_p\left(\frac{1}{RT}\right)^{c+d-(a+b)}$$

Rearranging,

$$K_p = K_c(RT)^{c+d-(a+b)}$$

Finally, if we let $\Delta n = c + d - (a + b)$, which is simply the sum of the stoichiometric coefficients of the gaseous products minus the sum of the stoichiometric coefficients of the gaseous reactants, we get the following general result:

$$K_p = K_c(RT)^{\Delta n} \qquad [14.2]$$

Notice that if the total number of moles of gas is the same after the reaction as before, then $\Delta n = 0$, and K_p is equal to K_c.

EXAMPLE 14.3 Relating K_p and K_c

Nitrogen monoxide, a pollutant in automobile exhaust, is oxidized to nitrogen dioxide in the atmosphere according to the following equation:

$$2\,NO(g) + O_2(g) \rightleftharpoons 2\,NO_2(g) \qquad K_p = 2.2 \times 10^{12} \text{ at } 25\,°C$$

Find K_c for this reaction.

Sort You are given K_p for the reaction and asked to find K_c.	**Given** $K_p = 2.2 \times 10^{12}$ **Find** K_c
Strategize Use Equation 14.2 to relate K_p and K_c.	**Equation** $K_p = K_c(RT)^{\Delta n}$
Solve Solve the equation for K_c. Compute Δn. Substitute the required quantities to compute K_c. The temperature must be in kelvins. The units are dropped when reporting K_c as described in the section that follows.	**Solution** $K_c = \dfrac{K_p}{(RT)^{\Delta n}}$ $\Delta n = 2 - 3 = -1$ $K_c = \dfrac{2.2 \times 10^{12}}{\left(0.08206\dfrac{L \cdot atm}{mol \cdot K} \times 298\,K\right)^{-1}}$ $= 5.4 \times 10^{13}$

Check The easiest way to check this answer is to substitute it back into Equation 14.2 and confirm that you get the original value for K_p.

$$K_p = K_c(RT)^{\Delta n}$$

$$= 5.4 \times 10^{13}\left(0.08206\frac{L \cdot atm}{mol \cdot K} \times 298\,K\right)^{-1}$$

$$= 2.2 \times 10^{12}$$

For Practice 14.3

Consider the following reaction and corresponding value of K_c:

$$H_2(g) + I_2(g) \rightleftharpoons 2\,HI(g) \qquad K_c = 6.2 \times 10^2 \text{ at } 25\,°C$$

What is the value of K_p at this temperature?

Units of K

Throughout this book, we express concentrations and partial pressures within the equilibrium constant in units of molarity and atmospheres, respectively. When expressing the value of the equilibrium constant, however, we have not included the units. Formally, the values of concentration or partial pressure that we substitute into the equilibrium constant expression are ratios of the concentration or pressure to a reference concentration (exactly 1 M) or a reference pressure (exactly 1 atm). For example, within the equilibrium constant expression, a pressure of 1.5 atm becomes

$$\frac{1.5 \text{ atm}}{1 \text{ atm}} = 1.5$$

Similarly, a concentration of 1.5 M becomes

$$\frac{1.5 \text{ M}}{1 \text{ M}} = 1.5$$

As long as concentration units are expressed in molarity for K_c and pressure units are expressed in atmospheres for K_p, we can skip this formality and simply enter the quantites directly into the equilibrium expression, dropping their corresponding units.

14.5 Heterogeneous Equilibria: Reactions Involving Solids and Liquids

Many chemical reactions involve pure solids or pure liquids as reactants or products. Consider, for example, the following reaction:

$$2\,CO(g) \rightleftharpoons CO_2(g) + C(s)$$

We might expect the expression for the equilibrium constant to be

$$K_c = \frac{[CO_2][C]}{[CO]^2} \quad \text{(incorrect)}$$

However, since carbon is a solid, its concentration is constant—it does not change. The concentration of a solid does not change because a solid does not expand to fill its container. Its concentration, therefore, depends only on its density, which is constant as long as *some* solid is present (Figure 14.6 ▼). Consequently, pure solids—those reactants or products labeled in the chemical equation with an (s)—are not included in the equilibrium expression. The correct equilibrium expression is therefore

$$K_c = \frac{[CO_2]}{[CO]^2} \quad \text{(correct)}$$

Similarly, the concentration of a pure liquid does not change. So, pure liquids—those reactants or products labeled in the chemical equation with an (ℓ)—are also excluded from the equilibrium expression. For example, what is the equilibrium expression for the following reaction?

$$CO_2(g) + H_2O(\ell) \rightleftharpoons H^+(aq) + HCO_3^-(aq)$$

Since $H_2O(\ell)$ is pure liquid, it is omitted from the equilibrium expression:

$$K_c = \frac{[H^+][HCO_3^-]}{[CO_2]}$$

A Heterogeneous Equilibrium

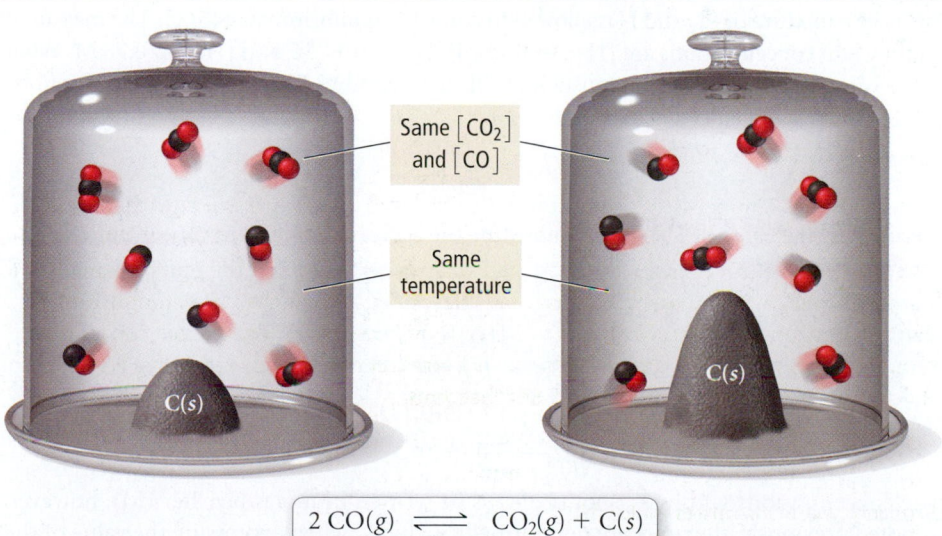

Same $[CO_2]$ and $[CO]$

Same temperature

C(s)

C(s)

$$2\,CO(g) \rightleftharpoons CO_2(g) + C(s)$$

▲ FIGURE 14.6 **Heterogeneous Equilibrium** The concentration of solid carbon (the number of atoms per unit volume) is constant as long as some solid carbon is present. The same is true for pure liquids. Thus, the concentrations of solids and pure liquids are not included in equilibrium constant expressions.

EXAMPLE 14.4 Writing Equilibrium Expressions for Reactions Involving a Solid or a Liquid

Write an expression for the equilibrium constant (K_c) for the following chemical equation.

$$CaCO_3(s) \rightleftharpoons CaO(s) + CO_2(g)$$

Solution

| Since $CaCO_3(s)$ and $CaO(s)$ are both solids, they are omitted from the equilibrium expression. | $K_c = [CO_2]$ |

For Practice 14.4

Write an equilibrium expression (K_c) for the following chemical equation.

$$4 HCl(g) + O_2(g) \rightleftharpoons 2 H_2O(\ell) + 2 Cl_2(g)$$

 Conceptual Connection 14.2 Heterogeneous Equilibria, K_p, and K_c

For which of the following reactions will $K_p = K_c$?

(a) $2 Na_2O_2(s) + 2 CO_2(g) \rightleftharpoons 2 Na_2CO_3(s) + O_2(g)$

(b) $NiO(s) + CO(g) \rightleftharpoons Ni(s) + CO_2(g)$

(c) $NH_4NO_3(s) \rightleftharpoons N_2O(g) + 2 H_2O(g)$

Answer: (b) Since Δn for gaseous reactants and products is zero, K_p will equal K_c.

14.6 Calculating the Equilibrium Constant from Measured Equilibrium Concentrations

The most direct way to obtain an experimental value for the equilibrium constant of a reaction is to measure the concentrations of the reactants and products in a reaction mixture at equilibrium. For example, consider the following reaction:

$$H_2(g) + I_2(g) \rightleftharpoons 2 HI(g)$$

> Since equilibrium constants depend on temperature, many equilibrium problems will state the temperature even though it has no formal part in the calculation.

Suppose a mixture of H_2 and I_2 is allowed to come to equilibrium at 445 °C. The measured equilibrium concentrations are $[H_2] = 0.11$ M, $[I_2] = 0.11$ M, and $[HI] = 0.78$ M. What is the value of the equilibrium constant at this temperature? The expression for K_c can be written from the balanced equation.

$$K_c = \frac{[HI]^2}{[H_2][I_2]}$$

To calculate the value of K_c, simply substitute the correct equilibrium concentrations into the expression for K_c.

$$K_c = \frac{[HI]^2}{[H_2][I_2]}$$

$$= \frac{0.78^2}{(0.11)(0.11)}$$

$$= 5.0 \times 10^1$$

The concentrations within K_c should always be written in moles per liter (M); however, as noted previously, the units are not normally included when expressing the value of the equilibrium constant, so that K_c is unitless.

For any reaction, the equilibrium *concentrations* of the reactants and products will depend on the initial concentrations (and in general vary from one set of initial concentrations to another). However, the equilibrium *constant* will always be the same at a given temperature, regardless of the initial concentrations. For example, Table 14.1 shows several different equilibrium concentrations of H_2, I_2, and HI, each from a different set of initial concentrations. Notice that the equilibrium constant is always the same, regardless of the initial concentrations. Notice also that, whether you start with only reactants or only products, the reaction reaches equilibrium concentrations in which the equilibrium constant is the same. In other words, no matter what the initial concentrations are, the reaction will always go in a direction so that the equilibrium concentrations—when substituted into the equilibrium expression—give the same constant, K.

TABLE 14.1 Initial and Equilibrium Concentrations for the Reaction
$H_2(g) + I_2(g) \rightleftharpoons 2\,HI(g)$ at 445 °C

Initial Concentrations			Equilibrium Concentrations			Equilibrium Constant
$[H_2]$	$[I_2]$	$[HI]$	$[H_2]$	$[I_2]$	$[HI]$	$K_c = \dfrac{[HI]^2}{[H_2][I_2]}$
0.50	0.50	0.0	0.11	0.11	0.78	$\dfrac{0.78^2}{(0.11)(0.11)} = 50$
0.0	0.0	0.50	0.055	0.055	0.39	$\dfrac{0.39^2}{(0.055)(0.055)} = 50$
0.50	0.50	0.50	0.165	0.165	1.17	$\dfrac{1.17^2}{(0.165)(0.165)} = 50$
1.0	0.50	0.0	0.53	0.033	0.934	$\dfrac{0.934^2}{(0.53)(0.033)} = 50$
0.50	1.0	0.0	0.033	0.53	0.934	$\dfrac{0.934^2}{(0.033)(0.53)} = 50$

In each of the entries in Table 14.1, and in the paragraph above, we calculated the equilibrium constant from values of the equilibrium concentrations of all the reactants and products. In most cases, however, we need only know the initial concentrations of the reactant(s) and the equilibrium concentration of any *one* reactant or product. The other equilibrium concentrations can be deduced from the stoichiometry of the reaction. For example, consider the following simple reaction:

$$A(g) \rightleftharpoons 2\,B(g)$$

Suppose that we have a reaction mixture in which the initial concentration of A is 1.00 M and the initial concentration of B is 0.00 M. When equilibrium is reached, the concentration of A is 0.75 M. Since [A] has changed by −0.25 M, we can deduce (based on the stoichiometry) that [B] must have changed by 2 × (+ 0.25 M) or + 0.50 M. We can summarize the initial conditions, the changes, and the equilibrium conditions in the following table:

	[A]	[B]
Initial	1.00	0.00
Change	−0.25	+0.50
Equilibrium	0.75	0.50

This type of table is often referred to as an ICE table (I = initial, C = change, E = equilibrium). To compute the equilibrium constant, we use the balanced equation to write an expression for the equilibrium constant and then substitute the equilibrium concentrations from the ICE table.

$$K = \frac{[B]^2}{[A]} = \frac{(0.50)^2}{(0.75)} = 0.33$$

In the examples that follow, we show the general procedure for solving these kinds of equilibrium problems in the left column and work two examples exemplifying the procedure in the center and right columns.

Procedure for Finding Equilibrium Constants from Experimental Concentration Measurements

To solve these types of problems, follow the procedure outlined below.

EXAMPLE 14.5 Finding Equilibrium Constants from Experimental Concentration Measurements

Consider the following reaction:

$$CO(g) + 2\,H_2(g) \rightleftharpoons CH_3OH(g)$$

A reaction mixture at 780 °C initially contains [CO] = 0.500 M and [H$_2$] = 1.00 M. At equilibrium, the CO concentration is found to be 0.15 M. Find the equilibrium constant for the reaction.

EXAMPLE 14.6 Finding Equilibrium Constants from Experimental Concentration Measurements

Consider the following reaction:

$$2\,CH_4(g) \rightleftharpoons C_2H_2(g) + 3\,H_2(g)$$

A reaction mixture at 1700 °C initially contains [CH$_4$] = 0.115 M. At equilibrium, the mixture contains [C$_2$H$_2$] = 0.035 M. What is the value of the equilibrium constant?

1. Using the balanced equation as a guide, prepare an ICE table showing the known initial concentrations and equilibrium concentrations of the reactants and products. Leave space in the middle of the table for determining the changes in concentration that occur during the reaction.

$$CO(g) + 2\,H_2(g) \rightleftharpoons CH_3OH(g)$$

	[CO]	[H$_2$]	[CH$_3$OH]
Initial	0.500	1.00	0.00
Change			
Equil	0.15		

$$2\,CH_4(g) \rightleftharpoons C_2H_2(g) + 3\,H_2(g)$$

	[CH$_4$]	[C$_2$H$_2$]	[H$_2$]
Initial	0.115	0.00	0.00
Change			
Equil		0.035	

2. For the reactant or product whose concentration is known both initially and at equilibrium, calculate the change in concentration that must have occurred.

$$CO(g) + 2\,H_2(g) \rightleftharpoons CH_3OH(g)$$

	[CO]	[H$_2$]	[CH$_3$OH]
Initial	0.500	1.00	0.00
Change	−0.35		
Equil	0.15		

$$2\,CH_4(g) \rightleftharpoons C_2H_2(g) + 3\,H_2(g)$$

	[CH$_4$]	[C$_2$H$_2$]	[H$_2$]
Initial	0.115	0.00	0.00
Change		+0.035	
Equil		0.035	

3. Use the change calculated in step 2 and the stoichiometric relationships from the balanced chemical equation to determine the changes in concentration of all other reactants and products. Since reactants are consumed during the reaction, the changes in their concentrations are negative. Since products are formed, the changes in their concentrations are positive.

$$CO(g) + 2\,H_2(g) \rightleftharpoons CH_3OH(g)$$

	[CO]	[H$_2$]	[CH$_3$OH]
Initial	0.500	1.00	0.00
Change	−0.35	−0.70 $2(-.15)$	+0.35
Equil	0.15		

$$2\,CH_4(g) \rightleftharpoons C_2H_2(g) + 3\,H_2(g)$$

	[CH$_4$]	[C$_2$H$_2$]	[H$_2$]
Initial	0.115	0.00	0.00
Change	−0.070	+0.035	+0.105 $3(.035)$
Equil		0.035	

4. Sum each column for each reactant and product to determine the equilibrium concentrations.

	[CO]	[H$_2$]	[CH$_3$OH]
Initial	0.500	1.00	0.00
Change	−0.35	−0.70	+0.35
Equil	0.15	0.30	0.35

	[CH$_4$]	[C$_2$H$_2$]	[H$_2$]
Initial	0.115	0.00	0.00
Change	−0.070	+0.035	+0.105
Equil	0.045	0.035	0.105

5. Use the balanced equation to write an expression for the equilibrium constant and substitute the equilibrium concentrations to compute K.

$$K_c = \frac{[CH_3OH]}{[CO][H_2]^2}$$

$$= \frac{0.35}{(0.15)(0.30)^2}$$

$$= 26$$

$$K_c = \frac{[C_2H_2][H_2]^3}{[CH_4]^2}$$

$$= \frac{(0.035)(0.105)^3}{0.045^2}$$

$$= 0.020$$

For Practice 14.5

The above reaction between CO and H_2 was carried out at a different temperature with initial concentrations of $[CO] = 0.27$ M and $[H_2] = 0.49$ M. At equilibrium, the concentration of CH_3OH was 0.11 M. Find the equilibrium constant at this temperature.

For Practice 14.6

The above reaction of CH_4 was carried out at a different temperature with an initial concentration of $[CH_4] = 0.087$ M. At equilibrium, the concentration of H_2 was 0.012 M. Find the equilibrium constant at this temperature.

14.7 The Reaction Quotient: Predicting the Direction of Change

When the reactants of a chemical reaction are mixed, they generally react to form products—we say that the reaction proceeds to the right (toward the products). The amount of products formed when equilibrium is reached depends on the magnitude of the equilibrium constant, as we have seen. However, what if a reaction mixture not at equilibrium contains both reactants *and products*? Can we predict the direction of change for such a mixture?

In order to compare the progress of a reaction to the equilibrium state of the reaction, we use a quantity called the *reaction quotient.* The definition of the reaction quotient takes the same form as the definition of the equilibrium constant, except that the reaction need not be at equilibrium. So, for the general reaction

$$aA + bB \rightleftharpoons cC + dD$$

we define the **reaction quotient** (Q_c) as the ratio—at any point in the reaction—of the concentrations of the products raised to their stoichiometric coefficients divided by the concentrations of the reactants raised to their stoichiometric coefficients. For gases with amounts measured in atmospheres, the reaction quotient simply uses the partial pressures in place of concentrations and is called Q_P.

$$Q_C = \frac{[C]^c[D]^d}{[A]^a[B]^b} \qquad Q_P = \frac{P_C^c P_D^d}{P_A^a P_B^b}$$

The difference between the reaction quotient and the equilibrium constant is that, at a given temperature, the equilibrium constant has only one value and it specifies the relative amounts of reactants and products *at equilibrium*. The reaction quotient, however, depends on the current state of the reaction and will have many different values as the reaction proceeds. For example, in a reaction mixture containing only reactants, the reaction quotient is zero ($Q_c = 0$):

$$Q_c = \frac{[0]^c[0]^d}{[A]^a[B]^b} = 0$$

In a reaction mixture containing only products, the reaction quotient is infinite ($Q_c = \infty$):

$$Q_c = \frac{[C]^c[D]^d}{[0]^a[0]^b} = \infty$$

In a reaction mixture containing both reactants and products, each at a concentration of 1 M, the reaction quotient is one ($Q_c = 1$):

$$Q_c = \frac{(1)^c(1)^d}{(1)^a(1)^b} = 1$$

Q, K, and the Direction of a Reaction

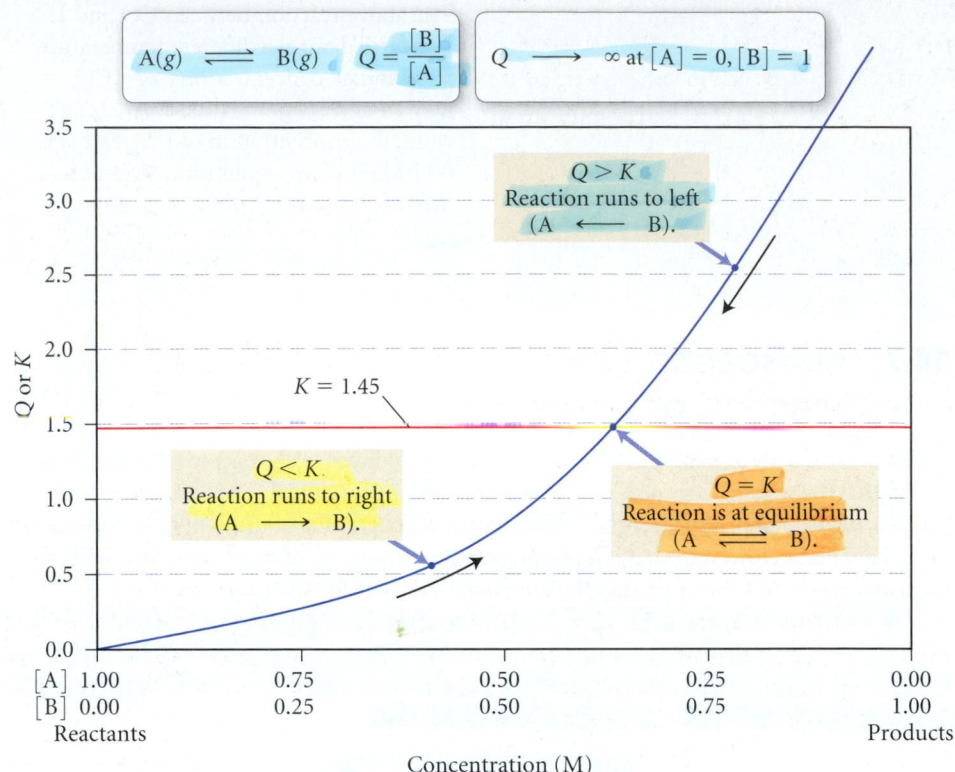

▲ **FIGURE 14.7 Q, K, and the Direction of a Reaction** The graph shows a plot of Q as a function of the concentrations of the reactants and products in a simple reaction $A \rightleftharpoons B$, in which $K = 1.45$ and the sum of the reactant and product concentrations is 1 M. Therefore, the far left of the graph represents pure reactant and the far right represents pure product. The very midpoint of the graph represents an equal mixture of A and B. When Q is less than K, the reaction moves in the forward direction ($A \longrightarrow B$). When Q is greater than K, the reaction moves in the reverse direction ($A \longleftarrow B$). When Q is equal to K, the reaction is at equilibrium.

The reaction quotient is useful because *the value of Q relative to K is a measure of the progress of the reaction toward equilibrium. At equilibrium, the reaction quotient is equal to the equilibrium constant.* For example, Figure 14.7 ▲ shows a plot of Q as a function of the concentrations of A and B for the simple reaction $A(g) \rightleftharpoons B(g)$, which has an equilibrium constant of $K = 1.45$. The following points are representative of three possible conditions.

Q	K	Predicted Direction of Reaction
0.55	1.45	To the right (toward products)
2.55	1.45	To the left (toward reactants)
1.45	1.45	No change (reaction at equilibrium)

For the first entry, Q is less than K and must therefore get larger as the reaction proceeds toward equilibrium. Q becomes larger as the reactant concentration decreases and the product concentration increases—the reaction proceeds to the right. For the second entry, Q is greater than K and must therefore get smaller as the reaction proceeds toward equilibrium. Q gets smaller as the reactant concentration increases and the product concentration decreases—the reaction proceeds to the left. In the third set of values, $Q = K$, implying that the reaction is at equilibrium—the reaction will not proceed in either direction.

Summarizing:

The reaction quotient (Q) is a measure of the progress of a reaction toward equilibrium.

➤ $Q < K$ Reaction goes to the right (toward products).
➤ $Q > K$ Reaction goes to the left (toward reactants).
➤ $Q = K$ Reaction is at equilibrium.

EXAMPLE 14.7 Predicting the Direction of a Reaction by Comparing Q and K

Consider the following reaction and its equilibrium constant:

$$I_2(g) + Cl_2(g) \rightleftharpoons 2\,ICl(g) \qquad K_p = 81.9$$

A reaction mixture contains $P_{I_2} = 0.114$ atm, $P_{Cl_2} = 0.102$ atm, and $P_{ICl} = 0.355$ atm. Is the reaction mixture at equilibrium? If not, in which direction will the reaction proceed?

Solution

To determine the progress of the reaction relative to the equilibrium state, calculate Q.	$$Q_p = \frac{P_{ICl}^2}{P_{I_2}P_{Cl_2}}$$ $$= \frac{0.355^2}{(0.114)(0.102)}$$ $$= 10.8$$
Compare Q to K.	$Q_p = 10.8;\ K_p = 81.9$ Since $Q_p < K_p$, the reaction is not at equilibrium and will proceed to the right.

For Practice 14.7

Consider the following reaction and its equilibrium constant:

$$N_2O_4(g) \rightleftharpoons 2\,NO_2(g) \qquad K_c = 5.85 \times 10^{-3}$$

A reaction mixture contains $[NO_2] = 0.0255$ M and $[N_2O_4] = 0.0331$ M. Calculate Q_c and determine the direction in which the reaction will proceed.

✦ Conceptual Connection 14.3 *Q* and *K*

For the reaction $N_2O_4(g) \rightleftharpoons 2\,NO_2(g)$, a reaction mixture at a certain temperature initially contains both N_2O_4 and NO_2 in their standard states (see the definition of standard state in Section 6.8). If $K_p = 0.15$, which of the following is true of the reaction mixture before any reaction occurs?

(a) $Q = K$; the reaction is at equilibrium.
(b) $Q < K$; the reaction will proceed to the right.
(c) $Q > K$; the reaction will proceed to the left.

Answer: (c) Since N_2O_4 and NO_2 are both in their standard states, they each have a partial pressure of 1.0 atm. Consequently, $Q_p = 1$. Since $K_p = 0.15$, $Q_p > K_p$, and the reaction proceeds to the left.

14.8 Finding Equilibrium Concentrations

In Section 14.6, we learned how to calculate an equilibrium constant given the equilibrium concentrations of the reactants and products. Just as commonly, however, we will want to calculate equilibrium concentrations of reactants or products given the equilibrium constant. These kinds of calculations are important because they allow us to calculate the amount of a reactant or product at equilibrium. For example, in a synthesis reaction, we might want to know how much of the product forms when the reaction reaches equilibrium. Or for the hemoglobin–oxygen equilibrium discussed in Section 14.1, we might want to know the concentration of oxygenated hemoglobin present under certain oxygen concentrations within the lungs or muscles.

We can divide these types of problems into the following two categories: (1) finding equilibrium concentrations when you are given the equilibrium constant and all but one of the equilibrium concentrations of the reactants and products; and (2) finding equilibrium concentrations when you are given the equilibrium constant and only initial concentrations. The second category of problem is more difficult than the first. We examine each separately.

Finding Equilibrium Concentrations When You Are Given the Equilibrium Constant and All but One of the Equilibrium Concentrations of the Reactants and Products

The equilibrium constant can be used to calculate the equilibrium concentration of one of the reactants or products, given the equilibrium concentrations of the others. To solve this type of problem, follow the general problem-solving procedure employed since the first chapter of this book.

EXAMPLE 14.8   Finding Equilibrium Concentrations When You Are Given the Equilibrium Constant and All but One Equilibrium Concentrations of the Reactants and Products

Consider the following reaction.

$$2\,COF_2(g) \rightleftharpoons CO_2(g) + CF_4(g) \qquad K_c = 2.00 \text{ at } 1000\,°C$$

In an equilibrium mixture, the concentration of COF_2 is 0.255 M and the concentration of CF_4 is 0.118 M. What is the equilibrium concentration of CO_2?

Sort You are given the equilibrium constant of a chemical reaction, together with the equilibrium concentrations of the reactant and one product. You are asked to find the equilibrium concentration of the other product.	**Given** $[COF_2] = 0.255\,M$ $[CF_4] = 0.118\,M$ $K_c = 2.00$ **Find** $[CO_2]$
Strategize You can calculate the concentration of the product by using the given quantities and the expression for K_c.	**Conceptual Plan** $\boxed{[COF_2], [CF_4], K_c} \longrightarrow \boxed{[CO_2]}$ $K_c = \dfrac{[CO_2][CF_4]}{[COF_2]^2}$
Solve Solve the equilibrium expression for $[CO_2]$ and then substitute in the appropriate values to compute it.	**Solution** $[CO_2] = K_c \dfrac{[COF_2]^2}{[CF_4]}$ $[CO_2] = 2.00\left(\dfrac{0.255^2}{0.118}\right)$ $= 1.10\,M$

Check You can check your answer by mentally substituting the given values of $[COF_2]$ and $[CF_4]$ as well as your calculated value for $[CO_2]$ back into the equilibrium expression.

$$K_c = \frac{[CO_2][CF_4]}{[COF_2]^2}$$

$[CO_2]$ was found to be roughly equal to one. $[COF_2]^2 \approx 0.06$ and $[CF_4] \approx 0.12$. Therefore K_c is approximately 2, as given in the problem.

For Practice 14.8

Diatomic iodine (I_2) decomposes at high temperature to form I atoms according to the following reaction:

$$I_2(g) \rightleftharpoons 2\,I(g) \qquad K_c = 0.011 \text{ at } 1200\,°C$$

In an equilibrium mixture, the concentration of I_2 is 0.10 M. What is the equilibrium concentration of I?

Finding Equilibrium Concentrations When You Are Given the Equilibrium Constant and Initial Concentrations or Pressures

More commonly, we know the equilibrium constant and only initial concentrations of reactants and need to find the *equilibrium concentrations* of the reactants or products. These kinds of problems are generally more involved than those we just examined and require a specific procedure to solve them. The procedure has some similarities to the one used in Examples 14.5 and 14.6 in that you must set up an ICE table showing the initial conditions, the changes, and the equilibrium conditions. However, unlike Examples 14.5 and 14.6, the changes in concentration are not known and must be represented with the variable x. For example, consider again the following simple reaction:

$$A(g) \rightleftharpoons 2\,B(g)$$

Suppose that, as before, we have a reaction mixture in which the initial concentration of A is 1.0 M and the initial concentration of B is 0.00 M. However, now we know the equilibrium constant, $K = 0.33$, and want to find the equilibrium concentrations. Set up the ICE table with the given initial concentrations and then *represent the unknown change in [A] with the variable x* as follows:

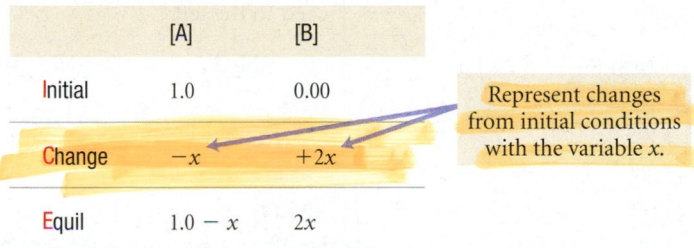

Notice that, due to the stoichiometry of the reaction, the change in [B] must be $+2x$. As before, each *equilibrium* concentration is simply the sum of the two entries above it in the ICE table. In order to find the equilibrium concentrations of A and B, we must find the value of the variable x. Since we know the value of the equilibrium constant, we can use the equilibrium expression to set up an equation in which x is the only variable.

$$K = \frac{[B]^2}{[A]} = \frac{(2x)^2}{1.0 - x} = 0.33$$

or simply,

$$\frac{4x^2}{1.0 - x} = 0.33$$

The above equation is a *quadratic* equation—it contains the variable x raised to the second power. In general, quadratic equations can be solved with the quadratic formula, which we introduce in Example 14.10. If the quadratic equation is a perfect square, however, it can be solved by simpler means, as shown in Example 14.9. For both of these examples, we give the general procedure in the left column and apply the procedure to the two different example problems in the center and right columns. Later in this section, we will also see that quadratic equations can often be simplified by making some approximations based on our chemical knowledge.

Procedure for Finding Equilibrium Concentrations from Initial Concentrations and the Equilibrium Constant	**EXAMPLE 14.9** Finding Equilibrium Concentrations from Initial Concentrations and the Equilibrium Constant	**EXAMPLE 14.10** Finding Equilibrium Concentrations from Initial Concentrations and the Equilibrium Constant
To solve these types of problems, follow the procedure outlined below.	Consider the following reaction. $N_2(g) + O_2(g) \rightleftharpoons 2\,NO(g)$ $\qquad K_c = 0.10$ (at 2000 °C) A reaction mixture (at 2000 °C) initially contains $[N_2] = 0.200$ M and $[O_2] = 0.200$ M. Find the equilibrium concentrations of the reactants and products at this temperature.	Consider the following reaction. $N_2O_4(g) \rightleftharpoons 2\,NO_2(g)$ $\qquad K_c = 0.36$ (at 100 °C) A reaction mixture at 100 °C initially contains $[NO_2] = 0.100$ M. Find the equilibrium concentrations of NO_2 and N_2O_4 at this temperature.

1. Using the balanced equation as a guide, prepare a table showing the known initial concentrations of the reactants and products. Leave room in the table for the changes in concentrations and for the equilibrium concentrations.

$N_2(g) + O_2(g) \rightleftharpoons 2\,NO(g)$

	$[N_2]$	$[O_2]$	$[NO]$
Initial	0.200	0.200	0.00
Change			
Equil			

$N_2O_4(g) \rightleftharpoons 2NO_2(g)$

	$[N_2O_4]$	$[NO_2]$
Initial	0.00	0.100
Change		
Equil		

2. Use the initial concentrations to calculate the reaction quotient (Q) for the initial concentrations. Compare Q to K and predict the direction in which the reaction will proceed.

$$Q_c = \frac{[NO]^2}{[N_2][O_2]} = \frac{0.00^2}{(0.200)(0.200)} = 0$$

$Q < K$, therefore the reaction will proceed to the right.

$$Q_c = \frac{[NO_2]^2}{[N_2O_4]} = \frac{(0.100)^2}{0.00} \quad \text{Products = 1}$$
$$= \infty \quad \text{Reactants 0}$$

$Q > K$, therefore the reaction will proceed to the left.

3. Represent the change in the concentration of one of the reactants or products with the variable x. Define the changes in the concentrations of the other reactants or products in terms of x. It is usually most convenient to let x represent the change in concentration of the reactant or product with the smallest stoichiometric coefficient.

$N_2(g) + O_2(g) \rightleftharpoons 2\,NO(g)$

	$[N_2]$	$[O_2]$	$[NO]$
Initial	0.200	0.200	0.00
Change	$-x$	$-x$	$+2x$
Equil			

$N_2O_4(g) \rightleftharpoons 2\,NO_2(g)$

	$[N_2O_4]$	$[NO_2]$
Initial	0.00	0.100
Change	$+x$	$-2x$
Equil		

4. Sum each column for each reactant and product to determine the equilibrium concentrations in terms of the initial concentrations and the variable x.

$N_2(g) + O_2(g) \rightleftharpoons 2\,NO(g)$

	$[N_2]$	$[O_2]$	$[NO]$
Initial	0.200	0.200	0.00
Change	$-x$	$-x$	$+2x$
Equil	$0.200 - x$	$0.200 - x$	$2x$

$N_2O_4(g) \rightleftharpoons 2\,NO_2(g)$

	$[N_2O_4]$	$[NO_2]$
Initial	0.00	0.100
Change	$+x$	$-2x$
Equil	x	$0.100 - 2x$

5. Substitute the expressions for the equilibrium concentrations (from step 4) into the expression for the equilibrium constant. Using the given value of the equilibrium constant, solve the expression for the variable x. In some cases, such as the first example here, you can simply take the square root of both sides of the expression to solve for x. In other cases, such as the second example here, you must solve a quadratic equation to find x.

Remember the quadratic formula:
$$ax^2 + bx + c = 0$$

$$x = \frac{-b \pm \sqrt{b^2 - 4ac}}{2a}$$

$$K_c = \frac{[NO]^2}{[N_2][O_2]}$$

$$= \frac{(2x)^2}{(0.200 - x)(0.200 - x)}$$

$$0.10 = \frac{(2x)^2}{(0.200 - x)^2}$$

$$\sqrt{0.10} = \frac{2x}{0.200 - x}$$

$$\sqrt{0.10}\,(0.200 - x) = 2x$$

$$\sqrt{0.10}\,(0.200) - \sqrt{0.10}\,x = 2x$$

$$0.063 = 2x + \sqrt{0.10}\,x$$

$$0.063 = 2.3x$$

$$x = 0.027$$

$$K_c = \frac{[NO_2]^2}{[N_2O_4]}$$

$$= \frac{(0.100 - 2x)^2}{x}$$

$$0.36 = \frac{0.0100 - 0.400x + 4x^2}{x}$$

$$0.36x = 0.0100 - 0.400x + 4x^2$$

$$4x^2 - 0.76x + 0.0100 = 0 \quad (quadratic)$$

$$x = \frac{-b \pm \sqrt{b^2 - 4ac}}{2a}$$

$$= \frac{-(-0.76) \pm \sqrt{(-0.76)^2 - 4(4)(0.0100)}}{2(4)}$$

$$= \frac{0.76 \pm 0.65}{8}$$

$$x = 0.176 \quad \text{or} \quad x = 0.014$$

6. Substitute x into the expressions for the equilibrium concentrations of the reactants and products (from step 4) and compute the concentrations. In cases where you solved a quadratic and have two values for x, choose the value for x that gives a physically realistic answer. For example, reject the value of x that results in any negative concentrations.

$$[N_2] = 0.200 - 0.027$$
$$= 0.173 \text{ M}$$

$$[O_2] = 0.200 - 0.027$$
$$= 0.173 \text{ M}$$

$$[NO] = 2(0.027)$$
$$= 0.054 \text{ M}$$

We reject the root $x = 0.176$ because it gives a negative concentration for NO_2. Using $x = 0.014$, we get the following concentrations:

$$[NO_2] = 0.100 - 2x$$
$$= 0.100 - 2(0.014) = 0.072 \text{ M}$$

$$[N_2O_4] = x$$
$$= 0.014 \text{ M}$$

7. Check your answer by substituting the computed equilibrium values into the equilibrium expression. The computed value of K should match the given value of K. Note that rounding errors could cause a difference in the least significant digit when comparing values of the equilibrium constant.

$$K_c = \frac{[NO]^2}{[N_2][O_2]}$$

$$= \frac{0.054^2}{(0.173)(0.173)} = 0.097$$

Since the computed value of K_c matches the given value (to within one digit in the least significant figure), the answer is valid.

$$K_c = \frac{[NO_2]^2}{[N_2O_4]} = \frac{0.072^2}{0.014}$$

$$= 0.37$$

Since the computed value of K_c matches the given value (to within one digit in the least significant figure), the answer is valid.

For Practice 14.9

The above reaction was carried out at a different temperature at which $K_c = 0.055$. This time, however, the reaction mixture started with only the product, $[NO] = 0.0100$ M, and no reactants. Find the equilibrium concentrations of N_2, O_2, and NO at equilibrium.

For Practice 14.10

The above reaction was carried out at the same temperature, but this time the reaction mixture initially contained only the reactant, $[N_2O_4] = 0.0250$ M, and no NO_2. Find the equilibrium concentrations of N_2O_4 and NO_2.

When the initial conditions are given in terms of partial pressures (instead of concentrations) and the equilibrium constant is given as K_p instead of K_c, use the same procedure, but substitute partial pressures for concentrations, as shown in the following example.

EXAMPLE 14.11 Finding Equilibrium Partial Pressures When You Are Given the Equilibrium Constant and Initial Partial Pressures

Consider the following reaction.

$$I_2(g) + Cl_2(g) \rightleftharpoons 2\,ICl(g) \qquad K_p = 81.9 \text{ (at 25 °C)}$$

A reaction mixture at 25 °C initially contains $P_{I_2} = 0.100$ atm, $P_{Cl_2} = 0.100$ atm, and $P_{ICl} = 0.100$ atm. Find the equilibrium partial pressures of I_2, Cl_2, and ICl at this temperature.

Solution

We follow the procedure outlined in Examples 14.5 and 14.6 (using the pressures in place of the concentrations) to solve the problem.

1. Using the balanced equation as a guide, prepare a table showing the known initial partial pressures of the reactants and products.

$$I_2(g) + Cl_2(g) \rightleftharpoons 2\,ICl(g)$$

	P_{I_2} (atm)	P_{Cl_2} (atm)	P_{ICl} (atm)
Initial	0.100	0.100	0.100
Change			
Equil			

2. Use the initial partial pressures to calculate the reaction quotient (Q). Compare Q to K and predict the direction in which the reaction will proceed.

$$Q_p = \frac{P_{ICl}^2}{P_{I_2} P_{Cl_2}} = \frac{0.100^2}{(0.100)(0.100)} = 1$$

$$K_p = 81.9 \text{ (given)}$$

$Q < K$, therefore the reaction will proceed to the right.

3. Represent the change in the partial pressure of one of the reactants or products with the variable x. Define the changes in the partial pressures of the other reactants or products in terms of x.

$$I_2(g) + Cl_2(g) \rightleftharpoons 2\,ICl(g)$$

	P_{I_2} (atm)	P_{Cl_2} (atm)	P_{ICl} (atm)
Initial	0.100	0.100	0.100
Change	$-x$	$-x$	$+2x$
Equil			

4. Sum each column for each reactant and product to determine the equilibrium partial pressures in terms of the initial partial pressures and the variable x.

$$I_2(g) + Cl_2(g) \rightleftharpoons 2\,ICl(g)$$

	P_{I_2} (atm)	P_{Cl_2} (atm)	P_{ICl} (atm)
Initial	0.100	0.100	0.100
Change	$-x$	$-x$	$+2x$
Equil	$0.100 - x$	$0.100 - x$	$0.100 + 2x$

5. Substitute the expressions for the equilibrium partial pressures (from step 4) into the expression for the equilibrium constant. Use the given value of the equilibrium constant to solve the expression for the variable x.

$$K_p = \frac{P_{ICl}^2}{P_{I_2} P_{Cl_2}}$$

$$= \frac{(0.100 + 2x)^2}{(0.100 - x)(0.100 - x)}$$

$$81.9 = \frac{(0.100 + 2x)^2}{(0.100 - x)^2} \qquad \text{(perfect square)}$$

$$\sqrt{81.9} = \frac{(0.100 + 2x)}{(0.100 - x)}$$

$$\sqrt{81.9}\,(0.100 - x) = 0.100 + 2x$$

	$\sqrt{81.9}\,(0.100) - \sqrt{81.9}\,x = 0.100 + 2x$
	$\sqrt{81.9}\,(0.100) - 0.100 = 2x + \sqrt{81.9}\,x$
	$0.805 = 11.05x$
	$x = 0.0729$
6. Substitute x into the expressions for the equilibrium partial pressures of the reactants and products (from step 4) and compute the partial pressures.	$P_{I_2} = 0.100 - 0.0729 = 0.027 \text{ atm}$ $P_{Cl_2} = 0.100 - 0.0729 = 0.027 \text{ atm}$ $P_{ICl} = 0.100 + 2(0.0729) = 0.246 \text{ atm}$
7. Check your answer by substituting the computed equilibrium partial pressures into the equilibrium expression. The computed value of K should match the given value of K.	$K_P = \dfrac{P_{ICl}^2}{P_{I_2}P_{Cl_2}} = \dfrac{0.246^2}{(0.027)(0.027)}$ $= 83$ Since the computed value of K_P matches the given value (within the uncertainty indicated by the significant figures), the answer is valid.

For Practice 14.11

The reaction between I_2 and Cl_2 (from Example 14.11) was carried out at the same temperature, but with the following initial partial pressures: $P_{I_2} = 0.150 \text{ atm}$, $P_{Cl_2} = 0.150 \text{ atm}$, $P_{ICl} = 0.00 \text{ atm}$. Find the equilibrium partial pressures of all three substances.

Simplifying Approximations in Working Equilibrium Problems

For some equilibrium problems of the type shown in the previous three examples, there is the possibility of using an approximation that makes solving the problem much easier without significant loss of accuracy. For example, if the equilibrium constant is relatively small, the reaction will not proceed very far to the right. If the initial reactant concentration is relatively large, we can therefore make the assumption that x is small relative to the initial concentration of reactant. To see how this approximation works, consider again the simple reaction $A \rightleftharpoons 2\,B$. Suppose that, as before, we have a reaction mixture in which the initial concentration of A is 1.0 M and the initial concentration of B is 0.0 M, and that we want to find the equilibrium concentrations. However, suppose that in this case the equilibrium constant is much smaller, say $K = 3.3 \times 10^{-5}$. The ICE table is identical to the one we set up previously:

	[A]	[B]
Initial	1.0	0.0
Change	$-x$	$+2x$
Equil	$1.0 - x$	$2x$

With the exception of the value of K, we end up with the exact quadratic equation that we had before: $K = \dfrac{[B]^2}{[A]} = \dfrac{(2x)^2}{1.0 - x} = 3.3 \times 10^{-5}$

or simply,

$$\dfrac{4x^2}{1.0 - x} = 3.3 \times 10^{-5}$$

This quadratic equation can be multiplied out and solved using the quadratic formula. However, since we know that K is small, we know that the reaction will not proceed very far toward products; therefore, x will also be small. If x is much smaller than 1.0, then $(1.0 - x)$ (the quantity in the denominator) can be approximated by (1.0).

$$\frac{2x^2}{(1.0 - x)} = 3.3 \times 10^{-5}$$

This approximation greatly simplifies the equation, which can then be easily solved for x as follows:

$$\frac{(2x)^2}{1.0} = 3.3 \times 10^{-5}$$

$$4x^2 = 3.3 \times 10^{-5}$$

$$x = \sqrt{\frac{3.3 \times 10^{-5}}{4}} = 0.0029$$

The validity of this approximation can be checked by comparing the computed value of x to the number it was subtracted from. The ratio of x to the number it is subtracted from should be less than 0.05 (or 5%) for the approximation to be valid. In this case, x was subtracted from 1.0, and therefore the ratio of the computed value of x to 1.0 is calculated as follows:

$$\frac{0.0029}{1.0} \times 100\% = 0.29\%$$

The approximation is therefore valid. In the two side-by-side examples that follow, we treat two nearly identical problems—the only difference is the initial concentration of the reactant. In Example 14.12, the initial concentration of the reactant is relatively large, the equilibrium constant is small, and the x is small approximation works well. In Example 14.13, however, the initial concentration of the reactant is much smaller, and even though the equilibrium constant is the same, the x is small approximation does not work (because the initial concentration is also small). In cases such as this, we have a couple of options to solve the problem. We can either go back and solve the equation exactly (using the quadratic formula, for example), or we can use the *method of successive approximations*, which is introduced in Example 14.13. In this method, we essentially solve for x as if it were small, and then substitute the value obtained back into the equation to solve for x again. This can be repeated until the computed value of x stops changing with each iteration, an indication that we have arrived at an acceptable value for x.

Note that the x is small approximation does not imply that x is zero. If that were the case, the reactant and product concentrations would not change from their initial values. The x is small approximation simply means that when x is added or subtracted to another number, it does not change that number by very much. For example, let us compute the value of the difference $1.0 - x$ when $x = 3.0 \times 10^{-4}$:

$$1.0 - x = 1.0 - 3.0 \times 10^{-4}$$

$$= 0.9997$$

$$= 1.0$$

Since the value of 1.0 is known only to two significant figures, subtracting the small x does not change the value at all. This situation is similar to weighing yourself on a bathroom scale with and without a penny in your pocket. Unless your scale is unusually precise, removing the penny from your pocket will not change the reading on the scale. This does not imply that the penny is weightless, only that its weight is small when compared to your weight. The weight of the penny can therefore be neglected in reading your weight with no detectable loss in accuracy.

Procedure for Finding Equilibrium Concentrations from Initial Concentrations in Cases with a Small Equilibrium Constant

To solve these types of problems, follow the procedure outlined below.

EXAMPLE 14.12 Finding Equilibrium Concentrations from Initial Concentrations in Cases with a Small Equilibrium Constant

Consider the following reaction for the decomposition of hydrogen disulfide:

$$2\,H_2S(g) \rightleftharpoons 2\,H_2(g) + S_2(g)$$
$$K_c = 1.67 \times 10^{-7} \text{ at } 800\,°C$$

A 0.500-L reaction vessel initially contains 0.0125 mol of H_2S at 800 °C. Find the equilibrium concentrations of H_2 and S_2.

EXAMPLE 14.13 Finding Equilibrium Concentrations from Initial Concentrations in Cases with a Small Equilibrium Constant

Consider the following reaction for the decomposition of hydrogen disulfide:

$$2\,H_2S(g) \rightleftharpoons 2\,H_2(g) + S_2(g)$$
$$K_c = 1.67 \times 10^{-7} \text{ at } 800\,°C$$

A 0.500-L reaction vessel initially contains 1.25×10^{-4} mol of H_2S at 800 °C. Find the equilibrium concentrations of H_2 and S_2.

---

1. Using the balanced equation as a guide, prepare a table showing the known initial concentrations of the reactants and products. (In these examples, you must first compute the concentration of H_2S from the given number of moles and volume.)

$$[H_2S] = \frac{0.0125 \text{ mol}}{0.500 \text{ L}} = 0.0250 \text{ M}$$

$$2\,H_2S(g) \rightleftharpoons 2\,H_2(g) + S_2(g)$$

	$[H_2S]$	$[H_2]$	$[S_2]$
Initial	0.0250	0.00	0.00
Change			
Equil			

$$[H_2S] = \frac{1.25 \times 10^{-4} \text{ mol}}{0.500 \text{ L}}$$
$$= 2.50 \times 10^{-4} \text{ M}$$

$$2\,H_2S(g) \rightleftharpoons 2\,H_2(g) + S_2(g)$$

	$[H_2S]$	$[H_2]$	$[S_2]$
Initial	2.50×10^{-4}	0.00	0.00
Change			
Equil			

---

2. Use the initial concentrations to calculate the reaction quotient (Q). Compare Q to K and predict the direction that the reaction will proceed.

By inspection, $Q = 0$; the reaction will proceed to the right.

By inspection, $Q = 0$; the reaction will proceed to the right.

---

3. Represent the change in the concentration of one of the reactants or products with the variable x. Define the changes in the concentrations of the other reactants or products with respect to x.

$$2\,H_2S(g) \rightleftharpoons 2\,H_2(g) + S_2(g)$$

	$[H_2S]$	$[H_2]$	$[S_2]$
Initial	0.0250	0.00	0.00
Change	$-2x$	$+2x$	$+x$
Equil			

$$2\,H_2S(g) \rightleftharpoons 2\,H_2(g) + S_2(g)$$

	$[H_2S]$	$[H_2]$	$[S_2]$
Initial	2.50×10^{-4}	0.00	0.00
Change	$-2x$	$+2x$	$+x$
Equil			

---

4. Sum each column for each reactant and product to determine the equilibrium concentrations in terms of the initial concentrations and the variable x.

$$2\,H_2S(g) \rightleftharpoons 2\,H_2(g) + S_2(g)$$

	$[H_2S]$	$[H_2]$	$[S_2]$
Initial	0.0250	0.00	0.00
Change	$-2x$	$+2x$	$+x$
Equil	$0.0250 - 2x$	$2x$	x

$$2\,H_2S(g) \rightleftharpoons 2\,H_2(g) + S_2(g)$$

	$[H_2S]$	$[H_2]$	$[S_2]$
Initial	2.50×10^{-4}	0.00	0.00
Change	$-2x$	$+2x$	$+x$
Equil	$2.50 \times 10^{-4} - 2x$	$2x$	x

---

5. Substitute the expressions for the equilibrium concentrations (from step 4) into the expression for the equilibrium constant. Use the given value of the equilibrium constant to solve the expression for the variable x.

$$K_c = \frac{[H_2]^2[S_2]}{[H_2S]^2}$$
$$= \frac{(2x)^2(x)}{(0.0250 - 2x)^2}$$

$$K_c = \frac{[H_2]^2[S_2]}{[H_2S]^2}$$
$$= \frac{(2x)^2 x}{(2.50 \times 10^{-4} - 2x)^2}$$

In this case, the resulting equation is cubic in x. Although cubic equations can be solved, the solutions are not usually simple. However, since the equilibrium constant is small, we know that the reaction does not proceed very far to the right. Therefore, x will be a small number and can be dropped from any quantities in which it is added to or subtracted from another number (as long as the number itself is not too small).

$$1.67 \times 10^{-7} = \frac{4x^3}{(0.0250 - 2x)^2}$$

$$1.67 \times 10^{-7} = \frac{4x^3}{(0.0250 - 2x)^2} \quad \text{x is small.}$$

$$1.67 \times 10^{-7} = \frac{4x^3}{6.25 \times 10^{-4}}$$

$$6.25 \times 10^{-4}(1.67 \times 10^{-7}) = 4x^3$$

$$x^3 = \frac{6.25 \times 10^{-4}(1.67 \times 10^{-7})}{4}$$

$$1.67 \times 10^{-7} = \frac{4x^3}{(2.50 \times 10^{-4} - 2x)^2}$$

$$1.67 \times 10^{-7} = \frac{4x^3}{(2.50 \times 10^{-4} - 2x)^2} \quad \text{x is small.}$$

$$1.67 \times 10^{-7} = \frac{4x^3}{6.25 \times 10^{-8}}$$

$$6.25 \times 10^{-8}(1.67 \times 10^{-7}) = 4x^3$$

$$x^3 = \frac{6.25 \times 10^{-8}(1.67 \times 10^{-7})}{4}$$

Check whether your approximation was valid by comparing the computed value of x to the number it was added to or subtracted from. The ratio of the two numbers should be less than 0.05 (or 5%) for the approximation to be valid. If approximation is not valid, proceed to step 5a.

$$x = 2.97 \times 10^{-4}$$

Checking the x is small approximation:

$$\frac{2.97 \times 10^{-4}}{0.0250} \times 100\% = 1.19\%$$

The x is small approximation is valid, proceed to step 6.

$$x = 1.38 \times 10^{-5}$$

Checking the x is small approximation:

$$\frac{1.38 \times 10^{-5}}{2.50 \times 10^{-4}} \times 100\% = 5.52\%$$

The approximation does not satisfy the < 5% rule (although it is close).

5a. If the approximation is not valid, you can either solve the equation exactly (either by hand or with your calculator), or use the method of successive approximations. In this case, we use the method of successive approximations.

Substitute the value obtained for x in step 5 back into the original cubic equation, but only at the exact spot where x was assumed to be negligible and then solve the equation for x again. Continue this procedure until the value of x obtained from solving the equation is the same as the one that is substituted into the equation.

$$1.67 \times 10^{-7} = \frac{4x^3}{(2.50 \times 10^{-4} - 2x)^2}$$

$$x = 1.38 \times 10^{-5}$$

$$1.67 \times 10^{-7} = \frac{4x^3}{(2.50 \times 10^{-4} - 2.76 \times 10^{-5})^2}$$

$$x = 1.27 \times 10^{-5}$$

If we substitute this value of x back into the cubic equation and solve it, we get $x = 1.28 \times 10^{-5}$, which is nearly identical to 1.27×10^{-5}. Therefore, we have arrived at the best approximation for x.

6. Substitue x into the expressions for the equilibrium concentrations of the reactants and products (from step 4) and compute the concentrations.

$$[H_2S] = 0.0250 - 2(2.97 \times 10^{-4})$$
$$= 0.0244 \, M$$

$$[H_2] = 2(2.97 \times 10^{-4})$$
$$= 5.94 \times 10^{-4} \, M$$

$$[S_2] = 2.97 \times 10^{-4} \, M$$

$$[H_2S] = 2.50 \times 10^{-4} - 2(1.28 \times 10^{-5})$$
$$= 2.24 \times 10^{-4} \, M$$

$$[H_2] = 2(1.28 \times 10^{-5})$$
$$= 2.56 \times 10^{-5} \, M$$

$$[S_2] = 1.28 \times 10^{-5} \, M$$

7. Check your answer by substituting the computed equilibrium values into the equilibrium expression. The computed value of K should match the given value of K. Note that the approximation method and rounding errors could cause a difference of up to about 5% when comparing values of the equilibrium constant.

$$K_c = \frac{(5.94 \times 10^{-4})^2 (2.97 \times 10^{-4})}{(0.0244)^2}$$
$$= 1.76 \times 10^{-7}$$

The computed value of K is close enough to the given value when we consider the uncertainty introduced by the approximation. Therefore the answer is valid.

$$K_c = \frac{(2.56 \times 10^{-5})^2 (1.28 \times 10^{-5})}{(2.24 \times 10^{-4})^2}$$
$$= 1.67 \times 10^{-7}$$

The computed value of K is equal to the given value. Therefore the answer is valid.

For Practice 14.12

The reaction in Example 14.12 was carried out at the same temperature with the following initial concentrations: $[H_2S] = 0.100\ M$, $[H_2] = 0.100\ M$, and $[S_2] = 0.00\ M$. Find the equilibrium concentration of $[S_2]$.

For Practice 14.13

The reaction in Example 14.13 was carried out at the same temperature with the following initial concentrations: $[H_2S] = 1.00 \times 10^{-4}\ M$, $[H_2] = 0.00\ M$, and $[S_2] = 0.00\ M$. Find the equilibrium concentration of $[S_2]$.

14.9 Le Châtelier's Principle: How a System at Equilibrium Responds to Disturbances

We have seen that a chemical system not in equilibrium tends to go toward equilibrium and that the concentrations of the reactants and products at equilibrium correspond to the equilibrium constant, K. What happens, however, when a chemical system already at equilibrium is disturbed? **Le Châtelier's principle** states that the chemical system will respond to minimize the disturbance.

| Pronounced "Le-sha-te-lyay."

> **Le Châtelier's principle:** When a chemical system at equilibrium is disturbed, the system shifts in a direction that minimizes the disturbance.

In other words, a system at equilibrium tends to maintain that equilibrium—it bounces back when disturbed.

We can understand Le Châtelier's principle by returning to our two neighboring countries analogy. Suppose the populations of Country A and Country B are at equilibrium. This means that the rate of people moving out of Country A and into Country B is equal to the rate of people moving into Country A and out of Country B, and the populations of the two countries are stable.

$$\text{Country A} \rightleftharpoons \text{Country B}$$

Now imagine disturbing the balance (Figure 14.8 on page 646). Suppose we add extra people to Country B. What happens? Since Country B suddenly becomes more crowded, the rate of people leaving Country B increases, because the pool of potential emigrants is now larger. The net flow of people is out of Country B and into Country A. We disturbed the equilibrium by adding more people to Country B, and people left Country B. In effect, the system responded by shifting in the direction that minimized the disturbance.

On the other hand, what happens if we add extra people to Country A? Since Country A suddenly gets more crowded, the rate of people leaving Country A goes up. The net flow of people is out of Country A and into Country B. We added people to Country A and the system responded by moving people out of Country A. Chemical systems behave similarly: When their equilibrium is disturbed, they react to counter the disturbance. We can disturb a system in chemical equilibrium in several different ways, including changing the concentration of a reactant or product, changing the volume or pressure, and changing the temperature. We consider each of these separately.

The two-country analogy should help you see the effects of disturbing a system in equilibrium—it should not be taken as an exact parallel.

Le Châtelier's Principle: An Analogy

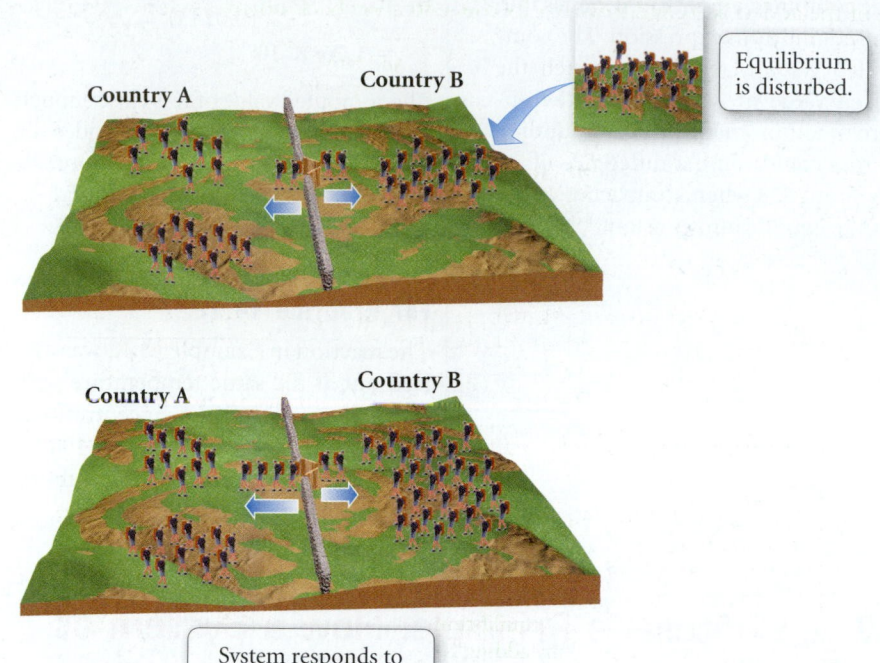

▶ **FIGURE 14.8 A Population Analogy for Le Châtelier's Principle** Adding population to Country B shifts the equilibrium to the left. People leave Country B (because it has become too crowded) and migrate to Country A until equilibrium is reestablished.

Equilibrium is disturbed.

System responds to minimize disturbance.

The Effect of a Concentration Change on Equilibrium

Consider the following reaction in chemical equilibrium:

$$N_2O_4(g) \rightleftharpoons 2\,NO_2(g)$$

Suppose we disturb the equilibrium by adding NO_2 to the equilibrium mixture (Figure 14.9 ▼). In other words, we increase the concentration of NO_2. What happens?

Le Châtelier's Principle: Changing Concentration

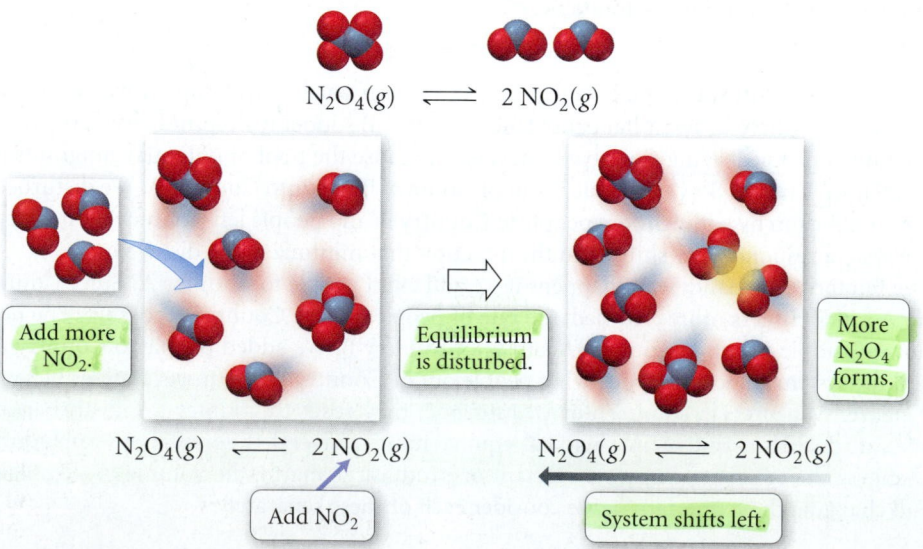

$$N_2O_4(g) \rightleftharpoons 2\,NO_2(g)$$

Add more NO_2.

Equilibrium is disturbed.

More N_2O_4 forms.

$$N_2O_4(g) \rightleftharpoons 2\,NO_2(g)$$

$$N_2O_4(g) \rightleftharpoons 2\,NO_2(g)$$

Add NO_2

System shifts left.

▲ **FIGURE 14.9 Le Châtelier's Principle: The Effect of a Concentration Change** Adding NO_2 causes the reaction to shift left, consuming some of the added NO_2 and forming more N_2O_4.

According to Le Châtelier's principle, the system shifts in a direction to minimize the disturbance. The reaction goes to the left (it proceeds in the reverse direction), consuming some of the added NO_2 and bringing its concentration back down, as shown graphically in Figure 14.10(a) ▼.

Add NO_2.

$$N_2O_4\,(g) \;\rightleftharpoons\; 2\,NO_2(g)$$

Reaction shifts left.

Le Châtelier's Principle: Graphical Representation

$$N_2O_4(g) \;\rightleftharpoons\; 2\,NO_2(g)$$

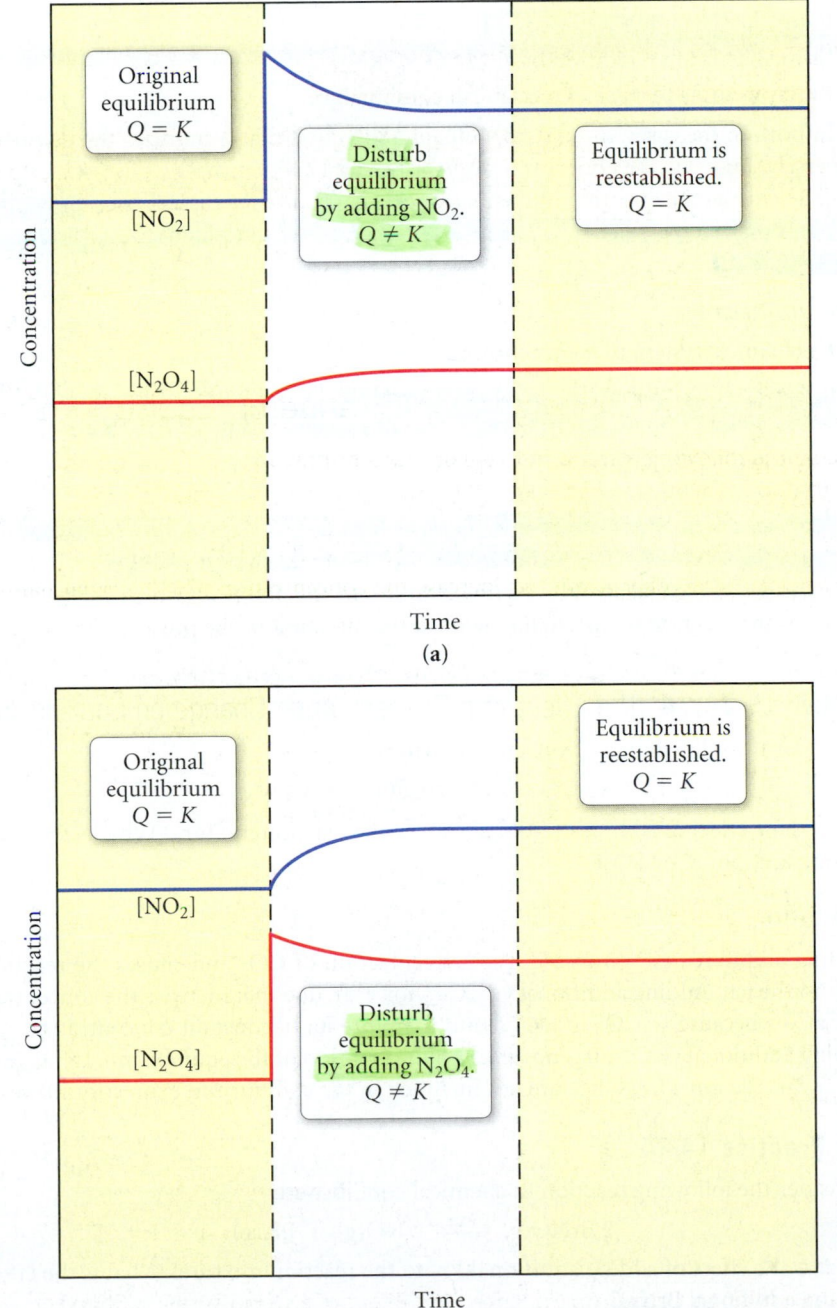

Time
(a)

Time
(b)

◀ **FIGURE 14.10 Le Châtelier's Principle: A Graphical Representation** The graph shows the concentrations of NO_2 and N_2O_4 for the reaction $N_2O_4(g) \longrightarrow 2\,NO_2(g)$ in three distinct stages of the reaction: initially at equilibrium (left); upon disturbance of the equilibrium by addition of more NO_2 (a) or N_2O_4 (b) to the reaction mixture (center); and upon reestablishment of equilibrium (right).

The reaction shifts to the left because the value of Q changes as follows:

- Before addition of NO_2: $Q = K$.
- Immediately after addition of NO_2: $Q > K$.
- Reaction shifts to left to reestablish equilibrium.

On the other hand, what happens if we add extra N_2O_4, increasing its concentration? In this case, the reaction shifts to the right, consuming some of the added N_2O_4 and bringing *its* concentration back down, as shown in Figure 14.10(b).

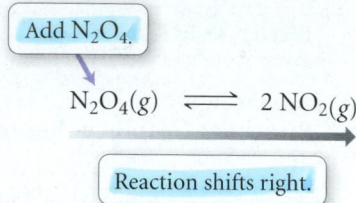

Add N_2O_4.

$$N_2O_4(g) \rightleftharpoons 2\,NO_2(g)$$

Reaction shifts right.

The reaction shifts to the right because the value of Q changes as follows:

- Before addition of N_2O_4: $Q = K$.
- Immediately after addition of N_2O_4: $Q < K$.
- Reaction shifts to right to reestablish equilibrium.

In both of the cases, the system shifts in a direction that minimizes the disturbance. Conversely, *lowering* the concentration of a reactant (which makes $Q > K$) causes the system to shift in the direction of the reactants to minimize the disturbance. Lowering the concentration of a product (which makes $Q < K$) causes the system to shift in the direction of products.

Summarizing,

If a chemical system is at equilibrium:

➤ *Increasing* the concentration of one or more of the *reactants* (which makes $Q < K$) causes the reaction to *shift to the right* (in the direction of the products).

➤ *Increasing* the concentration of one or more of the *products* (which makes $Q > K$) causes the reaction to *shift to the left* (in the direction of the reactants).

➤ *Decreasing* the concentration of one or more of the *reactants* (which makes $Q > K$) causes the reaction to *shift to the left* (in the direction of the reactants).

➤ *Decreasing* the concentration of one or more of the *products* (which makes $Q < K$) causes the reaction to *shift to the right* (in the direction of the products).

EXAMPLE 14.14   The Effect of a Concentration Change on Equilibrium

Consider the following reaction at equilibrium.

$$CaCO_3(s) \rightleftharpoons CaO(s) + CO_2(g)$$

What is the effect of adding additional CO_2 to the reaction mixture? What is the effect of adding additional $CaCO_3$?

Solution

Adding additional CO_2 increases the concentration of CO_2 and causes the reaction to shift to the left. Adding additional $CaCO_3$, however, does *not* increase the concentration of $CaCO_3$ because $CaCO_3$ is a solid and therefore has a constant concentration. Thus, adding additional $CaCO_3$ has no effect on the position of the equilibrium. (Note that, as we saw in Section 14.5, solids are not included in the equilibrium expression.)

For Practice 14.14

Consider the following reaction in chemical equilibrium:

$$2\,BrNO(g) \rightleftharpoons 2\,NO(g) + Br_2(g)$$

What is the effect of adding additional Br_2 to the reaction mixture? What is the effect of adding additional BrNO?

The Effect of a Volume (or Pressure) Change on Equilibrium

How does a system in chemical equilibrium respond to a volume change? Remember from Chapter 5 that changing the volume of a gas (or a gas mixture) results in a change in pressure. Remember also that pressure and volume are inversely related: a *decrease* in volume causes an *increase* in pressure, and an *increase* in volume causes a *decrease* in pressure. So, if the volume of a reaction mixture at chemical equilibrium is changed, the pressure changes and the system will shift in a direction to minimize that change. For example, consider the following reaction at equilibrium in a cylinder equipped with a moveable piston:

$$N_2(g) + 3 H_2(g) \rightleftharpoons 2 NH_3(g)$$

What happens if we push down on the piston, lowering the volume and raising the pressure (Figure 14.11a ▼)? How can the chemical system change so as to bring the pressure back down? Look carefully at the reaction coefficients. If the reaction shifts to the right, 4 mol of gas particles are converted to 2 mol of gas particles. From the ideal gas law ($PV = nRT$), we know that lowering the number of moles of a gas (n) results in a lower pressure (P). Therefore, the system shifts to the right, lowering the number of gas molecules and bringing the pressure back down, minimizing the disturbance.

Consider the same reaction mixture at equilibrium again. What happens if, this time, we pull *up* on the piston, *increasing* the volume (Figure 14.11b)? The higher volume results in a lower pressure and the system responds in such a way as to bring the pressure back up. It does this by shifting to the left, converting 2 mol of gas particles into 4 mol of gas particles, increasing the pressure and minimizing the disturbance.

In considering the effect of a change in volume, we are assuming that the change in volume is carried out at constant temperature.

Le Châtelier's Principle: Changing Pressure

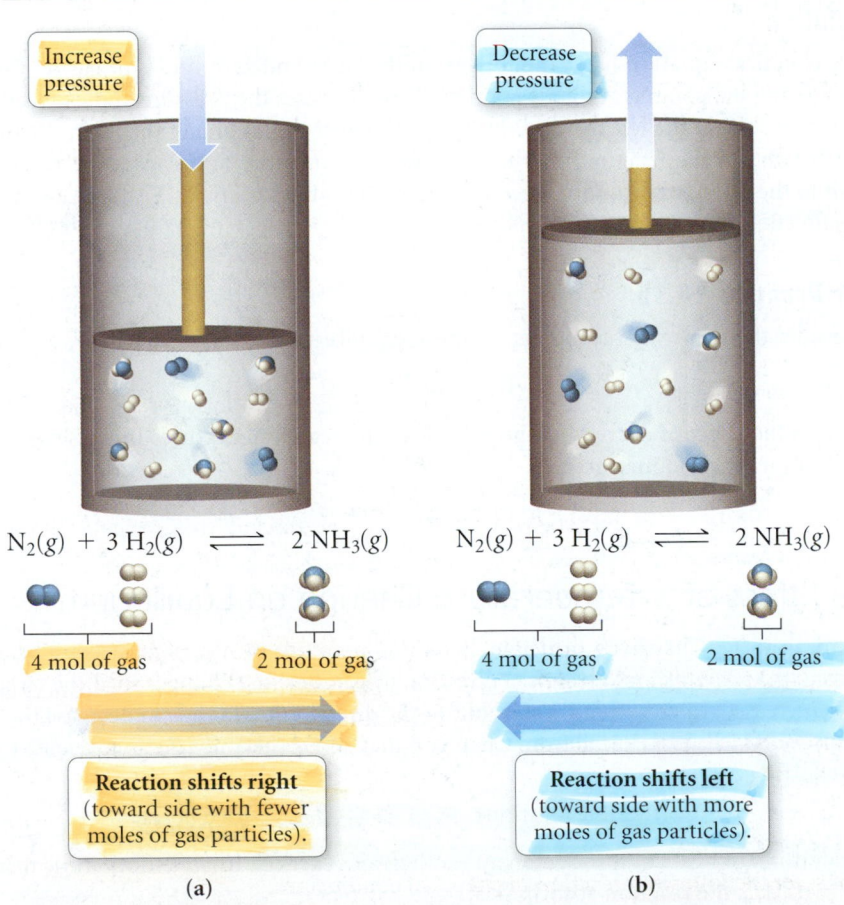

▲ FIGURE 14.11 Le Châtelier's Principle: The Effect of a Pressure Change (a) Decreasing the volume increases the pressure, causing the reaction to shift to the right (fewer moles of gas, lower pressure). (b) Increasing the volume reduces the pressure, causing the reaction to shift to the left (more moles of gas, higher pressure).

Consider again the same reaction mixture at equilibrium. What happens if, this time, we keep the volume the same, but increase the pressure *by adding an inert gas* to the mixture? Although the overall pressure of the mixture increases, the partial pressures of the reactants and products do not change. Consequently, there is no effect and the reaction does not shift in either direction.

Summarizing,

If a chemical system is at equilibrium:

▶ *Decreasing* the volume causes the reaction to shift in the direction that has *the fewer moles of gas particles.*

▶ *Increasing* the volume causes the reaction to shift in the direction that has *the greater number of moles of gas particles.*

▶ If a reaction has an equal number of moles of gas on both sides of the chemical equation, then a change in volume produces no effect on the equilibrium.

▶ Adding an inert gas to the mixture at a fixed volume has no effect on the equilibrium.

EXAMPLE 14.15 The Effect of a Volume Change on Equilibrium

Consider the following reaction at chemical equilibrium:

$$2\,KClO_3(s) \rightleftharpoons 2\,KCl(s) + 3\,O_2(g)$$

What is the effect of decreasing the volume of the reaction mixture? Increasing the volume of the reaction mixture? Adding an inert gas at constant volume?

Solution

The chemical equation has 3 mol of gas on the right and zero moles of gas on the left. Decreasing the volume of the reaction mixture increases the pressure and causes the reaction to shift to the left (toward the side with fewer moles of gas particles). Increasing the volume of the reaction mixture decreases the pressure and causes the reaction to shift to the right (toward the side with more moles of gas particles.) Adding an inert gas has no effect.

For Practice 14.15

Consider the following reaction at chemical equilibrium:

$$2\,SO_2(g) + O_2(g) \rightleftharpoons 2\,SO_3(g)$$

What is the effect of decreasing the volume of the reaction mixture? Increasing the volume of the reaction mixture?

The Effect of a Temperature Change on Equilibrium

According to Le Châtelier's principle, if the temperature of a system at equilibrium is changed, the system should shift in a direction to counter that change. So, if the temperature is increased, the reaction should shift in the direction that tends to decrease the temperature and vice versa. Recall from Chapter 6 that an exothermic reaction (negative ΔH) emits heat:

Exothermic reaction: $A + B \rightleftharpoons C + D + heat$

We can think of heat as a product in an exothermic reaction. In an endothermic reaction (positive ΔH), the reaction absorbs heat.

Endothermic reaction: $A + B + heat \rightleftharpoons C + D$

We can think of heat as a reactant in an endothermic reaction.

In considering the effect of a change in temperature, we are assuming that the heat is added (or removed) at constant pressure.

At constant pressure, raising the temperature of an *exothermic* reaction—think of this as adding heat—is similar to adding more product, causing the reaction to shift left. For example, the reaction of nitrogen with hydrogen to form ammonia is exothermic.

Add heat

$$N_2(g) + 3 H_2(g) \rightleftharpoons 2 NH_3(g) + heat$$

Reaction shifts left.
Smaller K

Raising the temperature of an equilibrium mixture of these three gases causes the reaction to shift left, absorbing some of the added heat and forming less products and more reactants. Note that, unlike adding additional NH_3 to the reaction mixture (which does *not* change the value of the equilibrium constant), *adding heat does change the value of the equilibrium constant*. The new equilibrium mixture will have more reactants and fewer products and therefore a smaller value of K.

Conversely, lowering the temperature causes the reaction to shift right, releasing heat and producing more products because the value of K has increased.

Remove heat

$$N_2(g) + 3 H_2(g) \rightleftharpoons 2NH_3(g) + heat$$

Rection shifts right.
Larger K

In contrast, for an *endothermic* reaction, raising the temperature—adding heat—causes the reaction to shift right to absorb the added heat. For example, the following reaction is endothermic.

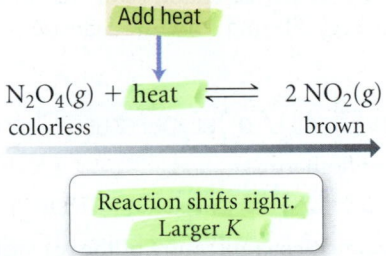

Add heat

$$N_2O_4(g) + heat \rightleftharpoons 2 NO_2(g)$$
colorless   brown

Reaction shifts right.
Larger K

Raising the temperature of an equilibrium mixture of these two gases causes the reaction to shift right, absorbing some of the added heat and producing more products because the value of K has increased. Since N_2O_4 is colorless and NO_2 is brown, the effects of changing the temperature of this reaction are easily seen (Figure 14.12 on page 652). On the other hand, lowering the temperature of a reaction mixture of these two gases causes the reaction to shift left, releasing heat, forming less products, and lowering the value of K.

Remove heat

$$N_2O_4(g) + heat \rightleftharpoons 2 NO_2(g)$$
colorless   brown

Reaction shifts left.
Smaller K

Le Châtelier's Principle: Changing Temperature

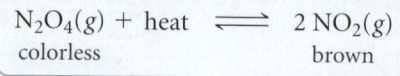

$$N_2O_4(g) + \text{heat} \rightleftharpoons 2\,NO_2(g)$$
colorless   brown

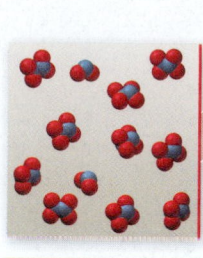

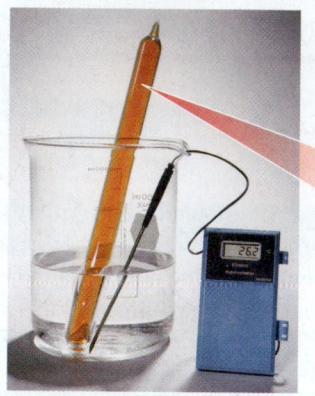
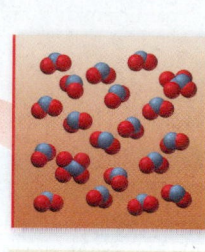

Lower temperature:
N_2O_4 favored

Higher temperature:
NO_2 favored

▲ **FIGURE 14.12 Le Châtelier's Principle: The Effect of a Temperature Change** Because the reaction is endothermic, raising the temperature causes a shift to the right, toward the formation of brown NO_2.

Summarizing:

In an *exothermic* chemical reaction, heat is a product.

➤ *Increasing* the temperature causes an exothermic reaction to *shift left* (in the direction of the reactants); the value of the equilibrium constant decreases.

➤ *Decreasing* the temperature causes an exothermic reaction to *shift right* (in the direction of the products); the value of the equilibrium constant increases.

In an *endothermic* chemical reaction, heat is a reactant.

➤ *Increasing* the temperature causes an endothermic reaction to *shift right* (in the direction of the products); the equilibrium constant increases.

➤ *Decreasing* the temperature causes an endothermic reaction to *shift left* (in the direction of the reactants); the equilibrium constant decreases.

Adding heat favors the endothermic direction. Removing heat favors the exothermic direction.

EXAMPLE 14.16   The Effect of a Temperature Change on Equilibrium

The following reaction is endothermic.

$$CaCO_3(s) \rightleftharpoons CaO(s) + CO_2(g)$$

What is the effect of increasing the temperature of the reaction mixture? Decreasing the temperature?

Solution

Since the reaction is endothermic, we can think of heat as a reactant:

$$\text{Heat} + CaCO_3(s) \rightleftharpoons CaO(s) + CO_2(g)$$

Raising the temperature is equivalent to adding a reactant, causing the reaction to shift to the right. Lowering the temperature is equivalent to removing a reactant, causing the reaction to shift to the left.

For Practice 14.16

The following reaction is exothermic.

$$2\,SO_2(g) + O_2(g) \rightleftharpoons 2\,SO_3(g)$$

What is the effect of increasing the temperature of the reaction mixture? Decreasing the temperature?

Chapter in Review

Key Terms

Section 14.2
reversible (619)
dynamic equilibrium (619)

Section 14.3
equilibrium constant (K) (622)
law of mass action (622)

Section 14.7
reaction quotient (Q) (633)

Section 14.9
Le Châtelier's principle (645)

Key Concepts

The Equilibrium Constant (14.1)

The relative concentrations of the reactants and the products at equilibrium are expressed by the equilibrium constant, K. The equilibrium constant measures how far a reaction proceeds toward products: a large K (much greater than 1) indicates a high concentration of products at equilibrium and a small K (much less than 1) indicates a low concentration of products at equilibrium.

Dynamic Equilibrium (14.2)

Most chemical reactions are reversible; they can proceed in either the forward or the reverse direction. When a chemical reaction is in dynamic equilibrium, the rate of the forward reaction equals the rate of the reverse reaction, so the net concentrations of reactants and products do not change. However, this does *not* imply that the concentrations of the reactants and the products are equal at equilibrium.

The Equilibrium Constant Expression (14.3)

The equilibrium constant expression is given by the law of mass action and is equal to the concentrations of the products, raised to their stoichiometric coefficients, divided by the concentrations of the reactants, raised to their stoichiometric coefficients. When the equation for a chemical reaction is reversed, multiplied, or added to another equation, K must be modified accordingly.

The Equilibrium Constant, K (14.4)

The equilibrium constant can be expressed in terms of concentrations (K_c) or in terms of partial pressures (K_p). These two constants are related by Equation 14.2. Concentration must always be expressed in units of molarity for K_c. Partial pressures must always be expressed in units of atmospheres for K_p.

States of Matter and the Equilibrium Constant (14.5)

The equilibrium constant expression contains only partial pressures or concentrations of reactants and products that exist as gases or solutes dissolved in solution. Pure liquids and solids are not included in the expression for the equilibrium constant.

Calculating K (14.6)

The equilibrium constant can be calculated from equilibrium concentrations or partial pressures by substituting measured values into the expression for the equilibrium constant (as obtained from the law of mass action). In most cases, the equilibrium concentrations of the reactants and products—and therefore the value of the equilibrium constant—can be calculated from the initial concentrations of the reactants and products and the equilibrium concentration of *just one* reactant or product.

The Reaction Quotient, Q (14.7)

The ratio of the concentrations (or partial pressures) of products raised to their stoichiometric coefficients to the concentrations of reactants raised to their stoichiometric coefficients *at any point in the reaction* is called the reaction quotient, Q. Like K, Q can be expressed in terms of concentrations (Q_c) or partial pressures (Q_p). At equilibrium, Q is equal to K; therefore, the direction in which a reaction will proceed can be determined by comparing Q to K. If $Q < K$, the reaction moves in the direction of the products; if $Q > K$, the reaction moves in the reverse direction.

Finding Equilibrium Concentrations (14.8)

There are two general types of problems in which K is given and one (or more) equilibrium concentrations can be found: (1) K, initial concentrations, and (at least) one equilibrium concentration are given; and (2) K and *only* initial concentrations are given. The first type is solved by rearranging the law of mass action and substituting given values. The second type is solved by using a variable x to represent the change in concentration.

Le Châtelier's Principle (14.9)

When a system at equilibrium is disturbed—by a change in the amount of a reactant or product, a change in volume, or a change in temperature—the system shifts in the direction that minimizes the disturbance.

Key Equations and Relationships

Expression for the Equilibrium Constant, K_c (14.3)

$$aA + bB \rightleftharpoons cC + dD$$

$$K = \frac{[C]^c[D]^d}{[A]^a[B]^b} \quad \text{(equilibrium concentrations only)}$$

Relationship between the Equilibrium Constant and the Chemical Equation (14.3)

1 If you reverse the equation, invert the equilibrium constant.

2 If you multiply the coefficients in the equation by a factor, raise the equilibrium constant to the same factor.

3 If you add two or more individual chemical equations to obtain an overall equation, multiply the corresponding equilibrium constants by each other to obtain the overall equilibrium constant.

Expression for the Equilibrium Constant, K_p (14.4)

$$aA + bB \rightleftharpoons cC + dD$$

$$K_p = \frac{P_C^c P_D^d}{P_A^a P_B^b} \quad \text{(equilibrium partial pressures only)}$$

Relationship between the Equilibrium Constants, K_c and K_p (14.4)

$$K_p = K_c(RT)^{\Delta n}$$

The Reaction Quotient, Q_c (14.7)

$$aA + bB \rightleftharpoons cC + dD$$

$$Q_c = \frac{[C]^c[D]^d}{[A]^a[B]^b} \quad \text{(concentrations at any point in the reaction)}$$

The Reaction Quotient, Q_p (14.7)

$$aA + bB \rightleftharpoons cC + dD$$

$$Q_P = \frac{P_C^c P_D^d}{P_A^a P_B^b} \quad \text{(partial pressures at any point in the reaction)}$$

Relationship of Q to the Direction of the Reaction (14.7)

$Q < K$ Reaction goes to the right.

$Q > K$ Reaction goes to the left.

$Q = K$ Reaction is at equilibrium.

Key Skills

Expressing Equilibrium Constants for Chemical Equations (14.3)
- Example 14.1 • For Practice 14.1 • Exercises 21, 22

Manipulating the Equilibrium Constant to Reflect Changes in the Chemical Equation (14.3)
- Example 14.2 • For Practice 14.2 • For More Practice 14.2 • Exercises 27–30

Relating K_p and K_c (14.4)
- Example 14.3 • For Practice 14.3 • Exercises 31, 32

Writing Equilibrium Expressions for Reactions Involving a Solid or a Liquid (14.5)
- Example 14.4 • For Practice 14.4 • Exercises 33, 34

Finding Equilibrium Constants from Experimental Concentration Measurements (14.6)
- Examples 14.5, 14.6 • For Practice 14.5, 14.6 • Exercises 35, 36, 41, 42

Predicting the Direction of a Reaction by Comparing Q and K (14.7)
- Example 14.7 • For Practice 14.7 • Exercises 45–48

Calculating Equilibrium Concentrations from the Equilibrium Constant and One or More Equilibrium Concentrations (14.8)
- Example 14.8 • For Practice 14.8 • Exercises 37–44

Finding Equilibrium Concentrations from Initial Concentrations and the Equilibrium Constant (14.8)
- Examples 14.9, 14.10 • For Practice 14.9, 14.10 • Exercises 51–56

Calculating Equilibrium Partial Pressures from the Equilibrium Constant and Initial Partial Pressures (14.8)
- Example 14.11 • For Practice 14.11 • Exercises 57, 58

Finding Equilibrium Concentrations from Initial Concentrations in Cases with a Small Equilibrium Constant (14.8)
- Examples 14.12, 14.13 • For Practice 14.12, 14.13 • Exercises 59, 60

Determining the Effect of a Concentration Change on Equilibrium (14.9)
- Example 14.14 • For Practice 14.14 • Exercises 61–64

Determining the Effect of a Volume Change on Equilibrium (14.9)
- Example 14.15 • For Practice 14.15 • Exercises 65, 66

Determining the Effect of a Temperature Change on Equilibrium (14.9)
- Example 14.16 • For Practice 14.16 • Exercises 67, 68

Exercises

Review Questions

1. Explain how a developing fetus in the womb gets oxygen.

2. What is dynamic equilibrium? Why is it called *dynamic*?

3. Give the general expression for the equilibrium constant of the following generic reaction: $aA + bB \rightleftharpoons cC + dD$.

4. What is the significance of the equilibrium constant? What does a large equilibrium constant tell us about a reaction? A small one?

5. What happens to the value of the equilibrium constant for a reaction if the reaction equation is reversed? Multiplied by a constant?

6. If two reactions sum to an overall reaction, and the equilibrium constants for the two reactions are K_1 and K_2, what is the equilibrium constant for the overall reaction?

7. Explain the difference between K_c and K_p. For a given reaction, how are the two constants related?

8. What units should be used when expressing concentrations or partial pressures in the equilibrium constant? What are the units of K_p and K_c? Explain.

9. Why are the concentrations of solids and liquids omitted from equilibrium expressions?

10. Does the value of the equilibrium constant depend on the initial concentrations of the reactants and products? Do the equilibrium concentrations of the reactants and products depend on their initial concentrations? Explain.

11. Explain how you might deduce the equilibrium constant for a reaction in which you know the initial concentrations of the reactants and products and the equilibrium concentration of only one reactant or product.

12. What is the definition of the reaction quotient (Q) for a reaction? What does Q measure?

13. What is the value of Q when each reactant and product is in its standard state? (See Section 6.8 for the definition of standard states.)

14. In what direction will a reaction proceed for each of the following conditions: (a) $Q < K$; (b) $Q > K$; and (c) $Q = K$?

15. Many equilibrium calculations involve finding the equilibrium concentrations of reactants and products given their initial concentrations and the equilibrium constant. Outline the general procedure used in solving these kinds of problems.

16. In equilibrium problems involving equilibrium constants that are small relative to the initial concentrations of reactants, we can often assume that the quantity x (which represents how far the reaction proceeds toward products) is small. When this assumption is made, the quantity x can be ignored when it is subtracted from a large number, but not when it is multiplied by a large number. In other words, $2.5 - x \approx 2.5$, but $2.5x \neq 2.5$. Explain why a small x can be ignored in the first case, but not in the second.

17. What happens to a chemical system at equilibrium when that equilibrium is disturbed?

18. What is the effect of a change in concentration of a reactant or product on a chemical reaction initially at equilibrium?

19. What is the effect of a change in volume on a chemical reaction (that includes gaseous reactants or products) initially at equilibrium?

20. What is the effect of a temperature change on a chemical reaction initially at equilibrium? How does the effect differ for an exothermic reaction compared to an endothermic one?

Problems by Topic

Equilibrium and the Equilibrium Constant Expression

21. Write an expression for the equilibrium constant of each of the following chemical equations.
    a. $SbCl_5(g) \rightleftharpoons SbCl_3(g) + Cl_2(g)$
    b. $2\,BrNO(g) \rightleftharpoons 2\,NO(g) + Br_2(g)$
    c. $CH_4(g) + 2\,H_2S(g) \rightleftharpoons CS_2(g) + 4\,H_2(g)$
    d. $2\,CO(g) + O_2(g) \rightleftharpoons 2\,CO_2(g)$

22. Find and fix each mistake in the following equilibrium constant expressions.
    a. $2\,H_2S(g) \rightleftharpoons 2\,H_2(g) + S_2(g)$

    $$K_{eq} = \frac{[H_2][S_2]}{[H_2S]}$$

    b. $CO(g) + Cl_2(g) \rightleftharpoons COCl_2(g)$

    $$K_{eq} = \frac{[CO][Cl_2]}{[COCl_2]}$$

23. When the following reaction comes to equilibrium, will the concentrations of the reactants or products be greater? Does the answer to this question depend on the initial concentrations of the reactants and products?

    $$A(g) + B(g) \rightleftharpoons 2\,C(g) \quad K_c = 1.4 \times 10^{-5}$$

24. Ethene (C_2H_4) can be halogenated by the following reaction:

    $$C_2H_4(g) + X_2(g) \rightleftharpoons C_2H_4X_2(g)$$

    where X_2 can be Cl_2 (green), Br_2 (brown), or I_2 (purple). Examine the three figures below representing equilibrium concentrations in this reaction at the same temperature for the three different halogens. Rank the equilibrium constants for these three reactions from largest to smallest.

(a)   (b)

(c)

25. H_2 and I_2 are combined in a flask and allowed to react according to the following reaction:

$$H_2(g) + I_2(g) \rightleftharpoons 2\,HI(g)$$

Examine the figures below (sequential in time) and answer the following questions:

a. Which figure represents the point at which equilibrium is reached?

b. How would the series of figures change in the presence of a catalyst?

c. Would the final figure (vi) have different amounts of reactants and products in the presence of a catalyst?

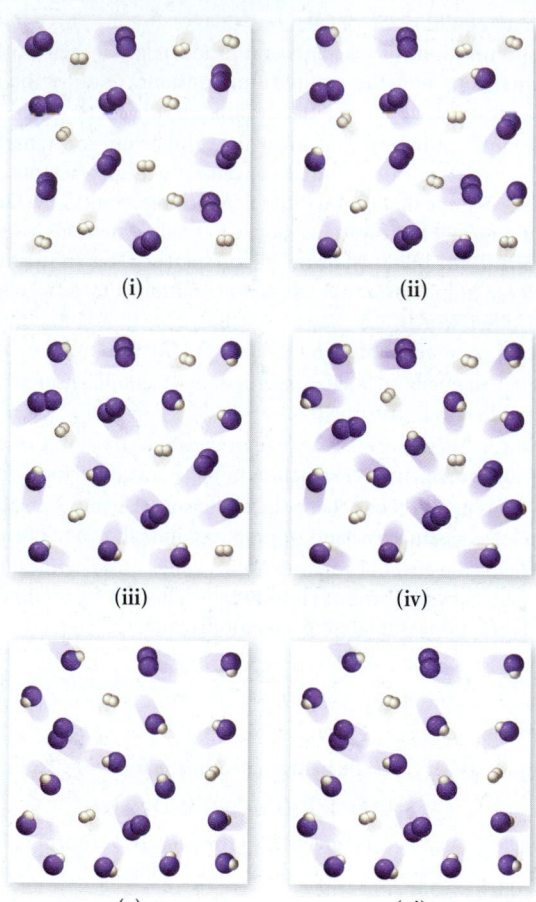

(i)   (ii)

(iii)   (iv)

(v)   (vi)

26. A chemist trying to synthesize a particular compound attempts two different synthesis reactions. The equilibrium constants for the two reactions are 23.3 and 2.2×10^4 at room temperature. However, upon carrying out both reactions for 15 minutes, the chemist finds that the reaction with the smaller equilibrium constant produced more of the desired product. Explain how this might be possible.

27. The reaction below has an equilibrium constant of $K_p = 2.26 \times 10^4$ at 298 K.

$$CO(g) + 2\,H_2(g) \rightleftharpoons CH_3OH(g)$$

Calculate K_p for each of the following reactions and predict whether reactants or products will be favored at equilibrium:

a. $CH_3OH(g) \rightleftharpoons CO(g) + 2\,H_2(g)$

b. $\frac{1}{2}CO(g) + H_2(g) \rightleftharpoons \frac{1}{2}CH_3OH(g)$

c. $2\,CH_3OH(g) \rightleftharpoons 2\,CO(g) + 4\,H_2(g)$

28. The reaction below has an equilibrium constant $K_p = 2.2 \times 10^6$ at 298 K.

$$2\,COF_2(g) \rightleftharpoons CO_2(g) + CF_4(g)$$

Calculate K_p for each of the following reactions and predict whether reactants or products will be favored at equilibrium:

a. $COF_2(g) \rightleftharpoons \frac{1}{2}CO_2(g) + \frac{1}{2}CF_4(g)$

b. $6\,COF_2(g) \rightleftharpoons 3\,CO_2(g) + 3\,CF_4(g)$

c. $2\,CO_2(g) + 2\,CF_4(g) \rightleftharpoons 4\,COF_2(g)$

29. Consider the following reactions and their respective equilibrium constants:

$$NO(g) + \tfrac{1}{2}Br_2(g) \rightleftharpoons NOBr(g) \qquad K_p = 5.3$$

$$2\,NO(g) \rightleftharpoons N_2(g) + O_2(g) \qquad K_p = 2.1 \times 10^{30}$$

Use these reactions and their equilibrium constants to predict the equilibrium constant for the following reaction:

$$N_2(g) + O_2(g) + Br_2(g) \rightleftharpoons 2\,NOBr(g)$$

30. Use the reactions below and their equilibrium constants to predict the equilibrium constant for the reaction, $2\,A(s) \rightleftharpoons 3\,D(g)$.

$$A(s) \rightleftharpoons \tfrac{1}{2}B(g) + C(g) \qquad K_1 = 0.0334$$

$$3\,D(g) \rightleftharpoons B(g) + 2\,C(g) \qquad K_2 = 2.35$$

K_p, K_c, and Heterogeneous Equilibria

31. Calculate K_c for each of the following reactions:

a. $I_2(g) \rightleftharpoons 2\,I(g) \quad K_p = 6.26 \times 10^{-22}$ (at 298 K)

b. $CH_4(g) + H_2O(g) \rightleftharpoons CO(g) + 3\,H_2(g)$

$$K_p = 7.7 \times 10^{24} \text{ (at 298 K)}$$

c. $I_2(g) + Cl_2(g) \rightleftharpoons 2\,ICl(g) \quad K_p = 81.9$ (at 298 K)

32. Calculate K_p for each of the following reactions:

a. $N_2O_4(g) \rightleftharpoons 2\,NO_2(g) \quad K_c = 5.9 \times 10^{-3}$ (at 298 K)

b. $N_2(g) + 3\,H_2(g) \rightleftharpoons 2\,NH_3(g)$

$$K_c = 3.7 \times 10^8 \text{ (at 298 K)}$$

c. $N_2(g) + O_2(g) \rightleftharpoons 2\,NO(g)$

$$K_c = 4.10 \times 10^{-31} \text{ (at 298 K)}$$

33. Write an equilibrium expression for each of the following chemical equations involving one or more solid or liquid reactants or products.

a. $CO_3^{2-}(aq) + H_2O(l) \rightleftharpoons HCO_3^-(aq) + OH^-(aq)$

b. $2\,KClO_3(s) \rightleftharpoons 2\,KCl(s) + 3\,O_2(g)$

c. $HF(aq) + H_2O(l) \rightleftharpoons H_3O^+(aq) + F^-(aq)$

d. $NH_3(aq) + H_2O(l) \rightleftharpoons NH_4^+(aq) + OH^-(aq)$

34. Find the mistake in the following equilibrium expression and fix it.

$$PCl_5(g) \rightleftharpoons PCl_3(l) + Cl_2(g)$$

$$K_{eq} = \frac{[PCl_3][Cl_2]}{[PCl_5]}$$

Relating the Equilibrium Constant to Equilibrium Concentrations and Equilibrium Partial Pressures

35. Consider the following reaction:

$$CO(g) + 2\,H_2(g) \rightleftharpoons CH_3OH(g)$$

An equilibrium mixture of this reaction at a certain temperature was found to have [CO] = 0.105 M, [H_2] = 0.114 M, and [CH_3OH] = 0.185 M. What is the value of the equilibrium constant (K_c) at this temperature?

36. Consider the following reaction:

$$NH_4HS(s) \rightleftharpoons NH_3(g) + H_2S(g)$$

An equilibrium mixture of this reaction at a certain temperature was found to have $[NH_3] = 0.278$ M and $[H_2S] = 0.355$ M. What is the value of the equilibrium constant (K_c) at this temperature?

37. Consider the following reaction:

$$N_2(g) + 3 H_2(g) \rightleftharpoons 2 NH_3(g)$$

Complete the following table. Assume that all concentrations are equilibrium concentrations in M.

T (K)	$[N_2]$	$[H_2]$	$[NH_3]$	$[K_c]$
500	0.115	0.105	0.439	___
575	0.110	___	0.128	9.6
775	0.120	0.140	___	0.0584

38. Consider the following reaction:

$$H_2(g) + I_2(g) \rightleftharpoons 2 HI(g)$$

Complete the following table. Assume that all concentrations are equilibrium concentrations in M.

T (°C)	$[H_2]$	$[I_2]$	$[HI]$	$[K_c]$
25	0.0355	0.0388	0.922	___
340	___	0.0455	0.387	9.6
445	0.0485	0.0468	___	50.2

39. Consider the following reaction:

$$2 NO(g) + Br_2(g) \rightleftharpoons 2 NOBr(g)$$
$$K_p = 28.4 \text{ at } 298 \text{ K}$$

In a reaction mixture at equilibrium, the partial pressure of NO is 118 torr and that of Br_2 is 176 torr. What is the partial pressure of NOBr in this mixture?

40. Consider the following reaction:

$$SO_2Cl_2(g) \rightleftharpoons SO_2(g) + Cl_2(g)$$
$$K_p = 2.91 \times 10^3 \text{ at } 298 \text{ K}$$

In a reaction at equilibrium, the partial pressure of SO_2 is 117 torr and that of Cl_2 is 255 torr. What is the partial pressure of SO_2Cl_2 in this mixture?

41. Consider the following reaction:

$$Fe^{3+}(aq) + SCN^-(aq) \rightleftharpoons FeSCN^{2+}(aq)$$

A solution is made containing an initial $[Fe^{3+}]$ of 1.0×10^{-3} M and an initial $[SCN^-]$ of 8.0×10^{-4} M. At equilibrium, $[FeSCN^{2+}] = 1.7 \times 10^{-4}$ M. Calculate the value of the equilibrium constant (K_c).

42. Consider the following reaction:

$$SO_2Cl_2(g) \rightleftharpoons SO_2(g) + Cl_2(g)$$

A reaction mixture is made containing an initial $[SO_2Cl_2]$ of 0.020 M. At equilibrium, $[Cl_2] = 1.2 \times 10^{-2}$ M. Calculate the value of the equilibrium constant (K_c).

43. Consider the following reaction:

$$H_2(g) + I_2(g) \rightleftharpoons 2 HI(g)$$

A reaction mixture in a 3.67-L flask at a certain temperature initially contains 0.763 g H_2 and 96.9 g I_2. At equilibrium, the flask contains 90.4 g HI. Calculate the equilibrium constant (K_c) for the reaction at this temperature.

44. Consider the following reaction:

$$CO(g) + 2 H_2(g) \rightleftharpoons CH_3OH(g)$$

A reaction mixture in a 5.19-L flask at a certain temperature contains 26.9 g CO and 2.34 g H_2. At equilibrium, the flask contains 8.65 g CH_3OH. Calculate the equilibrium constant (K_c) for the reaction at this temperature.

The Reaction Quotient and Reaction Direction

45. Consider the following reaction:

$$NH_4HS(s) \rightleftharpoons NH_3(g) + H_2S(g)$$

At a certain temperature, $K_c = 8.5 \times 10^{-3}$. A reaction mixture at this temperature containing solid NH_4HS has $[NH_3] = 0.166$ M and $[H_2S] = 0.166$ M. Will more of the solid form or will some of the existing solid decompose as equilibrium is reached?

46. Consider the following reaction:

$$2 H_2S(g) \rightleftharpoons 2 H_2(g) + S_2(g)$$
$$K_p = 2.4 \times 10^{-4} \text{ at } 1073 \text{ K}$$

A reaction mixture contains 0.112 atm of H_2, 0.055 atm of S_2, and 0.445 atm of H_2S. Is the reaction mixture at equilibrium? If not, in what direction will the reaction proceed?

47. Silver sulfate dissolves in water according to the following reaction:

$$Ag_2SO_4(s) \rightleftharpoons 2 Ag^+(aq) + SO_4^{2-}(aq)$$
$$K_c = 1.1 \times 10^{-5} \text{ at } 298 \text{ K}$$

A 1.5-L solution contains 6.55 g of dissolved silver sulfate. If additional solid silver sulfate is added to the solution, will it dissolve?

48. Nitrogen dioxide dimerizes according to the following reaction:

$$2 NO_2(g) \rightleftharpoons N_2O_4(g)$$
$$K_p = 6.7 \text{ at } 298 \text{ K}$$

A 2.25-L container contains 0.055 mol of NO_2 and 0.082 mol of N_2O_4 at 298 K. Is the reaction at equilibrium? If not, in what direction will the reaction proceed?

Finding Equilibrium Concentrations from Initial Concentrations and the Equilibrium Constant

49. Consider the following reaction and associated equilibrium constant:

$$aA(g) \rightleftharpoons bB(g) \qquad K_c = 2.0$$

Find the equilibrium concentrations of A and B for each of the following values of a and b. Assume that the initial concentration of A in each case is 1.0 M and that no B is present at the beginning of the reaction.

a. $a = 1; b = 1$ **b.** $a = 2; b = 2$

c. $a = 1; b = 2$

50. Consider the following reaction and associated equilibrium constant:

$$aA(g) + bB(g) \rightleftharpoons cC(g) \quad K_c = 4.0$$

Find the equilibrium concentrations of A, B, and C for each of the following values of a, b, and c. Assume that the initial concentrations of A and B are each 1.0 M and that no product is present at the beginning of the reaction.

a. $a = 1; b = 1; c = 2$
b. $a = 1; b = 1; c = 1$
c. $a = 2; b = 1; c = 1$ (set up equation for x; don't solve)

51. For the following reaction, $K_c = 0.513$ at 500 K.

$$N_2O_4(g) \rightleftharpoons 2 NO_2(g)$$

If a reaction vessel initially contains an N_2O_4 concentration of 0.0500 M at 500 K, what are the equilibrium concentrations of N_2O_4 and NO_2 at 500 K?

52. For the following reaction, $K_c = 255$ at 1000 K.

$$CO(g) + Cl_2(g) \rightleftharpoons COCl_2(g)$$

If a reaction mixture initially contains a CO concentration of 0.1500 M and a Cl_2 concentration of 0.175 M at 1000 K, what are the equilibrium concentrations of CO, Cl_2, and $COCl_2$ at 1000 K?

53. Consider the following reaction:

$$NiO(s) + CO(g) \rightleftharpoons Ni(s) + CO_2(g)$$
$$K_c = 4.0 \times 10^3 \text{ at } 1500 \text{ K}$$

If a mixture of solid nickel(II) oxide and 0.10 M carbon monoxide is allowed to come to equilibrium at 1500 K, what will be the equilibrium concentration of CO_2?

54. Consider the following reaction:

$$CO(g) + H_2O(g) \rightleftharpoons CO_2(g) + H_2(g)$$
$$K_c = 102 \text{ at } 500 \text{ K}$$

If a reaction mixture initially contains 0.125 M CO and 0.125 M H_2O, what will be the equilibrium concentration of each of the reactants and products?

55. Consider the following reaction:

$$HC_2H_3O_2(aq) + H_2O(l) \rightleftharpoons H_3O^+(aq) + C_2H_3O_2^-(aq)$$
$$K_c = 1.8 \times 10^{-5} \text{ at } 25 \text{ °C}$$

If a solution initially contains 0.210 M $HC_2H_3O_2$, what is the equilibrium concentration of H_3O^+ at 25 °C?

56. Consider the following reaction:

$$SO_2Cl_2(g) \rightleftharpoons SO_2(g) + Cl_2(g)$$
$$K_c = 2.99 \times 10^{-7} \text{ at } 227 \text{ °C}$$

If a reaction mixture initially contains 0.175 M SO_2Cl_2, what is the equilibrium concentration of Cl_2 at 227 °C?

57. Consider the following reaction:

$$Br_2(g) + Cl_2(g) \rightleftharpoons 2 BrCl(g)$$
$$K_p = 1.11 \times 10^{-4} \text{ at } 150 \text{ K}$$

A reaction mixture initially contains a Br_2 partial pressure of 755 torr and a Cl_2 partial pressure of 735 torr at 150 K. Calculate the equilibrium partial pressure of BrCl.

58. Consider the following reaction:

$$CO(g) + H_2O(g) \rightleftharpoons CO_2(g) + H_2(g)$$
$$K_p = 0.0611 \text{ at } 2000 \text{ K}$$

A reaction mixture initially contains a CO partial pressure of 1344 torr and a H_2O partial pressure of 1766 torr at 2000 K. Calculate the equilibrium partial pressures of each of the products.

59. Consider the following reaction:

$$A(g) \rightleftharpoons B(g) + C(g)$$

Find the equilibrium concentrations of A, B, and C for each of the following different values of K_c. Assume that the initial concentration of A in each case is 1.0 M and that the reaction mixture initially contains no products. Make any appropriate simplifying assumptions.

a. $K_c = 1.0$
b. $K_c = 0.010$
c. $K_c = 1.0 \times 10^{-5}$

60. Consider the following reaction:

$$A(g) \rightleftharpoons 2 B(g)$$

Find the equilibrium partial pressures of A and B for each of the following different values of K_p. Assume that the initial partial pressure of B in each case is 1.0 atm and that the initial partial pressure of A is 0.0 atm. Make any appropriate simplifying assumptions.

a. $K_c = 1.0$
b. $K_c = 1.0 \times 10^{-4}$
c. $K_c = 1.0 \times 10^5$

Le Châtelier's Principle

61. Consider the following reaction at equilibrium:

$$CO(g) + Cl_2(g) \rightleftharpoons COCl_2(g)$$

Predict whether the reaction will shift left, shift right, or remain unchanged upon each of the following disturbances:

a. $COCl_2$ is added to the reaction mixture.
b. Cl_2 is added to the reaction mixture.
c. $COCl_2$ is removed from the reaction mixture.

62. Consider the following reaction at equilibrium:

$$2 BrNO(g) \rightleftharpoons 2 NO(g) + Br_2(g)$$

Predict whether the reaction will shift left, shift right, or remain unchanged upon each of the following disturbances.

a. NO is added to the reaction mixture.
b. BrNO is added to the reaction mixture.
c. Br_2 is removed from the reaction mixture.

63. Consider the following reaction at equilibrium:

$$2 KClO_3(s) \rightleftharpoons 2 KCl(s) + 3 O_2(g)$$

Predict whether the reaction will shift left, shift right, or remain unchanged upon each of the following disturbances.

a. O_2 is removed from the reaction mixture.
b. KCl is added to the reaction mixture.
c. $KClO_3$ is added to the reaction mixture.
d. O_2 is added to the reaction mixture.

64. Consider the following reaction at equilibrium:

$$C(s) + H_2O(g) \rightleftharpoons CO(g) + H_2(g)$$

Predict whether the reaction will shift left, shift right, or remain unchanged upon each of the following disturbances.

a. C is added to the reaction mixture.
b. H_2O is condensed and removed from the reaction mixture.
c. CO is added to the reaction mixture.
d. H_2 is removed from the reaction mixture.

65. Each of the following reactions is allowed to come to equilibrium and then the volume is changed as indicated. Predict the effect (shift right, shift left, or no effect) of the indicated volume change.
   a. $I_2(g) \rightleftharpoons 2\,I(g)$ (volume is increased)
   b. $2\,H_2S(g) \rightleftharpoons 2\,H_2(g) + S_2(g)$ (volume is decreased)
   c. $I_2(g) + Cl_2(g) \rightleftharpoons 2\,ICl(g)$ (volume is decreased)

66. Each of the following reactions is allowed to come to equilibrium and then the volume is changed as indicated. Predict the effect (shift right, shift left, or no effect) of the indicated volume change.
   a. $CO(g) + H_2O(g) \rightleftharpoons CO_2(g) + H_2(g)$ (volume is decreased)
   b. $PCl_3(g) + Cl_2(g) \rightleftharpoons PCl_5(g)$ (volume is increased)
   c. $CaCO_3(s) \rightleftharpoons CaO(s) + CO_2(g)$ (volume is increased)

67. The following reaction is endothermic.
$$C(s) + CO_2(g) \rightleftharpoons 2\,CO(g)$$
   Predict the effect (shift right, shift left, or no effect) of increasing and decreasing the reaction temperature. How does the value of the equilibrium constant depend on temperature?

68. The following reaction is exothermic.
$$C_6H_{12}O_6(s) + 6\,O_2(g) \rightleftharpoons 6\,CO_2(g) + 6\,H_2O(g)$$
   Predict the effect (shift right, shift left, or no effect) of increasing and decreasing the reaction temperature. How does the value of the equilibrium constant depend on temperature?

69. Coal, which is primarily carbon, can be converted to natural gas, primarily CH_4, by the following exothermic reaction.
$$C(s) + 2\,H_2(g) \rightleftharpoons CH_4(g)$$
   Which of the following will favor CH_4 at equilibrium?
   a. adding more C to the reaction mixture
   b. adding more H_2 to the reaction mixture
   c. raising the temperature of the reaction mixture
   d. lowering the volume of the reaction mixture
   e. adding a catalyst to the reaction mixture
   f. adding neon gas to the reaction mixture

70. Coal can be used to generate hydrogen gas (a potential fuel) by the following endothermic reaction.
$$C(s) + H_2O(g) \rightleftharpoons CO(g) + H_2(g)$$
   If this reaction mixture is at equilibrium, predict whether each of the following will result in the formation of additional hydrogen gas, the formation of less hydrogen gas, or have no effect on the quantity of hydrogen gas.
   a. adding more C to the reaction mixture
   b. adding more H_2O to the reaction mixture
   c. raising the temperature of the reaction mixture
   d. increasing the volume of the reaction mixture
   e. adding a catalyst to the reaction mixture
   f. adding an inert gas to the reaction mixture

Cumulative Problems

71. Carbon monoxide replaces oxygen in oxygenated hemoglobin according to the following reaction:
$$HbO_2(aq) + CO(aq) \rightleftharpoons HbCO(aq) + O_2(aq)$$
   a. Use the following reactions and associated equilibrium constants at body temperature to find the equilibrium constant for the above reaction.
$$Hb(aq) + O_2(aq) \rightleftharpoons HbO_2(aq) \quad K_c = 1.8$$
$$Hb(aq) + CO(aq) \rightleftharpoons HbCO(aq) \quad K_c = 306$$
   b. Suppose that an air mixture becomes polluted with carbon monoxide at a level of 0.10%. Assuming the air contains 20.0% oxygen, and that the oxygen and carbon monoxide ratios that dissolve in the blood are identical to the ratios in the air, what would be the ratio of HbCO to HbO_2 in the bloodstream? Comment on the toxicity of carbon monoxide.

72. Nitrogen oxide is a pollutant in the lower atmosphere that irritates the eyes and lungs and leads to the formation of acid rain. Nitrogen oxide forms naturally in atmosphere according to the following endothermic reaction:
$$N_2(g) + O_2(g) \rightleftharpoons 2\,NO(g) \quad K_p = 4.1 \times 10^{-31} \text{ at 298 K.}$$
   Use the ideal gas law to calculate the concentrations of nitrogen and oxygen present in air at a pressure of 1.0 atm and a temperature of 298 K. Assume that nitrogen composes 78% of air by volume and that oxygen composes 21% of air. Find the "natural" equilibrium concentration of NO in air in units of molecules/cm^3. How would you expect this concentration to change in an automobile engine in which combustion is occurring?

73. Consider the following exothermic reaction:
$$C_2H_4(g) + Cl_2(g) \rightleftharpoons C_2H_4Cl_2(g)$$
   If you were a chemist trying to maximize the amount of $C_2H_4Cl_2$ produced, which of the following might you try? Assume that the reaction mixture reaches equilibrium.
   a. increasing the reaction volume
   b. removing $C_2H_4Cl_2$ from the reaction mixture as it forms
   c. lowering the reaction temperature
   d. adding Cl_2

74. Consider the following endothermic reaction:
$$C_2H_4(g) + I_2(g) \rightleftharpoons C_2H_4I_2(g)$$
   If you were a chemist trying to maximize the amount of $C_2H_4I_2$ produced, which of the following might you try? Assume that the reaction mixture reaches equilibrium.
   a. decreasing the reaction volume
   b. removing I_2 from the reaction mixture
   c. raising the reaction temperature
   d. adding C_2H_4 to the reaction mixture

75. Consider the following reaction:
$$H_2(g) + I_2(g) \rightleftharpoons 2\,HI(g)$$
   A reaction mixture at equilibrium at 175 K contains $P_{H_2} = 0.958$ atm, $P_{I_2} = 0.877$ atm, and $P_{HI} = 0.020$ atm. A second reaction mixture, also at 175 K, contains $P_{H_2} = P_{I_2} = 0.621$ atm, and $P_{HI} = 0.101$ atm. Is the second reaction at equilibrium? If not, what will be the partial pressure of HI when the reaction reaches equilibrium at 175 K?

76. Consider the following reaction:

$$2 H_2S(g) + SO_2(g) \rightleftharpoons 3 S(s) + 2 H_2O(g)$$

A reaction mixture initially containing 0.500 M H_2S and 0.500 M SO_2 was found to contain 0.0011 M H_2O at a certain temperature. A second reaction mixture at the same temperature initially contains $[H_2S] = 0.250$ M and $[SO_2] = 0.325$ M. Calculate the equilibrium concentration of H_2O in the second mixture at this temperature.

77. Ammonia can be synthesized according to the following reaction:

$$N_2(g) + 3 H_2(g) \rightleftharpoons 2 NH_3(g)$$
$$K_p = 5.3 \times 10^{-5} \text{ at } 725 \text{ K}$$

A 200.0-L reaction container initially contains 1.27 kg of N_2 and 0.310 kg of H_2 at 725 K. Assuming ideal gas behavior, calculate the mass of NH_3 (in g) present in the reaction mixture at equilibrium. What is the percent yield of the reaction under these conditions?

78. Hydrogen can be extracted from natural gas according to the following reaction:

$$CH_4(g) + CO_2(g) \rightleftharpoons 2 CO(g) + 2 H_2(g)$$
$$K_p = 4.5 \times 10^2 \text{ at } 825 \text{ K}$$

An 85.0-L reaction container initially contains 22.3 kg of CH_4 and 55.4 kg of CO_2 at 825 K. Assuming ideal gas behavior, calculate the mass of H_2 (in g) present in the reaction mixture at equilibrium. What is the percent yield of the reaction under these conditions?

79. The system described by the reaction

$$CO(g) + Cl_2(g) \rightleftharpoons COCl_2(g)$$

is at equilibrium at a given temperature when $P_{CO} = 0.30$ atm, $P_{Cl_2} = 0.10$ atm, and $P_{COCl_2} = 0.60$ atm. An additional pressure of $Cl_2(g) = 0.40$ atm is added. Find the pressure of CO when the system returns to equilibrium.

80. A reaction vessel at 27 °C contains a mixture of SO_2 (P = 3.00 atm) and O_2 (P = 1.00 atm). When a catalyst is added the reaction

$$2 SO_2(g) + O_2(g) \rightleftharpoons 2 SO_3(g)$$

takes place. At equilibrium the total pressure is 3.75 atm. Find the value of K_c.

81. At 70 K, CCl_4 decomposes to carbon and chlorine. The K_p for the decomposition is 0.76. Find the starting pressure of CCl_4 at this temperature that will produce a total pressure of 1.0 atm at equilibrium.

82. The equilibrium constant for the reaction $SO_2(g) + NO_2(g) \rightleftharpoons SO_3(g) + NO(g)$ is 3.0. Find the amount of NO_2 that must be added to 2.4 mol of SO_2 in order to form 1.2 mol of SO_3 at equilibrium.

83. A sample of $CaCO_3(s)$ is introduced into a sealed container of volume 0.654 L and heated to 1000 K until equilibrium is reached. The K_p for the reaction

$$CaCO_3(s) \rightleftharpoons CaO(s) + CO_2(g)$$

is 3.9×10^{-2} at this temperature. Calculate the mass of $CaO(s)$ that is present at equilibrium.

84. An equilibrium mixture contains N_2O_4, (P = 0.28 atm) and NO_2 (P = 1.1 atm) at 350 K. The volume of the container is doubled at constant temperature. Calculate the equilibrium pressures of the two gases when the system reaches a new equilibrium.

Challenge Problems

85. Consider the following reaction:

$$2 NO(g) + O_2(g) \rightleftharpoons 2 NO_2(g)$$

a. A reaction mixture at 175 K initially contains 522 torr of NO and 421 torr of O_2. At equilibrium, the total pressure in the reaction mixture is 748 torr. Calculate K_p at this temperature.

b. A second reaction mixture at 175 K initially contains 255 torr of NO and 185 torr of O_2. What is the equilibrium partial pressure of NO_2 in this mixture?

86. Consider the following reaction:

$$2 SO_2(g) + O_2(g) \rightleftharpoons 2 SO_3(g)$$
$$K_p = 0.355 \text{ at } 950 \text{ K}$$

A 2.75-L reaction vessel at 950 K initially contains 0.100 mol of SO_2 and 0.100 mol of O_2. Calculate the total pressure (in atmospheres) in the reaction vessel when equilibrium is reached.

87. Nitric oxide reacts with chlorine gas according to the following reaction:

$$2 NO(g) + Cl_2(g) \rightleftharpoons 2 NOCl(g)$$
$$K_p = 0.27 \text{ at } 700 \text{ K}$$

A reaction mixture initially contains equal partial pressures of NO and Cl_2. At equilibrium, the partial pressure of NOCl was measured to be 115 torr. What were the initial partial pressures of NO and Cl_2?

88. At a given temperature a system containing $O_2(g)$ and some oxides of nitrogen can be described by the following reactions:

$$2 NO(g) + O_2(g) \rightleftharpoons 2 NO_2(g) \quad K_p = 10^4$$
$$2NO_2(g) \rightleftharpoons N_2O_4(g) \quad K_p = 0.10$$

A pressure of 1 atm of $N_2O_4(g)$ is placed in a container at this temperature. Predict which, if any, component (other than N_2O_4) will be present at a pressure greater than 0.2 atm at equilibrium.

89. A sample of pure NO_2 is heated to 337 °C at which temperature it partially dissociates according to the equation

$$2\,NO_2(g) \rightleftharpoons 2\,NO(g) + O_2(g)$$

At equilibrium the density of the gas mixture is 0.520 g/L at 0.750 atm. Calculate K_c for the reaction.

90. When $N_2O_5(g)$ is heated it dissociates into $N_2O_3(g)$ and $O_2(g)$ according to the following reaction:

$$N_2O_5(g) \rightleftharpoons N_2O_3(g) + O_2(g) \quad K_c = 7.75 \text{ at a given temperature}$$

The $N_2O_3(g)$ dissociates to give $N_2O(g)$ and $O_2(g)$ according the following reaction:

$$N_2O_3(g) \rightleftharpoons N_2O(g) + O_2(g) \quad K_c = 4.00 \text{ at the same temperature}$$

When 4.00 mol of $N_2O_5(g)$ is heated in a 1.00-L reaction vessel to this temperature, the concentration of $O_2(g)$ at equilibrium is 4.50 mol/L. Find the concentrations of all the other species in the equilibrium system.

Conceptual Problems

91. The reaction $A(g) \rightleftharpoons 2\,B(g)$ has an equilibrium constant of $K_c = 1.0$ at a given temperature. If a reaction vessel contains equal initial amounts (in moles) of A and B, will the direction in which the reaction proceeds depend on the volume of the reaction vessel? Explain.

92. A particular reaction has an equilibrium constant of $K_p = 0.50$. A reaction mixture is prepared in which all the reactants and products are in their standard states. In which direction will the reaction proceed?

93. Consider the following reaction:

$$aA(g) \rightleftharpoons bB(g)$$

Each of the following entries in the table below represents equilibrium partial pressures of A and B under different initial conditions. What are the values of a and b in the reaction?

P_A (atm)	P_B (atm)
4.0	2.0
2.0	1.4
1.0	1.0
0.50	0.71
0.25	0.50

15

Acids and Bases

15.1 Heartburn

15.2 The Nature of Acids and Bases

15.3 Definitions of Acids and Bases

15.4 Acid Strength and the Acid Dissociation Constant (K_a)

15.5 Autoionization of Water and pH

15.6 Finding the [H_3O^+] and pH of Strong and Weak Acid Solutions

15.7 Base Solutions

15.8 The Acid–Base Properties of Ions and Salts

15.9 Polyprotic Acids

15.10 Acid Strength and Molecular Structure

15.11 Lewis Acids and Bases

15.12 Acid Rain

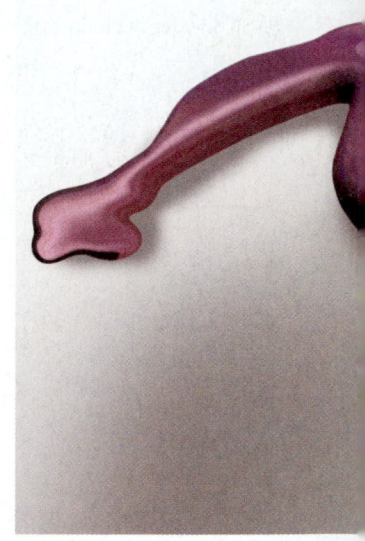

IN THIS CHAPTER, we apply the equilibrium concepts learned in the previous chapter to acid–base phenomena. Acids are common in many foods, such as limes, lemons, and vinegar, and in a number of consumer products, such as toilet cleaners and batteries. Bases are less common in foods but are found in a number of consumer products such as drain openers and antacids. We will examine three different models for acid–base behavior, all of which have different definitions of that behavior. In spite of their differences, the three models coexist, each being useful at explaining a particular range of acid–base phenomena. We will also learn how to calculate the acidity or basicity of solutions and define a useful scale, called the pH scale, to quantify acidity and basicity. These types of calculations often involve solving the kind of equilibrium problems that we examined in Chapter 14.

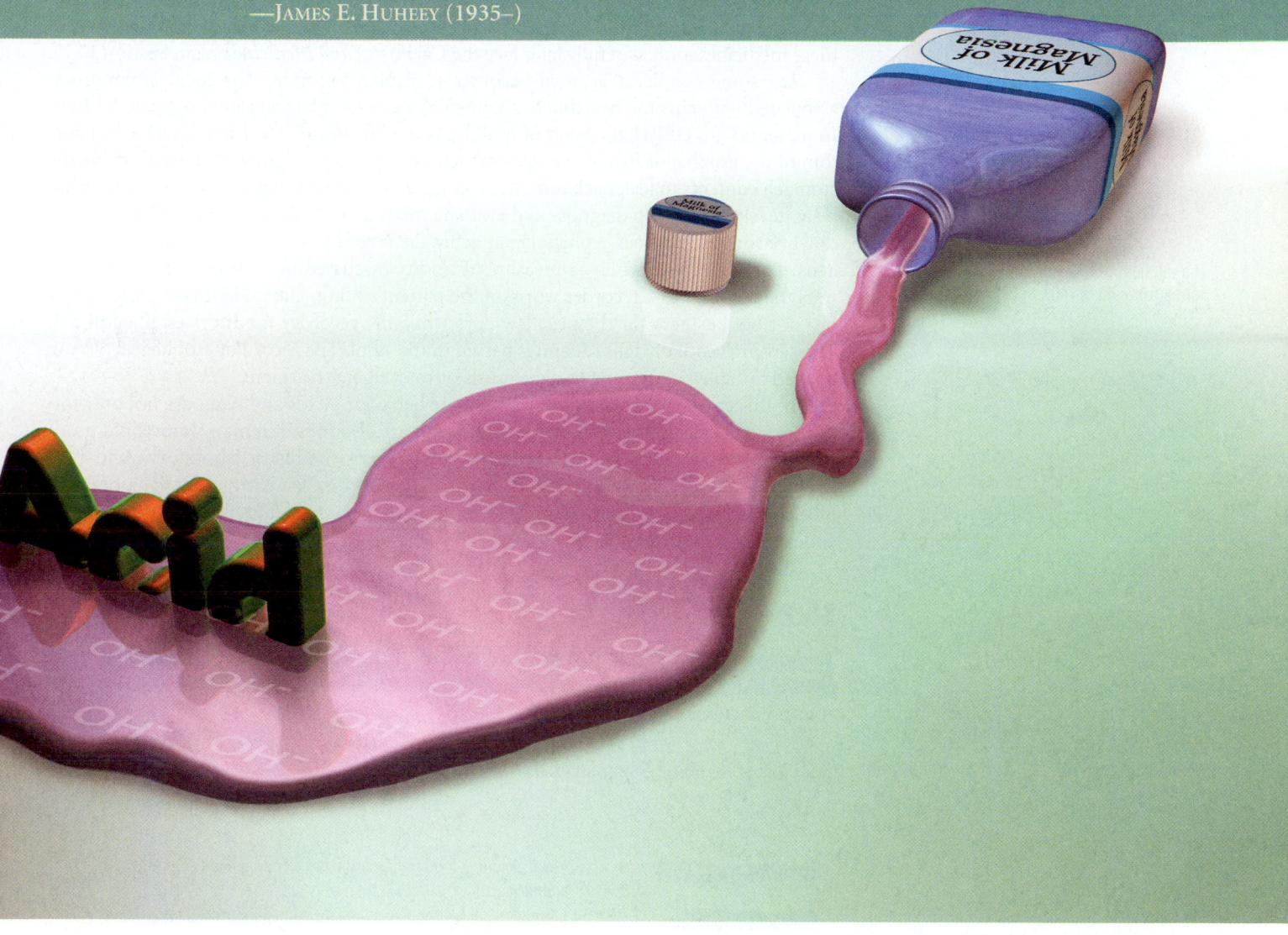

Milk of magnesia contains a base that can neutralize stomach acid and so relieve heartburn.

15.1 Heartburn

Heartburn is a painful burning sensation in the esophagus (the tube that joins the throat to the stomach) just below the chest. The pain is caused by hydrochloric acid (HCl), which is excreted in the stomach to kill microorganisms and activate enzymes that break down food. The hydrochloric acid can sometimes back up out of the stomach and into the esophagus, a phenomenon known as *acid reflux*. Recall from Section 4.8 that acids are substances that—by one definition that we will elaborate on shortly—produce H^+ ions in solution. When hydrochloric acid from the stomach comes in contact with the lining of the esophagus, the H^+ ions irritate the esophageal tissues, resulting in the burning sensation. Some of the acid can work its way into the lower throat and even the mouth, producing pain in the throat and a sour taste (characteristic of acids) in the mouth. Almost everyone experiences heartburn at some time, most likely after a large meal when the stomach is most full and the chances for reflux are greatest. Strenuous activity or lying in a horizontal position after a large meal increases the likelihood of stomach acid reflux and the resulting heartburn.

| Bases were first defined in Section 4.8.

The simplest way to relieve mild heartburn is to swallow repeatedly. Saliva contains the bicarbonate ion (HCO_3^-) that acts as a base and, when swallowed, neutralizes some of the acid in the esophagus. Later in this chapter, we will see just how bicarbonate acts as a base. Heartburn can also be treated with antacids such as Tums, milk of magnesia, or Mylanta. These over-the-counter medications contain more base than does saliva and therefore do a better job of neutralizing the esophageal acid. We will look at the bases in these medicines more carefully later (see the *Chemistry and Medicine* box in Section 15.7).

For some people, heartburn becomes a chronic problem. The medical condition associated with chronic heartburn is known as gastroesophageal reflux disease (GERD). In patients with GERD, the band of muscles (called the esophageal sphincter) at the bottom of the esophagus just above the stomach does not close tightly enough, allowing the stomach contents to leak back into the esophagus on a regular basis. A wireless sensor has been developed to help diagnose and evaluate treatment of GERD. Using a tube that goes down through the throat, a physician attaches the sensor to tissues in the patient's esophagus. The sensor reads pH—a measure of acidity discussed in Section 15.5—and transmits the readings to a recorder worn on the patient's body. The patient goes about his or her normal business for the next few days while the recorder monitors esophageal pH. The sensor eventually falls off and is passed in the stool. The record of esophageal pH can be read by a physician to make a diagnosis or to evaluate treatment.

| The concentration of stomach acid, [H_3O^+], varies from about 0.01 to 0.1 M.

In this chapter, we examine acid and base behavior. Acids and bases are not only important to our health (as we have just seen), but are also found in many household products, foods, medicines, and of course in nearly every chemistry laboratory. Acid–base chemistry is also central to much of biochemistry and molecular biology. The building blocks of proteins, for example, are amino acids and the molecules that carry the genetic code in DNA are bases.

15.2 The Nature of Acids and Bases

Acids have the following general properties: a sour taste; the ability to dissolve many metals; the ability to turn blue litmus paper red; and the ability to neutralize bases. Some common acids are listed in Table 15.1.

| For a review of naming acids, see Section 3.6.

TABLE 15.1 Some Common Acids	
Name	**Occurrence/Uses**
Hydrochloric acid (HCl)	Metal cleaning; food preparation; ore refining; main component of stomach acid
Sulfuric acid (H_2SO_4)	Fertilizer and explosives manufacturing; dye and glue production; automobile batteries; electroplating of copper
Nitric acid (HNO_3)	Fertilizer and explosives manufacturing; dye and glue production
Acetic acid ($HC_2H_3O_2$)	Plastic and rubber manufacturing; food preservative; active component of vinegar
Citric acid ($H_3C_6H_5O_3$)	Present in citrus fruits such as lemons and limes; used to adjust pH in foods and beverages
Carbonic acid (H_2CO_3)	Found in carbonated beverages due to the reaction of carbon dioxide with water
Hydrofluoric acid (HF)	Metal cleaning; glass frosting and etching
Phosphoric acid (H_3PO_4)	Fertilizer manufacture; biological buffering; preservative in beverages

| The formula for acetic acid can also be written as $CH_3COO\textbf{H}$.

Hydrochloric acid is found in most chemistry laboratories. It is used in industry to clean metals, to prepare and process some foods, and to refine metal ores. As we have just seen, hydrochloric acid is also the main component of stomach acid.

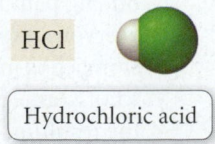

Hydrochloric acid

Sulfuric acid and nitric acid are also commonly used in the laboratory. They play major roles in the manufacture of fertilizers, explosives, dyes, and glues. Sulfuric acid, produced in larger quantities than any other chemical, is contained in most automobile batteries.

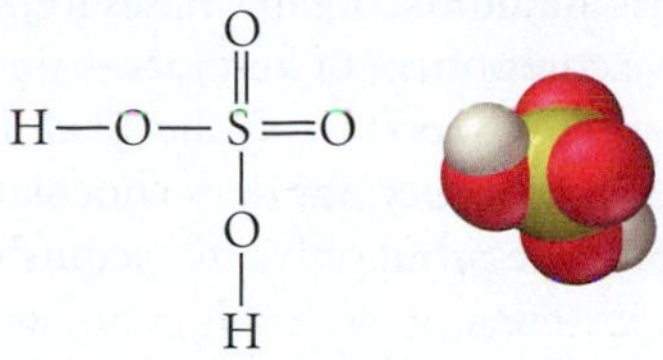

H_2SO_4

Sulfuric Acid

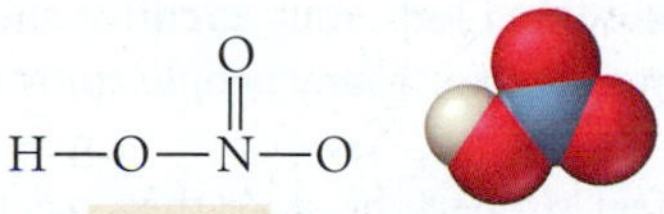

HNO_3

Nitric Acid

▲ The sour taste in vinegar is due to acetic acid.

Acetic acid is found in most people's homes as the active component of vinegar. It is also produced in improperly stored wines. The word vinegar originates from the French words *vin aigre*, which means sour wine. The presence of vinegar in wines is considered a serious fault, making the wine taste like salad dressing.

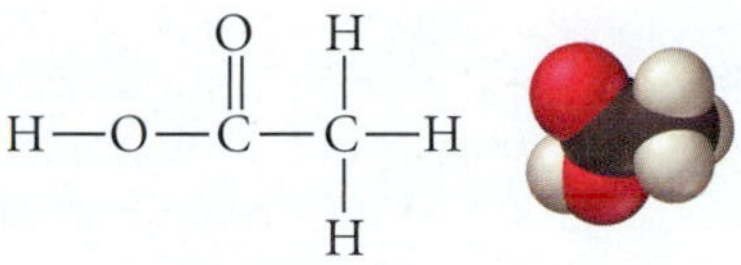

$HC_2H_3O_2$

Acetic acid

Acetic acid is an example of a **carboxylic acid**, an acid containing the following grouping of atoms:

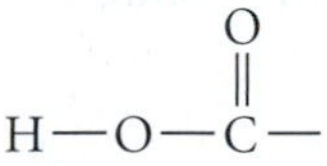

Carboxylic acid group

Carboxylic acids are often found in substances derived from living organisms. Other carboxylic acids include citric acid, the main acid in lemons and limes, and malic acid, an acid found in apples, grapes, and wine.

$H_3C_6H_5O_7$

$H_2C_4H_4O_5$

Citric acid

Malic acid

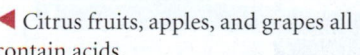

◀ Citrus fruits, apples, and grapes all contain acids.

Bases have the following general properties: a bitter taste; a slippery feel; the ability to turn red litmus paper blue; and the ability to neutralize acids. Because of their bitterness, bases are less common in foods than are acids. Our aversion to the taste of bases is probably an evolutionary adaptation to warn us against **alkaloids**, organic bases found in plants that are often poisonous. (For example, the active component of hemlock—the poisonous plant that caused the death of the Greek philosopher Socrates—is the alkaloid coniine.) Nonetheless, some foods, such as coffee and chocolate (especially dark chocolate), contain small amounts of base. Many people enjoy the bitterness, but only after acquiring the taste over time.

Bases feel slippery because they react with oils on the skin to form soaplike substances. Some household cleaning solutions, such as ammonia, are basic and have the characteristic slippery feel of a base. Bases turn red litmus paper blue; in the laboratory, litmus paper is routinely used to test the basicity of solutions.

Some common bases are listed in Table 15.2. Sodium hydroxide and potassium hydroxide are found in most chemistry laboratories. They are used in processing petroleum and cotton, and in soap and plastic manufacturing. Sodium hydroxide is the active ingredient in products such as Drano that work to unclog drains. Sodium bicarbonate can be found in most homes as baking soda and is also an active ingredient in many antacids.

Coffee is acidic overall, but bases present in coffee—such as caffeine—impart a bitter flavor.

▲ Many common household products and remedies contain bases.

TABLE 15.2 Common Bases

Name	Occurrence/Uses
Sodium hydroxide (NaOH)	Petroleum processing; soap and plastic manufacturing
Potassium hydroxide (KOH)	Cotton processing; electroplating; soap production; batteries
Sodium bicarbonate (NaHCO$_3$)	Antacid; ingredient of baking soda; source of CO$_2$
Sodium carbonate (Na$_2$CO$_3$)	Manufacture of glass and soap; general cleanser; water softener
Ammonia (NH$_3$)	Detergent; fertilizer and explosives manufacturing; synthetic fiber production

15.3 Definitions of Acids and Bases

What are the main characteristics of the molecules and ions that exhibit acid and base behavior? In this chapter, we examine three different definitions: the Arrhenius definition, the Brønsted–Lowry definition, and the Lewis definition. Why are there three definitions, and which one is correct? As Huheey notes in the quotation that opens this chapter, there really is no single "correct" definition. Rather, different definitions are convenient in different situations. The Lewis definition of acids and bases is discussed in Section 15.11; here we discuss the other two.

The Arrhenius Definition

In the 1880s, Swedish chemist Svante Arrhenius proposed the following molecular definitions of acids and bases:

Acid: A substance that produces H$^+$ ions in aqueous solution

Base: A substance that produces OH$^-$ ions in aqueous solution

For example, under the **Arrhenius definition**, HCl is an acid because it produces H$^+$ ions in solution (Figure 15.1 ◄):

$$HCl(aq) \longrightarrow H^+(aq) + Cl^-(aq)$$

Hydrogen chloride (HCl) is a covalent compound and does not contain ions. However, in water it *ionizes* completely to form H$^+(aq)$ ions and Cl$^-(aq)$ ions. The H$^+$ ions are highly reactive. In aqueous solution, they bond to water molecules in reactions such as the following:

$$H^+ + \;:\!\overset{\displaystyle H}{\underset{}{O}}\!:\!H \longrightarrow \left[\overset{\displaystyle H}{H\!:\!\overset{}{O}\!:\!H}\right]^+$$

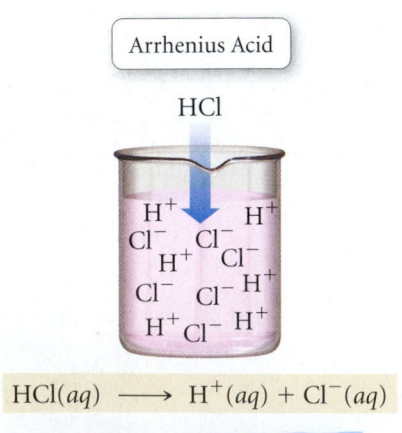

Arrhenius Acid

HCl

$$HCl(aq) \longrightarrow H^+(aq) + Cl^-(aq)$$

▲ FIGURE 15.1 Arrhenius Acid
An Arrhenius acid produces H$^+$ ions in solution.

The H_3O^+ ion is called the **hydronium ion**. In water, H^+ ions *always* associate with H_2O molecules to form hydronium ions and other associated species with the general formula $H(H_2O)_n^+$. For example, an H^+ ion can associate with two water molecules to form $H(H_2O)_2^+$, with three to form $H(H_2O)_3^+$, and so on. Chemists often use $H^+(aq)$ and $H_3O^+(aq)$ interchangeably, however, to mean the same thing—an H^+ ion that has been solvated in water.

NaOH is a base because it produces OH^- ions in solution (Figure 15.2 ▶):

$$NaOH(aq) \longrightarrow Na^+(aq) + OH^-(aq)$$

NaOH is an ionic compound and therefore contains Na^+ and OH^- ions. When NaOH is added to water, it *dissociates* or breaks apart into its component ions.

Under the Arrhenius definition, acids and bases naturally combine to form water, neutralizing each other in the process:

$$H^+(aq) + OH^-(aq) \longrightarrow H_2O(l)$$

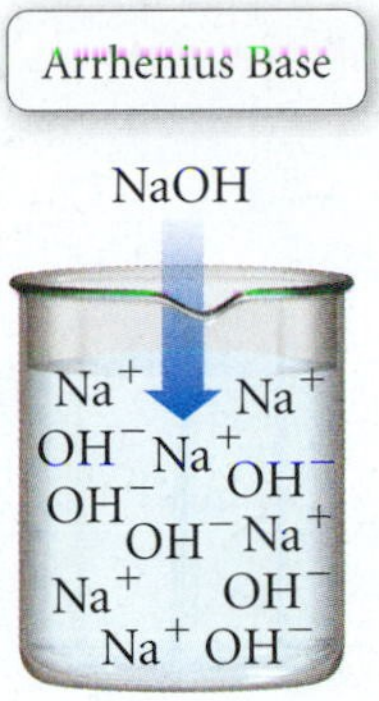

The Brønsted–Lowry Definition

A second, more widely applicable definition of acids and bases, called the **Brønsted–Lowry definition**, was introduced in 1923. This definition focuses on the *transfer of H^+ ions* in an acid–base reaction. Since an H^+ ion is a proton—a hydrogen atom without its electron—this definition focuses on the idea of a proton donor and a proton acceptor:

Acid: proton (H^+ ion) *donor*.

Base: proton (H^+ ion) *acceptor*.

Under this definition, HCl is an acid because, in solution, it donates a proton to water:

$$HCl(aq) + H_2O(l) \longrightarrow H_3O^+(aq) + Cl^-(aq)$$

This definition more clearly shows what happens to the H^+ ion from an acid—it associates with a water molecule to form H_3O^+ (a hydronium ion). The Brønsted–Lowry definition also works well with bases (such as NH_3) that do not inherently contain OH^- ions but that still produce OH^- ions in solution. In the Brønsted–Lowry definition, NH_3 is a base because it accepts a proton from water:

$$NH_3(aq) + H_2O(l) \rightleftharpoons NH_4^+(aq) + OH^-(aq)$$

In the Brønsted–Lowry definition, acids (proton donors) and bases (proton acceptors) always occur together. In the reaction between HCl and H_2O, HCl is the proton donor (acid) and H_2O is the proton acceptor (base):

$$\underset{\substack{\text{acid} \\ \text{(proton donor)}}}{HCl(aq)} + \underset{\substack{\text{base} \\ \text{(proton acceptor)}}}{H_2O(l)} \longrightarrow H_3O^+(aq) + Cl^-(aq)$$

In the reaction between NH_3 and H_2O, H_2O is the proton donor (acid) and NH_3 is the proton acceptor (base).

$$\underset{\substack{\text{base} \\ \text{(proton acceptor)}}}{NH_3(aq)} + \underset{\substack{\text{acid} \\ \text{(proton donor)}}}{H_2O(l)} \rightleftharpoons NH_4^+(aq) + OH^-(aq)$$

Notice that under the Brønsted–Lowry definition, some substances—such as water in the previous two equations—can act as acids *or* bases. Substances that can act as acids or bases are termed **amphoteric**. Notice also what happens when an equation representing Brønsted–Lowry acid–base behavior is reversed:

$$\underset{\substack{\text{acid} \\ \text{(proton donor)}}}{NH_4^+(aq)} + \underset{\substack{\text{base} \\ \text{(proton acceptor)}}}{OH^-(aq)} \rightleftharpoons NH_3(aq) + H_2O(l)$$

$$NaOH(aq) \longrightarrow Na^+(aq) + OH^-(aq)$$

▲ **FIGURE 15.2 Arrhenius Base**
An Arrhenius base produces OH^- ions in solution.

All Arrhenius acids and bases remain acids and bases under the Brønsted–Lowry definition. However, some Brønsted–Lowry acids and bases cannot naturally be classified as Arrhenius acids and bases.

In this reaction, NH_4^+ is the proton donor (acid) and OH^- is the proton acceptor (base). What was the base (NH_3) has become the acid (NH_4^+) and vice versa. NH_4^+ and NH_3 are often referred to as a **conjugate acid–base pair**, two substances related to each other by the transfer of a proton (Figure 15.3 ▼). Going back to the original forward reaction, we can identify the conjugate acid–base pairs as follows:

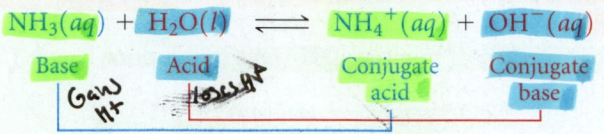

$$NH_3(aq) + H_2O(l) \rightleftharpoons NH_4^+(aq) + OH^-(aq)$$

Base   Acid   Conjugate acid   Conjugate base

In an acid–base reaction,

• A base accepts a proton and becomes a conjugate acid.
• An acid donates a proton and becomes a conjugate base.

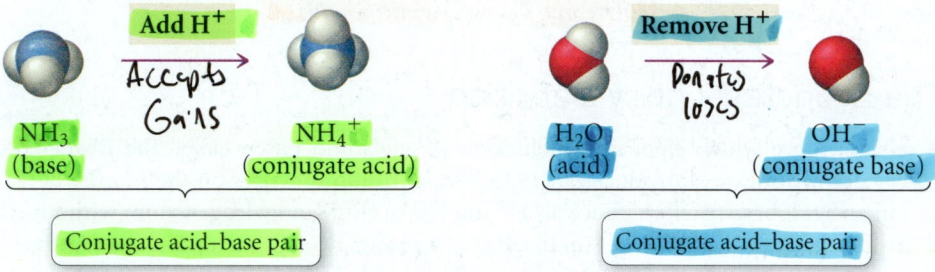

▶ **FIGURE 15.3 Conjugate Acid–base Pairs** A conjugate acid–base pair consists of two substances related to each other by the transfer of a proton.

EXAMPLE 15.1 Identifying Brønsted–Lowry Acids and Bases and Their Conjugates

In each of the following reactions, identify the Brønsted–Lowry acid, the Brønsted–Lowry base, the conjugate acid, and the conjugate base.

(a) $H_2SO_4(aq) + H_2O(l) \longrightarrow HSO_4^-(aq) + H_3O^+(aq)$
(b) $HCO_3^-(aq) + H_2O(l) \rightleftharpoons H_2CO_3(aq) + OH^-(aq)$

Solution

(a) Since H_2SO_4 donates a proton to H_2O in this reaction, it is the acid (proton donor). After H_2SO_4 donates the proton, it becomes HSO_4^-, the conjugate base. Since H_2O accepts a proton, it is the base (proton acceptor). After H_2O accepts the proton it becomes H_3O^+, the conjugate acid.

$$H_2SO_4(aq) + H_2O(l) \longrightarrow HSO_4^-(aq) + H_3O^+(aq)$$

$$H_2SO_4(aq) + H_2O(l) \longrightarrow HSO_4^-(aq) + H_3O^+(aq)$$

Acid   Base   Conjugate base   Conjugate acid

(b) Since H_2O donates a proton to HCO_3^- in this reaction, it is the acid (proton donor). After H_2O donates the proton, it becomes OH^-, the conjugate base. Since HCO_3^- accepts a proton, it is the base (proton acceptor). After HCO_3^- accepts the proton it becomes H_2CO_3, the conjugate acid.

$$HCO_3^-(aq) + H_2O(l) \rightleftharpoons H_2CO_3(aq) + OH^-(aq)$$

$$HCO_3^-(aq) + H_2O(l) \rightleftharpoons H_2CO_3(aq) + OH^-(aq)$$

Base   Acid   Conjugate acid   Conjugate base

For Practice 15.1

In each of the following reactions, identify the Brønsted–Lowry acid, the Brønsted–Lowry base, the conjugate acid, and the conjugate base.

(a) $C_5H_5N(aq) + H_2O(l) \rightleftharpoons C_5H_5NH^+(aq) + OH^-(aq)$
(b) $HNO_3(aq) + H_2O(l) \rightleftharpoons H_3O^+(aq) + NO_3^-(aq)$

> ### Conceptual Connection 15.1 Conjugate Acid–Base Pairs
>
> Which one of the following is not a conjugate acid–base pair?
>
> **(a)** $(CH_3)_3N$; $(CH_3)_3NH^+$ **(b)** H_2SO_4; H_2SO_3 **(c)** HNO_2; NO_2^-
>
> Answer: **(b)** H_2SO_4 and H_2SO_3 are two different acids, not a conjugate acid–base pair.

15.4 Acid Strength and the Acid Ionization Constant (K_a)

The strength of an electrolyte, first discussed in Section 4.5, is determined by the extent of the dissociation of the electrolyte into its component ions in solution. A *strong electrolyte* completely dissociates into ions in solution whereas a *weak electrolyte* only partially dissociates. Strong and weak acids are defined accordingly. A **strong acid** completely ionizes in solution whereas a **weak acid** only partially ionizes. In other words, the strength of an acid depends on the following equilibrium:

$$HA(aq) + H_2O(l) \rightleftharpoons H_3O^+(aq) + A^-(aq)$$

If the equilibrium lies far to the right, the acid is strong—it is virtually completely ionized. If the equilibrium lies to the left, the acid is weak—only a small percentage of the acid molecules are ionized. Of course, the range of acid strength is continuous, but for most purposes, the categories of strong and weak are useful.

Strong Acids

Hydrochloric acid (HCl) is an example of a strong acid.

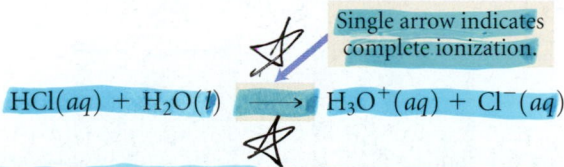

Single arrow indicates complete ionization.

$$HCl(aq) + H_2O(l) \longrightarrow H_3O^+(aq) + Cl^-(aq)$$

An HCl solution contains virtually no intact HCl; the HCl has essentially all ionized to form $H_3O^+(aq)$ and $Cl^-(aq)$ (Figure 15.4 ▼). A 1.0 M HCl solution will have an H_3O^+ concentration of 1.0 M. Abbreviating the concentration of H_3O^+ as $[H_3O^+]$, we say that a 1.0 M HCl solution has $[H_3O^+] = 1.0$ M.

A Strong Acid

When HCl dissolves in water, it ionizes completely.

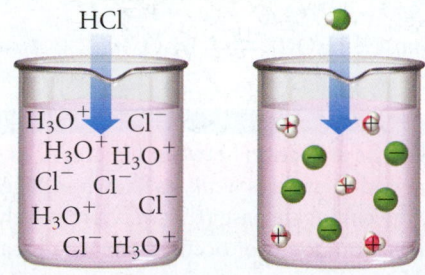

▲ **FIGURE 15.4 Ionization of a Strong Acid**   When HCl dissolves in water, it completely ionizes to form H_3O^+ and Cl^-. The solution contains virtually no intact HCl.

An ionizable proton is one that ionizes in solution. We discuss polyprotic acids in more detail in Section 15.9.

Table 15.3 lists the six important strong acids. The first five acids in the table are **monoprotic acids**, acids containing only one ionizable proton. Sulfuric acid is an example of a **diprotic acid**, an acid containing two ionizable protons.

TABLE 15.3 Strong Acids	
Hydrochloric acid (HCl)	Nitric acid (HNO₃)
Hydrobromic acid (HBr)	Perchloric acid (HClO₄)
Hydriodic acid (HI)	Sulfuric acid (H₂SO₄) (diprotic)

Weak Acids

The terms *strong* and *weak* acids are often confused with the terms *concentrated* and *dilute* acids. Can you articulate the difference between these terms?

In contrast to HCl, HF is an example of a weak acid, one that does not completely ionize in solution.

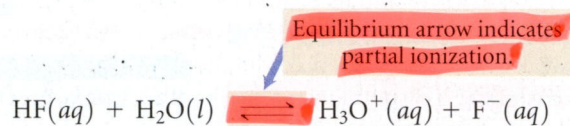

Equilibrium arrow indicates partial ionization.

$$HF(aq) + H_2O(l) \rightleftharpoons H_3O^+(aq) + F^-(aq)$$

An HF solution contains a lot of intact (or un-ionized) HF molecules; it also contains some $H_3O^+(aq)$ and $F^-(aq)$ (Figure 15.5 ◄). In other words, a 1.0 M HF solution has $[H_3O^+]$ that is much less than 1.0 M because only some of the HF molecules ionize to form H_3O^+.

► FIGURE 15.5 Ionization of a Weak Acid When HF dissolves in water, only a fraction of the dissolved molecules ionize to form H_3O^+ and F^-. The solution contains many intact HF molecules.

A Weak Acid

When HF dissolves in water, only a fraction of the molecules ionize.

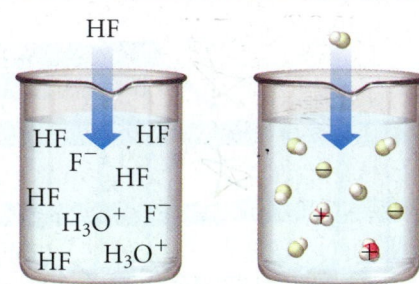

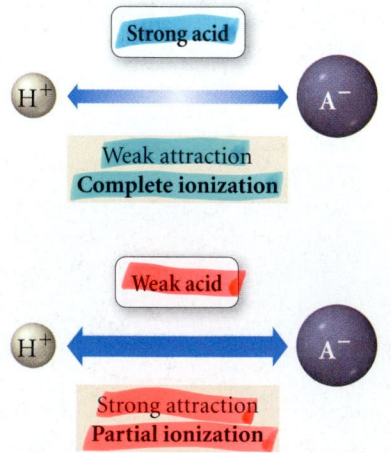

▲ FIGURE 15.6 Ionic Attraction and Acid Strength In a strong acid, the attraction between H^+ and A^- is weak, resulting in complete ionization. In a weak acid, the attraction between H^+ and A^- is strong, resulting in only partial ionization.

The degree to which an acid is strong or weak depends on the attraction between the anion of the acid (the conjugate base) and the hydrogen ion (relative to the attractions of these ions to water). Suppose HA is a generic formula for an acid. Then, the degree to which the following reaction proceeds in the forward direction depends on the strength of the attraction between H^+ and A^-.

$$\underset{\text{acid}}{HA(aq)} + H_2O(l) \rightleftharpoons H_3O^+(aq) + \underset{\text{conjugate base}}{A^-(aq)}$$

If the attraction between H^+ and A^- is *weak*, then the reaction favors the forward direction and the acid is *strong*. If the attraction between H^+ and A^- is *strong*, then the reaction favors the reverse direction and the acid is *weak*, as shown in Figure 15.6 ◄.

For example, in HCl, the conjugate base (Cl^-) has a relatively weak attraction to H^+, meaning that the reverse reaction does not occur to any significant extent. In HF, on the other hand, the conjugate base (F^-) has a greater attraction to H^+, meaning that the reverse reaction occurs to a significant degree. *In general, the stronger the acid, the weaker the conjugate base and vice versa*. This means that if the forward reaction (that of the acid) has high tendency to occur, then the reverse reaction (that of the conjugate base) has a low tendency to occur. Table 15.4 lists some common weak acids.

TABLE 15.4 Some Weak Acids

Hydrofluoric acid (HF)	Sulfurous acid (H_2SO_3) (*diprotic*)
Acetic acid ($HC_2H_3O_2$)	Carbonic acid (H_2CO_3) (*diprotic*)
Formic acid ($HCHO_2$)	Phosphoric acid (H_3PO_4) (*triprotic*)

The formulas for acetic acid and formic acid can also be written as CH_3COOH and $HCOOH$, respectively, to indicate that in these compounds the only H that ionizes is the one attached to an oxygen atom.

Notice that two of the weak acids in Table 15.4 are diprotic, meaning that they have two ionizable protons, and one is **triprotic (three ionizable protons)**. Let us return to sulfuric acid for a moment. Sulfuric acid is a diprotic acid that is strong in its first ionizable proton:

$$H_2SO_4(aq) + H_2O(l) \longrightarrow H_3O^+(aq) + HSO_4^-(aq)$$

but weak in its second ionizable proton:

$$HSO_4^-(aq) + H_2O(l) \rightleftharpoons H_3O^+(aq) + SO_4^{2-}(aq)$$

Sulfurous acid and carbonic acid are weak in both of their ionizable protons, and phosphoric acid is weak in all three of its ionizable protons. We discuss polyprotic acids in more detail in Section 15.9.

The Acid Ionization Constant (K_a)

The relative strengths of weak acids are quantified with the **acid ionization constant (K_a)**, which is simply the equilibrium constant for the ionization reaction of a weak acid. As we saw in Section 14.3, for the following two equivalent reactions:

$$HA(aq) + H_2O(l) \rightleftharpoons H_3O^+(aq) + A^-(aq)$$

$$HA(aq) \rightleftharpoons H^+(aq) + A^-(aq)$$

Sometimes K_a is also called the acid dissociation constant.

the equilibrium constant is

$$K_a = \frac{[H_3O^+][A^-]}{[HA]} = \frac{[H^+][A^-]}{[HA]}$$

Recall from Chapter 14 that the concentrations of pure solids or pure liquids are not included in the expression for K_c; therefore, $H_2O(l)$ is not included in the expression for K_a.

Since $[H_3O^+]$ is equivalent to $[H^+]$, both forms of the above expression are equal. Although the ionization constants for all weak acids are relatively small (otherwise the acid would not be a weak acid), they do vary in magnitude. The smaller the constant, the further to the left the equilibrium point for the ionization reaction lies, and the weaker the acid. Table 15.5 on page 672 lists the acid ionization constants for a number of common weak acids in order of decreasing acid strength.

Conceptual Connection 15.2 Conjugate Bases

Consider the following two acids and their K_a values:

HF	$K_a = 3.5 \times 10^{-4}$
HClO	$K_a = 2.9 \times 10^{-8}$

Which conjugate base, F^- or ClO^-, is stronger?

Answer: ClO^-, because the weaker the acid, the stronger the conjugate base.

TABLE 15.5   Acid Ionization Constants (K_a) for Some Monoprotic Weak Acids at 25°C

Acid	Formula	Structural Formula	Ionization Reaction	K_a
Chlorous acid	$HClO_2$	$H-O-Cl=O$	$HClO_2(aq) + H_2O(l) \rightleftharpoons$ $H_3O^+(aq) + ClO_2^-(aq)$	1.1×10^{-2}
Nitrous acid	HNO_2	$H-O-N=O$	$HNO_2(aq) + H_2O(l) \rightleftharpoons$ $H_3O^+(aq) + NO_2^-(aq)$	4.6×10^{-4}
Hydrofluoric acid	HF	$H-F$	$HF(aq) + H_2O(l) \rightleftharpoons$ $H_3O^+(aq) + F^-(aq)$	3.5×10^{-4}
Formic acid	$HCHO_2$	$H-O-\overset{\overset{\displaystyle O}{\|\|}}{C}-H$	$HCHO_2(aq) + H_2O(l) \rightleftharpoons$ $H_3O^+(aq) + CHO_2^-(aq)$	1.8×10^{-4}
Benzoic acid	$HC_7H_5O_2$	$H-O-\overset{\overset{\displaystyle O}{\|\|}}{C}-C_6H_5$	$HC_7H_5O_2(aq) + H_2O(l) \rightleftharpoons$ $H_3O^+(aq) + C_7H_5O_2^-(aq)$	6.5×10^{-5}
Acetic acid	$HC_2H_3O_2$	$H-O-\overset{\overset{\displaystyle O}{\|\|}}{C}-CH_3$	$HC_2H_3O_2(aq) + H_2O(l) \rightleftharpoons$ $H_3O^+(aq) + C_2H_3O_2^-(aq)$	1.8×10^{-5}
Hypochlorous acid	$HClO$	$H-O-Cl$	$HClO(aq) + H_2O(l) \rightleftharpoons$ $H_3O^+(aq) + ClO^-(aq)$	2.9×10^{-8}
Hydrocyanic acid	HCN	$H-C\equiv N$	$HCN(aq) + H_2O(l) \rightleftharpoons$ $H_3O^+(aq) + CN^-(aq)$	4.9×10^{-10}
Phenol	HC_6H_5O	$HO-C_6H_5$	$HC_6H_5O(aq) + H_2O(l) \rightleftharpoons$ $H_3O^+(aq) + C_6H_5O^-(aq)$	1.3×10^{-10}

15.5 Autoionization of Water and pH

We saw earlier that water acts as a base when it reacts with HCl and as an acid when it reacts with NH_3:

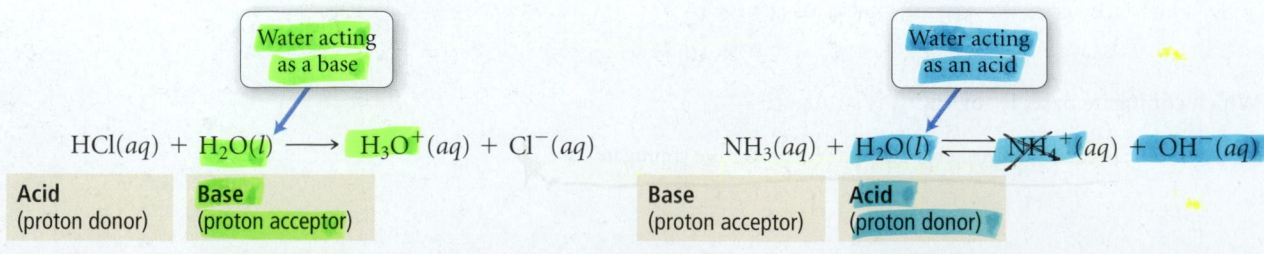

Water is *amphoteric*; it can act as either an acid or a base. Even in pure water, water acts as an acid and a base with itself, a process called **autoionization**:

> Water acting as both
> an acid and a base

$$H_2O(l) + H_2O(l) \rightleftharpoons H_3O^+(aq) + OH^-(aq)$$

Acid   **Base**
(proton donor)   (proton acceptor)

The autoionization reaction can also be written as follows:

$$H_2O(l) \rightleftharpoons H^+(aq) + OH^-(aq)$$

We can quantify the autoionization of water with the equilibrium constant for the autoionization reaction.

$$K_w = [H_3O^+][OH^-] = [H^+][OH^-]$$

This equilibrium constant is called the **ion product constant for water (K_w)** (sometimes called the *dissociation constant for water*). At 25 °C, $K_w = 1.0 \times 10^{-14}$. In pure water, since H_2O is the only source of these ions, the concentrations of H_3O^+ and OH^- are equal. Such a solution is said to be **neutral**. Since the concentrations are equal, they can be easily calculated from K_w.

$$[H_3O^+] = [OH^-] = \sqrt{K_w} = 1.0 \times 10^{-7} \quad \text{(in pure water at 25 °C)}$$

As you can see, in pure water, the concentrations of H_3O^+ and OH^- are *very small* (1.0×10^{-7} M) at room temperature.

An **acidic solution** contains an acid that creates additional H_3O^+ ions, causing $[H_3O^+]$ to increase. However, the *ion product constant still applies*:

$$[H_3O^+][OH^-] = K_w = 1.0 \times 10^{-14}$$

The concentration of H_3O^+ times the concentration of OH^- will always be 1.0×10^{-14} at 25 °C. If $[H_3O^+]$ increases, then $[OH^-]$ must decrease for the ion product constant to remain 1.0×10^{-14}. For example, if $[H_3O^+] = 1.0 \times 10^{-3}$, then $[OH^-]$ can be found by solving the ion product constant expression for $[OH^-]$:

$$(1.0 \times 10^{-3})[OH^-] = 1.0 \times 10^{-14}$$

$$[OH^-] = \frac{1.0 \times 10^{-14}}{1.0 \times 10^{-3}} = 1.0 \times 10^{-11} \text{ M}$$

In an acidic solution $[H_3O^+] > [OH^-]$.

A **basic solution** contains a base that creates additional OH^- ions, causing $[OH^-]$ to increase and $[H_3O^+]$ to decrease. For example, suppose $[OH^-] = 1.0 \times 10^{-2}$; then $[H_3O^+]$ can be found by solving the ion product constant expression for $[H_3O^+]$:

$$[H_3O^+](1.0 \times 10^{-2}) = 1.0 \times 10^{-14}$$

$$[H_3O^+] = \frac{1.0 \times 10^{-14}}{1.0 \times 10^{-2}} = 1.0 \times 10^{-12} \text{ M}$$

In a basic solution $[OH^-] > [H_3O^+]$.

Notice that changing $[H_3O^+]$ in an aqueous solution produces an inverse change in $[OH^-]$ and vice versa. We can summarize as follows:

- A *neutral solution* contains $[H_3O^+] = [OH^-] = 1.0 \times 10^{-7}$ M (at 25 °C).
- An *acidic solution* contains $[H_3O^+] > [OH^-]$.
- A *basic solution* contains $[OH^-] > [H_3O^+]$.
- In *all aqueous solutions* both H_3O^+ and OH^- are present, with $[H_3O^+][OH^-] = K_w = 1.0 \times 10^{-14}$ (at 25 °C).

EXAMPLE 15.2 Using K_w in Calculations

Calculate $[OH^-]$ at 25 °C for each of the following solutions and determine whether the solution is acidic, basic, or neutral.

(a) $[H_3O^+] = 7.5 \times 10^{-5}$ M   **(b)** $[H_3O^+] = 1.5 \times 10^{-9}$ M

(c) $[H_3O^+] = 1.0 \times 10^{-7}$ M

Solution

(a) To find $[OH^-]$ use the ion product constant. Substitute the given value for $[H_3O^+]$ and solve the equation for $[OH^-]$. Since $[H_3O^+] > [OH^-]$, the solution is acidic.	$[H_3O^+][OH^-] = K_w = 1.0 \times 10^{-14}$ $(7.5 \times 10^{-5})[OH^-] = 1.0 \times 10^{-14}$ $[OH^-] = \dfrac{1.0 \times 10^{-14}}{7.5 \times 10^{-5}} = 1.3 \times 10^{-10}$ M Acidic solution
(b) Substitute the given value for $[H_3O^+]$ and solve the acid ionization equation for $[OH^-]$. Since $[H_3O^+] < [OH^-]$, the solution is basic.	$(1.5 \times 10^{-9})[OH^-] = 1.0 \times 10^{-14}$ $[OH^-] = \dfrac{1.0 \times 10^{-14}}{1.5 \times 10^{-9}} = 6.7 \times 10^{-6}$ M Basic solution
(c) Substitute the given value for $[H_3O^+]$ and solve the acid ionization equation for $[OH^-]$. Since $[H_3O^+] = 1.0 \times 10^{-7}$ and $[OH^-] = 1.0 \times 10^{-7}$, the solution is neutral.	$(1.0 \times 10^{-7})[OH^-] = 1.0 \times 10^{-14}$ $[OH^-] = \dfrac{1.0 \times 10^{-14}}{1.0 \times 10^{-7}} = 1.0 \times 10^{-7}$ M Neutral solution

For Practice 15.2

Calculate $[H_3O^+]$ at 25 °C for each of the following solutions and determine whether the solution is acidic, basic, or neutral.

(a) $[OH^-] = 1.5 \times 10^{-2}$ M   **(b)** $[OH^-] = 1.0 \times 10^{-7}$ M

(c) $[OH^-] = 8.2 \times 10^{-10}$ M

The pH Scale: A Way to Quantify Acidity and Basicity

The pH scale is a compact way to specify the acidity of a solution. We define **pH** as follows:

$$pH = -\log[H_3O^+]$$

A solution with $[H_3O^+] = 1.0 \times 10^{-3}$ M (acidic) has a pH of:

$$pH = -\log[H_3O^+]$$
$$= -\log(1.0 \times 10^{-3})$$
$$= -(-3.00)$$
$$= 3.00$$

> The log of a number is the exponent to which 10 must be raised to obtain that number. Thus, $\log 10^1 = 1$; $\log 10^2 = 2$; $\log 10^{-1} = -1$; $\log 10^{-2} = -2$, etc. (see Appendix I).

Notice that the pH is reported to two *decimal places* here. This is because only the numbers to the right of the decimal point are significant in a logarithm. Since our original value for the concentration had two significant figures, the log of that number has two decimal places.

> When you take the log of a quantity, the result should have the same number of decimal places as the number of significant figures in the original quantity.

2 significant digits   2 decimal places

$$\log 1.0 \times 10^{-3} = 3.00$$

If the original number had three significant digits, the log would be reported to three decimal places:

3 significant digits   3 decimal places

$$\log(1.00 \times 10^{-3}) = 3.000$$

A solution with $[H_3O^+] = 1.0 \times 10^{-7}$ M (neutral) has a pH of

$$pH = -\log[H_3O^+]$$
$$= -\log(1.0 \times 10^{-7})$$
$$= -(-7.00)$$
$$= 7.00$$

In general, at 25 °C:

- pH < 7   The solution is *acidic*.
- pH > 7   The solution is *basic*.
- pH = 7   The solution is *neutral*.

Table 15.6 lists the pH of some common substances. Notice that, as we discussed in Section 15.2, many foods, especially fruits, are acidic and therefore have low pH values. Relatively few foods, however, are basic. The foods with the lowest pH values are limes and lemons, and they are among the sourest. Since the pH scale is a *logarithmic scale*, a change of 1 pH unit corresponds to a 10-fold change in H_3O^+ concentration (Figure 15.7 ▼). For example, a lime with a pH of 2.0 is 10 times more acidic than a plum with a pH of 3.0 and 100 times more acidic than a cherry with a pH of 4.0.

TABLE 15.6 The pH of Some Common Substances

Substance	pH
Gastric juice (human stomach)	1.0–3.0
Limes	1.8–2.0
Lemons	2.2–2.4
Soft drinks	2.0–4.0
Plums	2.8–3.0
Wines	2.8–3.8
Apples	2.9–3.3
Peaches	3.4–3.6
Cherries	3.2–4.0
Beers	4.0–5.0
Rainwater (unpolluted)	5.6
Human blood	7.3–7.4
Egg whites	7.6–8.0
Milk of magnesia	10.5
Household ammonia	10.5–11.5
4% NaOH solution	14

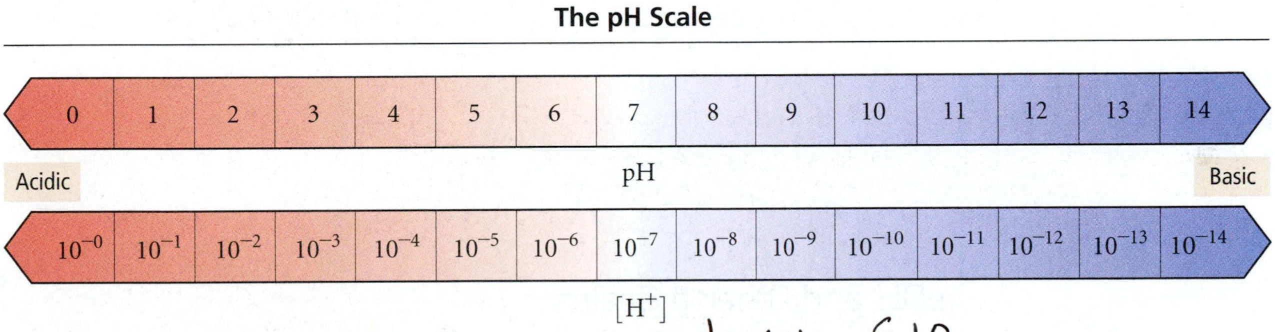

The pH Scale

▲ FIGURE 15.7 **The pH Scale** An increase of 1 on the pH scale corresponds to a decrease in $[H_3O^+]$ by a factor of 10.

EXAMPLE 15.3 Calculating pH from $[H_3O^+]$ or $[OH^-]$

Calculate the pH of each of the following solutions at 25 °C and indicate whether the solution is acidic or basic.

(a) $[H_3O^+] = 1.8 \times 10^{-4}$ M   **(b)** $[OH^-] = 1.3 \times 10^{-2}$ M

Solution

(a) To calculate pH, simply substitute the given $[H_3O^+]$ into the pH equation. Since pH < 7, this solution is acidic.

$$pH = -\log[H_3O^+]$$
$$= -\log(1.8 \times 10^{-4})$$
$$= -(-3.74)$$
$$= 3.74 \text{ (acidic)}$$

(b) First use K_w to find $[H_3O^+]$ from $[OH^-]$.

$$[H_3O^+][OH^-] = K_w = 1.0 \times 10^{-14}$$
$$[H_3O^+](1.3 \times 10^{-2}) = 1.0 \times 10^{-14}$$
$$[H_3O^+] = \frac{1.0 \times 10^{-14}}{1.3 \times 10^{-2}} = 7.7 \times 10^{-13} \text{ M}$$

Then simply substitute $[H_3O^+]$ into the pH expression to find pH.	$pH = -\log[H_3O^+]$
	$= -\log(7.7 \times 10^{-13})$
	$= -(-12.11)$
Since $pH > 7$, this solution is basic.	$= 12.11 \text{ (basic)}$

For Practice 15.3

Calculate the pH of each of the following solutions and indicate whether the solution is acidic or basic.

(a) $[H_3O^+] = 9.5 \times 10^{-9}\,M$ (b) $[OH^-] = 7.1 \times 10^{-3}\,M$

EXAMPLE 15.4   Calculating $[H_3O^+]$ from pH

Calculate the H_3O^+ concentration for a solution with a pH of 4.80.

Solution

To find the $[H_3O^+]$ from pH, start with the equation that defines pH. Substitute the given value of pH and then solve for $[H_3O^+]$. Since the given pH value was reported to two decimal places, the $[H_3O^+]$ is written to two significant figures. (Remember that $10^{\log x} = x$ (see Appendix I). Some calculators use an inv log key to represent this function.)	$pH = -\log[H_3O^+]$
	$4.80 = -\log[H_3O^+]$
	$-4.80 = \log[H_3O^+]$
	$10^{-4.80} = 10^{\log[H_3O^+]}$
	$10^{-4.80} = [H_3O^+]$
	$[H_3O^+] = 1.6 \times 10^{-5}\,M$

For Practice 15.4

Calculate the H_3O^+ concentration for a solution with a pH of 8.37.

pOH and Other p Scales

Notice that p is the mathematical function -log; thus, $pX = -\log X$.

The pOH scale is analogous to the pH scale but is defined with respect to $[OH^-]$ instead of $[H_3O^+]$.

$$pOH = -\log[OH^-]$$

A solution having an $[OH^-]$ of $1.0 \times 10^{-3}\,M$ (basic) has a pOH of 3.00. On the pOH scale, a pOH less than 7 is basic and a pOH greater than 7 is acidic. A pOH of 7 is neutral (Figure 15.8 ▼). We can derive a relationship between pH and pOH at 25 °C from the expression for K_w:

$$[H_3O^+][OH^-] = 1.0 \times 10^{-14}$$

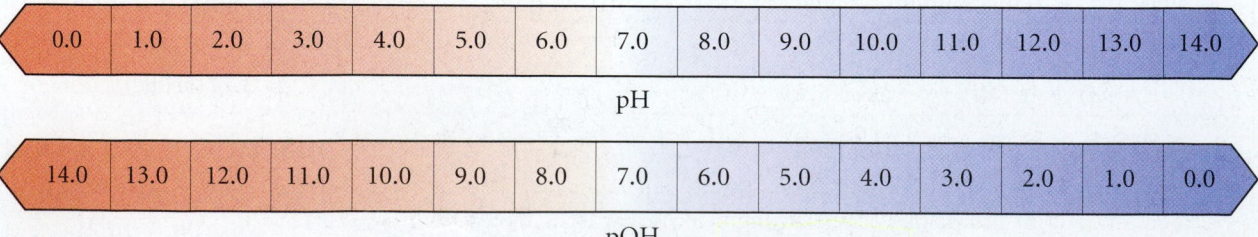

▲ FIGURE 15.8 pH and pOH

Chemistry and Medicine
Ulcers

An ulcer is a lesion that forms on the wall of the stomach or small intestine. Under normal circumstances, a thick layer of mucus lines the stomach wall and protects it from the hydrochloric acid

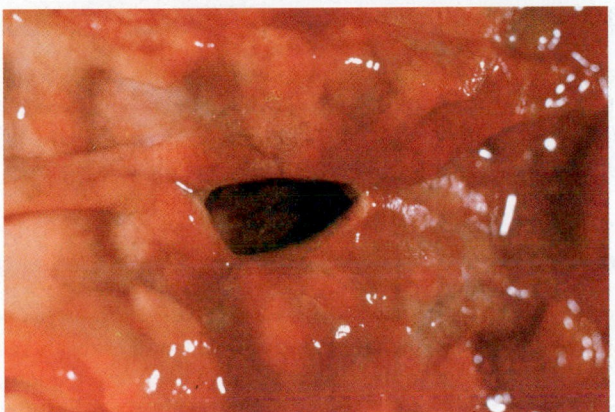

▲ An ulcer is a lesion in the stomach wall.

and other gastric juices in the stomach. When that mucous layer is damaged, however, stomach juices come in direct contact with the stomach wall and begin to digest it, resulting in an ulcer. The main symptom of an ulcer is a burning or gnawing pain in the stomach.

Acidic drugs, such as aspirin, and acidic foods, such as citrus fruits and pickling fluids, irritate ulcers. When consumed, these substances increase the acidity of the stomach, worsening the irritation to the stomach wall. On the other hand, antacids—which contain bases—relieve ulcers. Common antacids include Tums and milk of magnesia.

The causes of ulcers are complex. For many years, a stressful lifestyle and a rich diet were blamed. More recent research, however, has shown that a bacterial infection of the stomach lining is responsible for many ulcers. (The 2005 Nobel Prize in Physiology or Medicine was awarded to Barry J. Marshall and J. Robin Warren, both from Australia, for their discovery of the bacterial cause of ulcers.) Long-term use of some over-the-counter pain relievers, such as aspirin, is also believed to produce ulcers.

Question
Which dessert would be less likely to irritate an ulcer, key lime pie or meringue (made of egg whites)?

Taking the log of both sides, we get

$$\log([H_3O^+][OH^-]) = \log(1.0 \times 10^{-14})$$
$$\log[H_3O^+] + \log[OH^-] = -14.00$$
$$-\log[H_3O^+] - \log[OH^-] = 14.00$$
$$pH + pOH = 14.00$$

The sum of pH and pOH is always equal to 14.00 at 25 °C. Therefore, a solution with a pH of 3 has a pOH of 11.

Another common p scale is pK_a, defined as follows:

$$pK_a = -\log K_a$$

The pK_a of a weak acid is just another way to quantify its strength. The smaller the pK_a, the stronger the acid. For example, chlorous acid, with a K_a of 1.1×10^{-2}, has a pK_a of 1.96 and formic acid, with a K_a of 1.8×10^{-4}, has a pK_a of 3.74.

15.6 Finding the [H₃O⁺] and pH of Strong and Weak Acid Solutions

Finding the pH of a strong acid solution is relatively straightforward. In a solution containing a strong acid, there are two potential sources of H_3O^+, the strong acid itself and the ionization of water. If we let HA be a generic strong acid, the ionization reactions are as follows:

$$HA(aq) + H_2O(l) \longrightarrow H_3O^+(aq) + A^-(aq) \qquad \text{Strong}$$
$$H_2O(l) + H_2O(l) \rightleftharpoons H_3O^+(aq) + OH^-(aq) \qquad K_w = 1.0 \times 10^{-14}$$

Except in extremely dilute acid solutions, the ionization of water contributes a negligibly small amount of H_3O^+ compared to the strong acid. Recall from the previous section that autoionization in pure water produces an H_3O^+ concentration of 1.0×10^{-7} M. In a strong acid solution, the formation of H_3O^+ by the strong acid causes the autoionization

The only exceptions would be extremely dilute ($< 10^{-5}$ M) strong acid solutions.

equilibrium to shift left (as described by Le Châtelier's principle). Thus, in most strong acid solutions, the autoionization of water produces even less H_3O^+ than in pure water. The contribution of the autoionization reaction to the concentration of H_3O^+ can therefore be ignored.

Because strong acids, by definition, are completely ionized in solution, and because we can (in nearly all cases) ignore the contribution of the autoionization of water, *the concentration of H_3O^+ in a strong acid solution is simply equal to the concentration of the strong acid.* For example, a 0.10 M HCl solution has an H_3O^+ concentration of 0.10 M and a pH of 1.00.

$$0.10\ \text{M HCl} \implies [H_3O^+] = 0.10\ \text{M} \implies pH = -\log(0.10) = 1.00$$

Finding the pH of a weak acid solution is more complicated because the concentration of H_3O^+ is *not equal* to the concentration of the weak acid. For example, if you make solutions of 0.10 M HCl (a strong acid) and 0.10 M acetic acid (a weak acid) in the laboratory and measure the pH of each, you get the following results:

0.10 M HCl	pH = 1.00
0.10 M HC$_2$H$_3$O$_2$	pH = 2.87

The pH of the acetic acid solution is higher (it is less acidic) because it is a weak acid and therefore only partially ionizes. As was the case for solutions of strong acids, there are two potential sources of H_3O^+ in weak acid solutions, the weak acid itself and the ionization of water. If we let HA be a generic weak acid, the ionization reactions are as follows:

$$HA(aq) + H_2O(l) \rightleftharpoons H_3O^+(aq) + A^-(aq) \qquad K_a$$

$$H_2O(l) + H_2O(l) \rightleftharpoons H_3O^+(aq) + OH^-(aq) \qquad K_w = 1.0 \times 10^{-14}$$

The acid ionization constants (K_a) for weak acids vary from about 10^{-2} to 10^{-10} (see Table 15.5). Even the weakest of the weak acids has an ionization constant that is four orders of magnitude larger than that of water. Consequently, for most weak acid solutions, we can also ignore the contribution of the autoionization of water to the concentration of H_3O^+ and focus solely on the ionization of the weak acid itself.

Calculating the $[H_3O^+]$ formed by the ionization of a weak acid requires solving an equilibrium problem similar to those introduced in Chapter 14. Consider, for example, a 0.10 M solution of the generic weak acid HA with an acid ionization constant K_a. Since we can ignore the contribution of the autoionization of water, we simply have to determine the concentration of H_3O^+ formed by the following equilibrium:

$$HA(aq) + H_2O(l) \rightleftharpoons H_3O^+(aq) + A^-(aq) \qquad K_a$$

The ICE table was first introduced in Section 14.6.

We can summarize the initial conditions, the changes, and the equilibrium conditions in the following ICE table:

	[HA]	[H$_3$O$^+$]	[A$^-$]
Initial	0.10	≈ 0.00	0.00
Change	$-x$	$+x$	$+x$
Equilibrium	$0.10 - x$	x	x

The initial H_3O^+ concentration is listed as *approximately* zero because of the negligibly small contribution of H_3O^+ due to the autoionization of water (discussed previously). The variable x represents the amount of HA that ionizes. As discussed in Chapter 14, each *equilibrium* concentration is simply the sum of the two entries above it in the ICE table. In order to find the equilibrium concentration of H_3O^+, we must find the value of the variable x. We can use the equilibrium expression to set up an equation in which x is the only variable.

$$K_a = \frac{[H_3O^+][A^-]}{[HA]}$$

$$= \frac{x^2}{0.10 - x}$$

As in solving many other equilibrium problems, we arrive at a quadratic equation in *x*, which can be solved using the quadratic formula. In many cases, however, the *x is small* approximation discussed in Section 14.8 can be applied. Below, we give the general procedure for weak acid equilibrium problems in the left column and two examples of applying the procedure in the center and right columns. For both of these examples, the *x is small* approximation works well. In Example 15.7, we solve a problem in which the *x is small* approximation does not work. In such cases, we can solve the quadratic equation explicitly, or we can apply the method of successive approximations, also discussed in Section 14.8. Finally, in Example 15.8, we work a problem in which we must find the equilibrium constant of a weak acid from its pH.

Procedure for Finding the pH (or [H₃O⁺]) of a Weak Acid Solution	EXAMPLE 15.5 Finding the [H₃O⁺] of a Weak Acid Solution	EXAMPLE 15.6 Finding the pH of a Weak Acid Solution
To solve these types of problems, follow the procedure outlined below.	Find the $[H_3O^+]$ of a 0.100 M HCN solution.	Find the pH of a 0.200 M HNO_2 solution.

1. Write the balanced equation for the ionization of the acid and use it as a guide to prepare an ICE table showing the given concentration of the weak acid as its initial concentration. Leave room in the table for the changes in concentrations and for the equilibrium concentrations.

(Note that the H_3O^+ concentration is listed as approximately zero because the autoionization of water produces a negligibly small amount of H_3O^+).

$HCN(aq) + H_2O(l) \rightleftharpoons$
$\qquad H_3O^+(aq) + CN^-(aq)$

	[HCN]	[H₃O⁺]	[CN⁻]
Initial	0.100	≈ 0.00	0.00
Change			
Equil			

$HNO_2(aq) + H_2O(l) \rightleftharpoons$
$\qquad H_3O^+(aq) + NO_2^-(aq)$

	[HNO₂]	[H₃O⁺]	[NO₂⁻]
Initial	0.200	≈ 0.00	0.00
Change			
Equil			

2. Represent the change in the concentration of H_3O^+ with the variable *x*. Define the changes in the concentrations of the other reactants and products in terms of *x*, always keeping in mind the stoichiometry of the reaction.

$HCN(aq) + H_2O(l) \rightleftharpoons$
$\qquad H_3O^+(aq) + CN^-(aq)$

	[HCN]	[H₃O⁺]	[CN⁻]
Initial	0.100	≈ 0.00	0.00
Change	−x	+x	+x
Equil			

$HNO_2(aq) + H_2O(l) \rightleftharpoons$
$\qquad H_3O^+(aq) + NO_2^-(aq)$

	[HNO₂]	[H₃O⁺]	[NO₂⁻]
Initial	0.200	≈ 0.00	0.00
Change	−x	+x	+x
Equil			

3. Sum each column to determine the equilibrium concentrations in terms of the initial concentrations and the variable *x*.

$HCN(aq) + H_2O(l) \rightleftharpoons$
$\qquad H_3O^+(aq) + CN^-(aq)$

	[HCN]	[H₃O⁺]	[CN⁻]
Initial	0.100	≈ 0.00	0.00
Change	−x	+x	+x
Equil	0.100 − x	x	x

$HNO_2(aq) + H_2O(l) \rightleftharpoons$
$\qquad H_3O^+(aq) + NO_2^-(aq)$

	[HNO₂]	[H₃O⁺]	[NO₂⁻]
Initial	0.200	≈ 0.00	0.00
Change	−x	+x	+x
Equil	0.200 − x	x	x

4. Substitute the expressions for the equilibrium concentrations (from step 3) into the expression for the acid ionization constant (K_a). In many cases, you can make the approximation that *x is small* (as discussed in Section 14.8). **Substitute the value of the acid ionization constant (from Table 15.5) into the K_a expression and solve for *x*.**

$K_a = \dfrac{[H_3O^+][CN^-]}{[HCN]}$

$\quad = \dfrac{x^2}{0.100 - \cancel{x}} \quad (x \text{ is small})$

$4.9 \times 10^{-10} = \dfrac{x^2}{0.100}$

$\sqrt{4.9 \times 10^{-10}} = \sqrt{\dfrac{x^2}{0.100}}$

$4.9 \times 10^{-10} \times .100 = x^2$

$K_a = \dfrac{[H_3O^+][NO_2^-]}{[HNO_2]}$

$\quad = \dfrac{x^2}{0.200 - \cancel{x}} \quad (x \text{ is small})$

$4.6 \times 10^{-4} = \dfrac{x^2}{0.200}$

$\sqrt{4.6 \times 10^{-4}} = \sqrt{\dfrac{x^2}{0.200}}$

Confirm that the x is small approximation is valid by computing the ratio of x to the number it was subtracted from in the approximation. The ratio should be less than 0.05 (or 5%).	$x = \sqrt{(0.100)(4.9 \times 10^{-10})}$ $= 7.0 \times 10^{-6}$ $\dfrac{7.0 \times 10^{-6}}{0.100} \times 100\% = 7.0 \times 10^{-3}\%$ Therefore the approximation is valid.	$x = \sqrt{(0.200)(4.6 \times 10^{-4})}$ $= 9.6 \times 10^{-3}$ $\dfrac{9.6 \times 10^{-3}}{0.200} \times 100\% = 4.8\%$ Therefore the approximation is valid (but barely so).
5. Determine the H_3O^+ concentration from the computed value of x and compute the pH if necessary.	$[H_3O^+] = 7.0 \times 10^{-6}$ M (pH was not asked for in this problem.)	$[H_3O^+] = 9.6 \times 10^{-3}$ M $pH = -\log[H_3O^+]$ $= -\log(9.6 \times 10^{-3})$ $= 2.02$
6. Check your answer by substituting the computed equilibrium values into the acid ionization expression. The computed value of K_a should match the given value of K_a. Note that rounding errors and the x is small approximation could cause a difference in the least significant digit when comparing values of K_a.	$K_a = \dfrac{[H_3O^+][CN^-]}{[HCN]} = \dfrac{(7.0 \times 10^{-6})^2}{0.100}$ $= 4.9 \times 10^{-10}$ Since the computed value of K_a matches the given value, the answer is valid.	$K_a = \dfrac{[H_3O^+][NO_2^-]}{[HNO_2]} = \dfrac{(9.6 \times 10^{-3})^2}{0.200}$ $= 4.6 \times 10^{-4}$ Since the computed value of K_a matches the given value, the answer is valid.

For Practice 15.5

Find the H_3O^+ concentration of a 0.250 M hydrofluoric acid solution.

For Practice 15.6

Find the pH of a 0.0150 M acetic acid solution.

EXAMPLE 15.7 Finding the pH of a Weak Acid Solution in Cases Where the *x is small* Approximation Does Not Work

Find the pH of a 0.100 M $HClO_2$ solution.

Solution

1. Write the balanced equation for the ionization of the acid and use it as a guide to prepare an ICE table showing the given concentration of the weak acid as its initial concentration.

(Note that the H_3O^+ concentration is listed as approximately zero. Although a little H_3O^+ is present from the autoionization of water, this amount is negligibly small compared to the amount of H_3O^+ from the acid.)

$HClO_2(aq) + H_2O(l) \rightleftharpoons H_3O^+(aq) + ClO_2^-(aq)$

	$[HClO_2]$	$[H_3O^+]$	$[ClO_2^-]$
Initial	0.100	≈ 0.00	0.00
Change			
Equil			

2. Represent the change in $[H_3O^+]$ with the variable x. Define the changes in the concentrations of the other reactants and products in terms of x.

$HClO_2(aq) + H_2O(l) \rightleftharpoons H_3O^+(aq) + ClO_2^-(aq)$

	$[HClO_2]$	$[H_3O^+]$	$[ClO_2^-]$
Initial	0.100	≈ 0.00	0.00
Change	$-x$	$+x$	$+x$
Equil			

3. Sum each column to determine the equilibrium concentrations in terms of the initial concentrations and the variable x.

$HClO_2(aq) + H_2O(l) \rightleftharpoons H_3O^+(aq) + ClO_2^-(aq)$

	$[HClO_2]$	$[H_3O^+]$	$[ClO_2^-]$
Initial	0.100	≈ 0.00	0.00
Change	$-x$	$+x$	$+x$
Equil	$0.100 - x$	x	x

4.	Substitute the expressions for the equilibrium concentrations (from step 3) into the expression for the acid ionization constant (K_a). Make the *x is small* approximation and substitute the value of the acid ionization constant (from Table 15.5) into the K_a expression. Solve for x.	$K_a = \dfrac{[H_3O^+][ClO_2^-]}{[HNO_2]}$ $= \dfrac{x^2}{0.100 - x}$ (*x is small*) $0.011 = \dfrac{x^2}{0.100}$ $\sqrt{0.011} = \sqrt{\dfrac{x^2}{0.100}}$ $x = \sqrt{(0.100)(0.011)}$ $= 0.033$
	Check to see if the *x is small* approximation is valid by computing the ratio of x to the number it was subtracted from in the approximation. The ratio should be less than 0.05 (or 5%).	$\dfrac{0.033}{0.100} \times 100\% = 33\%$ Therefore, the *x is small* approximation is *not valid*.
4a.	If the *x is small* approximation is not valid, solve the quadratic equation explicitly or use the method of successive approximations to find x. In this case, we solve the quadratic equation.	$0.011 = \dfrac{x^2}{0.100 - x}$ $0.011(0.100 - x) = x^2$ $0.0011 - 0.011x = x^2$ $\underset{a}{x^2} + \underset{b}{0.011x} - \underset{c}{0.0011} = 0$ $x = \dfrac{-b \pm \sqrt{b^2 - 4ac}}{2a}$ $= \dfrac{-(0.011) \pm \sqrt{(0.011)^2 - 4(1)(-0.0011)}}{2(1)}$ $= \dfrac{-0.011 \pm 0.0672}{2}$ $x = -0.039 \ or \ x = 0.028$ Since x represents the concentration of H_3O^+, and since concentrations cannot be negative, we reject the negative root. $x = 0.028$
5.	Determine the H_3O^+ concentration from the computed value of x and compute the pH (if necessary).	$[H_3O^+] = 0.028 \, M$ $pH = -\log[H_3O^+]$ $= -\log 0.028$ $= 1.55$
6.	Check your answer by substituting the computed equilibrium values into the acid ionization expression. The computed value of K_a should match the given value of K_a. Note that rounding errors could cause a difference in the least significant digit when comparing values of K_a.	$K_a = \dfrac{[H_3O^+][ClO_2^-]}{[HClO_2]} = \dfrac{0.028^2}{0.100 - 0.028}$ $= 0.011$ Since the computed value of K_a matches the given value, the answer is valid.

For Practice 15.7

Find the pH of a 0.010 M HNO_2 solution.

EXAMPLE 15.8 Finding the Equilibrium Constant from pH

A 0.100 M weak acid (HA) solution has a pH of 4.25. Find K_a for the acid.

Solution

Use the given pH to find the equilibrium concentration of $[H_3O^+]$. Then write the balanced equation for the ionization of the acid and use it as a guide to prepare an ICE table showing all known concentrations.	$pH = -\log[H_3O^+]$ $4.25 = -\log[H_3O^+]$ $[H_3O^+] = 5.6 \times 10^{-5}$ M

$$HA(aq) + H_2O(l) \rightleftharpoons H_3O^+(aq) + A^-(aq)$$

	[HA]	[H₃O⁺]	[A⁻]
Initial	0.100	≈ 0.00	0.00
Change			
Equil		5.6×10^{-5}	

Use the equilibrium concentration of H_3O^+ and the stoichiometry of the reaction to predict the changes and equilibrium concentration for all species. For most weak acids, the initial and equilibrium concentrations of the weak acid (HA) will be equal because the amount that ionizes is usually very small compared to the initial concentration.	

$$HA(aq) + H_2O(l) \rightleftharpoons H_3O^+(aq) + A^-(aq)$$

	[HA]	[H₃O⁺]	[A⁻]
Initial	0.100	≈ 0.00	0.00
Change	-5.6×10^{-5}	$+5.6 \times 10^{-5}$	$+5.6 \times 10^{-5}$
Equil	$(0.100 - 5.6 \times 10^{-5})$ ≈ 0.100	5.6×10^{-5}	5.6×10^{-5}

Substitute the equilibrium concentrations into the expression for K_a and compute its value.	$K_a = \dfrac{[H_3O^+][A^-]}{[HA]}$ $= \dfrac{(5.6 \times 10^{-5})(5.6 \times 10^{-5})}{0.100}$ $= 3.1 \times 10^{-8}$

For Practice 15.8

A 0.175 M weak acid solution has a pH of 3.25. Find K_a for the acid.

◆ Conceptual Connection 15.3 The *x is small* Approximation

The initial concentration and K_a's of several weak acid (HA) solutions are listed below. For which of these is the *x is small* approximation *least* likely to work in finding the pH of the solution?

(a) initial $[HA] = 0.100$ M; $K_a = 1.0 \times 10^{-5}$

(b) initial $[HA] = 1.00$ M; $K_a = 1.0 \times 10^{-6}$

(c) initial $[HA] = 0.0100$ M; $K_a = 1.0 \times 10^{-3}$

(d) initial $[HA] = 1.0$ M; $K_a = 1.5 \times 10^{-3}$

Answer: (c) The validity of the *x is small* approximation depends on both the value of the equilibrium constant and the initial concentration—the closer that these are to one another, the less likely the approximation will be valid.

Percent Ionization of a Weak Acid

We can quantify the ionization of a weak acid based on the percentage of acid molecules that actually ionize. For instance, in Example 15.6, we found that a 0.200 M HNO_2 solution contains 9.6×10^{-3} M H_3O^+. We can define a useful quantity called the **percent ionization** of a weak acid as follows:

$$\text{Percent ionization} = \frac{\text{concentration of ionized acid}}{\text{initial concentration of acid}} \times 100\% = \frac{[H_3O^+]_{equil}}{[HA]_{init}} \times 100\%$$

Since the concentration of ionized acid is equal to the H_3O^+ concentration at equilibrium (for a monoprotic acid), we can simply use $[H_3O^+]_{equil}$ and $[HA]_{init}$ to compute the percent ionization. The 0.200 M HNO_2 solution therefore has the following percent ionization:

$$\% \text{ ionization} = \frac{[H_3O^+]_{equil}}{[HA]_{init}} \times 100\%$$

$$= \frac{9.6 \times 10^{-3} \text{ M}}{0.200 \text{ M}} \times 100\%$$

$$= 4.8\%$$

As you can see, the percent ionization is relatively small. In this case, even for a weak acid that has a relatively large K_a (HNO_2 has the second largest K_a in Table 15.5) the percent ionization indicates that less than five molecules out of one hundred ionize. For most other weak acids (with smaller K_a values) the percent ionization is even less.

In the example that follows, we calculate the percent ionization of a more concentrated HNO_2 solution. As you read through the example, notice the following important result: the computed H_3O^+ concentration is much greater (as we would expect for a more concentrated solution), but the *percent ionization* is actually smaller.

EXAMPLE 15.9 Finding the Percent Ionization of a Weak Acid

Find the percent ionization of a 2.5 M HNO_2 solution.

Solution

To find the percent ionization, you must find the equilibrium concentration of H_3O^+. Follow the procedure in Example 15.5, shown in condensed form here.

$$HNO_2(aq) + H_2O(l) \rightleftharpoons H_3O^+(aq) + NO_2^-(aq)$$

	[HNO₂]	[H₃O⁺]	[NO₂⁻]
Initial	2.5	≈ 0.00	0.00
Change	−x	+x	+x
Equil	2.5 − x	x	x

$$K_a = \frac{[H_3O^+][NO_2^-]}{[HNO_2]} = \frac{x^2}{2.5 - x} \quad (x \text{ is small})$$

$$4.6 \times 10^{-4} = \frac{x^2}{2.5}$$

$$x = 0.034$$

Therefore, $[H_3O^+] = 0.034$ M.

Use the definition of percent ionization to calculate it. (Since the percent ionization is less than 5%, the *x is small* approximation is valid.)

$$\% \text{ ionization} = \frac{[H_3O^+]_{equil}}{[HA]_{init}} \times 100\%$$

$$= \frac{0.034 \text{ M}}{2.5 \text{ M}} \times 100\%$$

$$= 1.4\%$$

For Practice 15.9

Find the percent ionization of a 0.250 M $HC_2H_3O_2$ solution.

Let us now summarize the results of Examples 15.6 and 15.9:

[HNO$_2$]	[H$_3$O$^+$]	Percent Dissociation
0.200	0.0096	4.8%
2.5	0.034	1.4%

Notice the following general result:

- The *equilibrium H$_3$O$^+$ concentration* of a weak acid *increases* with increasing initial concentration of the acid.
- The *percent ionization* of a weak acid *decreases* with increasing concentration of the acid.

In other words, as the concentration of a weak acid solution increases, the concentration of the hydronium ion also increases, but the increase is not linear. The H$_3$O$^+$ concentration increases more slowly than the concentration of the acid because as the acid concentration increases, a smaller fraction of weak acid molecules ionize.

We can rationalize this behavior by analogy with Le Châtelier's principle. Consider the following weak acid ionization equilibrium:

$$\text{HA(aq)} \rightleftharpoons \text{H}^+(aq) + \text{A}^-(aq)$$

<small>1 mol dissolved particles   2 mol dissolved particles</small>

If a weak acid solution initially at equilibrium is made more dilute, the system should (according to Le Châtelier's principle) respond to minimize the disturbance. The disturbance can be minimized by shifting to the right because the right side of the equation contains more particles in solution (2 mol versus 1 mol) than the left side. If the system shifts to the right, the percent ionization will be greater in the more dilute solution, as we observe.

 Conceptual Connection 15.4 Strong and Weak Acids

Which of the following solutions is most acidic (that is, has the lowest pH)?

(a) 1.0 M HCl   **(b)** 2.0 M HF

(c) A solution that is 1.0 M in HF and 1.0 M in HClO

Answer: **(a)** A weak acid solution will usually be less than 5% dissociated. Therefore, since HCl is the only strong acid, the 1.0 M solution is much more acidic than either a weak acid that is twice as concentrated or a combination of two weak acids with the same concentrations.

Mixtures of Acids

Finding the pH of a mixture of acids may seem difficult at first inspection. In many cases, however, the relative strengths of the acids in the mixture allow us to neglect the weaker acid and focus only on the stronger one. We consider two possible acid mixtures: a strong acid with a weak acid, and a weak acid with another weak acid.

A Strong Acid and a Weak Acid Consider a mixture that is 0.10 M in HCl and 0.10 M in HCHO$_2$. We now have three sources of H$_3$O$^+$ ions: the strong acid (HCl), the weak acid (HCHO$_2$), and the autoionization of water.

$$\text{HCl}(aq) + \text{H}_2\text{O}(l) \longrightarrow \text{H}_3\text{O}^+(aq) + \text{Cl}^-(aq) \qquad \text{Strong}$$

$$\text{HCHO}_2(aq) + \text{H}_2\text{O}(l) \rightleftharpoons \text{H}_3\text{O}^+(aq) + \text{CHO}_2^-(aq) \qquad K_a = 1.8 \times 10^{-4}$$

$$\text{H}_2\text{O}(l) + \text{H}_2\text{O}(l) \rightleftharpoons \text{H}_3\text{O}^+(aq) + \text{OH}^-(aq) \qquad K_w = 1.0 \times 10^{-14}$$

However, the complete ionization of HCl produces a significant concentration of H_3O^+ (0.10 M) which then suppresses the formation of additional H_3O^+ by the ionization of $HCHO_2$ or the autoionization of water. In other words, because of Le Châtelier's principle, the formation of H_3O^+ by the strong acid causes the weak acid to ionize even less than it would in the absence of the strong acid. To see this clearly, let us calculate $[H_3O^+]$ and $[CHO_2^-]$ in this solution.

If we neglect the contribution of $HCHO_2$ and the autoionization of water (which we will justify shortly), the concentration of H_3O^+ is equal to the initial concentration of HCl.

$$[H_3O^+] = [HCl] = 0.10\,M$$

To find $[CHO_2^-]$ we must solve an equilibrium problem. However, the initial concentration of H_3O^+ is not negligible (as it has been in all the other weak acid equilibrium problems that we have worked so far) because of the significant formation of H_3O^+ by the strong acid. Thus, the concentration of H_3O^+ formed by the strong acid becomes the initial concentration of H_3O^+ in the ICE table for $HCHO_2$ as shown here:

$$HCHO_2(aq) + H_2O(l) \rightleftharpoons H_3O^+(aq) + CHO_2^-(aq)$$

	$[HCHO_2]$	$[H_3O^+]$	$[CHO_2^-]$
Initial	0.10	0.10	0.00
Change	$-x$	$+x$	$+x$
Equil	$0.10 - x$	$0.10 + x$	x

We then use the equilibrium expression to set up an equation in which x is the only variable.

$$K_a = \frac{[H_3O^+][CHO_2^-]}{[HCHO_2]}$$

$$= \frac{(0.10 + x)x}{0.10 - x}$$

Since the equilibrium constant is small relative to the initial concentration of the acid, we can make the *x is small* approximation.

$$K_a = \frac{(0.10 + \cancel{x})x}{0.10 - \cancel{x}}$$

$$1.8 \times 10^{-4} = \frac{(0.10)x}{0.10}$$

$$x = 1.8 \times 10^{-4}$$

Checking the *x is small* approximation,

$$\frac{1.8 \times 10^{-4}}{0.10} \times 100\% = 0.18\%$$

We find that the approximation is valid. Therefore, $[CHO_2^-] = 1.8 \times 10^{-4}\,M$. We can now see why we are justified in neglecting the ionization of the weak acid ($HCHO_2$) in computing $[H_3O^+]$ for the mixture. Because of the stoichiometry of the weak acid ionization reaction, the contribution to the concentration of H_3O^+ by the weak acid must necessarily be equal to the concentration of CHO_2^- that we just calculated. Therefore, we have the following contributions to $[H_3O^+]$:

HCl contributes 0.10 M

$HCHO_2$ contributes 0.00018 M

As you can see, the amount of H_3O^+ contributed by $HCHO_2$ is completely negligible. Similarly, the amount of H_3O^+ contributed by the autoionization of water is even smaller and therefore also negligible.

A Mixture of Two Weak Acids When two weak acids are mixed, we again have three potential sources of H_3O^+ to consider: each of the two weak acids and the autoionization of water. However, if the K_a's of the two weak acids are sufficiently different in magnitude (if they differ by more than a factor of several hundred), then as long as the concentrations of the two acids are similar in magnitude (or the concentration of the stronger one is greater than that of the weaker), we can make the assumption that the weaker acid will not make a significant contribution to the concentration of H_3O^+. The reason we can make this assumption is similar to the reason we can make the same assumption in a mixture of strong acid and weak one: the stronger acid suppresses the ionization of the weaker one, in accordance with Le Châtelier's principle. Example 15.10 shows how to calculate the concentration of H_3O^+ in a mixture of two weak acids.

EXAMPLE 15.10 Mixtures of Weak Acids

Find the pH of a mixture that is 0.150 M in HF and 0.100 M in HClO.

Solution

The three possible sources of H_3O^+ ions are HF, HClO, and H_2O. Write the ionization equations for the three sources and their corresponding equilibrium constants. Since the equilibrium constant for the ionization of HF is about 12,000 times larger than that for the ionization of HClO, the contribution of HF to $[H_3O^+]$ is by far the greatest. We can therefore simply compute the $[H_3O^+]$ formed by HF and neglect the other two potential sources of H_3O^+.

$$HF(aq) + H_2O(l) \rightleftharpoons H_3O^+(aq) + F^-(aq) \qquad K_a = 3.5 \times 10^{-4}$$

$$HClO(aq) + H_2O(l) \rightleftharpoons H_3O^+(aq) + ClO^-(aq) \qquad K_a = 2.9 \times 10^{-8}$$

$$H_2O(l) + H_2O(l) \rightleftharpoons H_3O^+(aq) + OH^-(aq) \qquad K_w = 1.0 \times 10^{-14}$$

Write the balanced equation for the ionization of HF and use it as a guide to prepare an ICE table.

$$HF(aq) + H_2O(l) \rightleftharpoons H_3O^+(aq) + F^-(aq)$$

	[HF]	[H₃O]	[F⁻]
Initial	0.150	≈ 0.00	0.00
Change	−x	+x	+x
Equil	0.150 − x	x	x

Substitute the expressions for the equilibrium concentrations into the expression for the acid ionization constant (K_a). Since the equilibrium constant is small relative to the initial concentration of HF, you can make the x is small approximation. Substitute the value of the acid ionization constant (from Table 15.5) into the K_a expression and solve for x.

$$K_a = \frac{[H_3O^+][F^-]}{[HF]} = \frac{x^2}{0.150 - x} \qquad (x \text{ is small})$$

$$3.5 \times 10^{-4} = \frac{x^2}{0.150}$$

$$\sqrt{(0.150)(3.5 \times 10^{-4})} = \sqrt{x^2}$$

$$x = 7.2 \times 10^{-3}$$

Confirm that the x is small approximation is valid by computing the ratio of x to the number it was subtracted from in the approximation. The ratio should be less than 0.05 (or 5%).

$$\frac{7.2 \times 10^{-3}}{0.150} \times 100\% = 4.8\%$$

Therefore the approximation is valid (though barely so).

Determine the H_3O^+ concentration from the calculated value of x and find the pH.

$$[H_3O^+] = 7.2 \times 10^{-3} \text{ M}$$

$$pH = -\log(7.2 \times 10^{-3}) = 2.14$$

For Practice 15.10

Find the ClO^- concentration of the above mixture of HF and HClO.

15.7 Base Solutions

Strong Bases

By analogy with the definition of a strong acid, a **strong base** is a base that completely dissociates in solution. NaOH, for example, is a strong base:

$$NaOH(aq) \longrightarrow Na^+(aq) + OH^-(aq)$$

An NaOH solution contains no intact NaOH—it has all dissociated to form $Na^+(aq)$ and $OH^-(aq)$ (Figure 15.9 ▶). In other words, a 1.0 M NaOH solution will have $[OH^-] = 1.0$ M and $[Na^+] = 1.0$ M. The common strong bases are listed in Table 15.7.

As you can see, most strong bases are group 1A or group 2A metal hydroxides. The group 1A metal hydroxides are highly soluble in water and can form concentrated base solutions. The group 2A metal hydroxides, however, are only slightly soluble, a useful property for some applications (see the *Chemistry and Medicine* box in this section). Notice that the general formula for the group 2A metal hydroxides is $M(OH)_2$. When they dissolve, they produce 2 mol of OH^- per mole of the base. For example, $Sr(OH)_2$ dissociates as follows:

$$Sr(OH)_2(aq) \longrightarrow Sr^{2+}(aq) + 2\,OH^-(aq)$$

Unlike diprotic acids, which ionize in two steps, bases containing two OH^- ions dissociate in one step.

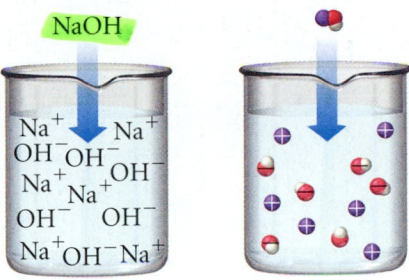

A Strong Base

▲ **FIGURE 15.9 Ionization of a Strong Base** When NaOH dissolves in water, it dissociates completely into Na^+ and OH^-. The solution contains virtually no intact NaOH.

TABLE 15.7 Strong Bases

Lithium hydroxide (LiOH)	Strontium hydroxide [$Sr(OH)_2$]
Sodium hydroxide (NaOH)	Calcium hydroxide [$Ca(OH)_2$]
Potassium hydroxide (KOH)	Barium hydroxide [$Ba(OH)_2$]

Weak Bases

A **weak base** is analogous to a weak acid. Unlike strong bases that contain OH^- and dissociate in water, the most common weak bases produce OH^- by accepting a proton from water, ionizing water to form OH^- according to the following general equation:

$$B(aq) + H_2O(l) \rightleftharpoons BH^+(aq) + OH^-(aq)$$

In this equation, B is simply a generic symbol for a weak base. Ammonia, for example, ionizes water as follows:

$$NH_3(aq) + H_2O(l) \rightleftharpoons NH_4^+(aq) + OH^-(aq)$$

The double arrow indicates that the ionization is not complete. An NH_3 solution contains mostly NH_3 with some NH_4^+ and OH^- (Figure 15.10 ▶). A 1.0 M NH_3 solution will have $[OH^-] < 1.0$ M. The extent of ionization of a weak base is quantified with the **base ionization constant, K_b**. For the general reaction in which a weak base ionizes water, K_b is defined as follows:

$$B(aq) + H_2O(l) \rightleftharpoons BH^+(aq) + OH^-(aq) \qquad K_b = \frac{[BH^+][OH^-]}{[B]}$$

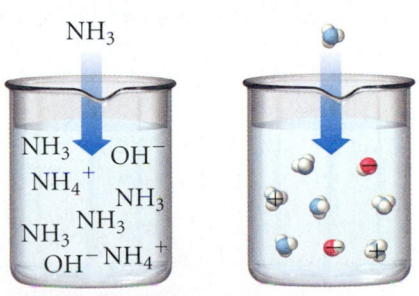

A Weak Base

▲ **FIGURE 15.10 Ionization of a Weak Base** When NH_3 dissolves in water, it partially ionizes water to form NH_4^+ and OH^-. Most of the NH_3 molecules in solution remain as NH_3.

By analogy with K_a, the smaller the value of K_b, the weaker the base. Table 15.8 lists some common weak bases, their ionization reactions, and values for K_b. The "p" scale can also be applied to K_b, so that $pK_b = -\log K_b$.

TABLE 15.8 Some Common Weak Bases

Weak Base	Ionization Reaction	K_b
Carbonate ion (CO_3^{2-})*	$CO_3^{2-}(aq) + H_2O(l) \rightleftharpoons HCO_3^-(aq) + OH^-(aq)$	1.8×10^{-4}
Methylamine (CH_3NH_2)	$CH_3NH_2(aq) + H_2O(l) \rightleftharpoons CH_3NH_3^+(aq) + OH^-(aq)$	4.4×10^{-4}
Ethylamine ($C_2H_5NH_2$)	$C_2H_5NH_2(aq) + H_2O(l) \rightleftharpoons C_2H_5NH_3^+(aq) + OH^-(aq)$	5.6×10^{-4}
Ammonia (NH_3)	$NH_3(aq) + H_2O(l) \rightleftharpoons NH_4^+(aq) + OH^-(aq)$	1.76×10^{-5}
Pyridine (C_5H_5N)	$C_5H_5N(aq) + H_2O(l) \rightleftharpoons C_5H_5NH^+(aq) + OH^-(aq)$	1.7×10^{-9}
Bicarbonate ion (HCO_3^-)* (or hydrogen carbonate)	$HCO_3^-(aq) + H_2O(l) \rightleftharpoons H_2CO_3(aq) + OH^-(aq)$	1.7×10^{-9}
Aniline ($C_6H_5NH_2$)	$C_6H_5NH_2(aq) + H_2O(l) \rightleftharpoons C_6H_5NH_3^+(aq) + OH^-(aq)$	3.9×10^{-10}

*The carbonate and bicarbonate ions must occur with a positively charged ion such as Na^+ that serves to balance the charge but does not have any part in the ionization reaction. For example, it is the bicarbonate ion that makes sodium bicarbonate ($NaHCO_3$) basic. We look more closely at ionic bases in Section 15.8.

All but two of the weak bases listed in Table 15.8 are either ammonia or *amines*, which can be thought of as ammonia with one or more hydrocarbon groups substituted for one or more hydrogen atoms. A common molecular feature of all these bases is a nitrogen atom with a lone pair (Figure 15.11 ▼). This lone pair acts as the proton acceptor that makes the substance a base, as shown in the following reactions for ammonia and methylamine:

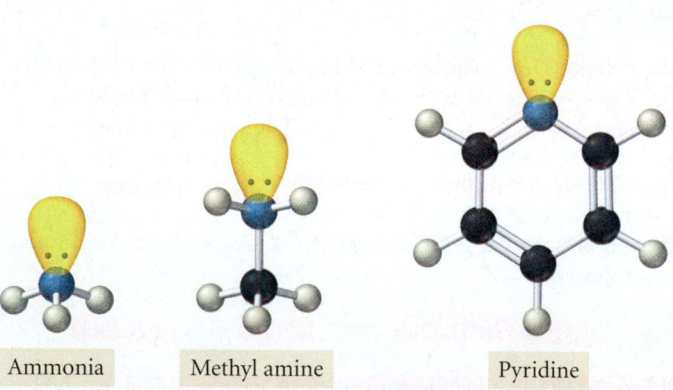

Ammonia   Methyl amine   Pyridine

▲ FIGURE 15.11 **Lone Pairs in Weak Bases** Many weak bases have a nitrogen atom with a lone pair that acts as the proton acceptor.

Finding the [OH⁻] and pH of Basic Solutions

Finding the [OH⁻] and pH of a strong base solution is relatively straightforward, as shown in the following example. As in calculating the [H_3O^+] in strong acid solutions, we can neglect the contribution of the autoionization of water to the [OH⁻] and focus solely on the strong base itself.

EXAMPLE 15.11 Finding the [OH⁻] and pH of a Strong Base Solution

What is the OH⁻ concentration and pH in each of the following solutions?

(a) 0.225 M KOH

(b) 0.0015 M $Sr(OH)_2$

Solution

(a) Since KOH is a strong base, it completely dissociates into K^+ and OH^- in solution. The concentration of OH^- will therefore be the same as the given concentration of KOH.

Use this concentration and K_w to find $[H_3O^+]$.

$$KOH(aq) \longrightarrow K^+(aq) + OH^-(aq)$$
$$[OH^-] = 0.225 \text{ M}$$
$$[H_3O^+][OH^-] = K_w = 1.00 \times 10^{-14}$$
$$[H_3O^+](0.225) = 1.00 \times 10^{-14}$$
$$[H_3O^+] = 4.44 \times 10^{-14} \text{ M}$$

$$\frac{1.0 \times 10^{-14}}{.225}$$

Then simply substitute $[H_3O^+]$ into the pH expression to find the pH.

$$pH = -\log[H_3O^+]$$
$$= -\log(4.44 \times 10^{-14})$$
$$= 13.353$$

(b) Since $Sr(OH)_2$ is a strong base, it completely dissociates into 1 mol of Sr^{2+} and 2 mol of OH^- in solution. The concentration of OH^- will therefore be twice the given concentration of $Sr(OH)_2$.

Use this concentration and K_w to find $[H_3O^+]$.

$$Sr^{2+}(aq) \longrightarrow Sr^{2+}(aq) + 2\,OH^-(aq)$$
$$[OH^-] = 2(0.0015) \text{ M}$$
$$= 0.0030 \text{ M}$$
$$[H_3O^+][OH^-] = K_w = 1.0 \times 10^{-14}$$
$$[H_3O^+](0.0030) = 1.0 \times 10^{-14}$$
$$[H_3O^+] = 3.3 \times 10^{-12} \text{ M}$$

Substitute $[H_3O^+]$ into the pH expression to find the pH.

$$pH = -\log[H_3O^+]$$
$$= -\log(3.3 \times 10^{-12})$$
$$= 11.48$$

For Practice 15.11

Find the $[OH^-]$ and pH of a 0.010 M $Ba(OH)_2$ solution.

Finding the $[OH^-]$ and pH of a *weak base* solution is analogous to finding the $[H_3O^+]$ and pH of a weak acid. We can neglect the contribution of the autoionization of water to the $[OH^-]$ and focus solely on the weak base itself. We find the contribution of the weak base by preparing an ICE table showing the relevant concentrations of all species and then use the base ionization constant expression to find the $[OH^-]$. The following example shows how to find the $[OH^-]$ and pH of a weak base solution.

EXAMPLE 15.12 Finding the [OH⁻] and pH of a Weak Base Solution

Find the $[OH^-]$ and pH of a 0.100 M NH_3 solution.

Solution

1. Write the balanced equation for the ionization of water by the base and use it as a guide to prepare an ICE table showing the given concentration of the weak base as its initial concentration. Leave room in the table for the changes in concentrations and for the equilibrium concentrations. (Note that we list the OH^- concentration as approximately zero. Although a little OH^- is present from the autoionization of water, this amount is negligibly small compared to the amount of OH^- formed by the base.)

$$NH_3(aq) + H_2O(l) \rightleftharpoons NH_4^+(aq) + OH^-(aq)$$

	[NH₃]	[NH₄⁺]	[OH⁻]
Initial	0.100	0.00	≈ 0.00
Change			
Equil			

2. Represent the change in the concentration of OH^- concentration with the variable x. Define the changes in the concentrations of the other reactants and products in terms of x.

$$NH_3(aq) + H_2O(l) \rightleftharpoons NH_4^+(aq) + OH^-(aq)$$

	$[NH_3]$	$[NH_4^+]$	$[OH^-]$
Initial	0.100	0.00	≈ 0.00
Change	$-x$	$+x$	$+x$
Equil			

3. Sum each column to determine the equilibrium concentrations in terms of the initial concentrations and the variable x.

$$NH_3(aq) + H_2O(l) \rightleftharpoons NH_4^+(aq) + OH^-(aq)$$

	$[NH_3]$	$[NH_4^+]$	$[OH^-]$
Initial	0.100	0.00	≈ 0.00
Change	$-x$	$+x$	$+x$
Equil	$0.100 - x$	x	x

4. Substitute the expressions for the equilibrium concentrations (from step 3) in to the expression for the base ionization constant.

In many cases, you can make the approximation that x is small (as discussed in Chapter 14).

Substitute the value of the base ionization constant (from Table 15.8) into the K_b expression and solve for x.

$$K_b = \frac{[NH_4^+][OH^-]}{[NH_3]}$$

$$= \frac{x^2}{0.100 - \cancel{x}} \quad (x \text{ is small})$$

$$1.76 \times 10^{-5} = \frac{x^2}{0.100}$$

$$\sqrt{1.76 \times 10^{-5}} = \sqrt{\frac{x^2}{0.100}}$$

$$x = \sqrt{(0.100)(1.76 \times 10^{-5})}$$

$$= 1.33 \times 10^{-3}$$

Confirm that the x is small approximation is valid by computing the ratio of x to the number it was subtracted from in the approximation. The ratio should be less than 0.05 (or 5%).

$$\frac{1.33 \times 10^{-3}}{0.100} \times 100\% = 1.33\%$$

Therefore the approximation is valid.

5. Determine the OH^- concentration from the computed value of x.

Use the expression for K_w to find $[H_3O^+]$.

Substitute $[H_3O^+]$ into the pH equation to find pH.

$$[OH^-] = 1.33 \times 10^{-3} \text{ M}$$

$$[H_3O^+][OH^-] = K_w = 1.00 \times 10^{-14}$$

$$[H_3O^+](1.33 \times 10^{-3}) = 1.00 \times 10^{-14}$$

$$[H_3O^+] = 7.52 \times 10^{-12} \text{ M}$$

$$pH = -\log[H_3O^+]$$

$$= -\log(7.52 \times 10^{-12})$$

$$= 11.124$$

For Practice 15.12

Find the $[OH^-]$ and pH of a 0.33 M methylamine solution.

15.8 The Acid–Base Properties of Ions and Salts

We have already seen examples of how some ions have basic properties. For example, the bicarbonate ion acts as a base according to the following equation:

$$HCO_3^-(aq) + H_2O(l) \rightleftharpoons H_2CO_3(aq) + OH^-(aq)$$

As we know, the bicarbonate ion does not form a stable compound by itself—charge neutrality requires that it pair with a cation (or counterion) to form an ionic compound, also called a *salt*. For example, the sodium salt of bicarbonate is sodium bicarbonate.

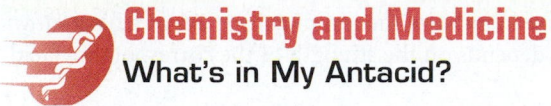

Chemistry and Medicine
What's in My Antacid?

In the opening section of this chapter, we discussed heartburn and its treatment with antacids. Some common antacids and their active ingredients include the following:

Amphogel	$Al(OH)_3$
Milk of magnesia	$Mg(OH)_2$
Maalox	$Mg(OH)_2$ and $Al(OH)_3$
Mylanta	$Mg(OH)_2$ and $Al(OH)_3$
Tums	$CaCO_3$

As you can see from this list, there are three main types of antacids: calcium-based, magnesium-based, and aluminum-based. The calcium-based antacids are suspected of causing acid rebound—which means that, although they initially relieve heartburn, they may also cause the stomach to produce more acid, resulting in a quick return of the symptoms. The aluminum- and magnesium-based antacids do not cause acid rebound, but have problems of their own. The aluminum-based antacids tend to cause constipation and the magnesium-based ones tend to cause diarrhea. (In fact, milk of magnesia can be taken as a laxative.) If you take repeated doses of these antacids, alternate between the two or choose one that contains both.

Notice the absence of group 1A metal hydroxides, such as KOH or NaOH, from the list of antacids. Why are those substances—which are completely soluble and act as strong bases—not used in antacids? The reason is that a solution containing sufficient KOH or NaOH to neutralize stomach acid would also burn the mouth and throat. In contrast, $Mg(OH)_2$

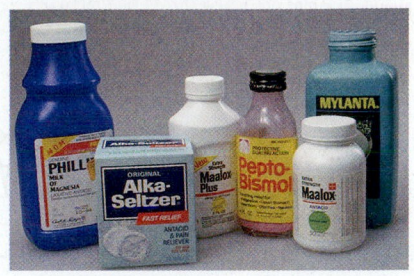

▶ Antacids contain a variety of bases designed to neutralize excess stomach acid.

and $Al(OH)_3$ are only slightly soluble. So liquid antacids containing these are actually *suspensions* of $Mg(OH)_2$ and $Al(OH)_3$—they are heterogeneous mixtures in which the solid is finely divided into the liquid. As a result, the concentration of OH^- in these suspensions is relatively small compared to what it would be with a group 1A metal hydroxide.

Initially, it might seem as though the relatively lower OH^- concentration would make the antacid much less effective. However, the solid $Mg(OH)_2$ or $Al(OH)_3$ continues to dissolve as the OH^- neutralizes stomach acid. For example, a suspension of magnesium hydroxide contains solid $Mg(OH)_2$ in equilibrium with dissolved Mg^{2+} and OH^- ions:

$$Mg(OH)_2(s) \rightleftharpoons Mg^{2+}(aq) + 2\,OH^-(aq)$$

As stomach acid is neutralized, OH^- is used up, and the equilibrium shifts to the right providing additional OH^- ions. In other words, a suspension of $Mg(OH)_2$ can provide a steady concentration of dissolved OH^- ions to neutralize stomach acid.

Question
Write chemical equations showing the reactions of each of the bases in the antacids discussed here with stomach acid (HCl).

Like all soluble salts, sodium bicarbonate dissociates in solution to form a sodium cation and bicarbonate anion:

$$NaHCO_3(s) \rightleftharpoons Na^+(aq) + HCO_3^-(aq)$$

The bicarbonate ion then acts as a weak base, ionizing water as just shown to form a basic solution. The sodium ion, however, has neither acidic nor basic properties (it does not ionize water), as we will see shortly. Consequently, the pH of a sodium bicarbonate solution is above 7. In this section, we look in general at the acid–base properties of salts and the ions they contain. Some salts are pH-neutral when put into water, others are acidic, and still others are basic, depending on their constituent anions and cations. In general, anions tend to form either *basic* or neutral solutions, while cations tend to form either *acidic* or neutral solutions.

Anions as Weak Bases

We can think of any anion as the conjugate base of an acid. For example, consider the following anions and their corresponding acids:

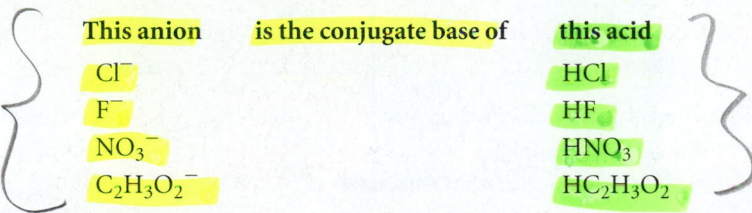

This anion	is the conjugate base of	this acid
Cl^-		HCl
F^-		HF
NO_3^-		HNO_3
$C_2H_3O_2^-$		$HC_2H_3O_2$

In general, the anion A^- is the conjugate base of the acid HA. Since virtually every anion can be envisioned as the conjugate base of an acid, the ion may itself act as a base. However, *not every anion acts as a base*—it depends on the strength of the corresponding acid. In general:

- An anion that is the conjugate base of a *weak acid* is itself a *weak base.*
- An anion that is the conjugate base of a *strong acid* is pH-*neutral* (forms solutions that are neither acidic nor basic).

For example, the Cl^- anion is the conjugate base of HCl, a strong acid. Therefore the Cl^- anion is pH-neutral (neither acidic nor basic). The F^- anion, however, is the conjugate base of HF, a weak acid. Therefore the F^- ion is itself a weak base and ionizes water according to the following reaction:

$$F^-(aq) + H_2O(l) \rightleftharpoons OH^-(aq) + HF(aq)$$

We can understand why the conjugate base of a weak acid is basic by asking ourselves why an acid is weak to begin with. Hydrofluoric acid is a weak acid because the following equilibrium lies to the left:

$$HF(aq) + H_2O(l) \rightleftharpoons H_3O^+(aq) + F^-(aq)$$

The equilibrium lies to the left because the F^- ion has a significant affinity for H^+ ions. Consequently, when F^- is put into water, its affinity for H^+ ions allows it to abstract H^+ ions from water molecules, thus acting as a weak base. In general, as shown in Figure 15.12 ▼, the weaker the acid, the stronger the conjugate base (as we saw in Section 15.4). In contrast, the conjugate base of a strong acid, such as Cl^-, does not act as a base because the following reaction lies far to the right:

$$HCl(aq) + H_2O(l) \longrightarrow H_3O^+(aq) + Cl^-(aq)$$

The reaction lies far to the right because the Cl^- ion has a very low affinity for H^+ ions. Consequently, when Cl^- is put into water, it does not abstract H^+ ions from water molecules.

	Acid	Base	
Strong	HCl	Cl^-	Neutral
	H_2SO_4	HSO_4^-	
	HNO_3	NO_3^-	
	H_3O^+	H_2O	
Weak	HSO_4^-	SO_4^{2-}	Weak
	H_2SO_3	HSO_3^-	
	H_3PO_4	$H_2PO_4^-$	
	HF	F^-	
	$HC_2H_3O_2$	$C_2H_3O_2^-$	
	H_2CO_3	HCO_3^-	
	H_2S	HS^-	
	HSO_3^-	SO_3^{2-}	
	$H_2PO_4^-$	HPO_4^{2-}	
	HCN	CN^-	
	NH_4^+	NH_3	
	HCO_3^-	CO_3^{2-}	
	HPO_4^{2-}	PO_4^{3-}	
	H_2O	OH^-	
Negligible	HS^-	S^{2-}	Strong
	OH^-	O^{2-}	

Acid Strength (left axis, increasing upward) • Base Strength (right axis, increasing downward)

▶ **FIGURE 15.12 Strength of Conjugate Acid–Base Pairs** The stronger an acid, the weaker is its conjugate base.

EXAMPLE 15.13 Determining Whether an Anion Is Basic or pH-Neutral

Classify each of the following anions as a weak base or pH-neutral (neither acidic nor basic):

(a) NO_3^- (b) NO_2^- (c) $C_2H_3O_2^-$

Solution

(a) From Table 15.3, we can see that NO_3^- is the conjugate base of a *strong* acid (HNO_3). NO_3^- is therefore pH-neutral.

(b) From Table 15.5 (or from its absence in Table 15.3), we know that NO_2^- is the conjugate base of a weak acid (HNO_2). NO_2^- is therefore a weak base.

(c) From Table 15.5 (or from its absence in Table 15.3), we know that $C_2H_3O_2^-$ is the conjugate base of a weak acid $(HC_2H_3O_2)$. $C_2H_3O_2^-$ is therefore a weak base.

For Practice 15.13

Classify each of the following anions as a weak base or pH-neutral:

(a) CHO_2^- (b) ClO_4^-

The pH of a solution containing an anion that acts as a weak base can be determined in a manner similar to that for any weak base solution. However, we need to know K_b for the anion acting as a base. As it turns out, K_b can be easily determined from K_a of the corresponding acid. Recall from Section 15.4 the expression for K_a for a generic acid HA:

$$HA(aq) + H_2O(l) \rightleftharpoons H_3O^+(aq) + A^-(aq)$$

$$K_a = \frac{[H_3O^+][A^-]}{[HA]}$$

Similarly, the expression for K_b for the conjugate base (A^-) is as follows:

$$A^-(aq) + H_2O(l) \rightleftharpoons OH^-(aq) + HA(aq) \quad reverse$$

$$K_b = \frac{[OH^-][HA]}{[A^-]}$$

If we multiply the expressions for K_a and K_b we get the following:

$$K_a \times K_b = \frac{[H_3O^+][A^-]}{[HA]} \frac{[OH^-][HA]}{[A^-]} = [H_3O^+][OH^-] = K_w$$

Or simply,

$$K_a \times K_b = K_w$$

The product of K_a for an acid and K_b for its conjugate base is simply K_w $(1.0 \times 10^{-14}$ at 25 °C). Consequently, we can find K_b for an anion acting as a base from the value of K_a for the corresponding acid. For example, for acetic acid $(HC_2H_3O_2)$, $K_a = 1.8 \times 10^{-5}$. The value of K_b for the conjugate base $(C_2H_3O_2^-)$ is therefore computed as follows:

$$K_a \times K_b = K_w$$

$$K_b = \frac{K_w}{K_a} = \frac{1.0 \times 10^{-14}}{1.8 \times 10^{-5}} = 5.6 \times 10^{-10}$$

Knowing K_b, we can find the pH of a solution containing an anion acting as a base, as shown in the following example.

EXAMPLE 15.14 Determining the pH of a Solution Containing an Anion Acting as a Base

Find the pH of a 0.100 M $NaCHO_2$ solution. The salt completely dissociates into $Na^+(aq)$ and $CHO_2^-(aq)$ and the Na^+ ion has no acid or base properties.

Solution

1. Since the Na^+ ion does not have any acidic or basic properties, we can ignore it. Write the balanced equation for the ionization of water by the basic anion and use it as a guide to prepare an ICE table showing the given concentration of the weak base as its initial concentration.

$$CHO_2^-(aq) + H_2O(l) \rightleftharpoons HCHO_2(aq) + OH^-(aq)$$

	[CHO$_2^-$]	[HCHO$_2$]	[OH$^-$]
Initial	0.100	0.00	≈ 0.00
Change			
Equil			

2. Represent the change in the concentration of OH^- with the variable x. Define the changes in the concentrations of the other reactants and products in terms of x.

$$CHO_2^-(aq) + H_2O(l) \rightleftharpoons HCHO_2(aq) + OH^-(aq)$$

	[CHO$_2^-$]	[HCHO$_2$]	[OH$^-$]
Initial	0.100	0.00	≈ 0.00
Change	$-x$	$+x$	$+x$
Equil			

3. Sum each column to determine the equilibrium concentrations in terms of the initial concentrations and the variable x.

$$CHO_2^-(aq) + H_2O(l) \rightleftharpoons HCHO_2(aq) + OH^-(aq)$$

	[CHO$_2^-$]	[HCHO$_2$]	[OH$^-$]
Initial	0.100	0.00	≈ 0.00
Change	$-x$	$+x$	$+x$
Equil	$0.100 - x$	x	x

4. Find K_b from K_a (for the conjugate acid).

$$K_a \times K_b = K_w$$

$$K_b = \frac{K_w}{K_a} = \frac{1.0 \times 10^{-14}}{1.8 \times 10^{-4}} = 5.6 \times 10^{-11}$$

Substitute the expressions for the equilibrium concentrations (from step 3) into the expression for K_b. In many cases, you can make the approximation that x is small (as discussed in Chapter 14).

$$K_b = \frac{[HCHO_2][OH^-]}{[CHO_2^-]}$$

$$= \frac{x^2}{0.100 - x}$$

Substitute the value of K_b into the K_b expression and solve for x.

$$5.6 \times 10^{-11} = \frac{x^2}{0.100}$$

$$x = 2.4 \times 10^{-6}$$

Confirm that the x is small approximation is valid by computing the ratio of x to the number it was subtracted from in the approximation. The ratio should be less than 0.05 (or 5%).

$$\frac{2.4 \times 10^{-6}}{0.100} \times 100\% = 0.0024\%$$

Therefore the approximation is valid.

5. Determine the OH^- concentration from the computed value of x.

$$[OH^-] = 2.4 \times 10^{-6} \text{ M}$$

Use the expression for K_w to find $[H_3O^+]$.

$$[H_3O^+][OH^-] = K_w = 1.0 \times 10^{-14}$$

$$[H_3O^+](2.4 \times 10^{-6}) = 1.0 \times 10^{-14}$$

$$[H_3O^+] = 4.2 \times 10^{-9} \text{ M}$$

Substitute $[H_3O^+]$ into the pH equation to find pH.

$$pH = -\log[H_3O^+]$$

$$= -\log(4.2 \times 10^{-9})$$

$$= 8.38$$

For Practice 15.14

Find the pH of a 0.250 M $NaC_2H_3O_2$ solution.

Cations as Weak Acids

In contrast to anions, which in some cases act as weak bases, cations can in some cases act as *weak acids*. We can generally divide cations into three categories: cations that are the counterions of strong bases; cations that are the conjugate acids of *weak* bases; and cations that are small, highly charged metals. We examine each individually.

Cations That Are the Counterions of Strong Bases Strong bases such as NaOH or Ca(OH)$_2$ generally contain hydroxide ions and a counterion. In solution, a strong base completely dissociates to form OH$^-$(aq) and the solvated counterion. Although these counterions interact with water molecules via ion–dipole forces, they do not ionize water and they do not contribute to the acidity or basicity of the solution. In general, therefore, *cations that are the counterions of strong bases are themselves pH-neutral* (they form solutions that are neither acidic nor basic). For example, Na$^+$, K$^+$, and Ca^{2+} are the counterions of the strong bases NaOH, KOH, and Ca(OH)$_2$ and are therefore themselves pH-neutral.

Cations That Are the Conjugate Acids of Weak Bases A cation can be formed from any nonionic weak base by adding a proton (H$^+$) to its formula. The cation will be the conjugate acid of the base. For example, consider the following cations and their corresponding weak bases.

This cation	is the conjugate acid of	this weak base
NH$_4^+$		NH$_3$
C$_2$H$_5$NH$_3^+$		C$_2$H$_5$NH$_2$
CH$_3$NH$_3^+$		CH$_3$NH$_2$

Any of these cations, with the general formula BH$^+$, will act as a weak acid according to the following general equation:

$$BH^+(aq) + H_2O(aq) \rightleftharpoons H_3O^+(aq) + B(aq)$$

In general, *a cation that is the conjugate acid of a weak base is a weak acid.*

The pH of a solution containing the conjugate acid of a weak base can be calculated just like that of any other weakly acidic solution. However, the value of K_a for the acid must be derived from K_b using the previously derived relationship $K_a \times K_b = K_w$.

Cations That Are Small, Highly Charged Metals *Small, highly charged metal cations such as Al^{3+} and Fe^{3+} form weakly acidic solutions.* For example, when Al^{3+} is dissolved in water, it becomes hydrated according to the following equation:

$$Al_3^+(aq) + 6\ H_2O(l) \longrightarrow Al(H_2O)_6^{3+}(aq)$$

The hydrated form of the ion then acts as a Brønsted–Lowry acid:

$$Al(H_2O)_6^{3+}(aq) + H_2O(aq) \rightleftharpoons Al(H_2O)_5(OH)^{2+} + H_3O^+(aq)$$

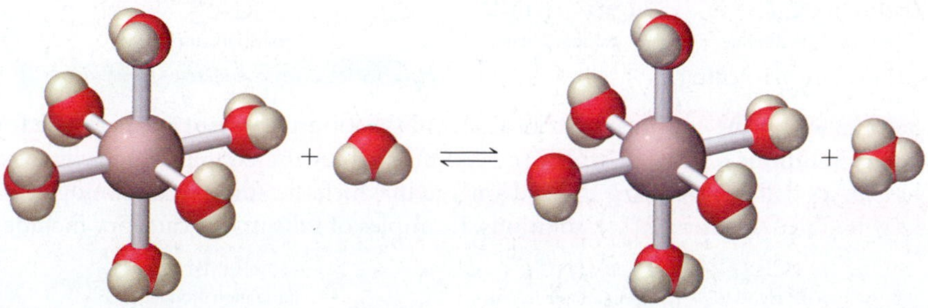

This type of behavior is not observed for alkali metals or alkaline earth metals, but is observed for the cations of most other metals.

EXAMPLE 15.15 Determining Whether a Cation Is Acidic or pH-Neutral

Classify each of the following cations as a weak acid or pH-neutral (neither acidic nor basic):

(a) $C_5H_5NH^+$ (b) Ca^{2+} (c) Cr^{3+}

Solution

(a) The $C_5H_5NH^+$ cation is the conjugate acid of a weak base. This ion is therefore a weak acid.

(b) The Ca^{2+} cation is the counterion of a strong base. This ion is therefore pH-neutral (neither acidic nor basic).

(c) The Cr^{3+} cation is a small, highly charged metal cation. It is therefore a weak acid.

For Practice 15.15

Classify each of the following cations as a weak acid or pH-neutral (neither acidic nor basic):

(a) Li^+ (b) $CH_3NH_3^+$ (c) Fe^{3+}

Classifying Salt Solutions as Acidic, Basic, or Neutral

Since salts contain both a cation and an anion, they can form acidic, basic, or neutral solutions when dissolved in water. The pH of the solution depends on the specific cation and anion involved. There are four different possibilities:

1. Salts in which neither the cation nor the anion acts as an acid or a base. (These salts form neutral solutions.)

2. Salts in which the anion acts as a base and the cation does not act as an acid. (These salts form basic solutions.)

3. Salts in which the cation acts as an acid and the anion does not act as a base. (These salts form acidic solutions.)

4. Salts in which the cation acts as an acid and the anion acts as a base. (These salts form solutions in which the pH depends on the relative strengths of the acid and the base.)

We examine examples of each individually.

1. **Salts in which neither the cation nor the anion acts as an acid or a base form pH-neutral solutions.** A salt in which the cation is the counterion of a strong base and in which the anion is the conjugate base of strong acid will form *neutral* solutions. Some examples of salts in this category include

$$\overset{+}{Na}\overset{-}{Cl} \qquad \overset{+}{Ca}(\overset{-}{NO_3})_2 \qquad \overset{+}{K}\overset{-}{Br}$$
sodium chloride   calcium nitrate   potassium bromide

Cations are pH-neutral   Anions are conjugate bases of *strong* acids.

2. **Salts in which the cation does not act as an acid and the anion acts as a base form basic solutions.** A salt in which the cation is the counterion of a strong base and in which the anion is the conjugate base of *weak* acid will form *basic* solutions. Examples of salts in this category include

$$NaF \qquad Ca(C_2H_3O_2)_2 \qquad KNO_2$$
sodium fluoride   calcium acetate   potassium nitrite

Cations are pH-neutral   Anions are conjugate bases of *weak* acids.

3. **Salts in which the cation acts as an acid and the anion does not act as a base form acidic solutions.** A salt in which the cation is either the conjugate acid of a weak base or a small, highly charged metal ion and in which the anion is the conjugate base of *strong* acid will form *acidic* solutions. Examples of salts in this category include

$$FeCl_3 \qquad Al(NO_3)_3 \qquad NH_4Br$$
iron(III) chloride   aluminum nitrate   ammonium bromide

Cations are conjugate acids of *weak* bases or small, highly charged metal ions.   Anions are conjugate bases of *strong* acids.

4. Salts in which the cation acts as an acid and the anion acts as a base form solutions in which the pH depends on the relative strengths of the acid and the base. A salt in which the cation is either the conjugate acid of a weak base or a small, highly charged metal ion and in which the anion is the conjugate base of a *weak* acid will form a solution in which the pH depends on the relative strengths of the acid and base. Examples of salts in this category include

FeF_3	$Al(C_2H_3O_2)_3$	NH_4NO_2
iron(III) fluoride	aluminum acetate	ammonium nitrite

Cations are conjugate acids of *weak* bases or small, highly charged metal ions.   Anions are conjugate bases of *weak* acids.

You can determine the overall pH of a solution containing one of these salts by comparing the K_a of the acid to the K_b of the base—the ion with the higher value of K dominates and determines whether the solution will be acidic or basic, as shown in part (e) of the following example. All these possibilities are summarized in Table 15.9.

TABLE 15.9 pH of Salt Solutions

		ANION	
		Conjugate base of strong acid	Conjugate base of weak acid
CATION	Conjugate acid of weak base	*Acidic*	*Depends on relative strengths*
	Small, highly charged metal ion	*Acidic*	*Depends on relative strengths*
	Counterion of strong base	*Neutral*	*Basic*

EXAMPLE 15.16 Determining the Overall Acidity or Basicity of Salt Solutions

Determine whether the solution formed by each of the following salts will be acidic, basic, or neutral:

(a) $SrCl_2$ **(b)** $AlBr_3$ **(c)** $CH_3NH_3NO_3$ **(d)** $NaCHO_2$ **(e)** NH_4F

Solution

(a) The Sr^{2+} cation is the counterion of a strong base $[Sr(OH)_2]$ and is therefore pH-neutral. The Cl^- anion is the conjugate base of a strong acid (HCl) and is therefore pH-neutral as well. The $SrCl_2$ solution will therefore be pH-neutral (neither acidic nor basic).

(b) The Al^{3+} cation is a small, highly charged metal ion (that is not an alkali metal or an alkaline earth metal) and is therefore a weak acid. The Br^- anion is the conjugate base of a strong acid (HBr) and is therefore pH-neutral. The $AlBr_3$ solution will therefore be acidic.

(c) The $CH_3NH_3^+$ ion is the conjugate acid of a weak base (CH_3NH_2) and is therefore acidic. The NO_3^- anion is the conjugate base of a strong acid (HNO_3) and is therefore pH-neutral. The $CH_3NH_3NO_3$ solution will therefore be acidic.

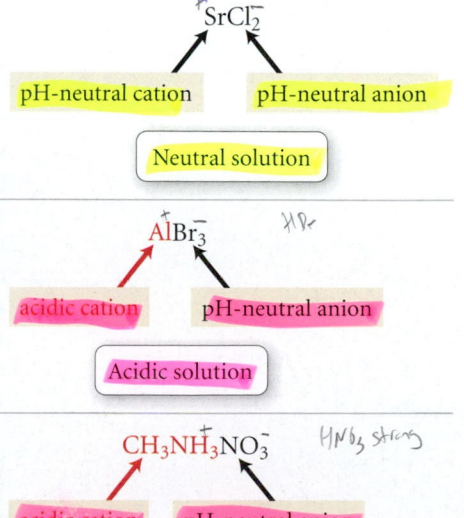

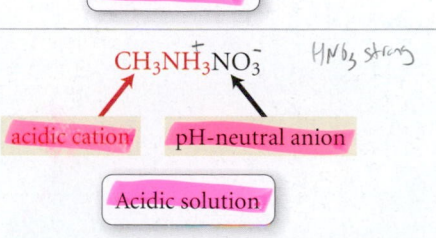

(d) The Na^+ cation is the counterion of a strong base and is therefore pH-neutral. The CHO_2^- anion is the conjugate base of a weak acid and is therefore basic. The $NaCHO_2$ solution will therefore be basic.

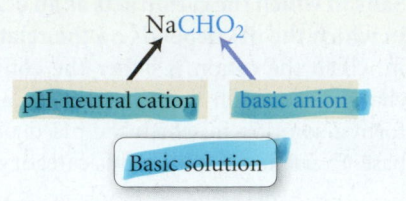

(e) The NH_4^+ ion is the conjugate acid of a weak base (NH_3) and is therefore acidic. The F^- ion is the conjugate base of a weak acid and is therefore basic. To determine the overall acidity or basicity of the solution, compare the values of K_a for the acidic cation and and K_b for the basic anion. Obtain each value of K from the conjugate by using $K_a \times K_b = K_w$.

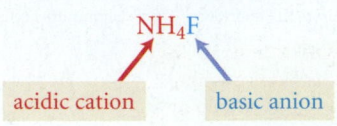

$$K_a(NH_4^+) = \frac{K_w}{K_b(NH_3)} = \frac{1.0 \times 10^{-14}}{1.76 \times 10^{-5}}$$
$$= 5.68 \times 10^{-10}$$

$$K_b(F^-) = \frac{K_w}{K_a(HF)} = \frac{1.0 \times 10^{-14}}{3.5 \times 10^{-4}}$$
$$= 2.9 \times 10^{-11}$$

Since K_a is greater than K_b, the solution is acidic.

$K_a > K_b$ | Acidic solution |

For Practice 15.16

Determine whether the solutions formed by each of the following salts will be acidic, basic, or neutral:

(a) $NaHCO_3$ **(b)** $CH_3CH_2NH_3Cl$ **(c)** KNO_3 **(d)** $Fe(NO_3)_3$

15.9 Polyprotic Acids

In Section 15.4, we learned that some acids, called polyprotic acids, contain two or more ionizable protons. For example, sulfurous acid (H_2SO_3) is a diprotic acid containing two ionizable protons and phosphoric acid (H_3PO_4) is a triprotic acid containing three ionizable protons. In general, a **polyprotic acid** ionizes in successive steps, each with its own K_a. For example, sulfurous acid ionizes as follows:

$$H_2SO_3(aq) \rightleftharpoons H^+(aq) + HSO_3^-(aq) \qquad K_{a_1} = 1.6 \times 10^{-2}$$
$$HSO_3^-(aq) \rightleftharpoons H^+(aq) + SO_3^{2-}(aq) \qquad K_{a_2} = 6.4 \times 10^{-8}$$

where K_{a_1} is the acid ionization constant for the first step and K_{a_2} is the acid ionization constant for the second step. Notice that K_{a_2} is smaller than K_{a_1}. This is true for all polyprotic acids and makes physical sense because the first proton must separate from a neutral molecule while the second must separate from an anion. The negative charge of the anion causes the positively charged proton to be held more tightly, making it more difficult to remove and resulting in a smaller value of K_a. Table 15.10 lists some common polyprotic acids and their acid ionization constants. Notice that in all cases, the values of K_a for each step become successively smaller. The value of K_{a_1} for sulfuric acid is listed as strong because, as we learned in Section 15.4, sulfuric acid is strong in the first step and weak in the second.

Finding the pH of Polyprotic Acid Solutions

Finding the pH of a polyprotic acid solution is simpler than might be imagined because, for most polyprotic acids, K_{a_1} is much larger than K_{a_2} (or K_{a_3} for triprotic acids). Therefore the amount of H_3O^+ contributed by the first ionization step is much larger than that contributed by the second or third ionization step. In addition, the production of H_3O^+

TABLE 15.10 Common Polyprotic Acids and Ionization Constants

Name (Formula)	Structure	K_{a_1}	K_{a_2}	K_{a_3}
Sulfuric Acid (H_2SO_4)		Strong	1.2×10^{-2}	
Oxalic Acid ($H_2C_2O_4$)		6.0×10^{-2}	6.1×10^{-5}	
Sulfurous Acid (H_2SO_3)		1.6×10^{-2}	6.4×10^{-8}	
Phosphoric Acid (H_3PO_4)		7.5×10^{-3}	6.2×10^{-8}	4.2×10^{-13}
Citric Acid ($H_3C_6H_5O_3$)		7.4×10^{-4}	1.7×10^{-5}	4.0×10^{-7}
Ascorbic Acid ($H_2C_6H_6O_6$)		8.0×10^{-5}	1.6×10^{-12}	
Carbonic Acid (H_2CO_3)		4.3×10^{-7}	5.6×10^{-11}	

by the first step inhibits additional production of H_3O^+ by the second step (because of Le Châtelier's principle.) Consequently, we can treat most polyprotic acid solutions as if the first step were the only one that contributes to the H_3O^+ concentration, as shown in the following example. A major exception is a dilute solution of sulfuric acid, which we cover in Example 15.18.

EXAMPLE 15.17 Finding the pH of a Polyprotic Acid Solution

Find the pH of a 0.100 M ascorbic acid ($H_2C_6H_6O_6$) solution.

Solution

To find the pH, you must find the equilibrium concentration of H_3O^+. Treat the problem as a weak acid pH problem with a single ionizable proton. The second proton contributes a negligible amount to the concentration of H_3O^+ and can therefore be ignored. Follow the procedure in Example 15.6, shown in condensed form here. Use K_{a_1} for ascorbic acid from Table 15.10.

$$H_2C_6H_6O_6(aq) + H_2O(l) \rightleftharpoons H_3O^+(aq) + HC_6H_6O_6^-(aq)$$

	[$H_2C_6H_6O_6$]	[H_3O^+]	[$HC_6H_6O_6^-$]
Initial	0.100	≈ 0.00	0.000
Change	$-x$	$+x$	$+x$
Equil	$0.100 - x$	x	x

$$K_{a_1} = \frac{[H_3O^+][HC_6H_6O_6^-]}{[H_2C_6H_6O_6]}$$

$$= \frac{x^2}{0.100 - x} \quad (x \text{ is small})$$

Confirm that the x *is small* approximation is valid by computing the ratio of x to the number it was subtracted from in the approximation. The ratio should be less than 0.05 (or 5%).

$$8.0 \times 10^{-5} = \frac{x^2}{0.100}$$

$$x = 2.8 \times 10^{-3}$$

$$\frac{2.8 \times 10^{-3}}{0.100} \times 100\% = 2.8\%$$

The approximation is valid. Therefore, $[H_3O^+] = 2.8 \times 10^{-3}$ M.

Compute the pH from H_3O^+ concentration.

$$pH = -\log(2.8 \times 10^{-3}) = 2.55$$

For Practice 15.17

Find the pH of a 0.050 M H_2CO_3 solution.

EXAMPLE 15.18 Dilute H_2SO_4 Solutions

Find the pH of a 0.0100 M sulfuric acid (H_2SO_4) solution.

Solution

Sulfuric acid is strong in its first ionization step and weak in its second. Begin by writing the equations for the two steps. As the concentration of an H_2SO_4 solution becomes smaller, the second ionization step becomes more significant because the percent ionization increases (as discussed in Section 15.6). Therefore, for a concentration of 0.0100 M, you can't simply neglect the H_3O^+ contribution from the second step, as you can for other polyprotic acids. You must calculate the H_3O^+ contributions from both steps.

$$H_2SO_4(aq) + H_2O(l) \longrightarrow H_3O^+(aq) + HSO_4^-(aq) \quad \text{Strong}$$

$$HSO_4^-(aq) + H_2O(l) \rightleftharpoons H_3O^+(aq) + SO_4^{2-}(aq) \quad K_{a_2} = 0.012$$

The $[H_3O^+]$ that results from the first ionization step is 0.0100 M (because the first step is strong). To determine the $[H_3O^+]$ formed by the second step, prepare an ICE table for the second step in which the initial concentration of H_3O^+ is 0.0100 M. The initial concentration of HSO_4^- must also be 0.0100 M (due to the stoichiometry of the ionization reaction).

	$[HSO_4^-]$	$[H_3O^+]$	$[SO_4^{2-}]$
Initial	0.0100	≈ 0.0100	0.000
Change	$-x$	$+x$	$+x$
Equil	$0.0100 - x$	$0.0100 + x$	x

Substitute the expressions for the equilibrium concentrations (from the table above) into the expression for K_{a_2}. In this case, you cannot make the x *is small* approximation because the equilibrium constant (0.012) *is not small* relative to the initial concentration (0.0100).

$$K_{a_2} = \frac{[H_3O^+][SO_4^{2-}]}{[HSO_4^-]}$$

$$= \frac{(0.0100 + x)x}{0.0100 - x}$$

Substitute the value of K_{a_2} and multiply out the expression to arrive at the standard quadratic form.

$$0.012 = \frac{0.0100x + x^2}{0.0100 - x}$$

$$0.012(0.0100 - x) = 0.0100x + x^2$$

$$0.00012 - 0.012x = 0.0100x + x^2$$

$$x^2 + 0.022x - 0.00012 = 0$$

| Solve the quadratic equation using the quadratic formula. | $$x = \frac{-b \pm \sqrt{b^2 - 4ac}}{2a}$$ $$= \frac{-(0.022) \pm \sqrt{(0.022)^2 - 4(1)(-0.00012)}}{2(1)}$$ $$= \frac{-0.022 \pm 0.031}{2}$$ $$x = -0.027 \quad \text{or} \quad x = 0.0045$$ Since x represents a concentration, and since concentrations cannot be negative, we reject the negative root. $$x = 0.0045$$ |
| Determine the H_3O^+ concentration from the computed value of x and compute the pH. Notice that the second step produces almost half as much H_3O^+ as the first step and cannot therefore be neglected. This will always be the case with dilute H_2SO_4 solutions. | $$[H_3O^+] = 0.0100 + x$$ $$= 0.0100 + 0.0045$$ $$= 0.0145 \text{ M}$$ $$pH = -\log[H_3O^+]$$ $$= -\log(0.0145)$$ $$= 1.84$$ |

For Practice 15.18

Find the pH and $[SO_4^{2-}]$ of a 0.0075 M sulfuric acid solution.

Finding the Concentration of the Anions for a Weak Diprotic Acid Solution

In some cases, we may want to know the concentrations of the anions formed by a polyprotic acid. For example, consider the following generic polyprotic acid H_2X and its ionization steps:

$$H_2X(aq) + H_2O(l) \rightleftharpoons H_3O^+(aq) + HX^-(aq) \qquad K_{a_1}$$
$$HX^-(aq) + H_2O(l) \rightleftharpoons H_3O^+(aq) + X^{2-}(aq) \qquad K_{a_2}$$

In Examples 15.17 and 15.18, we learned how to find the concentration of H_3O^+ for such a solution, which will be equal to the concentration of HX^-. However, what if we needed to find the concentration of X^{2-}? To find the concentration of X^{2-}, we simply use the concentration of HX^- and H_3O^+ (from the first ionization step) as the initial concentrations for the second ionization step. We then solve a second equilibrium problem using the second ionization equation and K_{a_2}, as shown in the following example.

EXAMPLE 15.19 Finding the Concentration of the Anions for a Weak Diprotic Acid Solution

Find the $[C_6H_6O_6^{2-}]$ of the 0.100 M ascorbic acid $(H_2C_6H_6O_6)$ solution in Example 15.17.

Solution

To find the $[C_6H_6O_6^{2-}]$, use the concentrations of $HC_6H_6O_6^-$ and H_3O^+ produced by the first ionization step (as calculated in Example 15.17) as the initial concentrations for the second step. Because of the 1:1 stoichiometry, $[HC_6H_6O_6^-] = [H_3O^+]$. Then solve an equilibrium problem for the second step similar to that of Example 15.6, shown in condensed form here. Use K_{a_2} for ascorbic acid from Table 15.10.

$$HC_6H_6O_6^-(aq) + H_2O(l) \rightleftharpoons H_3O^+(aq) + C_6H_6O_6^{2-}(aq)$$

	$[HC_6H_6O_6^-]$	$[H_3O^+]$	$[C_6H_6O_6^{2-}]$
Initial	2.8×10^{-3}	2.8×10^{-3}	0.000
Change	$-x$	$+x$	$+x$
Equil	$2.8 \times 10^{-3} - x$	$2.8 \times 10^{-3} + x$	x

$$K_{a_2} = \frac{[H_3O^+][C_6H_6O_6^{2-}]}{[HC_6H_6O_6^-]}$$

$$= \frac{(2.8 \times 10^{-3} + \cancel{x})x}{2.8 \times 10^{-3} - \cancel{x}} \qquad (x \text{ is small})$$

$$= \frac{(\cancel{2.8 \times 10^{-3}})x}{\cancel{2.8 \times 10^{-3}}}$$

$$x = K_{a_2} = 1.6 \times 10^{-12}$$

Since x is much smaller than 2.8×10^{-3}, the x is small approximation is valid. Therefore, $[C_6H_6O_6^{2-}] = 1.6 \times 10^{-12}$ M.

For Practice 15.19

Find the $[CO_3^{2-}]$ of the 0.050 M carbonic acid (H_2CO_3) solution in For Practice 15.17.

Notice from the results of the preceding example that the concentration of X^{2-} for a weak diprotic acid H_2X is simply equal to K_{a_2}. This result is general for all diprotic acids in which the *x is small* approximation is valid. Notice also that the concentration of H_3O^+ produced by the second ionization step of a diprotic acid is very small compared to the concentration produced by the first step, as shown in Figure 15.13 ▼.

Dissociation of a Polyprotic Acid

$$H_2C_6H_6O_6(aq) + H_2O(l) \rightleftharpoons H_3O^+(aq) + HC_6H_6O_6^-(aq)$$

$$[H_3O^+] = 2.8 \times 10^{-3} \text{ M}$$

$$HC_6H_6O_6^-(aq) + H_2O(l) \rightleftharpoons H_3O^+(aq) + C_6H_6O_6^{2-}(aq)$$

$$[H_3O^+] = 1.6 \times 10^{-12} \text{ M}$$

0.100 M $H_2C_6H_6O_6$

$$\text{Total } [H_3O^+] = 2.8 \times 10^{-3} \text{ M} + 1.6 \times 10^{-12} \text{ M}$$

$$= 2.8 \times 10^{-3} \text{ M}$$

▲ **FIGURE 15.13 Dissociation of a Polyprotic Acid** A 0.100 M $H_2C_6H_6O_6$ solution contains an H_3O^+ concentration of 2.8×10^{-3} M from the first step. The amount of H_3O^+ contributed by the second step is only 1.6×10^{-12} M, which is insignificant compared to the amount produced by the first step.

15.10 Acid Strength and Molecular Structure

We have learned that a Brønsted–Lowry acid is a proton (H^+) donor. However, we have not explored why some hydrogen-containing molecules act as acids while others do not, or why some acids are strong and others weak. For example, why is H_2S acidic while CH_4 is not? Or why is HF a weak acid while HCl is a strong acid? We will divide our discussion about these issues into two categories: binary acids (those containing hydrogen and only one other element) and oxyacids (those containing hydrogen bonded to an oxygen atom that is bonded to another element).

Binary Acids

Consider the bond between a hydrogen atom and some other generic element (which we will call Y):

$$H—Y$$

The factors affecting the ease with which this hydrogen will be donated (and therefore be acidic) are the *polarity* of the bond and the *strength* of the bond.

Bond Polarity Using the notation introduced in Chapter 9, the H—Y bond must be polarized as follows in order for the hydrogen atom to be acidic:

$$\overset{+}{\delta^+}H—Y^{\delta-}$$

This requirement makes physical sense because the hydrogen atom must be lost as a positively charged ion (H^+). Having a partial positive charge on the hydrogen atom therefore facilitates its loss. For example, consider the following three bonds and their corresponding dipole moments:

$$H—Li \quad\quad H—C \quad\quad H—F$$

Not acidic   Not acidic   Acidic

LiH is ionic with *the hydrogen having the negative charge*; therefore LiH is not acidic. The C—H bond is virtually nonpolar because the electronegativities of carbon and hydrogen are similar; therefore any C—H bonds in a compound will not be acidic. In contrast, the H—F bond is polar with the positive charge on the hydrogen atom. As we know from this chapter, HF is a weak acid. The partial positive charge on the hydrogen atom makes it easier for the hydrogen to be lost as an H^+ ion.

Bond Strength The strength of the H—Y bond also affects the strength of the corresponding acid. As you might expect, the stronger the bond, the weaker the acid, because the more tightly the hydrogen atom is held, the less likely it is to come off. We can see the effect of bond strength by comparing the bond strengths and acidities of the following hydrogen halides.

Acid	Bond Energy (kJ/mol)	Type of Acid
H—F	565	Weak
H—Cl	431	Strong
H—Br	364	Strong

HCl and HBr, with their weaker bonds, are both strong acids while H—F, with its stronger bond, is a weak acid. This is the case despite HF's greater bond polarity.

The Combined Effect of Bond Polarity and Bond Strength We can see the combined effect of bond polarity and bond strength by examining the trends in acidity of the group 6A and 7A hydrides (Figure 15.14 ▶). The hydrides become more acidic from left to right as the Y—H bond becomes more polar. The hydrides also become more acidic from top to bottom as the Y—H bond becomes weaker.

Oxyacids

Oxyacids contain a hydrogen atom bonded to an oxygen atom which is in turn bonded to some other atom (which we will call Y):

$$H—O—Y—\cdots$$

Y may or may not be bonded to other atoms. The factors affecting the ease with which this hydrogen will be donated (and therefore be acidic) are the *electronegativity of the element Y* and the *number of oxygen atoms attached to the element Y*.

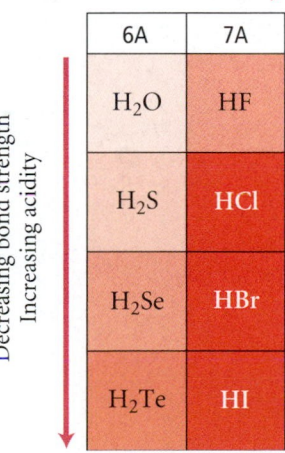

Increasing electronegativity
Increasing acidity

6A	7A
H_2O	HF
H_2S	HCl
H_2Se	HBr
H_2Te	HI

Decreasing bond strength / Increasing acidity

▲ **FIGURE 15.14 Acidity of the Group 6A and 7A Hydrides** From left to right, the hydrides become more acidic because the Y—H bond becomes more polar. From top to bottom, these hydrides become more acidic because the Y—H bond becomes weaker.

| Oxyacids are sometimes called oxoacids.

The Electronegativity of Y The more electronegative the element Y, the more it weakens and polarizes the O—H bond, resulting in greater acidity. We can see this effect by comparing the electronegativity of Y and the acid ionization constants of the following oxyacids:

Acid	Electronegativity of Y	K_a
H—O—I	2.5	2.3×10^{-11}
H—O—Br	2.8	2.0×10^{-9}
H—O—Cl	3.0	2.9×10^{-8}

Chlorine is the most electronegative of the three elements and the corresponding acid has the greatest K_a.

The Number of Oxygen Atoms Bonded to Y Oxyacids may contain additional oxygen atoms bonded to the element Y. Since these additional oxygen atoms are electronegative, they draw electron density away from the element Y, which in turn draws electron density away from the O—H bond, further weakening and polarizing it, and leading to increasing acidity. We can see this effect by comparing the following series of acid ionization constants:

Acid	Structure	K_a
$HClO_4$	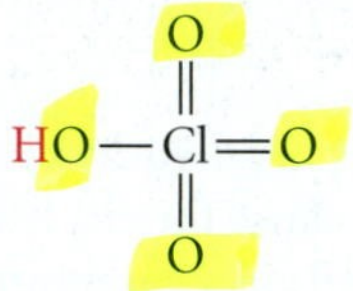 O=Cl with HO and two O (HO—Cl=O with O above and O below)	Strong
$HClO_3$	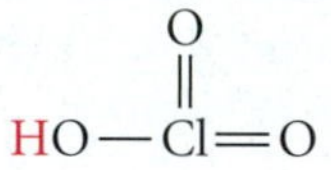 HO—Cl=O with O above	1
$HClO_2$	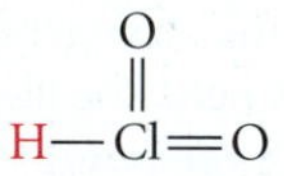 H—Cl=O with O above	1.1×10^{-2}
$HClO$	H—Cl=O	2.9×10^{-8}

The greater the number of oxygen atoms bonded to Y, the stronger the acid. On this basis we would predict that H_2SO_4 is a stronger acid than H_2SO_3 and that HNO_3 is stronger than HNO_2. As we have seen in this chapter, both H_2SO_4 and HNO_3 are strong acids, while H_2SO_3 and HNO_2 are weak acids, as predicted.

15.11 Lewis Acids and Bases

We began our definitions of acids and bases with the Arrhenius model. We then saw how the Brønsted–Lowry model expanded the range of substances that could be considered acids and bases by introducing the concept of a proton donor and proton acceptor. We now introduce a third model which further broadens the range of substances that can be considered acids. This third model is called the *Lewis model*, after G. N. Lewis, the American chemist who devised the electron-dot representation of chemical bonding (Section 9.1). While the Brønsted–Lowry model focuses on the transfer of a proton, the Lewis model focuses on the transfer of an electron pair. For example, consider the simple acid–base reaction between the H^+ ion and NH_3, shown here with Lewis structures:

$$H^+ + :NH_3 \longrightarrow [H:NH_3]^+$$

Bronsted-Lowry model focuses on the proton

Lewis model focuses on the electron pair

Under the Brønsted–Lowry model, the ammonia accepts a proton, thus acting as a base. Under the Lewis model, the ammonia acts as a base by *donating an electron pair*. The general definitions of acids and bases according to the Lewis model are as follows:

> **Lewis acid:** electron pair acceptor
>
> **Lewis base:** electron pair donor

Under the Lewis definition, H^+ in the above reaction is acting as an acid because it is accepting an electron pair from NH_3. In contrast, NH_3 is acting as a Lewis base because it is donating an electron pair to H^+.

Although the Lewis model does not significantly expand the view of a base—because all proton acceptors must have an electron pair to bind the proton–it significantly broadens the view of an acid. Under the Lewis model, a substance need not even contain hydrogen to be an acid. For example, consider the following gas-phase reaction between boron trifluoride and ammonia:

$$BF_3 + :NH_3 \longrightarrow F_3B:NH_3$$

Lewis acid   Lewis base   adduct

Boron trifluoride has an empty orbital that can accept the electron pair from ammonia and form the product (the product of a Lewis acid–base reaction is sometimes called an *adduct*). The above reaction demonstrates the following important property of Lewis acids:

> *A Lewis acid has an empty orbital (or can rearrange electrons to create an empty orbital) that can accept an electron pair.*

Consequently, the Lewis definition subsumes a whole new class of acids. We examine a few examples.

Molecules That Act as Lewis Acids

Since molecules with incomplete octets have empty orbitals, they can serve as Lewis acids. For example, both $AlCl_3$ and BCl_3 have incomplete octets:

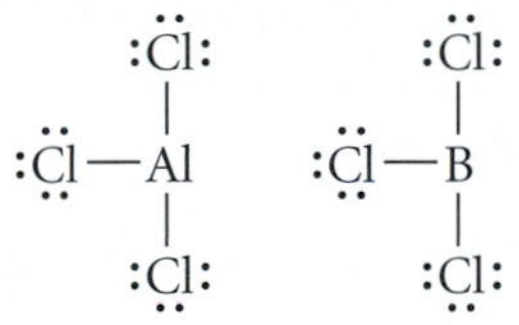

These both act as Lewis acids, as shown in the following reactions:

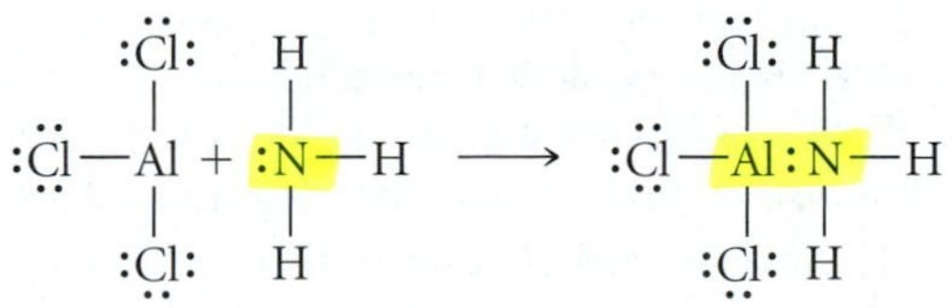

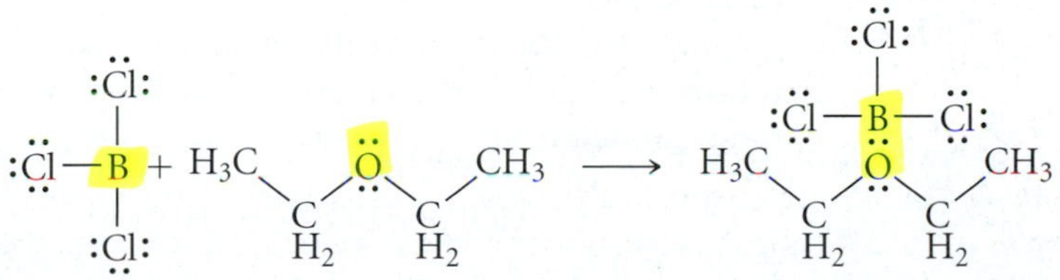

Some molecules that may not initially contain empty orbitals can rearrange their electrons to act as Lewis acids. For example, consider the following reaction between carbon dioxide and water:

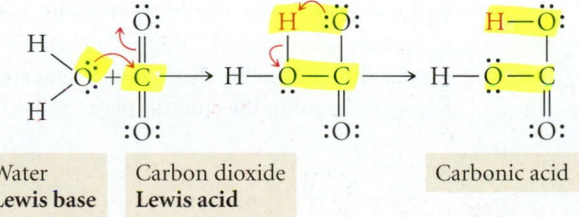

| Water | Carbon dioxide | Carbonic acid |
| **Lewis base** | **Lewis acid** | |

The electrons in the double bond on carbon move to the terminal oxygen atom, allowing carbon dioxide to act as a Lewis acid by accepting an electron pair from water. The molecule then undergoes a rearrangement in which the hydrogen atom shown in red bonds with the terminal oxygen atom instead of the internal one.

Cations That Act as Lewis Acids

Some cations, since they are positively charged and have lost some electrons, have empty orbitals that allow them to act as Lewis acids. For example, consider the hydration process of the Al^{3+} ion discussed in Section 15.8.

$$Al^{3+}(aq) + 6 \begin{bmatrix} H \\ | \\ :O: \\ | \\ H \end{bmatrix} (l) \longrightarrow \begin{bmatrix} H \\ | \\ Al \quad :O: \\ | \\ H \end{bmatrix}_6^{3+} (aq)$$

Lewis acid

Lewis base

The aluminum ion acts as a Lewis acid, accepting lone pairs from six water molecules to form the hydrated ion. Many other small, highly charged metal ions also act as Lewis acids.

15.12 Acid Rain

About 86% of U.S. energy comes from the combustion of fossil fuels, including petroleum, natural gas, and coal (Figure 15.15 ◄). Some fossil fuels, especially coal, contain small amounts of sulfur impurities. During combustion, these impurities react with oxygen to form SO_2. In addition, during combustion of any fossil fuel, nitrogen from the air reacts with oxygen to form NO_2. SO_2 and NO_2 emitted from fossil fuel combustion then react with water in the atmosphere to form sulfuric acid and nitric acid:

$$2\,SO_2 + O_2 + 2\,H_2O \longrightarrow 2\,H_2SO_4$$

$$4\,NO_2 + O_2 + 2\,H_2O \longrightarrow 4\,HNO_3$$

These acids combine with rain to form *acid rain*. The problem is greatest in the northeastern portion of the United States because many Midwestern power plants burn coal. The sulfur and nitrogen oxides produced by coal combustion in the Midwest are carried toward the Northeast by natural air currents, making rain in that portion of the country significantly acidic.

Rain is naturally somewhat acidic because of atmospheric carbon dioxide. Carbon dioxide combines with rainwater to form carbonic acid:

$$CO_2 + H_2O \longrightarrow H_2CO_3$$

However, carbonic acid is a relatively weak acid. Even rain that is saturated with CO_2 has a pH of only about 5.6, which is mildly acidic. However, when nitric acid and sulfuric acid mix with rain, the pH of the rain can fall below 4.3 (Figure 15.16 ◄). Remember that, because of the logarithmic nature of the pH scale, rain with a pH of 4.3 has an $[H_3O^+]$ that is 20 times greater than rain with a pH of 5.6. Rain that is this acidic has harmful effects on the environment.

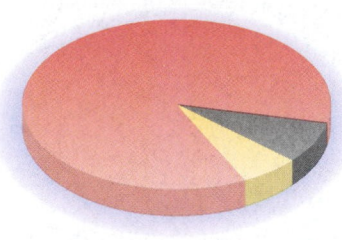

U.S. Energy Consumption by Source, 2004

■ Fossil Fuels: 86%
■ Nuclear: 8%
■ Renewable Energy: 6%

▲ FIGURE 15.15 **Sources of U.S. Energy**   About 86% of U.S. energy comes from fossil fuel combustion.

These equations represent simplified versions of the reactions that actually occur.

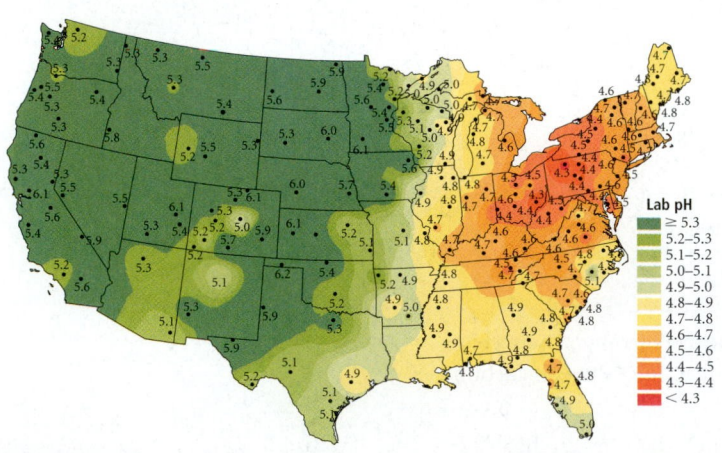

Lab pH
■ ≥ 5.3
■ 5.2–5.3
■ 5.1–5.2
■ 5.0–5.1
■ 4.9–5.0
■ 4.8–4.9
■ 4.7–4.8
■ 4.6–4.7
■ 4.5–4.6
■ 4.4–4.5
■ 4.3–4.4
■ < 4.3

▲ FIGURE 15.16 **Acid Rain**
Acid rain is a significant problem in the northeastern United States.

(a)   (b)

◀ FIGURE 15.17 The Effects of Acid Rain   (a) Buildings, gravestones, and statues damaged by acid rain are a common sight in the northeastern United States and in many other industrialized nations. (b) Forests have also suffered because some species of trees are highly susceptible to the effects of acid rain.

Effects of Acid Rain

Since acids dissolve metals, acid rain degrades metal structures. Bridges, railroads, and even automobiles can be damaged by acid rain. Since acids react with carbonates, acid rain also harms building materials that contain carbonates (CO_3^{2-}), including marble, cement, and limestone. Statues, buildings, and pathways in the Northeast show significant signs of acid rain damage (Figure 15.17a ▲).

Acid rain can also accumulate in lakes and rivers and affect aquatic life. Many lakes, especially those whose surrounding land contains significant amounts of limestone ($CaCO_3$), have the ability to neutralize acidic rain. In the Midwest, for example, limestone-rich soils prevent most lakes from becoming acidified. In the northeastern United States, however, the lack of limestone makes lakes more susceptible. In this part of the country, over 2000 lakes and streams have increased acidity levels due to acid rain. Aquatic plants, frogs, salamanders, and some species of fish are sensitive to acid levels and cannot survive in the acidified lakes. Trees can also be affected by acid rain because the acid removes nutrients from the soil, making it more difficult for trees to survive (Figure 15.17b ▲).

Acid Rain Legislation

Acid rain has been targeted by legislation. In 1990, the U.S. Congress passed amendments to the Clean Air Act specifically aimed at reducing acid rain. These amendments force electrical utilities—which are the most significant source of SO_2—to lower their SO_2 emissions gradually over time (Figure 15.18 ▶). The acidity of rain in the Northeast has already stabilized and should decrease in the coming years. Scientists expect many lakes, streams, and forests to recover once the pH of the rain returns to normal levels.

State-by-State SO₂ Emission Levels, 1990-2005

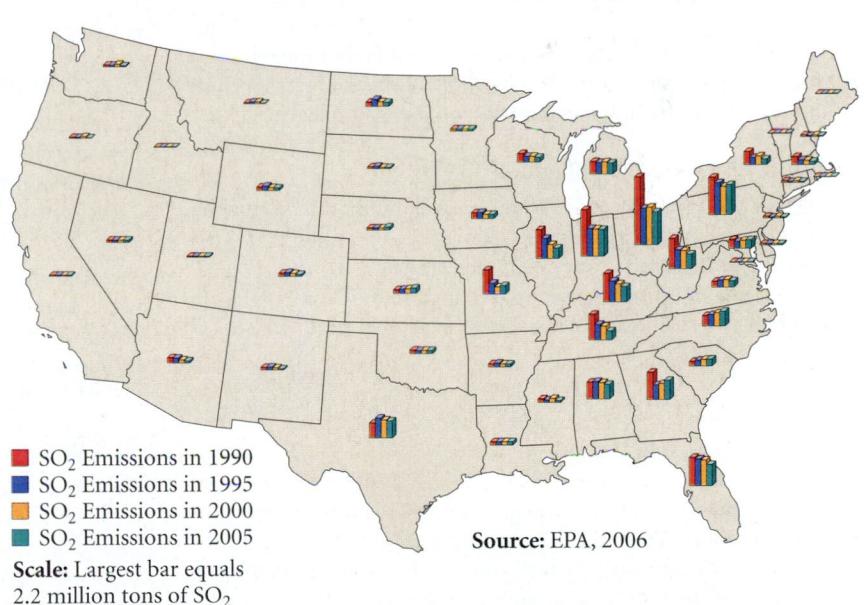

■ SO_2 Emissions in 1990
■ SO_2 Emissions in 1995
■ SO_2 Emissions in 2000
■ SO_2 Emissions in 2005

Scale: Largest bar equals 2.2 million tons of SO_2 emissions in Ohio, 1990

Source: EPA, 2006

▲ FIGURE 15.18 Emissions of Sulfur Dioxide   As a result of amendments to the Clean Air Act passed in 1990, SO_2 emissions have been decreasing and will continue to decrease in the coming years.

Chapter in Review

Key Terms

Section 15.2
carboxylic acid (665)
alkaloid (666)

Section 15.3
Arrhenius definitions
   (of acids and bases) (666)
hydronium ion (667)
Brønsted–Lowry definitions
   (of acids and bases) (667)
amphoteric (667)
conjugate acid–base pair (668)

Section 15.4
strong acid (669)
weak acid (669)
monoprotic acid (670)
diprotic acid (670)
triprotic acid (671)
acid ionization constant (K_a)
   (671)

Section 15.5
autoionization (673)
ion product constant for water
   (K_w) (673)

neutral (673)
acidic solution (673)
basic solution (673)
pH (674)

Section 15.6
percent ionization (683)

Section 15.7
strong base (687)
weak base (687)
base ionization constant (K_b)
   (687)

Section 15.9
polyprotic acid (698)

Section 15.11
Lewis acid (705)
Lewis base (705)

Key Concepts

Heartburn (15.1)

Hydrochloric acid from the stomach sometimes contacts the esophageal lining, resulting in irritation and burning, called heartburn. Heartburn is treated with antacids, bases that neutralize stomach acid. New medical methods are allowing better diagnosis and treatment for chronic heartburn.

The Nature of Acids and Bases (15.2)

Acids generally taste sour, dissolve metals, turn blue litmus paper red, and neutralize bases. Common acids include hydrochloric, sulfuric, nitric, and carboxylic acids. Bases generally taste bitter, feel slippery, turn red litmus paper blue, and neutralize acids. Common bases include sodium hydroxide, sodium bicarbonate, and potassium hydroxide.

Definitions of Acids and Bases (15.3)

The Arrhenius definition of acids and bases states that in an aqueous solution, an acid produces hydrogen ions and a base produces hydroxide ions. By the Brønsted–Lowry definition, an acid is a proton (hydrogen ion) donor and a base is a proton acceptor. In the Brønsted–Lowry definition two substances related by the transfer of a proton are called a conjugate acid–base pair.

Acid Strength and the Acid Dissociation Constant, K_a (15.4)

In a solution, a strong acid completely ionizes but a weak acid only partially ionizes. Generally, the stronger the acid, the weaker the conjugate base, and vice versa. The extent of dissociation of a weak acid is quantified by the acid dissociation constant, K_a, which is simply the equilibrium constant for the ionization of the weak acid.

Autoionization of Water and pH (15.5)

In an acidic solution, the concentration of hydrogen ions will always be greater than the concentration of hydroxide ions; however, $[H_3O^+]$ multiplied by $[OH^-]$ is always constant at a constant temperature. There are two types of logarithmic acid–base scales: pH and pOH. At 25 °C, the sum of the pH and pOH is always 14.

Finding the $[H_3O^+]$ and pH of Strong and Weak Acid Solutions (15.6)

In a strong acid solution, the hydrogen ion concentration equals the initial concentration of the acid. In a weak acid solution, the hydrogen ion concentration—which can be determined by solving an equilibrium problem—is lower than the initial acid concentration. The percent ionization of weak acids decreases as the acid (and hydrogen ion) concentration increases. In mixtures of two acids with large K_a differences, the concentration of hydrogen ions can usually be determined by considering only the stronger of the two acids.

Base Solutions (15.7)

A strong base dissociates completely while a weak base does not. The base ionization constant, K_b, describes the extent of ionization. Most weak bases produce hydroxide ions through the ionization of water.

Ions as Acids and Bases (15.8)

A cation will be a weak acid if it is the conjugate acid of a weak base; it will be neutral if it is the conjugate acid of a strong base. Conversely, an anion will be a weak base if it is the conjugate base of a weak acid; it will be neutral if it is the conjugate base of a strong acid. To calculate the pH of a basic anion, find K_b from K_w and K_a ($K_a \times K_b = K_w$).

Polyprotic Acids (15.9)

Polyprotic acids contain two or more ionizable protons. Generally, polyprotic acids ionize in successive steps, with the values of K_a becoming smaller for each step. In many cases, we can determine the $[H_3O^+]$ of a polyprotic acid solution by considering only the first ionization step. In those cases, the concentration of the acid anion formed in the second ionization step is equivalent to the value of K_{a_2}.

Acid Strength and Molecular Structure (15.10)

For binary acids, acid strength decreases with increasing bond energy, and increases with increasing bond polarity. Oxyacid strength increases with the electronegativity of the atoms bonded to the oxygen atom and also increases with the number of oxygen atoms in the molecule.

Lewis Acids and Bases (15.11)

The third model of acids and bases, the Lewis model, defines a base as an electron pair donor and an acid as an electron pair acceptor; therefore, an acid does not have to contain hydrogen. By this definition an acid can be a compound with an empty orbital—or one that will rearrange to make an empty orbital—or a cation.

Acid Rain (15.12)

The combustion of fossil fuels produces oxides of sulfur and nitrogen, which react with oxygen and water to form sulfuric and nitric acids. These acids then combine with rain to form acid rain. Acid rain corrodes man-made structures and also damages aquatic environments and forests. Environmental legislation has helped stabilize the amount of acid rain being produced.

Key Equations and Relationships

Note: In all of these equations $[H^+]$ is interchangeable with $[H_3O^+]$.

Expression for the Acid Ionization Constant, K_a (15.4)

$$K_a = \frac{[H_3O^+][A^-]}{[HA]}$$

The Ion Product Constant for Water, K_w (15.5)

$$K_w = [H_3O^+][OH^-] = 1.0 \times 10^{-14} \text{ (at 25 °C)}$$

Expression for the pH Scale (15.5)

$$pH = -\log[H_3O^+]$$

Expression for the pOH Scale (15.5)

$$pOH = -\log[OH^-]$$

Relationship between pH and pOH (15.5)

$$pH + pOH = 14.00$$

Expression for the pK_a Scale (15.5)

$$pK_a = -\log K_a$$

Expression for Percent Ionization (15.6)

$$\text{Percent ionization} = \frac{\text{concentration of ionized acid}}{\text{initial concentration of acid}} \times 100\%$$

$$= \frac{[H_3O^+]_{equil}}{[HA]_{init}} \times 100\%$$

Relationship between K_a, K_b, and K_w (15.8)

$$K_a \times K_b = K_w$$

Key Skills

Identifying Brønsted–Lowry Acids and Bases and Their Conjugates (15.3)
•Example 15.1 • For Practice 15.1 • Exercises 35, 36

Using K_w in Calculations (15.5)
•Example 15.2 • For Practice 15.2 • Exercises 47, 48

Calculating pH from $[H_3O^+]$ or $[OH^-]$ (15.5)
•Examples 15.3, 15.4 • For Practice 15.3, 15.4 • Exercises 49–52

Finding the pH of a Weak Acid Solution (15.6)
•Examples 15.5, 15.6, 15.7 • For Practice 15.5, 15.6, 15.7 • Exercises 59–64

Finding the Acid Ionization Constant from pH (15.6)
•Example 15.8 • For Practice 15.8 • Exercises 65, 66

Finding the Percent Ionization of a Weak Acid (15.6)
•Example 15.9 • For Practice 15.9 • Exercises 67–70

Mixtures of Weak Acids (15.6)
•Example 15.10 • For Practice 15.10 • Exercises 75, 76

Finding the $[OH^-]$ and pH of a Strong Base Solution (15.7)
•Example 15.11 • For Practice 15.11 • Exercises 77, 78

Finding the [OH⁻] and pH of a Weak Base Solution (15.7)
•Example 15.12   • For Practice 15.12   • Exercises 83, 84

Determining Whether an Anion Is Basic or Neutral (15.8)
•Example 15.13   • For Practice 15.13   • Exercises 89, 90

Determining the pH of a Solution Containing an Anion Acting as a Base (15.8)
•Example 15.14   • For Practice 15.14   • Exercises 91, 92

Determining Whether a Cation Is Acidic or Neutral (15.8)
•Example 15.15   • For Practice 15.15   • Exercises 93, 94

Determining the Overall Acidity or Basicity of Salt Solutions (15.8)
•Example 15.16   • For Practice 15.16   • Exercises 95, 96

Finding the pH of a Polyprotic Acid Solution (15.9)
•Example 15.17   • For Practice 15.17   • Exercises 105, 106

Finding the [H₃O⁺] in Dilute H₂SO₄ Solutions (15.9)
•Example 15.18   • For Practice 15.18   • Exercise 109

Finding the Concentration of the Anions for a Weak Diprotic Acid Solution (15.9)
•Example 15.19   • For Practice 15.19   • Exercises 107, 108

Exercises

Review Questions

1. What causes heartburn? What are some possible ways to alleviate heartburn?
2. What are the general physical and chemical properties of acids? Of bases?
3. What is a carboxylic acid? Give an example.
4. What is the Arrhenius definition of an acid? Of a base?
5. What is a hydronium ion? Does H^+ exist in solution by itself?
6. What is the Brønsted–Lowry definition of an acid? Of a base?
7. Why is there more than one definition of acid–base behavior? Which definition is the right one?
8. Describe amphoteric behavior and give an example.
9. What is a conjugate acid–base pair? Give an example.
10. Explain the difference between a strong acid and weak acid and give one example of each.
11. What are diprotic and triprotic acids? Give an example of each.
12. Give the definition of the acid ionization constant and explain its significance.
13. Write an equation for the autoionization of water and an expression for the ion product constant for water (K_w). What is the value of K_w at 25 °C?
14. What happens to the [OH⁻] of a solution when the [H₃O⁺] is increased? Decreased?
15. What is the definition of pH? What pH range is considered acidic? Basic? Neutral?
16. What is the definition of pOH? What pOH range is considered acidic? Basic? Neutral?
17. In most solutions containing a strong or weak acid, the autoionization of water can be neglected when computing [H₃O⁺]. Explain why this is so.

18. When computing [H₃O⁺] for weak acid solutions, we can often use the *x is small* approximation. Explain the nature of this approximation and why it is valid.
19. What is the percent ionization of an acid? Explain what happens to the percent ionization of a weak acid as a function of the concentration of the weak acid solution.
20. In computing [H₃O⁺] for a mixture of a strong acid and weak acid, the weak acid can often be neglected. Explain why this is so.
21. Write a generic equation showing how a weak base ionizes water.
22. How can you determine whether an anion will act as a weak base? Write a generic equation showing the reaction by which an anion, A^-, acts as a weak base.
23. What is the relationship between the acid ionization constant for a weak acid (K_a) and the base ionization constant for its conjugate base (K_b?)
24. What kinds of cations act as weak bases? Give some examples.
25. When calculating the [H₃O⁺] for a polyprotic acid, the second ionization step can often be neglected. Explain why this is so.
26. For a weak diprotic acid H_2X, what is the relationship between $[X^{2-}]$ and K_{a_2}? Under what conditions does this relationship exist?
27. For a binary acid, H—Y, what factors affect the relative ease with which the acid ionizes?
28. What factors affect the relative acidity of an oxyacid?
29. What is the Lewis definition of an acid? Of a base?
30. What is a general characteristic of a Lewis acid? Of a Lewis base?
31. What is acid rain? What causes it and where is the problem the greatest?
32. What are the main detrimental effects of acid rain? What is being done to address the problem of acid rain?

Problems by Topic

The Nature and Definitions of Acids and Base

33. Identify each of the following as an acid or a base and write a chemical equation showing how it is an acid or a base according to the Arrhenius definition.
    a. $HNO_3(aq)$ b. $NH_4^+(aq)$
    c. $KOH(aq)$ d. $HC_2H_3O_2(aq)$

34. Identify each of the following as an acid or a base and write a chemical equation showing how it is an acid or a base according to the Arrhenius definition.
    a. $NaOH(aq)$ b. $H_2SO_4(aq)$
    c. $HBr(aq)$ d. $Sr(OH)_2(aq)$

35. For each of the following, identify the Brønsted–Lowry acid, the Brønsted–Lowry base, the conjugate acid, and the conjugate base.
    a. $H_2CO_3(aq) + H_2O(l) \rightleftharpoons H_3O^+(aq) + HCO_3^-(aq)$
    b. $NH_3(aq) + H_2O(l) \rightleftharpoons NH_4^+(aq) + OH^-(aq)$
    c. $HNO_3(aq) + H_2O(l) \longrightarrow H_3O^+(aq) + NO_3^-(aq)$
    d. $C_5H_5N(aq) + H_2O(l) \rightleftharpoons C_5H_5NH^+(aq) + OH^-(aq)$

36. For each of the following, identify the Brønsted–Lowry acid, the Brønsted–Lowry base, the conjugate acid, and the conjugate base.
    a. $HI(aq) + H_2O(l) \longrightarrow H_3O^+(aq) + I^-(aq)$
    b. $CH_3NH_2(aq) + H_2O(l) \rightleftharpoons CH_3NH_3^+(aq) + OH^-(aq)$
    c. $CO_3^{2-}(aq) + H_2O(l) \rightleftharpoons HCO_3^-(aq) + OH^-(aq)$
    d. $HBr(aq) + H_2O(l) \longrightarrow H_3O^+(aq) + Br^-(aq)$

37. Write the formula for the conjugate base of each of the following acids.
    a. HCl   b. H_2SO_3 c. $HCHO_2$ d. HF

38. Write the formula for the conjugate acid of each of the following bases.
    a. NH_3 b. ClO_4^- c. HSO_4^- d. CO_3^{2-}

39. $H_2PO_4^-$ is amphoteric. Write equations that demonstrate both its acidic nature and its basic nature.

40. HCO_3^- is amphoteric. Write equations that demonstrate both its acidic nature and its basic nature.

Acid Strength and K_a

41. Classify each of the following acids as strong or weak. If the acid is weak, write an expression for the acid ionization constant (K_a).
    a. HNO_3 b. HCl   c. HBr   d. H_2SO_3

42. Classify each of the following acids as strong or weak. If the acid is weak, write an expression for the acid ionization constant (K_a).
    a. HF   b. $HCHO_2$ c. H_2SO_4 d. H_2CO_3

43. The following three diagrams represent three different solutions of the binary acid HA. Water molecules have been omitted for clarity and hydronium ions (H_3O^+) are represented by hydrogen ions (H^+). Rank the acids in order of decreasing acid strength.

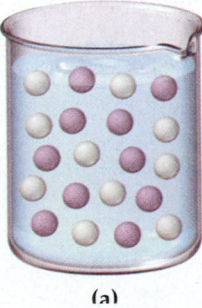
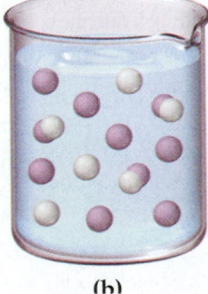
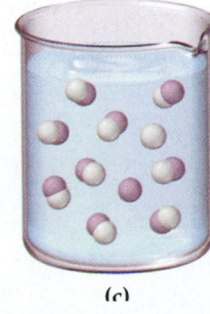

(a)   (b)   (c)

44. Rank the following solutions in order of decreasing $[H_3O^+]$: 0.10 M HCl; 0.10 M HF; 0.10 M HClO; 0.10 M HC_6H_5O.

45. Pick the stronger base from each of the following pairs:
    a. F^- or Cl^- b. NO_2^- or NO_3^-
    c. F^- or ClO^-

46. Pick the stronger base from each of the following pairs:
    a. ClO_4^- or ClO_2^- b. Cl^- or H_2O
    c. CN^- or ClO^-

Autoionization of Water and pH

47. Calculate $[OH^-]$ in each of the following aqueous solutions at 25 °C, and classify the solution as acidic or basic.
    a. $[H_3O^+] = 9.7 \times 10^{-9}$ M
    b. $[H_3O^+] = 2.2 \times 10^{-6}$ M
    c. $[H_3O^+] = 1.2 \times 10^{-9}$ M

48. Calculate $[H_3O^+]$ in each of the following aqueous solutions at 25 °C, and classify each solution as acidic or basic.
    a. $[OH^-] = 5.1 \times 10^{-4}$ M
    b. $[OH^-] = 1.7 \times 10^{-12}$ M
    c. $[OH^-] = 2.8 \times 10^{-2}$ M

49. Calculate the pH and pOH of each of the following solutions.
    a. $[H_3O^+] = 1.7 \times 10^{-8}$ M
    b. $[H_3O^+] = 1.0 \times 10^{-7}$ M
    c. $[H_3O^+] = 2.2 \times 10^{-6}$ M

50. Calculate $[H_3O^+]$ and $[OH^-]$ for each of the following solutions.
    a. pH = 8.55   b. pH = 11.23
    c. pH = 2.87

51. Complete the following table. (All solutions are at 25 °C.)

$[H_3O^+]$	$[OH^-]$	pH	Acidic or Basic
		3.15	
3.7×10^{-9}	___	___	___
___	___	11.1	___
___	1.6×10^{-11}	___	___

52. Complete the following table. (All solutions are at 25 °C.)

$[H_3O^+]$	$[OH^-]$	pH	Acidic or Basic
3.5×10^{-3}	___	___	___
___	3.8×10^{-7}	___	___
1.8×10^{-9}	___	___	___
___	___	7.15	___

53. Like all equilibrium constants, the value of K_w depends on temperature. At body temperature (37 °C), $K_w = 2.4 \times 10^{-14}$. What is the $[H_3O^+]$ and pH of pure water at body temperature?

54. The value of K_w increases with increasing temperature. Is the autoionization of water endothermic or exothermic?

Acid Solutions

55. For each of the following strong acid solutions, determine $[H_3O^+]$, $[OH^-]$, and pH.
  a. 0.15 M HCl
  b. 0.025 M HNO_3
  c. a solution that is 0.072 M in HBr and 0.015 M in HNO_3
  d. a solution that is 0.855% HNO_3 by mass (Assume a density of 1.01 g/mL for the solution.)

56. Determine the pH of each of the following solutions.
  a. 0.028 M HI
  b. 0.115 M $HClO_4$
  c. a solution that is 0.055 M in $HClO_4$ and 0.028 M in HCl
  d. a solution that is 1.85% HCl by mass (Assume a density of 1.01 g/mL for the solution.)

57. What mass of HI should be present in 0.250 L of solution to obtain a solution with each of the following pH's?
  a. pH = 1.25
  b. pH = 1.75
  c. pH = 2.85

58. What mass of $HClO_4$ should be present in 0.500 L of solution to obtain a solution with each of the following pH values?
  a. pH = 2.50
  b. pH = 1.50
  c. pH = 0.50

59. Determine the $[H_3O^+]$ and pH of a 0.100 M solution of benzoic acid.

60. Determine the $[H_3O^+]$ and pH of a 0.200 M solution of formic acid.

61. Determine the pH of an HNO_2 solution of each of the following concentrations. In which cases can you not make the simplifying assumption that x *is small*?
  a. 0.500 M
  b. 0.100 M
  c. 0.0100 M

62. Determine the pH of an HF solution of each of the following concentrations. In which cases can you not make the simplifying assumption that x *is small*?
  a. 0.250 M
  b. 0.0500 M
  c. 0.0250 M

63. If 15.0 mL of glacial acetic acid (pure $HC_2H_3O_2$) is diluted to 1.50 L with water, what is the pH of the resulting solution? The density of glacial acetic acid is 1.05 g/mL.

64. Calculate the pH of a formic acid solution that contains 1.35% formic acid by mass (Assume a density of 1.01 g/mL for the solution.)

65. A 0.185 M solution of a weak acid (HA) has a pH of 2.95. Calculate the acid ionization constant (K_a) for the acid.

66. A 0.115 M solution of a weak acid (HA) has a pH of 3.29. Calculate the acid ionization constant (K_a) for the acid.

67. Determine the percent ionization of a 0.125 M HCN solution.

68. Determine the percent ionization of a 0.225 M solution of benzoic acid.

69. Calculate the percent ionization of acetic acid solutions having the following concentrations.
  a. 1.00 M
  b. 0.500 M
  c. 0.100 M
  d. 0.0500 M

70. Calculate the percent ionization of formic acid solutions having the following concentrations.
  a. 1.00 M
  b. 0.500 M
  c. 0.100 M
  d. 0.0500 M

71. A 0.148 M solution of a monoprotic acid has a percent dissociation of 1.55%. Determine the acid ionization constant (K_a) for the acid.

72. A 0.085 M solution of a monoprotic acid has a percent dissociation of 0.59%. Determine the acid ionization constant (K_a) for the acid.

73. Find the pH and percent dissociation of each of the following HF solutions:
  a. 0.250 M
  b. 0.100 M
  c. 0.050 M

74. Find the pH and percent dissociation of a 0.100 M solution of a weak monoprotic acid having the following K_a values.
  a. $K_a = 1.0 \times 10^{-5}$
  b. $K_a = 1.0 \times 10^{-3}$
  c. $K_a = 1.0 \times 10^{-1}$

75. Find the pH of each of the following solutions of mixtures of acids.
  a. 0.115 M in HBr and 0.125 M in $HCHO_2$
  b. 0.150 M in HNO_2 and 0.085 M in HNO_3
  c. 0.185 M in $HCHO_2$ and 0.225 M in $HC_2H_3O_2$
  d. 0.050 M in acetic acid and 0.050 M in hydrocyanic acid

76. Find the pH of each of the following solutions of mixtures of acids.
  a. 0.075 M in HNO_3 and 0.175 M in $HC_7H_5O_2$
  b. 0.020 M in HBr and 0.015 M in $HClO_4$
  c. 0.095 M in HF and 0.225 M in HC_6H_5O
  d. 0.100 M in formic acid and 0.050 M in hypochlorous acid

Base Solutions

77. For each of the following strong base solutions, determine $[OH^-]$, $[H_3O^+]$, pH, and pOH.
  a. 0.15 M NaOH
  b. 1.5×10^{-3} M $Ca(OH)_2$
  c. 4.8×10^{-4} M $Sr(OH)_2$
  d. 8.7×10^{-5} M KOH

78. For each of the following strong base solutions, determine $[OH^-]$, $[H_3O^+]$, pH, and pOH.
  a. 8.77×10^{-3} M LiOH
  b. 0.0112 M $Ba(OH)_2$
  c. 1.9×10^{-4} M KOH
  d. 5.0×10^{-4} M $Ca(OH)_2$

79. Determine the pH of a solution that is 3.85% KOH by mass. Assume that the solution has density of 1.01 g/mL.

80. Determine the pH of a solution that is 1.55% NaOH by mass. Assume that the solution has density of 1.01 g/mL.

81. Write equations showing how each of the following weak bases ionizes water to form OH^-. Also write the corresponding expression for K_b.
  a. NH_3
  b. HCO_3^-
  c. CH_3NH_2

82. Write equations showing how each of the following weak bases ionizes water to form OH^-. Also write the corresponding expression for K_b.
  a. CO_3^{2-}
  b. $C_6H_5NH_2$
  c. $C_2H_5NH_2$

83. Determine the $[OH^-]$, pH, and pOH of a 0.15 M ammonia solution.

84. Determine the $[OH^-]$, pH, and pOH of a solution that is 0.125 M in CO_3^{2-}.

85. Caffeine ($C_8H_{10}N_4O_2$) is a weak base with a pK_b of 10.4. Calculate the pH of a solution containing a caffeine concentration of 455 mg/L.

86. Amphetamine ($C_9H_{13}N$) is a weak base with a pK_b of 4.2. Calculate the pH of a solution containing an amphetamine concentration of 225 mg/L.

87. Morphine is a weak base. A 0.150 M solution of morphine has a pH of 10.5. What is K_b for morphine?

88. A 0.135 M solution of a weak base has a pH of 11.23. Determine K_b for the base.

Acid–Base Properties of Ions and Salts

89. Which of the following anions act as weak bases in solution? For those anions that are basic, write an equation that shows how the anion acts as a base.
  a. Br^-
  b. ClO^-
  c. CN^-
  d. Cl^-

90. Classify each of the following anions as basic or neutral. For those anions that are basic, write an equation that shows how the anion acts as a base.
  a. $C_7H_5O_2^-$
  b. I^-
  c. NO_3^-
  d. F^-

91. Determine the $[OH^-]$ and pH of a solution that is 0.140 M in F^-.

92. Determine the $[OH^-]$ and pH of a solution that is 0.250 M in HCO_3^-.

93. Determine whether each of the following cations is acidic or pH-neutral. For those cations that are acidic, write an equation that shows how the cation acts as an acid.
  a. NH_4^+
  b. Na^+
  c. Co^{3+}
  d. $CH_2NH_3^+$

94. Determine whether each of the following cations is acidic or pH-neutral. For those cations that are acidic, write an equation that shows how the cation acts as an acid.
  a. Sr^{2+}
  b. Mn^{3+}
  c. $C_5H_5NH^+$
  d. Li^+

95. Determine whether each of the following salts will form a solution that is acidic, basic, or pH-neutral.
  a. $FeCl_3$
  b. NaF
  c. $CaBr_2$
  d. NH_4Br
  e. $C_6H_5NH_3NO_2$

96. Determine whether each of the following salts will form a solution that is acidic, basic, or pH-neutral.
  a. $Al(NO_3)_3$
  b. $C_2H_5NH_3NO_3$
  c. K_2CO_3
  d. RbI
  e. NH_4ClO

97. Arrange the following solutions in order of increasing acidity:

$$NaCl, NH_4Cl, NaHCO_3, NH_4ClO_2, NaOH$$

98. Arrange the following solutions in order of increasing basicity:

$$CH_3NH_3Br, KOH, KBr, KCN, C_5H_5NHNO_2$$

99. Determine the pH of each of the following solutions:
  a. 0.10 M NH_4Cl
  b. 0.10 M $NaC_2H_3O_2$
  c. 0.10 NaCl

100. Determine the pH of each of the following solutions:
  a. 0.20 M $KCHO_2$
  b. 0.20 M CH_3NH_3I
  c. 0.20 M KI

101. Calculate the concentration of all species in a 0.15 M KF solution.

102. Calculate the concentration of all species in a 0.225 M $C_6H_5NH_3Cl$ solution.

Polyprotic Acids

103. Write chemical equations and corresponding equilibrium expressions for each of the three ionization steps of phosphoric acid.

104. Write chemical equations and corresponding equilibrium expressions for each of the two ionization steps of carbonic acid.

105. Calculate the $[H_3O^+]$ and pH of each of the following polyprotic acid solutions:
  a. 0.350 M H_3PO_4
  b. 0.350 M $H_2C_2O_4$

106. Calculate the $[H_3O^+]$ and pH of each of the following polyprotic acid solutions:
  a. 0.125 M H_2CO_3
  b. 0.125 M $H_3C_6H_5O_7$

107. Calculate the concentration of all species in a 0.500 M solution of H_2SO_3.

108. Calculate the concentration of all species in a 0.155 M solution of H_2CO_3.

109. Calculate the $[H_3O^+]$ and pH of each of the following H_2SO_4 solutions. At approximately what concentration does the *x is small* approximation break down?
  a. 0.50 M
  b. 0.10 M
  c. 0.050 M

110. Consider a 0.10 M solution of a weak polyprotic acid (H_2A) with the possible values of K_{a_1} and K_{a_2} given below.
  a. $K_{a_1} = 1.0 \times 10^{-4}; K_{a_2} = 5.0 \times 10^{-5}$
  b. $K_{a_1} = 1.0 \times 10^{-4}; K_{a_2} = 1.0 \times 10^{-5}$
  c. $K_{a_1} = 1.0 \times 10^{-4}; K_{a_2} = 1.0 \times 10^{-6}$

Calculate the contributions to $[H_3O^+]$ from each ionization step. At what point can the contribution of the second step be neglected?

Molecular Structure and Acid Strength

111. Based on their molecular structure, pick the stronger acid from each of the following pairs of binary acids. Explain your reasoning.
  a. HF and HCl
  b. H_2O or HF
  c. H_2Se or H_2S

112. Based on molecular structure, arrange the following binary compounds in order of increasing acid strength. Explain your reasoning.

$$H_2Te, HI, H_2S, NaH$$

113. Based on their molecular structure, pick the stronger acid from each of the following pairs of oxyacids. Explain your reasoning.
  a. H_2SO_4 or H_2SO_3
  b. $HClO_2$ or HClO
  c. HClO or HBrO
  d. CCl_3COOH or CH_3COOH

114. Based on molecular structure, arrange the following sets of oxyacids in order of increasing acid strength. Explain your reasoning.

$$HClO_3, HIO_3, HBrO_3$$

115. Which is a stronger base, S^{2-} or Se^{2-}? Explain.

116. Which is a stronger base, PO_4^{3-} or AsO_4^{3-}? Explain.

Lewis Acids and Bases

117. Classify each of the following as either a Lewis acid or a Lewis base.
  a. Fe^{3+}
  b. BH_3
  c. NH_3
  d. F^-

118. Classify each of the following as either a Lewis acid or a Lewis base.
  a. $BeCl_2$
  b. OH^-
  c. $B(OH)_3$
  d. CN^-

119. Identify the Lewis acid and Lewis base from among the reactants in each of the following equations:
  a. $Fe^{3+}(aq) + 6 H_2O(l) \rightleftharpoons Fe(H_2O)_6^{3+}(aq)$
  b. $Zn^{2+}(aq) + 4 NH_3(aq) \rightleftharpoons Zn(NH_3)_4^{2+}(aq)$
  c. $(CH_3)_3N(g) + BF_3(g) \rightleftharpoons (CH_3)_3NBF_3(s)$

120. Identify the Lewis acid and Lewis base from among the reactants in each of the following equations:
  a. $Ag^+(aq) + 2 NH_3(aq) \rightleftharpoons Ag(NH_3)_2^+(aq)$
  b. $AlBr_3 + NH_3 \rightleftharpoons H_3NAlBr_3$
  c. $F^-(aq) + BF_3(aq) \rightleftharpoons BF_4^-(aq)$

Cumulative Problems

121. Consider the following molecular views of acid solutions. Based on the molecular view, determine whether the acid is weak or strong.

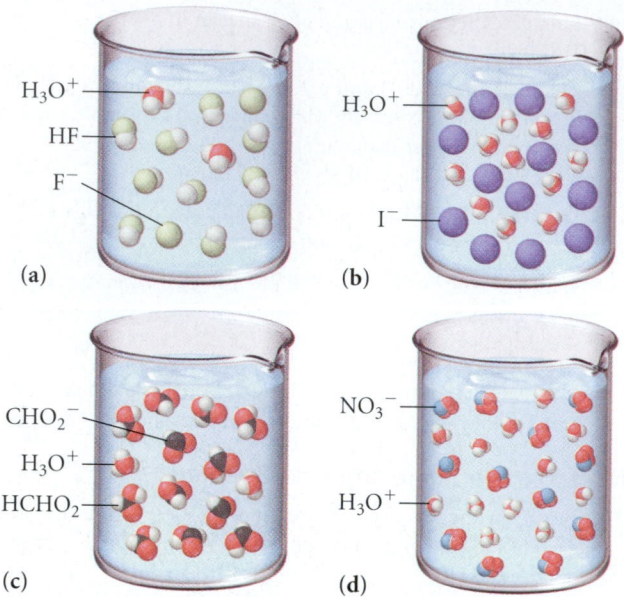

(a)   (b)

(c)   (d)

122. Consider the following molecular views of base solutions. Based on the molecular view, determine whether the base is weak or strong.

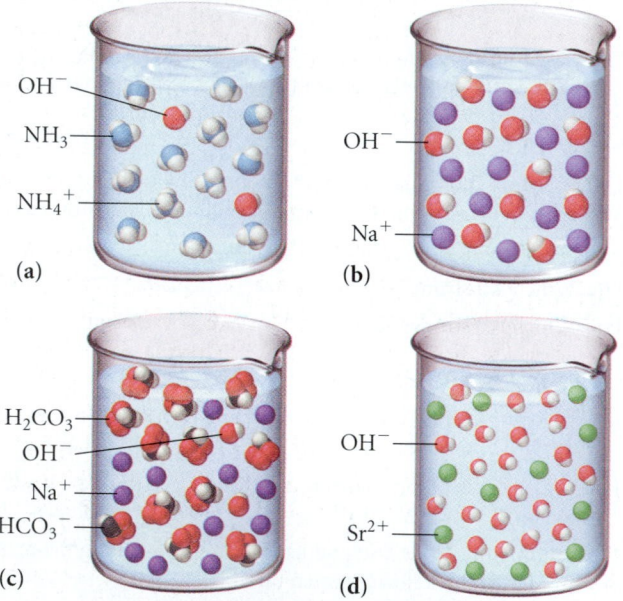

(a)   (b)

(c)   (d)

123. The binding of oxygen by hemoglobin in the blood involves the following equilibrium reaction:

$$HbH^+(aq) + O_2(aq) \rightleftharpoons HbO_2(aq) + H^+(aq)$$

In this equation, Hb is hemoglobin. The pH of normal human blood is highly controlled within a range of 7.35 to 7.45. Given the above equilibrium, why is this important? What would happen to the oxygen-carrying capacity of hemoglobin if blood became too acidic (a dangerous condition known as acidosis)?

124. Carbon dioxide dissolves in water according to the following equations:

$$CO_2(g) + H_2O(l) \rightleftharpoons H_2CO_3(aq)$$
$$H_2CO_3(aq) + H_2O(l) \rightleftharpoons HCO_3^-(aq) + H_3O^+(aq)$$

Carbon dioxide levels in the atmosphere have increased about 20% over the last century. Since Earth's oceans are exposed to atmospheric carbon dioxide, what effect might the increased CO_2 have on the pH of the world's oceans? What effect might this change have on the limestone structures (primarily $CaCO_3$) of coral reefs and marine shells?

125. Milk of magnesia is often taken to reduce the discomfort associated with acid stomach or heartburn. The recommended dose is 1 teaspoon, which contains 4.00×10^2 mg of $Mg(OH)_2$. What volume of an HCl solution with a pH of 1.3 can be neutralized by one dose of milk of magnesia? If the stomach contains 2.00×10^2 mL of pH 1.3 solution, will all the acid be neutralized? If not, what fraction is neutralized?

126. Lakes that have been acidified by acid rain can be neutralized by liming, the addition of limestone ($CaCO_3$). How much limestone (in kg) is required to completely neutralize a 4.3 billion liter lake with a pH of 5.5?

▲ Liming a lake.

127. Acid rain over the Great Lakes has a pH of about 4.5. Calculate the $[H_3O^+]$ of this rain and compare that value to the $[H_3O^+]$ of rain over the West Coast that has a pH of 5.4. How many times more concentrated is the acid in rain over the Great Lakes?

128. White wines tend to be more acidic than red wines. Find the $[H_3O^+]$ in a Sauvignon Blanc with a pH of 3.23 and a Cabernet Sauvignon with a pH of 3.64. How many times more acidic is the Sauvignon Blanc?

129. Common aspirin is acetylsalicylic acid, which has the structure shown below and a pK_a of 3.5.

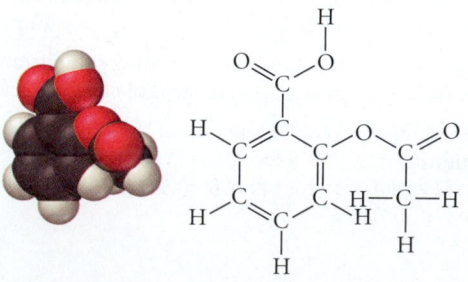

Calculate the pH of a solution in which one normal adult dose of aspirin (6.5×10^2 mg) is dissolved in 8.0 ounces of water.

130. The AIDS drug zalcitabine (also known as ddC) is a weak base with the structure shown below and a pK_b of 9.8.

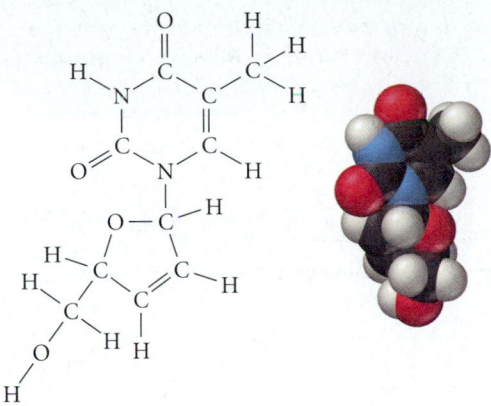

What percentage of the base is protonated in an aqueous zalcitabine solution containing 565 mg/L?

131. Determine the pH of each of the following solutions:
    a. 0.0100 M $HClO_4$
    b. 0.115 M $HClO_2$
    c. 0.045 M $Sr(OH)_2$
    d. 0.0852 M KCN
    e. 0.155 M NH_4Cl

132. Determine the pH of each of the following solutions:
    a. 0.0650 M HNO_3
    b. 0.150 M HNO_2
    c. 0.0195 M KOH
    d. 0.245 M CH_3NH_3I
    e. 0.318 M KC_6H_5O

133. Determine the pH of each of the following two-component solutions.
    a. 0.0550 M in HI and 0.00850 M in HF
    b. 0.112 M in NaCl and 0.0953 M in KF
    c. 0.132 M in NH_4Cl and 0.150 M HNO_3
    d. 0.0887 M in sodium benzoate and 0.225 M in potassium bromide
    e. 0.0450 M in HCl and 0.0225 M in HNO_3

134. Determine the pH of each of the following two-component solutions.
    a. 0.050 M KOH and 0.015 M $Ba(OH)_2$
    b. 0.265 M NH_4NO_3 and 0.102 M HCN
    c. 0.075 M RbOH and 0.100 M $NaHCO_3$
    d. 0.088 M $HClO_4$ and 0.022 M KOH
    e. 0.115 M NaClO and 0.0500 M KI

135. Write net ionic equations for the reactions that take place when aqueous solutions of the following substances are mixed:
    a. sodium cyanide and nitric acid
    b. ammonium chloride and sodium hydroxide
    c. sodium cyanide and ammonium bromide
    d. potassium hydrogen sulfate and lithium acetate
    e. sodium hypochlorite and ammonia

136. Morphine has the formula $C_{17}H_{19}NO_3$. It is a base and accepts one proton per molecule. It is isolated from opium. A 0.682-g sample of opium is found to require 8.92 mL of a 0.0116 M solution of sulfuric acid for neutralization. Assuming that morphine is the only acid or base present in opium, calculate the percent morphine in the sample of opium.

137. The pH of a 1.00 M solution of urea, a weak organic base, is 7.050. Calculate the K_a of protonated urea.

138. A solution is prepared by dissolving 0.10 mol of acetic acid and 0.10 mol of ammonium chloride in enough water to make 1.0 L of solution. Find the concentration of ammonia in the solution.

Challenge Problems

139. A student mistakenly calculates the pH of a 1.0×10^{-7} M HI solution to be 7.0. Explain why this calculation is incorrect and calculate the correct pH.

140. When 2.55 g of an unknown weak acid (HA) with a molar mass of 85.0 g/mol is dissolved in 250.0 g of water, the freezing point of the resulting solution is -0.257 °C. Calculate K_a for the unknown weak acid.

141. Calculate the pH of a solution that is 0.00115 M in HCl and 0.0100 M in $HClO_2$.

142. To what volume should 1 L of a solution of a weak acid HA be diluted to reduce the $[H^+]$ to one-half of that in the original solution?

143. HA, a weak acid, with $K_a = 1.0 \times 10^{-8}$, also forms the ion HA_2^-. The reaction is $HA(aq) + A^-(aq) \rightleftharpoons HA_2^-(aq)$ and its $K = 4.0$. Calculate the $[H^+]$, $[A^-]$, and $[HA_2^-]$ in a 1.0 M solution of HA.

144. Basicity in the gas phase can be defined as the proton affinity of the base, for example, $CH_3NH_2(g) + H^+(g) \rightleftharpoons CH_3NH_3^+(g)$. In the gas phase, $(CH_3)_3N$ is more basic than CH_3NH_2, while in solution the reverse is true. Account for this observation.

Conceptual Problems

145. Without doing any calculations, determine which of the following solutions would be most acidic.
    a. 0.0100 M in HCl and 0.0100 M in KOH
    b. 0.0100 M in HF and 0.0100 in KBr
    c. 0.0100 M in NH_4Cl and 0.0100 M in CH_3NH_3Br
    d. 0.100 M in NaCN and 0.100 M in $CaCl_2$

146. Without doing any calculations, determine which of the following solutions would be most basic.
    a. 0.100 M in NaClO and 0.100 Min NaF
    b. 0.0100 M in KCl and 0.0100 M in $KClO_2$
    c. 0.0100 M in HNO_3 and 0.0100 M in NaOH
    d. 0.0100 M in NH_4Cl and 0.0100 M in HCN

147. Rank the following in order of increasing acidity.

    CH_3COOH $CH_2ClCOOH$ $CHCl_2COOH$ CCl_3COOH

16

Aqueous Ionic Equilibrium

16.1 The Danger of Antifreeze

16.2 Buffers: Solutions That Resist pH Change

16.3 Buffer Effectiveness: Buffer Range and Buffer Capacity

16.4 Titrations and pH Curves

16.5 Solubility Equilibria and the Solubility Product Constant

16.6 Precipitation

16.7 Qualitative Chemical Analysis

16.8 Complex Ion Equilibria

WE HAVE ALREADY seen the importance of aqueous solutions, first in Chapters 4, 12, and 14, and most recently in Chapter 15 on acids and bases. We now turn our attention to two additional topics involving aqueous solutions: buffers (solutions that resist pH change) and solubility equilibria (the extent to which slightly soluble ionic compounds dissolve in water). Buffers are tremendously important in biology because nearly all physiological processes must occur within a narrow pH range. Solubility equilibria are related to the solubility rules that we learned in Chapter 4. But now we will find a more complicated picture: solids that we considered insoluble under the simple "solubility rules" are actually better described as being only very slightly soluble, as the quotation on the facing page suggests. Solubility equilibria are important in predicting not only solubility, but also precipitation reactions that might occur when aqueous solutions are mixed.

> *In the strictly scientific sense of the word, insolubility does not exist, and even those substances characterized by the most obstinate resistance to the solvent action of water may properly be designated as extraordinarily difficult of solution, not as insoluble.*
>
> —Otto N. Witt (1853–1915)

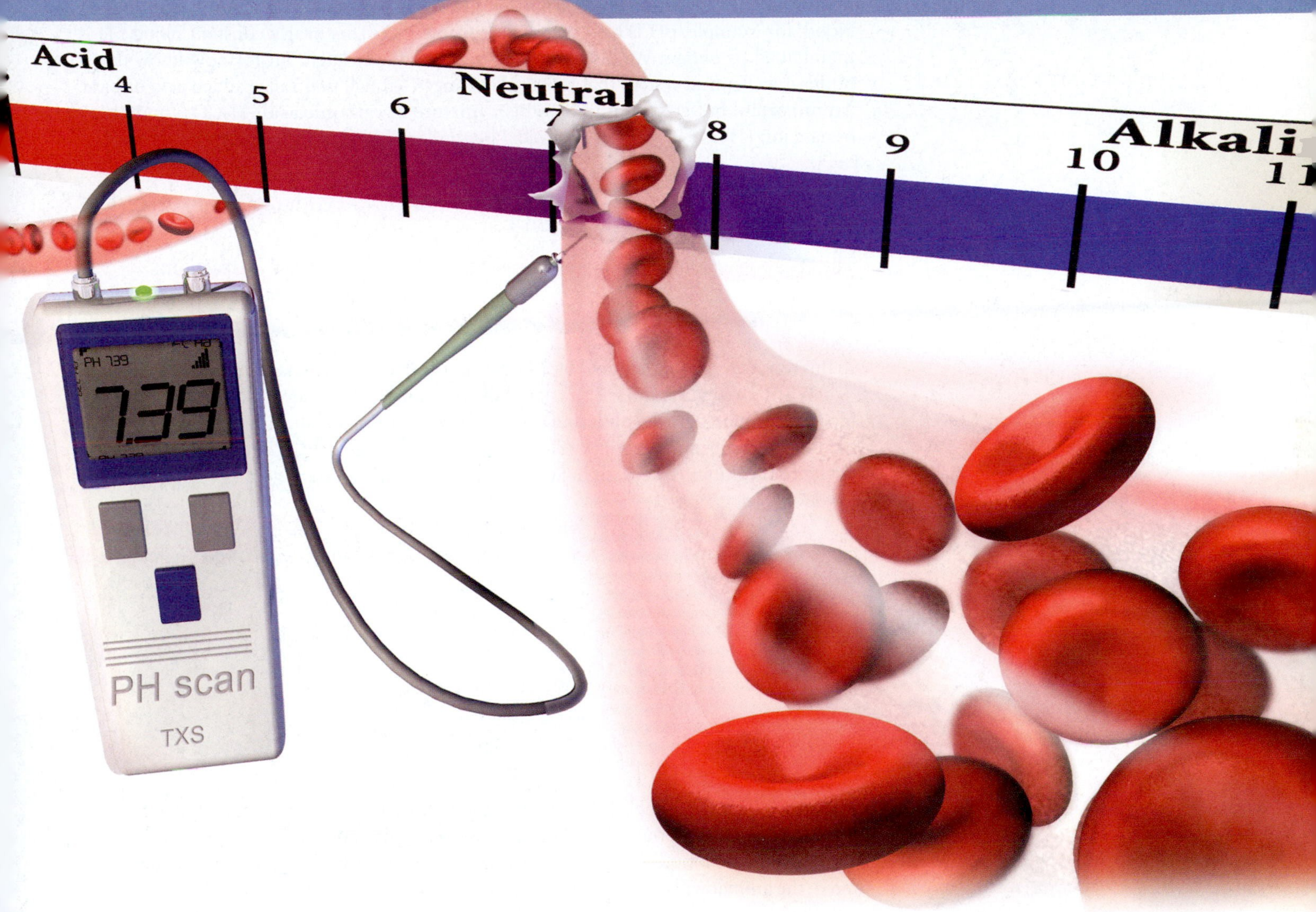

Acid

4 5 6 **Neutral** 7 8 9 10 **Alkali** 11

PH 7.39

7.39

PH scan

TXS

Human blood is held at nearly constant pH by the action of buffers, a main topic of this chapter.

16.1 The Danger of Antifreeze

Every year, thousands of dogs and cats die from consuming a common household product: antifreeze that was improperly stored or that leaked out of a car radiator. Most types of antifreeze used in cars are aqueous solutions of ethylene glycol ($HOCH_2CH_2OH$), an alcohol with the following structure:

Ethylene glycol has a somewhat sweet taste that often proves attractive to curious dogs and cats—and sometimes to young children, who are also vulnerable to this toxic compound. The first stage of ethylene glycol poisoning is a state resembling drunkenness.

Since the compound is an alcohol, it affects the brain of a dog or cat much as an alcoholic beverage would. Once ethylene glycol starts to be metabolized, however, a second and more deadly stage begins.

Ethylene glycol is oxidized in the liver to glycolic acid ($HOCH_2COOH$), which then enters the bloodstream. The acidity of blood is critically important, and tightly regulated, because many proteins require a narrow pH range for proper functioning. In human blood, for example, pH is held between 7.36 and 7.42. This nearly constant blood pH is maintained by *buffers*. We discuss buffers more carefully later, but for now know that a buffer is a chemical system that resists pH changes—it will neutralize added acid or base. An important buffer system in blood is a mixture of carbonic acid (H_2CO_3) and bicarbonate ion (HCO_3^-). The carbonic acid neutralizes added base according to the following reaction:

$$H_2CO_3(aq) + \underset{\text{added base}}{OH^-(aq)} \longrightarrow H_2O(l) + HCO_3^-(aq)$$

The bicarbonate ion neutralizes added acid according to the following reaction:

$$HCO_3^-(aq) + \underset{\text{added acid}}{H^+(aq)} \longrightarrow H_2CO_3(aq)$$

In this way, the carbonic acid and bicarbonate ion buffering system keeps blood pH constant.

When glycolic acid first enters the bloodstream, its tendency to lower the pH of the blood is countered by the buffering action of the bicarbonate ion. However, if the original quantities of consumed antifreeze are large enough, the glycolic acid overwhelms the capacity of the buffer (discussed in Section 16.3), causing blood pH to drop to dangerously low levels.

Low blood pH results in a condition called *acidosis*, in which the blood's ability to carry oxygen is reduced because the acid affects the equilibrium between hemoglobin (Hb) and oxygen:

$$\text{Excess } H^+$$

$$HbH^+(aq) + O_2(g) \rightleftharpoons HbO_2(aq) + H^+(aq)$$

$$\text{Shift left}$$

The excess acid causes the equilibrium to shift to the left, reducing hemoglobin's ability to carry oxygen. At this point, the cat or dog may begin hyperventilating in an effort to overcome the acidic blood's lowered oxygen-carrying capacity. If no treatment is administered, the animal will eventually go into a coma and die.

One treatment for ethylene glycol poisoning is the administration of ethyl alcohol (the alcohol found in alcoholic beverages). Because the two molecules are similar in structure, the liver enzyme that catalyzes the metabolism of ethylene glycol also acts on ethyl alcohol, but it has higher affinity for ethyl alcohol than for ethylene glycol. Consequently, the enzyme preferentially metabolizes ethyl alcohol, allowing the unmetabolized ethylene glycol to escape through the urine. If administered early, this treatment can save the life of a dog or cat that has consumed ethylene glycol.

16.2 Buffers: Solutions That Resist pH Change

Most solutions will rapidly change pH upon the addition of an acid or base. As we have just learned, however, a **buffer** resists pH change by neutralizing added acid or added base. Buffers contain significant amounts of both *a weak acid and its conjugate base* (or a weak base and its conjugate acid). For example, we saw that the buffer in blood was composed of carbonic acid (H_2CO_3) and its conjugate base, the bicarbonate ion (HCO_3^-). When additional base is added to a buffer, the weak acid reacts with the base, neutralizing it. When additional acid is added to a buffer, the conjugate base reacts with the acid, neutralizing it. In this way, a buffer can maintain a nearly constant pH.

Formation of a Buffer

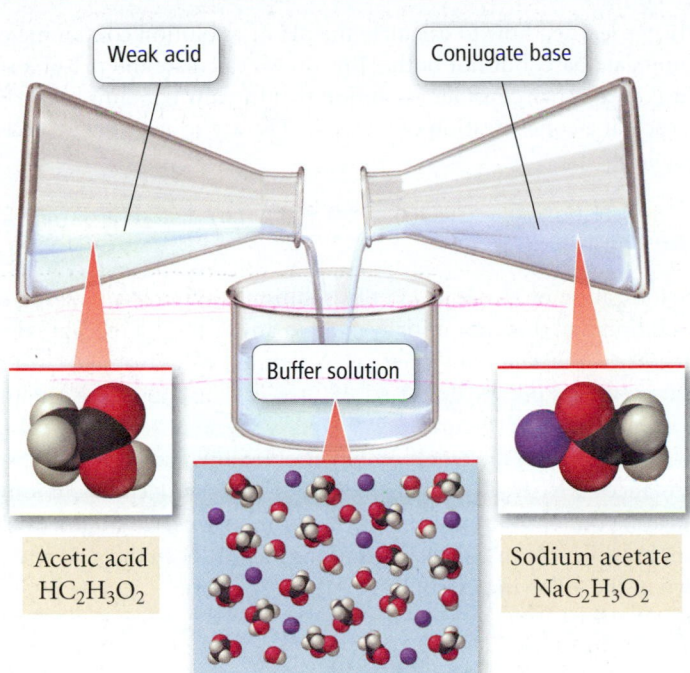

Weak acid

Conjugate base

Buffer solution

Acetic acid
$HC_2H_3O_2$

Sodium acetate
$NaC_2H_3O_2$

▲ **FIGURE 16.1 A Buffer Solution**   A buffer typically consists of a weak acid (which can neutralize added base) and its conjugate base (which can neutralize added acid).

A weak acid by itself, even though it partially ionizes to form some of its conjugate base, does not contain sufficient base to be a buffer. Similarly, a weak base by itself, even though it partially ionizes water to form some of its conjugate acid, does not contain sufficient acid to be a buffer. *A buffer contains significant amounts of both a weak acid and its conjugate base.* For example, a simple buffer can be made by dissolving acetic acid ($HC_2H_3O_2$) and sodium acetate ($NaC_2H_3O_2$) in water (Figure 16.1 ▲).

Suppose that a strong base, such as NaOH, were added to this solution. The acetic acid would neutralize the base according to the following reaction:

| $C_2H_3O_2^-$ is the conjugate base of $HC_2H_3O_2$.

$$NaOH(aq) + HC_2H_3O_2(aq) \longrightarrow H_2O(l) + NaC_2H_3O_2(aq)$$

As long as the amount of NaOH added was significantly less than the amount of $HC_2H_3O_2$ in solution, the buffer would neutralize the NaOH and the resulting pH change would be small. Suppose, on the other hand, that a strong acid, such as HCl, were added to the solution. Then the conjugate base, $NaC_2H_3O_2$, would neutralize the added HCl according to the following reaction:

$$HCl(aq) + NaC_2H_3O_2(aq) \longrightarrow HC_2H_3O_2(aq) + NaCl(aq)$$

As long as the amount of HCl added was significantly less than the amount of $NaC_2H_3O_2$ in solution, the buffer would neutralize the HCl and the resulting pH change would be small.

Summarizing:

➤ Buffers resist pH change.

➤ Buffers contain significant amounts of both a weak acid and its conjugate base.

➤ The weak acid neutralizes added base.

➤ The conjugate base neutralizes added acid.

Calculating the pH of a Buffer Solution

In Chapter 15, we learned how to calculate the pH of a solution containing either a weak acid or its conjugate base, but not both. How do we calculate the pH of a solution containing both? For example, consider a solution that initially contains both $HC_2H_3O_2$ and $NaC_2H_3O_2$, each at a concentration of 0.100 M. The acetic acid ionizes according to the following reaction:

$$HC_2H_3O_2(aq) + H_2O(l) \rightleftharpoons H_3O^+(aq) + C_2H_3O_2^-(aq)$$

Initial concentration: 0.100 M   0.100 M

Le Châtelier's principle was discussed in Section 14.9.

However, the ionization of $HC_2H_3O_2$ in this solution is suppressed compared to its ionization in a solution that does not initially contain any $C_2H_3O_2^-$, because the presence of $C_2H_3O_2^-$ *shifts the equilibrium to the left* (as we would expect from Le Châtelier's principle). In other words, the presence of the $C_2H_3O_2^-(aq)$ ion causes the acid to ionize even less than it normally would (Figure 16.2 ▼), resulting in a less acidic solution (higher pH). This effect is known as the **common ion effect**, so named because the solution contains two substances ($HC_2H_3O_2$ and $NaC_2H_3O_2$) that share a common ion ($C_2H_3O_2^-$). To find the pH of a buffer solution containing common ions, we simply work an equilibrium problem in which the initial concentrations include both the acid and its conjugate base, as shown in the following example.

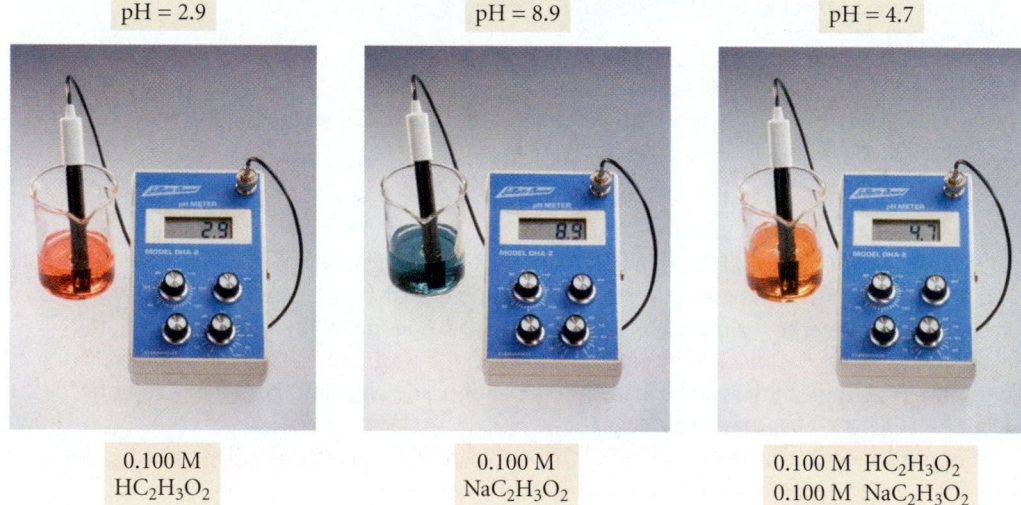

pH = 2.9	pH = 8.9	pH = 4.7
0.100 M $HC_2H_3O_2$	0.100 M $NaC_2H_3O_2$	0.100 M $HC_2H_3O_2$ 0.100 M $NaC_2H_3O_2$

▲ **FIGURE 16.2 The Common Ion Effect** The pH of a 0.100 M acetic acid solution is 2.9. The pH of a 0.10 M sodium acetate solution is 8.9. The pH of a solution that is 0.100 M in acetic acid and 0.100 M in sodium acetate is 4.7.

EXAMPLE 16.1 Calculating the pH of a Buffer Solution

Calculate the pH of a buffer solution that is 0.100 M in $HC_2H_3O_2$ and 0.100 M in $NaC_2H_3O_2$.

Solution

1. Write the balanced equation for the ionization of the acid and use it as a guide to prepare an ICE table showing the given concentrations of the acid and its conjugate base as the initial concentrations. Leave room in the table for the changes in concentrations and for the equilibrium concentrations.

$$HC_2H_3O_2(aq) + H_2O(l) \rightleftharpoons H_3O^+(aq) + C_2H_3O_2^-(aq)$$

	$[HC_2H_3O_2]$	$[H_3O^+]$	$[C_2H_3O_2^-]$
Initial	0.100	≈0.00	0.100
Change			
Equil			

2. Represent the change in the concentration of H_3O^+ with the variable x. Express the changes in the concentrations of the other reactants and products in terms of x.

$$HC_2H_3O_2(aq) + H_2O(l) \rightleftharpoons H_3O^+(aq) + C_2H_3O_2^-(aq)$$

	$[HC_2H_3O_2]$	$[H_3O^+]$	$[C_2H_3O_2^-]$
Initial	0.100	≈0.00	0.100
Change	$-x$	$+x$	$+x$
Equil			

3. Sum each column to determine the equilibrium concentrations in terms of the initial concentrations and the variable x.

$$HC_2H_3O_2(aq) + H_2O(l) \rightleftharpoons H_3O^+(aq) + C_2H_3O_2^-(aq)$$

	$[HC_2H_3O_2]$	$[H_3O^+]$	$[C_2H_3O_2^-]$
Initial	0.100	≈0.00	0.100
Change	$-x$	$+x$	$+x$
Equil	$0.100 - x$	x	$0.100 + x$

4. Substitute the expressions for the equilibrium concentrations (from step 3) into the expression for the acid ionization constant.

In most cases, you can make the approximation that x is small.

Substitute the value of the acid ionization constant (from Table 15.5) into the K_a expression and solve for x.

Confirm that x is small by computing the ratio of x to the number it was subtracted from in the approximation. The ratio should be less than 0.05 (or 5%). (See Sections 14.8 and 15.6 to review the *x is small* approximation.)

$$K_a = \frac{[H_3O^+][C_2H_3O^-]}{[HC_2H_3O_2]}$$

$$= \frac{x(0.100 + x)}{0.100 - x} \quad (x \text{ is small})$$

$$1.8 \times 10^{-5} = \frac{x(0.100)}{0.100}$$

$$x = 1.8 \times 10^{-5}$$

$$\frac{1.8 \times 10^{-5}}{0.100} \times 100\% = 0.018\%$$

Therefore the approximation is valid.

5. Determine the H_3O^+ concentration from the computed value of x and substitute into the pH equation to find pH.

$$[H_3O^+] = x = 1.8 \times 10^{-5} \text{ M}$$

$$pH = -\log[H_3O^+]$$

$$= -\log(1.8 \times 10^{-5})$$

$$= 4.74$$

For Practice 16.1

Calculate the pH of a buffer solution that is 0.200 M in $HC_2H_3O_2$ and 0.100 M in $NaC_2H_3O_2$.

For More Practice 16.1

Calculate the pH of the buffer that results from mixing 60.0 mL of 0.250 M $HCHO_2$ and 15.0 mL of 0.500 M $NaCHO_2$.

The Henderson–Hasselbalch Equation

Finding the pH of a buffer solution can be simplified by deriving an equation that relates the pH of the solution to the initial concentrations of the buffer components. Consider a buffer containing the generic weak acid HA and its conjugate base A$^-$. The acid ionization equation is as follows:

$$HA(aq) + H_2O(l) \rightleftharpoons H_3O^+(aq) + A^-(aq)$$

An expression for the concentration of H_3O^+ can be obtained from the acid ionization equation by solving it for $[H_3O^+]$.

$$K_a = \frac{[H_3O^+][A^-]}{[HA]}$$

$$[H_3O^+] = K_a \frac{[HA]}{[A^-]} \qquad [16.1]$$

If we make the same *x is small* approximation that we have normally made for weak acid or weak base equilibrium problems, *we can consider the equilibrium concentrations of HA and A^- to be essentially identical to the initial concentrations of HA and A^-* (see step 4 of Example 16.1). Therefore, to determine $[H_3O^+]$ for any buffer solution, we simply multiply K_a by the ratio of the concentrations of the acid and the conjugate base. For example, to find the $[H_3O^+]$ of the buffer in Example 16.1 (a solution that is 0.100 M in $HC_2H_3O_2$ and 0.100 M in $NaC_2H_3O_2$), we simply substitute the concentrations of $HC_2H_3O_2$ and $C_2H_3O_2^-$ into Equation 16.1 as follows:

$$[H_3O^+] = K_a \frac{[HC_2H_3O_2]}{[C_2H_3O_2^-]}$$

$$= K_a \frac{0.100}{0.100}$$

$$= K_a$$

In this solution, as in any buffer solution in which the concentrations of the acid and conjugate base are equal, $[H_3O^+]$ is simply equal to K_a.

We can derive an equation for the pH of a buffer by taking the logarithm of both sides of Equation 16.1 as follows:

$$[H_3O^+] = K_a \frac{[HA]}{[A^-]}$$

$$\log[H_3O^+] = \log\left(K_a \frac{[HA]}{[A^-]} \right)$$

$$\log[H_3O^+] = \log K_a + \log \frac{[HA]}{[A^-]} \qquad [16.2]$$

Multiplying both sides of Equation 16.2 by -1 and rearranging, we get the following:

$$-\log[H_3O^+] = -\log K_a - \log \frac{[HA]}{[A^-]}$$

$$-\log[H_3O^+] = -\log K_a + \log \frac{[A^-]}{[HA]}$$

Since $pH = -\log[H_3O^+]$ and since $pK_a = -\log K_a$, we obtain the following result:

$$pH = pK_a + \log \frac{[A^-]}{[HA]}$$

Since A^- is a weak base and HA is a weak acid, we can generalize the equation as follows:

$$pH = pK_a + \log \frac{[base]}{[acid]} \qquad [16.3]$$

where the base is the conjugate base of the acid or the acid is the conjugate acid of the base. This equation, known as the **Henderson–Hasselbalch equation**, allows us to easily calculate the pH of a buffer solution from the initial concentrations of the buffer components *as long as the x is small* approximation is valid. In the following example, we show how to find the pH of a buffer in two ways: in the left column we solve a common ion effect equilibrium problem similar to the one we solved in Example 16.1; in the right column we use the Henderson–Hasselbalch equation.

Recall that the variable *x* in a weak acid equilibrium problem represents the change in the initial acid concentration. The *x is small* approximation is valid because so little of the weak acid ionizes compared to its initial concentration.

Note that, as expected, the pH of a buffer increases with an increase in the amount of base relative to the amount of acid.

EXAMPLE 16.2   Calculating the pH of a Buffer Solution as an Equilibrium Problem and with the Henderson–Hasselbalch Equation

Calculate the pH of a buffer solution that is 0.050 M in benzoic acid ($HC_7H_5O_2$) and 0.150 M in sodium benzoate ($NaC_7H_5O_2$). For benzoic acid, $K_a = 6.5 \times 10^{-5}$.

Solution

Equilibrium Approach	Henderson–Hasselbalch Approach

Equilibrium Approach

Write the balanced equation for the ionization of the acid and use it as a guide to prepare an ICE table.

$$HC_7H_5O_2(aq) + H_2O(l) \rightleftharpoons H_3O^+(aq) + C_7H_5O_2^-(aq)$$

	$[HC_7H_5O_2]$	$[H_3O^+]$	$[C_7H_5O_2^-]$
Initial	0.050	≈0.00	0.150
Change	$-x$	$+x$	$+x$
Equil	$0.050 - x$	x	$0.150 + x$

Substitute the expressions for the equilibrium concentrations into the expression for the acid ionization constant. Make the *x is small* approximation and solve for *x*.

$$K_a = \frac{[H_3O^+][C_7H_5O_2^-]}{[HC_7H_5O_2]}$$

$$= \frac{x(0.150 + x)}{0.050 - x} \quad (x \text{ is small})$$

$$6.5 \times 10^{-5} = \frac{x(0.150)}{0.050}$$

$$x = 2.2 \times 10^{-5}$$

[handwritten: $6.5 \times 10^{-5} \times .050 = 3.25 \times 10^{-6}$ over $.150$; multiply]

Since $[H_3O^+] = x$, we compute pH as follows:

$$pH = -\log[H_3O^+]$$
$$= -\log(2.2 \times 10^{-5})$$
$$= 4.66$$

Henderson–Hasselbalch Approach

To find the pH of this solution, determine which component is the acid and which is the base and substitute their concentrations into the Henderson–Hasselbalch equation to compute pH.

$HC_7H_5O_2$ is the acid and $NaC_7H_5O_2$ is the base. Therefore, we compute the pH as follows:

$$pH = pK_a + \log\frac{[base]}{[acid]}$$

$$= -\log(6.5 \times 10^{-5}) + \log\frac{0.150}{0.050}$$

$$= 4.187 + 0.477$$

$$= 4.66$$

Confirm that the *x is small* approximation is valid by computing the ratio of *x* to the number it was subtracted from in the approximation. The ratio should be less than 0.05 (or 5%). (See Sections 14.8 and 15.6 to review the *x is small* approximation.)

$$\frac{2.2 \times 10^{-5}}{0.050} \times 100\% = 0.044\%$$

The approximation is valid.

Confirm that the *x is small* approximation is valid by calculating the $[H_3O^+]$ from the pH. Since $[H_3O^+]$ is formed by ionization of the acid, the calculated $[H_3O^+]$ has to be less than 0.05 (or 5%) of the initial concentration of the acid in order for the *x is small* approximation to be valid.

$$pH = 4.66 = -\log[H_3O^+]$$

$$[H_3O^+] = 10^{-4.66} = 2.2 \times 10^{-5} \text{ M}$$

$$\frac{2.2 \times 10^{-5}}{0.050} \times 100\% = 0.044\%$$

The approximation is valid.

For Practice 16.2

Calculate the pH of a buffer solution that is 0.250 M in HCN and 0.170 M in KCN. For HCN, $K_a = 4.9 \times 10^{-10}$ ($pK_a = 9.31$). Use both the equilibrium approach and the Henderson–Hasselbalch approach.

You may be wondering whether you should use the equilibrium approach or the Henderson–Hasselbalch equation when determining the pH of buffer solutions. The answer depends on the specific problem. In cases where the *x is small* approximation can be made, the Henderson–Hasselbalch equation is adequate. However, as you can see from Example 16.2, checking the *x is small* approximation is not as convenient with the Henderson–Hasselbalch equation (because the approximation is implicit). Thus, the equilibrium approach, although lengthier, gives you a better sense of the important quantities in the problem and the nature of the approximation. When first working buffer problems, use the equilibrium approach until you get a good sense for when the *x is small* approximation is adequate. Then, you can switch to the more streamlined approach in cases where the approximation applies (and only in those cases). In general, remember that the *x is small* approximation applies to problems in which both of the following are true: (a) the initial concentrations of acids (and/or bases) are not too dilute; and (b) the equilibrium constant is fairly small. Although the exact values depend on the details of the problem, for many buffer problems this means that the initial concentrations of acids and conjugate bases should be at least 10^2–10^3 times greater than the equilibrium constant (depending on the required accuracy).

Conceptual Connection 16.1 pH of Buffer Solutions

A buffer contains the weak acid HA and its conjugate base A^-. The weak acid has a pK_a of 4.82 and the buffer has a pH of 4.25. Which of the following is true of the relative concentrations of the weak acid and conjugate base in the buffer?

(a) $[HA] > [A^-]$ (b) $[HA] < [A^-]$ (c) $[HA] = [A^-]$

Which buffer component would you add to change the pH of the buffer to 4.72?

Answer: (a) Since the pH of the buffer is less than the pK_a of the acid, the buffer must contain more acid than base ($[HA] > [A^-]$). In order to raise the pH of the buffer from 4.25 to 4.72, you must add more of the weak base (adding a base will make the buffer solution more basic).

Calculating pH Changes in a Buffer Solution

When an acid or a base is added to a buffer, the buffer tends to resist a pH change. Nonetheless, the pH does change by a small amount. Calculating the pH change requires breaking up the problem into two parts: (1) a stoichiometry calculation (in which we calculate how the addition changes the relative amounts of acid and conjugate base); and (2) an equilibrium calculation (in which we calculate the pH based on the new amounts of acid and conjugate base). We demonstrate this calculation with a 1.0-L buffer solution that is 0.100 M in the generic acid HA and 0.100 M in its conjugate base A^-. How do we calculate the pH after addition of 0.025 mol of strong acid (H^+) to this buffer (assuming that the change in volume from adding the acid is negligible)?

The Stoichiometry Calculation As the added acid is neutralized, it converts a stoichiometric amount of the base into its conjugate acid through the following neutralization reaction (Figure 16.3a ▶):

$$H^+(aq) \; + \; A^-(aq) \longrightarrow HA(aq)$$

added acid   weak base in buffer

Neutralizing 0.025 mol of the strong acid (H^+) requires 0.025 mol of the weak base (A^-). Consequently, the amount of A^- *decreases* by 0.025 mol and the amount of HA *increases* by 0.025 mol because of the 1:1:1 stoichiometry of the neutralization reaction. We can track these changes in tabular form as follows:

It is best to work with amounts in moles instead of concentrations when tracking these changes, as explained below.

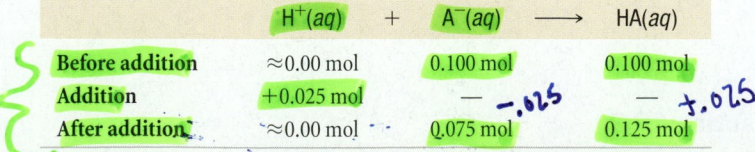

	$H^+(aq)$	+	$A^-(aq)$	⟶	HA(aq)
Before addition	≈0.00 mol		0.100 mol		0.100 mol
Addition	+0.025 mol		− -.025		− +.025
After addition	≈0.00 mol		0.075 mol		0.125 mol

Notice that this table *is not an ICE table*. It is simply a table that tracks the stoichiometric changes that occur during the neutralization of the added acid. We write ≈0.00 mol for the amount of H^+ because the amount is so small compared to the amounts of A^- and HA.

Action of a Buffer

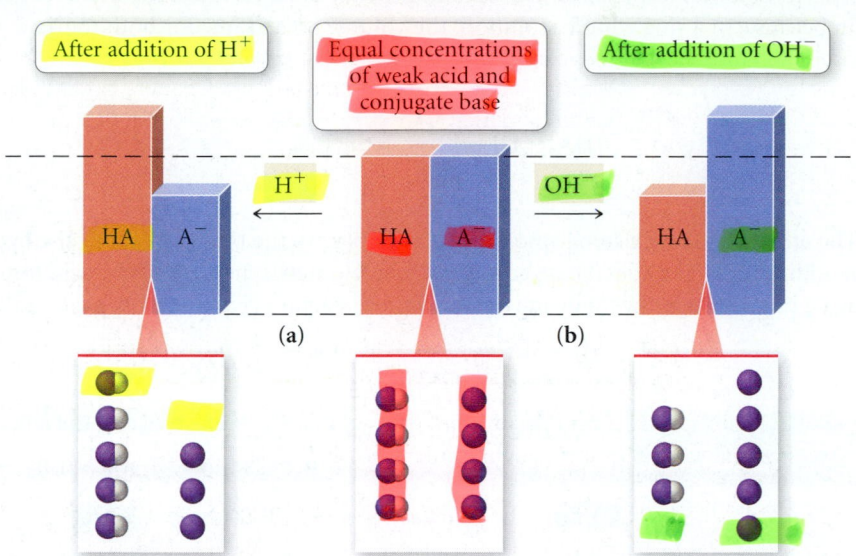

(a)   (b)

+ 4 product ↑
+ OH product ↓

◀ **FIGURE 16.3 Buffering Action**
(a) When an acid is added to a buffer, a stoichiometric amount of the weak base is converted to the conjugate acid.
(b) When a base is added to a buffer, a stoichiometric amount of the weak acid is converted to the conjugate base.

(Remember that weak acids ionize only to a small extent and that the presence of the common ion further suppresses the ionization.) The amount of H^+, of course, is not *exactly* zero, as we can see by completing the equilibrium part of the calculation.

The Equilibrium Calculation We have just seen that adding a small amount of acid to a buffer is equivalent to changing the initial concentrations of the acid and conjugate base present in the buffer (in this case, since the volume is 1.0 L, [HA] increased from 0.100 M to 0.125 M and [A⁻] decreased from 0.100 M to 0.075 M). Once these new initial concentrations are computed, the new pH can be calculated in the same way that we calculate the pH of any buffer: either by working a full equilibrium problem or by using the Henderson–Hasselbalch equation (see Examples 16.1 and 16.2). In this case, we work the full equilibrium problem. We begin by writing the balanced equation for the ionization of the acid and using it as a guide to prepare an ICE table. The initial concentrations for the ICE table are those calculated in the stoichiometry part of the calculation.

$$HA(aq) + H_2O(l) \rightleftharpoons H_3O^+(aq) + A^-(aq)$$

	[HA]	[H₃O⁺]	[A⁻]
			From stoichiometry calculation
Initial	0.125	≈0.00	0.075
Change	$-x$	$+x$	$+x$
Equil	$0.125 - x$	x	$0.075 + x$

We then substitute the expressions for the equilibrium concentrations into the expression for the acid ionization constant. As long as K_a is sufficiently small relative to the initial concentrations, we can make the *x is small* approximation and solve for x, which is equal to [H₃O⁺].

$$K_a = \frac{[H_3O^+][A^-]}{[HA]}$$

$$= \frac{x(0.075 + x)}{0.125 - x} \quad (x \text{ is small})$$

$$K_a = \frac{x(0.075)}{0.125} \quad \text{Flip CI's}$$

$$x = [H_3O^+] = K_a \frac{0.125}{0.075}$$

Once [H₃O⁺] is computed, pH can be calculated using $pH = -\log[H_3O^+]$.

Notice that, since the above expression for x contains a *ratio* of concentrations $[HA]/[A^-]$, the amounts of acid and base in moles may be substituted in place of concentration because, in a single buffer solution, the volume is the same for both the acid and the base. Therefore the volumes cancel:

$$[HA]/[A^-] = \frac{\dfrac{n_{HA}}{\cancel{V}}}{\dfrac{n_{A^-}}{\cancel{V}}} = n_{HA}/n_{A^-}$$

The effect of adding a small amount of strong base to the buffer is exactly the opposite of adding acid. The added base converts a stoichiometric amount of the acid into its conjugate base through the following neutralization reaction (Figure 16.3b, page 725):

$$OH^-(aq) + \underset{\text{weak acid in buffer}}{HA(aq)} \longrightarrow H_2O(l) + A^-(aq)$$
$$\underset{\text{added base}}{}$$

If we add 0.025 mol of OH^-, then the amount of A^- goes *up* by 0.025 mol and the amount of HA goes *down* by 0.025 mol as shown in the following table:

	$OH^-(aq)$	+	$HA(aq)$	$\longrightarrow$	$H_2O(l)$	+	$A^-(aq)$
Before addition	≈0.00 mol		0.100 mol				0.100 mol
Addition	+0.025 mol		—				—
After addition	≈0.00 mol		0.075 mol				0.125 mol

Summarizing, in calculating the pH of a buffer after adding small amounts of acid or base, remember the following:

➤ Adding a small amount of strong acid to a buffer converts a stoichiometric amount of the base to the conjugate acid.

➤ Adding a small amount of strong base to a buffer converts a stoichiometric amount of the acid to the conjugate base.

> The easiest way to remember these changes is relatively simple: adding acid creates more acid; adding base creates more base.

The following example and For Practice problems involve calculating pH changes in a buffer solution after small amounts of strong acid or strong base are added. As we have seen, these problems generally have two parts.

- Part I. Stoichiometry—using the stoichiometry of the neutralization equation to calculate the changes in the amounts (in moles) of the buffer components upon addition of the acid or base.
- Part II. Equilibrium—using the new amounts of buffer components to work an equilibrium problem to find pH. (For most buffers, this can also be done with the Henderson–Hasselbalch equation.)

EXAMPLE 16.3 Calculating the pH Change in a Buffer Solution After the Addition of a Small Amount of Strong Acid or Base

A 1.0-L buffer solution contains 0.100 mol $HC_2H_3O_2$ and 0.100 mol $NaC_2H_3O_2$. The value of K_a for $HC_2H_3O_2$ is 1.8×10^{-5}. Since the initial amounts of acid and conjugate base are equal, the pH of the buffer is simply equal to $pK_a = -\log(1.8 \times 10^{-5}) = 4.74$. Calculate the new pH after adding 0.010 mol of solid NaOH to the buffer. For comparison, compute the pH after adding 0.010 mol of solid NaOH to 1.0 L of pure water. (Ignore any small changes in volume that might occur upon addition of the base.)

Solution

Part I: Stoichiometry. The addition of the base converts a stoichiometric amount of acid to the conjugate base (adding base creates more base). Write an equation showing the neutralization reaction and then set up a table to track the changes.

	$OH^-(aq)$	+	$HC_2H_3O_2(aq)$	$\longrightarrow$	$H_2O(l)$	+	$C_2H_3O_2^-(aq)$
Before addition	≈0.00 mol		0.100 mol				0.100 mol
Addition	0.010 mol		—				—
After addition	≈0.00 mol		0.090 mol				0.110 mol

Part II: Equilibrium. Write the balanced equation for the ionization of the acid and use it as a guide to prepare an ICE table. Use the amounts of acid and conjugate base from part I as the initial amounts of acid and conjugate base in the ICE table.

$$HC_2H_3O_2(aq) + H_2O(l) \rightleftharpoons H_3O^+(aq) + C_2H_3O_2^-(aq)$$

	$[HC_2H_3O_2]$	$[H_3O^+]$	$[C_2H_3O_2^-]$
Initial	0.090	≈ 0.00	0.110
Change	$-x$	$+x$	$+x$
Equil	$0.090 - x$	x	$0.110 + x$

Substitute the expressions for the equilibrium concentrations of acid and conjugate base into the expression for the acid ionization constant. Make the *x is small* approximation and solve for *x*. Compute the pH from the value of *x*, which is equal to $[H_3O^+]$.

$$K_a = \frac{[H_3O^+][C_2H_3O_2^-]}{[HC_2H_3O_2]}$$

$$= \frac{x(0.110 + \cancel{x})}{0.090 - \cancel{x}} \quad (x\ is\ small)$$

$$1.8 \times 10^{-5} = \frac{x(0.110)}{0.090} \quad \rightarrow flip \quad \frac{.090}{x(.100)} = .8181 \times 1.8 \times 10^{-5}$$

$$x = [H_3O^+] = 1.47 \times 10^{-5}\ M$$

$$pH = -\log[H_3O^+]$$

$$= -\log(1.47 \times 10^{-5})$$

$$= 4.83$$

Confirm that *x is small* approximation is valid by computing the ratio of *x* to the smallest number it was subtracted from in the approximation. The ratio should be less than 0.05 (or 5%).

$$\frac{1.47 \times 10^{-5}}{0.090} \times 100\% = 0.016\%$$

The approximation is valid.

Part II: Equilibrium Alternative (using the Henderson–Hasselbalch equation). As long at the *x is small* approximation is valid, you can simply substitute the quantities of acid and conjugate base after the addition (from part I) into the Henderson–Hasselbalch equation and compute the new pH.

$$pH = pK_a + \log\frac{[base]}{[acid]}$$

$$= \left(-\log(1.8 \times 10^{-5})\right) + \left(\log\frac{0.110}{0.090}\right)$$

$$= 4.74 + 0.087$$

$$= 4.83$$

The pH of 1.0 L of water after adding 0.010 mol of NaOH is computed from the $[OH^-]$. For a strong base, $[OH^-]$ is just the number of moles of OH^- divided by the number of liters of solution.

$$[OH^-] = \frac{0.010\ mol}{1.0\ L} = 0.010\ M \quad \leftarrow OH\ added$$

$$pOH = -\log[OH^-] = -\log(0.010)$$

$$= 2.00$$

$$pOH + pH = 14.00$$

$$pH = 14.00 - pOH$$

$$= 14.00 - 2.00$$

$$= 12.00$$

Check Notice that the buffer solution changed from pH = 4.74 to pH = 4.83 upon addition of the base (a small fraction of a single pH unit). In contrast, the pure water changed from pH = 7.00 to pH = 12.00, five whole pH units (a factor of 10^5). Notice also that even the buffer solution got slightly more basic upon addition of a base, as we would expect. To check your answer, always make sure the pH goes in the direction you expect: adding base should make the solution more basic (higher pH); adding acid should make the solution more acidic (lower pH).

For Practice 16.3

Calculate the pH of the original buffer solution in Example 16.3 upon addition of 0.015 mol of NaOH to the original buffer.

For More Practice 16.3

Calculate the pH in the solution upon addition of 10.0 mL of 1.00 M HCl to the original buffer in Example 16.3.

Conceptual Connection 16.2 Adding Acid or Base to a Buffer

A buffer contains equal amounts of a weak acid and its conjugate base and has a pH of 5.25. Which of the following values would be reasonable to expect for the pH of the buffer after addition of a small amount of acid?

(a) 4.15   (b) 5.15   (c) 5.35   (d) 6.35

Answer: (b) Since acid is added to the buffer, the pH will become slightly lower (slightly more acidic). Answer (a) reflects too large a change in pH for a buffer, and answers (c) and (d) are in the wrong direction.

Buffers Containing a Base and Its Conjugate Acid

So far, we have seen examples of buffers composed of an acid and its conjugate base (where the conjugate base is an ion). However, a buffer can also be composed of a base and its conjugate acid (where the conjugate acid is an ion). For example, a solution containing significant amounts of both NH_3 and NH_4Cl will act as a buffer (Figure 16.4 ▼). The NH_3 is a weak base that neutralizes small amounts of added acid and the NH_4^+ ion is the conjugate acid that neutralizes small amounts of added base. We can calculate the pH of a solution such as this in the same way we did for a buffer containing a weak acid and its conjugate base. When using the Henderson–Hasselbalch equation, however, we must first compute pK_a for the conjugate acid of a weak base. Recall from Section 15.8 that for a conjugate acid–base pair, $K_a \times K_b = K_w$. Taking the negative logarithm of both sides of this equation, we get the following:

$$-\log(K_a \times K_b) = -\log K_w$$
$$-\log K_a - \log K_b = -\log K_w$$

Since $-\log K_a = pK_a$ and $-\log K_b = pK_b$, and substituting 10^{-14} for K_w, we get the following result (valid at 25 °C):

$$pK_a + pK_b = -\log 10^{-14}$$
$$pK_a + pK_b = 14 \qquad pK_b - 14 = pK_a \qquad [16.4]$$

Consequently, we can find pK_a of the conjugate acid by simply subtracting pK_b of the weak base from 14. The following example shows how to calculate the pH of a buffer composed of a weak base and its conjugate acid.

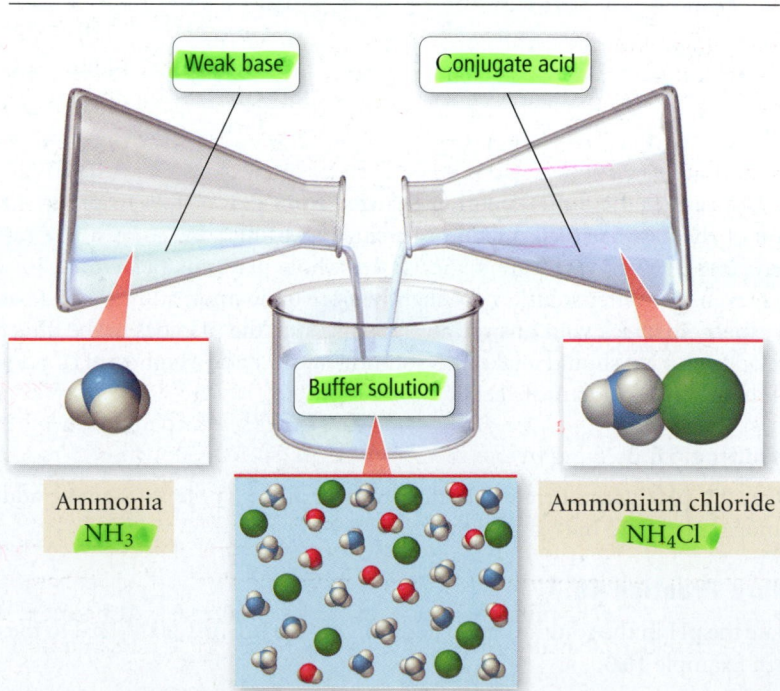

Formation of a Buffer

Weak base   Conjugate acid

Buffer solution

Ammonia
NH_3

Ammonium chloride
NH_4Cl

▶ FIGURE 16.4 Buffer Containing a Base A buffer can also consist of a weak base and its conjugate acid.

EXAMPLE 16.4 Using the Henderson–Hasselbalch Equation to Calculate the pH of a Buffer Solution Composed of a Weak Base and Its Conjugate Acid

Use the Henderson–Hasselbalch equation to calculate the pH of a buffer solution that is 0.50 M in NH_3 and 0.20 M in NH_4Cl. For ammonia, $pK_b = 4.75$.

Solution

Since K_b for $NH_3(1.8 \times 10^{-5})$ is much smaller than the initial concentrations in this problem, you can use the Henderson–Hasselbalch equation to calculate the pH of the buffer. However, you must first compute pK_a from pK_b by using Equation 16.4.

$$pK_a + pK_b = 14$$
$$pK_a = 14 - pK_b$$
$$= 14 - 4.75$$
$$= 9.25$$

Now, simply substitute the given quantities into the Henderson–Hasselbalch equation and compute pH.

$$pH = pK_a + \log \frac{[base]}{[acid]}$$
$$= 9.25 + \log \frac{0.50}{0.20}$$
$$= 9.25 + 0.40$$
$$= 9.65$$

For Practice 16.4

Calculate the pH of 1.0 L of the buffer in Example 16.4 upon addition of 0.010 mol of solid NaOH.

For More Practice 16.4

Calculate the pH of 1.0 L of the buffer in Example 16.4 upon addition of 30.0 mL of 1.0 M HCl.

16.3 Buffer Effectiveness: Buffer Capacity and Buffer Range

An effective buffer should neutralize moderate amounts of added acid or base. As we saw in the opening section of this chapter, however, a buffer can be destroyed by the addition of too much acid or too much base. What factors influence the effectiveness of a buffer? In this section, we look at two factors: *the relative amounts of the acid and conjugate base* and *the absolute concentrations of the acid and conjugate base.* We will then define the *capacity of a buffer* (how much added acid or base it can effectively neutralize) and the *range of a buffer* (the pH range over which a particular acid and its conjugate base can be effective).

Relative Amounts of Acid and Base

A buffer is most effective (most resistant to pH changes) when the concentrations of acid and conjugate base are equal. We will explore this idea by considering the behavior of a generic buffer composed of HA and A^- for which $pK_a = 5.00$. On page 730, we calculate the percent change in pH upon addition of 0.010 mol of NaOH for two different 1.0-liter solutions of this buffer system. Both solutions have exactly 0.20 mol of *total* acid and conjugate base. However, solution I has exactly equal amounts of acid and conjugate base (0.10 mol of each), while solution II has much more acid than conjugate base (0.18 mol HA and 0.020 mol A^-). The initial pH values of each solution can be calculated using the Henderson–Hasselbalch equation. Solution I has an initial pH of 5.00. Solution II has an initial pH of 4.05.

Solution I: 0.10 mol HA and 0.10 mol A⁻; initial pH = 5.00

OH⁻(aq) + HA(aq) ⟶ H₂O(l) + A⁻(aq)			
Before addition	≈0.00 mol	0.100 mol	0.100 mol
Addition	0.010 mol	—	—
After addition	≈0.00 mol	0.090 mol	0.110 mol

$$pH = pK_a + \log\frac{[base]}{[acid]}$$

$$= 5.00 + \log\frac{0.110}{0.090}$$

$$= 5.09$$

$$\% \text{ change} = \frac{5.09 - 5.00}{5.00} \times 100\%$$

$$= 1.8\%$$

Solution II: 0.18 mol HA and 0.020 mol A⁻; initial pH = 4.05

OH⁻(aq) + HA(aq) ⟶ H₂O(l) + A⁻(aq)			
Before addition	≈0.00 mol	0.18 mol	0.020 mol
Addition	0.010 mol	—	—
After addition	≈0.00 mol	0.17 mol	0.030 mol

$$pH = pK_a + \log\frac{[base]}{[acid]}$$

$$= 5.00 + \log\frac{0.030}{0.17}$$

$$= 4.25$$

$$\% \text{ change} = \frac{4.25 - 4.05}{4.05} \times 100\%$$

$$= 5.0\%$$

As you can see, the buffer with equal amounts of acid and conjugate base is more resistant to pH change and is therefore the more effective buffer. A buffer becomes less effective as the difference in the relative amounts of acid and conjugate base increases. As a guideline, we can say that an effective buffer must have a [base]/[acid] ratio in the range of 0.10 to 10. In other words, *the relative concentrations of acid and conjugate base should not differ by more than a factor of 10 in order for a buffer to be reasonably effective.*

Absolute Concentrations of the Acid and Conjugate Base

A buffer is most effective (most resistant to pH changes) when the concentrations of acid and conjugate base are highest. We will explore this idea by again considering a generic buffer composed of HA and A⁻ and a pK_a of 5.00. Below, we calculate the percent change in pH upon addition of 0.010 mol of NaOH for two 1.0-liter solutions of this buffer system. In this case, solution I is 10 times more concentrated in both the acid and the base than solution II. Both solutions have equal relative amounts of acid and conjugate base and therefore have the same initial pH of 5.00.

Solution I: 0.50 mol HA and 0.50 mol A⁻; initial pH = 5.00

OH⁻(aq) + HA(aq) ⟶ H₂O(l) + A⁻(aq)			
Before addition	≈0.00 mol	0.50 mol	0.50 mol
Addition	0.010 mol	—	—
After addition	≈0.00 mol	0.49 mol	0.51 mol

$$pH = pK_a + \log\frac{[base]}{[acid]}$$

$$= 5.00 + \log\frac{0.51}{0.49}$$

$$= 5.02$$

$$\% \text{ change} = \frac{5.02 - 5.00}{5.00} \times 100\%$$

$$= 0.4\%$$

Solution II: 0.050 mol HA and 0.050 mol A⁻; initial pH = 5.00

OH⁻(aq) + HA(aq) ⟶ H₂O(l) + A⁻(aq)			
Before addition	≈0.00 mol	0.050 mol	0.050 mol
Addition	0.010 mol	—	—
After addition	≈0.00 mol	0.040 mol	0.060 mol

$$pH = pK_a + \log\frac{[base]}{[acid]}$$

$$= 5.00 + \log\frac{0.060}{0.040}$$

$$= 5.18$$

$$\% \text{ change} = \frac{5.18 - 5.00}{5.00} \times 100\%$$

$$= 3.6\%$$

As you can see, the buffer with greater amounts of acid and conjugate base is most resistant to pH changes and therefore the more effective buffer. The more dilute the buffer components, the less effective the buffer.

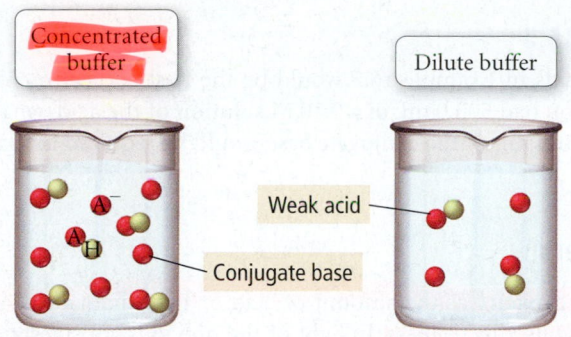

Concentrated buffer

Dilute buffer

Weak acid

Conjugate base

A concentrated buffer has more of the weak acid and its conjugate base than does a weak buffer. It can therefore neutralize more added acid or added base.

Buffer Range

In light of our guideline that the relative concentrations of acid and conjugate base should not differ by more than a factor of 10 in order for a buffer to be reasonably effective, we can calculate the pH range over which a particular acid and its conjugate base can be used to make an effective buffer. Since the pH of a buffer is given by the Henderson–Hasselbalch equation, we can calculate the outermost points of the effective range as follows:

Lowest pH for effective buffer occurs when the base is one-tenth as concentrated as the acid.

$$pH = pK_a + \log \frac{[\text{base}]}{[\text{acid}]}$$
$$= pK_a + \log 0.10$$
$$= pK_a - 1$$

Highest pH for effective buffer occurs when the base is 10 times as concentrated as the acid.

$$pH = pK_a + \log \frac{[\text{base}]}{[\text{acid}]}$$
$$= pK_a + \log 10$$
$$= pK_a + 1$$

Therefore, the effective range for a buffering system is one pH unit on either side of pK_a. For example, a buffering system with a pK_a for the weak acid of 5.0 can be used to prepare a buffer in the range of 4.0–6.0. The relative amounts of acid and conjugate base can be adjusted to achieve any pH within this range. As we noted earlier, however, the buffer would be most effective at pH 5.0, because the buffer components would be exactly equal at that pH.

EXAMPLE 16.5 Buffer Range

Which of the following acids would be the best choice to combine with its sodium salt to make a solution buffered at pH 4.25? For the best choice, calculate the ratio of the conjugate base to the acid required to attain the desired pH.

chlorous acid $(HClO_2)$ $pK_a = 1.95$ formic acid $(HCHO_2)$ $pK_a = 3.74$
nitrous acid (HNO_2) $pK_a = 3.34$ hypochlorous acid $(HClO)$ $pK_a = 7.54$

Solution

The best choice would be formic acid because its pK_a lies closest to the desired pH. The ratio of conjugate base (CHO_2^-) to acid $(HCHO_2)$ required can be calculated from the Henderson–Hasselbalch equation as follows:

$$pH = pK_a + \log \frac{[\text{base}]}{[\text{acid}]}$$
$$4.25 = 3.74 + \log \frac{[\text{base}]}{[\text{acid}]}$$
$$\log \frac{[\text{base}]}{[\text{acid}]} = 4.25 - 3.74$$
$$= 0.51$$
$$\frac{[\text{base}]}{[\text{acid}]} = 10^{0.51}$$
$$= 3.24$$

For Practice 16.5

Which of the acids in Example 16.5 would be the best choice to create a buffer with pH = 7.35? If you had 500.0 mL of a 0.10 M solution of the acid, what mass of the corresponding sodium salt of the conjugate base would be required to make the buffer?

Buffer Capacity

We define **buffer capacity** as the amount of acid or base that can be added to a buffer without destroying its effectiveness. In light of the previous considerations regarding the absolute concentrations of acid and conjugate base in an effective buffer, we can conclude that the *buffer capacity increases with increasing absolute concentrations of the buffer components*. The more concentrated the weak acid and conjugate base that compose the buffer are, the higher the buffer capacity. In addition, *overall buffer capacity increases as the relative concentrations of the buffer components become closer to each other*. As the ratio of the buffer components gets closer to 1, the *overall* capacity of the buffer (the ability to neutralize added acid *and* added base) becomes greater. In some cases, however, a buffer that must neutralize primarily added acid (or primarily added base) may be overweighed in one of the buffer components, as shown in the *Chemistry and Medicine* box in this section.

Chemistry and Medicine
Buffer Effectiveness in Human Blood

As we discussed in the opening section of this chapter, blood contains several buffering systems, one of the most important of which consists of carbonic acid and the carbonate ion. The concentrations of these buffer components in normal blood plasma are $[HCO_3^-] = 0.024$ M and $[H_2CO_3] = 0.0012$ M. The pK_a for carbonic acid at body temperature is 6.1. If we substitute these quantities into the Henderson–Hasselbalch equation, we compute the normal pH of blood as follows:

$$pH = pK_a + \log\frac{[base]}{[acid]}$$

$$= 6.1 = \log\frac{[HCO_3^-]}{[H_2CO_3]}$$

$$= 6.1 + \log\frac{0.024 \text{ M}}{0.0012 \text{ M}}$$

$$= 7.4$$

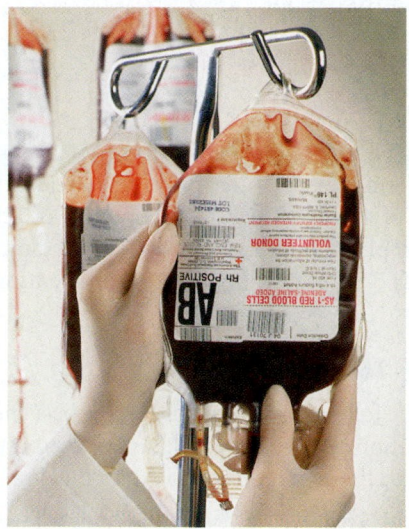

▲ Normal blood has a pH of 7.4.

Normal blood, therefore, has a pH of 7.4. Notice, however, that the concentration of the bicarbonate ion is 20 times higher than the concentration of carbonic acid and the pH of the buffer is more than one pH unit away from pK_a. Why?

The higher bicarbonate ion concentration in blood makes the buffer capacity of blood greater for acid than for base, which is necessary because the products of metabolism that enter blood are mostly acidic. For example, when we exercise, our bodies produce lactic acid ($HC_3H_5O_3$). The lactic acid enters the bloodstream and must be neutralized. The bicarbonate ion neutralizes the lactic acid according to the following equation:

$$HCO_3^-(aq) + HC_3H_5O_3(aq) \longrightarrow$$
$$H_2CO_3(aq) + C_3H_5O_3^-(aq)$$

An enzyme called carbonic anhydrase then catalyzes the conversion of carbonic acid into carbon dioxide and water:

$$H_2CO_3(aq) \rightleftharpoons CO_2(g) + H_2O(l)$$

The carbon dioxide is eliminated from the blood through breathing. When large amounts of lactic acid are produced, we must breathe faster to keep up with the need to eliminate carbon dioxide.

Question

A 70-kg person has a total blood volume of about 5.0 L. Given the carbonic acid and bicarbonate concentrations stated above, what volume (in mL) of 6.0 M HCl could be neutralized by blood without the blood pH dropping below 7.0 (which would result in death)?

Conceptual Connection 16.3 Buffer Capacity

A 1.0-L buffer solution is 0.10 M in HF and 0.050 M in NaF. Which of the following actions will destroy the buffer?

(a) adding 0.050 mol of HCl
(b) adding 0.050 mol of NaOH
(c) adding 0.050 mol of NaF
(d) None of the above

Answer: **(a)** Adding 0.050 mol of HCl will destroy the buffer because it will react with all of the NaF, leaving no conjugate base in the buffer mixture.

16.4 Titrations and pH Curves

We first examined acid–base titrations in Section 4.8. In an **acid–base titration**, a basic (or acidic) solution of unknown concentration is reacted with an acidic (or basic) solution of known concentration. The known solution is slowly added to the unknown one while the pH is monitored with either a pH meter or an **indicator** (a substance whose color depends on the pH). As the acid and base combine, they neutralize each other. At the **equivalence point**—the point in the titration when the number of moles of base is stoichiometrically equal to the number of moles of acid—the titration is complete. When the equivalence point is reached, neither reactant is in excess and the number of moles of the reactants are related by the reaction stoichiometry (Figure 16.5 ▼).

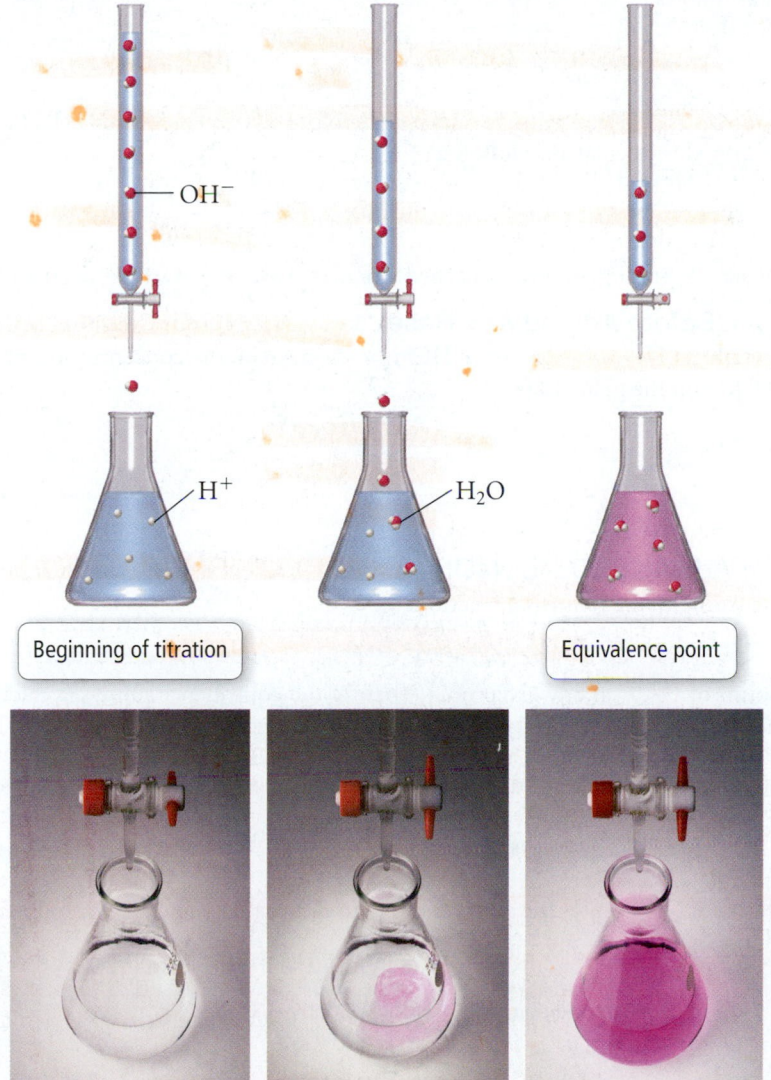

Beginning of titration

Equivalence point

◄ **FIGURE 16.5 Acid–Base Titration** As OH⁻ is added in a titration, it neutralizes the H⁺, forming water. At the equivalence point, the titration is complete.

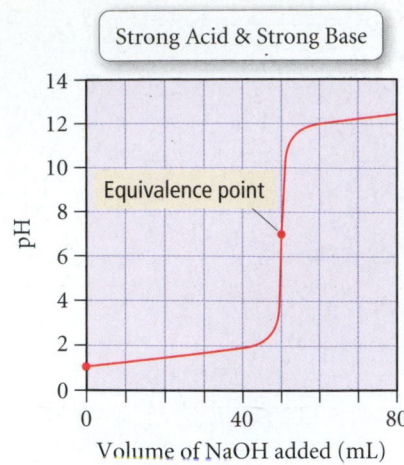

▲ FIGURE 16.6 Titration Curve: Strong Acid + Strong Base This curve represents the titration of 50.0 mL of 0.100 M HCl with 0.100 M NaOH.

In this section, we examine acid–base titrations more closely, concentrating on the pH changes that occur during the titration. A plot of the pH of the solution during a titration is known as a *titration curve* or *pH curve*. Figure 16.6 ◀ shows a pH curve for the titration of HCl with NaOH. Before any base is added to the solution, the pH is low (as expected for a solution of HCl). As the NaOH is added, the solution becomes less acidic as the NaOH begins to neutralize the HCl. The point of inflection in the middle of the curve is the equivalence point. Notice that the pH changes very quickly near the equivalence point (small amounts of added base cause large changes in pH). Beyond the equivalence point, the solution is basic because the HCl has been completely neutralized and excess base is being added to the solution. The exact shape of the pH curve depends on several factors, including the strength of the acid or base being titrated. We look at several combinations individually.

The Titration of a Strong Acid with a Strong Base

Consider the titration of 25.0 mL of 0.100 M HCl with 0.100 M NaOH. We will calculate the volume of base required to reach the equivalence point, as well as the pH at several points along the way.

Volume of NaOH Required to Reach the Equivalence Point The neutralization reaction that occurs during the titration is as follows:

$$HCl(aq) + NaOH(aq) \longrightarrow H_2O(l) + NaCl(aq)$$

The equivalence point is reached when the number of moles of base added equals the number of moles of acid initially in solution. We calculate the amount of acid initially in solution as follows:

$$\text{Initial mol HCl} = 0.0250 \, \cancel{L} \times \frac{0.100 \text{ mol}}{1 \, \cancel{L}} = 0.00250 \text{ mol HCl}$$

The amount of NaOH that must be added is therefore 0.00250 mol NaOH. The volume of NaOH required is calculated as follows:

$$\text{Volume NaOH solution} = 0.00250 \, \cancel{\text{mol}} \times \frac{1 \text{ L}}{0.100 \, \cancel{\text{mol}}} = 0.0250 \text{ L}$$

The equivalence point therefore is reached when 25.0 mL of NaOH has been added.

Initial pH (Before Adding Any Base) The initial pH of the solution is simply the pH of a 0.100 M HCl solution. Since HCl is a strong acid, the concentration of H_3O^+ is also 0.100 M and the pH is 1.00.

$$pH = -\log[H_3O^+]$$
$$= -\log(0.100)$$
$$= 1.00$$

pH After Adding 5.00 mL NaOH As NaOH is added to the solution, it neutralizes H_3O^+ according to the following reaction:

$$OH^-(aq) + H_3O^+(aq) \longrightarrow 2 H_2O(l)$$

The amount of H_3O^+ at any given point (before the equivalence point) is calculated by using the reaction stoichiometry—1 mol of NaOH neutralizes 1 mol of H_3O^+. The initial number of moles of H_3O^+ (as calculated above) is 0.00250 mol. The number of moles of NaOH added at 5.00 mL is as follows:

$$\text{mol NaOH added} = 0.00500 \, \cancel{L} \times \frac{0.100 \text{ mol}}{1 \, \cancel{L}} = 0.000500 \text{ mol NaOH}$$

The addition of OH^- causes the amount of H^+ to decrease as shown in the following table:

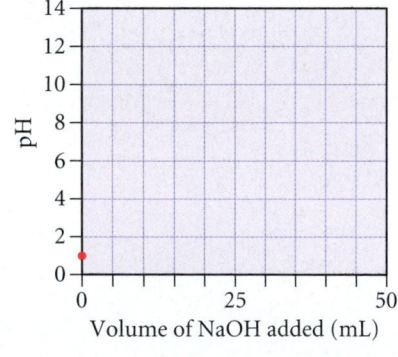

	$OH^-(aq) \, + \, H_3O^+(aq) \, \longrightarrow \, 2 H_2O(l)$	
Before addition	≈0.00 mol	0.00250 mol
Addition	0.000500 mol	—
After addition	≈0.00 mol	0.00200 mol

The H_3O^+ concentration is then computed by dividing the number of moles of H_3O^+ remaining by the *total volume* (initial volume plus added volume).

$$[H_3O^+] = \frac{0.00200 \text{ mol } H_3O^+}{\underbrace{0.0250 \text{ L}}_{\text{Initial volume}} + \underbrace{0.00500 \text{ L}}_{\text{Added volume}}} = 0.0667 \text{ M}$$

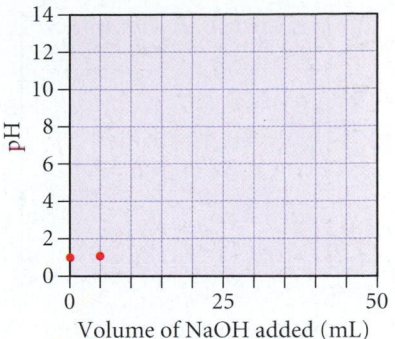

The pH is therefore 1.18:

$$pH = -\log 0.0667$$
$$= 1.18$$

pH's After Adding 10.0, 15.0, and 20.0 mL NaOH
As more NaOH is added, it further neutralizes the H_3O^+ in the solution. The pH at each of these points is calculated in the same way as it was after 5.00 mL of NaOH was added. The results are tabulated below.

Volume (mL)	pH
10.0	1.37
15.0	1.60
20.0	1.95

↑PH increase w ↑ in NaOH

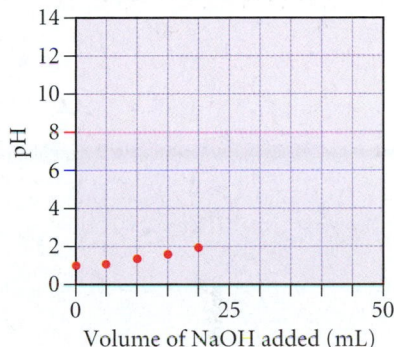

pH After Adding 25.0 mL NaOH (Equivalence Point)
The pH at the equivalence point of a strong acid–strong base titration will always be 7.00 (at 25 °C). At the equivalence point, the strong base has completely neutralized the strong acid. The only source of hydronium ions then becomes the ionization of water. The $[H_3O^+]$ at 25 °C from the ionization of water is 1.00×10^{-7} and the pH is therefore 7.00.

pH After Adding 30.00 mL NaOH
As NaOH is added beyond the equivalence point, it becomes the excess reagent. The amount of OH^- at any given point (past the equivalence point) is calculated by subtracting the initial amount of H_3O^+ from the amount of OH^- added. The number of moles of OH^- added at 30.00 mL is

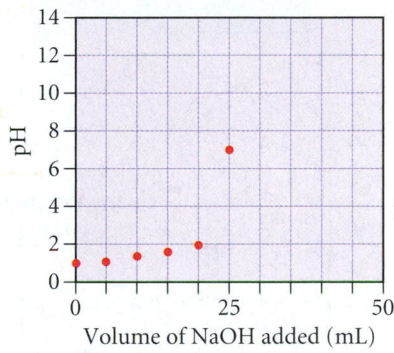

$$\text{mol } OH^- \text{ added} = 0.0300 \text{ L} \times \frac{0.100 \text{ mol}}{1 \text{ L}} = 0.00300 \text{ mol } OH^-$$

The number of moles of OH^- remaining after neutralization is shown in the following table:

	$OH^-(aq) + H_3O^+(aq) \longrightarrow 2 H_2O(l)$	
Before addition	≈0.00 mol	0.00250 mol
Addition	0.00300 mol	—
After addition	0.00050 mol	≈0.00 mol

The OH^- concentration is then computed by dividing the number of moles of OH^- remaining by the *total volume* (initial volume plus added volume).

$$[OH^-] = \frac{0.000500 \text{ mol } OH^-}{0.0250 \text{ L} + 0.0300 \text{ L}} = 0.00909 \text{ M}$$

We can then calculate the $[H_3O^+]$ and pH as follows:

$$[H_3O^+][OH^-] = 10^{-14}$$
$$[H_3O^+] = \frac{10^{-14}}{[OH^-]} = \frac{10^{-14}}{0.00909}$$
$$= 1.10 \times 10^{-12} \text{ M}$$
$$pH = -\log(1.10 \times 10^{-12})$$
$$= 11.96$$

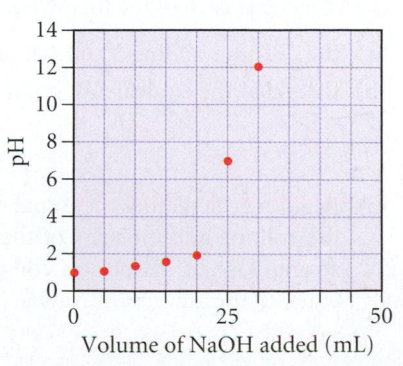

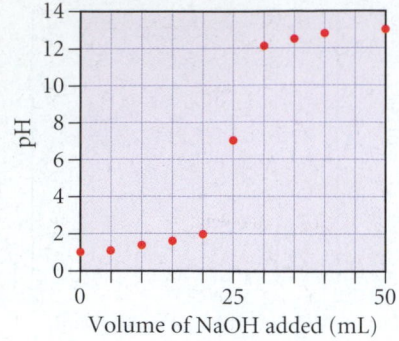

pH's After Adding 35.0, 40.0, and 50.0 mL NaOH
As more NaOH is added, it further increases the basicity of the solution. The pH at each of these points is calculated in the same way as it was after 30.00 mL of NaOH was added. The results are tabulated below.

Volume (mL)	pH
35.0	12.22
40.0	12.36
50.0	12.52

The Overall pH Curve
The overall pH curve for the titration of a strong acid with a strong base has the characteristic S-shape we just computed. The overall curve is shown here:

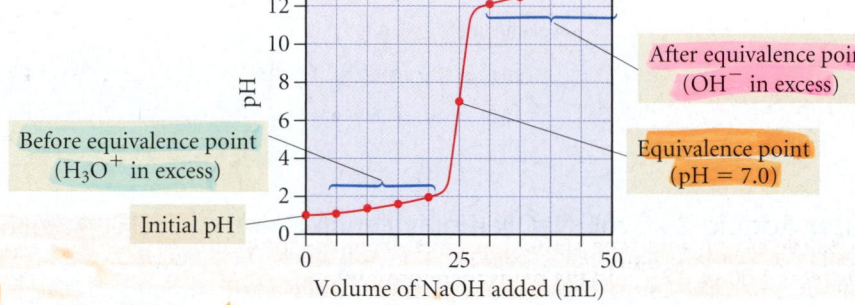

After equivalence point (OH^- in excess)

Before equivalence point (H_3O^+ in excess)

Equivalence point (pH = 7.0)

Initial pH

Summarizing:

> The initial pH is just the pH of the strong acid solution to be titrated.
> Before the equivalence point, H_3O^+ is in excess. Calculate the $[H_3O^+]$ by subtracting the number of moles of added OH^- from the initial number of moles of H_3O^+ and dividing by the *total* volume.
> At the equivalence point, neither reactant is in excess and the pH = 7.00.
> Beyond the equivalence point, OH^- is in excess. Calculate the $[OH^-]$ by subtracting the initial number of moles of H_3O^+ from the number of moles of added OH^- and dividing by the *total* volume.

The pH curve for the titration of a strong base with a strong acid is shown in Figure 16.7 ◄. Calculating the points along this curve is very similar to calculating the points along the pH curve for the titration of a strong acid with a strong base (which we just did). The main difference is that the curve starts basic and then turns acidic after the equivalence point (instead of vice versa).

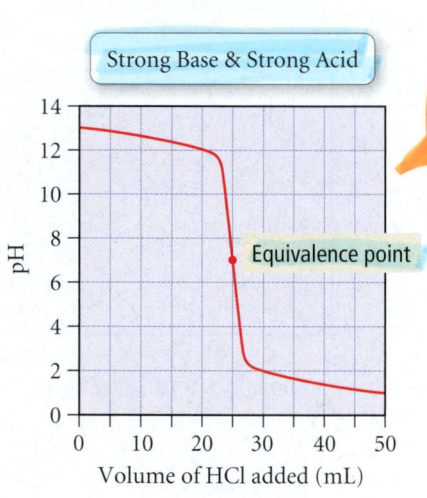

Strong Base & Strong Acid

Equivalence point

▲ **FIGURE 16.7 Titration Curve: Strong Base + Strong Acid** This curve represents the titration of 25.0 mL of 0.100 M NaOH with 0.100 M HCl.

EXAMPLE 16.6 Strong Acid–Strong Base Titration pH Curve

A 50.0-mL sample of 0.200 M sodium hydroxide is titrated with 0.200 M nitric acid. Calculate each of the following:

(a) the pH after adding 30.00 mL of HNO_3
(b) the pH at the equivalence point

Solution

(a) Begin by calculating the initial amount of NaOH (in moles) from the volume and molarity of the NaOH solution. Since NaOH is a strong base, it dissociates completely, so the amount of OH^- is equal to the amount of NaOH.

$$\text{moles NaOH} = 0.0500 \, \cancel{L} \times \frac{0.200 \, \text{mol}}{1 \, \cancel{L}}$$
$$= 0.0100 \, \text{mol}$$
$$\text{moles } OH^- = 0.0100 \, \text{mol}$$

Calculate the amount of HNO_3 (in moles) added at 30.0 mL from the molarity of the HNO_3 solution.

$$\text{mol } HNO_3 \text{ added} = 0.0300 \text{ L} \times \frac{0.200 \text{ mol}}{1 \text{ L}}$$

$$= 0.00600 \text{ mol } HNO_3$$

As HNO_3 is added to the solution, it neutralizes some of the OH^-. Calculate the number of moles of OH^- remaining by setting up a table based on the neutralization reaction that shows the amount of OH^- before the addition, the amount of H_3O^+ added, and the amounts left after the addition.

	$OH^-(aq)$ + $H_3O^+(aq)$ $\longrightarrow$ 2 $H_2O(l)$	
Before addition	0.0100 mol	≈0.00 mol
Addition	—	0.00600 mol
After addition	0.0040 mol	≈0.00 mol

Compute the OH^- concentration by dividing the amount of OH^- remaining by the *total volume* (initial volume plus added volume).

$$[OH^-] = \frac{0.0040 \text{ mol}}{0.0500 \text{ L} + 0.0300 \text{ L}}$$

$$= 0.0500 \text{ M}$$

Calculate the pOH from $[OH^-]$.

$$pOH = -\log 0.0500$$

$$= 1.30$$

Calculate the pH from the pOH using the equation pH + pOH = 14.

$$pH = 14 - pOH$$

$$= 14 - 1.30$$

$$= 12.70$$

(b) At the equivalence point, the strong base has completely neutralized the strong acid. The $[H_3O^+]$ at 25 °C from the ionization of water is 1.00×10^{-7} and the pH is therefore 7.00.

$$pH = 7.00$$

For Practice 16.6

Calculate the pH in the titration in Example 16.6 after adding a total of 60.0 mL of 0.200 M HNO_3.

The Titration of a Weak Acid with a Strong Base

Consider the titration of 25.0 mL of 0.100 M $HCHO_2$ with 0.100 M NaOH.

$$NaOH(aq) + HCHO_2(aq) \longrightarrow H_2O(l) + NaCHO_2(aq)$$

The concentrations and the volumes here are identical to those in our previous titration, in which we calculated the pH curve for the titration of a *strong* acid with a strong base. The only difference is that $HCHO_2$ is *weak* acid rather than a strong one. We begin our calculation by computing the volume required to reach the equivalence point of the titration.

Volume of NaOH Required to Reach the Equivalence Point From the stoichiometry of the equation, we can see that the equivalence point occurs when the amount (in moles) of added base equals the amount (in moles) of acid initially in solution.

$$\text{Initial mol } HCHO_2 = 0.0250 \text{ L} \times \frac{0.100 \text{ mol}}{1 \text{ L}} = 0.00250 \text{ mol } HCHO_2$$

Thus, the amount of NaOH that must be added is 0.00250 mol NaOH. The volume of NaOH required is therefore

$$\text{Volume NaOH solution} = 0.00250 \text{ mol} \times \frac{1 \text{ L}}{0.100 \text{ mol}} = 0.0250 \text{ L NaOH solution}$$

The equivalence point therefore occurs at 25.0 mL of added base. Notice that the volume of NaOH required to reach the equivalence point is identical to that required for a strong acid. *The volume at the equivalence point in an acid–base titration does not depend on whether the acid being titrated is a strong acid or a weak acid, but only on the amount (in moles) of acid present in solution before the titration begins and on the concentration of the added base.*

Initial pH (Before Adding Any Base) The initial pH of the solution is the pH of a 0.100 M $HCHO_2$ solution. Since $HCHO_2$ is a weak acid, we calculate the concentration of H_3O^+ and the pH by doing an equilibrium problem for the ionization of $HCHO_2$. The procedure for solving weak acid ionization problems is given in Examples 15.5 and 15.6. We show a highly condensed calculation below (K_a for $HCHO_2$ is 1.8×10^{-4}):

$$HCHO_2(aq) + H_2O(l) \rightleftharpoons H_3O^+(aq) + CHO_2^-(aq)$$

	[HCHO$_2$]	[H$_3$O$^+$]	[CHO$_2$$^-$]
Initial	0.100	≈0.00	0.00
Change	$-x$	$+x$	$+x$
Equil	$0.100 - x$	x	x

$$K_a = \frac{[H_3O^+][CHO_2^-]}{[HCHO_2]}$$

$$= \frac{x^2}{0.100 - x} \quad (x \text{ is small})$$

$$1.8 \times 10^{-4} = \frac{x^2}{0.100}$$

$$x = 4.24 \times 10^{-3}$$

Therefore, $[H_3O^+] = 4.24 \times 10^{-3}$ M.

$$pH = -\log(4.24 \times 10^{-3})$$

$$= 2.37$$

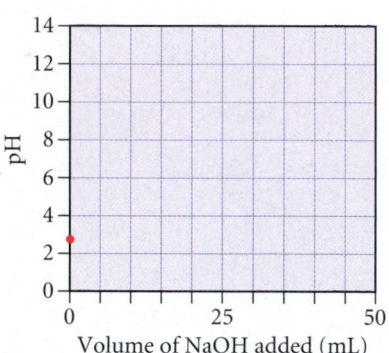

Notice that the pH begins at a higher value (less acidic) than it did for a strong acid of the same concentration, as we would expect because the acid is weak.

pH After Adding 5.00 mL NaOH When titrating a *weak acid* with a strong base, the added NaOH *converts a stoichiometric amount of the acid into its conjugate base.* As we calculated previously, 5.00 mL of the 0.100 M NaOH solution contains 0.000500 mol OH^-. When we add the 0.000500 mol OH^- to the weak acid solution, OH^- reacts stoichiometrically with $HCHO_2$, causing the amount of $HCHO_2$ to go *down* by 0.000500 mol and the amount of CHO_2^- to go *up* by 0.000500 mol. This is very similar to what happens when you add strong base to a buffer, and is summarized in the following table:

	$OH^-(aq)$	$+$	$HCHO_2(aq)$	$\longrightarrow$	$H_2O(l)$	$+$	$CHO_2^-(aq)$
Before addition	≈0.00 mol		0.00250 mol		—		0.00 mol
Addition	0.000500 mol		—		—		—
After addition	≈0.00 mol		0.00200 mol		—		0.000500 mol

Notice that, after the addition, the solution contains significant amounts of both an acid ($HCHO_2$) and its conjugate base (CHO_2^-)—*the solution is now a buffer.* To calculate the pH of a buffer (when the *x is small* approximation applies as it does here), we can use the Henderson–Hasselbalch equation and pK_a for $HCHO_2$ (which is 3.74).

$$pH = pK_a + \log\frac{[base]}{[acid]}$$

$$= 3.74 + \log\frac{0.000500}{0.00200}$$

$$= 3.74 - 0.60$$

$$= 3.14$$

pH's After Adding 10.0, 12.5, 15.0, and 20.0 mL NaOH

As more NaOH is added, it converts more $HCHO_2$ into CHO_2^-. We calculate the relative amounts of $HCHO_2$ and CHO_2^- at each of these volumes using the reaction stoichiometry, and then calculate the pH of the resulting buffer using the Henderson–Hasselbalch equation (as we did for the pH at 5.00 mL). The amounts of $HCHO_2$ and CHO_2^- (after addition of the OH^-) at each volume and the corresponding pH's are tabulated below.

Volume (mL)	mol HCHO$_2$	mol CHO$_2^-$	pH
10.0	0.00150	0.00100	3.56
12.5	0.00125	0.00125	3.74
15.0	0.00100	0.00150	3.92
20.0	0.00050	0.00200	4.34

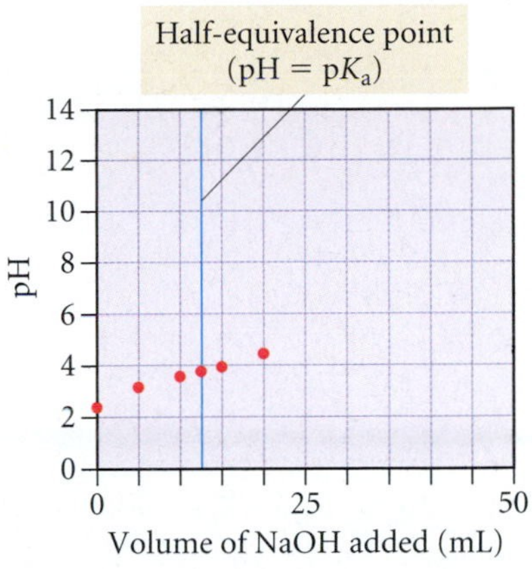

Notice that, as the titration proceeds, more of the $HCHO_2$ is converted to the conjugate base (CHO_2^-). Notice also that an added NaOH volume of 12.5 mL corresponds to exactly one-half of the equivalence point. At this volume, exactly one-half of the initial amount of $HCHO_2$ has been converted to CHO_2^-, resulting in *exactly equal amounts of weak acid and conjugate base.* For any buffer in which the amounts of weak acid and conjugate base are equal, the $pH = pK_a$ as shown here:

$$pH = pK_a + \log \frac{[base]}{[acid]}$$

If $[base] = [acid]$, then $[base]/[acid] = 1$.

$$pH = pK_a + \log 1$$
$$= pK_a + 0$$
$$= pK_a$$

pH After Adding 25.0 mL NaOH (Equivalence Point)

At the equivalence point, we have added 0.000250 mol of OH^- and have therefore converted all of the $HCHO_2$ into its conjugate base (CHO_2^-) as tabulated below.

	$OH^-(aq)$ +	$HCHO_2(aq)$ ⟶	$H_2O(l)$ +	$CHO_2^-(aq)$
Before addition	≈0.00 mol	0.00250 mol	—	0.00 mol
Addition	0.00250 mol	—	—	—
After addition	≈0.00 mol	≈0.00 mol	—	0.00250 mol

The solution is no longer a buffer (it no longer contains significant amounts of both a weak acid and its conjugate base). Instead, the solution is just that of an ion (CHO_2^-) acting as a weak base. We learned how to calculate the pH of solutions such as this in Section 15.8 (see Example 15.14) by solving an equilibrium problem involving the ionization of water by the weak base (CHO_2^-):

$$CHO_2^-(aq) + H_2O(l) \rightleftharpoons HCHO_2(aq) + OH^-(aq)$$

We calculate the initial concentration of CHO_2^- for the equilibrium problem by dividing the number of moles of CHO_2^- (0.00250 mol) by the *total* volume at the equivalence point (initial volume plus added volume).

Moles CHO_2^- at equivalence point

$$[CHO_2^-] = \frac{0.00250 \text{ mol}}{0.0250 \text{ L} + 0.0250 \text{ L}} = 0.050 \text{ M}$$

Initial volume   Added volume at equivalence point

Half-equivalence point
(pH = pK$_a$)

Since $pH = pK_a$ halfway to the equivalence point, titrations can be used to measure the pK_a of an acid.

We then proceed to solve the equilibrium problem as shown in condensed form below:

$$CHO_2^-(aq) + H_2O(l) \rightleftharpoons HCHO_2(aq) + OH^-(aq)$$

	$[CHO_2^-]$	$[HCHO_2]$	$[OH^-]$
Initial	0.0500	0.00	≈ 0.00
Change	$-x$	$+x$	$+x$
Equil	$0.0500 - x$	x	x

Before substituting into the expression for K_b, we find the value of K_b from K_a for formic acid ($K_a = 1.8 \times 10^{-4}$) and K_w:

$$K_a \times K_b = K_w$$

$$K_b = \frac{K_w}{K_a} = \frac{1.0 \times 10^{-14}}{1.8 \times 10^{-4}} = 5.6 \times 10^{-11}$$

Now we can substitute the equilibrium concentrations from the table above into the expression for K_b as follows:

$$K_b = \frac{[HCHO_2][OH^-]}{[CHO_2^-]}$$

$$= \frac{x^2}{0.0500 - x} \quad (x \text{ is small})$$

$$5.6 \times 10^{-11} = \frac{x^2}{0.0500}$$

$$x = 1.7 \times 10^{-6}$$

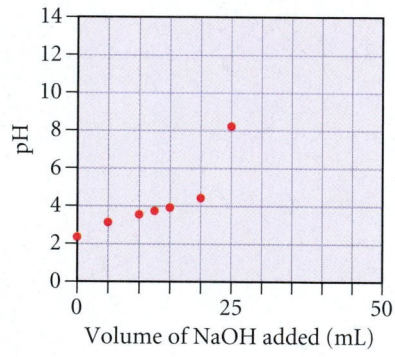

Remember that x represents the concentration of the hydroxide ion. We then calculate $[H_3O^+]$ and pH as follows:

$$[OH^-] = 1.7 \times 10^{-6} \text{ M}$$

$$[H_3O^+][OH^-] = K_w = 1.0 \times 10^{-14}$$

$$[H_3O^+](1.7 \times 10^{-6}) = 1.0 \times 10^{-14}$$

$$[H_3O^+] = 5.9 \times 10^{-9} \text{ M}$$

$$pH = -\log[H_3O^+]$$
$$= -\log(5.9 \times 10^{-9})$$
$$= 8.23$$

Notice that the pH at the equivalence point is *not* neutral but basic. *The titration of weak acid by a strong base will always have a basic equivalence point* because, at the equivalence point, all of the acid has been converted into its conjugate base, resulting in a weakly basic solution.

pH After Adding 30.00 mL NaOH   At this point in the titration, we have added 0.00300 mol of OH^-. NaOH has now become the excess reagent as shown in the following table:

	$OH^-(aq)$	$+ \ HCHO_2(aq)$	$\longrightarrow H_2O(l) \ +$	$CHO_2^-(aq)$
Before addition	≈ 0.00 mol	0.00250 mol	—	0.00 mol
Addition	0.00300 mol	—	—	—
After addition	0.00050 mol	≈ 0.00 mol	—	0.00250 mol

The solution becomes a mixture of a strong base (NaOH) and a weak base (CHO_2^-). The strong base completely overwhelms the weak base and the pH can be calculated by considering the strong base alone (as we did for the titration of a strong acid and a strong base). The OH^- concentration is then computed by dividing the amount of OH^- remaining by the *total volume* (initial volume plus added volume).

$$[OH^-] = \frac{0.00050 \text{ mol } OH^-}{0.0250 \text{ L} + 0.0300 \text{ L}} = 0.0091 \text{ M}$$

We can then calculate the $[H_3O^+]$ and pH as follows:

$$[H_3O^+][OH^-] = 10^{-14}$$

$$[H_3O^+] = \frac{10^{-14}}{[OH^-]} = \frac{10^{-14}}{0.0091} = 1.10 \times 10^{-12}\,M$$

$$pH = -\log(1.10 \times 10^{-12})$$

$$= 11.96$$

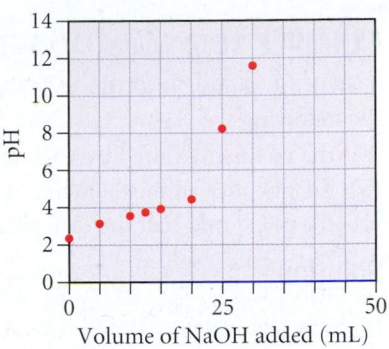

pH's After adding 35.0, 40.0, and 50.0 mL NaOH As more NaOH is added, the basicity of the solution increases further. The pH at each of these volumes is calculated in the same way as it was after 30.00 mL of NaOH was added. The results are tabulated below.

Volume (mL)	pH
35.0	12.22
40.0	12.36
50.0	12.52

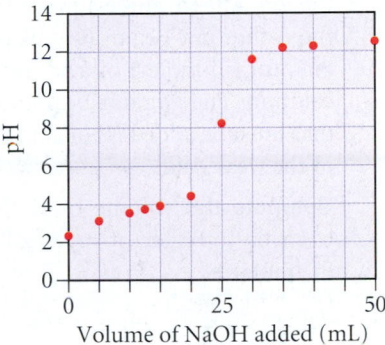

The Overall pH Curve The overall pH curve for the titration of a weak acid with a strong base has the characteristic S-shape similar to that for the titration of a strong acid with a strong base. The main difference is that the equivalence point pH is basic (not neutral). Notice that calculating the pH in different regions throughout the titration involves working different kinds of acid–base problems, all of which we have encountered before.

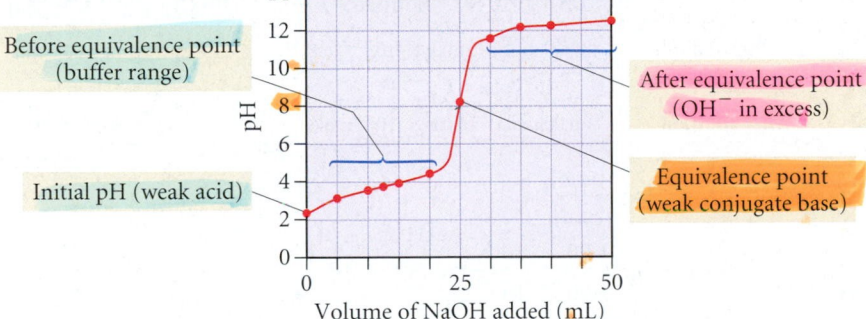

Before equivalence point (buffer range)

Initial pH (weak acid)

After equivalence point (OH^- in excess)

Equivalence point (weak conjugate base)

Volume of NaOH added (mL)

Summarizing:

▶ The initial pH is that of the weak acid solution to be titrated. Calculate the pH by working an equilibrium problem (similar to Examples 15.5 and 15.6) using the concentration of the weak acid as the initial concentration.

▶ Between the initial pH and the equivalence point, the solution becomes a buffer. Use the reaction stoichiometry to compute the amounts of each buffer component and then use the Henderson–Hasselbalch equation to compute the pH (as in Example 16.3).

▶ Halfway to the equivalence point, the buffer components are exactly equal and $pH = pK_a$.

▶ At the equivalence point, the acid has all been converted into its conjugate base. Calculate the pH by working an equilibrium problem for the ionization of water by the ion acting as a weak base (similar to Example 15.14). (Compute the concentration of the ion acting as a weak base by dividing the number of moles of the ion by the total volume at the equivalence point.)

▶ Beyond the equivalence point, OH^- is in excess. You can ignore the weak base and calculate the $[OH^-]$ by subtracting the initial number of moles of H_3O^+ from the number of moles of added OH^- and dividing by the *total* volume.

EXAMPLE 16.7   Weak Acid–Strong Base Titration pH Curve

A 40.0-mL sample of 0.100 M HNO_2 is titrated with 0.200 M KOH. Calculate each of the following:

(a) the volume required to reach the equivalence point
(b) the pH after adding 5.00 mL of KOH
(c) the pH at one-half the equivalence point

Solution

(a) The equivalence point occurs when the amount (in moles) of added base equals the amount (in moles) of acid initially in the solution. Begin by calculating the amount (in moles) of acid initially in the solution. The amount (in moles) of KOH that must be added is equal to the amount of the weak acid.

$$mol\ HNO_2 = 0.0400\ \cancel{L} \times \frac{0.100\ mol}{\cancel{L}}$$
$$= 4.00 \times 10^{-3}\ mol$$
$$mol\ KOH\ required = 4.00 \times 10^{-3}\ mol$$

Compute the volume of KOH required from the number of moles of KOH and the molarity.

$$volume\ KOH\ solution = 4.00 \times 10^{-3}\ \cancel{mol} \times \frac{1\ L}{0.200\ \cancel{mol}}$$
$$= 0.0200\ L\ KOH\ solution$$
$$= 20.0\ mL\ KOH\ solution$$

(b) Use the concentration of the KOH solution to compute the amount (in moles) of OH^- in 5.00 mL of the solution.

$$mol\ OH^- = 5.00 \times 10^{-3}\ \cancel{L} \times \frac{0.200\ mol}{1\ \cancel{L}}$$
$$= 1.00 \times 10^{-3}\ mol\ OH^-$$

Prepare a table showing the amounts of HNO_2 and NO_2^- before and after the addition of 5.00 mL KOH. The addition of the KOH stoichiometrically reduces the concentration of HNO_2 and increases the concentration of NO_2^-.

	$OH^-(aq)$	+	$HNO_2(aq)$	$\longrightarrow$	$H_2O(l)$	+	$NO_2^-(aq)$
Before addition	$\approx$0.00 mol		4.00×10^{-3} mol		—		0.00 mol
Addition	1.00×10^{-3} mol		—		—		—
After addition	$\approx$0.00 mol		3.00×10^{-3} mol		—		1.00×10^{-3} mol

Since the solution now contains significant amounts of a weak acid and its conjugate base, use the Henderson–Hasselbalch equation and pK_a for HNO_2 (which is 3.15) to calculate the pH of the solution.

$$pH = pK_a + \log \frac{[base]}{[acid]}$$
$$= 3.15 + \log \frac{1.00 \times 10^{-3}}{3.00 \times 10^{-3}}$$
$$= 3.15 - 0.48$$
$$= 2.67$$

(c) At one-half the equivalence point, the amount of added base is exactly one-half the initial amount of acid. The base converts exactly half of the HNO_2 into NO_2^-, resulting in equal amounts of the weak acid and its conjugate base. The pH is therefore equal to pK_a.

	$OH^-(aq)$	+	$HNO_2(aq)$	$\longrightarrow$	$H_2O(l)$	+	$NO_2^-(aq)$
Before addition	$\approx$0.00 mol		4.00×10^{-3} mol		—		0.00 mol
Addition	2.00×10^{-3} mol		—		—		—
After addition	$\approx$0.00 mol		2.00×10^{-3} mol		—		2.00×10^{-3} mol

$$pH = pK_a + \log \frac{[base]}{[acid]}$$
$$= 3.15 + \log \frac{2.00 \times 10^{-3}}{2.00 \times 10^{-3}}$$
$$= 3.15 + 0$$
$$= 3.15$$

For Practice 16.7

Determine the pH at the equivalence point for the titration between HNO_2 and KOH in Example 16.7.

The pH curve for the titration of a weak base with a strong acid is shown in Figure 16.8 ▼.

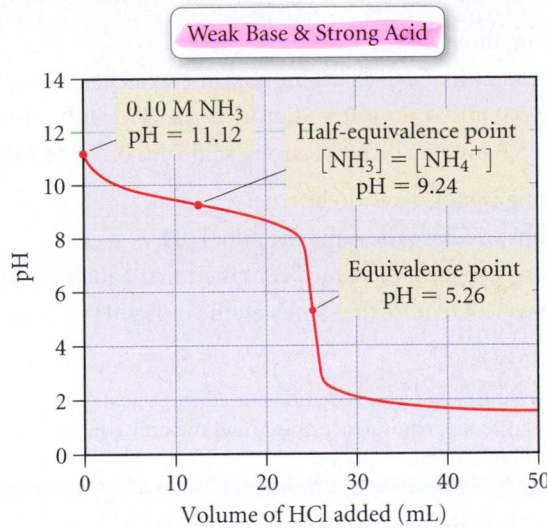

Weak Base & Strong Acid

0.10 M NH₃
pH = 11.12

Half-equivalence point
[NH₃] = [NH₄⁺]
pH = 9.24

Equivalence point
pH = 5.26

Volume of HCl added (mL)

◀ **FIGURE 16.8 Titration Curve: Weak Base with Strong Acid** Titration curve for the titration of 0.100 M NH₃ with 0.100 M HCl.

Calculating the points along this curve is very similar to calculating the points along the pH curve for the titration of a weak acid with a strong base (which we just did). The main differences are that the curve starts basic and has an acidic equivalence point.

Titration of a Polyprotic Acid

When a diprotic acid is titrated with a strong base, and if K_{a_1} and K_{a_2} are sufficiently different, the pH curve will have two equivalence points. For example, Figure 16.9 ▼ shows the pH curve for the titration of sulfurous acid (H_2SO_3) with sodium hydroxide. Recall from Section 15.9 that sulfurous acid ionizes in two steps as follows:

$$H_2SO_3(aq) \rightleftharpoons H^+(aq) + HSO_3^-(aq) \quad K_{a_1} = 1.6 \times 10^{-2}$$
$$HSO_3^-(aq) \rightleftharpoons H^+(aq) + SO_3^{2-}(aq) \quad K_{a_2} = 6.4 \times 10^{-8}$$

The first equivalence point in the titration curve represents the titration of the first proton while the second equivalence point represents the titration of the second proton. Notice that the volume required to reach the first equivalence point is identical to the volume required to the reach the second one because the number of moles H_2SO_3 in the first step determines the number of moles of HSO_3^- in the second step.

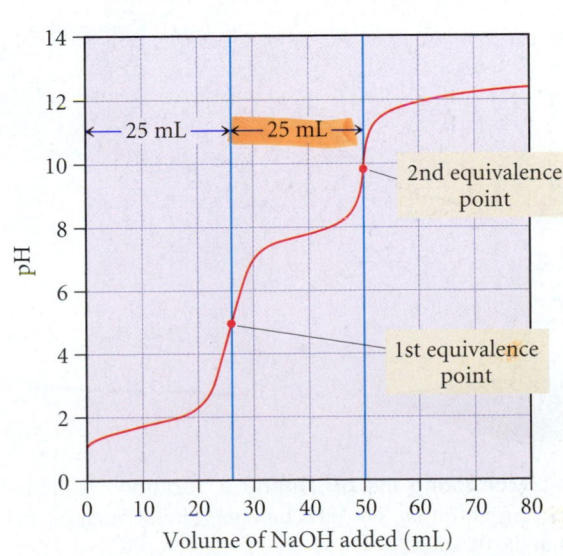

Titration of a Polyprotic Acid

25 mL — 25 mL

2nd equivalence point

1st equivalence point

Volume of NaOH added (mL)

◀ **FIGURE 16.9 Titration Curve: Diprotic Acid with Strong Base** This curve represents the titration of 25.0 mL of 0.100 M H_2SO_3 with 0.100 M NaOH.

Conceptual Connection 16.4 Acid–Base Titrations

Consider the following three titrations:

(i) the titration of 25.0 mL of a 0.100 M monoprotic weak acid with 0.100 M NaOH

(ii) the titration of 25.0 mL of a 0.100 M diprotic weak acid with 0.100 M NaOH

(iii) the titration of 25.0 mL of a 0.100 M strong acid with 0.100 M NaOH

Which of the following is most likely to be true?

(a) All three titrations will have the same the initial pH.

(b) All three titrations will have the same pH at their first equivalence point.

(c) All three titrations will require the same volume of NaOH to reach their first equivalence point.

Answer: (c) Since the volumes and concentrations of all three acids are the same, the volume of
NaOH required to reach the first equivalence point (and the only equivalence point for titrations i
and iii) will be the same for all three titrations.

Indicators: pH-Dependent Colors

The pH of a titration can be monitored with either a pH meter or an indicator. The direct monitoring of pH with a meter yields data like the pH curves we have examined
previously, allowing determination of the equivalence point from the pH curve itself, as
shown in Figure 16.10 ▼. With an indicator, however, we rely on the point where the indicator changes color—called the **endpoint**—to determine the equivalence point, as
shown in Figure 16.11 ▶. With the correct indicator, the endpoint of the titration
(indicated by the color change) will occur at the equivalence point (when the amount
of acid equals the amount of base).

An indicator is itself a weak organic acid that has a color different from that of its
conjugate base. For example, phenolphthalein (whose structure is shown in Figure 16.12 ▶)
is a common indicator whose acid form is colorless and conjugate base form is pink.

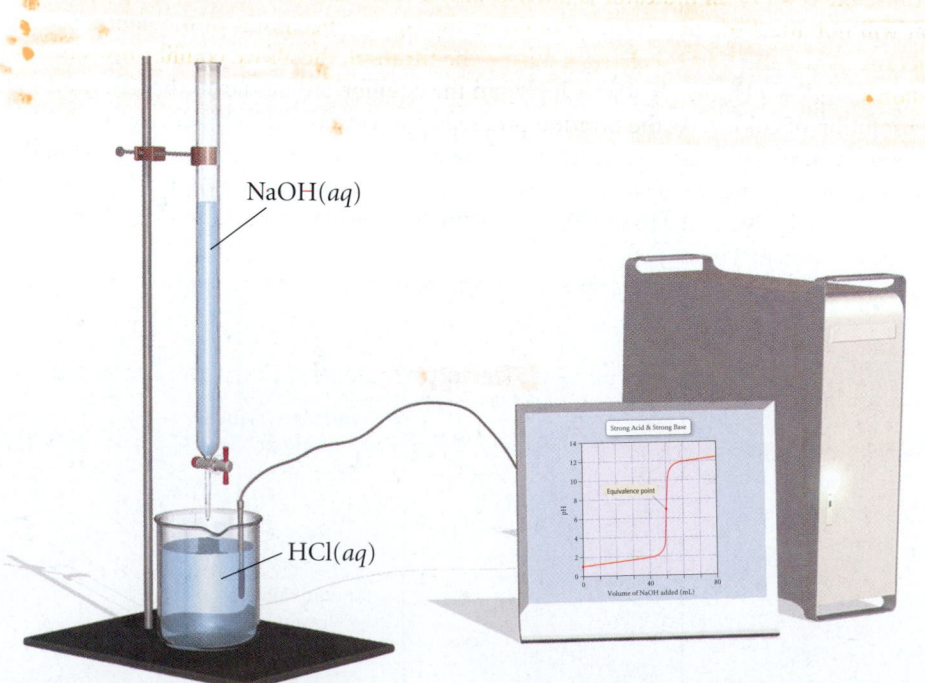

NaOH(aq)

HCl(aq)

▲ FIGURE 16.10 Monitoring the pH During a Titration   A pH meter can be used to
monitor the pH during a titration. The inflection point in the resulting pH curve signifies the
equivalence point in the titration.

Using an Indicator

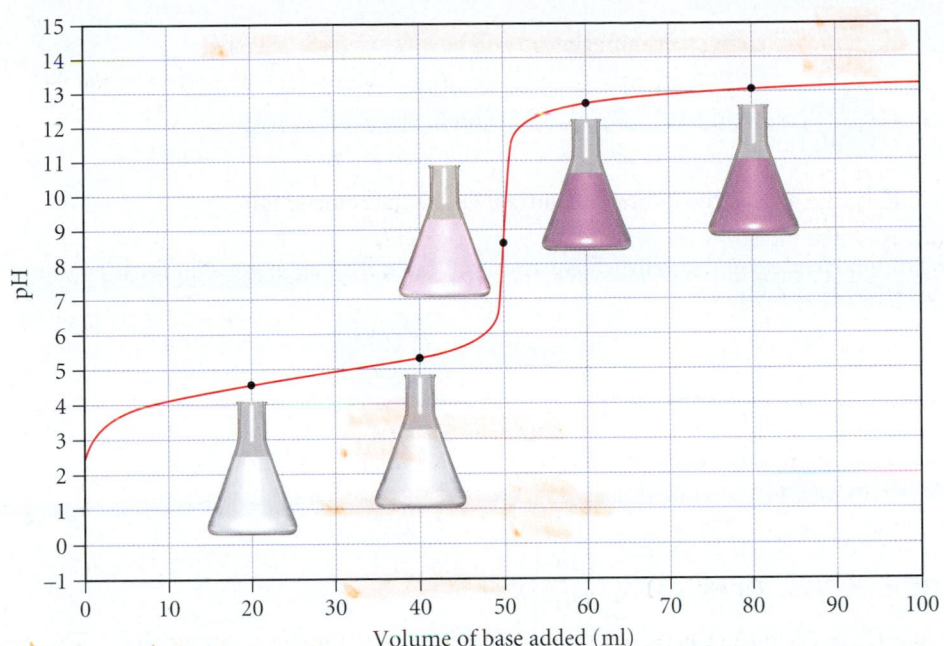

▲ **FIGURE 16.11** **Monitoring the Color Change During a Titration**   Titration of 0.100 M $HC_2H_3O_2$ with 0.100 M NaOH. The endpoint of a titration can be detected by a color change in an appropriate indicator (in this case, phenolphthalein).

If we let HIn represent the acid form of a generic indicator and In^- the conjugate base form, we have the following equilibrium:

$$\underset{\text{color 1}}{HIn(aq)} + H_2O(l) \rightleftharpoons H_3O^+(aq) + \underset{\text{color 2}}{In^-(aq)}$$

Because the color of an indicator is intense, only a small amount is required—an amount that will not affect the pH of the solution or the equivalence point of the neutralization reaction. When the $[H_3O^+]$ changes during the titration, the above equilibrium shifts in response. At low pH, the $[H_3O^+]$ is high and the equilibrium lies far to the left, resulting in a solution of color 1. As the titration proceeds, however, the $[H_3O^+]$ decreases, shifting the equilibrium to the right. Since the pH change is large near the equivalence point of the titration, there is a large change in $[H_3O^+]$ near the equivalence point. Provided that the correct indicator is chosen, there will also be a correspondingly large change in color. For the titration of a strong acid with a strong base, one drop of the base near the endpoint is usually enough to change the color of the indicator from color 1 to color 2.

Phenolphthalein, a Common Indicator

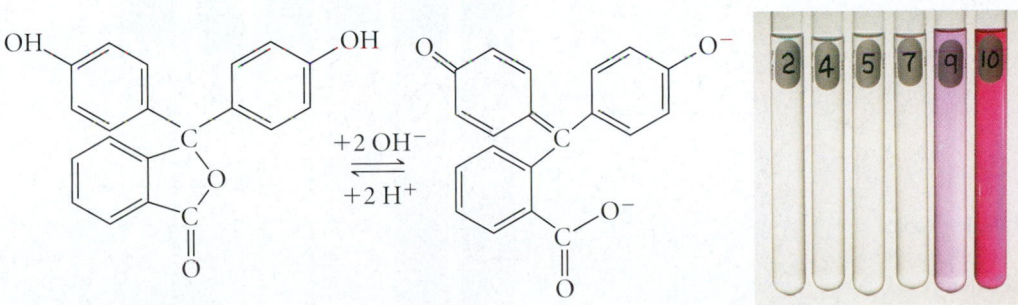

▲ **FIGURE 16.12** **Phenolphthalein**   Phenolphthalein, a weakly acidic compound, is colorless. Its conjugate base is pink.

The color of a solution containing an indicator depends on the relative concentrations of HIn and In^-. As a useful guideline, we can assume the following:

If $\dfrac{[In^-]}{[HIn]} = 1$, the indicator solution will be intermediate in color.

If $\dfrac{[In^-]}{[HIn]} > 10$, the indicator solution will be the color of In^-.

If $\dfrac{[In^-]}{[HIn]} < 0.1$, the indicator solution will be the color of HIn.

From the Henderson–Hasselbalch equation, we can derive an expression for the ratio of $[In^-]/[HIn]$ as follows:

$$pH = pK_a + \log\frac{[base]}{[acid]}$$

$$= pK_a + \log\frac{[In^-]}{[HIn]}$$

$$\log\frac{[In^-]}{[HIn]} = pH - pK_a$$

$$\frac{[In^-]}{[HIn]} = 10^{(pH-pK_a)}$$

Consider the following three pH values relative to pK_a and the corresponding colors of the indicator solution:

pH (relative to pK_a)	$[In^-]/[HIn]$ ratio	Color of Indicator Solution
$pH = pK_a$	$\dfrac{[In^-]}{[HIn]} = 10^0 = 1$	Intermediate color
$pH = pK_a + 1$	$\dfrac{[In^-]}{[HIn]} = 10^1 = 10$	Color of In^-
$pH = pK_a - 1$	$\dfrac{[In^-]}{[HIn]} = 10^{-1} = 0.10$	Color of HIn

When the pH of the solution equals the pK_a of the indicator, the solution will have an intermediate color. When the pH is 1 unit (or more) above pK_a, the indicator will be the color of In, and when the pH is 1 unit (or more) below pK_a, the indicator will be the color of HIn. As you can see, the indicator changes color within a range of two pH units centered at pK_a (Figure 16.13 ▼). Table 16.1 shows various indicators and their colors as a function of pH.

Indicator Color Change: Methyl Red

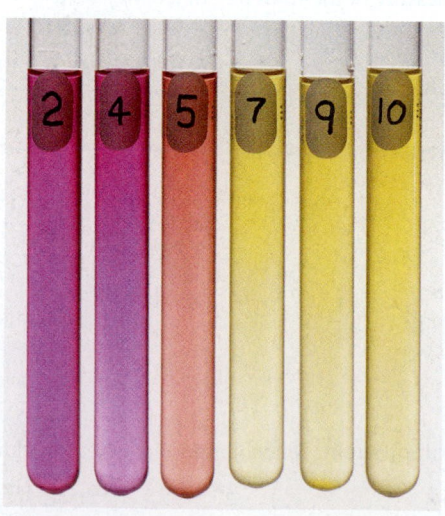

▶ FIGURE 16.13 Indicator Color Change   As this series of photos shows, an indicator (in this case, methyl red) generally changes color within a range of two pH units. (The pH's of the solutions are marked on the test tubes.)

TABLE 16.1 Ranges of Color Changes for Several Acid-Base Indicators

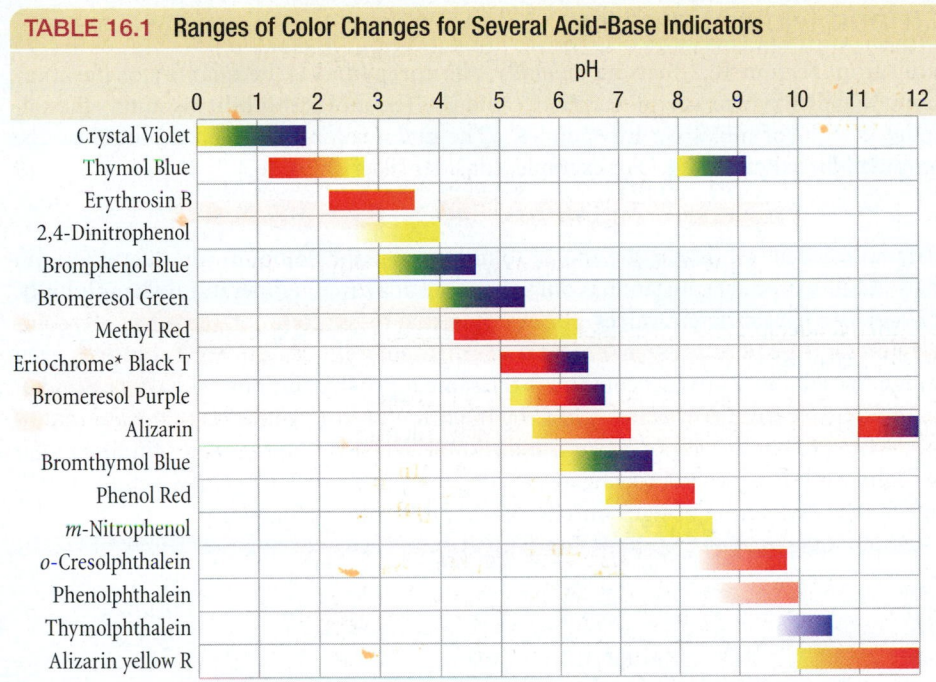

*Trademark of CIBA GEIGY CORP.

16.5 Solubility Equilibria and the Solubility Product Constant

Recall from Chapter 4 that a compound is considered *soluble* if it dissolves in water and *insoluble* if it does not. Recall also that, through the *solubility rules* (see Table 4.1), we classified ionic compounds simply as soluble or insoluble. Now we have the tools to examine *degrees* of solubility.

We can better understand the solubility of an ionic compound by applying the concept of equilibrium to the process of dissolution. For example, we can represent the dissolution of calcium fluoride in water with the following chemical equation:

$$CaF_2(s) \rightleftharpoons Ca^{2+}(aq) + 2\,F^-(aq)$$

The equilibrium expression for a chemical equation representing the dissolution of an ionic compound is called the **solubility product constant (K_{sp})**. For CaF_2, the expression of the solubility product constant is

$$K_{sp} = [Ca^{2+}][F^-]^2$$

Notice that, as we discussed in Section 14.5, solids are omitted from the equilibrium expression because the concentration of a solid is constant (it is determined by its density and does not change upon adding more solid).

The value of K_{sp} is a measure of the solubility of a compound. Table 16.2 lists the values of K_{sp} for a number of ionic compounds. A more complete listing can be found in Appendix IIC.

TABLE 16.2 Selected Solubility Product Constants (K_{sp})

Compound	Formula	K_{sp}	Compound	Formula	K_{sp}
Barium fluoride	BaF_2	2.45×10^{-5}	Lead(II) chloride	$PbCl_2$	1.17×10^{-5}
Barium sulfate	$BaSO_4$	1.07×10^{-10}	Lead(II) bromide	$PbBr_2$	4.67×10^{-6}
Calcium carbonate	$CaCO_3$	4.96×10^{-9}	Lead(II) sulfate	$PbSO_4$	1.82×10^{-8}
Calcium fluoride	CaF_2	1.46×10^{-10}	Lead(II) sulfide	PbS	9.04×10^{-29}
Calcium hydroxide	$Ca(OH)_2$	4.68×10^{-6}	Magnesium carbonate	$MgCO_3$	6.82×10^{-6}
Calcium sulfate	$CaSO_4$	7.10×10^{-5}	Magnesium hydroxide	$Mg(OH)_2$	2.06×10^{-13}
Copper(II) sulfide	CuS	1.27×10^{-36}	Silver chloride	$AgCl$	1.77×10^{-10}
Iron(II) carbonate	$FeCO_3$	3.07×10^{-11}	Silver chromate	Ag_2CrO_4	1.12×10^{-12}
Iron(II) hydroxide	$Fe(OH)_2$	4.87×10^{-17}	Silver bromide	$AgBr$	5.35×10^{-13}
Iron(II) sulfide	FeS	3.72×10^{-19}	Silver iodide	AgI	8.51×10^{-17}

K_{sp} and Molar Solubility

Recall from Section 12.2 that the *solubility* of a compound is the quantity of the compound that dissolves in a certain amount of liquid. The **molar solubility** is simply the solubility in units of moles per liter (mol/L). The molar solubility of a compound can be computed directly from K_{sp}. For example, consider silver chloride:

$$AgCl(s) \rightleftharpoons Ag^+(aq) + Cl^-(aq) \qquad K_{sp} = 1.77 \times 10^{-10}$$

First, notice that K_{sp} is *not* the molar solubility, but the solubility product constant. The solubility product constant has only one value at a given temperature. The solubility, however, can have different values. For example, due to the common ion effect, the solubility of AgCl in pure water is different from its solubility in an NaCl solution, even though the solubility product constant is the same for both solutions. Second, notice that the solubility of AgCl is directly related (by the reaction stoichiometry) to the amount of Ag^+ or Cl^- present in solution when equilibrium is reached. Consequently, finding molar solubility from K_{sp} involves solving an equilibrium problem. For AgCl, we set up an ICE table for the dissolution of AgCl into its ions in pure water as follows:

$$AgCl(s) \rightleftharpoons Ag^+(aq) + Cl^-(aq)$$

	$[Ag^+]$	$[Cl^-]$
Initial	0.00	0.00
Change	$+S$	$+S$
Equil	S	S

Alternatively, the variable x can be used in place of S, as it was for other equilibrium calculations.

We let S represent the concentration of AgCl that dissolves (which is the molar solubility), and then represent the concentrations of the ions formed in terms of S. In this case, for every 1 mol of AgCl that dissolves, 1 mol of Ag^+ and 1 mol of Cl^- are produced. Therefore, the concentrations of Ag^+ or Cl^- present in solution are simply equal to S. Substituting the equilibrium concentrations of Ag^+ and Cl^- into the expression for the solubility product constant, we get

$$\begin{aligned} K_{sp} &= [Ag^+][Cl^-] \\ &= S \times S \\ &= S^2 \end{aligned}$$

Therefore,

$$\begin{aligned} S &= \sqrt{K_{sp}} \\ &= \sqrt{1.77 \times 10^{-10}} \\ &= 1.33 \times 10^{-5}\ M \end{aligned}$$

So the molar solubility of AgCl is 1.33×10^{-5} mol per liter.

EXAMPLE 16.8 Calculating Molar Solubility from K_{sp}

Calculate the molar solubility of $PbCl_2$ in pure water.

Solution

Begin by writing the reaction by which solid $PbCl_2$ dissolves into its constituent aqueous ions and write the corresponding expression for K_{sp}.	$PbCl_2(s) \rightleftharpoons Pb^{2+}(aq) + 2\,Cl^-(aq)$ $K_{sp} = [Pb^{2+}][Cl^-]^2$

Use the stoichiometry of the reaction to prepare an ICE table, showing the equilibrium concentrations of Pb^{2+} and Cl^- relative to S, the amount of $PbCl_2$ that dissolves.

$$PbCl_2(s) \rightleftharpoons Pb^{2+}(aq) + 2\,Cl^-(aq)$$

	$[Pb^+]$	$[Cl^-]$
Initial	0.00	0.00
Change	$+S$	$+2S$
Equil	S	$2S$

Substitute the equilibrium expressions for $[Pb^{2+}]$ and $[Cl^-]$ from the previous step into the expression for K_{sp}.	$\begin{aligned} K_{sp} &= [Pb^{2+}][Cl^-]^2 \\ &= S(2S)^2 \\ &= 4S^3 \end{aligned}$
Solve for S and substitute the numerical value of K_{sp} (from Table 16.2) to compute S.	Therefore $$S = \sqrt[3]{\frac{K_{sp}}{4}}$$ $$S = \sqrt[3]{\frac{1.17 \times 10^{-5}}{4}}$$ $$= 1.43 \times 10^{-2}\,M$$

For Practice 16.8

Calculate the molar solubility of $Fe(OH)_2$ in pure water.

Chemistry in Your Day
Hard Water

In many parts of the United States, the water supply contains significant concentrations of $CaCO_3$ and $MgCO_3$. These salts dissolve into rainwater runoff as it flows through soils rich in $CaCO_3$ and $MgCO_3$, and are washed into lakes and reservoirs. Water containing these compounds is known as hard water. Hard water is not a health hazard because both calcium and magnesium are part of a healthy diet. However, their presence in water can be annoying. Because of the relatively low solubility of these ions water can easily become saturated with $CaCO_3$ and $MgCO_3$. A drop of water, for example, quickly becomes saturated with $CaCO_3$ and $MgCO_3$ as it evaporates. If evaporation continues, it will precipitate out some of its dissolved ions as salts. These precipitates show up as scaly deposits on faucets, sinks, or cookware. Washing cars or dishes with hard water leaves spots of $CaCO_3$ and $MgCO_3$ as these precipitate out of drying drops of water.

Water can be softened with a water softener. These devices replace the Ca^{2+} and Mg^{2+} ions present in hard water with other ions such as K^+ or Na^+. Since potassium and sodium salts are soluble, they do not form scaly deposits in the way that Ca^{2+} and Mg^{2+} ions do. However, when sodium is used to soften drinking water, the resulting water is high in sodium content, a disadvantage to those who must control their sodium intake due to high blood pressure.

▲ The water in reservoirs where the soil is rich in limestone (calcium carbonate) contains Ca^{2+} ions. Water containing dissolved $CaCO_3$ and $MgCO_3$ is called hard water. When hard water evaporates, it can leave deposits of these salts on the shores of lakes and reservoirs (above) and on plumbing fixtures (such as the pipe below).

Question

Use the K_{sp} values from Table 16.2 to calculate the molar solubility of $CaCO_3$ and $MgCO_3$. What mass of $CaCO_3$ (in grams) is in 5 L of water that is saturated with $CaCO_3$?

EXAMPLE 16.9 Calculating K_{sp} from Molar Solubility

The molar solubility of Ag_2SO_4 in pure water is 1.2×10^{-5} M. Calculate K_{sp}.

Solution

Begin by writing the reaction by which solid Ag_2SO_4 dissolves into its constituent aqueous ions and write the corresponding expression for K_{sp}.	$Ag_2SO_4(s) \rightleftharpoons 2\,Ag^+(aq) + SO_4^{2-}(aq)$ $K_{sp} = [Ag^+]^2[SO_4^{2-}]$

Use an ICE table to define $[Ag^+]$ and $[SO_4^{2-}]$ in terms of S, the amount of Ag_2SO_4 that dissolves.

$$Ag_2SO_4(s) \rightleftharpoons 2\,Ag^+(aq) + SO_4^{2-}(aq)$$

	$[Ag^+]$	$[SO_4^{2-}]$
Initial	0.00	0.00
Change	$+2S$	$+S$
Equil	$2S$	S

Substitute the expressions for $[Ag^+]$ and $[SO_4^{2-}]$ from the previous step into the expression for K_{sp}. Substitute the given value *of the molar solubility for S and compute* K_{sp}.

$$K_{sp} = [Ag^+]^2[SO_4^{2-}]$$
$$= (2S)^2 S$$
$$= 4S^3$$
$$= 4(1.2 \times 10^{-5})^3$$
$$= 6.9 \times 10^{-15}$$

For Practice 16.9

The molar solubility of AgBr in pure water is 7.3×10^{-7} M. Calculate K_{sp}.

K_{sp} and Relative Solubility

As we have just seen, molar solubility and K_{sp} are related, and one can be calculated from the other; however, you cannot always use the K_{sp} values of two different compounds directly to compare their relative solubilities. For example, consider the following compounds, their K_{sp} values, and their molar solubilities:

Compound	K_{sp}	Solubility
$Mg(OH)_2$	2.06×10^{-13}	3.72×10^{-5} M
$FeCO_3$	3.07×10^{-11}	5.54×10^{-6} M

Magnesium hydroxide has a smaller K_{sp} than iron(II) carbonate, but a higher molar solubility. Why? The relationship between K_{sp} and molar solubility depends on the stoichiometry of the dissociation reaction. Consequently, any direct comparison of K_{sp} values for different compounds can only be made if the compounds have the same dissociation stoichiometry. For example, consider the following compounds with the same dissociation stoichiometry, their K_{sp} values, and their molar solubilities:

Compound	K_{sp}	Solubility
$Mg(OH)_2$	2.06×10^{-13}	3.72×10^{-5} M
CaF_2	1.46×10^{-10}	3.32×10^{-4} M

In this case, magnesium hydroxide and calcium fluoride have the same dissociation stoichiometry (1 mol of each compound produces 3 mol of dissolved ions); therefore, the K_{sp} values can be directly compared as a measure of relative solubility.

The Effect of a Common Ion on Solubility

How is the solubility of an ionic compound affected when the compound is dissolved in a solution that already contains one of its ions? For example, what is the solubility of CaF_2 in a solution that is 0.100 M in NaF? The change in solubility can be explained by the common ion effect, which we first encountered in Section 16.2. We can represent the dissociation of CaF_2 in a 0.100 M NaF solution as follows:

$$\text{Common ion}$$
$$0.100 \text{ M } F^-(aq)$$

$$CaF_2(s) \rightleftharpoons Ca^{2+}(aq) + 2\,F^-(aq)$$

Equilibrium shifts left

In accordance with Le Châtelier's principle, the presence of the F^- ion in solution causes the equilibrium to shift to the left (compared to its position in pure water), which means that less CaF_2 dissolves—that is, its solubility is decreased. In general,

The solubility of an ionic compound is lower in a solution containing a common ion than in pure water.

We can calculate the exact value of the solubility by working an equilibrium problem in which the concentration of the common ion is accounted for in the initial conditions, as shown in the following example.

EXAMPLE 16.10 Calculating Molar Solubility in the Presence of a Common Ion

What is the molar solubility of CaF_2 in a solution containing 0.100 M NaF?

Solution

Begin by writing the reaction by which solid CaF_2 dissolves into its constituent aqueous ions and write the corresponding expression for K_{sp}.	$CaF_2(s) \rightleftharpoons Ca^{2+}(aq) + 2\,F^-(aq)$ $K_{sp} = [Ca^{2+}][F^-]^2$

Use the stoichiometry of the reaction to prepare an ICE table showing the initial concentration of the common ion. Fill in the equilibrium concentrations of Ca^{2+} and F^- relative to S, the amount of CaF_2 that dissolves.

$$CaF_2(s) \rightleftharpoons Ca^{2+}(aq) + 2\,F^-(aq)$$

	$[Ca^{2+}]$	$[F^-]$
Initial	0.00	0.100
Change	$+S$	$+2S$
Equil	S	$0.100 + 2S$

Substitute the equilibrium expressions for $[Ca^{2+}]$ and $[F^-]$ from the previous step into the expression for K_{sp}. Since K_{sp} is small, we can make the approximation that $2S$ is much less than 0.100 and will therefore be insignificant when added to 0.100 (this is similar to the *x is small* approximation that we have made for many equilibrium problems).

$$K_{sp} = [Ca^{2+}][F^-]^2$$
$$= S(0.100 + 2S)^2 \quad (S \text{ is small})$$
$$= S(0.100)^2$$

Solve for S and substitute the numerical value of K_{sp} (from Table 16.2) to compute S.

$$K_{sp} = S(0.100)^2$$
$$S = \frac{K_{sp}}{0.0100}$$
$$= \frac{1.46 \times 10^{-10}}{0.0100}$$
$$= 1.46 \times 10^{-8} \text{ M}$$

Note that the computed value of S is indeed small compared to 0.100, so our approximation is valid.

For comparison, the molar solubility of CaF_2 in pure water is 3.32×10^{-4} M, which means CaF_2 is over 20,000 times more soluble in water than in the NaF solution. (Confirm this for yourself by calculating the solubility in pure water from the value of K_{sp}.)

For Practice 16.10

Calculate the molar solubility of CaF_2 in a solution containing 0.250 M $Ca(NO_3)_2$.

The Effect of pH on Solubility

The pH of a solution can affect the solubility of a compound in that solution. For example, consider the dissociation of $Mg(OH)_2$, the active ingredient in milk of magnesia:

$$Mg(OH)_2(s) \rightleftharpoons Mg^{2+}(aq) + 2\,OH^-(aq)$$

The solubility of this compound is highly dependent on the pH of the solution into which it dissolves. If the pH is high, then the concentration of OH^- is high. In accordance with the common ion effect, this shifts the equilibrium to the left, lowering the solubility.

High $[OH^-]$

$$Mg(OH)_2(s) \rightleftharpoons Mg^{2+}(aq) + 2\,OH^-(aq)$$

Equilibrium shifts left

If the pH is low, then the concentration of $H_3O^+(aq)$ is high. As the $Mg(OH)_2$ dissolves, these H_3O^+ ions neutralize the newly dissolved OH^- ions, driving the reaction to the right.

H_3O^+ reacts with OH^-

$$Mg(OH)_2(s) \rightleftharpoons Mg^{2+}(aq) + 2\,OH^-(aq)$$

Equilibrium shifts right

Consequently, the solubility of $Mg(OH)_2$ in an acidic solution is higher than in a pH-neutral solution. In general,

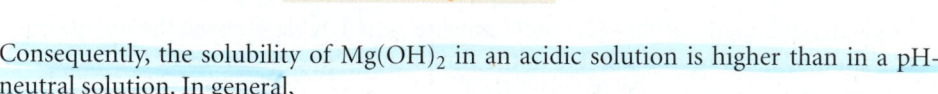

The solubility of an ionic compound with a strongly basic or weakly basic anion increases with increasing acidity (decreasing pH).

Common basic anions include OH^-, S^{2-}, and CO_3^{2-}. Therefore, hydroxides, sulfides, and carbonates are more soluble in acidic water than in pure water. Since rainwater is naturally acidic due to dissolved carbon dioxide, it can dissolve rocks high in limestone ($CaCO_3$) as it flows through the ground, sometimes resulting in huge underground caverns such as those at Carlsbad Caverns National Park in New Mexico. Dripping water saturated in $CaCO_3$ within the cave creates the dramatic mineral formations known as stalagmites and stalactites.

▲ Stalactites (which hang from the ceiling) and stalagmites (which grow up from the ground) are formed as calcium carbonate precipitates out of water evaporating in underground caves.

EXAMPLE 16.11 The Effect of pH on Solubility

Determine whether each of the following compounds will be more soluble in an acidic solution than in a neutral solution.

(a) BaF_2 (b) AgI   (c) $Ca(OH)_2$

Solution

(a) The solubility of BaF_2 will be greater in acidic solution because the F^- ion is a weak base. (F^- is the conjugate base of the weak acid HF, and is therefore a weak base.)

(b) The solubility of AgI will not be greater in acidic solution because the I^- is *not* a base. (I^- is the conjugate base of the *strong* acid HI, and is therefore pH-neutral.)

(c) The solubility of $Ca(OH)_2$ will be greater in acidic solution because the OH^- ion is a strong base.

For Practice 16.11

Which compound, $FeCO_3$ or $PbBr_2$, is more soluble in acid than in base? Why?

16.6 Precipitation

In Chapter 4, we learned that a precipitation reaction can occur upon the mixing of two solutions containing ionic compounds. The precipitation reaction occurs when one of the possible cross products—the combination of a cation from one solution and the anion from the other—is insoluble. As we have seen, however, the terms soluble and insoluble are extremes in a continuous range of solubility—many compounds are slightly soluble and even those that we categorized as insoluble in Chapter 4 actually have some limited degree of solubility (they have very small solubility product constants).

We can better understand precipitation reactions by revisiting a concept from Chapter 14—the reaction quotient (Q). The reaction quotient for the reaction by which an ionic compound dissolves is simply the product of the concentrations of the ionic components raised to their stoichiometric coefficients. For example, consider the reaction by which CaF_2 dissolves:

Na$_2$CrO$_4$
AgNO$_3$
Ag$_2$CrO$_4$

$$CaF_2(s) \rightleftharpoons Ca^{2+}(aq) + 2\,F^-(aq)$$

The reaction quotient for this reaction is

$$Q = [Ca^{2+}][F^-]^2$$

The difference between Q and K_{sp} is that K_{sp} is the value of this product *at equilibrium only*, whereas Q is the value of the product under any conditions. We can therefore use the value of Q to compare a solution containing any concentrations of the component ions to one that is at equilibrium.

For example, consider a solution of calcium fluoride in which Q is less than K_{sp}. Recall from Chapter 14 that if Q is less than K_{sp}, the reaction will proceed to the right (toward products). Consequently, if the solution contains any solid CaF_2, the CaF_2 will continue to dissolve. If all of the solid has already dissolved, the solution will simply remain as it is, containing less than the equilibrium amount of the dissolved ions. Such a solution is called an *unsaturated solution*. If more solid is added to an unsaturated solution, it will dissolve, as long as Q remains less than K_{sp}.

Now consider a solution in which Q is exactly equal to K_{sp}. In this case, the reaction is at equilibrium and will not proceed in either direction. Such a solution most likely contains at least a small amount of the solid in equilibrium with its component ions. However, the amount of solid may be too small to be visible. Such a solution is called a *saturated solution*.

Finally, consider a solution in which Q is greater than K_{sp}. In this case, the reaction will proceed to the left (toward the reactants) and solid calcium fluoride will form from the dissolved calcium and fluoride ions. In other words, the solid normally precipitates out of a solution in which Q is greater than K_{sp}. Under certain circumstances, however, Q can remain greater than K_{sp} for an unlimited period of time. Such a solution, called a

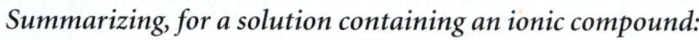

Seed crystal

Supersaturated solution of sodium acetate

Solid sodium acetate forming

► **FIGURE 16.14 Precipitation from a Supersaturated Solution** The excess solute in a supersaturated solution of sodium acetate will precipitate out if a small sodium acetate crystal is added.

supersaturated solution, is unstable and will form a precipitate when sufficiently disturbed. For example, Figure 16.14 ▲ shows a supersaturated solution of sodium acetate. When a small speck of solid sodium acetate is dropped into the solution, it triggers the precipitation reaction.

Summarizing, for a solution containing an ionic compound:

➤ If $Q < K_{sp}$, the solution is unsaturated. More of the solid ionic compound can dissolve in the solution.

➤ If $Q = K_{sp}$, the solution is saturated. The solution is holding the equilibrium amount of the dissolved ions and additional solid will not dissolve in the solution.

➤ If $Q > K_{sp}$, the solution is supersaturated. Under most circumstances, the excess solid will precipitate out of a supersaturated solution.

We can use Q to predict whether a precipitation reaction will occur upon the mixing of two solutions containing dissolved ionic compounds. For example, consider mixing a silver nitrate solution with a potassium iodide solution to form a mixture that is 0.010 M in $AgNO_3$ and 0.015 M in KI. Will a precipitate form in the newly mixed solution? From Chapter 4 we know that one of the cross products, KNO_3, is soluble and will therefore not precipitate. The other cross product, AgI, *may* precipitate if the concentrations of Ag^+ and I^- are high enough in the newly mixed solution: We simply compare Q to K_{sp} to find out. For AgI, $K_{sp} = 8.51 \times 10^{-17}$. For the newly mixed solution, $[Ag^+] = 0.010$ M and $[I^-] = 0.015$ M. We compute Q as follows:

$$Q = [Ag^+][I^-] = (0.010)(0.015) = 1.5 \times 10^{-4}$$

The value of Q is much greater than K_{sp}; therefore, AgI should precipitate out of the newly mixed solution.

EXAMPLE 16.12 Predicting Precipitation Reactions by Comparing Q and K_{sp}

A solution containing lead(II) nitrate is mixed with one containing sodium bromide to form a solution that is 0.0150 M in $Pb(NO_3)_2$ and 0.00350 M in NaBr. Will a precipitate form in the newly mixed solution?

Solution

First, determine the possible cross products and their K_{sp} values (Table 16.2). Any cross products that are soluble will *not* precipitate (see Table 4.1)	Possible cross products: $NaNO_3$ soluble $PbBr_2$ $K_{sp} = 4.67 \times 10^{-6}$

Compute Q and compare it to K_{sp}. A precipitate will only form if $Q > K_{sp}$.	$Q = [Pb^{2+}][Br^-]^2$ $= (0.0150)(0.00350)^2$ $= 1.84 \times 10^{-7}$ $Q < K_{sp}$; therefore no precipitate forms.

For Practice 16.12

The original solutions in Example 16.12 are concentrated through evaporation and then mixed again to form a solution that is now 0.0600 M in $Pb(NO_3)_2$ and 0.0158 M in NaBr. Will a precipitate form in this newly mixed solution?

Selective Precipitation

A solution may contain several different dissolved metal cations which can often be separated by **selective precipitation**, a process involving the addition of a reagent that forms a precipitate with one of the dissolved cations but not the others. For example, seawater contains dissolved magnesium and calcium cations with the concentrations $[Mg^{2+}] = 0.059$ M and $[Ca^{2+}] = 0.011$ M. We could separate these ions by adding a reagent that will precipitate one of the ions but not the other. The appropriate reagent must combine with both metal cations to form compounds having sufficiently different K_{sp} values. From Table 16.2, we find that $Mg(OH)_2$ has a K_{sp} of 2.06×10^{-13} and that $Ca(OH)_2$ has a K_{sp} of 4.68×10^{-6}. Consequently, a soluble hydroxide—such as KOH or NaOH—would make a good choice for the precipitating reagent. When an appropriate amount of KOH or NaOH is added to seawater, the hydroxide ion will cause the precipitation of $Mg(OH)_2$ (the compound with the lowest K_{sp}) but not $Ca(OH)_2$. Calculations associated with this selective precipitation are shown in the following examples. The main principle in these calculations is the comparison of Q to K_{sp} to determine the concentration that begins to trigger precipitation.

> The difference in K_{sp} values required for selective precipitation is a factor of at least 10^3.

EXAMPLE 16.13 Finding the Minimum Required Reagent Concentration for Selective Precipitation

As just discussed, the magnesium and calcium ions present in seawater ($[Mg^{2+}] = 0.059$ M and $[Ca^{2+}] = 0.011$ M) can be separated by selective precipitation with KOH. What minimum $[OH^-]$ is necessary to just begin to precipitate the Mg^{2+} ion?

Solution

The precipitation will just begin to occur when the value of Q for the precipitating compound just equals the value of K_{sp}. Set the expression for Q for magnesium hydroxide equal to the value of K_{sp}, and solve for $[OH^-]$. This will be the concentration above which $Mg(OH)_2$ just begins to precipitate.	$Q = [Mg^{2+}][OH^-]^2$ $= (0.059)[OH^-]^2$ When $Q = K_{sp}$, $(0.059)[OH^-]^2 = K_{sp} = 2.06 \times 10^{-13}$ $[OH^-]^2 = \dfrac{2.06 \times 10^{-13}}{0.059}$ $[OH^-] = 1.9 \times 10^{-6}$ M

For Practice 16.13

Suppose that the concentration of Mg^{2+} in the above solution was 0.025 M. What minimum $[OH^-]$ is necessary in this solution to just begin to precipitate the Mg^{2+} ion?

EXAMPLE 16.14 Finding the Concentrations of Ions Left in Solution After Selective Precipitation

Suppose we add potassium hydroxide to the solution in Example 16.13. When the $[OH^-]$ reaches 1.9×10^{-6} M (as we just calculated), magnesium hydroxide begins to precipitate out of solution. As we continue to add KOH, the magnesium hydroxide continues to precipitate. However, at some point, the $[OH^-]$ becomes high enough to begin to precipitate the calcium ions as well. What is the concentration of Mg^{2+} when Ca^{2+} just begins to precipitate?

Solution

First, find the OH^- concentration at which Ca^{2+} just begins to precipitate by writing the expression for Q for calcium hydroxide and substituting the concentration of Ca^{2+} from Example 16.13.	$Q = [Ca^{2+}][OH^-]^2$ $= (0.011)[OH^-]^2$
Then set the expression for Q equal to the value of K_{sp} for calcium hydroxide and solve for $[OH^-]$. This will be the concentration above which $Ca(OH)_2$ just begins to precipitate.	When $Q = K_{sp}$, $(0.011)[OH^-]^2 = K_{sp} = 4.68 \times 10^{-6}$ $[OH^-]^2 = \dfrac{4.68 \times 10^{-6}}{0.011}$ $[OH^-] = 2.06 \times 10^{-2}$ M
Next, find the concentration of Mg^{2+} when OH^- reaches the concentration you just calculated by writing the expression for Q *for magnesium hydroxide* and substituting the concentration of OH^- that you just calculated. Then set the expression for Q equal to the value of K_{sp} for magnesium hydroxide and solve for $[Mg^{2+}]$. This will be the concentration of Mg^{2+} that remains when $Ca(OH)_2$ just begins to precipitate.	$Q = [Mg^{2+}][OH^-]^2$ $= [Mg^{2+}](2.06 \times 10^{-2})^2$ When $Q = K_{sp}$, $[Mg^{2+}](2.06 \times 10^{-2})^2 = K_{sp} = 2.06 \times 10^{-13}$ $[Mg^{2+}] = \dfrac{2.06 \times 10^{-13}}{(2.06 \times 10^{-2})^2}$ $[Mg^{2+}] = 4.9 \times 10^{-10}$ M

As you can see from the results, the selective precipitation worked very well. The concentration of Mg^{2+} dropped from 0.059 M to 4.8×10^{-10} M before any calcium began to precipitate, which means that we separated 99.99% of the magnesium out of the solution.

For Practice 16.14

A solution is 0.085 M in Pb^{2+} and 0.025 M in Ag^+. (a) If selective precipitation is to be achieved using NaCl, what minimum concentration of NaCl is required to just begin to precipitate the ion that precipitates first? (b) What are the concentrations of both ions left in solution at the point where the second ion begins to precipitate?

16.7 Qualitative Chemical Analysis

Selective precipitation as discussed in Section 16.6 can be used in a systematic way to determine the metal ions present in an unknown solution. Such an analysis is known as **qualitative analysis**. The word *qualitative* means *involving quality or kind*. So qualitative analysis involves finding the *kind* of ions present in the solution. This stands in contrast to **quantitative analysis,** which is concerned with quantity, or the amounts of substances in a solution or mixture.

In the past, qualitative analysis by selective precipitation was used extensively to determine the metals present in a sample. This method of analysis—often dubbed *wet chemistry* because it involved the mixing of many aqueous solutions in the lab—has been replaced by more precise and less time-intensive instrumental techniques. Nonetheless, both for the sake of history and also because of the principles involved, we now examine a traditional qualitative analysis scheme. You may use such a scheme in your general chemistry laboratory as an exercise in qualitative analysis.

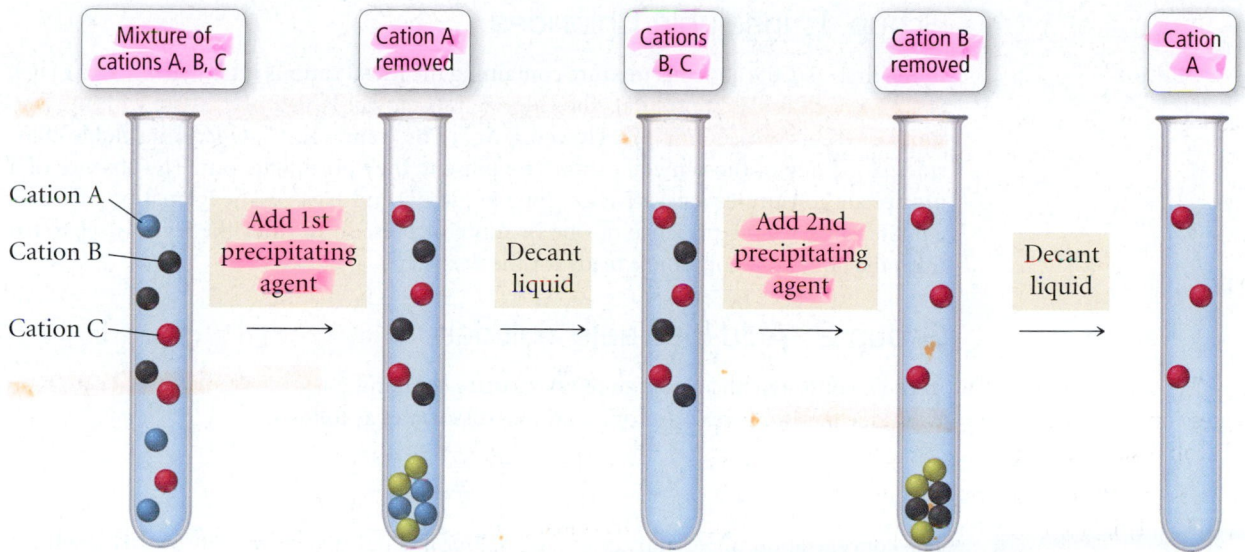

The basic idea behind qualitative analysis is simple. A sample containing a mixture of metal cations is subjected to the addition of several precipitating agents. At each step, some of the metal cations—those that form insoluble compounds with the precipitating agent—precipitate from the mixture and are separated out as solids. The remaining aqueous mixture is then subjected to the next precipitating agent, and so on (Figure 16.15 ▲).

A general qualitative analysis scheme is diagrammed in Figure 16.16 ▼. The scheme involves separating a mixture of the common ions into five groups by sequentially adding five different precipitating agents. After each precipitating agent is added, the mixture is put into a centrifuge to separate the solid from the liquid. The liquid is decanted for the next step, and the solid is set aside for subsequent analysis. We examine each group separately.

▲ **FIGURE 16.15 Qualitative Analysis** In qualitative analysis, specific ions are precipitated successively by the addition of appropriate reagents.

▼ **FIGURE 16.16** A General Qualitative Analysis Scheme

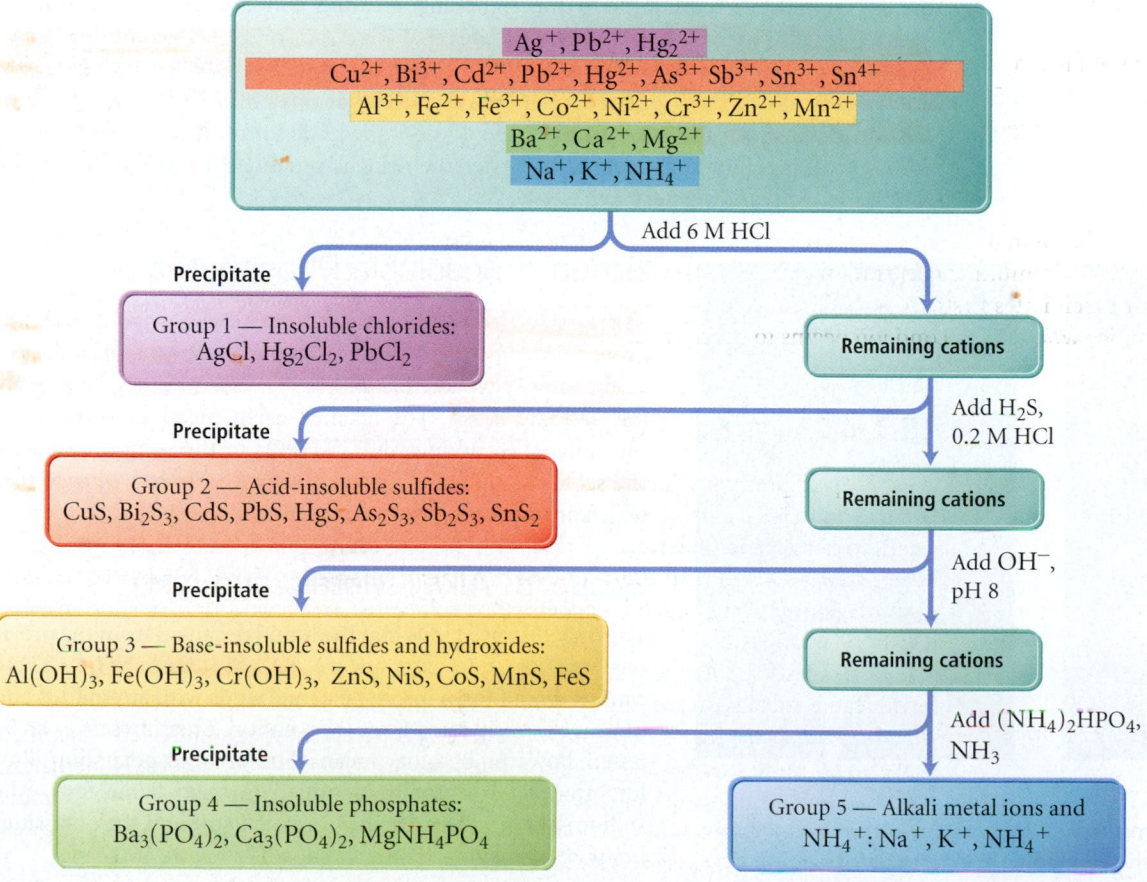

Group 1: Insoluble Chlorides

In the first step, the aqueous mixture containing the metal cations is treated with 6 M HCl. Since most chlorides are soluble, the chloride ions *do not form* a precipitate with the majority of the cations in mixture. However, Ag^+, Pb^{2+}, and Hg_2^{2+} *do form* insoluble chlorides. So, if any of those metal cations are present, they precipitate out. The absence of a precipitate constitutes a negative test for Ag^+, Pb^{2+}, and Hg_2^{2+}. The formation of a precipitate indicates the presence of one or more of these ions. After the solid is separated from the liquid, the solution is ready for the next step.

Group 2: Acid-Insoluble Sulfides

In the second step, the acidic aqueous mixture containing the remaining metal cations is treated with H_2S, a weak diprotic acid that dissociates as follows:

$$H_2S \rightleftharpoons H^+ + HS^-$$
$$HS^- \rightleftharpoons H^+ + S^{2-}$$

The concentration of S^{2-} ions in an H_2S solution is pH-dependent. At low pH (high H^+ concentration) the equilibria shift left, minimizing the amount of available S^{2-}. At high pH (low H^+ concentration) the equilibria shift right, maximizing the amount of available S^{2-}. At this stage, the solution is acidic (from the addition of HCl in the previous step), and the concentration of S^{2-} in solution is therefore relatively low. Only the most insoluble metal sulfides (those with the smallest K_{sp} values) will precipitate under these conditions. These include Hg^{2+}, Cd^{2+}, Bi^{3+}, Cu^{2+}, Sn^{4+}, As^{3+}, Pb^{2+}, and Sb^{3+}. If any of these metal cations are present, they precipitate out as sulfides. After the solid is separated from the liquid, the solution is ready for the next step.

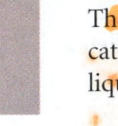

$(NH_4)_2S$ CdS Sb_2S_3 PbS

Group 3: Base-Insoluble Sulfides and Hydroxides

In the third step, the acidic aqueous mixture containing the remaining metal cations is treated with base and additional H_2S is added. The added base reacts with acid, shifting the H_2S ionization equilibria to the right and creating a higher S^{2-} concentration. This causes the precipitation of those sulfides that were too soluble to precipitate out in the previous step but not soluble enough to prevent precipitation with the higher sulfide ion concentration. The ions that precipitate as sulfides at this point (if they are present) are Fe^{2+}, Co^{2+}, Zn^{2+}, Mn^{2+}, and Ni^{2+}. In addition, the basic solution causes Cr^{3+}, Fe^{3+}, and Al^{3+} to precipitate as hydroxides. After the solid is separated from the liquid, the solution is ready for the next step.

Group 4: Insoluble Phosphates

At this stage, all of the cations have been precipitated except those belonging to the alkali metal family (group 1A in the periodic table) and the alkaline earth metal family (group 2A in the periodic table). The alkaline earth metal cations can be precipitated by adding $(NH_4)_2HPO_4$ to the solution, causing Mg^{2+}, Ca^{2+}, and Ba^{2+} to precipitate as metal phosphates, which are separated from the liquid.

Group 5: Alkali Metals and NH_4^+

The only dissolved ions that the liquid decanted from the previous step can now contain are Na^+, K^+, and NH_4^+. These cations do not form insoluble compounds with any anions and cannot be precipitated from the solution. Their presence can be tested, however, by other means. Sodium and potassium ions, for example, are usually identified through flame tests. The sodium ion produces a yellow-orange flame and the potassium ion produces a violet flame, as shown in Figure 16.17 ◄.

Sodium   Potassium

▲ **FIGURE 16.17 Flame Tests**   The sodium ion produces a yellow-orange flame. The potassium ion produces a violet flame.

By applying the above procedure, nearly two dozen metal cations can be separated from a solution initially containing all of them. Each of the groups can then be further analyzed to determine the specific ions present from that group. The specific procedures for these steps can be found in many general chemistry laboratory manuals.

16.8 Complex Ion Equilibria

We have learned about several different types of equilibria so far, including acid–base equilibria and solubility equilibria. We now turn to equilibria of another type, which involve primarily transition metal ions in solution. Transition metal ions tend to be good electron acceptors (or good Lewis acids). In aqueous solutions, water molecules can act as electron donors (or Lewis bases) to hydrate transition metal ions. For example, silver ions are hydrated by water in solution to form $Ag(H_2O)_2^+(aq)$. Chemists will often write $Ag^+(aq)$. as a shorthand notation for the hydrated silver ion, but the bare ion does not really exist by itself in solution.

Species such as $Ag(H_2O)_2^{2+}$ are known as *complex ions*. A **complex ion** contains a central metal ion bound to one or more *ligands*. A **ligand** is a neutral molecule or ion that acts as a Lewis base with the central metal ion. In $Ag(H_2O)_2^+$, water is acting as the ligand. If a stronger Lewis base is put into a solution containing $Ag(H_2O)_2^+$, the stronger Lewis base will displace the water in the complex ion. For example, ammonia reacts with $Ag(H_2O)_2^+$ according to the following reaction:

$$Ag(H_2O)_2^+(aq) + 2\,NH_3(aq) \rightleftharpoons Ag(NH_3)_2^+(aq) + 2\,H_2O(l)$$

For simplicity, water is often left out of the above equation and the reaction is written as follows:

$$Ag^+(aq) + 2\,NH_3(aq) \rightleftharpoons Ag(NH_3)_2^+(aq) \quad K_f = 1.7 \times 10^7$$

The equilibrium constant associated with the reaction for the formation of a complex ion, such as the one shown above, is called the **formation constant** (K_f). The expression for K_f is determined by the law of mass action, just as for any equilibrium constant. For $Ag(NH_3)_2^+$, the expression for K_f is written as follows:

$$K_f = \frac{[Ag(NH_3)_2^+]}{[Ag^+][NH_3]^2}$$

Notice that the value of K_f for $Ag(NH_3)_2^+$ is large, indicating that the formation of the complex ion is highly favored. Table 16.3 lists the formation constants for a number of common complex ions. You can see that, in general, the values of K_f are very large, indicating that the formation of complex ions is highly favored in each case. The following example shows how to use K_f in calculations.

We cover complex ions more thoroughly in Chapter 24. Here, we focus on the equilibria associated with their formation.

TABLE 16.3 Formation Constants of Selected Complex Ions in Water at 25 °C

Complex Ion	K_f	Complex Ion	K_f
$Ag(CN)_2^-$	1×10^{21}	$Cu(NH_3)_4^{2+}$	1.7×10^{13}
$Ag(NH_3)_2^+$	1.7×10^7	$Fe(CN)_6^{4-}$	1.5×10^{35}
$Ag(S_2O_3)_2^{3-}$	3.8×10^{13}	$Fe(CN)_6^{3-}$	2×10^{43}
AlF_6^{3-}	7×10^{19}	$Hg(CN)_4^{2-}$	1.8×10^{41}
$Al(OH)_4^-$	3×10^{33}	$HgCl_4^{2-}$	1.1×10^{16}
$CdBr_4^{2-}$	5.5×10^3	HgI_4^{2-}	2×10^{30}
CdI_4^{2-}	2×10^6	$Ni(NH_3)_6^{2+}$	2.0×10^8
$Cd(CN)_4^{2-}$	3×10^{18}	$Pb(OH)_3^-$	8×10^{13}
$Co(NH_3)_6^{3+}$	2.3×10^{33}	$Sn(OH)_3^-$	3×10^{25}
$Co(OH)_4^{2-}$	5×10^9	$Zn(CN)_4^{2-}$	2.1×10^{19}
$Co(SCN)_4^{2-}$	1×10^3	$Zn(NH_3)_4^{2+}$	2.8×10^9
$Cr(OH)_4^-$	8.0×10^{29}	$Zn(OH)_4^{2-}$	3×10^{15}
$Cu(CN)_4^{2-}$	1.0×10^{25}		

EXAMPLE 16.15   Complex Ion Equilibria

A 200.0-mL sample of a solution that is 1.5×10^{-3} M in $Cu(NO_3)_2$ is mixed with a 250.0-mL sample of a solution that is 0.20 M in NH_3. After the solution reaches equilibrium, what concentration of $Cu^{2+}(aq)$ remains?

Solution

<table>
<tr>
<td>Write the balanced equation for the complex ion equilibrium that occurs and look up the value of K_f in Table 16.3. Since this is an equilibrium problem, we will have to create an ICE table, which requires the initial concentrations of Cu^{2+} and NH_3. Calculate those concentrations from the given values.</td>
<td>

$Cu^{2+}(aq) + 4\,NH_3(aq) \rightleftharpoons Cu(NH_3)_4{}^{2+}(aq)$

$K_f = 1.7 \times 10^{13}$

$[Cu^{2+}]_{initial} = \dfrac{0.200\ \text{L} \times \dfrac{1.5 \times 10^{-3}\ \text{mol}}{1\ \text{L}}}{0.200\ \text{L} + 0.250\ \text{L}} = 6.7 \times 10^{-4}\ \text{M}$

$[NH_3]_{initial} = \dfrac{0.250\ \text{L} \times \dfrac{0.20\ \text{mol}}{1\ \text{L}}}{0.200\ \text{L} + 0.250\ \text{L}} = 0.11\ \text{M}$

</td>
</tr>
<tr>
<td>Construct an ICE table for the reaction and write down the initial concentrations of each species.</td>
<td>

$Cu^{2+}(aq) + 4\,NH_3(aq) \rightleftharpoons Cu(NH_3)_4{}^{2+}(aq)$

	$[Cu^{2+}]$	$[NH_3]$	$[Cu(NH_3)_4{}^{2+}]$
Initial	6.7×10^{-4}	0.11	0.0
Change			
Equil			

</td>
</tr>
<tr>
<td>Since the equilibrium constant is so large, and since the concentration of ammonia is much larger than the concentration of Cu^{2+}, we can assume that the reaction will be driven to the right so that most of the Cu^{2+} is consumed. Unlike previous ICE tables, where we let x represent the change in concentration in going to equilibrium, here we let x represent the small amount of Cu^{2+} that remains when equilibrium is reached.</td>
<td>

$Cu^{2+}(aq) + 4\,NH_3(aq) \rightleftharpoons Cu(NH_3)_4{}^{2+}(aq)$

	$[Cu^{2+}]$	$[NH_3]$	$[Cu(NH_3)_4{}^{2+}]$
Initial	6.7×10^{-4}	0.11	0.0
Change	$\approx(-6.7 \times 10^{-4})$	$\approx 4(-6.7 \times 10^{-4})$	$\approx(+6.7 \times 10^{-4})$
Equil	x	0.11	6.7×10^{-4}

</td>
</tr>
<tr>
<td>Substitute the expressions for the equilibrium concentrations into the expression for K_f and solve for x.</td>
<td>

$K_f = \dfrac{[Cu(NH_3)_6{}^{2+}]}{[Cu^{2+}][NH_3]^4}$

$= \dfrac{6.7 \times 10^{-4}}{x(0.11)^4}$

$x = \dfrac{6.7 \times 10^{-4}}{K_f(0.11)^4}$

$= \dfrac{6.7 \times 10^{-4}}{1.7 \times 10^{13}(0.11)^4}$

$= 2.7 \times 10^{-13}$

</td>
</tr>
<tr>
<td>Confirm that x is indeed small compared to the initial concentration of the metal cation.
The remaining Cu^{2+} is very small because the formation constant is very large.</td>
<td>

Since $x = 2.7 \times 10^{-13} \ll 6.7 \times 10^{-4}$; the approximation is valid. The remaining $[Cu^{2+}] = 2.7 \times 10^{-13}$ M.

</td>
</tr>
</table>

For Practice 16.15

A 125.0-mL sample of a solution that is 0.0117 M in $NiCl_2$ is mixed with a 175.0-mL sample of a solution that is 0.250 M in NH_3. After the solution reaches equilibrium, what concentration of $Ni^{2+}(aq)$ remains?

The Effect of Complex Ion Equilibria on Solubility

Recall from Section 16.5 that the solubility of an ionic compound with a strongly basic or weakly basic anion increases with increasing acidity because the acid reacts with the anion and drives the reaction to the right. Similarly, *the solubility of an ionic compound containing a metal cation that forms complex ions increases in the presence of Lewis bases that complex with the cation.* The most common Lewis bases that increase the solubility of metal cations are NH_3, CN^-, and OH^-. For example, silver chloride is only slightly soluble in pure water:

$$AgCl(s) \rightleftharpoons Ag^+(aq) + Cl^-(aq) \qquad K_{sp} = 1.77 \times 10^{-10}$$

However, adding ammonia increases its solubility dramatically because, as we saw previously in this section, the ammonia forms a complex ion with the silver cations:

$$Ag^+(aq) + 2\,NH_3(aq) \rightleftharpoons Ag(NH_3)_2{}^+(aq) \qquad K_f = 1.7 \times 10^7$$

The large value of K_f dramatically lowers the concentration of $Ag^+(aq)$ in solution and therefore drives the dissolution of $AgCl(s)$. The two above reactions can be added together as follows:

$$AgCl(s) \rightleftharpoons Ag^+(aq) + Cl^-(aq) \qquad K_{sp} = 1.77 \times 10^{-10}$$
$$Ag^+(aq) + 2\,NH_3(aq) \rightleftharpoons Ag(NH_3)_2{}^+(aq) \qquad K_f = 1.7 \times 10^7$$

$$AgCl(s) + 2\,NH_3(aq) \rightleftharpoons Ag(NH_3)_2{}^+(aq) + Cl^-(aq) \qquad K = K_{sp} \times K_f = 3.0 \times 10^{-3}$$

As we learned in Section 14.3, the equilibrium constant for a reaction that is the sum of two other reactions is simply the product of the equilibrium constants for the two other reactions. Therefore, adding ammonia changes the equilibrium constant for the dissolution of $AgCl(s)$ by a factor of $3.0 \times 10^{-3}/1.77 \times 10^{-10} = 1.7 \times 10^7$ (17 million), which makes the otherwise relatively insoluble $AgCl(s)$ quite soluble, as shown in Figure 16.18 (page 762).

 ## Conceptual Connection 16.5 Solubility and Complex Ion Equilibria

Which of the following compounds, when added to water, will be most likely to increase the solubility of CuS?

(a) NaCl **(b)** KNO_3 **(c)** NaCN **(d)** $MgBr_2$

Answer: **(c)** Only NaCN contains an anion (CN^-) that forms a complex ion with Cu^{2+} [from Table 16.3 we can see that $K_f = 1.0 \times 10^{25}$ for $Cu(CN)_4{}^{2-}$]. Therefore, the presence of CN^- will drive the dissolution reaction of CuS.

The Solubility of Amphoteric Metal Hydroxides

Many metal hydroxides are insoluble or only very slightly soluble in pH-neutral water. For example, $Al(OH)_3$ has $K_{sp} = 2 \times 10^{-32}$, which means that if you put $Al(OH)_3$ in water, the vast majority of it will settle to the bottom as an undissolved solid. All metal hydroxides, however, have a basic anion (OH^-) and therefore become more soluble in acidic solutions (see the previous subsection and Section 16.5). The metal hydroxides become more soluble because they can act as a base and react with $H_3O^+(aq)$. For example, $Al(OH)_3$ dissolves in acid according to the following reaction:

$$Al(OH)_3(s) + 3\,H_3O^+(aq) \longrightarrow Al^{3+}(aq) + 6\,H_2O(l)$$
Al(OH)₃ acts as a base in this reaction.

Interestingly, some metal hydroxides can also act as acids—they are *amphoteric*. The ability of an amphoteric metal hydroxide to act as an acid increases its solubility in basic solution. For example, $Al(OH)_3(s)$ dissolves in basic solution according to the following reaction:

$$Al(OH)_3(s) + OH^-(aq) \longrightarrow Al(OH)_4{}^-(aq)$$
Al(OH)₃ acts as an acid in this reaction.

Recall from Section 15.3 that a substance that can act as either an acid or a base is said to be amphoteric.

Complex Ion Formation

$$2\,NH_3(aq) + AgCl(s) \rightleftharpoons Ag(NH_3)_2{}^+(aq) + Cl^-(aq)$$

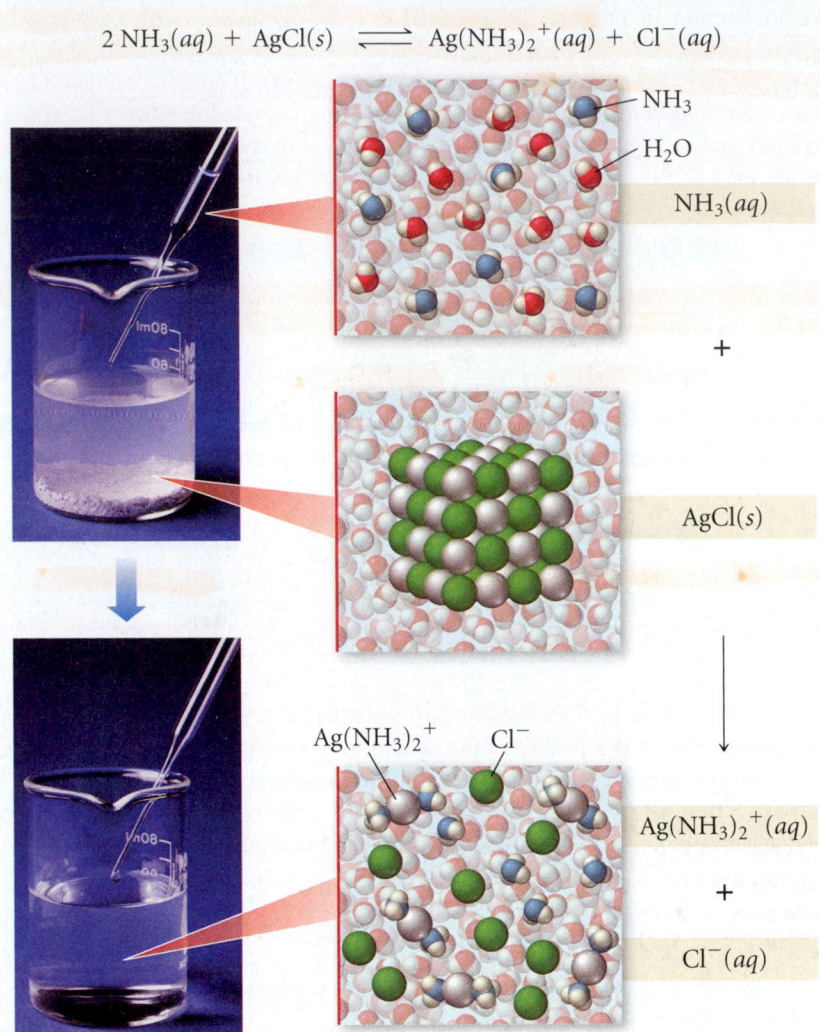

▶ FIGURE 16.18 Complex Ion Formation   Normally insoluble AgCl can be made soluble by the addition of NH_3, which forms a complex ion with Ag^+ and dissolves the AgCl.

Therefore $Al(OH)_3$ is soluble at high pH and soluble at low pH but *insoluble* in a pH-neutral solution.

We can observe the whole range of the pH-dependent solubility behavior of Al^{3+} by considering a hydrated aluminum ion in solution, beginning at an acidic pH. We know from Section 15.8 that Al^{3+} in solution is inherently acidic because it complexes with water to form $Al(H_2O)_6{}^{3+}(aq)$. The complex ion then acts as an acid by losing a proton from one of the complexed water molecules according to the reaction

$$Al(H_2O)_6{}^{3+}(aq) + H_2O(l) \rightleftharpoons Al(H_2O)_5(OH)^{2+}(aq) + H_3O^+(l)$$

Addition of base to the solution drives the reaction to the right and continues to remove protons from complexed water molecules as follows:

$$Al(H_2O)_5(OH)^{2+}(aq) + OH^-(aq) \rightleftharpoons Al(H_2O)_4(OH)_2{}^+(aq) + H_2O(l)$$

$$Al(H_2O)_4(OH)_2{}^+(aq) + OH^-(aq) \rightleftharpoons Al(H_2O)_3(OH)_3(s) + H_2O(l)$$
equivalent to $Al(OH)_3(s)$

The result of removing three protons from $Al(H_2O)_6{}^{3+}(aq)$ is the solid white precipitate $Al(H_2O)_3(OH)_3(s)$, which is more commonly written as $Al(OH)_3(s)$. The solution is now pH-neutral and the hydroxide is insoluble. Addition of more OH^- makes the solution basic and dissolves the solid precipitate as follows:

$$Al(H_2O)_3(OH)_3(s) + OH^-(aq) \rightleftharpoons Al(H_2O)_2(OH)_4{}^-(aq) + H_2O(l)$$

As the solution goes from acidic to neutral to basic, we see the solubility of Al^{3+} change accordingly, as illustrated in Figure 16.19 ▼.

The extent to which a metal hydroxide is capable of dissolving in both acid and base depends on the degree to which it is amphoteric. Cations that form amphoteric hydroxides include Al^{3+}, Cr^{3+}, Zn^{2+}, Pb^{2+}, and Sn^{2+}. Other metal hydroxides, such as those of Ca^{2+}, Fe^{2+}, and Fe^{3+}, are not amphoteric—they become soluble in acidic solutions, but not in basic ones.

pH-Dependent Solubility of an Amphoteric Hydroxide

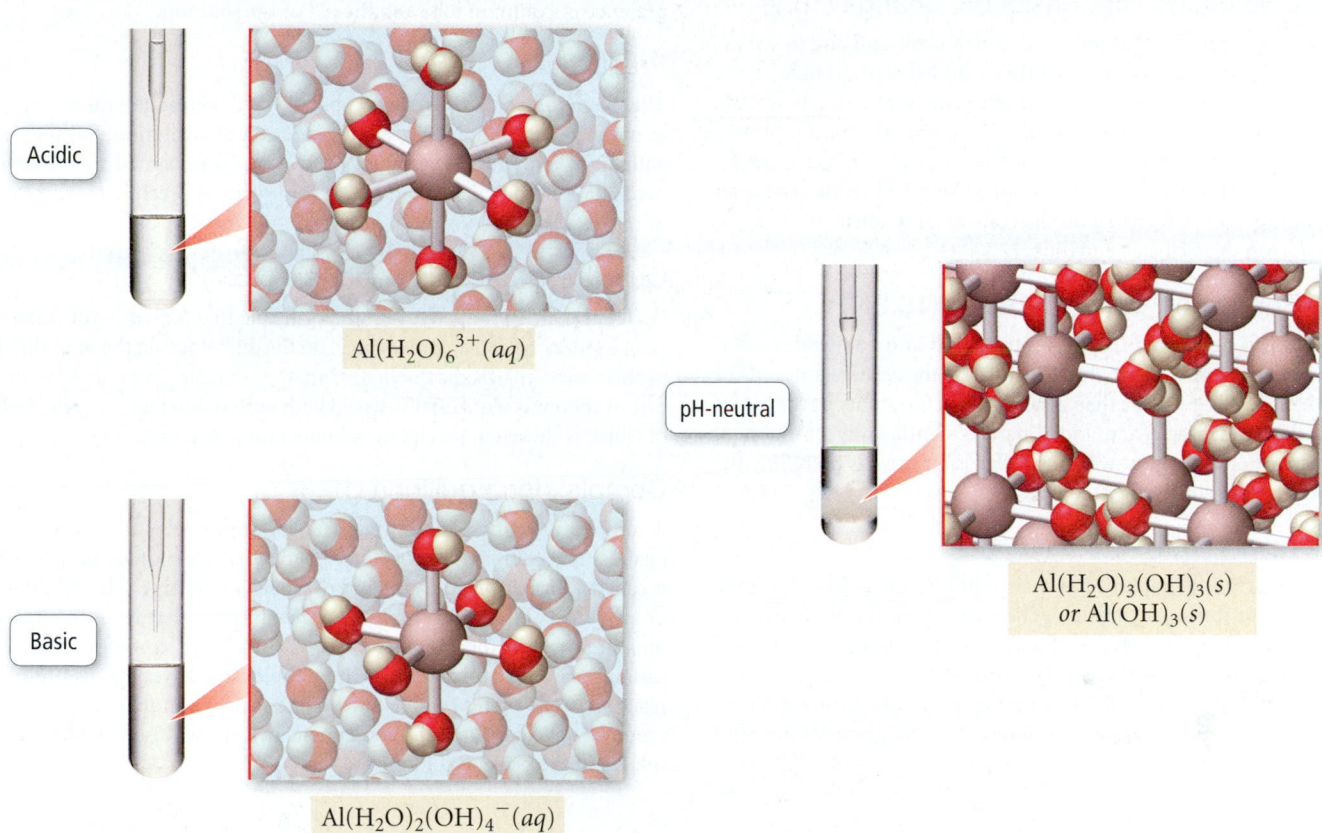

Acidic

$Al(H_2O)_6{}^{3+}(aq)$

pH-neutral

$Al(H_2O)_3(OH)_3(s)$
or $Al(OH)_3(s)$

Basic

$Al(H_2O)_2(OH)_4{}^-(aq)$

▲ FIGURE 16.19 **Solubility of an Amphoteric Hydroxide** Because aluminum hydroxide is amphoteric, its solubility is pH-dependent. At low pH, the formation of $Al(H_2O)_6{}^{3+}$ drives the dissolution. At neutral pH, insoluble $Al(OH)_3$ precipitates out of solution. At high pH, the formation of $Al(H_2O)_2(OH)_4{}^-$ drives the dissolution.

Chapter in Review

Key Terms

Section 16.2
buffer (718)
common ion effect (720)
Henderson–Hasselbalch
    equation (722)

Section 16.3
buffer capacity (732)

Section 16.4
acid–base titration (733)
indicator (733)
equivalence point (733)
endpoint (744)

Section 16.5
solubility product constant
    (K_{sp}) (747)
molar solubility (748)

Section 16.6
selective precipitation (755)

Section 16.7
qualitative analysis (756)
quantitative analysis (756)

Section 16.8
complex ion (759)
ligand (759)
formation constant (K_f) (759)

Key Concepts

The Dangers of Antifreeze (16.1)

Although the pH of human blood is closely regulated by buffers, their capacity to neutralize can be overwhelmed. For example, ethylene glycol, the main component of antifreeze, is metabolized by the liver into glycolic acid. The resulting acidity can exceed the buffering capacity of blood and cause acidosis, a serious condition that results in oxygen deprivation.

Buffers: Solutions That Resist pH Change (16.2)

Buffers contain significant amounts of both a weak acid and its conjugate base, enabling the buffer to neutralize added acid or added base. Adding a small amount of acid to a buffer converts a stoichiometric amount of base to the conjugate acid. Adding a small amount of base to a buffer converts a stoichiometric amount of the acid to the conjugate base. The pH of a buffer solution can be found either by solving an equilibrium problem, focusing on the common ion effect, or by using the Henderson–Hasselbalch equation.

Buffer Range and Buffer Capacity (16.3)

A buffer works best when the amounts of acid and conjugate base it contains are large and approximately equal. If the relative amounts of acid and base differ by more than a factor of 10, the ability of the buffer to neutralize added acid and added base is significantly diminished. The maximum pH range at which a buffer is effective is therefore one pH unit on either side of the acid's pK_a.

Titrations and pH Curves (16.4)

A titration curve is a graph of the change in pH versus added volume of acid or base during a titration. There are three types of titration curves, representing three types of acid–base reactions: a strong acid with a strong base, a weak acid with a strong base (or vice versa), and a polyprotic acid with a base. The equivalence point of a titration can be made visible by an indicator, a compound that changes color at a specific pH.

Solubility Equilibria and the Solubility Product Constant (16.5)

The solubility product constant (K_{sp}) is an equilibrium constant for the dissolution of an ionic compound in water. The molar solubility of an ionic compound can be determined from K_{sp} and vice versa. Although the value of K_{sp} is constant at a given temperature, the solubility of an ionic substance can depend on other factors such as the presence of common ions and the pH of the solution.

Precipitation (16.6)

The magnitude of K_{sp} can be compared with the reaction quotient, Q, in order to determine the relative saturation of a solution. Substances with sufficiently different values of K_{sp} can be separated by selective precipitation, in which an added reagent forms a precipitate with one of the dissolved cations but not others.

Qualitative Chemical Analysis (16.7)

Qualitative analysis operates on the principle that a mixture of cations can be separated and analyzed based on the differences in the solubilities of their salts. In a classic qualitative analysis scheme, an unknown mixture of cations is sequentially treated with several different reagents, each of which is chosen to precipitate a known subgroup of cations.

Complex Ion Equilibria (16.8)

A complex ion contains a central metal ion coordinated to two or more ligands. The equilibrium constant for the formation of a complex ion is called a formation constant and is usually quite large. The solubility of an ionic compound containing a metal cation that forms complex ions increases in the presence of Lewis bases that complex with the cation because the formation of the complex ion drives the dissolution reaction to the right. All metal hydroxides become more soluble in the presence of acids, but amphoteric metal hydroxides also become more soluble in the presence of bases.

Key Equations and Relationships

The Henderson–Hasselbalch Equation (16.2)

$$pH = pK_a + \log \frac{[\text{base}]}{[\text{acid}]}$$

Effective Buffer Range (16.3)

$$pH \text{ range} = pK_a \pm 1$$

The Relation between Q and K_{sp} (16.6)

If $Q < K_{sp}$, the solution is unsaturated. More of the solid ionic compound can dissolve in the solution.

If $Q = K_{sp}$, the solution is saturated. The solution is holding the equilibrium amount of the dissolved ions and additional solid will not dissolve in the solution.

If $Q > K_{sp}$, the solution is supersaturated. Under most circumstances, the solid will precipitate out of a supersaturated solution.

Key Skills

Calculating the pH of a Buffer Solution (16.2)

   • Example 16.1   • For Practice 16.1   • For More Practice 16.1   • Exercises 29, 30, 33, 34

Using the Henderson–Hasselbalch Equation to Calculate the pH of a Buffer Solution (16.2)

   • Example 16.2   • For Practice 16.2   • Exercises 37–42

Calculating the pH Change in a Buffer Solution After the Addition of a Small Amount of Strong Acid or Base (16.2)

• Example 16.3   • For Practice 16.3   • For More Practice 16.3   • Exercises 47–50

Using the Henderson–Hasselbalch Equation to Calculate the pH of a Buffer Solution Composed of a Weak Base and Its Conjugate Acid (16.2)

• Example 16.4   • For Practice 16.4   • For More Practice 16.4   • Exercises 37–40

Determining Buffer Range (16.3)

• Example 16.5   • For Practice 16.5   • Exercises 57, 58

Strong Acid–Strong Base Titration pH Curve (16.4)

• Example 16.6   • For Practice 16.6   • Exercises 67–70

Weak Acid–Strong Base Titration pH Curve (16.4)

• Example 16.7   • For Practice 16.7   • Exercises 65, 66, 71, 72, 75, 77–80

Calculating Molar Solubility from K_{sp} (16.5)

• Example 16.8   • For Practice 16.8   • Exercises 87, 88

Calculating K_{sp} from Molar Solubility (16.5)

• Example 16.9   • For Practice 16.9   • Exercises 89, 90, 92, 94

Calculating Molar Solubility in the Presence of a Common Ion (16.5)

• Example 16.10   • For Practice 16.10   • Exercises 95, 96

Determining the Effect of pH on Solubility (16.5)

• Example 16.11   • For Practice 16.11   • Exercises 97–100

Predicting Precipitation Reactions by Comparing Q and K_{sp} (16.6)

• Example 16.12   • For Practice 16.12   • Exercises 101–104

Finding the Minimum Required Reagent Concentration for Selective Precipitation (16.6)

• Example 16.13   • For Practice 16.13   • Exercises 105, 106

Finding the Concentrations of Ions Left in Solution After Selective Precipitation (16.6)

• Example 16.14   • For Practice 16.14   • Exercises 107, 108

Working with Complex Ion Equilibria (16.8)

• Example 16.15   • For Practice 16.15   • Exercises 109–112

Exercises

Review Questions

1. What is the pH range of human blood? How is human blood maintained in this pH range?

2. What is a buffer? How does a buffer work? How does it neutralize added acid? Added base?

3. What is the common ion effect?

4. What is the Henderson–Hasselbalch equation and why is it useful?

5. What is the pH of a buffer solution when the concentrations of both buffer components (the weak acid and its conjugate base) are equal? What happens to the pH when the buffer contains more of the weak acid than the conjugate base? More of the conjugate base than the weak acid?

6. Suppose that a buffer contains equal amounts of a weak acid and its conjugate base. What happens to the relative amounts of the weak acid and conjugate base upon the addition of a small amount of strong acid to the buffer? What happens upon addition of a small amount of strong base?

7. How do you use the Henderson–Hasselbalch equation to compute the pH of a buffer containing a base and its conjugate acid? Specifically, how do you obtain the correct value to use for pK_a?

8. What factors influence the effectiveness of a buffer? What are the characteristics of an effective buffer?

9. What is the effective pH range of a buffer (relative to the pK_a of the weak acid component)?

10. Describe an acid–base titration. What is the equivalence point?

11. The pH at the equivalence point of the titration of a strong acid with a strong base is 7.0. However, the pH at the equivalence point of the titration of a *weak* acid with a strong base is above 7.0. Explain why this is so.

12. The volume required to reach the equivalence point of an acid–base titration depends on the volume and concentration of the acid or base to be titrated and on the concentration of the acid or base used to do the titration. It does not, however, depend on the whether or not the acid or base being titrated is strong or weak. Explain why this is so.

13. Describe how you would calculate the following pH values in the titration of a strong acid with a strong base:
    a. initial pH
    b. pH before the equivalence point
    c. pH at the equivalence point
    d. pH beyond the equivalence point

14. Describe how you would calculate the following pH values in the titration of a weak acid with a strong base:
    a. initial pH
    b. pH before the equivalence point
    c. pH at one-half the equivalence point
    d. pH at the equivalence point
    e. pH beyond the equivalence point

15. The titration of a polyprotic acid with sufficiently different pK_a's displays two equivalence points. Explain why this is so.

16. In the titration of a polyprotic acid, the volume required to reach the first equivalence point is identical to the volume required to reach the second one. Explain why this is so.

17. Explain the difference between the endpoint and the equivalence point in a titration.

18. What is an indicator? Explain how an indicator can signal the equivalence point of a titration.

19. What is the solubility-product constant? Write a general expression for the solubility constant of a compound with the general formula A_mX_n.

20. What is molar solubility? Explain how to obtain the molar solubility of a compound from K_{sp}.

21. How does a common ion affect the solubility of a compound? More specifically, how is the solubility of a compound with the general formula AX different in a solution containing one of the common ions (A^+ or X^-) than it is in pure water? Explain.

22. How is the solubility of an ionic compound with a basic anion affected by pH? Explain.

23. For a given solution containing an ionic compound, explain the relationship between Q, K_{sp}, and the relative saturation of the solution.

24. What is selective precipitation? What are the conditions under which selective precipitation may occur?

25. What is qualitative analysis? How is *qualitative* analysis different from *quantitative* analysis?

26. What are the main groups in the general qualitative analysis scheme described in this chapter? Describe the steps and reagents necessary to identify each group.

Problems by Topic

The Common Ion Effect and Buffers

27. In which of the following solutions will HNO_2 ionize less than it would in pure water?
    a. 0.10 M NaCl   b. 0.10 M KNO_3
    c. 0.10 M NaOH   d. 0.10 M $NaNO_2$

28. A formic acid solution has a pH of 3.25. Which of the following substances will raise the pH of the solution upon addition? Explain.
    a. HCl   b. NaBr   c. $NaCHO_2$ d. KCl

29. Solve an equilibrium problem (using an ICE table) to calculate the pH of each of the following:
    a. a solution that is 0.15 M in $HCHO_2$ and 0.10 M in $NaCHO_2$
    b. a solution that is 0.12 M in NH_3 and 0.18 M in NH_4Cl

30. Solve an equilibrium problem (using an ICE table) to calculate the pH of each of the following:
    a. a solution that is 0.175 M in $HC_2H_3O_2$ and 0.110 M in $KC_2H_3O_2$
    b. a solution that is 0.195 M in CH_3NH_2 and 0.105 M in CH_3NH_3Br

31. Calculate the percent ionization of a 0.15 M benzoic acid solution in pure water and also in a solution containing 0.10 M sodium benzoate. Why is the percent ionization so different in the two solutions?

32. Calculate the percent ionization of a 0.13 M formic acid solution in pure water and also in a solution containing 0.11 M potassium formate. Explain the difference in percent ionization in the two solutions.

33. Solve an equilibrium problem (using an ICE table) to calculate the pH of each of the following solutions:
    a. 0.15 M HF
    b. 0.15 M NaF
    c. a mixture that is 0.15 M in HF and 0.15 M in NaF

34. Solve an equilibrium problem (using an ICE table) to calculate the pH of each of the following solutions:
    a. 0.18 M CH_3NH_2 b. 0.18 M CH_3NH_3Cl
    c. a mixture that is 0.18 M in CH_3NH_2 and 0.18 M in CH_3NH_3Cl

35. A buffer contains significant amounts of acetic acid and sodium acetate. Write equations showing how this buffer neutralizes added acid and added base.

36. A buffer contains significant amounts of ammonia and ammonium chloride. Write equations showing how this buffer neutralizes added acid and added base.

37. Use the Henderson–Hasselbalch equation to calculate the pH of each of the solutions in problem 29.

38. Use the Henderson–Hasselbalch equation to calculate the pH of each of the solutions in problem 30.

39. Use the Henderson–Hasselbalch equation to calculate the pH of a solution that is
  a. 0.125 M in HClO and 0.150 M in KClO
  b. 0.175 M in $C_2H_5NH_2$ and 0.150 M in $C_2H_5NH_3Br$
  c. 10.0 g of $HC_2H_3O_2$ and 10.0 g of $NaC_2H_3O_2$ in 150.0 mL of solution

40. Use the Henderson-Hasselbalch equation to calculate the pH of a solution that is
  a. 0.155 M in propanoic acid and 0.110 M in potassium propanoate
  b. 0.15 M in C_5H_5N and 0.10 M in C_5H_5NHCl
  c. 15.0 g of HF and 25.0 g of NaF in 125 mL of solution

41. Calculate the pH of the solution that results from each of the following mixtures:
  a. 50.0 mL of 0.15 M $HCHO_2$ with 75.0 mL of 0.13 M $NaCHO_2$
  b. 125.0 mL of 0.10 M NH_3 with 250.0 mL of 0.10 M NH_4Cl

42. Calculate the pH of the solution that results from each of the following mixtures:
  a. 150.0 mL of 0.25 M HF with 225.0 mL of 0.30 M NaF
  b. 175.0 mL of 0.10 M $C_2H_5NH_2$ with 275.0 mL of 0.20 M $C_2H_5NH_3Cl$.

43. Calculate the ratio of NaF to HF required to create a buffer with pH = 4.00.

44. Calculate the ratio of CH_3NH_2 to CH_3NH_3Cl concentration required to create a buffer with pH = 10.24.

45. What mass of sodium benzoate should be added to 150.0 mL of a 0.15 M benzoic acid solution in order to obtain a buffer with a pH of 4.25?

46. What mass of ammonium chloride should be added to 2.55 L of a 0.155 M NH_3 in order to obtain a buffer with a pH of 9.55?

47. A 250.0-mL buffer solution is 0.250 M in acetic acid and 0.250 M in sodium acetate.
  a. What is the initial pH of this solution?
  b. What is the pH after addition of 0.0050 mol of HCl?
  c. What is the pH after addition of 0.0050 mol of NaOH?

48. A 100.0-mL buffer solution is 0.175 M in HClO and 0.150 M in NaClO.
  a. What is the initial pH of this solution?
  b. What is the pH after addition of 150.0 mg of HBr?
  c. What is the pH after addition of 85.0 mg of NaOH?

49. For each of the following solutions, calculate the initial pH and the final pH after adding 0.010 mol of HCl.
  a. 500.0 mL of pure water
  b. 500.0 mL of a buffer solution that is 0.125 M in $HC_2H_3O_2$ and 0.115 M in $NaC_2H_3O_2$
  c. 500.0 mL of a buffer solution that is 0.155 M in $C_2H_5NH_2$ and 0.145 M in $C_2H_5NH_3Cl$

50. For each of the following solutions, calculate the initial pH and the final pH after adding 0.010 mol of NaOH.
  a. 250.0 mL of pure water
  b. 250.0 mL of a buffer solution that is 0.195 M in $HCHO_2$ and 0.275 M in $KCHO_2$
  c. 250.0 mL of a buffer solution that is 0.255 M in $CH_3CH_2NH_2$ and 0.235 M in $CH_3CH_2NH_3Cl$

51. A 350.0-mL buffer solution is 0.150 M in HF and 0.150 M in NaF. What mass of NaOH could this buffer neutralize before the pH rises above 4.00? If the same volume of the buffer was 0.350 M in HF and 0.350 M in NaF, what mass of NaOH could be handled before the pH rises above 4.00?

52. A 100.0-mL buffer solution is 0.100 M in NH_3 and 0.125 M in NH_4Br. What mass of HCl could this buffer neutralize before the pH fell below 9.00? If the same volume of the buffer were 0.250 M in NH_3 and 0.400 M in NH_4Br, what mass of HCl could be handled before the pH fell below 9.00?

53. Determine whether or not the mixing of each of the two solutions indicated below will result in a buffer.
  a. 100.0 mL of 0.10 M NH_3; 100.0 mL of 0.15 M NH_4Cl
  b. 50.0 mL of 0.10 M HCl; 35.0 mL of 0.150 M NaOH
  c. 50.0 mL of 0.15 M HF; 20.0 mL of 0.15 M NaOH
  d. 175.0 mL of 0.10 M NH_3; 150.0 mL of 0.12 M NaOH
  e. 125.0 mL of 0.15 M NH_3; 150.0 mL of 0.20 M NaOH

54. Determine whether or not the mixing of each of the two solutions indicated below will result in a buffer.
  a. 75.0 mL of 0.10 M HF; 55.0 mL of 0.15 M NaF
  b. 150.0 mL of 0.10 M HF; 135.0 mL of 0.175 M HCl
  c. 165.0 mL of 0.10 M HF; 135.0 mL of 0.050 M KOH
  d. 125.0 mL of 0.15 M CH_3NH_2; 120.0 mL of 0.25 M CH_3NH_3Cl
  e. 105.0 mL of 0.15 M CH_3NH_2; 95.0 mL of 0.10 M HCl

55. Blood is buffered by carbonic acid and the bicarbonate ion. Normal blood plasma is 0.024 M in HCO_3^- and 0.0012 M H_2CO_3 (pK_{a_1} for H_2CO_3 at body temperature is 6.1).
  a. What is the pH of blood plasma?
  b. If the volume of blood in a normal adult is 5.0 L, what mass of HCl could be neutralized by the buffering system in blood before the pH fell below 7.0 (which would result in death)?
  c. Given the volume from part (b), what mass of NaOH could be neutralized before the pH rose above 7.8?

56. The fluids within cells are buffered by $H_2PO_4^-$ and HPO_4^{2-}.
  a. Calculate the ratio of HPO_4^{2-} to $H_2PO_4^-$ required to maintain a pH of 7.1 within a cell.
  b. Could a buffer system employing H_3PO_4 as the weak acid and $H_2PO_4^-$ as the weak base be used as a buffer system within cells? Explain.

57. Which of the following buffer systems would be the best choice to create a buffer with pH = 7.20? For the best system, calculate the ratio of the masses of the buffer components required to make the buffer.

$HC_2H_3O_2/KC_2H_3O_2$	$HClO_2/KClO_2$
NH_3/NH_4Cl	$HClO/KClO$

58. Which of the following buffer systems would be the best choice to create a buffer with pH = 9.00? For the best system, calculate the ratio of the masses of the buffer components required to make the buffer.

HF/KF	HNO_2/KNO_2
NH_3/NH_4Cl	$HClO/KClO$

59. A 500.0-mL buffer solution is 0.100 M in HNO_2 and 0.150 M in KNO_2. Determine whether or not each of the following additions would exceed the capacity of the buffer to neutralize it.
  a. 250 mg NaOH   **b.** 350 mg KOH
  c. 1.25 g HBr   **d.** 1.35 g HI

60. A 1.0-L buffer solution is 0.125 M in HNO_2 and 0.145 M in $NaNO_2$. Determine the concentrations of HNO_2 and $NaNO_2$ after addition of each of the following:

    a. 1.5 g HCl   **b.** 1.5 g NaOH

    c. 1.5 g HI

Titrations, pH Curves, and Indicators

61. The graphs below labeled (a) and (b) show the titration curves for two equal-volume samples of monoprotic acids, one weak and one strong. Both titrations were carried out with the same concentration of strong base.

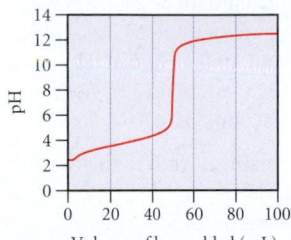

(a)

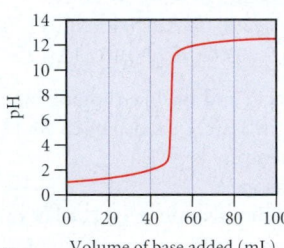

(b)

  (i)   What is the approximate pH at the equivalence point of each curve?

  (ii)   Which curve corresponds to the titration of the strong acid and which one to the titration of the weak acid?

62. Two 25.0-mL samples, one 0.100 M HCl and the other 0.100 M HF, were titrated with 0.200 M KOH. Answer each of the following questions regarding these two titrations.

    a. What is the volume of added base at the equivalence point for each titration?

    b. Predict whether the pH at the equivalence point for each titration will be acidic, basic, or neutral.

    c. Predict which titration curve will have the lower initial pH.

    d. Make a rough sketch of each titration curve.

63. Two 20.0-mL samples, one 0.200 M KOH and the other 0.200 M CH_3NH_2, were titrated with 0.100 M HI. Answer each of the following questions regarding these two titrations.

    a. What is the volume of added acid at the equivalence point for each titration?

    b. Predict whether the pH at the equivalence point for each titration will be acidic, basic, or neutral.

    c. Predict which titration curve will have the lower initial pH.

    d. Make a rough sketch of each titration curve.

64. The graphs below labeled (a) and (b) show, the titration curves for two equal-volume samples of bases, one weak and one strong. Both titrations were carried out with the same concentration of strong acid.

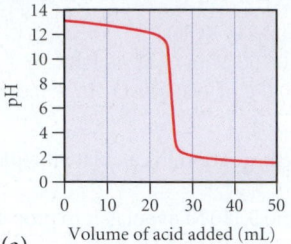

(a)

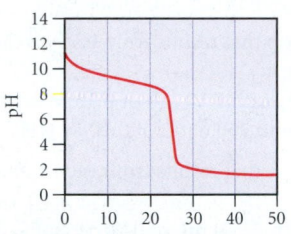

(b)

  (i)   What is the approximate pH at the equivalence point of each curve?

  (ii)   Which curve corresponds to the titration of the strong base and which one to the weak base?

65. Consider the following curve for the titration of a weak monoprotic acid with a strong base and answer each of the following questions.

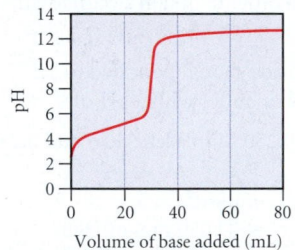

    a. What is the pH and what is the volume of added base at the equivalence point?

    b. At what volume of added base is the pH calculated by working an equilibrium problem based on the initial concentration and K_a of the weak acid?

    c. At what volume of added base does pH $=$ pK_a?

    d. At what volume of added base is the pH calculated by working an equilibrium problem based on the concentration and K_b of the conjugate base?

    e. Beyond what volume of added base is the pH calculated by focusing on the amount of excess strong based added?

66. Consider the following curve for the titration of a weak base with a strong acid and answer each of the following questions.

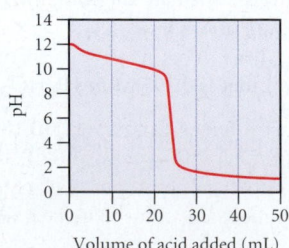

a. What is the pH and what is the volume of added acid at the equivalence point?
b. At what volume of added acid is the pH calculated by working an equilibrium problem based on the initial concentration and K_b of the weak base?
c. At what volume of added acid does pH $= 14 - pK_b$?
d. At what volume of added acid is the pH calculated by working an equilibrium problem based on the concentration and K_a of the conjugate acid?
e. Beyond what volume of added acid is the pH calculated by focusing on the amount of excess strong acid added?

67. Consider the titration of a 35.0-mL sample of 0.175 M HBr with 0.200 M KOH. Determine each of the following:
    a. the initial pH
    b. the volume of added base required to reach the equivalence point
    c. the pH at 10.0 mL of added base
    d. the pH at the equivalence point
    e. the pH after adding 5.0 mL of base beyond the equivalence point

68. A 20.0-mL sample of 0.125 M HNO_3 is titrated with 0.150 M NaOH. Calculate the pH for at least five different points throughout the titration curve and make a sketch of the curve. Indicate the volume at the equivalence point on your graph.

69. Consider the titration of a 25.0-mL sample of 0.115 M RbOH with 0.100 M HCl. Determine each of the following:
    a. the initial pH
    b. the volume of added acid required to reach the equivalence point
    c. the pH at 5.0 mL of added acid
    d. the pH at the equivalence point
    e. the pH after adding 5.0 mL of acid beyond the equivalence point

70. A 15.0-mL sample of 0.100 M $Ba(OH)_2$ is titrated with 0.125 M HCl. Calculate the pH for at least five different points throughout the titration curve and make a sketch of the curve. Indicate the volume at the equivalence point on your graph.

71. Consider the titration of a 20.0-mL sample of 0.105 M $HC_2H_3O_2$ with 0.125 M NaOH. Determine each of the following:
    a. the initial pH
    b. the volume of added base required to reach the equivalence point
    c. the pH at 5.0 mL of added base
    d. the pH at one-half of the equivalence point
    e. the pH at the equivalence point
    f. the pH after adding 5.0 mL of base beyond the equivalence point

72. A 30.0-mL sample of 0.165 M propanoic acid is titrated with 0.300 M KOH. Calculate the pH at each of the following volumes of added base: 0 mL, 5 mL, 10 mL, equivalence point, one-half equivalence point, 20 mL, 25 mL. Use your calculations to make a sketch of the titration curve.

73. Consider the titration of a 25.0-mL sample of 0.175 M CH_3NH_2 with 0.150 M HBr. Determine each of the following:
    a. the initial pH
    b. the volume of added acid required to reach the equivalence point
    c. the pH at 5.0 mL of added acid
    d. the pH at one-half of the equivalence point
    e. the pH at the equivalence point
    f. the pH after adding 5.0 mL of acid beyond the equivalence point

74. A 25.0-mL sample of 0.125 M pyridine is titrated with 0.100 M HCl. Calculate the pH at each of the following volumes of added acid: 0 mL, 10 mL, 20 mL, equivalence point, one-half equivalence point, 40 mL, 50 mL. Use your calculations to make a sketch of the titration curve.

75. Consider the following titration curves for two weak acids, both titrated with 0.100 M NaOH.

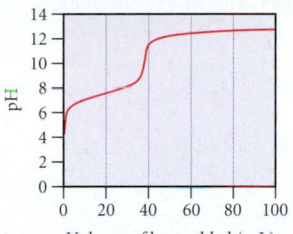

(a)

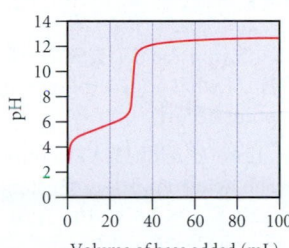

(b)

(i) Which of the two acid solutions is more concentrated?
(ii) Which of the two acids has the larger K_a?

76. Consider the following titration curves for two weak bases, both titrated with 0.100 M HCl.

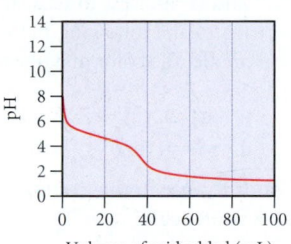

(a)

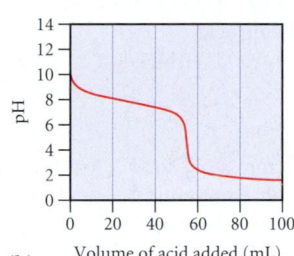

(b)

(i) Which of the two base solutions is more concentrated?
(ii) Which of the two bases has the larger K_b?

77. A 0.229-g sample of an unknown monoprotic acid was titrated with 0.112 M NaOH and the resulting titration curve is shown below. Determine the molar mass and pK_a of the acid.

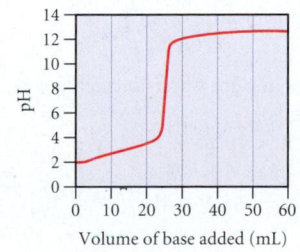

78. A 0.446-g sample of an unknown monoprotic acid was titrated with 0.105 M KOH and the resulting titration curve is shown below. Determine the molar mass and pK_a of the acid.

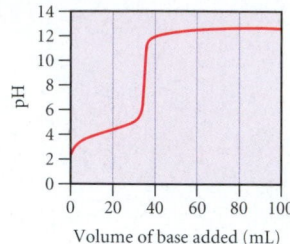

79. A 20.0 mL sample of 0.115 M sulfurous acid (H_2SO_3) solution was titrated with 0.1014 M KOH. At what added volume of base solution does each equivalence point occur?

80. A 20.0 mL sample of a 0.125 M diprotic acid (H_2A) solution was titrated with 0.1019 M KOH. The acid ionization constants for the acid are $K_{a_1} = 5.2 \times 10^{-5}$ and $K_{a_2} = 3.4 \times 10^{-10}$. At what added volume of base does each equivalence point occur?

81. Methyl red has a pK_a of 5.0 and is red in its acid form and yellow in its basic form. If several drops of this indicator are placed in a 25.0-mL sample of 0.100 M HCl, what color will the solution appear? If 0.100 M NaOH is slowly added to the HCl sample, in what pH range will the indicator change color?

82. Phenolphthalein has a pK_a of 9.7 and is colorless in its acid form and pink in its basic form. For each of the following pH's listed below, calculate $[In^-]/[HIn]$ and predict the color of a phenolphthalein solution.
    a. pH = 2.0   b. pH = 5.0
    c. pH = 8.0   d. pH = 11.0

83. Using Table 16.1, pick an indicator for use in the titration of the each of the following acids with a strong base.
    a. HF   b. HCl   c. HCN

84. Using Table 16.1, pick an indicator for use in the titration of each of the following bases with a strong acid.
    a. CH_3NH_2 b. NaOH   c. $C_6H_5NH_2$

Solubility Equilibria

85. Write balanced equations and expressions for K_{sp} for the dissolution of each of the following ionic compounds:
    a. $BaSO_4$ b. $PbBr_2$ c. Ag_2CrO_4

86. Write balanced equations and expressions for K_{sp} for the dissolution of each of the following ionic compounds:
    a. $CaCO_3$ b. $PbCl_2$ c. AgI

87. Use the K_{sp} values in Table 16.2 to calculate the molar solubility of each of the following compounds in pure water:
    a. AgBr   b. $Mg(OH)_2$ c. CaF_2

88. Use the K_{sp} values in Table 16.2 to calculate the molar solubility of each of the following compounds in pure water:
    a. CuS   b. Ag_2CrO_4 c. $Ca(OH)_2$

89. Use the given molar solubilities in pure water to calculate K_{sp} for each of the following compounds:
    a. NiS; molar solubility = 3.27×10^{-11} M
    b. PbF_2; molar solubility = 5.63×10^{-3} M
    c. MgF_2; molar solubility = 2.65×10^{-4} M

90. Use the given molar solubilities in pure water to calculate K_{sp} for each of the following compounds:
    a. $BaCrO_4$; molar solubility = 1.08×10^{-5} M
    b. Ag_2SO_3; molar solubility = 1.55×10^{-5} M
    c. $Pd(SCN)_2$; molar solubility = 2.22×10^{-8} M

91. Two compounds with general formulas AX and AX_2 have $K_{sp} = 1.5 \times 10^{-5}$. Which of the two compounds has the higher molar solubility?

92. Consider the compounds with the generic formulas listed below and their corresponding molar solubilities in pure water. Which compound will have the smallest value of K_{sp}?
    AX; molar solubility = 1.35×10^{-4} M
    AX_2; molar solubility = 2.25×10^{-4} M
    A_2X; molar solubility = 1.75×10^{-4} M

93. Use the K_{sp} value from Table 16.2 to calculate the solubility of iron (II) hydroxide in pure water in grams per 100.0 mL of solution.

94. The solubility of copper(I) chloride is 3.91 mg per 100.0 mL of solution. Calculate K_{sp} for CuCl.

95. Calculate the molar solubility of barium fluoride in each of the following:
    a. pure water   b. 0.10 M $Ba(NO_3)_2$
    c. 0.15 M NaF

96. Calculate the molar solubility of copper(II) sulfide in each of the following:
    a. pure water   b. 0.25 M $CuCl_2$
    c. 0.20 M K_2S

97. Calculate the molar solubility of calcium hydroxide in a solution buffered at each of the following pH's.
    a. pH = 4   b. pH = 7   c. pH = 9

98. Calculate the solubility (in grams per 1.00×10^2 mL of solution) of magnesium hydroxide in a solution buffered at pH = 10. How does this compare to the solubility of $Mg(OH)_2$ in pure water?

99. Determine whether or not each of the following compounds will be more soluble in acidic solution than in pure water. Explain.
    a. $BaCO_3$ b. CuS   c. AgCl   d. PbI_2

100. Determine whether or not each of the following compounds will be more soluble in acidic solution than in pure water. Explain.
    a. Hg_2Br_2 b. $Mg(OH)_2$ c. $CaCO_3$ d. AgI

Precipitation and Qualitative Analysis

101. A solution containing sodium fluoride is mixed with one containing calcium nitrate to form a solution that is 0.015 M in NaF and 0.010 M in $Ca(NO_3)_2$. Will a precipitate form in the mixed solution? If so, identify the precipitate.

102. A solution containing potassium bromide is mixed with one containing lead acetate to form a solution that is 0.013 M in KBr and 0.0035 M in $Pb(C_2H_3O_2)_2$. Will a precipitate form in the mixed solution? If so, identify the precipitate.

103. Predict whether or not a precipitate will form upon mixing 75.0 mL of a NaOH solution with pOH = 2.58 with 125.0 mL of a 0.018 M $MgCl_2$ solution. Identify the precipitate, if any.

104. Predict whether or not a precipitate will form upon mixing 175.0 mL of a 0.0055 M KCl solution with 145.0 mL of a 0.0015 M $AgNO_3$ solution. Identify the precipitate, if any.

105. Potassium hydroxide is used to precipitate each of the cations from their respective solution. Determine the minimum concentration of KOH required for precipitation to begin in each case.
  a. 0.015 M $CaCl_2$
  b. 0.0025 M $Fe(NO_3)_2$
  c. 0.0018 M $MgBr_2$

106. Determine the minimum concentration of the precipitating agent on the right to cause precipitation of the cation from the solution on the left.
  a. 0.035 M $BaNO_3$; NaF
  b. 0.085 M CaI_2; K_2SO_4
  c. 0.0018 M $AgNO_3$; RbCl

107. Consider a solution that is 0.010 M in Ba^{2+} and 0.020 M in Ca^{2+}.
  a. If sodium sulfate is used to selectively precipitate one of the cations while leaving the other cation in solution, which cation will precipitate first? What minimum concentration of Na_2SO_4 is required to cause the precipitation of the cation that precipitates first?
  b. What is the remaining concentration of the cation that precipitates first, when the other cation just begins to precipitate?

108. Consider a solution that is 0.022 M in Fe^{2+} and 0.014 M in Mg^{2+}.
  a. If potassium carbonate is used to selectively precipitate one of the cations while leaving the other cation in solution, which cation will precipitate first? What minimum concentration of K_2CO_3 is required to cause the precipitation of the cation that precipitates first?
  b. What is the remaining concentration of the cation that precipitates first, when the other cation just begins to precipitate?

Complex Ion Equilbria

109. A solution is made that is 1.1×10^{-3} M in $Zn(NO_3)_2$ and 0.150 M in NH_3. After the solution reaches equilibrium, what concentration of $Zn^{2+}(aq)$ remains?

110. A 120.0-mL sample of a solution that is 2.8×10^{-3} M in $AgNO_3$ is mixed with a 225.0-mL sample of a solution that is 0.10 M in NaCN. After the solution reaches equilibrium, what concentration of $Ag^+(aq)$ remains?

111. Use the appropriate values of K_{sp} and K_f to find the equilibrium constant for the following reaction:
$$FeS(s) + 6\,CN^-(aq) \rightleftharpoons Fe(CN)_6{}^{4-}(aq) + S^{2-}(aq)$$

112. Use the appropriate values of K_{sp} and K_f to find the equilibrium constant for the following reaction:
$$PbCl_2(s) + 3\,OH^-(aq) \rightleftharpoons Pb(OH)_3{}^-(aq) + 2\,Cl^-(aq)$$

Cumulative Problems

113. A 150.0-mL solution contains 2.05 g of sodium benzoate and 2.47 g of benzoic acid. Calculate the pH of the solution.

114. A solution is made by combining 10.0 mL of 17.5 M acetic acid with 5.54 g of sodium acetate and diluting to a total volume of 1.50 L. Calculate the pH of the solution.

115. A buffer is created by combining 150.0 mL of 0.25 M $HCHO_2$ with 75.0 mL of 0.20 M NaOH. Determine the pH of the buffer.

116. A buffer is created by combining 3.55 g of NH_3 with 4.78 g of HCl and diluting to a total volume of 750.0 mL. Determine the pH of the buffer.

117. A 1.0-L buffer solution initially contains 0.25 mol of NH_3 and 0.25 mol of NH_4Cl. In order to adjust the buffer pH to 8.75, should you add NaOH or HCl to the buffer mixture? What mass of the correct reagent should you add?

118. A 250.0-mL buffer solution initially contains 0.025 mol of $HCHO_2$ and 0.025 mol of $NaCHO_2$. In order to adjust the buffer pH to 4.10, should you add NaOH or HCl to the buffer mixture? What mass of the correct reagent should you add?

119. In analytical chemistry, bases used for titrations must often be standardized; that is, their concentration must be precisely determined. Standardization of sodium hydroxide solutions is often accomplished by titrating potassium hydrogen phthalate ($KHC_8H_4O_4$), also know as KHP, with the NaOH solution to be standardized.
  a. Write an equation for the reaction between NaOH and KHP.
  b. The titration of 0.5527 g of KHP required 25.87 mL of an NaOH solution to reach the equivalence point. What is the concentration of the NaOH solution?

120. A 0.5224-g sample of an unknown monoprotic acid was titrated with 0.0998 M NaOH. The equivalence point of the titration occurs at 23.82 mL. Determine the molar mass of the unknown acid.

121. A 0.25-mol sample of a weak acid with an unknown pK_a was combined with 10.0 mL of 3.00 M KOH and the resulting solution was diluted to 1.500 L. The measured pH of the solution was 3.85. What is the pK_a of the weak acid?

122. A 5.55-g sample of a weak acid with $K_a = 1.3 \times 10^{-4}$ was combined with 5.00 mL of 6.00 M NaOH and the resulting solution was diluted to 750 mL. The measured pH of the solution was 4.25. What is the molar mass of the weak acid?

123. A 0.552-g sample of ascorbic acid (vitamin C) was dissolved in water to a total volume of 20.0 mL and titrated with 0.1103 M KOH, and the equivalence point occurred at 28.42 mL. The pH of the solution at 10.0 mL of added base was 3.72. From this data, determine the molar mass and K_a for vitamin C.

124. Make a rough sketch of the titration curve from the previous problem by calculating the pH at the beginning of the titration, at one-half of the equivalence point, at the equivalence point, and at 5.0 mL beyond the equivalence point. Pick a suitable indicator for this titration from Table 16.1.

125. One of the main components of hard water is $CaCO_3$. When hard water evaporates, some of the $CaCO_3$ is left behind as a white mineral deposit. If a hard water solution is saturated with calcium carbonate, what volume of the solution has to evaporate to deposit 1.00×10^2 mg of $CaCO_3$?

126. Gout—a condition that results in joint swelling and pain—is caused by the formation of sodium urate ($NaC_5H_3N_4$) crystals within tendons, cartilage, and ligaments. Sodium urate will precipitate out of blood plasma when uric acid levels become abnormally high. This could happen as a result of eating too many rich foods and consuming too much alcohol, which is why gout is sometimes referred to as the "disease of kings." If the sodium concentration in blood plasma is 0.140 M, and K_{sp} for sodium urate is 5.76×10^{-8}, what minimum concentration of urate would result in the precipitation of sodium urate?

127. Pseudogout, a condition with symptoms similar to those of gout (see previous problem), is caused by the formation of calcium diphosphate ($Ca_2P_2O_7$) crystals within tendons, cartilage, and ligaments. Calcium diphosphate will precipitate out of blood plasma when diphosphate levels become abnormally high. If the calcium concentration in blood plasma is 9.2 mg/dL, and K_{sp} for calcium diphosphate is 8.64×10^{-13}, what minimum concentration of diphosphate results in the precipitation of calcium diphosphate?

128. Calculate the solubility of silver chloride in a solution that is 0.100 M in NH_3.

129. Calculate the solubility of copper(II) sulfide in a solution that is 0.150 M in NaCN.

130. Aniline, abbreviated ϕNH_2, where ϕ is C_6H_5, is an important organic base used in the manufacture of dyes. It has $K_b = 4.3 \times 10^{-10}$. In a certain manufacturing process it is necessary to keep the concentration of ϕNH_3^+ (its conjugate acid, called the anilinium ion) below 1.0×10^{-9} M in a solution that is 0.10 M in aniline. Find the concentration of NaOH necessary for this process.

131. The K_b of hydroxylamine, NH_2OH, is 1.10×10^{-8}. A buffer solution is prepared by mixing 100.0 mL of a 0.36 M hydroxylamine solution with 50.0 mL of a 0.26 M HCl solution. Find the pH of the resulting solution.

132. A 0.867-g sample of an unknown acid requires 32.2 mL of a 0.182 M barium hydroxide solution for neutralization. Assuming the acid is diprotic, calculate the molar mass of the acid.

133. A 25.0-mL volume of a sodium hydroxide solution requires 19.6 mL of a 0.189 M hydrochloric acid for neutralization. A 10.0-mL volume of a phosphoric acid solution requires 34.9 mL of the sodium hydroxide solution for complete neutralization. Calculate the concentration of the phosphoric acid solution.

Challenge Problems

134. Derive an equation similar to the Henderson–Hasselbalch equation for a buffer composed of a weak base and its conjugate acid. Instead of relating pH to pK_a and the relative concentrations of an acid and its conjugate base (as the Henderson–Hasselbalch equation does), the equation should relate pOH to pK_b and the relative concentrations of a base and its conjugate acid.

135. Since soap and detergent action is hindered by hard water, laundry formulations usually include water softeners—called builders—designed to remove hard water ions (especially Ca^{2+} and Mg^{2+}) from the water. A common builder used in North America is sodium carbonate. Suppose that the hard water used to do laundry contains 75 ppm $CaCO_3$ and 55 ppm $MgCO_3$ (by mass). What mass of Na_2CO_3 is required to remove 90.0% of these ions from 10.0 L of laundry water?

136. A 0.558-g sample of a diprotic acid with a molar mass of 255.8 g/mol is dissolved in water to a total volume of 25.0 mL. The solution is then titrated with a saturated calcium hydroxide solution.

a. Assuming that the pK_a values for each ionization step are sufficiently different to see two equivalence points, determine the volume of added base for the first and second equivalence points.

b. The pH after adding 25.0 mL of the base was 3.82. Find the value of K_{a_1}.

c. The pH after adding 20.0 mL past the first equivalence point was 8.25. Find the value of K_{a_2}.

137. When excess solid $Mg(OH)_2$ is shaken with 1.00 L of 1.0 M NH_4Cl solution, the resulting saturated solution has pH = 9.00. Calculate the K_{sp} of $Mg(OH)_2$.

138. What amount of solid NaOH must be added to 1.0 L of a 0.10 M H_2CO_3 solution to produce a solution with $[H^+] = 3.2 \times 10^{-11}$ M? There is no significant volume change as the result of the addition of the solid.

139. Calculate the solubility of $Au(OH)_3$ in (a) water and (b) 1.0 M nitric acid solution. ($K_{sp} = 5.5 \times 10^{-46}$.)

140. Calculate the concentration of I^- in a solution obtained by shaking 0.10 M KI with an excess of AgCl(s).

Conceptual Problems

141. Without doing any calculations, determine whether each of the following buffer solutions will have pH $=$ pK_a, pH $>$ pK_a, or pH $<$ pK_a. Assume that HA is a weak monoprotic acid.
  a. 0.10 mol HA and 0.050 mol of A$^-$ in 1.0 L of solution
  b. 0.10 mol HA and 0.150 mol of A$^-$ in 1.0 L of solution
  c. 0.10 mol HA and 0.050 mol of OH$^-$ in 1.0 L of solution
  d. 0.10 mol HA and 0.075 mol of OH$^-$ in 1.0 L of solution

142. A buffer contains 0.10 mol of a weak acid and 0.20 mol of its conjugate base in 1.0 L of solution. Determine whether or not each of the following additions exceed the capacity of the buffer.
  a. adding 0.020 mol of NaOH
  b. adding 0.020 mol of HCl
  c. adding 0.10 mol of NaOH
  d. adding 0.010 mol of HCl

143. Consider the following three solutions:
  (i) 0.10 M solution of a weak monoprotic acid
  (ii) 0.10 M solution of strong monoprotic acid
  (iii) 0.10 M solution of a weak diprotic acid

  Each solution is titrated with 0.15 M NaOH. Which of the following will be the same for all three solutions?
  a. the volume required to reach the final equivalence point
  b. the volume required to reach the first equivalence point
  c. the pH at the first equivalence point
  d. the pH at one-half the first equivalence point

144. Two monoprotic acid solutions (A and B) were titrated with identical NaOH solutions. The volume to reach the equivalence point for solution A was twice the volume required to reach the equivalence point for solution B, and the pH at the equivalence point of solution A was higher than the pH at the equivalence point for solution B. Which of the following is true of the two acids?
  a. The acid in solution A is more concentrated than in solution B and is also a stronger acid than that in solution B.
  b. The acid in solution A is less concentrated than in solution B and is also a weaker acid than that in solution B.
  c. The acid in solution A is more concentrated than in solution B and is also a weaker acid than that in solution B.
  d. The acid in solution A is less concentrated than in solution B and is also a stronger acid than that in solution B.

145. Describe the solubility of CaF_2 in each of the following solutions compared to its solubility in water.
  a. in a 0.10 M NaCl solution
  b. in a 0.10 M NaF solution
  c. in a 0.10 M HCl solution

17

Free Energy and Thermodynamics

17.1 Nature's Heat Tax: You Can't Win and You Can't Break Even

17.2 Spontaneous and Nonspontaneous Processes

17.3 Entropy and the Second Law of Thermodynamics

17.4 Heat Transfer and Changes in the Entropy of the Surroundings

17.5 Gibbs Free Energy

17.6 Entropy Changes in Chemical Reactions: Calculating ΔS°_{rxn}

17.7 Free Energy Changes in Chemical Reactions: Calculating ΔG°_{rxn}

17.8 Free Energy Changes for Nonstandard States: The Relationship between ΔG°_{rxn} and ΔG_{rxn}

17.9 Free Energy and Equilibrium: Relating ΔG°_{rxn} to the Equilibrium Constant (K)

THROUGHOUT THIS BOOK, we have examined and learned much about chemical and physical changes. For example, we have studied how fast chemical changes occur (kinetics) and how to predict how far they will go (through the use of equilibrium constants). We have learned that acids neutralize bases and that gases expand to fill their containers. We now turn to the following question: Why do these changes occur in the first place? What ultimately drives physical and chemical changes in matter? The answer may surprise you. The driving force behind chemical and physical change in the universe is a quantity called *entropy*, which is related to disorder or randomness. Nature tends toward that state in which energy is randomized to the greatest extent possible. Although it does not seem obvious at first glance, the freezing of water below 0 °C, the dissolving of a solid into a solution, the neutralization of an acid by a base, and even the development of a person from an embryo all increase the entropy in the universe. In this universe at least, entropy always increases.

Die Energie der Welt ist konstant. Die Entropie der Welt strebt einem Maximum zu. (The energy of the world is constant. The entropy of the world tends towards a maximum.)

—RUDOLF CLAUSIUS (1822–1888)

In this clever illusion, it seems that the water can perpetually flow through the canal. However, perpetual motion is forbidden by the laws of thermodynamics.

17.1 Nature's Heat Tax: You Can't Win and You Can't Break Even

Energy transactions are like gambling—you walk into the casino with your pockets full of cash and (if you keep gambling long enough) you walk out empty-handed. In the long run, you lose money in gambling because the casino takes a cut on each transaction. So it is with energy. Nature takes a cut—sometimes called nature's heat tax—on every energy transaction so that, in the end, energy is dissipated.

We learned in Chapter 6 that, according to the first law of thermodynamics, energy is conserved in chemical processes. When we burn gasoline to run a car, for example, the amount of energy produced by the chemical reaction does not vanish, nor does any new energy appear that was not present in the chemical bonds before the combustion. Some of

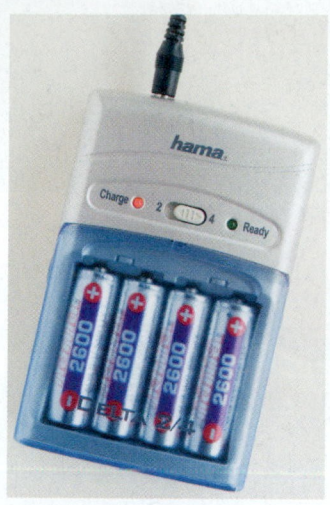

▲ A rechargeable battery will always require more energy to charge than the energy available for work during discharging because some energy must always be lost to the surroundings during the charging/discharging cycle.   ●

See the box entitled *Redheffer's Perpetual Motion Machine* in Section 6.2.

100 watts of electrical energy

95 watts dissipated as heat

5 watts emitted as light

▲ FIGURE 17.1 Energy Loss   In most energy transactions, some energy is lost to the surroundings, so that each transaction is only fractionally efficient.

the energy from the combustion reaction goes toward driving the car forward (about 20%), and the rest is dissipated into the surroundings as heat (just feel the engine). However, the total energy given off by the combustion reaction exactly equals the sum of the amount of energy driving the car forward and the amount dissipated as heat—energy is conserved. In other words, when it comes to energy, you can't win; you cannot create energy that was not there to begin with.

The picture becomes more interesting, however, when we consider the second law of thermodynamics. The second law—which we examine in more detail throughout this chapter—implies that not only can we not win in an energy transaction, but we cannot even break even. For example, consider a rechargeable battery.

Suppose that, upon using the fully charged battery for some application, the energy from the battery does 100 kJ of work. Then, recharging the battery to its original state will *necessarily* (according to the second law of thermodynamics) require *more than* 100 kJ of energy. Energy is not destroyed during the cycle of discharging and recharging the battery, but some energy must be lost to the surroundings in order for the process to occur at all. In other words, nature takes a *heat tax*, an unavoidable cut of every energy transaction.

The implications of the second law for energy use are significant. First of all, according to the second law, we cannot create a perpetual motion machine, that is, a machine that perpetually moves *without any energy input*. If the machine is to be in motion, it must pay the heat tax with each cycle of its motion—over time, it will therefore run down and stop moving.

Secondly, in most energy transactions, not only is the "heat tax" lost to the surroundings, but additional energy is also lost as heat because real-world processes do not achieve the theoretically possible maximum efficiency (Figure 17.1 ◀).

Consequently, the most efficient use of energy generally occurs with the smallest number of transactions. For example, heating your home with natural gas is generally cheaper and more efficient than heating it with electricity (Figure 17.2 ▶). Why? When you heat your home with natural gas, there is only one energy transaction—you burn the gas and the heat from the reaction warms the house. When you heat your home with electricity, however, several transactions occur. Most electricity is generated from the combustion of fossil fuels, so the fuel is burned and the heat from the reaction is used to boil water. The steam generated by the boiling water then turns a turbine on a generator to create electricity. The electricity must then travel from the power plant to your home, with some of the energy lost as heat during the trip. Finally, the electricity must run the heater that creates the heat. With each transaction, energy is lost to the surroundings, resulting in a less efficient use of energy than if you had burned natural gas directly.

◆ Conceptual Connection 17.1   Nature's Heat Tax and Diet

Advocates of a vegetarian diet argue that the amount of cropland required for one person to maintain a meat-based diet is about 6–10 times greater than the amount required for the same person to maintain a vegetarian diet. Use the concept of nature's heat tax to explain why this might be so.

Answer: A person subsisting on a vegetarian diet eats fruits and vegetables and metabolizes their energy-containing molecules. A person subsisting on a meat-based diet eats the meat of an animal such as a cow and metabolizes energy-containing molecules that were part of the cow. However, the cow synthesized its energy-containing molecules from compounds that it obtained by eating and digesting plants. Since breaking down and resynthesizing biological molecules requires energy—and since the cow also needs to extract some of the energy in its food to live—a meat-based diet adds additional energy transactions to the overall process of obtaining energy for life. Therefore, due to nature's heat tax, the meat-based diet is less efficient than the vegetarian diet.

Heating with Natural Gas

Heating with Electricity

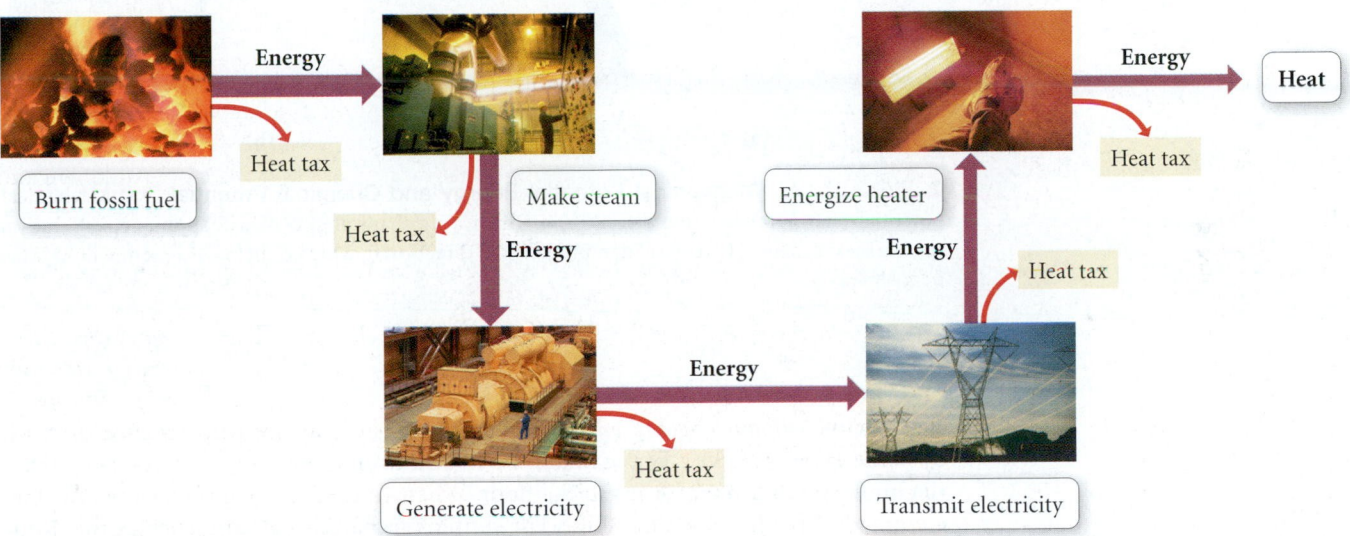

▲ **FIGURE 17.2 Heating with Gas versus Heating with Electricity**   When natural gas is used to heat a home, only a single energy transaction is involved, so the heat tax is minimized. When electricity is used to heat a home, a number of energy transactions are required, each of which involves some loss. The result is a much lower efficiency.

17.2 Spontaneous and Nonspontaneous Processes

A fundamental goal of thermodynamics is the prediction of *spontaneity*. For example, will rust spontaneously form when iron comes into contact with oxygen? Will water spontaneously decompose into hydrogen and oxygen? A **spontaneous process** is one

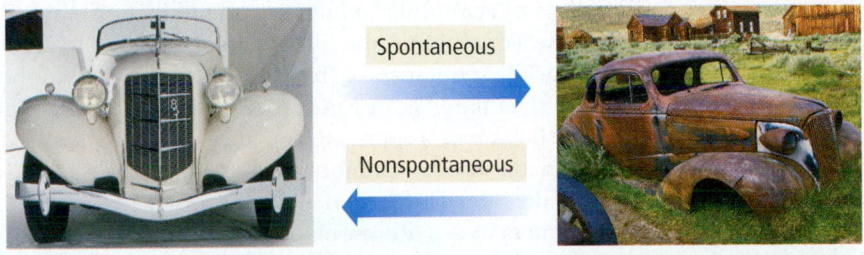

▲ Iron spontaneously rusts when it comes in contact with oxygen.

The Concept of Chemical Potential

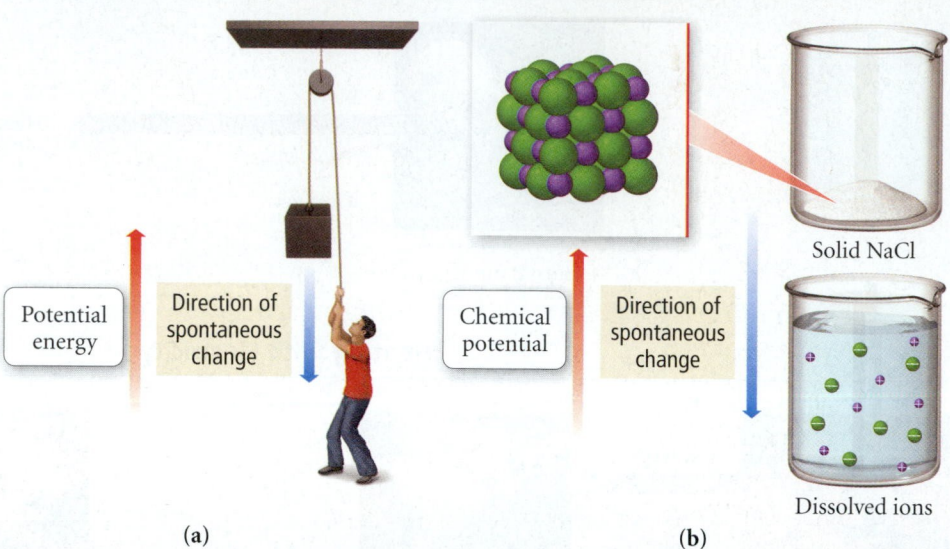

▲ **FIGURE 17.3 Mechanical Potential Energy and Chemical Potential** (**a**) Mechanical potential energy predicts the direction in which a mechanical system will spontaneously move. (**b**) We seek a chemical potential that predicts the direction in which a chemical system will spontaneously move.

that occurs *without ongoing outside intervention* (such as the performance of work by some external force). For example, when you drop a book in a gravitational field, the book spontaneously drops to the floor. When you place a ball on a slope, the ball spontaneously rolls down the slope. For simple mechanical systems, such as the dropping of a book or the rolling of a ball, the prediction of spontaneity is fairly intuitive. A mechanical system tends toward lowest potential energy, and this is usually easy to see (at least for *simple* mechanical systems). However, the prediction of spontaneity for chemical systems is not so intuitively obvious. We wish to develop a criterion for the spontaneity of chemical systems. In other words, we wish to develop a *chemical potential* that predicts the direction of a chemical system, much as mechanical potential energy predicts the direction of a mechanical system (Figure 17.3 ▲).

We must not, however, confuse the *spontaneity* of a chemical reaction with the *speed* of a chemical reaction. In thermodynamics, we study the *spontaneity* of a reaction—the direction in which and extent to which a chemical reaction proceeds. In kinetics, we study the *speed* of the reaction—how fast a reaction takes place (Figure 17.4 ◄). A reaction may be thermodynamically spontaneous but kinetically slow at a given temperature. For example, the conversion of diamond to graphite is thermodynamically spontaneous. But your diamonds will not become worthless anytime soon because the process is extremely slow kinetically. Although the rate of a spontaneous process can be increased by the use of a catalyst, a nonspontaneous process cannot be made spontaneous by the use of a catalyst. Catalysts affect only the rate of a reaction, not the spontaneity.

One last word about nonspontaneity—a nonspontaneous process is not *impossible.* The extraction of iron metal from iron ore is a nonspontaneous process; it does not happen if the iron ore is left to itself. However, it is not impossible. As we will see in a later section, a nonspontaneous process can be made spontaneous by coupling it to another process that is spontaneous, or by supplying energy from an external source. Iron can be separated from its ore if external energy is supplied, usually by means of another reaction (that is itself highly spontaneous).

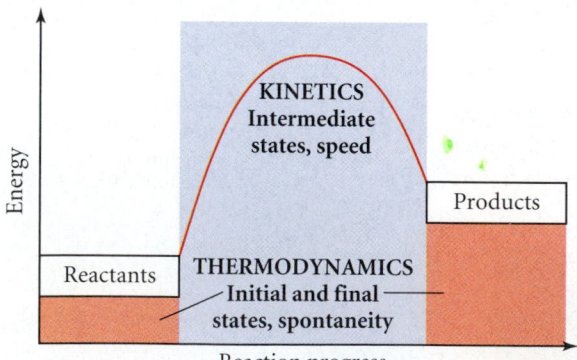

▲ **FIGURE 17.4 Thermodynamics and Kinetics** Thermodynamics deals with the relative chemical potentials of the reactants and products. It enables us to predict whether a reaction will be spontaneous, and to calculate how much work it can do. Kinetics deals with the chemical potential of intermediate states, and enables us to determine why a reaction is slow or fast.

▲ Even though graphite is thermodynamically more stable than diamond, the conversion of diamond to graphite is kinetically so slow that it does not occur at any measurable rate.

17.3 Entropy and the Second Law of Thermodynamics

The first candidate in our search for a chemical potential might be enthalpy, which we defined in Chapter 6. Perhaps, just as a mechanical system proceeds in the direction of lowest potential energy, so a chemical system might proceed in the direction of lowest enthalpy. If this were the case, all exothermic reactions would be spontaneous and all endothermic reactions would not. However, although *most* spontaneous reactions are exothermic, some spontaneous reactions are *endothermic*. For example, above 0 °C, ice spontaneously melts (an endothermic process). So enthalpy must not be the sole criterion for spontaneity.

See Section 6.5 for the definition of enthalpy.

We can learn more about the driving force behind chemical reactions by considering several processes (like ice melting) that involve an increase in enthalpy. What drives a spontaneous *endothermic* process? Consider each of the following processes:

• the melting of ice above 0 °C
• the evaporation of liquid water to gaseous water
• the dissolution of sodium chloride in water

Each of these processes is endothermic *and* spontaneous. Do they have anything in common? Notice that, in each process, disorder or randomness increases. In the melting of ice, the arrangement of the water molecules changes from a highly ordered one (in ice) to a somewhat disorderly one (in liquid water).

The use of the word *disorder* here is only analogous to our macroscopic notions of disorder. The definition of molecular disorder, which is covered shortly, is very specific.

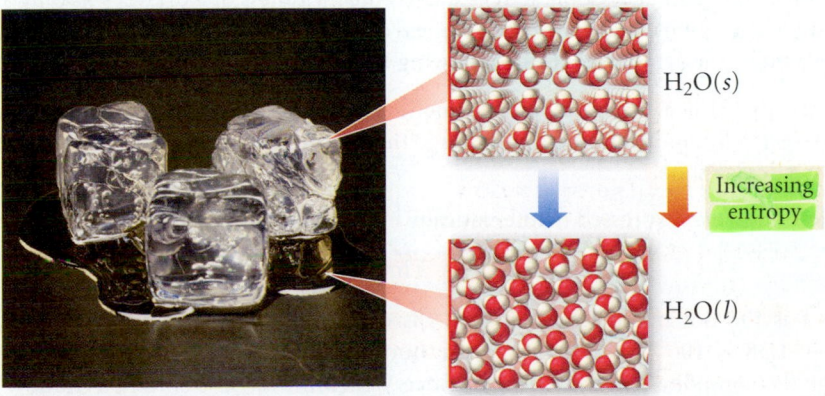

H₂O(s)

Increasing entropy

H₂O(l)

◀ When ice melts, the arrangement of water molecules changes from an orderly one to a more disorderly one.

In the evaporation of a liquid to a gas, the arrangement changes from a *somewhat* disorderly one (atoms or molecules in the liquid) to a *highly* disorderly one (atoms or molecules in the gas).

► When water evaporates, the arrangement of water molecules becomes still more disorderly.

In the dissolution of a salt into water, the arrangement again changes from an orderly one (in which the ions in the salt occupy regular positions in the crystal lattice) to a more disorderly one (in which the ions are randomly dispersed throughout the liquid water).

► When salt dissolves in water, the arrangement of the molecules and ions becomes more disorderly.

In all three of these processes, a quantity called *entropy*—related to disorder or randomness at the molecular level—increases.

Entropy

We have now hit upon the criterion for spontaneity in chemical systems: entropy. Informally, we can think of entropy as disorder or randomness. But the concept of disorder or randomness on the macroscopic scale—such as the messiness of a drawer—is only analogous to the concept of disorder or randomness on the molecular scale. Formally, **entropy,** abbreviated by the symbol S, has the following definition:

> Entropy (S) is a thermodynamic function that increases with the number of *energetically equivalent* ways to arrange the components of a system to achieve a particular state.

This definition was expressed mathematically by Ludwig Boltzmann as

$$S = k \ln W$$

▲ Boltzman's equation is engraved on his tombstone.

where k is the Boltzmann constant (the gas constant divided by Avogadro's number, $R/N_A = 1.38 \times 10^{-23}$ J/K) and W is the number of energetically equivalent ways to arrange the components of the system. Since W is unitless (it is simply a number), the units of entropy are joules per kelvin (J/K). We will talk about the significance of the units shortly. As you can see from the equation, as W increases, entropy increases.

Entropy, like enthalpy, is a *state function*—its value depends only on the state of the system, not on how the system got to that state. Therefore, for any process, *the change in entropy is just the entropy of the final state minus the entropy of the initial state.*

See the discussion of state functions in Section 6.2.

$$\Delta S = S_{\text{final}} - S_{\text{initial}}$$

Entropy determines the direction of chemical and physical change. *A chemical system proceeds in a direction that increases the entropy of the universe*—it proceeds in a direction that has the largest number of *energetically equivalent* ways to arrange its components. To better understand this tendency, let us examine the expansion of an ideal gas into a vacuum (a spontaneous process with no associated change in enthalpy). Consider a flask containing an ideal gas that is connected to another, evacuated, flask by a tube equipped with a stopcock. When the stopcock is opened, the gas spontaneously expands into the evacuated flask. Since the gas is expanding into a vacuum, the pressure against which it expands is zero, and therefore the work ($w = -P_{\text{ext}} \Delta V$) is also zero.

See the discussion of work done by an expanding gas in Section 6.3.

However, even though the total energy of the gas does not change during the expansion, the entropy does change. To see this, consider the following simplified system containing only four gas atoms.

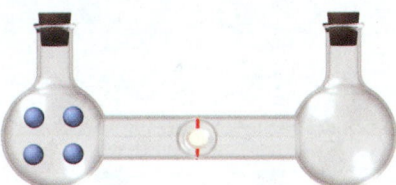

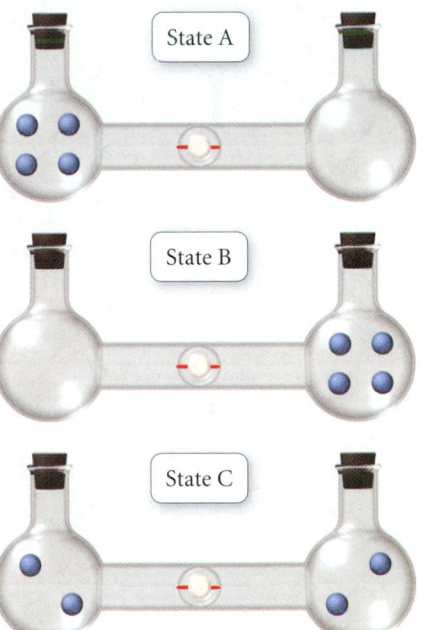

State A

State B

State C

When the stopcock is opened, there are several possible energetically equivalent final states that may result, each with the four atoms distributed in a different way. For example, there could be three atoms in the flask on the left and one in the flask on the right, or vice versa. For simplicity, we consider only the possibilities shown at right.

Since the energy of any one atom is the same in either flask, and since the atoms do not interact, all three states are energetically equivalent.

Now we ask the following question for each possible state: How many *internal arrangements* (sometimes called microstates) give rise to the same *external arrangement* (sometimes called the macrostate)? To keep track of the internal arrangements we label the atoms 1–4. However, since the atoms are all the same, there is no difference between them externally. For states A and B, there is only one internal arrangement that gives the specified external arrangement—atoms 1–4 on the left side or the right side, respectively.

External Arrangement (Macrostate)   Possible Internal Arrangements (Microstates)

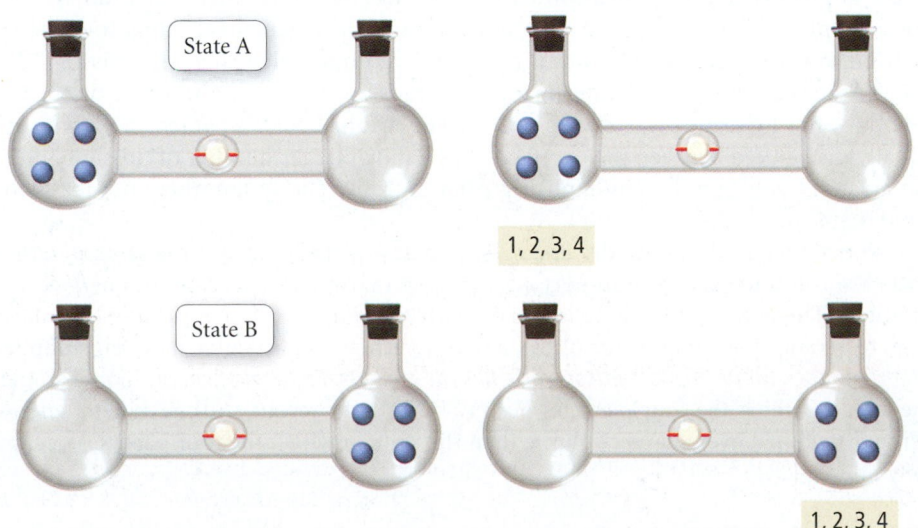

State A

1, 2, 3, 4

State B

1, 2, 3, 4

For state C, however, there are six possible internal arrangements that all give the same external arrangement (two atoms on each side).

External Arrangement (Macrostate)   **Possible Internal Arrangements (Microstates)**

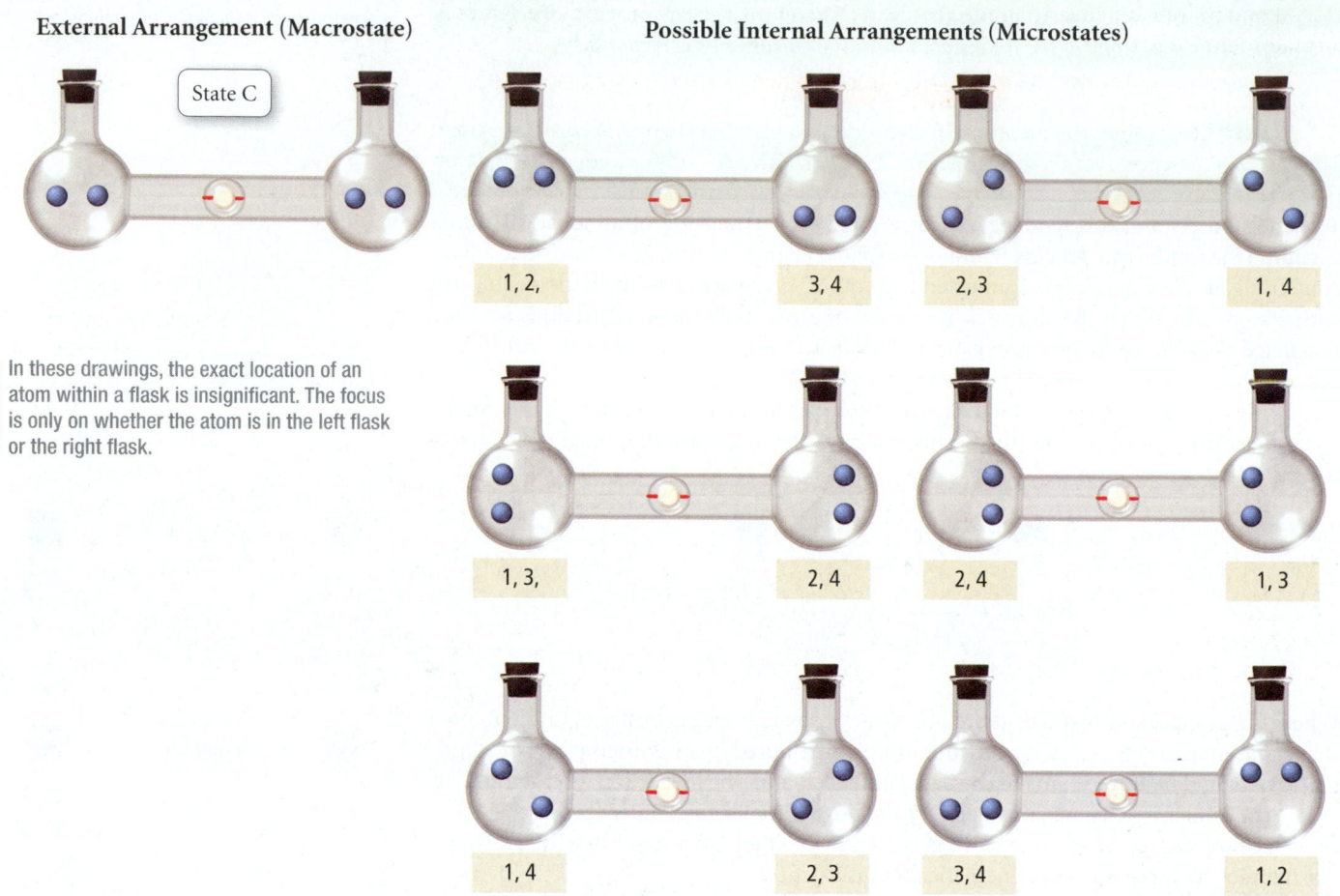

In these drawings, the exact location of an atom within a flask is insignificant. The focus is only on whether the atom is in the left flask or the right flask.

In other words, if the atoms are just randomly moving between the two flasks, the statistical probability of finding the atoms in state C is six times greater than the probability of finding the atoms in states A or B. Consequently, even for a simple system consisting of only four atoms, the atoms are most likely to be found in state C. State C has the greatest entropy—it has the greatest number of energetically equivalent ways to distribute its components.

As the number of atoms increases, the number of internal arrangements that leads to the atoms being equally distributed between the two flasks increases dramatically. For example, with 10 atoms, the number of internal arrangements leading to an equal distribution is 252 and with 20 atoms the number of internal arrangements is 184,756. Yet, the number of internal arrangements that leads to all of the atoms being on the left side does not increase—it is always only 1. The arrangement in which the atoms are equally distributed between the two flasks has a much larger number of possible internal arrangements and therefore much greater entropy. The system therefore tends toward that state.

For n particles, the number of ways to put r particles in one flask and $n - r$ particles in the other flask is $n!/[(n - r)!\, r!]$. For 10 atoms, $n = 10$ and $r = 5$.

Notice that, in the above discussion, the energy of the atom was the same in either flask. We can understand an important aspect of entropy by focusing on energy for a moment. The entropy of a state increases with the number of *energetically equivalent* ways to arrange the components of the system to achieve a particular state. This implies that *the state with the highest entropy also has the greatest dispersal of energy*. For example, the state in which the four particles occupy both flasks results in their kinetic energy being distributed over a larger volume than the state in which the four particles occupy only one flask.

Thus, at the heart of entropy is energy dispersal or energy randomization. *A state in which a given amount of energy is more highly dispersed (or more highly randomized) has more entropy than a state in which the same energy is more highly concentrated.*

Although we have already alluded to the *second law of thermodynamics*, we can now formally define it:

> **Second law of thermodynamics:** For any spontaneous process, the entropy of the *universe* increases ($\Delta S_{univ} > 0$).

The criterion for spontaneity is the entropy of the universe. Processes that increase the entropy of the universe—those that result in greater dispersal or randomization of energy—occur spontaneously. Processes that decrease the entropy of the universe do not occur spontaneously. Returning to the expansion of an ideal gas into a vacuum, the change in entropy in going from a state in which all of the atoms are in the left flask to the state in which the atoms are evenly distributed between both flasks is positive because the final state has a greater entropy than the initial state:

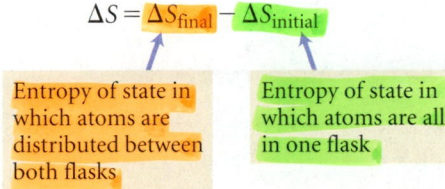

$$\Delta S = \Delta S_{final} - \Delta S_{initial}$$

Entropy of state in which atoms are distributed between both flasks

Entropy of state in which atoms are all in one flask

Since S_{final} is greater than $S_{initial}$, ΔS is positive and the process is spontaneous according to the second law.

Heat

The second law explains many phenomena not explained by the first law. For example, in Chapter 6, we learned that heat travels from a substance at higher temperature to one at lower temperature. If we drop an ice cube into water, heat travels from the water to the ice cube—the water cools and the ice warms (and eventually melts). Why? The first law would not prohibit some heat from flowing the other way— from the ice to the water. For example, the ice could lose 10 J of heat (cooling even more) and the water could gain 10 J of heat (warming even more). The first law of thermodynamics is not violated by such a heat transfer. Imagine putting ice into water only to have the water get warmer as it absorbed thermal energy from the ice! It will never happen. Why? Because heat transfer from cold to hot violates the second law of thermodynamics. According to the second law, energy is dispersed, not concentrated. The transfer of heat from a substance of higher temperature to one of lower temperature results in greater energy randomization—the energy that was concentrated in the hot substance becomes dispersed between the two substances. The second law describes this pervasive tendency.

✳ Conceptual Connection 17.2 Entropy

Consider the following changes in the distribution of six particles into three interconnected boxes. Which of the following has a positive ΔS?

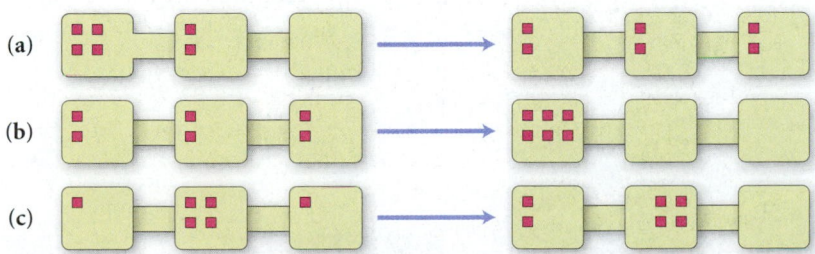

Answer: (a) The more spread out the particles are between the three boxes, the greater the entropy. Therefore, the entropy change is positive only in scheme **(a)**.

The Entropy Change Associated with a Change in State

The entropy of a sample of matter *increases* as it changes state from a solid to a liquid or from a liquid to a gas (Figure 17.5 ▼). We can informally think of this increase in entropy by analogy with macroscopic disorder. The gaseous state is more disorderly than the liquid state, which is in turn more disorderly than the solid state. More formally, however, the differences in entropy are related to the number of energetically equivalent ways of arranging the particles in each state—more in the gas than in the liquid, and more in the liquid than in the solid.

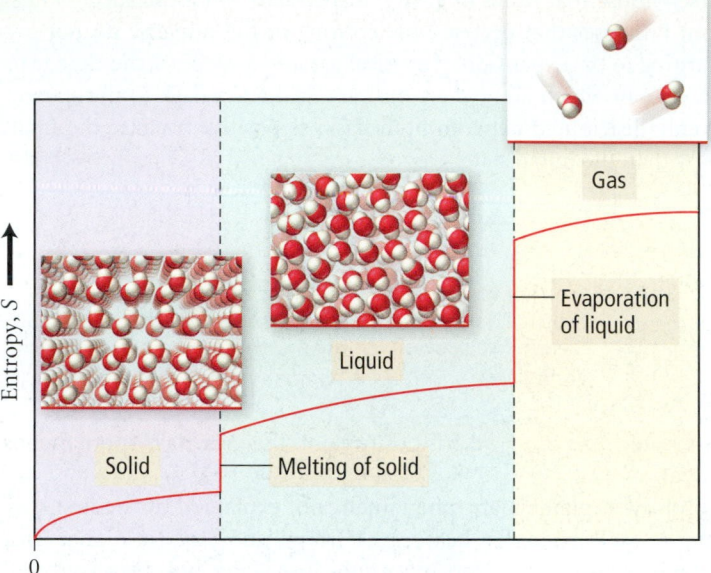

▶ **FIGURE 17.5 Entropy and Phase Change** Entropy increases in going from a solid to a liquid and in going from a liquid to a gas.

A gas has more energetically equivalent configurations because it has more ways to distribute its energy than a solid. The energy in a molecular solid consists largely of the vibrations of its molecules. If the same substance is vaporized, however, the energy can take the form of straight-line motions of the molecules (called translational energy) and rotations of the molecules (called rotational energy). In other words, when a solid vaporizes, there are new "places" to put energy (Figure 17.6 ▼). This increase in "places" to put energy results in more possible energetically equivalent configurations and therefore greater entropy.

As a result, we can easily predict the sign of ΔS for processes involving changes of state. In general, entropy increases ($\Delta S > 0$) for each of the following:

- the phase transition from a solid to a liquid
- the phase transition from a solid to a gas
- the phase transition from a liquid to a gas
- an increase in the number of moles of a gas during a chemical reaction

Additional "Places" for Energy

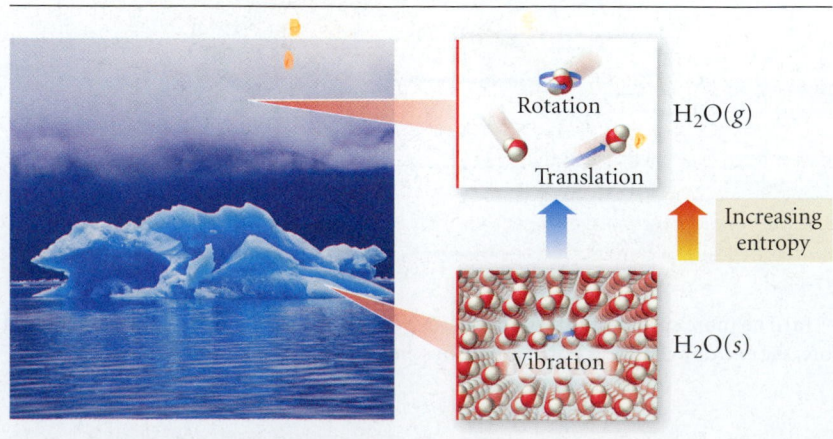

▶ **FIGURE 17.6 "Places" for Energy** In the solid phase, energy is contained largely in the vibrations between molecules. In the gas phase, energy can be contained in both the straight-line motion of molecules (translational energy) and the rotation of molecules (rotational energy).

EXAMPLE 17.1 Predicting the Sign of Entropy Change

Predict the sign of ΔS for each of the following processes:

(a) $H_2O(g) \longrightarrow H_2O(l)$

(b) Solid carbon dioxide sublimes.

(c) $2 N_2O(g) \longrightarrow 2 N_2(g) + O_2(g)$

Solution

(a) Since a gas has a greater entropy than a liquid, the entropy decreases and ΔS is therefore negative.

(b) Since a solid has a lower entropy than a gas, the entropy increases and ΔS is therefore positive.

(c) Since the number of moles of gas increases, the entropy increases and ΔS is therefore positive.

For Practice 17.1

Predict the sign of ΔS for each of the following processes:

(a) the boiling of water

(b) $I_2(g) \longrightarrow I_2(s)$

(c) $CaCO_3(s) \longrightarrow CaO(s) + CO_2(g)$

17.4 Heat Transfer and Changes in the Entropy of the Surroundings

We have now seen that the criterion for spontaneity is an increase in the entropy of the universe. However, you can probably think of several spontaneous processes in which entropy seems to decrease. For example, when water freezes at temperatures below 0 °C, the entropy of the water decreases, yet the process is spontaneous. Similarly, when water vapor in air condenses into fog on a cold night, the entropy of the water also decreases. Why are these processes spontaneous?

To answer this question, we must return to our statement of the second law: for any spontaneous process, the entropy *of the universe* increases ($\Delta S_{univ} > 0$). Even though the entropy *of the water* decreases during freezing and condensation, the entropy *of the universe* must somehow increase in order for these processes to be spontaneous. In Chapter 6, we found it helpful to distinguish between a thermodynamic system and its surroundings. The same distinction is helpful in our discussion of entropy. For the freezing of water, let us consider the water as the system. The surroundings are then the rest of the universe. Using these distinctions, ΔS_{sys} is the entropy change for the water itself, ΔS_{surr} is the entropy change for the surroundings, and ΔS_{univ} is the entropy change for the universe. The entropy change for the universe is just the sum of the entropy changes for the system and the surroundings:

$$\Delta S_{univ} = \Delta S_{sys} + \Delta S_{surr}$$

The second law states that the entropy of the universe must increase ($\Delta S_{univ} > 0$) for a process to be spontaneous. The entropy of the *system* could therefore decrease ($\Delta S_{sys} < 0$) as long as the entropy of the *surroundings* increases by a greater amount ($\Delta S_{surr} > -\Delta S_{sys}$), so that the overall entropy of the *universe* undergoes a net increase.

For liquid water freezing or water vapor condensing, we know that change in entropy for the system (ΔS_{sys}) is negative because the water becomes more orderly in both cases.

$$\Delta S_{univ} = \Delta S_{sys} + \Delta S_{surr}$$

Negative   Positive

For ΔS_{univ} to be positive, therefore, ΔS_{surr} must be *positive* and greater in absolute value (or magnitude) than ΔS_{sys}. But why does the freezing of ice or the condensation of water increase the entropy of the surroundings? Both processes are *exothermic*: they give off heat to the surroundings. If we think of entropy as the dispersal or randomization of energy, then *the release of heat energy by the system disperses that energy into the surroundings, increasing the entropy of the surroundings.* The freezing of water below 0 °C and the condensation of water vapor on a cold night both increase the entropy of the universe because the heat given off to the surroundings increases the entropy of the surroundings to a sufficient degree to overcome the entropy decrease in the water.

> Even though (as we saw earlier) enthalpy by itself cannot determine spontaneity, the increase in the entropy of the surroundings caused by the release of heat explains why exothermic processes are so *often* spontaneous.

Summarizing,

➤ An exothermic process increases the entropy of the surroundings.

➤ An endothermic process decreases the entropy of the surroundings.

The Temperature Dependence of ΔS_{surr}

We have just seen how the freezing of water increases the entropy of the surroundings by dispersing heat energy into the surroundings. However, we know that the freezing of water is not spontaneous at all temperatures. The freezing of water becomes *nonspontaneous* above 0 °C. Why? Because the magnitude of the increase in the entropy of the surroundings due to the dispersal of energy into the surroundings is *temperature dependent.*

$$\Delta S_{univ} = \Delta S_{sys} + \Delta S_{surr} \quad \text{(for water freezing)}$$

Negative   Positive, but magnitude depends on temperature

The greater the temperature, the smaller the increase in entropy for a given amount of energy dispersed into the surroundings. Recall that the units of entropy are joules per kelvin—that is, those of energy divided by those of temperature. Entropy is a measure of energy dispersal *per unit temperature*. The higher the temperature, the lower the amount of entropy for a given amount of energy dispersed. We can understand the temperature dependence of entropy changes due to heat flow with a simple analogy. Imagine that you have $1000 to give away. If you gave the $1000 to a rich man, the impact on his net worth would be negligible (because he already has so much money). If you gave the same $1000 to a poor man, however, his net worth would change substantially (because he has so little money). Similarly, if you disperse 1000 J of energy into surroundings that are hot, the entropy increase is small (because the impact of the 1000 J is small on surroundings that already contain a lot of energy). If you disperse the same 1000 J of energy into surroundings that are cold, however, the entropy increase is large (because the impact of the 1000 J is great on surroundings that contain little energy). Therefore, the impact of the heat released to the surroundings by the freezing of water depends on the temperature of the surroundings—the higher the temperature, the smaller the impact.

We can now see why the freezing of water is spontaneous at low temperature but nonspontaneous at high temperature.

$$\Delta S_{univ} = \Delta S_{sys} + \Delta S_{surr}$$

Negative   Positive and large at low temperature
Positive and small at high temperature

At low temperature, the decrease in entropy of the system (a negative quantity) is overcome by the large increase in the entropy of the surroundings (a positive quantity), resulting in a positive ΔS_{univ} and therefore a spontaneous process. At high temperature, however, the decrease in entropy of the system is not overcome by the increase in entropy of the surroundings (because the magnitude of the positive ΔS_{surr} is smaller at higher temperatures), resulting in a negative ΔS_{univ}; therefore, the freezing of water is not spontaneous at high temperature.

Quantifying Entropy Changes in the Surroundings

We have seen that when a system exchanges heat with the surroundings, it changes the entropy of the surroundings. At constant pressure, we can use q_{sys} to quantify the change in entropy for the surroundings (ΔS_{surr}). In general,

- A process that emits heat into the surroundings (q_{sys} negative) *increases* the entropy of the surroundings (positive ΔS_{surr}).
- A process that absorbs heat from the surroundings (q_{sys} positive) *decreases* the entropy of the surroundings (negative ΔS_{surr}).
- The magnitude of the change in entropy of the surroundings is proportional to the magnitude of q_{sys}.

We can summarize these three points with the following proportionality:

$$\Delta S_{surr} \propto -q_{sys} \qquad [17.1]$$

We have also seen that, for a given amount of heat exchanged with the surroundings, the magnitude of ΔS_{surr} is inversely proportional to the temperature. In general, the higher the temperature, the lower the magnitude of ΔS_{surr} for a given amount of heat exchanged:

$$\Delta S_{surr} \propto \frac{1}{T} \qquad [17.2]$$

Combining the proportionalities in Equations 17.1 and 17.2, we get the following general expression at constant temperature:

$$\Delta S_{surr} = \frac{-q_{sys}}{T}$$

For any chemical or physical process occurring at constant temperature and pressure, the entropy change of the surroundings is equal to the energy dispersed into the surroundings ($-\Delta H_{sys}$) divided by the temperature of the surroundings in kelvins.

From this equation, we can see why exothermic processes have a tendency to be spontaneous at low temperatures—they increase the entropy of the surroundings. As temperature increases, however, a given negative ΔH produces a smaller positive ΔS_{surr}; therefore, exothermicity becomes less of a determining factor for spontaneity as temperature increases.

Under conditions of constant pressure $q_{sys} = \Delta H_{sys}$; therefore,

$$\Delta S_{surr} = \frac{-\Delta H_{sys}}{T} \qquad \text{(constant } P, T\text{)} \qquad [17.3]$$

EXAMPLE 17.2 Calculating Entropy Changes in the Surroundings

Consider the combustion of propane gas:

$$C_3H_8(g) + 5\,O_2(g) \longrightarrow 3\,CO_2(g) + 4\,H_2O(g) \qquad \Delta H_{rxn} = -2044 \text{ kJ}$$

(a) Calculate the entropy change in the surroundings associated with this reaction occurring at 25 °C.

(b) Determine the sign of the entropy change for the system.

(c) Determine the sign of the entropy change for the universe. Will the reaction be spontaneous?

Solution

(a) The entropy change of the surroundings is given by Equation 17.3. Simply substitute the value of ΔH_{rxn} and the temperature in kelvins and compute ΔS_{surr}.	$T = 273 + 25 = 298 \text{ K}$ $\Delta S_{surr} = \dfrac{-\Delta H_{rxn}}{T}$ $= \dfrac{-(-2044 \text{ kJ})}{298 \text{ K}}$ $= +6.86 \text{ kJ/K}$ $= +6.86 \times 10^3 \text{ J/K}$
(b) Determine the number of moles of gas on each side of the reaction. An increase in the number of moles of gas implies a positive ΔS_{sys}.	$C_3H_8(g) + 5\,O_2(g) \longrightarrow 3\,CO_2(g) + 4\,H_2O(g)$ $\qquad$ 6 mol gas $\qquad\qquad\qquad\qquad$ 7 mol gas ΔS_{sys} is positive.

(c) The change in entropy of the universe is simply the sum of the entropy changes of the system and the surroundings. If the entropy changes of the system and surroundings are both the same sign, the entropy change for the universe will also have the same sign.	$\Delta S_{univ} = \Delta S_{sys} + \Delta S_{surr}$ Positive Positive Therefore, ΔS_{univ} is positive and reaction is spontaneous.

For Practice 17.2

Consider the following reaction between nitrogen and oxygen gas to form dinitrogen monoxide:

$$2\,N_2(g) + O_2(g) \longrightarrow 2\,N_2O(g) \qquad \Delta H_{rxn} = +163.2\,kJ$$

(a) Calculate the entropy change in the surroundings associated with this reaction occurring at 25 °C.

(b) Determine the sign of the entropy change for the system.

(c) Determine the sign of the entropy change for the universe. Will the reaction be spontaneous?

For More Practice 17.2

A reaction has $\Delta H_{rxn} = -107\,kJ$ and $\Delta S_{rxn} = 285\,J/K$. At what temperature is the change in entropy for the reaction equal to the change in entropy for the surroundings?

 Conceptual Connection 17.3 Entropy and Biological Systems

Biological systems seem to contradict the second law of thermodynamics. By taking energy from their surroundings and synthesizing large, complex biological molecules, plants and animals tend to concentrate energy, not disperse it. How can this be so?

Answer: Biological systems do not violate the second law of thermodynamics. The key to understanding this concept is that entropy changes in the system can be negative as long as the entropy change of the universe is positive. Biological systems can decrease their own entropy, but only at the expense of creating more entropy in the surroundings (which they do primarily by emitting heat generated by their metabolic processes). Thus, for any biological process, ΔS_{univ} is positive.

17.5 Gibbs Free Energy

Equation 17.3 gives us a relationship between the enthalpy change in a system and the entropy change in the surroundings. Recall that for any process the entropy change of the universe is the sum of the entropy change of the system and the entropy change of the surroundings:

$$\Delta S_{univ} = \Delta S_{sys} + \Delta S_{surr} \qquad [17.4]$$

Combining this equation with Equation 17.3 gives us the following relationship at constant termperature and pressure:

$$\Delta S_{univ} = \Delta S_{sys} - \frac{\Delta H_{sys}}{T} \qquad [17.5]$$

Notice that, by using Equation 17.5, we can calculate ΔS_{univ} while focusing only on the *system*. If we multiply Equation 17.5 by $-T$, we get the following expression:

$$-T\,\Delta S_{univ} = -T\,\Delta S_{sys} + \cancel{T}\frac{\Delta H_{sys}}{\cancel{T}}$$

$$= \Delta H_{sys} - T\,\Delta S_{sys} \qquad [17.6]$$

If we now drop the subscript *sys*—from now on ΔH and ΔS without subscripts will mean ΔH_{sys} and ΔS_{sys}—we get the following expression:

$$-T \Delta S_{univ} = \Delta H - T \Delta S \qquad [17.7]$$

The right hand side of Equation 17.7 represents the change in a thermodynamic function called the *Gibbs free energy*. The formal definition of **Gibbs free energy (G)** is

$$G = H - TS \qquad [17.8]$$

where H is enthalpy, T is the temperature in kelvins, and S is entropy. The *change* in Gibbs free energy, symbolized by ΔG, is therefore expressed as follows (at constant temperature):

$$\Delta G = \Delta H - T \Delta S \qquad [17.9]$$

If we combine Equations 17.7 and 17.9, we can understand the significance of ΔG:

$$\Delta G = -T \Delta S_{univ} \qquad \text{(constant } T, P) \qquad [17.10]$$

The change in Gibbs free energy for a process occurring at constant temperature and pressure is proportional to the negative of ΔS_{univ}. Since ΔS_{univ} is a criterion for spontaneity, ΔG is also a criterion for spontaneity (although opposite in sign). In fact, Gibbs free energy is sometimes called *chemical potential*, because it is analogous to mechanical potential energy discussed earlier. Just as mechanical systems tend toward lower potential energy, so chemical systems tend toward lower Gibbs free energy (or toward lower chemical potential) (Figure 17.7 ▼).

Summarizing (at constant temperature and pressure):

➤ ΔG is proportional to the negative of ΔS_{univ}.

➤ A decrease in Gibbs free energy ($\Delta G < 0$) corresponds to a spontaneous process.

➤ An increase in Gibbs free energy ($\Delta G > 0$) corresponds to a nonspontaneous process.

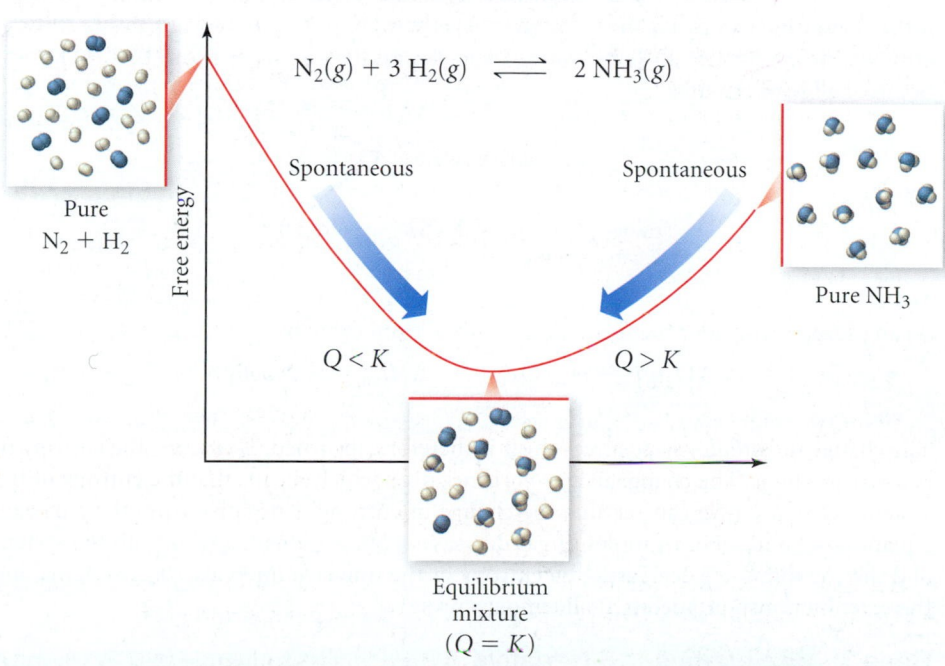

Free Energy Determines the Direction of Spontaneous Change

$$N_2(g) + 3\,H_2(g) \rightleftharpoons 2\,NH_3(g)$$

Pure $N_2 + H_2$

Spontaneous

Spontaneous

Pure NH_3

Free energy

$Q < K$

$Q > K$

Equilibrium mixture $(Q = K)$

▲ **FIGURE 17.7 Gibbs Free Energy** Gibbs free energy is also called chemical potential because it determines the direction of spontaneous change for chemical systems.

Notice that changes in Gibbs free energy can be computed solely with reference to the system. So, to determine whether a process is spontaneous, we simply have to find the change in *entropy* for the system (ΔS) and the change in *enthalpy* for the system (ΔH). We can then predict the spontaneity of the process at any temperature. In Chapter 6, we learned how to calculate changes in enthalpy (ΔH) for chemical reactions. In Section 17.6, we learn how to calculate changes in entropy (ΔS) for chemical reactions. We can then use those two quantities to compute changes in free energy (ΔG) for chemical reactions and therefore predict their spontaneity (Section 17.7). Before we move on to these matters, however, let us examine some examples that demonstrate how ΔH, ΔS, and T affect the spontaneity of chemical processes.

The Effect of ΔH, ΔS, and T on Spontaneity

Case 1: ΔH Negative, ΔS Positive If a reaction is exothermic ($\Delta H < 0$), and if the change in entropy for the reaction is positive ($\Delta S > 0$), then the change in free energy will be negative at all temperatures and the reaction will therefore be spontaneous at all temperatures.

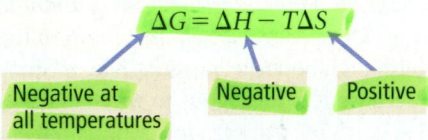

$$\Delta G = \Delta H - T\Delta S$$

Negative at all temperatures   Negative   Positive

As an example, consider the dissociation of N_2O:

$$2\,N_2O(g) \longrightarrow 2\,N_2(g) + O_2(g) \qquad \Delta H^\circ_{rxn} = -163.2 \text{ kJ/mol}$$
2 mol gas   3 mol gas

The change in *enthalpy* is negative—heat is emitted, increasing the entropy of the surroundings. The change in *entropy* for the reaction is positive, which means that the entropy of the system increases. (We can see that the change in entropy is positive from the balanced equation—the number of moles of gas increases.) Since the entropy of both the system and the surroundings increases, the entropy of the universe must also increase, making the reaction spontaneous at all temperatures.

Case 2: ΔH Positive, ΔS Negative If a reaction is endothermic ($\Delta H > 0$), and if the change in entropy for the reaction is negative ($\Delta S < 0$), then the change in free energy will be positive at all temperatures and the reaction will therefore be nonspontaneous at all temperatures.

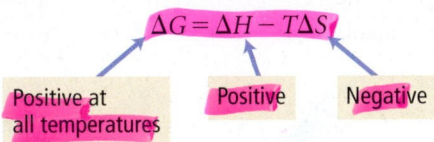

$$\Delta G = \Delta H - T\Delta S$$

Positive at all temperatures   Positive   Negative

As an example, consider the formation of ozone from oxygen:

$$3\,O_2(g) \longrightarrow 2\,O_3(g) \qquad \Delta H^\circ_{rxn} = +285.4 \text{ kJ}$$
3 mol gas   2 mol gas

The change in *enthalpy* is positive—heat is therefore absorbed, *decreasing* the entropy of the surroundings. The change in *entropy* is negative, which means that the entropy of the system decreases. (We can see that the change in entropy is negative from the balanced equation—the number of moles of gas decreases.) Since the entropy of both the system and the surroundings decreases, the entropy of the universe must also decrease, making the reaction nonspontaneous at all temperatures.

Case 3: ΔH Negative, ΔS Negative If a reaction is exothermic ($\Delta H < 0$), and if the change in entropy for the reaction is negative ($\Delta S < 0$), then the change in free energy will depend on temperature. The reaction will be spontaneous at low temperature, but nonspontaneous at high temperature.

Recall from Chapter 6 that ΔH° represents the standard enthalpy change. The definition of the standard state was first given in Section 6.8 and is summarized in Section 17.6.

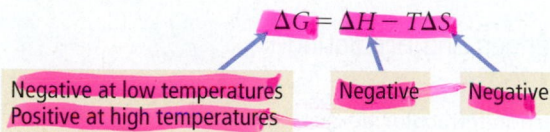

$$\Delta G = \Delta H - T\Delta S$$

Negative at low temperatures
Positive at high temperatures   Negative   Negative

As an example, consider the freezing of liquid water to form ice:

$$H_2O(l) \longrightarrow H_2O(s) \quad \Delta H° = -6.01 \text{ kJ}$$

The change in *enthalpy* is negative—heat is emitted, increasing the entropy of the surroundings. The change in *entropy* is negative, which means that the entropy of the system decreases. (We can see that the change in entropy is negative from the balanced equation because a liquid turns into a solid.)

Unlike the two previous cases, where the changes in *entropy* of the system and of the surroundings had the same sign, the changes in this case are opposite in sign. Therefore, the overall change in free energy depends on the relative magnitudes of the two changes. At a low enough temperature, the heat emitted into the surroundings causes a large entropy change in the surroundings, making the process spontaneous. At high temperature, the same amount of heat is dispersed into warmer surroundings, so the positive entropy change in the surroundings is smaller, resulting in a nonspontaneous process.

Case 4: ΔH Positive, ΔS Positive If a reaction is endothermic ($\Delta H > 0$), and if the change in entropy for the reaction is positive ($\Delta S > 0$), then the change in free energy will also depend on temperature. The reaction will be nonspontaneous at low temperature but spontaneous at high temperature.

$$\Delta G = \Delta H - T\Delta S$$

Positive at low temperatures
Negative at high temperatures   Positive   Positive

As an example, consider the vaporizing of liquid water to gaseous water:

$$H_2O(l) \longrightarrow H_2O(g) \quad \Delta H° = +40.7 \text{ kJ (at 100 °C)}$$

The change in *enthalpy* is positive—heat is absorbed from the surroundings, so the entropy of the surroundings decreases. The change in *entropy* is positive, which means that the entropy of the system increases. (We can see that the change in entropy is positive from the balanced equation because a liquid turns into a gas.) The changes in entropy of the system and the surroundings are again of opposite sign, only this time the entropy of the surroundings decreases while the entropy of the system increases. In cases such as this, high temperature favors spontaneity because the absorption of heat from the surroundings has less effect on the entropy of the surroundings as the temperature increases.

The results of this section are summarized in Table 17.1. Notice that when ΔH and ΔS have opposite signs, the spontaneity of the reaction does not depend on temperature. When ΔH and ΔS have the same sign, however, the spontaneity does depend on temperature. The temperature at which the reaction changes from being spontaneous to being nonspontaneous (or vice versa) is the temperature at which ΔG changes sign, which can be found by setting $\Delta G = 0$ and solving for T, as shown in part b of the following example.

TABLE 17.1 The Effect of ΔH, ΔS, and T on Spontaneity

ΔH	ΔS	Low Temperature	High Temperature	Example
−	+	Spontaneous ($\Delta G < 0$)	Spontaneous ($\Delta G < 0$)	$2 N_2O(g) \longrightarrow 2 N_2(g) + O_2(g)$
+	−	Nonspontaneous ($\Delta G > 0$)	Nonspontaneous ($\Delta G > 0$)	$3 O_2(g) \longrightarrow 2 O_3(g)$
−	−	Spontaneous ($\Delta G < 0$)	Nonspontaneous ($\Delta G > 0$)	$H_2O(l) \longrightarrow H_2O(s)$
+	+	Nonspontaneous ($\Delta G > 0$)	Spontaneous ($\Delta G < 0$)	$H_2O(l) \longrightarrow H_2O(g)$

EXAMPLE 17.3 Computing Gibbs Free Energy Changes and Predicting Spontaneity from ΔH and ΔS

Consider the following reaction for the decomposition of carbon tetrachloride gas:

$$CCl_4(g) \longrightarrow C(s, graphite) + 2\,Cl_2(g) \qquad \Delta H = +95.7\text{ kJ}; \Delta S = +142.2\text{ J/K}$$

(a) Calculate ΔG at 25 °C and determine whether the reaction is spontaneous.

(b) If the reaction is not spontaneous at 25 °C, determine at what temperature (if any) the reaction becomes spontaneous.

Solution

(a) Use Equation 17.9 to calculate ΔG from the given values of ΔH and ΔS. The temperature must be in kelvins. Also, *be sure to express both ΔH and ΔS in the same units (usually joules).*

$T = 273 + 25 = 298$ K

$$\Delta G = \Delta H - T\Delta S$$
$$= 95.7 \times 10^3\text{ J} - (298\text{ K})\,142.2\text{ J/K}$$
$$= 95.7 \times 10^3\text{ J} - 42.4 \times 10^3\text{ J}$$
$$= +53.3 \times 10^3\text{ J}$$

Therefore the reaction is not spontaneous.

(b) Since ΔS is positive, ΔG will become more negative with increasing temperature. To determine the temperature at which the reaction just becomes spontaneous, use Equation 17.9 to find the temperature at which ΔG changes from positive to negative (that is, set $\Delta G = 0$ and solve for T). The reaction will be spontaneous above this temperature.

$$\Delta G = \Delta H - T\Delta S$$
$$0 = 95.7 \times 10^3\text{ J} - (T)\,142.2\text{ J/K}$$
$$T = \frac{95.7 \times 10^3\text{ J}}{142.2\text{ J/K}}$$
$$= 673\text{ K}$$

For Practice 17.3

Consider the following reaction:

$$C_2H_4(g) + H_2(g) \longrightarrow C_2H_6(g) \qquad \Delta H = -137.5\text{ kJ}; \Delta S = -120.5\text{ J/K}$$

Calculate ΔG at 25 °C and determine whether the reaction is spontaneous. Does ΔG become more negative or more positive as the temperature increases?

 Conceptual Connection 17.4 ΔH, ΔS, and ΔG

Which of the following is true for the sublimation of dry ice (solid CO_2?)

(a) ΔH is positive, ΔS is positive, and ΔG is positive at low temperature and negative at high temperature.

(b) ΔH is negative, ΔS is negative, and ΔG is negative at low temperature and positive at high temperature.

(c) ΔH is negative, ΔS is positive, and ΔG is negative at all temperatures.

(d) ΔH is positive, ΔS is negative, and ΔG is positive at all temperatures.

Answer: (a) Since sublimation is endothermic (it requires energy to overcome the intermolecular forces that hold solid carbon dioxide together), ΔH is positive. Since the number of moles of gas increases when the solid turns into a gas, the entropy of the carbon dioxide increases and ΔS is positive. Since $\Delta G = \Delta H - T\Delta S$, ΔG is positive at low temperature and negative at high temperature.

17.6 Entropy Changes in Chemical Reactions: Calculating ΔS°_{rxn}

In Chapter 6, we learned how to calculate standard changes in enthalpy (ΔH°_{rxn}) for chemical reactions. We now turn to calculating standard changes in *entropy* for chemical reactions. Recall from Section 6.8 that we defined the standard enthalpy change for a reaction (ΔH°_{rxn}) as the change in enthalpy for a process in which all reactants and products are in their standard states. Recall also that the standard state of a substance is defined as follows:

- *For a Gas:* The standard state for a gas is the pure gas at a pressure of exactly 1 atm.

- *For a Liquid or Solid:* The standard state for a liquid or solid is the pure substance in its most stable form at a pressure of 1 atm and at the temperature of interest (often taken to be 25 °C).

- *For a Substance in Solution:* The standard state for a substance in solution is a concentration of exactly 1 M.

The standard state has recently been changed to a pressure of 1 bar, which is very close to 1 atm (1 atm = 1.013 bar). Both standards are now in common use.

We now define the **standard entropy change for a reaction (ΔS°_{rxn})** as the change in *entropy* for a process in which all reactants and products are in their standard states. Since entropy is a function of state, the standard change in entropy is therefore just the standard entropy of the products minus the standard entropy of the reactants.

$$\Delta S^\circ_{rxn} = S^\circ_{products} - S^\circ_{reactants}$$

But how do we find the standard entropies of the reactants and products? Recall from Chapter 6 that we defined *standard molar enthalpies of formation (ΔH°_f)* to use in computing in ΔH°_{rxn}. We now want to define **standard molar entropies (S°)** to use in computing ΔS°_{rxn}.

Standard Molar Entropies (S°) and the Third Law of Thermodynamics

In Chapter 6, we defined a *relative* zero for enthalpy. Recall that we assigned a value of zero to the standard enthalpy of formation for an element in its standard state. This was necessary because absolute values of enthalpy cannot be determined. In other words, for enthalpy, there is no absolute zero against which to measure all other values; therefore, we always have to rely on enthalpy changes from an arbitrarily assigned standard. For entropy, however, *there is an absolute zero*. The absolute zero of entropy is given by the *third law of thermodynamics*, which states the following:

> **Third law of thermodynamics: The entropy of a perfect crystal at absolute zero (0 K) is zero.**

A perfect crystal at a temperature of absolute zero has only one possible way ($W = 1$) to arrange its components (Figure 17.8 ▶). Therefore, based on Boltzmann's definition of entropy ($S = k \ln W$), its entropy is zero ($S = k \ln 1 = 0$).

All entropy values can then be measured against the absolute zero of entropy defined by the third law. Table 17.2 shows values of standard entropies at 25 °C for some selected substances. A more complete list can be found in Appendix IIB. Notice that standard entropy values are listed in units of joules per mole per kelvin (J/mol · K). The units of mole in the denominator is required because *entropy is an extensive property*—it depends on the amount of the substance.

Perfect crystal at 0 K
$W = 1 \quad S = 0$

▲ **FIGURE 17.8 Zero Entropy** A perfect crystal at 0 K has only one possible way to arrange for its components.

TABLE 17.2 Standard Molar Entropy Values ($S°$) for Selected Substances at 298 K

Substance	$S°$ (J/mol · K)	Substance	$S°$ (J/mol · K)	Substance	$S°$ (J/mol · K)
Gases		**Liquids**		**Solids**	
$H_2(g)$	130.7	$H_2O(l)$	70.0	$MgO(s)$	27.0
$Ar(g)$	154.8	$CH_3OH(l)$	126.8	$Fe(s)$	27.3
$CH_4(g)$	186.3	$Br_2(l)$	152.2	$Li(s)$	29.1
$H_2O(g)$	188.8	$C_6H_6(l)$	173.4	$Cu(s)$	41.6
$N_2(g)$	191.6			$Na(s)$	51.3
$NH_3(g)$	192.8			$K(s)$	64.7
$F_2(g)$	202.8			$NaCl(s)$	72.1
$O_2(g)$	205.2			$CaCO_3(s)$	91.7
$Cl_2(g)$	223.1			$FeCl_3(s)$	142.3
$C_2H_4(g)$	219.3				

At 25 °C, the standard entropy of any substance is a measure of the energy dispersed into one mole of that substance at 25 °C, which in turn depends on the number of "places" to put energy within the substance. The factors that affect the number of "places" to put energy—and therefore the standard entropy—include the state of the substance, the molar mass of the substance, the particular allotrope, the molecular complexity, and the extent of dissolution. We examine each of these separately.

| Some elements exist in two or more forms, called *allotropes*, within the same state.

Relative Standard Entropies: Gases, Liquids, and Solids

As we saw in Section 17.3, the entropy of a gas is generally greater than the entropy of a liquid, which is in turn greater than the entropy of a solid. We can easily see these trends in the tabulated values of standard entropies. For example, consider the relative standard entropies of liquid water and gaseous water at 25 °C.

	$S°$ (J/mol · K)
$H_2O(l)$	70.0
$H_2O(g)$	188.8

Gaseous water has a much greater standard entropy because, as we discussed in Section 17.3, it has more energetically equivalent ways to arrange its components, which in turn results in greater energy dispersal at 25 °C.

Relative Standard Entropies: Molar Mass

Consider the standard entropies of the noble gases at 25 °C:

	$S°$ (J/mol·K)	
$He(g)$	126.2	
$Ne(g)$	146.1	
$Ar(g)$	154.8	
$Kr(g)$	163.8	
$Xe(g)$	169.4	

The more massive the noble gas, the greater its entropy at 25 °C. A complete explanation of why entropy increases with increasing molar mass is beyond the scope of this book. Briefly, the energy states associated with the motion of heavy atoms are more closely spaced than those of lighter atoms. The more closely spaced energy states allow for greater dispersal of energy at a given temperature and therefore greater entropy. This trend holds only for elements in the same state. (The effect of a state change—from a liquid to a gas, for example—is far greater than the effect of molar mass.)

Relative Standard Entropies: Allotropes As mentioned previously, some elements can exist in two or more forms—called *allotropes*— that both have the same state. For example, the allotropes of carbon include diamond and graphite—both solid forms of carbon. Since the arrangement of atoms within these forms is different, their standard molar entropies are different. The molar entropies of the allotropes of carbon are as follows:

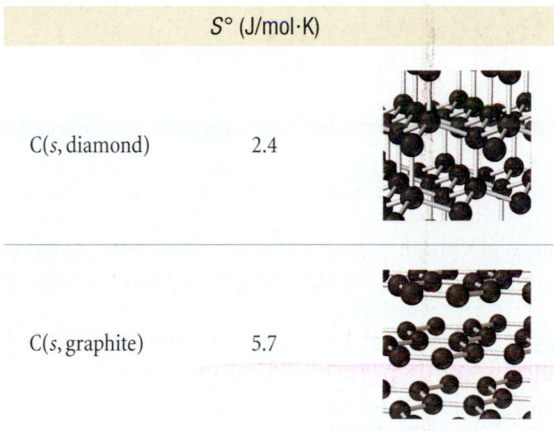

	S° (J/mol·K)	
C(s, diamond)	2.4	
C(s, graphite)	5.7	

In diamond the atoms are constrained by chemical bonds, which lock them in a highly restricted three-dimensional crystal structure. In graphite the atoms bond together in sheets, but the sheets have freedom to slide past each other. The less constrained structure of graphite results in more "places" to put energy and therefore greater entropy compared to diamond.

Relative Standard Entropies: Molecular Complexity For a given state of matter, entropy generally increases with increasing molecular complexity. For example, consider the standard entropies of the following two gases:

	Molar Mass (g/mol)	S° (J/mol · K)
Ar(g)	39.948	154.8
NO(g)	30.006	210.8

Ar has a greater molar mass than NO, yet it has less entropy at 25 °C. Why? Molecules generally have more "places" to put energy than do atoms. In a gaseous sample of argon, the only form that energy can take is the translational motion of the atoms. In a gaseous sample of NO, however, energy can take the form of translational motion, rotational motion, and (at high enough temperatures) vibrational motions of the molecules (Figure 17.9 ▼).

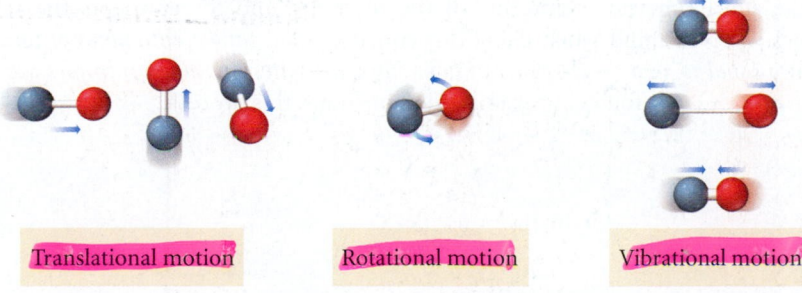

Translational motion	Rotational motion	Vibrational motion

◀ FIGURE 17.9 "Places" for Energy in Gaseous NO Energy can be contained in translational motion, rotational motion, and vibrational motion.

Therefore, for a given state, molecules will generally have a greater entropy than free atoms. Similarly, more complex molecules will generally have more entropy than simpler ones. For example, consider the standard entropies of each of the following:

	Molar Mass (g/mol)	$S°$ (J/mol · K)
$CO(g)$	28.01	197.7
$C_2H_4(g)$	28.05	219.3

These two substances have nearly the same molar mass, but the greater complexity of C_2H_4 compared to CO results in a greater molar entropy. When molecular complexity and molar mass both increase (as is often the case), molar entropy also increases, as shown by the following series:

	$S°$ (J/mol · K)
$NO(g)$	210.8
$NO_2(g)$	240.1
$N_2O_4(g)$	304.4

The increasing molecular complexity as you move down this list as well as the increasing molar mass results in more "places" to put energy and therefore greater entropy.

Relative Standard Entropies: Dissolution The dissolution of a crystalline solid into solution usually results in an increase in entropy. For example, consider the standard entropies of solid and aqueous potassium chlorate:

	$S°$ (J/mol · K)
$KClO_3(s)$	143.1
$KClO_3(aq)$	265.7

The standard entropies for aqueous solutions are for the solution in its standard state, which is defined as having a concentration of 1 M.

When solid potassium chlorate is dissolved in water, the energy that was concentrated within the crystal becomes dispersed throughout the entire solution. The greater energy dispersal results in greater entropy.

Calculating the Standard Entropy Change ($\Delta S°_{rxn}$) for a Reaction Since entropy is a state function, and since standard entropies for many common substances are tabulated, we can calculate the standard entropy change for a chemical reaction by computing the difference in entropy between the products and the reactants. More specifically,

To calculate $\Delta S°_{rxn}$, subtract the standard entropies of the reactants multiplied by their stoichiometric coefficients from the standard entropies of the products multiplied by their stoichiometric coefficients. In the form of an equation,

$$\Delta S°_{rxn} = \sum n_p S° \text{ (products)} - \sum n_r S° \text{ (reactants)} \qquad [17.11]$$

In Equation 17.11, n_p represents the stoichiometric coefficients of the products, n_r represents the stoichiometric coefficients of the reactants, and $S°$ represents the standard entropies. Keep in mind when using this equation that, *unlike enthalpies of formation, which are equal to zero for elements in their standard states, standard entropies are always nonzero at 25 °C.* The following example demonstrates this process.

EXAMPLE 17.4 Computing Standard Entropy Changes (ΔS°_{rxn})

Compute ΔS°_{rxn} for the following balanced chemical equation:

$$4\,NH_3(g) + 5\,O_2(g) \longrightarrow 4\,NO(g) + 6\,H_2O(g)$$

Solution

Begin by looking up (in Appendix IIB) the standard entropy for each reactant and product. Be careful always to note the correct state—(g), (l), (aq), or (s)—for each reactant and product.	**Reactant or product** $\quad$ **S° (in J/mol · K)** $NH_3(g)$ $\qquad$ 192.8 $O_2(g)$ $\qquad$ 205.2 $NO(g)$ $\qquad$ 210.8 $H_2O(g)$ $\qquad$ 188.8

Compute ΔS°_{rxn} by substituting these values into Equation 17.11. Remember to include the stoichiometric coefficients in your calculation.

$$\Delta S^\circ_{rxn} = \sum n_p S^\circ\,(\text{products}) - \sum n_r S^\circ\,(\text{reactants})$$

$$= [4(S^\circ_{NO(g)}) + 6(S^\circ_{H_2O(g)})] - [4(S^\circ_{NH_3(g)}) + 5(S^\circ_{O_2(g)})]$$

$$= [4(210.8\,J/K) + 6(188.8\,J/K)] - [4(192.8\,J/K) + 5(205.2\,J/K)]$$

$$= 1976.0\,J/K - 1797.2\,J/K$$

$$= 178.8\,J/K$$

Check Notice that ΔS°_{rxn} is positive, as you would expect for a reaction in which the number of moles of gas increases.

For Practice 17.4

Compute ΔS°_{rxn} for the following balanced chemical equation:

$$2\,H_2S(g) + 3\,O_2(g) \longrightarrow 2\,H_2O(g) + 2\,SO_2(g)$$

17.7 Free Energy Changes in Chemical Reactions: Calculating ΔG°_{rxn}

In the previous section, we learned how to calculate the standard change in entropy for a chemical reaction (ΔS°_{rxn}). However, the criterion for spontaneity is the **standard change in free energy** (ΔG°_{rxn}). In this section, we examine three ways to calculate the standard change in free energy for a reaction (ΔG°_{rxn}). In the first way, we calculate ΔH°_{rxn} and ΔS°_{rxn} from tabulated values of ΔH°_f and S°, and then use the relationship $\Delta G^\circ_{rxn} = \Delta H^\circ_{rxn} - T\Delta S^\circ_{rxn}$ to calculate ΔG°_{rxn}. In the second way, we use tabulated values of free energies of formation to calculate ΔG°_{rxn} directly. In the third way, we determine the free energy change for a stepwise reaction from the free energies of each of the steps. Finally, we look at what is "free" about free energy. Remember that ΔG°_{rxn} is extremely useful because it tells us about the spontaneity of a process. The more negative that ΔG°_{rxn} is, the more spontaneous the process (the further it will go toward products to reach equilibrium).

Calculating Free Energy Changes with $\Delta G^\circ_{rxn} = \Delta H^\circ_{rxn} - T\Delta S^\circ_{rxn}$

In Chapter 6 (Section 6.8), we learned how to use tabulated values of standard enthalpies of formation to calculate ΔH°_{rxn}. In the previous section, we learned how to use tabulated values of standard entropies to calculate ΔS°_{rxn}. We can use the values of ΔH°_{rxn} and ΔS°_{rxn} calculated in these ways to determine the standard free energy change for a reaction by using the following equation:

$$\Delta G^\circ_{rxn} = \Delta H^\circ_{rxn} - T\Delta S^\circ_{rxn} \qquad [17.12]$$

Since tabulated values of standard enthalpies of formation (ΔH°_f) and standard entropies (S°) are usually at 25 °C, the equation should (strictly speaking) be valid only when $T = 298$ K (25 °C). However, the changes in ΔH°_{rxn} and ΔS°_{rxn} over a limited temperature range are small when compared to the changes in the value of the temperature itself. Therefore, Equation 17.12 can be used to estimate changes in free energy at temperatures other than 25 °C.

EXAMPLE 17.5 Calculating the Standard Change in Free Energy for a Reaction using $\Delta G^\circ_{rxn} = \Delta H^\circ_{rxn} - T\Delta S^\circ_{rxn}$

One of the possible initial steps in the formation of acid rain is the oxidation of the pollutant SO_2 to SO_3 by the following reaction:

$$SO_2(g) + \tfrac{1}{2}O_2(g) \longrightarrow SO_3(g)$$

Calculate ΔG°_{rxn} at 25 °C and determine whether the reaction is spontaneous.

Solution

Begin by looking up (in Appendix IIB) the standard enthalpy of formation and the standard entropy for each reactant and product.

Reactant or product	ΔH°_f (kJ/mol)	S° (in J/mol·K)
$SO_2(g)$	−296.8	248.2
$O_2(g)$	0	205.2
$SO_3(g)$	−395.7	256.8

Calculate ΔH°_{rxn} using Equation 6.15.

$$\Delta H^\circ_{rxn} = \sum n_p \Delta H^\circ_f \,(\text{products}) - \sum n_r \Delta H^\circ_f \,(\text{reactants})$$

$$= [\Delta H^\circ_{f,\,SO_3(g)}] - \left[\Delta H^\circ_{f,\,SO_2(g)} + \tfrac{1}{2}(\Delta H^\circ_{f,\,O_2(g)}) \right]$$

$$= -395.7 \text{ kJ} - (-296.8 \text{ kJ} + 0.0 \text{ kJ})$$

$$= -98.9 \text{ kJ}$$

Calculate ΔS°_{rxn} using the Equation 17.11.

$$\Delta S^\circ_{rxn} = \sum n_p S^\circ(\text{products}) - \sum n_r S^\circ(\text{reactants})$$

$$= [S^\circ_{SO_3(g)}] - \left[S^\circ_{SO_2(g)} + \tfrac{1}{2}(S^\circ_{O_2(g)}) \right]$$

$$= 256.8 \text{ J/K} - \left[248.2 \text{ J/K} + \tfrac{1}{2}(205.2 \text{ J/K}) \right]$$

$$= -94.0 \text{ J/K}$$

Calculate ΔG°_{rxn} using the computed values of ΔH°_{rxn} and ΔS°_{rxn} and Equation 17.12. The temperature must be converted to kelvins.

$$T = 25 + 273 = 298 \text{ K}$$

$$\Delta G^\circ_{rxn} = \Delta H^\circ_{rxn} - T\Delta S^\circ_{rxn}$$

$$= -98.9 \times 10^3 \text{ J} - 298 \text{ K}\,(-94.0 \text{ J/K})$$

$$= -70.9 \times 10^3 \text{ J}$$

$$= -70.9 \text{ kJ}$$

The reaction is therefore spontaneous at this temperature, because ΔG_{rxn} is negative!

For Practice 17.5

Consider the oxidation of NO to NO_2:

$$NO(g) + \tfrac{1}{2}O_2(g) \longrightarrow NO_2(g)$$

Compute ΔG°_{rxn} at 25 °C and determine whether the reaction is spontaneous.

EXAMPLE 17.6 Estimating the Standard Change in Free Energy for a Reaction at a Temperature Other than 25 °C
Using $\Delta G^\circ_{rxn} = \Delta H^\circ_{rxn} - T\Delta S^\circ_{rxn}$

For the reaction in Example 17.5, estimate the value of ΔG°_{rxn} at 125 °C. Does the reaction become more or less spontaneous at this elevated temperature; that is, does the value of ΔG°_{rxn} become more negative (more spontaneous) or more positive (less spontaneous)?

Solution

Estimate ΔG°_{rxn} at the new temperature using the computed values of ΔH°_{rxn} and ΔS°_{rxn} from Example 17.5. For T, use the given temperature converted to kelvins. Make sure to use the same units for ΔH°_{rxn} and ΔS°_{rxn} (usually joules).

$$T = 125 + 273 = 398 \text{ K}$$

$$\begin{aligned} \Delta G^\circ_{rxn} &= \Delta H^\circ_{rxn} - T\Delta S^\circ_{rxn} \\ &= -98.9 \times 10^3 \text{ J} - 398 \text{ K} \, (-94.0 \text{ J/K}) \\ &= -61.5 \times 10^3 \text{ J} \\ &= -61.5 \text{ kJ} \end{aligned}$$

Since the value of ΔG°_{rxn} at this elevated temperature is less negative (or more positive) than the value of ΔG°_{rxn} at 25 °C (which is −70.9 kJ), the reaction is less spontaneous.

For Practice 17.6

For the reaction in For Practice 17.5, calculate the value of ΔG°_{rxn} at −55 °C. Does the reaction become more spontaneous (more negative ΔG°_{rxn}) or less spontaneous (more positive ΔG°_{rxn}) at the lower temperature?

Calculating ΔG°_{rxn} with Tabulated Values of Free Energies of Formation

Since ΔG°_{rxn} is the *change* in free energy for a chemical reaction—the difference in free energy between the products and the reactants—and since free energy is a state function, we can calculate ΔG°_{rxn} by simply subtracting the free energies of the reactants of the reaction from the free energies of the products of the reaction. Also, since we are interested only in *changes* in free energy (and not in absolute values of free energy itself), we are free to define the *zero* of free energy as conveniently as possible. By analogy with our definition of enthalpies of formation, we define the **free energy of formation (ΔG°_f)** as follows:

> The free energy of formation (ΔG°_f) is the change in free energy when 1 mol of a compound forms from its constituent elements in their standard states. The free energy of formation of pure elements in their standard states is zero.

We can then measure all changes in free energy relative to pure elements in their standard states. To calculate ΔG°_{rxn}, subtract the free energies of formation of the reactants multiplied by their stoichiometric coefficients from the free energies of formation of the products multiplied by their stoichiometric coefficients. In the form of an equation:

$$\Delta G^\circ_{rxn} = \sum n_p \Delta G^\circ_f (\text{products}) - \sum n_r \Delta G^\circ_f (\text{reactants}) \qquad [17.13]$$

In Equation 17.13, n_p represents the stoichiometric coefficients of the products, n_r represents the stoichiometric coefficients of the reactants, and ΔG_f° represents the standard free energies of formation. Keep in mind when using this equation that elements in their standard states have $\Delta G_f^\circ = 0$. Table 17.3 shows ΔG_f° values for some selected substances. A more complete list can be found in Appendix IIB. Notice that, by definition, elements have standard free energies of formation of zero. Notice also that most compounds have negative standard free energies of formation. This means that those compounds spontaneously form from their elements in their standard states. Compounds with positive free energies of formation do not spontaneously form from their elements and are therefore less commonly encountered.

The example that follows demonstrates the calculation of ΔG_{rxn}° from ΔG_f° values. Keep in mind, however, that this method of calculating ΔG_{rxn}° works only at the temperature for which the free energies of formation are tabulated, namely, 25 °C. Estimating ΔG_{rxn}° at other temperatures requires the use of $\Delta G_{rxn}^\circ = \Delta H_{rxn}^\circ - T\Delta S_{rxn}^\circ$, as demonstrated previously.

TABLE 17.3 Standard Molar Free Energies of Formation (ΔG_f°) for Selected Substances at 298 K

Substance	ΔG_f° (kJ/mol)	Substance	ΔG_f° (kJ/mol)
$H_2(g)$	0	$CH_4(g)$	−50.5
$O_2(g)$	0	$H_2O(g)$	−228.6
$N_2(g)$	0	$H_2O(l)$	−237.1
$C(s, graphite)$	0	$NH_3(g)$	−16.4
$C(s, diamond)$	2.900	$NO(g)$	+87.6
$CO(g)$	−137.2	$NO_2(g)$	+51.3
$CO_2(g)$	−394.4	$NaCl(s)$	−384.1

EXAMPLE 17.7 ΔG_{rxn}° from Standard Free Energies of Formation

Ozone in the lower atmosphere is a pollutant that can be formed by the following reaction involving the oxidation of unburned hydrocarbons:

$$CH_4(g) + 8\,O_2(g) \longrightarrow CO_2(g) + 2\,H_2O(g) + 4\,O_3(g)$$

Use the standard free energies of formation to determine ΔG_{rxn}° for this reaction at 25 °C.

Solution

Begin by looking up (in Appendix IIB) the standard free energies of formation for each reactant and product. Remember that the standard free energy of formation of a pure element in its standard state is zero.

Reactant/product	ΔG_f° (in kJ/mol)
$CH_4(g)$	−50.5
$O_2(g)$	0.0
$CO_2(g)$	−394.4
$H_2O(g)$	−228.6
$O_3(g)$	163.2

Compute ΔG_{rxn}° by substituting into Equation 17.13.

$$\Delta G_{rxn}^\circ = \sum n_p \Delta G_f^\circ(products) - \sum n_r \Delta G_f^\circ(reactants)$$

$$= [\Delta G_{f,CO_2(g)}^\circ + 2(\Delta G_{f,H_2O(g)}^\circ) + 4(\Delta G_{f,O_3(g)}^\circ)] - [\Delta G_{f,CH_4(g)}^\circ + 8(\Delta G_{f,O_2(g)}^\circ)]$$

$$= [-394.4\ kJ + 2(-228.6\ kJ) + 4(163.2\ kJ)] - [-50.5\ kJ + 8(0.0\ kJ)]$$

$$= -198.8\ kJ + 50.5\ kJ$$

$$= -148.3\ kJ$$

For Practice 17.7

One of the reactions occurring within a catalytic converter in the exhaust pipe of a car is the simultaneous oxidation of carbon monoxide and reduction of NO (both of which are harmful pollutants).

$$2\,CO(g) + 2\,NO(g) \longrightarrow 2\,CO_2(g) + N_2(g)$$

Use standard free energies of formation to find ΔG_{rxn}° for this reaction at 25 °C. Is the reaction spontaneous?

For More Practice 17.7

In For Practice 17.7, you calculated ΔG_{rxn}° for the simultaneous oxidation of carbon monoxide and reduction of NO using standard free energies of formation. Calculate ΔG_{rxn}° again at 25 °C, only this time use $\Delta G_{rxn}^{\circ} = \Delta H_{rxn}^{\circ} - T\Delta S_{rxn}^{\circ}$. How do the two values compare? Use your results to calculate ΔG_{rxn}° at 500.0 K and explain why you could not calculate ΔG_{rxn}° at 500.0 K using tabulated standard free energies of formation.

Determining ΔG_{rxn}° for a Stepwise Reaction from the Changes in Free Energy for Each of the Steps

Recall from Section 6.7 that, since enthalpy is a state function, ΔH_{rxn}° can be calculated for a stepwise reaction from the sum of the changes in enthalpy for each step according to Hess's law. Since free energy is also a state function, the same relationships that we covered in Chapter 6 for enthalpy also apply to free energy:

1. If a chemical equation is multiplied by some factor, then ΔG_{rxn} is also multiplied by the same factor.
2. If a chemical equation is reversed, then ΔG_{rxn} changes sign.
3. If a chemical equation can be expressed as the sum of a series of steps, then ΔG_{rxn} for the overall equation is the sum of the free energies of reactions for each step.

The following example illustrates the use of these relationship to calculate ΔG_{rxn}° for a stepwise reaction.

EXAMPLE 17.8   Determining ΔG_{rxn}° for a Stepwise Reaction

Find ΔG_{rxn}° for the following reaction:

$$3\,C(s) + 4\,H_2(g) \longrightarrow C_3H_8(g)$$

Use the following reactions with known ΔG's:

$$C_3H_8(g) + 5\,O_2(g) \longrightarrow 3\,CO_2(g) + 4\,H_2O(g) \quad \Delta G_{rxn}^{\circ} = -2074\,kJ$$
$$C(s) + O_2(g) \longrightarrow CO_2(g) \quad\quad\quad\quad\quad\quad \Delta G_{rxn}^{\circ} = -394.4\,kJ$$
$$2\,H_2(g) + O_2(g) \longrightarrow 2\,H_2O(g) \quad\quad\quad\quad \Delta G_{rxn}^{\circ} = -457.1\,kJ$$

Solution

To work this problem, manipulate the reactions with known ΔG_{rxn}°'s in such a way as to get the reactants of interest on the left, the products of interest on the right, and other species to cancel.

Since the first reaction has C_3H_8 as a reactant, and the reaction of interest has C_3H_8 as a product, reverse the first reaction and change the sign of ΔG_{rxn}°.	$3\,CO_2(g) + 4\,H_2O(g) \longrightarrow C_3H_8(g) + 5\,O_2(g)$	$\Delta G_{rxn}^{\circ} = +2074\,kJ$
The second reaction has C as a reactant and CO_2 as a product, just as required in the reaction of interest. However, the coefficient for C is 1, and in the reaction of interest, the coefficient for C is 3. Therefore, multiply this equation and its ΔG_{rxn}° by 3.	$3 \times [C(s) + O_2(g) \longrightarrow CO_2(g)]$	$\Delta G_{rxn}^{\circ} = 3 \times (-394.4\,kJ)$ $= -1183\,kJ$

The third reaction has $H_2(g)$ as a reactant, as required. However, the coefficient for H_2 is 2, and in the reaction of interest, the coefficient for H_2 is 4. Therefore multiply this reaction and its ΔG°_{rxn} by 2.	$2 \times [2\,H_2(g) + O_2(g) \longrightarrow 2\,H_2O(g)]$ $\Delta G^\circ_{rxn} = 2 \times (-457.1 \text{ kJ})$ $\phantom{2 \times [2\,H_2(g) + O_2(g) \longrightarrow 2\,H_2O(g)]\quad\Delta G^\circ_{rxn}\ } = -914.2$

Lastly, rewrite the three reactions after multiplying through by the indicated factors and show how they sum to the reaction of interest. ΔG°_{rxn} for the reaction of interest is then just the sum of the ΔG's for the steps.	$3\,CO_2(g) + 4\,H_2O(g) \longrightarrow C_3H_8(g) + 5\,O_2(g)$ $\Delta G^\circ_{rxn} = +2074 \text{ kJ}$ $3\,C(s) + 3\,O_2(g) \longrightarrow 3\,CO_2(g)$ $\Delta G^\circ_{rxn} = -1183 \text{ kJ}$ $4\,H_2(g) + 2\,O_2(g) \longrightarrow 4\,H_2O(g)$ $\Delta G^\circ_{rxn} = -914.2 \text{ kJ}$ $\overline{3\,C(s) + 4\,H_2(g) \longrightarrow C_3H_8(g)}$ $\Delta G^\circ_{rxn} = -23 \text{ kJ}$

For Practice 17.8

Find ΔG°_{rxn} for the following reaction:

$$N_2O(g) + NO_2(g) \longrightarrow 3\,NO(g)$$

Use the following reactions with known ΔG values:

$$2\,NO(g) + O_2(g) \longrightarrow 2\,NO_2(g) \qquad \Delta G^\circ_{rxn} = -71.2 \text{ kJ}$$
$$N_2(g) + O_2(g) \longrightarrow 2\,NO(g) \qquad \Delta G^\circ_{rxn} = +175.2 \text{ kJ}$$
$$2\,N_2O(g) \longrightarrow 2\,N_2(g) + O_2(g) \qquad \Delta G^\circ_{rxn} = -207.4 \text{ kJ}$$

Chemistry in Your Day

Making a Nonspontaneous Process Spontaneous

A process that is nonspontaneous can be made spontaneous by coupling it with another process that is highly spontaneous. For example, hydrogen gas is a potential future fuel because it can be used in a fuel cell (a type of battery in which the reactants are constantly supplied—see Chapter 18) to generate electricity. The main problem with switching to hydrogen is its source: Where can we get the vast amounts of hydrogen gas that would be needed to meet our world's energy needs? Earth's oceans and lakes, of course, contain vast amounts of hydrogen. However, that hydrogen is locked up in water molecules, and the decomposition of water into hydrogen and oxygen has a positive ΔG°_{rxn} and is therefore nonspontaneous.

$$H_2O(g) \longrightarrow H_2(g) + \tfrac{1}{2}O_2(g) \qquad \Delta G^\circ_{rxn} = +228.6 \text{ kJ}$$

The trick to obtaining hydrogen from water is to find another reaction with a highly negative ΔG°_{rxn} that might couple with the decomposition reaction to give an overall reaction with a negative ΔG°_{rxn}. For example, the oxidation of carbon monoxide to carbon dioxide has a large negative ΔG°_{rxn} and is therefore highly spontaneous.

$$CO(g) + \tfrac{1}{2}O_2(g) \longrightarrow CO_2(g) \qquad \Delta G^\circ_{rxn} = -257.2 \text{ kJ}$$

If we add the two reactions together, we get the following:

Nonspontaneous

$$H_2O(g) \longrightarrow H_2(g) + \tfrac{1}{2}O_2(g) \qquad \Delta G^\circ_{rxn} = +228.6 \text{ kJ}$$
$$CO(g) + \tfrac{1}{2}O_2(g) \longrightarrow CO_2(g) \qquad \Delta G^\circ_{rxn} = -257.2 \text{ kJ}$$
$$\overline{H_2O(g) + CO(g) \longrightarrow H_2(g) + CO_2(g)} \qquad \Delta G^\circ_{rxn} = -28.6 \text{ kJ}$$

Spontaneous

The reaction between water and carbon monoxide is therefore a spontaneous way to generate hydrogen gas.

The coupling of nonspontaneous reactions with highly spontaneous ones is also important in biological systems. The synthesis reactions that create the complex biological molecules (such as proteins and DNA) needed by living organisms, for example, are themselves nonspontaneous. However, living systems spontaneously grow and reproduce by coupling these nonspontaneous reactions to highly spontaneous ones. The main spontaneous reaction that ultimately drives the nonspontaneous ones is the metabolism of food. For example, the oxidation of glucose is highly spontaneous:

$$C_6H_{12}O_6(s) + 6\,O_2(g) \longrightarrow 6\,CO_2(g) + 6\,H_2O(l)$$
$$\Delta G^\circ_{rxn} = -2880 \text{ kJ}$$

Spontaneous reactions such as these ultimately drive the nonspontaneous reactions necessary to sustain life.

Why Free Energy Is "Free"

As we have discussed previously, we often want to use the energy released by a chemical reaction to do work. For example, in an automobile, we use the energy released by the combustion of gasoline to move the car forward. The change in free energy of a chemical reaction represents the maximum amount of energy available, or free, to do work (if ΔG°_{rxn} is negative). For many reactions, the amount of free energy is less than the change in enthalpy for the reaction. For example, consider the following reaction occurring at 25 °C:

$$C(s, graphite) + 2\,H_2(g) \longrightarrow CH_4(g)$$

$$\Delta H^\circ_{rxn} = -74.6\ kJ$$
$$\Delta S^\circ_{rxn} = -80.8\ J/K$$
$$\Delta G^\circ_{rxn} = -50.5\ kJ$$

The reaction is exothermic and gives off 74.6 kJ of heat energy. However, the maximum amount of energy available for useful work is only 50.5 kJ (Figure 17.10 ▶). Why? We can see that the change in entropy of the *system* is negative. Nevertheless, the reaction is spontaneous. This is possible only if some of the emitted heat goes to increase the entropy of the surroundings by an amount sufficient to make the change in entropy of the *universe* positive. The amount of energy available to do work (the free energy) is what is left after accounting for the heat that must be lost to the surroundings.

The change in free energy for a chemical reaction represents a *theoretical limit* as to how much work can be done by the reaction. In any *real* reaction, the amount of energy available to do work is even *less* than ΔG°_{rxn} because additional energy is lost to the surroundings as heat. In thermodynamics, a reaction that achieves the theoretical limit with respect to free energy is called a **reversible reaction**. A reversible reaction occurs infinitesimally slowly, and the free energy must be drawn out in infinitesimally small increments that exactly match the amount of energy that the reaction is producing in that increment (Figure 17.11 ▼).

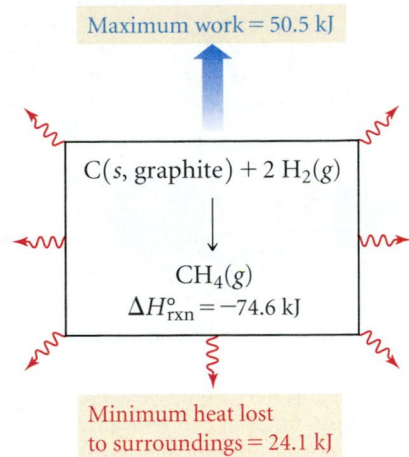

Maximum work = 50.5 kJ

$$C(s, graphite) + 2\,H_2(g)$$
$$\downarrow$$
$$CH_4(g)$$
$$\Delta H^\circ_{rxn} = -74.6\ kJ$$

Minimum heat lost to surroundings = 24.1 kJ

▲ **FIGURE 17.10 Free Energy**
Although the reaction produces 74.6 kJ of heat, only a maximum of 50.5 kJ is available to do work. The rest of the energy is lost to the surroundings.

The formal definition of a reversible reaction is as follows: *A reversible reaction is one that will change direction upon an infinitesimally small change in a variable (such as temperature or pressure) related to the reaction.*

Reversible Process

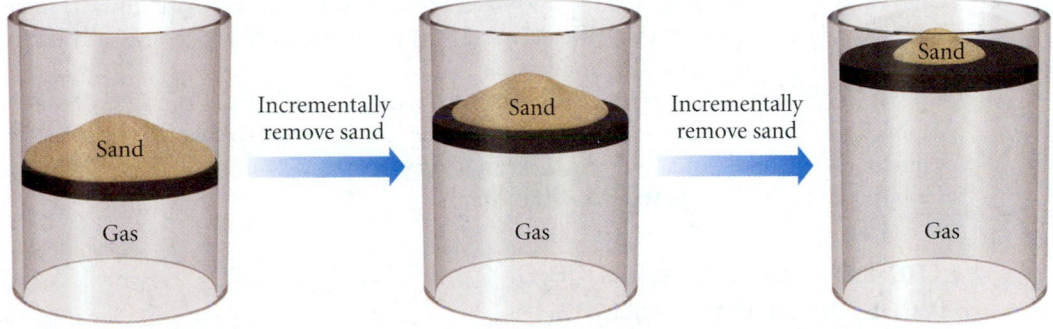

Weight of sand exactly matches pressure at each increment.

Incrementally remove sand

Incrementally remove sand

▲ **FIGURE 17.11 A Reversible Process**   In a reversible process, the free energy is drawn out in infinitesimally small increments that exactly match the amount of energy that the process is producing in that increment. In this case, grains of sand are removed one at a time, resulting in a series of small expansions in which the weight of sand almost exactly matches the pressure of the expanding gas. This process is close to reversible—each sand grain would need to have an infinitesimally small mass for the process to be fully reversible.

All real reactions, however, are **irreversible reactions** and therefore do not achieve the theoretical limit of available free energy. Let's return to our discharging battery from the opening section of this chapter as an example of this concept. A battery simply contains chemical reactants configured in such a way that, upon spontaneous reaction, it will produce an electrical current. The free energy released by the reaction can be harnessed to do work. For example, an electric motor can be wired to the battery and made to turn by

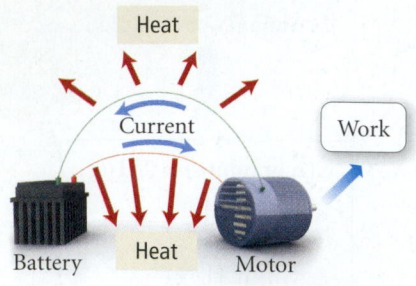

▲ **FIGURE 17.12 Energy Loss in a Battery** When current is drawn from a battery to do work, some energy is necessarily lost as heat due to resistance in the wire. Consequently, the quantity of energy required to recharge the battery will necessarily be more than the quantity of work done.

the flowing electrical current (Figure 17.12 ◄). However, owing to resistance in the wire, the flowing electrical current will also produce some heat, which is lost to the surroundings and is not available to do work. The amount of free energy lost as heat can be lowered by slowing down the rate of current flow. The slower the rate of current flow, the less free energy is lost as heat and the more is available to do work. However, only in the theoretical case of infinitesimally slow current flow will the maximum amount of work (equal to ΔG_{rxn}°) be done. Any real rate of current flow will result in some loss of energy as heat. This lost energy is the "heat tax" also discussed in the opening section of this chapter. Recharging the battery will necessarily require more energy than was obtained as work because some of the energy was lost as heat.

If the change in free energy of a chemical reaction is positive, then ΔG_{rxn}° represents *the minimum amount of energy required to make the reaction occur*. Again, ΔG_{rxn}° represents a theoretical limit. Making a real nonspontaneous reaction occur always requires more energy than the theoretical limit.

17.8 Free Energy Changes for Nonstandard States: The Relationship between ΔG_{rxn}° and ΔG_{rxn}

We have learned how to calculate the *standard* free energy change for a reaction (ΔG_{rxn}°). However, the standard free energy change applies only to a very narrow set of conditions, namely, those conditions in which the reactants and products are in their standard states. For example, consider the standard free energy change for the evaporation of liquid water to gaseous water:

$$H_2O(l) \rightleftharpoons H_2O(g) \qquad \Delta G_{rxn}^{\circ} = +8.59 \text{ kJ/mol}$$

The standard free energy change for this process is positive, so the process is nonspontaneous. However, if you spill water onto the floor under ordinary conditions, it spontaneously evaporates. Why? *Because ordinary conditions are not standard conditions* and ΔG_{rxn}° applies only to standard conditions. For a gas (such as the water vapor in the above reaction), standard conditions are those in which the pure gas is present at a partial pressure of 1 atmosphere. In a flask containing liquid water and water vapor under standard conditions ($P_{H_2O} = 1$ atm) at 25 °C the water would indeed not vaporize. In fact, since ΔG_{rxn}° would be negative for the reverse reaction, the reaction would spontaneously occur in reverse—water vapor would condense.

However, in open air under ordinary circumstances, the partial pressure of water vapor is much less than 1 atm. The conditions are not standard and therefore the value of ΔG_{rxn}° does not apply. For nonstandard conditions, we need to calculate ΔG_{rxn} (as opposed to ΔG_{rxn}°) to predict spontaneity.

▲ Spilled water spontaneously evaporates even though ΔG° for the vaporization of water is positive. Why?

The Free Energy Change of a Reaction under Nonstandard Conditions

The **free energy change of a reaction under nonstandard conditions (ΔG_{rxn})** can be calculated from ΔG_{rxn}° using the relationship

$$\Delta G_{rxn} = \Delta G_{rxn}^{\circ} + RT \ln Q \qquad\qquad [17.14]$$

where Q is the reaction quotient (defined in Section 14.7), T is the temperature in kelvins, and R is the gas constant in the appropriate units (8.314 J/mol · K). We demonstrate the use of this equation by applying it to the liquid–vapor water equilibrium under several different conditions, as shown in Figure 17.13 ▶ Note that by the law of mass action, for this equilibrium, $Q = P_{H_2O}$ (where the pressure is expressed in atmospheres).

$$H_2O(l) \longrightarrow H_2O(g) \qquad Q = P_{H_2O}$$

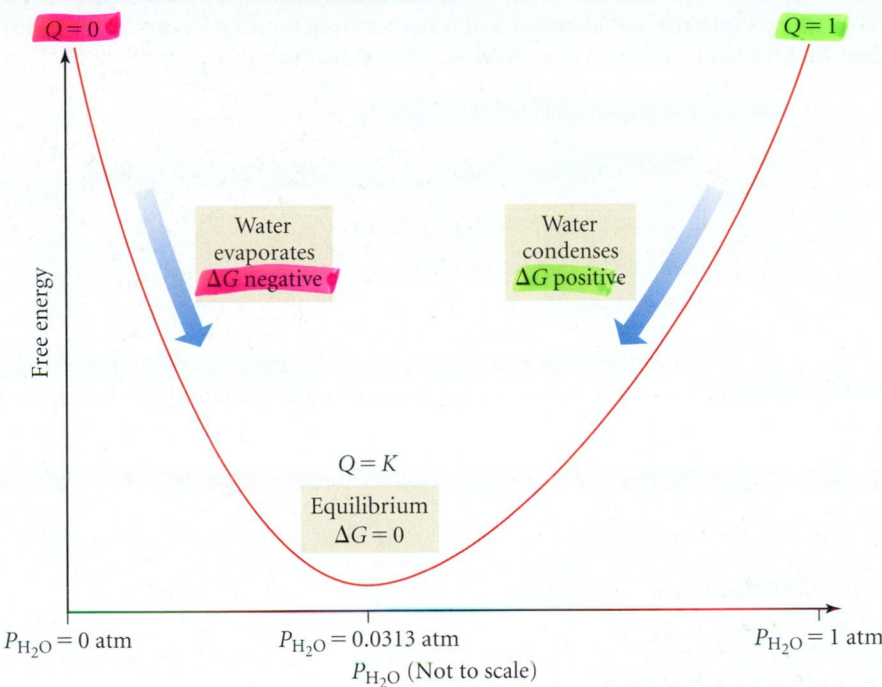

$$H_2O(l) \rightleftharpoons H_2O(g)$$

▲ FIGURE 17.13 Free Energy versus Pressure for Water   The free energy change for the vaporization of water is a function of pressure.

Standard Conditions Under standard conditions, $P_{H_2O} = 1$ atm and therefore $Q = 1$. Substituting, we get the following expression:

$$\Delta G_{rxn} = \Delta G^\circ_{rxn} + RT \ln Q$$
$$= +8.59 \text{ kJ/mol} + RT \ln(1)$$
$$= +8.59 \text{ kJ/mol}$$

Under standard conditions, Q will always be equal to 1, and (since $\ln(1) = 0$) the value of ΔG_{rxn} will therefore be equal to ΔG°_{rxn}, as expected. For the liquid–vapor water equilibrium, since $\Delta G^\circ_{rxn} > 0$, the reaction is not spontaneous in the forward direction but is spontaneous in the reverse direction. As stated previously, under standard conditions water vapor condenses into liquid water.

Equilibrium Conditions At 25.00 °C, liquid water is in equilibrium with water vapor at a pressure of 0.0313 atm; therefore $Q = K_p = 0.0313$. Substituting:

$$\Delta G_{rxn} = \Delta G^\circ_{rxn} + RT \ln(0.0313)$$
$$= +8.59 \text{ kJ/mol} + 8.314 \frac{J}{mol \cdot K}(298.15 \text{ K}) \ln(0.0313)$$
$$= +8.59 \text{ kJ/mol} + (-8.59 \times 10^3 \text{ J/mol})$$
$$= +8.59 \text{ kJ/mol} - 8.59 \text{ kJ/mol}$$
$$= 0$$

Under equilibrium conditions, the value of $RT \ln Q$ will always be equal in magnitude but opposite in sign to the value of ΔG°_{rxn}. Therefore, the value of ΔG_{rxn} will always be zero. Since $\Delta G_{rxn} = 0$, the reaction is not spontaneous in either direction, as expected for a reaction at equilibrium.

A water partial pressure of 5.00×10^{-3} atm corresponds to a relative humidity of 16% at 25 °C.

Other Nonstandard Conditions To calculate the value of ΔG_{rxn} under any other set of nonstandard conditions, simply compute Q and substitute the value into Equation 17.14. For example, the partial pressure of water vapor in the air on a dry (nonhumid) day might be 5.00×10^{-3} atm, so $Q = 5.00 \times 10^{-3}$. Substituting,

$$\Delta G_{rxn} = \Delta G^{\circ}_{rxn} + RT \ln(5.00 \times 10^{-3})$$

$$= +8.59 \, \text{kJ/mol} + 8.314 \frac{\text{J}}{\text{mol} \cdot \text{K}}(298 \, \text{K}) \ln(5.00 \times 10^{-3})$$

$$= +8.59 \, \text{kJ/mol} + (-13.1 \times 10^3 \, \text{J/mol})$$

$$= +8.59 \, \text{kJ/mol} - 13.1 \, \text{kJ/mol}$$

$$= -4.5 \, \text{kJ/mol}$$

Under these conditions, the value of $\Delta G_{rxn} < 0$, so the reaction is spontaneous in the forward direction, consistent with our experience of water evaporating when spilled on the floor.

EXAMPLE 17.9 Calculating ΔG_{rxn} under Nonstandard Conditions

Consider the following reaction at 298 K:

$$2 \, \text{NO}(g) + \text{O}_2(g) \longrightarrow 2 \, \text{NO}_2(g) \quad \Delta G^{\circ}_{rxn} = -71.2 \, \text{kJ}$$

Compute ΔG_{rxn} under the following conditions:

$$P_{\text{NO}} = 0.100 \, \text{atm}; \quad P_{\text{O}_2} = 0.100 \, \text{atm}; \quad P_{\text{NO}_2} = 2.00 \, \text{atm}$$

Is the reaction more or less spontaneous under these conditions than under standard conditions?

Solution

Use the law of mass action to calculate Q.	$Q = \dfrac{P^2_{\text{NO}_2}}{P^2_{\text{NO}} P_{\text{O}_2}}$ $= \dfrac{(2.00)^2}{(0.100)^2(0.100)}$ $= 4.00 \times 10^3$
Substitute Q, T, and ΔG°_{rxn} into Equation 17.14 to calculate ΔG_{rxn}. (Since the units of R include joules, write ΔG°_{rxn} in joules.)	$\Delta G_{rxn} = \Delta G^{\circ}_{rxn} + RT \ln Q$ $= -71.2 \times 10^3 \, \text{J} + 8.314 \frac{\text{J}}{\text{mol} \cdot \text{K}}(298 \, \text{K}) \ln(4.00 \times 10^3)$ $= -71.2 \times 10^3 \, \text{J} + 20.5 \times 10^3 \, \text{J}$ $= -50.7 \times 10^3 \, \text{J}$ $= -50.7 \, \text{kJ}$ The reaction is spontaneous under these conditions, but less spontaneous than it was under standard conditions (because ΔG_{rxn} is less negative than ΔG°_{rxn}).

Check The calculated result is consistent with what we would expect based on Le Châtelier's principle; increasing the concentration of the products and decreasing the concentration of the reactants relative to standard conditions should make the reaction less spontaneous than it was under standard conditions.

For Practice 17.9

Consider the following reaction at 298 K:

$$2\,H_2S(g) + SO_2(g) \longrightarrow 3\,S(s, \text{rhombic}) + 2\,H_2O(g) \quad \Delta G^\circ_{rxn} = -102\,kJ$$

Compute ΔG_{rxn} under the following conditions:

$$P_{H_2S} = 2.00\,\text{atm}; \; P_{SO_2} = 1.50\,\text{atm}; \; P_{H_2O} = 0.0100\,\text{atm}$$

Is the reaction more or less spontaneous under these conditions than under standard conditions?

Conceptual Connection 17.5 Free Energy Changes and Le Châtelier's Principle

According to Le Châtelier's principle and the dependence of free energy on reactant and product concentrations, which of the following statements is true? (Assume that both the reactants and products are gaseous.)

(a) A high concentration of reactants relative to products results in a more spontaneous reaction than one in which the reactants and products are in their standard states.

(b) A high concentration of products relative to reactants results in a more spontaneous reaction than one in which the reactants and products are in their standard states.

(c) A reaction in which the reactants are in standard states, but in which no products have formed, will have a ΔG_{rxn} that is more positive than ΔG°_{rxn}.

Answer: **(a)** A high concentration of reactants relative to products will lead to $Q < 1$, making the term $RT \ln Q$ in Equation 17.14 negative. ΔG_{rxn} will therefore be more negative than ΔG°_{rxn} and the reaction will therefore be more spontaneous.

17.9 Free Energy and Equilibrium: Relating ΔG°_{rxn} to the Equilibrium Constant (K)

We have learned throughout this chapter that ΔG°_{rxn} determines the spontaneity of a reaction when the reactants and products are in their standard states. In Chapter 14, we learned that the equilibrium constant (K) determines how far a reaction goes toward products, a measure of spontaneity. Therefore, as you might expect, the standard free energy change of a reaction and the equilibrium constant are related—the equilibrium constant becomes larger as the free energy change becomes more negative. In other words, if the reactants in a particular reaction undergo a large *negative* free energy change in going to products, then the reaction will have a large equilibrium constant, with products strongly favored at equilibrium. If, on the other hand, the reactants in a particular reaction undergo a large *positive* free energy change in going to products, then the reaction will have a small equilibrium constant, with reactants strongly favored at equilibrium.

We can derive a relationship between ΔG°_{rxn} and K from Equation 17.14. We know that at equilibrium $Q = K$ and $\Delta G_{rxn} = 0$. Making these substitutions,

$$\Delta G_{rxn} = \Delta G^\circ_{rxn} + RT \ln Q$$

$$0 = \Delta G^\circ_{rxn} + RT \ln K$$

$$\Delta G^\circ_{rxn} = -RT \ln K \qquad\qquad [17.15]$$

Notice that the relationship between ΔG_{rxn} and K is logarithmic—small changes in ΔG_{rxn} have a large effect on K.

We can better understand the relationship between ΔG_{rxn}° and K by considering the following ranges of values for K, as summarized in Figure 17.14 ▼.

- When $K < 1$, ln K is negative and ΔG_{rxn}° is therefore positive. Under standard conditions (when $Q = 1$) the reaction is spontaneous in the reverse direction.
- When $K > 1$, ln K is positive and ΔG_{rxn}° is therefore negative. Under standard conditions (when $Q = 1$) the reaction is spontaneous in the forward direction.
- When $K = 1$, ln K is zero and ΔG_{rxn}° is therefore zero. The reaction happens to be at equilibrium under standard conditions.

Free Energy and the Equilibrium Constant

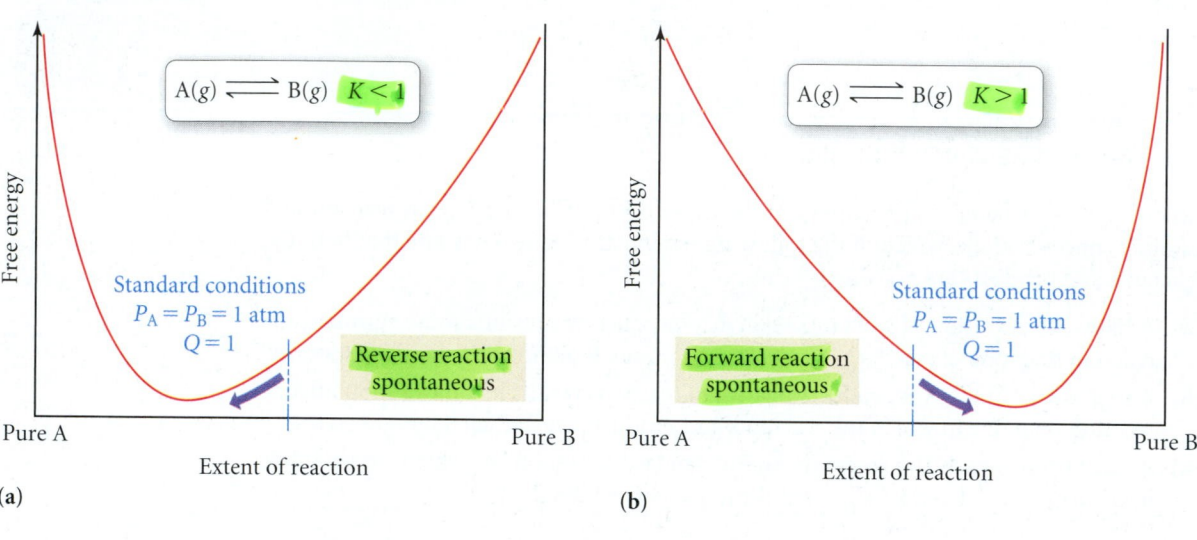

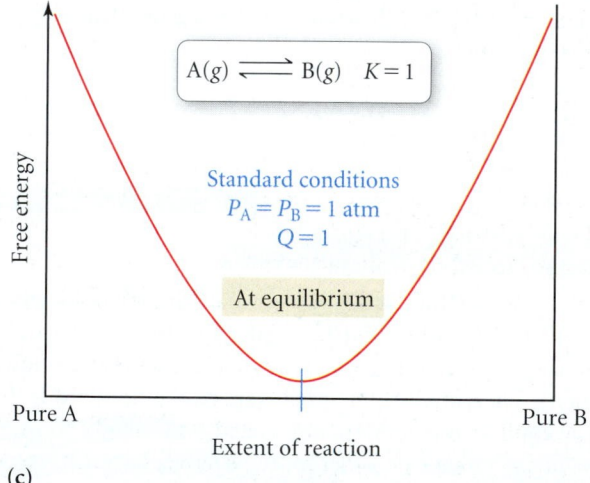

▶ **FIGURE 17.14 Free Energy and the Equilibrium Constant** (a) Free energy curve for a reaction with a small equilibrium constant. (b) Free energy curve for a reaction with a large equilibrium constant. (c) Free energy curve for a reaction in which $K = 1$.

EXAMPLE 17.10 The Equilibrium Constant and ΔG_{rxn}°

Use tabulated free energies of formation to calculate the equilibrium constant for the following reaction at 298 K:

$$N_2O_4(g) \rightleftharpoons 2\,NO_2(g)$$

Solution

Begin by looking up (in Appendix IIB) the standard free energies of formation for each reactant and product.

Reactant or product	ΔG_f° (in kJ/mol)
$N_2O_4(g)$	99.8
$NO_2(g)$	51.3

Compute ΔG°_{rxn} by substituting into Equation 17.13.	$\Delta G^\circ_{rxn} = \sum n_p \Delta G^\circ_f(\text{products}) - \sum n_r \Delta G^\circ_f(\text{reactants})$
	$= 2[\Delta G^\circ_f(NO_2)] - \Delta G^\circ_f(N_2O_4)$
	$= 2(51.3 \text{ kJ}) - 99.8 \text{ kJ}$
	$= 2.8 \text{ kJ}$
Compute K from ΔG°_{rxn} by solving Equation 17.15 for K and substituting the values of ΔG°_{rxn} and temperature.	$\Delta G^\circ_{rxn} = -RT \ln K$
	$\ln K = \dfrac{-\Delta G^\circ_{rxn}}{RT}$
	$= \dfrac{-2.8 \times 10^3 \text{ J/mol}}{8.314 \dfrac{\text{J}}{\text{mol} \cdot \text{K}}(298 \text{ K})}$
	$= -1.13$
	$K = e^{-1.13}$
	$= 0.32$

For Practice 17.10

Compute ΔG°_{rxn} at 298 K for the following reaction:

$$I_2(g) + Cl_2(g) \rightleftharpoons 2 \, ICl(g) \quad K_p = 81.9$$

The Temperature Dependence of the Equilibrium Constant

We now have an equation that relates the standard free energy change for a reaction (ΔG°_{rxn}) to the equilibrium constant for a reaction (K):

$$\Delta G^\circ_{rxn} = -RT \ln K \qquad [17.16]$$

We also have an equation for how the free energy change for a reaction (ΔG°_{rxn}) depends on temperature (T):

$$\Delta G^\circ_{rxn} = \Delta H^\circ_{rxn} - T \Delta S^\circ_{rxn} \qquad [17.17]$$

We can therefore combine these two equations to obtain an equation for how the equilibrium constant depends on temperature. Combining Equations 17.16 and 17.17, we get the following:

$$-RT \ln K = \Delta H^\circ_{rxn} - T \Delta S^\circ_{rxn} \qquad [17.18]$$

We can then divide both sides of Equation 17.18 by the quantity RT as follows:

$$-\ln K = \frac{\Delta H^\circ_{rxn}}{RT} - \frac{T \Delta S^\circ_{rxn}}{RT}$$

Cancelling and rearranging, we get the following important result:

$$\ln K = -\frac{\Delta H^\circ_{rxn}}{R}\left(\frac{1}{T}\right) + \frac{\Delta S^\circ_{rxn}}{R} \qquad [17.19]$$

$$y = \quad\quad mx \quad + \quad b$$

Equation 17.19 is in the form of a straight line. A plot of the natural log of the equilibrium constant ($\ln K$) versus the inverse of the temperature in kelvins ($1/T$) yields a straight line with a slope of $-\Delta H^\circ_{rxn}/R$ and a y-intercept of ΔS°_{rxn}. Such a plot is useful for obtaining thermodynamic data (namely, ΔH°_{rxn} and ΔS°_{rxn}) from measurements of K as a function of temperature. However, since ΔH°_{rxn} and ΔS°_{rxn} can themselves be slightly temperature dependent, this analysis works only over a relatively limited temperature range.

Chapter in Review

Key Terms

Section 17.2
spontaneous process (777)

Section 17.3
entropy (S) (780)
second law of thermodynamics (783)

Section 17.5
Gibbs free energy (G) (789)

Section 17.6
standard entropy change for a reaction (ΔS°_{rxn}) (793)
standard molar entropies (S°) (793)

third law of thermodynamics (793)

Section 17.7
standard free energy change (ΔG°_{rxn}) (797)
free energy of formation (ΔG°_f) (799)

reversible reaction (803)
irreversible reaction (803)
free energy change of a reaction under nonstandard conditions (ΔG_{rxn}) (804)

Key Concepts

Nature's Heat Tax: You Can't Win and You Can't Break Even (17.1)

The first law of thermodynamics states that energy can be neither created nor destroyed. The second law implies that for every energy transaction, some energy is lost to the surroundings; this lost energy is nature's "heat tax."

Spontaneous and Nonspontaneous Processes (17.2)

Both spontaneous and nonspontaneous processes can occur, but only spontaneous processes can take place without outside intervention. Thermodynamics is the study of the *spontaneity* of a reaction, *not* to be confused with kinetics, the study of the *rate* of a reaction.

Entropy and the Second Law of Thermodynamics (17.3)

The second law of thermodynamics states that for *any* spontaneous process, the entropy of the universe increases. Entropy (S) is proportional to the number of energetically equivalent ways in which the components of a system can be arranged and is a measure of energy dispersal per unit temperature. An example of a process in which entropy changes is a phase change, such as a change from a solid to a liquid.

Heat Transfer and Changes in the Entropy of the Surroundings (17.4)

For a process to be spontaneous, the total entropy of the universe (system plus surroundings) must increase. The entropy of the surroundings increases when the change in *enthalpy* of the system (ΔH_{sys}) is negative (i.e., for exothermic reactions). The change in entropy of the surroundings for a given ΔH_{sys} depends inversely on temperature— the greater the temperature, the lower the magnitude of ΔS_{surr}.

Gibbs Free Energy (17.5)

Gibbs free energy, G, is a thermodynamic function that is proportional to the negative of the change in the entropy of the universe. Therefore, a negative ΔG represents a spontaneous reaction and a positive ΔG represents a nonspontaneous reaction. The value of ΔG for a reaction can be calculated from the values of ΔH and ΔS for the *system* according to the equation $\Delta G = \Delta H - T\Delta S$.

Entropy Changes in Chemical Reactions: Calculating ΔS°_{rxn} (17.6)

The standard change in entropy for a reaction is calculated similarly to the standard change in enthalpy for a reaction: by subtracting the sum of the standard entropies of the reactants multiplied by their stoichiometric coefficients from the sum of the standard entropies of the products multiplied by their stoichiometric coefficients. In this equation, the standard entropies are *absolute*: an entropy of zero is determined by the third law of thermodynamics as the entropy of a perfect crystal at absolute zero. The absolute entropy of a substance depends on factors that affect the number of energetically equivalent arrangements of the substance; these include the phase, size, and molecular complexity of the substance.

Free Energy Changes in Chemical Reactions: Calculating ΔG°_{rxn} (17.7)

There are three methods of calculating ΔG°_{rxn}: (1) from ΔH° and ΔS°; (2) from free energies of formations (only at 25 °C); and (3) from the ΔG°'s of reactions that sum to the reaction of interest. The magnitude of a negative ΔG°_{rxn} represents the theoretical amount of energy available to do work while a positive ΔG°_{rxn} represents the minimum amount of energy required to make a nonspontaneous process occur.

Free Energy Changes for Nonstandard States: The Relationship between ΔG°_{rxn} and ΔG_{rxn} (17.8)

The value of ΔG°_{rxn} applies only to standard conditions, and most real conditions are not standard. Under nonstandard conditions, ΔG_{rxn} can be calculated from the equation $\Delta G_{rxn} = \Delta G^\circ_{rxn} + RT \ln Q$.

Free Energy and Equilibrium: Relating ΔG°_{rxn} to the Equilibrium Constant (K) (17.9)

Under standard conditions, the free energy change for a reaction is directly proportional to the negative of the natural log of the equilibrium constant, K; the more negative the free energy change (i.e., the more spontaneous the reaction), the larger the equilibrium constant. The temperature dependence of ΔG°_{rxn}, as given by $\Delta G^\circ = \Delta H^\circ - T\Delta S$, can be used to derive an expression for the temperature dependence of the equilibrium constant.

Key Equations and Relationships

The Definition of Entropy (17.3)
$$S = k \ln W \qquad k = 1.38 \times 10^{-23} \, \text{J/K}$$

Change in Entropy (17.3)
$$\Delta S = S_{\text{final}} - S_{\text{initial}}$$

Change in the Entropy of the Universe (17.4)
$$\Delta S_{\text{univ}} = \Delta S_{\text{sys}} + \Delta S_{\text{surr}}$$

Change in the Entropy of the Surroundings (17.4)
$$\Delta S_{\text{surr}} = \frac{-\Delta H_{\text{sys}}}{T} \quad (\text{constant } T, P)$$

Change in Gibb's Free Energy (17.5)
$$\Delta G = \Delta H - T\Delta S$$

The Relationship between Spontaneity and ΔH, ΔS, and T (17.5)

ΔH	ΔS	Low Temperature	High Temperature
−	+	Spontaneous	Spontaneous
+	−	Nonspontaneous	Nonspontaneous
−	−	Spontaneous	Nonspontaneous
+	+	Nonspontaneous	Spontaneous

Standard Change in Entropy (17.6)
$$\Delta S_{\text{rxn}}^{\circ} = \sum n_{\text{p}} S^{\circ}(\text{products}) - \sum n_{\text{r}} S^{\circ}(\text{reactants})$$

Methods for Calculating the Free Energy of Formation ($\Delta G_{\text{rxn}}^{\circ}$) (17.7)

1. $\Delta G_{\text{rxn}}^{\circ} = \Delta H_{\text{rxn}}^{\circ} - T\Delta S_{\text{rxn}}^{\circ}$

2. $\Delta G_{\text{rxn}}^{\circ} = \sum n_{\text{p}} \Delta G_{\text{f}}^{\circ}(\text{products}) - \sum n_{\text{r}} \Delta G_{\text{f}}^{\circ}(\text{reactants})$

3. $\Delta G_{\text{rxn(overall)}}^{\circ} = \Delta G_{\text{rxn(step 1)}}^{\circ} + \Delta G_{\text{rxn(step 2)}}^{\circ} + \Delta G_{\text{rxn(step 3)}}^{\circ} + \cdots$

The Relationship between $\Delta G_{\text{rxn}}^{\circ}$ and ΔG_{rxn} (17.8)
$$\Delta G_{\text{rxn}} = \Delta G_{\text{rxn}}^{\circ} + RT \ln Q \qquad R = 8.314 \, \text{J/mol} \cdot \text{K}$$

The Relationship between $\Delta G_{\text{rxn}}^{\circ}$ and K (17.9)
$$\Delta G_{\text{rxn}}^{\circ} = -RT \ln K$$

The Temperature Dependence of the Equilibrium Constant (17.9)
$$\ln K = -\frac{\Delta H_{\text{rxn}}^{\circ}}{R}\left(\frac{1}{T}\right) + \frac{\Delta S_{\text{rxn}}^{\circ}}{R}$$

Key Skills

Predicting the Sign of Entropy Change (17.3)

• Example 17.1   • For Practice 17.1   • Exercises 27, 28, 31–34, 37, 38

Calculating Entropy Changes in the Surroundings (17.4)

• Example 17.2   • For Practice 17.2   • For More Practice 17.2. Exercises 33–36

Computing Gibbs Free Energy Changes and Predicting Spontaneity from ΔH and ΔS (17.5)

• Example 17.3   • For Practice 17.3   • Exercises 39–44

Computing Standard Entropy Changes ($\Delta S_{\text{rxn}}^{\circ}$)(17.6)

• Example 17.4   • For Practice 17.4   • Exercises 51, 52

Calculating the Standard Change in Free Energy for a Reaction using $\Delta G_{\text{rxn}}^{\circ} = \Delta H_{\text{rxn}}^{\circ} - T\Delta S_{\text{rxn}}^{\circ}$ (17.7)

• Examples 17.5, 17.6   • For Practice 17.5, 17.6   • Exercises 55–58, 61, 62

Calculating $\Delta G_{\text{rxn}}^{\circ}$ from Standard Free Energies of Formation (17.7)

• Example 17.7   • For Practice 17.7   • For More Practice 17.7   • Exercises 59, 60

Determining $\Delta G_{\text{rxn}}^{\circ}$ for a Stepwise Reaction (17.7)

• Example 17.8   • For Practice 17.8   • Exercises 63, 64

Calculating ΔG_{rxn} under Nonstandard Conditions (17.8)

• Example 17.9   • For Practice 17.9   • Exercises 65–72

Relating the Equilibrium Constant and $\Delta G_{\text{rxn}}^{\circ}$ (17.9)

• Example 17.10   • For Practice 17.10   • Exercises 73–76

Exercises

Review Questions

1. What is the first law of thermodynamics and how does it relate to energy use?

2. What is nature's "heat tax" and how does it relate to energy use?

3. What is a perpetual motion machine? Is such a machine possible given the laws of thermodynamics?

4. Is it more efficient to heat your home with a natural gas furnace or an electric furnace? Explain.

5. What is a spontaneous process? Give an example.

6. Explain the difference between the spontaneity of a reaction (which depends on thermodynamics) and the speed at which the reaction occurs (which depends on kinetics). Can a catalyst make a nonspontaneous reaction spontaneous?

7. What is the precise definition of entropy? Explain the significance of entropy being a state function.

8. Explain why the entropy of a gas increases when it expands into a vacuum.

9. Explain the difference between macrostates (external arrangements of particles) and microstates (internal arrangements of particles).

10. Based on its fundamental definition, explain why entropy is a measure of energy dispersion.

11. Give the definition of the second law of thermodynamics. How does the second law explain why heat travels from a substance at higher temperature to one at lower temperature?

12. What happens to the entropy of a sample of matter when it changes state from a solid to a liquid? From a liquid to a gas?

13. Explain why water spontaneously freezes to form ice below $0\,°C$ even though the entropy of the water decreases during the phase transition. Why is the freezing of water not spontaneous above $0\,°C$?

14. Why do exothermic processes tend to be spontaneous at low temperatures? Why does their tendency toward spontaneity decrease with increasing temperature?

15. What is the significance of the change in Gibbs free energy (ΔG) for a reaction?

16. Predict the spontaneity of a reaction (and the temperature dependence of the spontaneity) for each of the following (for the system):
    a. ΔH negative, ΔS positive
    b. ΔH positive, ΔS negative
    c. ΔH negative, ΔS negative
    d. ΔH positive, ΔS positive

17. State the third law of thermodynamics and describe its significance.

18. Explain why the standard entropy of substance in the gas state is greater than its standard entropy in the liquid state.

19. How does the standard entropy of a substance depend on its molar mass? On its molecular complexity?

20. How is the standard entropy change for a reaction calculated from tables of standard entropies?

21. What are three different ways to calculate $\Delta G°$ for a reaction? Explain which way you would choose to calculate $\Delta G°$ for a reaction at a temperature other than $25\,°C$.

22. Why is free energy "free"?

23. Explain the difference between $\Delta G°$ and ΔG.

24. Explain why water spilled on the floor evaporates even though $\Delta G°$ for the evaporation process is positive at room temperature.

25. How do you calculate the change in free energy for a reaction under nonstandard conditions?

26. How is the value of $\Delta G°$ for a reaction related to the equilibrium constant for the reaction? What does a negative $\Delta G°$ for a reaction imply about K for the reaction? A positive $\Delta G°$?

Problems by Topic

Entropy, the Second Law of Thermodynamics, and the Direction of Spontaneous Change

27. Which of the following processes are spontaneous?
    a. the combustion of natural gas
    b. the extraction of iron metal from iron ore
    c. a hot drink cooling to room temperature
    d. drawing heat energy from the ocean's surface to power a ship

28. Which of the following processes are nonspontaneous? Are the nonspontaneous processes impossible?
    a. a bike going up a hill
    b. a meteor falling to Earth
    c. obtaining hydrogen gas from liquid water
    d. a ball rolling down a hill

29. Suppose a system of two particles, represented by circles, have the possibility of occupying energy states with 0, 10, or 20 J. Collectively, the particles must have 20 J of total energy. One way the two particles can distribute themselves is:

20 J _____
10 J __oo__
0 J   _____

Are there any other energetically equivalent configurations? If so, which configuration has the greatest entropy?

30. Suppose a system of three particles, represented by circles, have the possibility of occupying energy states with 0, 10, or 20 J. Collectively, the particles must have 30 J of total energy. One way the two particles can distribute themselves is shown below.

20 J _____
10 J _ooo_
0 J   _____

Are there any other energetically equivalent configurations? If so, which configuration has the greatest entropy?

31. Without doing any calculations, determine the sign of ΔS_{sys} for each of the following chemical reactions:

a. $2 KClO_3(s) \longrightarrow 2 KCl(s) + 3 O_2(g)$
b. $CH_2{=}CH_2(g) + H_2(g) \longrightarrow CH_3CH_3(g)$
c. $Na(s) + 1/2 Cl_2(g) \longrightarrow NaCl(s)$
d. $N_2(g) + 3 H_2(g) \longrightarrow 2 NH_3(g)$

32. Without doing any calculations, determine the sign of ΔS_{sys} for each of the following chemical reactions:
a. $Mg(s) + Cl_2(g) \longrightarrow MgCl_2(s)$
b. $2 H_2S(g) + 3 O_2(g) \longrightarrow 2 H_2O(g) + 2 SO_2(g)$
c. $2 O_3(g) \longrightarrow 3 O_2(g)$
d. $HCl(g) + NH_3(g) \longrightarrow NH_4Cl(s)$

33. Without doing any calculations, determine the sign of ΔS_{sys} and ΔS_{surr} for each of the chemical reactions below. In addition, predict under what temperatures (all temperatures, low temperatures, or high temperatures), if any, the reaction will be spontaneous.
a. $C_3H_8(g) + 5 O_2(g) \longrightarrow 3 CO_2(g) + 4 H_2O(g)$
$$\Delta H^\circ_{rxn} = -2044 \text{ kJ}$$
b. $N_2(g) + O_2(g) \longrightarrow 2 NO(g) \quad \Delta H^\circ_{rxn} = +182.6 \text{ kJ}$
c. $2 N_2(g) + O_2(g) \longrightarrow 2 N_2O(g) \quad \Delta H^\circ_{rxn} = +163.2 \text{ kJ}$
d. $4 NH_3(g) + 5 O_2(g) \longrightarrow 4 NO(g) + 6 H_2O(g)$
$$\Delta H^\circ_{rxn} = -906 \text{ kJ}$$

34. Without doing any calculations, determine the sign of ΔS_{sys} and ΔS_{surr} for each of the chemical reactions below. In addition, predict under what temperatures (all temperatures, low temperatures, or high temperatures), if any, the reaction will be spontaneous.
a. $2 CO(g) + O_2(g) \longrightarrow 2 CO_2(g) \quad \Delta H^\circ_{rxn} = -566.0 \text{ kJ}$
b. $2 NO_2(g) \longrightarrow 2 NO(g) + O_2(g) \quad \Delta H^\circ_{rxn} = +113.1 \text{ kJ}$
c. $2 H_2(g) + O_2(g) \longrightarrow 2 H_2O(g) \quad \Delta H^\circ_{rxn} = -483.6 \text{ kJ}$
d. $CO_2(g) \longrightarrow C(s) + O_2(g) \quad \Delta H^\circ_{rxn} = +393.5 \text{ kJ}$

35. Calculate ΔS_{surr} at the indicated temperature for a reaction having each of the following changes in enthalpy:
a. $\Delta H^\circ_{rxn} = -287 \text{ kJ}; 298 \text{ K}$
b. $\Delta H^\circ_{rxn} = -287 \text{ kJ}; 77 \text{ K}$
c. $\Delta H^\circ_{rxn} = +127 \text{ kJ}; 298 \text{ K}$
d. $\Delta H^\circ_{rxn} = +127 \text{ kJ}; 77 \text{ K}$

36. A reaction has $\Delta H_{rxn} = -127 \text{ kJ}$ and $\Delta S_{rxn} = 314 \text{ J/K}$. At what temperature is the change in entropy for the reaction equal to the change in entropy for the surroundings?

37. Given the values of ΔH°_{rxn}, ΔS°_{rxn}, and T below, determine ΔS_{univ} and predict whether or not the reaction will be spontaneous.
a. $\Delta H^\circ_{rxn} = -125 \text{ kJ}; \Delta S^\circ_{rxn} = +253 \text{ J/K}; T = 298 \text{ K}$
b. $\Delta H^\circ_{rxn} = +125 \text{ kJ}; \Delta S^\circ_{rxn} = -253 \text{ J/K}; T = 298 \text{ K}$
c. $\Delta H^\circ_{rxn} = -125 \text{ kJ}; \Delta S^\circ_{rxn} = -253 \text{ J/K}; T = 298 \text{ K}$
d. $\Delta H^\circ_{rxn} = -125 \text{ kJ}; \Delta S^\circ_{rxn} = -253 \text{ J/K}; T = 555 \text{ K}$

38. Given the values of ΔH°_{rxn}, ΔS°_{rxn}, and T below, determine ΔS_{univ} and predict whether or not the reaction will be spontaneous.
a. $\Delta H^\circ_{rxn} = +85 \text{ kJ}; \Delta S_{rxn} = +147 \text{ J/K}; T = 298 \text{ K}$
b. $\Delta H^\circ_{rxn} = +85 \text{ kJ}; \Delta S_{rxn} = +147 \text{ J/K}; T = 755 \text{ K}$
c. $\Delta H^\circ_{rxn} = +85 \text{ kJ}; \Delta S_{rxn} = -147 \text{ J/K}; T = 298 \text{ K}$
d. $\Delta H^\circ_{rxn} = -85 \text{ kJ}; \Delta S_{rxn} = +147 \text{ J/K}; T = 398 \text{ K}$

Standard Entropy Changes and Gibbs Free Energy

39. Calculate the change in Gibbs free energy for each of the sets of ΔH_{rxn}, ΔS_{rxn}, and T given in Problem 37. Predict whether or not the reaction will be spontaneous at the temperature indicated.

40. Calculate the change in Gibbs free energy for each of the sets of ΔH_{rxn}, ΔS_{rxn}, and T given in Problem 38. Predict whether or not the reaction will be spontaneous at the temperature indicated.

41. Calculate the free energy change for the following reaction at 25 °C. Is the reaction spontaneous?
$$C_3H_8(g) + 5 O_2(g) \longrightarrow 3 CO_2(g) + 4 H_2O(g)$$
$$\Delta H^\circ_{rxn} = -2217 \text{ kJ}; \Delta S^\circ_{rxn} = 101.1 \text{ J/K}$$

42. Calculate the free energy change for the following reaction at 25 °C. Is the reaction spontaneous?
$$2 Ca(s) + O_2(g) \longrightarrow 2 CaO(s)$$
$$\Delta H^\circ_{rxn} = -1269.8 \text{ kJ}; \Delta S^\circ_{rxn} = -364.6 \text{ J/K}$$

43. Fill in the blanks in the table below where both ΔH and ΔS refer to the system.

ΔH	ΔS	ΔG	Low Temperature	High Temperature
−	+	−	Spontaneous	___
−	−	Temperature dependent	___	___
+	+	___	___	Spontaneous
___	−	___	Nonspontaneous	Nonspontaneous

44. Predict the conditions (high temperature, low temperature, all temperatures, or no temperatures) under which each of the following reactions will be spontaneous.
a. $H_2O(g) \longrightarrow H_2O(l)$
b. $CO_2(s) \longrightarrow CO_2(g)$
c. $H_2(g) \longrightarrow 2 H(g)$
d. $2 NO_2(g) \longrightarrow 2 NO(g) + O_2(g)$ (endothermic)

45. How does the molar entropy of a substance change with increasing temperature?

46. What is the molar entropy of a pure crystal at 0 K? What is the significance of the answer to this question?

47. For each pair of substances, choose the one that you expect to have the higher standard molar entropy (S°) at 25 °C. Explain the reasons for your choice.
a. $CO(g); CO_2(g)$
b. $CH_3OH(l); CH_3OH(g)$
c. $Ar(g); CO_2(g)$
d. $CH_4(g); SiH_4(g)$
e. $NO_2(g); CH_3CH_2CH_3(g)$
f. $NaBr(s); NaBr(aq)$

48. For each pair of substances, choose the one that you expect to have the higher standard molar entropy (S°) at 25 °C. Explain the reasons for your choice.
a. $NaNO_3(s); NaNO_3(aq)$
b. $CH_4(g); CH_3CH_3(g)$
c. $Br_2(l); Br_2(g)$
d. $Br_2(g); F_2(g)$
e. $PCl_3(g); PCl_5(g)$
f. $CH_3CH_2CH_2CH_3(g); SO_2(g)$

49. Rank each of the following in order of increasing standard molar entropy (S°). Explain your reasoning.
a. $NH_3(g); Ne(g); SO_2(g); CH_3CH_2OH(g); He(g)$
b. $H_2O(s); H_2O(l); H_2O(g)$
c. $CH_4(g); CF_4(g); CCl_4(g)$

50. Rank each of the following in order of increasing standard molar entropy (S°). Explain your reasoning.
a. $I_2(g); F_2(g); Br_2(g); Cl_2(g)$
b. $H_2O(g); H_2O_2(g); H_2S(g)$
c. $C(s, \text{graphite}); C(s, \text{diamond}); C(s, \text{amorphous})$

51. Use data from Appendix IIB to calculate ΔS°_{rxn} for each of the reactions given below. In each case, try to rationalize the sign of ΔS°_{rxn}.
a. $C_2H_4(g) + H_2(g) \longrightarrow C_2H_6(g)$
b. $C(s) + H_2O(g) \longrightarrow CO(g) + H_2(g)$
c. $CO(g) + H_2O(g) \longrightarrow H_2(g) + CO_2(g)$
d. $2 H_2S(g) + 3 O_2(g) \longrightarrow 2 H_2O(l) + 2 SO_2(g)$

52. Use data from Appendix IIB to calculate ΔS°_{rxn} for each of the reactions given below. In each case, try to rationalize the sign of ΔS°_{rxn}.
a. $3 NO_2(g) + H_2O(l) \longrightarrow 2 HNO_3(aq) + NO(g)$
b. $Cr_2O_3(s) + 3 CO(g) \longrightarrow 2 Cr(s) + 3 CO_2(g)$
c. $SO_2(g) + \frac{1}{2} O_2(g) \longrightarrow SO_3(g)$
d. $N_2O_4(g) + 4 H_2(g) \longrightarrow N_2(g) + 4 H_2O(g)$

53. Find $\Delta S°$ for the formation of $CH_2Cl_2(g)$ from its gaseous elements in their standard states. Rationalize the sign of $\Delta S°$.

54. Find $\Delta S°$ for the reaction between nitrogen gas and fluorine gas to form nitrogen trifluoride gas. Rationalize the sign of $\Delta S°$.

55. Methanol burns in oxygen to form carbon dioxide and water. Write a balanced equation for the combustion of liquid methanol and calculate $\Delta H°_{rxn}$, $\Delta S°_{rxn}$, and $\Delta G°_{rxn}$ at 25 °C. Is the combustion of methanol spontaneous?

56. In photosynthesis, plants form glucose $(C_6H_{12}O_6)$ and oxygen from carbon dioxide and water. Write a balanced equation for photosynthesis and calculate $\Delta H°_{rxn}$, $\Delta S°_{rxn}$, and $\Delta G°_{rxn}$ at 25 °C. Is photosynthesis spontaneous?

57. For each of the following reactions, calculate $\Delta H°_{rxn}$, $\Delta S°_{rxn}$, and $\Delta G°_{rxn}$ at 25 °C and state whether or not the reaction is spontaneous. If the reaction is not spontaneous, would a change in temperature make it spontaneous? If so, should the temperature be raised or lowered from 25 °C ?
   a. $N_2O_4(g) \longrightarrow 2\,NO_2(g)$
   b. $NH_4Cl(s) \longrightarrow HCl(g) + NH_3(g)$
   c. $3\,H_2(g) + Fe_2O_3(s) \longrightarrow 2\,Fe(s) + 3\,H_2O(g)$
   d. $N_2(g) + 3\,H_2(g) \longrightarrow 2\,NH_3(g)$

58. For each of the following reactions, calculate $\Delta H°_{rxn}$, $\Delta S°_{rxn}$, and $\Delta G°_{rxn}$ at 25 °C and state whether or not the reaction is spontaneous. If the reaction is not spontaneous, would a change in temperature make it spontaneous? If so, should the temperature be raised or lowered from 25 °C ?
   a. $2\,CH_4(g) \longrightarrow C_2H_6(g) + H_2(g)$
   b. $2\,NH_3(g) \longrightarrow N_2H_4(g) + H_2(g)$
   c. $N_2(g) + O_2(g) \longrightarrow 2\,NO(g)$
   d. $2\,KClO_3(s) \longrightarrow 2\,KCl(s) + 3\,O_2(g)$

59. Use standard free energies of formation to calculate $\Delta G°$ at 25 °C for each of the reactions in Problem 57. How do the values of $\Delta G°$ calculated this way compare to those calculated from $\Delta H°$ and $\Delta S°$? Which of the two methods could be used to determine how $\Delta G°$ changes with temperature?

60. Use standard free energies of formation to calculate $\Delta G°$ at 25 °C for each of the reactions in Problem 58. How well do the values of $\Delta G°$ calculated this way compare to those calculated from $\Delta H°$ and $\Delta S°$? Which of the two methods could be used to determine how $\Delta G°$ changes with temperature?

61. Consider the following reaction:
$$2\,NO(g) + O_2(g) \longrightarrow 2\,NO_2(g)$$
Estimate $\Delta G°$ for this reaction at each of the following temperatures and predict whether or not the reaction will be spontaneous. (Assume that $\Delta H°$ and $\Delta S°$ do not change too much within the give temperature range.)
   a. 298 K   b. 715 K   c. 855 K

62. Consider the following reaction:
$$CaCO_3(s) \longrightarrow CaO(s) + CO_2(g)$$
Estimate $\Delta G°$ for this reaction at each of the following temperatures and predict whether or not the reaction will be spontaneous. (Assume that $\Delta H°$ and $\Delta S°$ do not change too much within the given temperature range.)
   a. 298 K   b. 1055 K   c. 1455 K

63. Determine $\Delta G°$ for the following reaction:
$$Fe_2O_3(s) + 3\,CO(g) \longrightarrow 2\,Fe(s) + 3\,CO_2(g)$$
Use the following reactions with known $\Delta G°_{rxn}$ values:

$2\,Fe(s) + \dfrac{3}{2}O_2(g) \longrightarrow Fe_2O_3(s)$ $\Delta G°_{rxn} = -742.2\,kJ$

$CO(g) + \dfrac{1}{2}O_2(g) \longrightarrow CO_2(g)$ $\Delta G°_{rxn} = -257.2\,kJ$

64. Calculate $\Delta G°_{rxn}$ for the following reaction:
$$CaCO_3(s) \longrightarrow CaO(s) + CO_2(g)$$
Use the following reactions and given $\Delta G°_{rxn}$ values:

$Ca(s) + CO_2(g) + \dfrac{1}{2}O_2(g) \longrightarrow CaCO_3(s)$ $\Delta G°_{rxn} = -734.4\,kJ$

$2\,Ca(s) + O_2(g) \longrightarrow 2\,CaO(s)$ $\Delta G°_{rxn} = -1206.6\,kJ$

Free Energy Changes, Nonstandard Conditions, and the Equilibrium Constant

65. Consider the sublimation of iodine at 25.0 °C:
$$I_2(s) \longrightarrow I_2(g)$$
   a. Find $\Delta G°_{rxn}$ at 25.0 °C.
   b. Find ΔG_{rxn} at 25.0 °C under the following nonstandard conditions:
      (i) $P_{I_2} = 1.00\,mmHg$
      (ii) $P_{I_2} = 0.100\,mmHg$
   c. Explain why iodine spontaneously sublimes in open air at 25.0 °C.

66. Consider the evaporation of methanol at 25.0 °C:
$$CH_3OH(l) \longrightarrow CH_3OH(g)$$
   a. Find $\Delta G°$ at 25.0 °C.
   b. Find ΔG at 25.0 °C under the following nonstandard conditions:
      (i) $P_{CH_3OH} = 150.0\,mmHg$
      (ii) $P_{CH_3OH} = 100.0\,mmHg$
      (iii) $P_{CH_3OH} = 10.0\,mmHg$
   c. Explain why methanol spontaneously evaporates in open air at 25.0 °C.

67. Consider the following reaction:
$$CH_3OH(g) \rightleftharpoons CO(g) + 2\,H_2(g)$$
Calculate ΔG for this reaction at 25 °C under the following conditions:
$$P_{CH_3OH} = 0.855\,atm$$
$$P_{CO} = 0.125\,atm$$
$$P_{H_2} = 0.183\,atm$$

68. Consider the following reaction:
$$CO_2(g) + CCl_4(g) \rightleftharpoons 2\,COCl_2(g)$$
Calculate ΔG for this reaction at 25 °C under the following conditions:
$$P_{CO_2} = 0.112\,atm$$
$$P_{CCl_4} = 0.174\,atm$$
$$P_{COCl_2} = 0.744\,atm$$

69. Use data from Appendix IIB to calculate the equilibrium constants at 25 °C for each of the following reactions.
   a. $2\,CO(g) + O_2(g) \rightleftharpoons 2\,CO_2(g)$
   b. $2\,H_2S(g) \rightleftharpoons 2\,H_2(g) + S_2(g)$

70. Use data from Appendix IIB to calculate the equilibrium constants at 25 °C for each of the following reactions. $\Delta G°_f$ for $BrCl(g)$ is $-1.0\,kJ/mol$.
   a. $2\,NO_2(g) \rightleftharpoons N_2O_4(g)$
   b. $Br_2(g) + Cl_2(g) \rightleftharpoons 2\,BrCl(g)$

71. Consider the following reaction:
$$CO(g) + 2\,H_2(g) \rightleftharpoons CH_3OH(g)$$
$$K_p = 2.26 \times 10^4 \text{ at } 25\,°C.$$
Calculate ΔG_{rxn} for the reaction at 25 °C under each of the following conditions:
   a. standard conditions   b. at equilibrium
   c. $P_{CH_3OH} = 1.0\,atm; P_{CO} = P_{H_2} = 0.010\,atm$

72. Consider the following reaction:

$$I_2(g) + Cl_2(g) \rightleftharpoons 2 ICl(g)$$

$$K_p = 81.9 \text{ at } 25 °C$$

Calculate ΔG_{rxn} for the reaction at 25 °C under each of the following conditions:
a. standard conditions
b. at equilibrium
c. $P_{ICl} = 2.55 \text{ atm}; P_{I_2} = 0.325 \text{ atm}; P_{Cl_2} = 0.221 \text{ atm}$

73. Estimate the value of the equilibrium constant at 525 K for each reaction in Problem 69.

74. Estimate the value of the equilibrium constant at 655 K for each reaction in Problem 70. (ΔH_f° for BrCl is 14.6 kJ/mol.)

75. Consider the following reaction:

$$H_2(g) + I_2(g) \rightleftharpoons 2 HI(g)$$

The data below show the equilibrium constant for this reaction measured at several different temperatures. Use the data to find ΔH_{rxn}° and ΔS_{rxn}° for the reaction.

Temperature	K_p
150 K	1.4×10^{-6}
175 K	4.6×10^{-4}
200 K	3.6×10^{-2}
225 K	1.1
250 K	15.5

76. Consider the following reaction:

$$2 NO(g) + O_2(g) \rightleftharpoons 2 NO_2(g)$$

The data below show the equilibrium constant for this reaction measured at several different temperatures. Use the data to find ΔH_{rxn}° and ΔS_{rxn}° for the reaction.

Temperature	K_p
170 K	3.8×10^{-3}
180 K	0.34
190 K	18.4
200 K	681

Cumulative Problems

77. Determine the sign of ΔS_{sys} in each of the following processes:
a. water boiling
b. water freezing
c.

78. Determine the sign of ΔS_{sys} for each of the following processes:
a. dry ice subliming
b. dew forming
c.

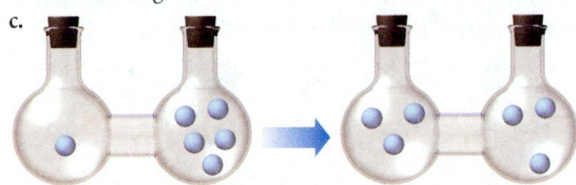

79. Our atmosphere is composed primarily of nitrogen and oxygen, which coexist at 25 °C without reacting to any significant extent. However, the two gases can react to form nitrogen monoxide according to the following reaction:

$$N_2(g) + O_2(g) \rightleftharpoons 2 NO(g)$$

a. Calculate ΔG° and K_p for this reaction at 298 K. Is the reaction spontaneous?
b. Estimate ΔG° at 2000 K. Does the reaction become more spontaneous with increasing temperature?

80. Nitrogen dioxide, a pollutant in the atmosphere, can combine with water to form nitric acid. One of the possible reactions is shown below. Calculate ΔG° and K_p for this reaction at 25 °C and comment on the spontaneity of the reaction.

$$3 NO_2(g) + H_2O(l) \longrightarrow 2 HNO_3(aq) + NO(g)$$

81. Ethene (C_2H_4) can be halogenated by the following reaction:

$$C_2H_4(g) + X_2(g) \rightleftharpoons C_2H_4X_2(g)$$

where X_2 can be Cl_2, Br_2, or I_2. Use the thermodynamic data given below to calculate ΔH°, ΔS°, ΔG°, and K_p for the halogenation reaction by each of the three halogens at 25 °C. Which reaction is most spontaneous? Least spontaneous? What is the main factor driving the difference in the spontaneity of the three reactions? Does higher temperature make the reactions more spontaneous or less spontaneous?

Compound	ΔH_f° (kJ/mol)	S° (J/mol · K)
$C_2H_4Cl_2(g)$	-129.7	308.0
$C_2H_4Br_2(g)$	-38.3	330.6
$C_2H_4I_2(g)$	$+66.5$	347.8

82. H_2 reacts with the halogens (X_2) according to the following reaction:

$$H_2(g) + X_2(g) \rightleftharpoons 2 HX(g)$$

where X_2 can be Cl_2, Br_2, or I_2. Use the thermodynamic data in Appendix IIB to calculate ΔH°, ΔS°, ΔG°, and K_p for the reaction between hydrogen and each of the three halogens. Which reaction is most spontaneous? Least spontaneous? What is the main factor driving the difference in the spontaneity of the three reactions? Does higher temperature make the reactions more spontaneous or less spontaneous?

83. Consider the following reaction occurring at 298 K:

$$N_2O(g) + NO_2(g) \rightleftharpoons 3 NO(g)$$

a. Show that the reaction is not spontaneous under standard conditions by calculating ΔG_{rxn}°.
b. If a reaction mixture contains only N_2O and NO_2 at partial pressures of 1.0 atm each, the reaction will be spontaneous until some NO forms in the mixture. What maximum partial pressure of NO builds up before the reaction ceases to be spontaneous?
c. Can the reaction be made more spontaneous by an increase or decrease in temperature? If so, what temperature is required to make the reaction spontaneous under standard conditions?

84. Consider the following reaction occurring at 298 K:

$$BaCO_3(s) \rightleftharpoons BaO(s) + CO_2(g)$$

    a. Show that the reaction is not spontaneous under standard conditions by calculating ΔG°_{rxn}.

    b. If $BaCO_3$ is placed in an evacuated flask, what partial pressure of CO_2 will be present when the reaction reaches equilibrium?

    c. Can the reaction be made more spontaneous by an increase or decrease in temperature? If so, what temperature is required to produce a carbon dioxide partial pressure of 1.0 atm?

85. Living organisms use energy from the metabolism of food to create an energy-rich molecule called adenosine triphosphate (ATP). The ATP then acts as an energy source for a variety of reactions that the living organism must carry out to survive. ATP provides energy through its hydrolysis, which can be symbolized as follows:

$$ATP(aq) + H_2O(l) \longrightarrow ADP(aq) + P_i(aq) \quad \Delta G^\circ_{rxn} = -30.5\,kJ$$

where ADP represents adenosine diphosphate and P_i represents an inorganic phosphate group (such as HPO_4^{2-}).

    a. Calculate the equilibrium constant, K, for the above reaction at 298 K.

    b. The free energy obtained from the oxidation (reaction with oxygen) of glucose ($C_6H_{12}O_6$) to form carbon dioxide and water can be used to re-form ATP by driving the above reaction in reverse. Calculate the standard free energy change for the oxidation of glucose and estimate the maximum number of moles of ATP that can be formed by the oxidation of one mole of glucose.

86. The standard free energy change for the hydrolysis of ATP was given in Problem 85. In a particular cell, the concentrations of ATP, ADP, and P_i are 0.0031 M, 0.0014 M, and 0.0048 M, respectively. Calculate the free energy change for the hydrolysis of ATP under these conditions. (Assume a temperature of 298 K.)

87. The following reactions are important ones in catalytic converters in automobiles. Calculate ΔG° for each at 298 K. Predict the effect of increasing temperature on the magnitude of ΔG°.

    a. $2\,CO(g) + 2\,NO(g) \longrightarrow N_2(g) + 2\,CO_2(g)$

    b. $5\,H_2(g) + 2\,NO(g) \longrightarrow 2\,NH_3(g) + 2\,H_2O(g)$

    c. $2\,H_2(g) + 2\,NO(g) \longrightarrow N_2(g) + 2\,H_2O(g)$

    d. $2\,NH_3(g) + 2\,O_2(g) \longrightarrow N_2O(g) + 3\,H_2O(g)$

88. Calculate ΔG° at 298 K for the following reactions and predict the effect on ΔG° of lowering the temperature.

    a. $NH_3(g) + HBr(g) \longrightarrow NH_4Br(s)$

    b. $CaCO_3(s) \longrightarrow CaO(s) + CO_2(g)$

    c. $CH_4(g) + 3\,Cl_2(g) \longrightarrow CHCl_3(g) + 3\,HCl(g)$

        (ΔG°_f for $CHCl_3(g)$ is -70.4 kJ/mol.)

89. All the oxides of nitrogen have positive values of ΔG°_f at 298 K, but only one common oxide of nitrogen has a positive ΔS°_f. Identify that oxide of nitrogen without reference to thermodynamic data and explain.

90. The trend in the ΔG°_f values of the hydrogen halides is to become less negative with increasing atomic number. The ΔG°_f of HI is slightly positive. On the other hand the trend in ΔS°_f is to become more positive with increasing atomic number. Explain these trends.

Challenge Problems

91. See the box in this chapter entitled *Chemistry in Your Day: Making a Nonspontaneous Process Spontaneous*. The hydrolysis of ATP, shown in Problem 85, is often used to drive nonspontaneous processes—such as muscle contraction and protein synthesis—in living organisms. The nonspontaneous process to be driven must be coupled to the ATP hydrolysis reaction. For example, suppose the nonspontaneous process is A + B $\longrightarrow$ AB (ΔG° positive). The coupling of a nonspontaneous reaction such as this one to the hydrolysis of ATP is often accomplished by the following mechanism.

$$A + ATP + H_2O \longrightarrow A{-}P_i + ADP$$
$$\underline{A{-}P_i + B \longrightarrow AB + P_i}$$
$$A + B + ATP + H_2O \longrightarrow AB + ADP + P_i$$

As long as ΔG°_{rxn} for the nonspontaneous reaction is less than 30.5 kJ, the reaction can be made spontaneous by coupling in this way to the hydrolysis of ATP. Suppose that ATP is to drive the reaction between glutamate and ammonia to form glutamine:

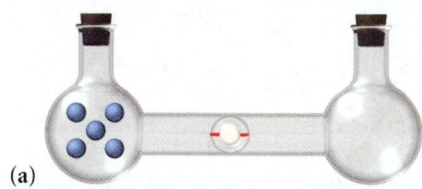

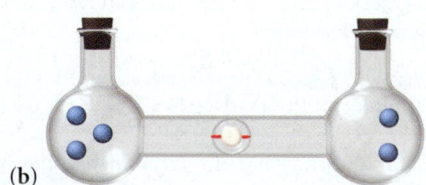

    a. Calculate K for the reaction between glutamate and ammonia. (The standard free energy change for the reaction is $+14.2$ kJ/mol. Assume a temperature of 298 K.)

    b. Write a set of reactions such as those shown above showing how the glutamate and ammonia reaction can couple with the hydrolysis of ATP. What is ΔG°_{rxn} and K for the coupled reaction?

92. Calculate the entropy of each of the following states and rank order the states in order of increasing entropy.

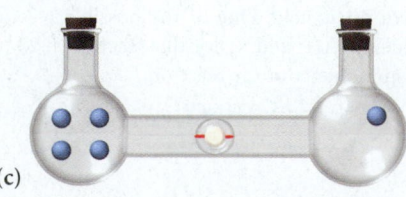

(a)

(b)

(c)

93. Suppose we redefine the standard state as $P = 2$ atm. Find the new standard ΔG_f° values of the following substances.
    a. $HCl(g)$ b. $N_2O(g)$ c. $H(g)$

    Explain the results in terms of the relative entropies of reactants and products of each reaction.

94. The ΔG for the freezing of $H_2O(l)$ at $-10\,^\circ C$ is -210 J/mol and the heat of fusion of ice at this temperature is 5610 J/mol. Find the entropy change of the universe when 1 mol of water freezes at $-10\,^\circ C$.

95. Consider the reaction that occurs during the Haber process:

$$N_2(g) + 3\,H_2(g) \longrightarrow 2\,NH_3(g)$$

    The equilibrium constant is 3.9×10^5 at 300 K and 1.2×10^{-1} at 500 K. Calculate ΔH_{rxn}° and ΔS_{rxn}° for this reaction.

96. The salt ammonium nitrate can follow three modes of decomposition: (a) to $HNO_3(g)$ and $NH_3(g)$, (b) to $N_2O(g)$ and $H_2O(g)$, and (c) to $N_2(g)$, $O_2(g)$, and $H_2O(g)$. Calculate ΔG_{rxn}° for each mode of decomposition at 298 K. Explain in light of these results how it is still possible to use ammonium nitrate as a fertilizer and the precautions that should be taken when it is used.

Conceptual Problems

97. Which is more efficient, a butane lighter or an electric lighter (such as can be found in most automobiles)? Explain.

98. Which of the following is true?
    a. A spontaneous reaction is always a fast reaction.
    b. A spontaneous reaction is always a slow reaction.
    c. The spontaneity of a reaction is not necessarily related to the speed of a reaction.

99. Which of the following processes is necessarily driven by an increase in the entropy of the surroundings?
    a. the condensation of water
    b. the sublimation of dry ice
    c. the freezing of water

100. Consider the following changes in the distribution of nine particles into three interconnected boxes. Which of the following has the most negative ΔS?

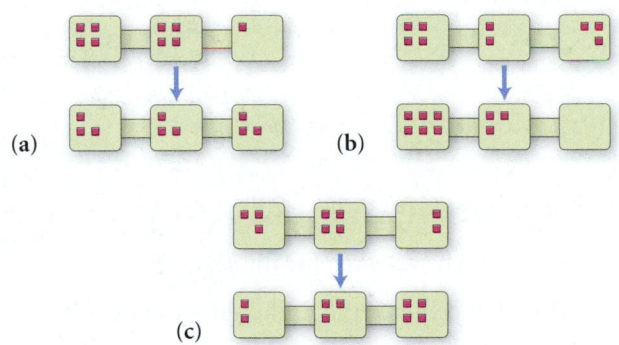

(a) (b) (c)

101. Which of the following statements is true?
    a. A reaction in which the entropy of the system increases can be spontaneous only if it is exothermic
    b. A reaction in which the entropy of the system increases can be spontaneous only if it is endothermic
    c. A reaction in which the entropy of the system decreases can be spontaneous only if it is exothermic

102. One of the following processes is spontaneous at 298 K. Which one?
    a. $H_2O(l) \longrightarrow H_2O(g, 1\text{ atm})$
    b. $H_2O(l) \longrightarrow H_2O(g, 0.10\text{ atm})$
    c. $H_2O(l) \longrightarrow H_2O(g, 0.010\text{ atm})$

103. The free energy change of the reaction $A(g) \longrightarrow B(g)$ is zero under certain conditions. The *standard* free energy change of the reaction is -42.5 kJ. Which of the following must be true about the reaction?
    a. The concentration of the product is greater than the concentration of the reactant.
    b. The reaction is at equilibrium.
    c. The concentration of the reactant is greater than the concentration of the product.

18

Electrochemistry

18.1 Pulling the Plug on the Power Grid

18.2 Balancing Oxidation–Reduction Equations

18.3 Voltaic (or Galvanic) Cells: Generating Electricity from Spontaneous Chemical Reactions

18.4 Standard Reduction Potentials

18.5 Cell Potential, Free Energy, and the Equilibrium Constant

18.6 Cell Potential and Concentration

18.7 Batteries: Using Chemistry to Generate Electricity

18.8 Electrolysis: Driving Nonspontaneous Chemical Reactions with Electricity

18.9 Corrosion: Undesirable Redox Reactions

THIS CHAPTER'S QUOTE BY MICHAEL FARADAY illustrates an important aspect of basic research (research for the sake of understanding how nature works). The Chancellor of the Exchequer (the British cabinet minister responsible for all financial matters) was questioning the value of Michael Faraday's investigations of electricity. The Chancellor wanted to know how the apparently esoteric research would ever be useful. Faraday responded in a way that the Chancellor would understand— the financial payoff. Today, of course, electricity is a fundamental form of energy that powers our entire economy. Although basic research does not always lead to useful applications, much of our technology has grown out of such research. The history of modern science shows that it is necessary first to understand nature (the goal of basic research) before you can harness nature. In this chapter, we seek to understand oxidation–reduction reactions (first introduced in Chapter 4) and how to harness them to generate electricity. The applications range from the batteries that power your flashlight to the fuel cells that may one day power our homes and automobiles.

Anode reaction:
$$2\,H_2(g) + 4\,OH^-(aq) \rightarrow 4\,H_2O(l) + 4\,e^-$$

Cathode reaction:
$$O_2(g) + \ldots + 4\,e^- \rightarrow 4\,OH^-(aq)$$

Overall reaction:
$$2\,H_2(g) + O_2(g) \rightarrow 2\,H_2O$$

Fuel Cell charge 100%
1 of 15
1:43
MENU

Fuel Cell
$$2H_2 + O_2 \rightarrow 2H_2O$$

The MP3 player shown here is powered by a hydrogen/oxygen fuel cell, a device that generates electricity from the reaction between hydrogen and oxygen to form water.

18.1 Pulling the Plug on the Power Grid

The power grid distributes centrally generated electricity throughout the country to homes and businesses. When you turn on a light or electrical appliance, electricity flows from the grid, through the wires in your home, and to the light or appliance. The electrical energy is then converted into light within the light bulb or into work within the appliance. The average U.S. household consumes about 1000 kilowatt-hours (kWh) of electricity per month. The local electrical utility, of course, monitors your electricity use and bills you for it.

In the future, you may have the option of disconnecting from the power grid. Several innovative companies are developing small, fuel-cell power plants—no bigger than a refrigerator—that sit next to a home and quietly generate enough electricity to power the home. The heat produced by the fuel cell's operation can be recaptured and used to heat water or the space within the home, eliminating the need for a hot-water heater and a

The kilowatt-hour is a unit of energy first introduced in Section 6.1.

▲ The energy produced by this fuel cell can power an entire house.

▲ General Motors recently demonstrated its HydroGen3—a five-passenger fuel-cell automobile with a top speed of 100 miles per hour and a range of 250 miles on one tank of fuel—to members of the U.S. Congress, who were encouraged to test drive the vehicle.

Review Section 4.9 on assigning oxidation states.

furnace. Similar fuel cells can also be used to power cars. Fuel cells are highly efficient and, although many obstacles are yet to be overcome, they may someday power our homes and cars while producing less pollution than fossil fuel combustion.

Fuel cells are based on oxidation–reduction reactions (see Section 4.9). The most common type of fuel cell—called the hydrogen–oxygen fuel cell—is based on the reaction between hydrogen and oxygen.

$$2 H_2(g) + O_2(g) \longrightarrow 2 H_2O(l)$$

In this reaction, hydrogen and oxygen form covalent bonds with one another. Recall that, according to Lewis theory, a single covalent bond is a shared electron pair. However, since oxygen is more electronegative than hydrogen, the electron pair in a hydrogen–oxygen bond is not *equally* shared, with oxygen getting the larger portion (see Section 9.6). In effect, oxygen has more electrons in H_2O than it does in elemental O_2—it has gained electrons in the reaction and has therefore been reduced.

In a direct reaction between hydrogen and oxygen, oxygen atoms gain the electrons directly from hydrogen atoms as the reaction proceeds. In a hydrogen–oxygen fuel cell, the same redox reaction occurs, but as we shall see in more detail in Section 18.3, the hydrogen and oxygen are separated, forcing the electrons to move through an external wire to get from hydrogen to oxygen. These moving electrons constitute an electrical current. In effect, fuel cells use the electron-gaining tendency of oxygen and the electron-losing tendency of hydrogen to force electrons to move through a wire, creating the electricity that can provide power for a home or an electric automobile. Smaller fuel cells can replace batteries and be used to power consumer electronics such as laptop computers, cell phones, and MP3 players (such as the one shown on p. 819). The generation of electricity from spontaneous redox reactions (such as a fuel cell) and the use of electricity to drive nonspontaneous redox reactions (such as those that occur in gold or silver plating) are both part of the field of electrochemistry, the subject of this chapter.

18.2 Balancing Oxidation–Reduction Equations

Recall from Section 4.9 that a fundamental definition of *oxidation* is the loss of electrons, and a fundamental definition of *reduction* is the gain of electrons. Recall also that oxidation–reduction reactions can be identified through changes in oxidation states: *oxidation corresponds to an increase in oxidation state and reduction corresponds to a decrease in oxidation state.* For example, consider the following reaction between calcium and water:

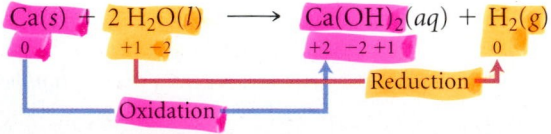

Since calcium increases in oxidation state from 0 to +2, it is oxidized. Since hydrogen decreases in oxidation state from +1 to 0, it is reduced.

Balancing redox reactions can be more complicated than balancing other types of reactions because both the mass (or number of each type of atom) and the *charge* must be balanced. Redox reactions occurring in aqueous solutions can be balanced by using a special procedure called the *half-reaction method of balancing*. In this procedure, the overall equation is broken down into two half-reactions: one for oxidation and one for reduction. The half-reactions are balanced individually and then added together. The steps differ slightly for reactions occurring in acidic and in basic solution. The following pair of examples demonstrate the method used for an acidic solution, and Example 18.3 demonstrates the method used for a basic solution.

Half-Reaction Method of Balancing Aqueous Redox Equations in Acidic Solution

General Procedure

EXAMPLE 18.1 Half-Reaction Method of Balancing Aqueous Redox Equations in Acidic Solution

EXAMPLE 18.1 Half-Reaction Method of Balancing Aqueous Redox Equations in Acidic Solution

Balance the following redox equation.

$$Al(s) + Cu^{2+}(aq) \longrightarrow Al^{3+}(aq) + Cu(s)$$

EXAMPLE 18.2 Half-Reaction Method of Balancing Aqueous Redox Equations in Acidic Solution

Balance the following redox equation.

$$Fe^{2+}(aq) + MnO_4^-(aq) \longrightarrow Fe^{3+}(aq) + Mn^{2+}(aq)$$

Step 1 *Assign oxidation states to all atoms and identify the substances being oxidized and reduced.*		
Step 2 *Separate the overall reaction into two half-reactions: one for oxidation and one for reduction.*	**Oxidation:** $Al(s) \longrightarrow Al^{3+}(aq)$ **Reduction:** $Cu^{2+}(aq) \longrightarrow Cu(s)$	**Oxidation:** $Fe^{2+}(aq) \longrightarrow Fe^{3+}(aq)$ **Reduction:** $MnO_4^-(aq) \longrightarrow Mn^{2+}(aq)$
Step 3 *Balance each half-reaction with respect to mass in the following order:* • *Balance all elements other than H and O.* • *Balance O by adding H_2O.* • *Balance H by adding H^+.*	All elements are balanced, so proceed to next step.	All elements are balanced, so proceed to balance H and O. $Fe^{2+}(aq) \longrightarrow Fe^{3+}(aq)$ $MnO_4^-(aq) \longrightarrow Mn^{2+}(aq) + \mathbf{4\,H_2O}(l)$ $\mathbf{8\,H^+(aq)} + MnO_4^-(aq) \longrightarrow$ $\qquad Mn^{2+}(aq) + 4\,H_2O(l)$
Step 4 *Balance each half-reaction with respect to charge by adding electrons. (The sum of the charges on both sides of the equation should be made equal by adding as many electrons as necessary.)*	$Al(s) \longrightarrow Al^{3+}(aq) + \mathbf{3\,e^-}$ $\mathbf{2\,e^-} + Cu^{2+}(aq) \longrightarrow Cu(s)$	$Fe^{2+}(aq) \longrightarrow Fe^{3+}(aq) + \mathbf{1\,e^-}$ $\mathbf{5\,e^-} + 8\,H^+(aq) + MnO_4^-(aq) \longrightarrow$ $\qquad Mn^{2+}(aq) + 4\,H_2O(l)$
Step 5 *Make the number of electrons in both half-reactions equal by multiplying one or both half-reactions by a small whole number.*	$\mathbf{2}[Al(s) \longrightarrow Al^{3+}(aq) + 3\,e^-]$ $2\,Al(s) \longrightarrow 2\,Al^{3+}(aq) + 6\,e^-$ $\mathbf{3}[2\,e^- + Cu^{2+}(aq) \longrightarrow Cu(s)]$ $6\,e^- + 3\,Cu^{2+}(aq) \longrightarrow 3\,Cu(s)]$	$\mathbf{5}[Fe^{2+}(aq) \longrightarrow Fe^{3+}(aq) + 1\,e^-]$ $5\,Fe^{2+}(aq) \longrightarrow 5\,Fe^{3+}(aq) + 5\,e^-$ $5\,e^- + 8\,H^+(aq) + MnO_4^-(aq) \longrightarrow$ $\qquad Mn^{2+}(aq) + 4\,H_2O(l)$
Step 6 *Add the two half-reactions together, canceling electrons and other species as necessary.*	$2\,Al(s) \longrightarrow 2\,Al^{3+}(aq) + \cancel{6\,e^-}$ $\cancel{6\,e^-} + 3\,Cu^{2+}(aq) \longrightarrow 3\,Cu(s)$ $\overline{}$ $2\,Al(s) + 3\,Cu^{2+}(aq) \longrightarrow$ $\qquad 2\,Al^{3+}(aq) + 3\,Cu(s)$	$5\,Fe^{2+}(aq) \longrightarrow 5\,Fe^{3+}(aq) + \cancel{5}\,e^-$ $\cancel{5}\,e^- + 8\,H^+(aq) + MnO_4^-(aq) \longrightarrow$ $\qquad Mn^{2+}(aq) + 4\,H_2O(l)$ $\overline{}$ $5\,Fe^{2+}(aq) + 8\,H^+(aq) + MnO_4^-(aq) \longrightarrow$ $\qquad 5\,Fe^{3+}(aq) + Mn^{2+}(aq) + 4\,H_2O(l)$

Step 7 *Verify that the reaction is balanced both with respect to mass and with respect to charge.*

Reactants	Products
2 Al	2 Al
3 Cu	3 Cu
+6 charge	+6 charge

Reactants	Products
5 Fe	5 Fe
8 H	8 H
1 Mn	1 Mn
4 O	4 O
+17 charge	+17 charge

For Practice 18.1

Balance the following redox reaction in acidic solution.

$$H^+(aq) + Cr(s) \longrightarrow H_2(g) + Cr^{3+}(aq)$$

For Practice 18.2

Balance the following redox reaction in acidic solution.

$$Cu(s) + NO_3^-(aq) \longrightarrow$$
$$Cu^{2+}(aq) + NO_2(g)$$

When a redox reaction occurs in basic solution, we balance the reaction in a similar manner, except that we must add an additional step to neutralize any H^+ with OH^-. The H^+ and the OH^- simply combine to form H_2O as shown in the following example.

EXAMPLE 18.3 Balancing Redox Reactions Occurring in Basic Solution

Balance the following equation occurring in basic solution:

$$I^-(aq) + MnO_4^-(aq) \longrightarrow I_2(aq) + MnO_2(s)$$

Solution

To balance redox reactions occuring in basic solution, we follow the half-reaction method outlined above, but add an extra step to neutralize the acid with OH^- as shown in step 3 below.

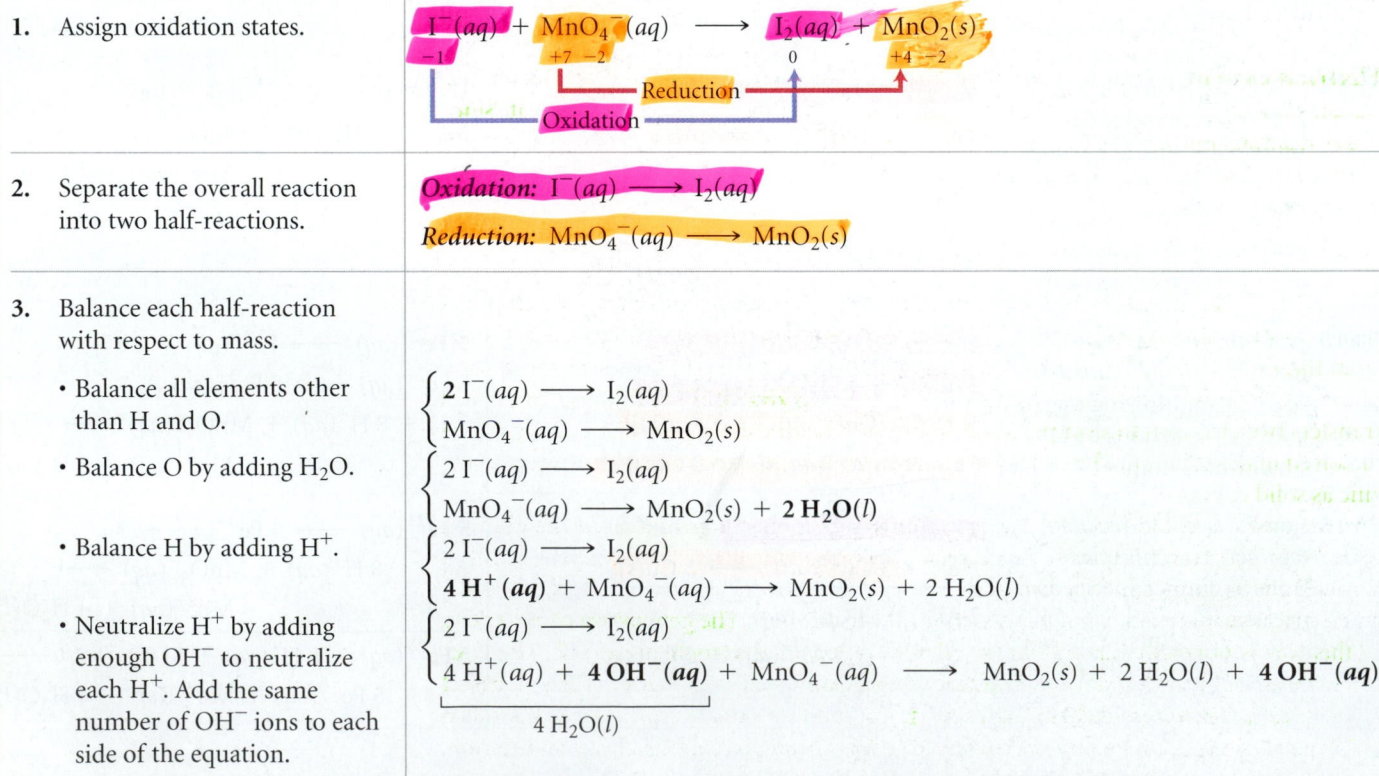

1. Assign oxidation states.

2. Separate the overall reaction into two half-reactions.

    Oxidation: $I^-(aq) \longrightarrow I_2(aq)$

    Reduction: $MnO_4^-(aq) \longrightarrow MnO_2(s)$

3. Balance each half-reaction with respect to mass.

    • Balance all elements other than H and O.

    $\begin{cases} 2\,I^-(aq) \longrightarrow I_2(aq) \\ MnO_4^-(aq) \longrightarrow MnO_2(s) \end{cases}$

    • Balance O by adding H_2O.

    $\begin{cases} 2\,I^-(aq) \longrightarrow I_2(aq) \\ MnO_4^-(aq) \longrightarrow MnO_2(s) + 2\,H_2O(l) \end{cases}$

    • Balance H by adding H^+.

    $\begin{cases} 2\,I^-(aq) \longrightarrow I_2(aq) \\ 4\,H^+(aq) + MnO_4^-(aq) \longrightarrow MnO_2(s) + 2\,H_2O(l) \end{cases}$

    • Neutralize H^+ by adding enough OH^- to neutralize each H^+. Add the same number of OH^- ions to each side of the equation.

    $\begin{cases} 2\,I^-(aq) \longrightarrow I_2(aq) \\ 4\,H^+(aq) + 4\,OH^-(aq) + MnO_4^-(aq) \longrightarrow MnO_2(s) + 2\,H_2O(l) + 4\,OH^-(aq) \end{cases}$

    $\underbrace{\qquad\qquad\qquad}_{4\,H_2O(l)}$

4.	Balance each half-reaction with respect to charge.	$2\,I^-(aq) \longrightarrow I_2(aq) + 2\,e^-$ $4\,H_2O(l) + MnO_4^-(aq) + 3\,e^- \longrightarrow MnO_2(s) + 2\,H_2O(l) + 4\,OH^-(aq)$
5.	Make the number of electrons in both half-reactions equal.	$3[2\,I^-(aq) \longrightarrow I_2(aq) + 2\,e^-]$ $6\,I^-(aq) \longrightarrow 3\,I_2(aq) + 6\,e^-$ $2[4\,H_2O(l) + MnO_4^-(aq) + 3\,e^- \longrightarrow MnO_2(s) + 2\,H_2O(l) + 4\,OH^-(aq)]$ $8\,H_2O(l) + 2\,MnO_4^-(aq) + 6\,e^- \longrightarrow 2\,MnO_2(s) + 4\,H_2O(l) + 8\,OH^-(aq)$
6.	Add the half-reactions together.	$6\,I^-(aq) \longrightarrow 3\,I_2(aq) + \cancel{6\,e^-}$ $\overset{4}{\cancel{8}}\,H_2O(l) + 2\,MnO_4^-(aq) + \cancel{6\,e^-} \longrightarrow 2\,MnO_2(s) + \cancel{4\,H_2O(l)} + 8\,OH^-(aq)$ $\overline{6\,I^-(aq) + 4\,H_2O(l) + 2\,MnO_4^-(aq) \longrightarrow 3\,I_2(aq) + 2\,MnO_2(s) + 8\,OH^-(aq)}$

7.	Verify that the reaction is balanced.	Reactants	Products
		6 I	6 I
		8 H	8 H
		2 Mn	2 Mn
		12 O	12 O
		−8 charge	−8 charge

For Practice 18.3

Balance the following redox reaction occurring in basic solution.

$$ClO^-(aq) + Cr(OH)_4^-(aq) \longrightarrow CrO_4^{2-}(aq) + Cl^-(aq)$$

18.3 Voltaic (or Galvanic) Cells: Generating Electricity from Spontaneous Chemical Reactions

Electrical current is simply the flow of electric charge. For example, electrons flowing through a wire or ions flowing through a solution both constitute electrical current. Since redox reactions involve the transfer of electrons from one substance to another, they have the potential to generate electrical current. For example, consider the following spontaneous redox reaction:

$$Zn(s) + Cu^{2+}(aq) \longrightarrow Zn^{2+}(aq) + Cu(s)$$

When Zn metal is placed in a Cu^{2+} solution, the greater tendency of zinc to lose electrons results in Zn being oxidized and Cu^{2+} being reduced—electrons are transferred directly from the Zn to the Cu^{2+} (Figure 18.1, page 824). Although the real process is more complicated, we can imagine that—on the atomic scale—a zinc atom within the zinc metal transfers two electrons to a copper ion in solution. The zinc atom then becomes a zinc ion dissolved in the solution. The copper ion accepts the two electrons and is deposited on the zinc as solid copper.

Suppose we could separate the zinc atoms and copper ions and force the electron transfer to occur another way—not directly from the zinc atom to the copper ion, but through a wire connecting the two half-reactions. The flowing electrons would constitute an electrical current and could be used to do electrical work. The generation of electricity in this way is normally carried out in a device called an **electrochemical cell**. The electrochemical cell just described is a **voltaic** (or **galvanic**) cell, one that *produces* electrical current from a *spontaneous* chemical reaction. A second type of electrochemical cell, called an **electrolytic cell**, *uses* electrical current to drive a *nonspontaneous* chemical reaction. We discuss voltaic cells in this section and electrolytic cells in Section 18.8.

A Spontaneous Redox Reaction: Zn + Cu²⁺

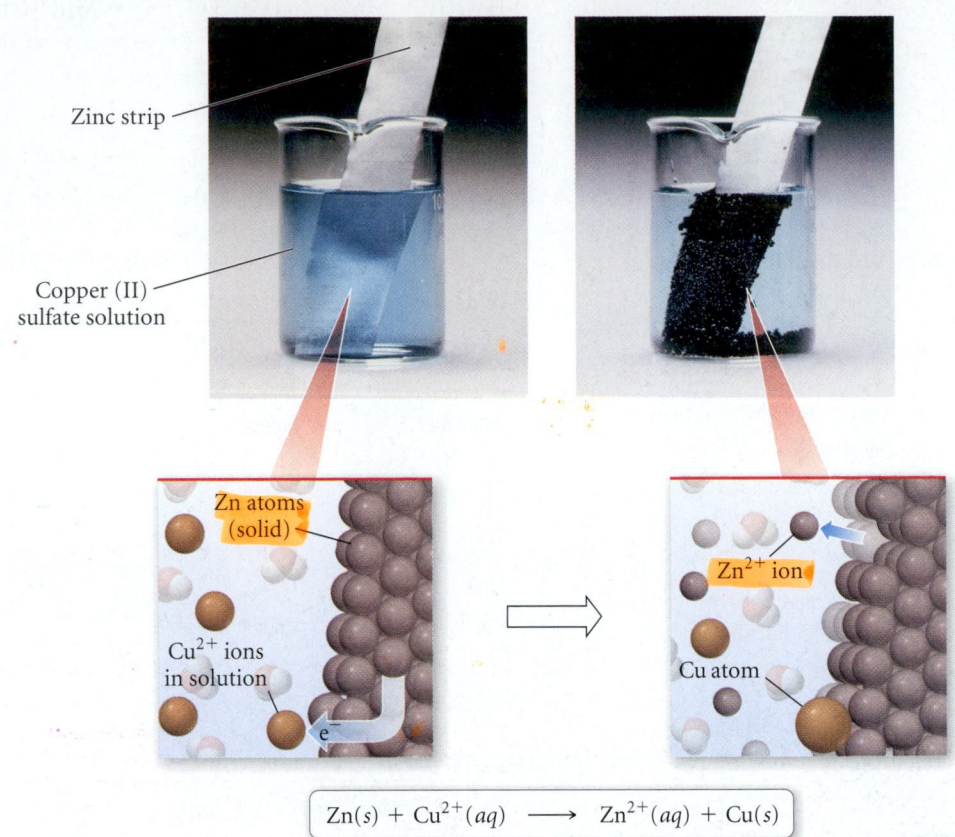

Zinc strip

Copper (II) sulfate solution

Zn atoms (solid)

Cu²⁺ ions in solution

e⁻

Zn²⁺ ion

Cu atom

$$Zn(s) + Cu^{2+}(aq) \longrightarrow Zn^{2+}(aq) + Cu(s)$$

▲ **FIGURE 18.1 A Spontaneous Oxidation–Reduction Reaction** When zinc is immersed in a solution containing copper ions, the zinc atoms transfer electrons to the copper ions. The zinc atoms are oxidized and dissolve in the solution. The copper ions are reduced and are deposited on the electrode.

For example, consider the voltaic cell in Figure 18.2 ▶. In this cell, a solid strip of zinc is placed in a $Zn(NO_3)_2$ solution to form a **half-cell**. Similarly, a solid strip of copper is placed in a $Cu(NO_3)_2$ solution to form a second half-cell. Then, the two half-cells are connected by attaching a wire from the zinc, through a lightbulb or other electrical device, to the copper. Since zinc loses electrons more easily than copper, electrons flow from the zinc through the wire to the copper. On the atomic scale, the zinc atoms are still oxidized to form zinc ions and the copper ions are still reduced to form copper atoms, but now the electrons flow through the wire to get from the zinc to the copper. The flowing electrons constitute an electrical current that lights the bulb.

In all electrochemical cells, the electrode where oxidation occurs is called the **anode** and the electrode where reduction occurs is called the **cathode.** In a voltaic cell, the anode is labeled with a negative (−) sign. The anode is negative because the oxidation reaction that occurs at the anode *releases electrons*. The cathode of a voltaic cell is labeled with a (+) sign. The cathode is positive because the reduction reaction that occurs at the cathode *takes up electrons.* Electrons flow from the anode to the cathode (from negative to positive) through the wires connecting the electrodes. As electrons flow out of the anode, positive ions (Zn^{2+} in the preceding example) form in the oxidation half-cell, resulting in a buildup of *positive charge* in the solution. As electrons flow into the cathode, positive ions (Cu^{2+} in the preceding example) are reduced at the reduction half-cell, resulting in a buildup of *negative charge* in the solution.

If the movement of electrons from anode to cathode were the only flow of charge, the flow would stop almost immediately. As positive ions accumulated at the anode, electrons would be attracted to the developing positive charge, preventing more electrons from

A Voltaic Cell

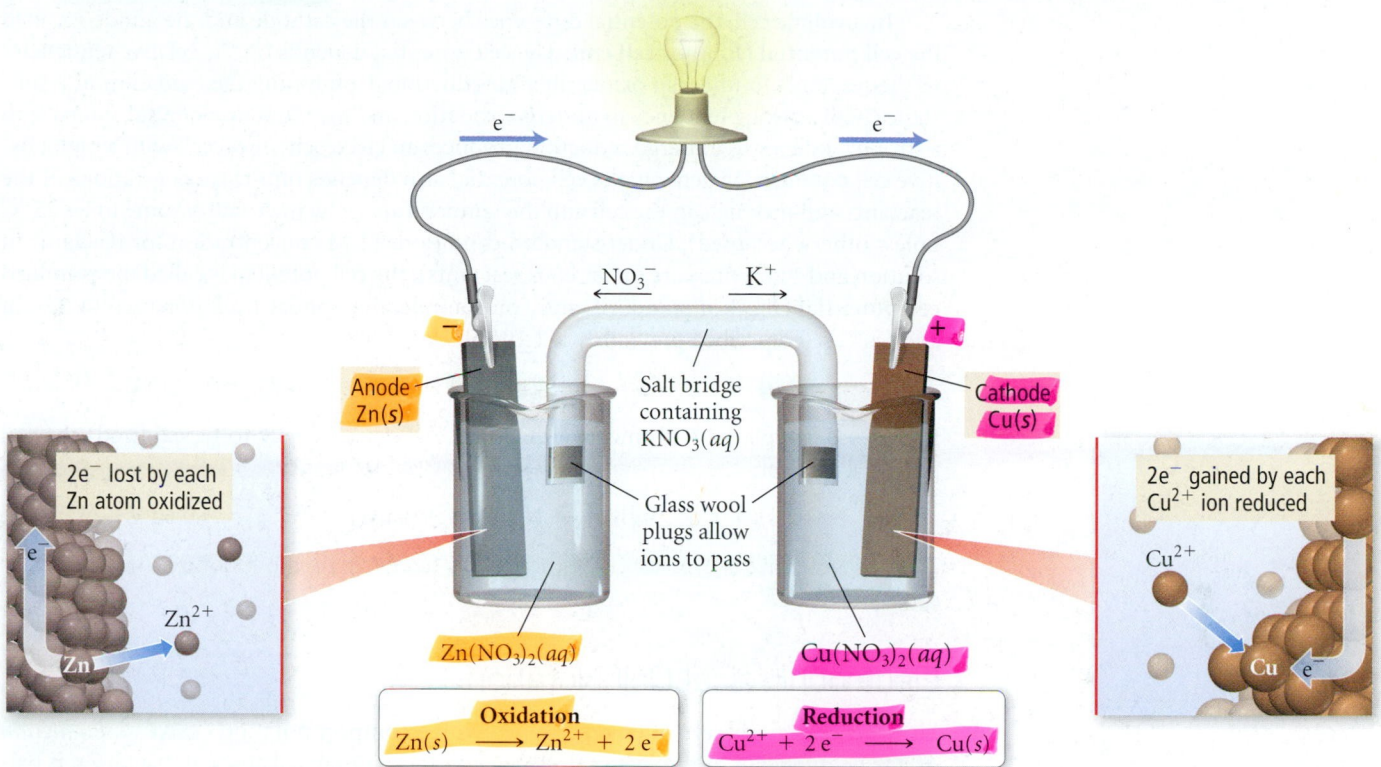

Oxidation
$$Zn(s) \longrightarrow Zn^{2+} + 2\,e^-$$

Reduction
$$Cu^{2+} + 2\,e^- \longrightarrow Cu(s)$$

flowing out of the anode. The cell needs a pathway by which counterions can flow between the half-cells without the solutions in the half-cells totally mixing. One such pathway is a **salt bridge**, an inverted, U-shaped tube that contains a strong electrolyte such as KNO_3 and connects the two half-cells (see Figure 18.2). The electrolyte is usually suspended in a gel and held within the tube by permeable stoppers. The salt bridge allows a flow of ions that neutralizes the charge buildup. *The negative ions within the salt bridge flow to neutralize the accumulation of positive charge at the anode, and the positive ions flow to neutralize the accumulation of negative charge at the cathode.* In other words, the salt bridge serves to complete the circuit, allowing electrical current to flow.

We can understand electrical current and why it flows by analogy with water current in a stream (Figure 18.3 ▼). The *rate of electrons flowing* through a wire is analogous to the *rate of water moving* through a stream. Electrical current is measured in units of **amperes (A)**. One ampere represents the flow of one coulomb (a measure of electrical charge) per second:

$$1\ A = 1\ C/s$$

Since an electron has a charge of 1.602×10^{-19} C, 1 A corresponds to the flow of 6.242×10^{18} electrons per second.

The *driving force* for electrical current is analogous to the driving force for water current. Water current is driven by a difference in gravitational potential energy—streams flow downhill, from higher to lower potential energy. The electrical current in a wire is also driven by a difference in potential energy called **potential difference**. *Potential difference is a measure of the difference in potential energy (usually in joules) per unit of charge (coulombs).* The SI unit of potential difference is the **volt (V)**, which is equal to one joule per coulomb:

$$1\ V = 1\ J/C$$

▲ **FIGURE 18.2 A Voltaic Cell**   The tendency of zinc to transfer electrons to copper results in a flow of electrons through the wire, lighting the bulb. The movement of electrons from the anode to the cathode creates a positive charge buildup at the anode and a negative charge buildup at the cathode. The flow of ions within the salt bridge neutralizes this charge buildup, allowing the reaction to continue.

| The ampere is often abbreviated as amp.

◄ **FIGURE 18.3 An Analogy for Electrical Current**   Just as water flows downhill in response to a difference in gravitational potential energy, electrons flow through a conductor in response to an electrical potential difference, creating an electrical current.

A large potential difference corresponds to a strong tendency for electron flow and is analogous to a steeply descending streambed. Potential difference, since it gives rise to the force that results in the motion of electrons, is also called **electromotive force (emf)**.

In a voltaic cell, the potential difference between the cathode and the anode is called the **cell potential (E_{cell})** or **cell emf**. The cell potential depends on the relative tendencies of the reactants to undergo oxidation and reduction. Combining the oxidation of a substance with a strong tendency to undergo oxidation and the reduction of a substance with a strong tendency to undergo reduction produces an electrochemical cell with a high positive cell potential. In general, the cell potential also depends on the concentrations of the reactants and products in the cell and the temperature (which we will assume to be 25 °C unless otherwise noted). Under standard conditions (1 M concentration for reactants in solution and 1 atm pressure for gaseous reactants), the cell potential is called the **standard cell potential ($E°_{cell}$)** or **standard emf**. For example, the standard cell potential in the Zn and Cu^{2+} cell described previously is 1.10 volts:

$$Zn(s) + Cu^{2+}(aq) \longrightarrow Zn^{2+}(aq) + Cu(s) \qquad E°_{cell} = +1.10 \text{ V}$$

If the zinc is replaced with nickel (which has a lower tendency to be oxidized) the cell potential is lower:

$$Ni(s) + Cu^{2+}(aq) \longrightarrow Ni^{2+}(aq) + Cu(s) \qquad E°_{cell} = +0.62 \text{ V}$$

The cell potential is a measure of the overall tendency of the redox reaction to occur spontaneously.

Electrochemical Cell Notation

Electrochemical cells are often represented with a compact notation called a *cell diagram* or *line notation*. For example, the electrochemical cell discussed above in which Zn is oxidized to Zn^{2+} and Cu^{2+} is reduced to Cu is represented as follows:

$$Zn(s)\,|\,Zn^{2+}(aq)\,\|\,Cu^{2+}(aq)\,|\,Cu(s)$$

In this representation,

- The oxidation half-reaction is always written on the left and the reduction on the right. A double vertical line, indicating the salt bridge, separates the two half-reactions.

- Substances in different phases are separated by a single vertical line, which represents the boundary between the phases.

- For some redox reactions, the reactants and products of one or both of the half-reactions may be in the same phase. In these cases, the reactants and products are simply separated from each other with a comma in the line diagram. Such cells use an inert electrode, such as platinum (Pt) or graphite, as the anode or cathode (or both).

Consider the following redox reaction in which Fe(s) is oxidized and $MnO_4^-(aq)$ is reduced:

$$5\,Fe(s) + 2\,MnO_4^-(aq) + 16\,H^+(aq) \longrightarrow 5\,Fe^{2+}(aq) + 2\,Mn^{2+}(aq) + 8\,H_2O(l)$$

The half-reactions for this overall reaction are as follows:

Oxidation: $Fe(s) \longrightarrow Fe^{2+}(aq) + 2\,e^-$

Reduction: $MnO_4^-(aq) + 5\,e^- + 8\,H^+(aq) \longrightarrow Mn^{2+}(aq) + 4\,H_2O(l)$

Notice that, in the reduction half-reaction, the principal species are all in the aqueous phase. However, the electron transfer needs an electrode (a solid surface) on which to occur. In this case, an inert platinum electrode is used, and the electron transfer takes place at its surface. Using line notation, we represent the electrochemical cell corresponding to the above reaction as follows:

$$Fe(s)\,|\,Fe^{2+}(aq)\,\|\,MnO_4^-(aq),\,H^+(aq),\,Mn^{2+}(aq)\,|\,Pt(s)$$

Inert Platinum Electrode

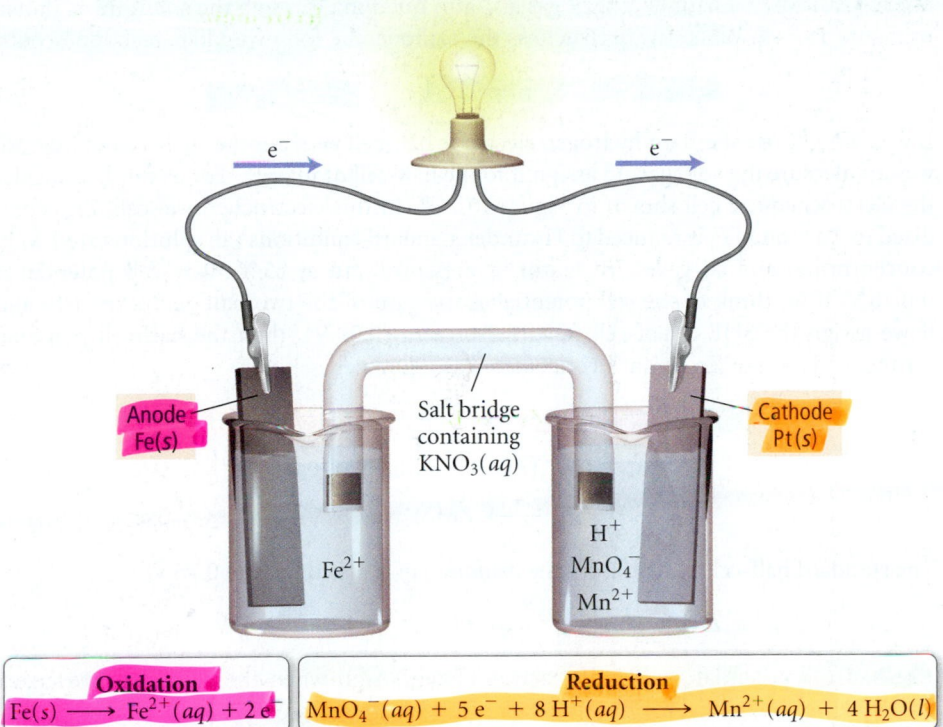

Oxidation

$$Fe(s) \longrightarrow Fe^{2+}(aq) + 2\,e^-$$

Reduction

$$MnO_4^-(aq) + 5\,e^- + 8\,H^+(aq) \longrightarrow Mn^{2+}(aq) + 4\,H_2O(l)$$

▲ **FIGURE 18.4 Inert Platinum Electrode**   When the participants in a half-reaction are all in the aqueous phase, a solid surface is needed for electron transfer to take place. In such cases an inert electrode of graphite or platinum is often used. In this electrochemical cell, an iron strip acts as the anode and a platinum strip acts as the cathode. Iron is oxidized at the anode and MnO_4^- is reduced at the cathode.

The $Pt(s)$ on the far right indicates the inert platinum electrode which acts as the cathode in this reaction, as shown in Figure 18.4 ▲.

18.4 Standard Reduction Potentials

As we have just seen, the standard cell potential (E°_{cell}) for an electrochemical cell depends on the specific half-reaction occurring in each of the half-cells. We can think of each half-cell as having its own individual potential, in which case E°_{cell} simply the sum of the two half-cell potentials. A simple analogy is two water tanks with different water levels connected by a common pipe, as shown in Figure 18.5 ▶. A half-reaction with a strong tendency to occur has a large positive half-cell potential and is analogous to a water tank with a high water level. A half-reaction with a lesser tendency to occur has a smaller half-cell potential and is analogous to a water tank with a lower water level. When the two tanks are connected by a common pipe, the water from the tank with the higher level flows into the one with the lower level. Similarly, when the two half-reactions are joined in an electrochemical cell, the one with the larger half-cell potential occurs in the forward direction and the one with the smaller potential occurs in the reverse direction.

The only problem with this analogy is that, unlike the water level in a tank, the half-cell potential cannot be measured directly—we can only measure the overall potential that occurs when two half-cells are combined in a whole cell. However, we can arbitrarily assign a potential of zero to a particular half-cell and then measure all other half-cell potentials relative to that zero.

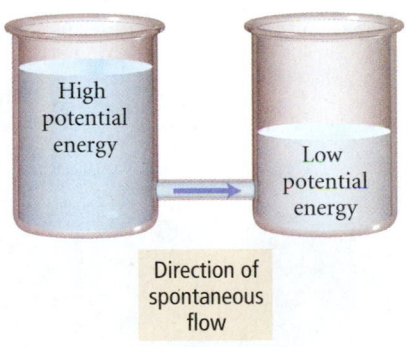

▲ **FIGURE 18.5 An Analogy for Half-Cell Potential**

Standard Hydrogen Electrode (SHE)

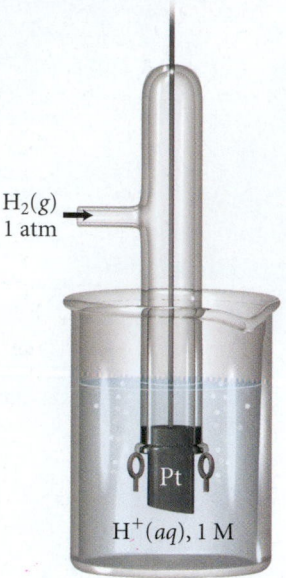

▲ **FIGURE 18.6 The Standard Hydrogen Electrode**   The standard hydrogen electrode (SHE) is arbitrarily assigned a half-cell potential of zero. All other half-cell potentials are then measured relative to the SHE.

The half-cell that is normally chosen to have a potential of zero is the **standard hydrogen electrode (SHE)** half-cell. This cell consists of an inert platinum electrode immersed in 1 M HCl with hydrogen gas at 1 atm bubbling through the solution, as shown in Figure 18.6 ◄. When the SHE acts as the cathode, the following half-reaction occurs:

$$2\,H^+(aq) + 2\,e^- \longrightarrow H_2(g) \qquad E^\circ_{red} = 0.00\ V$$

If we combine the standard hydrogen electrode half-cell with another half-cell of interest, we can measure the voltage and assign it to the half-cell of interest. For example, consider the electrochemical cell shown in Figure 18.7 ▼. In this electrochemical cell, Zn is oxidized to Zn^{2+} and H^+ is reduced to H_2 under standard conditions (all solutions are 1 M in concentration and all gases are 1 atm in pressure) and at 25 °C. The cell potential is +0.76 V. If we think of the cell potential as the sum of the two half-cell potentials, and if we assign the SHE a half-cell potential of zero (0.00 V), then the half-cell potential for the oxidation of Zn to Zn^{2+} is calculated as follows:

$$E^\circ_{cell} = E^\circ_{ox}(Zn \longrightarrow Zn^{2+}) + E^\circ_{red}(2\,H^+ \longrightarrow H_2)$$

$$0.76\ V = E^\circ_{ox}(Zn \longrightarrow Zn^{2+}) + 0.00\ V$$

$$E^\circ_{ox}(Zn \longrightarrow Zn^{2+}) = 0.76\ V$$

The standard half-cell potential for the *oxidation of zinc* is therefore 0.76 V.

$$Zn(s) \longrightarrow Zn^{2+}(aq) + 2\,e^- \qquad E^\circ_{ox} = 0.76\ V$$

The half-cell potential for a half-reaction changes sign when the reaction is reversed. So the half-cell potential for the *reduction of Zn^{2+}* is therefore −0.76 V.

$$Zn^{2+}(aq) + 2\,e^- \longrightarrow Zn(s) \qquad E^\circ_{red} = -0.76\ V$$

Measuring Half-Cell Potential with the SHE

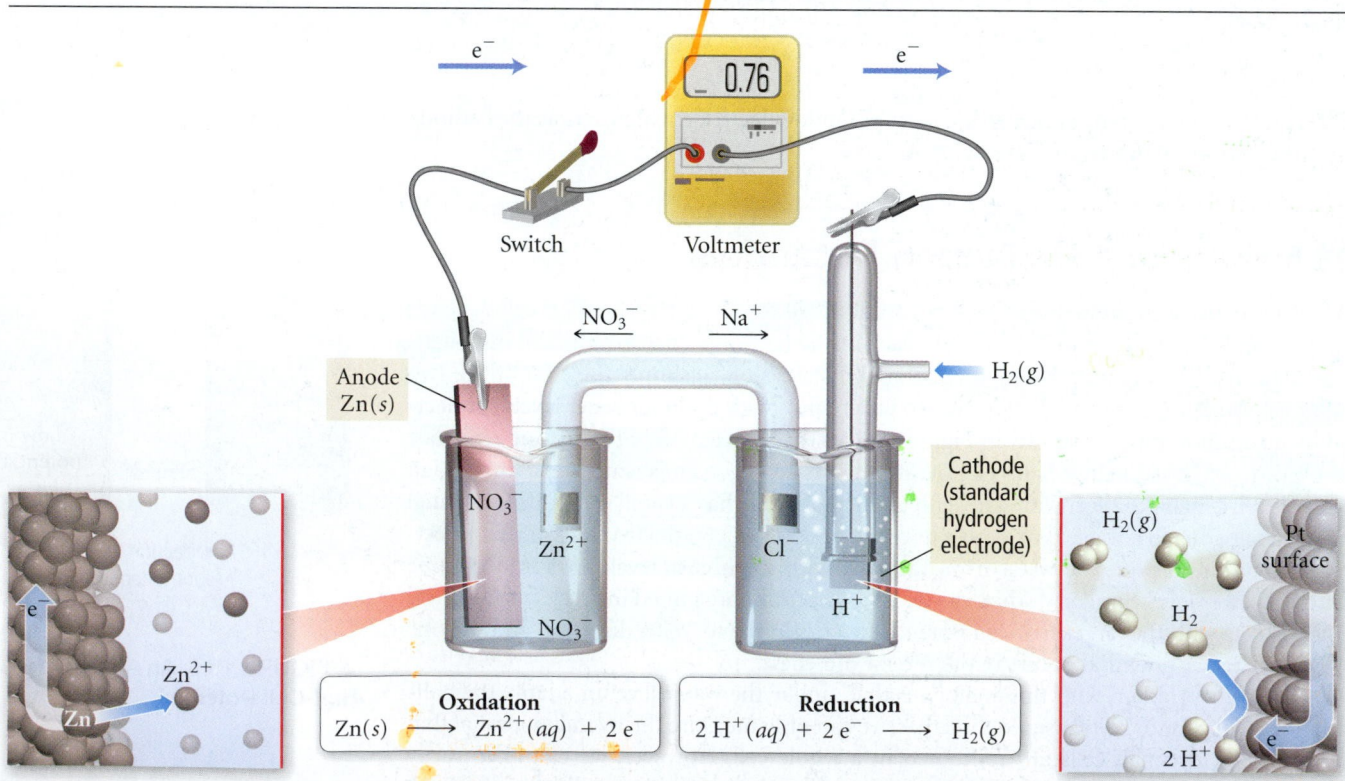

▲ **FIGURE 18.7 Measuring Half-Cell Potential**   Since the half-cell potential of the SHE is zero, the half-cell potential for the oxidation of Zn is equal to the cell potential.

Once we measure the half-cell potential of any particular half-reaction relative to the SHE, the half-cell potentials of other half-reactions can be measured with respect to it. For example, previously we saw that the standard cell potential for the $Zn(s)\,|\,Zn^{2+}(aq)\,||\,Cu^{2+}(aq)\,|\,Cu(s)$ cell was 1.10 V. Knowing that the half-cell potential for the oxidation of zinc is 0.76 V, we can calculate the half-cell potential for the reduction of Cu^{2+} as follows:

$$E^{\circ}_{cell} = E^{\circ}_{ox}(Zn \longrightarrow Zn^{2+}) + E^{\circ}_{red}(Cu^{2+} \longrightarrow Cu)$$

$$1.10\,V = 0.76\,V + E^{\circ}_{red}(Cu^{2+} \longrightarrow Cu)$$

$$E^{\circ}_{red}(Cu^{2+} \longrightarrow Cu) = 1.10\,V - 0.76\,V$$

$$= +0.34\,V$$

The overall cell potential for any electrochemical cell will always be the sum of the half-cell potential for the oxidation reaction and the half-cell potential for the reduction reaction.

$$E^{\circ}_{cell} = E^{\circ}_{ox} + E^{\circ}_{red}$$

By convention, however, standard half-cell potentials are listed only for *reduction* half-reactions, as shown in Table 18.1 on p. 830. The half-cell potential for any oxidation half-reaction can be obtained by simply changing the sign of the reduction potential listed in the table.

Summarizing the important concepts regarding standard half-cell potentials:

➤ The reduction potential of the standard hydrogen electrode (SHE) is exactly zero.

➤ Any half-cell with a greater tendency to undergo reduction than the SHE has a positive E°_{red}.

➤ Any half-cell with a lesser tendency to undergo reduction (or greater tendency to undergo oxidation) than the SHE has a negative E°_{red}.

➤ The cell potential of an electrochemical cell (E°_{cell}) can be obtained by summing the potentials of each half-cell. However, since half-cell potentials are listed as reduction half-reactions, the sign of the half-cell potential for the oxidation half-reaction must be changed ($E^{\circ}_{ox} = -E^{\circ}_{red}$).

A common mistake in calculating standard electrochemical cell potentials is to multiply the standard half-cell potentials by a constant to reflect the stoichiometry of the particular overall redox reaction. For example, consider the following half-reactions:

Oxidation: $Cr(s) \longrightarrow Cr^{3+}(aq) + 3\,e^-$ $\quad E^{\circ}_{ox} = -E^{\circ}_{red} = +0.73\,V$

Reduction: $Cl_2(g) + 2\,e^- \longrightarrow 2\,Cl^-(aq)$ $\quad E^{\circ}_{red} = +1.36\,V$

To obtain an overall balanced redox equation, we must multiply the oxidation half-reaction by two and the reduction half-reaction by three.

Oxidation: $\quad 2\,Cr(s) \longrightarrow 2\,Cr^{3+}(aq) + 6\,e^-$ $\quad\quad E^{\circ}_{ox} = +0.73\,V$

Reduction: $\quad 3\,Cl_2(g) + 6\,e^- \longrightarrow 6\,Cl^-(aq)$ $\quad\quad E^{\circ}_{red} = +1.36\,V$

$$2\,Cr(s) + 3\,Cl_2(g) \longrightarrow 2\,Cr^{3+}(aq) + 6\,Cl^-(aq) \quad\quad E^{\circ}_{cell} = 2.09\,V$$

Notice, however, that we **do not** multiply the half-cell potentials by the corresponding coefficients. Since standard reduction potentials are measured in units of V, and since a volt reflects potential energy difference *per unit charge*, the standard reduction potentials should not be multiplied by constants to reflect stoichiometry. In other words, standard electrode potentials are intensive properties analogous to quantities such as density—the density of a reactant or product does not change when you multiply the reaction by a constant and neither does the half-cell potential. The following example shows how to calculate the potential of an electrochemical cell from the standard reduction potentials of the half-reactions.

In an alternative method of calculating the overall cell potential, E°_{cell} is defined as the *difference* between the reducation potentials of the cathode and the anode as follows: $E^{\circ}_{cell} = E^{\circ}_{red,\,cathode} - E^{\circ}_{red,\,anode}$. Calculating the difference between the reduction potentials is numerically equivalent to changing the sign of $E^{\circ}_{red,\,anode}$ to determine E°_{ox} and then summing E°_{ox} and E°_{red} to obtain E°_{cell}.

TABLE 18.1 Standard Reduction Potentials at 25 °C

	Reduction Half-Reaction		$E°$ (V)	
Stronger oxidizing agent	$F_2(g) + 2\,e^-$	$\longrightarrow 2\,F^-(aq)$	2.87	Weaker reducing agent
	$H_2O_2(aq) + 2\,H^+(aq) + 2\,e^-$	$\longrightarrow 2\,H_2O(l)$	1.78	
	$PbO_2(s) + 4\,H^+(aq) + SO_4^{2-}(aq) + 2\,e^-$	$\longrightarrow PbSO_4(s) + 2\,H_2O(l)$	1.69	
	$MnO_4^-(aq) + 4\,H^+(aq) + 3\,e^-$	$\longrightarrow MnO_2(s) + 2\,H_2O(l)$	1.68	
	$MnO_4^-(aq) + 8\,H^+(aq) + 5\,e^-$	$\longrightarrow Mn^{2+}(aq) + 4\,H_2O(l)$	1.51	
	$Au^{3+}(aq) + 3\,e^-$	$\longrightarrow Au(s)$	1.50	
	$PbO_2(s) + 4\,H^+(aq) + 2\,e^-$	$\longrightarrow Pb^{2+}(aq) + 2\,H_2O(l)$	1.46	
	$Cl_2(g) + 2\,e^-$	$\longrightarrow 2\,Cl^-(aq)$	1.36	
	$Cr_2O_7^{2-}(aq) + 14\,H^+(aq) + 6\,e^-$	$\longrightarrow 2\,Cr^{3+}(aq) + 7\,H_2O(l)$	1.33	
	$O_2(g) + 4\,H^+(aq) + 4\,e^-$	$\longrightarrow 2\,H_2O(l)$	1.23	
	$MnO_2(s) + 4\,H^+(aq) + 2\,e^-$	$\longrightarrow Mn^{2+}(aq) + 2\,H_2O(l)$	1.21	
	$IO_3^-(aq) + 6\,H^+(aq) + 5\,e^-$	$\longrightarrow \frac{1}{2}\,I_2(aq) + 3\,H_2O(l)$	1.20	
	$Br_2(l) + 2\,e^-$	$\longrightarrow 2\,Br^-(aq)$	1.09	
	$VO_2^+(aq) + 2\,H^+(aq) + e^-$	$\longrightarrow VO^{2+}(aq) + H_2O(l)$	1.00	
	$NO_3^-(aq) + 4\,H^+(aq) + 3\,e^-$	$\longrightarrow NO(s) + 2\,H_2O(l)$	0.96	
	$ClO_2(g) + e^-$	$\longrightarrow ClO_2^-(aq)$	0.95	
	$Ag^+(aq) + e^-$	$\longrightarrow Ag(s)$	0.80	
	$Fe^{3+}(aq) + e^-$	$\longrightarrow Fe^{2+}(aq)$	0.77	
	$O_2(g) + 2\,H^+(aq) + 2\,e^-$	$\longrightarrow H_2O_2(aq)$	0.70	
	$MnO_4^-(aq) + e^-$	$\longrightarrow MnO_4^{2-}(aq)$	0.56	
	$I_2(s) + 2\,e^-$	$\longrightarrow 2\,I^-(aq)$	0.54	
	$Cu^+(aq) + e^-$	$\longrightarrow Cu(s)$	0.52	
	$O_2(g) + 2\,H_2O(l) + 4\,e^-$	$\longrightarrow 4\,OH^-(aq)$	0.40	
	$Cu^{2+}(aq) + 2\,e^-$	$\longrightarrow Cu(s)$	0.34	
	$SO_4^{2-}(aq) + 4\,H^+(aq) + 2\,e^-$	$\longrightarrow H_2SO_3(aq) + H_2O(l)$	0.20	
	$Cu^{2+}(aq) + e^-$	$\longrightarrow Cu^+(aq)$	0.16	
	$Sn^{4+}(aq) + 2\,e^-$	$\longrightarrow Sn^{2+}(aq)$	0.15	
	$2\,H^+(aq) + 2\,e^-$	$\longrightarrow H_2(g)$	0	
	$Fe^{3+}(aq) + 3\,e^-$	$\longrightarrow Fe(s)$	−0.036	
	$Pb^{2+}(aq) + 2\,e^-$	$\longrightarrow Pb(s)$	−0.13	
	$Sn^{2+}(aq) + 2\,e^-$	$\longrightarrow Sn(s)$	−0.14	
	$Ni^{2+}(aq) + 2\,e^-$	$\longrightarrow Ni(s)$	−0.23	
	$Cd^{2+}(aq) + 2\,e^-$	$\longrightarrow Cd(s)$	−0.40	
	$Fe^{2+}(aq) + 2\,e^-$	$\longrightarrow Fe(s)$	−0.45	
	$Cr^{3+}(aq) + e^-$	$\longrightarrow Cr^{2+}(aq)$	−0.50	
	$Cr^{3+}(aq) + 3\,e^-$	$\longrightarrow Cr(s)$	−0.73	
	$Zn^{2+}(aq) + 2\,e^-$	$\longrightarrow Zn(s)$	−0.76	
	$2\,H_2O(l) + 2\,e^-$	$\longrightarrow H_2(g) + 2\,OH^-(aq)$	−0.83	
	$Mn^{2+}(aq) + 2\,e^-$	$\longrightarrow Mn(s)$	−1.18	
	$Al^{3+}(aq) + 3\,e^-$	$\longrightarrow Al(s)$	−1.66	
	$Mg^{2+}(aq) + 2\,e^-$	$\longrightarrow Mg(s)$	−2.37	
	$Na^+(aq) + e^-$	$\longrightarrow Na(s)$	−2.71	
	$Ca^{2+}(aq) + 2\,e^-$	$\longrightarrow Ca(s)$	−2.76	
	$Ba^{2+}(aq) + 2\,e^-$	$\longrightarrow Ba(s)$	−2.90	
Weaker oxidizing agent	$K^+(aq) + e^-$	$\longrightarrow K(s)$	−2.92	Stronger reducing agent
	$Li^+(aq) + e^-$	$\longrightarrow Li(s)$	−3.04	

EXAMPLE 18.4 Calculating Standard Potentials for Electrochemical Cells from Standard Reduction Potentials of the Half-Reactions

Use tabulated standard half-cell potentials to calculate the standard cell potential for the following reaction occurring in an electrochemical cell at 25 °C. (The equation is balanced.)

$$Al(s) + NO_3^-(aq) + 4 H^+(aq) \longrightarrow Al^{3+}(aq) + NO(g) + 2 H_2O(l)$$

Solution

Begin by separating the reaction into oxidation and reduction half-reactions. [In this case, you can easily see that Al(s) is oxidized. In cases where it is not so clear, however, you many want to assign oxidation states to determine the correct half-reactions.]	Oxidation: $Al(s) \longrightarrow Al^{3+}(aq) + 3e^-$ Reduction: $NO_3^-(aq) + 4 H^+(aq) + 3e^- \longrightarrow NO(g) + 2 H_2O(l)$
Look up the standard half-cell potentials for each half-reaction. For the oxidation half-reaction, remember to change the sign of E_{red}° to obtain E_{ox}°. Add the half-cell reactions together to obtain the overall redox equation. Add the half-cell potentials together to obtain the overall standard cell potential.	Oxidation: $\qquad\qquad\qquad\qquad Al(s) \longrightarrow Al^{3+}(aq) + 3e^- \quad E_{ox}^\circ = -E_{red}^\circ = +1.66 V$ Reduction: $NO_3^-(aq) + 4 H^+(aq) + 3e^- \longrightarrow NO(g) + 2 H_2O(l) \quad E_{red}^\circ = 0.96 V$ $\qquad Al(s) + NO_3^-(aq) + 4 H^+(aq) \longrightarrow Al^{3+}(aq) + NO(g) + 2 H_2O(l)$ $E_{cell}^\circ = E_{ox}^\circ + E_{red}^\circ$ $\qquad = +1.66 V + 0.96 V$ $\qquad = 2.62 V$

For Practice 18.4

Use tabulated standard half-cell potentials to calculate the standard cell potential for the following reaction occurring in an electrochemical cell at 25 °C. (The equation is balanced.)

$$3 Pb^{2+}(aq) + 2 Cr(s) \longrightarrow 3 Pb(s) + 2 Cr^{3+}(aq)$$

Predicting the Spontaneous Direction of an Oxidation–Reduction Reaction

Table 18.1 lists half-reactions in order of *decreasing* reduction half-cell potential. Therefore, half-reactions near the top of the list—those having large *positive* reduction half-cell potentials—have a strong tendency to occur in the forward direction. In other words, *those substances at the top of the list have a strong tendency to be reduced and are therefore excellent oxidizing agents.* For example, the half-reaction at the top of the list is the reduction of fluorine gas. Since fluorine is very electronegative, it has a strong tendency to draw electrons to itself—it is readily reduced and oxidizes many other substances.

On the other hand, half-reactions near the bottom of the list—those having large *negative* reduction half-cell potentials—have a strong tendency to occur in the reverse direction. In other words, *those substances at the bottom of the list have a strong tendency to be oxidized and are therefore excellent reducing agents.* The half-reaction at the very bottom of the list is the reduction of lithium ions. Since lithium is a highly active metal, it has a strong tendency to lose electrons—it is readily oxidized and reduces many other substances.

Since half-reactions at the top of Table 18.1 have a tendency to occur in the forward direction, and half-reactions near the bottom of the table have a tendency to occur in the reverse direction, the relative positions of any two half-reactions in the table can be used to predict the spontaneity of the overall redox reaction. In general, *any reduction half-reaction will be spontaneous when paired with the reverse of a half-reaction below it in the table.*

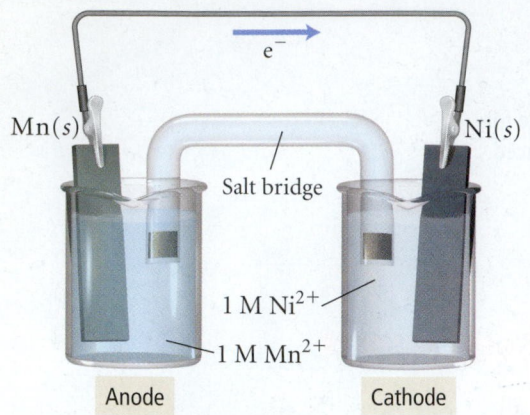

▲ FIGURE 18.8 Mn/Ni²⁺ Electrochemical Cell Since the reduction of Mn²⁺ lies below the reduction of Ni²⁺ in Table 18.1, the reduction of Ni²⁺ will be spontaneous when paired with the oxidation of Mn.

For example, consider the following two reduction half-reactions:

$$Ni^{2+}(aq) + 2\,e^- \longrightarrow Ni(s) \qquad E^{\circ}_{red} = -0.23\ V$$
$$Mn^{2+}(aq) + 2\,e^- \longrightarrow Mn(s) \qquad E^{\circ}_{red} = -1.18\ V$$

Since the reduction of Mn²⁺ lies below the reduction of Ni²⁺ on the table, the reduction of Ni²⁺ will be spontaneous when paired with the oxidation of Mn, as we can determine by summing the half-cell potentials:

$$Ni^{2+}(aq) + 2\,e^- \longrightarrow Ni(s) \qquad\qquad E^{\circ}_{red} = -0.23\ V$$
$$Mn(s) \longrightarrow Mn^{2+}(aq) + 2\,e^- \qquad E^{\circ}_{ox} = -E^{\circ}_{red} = +1.18\ V$$
$$\overline{Ni^{2+}(aq) + Mn(s) \longrightarrow Ni(s) + Mn^{2+}(aq) \quad E^{\circ}_{cell} = +0.95\ V}$$

A positive overall cell potential indicates a spontaneous reaction. The electrochemical cell corresponding to this redox reaction is shown in Figure 18.8 ◀. However, if we attempt to pair a reduction half-reaction on the table with the reverse of a half-reaction above it, we get negative cell potential, and the reaction is not spontaneous in the direction written. For example,

$$Ni^{2+}(aq) + 2\,e^- \longrightarrow Ni(s) \qquad\qquad E^{\circ}_{red} = -0.23\ V$$
$$Pb(s) \longrightarrow Pb^{2+}(aq) + 2\,e^- \qquad E^{\circ}_{ox} = -E^{\circ}_{red} = +0.13\ V$$
$$\overline{Ni^{2+}(aq) + Pb(s) \longrightarrow Ni(s) + Pb^{2+}(aq) \quad E^{\circ}_{cell} = -0.10\ V}$$

The reaction does not occur in the direction written because Ni(s) has a greater tendency to be oxidized than Pb(s). Consequently, the electrochemical cell runs in the reverse direction.

EXAMPLE 18.5 Predicting Spontaneous Redox Reactions and Sketching Electrochemical Cells

Without calculating E°_{cell}, predict whether each of the following redox reactions is spontaneous. If the reaction is spontaneous as written, make a sketch of the electrochemical cell in which the reaction could occur. If the reaction is not spontaneous as written, write an equation for the spontaneous direction in which the reaction would occur and make a sketch of the electrochemical cell in which the spontaneous reaction would occur. In your sketches, make sure to label the anode (which should be drawn on the left), the cathode, and the direction of electron flow.

(a) $Fe(s) + Mg^{2+}(aq) \longrightarrow Fe^{2+}(aq) + Mg(s)$

(b) $Fe(s) + Pb^{2+}(aq) \longrightarrow Fe^{2+}(aq) + Pb(s)$

Solution

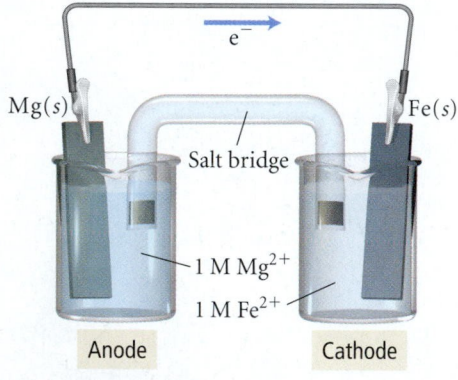

▲ FIGURE 18.9 Mg/Fe²⁺ Electrochemical Cell

(a) $Fe(s) + Mg^{2+}(aq) \longrightarrow Fe^{2+}(aq) + Mg(s)$

This reaction involves the reduction of Mg²⁺

$$Mg^{2+}(aq) + 2\,e^- \longrightarrow Mg(s)$$

paired with the reverse of a half-reaction *above it* in Table 18.1

$$Fe(s) \longrightarrow Fe^{2+}(aq) + 2\,e^-$$

Therefore, the reaction as written will *not be* spontaneous. However, the reverse reaction would be spontaneous.

$$Fe^{2+}(aq) + Mg(s) \longrightarrow Fe(s) + Mg^{2+}(aq)$$

The corresponding electrochemical cell is shown in Figure 18.9 ◀

(b) $Fe(s) + Pb^{2+}(aq) \longrightarrow Fe^{2+}(aq) + Pb(s)$

This reaction involves the reduction of Pb

$$Pb^{2+}(aq) + 2\,e^- \longrightarrow Pb(s)$$

paired with the reverse of a half-reaction *below it* in Table 18.1

$$Fe(s) \longrightarrow Fe^{2+}(aq) + 2\,e^-$$

Therefore, the reaction *will be* spontaneous as written. The corresponding electrochemical cell is shown in Figure 18.10 ◀.

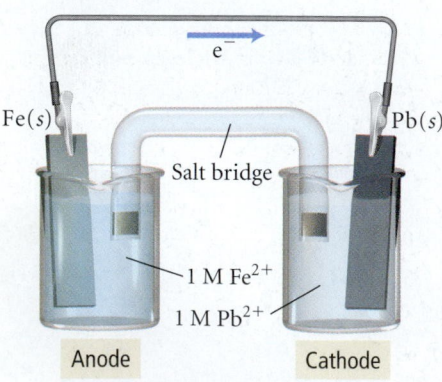

▲ FIGURE 18.10 Fe/Pb²⁺ Electrochemical Cell

For Practice 18.5

Will the following redox reactions be spontaneous under standard conditions?

(a) $Zn(s) + Ni^{2+}(aq) \longrightarrow Zn^{2+}(aq) + Ni(s)$

(b) $Zn(s) + Ca^{2+}(aq) \longrightarrow Zn^{2+}(aq) + Ca(s)$

 Conceptual Connection 18.1 Selective Oxidation

A solution contains both NaI and NaBr. Which of the following oxidizing agents could be added to the solution to selectively oxidize $I^-(aq)$ but not $Br^-(aq)$?

(a) Cl_2 (b) H_2O_2 (c) $CuCl_2$ (d) HNO_3

Answer: (d) The reduction of HNO_3 is below the reduction of Br_2 and above the reduction of I_2 in Table 18.1. Since any reduction half-reaction will be spontaneous when paired with the reverse of a half-reaction below it in the table, the reduction of HNO_3 will be spontaneous when paired with the oxidation of I^-, but will not be spontaneous when paired with the oxidation of Br^-.

Predicting Whether a Metal Will Dissolve in Acid

In Chapter 15, we learned that acids dissolve metals. Most acids dissolve metals by the reduction of H^+ ions to hydrogen gas and the corresponding oxidation of the metal to its ion. For example, if solid Zn is dropped into hydrochloric acid, the following reaction occurs.

$$2 H^+(aq) + 2e^- \longrightarrow H_2(g)$$
$$\underline{Zn(s) \longrightarrow Zn^{2+}(aq) + 2e^-}$$
$$Zn(s) + 2 H^+(aq) \longrightarrow Zn^{2+}(aq) + H_2(g)$$

$Zn(s) + 2 H^+(aq) \longrightarrow$

$Zn^{2+}(aq) + H_2(g)$

▲ When zinc is immersed in hydrochloric acid, the zinc is oxidized, forming ions that become solvated in the solution. Hydrogen ions are reduced, forming bubbles of hydrogen gas.

We observe the reaction as the dissolving of the zinc and the bubbling of hydrogen gas. The zinc is oxidized and the H^+ ions are reduced. Notice that this reaction involves the pairing of a reduction half-reaction (the reduction of H^+) with the reverse of a half-reaction below it on Table 18.1. Therefore, this reaction is spontaneous. What would happen, however, if we paired the reduction of H^+ with the oxidation of Cu? The reaction would not be spontaneous because it involves pairing the reduction of H^+ with the reverse of a half-reaction *above it* in the table. Consequently, copper does not react with H^+ and will not dissolve in acids such as HCl. In general, *metals whose reduction half-reactions lie below the reduction of H^+ to H_2 in Table 18.1 will dissolve in acids, while metals above it will not.*

An important exception to this rule is nitric acid (HNO_3), which can oxidize metals through the following reduction half-reaction:

$$NO_3^-(aq) + 4 H^+(aq) + 3 e^- \longrightarrow NO(g) + 2 H_2O(l) \qquad E^\circ_{red} = 0.96 V$$

Since this half-reaction is above the reduction of H^+ in Table 18.1, HNO_3 can oxidize metals (such as copper) that can't be oxidized by HCl.

Conceptual Connection 18.2 Metals Dissolving in Acids

Which of the following metals dissolves in HNO_3 but not in HCl?

(a) Fe   (b) Au   (c) Ag

Answer: (c) Ag falls *above* the half-reaction for the reduction of H^+ but *below* the half-reaction for the reduction of NO_3^- in Table 18.1.

18.5 Cell Potential, Free Energy, and the Equilibrium Constant

We have seen that a positive standard cell potential (E°_{cell}) corresponds to a spontaneous oxidation–reduction reaction. We also know (from Chapter 17) that the spontaneity of a reaction is determined by the sign of ΔG°. Therefore, E°_{cell} and ΔG° must be related. We also know from Section 17.9 that ΔG° for a reaction is related to the equilibrium constant (K) for the reaction. Since E°_{cell} and ΔG° are related, then E°_{cell} and K must also be related.

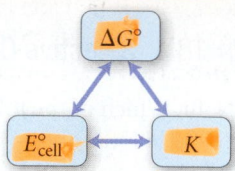

Before we look at the nature of each of these relationships in detail, let's consider the following generalizations.

For a spontaneous reaction (one that will proceed in the forward direction when all reactants and products are in their standard states):
- ΔG° is negative (<0)
- E°_{cell} is positive (>0)
- $K > 1$

For a nonspontaneous reaction (one that will proceed in the reverse direction when all reactants and products are in their standard states):
- ΔG° is positive (>0)
- E°_{cell} is negative (<0)
- $K < 1$

The Relationship between ΔG° and E°_{cell}

We can derive a relationship between ΔG° and E°_{cell} by briefly returning to the definition of potential difference given in Section 18.3—a potential difference is a measure of the difference of potential energy per unit charge (q):

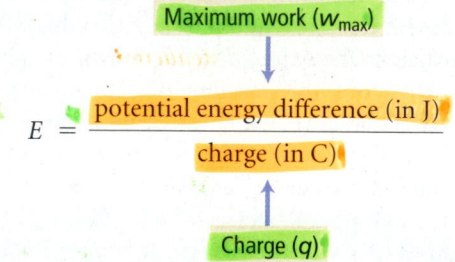

Since the potential energy difference represents the maximum amount of work that can be done by the system on the surroundings, we can write the following:

$$w_{max} = -qE^\circ_{cell} \tag{18.1}$$

The negative sign follows the convention used throughout this book that work done by the system on the surroundings is negative.

We can quantify the charge (q) that flows in an electrochemical reaction by using **Faraday's constant** (F), which represents the charge in coulombs of 1 mol of electrons.

$$F = \frac{96,485 \text{ C}}{\text{mol e}^-}$$

The total charge is therefore $q = nF$, where n is the number of moles of electrons from the balanced chemical equation and F is Faraday's constant. Substituting $q = nF$ into Equation 18.1,

$$w_{max} = -qE^\circ_{cell}$$
$$= -nFE^\circ_{cell} \tag{18.2}$$

Finally, recall from Chapter 17 that the standard change in free energy for a chemical reaction ($\Delta G°$) represents the maximum amount of work that can be done by the reaction. Therefore, $w_{max} = \Delta G°$. Making this substitution into Equation 18.2, we get the following important result:

$$\Delta G° = -nFE°_{cell} \qquad [18.3]$$

where $\Delta G°$ is the standard change in free energy for an electrochemical reaction, n is the number of moles of electrons transferred in the balanced equation, F is Faraday's constant, and $E°_{cell}$ is the standard cell potential. The following example shows how to apply this equation to calculate the standard free energy change for an electrochemical cell.

EXAMPLE 18.6 Relating $\Delta G°$ and $E°_{cell}$

Use the tabulated half-cell potentials to calculate $\Delta G°$ for the following reaction:

$$I_2(s) + 2\,Br^-(aq) \longrightarrow 2\,I^-(aq) + Br_2(l)$$

Is the reaction spontaneous?

Sort You are given a redox reaction and asked to find $\Delta G°$.	**Given** $I_2(s) + 2\,Br^-(aq) \longrightarrow 2\,I^-(aq) + Br_2(l)$ **Find** $\Delta G°$
Strategize Use the tabulated values of half-cell potentials to calculate $E°_{cell}$. Then use Equation 18.3 to calculate $\Delta G°$ from $E°_{cell}$.	**Conceptual Plan**

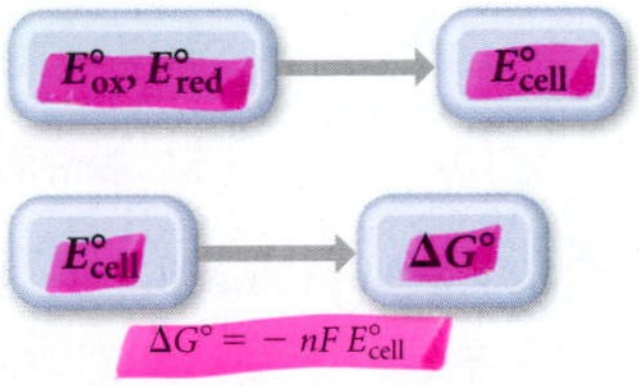

Solve Break the reaction up into oxidation and reduction half-reactions and find the standard half-cell potentials for each. Sum the half-cell potentials to find $E°_{cell}$.

Solution

Oxidation:	$2\,Br^-(aq) \longrightarrow Br_2(l) + 2\,e^-$	$E°_{ox} = -E°_{red} = -1.09\ V$
Reduction:	$I_2(s) + 2\,e^- \longrightarrow 2\,I^-(aq)$	$E°_{red} = 0.54\ V$
	$I_2(s) + 2\,Br^-(aq) \longrightarrow 2\,I^-(aq) + Br_2(l)$	$E°_{cell} = -0.55\ V$

Compute $\Delta G°$ from $E°_{cell}$. The value of n (the number of moles of electrons) corresponds to the number of electrons that are canceled in the half-reactions. Remember that $1\ V = 1\ J/C$.

$\Delta G° = -nFE°_{cell}$

$= -2\ mol\ e^- \left(\dfrac{96{,}485\ C}{mol\ e^-} \right) \left(-0.55\ \dfrac{J}{C} \right)$

$= +1.1 \times 10^5\ J$

Since $\Delta G°$ is positive, the reaction is not spontaneous under standard conditions.

Check The answer is in the correct units (joules) and seems reasonable in magnitude ($\approx 110\ kJ$) because we have seen (in Chapter 17) that values of $\Delta G°$ are typically in the range of plus or minus tens to hundreds of kilojoules. The sign is positive, as expected for a reaction in which $E°_{cell}$ is negative.

For Practice 18.6

Use tabulated half-cell potentials to calculate $\Delta G°$ for the following reaction:

$$2\,Na(s) + 2\,H_2O(l) \longrightarrow H_2(g) + 2\,OH^-(aq) + 2\,Na^+(aq)$$

Is the reaction spontaneous?

Conceptual Connection 18.3 Periodic Trends and the Direction of Spontaneity for Redox Reactions

Consider the result of Example 18.6. The calculation revealed that the reaction was not spontaneous. Based on conceptual reasoning, which of the following best explains why I_2 will not oxidize Br^-.

(a) Br is more electronegative than I; therefore Br^- would not be expected to give up an electron to I_2.

(b) I is more electronegative than Br; therefore I_2 would not be expected to give up an electron to Br^-.

(c) Br^- is in solution and I_2 is a solid. Solids do not gain electrons from substances in solution.

Answer: **(a)** Br is more electronegative than I. If the two atoms were in competition for the electron, the electron would go the more electronegative atom (Br). Therefore, I_2 does not spontaneously gain electrons from Br^-.

The Relationship between $E°_{cell}$ and K

We can derive a relationship between the standard cell potential ($E°_{cell}$) and the equilibrium constant for the redox reaction occurring in the cell (K) by returning to the relationship between $\Delta G°$ and K that we learned in Chapter 17. Recall from Section 17.9 that

$$\Delta G° = -RT \ln K \qquad [18.4]$$

By setting Equations 18.3 and 18.4 equal to each other, we get the following:

$$-nFE°_{cell} = -RT \ln K$$

$$E°_{cell} = \frac{RT}{nF} \ln K \qquad [18.5]$$

Equation 18.5 is usually simplified for use at 25 °C by making the following substitutions:

$$R = 8.314 \frac{J}{mol \cdot K}; T = 298.15\ K; F = \left(\frac{96,485\ C}{mol\ e^-}\right); \text{and } \ln K = 2.303 \log K$$

Substituting these into Equation 18.5, we get the following important result:

$$E°_{cell} = \frac{0.0592\ V}{n} \log K \qquad [18.6]$$

where $E°_{cell}$ is the standard cell potential, n is the number of moles of electrons transferred in the redox reaction, and K is the equilibrium constant for the balanced redox reaction at 25 °C.

EXAMPLE 18.7 Relating $E°_{cell}$ and K

Use the tabulated half-cell potentials to calculate K for the oxidation of copper by H^+:

$$Cu(s) + 2\ H^+(aq) \longrightarrow Cu^{2+}(aq) + H_2(g)$$

Sort You are given a redox reaction and asked to find K.	**Given** $Cu(s) + 2\ H^+(aq) \longrightarrow Cu^{2+}(aq) + H_2(g)$ **Find** K
Strategize Use the tabulated values of half-cell potentials to calculate $E°_{cell}$. Then use Equation 18.6 to calculate K from $E°_{cell}$.	**Conceptual Plan** 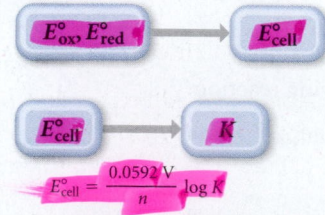

Solve	Solution
Break the reaction up into oxidation and reduction half-reactions and find the standard half-cell potentials for each. Sum the half-cell potentials to find $E°_{cell}$.	*Oxidation:* $\quad\quad\quad Cu(s) \longrightarrow Cu^{2+}(aq) + 2e^{-} \quad\quad E°_{ox} = -E°_{red} = -0.34\ V$ *Reduction:* $\quad 2\,H^{+}(aq) + 2e^{-} \longrightarrow H_2(g) \quad\quad\quad E°_{red} = 0.00\ V$ $\overline{\quad Cu(s) + 2\,H^{+}(aq) \longrightarrow Cu^{2+}(aq) + H_2(g) \quad E°_{cell} = -0.34\ V \quad}$

Compute K from $E°_{cell}$. The value of n (the number of moles of electrons) corresponds to the number of electrons that are canceled in the half-reactions.

$$E°_{cell} = \frac{0.0592\ V}{n}\log K$$

$$\log K = E°_{cell}\frac{n}{0.0592\ V}$$

$$\log K = -0.34\ V\ \frac{2}{0.0592\ V}$$

$$= -11.48$$

$$K = 10^{-11.48}$$

$$= 3.3 \times 10^{-12}$$

Check The answer has no units, as expected for an equilibrium constant. The magnitude of the answer is small, meaning that the reaction lies far to the left at equilibrium, as expected for a reaction in which $E°_{cell}$ is negative.

For Practice 18.7

Use the tabulated half-cell potentials to calculate K for the oxidation of iron by H^{+}:

$$2\,Fe(s) + 6\,H^{+}(aq) \longrightarrow 2\,Fe^{3+}(aq) + 3\,H_2(g)$$

Notice that the fundamental quantity in the above relationships is the standard change in free energy for a chemical reaction ($\Delta G°_{rxn}$). From that quantity, we can calculate both $E°_{cell}$ and K. The relationships between these three quantities can be summarized with the following diagram:

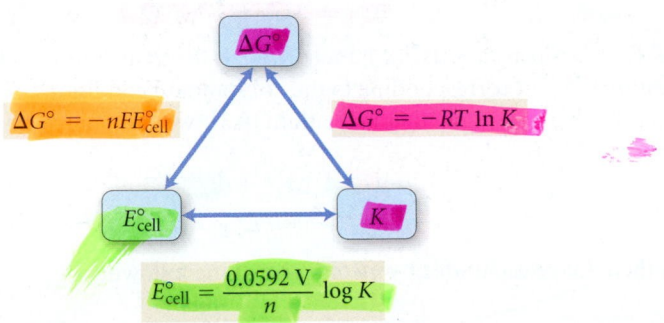

$$E°_{cell} = \frac{0.0592\ V}{n}\log K$$

18.6 Cell Potential and Concentration

We have learned how to find $E°_{cell}$ under standard conditions. For example, we know that when $[Cu^{2+}] = 1\ M$ and $[Zn^{2+}] = 1\ M$, the following reaction will produce a potential of 1.10 V.

$$Zn(s) + Cu^{2+}(aq, 1\ M) \longrightarrow Zn^{2+}(aq, 1\ M) + Cu(s) \quad E°_{cell} = 1.10\ V$$

However, what if $[Cu^{2+}] > 1\ M$ and $[Zn^{2+}] < 1\ M$? For example, how would the cell potential for the following conditions be different from the potential under standard conditions?

$$Zn(s) + Cu^{2+}(aq, 2\ M) \longrightarrow Zn^{2+}(aq, 0.010\ M) + Cu(s) \quad E_{cell} = ?$$

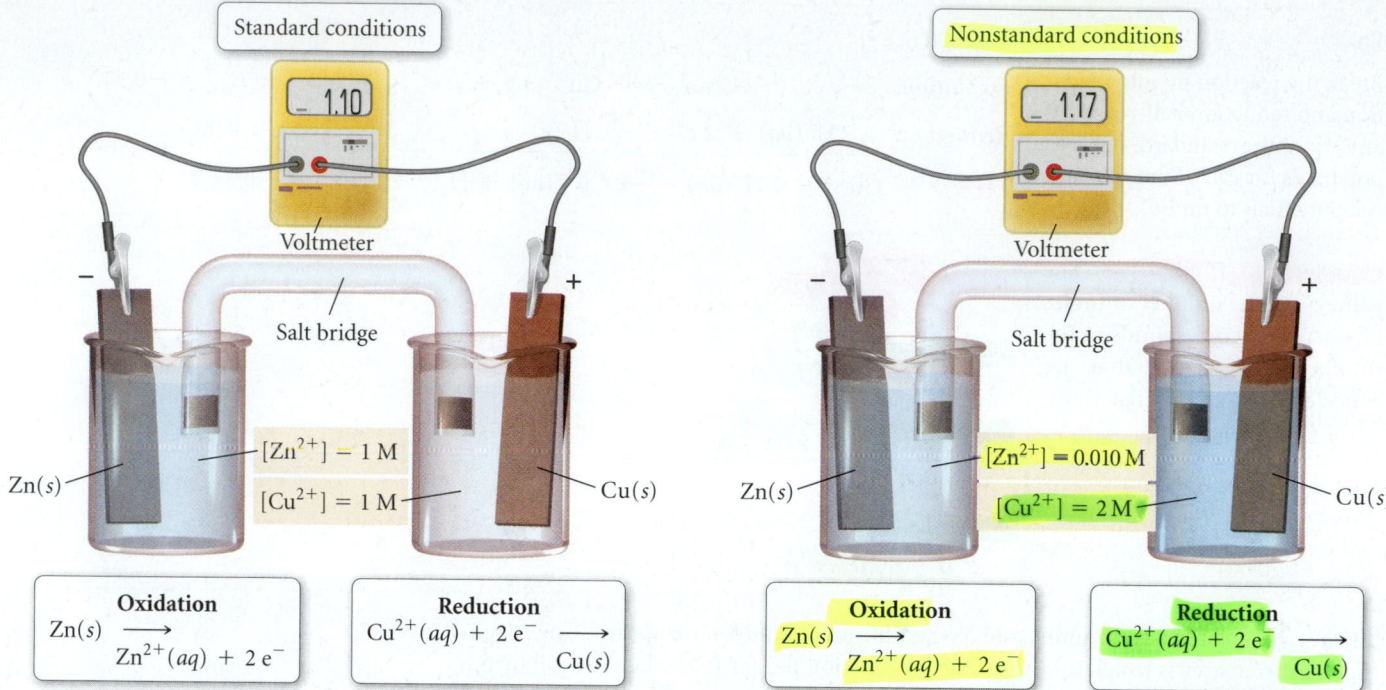

▲ FIGURE 18.11 Cell Potential and Concentration   This figure compares the Zn/Cu^{2+} electrochemical cell under standard and nonstandard conditions. In this case, the nonstandard conditions consist of a higher Cu^{2+} concentration ($[Cu^{2+}] > 1$ M) at the cathode and a lower Zn^{2+} concentration at the anode ($[Zn^{2+}] < 1$ M). According to Le Châtelier's principle, the forward reaction would therefore have a greater tendency to occur, resulting in a greater overall cell potential than the potential under standard conditions.

Since the concentration of a reactant is greater than standard conditions, and since the concentration of product is less than standard conditions, we could use Le Châtelier's principle to predict that the reaction would have an even stronger tendency to occur in the forward direction and that E_{cell} would therefore be greater than $+1.10$ V (Figure 18.11 ▲).

We can derive an exact relationship between E_{cell} (under nonstandard conditions) and E_{cell}° by considering the relationship between the change in free energy (ΔG) and the *standard* change in free energy (ΔG°) that we learned in Section 17.8:

$$\Delta G = \Delta G^\circ + RT \ln Q \qquad [18.7]$$

where R is the gas constant (8.314 J/mol·K), T is the temperature in kelvins, and Q is the equilibrium quotient corresponding to the nonstandard conditions. Since we know the relationship between ΔG and E_{cell} (Equation 18.3), we can substitute into Equation 18.7 as follows:

$$\Delta G = \Delta G^\circ + RT \ln Q$$
$$-nFE_{cell} = -nFE_{cell}^\circ + RT \ln Q$$

We can then divide each side by $-nF$ to arrive at the following:

$$E_{cell} = E_{cell}^\circ - \frac{RT}{nF} \ln Q \qquad [18.8]$$

As we have seen, R and F are constants; at $T = 25$ °C, $\frac{RT}{nF} \ln Q = \frac{0.0592 \text{ V}}{n} \log Q$. Substituting into Equation 18.8, we arrive at the **Nernst equation**:

$$E_{cell} = E_{cell}^\circ - \frac{0.0592 \text{ V}}{n} \log Q \qquad [18.9]$$

where E_{cell} is the cell potential in volts, E_{cell}° is the *standard* cell potential in volts, n is the number of moles of electrons transferred in the redox reaction, and Q is the reaction quotient. Notice that, under standard conditions, $Q = 1$, and (since $\log 1 = 0$) $E_{cell} = E_{cell}^\circ$, as expected. The following example shows how to calculate the cell potential under nonstandard conditions.

EXAMPLE 18.8 Calculating E_{cell} Under Nonstandard Conditions

An electrochemical cell is based on the following two half-reactions:

Oxidation: $Cu(s) \longrightarrow Cu^{2+}(aq, 0.010 \text{ M}) + 2 e^-$

Reduction: $MnO_4^-(aq, 2.0 \text{ M}) + 4 H^+(aq, 1.0 \text{ M}) + 3 e^- \longrightarrow MnO_2(s) + 2 H_2O(l)$

Compute the cell potential.

Sort You are given the half-reactions for a redox reaction and the concentrations of the aqueous reactants and products. You are asked to find the cell potential.	**Given** $[MnO_4^-] = 2.0 \text{ M}$; $[H^+] = 1.0 \text{ M}$; $[Cu^{2+}] = 0.010 \text{ M}$ **Find** E_{cell}
Strategize Use the tabulated values of half-cell potentials to calculate E°_{cell}. Then use Equation 18.9 to calculate E_{cell}.	**Conceptual Plan** $E^\circ_{ox}, E^\circ_{red} \longrightarrow E^\circ_{cell}$ $E^\circ_{cell}, [MnO_4^-], [H^+], [Cu^{2+}] \longrightarrow E_{cell}$ $E_{cell} = E^\circ_{cell} - \dfrac{0.0592 \text{ V}}{n} \log Q$
Solve Write the oxidation and reduction half-reactions, multiplying by the appropriate coefficients to cancel the electrons. Find the standard half-cell potentials for each and sum them to find E°_{cell}.	**Solution** **Ox:** $\qquad 3[Cu(s) \longrightarrow Cu^{2+}(aq) + 2e^-]$ $\qquad\qquad\qquad E^\circ_{ox} = -E^\circ_{red} = -0.34 \text{ V}$ **Red:** $\quad 2[MnO_4^-(aq) + 4 H^+(aq) + 3e^- \longrightarrow MnO_2(s) + 2 H_2O(l)]$ $\qquad\qquad\qquad E^\circ_{red} = 1.68 \text{ V}$ $\overline{3 Cu(s) + 2 MnO_4^-(aq) + 8 H^+(aq) \longrightarrow 3 Cu^{2+}(aq) + 2 MnO_2(s) + 4 H_2O(l)}$ $\qquad\qquad\qquad E^\circ_{cell} = 1.34 \text{ V}$
Compute E_{cell} from E°_{cell}. The value of n (the number of moles of electrons) corresponds to the number of electrons (in this case 6) that are canceled in the half-reactions. Determine Q based on the overall balanced equation and the given concentrations of the reactants and products. (Note that pure liquid water, solid MnO_2, and solid copper are omitted from the expression for Q.)	$E_{cell} = E^\circ_{cell} - \dfrac{0.0592 \text{ V}}{n} \log Q$ $= E^\circ_{cell} - \dfrac{0.0592 \text{ V}}{n} \log \dfrac{[Cu^{2+}]^3}{[MnO_4^-]^2 [H^+]^8}$ $= 1.34 \text{ V} - \dfrac{0.0592 \text{ V}}{6} \log \dfrac{(0.010)^3}{(2.0)^2 (1.0)^8}$ $= 1.34 \text{ V} - (-0.065 \text{ V})$ $= 1.41 \text{ V}$

Check The answer has the correct units (V). The value of E_{cell} is larger than E°_{cell}, as expected based on Le Châtelier's principle because one of the aqueous reactants has a concentration greater than standard conditions and the one aqueous product has a concentration less than standard conditions. Therefore the reaction would have a greater tendency to proceed toward products and have a greater cell potential.

For Practice 18.8

An electrochemical cell is based on the following two half-reactions:

Oxidation: $Ni(s) \longrightarrow Ni^{2+}(aq, 2.0 \text{ M}) + 2 e^-$

Reduction: $VO_2^+(aq, 0.010 \text{ M}) + 2 H^+(aq, 1.0 \text{ M}) + e^- \longrightarrow VO^{2+}(aq, 2.0 \text{ M}) + H_2O(l)$

Compute the cell potential.

From the above examples, and from Equation 18.9, we can conclude the following:

- When a redox reaction within a voltaic cell occurs under standard conditions, $Q = 1$; therefore $E_{cell} = E°_{cell}$.

$$E_{cell} = E°_{cell} - \frac{0.0592 \text{ V}}{n} \log Q$$

$$= E°_{cell} - \frac{0.0592 \text{ V}}{n} \log 1 \qquad \log 1 = 0$$

$$= E°_{cell}$$

- When a redox reaction within a voltaic cell occurs under conditions in which $Q < 1$, the greater concentration of reactants relative to products drives the reaction to the right, resulting in $E_{cell} > E°_{cell}$.
- When a redox reaction within an electrochemical cell occurs under conditions in which $Q > 1$, the greater concentration of products relative to reactants drives the reaction to the left, resulting in $E_{cell} < E°_{cell}$.
- When a redox reaction reaches equilibrium, $Q = K$. The redox reaction has no tendency to occur in either direction and $E_{cell} = 0$.

$$E_{cell} = E°_{cell} - \frac{0.0592 \text{ V}}{n} \log Q \qquad E°_{cell} \qquad \text{(see Equation 18.6)}$$

$$= E°_{cell} - \frac{0.0592 \text{ V}}{n} \log K$$

$$= E°_{cell} - E°_{cell}$$

$$= 0$$

This last point explains why batteries do not last forever—as the reactants are depleted, the reaction proceeds toward equilibrium and the potential tends toward zero.

 Conceptual Connection 18.4 Relating Q, K, E_{cell}, and $E°_{cell}$

In an electrochemical cell, $Q = 0.0010$ and $K = 0.10$. Which of the following is true of E_{cell} and $E°_{cell}$?

(a) E_{cell} is positive and $E°_{cell}$ is negative.   **(b)** E_{cell} is negative and $E°_{cell}$ is positive.

(c) Both E_{cell} and $E°_{cell}$ are positive.   **(d)** Both E_{cell} and $E°_{cell}$ are negative.

Answer: (a) Since $K < 1$, $E°_{cell}$ is negative (under standard conditions, the reaction is not spontaneous). Since $Q < K$, E_{cell} is positive (the reaction is spontaneous under the nonstandard conditions of the cell).

Concentration Cells

Since the cell potential depends not only on the half-reactions occurring in the cell, but also on the *concentrations* of the reactants and products in those half-reactions, it is possible to construct a voltaic cell in which both half-reactions are the same, but in which *a difference in concentration drives the current flow*. For example, consider the electrochemical cell shown in Figure 18.12 ▶, in which copper is oxidized at the anode and copper ions are reduced at the cathode. The second part of Figure 18.12 depicts this cell under nonstandard conditions, with $[Cu^{2+}] = 2.0$ M in one half-cell and $[Cu^{2+}] = 0.010$ M in the other:

$$Cu(s) + Cu^{2+}(aq, 2.0 \text{ M}) \longrightarrow Cu^{2+}(aq, 0.010 \text{ M}) + Cu(s)$$

A Concentration Cell

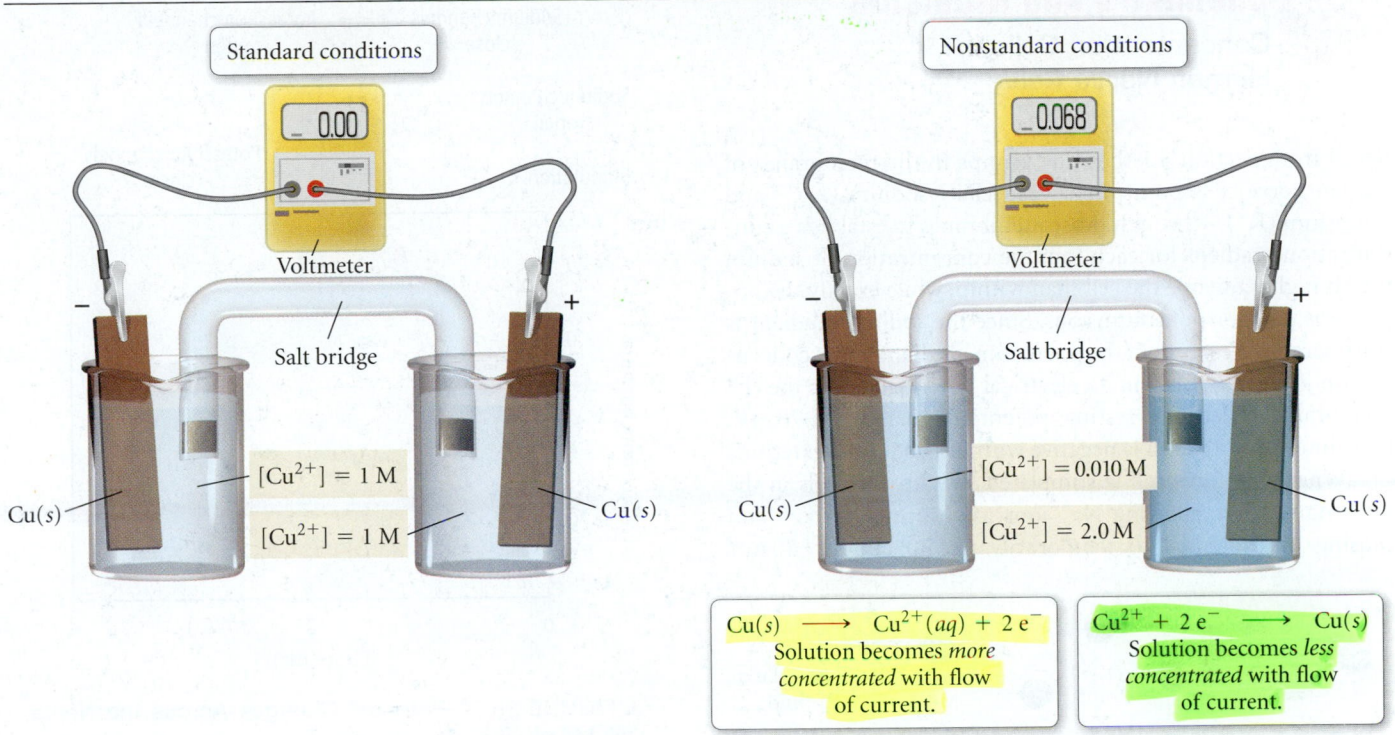

▲ **FIGURE 18.12 Cu/Cu²⁺ Concentration Cell** If the two half-cells have the same Cu^{2+} concentration, the cell potential is zero. If one half-cell has a greater Cu^{2+} concentration than the other, a spontaneous reaction occurs. In the reaction, Cu^{2+} ions in the more concentrated cell are reduced (to solid copper), while Cu^{2+} ions in the more dilute cell are formed (from solid copper). In effect, the concentration of copper ions in the two half-cells tends toward equality.

The half-reactions are identical and the *standard* cell potential is therefore zero.

$$Cu^{2+}(aq) + 2e^- \longrightarrow Cu(s) \qquad E^\circ_{red} = 0.34 \text{ V}$$
$$Cu(s) \longrightarrow Cu^{2+}(aq) + 2e^- \qquad E^\circ_{ox} = -E^\circ_{red} = -0.34 \text{ V}$$
$$\overline{Cu^{2+}(aq) + Cu(s) \longrightarrow Cu(s) + Cu^{2+}(aq) \qquad E^\circ_{cell} = +0.00 \text{ V}}$$

However, because of the different concentrations in the two half-cells, the cell potential must be calculated using the Nernst equation as follows:

$$E_{cell} = E^\circ_{cell} - \frac{0.0592 \text{ V}}{2} \log \frac{0.010}{2.0}$$
$$= 0.000 \text{ V} + 0.068 \text{ V}$$
$$= 0.068 \text{ V}$$

The cell produces a potential of 0.068 V. *Electrons spontaneously flow from the half-cell with the lower copper ion concentration to the half-cell with the higher copper ion concentration.* You can imagine a concentration cell in the same way you think about any concentration gradient. If you mix a concentrated solution of Cu^{2+} with a dilute solution, the Cu^{2+} ions flow from the concentrated solution to the dilute one. Similarly, in a concentration cell, the transfer of electrons *from* the dilute half-cell results in the forming of Cu^{2+} ions in the dilute half-cell. The electrons flow to the concentrated cell, where they react with Cu^{2+} ions and reduce them to $Cu(s)$. Therefore, *the flow of electrons has the effect of increasing the concentration of Cu^{2+} in the dilute cell and decreasing the concentration of Cu^{2+} in the concentrated half-cell.*

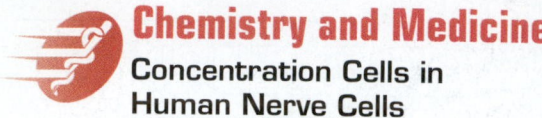

Chemistry and Medicine
Concentration Cells in Human Nerve Cells

Recall from Section 8.1 that tiny pumps in the membranes of human nerve cells pump ions—especially sodium (Na^+) and potassium (K^+)—through those membranes to establish a concentration gradient for each ion: the concentration of sodium ions is higher outside the cell than within, while exactly the opposite is true for potassium ions. Since the sodium gradient is considerably greater than the potassium gradient, these concentration gradients result in an electrical potential across the cell membrane, called the resting potential, of about -70 mV. (The interior of the cell is negative with respect to the exterior.)

When the nerve cell is stimulated, certain channels in the membrane open, allowing Na^+ ions to rush into the cell and causing the potential to temporarily rise to about $+30$ mV

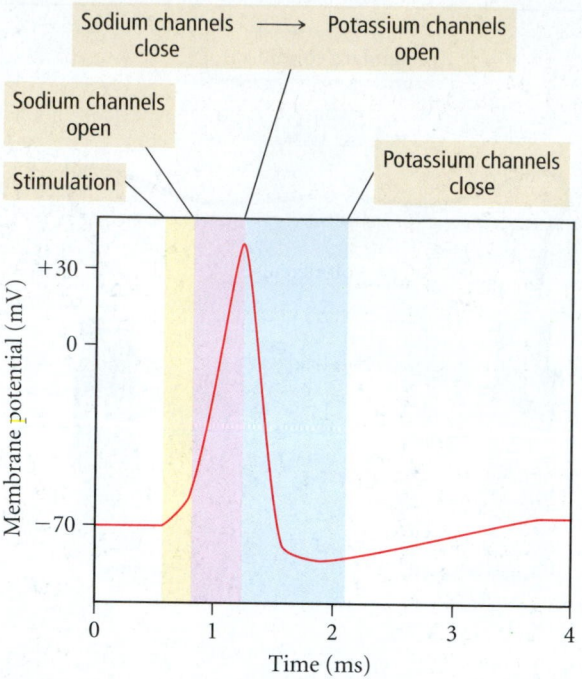

▲ **FIGURE 18.14 Potential Changes Across the Nerve Cell Membrane** The changes in ion concentrations that take place when a nerve cell is stimulated result in a spike in the electrochemical potential across the membrane.

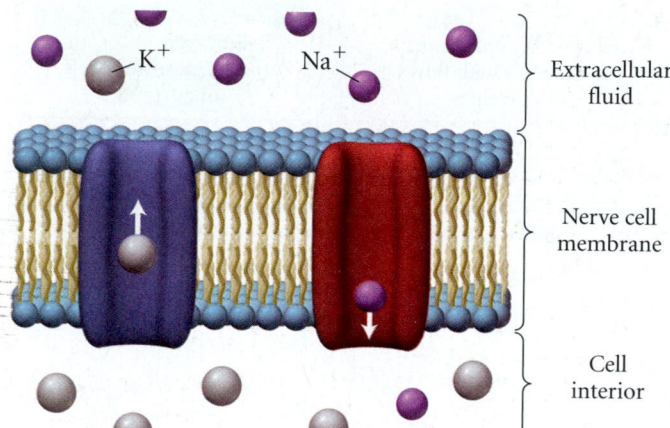

▲ **FIGURE 18.13 Concentration Changes in Nerve Cells** In a nerve cell at rest, the concentration of sodium ions is higher outside the cell than inside. The reverse is true for potassium ions. When a nerve cell is stimulated, sodium channels open and Na^+ ions flood into the cell. A fraction of a second later, the sodium channels close and potassium channels open, allowing K^+ ions to leave the cell.

(Figure 18.13 ◄). This is followed by the opening of other channels that allow K^+ ions to rush out of the cell, bringing the potential back down to near its resting potential. The result is a spike in the electrochemical potential across the membrane, which then provides the stimulus for a similar spike in the neighboring segment of the membrane (Figure 18.14 ▲). In this way, an electrical signal moves down the length of a nerve cell.

When the electrical signal reaches the end of the nerve cell, it triggers the release of a chemical neurotransmitter, which travels to the neighboring nerve cell and stimulates the same kind of electrochemical spike. In this way, neural signals can travel throughout the brain and nervous system of a human being.

18.7 Batteries: Using Chemistry to Generate Electricity

We have seen that we can combine the electron-losing tendency of one substance with the electron-gaining tendency of another to create electrical current in a voltaic cell. Batteries are simply voltaic cells conveniently packaged to act as portable sources of electricity. The actual oxidation and reduction reactions depend on the particular type of battery. In this section, we examine several different types.

Dry-Cell Batteries

Common flashlight batteries are called **dry-cell batteries** because they do not contain large amounts of liquid water. There are several familiar types of dry-cell batteries. The most inexpensive dry cells are composed of a zinc case that acts as the anode (Figure 18.15 ▶). The zinc is oxidized according to the following reaction.

Anode reaction: $Zn(s) \longrightarrow Zn^{2+}(aq) + 2e^-$ (oxidation)

The cathode is a carbon rod immersed in a moist paste of MnO_2 that also contains NH_4Cl. The MnO_2 is reduced to Mn_2O_3 according to the following reaction.

Cathode reaction: $2\,MnO_2(s) + 2\,NH_4^+(aq) + 2\,e^- \longrightarrow$
$$Mn_2O_3(s) + 2\,NH_3(g) + H_2O(l) \quad \text{(reduction)}$$

These two half-reactions produce a voltage of about 1.5 V. Two or more of these batteries can be connected in series (cathode-to-anode connection) to produce higher voltages.

The more common **alkaline batteries** (Figure 18.15b) employ slightly different half-reactions in a basic medium (therefore the name alkaline). In an alkaline battery, the reactions are as follows:

Anode reaction: $Zn(s) + 2\,OH^-(aq) \longrightarrow Zn(OH)_2(s) + 2e^-$

Cathode reaction: $2\,MnO_2(s) + 2\,H_2O(l) + 2e^- \longrightarrow$
$$2\,MnO(OH)(s) + 2\,OH^-(aq) \quad \text{(reduction)}$$

Overall reaction: $Zn(s) + 2\,MnO_2(s) + 2\,H_2O(l) \longrightarrow$
$$Zn(OH)_2(s) + 2\,MnO(OH)(s)$$

Alkaline batteries have a longer working life and a longer shelf life than their nonalkaline counterparts.

Lead–Acid Storage Batteries

The batteries in most automobiles are **lead–acid storage batteries**. These batteries consist of six electrochemical cells wired in series (Figure 18.16 ▼). Each cell produces 2 V for a total of 12 V. Each cell contains a porous lead anode where oxidation occurs and a lead(IV) oxide cathode where reduction occurs according to the following reactions:

Anode reaction: $Pb(s) + HSO_4^-(aq) \longrightarrow$
$$PbSO_4(s) + H^+(aq) + 2e^- \quad \text{(oxidation)}$$

Cathode reaction: $PbO_2(s) + HSO_4^-(aq) + 3\,H^+(aq) + 2e^- \longrightarrow$
$$PbSO_4(s) + 2\,H_2O(l) \quad \text{(reduction)}$$

Overall reaction: $Pb(s) + PbO_2(s) + 2\,HSO_4^-(aq) + 2\,H^+(aq) \longrightarrow$
$$2\,PbSO_4(s) + 2\,H_2O(l)$$

Both the anode and the cathode are immersed in sulfuric acid (H_2SO_4). As electrical current is drawn from the battery, both the anode and the cathode become coated with $PbSO_4(s)$. If the battery is run for a long time without recharging, too much $PbSO_4(s)$ develops on the surface of the electrodes and the battery goes dead. The lead–acid storage battery can be recharged, however, by running electrical current through it in reverse. The electrical current, which must come from an external source such as an alternator in a car, causes the preceding reaction to occur in reverse, converting the $PbSO_4(s)$ back to $Pb(s)$ and $PbO_2(s)$, recharging the battery.

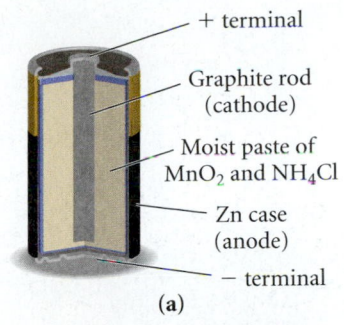

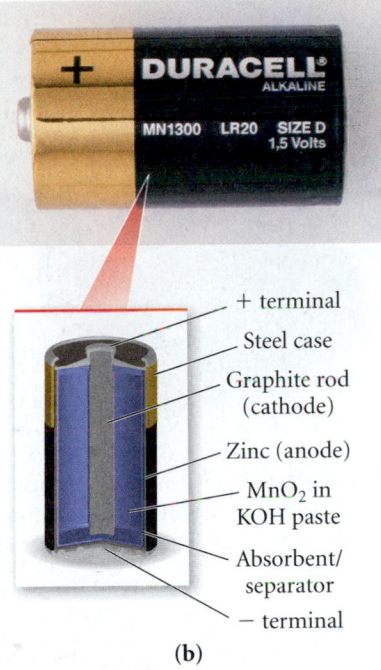

▲ FIGURE 18.15 Dry-Cell Batteries
(a) In a common dry-cell battery, the zinc case acts as the anode and a graphite rod immersed in a moist, slightly acidic paste of MnO_2 and NH_4Cl acts as the cathode. **(b)** The longer-lived alkaline batteries now in common use employ a graphite cathode immersed in a paste of MnO_2 and a base.

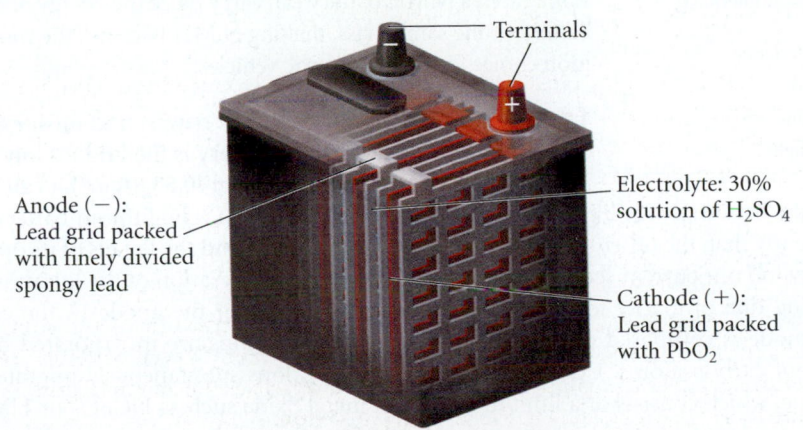

◄ FIGURE 18.16 Lead–Acid Storage Battery A lead–acid storage battery consists of six cells wired in series. Each cell contains a porous lead anode and a lead oxide cathode, both immersed in sulfuric acid.

Other Rechargeable Batteries

The ever-growing need to power electronic products such as laptops, cell phones, and digital cameras, as well as the growth in popularity of hybrid electric vehicles, has driven the need for efficient, long-lasting, rechargeable batteries. The most common types include the **nickel–cadmium (NiCad) battery**, the **nickel–metal hydride (NiMH) battery**, and the **lithium ion battery**.

The Nickel–Cadmium (NiCad) Battery Nickel–cadmium batteries consist of an anode composed of solid cadmium and a cathode composed of $NiO(OH)(s)$. The electrolyte is usually $KOH(aq)$. During operation, the cadmium is oxidized and the $NiO(OH)$ is reduced according to the following equations:

Anode reaction: $Cd(s) + 2\,OH^-(aq) \longrightarrow Cd(OH)_2(s) + 2\,e^-$

Cathode reaction: $2\,NiO(OH)(s) + 2\,H_2O(l) + 2\,e^- \longrightarrow$
$$2\,Ni(OH)_2(s) + 2\,OH^-(aq)$$

The overall reaction produces about 1.30 V. As current is drawn from the NiCad battery, solid cadmium hydroxide accumulates on the anode and solid nickel(II) hydroxide accumulates on the cathode. By running current in the opposite direction, the reactants can be regenerated from the products. A common problem in recharging NiCad and other rechargeable batteries is knowing when to stop. Once all of the products of the reaction are converted back to reactants, the charging process should ideally terminate—otherwise the electrical current will drive other, usually unwanted, reactions such as the electrolysis of water to form hydrogen and oxygen gas. These reactions will typically damage the battery and may sometimes even cause an explosion. Consequently, most commercial battery chargers have sensors designed to measure when the charging is complete. These sensors rely on the small changes in voltage or increases in temperature that occur once the products have all been converted back to reactants.

▲ Several types of batteries, including NiCad, NiMH, and lithium ion batteries, can be recharged by chargers that use household current.

The Nickel–Metal Hydride (NiMH) Battery Although NiCad batteries were the standard rechargeable battery for many years, they are being replaced by others, in part because of the toxicity of cadmium and the resulting disposal problems. One of these replacements is the nickel–metal hydride or NiMH battery. The NiMH battery uses the same cathode reaction as the NiCad battery but a different anode reaction. In the anode of a NiMH battery, hydrogen atoms held in a metal alloy are oxidized. If we let M represent the metal alloy, we can write the half-reactions as follows:

Anode reaction: $M \cdot H(s) + OH^-(aq) \longrightarrow M(s) + H_2O(l) + e^-$

Cathode reaction: $NiO(OH)(s) + H_2O(l) + e^- \longrightarrow Ni(OH)_2(s) + OH^-(aq)$

TABLE 18.2 Energy Density and Overcharge Tolerance of Several Rechargeable Batteries		
Battery Type	Energy Density (W·h/kg)	Overcharge Tolerance
NiCad	45–80	Moderate
NiMH	60–120	Low
Li ion	110–160	Low
Pb storage	30–50	High

In addition to being more environmentally friendly than NiCad batteries, NiMH batteries also have a greater energy density (energy content per unit battery mass), as summarized in Table 18.2. In some cases, a NiMH battery can carry twice the energy of a NiCad battery of the same mass, making NiMH batteries the most common choice for hybrid electric vehicles.

The Lithium Ion Battery The newest and most expensive common type of rechargeable battery is the lithium ion battery. Since lithium is the least dense metal ($0.53\ g/cm^3$) it can be used to make batteries with high energy densities (see Table 18.2). The lithium battery works differently than the other batteries we have examined so far, and the details of its operation are beyond our current scope. Briefly, we can think of the operation of the lithium battery as being due primarily to the motion of lithium ions from the anode to the cathode. The anode is composed of graphite into which lithium ions are incorporated between layers of carbon atoms. Upon discharge, the lithium ions spontaneously migrate to the cathode, which consists of a lithium transition metal oxide such as $LiCoO_2$ or $LiMn_2O_4$.

The transition metal is reduced during this process. Upon recharging, the transition metal is oxidized, forcing the lithium to migrate back into the graphite (Figure 18.17 ▶). The flow of lithium ions from the anode to the cathode causes a corresponding flow of electrons in the external circuit. Lithium ion batteries are commonly used in applications where light weight and high energy density are required. These include cell phones, laptop computers, and digital cameras.

Fuel Cells

We discussed the potential for *fuel cells* in the opening section of this chapter. Fuel cells may one day replace—or at least work in combination with—centralized power grid electricity. In addition, electric vehicles powered by fuel cells may one day replace vehicles powered by internal combustion engines. Fuel cells are like batteries, but the reactants must be constantly replenished. Normal batteries lose their ability to generate voltage with use because the reactants become depleted as electrical current is drawn from the battery. In a **fuel cell**, the reactants—the fuel—constantly flow through the battery, generating electrical current as they undergo a redox reaction.

The most common fuel cell is the hydrogen–oxygen fuel cell (Figure 18.18 ▼). In this cell, hydrogen gas flows past the anode (a screen coated with platinum catalyst) and undergoes oxidation:

Anode reaction: $2 H_2(g) + 4 OH^-(aq) \longrightarrow 4 H_2O(l) + 4 e^-$

Oxygen gas flows past the cathode (a similar screen) and undergoes reduction:

Cathode reaction: $O_2(g) + 2 H_2O(l) + 4 e^- \longrightarrow 4 OH^-(aq)$

The half-reactions sum to the following overall reaction:

Overall reaction: $2 H_2(g) + O_2(g) \longrightarrow 2 H_2O(l)$

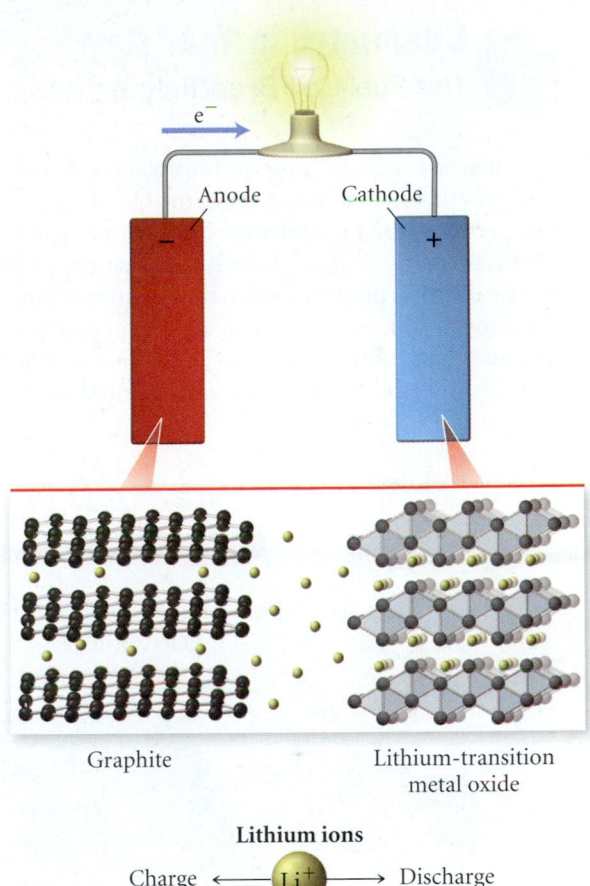

▲ **FIGURE 18.17 Lithium Ion Battery** In the lithium ion battery, the spontaneous flow of lithium ions from the graphite anode to the lithium transition metal oxide cathode causes a corresponding flow of electrons in the external circuit.

Hydrogen–Oxygen Fuel Cell

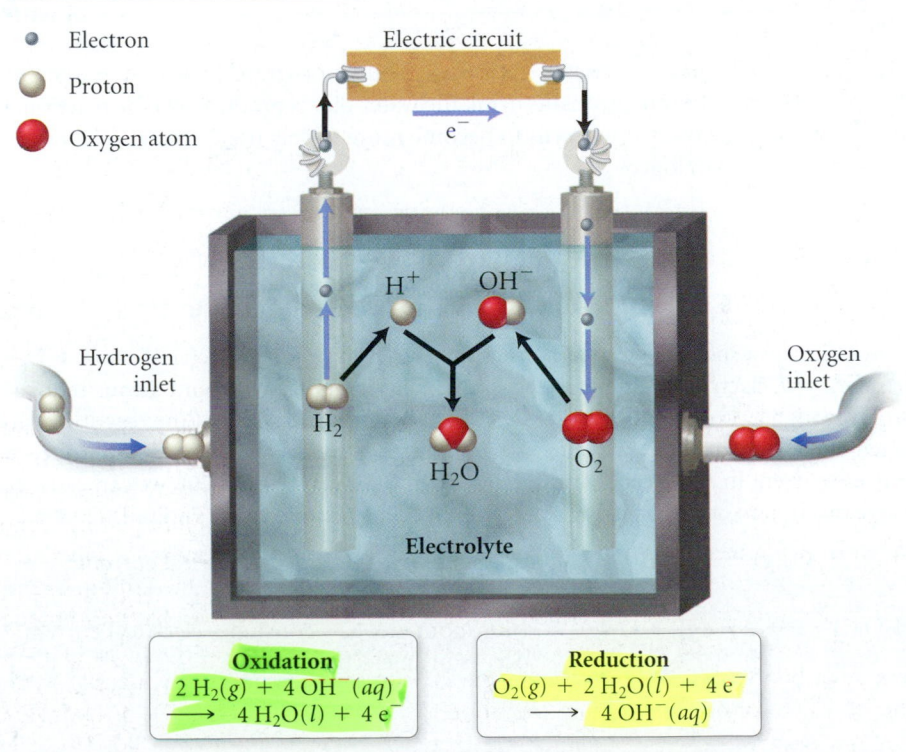

- Electron
- Proton
- Oxygen atom

Oxidation
$2 H_2(g) + 4 OH^-(aq)$
$\longrightarrow 4 H_2O(l) + 4 e^-$

Reduction
$O_2(g) + 2 H_2O(l) + 4 e^-$
$\longrightarrow 4 OH^-(aq)$

◀ **FIGURE 18.18 Hydrogen–Oxygen Fuel Cell** In this fuel cell, hydrogen and oxygen combine to form water.

Chemistry in Your Day
The Fuel-Cell Breathalyzer

Police often use a device called a breathalyzer to measure the amount of ethyl alcohol (C_2H_6O) in the bloodstream of a person suspected of driving under the influence of alcohol (Figure 18.19 ▼). Breathalyzers work because the quantity of ethyl alcohol in the breath is proportional to the quantity of ethyl alcohol in the bloodstream. One type of breathalyzer employs a fuel cell to measure the quantity of alcohol in the breath. When a suspect blows into the breathalyzer, ethyl alcohol is oxidized to acetic acid at the anode.

Anode: $CH_3CH_2OH(g) + 4\,OH^-(aq) \longrightarrow$
$\qquad\qquad$ ethyl alcohol
$$HC_2H_3O_2(g) + 3\,H_2O(l) + 4\,e^-$$
$\qquad\qquad\qquad$ acetic acid

At the cathode, oxygen is reduced.

Cathode: $O_2(g) + 2\,H_2O(l) + 4\,e^- \longrightarrow 4\,OH^-(aq)$

The overall reaction is simply the oxidation of ethyl alcohol to acetic acid and water.

Overall: $CH_3CH_2OH(g) + O_2(g) \longrightarrow$
$$HC_2H_3O_2(g) + H_2O(l)$$

The magnitude of electrical current produced depends on the quantity of alcohol in the breath. A higher current reveals a higher blood alcohol level. When calibrated correctly, the fuel-cell breathalyzer can precisely measure the blood alcohol level of a suspected drunk driver.

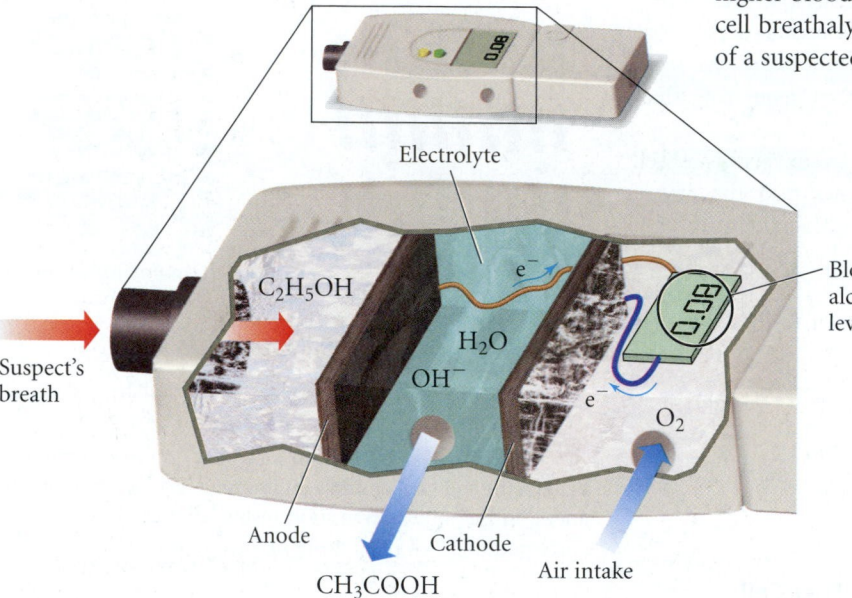

Electrolyte

C_2H_5OH

H_2O

OH^-

O_2

Suspect's breath

Anode

Cathode

Air intake

CH_3COOH

Blood alcohol level

◄ **FIGURE 18.19 Fuel-Cell Breathalyzer** The fuel-cell breathalyzer works by oxidizing ethyl alcohol in the breath to acetic acid. The electrical current that is produced is proportional to the concentration of ethyl alcohol in the breath.

Notice that the only product is water. In the space shuttle program, hydrogen–oxygen fuel cells provide electricity and astronauts drink the water that is produced by the reaction. In order for hydrogen-powered fuel cells to become more widely used, a practical source of hydrogen must be developed.

18.8 Electrolysis: Driving Nonspontaneous Chemical Reactions with Electricity

In a voltaic cell, a spontaneous redox reaction is used to produce electrical current. In an *electrolytic cell*, electrical current is used to drive an otherwise nonspontaneous redox reaction through a process called **electrolysis**. For example, we saw that the reaction of hydrogen with oxygen to form water is spontaneous and can be used to produce an electrical current in a fuel cell. However, by providing electrical current, we can cause the reverse reaction to occur, breaking water into hydrogen and oxygen (Figure 18.20 ▶).

$2\,H_2(g) + O_2(g) \longrightarrow 2\,H_2O(l)$ $\qquad$ (spontaneous—produces electrical current; occurs in a voltaic cell)

$2\,H_2O(l) \longrightarrow 2\,H_2(g) + O_2(g)$ $\qquad$ (nonspontaneous—consumes electrical current; occurs in an electrolytic cell)

Electrolysis of Water

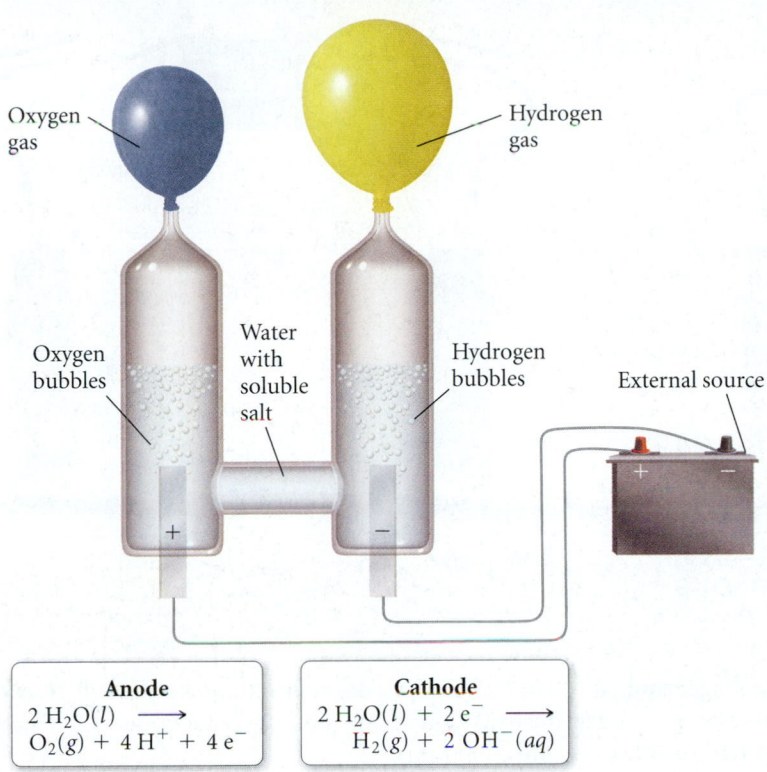

Anode	Cathode
$2 H_2O(l) \longrightarrow$ $O_2(g) + 4 H^+ + 4 e^-$	$2 H_2O(l) + 2 e^- \longrightarrow$ $H_2(g) + 2 OH^-(aq)$

▲ **FIGURE 18.20 Electrolysis of Water**   Electrical current can be used to decompose water into hydrogen and oxygen gas.

One of the problems associated with the widespread adoption of fuel cells is the scarcity of hydrogen. Where will the hydrogen to power these fuel cells come from? One possible answer is to obtain hydrogen from water through solar-powered electrolysis. In other words, a solar-powered electrolytic cell can be used to make hydrogen from water when the sun is shining. The hydrogen can then be converted back to water to generate electricity when needed. Hydrogen made in this way could also be used to power fuel-cell vehicles.

Electrolysis also has numerous other applications. For example, most metals are found in Earth's crust as metal oxides. Converting the oxides to pure metals requires the reduction of the metal, a nonspontaneous process. Electrolysis can be used to produce these metals. Thus, sodium can be produced by the electrolysis of molten sodium chloride (discussed in the following subsection). Electrolysis can also be used to plate metals onto other metals. For example, silver can be plated onto a less expensive metal using the electrolytic cell shown in Figure 18.21 ▶. In this cell, a silver electrode is placed in a solution containing silver ions. An electrical current then causes the oxidation of silver at the anode (replenishing the silver ions in solution) and the reduction of silver ions at the cathode (coating the less expensive metal with solid silver).

Anode reaction: $Ag(s) \longrightarrow Ag^+(aq) + e^-$

Cathode reaction: $Ag^+(aq) + e^- \longrightarrow Ag(s)$

Since the standard cell potential of the above reaction is zero, the reaction is not spontaneous under standard conditions. However, an external power source can be used to drive current flow and therefore cause the above reaction to occur.

Electrolytic Cell for Silver Plating

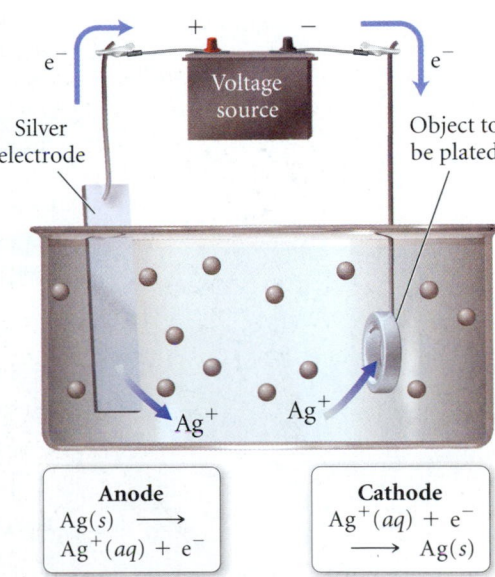

Anode	Cathode
$Ag(s) \longrightarrow$ $Ag^+(aq) + e^-$	$Ag^+(aq) + e^-$ $\longrightarrow Ag(s)$

▲ **FIGURE 18.21 Silver Plating**   Silver can be plated from a solution of silver ions onto metallic objects in an electrolytic cell.

Voltaic Cell

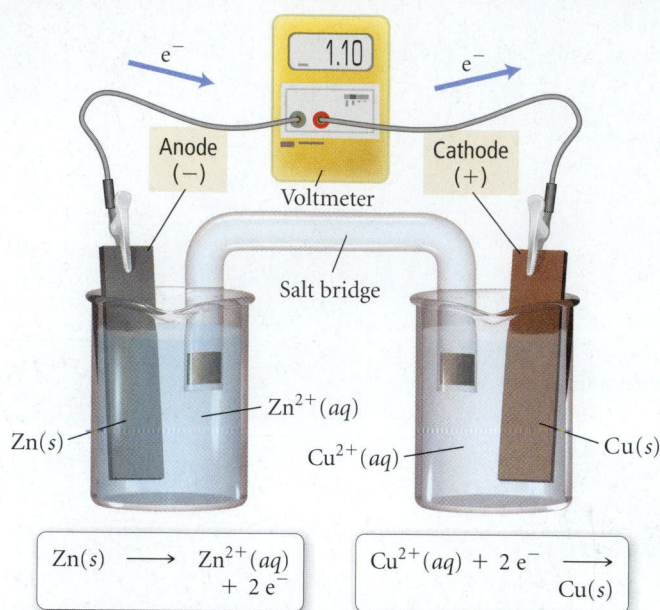

Electrolytic Cell

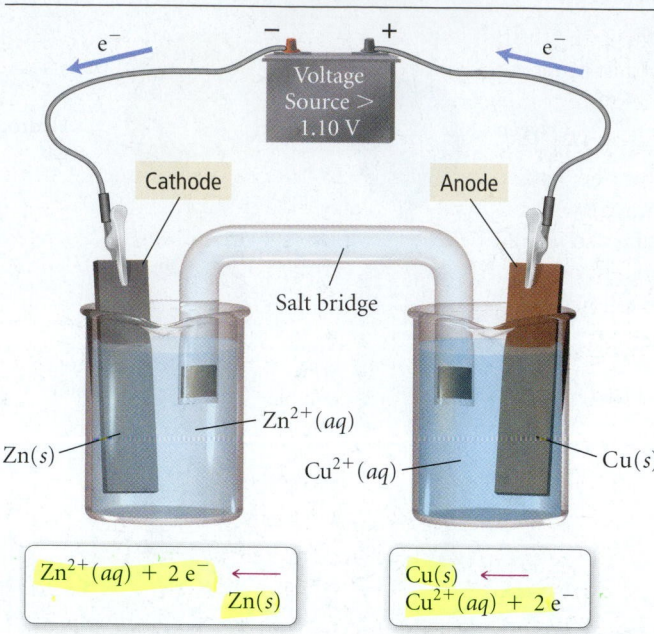

▲ FIGURE 18.22 **Voltaic versus Electrolytic Cells** In a Zn/Cu^{2+} voltaic cell, the reaction proceeds in the spontaneous direction. In a Zn^{2+}/Cu electrolytic cell, electrical current drives the reaction in the nonspontaneous direction.

The voltage required to cause electrolysis depends on the specific half-reactions. For example, we have seen that the oxidation of zinc and the reduction of Cu^{2+} produces a voltage of 1.10 V under standard conditions.

$$Cu^{2+}(aq) + 2e^- \longrightarrow Cu(s) \qquad E^{\circ}_{red} = 0.34\ V$$
$$\underline{Zn(s) \longrightarrow Zn^{2+}(aq) + 2e^- \qquad E^{\circ}_{ox} = -E^{\circ}_{red} = +0.76\ V}$$
$$Cu^{2+}(aq) + Zn(s) \longrightarrow Cu(s) + Zn^{2+}(aq) \qquad E^{\circ}_{cell} = +1.10\ V$$

If a power source producing *more than 1.10 V* is inserted into the voltaic cell, electrons can be forced to flow in the opposite direction, causing the reduction of Zn^{2+} and the oxidation of Cu, as shown in Figure 18.22 ▲. Notice that in the electrolytic cell, the anode has become the cathode (oxidation always occurs at the anode) and the cathode has become the anode.

In a *voltaic cell*, the anode is the source of electrons and is therefore labeled with a negative charge. The cathode draws electrons and is therefore labeled with a positive charge. In an *electrolytic cell*, however, the source of the electrons is the external power source. The external power source must *draw electrons away* from the anode; thus, the anode must be connected to the positive terminal of the battery (as shown in Figure 18.22). Similarly, the power source drives electrons toward the cathode (where they will be used in reduction), so the cathode must be connected to the *negative* terminal of the battery. The charge labels (+ and −) on an electrolytic cell are therefore opposite of what they are in a voltaic cell.

Summarizing,

In all electrochemical cells:

➤ Oxidation occurs at the anode.
➤ Reduction occurs at the cathode.

In voltaic cells:

➤ The anode is the source of electrons and has a negative charge (anode −).
➤ The cathode draws electrons and has a positive charge (cathode +).

In electrolytic cells:

➤ Electrons are drawn away from the anode, which must therefore be connected to the positive terminal of the external power source (anode +).
➤ Electrons are forced to the cathode, which must therefore be connected to the negative terminal of the power source (cathode −).

Predicting the Products of Electrolysis

Predicting the products of an electrolysis reaction is in some cases relatively simple and in other cases more complex. We cover the simpler cases first and follow with the more complex ones.

Pure Molten Salts Consider the electrolysis of a molten salt such as sodium chloride, shown in Figure 18.23 ▶. The species present in the cell are only Na^+ and Cl^-. The chloride ion cannot be further reduced (−1 is its lowest oxidation state), so it must be oxidized. The sodium ion cannot be further oxidized (+1 is its highest oxidation state), so it must be reduced. Thus we can write the half-reactions as follows:

Oxidation: $2\,Cl^-(l) \longrightarrow Cl_2(g) + 2e^-$

Reduction: $2\,Na^+(l) + 2e^- \longrightarrow 2\,Na(s)$

Overall: $2\,Na^+(l) + 2\,Cl^-(l) \longrightarrow 2\,Na(s) + Cl_2(g)$

Notice that, although the reaction as written is not spontaneous, it can be driven to occur in an electrolytic cell by an external power source. We can generalize as follows:

- In the electrolysis of a pure molten salt, the anion is oxidized and the cation is reduced.

Mixtures of Cations or Anions What if a molten salt contains more than one anion or cation? For example, suppose our electrolysis cell contained both NaCl and KCl. Which of the two cations would be reduced at the cathode? In order to answer this question, we must ask which of the two cations is most easily reduced. Although the values of half-cell potentials for aqueous solutions given in Table 18.1 do not apply to molten salts, the relative ordering of the reduction potentials does reflect the relative ease with which the metal cations are reduced. We can see from the table that the reduction of Na^+ is *above* the reduction of K^+ (that is, Na^+ has a less negative reduction potential); therefore, Na^+ is easier to reduce. Consequently, in a mixture of NaCl and KCl, Na^+ will have a greater tendency to be reduced at the cathode.

Similarly, what if a mixture of molten salts contained more than one anion? For example, in a mixture of NaBr and NaCl, which of the two anions would be oxidized at the cathode? The answer is similar: the anion that is most easily oxidized. Since the oxidation half-cell potential for the oxidation of Br^- is less negative than the oxidation half-cell potential for the oxidation of Cl^-, the bromide ion is more easily oxidized and so will be oxidized at the anode. In the electrolysis of a mixture of ions, therefore, we can generalize as follows:

- The cation that is most easily reduced (the one with the least negative, or most positive, reduction half-cell potential) is reduced first.
- The anion that is most easily oxidized (the one that has the least negative, or most positive, oxidation half-cell potential) is oxidized first.

Aqueous Solutions Electrolysis in an aqueous solution is complicated by the possibility of the electrolysis of water itself. Recall that water can be either oxidized or reduced according to the following half-reactions:

Oxidation: $2\,H_2O(l) \longrightarrow O_2(g) + 4\,H^+(aq) + 4\,e^-$ $E^\circ_{ox} = -E^\circ_{red} = -1.23\,V$ (standard conditions)

$\qquad\qquad\qquad\qquad\qquad\qquad\qquad\qquad\qquad\quad E_{ox} = -E_{red} = -0.82\,V\,([H^+] = 10^{-7}\,M)$

Reduction: $2\,H_2O(l) + 2\,e^- \longrightarrow H_2(g) + 2\,OH^-(aq)$ $E^\circ_{red} = -0.83\,V$ (standard conditions)

$\qquad\qquad\qquad\qquad\qquad\qquad\qquad\qquad\qquad\quad E_{red} = -0.41\,V\,([OH^-] = 10^{-7}\,M)$

The half-cell potentials under standard conditions are shown to the right of each half-reaction. However, in pure water at room temperature, the concentrations of H^+ and OH^- are not standard. The half-cell potentials for $[H^+] = 10^{-7}\,M$ and $[OH^-] = 10^{-7}\,M$ are shown in blue. Using those half-cell potentials, we can calculate E_{cell} for the electrolysis of water as follows:

$$E_{cell} = E_{ox} + E_{red} = -0.82\,V - 0.41\,V = -1.23\,V$$

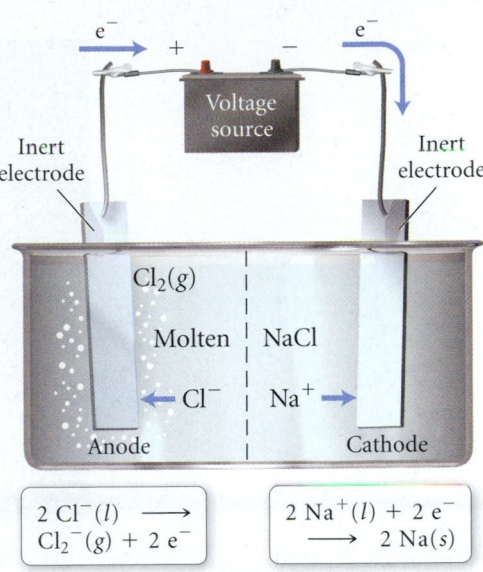

Electrolysis of a Molten Salt

$\begin{aligned} 2\,Cl^-(l) &\longrightarrow \\ Cl_2^-(g) + 2\,e^- & \end{aligned}$ $\begin{aligned} 2\,Na^+(l) + 2\,e^- \\ \longrightarrow 2\,Na(s) \end{aligned}$

▲ **FIGURE 18.23 Electrolysis of Molten NaCl** In the electrolysis of a pure molten salt, the anion (in this case Cl^-) is oxidized and the cation (in this case Na^+) is reduced.

Electrolysis of an Aqueous Salt Solution

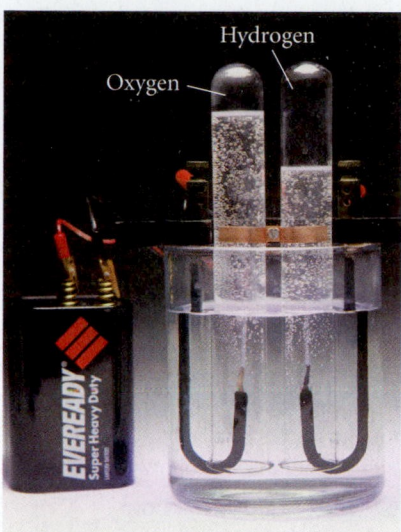

▲ Pure water is a poor conductor of electrical current, but the addition of an electrolyte allows electrolysis to take place, producing hydrogen and oxygen gas in a stoichiometric ratio.

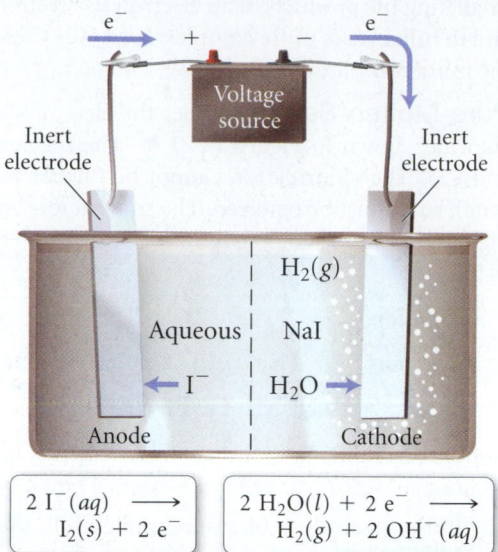

$$2\,I^-(aq) \longrightarrow I_2(s) + 2\,e^-$$

$$2\,H_2O(l) + 2\,e^- \longrightarrow H_2(g) + 2\,OH^-(aq)$$

▲ **FIGURE 18.24 Electrolysis of Aqueous NaI** In this cell, I^- is oxidized to I_2 at the anode and H_2O is reduced to H_2 at the cathode. Sodium ions are not reduced because their reduction potential is more negative than the reduction potential of water.

When a battery with a potential of several volts is connected to an electrolysis cell containing pure water, no reaction occurs because the concentration of ions in pure water is too low to conduct any significant electrical current. When an electrolyte such as Na_2SO_4 is added to the water, however, electrolysis occurs readily.

In any aqueous solution in which electrolysis is to take place, the electrolysis of water is also possible. For example, consider the electrolysis of a sodium iodide solution, as shown in Figure 18.24 ▲. In the electrolysis of *molten* NaI, we can easily predict that I^- is oxidized at the anode and that Na^+ is reduced at the cathode. In an aqueous solution, however, two different oxidation half-reactions are possible at the anode, the oxidation of I^- and the oxidation of water:

Oxidation: $\quad 2\,I^-(aq) \longrightarrow I_2(aq) + 2\,e^- \qquad E^\circ_{ox} = -E^\circ_{red} = -0.54\,V$

Oxidation: $\quad 2\,H_2O(l) \longrightarrow O_2(g) + 4\,H^+(aq) + 4\,e^-$

$$E_{ox} = -E_{red} = -0.82\,V\ ([H^+] = 10^{-7}\,M)$$

Similarly, two different reduction half-reactions are possible at the cathode, the reduction of Na^+ and the reduction of water:

Reduction: $\quad 2\,Na^+(aq) + 2\,e^- \longrightarrow 2\,Na(s) \qquad E^\circ_{red} = -2.71\,V$

Reduction: $\quad 2\,H_2O(l) + 2\,e^- \longrightarrow H_2(g) + 2\,OH^-(aq)$

$$E_{red} = -0.41\,V\ ([OH^-] = 10^{-7}\,M)$$

How do we know which reactions actually occur? In both cases, the answer is the same: *the half-reaction that occurs most easily (the one with the least negative, or most positive, half-cell potential) will be the one that actually occurs.* Therefore, of the two possible oxidation half-reactions, we can predict that the oxidation of I^- to I_2 will occur at the anode, and of the two possible reduction half-reactions, the reduction of H_2O to H_2 will occur at the cathode. Notice that Na^+ cannot be reduced in an aqueous solution—water will be reduced before Na^+. We can make the following generalization:

- The cations of active metals—those that are not easily reduced, such as Li^+, K^+, Na^+, Mg^{2+}, Ca^{2+}, and Al^{3+}—cannot be reduced from aqueous solutions by electrolysis because water is reduced at a lower voltage.

The Electrolysis of Aqueous Sodium Chloride and Overvoltage

An additional complication that must be considered when predicting the products of electrolysis is called *overvoltage*—an additional voltage that must be applied in order to get some nonspontaneous reactions to occur. We can demonstrate this concept by considering the electrolysis of a sodium chloride solution, shown in Figure 18.25 ▶. In order to predict the product of the electrolysis, we consider the two possible oxidation half-reactions,

Oxidation: $2\,Cl^-(aq) \longrightarrow Cl_2(g) + 2\,e^-$ $E^{\circ}_{ox} = -E^{\circ}_{red} = -1.36\,V$

Oxidation: $2\,H_2O(l) \longrightarrow O_2(g) + 4\,H^+(aq) + 4\,e^-$

$$E_{ox} = -E_{red} = -0.82\,V\,([H^+] = 10^{-7}\,M)$$

and the two possible reduction half-reactions,

Reduction: $2\,Na^+(aq) + 2\,e^- \longrightarrow 2\,Na(s)$ $E^{\circ}_{red} = -2.71\,V$

Reduction: $2\,H_2O(l) + 2\,e^- \longrightarrow H_2(g) + 2\,OH^-(aq)$

$$E_{red} = -0.41\,V\,([OH^-] = 10^{-7}\,M)$$

Since the oxidation of water has a lower half-cell potential than the oxidation of Cl^-, we would initially predict that water is oxidized at the anode and that water is reduced at the cathode. In other words, we initially predict that a sodium chloride solution would simply result in the electrolysis of water, producing oxygen gas at the anode and hydrogen gas at the cathode. If we construct such a cell, however, we find that, although hydrogen gas is indeed formed at the cathode (as predicted), oxygen gas is *not* formed at the anode—chlorine gas is formed instead. Why? The answer is that, even though the oxidation half-cell potential for the oxidation of water is $-0.82\,V$, the reaction actually requires a voltage greater than $0.82\,V$ in order to occur. (The reasons for this behavior are related to kinetic factors that our beyond our current scope.) This additional voltage, called the *overvoltage*, increases the voltage required for the oxidation of water to about $1.4\,V$. The result is that the chloride ion oxidizes more easily than water and $Cl_2(g)$ is therefore observed at the anode.

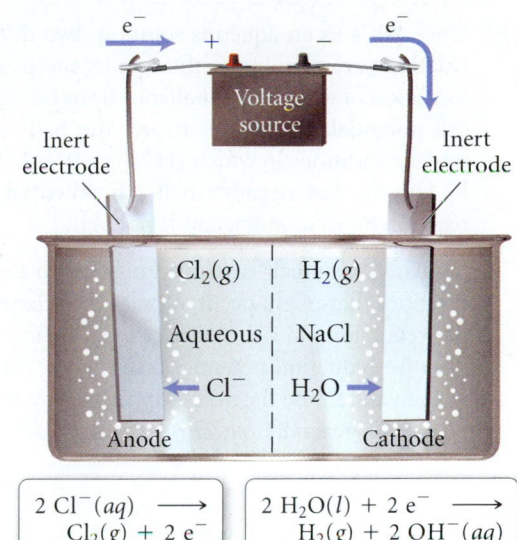

▲ FIGURE 18.25 Electrolysis of Aqueous NaCl: The Effect of Overvoltage Because of overvoltage, the anode reaction of this cell is the oxidation of Cl^- to Cl_2 gas rather than the oxidation of water to H^+ and O_2 gas.

EXAMPLE 18.9 Predicting the Products of Electrolysis Reactions

Predict the half-reaction occurring at the anode and the cathode for electrolysis of each of the following:

(a) a mixture of molten $AlBr_3$ and $MgBr_2$

(b) an aqueous solution of LiI

Solution

(a) In the electrolysis of a molten salt, the anion is oxidized and the cation is reduced. However, this mixture contains two cations. We start by writing the possible oxidation and reduction half-reactions that might occur.

Since Br^- is the only anion, we write the equation for its oxidation, which occurs at the anode.

At the cathode, both the reduction of Al^{3+} and the reduction of Mg^{2+} are possible. The one that actually occurs is the one that occurs most easily. Since the reduction of Al^{3+} has a less negative reduction half-cell potential in aqueous solution, we can assume that this ion is more easily reduced. Therefore, we would predict that the reduction of Al^{3+} occurs at the cathode.

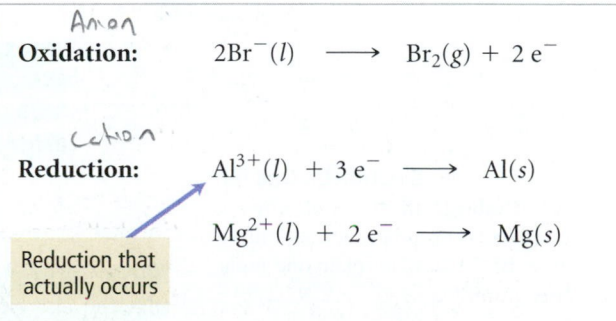

Anion

Oxidation: $2\,Br^-(l) \longrightarrow Br_2(g) + 2\,e^-$

cation

Reduction: $Al^{3+}(l) + 3\,e^- \longrightarrow Al(s)$

$Mg^{2+}(l) + 2\,e^- \longrightarrow Mg(s)$

Reduction that actually occurs

(b) Since LiI is in an aqueous solution, two different oxidation half-reactions are possible at the anode, the oxidation of I$^-$ and the oxidation of water. Write half-reactions for each including the half-cell potential. Remember to use the half-cell potential of water under conditions in which [H$^+$] = 10^{-7}M. Since the oxidation of I$^-$ has the less negative half-cell potential, it will be the half-reaction to occur at the anode.

Similarly, write half-reactions for the two possible reduction half-reactions that might occur at the cathode, the reduction of Li$^+$ and the reduction of water. Since the reduction of water has the less negative reduction potential (even when considering overvoltage, which would raise the necessary voltage by about 0.4–0.6 V), it will be the half-reaction to occur at the anode.

Oxidation that actually occurs
(least negative potential)

Oxidation:
$$2\,I^-(aq) \longrightarrow I_2(s) + 2\,e^-$$
$$E^\circ_{ox} = -E^\circ_{red} = -0.54\ V$$

Oxidation:
$$2\,H_2O(l) \longrightarrow O_2(g) + 4\,H^+(aq) + 4\,e^-$$
$$E_{ox} = -E_{red} = -0.82\ V\ ([H^+] = 10^{-7}\,M)$$

Reduction:
$$Li^+(aq) + e^- \longrightarrow Li(s)$$
$$E^\circ_{red} = -3.04\ V$$

Reduction:
$$2\,H_2O(l) + 2\,e^- \longrightarrow H_2(g) + 2\,OH^-(aq)$$
$$E_{red} = -0.41\ V([OH^-] = 10^{-7}\,M)$$

Reduction that actually occurs
(least negative potential)

For Practice 18.9

Predict the half-reactions occurring at the anode and the cathode for the electrolysis of aqueous Na$_2$SO$_4$.

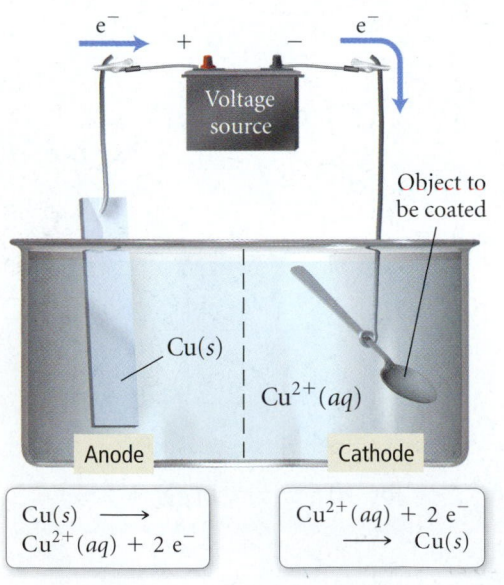

▲ **FIGURE 18.26 Electrolytic Cell for Copper Plating** In this cell, copper ions are plated onto other metals. It takes two moles of electrons to plate one mole of copper atoms.

Anode:
$$Cu(s) \longrightarrow Cu^{2+}(aq) + 2\,e^-$$

Cathode:
$$Cu^{2+}(aq) + 2\,e^- \longrightarrow Cu(s)$$

Stoichiometry of Electrolysis

In an electrolytic cell, electrical current is used to drive a particular chemical reaction. In a sense, the electrons act as a reactant and therefore have a stoichiometric relationship with the other reactants and products. Unlike ordinary reactants, for which we usually measure quantity as mass, for electrons we measure quantity as charge. For example, consider an electrolytic cell used to coat copper onto metals, as shown in Figure 18.26 ◀. The half-reaction by which copper is deposited onto the metal is as follows:

$$Cu^{2+}(aq) + 2\,e^- \longrightarrow Cu(s)$$

For every 2 mol of electrons that flow through cell, 1 mol of solid copper is plated. We can write the stoichiometric relationship as follows:

$$2\ mol\ e^- : 1\ mol\ Cu(s)$$

We can determine the number of moles of electrons that have flowed in a given electrolysis cell by measuring the total charge that has flowed through the cell, which in turn depends on the *magnitude* of the current and on the *time* that the current has run. Recall from Section 18.3 that the unit of current is the ampere:

$$1\ A = 1\,\frac{C}{s}$$

If we multiply the amount of current (in A) flowing through the cell by the time (in s) that the current flowed, we can find the total charge that passed through the cell in that time:

$$Current\left(\frac{C}{s}\right) \times time\,(s) = charge\ (C)$$

The relationship between charge and the number of moles of electrons is given by Faraday's constant, which, as we saw previously, corresponds to the charge in coulombs of 1 mol of electrons.

$$F = \frac{96,485 \text{ C}}{\text{mol e}^-}$$

These relationships can be used to solve problems involving the stoichiometry of electrolytic cells, as shown in the following example.

EXAMPLE 18.10   Stoichiometry of Electrolysis

Gold can be plated out of a solution containing Au^{3+} according to the following half-reaction:

$$Au^{3+}(aq) + 3\,e^- \longrightarrow Au(s)$$

What mass of gold (in grams) will be plated by the flow of 5.5 A of current for 25 minutes?

Sort You are given the half-reaction for the plating of gold, which shows the stoichiometric relationship between moles of electrons and moles of gold. You are also given the current and time. You are asked to find the mass of gold that will be deposited in that time.	**Given** 3 mol e⁻ : 1 mol Au 5.5 amps 25 min **Find** g Au
Strategize You need to find the amount of gold, which is related stoichiometrically to the number of electrons that have flowed through the cell. Begin with time in minutes and convert to seconds. Then, since current is a measure of charge per unit time, use the given current and the time to find the number of coulombs. You can then use Faraday's constant to calculate the number of moles of electrons and the stoichiometry of the reaction to find the number of moles of gold. Finally, use the molar mass of gold to convert to mass of gold.	**Conceptual Plan** 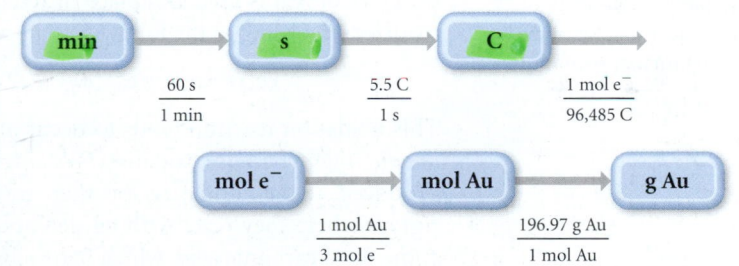
Solve Follow the conceptual plan to solve the problem, canceling units to arrive at mass of gold.	**Solution** $25 \text{ min} \times \dfrac{60 \text{ s}}{1 \text{ min}} \times \dfrac{5.5 \text{ C}}{1 \text{ s}} \times \dfrac{1 \text{ mol e}^-}{96,485 \text{ C}} \times \dfrac{1 \text{ mol Au}}{3 \text{ mol e}^-} \times \dfrac{196.97 \text{ g Au}}{1 \text{ mol Au}} = 5.6 \text{ g Au}$

Check The answer has the correct units (g Au). The magnitude of the answer is also reasonable if we consider that 10 amps of current for one hour is the equivalent of about 1/3 mol of electrons (check for yourself), which would produce 1/9 mol (or about 20 g) of gold.

For Practice 18.10

Silver can be plated out of a solution containing Ag^+ according to the following half-reaction:

$$Ag^+(aq) + e^- \longrightarrow Ag(s)$$

How much time (in minutes) would it take to plate 12 g of silver using a current of 3.0 A?

▲ A metal must usually be reduced to extract it from its ore. In corrosion, the metal is oxidized back to its more natural state.

▲ Aluminum is stable because its oxide forms a protective film over the underlying metal, preventing further oxidation.

18.9 Corrosion: Undesirable Redox Reactions

Corrosion is the (usually) gradual, nearly always undesired oxidation of metals that occurs when they are exposed to oxidizing agents in the environment. Notice from Table 18.1 that the reduction of oxygen in the presence of water has a half-cell potential of $+0.40$ V:

$$O_2(g) + 2\,H_2O(l) + 4\,e^- \longrightarrow 4\,OH^-(aq) \qquad E^\circ_{red} = +0.40\,V$$

In the presence of acid, the reduction of oxygen has an even greater half-cell potential of $+1.23$ V.

$$O_2(g) + 4\,H^+(aq) + 4\,e^- \longrightarrow 2\,H_2O(l) \qquad E^\circ_{red} = +1.23\,V$$

The reduction of oxygen, therefore, has a strong tendency to occur and can bring about the oxidation of other substances, especially metals. Notice that the half-reactions for the reduction of most metal ions lie *below* the half-reactions for the reduction of oxygen in Table 18.1. Consequently, the oxidation (or corrosion) of those metals will be spontaneous when paired with the reduction of oxygen. Corrosion is the opposite of the process by which metals are extracted from their ores. In extraction, the free metal is reduced out from its ore. In corrosion, the metal is oxidized.

Given the ease with which metals are oxidized in the presence of oxygen, acid, and water, why are metals used so frequently as building materials in the first place? Many metals form oxides that coat the surface of the metal and prevent further corrosion. For example, bare aluminum metal, with an oxidation half-cell potential of $+1.66$ V, is quickly oxidized in the presence of oxygen. However, the oxide that forms at the surface of aluminum is Al_2O_3. In its crystalline form, Al_2O_3 is sapphire, a highly inert and structurally solid substance. Consequently, the Al_2O_3 coating acts to protect the underlying aluminum metal, preventing further corrosion.

The oxides of iron, however, are not structurally stable, and tend to flake away from the underlying metal, exposing it to further corrosion. A significant part of the iron produced each year is used to replace rusted iron. Rusting is a redox reaction in which iron is oxidized according to the following half-reaction:

$$Fe(s) \longrightarrow Fe^{2+}(aq) + 2\,e^- \qquad E^\circ_{ox} = -E^\circ_{red} = +0.45\,V$$

This oxidation reaction tends to occur at defects on the surface of the iron—known as *anodic regions* because oxidation is occurring—as shown in Figure 18.27 ▼. The electrons produced at the anodic region then travel through the metal to areas called *cathodic regions* where they react with oxygen and H^+ ions dissolved in moisture. (The H^+ ions come from carbonic acid, which naturally forms in water from carbon dioxide in air.)

$$O_2(g) + 4\,H^+(aq) + 4\,e^- \longrightarrow 2\,H_2O(l) \qquad E^\circ_{red} = +1.23\,V$$

The overall reaction has a cell potential of $+1.67$ V and is therefore highly spontaneous.

$$2\,Fe(s) + O_2(g) + 4\,H^+(aq) \longrightarrow 2\,H_2O(l) + 2\,Fe^{2+}(aq) \qquad E^\circ_{cell} = +1.68\,V$$

▼ **FIGURE 18.27 Corrosion of Iron: Rusting** The oxidation of iron occurs at anodic regions on the metal surface. The iron ions migrate to cathodic regions, where they react with oxygen and water to form rust.

The Rusting of Iron

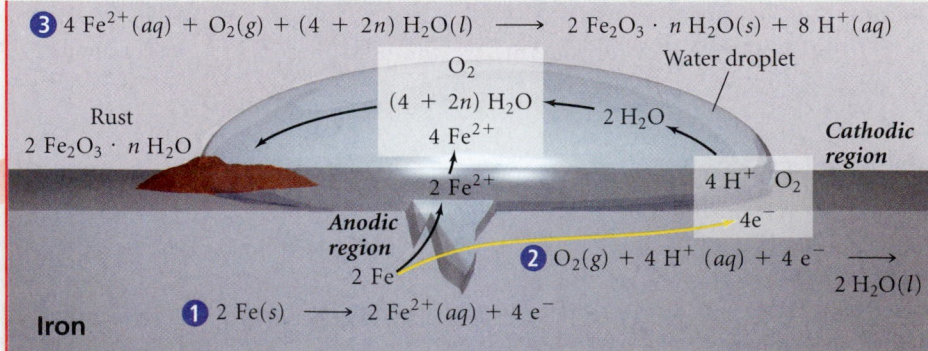

The Fe^{2+} ions formed in the anodic regions can migrate through moisture on the surface of the iron to cathodic regions, where they are further oxidized by reaction with more oxygen:

$$4\,Fe^{2+}(aq) + O_2(g) + (4 + 2n)\,H_2O(l) \longrightarrow 2\,Fe_2O_3 \cdot nH_2O(s) + 8\,H^+(aq)$$

rust

Rust is a hydrated form of iron(III) oxide whose exact composition depends on the conditions under which it forms.

Consider each of the following important components in the formation of rust:

- *Moisture must be present for rusting to occur.* The presence of water is necessary because water is a reactant in the last reaction, and also because charge (either electrons or ions) must be free to flow between the anodic and cathodic regions.
- *Additional electrolytes promote more rusting.* The presence of an electrolyte (such as sodium chloride) on the surface of iron promotes rusting because it enhances current flow. This is why cars rust so quickly in cold climates where roads are salted, or in areas directly adjacent to beaches where salt water mist is present.
- *The presence of acids promotes rusting.* Since H^+ ions are involved in the reduction of oxygen, lower pH enhances the cathodic reaction and leads to faster rusting.

▲ A scratch in paint often allows the rusting of the underlying iron.

Preventing Corrosion

Preventing the rusting of iron is a major industry. The most obvious way to prevent rust is to keep iron dry. Without water, the redox reaction cannot occur. Another way of preventing rust is to coat the iron with a substance that is impervious to water. Cars, for example, are painted and sealed to prevent rust. A scratch in the paint, however, can lead to rusting of the underlying iron.

Rust can also be prevented by placing a *sacrificial electrode* in electrical contact with the iron. The sacrificial electrode must be composed of a metal that oxidizes more easily than iron (that is, it must be below iron in Table 18.1). The sacrificial electrode then oxidizes in place of the iron (just as the more easily oxidizable species in a mixture will be the one to oxidize), protecting the iron from oxidation. Another way to protect iron from rusting is to coat it with a metal that oxidizes more easily than iron. Galvanized nails, for example, are coated with a thin layer of zinc. Since zinc has a more positive half-cell potential for oxidation, it will oxidize in place of the underlying iron (just as a sacrificial electrode does). The oxide of zinc is not crumbly and remains on the nail as a protective coating.

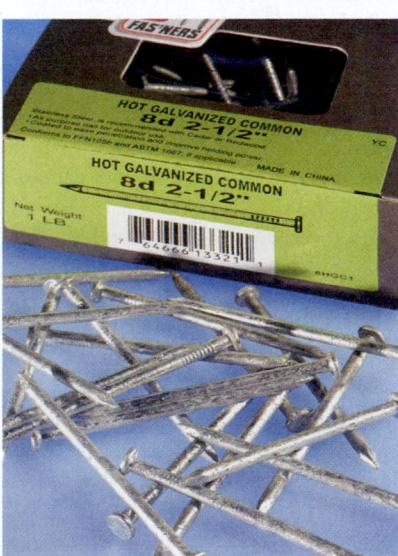

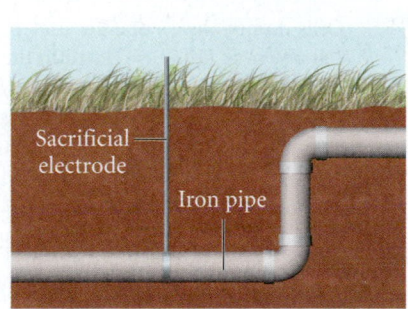

▲ If a metal more active than iron, such as magnesium or aluminum, is in electrical contact with iron, the metal rather than the iron will be oxidized. This principle underlies the use of sacrificial electrodes to prevent the corrosion of iron.

▲ In galvanized nails, a layer of zinc prevents the underlying iron from rusting. The zinc oxidizes in place of the iron, forming a protective layer of zinc oxide.

Conceptual Connection 18.5: Sacrificial Electrodes

Which of the following metals could not act as a sacrificial electrode for iron?

Zn, Mg, Mn, Cd

Answer: Cd. The half-cell potential for the oxidation of Fe is greater than the half-cell potential for the oxidation of Cd. Therefore, Fe will oxidize more easily than Cd.

Chapter in Review

Key Terms

Section 18.3

electrical current (823)
electrochemical cell (823)
voltaic (galvanic) cell (823)
electrolytic cell (823)
half-cell (824)
anode (824)
cathode (824)
salt bridge (825)
ampere (A) (825)
potential difference (825)

volt (V) (825)
electromotive force (emf) (826)
cell potential (cell emf) (E_{cell}) (826)
standard cell potential (standard emf) ($E°_{cell}$) (826)

Section 18.4

standard hydrogen electrode (SHE) (828)

Section 18.5

Faraday's constant (F) (834)

Section 18.6

Nernst equation (838)

Section 18.7

dry-cell battery (842)
alkaline battery (843)
lead–acid storage battery (843)
nickel–cadmium (NiCad) battery (844)

nickel–metal hydride (NiMH) battery (844)
lithium ion battery (844)
fuel cell (845)

Section 18.8

electrolysis (846)

Section 18.9

corrosion (854)

Key Concepts

Pulling the Plug on the Power Grid (18.1)

Oxidation–reduction reactions are reactions in which electrons are transferred. If the reactants of a redox reaction are separated and connected by an external wire, electrons flow through the wire. In the most common form of fuel cell, an electrical current is created in this way as hydrogen is oxidized and oxygen is reduced, with water the only product.

Balancing Oxidation–Reduction Equations (18.2)

Oxidation is the loss of electrons and corresponds to an increase in oxidation state; reduction is the gain of electrons and corresponds to a decrease in oxidation state. Redox reactions can be balanced by the half-reaction method, in which the oxidation and reduction reactions are balanced separately and then added. This method differs slightly for redox reactions in acidic and in basic solutions.

Voltaic (or Galvanic) Cells: Generating Electricity from Spontaneous Chemical Reactions (18.3)

A voltaic cell separates the reactants of a spontaneous redox reaction into two half-cells that are connected by a wire and a means to exchange ions, so that electricity is generated. The electrode where oxidation occurs is the anode and the electrode where reduction occurs is the cathode; electrons flow from the anode to the cathode. The rate of electrons flowing through a wire is measured in amperes (A), and the cell potential is measured in volts (V). A salt bridge is commonly used to allow ions to flow between the half-cell solutions, thereby preventing the build-up of charge. Cell diagram or line notation provides a technique for writing redox reactions concisely by separating the components of the reaction using lines or commas.

Standard Reduction Potentials (18.4)

The potentials of half-cells are measured in relation to that of a hydrogen electrode, which is assigned a half-cell potential of zero under standard conditions (solution concentrations of 1 M, gas pressures of 1 atm, and a temperature of 25 °C). Half-cell potentials are written as standard reduction potentials ($E°_{red}$). A species with a highly positive $E°_{red}$ is an excellent oxidizing agent; its reduction half-reaction will occur spontaneously when coupled with any half-reaction that has a lower $E°_{red}$, because the combined reaction will have a positive standard cell potential ($E°_{cell}$). For example, hydrochloric acid will dissolve any metal with an $E°_{red}$ lower than that of H^+.

Cell Potential, Free Energy, and the Equilibrium Constant (18.5)

In a spontaneous reaction, $E°_{cell}$ is positive, the change in free energy ($\Delta G°$) is negative, and the equilibrium constant (K) is greater than one. In a nonspontaneous reaction, $E°_{cell}$ is negative, $\Delta G°$ is positive, and K is less than one. Because $E°_{cell}$, $\Delta G°$ and K all relate to spontaneity, equations relating all three quantities can be derived.

Cell Potential and Concentration (18.6)

Cells do not always operate under standard conditions. The standard cell potential ($E°_{cell}$) is related to the cell potential (E_{cell}) by the Nernst equation, $E_{cell} = E°_{cell} - (0.0592 \text{ V}/n) \log Q$. As shown by this equation, E_{cell} is related to the reaction quotient (Q); since E_{cell} equals zero when Q equals K, a battery is depleted as the reaction proceeds toward equilibrium. In a concentration cell, the reactions at both electrodes are identical and electrons flow because of a difference in concentration. Nerve cells are a biological example of concentration cells.

Batteries: Using Chemistry to Generate Electricity (18.7)

Batteries are packaged voltaic cells. Dry-cell batteries, including alkaline batteries, do not contain large amounts of water. Rechargeable batteries, such as lead–acid storage, nickel–cadmium, nickel–metal hydride, and lithium ion batteries, allow the reaction to be reversed. Fuel cells are like batteries except that the reactants must be continually replenished.

Electrolysis: Driving Nonspontaneous Chemical Reactions with Electricity (18.8)

An electrolytic cell differs from a voltaic cell in that (1) an electrical charge is used to drive the reaction, and (2) although the anode is still the site of oxidation and the cathode the site of reduction they are represented with signs opposite those of a voltaic cell (anode +, cathode −). In electrolysis reactions, the anion is oxidized; if there is more than one anion, the anion with a less negative $E°_{ox}$ will be oxidized. Stoichiometry can be used to calculate the quantity of reactants consumed or products produced in an electrolytic cell.

Corrosion: Undesirable Redox Reactions (18.9)

Corrosion is the undesired oxidation of metal by environmental oxidizing agents. When some metals, such as aluminum, oxidize they form a stable compound that prevents further oxidation. Iron, however, does not form a structurally stable compound when oxidized and therefore rust flakes off and exposes more iron to corrosion. The corrosion of iron can be prevented by keeping water out of contact with metal, minimizing the presence of electrolytes and acids, or by coating the iron with a sacrificial electrode.

Key Equations and Relationships

Definition of an Ampere (18.3)

$$1\,A = 1\,C/s$$

Definition of a Volt (18.3)

$$1\,V = 1\,J/C$$

Standard Hydrogen Electrode (18.4)

$$2\,H^+(aq) + 2\,e^- \longrightarrow H_2(g) \qquad E^\circ_{red} = 0.00\,V$$

Equation for Cell Potential (18.4)

$$E^\circ_{cell} = E^\circ_{ox} + E^\circ_{red}$$

Relation between Half-Cell Potentials (18.4)

$$E^\circ_{ox} = -E^\circ_{red}$$

Relating ΔG° and E°_{cell} (18.5)

$$\Delta G^\circ = -nFE^\circ_{cell} \qquad F = \frac{96{,}485\,C}{mol\,e^-}$$

Relating E°_{cell} and K (18.5)

$$E^\circ_{cell} = \frac{0.0592\,V}{n} \log K$$

The Nernst Equation (18.6)

$$E_{cell} = E^\circ_{cell} - \frac{0.0592\,V}{n} \log Q$$

Key Skills

Half-Reaction Method of Balancing Aqueous Redox Equations in Acidic Solution (18.2)
 • Examples 18.1, 18.2 • For Practice 18.1, 18.2 • Exercises 37–40

Balancing Redox Reactions Occurring in Basic Solution (18.2)
 • Example 18.3 • For Practice 18.3 • Exercises 41, 42

Calculating Standard Potentials for Electrochemical Cells from Standard Reduction Potentials of the Half-Reactions (18.4)
 • Example 18.4 • For Practice 18.4 • Exercises 45, 46, 61, 62

Predicting Spontaneous Redox Reactions and Sketching Electrochemical Cells (18.4)
 • Example 18.5 • For Practice 18.5 • Exercises 43, 44, 47, 48, 51–54

Relating ΔG° and E°_{cell} (18.5)
 • Example 18.6 • For Practice 18.6 • Exercises 65, 66

Relating E°_{cell} and K (18.5)
 • Example 18.7 • For Practice 18.7 • Exercises 67–72

Calculating E_{cell} under Nonstandard Conditions (18.6)
 • Example 18.8 • For Practice 18.8 • Exercises 73–78

Predicting the Products of Electrolysis Reactions (18.8)
 • Example 18.9 • For Practice 18.9 • Exercises 91–96

Stoichiometry of Electrolysis (18.8)
 • Example 18.10 • For Practice 18.10 • Exercises 99–102

Exercises

Review Questions

1. Electrochemistry uses spontaneous redox reactions for what purpose?

2. In electrochemistry, what kind of reactions can be driven by electricity?

3. Give the basic definitions of oxidation and reduction and explain the basic procedure for balancing redox reactions.

4. Explain the difference between a voltaic (or galvanic) electrochemical cell and an electrolytic one.

5. What reaction (oxidation or reduction) occurs at the anode of a voltaic cell? What is the sign of the anode? Do electrons flow toward or away from the anode?

6. What reaction (oxidation or reduction) occurs at the cathode of a voltaic cell? What is the sign of the cathode? Do electrons flow toward or away from the cathode?

7. Explain the purpose of a salt bridge in an electrochemical cell.

8. What unit is used to measure the magnitude of electrical current? What unit is used to measure the magnitude of a potential difference? Explain the difference between electrical current and potential difference.

9. What is the definition of the standard cell potential (E°_{cell})? What does a large positive standard cell potential imply about the spontaneity of the redox reaction occurring in the cell? What does a negative standard cell potential imply about the reaction?

10. Describe the basic features of a cell diagram (or line notation) for an electrochemical cell.

11. Why do some electrochemical cells employ inert electrodes such as platinum?

12. Describe the standard hydrogen electrode and explain its use in determining half-cell potentials.

13. How is the cell potential of an electrochemical cell (E°_{cell}) related to the potentials of the half-cells?

14. Does a large positive half-cell reduction potential indicate a strong oxidizing agent or a strong reducing agent? What about a large negative half-cell reduction potential?

15. Can a spontaneous redox reaction be obtained by pairing any reduction half-reaction with one above it or with one below it in Table 18.1?

16. How can Table 18.1 be used to predict whether or not a metal will dissolve in HCl? In HNO_3?

17. Explain why E°_{cell}, ΔG°_{rxn}, and K are all interrelated.

18. Will a redox reaction with a small equilibrium constant ($K < 1$) have a positive or a negative E°_{cell}? Will it have a positive or a negative ΔG°_{rxn}?

19. How does E_{cell} depend on the concentrations of the reactants and products in the redox reaction occurring in the cell? What is the effect on E_{cell} of increasing the concentration of a reactant? Increasing the concentration of a product?

20. Use the Nernst equation to show that $E_{cell} = E^\circ_{cell}$ under standard conditions.

21. What is a concentration electrochemical cell?

22. What are the anode and cathode reactions in a common dry-cell battery? In an alkaline battery?

23. What are the anode and cathode reactions in a lead–acid storage battery? What happens when the battery is recharged?

24. What are the three common types of portable rechargeable batteries and how does each one work?

25. What is a fuel cell? What is the most common type of fuel cell and what reactions occur at its anode and cathode?

26. Explain how a fuel-cell breathalyzer works.

27. What are some applications of electrolysis?

28. The anode of an electrolytic cell must be connected to which terminal, positive or negative, of the power source?

29. What species is oxidized and what species reduced in the electrolysis of a pure molten salt?

30. If an electrolytic cell contains a mixture of species that can be oxidized, how can you decide which species will actually be oxidized? If it contains a mixture of species that can be reduced, how can you decide which one will actually be reduced?

31. Why does the electrolysis of an aqueous sodium chloride solution produce hydrogen gas at the cathode?

32. What is overvoltage in an electrochemical cell? Why is it important?

33. How is the amount of current flowing through an electrolytic cell related to the amount of product produced in the redox reaction?

34. What is corrosion? Why is corrosion only a problem with some metals (such as iron)?

35. Explain the role of each of the following in promoting corrosion: moisture, electrolytes, and acids.

36. How can the corrosion of iron be prevented?

Problems by Topic

Balancing Redox Reactions

37. Balance each of the following redox reactions occurring in acidic aqueous solution.

   a. $K(s) + Cr^{3+}(aq) \longrightarrow Cr(s) + K^+(aq)$

   b. $Al(s) + Fe^{2+}(aq) \longrightarrow Al^{3+}(aq) + Fe(s)$

   c. $BrO_3{}^-(aq) + N_2H_4(g) \longrightarrow Br^-(aq) + N_2(g)$

38. Balance each of the following redox reactions occurring in acidic aqueous solution.

   a. $Zn(s) + Sn^{2+}(aq) \longrightarrow Zn^{2+}(aq) + Sn(s)$

   b. $Mg(s) + Cr^{3+}(aq) \longrightarrow Mg^{2+}(aq) + Cr(s)$

   c. $MnO_4{}^-(aq) + Al(s) \longrightarrow Mn^{2+}(aq) + Al^{3+}(aq)$

39. Balance each of the following redox reactions occurring in acidic solution.

    a. $PbO_2(s) + I^-(aq) \longrightarrow Pb^{2+}(aq) + I_2(s)$

    b. $SO_3^{2-}(aq) + MnO_4^-(aq) \longrightarrow SO_4^{2-}(aq) + Mn^{2+}(aq)$

    c. $S_2O_3^{2-}(aq) + Cl_2(g) \longrightarrow SO_4^{2-}(aq) + Cl^-(aq)$

40. Balance each of the following redox reactions occurring in acidic solution.

    a. $I^-(aq) + NO_2^-(aq) \longrightarrow I_2(s) + NO(g)$

    b. $ClO_4^-(aq) + Cl^-(aq) \longrightarrow ClO_3^-(aq) + Cl_2(g)$

    c. $NO_3^-(aq) + Sn^{2+}(aq) \longrightarrow Sn^{4+}(aq) + NO(g)$

41. Balance each of the following redox reactions occurring in basic solution.

    a. $H_2O_2(aq) + ClO_2(aq) \longrightarrow ClO_2^-(aq) + O_2(g)$

    b. $Al(s) + MnO_4^-(aq) \longrightarrow MnO_2(s) + Al(OH)_4^-(aq)$

    c. $Cl_2(g) \longrightarrow Cl^-(aq) + ClO^-(aq)$

42. Balance each of the following redox reactions occurring in basic solution.

    a. $MnO_4^-(aq) + Br^-(aq) \longrightarrow MnO_2(s) + BrO_3^-(aq)$

    b. $Ag(s) + CN^-(aq) + O_2(g) \longrightarrow Ag(CN)_2^-(aq)$

    c. $NO_2^-(aq) + Al(s) \longrightarrow NH_3(g) + AlO_2^-(aq)$

Voltaic Cells, Standard Cell Potentials, and Direction of Spontaneity

43. Sketch a voltaic cell for each of the following overall redox reactions. Label the anode and cathode and indicate the half-reaction occurring at each electrode and the species present in each solution. Also indicate the direction of electron flow.

    a. $2 Ag^+(aq) + Pb(s) \longrightarrow 2 Ag(s) + Pb^{2+}(aq)$

    b. $2 ClO_2(g) + 2 I^-(aq) \longrightarrow 2 ClO_2^-(aq) + I_2(s)$

    c. $O_2(g) + 4 H^+(aq) + 2 Zn(s) \longrightarrow 2 H_2O(l) + 2 Zn^{2+}(aq)$

44. Sketch a voltaic cell for each of the following overall redox reactions. Label the anode and cathode and indicate the half-reaction occurring at each electrode and the species present in each solution. Also indicate the direction of electron flow.

    a. $Ni^{2+}(aq) + Mg(s) \longrightarrow Ni(s) + Mg^{2+}(aq)$

    b. $2 H^+(aq) + Fe(s) \longrightarrow H_2(g) + Fe^{2+}(aq)$

    c. $2 NO_3^-(aq) + 8 H^+(aq) + 3 Cu(s) \longrightarrow$
$$2 NO(g) + 4 H_2O(l) + 3 Cu^{2+}(aq)$$

45. Calculate the standard cell potential for each of the electrochemical cells in Problem 43.

46. Calculate the standard cell potential for each of the electrochemical cells in Problem 44.

47. Consider the following voltaic cell:

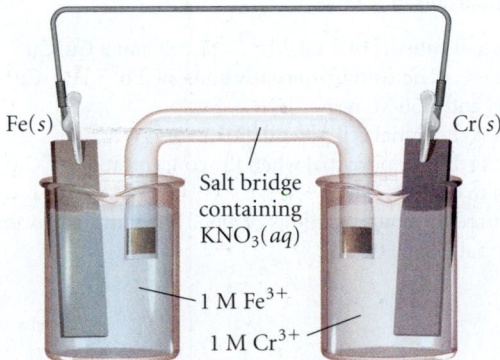

a. Determine the direction of electron flow and label the anode and the cathode.

b. Write a balanced equation for the overall reaction and calculate $E°_{cell}$.

c. Label each electrode as negative or positive.

d. Indicate the direction of anion and cation flow in the salt bridge.

48. Consider the following voltaic cell:

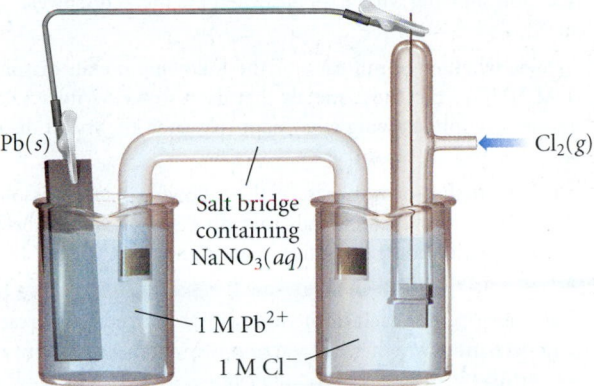

a. Determine the direction of electron flow and label the anode and the cathode.

b. Write a balanced equation for the overall reaction and calculate $E°_{cell}$.

c. Label each electrode as negative or positive.

d. Indicate the direction of anion and cation flow in the salt bridge.

49. Use line notation to represent each of the electrochemical cells in Problem 43.

50. Use line notation to represent each of the electrochemical cells in Problem 44.

51. Make a sketch of the voltaic cell represented with the following line notation. Write the overall balanced equation for the reaction and calculate $E°_{cell}$.

$$Sn(s)|Sn^{2+}(aq)||NO(g)|NO_3^-(aq), H^+(aq)|Pt(s)$$

52. Make a sketch of the voltaic cell represented with the following line notation. Write the overall balanced equation for the reaction and calculate $E°_{cell}$.

$$Mn(s)|Mn^{2+}(aq)||ClO_2^-(aq)|ClO_2(g)|Pt(s)$$

53. Which of the following redox reactions do you expect to occur spontaneously in the forward direction?

    a. $Ni(s) + Zn^{2+}(aq) \longrightarrow Ni^{2+}(aq) + Zn(s)$

    b. $Ni(s) + Pb^{2+}(aq) \longrightarrow Ni^{2+}(aq) + Pb(s)$

    c. $Al(s) + 3 Ag^+(aq) \longrightarrow Al^{3+}(aq) + 3 Ag(s)$

    d. $Pb(s) + Mn^{2+}(aq) \longrightarrow Pb^{2+}(aq) + Mn(s)$

54. Which of the following redox reactions do you expect to occur spontaneously in the reverse direction?

    a. $Ca^{2+}(aq) + Zn(s) \longrightarrow Ca(s) + Zn^{2+}(aq)$

    b. $2 Ag^+(aq) + Ni(s) \longrightarrow 2 Ag(s) + Ni^{2+}(aq)$

    c. $Fe(s) + Mn^{2+}(aq) \longrightarrow Fe^{2+}(aq) + Mn(s)$

    d. $2 Al(s) + 3 Pb^{2+}(aq) \longrightarrow 2 Al^{3+}(aq) + 3 Pb(s)$

55. Which metal could you use to reduce Mn^{2+} ions but not Mg^{2+} ions?

56. Which metal can be oxidized with an Sn^{2+} solution but not with an Fe^{2+} solution?

57. Decide whether or not each of the following metals dissolves in 1 M HCl. For those metals that do dissolve, write a balanced redox reaction showing what happens when the metal dissolves.
   a. Al   **b.** Ag   **c.** Pb

58. Decide whether or not each of the following metals dissolves in 1 M HCl. For those metals that do dissolve, write a balanced redox reaction showing what happens when the metal dissolves.
   a. Cu   **b.** Fe   **c.** Au

59. Decide whether or not each of the following metals dissolves in 1 M HNO_3 . For those metals that do dissolve, write a balanced redox reaction showing what happens when the metal dissolves.
   a. Cu   **b.** Au

60. Decide whether or not each of the following metals dissolves in 1 M HIO_3 . For those metals that do dissolve, write a balanced redox equation for the reaction that occurs.
   a. Au   **b.** Cr

61. Calculate E°_{cell} for each of the following balanced redox reactions and determine whether the reaction is spontaneous as written.
   a. $2\,Cu(s) + Mn^{2+}(aq) \longrightarrow 2\,Cu^+(aq) + Mn(s)$
   b. $MnO_2(s) + 4\,H^+(aq) + Zn(s) \longrightarrow$
$$Mn^{2+}(aq) + 2\,H_2O(l) + Zn^{2+}(aq)$$
   c. $Cl_2(g) + 2\,F^-(aq) \longrightarrow F_2(g) + 2\,Cl^-(aq)$

62. Calculate E°_{cell} for each of the following balanced redox reactions and determine whether the reaction is spontaneous as written.
   a. $O_2(g) + 2\,H_2O(l) + 4\,Ag(s) \longrightarrow 4\,OH^-(aq) + 4\,Ag^+(aq)$
   b. $Br_2(l) + 2\,I^-(aq) \longrightarrow 2\,Br^-(aq) + I_2(s)$
   c. $PbO_2(s) + 4\,H^+(aq) + Sn(s) \longrightarrow$
$$Pb^{2+}(aq) + 2\,H_2O(l) + Sn^{2+}(aq)$$

63. Which of the following metal cations is the best oxidizing agent?
   a. Pb^{2+} **b.** Cr^{3+} **c.** Fe^{2+} **d.** Sn^{2+}

64. Which of the following metals is the best reducing agent?
   a. Mn   **b.** Al   **c.** Ni   **d.** Cr

Cell Potential, Free Energy, and the Equilibrium Constant

65. Use tabulated half-cell potentials to calculate ΔG°_{rxn} for each of the following reactions at 25 °C.
   a. $Pb^{2+}(aq) + Mg(s) \longrightarrow Pb(s) + Mg^{2+}(aq)$
   b. $Br_2(l) + 2\,Cl^-(aq) \longrightarrow 2\,Br^-(aq) + Cl_2(g)$
   c. $MnO_2(s) + 4\,H^+(aq) + Cu(s) \longrightarrow$
$$Mn^{2+}(aq) + 2\,H_2O(l) + Cu^{2+}(aq)$$

66. Use tabulated half-cell potentials to calculate ΔG°_{rxn} for each of the following reactions at 25 °C.
   a. $2\,Fe^{3+}(aq) + 3\,Sn(s) \longrightarrow 2\,Fe(s) + 3\,Sn^{2+}(aq)$
   b. $O_2(g) + 2\,H_2O(l) + 2\,Cu(s) \longrightarrow$
$$4\,OH^-(aq) + 2\,Cu^{2+}(aq)$$
   c. $Br_2(l) + 2\,I^-(aq) \longrightarrow 2\,Br^-(aq) + I_2(s)$

67. Calculate the equilibrium constant for each of the reactions in Problem 65.

68. Calculate the equilibrium constant for each of the reactions in Problem 66.

69. Compute the equilibrium constant for the reaction between $Ni^{2+}(aq)$ and $Cd(s)$.

70. Compute the equilibrium constant for the reaction between $Fe^{2+}(aq)$ and $Zn(s)$.

71. Calculate ΔG°_{rxn} and E°_{cell} for a redox reaction with $n = 2$ that has an equilibrium constant of $K = 25$.

72. Calculate ΔG°_{rxn} and E°_{cell} for a redox reaction with $n = 3$ that has an equilibrium constant of $K = 0.050$.

Non-Standard Conditions and the Nernst Equation

73. A voltaic cell employs the following redox reaction:
$$Sn^{2+}(aq) + Mn(s) \longrightarrow Sn(s) + Mn^{2+}(aq)$$
Calculate the cell potential at 25 °C under each of the following conditions:
   a. standard conditions
   b. $[Sn^{2+}] = 0.0100$ M; $[Mn^{2+}] = 2.00$ M
   c. $[Sn^{2+}] = 2.00$ M; $[Mn^{2+}] = 0.0100$ M

74. A voltaic cell employs the following redox reaction:
$$2\,Fe^{3+}(aq) + 3\,Mg(s) \longrightarrow 2\,Fe(s) + 3\,Mg^{2+}(aq)$$
Calculate the cell potential at 25 °C under each of the following conditions:
   a. standard conditions
   b. $[Fe^{3+}] = 1.0 \times 10^{-3}$ M; $[Mg^{2+}] = 2.50$ M
   c. $[Fe^{3+}] = 2.00$ M; $[Mg^{2+}] = 1.5 \times 10^{-3}$ M

75. An electrochemical cell is based on the following two half-reactions:
Ox: $Pb(s) \longrightarrow Pb^{2+}(aq, 0.10\,M) + 2\,e^-$
Red: $MnO_4^-(aq, 1.50\,M) + 4\,H^+(aq, 2.0\,M) + 3\,e^- \longrightarrow$
$$MnO_2(s) + 2\,H_2O(l)$$
Compute the cell potential at 25 °C.

76. An electrochemical cell is based on the following two half-reactions:
Ox: $Sn(s) \longrightarrow Sn^{2+}(aq, 2.00\,M) + 2\,e^-$
Red: $ClO_2(g, 0.100\,atm) + e^- \longrightarrow ClO_2^-(aq, 2.00\,M)$
Compute the cell potential at 25 °C.

77. A voltaic cell consists of a Zn/Zn^{2+} half-cell and a Ni/Ni^{2+} half-cell at 25 °C. The initial concentrations of Ni^{2+} and Zn^{2+} are 1.50 M and 0.100 M, respectively.
   a. What is the initial cell potential?
   b. What is the cell potential when the concentration of Ni^{2+} has fallen to 0.500 M?
   c. What are the concentrations of Ni^{2+} and Zn^{2+} when the cell potential falls to 0.45 V?

78. A voltaic cell consists of a Pb/Pb^{2+} half-cell and a Cu/Cu^{2+} half-cell at 25 °C. The initial concentrations of Pb^{2+} and Cu^{2+} are 0.0500 M and 1.50 M, respectively.
   a. What is the initial cell potential?
   b. What is the cell potential when the concentration of Cu^{2+} has fallen to 0.200 M?
   c. What are the concentrations of Pb^{2+} and Cu^{2+} when the cell potential falls to 0.35 V?

79. Make a sketch of a concentration cell employing two Zn/Zn^{2+} half-cells. The concentration of Zn^{2+} in one of the half-cells is 2.0 M and the concentration in the other half-cell is 1.0×10^{-3} M. Label the anode and the cathode and indicate the half-reaction occurring at each electrode. Also show the direction of electron flow.

80. Consider the following concentration cell:

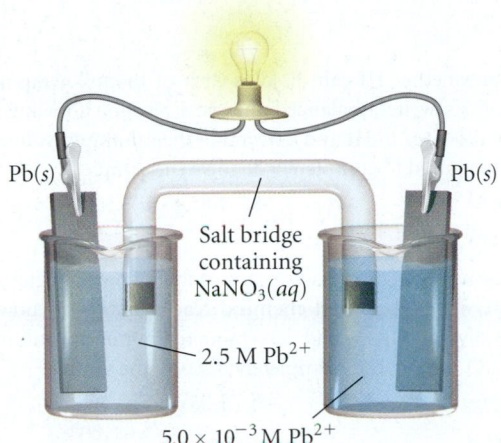

a. Label the anode and cathode.
b. Indicate the direction of electron flow.
c. Indicate what happens to the concentration of Pb^{2+} in each half cell.

81. A concentration cell consists of two Sn/Sn^{2+} half-cells. The cell has a potential of 0.10 V at 25 °C. What is the ratio of the Sn^{2+} concentrations in the two half-cells?

82. A Cu/Cu^{2+} concentration cell has a voltage of 0.22 V at 25 °C. The concentration of Cu^{2+} in one of the half-cells is 1.5×10^{-3} M. What is the concentration of Cu^{2+} in the other half-cell?

Batteries, Fuel Cells, and Corrosion

83. Determine the optimum mass ratio of Zn to MnO_2 in an alkaline battery.

84. What mass of lead sulfate is formed in a lead–acid storage battery when 1.00 g of Pb undergoes oxidation?

85. Use the tabulated values of ΔG_f° in Appendix IIB to calculate E_{cell}° for a fuel cell that employs the reaction between methane gas (CH_4) and oxygen to form carbon dioxide and gaseous water.

86. Use the tabulated values of ΔG_f° in Appendix IIB to calculate E_{cell}° for the fuel cell breathalyzer, which employs the reaction below. (ΔG_f° for $HC_2H_3O_2(g) = -374.2$ kJ/mol.)

$$CH_3CH_2OH(g) + O_2(g) \longrightarrow HC_2H_3O_2(g) + H_2O(g)$$

87. Which of the following metals, if coated onto iron, would prevent the corrosion of iron?
   a. Zn   b. Sn   c. Mn

88. Which of the following metals, if coated onto iron, would prevent the corrosion of iron?
   a. Mg   b. Cr   c. Cu

Electrolytic Cells and Electrolysis

89. Consider the following electrolytic cell:

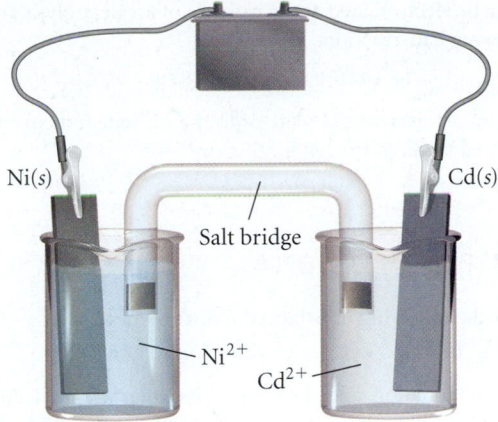

a. Label the anode and the cathode and indicate the half-reactions occurring at each.
b. Indicate the direction of electron flow.
c. Label the terminals on the battery as positive or negative and calculate the minimum voltage necessary to drive the reaction.

90. Draw an electrolytic cell in which Mn^{2+} is reduced to Mn and Sn is oxidized to Sn^{2+}. Label the anode and cathode, indicate the direction of electron flow, and write an equation for the half-reaction occurring at each electrode. What minimum voltage is necessary to drive the reaction?

91. Write equations for the half-reactions that occur in the electrolysis of molten potassium bromide.

92. What are the products obtained in the electrolysis of molten NaI?

93. Write equations for the half-reactions that occur in the electrolysis of a mixture of molten potassium bromide and molten lithium bromide.

94. What are the products obtained in the electrolysis of a molten mixture of KI and KBr?

95. Write equations for the half-reactions that occur at the anode and cathode for the electrolysis of each of the following aqueous solutions:
   a. NaBr(aq)   b. $PbI_2(aq)$
   c. $Na_2SO_4(aq)$

96. Write equations for the half-reactions that occur at the anode and cathode for the electrolysis of each of the following aqueous solutions:
   a. $Ni(NO_3)_2(aq)$ b. KCl(aq)
   c. $CuBr_2(aq)$

97. Make a sketch of an electrolysis cell that might be used to electroplate copper onto other metal surfaces. Label the anode and the cathode and show the reactions that occur at each.

98. Make a sketch of an electrolysis cell that might be used to electroplate nickel onto other metal surfaces. Label the anode and the cathode and show the reactions that occur at each.

99. Copper can be electroplated at the cathode of an electrolysis cell by the following half-reaction.

$$Cu^{2+}(aq) + 2\,e^- \longrightarrow Cu(s)$$

How much time would it take for 225 mg of copper to be plated at a current of 7.8 A?

100. Silver can be electroplated at the cathode of an electrolysis cell by the following half-reaction.

$$Ag^+(aq) + e^- \longrightarrow Ag(s)$$

What mass of silver would plate onto the cathode if a current of 5.8 A flowed through the cell for 55 min?

101. A major source of sodium metal is the electrolysis of molten sodium chloride. What magnitude of current is required to produce 1.0 kg of sodium metal in one hour?

102. What mass of aluminum metal can be produced per hour in the electrolysis of a molten aluminum salt by a current of 25 A?

Cumulative Problems

103. Consider the following unbalanced redox reaction.

$$MnO_4^-(aq) + Zn(s) \longrightarrow Mn^{2+}(aq) + Zn^{2+}(aq)$$

Balance the equation and determine the volume of a 0.500 M $KMnO_4$ solution required to completely react with 2.85 g of Zn.

104. Consider the following unbalanced redox reaction.

$$Cr_2O_7^{2-}(aq) + Cu(s) \longrightarrow Cr^{3+}(aq) + Cu^{2+}(aq)$$

Balance the equation and determine the volume of a 0.850 M $K_2Cr_2O_7$ solution required to completely react with 5.25 g of Cu.

105. Consider the following molecular views of an Al strip and Cu^{2+} solution. Draw a similar sketch showing what happens to the atoms and ions if the Al strip is submerged in the solution for a few minutes.

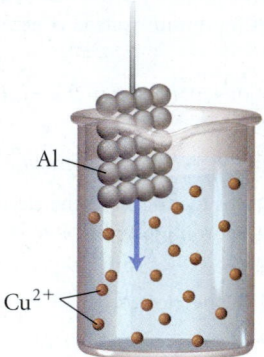

106. Consider the following molecular view of an electrochemical cell involving the following overall reaction.

$$Zn(s) + Ni^{2+}(aq) \longrightarrow Zn^{2+}(aq) + Ni(s)$$

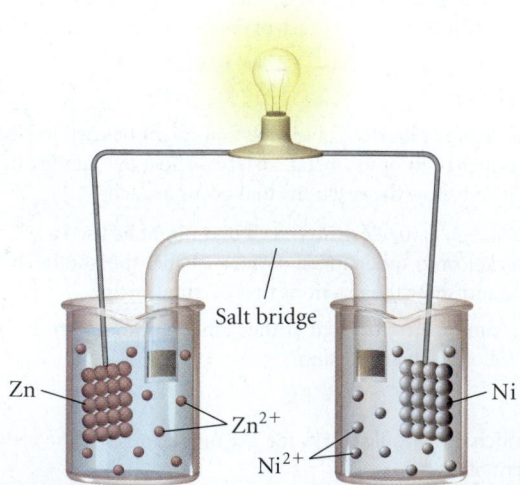

Draw a similar sketch showing how the cell might appear after it has generated a substantial amount of electrical current.

107. Determine whether HI can dissolve each of the following metal samples. If so, write a balanced chemical reaction showing how the metal dissolves in HI and determine the minimum volume of 3.5 M HI required to completely dissolve the sample.

　　a. 2.15 g Al　　　　　　　　**b.** 4.85 g Cu

　　c. 2.42 g Ag

108. Determine if HNO_3 can dissolve each of the following metal samples. If so, write a balanced chemical reaction showing how the metal dissolves in HNO_3 and determine the minimum volume of 6.0 M HNO_3 required to completely dissolve the sample.

　　a. 5.90 g Au　　　　　　　　**b.** 2.55 g Cu

　　c. 4.83 g Sn

109. The cell potential of the following electrochemical cell depends on the pH of the solution in the anode half-cell:

$$Pt(s)|H_2(g, 1\text{ atm})|H^+(aq, ?\text{ M})||Cu^{2+}(aq, 1.0\text{ M})|Cu(s)$$

What is the pH of the solution if E_{cell} is 355 mV?

110. The cell potential of the following electrochemical cell depends on the gold concentration in the cathode half-cell:

$$Pt(s)|H_2(g, 1.0\text{ atm})|H^+(aq, 1.0\text{ M})||Au^{3+}(aq, ?\text{ M})|Au(s)$$

What is the concentration of Au^{3+} in the solution if E_{cell} is 1.22 V?

111. A friend wants you to invest in his newly designed battery that produces 24 V in a single voltaic cell. Why should you be wary of investing in such a battery?

112. What voltage can theoretically be achieved in a battery in which lithium metal is oxidized and fluorine gas is reduced? Why might such a battery be difficult to produce?

113. A battery is constructed based on the oxidation of magnesium and the reduction of Cu^{2+}. The initial concentrations of Mg^{2+} and Cu^{2+} are 1.0×10^{-4} M and 1.5 M, respectively, in 1.0-liter half-cells.

　　a. What is the initial voltage of the battery?

　　b. What is the voltage of the battery after delivering 5.0 A for 8.0 h?

　　c. How long can the battery deliver 5.0 A before going dead?

114. A rechargeable battery is constructed based on a concentration cell constructed of two Ag/Ag^+ half-cells. The volume of each half-cell is 2.0 L and the concentrations of Ag^+ in the half-cells are 1.25 M and 1.0×10^{-3} M.

　　a. For how long can this battery deliver 2.5 A of current before it goes dead?

　　b. What mass of silver is plated onto the cathode by running at 3.5 A for 5.5 h?

　　c. Upon recharging, how long would it take to redissolve 1.00×10^2 g of silver at a charging current of 10.0 amps?

115. If a water electrolysis cell operates at a current of 7.8 A, how long will it take to generate 25.0 L of hydrogen gas at a pressure of 25.0 atm and a temperature of 25 °C?

116. When a suspected drunk driver blows 188 mL of his breath through the fuel-cell breathalyzer described in Section 18.7, the breathalyzer produces an average of 324 mA of current for 10 s. Assuming a pressure of 1.0 atm and a temperature of 25 °C, what percent (by volume) of the driver's breath is ethanol?

117. The K_{sp} of CuI is 1.1×10^{-12}. Find E_{cell} for the following cell.

$$Cu(s)|CuI(s)|I^-(aq)(1.0 \text{ M})||Cu^+(aq)(1.0 \text{ M})|Cu(s)$$

118. The K_{sp} of $Zn(OH)_2$ is 1.8×10^{-14}. Find E_{cell} for the following half-reaction.

$$Zn(OH)_2(s) + 2\,e^- \rightleftharpoons Zn(s) + 2\,OH^-(aq)$$

119. Calculate ΔG°_{rxn} and K for each of the following reactions:
    a. The disproportionation of $Mn^{2+}(aq)$ to $Mn(s)$ and $MnO_2(s)$ in acid solution at 25 °C.
    b. The disproportionation of $MnO_2(s)$ to $Mn^{2+}(aq)$ and $MnO_4^-(aq)$ in acid solution at 25 °C.

120. Calculate ΔG°_{rxn} and K for each of the following reactions
    a. The reaction of $Cr^{2+}(aq)$ with $Cr_2O_7^{2-}(aq)$ in acid solution to form $Cr^{3+}(aq)$.
    b. The reaction of $Cr^{3+}(aq)$ and $Cr(s)$ to form $Cr^{2+}(aq)$. [The reduction potential of $Cr^{2+}(aq)$ to $Cr(s)$ is -0.91 V.]

121. The molar mass of a metal (M) is 50.9 g/mol and it forms a chloride of unknown composition. Electrolysis of a sample of the molten chloride with a current of 6.42 A for 23.6 minutes produces 1.20 g of M at the cathode. Find the empirical formula of the chloride.

122. A metal forms the fluoride MF_3. Electrolysis of the molten fluoride by a current of 3.86 A for 16.2 minutes deposits 1.25 g of the metal. Calculate the molar mass of the metal.

Challenge Problems

123. Suppose a hydrogen–oxygen fuel-cell generator was used to produce electricity for a house. Use the balanced redox reactions and the standard cell potential to predict the volume of hydrogen gas (at STP) required each month to generate the electricity needed for a typical house. Assume the home uses 1.2×10^3 kWh of electricity per month.

124. A voltaic cell designed to measure $[Cu^{2+}]$ is constructed of a standard hydrogen electrode and a copper metal electrode in the Cu^{2+} solution of interest. If you wanted to construct a calibration curve for how the cell potential varies with the concentration of copper(II), what would you plot in order to obtain a straight line? What would be the slope of the line?

125. A thin layer of gold can be applied to another material by an electrolytic process. The surface area of an object to be gold plated is 49.8 cm^2 and the density of gold is 19.3 g/cm^3. A current of 3.25 A is applied to a solution that contains gold in the +3 oxidation state. Calculate the time required to deposit an even layer of gold 1.00×10^{-3} cm thick on the object.

126. To electrodeposit all the Cu and Cd from a solution of $CuSO_4$ and $CdSO_4$ required 1.20 F of electricity (1 F = 1 mol e$^-$). The mixture of Cu and Cd that was deposited had a mass of 50.36 g. What mass of $CuSO_4$ was present in the original mixture?

127. Sodium oxalate, $Na_2C_2O_4$, in solution is oxidized to CO_2 (g) by MnO_4^- which is reduced to Mn^{2+}. A 50.1-mL volume of a solution of MnO_4^- is required to titrate a 0.339-g sample of sodium oxalate. This solution of MnO_4^- is then used to analyze uranium-containing samples. A 4.62-g sample of a uranium-containing material requires 32.5 mL of the solution for titration. The oxidation of the uranium can be represented by the change $UO^{2+} \longrightarrow UO_2^{2+}$. Calculate the percentage of uranium in the sample.

Conceptual Problems

128. An electrochemical cell has a positive standard cell potential but a negative cell potential. What must be true of Q and K for the cell?
    a. $K > 1; Q > K$
    b. $K < 1; Q > K$
    c. $K > 1; Q < K$
    d. $K < 1; Q < K$

129. Which of the following oxidizing agents will oxidize Br^- but not Cl^-?
    a. $K_2Cr_2O_7$ (in acid)
    b. $KMnO_4$ (in acid)
    c. HNO_3

130. A redox reaction employed in an electrochemical cell has a negative ΔG°_{rxn}. Which of the following must be true?
    a. E°_{cell} is positive; $K < 1$
    b. E°_{cell} is positive; $K > 1$
    c. E_{cell} is negative; $K > 1$
    d. E°_{cell} is negative; $K < 1$

19

Radioactivity and Nuclear Chemistry

19.1 Diagnosing Appendicitis

19.2 The Discovery of Radioactivity

19.3 Types of Radioactivity

19.4 The Valley of Stability: Predicting the Type of Radioactivity

19.5 Detecting Radioactivity

19.6 The Kinetics of Radioactive Decay and Radiometric Dating

19.7 The Discovery of Fission: The Atomic Bomb and Nuclear Power

19.8 Converting Mass to Energy: Mass Defect and Nuclear Binding Energy

19.9 Nuclear Fusion: The Power of the Sun

19.10 Nuclear Transmutation and Transuranium Elements

19.11 The Effects of Radiation on Life

19.12 Radioactivity in Medicine and Other Applications

I N THIS CHAPTER, WE EXAMINE RADIOACTIVITY and nuclear chemistry, both of which are associated with changes occurring within the nuclei of atoms. Unlike ordinary chemical processes, in which elements retain their identity, nuclear processes result in the change of one element into another, often with the release of tremendous amounts of energy. Radioactivity has numerous applications, including the diagnosis and treatment of conditions such as cancer, thyroid disease, abnormal kidney and bladder function, and heart disease. Naturally occurring radioactivity allows us to estimate the age of fossils, rocks, and artifacts. Radioactivity also led to the discovery of nuclear fission, used for electricity generation and nuclear weapons. In this chapter, we will learn about radioactivity—how it was discovered, what it is, and how it is used.

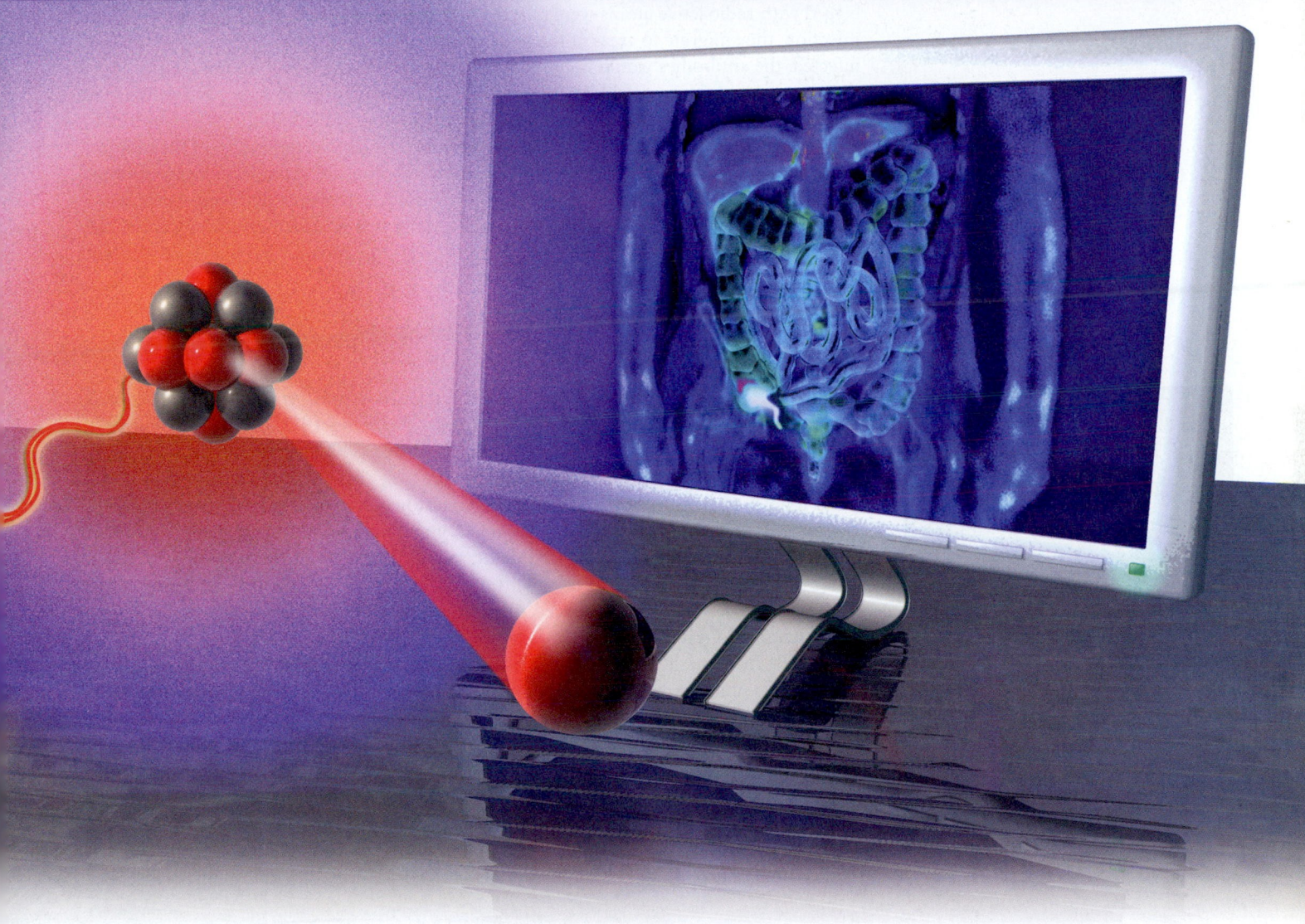

Antibodies tagged with radioactive atoms can be used to diagnose an infected appendix.

19.1 Diagnosing Appendicitis

One morning a few years ago I awoke with a dull pain on the lower right side of my abdomen that seemed to worsen by early afternoon. Since pain in this area can indicate appendicitis (inflammation of the appendix), and since I knew that appendicitis could be dangerous if left untreated, I went to the hospital emergency room. The doctor who examined me recommended a simple blood test to determine my white blood cell count. Patients with appendicitis usually have a high white blood cell count because the body is trying to fight the infection. However, the test was negative—I had a normal white blood cell count.

Although my symptoms were consistent with appendicitis, the negative blood test clouded the diagnosis. The doctor said that I could either have my appendix removed anyway (with the chance of its being healthy) or I could submit to another test that might confirm the appendicitis. I chose the additional test, which involved *nuclear medicine*, an area of medical practice that uses *radioactivity* to diagnose and treat disease. **Radioactivity** is the emission of subatomic particles or high-energy electromagnetic radiation by the nuclei of certain atoms. Such atoms are said to be **radioactive**. Most radioactive emissions are so energetic that they can pass through many types of matter (such as skin and muscle, in this case).

To perform the test, antibodies—naturally occurring molecules that fight infection— tagged with radioactive atoms were injected into my bloodstream. Since antibodies attack infection, they migrate to areas of the body where infection is present. If my appendix was infected, the antibodies would accumulate there. After waiting about an hour, I was taken to a room and laid on a table. A photographic film was inserted in a panel above me. Radioactivity is invisible to the eye, but it exposes photographic film. If my appendix had indeed been infected, it would have by then contained a high concentration of the radioactively tagged antibodies, and the film would show an exposure spot at the location of my appendix. In this procedure, my appendix was the radiation source that would expose the film. The test, however, was negative. No radioactivity was emanating from my appendix; it was healthy. After several hours, the pain in my abdomen subsided and I went home. I never did find out what caused the pain.

19.2 The Discovery of Radioactivity

Radioactivity was discovered in 1896 by a French scientist named Antoine-Henri Becquerel (1852–1908). Becquerel was interested in the newly discovered X-rays (see Chapter 7), which were a hot topic of physics research in his time. He hypothesized that X-rays were emitted in conjunction with **phosphorescence**, the long-lived *emission* of light that sometimes follows the absorption of light by certain atoms and molecules. Phosphorescence is probably most familiar to you as the *glow* in glow-in-the-dark toys. After such a toy is exposed to light, it reemits some of that light, usually at slightly longer wavelengths. If you turn off the room lights or put the toy in the dark, you can see the greenish glow of the emitted light. Becquerel hypothesized that the visible greenish glow was associated with the emission of X-rays (which are invisible).

To test his hypothesis, Becquerel placed crystals—composed of potassium uranyl sulfate, a compound known to phosphoresce—on top of a photographic plate wrapped in black cloth (Figure 19.1 ▼). He then exposed the crystals to sunlight. He knew the crystals phosphoresced because he could see the emitted light when he brought them back into the dark. If the crystals also emitted X-rays, the X-rays would pass through the black cloth and expose the underlying photographic plate. Becquerel performed the experiment several times and always got the same result—the photographic plate showed a dark exposure spot where the crystals had been. Becquerel believed his hypothesis was correct and presented the results—that phosphorescence and X-rays were linked—to the French Academy of Sciences.

▲ The greenish light emitted from glow-in-the-dark toys is phosphorescence.

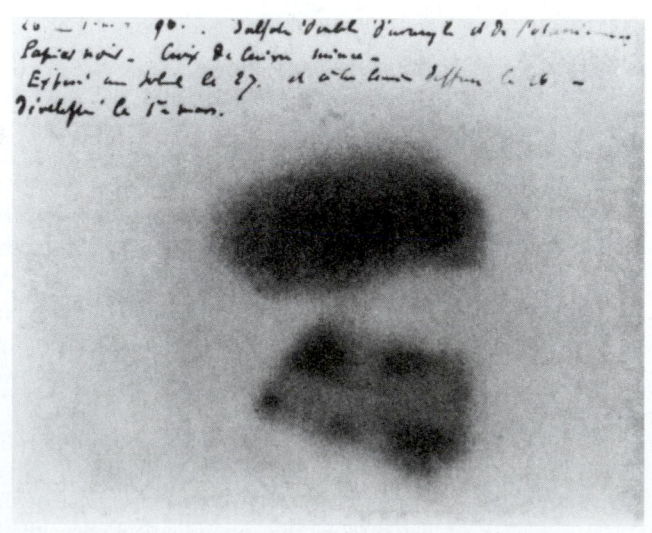

▶ **FIGURE 19.1 The Discovery of Radioactivity** This photographic plate (with Becquerel's original comments at top) played a key role in the discovery of radioactivity. Becquerel placed a uranium-containing compound on the plate (which was wrapped in black cloth to shield it from visible light). He found that the plate was darkened by some previously unknown form of penetrating radiation that was produced continuously, independently of phosphorescence.

Becquerel later retracted his results, however, when he discovered that a photographic plate with the same crystals showed a dark exposure spot even when the plate and the crystals were stored in a drawer and not exposed to sunlight. Becquerel realized that the crystals themselves were constantly emitting something that exposed the photographic plate, independent of whether or not they phosphoresced. Becquerel concluded that it was the uranium within the crystals that was the source of the emissions, and he called the emissions *uranic rays*.

Soon after Becquerel's discovery, a young graduate student named Marie Sklodowska Curie (1867–1934) (one of the first women in France to pursue doctoral work) decided to study uranic rays for her doctoral thesis. Her first task was to determine whether any other substances besides uranium (the heaviest known element at the time) emitted these rays. In her search, Curie discovered two new elements, both of which also emitted uranic rays. Curie named one of her newly discovered elements polonium, after her home country of Poland. The other element she named radium, because of its high level of radioactivity. Radium was so radioactive that it gently glowed in the dark and emitted significant amounts of heat. Since it was now clear that these rays were not unique to uranium, Curie changed the name of uranic rays to radioactivity. In 1903 Curie, her husband, Pierre Curie, and Becquerel were awarded the Nobel Prize in physics for the discovery of radioactivity. In 1911, Curie received a second Nobel Prize, this time in chemistry, for her discovery of the two new elements.

▲ Marie Curie, one of the first women in France to pursue a doctoral degree, was twice awarded the Nobel Prize, in 1903 and 1911. She is seen here with her daughters, about 1905. Irène (left) became a distinguished nuclear physicist in her own right, winning a Nobel Prize in 1935. Eve (right) wrote a highly acclaimed biography of her mother.

Element 96 is named curium in honor of Marie Curie and her contributions to our understanding of radioactivity.

19.3 Types of Radioactivity

While Curie focused her work on discovering the different kinds of radioactive elements, Ernest Rutherford and others focused on characterizing the radioactivity itself. These scientists found that the emissions were produced by the nuclei of radioactive atoms. Such nuclei were unstable and would spontaneously decompose, emitting small pieces of themselves to gain stability. These fragments were the radioactivity that Becquerel and Curie detected. Natural radioactivity can be divided into several different types, including *alpha (α) decay*, *beta (β) decay*, *gamma (γ) ray emission*, and *positron emission*. In addition, unstable atomic nuclei can sometimes attain greater stability by absorbing an atomic electron, a process called *electron capture*.

In order to understand these different types of radioactivity, we must briefly review the notation for symbolizing isotopes that was first introduced in Section 2.6. Recall that any isotope can be represented with the following notation:

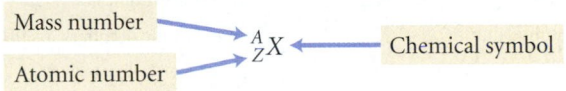

Mass number —→ $^A_Z X$ ←— Chemical symbol
Atomic number —→

▲ Radium, discovered by Marie Curie, is so radioactive that it glows visibly and emits heat.

Mass number (A) = the sum of the number of protons and number of neutrons in the nucleus

Atomic number (Z) = the number of protons in the nucleus

Thus, the number of neutrons in the nucleus (N) is $A - Z$.

$$N = A - Z$$

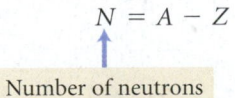

Number of neutrons

For example, the symbol $^{21}_{10}Ne$ represents the neon isotope containing 10 protons and 11 neutrons. The symbol $^{20}_{10}Ne$ represents the neon isotope containing 10 protons and 10 neutrons. Remember that most elements have several different isotopes. When discussing nuclear properties, a particular isotope (or species) of an element is often called a **nuclide**.

The main subatomic particles—protons, neutrons, and electrons—can all be represented with similar notation.

$$\text{Proton symbol } {}^1_1\text{p} \qquad \text{Neutron symbol } {}^1_0\text{n} \qquad \text{Electron symbol } {}^0_{-1}\text{e}$$

The 1 in the lower left of the proton symbol indicates 1 proton, and the 0 in the lower left corner of the neutron symbol represents 0 protons. The −1 in the lower left corner of the electron symbol is a bit different from the other atomic numbers, but it will make sense when we see it in the context of nuclear decay a bit later in this section.

Alpha (α) Decay

Alpha (α) decay occurs when an unstable nucleus emits a particle composed of two protons and two neutrons (Figure 19.2 ▼) Since two protons and two neutrons are identical to a helium-4 nucleus, the symbol for alpha radiation is the symbol for helium-4:

Alpha (α) particle ${}^4_2\text{He}$

Notice that when an element emits an alpha particle, the number of protons in its nucleus changes, transforming it into a different element. We symbolize this phenomenon with a **nuclear equation**, an equation that represents nuclear processes such as radioactivity. For example, the nuclear equation for the alpha decay of uranium-238 is

Parent nuclide   Daughter nuclide

$${}^{238}_{92}\text{U} \longrightarrow {}^{234}_{90}\text{Th} + {}^4_2\text{He}$$

The original atom is called the *parent nuclide* and the product of the decay is called the *daughter nuclide*. In this case, uranium-238 becomes thorium-234. Unlike a chemical reaction, in which elements retain their identity, a nuclear reaction often results in elements changing their identity. Like a chemical equation, however, a nuclear equation must be balanced. *The sum of the atomic numbers on both sides of a nuclear equation must be equal, and the sum of the mass numbers on both sides must also be equal.*

$${}^{238}_{92}\text{U} \longrightarrow {}^{234}_{90}\text{Th} + {}^4_2\text{He}$$

Reactants	Products
Sum of mass numbers = 238	Sum of mass numbers = 234 + 4 = 238
Sum of atomic numbers = 92	Sum of atomic numbers = 90 + 2 = 92

The identity and symbol of the daughter nuclide in any alpha decay can be deduced from the mass and atomic number of the parent nuclide. During alpha decay, the mass number decreases by 4 and the atomic number decreases by 2, as shown in the following example.

Alpha Decay

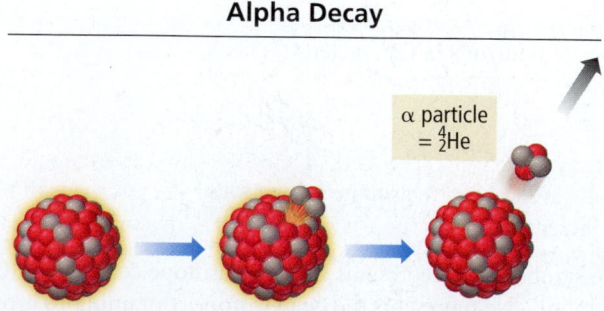

▲ **FIGURE 19.2 Alpha Decay**   In alpha decay, a nucleus emits a particle composed of two protons and two neutrons (a helium-4 nucleus).

Sidebar: As discussed in Section 19.4, nuclei are unstable when they are too large or when they contain an unbalanced ratio of neutrons to protons.

In nuclear chemistry, we are primarily interested in changes within the nucleus; therefore, the 2+ charge that we would normally write for a helium nucleus is omitted for an alpha particle.

EXAMPLE 19.1   Writing Nuclear Equations for Alpha Decay

Write a nuclear equation for the alpha decay of Ra-224.

Solution

Begin with the symbol for Ra-224 on the left side of the equation and the symbol for an alpha particle on the right side.	$^{224}_{88}Ra \longrightarrow ^{?}_{?}? + ^{4}_{2}He$
Equalize the sum of the mass numbers and the sum of the atomic numbers on both sides of the equation by writing the appropriate mass number and atomic number for the unknown daughter nuclide.	$^{224}_{88}Ra \longrightarrow ^{220}_{86}? + ^{4}_{2}He$
Using the periodic table, deduce the identity of the unknown daughter nuclide from the atomic number and write its symbol. Since the atomic number is 86, the daughter nuclide must be radon (Rn).	$^{224}_{88}Ra \longrightarrow ^{220}_{86}Rn + ^{4}_{2}He$

For Practice 19.1

Write a nuclear equation for the alpha decay of Po-216.

Alpha radiation is the 18-wheeler truck of radioactivity. The alpha particle is by far the most massive of all particles emitted by radioactive nuclei. Consequently, alpha radiation has the most potential to interact with and damage other molecules, including biological ones. Highly energetic radiation interacts with other molecules and atoms by ionizing them. If radiation ionizes molecules within the cells of living organisms, those molecules may undergo damaging chemical reactions, and the cell can die or begin to reproduce abnormally. The ability of radiation to ionize other molecules and atoms is called its **ionizing power**. Of all types of radioactivity, alpha radiation has the highest ionizing power.

However, alpha particles, because of their large size, also have the lowest **penetrating power**—the ability to penetrate matter. (Imagine a semi truck trying to get through a traffic jam.) In order for radiation to damage important molecules within living cells, it must penetrate into the cell. Alpha radiation does not easily penetrate into cells because it is stopped by a sheet of paper, by clothing, or even by air. Consequently, a low-level alpha emitter kept outside the body is relatively safe. If an alpha emitter is ingested, however, it becomes very dangerous because the alpha particles then have direct access to the molecules that compose organs and tissues.

Beta (β) Decay

Beta (β) decay occurs when an unstable nucleus emits an electron (Figure 19.3 ▶) How does a nucleus, which contains only protons and neutrons, emit an electron? The electron is formed when a neutron changes into a proton. In other words, in some unstable nuclei, a neutron will change into a proton and emit an electron in the process:

This kind of beta radiation is also called beta minus (β⁻) radiation due to its negative charge.

Beta decay   Neutron $\longrightarrow$ proton + emitted electron

The symbol for a beta (β) particle in a nuclear equation is

Beta (β) particle $^{0}_{-1}e$ •

Thus, beta decay can be represented as

$$^{1}_{0}n \longrightarrow ^{1}_{1}p + ^{0}_{-1}e$$

The −1 in the lower left corner reflects the charge of the electron, which is equivalent to an atomic number of −1 in a nuclear equation. When an atom emits a beta particle, its atomic number increases by 1 because it now has an additional proton. For example, the nuclear equation for the beta decay of radium-228 is

$$^{228}_{88}Ra \longrightarrow ^{228}_{89}Ac + ^{0}_{-1}e$$

Notice that the nuclear equation is still balanced—the sum of the mass numbers on both sides is equal and the sum of the atomic numbers on both sides is equal.

Beta Decay

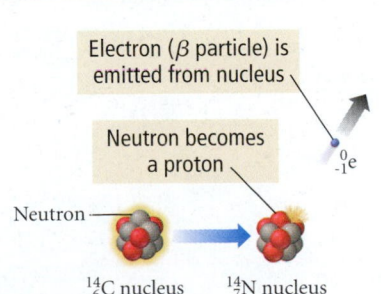

▲ **FIGURE 19.3 Beta Decay**   In beta decay, a neutron emits an electron and becomes a proton.

Beta radiation is the midsize car of radioactivity. Beta particles are much less massive than alpha particles and consequently have a lower ionizing power. However, because of their smaller size, beta particles have a higher penetrating power and require a sheet of metal or a thick piece of wood to stop them. Consequently, a beta emitter outside of the body poses a higher risk than an alpha emitter. Inside the body, however, the beta emitter does less damage than an alpha emitter.

Gamma (γ) Ray Emission

See Section 7.2 for a review of electromagnetic radiation.

Gamma (γ) ray emission is significantly different from alpha or beta radiation. Gamma radiation is *electromagnetic* radiation. Gamma rays are high-energy (short-wavelength) photons. The symbol for a gamma ray is

$$\text{Gamma } (\gamma) \text{ ray} \qquad {}^{0}_{0}\gamma$$

A gamma ray has no charge and no mass. When a gamma-ray photon is emitted from a radioactive atom, it does not change the mass number or the atomic number of the element. Gamma rays, however, are usually emitted in conjunction with other types of radiation. For example, the alpha emission of U-238 (discussed previously) is also accompanied by the emission of a gamma ray.

$$^{238}_{92}\text{U} \longrightarrow {}^{234}_{90}\text{Th} + {}^{4}_{2}\text{He} + {}^{0}_{0}\gamma$$

Gamma rays are the motorbikes of radioactivity. They have the lowest ionizing power, but the highest penetrating power. (Imagine a motorbike zipping through a traffic jam.) Stopping gamma rays requires several inches of lead shielding or thick slabs of concrete.

Positron Emission

Positron emission can be thought of as a type of beta emission and is sometimes referred to as beta plus emission (β^{+}).

Positron Emission

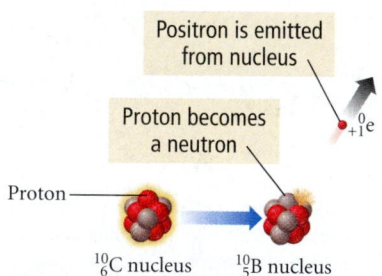

Positron is emitted from nucleus

Proton becomes a neutron

${}^{0}_{+1}\text{e}$

Proton

${}^{10}_{6}\text{C}$ nucleus ${}^{10}_{5}\text{B}$ nucleus

▲ **FIGURE 19.4 Positron Emission** In positron emission, a proton emits a positron and becomes a neutron.

Positron emission occurs when an unstable nucleus emits a positron (Figure 19.4 ◄). A **positron** is the *antiparticle* of the electron: that is, it has the same mass, but opposite charge. If a positron collides with an electron, the two particles annihilate each other, releasing energy in the form of gamma rays. In positron emission, a proton is converted into a neutron and emits a positron:

$$\text{Positron emission} \qquad \text{Proton} \longrightarrow \text{neutron} + \text{emitted positron}$$

The symbol for a positron in a nuclear equation is

$$\text{Positron} \qquad {}^{0}_{+1}\text{e}$$

Thus, positron emission can be represented as

$$^{1}_{1}\text{p} \longrightarrow {}^{1}_{0}\text{n} + {}^{0}_{+1}\text{e}$$

When an atom emits a positron, its atomic number *decreases* by 1 because it now has one fewer proton. For example, the nuclear equation for the positron emission of phosphorus-30 is

$$^{30}_{15}\text{P} \longrightarrow {}^{30}_{14}\text{Si} + {}^{0}_{+1}\text{e}$$

The identity and symbol of the daughter nuclide in any positron emission can be determined in a manner similar to that used for alpha and beta decay, as shown in Example 19.2. Positrons are similar to beta particles in their ionizing and penetrating power.

Electron Capture

Unlike the forms of radioactive decay described so far, electron capture involves a particle being *absorbed by* instead of *ejected from* an unstable nucleus. **Electron capture** occurs when a nucleus assimilates an electron from an inner orbital of its electron cloud. Like positron emission, the net effect of electron capture is the conversion of a proton into a neutron:

$$\text{Electron capture} \qquad \text{Proton} + \text{electron} \longrightarrow \text{neutron}$$

Thus, electron capture can be represented as

$$^{1}_{1}\text{p} + {}^{0}_{-1}\text{e} \longrightarrow {}^{1}_{0}\text{n}$$

When an atom undergoes electron capture, its atomic number decreases by 1 because it has one less proton. For example, the nuclear equation for electron capture in Ru-92 is as follows:

$$^{92}_{44}\text{Ru} + ^{0}_{-1}\text{e} \longrightarrow ^{92}_{43}\text{Tc}$$

The different kinds of radiation are summarized in Table 19.1

TABLE 19.1 Modes of Radioactive Decay

Decay Mode	Process	Change in: A	Z	N/Z*	Example
α	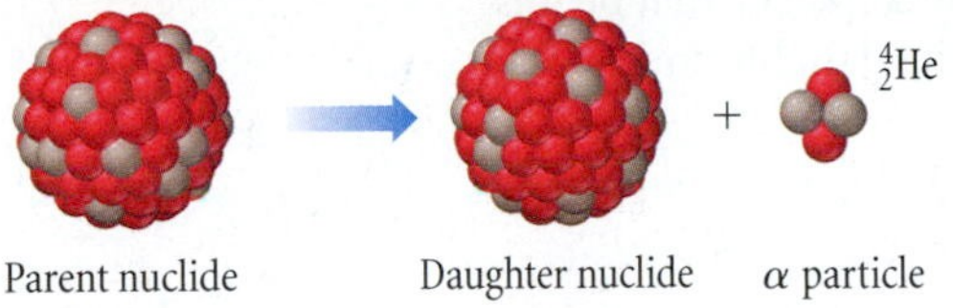 Parent nuclide → Daughter nuclide + $^{4}_{2}$He α particle	−4	−2	Increase	$^{238}_{92}\text{U} \longrightarrow ^{234}_{90}\text{Th} + ^{4}_{2}\text{He}$
β	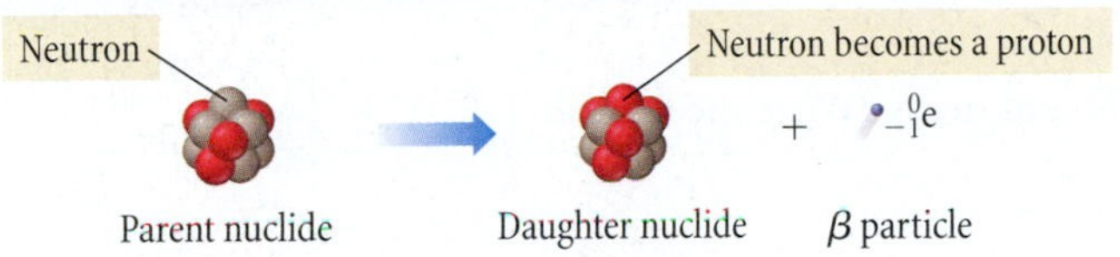 Neutron → Parent nuclide → Neutron becomes a proton + $^{0}_{-1}$e Daughter nuclide β particle	0	+1	Decrease	$^{228}_{88}\text{Ra} \longrightarrow ^{228}_{89}\text{Ac} + ^{0}_{-1}\text{e}$
γ	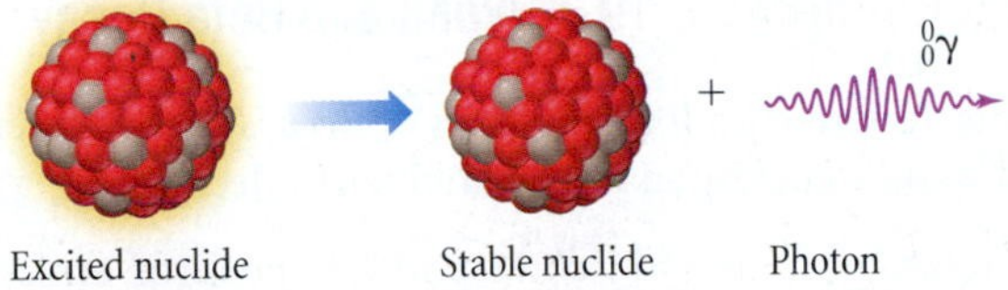 Excited nuclide → Stable nuclide + $^{0}_{0}\gamma$ Photon	0	0	None	$^{234}_{90}\text{Th} \longrightarrow ^{234}_{90}\text{Th} + ^{0}_{0}\gamma$
Positron emission	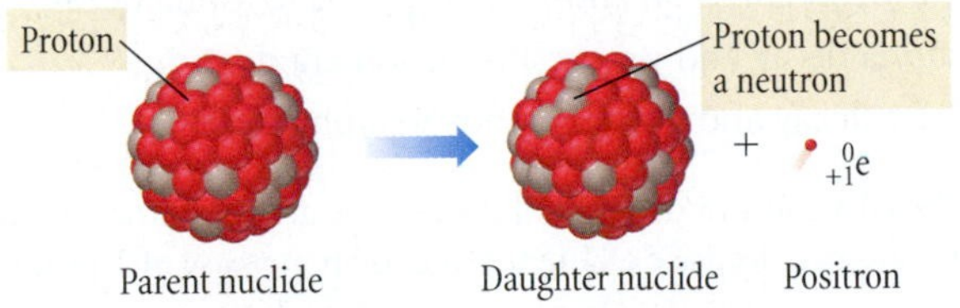 Proton → Parent nuclide → Proton becomes a neutron + $^{0}_{+1}$e Daughter nuclide Positron	0	−1	Increase	$^{30}_{15}\text{P} \longrightarrow ^{30}_{14}\text{Si} + ^{0}_{+1}\text{e}$
Electron capture	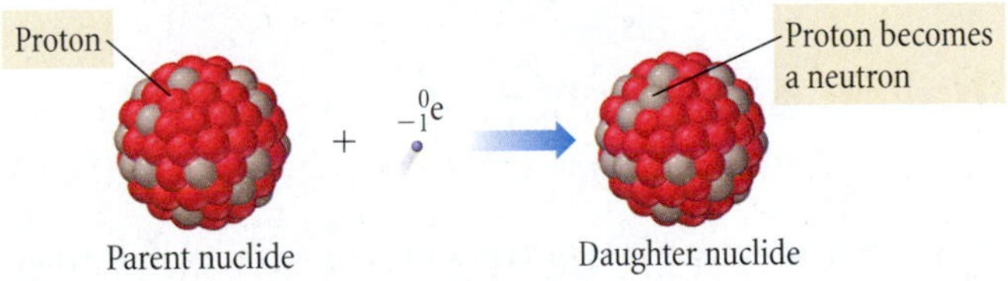 Proton → Parent nuclide + $^{0}_{-1}$e → Proton becomes a neutron Daughter nuclide	0	−1	Increase	$^{92}_{44}\text{Ru} + ^{0}_{-1}\text{e} \longrightarrow ^{92}_{43}\text{Tc}$

*Neutron-to-proton ratio

EXAMPLE 19.2 Writing Nuclear Equations for Beta Decay, Positron Emission, and Electron Capture

Write a nuclear equation for each of the following:

(a) beta decay in Bk-249   (b) positron emission in K-40

(c) electron capture in I-111

Solution

(a) In beta decay, the atomic number *increases* by 1 and the mass number remains unchanged.

The daughter nuclide is element number 98, californium.

$$^{249}_{97}\text{Bk} \longrightarrow ^{249}_{98}? + ^{0}_{-1}\text{e}$$

$$^{249}_{97}\text{Bk} \longrightarrow ^{249}_{98}\text{Cf} + ^{0}_{-1}\text{e}$$

(b) In positron emission, the atomic number *decreases* by 1 and the mass number remains unchanged.

The daughter nuclide is element number 18, argon.

$$^{40}_{19}\text{K} \longrightarrow \,^{40}_{18}? + \,^{0}_{+1}\text{e}$$

$$^{40}_{19}\text{K} \longrightarrow \,^{40}_{18}\text{Ar} + \,^{0}_{+1}\text{e}$$

(c) In electron capture, the atomic number also *decreases* by 1 and the mass number remains unchanged.

The daughter nuclide is element number 52, tellurium.

$$^{111}_{53}\text{I} + \,^{0}_{-1}\text{e} \longrightarrow \,^{111}_{52}?$$

$$^{111}_{53}\text{I} + \,^{0}_{-1}\text{e} \longrightarrow \,^{111}_{52}\text{Te}$$

For Practice 19.2

(a) Write three nuclear equations to represent the nuclear decay sequence that begins with the alpha decay of U-235 followed by a beta decay of the daughter nuclide and then another alpha decay.

(b) Write a nuclear equation for the positron emission of Na-22.

(c) Write a nuclear equation for electron capture in Kr-76.

For More Practice 19.2

Potassium-40 decays to produce Ar-40. What is the method of decay? Write the nuclear equation for this decay.

◆ Conceptual Connection 19.1 Alpha and Beta Decay

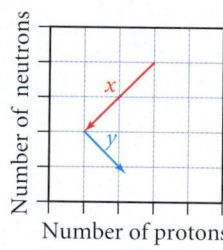

Consider the graphical representation of a series of decays shown at left. The arrows labeled *x* and *y* correspond to what kinds of decay?

(a) *x* corresponds to alpha decay and *y* corresponds to positron emission.

(b) *x* corresponds to positron emission and *y* corresponds to alpha decay.

(c) *x* corresponds to alpha decay and *y* corresponds to beta decay.

(d) *x* corresponds to beta decay and *y* corresponds to alpha decay.

Answer: **(c)** The arrow labeled *x* shows a decrease of 2 neutrons and 2 protons, indicative of alpha decay. The arrow labeled *y* shows a decrease of 1 neutron and an increase of 1 proton, indicative of beta decay.

19.4 The Valley of Stability: Predicting the Type of Radioactivity

So far, we have described various different types of radioactivity. But what causes a particular nuclide to be radioactive in the first place? And why do some nuclides decay via alpha decay, while others decay via beta decay or positron emission? The answers to these questions are not simple, but we can get a basic idea of the factors that influence the stability of the nucleus and the nature of its decay.

A nucleus is a collection of protons (positively charged) and neutrons (uncharged). We know, however, that positively charged particles such as protons repel one another. So what holds the nucleus together? The binding is provided by a fundamental force of physics known as the **strong force**. All nucleons—protons and neutrons—are attracted to one another by the strong force. However, the strong force acts only at very short distances. So we can think of the stability of a nucleus as a balance between the *repulsive* electrostatic force among protons and the *attractive* strong force among all nucleons. The neutrons in a nucleus therefore play an important role in stabilizing the nucleus because they attract other nucleons (through the strong force) but lack the repulsive force associated with positive charge. (It might seem that adding more neutrons would *always* lead to greater stability, so that the more neutrons the better. This is not the case, however, because protons and neutrons occupy energy levels in a nucleus that are similar to those occupied by electrons in an atom. As you add more neutrons, they must occupy increasingly higher energy levels within the nucleus. At some point, the energy payback from the strong nuclear force is not enough to compensate for the high energy state that the neutron must occupy.)

An important number in determining nuclear stability, therefore, is the *ratio* of neutrons to protons (N/Z). Figure 19.5 ▶ shows a plot of the number of neutrons versus the number of protons for all known stable nuclei. The green dots along the diagonal of the graph represent stable nuclei, and this region is known as the *valley (or island) of stability*. Notice that for the lighter elements, the N/Z ratio of stable isotopes is about one (equal numbers of neutrons and protons). For example, the most abundant isotope of carbon ($Z = 6$) is carbon-12, which contains 6 protons and 6 neutrons. However, beyond about $Z = 20$, the N/Z ratio of stable nuclei begins to get larger. For example, at $Z = 40$, stable nuclei have an N/Z ratio of about 1.25 and at $Z = 80$, the N/Z ratio reaches about 1.5. Above $Z = 83$, stable nuclei do not exist—bismuth ($Z = 83$) is the heaviest element with stable (nonradioactive) isotopes.

The type of radioactivity emitted by a nuclide depends in part on the N/Z ratio with the following possibilities:

N/Z too high: Nuclides that lie above the valley of stability have too many neutrons and will tend to convert neutrons to protons via beta decay.

N/Z too low: Nuclides that lie below the valley of stability have too many protons and will tend to convert protons to neutrons via positron emission or electron capture. (Alpha decay also raises the N/Z ratio for nuclides in which $N/Z > 1$, but the effect is smaller than for positron emission or electron capture.)

The following example shows how to use these considerations in predicting the mode of decay for a nucleus.

The Valley of Stability

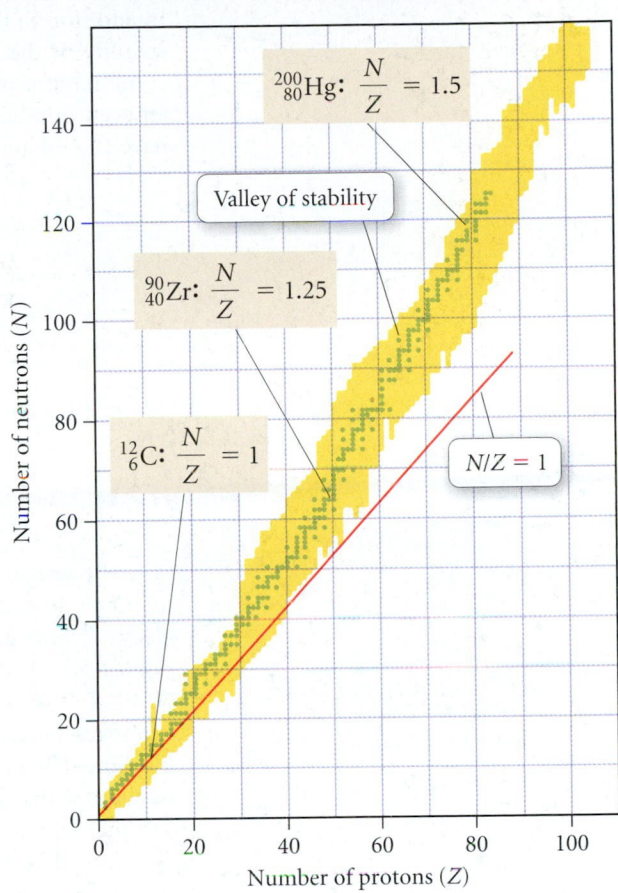

▲ **FIGURE 19.5 Stable and Unstable Nuclei**   A plot of N (the number of neutrons) versus Z (the number of protons) for all known stable nuclei—represented by green dots on this graph—shows that these nuclei cluster together in a region known as the valley (or island) of stability. Nuclei with an N/Z ratio that is too high tend to undergo beta decay. Nuclei with an N/Z ratio that is too low tend to undergo positron emission or electron capture.

EXAMPLE 19.3   Predicting the Type of Radioactive Decay

Predict whether each of the following nuclides is more likely to decay via beta decay or positron emission.

(a) Mg-28   **(b)** Mg-22   **(c)** Mo-102

Solution

(a) Magnesium-28 has 16 neutrons and 12 protons, so $N/Z = 1.33$. However, for $Z = 12$, we can see from Figure 19.5 that stable nuclei should have an N/Z ratio of about 1. Therefore, Mg-28 should undergo *beta decay*, resulting in the conversion of a neutron to a proton.

(b) Magnesium-22 has 10 neutrons and 12 protons, so $N/Z = 0.83$. Therefore, Mg-22 should undergo *positron emission*, resulting in the conversion of a proton to a neutron. (Electron capture would accomplish the same thing as positron emission, but in Mg-22, positron emission is the only decay mode observed.)

(c) Molybdenum-102 has 60 neutrons and 42 protons, so $N/Z = 1.43$. However, for $Z = 42$, we can see from Figure 19.5 that stable nuclei should have an N/Z ratio of about 1.3. Therefore, Mo-102 should undergo *beta decay*, resulting in the conversion of a neutron to a proton.

For Practice 19.3

Predict whether each of the following nuclides is more likely to decay via beta decay or positron emission.

(a) Pb-192   **(b)** Pb-212   **(c)** Xe-114

Magic Numbers

In addition to the N/Z ratio, the *actual number* of protons and neutrons also affects the stability of the nucleus. Table 19.2 shows the number of nuclei with different possible combinations of even or odd nucleons. Notice that a large number of stable nuclides have an even number of protons and an even number of neutrons. Only five stable nuclides have an odd and odd combination.

TABLE 19.2	Number of Stable Nuclides with Even and Odd Numbers of Nucleons	
Z	**N**	**Number of Nuclides**
Even	Even	157
Even	Odd	53
Odd	Even	50
Odd	Odd	5

The reason for this behavior is that nucleons occupy energy levels within the nucleus much as electrons occupy energy levels within an atom. Just as certain numbers of electrons have unique stability (in particular, the number of electrons associated with the noble gases: 2, 10, 18, 36, 54, etc.), so certain numbers of nucleons (N or Z = 2, 8, 20, 28, 50, 82, and N = 126) have unique stability. These numbers are often referred to as **magic numbers**. Nuclei containing a magic number of protons or neutrons are particularly stable. Since the magic numbers are even, this in part accounts for the abundance of stable nuclides with even numbers of nucleons. Moreover, nucleons also have a tendency to pair together (much as electrons pair together). This tendency and the resulting stability of paired nucleons also contribute to the abundance of stable nuclides with even numbers of nucleons.

Radioactive Decay Series

Atoms with $Z > 83$ are radioactive and decay in one or more steps involving mostly alpha and beta decay (with some gamma decay to carry away excess energy). For example, uranium (atomic number 92) is the heaviest naturally occurring element. Its most common isotope is U-238, an alpha emitter that decays to Th-234:

$$^{238}_{92}\text{U} \longrightarrow {}^{234}_{90}\text{Th} + {}^{4}_{2}\text{He}$$

The daughter nuclide, Th-234, is itself radioactive—it is a beta emitter that decays to Pa-234:

$$^{234}_{90}\text{Th} \longrightarrow {}^{234}_{91}\text{Pa} + {}^{0}_{-1}\text{e}$$

Protactinium-234 is also radioactive, decaying to U-234 via beta emission. Radioactive decay continues until a stable nuclide, Pb-206, is reached. The entire uranium-238 decay series is shown in Figure 19.6 ◀.

19.5 Detecting Radioactivity

The particles emitted by radioactive nuclei have a lot of energy and can therefore be detected with high sensitivity. In a radiation detector, the particles are detected through their interactions with atoms or molecules. The simplest radiation detectors are pieces of photographic film that become exposed when radiation passes through them. **Film-badge dosimeters**—which consist of photographic film held in a small case that is pinned to clothing—are standard for most people working with or near radioactive substances (Figure 19.7 ▶). These badges are collected and processed (or developed) regularly as a way to monitor a person's exposure. The more exposed the film has become in a given period of time, the more radioactivity the person has been exposed to.

A Decay Series

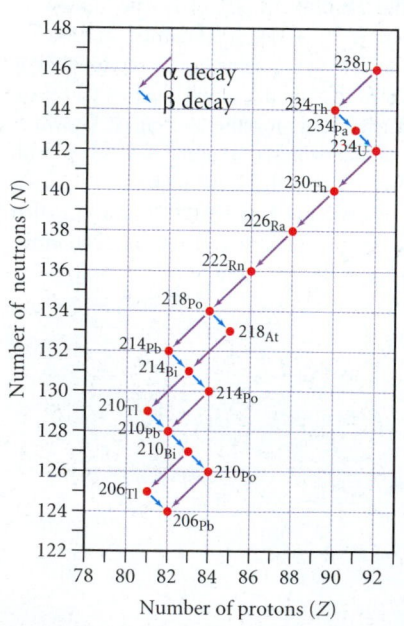

▲ FIGURE 19.6 The Uranium-238 Radioactive Decay Series Uranium-238 decays via a series of steps ending in Pb-206, which is stable. Each diagonal line to the left represents an alpha decay and each diagonal line to the right represents a beta decay.

Radioactivity can be instantly detected with devices such as a **Geiger-Müller counter** (Figure 19.8 ▼). In this instrument (commonly referred to simply as a Geiger counter), particles emitted by radioactive nuclei pass through an argon-filled chamber. The energetic particles create a trail of ionized argon atoms. An applied high voltage between a wire within the chamber and the chamber itself causes these newly formed ions to produce an electrical signal that can be detected on a meter or turned into an audible click. Each click corresponds to a radioactive particle passing through the argon gas chamber. This clicking is the stereotypical sound most people associate with a radiation detector.

A second type of device commonly used to detect radiation instantly is called a **scintillation counter**. In a scintillation counter, the radioactive emissions pass through a material (such as NaI or CsI) that emits ultraviolet or visible light in response to excitation by energetic particles. The radioactivity excites the atoms to a higher energy state. The atoms release this energy as light, which is then detected and turned into an electrical signal that can be read on a meter.

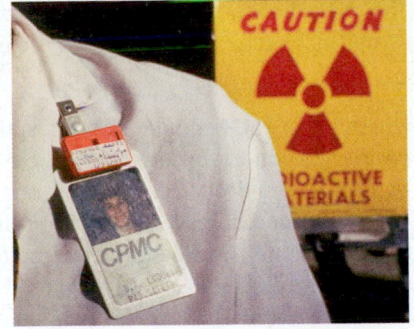

◀ **FIGURE 19.7 Film-Badge Dosimeter** A film-badge dosimeter consists simply of a piece of photographic film in a light-tight container. The film's exposure in a given time is proportional to the amount of radiation it receives.

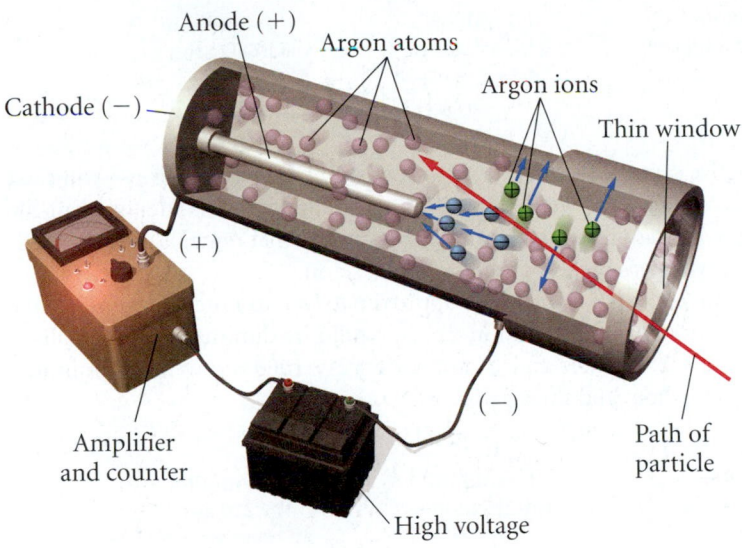

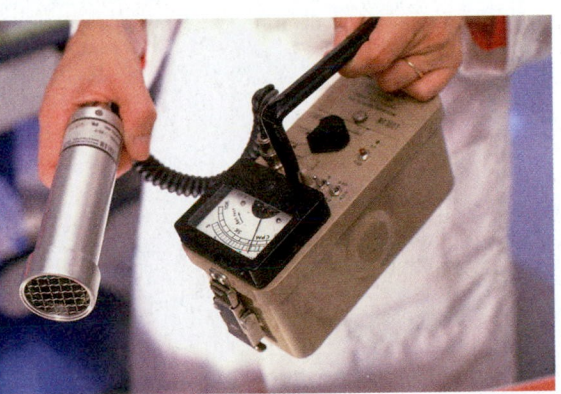

▲ **FIGURE 19.8 Geiger-Müller Counter** When ionizing radiation passes through the argon-filled chamber, it ionizes the atoms, giving rise to a brief, tiny pulse of electrical current that is transduced onto a meter or into an audible click.

19.6 The Kinetics of Radioactive Decay and Radiometric Dating

Radioactivity is a natural component of our environment. The ground beneath you most likely contains radioactive atoms that emit radiation into the air around you. The food you eat contains a residual quantity of radioactive atoms that are absorbed into your body fluids and incorporated into tissues. Small amounts of radiation from space make it through our atmosphere and constantly bombard Earth. Humans and other living organisms have evolved in this environment and have adapted to survive in it.

One reason for the radioactivity in our environment is the instability of all atomic nuclei beyond atomic number 83 (bismuth). Every element with more than 83 protons in its nucleus is unstable and therefore radioactive. In addition, some isotopes of elements with fewer than 83 protons are also unstable and radioactive. Radioactive nuclides *persist* in our environment because new ones are constantly being formed, and because many of the existing ones decay away only very slowly.

You may find it useful to review the discussion of first-order kinetics in Section 13.3.

All radioactive nuclei decay via first-order kinetics, so the rate of decay in a particular sample is directly proportional to the number of nuclei present:

$$\text{Rate} = kN$$

where N is the number of radioactive nuclei and k is the rate constant. Different radioactive nuclides decay into their daughter nuclides with different rate constants. Some nuclides decay quickly while others decay slowly.

The time it takes for one-half of the parent nuclides in a radioactive sample to decay to the daughter nuclides is called the *half-life*, and is identical to the concept of half-life for chemical reactions that we covered in Chapter 13. Thus, the relationship between the half-life of a nuclide and its rate constant is given by the same expression (Equation 13.19) that we derived for a first-order reaction in Section 13.4:

$$t_{1/2} = \frac{0.693}{k} \qquad [19.1]$$

Nuclides that decay quickly have short half-lives and large rate constants—they are considered very active (many decay events per unit time). Nuclides that decay slowly have long half-lives and are less active (fewer decay events per unit time). For example, thorium-232 is an alpha emitter with a half-life of 1.4×10^{10} years, or 14 billion years. If we start with a sample of Th-232 containing one million atoms, the sample would decay to $\frac{1}{2}$ million atoms in 14 billion years and then to $\frac{1}{4}$ million in another 14 billion years and so on.

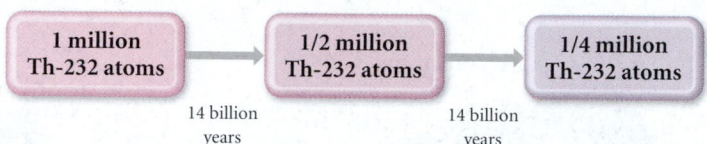

Notice that a radioactive sample does *not* decay to *zero* atoms in two half-lives—you can't add two half-lives together to get a "whole" life. The amount that remains after one half-life is always one-half of what was present at the start. The amount that remains after two half-lives is one-quarter of what was present at the start, and so on.

By contrast, radon-220 has a half-life of approximately 1 minute (Figure 19.9 ▼). If we had a 1-million-atom sample of radon-220, it would be diminished to $\frac{1}{4}$ million radon-220 atoms in just 2 minutes and to approximately 1000 atoms in 10 minutes. Table 19.3 lists several nuclides and their half-lives.

TABLE 19.3 Selected Nuclides and Their Half-Lives

Nuclide	Half-life	Type of Decay
$^{232}_{90}\text{Th}$	1.4×10^{10} yr	Alpha
$^{238}_{92}\text{U}$	4.5×10^{9} yr	Alpha
$^{14}_{6}\text{C}$	5730 yr	Beta
$^{220}_{86}\text{Rn}$	55.6 s	Alpha
$^{219}_{90}\text{Th}$	1.05×10^{-6} s	Alpha

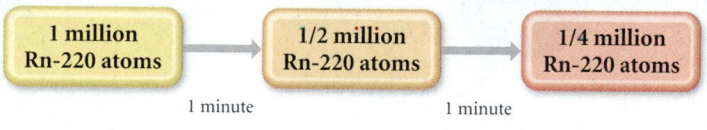

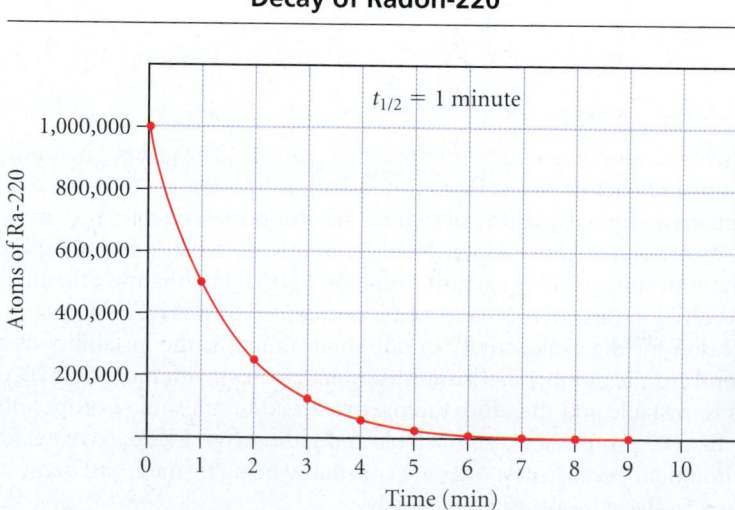

▶ **FIGURE 19.9 The Decay of Radon-220** Radon-220 decays with a half-life of approximately 1 minute.

Chemistry in the Environment
Environmental Radon

Radon—a radioactive gas—is one of the products of the radioactive decay series of uranium. Wherever there is uranium in the ground, there is likely to be radon seeping up into the air. Because radon is a gas, it can be inhaled into the lungs, where it decays and increases the risk of lung cancer. Its radioactive daughter nuclides, attached to dust particles, can also enter the lungs, with similar effects. The radioactive decay of radon is by far the single greatest source of human exposure to radiation.

Homes built in areas with significant uranium deposits in the ground pose the greatest risk. These homes can accumulate radon levels that are above what the U.S. Environmental Protection Agency (EPA) considers safe. The risk is especially high for smokers, who already have an increased risk of lung cancer because of carcinogens in cigarette smoke. Simple test kits are available to test indoor air and determine radon levels. Very high indoor radon levels require the installation of a ventilation system to purge radon from the house. Lower levels can be vented by keeping windows and doors open.

▶ The areas in red are at greatest risk for radon contamination.

Question

Suppose that a house contains 1.80×10^{-3} mol of radon-222 (which has a half-life of 3.8 days). If no new radon entered the house, how long would it take for the radon to decay to 4.50×10^{-4} mol?

EPA Map of Radon Zones

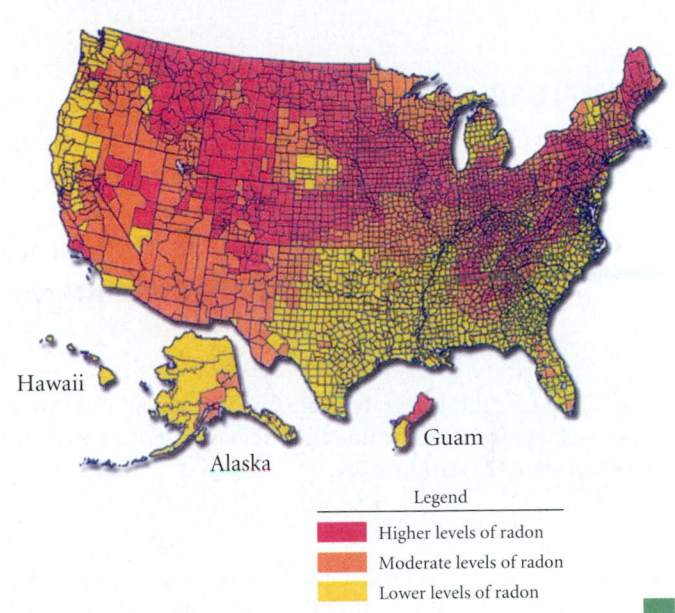

Hawaii

Alaska

Guam

Legend

█ Higher levels of radon
█ Moderate levels of radon
█ Lower levels of radon

Conceptual Connection 19.2 Half-Life

Consider the following graph representing the decay of a radioactive nuclide:

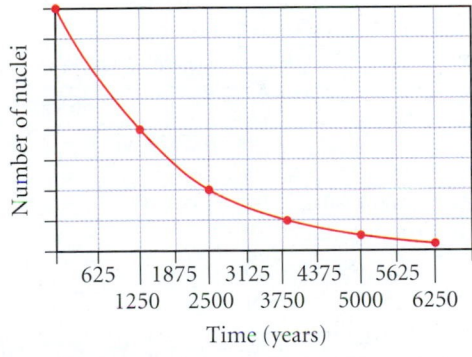

What is the half-life of the nuclide?

(a) 625 years   (b) 1250 years   (c) 2500 years   (d) 3125 years

Answer: (b) The half-life is the time it takes for the number of nuclei to decay to one-half of their original number.

The Integrated Rate Law

In Chapter 13, we learned that for first-order chemical reactions, the concentration of a reactant as a function of time is given by the integrated rate law:

$$\ln \frac{[A]_t}{[A]_0} = -kt \qquad\qquad [19.2]$$

Since nuclear decay follows first-order kinetics, we can substitute the number of nuclei for concentration to arrive at the equation

$$\ln \frac{N_t}{N_0} = -kt \qquad \text{[19.3]}$$

where N_t is the number of radioactive nuclei at time t and N_0 is the initial number of radioactive nuclei. The following example demonstrates how to use this equation.

EXAMPLE 19.4 Radioactive Decay Kinetics

Plutonium-236 is an alpha emitter with a half-life of 2.86 years. If a sample initially contains 1.35 mg of Pu-236, what mass of Pu-236 will be present after 5.00 years?

Sort You are given the initial mass of Pu-236 in a sample and asked to find the mass after 5.00 years.

Given $m_{Pu\text{-}236}(\text{initial}) = 1.35$ mg;
$t = 5.00$ yr; $t_{1/2} = 2.86$ yr

Find $m_{Pu\text{-}236}(\text{final})$

Strategize Use the integrated rate law (Equation 19.3) to solve this problem. However, you must determine the value of the rate constant (k) from the half-life expression (Equation 19.1).

Conceptual Plan

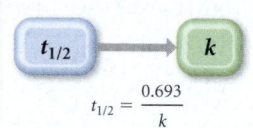

$$t_{1/2} = \frac{0.693}{k}$$

Use the value of the rate constant, the initial mass of Pu-236, and the time along with integrated rate law to find the final mass of Pu-236. Since the mass of the Pu-236 ($m_{Pu\text{-}236}$) is directly proportional to the number of atoms (N), and since the integrated rate law contains the ratio (N_t/N_0), the initial and final masses can be substituted for the initial and final number of atoms.

$$k, m_{Pu\text{-}236}(\text{initial}), t \longrightarrow m_{Pu\text{-}236}(\text{final})$$

$$\ln \frac{N_t}{N_0} = -kt$$

Solve Follow your plan. Begin by finding the rate constant from the half-life.

Solution

$$t_{1/2} = \frac{0.693}{k}$$

$$k = \frac{0.693}{t_{1/2}} = \frac{0.693}{2.86 \text{ yr}}$$

$$= 0.2423/\text{yr}$$

Solve the integrated rate law for N_t and substitute the values of the rate constant, the initial mass of Pu-236, and the time into the solved equation. Compute the final mass of Pu-236.

$$\ln \frac{N_t}{N_0} = -kt$$

$$\frac{N_t}{N_0} = e^{-kt}$$

$$N_t = N_0 e^{-kt}$$

$$N_t = 1.35 \text{ mg} \left[e^{-(0.2423/\text{yr})(5.00\text{ yr})} \right]$$

$$N_t = 0.402 \text{ mg}$$

Check The units of the answer (mg) are correct. The magnitude of the answer (0.402 mg) is about one-third of the original mass (1.35 mg) which seems reasonable given that the amount of time is between one and two half-lives. (One half-life would result in one-half of the original mass and two half-lives would result in one-fourth of the original mass.)

For Practice 19.4

How long would it take for the 1.35-mg sample of Pu-236 in Example 19.4 to decay to 0.100 mg?

Since radioactivity is a first-order process, the rate of decay is linearly proportional to the number of nuclei in the sample. Therefore, as shown below, the initial rate of decay (rate$_0$) and the rate of decay at time t (rate$_t$) can also be used in the integrated rate law.

$$\text{Rate}_t = kN_t \qquad \text{Rate}_0 = kN_0$$

$$\frac{N_t}{N_0} = \frac{\text{rate}_t/k}{\text{rate}_0/k} = \frac{\text{rate}_t}{\text{rate}_0}$$

Substituting into Equation 19.3, we get the following result:

$$\ln \frac{\text{rate}_t}{\text{rate}_0} = -kt \qquad\qquad [19.4]$$

We can use Equation 19.4 to predict how the rate of decay of a radioactive sample will change with time or how much time has passed based on how the rate has changed (see examples later in this section).

The presence of radioactive isotopes in our environment, and their predictable decay with time, can therefore be used to estimate the age of rocks or artifacts containing those isotopes. The technique is known as **radiometric dating**, and we examine two different types individually.

Radiocarbon Dating: Using Radioactivity to Measure the Age of Fossils and Artifacts

Radiocarbon dating, a technique devised in 1949 by Willard Libby at the University of Chicago, is used by archeologists, geologists, anthropologists, and other scientists to estimate the ages of fossils and artifacts. For example, in 1947, young shepherds searching for a stray goat near the Dead Sea (east of Jerusalem) entered a cave and discovered ancient scrolls stuffed into jars. These scrolls—now named the Dead Sea Scrolls—are 2000-year-old texts of the Hebrew Bible, predating other previously known manuscripts by almost a thousand years.

Libby received the Nobel Prize in 1960 for the development of radiocarbon dating.

◄ The Dead Sea Scrolls are 2000-year-old biblical manuscripts. Their age was determined by radiocarbon dating.

The Dead Sea Scrolls, like other ancient artifacts, contain a radioactive signature that reveals their age. This signature results from the presence of carbon-14 (which is radioactive) in the environment. Carbon-14 is constantly formed in the upper atmosphere by the neutron bombardment of nitrogen:

$$^{14}_{7}\text{N} + ^{1}_{0}\text{n} \longrightarrow ^{14}_{6}\text{C} + ^{1}_{1}\text{H}$$

Carbon-14 then decays back to nitrogen by beta emission with a half-life of 5,730 years.

$$^{14}_{6}\text{C} \longrightarrow ^{14}_{7}\text{N} + ^{0}_{-1}\text{e} \qquad t_{1/2} = 5{,}730 \text{ yr}$$

The continuous formation of carbon-14 in the atmosphere and its continuous decay back to nitrogen-14 produces a nearly constant equilibrium amount of atmospheric carbon-14, which is oxidized to carbon dioxide and incorporated into plants by photosynthesis. The C-14 then makes its way up the food chain and ultimately into all living organisms. As a result, all living plants, animals, and humans contain the same ratio of carbon-14 to carbon-12 ($^{14}\text{C}:^{12}\text{C}$) as is found in the atmosphere. When a living organism dies,

▲ Western bristlecone pine trees can live up to 5,000 years, and their age can be precisely determined by counting the annual rings in their trunks. They can therefore be used to calibrate the timescale for radiocarbon dating.

however, it stops incorporating new carbon-14 into its tissues. The $^{14}C : {}^{12}C$ ratio then decreases with a half-life of 5,730 years. Since many artifacts, including the Dead Sea Scrolls, are made from materials that were once living—such as papyrus, wood, or other plant and animal derivatives—the $^{14}C : {}^{12}C$ ratio in these artifacts indicates their age. For example, suppose an ancient artifact has a $^{14}C : {}^{12}C$ ratio that is 25% of that found in living organisms. How old is the artifact? Since it contains one-quarter as much carbon-14 as a living organism, it must be two half-lives or 11,460 years old.

The accuracy of carbon-14 dating can be checked against objects whose ages are known from historical sources. These kinds of comparisons have revealed that ages obtained from C-14 dating could deviate from the actual ages by up to about 5%. For a 6,000-year-old object, that would result in an error of about 300 years. The reason for the deviations is the variance of atmospheric C-14 levels over time.

In order to make C-14 dating more accurate, scientists have studied the carbon-14 content of western bristlecone pine trees, which can live up to 5,000 years. The tree trunk contains growth rings corresponding to each year of the tree's life, and the wood laid down in each ring incorporates carbon derived from the carbon dioxide in the atmosphere at that time. The rings thus provide a record of the historical atmospheric carbon-14 content. In addition, the rings of living trees can be correlated with the rings of dead trees, allowing the record to be extended back about 11,000 years. Using the data from the bristlecone pine, the 5% deviations from historical dates can be corrected. In effect, the known ages of bristlecone pine trees are used to calibrate C-14 dating, resulting in more accurate results. The maximum age that can be estimated from carbon-14 dating is about 50,000 years— beyond that, the amount of carbon-14 becomes too low to measure accurately.

EXAMPLE 19.5 Radiocarbon Dating

A skull believed to belong to an ancient human being is found to have a carbon-14 decay rate of 4.50 disintegrations per minute per gram of carbon (4.50 dis/min · g C). If living organisms have a decay rate 15.3 dis/min · g C, how old is the skull? (The decay rate is directly proportional to the amount of carbon-14 present).

Sort You are given the current rate of decay for the skull and the assumed initial rate. You are asked to find the age of the skull, which is the time that passed in order for the rate to have reached its current value.	**Given** $\text{rate}_t = 4.50 \text{ dis/min} \cdot \text{g C}$; $\quad\quad\quad \text{rate}_0 = 15.3 \text{ dis/min} \cdot \text{g C}$; **Find** t
Strategize Use the expression for half-life (Equation 19.1) to find the rate constant (k) from the half-life for C-14, which is 5,730 yr (Table 19.3).	**Conceptual Plan** $\boxed{t_{1/2}} \longrightarrow \boxed{k}$ $t_{1/2} = \dfrac{0.693}{k}$
Use the value of the rate constant and the initial and current rates to find t from the integrated rate law (Equation 19.4).	$\boxed{k, \text{rate}_t, \text{rate}_0} \longrightarrow \boxed{t}$ $\ln \dfrac{\text{rate}_t}{\text{rate}_0} = -kt$
Solve Follow your plan. Begin by finding the rate constant from the half-life.	**Solution** $t_{1/2} = \dfrac{0.693}{k}$ $k = \dfrac{0.693}{t_{1/2}} = \dfrac{0.693}{5730 \text{ yr}}$ $= 1.209 \times 10^{-4}/\text{yr}$

Chemistry in Your Day
Radiocarbon Dating and the Shroud of Turin

The Shroud of Turin—kept in the cathedral of Turin in Italy—is an old linen cloth that bears a mysterious image. Many people have interpreted the image as that of a man who appears to have been crucified. The image becomes clearer if the shroud is photographed and viewed as a negative. It has been suggested that the shroud is the original burial cloth of Jesus, miraculously imprinted with his image. In 1988, three independent laboratories were chosen by the Roman Catholic Church to perform radiocarbon dating on the shroud. The laboratories took samples of the cloth and measured the carbon-14 content. The three independent laboratories all got similar results—the shroud was made from linen originating in about A.D. 1325. Although some have disputed the results (and continue to do so), and although no scientific test is 100% reliable, newspapers around the world announced that the Shroud could not date to biblical times.

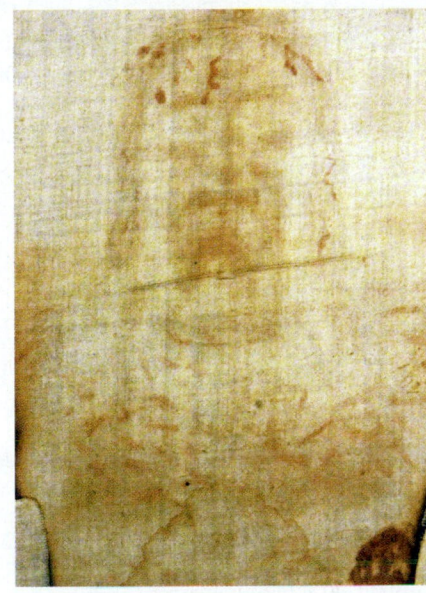

◀ The linen cloth known as the Shroud of Turin bears the image of a man believed by some to be Jesus.

Substitute the rate constant and the initial and current rates into the integrated rate law and solve for t.	$\ln \dfrac{\text{rate}_t}{\text{rate}_0} = -kt$

$$t = -\frac{\ln \dfrac{\text{rate}_t}{\text{rate}_0}}{k} = -\frac{\ln \dfrac{4.50 \; \text{dis/min} \cdot \text{g C}}{15.3 \; \text{dis/min} \cdot \text{g C}}}{1.209 \times 10^{-4}/\text{yr}}$$

$$= 1.0 \times 10^4 \; \text{yr}$$

Check   The units of the answer (yr) are correct. The magnitude of the answer is about 10,000 years, which is a little less than two half-lives. This value is reasonable given that two half-lives would result in a decay rate of about 3.8 dis/min · g C.

For Practice 19.5

An ancient scroll is claimed to have originated from Greek scholars in about 500 B.C. A measure of its carbon-14 decay rate gives a value that is 89% of that found in living organisms. How old is the scroll and could it be authentic?

Uranium/Lead Dating

Radiocarbon dating is limited to measuring the ages of objects that were once living and that are relatively young (<50,000 years). Other radiometric dating techniques can measure the ages of prehistoric objects that were never alive. The most dependable technique for these purposes relies on the ratio of uranium-238 to lead-206 within igneous rocks (rocks of volcanic origin). This technique measures the time that has passed since the rock solidified (at which point the "radiometric clock" was reset).

Since U-238 decays into Pb-206 with a half-life of 4.5×10^9 years, the relative amounts of U-238 and Pb-206 in a uranium-containing rock reveal its age. For example, if a rock originally contained U-238 and currently contains equal amounts of U-238 and Pb-206, it would be 4.5 billion years old, assuming that the rock did not contain any Pb-206 when it was formed. The latter assumption can be tested because the lead that results from the decay of uranium has a different isotopic composition than the lead that would have been deposited in rocks at the time of their formation. The following example shows how the relative amounts of Pb-206 and U-238 in a rock can be used to estimate its age.

EXAMPLE 19.6 Uranium/Lead Dating to Estimate the Age of a Rock

A meteor is found to contain 0.556 g of Pb-206 to every 1.00 g of U-238. Assuming that the meteor did not contain any Pb-206 at the time of its formation, determine the age of the meteor. Uranium-238 decays to lead-206 with a half-life of 4.5 billion years.

Sort You are given the current masses of Pb-206 and U-238 in a rock and asked to find its age. You are also given the half-life of U-238.

Given $m_{U\text{-}238} = 1.00$ g; $m_{Pb\text{-}206} = 0.556$ g; $t_{1/2} = 4.5 \times 10^9$ yr

Find t

Strategize Use the integrated rate law (Equation 19.3) to solve this problem. However, you must first determine the value of the rate constant (k) from the half-life expression (Equation 19.1).

Before substituting into the integrated rate law, you will also need the ratio of the current amount of U-238 to the original amount (N_t/N_0). The current mass of uranium is simply 1.00 g. The initial mass includes the current mass (1.00 g) plus the mass that has decayed into lead-206, which can be found from the current mass of Pb-206.

Use the value of the rate constant and the initial and current amounts of U-238 along with integrated rate law to find t.

Conceptual Plan

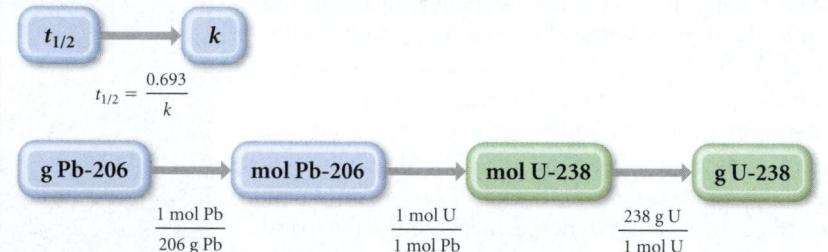

$$t_{1/2} \xrightarrow{} k$$
$$t_{1/2} = \frac{0.693}{k}$$

$$\text{g Pb-206} \xrightarrow{} \text{mol Pb-206} \xrightarrow{} \text{mol U-238} \xrightarrow{} \text{g U-238}$$
$$\frac{1 \text{ mol Pb}}{206 \text{ g Pb}} \qquad \frac{1 \text{ mol U}}{1 \text{ mol Pb}} \qquad \frac{238 \text{ g U}}{1 \text{ mol U}}$$

$$k, N_t, N_0 \xrightarrow{} t$$
$$\ln \frac{N_t}{N_0} = -kt$$

Solve Follow your plan. Begin by finding the rate constant from the half-life.

Solution

$$t_{1/2} = \frac{0.693}{k}$$

$$k = \frac{0.693}{t_{1/2}} = \frac{0.693}{4.5 \times 10^9 \text{ yr}}$$

$$= 1.54 \times 10^{-10} /\text{yr}$$

Determine the mass in grams of U-238 that would have been required to form the given mass of Pb-206.

$$0.556 \text{ g Pb-206} \times \frac{1 \text{ mol Pb-206}}{206 \text{ g Pb-206}} \times \frac{1 \text{ mol U-238}}{1 \text{ mol Pb-206}} \times \frac{238 \text{ g U-238}}{1 \text{ mol U-238}} = 0.6424 \text{ g U-238}$$

Substitute the rate constant and the initial and current masses of U-238 into the integrated rate law and solve for t. (The initial mass of U-238 is simply the sum of the current mass and the mass that would have been required to form the given mass of Pb-206.)

$$\ln \frac{N_t}{N_0} = -kt$$

$$t = -\frac{\ln \dfrac{N_t}{N_0}}{k} = -\frac{\ln \dfrac{1.00 \text{ g}}{1.00 \text{ g} + 0.6424 \text{ g}}}{1.54 \times 10^{-10}/\text{yr}}$$

$$= 3.2 \times 10^9 \text{ yr}$$

Check The units of the answer (yr) are correct. The magnitude of the answer is about 3.2 billion years, which is less than one half-life. This value is reasonable given that less than half of the uranium has decayed into lead.

For Practice 19.6

A rock is found to contain a Pb-206 to U-238 mass ratio of 0.145 : 1.00. Assuming that the rock did not contain any Pb-206 at the time of its formation, determine the age of the rock.

The Age of Earth The uranium/lead radiometric dating technique as well as other radiometric dating techniques (such as the decay of potassium-40 to argon-40) have been widely used to measure the ages of rocks on Earth and have produced highly consistent results. Rocks with ages greater than 3.5 billion years have been found on every continent. The oldest rocks have an age of approximately 4.0 billion years, establishing a lower bound for Earth's age (Earth must be at least as old as its oldest rocks). The ages of about 70 meteorites that have struck Earth have also been extensively studied and have been found to be about 4.5 billion years old. Since the meteorites were formed at the same time as our solar system (which includes Earth), the best estimate for Earth's age is therefore about 4.5 billion years. That age is consistent with the estimated age of our universe— about 13.7 billion years.

> The age of the universe is estimated from its expansion rate, which can be measured by examining changes in the wavelength of light from distant galaxies.

19.7 The Discovery of Fission: The Atomic Bomb and Nuclear Power

In the mid-1930s Enrico Fermi (1901–1954), an Italian physicist, tried to synthesize a new element by bombarding uranium—the heaviest known element at that time—with neutrons. Fermi hypothesized that if a neutron were to be incorporated into the nucleus of a uranium atom, the nucleus might undergo beta decay, converting a neutron into a proton. If that happened, a new element, with atomic number 93, would be synthesized for the first time. The nuclear equation for the process is

> The element with atomic number 100 is named fermium in honor of Enrico Fermi.

$$^{238}_{92}\text{U} + {}^{1}_{0}\text{n} \longrightarrow {}^{239}_{92}\text{U} \longrightarrow {}^{239}_{93}\text{X} + {}^{0}_{-1}\text{e}$$

Neutron   Newly synthesized element

Fermi performed the experiment and detected the emission of beta particles. However, his results were inconclusive. Had he synthesized a new element? Fermi never chemically examined the products to determine their composition and therefore could not say with certainty that he had.

Three researchers in Germany—Lise Meitner (1878–1968), Fritz Strassmann (1902–1980), and Otto Hahn (1879–1968)—repeated Fermi's experiments, but then performed careful chemical analysis of the products. What they found in the products— several elements *lighter* than uranium—would change the world forever. On January 6, 1939, Meitner, Strassmann, and Hahn reported that the neutron bombardment of uranium resulted in **nuclear fission**—the splitting of the uranium atom. The nucleus of the neutron-bombarded uranium atom had been split into barium, krypton, and other smaller products. They also realized that the process emitted enormous amounts of energy. A nuclear equation for a fission reaction, showing how uranium breaks apart into the daughter nuclides, is shown here.

▲ Lise Meitner in Otto Hahn's Berlin laboratory. Together with Hahn and Fritz Strassmann, Meitner determined that U-235 could undergo nuclear fission.

$$^{235}_{92}\text{U} + {}^{1}_{0}\text{n} \longrightarrow {}^{140}_{56}\text{Ba} + {}^{93}_{36}\text{Kr} + 3\,{}^{1}_{0}\text{n} + \text{energy}$$

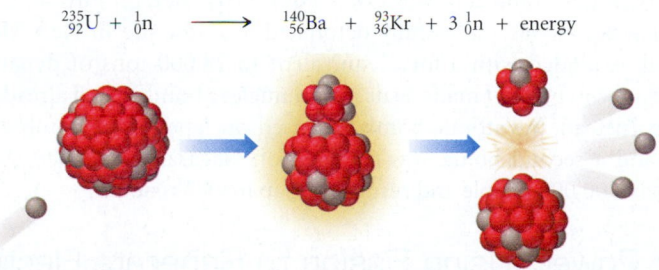

> The element with atomic number 109 is named meitnerium in honor of Lise Meitner.

Notice that the initial uranium atom is the U-235 isotope, which constitutes less than 1% of all naturally occurring uranium. U-238, the most abundant uranium isotope, does not undergo fission. Notice also that the process produces three neutrons, which have the potential to initiate fission in three other U-235 atoms.

Fission Chain Reaction

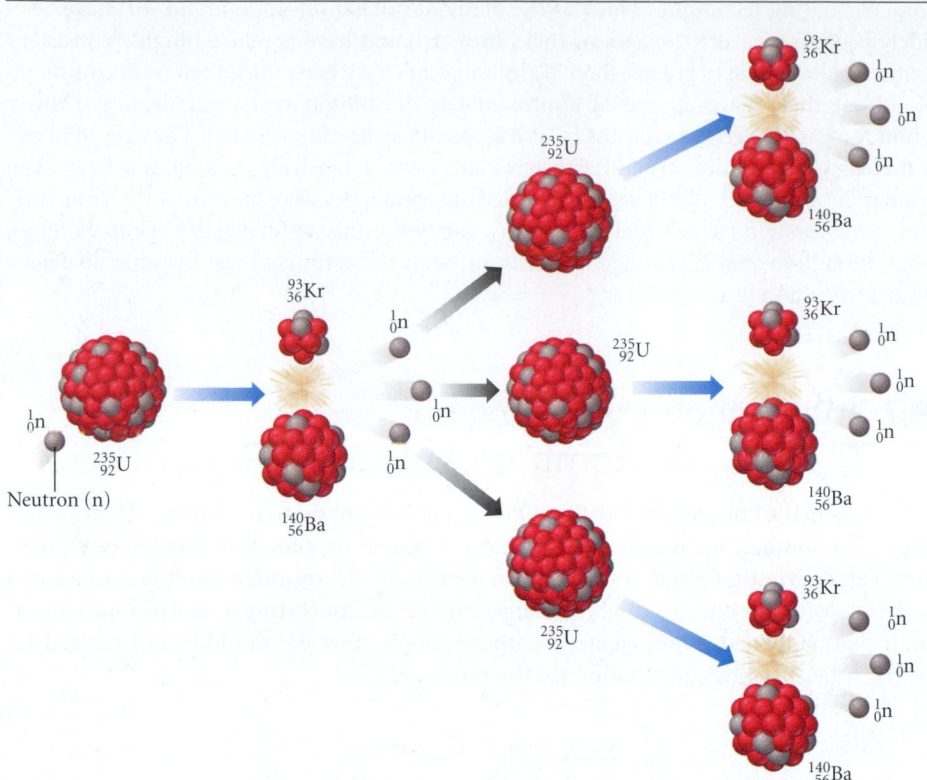

▶ **FIGURE 19.10 A Self-Amplifying Chain Reaction** The fission of one U-235 nucleus emits neutrons which can then initiate fission in other U-235 nuclei, resulting in a chain reaction that can release enormous amounts of energy.

▲ On July 16, 1945, in the New Mexico desert, the world's first atomic bomb was detonated. It had the power of 18,000 tons of dynamite.

Scientists quickly realized that a sample rich in U-235 could undergo a **chain reaction** in which neutrons produced by the fission of one uranium nucleus would induce fission in other uranium nuclei (Figure 19.10 ▲). The result would be a self-amplifying reaction capable of producing an enormous amount of energy—an atomic bomb. However, to make a bomb, a **critical mass** of U-235—enough U-235 to produce a self-sustaining reaction—would be necessary. Fearing that Nazi Germany would develop such a bomb, several U.S. scientists persuaded Albert Einstein, the most famous scientist of the time, to write a letter to President Franklin Roosevelt warning of this possibility (see p. 885). Einstein wrote, ". . . and it is conceivable—though much less certain—that extremely powerful bombs of a new type may thus be constructed. A single bomb of this type, carried by boat and exploded in a port, might very well destroy the whole port together with some of the surrounding territory."

Roosevelt was convinced by Einstein's letter, and in 1941 he assembled the resources to begin the costliest scientific project ever attempted. The top-secret endeavor was called the *Manhattan Project* and its main goal was to build an atomic bomb before the Germans did. The project was led by physicist J. R. Oppenheimer (1904–1967) at a high-security research facility in Los Alamos, New Mexico. Four years later, on July 16, 1945, the world's first nuclear weapon was successfully detonated at a test site in New Mexico. The first atomic bomb exploded with a force equivalent to 18,000 tons of dynamite. Ironically, the Germans—who had *not* made a successful nuclear bomb—had already been defeated by this time. Instead, the atomic bomb was used on Japan. One bomb was dropped on Hiroshima and a second bomb was dropped on Nagasaki. Together, the bombs killed approximately 200,000 people and resulted in Japan's surrender.

Nuclear Power: Using Fission to Generate Electricity

Nuclear reactions, such as fission, generate enormous amounts of energy. In a nuclear bomb, the energy is released all at once. However, the energy can also be released more slowly and used for peaceful purposes such as electricity generation. In the United States, about 20% of electricity is generated by nuclear fission. In some other countries, as much

Albert Einstein
Old Grove Rd.
Nassau Point
Peconic, Long Island

August 2nd, 1939

F.D. Roosevelt,
President of the United States,
White House
Washington, D.C.

Sir:

Some recent work by E.Fermi and L. Szilard, which has been communicated to me in manuscript, leads me to expect that the element uranium may be turned into a new and important source of energy in the immediate future. Certain aspects of the situation which has arisen seem to call for watchfulness and, if necessary, quick action on the part of the Administration. I believe therefore that it is my duty to bring to your attention the following facts and recommendations:

In the course of the last four months it has been made probable - through the work of Joliot in France as well as Fermi and Szilard in America - that it may become possible to set up a nuclear chain reaction in a large mass of uranium, by which vast amounts of power and large quantities of new radium-like elements would be generated. Now it appears almost certain that this could be achieved in the immediate future.

This new phenomenon would also lead to the construction of bombs, and it is conceivable - though much less certain - that extremely powerful bombs of a new type may thus be constructed. A single bomb of this type, carried by boat and exploded in a port, might very well destroy the whole port together with some of the surrounding territory. However, such bombs might very well prove to be too heavy for transportation by air.

◀ Einstein's letter (part of which is shown here) helped persuade Franklin Roosevelt to begin research into the building of a fission bomb.

as 70% of electricity is generated by nuclear fission. To get an idea of the amount of energy released during fission, imagine a hypothetical nuclear-powered car. Suppose the fuel for such a car was a uranium cylinder about the size of a pencil. How often would you have to refuel the car? The energy content of the uranium cylinder would be equivalent to about 1000 twenty-gallon tanks of gasoline. If you refuel your gasoline-powered car once a week, your nuclear-powered car could go 1000 weeks—almost 20 years—before refueling.

Similarly, a nuclear-powered electricity generation plant can produce a lot of electricity from a small amount of fuel. Such plants exploit the heat created by fission. The heat is used to boil water and make steam, which then turns the turbine on a generator to produce electricity (Figure 19.11, page 886). The fission reaction itself occurs in the nuclear core of the power plant. The core consists of uranium fuel rods—enriched to about 3.5% U-235—interspersed between retractable neutron-absorbing control rods. When the control rods are fully retracted from the fuel rod assembly, the chain reaction can occur. When the control rods are fully inserted into the fuel assembly, however, they absorb the neutrons that would otherwise induce fission, shutting down the chain reaction. By retracting or inserting the control rods, the operator can increase or decrease the rate of fission. In this way, the fission reaction is controlled to produce the right amount of heat needed for electricity generation. In case of a power failure, the control rods automatically drop into the fuel rod assembly, shutting down the fission reaction.

A typical nuclear power plant generates enough electricity for a city of about 1 million people and uses about 50 kg of fuel per day. In contrast, a coal-burning power plant uses about 2,000,000 kg of fuel to generate the same amount of electricity. Furthermore, a

Nuclear Reactor

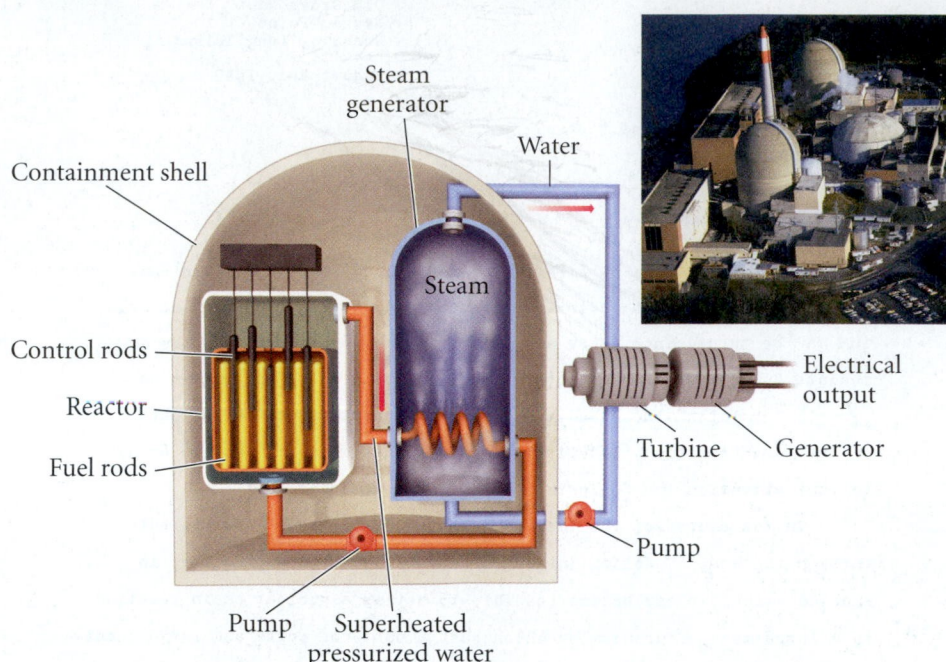

▲ **FIGURE 19.11 A Nuclear Reactor** The fission of U-235 in the core of a nuclear power plant generates heat that is used to create steam, which is used to turn a turbine on an electrical generator. The control rods can be raised or lowered to control the fission reaction. (Note that the water carrying heat away from the reactor core is contained within its own pipes and does not come into direct contact with the steam that drives the turbines.)

Reactor cores in the United States are not made of graphite and could not burn in the way that the Chernobyl core did.

▲ In 1986, the reactor core at Chernobyl (in what is now Ukraine) overheated, exploded, and destroyed part of the containment structure. The release of radioactive nuclides into the environment forced the government to relocate over 335,000 people. It is estimated that there may eventually be several thousand additional cancer deaths among the exposed populations.

nuclear power plant generates no air pollution and no greenhouse gases. A coal-burning power plant, in contrast, emits pollutants such as carbon monoxide, nitrogen oxides, and sulfur oxides. Coal-burning power plants also emit carbon dioxide, a greenhouse gas.

Nuclear power generation, however, is not without problems. Foremost among them is the danger of nuclear accidents. In spite of safety precautions, the fission reaction occurring in nuclear power plants can overheat. The most famous example of this occurred in Chernobyl, in the former Soviet Union, on April 26, 1986. Operators of the plant were performing an experiment designed to reduce maintenance costs. In order to perform the experiment, however, many of the safety features of the reactor core were disabled. The experiment failed with disastrous results. The nuclear core, composed partly of graphite, overheated and began to burn. The accident caused 31 immediate deaths and produced a fire that scattered radioactive debris into the atmosphere, making much of the surrounding land (within about a 20-mile radius) uninhabitable. It is important to understand, however, that a nuclear power plant *cannot* become a nuclear bomb. The uranium fuel used in electricity generation is not sufficiently enriched in U-235 to produce a nuclear detonation. It is also important to understand that U.S. nuclear power plants have additional safety features to prevent similar accidents. For example, U.S. nuclear power plants have large containment structures designed to contain radioactive debris in the event of an accident.

A second problem associated with nuclear power is waste disposal. Although the amount of fuel used in electricity generation is small compared to other fuels, the products of the reaction are radioactive and have long half-lives. What do we do with this waste? Currently, in the United States, nuclear waste is stored on site at the nuclear power plants. However, a permanent disposal site is being developed in Yucca Mountain, Nevada. The site is scheduled to be operational in 2010, but political resistance may alter current plans.

19.8 Converting Mass to Energy: Mass Defect and Nuclear Binding Energy

We have seen that nuclear fission produces large amounts of energy. But where does the energy come from? We can answer this question by carefully examining the masses of the reactants and products in the fission equation from Section 19.7.

$$^{235}_{92}\text{U} + ^{1}_{0}\text{n} \longrightarrow ^{140}_{56}\text{Ba} + ^{93}_{36}\text{Kr} + 3\,^{1}_{0}\text{n}$$

Mass Reactants		Mass Products	
$^{235}_{92}\text{U}$	235.04392 amu	$^{140}_{56}\text{Ba}$	139.910581 amu
$^{1}_{0}\text{n}$	1.00866 amu	$^{93}_{36}\text{Kr}$	92.931130 amu
		$3\,^{1}_{0}\text{n}$	3(1.00866) amu
Total	236.05258 amu		235.86769 amu

Notice that the products of the nuclear reaction have less mass than the reactants. The missing mass is converted to energy. In Chapter 1, we learned that matter is conserved in chemical reactions. In nuclear reactions, however, matter can be converted to energy. The relationship between the amount of matter that is lost and the amount of energy formed is given by Einstein's famous equation relating the two quantities,

$$E = mc^2$$

where E is the energy produced, m is the mass lost, and c is the speed of light. For example, in the above fission reaction, we can calculate the quantity of energy produced as follows:

$$\text{Mass lost } (m) = 236.05258 \text{ amu} - 235.86769 \text{ amu}$$

$$= 0.18489 \text{ amu} \times \frac{1.66054 \times 10^{-27} \text{ kg}}{1 \text{ amu}}$$

$$= 3.0702 \times 10^{-28} \text{ kg}$$

$$\text{Energy produced } (E) = mc^2$$

$$= 3.0702 \times 10^{-28} \text{ kg} \,(2.9979 \times 10^8 \text{ m/s})^2$$

$$= 2.7593 \times 10^{-11} \text{ J}$$

The result is the energy produced when one nucleus of U-235 undergoes fission. To compare to a chemical reaction, we can calculate the energy produced *per mole* of U-235:

$$2.7593 \times 10^{-11} \frac{\text{J}}{\text{U-235 atom}} \times \frac{6.0221 \times 10^{23} \text{ U-235 atoms}}{1 \text{ mol U-235}}$$

$$= 1.6617 \times 10^{13} \text{ J/mol U-235}$$

The energy produced by the fission of 1 mol of U-235 is therefore about 17 billion kJ. In contrast, a highly exothermic chemical reaction might produce 1000 kJ per mole of reactant. Fission therefore produces over a million times more energy per mole than chemical processes.

Mass Defect

The formation of a stable nucleus from its component particles can be viewed as a nuclear reaction in which mass is converted to energy. For example, consider the formation of helium-4 from its components:

$$2\,^{1}_{1}\text{H} + 2\,^{1}_{0}\text{n} \longrightarrow ^{4}_{2}\text{He}$$

Mass Reactants		Mass Products	
$2\,^{1}_{1}\text{H}$	2(1.00783) amu	$^{4}_{2}\text{He}$	4.00260 amu
$2\,^{1}_{0}\text{n}$	2(1.00866) amu		
Total	4.03298 amu		4.00260 amu

In a chemical reaction, there are also mass changes associated with the emission or absorption of energy. Because the energy involved in chemical reactions is so much smaller than that of nuclear reactions, however, these mass changes are completely negligible.

The electrons are contained on the left side in the two $^{1}_{1}\text{H}$, and on the right side in $^{4}_{2}\text{He}$. If you write the equation using only two protons on the left ($^{1}_{1}\text{p}$), you must also add two electrons to the left.

A helium-4 atom has less mass than the sum of the masses of its separate components. This difference in mass, known as the **mass defect**, exists in all stable nuclei. The energy corresponding to the mass defect—obtained by substituting the mass defect into the equation $E = mc^2$—is known as the **nuclear binding energy**, the amount of energy that would be required to break apart the nucleus into its component nucleons.

> An electron volt is defined as the kinetic energy of an electron that has been accelerated through a potential difference of 1 V.

Although chemists often report energies in joules, nuclear physicists often use the electron volt (eV) or megaelectron volt (MeV): 1 MeV $= 1.602 \times 10^{-13}$ J. Unlike energy in joules, which is usually reported per mole, energy in electron volts is usually reported per nucleus. A particularly useful conversion for calculating and reporting nuclear binding energies is the relationship between amu (mass units) and MeV (energy units):

$$1 \text{ amu} = 931.5 \text{ MeV}$$

In other words, a mass defect of 1 amu, when substituted into the equation $E = mc^2$, gives an energy of 931.5 MeV. Using this conversion factor, we can easily calculate the binding energy of the helium nucleus as follows:

$$\text{Mass defect} = 4.03298 \text{ amu} - 4.00260 \text{ amu}$$
$$= 0.03038 \text{ amu}$$

$$\text{Nuclear binding energy} = 0.03038 \text{ amu} \times \frac{931.5 \text{ MeV}}{1 \text{ amu}}$$
$$= 28.30 \text{ MeV}$$

So the binding energy of the helium nucleus is 28.30 MeV. In order to compare the binding energy of one nucleus to that of another, we calculate the *binding energy per nucleon*, which is simply the nuclear binding energy of a nuclide divided by the number of nucleons in the nuclide. For helium-4, we calculate the binding energy per nucleon as follows:

$$\text{Binding energy per nucleon} = \frac{28.30 \text{ MeV}}{4 \text{ nucleons}}$$
$$= 7.075 \text{ MeV per nucleon}$$

The binding energy per nucleon for other nuclides can be calculated in the same way. For example, the nuclear binding energy of carbon-12 is 7.680 MeV per nucleon. Since the binding energy per nucleon of carbon-12 is greater than that of helium-4, we can conclude the carbon-12 nuclide is more *stable* (it has lower potential energy).

EXAMPLE 19.7 Mass Defect and Nuclear Binding Energy

Calculate the mass defect and nuclear binding energy per nucleon (in MeV) for C-16, a radioactive isotope of carbon with a mass of 16.014701 amu.

Solution

Compute the mass defect as the difference between the mass of one C-16 atom and the sum of the masses of 6 hydrogen atoms and 10 neutrons.	$\text{Mass defect} = 6(\text{mass } {}_1^1\text{H}) + 10(\text{mass } {}_0^1\text{n}) - \text{mass } {}_6^{16}\text{C}$ $= 6(1.00783) \text{ amu} + 10(1.00866) \text{ amu} - 16.014701 \text{ amu}$ $= 0.118879 \text{ amu}$
Compute the nuclear binding energy by converting the mass defect (in amu) into MeV. (Use 1 amu = 931.5 MeV.)	$0.118879 \text{ amu} \times \dfrac{931.5 \text{ MeV}}{\text{amu}} = 110.74 \text{ MeV}$
Compute the nuclear binding energy per nucleon by dividing by the number of nucleons in the nucleus.	$\text{Nuclear binding energy per nucleon} = \dfrac{110.74 \text{ MeV}}{16 \text{ nucleons}}$ $= 6.921 \text{ MeV/nucleon}$

For Practice 19.7

Calculate the mass defect and nuclear binding energy per nucleon (in MeV) for U-238, which has a mass of 238.050784 amu.

The Curve of Binding Energy

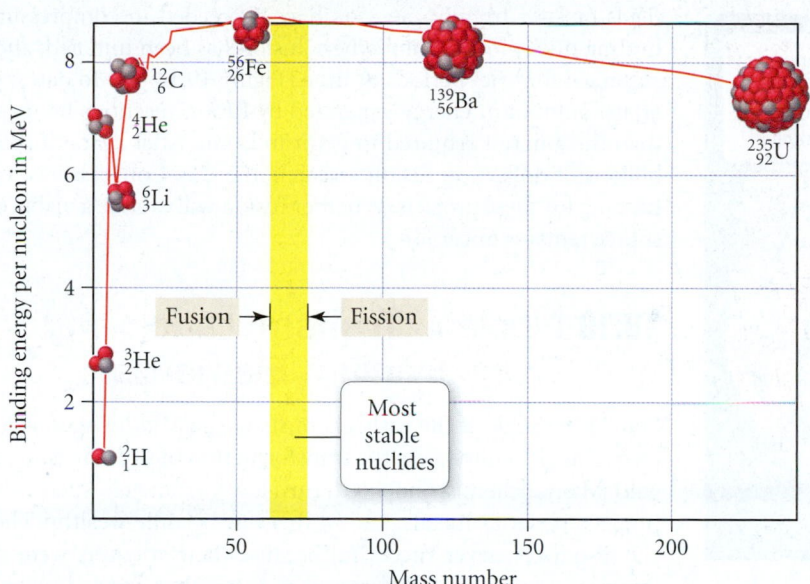

◀ **FIGURE 19.12 Nuclear Binding Energy per Nucleon** The nuclear binding energy per nucleon (a measure of the stability of a nucleus) reaches a maximum at Fe-56. Consequently, energy can be obtained either by breaking a heavy nucleus up into lighter ones (fission) or by combining lighter nuclei into heavier ones (fusion).

Figure 19.12 ▲ shows the binding energy per nucleon plotted as a function of mass number (A). Notice that the binding energy per nucleon is relatively low for small mass numbers and increases until about $A = 60$, where it reaches a maximum. This means that nuclides with mass numbers of about 60 are among the most stable. Beyond $A = 60$, the binding energy per nucleon decreases again. Figure 19.12 clearly shows why nuclear fission is a highly exothermic process. When a heavy nucleus, such as U-235, breaks up into smaller nuclei, such as Ba-140 and Kr-93, the binding energy per nucleon increases. This is analogous to a chemical reaction in which weak bonds break and strong bonds form. In either case, the process is exothermic. Notice also, however, that Figure 19.12 reveals that the combining of two lighter nuclei (below $A = 60$) to form a heavier nucleus should also be exothermic. This process is called *nuclear fusion*.

19.9 Nuclear Fusion: The Power of the Sun

As we have learned, nuclear fission is the *splitting* of a heavy nucleus to form two or more lighter ones. **Nuclear fusion**, in contrast, is the *combination* of two light nuclei to form a heavier one. Both fusion and fission emit large amounts of energy because, as we have just seen, they both form daughter nuclides with greater binding energies per nucleon than the parent nuclides. Nuclear fusion is the energy source of stars, including our sun. In stars, hydrogen atoms fuse together to form helium atoms, emitting energy in the process.

Nuclear fusion is also the basis of modern nuclear weapons called hydrogen bombs. A modern hydrogen bomb has up to 1000 times the explosive force of the first atomic bombs. These bombs employ the following fusion reaction:

$$\,^{2}_{1}H + \,^{3}_{1}H \longrightarrow \,^{4}_{2}He + \,^{1}_{0}n$$

In this reaction, deuterium (the isotope of hydrogen with one neutron) and tritium (the isotope of hydrogen with two neutrons) combine to form helium-4 and a neutron (Figure 19.13 ▶). Because fusion reactions require two positively charged nuclei (which repel each other) to fuse together, extremely high temperatures are required. In a hydrogen bomb, a small fission bomb is detonated first, providing temperatures high enough for fusion to proceed.

Nuclear fusion has been intensely investigated as a way to produce electricity. Because of the higher energy production—fusion provides about 10 times more energy per gram of fuel than does fission—and because the products of the reaction are less problematic than those of fission, fusion holds promise as a future energy source. However, in spite of intense efforts, the generation of electricity by fusion remains elusive. One of the main

Deuterium-Tritium Fusion Reaction

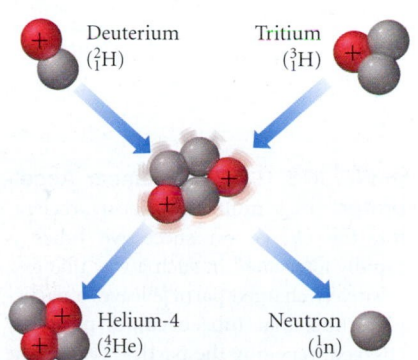

▲ **FIGURE 19.13 A Nuclear Fusion Reaction** In this particular reaction, two heavy isotopes of hydrogen, deuterium (hydrogen-2) and tritium (hydrogen-3), fuse to form helium-4 and a neutron.

Tokamak Fusion Reactor

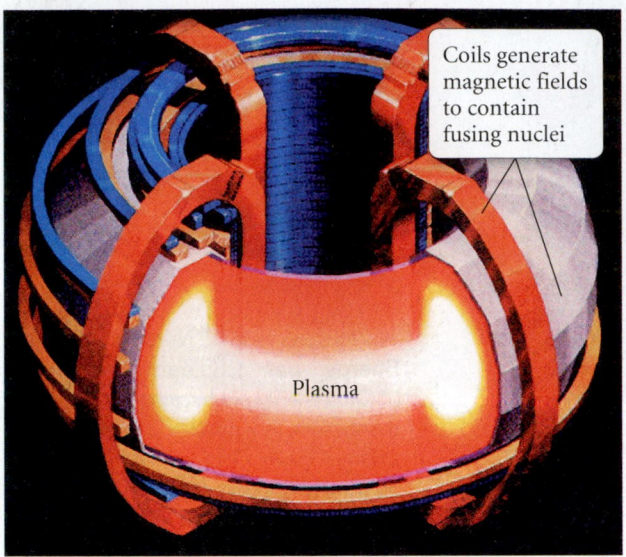

Coils generate magnetic fields to contain fusing nuclei

Plasma

▲ **FIGURE 19.14 Tokamak Fusion Reactor** A tokamak uses powerful magnetic fields to confine nuclear fuel at the enormous temperatures needed for fusion.

▲ The Joliot-Curies won the 1935 Nobel Prize in Chemistry for their work on nuclear transmutation.

problems is the high temperature required for fusion to occur—no material can withstand those temperatures. Using powerful magnetic fields or laser beams, scientists have succeeded in compressing and heating nuclei to the point where fusion has been initiated, and even sustained for brief periods of time (Figure 19.14 ◄). To date, however, the amount of energy generated by fusion reactions has been less than the amount required to get it to occur. After years of spending billions of dollars on fusion research, the U.S. Congress has reduced funding for these projects. Whether fusion will ever be a viable energy source remains uncertain.

19.10 Nuclear Transmutation and Transuranium Elements

One of the goals of the early chemists of the Middle Ages, who were known as alchemists, was the transformation of ordinary metals into gold. Many alchemists hoped to turn low-cost metals, such as lead or tin, into precious metals, and in this way become wealthy. These alchemists were never successful because their attempts were merely chemical—they mixed different metals together or tried to get them to react with other substances in order to turn them into gold. In a chemical reaction, an element retains its identity, so a less valuable metal—such as lead—will always remain lead, even though it forms a compound with another element.

Nuclear reactions, however, result in the transformation of one element into another, a process known as **transmutation**. We have already seen how this is true in radioactive decay, in fission, and in fusion. In addition, other nuclear reactions that transmute elements are possible. For example, in 1919 Ernest Rutherford bombarded nitrogen-17 with alpha particles to form oxygen:

$$^{14}_{7}\text{N} + {}^{4}_{2}\text{He} \longrightarrow {}^{17}_{8}\text{O} + {}^{1}_{1}\text{H}$$

Irène Joliot-Curie (daughter of Marie Curie) and her husband Frédéric bombarded aluminum-27 with alpha particles to form phosphorus:

$$^{27}_{13}\text{Al} + {}^{4}_{2}\text{He} \longrightarrow {}^{30}_{15}\text{P} + {}^{1}_{0}\text{n}$$

In the 1930s, scientists began building devices that could accelerate particles to high velocities, opening the door to even more possibilities. These devices are generally of two types, the **linear accelerator** and the **cyclotron**.

In a single-stage linear accelerator, a charged particle such as a proton is accelerated in an evacuated tube. The accelerating force is provided by a potential difference between the ends of the tube. In multistage linear accelerators, such as the Stanford Linear Accelerator (SLAC) at Stanford University, a series of tubes of increasing length are connected to a source of alternating voltage, as shown in Figure 19.15 ▼. The voltage alternates in such a way that, as a positively charged particle leaves a particular tube, that tube becomes

▶ **FIGURE 19.15 The Linear Accelerator** In a multistage linear accelerator, the charge on successive tubes is rapidly alternated in such a way that as a positively charged particle leaves a particular tube, that tube becomes positively charged, repelling the particle toward the next tube. At the same time, the tube that the particle is now approaching becomes negatively charged, pulling the particle toward it. This process is repeated through a number of tubes until the particle has been accelerated to a high velocity.

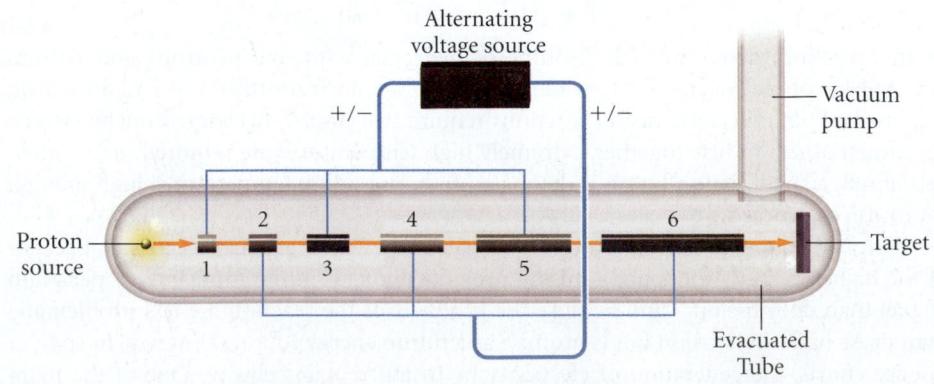

Alternating voltage source

+/−   +/−

Vacuum pump

Proton source

2   4   6

1   3   5

Target

Evacuated Tube

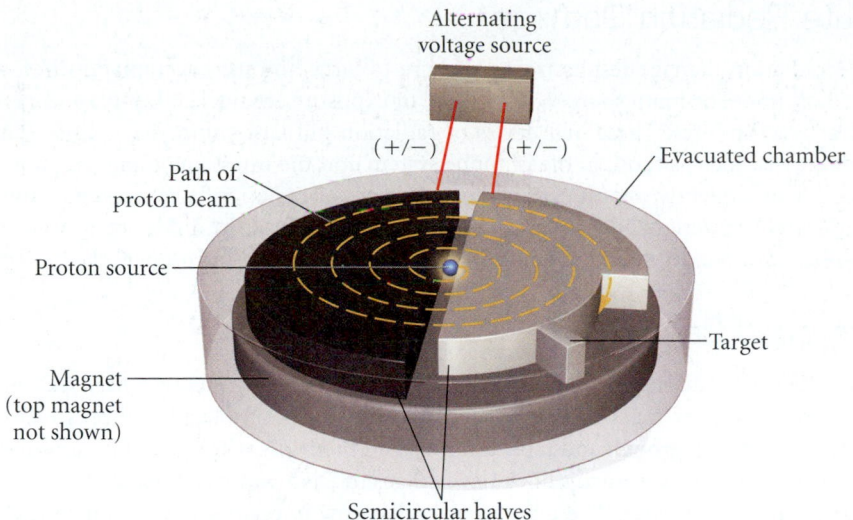

▲ FIGURE 19.16 The Cyclotron   In a cyclotron, two semicircular D-shaped structures are subject to an alternating voltage. A charged particle, starting from a point between the two, is accelerated back and forth between them, while additional magnets cause the particle to move in a spiral path.

positively charged, repelling the particle to the next tube. At the same time, the tube the particle is now approaching becomes negatively charged, pulling the particle toward it. This continues throughout the linear accelerator, allowing the particle to be accelerated to velocities up to 90% of the speed of light.

In a cyclotron, a similarly alternating voltage is used to accelerate a charged particle, only this time the alternating voltage is applied between the two semicircular halves of the cyclotron (Figure 19.16 ▲). A charged particle originally in the middle of the two semicircles is accelerated back and forth between them. Additional magnets cause the particle to move in a spiral path. As the charged particle spirals out from the center, it gains speed and eventually exits the cyclotron aimed at the target.

With linear accelerators or cyclotrons, all sorts of nuclear transmutations can be achieved. In this way, scientists have made nuclides that don't normally exist in nature. For example, uranium-238 can be made to collide with carbon-12 to form an element with atomic number 98:

$$^{238}_{92}\text{U} + {}^{12}_{6}\text{C} \longrightarrow {}^{244}_{98}\text{Cf} + 6\,{}^{1}_{0}\text{n}$$

This element was named californium (Cf) because it was first produced (by a slightly different nuclear reaction) at the University of California at Berkeley. Many other nuclides with atomic numbers larger than that of uranium have been synthesized since the 1940s. These synthetic elements—called transuranium elements—have been added to modern versions of the periodic table.

▲ The Stanford Linear Accelerator (top) is located at Stanford University in California. The Fermilab National Accelerator Laboratory complex in Batavia, Illinois (bottom), includes two cyclotrons in a figure-8 configuration.

Most synthetic elements are unstable and have very short half-lives. Some exist for only fractions of a second after they are made.

 ### Conceptual Connection 19.3   Nuclear Transformations

Californium-252 is bombarded with a boron-10 nucleus to produce another nuclide and six neutrons. What nuclide is formed?

Answer: Lawrencium-256

19.11 The Effects of Radiation on Life

As we discussed in Section 19.2, the energy associated with radioactivity can ionize molecules. When radiation ionizes important molecules in living cells, problems can develop. The ingestion of radioactive materials—especially alpha and beta emitters—is particularly dangerous because the radioactivity is then inside the body and can do even more damage. The effects of radiation can be divided into three different types: acute radiation damage, increased cancer risk, and genetic effects.

Acute Radiation Damage

Acute radiation damage results from exposure to large amounts of radiation in a short period of time. The main sources of this kind of exposure are nuclear bombs and exposed nuclear reactor cores. These high levels of radiation kill large numbers of cells. Rapidly dividing cells, such as those in the immune system and the intestinal lining, are most susceptible. Consequently, people exposed to high levels of radiation have weakened immune systems and a lowered ability to absorb nutrients from food. In milder cases recovery is possible with time. In more extreme cases death results, often from unchecked infection.

Increased Cancer Risk

| DNA and its function in the body are explained in more detail in Chapter 21.

Lower doses of radiation over extended periods of time can increase cancer risk. Radiation increases cancer risk because it can damage DNA, the molecules in cells that carry instructions for cell growth and replication. When the DNA within a cell is damaged, the cell normally dies. Occasionally, however, changes in DNA cause cells to grow abnormally and to become cancerous. These cancerous cells grow into tumors that can spread and, in some cases, cause death. Cancer risk increases with increasing radiation exposure. However, cancer is so prevalent and has so many convoluted causes that it is difficult to determine an exact threshold for increased cancer risk from radiation exposure.

Genetic Defects

Another possible effect of radiation exposure is genetic defects in offspring. If radiation damages the DNA of reproductive cells—such as eggs or sperm—then the offspring that develop from those cells may have genetic abnormalities. Genetic defects of this type have been observed in laboratory animals exposed to high levels of radiation. However, such genetic defects—with a clear causal connection to radiation exposure—have yet to be verified in humans, even in studies of Hiroshima survivors.

Measuring Radiation Exposure

We can measure radiation exposure in a number of different ways. For example, we can simply measure the number of decay events to which a person is exposed. The unit used for this type of exposure measurement is called the *curie* (Ci), defined as 3.7×10^{10} decay events per second. A person exposed to a curie of radiation from an alpha emitter is being bombarded by 3.7×10^{10} alpha particles per second. However, we already know that different kinds of radiation produce different effects. For example, we know that alpha radiation has a much greater ionizing power than beta radiation. Consequently, a certain number of alpha decays occurring within a person's body (due to the ingestion of an alpha emitter) would do more damage than the same number of beta decays. If the alpha emitter and beta emitter were external to the body, however, the radiation from the alpha emitter would largely be stopped by clothing or the skin (due to the low penetrating power of alpha radiation), while the radiation from the beta emitter could penetrate the skin and therefore cause more damage.

A better way to assess radiation exposure is to measure the amount of energy actually absorbed by body tissue. The units used for this type of exposure measurement are called the *gray* (Gy), which corresponds to 1 J of energy absorbed per kilogram of body tissue, and the *rad* (for radiation absorbed dose), which corresponds to 0.01 Gy.

$$1 \text{ gray (Gy)} = 1 \text{ J/kg body tissue}$$

$$1 \text{ rad} = 0.01 \text{ J/kg body tissue}$$

Although these units measure the actual energy absorbed by bodily tissues, they still do not account for the amount of damage to biological molecules caused by that energy absorption, which differs from one type of radiation to another and from one type of biological tissue to another. For example, when a gamma ray passes through biological tissue, the energy absorbed is spread out over the long distance that the radiation travels through the body, resulting in a low ionization density within the tissue. When an alpha particle passes through biological tissue, however, the energy is absorbed over a much shorter distance, resulting in a much higher ionization density. The higher ionization density results in greater damage, even though the amount of energy absorbed might be the same.

Consequently, a correction factor, called the **biological effectiveness factor**, or **RBE** (for *Relative Biological Effectiveness*), is usually multiplied by the dose in rads to obtain the dose in a unit called the **rem** for roentgen *equivalent man*.

> A *roentgen* is defined as the amount of radiation that produces 2.58×10^{-4} C of charge per kg of air.

$$\text{Dose in rads} \times \text{biological effectiveness factor} = \text{dose in rems}$$

The biological effectiveness factor for alpha radiation, for example, is much higher than that for gamma radiation.

On average, each of us is exposed to approximately 360 mrem of radiation per year from sources shown in Table 19.4. The majority of this exposure comes from natural sources, especially radon, one of the products in the uranium decay series. As you can see from Table 19.4, however, some medical procedures also involve exposure levels similar to those received from natural sources.

TABLE 19.4   Exposure by Source for Persons Living in the United States

Source	Dose
Natural Radiation	
A 5-hour jet airplane ride	2.5 mrem/trip (0.5 mrem/hr at 39,000 feet) (Whole body dose)
Cosmic radiation from outer space	27 mrem/yr (whole body dose)
Terrestrial radiation	28 mrem/yr (whole body dose)
Natural radionuclides in the body	35 mrem/yr (whole body dose)
Radon gas	200 mrem/yr (lung dose)
Diagnostic Medical Procedures	
Chest X-ray	8 mrem (whole body dose)
Dental X-rays (panoramic)	30 mrem (skin dose)
Dental X-rays (two bitewings)	80 mrem (skin dose)
Mammogram	138 mrem per image
Barium enema (X-ray portion only)	406 mrem (bone marrow dose)
Upper gastrointestinal tract	244 mrem (X-ray portion only) (bone marrow dose)
Thallium heart scan	500 mrem (whole body dose)
Consumer Products	
Building materials	3.5 mrem/year (whole body dose)
Luminous watches (H-3 and Pm-147)	0.04–0.1 mrem/year (whole body dose)
Tobacco products (to smokers of 30 cigarettes per day)	16,000 mrem/year (bronchial epithelial dose)

Source: **Department of Health and Human Services,** *National Institutes of Health.*

It takes much more than the average radiation dose to produce significant health effects in humans. The first measurable effect, a decreased white blood cell count, occurs at instantaneous exposures of approximately 20 rem (Table 19.5). Exposures of 100 rem produce a definite increase in cancer risk, and exposures of over 500 rem often result in death.

TABLE 19.5   Effects of Radiation Exposure

Approximate Dose (rem)	Probable Outcome
20–100	Decreased white blood cell count; possible increase in cancer risk
100–400	Radiation sickness including vomiting and diarrhea; skin lesions; increase in cancer risk
500	Death (often within 2 months)
1000	Death (often within 2 weeks)
2000	Death (within hours)

Conceptual Connection 19.4 **Radiation Exposure**

Suppose a person ingests equal amounts of two nuclides, both of which are beta emitters (of roughly equal energy). Nuclide A has a half-life of 8.5 hours and Nuclide B has a half-life of 15.0 hours. Both nuclides are eliminated from the body within 24 hours of ingestion. Which of the two nuclides produces the greater radiation exposure?

Answer: Nuclide A. Because nuclide A has a shorter half-life, more of the nuclides will decay, and therefore produce radiation, before they exit the body.

19.12 Radioactivity in Medicine and Other Applications

Radioactivity is often perceived as dangerous; however, it is also immensely useful to physicians in the diagnosis and treatment of disease and has numerous other valuable applications. The use of radioactivity in medicine can be broadly divided into *diagnostic techniques* (used to diagnose disease) and *therapeutic techniques* (used to treat disease).

Diagnosis in Medicine

The use of radioactivity in diagnosis usually involves a **radiotracer**, a radioactive nuclide that has been attached to a compound or introduced into a mixture in order to track the movement of the compound or mixture within the body. Tracers are useful in the diagnosis of disease because of two main factors: (1) the sensitivity with which radioactivity can be detected, and (2) the identical chemical behavior of a radioactive nucleus and its nonradioactive counterpart. For example, the thyroid gland naturally concentrates iodine. When a patient is given small amounts of iodine-131 (a radioactive isotope of iodine), the radioactive iodine accumulates in the thyroid, just as nonradioactive iodine does. However, the radioactive iodine emits radiation, which can then be detected with great sensitivity and used to measure the rate of iodine uptake by the thyroid, and to image the gland.

Since different elements are taken up preferentially by different organs or tissues, various radiotracers can be used to monitor metabolic activity and image a variety of organs and structures, including the kidneys, heart, brain, gallbladder, bones, and arteries, as shown in Table 19.6. Radiotracers can also be used to locate infections or cancers within the body. To locate an infection, antibodies are "labeled" with a radioactive nuclide, such as technetium-99m (where "m" means metastable), and administered to the patient. The antibodies then aggregate at the infected site, as described in the opening section of this chapter. Cancerous tumors can be detected because they naturally concentrate phosphorus. When a patient is given phosphorus-32 (a radioactive isotope of phosphorus) or a phosphate compound incorporating another radioactive isotope such as Tc-99m, the tumors concentrate the radioactive substance and become sources of radioactivity that can be detected (Figure 19.17 ◀).

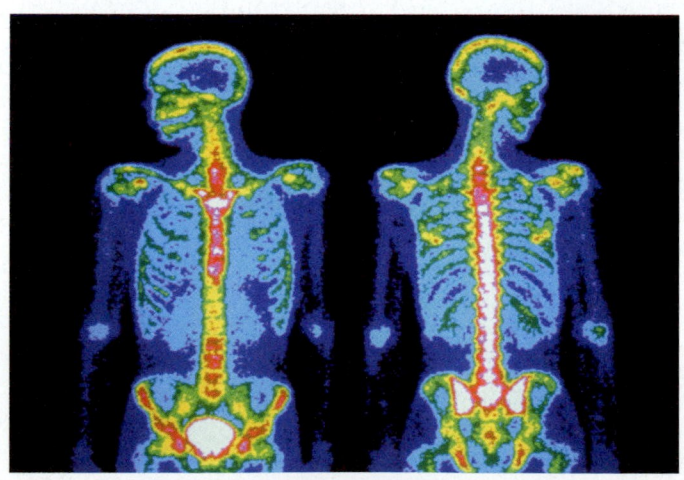

▲ FIGURE 19.17 A Bone Scan
These images, front and rear views of the human body, were created by the gamma ray emissions of Tc-99m. Such scans are often used to locate cancer that has metastasized to the bones from a primary tumor elsewhere.

TABLE 19.6 Common Radiotracers

Nuclide	Type of Emission	Half-Life	Part of Body Studied
Technetium-99m	Gamma (primarily)	6.01 hours	Various organs, bones
Iodine-131	Beta	8.0 days	Thyroid
Iron-59	Beta	44.5 days	Blood, spleen
Thallium-201	Electron capture	3.05 days	Heart
Fluorine-18	Positron emission	1.83 hours	PET studies of heart, brain
Phosphorus-32	Beta	14.3 days	Tumors in various organs

A specialized imaging technique known as **positron emission tomography (PET)** employs positron-emitting nuclides, such as fluorine-18, synthesized in cyclotrons. The fluorine-18 is attached to a metabolically active substance such as glucose and administered to the patient. As the glucose travels through the bloodstream and to the heart and brain, it carries the radioactive fluorine, which decays with a half-life of just under 2 hours. When a fluorine-18 nuclide decays, it emits a positron which immediately combines with any nearby electrons. Since a positron and an electron are antiparticles, they annihilate one other, producing two gamma rays that travel in exactly opposing directions. The gamma rays are detected by an array of detectors that can locate the point of origin with great accuracy. The result is a set of highly detailed images that show both the rate of glucose metabolism and structural features of the imaged organ (Figure 19.18 ▶).

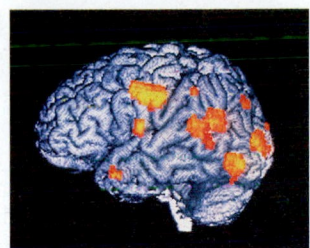

▲ **FIGURE 19.18 A PET Scan** The colored areas indicate regions of high metabolic activity in the brain of a schizophrenic patient experiencing hallucinations.

Radiotherapy in Medicine

Because radiation kills cells, and because it is particularly effective at killing rapidly dividing cells, it is often used as a therapy for cancer (cancer cells reproduce much faster than normal cells). Gamma rays are focused on internal tumors to kill them. The gamma ray beam is usually moved in a circular path around the tumor (Figure 19.19 ▶), maximizing the exposure of the tumor while minimizing the exposure of the surrounding healthy tissue. Nonetheless, cancer patients receiving such treatment usually develop the symptoms of radiation sickness, which include vomiting, skin burns, and hair loss.

Some people wonder why radiation—which is known to cause cancer—can also be used to treat cancer. The answer lies in risk analysis. A cancer patient is normally exposed to radiation doses of about 100 rem. Such a dose increases cancer risk by about 1%. However, if the patient has a 100% chance of dying from the cancer that he already has, such a risk becomes acceptable, especially since there is a significant chance of curing the cancer.

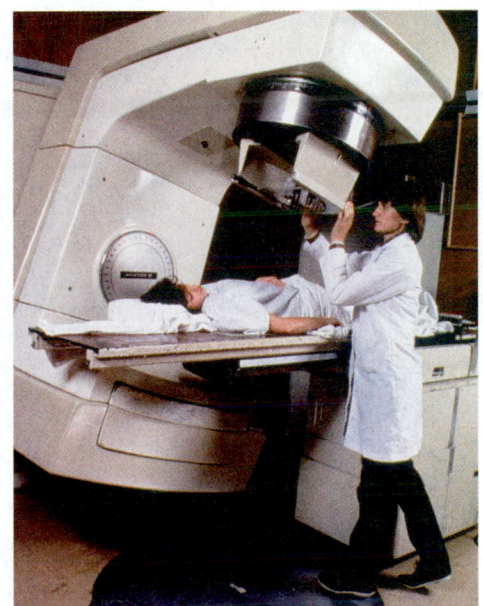

▲ **FIGURE 19.19 Radiotherapy for Cancer** This treatment involves exposing a malignant tumor to gamma rays generated by nuclides such as cobalt-60. The beam is moved in a circular pattern around the tumor to maximize exposure of the tumor to radiation while minimizing the exposure of healthy tissues.

Other Applications

Radioactivity is often used to kill microorganisms. For example, physicians use radiation to sterilize medical devices that are to be surgically implanted. The radiation kills bacteria that might otherwise lead to infection. Similarly, radiation can be used to kill bacteria and parasites in foods. Like the pasteurization of milk, the irradiation of foods makes them safer to consume and gives them a longer shelf life (Figure 19.20 ▼). For example, the irradiation of raw meat and poultry kills *E. coli* and *Salmonella*, bacteria that can lead to serious illness and even death when consumed. The irradiation of food does not, however, make the food itself radioactive, nor does it decrease the nutritional value of the food. In the United States, the irradiation of many different types of foods—including beef, poultry, potatoes, flour, and fruit—has been approved by the U.S. Food and Drug Administration (FDA) and the U.S. Department of Agriculture (USDA).

Radioactivity can also be used to control the populations of harmful insects. For example, fruit flies can be raised in large numbers in captivity and sterilized with radiation. When these fruit flies are released, they mate with wild fruit flies but do not produce offspring. The

◀ **FIGURE 19.20 Irradiation of Food** Irradiation kills microbes that cause food to decay, allowing for longer and safer storage. The food is not made radioactive and its properties are unchanged in the process. These strawberries were picked at the same time, but those on the left were irradiated before storage.

efforts of the wild fruit flies, which might otherwise lead to reproduction, are wasted and the next generation of flies is smaller than it would otherwise have been. Similar strategies have been employed to control the populations of disease-carrying mosquitoes.

Chapter in Review

Key Terms

Section 19.1
radioactivity (866)
radioactive (866)

Section 19.2
phosphorescence (866)

Section 19.3
nuclide (867)
alpha (α) decay (868)
alpha (α) particle (868)
nuclear equation (868)
ionizing power (869)
penetrating power (869)
beta (β) decay (869)

beta (β) particle (869)
gamma (γ) ray emission (870)
gamma (γ) ray (870)
positron emission (870)
positron (870)
electron capture (870)

Section 19.4
strong force (872)
magic numbers (874)

Section 19.5
film-badge dosimeter (874)
Geiger-Müller counter (875)
scintillation counter (875)

Section 19.6
radiometric dating (879)
radiocarbon dating (879)

Section 19.7
nuclear fission (883)
chain reaction (884)
critical mass (884)

Section 19.8
mass defect (888)
nuclear binding energy (888)

Section 19.9
nuclear fusion (889)

Section 19.10
transmutation (890)
linear accelerator (890)
cyclotron (890)

Section 19.11
biological effectiveness factor
  (RBE) (893)
rem (893)

Section 19.12
radiotracer (894)
positron emission tomography
  (PET) (895)

Key Concepts

Diagnosing Appendicitis (19.1)

Radioactivity is the emission of subatomic particles or energetic electromagnetic radiation by the nuclei of certain atoms. Because some of these emissions can pass through matter, radioactivity is useful in medicine and many other areas of study. For example, antibodies can be radioactively tagged and injected into a person's body; if an infection is present, the radioactivity accumulates at the site of infection, allowing imaging and diagnosis of the infected organ.

The Discovery of Radioactivity (19.2)

Radioactivity was discovered by Antoine-Henri Becquerel when he found that uranium causes a photographic exposure in the absence of light. Marie Sklodowska Curie later found that this phenomenon was not unique to uranium, and she began calling the rays that produced the exposure radioactivity. In her search, Curie discovered two new elements, polonium and radium.

Types of Radioactivity (19.3)

The major types of natural radioactivity are alpha (α) decay, beta (β) decay, gamma (γ) ray emission, and positron emission. Alpha radiation consists of helium nuclei. Beta particles are electrons. Gamma rays are electromagnetic radiation of very high energy. Positrons are the antiparticles of electrons. In addition, a nucleus may absorb one of its orbital electrons (electron capture). Each radioactive process can be represented with a nuclear equation, in which the parent nuclide changes into the daughter nuclide. In a nuclear equation, although the specific types of atoms may not balance, the atomic numbers and mass numbers must. Each type of radioactivity has a different ionizing and penetrating power. These values are inversely related; a particle with a higher ionizing power has a lower penetrating power. Alpha particles are the most massive and they therefore have the highest ionizing power, followed by beta particles and positrons, which are equivalent in their ionizing power. Gamma rays have the lowest ionizing power.

The Valley of Stability: Predicting the Type of Radioactivity (19.4)

The stability of a nucleus, and therefore the probability that it will undergo radioactive decay, depends largely on two factors. The first is the ratio of neutrons to protons (N/Z), because neutrons provide a strong force which overcomes the electromagnetic repulsions between the positive protons. This ratio is one for smaller elements, but becomes greater than one for larger elements. The second factor in nuclei stability is a concept known as magic numbers; certain numbers of nucleons are more stable than others.

Detecting Radioactivity (19.5)

Radiation detectors are used to determine the quantity of radioactivity in an area or sample. Film-badge dosimeters utilize photographic film for that purpose; however, such detectors do not provide an instantaneous response. Two detectors that instantly register the amount of radiation are the Geiger-Müller counter, which uses the ionization of argon by radiation to produce an electrical signal, and the scintillation counter, which uses the emission of light induced by radiation.

The Kinetics of Radioactive Decay and Radiometric Dating (19.6)

All radioactive elements decay according to first-order kinetics (Chapter 13); the half-life equation and the integrated rate law for radioactive decay are derived from the first-order rate laws. The kinetics of radioactive decay can be used to date objects and artifacts. The age of materials that were once part of living organisms can be measured by carbon-14 dating. The age of ancient rocks and even Earth itself can be determined by uranium/lead dating.

The Discovery of Fission: The Atomic Bomb and Nuclear Power (19.7)

Fission is the splitting of an atom, such as uranium-235, into two atoms of lesser atomic weight. Because the fission of one uranium-235 atom releases enormous amounts of energy, and produces neutrons that can split other uranium-235 atoms, the energy from these collective reactions can be harnessed in an atomic bomb or nuclear reactor. Nuclear power produces no pollution and requires little mass to release lots of energy; however, there is always a danger of accidents, and it is difficult to dispose of nuclear waste.

Converting Mass to Energy: Mass Defect and Nuclear Binding Energy (19.8)

In a nuclear fission reaction, a substantial amount of mass is converted into energy. The difference in mass between the products and the reactants is called the mass defect and the corresponding energy, calculated from Einstein's equation $E = mc^2$, is the nuclear binding energy. The stability of the nucleus is determined by the binding energy per nucleon, which increases up to mass number 60 and then decreases.

Nuclear Fusion: The Power of the Sun (19.9)

Stars produce their energy by a process that is the opposite of fission: nuclear fusion, the combination of two light nuclei to form a heavier one. Modern nuclear weapons employ fusion. Although fusion has been examined as a possible method to produce electricity, experiments with hydrogen fusion have thus far been more costly than productive.

Nuclear Transmutation and Transuranium Elements (19.10)

Nuclear transmutation, the changing of one element to another element, has been used to create the transuranium elements, elements with atomic numbers greater than that of uranium. Two devices are most commonly used to accelerate particles to the high speeds necessary for transmutation reactions: the linear accelerator and the cyclotron. Both use alternating voltage to propel particles by electromagnetic forces.

The Effects of Radiation on Life (19.11)

The effects of radiation can be grouped into three categories. Acute radiation damage is caused by a large exposure to radiation for a short period of time. Lower radiation exposures may result in increased cancer risk because of damage to DNA. Genetic defects are caused by damage to the DNA of reproductive cells. The most effective unit of measurement for the amount of radiation absorbed is the rem, which takes into account the different penetrating and ionizing powers of the various types of radiation.

Radioactivity in Medicine (19.12)

Radioactivity is central to the diagnosis of medical problems by means of radiotracers and positron emission tomography (PET). Both of these techniques can provide data about the appearance and metabolic activity of an organ, or help locate a tumor. Radiation is also used to treat cancer because it can kill cells. Although this treatment has unpleasant side effects and increases the risk of a new cancer developing, the risk is usually acceptable compared to the risk of death from an established cancer. Radiation can also be used to kill bacteria in foods and to control harmful insect populations.

Key Equations and Relationships

The First-Order Rate Law (19.6)
$$\text{Rate} = kN$$

The Half-Life Equation (19.6)
$$t_{1/2} = \frac{0.693}{k} \qquad k = \text{rate constant}$$

The Integrated Rate Law (19.6)
$$\ln \frac{N_t}{N_0} = -kt \qquad N_t = \text{number of radioactive nuclei at time } t$$
$$N_0 = \text{initial number of radioactive nuclei}$$

Einstein's Energy–Mass Equation (19.8)
$$E = mc^2$$

Key Skills

Writing Nuclear Equations for Alpha Decay (19.3)
   • Example 19.1   • For Practice 19.1   • Exercises 33–38

Writing Nuclear Equations for Beta Decay, Positron Emission, and Electron Capture (19.3)
   • Example 19.2   • For Practice 19.2   • For More Practice 19.2   • Exercises 33–38

Predicting the Type of Radioactive Decay (19.4)
   • Example 19.3   • For Practice 19.3   • Exercises 43, 44

Using Radioactive Decay Kinetics (19.6)
   • Example 19.4   • For Practice 19.4   • Exercises 47–54

Using Radiocarbon Dating (19.6)
   • Example 19.5   • For Practice 19.5   • Exercises 55, 56

Using Uranium/Lead Dating to Estimate the Age of a Rock (19.6)
   • Example 19.6   • For Practice 19.6   • Exercises 57, 58

Determining the Mass Defect and Nuclear Binding Energy (19.8)
   • Example 19.7   • For Practice 19.7   • Exercises 67–74

Exercises

Review Questions

1. What is radioactivity? How and by whom was radioactivity discovered?

2. Explain the role of Marie Curie in the discovery of radioactivity.

3. Define A, Z, and X in the following notation used to specify a nuclide: $^A_Z X$.

4. Use the notation from Question 3 to write symbols for a proton, a neutron, and an electron.

5. What is an alpha particle? What happens to the mass number and atomic number of a nuclide that emits an alpha particle?

6. What is a beta particle? What happens to the mass number and atomic number of a nuclide that emits a beta particle?

7. What is a gamma ray? What happens to the mass number and atomic number of a nuclide that emits a gamma ray?

8. What is a positron? What happens to the mass number and atomic number of a nuclide that emits a positron?

9. Describe the process of electron capture. What happens to the mass number and atomic number of a nuclide that undergoes electron capture?

10. List alpha particles, beta particles, positrons, and gamma rays in order of each of the following: (a) increasing ionizing power; (b) increasing penetrating power.

11. Explain why the ratio of neutrons to protons (N/Z) is important in determining nuclear stability.

12. Explain how you might use the N/Z ratio of a nuclide to predict the kind of radioactive decay that it might undergo.

13. What are magic numbers? How are they important in determining the stability of a nuclide?

14. Explain the basic way that each of the following detect radioactivity: (a) film-badge dosimeter; (b) Geiger-Müller counter; and (c) scintillation counter.

15. Explain the concept of half-life with respect to radioactive nuclides. What rate law is characteristic of radioactivity?

16. Explain the main concepts behind the technique of radiocarbon dating. How can radiocarbon dating be corrected for changes in atmospheric concentrations of C-14? What range of ages can be reliably determined by C-14 dating?

17. Explain how the uranium to lead ratio in a rock can be used to estimate its age. How does this dating technique also give an estimate for Earth's age? How old is Earth according to this dating method?

18. Describe fission. Include the concepts of chain reaction and critical mass in your description. How and by whom was fission discovered?

19. What was the Manhattan Project? Briefly describe its development and culmination.

20. Explain how fission can be used to generate electricity.

21. Describe the advantages and disadvantages of using fission to generate electricity.

22. The products of a nuclear reaction usually have a different mass than the reactants. Explain why this is so.

23. Explain the concepts of mass defect and nuclear binding energy. At what mass number does the nuclear binding energy per nucleon peak? What is the significance of this?

24. What is fusion? Why can fusion and fission both produce energy? Explain.

25. What are some of the problems associated with using fusion to generate electricity?

26. Explain transmutation and give one or two examples.

27. How does a linear accelerator work? For what purpose is it used?

28. Explain the basic principles by which a cyclotron functions.

29. How does radiation affect living organisms? What kinds of effects are possible?

30. Explain why different kinds of radiation produce different effects in biological tissue, even though the amount of radiation exposure may be the same.

31. Explain the significance of the biological effectiveness factor in measuring radiation exposure. What types of radiation would you expect to have the highest biological effectiveness factor?

32. Describe some of the medical uses, both in diagnosis and in treatment of disease, of radioactivity.

Problems by Topic

Radioactive Decay and Nuclide Stability

33. Write a nuclear equation for the indicated decay of each of the following nuclides.
    a. U-234 (alpha)
    b. Th-230 (alpha)
    c. Pb-214 (beta)
    d. N-13 (positron emission)
    e. Cr-51 (electron capture)

34. Write a nuclear equation for the indicated decay of each of the following nuclides.
    a. Po-210 (alpha)
    b. Ac-227 (beta)
    c. Tl-207 (beta)
    d. O-15 (positron emission)
    e. Pd-103 (electron capture)

35. Write a partial decay series for Th-232 undergoing the following sequential decays: $\alpha, \beta, \beta, \alpha$.

36. Write a partial decay series for Rn-220 undergoing the following sequential decays: $\alpha, \alpha, \beta, \alpha$.

37. Fill in the missing particles in each of the following nuclear equations.

    a. $\underline{\quad} \longrightarrow {}^{217}_{85}At + {}^4_2He$

    b. ${}^{241}_{94}Pu \longrightarrow {}^{241}_{95}Am + \underline{\quad}$

    c. ${}^{19}_{11}Na \longrightarrow {}^{19}_{10}Ne + \underline{\quad}$

    d. ${}^{75}_{34}Se + \underline{\quad} \longrightarrow {}^{75}_{33}As$

38. Fill in the missing particles in each of the following nuclear equations.

    a. ${}^{241}_{95}Am \longrightarrow {}^{237}_{93}Np + \underline{\quad}$

    b. $\underline{\quad} \longrightarrow {}^{233}_{92}U + {}^0_{-1}e$

    c. ${}^{237}_{93}Np \longrightarrow \underline{\quad} + {}^4_2He$

    d. ${}^{75}_{35}Br \longrightarrow \underline{\quad} + {}^0_{+1}e$

39. Determine whether or not each of the following nuclides is likely to be stable. State your reasons.
    a. Mg-26   b. Ne-25   c. Co-51   d. Te-124

40. Determine whether or not each of the following nuclides is likely to be stable. State your reasons.
    a. Ti-48   b. Cr-63   c. Sn-102   d. Y-88

41. The first six elements of the first transition series have the following number of stable isotopes:

Element	Number of Stable Isotopes
Sc	1
Ti	5
V	1
Cr	3
Mn	1
Fe	4

    Explain why Sc, V, and Mn each has only one stable isotope while the other elements have several.

42. Neon and magnesium each has three stable isotopes while sodium and aluminum each has only one. Explain why this might be so.

43. Predict a likely mode of decay for each of the following unstable nuclides.
    a. Mo-109   b. Ru-90   c. P-27   d. Rn-196

44. Predict a likely mode of decay for each of the following unstable nuclides.
    a. Sb-132   b. Te-139   c. Fr-202   d. Ba-123

45. Which one of each of the following pair of nuclides would you expect to have the longest half-life?
    a. Cs-113 or Cs-125   b. Fe-62 or Fe-70

46. Which one of each of the following pair of nuclides would you expect to have the longest half-life?
    a. Cs-149 or Cs-139   b. Fe-45 or Fe-52

The Kinetics of Radioactive Decay and Radiometric Dating

47. One of the nuclides in spent nuclear fuel is U-235, an alpha emitter with a half-life of 703 million years. How long would it take for the amount of U-235 to reach one-eighth of its initial amount?

48. A patient is given 0.050 mg of technetium-99m, a radioactive isotope with a half-life of about 6.0 hours. How long until the radioactive isotope decays to 6.3×10^{-3} mg? (Assume no excretion of the nuclide from the body.)

49. A radioactive sample contains 1.55 g of an isotope with a half-life of 3.8 days. What mass of the isotope will remain after 5.5 days? (Assume no excretion of the nuclide from the body.)

50. At 8:00 A.M., a patient receives a 58-mg dose of I-131 to obtain an image of her thyroid. If the nuclide has a half-life of 8 days, what mass of the nuclide remains in the patient at 5:00 P.M. the next day?

51. A sample of F-18 has an initial decay rate of 1.5×10^5/s. How long will it take for the decay rate to fall to 1.0×10^2/s? (F-18 has a half-life of 1.83 hours.)

52. A sample of Tl-201 has an initial decay rate of 5.88×10^4/s. How long will it take for the decay rate to fall to 55/s? (Tl-201 has a half-life of 3.042 days.)

53. A wooden boat discovered just south of the Great Pyramid in Egypt has a carbon-14/carbon-12 ratio that is 72.5% of that found in living organisms. How old is the boat?

54. A layer of peat beneath the glacial sediments of the last ice age has a carbon-14/carbon-12 ratio that is 22.8% of that found in living organisms. How long ago was this ice age?

55. An ancient skull has a carbon-14 decay rate of 0.85 disintegrations per minute per gram of carbon (0.85 dis/min · g C). How old is the skull? (Assume that living organisms have a carbon-14 decay rate of 15.3 dis/min · g C and that carbon-14 has a half-life of 5730 yr.)

56. A mammoth skeleton has a carbon-14 decay rate of 0.48 disintegrations per minute per gram of carbon (0.48 dis/min · g C). When did the mammoth live? (Assume that living organisms have a carbon-14 decay rate of 15.3 dis/min · g C and that carbon-14 has a half-life of 5730 yr.)

57. A rock from Australia was found to contain 0.438 g of Pb-206 to every 1.00 g of U-238. Assuming that the rock did not contain any Pb-206 at the time of its formation, how old is the rock?

58. A meteor has a Pb-206:U-238 mass ratio of 0.855:1.00. What is the age of the meteor? (Assume that the meteor did not contain any Pb-206 at the the time of its formation.)

Fission, Fusion, and Transmutation

59. Write a nuclear reaction for the neutron-induced fission of U-235 to form Xe-144 and Sr-90. How many neutrons are produced in the reaction?

60. Write a nuclear reaction for the neutron-induced fission of U-235 to produce Te-137 and Zr-97. How many neutrons are produced in the reaction?

61. Write a nuclear equation for the fusion of two H-2 atoms to form He-3 and one neutron.

62. Write a nuclear equation for the fusion of H-3 with H-1 to form He-4.

63. A breeder nuclear reactor is a reactor in which nonfissile U-238 is converted into fissile Pu-239. The process involves bombardment of U-238 by neutrons to form U-239 which then undergoes two sequential beta decays. Write nuclear equations to represent this process.

64. Write a series of nuclear equations to represent the bombardment of Al-27 with a neutron to form a product that then undergoes an alpha decay followed by a beta decay.

65. Rutherfordium-257 was synthesized by bombarding Cf-249 with C-12. Write a nuclear equation for this reaction.

66. Element 107, now named bohrium, was synthesized by German researchers by colliding bismuth-209 with chromium-54 to form a bohrium isotope and one neutron. Write a nuclear equation to represent this reaction.

Energetics of Nuclear Reactions, Mass Defect, and Nuclear Binding Energy

67. If 1.0 g of matter were converted to energy, how much energy would be formed?

68. A typical home uses approximately 1.0×10^3 kWh of energy per month. If the energy came from a nuclear reaction, what mass would have to be converted to energy per year to meet the energy needs of the home?

69. Calculate the mass defect and nuclear binding energy per nucleon of the each of the nuclides indicated below.
    a. O-16 (atomic mass = 15.994915 amu)
    b. Ni-58 (atomic mass = 57.935346 amu)
    c. Xe-129 (atomic mass = 128.904780 amu)

70. Calculate the mass defect and nuclear binding energy per nucleon of the each of the nuclides indicated below.
    a. Li-7 (atomic mass = 7.016003 amu)
    b. Ti-48 (atomic mass = 47.947947 amu)
    c. Ag-107 (atomic mass = 106.905092 amu)

71. Calculate the quantity of energy produced per gram of U-235 (atomic mass = 235.043922 amu) for the neutron-induced fission of U-235 to form Xe-144 (atomic mass = 143.9385 amu) and Sr-90 (atomic mass =89.907738 amu) (discussed in Problem 59).

72. Calculate the quantity of energy produced per mole of U-235 (atomic mass = 235.043922 amu) for the neutron-induced fission of U-235 to produce Te-137 (atomic mass = 136.9253 amu) and Zr-97 (atomic mass = 96.910950 amu) (discussed in Problem 60).

73. Calculate the quantity of energy produced per gram of reactant for the fusion of two H-2 (atomic mass = 2.014102 amu) atoms to form He-3 (atomic mass = 3.016029 amu) and one neutron (discussed in Problem 61).

74. Calculate the quantity of energy produced per gram of reactant for the fusion of H-3 (atomic mass = 3.016049 amu) with H-1 (atomic mass =1.007825 amu) to form He-4 (atomic mass = 4.002603 amu) (discussed in Problem 62).

Effects and Applications of Radioactivity

75. A 75-kg human is exposed to 32.8 rad of radiation. How much energy was absorbed by the person's body? Compare this energy to the amount of energy absorbed by a person's body if they jumped from a chair to the floor (assume that all of the energy from the fall is absorbed by the person).

76. If a 55-gram laboratory mouse is exposed to 20.5 rad of radiation, how much energy was absorbed by the mouse's body?

77. PET studies require fluorine-18, which is produced in a cyclotron and decays with a half-life of 1.83 hours. Assuming that the F-18 can be transported at 60.0 miles/hour, how close must the hospital be to the cyclotron if 65% of the F-18 produced is to make it to the hospital?

78. Suppose a patient is given 155 mg of I-131, a beta emitter with a half-life of 8.0 days. Assuming that none of the I-131 is eliminated from the person's body in the first 4.0 hours of treatment, what is the exposure (in Ci) during those first four hours?

Cumulative Problems

79. Complete each of the following nuclear equations and calculate the energy change (in J/mol of reactant) associated with each one. (Be-9 = 9.012182 amu, Bi-209 = 208.980384 amu, He-4 = 4.002603 amu, Li-6 = 6.015122 amu, Ni-64 = 63.927969 amu, Rg-272 = 272.1535 amu, Ta-179 = 178.94593 amu, and W-179 = 178.94707 amu)

    a. _____ + $^9_4$Be $\longrightarrow$ $^6_3$Li + $^4_2$He
    b. $^{209}_{83}$Bi + $^{64}_{28}$Ni $\longrightarrow$ $^{272}_{111}$Rg + _____
    c. $^{179}_{74}$W + _____ $\longrightarrow$ $^{179}_{73}$Ta

80. Complete each of the following nuclear equations and calculate the energy change (in J/mol of reactant) associated with each one. (Al-27 = 26.981538 amu, Am-241 = 241.056822 amu, He-4 = 4.002603 amu, Np-237 = 237.048166 amu, P-30 = 29.981801 amu, S-32 = 31.972071 amu, and Si-29 = 28.976495 amu.)

    a. $^{27}_{13}$Al + $^4_2$He $\longrightarrow$ $^{30}_{15}$P + _____
    b. $^{32}_{16}$S + _____ $\longrightarrow$ $^{29}_{14}$Si + $^4_2$He
    c. $^{241}_{95}$Am $\longrightarrow$ $^{237}_{93}$Np + _____

81. Write a nuclear equation for the most likely mode of decay for each of the following unstable nuclides:
    a. Ru-114   b. Ra-216   c. Zn-58   d. Ne-31

82. Write a nuclear equation for the most likely mode of decay for each of the following unstable nuclides:
    a. Kr-74   b. Th-221   c. Ar-44   d. Nb-85

83. Bismuth-210 is beta emitter with a half-life of 5.0 days. If a sample contains 1.2 g of Bi-210 (atomic mass = 209.984105 amu), how many beta emissions would occur in 13.5 days? If a person's body intercepts 5.5% of those emissions, to what dose of radiation (in Ci) is the person exposed?

84. Polonium-218 is an alpha emitter with a half-life of 3.0 minutes. If a sample contains 55 mg of Po-218 (atomic mass = 218.008965 amu), how many alpha emissions would occur in 25.0 minutes? If the polonium were ingested by a person, to what dose of radiation (in Ci) would the person be exposed?

85. Radium-226 (atomic mass = 226.025402 amu) decays to radon-224 (a radioactive gas) with a half-life of 1.6×10^3 years. What volume of radon gas (at 25.0 °C and 1.0 atm) is produced by 25.0 g of radium in 5.0 days? (Report your answer to two significant digits.)

86. In one of the neutron-induced fission reactions of U-235 (atomic mass = 235.043922 amu), the products are Ba-140 and Kr-93 (a radioactive gas). What volume of Kr-93 (at 25.0 °C and 1.0 atm) is produced when 1.00 g of U-235 undergoes this fission reaction?

87. When a positron and an electron annihilate one another, the resulting mass is completely converted to energy.
    a. Calculate the energy associated with this process in kJ/mol.
    b. If all of the energy from the annihilation were carried away by two gamma ray photons, what would be the wavelength of the photons?

88. A typical nuclear reactor produces about 1.0 MW of power per day. What is the minimum rate of mass loss required to produce this much energy?

89. Find the binding energy in an atom of ^{3}He, which has a mass of 3.016030 amu.

90. The overall hydrogen burning reaction in stars can be represented as the conversion of four protons to one α particle. Use the data for the mass of H-1 and He-4 to calculate the energy released by this process.

91. The nuclide ^{247}Es can be made by bombardment of ^{238}U in a reaction that emits five neutrons. Identify the bombarding particle.

92. The nuclide ^{6}Li reacts with ^{2}H to form two identical particles. Identify the particles.

93. The half-life of ^{238}U is 4.5×10^9 yr. A sample of rock of mass 1.6 g is found to produce 29 dis/s. Assuming all the radioactivity is due to ^{238}U, find the percent by mass of ^{238}U in the rock.

94. The half-life of ^{232}Th is 1.4×10^{10} yr. Find the number of disintegrations per hour emitted by 1.0 mol of ^{232}Th in 1 minute.

Challenge Problems

95. The space shuttle carries about 72,500 kg of solid aluminum fuel, which is oxidized with ammonium perchlorate according to the following reaction:

$$10 \, Al(s) + 6 \, NH_4ClO_4(s) \longrightarrow$$
$$4 \, Al_2O_3(s) + 2 \, AlCl_3(s) + 12 \, H_2O(g) + 3 \, N_2(g)$$

The space shuttle also carries about 608,000 kg of oxygen (which reacts with hydrogen to form gaseous water).

  a. Assuming that aluminum and oxygen are the limiting reactants, determine the total energy produced by these fuels. (ΔH_f° for solid ammonium perchlorate is -295 kJ/mol.)

  b. Suppose that a future space shuttle was powered by matter–antimatter annihilation. The matter could be normal hydrogen (containing a proton and an electron) and the antimatter could be antihydrogen (containing an antiproton and a positron). What mass of antimatter would be required to produce the energy equivalent of the aluminum and oxygen fuel currently carried on the space shuttle?

96. Suppose that an 85.0-gram laboratory animal ingested 10.0 mg of a substance that contained 2.55% by mass Pu-239, an alpha emitter with a half-life of 24,110 years.

  a. What is the animal's initial radiation exposure in Curies?

  b. If all of the energy from the emitted alpha particles is absorbed by the animal's tissues, and if the energy of each emission is 7.77×10^{-12} J what is the dose in rads to the animal in the first 4.0 hours following the ingestion of the radioactive material? Assuming a biological effectiveness factor of 20, what is the 4.0-hour dose in rems?

97. In addition to the natural radioactive decay series that begins with U-238 and ends with Pb-206, there are natural radioactive decay series that begin with U-235 and Th-232. Both of these series end with nuclides of Pb. Predict the likely end product of each series and the number of α decay steps that occur.

Conceptual Problems

98. Closely examine the following diagram representing the beta decay of fluorine-21 and draw in the missing nucleus.

$${}^{21}_{9}F \longrightarrow \quad + \quad {}^{0}_{-1}e$$

99. Approximately how many half-lives must pass for the amount of radioactivity in a substance to decrease to below 1% of its initial level?

100. A person is exposed for three days to identical amounts of two different nuclides that emit positrons of roughly equal energy. The half-life of nuclide A is 18.5 days and the half-life of nuclide B is 255 days. Which of the two nuclides poses the greater health risk?

101. Identical amounts of two different nuclides, an alpha emitter and a gamma emitter, with roughly equal half-lives are spilled in a building adjacent to your bedroom. Which of the two nuclides poses the greater health threat to you while you sleep in your bed? If you accidentally wander into the building and ingest equal amounts of the two nuclides, which one poses the greater health threat?

102. Drugstores in many areas now carry tablets, under such trade names as Iosat and NoRad, designed to be taken in the event of an accident at a nuclear power plant or a terrorist attack that release radioactive material. These tablets contain potassium iodide (KI). Can you explain the nature of the protection that they provide? (Hint: see the label in the photo.)

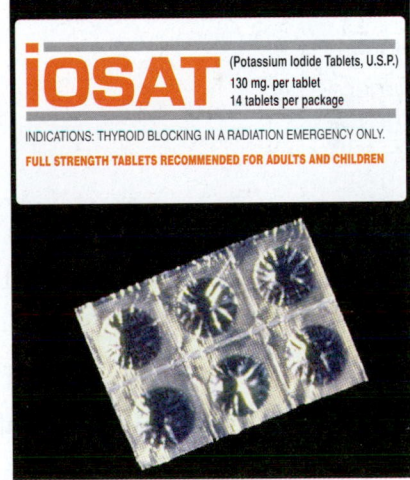

20

Organic Chemistry

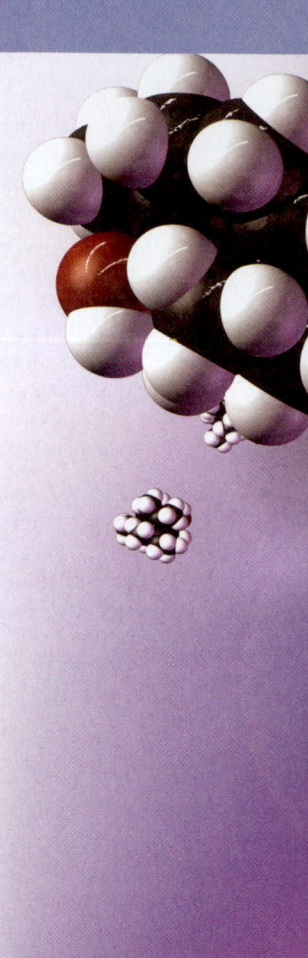

20.1 Fragrances and Odors

20.2 Carbon: Why It Is Unique

20.3 Hydrocarbons: Compounds Containing Only Carbon and Hydrogen

20.4 Alkanes: Saturated Hydrocarbons

20.5 Alkenes and Alkynes

20.6 Hydrocarbon Reactions

20.7 Aromatic Hydrocarbons

20.8 Functional Groups

20.9 Alcohols

20.10 Aldehydes and Ketones

20.11 Carboxylic Acids and Esters

20.12 Ethers

20.13 Amines

20.14 Polymers

O RGANIC CHEMISTRY IS THE STUDY of carbon-containing compounds. Since life has organized itself around organic compounds, organic chemistry is critical to the study of life. Carbon is unique in the sheer number of compounds that it can form. Millions of organic compounds are known, and new ones are discovered every day. Carbon is also unique in the diversity of compounds that it can form. In most cases, a fixed number of carbon atoms can combine with a fixed number of atoms of another element to form a large number of different compounds. For example, 10 carbon atoms and 22 hydrogen atoms can form 75 distinctly different compounds. In other words, with carbon as the backbone, nature can take the same combination of atoms and bond them together in slightly different ways to produce a huge diversity of substances. It is not surprising that life is based on the chemistry of carbon because life needs diversity to exist, and organic chemistry is nothing if it is not diverse. In this chapter, we peer into Friedrich Wöhler's "primeval tropical forest" (see the quotation on the facing page) and will discover the most remarkable things.

Organic chemistry just now is enough to drive one mad. It gives one the impression of a primeval, tropical forest full of the most remarkable things. . . .

—FRIEDRICH WÖHLER (1800–1882)

About half of all men's colognes contain at least some patchouli alcohol ($C_{15}H_{26}O$), an organic compound (pictured here) derived from the patchouli plant. Patchouli alcohol has a pungent, musty, earthy fragrance.

20.1 Fragrances and Odors

Have you ever ridden an elevator with someone wearing too much perfume? Or accidentally gotten too close to a skunk? Or gotten a whiff of rotting fish? What causes these fragrances and odors? When we inhale certain molecules called odorants, they bind with olfactory receptors in our noses. This interaction sends a nerve signal to the brain that we experience as a smell. Some smells, such as that of perfume, are pleasant (when not overdone). Other smells, such as that of the skunk or rotting fish, are unpleasant. Our sense of smell helps us identify food, people, and other organisms, and alerts us to dangers such as polluted air or spoiled food. Smell (or olfaction) is one way we probe the environment around us.

CH₃CH=CHCH₂SH
2-Butene-1-thiol

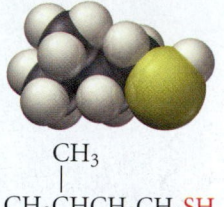

CH_3
|
$CH_3CHCH_2CH_2SH$
3-Methyl-1-butanethiol

▲ The smell of skunk is due primarily to the molecules shown here.

Odorants, if they are to reach our noses, must be volatile. However, many volatile substances have no scent at all. Nitrogen, oxygen, water, and carbon dioxide molecules, for example, are constantly passing through our noses, yet they produce no smell because they do not bind to olfactory receptors. Most common smells are caused by **organic molecules**, molecules containing carbon combined with several other elements including hydrogen, nitrogen, oxygen, and sulfur. For example, organic molecules are responsible for the smells of roses, vanilla, cinnamon, almond, jasmine, body odor, and rotting fish. When you wander into a rose garden, you experience the sweet smell caused in part by geraniol, an organic compound emitted by roses. Men's colognes often contain patchouli alcohol, an earthy-smelling organic compound that can be extracted from the patchouli plant. If you have been in the vicinity of skunk spray (or have been unfortunate enough to be sprayed yourself), you are familiar with the unpleasant smell caused primarily by 2-butene-1-thiol and 3-methyl-1-butanethiol, two particularly odoriferous compounds present in the secretion that skunks use to defend themselves.

The study of compounds containing carbon combined with one or more of the elements mentioned previously (that is, hydrogen, nitrogen, oxygen, and sulfur), including their properties and their reactions, is known as **organic chemistry**. Besides composing much of what we smell, organic compounds are prevalent in foods, drugs, petroleum products, and pesticides. Organic chemistry is also the basis for living organisms. Life has evolved based on carbon-containing compounds, making organic chemistry of utmost importance to any person interested in understanding living organisms.

20.2 Carbon: Why It Is Unique

Why did life evolve based on the chemistry of carbon? Why is life not based on some other element? The answer may not be simple, but we know that life—in order to exist—must have complexity. It is also clear that carbon chemistry is complex. The number of compounds containing carbon is greater than the number of compounds of all the rest of the elements combined. There are several reasons for this behavior including carbon's ability to form four covalent bonds, its ability to form double and triple bonds, and its tendency to *catenate* (that is, to form chains).

Carbon's Tendency to Form Four Covalent Bonds Carbon—with its four valence electrons—can form four covalent bonds. Consider the Lewis structure and space-filling models of two simple carbon compounds, methane and ethane.

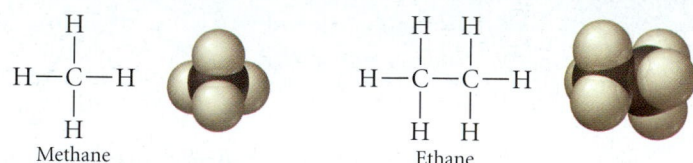

The geometry about a carbon atom forming four single bonds is tetrahedral, as shown above for methane. Carbon's ability to form four bonds, and to form those bonds with a number of different elements, results in the potential to form many different compounds. As you learn to draw structures for organic compounds, always remember to draw carbon with four bonds.

Carbon's Ability to Form Double and Triple Bonds Carbon atoms can also form double bonds (trigonal planar geometry) and triple bonds (linear geometry), adding even more diversity to the number of compounds that carbon can form.

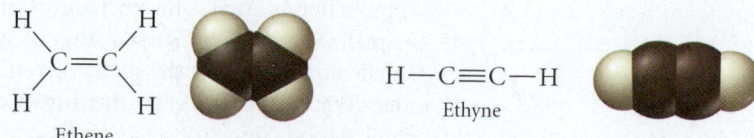

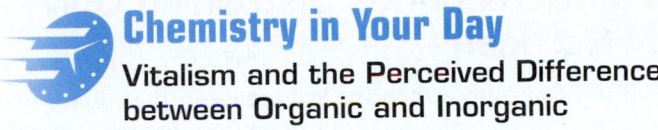

Chemistry in Your Day
Vitalism and the Perceived Difference between Organic and Inorganic

By the end of the eighteenth century, chemists had learned that compounds could be broadly divided into two categories: organic and inorganic. Organic compounds came from living things. Sugar—obtained from sugarcane or the sugar beet—is a common example of an organic compound. Inorganic compounds, in contrast, came from the earth. Salt—mined from the ground or extracted from ocean water—is a common example of an inorganic compound.

Organic and inorganic compounds were different, not only in their origin, but also in their properties. Organic compounds were easily decomposed. Sugar, for example, easily decomposes into carbon and water when heated. (Think of the last time you burned something sugary in a pan or in the oven.) Inorganic compounds, however, were more difficult to decompose. Salt must be heated to very high temperatures before it decomposes. Even more curious to these early chemists was their inability to synthesize a single organic compound in the laboratory. Many inorganic compounds could be easily synthesized, but organic compounds could not.

The origin and properties of organic compounds led early chemists to postulate that organic compounds were unique to living organisms. They hypothesized that living organisms contained a *vital force*—a mystical or supernatural power—that allowed them to produce organic compounds. They thought that it was impossible to produce an organic compound outside of a living organism because the vital force was not present.

This belief—which became known as *vitalism*—explained why no chemist had succeeded in synthesizing an organic compound in the laboratory.

An experiment performed in 1828 by German chemist Friedrich Wöhler (1800–1882) marked the beginning of the end of vitalism. Wöhler heated ammonium cyanate (an inorganic compound) and formed urea (an organic compound).

$$NH_4OCN \xrightarrow{\text{heat}} H_2NCONH_2$$
ammonium cyanate   urea

Urea was a known organic compound that had previously been isolated only from urine. Although it was not realized at the time, Wöhler's simple experiment was a key step in opening all of life to scientific investigation. He showed that the compounds composing living organisms—like all compounds—followed scientific laws and could be studied and understood. Today, known organic compounds number in the millions, and modern organic chemistry is a vast field that produces substances as diverse as drugs, petroleum products, and plastics.

Freidrich Wöhler

▶ The synthesis of urea in 1828 by German chemist Friedrich Wöhler marked the beginning of the end for vitalism.

Urea

In contrast, silicon (the element in the periodic table with properties closest to that of carbon) does not readily form double or triple bonds because the greater size of silicon atoms results in a Si — Si bond that is too long for much overlap between nonhybridized *p* orbitals.

Carbon's Tendency to Catenate Carbon, more than any other element, can bond to itself to form chain, branched, and ring structures.

Propane   Isobutane   Cyclohexane

Although other elements can form chains, none beats carbon at this ability. Silicon, for example, can form chains with itself. However, silicon's affinity for oxygen (the Si — O bond is 142 kJ/mol stronger than the Si — Si bond) coupled with the prevalence of oxygen in our atmosphere means that silicon–silicon chains are readily oxidized to form silicates (the silicon–oxygen compounds that compose a significant proportion of minerals). By contrast, the C — C bond (347 kJ/mol) and the C — O bond (359 kJ/mol) are nearly the same strength, allowing carbon chains to exist relatively peacefully in an oxygen-rich environment. In other words, silicon's affinity for oxygen robs it of the rich diversity that catenation provides to carbon.

20.3 Hydrocarbons: Compounds Containing Only Carbon and Hydrogen

Hydrocarbons—compounds that contain only carbon and hydrogen—are the simplest organic compounds. However, because of the uniqueness of carbon (just discussed), many different kinds of hydrocarbons exist. Hydrocarbons are commonly used as fuels. Candle wax, oil, gasoline, liquid propane (LP) gas, and natural gas are all composed of hydrocarbons. Hydrocarbons are also the starting materials in the synthesis of many different consumer products including fabrics, soaps, dyes, cosmetics, drugs, plastic, and rubber.

As shown in Figure 20.1 ▼, hydrocarbons can be classified into four different types: **alkanes**, **alkenes**, **alkynes**, and **aromatic hydrocarbons**. Alkanes, alkenes, and alkynes— also called **aliphatic hydrocarbons**—are differentiated based on the kinds of bonds between carbon atoms. As shown in Table 20.1, alkanes have only single bonds between carbon atoms, alkenes have a double bond, and alkynes have a triple bond.

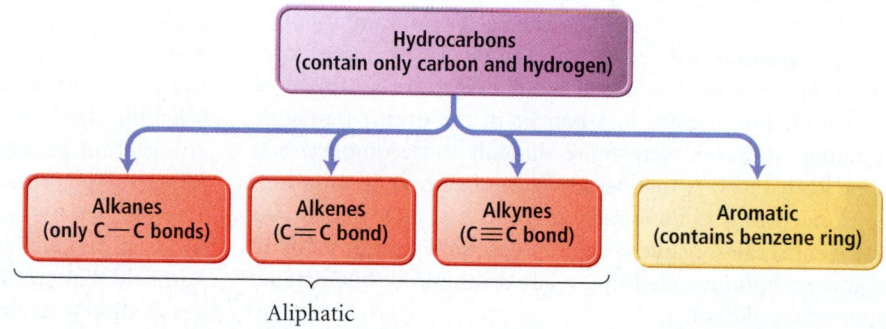

Hydrocarbons
(contain only carbon and hydrogen)

| Alkanes (only C—C bonds) | Alkenes (C=C bond) | Alkynes (C≡C bond) | Aromatic (contains benzene ring) |

Aliphatic

▲ **FIGURE 20.1** Four Types of Hydrocarbons

TABLE 20.1	Alkanes, Alkenes, Alkynes		
Type of Hydrocarbon	Type of bonds	Generic Formula*	Example
Alkane	All single	C_nH_{2n+2}	Ethane
Alkenes	One (or more) double	C_nH_{2n}	Ethene
Alkynes	One (or more) triple	C_nH_{2n-2}	Ethyne

* n is the number of carbon atoms. These formulas apply only to noncyclic structures containing no more than one multiple bond.

Drawing Hydrocarbon Structures

Throughout this book, we have relied primarily on molecular formulas as the simplest way to represent compounds. In organic chemistry, however, molecular formulas are insufficient because, as we have already discussed, the same atoms can bond together in different ways to form different compounds. For example, consider an alkane with 4 carbon atoms and 10 hydrogen atoms. Two different structures, named butane and isobutane, are possible.

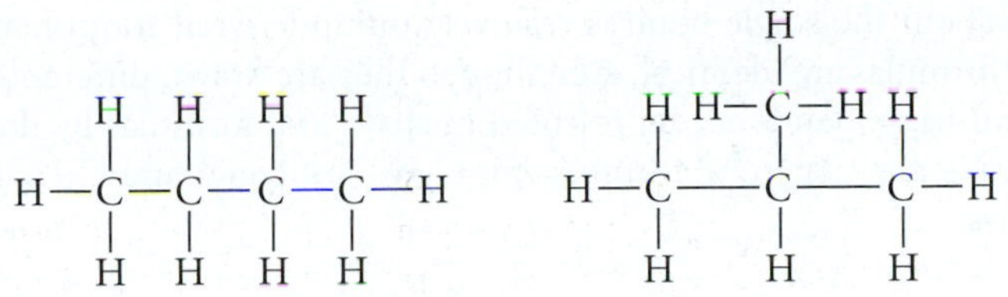

Butane and isobutane are **structural isomers**, molecules with the same molecular formula but different structures. Because of their different structures, they have different properties—they are indeed different compounds. Isomerism is ubiquitous in organic chemistry. Butane has 2 structural isomers. Pentane (C_5H_{12}) has 3, hexane (C_6H_{14}) has 5, and decane ($C_{10}H_{22}$) has 75!

The structure of a particular hydrocarbon is represented with a **structural formula**, a formula that shows not only the numbers of each kind of atoms, but also how the atoms are bonded together. Organic chemists use several different kinds of structural formulas. For example, we can represent butane and isobutane in each of the following ways:

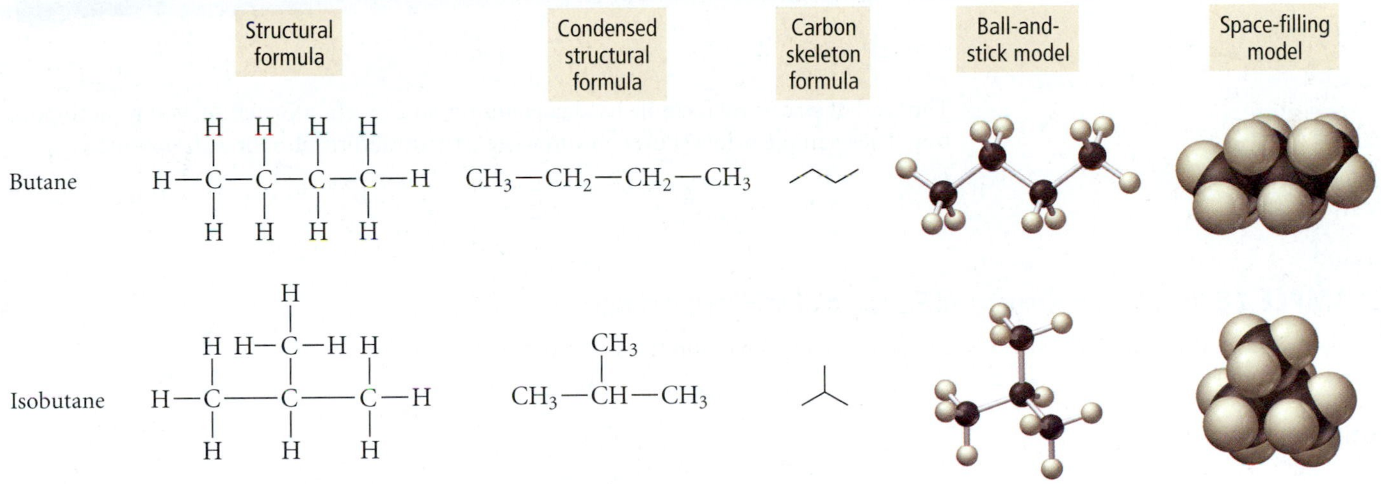

Structural formula	Condensed structural formula	Carbon skeleton formula	Ball-and-stick model	Space-filling model

Butane $CH_3-CH_2-CH_2-CH_3$

Isobutane $CH_3-CH-CH_3$ with CH_3

The structural formula shows all of the carbon and hydrogen atoms in the molecule and how they are bonded together. The condensed structural formula groups the hydrogen atoms together with the carbon atom to which they are bonded. Condensed structural formulas may show some of the bonds (as above) or none at all. For example, the condensed structural formula for butane can also be written as $CH_3CH_2CH_2CH_3$. The carbon skeleton formula (also called a line formula) shows the carbon–carbon bonds only as lines. Each end or bend of a line represents a carbon atom bonded to as many hydrogen atoms as necessary to form a total of four bonds.

Note that structural formulas are generally not three-dimensional representations of the molecule—as space-filling or ball-and-stick models are—but rather two-dimensional representations that show how atoms are bonded together. As such, the most important feature of a structural formula is the *connectivity* of the atoms, not the exact way the formula is drawn. For example, consider the following two condensed structural formulas for butane and the corresponding space-filling models below them:

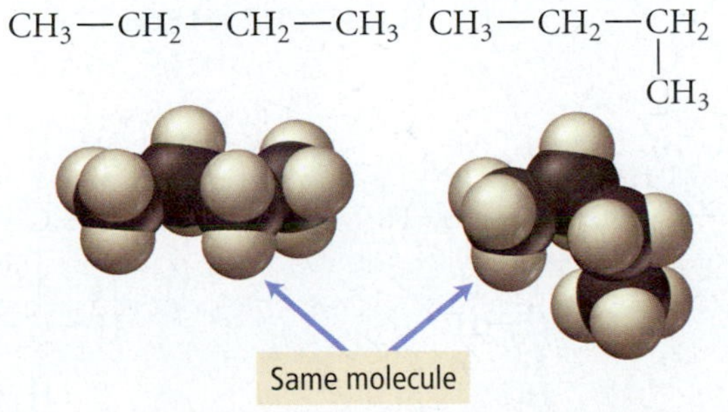

$CH_3-CH_2-CH_2-CH_3$ $CH_3-CH_2-CH_2$ with CH_3

Same molecule

Since rotation about the single bond is relatively unhindered at room temperature, the two structural formulas are identical, even though they are drawn differently.

Double and triple bonds are represented in structural formulas by double or triple lines. For example, the structural formulas for C_3H_6 (propene) and C_3H_4 (propyne) are drawn as follows:

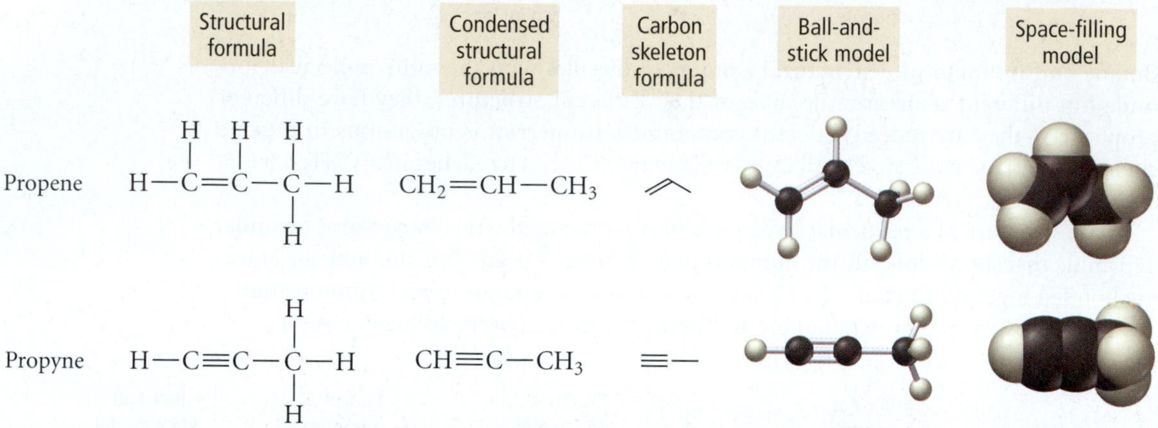

	Structural formula	Condensed structural formula	Carbon skeleton formula	Ball-and-stick model	Space-filling model

Propene
$$\begin{array}{c} H \quad H \quad H \\ | \quad\ | \quad\ | \\ H-C=C-C-H \\ | \\ H \end{array}$$
$CH_2{=}CH{-}CH_3$

Propyne
$$\begin{array}{c} H \\ | \\ H-C{\equiv}C-C-H \\ | \\ H \end{array}$$
$CH{\equiv}C{-}CH_3$

The kind of structural formula used depends on how much information you want to portray. The example below shows how to write structural formulas for a compound.

EXAMPLE 20.1 Writing Structural Formulas for Hydrocarbons

Write the structural formulas and carbon skeleton formulas for the five isomers of C_6H_{12} (hexane).

Solution

To start, draw the carbon backbone of the straight-chain isomer.	C—C—C—C—C—C
Next, determine the carbon backbone structure of the other isomers by arranging the carbon atoms in four other unique ways.	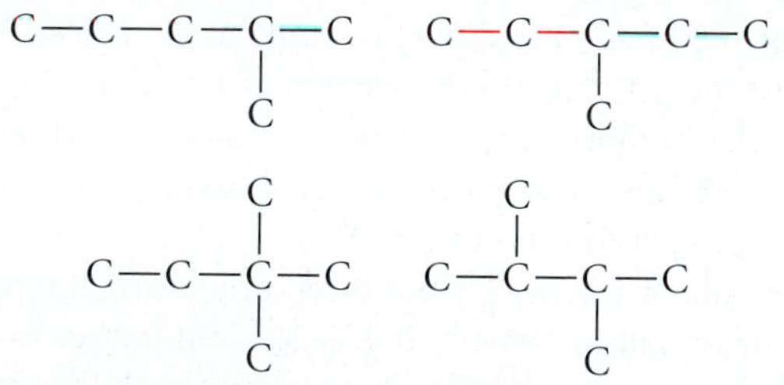 (carbon skeleton formulas)
Fill in all the hydrogen atoms so that each carbon has four bonds.	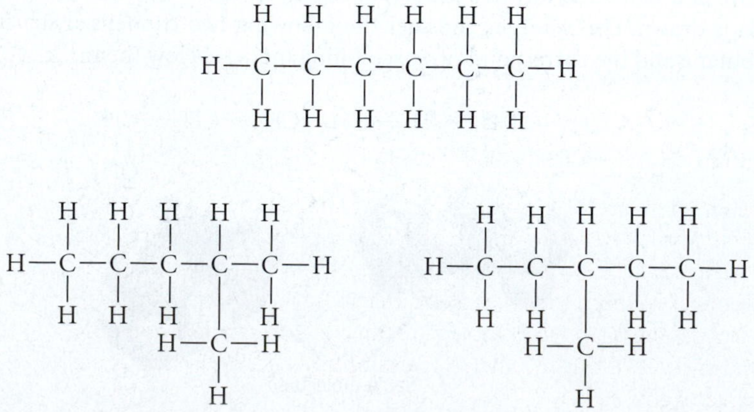 (structural formulas)

$$\begin{array}{cccc} C-C-C-C-C & \quad & C-C-C-C-C \\ \quad\ | & & \quad\ | \\ \quad\ C & & \quad\ C \end{array}$$

$$\begin{array}{cccc} C & & C \\ | & & | \\ C-C-C-C & \quad & C-C-C-C \\ | & & | \\ C & & C \end{array}$$

$$\begin{array}{c} H \quad H \quad H \quad H \quad H \quad H \\ | \quad\ | \quad\ | \quad\ | \quad\ | \quad\ | \\ H-C-C-C-C-C-C-H \\ | \quad\ | \quad\ | \quad\ | \quad\ | \quad\ | \\ H \quad H \quad H \quad H \quad H \quad H \end{array}$$

$$\begin{array}{c} H \quad H \quad H \quad H \quad H \\ | \quad\ | \quad\ | \quad\ | \quad\ | \\ H-C-C-C-C-C-H \\ | \quad\ | \quad\ | \quad\ \ | \\ H \quad H \quad H \quad\ \ H \\ \qquad\qquad H-C-H \\ \qquad\qquad\ \ | \\ \qquad\qquad\ \ H \end{array}$$

$$\begin{array}{c} H \quad H \quad H \quad H \quad H \\ | \quad\ | \quad\ | \quad\ | \quad\ | \\ H-C-C-C-C-C-H \\ | \quad\ | \quad\ | \quad\ | \quad\ | \\ H \quad H \quad\ \ H \quad H \\ \qquad\quad H-C-H \\ \qquad\quad\ \ | \\ \qquad\quad\ \ H \end{array}$$

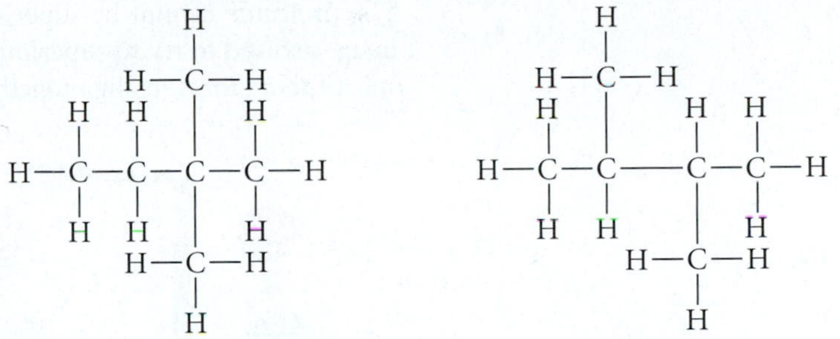

Write the carbon skeleton formulas by using lines to represent each carbon–carbon bond. Remember that each end or bend represents a carbon atom.

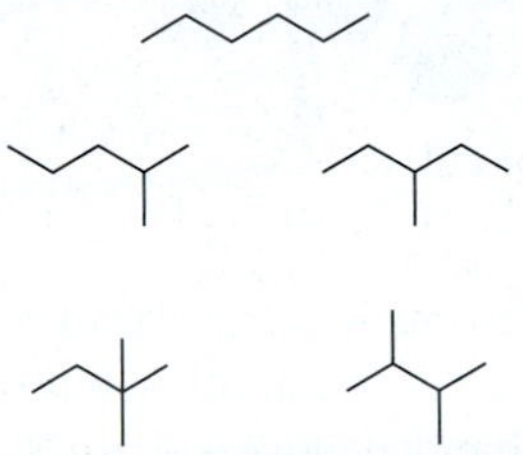

For Practice 20.1

Write the structural formulas and carbon skeleton formulas for the three isomers of C_5H_{12} (pentane).

⬥ Conceptual Connection 20.1 Organic Structures

Which of the following structures is an *isomer* of $CH_3-CH-CH_2-CH-CH_3$ (and not just the same structure)?

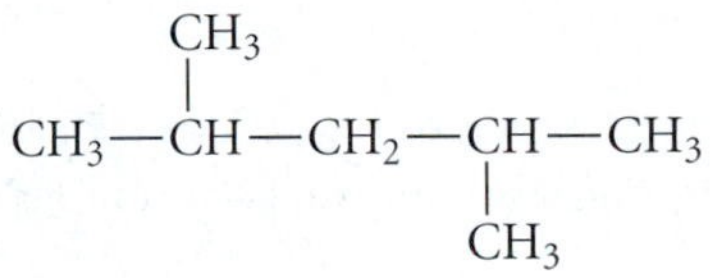

with substituents CH_3 on the second carbon and CH_3 on the fourth carbon

(a) 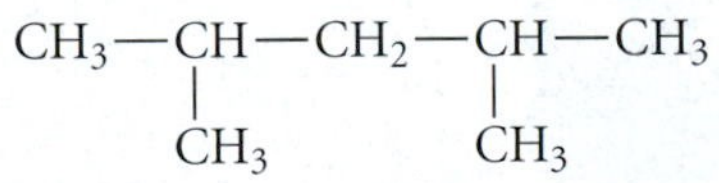 $CH_3-CH-CH_2-CH-CH_3$ with CH_3, CH_3 substituents

(b) 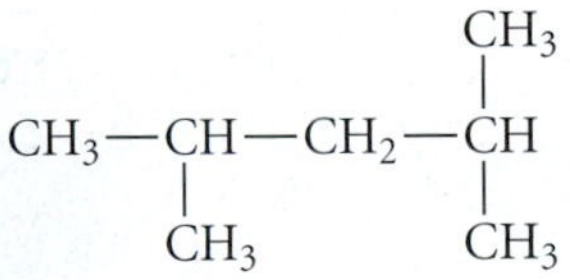 $CH_3-CH-CH_2-CH$ with CH_3, CH_3, and CH_3 substituents

(c) 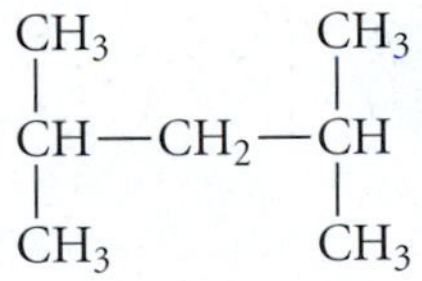 $CH-CH_2-CH$ with CH_3, CH_3, CH_3, CH_3 substituents

(d) 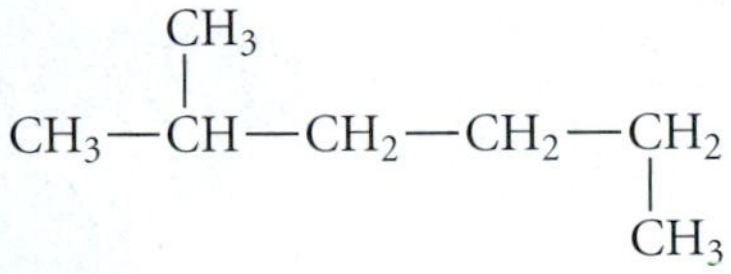 $CH_3-CH-CH_2-CH_2-CH_2$ with CH_3 and CH_3 substituents

Answer: **(d)** The others are just the same structure drawn in slightly different ways.

Stereoisomerism and Optical Isomerism

Stereoisomers are molecules in which the atoms have the same connectivity, but have a different spatial arrangement. Stereoisomers can themselves be of two types: geometric (or cis–trans) isomers, and optical isomers. We discuss geometric isomers in Section 20.5. **Optical isomers** are two molecules that are nonsuperimposable mirror images of one another. For example, consider the molecule shown at right:

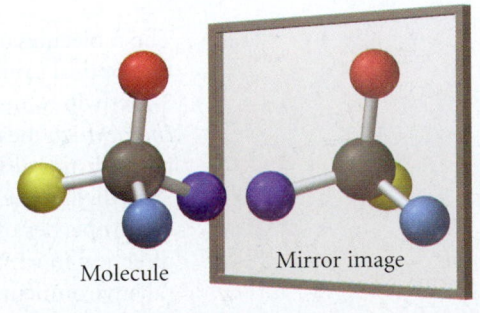

Molecule   Mirror image

The molecule cannot be superimposed onto its mirror image. If you swing the mirror image around to try to superimpose the two, you find that there is no way to get all four substituent atoms to align together.

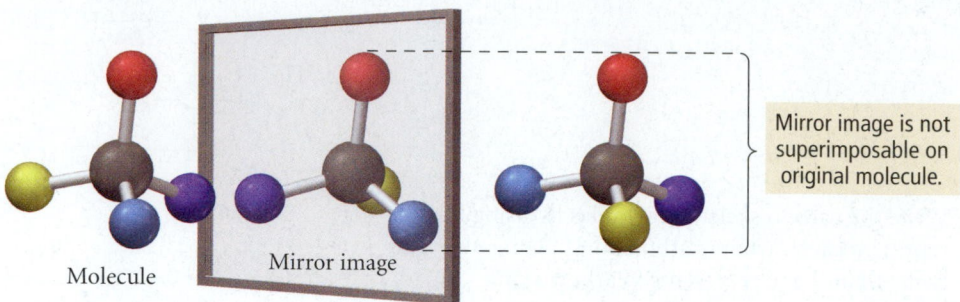

Molecule   Mirror image

Mirror image is not superimposable on original molecule.

Optical isomers are similar to your right and left hands (Figure 20.2 ▼). The two are mirror images of one another, but you cannot superimpose one on the other. For this reason, a right-handed glove will not fit on your left hand and vice versa.

► FIGURE 20.2 Mirror Images
Your left and right hands are nonsuperimposable mirror images, just as are optical isomers.

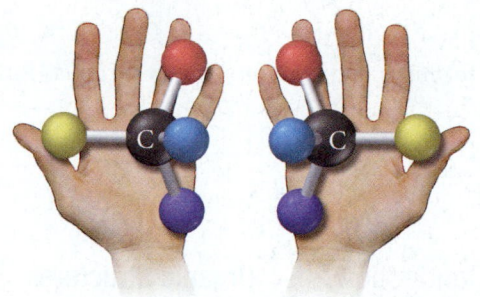

Any carbon atom with four different substituents in a tetrahedral arrangement will exhibit optical isomerism. For example, consider 3-methylhexane:

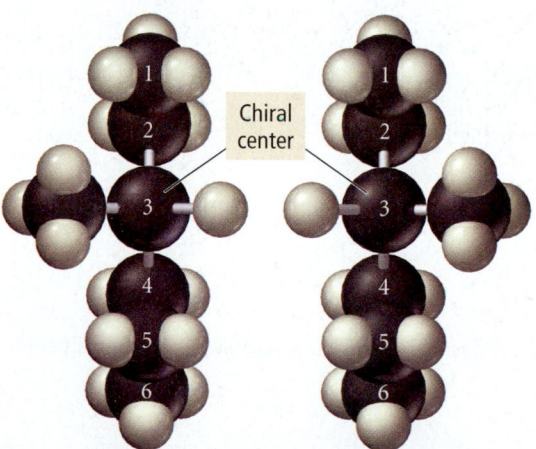

Chiral center

Optical isomers of 3-methylhexane

The molecules on the left and right are nonsuperimposable mirror images and are therefore optical isomers of one other; they are also called **enantiomers**. Any molecule, such as 3-methylhexane, that exhibits optical isomerism is said to be **chiral**, from the Greek word *cheir*, which means "hand." Some of the physical and chemical properties of entantiomers are indistinguishable from one another. For example, both of the optical isomers of 3-methylhexane have identical freezing points, melting points, and densities. However, the properties of enantiomers differ from one another in two important ways: (1) in the direction in which they rotate polarized light and (2) in their chemical behavior in a chiral environment.

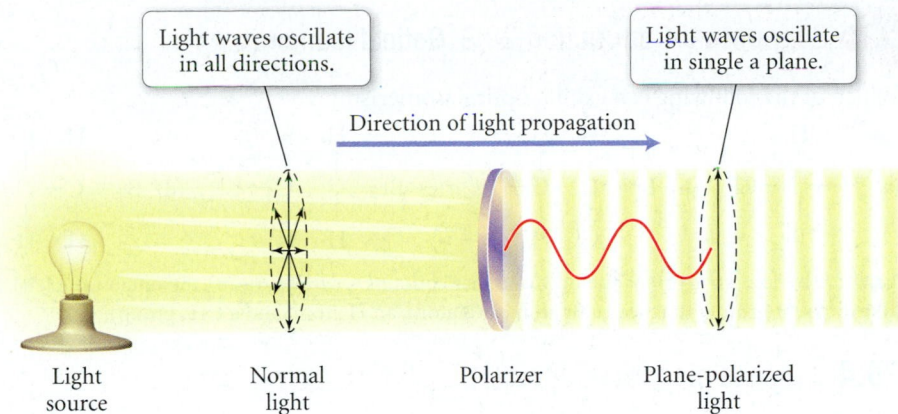

▲ **FIGURE 20.3 Plane-Polarized Light**   The electric field of plane-polarized light oscillates in one plane.

Rotation of Polarized Light *Plane-polarized light* is light whose electric field waves oscillate in only one plane as shown in Figure 20.3 ▲. When a beam of plane-polarized light is directed through a sample containing only one of two optical isomers, the plane of polarization of the light is rotated as shown in Figure 20.4 ▼. One of the two optical isomers rotates the polarization of the light clockwise and is called the **dextrorotatory** isomer (or the *d* isomer). The other isomer rotates the polarization of the light counterclockwise and is called the **levorotatory** isomer (or the *l* isomer). An equimolar mixture of both optical isomers does not rotate the polarization of light at all and is called a **racemic mixture**.

Dextrorotatory means turning clockwise or to the right. *Levorotatory* means turning counterclockwise or to the left.

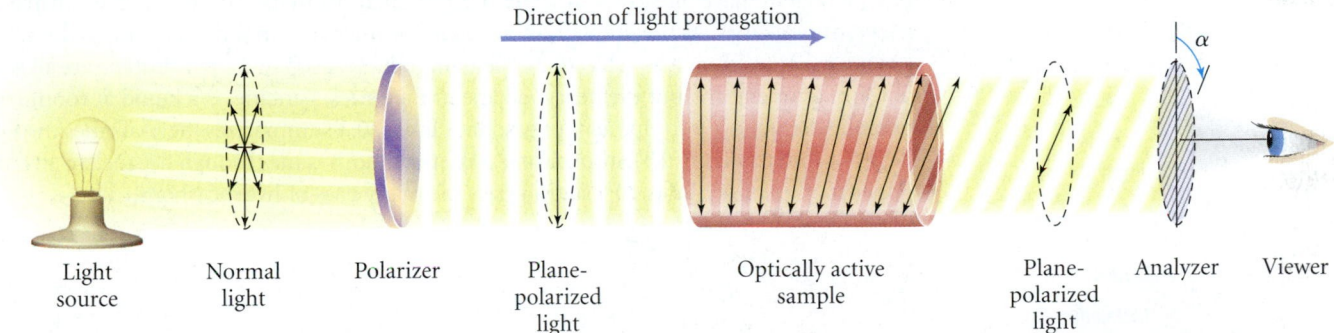

▲ **FIGURE 20.4 Rotation of Plane-Polarized Light**   Plane-polarized light is rotated as it passes through a sample containing only one of two optical isomers.

Chemical Behavior in a Chiral Environment Optical isomers will exhibit different chemical behavior when they are in a chiral environment. For example, enzymes are large biological molecules that catalyze reactions in living organisms. They provide chiral environments that distinguish between the two optical isomers. Consider the following simplified picture of two enantiomers in a chiral environment.

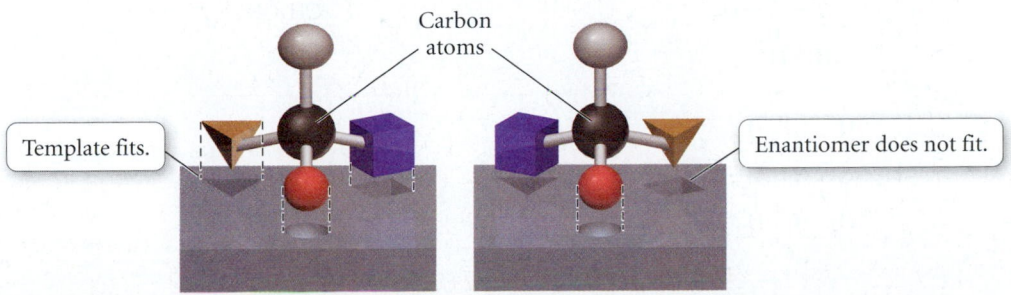

One of the enantiomers fits the template, but the other does not, no matter how it is rotated. In a similar manner, an enzyme will catalyze the reaction of one enantiomer because it fits the "template" but not the other.

Conceptual Connection 20.2 Optical Isomers

Which of the following can exhibit optical isomerism?

(a) $H-\overset{\displaystyle H}{\underset{\displaystyle Br}{\overset{|}{\underset{|}{C}}}}-Cl$ (b) $Br-\overset{\displaystyle H}{\underset{\displaystyle Cl}{\overset{|}{\underset{|}{C}}}}-\overset{\displaystyle H}{\underset{\displaystyle H}{\overset{|}{\underset{|}{C}}}}-H$ (c) $Br-\overset{\displaystyle H}{\underset{\displaystyle H}{\overset{|}{\underset{|}{C}}}}-\overset{\displaystyle H}{\underset{\displaystyle H}{\overset{|}{\underset{|}{C}}}}-Cl$ (d) $Br-\overset{\displaystyle H}{\underset{\displaystyle H}{\overset{|}{\underset{|}{C}}}}-\overset{\displaystyle Cl}{\underset{\displaystyle H}{\overset{|}{\underset{|}{C}}}}-Cl$

Answer: (b) This structure is the only one that contains a carbon atom (the one on the left) with four different groups attached (a Br atom, a Cl atom, an H atom, and a CH_3 group).

20.4 Alkanes: Saturated Hydrocarbons

As we have seen, alkanes are hydrocarbons containing only single bonds. Alkanes are often called **saturated hydrocarbons** because they are saturated (loaded to capacity) with hydrogen. The simplest hydrocarbons are methane (CH_4), the main component of natural gas, ethane (C_2H_6), a minority component in natural gas, and propane (C_3H_8), the main component of liquid propane (LP) gas.

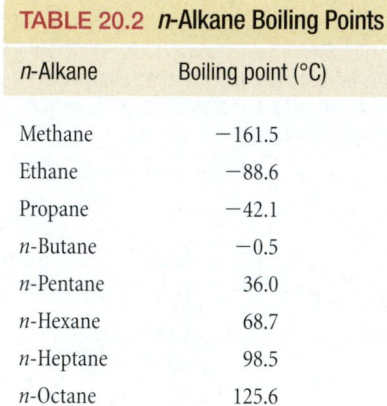

Methane   Ethane   Propane

For alkanes containing four or more carbon atoms, branching of the carbon chain becomes possible (as we have already seen). The straight-chain isomers are often called normal alkanes, or n-alkanes. As the number of carbon atoms increases in the n-alkanes, so does their boiling point (as shown in Table 20.2). Methane, ethane, propane, and n-butane are all gases at room temperature, but the next n-alkane in the series, pentane, is a liquid at room temperature. Pentane is a component of gasoline. Table 20.3 summarizes the n-alkanes through decane, which contains 10 carbon atoms. Like pentane, hexane through decane are all components of gasoline. Table 20.4 summarizes the many uses of hydrocarbons.

TABLE 20.2 n-Alkane Boiling Points

n-Alkane	Boiling point (°C)
Methane	−161.5
Ethane	−88.6
Propane	−42.1
n-Butane	−0.5
n-Pentane	36.0
n-Hexane	68.7
n-Heptane	98.5
n-Octane	125.6

TABLE 20.3 n-Alkanes

n	Name	Molecular Formula C_nH_{2n+2}	Structural Formula	Condensed Structural Formula
1	Methane	CH_4	$H-\overset{\displaystyle H}{\underset{\displaystyle H}{\overset{|}{\underset{|}{C}}}}-H$	CH_4
2	Ethane	C_2H_6	$H-\overset{\displaystyle H}{\underset{\displaystyle H}{\overset{|}{\underset{|}{C}}}}-\overset{\displaystyle H}{\underset{\displaystyle H}{\overset{|}{\underset{|}{C}}}}-H$	CH_3CH_3
3	Propane	C_3H_8	$H-\overset{\displaystyle H}{\underset{\displaystyle H}{\overset{|}{\underset{|}{C}}}}-\overset{\displaystyle H}{\underset{\displaystyle H}{\overset{|}{\underset{|}{C}}}}-\overset{\displaystyle H}{\underset{\displaystyle H}{\overset{|}{\underset{|}{C}}}}-H$	$CH_3CH_2CH_3$
4	n-Butane	C_4H_{10}	$H-\overset{\displaystyle H}{\underset{\displaystyle H}{\overset{|}{\underset{|}{C}}}}-\overset{\displaystyle H}{\underset{\displaystyle H}{\overset{|}{\underset{|}{C}}}}-\overset{\displaystyle H}{\underset{\displaystyle H}{\overset{|}{\underset{|}{C}}}}-\overset{\displaystyle H}{\underset{\displaystyle H}{\overset{|}{\underset{|}{C}}}}-H$	$CH_3CH_2CH_2CH_3$

| 5 | n-Pentane | C_5H_{12} | 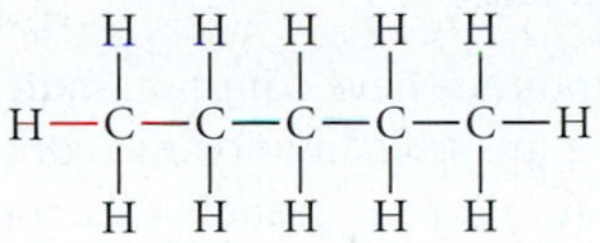 | $CH_3CH_2CH_2CH_2CH_3$ |

| 6 | n-Hexane | C_6H_{14} | 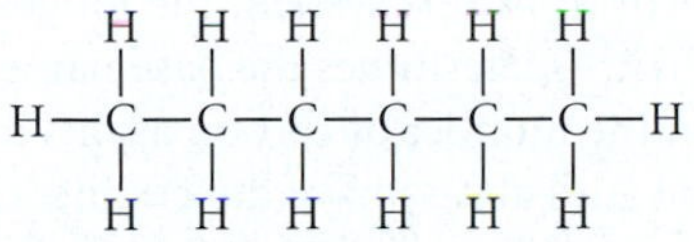 | $CH_3CH_2CH_2CH_2CH_2CH_3$ |

| 7 | n-Heptane | C_7H_{16} | 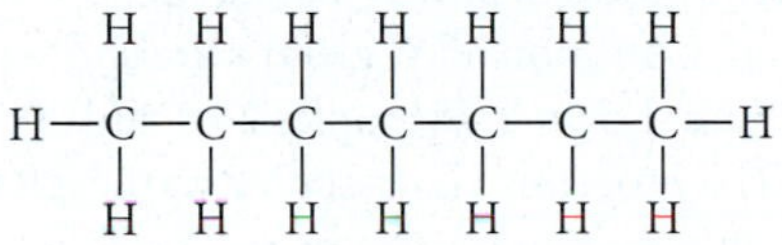 | $CH_3CH_2CH_2CH_2CH_2CH_2CH_3$ |

| 8 | n-Octane | C_8H_{18} | 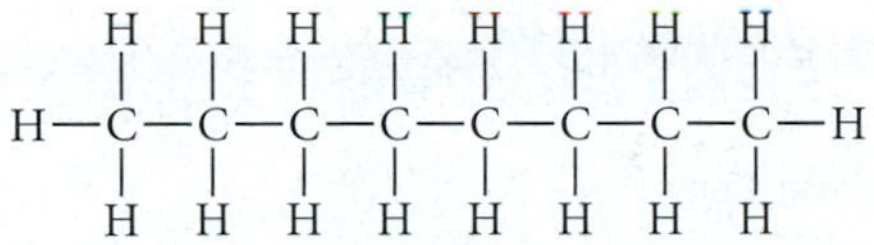 | $CH_3CH_2CH_2CH_2CH_2CH_2CH_2CH_3$ |

| 9 | n-Nonane | C_9H_{20} | 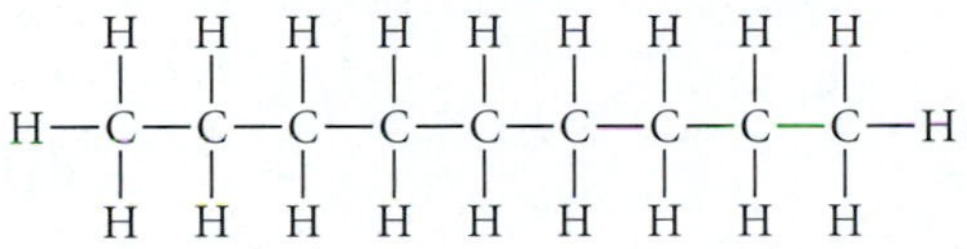 | $CH_3CH_2CH_2CH_2CH_2CH_2CH_2CH_2CH_3$ |

| 10 | n-Decane | $C_{10}H_{22}$ | 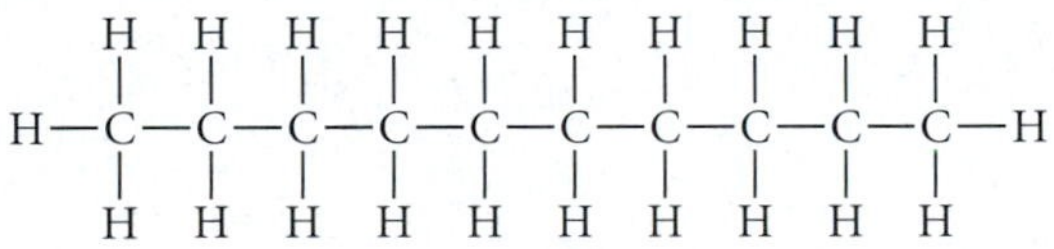 | $CH_3CH_2CH_2CH_2CH_2CH_2CH_2CH_2CH_2CH_3$ |

TABLE 20.4 Uses of Hydrocarbons

Number of Carbon Atoms	State	Major Uses
1–4	Gas	Heating fuel, cooking fuel
5–7	Low-boiling liquids	Solvents, gasoline
6–18	Liquids	Gasoline
12–24	Liquids	Jet fuel, portable-stove fuel
18–50	High-boiling liquids	Diesel fuel, lubricants, heating oil
50+	Solids	Petroleum jelly, paraffin wax

Naming Alkanes

Many organic compounds have common names that can be learned only through familiarity. Because there are so many organic compounds, however, a systematic method of
nomenclature is required. In this book, we adopt the nomenclature system recommended
by the IUPAC (International Union of Pure and Applied Chemistry), which is used
throughout the world. In this system, the longest continuous chain of carbon atoms—
called the base chain—determines the base name of the compound. The root of the base
name depends on the number of carbon atoms in the base chain, as shown in Table 20.5.
Base names for alkanes always have the ending -*ane*. Groups of carbon atoms branching
off the base chain are called alkyl groups and are named as substituents. A substituent is
simply an atom or group of atoms that has been substituted for a hydrogen atom in an organic compound. Common alkyl groups are shown in Table 20.6.

The procedure shown in Examples 20.2 and 20.3 will allow you to systematically name
many alkanes. The procedure is presented in the left column and two examples of applying
the procedure are shown in the center and right columns.

TABLE 20.5 Prefixes for Base Names of Alkane Chains	
Number of Carbon Atoms	Prefix
1	meth-
2	eth-
3	prop-
4	but-
5	pent-
6	hex-
7	hept-
8	oct-
9	non-
10	dec-

TABLE 20.6 Common Alkyl Groups	
Condensed Structural Formula	Name
$-CH_3$	Methyl
$-CH_2CH_3$	Ethyl
$-CH_2CH_2CH_3$	Propyl
$-CH_2CH_2CH_2CH_3$	Butyl

$-CHCH_3$
     $\mid$
     CH_3
	Isopropyl

$-CH_2CHCH_3$
        $\mid$
        CH_3
	Isobutyl

$-CHCH_2CH_3$
    $\mid$
    CH_3
	sec-Butyl

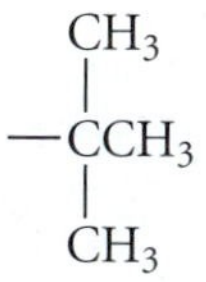 CH_3
     $\mid$
 $-CCH_3$
     $\mid$
     CH_3
	tert-Butyl

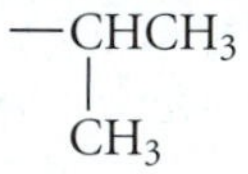

Procedure for Naming Alkanes

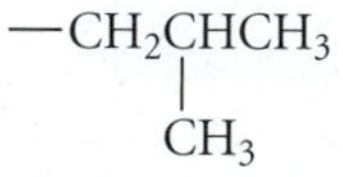

EXAMPLE 20.2
Naming Alkanes

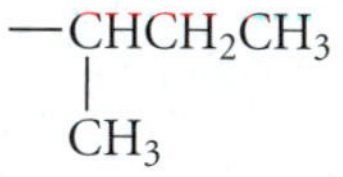

Name the following alkane.

$$CH_3-CH_2-CH-CH_2-CH_3$$
$$\mid$$
$$CH_2$$
$$\mid$$
$$CH_3$$

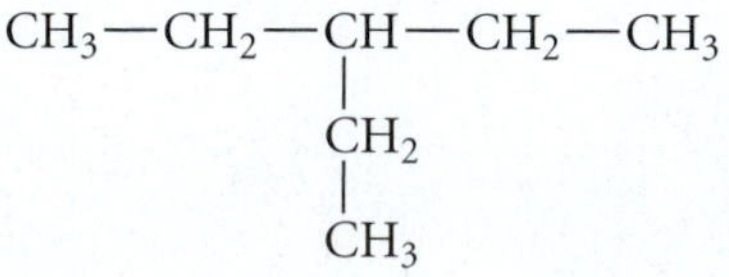

EXAMPLE 20.3
Naming Alkanes

Name the following alkane.

$$CH_3-CH-CH_2-CH-CH_2-CH_2-CH-CH_3$$
$$\mid \qquad\qquad \mid \qquad\qquad\qquad \mid$$
$$CH_3 \qquad\quad CH_2 \qquad\qquad\quad CH_3$$
$$\qquad\qquad\quad \mid$$
$$\qquad\qquad\quad CH_3$$

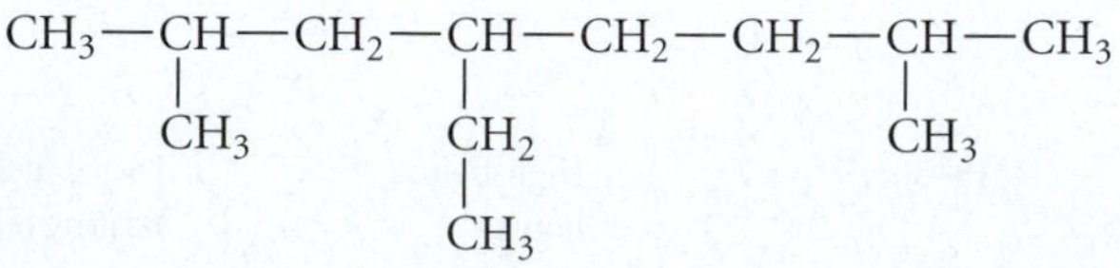

	Solution	**Solution**
1. Count the number of carbon atoms in the longest continuous carbon chain to determine the base name of the compound. Find the prefix corresponding to this number of atoms in Table 20.5 and add the ending -*ane* to form the base name.	This compound has five carbon atoms in its longest continuous chain. 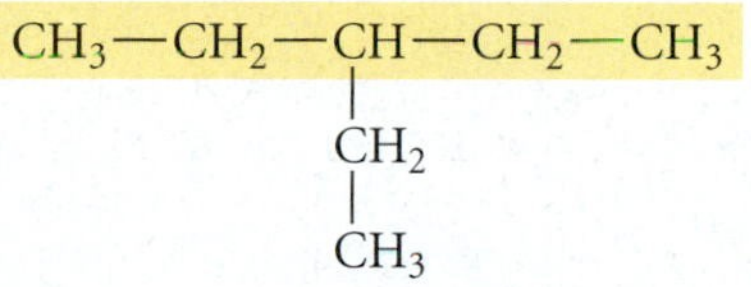 $CH_3-CH_2-CH-CH_2-CH_3$ 　　　　　　│ 　　　　　　CH_2 　　　　　　│ 　　　　　　CH_3 The correct prefix from Table 20.5 is *pent-*. The base name is pentane.	This compound has eight carbon atoms in its longest continuous chain. 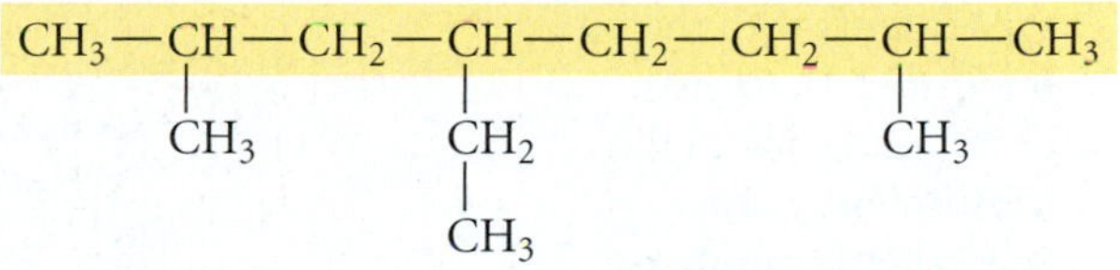 $CH_3-CH-CH_2-CH-CH_2-CH_2-CH-CH_3$ 　　　│　　　　　│　　　　　　　　│ 　　　CH_3　　　CH_2　　　　　　　CH_3 　　　　　　　　　│ 　　　　　　　　　CH_3 The correct prefix from Table 20.5 is *oct-*. The base name is octane.
2. Consider every branch from the base chain to be a substituent. Name each substituent according to Table 20.6.	This compound has one substituent named *ethyl*. 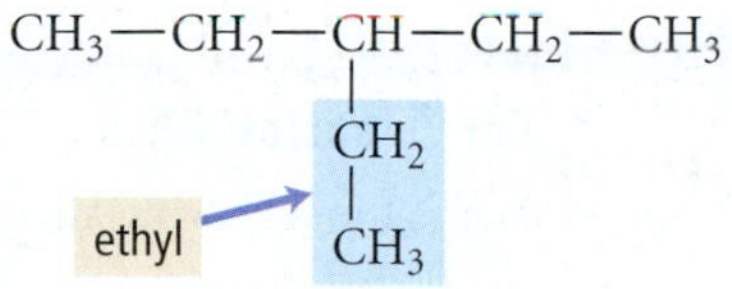 $CH_3-CH_2-CH-CH_2-CH_3$ 　　　　　　│ 　　　　　　CH_2 　ethyl　→　CH_3	This compound has one substituent named *ethyl* and two named *methyl*. $CH_3-CH-CH_2-CH-CH_2-CH_2-CH-CH_3$ 　　　│　　　　│　　　　　　　│ 　　　CH_3　　CH_2　←ethyl　CH_3 　　　　　　　│ 　　　　　　　CH_3 　　　　　　　methyl 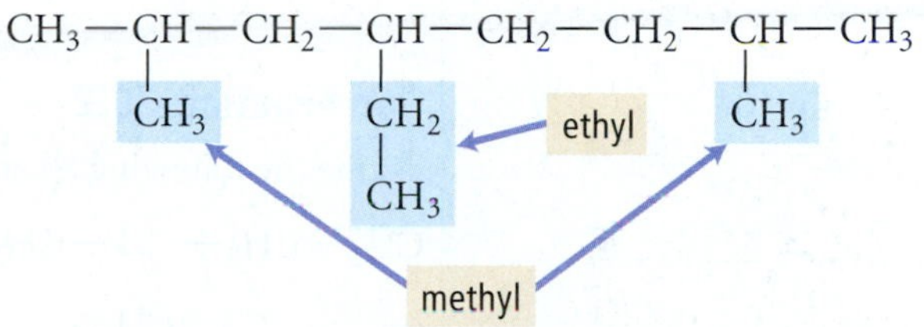
3. Beginning with the end closest to the branching, number the base chain and assign a number to each substituent. (If two substituents occur at equal distances from each end, go to the next substituent to determine from which end to start numbering.)	The base chain is numbered as follows: 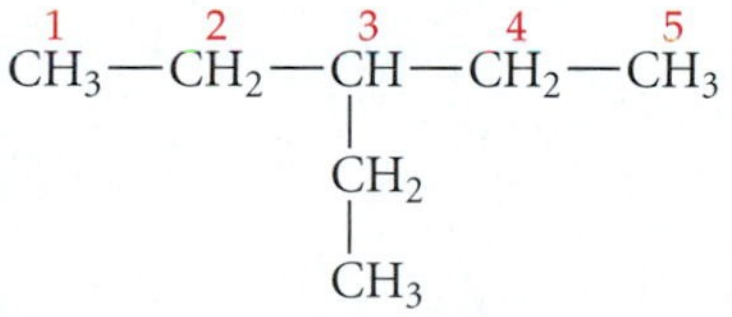 　1　　　2　　　3　　4　　　5 $CH_3-CH_2-CH-CH_2-CH_3$ 　　　　　　│ 　　　　　　CH_2 　　　　　　│ 　　　　　　CH_3 The ethyl substituent is assigned the number 3.	The base chain is numbered as follows: 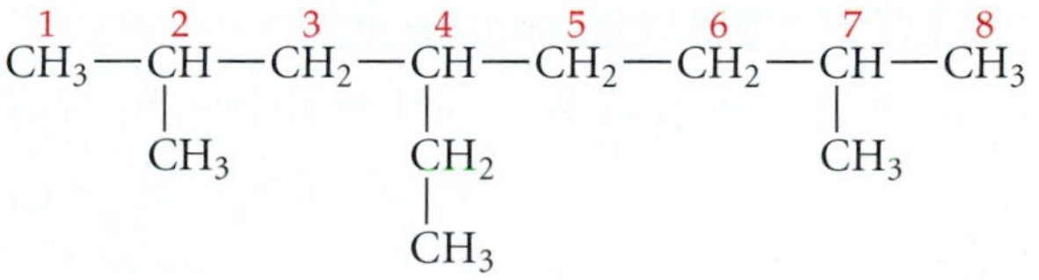 　1　　2　　3　　4　　5　　6　　7　　8 $CH_3-CH-CH_2-CH-CH_2-CH_2-CH-CH_3$ 　　　│　　　　│　　　　　　　│ 　　　CH_3　　CH_2　　　　　　CH_3 　　　　　　　│ 　　　　　　　CH_3 The ethyl substituent is assigned the number 4 and the two methyl substituents are assigned the numbers 2 and 7.
4. Write the name of the compound in the following format: (substituent number)-(substituent name)(base name) If there are two or more substituents, give each one a number and list them alphabetically with hyphens between words and numbers.	The name of the compound is 　　3-ethylpentane	The basic form of the name of the compound is 　　4-ethyl-2,7-methyloctane Ethyl is listed before methyl because substituents are listed in alphabetical order.

5. If a compound has two or more identical substituents, designate the number of identical substituents with the prefix *di-* (2), *tri-* (3), or *tetra-* (4) before the substituent's name. Separate the numbers indicating the positions of the substituents relative to each other with a comma. The prefixes are not taken into account when alphabetizing.	Does not apply to this compound.	This compound has two methyl substituents; therefore, the name of the compound is 4-ethyl-2,7-dimethyloctane

For Practice 20.2

Name the following alkane.

$$CH_3-CH_2-CH-CH_2-CH_2-CH_3$$
$$\qquad\qquad\;\; |$$
$$\qquad\qquad\; CH_3$$

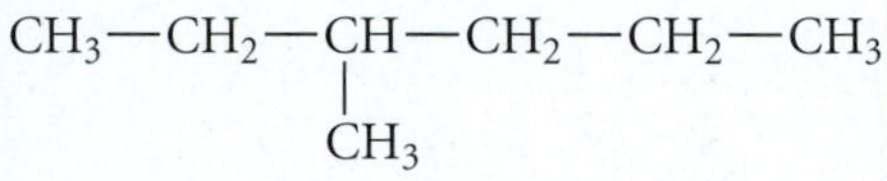

For Practice 20.3

Name the following alkane.

$$CH_3-CH_2-CH-CH_2-CH-CH_2-CH_3$$
$$\qquad\qquad\;\; |\qquad\qquad\; |$$
$$\qquad\qquad\; CH_3\qquad\quad CH_3$$

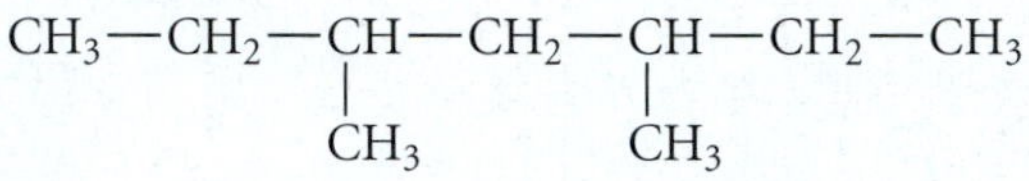

EXAMPLE 20.4 Naming Alkanes

Name the following alkane: $CH_3-CH-CH_2-CH-CH_3$
$\qquad\qquad\qquad\qquad\qquad\qquad\quad\; | \qquad\qquad\; |$
$\qquad\qquad\qquad\qquad\qquad\qquad\; CH_3 \qquad\; CH_3$

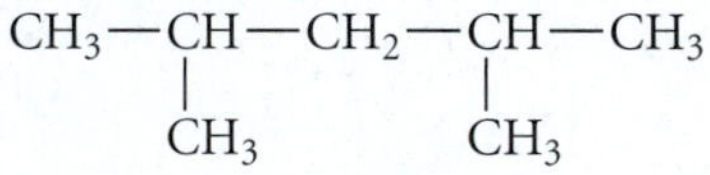

Solution

1. The longest continuous carbon chain has five atoms. Therefore the base name is pentane.	$CH_3-CH-CH_2-CH-CH_3$ $\qquad\quad | \qquad\qquad\; |$ $\qquad\; CH_3 \qquad\quad CH_3$
2. This compound has two substituents, both of which are named methyl.	$CH_3-CH-CH_2-CH-CH_3$ $\qquad\quad | \qquad\qquad\; |$ $\qquad\; CH_3 \qquad\quad CH_3$ methyl
3. Since both substituents are equidistant from the ends, it does not matter from which end you start numbering.	$\overset{1}{C}H_3-\overset{2}{C}H-\overset{3}{C}H_2-\overset{4}{C}H-\overset{5}{C}H_3$ $\qquad\quad | \qquad\qquad\; |$ $\qquad\; CH_3 \qquad\quad CH_3$
4, 5. Use the general form for the name:	2,4-dimethylpentane

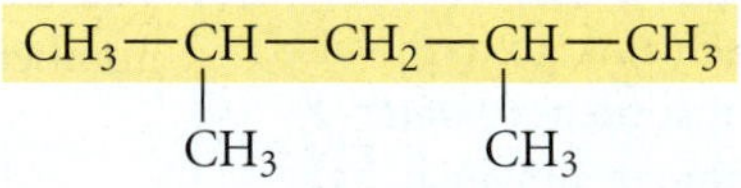

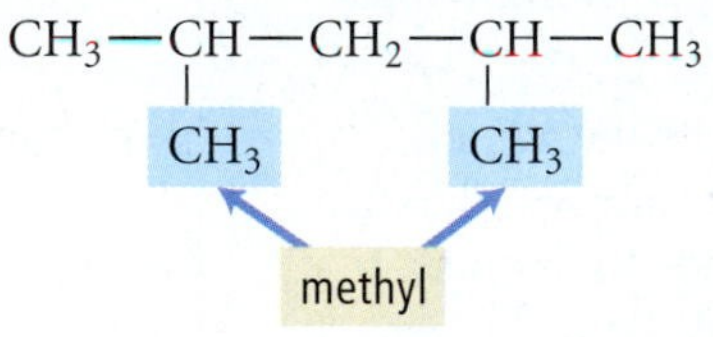

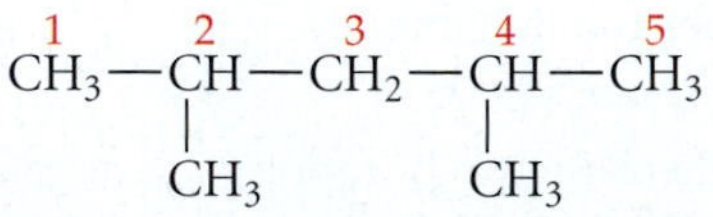

   (substituent number)-(substituent name)(base name)

Since this compound contains two identical substituents, step 5 from the naming procedure applies and we use the prefix *di-*. We also indicate the position of each substituent with a number separated by a comma.

For Practice 20.4

Name the following alkane: $CH_3-CH-CH_2-CH-CH-CH_3$
$\qquad\qquad\qquad\qquad\qquad\qquad\quad | \qquad\qquad | \quad\; |$
$\qquad\qquad\qquad\qquad\qquad\qquad CH_3 \qquad CH_3\; CH_3$

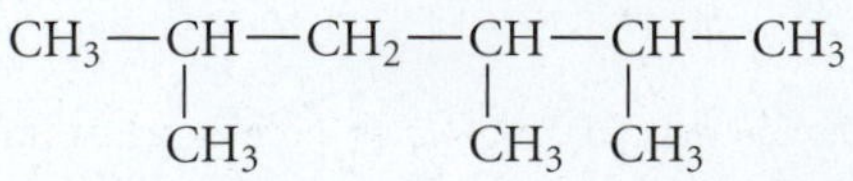

20.5 Alkenes and Alkynes

Alkenes are hydrocarbons containing at least one double bond between carbon atoms, and alkynes contain at least one triple bond. Because of the double or triple bond, alkenes and alkynes have fewer hydrogen atoms than the corresponding alkane and are therefore called **unsaturated hydrocarbons**—they are not loaded to capacity with hydrogen. As we learned earlier, noncyclic alkenes have the formula C_nH_{2n} and noncyclic alkynes have the formula C_nH_{2n-2}. The simplest alkene is ethene (C_2H_4), also called ethylene.

The formulas shown here for alkenes and alkynes assume only one multiple bond.

Ethene or ethylene	C_2H_4	$\overset{H}{\underset{H}{\diagup}}C=C\overset{H}{\underset{H}{\diagdown}}$	
	Formula	Structural formula	Space-filling model

The geometry about each carbon atom in ethene is trigonal planar, making ethene a flat, rigid molecule. Ethene is a ripening agent in fruit. For example, when a banana within a cluster of bananas begins to ripen, it emits ethene. The ethene then causes other bananas in the cluster to ripen. Banana farmers usually pick bananas green for ease of shipping. When the bananas arrive at their destination, they are often "gassed" with ethene to initiate ripening so that they will be ready to sell. The names and structures of several other alkenes are shown in Table 20.7. Most of them do not have familiar uses other than their presence as minority components of fuels.

TABLE 20.7 Alkenes

n	Name	Molecular Formula C_nH_{2n}	Structural Formula	Condensed Structural Formula
2	Ethene	C_2H_4	$\overset{H}{\underset{H}{\diagup}}C=C\overset{H}{\underset{H}{\diagdown}}$	$CH_2{=}CH_2$
3	Propene	C_3H_6	$\overset{H}{\underset{H}{\diagup}}C=C-C-H$	$CH_2{=}CHCH_3$
4	1-Butene*	C_4H_8	$\overset{H}{\underset{H}{\diagup}}C=C-C-C-H$	$CH_2{=}CHCH_2CH_3$
5	1-Pentene*	C_5H_{10}	$\overset{H}{\underset{H}{\diagup}}C=C-C-C-C-H$	$CH_2{=}CHCH_2CH_2CH_3$
6	1-Hexene*	C_6H_{12}	$\overset{H}{\underset{H}{\diagup}}C=C-C-C-C-C-H$	$CH_2{=}CHCH_2CH_2CH_2CH_3$

* These alkenes have one or more isomers depending on the position of the double bond. The isomers shown here have the double bond in the 1 position, meaning the first carbon–carbon bond of the chain.

The simplest alkyne is ethyne, C_2H_2, also called acetylene.

Ethyne or acetylene C_2H_2 $H—C{\equiv}C—H$

Formula   Structural formula   Space-filling model

The geometry about each carbon atom in ethyne is linear, making ethyne a linear molecule. Ethyne (or acetylene) is commonly used as fuel for welding torches. The names and structures of several other alkynes are shown in Table 20.8. Like alkenes, the alkynes do not have familiar uses other than their presence as minority components of gasoline.

▶ Welding torches often burn ethyne in pure oxygen to produce the very hot flame needed for melting metals.

TABLE 20.8   Alkynes

n	Name	Molecular Formula C_nH_{2n-2}	Structural Formula	Condensed Structural Formula
2	Ethyne	C_2H_2	$H—C{\equiv}C—H$	$CH{\equiv}CH$
3	Propyne	C_3H_4	$H—C{\equiv}C—\overset{\displaystyle H}{\underset{\displaystyle H}{C}}—H$	$CH{\equiv}CCH_3$
4	1-Butyne*	C_4H_6	$H—C{\equiv}C—\overset{\displaystyle H}{\underset{\displaystyle H}{C}}—\overset{\displaystyle H}{\underset{\displaystyle H}{C}}—H$	$CH{\equiv}CCH_2CH_3$
5	1-Pentyne*	C_5H_8	$H—C{\equiv}C—\overset{\displaystyle H}{\underset{\displaystyle H}{C}}—\overset{\displaystyle H}{\underset{\displaystyle H}{C}}—\overset{\displaystyle H}{\underset{\displaystyle H}{C}}—H$	$CH{\equiv}CCH_2CH_2CH_3$
6	1-Hexyne*	C_6H_{10}	$H—C{\equiv}C—\overset{\displaystyle H}{\underset{\displaystyle H}{C}}—\overset{\displaystyle H}{\underset{\displaystyle H}{C}}—\overset{\displaystyle H}{\underset{\displaystyle H}{C}}—\overset{\displaystyle H}{\underset{\displaystyle H}{C}}—H$	$CH{\equiv}CCH_2CH_2CH_2CH_3$

* These alkynes have one or more isomers depending on the position of the triple bond. The isomers shown here have the triple bond in the 1 position, meaning the first carbon–carbon bond of the chain.

Naming Alkenes and Alkynes

Alkenes and alkynes are named in the same way as alkanes with the following exceptions.

- The base chain is the longest continuous carbon chain *that contains the double or triple bond*.
- The base name has the ending *-ene* for alkenes and *-yne* for alkynes.
- The base chain is numbered to give the double or triple bond the lowest possible number.
- A number indicating the position of the double or triple bond (lowest possible number) is inserted just before the base name.

For example, the alkene and alkyne shown here are named as follows:

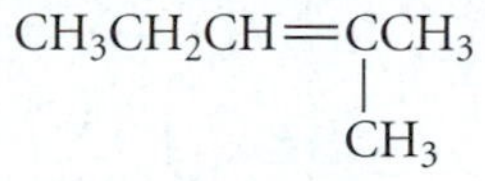

$$CH_3CH_2CH{=}CCH_3$$
$$|$$
$$CH_3$$
2-Methyl-2-pentene

$$CH{\equiv}CCH_2CH_3$$
1-butyne

EXAMPLE 20.5 Naming Alkenes and Alkynes

Name the following compounds:

(a)
$$CH_3$$
$$|$$
$$CH_3{-}C{=}C{-}CH_2{-}CH_3$$
$$|$$
$$CH_2$$
$$|$$
$$CH_3$$

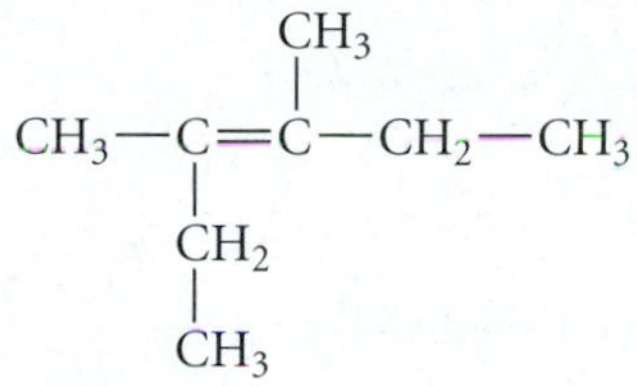

(b)
$$CH_3$$
$$|$$
$$CH_3{-}CH$$
$$|$$
$$CH_3{-}CH{-}CH{-}C{\equiv}CH$$
$$|$$
$$CH_3$$

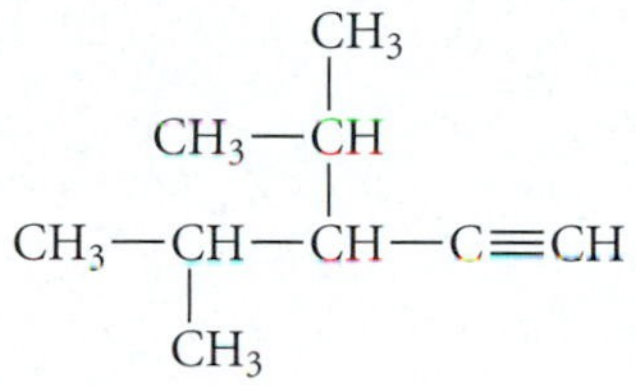

Solution

(a) 1. The longest continuous carbon chain containing the double bond has six carbon atoms. The base name is therefore *hexene*.	$$CH_3$$ $$\|$$ $$CH_3{-}C{=}C{-}CH_2{-}CH_3$$ $$\|$$ $$CH_2$$ $$\|$$ $$CH_3$$
2. The two substituents are both methyl.	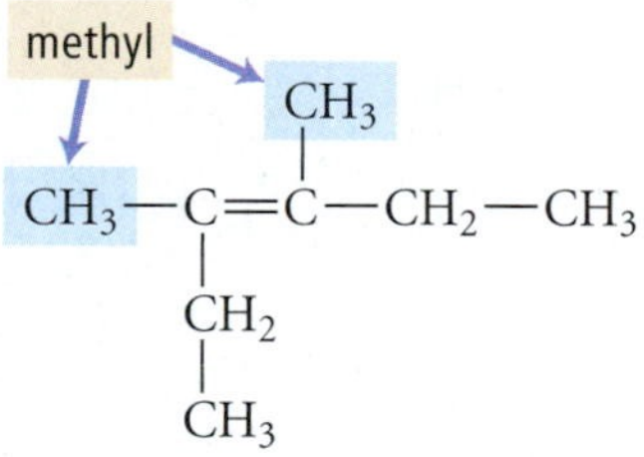 methyl $$CH_3$$ $$\|$$ $$CH_3{-}C{=}C{-}CH_2{-}CH_3$$ $$\|$$ $$CH_2$$ $$\|$$ $$CH_3$$
3. One of the exceptions to naming alkenes is to number the chain so that the *double bond* has the lowest number. In this case, the double bond is equidistant from the ends. The double bond is assigned the number 3. They two methyl groups are therefore at positions 3 and 4.	$$CH_3$$ $$\|$$ $$CH_3{-}C{=}C{-}CH_2{-}CH_3$$ $$\quad 3\quad 4\quad 5\quad\quad 6$$ $$\|$$ $$2\,CH_2$$ $$\|$$ $$CH_3$$ $$\quad 1$$
4, 5. Use the general form for the name:	3,4-dimethyl-3-hexene

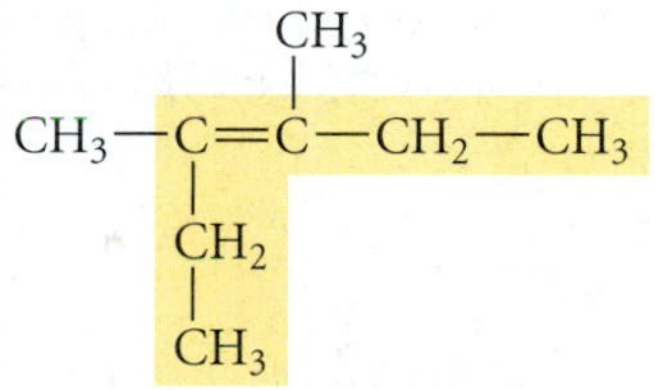

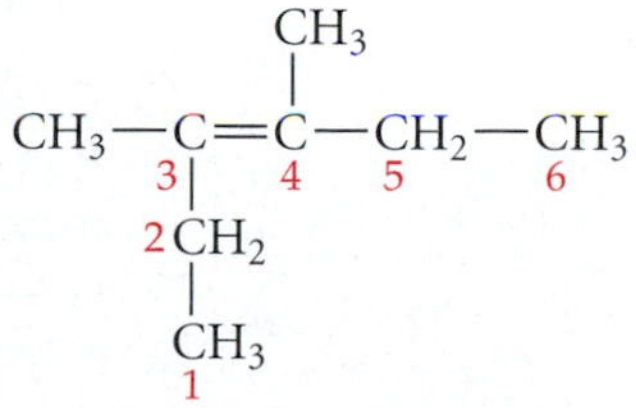

(substituent number)-(substituent name)(base name)

Since this compound contains two identical substituents, step 5 of the naming procedure applies, so use the prefix *di-*. In addition, indicate the position of each substituent with a number separated by a comma. Since this compound is an alkene, specify the position of the double bond, isolated by hyphens, just before the base name.

(b) 1. The longest continuous carbon chain containing the triple bond is five carbons long; therefore the base name is *pentyne*.

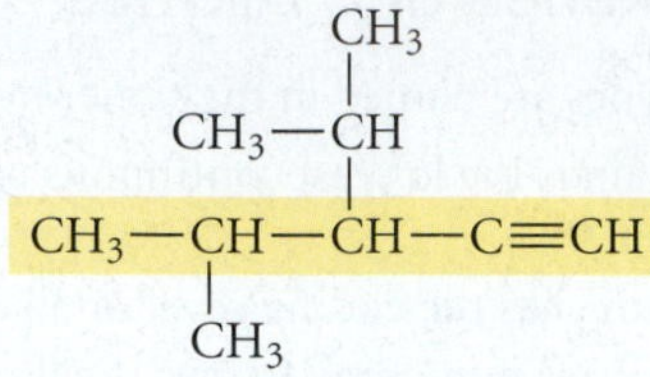

$$CH_3-CH-CH-C\equiv CH$$

with CH_3-CH (bearing CH_3) and CH_3

2. There are two substituents; one is a methyl group and the other an isopropyl group.

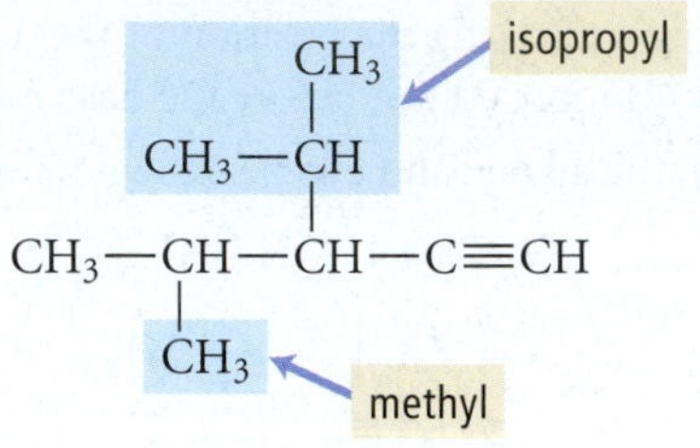

isopropyl
methyl

3. Number the base chain, giving the triple bond the lowest number (1). The isopropyl and methyl groups are therefore given the numbers 3 and 4, respectively.

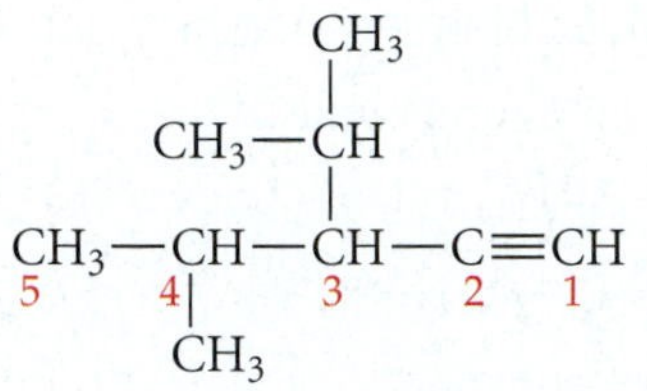

$$\underset{5}{CH_3}-\underset{4}{CH}-\underset{3}{CH}-\underset{2}{C}\equiv\underset{1}{CH}$$

4. Use the general form for the name:

(substituent number)-(substituent name)(base name)

Since there are two substituents, list both of them in alphabetical order. Since this compound is an alkyne, specify the position of the triple bond with a number isolated by hyphens just before the base name.

3-isopropyl-4-methyl-1-pentyne

For Practice 20.5

Name the following compounds:

(a) 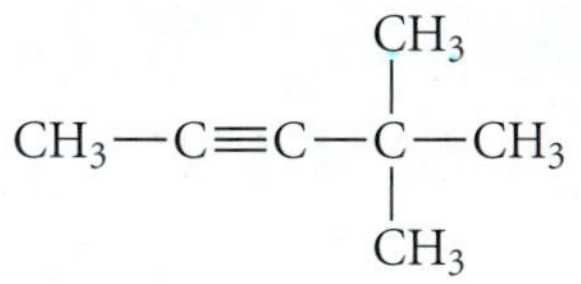 $CH_3-C\equiv C-\overset{\displaystyle CH_3}{\underset{\displaystyle CH_3}{C}}-CH_3$

(b) 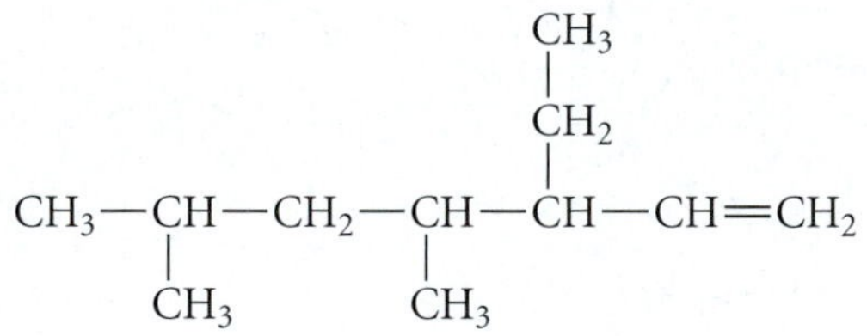 $CH_3-CH-CH_2-CH-\overset{\displaystyle CH_3}{\underset{\displaystyle CH_3}{CH}}-CH=CH_2$ with CH_3 and CH_3 substituents

Geometric (Cis–Trans) Isomerism in Alkenes

A major difference between a single bond and a double bond is the degree to which rotation occurs about the bond. As discussed in Section 10.7, rotation about a double bond is highly restricted due to the overlap between unhybridized p orbitals on the adjacent carbon atoms. Consider, for example, the difference between 1,2-dichloroethane and 1,2-dichloroethene:

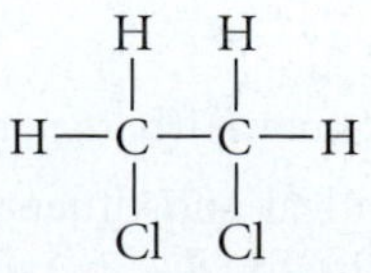

$$H-\overset{\displaystyle H}{\underset{\displaystyle Cl}{C}}-\overset{\displaystyle H}{\underset{\displaystyle Cl}{C}}-H$$

1, 2-Dichloroethane

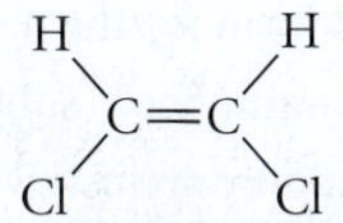

1, 2-Dichloroethene

The hybridization of the carbon atoms in 1,2-dichloroethane is sp^3, resulting in relatively free rotation about the sigma single bond. Consequently, there is no difference between the following two structures at room temperature because they quickly interconvert:

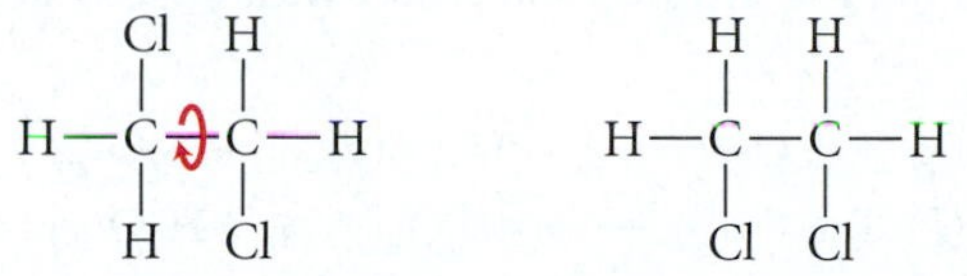

In contrast, rotation about the double bond (sigma + pi) in 1,2-dichloroethene is restricted, so that, at room temperature, 1,2-dichloroethene can exist in two isomeric forms:

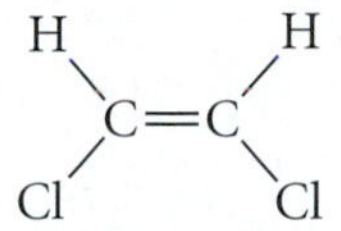

cis-1,2-Dichloroethene

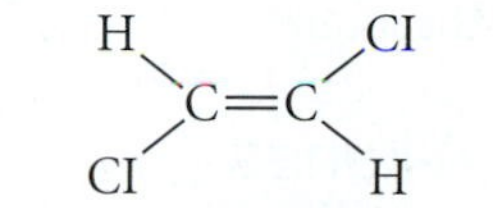

trans-1,2-Dichloroethene

These two forms of 1,2-dichloroethene are indeed different compounds with different properties as shown in Table 20.9. This kind of isomerism is a type of stereoisomerism (see Section 20.3) called **geometric** (or **cis–trans**) **isomerism**. We distinguish between the two isomers with the designations cis (meaning "same side") and trans (meaning "opposite sides"). Cis–trans isomerism is common in alkenes. As another example, consider cis- and trans-2-butene.

Like the two isomers of 1,2-dichloroethene, these two isomers have different physical properties. For example, cis-2-butene boils at 3.7 °C, and trans-2-butene boils at 0.9 °C.

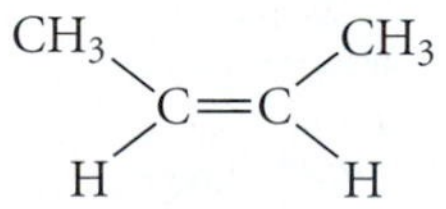

cis-2-Butene

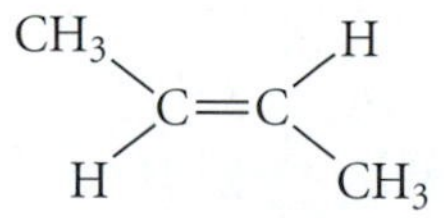

trans-2-Butene

TABLE 20.9 Physical Properties of cis- and trans-1,2-Dichloroethene

Name	Structure	Space-filling Model	Density (g/mL)	Melting Point (°C)	Boiling Point (°C)
cis-1,2-Dichloroethene	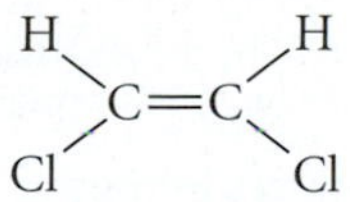		1.284	−80.5	60.1
trans-1,2-Dichloroethene	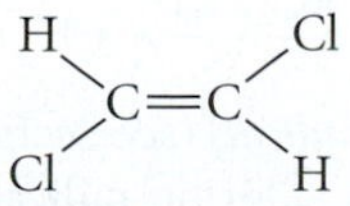	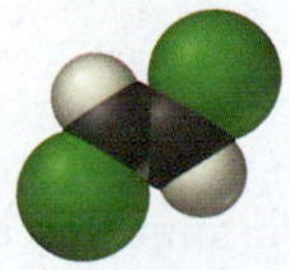	1.257	−49.4	47.5

20.6 Hydrocarbon Reactions

One of the most common hydrocarbon reactions is combustion, the burning of hydrocarbons in the presence of oxygen. Alkanes, alkenes, and alkynes all undergo combustion. In a combustion reaction, the hydrocarbon reacts with oxygen to form carbon dioxide and water.

$$CH_3CH_2CH_3(g) + 5 O_2(g) \longrightarrow 3 CO_2(g) + 4 H_2O(g) \quad \text{Alkane combustion}$$

$$CH_2{=}CHCH_2CH_3(g) + 6 O_2(g) \longrightarrow 4 CO_2(g) + 4 H_2O(g) \quad \text{Alkene combustion}$$

$$CH{\equiv}CCH_3(g) + 4 O_2(g) \longrightarrow 3 CO_2(g) + 2 H_2O(g) \quad \text{Alkyne combustion}$$

Hydrocarbon combustion reactions are highly exothermic and are commonly used to warm homes and buildings, to generate electricity, and to power the engines of cars, ships, and airplanes. Approximately 90% of energy produced in the United States is generated by hydrocarbon combustion.

Reactions of Alkanes

In addition to combustion, alkanes also undergo substitution reactions, in which one or more hydrogen atoms on an alkane are replaced by one or more other atoms. The most common substitution reaction is halogen substitution. For example, methane reacts with chlorine gas in the presence of heat or light to form chloromethane:

$$\underset{\text{Methane}}{CH_4(g)} + \underset{\text{Chlorine}}{Cl_2(g)} \xrightarrow{\text{heat or light}} \underset{\text{Chloromethane}}{CH_3Cl(g)} + HCl(g)$$

Ethane reacts with chlorine gas to form chloroethane:

$$\underset{\text{Ethane}}{CH_3CH_3(g)} + \underset{\text{Chlorine}}{Cl_2(g)} \xrightarrow{\text{heat or light}} \underset{\text{Chloroethane}}{CH_3CH_2Cl(g)} + HCl(g)$$

Multiple halogenation reactions can occur because halogens can replace more than one of the hydrogen atoms on an alkane. For example, chloromethane can continue to react with chlorine as follows:

$$\underset{\text{Chloromethane}}{CH_3Cl(g)} + \underset{\text{Chlorine}}{Cl_2(g)} \xrightarrow{\text{heat or light}} \underset{\substack{\text{Dichloromethane} \\ \text{(also known as} \\ \text{methylene chloride)}}}{CH_2Cl_2(g)} + HCl(g)$$

$$\underset{\text{Dichloromethane}}{CH_2Cl_2(g)} + \underset{\text{Chlorine}}{Cl_2(g)} \xrightarrow{\text{heat or light}} \underset{\substack{\text{Trichloromethane} \\ \text{(also known as} \\ \text{chloroform)}}}{CHCl_3(g)} + HCl(g)$$

$$\underset{\text{Trichloromethane}}{CHCl_3(g)} + \underset{\text{Chlorine}}{Cl_2(g)} \xrightarrow{\text{heat or light}} \underset{\substack{\text{Tetrachloromethane} \\ \text{(also known as} \\ \text{carbon tetrachloride)}}}{CCl_4(g)} + HCl(g)$$

The general form for halogen substitution reactions is

$$\underset{\text{Alkane}}{R{-}H} + \underset{\text{Halogen}}{X_2} \xrightarrow{\text{heat or light}} \underset{\text{Haloalkane}}{R{-}X} + \underset{\substack{\text{Hydrogen} \\ \text{halide}}}{HX}$$

The halogenation of hydrocarbons requires initiation with heat or light, which causes the chlorine–chlorine bond to break:

$$Cl{-}Cl \xrightarrow{\text{heat or light}} Cl\cdot + Cl\cdot$$

The resulting chlorine atoms are *free radicals* (see Section 9.9), as indicated by the dot that represents an unpaired electron. The chlorine radical is highly reactive and attacks the $C{-}H$ bond in hydrocarbons. The reaction then proceeds by the following mechanism:

$$Cl\cdot + R{-}H \longrightarrow R\cdot + HCl$$

$$R\cdot + Cl_2 \longrightarrow R{-}Cl + Cl\cdot$$

Notice that a chlorine free radical is produced as a product of the last step. This free radical can go on to react again, unless it encounters another chlorine free radical, in which case it reacts with it to re-form Cl_2.

Reactions of Alkenes and Alkynes

Alkenes and alkynes undergo addition reactions in which molecules add across the multiple bond. For example, ethene reacts with chlorine gas to form dichloroethane.

$$H_2C=CH_2 + Cl-Cl \longrightarrow H-\underset{Cl}{\underset{|}{C}}H_2-\underset{Cl}{\underset{|}{C}}H_2-H$$

Notice that the addition of chlorine converts the carbon–carbon double bond into a single bond because each carbon atom now has a new bond to a chlorine atom. Alkenes and alkynes can also add hydrogen in hydrogenation reactions. For example, in the presence of an appropriate catalyst, propene reacts with hydrogen gas to form propane.

$$H_3C-CH=CH_2 + H-H \xrightarrow{\text{catalyst}} H_3C-CH_2-CH_3$$

Hydrogenation reactions convert unsaturated hydrocarbons into saturated hydrocarbons. Hydrogenation reactions are also used to convert unsaturated vegetable oils into saturated fats. Most vegetable oils are unsaturated because their carbon chains contain double bonds. The double bonds put bends into the carbon chains that result in less efficient packing of molecules; the result is that vegetable oils are liquids at room temperature while saturated fats are solids at room temperature. When hydrogen is added to the double bonds of vegetable oil, the unsaturated fat is converted into a saturated fat, turning the liquid oil into a solid at room temperature. The words "partially hydrogenated vegetable oil" on a label indicate a food product that contains saturated fats made via hydrogenation reactions.

Alkenes can also add unsymmetrical reagents across the double bond. For example, ethene reacts with hydrogen chloride to form chloroethane:

$$H_2C=CH_2 + HCl \longrightarrow H-\underset{H}{\underset{|}{C}}H_2-\underset{Cl}{\underset{|}{C}}H_2-H$$
Chloroethane

▲ Partially hydrogenated vegetable oil is a saturated fat that is made by hydrogenating unsaturated fats.

If the alkene is also unsymmetrical, then the addition of an unsymmetrical reagent leads to the potential for two different products. For example, when HCl adds to propene, two products are possible:

$$CH_3-\underset{H}{\underset{|}{C}}H-\underset{Cl}{\underset{|}{C}}H-H \quad \text{1-chloropropane}$$
(not observed)

$$CH_3-CH=CH_2 + HCl$$

$$CH_3-\underset{Cl}{\underset{|}{C}}H-\underset{H}{\underset{|}{C}}H-H \quad \text{2-chloropropane}$$

When this reaction is carried out in the lab, however, only the 2-chloropropane is observed to form. The product of the addition of an unsymmetrical reagent to an unsymmetrical alkene can be predicted with Markovnikov's rule, which states the following:

> **When a polar reagent is added to an unsymmetrical alkene, the positive end (the least electronegative part) of the reagent adds to the carbon atom that has the most hydrogen atoms.**

In most reactions of this type, the positive end of the reagent is hydrogen; therefore, the hydrogen atom goes to the carbon atom that already contains the most hydrogen atoms.

EXAMPLE 20.6 Alkene Addition Reactions

Determine the products of the following reactions:

(a) $CH_3CH_2CH{=}CH_2 + Br_2 \longrightarrow$ (b) $CH_3CH_2CH{=}CH_2 + HBr \longrightarrow$

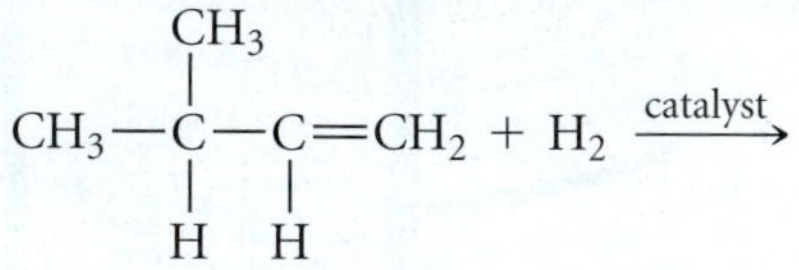

Solution

(a) The reaction of 1-butene with bromine is an example of a symmetric addition. The bromine adds across the double bond so that each carbon forms a single bond to a bromine atom.

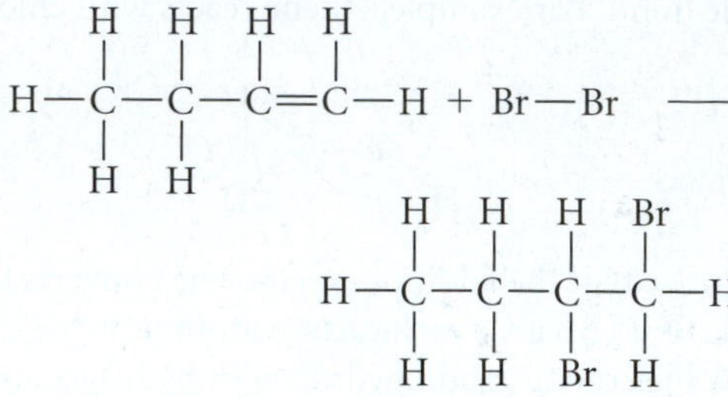

(b) The reaction of 1-butene with hydrogen bromide is an example of an unsymmetrical addition. Use Markovnikov's rule to predict which carbon the hydrogen will bond with and which carbon the bromine atom will bond with. Markovnikov's rule simply stated says that the hydrogen will add to the carbon which already has the most hydrogens, or the end carbon in this case.

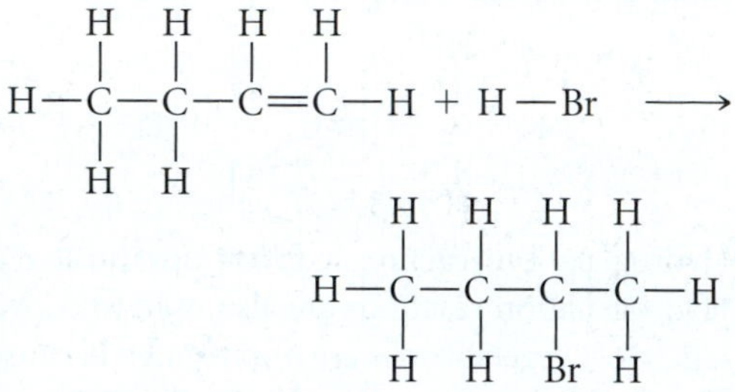

For Practice 20.6

Determine the products of the following reactions:

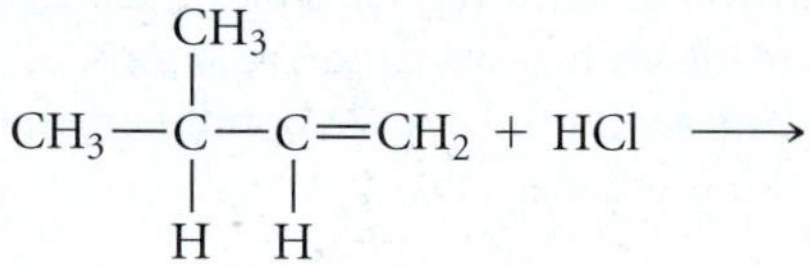

20.7 Aromatic Hydrocarbons

As you might imagine, determining the structure of organic compounds has not always been easy. In the mid-1800s chemists were trying to determine the structure of a particularly stable organic compound named benzene (C_6H_6). In 1865, Friedrich August Kekulé (1829–1896) had a dream in which he envisioned chains of carbon atoms as snakes. The snakes danced before him, and one of them twisted around and bit its tail. Based on that vision, Kekulé proposed the following structure for benzene.

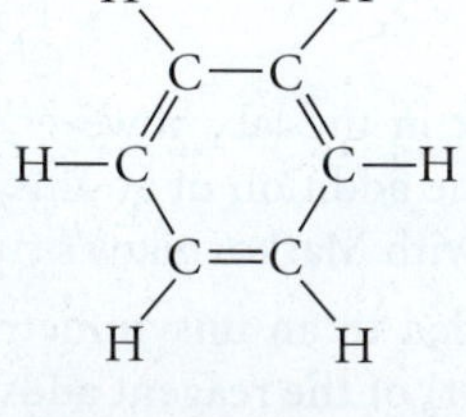

This structure shows alternating single and double bonds. When we examine the bond lengths in benzene, however, we find that all the bonds are the same length. The structure of benzene is better represented by the following resonance structures.

Resonance structures were defined in Section 9.8. Recall that the actual structure of a molecule represented by resonance structures is intermediate between the two resonance structures and is called a *resonance hybrid*.

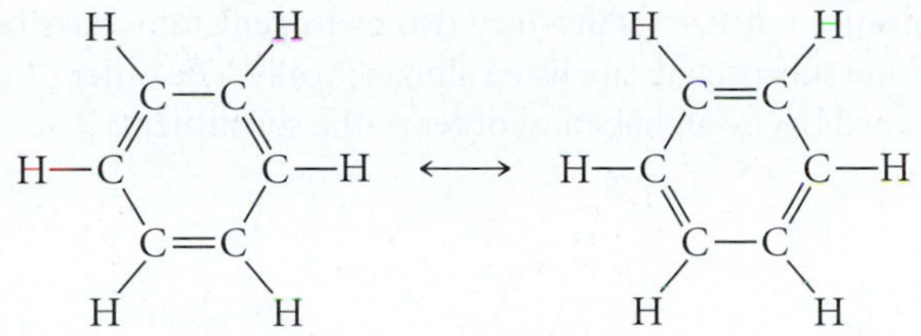

The true structure of benzene is a hybrid of the two resonance structures. Benzene is often represented with the following shorthand notation.

The ring represents the delocalized π electrons which occupy the molecular orbital shown on the right. When drawing benzene rings, either by themselves or as parts of other compounds, organic chemists often use either this diagram or just one of the resonance structures with alternating double bonds. Both representations, however, mean the same thing—a benzene ring.

The ring structure of benzene occurs in many organic compounds. An atom or group of atoms can be substituted for one or more of the six hydrogen atoms on the ring to form substituted benzenes. The following are two examples of substituted benzenes.

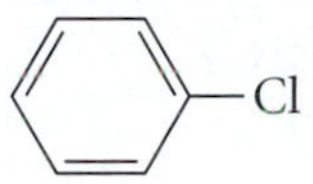

Chlorobenzene 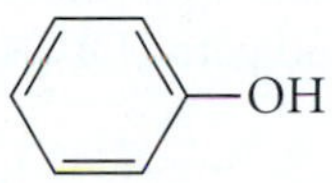   Phenol

Since many compounds containing benzene rings have pleasant aromas, benzene rings are also called aromatic rings, and compounds containing them are called aromatic compounds. For example, the pleasant smells of cinnamon, vanilla, and jasmine are all caused by aromatic compounds.

Naming Aromatic Hydrocarbons

Monosubstituted benzenes—benzenes in which only one of the hydrogen atoms has been substituted—are often named as derivatives of benzene.

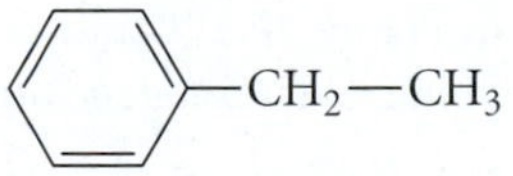

Ethylbenzene 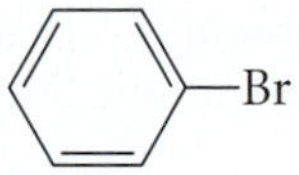   Bromobenzene

These names have the following general form:

(name of substituent) benzene

However, many monosubstituted benzenes have common names that can only be learned through familiarity. The following are four examples:

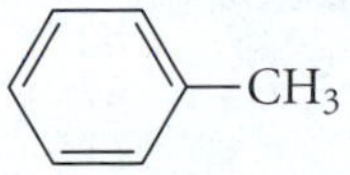

Toluene 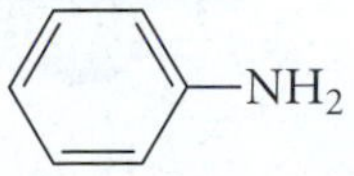   Aniline 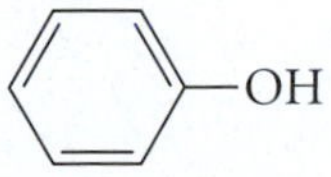   Phenol 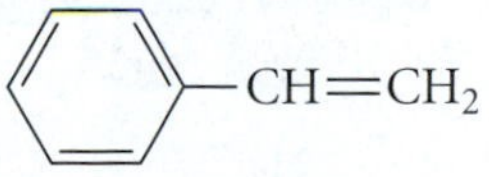   Styrene

Some substituted benzenes, especially those with large substituents, are named by treating the benzene ring as the substituent. In these cases, the benzene substituent is called a **phenyl group**.

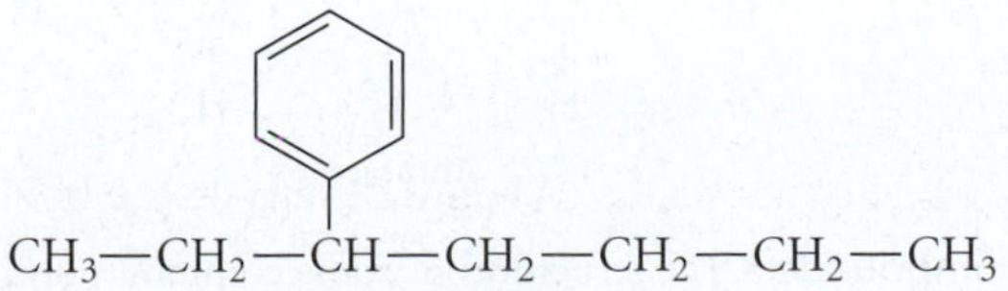

$CH_3-CH_2-CH-CH_2-CH_2-CH_2-CH_3$ $CH_2=CH-CH_2-CH-CH_2-CH_3$

3-Phenylheptane 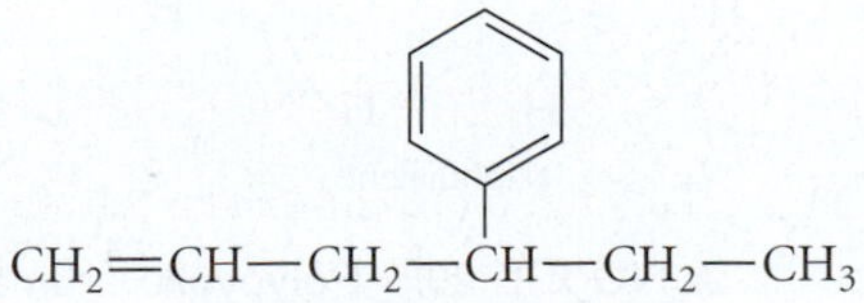   4-Phenyl-1-hexene

Disubstituted benzenes—benzenes in which two hydrogen atoms have been substituted—are numbered and the substituents are listed alphabetically. The order of numbering on the ring is also determined by the alphabetical order of the substituents.

1-Chloro-3-ethylbenzene   1-Bromo-2-chlorobenzene

When the two substituents are identical, use the prefix *di-*.

1,2-Dichlorobenzene   1,3-Dichlorobenzene   1,4-Dichlorobenzene

Also in common use, in place of numbering, are the prefixes ortho (1,2 disubstituted), meta (1,3 disubstituted), and para (1,4 disubstituted).

ortho-Dichlorobenzene   *meta*-Dichlorobenzene   *para*-Dichlorobenzene
or   or   or
o-Dichlorobenzene   *m*-Dichlorobenzene   *p*-Dichlorobenzene

Compounds containing fused aromatic rings are called polycyclic aromatic hydrocarbons. Some common examples are shown in Figure 20.5 ▼ and include naphthalene, the substance that composes mothballs, and pyrene, a carcinogen found in cigarette smoke.

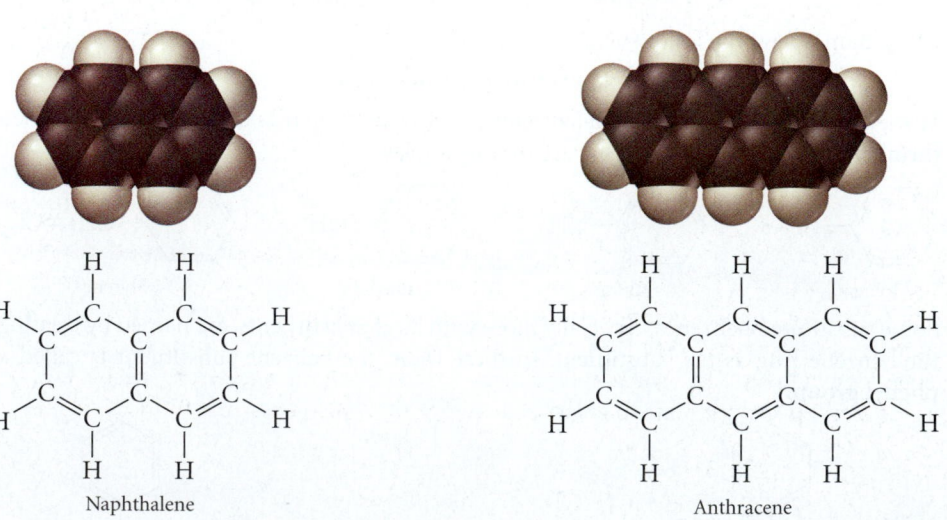

Naphthalene   Anthracene

▲ **FIGURE 20.5 Polycyclic aromatic compounds**   The structures of some common polycyclic aromatic compounds contain fused rings.

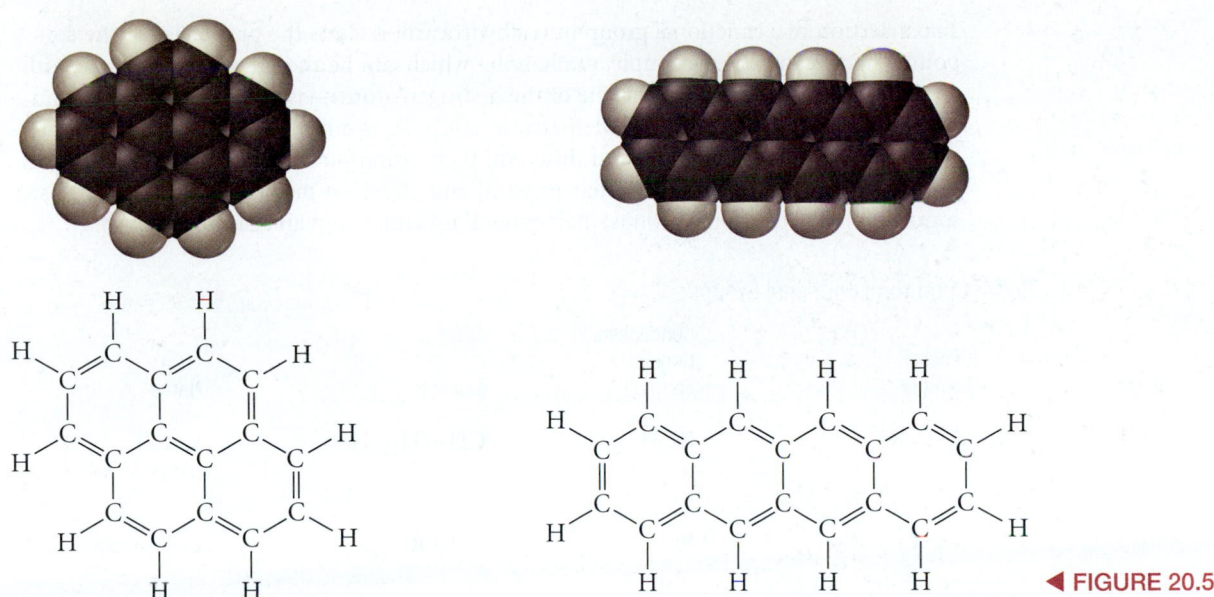

Pyrene

Tetracene

◀ FIGURE 20.5
Continued

Reactions of Aromatic Compounds

We might expect benzene to react similarly to alkenes, readily undergoing addition reactions across its double bonds. However, because of electron delocalization around the ring and the resulting greater stability, benzene does not readily undergo addition reactions. Instead, benzene undergoes substitution reactions in which the hydrogen atoms are replaced by other atoms or groups of atoms as shown in the following examples:

Chlorobenzene

> The substances shown over the arrows are catalysts needed to increase the rate of the reaction.

20.8 Functional Groups

Most other families of organic compounds can be thought of as hydrocarbons with a functional group—a characteristic atom or group of atoms—inserted into the hydrocarbon. A group of organic compounds with the same functional group forms a family. For example, the members of the family of alcohols have an —OH functional group and the general formula R—OH, where R represents a hydrocarbon group. Some specific examples of alcohols include methanol and isopropyl alcohol (also known as rubbing alcohol).

Methanol

Isopropyl alcohol

The insertion of a functional group into a hydrocarbon alters the properties of the compound significantly. For example, methanol—which can be thought of as methane with an —OH group substituted for one of the hydrogen atoms—is a polar, hydrogen-bonded liquid at room temperature. Methane, in contrast, is a nonpolar gas. Although each member of a family is unique and different, their common functional group also gives them some similarities in both their physical and chemical properties. Table 20.10 lists some common functional groups, their general formulas, and an example of each.

TABLE 20.10   Some Common Functional Groups

Family	General Formula*	Condensed General Formula	Example	Name
Alcohols	R—OH	ROH	CH_3CH_2OH	Ethanol (ethyl alcohol)
Ethers	R—O—R	ROR	CH_3OCH_3	Dimethyl ether
Aldehydes	$R-\overset{\overset{\displaystyle O}{\|\|}}{C}-H$	RCHO	$CH_3-\overset{\overset{\displaystyle O}{\|\|}}{C}-H$	Ethanal (acetaldehyde)
Ketones	$R-\overset{\overset{\displaystyle O}{\|\|}}{C}-R$	RCOR	$CH_3-\overset{\overset{\displaystyle O}{\|\|}}{C}-CH_3$	Propanone (acetone)
Carboxylic acids	$R-\overset{\overset{\displaystyle O}{\|\|}}{C}-OH$	RCOOH	$CH_3-\overset{\overset{\displaystyle O}{\|\|}}{C}-OH$	Ethanoic acid (acetic acid)
Esters	$R-\overset{\overset{\displaystyle O}{\|\|}}{C}-OR$	RCOOR	$CH_3-\overset{\overset{\displaystyle O}{\|\|}}{C}-OCH_3$	Methyl acetate
Amines	$R-\overset{\overset{\displaystyle R}{\|}}{N}-R$	R_3N	$CH_3CH_2-\overset{\overset{\displaystyle H}{\|}}{N}-H$	Ethylamine

*In ethers, ketones, esters, and amines, the two R groups may be the same or different.

20.9 Alcohols

As we just discussed, **alcohols** are organic compounds containing the —OH functional group, or **hydroxyl group**, and they have the general formula R—OH. In addition to methanol and isopropyl alcohol (shown previously), other common alcohols include the following:

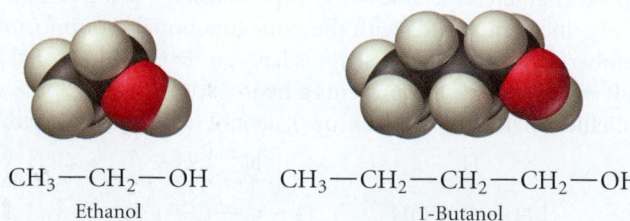

CH_3-CH_2-OH
Ethanol

$CH_3-CH_2-CH_2-CH_2-OH$
1-Butanol

Naming Alcohols

Alcohols are named like alkanes with the following differences:

- The base chain is the longest continuous carbon chain that contains the —OH functional group.
- The base name has the ending -ol.
- The base chain is numbered to give the —OH group the lowest possible number.
- A number indicating the position of the —OH group is inserted just before the base name. For example,

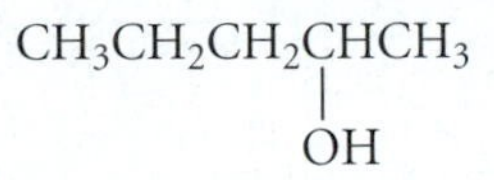

$$CH_3CH_2CH_2\underset{\underset{OH}{|}}{C}HCH_3 \qquad \underset{\underset{OH}{|}}{C}H_2-CH_2-\underset{\underset{CH_3}{|}}{C}H-CH_3$$

2-Pentanol 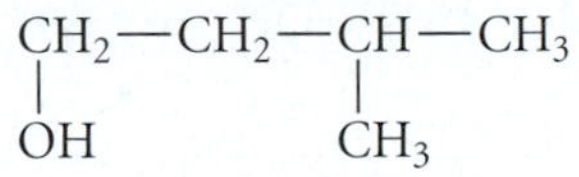   3-Methyl-1-butanol

About Alcohols

Among the most familiar alcohols is ethanol, the alcohol in alcoholic beverages. Ethanol is most commonly formed by the yeast fermentation of sugars, such as glucose, from fruits and grains.

$$\underset{\text{Glucose}}{C_6H_{12}O_6} \xrightarrow{\text{yeast}} 2\ \underset{\text{Ethanol}}{CH_3CH_2OH} + 2\ CO_2$$

Alcoholic beverages contain ethanol, water, and a few other components that give flavor and color. Beer usually contains 3–6% ethanol. Wine contains about 12–14% ethanol, and spirits—beverages like whiskey, rum, or tequila—range from 40% to 80% ethanol, depending on their proof. The proof of an alcoholic beverage is twice the percentage of its ethanol content, so an 80-proof whiskey contains 40% ethanol. Ethanol is also used as a gasoline additive because it increases the octane rating of gasoline and fosters its complete combustion, reducing the levels of certain pollutants such as carbon monoxide and the precursors of ozone.

Isopropyl alcohol (or 2-propanol) can be purchased at any drug store as rubbing alcohol. It is commonly used as a disinfectant for wounds and to sterilize medical instruments. Isopropyl alcohol should never be consumed internally, as it is highly toxic. Four ounces of isopropyl alcohol can cause death. A third common alcohol is methanol, also called wood alcohol. Methanol is commonly used as a laboratory solvent and as a fuel additive. Like isopropyl alcohol, methanol is toxic and should never be consumed.

Alcohol Reactions

Alcohols undergo a number of reactions including substitution, elimination (or dehydration), and oxidation. Alcohols also react with active metals to form strong bases.

Substitution Alcohols react with acids such as HBr to form halogenated hydrocarbons as follows.

$$ROH + HBr \longrightarrow R-Br + H_2O$$

The halogen replaces the hydroxyl group on the alcohol. For example, ethanol reacts with hydrobromic acid to form bromoethane and water.

$$CH_3CH_2OH + HBr \longrightarrow CH_3CH_2Br + H_2O$$

Elimination (or Dehydration) In the presence of concentrated acids such as H_2SO_4, alcohols react to eliminate water, forming an alkene. For example, ethanol eliminates water to form ethene according to the following reaction:

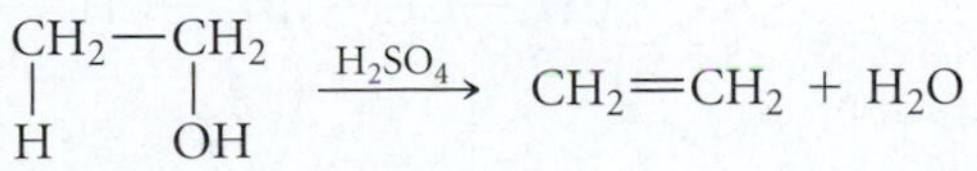

$$\underset{\underset{H}{|}}{C}H_2-\underset{\underset{OH}{|}}{C}H_2 \xrightarrow{H_2SO_4} CH_2{=}CH_2 + H_2O$$

Oxidation In organic chemistry, we think of oxidation and reduction from the point of view of the carbon atoms in the organic molecule. Thus, oxidation is the gaining of oxygen or the losing of hydrogen by a carbon atom. Reduction is then the loss of oxygen or the gaining of hydrogen by a carbon atom. We can draw the following series showing relative states of oxidation:

$$\text{Oxidation} \longrightarrow$$

$$CH_3{-}CH_3 \xrightarrow[\text{gain O}]{} CH_3{-}CH_2{-}OH \xrightarrow[\text{lose H}]{} CH_3{-}\overset{\displaystyle O}{\underset{\|}{C}}{-}H \xrightarrow[\text{gain O}]{} CH_3{-}\overset{\displaystyle O}{\underset{\|}{C}}{-}OH$$

$$\text{Ethane} \qquad \text{Ethanol} \qquad \text{Aldehyde} \qquad \text{Carboxylic acid}$$

$$\longleftarrow \text{Reduction}$$

Under this scenario, an alcohol is a partially oxidized hydrocarbon; it can be further oxidized to form an aldehyde or carboxylic acid, or it can be reduced to form a hydrocarbon (but this is rare). For example, ethanol can be oxidized to acetic acid according to the following reaction:

$$CH_3CH_2OH \xrightarrow[H_2SO_4]{Na_2Cr_2O_7} CH_3COOH$$

Reaction with Active Metals Alcohols react with active metals, such as sodium, much as water does. For example, methanol reacts with sodium to form *sodium methoxide* and hydrogen gas:

$$CH_3OH + Na \longrightarrow CH_3ONa + \tfrac{1}{2}H_2$$

$$\text{Sodium methoxide}$$

The reaction of *water* with sodium produces *sodium hydroxide* and hydrogen gas.

$$H_2O + Na \longrightarrow NaOH + \tfrac{1}{2}H_2$$

In both cases, a strong base is formed (OH^- for water and CH_3O^- for methanol).

EXAMPLE 20.7 Alcohol Reactions

Determine the type of reaction (that is, substitution, dehydration, oxidation, or reaction with an active metal) occurring in each case, and write formulas for the products.

(a) $CH_3{-}\overset{\displaystyle CH_3}{\underset{\displaystyle |}{CH}}{-}CH_2{-}CH_2{-}OH + HBr \longrightarrow$

(b) $CH_3{-}\overset{\displaystyle CH_3}{\underset{\displaystyle |}{CH}}{-}CH_2{-}CH_2{-}OH \xrightarrow[H_2SO_4]{Na_2Cr_2O_7}$

Solution

(a) An alcohol reacting with an acid is an example of a *substitution reaction*. The product of the substitution reaction is a halogenated hydrocarbon and water.

$CH_3{-}\overset{\displaystyle CH_3}{\underset{\displaystyle |}{CH}}{-}CH_2{-}CH_2{-}OH + HBr \longrightarrow$

$CH_3{-}\overset{\displaystyle CH_3}{\underset{\displaystyle |}{CH}}{-}CH_2{-}CH_2{-}Br + H_2O$

(b) An alcohol in solution with sodium dichromate and acid will undergo an *oxidation reaction*. The product of the oxidation reaction is a carboxylic acid functional group instead of the alcohol.

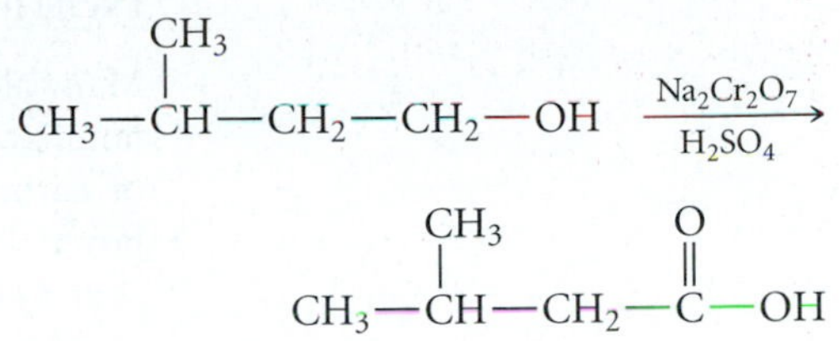

$$CH_3-CH(CH_3)-CH_2-CH_2-OH \xrightarrow[H_2SO_4]{Na_2Cr_2O_7}$$

$$CH_3-CH(CH_3)-CH_2-C(=O)-OH$$

For Practice 20.7

Determine the type of reaction (that is, substitution, dehydration, oxidation, or reaction with an active metal) occurring in each case, and write formulas for the products.

(a) $CH_3CH_2OH + Na \longrightarrow$

(b) 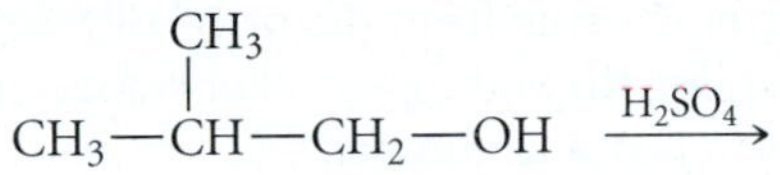 $CH_3-CH(CH_3)-CH_2-OH \xrightarrow{H_2SO_4}$

20.10 Aldehydes and Ketones

Aldehydes and **ketones** have the following general formulas:

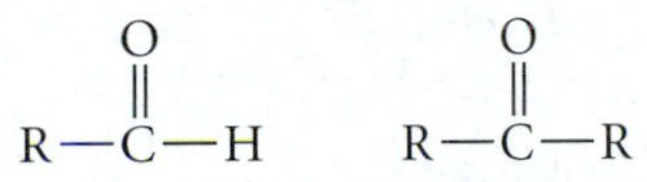

$$R-C(=O)-H \quad R-C(=O)-R$$
Aldehyde   Ketone

The condensed structural formula for aldehydes is RCHO and that for ketones is RCOR.

Both aldehydes and ketones contain a **carbonyl group:**

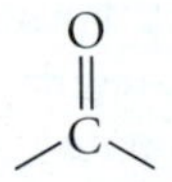

$$C=O$$

Ketones have an R group attached to both sides of the carbonyl, while aldehydes have one R group and a hydrogen atom. (An exception is formaldehyde, which is an aldehyde with two H atoms attached to the carbonyl group.)

Formaldehyde or methanal

Other common aldehydes and ketones are shown in Figure 20.6 ▼.

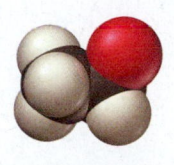
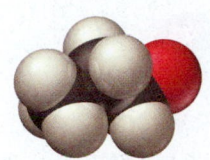

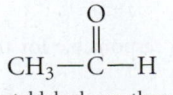
$CH_3-C(=O)-H$
Acetaldehyde or ethanal

$CH_3-CH_2-C(=O)-H$
Propanal

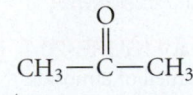
$CH_3-C(=O)-CH_3$
Acetone or propanone

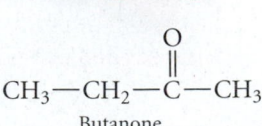
$CH_3-CH_2-C(=O)-CH_3$
Butanone

▲ FIGURE 20.6 **Common Aldehydes and Ketones**

Naming Aldehydes and Ketones

Many aldehydes and ketones have common names that can be learned only by becoming familiar with them. Simple aldehydes are systematically named according to the number of carbon atoms in the longest continuous carbon chain that contains the carbonyl group. Form the base name from the name of the corresponding alkane by dropping the *-e* and adding the ending *-al*.

$$CH_3-CH_2-CH_2-\overset{\displaystyle O}{\overset{\displaystyle \|}{C}}-H$$
Butanal

$$CH_3-CH_2-CH_2-CH_2-\overset{\displaystyle O}{\overset{\displaystyle \|}{C}}-H$$
Pentanal

Simple ketones are systematically named according to the longest continuous carbon chain containing the carbonyl group. Form the base name from the name of the corresponding alkane by dropping the letter *-e* and adding the ending *-one*. For ketones, number the chain to give the carbonyl group the lowest possible number.

$$CH_3-CH_2-CH_2-\overset{\displaystyle O}{\overset{\displaystyle \|}{C}}-CH_3$$
2-Pentanone

$$CH_3-CH_2-CH_2-\overset{\displaystyle O}{\overset{\displaystyle \|}{C}}-CH_2-CH_3$$
3-Hexanone

About Aldehydes and Ketones

The most familiar aldehyde is probably formaldehyde, shown earlier in this section. Formaldehyde is a gas with a pungent odor. It is often mixed with water to make formalin, a preservative and disinfectant. Formaldehyde is also found in wood smoke, which is one reason that smoking foods preserves them—the formaldehyde kills bacteria. Aromatic aldehydes, those that also contain an aromatic ring, have pleasant aromas. For example, vanillin is responsible for the smell of vanilla, cinnamaldehyde is the sweet-smelling component of cinnamon, and benzaldehyde accounts for the smell of almonds (Figure 20.7 ▼).

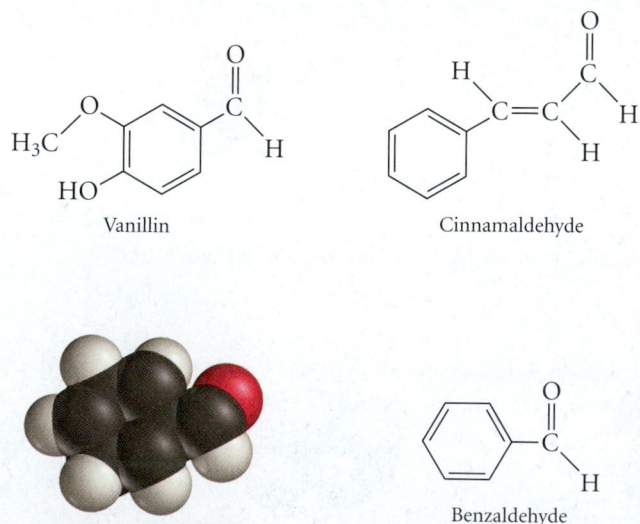

Vanillin

Cinnamaldehyde

Benzaldehyde

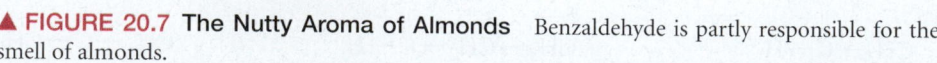

▲ FIGURE 20.7 **The Nutty Aroma of Almonds** Benzaldehyde is partly responsible for the smell of almonds.

The most familiar ketone is acetone, the main component of nail polish remover. Many ketones also have pleasant aromas. For example, carvone is largely responsible for the smell of spearmint, 2-heptanone (among other compounds) for the smell of cloves, and ionone for the smell of raspberries (Figure 20.8 ▼).

$$CH_3-CH_2-CH_2-CH_2-CH_2-\overset{\displaystyle O}{\overset{\|}{C}}-CH_3$$

2-Heptanone

Carvone

Ionone

▲ **FIGURE 20.8 The Fragrance of Raspberries**   Ionone is partly responsible for the smell of raspberries.

Aldehyde and Ketone Reactions

Aldehydes and ketones can be formed by the *oxidation* of alcohols. For example, ethanol can be oxidized to ethanal, and 2-propanol can be oxidized to 2-propanone (or acetone) as follows:

Ethanol $\xrightarrow{\text{oxidation}}$ Ethanal

2-Propanol $\xrightarrow{\text{oxidation}}$ 2-Propanone

In the reverse reaction, an aldehyde or ketone is reduced to an alcohol. For example, 2-butanone can be reduced to 2-butanol in the presence of a reducing agent:

2-Butanone $\xrightarrow{\text{reduction}}$ 2-Butanol

Formaldehyde

Acetaldehyde

Acetone

▲ FIGURE 20.9 **Charge Density Plots of the Carbonyl Group** The carbonyl group is highly polar, as shown in these plots of electrostatic potential.

The carbonyl group in aldehydes and ketones is unsaturated, much like the double bond in an alkene. Therefore, the most common reactions of aldehydes and ketones are **addition reactions**. However, in contrast to the carbon–carbon double bond in alkenes, which is nonpolar, the double bond in the carbonyl group is highly polar (Figure 20.9 ▲). Consequently, additions across the double bond result in the more electronegative part of the reagent adding to the carbon atom and the less electronegative part (often hydrogen) adding to the oxygen atom. For example, HCN adds across the carbonyl double bond in formaldehyde as follows:

20.11 Carboxylic Acids and Esters

Carboxylic acids and **esters** have the following general formulas:

$$
\underset{\text{Carboxylic acid}}{R-\overset{\overset{\displaystyle O}{\|}}{C}-OH} \qquad \underset{\text{Ester}}{R-\overset{\overset{\displaystyle O}{\|}}{C}-OR}
$$

The structures of some common carboxylic acids and esters are shown in Figure 20.10 ▼.

> The condensed structural formula for carboxylic acids is RCOOH and that for esters is RCOOR.

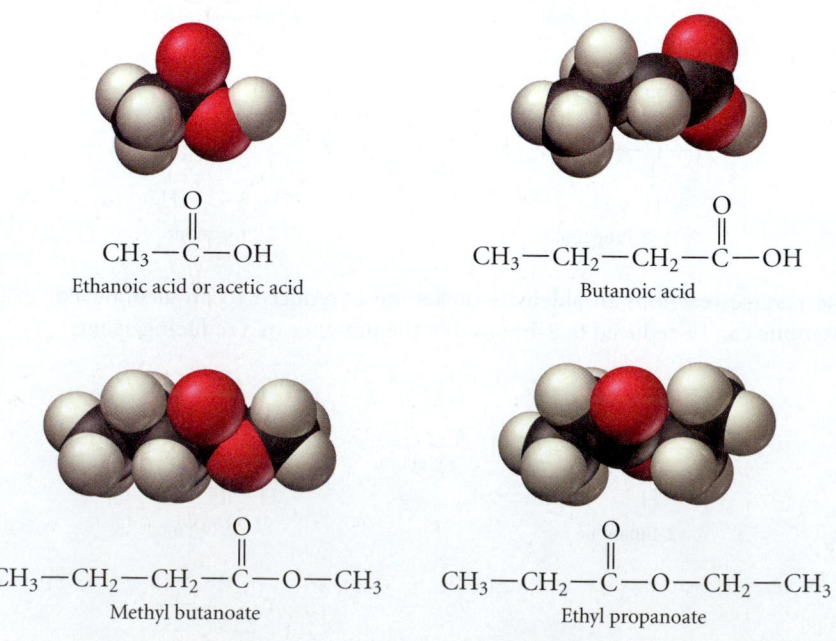

$$
\underset{\text{Ethanoic acid or acetic acid}}{CH_3-\overset{\overset{\displaystyle O}{\|}}{C}-OH}
$$

$$
\underset{\text{Butanoic acid}}{CH_3-CH_2-CH_2-\overset{\overset{\displaystyle O}{\|}}{C}-OH}
$$

$$
\underset{\text{Methyl butanoate}}{CH_3-CH_2-CH_2-\overset{\overset{\displaystyle O}{\|}}{C}-O-CH_3}
$$

$$
\underset{\text{Ethyl propanoate}}{CH_3-CH_2-\overset{\overset{\displaystyle O}{\|}}{C}-O-CH_2-CH_3}
$$

▶ FIGURE 20.10 **Common Carboxylic Acids and Esters**

Naming Carboxylic Acids and Esters

Carboxylic acids are systematically named according to the number of carbon atoms in the longest chain containing the —COOH functional group. Form the base name by dropping the -e from the name of the corresponding alkane and adding the ending –oic acid.

$$CH_3—CH_2—\overset{\overset{\displaystyle O}{\|}}{C}—OH$$
Propanoic acid

$$CH_3—CH_2—CH_2—CH_2—\overset{\overset{\displaystyle O}{\|}}{C}—OH$$
Pentanoic acid

Esters are systematically named as if they were derived from a carboxylic acid by replacing the H on the OH with an alkyl group. The R group from the parent acid forms the base name of the compound. Change the -ic on the name of the corresponding carboxylic acid to -ate, and drop acid. The R group that replaced the H on the carboxylic acid is named as an alkyl group with the ending -yl, as shown in the following examples.

$$CH_3—CH_2—\overset{\overset{\displaystyle O}{\|}}{C}—OH$$
Propanoic acid

$$CH_3—CH_2—CH_2—CH_2—\overset{\overset{\displaystyle O}{\|}}{C}—OH$$
Pentanoic acid

$$CH_3—CH_2—\overset{\overset{\displaystyle O}{\|}}{C}—OCH_3$$
Methyl propanoate

$$CH_3—CH_2—CH_2—CH_2—\overset{\overset{\displaystyle O}{\|}}{C}—OCH_2CH_3$$
Ethyl pentanoate

About Carboxylic Acids and Esters

Like all acids, carboxylic acids taste sour. The most familiar carboxylic acid is ethanoic acid, which is better known by its common name, acetic acid. Acetic acid is the active ingredient in vinegar. It can be formed by the oxidation of ethanol, which is why wines left open to air become sour. Some yeasts and bacteria also form acetic acid when they metabolize sugars in bread dough. These are often added to bread dough to make sourdough bread. Other common carboxylic acids include methanoic acid (usually called formic acid), present in bee stings and ant bites; lactic acid, present in muscles after intense exercise and causing soreness; and citric acid, present in limes, lemons, and oranges (see Figure 20.11 ▼).

$$H—\overset{\overset{\displaystyle O}{\|}}{C}—OH$$
Formic or methanoic acid

$$CH_2—\overset{\overset{\displaystyle OH}{|}}{CH}—\overset{\overset{\displaystyle O}{\|}}{C}—OH$$
Lactic acid

$$HO—\overset{\overset{\displaystyle O}{\|}}{C}—\underset{\underset{\displaystyle CH_2—\overset{\overset{\displaystyle }{|}}{C}—OH}{\underset{\displaystyle O}{\|}}}{\overset{\overset{\displaystyle CH_2—\overset{\overset{\displaystyle O}{\|}}{C}—OH}{|}}{C}}—OH$$
Citric acid

▲ **FIGURE 20.11 The Tart Taste of Limes**   Citric acid is partly responsible for the sour taste of limes.

Esters are best known for their sweet smells. For example, methyl butanoate is largely responsible for the smell and taste of apples, and ethyl butanoate is largely responsible for the smell and taste of pineapples (see Figure 20.12 ▼).

$$CH_3-CH_2-CH_2-\overset{\overset{\displaystyle O}{\|}}{C}-O-CH_3$$
Methyl butanoate

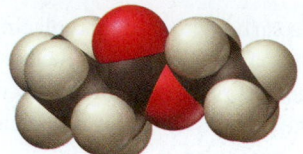

$$CH_3-CH_2-CH_2-\overset{\overset{\displaystyle O}{\|}}{C}-O-CH_2-CH_3$$
Ethyl butanoate

▲ FIGURE 20.12 **The Aroma of Pineapple**   Ethyl butanoate is partly responsible for the aroma of pineapples.

Carboxylic Acid and Ester Reactions

Carboxylic acids act as weak acids in solution according to the following equation:

$$RCOOH(aq) + H_2O(l) \rightleftharpoons H_3O^+(aq) + RCOO^-(aq)$$

Like all acids, carboxylic acids react with strong bases via neutralization reactions. For example, propanoic acid reacts with sodium hydroxide to form sodium propanoate and water:

$$CH_3CH_2COOH(aq) + NaOH(aq) \longrightarrow CH_3CH_2COO^-Na^+(aq) + HOH(l)$$

A carboxylic acid reacts with an alcohol to form an ester via a **condensation reaction**, a reaction in which two (or more) organic compounds are joined, often with the loss of water (or some other small molecule).

$$\underset{\text{Acid}}{R-\overset{\overset{\displaystyle O}{\|}}{C}-OH} + \underset{\text{Alcohol}}{HO-R'} \xrightarrow{H_2SO_4} \underset{\text{Ester}}{R-\overset{\overset{\displaystyle O}{\|}}{C}-O-R'} + \underset{\text{Water}}{H_2O}$$

An important example of this reaction is the formation of acetylsalicylic acid (aspirin) from ethanoic acid (acetic acid) and salicylic acid (originally obtained from the bark of the willow tree).

Acetic acid   Salicylic acid   Acetylsalicylic acid

Strong heating of a carboxylic acid causes it to undergo a condensation reaction with itself to form an acid anhydride (anhydride means "without water").

$$RCOOH(aq) + HOOCR(aq) \longrightarrow RCOOOCR(aq) + HOH(aq)$$
Acid anhydride

Addition of water to an acid anhydride reverses the above reaction and regenerates the carboxylic acid molecules.

Conceptual Connection 20.3 Oxidation

Arrange the following compounds from least oxidized to most oxidized.

(a)
$$CH_3-\overset{\overset{\displaystyle O}{\|}}{C}-CH_3$$

(b)
$$CH_3-\overset{\overset{\displaystyle O}{\|}}{C}-O-CH_3$$

(c) $CH_3-CH_2-CH_3$

(d)
$$CH_3-\underset{\underset{\displaystyle OH}{|}}{CH}-CH_3$$

Answer: (**c, d, a, b**) Oxidation includes the gain of oxygen or the loss of hydrogen.

20.12 Ethers

Ethers are organic compounds with the general formula ROR. The R groups may be the same or different. Some common ethers are shown in Figure 20.13 ▼

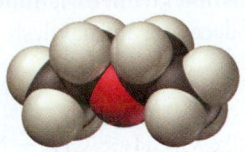

$$CH_3-O-CH_3$$
Dimethyl ether

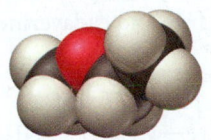

$$CH_3-O-CH_2-CH_3$$
Ethyl methyl ether

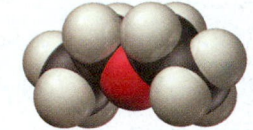

$$CH_3-CH_2-O-CH_2-CH_3$$
Diethyl ether

▲ **FIGURE 20.13 Ethers**

Naming Ethers

Common names for ethers have the following format:

(R group 1) (R group 2) ether

If the two R groups are different, use each of their names in alphabetical order. If the two R groups are the same, use the prefix *di-*. Some examples include

$$H_3C-CH_2-CH_2-O-CH_2-CH_2-CH_3$$
Dipropyl ether

$$H_3C-CH_2-O-CH_2-CH_2-CH_3$$
Ethyl propyl ether

About Ethers

The most common ether is diethyl ether. Diethyl ether is a common laboratory solvent because of its ability to dissolve many organic compounds and because of its low boiling point (34.6 °C). The low boiling point allows for easy removal of the solvent when necessary. Diethyl ether was also used as a general anesthetic for many years. When inhaled, diethyl ether depresses the central nervous system, causing unconsciousness and insensitivity to pain. Its use as an anesthetic, however, has decreased in recent years because other compounds have the same anesthetic effect with fewer side effects (such as nausea).

20.13 Amines

An amine is an organic compound that contains nitrogen. The simplest nitrogen-containing compound is ammonia (NH_3). **Amines** are derivatives of ammonia with one or more of the hydrogen atoms replaced by alkyl groups. Like ammonia, amines are weak bases. They are systematically named according to the hydrocarbon groups attached to the nitrogen and given the ending -*amine*.

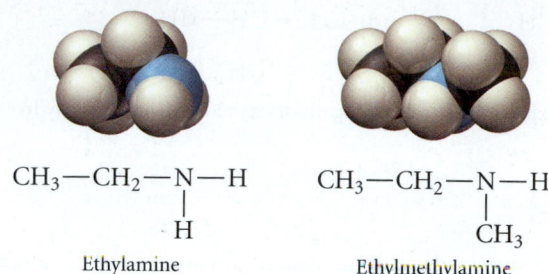

$$CH_3—CH_2—\underset{\underset{\displaystyle H}{|}}{N}—H$$

Ethylamine

$$CH_3—CH_2—\underset{\underset{\displaystyle CH_3}{|}}{N}—H$$

Ethylmethylamine

Amines are most commonly known for their awful odors. When a living organism dies, bacteria that feast on its proteins emit amines. For example, trimethylamine causes the smell of rotten fish, and cadaverine causes the smell of decaying animal flesh.

$$CH_3—\overset{\overset{\displaystyle CH_3}{|}}{N}—CH_3$$

Trimethylamine

$$NH_2—CH_2—CH_2—CH_2—CH_2—CH_2—NH_2$$

Cadaverine

Amine Reactions

Just as carboxylic acids act as weak acids, so amines act as weak bases as follows:

$$RNH_2(aq) + H_2O(l) \rightleftharpoons RNH_3^+(aq) + OH^-(aq)$$

Like all bases, amines react with strong acids to form salts called ammonium salts. For example, methylamine reacts with hydrochloric acid to form methylammonium chloride:

$$CH_3NH_2(aq) + HCl(aq) \longrightarrow CH_3NH_3^+Cl^-(aq)$$
$$\text{Methylammonium chloride}$$

An important amine reaction which we will see again in Chapter 21 is the condensation reaction between a carboxylic acid and an amine.

$$CH_3COOH(aq) + HNHR(aq) \longrightarrow CH_3CONHR(aq) + HOH(l)$$

This reaction is responsible for the formation of proteins from amino acids (see Section 21.4).

20.14 Polymers

Polymers are long, chainlike molecules composed of repeating units called monomers. In Chapter 21, we will learn about natural polymers such as starches, proteins, and DNA, which play important roles in living organisms. In this section, we learn about synthetic polymers, which compose many frequently encountered plastic products such as PVC tubing, styrofoam coffee cups, nylon rope, and plexiglass windows. Polymer materials are common in our everyday lives since they are found in everything from computers to toys to packaging materials. The simplest synthetic polymer is probably polyethylene. The polyethylene monomer is ethene (also called ethylene).

$$C_2H{=}CH_2$$

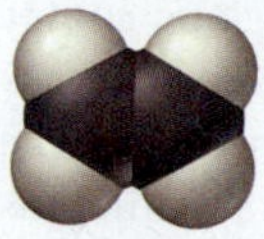

Monomer

Ethene or ethylene

Ethene monomers can react with each other, breaking the double bond between carbons and adding together to make a long polymer chain:

$$\cdots CH_2-CH_2-CH_2-CH_2-CH_2-CH_2-CH_2-CH_2-CH_2\cdots$$

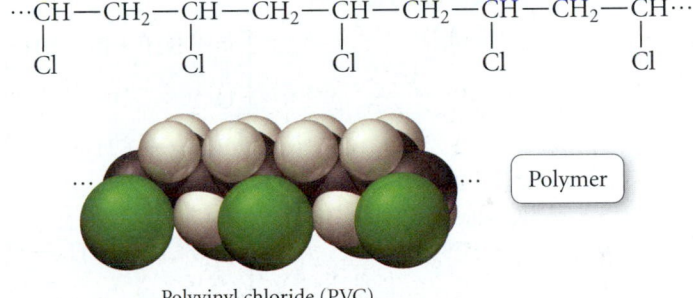

Polyethylene

Polyethylene is the plastic that is used for soda bottles, juice containers, and garbage bags (Figure 20.14 ▼). It is an example of an **addition polymer**, a polymer in which the monomers simply link together without the elimination of any atoms.

An entire class of polymers can be thought of as substituted polyethylenes. For example, polyvinyl chloride (PVC)—the plastic used to make certain kinds of pipes and plumbing fixtures—is composed of monomers in which a chlorine atom has been substituted for one of the hydrogen atoms in ethene (Figure 20.15 ▼). These monomers react together to form PVC.

$$\underset{\underset{Cl}{|}}{HC}=CH_2$$

Monomer

Chloroethene

$$\cdots\underset{\underset{Cl}{|}}{CH}-CH_2-\underset{\underset{Cl}{|}}{CH}-CH_2-\underset{\underset{Cl}{|}}{CH}-CH_2-\underset{\underset{Cl}{|}}{CH}-CH_2-\underset{\underset{Cl}{|}}{CH}\cdots$$

Polymer

Polyvinyl chloride (PVC)

▲ FIGURE 20.14 **Polyethylene**
Polyethylene is the plastic from which soda and juice bottles are made.

▲ FIGURE 20.15 **Polyvinyl Chloride**
Polyvinyl chloride is used for many plastic plumbing supplies, such as pipes and connectors.

Table 20.11 shows several other substituted polyethylene polymers.

TABLE 20.11   Polymers of Commercial Importance

Polymer	Structure	Uses
Addition Polymers		
Polyethylene	$-(CH_2-CH_2)_n$	Films, packaging, bottles
Polypropylene	$\left[\begin{array}{c} CH_2-CH_2 \\ \mid \\ CH_3 \end{array}\right]_n$	Kitchenware, fibers, appliances
Polystyrene	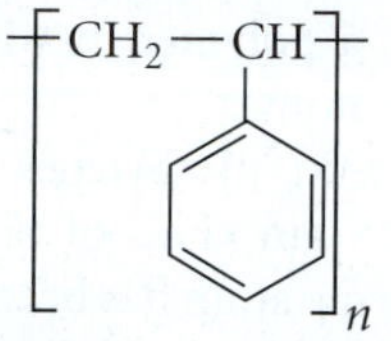 $\left[\begin{array}{c} CH_2-CH \\ \mid \\ C_6H_5 \end{array}\right]_n$	Packaging, disposable food containers, insulation
Polyvinyl chloride	$\left[\begin{array}{c} CH_2-CH \\ \mid \\ Cl \end{array}\right]_n$	Pipe fittings, clear film for meat packaging
Condensation Polymers		
Polyurethane	$\left[\begin{array}{c} C-NH-R-NH-C-O-R'-O \\ \parallel \qquad\qquad\quad \parallel \\ O \qquad\qquad\qquad\quad O \end{array}\right]_n$ R, R' = $-CH_2-CH_2-$ (*for example*)	"Foam" furniture stuffing, spray-on insulation, automotive parts, footwear, water-protective coatings
Polyethylene terephthalate (a polyester)	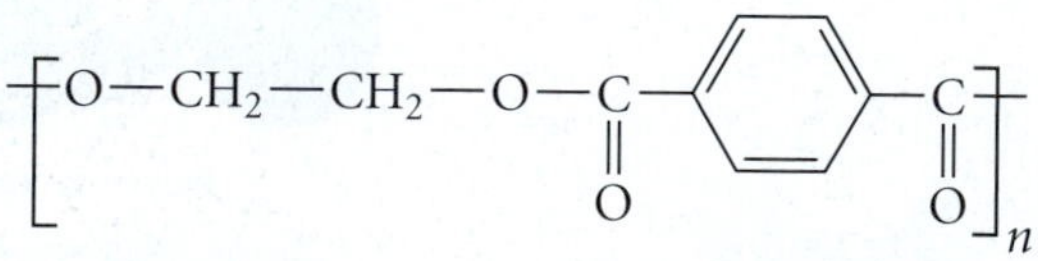 $\left[\begin{array}{c} O-CH_2-CH_2-O-C-\!\!\!\bigcirc\!\!\!-C \\ \parallel\qquad\qquad\qquad\parallel \\ O\qquad\qquad\qquad\ O \end{array}\right]_n$	Tire cord, magnetic tape, apparel, soda bottles
Nylon 6,6	$\left[\begin{array}{c} NH-(CH_2)_6-NH-C-(CH_2)_4-C \\ \parallel\qquad\qquad\quad\parallel \\ O\qquad\qquad\ O \end{array}\right]_n$	Home furnishings, apparel, carpet fibers, fish line, polymer blends

Some polymers—called copolymers—consist of two different kinds of monomers. For example, the monomers that compose nylon 6,6 are hexamethylenediamine and adipic acid. These two monomers add together via a condensation reaction as follows:

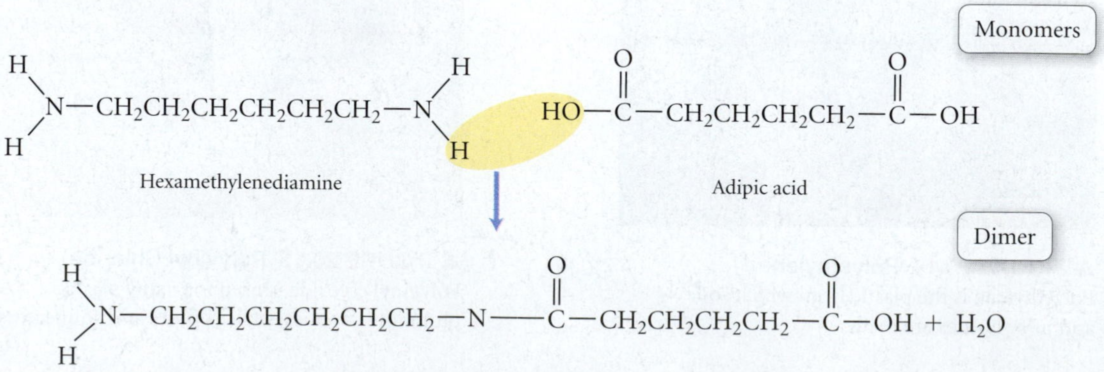

Monomers

$$\begin{array}{ccc} \overset{H}{\underset{H}{\diagup}}N-CH_2CH_2CH_2CH_2CH_2CH_2-N\overset{H}{\underset{H}{\diagdown}} & & \overset{O}{\overset{\parallel}{HO-C}}-CH_2CH_2CH_2CH_2-\overset{O}{\overset{\parallel}{C}}-OH \end{array}$$

Hexamethylenediamine   Adipic acid

Dimer

$$\overset{H}{\underset{H}{\diagup}}N-CH_2CH_2CH_2CH_2CH_2CH_2-N-\overset{O}{\overset{\parallel}{C}}-CH_2CH_2CH_2CH_2-\overset{O}{\overset{\parallel}{C}}-OH + H_2O$$

Chemistry in Your Day
Kevlar

In 1965, Stephanie Kwolek, working for DuPont to develop new polymer fibers, noticed an odd cloudy product from a polymerization reaction. Some researchers might have rejected the product, but Kwolek insisted on examining its properties more carefully. The results were astonishing—when the polymer was spun into a fiber, it was stronger than any other fiber known before. Kwolek had discovered Kevlar, a material that is pound for pound five times stronger than steel.

Kevlar is a condensation polymer containing aromatic rings and amide linkages:

The polymeric chains within Kevlar crystallize in a parallel arrangement (like dry spaghetti noodles in a box), with strong cross-linking between neighboring chains due to hydrogen bonding. The hydrogen bonding occurs between the N—H groups on one chain and the C=O groups on neighboring chains:

This structure is responsible for Kevlar's high strength and its other properties, including fire resistance and chemical resistance (for example, resistance to attack by acids).

Today, DuPont sells hundreds of millions of dollars' worth of Kevlar every year. Kevlar is particularly well-known for its use in bulletproof vests (Figure 20.16). With this application alone, Kwolek's discovery has saved thousands of lives. In addition, Kevlar is used to make helmets, radial tires, brake pads, racing sails, suspension bridge cables, skis, and high-performance hiking and camping gear.

▲ FIGURE 20.16 Kevlar   Kevlar is used to make bulletproof vests.

Question

Examine the structure of the Kevlar polymer above. Knowing that the polymer is a condensation polymer, draw the structures of the monomers before the condensation reaction.

The product that forms between the reaction of two monomers is called a **dimer**. The polymer (nylon 6,6) forms as the dimer continues to add more monomers. Polymers that eliminate an atom or a small group of atoms during polymerization are called **condensation polymers**. Nylon 6,6 and other similar nylons can be drawn into fibers and used to make consumer products such as panty hose, carpet fibers, and fishing line. Table 20.11 shows other condensation polymers.

Chapter in Review

Key Terms

Section 20.1

organic molecule (903)
organic chemistry (904)

Section 20.3

alkane (906)
alkene (906)

alkyne (906)
aromatic hydrocarbon (906)
aliphatic hydrocarbon (906)

structural isomers (907)
structural formula (907)
stereoisomers (909)

optical isomers (909)
enantiomers (910)
chiral (910)
dextrorotatory (911)
levorotatory (911)
racemic mixture (911)

Section 20.4
saturated hydrocarbon (912)

Section 20.5
unsaturated hydrocarbon (917)

geometric (cis–trans) isomerism (921)

Section 20.7
phenyl group (925)
disubstituted benzene (926)

Section 20.9
hydroxyl group (928)
alcohol (928)

Section 20.10
aldehyde (931)

ketone (931)
carbonyl group (931)
addition reaction (934)

Section 20.11
carboxylic acid (934)
esters (934)
condensation reaction (936)

Section 20.12
ether (937)

Section 20.13
amine (938)

Section 20.14
addition polymer (939)
dimer (941)
condensation polymer (941)

Key Concepts

Fragrances and Odors (20.1)

Organic chemistry is the study of organic compounds, those that contain carbon (and other elements including hydrogen, oxygen, and nitrogen). These compounds produce many common odors.

Carbon (20.2)

Carbon forms more compounds than all the other elements combined for several reasons. Its four valence electrons (combined with its size) allow carbon to form four bonds (in the form of single, double, or triple bonds). Carbon also has the capacity to catenate, to form long chains, because of the strength of the carbon–carbon bond.

Before the properties of carbon were researched, some scientists believed that organic compounds had a vital force and could therefore never be synthesized in the laboratory. In 1828, Friedrich Wöhler disproved this theory by synthesizing urea from an inorganic compound.

Hydrocarbons (20.3)

Organic compounds containing only carbon and hydrogen are called hydrocarbons, most commonly known as the key components of our world's fuels. Hydrocarbons can be divided into four different types: alkanes, alkenes, alkynes, and aromatic hydrocarbons.

Stereoisomers are molecules with the same atoms bonded in the same order, but arranged differently in space. Optical isomerism, a type of stereoisomerism, occurs when two molecules are nonsuperimposable mirror images of one another.

Alkanes (20.4)

Alkanes are saturated hydrocarbons—they contain only single bonds and can therefore be represented by the generic formula C_nH_{2n+2}. Alkane names always end in -ane.

Alkenes and Alkynes (20.5)

Alkenes and alkynes are unsaturated hydrocarbons—they contain double bonds (alkenes) or triple bonds (alkynes) and are represented by the generic formula C_nH_{2n} and C_nH_{2n-2}, respectively. Alkene names always end in -ene and alkynes end in -yne.

Because rotation about a double bond is severely restricted, geometric (or cis–trans) isomerism occurs in alkenes.

Hydrocarbon Reactions (20.6)

The most common hydrocarbon reaction is probably combustion in which hydrocarbons react with oxygen to form carbon dioxide and water; this reaction is exothermic and is used to provide most of our society's energy. Alkanes can also undergo substitution reactions, where heat or light causes another atom, commonly a halogen such as bromine, to be substituted for a hydrogen atom. Unsaturated hydrocarbons undergo addition reactions. If the addition reaction is between two unsymmetrical molecules, Markovnikov's rule predicts that the positive end of the polar reagent will add to the carbon with the most hydrogen atoms.

Aromatic Hydrocarbons (20.7)

Aromatic hydrocarbons contain six-membered benzene rings represented with alternating single and double bonds that become equivalent through resonance. These compounds are called aromatic because they often produce pleasant fragrances. Because of the stability of the aromatic ring, benzene is more stable than a straight-chain alkene, and it undergoes substitution rather than addition reactions.

Functional Groups (20.8)

Characteristic groups of atoms, such as hydroxyl ($-OH$), are called functional groups. Molecules that contain the same functional group have similar chemical and physical properties, and are referred to as families.

Alcohols (20.9)

The family of alcohols contains the $-OH$ group and is named with the suffix -ol. Alcohols are commonly used in gasoline, in alcoholic beverages, and in sterilization procedures. Alcohols undergo substitution reactions, in which a substituent such as a halogen replaces the hydroxyl group. They also undergo elimination reactions, in which water is eliminated across a bond to form an alkene, and oxidation or reduction reactions. Alcohols also react with active metals to form alkoxide ions and hydrogen gas.

Aldehydes and Ketones (20.10)

Aldehydes and ketones both contain a carbonyl group (a carbon atom double-bonded to oxygen); in aldehydes, this group is at the end of a carbon chain while in ketones, it is between two other carbon atoms. Aldehydes are named with the suffix -al and ketones with the suffix -one. Formaldehyde is used as a preservative, while acetone is an organic solvent. A carbonyl can be formed by the oxidation of an alcohol or reverted to an alcohol by reduction. Like alkenes, carbonyls undergo addition reactions; however, because the carbon–oxygen bond is highly polar, the electronegative component of the reagent always adds to the carbon atom, and the less electronegative part adds to the oxygen.

Carboxylic Acids and Esters (20.11)

Carboxylic acids contain a carbonyl group and a hydroxide on the same carbon and are named with the suffix -oic acid. Esters contain a carbonyl group bonded to an oxygen atom which is in turn bonded to an R group; they are named with the suffix -oate. Carboxylic acids taste sour, such as acetic acid in vinegar, while esters smell sweet. Carboxylic acids react as weak acids but can also form esters through condensation reactions with alcohols.

Ethers (20.12)

The family of ethers contains an oxygen atom between two R groups. Ethers are named with the ending *-yl ether*. A common ether, diethyl ether, is used as a solvent in many organic experiments.

Amines (20.13)

Amines are organic compounds that contain nitrogen and are named with the suffix *-amine*. They are known for their terrible odors, such as the smell of decaying animal flesh produced by cadaverine. Amines act as weak bases and will produce a salt when mixed with a strong acid.

The combination of an amine with a carboxylic acid leads to a condensation reaction; this reaction is used by our bodies to produce proteins from amino acids.

Polymers (20.14)

Polymers are long, chainlike molecules that consist of repeating units called monomers. Polymers can be natural or synthetic; an example of a common synthetic polymer is polyethylene, the plastic in soda bottles and garbage bags. Polyethylene is an addition polymer, a polymer formed without the elimination of any atoms. Condensation polymers, such as nylon, are formed by the elimination of small groups of atoms.

Key Equations and Relationships

Halogen Substitution Reactions in Alkanes (20.6)

$$\underset{\text{alkane}}{R-H} + \underset{\text{halogen}}{X_2} \xrightarrow{\text{heat or light}} \underset{\text{haloalkane}}{R-X} + \underset{\text{hydrogen halide}}{HX}$$

Common Functional Groups (20.8)

Family	General Formula	Condensed General Formula	Example	Name
Alcohols	R—OH	ROH	CH_3CH_2OH	Ethanol (ethyl alcohol)
Ethers	R—O—R	ROR	CH_3OCH_3	Dimethyl ether
Aldehydes	R—C(=O)—H	RCHO	H_3C—C(=O)—H	Ethanal (acetaldehyde)
Ketones	R—C(=O)—R	RCOR	H_3C—C(=O)—CH_3	Propanone (acetone)
Carboxylic acids	R—C(=O)—OH	RCOOH	H_3C—C(=O)—OH	Acetic acid
Esters	R—C(=O)—OR	RCOOR	H_3C—C(=O)—OCH_3	Methyl acetate
Amines	R—N(R)—R	R_3N	H_3CH_2C—N(H)—H	Ethylamine

Alcohol Reactions (20.9)

Substitution $ROH + HBr \longrightarrow R-Br + H_2O$

Oxidation

$$\underset{\text{Alcohol}}{R-CH_2-\underset{\underset{OH}{|}}{C}H_2} \xrightarrow{\text{lose H}} \underset{\text{Aldehyde}}{R-CH_2-\underset{\underset{O}{\|}}{C}H}$$

Elimination $\underset{\underset{OH}{|}}{R-CH_2-CH_2} \xrightarrow{H_2SO_4} R-CH=CH_2 + H_2O$

Carboxylic Acid Condensation Reactions (20.11)

$$\underset{\text{Acid}}{R-\overset{\overset{O}{\|}}{C}-OH} + \underset{\text{Alcohol}}{HO-R'} \xrightarrow{H_2SO_4} \underset{\text{Ester}}{R-\overset{\overset{O}{\|}}{C}-O-R'} + \underset{\text{Water}}{H_2O}$$

Amine Acid–Base Reactions (20.13)

$$RNH_2(aq) + H_2O(l) \longrightarrow RNH_3^+(aq) + OH^-(aq)$$

Amine–Carboxylic Acid Condensation Reactions (20.13)

$$CH_3COOH(aq) + HNHR(aq) \longrightarrow CH_3CONHR(aq) + HOH(l)$$

Key Skills

Writing Structural Formulas for Hydrocarbons (20.3)
- Example 20.1 • For Practice 20.1 • Exercises 37, 38

Naming Alkanes (20.4)
- Examples 20.2, 20.3, 20.4 • For Practice 20.2, 20.3, 20.4 • Exercises 43, 44

Naming Alkenes and Alkynes (20.5)
- Example 20.5 • For Practice 20.5 • Exercises 53–56

Writing Reactions: Addition Reactions (20.6)
- Example 20.6 • For Practice 20.6 • Exercises 59–62

Writing Reactions: Alcohols (20.9)
- Example 20.7 • For Practice 20.7 • Exercises 75, 76

Exercises

Review Questions

1. What kinds of molecules often impact our sense of smell?

2. What is organic chemistry?

3. What is unique about carbon and carbon-based compounds? Why did life evolve around carbon?

4. Why does carbon form such a large diversity of compounds?

5. Why does silicon not exhibit the great diversity of compounds that carbon does?

6. Describe the geometry and hybridization about a carbon atom that forms:
   a. four single bonds
   b. two single bonds and one double bond
   c. one single bond and one triple bond

7. What are hydrocarbons? What are their main uses?

8. What are the main classifications of hydrocarbons? What are their generic molecular formulas?

9. Explain the differences between a structural formula, a condensed structural formula, a carbon skeleton formula, a ball-and-stick model, and a space-filling model.

10. What are structural isomers? How do the properties of structural isomers differ from one another?

11. What are optical isomers? How do the properties of optical isomers differ from one another?

12. Define each of the following terms related to optical isomerism: enantiomers, chiral, dextrorotatory, levorotatory, racemic mixture.

13. What is the difference between saturated and unsaturated hydrocarbons?

14. What are the main differences in the way that alkanes, alkenes, and alkynes are named?

15. Explain geometric isomerism in alkenes. How do the properties of geometric isomers differ from one another?

16. What are hydrocarbon combustion reactions? Give an example.

17. What kinds of reactions are common to alkanes? Give an example of each.

18. Describe each of the following kinds of reactions.
    a. substitution reaction   b. addition reaction
    c. elimination reaction

19. What kinds of reactions are common to alkenes? Give an example of each.

20. Explain Markovnikov's rule and give an example of a reaction to which it applies.

21. What is the structure of benzene? What are the different ways in which this structure is represented?

22. What kinds of reactions are common to aromatic compounds? Give an example of each.

23. What is a functional group? Give some examples.

24. What is the generic structure of alcohols? Write the structures of two specific alcohols.

25. Explain oxidation and reduction with respect to organic compounds.

26. What kinds of reactions are common to alcohols? Give an example of each.

27. What are the generic structures for aldehydes and ketones? Write a structure for a specific aldehyde and ketone.

28. What kind of reactions are common to aldehydes and ketones? Give an example of each.

29. What are the generic structures for carboxylic acids and esters? Write a structure for a specific carboxylic acid and ester.

30. What kind of reactions are common to carboxylic acids and esters? Give an example of each.

31. What is the generic structure of ethers? Write the structures of two specific ethers.

32. What is the generic structure of amines? Write the structures of two specific amines.

33. What is a polymer? What is the difference between a polymer and a copolymer?

34. What is the difference between an addition polymer and a condensation polymer?

Problems by Topic

Hydrocarbons

35. Based on the molecular formula, determine whether each of the following is an alkane, alkene, or alkyne. (Assume that the hydrocarbons are noncyclical and there is no more than one multiple bond.)
  a. C_5H_{12} **b.** C_3H_6 **c.** C_7H_{12} **d.** $C_{11}H_{22}$

36. Based on the molecular formula, determine whether each of the following is an alkane, alkene, or alkyne. (Assume that the hydrocarbons are noncyclical and there is no more than one multiple bond.)
  a. C_8H_{16} **b.** C_4H_6 **c.** C_7H_{16} **d.** C_2H_2

37. Write structural formulas for each of the nine structural isomers of heptane.

38. Write structural formulas for any 6 of the 18 structural isomers of octane.

39. Determine whether the following molecules will exhibit optical isomerism:
  a. CCl_4

  b. 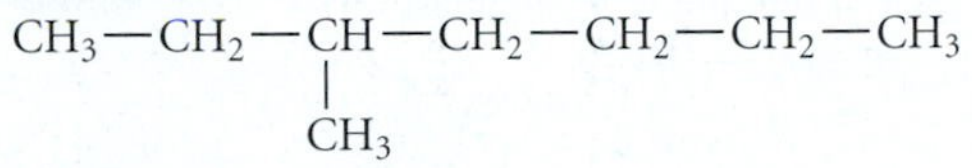 $CH_3-CH_2-CH-CH_2-CH_2-CH_2-CH_3$
                         $|$
                         CH_3

  c.

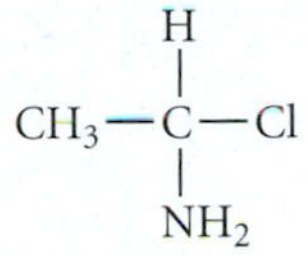

  d. $CH_3CHClCH_3$

40. Determine whether the following molecules will exhibit optical isomerism:
  a. $CH_3CH_2CHClCH_3$ **b.** $CH_3CCl_2CH_3$

  c.

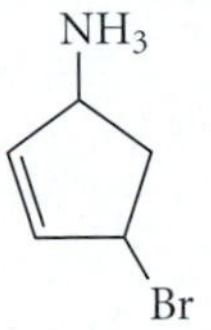

  d.

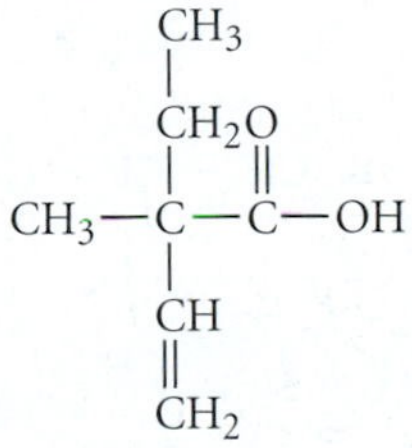

41. Determine whether the following pairs are the same molecules or enantiomers:

  a.

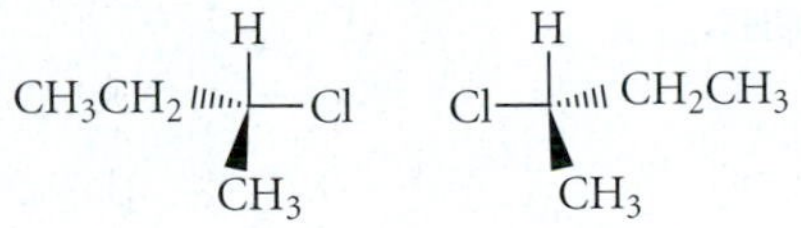

  b.

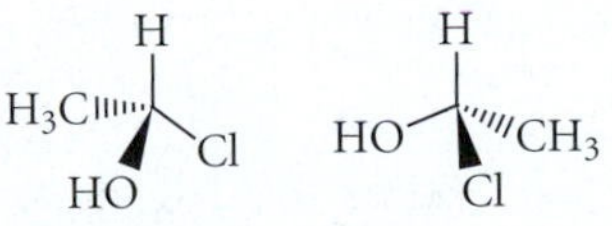

  c.

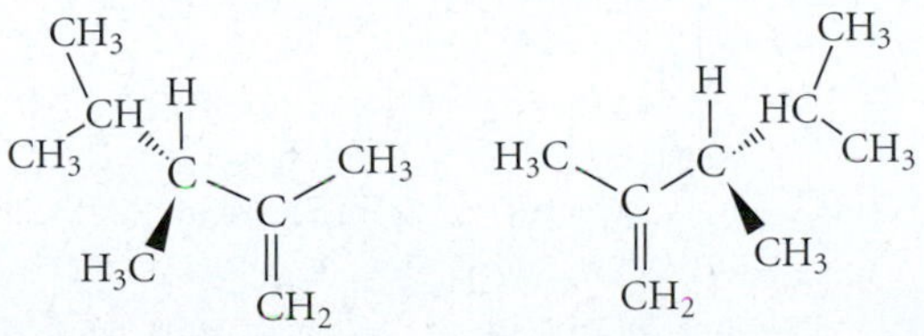

42. Determine whether the following pairs are the same molecules or enantiomers:

  a.

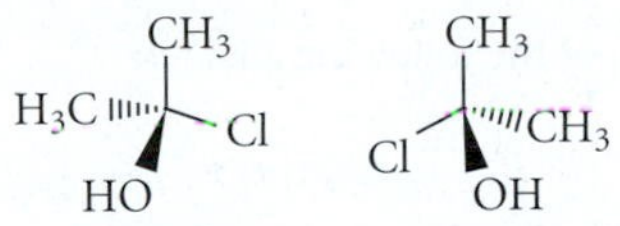

  b.

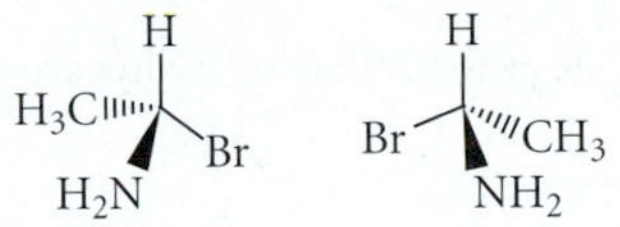

  c.

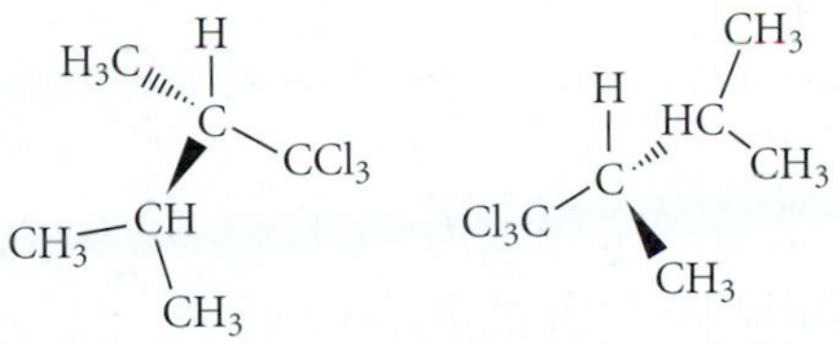

Alkanes

43. Name each of the following alkanes:
  a. $CH_3-CH_2-CH_2-CH_2-CH_3$

  b. 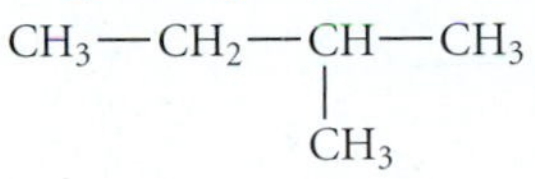 $CH_3-CH_2-CH-CH_3$
                      $|$
                      CH_3

  c.

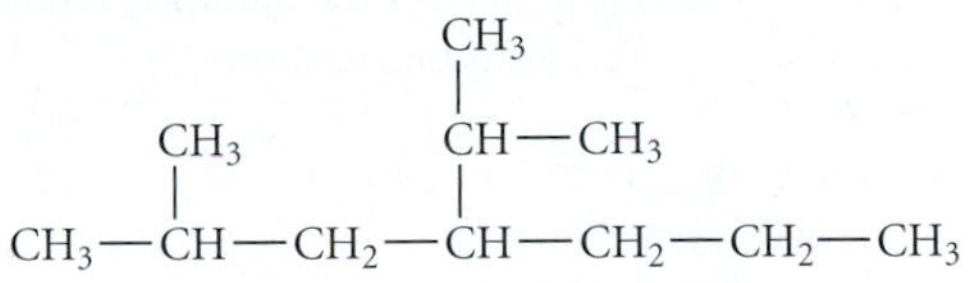

  d.

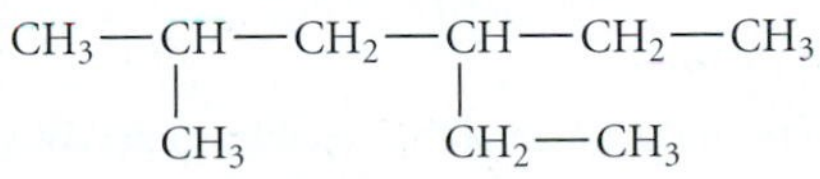

44. Name each of the following alkanes:
  a. 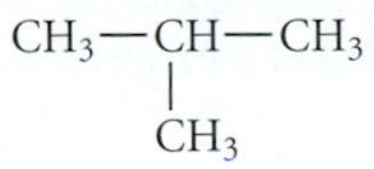 $CH_3-CH-CH_3$
             $|$
             CH_3

  b.

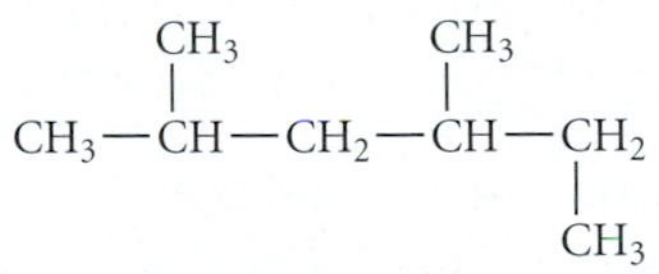

  c.

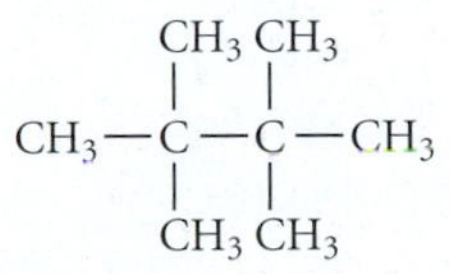

  d.

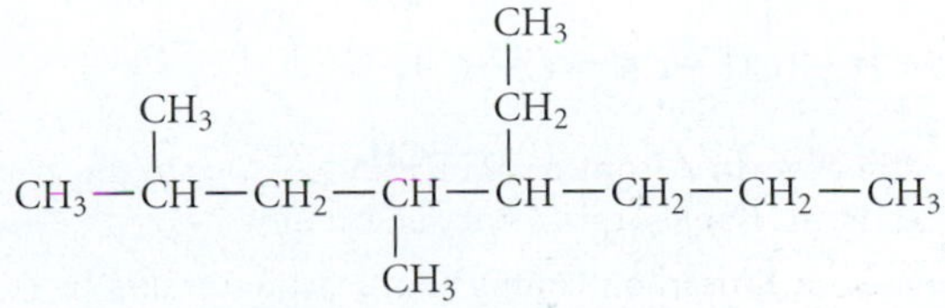

45. Draw a structure for each of the following alkanes:
  a. 3-ethylhexane
  b. 3-ethyl-3-methylpentane
  c. 2,3-dimethylbutane
  d. 4,7-diethyl-2,2-dimethylnonane

46. Draw a structure for each of the following alkanes:
  a. 2,2-dimethylpentane
  b. 3-isopropylheptane
  c. 4-ethyl-2,2-dimethylhexane
  d. 4,4-diethyloctane

47. Complete and balance each of the following hydrocarbon combustion reactions:
  a. $CH_3CH_2CH_3 + O_2 \longrightarrow$
  b. $CH_3CH_2CH{=}CH_2 + O_2 \longrightarrow$
  c. $CH{\equiv}CH + O_2 \longrightarrow$

48. Complete and balance each of the following hydrocarbon combustion reactions:
  a. $CH_3CH_2CH_2CH_3 + O_2 \longrightarrow$
  b. $CH_2{=}CHCH_3 + O_2 \longrightarrow$
  c. $CH{\equiv}CCH_2CH_3 + O_2 \longrightarrow$

49. What are all the possible products for the following alkane substitution reactions? (Assume monosubstitution.)
  a. $CH_3CH_3 + Br_2 \longrightarrow$
  b. $CH_3CH_2CH_3 + Cl_2 \longrightarrow$
  c. $CH_2Cl_2 + Br_2 \longrightarrow$
  d.
$$CH_3{-}\underset{\underset{CH_3}{|}}{CH}{-}CH_3 + Cl_2 \longrightarrow$$

50. What are all the possible products for the following alkane substitution reactions? (Assume monosubstitution.)
  a. $CH_4 + Cl_2 \longrightarrow$
  b. $CH_3CH_2Br + Br_2 \longrightarrow$
  c. $CH_3CH_2CH_2CH_3 + Cl_2 \longrightarrow$
  d. $CH_3CHBr_2 + Br_2 \longrightarrow$

Alkenes and Alkynes

51. Write structural formulas for all of the possible isomers of *n*-hexene that can be formed by moving the position of the double bond.

52. Write structural formulas for all of the possible isomers of *n*-pentyne that can be formed by moving the position of the triple bond.

53. Name each of the following alkenes:
  a. $CH_2{=}CH{-}CH_2{-}CH_3$

  b.
$$CH_3{-}\underset{\underset{CH_3}{|}}{CH}{-}\underset{\underset{CH_3}{|}}{C}{=}CH{-}CH_3$$

  c.
$$CH_2{=}HC{-}\underset{\underset{\underset{\underset{CH_3}{|}}{CH}}{|}}{CH}{-}CH_2{-}CH_2{-}CH_3$$

  d.
$$CH_3{-}\underset{\underset{CH_3}{|}}{CH}{-}CH{=}\underset{\underset{\underset{\underset{CH_3}{|}}{CH_2}}{|}}{C}{-}CH_3$$

54. Name each of the following alkenes:
  a. $CH_3{-}CH_2{-}CH{=}CH{-}CH_2{-}CH_3$

  b.
$$CH_3{-}\underset{\underset{CH_3}{|}}{CH}{-}CH{=}CH{-}CH_3$$

  c.
$$CH_3{-}\underset{\underset{CH_3}{|}}{CH}{-}CH{=}\underset{\underset{\underset{\underset{CH_3}{|}}{CH_2}}{|}}{C}{-}\underset{\underset{CH_3}{|}}{CH}{-}CH_3$$

  d.
$$CH_3{-}\underset{\underset{CH_3}{|}}{\overset{\overset{CH_3}{|}}{C}}{-}CH{=}C{-}CH_2{-}CH_3$$

55. Name each of the following alkynes:
  a. $CH_3{-}C{\equiv}C{-}CH_3$

  b.
$$CH_3{-}C{\equiv}C{-}\underset{\underset{CH_3}{|}}{\overset{\overset{CH_3}{|}}{C}}{-}CH_2{-}CH_3$$

  c.
$$CH{\equiv}C{-}\underset{\underset{\underset{\underset{CH_3}{|}}{CH{-}CH_3}}{|}}{CH}{-}CH_2{-}CH_2{-}CH_3$$

  d.
$$CH_3{-}\underset{\underset{\underset{\underset{CH_3}{|}}{CH_2}}{|}}{CH}{-}C{\equiv}C{-}\underset{\underset{\underset{\underset{CH_3}{|}}{CH_2}}{|}}{\overset{\overset{CH_3}{|}}{CH}}{-}CH_2$$

56. Name each of the following alkynes:
  a.
$$CH{\equiv}C{-}\underset{\underset{CH_3}{|}}{CH}{-}CH_3$$

  b.
$$CH_3{-}C{\equiv}C{-}\underset{\underset{CH_3}{|}}{CH}{-}\overset{\overset{CH_3}{|}}{CH}{-}CH_2{-}CH_3$$

  c.
$$CH{\equiv}C{-}\underset{\underset{\underset{\underset{CH_3}{|}}{CH_2}}{|}}{\overset{\overset{CH_3}{|}}{C}}{-}CH_2{-}CH_3$$

  d.
$$CH_3{-}C{\equiv}C{-}\underset{\underset{\underset{\underset{CH_3}{|}}{CH_2}}{|}}{CH}{-}\underset{\underset{CH_3}{|}}{\overset{\overset{CH_3}{|}}{C}}{-}CH_3$$

57. Provide correct structures for each of the following:
   a. 4-octyne
   b. 3-nonene
   c. 3,3-dimethyl-1-pentyne
   d. 5-ethyl-3,6-dimethyl-2-heptene

58. Provide correct structures for each of the following:
   a. 2-hexene
   b. 1-heptyne
   c. 4,4-dimethyl-2-hexene
   d. 3-ethyl-4-methyl-2-pentene

59. What are the products of the following alkene addition reactions?
   a. $CH_3-CH=CH-CH_3 + Cl_2 \longrightarrow$

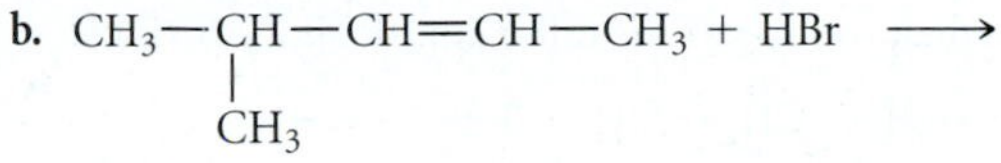

   b. $CH_3-\underset{\underset{CH_3}{|}}{CH}-CH=CH-CH_3 + HBr \longrightarrow$

   c. $CH_3-CH_2-CH=CH-CH_3 + Br_2 \longrightarrow$

   d. $CH_3-\underset{\underset{CH_3}{|}}{CH}-CH=\overset{\overset{CH_3}{|}}{C}-CH_3 + HCl \longrightarrow$

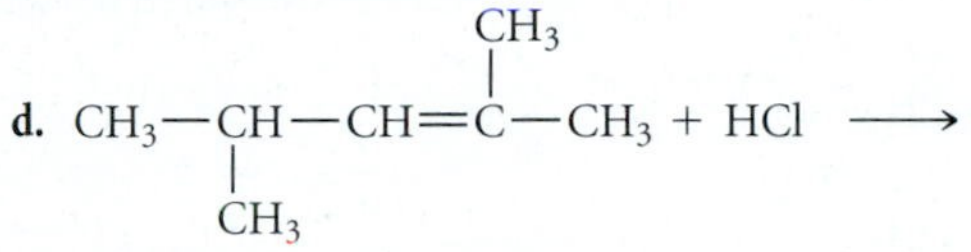

60. What are the products of the following alkene addition reactions?
   a. $CH_3-\underset{\underset{CH_3}{|}}{CH}-CH=CH_2 + Br_2 \longrightarrow$

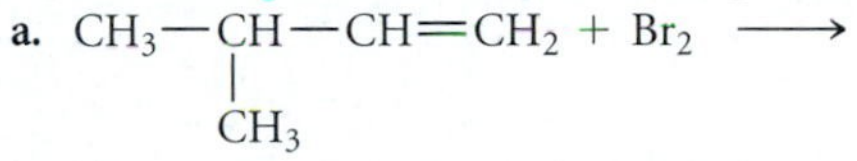

   b. $CH_2=CH-CH_3 + Cl_2 \longrightarrow$

   c. $CH_3-\underset{\underset{CH_3}{|}}{\overset{\overset{CH_3}{|}}{C}}-CH=CH_2 + HCl \longrightarrow$

   d. $CH_3-\underset{\underset{CH_2-CH_3}{|}}{CH}-CH=\overset{\overset{CH_3}{|}}{C}-CH_3 + HBr \longrightarrow$

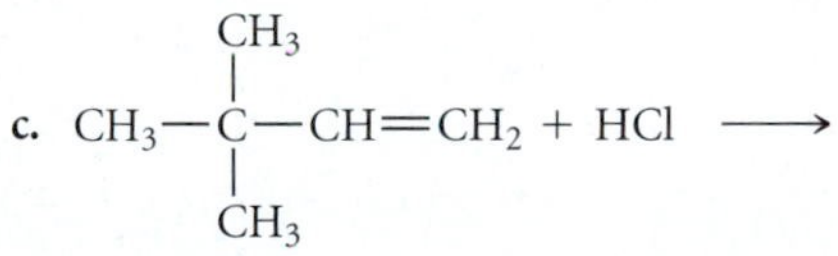

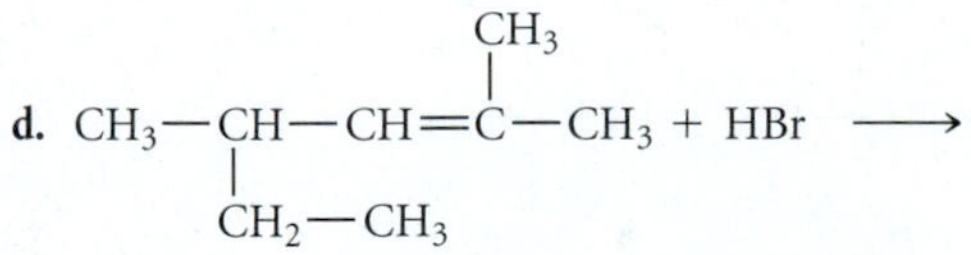

61. Complete the following hydrogenation reactions:
   a. $CH_2=CH-CH_3 + H_2 \longrightarrow$

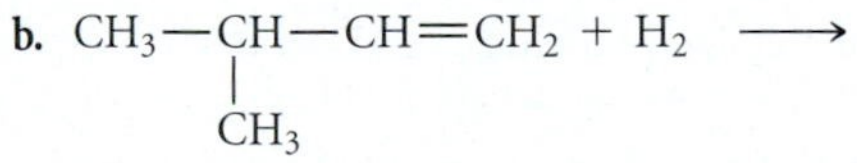

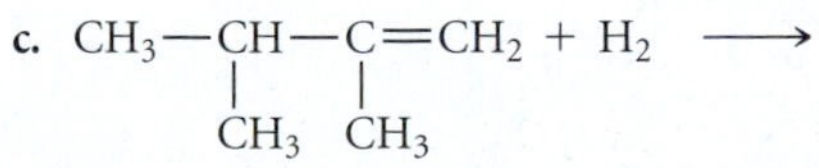

   b. $CH_3-\underset{\underset{CH_3}{|}}{CH}-CH=CH_2 + H_2 \longrightarrow$

   c. $CH_3-\underset{\underset{CH_3}{|}}{CH}-\overset{\overset{CH_3}{|}}{C}=CH_2 + H_2 \longrightarrow$

62. Complete the following hydrogenation reactions:
   a. $CH_3-CH_2-CH=CH_2 + H_2 \longrightarrow$

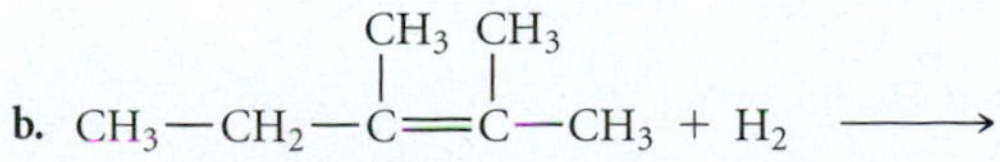

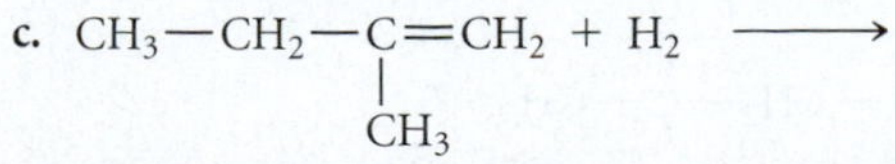

   b. $CH_3-CH_2-\underset{\underset{CH_3}{|}}{\overset{\overset{CH_3}{|}}{C}}=\overset{\overset{CH_3}{|}}{C}-CH_3 + H_2 \longrightarrow$

   c. $CH_3-CH_2-\underset{\underset{CH_3}{|}}{C}=CH_2 + H_2 \longrightarrow$

Aromatic Hydrocarbons

63. Name each of the following monosubstituted benzenes:

   a.   **b.**   **c.**

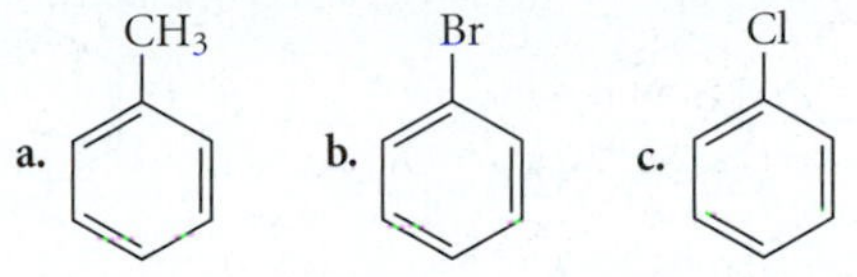

64. Name each of the following monosubstituted benzenes:

   a.   **b.**   **c.**

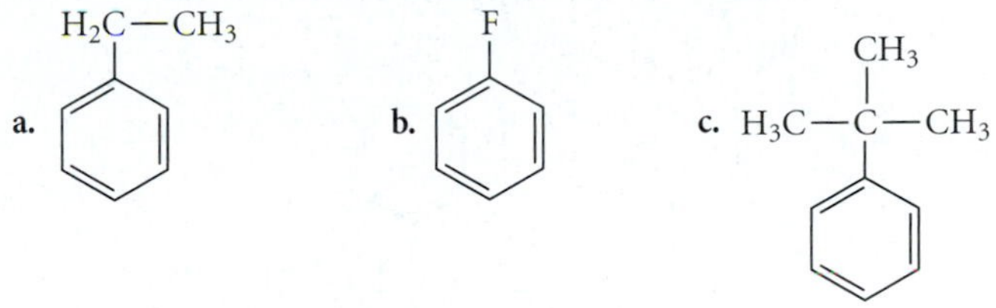

65. Name each of the following compounds in which the benzene ring is best treated as a substituent:

   a. $CH_3-CH-CH_2-\underset{\underset{CH_2-CH_3}{|}}{CH}-CH_2-\overset{\overset{C_6H_5}{|}}{CH}-CH_2-CH_3$
   with $\underset{CH_3}{|}$ on second carbon

   b. $CH_3-\underset{\underset{C_6H_5}{|}}{CH}-CH=CH-CH_2-CH_2-CH_2-CH_3$

   c. $CH_3-C\equiv C-CH-\underset{\underset{CH_3}{|}}{CH}-\underset{\underset{CH_3}{|}}{CH}-CH_2-CH_3$
   with phenyl

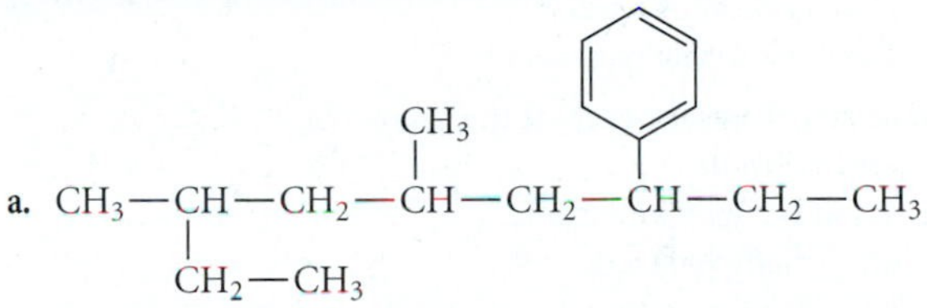

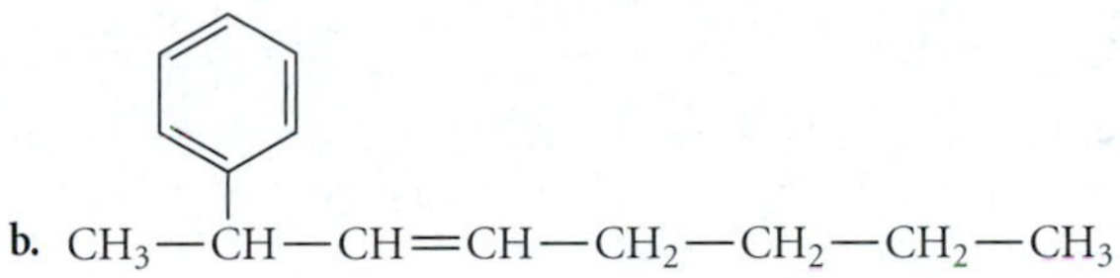

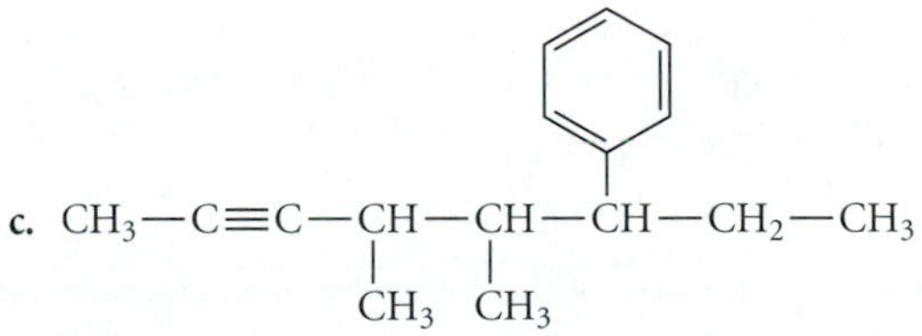

66. Name each of the following compounds in which the benzene ring is best treated as a substituent:

   a. $H_3C-CH_2-\overset{\overset{C_6H_5}{|}}{CH}-\underset{\underset{H_3C}{|}}{CH}-CH_2-CH_3$

   b. $\overset{\overset{C_6H_5}{|}}{CH_2}-CH_2-CH_2-CH_2-C\equiv C-CH_3$

   c. $CH_3-CH-\underset{\underset{CH_3}{|}}{CH}-\overset{\overset{CH_3}{|}}{C}=\overset{\overset{C_6H_5}{|}}{CH}-\underset{\underset{CH_3}{|}}{CH}-CH_2-CH_3$

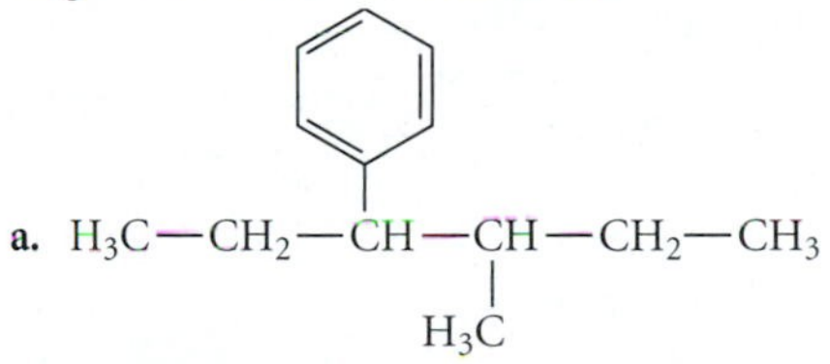

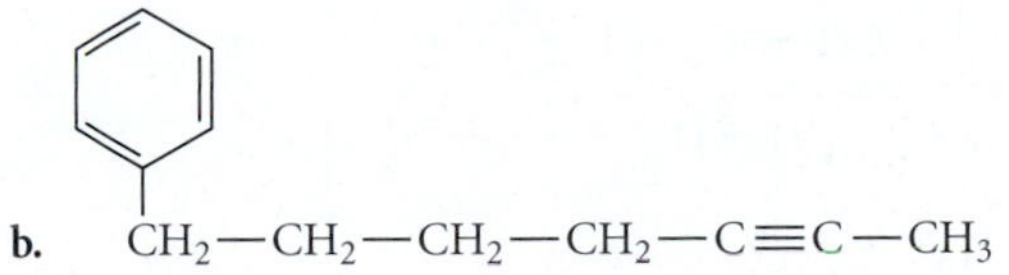

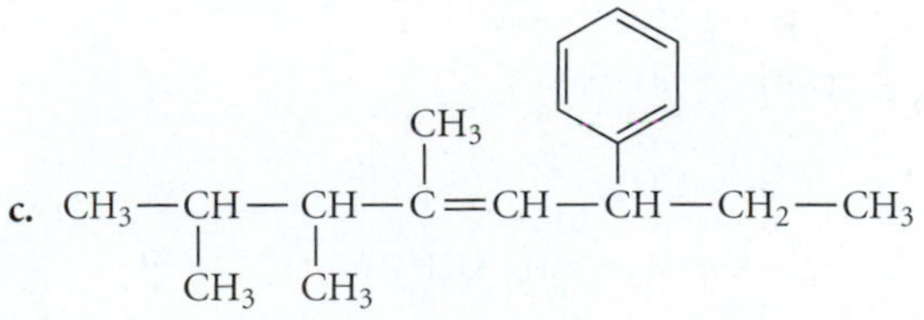

67. Name each of the following disubstituted benzenes:

a. 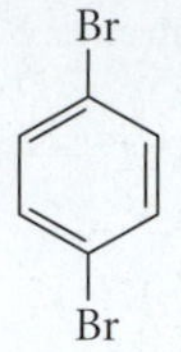 *(structure: benzene ring with Br at top and Br at bottom — para-dibromobenzene)*

b. 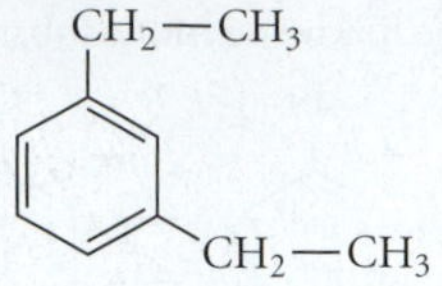 *(structure: benzene ring with CH_2-CH_3 at top and CH_2-CH_3 at lower position — meta)*

c. 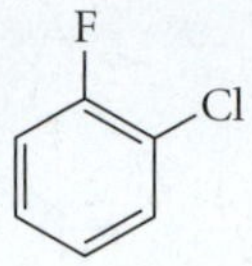 *(structure: benzene ring with F at top and Cl at adjacent position)*

68. Name each of the following disubstituted benzenes:

a. 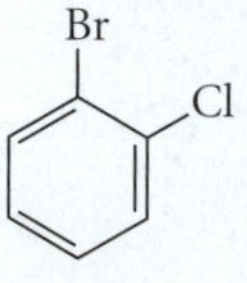 *(structure: benzene ring with Br and Cl on adjacent carbons)*

b. 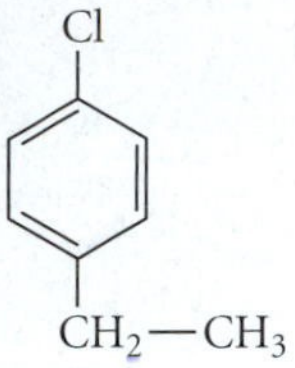 *(structure: benzene ring with Cl at top and CH_2-CH_3 at bottom — para)*

c. 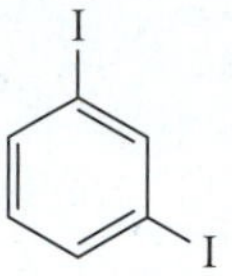 *(structure: benzene ring with I at top and I at meta position)*

69. Draw structures for each of the following:
   a. isopropylbenzene
   b. *meta*-dibromobenzene
   c. 1-chloro-4-methylbenzene

70. Draw structures for each of the following:
   a. ethylbenzene
   b. 1-iodo-2-methylbenzene
   c. *para*-diethylbenzene

71. What are the products of the following aromatic substitution reactions?

a. 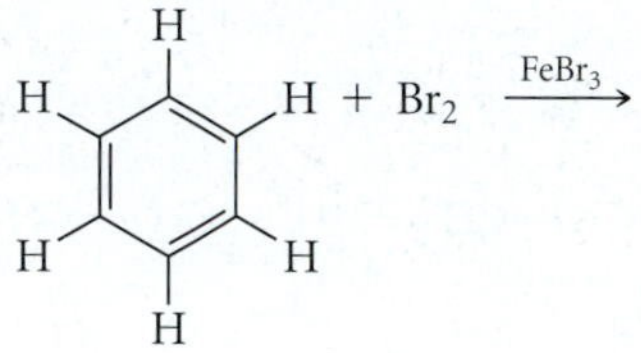 *(benzene with all H shown)* $+ Br_2 \xrightarrow{FeBr_3}$

b. 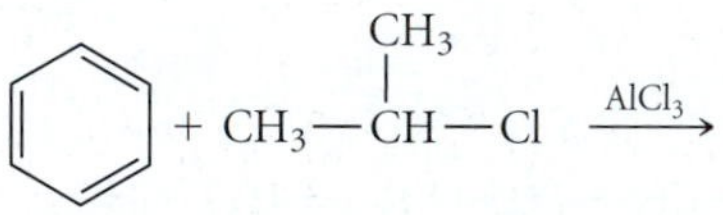 *(benzene ring)* $+ CH_3-\overset{\overset{\displaystyle CH_3}{|}}{CH}-Cl \xrightarrow{AlCl_3}$

72. What are the products of the following aromatic substitution reactions?

a. 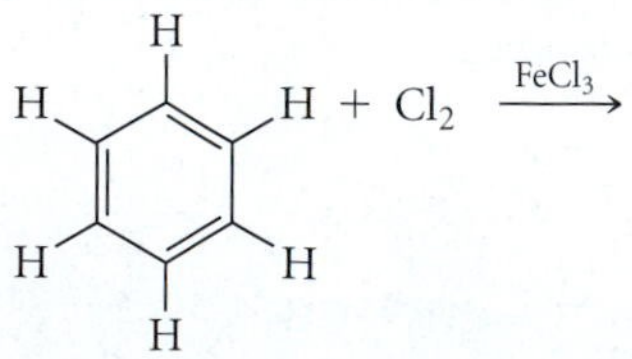 *(benzene with all H shown)* $+ Cl_2 \xrightarrow{FeCl_3}$

b. 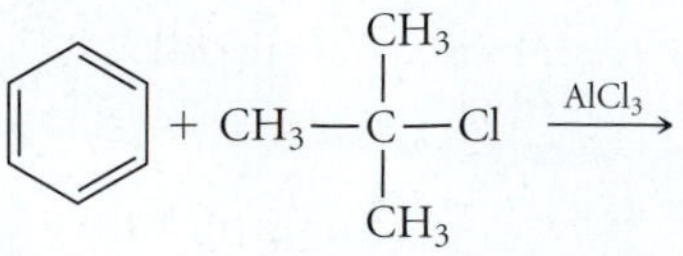 *(benzene ring)* $+ CH_3-\overset{\overset{\displaystyle CH_3}{|}}{\underset{\underset{\displaystyle CH_3}{|}}{C}}-Cl \xrightarrow{AlCl_3}$

Alcohols

73. Name each of the following alcohols:

a. $CH_3-CH_2-CH_2-OH$

b. 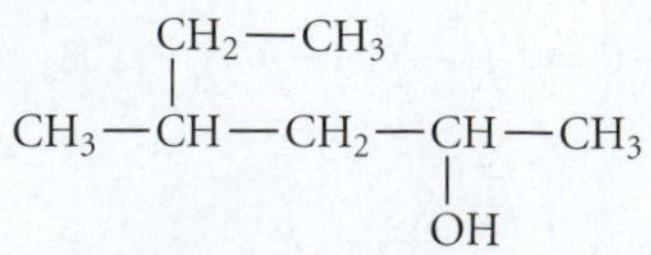 $CH_3-\overset{\overset{\displaystyle CH_2-CH_3}{|}}{CH}-CH_2-\underset{\underset{\displaystyle OH}{|}}{CH}-CH_3$

c. 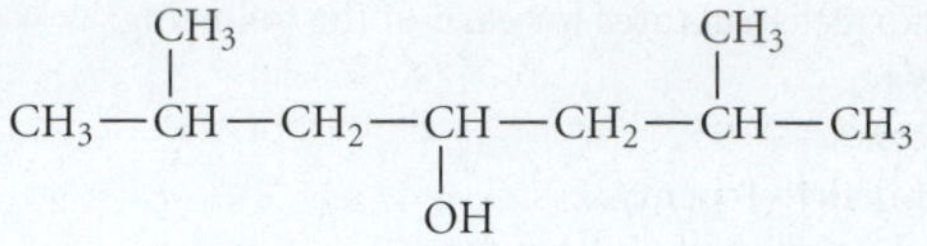 $CH_3-\overset{\overset{\displaystyle CH_3}{|}}{CH}-CH_2-CH-CH_2-\overset{\overset{\displaystyle CH_3}{|}}{CH}-CH_3$, with OH below the middle CH

d. 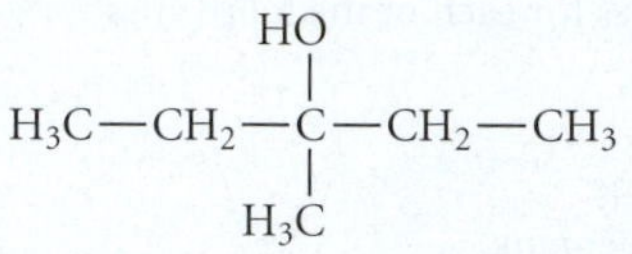 $H_3C-CH_2-\overset{\overset{\displaystyle HO}{|}}{\underset{\underset{\displaystyle H_3C}{|}}{C}}-CH_2-CH_3$

74. Draw a structure for each of the following alcohols:
   a. 2-butanol   b. 2-methyl-1-propanol
   c. 3-ethyl-1-hexanol d. 2-methyl-3-pentanol

75. What are the products of the following alcohol reactions?

a. $CH_3-CH_2-CH_2-OH + HBr \longrightarrow$

b. 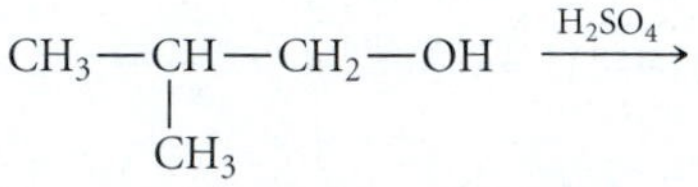 $CH_3-\overset{\overset{\displaystyle }{}}{CH}-CH_2-OH \xrightarrow{H_2SO_4}$ with CH_3 below CH

c. $CH_3-CH_2-OH + Na \longrightarrow$

d. 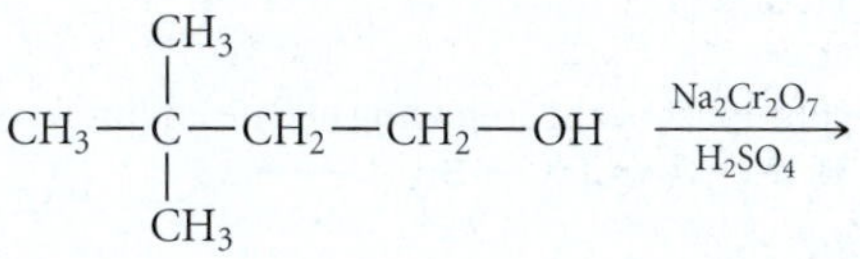 $CH_3-\overset{\overset{\displaystyle CH_3}{|}}{\underset{\underset{\displaystyle CH_3}{|}}{C}}-CH_2-CH_2-OH \xrightarrow[H_2SO_4]{Na_2Cr_2O_7}$

76. What are the products of the following alcohol reactions?

a. 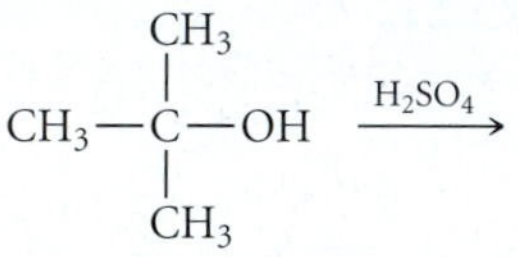 $CH_3-\overset{\overset{\displaystyle CH_3}{|}}{\underset{\underset{\displaystyle CH_3}{|}}{C}}-OH \xrightarrow{H_2SO_4}$

b. 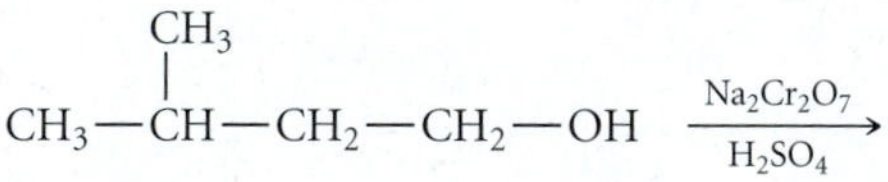 $CH_3-\overset{\overset{\displaystyle CH_3}{|}}{CH}-CH_2-CH_2-OH \xrightarrow[H_2SO_4]{Na_2Cr_2O_7}$

c. $CH_3-CH_2-OH + HCl \longrightarrow$

d. 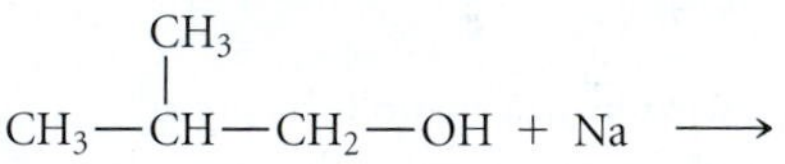 $CH_3-\overset{\overset{\displaystyle CH_3}{|}}{CH}-CH_2-OH + Na \longrightarrow$

Aldehydes and Ketones

77. Name each of the following aldehydes and ketones:

a. 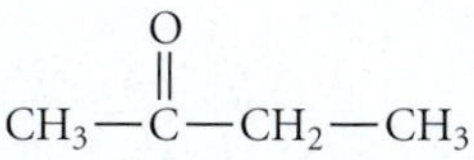 $CH_3-\overset{\overset{\displaystyle O}{||}}{C}-CH_2-CH_3$

b. $CH_3-CH_2-CH_2-CH_2-\overset{\overset{\displaystyle O}{||}}{CH}$

c. $CH_3-\overset{\overset{\displaystyle CH_3}{|}}{\underset{\underset{\displaystyle CH_3}{|}}{C}}-CH_2-\overset{\overset{\displaystyle CH_3}{|}}{CH}-CH_2-\overset{\overset{\displaystyle O}{||}}{C}-H$

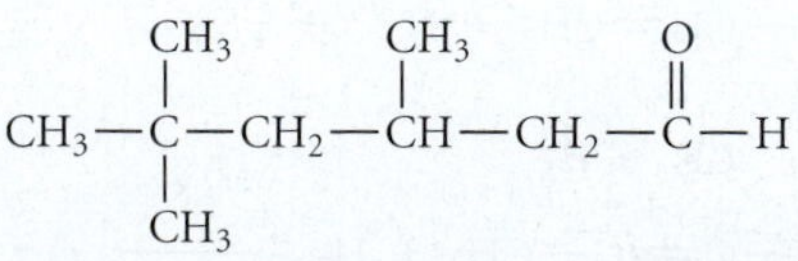

d. 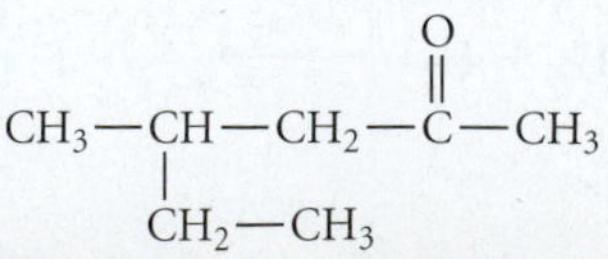 $CH_3-\underset{\underset{\displaystyle CH_2-CH_3}{|}}{CH}-CH_2-\overset{\overset{\displaystyle O}{||}}{C}-CH_3$

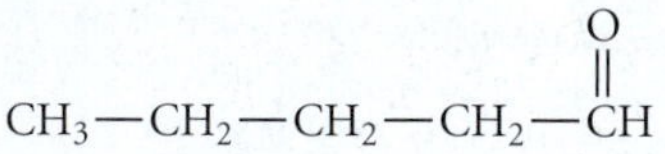

78. Draw the structure of each of the following aldehydes and ketones:
    a. hexanal
    b. 2-pentanone
    c. 2-methylbutanal
    d. 4-heptanone

79. Determine the product of the following addition reaction:

$$CH_3-CH_2-CH_2-\overset{\overset{\displaystyle O}{\|}}{C}H \ + \ H-C\equiv N \ \xrightarrow{NaCN}$$

80. Determine the product of the following addition reaction:

$$CH_3-\overset{\overset{\displaystyle O}{\|}}{C}-CH_2-CH_3 \ + \ HCN \ \xrightarrow{NaCN}$$

Carboxylic Acids and Esters

81. Name each of the following carboxylic acids and esters:

    a. $CH_3-CH_2-CH_2-\overset{\overset{\displaystyle O}{\|}}{C}-O-CH_3$

    b. $CH_3-CH_2-\overset{\overset{\displaystyle O}{\|}}{C}-OH$

    c. $CH_3-\underset{\underset{\displaystyle CH_3}{|}}{C}H-CH_2-CH_2-CH_2-\overset{\overset{\displaystyle O}{\|}}{C}-OH$

    d. $CH_3-CH_2-CH_2-CH_2-\overset{\overset{\displaystyle O}{\|}}{C}-O-CH_2-CH_3$

82. Draw the structure of each of the following carboxylic acids and esters:
    a. pentanoic acid
    b. methyl hexanoate
    c. 3-ethylheptanoic acid
    d. butyl ethynoate

83. Determine the products of the following carboxylic acid reactions:

    a. $CH_3-CH_2-CH_2-CH_2-\overset{\overset{\displaystyle O}{\|}}{C}-OH \ +$
    $CH_3-CH_2-OH \ \xrightarrow{H_2SO_4}$

    b. (cyclic structure) $\xrightarrow{\text{Heat}}$

84. Determine the products of the following carboxylic acid reactions:

    a. $CH_3-CH_2-\overset{\overset{\displaystyle O}{\|}}{C}-OH \ + \ NaOH \ \longrightarrow$

    b. $CH_3-CH_2-CH_2-\overset{\overset{\displaystyle O}{\|}}{C}-OH \ +$
    $CH_3-CH_2-CH_2-OH \ \xrightarrow{H_2SO_4}$

Ethers

85. Name each of the following ethers:
    a. $CH_3-CH_2-CH_2-O-CH_2-CH_3$
    b. $CH_3-CH_2-CH_2-CH_2-CH_2-O-CH_2-CH_3$
    c. $CH_3-CH_2-CH_2-O-CH_2-CH_2-CH_3$
    d. $CH_3-CH_2-O-CH_2-CH_2-CH_2-CH_3$

86. Draw a structure for each of the following ethers:
    a. ethyl propyl ether
    b. dibutyl ether
    c. methyl hexyl ether
    d. dipentyl ether

Amines

87. Name the following amines:
    a. $CH_3-CH_2-\underset{\underset{\displaystyle H}{|}}{N}-CH_2-CH_3$

    b. $CH_3-CH_2-CH_2-\underset{\underset{\displaystyle H}{|}}{N}-CH_3$

    c. $CH_3-CH_2-CH_2-\underset{\underset{\displaystyle H}{\overset{\overset{\displaystyle CH_3}{|}}{N}}}{}-CH_2-CH_2-CH_2-CH_3$

88. Draw structures for the following amines:
    a. isopropylamine
    b. triethylamine
    c. butylethylamine

89. Classify the following amine reactions as acid–base or condensation reactions; determine the products.
    a. $CH_3NHCH_3 + HCl \longrightarrow$
    b. $CH_3CH_2NH_2 + CH_3CH_2COOH \longrightarrow$
    c. $CH_3NH_2 + H_2SO_4 \longrightarrow$

90. Determine the products of the following amine reactions:
    a. $N(CH_2CH_3)_3 + HNO_3 \longrightarrow$

    b. $CH_3-\underset{\underset{\displaystyle CH_3}{|}}{\overset{\overset{\displaystyle H}{|}}{N}}-CH-CH_3 + HCN \longrightarrow$

    c. $CH_3-\underset{\underset{\displaystyle CH_3}{|}}{\overset{\overset{\displaystyle H}{|}}{N}}-CH-CH_3 +$
    $CH_3-\underset{\underset{\displaystyle CH_3}{|}}{C}H-CH_2-\overset{\overset{\displaystyle O}{\|}}{C}-OH \longrightarrow$

Polymers

91. Teflon is an addition polymer formed from the following monomer. Draw the structure of the polymer.

$$\overset{\displaystyle F \qquad F}{\underset{\displaystyle F \qquad F}{C=C}}$$

92. Saran, the polymer used to make saran wrap, is an addition polymer formed from the following two monomers, vinylidene chloride and vinyl chloride. Draw the structure of the polymer. (Hint: The monomers alternate.)

$$\overset{\displaystyle H \qquad Cl}{\underset{\displaystyle H \qquad Cl}{C=C}} \qquad \overset{\displaystyle H \qquad H}{\underset{\displaystyle H \qquad Cl}{C=C}}$$

Vinylidene chloride   Vinyl chloride

93. One kind of polyester is a condensation copolymer formed between terephthalic acid and ethylene glycol. Draw the structure of the dimer and circle the ester functional group. [Hint: Water (circled) is eliminated when the bond between the monomers forms.]

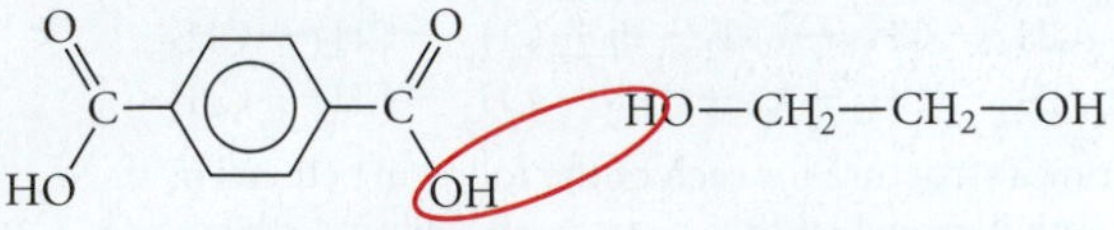

Terephthalic acid   Ethylene glycol

94. Nomex, a condensation copolymer used by firefighters because of its flame-resistant properties, is formed between isophthalic acid and *m*-aminoaniline. Draw the structure of the dimer. (Hint: Water is eliminated when the bond between the monomers forms.)

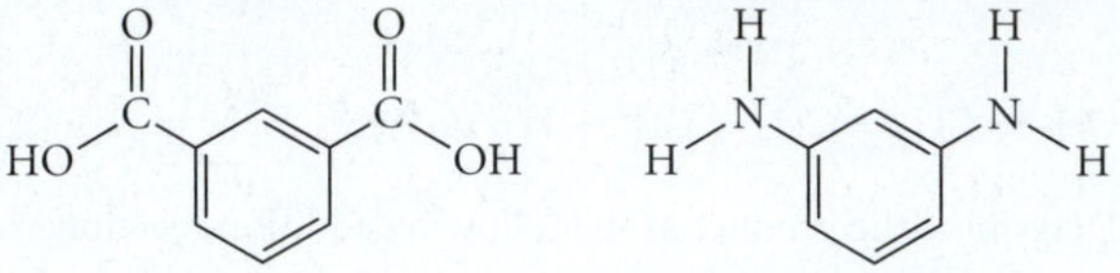

Isophthalic acid   *m*-Aminoaniline

Cumulative Problems

95. Identify each of the following organic compounds as an alkane, alkene, alkyne, aromatic hydrocarbon, alcohol, ether, aldehyde, ketone, carboxylic acid, ester, or amine, and provide a name for the compound:

a. 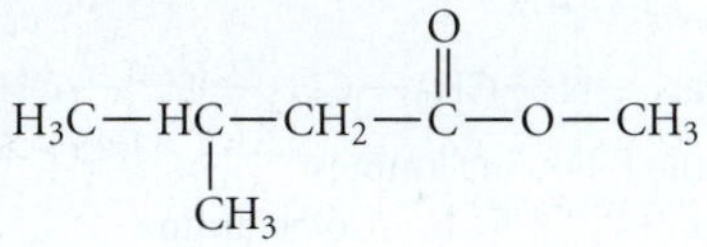 $H_3C-HC-CH_2-\overset{\overset{\displaystyle O}{\|}}{C}-O-CH_3$
   with CH_3 branch

b. $CH_3-CH_2-\overset{\overset{\displaystyle CH_3}{|}}{CH}-CH_2-O-CH_2-CH_3$

c. aromatic ring with H_2C-CH_3 and CH_3 substituents

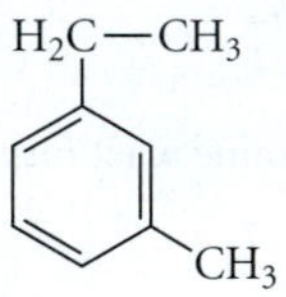

d. 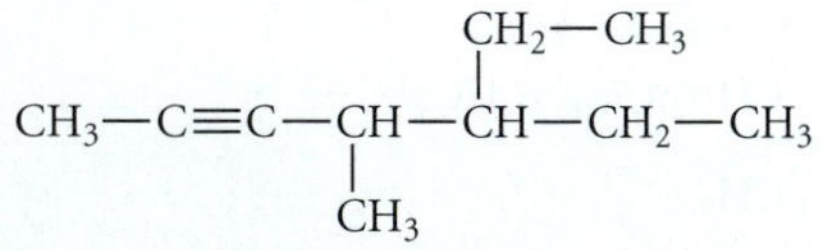 $CH_3-C\equiv C-\overset{\overset{\displaystyle CH_2-CH_3}{|}}{CH}-\overset{\underset{\displaystyle CH_3}{|}}{CH}-CH_2-CH_3$

e. 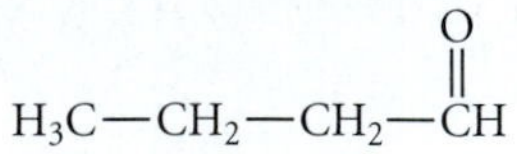 $H_3C-CH_2-CH_2-\overset{\overset{\displaystyle O}{\|}}{CH}$

f. 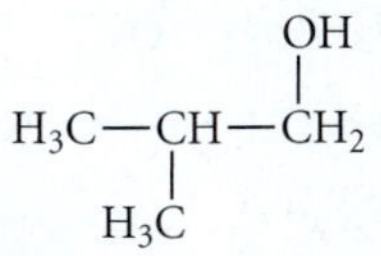 $H_3C-\overset{\overset{\displaystyle OH}{|}}{CH}-CH_2$
   with H_3C branch

96. Identify each of the following organic compounds as an alkane, alkene, alkyne, aromatic hydrocarbon, alcohol, ether, aldehyde, ketone, carboxylic acid, ester, or amine, and provide a name for the compound:

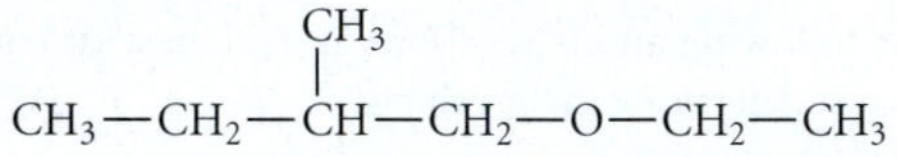

a. 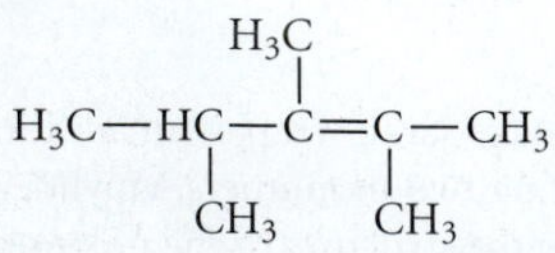 $H_3C-HC-\overset{\overset{\displaystyle H_3C}{|}}{C}=C-CH_3$
   with CH_3 and CH_3 branches

b. $CH_3-\overset{\overset{\displaystyle CH_3}{|}}{\underset{\underset{\displaystyle CH_3}{|}}{C}}-CH_2-\overset{\overset{\displaystyle CH_3}{|}}{CH}-CH_2-CH_3$

c. $CH_3-CH_2-\overset{\overset{\displaystyle CH_3}{|}}{CH}-CH_2-\overset{\overset{\displaystyle O}{\|}}{C}-OH$

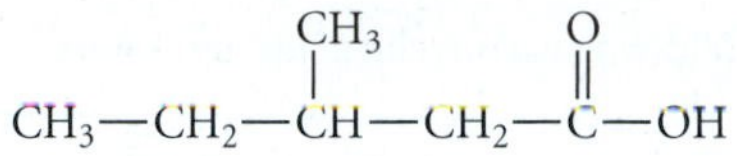

d. 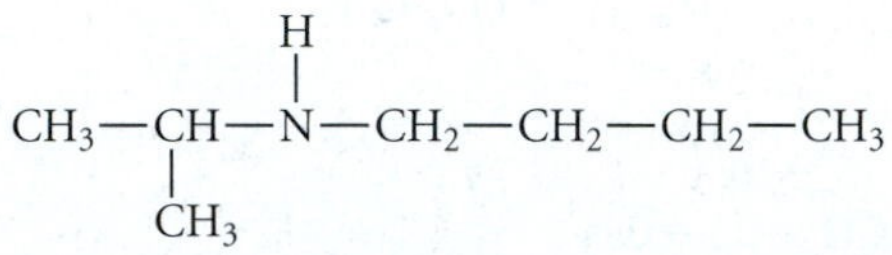 $CH_3-\overset{\overset{\displaystyle H}{|}}{CH}-N-CH_2-CH_2-CH_2-CH_3$
   with CH_3 branch

e. 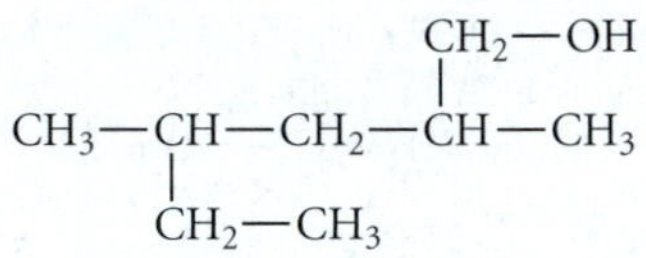 $CH_3-\overset{\overset{\displaystyle CH_2-OH}{|}}{CH}-CH_2-\overset{\underset{\displaystyle CH_2-CH_3}{|}}{CH}-CH_3$

f. 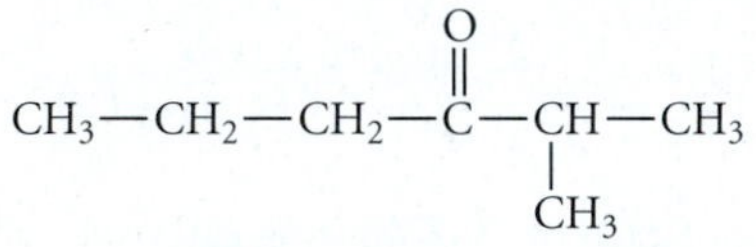 $CH_3-CH_2-CH_2-\overset{\overset{\displaystyle O}{\|}}{C}-\overset{\underset{\displaystyle CH_3}{|}}{CH}-CH_3$

97. Name each of the following compounds:

a. 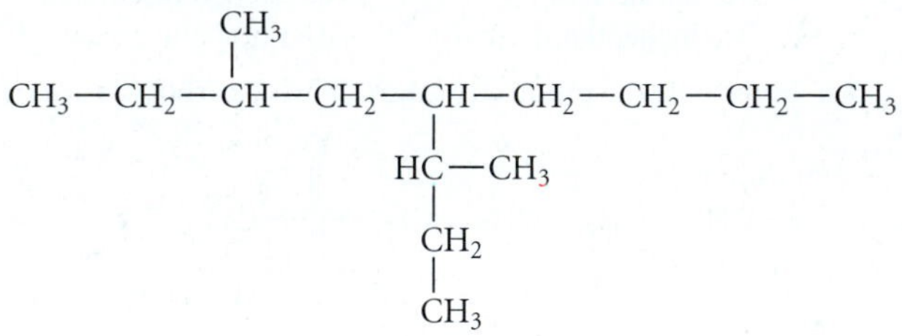 $CH_3-CH_2-\overset{\overset{\displaystyle CH_3}{|}}{CH}-CH_2-\overset{\underset{\displaystyle HC-CH_3}{|}}{CH}-CH_2-CH_2-CH_2-CH_3$
   with $HC-CH_3$, CH_2, CH_3 branch chain

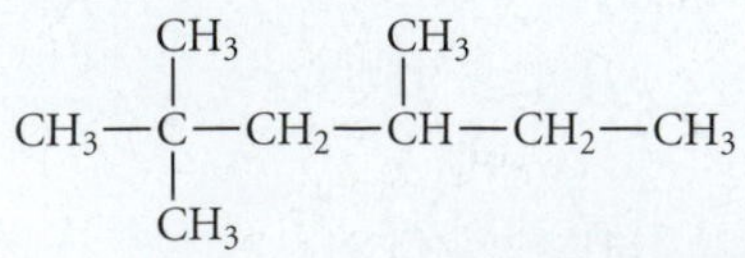

b. 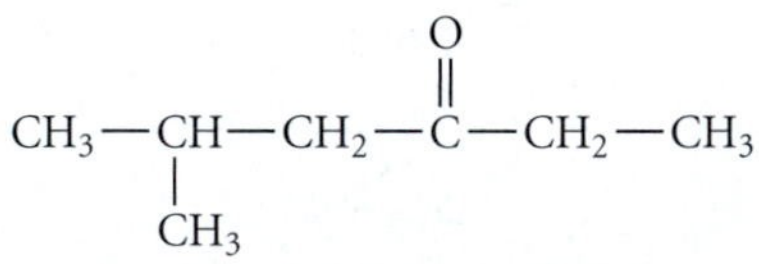 $CH_3-\overset{\underset{\displaystyle CH_3}{|}}{CH}-CH_2-\overset{\overset{\displaystyle O}{\|}}{C}-CH_2-CH_3$

c. $CH_3-\overset{\overset{\displaystyle OH}{|}}{CH}-\overset{\underset{\displaystyle CH_3}{|}}{CH}-CH_3$

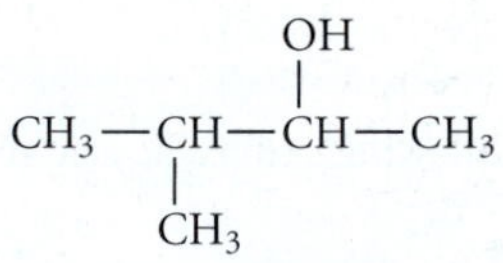

d. 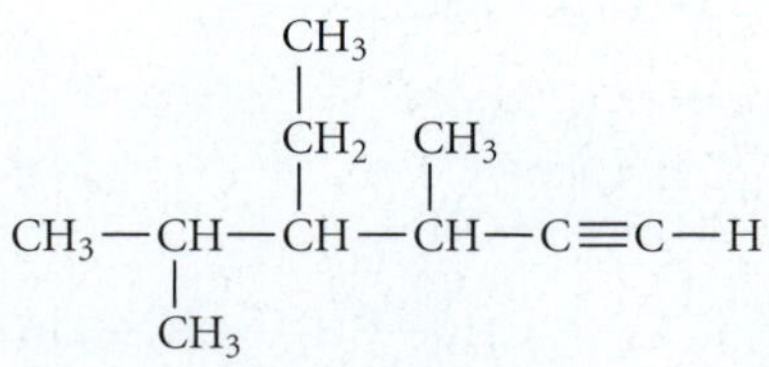 $CH_3-\overset{\underset{\displaystyle CH_3}{|}}{CH}-\overset{\overset{\displaystyle CH_3}{|}\overset{\displaystyle CH_2}{|}}{CH}-CH-C\equiv C-H$

98. Name each of the following compounds:

a. 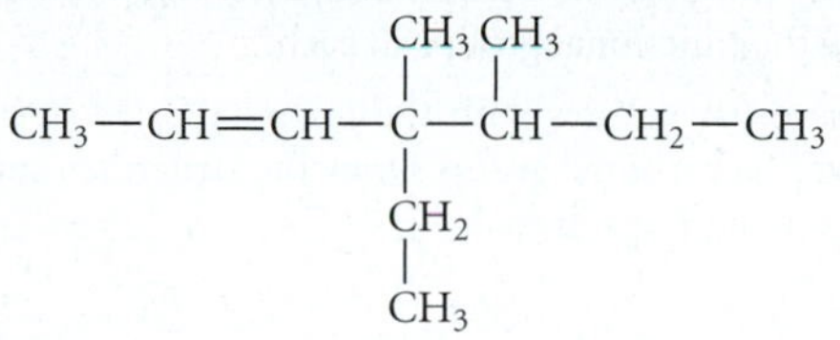 $CH_3-CH=CH-\underset{\underset{\underset{CH_3}{|}}{\underset{CH_2}{|}}{\overset{\overset{CH_3}{|}}{C}}-\underset{\underset{CH_3}{|}}{\overset{\overset{CH_3}{|}}{CH}}-CH_2-CH_3$

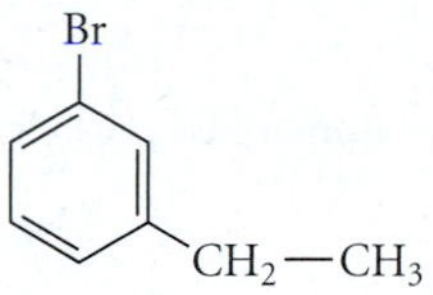

b. [structure: benzene ring with Br and CH_2-CH_3]

c. $CH_3-CH_2-\underset{\underset{CH_3}{|}}{CH}-CH_2-\overset{\overset{O}{||}}{C}-O-\underset{\underset{CH_3}{|}}{CH}-CH_3$

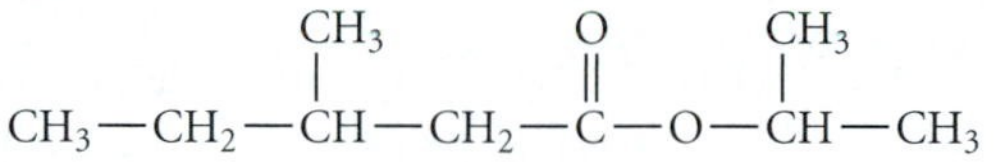

d. 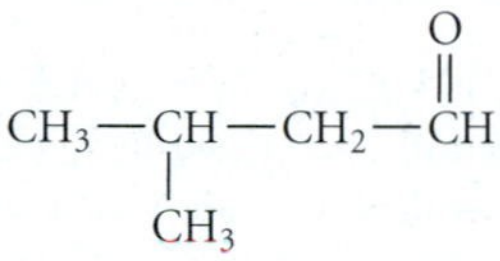 $CH_3-\underset{\underset{CH_3}{|}}{CH}-CH_2-\overset{\overset{O}{||}}{CH}$

99. For each of the following, determine whether the two structures are isomers or the same molecule drawn in two different ways:

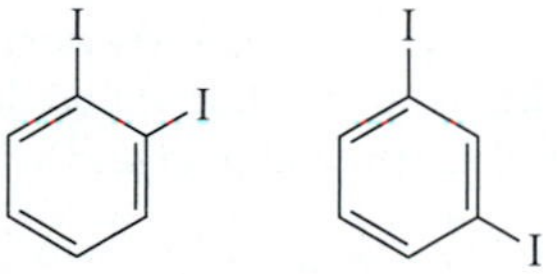

a. 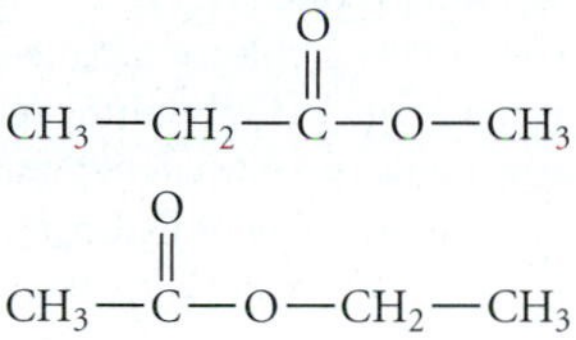 $CH_3-CH_2-\overset{\overset{O}{||}}{C}-O-CH_3$

   $CH_3-\overset{\overset{O}{||}}{C}-O-CH_2-CH_3$

b. [two benzene ring structures with I substituents]

c. $CH_3-\underset{\underset{CH_3}{|}}{HC}-\underset{\underset{CH_3}{|}}{CH}-CH_3$

   $CH_3-\underset{\underset{CH_3}{|}}{HC}-\underset{\underset{CH_3}{|}}{CH}-CH_3$

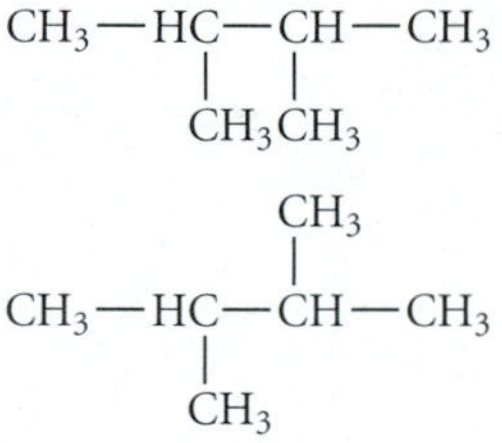

100. For each of the following, determine whether the two structures are isomers or the same molecule drawn two different ways:

a. 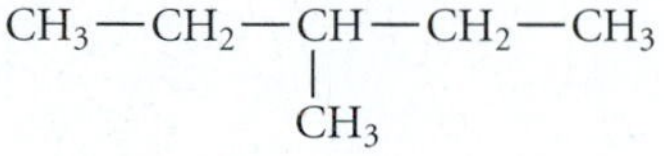 $CH_3-CH_2-\underset{\underset{CH_3}{|}}{CH}-CH_2-CH_3$

   $CH_3-CH_2-\underset{\underset{\underset{CH_3}{|}}{CH_2}}{\underset{|}{CH}}-CH_3$

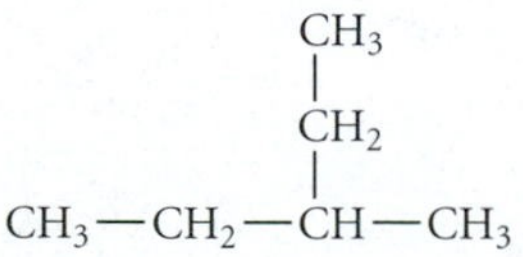

b. 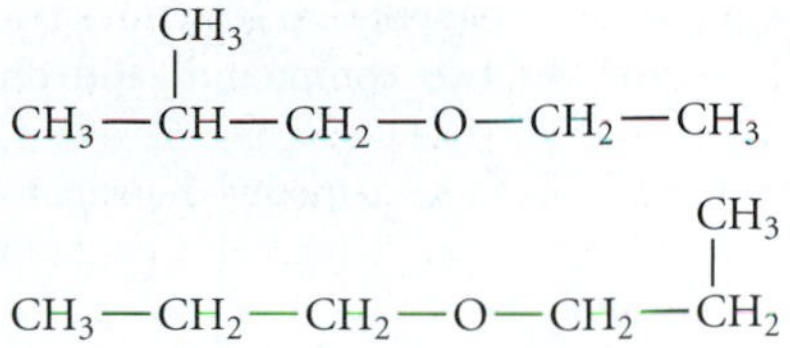 $CH_3-\underset{\underset{CH_3}{|}}{CH}-CH_2-O-CH_2-CH_3$

   $CH_3-CH_2-CH_2-O-CH_2-\underset{\underset{CH_3}{|}}{CH_2}$

c. 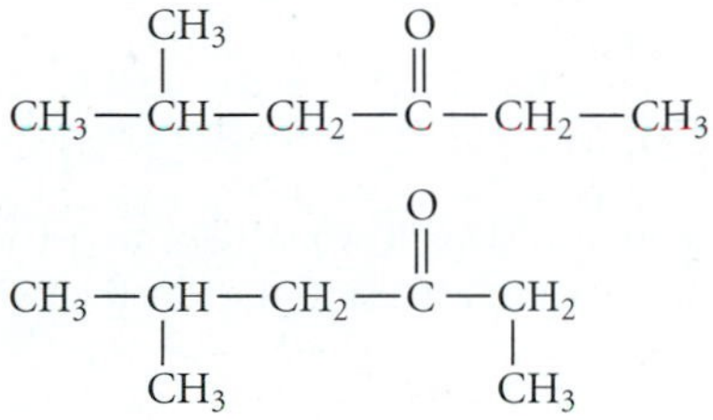 $CH_3-\underset{\underset{CH_3}{|}}{CH}-CH_2-\overset{\overset{O}{||}}{C}-CH_2-CH_3$

   $CH_3-\underset{\underset{CH_3}{|}}{CH}-CH_2-\overset{\overset{O}{||}}{C}-\underset{\underset{CH_3}{|}}{CH_2}$

101. What is the minimum amount of hydrogen gas, in grams, required to completely hydrogenate 15.5 kg of 2-butene?

102. How many kilograms of CO_2 are produced by the complete combustion of 3.8 kg of *n*-octane?

103. Classify the following organic reactions as combustion, alkane substitution, alkene addition or hydrogenation, aromatic substitution, or alcohol substitution, elimination, or oxidation.

a. $2\ CH_3CH=CH_2 + 9\ O_2 \longrightarrow 6\ CO_2 + 6\ H_2O$

b. $CH_3CH_2CH_3 + Cl_2 \longrightarrow CH_3CH_2CH_2Cl + HCl$

c. 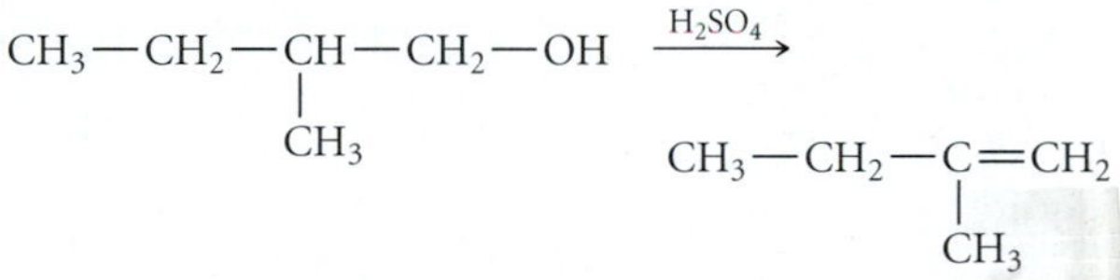 $CH_3-CH_2-\underset{\underset{CH_3}{|}}{CH}-CH_2-OH \xrightarrow{H_2SO_4}$

   $CH_3-CH_2-\underset{\underset{CH_3}{|}}{C}=CH_2$

d. 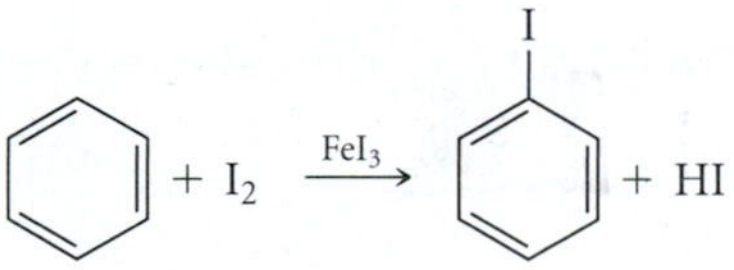 [benzene ring] $+ I_2 \xrightarrow{FeI_3}$ [benzene ring with I] $+ HI$

104. Determine the products of the following reactions:

a. 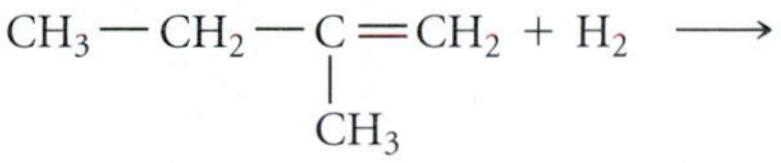 $CH_3-CH_2-\underset{\underset{CH_3}{|}}{C}=CH_2 + H_2 \longrightarrow$

b. $CH_3-CH_2-CH_2-CH_2-OH + HCl \longrightarrow$

c. 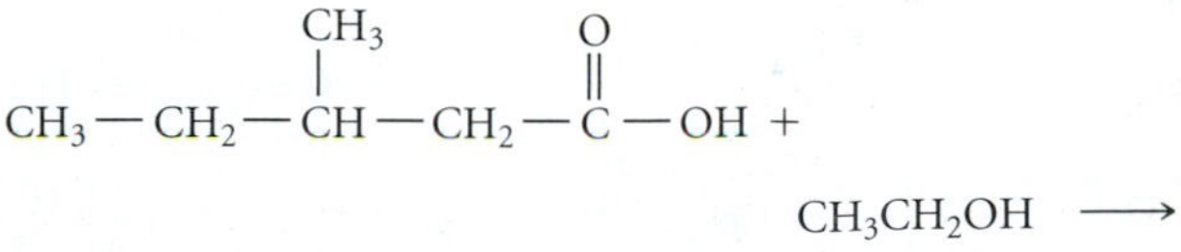 $CH_3-CH_2-\underset{\underset{CH_3}{|}}{CH}-CH_2-\overset{\overset{O}{||}}{C}-OH +$

   $CH_3CH_2OH \longrightarrow$

d. 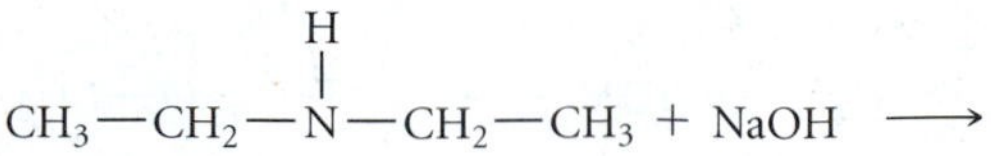 $CH_3-CH_2-\underset{\underset{H}{|}}{N}-CH_2-CH_3 + NaOH \longrightarrow$

105. Draw a structure corresponding to each of the following names and indicate those structures that can exist as stereoisomers.

a. 3-methyl-1-pentene   b. 3,5-dimethyl-2-hexene

c. 3-propyl-2-hexene

106. Two of the following names correspond to structures that display stereoisomerism. Identify the two compounds and draw their structures.

a. 3-methyl-3-pentanol   b. 2-methyl-2-pentanol
c. 3-methyl-2-pentanol   d. 2-methyl-3-pentanol
e. 2,4-dimethyl-3-pentanol

107. There are 11 structures (ignoring stereoisomerism) with the formula C_4H_8O that have no carbon branches. Draw the structures and identify the functional groups in each.

108. There are seven structures with the formula C_3H_7NO in which the O is part of a carbonyl group. Draw the structures and identify the functional groups in each.

Challenge Problems

109. Determine the one or two steps it would take to get from the starting material to the product using the reactions found in this chapter.

a.

b.

$$CH_3-\underset{\underset{CH_2-CH_3}{|}}{CH}-CH_2-\underset{\underset{CH_2-OH}{|}}{CH}-CH_3 \longrightarrow$$

$$CH_3-\underset{\underset{CH_2-CH_3}{|}}{CH}-CH_2-\underset{\underset{\underset{CH_2-CH_3}{|}}{|}}{\overset{\overset{CH_2}{\|}}{C}}-CH_3$$

c. $CH_3-CH_2-\underset{\underset{CH_3}{|}}{C}=CH_2 \longrightarrow CH_3-CH_2-\underset{\underset{CH_3}{|}}{\overset{\overset{Br}{|}}{C}}-CH_3$

110. Given the following synthesis of ethyl 3-chloro-3-methylbutanoate, fill in the missing intermediates or reactants.

$$CH_3-\underset{\underset{CH_3}{|}}{CH}-CH_2-CH_2-OH \xrightarrow{(a)} (b)$$

$$\xrightarrow{(c)} CH_3-\underset{\underset{CH_3}{|}}{CH}-CH_2-\overset{\overset{O}{\|}}{C}-O-CH_2-CH_3$$

$$\xrightarrow{(d)} CH_3-\underset{\underset{CH_3}{|}}{\overset{\overset{Cl}{|}}{C}}-CH_2-\overset{\overset{O}{\|}}{C}-O-CH_2-CH_3$$

111. For the chlorination of propane two isomers are possible:

$$CH_3CH_2CH_3 + Cl_2 \longrightarrow$$

$$CH_3-CH_2-CH_2-Cl + CH_3-\underset{\underset{Cl}{|}}{\overset{}{C}H}-CH_3$$
$$\text{1-chloropropane} \qquad \text{2-chloropropane}$$

Propane has six hydrogen atoms on terminal carbon atoms—called primary (1°) hydrogen atoms—and two hydrogen atoms on the interior carbon atom—called secondary (2°) hydrogen atoms.

a. If the two different types of hydrogen atoms were equally reactive, what ratio of 1-chloropropane to 2-chloropropane would we expect as monochlorination products?

b. The result of a reaction yields 55% 2-chloropropane and 45% 1-chloropropane. What can we conclude about the relative reactivity of the two different kinds of hydrogen atoms? Determine a ratio of the reactivity of one type of hydrogen atom to the other.

112. There are two isomers of C_4H_{10}. Suppose that each isomer is treated with Cl_2 and the products that have the composition $C_4H_8Cl_2$ are isolated. Find the number of different products that form from each of the original C_4H_{10} compounds. Do not consider optical isomerism.

113. Identify the compounds that were formed in the previous problem that are chiral.

Conceptual Problems

114. Pick the more oxidized structure from each pair:

a. 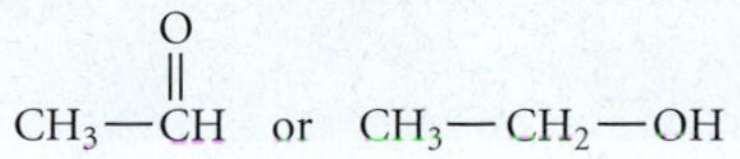 $CH_3-\overset{\overset{\displaystyle O}{\|}}{CH}$ or CH_3-CH_2-OH

b. CH_3-CH_2-OH or CH_3-CH_3

c. 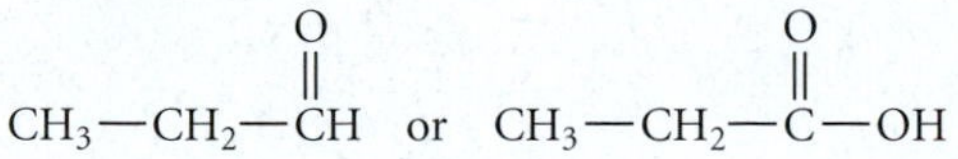 $CH_3-CH_2-\overset{\overset{\displaystyle O}{\|}}{CH}$ or $CH_3-CH_2-\overset{\overset{\displaystyle O}{\|}}{C}-OH$

115. Draw the structure and name a compound with the formula C_8H_{18} that forms only one product with the formula $C_8H_{17}Br$ when it is treated with Br_2.

116. Determine whether each of the following structures is chiral.

a. 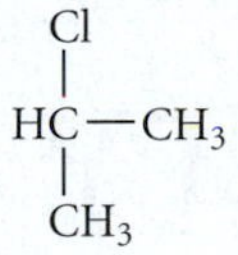 $\overset{\overset{\displaystyle Cl}{|}}{\underset{\underset{\displaystyle CH_3}{|}}{HC}}-CH_3$

b. 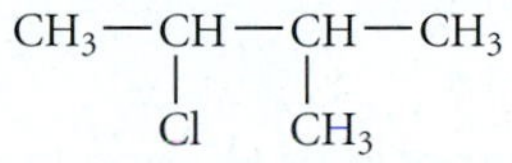 $CH_3-\overset{\overset{\displaystyle}{\underset{\underset{\displaystyle Cl}{|}}{CH}}}-\overset{\overset{\displaystyle}{\underset{\underset{\displaystyle CH_3}{|}}{CH}}}-CH_3$

c. CH_3-CH_2-OH

d. 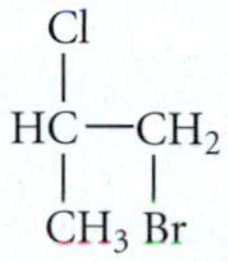 $\overset{\overset{\displaystyle Cl}{|}}{\underset{\underset{\displaystyle CH_3}{|}}{HC}}-\overset{\overset{\displaystyle}{\underset{\underset{\displaystyle Br}{|}}{CH_2}}}{}$

21

Biochemistry

21.1 Diabetes and the Synthesis of Human Insulin

21.2 Lipids

21.3 Carbohydrates

21.4 Proteins and Amino Acids

21.5 Protein Structure

21.6 Nucleic Acids: Blueprints for Proteins

21.7 DNA Replication, the Double Helix, and Protein Synthesis

I N THE PREVIOUS CHAPTER, we examined organic chemistry and learned about the different types of organic compounds and their structures and chemistry. In this chapter, we turn to biochemical compounds, those organic compounds important in living organisms. Biochemistry—the area at the interface between chemistry and biology that strives to understand living organisms at the molecular level—exploded in the second half of the twentieth century. That explosion began with the discovery of the structure of DNA in 1953 by James D. Watson and Francis H. C. Crick and continues to this day, most recently marked by the 2003 completion of the Human Genome Project, which succeeded in mapping the 3 billion base pairs within the DNA of humans. The benefits of biochemistry to humankind are numerous, ranging from a better understanding of illnesses and better drugs to cure them to a better understanding of ourselves and our origins.

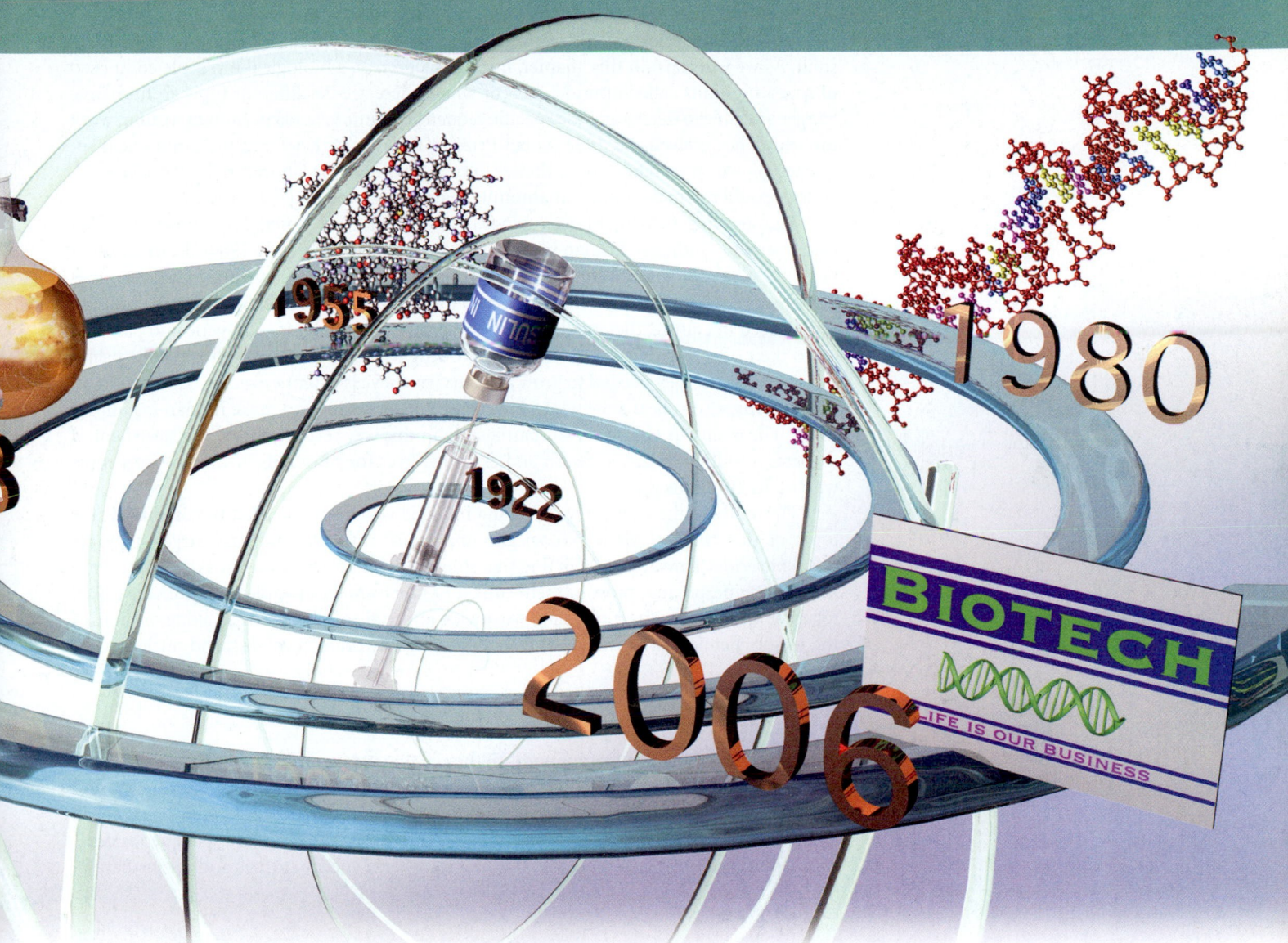

> *We've discovered the secret of life.*
> —FRANCIS H. C. CRICK (1916–2004)

The above image depicts a timeline marking the important events in the treatment of diabetes.

21.1 Diabetes and the Synthesis of Human Insulin

Diabetes is a serious disease afflicting over 16 million people in the United States alone. Today, it is a chronic but generally manageable ailment; at the beginning of the twentieth century, however, it was often fatal. The most dangerous form of this condition, type 1 diabetes, develops when the pancreas does not make enough *insulin*, a protein that promotes the absorption of glucose from the blood into cells, where glucose is used for energy. Consequently, diabetics have high blood-sugar levels that can lead to a number of complications, including heart disease, blindness, and kidney failure. Before 1922, diabetics could control their blood-sugar levels only through diet, but this was often not enough to overcome the disease.

Important advances throughout the twentieth century dramatically changed the prognosis for diabetics. The initial breakthrough came in 1922, when researchers first injected insulin from animal sources into a hospitalized diabetic. The insulin worked, resulting in a nearly complete recovery for the patient. Within a year, insulin harvested from the pancreases of slaughtered pigs and cattle was made widely available, and, for many patients, diabetes became a long-term manageable disease. However, insulin taken from pigs and cattle is not identical to human insulin, and some patients could not tolerate the animal insulin as well as others.

In 1955, Frederick Sanger discovered the detailed chemical structure of human insulin. As we will learn in this chapter, insulin is a *protein*, a biological molecule composed of repeating units called amino acids (of which there are 20 different types in humans). Sanger was able to determine the specific sequence of amino acids in human insulin, work for which he received the 1958 Nobel Prize in Chemistry. Knowing the amino acid sequence allowed researchers to synthesize human insulin in the laboratory by 1963. However, they could not make sufficient amounts to meet the needs of diabetics.

The growing field of biotechnology, however, allowed a fledgling company called Genentech to synthesize human insulin on a large scale by the early 1980s. Researchers at Genentech were able to insert the human *gene* for insulin—the blueprint that determines how insulin is synthesized in humans—into the DNA of bacterial cells. When the bacteria reproduced in culture, they made copies of the inserted human insulin gene and passed it on to their offspring. Furthermore, as the growing bacterial culture synthesized the bacterial proteins that they needed to grow and survive, they also synthesized human insulin. In other words, researchers at Genentech were able to get bacteria to make human insulin for them! The ability to synthesize human insulin in this way revolutionized the treatment of diabetes, resulting in better health and extended lives for hundreds of millions of people who suffer from this disease.

The study of the chemistry occurring in living organisms is called **biochemistry**, the topic of this chapter. Many biologically important molecules are very large—they are *macromolecules*. However, we will find that studying their structures is not as difficult as you might imagine, because most of them consist of much smaller, simpler components linked together into long chains (polymers). We divide our study of biochemistry along the lines of the major chemical components of cells: lipids, carbohydrates, proteins, and nucleic acids.

21.2 Lipids

Lipids are those chemical components of the cell that are insoluble in water but soluble in nonpolar solvents. Lipids include fatty acids, fats, oils, phospholipids, glycolipids, and steroids. Their insolubility in water makes lipids ideal as the structural components of cell membranes, which must separate the aqueous interior of the cell from its aqueous environment in the body. Lipids are also used for long-term energy storage and for insulation.

Fatty Acids

One type of lipid is the **fatty acid**, which is a carboxylic acid (see Section 20.11) with a long hydrocarbon tail. The general structure for a fatty acid is

$$R-\overset{\overset{\displaystyle O}{\|}}{C}-OH$$

Fatty acid general structure

where R represents a hydrocarbon chain containing 3 to 19 carbon atoms. Fatty acids differ only in their R group. A common fatty acid is myristic acid, which has the R group $CH_3(CH_2)_{12}$.

$$H_3C-\underset{CH_2}{\overset{CH_2}{}}-\underset{CH_2}{\overset{CH_2}{}}-\underset{CH_2}{\overset{CH_2}{}}-\underset{CH_2}{\overset{CH_2}{}}-\underset{CH_2}{\overset{CH_2}{}}-\underset{CH_2}{\overset{CH_2}{}}-\overset{\overset{\displaystyle O}{\|}}{C}-OH$$

Myristic acid

Myristic acid, which occurs in butterfat and in coconut oil, is an example of a *saturated* fatty acid: its carbon chain has no double bonds. Other fatty acids—called *monounsaturated* or *polyunsaturated* fatty acids—have one or more double bonds in their carbon chains. For example, oleic acid—found in olive oil, peanut oil, and human fat—is a monounsaturated fatty acid.

$$H_3C-CH_2-CH_2-CH_2-CH_2-CH_2-CH_2-CH=CH-CH_2-CH_2-CH_2-CH_2-CH_2-C(=O)-OH$$

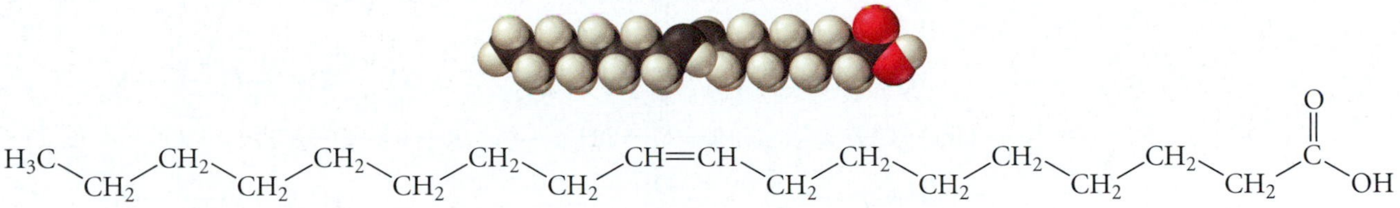

Oleic acid

The long hydrocarbon tails of fatty acids make them insoluble in water. Table 21.1 contains a list of several different fatty acids, some common sources for each, and their melting points. Notice that the melting point increases with increasing length of the carbon chain. The longer the chains, the greater the dispersion forces between adjacent molecules in the solid state, and these forces must be overcome in order for the compound to melt. Melting point temperature decreases with the presence of double bonds.

Dispersion forces are discussed in Section 11.3.

TABLE 21.1 Fatty Acids

Saturated Fatty Acids

Name	Number of Carbons	mp (°C)	Structure	Sources
Butyric acid	4	−7.9	$CH_3CH_2CH_2COOH$	Milk fat
Caproic acid	10	31	$CH_3(CH_2)_8COOH$	Milk fat, whale oil
Myristic acid	14	59	$CH_3(CH_2)_{12}COOH$	Butterfat, coconut oil
Palmitic acid	16	64	$CH_3(CH_2)_{14}COOH$	Beef fat, butterfat
Stearic acid	18	70	$CH_3(CH_2)_{16}COOH$	Beef fat, butterfat

Unsaturated Fatty Acids

Name	Number of Carbons	Number of Double Bonds	mp (°C)	Structure	Sources
Oleic acid	18	1	4	$CH_3(CH_2)_7CH=CH(CH_2)_7COOH$	Olive oil, peanut oil
Linoleic acid	18	2	−5	$CH_3(CH_2)_4(CH=CHCH_2)_2(CH_2)_6COOH$	Linseed oil, corn oil
Linolenic acid	18	3	−11	$CH_3CH_2(CH=CHCH_2)_3(CH_2)_6COOH$	Linseed oil, corn oil

For example, stearic acid and oleic acid have the same number of carbon atoms, but stearic acid melts at 70 °C and oleic acid melts at 4 °C. The double bond puts a "kink" in the carbon chain that makes it more difficult for neighboring molecules to interact over the entire length of the chain (Figure 21.1 ▼), thus lowering the melting point.

The Effect of Unsaturation

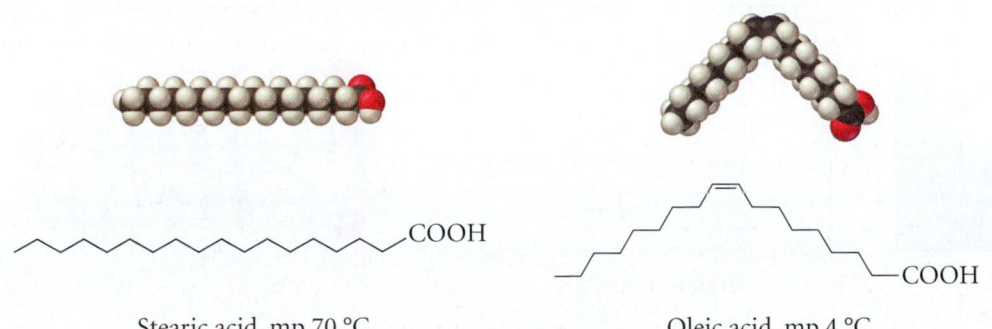

Stearic acid, mp 70 °C   Oleic acid, mp 4 °C

◄ FIGURE 21.1 **The Effect of Unsaturation** A double bond puts a bend in the carbon chain of a fatty acid or fat that makes it more difficult for neighboring molecules to interact over the entire length of the carbon chain, thus lowering

The general structure of esters is discussed in Section 20.11.

Fats and Oils

Fats and oils are **triglycerides**, triesters composed of glycerol with three fatty acids attached. Triglycerides form by the reaction of glycerol with three fatty acids:

$$
\begin{array}{ccc}
& R-\overset{\displaystyle O}{\overset{\|}{C}}-OH & \\
H_2C-OH & O & H_2C-O-\overset{\displaystyle O}{\overset{\|}{C}}-R \\
HO-CH \quad + \quad R-\overset{\|}{C}-OH & \longrightarrow & R-\overset{O}{\overset{\|}{C}}-O-CH \quad O \quad + 3H_2O \\
H_2C-OH & O & H_2C-O-\overset{\|}{C}-R \\
& R-\overset{\|}{C}-OH &
\end{array}
$$

Ester linkage

Glycerol   3 Fatty acids   Triglyceride

The bonds that join the glycerol to the fatty acids are called **ester linkages**. For example, tristearin—the main component of beef fat—is formed from the reaction of glycerol and three stearic acid molecules (Figure 21.2 ▼). If the fatty acids in a triglyceride are saturated, the triglyceride is called a **saturated fat** and tends to be solid at room temperature. Triglycerides from warm-blooded animals (for example, lard) are generally saturated.

$$
\begin{array}{c}
H_2C-OH \\
HO-CH \quad + \quad 3 \\
H_2C-OH
\end{array}
\left[\begin{array}{c}
\qquad\qquad\qquad\qquad\qquad\qquad\qquad O \\
CH_2\ CH_2\ CH_2\ CH_2\ CH_2\ CH_2\ CH_2\ CH_2\ \overset{\|}{C} \\
H_3C \quad CH_2\ CH_2\ CH_2\ CH_2\ CH_2\ CH_2\ CH_2\ CH_2\ OH
\end{array} \right]
$$

Glycerol   3 Stearic acid molecules

$$\longrightarrow$$

$$
\begin{array}{c}
H_2C-O-\overset{O}{\overset{\|}{C}} \quad CH_2\ CH_2\ CH_2\ CH_2\ CH_2\ CH_2\ CH_2\ CH_2 \\
\qquad\qquad\qquad CH_2\ CH_2\ CH_2\ CH_2\ CH_2\ CH_2\ CH_2\ CH_2\ CH_3 \\
HC-O-\overset{O}{\overset{\|}{C}} \quad CH_2\ CH_2\ CH_2\ CH_2\ CH_2\ CH_2\ CH_2\ CH_2 \quad + 3H_2O \\
\qquad\qquad\qquad CH_2\ CH_2\ CH_2\ CH_2\ CH_2\ CH_2\ CH_2\ CH_2\ CH_3 \\
H_2C-O-\overset{O}{\overset{\|}{C}} \quad CH_2\ CH_2\ CH_2\ CH_2\ CH_2\ CH_2\ CH_2\ CH_2 \\
\qquad\qquad\qquad CH_2\ CH_2\ CH_2\ CH_2\ CH_2\ CH_2\ CH_2\ CH_2\ CH_3
\end{array}
$$

Tristearin

▶ **FIGURE 21.2 The Formation of Tristearin** The reaction between glycerol and stearic acid forms tristearin.

If, however, the fatty acids in a triglyceride are unsaturated, the triglyceride is called an **unsaturated fat**, or an oil, and tends to be liquid at room temperature. Triglycerides from plants (for example, olive oil, corn oil, or canola oil) or from cold-blooded animals (for example, fish oil) are generally unsaturated.

Tristearin: A Saturated Fat

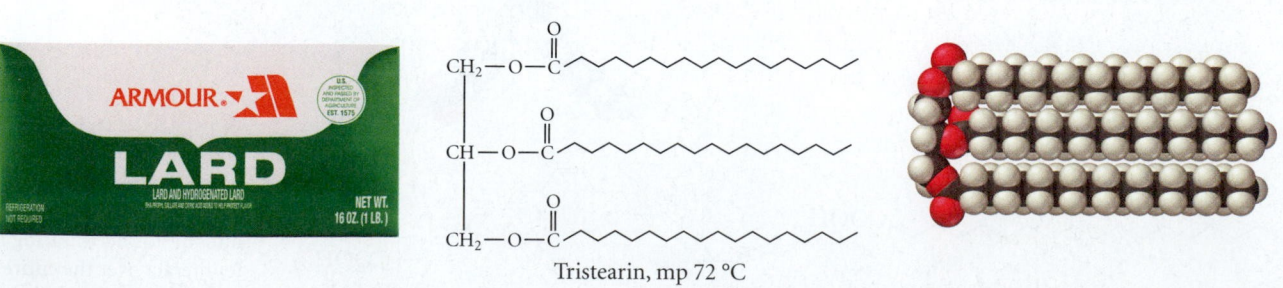

Tristearin, mp 72 °C

▲ Tristearin is a triglyceride found in lard—it is a saturated fat.

Triolein: A Monounsaturated Fat

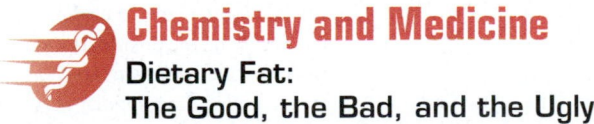

Triolein, mp −4 °C

▲ Triolein is a triglyceride found in olive oil—it is a monounsaturated fat.

Most of the fats and oils in our diet are triglycerides. During digestion, triglycerides are broken down into fatty acids, glycerol, monoglycerides, and diglycerides. These products pass through the intestinal wall and then reassemble into triglycerides before they are absorbed into the blood. This process, however, is slower than the digestion of other food types, and for this reason eating fats and oils gives a lasting feeling of fullness.

The effect of fats and oils on health has been widely debated. Some diets call for a drastic reduction of daily intake of fats and oils, whereas others actually call for an *increase* in fats and oils. The U.S. Food and Drug Administration (FDA) has tended to recommend a moderate consumption of fats and oils, less than 30% of total caloric intake.

Chemistry and Medicine
Dietary Fat: The Good, the Bad, and the Ugly

Today, the debate about fat has focused as much on the *kind* of fats that should be included in a healthy diet as on their quantity. Nutrition experts generally agree that unsaturated fats from plants and fish are better for you than the saturated fats from warm-blooded animals. However, unsaturated fats are liquids at room temperature and do not have the creamy, melt-in-your-mouth feel of saturated fats (try making chocolate chip cookies with vegetable oil). Consequently, producers started using hydrogenation reactions (see Section 20.6) to convert polyunsaturated vegetable oil into a saturated fat (known as vegetable shortening). The vegetable shortening, even though it was highly saturated, was supposed to be better for your health than lard or butter because it came from plant sources and did not contain cholesterol.

More recent research, however, has raised serious concerns about saturated fats in general (regardless of their source) and in particular about the presence of "trans" fats in vegetable shortening. The hydrogenation process converts cis double bonds (see Section 20.5) between carbon atoms in oils into single bonds. However, the catalyst and the temperatures used also cause small amounts of the reverse reaction, the conversion of single bonds into double bonds. The double bonds that form in this way can be either the naturally occurring cis double bond or an unnatural trans double bond (Figure 21.3▼). So hydrogenated vegetable oil not only has fewer double bonds overall, but some of the double bonds are trans, which is why the fats

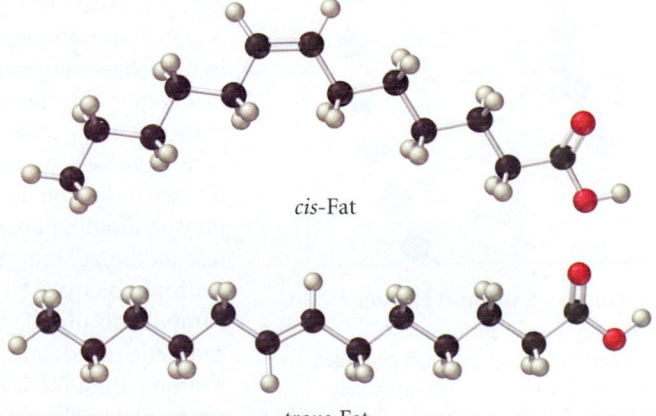

cis-Fat

trans-Fat

▶ FIGURE 21.3 Cis and Trans Fats The hydrogenation process introduces some trans double bonds into fats. The double bonds in

are called trans fats. Consumption of too much trans fat, even more than consumption of too much saturated fat, raises the risk of coronary artery disease. Thus trans fat has become the target of many health scientists and consumer groups. In response, the FDA has promulgated regulations that require the nutrition labels on foods to include the amount of trans fat as well as saturated and unsaturated fat.

Servings Per Package: 28	
Amount Per Serving	
Calories 140 Calories from Fat 70	
	% Daily Value*
Total Fat 7 g	**11%**
Saturated Fat 2 g	10%
Trans Fat 0 g	
Cholesterol 0 mg	**0%**
Sodium 150 mg	**6%**
Total Carbohydrate 17 g	**6%**
Dietary Fiber 1 g	4%
Sugars 0 g	

Phosphatidyl Choline: A Phospholipid

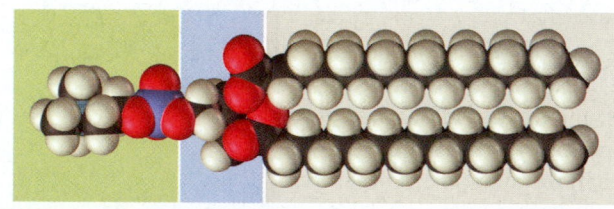

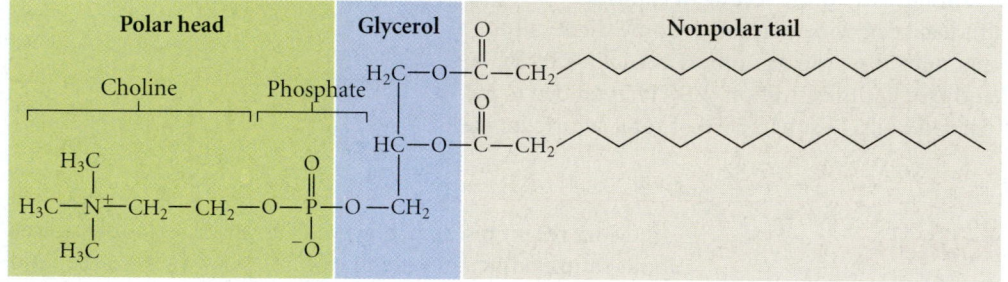

Polar head   Choline   Phosphate   Glycerol   Nonpolar tail

Polar head   Nonpolar tails

▲ **FIGURE 21.5 Schematic for Phospholipid or Glycolipid** Phospholipids and glycolipids can be represented as a circle (representing the polar part of the molecule) with two long tails (representing the nonpolar part of the molecule).

▲ **FIGURE 21.4 Phosphatidylcholine** Phosphatidylcholine is a phospholipid. The structure is similar to a triglyceride except that one of the fatty acid groups has been replaced with a phosphate group.

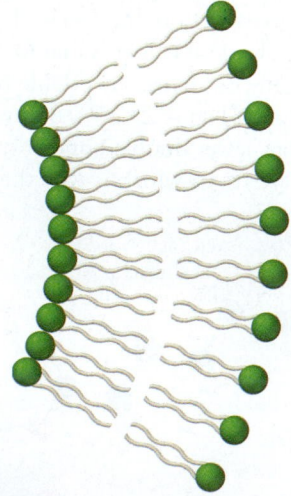

▲ **FIGURE 21.6 Lipid Bilayer** Lipid bilayers are composed of phospholipids or glycolipids arranged in a structure that encapsulates cells and many cellular structures.

Other Lipids

Other lipids found in cells include phospholipids, glycolipids, and steroids. **Phospholipids** have the same basic structure as triglycerides, except that one of the fatty acid groups is replaced with a phosphate group. Unlike a fatty acid, which is nonpolar, the phosphate group is polar. The phospholipid molecule therefore has a polar region and a nonpolar region. For example, consider the structure of phosphatidylcholine, a phospholipid found in the cell membranes of higher animals (Figure 21.4 ▲). The polar part of the molecule is *hydrophilic* (has a strong affinity for water) while the nonpolar part is *hydrophobic* (is repelled by water). **Glycolipids** have similar structures and properties. The nonpolar section of a glycolipid is composed of a fatty acid chain and a hydrocarbon chain. However, the polar section is a sugar molecule such as glucose. A schematic way to represent phospholipids or glycolipids is as a circle with two long tails (Figure 21.5 ▲). The circle represents the polar, hydrophilic part of the molecule and the tails represent the nonpolar, hydrophobic parts. The structure of phospholipids and glycolipids makes them ideal as components of cell membranes, where the polar parts can interact with the aqueous environments inside and outside the cell and the nonpolar parts interact with each other, forming a double-layered structure called a **lipid bilayer** (Figure 21.6 ◄). Lipid bilayers encapsulate cells and many cellular structures.

Steroids are lipids with the following four-ring structure:

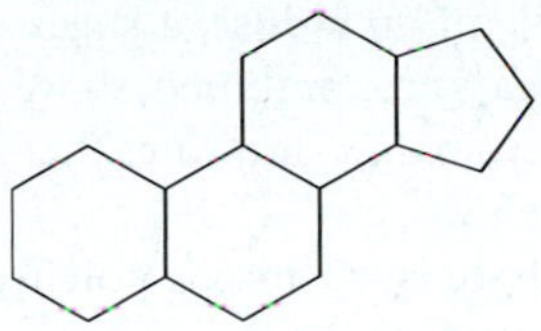

Some common steroids include cholesterol, testosterone, and β-estradiol:

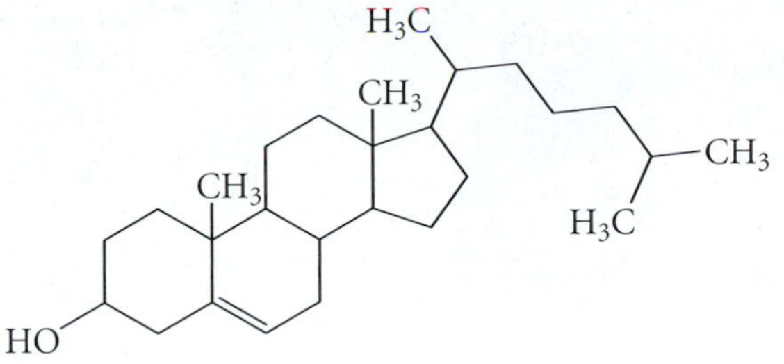

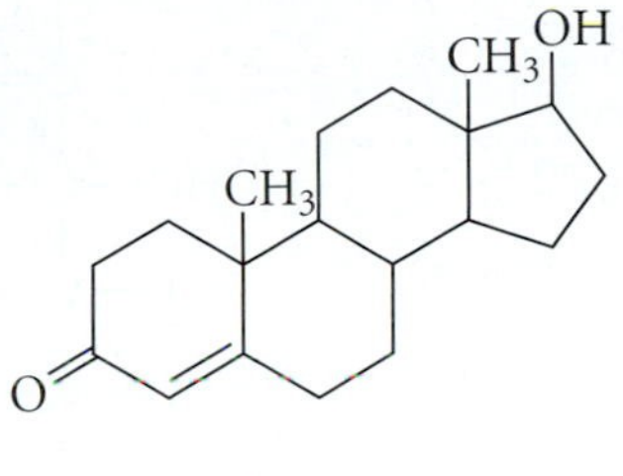

Testosterone

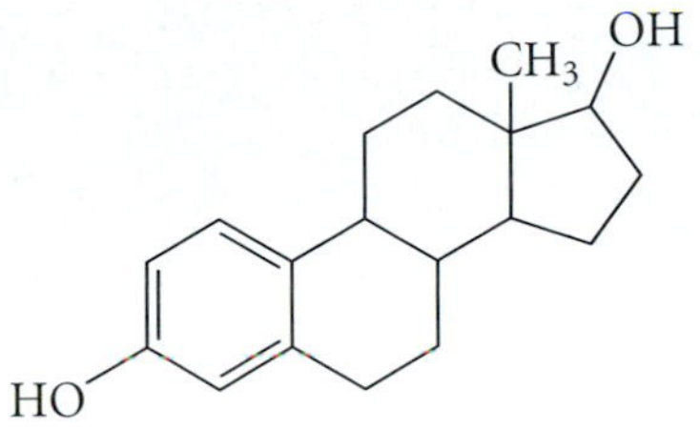

β-Estradiol

Cholesterol

Although cholesterol has a bad reputation, it serves many important functions in the body. Like phospholipids and glycolipids, cholesterol is part of cell membranes. Cholesterol also serves as a starting material (or precursor) for the body to synthesize other steroids such as testosterone, a principal male hormone, and β-estradiol, a principal female hormone.

21.3 Carbohydrates

Carbohydrates are responsible for short-term storage of energy in living organisms, and they make up the main structural components of plants. Carbohydrates—as their name, which means carbon and water, implies—often have the general formula $(CH_2O)_n$. Structurally, we identify **carbohydrates** as polyhydroxy aldehydes or ketones. For example, glucose, with the formula $C_6H_{12}O_6$, has the following structure:

> The name should not be taken literally—the hydrogen and oxygen in carbohydrates are not bonded together as they are in water.

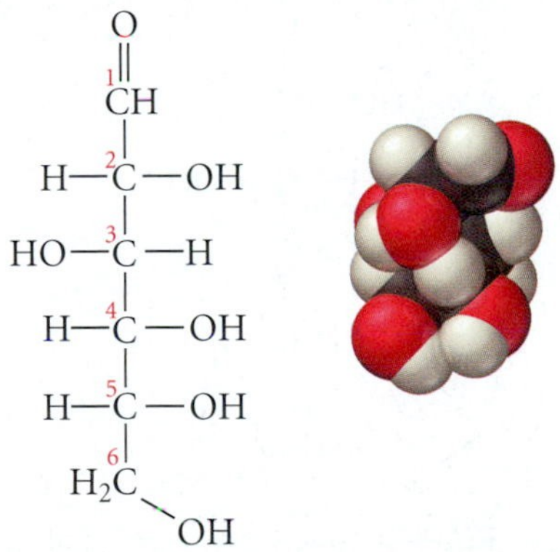

Glucose

Glucose is a six-carbon aldehyde (that is, it contains the —CHO group) with —OH groups on five of the six carbon atoms. The many —OH groups make glucose soluble in water (and therefore in blood), which is important in the role of glucose as the primary fuel of cells. Glucose is easily transported in the bloodstream and is soluble within the aqueous interior of a cell. Carbohydrates can be broadly classified as simple carbohydrates (or simple sugars) and complex carbohydrates.

> As we learned in Section 20.10, aldehydes have the general structure RCHO, and ketones have the general structure RCOR.

Simple Carbohydrates: Monosaccharides and Disaccharides

Monosaccharides—meaning "one sugar"—are the simplest carbohydrates. Monosaccharides contain between three and eight carbon atoms and have only one aldehyde or ketone functional group. The general names for monosaccharides have a prefix that indicates the number of carbon atoms, followed by the suffix -ose. The most common carbohydrates in living organisms are pentoses and hexoses.

> 3-carbon sugar: triose
> 4-carbon sugar: tetrose
> 5-carbon sugar: pentose
> 6-carbon sugar: hexose
> 7-carbon sugar: heptose
> 8-carbon sugar: octose

Glucose, whose structure we saw previously, is an example of a **hexose**, a six-carbon sugar. Glucose is also an example of an **aldose**, a sugar with an aldehyde group. Often, these two ways of designating sugars are combined, so we say that glucose is an *aldohexose* (*aldo-* indicates that it is an aldehyde; *-hex-* indicates that it has six carbon atoms; and *-ose* indicates that it is a carbohydrate).

Another common carbohydrate is fructose, a polyhydroxy *ketone* with the following structure:

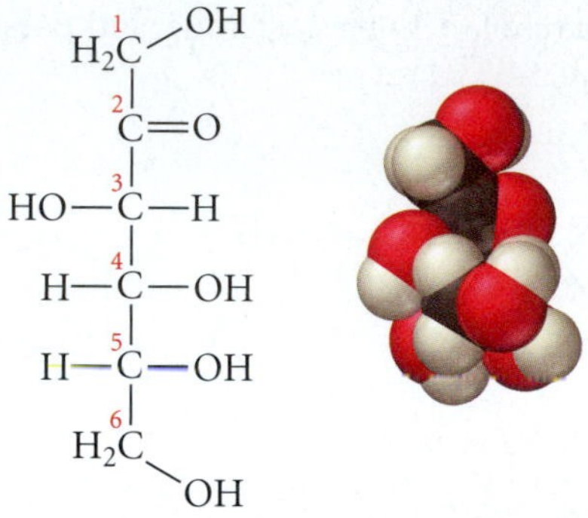

Fructose

Glucose and fructose are structural isomers—they both have the same formula ($C_6H_{12}O_6$), but they have different structures. Fructose is an example of a **ketose**, a sugar that is a ketone. Since fructose has six carbon atoms, it is a *ketohexose*. Fructose, often called fruit sugar, is found in many fruits and vegetables and is a major component of honey.

EXAMPLE 21.1   Carbohydrates and Optical Isomerism

Closely examine the structure of glucose below. Does glucose exhibit optical isomerism (discussed in Section 20.3)? If so, which carbon atoms are chiral?

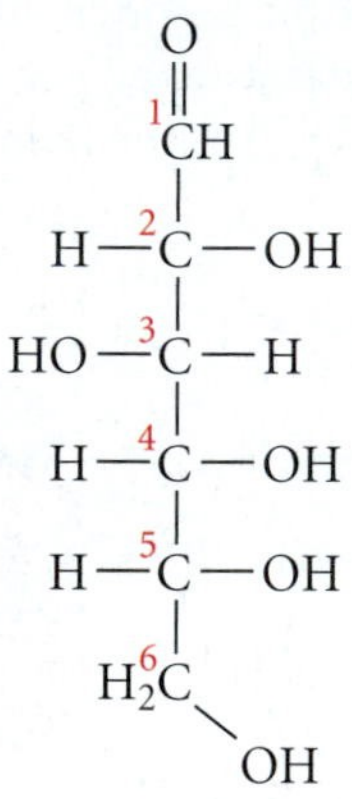

Glucose

Solution

Any carbon atom with four different substituents attached to it will be chiral. Glucose has four chiral carbon atoms (those labeled 2, 3, 4, and 5) and therefore exhibits optical isomerism.

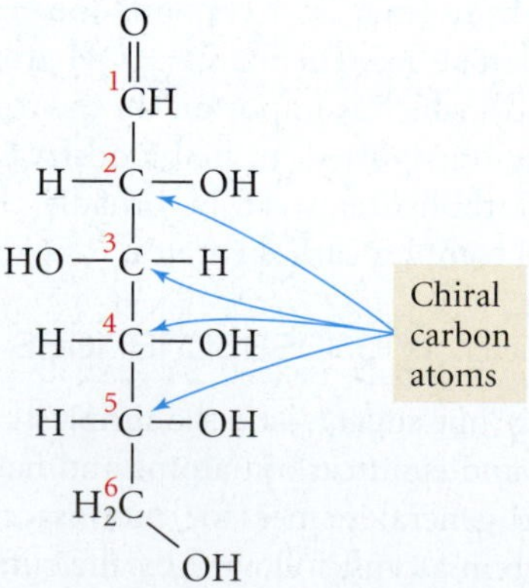

Glucose

Variations in the positions of the —OH and —H groups on these carbon atoms results in a number of different possible isomers for glucose. For example, switching the relative positions of the —OH and —H group on the carbon atom closest to the carbonyl group results in mannose, an optical isomer of glucose.

For Practice 21.1

Examine the structure of fructose (p. 962). Is fructose optically active? How many of the carbon atoms in fructose are chiral?

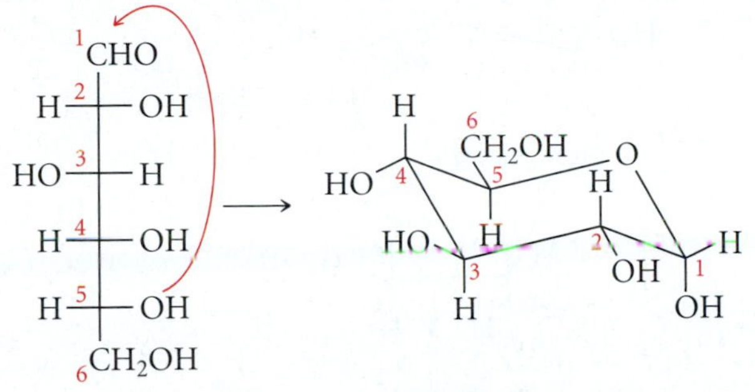

▲ **FIGURE 21.7 Intramolecular Reaction of Glucose to Form a Ring**
The alcohol group on C5 in glucose reacts with the carbonyl group (C1) to form a closed ring.

Most five- and six-carbon monosaccharides undergo an intramolecular reaction that converts their straight carbon chain into a ring. For example, in glucose, the alcohol group on C5 reacts with the carbonyl group (C1) as shown in Figure 21.7 ▲ to form the following ring structure:

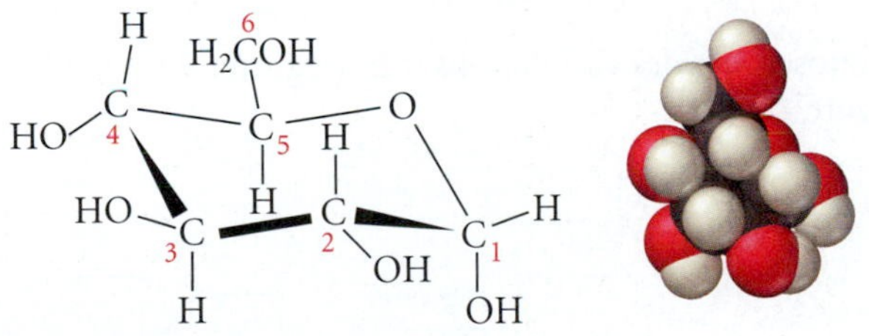

In a glucose solution, the vast majority of the molecules are in their ring form. However, the molecules in ring form exist in equilibrium with a small fraction in the open-chain form. Other common monosaccharides, in their ring form, include fructose (discussed previously in its straight-chain form) and galactose.

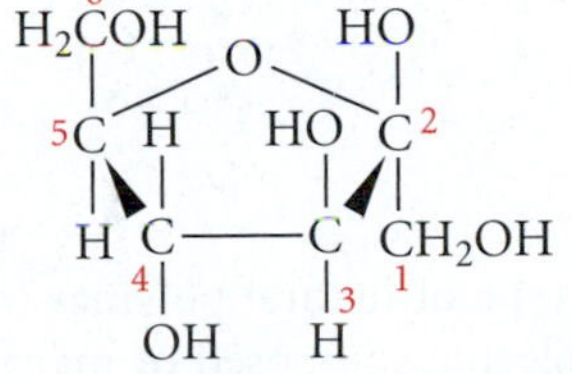

Fructose

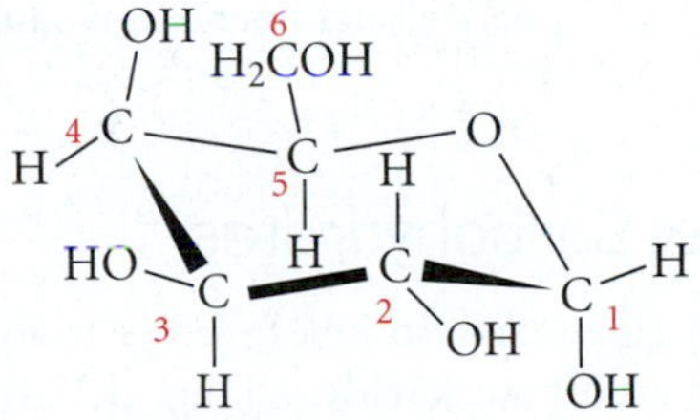

Galactose

Galactose, also known as brain sugar, is a hexose usually found combined with other monosaccharides in disaccharides such as lactose (see next paragraph). Galactose also occurs within the brain and nervous system of most animals. The difference between galactose and glucose lies only in the stereochemistry at C4. Notice that in galactose, the —OH group is roughly perpendicular to the plane of the ring while in glucose it is roughly in the same plane as the ring.

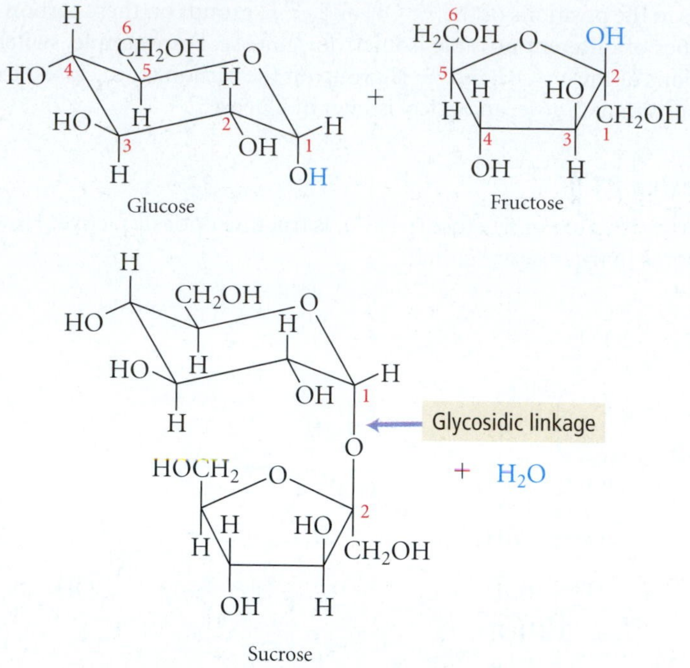

▶ **FIGURE 21.8 Formation of a Glycosidic Linkage** Glucose and fructose can join, eliminating water and forming a glycosidic linkage that results in the disaccharide sucrose, commonly known as table sugar.

Two monosaccharides can link together via a **glycosidic linkage** to form a **disaccharide**, a carbohydrate that can be decomposed into two simpler sugars. For example, glucose and fructose can be joined to form sucrose, commonly known as table sugar (Figure 21.8 ▲). When we eat disaccharides, the link between individual monosaccharides is broken during digestion by **hydrolysis**, the splitting of a chemical bond with water resulting in the addition of H and OH to the products.

$$R—R + HOH \longrightarrow R—H + ROH$$

The resultant monosaccharides can then pass through the intestinal wall and enter the bloodstream (Figure 21.9 ▼).

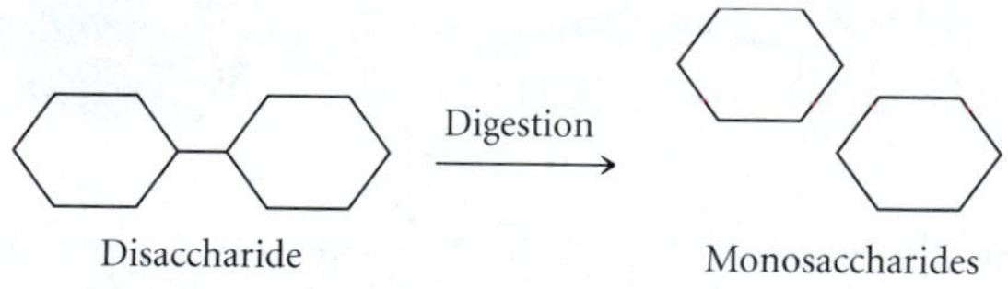

▲ **FIGURE 21.9 Hydrolysis of Disaccharides** A disaccharide is hydrolyzed during digestion. The resultant monosaccharides can then pass through the intestinal wall and enter the bloodstream.

Complex Carbohydrates

Monosaccharides can also link together to form a type of natural polymer (or biopolymer) called a **polysaccharide**, a long, chainlike molecule composed of many monosaccharide units bonded together. Polysaccharides are known as **complex carbohydrates** because of their long chains of sugars. The most common polysaccharides are *cellulose*, *starch*, and *glycogen*, all three of which are composed of repeating glucose units. The main difference among them lies in the way the units are bonded together. In cellulose, the oxygen atoms are roughly parallel with the planes of the rings. This is called a β-glycosidic linkage. In starch and glycogen, the oxygen atoms joining neighboring glucose units point down relative to the planes of the rings. This is called an α-glycosidic linkage.

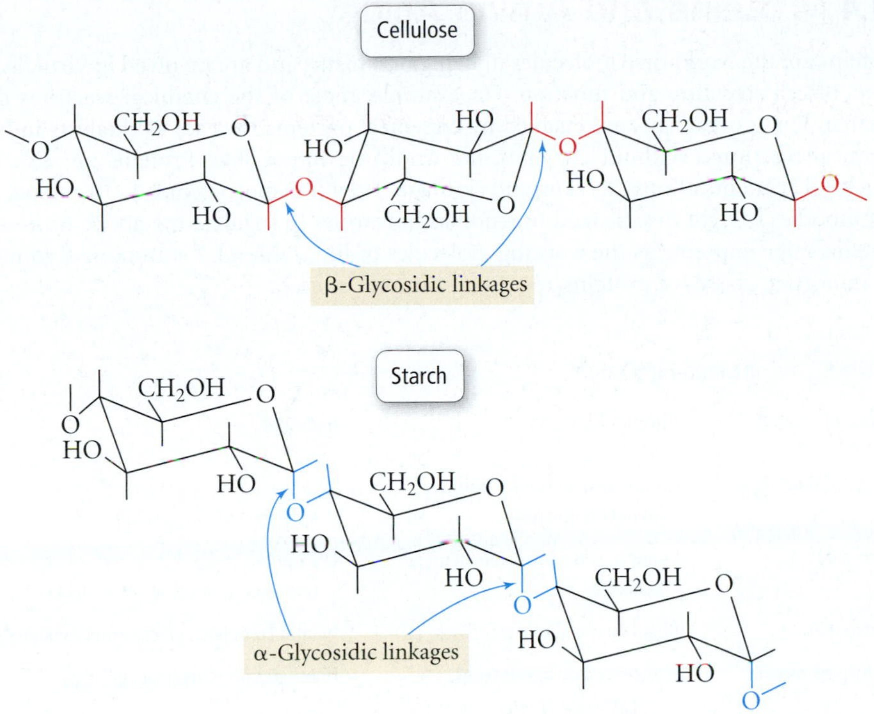

Cellulose

β-Glycosidic linkages

Starch

α-Glycosidic linkages

◄ The main difference between starch and cellulose lies in the way the units are bonded together.

Cellulose The main structural component of plants, **cellulose**, is the most abundant organic substance on Earth. It consists of glucose units bonded together by β-glycosidic linkages. This structure allows neighboring cellulose molecules to form multiple hydrogen bonds with one another, resulting in the rigid and structurally stable properties we associate with wood and fiber. Humans lack the enzyme required to digest cellulose. When we eat cellulose (usually called fiber when it is present in foods), it passes right through the intestine, providing bulk to stools and preventing constipation. Some bacteria, however, do have the enzyme required to metabolize cellulose into its component glucose units. These bacteria are found in the guts of termites and ruminants such as cows, allowing them to extract caloric content from cellulose.

Starch The main energy storage medium for plants is **starch**, the soft, pliable substance abundant in potatoes and grains. Starch is composed of two slightly different polysaccharides, *amylose* and *amylopectin*. Both are composed of glucose units bonded together by α-glycosidic linkages, but amylopectin contains branches in the chains. When we digest starch, the link between individual glucose units is broken by hydrolysis, allowing glucose molecules to pass through the intestinal wall and into the bloodstream (Figure 21.10 ▼).

Glycogen The structure of **glycogen** is similar to that of amylopectin, but the chain is even more highly branched. Animals use glycogen to store glucose in the muscles. Glycogen's highly branched structure leaves many end groups that can be quickly hydrolyzed to meet energy needs. When muscles become depleted of glycogen, muscle movement and exercise become much more difficult. Marathon runners often "hit the wall" at about mile 20 because they have depleted most of the glycogen from their muscles.

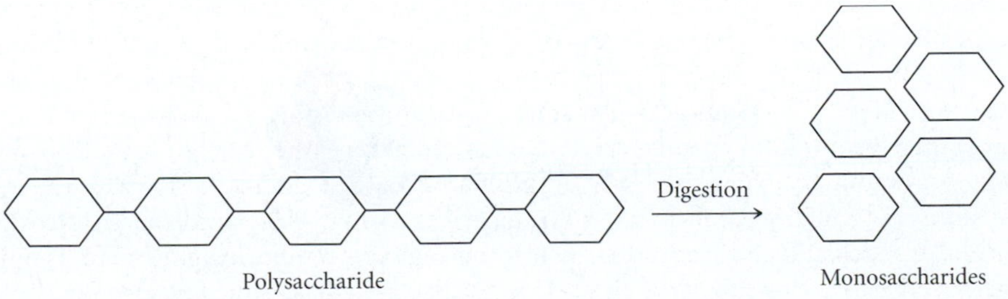

Polysaccharide   Digestion →   Monosaccharides

◄ **FIGURE 21.10 Hydrolysis of Polysaccharides** A polysaccharide is hydrolyzed during digestion. The resultant monosaccharides can then pass through the intestinal wall and enter the bloodstream.

21.4 Proteins and Amino Acids

Proteins are the workhorse molecules in living organisms and are involved in virtually all facets of cell structure and function. For example, most of the chemical reactions that occur in living organisms are enabled by **enzymes**, proteins that act as catalysts in biochemical reactions. Without enzymes, life would be impossible. Proteins are also the structural elements of muscle, skin, and cartilage. They transport oxygen in the blood, act as antibodies to fight disease, and function as hormones to regulate metabolic processes. Proteins reign supreme as the working molecules of life. Table 21.2 summarizes some of the important classes of proteins and gives examples of each.

TABLE 21.2 Protein Functions

Class of Protein	Primary Function	Example
Structural proteins	Compose structures within living organisms	Collagen (skin, tendon, cartilage), keratin (hair, fingernails)
Enzymes	Catalyze and control biochemical reactions	DNA polymerase (involved in replication of DNA)
Hormones	Regulate metabolic processes	Insulin (regulates glucose metabolism)
Transport proteins	Transport substances from one place to another	Hemoglobin (transports oxygen)
Storage proteins	Provide source of essential nutrients	Casein (protein in mammalian milk)
Contractile and motile proteins	Mediate motion and muscle contraction	Actin and myosin (provide muscle contraction)
Protective proteins	Protect and defend cells	Antibodies (neutralize infectious agents)

Amino Acids: The Building Blocks of Proteins

In a protein, an R group does not necessarily mean a pure alkyl group. See Table 21.3 for possible R groups.

Proteins are polymers of amino acids. **Amino acids** contain a carbon atom—called the α-carbon—bonded to four different groups: an amine group, an R group (also called a side chain), a carboxylic acid group, and a hydrogen atom:

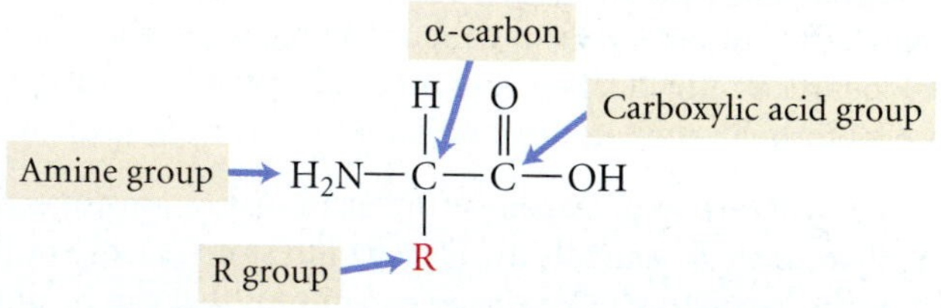

Amino acid general structure

Amino acids differ from each other only in their R group. A simple amino acid is alanine, which has a methyl group (CH_3) as its R group:

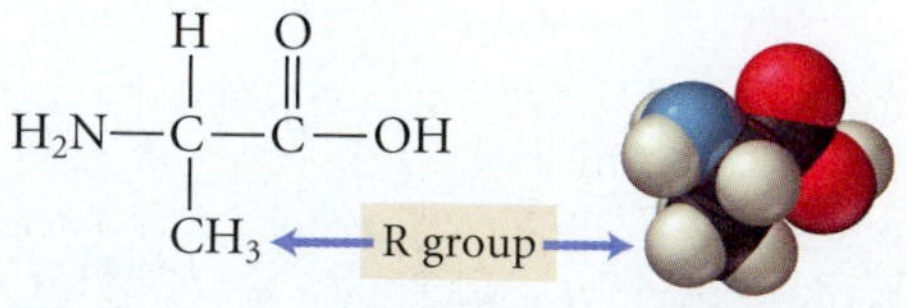

Alanine

Other amino acids include glycine (R = H), phenylalanine (R = $CH_2C_6H_6$), serine (R = CH_2OH), aspartic acid (R = CH_2COOH), and lysine (R = $CH_2CH_2CH_2CH_2NH_2$).

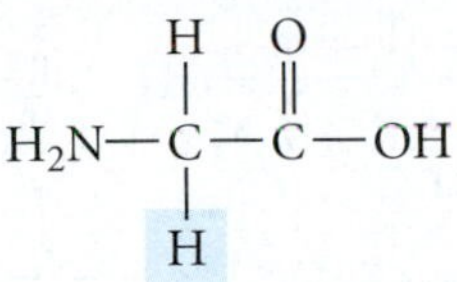

Glycine

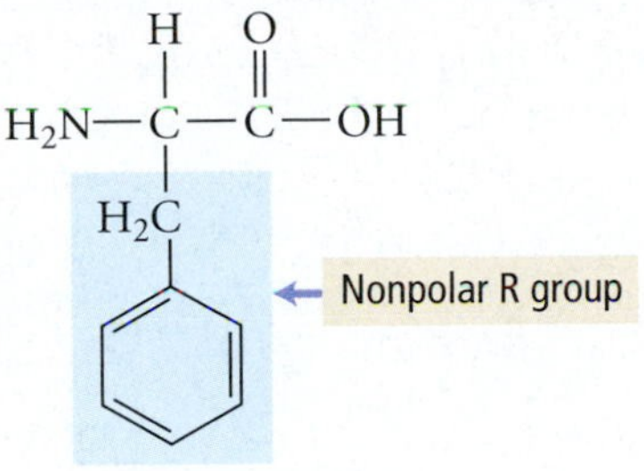

Phenylalanine ← Nonpolar R group

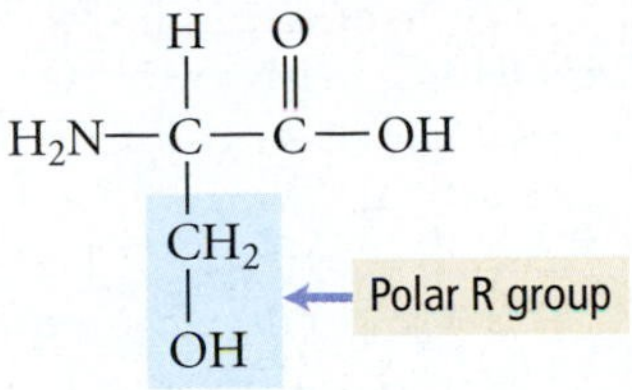

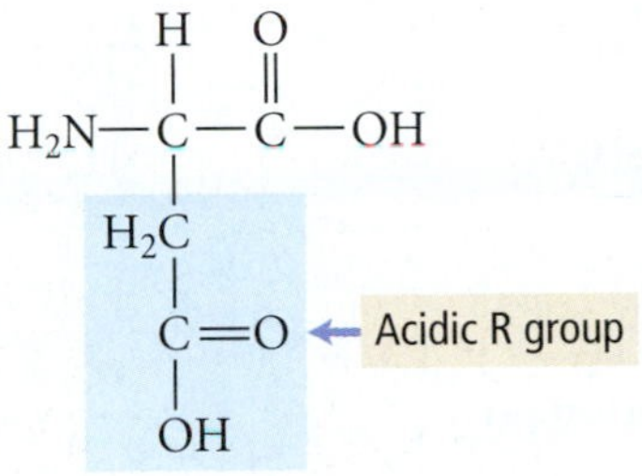

Serine ← Polar R group

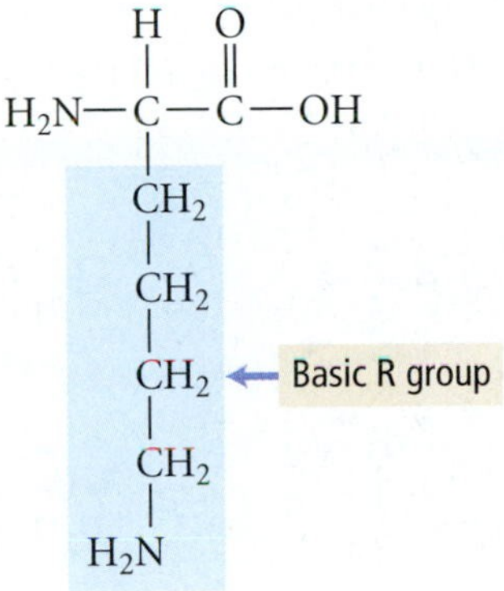

Aspartic acid ← Acidic R group

Lysine ← Basic R group

The R groups, or side chains, differ chemically from one amino acid to another. For example, phenylalanine has a large nonpolar R group, whereas serine has a polar one. Aspartic acid has an acidic R group, whereas lysine, since it contains nitrogen, has a basic one. When amino acids are strung together to make a protein, the chemical properties of the R groups determine the structure and properties of the protein. Table 21.3 on p. 968 shows the most common amino acids in proteins and their three-letter abbreviations.

Since all amino acids (except glycine) contain four different groups attached to a tetrahedral carbon (the α-carbon), all amino acids are chiral about that carbon. The amino acids that compose naturally occurring proteins are the L-enantiomers, and are thus called L-amino acids. Why life on Earth settled on this enantiomer over the other is an interesting question that remains to be answered. (It seems just as likely that life could have used the D-enantiomer.)

Although we usually write the structures of amino acids as neutral, their actual structure is ionic and depends on pH. In general, amino acids undergo an intramolecular acid–base reaction to form a *dipolar ion*, or *zwitterion*:

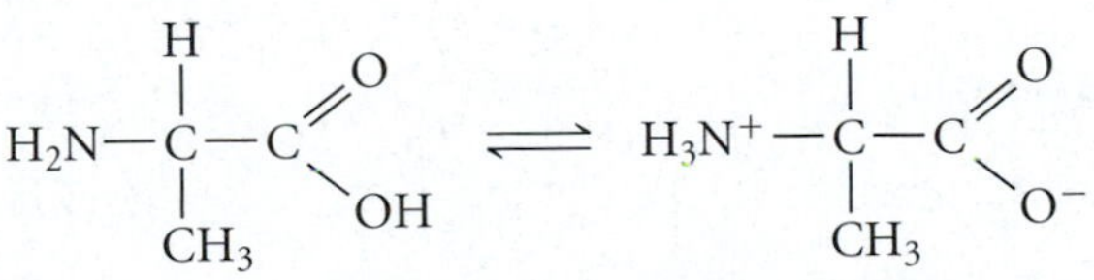

Dipolar ion

At room temperature this equilibrium lies far to the right. Since one side of the dipolar ion is positively charged and the other negatively charged, amino acids are highly polar and soluble in water. They also have fairly high melting points (usually > 200 °C). In addition, the intramolecular acid–base reaction makes amino acids less acidic and less basic than most carboxylic acids and amines, respectively.

TABLE 21.3 Common Amino Acids

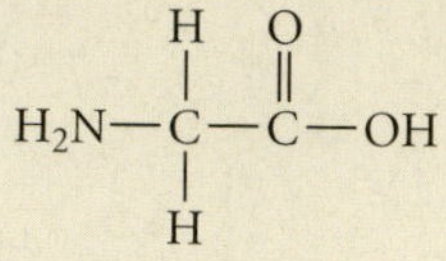

Glycine (Gly)

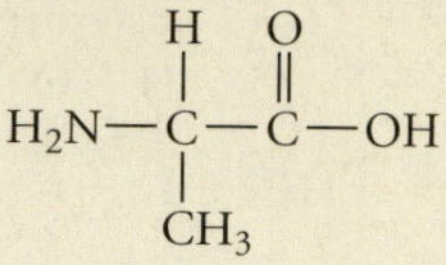

Alanine (Ala)

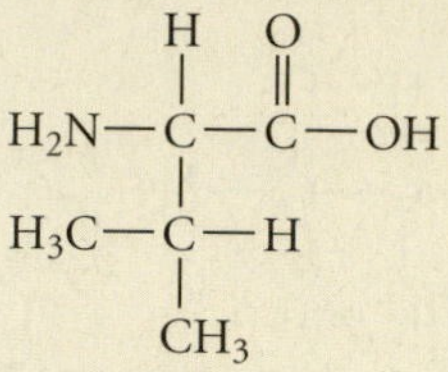

Valine (Val)

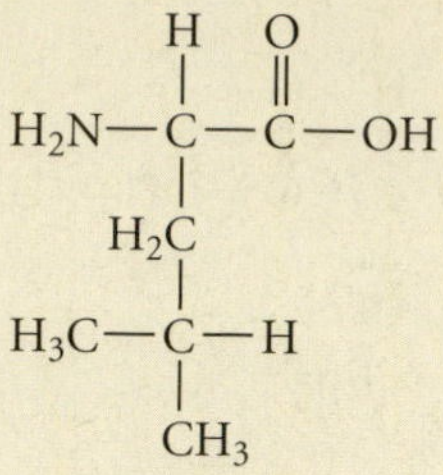

Leucine (Leu)

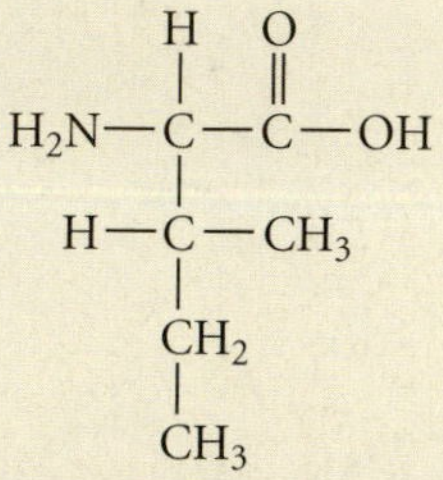

Isoleucine (Ile)

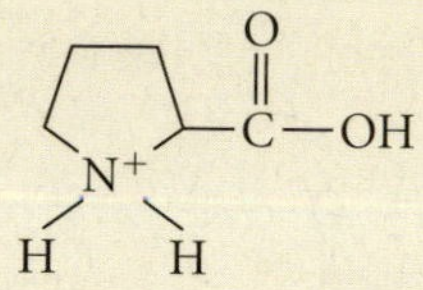

Proline (Pro)

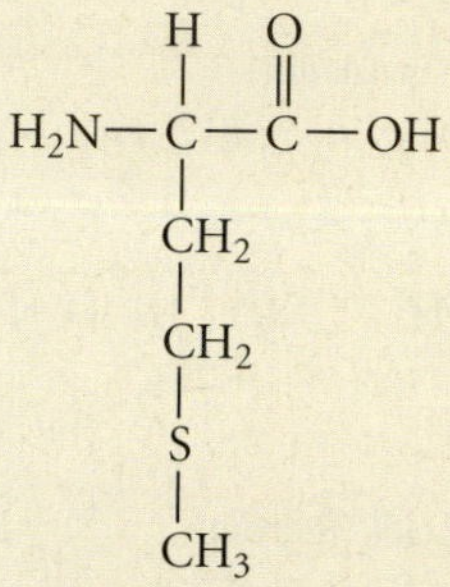

Methionine (Met)

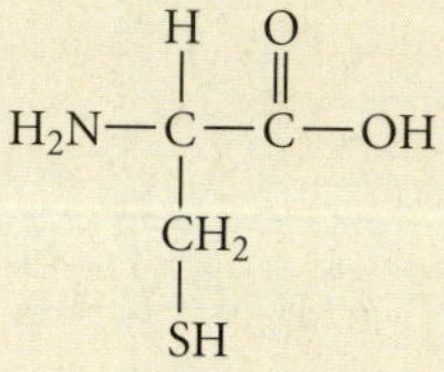

Cysteine (Cys)

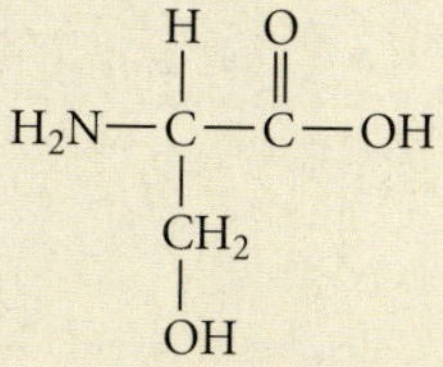

Serine (Ser)

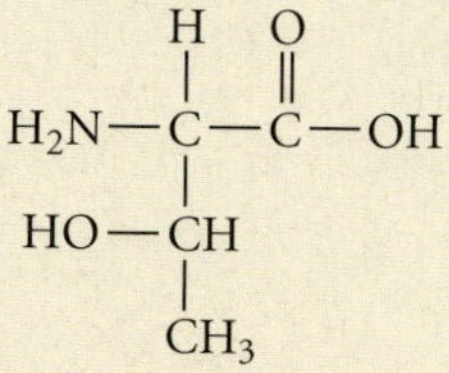

Threonine (Thr)

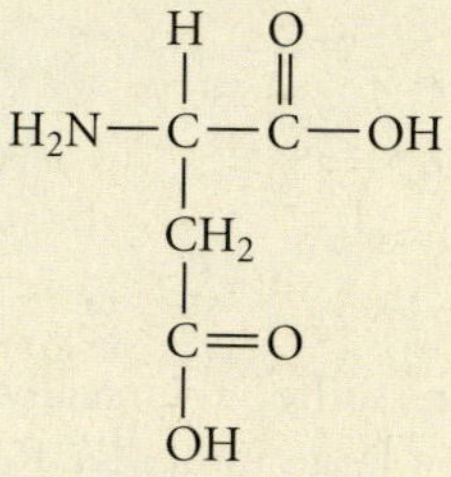

Aspartic acid (Asp)

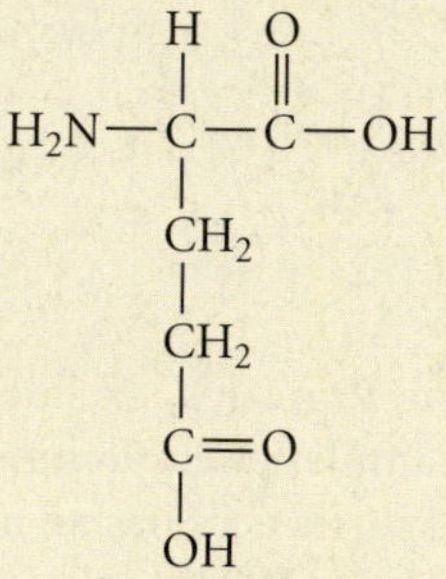

Glutamic acid (Glu)

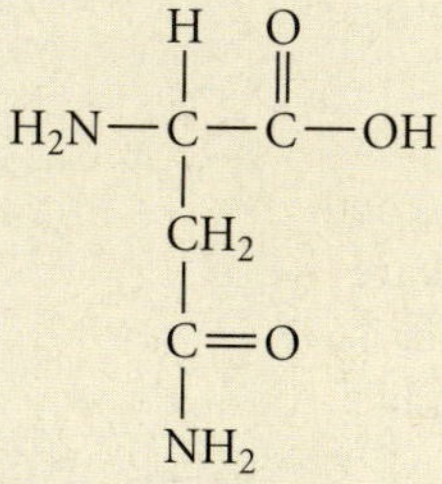

Asparagine (Asn)

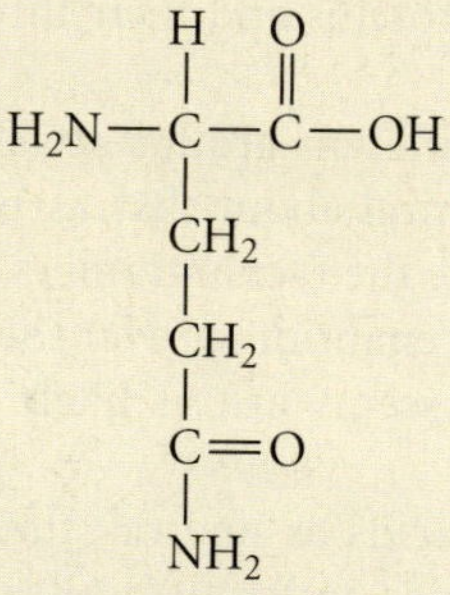

Glutamine (Gln)

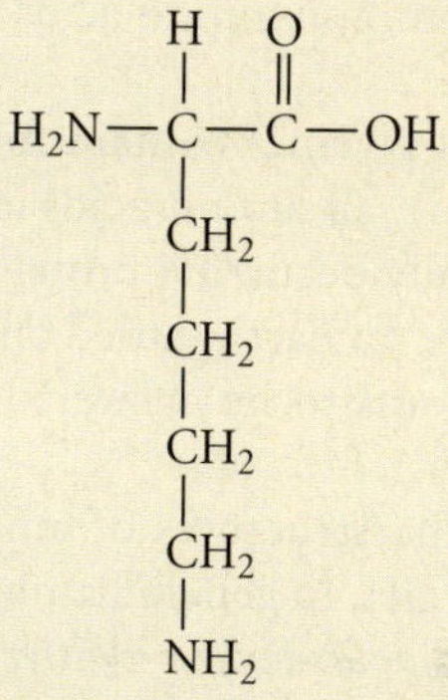

Lysine (Lys)

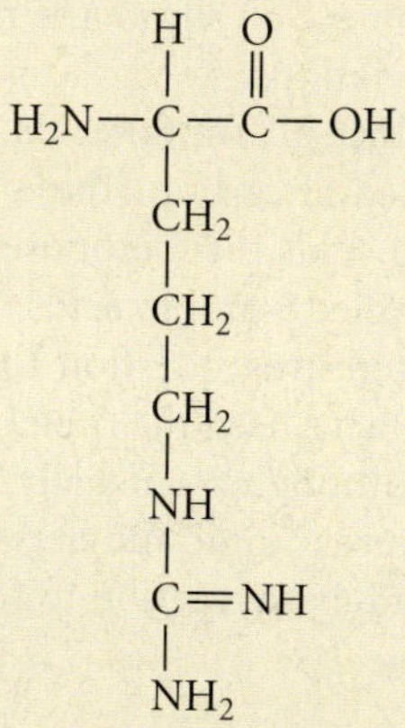

Arginine (Arg)

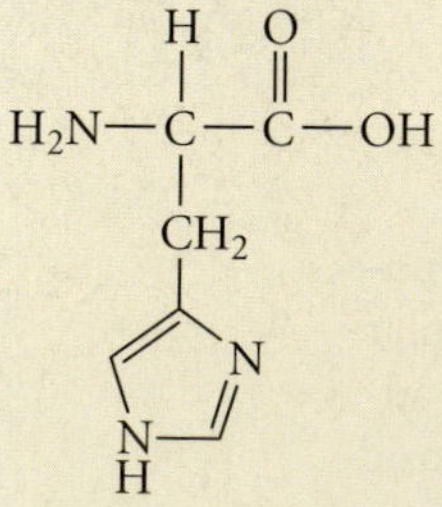

Histidine (His)

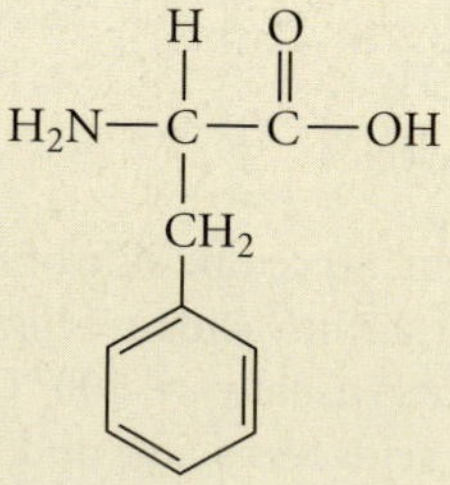

Phenylalanine (Phe)

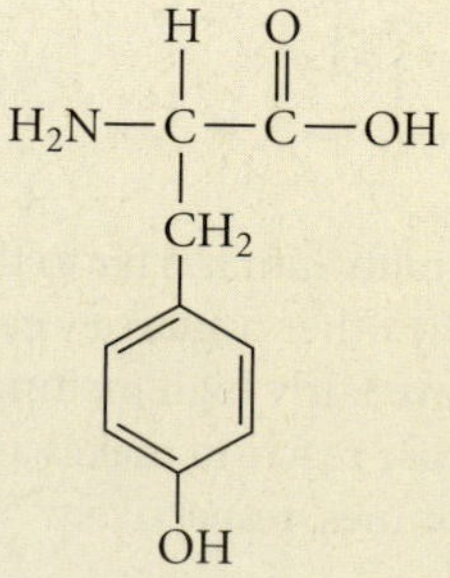

Tyrosine (Tyr)

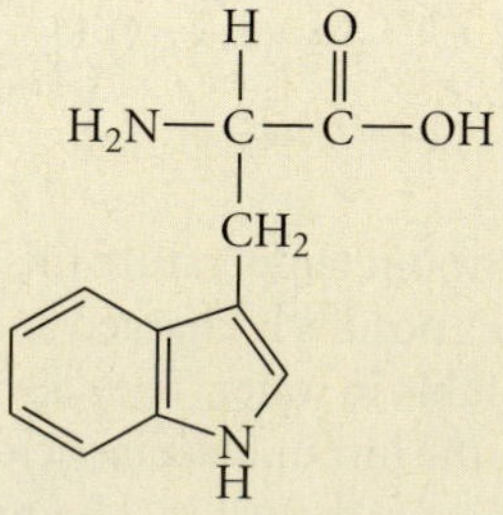

Tryptophan (Trp)

Peptide Bonding between Amino Acids

Amino acids link together through the reaction of the amine end of one amino acid with the carboxylic end of another.

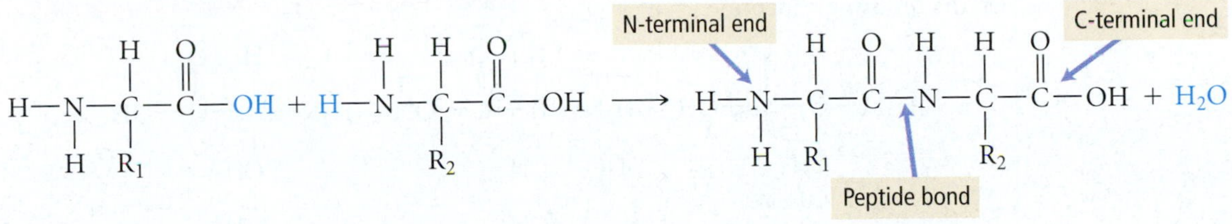

The resulting bond is called a **peptide bond**, and the resulting molecule—two amino acids linked together—is called a **dipeptide**. When two or more amino acids are linked in this way, the resulting molecule has two distinct ends: an amino terminal (or N-terminal end) and a carboxyl terminal (or C-terminal end). A *tripeptide* is simply three amino acids joined by peptide bonds; a *tetrapeptide* is four, and so on. Short chains of amino acids are generally called *oligopeptides*, and longer chains (more than 20) are called **polypeptides**. Functional proteins usually contain one or more polypeptide chains with each chain consisting of hundreds or even thousands of amino acids joined by peptide bonds.

The formation of a peptide bond is an example of a condensation reaction (see Section 20.11).

EXAMPLE 21.2 Peptide Bonds

Show the reaction by which valine, cysteine, and phenylalanine link (in that order) via peptide bonds. Make valine N-terminal and identify the N-terminal and C-terminal ends in the resulting tripeptide.

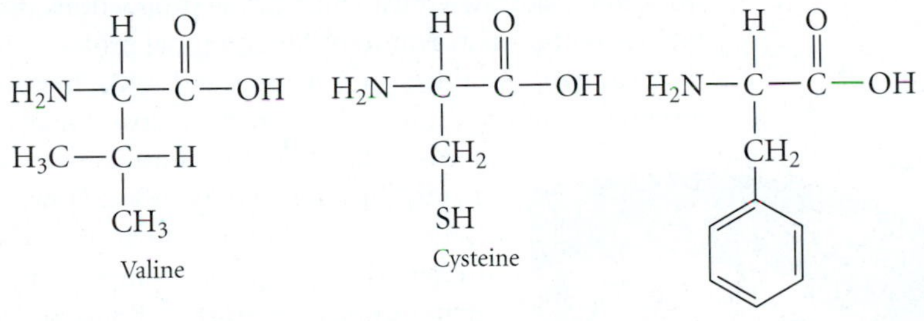

Solution

Peptide bonds are formed when the carboxylic end of one amino acid reacts with the amine end of another amino acid.

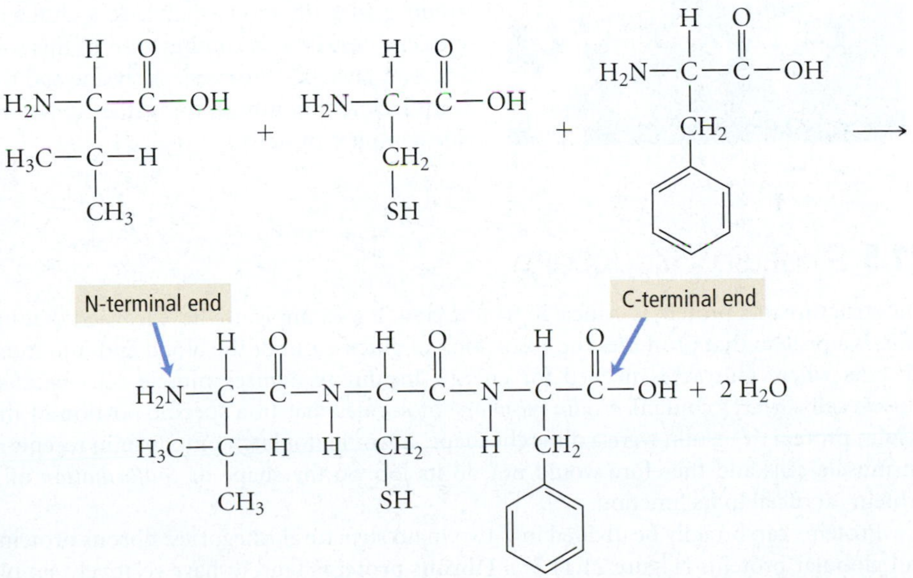

For Practice 21.2

Show the reaction by which alanine, threonine, and serine link (in that order) via peptide bonds. Make alanine N-terminal and identify the N-terminal and C-terminal ends in the resulting tripeptide.

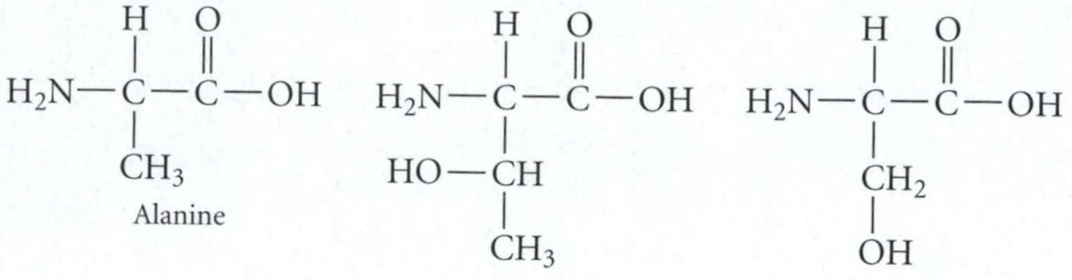

$$H_2N-\overset{\overset{\displaystyle H}{|}}{\underset{\underset{\displaystyle CH_3}{|}}{C}}-\overset{\overset{\displaystyle O}{\|}}{C}-OH \qquad H_2N-\overset{\overset{\displaystyle H}{|}}{\underset{\underset{\underset{\displaystyle CH_3}{|}}{\underset{\displaystyle HO-CH}{|}}}{C}}-\overset{\overset{\displaystyle O}{\|}}{C}-OH \qquad H_2N-\overset{\overset{\displaystyle H}{|}}{\underset{\underset{\underset{\displaystyle OH}{|}}{\underset{\displaystyle CH_2}{|}}}{C}}-\overset{\overset{\displaystyle O}{\|}}{C}-OH$$

Alanine   Threonine   Serine

Conceptual Connection 21.1 Peptides

How many different tripeptides can be formed from the three amino acids listed below? (The amino acids are indicated using the three-letter amino acid abbreviations in Table 21.3.)

Ser, Ala, Gly

Answer: Six possible tripeptides can be made from three amino acids. They are (1) Ser-Gly-Ala; (2) Gly-Ala-Ser; (3) Ala-Ser-Gly; (4) Ala-Gly-Ser; (5) Ser-Ala-Gly; (6) Gly-Ser-Ala. Notice that bonding the amino acids in reverse order results in a different molecule because the N-terminal and the C-terminal ends reside on different amino acids. For example, in Ser-Gly-Ala, the N-terminal amino acid is Ser (conventionally drawn on the left) and the C-terminal side is Ala. In Ala-Gly-Ser, however, the N-terminal amino acid is Ala and the C-terminal one is Ser.

Chemistry and Medicine
The Essential Amino Acids

Of the 20 amino acids necessary for protein synthesis in humans, about half can be synthesized in the body. The other 10, called *essential amino acids* (see Table 21.4), must come from food, and they must come in the proportions that match the needs of protein synthesis. Many animal proteins, especially those in dairy products such as eggs and milk, contain all the essential amino acids in proportions that closely match human needs. They are therefore known as high-quality, or complete, proteins. However, most plant proteins do not contain the 10 essential amino acids in the right proportions. For example, grains are short on lysine and beans or lentils are short on methionine. Since different types of plants are generally deficient in different amino acids, a vegetarian can obtain complete protein by matching foods lacking in one amino acid with foods that lack a different one. The best-known combination of this sort is rice and beans—the rice lacks lysine and the beans lack methionine, but together they provide complete protein.

TABLE 21.4 Essential Amino Acids

Phenylalanine	Valine
Tryptophan	Threonine
Isoleucine	Methionine
Histidine	Arginine
Lysine	Leucine

21.5 Protein Structure

The structure of a protein is critical to its function. For example, we have learned that insulin is a protein that promotes the absorption of glucose out of the blood and into muscle cells where glucose is needed for energy. Insulin recognizes muscle cells because muscle cell surfaces contain *insulin receptors*, molecules that fit a specific portion of the insulin protein. If insulin were a different shape, it would not latch onto insulin receptors on muscle cells and therefore would not do its job. So the shape or *conformation* of a protein is crucial to its function.

Proteins can broadly be divided into two main structural categories: fibrous proteins and globular proteins (Figure 21.11 ▶). **Fibrous proteins** tend to have relatively simple

Fibrous and Globular Proteins

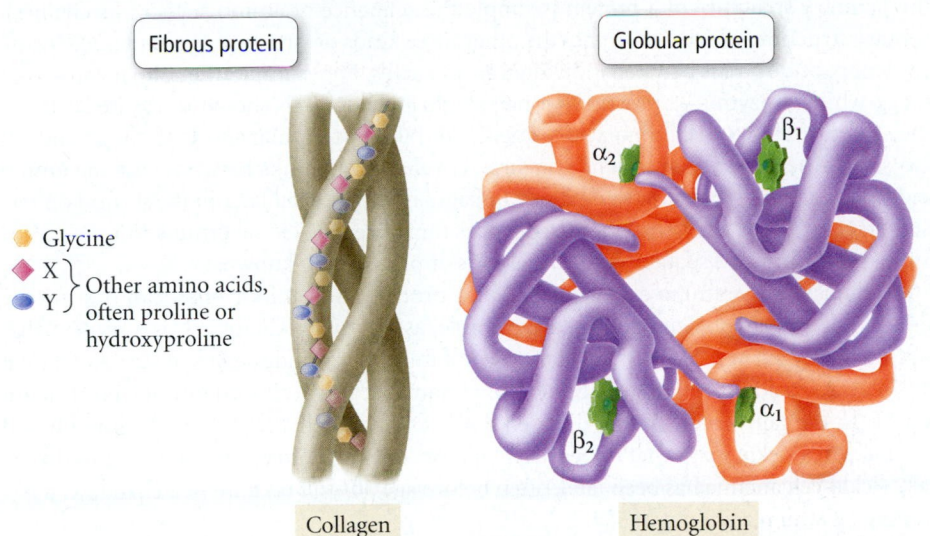

Collagen

Hemoglobin

◀ **FIGURE 21.11 Fibrous and Globular Proteins** Proteins can broadly be divided into fibrous proteins (which have relatively simple linear structures) and globular proteins (which have more complex three-dimensional structures).

linear structures and tend to be insoluble in aqueous solutions. They serve primarily structural functions within living organisms. Collagen and keratin, for example, are both fibrous proteins (see Table 21.2). **Globular proteins** tend to have more complex structures that are highly complicated and specific but often roughly spherical in overall shape. Globular proteins are generally structured so that polar side chains on amino acids are oriented toward the exterior of the protein, while nonpolar side chains are oriented toward the interior of the protein. Consequently, globular proteins tend to be soluble in water, but they maintain a nonpolar environment within the protein that excludes water. Hemoglobin and insulin, for example, are both globular proteins.

Protein structure can be analyzed at four levels: primary structure, secondary structure, tertiary structure, and quaternary structure (Figure 21.12 ▼). We examine each of these categories separately.

Levels of Protein Structure

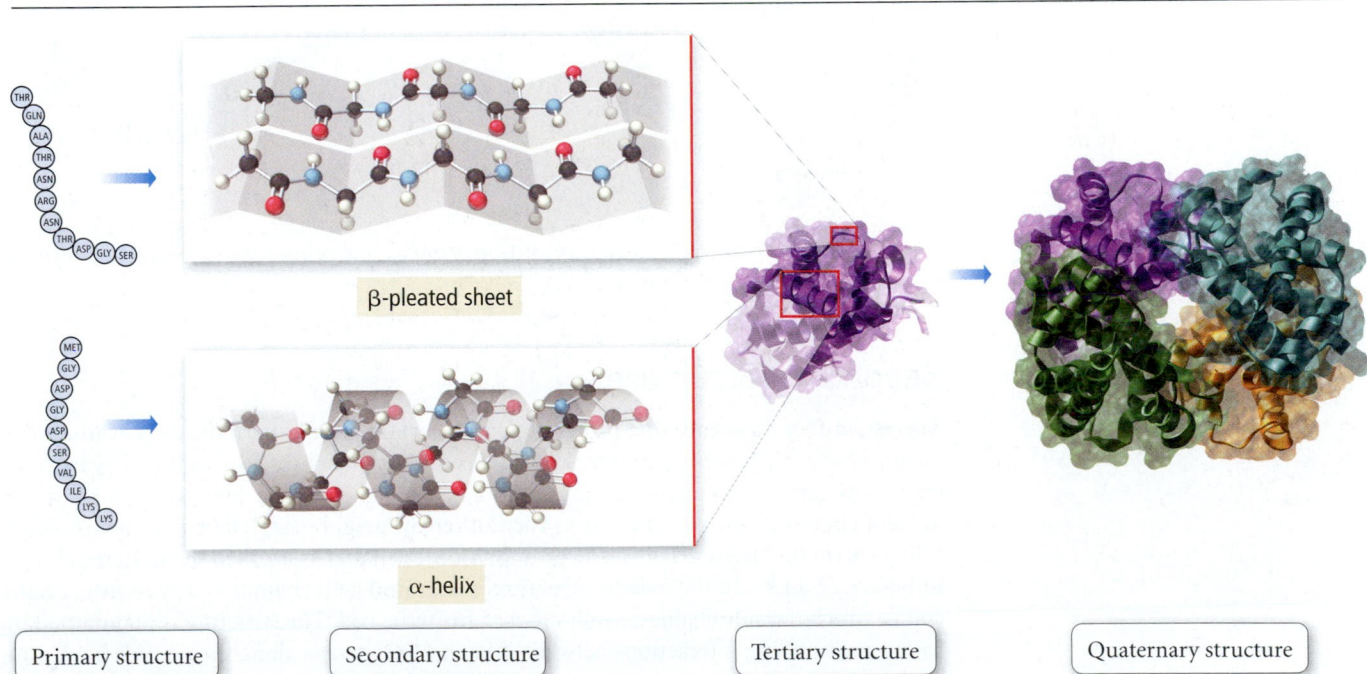

β-pleated sheet

α-helix

Primary structure

Secondary structure

Tertiary structure

Quaternary structure

▲ **FIGURE 21.12 Levels of Protein Structure** Protein structure can be analyzed at four levels: primary, secondary, tertiary and quaternary.

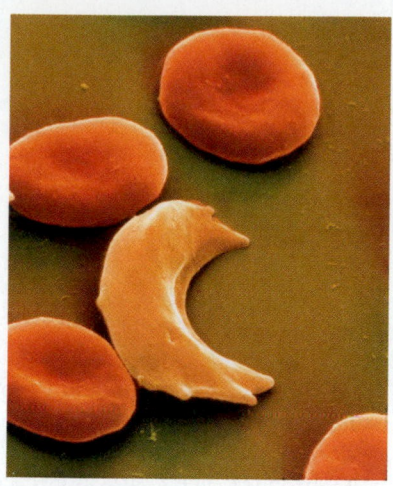

▲ The genetic disease known as sickle-cell anemia results in red blood cells with a sickle shape. These cells impede circulation of blood, causing damage to major organs.

Primary Structure

The **primary structure** of a protein is simply the sequence of amino acids in its chain(s). Primary structure, which determines the other three kinds of structure, is maintained by the covalent peptide bonds between individual amino acids. For example, the primary structure of egg-white lysozyme—a protein that helps fight infection—is shown in Figure 21.13 ▼. The structure shows the amino acid sequence and the N-terminal and C-terminal ends. It also shows the presence of *disulfide linkages*, covalent cross-links between cysteine amino acids in the polymer. We discuss disulfide linkages in more detail later in the section on tertiary structure. The first amino acid sequences for proteins were determined in the 1950s. Today, the amino acid sequences for thousands of proteins are known.

Changes in the amino acid sequence of a protein, even minor ones, can destroy the function of a protein. Hemoglobin, for example, as we saw in Chapter 1, is a protein that transports oxygen in the blood. It is composed of four polypeptide chains made up of a total of 574 amino acid units. The replacement of glutamic acid by valine in just one position on two of these chains results in the disease known as sickle-cell anemia, in which red blood cells take on a sickle shape that impedes circulation, causing damage to major organs. In the past, sickle-cell anemia has been fatal, often before age 30—all because of a change in a few atoms of 2 amino acids out of 574.

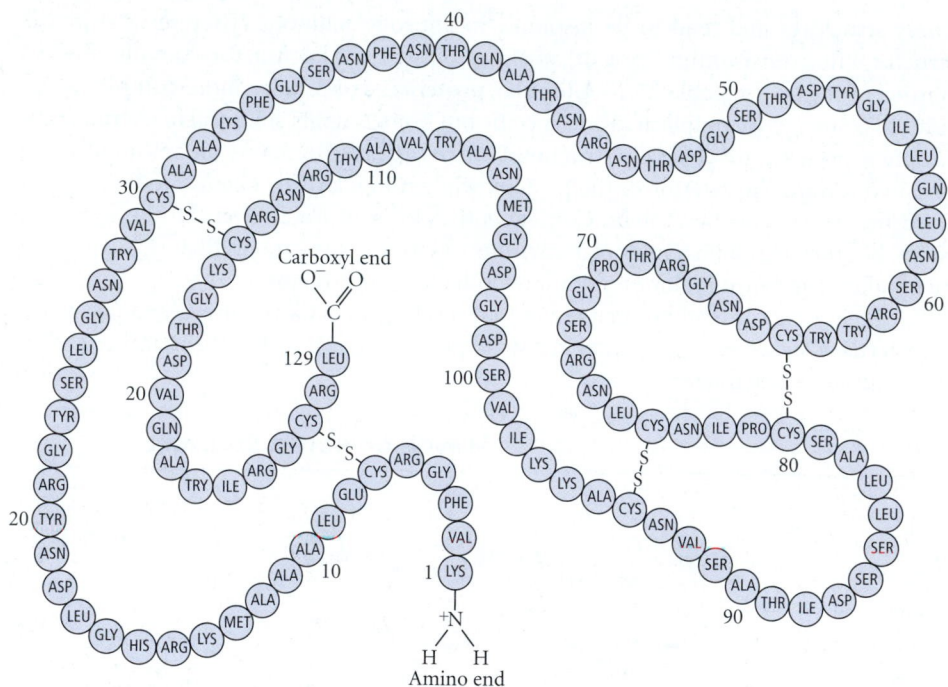

▲ **FIGURE 21.13 Primary Structure of Egg-White Lysozyme**   Primary structure refers to the sequence of amino acids in a protein.

Secondary Structure

The **secondary structure** of a protein refers to certain regular periodic or repeating patterns in the arrangement of protein chains. Secondary structure is maintained by interactions between amino acids that are fairly close together in the linear sequence of the protein chain or that are adjacent to each other on neighboring chains or chains which fold back on themselves. The most common of these patterns is called the **α-helix**, shown in Figure 21.14 ▶. In the α-helix structure, the amino acid chain is wrapped into a tight coil in which the side chains extend outward from the coil. The structure is maintained by hydrogen-bonding interactions between NH and CO groups along the peptide backbone of the protein. Some proteins—such as keratin, which composes hair—have the α-helix pattern throughout their entire chain. Other proteins have very little or no α-helix pattern in their chain. It depends on the particular protein.

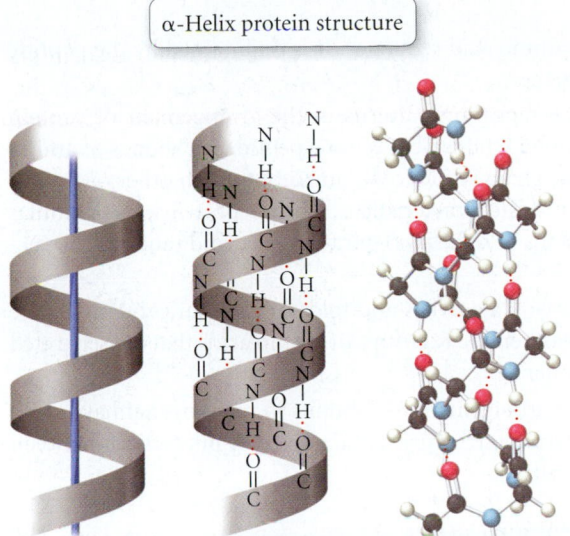

α-Helix protein structure

◀ FIGURE 21.14 The α-Helix Structure The α-helix is an example of secondary protein structure.

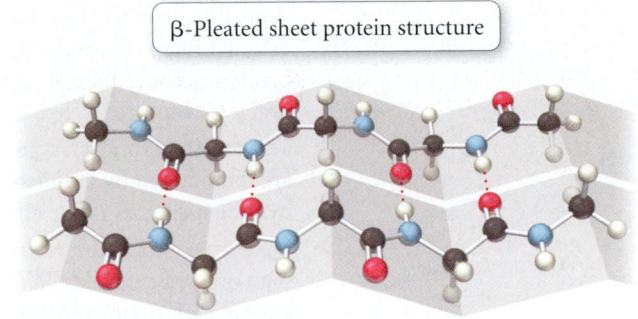

β-Pleated sheet protein structure

▲ FIGURE 21.15 The β-Pleated Sheet Structure The β-pleated sheet is an example of secondary protein structure.

A second common pattern in the secondary structure of proteins is called the **β-pleated sheet** (Figure 21.15 ▲). In this structure, the chain is extended (as opposed to coiled) and forms a zigzag pattern. The peptide backbones of neighboring chains interact with one another through hydrogen bonding to form zigzag shaped sheets. Some proteins—such as silk—have the β-pleated sheet structure throughout their entire chain. Since the protein chains in the β-pleated sheet are fully extended, silk is inelastic. Many proteins have some sections that are β-pleated sheet, other sections that are α-helical, and still other sections that have less regular patterns called **random coils**.

Tertiary Structure

The **tertiary structure** of a protein consists of the large-scale bends and folds due to interactions between the R groups of amino acids that are separated by large distances in the linear sequence of the protein chain. These interactions, shown in Figure 21.16 ▶, include hydrogen bonding, disulfide linkages (covalent bonds between cysteine amino acids), hydrophobic interactions (attractions between large, nonpolar side chains), and salt bridges (acid–base interactions between acidic and basic side chains). Fibrous proteins generally lack tertiary structure; they simply extend in a long continuous chain with some secondary structure. Globular proteins, by contrast, fold in on themselves, forming complex globular shapes rich in tertiary structure.

Quaternary Structure

Some proteins—called *monomeric* proteins—are composed of only one polypeptide chain. However, many proteins—called *multimeric* proteins—are composed of several polypeptide chains, called subunits. We just saw, for example, that hemoglobin is composed of four such subunits. The way that subunits fit together in a multimeric protein is called the **quaternary structure** of the protein. Quaternary structure is maintained by the same types of interactions that maintain tertiary structure, but the interactions are between amino acids on different subunits.

Interactions That Maintain Tertiary Structure

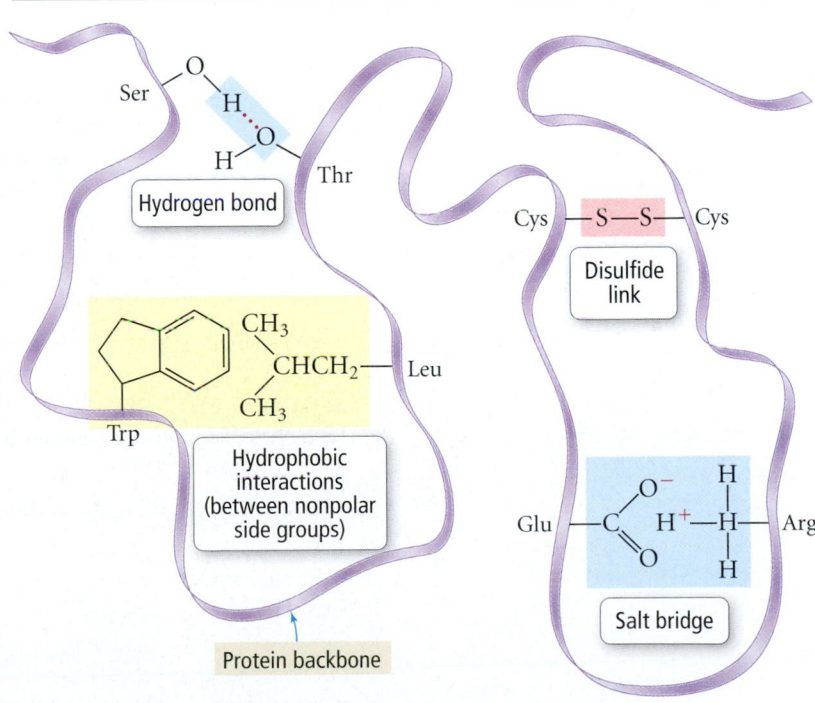

▲ FIGURE 21.16 Interactions within Proteins The tertiary structure of a protein is maintained by interactions between the R groups of amino acids that are separated by large distances in the linear sequence of the protein chain.

Summarizing:

➤ Primary structure is simply the amino acid sequence. It is maintained by the peptide bonds that hold amino acids together.

➤ Secondary structure refers to the repeating patterns in the arrangement of protein chains. These are maintained by interactions between the peptide backbones of amino acids that are close together in the chain sequence or adjacent to each other on neighboring chains. Secondary structure is characteristic of fibrous proteins, but globular proteins also frequently have regions of α-helix, β-pleated sheet, and random coil secondary structure.

➤ Tertiary structure refers to the large-scale twists and folds of globular proteins. These are maintained by interactions between the R groups of amino acids that are separated by long distances in the chain sequence.

➤ Quaternary structure refers to the arrangement of subunits in proteins that have more than one polypeptide chain. Quaternary structure is maintained by interactions between amino acids on the different subunits.

21.6 Nucleic Acids: Blueprints for Proteins

We have seen that the amino acid sequence in a protein determines the protein's structure and function. If the amino acid sequence is incorrect, the protein is unlikely to function properly. How do the cells in living organisms synthesize the many thousands of different proteins required, each with the correct amino acid sequence? The answer lies in nucleic acids, molecules that serve as blueprints for making proteins. Nucleic acids employ a chemical code to specify the correct amino acid sequences for proteins. Nucleic acids can be broadly divided into two types: deoxyribonucleic acid, or DNA, which exists primarily in the nucleus of the cell; and ribonucleic acid, or RNA, which exists throughout the cell.

The Basic Structure of Nucleic Acids

Like proteins, nucleic acids are polymers. The individual units composing nucleic acids are called **nucleotides**. Each nucleotide has three parts: a sugar, a base, and a phosphate group, which serves as a link between sugars (Figure 21.17 ▼).

DNA: Basic Structure

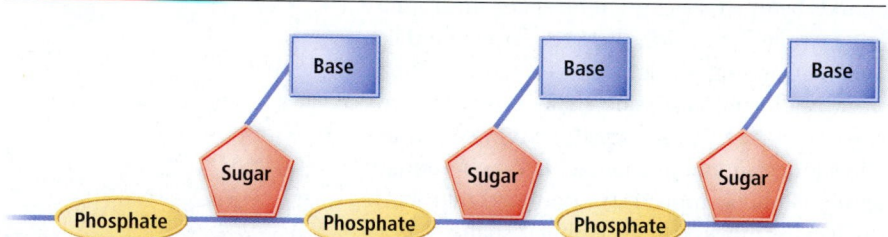

▲ **FIGURE 21.17 DNA Structure** DNA is composed of repeating units called nucleotides. Each nucleotide contains a sugar, a base, and a phosphate group.

Sugars In DNA, the sugar is deoxyribose, whereas in RNA the sugar is ribose.

Deoxyribose

Ribose

The sugar attaches to a base at C1 and attaches to phosphate groups at C3 and C5. When a base is attached to the sugar, the numbers of the carbon atoms in the sugar ring are primed to distinguish them from the carbon atoms on the bases (which are not primed). For example, C5 becomes C5′ and C3 becomes C3′.

Bases Every nucleotide in DNA has the same sugar, but one of four different bases. In DNA, the four bases are adenine (A), cytosine (C), guanine (G), and thymine (T). Each of these bases bonds to the sugar via the circled nitrogen atom.

Purine Bases

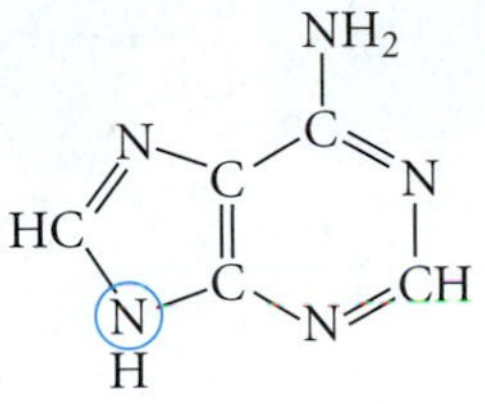

Adenine 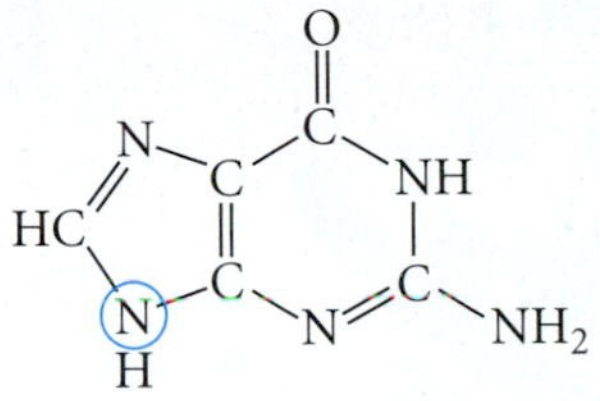   Guanine

Pyrimidine Bases

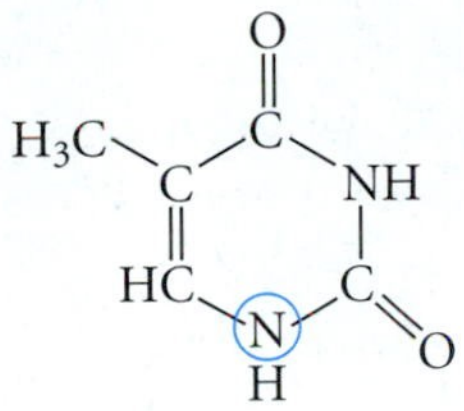

Thymine 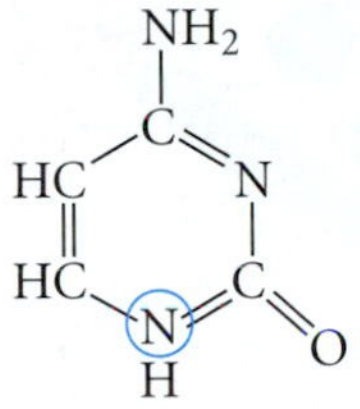   Cytosine

Adenine and guanine are called *purine* bases because they resemble the bicyclic compound purine. Cytosine and thymine are called *pyrimidine* bases because they resemble the monocyclic compound pyrimidine.

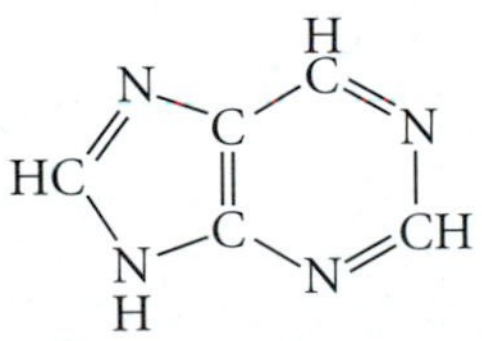

Purine 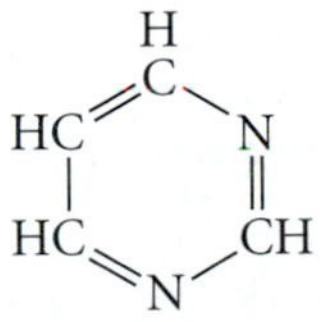   Pyrimidine

In RNA, the base uracil (U), also a pyridimine base, replaces thymine. Uracil and thymine differ only in a methyl (—CH_3) group.

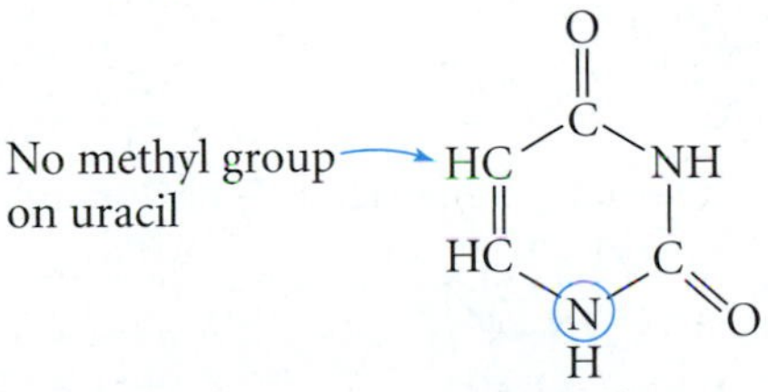

No methyl group on uracil

Uracil

The bases in nucleic acids are **complementary**—that is, they are capable of precise pairing. Each pyrimidine base pairs with one purine base via specific hydrogen bonds that occur between the two bases. More specifically, guanine pairs with cytosine through three hydrogen bonds and adenine pairs with thymine through two hydrogen bonds as shown in Figure 21.18 on page 976. In other words, guanine is complementary to cytosine and adenine is complementary to thymine. This base pairing is central to DNA replication, as described in Section 21.7.

Phosphate Links The sugar units in nucleic acids link together by phosphate groups, which bind to C5′ and C3′ of the sugar as shown in Figure 21.19 on page 976. Note that a nucleic acid molecule will have two distinct ends, referred to as the 5′ end and the 3′ end.

Base-pairing in DNA

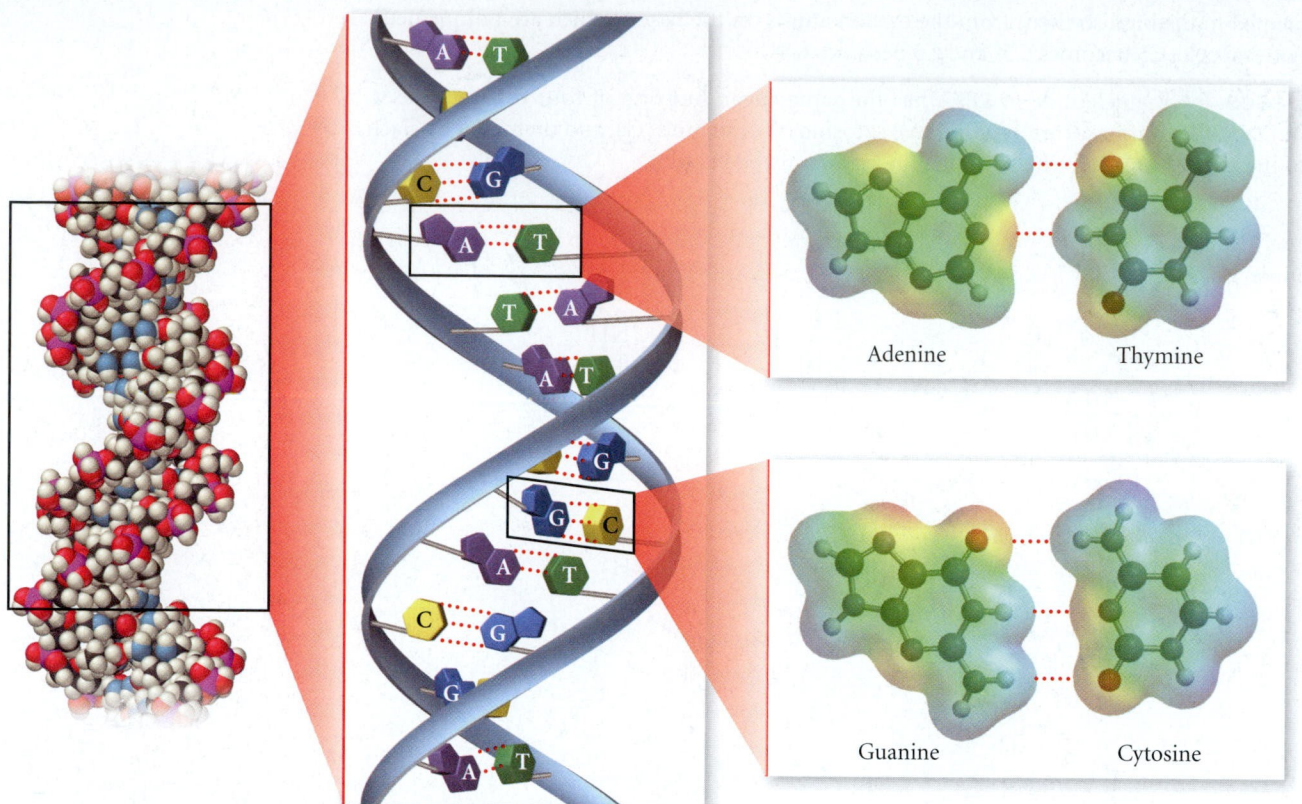

▲ **FIGURE 21.18 Base Pairing in DNA** The bases in nucleic acids are complementary. Each pyrimidine base pairs with one purine base (G with C, A with T) via specific hydrogen bonds that occur between the two bases.

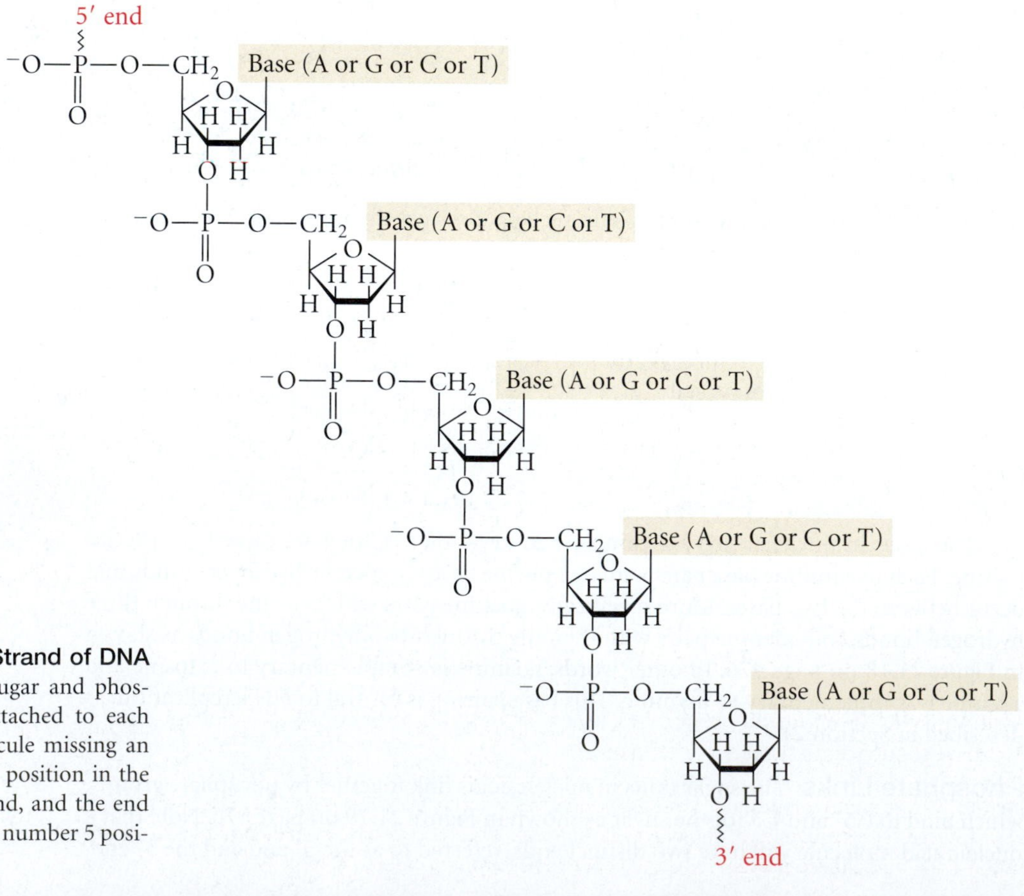

▶ **FIGURE 21.19 Short Strand of DNA**
DNA contains alternating sugar and phosphate groups with bases attached to each sugar. The end of the molecule missing an attachment at the number 3 position in the sugar ring is called the 3′ end, and the end missing an attachment at the number 5 position is called the 5′ end.

The Genetic Code

The order of bases in a nucleic acid chain specifies the order of amino acids in a protein. However, since there are only four bases, and since together they must have the ability to code for 20 different amino acids, a single base cannot code for a single amino acid. It takes a sequence of three bases—called a **codon**—to code for one amino acid (Figure 21.20 ▼). The genetic code—the code that identifies the amino acid specified by a particular codon—was worked out in 1961. It is nearly universal, meaning that the same codons specify the same amino acids in nearly all organisms. For example, in DNA the sequence AGT codes for the amino acid serine and the sequence ACC codes for the amino acid threonine. It does not matter if you are a rat, a bacterium, or a human—the code is the same.

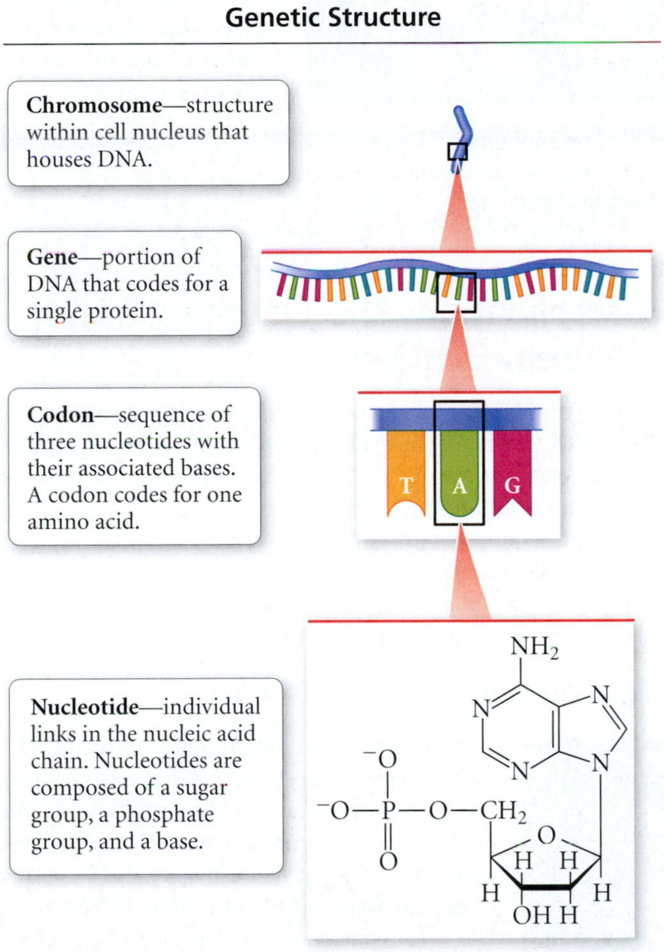

Genetic Structure

Chromosome—structure within cell nucleus that houses DNA.

Gene—portion of DNA that codes for a single protein.

Codon—sequence of three nucleotides with their associated bases. A codon codes for one amino acid.

Nucleotide—individual links in the nucleic acid chain. Nucleotides are composed of a sugar group, a phosphate group, and a base.

▲ **FIGURE 21.20 Genetic Structure** The hierarchical structure of genetic information is as follows: chromosome, gene, codon, and nucleotide.

Conceptual Connection 21.2 The Genetic Code

Assuming you have four different bases, how many amino acids could you code for with a two-base sequence? A three-base sequence?

Answer: The number of unique two-base sequences of four bases is $4^2 = 16$. The number of unique three-base sequences of four bases is $4^3 = 64$. Thus, a two-base system could code for 16 amino acids, and a three-base system could code for 64 amino acids.

In addition to having a codon for each amino acid, genes also contain additional coding that signals, for example, where the gene begins and where it ends.

A **gene** is a sequence of codons within a DNA molecule that codes for a single protein. Since proteins vary in size from a few dozen to thousands of amino acids, genes vary in length from dozens to thousands of codons. For example, egg-white lysozyme (Figure 21.13, p. 972) is composed of 129 amino acids. The lysozyme gene, then, must contain 129 codons—one for each amino acid in the lysozyme protein. Each codon is like a three-letter word that specifies one amino acid. String the correct number of codons together in the correct sequence, and you have a gene, the instructions for the amino acid sequence in a protein. Genes are contained in structures called **chromosomes** (Figure 21.21 ◄). There are 46 chromosomes in the nuclei of human cells.

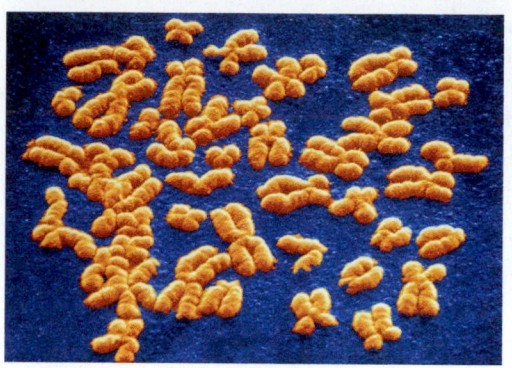

▲ FIGURE 21.21 **Chromosomes** Genes are contained in structures called chromosomes. Most human cells contain 46 chromosomes.

21.7 DNA Replication, the Double Helix, and Protein Synthesis

Most of the cells in our bodies contain all of the genes required to make all of the proteins that we need—the DNA within any one cell is *complete*. However, any particular cell does not express all those genes; it does not synthesize all those proteins. Cells synthesize only the proteins that are important to their function. For example, a pancreatic cell expresses the insulin gene within its nucleus to synthesize insulin. By contrast, pancreatic cells do not express the gene for keratin (the protein in hair), even though the keratin gene is also contained in their nuclei. The cells in our scalp, however, which also have both insulin and keratin genes in their nuclei, synthesize keratin but not insulin.

DNA Replication and the Double Helix

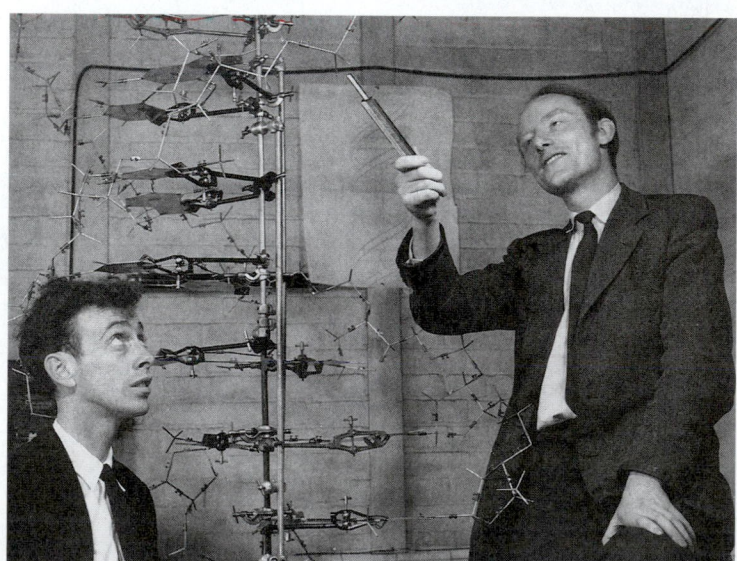

▲ FIGURE 21.22 **Watson and Crick** James Watson and Francis Crick discovered the structure of DNA, including the double helix and the pairing of complementary bases.

The human body contains on the order of 10^{13} cells, most of which have complete copies of the DNA that originally was present in only one cell (the fertilized egg). When a cell divides, it makes complete copies of its DNA for each daughter cell. The ability of DNA to copy itself is related to its structure, discovered in 1953 by James D. Watson and Francis H. C. Crick. Watson and Crick, aided by evidence from X-ray diffraction photos (see Section 11.10), determined that DNA exists as two complementary strands wound around each other in a double helix (Figure 21.22 ◄). The strands are antiparallel, so that one runs $3' \longrightarrow 5'$ while the other runs $5' \longrightarrow 3'$. The bases on each DNA strand are directed toward the interior of the helix, where they hydrogen bond to their complementary bases on the other strand. For example, if a section of DNA contains the bases

A T G A A T C C G A C

the complementary strand would have the sequence

T A C T T A G G C T G

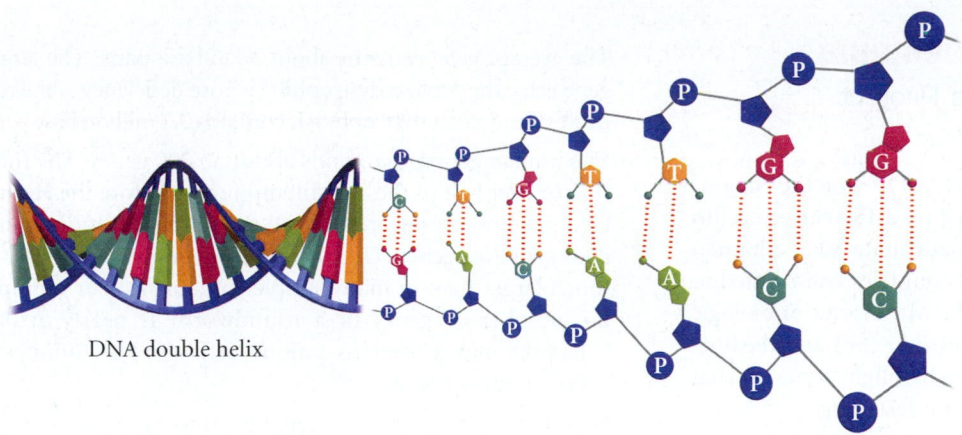

▲ **FIGURE 21.23 DNA Double Helix**   The two complementary strands of DNA wrap around one another to form a double helix.

As we saw earlier, A pairs only with T, and C pairs only with G. The two complementary strands are tightly wrapped into a helical coil, the famous DNA double helix structure (Figure 21.23 ▲).

When a cell is about to divide, the DNA unwinds and the hydrogen bonds joining the complementary bases break (Figure 21.24 ▼), forming two daughter strands. With the help of the enzyme DNA polymerase, a complement to each daughter strand—with the correct complementary bases in the correct sequence—is formed. The hydrogen bonds between the strands then re-form, resulting in two complete copies of the original DNA, one for each daughter cell.

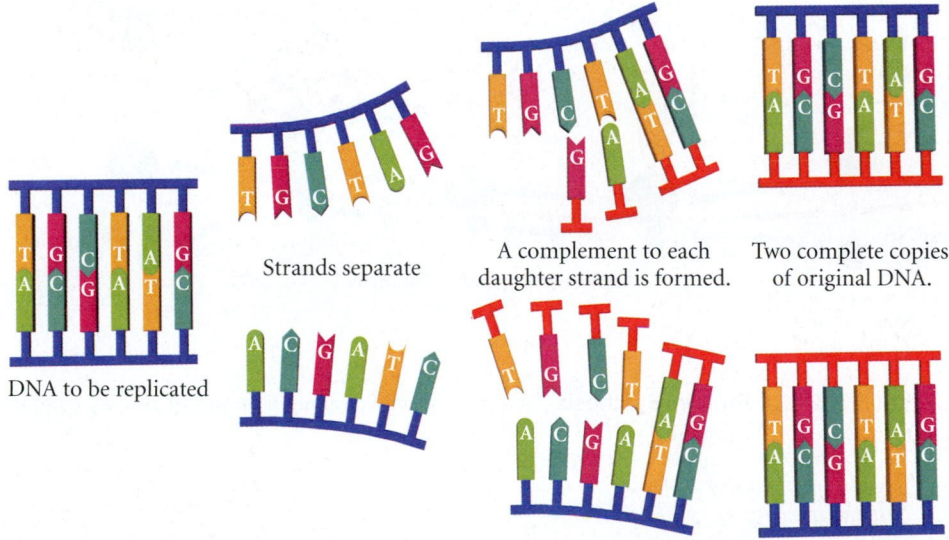

Strands separate

A complement to each daughter strand is formed.

Two complete copies of original DNA.

DNA to be replicated

▲ **FIGURE 21.24 DNA Replication**   When a cell is about to divide, its DNA unwinds. With the help of the enzyme DNA polymerase, a complement to each daughter strand is formed, resulting in two complete copies of the original DNA.

Protein Synthesis

Living organisms must continually synthesize thousands of proteins to survive, each when it is needed, and in the quantities required. When a cell needs to make a particular protein, the gene—the section of the DNA that codes for that particular protein—unravels. Complementary copies of that gene are then synthesized (or transcribed) as single strands of messenger RNA (or mRNA). The mRNA moves out of the cell's nucleus to structures in the cytoplasm called *ribosomes*. At the ribosome, protein synthesis occurs. The ribosome

Chemistry and Medicine
The Human Genome Project

In 1990, the U.S. Department of Energy (DOE) and the National Institutes of Health (NIH) embarked on a 15-year project to map the human genome, all of the genetic material of a human being. Over 2500 researchers from 18 countries contributed to this research, which has been called the Mt. Everest of biology. An initial draft of the map was completed in 2001 and the final draft was completed in 2003. Below, we highlight some of what has been learned through this massive undertaking.

- The human genome contains 3165 million nucleotide base pairs.

- The average gene contains about 3000 base pairs. The largest gene is for the protein dystrophin (whose deficiency is the root cause of muscular dystrophy); it contains 2.4 million base pairs.

- The human genome contains about 30,000 genes. The function of over half of these is still unknown. Before the Human Genome Project, researchers had estimated that humans had about 100,000 genes. The number of genes in humans is not much larger than in many simpler organisms. For example, the number of genes in a roundworm is nearly 20,000. Whatever makes humans unique, it is not the number of genes in our genome.

- Less than 2% of our DNA actually consists of genes. These genes are aggregated in seemingly random areas within the genome, with vast expanses of noncoding DNA between the

moves along the mRNA chain that codes for the protein, "reading" the sequence of codons. At each codon, the specified amino acid is brought into place and a peptide bond is formed with the previous amino acid (Figure 21.25 ▼). As the ribosome moves along the mRNA, the protein is formed. All of this is orchestrated by enzymes that catalyze the necessary reactions.

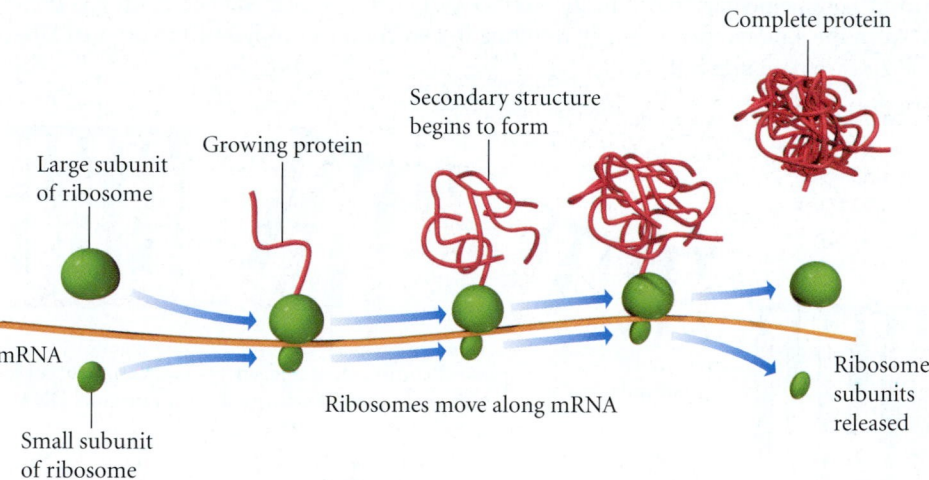

▲ FIGURE 21.25 **Protein Synthesis**   A ribosome moves along a strand of mRNA, joining amino acids to form a protein.

Summarizing:

➤ DNA contains the code for the sequence of amino acids in proteins.

➤ A codon—three nucleotides with their bases—codes for one amino acid.

➤ A gene—a sequence of codons—codes for one protein.

➤ Genes are contained in structures called chromosomes that occur within cells. Humans have 46 chromosomes in the nuclei of their cells.

➤ When a human cell divides, each daughter cell receives a complete copy of the DNA—all 46 chromosomes.

➤ When a cell synthesizes a protein, the base sequence of the gene which codes for that protein is transferred to mRNA. mRNA then moves to a ribosome, where the amino acids are linked in the correct sequence to synthesize the protein. The general sequence of information flow is

$$DNA \longrightarrow RNA \longrightarrow PROTEIN$$

coding regions. This stands in contrast to other organisms, which tend to have more uniform distribution of genes throughout their genome.

- The order of DNA base pairs is 99.9% identical in all humans.
- About 1.4 million single base-pair differences (called SNPs for single-nucleotide polymorphisms) have been identified in the human genome. Understanding SNPs can help in identifying individuals who are susceptible to certain diseases. Knowledge of SNPs may also allow physicians to tailor drugs to match individuals.

Knowledge of the human genome is expected to lead to the development of new therapies in several ways. First, knowledge of genes can lead to smart drug design. Instead of developing drugs by trial and error (the current procedure for many

drugs), knowledge of a specific gene will allow scientists to design drugs to carry out a specific function related to that gene or its protein product. Second, human genes themselves can provide the blueprint for the production of certain types of drugs, either in the laboratory or by other organisms. For example, we have seen that insulin can be made by inserting the insulin gene into bacteria, which then synthesize the needed drug. Intriguingly, it may even be possible to replace abnormal or missing genes in the cells of diseased patients. Such gene therapies are still in the early stages of development, but they may eventually give us a powerful new tool for combating inherited diseases.

Although the completion of the Human Genome Project may seem like the end, it is really just the beginning. Thousands of studies in the coming years will rely on the data obtained through this endeavor.

Chapter in Review

Key Terms

Section 21.1
biochemistry (956)

Section 21.2
lipid (956)
fatty acid (956)
triglyceride (958)
ester linkage (958)
saturated fat (958)
unsaturated fat (958)
phospholipid (960)
glycolipid (960)
lipid bilayer (960)
steroid (961)

Section 21.3
carbohydrate (961)
monosaccharide (961)
hexose (962)
aldose (962)
ketose (962)
glycosidic linkage (964)
disaccharide (964)
hydrolysis (964)
polysaccharide (964)
complex carbohydrate (964)
cellulose (965)
starch (965)
glycogen (965)

Section 21.4
enzyme (966)
amino acid (966)
peptide bond (969)
dipeptide (969)
polypeptide (969)

Section 21.5
fibrous protein (971)
globular protein (971)
primary structure (972)
secondary structure (972)
α-helix (972)
β-pleated sheet (973)
random coil (973)

tertiary structure (973)
quaternary structure (973)

Section 21.6
nucleotide (974)
complementary (975)
codon (977)
gene (978)
chromosome (978)

Key Concepts

Diabetes and the Synthesis of Human Insulin (21.1)

Diabetes is a chronic illness that occurs when the pancreas cannot make enough insulin, a protein that promotes the absorption of glucose into cells. The chemical structure of insulin was discovered in 1955 by Frederick Sanger, who made it possible for insulin to be synthesized in the laboratory. Eventually, the human gene that codes for the production of insulin was inserted into bacteria, which were then able to produce enough insulin to supply diabetics. This example demonstrates the usefulness of biochemistry, the study of the chemicals that compose living organisms.

Lipids (21.2)

Lipids are biological chemicals that are nonpolar and, therefore, insoluble in water. They are used to compose cell membranes, for energy storage, and for insulation. A type of lipid called a fatty acid is a carboxylic acid with a long hydrocarbon chain. Fatty acids can be saturated, meaning they contain the maximum number of hydrogen atoms, or unsaturated, meaning they contain one or more carbon–carbon double bonds. Satu-

rated fatty acids experience greater molecular forces, making them solid at room temperature, while unsaturated fatty acids are liquids. Fats and oils are triglycerides, triesters composed of glycerol bonded by ester linkages to three fatty acids. Like fatty acids, triglycerides can be saturated (fats) or unsaturated (oils). Other lipids include phospholipids, glycerol bonded to two nonpolar fatty acids and a polar phosphate group, used in animal cell membranes; glycolipids, similar to phospholipids but with a sugar molecule as its polar head; and steroids, four-ringed lipids that include cholesterol and sex hormones.

Carbohydrates (21.3)

Carbohydrates are polyhydroxy aldehydes or ketones and generally have the formula $(CH_2O)_n$. They are used for short-term energy storage and plant structure composition. Monosaccharides, the simplest carbohydrates, contain three to eight carbons with one aldehyde or ketone functional group. Glucose, an example of a hexose, can exist both in a linear form and a ring form. Two monosaccharides can combine to form a disaccharide. For example, glucose and fructose can be combined to form sucrose. The glycosidic linkages that connect the two monosaccharides can be broken during digestion by hydrolysis. Polysaccharides are polymers of monosaccharides known as complex carbohydrates. They include cellulose, also called fiber, the main structural component of plants; starch, an energy storage compound found in potatoes and grains; and glycogen, used by animals to store glucose in the muscles.

Proteins and Amino Acids (21.4)

Proteins are polymers of amino acids and are used in a variety of biological functions including structure composition, metabolic regulation, and muscle contraction. Enzymes are particularly important proteins used to catalyze biochemical reactions in cells. Amino acids contain a carbon atom bonded to an amine group, a carboxylic acid group, a hydrogen atom, and an R group. There are 20 amino acids in humans and they all differ only in their R group. Amino acids form peptide bonds between the amine end of one amino acid with the carboxylic end of another, creating dipeptides, tripeptides, etc., or simply polypeptides, large examples of which are called proteins.

Protein Structure (21.5)

Protein structure and shape are critical for protein function. Proteins can be broadly divided into two structural categories. Fibrous proteins are generally linear, insoluble structures that serve structural functions. Globular proteins fold into roughly spherical conformations with nonpolar side chains oriented to the interior while polar side chains orient to the exterior, making them soluble in water. Protein structure can be analyzed at four levels. The primary structure is simply the sequence of the amino acid chain. The secondary structure refers to certain regular repeating patterns in the arrangement of protein chains, such as α-helix and β-pleated sheet patterns. The tertiary structure refers to large-scale bends and folds due to interactions between the R groups of amino acids such as hydrogen bonding, disulfide linkages, hydrophobic interactions, and salt bridges. Quaternary structure shows the way that monomeric subunits fit together in a multimeric protein that has more than one polypeptide chain.

Nucleic Acids (21.6)

Nucleic acids, such as DNA and RNA, are the chemical blueprints used to synthesize proteins. They are polymers of nucleotides, which are themselves composed of a sugar, a base, and a phosphate group. The bases of DNA are adenine (A), cytosine (C), guanine (G), and thymine (T), which are subject to complementary pairing: each pyrimidine base combines with one purine base. Phosphate links bind the C5′ carbon of one sugar with the C3′ carbon of another sugar to make the polymeric chain. The order of the bases in a nucleic acid chain specifies the order of amino acids in a protein. Each amino acid is coded by a codon, a sequence of three bases. A gene is a sequence of codons that codes for a specific protein. Genes, in turn, make up structures called chromosomes.

DNA Replication, the Double Helix, and Protein Synthesis (21.7)

Though the DNA code is complete in any cell in the body, only certain cells express certain genes. DNA is complete in every cell due to its ability to replicate. DNA exists as two complementary strands wound around each other in a double helix. The strands are antiparallel and the bases of each strand face the interior and hydrogen bond to their complements on the other strand. A always binds to T and C always binds to G. In order to replicate, a DNA strand divides and an enzyme called DNA polymerase creates the complement of each of the divided strands, thereby making two copies of the original DNA molecule. To synthesize proteins, the section of DNA that codes for that gene is unraveled. Messenger RNA (mRNA) is synthesized as a copy of the gene. The mRNA combines with ribosomes, structures that "read" the mRNA code and synthesize the correct sequence of amino acids.

Key Skills

Recognizing and Working with the Basic Structures of Lipids (21.2)
- Exercises 31–34

Recognizing and Working with the Basic Structures of Carbohydrates (21.3)
- Exercises 37–40

Identifying Chiral Carbon Atoms in Carbohydrates (21.3)
- Example 21.1   • For Practice 21.1   • Exercises 41, 42, 82

Drawing Structures for Amino Acids and Peptide Bonds (21.4)
- Example 21.2   • For Practice 21.2   • Exercises 49–58

Recognizing Levels of Protein Structure (21.5)
- Exercises 59–62

Recognizing Nucleic Acids and Nucleotides (21.6)
- Exercises 63, 64

Exercises

Review Questions

1. What is biochemistry? What are the significant advances in biochemistry that have helped diabetics?
2. What is a lipid? What roles do lipids play in living organisms?

3. What is a fatty acid? Draw the structure of a fatty acid.
4. What is the effect of double bonds within the hydrocarbon chain of a fatty acid?

5. What are triglycerides? Draw a structure of a triglyceride.

6. Explain the difference, both in terms of structure and in terms of properties, between a saturated fat and an unsaturated fat.

7. Describe the basic structure of phospholipids and glycolipids. What functions do these lipids have in living organisms?

8. What is a steroid? What are some functions of steroids?

9. What are carbohydrates? What are the main functions of carbohydrates in living organisms?

10. What is the difference between monosaccharides and disaccharides? Between aldoses and ketoses?

11. What is the difference between simple and complex carbohydrates?

12. Explain the differences among cellulose, starch, and glycogen. Describe the function of each.

13. What roles do proteins play in living organisms? Give some specific examples.

14. Describe the basic structure of an amino acid. How are amino acids linked together to form proteins?

15. How are the properties of the R groups in amino acids related to the properties of proteins?

16. Explain why amino acids are chiral.

17. Draw the structure of a neutral amino acid and its dipolar ion.

18. Draw the structure of any two amino acids and show how they link together to form a dipeptide.

19. What are the essential amino acids? What does it mean for an amino acid to be essential?

20. Explain the difference between fibrous proteins and globular proteins.

21. Describe the various levels of protein structure (primary, secondary, tertiary, and quaternary).

22. What types of interactions or bonds maintain each of the structures listed in the previous problem?

23. Describe the secondary structures known as α-helix and β-pleated sheet.

24. What is the function of nucleic acids in living organisms?

25. What is the basic structure of a nucleic acid?

26. The bases in nucleic acids are *complementary*. What does this mean?

27. What is a codon? A gene? A chromosome?

28. Do most cells contain complete copies of an organism's DNA? Do most cells express all of the genes contained in their DNA?

29. Explain the mechanism by which DNA is replicated.

30. Explain the mechanism by which proteins are synthesized from the information contained within DNA.

Problems by Topic

Lipids

31. Which of the following molecules is a lipid? If the molecule is a lipid, determine the kind of lipid. If it is a fatty acid or a triglyceride, classify it as saturated or unsaturated.

a. $CH_3-CH_2-CH_2-CH_2-CH_2-CH_2-CH_2-CH_2-CH_2-CH_3$

b.

c.

d.

32. Which of the following molecules is a lipid? If the molecule is a lipid, determine the kind of lipid. If it is a fatty acid or a triglyceride, classify it as saturated or unsaturated.

a.

b. $CH_3-CH_2-CH_2-CH_2-CH_2-CH_2-CH_2-CH_2-CH_2-CH_2-CH_2-CH_3$

c.

d.

33. Determine whether or not each of the following is a fatty acid. If it is a fatty acid, classify it as saturated, monounstaturated, or polyunsaturated.

a.

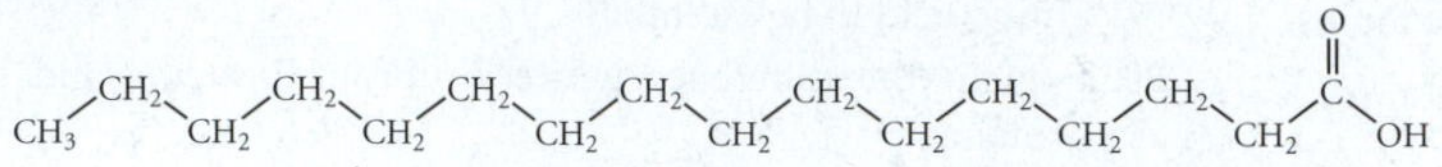

b.

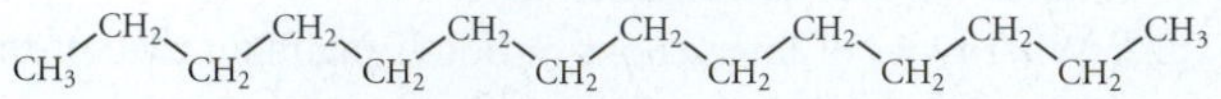

c.

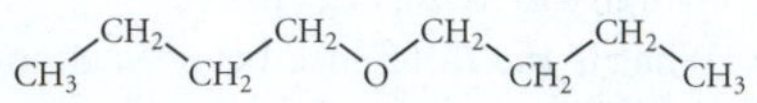

d.

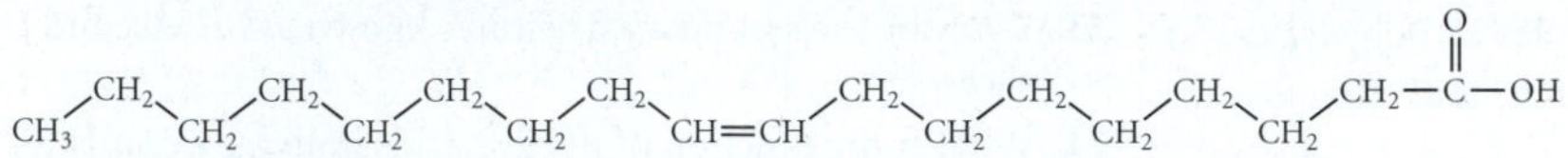

34. Which of the following fatty acids is most likely to be a solid at room temperature?

a. 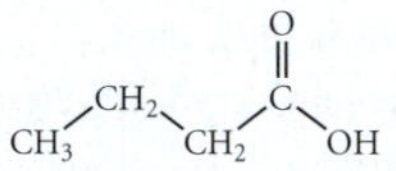   **b.**

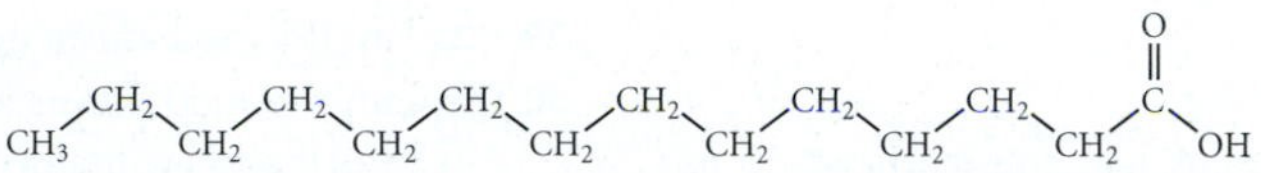

c.

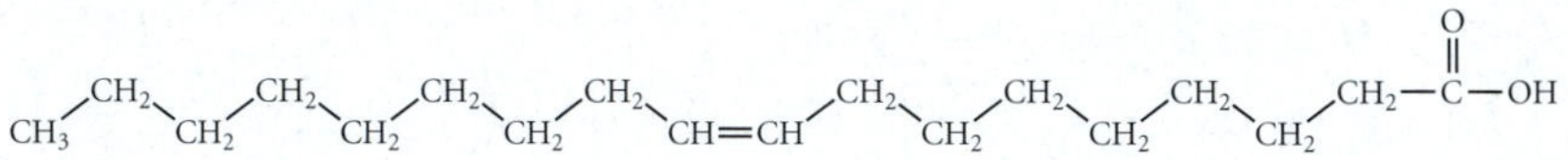

d.

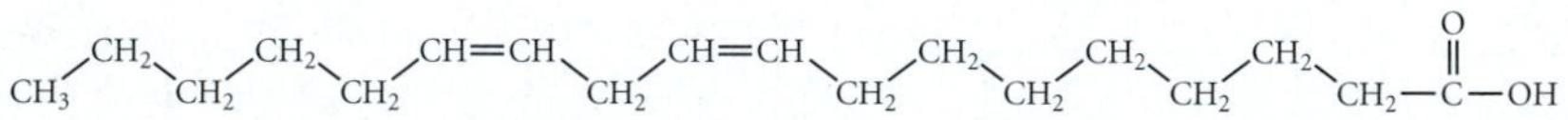

35. Draw structures showing the reaction of glycerol with linoleic acid to form the triglyceride trilinolean. Would you expect this triglyceride to be a fat or an oil?

36. Draw structures showing the reaction of glycerol with myristic acid to form the triglyceride trimyristin. Would you expect this triglyceride to be a fat or an oil?

Carbohydrates

37. Which of the following molecules is a carbohydrate? If the molecule is a carbohydrate, classify it as a monosaccharide, disaccharide, or trisaccharide.

a.

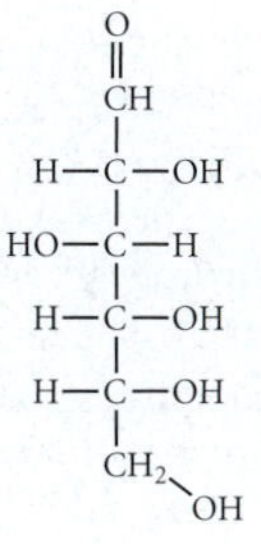

b.

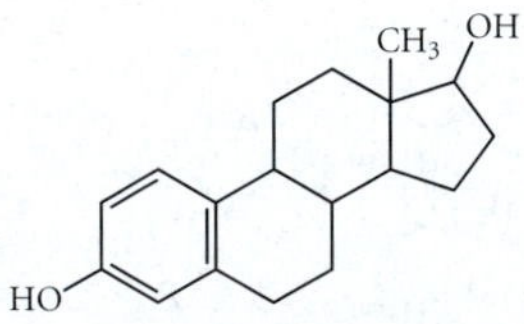

c.

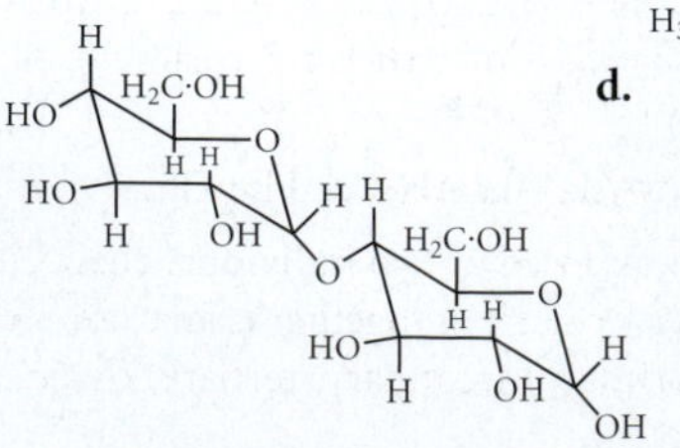

d.

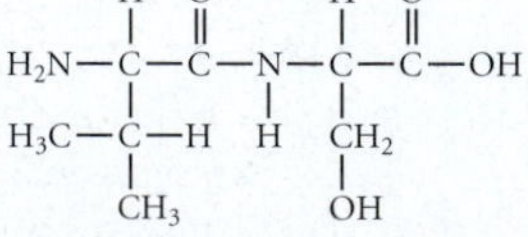

38. Which of the following molecules is a carbohydrate? If the molecule is a carbohydrate, classify it as a monosaccharide, disaccharide, or trisaccharide.

a.

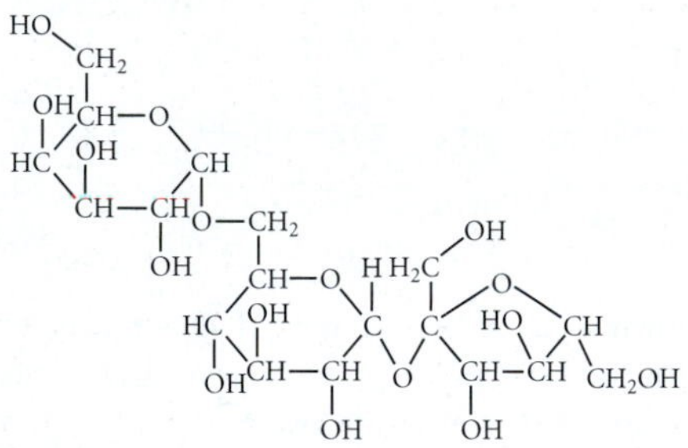

b.

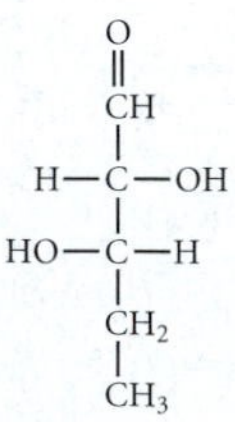

c.

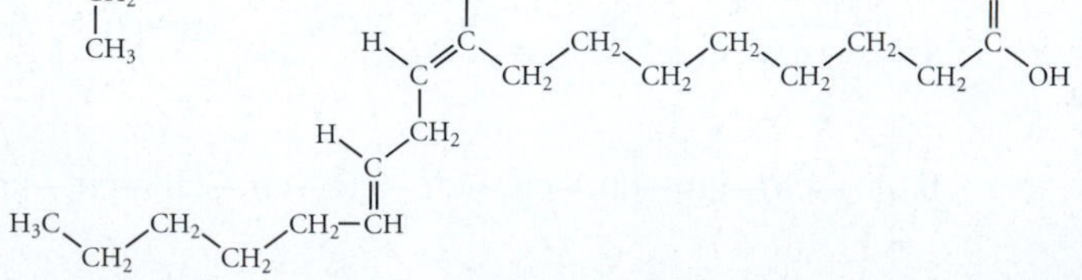

d.

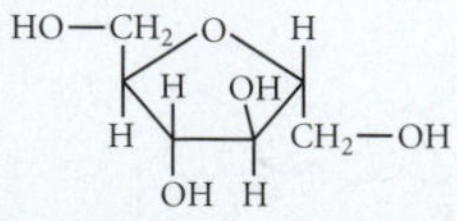

39. Classify each of the following as an aldose or a ketose. Also classify each as a triose, tetrose, pentose, etc.

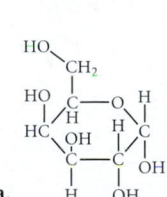

a.

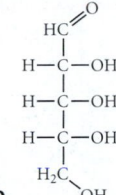

b.

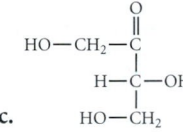

c.

d.

40. Classify each of the following as an aldose or a ketose. Also classify each as a triose, tetrose, pentose, etc.

a.

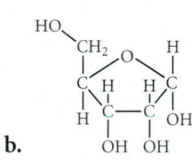

b.

c.

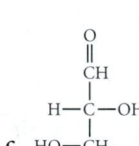

d.

41. How many chiral carbon atoms in each of the structures in Problem 39.

42. How many chiral carbon atoms in each of the structures in Problem 40.

43. Draw structures for the straight-chain and ring forms of glucose.

44. Draw structures for the straight-chain and ring forms of fructose.

45. Draw the products that result from the hydrolysis of the following carbohydrate:

46. Draw the products that result from the hydrolysis of the following carbohydrate:

47. Draw the structure of sucrose. Label the glucose and fructose rings in this disaccharide.

48. Lactose is a disaccharide of glucose and galactose. Draw the structure of lactose.

Amino Acids and Proteins

49. Draw each of the following amino acids in their dipolar ion form:
   a. Thr   **b.** Ala   **c.** Leu   **d.** Lys

50. Draw each of the following amino acids in their dipolar ion form:
   a. Val   **b.** Phe   **c.** Tyr   **d.** Cys

51. Draw the structures of the two enantiomers of alanine.

52. Draw the structures of the two entantiomers of cysteine.

53. How many different tripeptides can be made from one molecule of each of the following amino acids: serine, glycine, and cysteine? Give the amino acid sequence of each one.

54. How many dipeptides can be made from leucine and serine? Give the amino acid sequence for each one.

55. Show the reaction by which serine and tyrosine form a peptide bond.

56. Show the reaction by which valine and asparagine form a peptide bond.

57. Draw structures for each of the following tripeptides.
   a. Gln-Met-Cys   **b.** Ser-Leu-Cys
   c. Cys-Leu-Ser

58. Draw structures for each of the following tetrapeptides.
   a. Ser-Ala-Leu-Cys   **b.** Gln-Met-Cys-Gly
   c. Gly-Cys-Met-Gln

59. A phenylalanine amino acid on a protein strand undergoes hydrophobic interactions with another phenylalanine amino acid that is 26 amino acid units away. The resulting fold in the protein is an example of what kind of structure? (primary, secondary, tertiary, or quaternary)

60. An amino acid on a protein strand forms a hydrogen bond to another amino acid that is 4 amino acid units away. The next amino acid on the chain does the same, hydrogen bonding to an amino acid that is 4 amino acids away from it. This pattern repeats itself over a significant part of the protein chain. The resulting pattern in the protein is an example of what kind of structure (primary, secondary, tertiary, or quaternary)?

61. The following is the amino acid sequence in one section of a protein. It represents what kind of structure? (primary, secondary, tertiary, or quaternary)

-Lys-Glu-Thr-Ala-Ala-Ala-Lys-Phe-Glu-

62. A dimeric protein is composed of two individual chains of amino acids. The way these two chains fit together is an example of what kind of structure? (primary, secondary, tertiary, or quaternary)

Nucleic Acids

63. Which of the following is a nucleotide? For each nucleotide, identify the base as A, T, C, or G.

a.

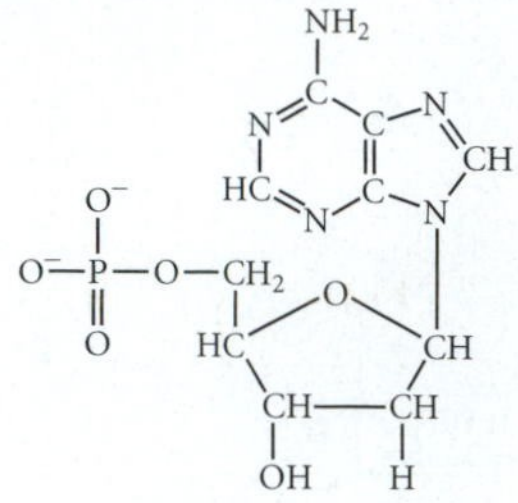

b.

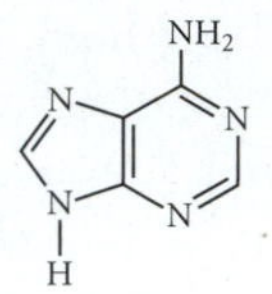

c.

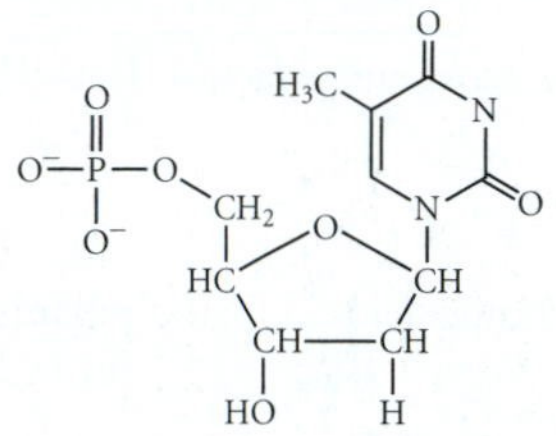

d.

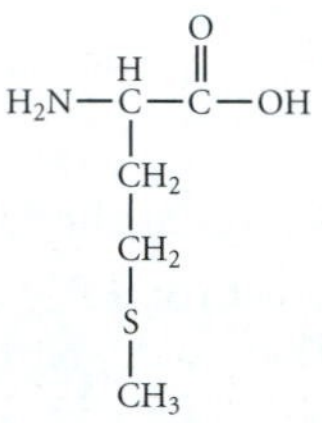

64. Which of the following is a nucleotide? For each nucleotide, identify the base as A, T, C, or G.

a.

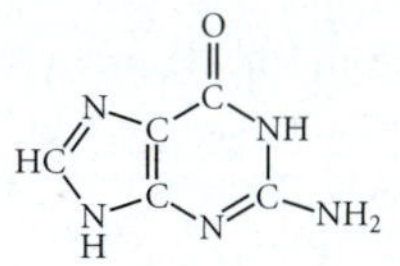

b.

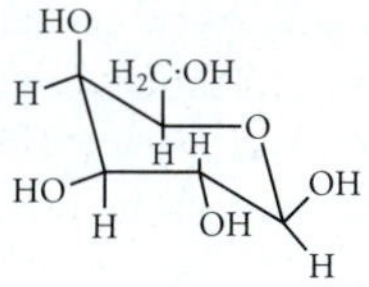

c.

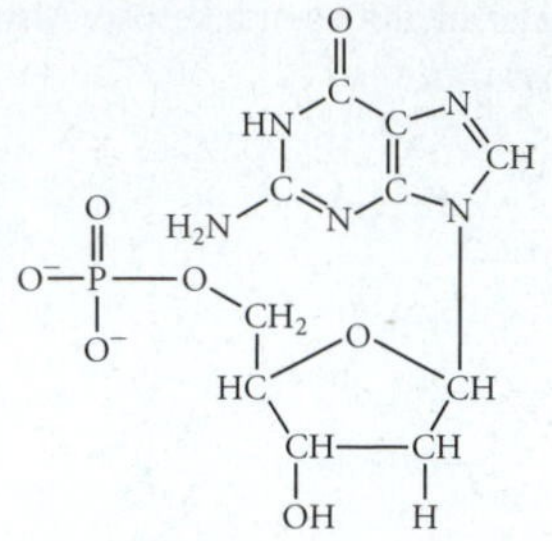

d.

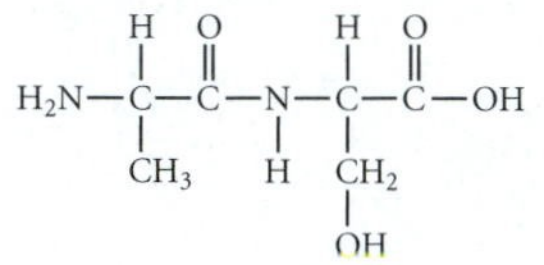

65. Draw the structures of the two purine bases present in nucleic acids.

66. Draw the structures of the three pyrimidine bases present in nucleic acids.

67. Draw the DNA strand that is complementary to the following DNA strand.

T G T A C G C

68. Draw the DNA strand that is complementary to the following DNA strand.

A T G A C T G

69. A monomeric protein contains 154 amino acids. How many codons are required to code for these amino acids? How many nucleotides?

70. A dimeric protein contains 142 amino acids in one strand and 148 in the other. How many codons are required to code for these amino acids? How many nucleotides?

Cumulative Problems

71. Name the class of biochemical compound that contains each of the following types of linkages.
    a. peptide bonds
    b. glycosidic linkage
    c. ester linkage

72. Name the type of polymer associated with each of the following monomers.
    a. nucleotide
    b. amino acid
    c. saccharide

73. What is the difference between a codon and a nucleotide? A codon and a gene?

74. What is the difference between a fatty acid and a triglyceride? A triglyceride and a phospholipid?

75. The amino acid alanine has the following condensed structural formula:

$$NH_2CH(CH_3)COOH$$

Determine the VSEPR geometry about each internal atom and make a three-dimensional sketch of the molecule.

76. The amino acid serine has the following condensed structural formula:

$$NH_2CH(CH_2OH)COOH$$

Determine the VSEPR geometry about each internal atom and make a three-dimensional sketch of the molecule.

77. Which amino acids in Table 21.3 are most likely to be involved in hydrophobic interactions?

78. Sickle-cell anemia is caused by a genetic defect that substitutes valine for glutamic acid at one position in two of the four chains of the hemoglobin protein. The result is a decrease in the water solubility of hemoglobin. Examine the structures of valine and glutamic acid and explain why this might be so.

79. Determining the amino acid sequence in a protein usually involves treating the protein with various reagents that break up the protein into smaller fragments which can be individually sequenced. Treating a particular 11-amino acid polypeptide with one reagent produced the following fragments:

Ala-Leu-Phe-Gly-Asn-Lys   Trp-Glu-Cys   Gly-Arg

Treating the same polypeptide with a different reagent produced the following fragments:

Glu-Cys   Gly-Asn-Lys-Trp   Gly-Arg-Ala-Leu-Phe

What is the amino acid sequence of the polypeptide?

80. Treating a particular polypeptide with one reagent (as described in the previous problem) produced the following fragments:

Gly-Glu-Ser-Lys   Trp-Arg   Leu-Thr-Ala-Trp

Treating the same polypeptide with a different reagent produced the following fragments:

Gly-Glu   Thr-Ala-Trp   Ser-Lys-Trp-Arg-Leu

What is the amino acid sequence of the polypeptide?

Challenge Problems

81. One way to fight viral infections is to prevent viruses from replicating their DNA. Without DNA replication, the virus cannot multiply. Some viral drug therapies cause the introduction of *fake* nucleotides into cells. When the virus uses one of these fake nucleotides in an attempt to replicate its DNA, the fake nucleotide doesn't work and DNA replication is halted. For example, azidothymidine (AZT), a drug used to fight the human immunodeficiency virus (HIV) that causes AIDS, results in the introduction of the following fake thymine-containing nucleotide into cells. Examine the structures of the real nucleotide and the AZT fake nucleotide. Can you propose a mechanism for how this fake nucleotide might halt DNA replication?

AZT nucleotide

Actual nucleotide

82. Draw the following molecules and then identify the chiral centers within them.
   a. ribose   **b.** galactose
   c. 5-deoxyribose (Hint: the 5 indicates that the oxygen is removed from the 5th carbon)

83. Glucose transport across the red blood cell membranes (erythrocyte membrane) is a well-studied transport system. In one laboratory project the following data were obtained for glucose transport.

[Glucose]$_{outside}$ (mM)	Rate of Glucose Entry (μM/min)
0.5	12
1.0	19
2.0	27
3.0	32
4.0	35

The kinetics of glucose transport through the membrane follows the Michaelis–Menten equation:

$$V_0 = \frac{V_{max} \, [\text{glucose}]}{K_t + [\text{glucose}]}$$

V_0 = rate of glucose entry

V_{max} = maximum rate (the point at which addition of glucose has no effect on the rate)

K_t = transport constant

The Michaelis–Menten equation can be rearranged so that a plot $1/V_0$ versus $1/[\text{glucose}]$ produces a straight line. Rearrange the equation and plot the data in order to determine K_t and V_{max} for glucose transport across the erythrocyte membrane.

84. Eukaryotic DNA is equipped with special ends called telomers. Telomers comprise hexanucleotide sequences that repeat at the ends of the DNA. For example, human DNA is equipped with repeating AGGGTT sequences. Functionally, telomers protect the ends of chromosomes from being treated as a broken piece of DNA needing repair. Interestingly, telomers are cut off each time the DNA is replicated, indicating a possible cellular clock that allows only a certain number of cellular replications. Telomerase is the enzyme which catalyzes the synthesis of telomers. Telomerase is found in limited quantities within certain cells such as fetal tissue, adult male germ cells, and stem cells. It is also found in over 85% of tumor cells. It is speculated that the telomerase activity may be linked to cancer. Give an explanation for why telomerase activity could be associated with cancer and speculate on ways in which cancer treatments in the future may capitalize on research on this enzyme.

Conceptual Problems

85. How many different tetrapeptides can form from four different amino acids?

86. Could the genetic code have been based on just three-bases and three-base codons? Explain why or why not. (Assume that the code must accommodate 20 different amino acids.)

87. The genetic code is random, which means that a particular codon could have coded for a different amino acid. The genetic code is also nearly universal, meaning that it is the same code in nearly all organisms (and in the few where it differs, it does so only slightly). If scientists ever find life on another planet, they will be curious to know its genetic code. What would a completely different genetic code indicate about the origin of the life-form? What would a genetic code identical to terrestrial life indicate?

22

Chemistry of the Nonmetals

22.1   Insulated Nanowires

22.2   The Main-Group Elements: Bonding and Properties

22.3   Silicates: The Most Abundant Matter in Earth's Crust

22.4   Boron: An Interesting Group 3A Element and Its Remarkable Structures

22.5   Carbon, Carbides, and Carbonates

22.6   Nitrogen and Phosphorus: Essential Elements for Life

22.7   Oxygen

22.8   Sulfur: A Dangerous but Useful Element

22.9   Halogens: Reactive Elements with High Electronegativity

THROUGHOUT THIS BOOK, YOU HAVE been introduced to many chemical topics, but you still may not know the composition of some everyday objects such as a drinking glass, a computer chip, or even rocks and soil. In this chapter and the following two chapters, we explore the descriptive chemistry of the nonmetals and metals. These descriptions are part of a branch of chemistry called inorganic chemistry. We begin our exploration of descriptive inorganic chemistry in this chapter with the chemistry of some of the main-group elements. The main-group elements are gathered together because their valence electrons occupy only *s* or *p* orbitals; however, their properties vary greatly. The main-group elements include metals, nonmetals, and semimetals, and they may be solids, liquids, or gases at room temperature. This great diversity of properties, bonding, and structures cannot be adequately described in a single chapter. Therefore, this chapter focuses on only a few main-group elements (silicon, boron, carbon, nitrogen, phosphorus, oxygen, sulfur, and the halogens) and their compounds to illustrate the diversity within the group.

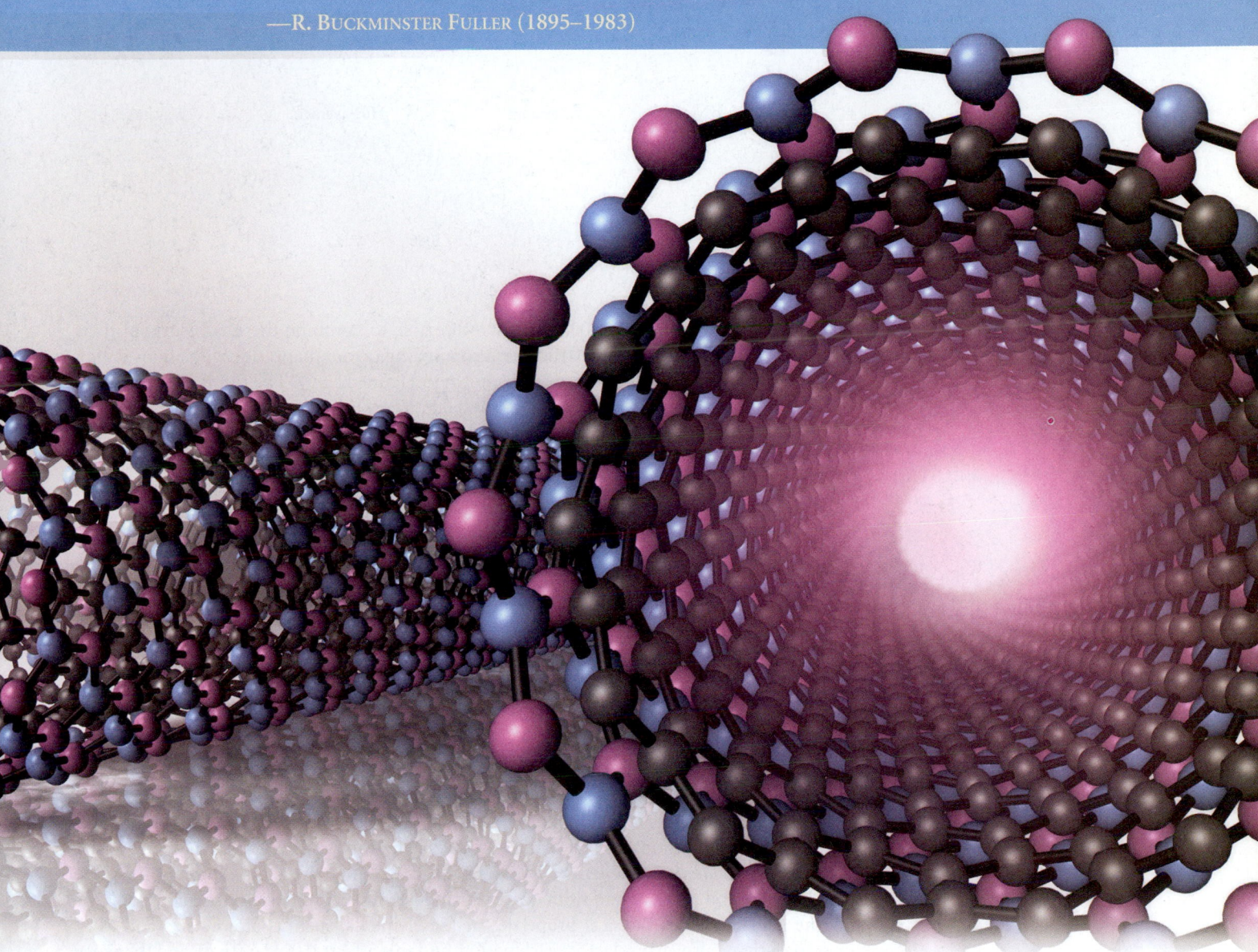

What one learns in chemistry is that Nature wrote all the rules of structuring; man does not invent chemical structuring rules; he only discovers the rules. All the chemist can do is find out what Nature permits, and any substances that are thus developed or discovered are inherently natural.

—R. BUCKMINSTER FULLER (1895–1983)

Scientists have recently been able to synthesize carbon nanotubes with an insulating boron nitride sheath. These structures are like electrical wires that are 100,000 times thinner than a human hair.

22.1 Insulated Nanowires

In 1991, scientists discovered carbon nanotubes, the long, thin, hollow cylinders of carbon atoms depicted on the cover of this book and discussed in some detail in Section 22.5. In the late 1990s, scientists discovered that similar tubes can be made from boron nitride. Boron nitride contains BN units that are isoelectronic with carbon in the sense that each BN unit contains eight valence electrons, or four per atom (just like carbon). Also the size and electronegativity of a carbon atom are almost equal to the average of those properties for a boron atom and a nitrogen atom, as shown in Table 22.1. Therefore, BN forms a number of structures that are similar to those formed by carbon, including nanotubes.

An important difference between boron nitride and carbon nanotubes is their conductivity. Carbon nanotubes conduct electrical current, while the boron nitride tubes are

TABLE 22.1   Properties of BN and C

Element	Atomic Radius (pm)	Ionization Energy (kJ/mol)	Electronegativity
B	85	800	2
BN	77.5 average	1101 average	2.5 average
N	70	1402	3
C	77	1086	2.5

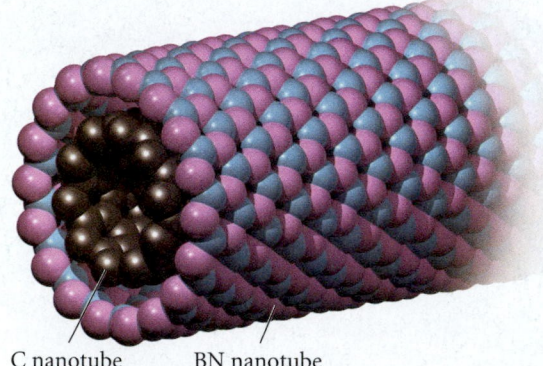

C nanotube   BN nanotube

▲ FIGURE 22.1 **Boron Nitride Nanotube** The model represents an insulating BN nanotube filled with a conducting carbon nanotube.

insulating. In 2003, scientists were able to combine these two sorts of nanotubes into one structure: a conducting carbon nanotube with an insulating boron nitride sheath, as shown in Figure 22.1 ◄. The result is an insulated conducting wire that is 100,000 times thinner than a human hair. Such thin wires may someday be used in computers and other electronic devices, allowing them to continue to become smaller and more efficient.

The more we learn about the structures and reactivities of known materials, the better equipped we are to discover new materials and applications. Even though it may seem that most inorganic compounds have already been discovered and analyzed, new materials, with immense impacts on our society, are constantly being discovered. As Buckminster Fuller states in this chapter's opening quote, we continue to find "what Nature permits." In some cases, what nature permits turns out to be extremely useful to society.

22.2 The Main-Group Elements: Bonding and Properties

The **main-group elements** are identified as a group by their valence electrons and their electron configurations. In this chapter, we focus on groups 3A–7A, the major part of the *p* block in the periodic table. The *p* orbitals fill across any row of this section of the periodic table, and they contain from one electron in group 3A to five electrons in group 7A (the halogens). The physical properties of the elements, such as atomic size and electronegativity, also change across each period, and this affects their reactivity and the types of compounds they form.

1A																	8A
1 H	2A											3A	4A	5A	6A	7A	2 He
3 Li	4 Be											5 B	6 C	7 N	8 O	9 F	10 Ne
11 Na	12 Mg	3B	4B	5B	6B	7B	8B	8B	8B	1B	2B	13 Al	14 Si	15 P	16 S	17 Cl	18 Ar
19 K	20 Ca	21 Sc	22 Ti	23 V	24 Cr	25 Mn	26 Fe	27 Co	28 Ni	29 Cu	30 Zn	31 Ga	32 Ge	33 As	34 Se	35 Br	36 Kr
37 Rb	38 Sr	39 Y	40 Zr	41 Nb	42 Mo	43 Tc	44 Ru	45 Rh	46 Pd	47 Ag	48 Cd	49 In	50 Sn	51 Sb	52 Te	53 I	54 Xe
55 Cs	56 Ba	57 La	72 Hf	73 Ta	74 W	75 Re	76 Os	77 Ir	78 Pt	79 Au	80 Hg	81 Tl	82 Pb	83 Bi	84 Po	85 At	86 Rn
87 Fr	88 Ra	89 Ac	104 Rf	105 Db	106 Sg	107 Bh	108 Hs	109 Mt	110 Ds	111 Rg	112		114		116		

Atomic Size and Types of Bonds

Recall from Section 8.6 that the atomic radius of the main-group elements becomes smaller as you move to the right across any row in the periodic table because of the increasing effective nuclear charge. This increasing effective charge results in smaller radii, increasing electronegativity, and increasing ionization energy as you move to the right across the periods. Consequently, as we have seen since the early chapters of this book, the nonmetals on the right side of the periodic table tend to form anions in ionic compounds. They are easily reduced, gaining electrons to completely fill their p orbitals and to attain a noble gas electron configuration. These elements are often used in reactions as *oxidizing agents*—they oxidize other substances while they are themselves reduced. The smallest halogens and the elements in the oxygen group are the strongest oxidizing agents in the p block.

Elements near the center of the p block have fewer p electrons and do not usually fill the p orbitals by forming anions; instead they share electrons, forming covalent compounds. This type of reactivity can be seen in the vast array of covalent compounds formed by the smaller elements in the carbon and nitrogen groups. The main-group elements on the far left of the p block have only one p electron and can form cations in ionic compounds and electron-deficient species (species with an incomplete octet) in covalent compounds.

We can see that, as we move to the right across any row in the p block, the type of bonding changes as the elements become less metallic. Recall also (from Section 8.8) that metallic character increases going down each column. The diagonal group of metalloid elements stretching from boron to astatine divides the main-group elements. To the left of this diagonal, the elements are metals that form cations and metallic compounds; to the right, the elements are nonmetals that form anions and covalent compounds.

The vast range in elemental properties, from those of metallic elements such as thallium and lead that have very low electronegativity values of 1.8 and 1.9 (respectively) to those of the nonmetallic elements such as oxygen and fluorine that have the highest electronegativity values of 3.5 and 4.0 (respectively), provides for the great chemical diversity of the elements in the p block. Metals, alloys, simple covalent compounds, enormous covalent network compounds, simple binary ionic compounds, and complex chain and layered ionic compounds can all be found in the chemistry of these elements.

See Section 8.6 for a more thorough discussion of the periodic trends and exceptions in these properties of the elements.

Metals, Nonmetals, and Metalloids

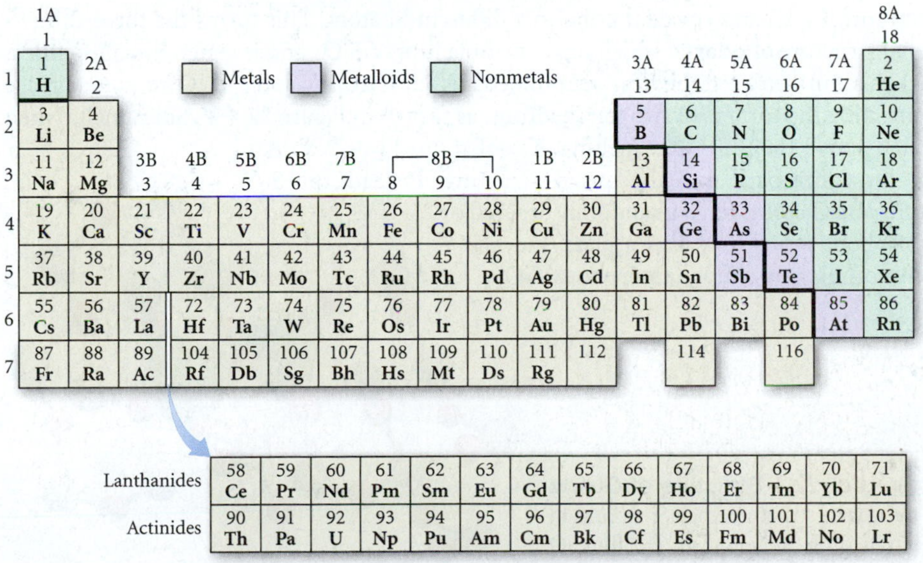

22.3 Silicates: The Most Abundant Matter in Earth's Crust

The most abundant elements in Earth's crust are oxygen (45–50% by mass) and silicon (about 28% by mass). The only other elements that individually comprise more than 1% of the crust's mass are aluminum, iron, calcium, magnesium, sodium, and potassium (as shown in Figure 22.2 ▼), and most of these are found in silicon and oxygen compounds. So, in order to understand most of the matter on Earth's surface, we must understand silicon and oxygen compounds.

▶ FIGURE 22.2 **Major Elements in Earth's Crust** The major components of Earth's crust are oxygen and silicon. Only a few other elements compose more than 1% of the crust.

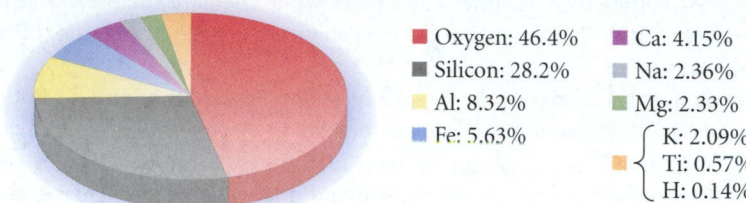

- Oxygen: 46.4%
- Silicon: 28.2%
- Al: 8.32%
- Fe: 5.63%
- Ca: 4.15%
- Na: 2.36%
- Mg: 2.33%
- { K: 2.09%
  Ti: 0.57%
  H: 0.14%

Silicates are covalent atomic solids (see Section 11.12) that contain silicon, oxygen, and various metal atoms. Silicates are found in rocks, clays, and soils. Their great diversity illustrates again a theme we have encoutered since Chapter 1 of this book: the properties of substances are determined by their atomic and molecular structures. The structures of silicates, therefore, determine their properties—and since these structures are varied, their properties are also varied. Some silicates form strong three-dimensional materials, while others break into sheets, and still others are highly fibrous. We now examine more closely several of these structures.

Quartz and Glass

Silicon and oxygen form a network covalent structure in which a silicon atom bonds to four oxygen atoms, forming a tetrahedral shape with the silicon atom in the middle and the four oxygen atoms at the corners of the tetrahedron, as shown in Figure 22.3 ◀. In this structure, the silicon atom bonds to each oxygen atom with a single covalent sigma bond. In contrast to carbon, which often bonds to oxygen with a double bond (one sigma and one pi bond), silicon forms only single bonds with oxygen, because the silicon atom is too large for there to be substantial overlap between the *p* orbitals on the two atoms. The silicon atom in this structure, by bonding to four oxygen atoms, obtains a complete octet. However, each oxygen atom is one electron short of an octet. Therefore, each O atom forms a second covalent bond to a different Si atom. This forms the three-dimensional structure of **quartz**, which has a formula unit of SiO_2 and is generally called **silica**. Each Si atom is in a tetrahedron surrounded by four O atoms, and each O atom is a bridge connecting the corners of two tetrahedrons, as shown in Figure 22.4 ▼. Silica melts when heated above 1500 °C. After melting, if cooled quickly, silica does not crystallize back into the quartz structure. Instead, the Si atoms and O atoms form a randomly ordered or amorphous structure called a glass. Indeed, common glass is simply amorphous SiO_2.

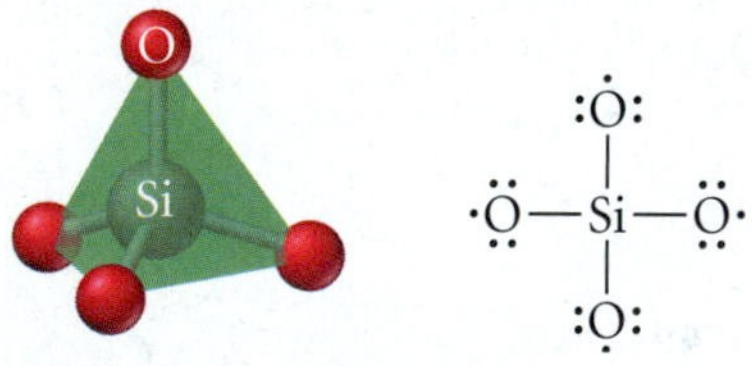

▲ FIGURE 22.3 **SiO₄ Tetrahedron** In an SiO₄ tetrahedron, silicon occupies the center of the tetrahedron and one oxygen atom occupies each corner.

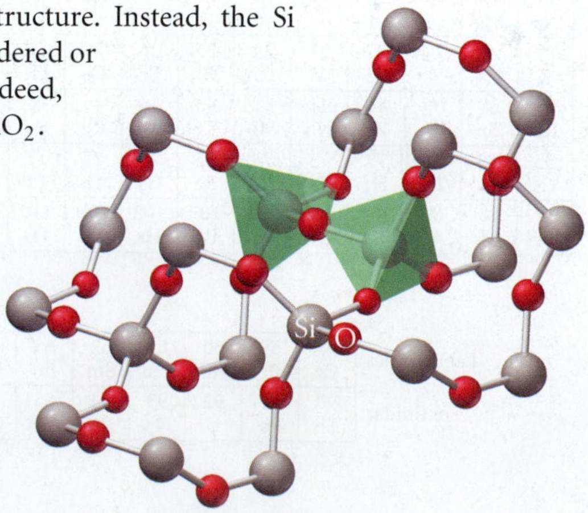

▲ FIGURE 22.4 **Structure of Quartz** In the quartz structure, each Si atom is in a tetrahedron surrounded by four O atoms, and each O atom is a bridge connecting the corners of two tetrahedrons.

Aluminosilicates

Aluminosilicates are a family of compounds in which aluminum atoms substitute for silicon atoms in some of the silicon lattice sites of the silica structure. Since the aluminum ion has only three valence electrons (in contrast to the four valence electrons of silicon), an individual SiO_2 unit becomes AlO_2^- upon substitution of aluminum. The negative charge is balanced by a positive counterion. A common aluminosilicate is feldspar, an example of which is the mineral albite. In albite, one-fourth of the Si atoms are replaced by Al atoms, with Na^+ ions providing the necessary balancing positive charge. The formula for albite is $Na(AlSi_3O_8)$, but it can be written as $Na(AlO_2)(SiO_2)_3$ to illustrate the substitution of Al for Si.

EXAMPLE 22.1 Determining the Composition of an Aluminosilicate

Anorthite is a crystal in which Al atoms substitute for one-half of the Si atoms and the charge is balanced by Ca^{2+} ions. Write the formula for anorthite.

Solution

The AlO_2^- unit substitutes for one-half of the SiO_2 units; therefore, the formula has equal numbers of AlO_2^- and SiO_2 units. Every AlO_2^- ion in the formula must be balanced by a corresponding positive charge. Since Ca^{2+} has a 2+ charge, it can balance two AlO_2^- units. Therefore, the formula for anorthite is $Ca(Al_2Si_2O_8)$ or $Ca(AlO_2)_2(SiO_2)_2$.

For Practice 22.1

Orthoclase is a crystal in which Al^{3+} substitutes for one-fourth of the Si^{4+} ions and the charge is balanced by K^+ ions. Write the formula for orthoclase.

Individual Silicate Units, Silicate Chains, and Silicate Sheets

In many silicate compounds, the oxygen atoms are not connected to two silicon atoms to form the neutral compound that is found in quartz. Instead, the oxygen atoms gain electrons from metal atoms to form polyatomic anions, such as SiO_4^{4-}. The positively charged metal ions then bond to the negatively charged silicon oxide. In these minerals, the SiO_4 tetrahedrons are either alone, in chains, or in sheets.

When the tetrahedrons are alone (not bonded to other tetrahedrons), they form the SiO_4^{4-} polyatomic anion, (which has four extra electrons that satisfy the octet rule for the four oxygen atoms). These types of silicates are known as **orthosilicates** (or nesosilicates) and require cations that have a total charge of 4+ to neutralize the negative charge. The cations could be of a single metal, such as Zn^{2+} in Zn_2SiO_4 (the mineral willemite), or they could be a mixture of different metals, such as the family of crystals called olivines [$(Mg,Fe)_2SiO_4$], where the Mg^{2+} and Fe^{2+} ions can exist in variable proportion, providing a total charge of 4+. All of these compounds are held together by the ionic bonding between the metal cations and SiO_4^{4-} polyatomic anions.

The silicate tetrahedrons can also form structures called **pyrosilicates** (or sorosilicates) in which two tetrahedrons share one corner (as shown in Figure 22.5 ▼), forming the disilicate ion, which has the formula $Si_2O_7^{6-}$. This group requires cations that balance the 6− charge on $Si_2O_7^{6-}$. Again, these cations can be the same metal ions or a mixture of different metal ions. For example, in the mineral hardystonite ($Ca_2ZnSi_2O_7$), two Ca^{2+} ions and one Zn^{2+} ion together provide the 6+ charge.

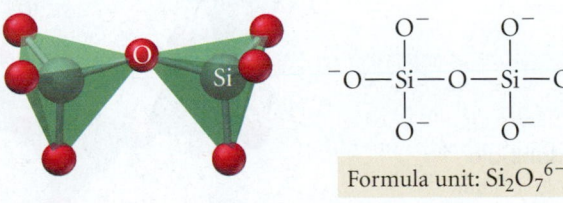

Formula unit: $Si_2O_7^{6-}$

◄ **FIGURE 22.5 Pyrosilicate Structure** In pyrosilicates, the silicate tetrahedrons share one corner, forming $Si_2O_7^{6-}$ units. Pyrosilicates are also called sorosilicates.

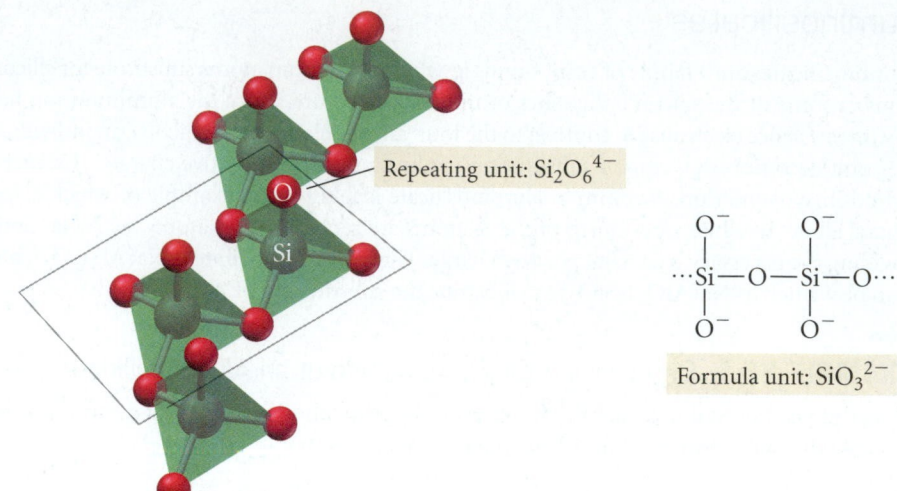

Repeating unit: $Si_2O_6^{4-}$

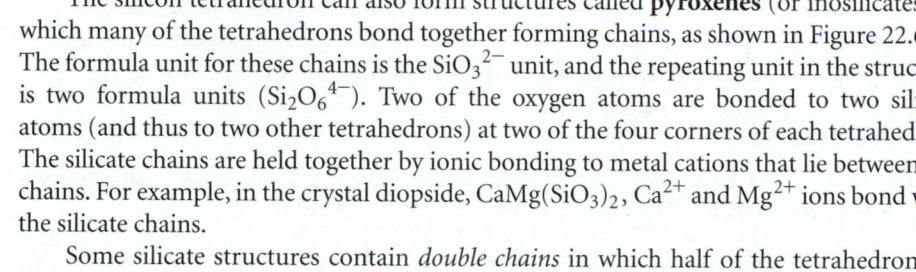

Formula unit: SiO_3^{2-}

▶ **FIGURE 22.6 Pyroxene Structure**
In pyroxenes, chains of silicate tetrahedrons are formed. Pyroxenes are also called inosilicates.

▲ The fibrous texture of asbestos is due to silicate double chains of the amphibole structure.

The silicon tetrahedron can also form structures called **pyroxenes** (or inosilicates) in which many of the tetrahedrons bond together forming chains, as shown in Figure 22.6 ▲. The formula unit for these chains is the SiO_3^{2-} unit, and the repeating unit in the structure is two formula units ($Si_2O_6^{4-}$). Two of the oxygen atoms are bonded to two silicon atoms (and thus to two other tetrahedrons) at two of the four corners of each tetrahedron. The silicate chains are held together by ionic bonding to metal cations that lie between the chains. For example, in the crystal diopside, $CaMg(SiO_3)_2$, Ca^{2+} and Mg^{2+} ions bond with the silicate chains.

Some silicate structures contain *double chains* in which half of the tetrahedrons of one chain are bonded to tetrahedrons in another chain through oxygen atoms. The minerals with double silicate chains are called amphiboles, and the repeating unit in the crystal is $Si_4O_{11}^{6-}$, as shown in Figure 22.7 ▼. Half of the tetrahedrons are bonded by two of the four corner O atoms, and half of the tetrahedrons are bonded by three of the four corners, bonding the two chains together. The bonding within the double chains is very strong, but the bonding between the double chains is not so strong, often resulting in fibrous minerals such as asbestos. An example of an asbestos-type mineral is tremolite, $Ca_2(OH)_2Mg_5(Si_4O_{11})_2$. In this crystal, hydroxide ions that are bonded to some of the metal cations help balance the charge between the cations and the anionic silicate chains.

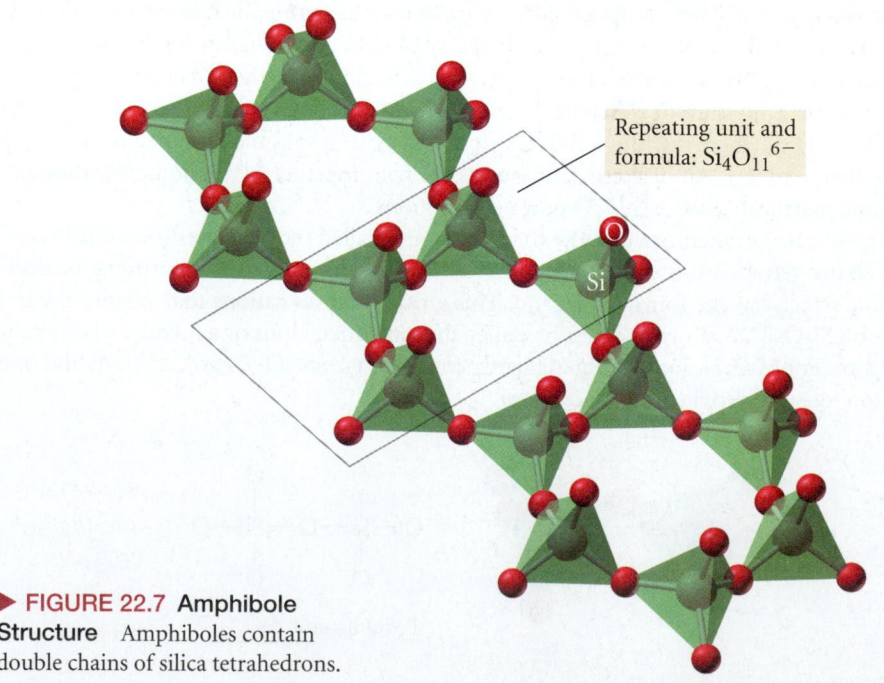

Repeating unit and formula: $Si_4O_{11}^{6-}$

▶ **FIGURE 22.7 Amphibole Structure** Amphiboles contain double chains of silica tetrahedrons.

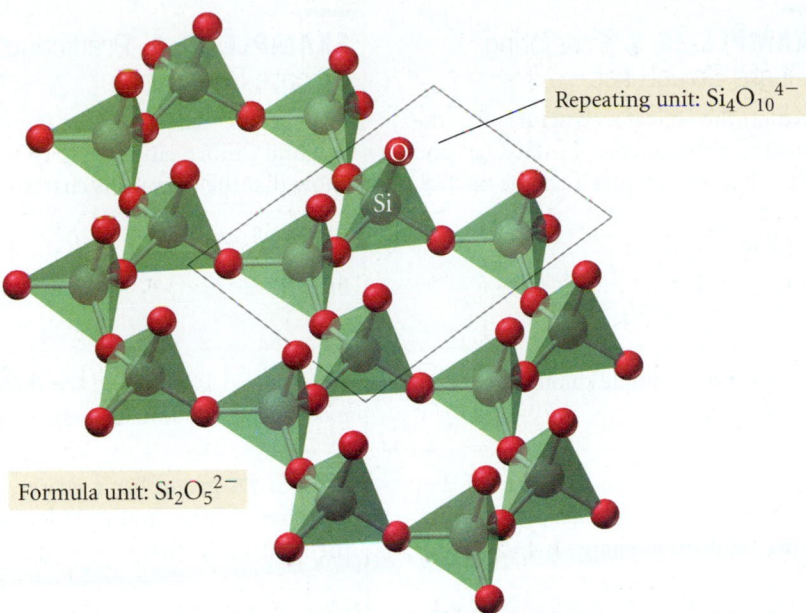

Repeating unit: $Si_4O_{10}{}^{4-}$

O

Si

Formula unit: $Si_2O_5{}^{2-}$

▲ **FIGURE 22.8 Phyllosilicate Structure** In phyllosilicates, three of the four oxygens are bonded to two silicon atoms, forming sheets of silica tetrahedrons.

When three of the four oxygen atoms are bonded between the silicate tetrahedrons, the sheet structures shown in Figure 22.8 ▲ are formed. These compounds are called phyllosilicates and have a formula unit of $Si_2O_5{}^{2-}$. These sheets of tetrahedral silicates are bonded together by metal cations that lie between the sheets. For example, the mineral *talc*, $Mg_3(Si_2O_5)_2(OH)_2$, is a phyllosilicate. The weak interactions between silicate sheets gives talc its slippery feel (much as do the weak interactions between sheets of carbon atoms in graphite). Table 22.2 summarizes the different kinds of silicate structures.

▲ The flaky texture of mica is due to silicate sheets of the phyllosilicate structure.

EXAMPLE 22.2 Composition and Charge Balance of Silicates

The silicate chrysotile is an amphibole with the formula $Mg_6Si_4O_{11}(OH)_x$. Use charge balancing to calculate the value of x in the formula.

Solution

The silicate unit for amphiboles is $Si_4O_{11}{}^{6-}$. There are six Mg^{2+} ions in this formula for a total charge of $12+$. To balance, another $6-$ charge needs to be added to the $6-$ charge for the silicate. Therefore, 6 OH^- ions need to be in the formula for chrysotile, giving a formula of $Mg_6Si_4O_{11}(OH)_6$.

For Practice 22.2

Use charge balancing to calculate how many hydroxide ions there are in the formula of the mineral pyrophyllite, $Al_2(Si_2O_5)_2(OH)_x$.

TABLE 22.2 Types of Silicate Structures

Tetrahedrons	Shared Vertices	Formula Unit	Si : O Ratio	Class Name	Example
Single tetrahedron	0	$SiO_4{}^{4-}$	1 : 4	Orthosilicates, nesosilicates	Olivine, Mg_2SiO_4
Double tetrahedron	1	$Si_2O_7{}^{6-}$	2 : 7	Pyrosilicates, sorosilicates	Hardystonite, $Ca_2ZnSi_2O_7$
Single chain	2	$SiO_3{}^{2-}$	1 : 3	Pyroxenes, inosilicates	Jadeite, $NaAl(SiO_3)_2$
Double chain	2 and 3	$Si_4O_{11}{}^{6-}$	4 : 11	Amphiboles	Tremolite, $Ca_2(OH)_2Mg_5(Si_4O_{11})_2$
Sheet	3	$Si_2O_5{}^{2-}$	2 : 5	Phyllosilicates	Talc, $Mg_3(Si_2O_5)_2(OH)_2$
Network covalent	4	SiO_2	1 : 2	Silicas, tectosilicates	Quartz, SiO_2
Network covalent	4	$AlSi_3O_8{}^-$ or $Al_2Si_2O_8{}^{2-}$	Variable	Feldspars	Albite, $NaAlSi_3O_8$

Procedure for Predicting Types of Silicate Structure and Accounting for Charge Balance	**EXAMPLE 22.3** Predicting Silicate Structures Predict the silicate structure for the mineral spudomene, $LiAlSi_2O_6$, and show that the formula is charge neutral.	**EXAMPLE 22.4** Predicting Silicate Structures Predict the silicate structure for the mineral thortveitite, $Sc_2Si_2O_7$, and show that the formula is charge neutral.
	Solution	**Solution**
Determine the ratio of Si to O in the formula.	$Si:O = 1:3$	$Si:O = 2:7$
Match the Si:O ratio to the type of silicate in Table 22.2.	A 1:3 ratio is a single chain, a pyroxene (or inosilicate).	A 2:7 ratio is a double tetrahedron, a pyrosilicate (or sorosilicate).
Determine the total anion charge.	Each SiO_3 group has a charge of 2−, and there are two groups per formula, so the total anion charge is 4−.	Each Si_2O_7 group has a charge of 6−, and there is one group per formula, so the total anion charge is 6−.
Determine the total cation charge and show that it matches the total anion charge.	The Li^+ cation has a 1+ charge and the Al^{3+} cation has a 3+ charge for a total of 4+, which matches the anion charge.	Each scandium cation has a charge of 3+ for a total of 6+, which matches the anion charge.
	For Practice 22.3 Predict the silicate structure for the mineral phenakite, Be_2SiO_4, and show that the formula is charge neutral.	**For Practice 22.4** Predict the silicate structure for the mineral diopside, $CaMgSi_2O_6$, and show that the formula is charge neutral.

22.4 Boron: An Interesting Group 3A Element and Its Remarkable Structures

The group 3A elements all have a filled *s* sublevel and one electron in the *p* sublevel. This electron configuration does not allow these main-group elements, especially boron, to easily attain a full octet. Most of the elements in the group are metals; however, because of its small size and higher electronegativity, boron behaves as a semimetal. These characteristics give boron some special properties that result in a wide array of different structures not common to most elements.

Elemental Boron

The structure of elemental boron is complex; there are at least five different allotropes (different structures with the same elemental composition). The structure of each allotrope is based on an icosahedron (Figure 22.9 ◀), a geometrical shape that is very roughly spherical and contains 20 triangular faces joined at 12 vertices. Twelve boron atoms occupy the twelve vertices. The different allotropes connect the icosahedrons in different ways and also have boron atoms outside the icosahedrons, bridging the icosahedrons together.

Boron is rare in Earth's crust, making up less than 0.001% by mass. Yet because it occurs in high concentrations at various deposits around the world, it can be mined and used in large quantities. The largest deposit is found at an old, volcanically active site in Boron, California. Naturally occurring boron is always found in compounds, and it is almost always bonded to oxygen. Among the major sources of boron are the sodium borates, which include borax, $Na_2[B_4O_5(OH)_4] \cdot 8\,H_2O$, and kernite, $Na_2[B_4O_6(OH)_2] \cdot 3H_2O$. Another major source of boron is calcium borate, or colemanite, $Ca_2B_6O_{11} \cdot 5H_2O$. In all of these compounds, boron is bonded in polyatomic anions.

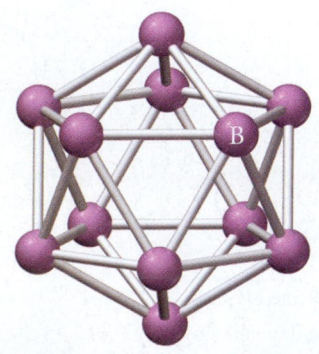

▲ **FIGURE 22.9** B_{12} **Icosahedron** An icosahedron contains 20 triangular faces that are connected at 12 vertices. Elemental boron forms several different structures having icosahedron units bonded together in different arrangements.

The main use for boron today is in glass manufacture. Adding boron oxide to silicon oxide glass changes the thermal expansion of the glass, which is important if glassware is intended to be heated. When glass is heated, the outer edge of the glass warms and expands more quickly than the inner edge, creating stress that can result in cracking. Adding boron oxide to glass reduces its thermal expansion, allowing the glass (called borosilicate glass or Pyrex) to be heated without cracking.

Elemental boron is also used in the nuclear energy industry. Boron readily absorbs neutrons and is therefore used in the control rods of nuclear reactors. When the nuclear reaction needs to be slowed down, the rods are inserted into the reactor to absorb the neutrons (see Section 19.7).

Boron–Halogen Compounds: Trihalides

Boron forms many covalently bonded compounds in which boron atoms bond to each other. In some ways, these compounds are similar to those in which carbon covalently bonds to itself; however, the structures are different because boron is less electronegative and has only three valence electrons. As we saw earlier in this section, boron has a tendency to form polyhedral cluster structures such as the icosahedral structure of elemental boron. Boron also tends to form electron-deficient compounds (compounds in which boron lacks an octet).

Boron halides have the general formula BX_3 and have a trigonal planar structure.

The bonds in the boron trihalides are stronger and shorter than a typical single bond, which can be explained using valence bond theory and hybridization. The boron atom uses sp^2 hybridized orbitals to form sigma bonds with the three halogen atoms. Because boron's three valence electrons are used to form the sigma bonds, the third p orbital of boron is an empty orbital that is perpendicular to the trigonal plane of the molecule. Each halogen atom has a filled p orbital, also perpendicular to the trigonal plane of the molecule. The empty p orbital on the boron atom can overlap with the full p orbitals on the halogens, forming a coordinate-covalent type bond that, like a normal double bond, is shorter and stronger than a single bond.

The boron trihalides are strong Lewis acids. For example, BF_3 reacts with NH_3 according to the following Lewis acid–base reaction:

$$BF_3(g) + :NH_3(g) \longrightarrow F_3B:NH_3(s)$$

Boron trihalides are used as Lewis acids in many organic reactions, such as those in which alcohols or acids are converted to esters. In water, the trihalides hydrolyze to form acidic solutions according to the following reactions:

$$BF_3 + H_2O \longrightarrow BF_3 \cdot H_2O \longrightarrow BF_3OH^-(aq) + H^+(aq)$$
$$4\,BF_3 + 3\,H_2O \longrightarrow 3\,H^+ + 3\,BF_4^-(aq) + B(OH)_3(aq)$$
$$BCl_3 + 3\,H_2O \longrightarrow B(OH)_3(aq) + 3\,H^+(aq) + 3\,Cl^-(aq)$$

Boron–Oxygen Compounds

Boron forms very strong bonds with oxygen in structures that contain trigonal BO_3 structures. The formula for the crystalline structure of boron and oxygen is B_2O_3. In this compound, the trigonal BO_3 structures hook together to form interlocking B_6O_6 hexagonal rings, as shown in Figure 22.10 ▶. Each hexagonal ring has a boron atom at the six corners and an oxygen atom in the middle of each of the six sides. The compound B_2O_3 melts at 450 °C. If the molten B_2O_3 cools quickly, it forms a glass. The glass still contains many interlocking B_6O_6 hexagonal rings but lacks the long-range order of the crystal. Molten boron oxide dissolves many metal oxides and silicon oxide to form glasses of many different compositions.

▼ FIGURE 22.10 **B_2O_3 Structure**
Crystalline B_2O_3 consists of BO_3 trigonal structures that form hexagonal rings of B_6O_6.

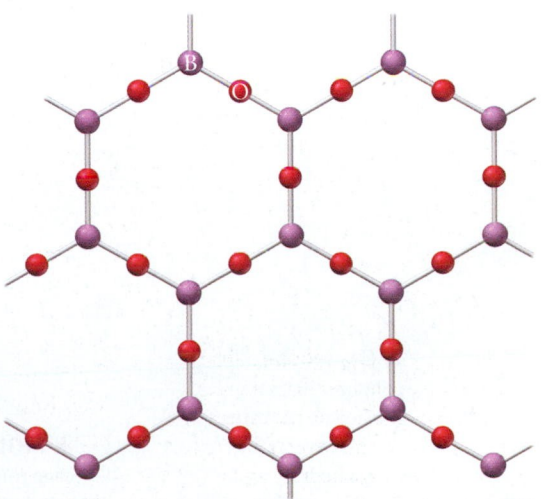

Boron–Hydrogen Compounds: Boranes

Compounds composed of boron and hydrogen, called **boranes**, form many unique cluster, cagelike, and netlike structures. ***closo*-Boranes** have the formula $B_nH_n^{2-}$ and form fully closed polyhedrons with triangular sides; two of these structures are shown in Figure 22.11 ▼. Each of the vertices in the polyhedrons is a boron atom with an attached hydrogen atom. The *closo*-borane with the formula $B_{12}H_{12}^{2-}$ forms the full icosohedral shape, as does elemental boron, but with the added hydrogen atoms.

closo-Boranes

$B_6H_6^{2-}$ $B_7H_7^{2-}$

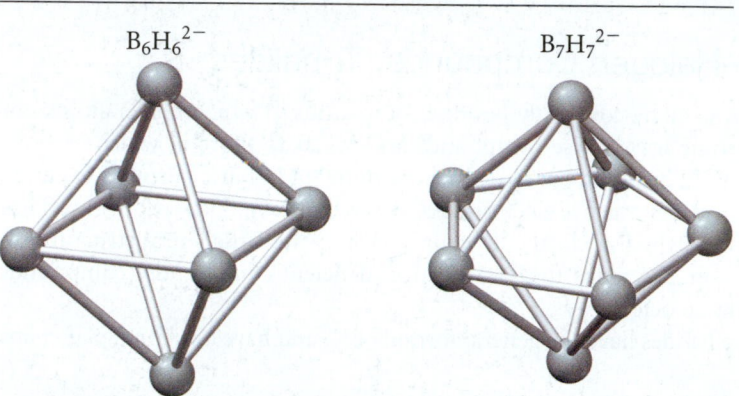

▲ **FIGURE 22.11** *closo*-**Borane Structures**   *closo*-Borane structures form closed polyhedrons with triangular faces. In this figure, each sphere represents a BH unit. $B_6H_6^{2-}$ has an octahedral (square bipyramidal) shape. $B_7H_7^{2-}$ has a pentagonal bipyramidal shape.

If the borane polyhedron is missing one or more boron atoms, extra hydrogen atoms attach to the structure and make the borane neutral. Over 35 structurally different neutral boranes have been identified, ranging from B_2H_6 to $B_{20}H_{26}$. These neutral boranes can be classified on the basis of their different chemical formulas. The ***nido*-Boranes**, named from the Latin word for *net*, have the formula B_nH_{n+4}. They consist of a cage of boron atoms missing one corner. The ***arachno*-Boranes**, named from the Greek word for *web*, have the formula B_nH_{n+6}. They consist of a cage of boron atoms that is missing two or three corners. Examples of a *nido*- and an *arachno*-borane are shown in Figure 22.12 ▼.

We have seen that boranes form interesting structures, but they are also valuable as catalysts in organic reactions. For example, adding an alkene to a diborane forms an alkane bonded to the boron atom. The alkane can be cleaved from the boron, resulting in a net hydrogenation reaction that can be carried out under mild conditions.

$$B_2H_6(g) + 6\ CH_2{=}CHCH_3(g) \longrightarrow 2\ B(CH_2CH_2CH_3)_3(l)$$

A *nido*-borane   An *arachno*-borane

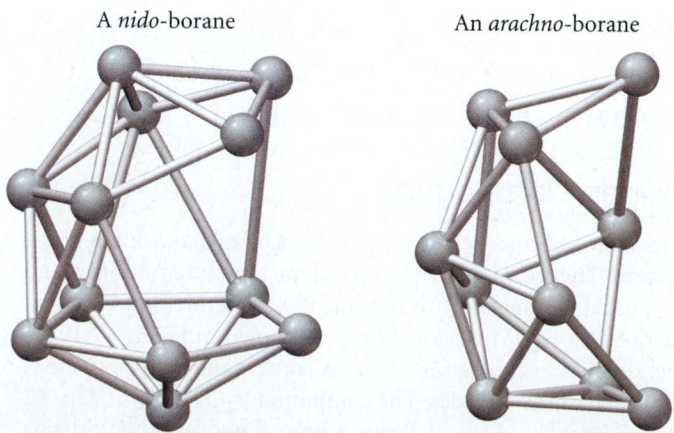

▲ **FIGURE 22.12** *nido*- and *arachno*-**Boranes**   The *nido*-borane structure forms a cage missing one boron atom from a corner. The *arachno*-borane structure forms a net missing more than one boron from a corner.

22.5 Carbon, Carbides, and Carbonates

The group 4A elements exhibit the most versatile bonding of all elements. As we saw in Chapter 20, carbon has the ability to bond with other carbon atoms and with a few other elements to form the great variety of organic compounds. From these come the molecules of life that we examined in Chapter 21. Here we focus on elemental carbon and those compounds of carbon which are known as *inorganic* (rather than organic).

Carbon

Elemental carbon is found in several different forms. Two well-known naturally occurring crystalline forms of carbon are **graphite** and **diamond**. Graphite deposits are found mostly in mines in East Asia and Canada. Graphite's structure, shown in Figure 22.13 ▼, consists of flat sheets of carbon atoms bonded together as interconnected hexagonal rings. Although the covalent bonds within the sheets are strong, the interactions between the sheets are weak, allowing the layers of graphite to slip easily past each other, making graphite a good lubricant. The electrons in the extended pi bonding network within a sheet make graphite a good electrical conductor in the direction of the plane of the sheets. Because of its relative stability and electrical properties, graphite is used for electrodes in electrochemical applications and for heating elements in furnaces.

The density of graphite is 2.2 g/cm^3. Under high pressure, the carbon atoms in graphite rearrange to form diamond, which has a higher density of 3.5 g/cm^3. Diamonds form naturally when carbon is exposed to high pressures deep underground. Through movements in Earth's crust, the diamonds are brought near the surface. Most diamonds are found in Africa, mainly in the Congo region and in South Africa. The first synthetic diamonds were produced in the 1940s, using pressures of 50,000 atm and a temperature of 1600 °C. The diamond structure, shown in Figure 22.14 ▼, consists of carbon atoms connected to four other carbon atoms at the corners of a tetrahedron. This bonding extends throughout three dimensions, making giant molecules described as network covalent solids (see Section 11.12).

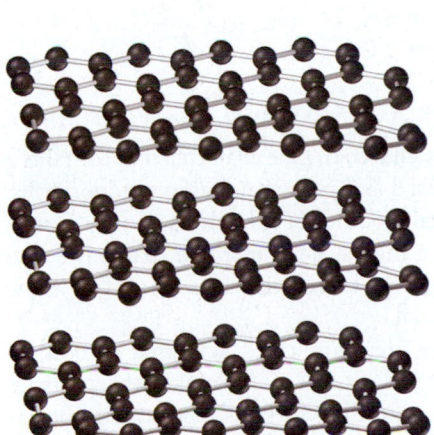

▲ FIGURE 22.13 Graphite Structure
The carbon atoms in graphite bond strongly within the plane of the carbon atoms but bond weakly between the sheets.

▲ FIGURE 22.14 Diamond Structure
The diamond structure is composed of carbon atoms connected to four other carbon atoms at the corners of a tetrahedron.

✦ Conceptual Connection 22.1 Phase Changes and Pressure

Why do high pressures favor the formation of diamond from graphite?

Answer: An increase in pressure favors the denser phase, in this case diamond.

Diamond is very hard and is an excellent conductor of heat. Consequently, the largest use of diamonds is in abrasives and cutting tools. Small diamonds are used at the cutting edge of tools, making the edges much harder and giving them a longer life. Natural diamonds are used as gems for their brilliance and relative inertness.

Carbon is also naturally found in some noncrystalline forms. **Coal** is the product of the decomposition of ancient plant material that has been buried for millions of years, during which time it undergoes a process called carbonization. The reaction, which occurs under high pressure in the presence of water and the absence of air, removes most of the hydrogen and oxygen (which are lost as volatile gases such as methane and water) from the original organic compounds that composed the plant. The resulting coal contains a mixture of various hydrocarbons and carbon-rich particles. It is extensively mined and used as an energy source throughout the world. Coal types are classified by the amount of carbon and other elements that they contain, as shown in Table 22.3.

TABLE 22.3 Approximate Composition of the Main Types of Coal

Type of Coal	Free C (mol %)	Total C (mol %)	H (mol %)	O (mol %)	S (mol %)
Lignite	22	71	4	23	1
Bituminous	60	80	6	8	5
Anthracite	88	93	3	3	1

Among the types of coal listed in Table 22.3, anthracite has the highest carbon content and consequently yields the most energy per mass when burned. Bituminous coal also contains a relatively high amount of carbon but has in addition high levels of sulfur, which increases the formation of sulfur oxides, pollutants that result in acid rain (see Sections 3.6 and 15.12).

Heating coal in the absence of air forms a solid called **coke** that is composed mainly of carbon and ash. Coke is used in the steel industry for the reduction of iron ore to iron. In a blast furnace, the carbon in the coke is oxidized to form carbon monoxide, which is used to reduce the iron in iron oxide according to the following reactions:

$$O_2(g) + \underset{\text{coke}}{C(s)} \longrightarrow CO_2(g)$$

$$CO_2(g) + \underset{\text{coke}}{C(s)} \longrightarrow 2\,CO(g)$$

$$Fe_2O_3(s) + 3\,CO(g) \longrightarrow 2\,Fe(s) + 3\,CO_2(g)$$

Heating wood in the absence of air produces **charcoal**. Like coal, charcoal contains a high amount of amorphous free carbon and is used as a common fuel in outdoor grills. Charcoal retains the general shape of the original wood even though its density is substantially decreased. Consequently, charcoal has a high surface area that makes it useful for filtration. The impurities in a liquid or gas are adsorbed on the charcoal surface as the liquid or gas flows through the pores in the charcoal.

Very fine carbon particles with high surface areas are called **activated carbon**, or *activated charcoal*. The large surface area, greater than $10^3 \, m^2/g$, makes the particles extremely efficient at adsorbing other molecules onto their surfaces. Activated carbon is made by heating amorphous carbon in steam, which breaks the grains into smaller sizes and removes any other materials adsorbed on the surface. Activated carbon is used as a gas filter to remove impurities in gas streams and as a decolorizing agent, removing impurities that discolor organic products such as sugar or wheat flour.

Soot is an amorphous form of carbon that is produced during the incomplete combustion of hydrocarbons, as indicated by blue or black smoke. Toxic carbon monoxide is also formed in the process.

$$H_xC_y(s) + O_2(g) \longrightarrow H_2O(g) + CO_2(g) + CO(g) + \underset{\text{soot}}{C(s)}$$

▲ The black color of automobile tires is from the carbon black that is added to strengthen the tires and maintain flexibility.

Carbon black, a fine, powdered form of carbon, is a component of soot. Over a million tons of carbon black, a strengthener for rubber tires, are used in manufacturing each day. The black color of automobile tires is due to the several kilograms of carbon black within each tire, over 25% of the mass of the typical tire.

In the 1980s a new form of carbon was discovered when a powerful laser was aimed at a graphite surface. The structure of this new form of carbon was soccer-ball-shaped clusters of C_{60} that contained five- and six-membered carbon rings wrapped into a 20-sided icosahedral structure, as shown in Figure 22.15 ▼. The compound was named *buckminsterfullerene*, honoring R. Buckminster Fuller, a twentieth-century engineer and architect who advocated the construction of buildings using the structurally strong geodesic dome shape that he had patented.

Carbon clusters similar to C_{60} but containing from 36 to over 100 carbon atoms have also been identified; they are generally called **fullerenes**, and nicknamed *buckyballs*. Fullerenes are black solids in which the carbon clusters are held to one another by dispersion forces. They are somewhat soluble in nonpolar solvents, and the different fullerenes form solutions of different colors.

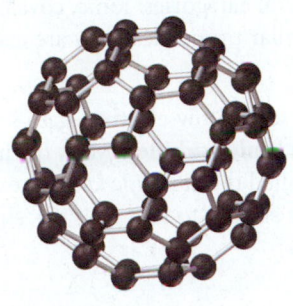

C_{60}

◀ **FIGURE 22.15** C_{60} **and a Geodesic Dome** The C_{60} structure resembles Buckminster Fuller's geodesic dome.

Long structures, called **nanotubes**, consisting of interconnected C_6 rings have also been developed and are featured on the front cover of this book. The first nanotubes discovered consisted of tubes with double walls of C_6 rings with closed ends. However, the ends of the tubes can be opened when they are heated under the proper conditions. Salts and organometallic compounds can be introduced into the nanotubes, and some compounds that are generally not stable can also be formed inside nanotubes. These materials open a new synthetic route to making novel chemicals.

Today, two general types of nanotubes can be produced: (1) single-walled nanotubes (SWNT), which have one layer of interconnected C_6 rings forming the walls of the nanotubes, and (2) multiwalled nanotubes (MWNT), which have concentric layers of interconnected C_6 rings forming the walls of the nanotubes. Both types of nanotubes are shown in Figure 22.16 ▼.

(a) Single-walled nanotube (SWNT)

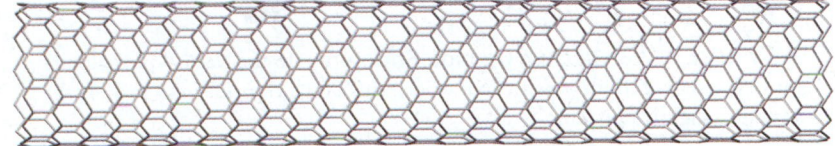

(b) Multiwalled nanotube (MWNT)

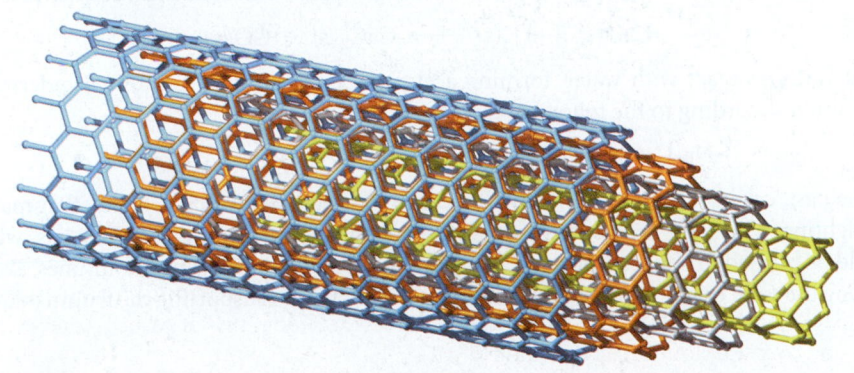

◀ **FIGURE 22.16** **Carbon nanotubes** Carbon nanotubes can be formed with **(a)** a single wall of carbon atoms or **(b)** multiple walls of carbon atoms.

▲ FIGURE 22.17 The Nanocar
This nanocar has buckyballs for wheels and can actually roll across an atomic surface.

Nanotubes are 100 times stronger than steel and only one-sixteenth as dense. Consequently, carbon nanotubes are used commercially for lightweight applications that require strength, such as golf clubs and bicycle frames. When the nanotubes are lined up parallel to one another, a bundle of the tubes can form a "wire" with very low electrical resistance. These tiny wires open up the possibility of making incredibly small electronic devices. Other applications include using nanotubes and buckyballs to make nanomachines. For example, Figure 22.17 ◀ shows a nanocar that has buckyballs for wheels. The car can actually roll across an atomic surface and is so small that 20,000 of them laid end to end would span the thickness of a human hair.

Carbides

Binary compounds composed of carbon combined with less electronegative elements are called **carbides**. Carbides are classified into three general categories: ionic, covalent, and metallic. All three types of carbides have generally similar properties; they are extremely hard materials with high melting points.

Ionic Carbides Compounds composed of carbon and a low-electronegativity metal such as an alkali metal or an alkaline earth metal are **ionic carbides**. Most of the ionic carbides contain the dicarbide ion, C_2^{2-}, commonly called the *acetylide ion*. For example, calcium carbide has the formula CaC_2 and a structure similar to that of NaCl, shown in Figure 22.18 ▼.

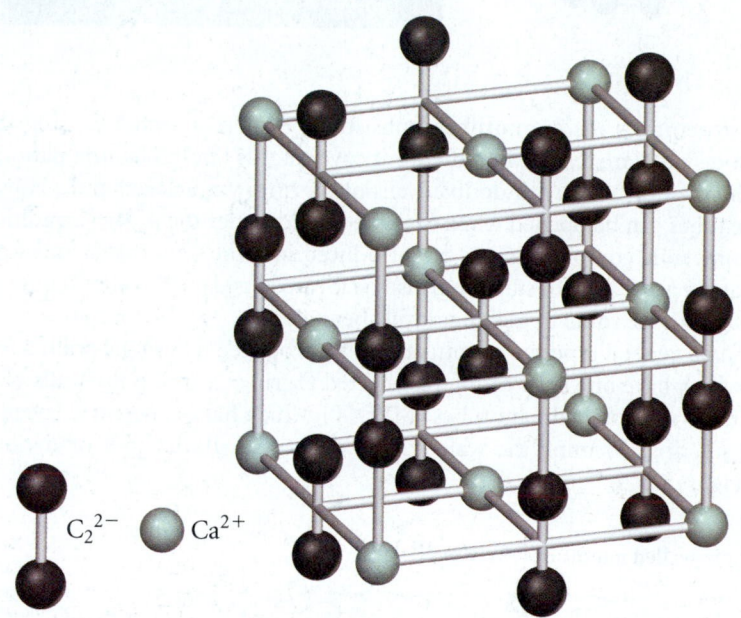

C_2^{2-} Ca^{2+}

▲ FIGURE 22.18 Calcium Carbide Structure   In the NaCl-type structure for CaC_2, the dicarbide ions are in the positions of the chloride ions, making the structure slightly noncubic.

Calcium carbide is formed by the reaction of calcium oxide with coke in an electric furnace.

$$CaO(s) + 3\,C(s) \longrightarrow CaC_2(s) + CO(g)$$

Ionic carbides react with water, forming acetylene. For example, sodium carbide reacts with water according to the following reaction:

$$Na_2C_2(s) + 2\,H_2O(l) \longrightarrow 2\,NaOH(aq) + C_2H_2(g)$$

In the past, calcium carbide was used as a source of acetylene (which is highly flammable) for lighting. The solid CaC_2 was made to react with water, releasing acetylene gas, which would be burned in applications such as automobile headlights and lamps for mines. Transporting the solid calcium carbide was more convenient than transporting the flammable gas.

Covalent Carbides Compounds composed of carbon and low-electronegativity *non-metals* or *metalloids* are **covalent carbides**. The most important covalent carbide is silicon carbide (SiC), a very hard material. Over 500,000 tons of silicon carbide are produced annually, mostly for use as an abrasive material in the cutting and polishing of metals. In a process analogous to the formation of calcium carbide, silicon carbide is formed by the reaction of silicon oxide with coke at high temperatures.

$$SiO_2(s) + 3\,C(s) \longrightarrow SiC(s) + 2\,CO(g)$$

Recently, a gem-quality form of SiC, called *moissanite*, has been developed. Moissanite is described as being more brilliant than all other gems, including diamonds. Yet moissanite costs much less than diamond and is consequently sold as a diamond substitute (much like the more common diamond substitute cubic zirconia, ZrO_2). Moissanite was first identified in small particles at the Diablo Canyon meteorite impact crater in Arizona, and is therefore sometimes advertised as a gift from the stars.

Metallic Carbides Compounds composed of carbon and metals that have a metallic lattice with holes small enough to fit carbon atoms are **metallic carbides**. Metallic carbides retain many of their metallic properties, such as high electrical conductivity, but they are stronger, harder, and less malleable than the corresponding metals. Adding carbon to steel, for example, increases its hardness by forming regions of cementite (Fe_3C) in the steel matrix. Tungsten carbide (WC) is a metallic carbide used in cutting tools.

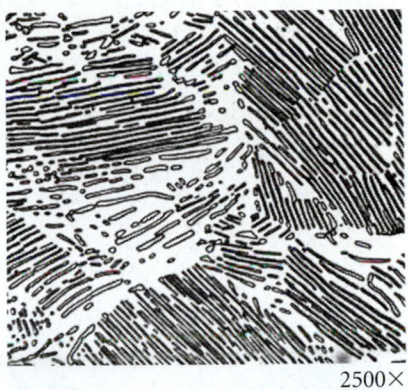

2500×

▲ This micrograph shows cementite (dark regions) in steel.

Carbon Oxides

Carbon forms two stable oxides, carbon monoxide and carbon dioxide. Our atmosphere contains about 0.04% carbon dioxide by volume. Plants use atmospheric carbon dioxide to produce sugars during photosynthesis.

$$6\,CO_2(g) + 6\,H_2O(g) \longrightarrow C_6H_{12}O_6(s) + 6\,O_2(g)$$

Carbon dioxide is returned to the atmosphere by animal respiration, plant and animal decay, and (in modern history) fossil fuel combustion. Because carbon dioxide is highly soluble in water, the oceans of the world act as a reservoir for CO_2, keeping the amount of CO_2 in the atmosphere generally stable. As we saw in Section 6.9, however, the increase in the combustion of fossil fuels in the last century has increased the amount of CO_2 in the atmosphere by about 25%.

Recall from Section 11.8 that CO_2 has a triple point at $-57\,°C$ and 5.1 atm. At atmospheric pressure, therefore, the liquid phase of CO_2 does not exist. Solid carbon dioxide sublimes directly to the gas phase when heated, which is why solid CO_2 is often called "dry ice."

Carbon monoxide (CO) is a colorless, odorless, and tasteless gas. The boiling point of carbon monoxide is $-192\,°C$ at atmospheric pressure, and CO is only very slightly soluble in water. As we saw at the very beginning of this book (see Section 1.1), carbon monoxide is toxic because it interferes with the ability of hemoglobin to bind oxygen. Unlike carbon dioxide, which is very stable, carbon monoxide is relatively reactive and can be used as a reducing agent. For example, carbon monoxide reacts with oxygen and metal oxides to form carbon dioxide.

$$2\,CO(g) + O_2(g) \longrightarrow 2\,CO_2(g)$$
$$CO(g) + CuO(s) \longrightarrow CO_2(g) + Cu(s)$$

Carbon monoxide also reduces many nonmetals, producing compounds with the reduced form of the nonmetal.

$$CO(g) + Cl_2(g) \longrightarrow COCl_2(g)$$
$$CO(g) + S(s) \longrightarrow COS(g)$$

The product of the first reaction listed above is phosgene ($COCl_2$), also known as carbonyl chloride, a poisonous gas that was used in World War I as a chemical weapon. Phosgene is now an important industrial chemical used in the production of polycarbonates. The product of the second reaction, carbonyl sulfide (COS), is a fungicide.

Carbonates

When carbon dioxide dissolves in water, it forms carbonic acid (H_2CO_3). As a weak acid, carbonic acid partially ionizes into hydrogen carbonate (or bicarbonate) and carbonate.

$$CO_2(aq) + H_2O(l) \rightleftharpoons \underset{\text{carbonic acid}}{H_2CO_3(aq)}$$

$$H_2CO_3(aq) \rightleftharpoons H^+(aq) + \underset{\text{hydrogen carbonate}}{HCO_3^-(aq)} \rightleftharpoons 2\,H^+(aq) + \underset{\text{carbonate}}{CO_3^{2-}(aq)}$$

As we saw in Section 12.4, the solubility of carbon dioxide, like that of other gases, increases with increasing pressure. Carbon dioxide under high pressure is used to carbonate soft drinks. Under most conditions, less than 0.5% of the dissolved carbon dioxide reacts with water to form carbonic acid. This leaves most of the carbon dioxide as dissolved gas molecules so the soft drink does not acquire much of a sour acidic taste.

The hydrated crystal of sodium carbonate, $Na_2CO_3 \cdot 10\,H_2O$, is known as **washing soda**. It can be heated to remove the water and form the stable anhydrous sodium carbonate, Na_2CO_3. All the alkali metal ions form stable carbonates, even when heated. The carbonates all make basic solutions when added to water because the carbonate ions readily ionize water (as described more fully in Section 15.8).

$$Na_2CO_3(s) \longrightarrow 2\,Na^+(aq) + CO_3^{2-}(aq)$$

$$CO_3^{2-}(aq) + H_2O(l) \longrightarrow HCO_3^-(aq) + OH^-(aq)$$

Sodium bicarbonate ($NaHCO_3$) is known as baking soda. When heated, baking soda gives off carbon dioxide gas, which is why it is used in baking to help raise dough.

$$2\,NaHCO_3(s) \longrightarrow Na_2CO_3(s) + H_2O(l) + CO_2(g)$$

Baking *powder* is a mixture of $NaHCO_3$ and an acid. The two components of the mixture are kept from reacting by a starch filler. When water is added to the mixture, however, the two components dissolve and react, producing the carbon dioxide that forms pockets of gas in baked products. You can perform a simple test to determine if baking powder is still good (that is, whether the acid has not already slowly reacted with the sodium bicarbonate) by pouring some boiling water over a small sample of the baking powder. If the hot water produces bubbles, then the baking powder is still active. Alka-Seltzer is another common product that uses sodium bicarbonate, in this case mixed with citric acid and aspirin. When put in water, the acid and carbonate react to produce carbon dioxide, giving the familiar fizz.

▲ Alka-Seltzer contains sodium bicarbonate mixed with citric acid and aspirin. When put in water, the acid and carbonate react to produce the fizz.

◆ Conceptual Connection 22.2 Carbonate Solubility

As we saw in Chapter 4, the carbonates of metal ions other than group 1A are insoluble in water. Which of the following would increase their solubility?

(a) adding acid to the solution

(b) adding base to the solution

(c) increasing the amount of the solid carbonate in the solution

Answer: **(a)** Since the carbonate ion is basic, adding acid to the solution drives the dissolution reaction to the right because the acid reacts with the carbonate ion. (Recall from Section 16.5 that the solubility of an ionic compound with a basic anion increases with increasing acidity.)

22.6 Nitrogen and Phosphorus: Essential Elements for Life

The group 5A elements range from the nonmetallic nitrogen and phosphorus to metallic bismuth. Both nitrogen and phosphorus are classified as nonmetals because they are nonconductors and form acidic oxides. They both have s^2p^3 electron configurations, and yet their chemical properties are very different. Phosphorus is a much larger and less electronegative atom and also has d orbitals available for bonding.

Elemental Nitrogen and Phosphorus

Nitrogen and phosphorus have been known for many years. Nitrogen was identified in 1772 and phosphorus in 1669. Elemental nitrogen is a diatomic gas that makes up about 78% of Earth's atmosphere by volume (see Section 5.6). To obtain elemental nitrogen, air is cooled to below $-196\,°C$, which liquefies it. When the liquid air is warmed slightly, the nitrogen boils off, leaving liquid oxygen (which boils at the higher temperature of $-183\,°C$). Passing the vaporized gas over hot copper metal purifies the nitrogen by removing residual oxygen, which reacts with the copper to form CuO. Nitrogen gas can also be separated from the other atmospheric gases by passing air through certain silicate materials called zeolites, which have channels of just the right diameter to separate gas molecules of different size. Some mineral sources for nitrogen are saltpeter (KNO_3) and Chile saltpeter ($NaNO_3$).

As we first saw in Section 9.5, nitrogen molecules contain a triple bond between the two N atoms. The strength of the triple bond makes N_2 very stable, and many chemical attempts to break the bond have not been commercially successful. When nitrogen gas is heated with oxygen or hydrogen, nitric oxide (NO) or ammonia (NH_3), respectively, form with low yields. When nitrogen gas is heated with active metals, metal nitrides form. Beyond this, however, nitrogen gas is relatively unreactive.

The stability of elemental nitrogen makes it useful as a protective atmosphere to prevent oxidation in many industrial processes. For example, industrial furnaces use a nitrogen atmosphere to anneal products made of metal, and chemical reactions sensitive to oxygen are carried out in a nitrogen atmosphere. Nitrogen is also used to preserve a variety of foods.

Elemental phosphorus was first isolated by accident from urine when Henning Brand, a seventeenth-century physician and alchemist from Hamburg, Germany, was distilling urine in an ill-informed attempt to obtain gold from the golden liquid. The elemental form of phosphorus that he obtained instead was a white, waxy, flammable solid called **white phosphorus**. White phosphorus is highly toxic to humans. For over a hundred years, the phosphorus-containing compounds in urine were the main source for producing elemental phosphorus. Today, however, phosphorus is obtained from a calcium phosphate mineral called apatite [$Ca_3(PO_4)_2$]. The mineral is heated with sand and coke in an electric furnace.

$$2\,Ca_3(PO_4)_2(s) + 6\,SiO_2(s) + 10\,C(s) \longrightarrow P_4(g) + 6\,CaSiO_3(l) + 10\,CO(g)$$

apatite   sand   coke   white phosphorus

The desired product, white phosphorus, spontaneously burns in air and is normally stored under water to prevent contact with air.

White phosphorus consists of P_4 molecules in a tetrahedral shape, with the phosphorus atoms at the corners of the tetrahedron, as shown in Figure 22.19 ▶. The bond angles between the three P atoms on any one face of the tetrahedron is small (60°) and strained, making the P_4 molecule unstable and reactive.

When heated to about 300 °C in the absence of air, white phosphorus slightly changes its structure to a different allotrope called **red phosphorus**, which is amorphous. The general structure of red phosphorus is similar to that of white phosphorus, except that one of the bonds between two phosphorus atoms in the tetrahedron is broken, as shown in Figure 22.20 ▼. These two phosphorus atoms then link to other phosphorus atoms, making chains that can vary in structure.

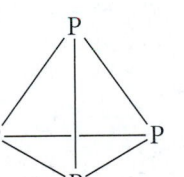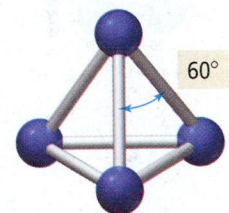

▲ FIGURE 22.19 **White Phosphorus** The small bond angle of 60° between the phosphorus atoms at the corners of the tetrahedron puts a great strain on the structure and makes the P_4 molecule unstable.

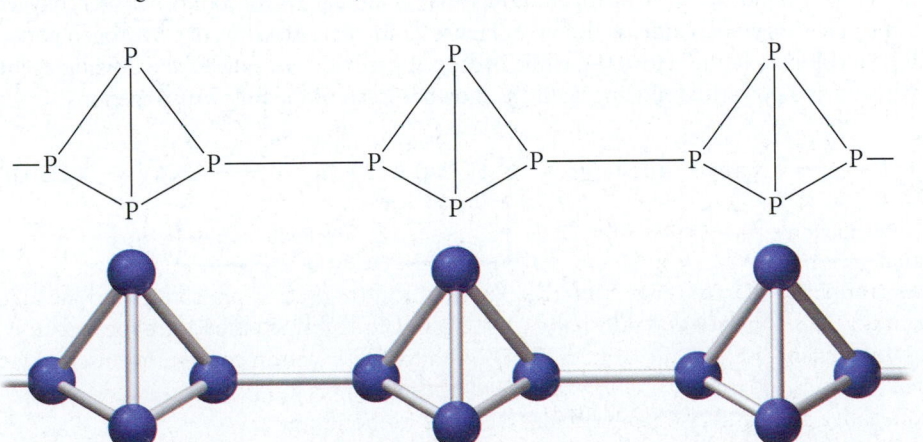

◀ FIGURE 22.20 **Red Phosphorus** Red phosphorus consists of chains of phosphorus atoms that form amorphous structures.

Red phosphorus is neither as reactive nor as toxic as white phosphorus, and even though it is still flammable, it can be stored in air. Red phosphorus is used commercially in applications such as match heads. Rubbing the match head onto a surface produces enough heat (through friction) to ignite the phosphorus. Today, most strike-anywhere matches use the phosphorus compound tetraphosphorus trisulfide (P_4S_3) and an oxidizing agent, potassium chlorate ($KClO_3$).

A third allotrope of phosphorus is **black phosphorus**. Black phosphorus is obtained by heating white phosphorus under pressure. This form of phosphorus is the most thermodynamically stable form, and therefore the least reactive. Black phosphorus has a layered structure similar to that of graphite.

Nitrogen Compounds

Nitrogen, with a valence electron configuration of $2s^2 2p^3$, can gain three electrons or lose five electrons to obtain an octet. Nitrogen forms many covalent compounds with oxidation states from -3 to $+5$, as shown in Table 22.4.

Nitrogen Hydrides The most common nitrogen hydride is **ammonia** (NH_3), the strong-smelling compound in which nitrogen displays its lowest oxidation state (-3). Ammonia is important because of its reaction with sulfuric acid (or phosphoric acid) to produce ammonium salts for fertilizers.

$$2\,NH_3(g) + H_2SO_4(aq) \longrightarrow (NH_4)_2SO_4(aq)$$

For hundreds of years, natural biological materials such as animal manure were used as nitrogen-containing fertilizers. In the 1800s, however, the nitrogen-bearing nitrate mineral $NaNO_3$ was discovered in Chile (and named Chile saltpeter). This nitrate mineral became an important source of fertilizer and made the country of Chile very wealthy; yet it was a limited source.

The obvious *unlimited* source of nitrogen is the N_2 gas in the atmosphere, but the strong triple bond in elemental nitrogen renders it unusable by plants. In order to be used as fertilizer, elemental nitrogen has to be *fixed*, which means that it has to be converted into a nitrogen-containing compound such as NH_3. However, the direct reaction of nitrogen gas with hydrogen gas to form ammonia is very slow and produces low yields of ammonia under normal conditions.

$$N_2(g) + 3\,H_2(g) \rightleftharpoons 2\,NH_3(g)$$

In the early 1900s, the German chemist Fritz Haber studied the equilibrium conditions for this reaction and showed that high pressures and lower temperatures favored the product. With new techniques for carrying out the reaction at a higher pressure, and using a catalyst to increase the reaction rate, the industrial process for producing ammonia from nitrogen gas and hydrogen gas—now called the **Haber-Bosch process**—became practical by the middle 1930s. This process is now the main industrial process for making ammonia and fixing nitrogen for many uses, including fertilizers and explosives.

Hydrazine (N_2H_4) is another nitrogen and hydrogen compound in which nitrogen has a negative oxidation state (-2). Hydrazine is the nitrogen analog of hydrogen peroxide; it has a bond between nitrogen atoms that is similar to the bond between oxygen atoms in hydrogen peroxide, as shown in Figure 22.21 ◀. Hydrazine, like hydrogen peroxide, is a colorless liquid. However, while hydrogen peroxide is a powerful oxidizing agent, hydrazine is a powerful reducing agent, as shown in each of the following reactions:

TABLE 22.4 Oxidation States of Various Nitrogen Compounds

Nitrogen-Containing Compound	Oxidation State
NH_3	-3
N_2H_4	-2
H_2NOH	-1
HN_3	$-\frac{1}{3}$
N_2	0
N_2O	$+1$
NO	$+2$
N_2O_3, NF_3	$+3$
NO_2, N_2O_4	$+4$
N_2O_5, HNO_3	$+5$

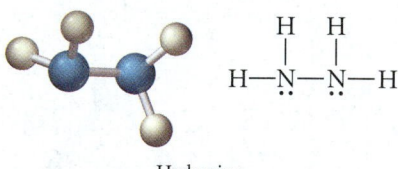

Hydrazine

Hydrogen peroxide

▲ **FIGURE 22.21 Hydrazine and Hydrogen Peroxide** Hydrazine forms a structure similar to hydrogen peroxide with an N—N bond in the place of the O—O bond.

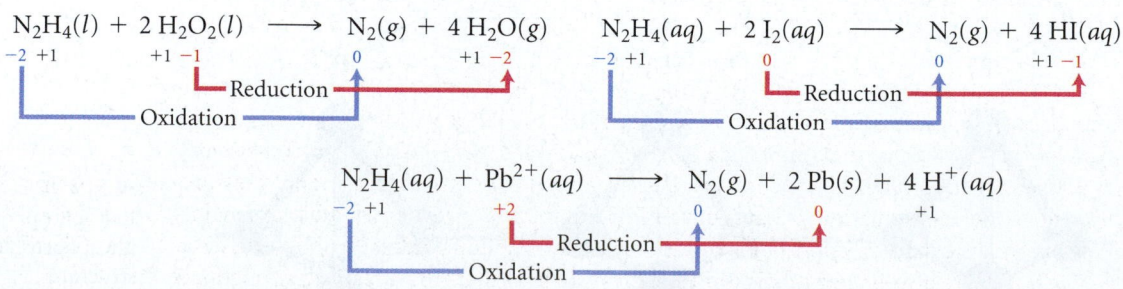

The oxidation state of each atom is shown directly below its symbol. Notice that in each reaction, nitrogen is oxidized and causes the reduction of the other reactant.

Hydrogen azide (HN_3) is a nitrogen and hydrogen compound with a higher nitrogen-to-hydrogen ratio than ammonia or hydrazine. Ammonia and hydrazine are both basic:

$$N_2H_4(aq) + H_2O(l) \longrightarrow N_2H_5^{2+}(aq) + OH^-(aq)$$
$$NH_3(aq) + H_2O(l) \longrightarrow NH_4^+(aq) + OH^-(aq)$$

Hydrogen azide, by contrast, is acidic, ionizing in water to form the azide ion (N_3^-) as follows:

$$HN_3(aq) + H_2O(l) \longrightarrow H_3O^+(aq) + N_3^-(aq)$$

The N_3^- ion can be represented with the following resonance structures (formal charges indicated in red):

$$\left[:\ddot{N}=N=\ddot{N}: \right]^- \qquad \left[:N\equiv N-\ddot{\ddot{N}}: \right]^- \qquad \left[\cdot\ddot{N}=\ddot{N}-\ddot{N}\cdot \right]^-$$

Since the rightmost structure has the least amount of formal charge, it contributes most to the hybrid structure.

Hydrogen azide is thermodynamically unstable compared to its constituent elements and reacts explosively to produce hydrogen and nitrogen gas.

$$2\,HN_3(l) \longrightarrow H_2(g) + 3\,N_2(g)$$

The sodium azide salt is a stable solid at room temperature, but at elevated temperatures, or with a spark, it quickly forms elemental sodium and nitrogen gas.

$$2\,NaN_3(s) \longrightarrow 2\,Na(l) + 3\,N_2(g)$$

The large volume of N_2 gas that can be produced from a small volume of $NaN_3(s)$ is the basis for air bags in automobiles. However, pure sodium azide would also form liquid sodium, which would be dangerous because of its high reactivity. Therefore, other components, such as KNO_3 and SiO_2, are added to the mixture to react with the liquid sodium as follows:

$$10\,Na(l) + 2\,KNO_3(s) \longrightarrow K_2O(s) + 5\,Na_2O(s) + N_2(g)$$
$$2\,K_2O(s) + SiO_2(s) \longrightarrow K_4SiO_4(s)$$
$$2\,Na_2O(s) + SiO_2(s) \longrightarrow Na_4SiO_4(s)$$

The overall reaction thus produces the large volume of nitrogen gas required to fill the air bag quickly, along with harmless potassium and sodium silicates.

Nitrogen Oxides Under certain conditions, especially high temperatures, nitrogen is oxidized by oxygen to form a number of different oxides. For example, during lightning storms, nitrogen monoxide (NO) gas is formed in the upper atmosphere.

$$N_2(g) + O_2(g) \xrightarrow{\text{lightning}} 2\,NO(g)$$

Other nitrogen oxides, such as nitrogen dioxide and dinitrogen trioxide, can be formed from the further oxidation of nitrogen monoxide.

$$2\,NO(g) + O_2(g) \longrightarrow 2\,NO_2(g)$$
$$NO(g) + NO_2(g) \longrightarrow N_2O_3(l)$$

All nitrogen oxides are thermodynamically unstable and will eventually decompose into their constituent elements or react to form more stable compounds. However, many of these reactions are kinetically slow, so that some nitrogen oxides can persist for long periods of time.

The most important nitrogen oxide, because of its significance in biological systems, is probably nitrogen monoxide (NO), also called nitric oxide. In 1987, nitrogen monoxide was named molecule of the year by the journal *Science* because of a number of discoveries related to its biological functions. For example, NO helps control blood pressure through blood vessel dilation; it is important in memory and digestion; and it plays major roles in inducing male erections and female uterine contractions. Adjusting NO levels is a key part of the relatively new medications (such as Viagra) that treat erectile dysfunction.

Earlier in this section, we saw that NO is formed in the atmosphere by lightning. In Sections 3.6 and 15.12 we saw that NO and NO_2 are also formed as by-products of fossil fuel combustion and are among the major precursors of acid rain.

Nitrogen monoxide and nitrogen dioxide are both reactive free radicals (they contain unpaired electrons). At low temperature, two NO_2 molecules dimerize to form N_2O_4, pairing the two lone electrons. If N_2O_4 is heated, however, it decomposes back to NO_2. Consequently, the equilibrium between NO_2 and N_2O_4 is highly temperature-dependent, as we saw in Section 14.9.

$$\underset{\text{colorless}}{N_2O_4(g)} \xrightarrow{\text{heat}} \underset{\text{reddish brown}}{2\,NO_2(g)}$$

Dinitrogen monoxide (N_2O), also called nitrous oxide, is a good oxidizing agent. It can support the combustion of active metals.

$$Mg(s) + N_2O(g) \longrightarrow MgO(s) + N_2(g)$$

Dinitrogen monoxide is unstable when heated, decomposing into nitrogen and oxygen gas.

$$2\,N_2O(g) \xrightarrow{\text{heat}} 2\,N_2(g) + O_2(g)$$

Dinitrogen monoxide (often called "nitrous" or laughing gas) is commonly used as an anesthetic by dentists and to pressurize food dispensers (such as whipped-cream dispensers). Commercially, N_2O is produced by the decomposition of ammonium nitrate.

$$NH_4NO_3(aq) \xrightarrow{\text{heat}} N_2O(g) + 2\,H_2O(l)$$

Nitric Acid, Nitrates, and Nitrides Nitric acid is a very important commercial product of nitrogen. In an electric furnace, nitric acid can be formed from nitrogen and oxygen gas.

$$2\,N_2(g) + 5\,O_2(g) + 2\,H_2O(g) \longrightarrow 4\,HNO_3(g)$$

This reaction is thermodynamically favored, but it is kinetically very slow. (What would happen to our atmosphere if this reaction were thermodynamically *and* kinetically favored?) Because this reaction is slow, a more efficient process, called the **Ostwald process**, is used for commercial preparation of nitric acid.

The first step of the Ostwald process is to pass ammonia gas over hot metal gauze at 600–700 °C to form NO gas. The gauze is made from metals such as platinum and rhodium that are good catalysts for this reaction.

$$4\,NH_3(g) + 5\,O_2(g) \xrightarrow{\text{catalyst}} 4\,NO(g) + 6\,H_2O(g)$$

Next, additional oxygen is added to oxidize the NO to NO_2 gas, which is then passed through a water spray to form nitric acid.

$$3\,NO_2(g) + H_2O(l) \longrightarrow 2\,HNO_3(l) + NO(g)$$

These steps are similar to the natural process that forms acid rain from NO and NO_2 gas in the atmosphere, but under controlled conditions. The NO gas made during this last step can be recycled back to form more NO_2 and eventually more HNO_3.

Nitric acid is a strong acid that completely ionizes in water. Concentrated nitric acid is 70% nitric acid by mass, or 16 M. A small fraction of the HNO_3 in a bottle of concentrated nitric acid will react with water to form NO_2, a reddish brown gas that, in small amounts, gives the acid its characteristic pale yellow color.

$$4\,HNO_3(aq) \longrightarrow 4\,NO_2(g) + O_2(g) + 2\,H_2O(l)$$

The main commercial uses of nitric acid are in the production of fertilizers and explosives. Over a million tons of ammonium nitrate fertilizer is produced annually by the reaction between ammonia and nitric acid.

$$NH_3(g) + HNO_3(aq) \longrightarrow NH_4NO_3(aq)$$

Besides being a good fertilizer, ammonium nitrate (as well as some other nitrates) are also good explosives. Ammonium nitrate explodes according to the following reaction:

$$2\,NH_4NO_3(s) \xrightarrow{\text{heat}} 2\,N_2O(g) + 4\,H_2O(g) \longrightarrow 2\,N_2(g) + O_2(g) + 4\,H_2O(g)$$

Metal nitrates are responsible for the various colors seen in fireworks displays. The different metal ions emit different colors as the nitrate explodes in air. For example, copper nitrate produces a green-colored light according to the following reaction:

$$2\,Cu(NO_3)_2(s) \xrightarrow{heat} 2\,CuO(s) + 4\,NO_2(g) + O_2(g) + green\ light$$

As we learned in Chapter 4, nitrates are very soluble. For reactions that need soluble metal cations, a nitrate compound is a good source for the cation without interference from the anion.

Nitrites are compounds containing the nitrite ion (NO_2^-). Sodium nitrite is used as a food preservative because it kills *Clostridium botulinum* bacteria, the cause of botulism, and because it keeps meat from discoloring (or browning) when it is exposed to air. Recently there has been some concern over this practice, both because it hides the true age of the meat and because the nitrites can react with amines in the meat to form compounds called nitrosamines, which are suspected cancer-causing agents. However, there is currently no evidence that nitrites at levels currently used in meats increase cancer risk in humans.

Phosphorus Compounds

Phosphorus has a valence electron configuration of $3s^2 3p^3$, similar to that of nitrogen. Phosphorus also forms many compounds with oxidation states ranging from -3 through $+5$. The most stable compounds have the $+5$ oxidation state.

Phosphine Phosphine (PH_3) is a colorless, poisonous gas that smells like decaying fish and has an oxidation state of -3 for phosphorus. Since phosphorus is less electronegative than nitrogen, phosphine is less polar than ammonia. Phosphine can be formed from the hydrolysis of metal phosphides.

$$Ca_3P_2(s) + 6\,H_2O(l) \longrightarrow 2\,PH_3(g) + 3\,Ca(OH)_2(aq)$$

The disproportionation of white phosphorus in a basic solution can also produce phosphine.

$$2\,P_4 + 3\,OH^- + 9\,H_2O \longrightarrow 5\,PH_3 + 3\,H_2PO_4^-$$

When heated, phosphine decomposes to phosphorus and hydrogen.

$$4\,PH_3(g) \longrightarrow P_4(s) + 6\,H_2(g)$$

Disproportionation is a reaction in which an element is both reduced and oxidized during the same reaction. In this equation the phosphorus in P_4 is both oxidized and reduced. Phosphorus has an oxidation number of 0 in P_4 and is reduced to -3 in PH_3 and oxidized to $+5$ in $H_2PO_4^-$.

Like ammonia, phosphine can form phosphonium compounds such as PH_4Cl and PH_4I. Unlike ammonia, however, phosphine is not basic in aqueous solution.

Phosphorus Halides When phosphorus reacts with the halogens, it forms phosphorus halides, the most important of which generally have the formulas PX_3 and PX_5.

$$P_4(s) + 6\,Cl_2(g) \longrightarrow 4\,PCl_3(l)$$

$$P_4(s) + 10\,Cl_2(g) \longrightarrow 4\,PCl_5(s) \qquad (with\ excess\ chlorine)$$

Phosphorus halides react with water to form phosphoric acid and the corresponding hydrogen halide. For example, PCl_3 reacts with water as follows:

$$PCl_3(l) + 3\,H_2O(l) \longrightarrow H_3PO_3(aq) + 3\,HCl(aq)$$

Reaction of PCl_3 with oxygen at room temperature forms phosphorus oxychloride.

$$2\,PCl_3(l) + O_2(g) \longrightarrow 2\,POCl_3(l)$$

Other phosphorus oxyhalides can be formed from reactions of $POCl_3$ with metal fluorides or iodides.

$$POCl_3(l) + 3\,NaI(s) \longrightarrow POI_3(g) + 3\,NaCl(s)$$

The phosphorus halides and oxyhalides are important compounds in organic chemistry and serve as starting materials for the production of many phosphorus-containing compounds. Many of the key compounds in pesticides, oil additives, fire retardants for clothing, and surfactants, for example, are commercially made from phosphorus oxyhalides.

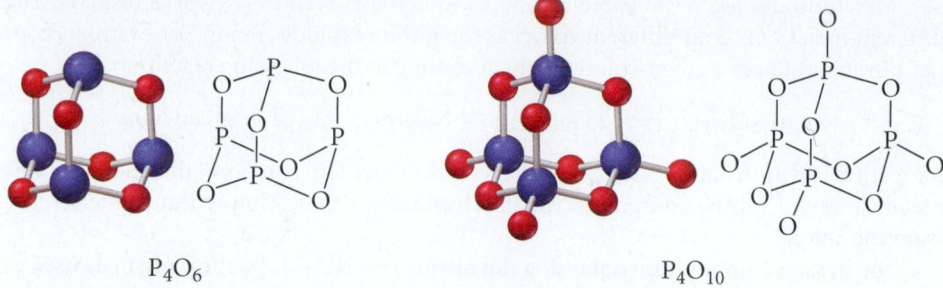

▲ FIGURE 22.22 Tetraphosphorus Hexaoxide and Decaoxide, P_4O_6 and P_4O_{10}
The P_4O_6 structure has the P atoms at the corners of a tetrahedron and the O atoms on the edges.
The P_4O_{10} structure has O atoms also bonded to the P atoms at the corners.

Phosphorus Oxides White phosphorus reacts directly with oxygen to form phosphorus oxides, as in the following reaction:

$$P_4(s) + 5\,O_2(g) \longrightarrow P_4O_{10}(s)$$

The product, however, depends upon the amount of oxygen. Tetraphosphorus hexaoxide, $P_4O_6(s)$, forms when the amount of oxygen is limited, and tetraphosphorus decaoxide, $P_4O_{10}(s)$ is formed when greater amounts of oxygen are available, as shown above.

Phosphorus oxides form interesting cage structures, as shown in Figure 22.22 ▲. The P_4O_6 structure can be pictured as a tetrahedron with a phosphorus atom at each of the vertices and an oxygen atom between each pair of phosphorus atoms. The P_4O_{10} structure has four additional oxygen atoms bonded to each phosphorus atom at the vertices of the tetrahedron.

Phosphoric Acid and Phosphates Phosphoric acid and phosphates are among the most important phosphorus-containing compounds. Phosphoric acid is a colorless solid that melts at 42 °C. Concentrated phosphoric acid is 85% phosphoric acid by mass, or 14.7 M. Phosphoric acid is produced from the oxidation of white phosphorus to tetraphosphorus decaoxide (see reaction above), which is then reacted with water.

$$P_4O_{10}(s) + 6\,H_2O(l) \longrightarrow 4\,H_3PO_4(aq)$$

This method produces a very pure phosphoric acid. A less pure product can be formed from the reaction of calcium phosphate (a mineral source of phosphate) with concentrated sulfuric acid.

$$Ca_3(PO_4)_2(s) + 3\,H_2SO_4(aq) \longrightarrow 3\,CaSO_4(s) + 2\,H_3PO_4(aq)$$

One direct use of phosphoric acid is rust removal. In steel production, thick steel slabs must be heated and rolled into thinner ones. During this process, the hot steel is exposed to air, which oxidizes the surface. To remove this rust, the thin steel sheets pass through phosphoric or hydrochloric acid baths, which dissolve the rust from the metal.

A major use of phosphoric acid is for fertilizer production. In the past, phosphorus-containing materials such as fish, bones, and bat guano were used as fertilizer. Sulfuric acid can decompose bones to make phosphorus compounds that are more readily used by plants. Today many different phosphorus compounds, based on phosphoric acid chemistry, have been developed specifically as fertilizers for various types of plants.

Sodium phosphate compounds produced from phosphoric acid are used as additives in detergents. Compounds such as sodium pyrophosphate ($Na_4P_2O_7$) and sodium tripolyphosphate ($Na_5P_3O_{10}$) can remove metal ions such as Ca^{2+} and Mg^{2+} from hard water, increasing the effectiveness of the detergent and preventing scum rings on sinks and tubs. However, phosphate compounds in detergents are being replaced by other compounds because of the ecological problems—primarily the overfertilization of algae in bodies of water—associated with the phosphates.

Phosphoric acid and phosphates are also important chemicals in the food industry. Phosphoric acid is a soft drink additive. At a low concentration, phosphoric acid is nontoxic and adds a tart, acidic taste to soft drinks. It also prevents bacterial growth in the soda. Table 22.5 summarizes the uses of phosphates in the food industry.

TABLE 22.5 Uses of Phosphates in the Food Industry	
Phosphoric acid, H_3PO_4	Flavor agent in soda, yeast nutrient
Sodium dihydrogen phosphate (also Sodium phosphate monobasic), NaH_2PO_4	Emulsifier, pH buffering agent
Sodium hydrogen phosphate (also Sodium phosphate dibasic), Na_2HPO_4	Baking powder, fermentation auxiliary
Sodium hexametaphosphate, $(NaPO_3)_6$	Preservative, pH buffering agent
Sodium trimetaphosphate, $(NaPO_3)_3$	Starch modifier, juice dispersant
Iron (III) pyrophosphate nonahydrate $Fe_4(P_2O_7)_3 \cdot 9H_2O$	Nutritional supplement
Sodium monofluorophosphate, Na_2PO_3F	Fluoride source for toothpaste
Pyrophosphate, $P_2O_7{}^{4-}$	Tartar control for toothpaste

22.7 Oxygen

The group 6A elements have an s^2p^4 valence electron configuration and a strong attraction for electrons. They can obtain a full octet by gaining only two more electrons. Because of its small size, oxygen is a much stronger oxidizing agent than the rest of the group 6A elements. Oxygen has the second highest electronegativity of any element (3.5), while the rest of the 6A elements range from 2.5 to 2.0 as you move down the column. Because of its high abundance (almost half the mass of Earth's crust is composed of oxygen) and its high reactivity, oxygen is found in many common compounds, including metal oxides, carbonates, silicates, hydrates, and water. Oxygen is also critical for life since the oxidation of biomolecules by oxygen provides energy for most living systems on Earth.

Elemental Oxygen

Oxygen exists naturally as a colorless, odorless, diatomic, nonpolar gas. It condenses to a pale blue liquid at $-183\ °C$. Oxygen is slightly soluble in water (0.04 g in 1 L or 0.001 M at 25 °C). This rather low concentration of oxygen is still enough to support life in aquatic environments. A few types of living systems that dwell deep in the ocean near vents that exude sulfur-containing fumes base their life processes on sulfur chemistry rather than oxygen chemistry.

Today, about 21% of Earth's atmosphere is composed of O_2, but this was not always the case. Earth's early atmosphere was reducing (rather than oxidizing) and contained hydrogen, methane, ammonia, and carbon dioxide. About 2.7 billion years ago, cyanobacteria (blue-green algae) began to convert the carbon dioxide and water to oxygen by photosynthesis. It took hundreds of millions of years to reach the present oxygen composition.

The credit for the discovery of oxygen is given to Joseph Priestley, an English scientist and minister. In 1774, he isolated oxygen by focusing sunlight on mercury(II) oxide and collecting the gas that was released as the red powder oxide formed liquid mercury. He tested the gas by using it to make a candle burn more brightly. He carried out a number of experiments with oxygen over the years, including bravely breathing his newfound gas. As we saw in Chapter 1, Antoine Lavoisier is credited with recognizing that oxygen is necessary for combustion. He described combustion as the reaction of a substance with oxygen (and not the loss of phlogiston). These discoveries and explanations were important steps in the development of modern chemistry.

Oxygen is one of the most abundantly produced industrial chemicals. The major production method is the fractionation of air. Air is cooled until its components liquefy. Then the air is warmed, and components such as N_2 and Ar are separated, leaving oxygen behind. Most commercial oxygen is stored and transported as a gas in tanks under high pressure. Another method for the production of oxygen is the electrolysis of water. Passing an electric current through water containing a small amount of an electrolyte will form hydrogen gas at the cathode and oxygen gas at the anode (see Section 18.8). Because of the large amount of electricity needed, however, electrolysis is not a cost-efficient method for oxygen production.

$$2\ H_2O(l) \xrightarrow{\text{electric current}} 2\ H_2(g)\ +\ O_2(g)$$

▲ Black smokers are vents found under the ocean that provide energy based on sulfur chemistry for life dwelling near the vents.

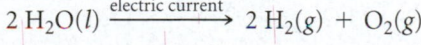

In the laboratory, oxygen can be produced by the heating and decomposition of metal oxides and other oxygen-containing compounds. The oxides of mercury, silver, and gold lose all their oxygen when heated, while the oxides of other metals, such as barium, lose only some of their oxygen.

$$2\ HgO(s) \xrightarrow{\text{heat}} 2\ Hg(l)\ +\ O_2(g)$$

$$2\ BaO_2(s) \xrightarrow{\text{heat}} 2\ BaO(s)\ +\ O_2(g)$$

Metal nitrates and chlorates can also be heated to yield oxygen. Use of a catalyst, such as manganese oxide or iron oxide, can make these reactions very fast and dangerous.

$$2\ NaNO_3(s) \xrightarrow{\text{heat}} 2\ NaNO_2(s)\ +\ O_2(g)$$

$$2\ KClO_3(s) \xrightarrow[\text{catalyst}]{\text{heat}} 2\ KCl\ (s)\ +\ 3\ O_2(g)$$

Uses for Oxygen

The greatest industrial use for oxygen is the enrichment of the air in a blast furnace for the conversion of high-carbon iron to steel. Large quantities of oxygen are also used in oxyhydrogen or oxyacetylene torches for the cutting of metals. Oxygen is also used to create artificial air for use underwater, for high-altitude travel, and in safety equipment.

Oxygen is important in the treatment of a number of medical conditions, such as acute and chronic lung diseases and heart disorders. Generally, patients use masks or nasal catheters to deliver a higher oxygen level from a tank of compressed oxygen. New technology, however, has developed portable oxygen concentrators that use molecular sieves to separate and concentrate oxygen from air. Hyperbaric oxygen therapy is the application of high oxygen levels to patients with skin wounds, such as those with skin grafts or hard-to-heal wounds associated with diabetes. The high oxygen level kills anaerobic bacteria that can infect such wounds.

Oxides

As a strong oxidizing agent, oxygen reacts with most other elements to form oxides. Oxides can be classified according to the oxidation state of oxygen in the oxide, as shown in Table 22.6. The type of oxide that forms depends on the size and charge of the metal. Regular oxides are more stable for the smaller ions with a higher charge. Superoxides are more stable for the larger ions with a smaller charge.

Oxygen also reacts with many nonmetals to form covalent compounds. Many of these nonmetals form several different binary oxides. For example, we have already seen that carbon forms CO and CO_2 and that nitrogen forms N_2O, NO, N_2O_3, NO_2, N_2O_4, and N_2O_5.

TABLE 22.6 Types of Oxides

Class	Ion	Oxidation State of O	Example
Oxide	O^{2-}	-2	Li_2O, MgO
Peroxide	O_2^{2-}	-1	Na_2O_2, BaO_2
Superoxide	O_2^{-}	$-\frac{1}{2}$	RbO_2, CsO_2

Ozone

Ozone (O_3), an allotrope of oxygen, is a toxic blue diamagnetic gas with a strong odor. The smell can be detected at levels as low as 0.01 ppm and is often noticed in electrical storms or near electrical equipment because ozone is formed from O_2 by an electrical discharge. Ozone is denser than O_2 and condenses to a deep blue liquid at $-112\ ^\circ C$. Ozone is naturally made by the irradiation of O_2 with ultraviolet light in the upper atmosphere.

$$3\ O_2(g) \xrightarrow{\text{UV radiation}} 2\ O_3(g)$$

Ozone can also be produced by passing O_2 gas through an electric field. The volume of gas decreases as the O_2 is converted to O_3. Ozone is produced industrially by the electrolysis of cold concentrated sulfuric acid.

Ozone is thermodynamically unstable and decomposes spontaneously to oxygen.

$$2 O_3(g) \longrightarrow 3 O_2(g)$$

The commercial use of ozone is as a strong oxidizing agent. For example, ozone can oxidize NO_2 to N_2O_5 or PbS to $PbSO_4$.

$$2 NO_2(g) + O_3(g) \longrightarrow N_2O_5(g) + O_2(g)$$

$$PbS(s) + 4 O_3(g) \longrightarrow PbSO_4(s) + 4 O_2(g)$$

Ozone kills bacteria and is used as a disinfectant to purify water. It is an environmentally safe replacement for chlorine in water-purification plants because the only by-product is O_2. However, since ozone naturally decomposes, it must constantly be replenished, an economic drawback that limits its use.

Ozone gets into the air we breathe because, as we saw in Section 5.11, ozone is formed from a by-product of the combustion that occurs in automobile engines. Since ozone is a strong oxidizing agent, it is also a harmful substance. In the lower atmosphere, ozone causes a number of problems, including damage to the lungs and skin, stinging of the eyes, and damage to most plant and animal tissues. Ozone also reacts with many types of plastic and rubber materials, causing them to become brittle and to crack.

As we also saw in Section 5.11, ozone in the upper atmosphere is important in the absorption of harmful ultraviolet radiation from the sun. The ozone absorbs the UV radiation and breaks apart to O_2 and O. The oxygen atom will often react with another O_2 molecule to re-form the ozone. In this cycle, each ozone molecule absorbs many ultraviolet photons. Recall also from Section 5.11 that the ozone layer has been depleted by chlorofluorocarbons (CFCs). However, legislation has banned CFCs, in the hope that the reduction of CFCs will help the ozone layer to recover.

22.8 Sulfur: A Dangerous but Useful Element

Like oxygen, sulfur is a nonmetal that belongs to the 6A family. However, sulfur's $3p$ orbitals extend farther out from the nucleus than do oxygen's $2p$ orbitals. Consequently, sulfur is larger and is a much weaker oxidizing agent than oxygen. Unlike oxygen, which forms compounds with only negative oxidation states, sulfur forms compounds with both negative and positive oxidation states. Sulfur, selenium, and tellurium generally form covalent compounds with a +4 or +6 oxidation state, forming anions only when bonding with highly electropositive metals. Sulfur is also much less abundant than oxygen, yet still composes about 0.06% of the mass of Earth's crust.

Elemental Sulfur

Elemental sulfur is found in a few natural deposits, mostly deep underground in the Gulf Coast area of the United States and in Eastern Europe. The deposits are believed to have formed as a by-product of anaerobic bacteria that have decomposed sulfur-containing minerals over many years. The process to recover the sulfur, called the **Frasch process**, is diagrammed in Figure 22.23 ▶. Superheated water and compressed air are run down long pipes into the sulfur deposits.

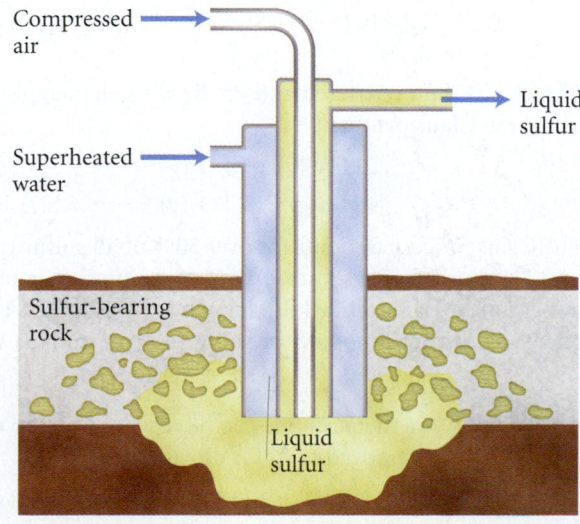

◀ FIGURE 22.23 **The Frasch Process**
The Frasch process extracts molten sulfur from the ground by forcing superheated water into beds of deposited solid sulfur.

▲ FIGURE 22.24 **Sulfur Deposits**   Some sulfur deposits can also be found at Earth's surface where volcanic activity has brought the element to the surface.

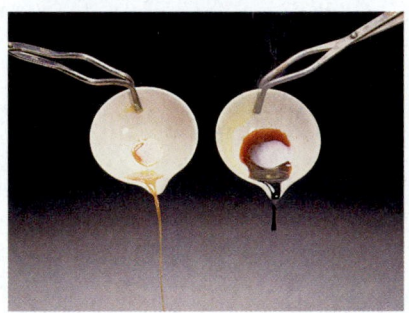

▲ Molten sulfur below 150 °C (left) and nearing 180 °C (right).

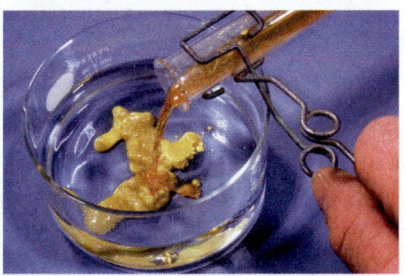

▲ FIGURE 22.25 **Quenching Liquid Sulfur**   When hot liquid sulfur is poured into cold water, the sulfur will quench into an amorphous solid.

The hot water melts the sulfur, which is pushed up to the surface along with the hot water through even larger pipes. Some sulfur deposits can also be found at Earth's surface where volcanic activity has made the sulfur accessible (Figure 22.24 ▲), and often in hot springs.

Sulfur has several allotropes, but the most common naturally occurring one is composed of an S_8 ring structure called cyclooctasulfur. Most of the different allotropes of sulfur are also ring structures with rings ranging in size from S_6 to S_{20}. When heated above its melting point of 112 °C, cyclooctasulfur forms a straw-yellow liquid with low viscosity. Above 150 °C, the rings begin to break, and the sulfur becomes a darker, more viscous liquid as the broken rings entangle one another. The color is darkest at 180 °C and the liquid becomes very viscous, pouring only very slowly, as can be seen in the photo at left. Above this temperature, however, the intermolecular forces between the S_8 chains weaken, and the liquid becomes less viscous again. If the hot liquid is poured into cold water, the sulfur will quench into an amorphous solid, as shown in Figure 22.25 ◄. Initially, this amorphous material is flexible like a plastic, but it eventually hardens into a brittle solid.

The Frasch process, while important, accounts for less than one-third of the world's sulfur production because sulfur is a by-product of a number of other industrial processes and can be recovered from them. For example, dihydrogen sulfide (H_2S), which is commonly called hydrogen sulfide, is a component of natural gas. The H_2S can be separated from the other components by passing the gas through organic solvents such as ethanolamine. The H_2S dissolves in the organic solvent as follows:

$$HOC_2H_4NH_2(l) + H_2S(g) \longrightarrow HOC_2H_4NH_3^+(\text{solvent}) + HS^-(\text{solute})$$
$$\underset{\text{ethanolamine}}{}$$

The H_2S is then recovered and oxidized to elemental sulfur through a two-step process called the **Claus process**.

$$2\,H_2S(g) + 3\,O_2(g) \longrightarrow 2\,SO_2(g) + 2\,H_2O(g)$$
$$4\,H_2S(g) + 2\,SO_2(g) \longrightarrow 6\,S(s) + 4\,H_2O(g)$$

The Claus process accounts for over 50% of all sulfur produced.

The third major source of sulfur production is metal sulfide minerals, such as the mineral iron(II) disulfide (also known as iron pyrite). Roasting iron pyrite in the absence of air will disproportionate the sulfur to form iron(II) sulfide and elemental sulfur.

$$2\,FeS_2(s) \xrightarrow{\text{heat}} 2\,FeS(s) + S_2(g)$$

Alternatively, the metal sulfide can be roasted in air to oxidize the metal sulfide. The sulfur can then be removed as sulfur dioxide.

$$2\,ZnS(s) + 3\,O_2(g) \longrightarrow 2\,ZnO(s) + 2\,SO_2(g)$$

Hydrogen Sulfide and Metal Sulfides

Hydrogen sulfide, which we just saw is a component of natural gas, is toxic. Hydrogen sulfide is formed by the reactions of anaerobic bacteria on organic substances. It is found in rotting vegetation and bogs, which are natural sources of H_2S for the atmosphere. Fortunately, humans can detect the odor of H_2S (a rotten egg smell) at concentrations as low as 0.02 ppm, which pose no threat to human health. Levels as low as 10 ppm, however, can cause nausea, and 100 ppm can cause death. However, the smell of H_2S becomes more difficult to detect at high levels because the H_2S also has an anesthetic effect that dulls the sense of smell. Consequently, although faint H_2S odor is no cause for concern, the sudden onset of strong H_2S odor is a reason to move quickly to fresh air.

We may initially think that hydrogen sulfide (H_2S) might be similar to water (H_2O), but it is not. Water has a larger bond angle (104.5°) than hydrogen sulfide (92.5°) and is much more polar. Because of its polarity, water forms strong hydrogen bonds, but hydrogen sulfide does not. In addition, the O—H bond is much stronger than the S—H bond. These differences result in a much lower boiling point and greater reactivity for hydrogen sulfide compared to water. Water is a stable molecule in the presence of air and oxygen. Hydrogen sulfide can burn in air, reacting with oxygen to form elemental sulfur or sulfur oxides.

EXAMPLE 22.5   Balancing of and Assigning Oxidation States to Sulfur Reactions

Write a balanced equation for the reaction of O_2 and H_2S to form elemental S (in the form of S_8). Identify the change of the oxidation state for S.

Solution

Write the skeletal equation. The products are elemental S_8 and H_2O.	$H_2S(g) + O_2(g) \longrightarrow H_2O(g) + S_8(s)$
Since hydrogen is initially balanced, balance S first, followed by H and O.	$8\,H_2S(g) + 4\,O_2(g) \longrightarrow 8\,H_2O(g) + S_8(s)$
Assign oxidation states to each element (see Section 4.9).	$\underset{+1\ -2}{8\,H_2S(g)} + \underset{0}{4\,O_2(g)} \longrightarrow \underset{+1\ -2}{8\,H_2O(g)} + \underset{0}{S_8(s)}$ $\longmapsto$ Oxidation $\longrightarrow$ The oxidation state of S changes from -2 in H_2S to 0 in S_8; therefore, S is oxidized.

For Practice 22.5

Write a balanced equation for reaction of oxygen with H_2S to form SO_2. Identify the change of the oxidation state for S.

Hydrogen sulfide can be formed from the reactions of metal sulfides with hydrochloric acid.

$$FeS(s) + 2\,HCl(aq) \longrightarrow FeCl_2(s) + H_2S(g)$$

Only a few of the metal sulfides, those with group 1A and 2A metals and Al, are very soluble in water. Some common metal sulfides and their solubility product constants are listed in Table 22.7 (on page 1016). The low solubility of these sulfides allows the use of H_2S as a good analytical method to determine whether metal ions are present in a solution. Sodium sulfide is used to precipitate toxic metals from industrial waste sources.

Metal sulfides have a number of industrial uses, mostly based on their toxicity to bacteria. For example, SeS_2 is used as a shampoo additive to kill bacteria and stop dandruff, and As_2S_3 is used to kill internal and skin parasites.

TABLE 22.7 Common Metal Sulfides			
Sulfide	Formula	Common Name	K_{sp} (at 25 °C)
Iron(II) disulfide	FeS_2	Pyrite	3.72×10^{-19}
Zinc sulfide	ZnS	Sphalerite	2.0×10^{-22}
Lead(II) sulfide	PbS	Galena	9.04×10^{-29}
Mercury(II) sulfide	HgS	Cinnabar	1.6×10^{-54}

Sulfur Dioxide

Sulfur dioxide is another toxic sulfur compound. Under standard conditions, it is a color-less, dense gas with an acidic taste. The acidic taste is from the reaction of the gas with the water in your mouth.

$$SO_2(g) + H_2O(l) \longrightarrow H_2SO_3(aq)$$

Sulfur dioxide is produced naturally by volcanic activity in which sulfides are oxidized at the high volcanic temperatures. Sulfur dioxide is also a pollutant that is formed in many industrial processes, such as coal and oil combustion and metal extraction. As we have seen, when the sulfur dioxide is emitted into the air, it reacts with oxygen and water to produce acid rain.

$$2\,SO_2(g) + 2\,H_2O(g) + O_2(g) \longrightarrow 2\,H_2SO_4(aq)$$

Sulfuric acid as a pollutant is destructive to living plants, animals, and man-made struc-tures (see Section 15.12).

To prevent the emission of SO_2 into the atmosphere, industrial processes "scrub" the gas to remove the SO_2. The exhaust gas flows through stacks lined with calcium carbon-ate that, when heated, captures sulfur oxides in calcium sulfate dust.

$$CaCO_3(s) \xrightarrow{\text{heat}} CaO(s) + CO_2(g)$$

$$2\,CaO(s) + 2\,SO_2(g) + O_2(g) \longrightarrow 2\,CaSO_4(s)$$

The $CaSO_4$ dust can be collected and disposed of. New uses for the tons of waste $CaSO_4$, such as fireproof insulation, are needed.

One use for SO_2 again capitalizes on its toxicity. Fruits and other vegetation are sprayed with a solution containing SO_2 to kill mold and preserve the fruit. As a result, the fruits and vegetables can be shipped throughout the world, making it possible for inhabi-tants of the Northern Hemisphere to have summer fruits (from the Southern Hemi-sphere) in the winter.

Sulfuric Acid

The most important use of sulfur and its compounds is the production of sulfuric acid. In fact, sulfuric acid is the most abundantly produced chemical in the world because it is a strong acid, a strong oxidizing agent, and a good dehydrating agent. It is also plentiful and inexpensive. Sulfuric acid has many uses in fertilizer, color dyes, petrochemicals, paints, plastics, explosives, battery, steel, and detergent industries, to name a few.

Pure H_2SO_4 melts at 10.4 °C and boils at 337 °C. At room temperature, it is an oily, dense liquid. Sulfuric acid reacts vigorously and exothermically with water. Pure or con-centrated H_2SO_4 must be added very slowly to water to avoid rapid heating, boiling, and splattering.

▲ FIGURE 22.26 Dehydration of Sucrose Sulfuric acid dehydrates sucrose by removing the hydrogen and oxygen as water molecules and leaving carbon behind. The porous carbon foam is formed because the reaction is very exothermic.

$$H_2SO_4(l) \xrightarrow{H_2O(l)} H_2SO_4(aq) \quad \text{highly exothermic}$$

The strong attraction between sulfuric acid and water makes sulfuric acid a very strong dehydrating agent. As shown in Figure 22.26 ◄, the affinity for water is strong enough to decompose some organic materials, such as sucrose.

$$C_{12}H_{22}O_{11}(s) + H_2SO_4(l) \longrightarrow 12\,C(s) + 11\,H_2O(g) + H_2SO_4(aq)$$

Sulfuric acid is produced industrially by a method known as the **contact process**, developed in the early twentieth century. In this method, elemental sulfur is first heated in air to form SO_2 gas, which is then heated in contact with a V_2O_5 catalyst to form SO_3 gas.

$$S(g) + O_2(g) \xrightarrow{\text{heat}} SO_2(g)$$

$$2\ SO_2(g) + O_2(g) \xrightarrow{V_2O_5 \text{ catalyst}} 2\ SO_3(g)$$

The SO_3 gas is then absorbed into concentrated sulfuric acid, producing a dense form of sulfuric acid called oleum, $H_2S_2O_7$, which gives H_2SO_4 when dissolved in water.

$$SO_3(g) + H_2SO_4(l) \longrightarrow H_2S_2O_7(l)$$

$$H_2S_2O_7(l) + H_2O(l) \longrightarrow 2\ H_2SO_4(aq)$$

As we have already seen, sulfuric acid is used in the production of fertilizer, which consumes a significant amount of the sulfuric acid produced.

22.9 Halogens: Reactive Elements with High Electronegativity

The halogens are all one electron short of a noble gas electron configuration. They are the most electronegative elements in their respective periods and are therefore very reactive. They are not naturally found in their elemental form but always occur within compounds. The source of most of the halogens (except fluorine) is the dissolved salts present in seawater. The major sources for fluorine are several different minerals, including fluorspar (CaF_2) and fluoroapatite [$Ca_{10}F_2(PO_4)_6$].

We have already seen some of the properties of the halogens, especially those that exhibit periodic trends, in Section 8.9. For example, the atomic radius of the halogens increases regularly from fluorine to iodine, as shown in Table 22.8. Due partly to its small size, fluorine has the highest electronegativity of all elements, and is always found in oxidation states of −1 or 0. The other halogens can be found with oxidation states ranging from −1 to +7. The positive oxidation states occur when the halogen bonds to more electronegative elements such as fluorine or oxygen.

TABLE 22.8 Selected Properties of the Halogens

Element	Melting Point (°C)	Boiling Point (°C)	Atomic Radius (pm)	Electronegativity
Fluorine	−219	−188	72	4.0
Chlorine	−101	−34	99	3.0
Bromine	−7	60	113	2.8
Iodine	114	185	133	2.5

EXAMPLE 22.6 Determining the Oxidation State of Halogens in Compounds

Calculate the oxidation state for Cl in the following compounds:

(a) ClO_3^- **(b)** HClO   **(c)** Cl_2

Solution

(a) For ClO_3^- each O atom has an oxidation state of −2 for a total of −6 for the three O atoms. Therefore, the Cl atom has to be +5 for the sum of the oxidation states to equal the charge of the ion (5 − 6 = −1).

(b) For HClO, the O atom has an oxidation state of −2 and the H atom has an oxidation state of +1. Therefore the Cl atom has to be +1 to have a neutral charge (1 + 1 − 2 = 0).

(c) For Cl_2 the Cl atoms are in their elemental state, so they have an oxidation state of 0.

For Practice 22.6

Calculate the oxidation state of Cl in ClO_4^- and Cl^-.

Elemental Fluorine and Hydrofluoric Acid

Fluorine is the most reactive element and forms binary compounds with all elements except He, Ne, and Ar. Fluorine even forms compounds with some of the noble gases, producing species such as XeF_2, XeF_6, and $XeOF_4$. The high reactivity of fluorine is related to several factors. First, the F—F bond is among the weakest halogen–halogen bonds, as shown in Table 22.9. When a halogen reacts with other substances, the halogen–halogen bond must break. The energy required to break that bond is small for F_2, allowing the resulting reaction to be more exothermic. Second, the small size of fluorine results in a high lattice energy for the ionic compounds that it forms. The high lattice energy means the compounds are very stable.

TABLE 22.9 Comparison of Halogen X—X Bond Energy

Halogen	F—F	Cl—Cl	Br—Br	I—I
Bond energy (kJ/mol)	159	243	193	151

The high reactivity of fluorine can be illustrated by its ability to burn many substances, such as iron and sulfur, that do not readily burn with oxygen.

$$Fe(s) + F_2(g) \longrightarrow FeF_2(s)$$

$$S(s) + 3\,F_2(g) \longrightarrow SF_6(g)$$

Fluorine gas even reacts with asbestos and glass, two materials commonly used as containers for reactive substances. Consequently, fluorine is normally held in metal containers composed of iron, copper, or nickel. These metals also react with fluorine, but a thin layer of the product coats the surface of the metal, protecting the underlying metal from further reaction.

Elemental fluorine is produced from the electrolysis of hydrofluoric acid, forming F_2 and H_2 gases.

Oxidation: $\qquad\qquad\qquad 2\,F^-(g) \longrightarrow F_2(g) + 2\,e^-$

Reduction: $\qquad\qquad 2\,H^+(g) + 2\,e^- \longrightarrow H_2(g)$

Gaseous hydrogen fluoride can be obtained from the reaction of the mineral fluorspar (CaF_2) with sulfuric acid.

$$CaF_2(s) + H_2SO_4(l) \longrightarrow 2\,HF(g) + CaSO_4(s)$$

In its solid form, HF has a crystal structure that contains zigzag chains of alternating H and F atoms. In aqueous solutions, HF is a weak acid ($K_a = 6.8 \times 10^{-4}$ for HF at 298 K). Like all anions in aqueous solution, the F^- ions from ionized HF are solvated by water molecules. However, the F^- ions can also associate with other HF molecules to form HF_2^-.

$$F^-(aq) + HF(aq) \longrightarrow HF_2^-(aq)$$

The structure of HF_2^-, shown below, is unique because it contains a bridging hydrogen atom (a hydrogen atom that essentially forms two bonds).

Hydrofluoric acid is a strong oxidizing agent and reacts with glass according to the following reactions:

$$SiO_2(s) + 4\,HF(aq) \longrightarrow SiF_4(g) + 2\,H_2O(l)$$

or

$$SiO_2(s) + 6\,HF(aq) \longrightarrow SiF_6^{2-}(aq) + 2\,H^+(aq) + 2\,H_2O(l)$$

As a result, hydrofluoric acid cannot be held in a glass container and is generally stored in plastic. The ability of HF to react with glass makes it useful in etching glass. The parts of the glass to be etched are left exposed, and the rest is masked with a nonreactive substance such as plastic. The surface is then exposed to hydrofluoric acid and the nonmasked glass etches away, leaving the desired pattern. Hydrofluoric acid is particularly dangerous because it quickly penetrates into tissues, damaging internal organs and bones. Direct exposure of just 2% of body surface area to concentrated hydrofluoric acid can be fatal.

Elemental Chlorine

Historically, the main source for chlorine has been seawater. Electrolysis of NaCl in seawater produces Cl_2 gas and H_2 gas.

$$2\,NaCl(aq) + 2\,H_2O(l) \xrightarrow{\text{electricity}} Cl_2(g) + 2\,NaOH(aq) + H_2(g)$$

Today much Cl_2 gas is produced and collected as a by-product of the various metal processing methods, such as in reduction of metal chlorides to form metals.

$$MgCl_2(l) \xrightarrow{\text{electricity}} Mg(s) + Cl_2(g)$$

Halogen Compounds

Halogens form ionic compounds with most metals; they form covalent compounds with many nonmetals, and they can bond with themselves to form interhalogen compounds. Here we look at some of these halogen compounds.

Interhalogen Compounds Covalent compounds composed of two different halogens are known as **interhalogen compounds** or *interhalides*. The general formula of these compounds is AB_n, where A is the larger halogen, B is the smaller halogen, and n can be 1, 3, 5, or 7. The smaller halogens surround the larger halogen in the AB_3 compounds. The only known AB_5 compounds contain fluorine as the smaller halogen (because the other halogens are too large to fit five of them around another halogen). The compound IF_7 is the only known interhalogen compound with $n = 7$. The large size of the iodine atom allows seven small fluorine atoms to surround it. The interhalides can be formed by allowing the elemental halogens to react with one another. Interhalide polyatomic ions, such as ICl_2^+ and ICl_4^-, also exist.

EXAMPLE 22.7 Formation of Interhalogen Compounds

Write a balanced equation for the formation of ClF_3 gas from the elemental halogens.

Solution

The unbalanced reaction is

$$Cl_2(g) + F_2(g) \longrightarrow ClF_3(g)$$

At least 2 ClF_3 molecules are formed from each Cl_2 molecule; therefore, add a 2 before ClF_3 and then a 3 before F_2. The balanced equation is

$$Cl_2(g) + 3\,F_2(g) \longrightarrow 2\,ClF_3(g)$$

For Practice 22.7

Write a balanced equation for the formation of IF_5 gas from the elemental halogens.

The geometry of the interhalides can be predicted from the valence shell electron pair repulsion model (VSEPR). Because the halogens do not form double or triple bonds, the shape of these compounds is relatively straightforward to determine, as shown in the following examples.

Steps in Determining the Shape of Interhalogen Compounds	**EXAMPLE 22.8** Molecular Shapes of Interhalogen Compounds Determine the molecular geometry of IBr_2^-.	**EXAMPLE 22.9** Molecular Shapes of Interhalogen Compounds Determine the molecular geometry of IF_7.
	Solution	Solution
Identify the central atom.	I	I
Draw the Lewis structure (see Section 9.5).	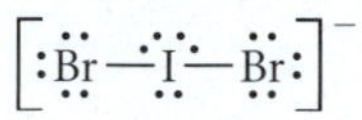 $$\left[:\ddot{B}r - \ddot{I} - \ddot{B}r: \right]^-$$	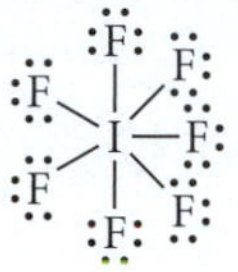
Count the number of bonds and lone pairs on the central atom.	The I atom has two bonds and three lone pairs for a total of five electron groups.	The I atom has seven bonds on the central atom for a total of seven electron groups.
Determine the electron geometry from the number of electron groups (see Section 10.4).	With five electron groups, the electron geometry is *trigonal bipyramidal*.	With seven electron groups, the electron geometry is *pentagonal bipyramidal*.
Determine the molecular shape from the number of bonds and lone pairs.	With two bonds the molecular geometry is *linear* because the three lone pairs will occupy equatorial positions.  	With seven bonds the molecular geometry will also be *pentagonal bipyramidal*. 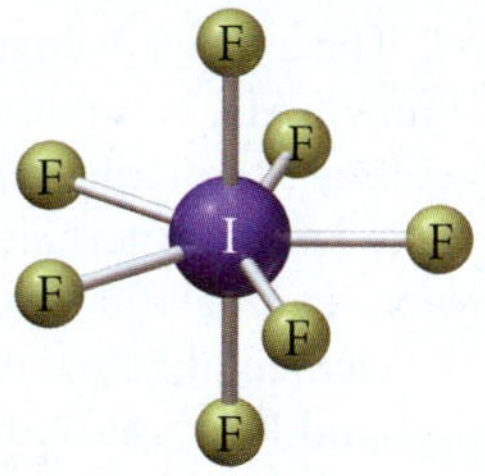

For Practice 22.8

Determine the electron geometry and molecular geometry of ICl_2^+.

For Practice 22.9

Determine the electron geometry and molecular geometry of BrF_5.

The only industrially useful interhalide is ClF_3, used in the nuclear energy industry to produce $UF_6(g)$, which is then used to separate ^{235}U (<1% of naturally occurring uranium) from ^{238}U. Uranium ores react with gaseous hydrogen fluoride to form $UF_4(s)$, which is then reacted with $ClF_3(g)$ to form the gaseous uranium compound.

$$UO_2(s) + 4\,HF(g) \longrightarrow UF_4(s) + 2\,H_2O(g)$$

$$UF_4(s) + ClF_3(g) \longrightarrow UF_6(g) + ClF(g)$$

Because of the difference in mass between the two isotopes, $^{238}UF_6$ effuses more slowly than $^{235}UF_6$ (see Section 5.9 for a description of effusion). The mixture of gases is allowed to flow through barriers with very small pores. Since $^{235}UF_6$ effuses more quickly, the initial flow of gas that exits the pores is enriched in $^{235}UF_6$. By repeating the process, the two isotopes can be nearly completely separated.

Halogen Oxides Most halogen oxides are unstable and many are explosive. A unique halogen oxide is OF_2, because oxygen usually has a negative oxidation state in its compounds but has a +2 oxidation state in OF_2 (due to the high electronegativity of fluorine).

A number of different chlorine oxides are known, including Cl_2O, ClO_2, Cl_2O_6, and Cl_2O_7. Chlorine dioxide is a powerful oxidizing agent used to bleach flour and wood pulp (to make white paper). Because ClO_2 is explosive, the gas is diluted with CO_2 or N_2 for safety. Some water treatment plants are now using ClO_2 for water disinfection in place of Cl_2. ClO_2 is produced by the reduction of sodium chlorite with Cl_2 or by the reduction of sodium chlorate with hydrochloric acid.

$$2\ NaClO_2(aq) + Cl_2(g) \longrightarrow 2\ NaCl(aq) + 2\ ClO_2(g)$$

$$2\ NaClO_3(aq) + 4\ HCl(aq) \longrightarrow 2\ ClO_2(g) + Cl_2(g) + 2\ H_2O(l) + 2\ NaCl(aq)$$

EXAMPLE 22.10 Indentifying Changes in Oxidation States.

Identify the change of oxidation state for Cl in the production of chlorine dioxide from sodium chlorite. Identify the oxidizing agent and the reducing agent.

Solution

First determine the oxidation state of Cl in each compound.

$$2\ \underset{+1\ +3\ -2}{NaClO_2(aq)} + \underset{0}{Cl_2(g)} \longrightarrow 2\ \underset{+1\ -1}{NaCl(aq)} + 2\ \underset{+4\ -2}{ClO_2(g)}$$

The Cl in the $NaClO_2$ was oxidized from +3 to +4 by the oxidizing agent Cl_2, which was reduced from 0 to −1 by the reducing agent $NaClO_2$.

For Practice 22.10

Identify the change of oxidation state for Cl in the production of chlorine dioxide from sodium chlorate. Identify the oxidizing agent and the reducing agent.

Chapter in Review

Key Terms

Section 22.2

main-group elements (990)

Section 22.3

silicates (992)
quartz (992)
silica (992)
aluminosilicates (993)
orthosilicates (993)
pyrosilicates (993)
pyroxenes (994)

Section 22.4

boranes (998)
closo-borane (998)

nido-boranes (998)
arachno-boranes (998)

Section 22.5

graphite (999)
diamond (999)
coal (1000)
coke (1000)
charcoal (1000)
activated carbon (1000)
soot (1000)
carbon black (1000)
fullerenes (1000)
nanotubes (1001)
carbides (1002)

ionic carbides (1002)
covalent carbides (1003)
metallic carbides (1003)
washing soda (1004)

Section 22.6

white phosphorus (1005)
red phosphorus (1005)
black phosphorus (1006)
ammonia (1006)
Haber–Bosch process (1006)
hydrazine (1006)
hydrogen azide (1007)
Ostwald process (1008)
phosphine (1009)

Section 22.7

ozone (1012)

Section 22.8

Frasch process (1013)
Claus process (1014)
contact process (1017)

Section 22.9

interhalogen compounds (1019)

Key Concepts

Bonding and Properties (22.2)

Main-group elements are defined by their electron configurations and their placement in the periodic table. The properties of the main-group elements show great diversity. Metals, nonmetals, and metalloids are all found among the main-group elements. Some of the main-group elements tend to form covalent bonds while others form ionic bonds.

The Most Common Matter: Silicates (23.3)

Silicates are covalent atomic solids that contain silicon, oxygen, and various metal atoms. Silicates are found in rocks, clays, and soils. Four oxygen atoms bond to silicon, forming a negatively charged polyatomic anion that has a structure with a tetrahedral shape. Various metal ions within the structure balance the charge of the compound. The

SiO_4 tetrahedrons can link to form chains, double chains, sheets, or even extended three-dimensional structures. The properties of the silicates depend on the connections between the silicate tetrahedrons. Because of the wide variety of combinations of tetrahedron connections and the many different metal ions that fit within the structure, an enormous variety of different silicate minerals exist in nature, making the silicate materials the most common structures found on Earth.

Boron (22.4)

Because of its small size and high electronegativity, boron behaves as a metalloid. The structure of elemental boron consists of icosahedron structures bonded together in various ways. Boron tends to form electron-deficient compounds. Compounds of boron and hydrogen form cluster compounds resembling spheres, cages, and nets.

Carbon (22.5)

Organic chemistry is based on the chemistry of carbon, yet carbon is also very important in many inorganic compounds and applications. Graphite and diamond are two well-known structures of elemental carbon, but many other forms of elemental carbon, such as carbon black, coke, and the newly discovered fullerenes have many industrial applications. Important types of inorganic carbon compounds include carbides and carbonates. Carbon can form carbides with metallic, covalent, or ionic properties. Carbon and its oxygen compounds are intimately involved in the functions of life.

Nitrogen and Phosphorus (22.6)

Nitrogen and phosphorus are two elements that have been known for over 200 years. Nitrogen and phosphorus both form compounds with oxidation numbers ranging from -3 up to $+5$. Both nitrogen and phosphorus compounds are very important for plant growth and their most important use is as fertilizers. The strong triple bond between nitrogen atoms in N_2 makes nitrogen from the atmosphere inaccessible for use by most plants in nature, so ingenious chemical processes have been devised to make nitrogen compounds that are more available to plants.

Oxygen (22.7)

Oxygen is the most common element on Earth. It is found in the atmosphere as the elemental gas and as many oxide gases. Oxygen is found in the ocean within water and in Earth's crust as silicate and oxide compounds. Oxygen is a strong oxidizing agent and forms compounds with $-\frac{1}{2}$, -1, or -2 oxidation states. Ozone, O_3, is a helpful molecule in the upper atmosphere, where it absorbs harmful ultraviolet radiation, but a harmful molecule at Earth's surface.

Sulfur (22.8)

More sulfuric acid is produced than any other chemical. Most of the sulfuric acid is used to make fertilizers. Other uses take advantage of its strong oxidation and dehydration properties. Elemental sulfur has several allotropes ranging from ring structures to chain structures and amorphous materials, depending upon the temperature.

Halogens (22.9)

The halogens are the most electronegative elements, so they are always found as compounds, usually ionic. When they bond with other electronegative elements, however, they can form covalent compounds. Fluorine has special chemical properties because it is the most electronegative element and is very small, making it a very strong oxidizing agent. Interhalogen compounds are formed between two halogens, with the larger halogen as the central atom of the structure.

Key Skills

Determining the Composition, Charge Balance, and Type of Silicates and Aluminosilicates (22.3)
  • Examples 22.1, 22.2, 22.3, 22.4 • For Practice 22.1, 22.2, 22.3, 22.4 • Exercises 17–26

Writing Equations for Reactions and Assigning Oxidation States (22.8, 22.9)
  • Examples 22.5, 22.6, 22.10 • For Practice 22.5, 22.6, 22.10 • Exercises 43–46, 55–58, 73, 74, 77, 78

Formation of Interhalogen Compounds and Their Structures (22.9)
  • Examples 22.7, 22.8, 22.9 • For Practice 22.7, 22.8, 22.9 • Exercises 75, 76, 80

Exercises

Review Questions

1. Why does BN form compounds similar to those of elemental carbon?

2. What is the main characteristic that determines whether or not an element is identified as a main-group element?

3. Does the metallic characteristic of a main-group element increase or decrease down a family? Explain why.

4. Why does silicon form only single bonds with oxygen but carbon, which is in the same family as silicon, form double bonds with oxygen in many compounds?

5. What is the difference between SiO_2 that is cooled slowly and SiO_2 that is cooled quickly?

6. What is the difference between a rock and a mineral?

7. Briefly define each of the following:
   a. orthosilicate   b. amphibole
   c. pyroxene   d. pyrosilicate
   e. feldspar

8. Why is boron oxide often added to silica glass?

9. Why does boron form electron-deficient bonds? Give an example.

10. Name three allotropes of crystalline carbon.

11. Explain why solid CO_2 is called dry ice.

12. Nitric acid and phosphoric acid are two major chemical products of the chemical industry. Describe some of the uses for these chemicals.

13. Give the typical concentration of oxygen in dry air.

14. Describe how nitrogen can be separated from the other components of air.

15. Earth's atmosphere originally did not contain oxygen. Explain how the atmosphere gained oxygen.

16. Name a benefit, a hazard, and a useful commercial application of ozone.

Problems by Topic

Silicates: The Most Abundant Matter in Earth's Crust

17. Silicon bonds to oxygen to form a tetrahedral shape in both the network covalent silica compound, SiO_2, and in ionic silicate compounds. What is the oxidation state of Si in each of these structures?
    a. silica compound, SiO_2 b. orthosilicates, SiO_4^{4-}
    c. pyrosilicates, $Si_2O_7^{6-}$

18. What is the oxidation state of Si in each of these structures?
    a. pyroxenes, SiO_3^{2-}
    b. amphiboles, $Si_4O_{11}^{6-}$
    c. phyllosilicates, $Si_2O_5^{2-}$

19. In the orthosilicate garnet, the formula unit has three SiO_4^{4-} units and is balanced by Ca^{2+} and Al^{3+} cations. Determine the formula unit of garnet.

20. In the pyroxene kanoite, the formula unit has two SiO_3^{2-} units and is balanced by manganese and magnesium ions. Determine the formula unit of kanoite. Assume that the oxidation state of Mn is +2.

21. Kaolin is a clay material that is a phyllosilicate. Use charge balancing to determine how many hydroxide ions are in the formula for kaolin, $Al_2Si_2O_5(OH)_x$.

22. Tremolite is a double-chain silicate in the amphibole class. Use charge balancing to determine how many hydroxide ions are in the formula for tremolite, $Ca_2Mg_5Si_8O_{22}(OH)_x$.

23. How are the silica tetrahedrons linked for $ZrSiO_4$? To what class of silicates does this compound belong?

24. How are the silica tetrahedrons linked for $CaSiO_3$? To what class of silicates does this compound belong?

25. Predict the structure and give the charges on the cations in one of the minerals in the hornblende family, $Ca_2Mg_4FeSi_7AlO_{22}(OH)_2$.

26. Predict the structure and give the charges on the cations in the mineral hedenbergite, $CaFeSi_2O_6$.

Boron, an Interesting Group 3A Element

27. A major source of boron is the mineral kernite, $Na_2[B_4O_5(OH)_4] \cdot 3\,H_2O$. Calculate how many grams of boron can be produced from 1.0×10^3 kg of a kernite-bearing ore if the ore contains 0.98% kernite by mass and the process has a 65% yield.

28. A mineral of boron that is not very common is ulexite, $NaCaB_5O_9 \cdot 8\,H_2O$. Calculate how many grams of boron can be produced from 5.00×10^2 kg of ulexite-bearing ore if the ore contains 0.032% ulexite by mass and the process has an 88% yield.

29. Explain why the bond angles in BCl_3 and NCl_3 are different.

30. Explain why the bond between boron and Cl in the molecule BCl_3 is shorter than would be expected from a single B—Cl bond.

31. Predict the number of vertices and faces on the following closo-boranes:
    a. $B_6H_6^{2-}$ b. $B_{12}H_{12}^{2-}$

32. Predict the number of vertices and faces on the following closo-boranes:
    a. $B_4H_4^{2-}$ b. $B_9H_9^{2-}$

33. Describe the differences between a closo-borane, a nido-borane, and an arachno-borane.

34. Describe how boron is used in the nuclear industry.

Carbon, Carbides, and Carbonates

35. Explain why the graphite structure of carbon allows graphite to be used as a lubricant, but the diamond structure of carbon does not.

36. Explain why graphite can conduct electricity, but diamond does not.

37. Describe the difference between regular charcoal and activated charcoal.

38. Explain why the structure of charcoal allows carbon to act as a good filter while the diamond structure does not.

39. Describe the difference between an ionic carbide and a covalent carbide. State which types of atoms will form these carbides with carbon.

40. Silicon carbide can be produced by heating silicone polymers, forming methane gas, hydrogen gas, and silicon carbide. Balance the reaction of heating $[(CH_3)_2Si]_8$ to form silicon carbide.

41. Using the phase diagram given in Section 11.8 state what happens to the phase of CO_2 during the following processes.
    a. reducing the pressure on solid CO_2 that is at $-80\ °C$
    b. decreasing the temperature on CO_2 gas that is held at a pressure of 20 atm
    c. increasing the temperature on solid CO_2 that is held at a pressure of 0.8 atm

42. Using the phase diagram given in Section 11.8 state what happens to the phase of CO_2 during the following processes.
    a. reducing the temperature from the critical point
    b. increasing the pressure on CO_2 gas that is held at a temperature of $-50\ °C$
    c. increasing the temperature on solid CO_2 that is held at a pressure of 20 atm

43. Predict the products for the following reactions and write a balanced equation.
    a. $CO(g) + CuO(s)$ b. $SiO_2(s) + C(s)$
    c. $S(s) + CO(g)$

44. Predict the products for the following reactions and write a balanced equation.
    a. $CO(g) + Cl_2(g)$ b. $CO_2(g) + Mg(s)$
    c. $S(s) + C(s)$

45. Give the oxidation state for carbon in:
    a. CO   b. CO_2 c. C_3O_2

46. Write a balanced reaction for the gas release reaction of Alka-Seltzer, sodium bicarbonate with citric acid, $C_6H_8O_7$. (The acid is a triprotic acid and consists of a chain of three carbon atoms each with a carboxylic acid group, COOH.)

Nitrogen and Phosphorus: Essential Elements for Life

47. Described what is meant by fixing nitrogen.

48. Explain why the diatomic nitrogen atom is unusable by most plants. Where do plants get nitrogen?

49. Describe the differences in the allotropes of white and red phosphorus. Explain why red phosphorus is more stable.

50. Describe how red and black phosphorus can be made from white phosphorus.

51. Saltpeter and Chile saltpeter are two important mineral sources for nitrogen. Calculate the mass percent of nitrogen in both minerals.

52. Apatite is a main mineral source for the production of phosphorus. Calculate the atomic and mass percent of P in the mineral apatite.

53. Using the tables in Appendix IIB, show whether or not hydrogen azide is unstable at room temperature compared to its elements, H_2 and N_2. Do any temperatures exist where hydrogen azide is stable?

54. Using the tables in Appendix IIB, show if dinitrogen monoxide is unstable at room temperature compared to its elements, O_2 and N_2. Do any temperatures exist where dinitrogen monoxide is stable?

55. Predict the products for the following reactions and write a balanced equation.
    a. $NH_4NO_3(aq)$ + heat   b. $NO_2(g) + H_2O(l)$
    c. $PCl_3(l) + O_2(g)$

56. Predict the products for the following reactions and write a balanced equation.
    a. $NO(g) + NO_2(g)$ b. heating PH_3
    c. $P_4(s) + 5 O_2(g)$

57. Rank the following nitrogen ions from the one with N in the highest oxidation state to the one with N in the lowest.
    $$N_3^-, N_2H_5^+, NO_3^-, NH_4^+, NO_2^-$$

58. Determine the oxidation state of N in the compounds of the following reaction for the formation of nitric acid. Identify the oxidizing agent and the reducing agent.
    $$3 NO_2(g) + H_2O(l) \longrightarrow 2 HNO_3(l) + NO(g)$$

59. Draw Lewis structures for the phosphorus halides PCl_3 and PCl_5. Describe their VSEPR shape.

60. Dinitrogen pentoxide is an ionic compound formed from the ions NO_2^+ and NO_3^-. Give the oxidation state of N in each ion and the VSEPR shape for each of the ions.

61. Ammonium carbonate is produced from the reaction of urea, $CO(NH_2)_2$, with water. Write a balanced equation for this reaction and determine how much urea is needed to produce 23 g of ammonium carbonate.

62. Explain why phosphine, PH_3, is less polar than ammonia.

63. Reacting oxygen with white phosphorus can form either P_4O_6 or P_4O_{10}. Explain the different conditions that will determine which product is formed.

64. P_4O_{10} is one of the most effective drying agents, having the ability to extract water from other molecules. The P_4O_{10} forms phosphoric acid. Write balanced reactions for the reaction of P_4O_{10} with:
    a. HNO_3, forming N_2O_5 b. H_2SO_4, forming SO_3

Oxygen

65. Name the major source for the element oxygen and describe how it is produced.

66. Explain why either greatly decreasing or increasing the percentage of oxygen in the atmosphere would be dangerous.

67. Identify the following molecules as oxides, peroxides, or superoxides:
    a. LiO_2 b. CaO c. K_2O_2

68. Identify the following molecules as oxides, peroxides, or superoxides:
    a. MgO b. Na_2O_2 c. CsO_2

Sulfur: A Dangerous but Useful Element

69. Explain why the viscosity of liquid sulfur increases with increasing temperature initially, but then decreases upon further increases in temperature.

70. Sulfur dioxide is a toxic sulfur compound. List one natural source and one industrial source that produces SO_2.

71. Calculate the maximum mass (in grams) of the following metal sulfides that can be dissolved in 1.0 L of a solution that is 5.00×10^{-5} M in Na_2S.
    a. PbS b. ZnS

72. A coal source contains 1.1% sulfur by mass. If 2.0×10^4 kg of coal is burned and forms oxides, calculate the mass of $CaSO_4(s)$ that is produced from "scrubbing" the SO_2 pollutant out of the exhaust gas. Assume that all of the sulfur in the coal is converted to calcium sulfate.

73. Write the equation for the reaction of roasting iron pyrite in the absence of air to form elemental sulfur. Calculate the volume of S_2 gas that can be produced from roasting 5.5 kg of iron pyrite. Assume that all of the sulfur in the iron pyrite is converted to S_2 gas. (Assume STP to calculate the gas volume.)

74. Write an overall reaction from the two steps in the Claus process. Calculate the volume of H_2S gas needed to produce 1.0 kg of $S(s)$. (Assume STP to calculate the gas volume.)

Halogens

75. Determine the oxidation state of Xe and give the VSEPR structure for the following compounds:
    a. XeF_2 b. XeF_6 c. $XeOF_4$

76. Describe the shapes of the following halogen compounds:
    a. BrF_4^- b. IF_3 c. BrO_2^- d. ClO_4^-

77. When chlorine is bubbled through a colorless aqueous solution containing bromide ions, the solution turns red and the chlorine is reduced. Write a balanced equation for this reaction and identify the oxidizing and reducing agent.

78. Carbon tetrachloride is produced by passing chlorine gas over carbon disulfide in the presence of a catalyst. The reaction also produces S_2Cl_2. Write a balanced reaction and identify which element is being oxidized and which element is being reduced.

79. If 55 g of $SiO_2(s)$ glass is placed into 111 L of 0.032 M HF, is there enough HF to dissolve all of the glass? Determine which substance is the limiting reagent and calculate how much of the other reagent is left if the reaction proceeds to completion.

80. Describe the difference in the types of bonds (single, double, triple) and the shapes of the following two iodine ions: ICl_4^- and IO_4^-.

Cumulative Problems

81. From the compositions of lignite and bituminous coal, calculate the mass of sulfuric acid that could potentially form as acid rain from burning 1.00×10^2 kg of each type of coal.

82. Calculate the volume of CO_2 released from heating and decomposing 88 g of sodium bicarbonate. (Assume standard pressure and temperature.)

83. All of the halogens can form oxoacids. The perhalic acids have the general formula of HXO_4. Explain why $HClO_4$ is a much stronger acid than HIO_4.

84. The halogens can form oxoacids with different amounts of oxygen. Explain why $HClO_4$ is a stronger acid than $HClO_2$.

85. Determine the ratio of effusion rates of HCl compared to each of the following gases:
    a. Cl_2　　　b. HF　　　c. HI

86. Calculate the ratio of effusion rates for the following pairs of gases:
    a. $^{238}UF_6$ and ClF　　　b. $^{238}UF_6$ and $^{235}UF_6$

87. Sodium peroxide is a very powerful oxidizing agent. Balance the reaction of sodium peroxide with elemental iron to give sodium oxide and Fe_3O_4.

88. Sulfur dioxide is a reducing agent and when it is bubbled through an aqueous solution containing Br_2, a red-colored solution, it reduces the bromine to colorless bromide ions and forms sulfuric acid. Write a balanced equation for this reaction and identify the oxidizing and reducing agent.

89. Using the molecular orbital model for a diatomic molecule, explain the different bond lengths for the ions of oxygen. Also state which ion is diamagnetic

Ion	O—O Bond Length (pm)
O_2^+	112
O_2	121
O_2^-	133
O_2^{2-}	149

90. The *closo*-borane with the formula $B_6H_6^{2-}$ has the six B atoms at vertices, forming an octahedron structure with eight faces. The formula for the number of sides is $2n - 4$, where n is the number of boron atoms. Determine the number of vertices and faces for the following *closo*-boranes.
    a. $B_4H_4^{2-}$　　　b. $B_{12}H_{12}^{2-}$

Challenge Problems

91. Calculate the standard enthalpy of reaction for reducing the different forms of iron oxide to iron metal and CO_2 from the reaction of the oxide with CO. Identify which reaction is the most exothermic per mole of iron and explain why.
    a. Fe_3O_4　　　b. FeO　　　c. Fe_2O_3

92. Balance the equation for the production of acetylene (C_2H_2) from the reaction of calcium carbide with water. If the acetylene were burned to form water and carbon dioxide, how many kilojoules of energy would be produced from the complete reaction of 18 g of calcium carbide?

93. Carbon suboxide, C_3O_2, is a linear molecule with the two oxygen atoms at the end and double bonds between each carbon and oxygen atom.
    a. Draw the Lewis structure for C_3O_2.
    b. State the type of hybridization of each carbon atom.
    c. Calculate the heat of reaction for the reaction of carbon suboxide with water to form malonic acid ($HO_2CCH_2CO_2H$). Hint: Each end of malonic acid has a carbon double bonded to an oxygen and bonded to a hydroxide.

94. Calcium carbonate is insoluble in water. Yet, it dissolves in an acidic solution. Calculate the standard enthalpy, entropy, and Gibbs free energy change for the reaction between solid calcium carbonate and hydrochloric acid. What drives the reaction, the enthalpy change or the entropy change?

95. When hydrazine is dissolved in water it acts like a base.
    $$N_2H_4(aq) + H_2O(l) \rightleftharpoons N_2H_5^+(aq) + OH^-(aq)$$
    $$K_{b_1} = 8.5 \times 10^{-7}$$
    $$N_2H_5^+(aq) + H_2O(l) \rightleftharpoons N_2H_6^{2+}(aq) + OH^-(aq)$$
    $$K_{b_2} = 8.9 \times 10^{-16}$$
    a. Calculate the K_b for the overall reaction of hydrazine forming $N_2H_6^{2+}$.
    b. Calculate K_{a_1} for $N_2H_5^+$.
    c. Calculate the concentration of hydrazine and both cations in a solution buffered at a pH of 8.5 for a solution that was made by dissolving 0.012 mol of hydrazine in 1 L of water.

96. Solid fuel in the booster rockets for spacecraft consists of aluminum powder as the fuel and ammonium perchlorate as the oxidizing agent:
    $$3\ NH_4ClO_4(s) + 3\ Al(s) \longrightarrow$$
    $$Al_2O_3(s) + AlCl_3(s) + 6\ H_2O(g) + 3\ NO(g).$$
    If a rocket launch burns 2.200×10^3 kg of aluminum, calculate the energy produced in joules. Calculate the volume of the gas produced assuming it was cooled back to 298 K at 1 atm. The standard enthalpy of formation of solid ammonium perchlorate is -295.3 kJ/mol.

Conceptual Problems

97. Explain why fine particles of activated charcoal can absorb more (as a filter) than large briquettes of charcoal.

98. The two major components of the atmosphere are the diatomic molecules of nitrogen and oxygen. Explain why pure nitrogen can be used as a protective atmosphere, but pure oxygen is much more reactive.

99. Explain why nitrogen can form compounds with many different oxidation numbers.

100. Describe how sodium dihydrogen phosphate can be used as a pH buffering agent.

101. Explain why H_2S has a different bond angle and is much less reactive than H_2O.

102. Explain why fluorine is found only with the oxidation state of -1 or 0, while the other halogens can be found in compounds with other oxidation states.

103. Why do some substances burn in fluorine gas even if they do not burn in oxygen gas?

104. Explain why SO_2 is used as a reducing agent but SO_3 is not.

105. Explain why the interhalogen molecule $BrCl_3$ exists but $ClBr_3$ does not.

23

Metals and Metallurgy

23.1 Vanadium: A Problem and an Opportunity

23.2 The General Properties and Natural Distribution of Metals

23.3 Metallurgical Processes

23.4 Metal Structures and Alloys

23.5 Sources, Properties, and Products of Some of the 3d Transition Metals

IF YOU LOOK AROUND YOUR HOUSE, classroom, or neighborhood, you will notice that much of what you see is made of metal or at least has some metal parts. Can you imagine life without metals? Without metals, we would have no skyscrapers with their rigid framework of steel beams, no automobiles with their metal engines and bodies, and no electricity transmitted through copper and aluminum wire. Metallurgy is in some ways both a very old and a very new science. The roots of chemistry are found in the distant past, as our primitive ancestors searched for and processed metals. Of the three prehistoric ages of humankind—the Stone Age, the Bronze Age, and the Iron Age—two are named after our quest for metals. But most metals are not found as elements on Earth; instead they are found in compounds, often scattered within other compounds. The mining, separating, and refining of the vast array of metals that are known today is a fairly new science. Many modern metals were not available even as recently as 200 years ago. In the middle of the nineteenth century, gold was cheaper than aluminum because aluminum was so difficult to refine, and titanium, which today is important in the aerospace industry, could not even be manufactured. In this chapter, we cover the area of chemistry known as metallurgy.

The metals are not presented immediately to the hand of man . . . but they are, for the most part, buried in darkness, in the bowels of the earth, where they are so much disguised, by combination and mixture with other substances, that they often appear entirely unlike themselves.

—WILLIAM HENRY (1774–1836)

Oil can be contaminated with vanadium, which appears to have been the metal used for oxygen transport by some ancient life forms.

23.1 Vanadium: A Problem and an Opportunity

Recently, members of a university's chemistry and engineering departments met with representatives of an oil company to discuss how the faculty and students could cooperate on some industrial research. A major topic was vanadium contamination in oil, a real problem of continuing concern. Because vanadium is toxic, the U.S. government regulates the quantity of vanadium that can be emitted by industrial processes, so oil companies have to figure out environmentally acceptable ways to remove vanadium from oil.

Vanadium is a rare element, making up only about 0.015% of Earth's crust (by mass). It is a soft white metal with high ductility, which means that it can be easily drawn into narrow wires and thin sheets. Because of the high reactivity of pure vanadium, this element is

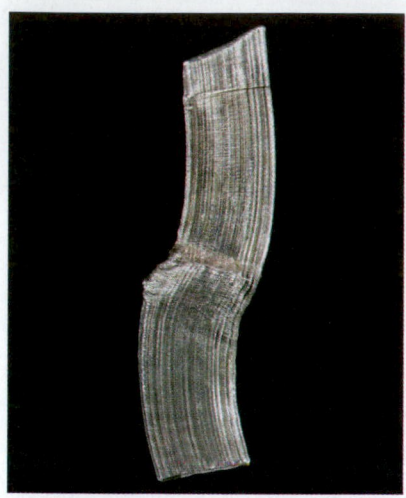

▲ Vanadium is a soft white metal.

naturally found within compounds. However, vanadium compounds are not normally found in the rocks or soils near oil fields, so why would this rare, reactive metal occur in oil? The answer lies in the biology of ancient life-forms. Although most modern animals use iron in hemoglobin to facilitate the transport of oxygen through their bloodstreams (see Section 1.1 and 14.1), and a few (such as lobsters) use copper for this function, it appears that some extinct animals used vanadium. Even today, one group of simple marine organisms, the tunicates or sea squirts, are suspected of using vanadium compounds for oxygen transport. Therefore, it appears that the source of vanadium in some crude oil is the very animals from which the oil was formed.

During the refining process, crude oil is heated to carry out the reactions that form the different petroleum products. If some of the vanadium-containing oil is burned during heating, the vanadium forms vanadate compounds, ionic compounds with vanadium oxide polyatomic ions such as VO_4^{3-} and VO_3^-. These compounds, which have low melting points, can dissolve the protective oxide coatings on stainless steel containers, allowing the steel to corrode and ruining the containers in which the oil is stored. If the vanadium compounds are not removed from the oil before storage, the problem is passed along to the customer. In addition, vanadium and many of its compounds are toxic, and so oil companies cannot simply remove the vanadium compounds and dump them into the environment.

The presence of vanadium in crude oil could, however, turn out to be profitable for the oil companies. If vanadium could be economically recovered from oil, then oil could become a major source of vanadium, a valuable metal with several important industrial uses including the production of iron alloys and sulfuric acid.

The recovery of vanadium from oil involves the field of metallurgy, the topic of this chapter. **Metallurgy** includes all the processes associated with mining, separating, and refining metals and the subsequent production of pure metals and mixtures of metals called *alloys* (which we will define in Section 23.4). In this chapter we explore the general properties and natural distribution of metals; several different categories of metallurgical processes, including pyrometallurgy, hydrometallurgy, electrometallurgy, and powder metallurgy; the structures and alloys formed by metals; and the metallic products and applications for several selected materials.

23.2 The General Properties and Natural Distribution of Metals

Metals share a number of properties. All metals are opaque (you cannot see through them), and they are good conductors of heat and electricity. They generally have high malleability (the ability to be bent or hammered into desired forms) and ductility (the ability to be drawn into wires). These properties can be explained by the bonding theories that we have already examined for metals: the electron sea model and band theory. In the electron sea model, discussed in Section 9.11, each metal atom within a metal sample donates one or more electrons to an *electron sea*, which can then flow within the metal. In band theory, presented in Section 11.13, the atomic orbitals of the metal atoms are combined, forming *bands* that are delocalized over the entire crystalline solid; electrons can move freely within these bands. The mobile electrons in both of these models give metals many of their shared properties.

Each metal, however, is also unique, and even their shared properties can vary within a range. For example, some metals, such as copper, silver, and aluminum, have much higher thermal and electrical conductivities than do other metals. Table 23.1 shows the thermal conductivity and electrical resistivity (a low resistivity corresponds to a high conductivity) of a few metals. Notice that both properties can vary by factors of 10 or more. Most metals are very strong and are malleable and ductile. But lead is a soft, weak metal, and chromium is a brittle metal that will not bend without breaking. When we think about the shared properties of metals, therefore, we must allow for a range of properties as well as some unique ones.

TABLE 23.1	Thermal Conductivity and Electrical Resistivity of Several Metals	
Metal	Thermal Conductivity (W/cm · K)	Electrical Resistivity ($\mu\Omega$ · cm)
Ag, silver	4.29	1.59
Cu, copper	4.01	1.67
Fe, iron	0.804	9.71
V, vanadium	0.307	24.8

Over 75% of the elements in the periodic table are metals, yet metals make up only about 25% of the mass of Earth's crust (see Figure 22.2 for the composition of Earth's crust). Earth's core is thought to be composed of iron and nickel, but because the core is so far from the surface, these metals are not accessible. The most abundant metal on Earth is aluminum, a main-group metal. Several alkali and alkaline earth metals (calcium, sodium, potassium, and magnesium) make up more than 1% of Earth's crust. Iron, which makes up about 5% of the crust, is the only transition metal that accounts for more than 0.1% of the crust. Of the first-row transition metals, titanium, chromium, iron, nickel, copper, and zinc are all plentiful enough to be important industrial materials.

Only a few metals are found naturally as elements; these include nickel, copper, palladium, silver, platinum, and gold. Because of their low reactivity, these metals are often referred to as the "noble metals." They are often concentrated within mountainous or volcanic regions and are found as small isolated veins of the metal within a rock matrix.

Most of the rest of the metals occur naturally in positive oxidation states within mineral deposits. **Minerals** are homogeneous, naturally occurring, crystalline inorganic solids. A rock that contains a high concentration of a specific mineral is called an **ore**. Metallurgical processes are designed to separate useful minerals from other, nonuseful material in the ore.

The main source for the alkali metals are chloride minerals, such as halite (sodium chloride) and sylvite (potassium chloride). Halite can be found in large deposits from dried ancient oceans or can be precipitated by evaporating ocean water. The main source for some metals are oxide minerals, such as hematite (Fe_2O_3), rutile (TiO_2), and cassiterite (SnO_2). The ores containing these minerals are unevenly distributed throughout the Earth. For example, in North America there are large deposits of hematite but no substantial deposits of cassiterite. Sulfides are the most important minerals for many metals, such as galena (PbS), cinnabar (HgS), sphalerite (ZnS), and molybdenite (MoS_2).

The main mineral source for some metals can be complex. For example, the main *mineral* sources for vanadium are vanadinite [$Pb_5(VO_4)_3Cl$] and carnotite [$K_2(UO_2)_2(VO_4)_2 \cdot 3\,H_2O$], which is also a main mineral source for uranium. Vanadium, as we have seen, is also found in crude oil. There are no specific minerals that contain the metal radium. Yet radium can substitute for uranium within uranium-containing minerals. Therefore, the small amount of radium that replaces uranium in carnotite is also the major source of radium. Tantalum and niobium were named after the Greek god Tantalus and his daughter Niobe. These two metals are always found together in mixed deposits of columbite [$Fe(NbO_3)_2$] and tantalite [$Fe(TaO_3)_2$]. The minerals are mined together and the elements are later separated by fractional crystallization.

▲ Gold is one of the few metals that can be found as an element in nature, often in veins like the one shown above.

▲ Most NaCl salt is produced by trapping ocean water in shallow basins and letting the water evaporate, leaving salt.

23.3 Metallurgical Processes

To obtain elemental metals, the mined material must first be physically *separated* into its metal-containing and nonmetal-containing components. Then, the elemental metal must be extracted from the compounds in which it is found, a process called **extractive metallurgy**. In this section, we look at several metallurgical processes, including pyrometallurgy, hydrometallurgy, electrometallurgy, and powder metallurgy. Finally, the metal must be **refined**, a process in which the crude material is purified.

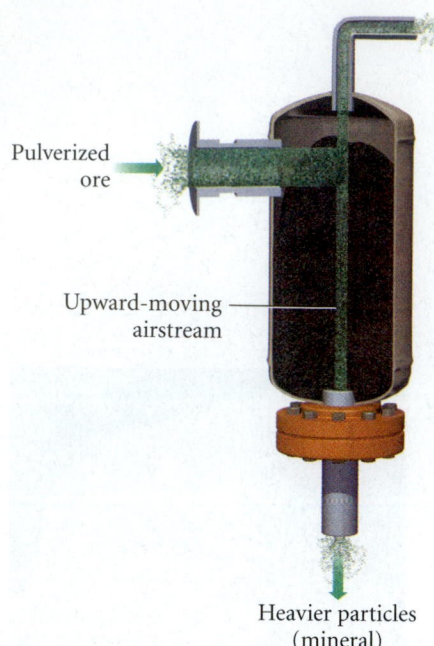

Lighter
particles
(gangue)

Pulverized
ore

Upward-moving
airstream

Heavier particles
(mineral)

▲ FIGURE 23.1 Separation by Air
An industrial cyclone can separate the
crushed light particles of gangue from the
heavier metal-bearing particles.

Pyrometallurgy contains the Greek stem
pyro, meaning fire or heat.

Hydrated compounds are discussed in
Section 3.5.

Separation

The first step in processing metal-containing ores is to crush the ore into smaller particles. The particles that contain the minerals are then separated from the undesired material, called **gangue** (pronounced "gang"), usually by physical methods. For example, in some cases, a strong air current is all that is needed to separate the different particles. A cyclone of wind can lift the gangue away from the metal-containing particles, as shown in Figure 23.1 ◄. If the minerals are magnetic, magnets can be used to separate the minerals. Electrostatic forces can even be used to separate polar minerals from nonpolar gangue.

Sometimes solutions are used to separate the minerals from the gangue, as shown in Figure 23.2 ▼. A wetting agent (or detergent) that preferentially attaches to the mineral surfaces can be added to a mixture of the mined material and water. Air is then blown through the mixture, forming bubbles and a froth. Because of the wetting agent, the minerals segregate into the froth where they can be collected. The gangue separates into the solution.

Pyrometallurgy

Once the mineral is separated from the gangue, the elemental metal must be extracted from the mineral. There are several techniques by which this separation can be achieved. In **pyrometallurgy**, heat is used to extract a metal from its mineral. Different heating conditions have different effects on the mineral.

Calcination is the heating of an ore in order to decompose it and drive off a volatile product. For example, when carbonate minerals are heated, carbon dioxide is driven off, as shown in the following examples:

$$PbCO_3(s) \xrightarrow{\text{heat}} PbO(s) + CO_2(g)$$

$$4\,FeCO_3(s) + O_2(g) \xrightarrow{\text{heat}} 2\,Fe_2O_3(s) + 4\,CO_2(g)$$

Many minerals are found in a hydrated form (that is, they contain water). Calcination can also drive off water, as in the following reaction:

$$Fe_2O_3 \cdot 2\,Fe(OH)_3(s) \xrightarrow{\text{heat}} 2\,Fe_2O_3(s) + 3\,H_2O(g)$$

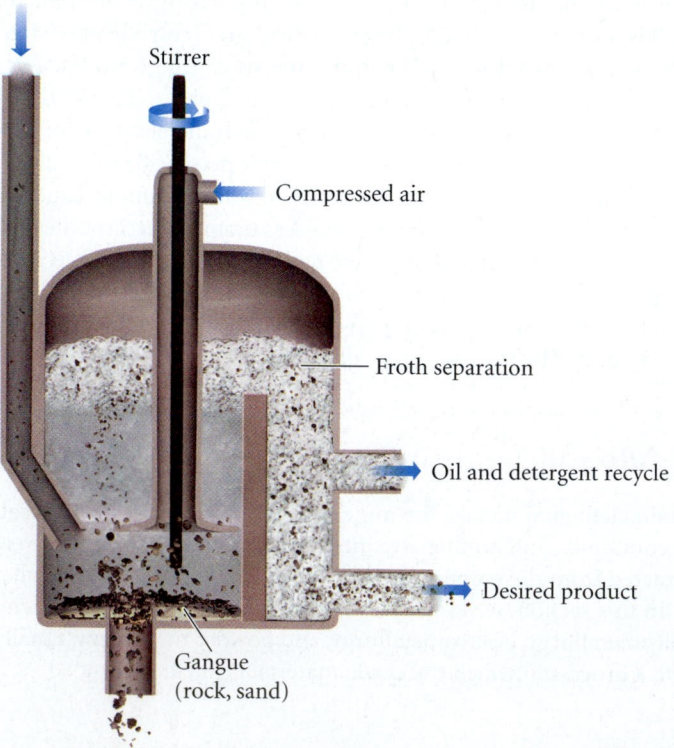

Water/ore/oil/detergent mixture

Stirrer

Compressed air

Froth separation

Oil and detergent recycle

Desired product

Gangue
(rock, sand)

▶ FIGURE 23.2 Separation by
Using a Solution   A wetting agent
helps the minerals attach to the froth in a
bubbled solution. The minerals in the
froth are separated from the gangue remaining in the solution.

Heating that causes a chemical reaction between the furnace atmosphere and the mineral is called **roasting**. Roasting is particularly important in processing sulfide ores. The ores are heated in the presence of oxygen, thereby converting the sulfide into an oxide and emitting sulfur dioxide. For example, the roasting of lead(II) sulfide occurs by the following reaction:

$$2\,PbS(s) + 3\,O_2(g) \xrightarrow{\text{heat}} 2\,PbO(s) + 2\,SO_2(g)$$

In some cases, especially with the less active metals such as mercury, roasting the sulfide produces the pure metal.

$$HgS(s) + O_2(g) \xrightarrow{\text{heat}} Hg(g) + SO_2(g)$$

When roasting forms a liquid product, which often helps in separation, it is called **smelting**. Consider for example the smelting of zinc oxide:

$$ZnO(s) + C(s) \xrightarrow{\text{heat}} Zn(l) + CO(g)$$

The gaseous carbon monoxide separates from the liquid zinc, allowing the metal to be recovered. In some cases, a *flux* must be added to the mixture during smelting to help separate the two materials. The **flux** is a material that will react with the gangue to form a substance with a low melting point. For example, oxides of silicon within gangue can be liquefied by reaction with calcium carbonate according to the following reaction:

$$\underset{\text{gangue}}{SiO_2(s)} + \underset{\text{flux}}{CaCO_3(s)} \longrightarrow CO_2(g) + \underset{\text{slag}}{CaSiO_3(l)}$$

The waste liquid solution that is formed between the flux and gangue is usually a silicate material called a **slag**. The liquid metal and the liquid slag have different densities and therefore separate. Holes are tapped at different heights into the side of the container holding the liquid metal and slag, allowing the more dense liquid to flow out of the lower tap holes and the less dense liquid to flow out of the higher tap holes.

Hydrometallurgy

The use of an aqueous solution to extract metals from their ores is known as **hydrometallurgy**. One of the earliest examples of hydrometallurgy is a process used to obtain gold. Gold can be found in its elemental state, but it often occurs as very small particles mixed with other substances. The gold can be separated out of the mixture by selectively dissolving it into solution, a process called **leaching**. Solid gold reacts with sodium cyanide to form a soluble gold complex.

$$4\,Au(s) + 8\,CN^-(aq) + O_2(g) + 2\,H_2O(l) \longrightarrow 4\,Au(CN)_2^-(aq) + 4\,OH^-(aq)$$

The impurities are then filtered out of the solution and the gold is reduced back to elemental gold with a reactive metal such as zinc:

$$2\,Au(CN)_2^-(aq) + Zn(s) \longrightarrow Zn(CN)_4^{2-}(aq) + 2\,Au(s)$$

This process for obtaining gold has been practiced for many years and has often resulted in the contamination of streams and rivers with cyanide. New alternatives, using the thiosulfate ion ($S_2O_3^-$), are being investigated to replace it.

Different acid, base, and salt solutions are sometimes used to selectively separate out metal-bearing minerals. For example, sulfuric acid is used to separate the copper and iron from the mineral chalcopyrite, $CuFeS_2$, and a sodium chloride solution is used to separate the lead from the insoluble mineral anglesite, $PbSO_4$.

$$2\,CuFeS_2(s) + H_2SO_4(aq) + 4\,O_2(g) \longrightarrow 2\,CuSO_4(aq) + Fe_2O_3(s) + 3\,S(s) + H_2O(l)$$

$$PbSO_4(s) + 4\,NaCl(aq) \longrightarrow Na_2(PbCl_4)(aq) + Na_2SO_4(aq)$$

Hydrometallurgy is often more economical than pyrometallurgy, because of the high energy costs associated with reaching the elevated temperatures needed for calcination and roasting.

See Section 18.8 for a description of electrolysis.

Electrometallurgy

In **electrometallurgy**, electrolysis is used to produce metals from their compounds. A well-known electrometallurgical process is the production of aluminum metal using the Hall process. The main source of aluminum is bauxite, a hydrated oxide with the formula $Al_2O_3 \cdot n\,H_2O$. A hydrometallurgical process, called the *Bayer process*, separates the bauxite from the iron and silicon oxide with which it is usually found. In this process, the bauxite is digested with a hot concentrated aqueous NaOH solution under high pressure. The aluminum oxide dissolves, leaving the other oxides as solids.

$$Al_2O_3 \cdot n\,H_2O(s) + 2\,OH^-(aq) + 2\,H_2O(l) \longrightarrow 2\,Al(OH)_4{}^-(aq)$$

The basic aluminum solution is separated from the oxide solids, and then the aluminum oxide is precipitated out of solution by neutralizing it [recall the pH dependence of the solubility of $Al(OH)_3$ from Section 16.8]. Calcination of the precipitate at temperatures greater than 1000 °C yields anhydrous alumina (Al_2O_3). Electrolysis is then used to reduce the aluminum out of the aluminum oxide. Because Al_2O_3 melts at such a high temperature (greater than 2000 °C), however, the electrolysis cannot be carried out on molten Al_2O_3. Instead, in the Hall process the Al_2O_3 is dissolved into molten cryolite (Na_3AlF_6), and graphite rods are used as electrodes to carry out the electrolysis in the liquid mixture, as illustrated in Figure 23.3 ▼. The carbon that composes the graphite electrodes is oxidized by the dissolved oxygen ions in the molten salt and converted to carbon dioxide. The aluminum ions dissolved in the molten salt are reduced to molten aluminum, which sinks down to the bottom of the cell and is removed.

Oxidation: $C(s) + 2\,O^{2-}(\text{dissolved}) \longrightarrow CO_2(g) + 4\,e^-$

Reduction: $3\,e^- + Al^{3+}(\text{dissolved}) \longrightarrow Al(l)$

Another important use of electrometallurgy is the refinement of copper. The copper must first, however, be extracted from its minerals. The most abundant copper source is the mineral chalcopyrite, $CuFeS_2$. First, the mineral is converted to CuS by roasting the chalcopyrite in air. During this process, the iron also forms oxides and sulfides. Silica is added to form an iron silicate slag, which is then removed. The remaining copper sulfide is reduced with oxygen to form copper metal, but the metal is not very pure. Electrolysis is used to refine (or purify) the copper.

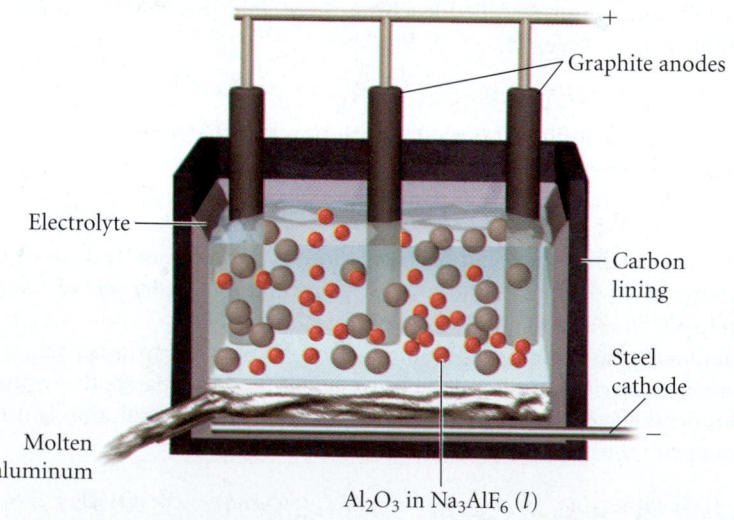

Graphite anodes

Electrolyte

Carbon lining

Steel cathode

Molten aluminum

Al_2O_3 in Na_3AlF_6 (l)

▲ **FIGURE 23.3 The Hall Process**   Using the Hall process, aluminum metal is produced by reducing the aluminum ions in alumina. Graphite electrodes act as reducing agents as the carbon is oxidized to form carbon dioxide by the dissolved oxide ions in the molten cryolite.

In the electrolysis cell used to purify copper, both the anode and the cathode are made of copper, as shown in Figure 23.4 ▶. The anode is the impure copper (to be purified) and the cathode is a thin sheet of pure copper. As the current flows through the cell, the copper from the anode oxidizes and dissolves in a copper sulfate solution. It then plates out as pure copper on the cathode. The impurities in the copper anode separate from the copper during electrolysis because, even though the more active metals also oxidize from the anode, they stay in solution and do not plate out on the cathode. The less active metals do not oxidize at all and simply fall to the bottom of the cell as the copper is dissolved from the anode. The sludge at the bottom of the electrolysis cell contains many precious metals, including gold and silver. About one-quarter of the U.S. production of silver is from the impurities recovered from the refinement of copper.

$$Cu^{2+} + 2e^- \longrightarrow Cu$$
Cathode

$$Cu \longrightarrow Cu^{2+} + 2e^-$$
Anode

| Thin sheet of pure copper | Anode sludge | $CuSO_4$, H_2SO_4 solution | Slab of impure copper |

▲ **FIGURE 23.4 Copper Electrolysis Cell**   Copper is refined by electrolysis. The impure copper is oxidized at the anode and then reduced to form pure metal on the cathode. Many precious metals from the impure copper collect in the sludge at the bottom of the electrolysis cell.

Powder Metallurgy

Powder metallurgy, first developed in the 1920s, is a process by which metallic components are made from powdered metal. In powder metallurgy, micron-sized metal particles are pressed together under high pressures to form the desired component. The component is then heated (sintered). The sintering process occurs below the melting point of the powder, but at a temperature that is high enough for the particles to fuse together, strengthening the metal and increasing its density.

Originally, iron powder from mill scrap was used in powder metallurgy. The scrap was primarily iron oxide that would fall off the steel as it was being milled. The iron oxide dust would be heated in a hydrogen atmosphere to reduce the oxide to iron particles. The powdered metal is called *iron sponge* because of the many holes formed in the particles when the oxygen escapes.

In the 1960s a new method for the development of powdered metal, called water atomization, was introduced by the A. O. Smith Company in the United States. Pure metal is melted and a small stream of the liquid is allowed to flow from the bottom of the container of molten metal. A high-pressure blast of cold water hits the stream, breaking it into small droplets that quickly solidify. The powdered metal made in this way results in more smooth and dense particles than the sponge powder from oxide scrap.

Products made by powder metallurgy are an important part of the metal industry. In 2002, over 700,000 tons of iron parts were produced this way, with over 500 million pounds of powdered iron parts being used in the auto industry. The average automobile has about 40 pounds of these metal parts. Beyond powdered iron metal, many copper, bronze, carbide, and brass parts are made through powder metallurgical processes.

Powder metallurgy offers several advantages over traditional casting or milling of metal. For example, there is no scrap involved in making metal components through powder metallurgy—the part can be pressed directly into the desired shape. Intricate teeth on gears and multiple holes can be designed into the press and therefore do not have to be machined after production. Making cast metal objects from metals with high melting points, such as molybdenum and tungsten, can be difficult because of the high temperatures necessary to melt the metal. Using the powder avoids the need for high temperatures.

▲ These metal products were all made by powder metallurgy.

23.4 Metal Structures and Alloys

The structures of metals can be described as the closest packing of spheres, first discussed in Section 11.11. Elemental metals generally crystallize in one of the basic types of crystal lattices, including face-centered cubic, body-centered cubic, and hexagonal closest

packed. The crystal structure of a metal may change, however, as a function of temperature and pressure. Table 23.2 shows the crystal structures for the 3*d* transition metals at
atmospheric pressure.

TABLE 23.2 **The Crystal Structures of the 3*d* Elements**		
Metal	Natural Crystal Structure at Room Temperature	Other Crystal Structures at Different Temperatures and Pressures
Sc	Hexagonal closest packed	Face-centered cubic, body-centered cubic
Ti	Hexagonal closest packed	Body-centered cubic above 882 °C
V	Body-centered cubic	
Cr	Body-centered cubic	Hexagonal closest packed
Mn	Alpha complex body-centered cubic form	Beta simple cubic form above 727 °C
		Face-centered cubic above 1095 °C
		Body-centered cubic above 1133 °C
Fe	Body-centered cubic	Face-centered cubic above 909 °C
		Body-centered cubic above 1403 °C
Co	Hexagonal closest packed	Face-centered cubic above 420 °C
Ni	Face-centered cubic	
Cu	Face-centered cubic	
Zn	Hexagonal closest packed	

Alloys

An **alloy** is a metallic material that contains more than one type of element. Some alloys
are simply solid solutions, while others are specific compounds with definite ratios of the
component elements. Alloys have metallic properties, but they can consist of either two or
more metals or a metal and a nonmetal. Alloys can be broadly classified as substitutional
or interstitial. In a **substitutional alloy**, one metal atom substitutes for another in the
crystal structure. The crystal structure may either stay the same upon the substitition, or
it may change to accommodate the differences between the atoms. In an **interstitial alloy**,
small, usually nonmetallic atoms fit in between the metallic atoms of a crystal. The alloy
still maintains its metallic properties with these interstitial atoms in the structure.

Substitutional Alloys: Miscible Solid Solutions

In order for two metals to form a substitutional alloy, the radii of the two metal atoms
must be similar, usually within 15% of each other. For example, the atomic radii of copper and nickel are both 135 pm and both of the elements form the face-centered cubic
structure. Thus, either metal can easily replace the other in the metal crystal structure.

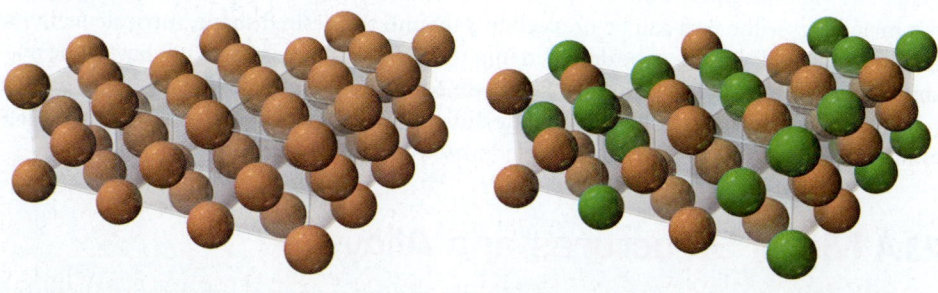

▲ Copper crystallizes in the face-centered
structure shown here.

▲ In a nickel and copper alloy, nickel atoms
simply substitute for some of the copper atoms.

Figure 23.5 ▶, shows a phase diagram for a copper and nickel alloy. This phase diagram, called a *binary* phase diagram, is different from those that we discussed in Section 11.8, which show the phases of a pure substance at different pressures and temperatures. This diagram shows the different phases for a mixture at different *compositions* and temperatures. The *x*-axis shows the composition (in this case, the mole percent of nickel in the alloy, with the left side representing pure copper and the right side representing pure nickel). The *y*-axis shows the temperature.

Pure copper melts at 1084 °C, as indicated by the change from solid to liquid at 0% nickel. Pure nickel melts at 1455 °C, as indicated by the change from solid to liquid at 100% nickel. The area on the diagram above the line connecting the melting points of copper and nickel represents a liquid solution of the two metals. The area below that line represents a solid solution of the two metals. Any ratios of copper and nickel can form the face-centered cubic structure with the copper and nickel atoms fitting in the crystal.

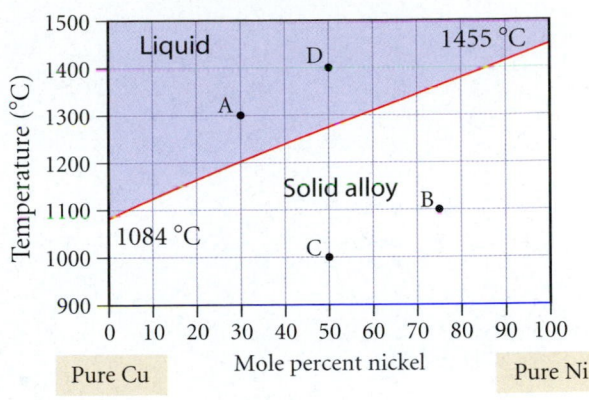

▲ **FIGURE 23.5 Cu–Ni Phase Diagram** Because copper and nickel have similar crystal structures and similar size, they form a miscible solid solution. A solution forms at all compositions from pure copper to pure nickel.

Guidelines for Interpreting a Phase Diagram	**EXAMPLE 23.1** Alloy Compositions in a Miscible Solid Solution	**EXAMPLE 23.2** Alloy Compositions in a Miscible Solid Solution
	Determine the composition and phase present at point A on Figure 23.5.	Determine the composition and phase present at point B on Figure 23.5.
Obtain the temperature and composition of the alloy.	Point A represents 30 mol % Ni at 1300 °C.	Point B represents 75 mol % Ni at 1100 °C.
Identify the phase.	The point is above the melting point line, so the phase is liquid.	The point is below the melting point line, so the phase is solid.
Identify the amount of copper and nickel in the phase.	This liquid phase is made up of 30 mol % Ni, so it is 70 mol % Cu.	This solid phase is made up of 75 mol % Ni, so it is 25 mol % Cu.
	For Practice 23.1 Determine the composition and phase present at point C on Figure 23.5.	**For Practice 23.2** Determine the composition and phase present at point D on Figure 23.5.

The chromium and vanadium phase diagram in Figure 23.6, on page 1036, shows that complete mixture also occurs between Cr and V atoms in the crystal. Both Cr and V form the body-centered cubic crystal, and the atoms are close in size. Yet the phase diagram reveals an important difference compared to the Cu and Ni alloy: the melting temperature does not vary in a uniform way from the lower melting point of Cr to the higher melting point of V. The alloys with compositions that are intermediate between the two pure metals melt at lower temperatures than either pure metal. The lowest melting point is 1750 °C at a composition of 30 mol % vanadium. Solid compositions that melt at temperatures lower than either of the pure metals, as well as compositions that melt at temperatures higher than either of the two metals, are common in these types of phase diagrams.

Alloys with Limited Solubility

Nickel crystallizes in the face-centered cubic structure and chromium in the body-centered cubic structure. Because of their different structures, these two metals cannot form a miscible solid solution throughout the entire composition range. At some intermediate

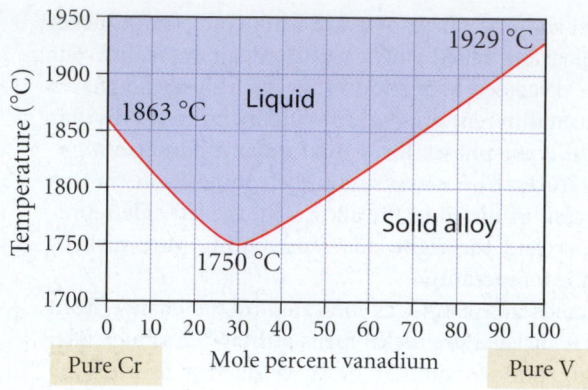

▲ **FIGURE 23.6 Cr–V Phase Diagram** In a binary phase diagram of two metals that form a miscible solid solution, an intermediate composition can have either the highest or lowest melting point. In the chromium and vanadium phase diagram an intermediate composition has the lowest melting point.

composition, the structure has to change. Figure 23.7 ▼ shows the nickel and chromium phase diagram from about 700 °C up to 1900 °C. Notice that the diagram has two different solid phases: face-centered cubic and body-centered cubic. From pure nickel (0 mol % chromium) to about 40–50 mol % chromium, the structure is face-centered cubic. In this structure, Cr atoms substitute for Ni atoms in the face-centered cubic structure of nickel. However, beyond a certain percentage of chromium (which depends on temperature), the structure is no longer stable. At 700 °C, about 40 mol % Cr can fit in the crystal, but at 1200 °C about 50 mol % Cr can fit in. Adding additional Cr beyond these points results in a different phase.

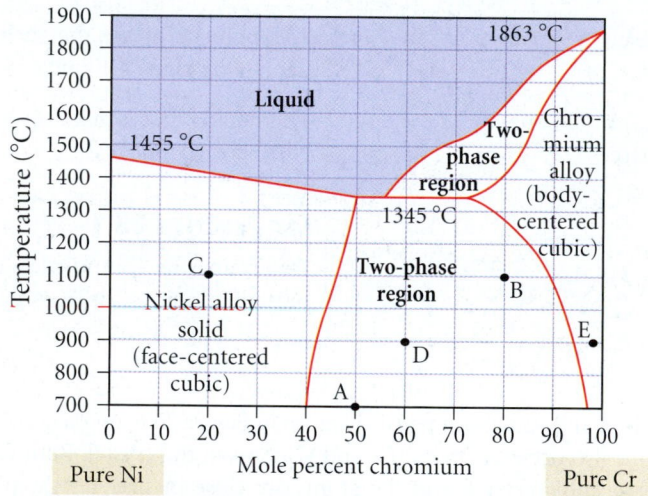

▲ **FIGURE 23.7 Cr–Ni Phase Diagram** Because chromium and nickel form solids with different crystal structures, the two metals do not form solid solutions at all compositions. A two-phase region exists at compositions between the possible compositions of the two different structures. In the two-phase region, both crystal structures coexist in equilibrium.

At the other end of the diagram (nearly pure chromium), and at 700 °C, only a small amount of nickel can be substituted into the body-centered cubic structure of the chromium. As the temperature rises to 1300 °C, however, about 20 mol % nickel can be accommodated into the chromium structure. The region in between the two phases is called a **two-phase region**. At these compositions, the two phases (nickel-rich face-centered cubic and chromium-rich body-centered cubic) exist together. The amount of each phase depends upon the composition of the alloy.

The composition and relative amounts of the two different phases that coexist in a two-phase region can be determined from a phase diagram. Point A on Figure 23.7 in the Cr–Ni phase diagram represents 50% composition at 700 °C, and both of the phases are present. Some of the Cr atoms have substituted into the nickel-rich face-centered cubic structure, but there is too much Cr to all fit into the crystal. So the leftover Cr atoms form the chromium-rich body-centered cubic structure with a small number of Ni atoms in the crystal.

The two crystals that exist at 700 °C are (1) the nickel-rich face-centered cubic structure with 40 mol % Cr; and (2) the chromium-rich body-centered cubic structure with 5 mol % Ni. The 50% composition on the phase diagram has just slightly more Cr atoms than can fit into the nickel structure. So most of the crystals in the two-phase region are the nickel-rich face-centered cubic structure with only a small amount of the chromium-rich body-centered cubic structure, as determined by a method called the *lever rule*. The **lever rule** tells us that in a two-phase region, whichever phase is closest to the composition of the alloy is the more abundant phase. In this example, the 50 mol % Cr composition is closer to the 40 mol % Cr composition of the nickel-rich face-centered cubic phase than the 95 mol % Cr composition of the chromium-body-centered cubic phase, so more face-centered cubic crystals are present than body-centered cubic crystals.

Process for Interpreting the Phases and Compositions in a Phase Diagram as Described by the Lever Rule	**EXAMPLE 23.3** Alloy Compositions in a Solid Solution with Limited Solubility Determine the composition, relative amounts, and phases present at point B on Figure 23.7.	**EXAMPLE 23.4** Alloy Compositions in a Solid Solution with Limited Solubility Determine the composition, relative amounts, and phases present at point C on Figure 23.7.
Obtain the temperature and composition of the alloy.	Point B represents 80 mol % Cr at 1100 °C.	Point C represents 20 mol % Cr at 1100 °C.
Identify the phase(s).	Point B is located in the two-phase region, consisting of the nickel-rich face-centered cubic structure and the chromium-rich body-centered cubic structure.	Point C is located in the one-phase region of the nickel-rich face-centered cubic structure.
Identify the amount of Ni and Cr in the phases.	The nickel-rich face-centered cubic structure has 45 mol % Cr and 55 mol % Ni and the chromium rich body-centered cubic structure has 90 mol % Cr and 10 mol % Ni.	The nickel-rich face-centered cubic structure at this point has 20 mol % Cr and 80 mol % Ni.
Identify the relative amounts of the phases.	At a composition of 80 mol % Cr, the composition is closer to the Cr-rich body-centered cubic phase, so there is more of this phase than there is of the Ni-rich face-centered cubic phase.	There is only one phase. It is 100 mol % of the Ni face-centered cubic phase.
	For Practice 23.3 Determine the composition, relative amounts, and phases present at point D on Figure 23.7.	**For Practice 23.4** Determine the composition, relative amounts, and phases present at point E on Figure 23.7.

Interstitial Alloys

In contrast to substitutional alloys, where one metal atom substitutes for another in the lattice, interstitial alloys contain atoms of one kind that fit into the holes, or interstitial sites, of the crystal structure of the other. In metals with the interstitial elements hydrogen, boron, nitrogen, or carbon, the alloy that results retains its metallic properties.

Closest packed crystal structures have two different types of holes between the atoms in the crystalline lattice. An **octahedral hole**, shown in Figure 23.8 ◀, exists in the middle of six atoms on two adjacent closest packed sheets of metal atoms. The hole is located directly above the center of three closest packed metal atoms in one sheet and below the three metal atoms in the adjacent sheet. This configuration of metal atoms is the same as a regular octahedral configuration that can be described as four atoms in a square plane with one atom above and one atom below the square. The size of an octahedral hole can be therefore calculated by determining the size of a hole on a square of four atoms in a plane, as shown in Figure 23.9 ◀. Any of the four corners of the square is the 90° angle of a right triangle formed from the adjacent two sides and the diagonal line that runs across the center of the square. According to the Pythagorean theorem, the length of the diagonal line (c) is related to the length of the sides (a and b) as follows:

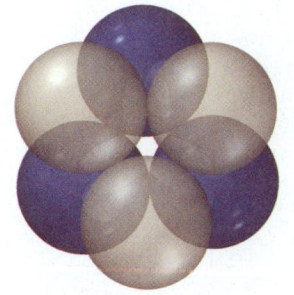

▲ **FIGURE 23.8** Octahedral Holes in Closest Packed Crystals Octahedral holes are found in a closest packed structure. The octahedral hole is surrounded by six of the atoms in the closest packed structure.

$$c^2 = a^2 + b^2 \qquad [23.1]$$

If we let r_m equal the metal atom radius, the lengths of the sides are $2r_m$. We can therefore substitute $2r_m$ for both a and b:

$$c^2 = (2r_m)^2 + (2r_m)^2$$
$$c^2 = 8r_m^2$$
$$c = 2.828 r_m \qquad [23.2]$$

We can see from Figure 23.9 that length of the diagonal (c) is twice the radius of the atoms plus twice the radius of the hole:

$$c = 2r_m + 2r_{hole} \qquad [23.3]$$

Combining Equations 23.2 and 23.3, we get the following important result:

$$2.828 r_m = 2r_m + 2r_{hole}$$
$$0.828 r_m = 2r_{hole}$$
$$r_{hole} = 0.414 r_m \qquad [23.4]$$

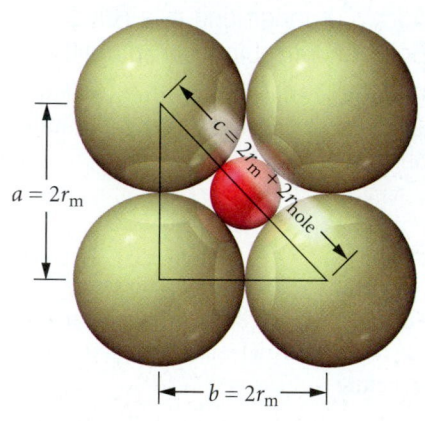

▲ **FIGURE 23.9** A Different View of an Octahedral Hole An octahedral hole can be viewed as the area in the middle of a square plane of atoms, with one additional atom above the hole and one additional atom below the hole, accounting for the six close atoms. The diagonal of the square is equal to the radius of the two corner atoms plus the diameter of the hole. The length of the diagonal is related to the lengths of the sides of the square by the Pythagorean theorem.

The octahedral hole, surrounded by six metal atoms, has a radius that is 41.4% of the metal atom radius. By contrast, the hole in the center of a cube in the simple cubic structure has a radius that is 73% of the metal atom radius. The *number* of octahedral holes in a closest packed structure is equal to the number of metal atoms.

The second type of interstitial hole in a closest packed structure is a **tetrahedral hole**, which is formed directly above the center point of three closest packed metal atoms in one plane and below a fourth metal atom located directly above the center point in the adjacent plane, as shown in Figure 23.10 ▶. The number of tetrahedral holes in a closest packed structure is equal to *twice* the number of metal atoms. Because this hole is surrounded by only four atoms, the hole is smaller than the octahedral interstitial hole. Geometric considerations similar to those used previously for the octahedral hole show that the tetrahedral hole has radius that is 23% of the metal atom radius.

Interstitial alloys form when the small nonmetallic atoms fit within the octahedral or tetrahedral holes of the crystalline lattice of the metal. The formulas for these alloys depend both on the type of hole occupied by the nonmetallic atom and on the fraction of holes occupied. For example, titanium and carbon form an alloy with a closest packed structure for titanium in which all of the octahedral holes are filled with carbon atoms. Since the number of octahedral holes in a closest packed structure is equal to the number of atoms in the structure, the ratio of carbon atoms to titanium atoms must be 1:1 and

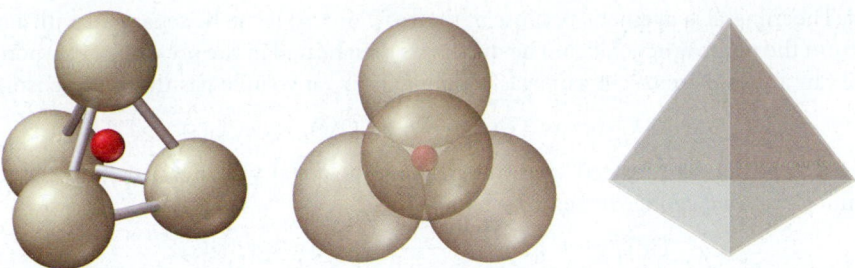

▲ **FIGURE 23.10 A Tetrahedral Hole in a Closest Packed Crystal** Tetrahedral holes are found in a closest packed structure. The tetrahedral hole is surrounded by four of the atoms in the closest packed structure.

the corresponding formula is therefore TiC. In the compound formed between molybdenum and nitrogen, by contrast, only one-half of the octahedral holes in the closest packed structure of Mo are filled with N. Therefore, the formula for this compound must be Mo_2N. Table 23.3 shows the formulas and relative number of holes filled for several different interstitial alloys.

TABLE 23.3 Formulas of Several Interstitial Alloys

Compound	Type of Interstitial Hole Occupied	Fraction of Holes Occupied	Formula
Titanium carbide	Octahedral	All	TiC
Molybdenum nitride	Octahedral	One-half	Mo_2N
Tungsten nitride	Octahedral	One-half	W_2N
Manganese nitride	Octahedral	One-quarter	Mn_4N
Palladium hydride	Tetrahedral	One-quarter	Pd_2H
Titanium hydride	Tetrahedral	All	TiH_2

❧ Conceptual Connection 23.1 Interstitial Alloys

An interstitial alloy contains a nonmetal (X) that occupies one-eighth of the tetrahedral holes of the closest packed structure of the metal (M). What is the formula of the alloy?

Answer: M_4X. Since there are twice as many tetrahedral holes as metal atoms in a closest packed structure, and since one-eighth of them are occupied by nonmetal atoms, there must be one-quarter as many nonmetal atoms as metal atoms.

23.5 Sources, Properties, and Products of Some of the 3*d* Transition Metals

In this section we examine major sources, interesting properties, and important products of several of the 3*d* transition metals, specifically, titanium, chromium, manganese, cobalt, copper, nickel, and zinc. We also survey the different metallurgical methods used to separate and refine them. The variety of uses for these metals demonstrates their varied properties.

Titanium

Titanium is the ninth most abundant element in Earth's crust, and the fourth most abundant metal. Titanium was discovered in 1791, but the pure metal was not isolated until 1910. The principal minerals of titanium are rutile (TiO_2) and ilmenite ($FeTiO_3$). Another source of titanium is coal ash, the residue from burned coal. The black shiny mineral ilmenite is found in granite deposits along the North Atlantic coast, often within silica sand. The mineral is magnetic because of the Fe^{2+} ions, so it can be separated with a magnet from the nonmagnetic silica. The mineral is then heated in the presence of carbon and under an atmosphere of chlorine gas, forming $TiCl_4$, a volatile gas that can be isolated.

$$FeTiO_3(s) + 3\,Cl_2(g) + 3\,C(s) \longrightarrow 3\,CO(g) + FeCl_2(s) + TiCl_4(g)$$

The $TiCl_4$ gas is then reacted with hot magnesium metal turnings, forming elemental titanium as a solid sponge material.

$$TiCl_4(g) + 2\,Mg(s) \longrightarrow 2\,MgCl_2(l) + Ti(s)$$

Titanium is very reactive, readily oxidizing in the presence of oxygen and even nitrogen. Consequently, the elemental titanium is **arc-melted**—a method in which the solid metal is melted with an arc from a high-voltage electric source in a controlled atmosphere to prevent oxidation—and then collected in a water-cooled copper pot. Because of titanium's high reactivity, any further processing must be done under a protective atmosphere of an inert gas, such as Ar, to prevent oxidation.

Despite its reactivity, solid titanium is highly resistant to corrosion in air, acid, and seawater because it quickly reacts with oxygen to form an oxide that coats the surface, preventing further oxidation of the underlying metal. Consequently, titanium is often used in the production of ship components such as propeller shafts and rigging. Titanium is also very strong and light; it is stronger than steel but less than half as dense. Although titanium is more dense than aluminum, it is twice as strong. For these reasons, titanium is used in the airline industry for the production of jet engine parts. When titanium is alloyed with 5% aluminum and trace amounts of Fe, Cr, and Mo, the resulting metal is able to retain strength under higher temperature.

The largest use of titanium, however, is as titanium dioxide (TiO_2), which forms a clear crystal but a brilliant white powder. Most white paints use TiO_2 as the pigment. Titanium dioxide is made by reacting sulfuric acid with ilmenite, which dissolves the titanium into solution.

$$FeTiO_3(s) + 3\,H_2SO_4(l) \longrightarrow FeSO_4(aq) + Ti(SO_4)_2(aq) + 3\,H_2O(l)$$

Neutralizing the solution with strong base forms titanium oxide.

$$Ti^{4+}(aq) + 4\,OH^-(aq) \longrightarrow TiO_2(s) + 2\,H_2O(l)$$

Calcination of the oxide dries it to form TiO_2 rutile crystals. Large rutile crystals are sometimes used as gems because they resemble diamonds.

Chromium

The name chromium comes from the Greek root *chroma*, which means color. The different compounds of chromium are brightly colored, as tabulated in Table 23.4 and shown in Figure 23.11 ▶. The main ore source of chromium is chromite, $FeCr_2O_4$.

TABLE 23.4 The Colors of Various Chromium Compounds

Compound	Color
Chromates (CrO_4^{2-})	Yellow
Chromium(II) iodide	Red-brown
Chromium(III) iodide	Green-black
Chromium(II) chloride	White
Chromium(III) chloride	Violet
Dichromates ($Cr_2O_7^{2-}$)	Orange
Chromium(III) oxide	Deep green
Chrome alum	Purple
Chromium(VI) oxychloride (CrO_2Cl_2)	Dark red
Chromium(II) acetate	Red

There are no appreciable sources of chromium ores found in the United States. Chromium metal is produced by reducing the ore with aluminum.

$$3\,FeCr_2O_4(s) + 8\,Al(s) \longrightarrow 4\,Al_2O_3(s) + 6\,Cr(s) + 3\,Fe(s)$$

Metallic chromium is a white, hard, lustrous, and brittle metal. The metal easily dissolves in acids such as hydrochloric acid and sulfuric acid, but it does not dissolve in nitric acid. The main use of chromium is to produce steel alloys called *stainless steels.* Reducing chromite with carbon produces ferrochrome, an alloy that is added to steel.

$$FeCr_2O_4(s) + 4\,C(s) \xrightarrow{\text{heat}} Fe_xCr_y(s) + 4\,CO(g)$$

The chromium reacts with any oxygen in the steel to protect the iron from rusting. Chromium compounds have also been used extensively in metal coatings such as paints, because the chromium helps to rustproof the underlying metal, and in wood preservatives, because these compounds kill many of the bacteria and molds that rot wood. Chromium compounds are being used in fewer applications today, however, because of their toxicity and potential carcinogenicity. Nonetheless, because of their great corrosion resistance, chromate coatings are still used on large outdoor steel structures such as bridges, and chromate paints are used to mark streets (on the pavement) and street signs.

Because chromium has the electron configuration [Ar] $4s^1 3d^5$, with six orbitals available for bonding, it can have oxidation states from +1 to +6. Low-oxidation-state chromium exists primarily as the *cation* in salt compounds such as $Cr(NO_3)_3$ and $CrCl_3$. High-oxidation-state chromium can be found within the polyatomic *anions* of salts. The most important compounds with chromium in the +6 oxidation state are the chromate and dichromate compounds. In an acidic solution (below pH 6), the orange-red dichromate ion $Cr_2O_7^{2-}$ is more stable. In more basic solutions (above pH 6), the yellow chromate ion CrO_4^{2-} dominates. Adding acid to a solution containing the chromate ion produces the dichromate ion.

▲ **FIGURE 23.11 Chromium Compounds** Chromium compounds tend to be brightly colored.

$$2\,H^+(aq) + 2\,CrO_4^{2-}(aq) \longrightarrow Cr_2O_7^{2-}(aq) + H_2O(l)$$

The chromate ion has a tetrahedral arrangement of oxygen atoms around the chromium atom. The dichromate ion has one bridging oxygen between the two tetrahedra around the chromium ions.

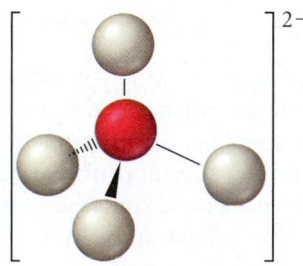

Chromate (CrO_4^{2-})

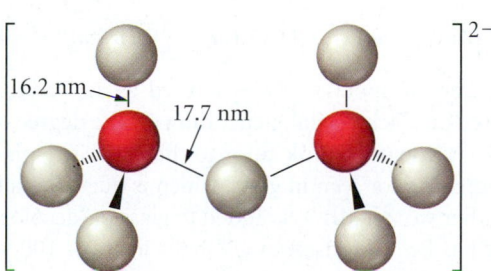

16.2 nm   17.7 nm

Dichromate $(Cr_2O_7^{2-})$

The high-oxidation-state chromates and dichromates are very strong oxidizing agents (that is, they are easily reduced). Consequently, they are used as coatings to prevent oxidation of other metal surfaces. The chromates react with the atoms on the surface of the other metals, forming a strongly bonded gel-like film. The film is nonmetallic and bonds very effectively with paint and resins that are applied over the film. The oxidizing

▲ Ammonium dichromate burns without needing additional oxygen.

power of dichromate is demonstrated by the ability of ammonium dichromate to sustain combustion without any additional oxygen. Once ignited, ammonium dichromate burns with visible flames, giving off a smoke of green chromium oxide dust.

$$(NH_4)_2Cr_2O_7(s) \xrightarrow{\text{heat}} 4\,H_2O(g) + N_2(g) + Cr_2O_3(s)$$

Manganese

Manganese, with the electron configuration $[Ar]\,4s^2 3d^5$, exhibits the widest range of oxidation states, from $+1$ to $+7$. The most common natural sources of manganese are pyrolusite (MnO_2), hausmannite (Mn_3O_4), and rhodochrosite ($MnCO_3$) minerals. Calcination of rhodochrosite produces manganese(IV) oxide.

$$MnCO_3(s) + {}^1\!/_2\,O_2(g) \xrightarrow{\text{heat}} MnO_2(s) + CO_2(g)$$

Then the manganese(IV) oxide or pyrolusite minerals are reacted with active metals such as Al or Na to produce the elemental metal.

$$3\,MnO_2(s) + 4\,Al(s) \longrightarrow 3\,Mn(s) + 2\,Al_2O_3(s)$$

Pyrolusite, however, is often found as an impure mineral, containing mixtures of MnO and Fe_2O_3. Heating the mineral in the presence of carbon can reduce the mineral to form an alloy of manganese and iron, ferromanganese.

$$MnO_2(s) + MnO(s) + Fe_2O_3(s) + n\,C(s) \longrightarrow Mn_xFe_yC_z(s) + n\,CO(g)$$

The ferromanganese generally contains about 5–6% carbon and is used as an alloying material in steel. Manganese is usually added to steel alloys to change the physical properties of the steel. For example, added manganese makes the steel easier to deform at high temperatures. Therefore, adding manganese to steel helps the rolling and forging steps of steel production. Steel alloys containing about 12% manganese are used for military armor and industrial applications such as bulldozer blades. Manganese is also used to strengthen copper, aluminum, and magnesium alloys.

Manganese is a reactive metal that dissolves in most acids. When heated in the presence of air, it forms the various manganese oxides, including MnO, Mn_3O_4, and MnO_2. When heated in pure oxygen, the high-oxidation-state oxide Mn_2O_7 is also formed. At high oxidation states, the manganese compounds are good oxidizing agents. Dissolving MnO_2 into a solution of hydrochloric acid oxidizes the chloride, producing chlorine gas.

$$MnO_2(s) + 4\,HCl(aq) \longrightarrow 2\,Cl_2(g) + 2\,H_2O(l) + Mn^{2+}(aq)$$

The permanganate ion (MnO_4^-) has an oxidation state of $+7$ for Mn and is also an important oxidizing agent. The permanganate ion can even oxidize hydrogen peroxide, which is itself generally used as an oxidizing agent.

$$2\,MnO_4^-(aq) + 3\,H_2O_2(aq) + 2\,H^+(aq) \longrightarrow 2\,MnO_2(s) + 3\,O_2(g) + 4\,H_2O(g)$$

The compound MnO_2 is used as an additive to glass. By itself, MnO_2 is either a brown or black crystal, depending on the degree of hydration. Yet, when added to a silica glass, it imparts a pink color to the glass. The pink color is useful because it counteracts the green color seen in glass, which is due to small concentrations of impure iron oxides. In other words, MnO_2 is added to glass to "decolorize" it. Old manganese-containing glass that has been exposed to UV light for over 100 years develops a slight purple tint. This purple color is due to the oxidation of MnO_2 in the glass to Mn(VII) oxides.

Cobalt

Cobalt ore is often found within the ores of other metals, such as iron, nickel, lead, and silver. Cobalt's most common ores are sulfide minerals, such as cobaltite (CoAsS), which is collected as a by-product in the extraction processes of the other metal ores. There are no large deposits of cobalt ores in the United States, but recent surveys have indicated cobalt-rich deposits in the Pacific Ocean near the Hawaiian Islands.

Cobalt, like iron and nickel, is **ferromagnetic** and is important in the production of magnets. Like the paramagnetic materials described in Section 8.7, the atoms in ferromagnetic materials contain unpaired electrons. In ferromagnetic materials, however, these electrons can all align with their spins oriented in a the same direction, resulting in permanent magnetic field.

Magnets are becoming more important in industrial and military applications. In the United States, concerns have arisen over the lack of domestic sources of cobalt for these applications. New and stronger magnetic materials that do not require cobalt have been developed, but these require neodymium, which is found mostly in China. In the late 1990s the last U.S. producer of these magnets moved its facilities to China, closer to the source of the raw materials.

Cobalt is also an important additive to high-strength steels. Carbaloy, a mixture composed primarily of cobalt metal with grains of tungsten carbide, is a very tough material. The strength of the cobalt and hardness of the carbides make this a good material for industrial cutting and abrasion applications. In addition, cobalt forms many compounds with brilliant blue colors and has been used in making pigments and inks. Cobalt compounds are also essential for health, because cobalt is part of vitamin B_{12}. Deficiencies of B_{12} can cause anemia.

Copper

Because copper can be found in its elemental form, it was one of the first elements to be isolated and used by humans—copper products have been employed for over 10,000 years. Ancient civilizations used copper to form tools. The earliest known artifacts produced by the smelting of copper are at a site in Tepe Yahya, Iran, dating from about 3800 B.C. The discovery of **bronze**, a copper and tin alloy, improved toolmaking because the bronze alloys had increased strength, wear resistance, and corrosion resistance.

Important ores of copper are chalcopyrite ($CuFeS_2$) and malachite [$Cu_2(OH)_2CO_3$]. Often ores of copper can be found near deposits of elemental copper, giving early humans an easy way of finding copper ores. The electrometallurgical production of pure copper from chalcopyrite was previously described in Section 23.3.

The high natural abundance and generally high concentration of copper and copper ores makes copper an economical choice for many industrial applications. Today over 40% of copper products are made from recycled copper. The ease and low cost of recycling copper has kept copper an important industrial metal.

The high conductivity of copper is second only to that of silver, making copper the most important metal for wires in the electrical industry. Along with iron in steel, copper is now one of the most widely used metals being found in electrical motors and devices, the electrical wiring that snakes though the walls of houses, and the electrical transmission network linking power sources to homes and industries all over the world.

Because of its high heat conductivity, copper is also used in applications as a heat exchange material—for example, in car radiators. In addition, copper is used to make pipes for water distribution. Copper pipes can be easily connected to each other with watertight seals by soldering. Copper displaced lead for use as water pipes because of the toxicity of lead, but advancing technology keeps presenting new options. Today, because of its lower price and lighter weight, plastic is used for many water pipes.

Copper has a distinctive reddish color that can be polished to a beautiful metallic luster; therefore it has been used in architecture, as a decorative metal in jewelry, and as a material

▲ Copper wiring is used to transmit electricity throughout the United States.

▲ Copper is used in a variety of applications.

for sculpture. For example, copper is used as a decorative sheathing for domes and art work on top of buildings. When exposed to the atmosphere and rain, some of the copper begins to oxidize, forming a number of different compounds, such as malachite and brochanite [$Cu_4SO_4(OH)_6$], that have beautiful blue or green colors. Copper roofing shingles can be bought as polished copper sheets or as sheets that have already been oxidized to exhibit a beautiful weathered *patina* (a coating that comes with age and use). Chemical compounds can speed up the aging process to produce the patina, without the 20 years that would be required for the patina to develop naturally.

Even though copper has many useful properties, it is not a very strong metal; therefore, alloys of copper with improved strength have been developed. Bronze, as we have seen, was one of the first alloys ever produced. This alloy of copper and tin has been used for thousands of years because it could be made in the low heat of a Stone Age campfire. **Brass**, another widely used alloy, contains copper and zinc. Many brass and bronze alloys also contain other metals to achieve certain physical properties. Some of the most important applications of bronze and brass are for plumbing fixtures, bearings, and art decorations. In addition, the tendency of brass and bronze alloys (unlike those of iron) not to spark when struck make them useful in applications where a spark could be dangerous.

Nickel

Most of the world's nickel comes from deposits in Ontario, Canada. These deposits are believed to have been formed by a meteorite impact (most meteorites have high nickel content). The nickel occurs as a sulfide compound mixed with copper and iron sulfides. To produce nickel, the sulfides are roasted in air to form metal oxides, and then they are reduced to the elemental metals with carbon. The metal mixture is then heated in the presence of carbon monoxide, forming nickel carbonyl, which has a boiling point of 43 °C and can therefore be collected as a gas.

$$Ni(s) + 4\,CO(g) \xrightarrow{\text{heat}} Ni(CO)_4(g)$$

When the nickel carbonyl is heated past 200 °C, it decomposes back to nickel metal and carbon monoxide. This way of refining nickel is called the *Mond process*.

Nickel metal is not very reactive and is resistant to corrosion. It has properties similar to those of platinum and palladium. Consequently, nickel is used as an alloying metal in the production of stainless steels. Many nickel alloys are used for applications where corrosion resistance is important. For example, the alloy Monel contains 72% Ni, 25% Cu, and 3% Fe and is resistant to reaction with most chemicals. Monel will not react even with fluorine gas at room temperature. Nickel–steel alloys are used for armor plates, and elemental nickel is often plated onto other metals as a protective coating.

Zinc

Elemental zinc was officially discovered in Europe in 1746 when calamine (zinc silicate) was reduced with charcoal to produce the metal. However, zinc had been used for many hundreds of years before this discovery because zinc ores were used, along with copper ores, to form copper–zinc brass alloys. The main sources of zinc are ores composed of sphalerite (ZnS), smithsonite ($ZnCO_3$), and an oxide mixture of zinc, iron, and manganese called franklinite. These ores are roasted to form the oxides of the metals and then reduced with carbon to produce the elemental metals. Zinc combines with a number of different metals to form useful alloys. As we have seen, the combination with copper produces the brass family of alloys. The combination of zinc and nickel with copper produces alloys with a silver color called German or silver brass. Zinc is also used in solder alloys with low melting points.

Zinc is commonly used as an anticorrosion coating for steel products in a process called *galvanizing*. Galvanizing a steel object (such as a nail) involves simply dipping the object into a molten bath of zinc. The zinc, which is more reactive than the iron in steel, preferentially oxidizes, forming a tough protective coating. Zinc compounds are also used to coat steel before applying other coatings. The zinc compounds, such as zinc phosphate, adhere strongly to the steel surface, forming rough crystals onto which other coatings, such as paint, adhere very well.

Zinc compounds are also used in paint intended for steel. If the paint gets scratched, the underlying metal can start to rust. The zinc ions in the paint migrate toward the defect and form a protective zinc oxide coating that prevents further corrosion. Generally, zinc and zinc compounds have been considered safe. Zinc additives in coatings have replaced many of the chromium and lead additives (both of which are toxic) that were previously used. Today, however, even zinc additives in coatings are being studied for environmental hazards. In Europe, all substances that contain zinc compounds must be labeled as potential polluters of environmental water. More environmentally safe organic compounds have been developed to replace the metallic anticorrosion additives, but these compounds are more expensive to produce. Opportunities abound for chemists to develop needed products that are both environmentally safe and economically viable.

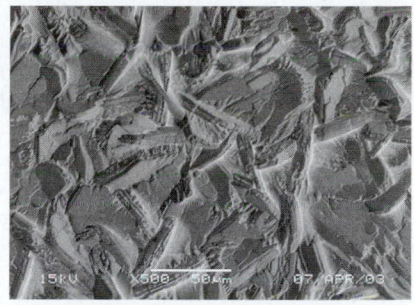

▲ Zinc phosphate adheres strongly to steel surfaces, forming rough crystals onto which other coatings, such as paint, can be applied.

Chapter in Review

Key Terms

Section 23.1
metallurgy (1028)

Section 23.2
minerals (1029)
ore (1029)

Section 23.3
extractive metallurgy (1029)

refining (1029)
gangue (1030)
pyrometallurgy (1030)
calcination (1030)
roasting (1031)
smelting (1031)
flux (1031)
slag (1031)
hydrometallurgy (1031)

leaching (1031)
electrometallurgy (1032)
powder metallurgy (1033)

Section 23.4
alloy (1034)
substitutional alloy (1034)
interstitial alloy (1034)
two-phase region (1036)

lever rule (1037)
octahedral hole (1038)
tetrahedral hole (1038)

Section 23.5
arc-melting (1040)
ferromagnetic (1043)
bronze (1044)
brass (1044)

Key Concepts

General Properties (23.2)

Metals have many common physical properties, such as high conductivity of electricity and heat and high malleability and ductility. However, these properties vary among the different metals. For example, some metals are very malleable, while other metals are more brittle.

Natural Distribution of Metals (23.2)

Metals are unevenly distributed throughout Earth's crust. Only a few metals individually make up more than 1% of the crust, with all of the metals making up about 25% of the crust's mass. Few metals are found in their elemental state; most are found in ores, rocks that contain a high concentration of metal-containing minerals. Most metal-containing minerals are oxides, sulfides, chlorides, carbonates, or more complex compounds.

Metallurgical Processes (23.3)

To be useful, metals have to be separated from the gangue, the nonuseful part of the ores, reduced to the elemental metals, and refined to reach higher purity. Extractive metallurgy is the general term for the processes used to separate the metal from the ore. Among these processes are pyrometallurgy, hydrometallurgy, and electrometallurgy. A new method of forming metal components from micron-sized metal particles is called powder metallurgy.

Phase Diagrams (23.4)

The structure of metals can be described as the packing of spheres, including body-centered cubic and the closest packed types: face-centered cubic and hexagonal closest packed. When two types of metal atoms bond together, they form an alloy. A binary phase diagram is a graphical representation of the crystal types of alloys present at different compositions and temperatures. If the two metals are similar in size and crystallize with the same crystal structure, they will most likely form a miscible solid solution, which means that they can form an alloy at any composition ratio. If the two metals are dissimilar in size or crystal structure, the solubility of one atom in the other's crystal structure is often limited. At certain compositions two different crystals can coexist in equilibrium; this is called a two-phase region. The lever rule determines which phase is present in a greater proportion.

Types of Alloys (23.4)

There are generally two different types of alloys, substitutional and interstitial. A substitutional alloy is a mixture in which one type of metal atom replaces another type of metal atom in the crystal structure. In an interstitial alloy, one type of atom (either a metal or nonmetal) fits into the interstitial holes within the crystal structure of the metal. Interstitial alloys can be made with different atoms filling different fractions of the different types of holes.

Key Equations and Relationships

Lever Rule (23.4)

In a two-phase region on a phase diagram, two different crystal structures coexist in equilibrium. Whichever phase has a composition closer to the overall composition of the alloy is the phase present in the larger relative amount.

Key Skills

Using a Phase Diagram to Determine the Composition, Relative Amounts, and Phases Present (23.4)
- Examples 23.1, 23.2, 23.3, 23.4   • For Practice 23.1, 23.2, 23.3, 23.4   • Exercises 35–38

Determining the Composition of an Interstitial Alloy from the Occupancy of the Interstial Holes (23.4)
- Exercises 41, 42

Exercises

Review Questions

1. What explains the presence of vanadium in oil sources?

2. Name three categories of metallurgical processes.

3. Why is Ni not considered a commonly found metal even though it composes over 2% of the total mass of Earth? Only Fe and Mg have a higher percent composition of Earth's total mass.

4. Metal elements are found in both minerals and ores. Describe the difference between a mineral and an ore.

5. Ores contain minerals and gangue. Describe the difference between a mineral and gangue.

6. Calcining, roasting and smelting are three pyrometallurgical processes. Describe the differences among the processes.

7. What compound has been historically used to leach gold from gold ores? Why should this process be discontinued?

8. Name three benefits of making metal components from the powder metallurgical process.

9. Describe the difference between body-centered cubic and face-centered cubic structures.

10. Describe the difference between a substitutional alloy and interstitial alloy.

11. Describe why copper was one of the first metals that humans used early in history.

12. Describe why bronze was one of the first alloys that humans used early in history.

13. Both brass and bronze contain copper. Describe the difference between these two alloys.

14. Name the properties of copper that make it useful for electric wires and the properties that make it useful for water pipes.

Problems by Topic

The General Properties and Natural Distribution of Metals

15. Describe three typical properties of metals.

16. Describe whether each of the following are generally higher or lower for metals compared to nonmetals.
a. thermal conductivity
b. electrical resistivity
c. transparency
d. ductility

17. List four metal elements that compose more than 1% of Earth's crust.

18. List four metal elements that are found as elements in their natural state.

19. Give the name and formula of an important mineral source for the metals Fe, Hg, V, and Nb.

20. Give the name and formula of an important mineral source for the metals Ti, Zn, U, and Ta.

Metallurgical Processes

21. Two ores of magnesium are $MgCO_3$ and $Mg(OH)_2$. Write balanced equations for the calcinations of these two minerals to form MgO.

22. Two ores of copper are CuO and CuS. Write balanced equations for the roasting of CuO with C to form Cu metal and the roasting of CuS with O_2 to form CuO.

23. To help separate the desired material from waste material a flux is often added during smelting. Give the definition of a flux and identify the flux in the following reaction:

$$SiO(s) + MgO(s) \longrightarrow MgSiO_3(l)$$

24. Give the definition of a slag and identify the slag in the following reaction:

$$SiO_2(s) + MgO(s) \longrightarrow MgSiO_3(l)$$

25. Give a general description of how hydrometallurgy is used in extracting metals from ores.

26. Give a general description of how electrometallurgy is used in extracting metals from ores.

27. How is Al_2O_3 separated from other oxides using the Bayer process? What is the soluble form of aluminum that is formed during the Bayer process from Al_2O_3?

28. In the purification of copper using an electrochemical cell, which electrode contains the pure copper and which electrode has the impure copper? Explain how gold is obtained from this process.

29. Describe the difference between sponge powdered iron and water-atomized powdered iron.

30. Describe the difference in the processing of sponge powdered iron and water-atomized powdered iron.

Metal Structures and Alloys

31. Determine the composition of each of the following vanadium alloys.
a. One-half of the V atoms are replaced by Cr atoms.
b. One-fourth of the V atoms are replaced by Fe atoms.
c. One-fourth of the V atoms are replaced by Cr atoms and one-fourth of the V atoms are replaced by Fe atoms.

32. Determine the composition of each of the following cobalt alloys.
a. One-third of the Co atoms are replaced by Zn atoms.
b. One-eighth of the Co atoms are replaced by Ti atoms.
c. One-third of the Co atoms are replaced by Zn atoms and one-sixth of the Co atoms are replaced by Ti atoms.

33. Using Table 23.2, explain why it would be reasonable to expect that Cr and Fe would form miscible alloys.

34. Using Table 23.2, explain why it would be reasonable to expect that Co and Cu would not form miscible alloys.

35. Determine the composition and phases present at points A and B on the Cr–Fe phase diagram.

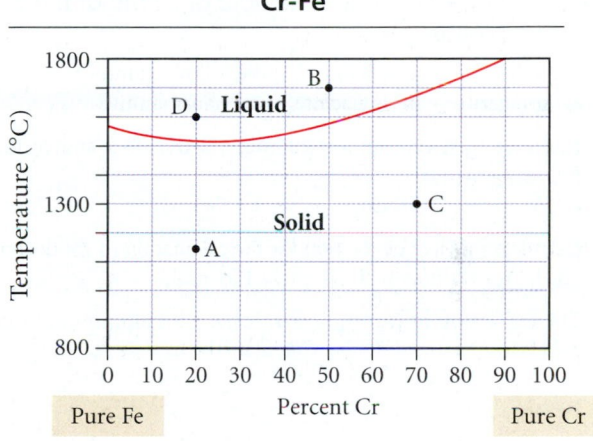

36. Determine the composition and phase present at points C and D on the Cr–Fe phase diagram (see previous problem for diagram).

37. Determine the composition, relative amounts, and phases present at points A and B on the Co–Cu phase diagram.

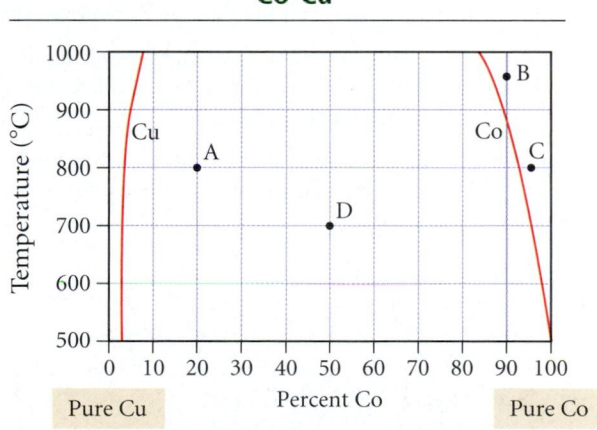

38. Determine the composition, relative amounts and phases present at points C and D on the Co–Cu phase diagram (see previous problem for phase diagram).

39. The elements Mn and Si are added to steel to improve its properties in electric motors and some C is often found as a detrimental impurity. Which of these elements would fill interstitial holes in the Fe lattice and which would substitute for the Fe in the lattice?

40. The elements Si and P are added to steel to improve its properties in electric motors, and some N is often found as a detrimental impurity. Which of these elements would fill interstitial holes in the Fe lattice and which would substitute for the Fe in the lattice?

41. Determine the formula for the following interstitial alloys:
    a. Nitrogen occupies one-half of the octahedral sites of a closest packed Mo structure.
    b. Hydrogen occupies all of the tetrahedral sites of a Cr closest packed structure.

42. Determine the formula for the following interstitial alloys:
    a. Nitrogen occupies one-fourth of the octahedral sites of a closest packed Fe structure.
    b. Hydrogen occupies one-half of the tetrahedral sites of a Ti closest packed structure.

Sources, Properties, and Products of Some of the 3d Transition Metals

43. Identify which metal is found in the following minerals:
    a. sphalerite   b. malachite   c. hausmannite

44. Name at least one important mineral that is a source for the following metals:
    a. Fe   b. Co   c. Cr

45. Calculate the heat of reaction for the calcination of rhodochrosite. (ΔH_f°) for rhodochrosite is -894.1 kJ/mol.)

46. The extraction of Mn from pyrolusite with aluminum produces pure Mn metal and Al_2O_3. Calculate the heat of reaction.

47. Describe the effects of adding Cr to steel and give a use for chromium–steel alloys.

48. Describe the effects of adding Mn to steel and give a use for manganese–steel alloys.

49. Calculate the mole percent and mass percent of Ti in the minerals rutile and ilmenite.

50. Calculate the mole percent and mass percent of Mn in the minerals pyrolusite and rhodochrosite.

51. Explain why it is important to use an inert atmosphere to surround the metal when arc-melting titanium.

52. Titanium is a very reactive metal. Explain why titanium has a high corrosion resistance to seawater and can be used for the production of ship components.

53. Which compound of Ti is the most important industrial product of titanium metal? Describe an application for this compound.

54. Describe how Zn in used to protect the surface of steel products. Give the name of this process.

55. Describe the Bayer process.

56. Describe the Mond process.

57. Which metals are found in carbaloy steel?

58. Which metals are found in monel steel?

Cumulative Problems

59. After 2.0×10^4 kg of an ore that contains 0.051% ilmenite is mined, the percent yield from extracting and refining the metals from the mineral is 87% for the iron and 63% for the titanium. Calculate the mass of iron and titanium produced from the ore.

60. Calculate the mass of aluminum metal that is needed to produce Cr metal from 5.00×10^2 g of chromite. Calculate how many grams of Cr metal are produced from 5.00×10^2 g of chromite.

61. State how many lattice atoms surround a tetrahedral hole and an octahedral hole in a closest packed structure. Describe which hole site is larger and explain why.

62. Describe why the crystal structure and atomic size of the two elements are important factors in a two-component phase diagram.

63. Explain why Mn will form compounds with higher oxidation states than Cr.

64. When MnO_2 is added to silica glass, the glass has a pink color. Explain why MnO_2 is added to the glass.

65. Co, Fe, and Ni are ferromagnetic. Describe the difference between ferromagnetic and paramagnetic.

66. Is the chromate ion or dichromate ion more stable in an acidic solution? Which ion has a higher Cr : O ratio? Which ion has a higher oxidation state for Cr?

Challenge Problems

67. Iron powder was placed into a tall cylinder-shaped die and pressed from the top and bottom to make a cylinder-shaped pressed part with a height of 5.62 cm and radius of 4.00 cm. The density of the iron powder before it was pressed was 2.41 g/mL. The density of a pressed iron part is 6.85 g/mL. The density of pure solid iron is 7.78 g/mL.
   a. Calculate the original height of the powder before it was pressed.
   b. Calculate the theoretical height of the pressed component if it could be pressed to the same density of pure iron.
   c. Calculate the percentage of the component that is composed of voids between the iron particles.

68. When a part is made from pressing together powdered metal, no metal is wasted, compared to metal that is scrapped from cutting a metal part from a solid metal plate. If a circular part with a diameter of 10.0 cm is made from an original shape of a square with a side length of 10.0 cm, calculate the percentage of the metal that is thrown away as scrap. If the circular shape also has a circular hole with a diameter of 6.0 cm, calculate the percentage of the metal that is thrown away as scrap.

69. Hydrogen can be in both the octahedral and tetrahedral holes for lanthanum. Determine the percentage of the holes that are filled if the formula is $LaH_{2.76}$.

Conceptual Problems

70. Explain why metals such as Ni and Co are economical to mine and use in industrial processes even though they have a very low natural abundance in Earth's crust.

71. Explain why metals such as Au and Ag are found in their elemental states in nature, but metals like Na and Ca are always found in compounds in nature.

24

Transition Metals and Coordination Compounds

24.1 The Colors of Rubies and Emeralds

24.2 Properties of Transition Metals

24.3 Coordination Compounds

24.4 Structure and Isomerization

24.5 Bonding in Coordination Compounds

24.6 Applications of Coordination Compounds

I N THIS CHAPTER, WE EXAMINE the chemistry of the transition metals and an important class of their compounds called coordination compounds. We will see that coordination compounds form all of the types of isomers that we have studied so far, as well as some new types. As we study the transition metals, we will draw on much of what we learned in Chapters 7 and Chapter 8 about electronic structure and periodicity. We will also briefly revisit valence bond theory to explain bonding in coordination compounds, but we will quickly shift to a different theory—called crystal field theory—that better explains many of the properties of these compounds. Transition metals and coordination compounds are important, not only because of their interesting chemistry, but because of their numerous applications. For example, coordination compounds are the basis for a number of therapeutic drugs, chemical sensors, and coloring agents. In addition, many biological molecules contain transition elements in arrangements that are similar to coordination compounds. For example, the oxygen-carrying site on hemoglobin is an iron ion bonded partly to a flat molecule called a porphyrin and partly to an amino acid in the hemoglobin molecule. An oxygen molecule reversibly bonds to the iron and is transported throughout the body by blood flow.

The red color of ruby is caused by a splitting of the d-orbital energy levels in Cr^{3+} by the host crystal.

24.1 The Colors of Rubies and Emeralds

Rubies are deep red and emeralds are brilliant green, yet the color of both gemstones is caused by the same ion, Cr^{3+}. The difference lies in the crystal that is hosting the ion. Rubies are crystals of aluminum oxide (Al_2O_3) in which about 1% of the Al^{3+} ions are replaced by Cr^{3+} ions. Emeralds, by contrast, are crystals of beryllium aluminum silicate [$Be_3Al_2(SiO_3)_6$] in which a similar percentage of the Al^{3+} ions are replaced by Cr^{3+}. The imbedded Cr^{3+} ion is red in the aluminum oxide crystal, but green in the beryllium aluminum silicate crystal. Why?

The answer to this question lies in the effect that the host crystal has on the energies of the atomic orbitals in Cr^{3+}. Atoms in the crystal create a field around the ion—sometimes called the *crystal field*—that splits the five normally degenerate *d* orbitals into two or more levels. The color of the gemstone is caused by electron transitions between these levels.

▲ **Ruby and Emerald** The red color of ruby and the green color of emerald are both caused by Cr^{3+}.

In rubies, the crystal field is stronger (and the corresponding splitting of the *d* orbitals greater) than it is in emeralds. Recall from Chapter 7 that the color of a substance depends on the colors *absorbed* by that substance, which in turn depends on the energy differences between the orbitals involved in the absorption. The greater splitting in ruby results in a greater energy difference between the *d* orbitals, and consequently the absorption of a different color of light than in emerald.

The colors of several other gemstones are also caused by the splitting of the *d* orbitals in transition metal ions imbedded within a host crystal. For example, the red in garnet [$Mg_3Al_2(SiO_4)_3$ host crystal] and the yellow-green of peridot (Mg_2SiO_4 host crystal) are both caused by transitions between *d* orbitals in Fe^{2+}. Similarly, the blue in turquoise, $[Al_6(PO_4)_4(OH)_8 \cdot 4H_2O]^{2-}$ as a host crystal is caused by transitions between the *d* orbitals in Cu^{2+}.

In this chapter, we examine the properties of the transition metals and their ions more closely. We also examine the properties of a common type of transition metal compound, called a coordination compound, that we first encountered in Chapter 16 (see Section 16.8). In a coordination compound, bonds to a central metal ion split the *d* orbitals much as they are split in the crystals of gemstones. The theory that explains these splittings and the corresponding colors is called crystal field theory, which we will also examine in this chapter.

▲ **Garnet, Peridot, and Turquoise** The red in garnet and the yellow-green of peridot are both caused by Fe^{2+}. The blue of turquoise is caused by Cu^{2+}.

24.2 Properties of Transition Metals

Transition metals, the elements in the *d* block of the periodic table, are an interesting study in similarities and differences. When their properties are compared with the varied properties of the main-group elements, they seem markedly similar. For example, almost all transition metals have moderate to high densities, good electrical conductivity, high melting points, and moderate to extreme hardness. Their similar properties are related to their similar electron configurations: they all have electrons in *d* orbitals that can be involved in metallic bonding. In spite of their similarities, however, each element is also unique, and they exhibit a wide variety of chemical behavior. Before we examine some of the periodic properties of the transition metals, we review the electron configurations of these elements, first discussed in Chapters 7 and 8.

Electron Configurations

Recall from Section 8.4 that as you move to the right across a row of transition elements, electrons are added to $(n - 1)d$ orbitals (where *n* is the row number in the periodic table and also the quantum number of the highest occupied principal level). For example, as you move across the fourth-period transition metals, electrons are added to the $3d$ orbitals, as shown in Table 24.1.

In general, the ground state electron configuration for the first two rows of transition elements is [noble gas] $ns^2(n - 1)d^x$ and that in the third and fourth rows is [noble gas] $ns^2(n - 2)f^{14}(n - 1)d^x$, where *x* ranges from 1 to 10. Recall from Section 8.4, however, that because the *ns* and $(n - 1)d$ sublevels are close in energy, many exceptions occur. For example, in the first transition series of the *d* block, the outer configuration is $4s^2 3d^x$ with two exceptions: Cr is $4s^1 3d^5$ and Cu is $4s^1 3d^{10}$. This behavior is related to the closely spaced $3d$ and $4s$ energy levels and the stability associated with a half-filled or completely filled d sublevel.

Recall from Section 8.7 that the transition metals form ions by losing electrons from the ns orbital *before* losing electrons from the $(n-1)d$ orbitals. For example, Fe^{2+} has an electron configuration of [Ar] $3d^6$, having lost both of the $4s$ electrons to form the 2+ charge. The following examples review the writing of electron configurations for transition metals and their ions.

TABLE 24.1 First–Row Transition Metal Orbital Occupancy

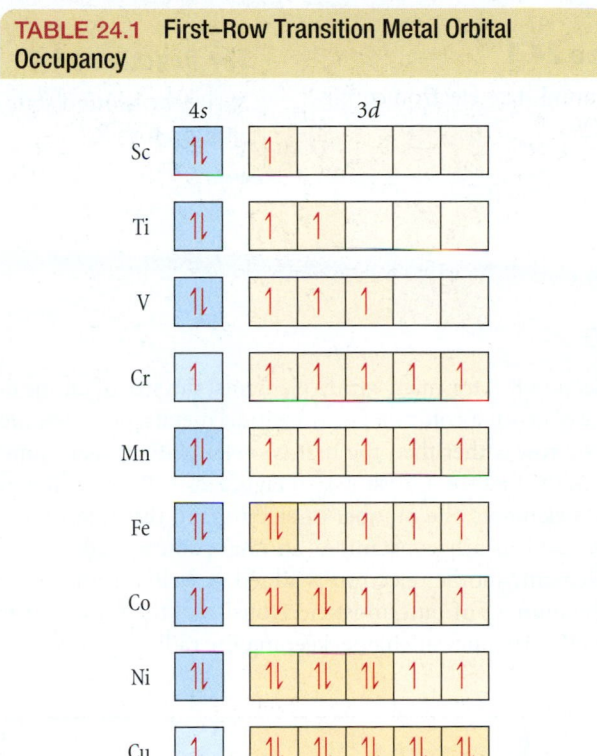

Steps in Writing Electron Configurations	EXAMPLE 24.1 Writing Electron Configurations for Transition Metals	EXAMPLE 24.2 Writing Electron Configurations for Transition Metals
	Write the ground state electron configuration for Zr.	Write the ground state electron configuration for Co^{3+}.
Identify the noble gas that precedes the element and put it in square brackets.	Solution [Kr]	Solution [Ar]
Count down the periods to determine the outer principal quantum level—this is the quantum level for the s orbital. Subtract one to obtain the quantum level for the d orbital. If the element is in the third or fourth transition series, include $(n-2)f^{14}$ electrons in the configuration.	Zr is in the fifth period so the orbitals we use are [Kr] $5s4d$	Co is in the fourth period so the orbitals we use are [Ar] $4s3d$

| Count across the row to see how many electrons are in the neutral atom and fill the orbitals accordingly. | Zr has four more electrons than Kr. $$[Kr]\ 5s^2 4d^2$$ | Co has nine more electrons than Ar. $$[Ar]\ 4s^2 3d^7$$ |
| For an ion, remove the required number of electrons, first from the *s* and then from the *d* orbitals. | | Co^{3+} has lost three electrons relative to the Co atom. $$[Ar]\ 4s^0 3d^6 \quad or \quad [Ar]\ 3d^6$$ |

For Practice 24.1

Write the ground state electron configuration for Os.

For Practice 24.2

Write the ground state electron configuration for Nb^{2+}.

Atomic Size

As discussed in Section 8.6, for main-group elements, the size of atoms decreases across a period and increases down a column. For transition metals, however, there is little variation in size across a row (other than the first two elements in each transition metal row, such as Sc and Ti in the first row), as shown in Figure 24.1 ▼. The difference is that, across a row of transition elements, the number of electrons in the outermost principal energy level (highest *n* value) is nearly constant. As another proton is added to the nucleus with each successive element, another electron is added as well, but the electron goes into an $n-1$ orbital. The number of outermost electrons stays the same and they experience a roughly constant effective nuclear charge, keeping the radius approximately constant.

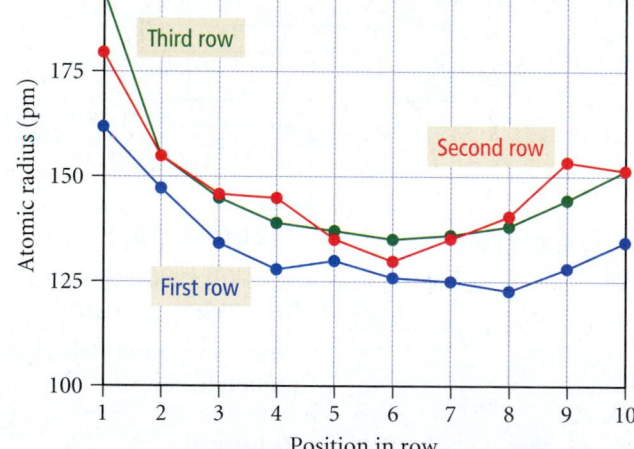

▶ **FIGURE 24.1 Trends in Atomic Radius** With the exception of a decrease in radius from the first to the second element, there is only a small variation in atomic radius across a row. There is a small and expected increase in radius from the first to the second transition row but virtually no difference in radius from the second to the third.

Looking down a group, we see a small but expected increase in size from the first transition metal row to the second, but the size of elements in the third row is about the same as it is for those in the second row. This pattern is also different from that of the main-group elements, especially when we consider that in any given column, the third transition row has 32 more electrons than the second row. The reason that the third transition row elements are not larger is because 14 of the 32 electrons have gone into a $(n-2)f$ sublevel, and while electrons in *f* orbitals are in lower principal quantum levels, they are not very effective at shielding the outer electrons from nuclear charge. Consequently, the outer electrons are held more tightly by the nucleus, offsetting the typical increase in size between the periods—an effect called the **lanthanide contraction**.

▶ Conceptual Connection 24.1 Atomic Size

Which element has the larger atomic radius, Fe or W?

Answer: The element W has the larger radius because it is in the third transition row and Fe is in the first. Atomic radii increase from the first to the second transition row and stay roughly constant from the second to the third.

Ionization Energy

First ionization energy values of transition elements follow the expected main-group periodic trend and slowly increase across a row (Figure 24.2 ▼), but the increase is smaller than for main-group elements. As we move down a group, we see that the third transition row generally has a higher ionization energy than do the first two rows, a trend counter to that observed in the main-group elements. In the transition elements the charge of the nucleus increases substantially from one row to the next, but there is only a small increase in atomic size between the first and second rows, and no increase in size between the second and third row. The outer electrons are therefore held more tightly in the third transition row than in the first two rows.

Electronegativity

The electronegativity values of the transition metals, like their ionization energies, follow the main-group trend and slowly increase across a row, as shown in Figure 24.3 ▼. The increase is smaller than the increase that occurs in the main-group elements, but that is not unexpected given the similarity in the sizes of the atoms. The trend in electronegativity values down a group is another example of the transition metals behaving differently from the main-group elements. The electronegativity values generally increase from the first transition row to the second, but then there is no further increase for the third row. In the main-group elements, by contrast, we see a *decrease* in electronegativity down a group. The difference is again caused by the relatively small change in atomic size as you move down a column for the transition elements, accompanied by a large increase in nuclear charge. One of the heaviest metals, gold (Au), is also the most electronegative metal. Its electronegativity value (EN = 2.4) is even higher than that of some nonmetals (EN of P is 2.1), and compounds of an Au^- ion have been observed.

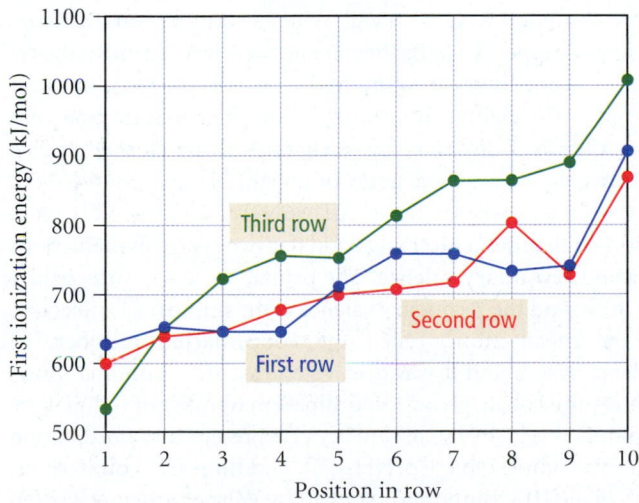

▲ **FIGURE 24.2 Trends in First Ionization Energy** First ionization energy generally increases across a row, following the main-group trend. However, in contrast to the main-group trend, the third transition row has a greater ionization energy than the first and second rows.

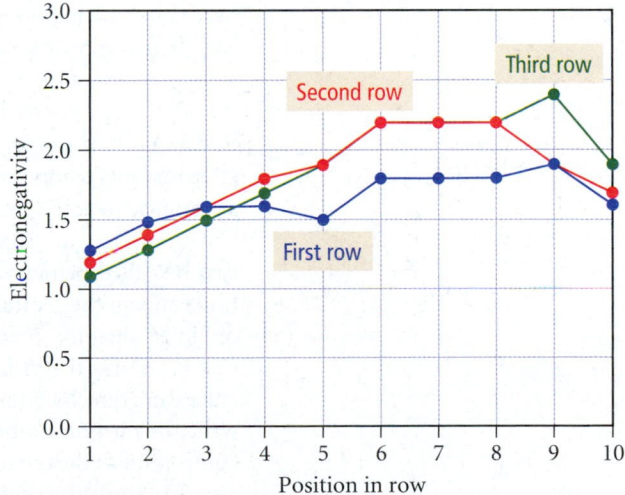

▲ **FIGURE 24.3 Trends in Electronegativity** The electronegativity of the transition elements generally increases across a row, following the main-group trend. However, in contrast to main-group trend, electronegativity increases from the first transition row to the second. There is little electronegativity difference between the second and third transition rows.

Oxidation States

Unlike main-group metals, which tend to exhibit only one oxidation state, the transition metals often exhibit a variety of oxidation states, as shown in Figure 24.4 ▼. The highest oxidation state for a transition metal is +7 for manganese (Mn). The electron configuration of manganese in this oxidation state corresponds to the loss of all the electrons in the $4s$ and $3d$ orbitals, leaving a noble gas electron configuration ([Ar]). This is the same configuration we get for all of the highest oxidation states of the elements to the left of Mn. To the right of manganese, the oxidation states are all lower, mostly +2 or +3. A +2 oxidation state for a transition metal is not surprising, since $4s$ electrons are readily lost.

Metals in high oxidation states, such as +7, exist only when the metal is bound to a highly electronegative element such as oxygen; they do not exist as bare ions.

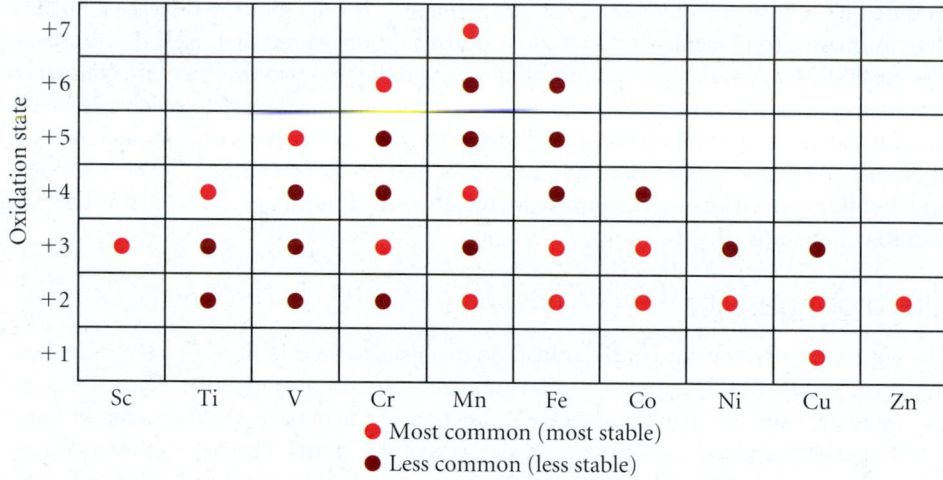

▲ FIGURE 24.4 **First-Row Transition Metal Oxidation States** The transition metals exhibit many more oxidation states than the main-group elements. These oxidation states range from +7 to +1.

24.3 Coordination Compounds

Lewis acid and base theory was covered in Section 15.11.

We saw at the end of Chapter 16 that transition metals tend to form *complex ions*. A **complex ion** contains a central metal ion bound to one or more *ligands*. A **ligand** is a Lewis base (or electron donor) that forms a bond with the metal. When a complex ion combines with one or more *counterions* (ions of opposite charge that are not acting as ligands), the resulting neutral compound is called a **coordination compound**. The first coordination compounds were discovered in the early eighteenth century, but their nature was not understood for nearly 200 years after their discovery. Swiss chemist Alfred Werner studied coordination compounds extensively—especially a series of cobalt(III) compounds with ammonia, whose formulas were then written as $CoCl_3 \cdot 6NH_3$, $CoCl_3 \cdot 5NH_3$ and, $CoCl_3 \cdot 4NH_3$. In 1893, he proposed that the central metal ion has two types of interactions that he called **primary valence** and **secondary valence**. The primary valence is the oxidation state on the central metal atom, and the secondary valence is the number of molecules or ions directly bound to the metal atom, called the **coordination number**. In $CoCl_3 \cdot 6NH_3$ the primary valence is +3, and it was discovered that the ammonia molecules were directly bound to the central cobalt, giving a coordination number of 6. Today we write the formula of this compound as $[Co(NH_3)_6]Cl_3$ to better represent the coordination compound as the combination of a complex ion, $Co(NH_3)_6^{3+}$, and three Cl^- counterions.

The formulas of the other cobalt(III) compounds studied by Werner are now written as $[Co(NH_3)_5Cl]Cl_2$ and $[Co(NH_3)_4Cl_2]Cl$. In these two cases, the complex ions are $Co(NH_3)_5Cl^{2+}$ (with two Cl^- counterions) and $Co(NH_3)_4Cl_2^+$ (with one Cl^-) counterion, respectively. With this series of compounds, Werner demonstrated that the Cl^- could replace NH_3 in the secondary valence. In other words, Cl^- could act as a counterion, or it could bond directly to the metal as part of the complex ion.

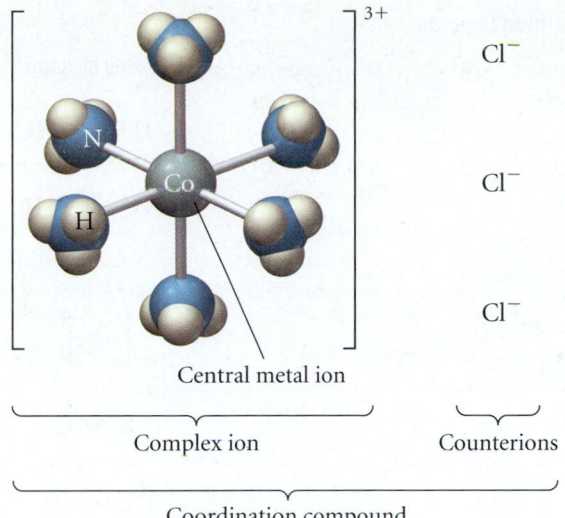

▲ **Complex Ion and Coordination Compound** A co-ordination compound contains a complex ion and corresponding counterions. The complex ion contains a central metal atom coordinated to several ligands. The particular compound shown here is $[Co(NH_3)_6]Cl_3$.

The complex ion itself contains the metal ion in the center and the ligands—which can be neutral molecules or ions—arranged around it. We can think of the metal–ligand complex as a Lewis acid–base adduct (see Section 15.11) because the bond is formed when the ligand donates a pair of electrons to an empty orbital on the metal. For example, consider the reaction between the silver ion and ammonia:

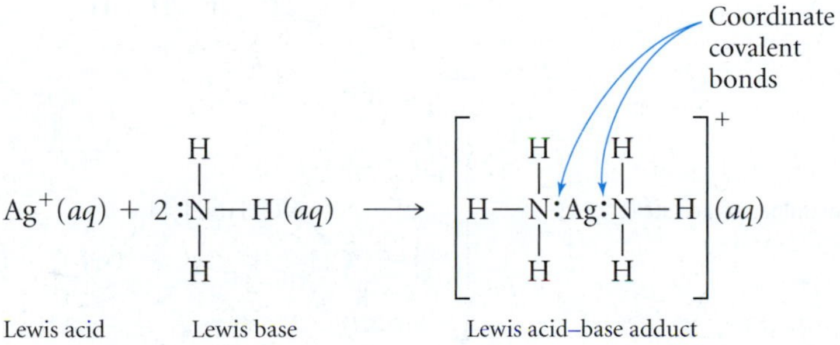

A bond of this type, which we first encountered in Section 10.6, is often referred to as a **coordinate covalent bond**. Ligands are therefore good Lewis bases and have at least one pair of electrons to donate to, and bond with, the central metal ion. Table 24.2 contains a number of common ligands.

Ligands that donate only one electron pair to the central metal are called **monodentate**. Some ligands, however, have the ability to donate two pairs of electrons (from two different atoms) to the metal; these are called **bidentate**. Examples of bidentate ligands include the oxalate ion (abbreviated ox) and the ethylenediamine molecule (abbreviated en).

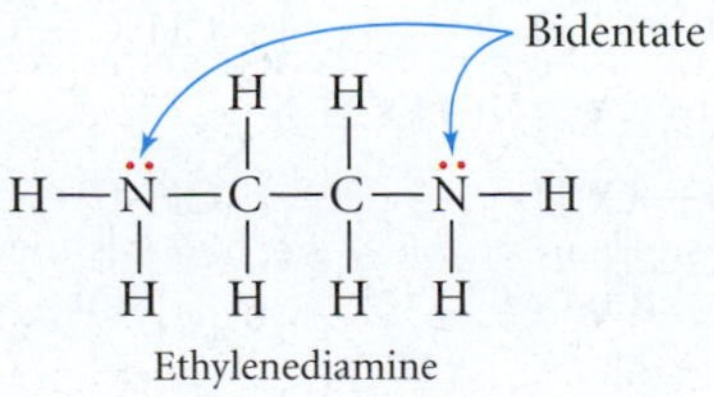

Ethylenediamine

TABLE 24.2   Common Ligands

Name	Lewis diagram
Water	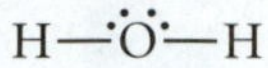 $H\!-\!\ddot{\text{O}}\!-\!H$
Ammonia	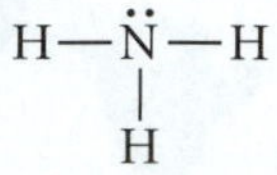 $H\!-\!\ddot{\text{N}}\!-\!H$, with H below
Chloride ion	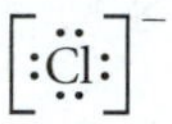 $\left[:\!\ddot{\text{Cl}}\!:\right]^{-}$
Carbon monoxide	 $:\!C\!\equiv\!O:$
Cyanide ion	 $\left[:\!C\!\equiv\!N\!:\right]^{-}$
Thiocyanate ion	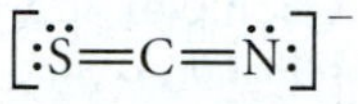 $\left[:\!\ddot{\text{S}}\!=\!C\!=\!\ddot{\text{N}}\!:\right]^{-}$
Oxalate ion (ox)	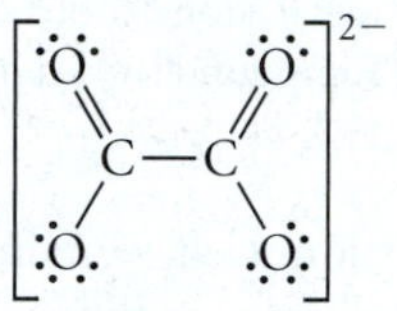 (oxalate Lewis structure) $^{2-}$
Ethylenediamine (en)	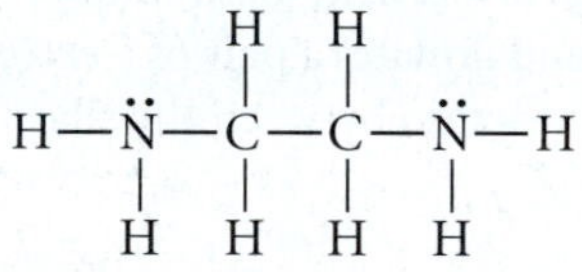 $H\!-\!\ddot{\text{N}}\!-\!C\!-\!C\!-\!\ddot{\text{N}}\!-\!H$
Ethylenediaminetetraacetate (EDTA)	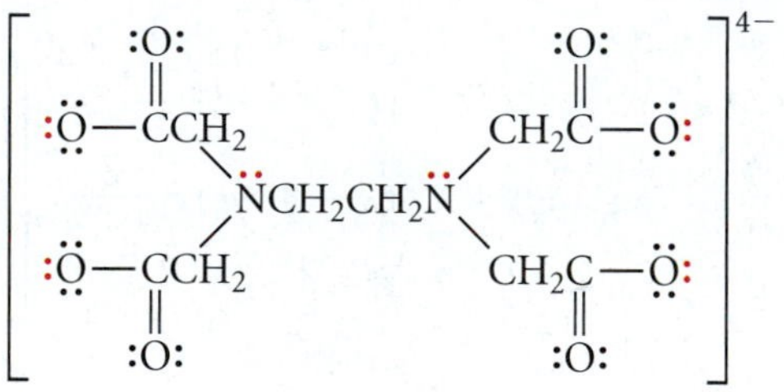 (EDTA Lewis structure) $^{4-}$

The ethylenediamine ligand bonded to Co^{3+} is shown in Figure 24.5(a) ▶. Some ligands, called **polydentate** ligands, can donate even more than two electron pairs (from more than two atoms) to the metal. The most common polydentate ligand is the ethylene-diaminetetraacetate ion ($EDTA^{4-}$).

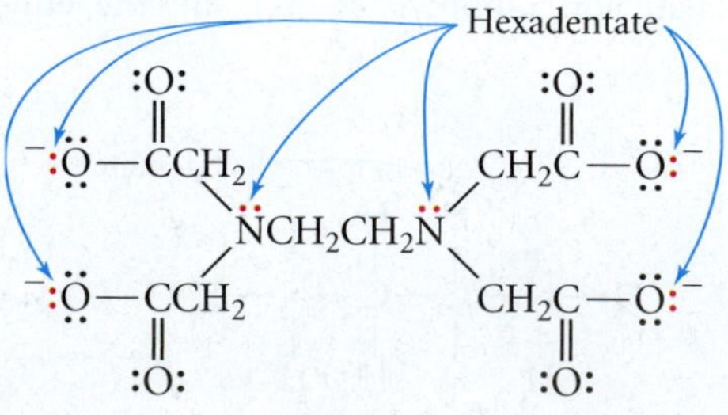

EDTA^{4-}

Bidentate and Polydentate Ligands Coordinated to Co (III)

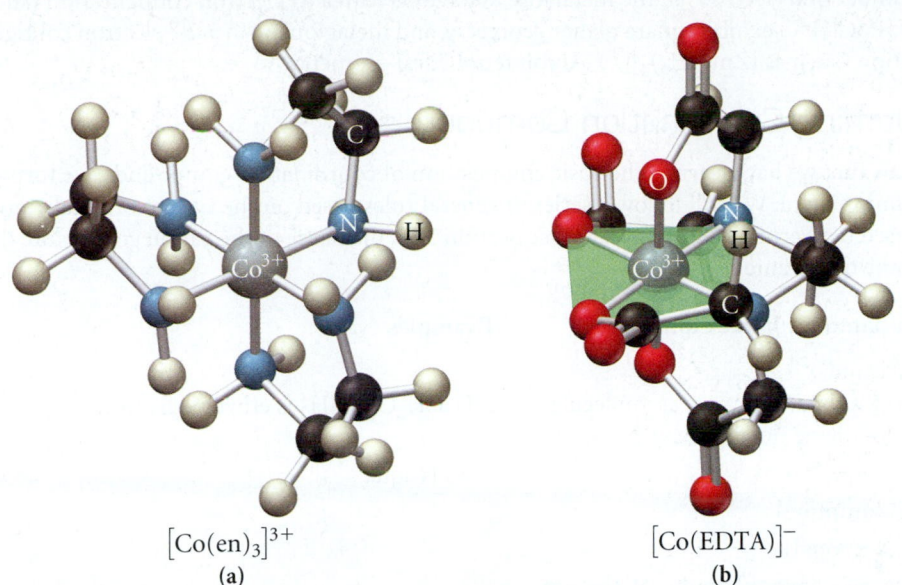

$$[Co(en)_3]^{3+}$$
(a)

$$[Co(EDTA)]^-$$
(b)

◀ **FIGURE 24.5 Bidentate and Polydentate Ligands Coordinated to Co(III)** **(a)** Ethylenediamine is a bidentate ligand; **(b)** EDTA is a hexadentate ligand.

The EDTA ligand can wrap itself completely around the metal, donating up to six pairs of electrons, as shown in Figure 24.5(b). A complex ion that contains either a bidentate or polydentate ligand is called a **chelate** (pronounced "key-late"), and the coordinating ligand is known as a **chelating agent**.

A survey of many coordination compounds shows that coordination numbers can vary from as low as 2 to as high as 12. The most common coordination numbers are 6, as occurs in $[Co(NH_3)_6]^{3+}$, and 4, as occurs in $[PdCl_4]^{2-}$. Coordination numbers greater than 6 are rarely observed for the first-row transition metals. Typically, only 1+ metal ions have a coordination number as low as 2, as occurs in $[Ag(NH_3)_2]^+$. Odd coordination numbers are known, but they are rare.

The common geometries of complex ions, shown in Table 24.3, depend in part on their coordination number. A coordination number of 2 results in a linear geometry, and a coordination number of 6 results in an octahedral geometry. A coordination number of

TABLE 24.3 Common Geometries of Complex Ions

Coordination Number	Shape	Model	Example
2	Linear		$[Ag(NH_3)_2]^+$
4	Square planar		$[PdCl_4]^{2-}$
4	Tetrahedral		$[Zn(NH_3)_4]^{2+}$
6	Octahedral		$[Fe(H_2O)_6]^{3+}$

4 can have either a tetrahedral geometry or a square planar geometry, depending on the number of d electrons in the metal ion. Metal ions with a d^8 electron configuration (such as $[PdCl_4]^{2-}$) exhibit square planar geometry, and metal ions with a d^{10} electron configuration (such as $[Zn(NH_3)_4]^{2+}$) exhibit tetrahedral geometry.

Naming Coordination Compounds

Now that we have learned the basic composition of coordination compounds, we turn to naming them. We will follow a series of general rules based on the system originally proposed by Werner. As with all salts (see Section 3.5), the name of the cation goes before the name of the anion.

Guidelines for naming complex ions

Examples

1. Name the ligands.
   - Neutral ligands are named as molecules with the following notable exceptions.

     $NH_2CH_2CH_2NH_2$ is ethylenediamine.

     H_2O (aqua)

     H_2O is aqua.

     NH_3 (ammine)

     CO (carbonyl)

   - Anionic ligands have the name of the ion with the endings modified as follows:

     -*ide* becomes -o

     Cl^- is chloro.

     -*ate* becomes -ato

     SO_4^{2-} is sulfato.

     -*ite* becomes -ito

     SO_3^{2-} is sulfito.

   Table 24.4 lists the names of some common ligands.

2. List the names of the ligands in alphabetical order before the name of the metal cation.

   Ammine (NH_3) is listed before chloro (Cl^-) which is listed before nitrito (NO_2^-).

3. Use a prefix to indicate how many ligands (when there is more than one of a particular type) as follows: *di-* (2), *tri-* (3), *tetra-* (4), *penta-* (5), or *hexa-* (6).

   Trichloro indicates three Cl^- ligands.

   Tetraammine indicates four NH_3 ligands.

   Tris(ethylenediamine) indicates three ethylene-diamine ligands.

   If the name of the ligand already contains a prefix, such as ethylenediamine, put parentheses around the ligand name and use *bis-* (2), *tris-* (3), or *tetrakis-* (4) to indicate the number.

   Prefixes do not affect the order in which the ligands are listed.

4. Name the metal.

   a. When the complex ion is a cation, use the name of the metal followed by the oxidation state written with a roman numeral.

   In cations:

   Co^{3+} is cobalt(III).

   Pt^{2+} is platinum(II).

   Cu^+ is copper(I).

   b. If the complex ion is an anion, drop the ending of the metal and add -*ate* followed by the oxidation state written with a Roman numeral. Some metals use the Latin root with the -*ate* ending. Table 24.5 lists the names for some common metals in anionic complexes.

   In anions:

   Co^{3+} is cobaltate(III).

   Pt^{2+} is platinate(II).

   Cu^+ is cuprate(I).

5. Write the entire name of the complex ion by listing the ligands first followed by the metal.

   $[Pt(NH_3)_2Cl_4]^{2-}$ is diamminetetrachloroplatinate(II).

   $[Co(NH_3)_6]^{3+}$ is hexaamminecobalt(III).

TABLE 24.4 Names and Formulas of Common Ligands

Ligand	Name in Complex Ion
Anions	
Bromide, Br^-	Bromo
Chloride, Cl^-	Chloro
Hydroxide, OH^-	Hydroxo
Cyanide, CN^-	Cyano
Nitrite, NO_2^-	Nitro
Oxalate, $C_2O_4^{2-}$ (ox)	Oxalato
Neutral molecules	
Water, H_2O	Aqua
Ammonia, NH_3	Ammine
Carbon monoxide CO	Carbonyl
Ethylenediamine (en)	Ethylenediamine
Ethylenediaminetetraacetate (EDTA)	Ethylenediaminetetraacetato

TABLE 24.5 Names of Common Metals when Found in Anionic Complex Ions

Metal	Name in Anionic Complex
Chromium	Chromate
Cobalt	Cobaltate
Copper	Cuprate
Gold	Aurate
Iron	Ferrate
Lead	Plumbate
Manganese	Manganate
Molybdenum	Molybdate
Nickel	Nickelate
Platinum	Platinate
Silver	Argentate
Tin	Stannate
Zinc	Zincate

When we write the *formula* of a complex ion, the symbol of the metal is written first, followed by neutral molecules and then anions. If there is more than one anion or neutral molecule acting as a ligand, then list them in alphabetical order based on the chemical symbol.

Steps in Naming Coordination Compounds	EXAMPLE 24.3 Naming Coordination Compounds Name the following: $[Cr(H_2O)_5Cl]Cl_2$.	EXAMPLE 24.4 Naming Coordination Compounds Name the following: $K_3[Fe(CN)_6]$.
	Solution	**Solution**
Identify the cation and anion and first name the simple ion (i.e., not the complex one).	$[Cr(H_2O)_5Cl]^{2+}$ is a complex cation. Cl^- is chloride.	K^+ is potassium. $[Fe(CN)_6]^{3-}$ is a complex anion.
List the ligand names in alphabetical order and give each a name.	H_2O is aqua. Cl^- is chloro.	CN^- is cyano.
Name the metal ion.	Cr^{3+} is chromium(III).	Fe^{3+} is ferrate(III) because the complex is anionic.
Name the complex ion by adding prefixes to indicate the number of each ligand followed by the name of each ligand followed by the name of the metal ion.	$[Cr(H_2O)_5Cl]^{2+}$ is pentaaquachlorochromium(III).	$[Fe(CN)_6]^{3-}$ is hexacyanoferrate(III).
Name the compound by writing the name of the cation before the anion. The only space is between ion names.	$[Cr(H_2O)_5Cl]Cl_2$ is pentaaquachlorochromium(III) chloride.	$K_3[Fe(CN)_6]$ is potassium hexacyanoferrate(III).
	For Practice 24.3 Name the following: $[Mn(CO)(NH_3)_5]SO_4$.	**For Practice 24.4** Name the following: $Na_2[PtCl_4]$.

24.4 Structure and Isomerization

Isomerism is common in coordination compounds. We can broadly divide the isomerism observed in coordination compounds into two categories, each with subcategories, as shown in Figure 24.6 ▼. **Structural isomers** are those in which atoms are connected to one another in different ways, whereas **stereoisomers** are those in which atoms are connected in the same way but the ligands have a different spatial arrangement about the metal atom.

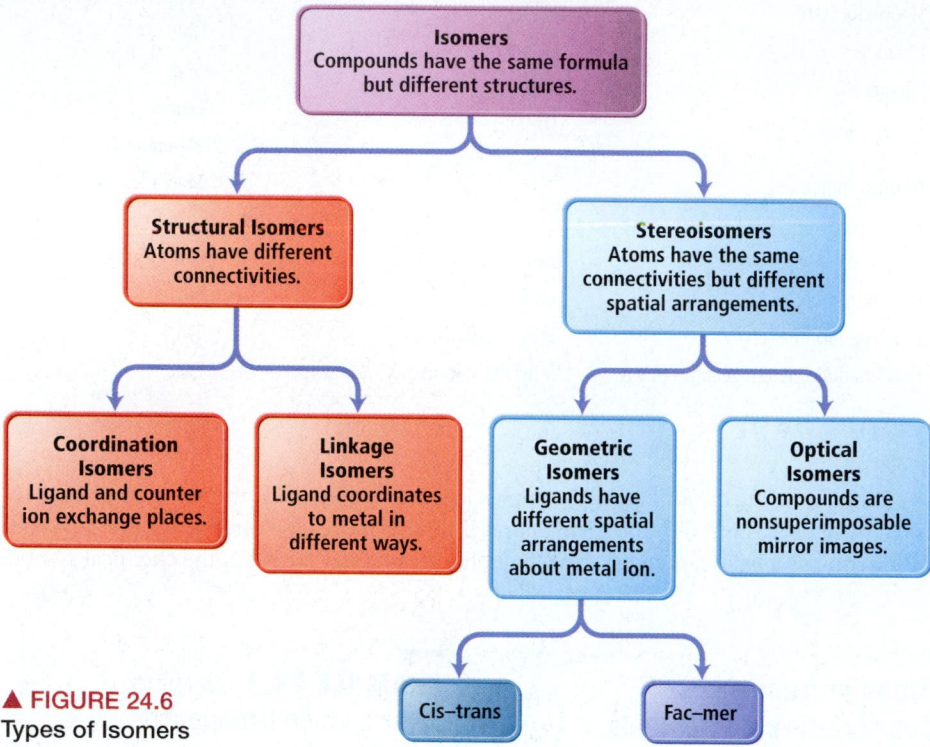

▲ **FIGURE 24.6**
Types of Isomers

Structural Isomerism

The broad category of structural isomers can be further broken down into two types: coordination isomers and linkage isomers. **Coordination isomers** occur when a coordinated ligand exchanges places with the uncoordinated counterion. For example, there are two compounds that have the general formula $Co(NH_3)_5BrCl$. One has the bromine coordinated to the metal and chloride as a counterion, pentaamminebromocobalt(II) chloride, $[Co(NH_3)_5Br]Cl$, and the second has the chlorine coordinated to the metal and bromide as the counterion, pentaamminechlorocobalt(II) bromide ($[Co(NH_3)_5Cl]Br$).

Some ligands can coordinate to the metal in different ways, leading to **linkage isomers**. For example, the nitrite ion (NO_2^-) has a lone pair on the N atom and also lone pairs on the O atoms—either of the two atoms can form coordinate covalent bonds with the metal. When the nitrite ion coordinates through the N atom it is called a nitro ligand and is represented as NO_2^-, but when it coordinates through the O atom, it is called nitrito and is usually represented as ONO^-. An example of linkage isomerization can be seen in the yellow-orange complex ion pentaamminenitrocobalt(III), $[Co(NH_3)_5NO_2]^{2+}$, contrasted with the red-orange complex ion pentaamminenitritocobalt(III), $[Co(NH_3)_5ONO]^{2+}$, as shown in Figure 24.7 ▶. Other ligands capable of linkage isomerization as well as their names are listed in Table 24.6.

Stereoisomerism

The broad category of stereoisomers can also be further broken down into two types: geometric isomers and optical isomers. **Geometric isomers** result when the ligands bonded to the metal have a different spatial arrangement. One type of geometric isomerism, as we saw in Section 20.5, is cis–trans isomerism, which in complex ions occurs in square planar

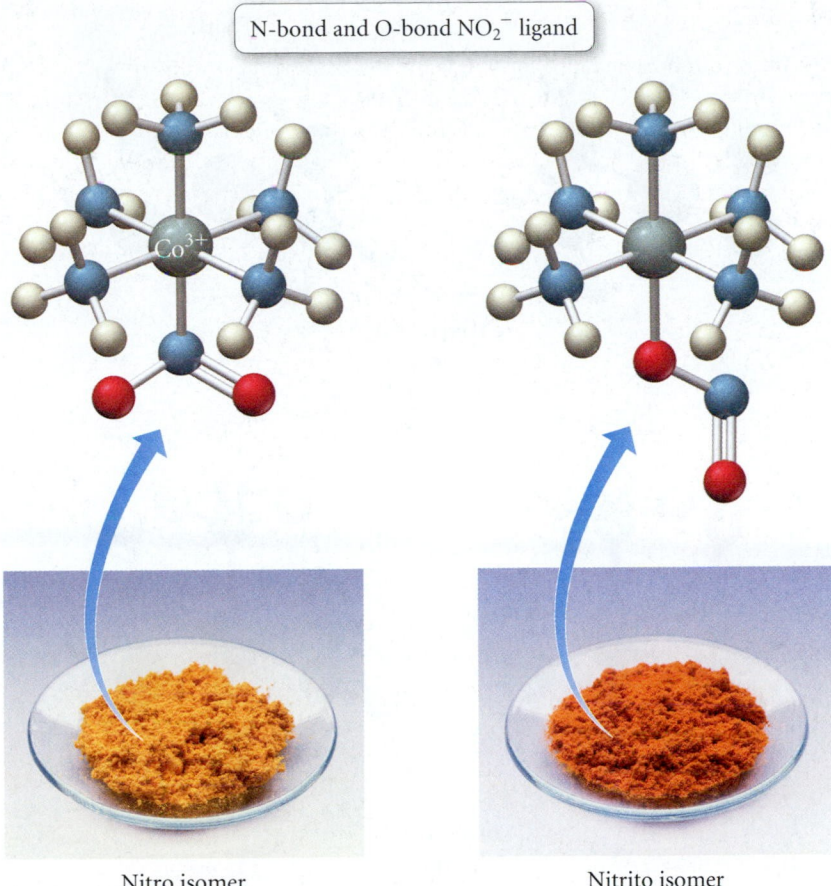

Nitro isomer   Nitrito isomer

▲ **FIGURE 24.7 Linkage Isomers**   In $[Co(NH_3)_5NO_2]^{2+}$, the NO_2 ligand bonds to the central metal atom through the nitrogen atom. In $[Co(NH_3)_5ONO]^{2+}$, the NO_2 ligand bonds through the oxygen atoms. The different isomers have different colors.

TABLE 24.6   Ligands Capable of Linkage Isomerization

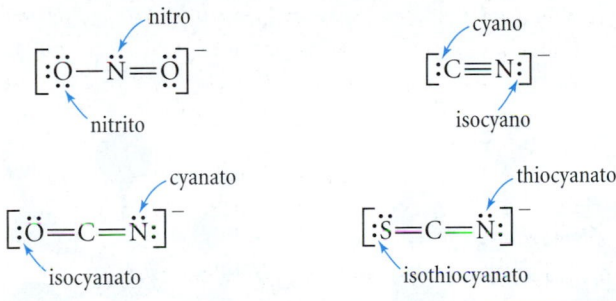

complexes of the general formula MA_2B_2 or octahedral complexes of the general formula MA_4B_2. For example, cis–trans isomerism can be seen in the square planar complex $Pt(NH_3)_2Cl_2$. Figure 24.8(a), on page 1064, shows the two distinct ways in which the ligands can be oriented around the metal. In one complex, the Cl^- ligands are next to each other on one side of the molecule—this is the cis isomer. In the other complex, the Cl^- ligands are on opposite sides of the molecule—this is the trans isomer. Geometric isomerism also exists in the octahedral complex ion $[Co(NH_3)_4Cl_2]^+$. As shown in Figure 24.8(b), there are two distinct ways to arrange the ligands around the metal, one with the Cl^- ligands on the same side (the cis isomer) and another with the Cl^- ligands on opposite sides of the metal (the trans isomer). Note that cis–trans isomerism does not occur in tetrahedral complexes because all bond angles around the metal are 109.5°, and each corner of a tetrahedron can be considered to be adjacent to all three other corners.

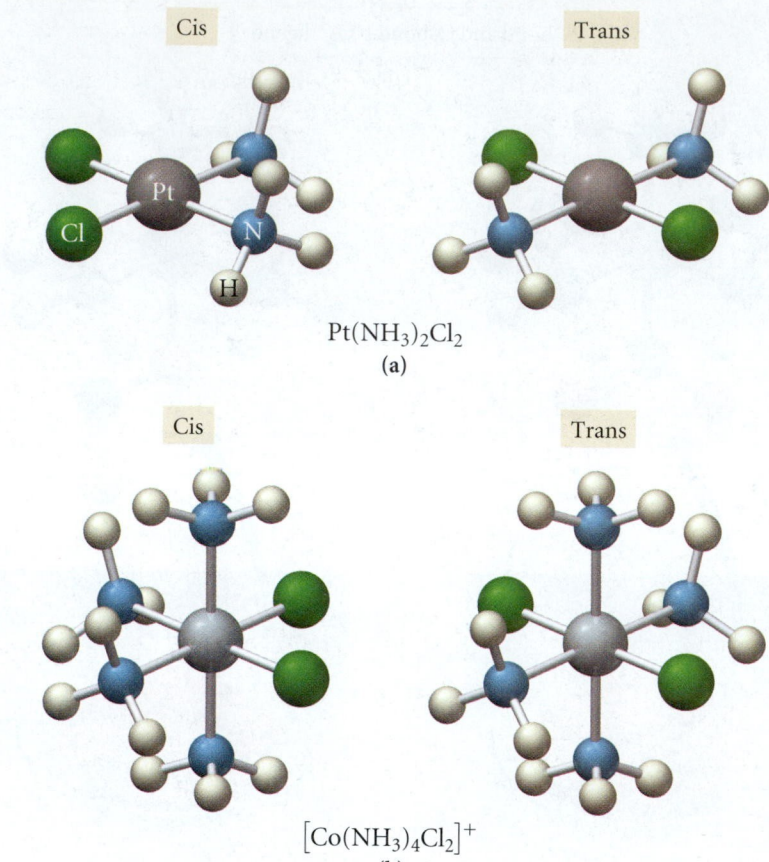

Cis   **Trans**

$Pt(NH_3)_2Cl_2$
(a)

Cis   **Trans**

$[Co(NH_3)_4Cl_2]^+$
(b)

▶ FIGURE 24.8 **Cis–trans Isomerism** (a) Cis–trans isomerism in square planar $Pt(NH_3)_2Cl_2$. In the cis isomer, the Cl^- ligands are next to each other on one side of the molecule. In the trans isomer, the Cl^- ligands are on opposite sides of the molecule. (b) Cis–trans isomerism in octahedral $[Co(NH_3)_4Cl_2]^+$. In the cis isomer, the Cl^- ligands are on the same side. In the trans isomer, the Cl^- ligands are on opposite sides.

Another type of geometric isomerism is fac–mer isomerism, which occurs in octahedral complexes of the general formula MA_3B_3. For example, in $Co(NH_3)_3Cl_3$, there are two distinct ways to arrange the ligands around the metal (see Figure 24.9 ▼). In the fac isomer the three Cl^- ligands are all on one side of the molecule and make up one face of the octahedron (fac is short for facial). In the mer isomer the three ligands inscribe an arc around the middle of the octahedron (mer is short for meridian).

Fac   **Mer**

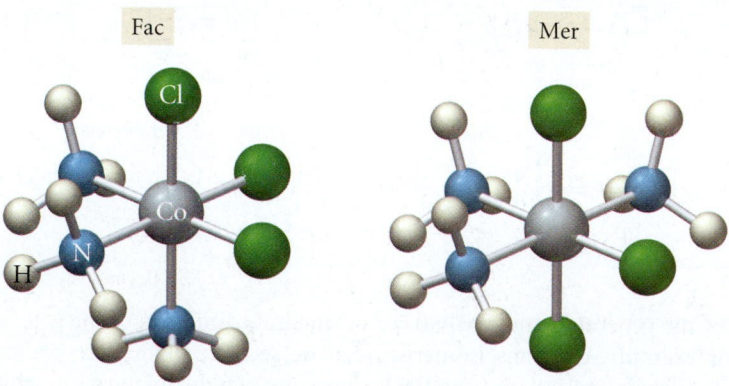

▶ FIGURE 24.9 **Fac–Mer Isomerism in $Co(NH_3)_3Cl_3$** In the fac isomer, the three Cl^- ligands are all on one side of the molecule and make up one face of the octahedron. In the mer isomer, the three ligands inscribe an arc around the middle (or meridian) of the octahedron.

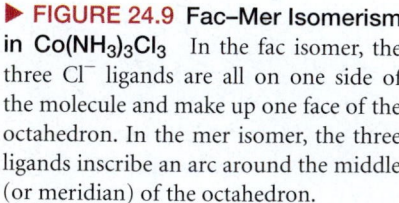

Procedure for Identifying and Drawing Geometric Isomers

EXAMPLE 24.5 Identifying and Drawing Geometric Isomers

Draw the structures and label the type for all the isomers of $[Co(en)_2Cl_2]^+$.

EXAMPLE 24.6 Identifying and Drawing Geometric Isomers

Draw the structures and label the type for all the isomers of $[Ni(CN)_2Cl_2]^{2-}$.

	Solution	**Solution**
Identify the coordination number and the geometry around the metal.	The ethylenediamine (en) ligand is bidentate so each occupies two coordination sites. Each Cl^- is monodentate, occupying one site. The total coordination number is 6 so this must be an octahedral complex.	All the ligands are monodentate so the total coordination number is 4. Ni^{2+} is a d^8 electronic configuration so we expect a square planar complex.
Identify if this will be cis–trans or fac–mer isomerism.	With ethylenediamine occupying four sites and Cl^- occupying two sites, we fit the general formula MA_4B_2, leading to cis–trans isomers.	Square planar complexes can only have cis–trans isomers.
Draw and label the two isomers.	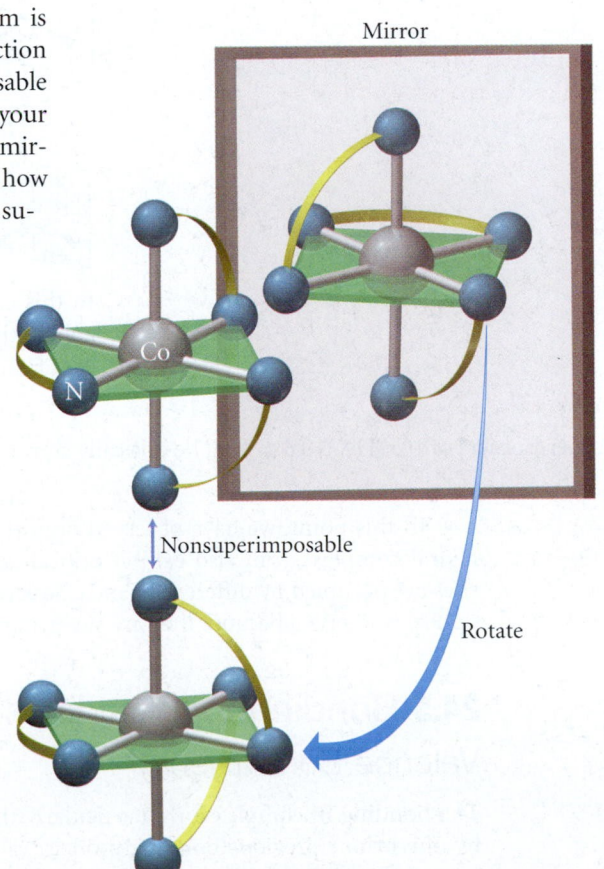	

For Practice 24.5

Draw the structures and label the type for all the isomers of $[Cr(H_2O)_3Cl_3]^+$.

For Practice 24.6

Draw the structures and label the type for all the isomers of $[Co(NH_3)_2Cl_2(ox)]^-$.

The second category of stereoisomerism is optical isomerism. As we discussed in Section 20.3, **optical isomers** are nonsuperimposable mirror images of one another. If you hold your right hand up to a mirror, the image in the mirror looks like your left hand. No matter how you rotate or flip your left hand, you cannot superimpose it on your right hand. As discussed in Section 20.3, molecules or ions that exhibit this quality are known as being *chiral*. The isomers are called *enantiomers*, and they exhibit the property of optical activity (the rotation of polarized light). For example, as shown in Figure 24.10 ▶, the complex ion $[Co(en)_3]^{3+}$ is nonsuperimposable on its mirror image, so it is a chiral complex.

▶ **FIGURE 24.10 Optical Isomerism in $[Co(en)_3]^{3+}$** The mirror images of $[Co(en)_3]^{3+}$ are not superimposable. (The connected nitrogen atoms represent the ethylenediamine ligand.)

EXAMPLE 24.7 Recognizing and Drawing Optical Isomers

Determine whether the cis or trans isomers shown in Example 24.5 will be optically active.

Solution

Draw the trans isomer of $[Co(en)_2Cl_2]^+$ and its mirror image. Check to see if they are superimposable by rotating one isomer by 180°.

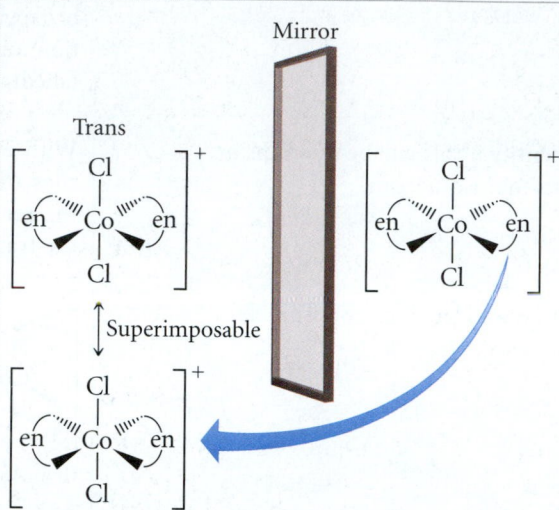

In this case the two are identical so there is no optical activity.

Draw the cis isomer and its mirror image. Check to see if they are superimposable by rotating one isomer by 180°.

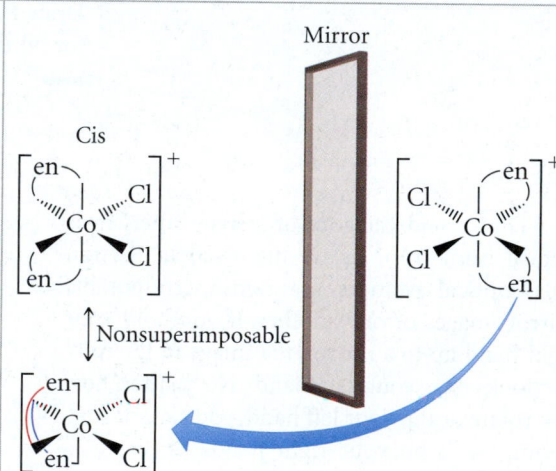

In this case the two structures are not superimposable so the cis isomer will exhibit optical activity.

For Practice 24.7

Determine whether the fac or mer isomers of $[Cr(H_2O)_3Cl_3]^+$ will be optically active.

To this point, we have observed optical isomerism in octahedral complexes. Tetrahedral complexes can also exhibit optical isomerism, but only if all four coordination sites are occupied by different ligands. Square planar complexes do not normally exhibit optical isomerism because they are superimposable on their mirror images.

24.5 Bonding in Coordination Compounds

Valence Bond Theory

The bonding in complex ions, particularly the geometries of the ions, can be described by one of our previous bonding models, valence bond theory (see Chapter 10). Recall from Section 10.6 that in valence bond theory, a coordinate covalent bond is the overlap between a completely filled atomic orbital and an empty atomic orbital. In complex ions, the filled orbital is on the ligand, and the empty orbital is on the metal ion. The metal ion orbitals are hybridized according to the geometry of the complex ion. The common hybridization schemes are shown in Figure 24.11 ▼. An octahedral complex ion requires six empty orbitals in an octahedral arrangement on the metal ion. A full set of d^2sp^3 hybrid orbitals results in the exact orbitals needed for this geometry. A set of sp^3 hybrid orbitals results in a tetrahedral arrangement of orbitals; a set of dsp^2 hybrid orbitals results in a square planar arrangement; and a set of sp hybrid orbitals results in a linear arrangement of orbitals. In each case, the coordinate covalent bond is formed by the overlap between the orbitals on the ligands and the hybridized orbitals on the metal ion.

Geometry	Hybridization	Orbitals
Linear	sp	
Tetrahedral	sp^3	
Square planar	dsp^2	
Octahedral	d^2sp^3	

▲ FIGURE 24.11 **Common Hybridization Schemes in Complex Ions** The valence bond model hybridization schemes can be deduced from the geometry of the complex ion.

Crystal Field Theory

Valence bond theory, while useful for describing the geometries of the complex ions, cannot explain other properties such as color and magnetism. Crystal field theory (CFT), a bonding model for transition metal complexes, can account for these properties. To illustrate the basic principles of CFT, we examine the central metal atom's d orbitals in an octahedral complex.

Octahedral Complexes

The basic premise of CFT is that complex ions form because of attractions between the electrons on the ligands and the positive charge on the metal ion. However, the electrons on the ligands also repel the electrons in the *unhybridized* metal d orbitals. CFT focuses on these repulsions. Figure 24.12 on page 1068 shows how the ligand positions superimpose on the d orbitals in an octahedral complex. Notice that the ligands in an octahedral complex are located in the same space as the lobes of the $d_{x^2-y^2}$ and d_{z^2} orbitals. The repulsions between electron pairs in the ligands and any potential electrons in the d orbitals result in an increase in the energies of these orbitals. In contrast, the d_{xy}, d_{xz}, and d_{yz} orbitals lie *between* the axes and have nodes directly on the axes, which results in less repulsion and lower energies for these three orbitals. In other words, the d orbitals—which are degenerate in the bare metal ion—are split into higher and lower energy levels because of

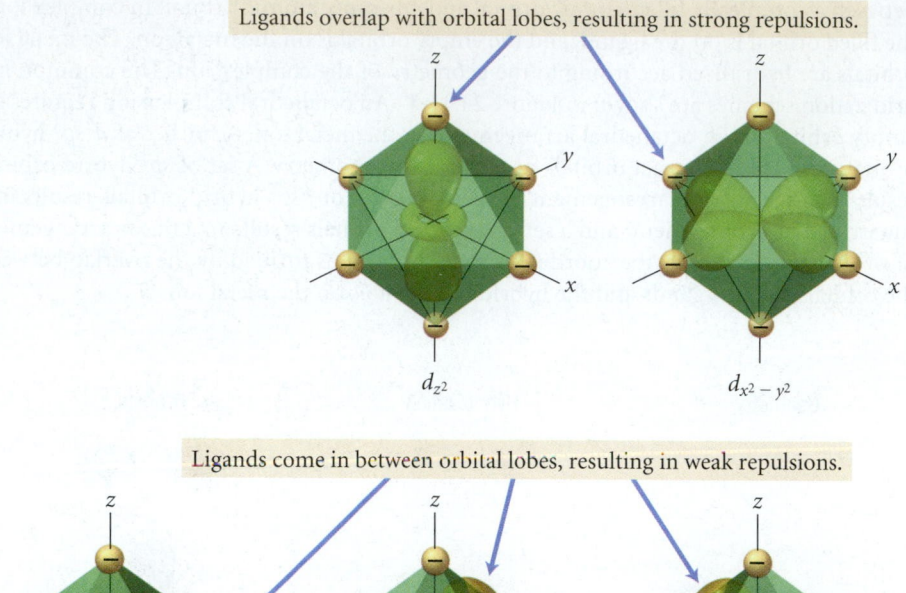

Ligands overlap with orbital lobes, resulting in strong repulsions.

d_{z^2} $d_{x^2-y^2}$

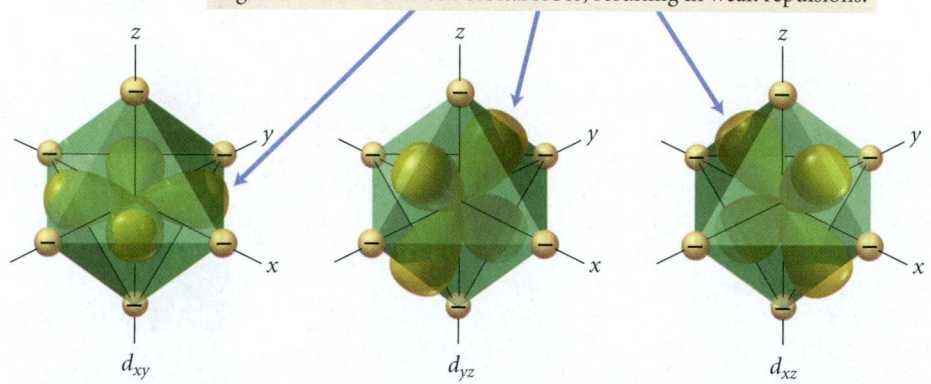

Ligands come in between orbital lobes, resulting in weak repulsions.

d_{xy} d_{yz} d_{xz}

▲ **FIGURE 24.12 Relative Positions of** *d* **Orbitals and Ligands in an Octahedral Complex**
The ligands in an octahedral complex (represented here as spheres of negative charge) interact most strongly with the d_{z^2} and $d_{x^2-y^2}$ orbitals.

the spatial arrangement of the ligands (as shown in Figure 24.13 ▼). The difference in energy between these split *d* orbitals is known as the crystal field splitting energy (Δ). The magnitude of the splitting depends on the particular complex. In **strong-field complexes**, the splitting is large; and in **weak-field complexes**, the splitting is small.

▶ **FIGURE 24.13** *d* **Orbital Splitting in an Octahedral Field**   The otherwise degenerate *d* orbitals are split into two energy levels by the ligands in an octahedral complex ion.

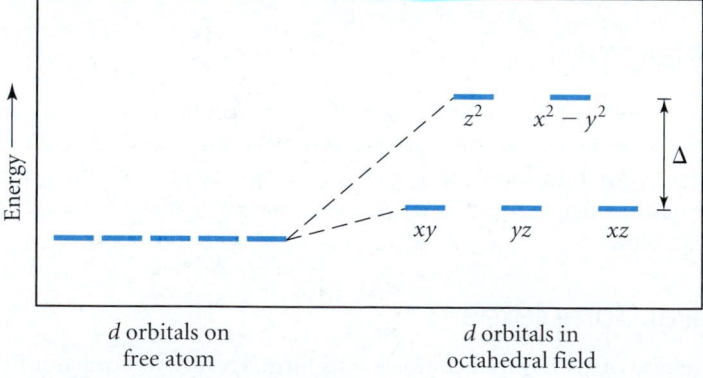

d orbitals on free atom

d orbitals in octahedral field

The Color of Complex Ions and Crystal Field Strength

We saw in the opening section of this chapter that transition metals in host crystals often show brilliant colors because of the crystal field splitting of their *d* orbitals. Solutions of complex ions show brilliant colors because of similar splittings. For example, an $[Fe(CN)_6]^{3-}$ solution is deep red, and an $[Ni(NH_3)_6]^{2+}$ solution is blue (see Figure 24.14 ▶). Recall from Chapter 7 that the color of an object is related to the absorption of light energy by its electrons. If a substance absorbs all of the visible wavelengths, it will appear black. If it reflects all the wavelengths (absorbs no light), it will appear white. A substance will appear to

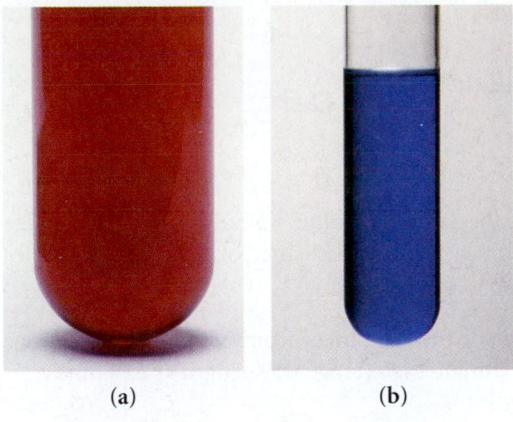

▲ FIGURE 24.14 Colors of Complex Ions
(a) The complex ion $[Fe(CN)_6]^{3-}$ forms a deep red solution, and (b) $[Ni(NH_3)_6]^{2+}$ is blue.

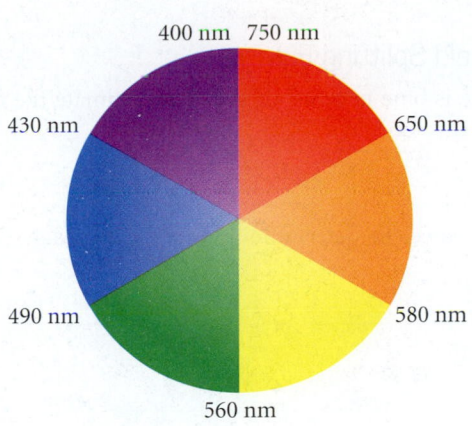

400 nm 750 nm
430 nm   650 nm
490 nm   580 nm
560 nm

▲ FIGURE 24.15 The Color Wheel
Colors across from one another on the color wheel are said to be complementary. A substance that absorbs a color on the wheel will appear to be the complementary color.

be a particular color if it absorbs most visible light but reflects the wavelengths associated with that color. A substance can also appear to be a particular color if it reflects most wavelengths but absorbs the *complementary color* on a color wheel (see Figure 24.15 ▲). For example, a substance that absorbs green light (the complement of red) will appear red. A solution of $[Ti(H_2O)_6]^{3+}$ is purple because it absorbs strongly between 490 and 580 nm, the yellow-green region of the visible spectrum.

The easiest way to measure the energy difference between the *d* orbitals in a complex ion is to use spectroscopy to find the wavelength of light absorbed when an electron makes a transition from the lower energy *d* orbitals to the higher energy ones. With that information we can calculate the crystal field splitting energy, Δ:

$$E_{photon} = h\nu = hc/\lambda = \Delta$$

For example, consider the $[Ti(H_2O)_6]^{3+}$ absorption spectrum shown in Figure 24.16 ▼. The maximum absorbance is at 498 nm. Using this wavelength, we calculate Δ as follows:

$$\Delta = hc/\lambda = (6.626 \times 10^{-34}\,\text{J}\cdot\text{s})(3.00 \times 10^8\,\text{m/s})/(498\,\text{nm} \times 1 \times 10^{-9}\,\text{m/nm})$$

$$\Delta = 3.99 \times 10^{-19}\,\text{J}$$

This energy corresponds to a single $[Ti(H_2O)_6]^{3+}$ ion. We can easily convert to kilojoules per mole:

$$\Delta = (3.99 \times 10^{-19}\,\text{J/ion})(6.02 \times 10^{23}\,\text{ion/mol})(1\,\text{kJ}/1000\,\text{J}) = 240\,\text{kJ/mol}$$

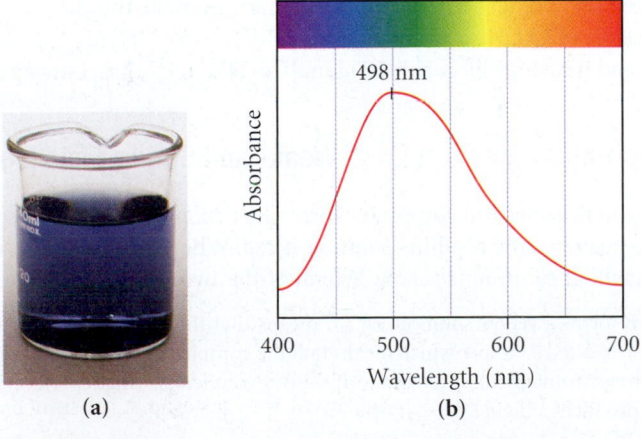

498 nm

Absorbance

400   500   600   700
Wavelength (nm)

(a)   (b)

◀ FIGURE 24.16 The Color and Absorption Spectrum of $[Ti(H_2O)_6]^{3+}$
(a) A solution containing $[Ti(H_2O)_6]^{3+}$ is purple. (b) The absorption spectrum of $[Ti(H_2O)_6]^{3+}$ extends across the green-yellow region of the spectrum.

EXAMPLE 24.8 Crystal Field Splitting Energy

The complex ion $[Cu(NH_3)_6]^{2+}$ is blue in aqueous solution. Estimate the crystal field splitting energy (in kJ/mol) for this ion.

Solution

Begin by consulting the color wheel to determine approximately what wavelength is being absorbed.	Since the solution is blue, we can deduce that orange light must be absorbed since orange is the complementary color to blue.
Estimate the absorbed wavelength.	The color orange ranges from 580 to 650 nm, so we can estimate the average wavelength as 615 nm.
Calculate the energy corresponding to this wavelength, using $E = hc/\lambda$. This energy will correspond to Δ.	$$E = \frac{(6.626 \times 10^{-34}\,\text{J} \cdot \text{s})(3.00 \times 10^{8}\,\text{m/s})}{(615\,\text{nm})(1 \times 10^{-9}\,\text{m/nm})}$$ $$E = 3.23 \times 10^{-19}\,\text{J} = \Delta$$
Convert J/ion into kJ/mol.	$$\Delta = \frac{(3.23 \times 10^{-19}\,\text{J/ion})(6.02 \times 10^{23}\,\text{ion/mol})}{(1000\,\text{J/kJ})}$$ $$\Delta = 195\,\text{kJ/mol}$$

For Practice 24.8

The complex ion $[Co(NH_3)_5NO_2]^{2+}$ is yellow. Estimate the crystal field splitting energy (in kJ/mol) for this ion.

The magnitude of the crystal field splitting in a complex ion—and, therefore whether it is a strong-field or weak-field complex—depends in large part on the ligands attached to the central metal ion. Spectroscopic studies of various ligands attached to the same metal allow us to arrange different ligands in order of their ability to split the *d* orbitals. This list is known as the *spectrochemical series* and is arranged from ligands that result in the largest Δ to those that result in the smallest:

$$CN^- > NO_2^- > en > NH_3 > H_2O > OH^- > F^- > Cl^- > Br^- > I^-$$

large Δ small Δ

typically strong-field ligands typically weak-field ligands

Ligands that produce large values of Δ are known as *strong-field ligands* and those that give small values of Δ are known as *weak-field ligands*.

The metal ion can also have an effect on the magnitude of Δ. If we examine different metal ions with the same ligand, we find that Δ increases as the charge on the metal ion increases. The greater charge on the metal draws the ligands closer, causing greater repulsion with the *d* orbitals and therefore a larger Δ. An example of this behavior occurs in the complex ions between NH_3 (a ligand in the middle of the spectrochemical series) and the +2 or +3 oxidation states of Co. Hexaamminecobalt(II) ion, $[Co(NH_3)_6]^{2+}$, has a weak crystal field (small Δ) and hexaamminecobalt(III) ion, $[Co(NH_3)_6]^{3+}$, has a strong field (large Δ).

 Conceptual Connection 24.2 Weak- and Strong-Field Ligands

Two ligands, A and B, both form complexes with a particular metal ion. When the metal ion complexes with ligand A, the resulting solution is red. When the metal ion complexes with ligand B, the resulting solution is yellow. Which of the two ligands produces the larger Δ?

Answer: Ligand B forms a yellow solution, which means that the complex absorbs in the violet region. Ligand A forms a red solution, which means that the complex absorbs in the green. Since the violet region of the electromagnetic spectrum is of shorter wavelength (higher energy) than the green region, ligand B produces a higher Δ.

Magnetic Properties

The strength of the crystal field potentially affects the magnetic properties of a transition metal complex. Recall that, according to Hund's rule, electrons occupy degenerate orbitals singly as long as an empty orbital is available. When the energies of the d orbitals are split by ligands, the lower energy orbitals fill first. Once they are half-filled, the next electron can either (1) pair with an electron in one of the lower energy half-filled orbitals by overcoming the electron–electron repulsion associated with having two electrons in the same orbital or (2) go into an empty orbital of higher energy by overcoming the energy difference between the orbitals—in this case, the crystal field splitting energy, Δ. The magnitude of Δ compared to the electron–electron repulsions determines the d orbital occupancy and therefore the number of unpaired electrons.

Let us compare two iron(II) complexes to see the difference in behavior under strong- and weak-field conditions. $[Fe(CN)_6]^{4-}$ is known to be diamagnetic and $[Fe(H_2O)_6]^{2+}$ is known to be paramagnetic. Both of these complexes contain Fe^{2+}, which has an electron configuration of $[Ar]\ 3d^6$. In the case of $[Fe(CN)_6]^{4-}$, CN^- is a strong-field ligand that generates a large Δ, so it will take more energy to occupy the higher energy level than it does to pair the electrons in the lower energy level. The result is that all six electrons are paired and the compound is diamagnetic, as shown at right.

In $[Fe(H_2O)6]^{2+}$, H_2O is a weak-field ligand that generates a small Δ, so the electron pairing energy is greater than Δ. The result is that the first five electrons occupy the five d orbitals singly and only the sixth pairs up, resulting in a paramagnetic compound with four unpaired electrons, as shown at right.

In general, complexes with strong-field ligands have fewer unpaired electrons relative to the free metal ion, and are therefore called **low-spin complexes**. Complexes with weak-field ligands, in contrast, have the same number of unpaired electrons as the free metal ion and are called **high-spin complexes**.

When we examine the orbital diagrams of the d^1 through d^{10} metal ions in octahedral complexes, we find that only d^4, d^5, d^6, and d^7 metal ions have low- and high-spin possibilities. Since there are three lower energy d orbitals, the d^1, d^2, and d^3 metal ions will always have unpaired electrons, independent of Δ. In the d^8, d^9, and d^{10} metal ions, the three lower energy orbitals are completely filled, so the remaining electrons fill the two higher orbitals (as expected by Hund's rule), also independent of Δ.

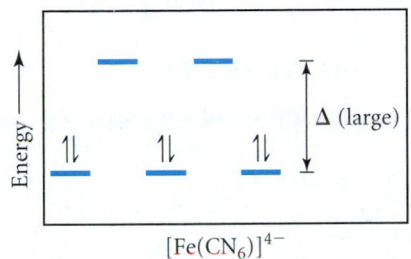

$[Fe(CN_6)]^{4-}$

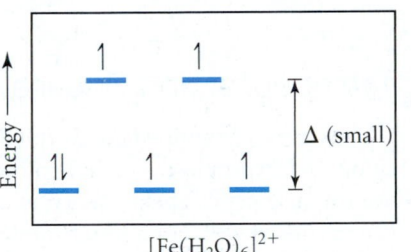

$[Fe(H_2O)_6]^{2+}$

Recall from Section 8.7 that a paramagnetic species contains unpaired electrons and a diamagnetic one does not.

Guidelines for Determining the Number of Unpaired Electrons in Octahedral Complexes	**EXAMPLE 24.9** High- and Low-Spin Octahedral Complexes	**EXAMPLE 24.10** High- and Low-Spin Octahedral Complexes
	How many unpaired electrons would you expect for the complex ion $[CoF_6]^{3-}$?	How many unpaired electrons would you expect for the complex ion $[Co(NH_3)_5NO_2]^{2+}$?
Begin by determining the charge and number of d electrons on the metal.	**Solution** The metal is Co^{3+} and has a d^6 electronic configuration.	**Solution** The metal is Co^{3+} and has a d^6 electronic configuration.
Look at the spectrochemical series to determine whether the ligand is a strong-field or a weak-field ligand.	F^- is a weak-field ligand, so Δ will be relatively small.	NH_3 and NO_2^- are both strong-field ligands, so Δ will be relatively large.

Decide if the complex will be high- or low-spin and draw the electron configuration.	Weak-field ligands yield high-spin configurations.	Strong-field ligands yield low-spin configurations.
Count the unpaired electrons.	This configuration has four unpaired electrons.	This configuration has no unpaired electrons.

For Practice 24.9

How many unpaired electrons would you expect for the complex ion $[FeCl_6]^{3-}$?

For Practice 24.10

How many unpaired electrons would you expect for the complex ion $[Co(CN)_6]^{4-}$?

Tetrahedral and Square Planar Complexes

So far, we have examined the d orbital energy changes for octahedral complexes, but transition metal complexes can have other geometries, such as tetrahedral and square planar. We can use crystal field theory to determine the d orbital splitting pattern for these geometries as well. For a tetrahedral complex, the d orbital splitting pattern is the opposite of the octahedral splitting pattern: three d orbitals (d_{xy}, d_{xz}, and d_{yz}) are higher in energy and two d orbitals ($d_{x^2-y^2}$ and d_{z^2}) are lower in energy, as shown in Figure 24.17 ▼. Almost all tetrahedral complexes are high-spin because of reduced ligand–metal interactions. The d orbitals in a tetrahedral complex are interacting with only four ligands, as opposed to six in the octahedral complex, so the value of Δ is generally smaller.

A square planar complex gives us the most complex splitting pattern of the three geometries we have examined, as shown in Figure 24.18 ▶. As we discussed previously, square planar complexes occur in d^8 metal ions, such as Pt^{2+}, Pd^{2+}, Ir^+, or Au^{3+}, and in nearly all cases they are low-spin.

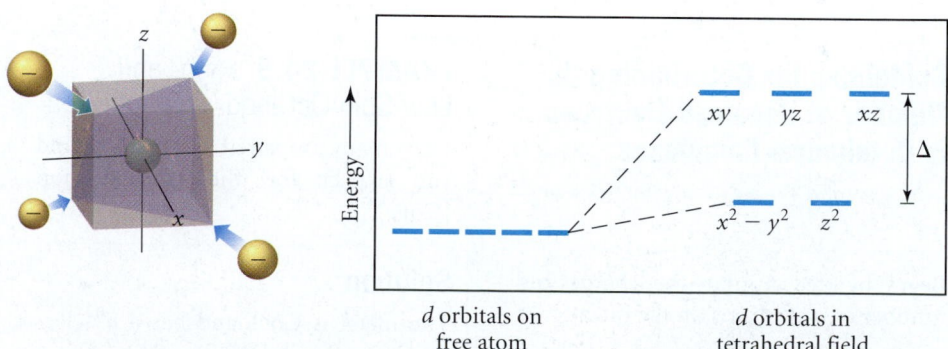

▲ FIGURE 24.17 **Splitting of d Orbitals by a Tetrahedral Ligand Geometry** In tetrahedral complexes, the splitting of the d orbitals has a pattern that is the opposite of the octahedral splitting pattern. The d_{xy}, d_{yz}, and d_{xz} orbitals are higher in energy than the d_{z^2} and $d_{x^2-y^2}$ orbitals.

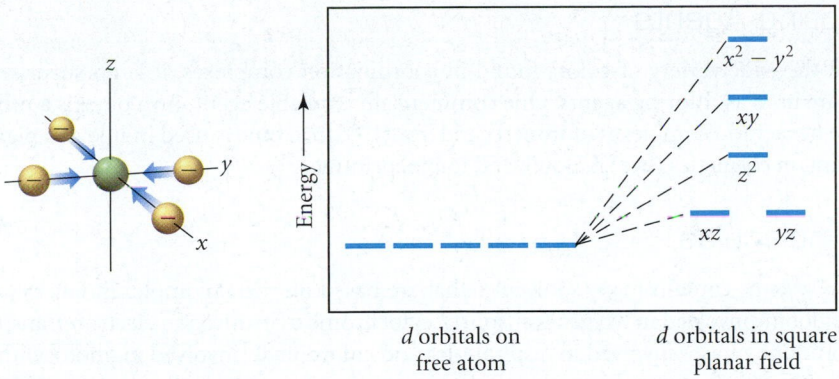

▲ **FIGURE 24.18 Splitting of *d* Orbitals by a Square Planar Ligand Geometry**
Square planar complexes produce the *d* orbital energy pattern shown here.

24.6 Applications of Coordination Compounds

Coordination compounds are commonly found in living systems, in industry, and even in household products. In Chapter 23, we saw how both silver and gold metals are extracted from their respective ores using cyanide complexes and how nickel metal is extracted by forming the gaseous carbonyl complex, $Ni(CO)_4$. In this section we will describe a few other applications of coordination compounds.

Chelating Agents

Earlier in this chapter, we mentioned the chelating agent ethylenediaminetetraacetate ion ($EDTA^{4-}$). This ligand has lone pairs on six different donor atoms that can interact with a metal ion, so this ligand makes very stable metal complexes. One use of EDTA is in the medical treatment of heavy metal poisoning such as lead poisoning. The patient is given $[Ca(EDTA)]^{2-}$ and since the lead complex ($K_f = 2 \times 10^{18}$) is more stable than the calcium complex ($K_f = 4 \times 10^{10}$), the lead displaces the calcium. The body then excretes the lead complex and leaves behind the calcium, which is nontoxic (and is in fact a nutrient).

Chemical Analysis

Some ligands are selective in their binding, preferring specific metal ions, so they can be used in chemical analysis. For example, dimethylglyoxime (dmg) is used to chemically analyze a sample for Ni^{2+} or Pd^{2+}. In the presence of Ni^{2+}, an insoluble red precipitate forms, and in the presence of Pd^{2+}, an insoluble yellow precipitate forms. Another ligand used in chemical analysis is the SCN^- ligand, which is used to test for Co^{2+} or Fe^{3+}. In the presence of Co^{2+} a blue solution forms, and in the presence of Fe^{3+} a deep red solution forms (see Figure 24.19 ▼).

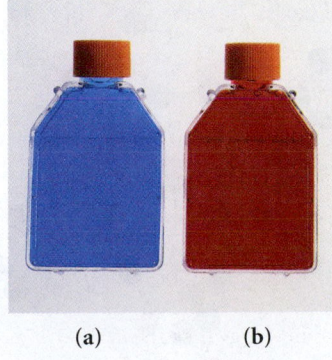

(a)   **(b)**

◀ **FIGURE 24.19 Chemical Analysis with SCN^-** (a) Blue indicates Co^{2+}. (b) Red indicates Fe^{3+}.

Coloring Agents

Given the wide variety of colors found in coordination complexes, it is no surprise that they are used as coloring agents. One commercially available agent, iron blue, is a mixture of the hexacyano complexes of iron(II) and iron(III). Iron blue is used in ink, as a pigment in paint, in cosmetics (eye shadow), and in blueprinting.

Biomolecules

Living systems contain many molecules that are based on metal complexes. For example, hemoglobin (involved in oxygen transport), cytochrome c (involved in electron transport), carbonic anhydrase (involved in respiration), and chlorophyll (involved in photosynthesis) all have coordinated metal ions that are critical to their structure and function. Table 24.7 summarizes the biological significance of many of the other first-row transition metals.

TABLE 24.7 Transition Metals and Some of Their Functions in the Human Body	
Transition Metal	Biological Function
Chromium	Works with insulin to control utilization of glucose
Manganese	Fat and carbohydrate synthesis
Molybdenum	Involved in hemoglobin synthesis
Iron	Oxygen transport
Copper	Involved in hemoglobin synthesis
Zinc	Involved in cell reproduction and tissue growth; part of more than 70 enzymes; assists in the utilization of carbohydrate, protein, and fat

Hemoglobin and Cytochrome C In hemoglobin and in cytochrome c, an iron complex called a heme is connected to a protein, as shown below. A heme is an iron ion coordinated to a flat, polydentate ligand called a porphyrin (Figure 24.20 ▶). The porphyrin ligand has a planar ring structure with four nitrogen atoms that can coordinate to the metal ion. Different porphyrins have different substituent groups connected around the outside of the ring.

In hemoglobin, the iron complex is octahedral, with the four nitrogen atoms of the porphyrin in a square planar arrangement around the metal. A nitrogen atom from a nearby amino acid of the protein occupies the fifth coordination site, and either O_2 or

Hemoglobin was discussed in Sections 1.1 and 14.1.

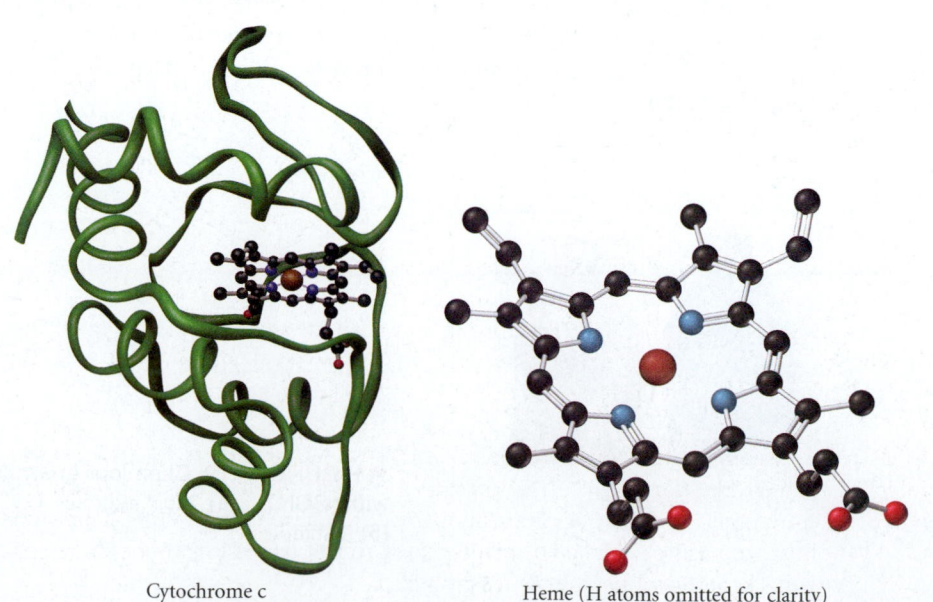

Cytochrome c   Heme (H atoms omitted for clarity)

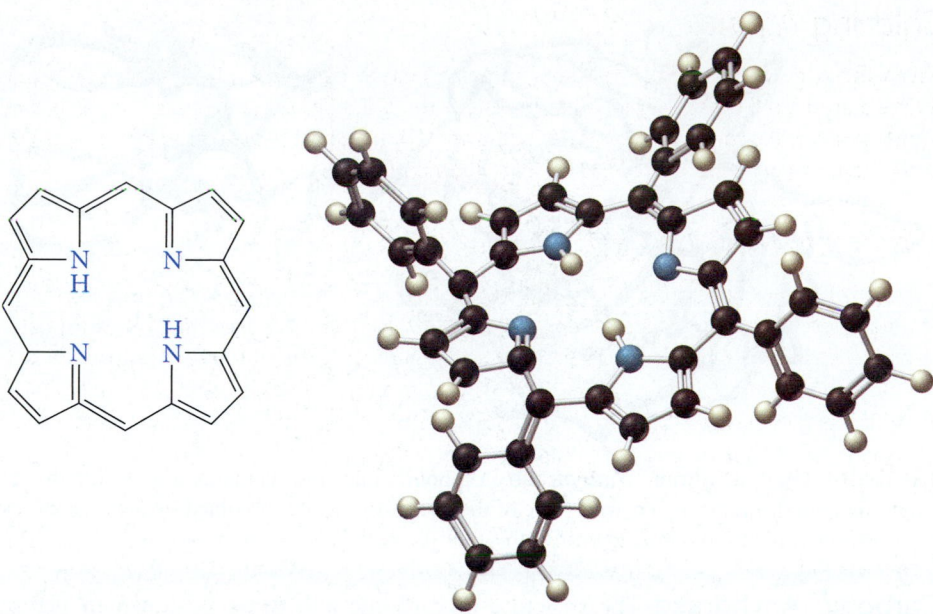

▲ **FIGURE 24.20 Porphyrin** A porphyrin has four nitrogen atoms that can coordinate to a central metal atom.

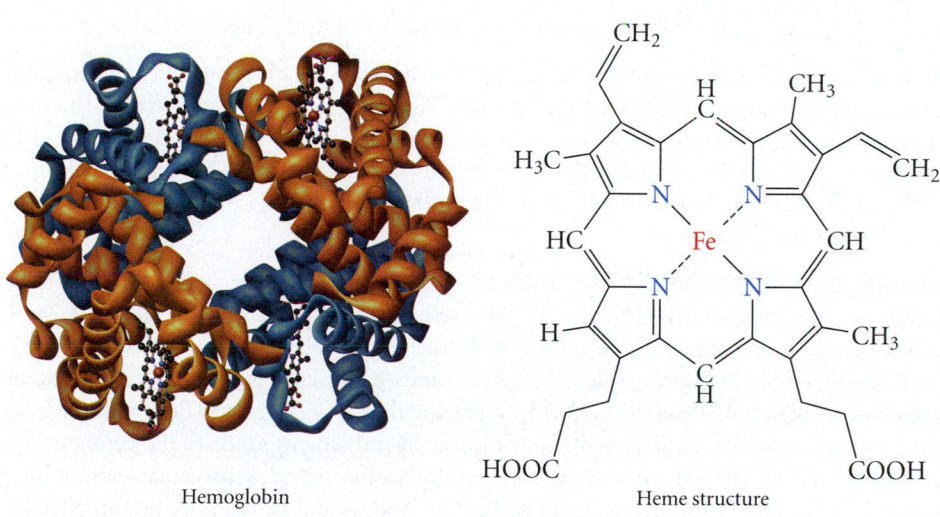

Hemoglobin   Heme structure

▲ **FIGURE 24.21 Hemoglobin** In hemoglobin, the iron complex is octahedral, with the four nitrogen atoms of the porphyrin in a square planar arrangement around the metal. A nitrogen atom from a nearby amino acid of the protein occupies the fifth coordination site, and either O_2 or H_2O occupies the last coordination site.

H_2O occupies the last coordination site (see Figure 24.21 ▲). In the lungs, where the oxygen content is high, the hemoglobin coordinates to an O_2 molecule. The hemoglobin is then carried by the bloodstream to areas throughout the body that are depleted in oxygen, where oxygen is released and replaced by a water molecule. The hemoglobin then travels back to the lungs to repeat the cycle.

Chlorophyll Chlorophyll, shown in Figure 24.22 ▶, is another porphyrin-based biomolecule, but here the porphyrin is not surrounded by a protein, and the coordinated metal is magnesium (which is not a transition metal). Chlorophyll is essential for the process of photosynthesis performed by plants, in which light energy from the sun is converted to chemical energy to fuel the plant's growth.

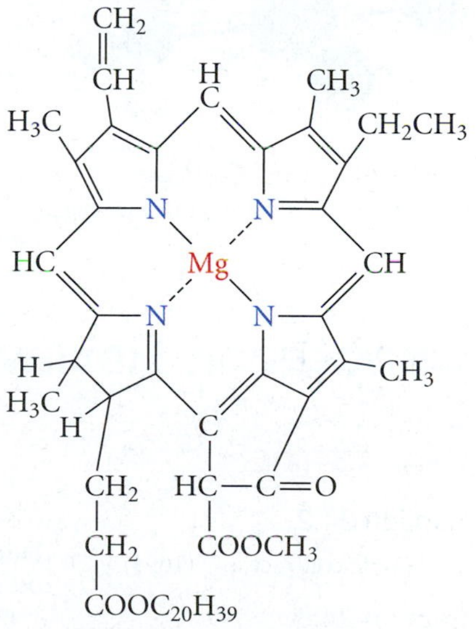

▲ **FIGURE 24.22 Chlorophyll** Chlorophyll, involved in photosynthesis in plants, contains magnesium coordinated to a porphyrin.

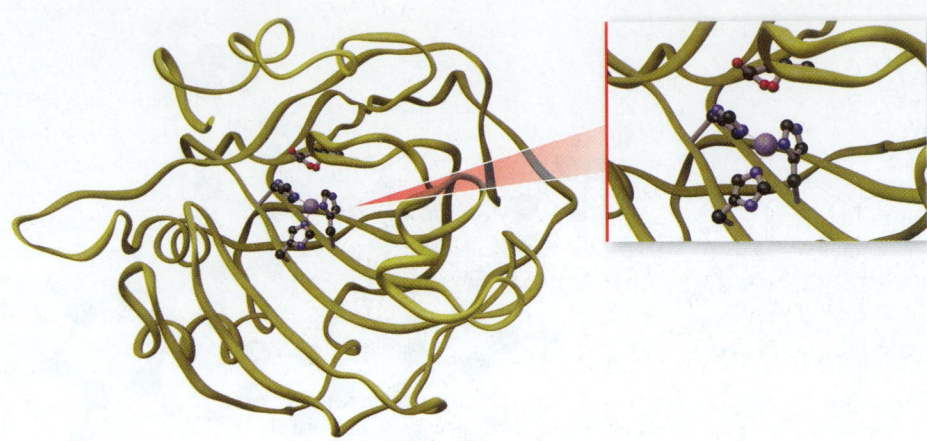

▲ FIGURE 24.23 **Carbonic Anhydrase**   Carbonic anhydrase contains a zinc ion that is bound in a tetrahedral complex, with three of the coordination sites occupied by nitrogen atoms from surrounding amino acids and the fourth site available to bind a water molecule.

Carbonic Anyhdrase   The structure of carbonic anhydrase is shown in Figure 24.23 ▲. The zinc ion is bound in a tetrahedral complex, with three of the coordination sites occupied by nitrogen atoms from surrounding amino acids and the fourth site available to bind a water molecule. Carbonic anhydrase catalyzes the reaction between water and CO_2 in respiration:

$$H_2O(l) + CO_2(g) \rightleftharpoons H^+(aq) + HCO_3^-(aq)$$

A water molecule alone is not acidic enough to react with a CO_2 molecule at a sufficient rate. When the water molecule is bound to the zinc ion in carbonic anhydrase, the positive charge on the metal draws electron density from the O—H bond and the H_2O becomes more acidic—sufficiently so to readily lose a proton. The resulting bound OH^- can easily react with a CO_2 molecule, and the reaction is therefore much faster than the uncatalyzed version.

Drugs and Therapeutic Agents   In the mid-1960s researchers found that the platinum(II) complex *cis*-[Pt(NH$_3$)$_2$Cl$_2$], known as cisplatin and shown in Figure 24.24 ◄, was an effective anticancer agent. Interestingly, the closely related geometric isomer *trans*-[Pt(NH$_3$)$_2$Cl$_2$] has little or no effect on cancer tumors. Cisplatin is believed to function by attaching itself to the cancer cell's DNA by replacing the Cl^- ligands with donor atoms from the DNA strands. The cis arrangement of the Cl^- ligands corresponds to the geometry required to bind to the DNA strands. The trans isomer, although closely related, cannot bind properly due to the trans arrangement of the Cl^- ligands and is therefore not an effective agent. Cisplatin and other closely related platinum(II) complexes are still used today in the chemotherapy for certain types of cancer and are among the most effective anticancer agents available in these cases.

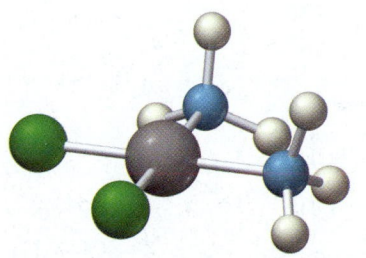

▲ FIGURE 24.24 **Cisplatin**
Cisplatin is an effective anticancer agent.

Chapter in Review

Key Terms

Section 24.2
lanthanide contraction (1054)

Section 24.3
complex ion (1056)
ligand (1056)
coordination compound
    (1056)
primary valence (1056)

secondary valence (1056)
coordination number (1056)
coordinate covalent bond
    (1057)
monodentate (1057)
bidentate (1057)
polydentate (1058)
chelate (1059)
chelating agent (1059)

Section 24.4
structural isomers (1062)
stereoisomers (1062)
coordination isomers (1062)
linkage isomers (1062)
geometric isomers (1062)
optical isomers (1065)

Section 24.5
strong-field complex (1068)
weak-field complex (1068)
low-spin complex (1071)
high-spin complex (1071)

Key Concepts

Electron Configurations (24.2)

As we work across a row of transition elements, we add electrons to the $(n - 1)d$ orbitals, resulting in a general electron configuration for first- and second-row transition elements of [noble gas] $ns^2(n - 1)d^x$ and for the third and fourth rows of [noble gas] $ns^2(n - 2)f^{14}(n - 1)d^x$, where x ranges from 1 to 10. A transition element forms a cation by losing electrons from the ns orbitals before losing electrons from the $(n - 1)d$ orbitals.

Periodic Trends (24.2)

As we look at the variations in atomic size, ionization energy, and electronegativity across a row in the periodic table, we find behavior similar to what we observe for main-group elements. Looking down a group, however, we see that atomic size increases from the first row to the second, but size in third row is no larger than in the second because of the lanthanide contraction. This effect causes the trend down a group for ionization energy and electronegativity to be the opposite of what we see for the rest of the periodic table.

Composition and Naming of Coordination Compounds (24.3)

A coordination compound is made up of a complex ion and a counterion. A complex ion contains a central metal ion bound to one or more ligands. The number of ligands directly bound to the metal ion is called the coordination number. The ligand forms a coordinate covalent bond to the metal ion by donating a pair of electrons to an empty orbital on the metal. Ligands that donate a single pair of electrons are termed monodentate. A ligand that donates two pairs of electrons is termed bidentate, and a ligand that donates more than two pairs is termed polydentate. In naming coordination compounds, we use the name of the cation followed by the name of the anion. To name a complex ion we use the guidelines outlined in Section 24.3.

Types of Isomers (24.4)

We can broadly divide the isomerism observed in coordination compounds into two categories: structural isomers, in which atoms are connected differently to one another, and stereoisomers, in which atoms are connected in the same way but the ligands have a different spatial arrangement about the metal atom. These broad categories can each be further broken down into two types. Structural isomers can be either coordination isomers (a coordinated ligand exchanges places with an uncoordinated counterion,) or linkage isomers (a particular ligand has the ability to coordinate to the metal in different ways). Stereoisomers can be either geometric isomers (the ligands bonded to the metal have a different spatial arrangement relative to each other, leading to cis–trans or fac–mer isomers) or optical isomers (nonsuperimposable mirror images of one another).

Crystal Field Theory (24.5)

Crystal field theory is a bonding model for transition metal complex ions. The model describes how the degeneracy of the d orbitals is broken by the repulsive forces between the electrons on the ligands around the metal ion and the d orbitals in the metal ion. The energy difference between the split d orbitals is called the crystal field splitting energy (Δ). The magnitude of Δ depends in large part on the ligands bound to the metal. When we examine the magnetic properties of complex ions we find that octahedral complexes with a d^4, d^5, d^6, or d^7 metal ion can have two possible electronic configurations with different numbers of unpaired electrons. The first, called high-spin, has the same number of unpaired electrons as the free metal ion and is usually the result of a weak crystal field. The second, called low-spin, has fewer unpaired electrons than the free metal ion and is usually the result of a strong crystal field.

Key Equations and Relationships

Crystal Field Splitting Energy (24.5)

$\Delta = hc/\lambda$ (where λ is the wavelength of maximum absorption)

Key Skills

Writing Electronic Configurations for Transition Metals and Their Ions (24.2)

- Examples 24.1, 24.2   • For Practice 24.1, 24.2   • Exercises 19, 20, 57, 58

Naming Coordination Compounds (24.3)

- Examples 24.3, 24.4   • For Practice 24.3, 24.4   • Exercises 23–28

Recognizing and Drawing Geometric Isomers (24.4)

- Examples 24.5, 24.6   • For Practice 24.5, 24.6   • Exercise 35–40, 61, 62

Recognizing and Drawing Optical Isomers (24.4)

- Example 24.7   • For Practice 24.7   • Exercises 39, 40, 61, 62

Calculating Crystal Field Splitting Energy (24.5)

- Example 24.8   • For Practice 24.8   • Exercises 43–46

Recognizing and Predicting High-Spin and Low-Spin Octahedral Complex Ions (24.5)

- Examples 24.9, 24.10   • For Practice 24.9, 24.10   • Exercises 49–52, 65

Exercises

Review Questions

1. When a transition metal atom forms an ion, which electrons are lost first?

2. Explain why transition metals exhibit multiple oxidation states instead of a single oxidation state (like most of the main-group metals).

3. Why is the +2 oxidation state so common for transition metals?

4. Explain why atomic radii of elements in the third row of the transition metals are no larger than those of elements in the second row.

5. Gold is the most electronegative transition metal. Explain.

6. Briefly define each of the following:
   a. coordination number   b. ligand
   c. bidentate and polydentate   d. complex ion
   e. chelating agent

7. Using the Lewis acid–base definition, how would you categorize a ligand? How would you categorize a transition metal ion?

8. Explain the differences between each of the following pairs:
   a. structural isomer and stereoisomer

   b. linkage isomer and coordination isomer
   c. geometric isomer and optical isomer

9. Which of the following structures for a complex ion can exhibit cis–trans isomerism: linear, tetrahedral, square planar, octahedral?

10. How can you tell whether a complex ion will be optically active?

11. Explain the differences between weak-field and strong-field metal complexes.

12. Explain why compounds of Sc^{3+} are colorless, but compounds of Ti^{3+} are colored.

13. Explain why compounds of Zn^{2+} are white, but compounds of Cu^{2+} are often blue or green.

14. Explain the differences between high-spin and low-spin metal complexes.

15. Explain why almost all tetrahedral complexes are high-spin.

16. Explain how crystal field theory accounts for why so many transition metal compounds are colored.

Problems by Topic

Properties of Transition Metals

17. Write the ground state electron configuration for the following species:
   a. Ni, Ni^{2+} b. Mn, Mn^{4+}
   c. Y, Y^+ d. Ta, Ta^{2+}

18. Write the ground state electron configuration for the following species:
   a. Zr, Zr^{2+} b. Co, Co^{2+}
   c. Tc, Tc^{3+} d. Os, Os^{4+}

19. Determine the highest possible oxidation state for the following:
   a. V   b. Re   c. Pd

20. Which first-row transition metal(s) has the following highest possible oxidation state?
   a. +3   b. +7   c. +4

Coordination Compounds

21. Determine the oxidation state and coordination number of the metal ion in each of the following:
   a. $[Cr(H_2O)_6]^{3+}$ b. $[Co(NH_3)_3Cl_3]^-$
   c. $[Cu(CN)_4]^{2-}$ d. $[Ag(NH_3)_2]^+$

22. Determine the oxidation state and coordination number of the metal ion in each of the following:
   a. $[Co(NH_3)_5Br]^{2+}$ b. $[Fe(CN)_6]^{4-}$
   c. $[Co(ox)_3]^{4-}$ d. $[PdCl_4]^{2-}$

23. Name the following:
   a. $[Cr(H_2O)_6]^{3+}$
   b. $[Cu(CN)_4]^{2-}$
   c. $[Fe(NH_3)_5Br]SO_4$
   d. $[Co(H_2O)_4(NH_3)(OH)]Cl_2$

24. Name the following:
   a. $[Cu(en)_2]^{2+}$
   b. $[Mn(CO)_3(NO_2)_3]^{2+}$
   c. $Na[Cr(H_2O)_2(ox)_2]$
   d. $[Co(en)_3][Fe(CN)_6]$

25. Write the correct formula for the following:
   a. hexaamminechromium(III) ion
   b. potassium hexacyanoferrate(III)
   c. ethylenediaminedithiocyanatocopper(II)
   d. tetraaquaplatinum(II) hexachloroplatinate(IV)

26. Write the correct formula for the following:
   a. hexaaquanickel (II) chloride
   b. pentacarbonylchloromanganese(I)
   c. ammonium diaquatetrabromovanadate(III)
   d. tris(ethylenediamine)cobalt(III) trioxalatoferrate(III)

27. Write the formula and give the name of each of the following:
   a. a complex ion with Co^{3+} as the central ion and three NH_3 molecules and three CN^- ions as ligands
   b. a complex ion with Cr^{3+} as the central ion and a coordination number of 6 with ethylenediamine ligands

28. Write the formula and give the name of each of the following:

    a. a complex ion with four water molecules and two ONO^- ions connected to an Fe(III) ion

    b. a coordination compound made of two complex ions: one a complex of V(III) with two ethylenediamine molecules and two Cl^- ions as ligands and the other a complex of Ni(II) having a coordination number of 4 with Cl^- ions as ligands

Structure and Isomerism

29. Draw two linkage isomers of $[Mn(NH_3)_5(NO_2)]^{2+}$.

30. Draw two linkage isomers of $[PtCl_3(SCN)]^{2-}$.

31. Write the formulas and names for the coordination isomers of $[Fe(H_2O)_6]Cl_2$.

32. Write the formulas and names for the coordination isomers of $[Co(en)_3][Cr(ox)_3]$.

33. Which of the following complexes will exhibit geometric isomerism?

    a. $[Cr(NH_3)_5(OH)]^{2+}$

    b. $[Cr(en)_2Cl_2]^+$

    c. $[Cr(H_2O)(NH_3)_3Cl_2]^+$

    d. $[Pt(NH_3)Cl_3]^-$

    e. $[Pt(H_2O)_2(CN)_2]$

34. Which of the following complexes will exhibit geometric isomerism?

    a. $[Co(H_2O)_2(ox)_2]^-$

    b. $[Co(en)_3]^{3+}$

    c. $[Co(H_2O)_2(NH_3)_2(ox)]^+$

    d. $[Ni(NH_3)_2(en)]^{2+}$

    e. $[Ni(CO)_2Cl_2]$

35. If W, X, Y, and Z are different monodentate ligands, how many geometric isomers are there for the following?

    a. square-planar $[NiWXYZ]^{2+}$

    b. tetrahedral $[ZnWXYZ]^{2+}$

36. How many geometric isomers are there for the following?

    a. $[Fe(CO)_3Cl_3]$

    b. $[Mn(CO)_2Cl_2Br_2]^+$

37. Draw the structures and label the type for all the isomers of each of the following:

    a. $[Cr(CO)_3(NH_3)_3]^{3+}$

    b. $[Pd(CO)_2(H_2O)Cl]^+$

38. Draw the structures and label the type for all the isomers of each of the following:

    a. $[Fe(CO)_4Cl_2]^+$

    b. $[Pt(en)Cl_2]$

39. Determine if either isomer of $[Cr(NH_3)_2(ox)_2]^-$ is optically active.

40. Determine if either isomer of $[Fe(CO)_3Cl_3]$ is optically active.

Bonding in Coordination Compounds

41. Draw the octahedral crystal field splitting diagram for the following metal ions:

    a. Zn^{2+}

    b. Fe^{3+} (high- and low-spin)

    c. V^{3+}

    d. Co^{2+} (high-spin)

42. Draw the octahedral crystal field splitting diagram for the following metal ions:

    a. Cr^{3+}

    b. Cu^{2+}

    c. Mn^{3+} (high- and low-spin)

    d. Fe^{2+} (low-spin)

43. The $[CrCl_6]^{3-}$ ion has a maximum in its absorption spectrum at 735 nm. Calculate the crystal field splitting energy (in kJ/mol) for this ion.

44. The absorption spectrum of the complex ion $[Rh(NH_3)_6]^{3+}$ has maximum absorbance at 295 nm. Calculate the crystal field splitting energy (in kJ/mol) for this ion.

45. Three complex ions of cobalt(III), $[Co(CN)_6]^{3-}$, $[Co(NH_3)_6]^{3+}$, and $[CoF_6]^{3-}$, are found to absorb light at wavelengths of (in no particular order) 290 nm, 440 nm, and 770 nm. Match each complex ion to the appropriate wavelength absorbed. What color would you expect each solution to be?

46. Three bottles of aqueous solutions are discovered in an abandoned lab. The solutions are green, yellow, and purple. It is known that three complex ions of chromium(III) were commonly used in that lab: $[Cr(H_2O)_6]^{3+}$, $[Cr(NH_3)_6]^{3+}$, and $[Cr(H_2O)_4Cl_2]^+$. What is the likely identity of each of the solutions?

47. The $[Mn(NH_3)_6]^{2+}$ ion is paramagnetic with five unpaired electrons. Is the NH_3 ligand inducing a strong or weak field?

48. The complex $[Fe(H_2O)_6]^{2+}$ is paramagnetic. Is the H_2O ligand inducing a strong or weak field?

49. How many unpaired electrons would you expect for the following complex ions?

    a. $[RhCl_6]^{3-}$ **b.** $[Co(OH)_6]^{4-}$

    c. cis-$[Fe(en)_2(NO_2)_2]^+$

50. How many unpaired electrons would you expect for the following complex ions?

    a. $[Cr(CN)_6]^{4-}$ **b.** $[MnF_6]^{4-}$

    c. $[Ru(en)_3]^{2+}$

51. How many unpaired electrons would you expect for the complex ion $[CoCl_4]^{2-}$ if it is a tetrahedral shape?

52. The complex ion $[PdCl_4]^{2-}$ is known to be diamagnetic. Use this information to determine if it is a tetrahedral or square planar structure.

Applications of Coordination Compounds

53. What structural feature do hemoglobin, cytochrome c, and chlorophyll have in common?

54. Identify the central metal atom in the following:

    a. hemoglobin   **b.** carbonic anhydrase

    c. chlorophyll   **d.** iron blue

55. Hemoglobin in our bodies exists in two predominant forms. One form, known as oxyhemoglobin, has O_2 bound to the iron and the other, known as deoxyhemoglobin, has a water molecule bound instead. Oxyhemoglobin is a low-spin complex that gives arterial blood its red color, and deoxyhemoglobin is a high-spin complex that gives venous blood its blue color. Explain these observations in terms of crystal field splitting. Would you categorize O_2 as a strong- or weak-field ligand?

56. Carbon monoxide and the cyanide ion are both toxic because they bind more strongly than oxygen to the iron in hemoglobin (Hb).

$$Hb + O_2 \rightleftharpoons HbO_2 \qquad K = 2 \times 10^{12}$$
$$Hb + CO \rightleftharpoons HbCO \qquad K = 1 \times 10^{14}$$

Calculate the equilibrium constant value for the reaction:

$$HbO_2 + CO \rightleftharpoons HbCO + O_2$$

Does the equilibrium favor reactants or products?

Cumulative Problems

57. In Chapter 8 we learned that Cr and Cu were exceptions to the normal orbital filling, resulting in a $[Ar]\ 4s^1 3d^x$ configuration. Write the ground state electron configuration for the following species:
  a. $Cr, Cr^+, Cr^{2+}, Cr^{3+}$ **b.** Cu, Cu^+, Cu^{2+}

58. In the second row of the transition metals there are five elements that do not follow the normal orbital filling. Five—Nb, Mo, Ru, Rh, and Ag—result in a $[Kr]\ 5s^1 4d^x$ configuration and Pd results in a $[Kr]\ 4d^{10}$ configuration. Write the ground state electron configuration for the following species:
  a. Mo, Mo^+, Ag, Ag^+ **b.** Ru, Ru^{3+}
  c. Rh, Rh^{2+} **d.** Pd, Pd^+, Pd^{2+}

59. Draw the Lewis diagrams for the following ligands. Indicate the lone pair(s) that may be donated to the metal. Indicate any you expect to be bidentate or polydentate.
  a. NH_3 **b.** SCN^- **c.** H_2O

60. Draw the Lewis diagrams for the following ligands. Indicate the lone pair(s) that may be donated to the metal. Indicate any you expect to be bidentate or polydentate.
  a. CN^-
  b. bipyridyl (bipy), which has the following structure:

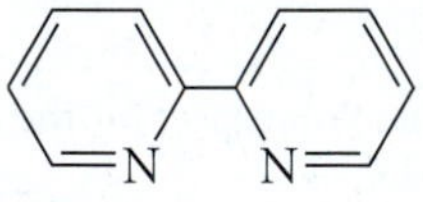

  c. NO_2^-

61. List all the different formulas for an octahedral complex made from a metal (M) and three different ligands (A, B, and C). Describe any isomers for each complex.

62. Amino acids, such as glycine (gly), can form complexes with the trace metal ions found in the bloodstream. Glycine, whose structure is shown below, acts as a bidentate ligand coordinating with the nitrogen atom and one of the oxygen atoms.

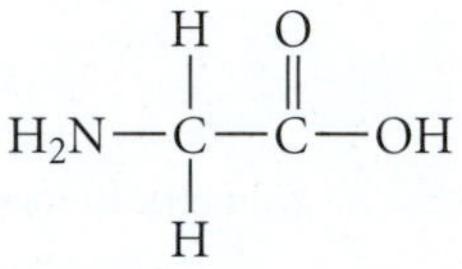

Draw all the possible isomers of
  a. square planar $[Ni(gly)_2]$
  b. tetrahedral $[Zn(gly)_2]$
  c. octahedral $[Fe(gly)_3]$

63. Oxalic acid solutions are used to remove rust stains. Draw a complex ion that is likely responsible for this effect. Does it have any isomers?

64. W, X, Y, and Z are different monodentate ligands.
  a. Will the square planar $[NiWXYZ]^{2+}$ be optically active?
  b. Will the tetrahedral $[ZnWXYZ]^{2+}$ be optically active?

65. Hexacyanomanganate(III) ion is a low-spin complex. Draw the crystal field splitting diagram with electrons filled in appropriately. Is this complex paramagnetic or diamagnetic?

66. Determine the color and approximate wavelength absorbed most strongly by each of the following solutions.
  a. blue solution   **b.** red solution
  c. yellow solution

Challenge Problems

67. When a solution of $PtCl_2$ is allowed to react with the ligand trimethylphosphine, $P(CH_3)_3$, two compounds are produced. Both compounds give the same elemental analysis: 46.7% Pt; 17.0% Cl; 14.8% P; 17.2% C; 4.34% H. Determine the formulas, draw the structures, and give the systematic names for both compounds.

68. Draw a crystal field splitting diagram for a trigonal planar complex ion. Assume the plane of the molecule is perpendicular to the z axis.

69. Draw a crystal field splitting diagram for a trigonal bipyramidal complex ion. Assume the axial positions are on the z axis.

70. Explain why $[Ni(NH_3)_4]^{2+}$ is paramagnetic, while $[Ni(CN)_4]^{2-}$ is diamagnetic.

71. Sulfide (S^{2-}) salts are notoriously insoluble in aqueous solution.
   a. Calculate the molar solubility of nickel(II) sulfide in water. $K_{sp}(NiS) = 3 \times 10^{-16}$

   b. Nickel(II) ions form a complex ion in the presence of ammonia with a formation constant (K_f) of 2.0×10^8:
   $Ni^{2+} + 6 NH_3 \rightleftharpoons [Ni(NH_3)_6]^{2+}$ Calculate the molar solubility of NiS in 3.0 M NH_3.

   c. Explain any differences in the answers to parts **a** and **b**.

Conceptual Problems

72. Two ligands, A and B, both form complexes with a particular metal ion. When the metal ion complexes with ligand A, the resulting solution is green. When the metal ion complexes with ligand B, the resulting solution is violet. Which of the two ligands results in the largest Δ?

73. Which element has the higher ionization energy, Cu or Au?

74. The complexes of Fe^{3+} have magnetic properties that depend on whether the ligands are strong or weak field. Explain why this observation supports the idea that electrons are lost from the $4s$ orbital before the $3d$ orbitals in the transition metals.

Appendix I: Common Mathematical Operations in Chemistry

A. Scientific Notation

A number written in scientific notation consists of a **decimal part**, a number that is usually between 1 and 10, and an **exponential part**, 10 raised to an **exponent**, n.

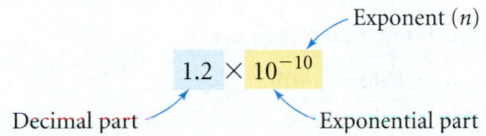

Exponent (n)

1.2×10^{-10}

Decimal part   Exponential part

Each of the following numbers is written in both scientific and decimal notation.

$1.0 \times 10^5 = 100{,}000$ $1.0 \times 10^{-5} = 0.000001$

$6.7 \times 10^3 = 6700$ $6.7 \times 10^{-3} = 0.0067$

A positive exponent means 1 multiplied by 10 n times.

$10^0 = 1$

$10^1 = 1 \times 10$

$10^2 = 1 \times 10 \times 10 = 100$

$10^3 = 1 \times 10 \times 10 \times 10 = 1000$

A negative exponent ($-n$) means 1 divided by 10 n times.

$$10^{-1} = \frac{1}{10} = 0.1$$

$$10^{-2} = \frac{1}{10 \times 10} = 0.01$$

$$10^{-3} = \frac{1}{10 \times 10 \times 10} = 0.001$$

To convert a number to scientific notation, we move the decimal point to obtain a number between 1 and 10 and then multiply by 10 raised to the appropriate power. For example, to write 5983 in scientific notation, we move the decimal point to the left three places to get 5.983 (a number between 1 and 10) and then multiply by 1000 to make up for moving the decimal point.

$5983 = 5.983 \times 1000$

Since 1000 is 10^3, we write

$5983 = 5.983 \times 10^3$

We can do this in one step by counting how many places we move the decimal point to obtain a number between 1 and 10 and then writing the decimal part multiplied by 10 raised to the number of places we moved the decimal point.

$$5983 = 5.983 \times 10^3$$

If the decimal point is moved to the left, as in the previous example, the exponent is positive. If the decimal is moved to the right, the exponent is negative.

$$0.00034 = 3.4 \times 10^{-4}$$

To express a number in scientific notation:

1. **Move the decimal point to obtain a number between 1 and 10.**
2. **Write the result from step 1 multiplied by 10 raised to the number of places you moved the decimal point.**
   - *The exponent is positive if you moved the decimal point to the left.*
   - *The exponent is negative if you moved the decimal point to the right.*

Consider the following additional examples:

$$290{,}809{,}000 = 2.90809 \times 10^8$$
$$0.000000000070 \text{ m} = 7.0 \times 10^{-11} \text{ m}$$

Multiplication and Division

To multiply numbers expressed in scientific notation, multiply the decimal parts and add the exponents.

$$(A \times 10^m)(B \times 10^n) = (A \times B) \times 10^{m+n}$$

To divide numbers expressed in scientific notation, divide the decimal parts and subtract the exponent in the denominator from the exponent in the numerator.

$$\frac{(A \times 10^m)}{(B \times 10^n)} = \left(\frac{A}{B}\right) \times 10^{m-n}$$

Consider the following example involving multiplication:

$$(3.5 \times 10^4)(1.8 \times 10^6) = (3.5 \times 1.8) \times 10^{4+6}$$
$$= 6.3 \times 10^{10}$$

Consider the following example involving division:

$$\frac{(5.6 \times 10^7)}{(1.4 \times 10^3)} = \left(\frac{5.6}{1.4}\right) \times 10^{7-3}$$
$$= 4.0 \times 10^4$$

Addition and Subtraction

To add or subtract numbers expressed in scientific notation, rewrite all the numbers so that they have the same exponent, then add or subtract the decimal parts of the numbers. The exponents remained unchanged.

$$\begin{array}{r} A \times 10^n \\ \pm B \times 10^n \\ \hline (A \pm B) \times 10^n \end{array}$$

Notice that the numbers *must have* the same exponent. Consider the following example involving addition:

$$4.82 \times 10^7$$
$$+3.4 \times 10^6$$

First, express both numbers with the same exponent. In this case, we rewrite the lower number and perform the addition as follows:

$$4.82 \times 10^7$$
$$+0.34 \times 10^7$$
$$\overline{5.16 \times 10^7}$$

Consider the following example involving subtraction.

$$7.33 \times 10^5$$
$$-1.9 \times 10^4$$

First, express both numbers with the same exponent. In this case, we rewrite the lower number and perform the subtraction as follows:

$$7.33 \times 10^5$$
$$-0.19 \times 10^5$$
$$\overline{7.14 \times 10^5}$$

Powers and Roots

To raise a number written in scientific notation to a power, raise the decimal part to the power and multiply the exponent by the power:

$$(4.0 \times 10^6)^2 = 4.0^2 \times 10^{6 \times 2}$$
$$= 16 \times 10^{12}$$
$$= 16 \times 10^{13}$$

To take the nth root of a number written in scientific notation, take the nth root of the decimal part and divide the exponent by the root:

$$(4.0 \times 10^6)^{1/3} = 4.0^{1/3} \times 10^{6/3}$$
$$= 1.6 \times 10^2$$

B. Logarithms

Common (or Base 10) Logarithms

The common or base 10 logarithm (abbreviated log) of a number is the exponent to which 10 must be raised to obtain that number. For example, the log of 100 is 2 because 10 must be raised to the second power to get 100. Similarly, the log of 1000 is 3 because 10 must be raised to the third power to get 1000. The logs of several multiples of 10 are shown below.

$$\log 10 = 1$$
$$\log 100 = 2$$
$$\log 1000 = 3$$
$$\log 10,000 = 4$$

Because $10^0 = 1$ by definition, $\log 1 = 0$..

The log of a number smaller than one is negative because 10 must be raised to a negative exponent to get a number smaller than one. For example, the log of 0.01 is -2 because 10 must be raised to -2 to get 0.01. Similarly, the log of 0.001 is -3 because 10 must be raised to -3 to get 0.001. The logs of several fractional numbers are shown below.

$$\log 0.1 = -1$$
$$\log 0.01 = -2$$
$$\log 0.001 = -3$$
$$\log 0.0001 = -4$$

The logs of numbers that are not multiples of 10 can be computed on your calculator. See your calculator manual for specific instructions.

Inverse Logarithms

The inverse logarithm or invlog function is exactly the opposite of the log function. For example, the log of 100 is 2 and the inverse log of 2 is 100. The log function and the invlog function undo one another.

$$\log 100 = 2$$
$$\text{invlog } 2 = 100$$
$$\text{invlog}(\log 100) = 100$$

The inverse log of a number is simply 10 rasied to that number.

$$\text{invlog } x = 10^x$$
$$\text{invlog } 3 = 10^3 = 1000$$

The inverse logs of numbers can be computed on your calculator. See your calculator manual for specific instructions.

Natural (or Base *e*) Logarithms

The natural (or base e) logarithm (abbreviated ln) of a number is the exponent to which e (which has the value of 2.71828…) must be raised to obtain that number. For example, the ln of 100 is 4.605 because e must be raised to 4.605 to get 100. Similarly, the ln of 10.0 is 2.303 because e must be raised to 2.303 to get 10.0.

The inverse natural logarithm or invln function is exactly the opposite of the ln function. For example, the ln of 100 is 4.605 and the inverse ln of 4.605 is 100. The inverse ln of a number is simply e rasied to that number.

$$\text{invln } x = e^x$$
$$\text{invln } 3 = e^3 = 20.1$$

The invln of a number can be computed on your calculator. See your calculator manual for specific instructions.

Mathematical Operations Using Logarithms

Because logarithms are exponents, mathematical operations involving logarithms are similar to those involving exponents as follows:

$$\log(a \times b) = \log a + \log b \qquad \ln(a \times b) = \ln a + \ln b$$

$$\log \frac{a}{b} = \log a - \log b \qquad \ln \frac{a}{b} = \ln a - \ln b$$

$$\log a^n = n \log a \qquad \ln a^n = n \ln a$$

C. Quadratic Equations

A quadratic equation contains at least one term in which the variable x is raised to the second power (and no terms in which x is raised to a higher power). A quadratic equation has the following general form:

$$ax^2 + bx + c = 0$$

A quadratic equation can be solved for x using the quadratic formula:

$$x = \frac{-b \pm \sqrt{b^2 - 4ac}}{2a}$$

Quadratic equations are often encountered when solving equilibrium problems. Below we show how to use the quadratic formula to solve a quadratic equation for x.

$$3x^2 - 5x + 1 = 0 \quad (quadratic\ equation)$$

$$x = \frac{-b \pm \sqrt{b^2 - 4ac}}{2a}$$

$$= \frac{-(-5) \pm \sqrt{(-5)^2 - 4(3)(1)}}{2(3)}$$

$$= \frac{5 \pm 3.6}{6}$$

$$x = 1.43 \quad or \quad x = 0.233$$

As you can see, the solution to a quadratic equation usually has two values. In any real chemical system, one of the values can be eliminated because it has no physical significance. (For example, it may correspond to a negative concentration, which does not exist.)

D. Graphs

Graphs are often used to visually show the relationship between two variables. For example, in Chapter 5 we show the following relationship between the volume of a gas and its pressure:

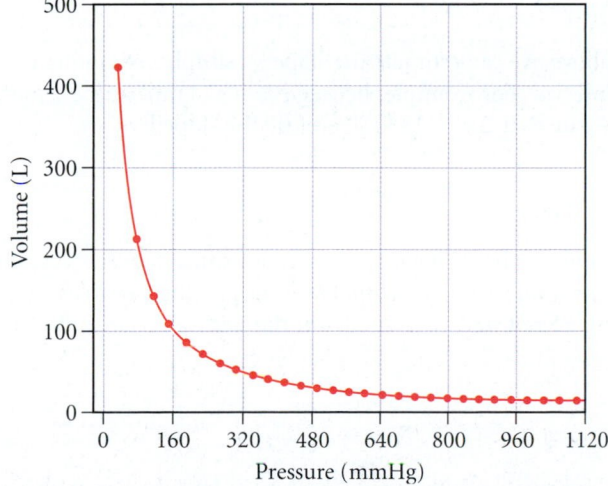

Volume versus Pressure   A plot of the volume of a gas sample—as measured in a J-tube—versus pressure. The plot shows that volume and pressure are inversely related.

The horizontal axis is the x-axis and is normally used to show the independent variable. The vertical axis is the y-axis and is normally used to show how the other variable (called the dependent variable) varies with a change in the independent variable. In this case, the graph shows that as the pressure of a gas sample increases, its volume decreases.

Many relationships in chemistry are *linear*, which means that if you change one variable by a factor of n the other variable will also change by a factor of n. For example, the volume of a gas is linearly related to the number of moles of gas. When two quantities are linearly related, a graph of one versus the other produces a straight line. For example, the graph below shows how the volume of an ideal gas sample depends on the number of moles of gas in the sample:

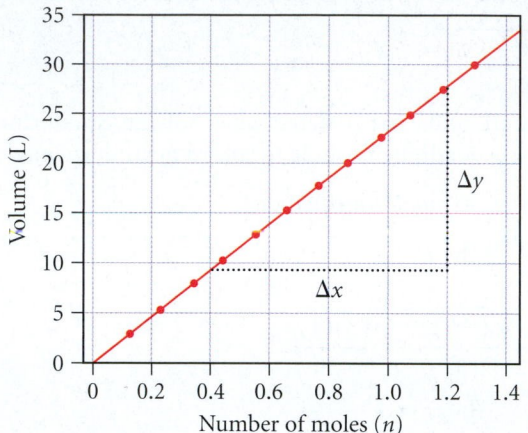

Volume versus Number of Moles   The volume of a gas sample increases linearly with the number of moles of gas in the sample.

A linear relationship between any two variables x and y can be expressed by the following equation:

$$y = mx + b$$

where m is the slope of the line and b is the y-intercept. The slope is the change in y divided by the change in x.

$$m = \frac{\Delta y}{\Delta x}$$

For the graph above, we can estimate the slope by simply estimating the changes in y and x for a given interval. For example, between $x = 0.4$ mol and 1.2 mol, $\Delta x = 0.80$ mol and we can estimate that $\Delta y = 18$ L. Therefore the slope is

$$m = \frac{\Delta y}{\Delta x} = \frac{18\text{ L}}{0.80\text{ mol}} = 23\text{ mol/L}$$

In several places in this book, logarithmic relationships between variables can be plotted in order to obtain a linear relationship. For example, the variables $[A]_t$ and t in the following equation are not linearly related, but the natural logarithm of $[A]_t$ and t are linearly related.

$$\ln[A]_t = -kt + \ln[A]_0$$
$$y = mx + b$$

A plot of $\ln[A]_t$ versus t will therefore produce a straight line with slope $= -k$ and y-intercept $= \ln[A]_0$.

Appendix II: Useful Data

A. Atomic Colors

Atomic number:	1	4	5	6	7	8	9
Atomic symbol:	H	Be	B	C	N	O	F

Atomic number:	11	12	14	15	16	17	19
Atomic symbol:	Na	Mg	Si	P	S	Cl	K

Atomic number:	20	29	30	35	53	54
Atomic symbol:	Ca	Cu	Zn	Br	I	Xe

B. Standard Thermodynamic Quantities for Selected Substances at 25 °C

Substance	ΔH_f° (kJ/mol)	ΔG_f° (kJ/mol)	S° (J/mol·K)
Aluminum			
Al(s)	0	0	28.32
Al(g)	330.0	289.4	164.6
Al^{3+}(aq)	−538.4	−483	−325
AlCl$_3$(s)	−704.2	−628.8	109.3
Al$_2$O$_3$(s)	−1675.7	−1582.3	50.9
Barium			
Ba(s)	0	0	62.5
Ba(g)	180.0	146.0	170.2
Ba^{2+}(aq)	−537.6	−560.8	9.6
BaCO$_3$(s)	−1213.0	−1134.4	112.1
BaCl$_2$(s)	−855.0	−806.7	123.7
BaO(s)	−548.0	−520.3	72.1
Ba(OH)$_2$(s)	−944.7		

Substance	ΔH_f° (kJ/mol)	ΔG_f° (kJ/mol)	S° (J/mol·K)
BaSO$_4$(s)	−1473.2	−1362.2	132.2
Beryllium			
Be(s)	0	0	9.5
BeO(s)	−609.4	−580.1	13.8
Be(OH)$_2$(s)	−902.5	−815.0	45.5
Bismuth			
Bi(s)	0	0	56.7
BiCl$_3$(s)	−379.1	−315.0	177.0
Bi$_2$O$_3$(s)	−573.9	−493.7	151.5
Bi$_2$S$_3$(s)	−143.1	−140.6	200.4
Boron			
B(s)	0	0	5.9
B(g)	565.0	521.0	153.4

(continued on the next page)

Substance	ΔH_f° (kJ/mol)	ΔG_f° (kJ/mol)	S° (J/mol·K)
$BCl_3(g)$	−403.8	−388.7	290.1
$BF_3(g)$	−1136.0	−1119.4	254.4
$B_2H_6(g)$	36.4	87.6	232.1
$B_2O_3(s)$	−1273.5	−1194.3	54.0
$H_3BO_3(s)$	−1094.3	−968.9	90.0
Bromine			
$Br(g)$	111.9	82.4	175.0
$Br_2(l)$	0	0	152.2
$Br_2(g)$	30.9	3.1	245.5
$Br^-(aq)$	−121.4	−102.8	80.71
$HBr(g)$	−36.3	−53.4	198.7
Cadmium			
$Cd(s)$	0	0	51.8
$Cd(g)$	111.8	77.3	167.7
$Cd^{2+}(aq)$	−75.9	−77.6	−73.2
$CdCl_2(s)$	−391.5	−343.9	115.3
$CdO(s)$	−258.4	−228.7	54.8
$CdS(s)$	−161.9	−156.5	64.9
$CdSO_4(s)$	−933.3	−822.7	123.0
Calcium			
$Ca(s)$	0	0	41.6
$Ca(g)$	177.8	144.0	154.9
$Ca^{2+}(aq)$	−542.8	−553.6	−53.1
$CaC_2(s)$	−59.8	−64.9	70.0
$CaCO_3(s)$	−1207.6	−1129.1	91.7
$CaCl_2(s)$	−795.4	−748.8	108.4
$CaF_2(s)$	−1228.0	−1175.6	68.5
$CaH_2(s)$	−181.5	−142.5	41.4
$Ca(NO_3)_2(s)$	−938.2	−742.8	193.2
$CaO(s)$	−634.9	−603.3	38.1
$Ca(OH)_2(s)$	−985.2	−897.5	83.4
$CaSO_4(s)$	−1434.5	−1322.0	106.5
$Ca_3(PO_4)_2(s)$	−4120.8	−3884.7	236.0
Carbon			
$C(s, \text{graphite})$	0	0	5.7
$C(s, \text{diamond})$	1.88	2.9	2.4
$C(g)$	716.7	671.3	158.1
$CH_4(g)$	−74.6	−50.5	186.3
$CH_3Cl(g)$	−81.9	−60.2	234.6
$CH_2Cl_2(g)$	−95.4		270.2
$CH_2Cl_2(l)$	−124.2	−63.2	177.8
$CHCl_3(l)$	−134.1	−73.7	201.7
$CCl_4(g)$	−95.7	−62.3	309.7
$CCl_4(l)$	−128.2	−66.4	216.4
$CH_2O(g)$	−108.6	−102.5	218.8
$CH_2O_2(\text{l, formic acid})$	−425.0	−361.4	129.0
$CH_3NH_2(\text{g, methylamine})$	−22.5	32.7	242.9
$CH_3OH(l)$	−238.6	−166.6	126.8
$CH_3OH(g)$	−201.0	−162.3	239.9
$C_2H_2(g)$	227.4	209.9	200.9
$C_2H_4(g)$	52.4	68.4	219.3

Substance	ΔH_f° (kJ/mol)	ΔG_f° (kJ/mol)	S° (J/mol·K)
$C_2H_6(g)$	−84.68	−32.0	229.2
$C_2H_5OH(l)$	−277.6	−174.8	160.7
$C_2H_5OH(g)$	−234.8	−167.9	281.6
$C_2H_3Cl(g, \text{vinyl chloride})$	37.2	53.6	264.0
$C_2H_4Cl_2(l, \text{dichloroethane})$	−166.8	−79.6	208.5
$C_2H_4O(g, \text{acetaldehyde})$	−166.2	−133.0	263.8
$C_2H_4O_2(l, \text{acetic acid})$	−484.3	−389.9	159.8
$C_3H_8(g)$	−103.85	−23.4	270.3
$C_3H_6O(l, \text{acetone})$	−248.4	−155.6	199.8
$C_3H_7OH(l, \text{isopropanol})$	−318.1		181.1
$C_4H_{10}(l)$	−147.3	−15.0	231.0
$C_4H_{10}(g)$	−125.7	−15.71	310.0
$C_6H_6(l)$	49.1	124.5	173.4
$C_6H_5NH_2(l, \text{aniline})$	31.6	149.2	191.9
$C_6H_5OH(s, \text{phenol})$	−165.1	−50.4	144.0
$C_6H_{12}O_6(s, \text{glucose})$	−1273.3	−910.4	212.1
$C_8H_{18}(l)$	−250.1		
$C_{10}H_8(s, \text{naphthalene})$	78.5	201.6	167.4
$C_{12}H_{22}O_{11}(s, \text{sucrose})$	−2226.1	−1544.3	360.24
$CO(g)$	−110.5	−137.2	197.7
$CO_2(g)$	−393.5	−394.4	213.8
$CO_2(aq)$	−413.8	−386.0	117.6
$CO_3^{2-}(aq)$	−677.1	−527.8	−56.9
$HCO_3^-(aq)$	−692.0	−586.8	91.2
$H_2CO_3(aq)$	−699.7	−623.2	187.4
$CN^-(aq)$	151	166	118
$HCN(l)$	108.9	125.0	112.8
$HCN(g)$	135.1	124.7	201.8
$CS_2(l)$	89.0	64.6	151.3
$CS_2(g)$	116.7	67.1	237.8
$COCl_2(g)$	−219.1	−204.9	283.5
$C_{60}(s)$	2327.0	2302.0	426.0
Cesium			
$Cs(s)$	0	0	85.2
$Cs(g)$	76.5	49.6	175.6
$Cs^+(aq)$	−258.0	−292.0	132.1
$CsBr(s)$	−400	−387	117
$CsCl(s)$	−438	−414	101.2
$CsF(s)$	−553.5	−525.5	92.8
$CsI(s)$	−342	−337	127
Chlorine			
$Cl(g)$	121.3	105.3	165.2
$Cl_2(g)$	0	0	223.1
$Cl^-(aq)$	−167.1	−131.2	56.6
$HCl(g)$	−92.3	−95.3	186.9
$HCl(aq)$	−167.2	−131.2	56.5
$ClO_2(g)$	102.5	120.5	256.8
$Cl_2O(g)$	80.3	97.9	266.2
Chromium			
$Cr(s)$	0	0	23.8
$Cr(g)$	396.6	351.8	174.5
$Cr^{3+}(aq)$	−1971		

Substance	ΔH_f° (kJ/mol)	ΔG_f° (kJ/mol)	S° (J/mol·K)
$CrO_4^{2-}(aq)$	−872.2	−717.1	44
$Cr_2O_3(s)$	−1139.7	−1058.1	81.2
$Cr_2O_7^{2-}(aq)$	−1476	−1279	238
Cobalt			
$Co(s)$	0	0	30.0
$Co(g)$	424.7	380.3	179.5
$CoO(s)$	−237.9	−214.2	53.0
$Co(OH)_2(s)$	−539.7	−454.3	79.0
Copper			
$Cu(s)$	0	0	33.2
$Cu(g)$	337.4	297.7	166.4
$Cu^+(aq)$	51.9	50.2	−26
$Cu^{2+}(aq)$	64.9	65.5	−98
$CuCl(s)$	−137.2	−119.9	86.2
$CuCl_2(s)$	−220.1	−175.7	108.1
$CuO(s)$	−157.3	−129.7	42.6
$CuS(s)$	−53.1	−53.6	66.5
$CuSO_4(s)$	−771.4	−662.2	109.2
$Cu_2O(s)$	−168.6	−146.0	93.1
$Cu_2S(s)$	−79.5	−86.2	120.9
Fluorine			
$F(g)$	79.38	62.3	158.75
$F_2(g)$	0	0	202.79
$F^-(aq)$	−335.35	−278.8	−13.8
$HF(g)$	−273.3	−275.4	173.8
Gold			
$Au(s)$	0	0	47.4
$Au(g)$	366.1	326.3	180.5
Helium			
$He(g)$	0	0	126.2
Hydrogen			
$H(g)$	218.0	203.3	114.7
$H^+(aq)$	0	0	0
$H^+(g)$	1536.3	1517.1	108.9
$H_2(g)$	0	0	130.7
Iodine			
$I(g)$	106.76	70.2	180.79
$I_2(s)$	0	0	116.14
$I_2(g)$	62.42	19.3	260.69
$I^-(aq)$	−56.78	−51.57	106.45
$HI(g)$	26.5	1.7	206.6
Iron			
$Fe(s)$	0	0	27.3
$Fe(g)$	416.3	370.7	180.5
$Fe^{2+}(aq)$	−87.9	−84.94	113.4
$Fe^{3+}(aq)$	−47.69	−10.54	293.3
$FeCO_3(s)$	−740.6	−666.7	92.9
$FeCl_2(s)$	−341.8	−302.3	118.0
$FeCl_3(s)$	−399.5	−334.0	142.3

Substance	ΔH_f° (kJ/mol)	ΔG_f° (kJ/mol)	S° (J/mol·K)
$FeO(s)$	−272.0	−255.2	60.75
$Fe(OH)_3(s)$	−823.0	−696.5	106.7
$FeS_2(s)$	−178.2	−166.9	52.9
$Fe_2O_3(s)$	−824.2	−742.2	87.4
$Fe_3O_4(s)$	−1118.4	−1015.4	146.4
Lead			
$Pb(s)$	0	0	64.8
$Pb(g)$	195.2	162.2	175.4
$Pb^{2+}(aq)$	0.92	−24.4	18.5
$PbBr_2(s)$	−278.7	−261.9	161.5
$PbCO_3(s)$	−699.1	−625.5	131.0
$PbCl_2(s)$	−359.4	−314.1	136.0
$PbI_2(s)$	−175.5	−173.6	174.9
$Pb(NO_3)_2(s)$	−451.9		
$PbO(s)$	−217.3	−187.9	68.7
$PbO_2(s)$	−277.4	−217.3	68.6
$PbS(s)$	−100.4	−98.7	91.2
$PbSO_4(s)$	−920.0	−813.0	148.5
Lithium			
$Li(s)$	0	0	29.1
$Li(g)$	159.3	126.6	138.8
$Li^+(aq)$	−278.47	−293.3	12.24
$LiBr(s)$	−351.2	−342.0	74.3
$LiCl(s)$	−408.6	−384.4	59.3
$LiF(s)$	−616.0	−587.7	35.7
$LiI(s)$	−270.4	−270.3	86.8
$LiNO_3(s)$	−483.1	−381.1	90.0
$LiOH(s)$	−487.5	−441.5	42.8
$Li_2O(s)$	−597.9	−561.2	37.6
Magnesium			
$Mg(s)$	0	0	32.7
$Mg(g)$	147.1	112.5	148.6
$Mg^{2+}(aq)$	−467.0	−455.4	−137
$MgCl_2(s)$	−641.3	−591.8	89.6
$MgCO_3(s)$	−1095.8	−1012.1	65.7
$MgF_2(s)$	−1124.2	−1071.1	57.2
$MgO(s)$	−601.6	−569.3	27.0
$Mg(OH)_2(s)$	−924.5	−833.5	63.2
$MgSO_4(s)$	−1284.9	−1170.6	91.6
$Mg_3N_2(s)$	−461	−401	88
Manganese			
$Mn(s)$	0	0	32.0
$Mn(g)$	280.7	238.5	173.7
$Mn^{2+}(aq)$	−219.4	−225.6	−78.8
$MnO(s)$	−385.2	−362.9	59.7
$MnO_2(s)$	−520.0	−465.1	53.1
$MnO_4^-(aq)$	−529.9	−436.2	190.6
Mercury			
$Hg(l)$	0	0	75.9
$Hg(g)$	61.4	31.8	175.0

(continued on the next page)

Substance	ΔH_f° (kJ/mol)	ΔG_f° (kJ/mol)	S° (J/mol·K)
$Hg^{2+}(aq)$	170.21	164.4	−36.19
$Hg_2^{2+}(aq)$	166.87	153.5	65.74
$HgCl_2(s)$	−224.3	−178.6	146.0
$HgO(s)$	−90.8	−58.5	70.3
$HgS(s)$	−58.2	−50.6	82.4
$Hg_2Cl_2(s)$	−265.4	−210.7	191.6
Nickel			
$Ni(s)$	0	0	29.9
$Ni(g)$	429.7	384.5	182.2
$NiCl_2(s)$	−305.3	−259.0	97.7
$NiO(s)$	−239.7	−211.7	37.99
$NiS(s)$	−82.0	−79.5	53.0
Nitrogen			
$N(g)$	472.7	455.5	153.3
$N_2(g)$	0	0	191.6
$NF_3(g)$	−132.1	−90.6	260.8
$NH_3(g)$	−45.9	−16.4	192.8
$NH_3(aq)$	−80.29	−26.50	111.3
$NH_4^+(aq)$	−133.26	−79.31	111.17
$NH_4Br(s)$	−270.8	−175.2	113.0
$NH_4Cl(s)$	−314.4	−202.9	94.6
$NH_4CN(s)$	0.4		
$NH_4F(s)$	−464.0	−348.7	72.0
$NH_4HCO_3(s)$	−849.4	−665.9	120.9
$NH_4I(s)$	−201.4	−112.5	117.0
$NH_4NO_3(s)$	−365.6	−183.9	151.1
$NH_4NO_3(aq)$	−339.9	−190.6	259.8
$HNO_3(g)$	−133.9	−73.5	266.9
$HNO_3(aq)$	−207	−110.9	146
$NO(g)$	91.3	87.6	210.8
$NO_2(g)$	33.2	51.3	240.1
$NO_3^-(aq)$	−206.85	−110.2	146.70
$NOBr(g)$	82.2	82.4	273.7
$NOCl(g)$	51.7	66.1	261.7
$N_2H_4(l)$	50.6	149.3	121.2
$N_2H_4(g)$	95.4	159.4	238.5
$N_2O(g)$	81.6	103.7	220.0
$N_2O_4(l)$	−19.5	97.5	209.2
$N_2O_4(g)$	11.1	99.8	304.4
$N_2O_5(s)$	−43.1	113.9	178.2
$N_2O_5(g)$	13.3	117.1	355.7
Oxygen			
$O(g)$	249.2	231.7	161.1
$O_2(g)$	0	0	205.2
$O_3(g)$	142.7	163.2	238.9
$OH^-(aq)$	−230.02	−157.3	−10.90
$H_2O(l)$	−285.8	−237.1	70.0
$H_2O(g)$	−241.8	−228.6	188.8
$H_2O_2(l)$	−187.8	−120.4	109.6
$H_2O_2(g)$	−136.3	−105.6	232.7

Substance	ΔH_f° (kJ/mol)	ΔG_f° (kJ/mol)	S° (J/mol·K)
Phosphorus			
$P(s, white)$	0	0	41.1
$P(s, red)$	−17.6	−12.1	22.8
$P(g)$	316.5	280.1	163.2
$P_2(g)$	144.0	103.5	218.1
$P_4(g)$	58.9	24.4	280.0
$PCl_3(l)$	−319.7	−272.3	217.1
$PCl_3(g)$	−287.0	−267.8	311.8
$PCl_5(s)$	−443.5		
$PCl_5(g)$	−374.9	−305.0	364.6
$PF_5(g)$	−1594.4	−1520.7	300.8
$PH_3(g)$	5.4	13.5	210.2
$POCl_3(l)$	−597.1	−520.8	222.5
$POCl_3(g)$	−558.5	−512.9	325.5
$PO_4^{3-}(aq)$	−1277.4	−1018.7	−220.5
$HPO_4^{2-}(aq)$	−1292.1	−1089.2	−33.5
$H_2PO_4^-(aq)$	−1296.3	−1130.2	90.4
$H_3PO_4(s)$	−1284.4	−1124.3	110.5
$H_3PO_4(aq)$	−1288.3	−1142.6	158.2
$P_4O_6(s)$	−1640.1		
$P_4O_{10}(s)$	−2984	−2698	228.9
Platinum			
$Pt(s)$	0	0	41.6
$Pt(g)$	565.3	520.5	192.4
Potassium			
$K(s)$	0	0	64.7
$K(g)$	89.0	60.5	160.3
$K^+(aq)$	−252.14	−283.3	101.2
$KBr(s)$	−393.8	−380.7	95.9
$KCN(s)$	−113.0	−101.9	128.5
$KCl(s)$	−436.5	−408.5	82.6
$KClO_3(s)$	−397.7	−296.3	143.1
$KClO_4(s)$	−432.8	−303.1	151.0
$KF(s)$	−567.3	−537.8	66.6
$KI(s)$	−327.9	−324.9	106.3
$KNO_3(s)$	−494.6	−394.9	133.1
$KOH(s)$	−424.6	−379.4	81.2
$KOH(aq)$	−482.4	−440.5	91.6
$KO_2(s)$	−284.9	−239.4	116.7
$K_2CO_3(s)$	−1151.0	−1063.5	155.5
$K_2O(s)$	−361.5	−322.1	94.14
$K_2O_2(s)$	−494.1	−425.1	102.1
$K_2SO_4(s)$	−1437.8	−1321.4	175.6
Rubidium			
$Rb(s)$	0	0	76.8
$Rb(g)$	80.9	53.1	170.1
$Rb^+(aq)$	−251.12	−283.1	121.75
$RbBr(s)$	−394.6	−381.8	110.0
$RbCl(s)$	−435.4	−407.8	95.9
$RbClO_3(s)$	−392.4	−292.0	152

Substance	ΔH_f° (kJ/mol)	ΔG_f° (kJ/mol)	S° (J/mol·K)
RbF(s)	−557.7		
RbI(s)	−333.8	−328.9	118.4
Scandium			
Sc(s)	0	0	34.6
Sc(g)	377.8	336.0	174.8
Selenium			
Se(s, gray)	0	0	42.4
Se(g)	227.1	187.0	176.7
H$_2$Se(g)	29.7	15.9	219.0
Silicon			
Si(s)	0	0	18.8
Si(g)	450.0	405.5	168.0
SiCl$_4$(l)	−687.0	−619.8	239.7
SiF$_4$(g)	−1615.0	−1572.8	282.8
SiH$_4$(g)	34.3	56.9	204.6
SiO$_2$(s, quartz)	−910.7	−856.3	41.5
Si$_2$H$_6$(g)	80.3	127.3	272.7
Silver			
Ag(s)	0	0	42.6
Ag(g)	284.9	246.0	173.0
Ag$^+$(aq)	105.79	77.11	73.45
AgBr(s)	−100.4	−96.9	107.1
AgCl(s)	−127.0	−109.8	96.3
AgF(s)	−204.6	−185	84
AgI(s)	−61.8	−66.2	115.5
AgNO$_3$(s)	−124.4	−33.4	140.9
Ag$_2$O(s)	−31.1	−11.2	121.3
Ag$_2$S(s)	−32.6	−40.7	144.0
Ag$_2$SO$_4$(s)	−715.9	−618.4	200.4
Sodium			
Na(s)	0	0	51.3
Na(g)	107.5	77.0	153.7
Na$^+$(aq)	−240.34	−261.9	58.45
NaBr(s)	−361.1	−349.0	86.8
NaCl(s)	−411.2	−384.1	72.1
NaCl(aq)	−407.2	−393.1	115.5
NaClO$_3$(s)	−365.8	−262.3	123.4
NaF(s)	−576.6	−546.3	51.1
NaHCO$_3$(s)	−950.8	−851.0	101.7
NaHSO$_4$(s)	−1125.5	−992.8	113.0
NaI(s)	−287.8	−286.1	98.5
NaNO$_3$(s)	−467.9	−367.0	116.5
NaNO$_3$(aq)	−447.5	−373.2	205.4
NaOH(s)	−425.8	−379.7	64.4
NaOH(aq)	−470.1	−419.2	48.2
NaO$_2$(s)	−260.2	−218.4	115.9
Na$_2$CO$_3$(s)	−1130.7	−1044.4	135.0
Na$_2$O(s)	−414.2	−375.5	75.1
Na$_2$O$_2$(s)	−510.9	−447.7	95.0

Substance	ΔH_f° (kJ/mol)	ΔG_f° (kJ/mol)	S° (J/mol·K)
Na$_2$SO$_4$(s)	−1387.1	−1270.2	149.6
Na$_3$PO$_4$(s)	−1917	−1789	173.8
Strontium			
Sr(s)	0	0	55.0
Sr(g)	164.4	130.9	164.6
Sr^{2+}(aq)	−545.51	−557.3	−39
SrCl$_2$(s)	−828.9	−781.1	114.9
SrCO$_3$(s)	−1220.1	−1140.1	97.1
SrO(s)	−592.0	−561.9	54.4
SrSO$_4$(s)	−1453.1	−1340.9	117.0
Sulfur			
S(s, rhombic)	0	0	32.1
S(s, monoclinic)	0.3	0.096	32.6
S(g)	277.2	236.7	167.8
S$_2$(g)	128.6	79.7	228.2
S$_8$(g)	102.3	49.7	430.9
S^{2-}(aq)	41.8	83.7	22
SF$_6$(g)	−1220.5	−1116.5	291.5
HS$^-$(aq)	−17.7	12.4	62.0
H$_2$S(g)	−20.6	−33.4	205.8
H$_2$S(aq)	−39.4	−27.7	122
SOCl$_2$(l)	−245.6		
SO$_2$(g)	−296.8	−300.1	248.2
SO$_3$(g)	−395.7	−371.1	256.8
SO$_4^{2-}$(aq)	−909.3	−744.6	18.5
HSO$_4^-$(aq)	−886.5	−754.4	129.5
H$_2$SO$_4$(l)	−814.0	−690.0	156.9
H$_2$SO$_4$(aq)	−909.3	−744.6	18.5
S$_2$O$_3^{2-}$(aq)	−648.5	−522.5	67
Tin			
Sn(s, white)	0	0	51.2
Sn(s, gray)	−2.1	0.1	44.1
Sn(g)	301.2	266.2	168.5
SnCl$_4$(l)	−511.3	−440.1	258.6
SnCl$_4$(g)	−471.5	−432.2	365.8
SnO(s)	−280.7	−251.9	57.2
SnO$_2$(s)	−577.6	−515.8	49.0
Titanium			
Ti(s)	0	0	30.7
Ti(g)	473.0	428.4	180.3
TiCl$_4$(l)	−804.2	−737.2	252.3
TiCl$_4$(g)	−763.2	−726.3	353.2
TiO$_2$(s)	−944.0	−888.8	50.6
Tungsten			
W(s)	0	0	32.6
W(g)	849.4	807.1	174.0
WO$_3$(s)	−842.9	−764.0	75.9

(continued on the next page)

Substance	ΔH_f° (kJ/mol)	ΔG_f° (kJ/mol)	S° (J/mol·K)
Uranium			
U(s)	0	0	50.2
U(g)	533.0	488.4	199.8
$UF_6(s)$	−2197.0	−2068.5	227.6
$UF_6(g)$	−2147.4	−2063.7	377.9
$UO_2(s)$	−1085.0	−1031.8	77.0
Vanadium			
V(s)	0	0	28.9
V(g)	514.2	754.4	182.3

Substance	ΔH_f° (kJ/mol)	ΔG_f° (kJ/mol)	S° (J/mol·K)
Zinc			
Zn(s)	0	0	41.6
Zn(g)	130.4	94.8	161.0
$Zn^{2+}(aq)$	−153.39	−147.1	−109.8
$ZnCl_2(s)$	−415.1	−369.4	111.5
ZnO(s)	−350.5	−320.5	43.7
ZnS(s, zinc blende)	−206.0	−201.3	57.7
$ZnSO_4(s)$	−982.8	−871.5	110.5

C. Aqueous Equilibrium Constants

1. Dissociation Constants for Acids at 25 °C

Name	Formula	K_{a_1}	K_{a_2}	K_{a_3}
Acetic	$HC_2H_3O_2$	1.8×10^{-5}		
Acetylsalicylic	$HC_9H_7O_4$	3.3×10^{-4}		
Adipic	$H_2C_6H_8O_4$	3.9×10^{-5}	3.9×10^{-6}	
Arsenic	H_3AsO_4	5.5×10^{-3}	1.7×10^{-7}	5.1×10^{-12}
Arsenous	H_3AsO_3	5.1×10^{-10}		
Ascorbic	$H_2C_6H_6O_6$	8.0×10^{-5}	1.6×10^{-12}	
Benzoic	$HC_7H_5O_2$	6.5×10^{-5}		
Boric	H_3BO_3	5.4×10^{-10}		
Butanoic	$HC_4H_7O_2$	1.5×10^{-5}		
Carbonic	H_2CO_3	4.3×10^{-7}	5.6×10^{-11}	
Chloroacetic	$HC_2H_2O_2Cl$	1.4×10^{-3}		
Chlorous	$HClO_2$	1.1×10^{-2}		
Citric	$H_3C_6H_5O_7$	7.4×10^{-4}	1.7×10^{-5}	4.0×10^{-7}
Cyanic	HCNO	2×10^{-4}		
Formic	$HCHO_2$	1.8×10^{-4}		
Hydrazoic	HN_3	2.5×10^{-5}		
Hydrocyanic	HCN	4.9×10^{-10}		
Hydrofluoric	HF	3.5×10^{-4}		
Hydrogen chromate ion	$HCrO_4^-$	3.0×10^{-7}		
Hydrogen peroxide	H_2O_2	2.4×10^{-12}		
Hydrogen selenate ion	$HSeO_4^-$	2.2×10^{-2}		
Hydrosulfuric	H_2S	8.9×10^{-8}	1×10^{-19}	
Hydrotelluric	H_2Te	2.3×10^{-3}	1.6×10^{-11}	

Name	Formula	K_{a_1}	K_{a_2}	K_{a_3}
Hypobromous	HBrO	2.8×10^{-9}		
Hypochlorous	HClO	2.9×10^{-8}		
Hypoiodous	HIO	2.3×10^{-11}		
Iodic	HIO_3	1.7×10^{-1}		
Lactic	$HC_3H_5O_3$	1.4×10^{-4}		
Maleic	$H_2C_4H_2O_4$	1.2×10^{-2}	5.9×10^{-7}	
Malonic	$H_2C_3H_2O_4$	1.5×10^{-3}	2.0×10^{-6}	
Nitrous	HNO_2	4.6×10^{-4}		
Oxalic	$H_2C_2O_4$	5.9×10^{-2}	6.4×10^{-5}	
Paraperiodic	H_5IO_6	2.8×10^{-2}	5.3×10^{-9}	
Phenol	HC_6H_5O	1.3×10^{-10}		
Phosphoric	H_3PO_4	7.5×10^{-3}	6.2×10^{-8}	4.2×10^{-13}
Phosphorous	H_3PO_3	5×10^{-2}	2.0×10^{-7}	
Propanoic	$HC_3H_5O_2$	1.3×10^{-5}		
Pyruvic	$HC_3H_3O_3$	4.1×10^{-3}		
Pyrophosphoric	$H_4P_2O_7$	1.2×10^{-1}	7.9×10^{-3}	2.0×10^{-7}
Selenous	H_2SeO_3	2.4×10^{-3}	4.8×10^{-9}	
Succinic	$H_2C_4H_4O_4$	6.2×10^{-5}	2.3×10^{-6}	
Sulfuric	H_2SO_4	Strong acid	1.2×10^{-2}	
Sulfurous	H_2SO_3	1.7×10^{-2}	6.4×10^{-8}	
Tartaric	$H_2C_4H_4O_6$	1.0×10^{-3}	4.6×10^{-5}	
Trichloroacetic	$HC_2Cl_3O_2$	2.2×10^{-1}		
Trifluoroacetic acid	$HC_2F_3O_2$	3.0×10^{-1}		

2. Dissociation Constants for Hydrated Metal Ions at 25 °C

Cation	Hydrated Ion	K_a
Al^{3+}	$Al(H_2O)_6^{3+}$	1.4×10^{-5}
Be^{2+}	$Be(H_2O)_6^{2+}$	3×10^{-7}
Co^{2+}	$Co(H_2O)_6^{2+}$	1.3×10^{-9}
Cr^{3+}	$Cr(H_2O)_6^{3+}$	1.6×10^{-4}
Cu^{2+}	$Cu(H_2O)_6^{2+}$	3×10^{-8}
Fe^{2+}	$Fe(H_2O)_6^{2+}$	3.2×10^{-10}

Cation	Hydrated Ion	K_a
Fe^{3+}	$Fe(H_2O)_6^{3+}$	6.3×10^{-3}
Ni^{2+}	$Ni(H_2O)_6^{2+}$	2.5×10^{-11}
Pb^{2+}	$Pb(H_2O)_6^{2+}$	3×10^{-8}
Sn^{2+}	$Sn(H_2O)_6^{2+}$	4×10^{-4}
Zn^{2+}	$Zn(H_2O)_6^{2+}$	2.5×10^{-10}

3. Dissociation Constants for Bases at 25 °C

Name	Formula	K_b
Ammonia	NH_3	1.76×10^{-5}
Aniline	$C_6H_5NH_2$	3.9×10^{-10}
Bicarbonate ion	HCO_3^-	1.7×10^{-9}
Carbonate ion	CO_3^{2-}	1.8×10^{-4}
Codeine	$C_{18}H_{21}NO_3$	1.6×10^{-6}
Diethylamine	$(C_2H_5)_2NH$	6.9×10^{-4}
Dimethylamine	$(CH_3)_2NH$	5.4×10^{-4}
Ethylamine	$C_2H_5NH_2$	5.6×10^{-4}
Ethylenediamine	$C_2H_8N_2$	8.3×10^{-5}
Hydrazine	H_2NNH_2	1.3×10^{-6}
Hydroxylamine	$HONH_2$	1.1×10^{-8}

Name	Formula	K_b
Ketamine	$C_{13}H_{16}ClNO$	3×10^{-7}
Methylamine	CH_3NH_2	4.4×10^{-4}
Morphine	$C_{17}H_{19}NO_3$	1.6×10^{-6}
Nicotine	$C_{10}H_{14}N_2$	1.0×10^{-6}
Piperidine	$C_5H_{10}NH$	1.33×10^{-3}
Propylamine	$C_3H_7NH_2$	3.5×10^{-4}
Pyridine	C_5H_5N	1.7×10^{-9}
Strychnine	$C_{21}H_{22}N_2O_2$	1.8×10^{-6}
Triethylamine	$(C_2H_5)_3N$	5.6×10^{-4}
Trimethylamine	$(CH_3)_3N$	6.4×10^{-5}

4. Solubility Product Constants for Compounds at 25 °C

Compound	Formula	K_{sp}
Aluminum hydroxide	$Al(OH)_3$	1.3×10^{-33}
Aluminum phosphate	$AlPO_4$	9.84×10^{-21}
Barium carbonate	$BaCO_3$	2.58×10^{-9}
Barium chromate	$BaCrO_4$	1.17×10^{-10}
Barium fluoride	BaF_2	2.45×10^{-5}
Barium hydroxide	$Ba(OH)_2$	5.0×10^{-3}
Barium oxalate	BaC_2O_4	1.6×10^{-6}
Barium phosphate	$Ba_3(PO_4)_2$	6×10^{-39}
Barium sulfate	$BaSO_4$	1.07×10^{-10}
Cadmium carbonate	$CdCO_3$	1.0×10^{-12}
Cadmium hydroxide	$Cd(OH)_2$	7.2×10^{-15}
Cadmium sulfide	CdS	8×10^{-28}
Calcium carbonate	$CaCO_3$	4.96×10^{-9}
Calcium chromate	$CaCrO_4$	7.1×10^{-4}
Calcium fluoride	CaF_2	1.46×10^{-10}
Calcium hydroxide	$Ca(OH)_2$	4.68×10^{-6}
Calcium hydrogen phosphate	$CaHPO_4$	1×10^{-7}
Calcium oxalate	CaC_2O_4	2.32×10^{-9}
Calcium phosphate	$Ca_3(PO_4)_2$	2.07×10^{-33}
Calcium sulfate	$CaSO_4$	7.10×10^{-5}
Chromium(III) hydroxide	$Cr(OH)_3$	6.3×10^{-31}
Cobalt(II) carbonate	$CoCO_3$	1.0×10^{-10}
Cobalt(II) hydroxide	$Co(OH)_2$	5.92×10^{-15}
Cobalt(II) sulfide	CoS	5×10^{-22}
Copper(I) bromide	$CuBr$	6.27×10^{-9}
Copper(I) chloride	$CuCl$	1.72×10^{-7}
Copper(I) cyanide	$CuCN$	3.47×10^{-20}
Copper(II) carbonate	$CuCO_3$	2.4×10^{-10}
Copper(II) hydroxide	$Cu(OH)_2$	2.2×10^{-20}
Copper(II) phosphate	$Cu_3(PO_4)_2$	1.40×10^{-37}
Copper(II) sulfide	CuS	1.27×10^{-36}
Iron(II) carbonate	$FeCO_3$	3.07×10^{-11}
Iron(II) hydroxide	$Fe(OH)_2$	4.87×10^{-17}
Iron(II) sulfide	FeS	3.72×10^{-19}

Compound	Formula	K_{sp}
Iron(III) hydroxide	$Fe(OH)_3$	2.79×10^{-39}
Lanthanum fluoride	LaF_3	2×10^{-19}
Lanthanum iodate	$La(IO_3)_3$	7.50×10^{-12}
Lead(II) bromide	$PbBr_2$	4.67×10^{-6}
Lead(II) carbonate	$PbCO_3$	7.40×10^{-14}
Lead(II) chloride	$PbCl_2$	1.17×10^{-5}
Lead(II) chromate	$PbCrO_4$	2.8×10^{-13}
Lead(II) fluoride	PbF_2	3.3×10^{-8}
Lead(II) hydroxide	$Pb(OH)_2$	1.43×10^{-20}
Lead(II) iodide	PbI_2	9.8×10^{-9}
Lead(II) phosphate	$Pb_3(PO_4)_2$	1×10^{-54}
Lead(II) sulfate	$PbSO_4$	1.82×10^{-8}
Lead(II) sulfide	PbS	9.04×10^{-29}
Magnesium carbonate	$MgCO_3$	6.82×10^{-6}
Magnesium fluoride	MgF_2	5.16×10^{-11}
Magnesium hydroxide	$Mg(OH)_2$	2.06×10^{-13}
Magnesium oxalate	MgC_2O_4	4.83×10^{-6}
Manganese(II) carbonate	$MnCO_3$	2.24×10^{-11}
Manganese(II) hydroxide	$Mn(OH)_2$	1.6×10^{-13}
Manganese(II) sulfide	MnS	2.3×10^{-13}
Mercury(I) bromide	Hg_2Br_2	6.40×10^{-23}
Mercury(I) carbonate	Hg_2CO_3	3.6×10^{-17}
Mercury(I) chloride	Hg_2Cl_2	1.43×10^{-18}
Mercury(I) chromate	Hg_2CrO_4	2×10^{-9}
Mercury(I) cyanide	$Hg_2(CN)_2$	5×10^{-40}
Mercury(I) iodide	Hg_2I_2	5.2×10^{-29}
Mercury(II) hydroxide	$Hg(OH)_2$	3.1×10^{-26}
Mercury(II) sulfide	HgS	1.6×10^{-54}
Nickel(II) carbonate	$NiCO_3$	1.42×10^{-7}
Nickel(II) hydroxide	$Ni(OH)_2$	5.48×10^{-16}
Nickel(II) sulfide	NiS	3×10^{-20}
Silver bromate	$AgBrO_3$	5.38×10^{-5}
Silver bromide	$AgBr$	5.35×10^{-13}
Silver carbonate	Ag_2CO_3	8.46×10^{-12}

(continued on the next page)

Compound	Formula	K_{sp}
Silver chloride	AgCl	1.77×10^{-10}
Silver chromate	Ag_2CrO_4	1.12×10^{-12}
Silver cyanide	AgCN	5.97×10^{-17}
Silver iodide	AgI	8.51×10^{-17}
Silver phosphate	Ag_3PO_4	8.89×10^{-17}
Silver sulfate	Ag_2SO_4	1.20×10^{-5}
Silver sulfide	Ag_2S	6×10^{-51}
Strontium carbonate	$SrCO_3$	5.60×10^{-10}
Strontium chromate	$SrCrO_4$	3.6×10^{-5}

Compound	Formula	K_{sp}
Strontium phosphate	$Sr_3(PO_4)_2$	1×10^{-31}
Strontium sulfate	$SrSO_4$	3.44×10^{-7}
Tin(II) hydroxide	$Sn(OH)_2$	5.45×10^{-27}
Tin(II) sulfide	SnS	1×10^{-26}
Zinc carbonate	$ZnCO_3$	1.46×10^{-10}
Zinc hydroxide	$Zn(OH)_2$	3×10^{-17}
Zinc oxalate	ZnC_2O_4	2.7×10^{-8}
Zinc sulfide	ZnS	2×10^{-25}

5. Complex Ion Formation Constants in Water at 25 °C

Complex Ion	K_f
$[Ag(CN)_2]^-$	1×10^{21}
$[Ag(EDTA)]^{3-}$	2.1×10^7
$[Ag(en)_2]^+$	5.0×10^7
$[Ag(NH_3)_2]^+$	1.7×10^7
$[Ag(SCN)_4]^{3-}$	1.2×10^{10}
$[Ag(S_2O_3)_2]^-$	2.8×10^{13}
$[Al(EDTA)]^-$	1.3×10^{16}
$[AlF_6]^{3-}$	7×10^{19}
$[Al(OH)_4]^-$	3×10^{33}
$[Al(ox)_3]^{3-}$	2×10^{16}
$[CdBr_4]^{2-}$	5.5×10^3
$[Cd(CN)_4]^{2-}$	3×10^{18}
$[CdCl_4]^{2-}$	6.3×10^2
$[Cd(en)_3]^{2+}$	1.2×10^{12}
$[CdI_4]^{2-}$	2×10^6
$[Co(EDTA)]^{2-}$	2.0×10^{16}
$[Co(EDTA)]^-$	1×10^{36}
$[Co(en)_3]^{2+}$	8.7×10^{13}
$[Co(en)_3]^{3+}$	4.9×10^{48}
$[Co(NH_3)_6]^{2+}$	1.3×10^5
$[Co(NH_3)_6]^{3+}$	2.3×10^{33}
$[Co(OH)_4]^{2-}$	5×10^9
$[Co(ox)_3]^{4-}$	5×10^9
$[Co(ox)_3]^{3-}$	1×10^{20}
$[Co(SCN)_4]^{2-}$	1×10^3
$[Cr(EDTA)]^-$	1×10^{23}
$[Cr(OH)_4]^-$	8.0×10^{29}
$[CuCl_3]^{2-}$	5×10^5
$[Cu(CN)_4]^{2-}$	1.0×10^{29}
$[Cu(EDTA)]^{2-}$	5×10^{18}
$[Cu(en)_2]^{2+}$	1×10^{20}
$[Cu(NH_3)_4]^{2+}$	1.7×10^{13}
$[Cu(ox)_2]^{2-}$	3×10^8
$[Fe(CN)_6]^{4-}$	1.5×10^{35}

Complex Ion	K_f
$[Fe(CN)_6]^{3-}$	2×10^{43}
$[Fe(EDTA)]^{2-}$	2.1×10^{14}
$[Fe(EDTA)]^-$	1.7×10^{24}
$[Fe(en)_3]^{2+}$	5.0×10^9
$[Fe(ox)_3]^{4-}$	1.7×10^5
$[Fe(ox)_3]^{3-}$	2×10^{20}
$[Fe(SCN)]^{2+}$	8.9×10^2
$[Hg(CN)_4]^{2-}$	1.8×10^{41}
$[HgCl_4]^{2-}$	1.1×10^{16}
$[Hg(EDTA)]^{2-}$	6.3×10^{21}
$[Hg(en)_2]^{2+}$	2×10^{23}
$[HgI_4]^{2-}$	2×10^{30}
$[Hg(ox)_2]^{2-}$	9.5×10^6
$[Ni(CN)_4]^{2-}$	2×10^{31}
$[Ni(EDTA)]^{2-}$	3.6×10^{18}
$[Ni(en)_3]^{2+}$	2.1×10^{18}
$[Ni(NH_3)_6]^{2+}$	2.0×10^8
$[Ni(ox)_3]^{4-}$	3×10^8
$[PbCl_3]^-$	2.4×10^1
$[Pb(EDTA)]^{2-}$	2×10^{18}
$[PbI_4]^{2-}$	3.0×10^4
$[Pb(OH)_3]^-$	8×10^{13}
$[Pb(ox)_2]^{2-}$	3.5×10^6
$[Pb(S_2O_3)_3]^{4-}$	2.2×10^6
$[PtCl_4]^{2-}$	1×10^{16}
$[Pt(NH_3)_6]^{2+}$	2×10^{35}
$[Sn(OH)_3]^-$	3×10^{25}
$[Zn(CN)_4]^{2-}$	2.1×10^{19}
$[Zn(EDTA)]^{2-}$	3×10^{16}
$[Zn(en)_3]^{2+}$	1.3×10^{14}
$[Zn(NH_3)_4]^{2+}$	2.8×10^9
$[Zn(OH)_4]^{2-}$	2×10^{15}
$[Zn(ox)_3]^{4-}$	1.4×10^8

D. Standard Reduction Half-Cell Potentials at 25 °C

Half-Reaction	$E°$ (V)
$F_2(g) + 2\,e^- \longrightarrow 2\,F^-(aq)$	2.87
$O_3(g) + 2\,H^+(aq) + 2\,e^- \longrightarrow O_2(g) + H_2O(l)$	2.08
$Ag^{2+}(aq) + e^- \longrightarrow Ag^+(aq)$	1.98
$Co^{3+}(aq) + e^- \longrightarrow Co^{2+}(aq)$	1.82
$H_2O_2(aq) + 2\,H^+(aq) + 2\,e^- \longrightarrow 2\,H_2O(l)$	1.78
$PbO_2(s) + 4\,H^+(aq) + SO_4^{2-}(aq) + 2\,e^- \longrightarrow$ $PbSO_4(s) + 2\,H_2O(l)$	1.69
$MnO_4^-(aq) + 4\,H^+(aq) + 3\,e^- \longrightarrow MnO_2(s) + 2\,H_2O(l)$	1.68
$2\,HClO(aq) + 2\,H^+(aq) + 2\,e^- \longrightarrow Cl_2(g) + 2\,H_2O(l)$	1.61
$MnO_4^-(aq) + 8\,H^+(aq) + 5\,e^- \longrightarrow Mn^{2+}(aq) + 4\,H_2O(l)$	1.51
$Au^{3+}(aq) + 3\,e^- \longrightarrow Au(s)$	1.50
$2\,BrO_3^-(aq) + 12\,H^+(aq) + 10\,e^- \longrightarrow Br_2(l) + 6\,H_2O(l)$	1.48
$PbO_2(s) + 4\,H^+(aq) + 2\,e^- \longrightarrow Pb^{2+}(aq) + 2\,H_2O(l)$	1.46
$Cl_2(g) + 2\,e^- \longrightarrow 2\,Cl^-(aq)$	1.36
$Cr_2O_7^{2-}(aq) + 14\,H^+(aq) + 6\,e^- \longrightarrow 2\,Cr^{3+}(aq) + 7\,H_2O(l)$	1.33
$O_2(g) + 4\,H^+(aq) + 4\,e^- \longrightarrow 2\,H_2O(l)$	1.23
$MnO_2(s) + 4\,H^+(aq) + 2\,e^- \longrightarrow Mn^{2+}(aq) + 2\,H_2O(l)$	1.21
$IO_3^-(aq) + 6\,H^+(aq) + 5\,e^- \longrightarrow \frac{1}{2}I_2(aq) + 3\,H_2O(l)$	1.20
$Br_2(l) + 2\,e^- \longrightarrow 2\,Br^-(aq)$	1.09
$AuCl_4^-(aq) + 3\,e^- \longrightarrow Au(s) + 4\,Cl^-(aq)$	1.00
$VO_2^+(aq) + 2\,H^+(aq) + e^- \longrightarrow VO^{2+}(aq) + H_2O(l)$	0.99
$HNO_2(aq) + H^+(aq) + e^- \longrightarrow NO(g) + 2\,H_2O(l)$	0.98
$NO_3^-(aq) + 4\,H^+(aq) + 3\,e^- \longrightarrow NO(g) + 2\,H_2O(l)$	0.96
$ClO_2(g) + e^- \longrightarrow ClO_2^-(aq)$	0.95
$2\,Hg^{2+}(aq) + 2\,e^- \longrightarrow 2\,Hg_2^{2+}(aq)$	0.92
$Ag^+(aq) + e^- \longrightarrow Ag(s)$	0.80
$Hg_2^{2+}(aq) + 2\,e^- \longrightarrow 2\,Hg(l)$	0.80
$Fe^{3+}(aq) + e^- \longrightarrow Fe^{2+}(aq)$	0.77
$PtCl_4^{2-}(aq) + 2\,e^- \longrightarrow Pt(s) + 4\,Cl^-(aq)$	0.76
$O_2(g) + 2\,H^+(aq) + 2\,e^- \longrightarrow H_2O_2(aq)$	0.70
$MnO_4^-(aq) + e^- \longrightarrow MnO_4^{2-}(aq)$	0.56
$I_2(s) + 2\,e^- \longrightarrow 2\,I^-(aq)$	0.54
$Cu^+(aq) + e^- \longrightarrow Cu(s)$	0.52
$O_2(g) + 2\,H_2O(l) + 4\,e^- \longrightarrow 4\,OH^-(aq)$	0.40
$Cu^{2+}(aq) + 2\,e^- \longrightarrow Cu(s)$	0.34

Half-Reaction	$E°$ (V)
$BiO^+(aq) + 2\,H^+(aq) + 3\,e^- \longrightarrow Bi(s) + H_2O(l)$	0.32
$Hg_2Cl_2(s) + 2\,e^- \longrightarrow 2\,Hg(l) + 2\,Cl^-(aq)$	0.27
$AgCl(s) + e^- \longrightarrow Ag(s) + Cl^-(aq)$	0.22
$SO_4^{2-}(aq) + 4\,H^+(aq) + 2\,e^- \longrightarrow H_2SO_3(aq) + H_2O(l)$	0.20
$Cu^{2+}(aq) + e^- \longrightarrow Cu^+(aq)$	0.16
$Sn^{4+}(aq) + 2\,e^- \longrightarrow Sn^{2+}(aq)$	0.15
$S(s) + 2\,H^+(aq) + 2\,e^- \longrightarrow H_2S(g)$	0.14
$AgBr(s) + e^- \longrightarrow Ag(s) + Br^-(aq)$	0.071
$2\,H^+(aq) + 2\,e^- \longrightarrow H_2(g)$	0.00
$Fe^{3+}(aq) + 3\,e^- \longrightarrow Fe(s)$	−0.036
$Pb^{2+}(aq) + 2\,e^- \longrightarrow Pb(s)$	−0.13
$Sn^{2+}(aq) + 2\,e^- \longrightarrow Sn(s)$	−0.14
$AgI(s) + e^- \longrightarrow Ag(s) + I^-(aq)$	−0.15
$N_2(g) + 5\,H^+(aq) + 4\,e^- \longrightarrow N_2H_5^+(aq)$	−0.23
$Ni^{2+}(aq) + 2\,e^- \longrightarrow Ni(s)$	−0.23
$Co^{2+}(aq) + 2\,e^- \longrightarrow Co(s)$	−0.28
$PbSO_4(s) + 2\,e^- \longrightarrow Pb(s) + SO_4^{2-}(aq)$	−0.36
$Cd^{2+}(aq) + 2\,e^- \longrightarrow Cd(s)$	−0.40
$Fe^{2+}(aq) + 2\,e^- \longrightarrow Fe(s)$	−0.45
$2\,CO_2(g) + 2\,H^+(aq) + 2\,e^- \longrightarrow H_2C_2O_4(aq)$	−0.49
$Cr^{3+}(aq) + e^- \longrightarrow Cr^{2+}(aq)$	−0.50
$Cr^{3+}(aq) + 3\,e^- \longrightarrow Cr(s)$	−0.73
$Zn^{2+}(aq) + 2\,e^- \longrightarrow Zn(s)$	−0.76
$2\,H_2O(l) + 2\,e^- \longrightarrow H_2(g) + 2\,OH^-(aq)$	−0.83
$Mn^{2+}(aq) + 2\,e^- \longrightarrow Mn(s)$	−1.18
$Al^{3+}(aq) + 3\,e^- \longrightarrow Al(s)$	−1.66
$H_2(g) + 2\,e^- \longrightarrow 2\,H^-(aq)$	−2.23
$Mg^{2+}(aq) + 2\,e^- \longrightarrow Mg(s)$	−2.37
$La^{3+}(aq) + 3\,e^- \longrightarrow La(s)$	−2.38
$Na^+(aq) + e^- \longrightarrow Na(s)$	−2.71
$Ca^{2+}(aq) + 2\,e^- \longrightarrow Ca(s)$	−2.76
$Ba^{2+}(aq) + 2\,e^- \longrightarrow Ba(s)$	−2.90
$K^+(aq) + e^- \longrightarrow K(s)$	−2.92
$Li^+(aq) + e^- \longrightarrow Li(s)$	−3.04

E. Vapor Pressure of Water at Various Temperatures

T (°C)	P (torr)	T (°C)	P (torr)	T (°C)	P (torr)	T (°C)	P (torr)
0	4.58	21	18.65	35	42.2	92	567.0
5	6.54	22	19.83	40	55.3	94	610.9
10	9.21	23	21.07	45	71.9	96	657.6
12	10.52	24	22.38	50	92.5	98	707.3
14	11.99	25	23.76	55	118.0	100	760.0
16	13.63	26	25.21	60	149.4	102	815.9
17	14.53	27	26.74	65	187.5	104	875.1
18	15.48	28	28.35	70	233.7	106	937.9
19	16.48	29	30.04	80	355.1	108	1004.4
20	17.54	30	31.82	90	525.8	110	1074.6

Appendix III: Answers to Selected Exercises

Chapter 1

33. a. theory   **b.** observation
   c. law   **d.** observation

35. Several answers possible.

37. a. mixture, homogeneous
   b. pure substance, compound
   c. pure substance, element
   d. mixture, heterogeneous

39.

Substance	Pure or Mixture	Type
Aluminum	Pure	Element
Apple juice	Mixture	Homogeneous
Hydrogen peroxide	Pure	Compound
Chicken soup	Mixture	Heterogeneous

41. a. pure substance, compound   **b.** mixture, heterogeneous
   c. mixture, homogeneous   **d.** pure substance, element

43. physical, chemical, physical, physical, physical

45. a. chemical   **b.** physical
   c. physical   **d.** chemical

47. a. chemical   **b.** physical
   c. chemical   **d.** chemical

49. a. physical   **b.** chemical
   c. physical

51. a. $0\,°C$ **b.** $-321\,°F$
   c. $-78.3\,°F$ **d.** $310.2\,K$

53. $-62.2\,°C$, $210.9\,K$

55. a. 1.2 nm   **b.** 22 fs
   c. 1.5 Gg   **d.** 3.5 ML

57.

1245 kg	1.245×10^6 g	1.245×10^9 mg
515 km	5.15×10^6 dm	5.15×10^7 cm
122.355 s	1.22355×10^5 ms	0.122355 ks
3.345 kJ	3.345×10^3 J	3.345×10^6 mJ

59. 10,000 1-cm squares

61. no

63. $1.26\,g/cm^3$

65. a. 463 g   **b.** 3.7 L

67. a. 73.0 mL   **b.** $88.2\,°C$ **c.** 645 mL

69. a. 1,0̲50,501   **b.** 0.00̲20
   c. 0.0000000000000̲0 2   **d.** 0.00̲1090

71. a. 3   **b.** ambiguous, without more
information assume 3 significant figures
   c. 3   **d.** 5
   e. ambiguous, without more information assume
1 significant figure

73. a. not exact   **b.** exact   **c.** not exact   **d.** exact

75. a. 156.9   **b.** 156.8   **c.** 156.8   **d.** 156.9

77. a. 1.84   **b.** 0.033
   c. 0.500   **d.** 34

79. a. 41.4   **b.** 133.5
   c. 73.0   **d.** 0.42

81. a. 391.3   **b.** 1.1×10^4
   c. 5.96   **d.** 5.93×10^4

83. a. 60.6 in   **b.** 3.14×10^3 g
   c. 3.7 qt   **d.** 4.29 in

85. 5.0×10^1 min

87. 33 mi/gal

89. a. $1.95 \times 10^{-4}\,km^2$ **b.** $1.95 \times 10^4\,dm^2$
   c. $1.95 \times 10^6\,cm^2$

91. $0.680\,mi^2$

93. 0.95 mL

95. 3.1557×10^7 s/solar year

97. a. extensive   **b.** intensive   **c.** intensive
   d. intensive   **e.** extensive

99. $-34°$

101. $F = kg(m/s^2) = N$ (for newton), kN, pN

103. a. 2.2×10^{-6} **b.** 0.0159
   c. 6.9×10^4

105. a. mass of can of gold $= 1.9 \times 10^4$ g
     mass of can of sand $= 3.0 \times 10^3$ g
   b. Yes, the thief sets off the trap because the can of sand is
lighter than the gold cylinder.

107. $21\,in^3$

109. $7.6\,g/cm^3$

111. 3.11×10^5 lb

113. 3.3×10^2 km

115. 6.8×10^{-15}

117. $7.3 \times 10^{11}\,g/cm^3$

119. a. $1.6 \times 10^4\,nm^3$ **b.** 1.3×10^{-18} g oxygen
   c. 1.7×10^2 g oxygen   **d.** 1.3×10^{20} nanocontainers
   e. 2.0 L, not feasible when total blood volume in an adult is 5 L
(this is almost a 40% increase in blood volume)

121. (c)

123. substance A

125. a. law   **b.** theory   **c.** observation   **d.** law

Chapter 2

31. 13.5 g

33. These results are not consistent with the law of definite proportions because sample 1 is composed of 11.5 parts Cl to 1 part C and sample 2 is composed of 9.05 parts Cl to 1 part C. The law of definite proportions states that a given compound always contains exactly the same proportion of elements by mass.

35. 23.8 g

37. For the law of multiple proportions to hold, the ratio of the masses of O combining with 1 g of Os in the compound should be a small whole number. $0.3369/0.168 = 2.00$

39. Sample 1: $1.00 \, g \, O_2/1.00 \, g \, S$; sample 2: $1.50 \, g \, O_2/1.00 \, g \, S$
Sample 2/sample 1 $= 1.50/1.00 = 1.50$
3 O atoms/2 O atoms $= 1.5$

41. a. not consistent
b. consistent: Dalton's atomic theory states that the atoms of a given element are identical.
c. consistent: Dalton's atomic theory states that atoms combine in simple whole-number ratios to form compounds.
d. not consistent

43. a. consistent: Rutherford's nuclear model states that the atom is largely empty space.
b. consistent: Rutherford's nuclear model states that most of the atom's mass is concentrated in a tiny region called the nucleus.
c. not consistent   **d.** not consistent

45. $-2.3 \times 10^{-19} \, C$

47. 9.4×10^{13} excess electrons, $8.5 \times 10^{-17} \, kg$

49. a, b, c

51. $1.83 \times 10^3 \, e^-$

53. a. $^{23}_{11}Na$ **b.** $^{16}_{8}O$
c. $^{27}_{13}Al$ **d.** $^{127}_{53}I$

55. a. $7 \, ^1_1p$ and $7 \, ^0_1n$ **b.** $11 \, ^1_1p$ and $12 \, ^0_1n$
c. $86 \, ^1_1p$ and $136 \, ^0_1n$ **d.** $82 \, ^1_1p$ and $126 \, ^0_1n$

57. $6 \, ^1_1p$ and $8 \, ^1_0n$, $^{14}_6C$

59. a. $28 \, ^1_1p$ and $26 \, e^-$ **b.** $16 \, ^1_1p$ and $18 \, e^-$
c. $35 \, ^1_1p$ and $36 \, e^-$ **d.** $24 \, ^1_1p$ and $21 \, e^-$

61. a. $2-$ **b.** $1+$ **c.** $3+$ **d.** $1+$

63.

Symbol	Ion Formed	Number of Electrons in Ion	Number of Protons in Ion
Ca	Ca^{2+}	18	20
Be	Be^{2+}	2	4
Se	Se^{2-}	36	34
In	In^{3+}	46	49

65. a. sodium, metal   **b.** magnesium, metal
c. bromine, nonmetal   **d.** nitrogen, nonmetal
e. arsenic, metalloid

67. a, b

69. a. alkali metal   **b.** halogen
c. alkaline earth metal   **d.** alkaline earth metal
e. noble gas

71. Cl and F because they are in the same group or family. Elements in the same group or family have similar chemical properties.

73. 85.47 amu, mass spectrum will have a peak at 86.91 amu and another peak about 2.5 times larger at 84.91 amu

75. The fluorine-19 isotope must have a large percent abundance, which would make fluorine produce a large peak at this mass. Chlorine has two isotopes (Cl-35 and Cl-37). The atomic mass is simply the weighted average of these two, which means that there is no chlorine isotope with a mass of 35.45 amu.

77. 121.8 amu, Sb

79. 2.3×10^{24} atoms

81. a. 0.295 mol Ar   **b.** 0.0543 mol Zn
c. 0.144 mol Ta   **d.** 0.0304 mol Li

83. 2.11×10^{22} atoms

85. a. 1.01×10^{23} atoms   **b.** 6.78×10^{21} atoms
c. 5.39×10^{21} atoms   **d.** 5.6×10^{20} atoms

87. 2.6×10^{21} atoms

89. $3.239 \times 10^{-22} \, g$

91. 1.50 g

93. C_2O_3

95. $4.82241 \times 10^7 \, C/kg$

97. ^{237}Pa, ^{238}U, ^{239}Np, ^{240}Pu, ^{235}Ac, ^{234}Ra, etc.

99.

Symbol	Z	A	#p	$\#e^-$	#n	Charge
O	8	16	8	10	8	$2-$
Ca^{2+}	20	40	20	18	20	$2+$
Mg^{2+}	12	25	12	10	13	$2+$
N^{3-}	7	14	7	10	7	$3-$

101. $V_n = 8.2 \times 10^{-8} \, pm^3$, $V_a = 1.4 \times 10^6 \, pm^3$, $5.9 \times 10^{-12} \%$

103. 6.022×10^{21} dollars total, 9.3×10^{11} dollars per person, billionaires

105. 15.985 amu

107. 4.76×10^{24} atoms

109. $Li - 6 = 7.494\%$, $Li - 7 = 92.506\%$

111. 1×10^{78} atoms/universe

113. c. The law of multiple proportions states that when two elements form different compounds, the masses of element B that combine with 1 gram of element A can be expressed in whole number ratios.

115. a. contains the greatest amount of an element in moles and **c.** contains the greatest mass of an element.

Chapter 3

23. a. 3 Ca, 2 P, 8 O   **b.** 1 Sr, 2 Cl
c. 1 K, 1 N, 3 O   **d.** 1 Mg, 2 N, 4 O

25. a. NH_3 **b.** C_2H_6 **c.** SO_3

27. a. atomic   **b.** molecular
c. atomic   **d.** molecular

29. a. molecular   **b.** ionic
c. ionic   **d.** molecular

31. a. molecular element   **b.** molecular compound
c. atomic element

33. a. MgS   **b.** BaO   **c.** $SrBr_2$ **d.** $BeCl_2$

35. a. $Ba(OH)_2$ **b.** $BaCrO_4$ **c.** $Ba_3(PO_4)_2$ **d.** $Ba(CN)_2$

37. a. magnesium nitride   **b.** potassium fluoride
c. sodium oxide   **d.** lithium sulfide

39. a. tin(IV) chloride   **b.** lead(II) iodide
c. iron(III) oxide   **d.** copper(II) iodide

41. a. tin(II) oxide   **b.** chromium(III) sulfide
c. rubidium iodide   **d.** barium bromide

43. a. copper(I) nitrite   **b.** magnesium acetate
c. barium nitrate   **d.** lead(II) acetate
e. potassium chlorate   **f.** lead(II) sulfate

45. a. $NaHSO_3$ **b.** $LiMnO_4$
c. $AgNO_3$ **d.** K_2SO_4
e. $RbHSO_4$ **f.** $KHCO_3$

47. a. cobalt(II) sulfate heptahydrate   **b.** $IrBr_3 \cdot 4H_2O$
c. Magnesium bromate hexahydrate   **d.** $K_2CO_3 \cdot 2H_2O$

49. a. carbon monoxide   **b.** nitrogen triiodide
c. silicon tetrachloride   **d.** tetranitrogen tetraselenide
e. diiodine pentaoxide

51. a. PCl_3 **b.** ClO **c.** S_2F_4 **d.** PF_5
e. P_2S_5

53. a. hydroiodic acid   **b.** nitric acid
c. carbonic acid   **d.** acetic acid

55. a. HF **b.** HBr
c. H_2SO_3

57. a. 46.01 amu   **b.** 58.12 amu
c. 180.16 amu   **d.** 238.03 amu

59. a. 2.2×10^{23} molecules   **b.** 7.06×10^{23} molecules
c. 4.16×10^{23} molecules   **d.** 1.09×10^{23} molecules

61. 2.992×10^{-23} g

63. 0.10 mg

65. a. 74.87% C   **b.** 79.88% C
c. 92.24% C   **d.** 37.23% C

67. NH_3: 82.27% N
$CO(NH_2)_2$: 46.65% N
NH_4NO_3: 35.00% N
$(NH_4)_2SO_4$: 21.20% N
NH_3 has the highest N content

69. 20.8 g F

71. 196 μg KI

73. a. 2:1   **b.** 4:1   **c.** 6:2:1

75. a. 0.885 mol H   **b.** 5.2 mol H
c. 29 mol H   **d.** 33.7 mol H

77. a. 3.3 g Na   **b.** 3.6 g Na   **c.** 1.4 g Na   **d.** 1.7 g Na

79. a. Ag_2O **b.** $Co_3As_2O_8$ **c.** $SeBr_4$

81. a. C_5H_7N **b.** $C_4H_5N_2O$

83. NCl_3

85. a. $C_{12}H_{14}N_2$ **b.** $C_6H_3Cl_3$ **c.** $C_{10}H_{20}N_2S_4$

87. CH_2

89. C_2H_4O

91. $2 SO_2(g) + O_2(g) + 2 H_2O(l) \longrightarrow 2 H_2SO_4(aq)$

93. $2 Na(s) + 2 H_2O(l) \longrightarrow H_2(g) + 2 NaOH(aq)$

95. $C_{12}H_{22}O_{11}(s) + H_2O(l) \longrightarrow 4 C_2H_5OH(aq) + 4 CO_2(g)$

97. a. $PbS(s) + 2 HBr(aq) \longrightarrow PbBr_2(s) + H_2S(g)$
b. $CO(g) + 3 H_2(g) \longrightarrow CH_4(g) + H_2O(l)$
c. $4 HCl(aq) + MnO_2(s) \longrightarrow$
$MnCl_2(aq) + 2 H_2O(l) + Cl_2(g)$
d. $C_5H_{12}(l) + 8 O_2(g) \longrightarrow 5 CO_2(g) + 6 H_2O(g)$

99. a. $2 CO_2(g) + CaSiO_3(s) + H_2O(l) \longrightarrow$
$SiO_2(s) + Ca(HCO_3)_2(aq)$
b. $2 Co(NO_3)_3(aq) + 3 (NH_4)_2S(aq) \longrightarrow$
$Co_2S_3(s) + 6 NH_4NO_3(aq)$
c. $Cu_2O(s) + C(s) \longrightarrow 2 Cu(s) + CO(g)$
d. $H_2(g) + Cl_2(g) \longrightarrow 2 HCl(g)$

101. a. inorganic   **b.** organic
c. organic   **d.** inorganic

103. a. alkene   **b.** alkane
c. alkyne   **d.** alkane

105. a. $CH_3CH_2CH_3$ **b.** propane
c. $CH_3CH_2CH_2CH_2CH_2CH_2CH_2CH_3$ **d.** pentane

107. a. functionalized hydrocarbon, alcohol   **b.** hydrocarbon
c. functionalized hydrocarbon, ketone
d. functionalized hydrocarbon, amine

109. 1.50×10^{24} molecules EtOH

111. a. K_2CrO_4, 40.27% K, 26.78% Cr, 32.95% O
b. $Pb_3(PO_4)_2$, 76.60% Pb, 7.63% P, 15.77% O

c. H_2SO_3, 2.46% H, 39.07% S, 58.47% O
d. $CoBr_2$, 26.94% Co, 73.06% Br

113. 1.80×10^2 g Cl/yr

115. M = Fe

117. estradiol = $C_{18}H_{24}O_2$

119. $C_{18}H_{20}O_2$

121. $7 H_2O$

123. C_6H_9BrO

125. 1.87×10^{21} atoms

127. 92.93 amu

129. 0.224 g

131. 22.0% by mass

133. 1.6×10^7 kg Cl

135. 7.8×10^3 kg rock

137. a. O   **b.** N
c. O   **d.** N

139. A chemical formula does not express a mass ratio, it expresses a molar ratio. The statement should be: "When a chemical equation is balanced, the number of atoms of each type on both sides of the equation will be equal."

141. O, S, H

Chapter 4

25. $2 C_6H_{14}(g) + 19 O_2(g) \longrightarrow$
$12 CO_2(g) + 14 H_2O(g)$, 47 mol O_2

27. a. 2.6 mol   **b.** 12 mol
c. 0.194 mol   **d.** 28.7 mol

29.

mol SiO_2	mol C	mol SiC	mol CO
3	9	3	6
2	6	2	4
5	15	5	10
2.8	8.4	2.8	5.6
0.517	1.55	0.517	1.03

31. a. 9.3 g HBr, 0.12 g H_2

33. a. 3.8 g   **b.** 4.5 g
c. 4.1 g   **d.** 4.7 g

35. a. 4.42 g HCl   **b.** 8.25 g HNO_3
c. 4.24 g H_2SO_4

37. a. Na   **b.** Na
c. Br_2 **d.** Na

39. 3 molecules Cl_2

41. a. 2 mol   **b.** 7 mol
c. 9.40 mol

43. a. 2.5 g   **b.** 31.1 g   **c.** 1.16 g

45. limiting reactant: Pb^{2+}, theoretical yield: 34.5 g $PbCl_2$, percent yield: 85.3%

47. limiting reactant: NH_3, theoretical yield: 240.5 kg CH_4N_2O, percent yield: 70.01%

49. a. 1.5 M LiCl   **b.** 0.116 M $C_6H_{12}O_6$
c. 5.05×10^{-3} M NaCl

51. a. 1.3 mol   **b.** 1.5 mol   **c.** 0.211 mol

53. 37 g

55. 0.27 M

57. 6.0 L

59. 37.1 mL

61. 2.1 L

63. a. yes   **b.** no   **c.** yes   **d.** no

65. a. soluble Ag^+, NO_3^- **b.** soluble Pb^{2+}, $C_2H_3O_2^-$
   c. soluble K^+, NO_3^- **d.** soluble NH_4^+, S^{2-}

67. a. NO REACTION   **b.** NO REACTION
   c. $CrBr_2(aq) + Na_2CO_3(aq) \longrightarrow CrCO_3(s) + 2\,NaBr(aq)$
   d. $3\,NaOH(aq) + FeCl_3(aq) \longrightarrow Fe(OH)_3(s) + 3\,NaCl(aq)$

69. a. $K_2CO_3(aq) + Pb(NO_3)_2(aq) \longrightarrow$
$$PbCO_3(s) + 2\,KNO_3(aq)$$
   b. $Li_2SO_4(aq) + Pb(C_2H_3O_2)_2(aq) \longrightarrow$
$$PbSO_4(s) + 2\,LiC_2H_3O_3(aq)$$
   c. $Cu(NO_3)_2(aq) + MgS(aq) \longrightarrow$
$$CuS(s) + Mg(NO_3)_2(aq)$$
   d. NO REACTION

71. a. Complete:
$$H^+(aq) + Cl^-(aq) + Li^+(aq) + OH^-(aq) \longrightarrow$$
$$H_2O(l) + Li^+(aq) + Cl^-(aq)$$
Net: $H^+(aq) + OH^-(aq) \longrightarrow H_2O(l)$
   b. Complete:
$$Mg^{2+}(aq) + S^{2-}(aq) + Cu^{2+}(aq) + 2\,Cl^-(aq) \longrightarrow$$
$$CuS(s) + Mg^{2+}(aq) + 2\,Cl^-(aq)$$
Net: $Cu^{2+}(aq) + S^{2-}(aq) \longrightarrow CuS(s)$
   c. Complete:
$$Na^+(aq) + OH^-(aq) + H^+(aq) + NO_3^-(aq) \longrightarrow$$
$$H_2O(l) + Na^+(aq) + NO_3^-(aq)$$
Net: $H^+(aq) + OH^-(aq) \longrightarrow H_2O(l)$
   d. Complete:
$$6\,Na^+(aq) + 2\,PO_4^{3-}(aq) + 3\,Ni^{2+}(aq) + 6\,Cl^-(aq) \longrightarrow$$
$$Ni_3(PO_4)_2(s) + 6\,Na^+(aq) + 6\,Cl^-(aq)$$
Net: $3\,Ni^{2+}(aq) + 2\,PO_4^{3-}(aq) \longrightarrow Ni_3(PO_4)_2(s)$

73. Complete:
$$Hg_2^{2+}(aq) + 2\,NO_3^-(aq) + 2\,Na^+(aq) + 2\,Cl^-(aq) \longrightarrow$$
$$Hg_2Cl_2(s) + 2\,Na^+(aq) + 2\,NO_3^-(aq)$$
Net: $Hg_2^{2+}(aq) + 2\,Cl^-(aq) \longrightarrow Hg_2Cl_2(s)$

75. Molecular: $HBr(aq) + KOH(aq) \longrightarrow H_2O(l) + KBr(aq)$
Net ionic: $H^+(aq) + OH^-(aq) \longrightarrow H_2O(l)$

77. a. $H_2SO_4(aq) + Ca(OH)_2(aq) \longrightarrow 2\,H_2O(l) + CaSO_4(s)$
   b. $HClO_4(aq) + KOH(aq) \longrightarrow H_2O(l) + KClO_4(aq)$
   c. $H_2SO_4(aq) + 2\,NaOH(aq) \longrightarrow 2\,H_2O(l) + Na_2SO_4(aq)$

79. 0.337 M

81. a. $2\,HBr(aq) + NiS(s) \longrightarrow H_2S(g) + NiBr_2(aq)$
   b. $NH_4I(aq) + NaOH(aq) \longrightarrow$
$$H_2O(l) + NH_3(g) + NaI(aq)$$
   c. $2\,HBr(aq) + Na_2S(aq) \longrightarrow H_2S(g) + 2\,NaBr(aq)$
   d. $2\,HClO_4(aq) + Li_2CO_3(aq) \longrightarrow$
$$H_2O(l) + CO_2(g) + 2\,LiClO_4(aq)$$

83. a. Ag: 0   **b.** Ag: +1
   c. Ca: +2, F: −1   **d.** H: +1, S: −2
   e. C: +4, O: −2   **f.** Cr: +6, O: −2

85. a. +2   **b.** +6   **c.** +3

87. a. redox reaction, oxidizing agent: O_2, reducing agent: Li
   b. redox reaction, oxidizing agent: Fe^{2+}, reducing agent: Mg
   c. not a redox reaction   **d.** not a redox reaction

89. a. $S(s) + O_2(g) \longrightarrow SO_2(g)$
   b. $2\,C_3H_6(g) + 9\,O_2(g) \longrightarrow 6\,CO_2(g) + 6\,H_2O(g)$
   c. $2\,Ca(s) + O_2(g) \longrightarrow 2\,CaO(g)$
   d. $C_5H_{12}S(l) + 9\,O_2(g) \longrightarrow$
$$5\,CO_2(g) + SO_2(g) + 6\,H_2O(g)$$

91. 3.32 M

93. 1.1 g

95. 3.1 kg

97. limiting reactant: $C_7H_6O_3$, theoretical yield: 1.63 g $C_9H_8O_4$, percent yield: 74.8%

99. b.

101. a. $2\,HCl(aq) + Hg_2(NO_3)_2(aq) \longrightarrow$
$$Hg_2Cl_2(s) + 2\,HNO_3(aq)$$
   b. $KHSO_3(aq) + HNO_3(aq) \longrightarrow$
$$H_2O(l) + SO_2(g) + KNO_3(aq)$$
   c. $2\,NH_4Cl(aq) + Pb(NO_3)_2(aq) \longrightarrow$
$$PbCl_2(s) + 2\,NH_4NO_3(aq)$$
   d. $2\,NH_4Cl(aq) + Ca(OH)_2(aq) \longrightarrow$
$$2\,NH_3(g) + 2\,H_2O(g) + CaCl_2(aq)$$

103. 22 g

105. 6.9 g

107. $NaNO_3$ is more economical

109. Br is the oxidizing agent, Au is the reducing agent, 38.8 g $KAuF_4$

111. Ca^{2+} and Cu^{2+} present in the original solution
Net ionic for first precipitate:
$Ca^{2+}(aq) + SO_4^{2-}(aq) \longrightarrow CaSO_4(s)$
Net ionic for second precipitate:
$Cu^{2+}(aq) + CO_3^{2-}(aq) \longrightarrow CuCO_3(s)$

113. 3.4×10^4 kg

115. 2.0 mg

117. 96.9 g

119. d. The mass ratio must be at least $(4 \times 39.09)/32 = 4.88$ or K will be the limiting reactant. The mass ratio of 1.5/0.38 is 3.94, which is less than 4.88.

121. a. Add 0.5 mol solute
   b. Add 1 L solvent
   c. Add 0.33 L solvent

Chapter 5

29. a. 0.832 atm   **b.** 632 mmHg
   c. 12.2 psi   **d.** 8.43×10^4 Pa

31. a. 809.0 mmHg   **b.** 1.064 atm
   c. 809.0 torr   **d.** 107.9 kPa

33. a. 827 mmHg   **b.** 711 mmHg

35. 5.70×10^2 mmHg

37. 58.9 mL

39. 4.33 L

41. 3.0 L

43. 2.1 mol

45. Yes, the final gauge pressure is 43.5 psi which exceeds the maximum rating.

47. 16.2 L

49. b

51. 4.76 atm

53. 11.1 L

55. 9.43 g/L

57. 44.0 g/mol

59. 4.00 g/mol

61. $P_{tot} = 0.866$ atm, $mass_{N_2} = 0.514$ g, $mass_{O_2} = 0.224$ g, $mass_{He} = 0.0473$ g

63. 1.84 atm

65. $\chi_{N_2} = 0.627$, $\chi_{O_2} = 0.373$, $P_{N_2} = 0.687$ atm, $P_{O_2} = 0.409$ atm

67. $P_{H_2} = 0.921$ atm, $mass_{H_2} = 0.0539$ g

69. 7.47×10^{-2} g

71. 38 L

73. $V_{H_2} = 48.2$ L, $V_{CO} = 24.1$ L

75. 22.8 g NaN_3

77. 60.5%

79. a. yes   **b.** no

c. No. Even though the argon atoms are more massive than the helium atoms, both have the same kinetic energy at a given temperature. The argon atoms therefore move more slowly, and so exert the same pressure as the helium atoms.

d. He

81. F_2: $u_{rms} = 442$ m/s, $KE_{avg} = 3.72 \times 10^3$ J;

Cl_2: $u_{rms} = 324$ m/s, $KE_{avg} = 3.72 \times 10^3$ J;

Br_2: $u_{rms} = 216$ m/s, $KE_{avg} = 3.72 \times 10^3$ J;

rankings: u_{rms}: $Br_2 < Cl_2 < F_2$, KE_{avg}: $Br_2 = Cl_2 = F_2$,

rate of effusion: $Br_2 < Cl_2 < F_2$

83. rate $^{238}UF_6$/rate $^{235}UF_6 = 0.99574$

85. krypton

87. A has the higher molar mass, B has the higher rate of effusion.

89. That the volume of gas particles is small compared to the space between them breaks down under conditions of high pressure. At high pressure the particles themselves occupy a significant portion of the total gas volume.

91. 0.05826 L (ideal); 0.0708 L (V.D.W.); Difference because of high pressure, at which Ne no longer acts ideally.

93. 97.7%

95. 27.8 g/mol

97. C_4H_{10}

99. 4.70 L

101. $2 \, HCl(aq) + K_2S(s) \longrightarrow H_2S(g) + 2 \, KCl(aq)$, 0.191 g $K_2S(s)$

103. 11.7 L

105. $mass_{air} = 8.56$ g, $mass_{He} = 1.20$ g, mass difference = 7.36 g

107. 4.76 L/s

109. total force = 6.15×10^3 pounds; no, the can cannot withstand this force

111. 5.8×10^3 balloons

113. 4.0 cm

115. 77.7%

117. 7.3×10^{-3} mol

119. 311 K

121. 5.0 g

123. C_3H_8

125. $P_{CH_4} = 7.30 \times 10^{-2}$ atm, $P_{O_2} = 4.20 \times 10^{-1}$ atm,

$P_{NO} = 2.79 \times 10^{-3}$ atm, $P_{CO_2} = 5.03 \times 10^{-3}$ atm,

$P_{H_2O} = 5.03 \times 10^{-3}$ atm, $P_{NO_2} = 2.51 \times 10^{-2}$ atm,

$P_{OH} = 1.01 \times 10^{-2}$ atm, $P_{tot} = 0.533$ atm

127. 0.42

129. Because helium is less dense than air, the balloon moves in a direction opposite the direction the air inside the car is moving due to the acceleration and deceleration of the car.

131. −29%

133. a. false   **b.** false   **c.** false   **d.** true

Chapter 6

31. a. 8.48×10^3 cal   **b.** 4.289×10^6 J

c. 8.48×10^4 cal   **d.** 4.50×10^8 J

33. a. 9.017×10^6 J   **b.** 9.017×10^3 kJ

c. 2.50 kWh

35. d.

37. a. heat, +   **b.** work, −   **c.** heat, +

39. -5.40×10^2 kJ

41. 311 kJ

43. The drinks that went into cooler B had more thermal energy than the refrigerated drinks that went into cooler A. The temperature difference between the drinks in cooler B and the ice was greater than the difference between the drinks and the ice in cooler A. More thermal energy was exchanged between the drinks and the ice in cooler B, which resulted in more melting.

45. 4.70×10^5 J

47. a. 7.6×10^2 °C   **b.** 4.3×10^2 °C

c. 1.3×10^2 °C   **d.** 49 °C

49. -2.8×10^2 J

51. 489 J

53. $\Delta E = -3463$ J, $\Delta H = -3452$ kJ

55. a. exothermic, −   **b.** endothermic, +

c. exothermic, −

57. -4.30×10^3 kJ

59. 9.5×10^2 g CO_2

61. Measurement B corresponds to conditions of constant pressure. Measurement A corresponds to conditions of constant volume. When a fuel is burned under constant pressure some of the energy released does work on the atmosphere by expanding against it. Less energy is manifest as heat due to this work. When a fuel is burned under constant volume, all of the energy released by the combustion reaction is evolved as heat.

63. -6.3×10^3 kJ/mol

65. -1.6×10^5 J

67. a. $-\Delta H_1$ **b.** $2 \, \Delta H_1$

c. $-\frac{1}{2}\Delta H_1$

69. -23.9 kJ

71. 87.8 kJ

73. a. $\frac{1}{2} N_2(g) + \frac{3}{2} H_2(g) \longrightarrow NH_3(g)$, $\Delta H_f^\circ = -45.9$ kJ/mol

b. $C(s, graphite) + O_2(g) \longrightarrow$

$CO_2(g)$, $\Delta H_f^\circ = -393.5$ kJ/mol

c. $2 \, Fe(s) + 3/2 \, O_2(g) \longrightarrow$

$Fe_2O_3(s)$, $\Delta H_f^\circ = -824.2$ kJ/mol

d. $C(s, graphite) + 2 \, H_2(g) \longrightarrow$

$CH_4(g)$, $\Delta H_f^\circ = -74.6$ kJ/mol

75. -382.1 kJ/mol

77. a. -137.1 kJ   **b.** -41.2 kJ

c. -137 kJ   **d.** 290.7 kJ

79. $6 \, CO_2(g) + 6 \, H_2O(l) \longrightarrow$

$C_6H_{12}O_6(s) + 9 \, O_2(g)$, $\Delta H_{rxn}^\circ = 2803$ kJ

81. -401.6 kJ/mol

83. a. 5.49 g CO_2 **b.** 5.96 g CO_2

c. 6.94 g CO_2

Natural gas, $CH_4(g)$, contributes the least to global warming by producing the least $CO_2(g)$ per kJ of heat produced.

85. 2×10^{13} kg CO_2 produced per year, 150 years

87. $\Delta E = -1.7$ J, $q = -0.5$ J, $w = -1.2$ J

89. 78 g

91. $\Delta H = 6.0$ kJ/mol, 1.1×10^2 g

93. 26.3 °C

95. palmitic acid: 9.9378 Cal/g, sucrose: 3.938 Cal/g, fat contains more Cal/g than sugar

97. $\Delta H = \Delta E + nR \, \Delta T$

99. 5.7 Cal/g

101. $\Delta E = 0, \Delta H = 0, q = -w = 3.0 \times 10^3$ J

103. 7.3×10^3 g H_2SO_4

105. 7.2×10^2 g

107. 78.2 °C

109. $C_v = \dfrac{3}{2}R, C_p = \dfrac{5}{2}R$

111. d.

113. a. At constant pressure, heat can be added and work can be done on the system. $\Delta E = q + w$, therefore $q = \Delta E - w$.

115. The aluminum is cooler because it has a lower heat capacity (specific heat).

117. The reactants lose energy by breaking chemical bonds.

Chapter 7

37. 499 s

39. (i) d, c, b, a

(ii) a, b, c, d

41. a. 4.74×10^{14} Hz   b. 5.96×10^{14} Hz

c. 5.8×10^{18} Hz

43. a. 3.14×10^{-19} J   b. 3.95×10^{-19} J

c. 3.8×10^{-15} J

45. 1.25×10^{16} photons

47. a. 79.8 kJ/mol   b. 239 kJ/mol

c. 798 kJ/mol

49.

51. 6.33 nm

53. 1.1×10^{-34} m. The wavelength of a baseball is negligible with respect to its size.

55. 2s

57. a. $l = 0$ b. $l = 0, 1$

c. $l = 0, 1, 2$ d. $l = 0, 1, 2, 3$

59. c.

61. See Figures 7.25 and 7.26. The 2s and 3p orbitals would, on average, be farther from the nucleus and have more nodes than the 1s and 2p orbitals.

63. $n = 1$

65. $2p \longrightarrow 1s$

67. a. 122 nm, UV   b. 103 nm, UV

c. 486 nm, visible   d. 434 nm, visible

69. $n = 2$

71. 344 nm

73. 6.4×10^{17} photons/s

75. 0.0547 nm

77. 91.2 nm

79. a. 4   b. 9   c. 16

81. $n = 4 \longrightarrow n = 3, n = 5 \longrightarrow n = 3,$
$n = 6 \longrightarrow n = 3$, respectively

83. 4.84×10^{14} s^{-1}

85. 11 m

87. a. $E_1 = 2.51 \times 10^{-18}$ J, $E_2 = 1.00 \times 10^{-17}$ J,
$E_3 = 2.26 \times 10^{-17}$ J

b. 26.5 nm, UV; 15.8 nm, UV

89.

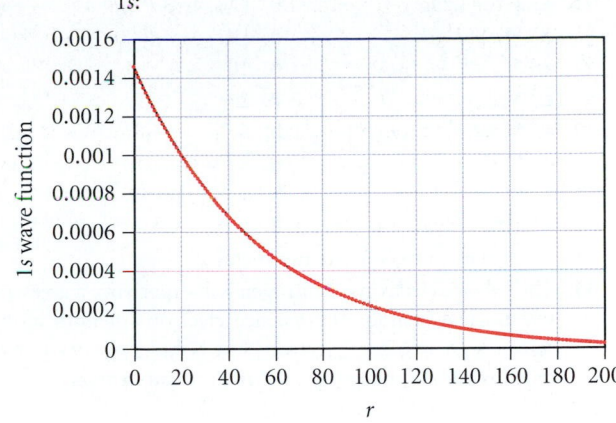

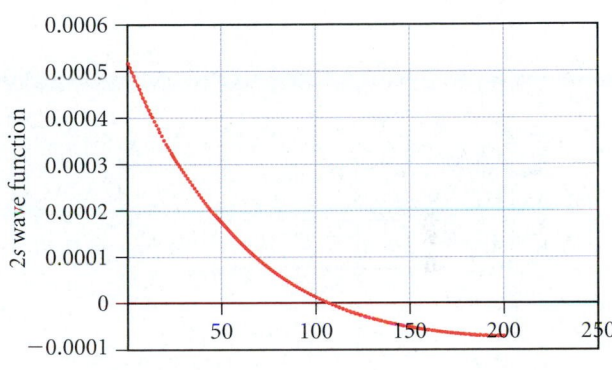

The plot for the 2s wave function extends below the x-axis. The x-intercept represents the radial node of the orbital.

91. 7.39×10^5 m/s

93. In the Bohr model, electrons exist in specific orbits encircling the atom. In the quantum mechanical model, electrons exist in orbitals that are really probability density maps of where the electron is likely to be found. The Bohr model is inconsistent with Heisenberg's uncertainty principle.

95. a. yes   b. no   c. yes   d. no

Chapter 8

41. a. $1s^2 2s^2 2p^6 3s^2 3p^3$ b. $1s^2 2s^2 2p^2$

c. $1s^2 2s^2 2p^6 3s^1$ d. $1s^2 2s^2 2p^6 3s^2 3p^6$

43. a.

| 1s | 2s | | 2p | |
| 1↑↓ | 1↑↓ | 1↑ | 1↑ | 1↑ |

b.

| 1s | 2s | | 2p | |
| 1↑↓ | 1↑↓ | 1↑↓ | 1↑↓ | 1↑ |

c.

| 1s | 2s | | 2p | | 3s |
| 1↑↓ | 1↑↓ | 1↑↓ | 1↑↓ | 1↑↓ | 1↑↓ |

d.

| 1s | 2s | | 2p | | 3s | | 3p | |
| 1↑↓ | 1↑↓ | 1↑↓ | 1↑↓ | 1↑↓ | 1↑↓ | 1↑ | | |

45. a. $[Ne] 3s^2 3p^3$ **b.** $[Ar] 4s^2 3d^{10} 4p^2$
c. $[Kr] 5s^2 4d^2$ **d.** $[Kr] 5s^2 4d^{10} 5p^5$
47. a. 1   **b.** 10
c. 5   **d.** 2
49. a. V, As   **b.** Se
c. V   **d.** Kr
51. a. 2   **b.** 1
c. 10   **d.** 6
53. reactive metal: **a**, reactive nonmetal: **c**
55. The valence electrons of nitrogen will experience a greater effective nuclear charge. The valence electrons of both atoms are screened by two core electrons but N has a greater number of protons and therefore a greater net nuclear charge.
57. a. 1+   **b.** 2+
c. 6+   **d.** 4+
59. a. In   **b.** Si
c. Pb   **d.** C
61. F, S, Si, Ge, Ca, Rb
63. a. $[Ne]$ **b.** $[Kr]$
c. $[Kr]$ **d.** $[Ar] 3d^6$
e. $[Ar] 3d^9$
65. a. $[Ar]$ Diamagnetic

b. 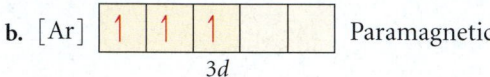 Paramagnetic

c. 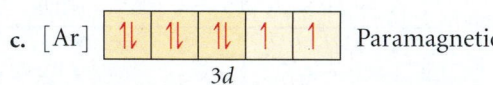 Paramagnetic

d. 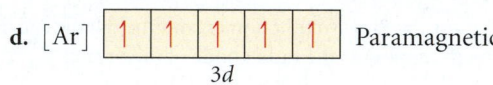 Paramagnetic

67. a. Li   **b.** I^-
c. Cr   **d.** O^{2-}
69. O^{2-}, F^-, Ne, Na^+, Mg^{2+}
71. a. Br   **b.** Na
c. cannot tell based on periodic trends   **d.** P
73. In, Si, N, F
75. a. second and third   **b.** fifth and sixth
c. sixth and seventh   **d.** first and second
77. a. Na   **b.** S
c. C   **d.** F
79. a. Sr   **b.** Bi
c. cannot tell based on periodic trends   **d.** As
81. S, Se, Sb, In, Ba, Fr
83. $Sr(s) + I_2(g) \longrightarrow SrI_2(s)$
85. $2 Li(s) + 2 H_2O(l) \longrightarrow 2 Li^+(aq) + 2 OH^-(aq) + H_2(g)$
87. $H_2(g) + Br_2(g) \longrightarrow 2 HBr(g)$
89. Br: $1s^2 2s^2 2p^6 3s^2 3p^6 4s^2 3d^{10} 4p^5$
Kr: $1s^2 2s^2 2p^6 3s^2 3p^6 4s^2 3d^{10} 4p^6$
Krypton's outer electron shell is filled, giving it chemical stability. Bromine is missing an electron from its outer shell and subsequently has a high electron affinity. Bromine tends to be easily reduced by gaining an electron, giving the bromide ion stability due to the filled p subshell which corresponds to krypton's chemically stable electron configuration.
91. V: $[Ar] 4s^2 3d^3$
V^{3+}: $[Ar] 3d^2$

Both V and V^{3+} contain unpaired electrons in their $3d$ orbitals.
93. A substitute for K^+ would need to exhibit a 1+ electric charge and have similar mass and atomic radius. Na^+ and Rb^+ would not be good substitutes because their radii are significantly smaller and larger, respectively. Based on mass, Ca^+ and Ar^+ are the closest to K^+. Because the first ionization energy of Ca^+ is closest to that of K^+, Ca^+ is the best choice for a substitute. The difficulty lies in Ca's low second ionization energy, making it easily oxidized.
95. Si, Ge
97. a. N: $[He] 2s^2 2p^3$, Mg: $[Ne] 3s^2$, O: $[He] 2s^2 2p^4$, F: $[He] 2s^2 2p^5$, Al: $[Ne] 3s^2 3p^1$
b. Mg, Al, N, O, F   **c.** Al, Mg, O, N, F
d. Aluminum's first ionization energy is lower than Mg because its $3p$ electron is shielded by the $3s$ orbital. Oxygen's first ionization energy is lower than that of N because its fourth $2p$ electron experiences electron–electron repulsion by the other electron in its orbital.
99. For main-group elements, atomic radii decrease across a period because the addition of a proton in the nucleus and an electron in the outermost energy level increases Z_{eff}. This does not happen in the transition metals because the electrons are added to the $n_{highest-1}$ orbital and the Z_{eff} stays roughly the same.
101. Noble gases are exceptionally unreactive due to the stability of their completely filled outer quantum levels and their high ionization energies. The ionization energies of Kr, Xe, and Rn are low enough to form some compounds.
103. 6A: $ns^2 np^4$, 7A: $ns^2 np^5$, group 7A elements require only one electron to achieve a noble gas configuration. Since group 6A elements require two electrons, their affinity for one electron is less negative, because one electron will merely give them an np^5 configuration.
105. 85
107. a. $d_{Ar} \approx 2$ g/L, $d_{Xe} \approx 6.5$ g/L   **b.** $d_{118} \approx 13$ g/L
c. mass $= 3.35 \times 10^{-23}$ g/Ne atom, density of Ne atom $= 2.3 \times 10^4$ g/L. The separation of Ne atoms relative to their size is immense.
d. Kr: 2.69×10^{22} atoms/L, Ne: 2.69×10^{22} atoms/L. It seems Ar will also have 2.69×10^{22} atoms/L. $d_{Ar} = 1.78$ g/L. This corresponds to accepted values.
109. Density increases to the right, because, though electrons are added successively across the period, they are added to the $3d$ subshell which is not a part of the outermost principal energy level. As a result, the atomic radius does not increase significantly across the period while mass does.

111.

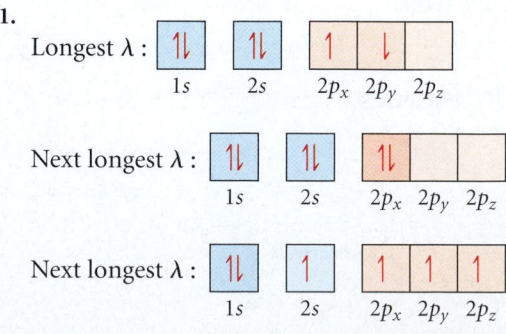

113. 168, noble gas

115. A relatively high effective nuclear charge is found in gallium with its completed $3d$ subshell and in thallium with its completed $4f$ subshell, accounting for the relatively high first ionization energies of these elements.

117. The second electron affinity requires the addition of an electron to something that is already negatively charged. The monoanions of both of these elements have relatively high electron density in a relatively small volume. As we shall see in Chapter 9 the dianions of these elements do exist in many compounds because they are stabilized by chemical bonding.

119. **a.** any group 6A element
   b. any group 5A element
   c. any group 1A element

121. $5p : n = 5$
   $l = 1$
   $m_l = 1, 0, \text{ or } -1$
   $m_s = \frac{1}{2} \text{ or } -\frac{1}{2}$
   $6d : n = 6$
   $l = 2$
   $m_l = 2, 1, 0, 1, -2$
   $m_s = \frac{1}{2} \text{ or } -\frac{1}{2}$

Chapter 9

37. $1s^2 2s^2 2p^3$ $\cdot \ddot{\text{N}} :$

39. **a.** $\cdot \dot{\text{Al}} \cdot$ **b.** Na^+

   c. $\cdot \ddot{\text{Cl}} :$ **d.** $\left[: \ddot{\text{Cl}} : \right]^-$

41. **a.** $\text{Na}^+ \left[: \ddot{\text{F}} : \right]^-$ **b.** $\text{Ca}^{2+} \left[: \ddot{\text{O}} : \right]^{2-}$

   c. $\text{Sr}^{2+} \, 2\left[: \ddot{\text{Br}} : \right]^-$ **d.** $2\,\text{K}^+ \left[: \ddot{\text{O}} : \right]^{2-}$

43. **a.** SrSe   **b.** BaCl_2
   c. Na_2S **d.** Al_2O_3

45. As the size of the alkaline earth metal ions increases, so does the distance between the metal cations and oxygen anions. Therefore, the magnitude of the lattice energy decreases accordingly because the potential energy decreases as the distance increases.

47. One factor of lattice energy is the product of the charges of the two ions. The product of the ion charges for CsF is -1 while that for BaO is -4. Because this product is four times greater, the lattice energy is also four times greater.

49. -708 kJ/mol

51. **a.** H:H, filled duets, 0 formal charge on both atoms

   b. $: \ddot{\text{Cl}} : \ddot{\text{Cl}} :$, filled octets, 0 formal charge on both atoms

   c. $\ddot{\text{O}} = \ddot{\text{O}}$, filled octets, 0 formal charge on both atoms

   d. $: \text{N} \equiv \text{N} :$, filled octets, 0 formal charge on both atoms

53. **a.** $\text{H} - \ddot{\text{P}} - \text{H}$ with H below   **b.** $: \ddot{\text{S}} - \ddot{\text{Cl}} :$ with $: \ddot{\text{Cl}} :$ below

   c. $\text{H} - \ddot{\text{I}} :$ **d.** $\text{H} - \overset{\text{H}}{\underset{\text{H}}{\text{C}}} - \text{H}$

55. **a.** pure covalent   **b.** polar covalent
   c. pure covalent   **d.** ionic bond

57. $: \text{C} = \text{O} :$ (with dipole arrow), 25%

59. **a.** $: \ddot{\text{I}} - \overset{:\ddot{\text{I}}:}{\underset{:\ddot{\text{I}}:}{\text{C}}} - \ddot{\text{I}} :$ **b.** $: \text{N} \equiv \text{N} - \ddot{\text{O}} :$

   c. $\text{H} - \overset{\text{H}}{\underset{\text{H}}{\text{Si}}} - \text{H}$ **d.** $: \ddot{\text{Cl}} - \overset{:\text{O}:}{\text{C}} - \ddot{\text{Cl}} :$ (double bond C=O)

   e. $\text{H} - \overset{:\ddot{\text{O}} - \text{H}}{\underset{\text{H}}{\text{C}}} - \text{H}$ **f.** $\left[: \ddot{\text{O}} - \text{H} \right]^-$

   g. $\left[: \ddot{\text{Br}} - \ddot{\text{O}} : \right]^-$

61. **a.** $: \overset{-1}{\ddot{\text{O}}} - \overset{+1}{\text{Se}} = \overset{0}{\ddot{\text{O}}} \longleftrightarrow : \overset{0}{\ddot{\text{O}}} = \overset{+1}{\text{Se}} - \overset{-1}{\ddot{\text{O}}} :$

   b. resonance structures of carbonate ion $[\text{CO}_3]^{2-}$ with formal charges shown

   c. $\left[: \ddot{\text{Cl}} - \overset{-1}{\ddot{\text{O}}} : \right]^-$ ($\overset{0}{\text{Cl}}$)

   d. $\left[: \overset{0}{\ddot{\text{O}}} = \overset{0}{\text{N}} - \overset{-1}{\ddot{\text{O}}} : \right]^- \longleftrightarrow \left[: \overset{-1}{\ddot{\text{O}}} - \overset{0}{\text{N}} = \overset{0}{\ddot{\text{O}}} : \right]^-$

63. $\text{H} - \overset{\text{H}}{\text{C}} = \overset{0}{\ddot{\text{S}}}$ ($\overset{0}{\text{C}}$) $\text{H} - \overset{\text{H}}{\underset{+2}{\text{S}}} = \overset{-2}{\ddot{\text{C}}}$, H_2CS is the better structure

65. $: \text{O} \equiv \text{C} - \ddot{\text{O}} :$ does not provide a significant contribution to the resonance hybrid as is has a $+1$ formal charge on a very electronegative atom (oxygen).

67. **a.** B bonded to three Cl: $: \ddot{\text{Cl}}$, $: \ddot{\text{Cl}} :$, $: \ddot{\text{Cl}} :$

   b. $\ddot{\text{O}} = \ddot{\text{N}} - \ddot{\text{O}} : \longleftrightarrow : \ddot{\text{O}} - \ddot{\text{N}} = \ddot{\text{O}}$

   c. B bonded to three H: $\overset{\text{H}}{\text{B}}$, H, H

69. a.

$$\left[\begin{array}{c} :O:^0 \\ ^{-1}:O-\underset{0}{P}-O:^{-1} \\ :O:^{-1} \end{array}\right]^{3-} \longleftrightarrow \left[\begin{array}{c} :O:^{-1} \\ ^{-1}:O-\underset{0}{P}=O^0 \\ :O:^{-1} \end{array}\right]^{3-} \longleftrightarrow$$

$$\left[\begin{array}{c} :O:^{-1} \\ ^{-1}:O-\underset{0}{P}-O:^{-1} \\ :O:^0 \end{array}\right]^{3-} \longleftrightarrow \left[\begin{array}{c} :O:^{-1} \\ ^0O=\underset{0}{P}-O:^{-1} \\ :O:^{-1} \end{array}\right]^{3-}$$

b. $\left[:C\underset{0}{\equiv}\underset{-1}{N}:\right]^{-}$

c.

$$\left[\begin{array}{c} :O:^0 \\ :O-S-O: \\ {}_{-1}\quad 0\quad {}_{-1} \end{array}\right]^{2-} \longleftrightarrow \left[\begin{array}{c} :O:^{-1} \\ :O-S=O: \\ {}_{-1}\quad 0\quad 0 \end{array}\right]^{2-} \longleftrightarrow$$

$$\left[\begin{array}{c} :O:^{-1} \\ O=S-O: \\ 0\quad 0\quad {}_{-1} \end{array}\right]^{2-}$$

d. $\left[\begin{array}{c} :O-Cl-O: \\ {}_{-1}\quad {}_{+1}\quad {}_{-1} \end{array}\right]^{-}$

71. a.

$$\begin{array}{c} :F: \\ :F-P-F: \\ :F: \quad :F: \end{array}$$

b. $\left[:I-I-I:\right]^{-}$

c.

$$\begin{array}{c} :F: \\ :F-S-F: \\ :F: \end{array}$$

d.

$$\begin{array}{c} :F: \\ :F-Ge-F: \\ :F: \end{array}$$

73. H_3CCH_3, H_2CCH_2, $HCCH$

75. -128 kJ

77. -614 kJ

79. a.

$$\begin{array}{c} :I: \\ :I-B-I: \end{array}$$

b. $2\,K^+\left[:S:\right]^{2-}$

c.

$$\begin{array}{c} :O: \\ \| \\ H-C-F: \end{array}$$

d.

$$\begin{array}{c} :Br: \\ | \\ :Br-P-Br: \end{array}$$

81. a.

$$Ba^{2+}\left[\begin{array}{c} :O: \\ \| \\ :O-C-O: \end{array}\right]^{2-} \longleftrightarrow$$

$$Ba^{2+}\left[\begin{array}{c} :O: \\ | \\ :O-C=O: \end{array}\right]^{2-} \longleftrightarrow Ba^{2+}\left[\begin{array}{c} :O: \\ | \\ O=C-O: \end{array}\right]^{2-}$$

b. $Ca^{2+}\,2\left[:O-H\right]^{-}$

c.

$$K^+\left[\begin{array}{c} :O: \\ :O-N-O: \end{array}\right]^{-} \longleftrightarrow K^+\left[\begin{array}{c} :O: \\ :O-N=O: \end{array}\right]^{-} \longleftrightarrow$$

$$K^+\left[\begin{array}{c} :O: \\ O=N-O: \end{array}\right]^{-}$$

d. $Li^+\left[:I-O:\right]^{-}$

83. a.

$$\begin{array}{cc} H & H \\ | & | \\ H-C-C-H \\ | & | \\ H-C-C-H \\ | & | \\ H & H \end{array}$$

b.

$$\begin{array}{cc} H\quad\quad H \\ C=C \\ C=C \\ H\quad\quad H \end{array} \longleftrightarrow \begin{array}{cc} H\quad\quad H \\ C-C \\ C-C \\ H\quad\quad H \end{array}$$

c.

cyclohexane structure

d.

benzene resonance structures

85. CH_2O_2,

$$\begin{array}{c} :O: \\ \| \\ H-C-O-H \end{array}$$

87. The reaction is exothermic due to the energy released when the Al_2O_3 lattice forms.

89.

$$\underset{0}{H}-\underset{0}{O}-\underset{+1}{N}-\underset{-1}{O}: \overset{:O:^0}{} \longleftrightarrow \underset{0}{H}-\underset{0}{O}-\underset{+1}{N}=\underset{0}{O}: \overset{:O:^{-1}}{} \longleftrightarrow$$

Most important

$$\underset{0}{H}-\underset{+1}{O}=\underset{+1}{N}-\underset{-1}{O}: \overset{:O:^{-1}}{}$$

91.
$$\left[\ddot{\ddot{C}}=N=\ddot{O}\right]^{-} \longleftrightarrow \left[:C\equiv\ddot{N}-\ddot{O}:\right]^{-}$$
$$\quad\; {-2}\quad{+1}\quad 0 \qquad\qquad\; {-1}\quad{+1}\quad{-1}$$

The fulminate ion is less stable because nitrogen is more electronegative than carbon and should therefore be terminal to accomodate the negative formal charge.

93.

Nonpolar — :S: — Polar

H—C—N—H

Nonpolar   Polar

95. a. $\left[\ddot{O}=\ddot{O}\cdot\right]^{-}$ **b.** $\left[\cdot\ddot{\ddot{O}}:\right]^{-}$

c. $:\ddot{O}-H$ **d.** $H-\overset{\displaystyle H}{\underset{\displaystyle H}{C}}-\ddot{O}-\ddot{O}\cdot$

97. $\Delta H_{rxn(H_2)} = -243\,kJ/mol = -121\,kJ/g$
$\Delta H_{rxn(CH_4)} = -802\,kJ/mol = -50.0\,kJ/g$
CH_4 yields more energy per mole while H_2 yields more energy per gram.

99. a. $\overset{\displaystyle :O:}{\ddot{O}=Cl}-\ddot{O}-\overset{\displaystyle :O:}{Cl=\ddot{O}}$
  b. $H-\overset{\displaystyle :O:}{\underset{\displaystyle :O:}{\overset{\displaystyle \|}{\underset{\displaystyle |}{P}}}}-\ddot{O}-H$, H

c. $H-\ddot{O}-\overset{\displaystyle :O:}{\underset{\displaystyle :O:}{\overset{\displaystyle \|}{\underset{\displaystyle |}{As}}}}-\ddot{O}-H$, H

101. $Na^+F^-, Na^+O^{2-}, Mg^{2+}F^-, Mg^{2+}O^{2-}, Al^{3+}O^{2-}$

103. $\ddot{O}=\ddot{S}=\ddot{O} + :\ddot{O}-H \longrightarrow H-\ddot{O}-\overset{\displaystyle :O:}{\overset{\displaystyle \|}{S}}-\ddot{O}:$

$H-\ddot{O}-\overset{\displaystyle :O:}{\overset{\displaystyle \|}{S}}-\ddot{O}: + \ddot{O}=\ddot{O} \longrightarrow$

$:\ddot{O}-\overset{\displaystyle :O:}{\overset{\displaystyle \|}{S}}-\ddot{O}: + H-\ddot{O}-\ddot{O}\cdot$

$:\ddot{O}-\overset{\displaystyle :O:}{\overset{\displaystyle \|}{S}}-\ddot{O}: + H-\ddot{O}-H \longrightarrow :\ddot{O}-\overset{\displaystyle H\;\; :O}{\underset{\displaystyle H-\ddot{O}:}{\overset{\displaystyle |\;\;\|}{S}}}=O:$

$\Delta H_{rxn} = -172\,kJ$

105. $r_{HCl} = 113\,pm$
$r_{HF} = 84\,pm$
These values are close to the accepted values.

107. $:P\overset{\displaystyle \ddot{P}}{\underset{\displaystyle \ddot{P}}{<\!\!\!|\!\!\!>}}P:$

109. 126 kJ/mol

111. The compounds are energy rich because a great deal of energy is released when these compounds undergo a reaction that breaks weak bonds and forms strong ones.

113. The theory is successful because it allows us to predict and account for many chemical observations. The theory is limited because electrons cannot be treated as localized "dots."

Chapter 10

29. 4

31. a. 4 e⁻ groups, 4 bonding groups, 0 lone pairs
   b. 5 e⁻ groups, 3 bonding groups, 2 lone pairs
   c. 6 e⁻ groups, 5 bonding groups, 1 lone pair

33. a. e⁻ geometry: tetrahedral
   molecular geometry: trigonal pyramidal
   idealized bond angle: 109.5°, deviation
   b. e⁻ geometry: tetrahedral
   molecular geometry: bent
   idealized bond angle: 109.5°, deviation
   c. e⁻ geometry: tetrahedral
   molecular geometry: tetrahedral
   idealized bond angle: 109.5°, deviation (due to large size of Cl compared to H)
   d. e⁻ geometry: linear
   molecular geometry: linear
   idealized bond angle: 180°

35. H_2O has a smaller bond angle due to lone pair–lone pair repulsions, the strongest electron group repulsion.

37. a. seesaw, S (with F atoms)   **b.** T-shape, $Cl-F$ (with F atoms)

c. linear, $F-I-F$ **d.** square planar, $Br-I-Br$ (with Br atoms)

39. a. linear, $H-C\equiv C-H$ **b.** Trigonal planar,
$\overset{\displaystyle H}{\underset{\displaystyle H}{}}C=C\overset{\displaystyle H}{\underset{\displaystyle H}{}}$

c. tetrahedral, $H\cdots C-C\cdots H$ (with H atoms)

41. a. The lone pair will cause lone pair–bonding pair repulsions, pushing the three bonding pairs out of the same plane. The correct molecular geometry is trigonal pyramidal.
   b. The lone pair should take an equatorial position to minimize 90° bonding pair interactions. The correct molecular geometry is seesaw.
   c. The lone pairs should take positions on opposite sides of the central atom to reduce lone pair–lone pair interactions. The correct molecular geometry is square planer.

43. a. C: tetrahedral   **b.** C's: tetrahedral
    O: bent   O: bent

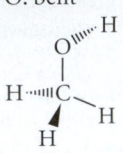

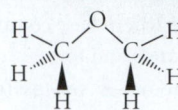

c. O's: bent

45. The vectors of the polar bonds in both CO_2 and CCL_4 oppose each other with equal magnitude and sum to 0.

47. PF_3, polar
SBr_2, nonpolar
$CHCl_3$, polar
CS_2, nonpolar

49. a. polar   **b.** polar
   c. polar   **d.** nonpolar

51. a. 0   **b.** 3
   c. 1

53. P: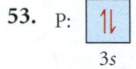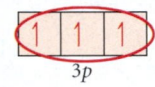

H₁:

Expected bond angle = 90°

Valence bond theory is compatible with experimentally determined bond angle of 93.3° without hybrid orbitals.

H₂:

H₃:

55.

57. sp^2

59. a. sp^3

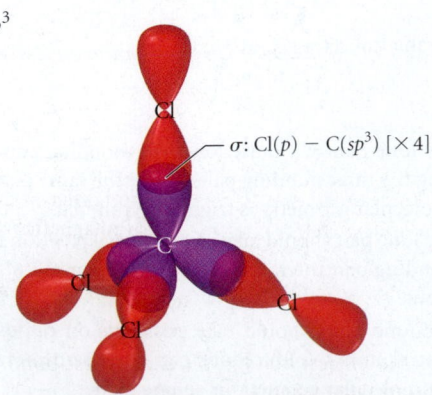

b. sp^3

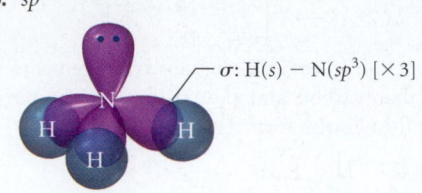

σ: H(s) − N(sp³) [×3]

c. sp^3

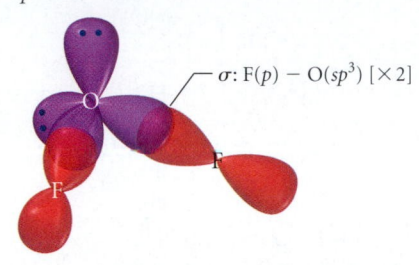

σ: F(p) − O(sp³) [×2]

d. sp

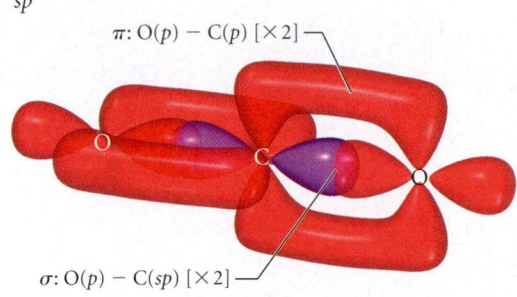

π: O(p) − C(p) [×2]

σ: O(p) − C(sp) [×2]

61. a. sp^2

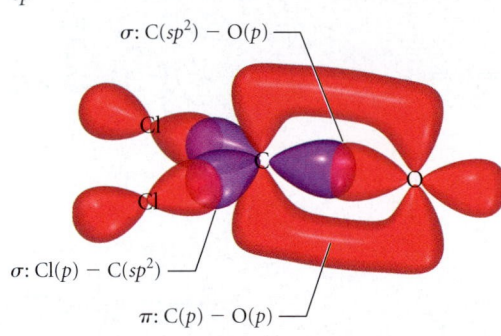

σ: C(sp²) − O(p)

σ: Cl(p) − C(sp²)

π: C(p) − O(p)

b. sp^3d^2

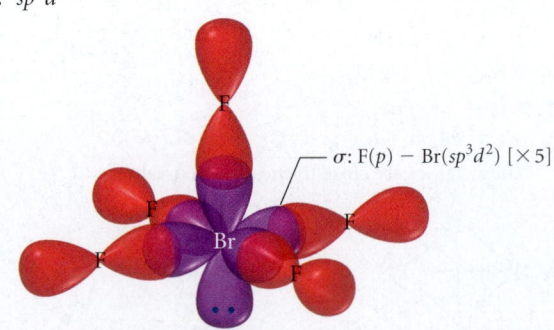

σ: F(p) − Br(sp³d²) [×5]

σ: Cl(p) − C(sp³) [×4]

c. sp^3d

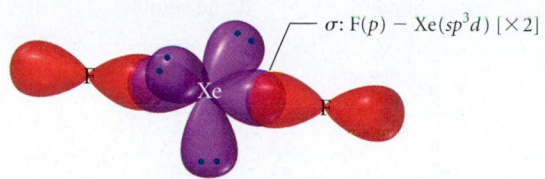

σ: F(p) − Xe(sp^3d) [×2]

d. sp^3d

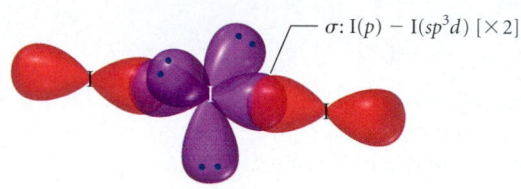

σ: I(p) − I(sp^3d) [×2]

63. a. N's: sp^2

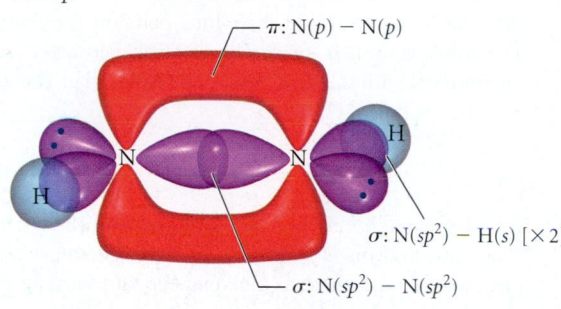

π: N(p) − N(p)

σ: N(sp^2) − H(s) [×2]

σ: N(sp^2) − N(sp^2)

b. N's: sp^3

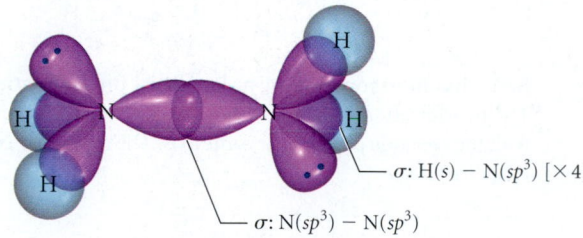

σ: H(s) − N(sp^3) [×4]

σ: N(sp^3) − N(sp^3)

c. C: sp^3
N: sp^3

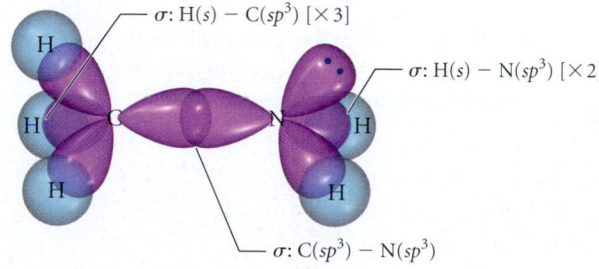

σ: H(s) − C(sp^3) [×3]

σ: H(s) − N(sp^3) [×2]

σ: C(sp^3) − N(sp^3)

65.

H H O
| | ∥
H−C−C−C−O−H
| |
H N
| |
H H

sp^3 sp^2
sp^3

67.

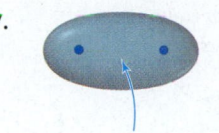

Constructive interference

69. Be_2^+ $\underline{1}$ σ_{2s}^* Be_2^-

$\underline{1\!\downarrow}$ σ_{2s} $\underline{1}$ σ_{2p}

$\underline{1\!\downarrow}$ σ_{2s}^*

$\underline{1\!\downarrow}$ σ_{2s}

bond order $Be_2^+ = 1/2$
bond order $Be_2^- = 1/2$
Both will exist in gas phase.

71.   Bonding

Antibonding

73. a.

_____		σ_{2p}^*
_____	_____	π_{2p}^*
_____		σ_{2p}
_____	_____	π_{2p}
$\underline{1\!\downarrow}$		σ_{2s}^*
$\underline{1\!\downarrow}$		σ_{2s}

bond order = 0
diamagnetic

b.

_____		σ_{2p}^*
_____	_____	π_{2p}^*
_____		σ_{2p}
$\underline{1}$	$\underline{1}$	π_{2p}
$\underline{1\!\downarrow}$		σ_{2s}^*
$\underline{1\!\downarrow}$		σ_{2s}

bond order = 1
paramagnetic

c.

_____		σ_{2p}^*
_____	_____	π_{2p}^*
_____		σ_{2p}
$\underline{1\!\downarrow}$	$\underline{1\!\downarrow}$	π_{2p}
$\underline{1\!\downarrow}$		σ_{2s}^*
$\underline{1\!\downarrow}$		σ_{2s}

bond order = 2
diamagnetic

d.

_____		σ_{2p}^*
_____	_____	π_{2p}^*
$\underline{1}$		σ_{2p}
$\underline{1\!\downarrow}$	$\underline{1\!\downarrow}$	π_{2p}
$\underline{1\!\downarrow}$		σ_{2s}^*
$\underline{1\!\downarrow}$		σ_{2s}

bond order = 2.5
paramagnetic

75. a. not stable   **b.** not stable
c. stable   **d.** not stable

77. C_2^- has the highest bond order, the highest bond energy, and the shortest bond length.

79.

$$\underline{\qquad} \quad \sigma_{2p}^{*}$$
$$\underline{\qquad} \quad \underline{\qquad} \quad \pi_{2p}^{*}$$
$$\underline{\uparrow\downarrow} \quad \underline{\uparrow\downarrow} \quad \pi_{2p}$$
$$\underline{\uparrow\downarrow} \quad \sigma_{2p}$$
$$\underline{\uparrow\downarrow} \quad \sigma_{2s}^{*}$$
$$\underline{\uparrow\downarrow} \quad \sigma_{2s}$$

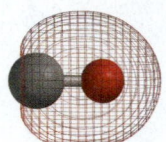

bond order = 3

81. a.

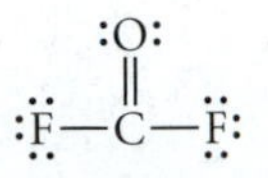

:O:
‖
:F̈—C—F̈:

trigonal planar
polar
C: sp^2

b. :C̈l—S̈—S̈—C̈l:

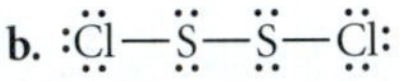

bent polar
S's sp^3

c.

   :F̈:
   |
:F̈—S—F̈:
   |
   :F̈:

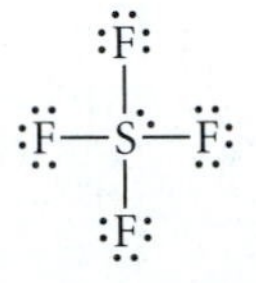

seesaw polar
S: sp^3d

83. a.

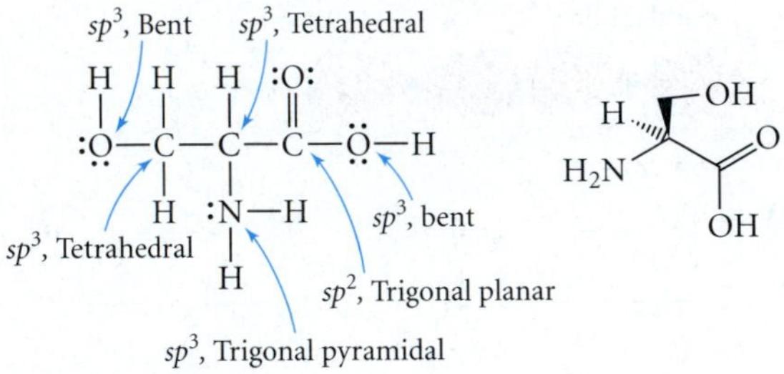

sp^3, Bent sp^3, Tetrahedral

H H H :O:
| | | ‖
:Ö—C—C—C—Ö—H
| |
H :N—H sp^3, bent
     |
sp^3, Tetrahedral H

sp^3, Trigonal pyramidal sp^2, Trigonal planar

H⸍⸍⸍⸍—OH

H₂N   O

       OH

b.

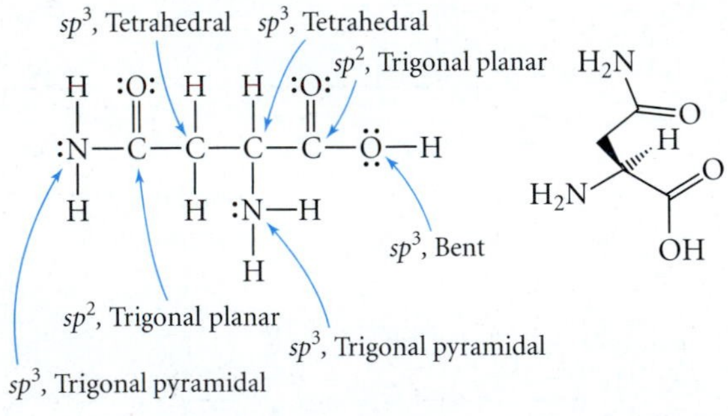

sp^3, Tetrahedral sp^3, Tetrahedral
          sp^2, Trigonal planar H₂N

H :O: H H :O:
| ‖ | | ‖
:N—C—C—C—C—Ö—H
| | :N—H
H   H |
           H
     sp^3, Bent

sp^2, Trigonal planar
       sp^3, Trigonal pyramidal
sp^3, Trigonal pyramidal

H₂N
      O
        H
H₂N
     O
      OH

c.

sp^3, Bent sp^3, Tetrahedral

H H H :O: sp^3, Bent
| | | ‖
:S—C—C—C—Ö—H
| | |
H H :N—H
sp^3, Tetrahedral | sp^2, Trigonal planar
       H

sp^3, Trigonal pyramidal

         SH

H
 N⸍⸍⸍⸍ O
 | H OH
 H

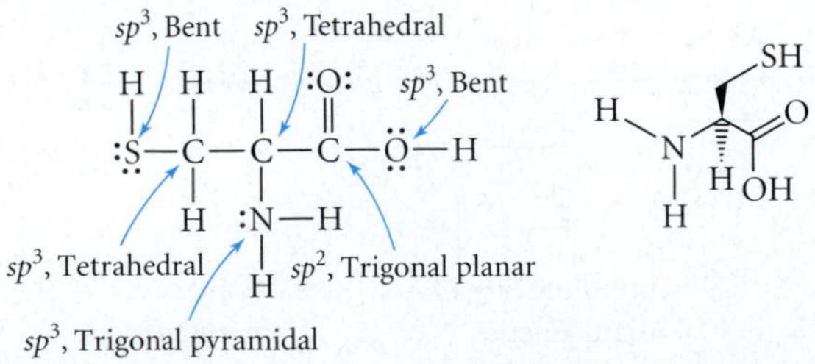

85. s bonds: 25
p bonds: 4
lone pairs: on O's and N (without methyl group): sp^2 orbitals
           on N's (with methyl group): sp^3 orbitals

87. a. water soluble   **b.** fat soluble
   c. water soluble   **d.** fat soluble

89.

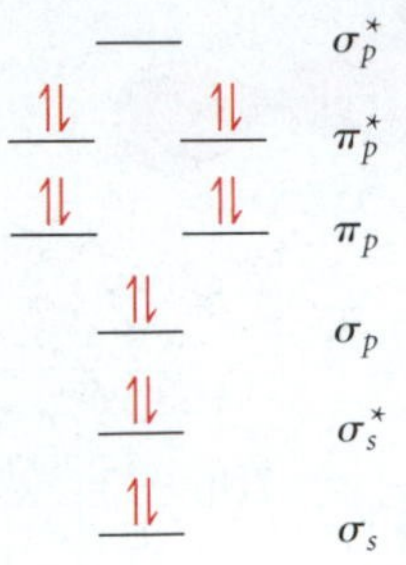

$$\underline{\qquad} \quad \sigma_{p}^{*}$$
$$\underline{\uparrow\downarrow} \quad \underline{\uparrow\downarrow} \quad \pi_{p}^{*}$$
$$\underline{\uparrow\downarrow} \quad \underline{\uparrow\downarrow} \quad \pi_{p}$$
$$\underline{\uparrow\downarrow} \quad \sigma_{p}$$
$$\underline{\uparrow\downarrow} \quad \sigma_{s}^{*}$$
$$\underline{\uparrow\downarrow} \quad \sigma_{s}$$

bond order = 1

91. BrF, unhybridized, linear

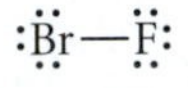

:B̈r—F̈:

BrF_2^{-} has two bonds and three lone pairs on the central atom. The hybridization is sp^3d. The electron geometry is trigonal bipyramidal with the three lone pairs equatorial. The molecular geometry is linear.

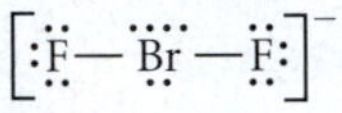

$$\left[:\ddot{F}—\ddot{B}r—\ddot{F}: \right]^{-}$$

BrF_3 has three bonds and two lone pairs on the central atom. The hybridization is sp^3d. The electron geometry is trigonal bipyramidal with the two lone pairs equatorial. The molecular geometry is T-shaped.

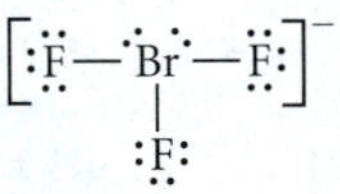

$$\left[:\ddot{F}—\ddot{B}r—\ddot{F}: \atop | \atop :\ddot{F}: \right]^{-}$$

BrF_4^{-} has four bonds and two lone pairs on the central atom. The hybridization is sp^3d^2. The electron geometry is octahedral with the two lone pairs on the same axis. The molecular geometry is square planar.

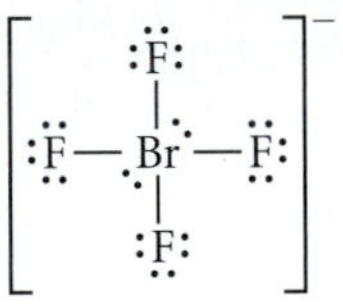

$$\left[\begin{array}{c} :\ddot{F}: \\ | \\ :\ddot{F}—Br—\ddot{F}: \\ | \\ :\ddot{F}: \end{array} \right]^{-}$$

BrF_5 has five bonds and one lone pair on the central atom. The hybridization is sp^3d^2. The electron geometry is octahedral. The molecular geometry is square pyramidal.

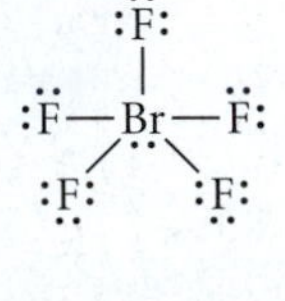

       :F̈:
       |
:F̈—Br—F̈:
   :F̈:   :F̈:

93. According to valence bond theory, CH_4, NH_3, and H_2O are all sp^3 hybridized. This hybridization results in a tetrahedral electron group configuration with a 109.5° bond angle. NH_3 and H_2O deviate from this idealized bond angle because their lone electron pairs exist in their own sp^3 orbitals. The presence of lone pairs lowers the tendency for the central atom's orbitals to hybridize. As a result, as lone pairs are added, the bond angle moves further from the 109.5° hybrid angle and closer to the 90° unhybridized angle.

95.

___	___	Antibonding
	___	Antibonding
⇅		Nonbonding
⇅	⇅	Bonding
	⇅	Bonding

NH_3 is stable due to its bond order of 3.

97. In NO_2^+, the central N has two electron groups, so the hybridization is *sp* and the ONO angle is 180°. In NO_2^- the central N has three electron groups, two bonds and one lone pair. The ideal hybridization is sp^2 but the ONO bond angle should close down a bit because of the lone pair. A bond angle around 115° is a good guess. In NO_2 there are three electron groups, but one group is a lone pair. Again the ideal hybridization would be sp^2, but since one unpaired electron must be much smaller than a lone pair or even a bonding pair, we predict that the ONO bond angle will spread and be greater than 120°. As a guess the angle is probably significantly greater than 120°.

$$\left[\ddot{O}=N=\ddot{O}\right]^+$$

$$\left[\ddot{O}=\ddot{N}-\ddot{O}\colon\right]$$

$$\ddot{O}=\dot{N}=\ddot{O}\colon$$

99. **a.** This is the best.

b. This statement is similar to **a.** but leaves out non-bonding lone-pair electron groups.

c. Molecular geometries are not determined by overlapping orbitals, but rather by the number and type pf electron groups around each central atom.

101. Lewis theory defines a single bond, double bond, and triple bond as a sharing of two electrons, four electrons, and six electrons respectively between two atoms. Valence bond theory defines a single bond as a sigma overlap of two orbitals, a double bond as a single sigma bond combined with a pi bond, and a triple bond as a double bond with an additional pi bond. Molecular orbital theory defines a single bond, double bond, and triple bond as a bond order of 1, 2, or 3 respectively between two atoms.

Chapter 11

49. **a.** dispersion **b.** dispersion, dipole–dipole

c. dispersion

d. dispersion, dipole–dipole, hydrogen bonding

e. dispersion

f. dispersion, dipole–dipole, hydrogen bonding

g. dispersion, dipole–dipole **h.** dispersion

51. **a, b, c, d.** Boiling point increases with increasing intermolecular forces. The molecules increase in their intermolecular forces as follows: **a**, dispersion forces; **b**, stronger dispersion forces (broader electron cloud); **c**, dispersion forces and dipole–dipole interactions; **d**, dispersion forces, dipole–dipole interactions, and hydrogen bonding.

53. **a.** CH_3OH, hydrogen bonding

b. CH_3CH_2OH, hydrogen bonding

c. CH_3CH_3, greater mass, broader electron cloud causes greater dispersion forces

55. **a.** Br_2, smaller mass results in weaker dispersion forces

b. H_2S, lacks hydrogen bonding

c. PH_3, lacks hydrogen bonding

57. **a.** not homogeneous

b. homogeneous, dispersion, dipole–dipole, hydrogen bonding, ion–dipole

c. homogeneous, dispersion

d. homogeneous, dispersion, dipole–dipole, hydrogen bonding

59. Water. Surface tension increases with increasing intermolecular forces, and water can hydrogen bond while acetone cannot.

61. compound A

63. When the tube is clean, water experiences adhesive forces with glass that are stronger that its cohesive forces, causing it to climb the surface of a glass tube. Water does not experience strong intermolecular forces with oil, so if the tube is coated in oil, the water's cohesive forces will be greater and it will not be attracted to the surface of the tube.

65. The water in the 12-cm dish will evaporate more quickly. The vapor pressure does not change but the surface area does. The water in the dish evaporates more quickly because the greater surface area allows for more molecules to obtain enough energy at the surface and break free.

67. Water is more volatile than vegetable oil. When the water evaporates, the endothermic process results in cooling.

69. 0.423 L

71. 86 °C

73. ΔH_{vap} = 24.7 kJ/mol, bp = 239 K

75. 41 torr

77. 15.9 kJ

79. 2.7 °C

81. 30.5 kJ

83. **a.** solid **b.** liquid

c. gas **d.** supercritical fluid

e. solid/liquid **f.** liquid/gas

g. solid/liquid/gas

85.

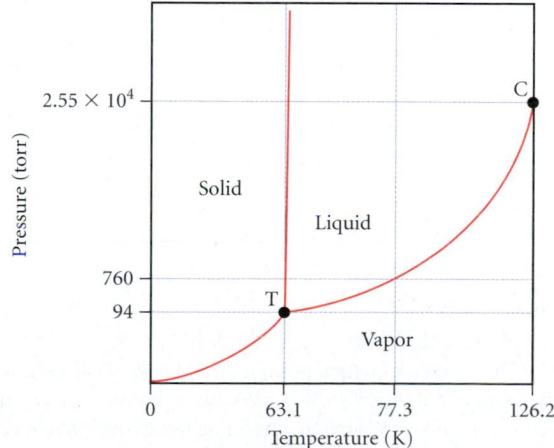

N_2 has a stable liquid phase at 1 atm.

87. **a.** 0.027 mmHg **b.** rhombic

89. Water has strong intermolecular forces. It is polar and experiences hydrogen bonding.

91. Water's exceptionally high specific heat capacity has a moderating effect on Earth's climate. Also, its high ΔH_{vap} causes water evaporation and condensation to have a strong effect on temperature.

93. 162 pm

95. a. 1   **b.** 2

   c. 4

97. $l = 393$ pm, $d = 21.3$ g/cm^3

99. 134.5 pm

101. 6.0×10^{23} atoms/mol

103. a. atomic   **b.** molecular

   c. ionic   **d.** atomic

105. LiCl(s). The other three solids are held together by intermolecular forces while LiCl is held together by stronger coulombic interactions between the cations and anions of the crystal lattice.

107. a. TiO$_2$(s), ionic solid

   b. SiCl$_4$(s), larger, stronger dispersion forces

   c. Xe(s), larger, stronger dispersion forces

   d. CaO, ions have greater charge, and therefore stronger coulombic forces

109. TiO$_2$

111. Cs: $1(1) = 1$

   Cl: $8(1/8) = 1$

   1:1

   CsCl

   Ba: $8(1/8) + 6(1/2) = 4$

   Cl: $8(1) = 8$

   $4:8 = 1:2$

   BaCl$_2$

113. a.

115. a. p-type   **b.** n-type

117. The general trend is that melting point increases with increasing mass. This is due to the fact that the electrons of the larger molecules are held more loosely and a stronger dipole moment can be induced more easily. HF is the exception to the rule. It has a relatively high melting point due to hydrogen bonding.

119. yes, 1.22 g

121. gas $\longrightarrow$ liquid $\longrightarrow$ solid

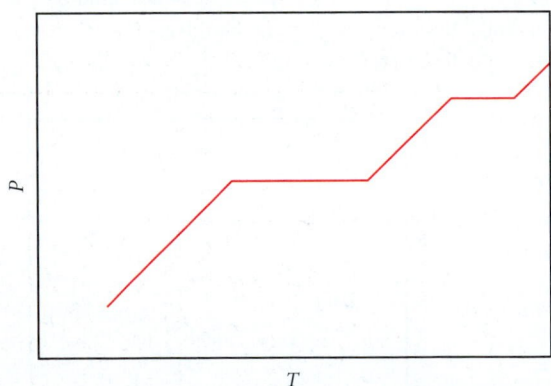

123. 26 °C

125. 3.4×10^3 g H$_2$O

127. CsCl has a higher melting point than AgI because of its higher coordination number. In CsCl, one anion bonds to eight cations (and vice versa) while in AgI, one anion bonds only to four cations.

129. a. 4r

   b. $c^2 = a^2 + b^2$ $c = 4r, a = l, b = l$

   $(4r)^2 = l^2 + l^2$

   $16r^2 = 2l^2$

   $8r^2 = l^2$

   $l = \sqrt{8r^2}$

   $l = 2\sqrt{2}r$

131. 8 atoms/unit

133. a. CO$_2$(s) $\longrightarrow$ CO$_2$(g) at 195 K

   b. CO$_2$(s) $\longrightarrow$ triple point at 216 K $\longrightarrow$ CO$_2$(g) just above 216 K

   c. CO$_2$(s) $\longrightarrow$ CO$_2$(l) at somewhat above 216 K $\longrightarrow$ CO$_2$(g) at around 250 K

   d. CO$_2$(s) $\longrightarrow$ CO$_2$(g) $\longrightarrow$ supercritical fluid

135. 2.00 g/cm^3

137. Decreasing the pressure will decrease the temperature of liquid nitrogen. Because the nitrogen is boiling, its temperature must be constant at a given pressure. As the pressure decreases, the boiling point decreases, and therefore so does the temperature. If the pressure drops below the pressure of the triple point, the phase change will shift from vaporization to sublimation and the liquid nitrogen will become solid.

139. body diagonal $= \sqrt{6}\,r$, radius $= (\sqrt{3} - \sqrt{2})r/\sqrt{2} = 0.2247r$

141. This is a valid criticism because icebergs displace the same volume of water regardless of whether or not they melt. The melting of ice sheets on Antarctica would increase ocean levels because that ice is not currently displacing any water.

143. A

145. $\Delta H_{sub} = \Delta H_{vap} + \Delta H_{fus}$

147. The vats of water will prevent the cellar from reaching freezing temperatures. Because the heat capacity of water is so high, the water will lose a lot of heat before the temperature is low enough to freeze everything.

Chapter 12

29. a. hexane, toluene, or CCl$_4$; dispersion forces

   b. water, methanol; dispersion, dipole–dipole, hydrogen bonding

   c. hexane, toluene, or CCl$_4$; dispersion forces

   d. water, acetone, methanol, ethanol; dispersion, ion–dipole

31. HOCH$_2$CH$_2$CH$_2$OH

33. a. water; dispersion, dipole–dipole, hydrogen bonding

   b. hexane; dispersion

   c. water; dispersion, dipole–dipole

   d. water; dispersion, dipole–dipole, hydrogen bonding

35. a. endothermic

   b. The lattice energy is greater in magnitude than the heat of hydration.

   c.

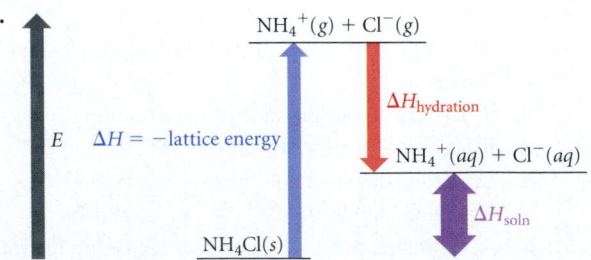

   d. The solution forms because chemical systems tend toward greater entropy.

37. -797 kJ/mol

39. $\Delta H_{soln} = -6 \times 10^1$ kJ/mol, -7 kJ of energy evolved

41. unsaturated

43. About 31 g will precipitate.

45. Boiling water releases any O_2 dissolved in it. The solubility of gases decreases with increasing temperature.

47. As pressure increases, nitrogen will more easily dissolve in blood. To reverse this process, divers should ascend to lower pressures.

49. 1.1 g

51. 2.28 M, 2.4 m, 12.3% by mass

53. 319 mL

55. 1.6×10^2 g Ag

57. 1.4×10^4 g

59. Add water to 7.31 mL of concentrated solution until a total volume of 1.15 L is acquired.

61. **a.** Add water to 3.73 g KCl to a volume of 100 mL.
   b. Add 3.59 g KCl to 96.41 g H_2O.
   c. Add 5.0 g KCl to 95 g H_2O.

63. **a.** 0.417 M   **b.** 0.444 m **c.** 7.41% by mass
   d. 0.00794   **e.** 0.794% by mole

65. The level has decreased more in the beaker filled with pure water. The dissolved salt in the seawater decreases the vapor pressure and subsequently lowers the rate of vaporization.

67. 30.4 torr

69. 23.0 torr

71. **a.** $P_{hep} = 24.4$ torr, $P_{oct} = 5.09$ torr   **b.** 29.5 torr
   c. 80.8% heptane by mass, 19.2% octane by mass
   d. The vapor is richer in the more volatile component.

73. $P_{chl} = 51.9$ torr, $P_{ace} = 274$ torr, $P_{tot} = 326$ torr. The solution is not ideal. The chloroform–acetone interactions are stronger than the chloroform–chloroform and acetone-acetone interactions.

75. freezing point (fp) $= -1.27\,°C$, bp $= 100.349\,°C$

77. 1.8×10^2 g/mol

79. 26.1 atm

81. 6.36×10^3 g/mol

83. **a.** fp $= -0.558\,°C$, bp $= 100.154\,°C$
   b. fp $= -1.98\,°C$, bp $= 100.546\,°C$
   c. fp $= -2.5\,°C$, bp $= 100.70\,°C$

85. **a.** $-0.632\,°C$ **b.** 5.4 atm   **c.** $100.18\,°C$

87. 3.4

89. Chloroform is polar and has stronger solute–solvent interactions than nonpolar carbon tetrachloride.

91. $\Delta H_{soln} = 51$ kJ/mol, $-8.7\,°C$

93. 2.2×10^{-3} M/atm

95. 1.3×10^4 L

97. 0.24 g

99. $-24\,°C$

101. **a.** 1.1% by mass/V **b.** 1.58% by mass/V
    c. 5.3% by mass/V

103. 2.484

105. 0.227 atm

107. x_{CHCl_3}(original) $= 0.657$, P_{CHCl_3}(condensed) $= 0.346$ atm

109. 6.4×10^{-3} L

111. 22.4% glucose by mass, 77.6% sucrose by mass

113. $P_{iso} = 0.131$ atm, $P_{pro} = 0.068$ atm. The major intermolecular attractions are between the OH groups. The OH group at the end of the chain in propyl alcohol is more accessible than the one in the middle of the chain in isopropyl alcohol. In addition, the molecular shape of propyl alcohol is a straight chain of carbon atoms, while that of isopropyl alcohol is a branched chain and is more like a ball. The contact area between two ball-like objects is smaller than that of two chain-like objects. The smaller contact area in isopropyl alcohol means the molecules don't attract each other as

strongly as do those of propyl alcohol. As a result of both of these factors, the vapor pressure of isopropyl alcohol is higher.

115. **a.** The two substances mix because their intermolecular forces between themselves are roughly equal to the forces between each other.
    b. 0
    c. ΔH_{solute} and $\Delta H_{solvent}$ are positive, ΔH_{mix} is negative and equals the sum of the first two.

117. **d.**

119. The balloon not only loses He, it also takes in N_2 and O_2 from the air (due to the tendency towards mixing), increasing the density of the balloon.

Chapter 13

25. **a.** Rate $= -\dfrac{1}{2}\dfrac{\Delta[HBr]}{\Delta t} = \dfrac{\Delta[H_2]}{\Delta t} = \dfrac{\Delta[Br_2]}{\Delta t}$
   b. 1.5×10^{-3} M/s
   c. 0.011 mol Br_2

27. **a.** Rate $= -\dfrac{1}{2}\dfrac{\Delta[A]}{\Delta t} = -\dfrac{\Delta[B]}{\Delta t} = \dfrac{1}{3}\dfrac{\Delta[C]}{\Delta t}$
   b. $\dfrac{\Delta[B]}{\Delta t} = -0.0500$ M/s, $\dfrac{\Delta[C]}{\Delta t} = 0.150$ M/s

29. **a.** $0 \longrightarrow 10$ s: Rate $= 8.7 \times 10^{-3}$ M/s
   $40 \longrightarrow 50$ s: Rate $= 6.0 \times 10^{-3}$ M/s
   b. 1.4×10^{-2} M/s

31. **a.** (i) 1.0×10^{-2} M/s   (ii) 8.5×10^{-3} M/s   (iii) 0.013 M/s
   b.

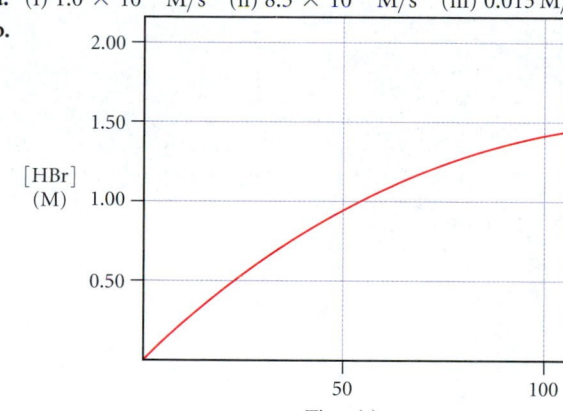

33. **a.** first order
   b.

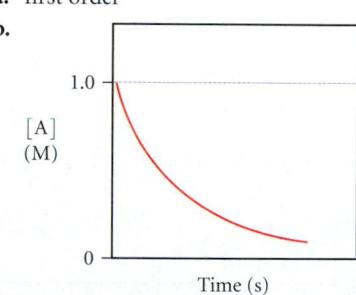

   c. Rate $= k[A]^1$, $k = 0.010$ s^{-1}

35. **a.** s^{-1} **b.** M^{-1} s^{-1}
   c. M · s^{-1}

37. **a.** Rate $= k[A][B]^2$ **b.** third order
   c. 2   **d.** 4
   e. 1   **f.** 8

39. second order, Rate $= 5.25$ M^{-1} s^{-1}[A]^2

41. Rate $= k[NO_2][F_2]$, $k = 2.57$ M^{-1} s^{-1}, second order

43. **a.** zero order   **b.** first order   **c.** second order

45. second order, $k = 2.25 \times 10^{-2} \, M^{-1} \, s^{-1}$, [AB] at 25 s = 0.619 M

47. first order, $k = 1.12 \times 10^{-2} \, s^{-1}$, Rate = $2.8 \times 10^{-3} \, M/s$

49. a. $4.5 \times 10^{-3} \, s^{-1}$ **b.** Rate = $4.5 \times 10^{-3} \, s^{-1}[A]$

    c. $1.5 \times 10^2 \, s$ **d.** [A] = 0.0908 M

51. a. $4.88 \times 10^3 \, s$ **b.** $9.8 \times 10^3 \, s$

    c. $1.8 \times 10^3 \, s$ **d.** 0.146 M at 200 s, 0.140 M at 500 s

53. 6.8×10^8 yrs; 1.8×10^{17} atoms

55.

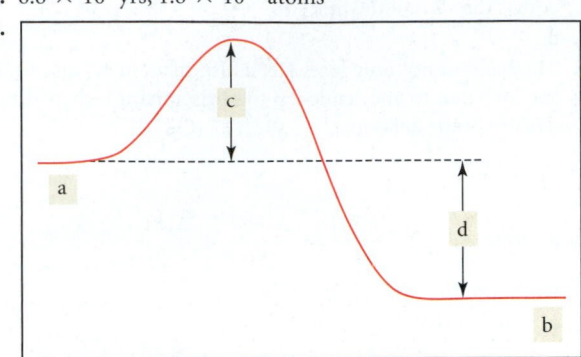

57. $17 \, s^{-1}$

59. $E_a = 251 \, kJ/mol$, $A = 7.93 \times 10^{11} \, s^{-1}$

61. $E_a = 23.0 \, kJ/mol$, $A = 8.05 \times 10^{10} \, s^{-1}$

63. a. $122 \, kJ/mol$ **b.** $0.101 \, s^{-1}$

65. $47.85 \, kJ/mol$

67. a.

69. The mechanism is valid.

71. a. $Cl_2(g) + CHCl_3(g) \longrightarrow HCl(g) + CCl_4(g)$

    b. $Cl(g)$, $CCl_3(g)$ **c.** Rate = $k[Cl_2]^{1/2}[CHCl_3]$

73. Heterogeneous catalysts require a large surface area because catalysis can only happen at the surface. A greater surface area means greater opportunity for the substrate to react, which results in a faster reaction.

75. 10^{12}

77. a. first order, $k = 0.0462 \, hr^{-1}$

    b. 15 hr   **c.** $5.0 \times 10^1 \, hr$

79. 0.0531 M/s

81. 219 torr

83. $1 \times 10^{-7} \, s$

85. a. 2

    b.

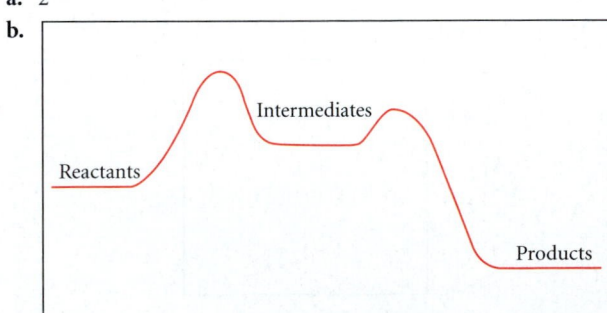

    c. first step   **d.** exothermic

87. a. 5.41 s   **b.** 2.2 s for 25%, 5.4 s for 50%

    c. 0.28 at 10 s, 0.077 at 20 s

89. a. $E_a = 89.5 \, kJ/mol$, $A = 4.22 \times 10^{11} \, s^{-1}$

    b. $2.5 \times 10^{-5} \, M^{-1} \, s^{-1}$ **c.** $6.0 \times 10^{-4} \, M/s$

91. a. No

    b. No bond is broken and the two radicals attract each other.

    c. Formation of diatomic gases from atomic gases.

93. 1.35×10^4 years

95. a. All are valid. For each, all steps sum to overall reaction and the predicted rate law is consistent with experimental data.

    b. Buildup of $I(g)$ and/or $H_2I(g)$.

97. a. 0%   **b.** 25%   **c.** 33%

99. 174 kJ

101. a. second order

    b. $CH_3NC + CH_3NC \underset{k_2}{\overset{k_1}{\rightleftharpoons}} CH_3NC^* + CH_3NC$ (fast)

    $CH_3NC^* \xrightarrow{k_3} CH_3CN$ (slow)

    Rate = $k_3[CH_3NC^*]$

    $k_1[CH_3NC]^2 = k_2[CH_3NC^*][CH_3NC]$

    $[CH_3NC^*] = \dfrac{k_1}{k_2}[CH_3NC]$

    Rate = $k_3 \times \dfrac{k_1}{k_2}[CH_3NC]$

    Rate = $k[CH_3NC]$

103. a. 48 s   **b.** $1.5 \times 10^2 \, s$ **c.** 0.179 atm

105. $k = 3.20 \times 10^{-4} \, s^{-1}$

107. B is first order and A is second order.

B will be linear if you plot ln[B] vs. time, A will be linear if you plot 1/[A] vs. time.

Chapter 14

21. a. $K = \dfrac{[SbCl_3][Cl_2]}{[SbCl_5]}$ **b.** $K = \dfrac{[NO]^2[Br_2]}{[BrNO]^2}$

    c. $K = \dfrac{[CS_2][H_2]^4}{[CH_4][H_2S]^2}$ **d.** $K = \dfrac{[CO_2]^2}{[CO]^2[O_2]}$

23. The concentration of the reactants will be greater. No, this is not dependent on initial concentrations; it is dependent on the value of K_c.

25. a. figure v

    b. The change in the decrease of reactants and increase of products would be faster.

    c. No, catalysts affect kinetics, not equilibrium.

27. a. 4.42×10^{-5}, reactants favored

    b. 1.50×10^2, products favored

    c. 1.96×10^{-9}, reactants favored

29. 1.3×10^{-29}

31. a. 2.56×10^{-23} **b.** 1.3×10^{22} **c.** 81.9

33. a. $K_c = \dfrac{[HCO_3^-][OH^-]}{[CO_3^{2-}]}$ **b.** $K_c = [O_2]^3$

    c. $K_c = \dfrac{[H_3O^+][F^-]}{[HF]}$ **d.** $K_c = \dfrac{[NH_4^+][OH^-]}{[NH_3]}$

35. 136

37.

$T \, (K)$	$[N_2]$	$[H_2]$	$[NH_3]$	K_c
500	0.115	0.105	0.439	1.45×10^{-3}
575	0.110	0.249	0.128	9.6
775	0.120	0.140	4.39×10^{-3}	0.0584

39. 303 torr

41. 3.3×10^2

43. 764

45. More solid will form.

47. Additional solid will not dissolve.

49. a. [A] = 0.33 M, [B] = 0.67 M

    b. [A] = 0.41 M, [B] = 0.59 M

    c. [A] = 0.50 M, [B] = 1.0 M

51. $[N_2O_4] = 0.0115 \, M$, $[NO_2] = 0.0770 \, M$

53. 0.10 M

55. 1.9×10^{-3} M

57. 7.84 torr

59. a. [A] = 0.38 M, [B] = 0.62 M, [C] = 0.62 M
   b. [A] = 0.90 M, [B] = 0.095 M, [C] = 0.095 M
   c. [A] = 1.0 M, [B] = 3.2×10^{-3} M, [C] = 3.2×10^{-3} M

61. a. shift left   **b.** shift right
   c. shift right

63. a. shift right   **b.** no effect
   c. no effect   **d.** shift left

65. a. shift right   **b.** shift left
   c. no effect

67. Increase temperature $\longrightarrow$ shift right, decrease temperature $\longrightarrow$ shift left. Increasing the temperature will increase the equilibrium constant.

69. b, d

71. a. 1.7×10^2
   b. $\dfrac{[Hb-CO]}{[Hb-O_2]} = 0.85$ or 17/20

CO is highly toxic, as it blocks O_2 uptake by hemoglobin. CO at a level of 0.1% will replace nearly half of the O_2 in blood.

73. b, c, d

75. 0.0144 atm

77. 3.1×10^2 g, 20% yield

79. 0.12 atm

81. 0.72 atm

83. 0.017 g

85. a. 29.3   **b.** 86.3 torr

87. $P_{NO} = P_{Cl_2} = 429$ torr

89. 2.52×10^{-3}

91. Yes, because the volume affects Q.

93. a = 1, b = 2

Chapter 15

33. a. acid, $HNO_3(aq) \longrightarrow H^+(aq) + NO_3^-(aq)$
   b. acid, $NH_4^+(aq) \rightleftharpoons H^+(aq) + NH_3(aq)$
   c. base, $KOH(aq) \longrightarrow K^+(aq) + OH^-(aq)$
   d. acid, $HC_2H_3O_2(aq) \rightleftharpoons H^+(aq) + C_2H_3O_2^-(aq)$

35. a. $H_2CO_3(aq) + H_2O(l) \rightleftharpoons H_3O^+(aq) + HCO_3^-(aq)$
      acid   base   conj. acid   conj. base
   b. $NH_3(aq) + H_2O(l) \rightleftharpoons NH_4^+(aq) + OH^-(aq)$
      base   acid   conj. acid   conj. base
   c. $HNO_3(aq) + H_2O(l) \longrightarrow H_3O^+(aq) + NO_3^-(aq)$
      acid   base   conj. acid   conj. base
   d. $C_5H_5N(aq) + H_2O(l) \rightleftharpoons C_5H_5NH^+(aq) + OH^-(aq)$
      base   acid   conj. acid   conj. base

37. a. Cl^- **b.** HSO_3^-
   c. CHO_2^- **d.** F^-

39. $H_2PO_4^-(aq) + H_2O(l) \rightleftharpoons HPO_4^{2-}(aq) + H_3O^+(aq)$
   $H_2PO_4^-(aq) + H_2O(l) \rightleftharpoons H_3PO_4(aq) + OH^-(aq)$

41. a. strong   **b.** strong
   c. strong   **d.** weak, $K_a = \dfrac{[H_3O^+][HSO_3^-]}{[H_2SO_3]}$

43. a, b, c

45. a. F^- **b.** NO_2^- **c.** ClO^-

47. a. 1.0×10^{-6}, basic   **b.** 4.5×10^{-9}, acidic
   c. 8.3×10^{-6}, basic

49. a. pH = 7.77, pOH = 6.23   **b.** pH = 7.00, pOH = 7.00
   c. pH = 5.66, pOH = 8.34

51.

$[H_3O^+]$	$[OH^-]$	pH	Acidic or Basic
7.1×10^{-4}	1.4×10^{-11}	3.15	Acidic
3.7×10^{-9}	2.7×10^{-6}	8.43	Basic
7.9×10^{-12}	1.3×10^{-3}	11.1	Basic
6.3×10^{-4}	1.6×10^{-11}	3.20	Acidic

53. $[H_3O^+] = 1.5 \times 10^{-7}$ M, pH = 6.81

55. a. $[H_3O^+] = 0.15$ M, $[OH^-] = 6.7 \times 10^{-14}$ M, pH = 0.82
   b. $[H_3O^+] = 0.025$ M, $[OH^-] = 4.0 \times 10^{-13}$ M, pH = 1.60
   c. $[H_3O^+] = 0.087$ M, $[OH^-] = 1.1 \times 10^{-13}$ M, pH = 1.06
   d. $[H_3O^+] = 0.137$ M, $[OH^-] = 7.30 \times 10^{-14}$ M, pH = 0.863

57. a. 1.8 g   **b.** 0.57 g   **c.** 0.045 g

59. $[H_3O^+] = 2.5 \times 10^{-3}$ M, pH = 2.59

61. a. 1.82 (approximation valid)   **b.** 2.18 (approximation breaks down)   **c.** 2.72 (approximation breaks down)

63. 2.75

65. 6.8×10^{-6}

67. 0.0063%

69. a. 0.42%   **b.** 0.60%
   c. 1.3%   **d.** 1.9%

71. 3.61×10^{-5}

73. a. pH = 2.03, percent ionization = 3.7%
   b. pH = 2.24, percent ionization = 5.7%
   c. pH = 2.40, percent ionization = 8.0%

75. a. 0.939   **b.** 1.07
   c. 2.19   **d.** 3.02

77. a. $[OH^-] = 0.15$ M, $[H_3O^+] = 6.7 \times 10^{-14}$ M, pH = 13.17, pOH = 0.83
   b. $[OH^+] = 0.003$ M, $[H_3O^+] = 3.3 \times 10^{-12}$ M, pH = 11.48, pOH = 2.52
   c. $[OH^-] = 9.6 \times 10^{-4}$ M, $[H_3O^+] = 1.0 \times 10^{-11}$ M, pH = 10.98, pOH = 3.02
   d. $[OH^-] = 8.7 \times 10^{-5}$ M, $[H_3O^+] = 1.1 \times 10^{-10}$ M, pH = 9.93, pOH = 4.07

79. 13.842

81. a. $NH_3(aq) + H_2O(l) \rightleftharpoons NH_4^+(aq) + OH^-(aq)$, $K_b = \dfrac{[NH_4^+][OH^-]}{[NH_3]}$
   b. $HCO_3^-(aq) + H_2O(l) \rightleftharpoons H_2CO_3(aq) + OH^-(aq)$, $K_b = \dfrac{[H_2CO_3][OH^-]}{[HCO_3^-]}$
   c. $CH_3NH_2(aq) + H_2O(l) \rightleftharpoons CH_3NH_3^+(aq) + OH^-(aq)$, $K_b = \dfrac{[CH_3NH_3^+][OH^-]}{[CH_3NH_2]}$

83. $[OH^-] = 1.6 \times 10^{-3}$ M, pOH = 2.79, pH = 11.21

85. 7.48

87. 6.7×10^{-7}

89. a. neutral
   b. basic, $ClO^-(aq) + H_2O(l) \rightleftharpoons HClO(aq) + OH^-(aq)$
   c. basic, $CN^-(aq) + H_2O(l) \rightleftharpoons HCN(aq) + OH^-(aq)$
   d. neutral

91. $[OH^-] = 2.0 \times 10^{-6}$ M, pH = 8.30

93. a. acidic, $NH_4^+(aq) + H_2O(l) \rightleftharpoons NH_3(aq) + H_3O^+(aq)$
   b. neutral
   c. acidic, $Co(H_2O)_6^{3+}(aq) + H_2O(l) \rightleftharpoons$
                  $Co(H_2O)_5(OH)^{2+}(aq) + H_3O^+(aq)$
   d. acidic, $CH_2NH_3^+(aq) + H_2O(l) \rightleftharpoons$
                  $CH_2NH_2(aq) + H_3O^+(aq)$

95. a. acidic   **b.** basic
   c. neutral   **d.** acidic
   e. acidic

97. $NaOH, NaHCO_3, NaCl, NH_4ClO_2, NH_4Cl$

99. a. 5.13   **b.** 8.87   **c.** 7.0

101. $[K^+] = 0.15\,M, [F^-] = 0.15\,M, [HF] = 2.1 \times 10^{-6}\,M,$
    $[OH^-] = 2.1 \times 10^{-6}\,M; [H_3O^+] = 4.8 \times 10^{-9}\,M$

103. $H_3PO_4(aq) + H_2O(l) \rightleftharpoons H_2PO_4^-(aq) + H_3O^+(aq),$

$$K_{a_1} = \frac{[H_3O^+][H_2PO_4^-]}{[H_3PO_4]}$$

$H_2PO_4^-(aq) + H_2O(l) \rightleftharpoons HPO_4^{2-}(aq) + H_3O^+(aq),$

$$K_{a_2} = \frac{[H_3O^+][HPO_4^{2-}]}{[H_2PO_4^-]}$$

$HPO_4^{2-}(aq) + H_2O(l) \rightleftharpoons PO_4^{3-}(aq) + H_3O^+(aq),$

$$K_{a_3} = \frac{[H_3O^+][PO_4^{3-}]}{[HPO_4^{2-}]}$$

105. a. $[H_3O^+] = 0.048\,M, pH = 1.32$
   b. $[H_3O^+] = 0.12\,M, pH = 0.92$

107. $[H_2SO_3] = 0.418\,M$
    $[HSO_3^-] = 0.082\,M$
    $[SO_3^{2-}] = 6.4 \times 10^{-8}\,M$
    $[H_3O^+] = 0.082\,M$

109. a. $[H_3O^+] = 0.50\,M, pH = 0.30$
   b. $[H_3O^+] = 0.11\,M, pH = 0.96$ (*x is small* approximation breaks down)
   c. $[H_3O^+] = 0.059\,M, pH = 1.23$

111. a. HCl, weaker bond   **b.** HF, bond polarity
   c. H_2Se, weaker bond

113. a. H_2SO_4, more oxygen atoms bonded to S
   b. $HClO_2$, more oxygen atoms bonded to Cl
   c. HClO, Cl has higher electronegativity
   d. CCl_3COOH, Cl has higher electronegativity

115. S^{2-}, its conjugate acid (H_2S), is a weaker acid than H_2S

117. a. Lewis acid   **b.** Lewis acid
   c. Lewis base   **d.** Lewis base

119. a. acid: Fe^{3+}, base: H_2O **b.** acid: Zn^{2+}, base: NH_3
   c. acid: BF_3, base: $(CH_3)_3N$

121. a. weak   **b.** strong
   c. weak   **d.** strong

123. If blood became acidic, the H^+ concentration would increase. According to Le Châtelier's principle, equilibrium would be shifted to the left and the concentration of oxygenated Hb would decrease.

125. All acid will be neutralized.

127. $[H_3O^+]$(Great Lakes) $= 3 \times 10^{-5}\,M$,
    $[H_3O^+]$(West Coast) $= 4 \times 10^{-6}\,M$. The rain over the Great Lakes is about 8 times more concentrated.

129. 2.7

131. a. 2.000   **b.** 1.52
   c. 12.95   **d.** 11.12
   e. 5.03

133. a. 1.260   **b.** 8.22

   c. 0.824   **d.** 8.57
   e. 1.171

135. a. $CN^-(aq) + H^+(aq) \rightleftharpoons HCN(aq)$
   b. $NH_4^+(aq) + OH^-(aq) \rightleftharpoons NH_3(aq) + H_2O(l)$
   c. $CN^-(aq) + NH_4^+(aq) \rightleftharpoons HCN(aq) + NH_3(aq)$
   d.
   $HSO_4^-(aq) + C_2H_3O_2^-(aq) \rightleftharpoons SO_4^{2-}(aq) + HC_2H_3O_2(aq)$
   e. no reaction between the major species

137. 0.794

139. 6.79

141. 2.14

143. $[A^-] = 4.5 \times 10^{-5}\,M$
    $[H^+] = 2.2 \times 10^{-4}\,M$
    $[HA_2^-] = 1.8 \times 10^{-4}\,M$

145. b.

147. $CH_3COOH < CH_2ClCOOH < CHCl_2COOH <$
    CCl_3COOH

Chapter 16

27. d.

29. a. 3.57   **b.** 9.08

31. pure water: 2.1%, in $NaC_7H_5O_2$: 0.065%. The percent ionization in the sodium benzoate solution is much smaller because the presence of the benzoate ion shifts the equilibrium to the left.

33. a. 2.14   **b.** 8.32   **c.** 3.46

35. $HCl + NaC_2H_3O_2 \longrightarrow HC_2H_3O_2 + NaCl$
    $NaOH + HC_2H_3O_2 \longrightarrow NaC_2H_3O_2 + H_2O$

37. a. 3.57   **b.** 9.07

39. a. 7.62   **b.** 10.82   **c.** 4.61

41. a. 3.86   **b.** 8.95

43. 3.5

45. 3.7 g

47. a. 4.74   **b.** 4.68   **c.** 4.81

49. a. initial 7.00   **b.** initial 4.71   **c.** initial 10.78
     after 1.70   after 4.56   after 10.66

51. 1.2 g; 2.7 g

53. a. yes   **b.** no   **c.** yes
   d. no   **e.** no

55. a. 7.4   **b.** 0.3 g   **c.** 0.14 g

57. KClO/HClO $= 0.79$

59. a. does not exceed capacity   **b.** does not exceed capacity
   c. does not exceed capacity   **d.** does not exceed capacity

61. i. (a) pH $= 8$, (b) pH $= 7$ **ii.** (a) weak acid, (b) strong acid

63. a. 40.0 mL HI for both   **b.** KOH: neutral, CH_3NH_2: acidic
   c. CH_3NH_2 **d.** Titration of KOH with HI:

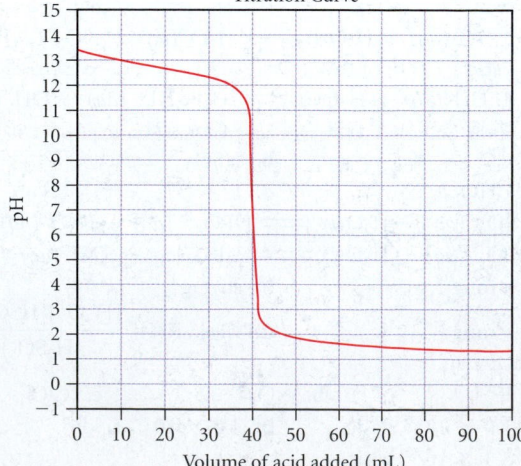

Titration Curve

Titration of CH_3NH_2 with HI:

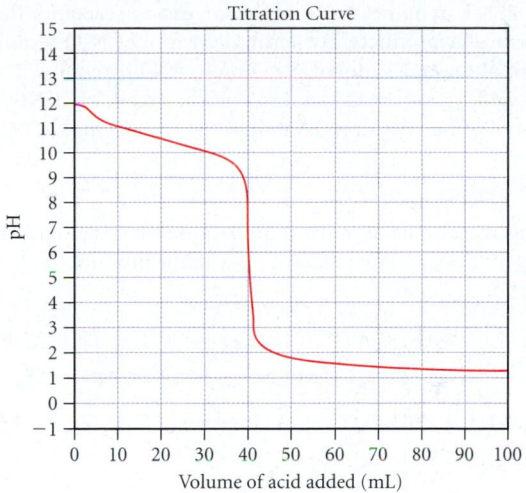

Titration Curve

65. a. pH = 9, added base = 30 mL
    b. 0 mL   c. 15 mL   d. 30 mL   e. 30 mL
67. a. 0.757   b. 30.6 mL   c. 1.038
    d. 7   e. 12.15
69. a. 13.06   b. 28.8 mL   c. 12.90
    d. 7   e. 2.07
71. a. 2.86   b. 16.8 mL   c. 4.37
    d. 4.74   e. 8.75   f. 12.17
73. a. 11.94   b. 29.2 mL   c. 11.33
    d. 10.64   e. 5.87   f. 1.90
75. i. (a)   ii. (b)
77. pK_a = 3, 82 g/mol
79. First equivalence: 22.7 mL
    Second equivalence: 45.4 mL
81. The indicator will appear red. The pH range is 4 to 6.
83. a. phenol red, *m*-nitrophenol
    b. alizarin, bromothymol blue, phenol red
    c. alizarin yellow R
85. a. $BaSO_4(s) \rightleftharpoons$
    $Ba^{2+}(aq) + SO_4^{2-}(aq), K_{sp} = [Ba^{2+}][SO_4^{2-}]$
    b. $PbBr_2(s) \rightleftharpoons Pb^{2+}(aq) + 2 Br^-(aq), K_{sp} = [Pb^{2+}][Br^-]^2$
    c. $Ag_2CrO_4(s) \rightleftharpoons 2 Ag^+(aq) + CrO_4^{2-}(aq),$
    $K_{sp} = [Ag^+]^2[CrO_4^{2-}]$
87. a. 7.31×10^{-7} M   b. 3.72×10^{-5} M   c. 3.32×10^{-4} M
89. a. 1.07×10^{-21} b. 7.14×10^{-7} c. 7.44×10^{-11}
91. AX_2
93. 2.07×10^{-5} g/100 mL
95. a. 0.0183 M   b. 0.00755 M   c. 0.00109 M
97. a. 5×10^{14} M   b. 5×10^8 M   c. 5×10^4 M
99. a. more soluble, CO_3^{2-} is basic b. more soluble, S^{2-} is basic
    c. not, neutral   d. not, neutral
101. precipitate will form, CaF_2
103. precipitate will form, $Mg(OH)_2$
105. a. 0.018 M   b. 1.4×10^{-7} M   c. 1.1×10^{-5} M
107. a. $BaSO_4$, 1.1×10^{-8} M   b. 3.0×10^{-8} M
109. 8.7×10^{-10} M
111. 5.6×10^{16}
113. 4.03
115. 3.57
117. HCl, 4.7 g

119. a. $NaOH(aq) + KHC_8H_4O_4(aq) \longrightarrow Na^+(aq) +$
    $K^+(aq) + C_8H_4O_4^{2-}(aq) + H_2O(l)$
    b. 0.1046 M
121. 4.73
123. 176 g/mol; 1.0×10^{-4}
125. 14.2 L
127. 1.6×10^{-7} M
129. 8.0×10^{-8} M
131. 6.3
133. 0.172 M
135. 51.6 g
137. 1.8×10^{-11} (based on this data)
139. a. 5.5×10^{-25} M   b. 5.5×10^{-4} M
141. a. pH $<$ pK_a b. pH $>$ pK_a
    c. pH $=$ pK_a d. pH $>$ pK_a
143. b.
145. a. no difference   b. less soluble   c. more soluble

Chapter 17

27. a, c
29. Yes. Putting one particle in the 20 J state and the other in the 0 J state has the greater entropy.
31. a. $\Delta S > 0$ b. $\Delta S < 0$
    c. $\Delta S < 0$ d. $\Delta S < 0$
33. a. $\Delta S_{sys} > 0, \Delta S_{surr} > 0$, spontaneous at all temperatures
    b. $\Delta S_{sys} < 0, \Delta S_{surr} < 0$, nonspontaneous at all temperatures
    c. $\Delta S_{sys} < 0, \Delta S_{surr} < 0$, nonspontaneous at all temperatures
    d. $\Delta S_{sys} > 0, \Delta S_{surr} > 0$, spontaneous at all temperatures
35. a. 963 J/K   b. 3.73×10^3 J/K
    c. -426 J/K   d. -1.65×10^3 J/K
37. a. 672 J/K, spontaneous   b. -672 J/K, nonspontaneous
    c. 166 J/K, spontaneous   d. -28 J/K, nonspontaneous
39. a. -2.00×10^2 kJ, spontaneous
    b. $- 2.00 \times 10^2$ kJ, nonspontaneous
    c. -5.0×10^1 kJ, spontaneous   d. 15 kJ, nonspontaneous
41. -2.247×10^6 J, spontaneous
43.

ΔH	ΔS	ΔG	Low Temperature	High Temperature
$-$	$+$	$-$	Spontaneous	Spontaneous
$-$	$-$	Temperature dependent	Spontaneous	Nonspontaneous
$+$	$+$	Temperature dependent	Nonspontaneous	Spontaneous
$+$	$-$	$+$	Nonspontaneous	Nonspontaneous

45. It increases.
47. a. $CO_2(g)$, greater molar mass and complexity
    b. $CH_3OH(g)$, gas phase
    c. $CO_2(g)$, greater molar mass and complexity
    d. $SiH_4(g)$, greater molar mass
    e. $CH_3CH_2CH_3(g)$, greater molar mass and complexity
    f. NaBr(aq), aqueous
49. a. He, Ne, SO_2, NH_3, CH_3CH_2OH. From He to Ne there is an increase in molar mass, beyond that, the molecules increase in complexity.
    b. $H_2O(s), H_2O(l), H_2O(g)$; increase in entropy in going from solid to liquid to gas phase.
    c. CH_4, CF_4, CCl_4; increasing entropy with increasing molar mass.

51. a. -120.8 J/K, decrease in moles of gas

  b. 133.9 J/K, increase in moles of gas

  c. -42.0 J/K, small change because moles of gas stay constant

  d. -390.8 J/K, decrease in moles of gas

53. -89.3 J/K, decrease in moles of gas

55. $\Delta H^\circ_{rxn} = -638.5$ kJ, $\Delta S^\circ_{rxn} = 259.4$ J/K, $\Delta G^\circ_{rxn} = -7.158 \times 10^3$ kJ; yes

57. a. $\Delta H^\circ_{rxn} = 55.3$ kJ, $\Delta S^\circ_{rxn} = 175.8$ J/K, $\Delta G^\circ_{rxn} = 2.9 \times 10^3$ J/mol; nonspontaneous, becomes spontaneous at high temperatures

  b. $\Delta H^\circ_{rxn} = 176.2$ kJ, $\Delta S^\circ_{rxn} = 285.1$ J/K, $\Delta G^\circ_{rxn} = 91.2$ kJ; nonspontaneous, becomes spontaneous at high temperatures

  c. $\Delta H^\circ_{rxn} = 98.8$ kJ, $\Delta S^\circ_{rxn} = 141.5$ J/K, $\Delta G^\circ_{rxn} = 56.6$ kJ; nonspontaneous, becomes spontaneous at high temperatures

  d. $\Delta H^\circ_{rxn} = -91.8$ kJ, $\Delta S^\circ_{rxn} = -198.1$ J/K, $\Delta G^\circ_{rxn} = -32.8$ kJ; spontaneous

59. a. 2.8 kJ   **b.** 91.2 kJ

  c. 56.4 kJ   **d.** -32.8 kJ

  Values are comparable. The method using ΔH° and ΔS° can be used to determine how ΔG° changes with temperature.

61. a. -72.5 kJ, nonspontaneous   **b.** -11.4 kJ, spontaneous

  c. 9.1 kJ, nonspontaneous

63. -29.4 kJ

65. a. 19.3 kJ   **b.** (i) 2.9 kJ

                             (ii) -2.9 kJ

  c. The partial pressure of iodine is very low.

67. 11.9 kJ

69. a. 1.48×10^{90} **b.** 2.09×10^{-26}

71. a. -24.8 kJ   **b.** 0   **c.** 9.4 kJ

73. a. 4.32×10^{26}

  b. 1.51×10^{-13}

75. $\Delta H^\circ = 50.6$ kJ

  $\Delta S^\circ = 226$ J$\cdot$K

77. a. $+$ **b.** $-$ **c.** $-$

79. a. $\Delta G^\circ = 175.2$ kJ, $K = 1.95 \times 10^{-31}$, nonspontaneous

  b. 133 kJ, yes

81. Cl_2: $\Delta H^\circ_{rxn} = -182.1$ kJ, $\Delta S^\circ_{rxn} = -134.4$ J/K,

  $\Delta G^\circ_{rxn} = -142.0$ kJ $K = 7.94 \times 10^{24}$

  Br_2: $\Delta H^\circ_{rxn} = -121.6$ kJ, $\Delta S^\circ_{rxn} = -134.2$ J/K,

  $\Delta G^\circ_{rxn} = -81.6$ kJ $K = 2.01 \times 10^{14}$

  I_2: $\Delta H^\circ_{rxn} = -48.3$ kJ, $\Delta S^\circ_{rxn} = -132.2$ J/K,

  $\Delta G^\circ_{rxn} = -8.9$ kJ $K = 36.3$

  Cl_2 is the most spontaneous, I_2 is the least. Spontaneity is determined by the standard enthalpy of formation of the dihalogenated ethane. Higher temperatures make the reactions less spontaneous.

83. a. 107.8 kJ   **b.** 5.0×10^{-7} atm

  c. spontaneous at higher temperatures, $T = 923.4$ K

85. a. 2.22×10^5 **b.** 94.4 mol

87. a. $\Delta G^\circ = -689.6$ kJ, ΔG° becomes less negative

  b. $\Delta G^\circ = -665.2$ kJ, ΔG° becomes less negative

  c. $\Delta G^\circ = -632.4$ kJ, ΔG° becomes less negative

  d. $\Delta G^\circ = -549.3$ kJ, ΔG° becomes less negative

89. With one exception, the formation of any oxide of nitrogen at 298 K requires more moles of gas as reactants than are formed as products. For example, 1 mol of N_2O requires 0.5 mol of O_2 and 1 mol of N_2, 1 mol of N_2O_3 requires 1 mol of N_2 and 1.5 mol of O_2, and so on. The exception is NO, where 1 mol of NO requires 0.5 mol of O_2 and 0.5 mol of N_2:

$$\frac{1}{2}N_2(g) + \frac{1}{2}O_2(g) \longrightarrow NO(g)$$

This reaction has a positive ΔS because what is essentially mixing of the N and O has taken place in the product.

91. a. 3.24×10^{-3}

  b.

  $NH_3 + ATP + H_2O \longrightarrow NH_3{-}P_i + ADP$

  $NH_3{-}P_i + C_5H_8O_4N^- \longrightarrow C_5H_9O_3N_2 + P_i + H_2O$

  $NH_3 + C_5H_8O_4N^- + ATP \longrightarrow C_5H_9O_3N_2 + ADP + P_i$

  $\Delta G^\circ = -16.3$ kJ, $K = 7.20 \times 10^2$

93. a. -95.3 kJ/mol. Since the number of moles of reactants and products are the same, the decrease in volume affects the entropy of both equally, so there is no change in ΔG.

  b. 102.8 J/mol. The entropy of the reactants (1.5 mol) is decreased more than the entropy of the product (1 mol). Since the product is relatively more favored at lower volume, ΔG is less positive.

  c. 204.2 kJ/mol. The entropy of the product (1 mol) is decreased more than the entropy of the reactant (0.5 mol). Since the product is relatively less favored, ΔG is more positive.

95. $\Delta H^\circ = -93$ kJ, $\Delta S^\circ = -2.0 \times 10^2$ J/K

97. A butane lighter is more efficient because the energy from the combustion is being immediately used. There is more than one energy transition involved in an electric lighter and more energy is lost (and wasted) in those transitions.

99. a and c

101. c

103. a and b

Chapter 18

37. a. $3 K(s) + Cr^{3+}(aq) \longrightarrow Cr(s) + 3 K^+(aq)$

  b. $2 Al(s) + 3 Fe^{2+}(aq) \longrightarrow 2 Al^{3+}(aq) + 3 Fe(s)$

  c. $2 BrO_3^-(aq) + 3 N_2H_4(g) \longrightarrow$
  $\qquad\qquad 2 Br^-(aq) + 3 N_2(g) + 6 H_2O(l)$

39. a. $PbO_2(s) + 2 I^-(aq) + 4 H^+(aq) \longrightarrow$
  $\qquad\qquad Pb^{2+}(aq) + I_2(s) + 2 H_2O(l)$

  b. $5 SO_3^{2-}(aq) + 2 MnO_4^-(aq) + 6 H^+(aq) \longrightarrow$
  $\qquad\qquad 5 SO_4^{2-}(aq) + 2 Mn^{2+}(aq) + 3 H_2O(l)$

  c. $S_2O_3^{2-}(aq) + 4 Cl_2(g) + 5 H_2O(l) \longrightarrow$
  $\qquad\qquad 2 SO_4^{2-}(aq) + 8 Cl^-(aq) + 10 H^+(aq)$

41. a. $H_2O_2(aq) + 2 ClO_2(aq) + 2 OH^-(aq) \longrightarrow$
  $\qquad\qquad O_2(g) + 2 ClO_2^-(aq) + 2 H_2O(l)$

  b. $Al(s) + MnO_4^-(aq) + 2 H_2O(l) \longrightarrow$
  $\qquad\qquad Al(OH)_4^-(aq) + MnO_2(s)$

  c. $Cl_2(g) + 2 OH^-(aq) \longrightarrow Cl^-(aq) + ClO^-(aq) + H_2O(l)$

43. a.

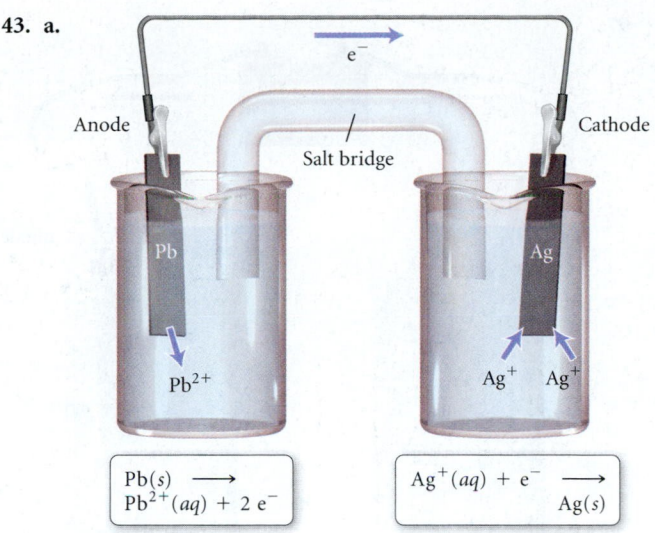

$$Pb(s) \longrightarrow Pb^{2+}(aq) + 2\,e^-$$

$$Ag^+(aq) + e^- \longrightarrow Ag(s)$$

b.

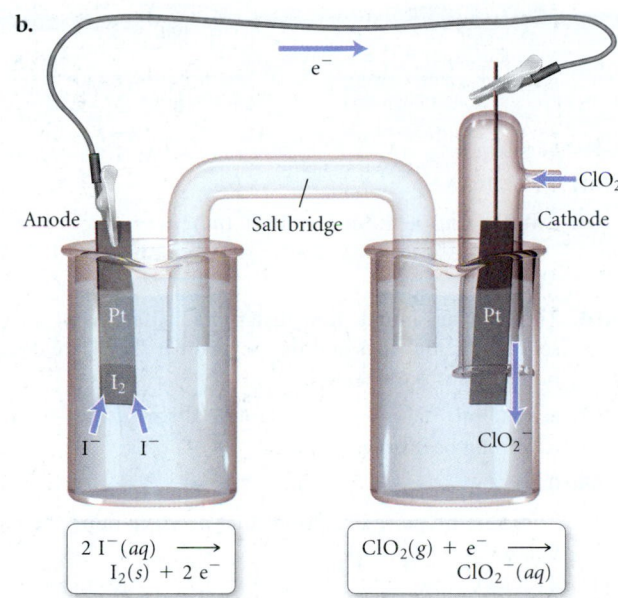

$$2\,I^-(aq) \longrightarrow I_2(s) + 2\,e^-$$

$$ClO_2(g) + e^- \longrightarrow ClO_2^-(aq)$$

c.

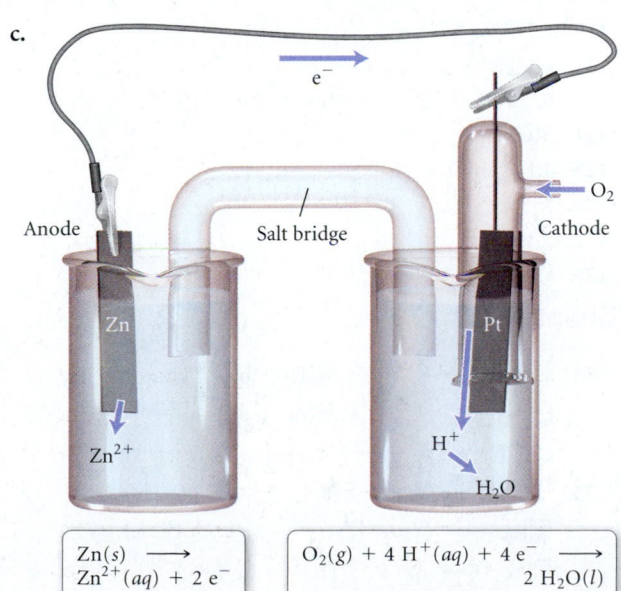

$$Zn(s) \longrightarrow Zn^{2+}(aq) + 2\,e^-$$

$$O_2(g) + 4\,H^+(aq) + 4\,e^- \longrightarrow 2\,H_2O(l)$$

45. a. 0.93 V   **b.** 0.41 V   **c.** 1.99 V

47. a, c, d.

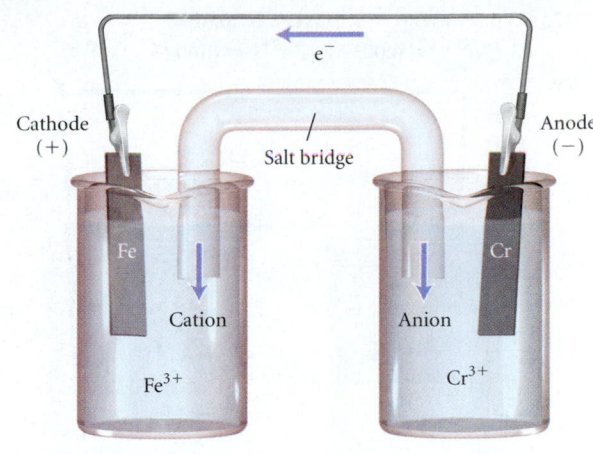

b. $Cr(s) + Fe^{3+}(aq) \longrightarrow Cr^{3+}(aq) + Fe(s),\ E^\circ_{cell} = 0.69$ V

49. a. $Pb(s)|Pb^{2+}(aq)||Ag^+(aq)|Ag(s)$

    b. $Pt(s), I_2(s)|I^-(aq)||ClO_2^-(aq)|ClO_2(g)|Pt(s)$

    c. $Zn(s)|Zn^{2+}(aq)||H_2O(l)|H^+(aq)|O_2(g)|Pt(s)$

51.

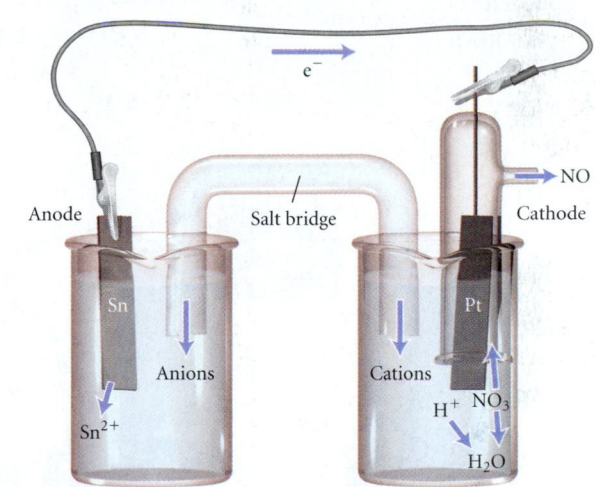

$$3\,Sn(s) + 2\,NO_3^-(aq) + 8\,H^+(aq) \longrightarrow$$
$$3\,Sn^{2+}(aq) + 2\,NO(g) + 4\,H_2O(l),\ E^\circ_{cell} = 1.10\ \text{V}$$

53. b, c

55. aluminum

57. a. yes, $2\,Al(s) + 6\,H^+(aq) \longrightarrow 2\,Al^{3+}(aq) + 3\,H_2(g)$

    b. no

    c. yes, $Pb(s) + 2\,H^+(aq) \longrightarrow Pb^{2+}(aq) + H_2(g)$

59. a. yes, $3\,Cu(s) + 2\,NO_3^-(aq) + 8\,H^+(aq) \longrightarrow$
$$3\,Cu^{2+}(aq) + 2\,NO(g) + 4\,H_2O(l)$$

    b. no

61. a. −1.70 V, nonspontaneous   **b.** 1.97 V, spontaneous

    c. −1.51, nonspontaneous

63. a

65. a. −432 kJ   **b.** 52 kJ   **c.** -1.7×10^2 kJ

67. a. 5.31×10^{75} **b.** 7.7×10^{-10} **c.** 6.3×10^{29}

69. 5.54×10^5

71. $\Delta G^\circ = -7.97$ kJ, $E^\circ_{cell} = 0.041$ V

73. a. 1.04 V   **b.** 0.97 V   **c.** 1.11 V

75. 1.87 V

77. a. 0.56 V   **b.** 0.52 V

  c. $[Ni^{2+}] = 0.003$ M, $[Zn^{2+}] = 1.60$ M

79.

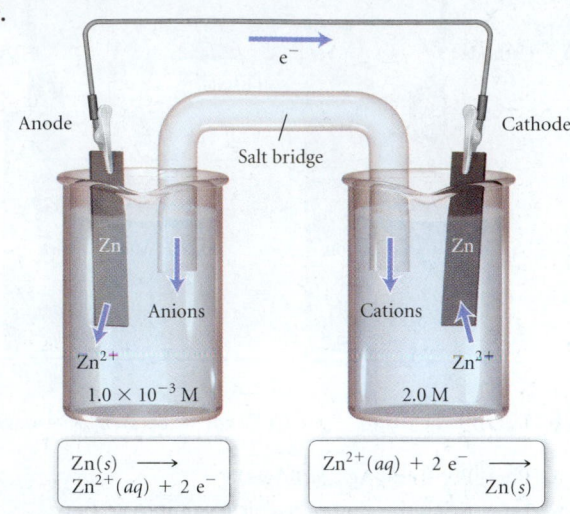

$$\begin{array}{c} Zn(s) \longrightarrow \\ Zn^{2+}(aq) + 2\,e^- \end{array}$$ $$\begin{array}{c} Zn^{2+}(aq) + 2\,e^- \longrightarrow \\ Zn(s) \end{array}$$

81. $\dfrac{[Sn^{2+}](ox)}{[Sn^{2+}](red)} = 4.2 \times 10^{-4}$

83. 0.3762

85. 1.038 V

87. a, c

89.

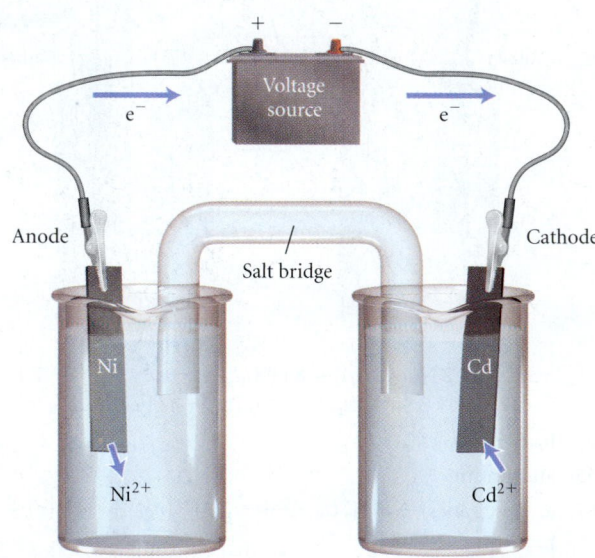

minimum voltage = 0.17 V

91. oxidation: $2\,Br^-(l) \longrightarrow Br_2(g) + 2\,e^-$

  reduction: $K^+(l) + e^- \longrightarrow K(l)$

93. oxidation: $2\,Br^-(l) \longrightarrow Br_2(g) + 2\,e^-$

  reduction: $K^+(l) + e^- \longrightarrow K(l)$

95. a. anode: $2\,Br^- \longrightarrow Br_2(l) + 2\,e^-$

  cathode: $2\,H_2O(l) + 2\,e^- \longrightarrow H_2(g) + 2\,OH^-(aq)$

  b. anode: $2\,I^-(aq) \longrightarrow I_2(s) + 2\,e^-$

  cathode: $Pb^{2+}(aq) + 2\,e^- \longrightarrow Pb(s)$

  c. anode: $2\,H_2O(l) \longrightarrow O_2(g) + 4\,H^+(aq) + 4\,e^-$

  cathode: $2\,H_2O(l) + 2\,e^- \longrightarrow H_2(g) + 2\,OH^-(aq)$

97.

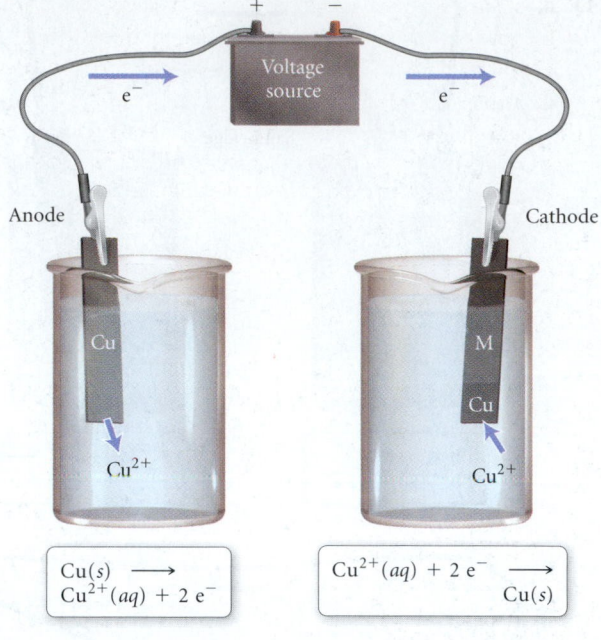

$$\begin{array}{c} Cu(s) \longrightarrow \\ Cu^{2+}(aq) + 2\,e^- \end{array}$$ $$\begin{array}{c} Cu^{2+}(aq) + 2\,e^- \longrightarrow \\ Cu(s) \end{array}$$

99. 88 s

101. 1.2×10^3 A

103. $2\,MnO_4^-(aq) + 5\,Zn(s) + 16\,H^+(aq) \longrightarrow$

  $2\,Mn^{2+}(aq) + 5\,Zn^{2+}(aq) + 8\,H_2O(l)$

  34.9 mL

105. The drawing should show that several Al atoms dissolve into solution as Al^{3+} ions and that several Cu^{2+} ions are deposited on the Al surface as solid Cu.

107. a. 68.3 mL   **b.** cannot be dissolved

  c. cannot be dissolved

109. 0.25

111. There are no paired reactions that produce more than about 5 or 6 V.

113. a. 2.83 V   **b.** 2.71 V   **c.** 16 hr

115. 176 hr

117. 0.71 V

119. a. $\Delta G° = 461$ kJ, $K = 1.4 \times 10^{-81}$

  b. $\Delta G° = 2.7 \times 10^2$ kJ, $K = 2.0 \times 10^{-48}$

121. MCl_4

123. 4.1×10^5 L

125. 435 s

127. 8.39% U

129. a

Chapter 19

33. a. $^{234}_{92}U \longrightarrow\ ^4_2He +\ ^{230}_{90}Th$ **b.** $^{230}_{90}Th \longrightarrow\ ^4_2He +\ ^{226}_{88}Ra$

  c. $^{214}_{82}Pb \longrightarrow\ ^0_{-1}e +\ ^{214}_{83}Bi$ **d.** $^{13}_{7}N \longrightarrow\ ^0_{+1}e +\ ^{13}_{6}C$

  e. $^{51}_{24}Cr +\ ^0_{-1}e \longrightarrow\ ^{51}_{23}V$

35. $^{232}_{90}Th \longrightarrow\ ^4_2He +\ ^{228}_{88}Ra$

  $^{228}_{88}Ra \longrightarrow\ ^0_{-1}e +\ ^{228}_{89}Ac$

  $^{228}_{89}Ac \longrightarrow\ ^0_{-1}e +\ ^{228}_{90}Th$

  $^{228}_{90}Th \longrightarrow\ ^4_2He +\ ^{224}_{88}Ra$

37. a. $^{221}_{87}Fr$ **b.** $^0_{-1}e$ **c.** $^0_{+1}e$ **d.** $^0_{-1}e$

39. a. stable, N/Z ratio is close to 1, acceptable for low Z atoms
  b. not stable, N/Z ratio much too high for low Z atom
  c. not stable, N/Z ratio is less than 1, much too low
  d. stable, N/Z ratio is acceptable for this Z

41. Sc, V, and Mn, each have odd numbers of protons. Atoms with an odd number of protons typically have less stable isotopes than those with an even number of protons.

43. a. beta decay **b.** positron emission
  c. positron emission **d.** positron emission

45. a. Cs-125 **b.** Fe-62

47. 2.11×10^9 yr

49. 0.57 g

51. 19 hr

53. 2.66×10^3 yr

55. 2.4×10^4 yr

57. 2.7×10^9 yr

59. $^{235}_{92}U + ^{1}_{0}n \longrightarrow ^{144}_{54}Xe + ^{90}_{38}Sr + 2\,^{1}_{0}n$

61. $^{2}_{1}H + ^{2}_{1}H \longrightarrow ^{3}_{2}He + ^{1}_{0}n$

63. $^{238}_{92}U + ^{1}_{0}n \longrightarrow ^{239}_{92}U$
  $^{239}_{92}U \longrightarrow ^{239}_{93}Np + ^{0}_{-1}e$
  $^{239}_{93}Np \longrightarrow ^{239}_{94}Pu + ^{0}_{-1}e$

65. $^{249}_{98}Cf + ^{12}_{6}C \longrightarrow ^{257}_{104}Rf + 4\,^{1}_{0}n$

67. 9.0×10^{13} J

69. a. mass defect = 0.13701 amu
  binding energy = 7.976 MeV/nucleon
  b. mass defect = 0.54369 amu
  binding energy = 8.732 MeV/nucleon
  c. mass defect = 1.16754 amu
  binding energy = 8.437 MeV/nucleon

71. 7.228×10^{10} J/g U-235

73. 7.84×10^{10} J/g H-2

75. radiation: 25.5 J, fall: 370 J

77. 68 mi

79. a. $^{1}_{1}p + ^{9}_{4}Be \longrightarrow ^{6}_{3}Li + ^{4}_{2}He$
  1.03×10^{11} J/mol
  b. $^{209}_{83}Bi + ^{64}_{28}Ni \longrightarrow ^{272}_{111}Rg + ^{1}_{0}n$
  1.141×10^{13} J/mol
  c. $^{179}_{74}W + ^{0}_{-1}e \longrightarrow ^{179}_{73}Ta$
  7.59×10^{10} J/mol

81. a. $^{114}_{44}Ru \longrightarrow ^{0}_{-1}e + ^{114}_{45}Rh$
  b. $^{216}_{88}Ra \longrightarrow ^{0}_{+1}e + ^{216}_{87}Fr$
  c. $^{58}_{30}Zn \longrightarrow ^{0}_{+1}e + ^{58}_{29}Cu$
  d. $^{31}_{10}Ne \longrightarrow ^{0}_{-1}e + ^{31}_{11}Na$

83. 2.9×10^{21} beta emissions, 3700 Ci

85. 1.6×10^{-5} L

87. a. 4.93×10^7 kJ/mol **b.** 2.42 pm

89. 7.72 MeV

91. ^{14}N

93. 0.15%

95. a. 1.164×10^{10} kJ **b.** 0.1295 g

97. U-235 forms Pb-207 in seven a-decays and four b-decays and Th-232 forms Pb-208 in six a-decays and four b-decays.

99. 7

101. The gamma emitter is the greater health threat while you sleep. The alpha emitter is the greater health threat if ingested.

Chapter 20

35. a. alkane **b.** alkene **c.** alkyne **d.** alkene

37. $CH_3-CH_2-CH_2-CH_2-CH_2-CH_2-CH_3$

$CH_3-CH-CH_2-CH_2-CH_2-CH_3$ with CH_3 branch

$CH_3-CH_2-CH-CH_2-CH_2-CH_3$ with CH_3 branch

$CH_3-CH-CH-CH_2-CH_3$ with CH_3 and CH_3 branches

$CH_3-CH_2-C-CH_2-CH_3$ with two CH_3 branches

$H_3C-C-CH_2-CH_2-CH_3$ with two CH_3 branches

$H_3C-CH-CH_2-CH-CH_3$ with CH_3 and CH_3 branches

$H_3C-C-CH-CH_3$ with CH_3 branches

$H_3C-CH_2-CH-CH_2-CH_3$ with CH_2-CH_3 branch

39. a. no **b.** yes **c.** yes **d.** no

41. a. enantiomers **b.** same **c.** enantiomers

43. a. pentane **b.** 2-methylbutane
  c. 4-isopropyl-2-methylheptane
  d. 4-ethyl-2-methylhexane

45. a. $CH_3-CH_2-CH-CH_2-CH_2-CH_3$ with CH_2-CH_3 branch
  b. $CH_3-CH_2-C-CH_2-CH_3$ with CH_3 and CH_2-CH_3 branches
  c. $CH_3-CH-CH-CH_3$ with CH_3 and CH_3 branches

d.
$$CH_3-\underset{\underset{CH_3}{|}}{\overset{\overset{CH_3}{|}}{C}}-CH_2-\underset{\underset{CH_2-CH_3}{|}}{CH}-CH_2-CH_2-\overset{\overset{CH_2-CH_3}{|}}{CH}-CH_2-CH_3$$

47. a. $CH_3CH_2CH_3 + 5\,O_2 \longrightarrow 3\,CO_2 + 4\,H_2O$
  b. $CH_3CH_2CH{=}CH_2 + 6\,O_2 \longrightarrow 4\,CO_2 + 4\,H_2O$
  c. $2\,CH{\equiv}CH + 5\,O_2 \longrightarrow 4\,CO_2 + 2\,H_2O$

49. a. CH_3CH_2Br **b.** $CH_3CH_2CH_2Cl,\ CH_3CHClCH_3$
  c. $CHCl_2Br$
  d.
$$CH_3-\underset{\underset{CH_3}{|}}{\overset{\overset{H}{|}}{C}}-CH_2-Cl$$
$$CH_3-\underset{\underset{CH_3}{|}}{\overset{\overset{Cl}{|}}{C}}-CH_3$$

51. $CH_2{=}CH-CH_2-CH_2-CH_2-CH_3$
  $CH_3-CH{=}CH-CH_2-CH_2-CH_3$
  $CH_3-CH_2-CH{=}CH-CH_2-CH_3$

53. a. 1-butene
  b. 3,4-dimethyl-2-pentene
  c. 3-isopropyl-1-hexene
  d. 2,4-dimethyl-3-hexene

55. a. 2-butyne
  b. 4,4-dimethyl-2-hexyne
  c. 3-isopropyl-1-hexyne
  d. 3,6-dimethyl-4-nonyne

57. a. $CH_3-CH_2-CH-C{\equiv}C-CH_2-CH_2-CH_3$
  b.
$$CH_3-CH_2-\overset{\overset{\|}{}}{CH}$$
$$CH-CH_2-CH_2-CH_2-CH_2-CH_3$$

  c.
$$CH{\equiv}C-\underset{\underset{CH_3}{|}}{\overset{\overset{CH_3}{|}}{C}}-CH_2-CH_2-CH_3$$

  d.
$$CH_3-CH{=}\overset{\overset{CH_3}{|}}{C}-CH_2-\underset{\underset{CH_2-CH_3}{|}}{CH}-\overset{\overset{CH_3}{|}}{CH}-CH_3$$

59. a.
$$CH_3-\underset{\underset{Cl}{|}}{CH}-\underset{\underset{Cl}{|}}{CH}-CH_3$$
  b.
$$CH_3-\underset{\underset{CH_3}{|}}{CH}-CH_2-\underset{\underset{Br}{|}}{CH}-CH_3\ +$$
$$CH_3-\underset{\underset{CH_3}{|}}{CH}-\underset{\underset{Br}{|}}{CH}-CH_2-CH_3$$

c.
$$CH_3-CH_2-\underset{\underset{Br}{|}}{CH}-\underset{\underset{Br}{|}}{CH}-CH_3$$

d.
$$CH_3-\underset{\underset{CH_3}{|}}{CH}-CH_2-\underset{\underset{Cl}{|}}{\overset{\overset{CH_3}{|}}{C}}-CH_3$$

61. a. $CH_2{=}CH-CH_3 + H_2 \longrightarrow CH_3-CH_2-CH_3$

  b.
$$CH_3-\underset{\underset{CH_3}{|}}{CH}-CH{=}CH_2 + H_2 \longrightarrow$$
$$CH_3-\underset{\underset{CH_3}{|}}{CH}-CH_2-CH_3$$

  c.
$$CH_3-\underset{\underset{CH_3}{|}}{CH}-\overset{\overset{CH_3}{|}}{C}{=}CH_2 + H_2 \longrightarrow$$
$$CH_3-\underset{\underset{CH_3}{|}}{CH}-\underset{\underset{CH_3}{|}}{CH}-CH_3$$

63. a. methylbenzene or toluene
  b. bromobenzene   **c.** chlorobenzene

65. a. 3,5-dimethyl-7-phenylnonane
  b. 2-phenyl-3-octene
  c. 4,5-dimethyl-6-phenyl-2-octyne

67. a. 1,4-dibromobenzene or p-dibromobenzene
  b. 1,3-diethylbenzene or m-diethylbenzene
  c. 1-chloro-2-fluorobenzene or o-chlorofluorobenzene

69. a.
$$CH_3-CH-CH_3$$
  b. Br
  c. Cl

71. a.
  $+\ HBr$
  b. $CH_3-CH-CH_3$
  $+\ HCl$

73. a. 1-propanol   **b.** 4-methyl-2-hexanol
  c. 2,6-dimethyl-4-heptanol   **d.** 3-methyl-3-pentanol

75. a. $CH_3CH_2CH_2Br + H_2O$ **b.** $CH_3-\overset{\overset{CH_3}{|}}{C}{=}CH_2 + H_2O$
  c. $CH_3CH_2ONa + \frac{1}{2}H_2$ **d.** $CH_3-\underset{\underset{CH_3}{|}}{\overset{\overset{CH_3}{|}}{C}}-CH_2-\overset{\overset{O}{\|}}{C}-OH$

77. a. butanone   **b.** pentanal
  c. 3,5,5-trimethylhexanal **d.** 4-methyl-2-hexanone

79. 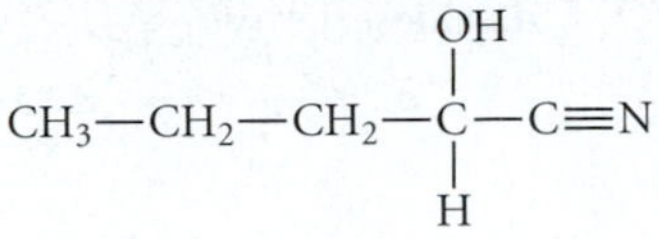
$$CH_3-CH_2-CH_2-\overset{\overset{\displaystyle OH}{|}}{\underset{\underset{\displaystyle H}{|}}{C}}-C\equiv N$$

81. a. methylbutanoate   **b.** propanoic acid
  c. 5-methylhexanoic acid **d.** ethylpentanoate

83. a. 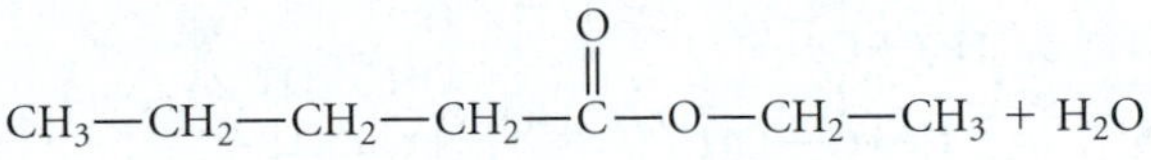
$$CH_3-CH_2-CH_2-CH_2-\overset{\overset{\displaystyle O}{\|}}{C}-O-CH_2-CH_3 + H_2O$$

b.

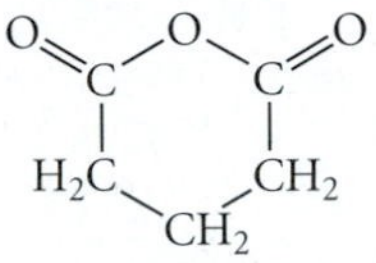

+ H₂O

85. a. ethyl propyl ether   **b.** ethyl pentyl ether
  c. dipropyl ether   **d.** butyl ethyl ether

87. a. diethylamine   **b.** methylpropylamine
  c. butylmethylpropylamine

89. a. acid–base, $(CH_3)_2NH_2^+(aq) + Cl^-(aq)$
  b. condensation, $CH_3CH_2CONHCH_2CH_3(aq) + H_2O$
  c. acid–base, $CH_3NH_3^+(aq) + HSO_4^-(aq)$

91. 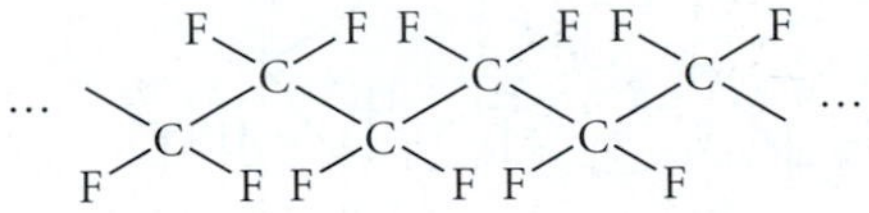 ...

93.

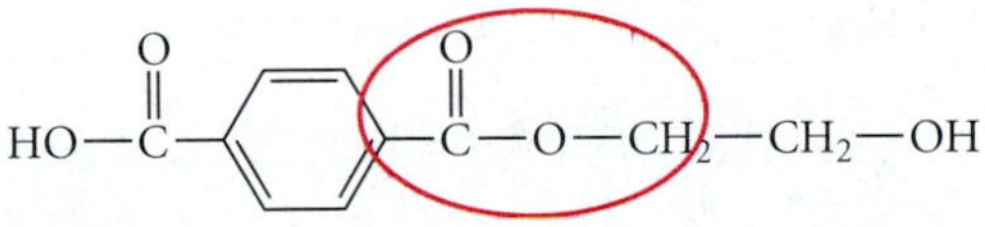

95. a. ester, methyl 3-methylbutanoate
  b. ether, ethyl 2-methylbutyl ether
  c. aromatic, 1-ethyl-3-methylbenzene or *m*-ethylmethylbenzene
  d. alkyne, 5-ethyl-4-methyl-2-heptyne
  e. aldehyde, butanal   **f.** alcohol, 2-methyl-1-propanol

97. a. 5-isobutyl-3-methylnonane
  b. 5-methyl-3-hexanone
  c. 3-methyl-2-butanol
  d. 4-ethyl-3,5-dimethyl-1-hexyne

99. a. isomers   **b.** isomers   **c.** same

101. 558 g

103. a. combustion   **b.** alkane substitution
  c. alcohol elimination   **d.** aromatic substitution

105. a. 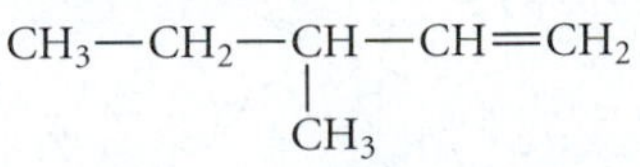
$$CH_3-CH_2-\overset{\overset{\displaystyle }{|}}{\underset{\underset{\displaystyle CH_3}{|}}{CH}}-CH=CH_2$$
Can exist as a stereoisomer

b. 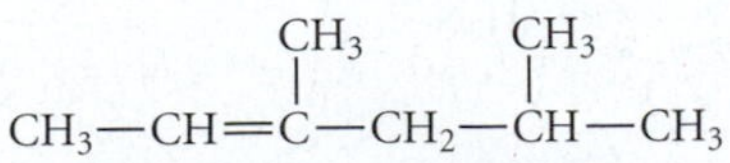
$$CH_3-CH=\overset{\overset{\displaystyle CH_3}{|}}{C}-CH_2-\overset{\overset{\displaystyle CH_3}{|}}{CH}-CH_3$$
Can exist as a stereoisomer

c. 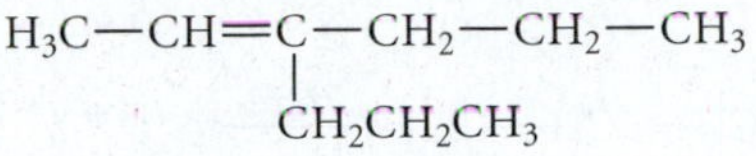
$$H_3C-CH=\overset{\overset{\displaystyle }{|}}{\underset{\underset{\displaystyle CH_2CH_2CH_3}{|}}{C}}-CH_2-CH_2-CH_3$$
Can exist as a stereoisomer

107. 1. 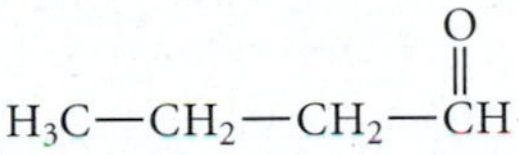
$$H_3C-CH_2-CH_2-\overset{\overset{\displaystyle O}{\|}}{CH}$$
Aldehyde

2. 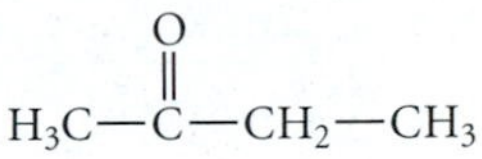
$$H_3C-\overset{\overset{\displaystyle O}{\|}}{C}-CH_2-CH_3$$
Ketone

3. $H_3C-CH=CH-O-CH_3$
Alkene, ether

4. $H_2C=CH-O-CH_2-CH_3$
Alkene, ether

5. $H_2C=CH-CH_2-O-CH_3$
Alkene, ether

6. $H_3C-CH=CH-CH_2-OH$
Alkene, alcohol

7. 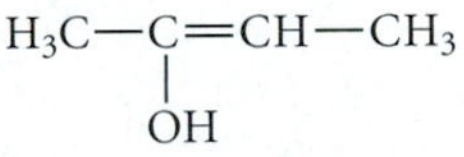
$$H_3C-\overset{\overset{\displaystyle }{|}}{\underset{\underset{\displaystyle OH}{|}}{C}}=CH-CH_3$$
Alkene, alcohol

8. 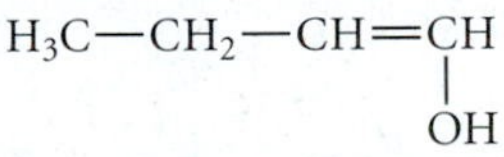
$$H_3C-CH_2-CH=\overset{\overset{\displaystyle }{|}}{\underset{\underset{\displaystyle OH}{|}}{CH}}$$
Alkene, alcohol

9. 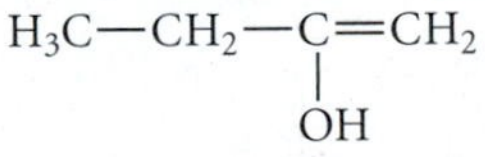
$$H_3C-CH_2-\overset{\overset{\displaystyle }{|}}{\underset{\underset{\displaystyle OH}{|}}{C}}=CH_2$$
Alkene, alcohol

10. 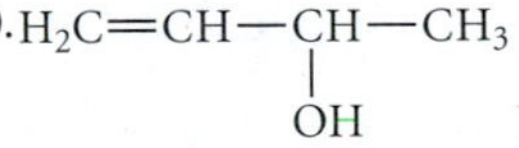
$$H_2C=CH-\overset{\overset{\displaystyle }{|}}{\underset{\underset{\displaystyle OH}{|}}{CH}}-CH_3$$
Alkene, alcohol

11. $H_2C=CH-CH_2-CH_2-OH$
Alkene, alcohol

109. a.

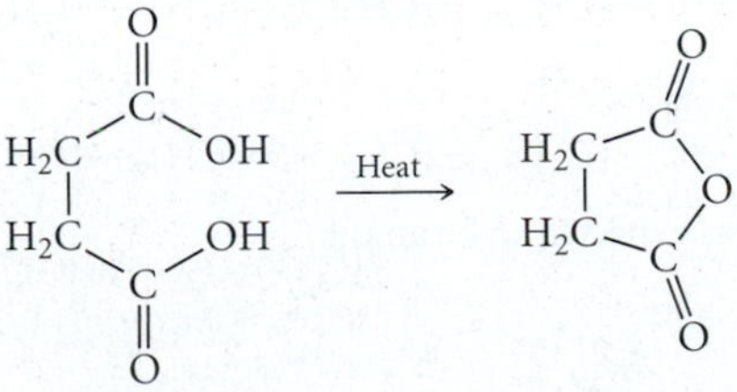

b.

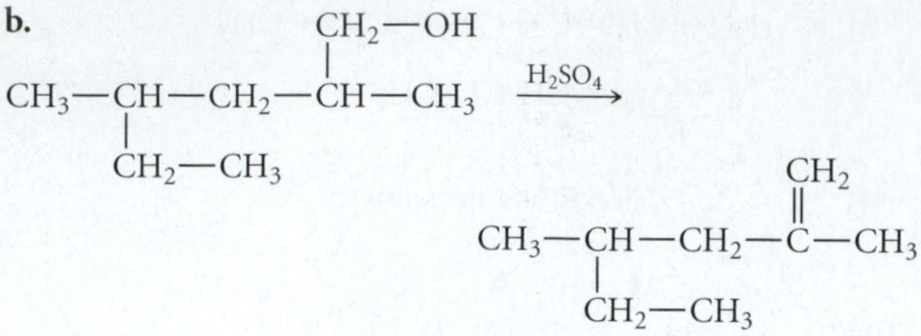

$$CH_3-CH-CH_2-CH-CH_3 \xrightarrow{H_2SO_4}$$
with CH_2-OH above the second carbon and CH_2-CH_3 below the second carbon

$$CH_3-CH-CH_2-\overset{\overset{\displaystyle CH_2}{\|}}{C}-CH_3$$
with CH_2-CH_3 below

c.

$$CH_3-CH_2-C=CH_2 + HBr \longrightarrow CH_3-CH_2-\overset{Br}{\underset{CH_3}{C}}-CH_3$$
with CH_3 below the left carbon

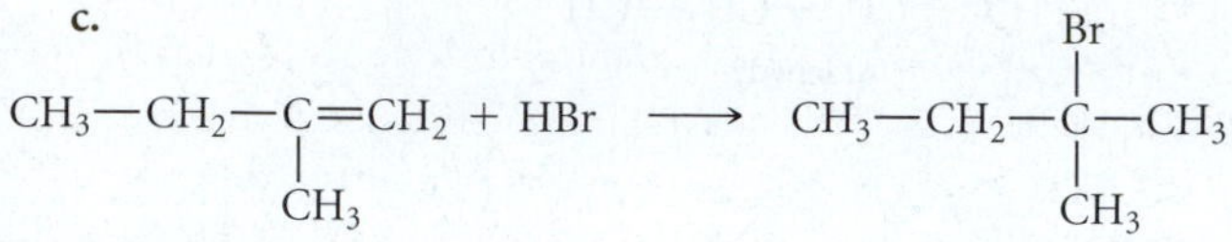

111. a. 3:1

b. 2° hydrogen atoms are more reactive. The reactivity of 2° hydrogens to 1° hydrogens is 11:3.

113.
$$CH_3-\overset{\overset{\displaystyle Cl\ \ Cl}{|\ \ \ |}}{\underset{\underset{\displaystyle H\ \ H}{|\ \ \ |}}{C-C}}-CH_3$$
Chiral

$$Cl-CH_2-\overset{Cl}{\underset{}{CH}}-CH_2-CH_3$$
Chiral

$$Cl-CH_2-CH_2-\overset{Cl}{\underset{}{CH}}-CH_3$$
Chiral

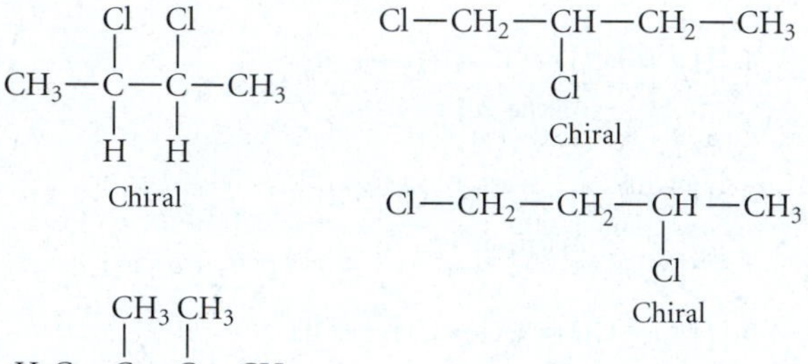

115. 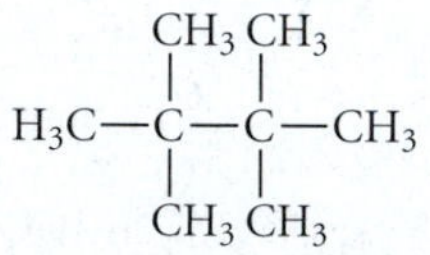
$$H_3C-\overset{\overset{\displaystyle CH_3\ CH_3}{|\ \ \ \ |}}{\underset{\underset{\displaystyle CH_3\ CH_3}{|\ \ \ \ |}}{C-C}}-CH_3$$
2,2,3,3-tetramethylbutane

Chapter 21

31. c. saturated fatty acid; **d.** steroid

33. a. saturated fatty acid
 b. not a fatty acid
 c. not a fatty acid
 d. monounsaturated fatty acid

35.

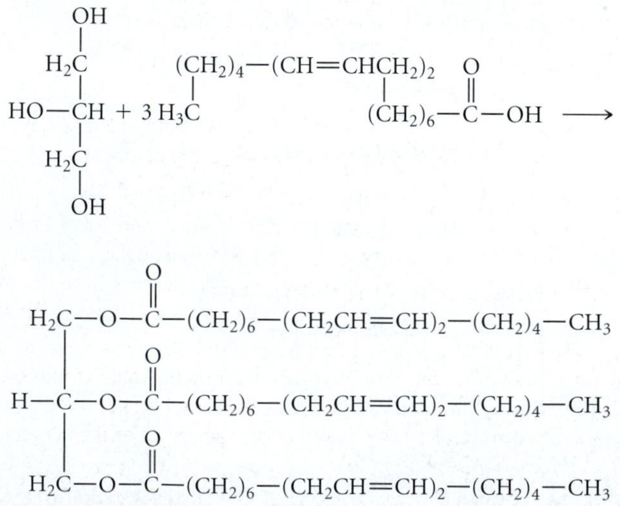

$$H_2C-OH$$
$$HO-CH + 3\ H_3C-(CH_2)_4-(CH=CHCH_2)_2-(CH_2)_6-\overset{\overset{\displaystyle O}{\|}}{C}-OH \longrightarrow$$
$$H_2C-OH$$

$$H_2C-O-\overset{\overset{\displaystyle O}{\|}}{C}-(CH_2)_6-(CH_2CH=CH)_2-(CH_2)_4-CH_3$$
$$H-C-O-\overset{\overset{\displaystyle O}{\|}}{C}-(CH_2)_6-(CH_2CH=CH)_2-(CH_2)_4-CH_3$$
$$H_2C-O-\overset{\overset{\displaystyle O}{\|}}{C}-(CH_2)_6-(CH_2CH=CH)_2-(CH_2)_4-CH_3$$

Triglyceride is expected to be an oil.

37. a. monosaccharide; **c.** disaccharide

39. a. aldose, hexose   **b.** aldose, pentose
 c. ketose, tetrose   **d.** aldose, tetrose

41. a. 5   **b.** 3   **c.** 1   **d.** 3

43.

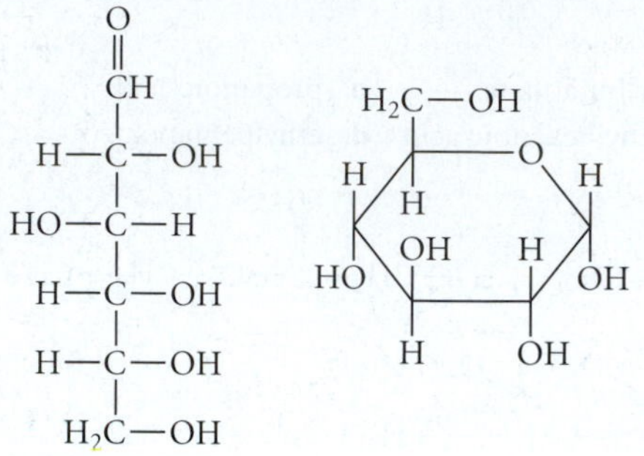

$$\overset{\overset{\displaystyle O}{\|}}{\underset{}{CH}}$$
$$H-C-OH$$
$$HO-C-H$$
$$H-C-OH$$
$$H-C-OH$$
$$H_2C-OH$$

(haworth ring structure with H_2C-OH, O, H, OH, HO, OH, H, OH)

45.

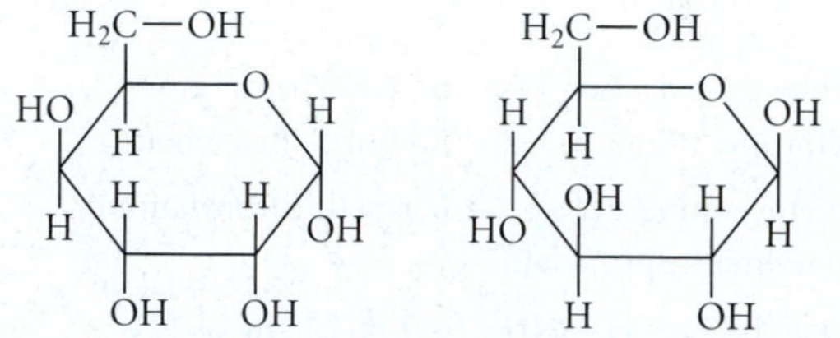

(two ring structures with H_2C-OH, HO, H, OH, OH labels)

47.

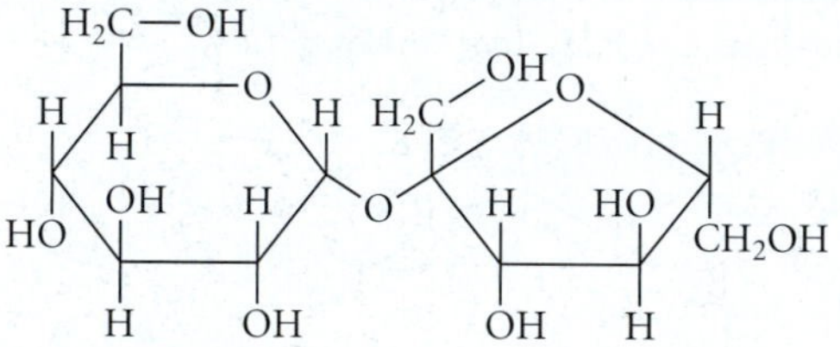

(disaccharide structure)

Glucose   Fructose

49. a.
$$H_3N^+-CH-\overset{\overset{\displaystyle O}{\|}}{C}-O^-$$
$$HO-CH$$
$$CH_3$$

b.
$$H_3N^+-CH-\overset{\overset{\displaystyle O}{\|}}{C}-O^-$$
$$CH_3$$

c.
$$H_3N^+-CH-\overset{\overset{\displaystyle O}{\|}}{C}-O^-$$
$$CH_2$$
$$H_3C-CH$$
$$CH_3$$

d.
$$H_3N^+-CH-\overset{\overset{\displaystyle O}{\|}}{C}-O^-$$
$$CH_2$$
$$CH_2$$
$$S$$
$$CH_3$$

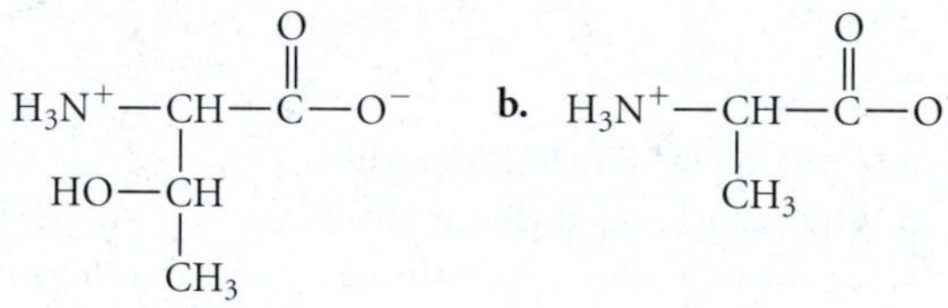

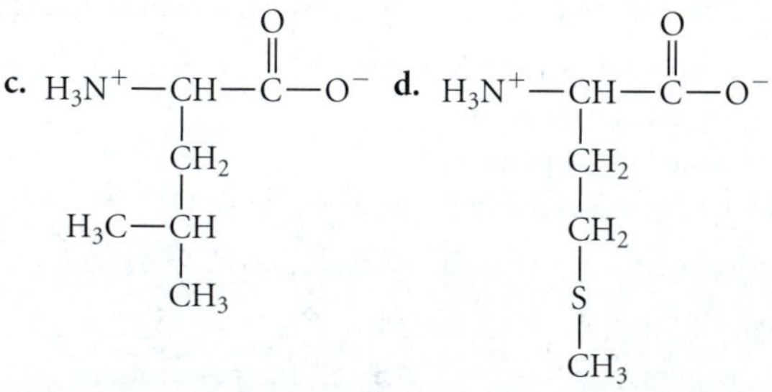

51.
$$H_2N\cdots\overset{\overset{\displaystyle H\ \ O}{|\ \ \|}}{\underset{\underset{\displaystyle H_3C}{}}{C}}-C-OH$$
     $$H_3C\cdots\overset{\overset{\displaystyle H\ \ O}{|\ \ \|}}{\underset{\underset{\displaystyle H_2N}{}}{C}}-C-OH$$

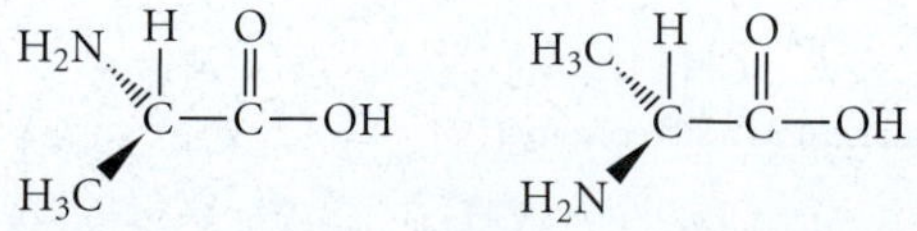

53. 6, SerGlyCys, SerCysGly, GlySerCys, GlyCysSer, CysSerGly, CysGlySer

55.

$$H_2N-\overset{\overset{\displaystyle H}{|}}{\underset{\underset{\displaystyle CH_2}{|}}{C}}-\overset{\overset{\displaystyle O}{\|}}{C}-OH \;+\; H_2N-\overset{\overset{\displaystyle H}{|}}{\underset{\underset{\displaystyle CH_2}{|}}{C}}-\overset{\overset{\displaystyle O}{\|}}{C}-OH \longrightarrow$$

with $-OH$ on the first CH_2 and a benzene ring with OH on the second.

$$H_2N-\overset{H}{\underset{CH_2}{C}}-\overset{O}{C}-NH-\overset{H}{\underset{CH_2}{C}}-\overset{O}{C}-OH$$

(with OH on first CH_2, benzene ring with OH on second)

$+ H_2O$

57. a.

$$H_2N-\overset{H}{\underset{\substack{CH_2 \\ CH_2 \\ C=O \\ NH_2}}{C}}-\overset{O}{C}-NH-\overset{H}{\underset{\substack{CH_2 \\ CH_2 \\ S \\ CH_3}}{C}}-\overset{O}{C}-NH-\overset{H}{\underset{SH}{C}}-\overset{O}{C}-OH$$

b.

$$H_2N-\overset{H}{\underset{\substack{CH_2 \\ OH}}{C}}-\overset{O}{C}-NH-\overset{H}{\underset{\substack{CH_2 \\ H_3C-CH \\ CH_3}}{C}}-\overset{O}{C}-NH-\overset{H}{\underset{SH}{C}}-\overset{O}{C}-OH$$

c.

$$H_2N-\overset{H}{\underset{\substack{CH_2 \\ SH}}{C}}-\overset{O}{C}-NH-\overset{H}{\underset{\substack{CH_2 \\ H_3C-CH \\ CH_3}}{C}}-\overset{O}{C}-NH-\overset{H}{\underset{\substack{CH_2 \\ OH}}{C}}-\overset{O}{C}-OH$$

59. tertiary

61. primary

63. a. A; **c.** T

65.

67. A C A T G C G

69. 154 codons, 462 nucleotides

71. a. protein   **b.** carbohydrate   **c.** lipid

73. A codon is composed of three nucleotides. A codon codes for a specific amino acid while a gene codes for an entire protein.

75.

$$H-\overset{H}{\underset{}{N}}-\overset{H}{\underset{CH_3}{C}}-\overset{O}{C}-OH$$

Trigonal planar (arrow to O); Trigonal pyramidal (arrow to N); Bent (arrow to OH); Tetrahedral (arrow to C)

77. valine, leucine, isoleucine, phenylalanine

79. Gly-Arg-Ala-Leu-Phe-Gly-Asn-Lys-Trp-Glu-Cys

81. When the fake thymine nucleotide is added to the replicating DNA, the chain cannot continue to form because the $-N\!=\!N^+\!=\!NH$ group on the sugar prevents future phosphate linkages.

83. $V_{max} = 47.6$, $K_t = 1.68$

85. 24

87. If the genetic code was the same, it would indicate that life on Earth and life on the other planet had shared origins. If the code were different, it would indicate that life on the other planet originated independently of life on Earth.

Chapter 22

17. a. +4   **b.** +4   **c.** +4

19. $Ca_3Al_2(SiO_4)_3$

21. 4

23. tetrahedrons stand alone, orthosilicates

25. amphibole or double-chain structure; $Ca^{2+}, Mg^{2+}, Fe^{2+}, Al^{3+}$

27. 950 g

29. NCl_3 has a lone pair that BCl_3 lacks, giving it a trigonal pyramidal shape, as opposed to BCl_3's trigonal planar shape.

31. a. 6 vertices, 8 faces   **b.** 12 vertices, 20 faces

33. *closo*-Boranes have the formula $B_nH_n^{2-}$ and form fully closed polyhedra, *nido*-boranes have the formula B_nH_{n+4} and consist of a cage missing a corner, and *arachno*-boranes have the formula B_nH_{n+6} and consist of a cage missing two or three corners.

35. Graphite consists of covalently bonded sheets that are held to each other by weak interactions, allowing them to slip past each other. Diamond is not a good lubricant because it is an extremely strong network covalent solid, where all of the carbon atoms are covalently bonded.

37. Activated charcoal consists of fine particles, rather than a lump of charcoal, and subsequently has a much higher surface area.

39. Ionic carbides are composed of carbon, generally in the form of the carbide ion, C_2^{2-}, and low-electronegativity metals, such as the alkali and alkaline earth metals. Covalent carbides are composed of carbon and low-electronegativity nonmetals or metalloids, such as silicon.

41. a. solid $\longrightarrow$ gas   **b.** gas $\longrightarrow$ liquid $\longrightarrow$ solid

    c. solid $\longrightarrow$ gas

43. a. $CO(g) + CuO(s) \longrightarrow CO_2(g) + Cu(s)$

    b. $SiO_2(s) + 3\,C(s) \longrightarrow SiC(s) + 2\,CO(g)$

    c. $S(s) + CO(g) \longrightarrow COS(g)$

45. a. $+2$ **b.** $+4$ **c.** $+4/3$

47. Fixing nitrogen refers to converting N_2 to a nitrogen-containing compound.

49. White phosphorus consists of P_4 molecules in a tetrahedral shape with the atoms at the corners of the tetrahedron. This allotrope is unstable because of the strain from the bond angles. Red phosphorus is much more stable because one bond of the tetrahedron is broken, allowing the phosphorus atoms to make chains with bond angles that are less strained.

51. saltpeter: 13.86% N by mass

    Chile saltpeter: 16.48% N by mass

53. HN_3 has a positive ΔG_f° meaning that it spontaneously decomposes into H_2 and N_2 at room temperature. There are no temperatures at which HN_3 will be stable. ΔH_f is positive and ΔS_f is negative, so ΔG_f will always be negative.

55. a. $NH_4NO_3(aq) + \text{heat} \longrightarrow N_2O(g) + 2\,H_2O(l)$

    b. $3\,NO_2(g) + H_2O(l) \longrightarrow 2\,HNO_3(l) + NO(g)$

    c. $2\,PCl_3(l) + O_2(g) \longrightarrow 2\,POCl_3(l)$

57. NO_3^-, NO_2^-, N_3^-, $N_2H_5^+$, NH_4^+

59.

Trigonal pyramidal   Trigonal bipyramidal

61. $CO(NH_2)_2 + 2\,H_2O \longrightarrow (NH_4)_2CO_3$

    14 g

63. P_4O_6 forms if there is only a limited amount of oxygen available, while P_4O_{10} will form with greater amounts of oxygen.

65. The major source of oxygen is the fractionation of air by which air is cooled and liquefied and oxygen is separated from the other components.

67. a. superoxide   **b.** oxide   **c.** peroxide

69. Initially, liquid sulfur becomes less viscous when heated because the S_8 rings have greater thermal energy which overcomes intermolecular forces. Above 150 °C the rings break and the broken rings entangle one another, causing greater viscosity.

71. a. 4.3×10^{-22} g   **b.** 4.0×10^{-19} g

73. $2\,FeS_2(s) \xrightarrow{\text{heat}} 2\,FeS(s) + S_2(g)$

    510 L

75. a. +2, linear   **b.** +6, octahedral   **c.** +6, square pyramidal

77. $Cl_2(g) + 2\,Br^-(aq) \longrightarrow 2\,Cl^-(aq) + Br_2(l)$

    Oxidizing agent: Cl_2

    Reducing agent: Br^-

79. No, there is not enough HF to dissolve all of the SiO_2.

    HF is the limiting reagent. 1.6 g SiO_2.

81. 8 kg from lignite, 40 kg from bituminous

83. Chlorine is much more electronegative than iodine, allowing it to withdraw an electron and ionize in solution much more easily.

85. a. $\text{rate}_{HCl}/\text{rate}_{Cl_2} = 1.395$ **b.** $\text{rate}_{HCl}/\text{rate}_{HF} = 0.7407$

    c. $\text{rate}_{HCl}/\text{rate}_{HI} = 0.5339$

87. $4\,Na_2O_2 + Fe + 2\,e^- \longrightarrow 4\,Na_2O + FeO_4^{2-}$

89. The bond length of the O_2 species increases as electrons are added because they are added to the p^* antibonding orbital. O_2^{2-} is diamagnetic.

91. a. -13.6 kJ/mol   **b.** -11.0 kJ/mol   **c.** -24.8 kJ/mol

    Fe_2O_3 is the most exothermic because it has the highest oxidation state and is therefore able to oxidize the most CO per mol Fe.

93. a. $\ddot{O}{=}C{=}C{=}C{=}\ddot{O}$ **b.** sp **c.** -92 kJ/mol

95. a. 7.6×10^{-22}

    b. 1.2×10^{-8}

    c. $[N_2H_4] = 0.009$ M, $[N_2H_5^+] = 0.0025$ M, $[N_2H_6^{2+}] = 7.0 \times 10^{-13}$ M

97. Fine particles of activated charcoal have a much greater total surface area than does a charcoal briquette, allowing it to interact with more impurities.

99. Nitrogen can either gain three electrons or lose five electrons to obtain an octet, giving it the possibility of an oxidation state from -3 to $+5$.

101. H_2S has a different bond angle than H_2O because it is larger and less polar, and because S is less electronegative than O. H_2S is more reactive because the O—H bond is much stronger than the S—H bond.

103. Fluorine has a greater electronegativity than oxygen and is more reactive.

105. Cl is small enough to fit around Br but Br is too large to fit around Cl and form stable bonds.

Chapter 23

15. Metals are typically opaque, are good conductors of heat and electricity, and are ductile and malleable, meaning they can be drawn into wires and flattened into sheets.

17. aluminum, iron, calcium, magnesium, sodium, potassium

19. Fe: hematite (Fe_2O_3), magnetite (Fe_3O_4)

    Hg: cinnabar (HgS)

    V: vanadite $[Pb_5(VO_4)Cl]$, carnotite $[K_2(UO_2)_2(VO_4)_2 \cdot 3\,H_2O]$

    Nb: columbite $[Fe(NbO_3)_2]$

21. $MgCO_3(s) + \text{heat} \longrightarrow MgO(s) + CO_2(g)$

    $Mg(OH)_2(s) + \text{heat} \longrightarrow MgO(s) + H_2O(g)$

23. The flux is a material that will react with the gangue to form a substance with a low melting point. MgO is the flux.

25. Hydrometallurgy is used to separate metals from ores by selectively dissolving the metal in a solution, filtering out impurities, and then reducing the metal to its elemental form.

27. The Bayer process is a hydrometallurgical process by which Al_2O_3 is selectively dissolved, leaving other oxides as solids. The soluble form of aluminum is $Al(OH)_4^-$.

29. Sponge powdered iron contains many small holes in the iron particles due to the escaping of the oxygen when the iron is reduced. Water atomized powdered iron has much more smooth and dense particles as the powder is formed from molten iron.

31. **a.** 50% Cr, 50% V by moles; 50.5% Cr, 49.5% V by mass
    b. 25% Fe, 75% V by moles; 26.8% Fe, 73.2% V by mass
    c. 25% Cr, 25% Fe, 50% V by moles; 24.8% Cr, 26.6% Fe, 48.6% V by mass

33. Cr and Fe are very close to each other in mass, so their respective atomic radii are probably close enough to form an alloy. Also, they both form body-centered cubic structures.

35. A: solid, 20% Cr, 80% Fe
    B: liquid, 50% Cr, 50% Fe

37. A: solid (20% Co and 80% Cu overall. Two phases; one is the Cu structure with 4% Co, and the other is the Co structure with 7% Cu. There will be more of the Cu structure).
    B: solid (Co structure), 90% Co, 10% Cu

39. C would fill interstitial holes; Mn and Si would substitute for Fe.

41. **a.** Mo_2N **b.** CrH_2

43. **a.** zinc   **b.** copper   **c.** manganese

45. -19.4 kJ/mol

47. When Cr is added to steel it reacts with oxygen in steel to prevent it from rusting. A Cr steel alloy would be used in any situation where the steel might be easily oxidized, such as when it comes in contact with water.

49. rutile: 33.3% Ti by moles, 59.9% Ti by mass
    ilmenite: 20.0% Ti by moles, 31.6% Ti by mass

51. Titanium must be arc-melted in an inert atmosphere because the high temperature and flow of electrons would cause the metal to oxidize in a normal atmosphere.

53. TiO_2 is the most important industrial product of titanium and it is often used as a pigment in white paint.

55. The Bayer process is a hydrometallurgical process used to separate Al_2O_3 from other oxides. The Al_2O_3 is selectively dissolved by hot, concentrated NaOH. The other oxides are removed as solids and the Al_2O_3 precipitates out of solution when the solution is neutralized.

57. cobalt and tungsten

59. 3.3 kg Fe, 2.0 kg Ti

61. Four atoms surround a tetrahedral hole and six atoms surround an octahedral hole. The octahedral hole is larger because it is surrounded by a greater number of atoms.

63. Mn has one more electron orbital available for bonding than does chromium.

65. Ferromagnetic atoms, like paramagnetic ones, have unpaired electrons. However, in ferromagnetic atoms, these electrons align with their spin oriented in the same direction, resulting in a permanent magnetic field.

67. **a.** 16.0 cm   **b.** 4.95 cm   **c.** 14%

69. 92%

71. Au and Ag are found in elemental form because of their low reactivity. Na and Ca are group 1 and group 2 metals, respectively, and are highly reactive as they readily lose their valence electrons to obtain octets.

Chapter 24

17. **a.** $[Ar]4s^23d^8$, $[Ar]3d^8$
    b. $[Ar]4s^23d^5$, $[Ar]3d^3$
    c. $[Kr]5s^24d^1$, $[Kr]5s^14d^1$
    d. $[Xe]6s^24f^{14}5d^3$, $[Xe]4f^{14}5d^3$

19. **a.** $+5$ **b.** $+7$ **c.** $+4$

21. **a.** $+3, 6$ **b.** $+2, 6$ **c.** $+2, 4$ **d.** $+1, 2$

23. **a.** hexaaquachromium(III)
    b. tetracyanocuprate(II)
    c. pentaaminebromoiron(III) sulfate
    d. aminetetraaquahydroxycobalt(III) chloride

25. **a.** $[Cr(NH_3)_6]^{3+}$ **b.** $K_3[Fe(CN)_6]$
    c. $[Cu(en)(SCN)_2]$ **d.** $[Pt(H_2O)_4][PtCl_6]$

27. **a.** $[Co(NH_3)_3(CN)_3]$, triaminetricyanocobalt(III)
    b. $[Cr(en)_3]^{3+}$, tris(ethylenediamine)chromium(III)

29.

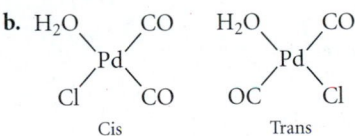

31. $[Fe(H_2O)_5Cl]Cl \cdot H_2O$, pentaaquachloroiron(II) chloride monohydrate
    $[Fe(H_2O)_4Cl_2] \cdot 2H_2O$, tetraaquadichloroiron(II) dihydrate

33. b, c, e

35. **a.** 3   **b.** 1

37. **a.**

Fac   Mer

b.

Cis   Trans

39. *cis* isomer is optically active

41. **a.**
    ⇅ ⇅
    ⇅ ⇅ ⇅

b.
    ↑ ↑   __ __
    ↑ ↑ ↑   ⇅ ⇅ ↑

c.
    __ __
    ↑ ↑

d.
    ↑ ↑
    ⇅ ⇅ ↑

43. 163 kJ/mol

45. $[Co(CN)_6]^{3-} \longrightarrow$ 290 nm, colorless

$[Co(NH_3)_6]^{3+} \longrightarrow$ 440 nm, yellow

$[CoF_6]^{3-} \longrightarrow$ 770 nm, green

47. weak

49. a. 4   **b.** 3   **c.** 5

51. 3

53. porphyrin

55. Water is a weak field ligand that forms a high-spin complex with hemoglobin. Because deoxyhemoglobin is weak field it absorbs large wavelength light and appears blue. Oxyhemoglobin is a low-spin complex and absorbs small wavelength light, so O_2 must be a strong field ligand.

57. a. $[Ar]4s^1 3d^5$, $[Ar]3d^5$, $[Ar]3d^4$, $[Ar]3d^3$

b. $[Ar]4s^1 3d^{10}$, $[Ar]3d^{10}$, $[Ar]3d^9$

59. a. 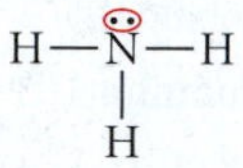 H—N̈—H
               |
               H

b. $\left[\ddot{\text{S}}=\text{C}=\ddot{\text{N}}\right]$ **c.** Ö with H   H

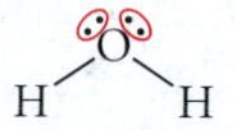

61. $[MA_2B_2C_2]$ all cis; A trans and B and C cis; B trans and A and C cis; C trans and A and B cis; all trans.

$[MA_2B_3C]$ will have fac–mer isomers.

$[MAB_2C_3]$ will have fac–mer isomers.

$[MAB_3C_2]$ will have fac–mer isomers.

$[MA_3B_2C]$ will have fac–mer isomers.

$[MA_2BC_3]$ will have fac–mer isomers.

$[MA_3BC_2]$ will have fac–mer isomers.

$[MABC_2]$ will have AB cis–trans isomers.

$[MAB_4C]$ will have AC cis–trans isomers.

$[MA_4BC]$ will have BC cis–trans isomers.

$[MABC_4]$ will have AB cis–trans isomers.

63. 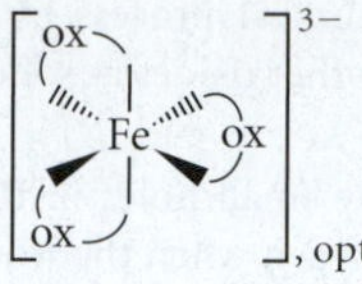 $\left[\begin{array}{c} \text{ox} \\ \text{Fe} \text{ ox} \\ \text{ox} \end{array}\right]^{3-}$, optical isomers

65. 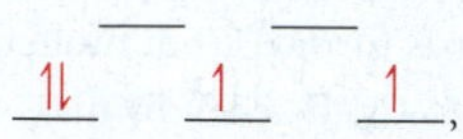 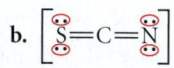, paramagnetic

67.

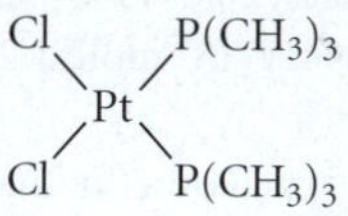

Cl   P(CH₃)₃
    \ /
    Pt
    / \
Cl   P(CH₃)₃

cis-dichlorobis (trimethyl phospine) platinum(II)

Cl   P(CH₃)₃
    \ /
    Pt
    / \
(CH₃)₃P   Cl

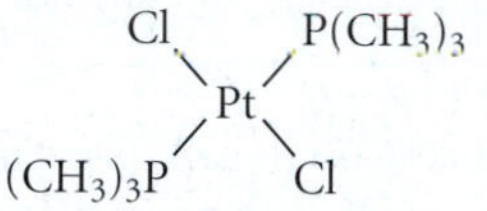

trans-dichlorobis (trimethyl phospine) platinum(II)

69. ____ d_{z^2}

____ ____ $d_{x^2-y^2}$ and d_{xy}

____ ____ d_{xz} and d_{yz}

71. a. 2×10^{-8} M

b. 6.6×10^{-3} M

c. NiS will dissolve more easily in the ammonia solution because the formation of the complex ion is favorable, removing Ni^{2+} ions from the solution allowing more NiS to dissolve.

73. Au

Appendix IV: Answers to In-Chapter Practice Problems

Chapter 1

1.1. **a.** The composition of the copper is not changing, thus, being hammered flat is a physical change that signifies a physical property.
b. The dissolution and color change of the nickel indicate that it is undergoing a chemical change and exhibiting a chemical property.
c. Vaporization is a physical change indicative of a physical property.
d. When a match ignites, a chemical change begins as the match reacts with oxygen to form carbon dioxide and water. Flammability is a chemical property.

1.2. **a.** 29.8 °C
b. 303.0 K

1.3. 21.4 g/cm^3 This matches the density of platinum.

1.3. For More Practice

4.50 g/cm^3 The metal is titanium.

1.4. The thermometer shown has markings every 1 °F; thus, the first digit of uncertainty is 0.1. The answer is 103.4 °F.

1.5. **a.** Each figure in this number is significant by rule 1: three significant figures.
b. This is a defined quantity that has an unlimited number of significant figures.
c. Both 1's are significant (rule 1) and the interior zero is significant as well (rule 2): three significant figures.
d. Only the two 9's are significant, the leading zeroes are not (rule 3): two significant figures.
e. There are five significant figures because the 1, 4, and 5 are nonzero (rule 1) and the trailing zeroes are after a decimal point so they are significant as well (rule 4).
f. The number of significant figures is ambiguous because the trailing zeroes occur before an implied decimal point (rule 4). Assume two significant figures.

1.6. **a.** 0.381
b. 121.0
c. 1.174
d. 8

1.7. 3.15 yd

1.8. 2.446 gal

1.9. 1.61 × 10^6 cm^3

1.9. For More Practice

3.23 × 10^3 kg

1.10. 1.03 kg

1.10. For More Practice

2.9 × 10^{-2} cm^3

1.11. 0.855 cm

1.12. 2.70 g/cm^3

Chapter 2

2.1. For the first sample:

$$\frac{\text{mass of oxygen}}{\text{mass of carbon}} = \frac{17.2 \text{ g O}}{12.9 \text{ g C}} = 1.33 \text{ or } 1.33 : 1$$

For the second sample:

$$\frac{\text{mass of oxygen}}{\text{mass of carbon}} = \frac{10.5 \text{ g O}}{7.88 \text{ g C}} = 1.33 \text{ or } 1.33 : 1$$

The ratios of oxygen to carbon are the same in the two samples of carbon monoxide, so these results are consistent with the law of definite proportions.

2.2. $\dfrac{\text{mass of hydrogen to 1 g of oxygen in hydrogen peroxide}}{\text{mass of hydrogen to 1 g of oxygen in water}} = \dfrac{0.250}{0.125} = 2.00$

The ratio of the mass of hydrogen from one compound to the mass of hydrogen in the other is equal to 2. This is a simple whole number and therefore consistent with the law of multiple proportions.

2.3. **a.** $Z = 6$, $A = 13$, $^{13}_{6}\text{C}$

b. 19 protons, 20 neutrons

2.4. **a.** N^{3-}

b. Rb^{+}

2.5. 24.31 amu

2.5. For More Practice

70.92 amu

2.6. 4.65×10^{-2} mol Ag

2.7. 0.563 mol Cu

2.7. For More Practice

22.6 g Ti

2.8. 1.3×10^{22} C atoms

2.8. For More Practice

6.87 g W

2.9. $l = 1.72$ cm

2.9. For More Practice

2.90×10^{24} Cu atoms

Chapter 3

3.1. **a.** C_5H_{12}

b. HgCI

c. CH_2O

3.2. **a.** molecular element

b. molecular compound

c. atomic element

d. ionic compound

e. ionic compound

3.3. K_2S

3.4. AlN

3.5. silver nitride

3.5. For More Practice
Rb_2S

3.6. iron(II) sulfide

3.6. For More Practice
RuO_2

3.7. tin(II) chlorate

3.7. For More Practice
$Co_3(PO_4)_2$

3.8. dinitrogen pentoxide

3.8. For More Practice
PBr_3

3.9. hydrofluoric acid

3.10. nitrous acid

3.10. For More Practice
$HClO_4$

3.11. 164.10 amu

3.12. 5.839×10^{20} $C_{13}H_{18}O_2$ molecules

3.12. For More Practice
1.06 g H_2O

3.13. 53.29%

3.13. For More Practice
74.19% Na

3.14. 83.9 g Fe_2O_3

3.14. For More Practice
8.6 g Na

3.15. 4.0 g O

3.15. For More Practice
3.60 g C

3.16. CH_2O

3.17. $C_{13}H_{18}O_2$

3.18. C_6H_6

3.18. For More Practice
$C_2H_8N_2$

3.19. C_2H_5

3.20. C_2H_4O

3.21. $SiO_2(s) + 3\,C(s) \longrightarrow SiC(s) + 2\,CO(g)$

3.22. $2\,C_2H_6(g) + 7\,O_2(g) \longrightarrow 4\,CO_2(g) + 6\,H_2O(g)$

Chapter 4

4.1. 4.08 g HCl

4.2. 22 kg HNO_3

4.3. H_2 is the limiting reagent, since it produces the least amount of NH_3. Therefore, 29.4 kg NH_3 is the theoretical yield.

4.4. CO is the limiting reagent, since it only produces 114 g Fe. Therefore, 114 g Fe is the theoretical yield: percentage yield = 63.4% yield

4.5. 0.214 M $NaNO_3$

4.5. For More Practice

44.6 g KBr

4.6. 402 g $C_{12}H_{22}O_{11}$

4.6. For More Practice

221 mL of KCl solution

4.7. 667 mL

4.7. For More Practice

0.105 L

4.8. 51.4 mL HNO_3 solution

4.8. For More Practice

0.170 g CO_2

4.9. **a.** Insoluble.
  b. Insoluble.
  c. Soluble.
  d. Soluble.

4.10. $NH_4Cl(aq) + Fe(NO_3)_3(aq) \longrightarrow$ NO REACTION

4.11. $2\,NaOH(aq) + CuBr_2(aq) \longrightarrow Cu(OH)_2(s) + 2\,NaBr(aq)$

4.12. $2\,H^+(aq) + 2\,I^-(aq) + Ba^{2+}(aq) + 2\,OH^-(aq) \longrightarrow 2\,H_2O(\ell) + Ba^{2+}(aq) + 2\,I^-(aq)$

  $H^+(aq) + OH^-(aq) \longrightarrow H_2O(\ell)$

4.12. For More Practice

$2\,Ag^+(aq) + 2\,NO_3^-(aq) + Mg^{2+}(aq) + 2\,Cl^-(aq) \longrightarrow$

$2\,AgCl(s) + Mg^{2+}(aq) + 2\,NO_3^-(aq)$

$Ag^+(aq) + Cl^-(aq) \longrightarrow AgCl(s)$

4.13. $H_2SO_4(aq) + 2\,LiOH(aq) \longrightarrow 2\,H_2O(l) + Li_2SO_4(aq)$

$H^+(aq) + OH^-(aq) \longrightarrow H_2O(aq)$

4.14. 9.03×10^{-2} M H_2SO_4

4.14. For More Practice

24.5 mL NaOH solution

4.15. $2\,HBr(aq) + K_2SO_3(aq) \longrightarrow H_2O(l) + SO_2(g) + 2\,KBr\,(aq)$

4.15. For More Practice

$2\,H^+(aq) + S^{2-}(aq) \longrightarrow H_2S(g)$

4.16. **a.** Cr = 0.
  b. Cr^{3+} = +3.

    c. $Cl^- = -1, C = +4$.

    d. $Br = -1, Sr = +2$.

    e. $O = -2, S = +6$.

    f. $O = -2, N = +5$.

4.17. Sn is oxidized and N is reduced.

4.17. For More Practice

    b. Reaction b is the only redeox reaction. Al is oxidized and O is reduced.

4.18. **a.** This is a redox reaction in which Li is the reducing agent (it is oxidized) and Cl_2 is the oxidizing reagent (it is reduced).

    b. This is a redox reaction in which Al is the reducing agent and Sn^{2+} is the oxidizing agent.

    c. This is not a redox reaction because no oxidation states change.

    d. This is a redox reaction in which C is the reducing agent and O_2 is the oxidizing agent.

4.19. $2 C_2H_5SH(l) + 9 O_2(g) \longrightarrow 4 CO_2(g) + 2 SO_2(g) + 6 H_2O(g)$

Chapter 5

5.1. 15.0 psi

5.1. For More Practice

    80.6 kPa

5.2. 2.1 atm at a depth of approximately 11 m.

5.3. 123 mL

5.4. 11.3 L

5.5. 1.63 atm, 23.9 psi

5.6. 16.1 L

5.6. For More Practice

    976 mmHg

5.7. $d = 4.91$ g/L

5.7. For More Practice

    44.0 g/mol

5.8. 70.7 g/mol

5.9. 0.0610 mol H_2

5.10. 4.2 atm

5.11. 12.0 mg H_2

5.12. 82.3 g Ag_2O

5.12. For More Practice

    7.10 g Ag_2O

5.13. 6.53 L O_2

5.14. $u_{rms} = 238$ m/s

5.15. $\dfrac{rate_{H_2}}{rate_{Kr}} = 6.44$

Chapter 6

6.1. $\Delta E = 71$ J

6.2. $C_s = 0.38 \dfrac{J}{g \cdot ^\circ C}$

The specific heat capacity of gold is 0.128 J/g·°C; therefore the rock cannot be pure gold.

6.2. For More Practice

$T_f = 42.1 \,^\circ C$

6.3. $-122 \, J$

6.3. For More Practice

$\Delta E = -998 \, J$

6.4. $\Delta E_{rxn} = -3.91 \times 10^3 \dfrac{kJ}{mol \, C_6H_{14}}$

6.4. For More Practice

$C_{cal} = 4.55 \dfrac{kJ}{^\circ C}$

6.5. **a.** endothermic, positive ΔH.
b. endothermic, positive ΔH.
c. exothermic, positive ΔH.

6.6. $-2.06 \times 10^3 \, kJ$

6.6. For More Practice

33 g C_4H_{10}
99 g CO_2

6.7. $\Delta H_{rxn} = -68 \, kJ$

6.8. $N_2O(g) + NO_2(g) \longrightarrow 3 \, NO(g)$, $\Delta H_{rxn} = +157.6 \, kJ$

6.8. For More Practice

$3 \, H_2(g) + O_3(g) \longrightarrow 3 \, H_2O(g)$, $\Delta H = -868.1 \, kJ$

6.9. **a.** $Na(s) + \dfrac{1}{2} Cl_2(g) \longrightarrow NaCl(s)$, $\Delta H_f^\circ = -411.2 \, kJ/mol$

b. $Pb(s) + N_2(g) + 3 \, O_2(g) \longrightarrow Pb(NO_3)_2(s)$, $\Delta H_f^\circ = -451.9 \, kJ/mol$

6.10. $\Delta H_{rxn}^\circ = -851.5 \, kJ$

6.11. $\Delta H_{rxn}^\circ = -1648.4 \, kJ$

111 kJ emitted (-111 kJ)

6.12. $1.2 \times 10^2 \, kg \, CO_2$

Chapter 7

7.1. $5.83 \times 10^{14} \, s^{-1}$

7.2. 2.64×10^{20} photons

7.2. For More Practice

435 nm

7.3. **a.** blue<green<red.
b. red<green<blue.
c. red<green<blue.

7.4. $6.1 \times 10^6 \, m/s$

7.5. For the 5d orbitals:
$n = 5$
$l = 2$

$m_l = -2, -1, 0, 1, 2$
The 5 integer values for m_l signifies that there are five 5*d* orbitals.

7.6. a. *l* cannot equal 3 if $n = 3$. $l = 2$
    b. m_l cannot equal -2 if $l = -1$. Possible values for $m_l = -1, 0,$ or 1
    c. *l* cannot be 1 if $n = 1$. $l = 0$.

7.7. 397 nm

7.7. For More Practice
    $n = 1$

Chapter 8

8.1. a. Cl $1s^2 2s^2 2p^6 3s^2 3p^5$ or [Ne] $3s^2 3p^5$
    b. Si $1s^2 2s^2 2p^6 3s^2 3p^2$ or [Ne] $3s^2 3p^2$
    c. Sr $1s^2 2s^2 2p^6 3s^2 3p^6 4s^2 3d^{10} 4p^6 5s^2$ or [Kr] $5s^2$
    d. O $1s^2 2s^2 2p^4$ or [He] $2s^2 p^4$

8.2. There are no unpaired electrons.

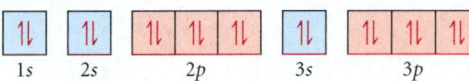

8.3. $1s^2 2s^2 2p^6 3s^2 3p^3$ or [Ne] $3s^2 3p^3$. The five electrons in the $3s^2 3p^3$ orbitals are the valence electrons, while the 10 electrons in the $1s^2 2s^2 2p^6$ orbitals belong to the core.

8.4. Bi   [Xe] $6s^2 4f^{14} 5d^{10} 6p^3$

8.4. For More Practice
    I   [Kr] $5s^2 4d^{10} 5p^5$

8.5. a. Sn
    b. cannot predict
    c. W
    d. Se

8.5. For More Practice
    Rb > Ca > Si > S > F.

8.6. a. [Ar] $4s^0 3d^7$. Co^{2+} is paramagnetic.

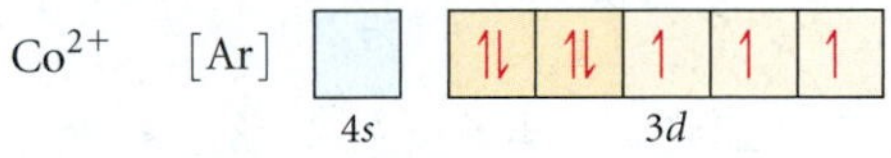

    Co^{2+} [Ar]

    b. [He] $2s^2 2p^6$. N^{3-} is diamagnetic.

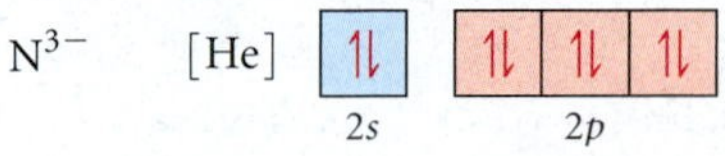

    N^{3-} [He]

    c. [Ne] $3s^2 3p^6$. Ca^{2+} is diamagnetic.

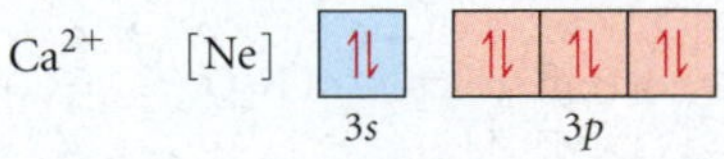

    Ca^{2+} [Ne]

8.7. a. K

  b. F^-

  c. Cl^-

8.7. For More Practice

  $Cl^- > Ar > Ca^{2+}$.

8.8. a. I

  b. Ca

  c. cannot predict

  d. F

8.8. For More Practice

  $F > S > Si > Ca > Rb$.

8.9. a. Sn

  b. cannot predict based on simple trends (Po is larger)

  c. Bi

  d. B

8.9. For More Practice

  $Cl < Si < Na < Rb$.

8.10. a. $2\,Al(s) + 3\,Cl_2(g) \longrightarrow 2\,AlCl_3(s)$

  b. $2\,Li(s) + 2\,H_2O(l) \longrightarrow 2\,Li^+(aq) + 2\,OH^-(aq) + H_2(g)$

  c. $H_2(g) + Br_2(l) \longrightarrow 2\,HBr(g)$

Chapter 9

9.1. Mg_3N_2

9.2. $KI < LiBr < CaO$.

9.2. For More Practice

  $MgCl_2$

9.3. a. pure covalent

  b. ionic

  c. polar covalent

9.4. $:C\!\equiv\!O:$

9.5.
$$
\begin{array}{c}
:\!O\!: \\
\|\\
H\!-\!C\!-\!H
\end{array}
$$

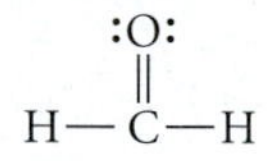

9.6. $\left[:\!\ddot{C}l\!:\!\ddot{O}\!:\right]^-$

9.7. $\left[:\!\ddot{O}\!=\!\ddot{N}\!-\!\ddot{O}\!:\right]^- \longleftrightarrow \left[:\!\ddot{O}\!-\!\ddot{N}\!=\!\ddot{O}\!:\right]^-$

9.8.

Structure	A			B			C		
	$:\!\ddot{N}\!=\!N\!=\!\ddot{O}\!:$			$:N\!\equiv\!N\!-\!\ddot{O}\!:$			$:\!\ddot{N}\!-\!N\!\equiv\!O:$		
number of valence e^-	5	5	6	5	5	6	5	5	6
number of lone pair e^-	-4	-0	-4	-2	-0	-6	-6	-0	-2
$\frac{1}{2}$ (number of bonding e^2)	-2	-4	-2	-3	-4	-1	-1	-4	-3
Formal charge	-1	$+1$	0	0	$+1$	-1	-2	$+1$	$+1$

Structure B contributes the most to the correct overall structure of N_2O.

9.8. For More Practice

The nitrogen is $+1$, the singly bonded oxygen atoms are -1, and the double-bonded oxygen atom has no formal charge.

9.9.

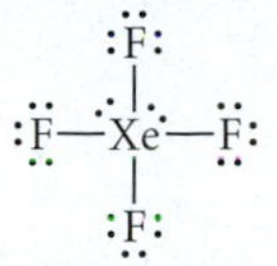

$$:\ddot{F}:$$
$$|$$
$$:\ddot{F}-\ddot{Xe}-\ddot{F}:$$
$$|$$
$$:\ddot{F}:$$

9.9. For More Practice

$$H$$
$$|$$
$$:\ddot{O}:$$
$$|$$
$$H-\ddot{O}-P^0-\ddot{O}-H$$
$$\|$$
$$:\ddot{O}:^0$$

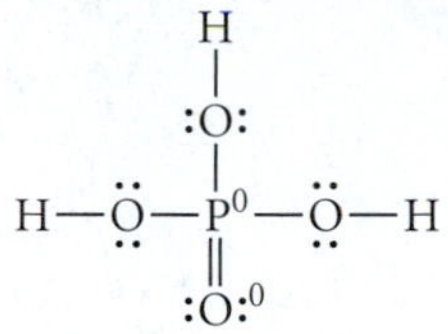

9.10. $CH_3OH(g) + \dfrac{3}{2} O_2(g) \longrightarrow CO_2(g) + 2\,H_2O(g).$

$\Delta H_{rxn} = -641 \text{ kJ}$

9.10. For More Practice

$\Delta H_{rxn} = -8.0 \times 10^1 \text{ kJ}$

Chapter 10

10.1. tetrahedral

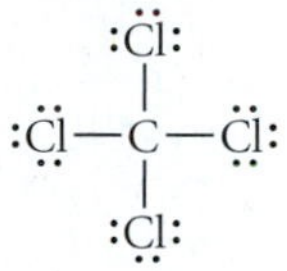

$$:\ddot{Cl}:$$
$$|$$
$$:\ddot{Cl}-C-\ddot{Cl}:$$
$$|$$
$$:\ddot{Cl}:$$

10.2. bent

10.3. linear

10.4.

Atom	Number of Electron Groups	Number of Lone Pairs	Molecular Geometry
Carbon (left)	4	0	Tetrahedral
Carbon (right)	3	0	Trigonal planar
Oxygen	4	2	Bent

10.5. The molecule is nonpolar.

10.6. The xenon atom has six electron groups and therefore has an octahedral electron geometry. An octahedral electron geometry corresponds to sp^3d^2 hybridization (refer to Table 10.3).

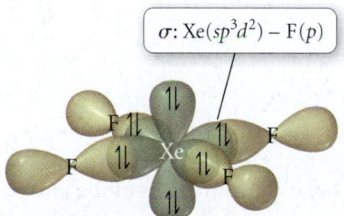

10.7. Since there are only two electron groups around the central atom (C), the electron geometry is linear. According to Table 10.3, the corresponding hybridization on the carbon atom is sp.

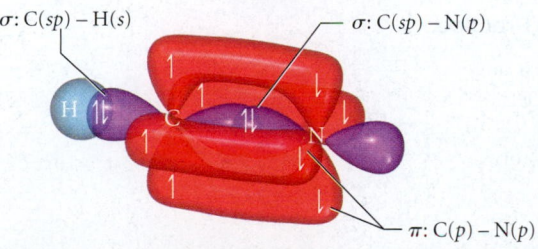

10.8. Since there are only two electron groups about the central atom (C) the electron geometry is linear. The hybridization on C is *sp* (refer to Table 10.3).

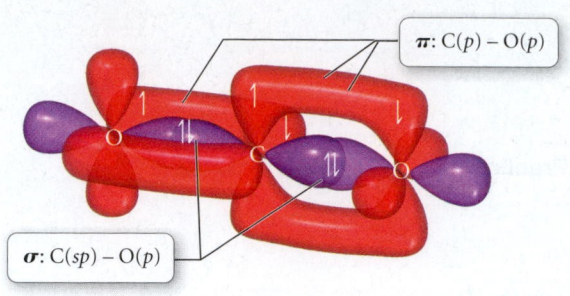

π: C(*p*) – O(*p*)

σ: C(*sp*) – O(*p*)

10.8. For More Practice

There are five electron groups about the central atom (I); therefore the electron geometry is trigonal bipyramidal and the corresponding hybridization of I is sp^3d (refer to Table 10.3).

10.9. H_2^+ bond order $= +\dfrac{1}{2}$

Since the bond order is positive, the H_2^+ ion should be stable; however, the bond order of H_2^+ is lower than the bond order of H_2 (bond order = 1). Therefore, the bond in H_2^+ is weaker than in H_2.

10.10. The bond order of N_2^+ is 2.5, which is lower than that of the N_2 molecule (bond order = 3), therefore the bond is weaker. The MO diagram shows that the N_2^+ ion has one unpaired electron and is therefore paramagnetic.

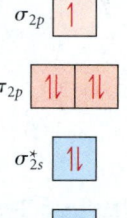

σ_{2p} $\boxed{1}$

π_{2p} $\boxed{1\!\downarrow}$ $\boxed{1\!\downarrow}$

σ_{2s}^* $\boxed{1\!\downarrow}$

σ_{2s} $\boxed{1\!\downarrow}$

10.10. For More Practice

The bond order of Ne_2 is 0, which indicates that dineon does not exist.

10.11. The bond order of NO is +2.5. The MO diagram shows that the NO ion has one unpaired electron and is therefore paramagnetic.

Chapter 11

11.1. b, c

11.2. HF has a higher boiling point than HCl because, unlike HCl, HF is able to form hydrogen bonds. The hydrogen bond is the strongest of the intermolecular forces and requires more energy to break.

11.3. 5.83×10^3 kJ

11.3. For More Practice

49 °C

11.4. 33.8 kJ/mol

11.5. 7.04×10^3 torr

11.6. 29.4°

11.7. $7.18 \dfrac{g}{cm^3}$

Chapter 12

12.1. a. not soluble
    b. soluble
    c. not soluble
    d. not soluble

12.2. $2.7 \times 10^{-4}\,M$

12.3. $42.5\,g\ C_{12}H_{22}O_{11}$

12.3. For More Practice

    $3.3 \times 10^4\,L$

12.4. a. $M = 0.415\,M$
    b. $m = 0.443\,m$
    c. % by mass $= 13.2\%$
    d. $x_{C_{12}H_{22}O_{11}} = 0.00793$
    e. mole percent $= 0.793\%$

12.5. 22.5 torr

12.5. For More Practice
    0.144

12.6. 0.014 mol NaCl

12.7. a. $P_{benzene} = 26.6\,torr$
        $P_{toluene} = 20.4\,torr$
    b. 47.0 torr
    c. 52.5% benzene; 47.5% toluene
    The vapor will be richer in the more volatile component, which in this case is benzene.

12.8. $T_f = -4.8\,°C$

12.9. 101.84 °C

12.10. 11.8 atm

12.11. $-0.60\,°C$

Chapter 13

13.1. $\dfrac{\Delta[H_2O_2]}{\Delta t} = -4.40 \times 10^{-3}\,M/s$

    $\dfrac{\Delta[I_3^-]}{\Delta t} = 4.40 \times 10^{-3}\,M/s$

13.2. a. Rate $= k[CHCl_3][Cl_2]^{1/2}$. (Fractional-order reactions are not common but are occasionally observed.)
    b. $3.5\,M^{-1/2} \cdot s^{-1}$

13.3. $5.78 \times 10^{-2}\,M$

13.4. 0.0277 M

13.5. $1.64 \times 10^{-3}\,M$

13.6. 79.2 s

13.7. $2.07 \times 10^{-5}\dfrac{L}{mol \cdot s}$

13.8. $6.13 \times 10^{-4} \dfrac{\text{L}}{\text{mol} \cdot \text{s}}$

13.9. $2 \text{ A} + \text{B} \longrightarrow \text{A}_2\text{B}$

Rate $= k[\text{A}]^2$

Chapter 14

14.1. $K = \dfrac{[\text{CO}_2]^3[\text{H}_2\text{O}]^4}{[\text{C}_3\text{H}_8][\text{O}_2]^5}$

14.2. 2.1×10^{-13}

14.2. **For More Practice**

1.4×10^2

14.3. 6.2×10^2

14.4. $K_c = \dfrac{[\text{Cl}_2]^2}{[\text{HCl}]^4[\text{O}_2]}$

14.5. 9.4

14.6. 1.1×10^{-6}

14.7. $Q_c = 0.0196$
Reaction proceeds to the left.

14.8. 0.033 M

14.9. $[\text{N}_2] = 4.45 \times 10^{-3} \text{ M}$
$[\text{O}_2] = 4.45 \times 10^{-3} \text{ M}$
$[\text{NO}] = 1.1 \times 10^{-3} \text{ M}$

14.10. $[\text{N}_2\text{O}_4] = 0.005 \text{ M}$
$[\text{NO}_2] = 0.041 \text{ M}$

14.11. $P_{\text{I}_2} = 0.027 \text{ atm}$
$P_{\text{Cl}_2} = 0.027 \text{ atm}$
$P_{\text{ICl}} = 0.246 \text{ atm}$

14.12. $1.67 \times 10^{-7} \text{ M}$

14.13. $6.78 \times 10^{-6} \text{ M}$

14.14. Adding Br_2 increases the concentration of Br_2, causing a shift to the left (away from the Br_2). Adding BrNO increases the concentration of BrNO, causing a shift to the right.

14.15. Decreasing the volume causes the reaction to shift right. Increasing the volume causes the reaction to shift left.

14.16. If we increase the temperature, the reaction shifts to the left. If we decrease the temperature, the reaction shifts to the right.

Chapter 15

15.1. **a.** H_2O donates a proton to $\text{C}_5\text{H}_5\text{N}$, making it the acid. The conjugate base is therefore OH^-. Since $\text{C}_5\text{H}_5\text{N}$ accepts the proton, it is the base and becomes the conjugate acid $\text{C}_5\text{H}_5\text{NH}^+$.
b. Since HNO_3 donates a proton to H_2O, it is the acid, making NO_3^- the conjugate base. Since H_2O is the proton acceptor, it is the base and becomes the conjugate acid, H_3O^+.

15.2. **a.** $[\text{H}_3\text{O}^+] = 6.7 \times 10^{-13} \text{ M}$
Since $[\text{H}_3\text{O}^+] < [\text{OH}^-]$, the solution is basic.

   b. $[H_3O^+] = 1.0 \times 10^{-7} M$
   Neutral solution.
   c. $[H_3O^+] = 1.2 \times 10^{-5} M$
   Since $[H_3O^+] > [OH^-]$, the solution is acidic.

15.3. **a.** 8.02 (basic)
   b. 11.85 (basic)

15.4. $4.3 \times 10^{-9} M$

15.5. $9.4 \times 10^{-3} M$

15.6. 3.28

15.7. 2.72

15.8. 1.8×10^{-6}

15.9. 0.85%

15.10. $4.0 \times 10^{-7} M$

15.11. $[OH^-] = 0.020 M$
   pH = 12.30

15.12. $[OH^-] = 1.2 \times 10^{-2} M$
   pH = 12.08

15.13. **a.** weak base
   b. pH-neutral

15.14. 9.07

15.15. **a.** pH-neutral
   b. weak acid
   c. weak acid

15.16. **a.** basic
   b. acidic
   c. pH-neutral
   d. acidic

15.17. 3.83

15.18. $[SO_4^{2-}] = 0.00386 M$
   pH = 1.945

15.19. $5.6 \times 10^{-11} M$

Chapter 16

16.1. 4.44

16.1. For More Practice
   3.44

16.2. 9.14

16.3. 4.87

16.3. For More Practice
   4.65

16.4. 9.68

16.4. For More Practice
   9.56

16.5. hypochlorous acid (HClO); 2.4 g NaClO

16.6. 1.74

16.7. 7.99

16.8. 2.30×10^{-6} M

16.9. 5.3×10^{-13}

16.10. 1.21×10^{-5} M

16.11. $FeCO_3$ will be more soluble in an acidic solution than $PbBr_2$ because the CO_3^{2-} ion is a basic anion, whereas Br^- is the conjugate base of a strong acid (HBr) and is therefore pH-neutral.

16.12. $Q > K_{sp}$; therefore, a precipitate forms.

16.13. 2.9×10^{-6} M

16.14. **a.** AgCl precipitates first; $[NaCl] = 7.08 \times 10^{-9}$M

b. $[Ag^+]$ is 1.5×10^{-8} M when $PbCl_2$ begins to precipitate, and $[Pb^{2+}]$ is 0.085 M

16.15. 9.6×10^{-6} M.

Chapter 17

17.1. **a.** positive
b. negative
c. positive

17.2. **a.** -548 J/K
b. ΔS_{sys} is negative.
c. ΔS_{univ} is negative, and the reaction is not spontaneous.

17.2. **For More Practice**
375 K

17.3. $\Delta G = -101.6 \times 10^3$ J
Therefore the reaction is spontaneous. Since both ΔH and ΔS are negative, as the temperature increases ΔG will become more positive.

17.4. -153.2 J/K

17.5. $\Delta G_{rxn}^{\circ} = -36.3$ kJ
Since ΔG_{rxn}° is negative, the reaction is spontaneous at this temperature.

17.6. $\Delta G_{rxn}^{\circ} = -42.1$ kJ
Since the value of ΔG_{rxn}° at the lowered temperature is more negative (or less positive) (which is -36.3 kJ), the reaction is more spontaneous.

17.7. $\Delta G_{rxn}^{\circ} = -689.6$ kJ
Since ΔG_{rxn}° is negative, the reaction is spontaneous at this temperature.

17.7. **For More Practice**
$\Delta G_{rxn}^{\circ} = -689.7$ kJ (at 25°)
The value calculated for ΔG_{rxn}° from the tabulated values (-689.6 kJ) is the same, to within 1 in the least significant digit, as the value calculated using the equation for ΔG_{rxn}° .
$\Delta G_{rxn}^{\circ} = -649.7$ kJ (at 500.0 K)
You could not calculate ΔG_{rxn}° at 500.0 K using tabulated ΔG_f° values because the tabulated values of free energy are calculated at a standard temperature of 298 K, much lower than 500 K.

17.8. $+107.1$ kJ

17.9. $\Delta G_{rxn} = -129$ kJ

The reaction is more spontaneous under these conditions than under standard conditions because ΔG_{rxn} is more negative than ΔG°_{rxn} .

17.10. -10.9 kJ

Chapter 18

18.1. $2 \, Cr(s) + 6 \, H^{+}(aq) \longrightarrow 2 \, Cr^{3+}(aq) + 3 \, H_2(g)$

18.2. $Cu(s) + 4 \, H^{+}(aq) + 2 \, NO_3^{-}(aq) \longrightarrow Cu^{2+}(aq) + 2 \, NO_2(g) + 2 \, H_2O(l)$

18.3. $3 \, ClO^{-}(aq) + 2 \, Cr(OH)_4^{-}(aq) + 2 \, OH^{-}(aq) \longrightarrow 3 \, Cl^{-}(aq) + 2 \, CrO_4^{2-}(aq) + 5 \, H_2O(l)$

18.4. $+0.60$ V

18.5. **a.** The reaction *will* be spontaneous as written.
b. The reaction *will not* be spontaneous as written.

18.6. $\Delta G^{\circ} = -3.63 \times 10^5$ J

Since ΔG° is negative, the reaction is spontaneous.

18.7. 4.5×10^3

18.8. 1.09 V

18.9. *Anode:* $2 \, H_2O(l) \longrightarrow O_2(g) + 4 \, H^{+}(aq) + 4 \, e^{-}$
Cathode: $2 \, H_2O(l) + 2 \, e^{-} \longrightarrow H_2(g) + 2 \, OH^{-}(aq)$

18.10. 6.0×10^1 min

Chapter 19

19.1. $^{216}_{84}Po \longrightarrow {}^{212}_{82}Pb + {}^{4}_{2}He$

19.2. a. $^{235}_{92}U \longrightarrow {}^{231}_{90}Th + {}^{4}_{2}He$

$^{231}_{90}Th \longrightarrow {}^{231}_{91}Pa + {}^{0}_{-1}e$

$^{231}_{91}Pa \longrightarrow {}^{227}_{89}Ac + {}^{4}_{2}He$

b. $^{22}_{11}Na \longrightarrow {}^{22}_{10}Ne + {}^{0}_{+1}e$

c. $^{76}_{36}Kr + {}^{0}_{-1}e \longrightarrow {}^{76}_{35}Br$

19.2. For More Practice

Positron emission $\left(^{40}_{19}K \longrightarrow {}^{40}_{18}Ar + {}^{0}_{+1}e \right)$ or electron capture $\left(^{40}_{19}K + {}^{0}_{+1}e \longrightarrow {}^{40}_{18}Ar \right)$

19.3. a. positron emission
b. beta decay
c. positron emission

19.4. 10.7 yr

19.5. $t = 964$ yr

No, the C-14 content suggests that the scroll is from about A.D. 1000, not 500 B.C.

19.6. 1.0×10^9 yr

19.7. Mass defect $= 1.934$ amu

Nuclear binding energy $= 7.569$ MeV/nucleon

Chapter 20

20.1.

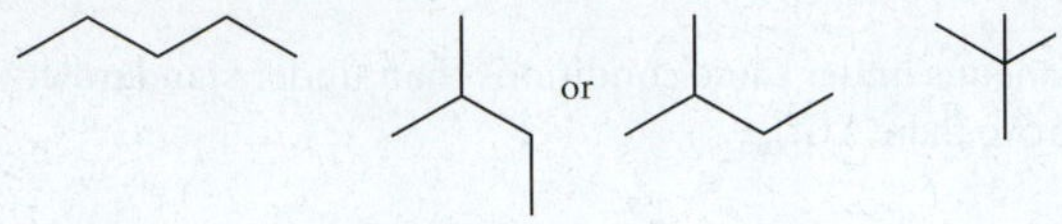

20.2. 3-methylhexane

20.3. 3,5-dimethylheptane

20.4. 2,3,5-trimethylhexane

20.5. **a.** 4,4-dimethyl-2-pentyne
b. 3-ethyl-4,6-dimethyl-1-heptene

20.6. **a.** 2-methylbutane

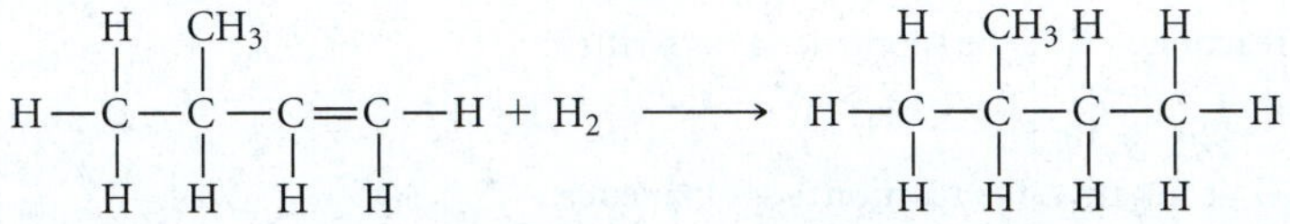

b. 2-chloro-3-methylbutane

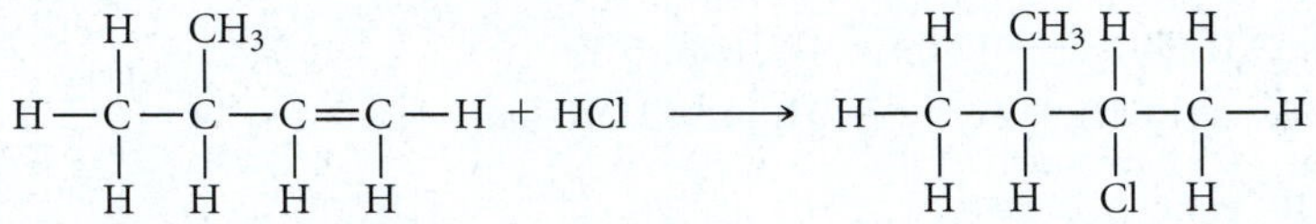

20.7. **a.** Alcohol reacting with an active metal.

$$CH_3CH_2OH + Na \longrightarrow CH_3CH_2ONa + \frac{1}{2}H_2$$

b. dehydration reaction

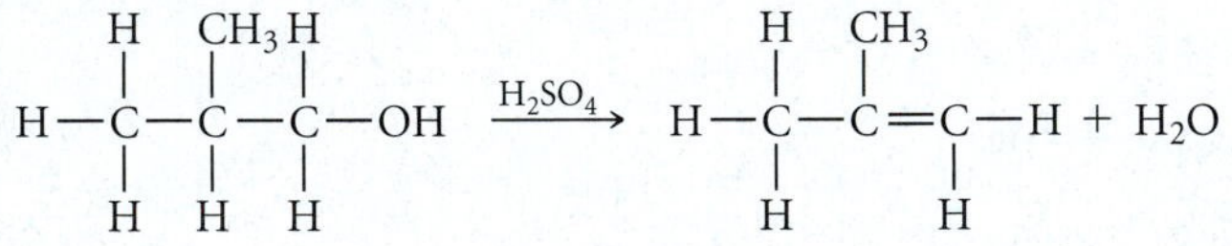

Chapter 21

21.1. Fructose is optically active. It contains three chiral carbons.

21.2.

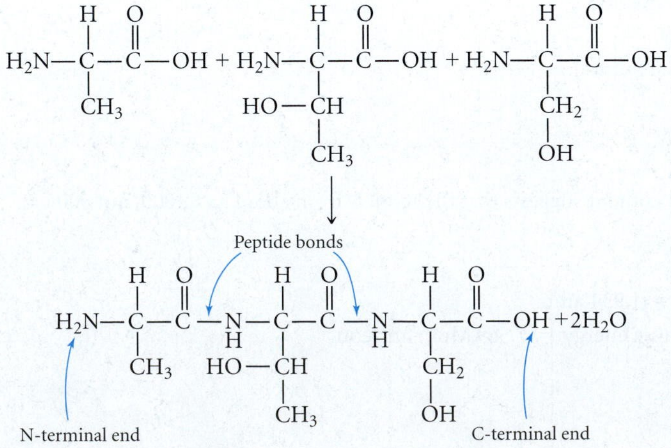

Chapter 22

22.1. $KAlSi_3O_8$

22.2. $x = 2$

22.3. Orthosilicate (or neosilicate): Each of the two Be ions has a charge of 2+ for a total of 4+, and the SiO_4 unit has a charge of 4−.

22.4. Inosilicate (or pyroxene): Ca and Mg each have a charge of 2+ for a total of 4+, and the Si_2O_6 unit has a charge of 4− (two SiO_3^{2-} units).

22.5. $2\,H_2S(g) + 3\,O_2(g) \longrightarrow 2\,H_2O(g) + SO_2(g)$
S changes from the −2 to +4 oxidation state

22.6. The oxidation state for Cl is +7 in ClO_4^- and −1 in Cl^-.

22.7. $I_2(s) + 5\,F_2(g) \longrightarrow 2\,IF_5(g)$

22.8. The electron geometry is tetrahedral and the shape is bent for ICl_2^+.

22.9. The electron geometry is octahedral for BrF_5, and the molecular geometry is square pyramidal.

22.10. The oxidation number changes from −1 to 0 for the oxidation of the Cl in HCl to Cl_2 and from +5 to +4 for the reduction of the Cl in $NaClO_3$ to ClO_2. The oxidizing agent is $NaClO_3$, and the reducing agent is HCl.

Chapter 23

23.1. At 50 mol % Ni and 1000 °C this is a solid phase with half of the atoms each Ni and Cu.

23.2. At 50 mol % Ni and 1400 °C this is a liquid phase with half of the atoms each Ni and Cu.

23.3. At 900 °C and 60 mol % Cr this is a two-phase region with more Ni-rich face-centered cubic crystals than Cr-rich body-centered cubic crystals. The Ni-rich phase is about 42 mol % Cr and 58 mol % Ni. The Cr-rich phase is about 94 mol % Cr and 6 mol % Ni.

23.4. At 900 °C and 98 mol % Cr this is a single-phase region with 100 mol % of the Cr-rich body-centered cubic crystals which contains 2% Ni.

Chapter 24

24.1. $[Xe]6s^2 4f^{14} 5d^6$

24.2. $[Kr]\,5s^0 4d^3$ or $[Kr]\,4d^3$

24.3. pentaamminecarbonylmanganese(II) sulfate

24.4. sodium tetrachloroplatinate(II)

24.5. The complex ion $[Cr(H_2O)_3Cl_3]^+$ fits the general formula MA_3B_3, which results in fac and mer isomers.

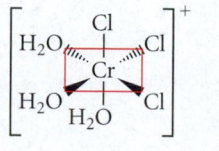

Fac 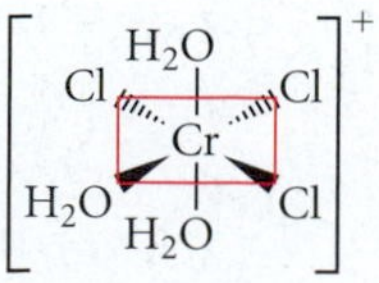   Mer

24.6. The oxalate ligand is a small bidentate ligand so it will have to occupy two adjacent (cis) positions of the octahedron. There are three ways to arrange the two NH_3 and two Cl^- ligands in the four remaining positions. One has both NH_3 and both Cl^- in cis positions (cis isomer). Another has the NH_3 ligands in a trans arrangement with both Cl^- in cis positions (*trans*-ammine isomer). The third has both NH_3 ligands cis and the Cl^- ligands trans (*trans*-chloro isomer).

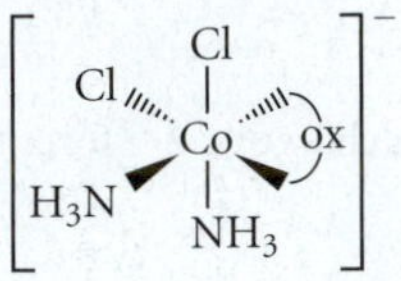

Cis isomer

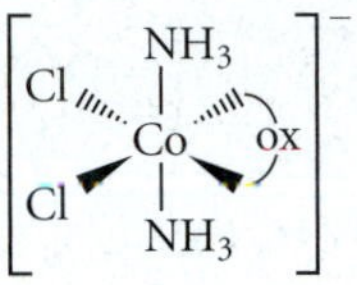

Trans (in NH_3) 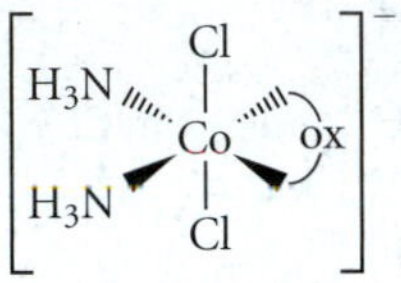   Trans (in Cl^-)

24.7. Both the fac and mer isomers are superimposable (by rotating 180°) on their mirror images, so neither one is optically active.

24.8. 288 kJ/mol

24.9. 5 unpaired electrons

24.10. 1 unpaired electron

Glossary

accuracy A term that refers to how close a measured value is to the actual value. (1.7)

acid A molecular compound that is able to donate an H^+ ion (proton) when dissolved in water, thereby increasing the concentration of H^+. (3.6)

acid ionization constant (K_a) The equilibrium constant for the ionization reaction of a weak acid; used to compare the relative strengths of weak acids. (15.4)

acid–base reaction (neutralization reaction) A reaction in which an acid reacts with a base and the two neutralize each other, producing water. (4.8)

acid–base titration A laboratory procedure in which a basic (or acidic) solution of unknown concentration is reacted with an acidic (or basic) solution of known concentration, in order to determine the concentration of the unknown. (16.4)

acidic solution A solution containing an acid that creates additional H_3O^+ ions, causing $[H_3O^+]$ to increase. (15.4)

activated carbon Very fine carbon particles with high surface area. (22.5)

activated complex (transition state) A high-energy intermediate state between reactant and product. (13.5)

activation energy An energy barrier in a chemical reaction that must be overcome for the reactants to be converted into products. (13.5)

active site The specific area of an enzyme at which catalysis occurs. (13.7)

actual yield The amount of product actually produced by a chemical reaction. (4.3)

addition polymer A polymer in which the monomers simply link together without the elimination of any atoms. (20.14)

addition reaction A type of organic reaction in which two substituents are added across a double bond. (20.10)

alcohol A member of the family of organic compounds that contain a hydroxyl functional group ($-OH$). (3.11, 20.9)

aldehyde A member of the family of organic compounds that contain a carbonyl functional group ($C=O$) bonded to two R group, one of which is a hydrogen atom. (20.10)

aldose A sugar that is an aldehyde. (21.3)

aliphatic hydrocarbons Organic compounds in which carbon atoms are joined in straight or branched chains. (20.3)

alkali metals Highly reactive metals in group 1A of the periodic table. (2.7)

alkaline battery A dry-cell battery that employs slightly different half-reactions in a basic medium. (18.7)

alkaline earth metals Fairly reactive metals in group 2A of the periodic table. (2.7)

alkaloid Organic bases found in plants; they are often poisonous. (15.2)

alkane A hydrocarbon containing only single bonds. (3.11)

alkene A hydrocarbon containing one or more carbon–carbon double bonds. (3.11)

alkyne A hydrocarbon containing one or more carbon–carbon triple bonds. (3.11)

alloy A metallic material that contains more than one element. (23.4)

alpha (α) decay The form of radioactive decay that occurs when an unstable nucleus emits a particle composed of two protons and two neutrons. (19.3)

alpha (α) particle A low-energy particle released during alpha decay; equivalent to a He-4 nucleus. (19.3)

α-helix A pattern in the secondary structure of a protein that occurs when the amino acid chain is wrapped tightly in a coil with the side chains extending outward. (21.5)

aluminosilicates Members of a family of compounds in which aluminum atoms substitute for silicon atoms in some of the silicon lattice sites of the silica structure. (22.3)

amino acids Organic compounds that contain a carbon atom, called the α-carbon, bonded to four different groups: an amine group, an R group, a carboxylic acid group, and a hydrogen atom. (21.4)

ammonia NH_3, the strong smelling compound in which nitrogen displays its lowest oxidation state (-3); (22.6)

amorphous solid A solid in which atoms or molecules do not have any long-range order. (1.3, 11.2)

ampere (A) The SI unit for electrical current; 1 A $=$ 1 C/s. (18.3)

amphoteric Able to act as either an acid or a base. (15.3)

amplitude The vertical height of a crest (or depth of a trough) of a wave; a measure of wave intensity. (7.2)

angular momentum quantum number (l) An integer that determines the shape of an orbital. (7.5)

anion A negatively charged ion. (2.6)

anode The electrode in an electrochemical cell where oxidation occurs; electrons flow away from the anode. (18.3)

antibonding orbital A molecular orbital that is higher in energy than any of the atomic orbitals from which it was formed. (10.8)

aqueous solution A solution in which water acts as the solvent. (4.4, 12.2)

***arachno*-boranes** Boranes with the formula B_nH_{n+6}, consisting of a cage of boron atoms that is missing two or three corners. (22.4)

arc-melting A method in which the solid metal is melted with an arc from a high-voltage electric source in a controlled atmosphere to prevent oxidation. (23.5)

Arrhenius definitions (of acids and bases) The definitions of an acid as a substance that produces H^+ ions in aqueous solution and a base as a substance that produces OH^- ions in aqueous solution. (4.8, 15.3)

Arrhenius equation An equation which relates the rate constant of a reaction to the temperature, the activation energy, and the frequency factor; $k = Ae^{\frac{-E_a}{RT}}$. (13.5)

Arrhenius plot A plot of the natural log of the rate constant ($\ln k$) versus the inverse of the temperature in kelvins ($1/T$) that yields a straight line with a slope of $-E_a/R$ and a y-intercept of $\ln A$. (13.5)

atmosphere (atm)   A unit of pressure based on the average pressure of air at sea level; 1 atm = 101,325 Pa. (5.2)

atom   A submicroscopic particle that constitutes the fundamental building block of ordinary matter; the smallest identifiable unit of an element. (1.1)

atomic element   Those elements that exist in nature with single atoms as their basic units. (3.4)

atomic mass (atomic weight)   The average mass in amu of the atoms of a particular element based on the relative abundance of the various isotopes; it is numerically equivalent to the mass in grams of one mole of the element. (2.8)

atomic mass unit (amu)   A unit used to express the masses of atoms and subatomic particles, defined as $1/12^{th}$ the mass of a carbon atom containing six protons and six neutrons. (2.6)

atomic number (Z)   The number of protons in an atom; the atomic number defines the element. (2.6)

atomic solids   Solids whose composite units are atoms; they include nonbonding atomic solids, metallic atomic solids, and network covalent solids. (11.12)

atomic theory   The theory that each element is composed of tiny indestructible particles called atoms, that all atoms of a given element have the same mass and other properties, and that atoms combine in simple, whole-number ratios to form compounds. (1.2, 2.3)

aufbau principle   The principle that indicates the pattern of orbital filling in an atom. (8.3)

autoionization   The process by which water acts as an acid and a base with itself. (15.4)

Avogadro's law   The law that states that the volume of a gas is directly proportional to its amount in moles ($V \propto n$). (5.3)

Avogadro's number   The number of ^{12}C atoms in exactly 12 g of ^{12}C; equal to 6.0221421×10^{23}. (2.9)

balanced   see *chemical equation* (3.10)

ball and stick model   A representation of the arrangement of atoms in a molecule that shows how the atoms are bonded to each other and the overall shape of the molecule. (3.3)

band gap   An energy gap that exists between the valence band and conduction band of semiconductors and insulators. (11.13)

band theory   A model for bonding in atomic solids that comes from molecular orbital theory in which atomic orbitals combine and become delocalized over the entire crystal. (11.13)

barometer   An instrument used to measure atmospheric pressure. (5.2)

base ionization constant (K_b)   The equilibrium constant for the ionization reaction of a weak base; used to compare the relative strengths of weak bases. (15.7)

basic solution   A solution containing a base that creates additional OH^- ions, causing the $[OH^-]$ to increase. (15.4)

beta (β) decay   The form of radioactive decay that occurs when an unstable nucleus emits an electron. (19.3)

beta (β) particle   A medium-energy particle released during beta decay; equivalent to an electron. (19.3)

β-pleated sheet   A pattern in the secondary structure of a protein that occurs when the amino acid chain is extended and forms a zigzag pattern. (21.5)

bidentate   Describes ligands that donate two electron pairs to the central metal. (24.3)

bimolecular   An elementary step in a reaction that involves two particles, either the same species or different, that collide and go on to form products. (13.6)

binary acid   An acid composed of hydrogen and a nonmetal. (3.6)

binary compound   A compound that contains only two different elements. (3.5)

biochemistry   The study of the chemistry occurring in living organisms. (21.1)

biological effectiveness factor (RBE)   A correction factor multiplied by the dose of radiation exposure in rad to obtain the dose rem. (19.11)

black phosphorus   An allotrope of phosphorus with a structure similar to that of graphite; the most thermodynamically stable form. (22.6)

body-centered cubic   A unit cell that consists of a cube with one atom at each corner and one atom at the center of the cube. (11.11)

boiling point   The temperature at which the vapor pressure of a liquid equals the external pressure. (11.5)

boiling point elevation   The effect of a solute that causes a solution to have a higher boiling point than the pure solvent. (12.7)

bomb calorimeter   A piece of equipment designed to measure ΔE_{rxn} for combustion reactions at constant volume. (6.4)

bond energy   The energy required to break 1 mol of the bond in the gas phase. (9.10)

bond length   The average length of a bond between two particular atoms in a variety of compounds. (9.10)

bond order   For a molecule, the number of electrons in bonding orbitals minus the number of electrons in nonbonding orbitals divided by two; a positive bond order implies that the molecule is stable. (10.8)

bonding orbital   A molecular orbital that is lower in energy than any of the atomic orbitals from which it was formed. (10.8)

bonding pair   A pair of electrons shared between two atoms. (9.5)

boranes   Compounds composed of boron and hydrogen. (22.4)

Born–Haber cycle   A hypothetical series of steps based on Hess's law that represents the formation of an ionic compound from its constituent elements. (9.4)

Boyle's law   The law that states that volume of a gas is inversely proportional to its pressure ($V \propto \frac{1}{P}$). (5.3)

brass   A widely used alloy that contains copper and zinc. (23.5)

Brønsted-Lowry definitions (of acids and bases)   The definitions of an acid as a proton (H^+ ion) donor and a base as a proton acceptor. (15.3)

bronze   An alloy of copper and tin has been used for thousands of years. (23.5)

buffer   A solution containing significant amounts of both a weak acid and its conjugate base (or a weak base and its conjugate acid) that resists pH change by neutralizing added acid or added base (16.2)

buffer capacity   The amount of acid or base that can be added to a buffer without destroying its effectiveness. (16.3)

calcination   The heating of an ore in order to decompose it and drive off a volatile product. (23.3)

calorie (cal)   A unit of energy defined as the amount of energy required to raise one gram of water 1 °C; equal to 4.184 J. (6.1)

Calorie (Cal)   Shorthand notation for the kilocalorie (kcal), or 1000 calories; also called the nutritional calorie, the unit of energy used on nutritional labels. (6.1)

calorimetry   The experimental procedure used to measure the heat evolved in a chemical reaction. (6.4)

capillary action   The ability of a liquid to flow against gravity up a narrow tube due to adhesive and cohesive forces. (11.4)

carbohydrate   A polyhydroxyl aldehyde or ketone. (21.3)

carbon black   A fine powdered form of carbon. (22.5)

carbonyl group   A functional group consisting of a carbon atom double-bonded to an oxygen atom (C=O). (20.10)

carboxylic acid   An organic acid containing the functional group —COOH. (15.2, 20.11)

catalyst   A substance that is not consumed in a chemical reaction, but increases the rate of the reaction by providing an alternate mechanism in with the rate-determining step has a smaller activation energy. (13.7)

cathode   The electrode in an electrochemical cell where reduction occurs; electrons flow toward the cathode. (18.3)

cathode rays   A stream of electrons produced when a high electrical voltage is applied between two electrodes within a partially evacuated tube. (2.4)

cation   A positively charged ion. (2.6)

cell potential (cell emf) (E_{cell})   The potential difference between the cathode and the anode in an electrochemical cell. (18.3)

cellulose   A polysaccharide that consists of glucose units bonded together by β-glycosidic linkages; the main structural component of plants, and the most abundant organic substance on earth. (21.3)

Celsius (°C) scale   The temperature scale most often used by scientists (and by most countries other than the United States), on which pure water freezes at 0 °C and boils at 100 °C (at atmospheric pressure). (1.6)

chain reaction   A series of reactions in which previous reactions cause future ones; in a fission bomb, neutrons produced by the fission of one uranium nucleus induce fission in other uranium nuclei. (19.7)

charcoal   A fuel similar to coal made by heating wood in the absence of air. (22.5)

Charles's law   The law that states that the volume of a gas is directly proportional to its temperature ($V \propto T$). (5.3)

chelate   A complex ion that contains either a bi- or polydentate ligand. (24.3)

chelating agent   The coordinating ligand of a chelate. (24.3)

chemical bond   The sharing or transfer of electrons to attain stable electron configurations for the bonding atoms. (9.3)

chemical change   A change that alters the molecular composition of a substance; see also *chemical reaction*. (1.4)

chemical energy   The energy associated with the relative positions of electrons and nuclei in atoms and molecules (6.1)

chemical equation   A symbolic representation of a chemical reaction; a balanced equation contains equal numbers of the atoms of each element on both sides of the equation. (3.10)

chemical formula   A symbolic representation of a compound which indicates the elements present in the compound and the relative number of atoms of each. (3.3)

chemical property   A property that a substance displays only by changing its composition via a chemical change. (1.4)

chemical reaction   A process by which one or more substances are converted to one or more different substances; see also *chemical change*. (3.10)

chemical symbol   A one- or two-letter abbreviation for an element that is listed directly below its atomic number on the periodic table. (2.6)

chemistry   The science that seeks to understand the behavior of matter by studying the behavior of atoms and molecules. (1.1)

chiral molecule   A molecule that is not superimposable on its mirror image, and thus exhibits optical isomerism. (20.3)

chromosome   The DNA-containing structures that occur in the nuclei of living cells. (21.6)

cis-trans isomerism   Another term for geometric isomerism; cis-isomers have the same functional group on the same side of a bond and trans-isomers have the same functional group on opposite sides of a bond.(20.5)

Claus process   A industrial process for obtaining sulfur through the oxidation of hydrogen sulfide. (22.8)

Clausius-Clapeyron equation   An equation that displays the exponential relationship between vapor pressure and temperature;
$$\ln (P_{vap}) = \frac{-\Delta H_{vap}}{R}\left(\frac{1}{T}\right) + \ln \beta \quad (11.5)$$

closo-**boranes**   Boranes that have the formula $B_{12}H_{12}{}^{2-}$ and form the full icosohedral shape. (22.4)

coal   A solid, black fuel with high carbon content, the product of the decomposition of ancient plant material. (22.5)

codon   A sequence of three bases in a nucleic acid that codes for one amino acid. (21.6)

coffee-cup calorimeter   A piece of equipment designed to measure ΔH_{rxn} for reactions at constant pressure. (6.5)

coke   A solid formed by heating coal in the absence of air that consists primarily of carbon and ash. (22.5)

colligative property   A property that depends on the amount of a solute but not on the type. (12.7)

collision model   A model of chemical reactions in which a reaction occurs after a sufficiently energetic collision between two reactant molecules. (13.5)

colloidal dispersion (colloid)   A mixture in which a dispersed substance is finely dived but not truly dissolved in a dispersing medium. (12.8)

combustion analysis   A method of obtaining empirical formulas for unknown compounds, especially those containing carbon and hydrogen, by burning a sample of the compound in pure oxygen and analyzing the products of the combustion reaction. (3.9)

combustion reaction   A type of chemical reaction in which a substance combines with oxygen to form one or more oxygen-containing compounds; the reaction often causes the evolution of heat and light in the form of a flame. (3.10)

common ion effect   The tendency for a common ion to decrease the solubility of an ionic compound or to decrease the ionization of a weak acid or weak base. (16.2)

common name   A traditional name of a compound that gives little or no information about its chemical structure; for example, the common name of $NaHCO_3$ is "baking soda." (3.5)

complementary   Capable of precise pairing; in particular, the bases of nucleic acids. (21.6)

complementary properties   Those properties that exclude one another, i.e. the more you know about one, the less you know about the other. For example, the wave nature and particle nature of the electron are complementary. (7.4)

complete ionic equation   An equation which lists individually all of the ions present as either reactants or products in a chemical reaction. (4.7)

complex carbohydrate   Another term for a polysaccharide based on the fact that it is made up of many simple sugars. (21.3)

complex ion   An ion that contains a central metal ion bound to one or more ligands. (16.8, 24.3)

compound   A substance composed of two or more elements in fixed, definite proportions. (1.3)

concentrated solution   A solution that contains a large amount of solute relative to the amount of solvent. (4.4, 12.5)

condensation   The phase transition from gas to liquid. (11.5)

condensation polymer   A polymer formed by elimination of an atom or small group of atoms (usually water) between pairs of monomers during polymerization. (20.14)

condensation reaction   A reaction in which two or more organic compounds are joined, often with the loss of water or some other small molecule. (20.11)

conjugate acid–base pair   Two substances related to each other by the transfer of a proton. (15.3)

constructive interference   The interaction of waves from two sources that align with overlapping crests, resulting in a wave of greater amplitude. (7.2)

contact process   An industrial method for the production of sulfuric acid. (22.8)

conversion factor   A factor used to convert between two different units; a conversion factor can be constructed from any two quantities known to be equivalent. (1.8)

coordinate covalent bond   The bond formed when a ligand donates electrons to an empty orbital of a metal in a complex ion. (24.3)

coordination compound   A neutral compound made when a complex ion combines with one or more counterions. (24.3)

coordination isomers   Isomers of complex ions that occur when a coordinated ligand exchanges places with the uncoordinated counterion. (24.4)

coordination number (secondary valence)   The number of molecules or ions directly bound to the metal atom in a complex ion. (24.3)

coordination number   The number of atoms with which each atom in a crystal lattice is in direct contact (11.11)

core electrons   Those electrons in a complete principal energy level and those in complete d and f sublevels. (8.4)

corrosion   The gradual, nearly always undesired oxidation of metals that occurs when they are exposed to oxidizing agents in the environment. (18.9)

covalent bond   A chemical bond in which two atoms share electrons that interact with the nuclei of both atoms, lowering the potential energy of each through electrostatic interactions. (3.2, 9.2)

covalent carbides   Binary compounds composed of carbon combined with low-electronegativity nonmetals or metalloids. (22.5)

covalent radius (bonding atomic radius)   Defined in nonmetals as one-half the distance between two atoms bonded together, and in metals as one-half the distance between two adjacent atoms in a crystal of the metal. (8.6)

critical mass   The necessary amount of a radioactive isotope required to produce a self-sustaining fission reaction. (19.7)

critical point   The temperature and pressure above which a supercritical fluid exists. (11.8)

critical pressure   The pressure required to bring about a transition to a liquid at the critical temperature. (11.5)

critical temperature   The temperature above which a liquid cannot exist, regardless of pressure. (11.5)

crystalline lattice   The regular arrangement of atoms in a crystalline solid. (11.11)

crystalline solid (crystal)   A solid in which atoms, molecules, or ions are arranged in patterns with long-range, repeating order. (1.3, 11.2)

cubic closest packing   A closest-packed arrangement in which the third layer of atoms is offset from the first; the same structure as the face-centered cubic. (11.11)

cyclotron   A particle accelerator in which a charged particle is accelerated in an evacuated ring-shaped tube by an alternating voltage applied to each semi-circular half of the ring. (19.10)

Dalton's law of partial pressures   The law stating that the sum of the partial pressures of the components in a gas mixture must equal the total pressure. (5.6)

de Broglie relation   The observation that the wavelength of a particle is inversely proportional to its momentum

$$\lambda = \frac{h}{mv}\ (7.4)$$

decanting   A method of separating immiscible liquids by pouring the top layer into another container. (1.3)

degenerate   A term describing two or more electron orbitals with the same value of n that have the same energy. (8.3)

density (d)   The ratio of an object's mass to its volume. (1.6)

deposition   The phase transition from gas to solid. (11.6)

derived unit   A unit that is a combination of other base units. For example, the SI unit for speed is meters per second (m/s), a derived unit. (1.6)

destructive interference   The interaction of waves from two sources aligned so that the crest of one overlaps the trough of the other, resulting in cancellation. (7.2)

deterministic   A characteristic of the classical laws of motion, which imply that present circumstances determine future events. (7.4)

dextrorotatory   Capable of rotating the plane of polarization of light clockwise. (20.3)

diamagnetic   The state of an atom or ion that contains only paired electrons and is, therefore, slightly repelled by an external magnetic field. (8.7, 10.8)

diamond   An elemental form of carbon with a crystal structure that consists of carbon atoms connected to four other carbon atoms at the corners of a tetrahedron, creating a strong network covalent solid. (22.5)

diffraction   The phenomena by which a wave emerging from an aperture spreads out to form a new wave front. (7.2)

diffusion   The process by which a gas spreads through a space occupied by another gas. (5.9)

dilute solution   A solution that contains a very small amount of solute relative to the amount of solvent. (4.4, 12.5)

dimensional analysis   The use of units as a guide to solving problems. (1.8)

dimer   The product that forms from the reaction of two monomers. (20.14)

diode   A device that allows the flow of electrical current in only one direction. (11.13)

dipeptide   Two amino acids linked together. (21.4)

dipole moment   A measure of the separation of positive and negative charge in a molecule. (9.6)

dipole–dipole force   An intermolecular force exhibited by polar molecules that results from the uneven charge distribution. (11.3)

diprotic acid   An acid that contains two ionizable protons. (4.8, 15.4)

disaccharide   A carbohydrate composed of two monosaccharides. (21.3)

dispersion force (London force)   An intermolecular force exhibited by all atoms and molecules that results from fluctuations in the electron distribution. (11.3)

distillation   The process by which mixtures of miscible liquids are separated by heating the mixture to boil off the more volatile liquid. The vaporized component is then recondensed and collected in a separate flask. (1.3)

disubstituted benzene   A benzene in which two hydrogen atoms have been replaced by other atoms. (20.7)

double bond   The bond that forms when two electrons are shared between two atoms. (9.5)

dry-cell battery   A battery that does not contain a large amount of liquid water, often using the oxidation of zinc and the reduction of MnO_2 to provide the electrical current. (18.7)

duet   A Lewis structure with two dots, signifying a filled outer electron shell for the elements H and He. (9.3)

dynamic equilibrium   The point at which the rate of the reverse reaction or process equals the rate of the forward reaction or process. (11.5, 12.4, 14.2)

effective nuclear charge (Z_{eff})   The actual nuclear charge experienced by an electron, defined as the charge of the nucleus plus the charge of the shielding electrons. (8.3)

effusion   The process by which a gas escapes from a container into a vacuum through a small hole. (5.9)

electrical charge   A fundamental property of certain particles that causes them to experience a force in the presence of electric fields. (2.4)

electrical current   The flow of electric charge. (18.3)

electrochemical cell   A device in which a chemical reaction either produces or is carried out by an electrical current. (18.3)

electrolysis   The process by which electrical current is used to drive an otherwise nonspontaneous redox reaction. (18.8)

electrolyte   A substance that dissolves in water to form solutions that conduct electricity. (4.5)

electrolytic cell   An electrochemical cell which uses electrical current to drive a nonspontaneous chemical reaction. (18.3)

electromagnetic radiation   A form of energy embodied in oscillating electric and magnetic fields. (7.2)

electromagnetic spectrum   The range of the wavelengths of all possible electromagnetic radiation. (7.2)

electrometallurgy   The use of electrolysis to produce metals from their compounds. (23.3)

electromotive force (emf)   The force that results in the motion of electrons due to a difference in potential. (18.3)

electron   A negatively charged, low mass particle found outside the nucleus of all atoms that occupies most of the atom's volume but contributes almost none of its mass. (2.4)

electron affinity (EA)   The energy change associated with the gaining of an electron by an atom in its gaseous state. (8.7)

electron capture   The form of radioactive decay that occurs when a nucleus assimilates an electron from an inner orbital. (19.3)

electron configuration   A notation that shows the particular orbitals that are occupied by electrons in an atom. (8.3)

electron geometry   The geometrical arrangement of electron groups in a molecule. (10.3)

electron groups   A general term for lone pairs, single bonds, multiple bonds, or lone electrons in a molecule. (10.2)

electron spin   A fundamental property of electrons; spin can have a value of $\pm \frac{1}{2}$. (8.3)

electronegativity   The ability of an atom to attract electrons to itself in a covalent bond. (9.6)

element   A substance that cannot be chemically broken down into simpler substances. (1.3)

elementary step   An individual step in a reaction mechanism. (13.6)

emission spectrum   The range of wavelengths emitted by a particular element; used to identify the element. (7.3)

empirical formula   A chemical formula that shows the simplest whole number ratio of atoms in the compound. (3.3)

empirical formula molar mass   The sum of the masses of all the atoms in an empirical formula. (3.9)

enantiomers(optical isomers)   Two molecules that are nonsuperimposable mirror images of one another. (20.3, 24.4)

endothermic reaction   A chemical reaction that absorbs heat from its surroundings; for an endothermic reaction, $\Delta H > 0$. (6.5)

endpoint   The point of pH change where an indicator changes color (16.4)

energy   The capacity to do work. (1.5, 6.1)

English system   The system of units used in the United States and various other countries in which the inch is the unit of length, the pound is the unit of force, and the ounce is the unit of mass. (1.6)

enthalpy (H)   The sum of the internal energy of a system and the product of its pressure and volume; the energy associated with the breaking and forming of bonds in a chemical reaction. (6.5)

entropy   A thermodynamic function that is proportional to the number of energetically equivalent ways to arrange the components of a system to achieve a particular state; a measure of the energy randomization or energy dispersal in a system. (12.2, 17.3)

enzyme   A biochemical catalyst made of protein that increases the rates of biochemical reactions. (13.7, 21.4)

equilibrium constant (K)   The ratio, at equilibrium, of the concentrations of the products of a reaction raised to their stoichiometric coefficients to the concentrations of the reactants raised to their stoichiometric coefficients. (14.3)

equivalence point   The point in a titration at which the added solute completely reacts with the solute present in the solution; for acid–base titrations, the point at which the amount of acid is stoichiometrically equal to the amount of base in solution. (4.8, 16.4)

ester   A family of organic compounds with the general structure R—COO—R. (20.11)

ester linkage   The bonds that form between a carboxylic acid and an alcohol to form an ester, such as those in triglycerides. (21.2)

ether   A member of the family of organic compounds of the form R—O—R'. (20.12)

exact numbers   Numbers that have no uncertainty and thus do not limit the number of significant figures in any calculation. (1.7)

exothermic reaction   A chemical reaction that releases heat to its surroundings; for an exothermic reaction, $\Delta H < 0$. (6.5)

experiment   A highly controlled procedure designed to generate observations that may support a hypothesis or prove it wrong. (1.2)

exponential factor   A number between 0 and 1 that represents the fraction of molecules that have enough energy to make it over the activation barrier on a given approach. (13.5)

extensive property   A property that depends on the amount of a given substance, such as mass. (1.6)

extractive metallurgy   The process by which an elemental metal must be extracted from the compounds in which it is found. (23.3)

face-centered cubic   A crystal structure whose unit cell consists of a cube with one atom at each corner and one atom in the center of every face. (11.11)

Fahrenheit (°F) scale   The temperature scale that is most familiar in the United States, on which pure water freezes at 32 °F and boils at 212 °F. (1.6)

family   A group of organic compounds with the same functional group. (3.11)

family (group)   Columns within the main group elements in the periodic table that contain elements that exhibit similar chemical properties. (2.7)

Faraday's constant (F)   The charge in coulombs of 1 mol of electrons: $F = \dfrac{96,485 \text{ C}}{\text{mol e}^-}$. (18.5)

fatty acid   A carboxylic acid with a long hydrocarbon tail. (21.2)

ferromagnetic   The state of an atom or ion that is very strongly attracted by an external magnetic field. (23.5)

fertilizer   A material containing large amounts of nitrogen or phosphorus that is used to increase plant growth. (22.6)

fibrous protein   A protein with a relatively linear structure; fibrous proteins tend to be insoluble in aqueous solutions. (21.5)

film-badge dosimeter   A device for monitoring exposure to radiation consisting of photographic film held in a small case that is pinned to clothing (19.5)

filtration   A procedure used to separate a mixture composed of an insoluble solid and a liquid by pouring it through filter paper or some other porous membrane or layer. (1.3)

first law of thermodynamics   The law stating that the total energy of the universe is constant. (6.2)

flux   In pyrometallurgy, material that will react with the gangue to form a substance with a low melting point. (23.3)

formal charge   The charge that an atom in a Lewis structure would have if all the bonding electrons were shared equally between the bonded atoms. (9.8)

formation constant (K_f)   The equilibrium constant associated with reactions for the formation of complex ions. (16.8)

formula mass   The average mass of a molecule of a compound in amu. (3.7)

formula unit   The smallest, electrically neutral collection of ions in an ionic compound. (3.4)

Frasch process   An industrial process for the recovery of sulfur that uses superheated water to liquefy sulfur deposits in Earth's crust and bring the molten sulfur to the surface. (22.8)

free energy of formation (ΔG_f°)   The change in free energy when 1 mol of a compound forms from its constituent elements in their standard states. (17.7)

free radical   A molecule or ion with an odd number of electrons in its Lewis structure. (9.9)

freezing   The phase transition from liquid to solid. (11.6)

freezing point depression   The effect of a solute that causes a solution to have a lower melting point than the pure solvent. (12.7)

frequency (ν)   For waves, the number of cycles (or complete wavelengths) that pass through a stationary point in one second. (7.2)

frequency factor   The number of times that reactants approach the activation energy per unit time. (13.5)

fuel cell   A voltaic cell that uses the oxidation of hydrogen and the reduction of oxygen, forming water, to provide electrical current. (18.7)

fullerenes   Carbon clusters, such as C_{60}, bonded in roughly spherical shapes containing from 36 to over 100 carbon atoms. (22.5)

functional group   A characteristic atom or group of atoms that imparts certain chemical properties to an organic compound. (3.11)

gamma (γ) rays   The form of electromagnetic radiation with the shortest wavelength and highest energy. (7.2, 19.3)

gamma (γ) ray emission   The form of radioactive decay that occurs when an unstable nucleus emits extremely high frequency electromagnetic radiation. (19.3)

gangue   The undesirable minerals that are separated from specific ores. (23.3)

gas   A state of matter in which atoms or molecules have a great deal of space between them and are free to move relative to one another; lacking a definite shape or volume, a gas conforms to those of its container. (1.3)

gas-evolution reaction   A reaction in which two aqueous solutions are mixed and a gas forms, resulting in bubbling. (4.8)

Geiger-Müller counter   A device used to detect radioactivity that uses argon atoms that become ionized in the presence of energetic particles to produce an electrical signal. (19.5)

gene   A sequence of codons within a DNA molecule that codes for a single protein. (21.6)

geometric isomerism   A form of stereoisomerism involving the orientation of functional groups in a molecule that contains bonds incapable of rotating. (20.5)

geometric isomers   For complex ions, isomers that result when the ligands bonded to the metal have a different spatial arrangement. (24.4)

Gibbs free energy (G)   A thermodynamic state function related to enthalpy and entropy by the equation $G = H - TS$; chemical systems tend towards lower Gibbs free energy, also called the *chemical potential*. (17.5)

globular protein   A protein that folds into a roughly spherical shape so that its polar side chains are oriented outward and its nonpolar side chains toward the interior; globular proteins tend to be soluble in water. (21.5)

glycogen   A highly branched form of starch. (21.3)

glycolipid   A triglyceride composed of a fatty acid, a hydrocarbon chain, and a sugar molecule as the polar section. (21.2)

glycosidic linkage   A bond between carbohydrates that results from a dehydration reaction. (21.3)

graphite   An elemental form of carbon consisting of flat sheets of carbon atoms, bonded together as interconnected hexagonal rings held together by intermolecular forces, that can easily slide past each other. (22.5)

Haber-Bosch process   The industrial process for producing ammonia from nitrogen gas and hydrogen gas. (22.6)

half-cell   One half of an electrochemical cell where either oxidation or reduction occurs. (18.3)

half-life ($t_{1/2}$)   The time required for the concentration of a reactant or the amount of a radioactive isotope to fall to one-half of its initial value. (13.4)

halogens   Highly reactive nonmetals in group 7A of the periodic table. (2.7)

heat (q)   The flow of energy caused by a temperature difference (6.1)

heat capacity (C)   The quantity of heat required to change a system's temperature by 1 °C. (6.3)

heat of fusion (ΔH_{fus})   The amount of heat required to melt 1 mole of a solid. (11.6)

heat of hydration ($\Delta H_{hydration}$)   The enthalpy change that occurs when 1 mole of gaseous solute ions are dissolved in water. (12.3)

heat of reaction (ΔH_{rxn})   The enthalpy change for a chemical reaction. (6.5)

heat of vaporization (ΔH_{vap})   The amount of heat required to vaporize one mole of a liquid to a gas. (11.5)

Heisenberg's uncertainty principle   The principle stating that due to the wave-particle duality, it is fundamentally impossible to precisely determine both the position and velocity of a particle at a given moment in time. (7.4)

Henderson-Hasselbalch equation   An equation used to easily calculate the pH of a buffer solution from the initial concentrations of the buffer components, assuming that the "*x is small*" approximation is valid: $pH = pK_a + \log \frac{[\text{base}]}{[\text{acid}]}$ (16.2)

Henry's law   An equation that expresses the relationship between solubility of a gas and pressure: $S_{gas} = k_H P_{gas}$ (12.4)

Hess's law   The law stating that if a chemical equation can be expressed as the sum of a series of steps, then ΔH_{rxn} for the overall equation is the sum of the heats of reactions for each step (6.7)

heterogeneous catalysis   Catalysis in which the catalyst and the reactants exist in different phases. (13.7)

heterogeneous mixture   A mixture in which the composition varies from one region to another. (1.3)

hexagonal closest packing   A closest packed arrangement in which the atoms of the third layer align exactly over those in the first layer. (11.11)

hexose   A six-carbon sugar. (21.3)

high-spin complex   A complex ion with weak field ligands that have the same number of unpaired electrons as the free metal ion. (24.5)

homogeneous catalysis   Catalysis in which the catalyst exists in the same phase as the reactants. (13.7)

homogeneous mixture   A mixture with the same composition throughout. (1.3)

Hund's rule   The principle stating that when electrons fill degenerate orbitals they first fill them singly with parallel spins. (8.3)

hybrid orbitals   Orbitals formed from the combination of standard atomic orbitals that correspond more closely to the actual distribution of electrons in a chemically bonded atom. (10.7)

hybridization   A mathematical procedure in which standard atomic orbitals are combined to form new, hybrid orbitals. (10.7)

hydrate   An ionic compound that contains a specific number of water molecules associated with each formula unit. (3.5)

hydrazine N_2H_4, a nitrogen and hydrogen compound in which nitrogen has a negative oxidation state (-2). (22.6)

hydrocarbon   An organic compound that contains only carbon and hydrogen. (3.11)

hydrogen azide   A nitrogen and hydrogen compound with a higher hydrogen-to-nitrogen ratio than ammonia or hydrazine. (22.6)

hydrogen bond   A strong dipole–dipole attractive force between a hydrogen bonded to O, N, or F and one of these electronegative atoms on a neighboring molecule. (11.3)

hydrogenation   The catalyzed addition of hydrogen to alkene double bonds to make single bonds. (13.7)

hydrolysis   The splitting of a chemical bond with water, resulting in the addition of H and OH to the products. (21.3)

hydrometallurgy   The use of an aqueous solution to extract metals from their ores. (23.3)

hydronium ion H_3O^+, the ion formed from the association of a water molecule with an H^+ ion donated by an acid. (4.8, 15.3)

hypothesis   A tentative interpretation or explanation of an observation. A good hypothesis is *falsifiable*. (1.2)

hypoxia   A physiological condition caused by low levels of oxygen, marked by dizziness, headache, and shortness of breath and eventually unconsciousness or even death in severe cases. (5.6)

ideal gas constant   The proportionality constant of the ideal gas law, R, equal to 8.314 J/mol · K or 0.08206 L · atm/mol · K. (5.4)

ideal gas law   The law that combines the relationships of Boyle's, Charles's, and Avogadro's laws into one comprehensive equation of state with the proportionality constant R in the form $PV = nRT$. (5.4)

ideal solution   A solution that follows Raoult's law at all concentrations for both solute and solvent. (12.6)

indeterminacy   The principle that present circumstances do not necessarily determine future evens in the quantum-mechanical realm. (7.4)

indicator   A dye whose color depends on the pH of the solution it is dissolved in; often used to detect the endpoint of a titration. (4.8, 16.4)

infrared (IR) radiation   Electromagnetic radiation emitted from warm objects, with wavelengths slightly larger than those of visible light. (7.2)

insoluble   Incapable of dissolving in water or being extremely difficult of solution. (4.5)

integrated rate law   A relationship between the concentrations of the reactants in a chemical reaction and time. (13.4)

intensive property   A property such as density that is independent of the amount of a given substance. (1.6)

interference   The superposition of two or more waves overlapping in space, resulting in either an increase in amplitude (constructive interference) or a decrease in amplitude (destructive interfence). (7.2)

interhalogen compounds   A class of covalent compounds that contain two different halogens. (22.9)

internal energy (E)   The sum of the kinetic and potential energies of all of the particles that compose a system. (6.2)

International System of Units (SI)   The standard unit system used by scientists, based on the metric system. (1.6)

interstitial alloy   An alloy in which small, usually nonmetallic atoms fit between the metallic atoms of a crystal. (23.4)

ion   An atom or molecule with a net charge caused by the loss or gain of electrons. (2.6)

ion product constant for water (K_w)   The equilibrium constant for the autoionization of water. (15.5)

ion–dipole force   An intermolecular force between an ion and the oppositely charged end of a polar molecule. (11.3)

ionic bond   A chemical bond formed between two oppositely charged ions, generally a metallic cation and a nonmetallic anion, that are attracted to one another by electrostatic forces. (3.2, 9.2)

ionic carbides   Binary compounds composed of carbon combined with low-electronegativity metals. (22.5)

ionic compound   A compound composed of cations and anions bound together by electrostatic attraction. (3.4)

ionic solids   Solids whose composite units are ions; they generally have high melting points. (11.12)

ionization energy (IE)   The energy required to remove an electron from an atom or ion in its gaseous state. (8.7)

ionizing power   The ability of radiation to ionize other molecules and atoms. (19.3)

irreversible reaction   A reaction that does not achieve the theoretical limit of available free energy. (17.7)

isotopes   Atoms of the same element with the same number of protons but different numbers of neutrons and consequently different masses. (2.6)

joule (J)   The SI unit for energy: equal to $1 \text{ kg} \cdot \text{m}^2/\text{s}^2$. (6.1)

kelvin (K)   The SI standard unit of temperature. (1.6)

Kelvin scale   The temperature scale that assigns 0 K (-273 °C or -459 °F) to the coldest temperature possible, absolute zero, the temperature at which molecular motion virtually stops. 1 K = 1°C. (1.6)

ketone   A member of the family of organic compounds that contain a carbonyl functional group (C = O) bonded to two R group, neither of which is a hydrogen atom. (20.10)

ketose   A sugar that is a ketone. (21.3)

kilogram (kg)   The SI standard unit of mass defined as the mass of a block of metal kept at the International Bureau of Weights and Measures at Sèvres, France. (1.6)

kilowatt-hour (kWh)   An energy unit used primarily to express large amounts of energy produced by the flow of electricity; equal to 3.60×10^6 J. (6.1)

kinetic energy   The energy associated with motion of an object. (1.5, 6.1)

kinetic molecular theory   A model of an ideal gas as a collection of point particles in constant motion undergoing completely elastic collisions. (5.8)

lanthanide contraction   The trend toward leveling off in size of the atoms in the third and fourth transition rows due to the ineffective shielding of the *f* sublevel electrons. (24.2)

lattice energy   The energy associated with forming a crystalline lattice from gaseous ions. (9.4)

law   see *scientific law*

law of conservation of energy   A law stating that energy can neither be created nor destroyed, only converted from one form to another. (1.5, 6.1)

law of conservation of mass   A law stating that matter is neither created nor destroyed in a chemical reaction. (1.2)

law of definite proportions   A law stating that all samples of a given compound have the same proportions of their constituent elements. (2.3)

law of mass action   The relationship between the balanced chemical equation and the expression of the equilibrium constant. (14.3)

law of multiple proportions   A law stating that when two elements (A and B) form two different compounds, the masses of element B that combine with one gram of element A can be expressed as a ratio of small whole numbers. (2.3)

Le Châtelier's principle   The principle stating that when a chemical system at equilibrium is disturbed, the system shifts in a direction that minimizes the disturbance. (14.9)

leaching   The process by which a metal is separated out of a mixture by selectively dissolving it into solution. (23.3)

lead-acid storage battery   A battery that uses the oxidation of lead and the reduction of lead(IV) oxide in sulfuric acid to provide electrical current. (18.7)

lever rule   The rule that states that in a two-phase region, whichever phase is closest to the composition of the alloy is the more abundant phase. (23.4)

levorotatory   Capable of rotating the polarization of light counterclockwise. (20.3)

Lewis acid   An atom, ion, or molecule that is an electron pair acceptor. (15.11)

Lewis base   An atom, ion, or molecule that is an electron pair donor. (15.11)

Lewis electron-dot structures (Lewis structures)   A drawing that represents chemical bonds between atoms as shared or transferred electrons; the valence electrons of atoms are represented as dots. (9.1)

Lewis theory   A simple model of chemical bonding using diagrams that represent bonds between atoms as lines or pairs of dots. In this theory, atoms bond together to obtain stable octets (8 valence electrons). (9.1)

ligand   A neutral molecule or an ion that acts as a Lewis base with the central metal ion in a complex ion. (16.8, 24.3)

limiting reactant   The reactant that has the smallest stoichiometric amount in a reactant mixture and consequently limits the amount of product in a chemical reaction. (4.3)

linear accelerator   A particle accelerator in which a charged particle is accelerated in an evacuated tube by a potential difference between the ends of the tube or by alternating charges in sections of the tube. (19.10)

linear geometry   The molecular geometry of three atoms with a 180° bond angle due to the repulsion of two electron groups. (10.2)

linkage isomers   Isomers of complex ions that occur when some ligands coordinate to the metal in different ways. (24.4)

lipid   A member of the class of biochemical compounds that are insoluble in water but soluble in nonpolar solvents; include fatty acids, triglycerides, and steroids. (21.2)

lipid bilayer   A double-layered structure made of phospholipids or glycolipids, in which the polar heads of the molecules interact with the environment and the nonpolar tails interact with each other; a component of many cellular membranes. (21.2)

liquid   A state of matter in which atoms or molecules pack about as closely as they do in solid matter but are free to move relative to each other, giving a fixed volume but not a fixed shape. (1.3)

liter (L)   A unit of volume equal to 1000 cm^3 or 1.057 qt. (1.6)

lithium-ion battery   A battery that produces electrical current in the form of motion of lithium ions from the anode to the cathode. (18.7)

lone pair   A pair of electrons associated with only one atom. (9.5)

low-spin complex   A complex ion with strong field ligands that have fewer unpaired electrons than the free metal ion. (24.5)

magic numbers   Certain numbers of nucleons (N or Z = 2, 8, 20, 28, 50, 82, and N = 126) that confer unique stability. (19.4)

magnetic quantum number (m_l)   An integer that specifies the orientation of an orbital. (7.5)

main-group elements   Those elements found in the *s* or *p* blocks of the periodic table, whose properties tend to be predictable based on their position in the table. (2.7, 22.2)

manometer   An instrument used to determine the pressure of a gaseous sample, consisting of a liquid-filled U-shaped tube with one end exposed to the ambient pressure and the other end connected to the sample. (5.2)

mass   A measure of the quantity of matter making up an object. (1.6)

mass defect   The difference in mass between the nucleus of an atom and the sum of the separated particles that make up that nucleus. (19.8)

mass number (*A*)   The sum of the number of protons and neutrons in an atom. (2.6)

mass percent composition (mass percent)   An element's percentage of the total mass of a compound containing the element. (3.8)

mass spectrometry   An experimental method of determining the precise mass and relative abundance of isotopes in a given sample using an instrument called a *mass spectrometer*. (2.8)

matter   Anything that occupies space and has mass. (1.3)

mean free path   The average distance that a molecule in a gas travels between collisions. (5.9)

melting (fusion)   The phase transition from solid to liquid. (11.6)

melting point   The temperature at which the molecules of a solid have enough thermal energy to overcome intermolecular forces and become a liquid. (11.6)

metals   A large class of elements that are generally good conductors of heat and electricity, malleable, ductile, lustrous, and tend to lose electrons during chemical changes. (2.7)

metallic atomic solids   Atomic solids held together by metallic bonds; they have variable melting points. (11.12)

metallic bonding   The type of bonding that occurs in metal crystals, in which metal atoms donate their electrons to an electron sea, delocalized over the entire crystal lattice. (9.2)

metallic carbides   Binary compounds composed of carbon combined with metals that have a metallic lattice with holes small enough to fit carbon atoms. (22.5)

metalloids   A category of elements found on the boundary between the metals and nonmetals of the periodic table, with properties intermediate between those of both groups; also called *semimetals*. (2.7)

metallurgy   The part of chemistry that includes all the processes associated with mining, separating, and refining metals and the subsequent production of pure metals and mixtures of metals called alloys. (23.1)

meter (m)   The SI standard unit of length; equivalent to 39.37 inches. (1.6)

metric system   The system of measurements used in most countries in which the meter is the unit of length, the kilogram is the unit of mass, and the second is the unit of time. (1.6)

microwaves   Electromagnetic radiation with wavelengths slightly longer than those of infrared radiation; used for radar and in microwave ovens. (7.2)

milliliter (mL)   A unit of volume equal to 10^{-3} L or 1 cm^3. (1.6)

millimeter of mercury (mmHg)   A common unit of pressure referring to the air pressure required to push a column of mercury to a height of 1 mm in a barometer; 760 mmHg = 1 atm. (5.2)

minerals   are homogenous, naturally occurring, crystalline inorganic solids. (23.2)

miscibility   The ability to mix without separating into two phases. (11.3)

miscible   The ability of two or more substances to be soluble in each other in all proportions. (12.2)

mixture   A substance composed of two or more different types of atoms or molecules that can be combined in variable proportions. (1.3)

molality (*m*)   A means of expressing solution concentration as the number of moles of solute per kilogram of solvent. (12.5)

molar heat capacity   The amount of heat required to raise the temperature of one mole of a substance by 1 °C. (6.3)

molar mass   The mass in grams of one mole of atoms of an element; numerically equivalent to the atomic mass of the element in amu. (2.9)

molar solubility   The solubility of a compound in units of moles per liter. (16.5)

molar volume   The volume occupied by one mole of a gas; the molar volume of an ideal gas at STP is 22.4 L. (5.5)

molarity (M)   A means of expressing solution concentration as the number of moles of solute per liter of solution. (4.4, 12.5)

mole (mol)   A unit defined as the amount of material containing 6.0221421×10^{23} (Avogadro's number) particles. (2.9)

mole fraction (χ_A)   The number of moles of a component in a mixture divided by the total number of moles in the mixture. (5.6)

mole fraction (χ_{solute})   A means of expressing solution concentration as the number of moles of solute per moles of solution. (12.5)

mole percent   A means of expressing solution concentration as the mole fraction multiplied by 100%. (12.5)

molecular compound   Compounds composed of two or more covalently bonded nonmetals. (3.4)

molecular element   Those elements that exist in nature with diatomic or polyatomic molecules as their basic unit. (3.4)

molecular equation   An equation showing the complete neutral formula for each compound in a reaction. (4.7)

molecular formula   A chemical formula that shows the actual number of atoms of each element in a molecule of a compound. (3.3)

molecular geometry   The geometrical arrangement of atoms in a molecule. (10.3)

molecular orbital theory   An advanced model of chemical bonding in which electrons reside in molecular orbitals delocalized over the entire molecule. In the simplest version, the molecular orbitals are simply linear combinations of atomic orbitals. (10.8)

molecular solids   Solids whose composite units are molecules; they generally have low melting points. (11.12)

molecularity   The number of reactant particles involved in an elementary step. (13.6)

molecule   Two or more atoms joined chemically in a specific geometrical arrangement. (1.1)

monodentate   Describes ligands that donate only one electron pair to the central metal. (24.3)

monoprotic acid   An acid that contains only one ionizable proton. (15.4)

monosaccharide   The simplest carbohydrates, with 3 to 8 carbon atoms and only one aldehyde or ketone group. (21.3)

nanotubes   Long, tubular structures consisting of interconnected C$_6$ rings. (22.5)

natural abundance   The relative percentage of a particular isotope in a naturally occurring sample with respect to other isotopes of the same element. (2.6)

Nernst equation   The equation relating the cell potential of an electrochemical cell to the standard cell potential and the reaction quotient; $E_{cell} = E^{\circ}_{cell} - \dfrac{0.0592 \text{ V}}{n} \log Q$. (18.6)

net ionic equation   An equation that shows only the species that actually change during the reaction. (4.7)

network covalent atomic solids   Atomic solids held together by covalent bonds; they have high melting points. (11.12)

neutral   The state of a solution where the concentrations of H_3O^+ and OH^- are equal. (15.4)

neutron   An electrically neutral subatomic particle found in the nucleus of an atom, with a mass almost equal to that of a proton. (2.5)

nickel-cadmium (NiCad) battery   A battery that consists of an anode composed of solid cadmium and a cathode composed of $NiO(OH)(s)$ in a KOH solution. (18.7)

nickel-metal-hydride (NiMH) battery   A battery that uses the same cathode reaction as the NiCad battery but a different anode reaction, the oxidation of hydrogens in a metal alloy. (18.7)

***nido*-boranes**   Boranes that have the formula B_nH_{n+4} and consist of a cage of boron atoms missing one corner. (22.4)

nitrogen narcosis   A physiological condition caused by an increased partial pressure of nitrogen, resulting in symptoms similar to those of intoxication. (5.6)

noble gases   The group 8A elements, which are largely unreactive (inert) due to their stable filled p orbitals. (2.7)

node   A point where the wave function (ψ), and therefore the probability density (ψ^2) and radial distribution function, all go through zero (7.6)

nonbonding atomic solids   Atomic solids held together by dispersion forces; they have low melting points. (11.12)

nonbonding orbital   An orbital whose electrons remain localized on an atom. (10.8)

nonelectrolyte   A compound that does not dissociate into ions when dissolved in water. (4.5)

nonmetal   A class of elements that tend to be poor conductors of heat and electricity and usually gain electrons during chemical reactions. (2.7)

nonvolatile   Not easily vaporized. (11.5)

normal boiling point   The temperature at which the vapor pressure of a liquid equals 1 atm. (11.5)

n-type semiconductor   A semiconductor that employs negatively charged electrons in the conduction band as the charge carriers. (11.13)

nuclear binding energy   The amount of energy that would be required to break apart the nucleus into its component nucleons. (19.8)

nuclear equation   An equation that represents nuclear processes such as radioactivity. (19.3)

nuclear fission   The splitting of the nucleus of an atom, resulting in a tremendous release of energy. (19.7)

nuclear fusion   The combination of two light nuclei to form a heavier one. (19.9)

nuclear theory   The theory that most of the atom's mass and all of its positive charge is contained in a small, dense nucleus. (2.5)

nucleotides   The individual units composing nucleic acids; each consists of a phosphate group, a sugar, and a nitrogenous base. (21.6)

nucleus   The very small, dense core of the atom that contains most of the atom's mass and all of its positive charge; it is composed of protons and neutrons. (2.5)

nuclide   A particular isotope of an atom. (19.3)

octahedral arrangement   The molecular geometry of seven atoms with 90° bond angles. (10.2)

octahedral hole   A space that exists in the middle of six atoms on two adjacent close-packed sheets of atoms in a crystal lattice. (23.4)

octet   A Lewis structure with eight dots, signifying a filled outer electron shell for s and p block elements. (9.3)

octet rule   The tendency for most bonded atoms to possess or share eight electron in their outer shell to obtain stable electron configurations and lower their potential energy. (9.3)

optical isomers   Two molecules that are nonsuperimposable mirror images of one another. (20.3, 24.4)

orbital   A probability distribution map, based on the quantum mechanical model of the atom, used to describe the likely position of an electron in an atom; also an allowed energy state for an electron. (7.5)

orbital diagram   A diagram which gives information similar to an electron configuration, but symbolizes an electron as an arrow in a box representing an orbital, with the arrow's direction denoting the electron's spin. (8.3)

ore   A rock that contains a high concentration of a specific mineral. (23.2)

organic chemistry   The study of carbon-based compounds. (20.1)

organic molecule   A molecule containing carbon combined with several other elements including hydrogen, nitrogen, oxygen, or sulfur. (20.1)

orthosilicates   Silicates in which tetrahedral SO_4^{4-} ions stand alone. (22.3)

osmosis   The flow of solvent from a solution of lower solute concentration to one of higher solute concentration. (12.7)

osmotic pressure   The pressure required to stop osmotic flow. (12.7)

Ostwald process   A industrial process used for commercial preparation of nitric acid. (22.6)

overall order   The sum of the orders of all reactants in a chemical reaction. (13.3)

oxidation   The loss of one or more electrons; also the gaining of oxygen or the loss of hydrogen. (4.9)

oxidation state (oxidation number)   A positive or negative whole number that represents the "charge" an atom in a compound would have if all shared electrons were assigned to the atom with a greater attraction for those electrons. (4.9)

oxidation-reduction (redox) reaction   Reactions in which electrons are transferred from one reactant to another and the oxidation states of certain atoms are changed. (4.9)

oxidizing agent   A substance that causes the oxidation of another substance; an oxidizing agent gains electrons and is reduced. (4.9)

oxyacid   An acid composed of hydrogen and an oxyanion. (3.6)

oxyanion   A polyatomic anion containing a nonmetal covalently bonded to one or more oxygen atoms. (3.5)

oxygen toxicity   A physiological condition caused by an increased level of oxygen in the blood, resulting in muscle twitching, tunnel vision, and convulsions. (5.6)

ozone O_3, an allotrope of oxygen that is a toxic blue diamagnetic gas with a strong odor. (22.7)

packing efficiency   The percentage of volume of a unit cell occupied by the atoms, assumed to be spherical. (11.11)

paramagnetic   The state of an atom or ion that contains unpaired electrons and is, therefore, attracted by an external magnetic field. (8.7, 10.8)

partial pressure (P_n)   The pressure due to any individual component in a gas mixture. (5.6)

parts by mass   A unit for expressing solution concentration as the mass of the solute divided by the mass of the solution multiplied by a multiplication factor. (12.5)

parts by volume   A unit for expressing solution concentration as the volume of the solute divided by the volume of the solution multiplied by a multiplication factor. (12.5)

parts per billion (ppb)   A unit for expressing solution concentration in parts by mass where the multiplication factor is 10^9. (12.5)

parts per million (ppm)   A unit for expressing solution concentration in parts by mass where the multiplication factor is 10^6. (12.5)

pascal (Pa)   The SI unit of pressure, defined as $1 \ N/m^2$. (5.2)

Pauli exclusion principle   The principle that no two electrons in an atom can have the same four quantum numbers. (8.3)

penetrating power   The ability of radiation to penetrate matter. (19.3)

penetration   The phenomenon of some higher-level atomic orbitals having significant amounts of probability within the space occupied by orbitals of lower energy level. For example, the 2s orbital penetrates into the 1s orbital. (8.3)

peptide bond   The bond that forms between the amine end of one amino acid and the carboxylic end of another. (21.4)

percent by mass   A unit for expressing solution concentration in parts by mass with a multiplication factor of 100%. (12.5)

percent ionic character   The ratio of a bond's actual dipole moment to the dipole moment it would have if the electron were transferred completely from one atom to the other, multiplied by 100%. (9.6)

percent ionization   The concentration of ionized acid in a solution divided by the initial concentration of acid multiplied by 100%. (15.6)

percent yield   The percentage of the theoretical yield of a chemical reaction that is actually produced; the ratio of the actual yield to the theoretical yield multiplied by 100%. (4.3)

periodic law   A law based on the observation that when the elements are arranged in order of increasing mass, certain sets of properties recur periodically. (2.7)

periodic property   A property of an element that is predictable based on an element's position in the periodic table. (8.1)

permanent dipole   A permanent separation of charge; a molecule with a permanent dipole always has a slightly negative charge at one end and a slightly positive charge at the other. (11.3)

pH   The negative log of the concentration of H_3O^+ in a solution; the pH scale is a compact way to specify the acidity of a solution. (15.4)

phase diagram   A map of the phase of a substance as a function of pressure and temperature. (11.8)

phenyl group   A benzene ring treated as a substituent. (20.7)

phosphine PH_3, a colorless, poisonous gas that smells like decaying fish and has an oxidation state of -3 for phosphorus. (22.6)

phospholipid   Compound similar in structure to a triglyceride but with one fatty acid replaced by a phosphate group. (21.2)

phosphorescence   The long-lived emission of light that sometimes follows the absorption of light by certain atoms and molecules. (19.2)

photoelectric effect   The observation that many metals emit electrons when light falls upon them. (7.2)

photon (quantum)   The smallest possible packet of electromagnetic radiation with an energy equal to $h\nu$. (7.2)

physical change   A change that alters only the state or appearance of a substance but not its chemical composition. (1.4)

physical property   A property that a substance displays without changing its chemical composition. (1.4)

pi (π) bond   The bond that forms between two p orbitals that overlap side to side. (10.7)

p-n junctions   Tiny areas in electronic circuits that have p-type semiconductors on one side and n-type on the other. (11.13)

polar covalent bond   A covalent bond between two atoms with significantly different electronegativities, resulting in an uneven distribution of electron density. (9.6)

polyatomic ion   An ion composed of two or more atoms. (3.5)

polydentate   Describes ligands that donate more than one electron pair to the central metal. (24.3)

polypeptide   A chain of amino acids joined together by peptide bonds. (21.4)

polyprotic acid   An acid that contains more than one ionizable proton and releases them sequentially. (4.8, 15.9)

polysaccharide   A long, chainlike molecule composed of many monosaccharide units bonded together. (21.3)

positron   The particle released in positron emission; equal in mass to an electron but opposite in charge. (19.3)

positron emission   The form of radioactive decay that occurs when an unstable nucleus emits a positron. (19.3)

positron emission tomography (PET)   A specialized imaging technique that employs positron-emitting nuclides, such as fluorine-18, as a radiotracer. (19.12)

potential difference   A measure of the difference in potential energy (usually in joules) per unit of charge (coulombs). (18.3)

potential energy   The energy associated with the position or composition of an object. (1.5, 6.1)

powder metallurgy   A process by which metallic components are made from powdered metal. (23.3)

precipitate   A solid, insoluble ionic compound that forms in, and separates from, a solution. (4.6)

precipitation reaction   A reaction in which a solid, insoluble product forms upon mixing two solutions. (4.6)

precision   A term that refers to how close a series of measurements are to one another or how reproducible they are. (1.7)

prefix multipliers   Multipliers that change the value of the unit by powers of 10. (1.6)

pressure   A measure of force exerted per unit area; in chemistry, most commonly the force exerted by gas molecules as they strike the surfaces around them. (5.1)

pressure-volume work   The work that occurs when a volume change takes place against an external pressure. (6.3)

primary structure   The sequence of amino acids in a protein chain. (21.5)

primary valence   The oxidation state on the central metal atom in a complex ion. (24.3)

principal level (shell)   The group of orbitals with the same value of n. (7.5)

principal quantum number (n)   An integer that specifies the overall size and energy of an orbital. The higher the quantum number n, the greater the average distance between the electron and the nucleus and the higher its energy. (7.5)

probability density   The probability (per unit volume) of finding the electron at a point in space as expressed by a three-dimensional plot of the wave function squared (ψ^2) (7.6)

products   The substances produced in a chemical reaction; they appear on the right-hand side of a chemical equation. (3.10)

proton   A positively charged subatomic particle found in the nucleus of an atom. (2.5)

p-type semiconductor   A semiconductor that employs positively charged "holes" in the valence band as the charge carriers. (11.13)

pure substance   A substance composed of only one type of atom or molecule. (1.3)

pyrometallurgy   A technique of extractive metallurgy in which heat is used to extract a metal from its mineral. (23.3)

pyrosilicates   Silicates in which two SO_4^{4-} tetrahedral ions share a corner. (22.3)

pyroxenes   Silicates in which SO_4^{4-} tetrahedral ions bond together to form chains. (22.3)

qualitative analysis   A systematic way to determine the ions present in an unknown solution. (16.7)

quantitative analysis   A systematic way to determine the amounts of substances in a solution or mixture. (16.7)

quantum number   One of four interrelated numbers that determine the shape and energy of orbitals, as specified by a solution of the Schrödinger equation. (7.5)

quantum-mechanical model   A model that explains the behavior of absolutely small particles such as electrons and photons. (7.1)

quartz   A silicate crystal which has a formula unit of SiO_2. (22.3)

quaternary structure   The way that subunits fit together in a multimeric protein. (21.5)

racemic mixture   An equimolar mixture of two optical isomers that does not rotate the plane of polarization of light at all. (20.3)

radio waves   The form of electromagnetic radiation with the longest wavelengths and smallest energy. (7.2)

radioactive   The state of those unstable atoms that emit subatomic particles or high-energy electromagnetic radiation. (19.1)

radioactivity   The emission of subatomic particles or high-energy electromagnetic radiation by the unstable nuclei of certain atoms. (2.5, 19.1)

radiocarbon dating   A form of radiometric dating based on the C-14 isotope. (19.6)

radiometric dating   A technique used to estimate the age of rocks, fossils, or artifacts that depends on the presence of radioactive isotopes and their predictable decay with time. (19.6)

radiotracer   A radioactive nuclide that has been attached to a compound or introduced into a mixture in order to track the movement of the compound or mixture within the body. (19.12)

random coils   Sections of a protein's secondary structure that have less regular patterns than α-helixes or β-pleated sheets. (21.5)

random error   Error that has equal probability of being too high or too low. (1.7)

Raoult's law   An equation used to determine the vapor pressure of a solution; $P_{soln} = X_{solv}P_{solv}^\circ$. (12.6)

rate constant (k)   A constant of proportionality in the rate law. (13.3)

rate law   A relationship between the rate of a reaction and the concentration of the reactants. (13.3)

rate-determining step   The step in a reaction mechanism that occurs much more slowly than any of the other steps. (13.6)

reactants   The starting substances of a chemical reaction; they appear on the left-hand side of a chemical equation. (3.10)

reaction intermediates   Species that are formed in one step of a reaction mechanism and consumed in another. (13.6)

reaction mechanism   A series of individual chemical steps by which an overall chemical reaction occurs. (13.6)

reaction order (n)   A value in the rate law that determines how the rate depends on the concentration of the reactants. (13.3)

reaction quotient (Q_c)   The ratio, at any point in the reaction, of the concentrations of the products of a reaction raised to their stoichiometric coefficients to the concentrations of the reactants raised to their stoichiometric coefficients. (14.7)

recrystallization   A technique used to purify solids in which the solid is put into hot solvent until the solution is saturated; when the solution cools, the purified solute comes out of solution. (12.4)

red phosphorus   An allotrope of phosphorus similar in structure to white phosphorus but with one of the bonds between two phosphorus atoms in the tetrahedron broken; red phosphorus is more stable than white. (22.6)

reducing agent   A substance that causes the reduction of another substance; a reducing agent loses electrons and is oxidized. (4.9)

reduction   The gaining of one or more electrons; also the gaining of hydrogen or the loss of oxygen. (4.9)

refine   To purify, particularly a metal. (23.3)

refining   A process in which the crude material is purified. (23.3)

rem   A unit of the dose of radiation exposure that stands for roentgen equivalent man, where a roentgen is defined as the amount of radiation that produces 2.58×10^{-4} C of charge per kg of air. (19.11)

resonance hybrid   The actual structure of a molecule that is intermediate between two or more resonance structures. (9.8)

resonance structures   Two or more valid Lewis structures that are shown with double-headed arrows between them to indicate that the actual structure of the molecule is intermediate between them. (9.8)

reversible   As applied to a reaction, the ability to proceed in either the forward or the reverse direction. (14.2)

reversible reaction   A reaction that achieves the theoretical limit with respect to free energy and will change direction upon an infinitesimally small change in a variable (such as temperature or pressure) related to the reaction. (17.7)

roasting   Heating that causes a chemical reaction between a furnace atmosphere and a mineral in order to process ores. (23.3)

salt   An ionic compound formed in a neutralization reaction by the replacement of an H^+ ion from the acid with a cation from the base. (4.8)

salt bridge   An inverted, U-shaped tube containing a strong electrolyte such as KNO_3 that connects the two half-cells, allowing a flow of ions that neutralizes the charge build-up. (18.3)

saturated fat   A triglyceride with no double bonds in the hydrocarbon chain; saturated fats tend to be solid at room temperature. (21.2)

saturated hydrocarbon   A hydrocarbon containing no double bonds in the carbon chain. (20.4)

saturated solution   A solution in which the dissolved solute is in dynamic equilibrium with any undissolved solute; any added solute will not dissolve. (12.4)

scientific law   A brief statement or equation that summarizes past observations and predicts future ones. (1.2)

scientific method   An approach to acquiring knowledge about the natural world that begins with observations and leads to the formation of testable hypotheses. (1.2)

scintillation counter   A device for the detection of radioactivity using a material that emits ultraviolet or visible light in response to excitation by energetic particles. (19.5)

second (s)   The SI standard unit of time, defined as the duration of 9,192,631,770 periods of the radiation emitted from a certain transition in a cesium-133 atom. (1.6)

second law of thermodynamics   A law stating that for any spontaneous process, the entropy of the universe increases ($\Delta S_{univ} > 0$). (17.3)

secondary structure   The regular periodic or repeating patterns in the arrangement of protein chains. (21.5)

secondary valence The number of molecules or ions directly bound to the metal atom in a complex ion; also called the *coordination number*. (24.3)

seesaw The molecular geometry of a molecule with trigonal bipyramidal electron geometry and one lone pair in an axial position. (10.3)

selective precipitation A process involving the addition of a reagent to a solution that forms a precipitate with one of the dissolved ions but not the others. (16.6)

semiconductor A material with intermediate electrical conductivity that can be changed and controlled. (2.7)

semipermeable membrane A membrane that selectively allows some substances to pass through but not others. (12.7)

shielding The effect on an electron of repulsion by electrons in lower-energy orbitals that screen it from the full effects of nuclear charge. (8.3)

sigma (σ) bond The resulting bond that forms between a combination of any two s, p, or hybridized orbitals that overlap end to end. (10.7)

significant figures (significant digits) In any reported measurement, the non-place-holding digits that indicate the precision of the measured quantity. (1.7)

silica A silicate crystal which has a formula unit of SiO_2, also called *quartz*. (22.3)

silicates Covalent atomic solids that contain silicon, oxygen, and various metal atoms. (22.3)

simple cubic A unit cell that consists of a cube with one atom at each corner. (11.11)

slag In pyrometallurgy, the waste liquid solution that is formed between the flux and gangue; usually a silicate material. (23.3)

smelting A form of roasting in which the product is liquefied, which aids in the separation. (23.3)

solid A state of matter in which atoms or molecules are packed close to one another in fixed locations with definite volume. (1.3)

solubility The amount of a substance that will dissolve in a given amount of solvent. (12.2)

solubility product constant (K_{sp}) The equilibrium expression for a chemical equation representing the dissolution of a slightly to moderately soluble ionic compound. (16.5)

soluble Able to dissolve to a significant extent, usually in water. (4.5)

solute The minority component of a solution. (4.4, 12.1)

solution A homogenous mixture of two substances. (4.4, 12.1)

solvent The majority component of a solution. (4.4, 12.1)

space-filling molecular model A representation of a molecule that shows how the atoms fill the space between them. (3.3)

specific heat capacity (C_s) The amount of heat required to raise the temperature of 1 g of a substance by 1 °C. (6.3)

spectator ion Ions in a complete ionic equation that do not participate in the reaction and therefore remain in solution. (4.7)

spin quantum number, m_s The fourth quantum number, which denotes the electron's spin as either $1/2$ (up arrow) or $-1/2$ (down arrow). (8.3)

spontaneous process A process that occurs without ongoing outside intervention. (17.2)

square planar The molecular geometry of a molecule with octahedral electron geometry and two lone pairs. (10.3)

square pyramidal The molecular geometry of a molecule with octahedral electron geometry and one lone pair. (10.3)

standard cell potential (standard emf) ($E°_{cell}$) The cell potential for a system in standard states (solute concentration of 1 M and gaseous reactant partial pressure of 1 atm). (18.3)

standard change in free energy ($\Delta G°_{rxn}$) The change in free energy for a process when all reactants and products are in their standard states. (17.7)

standard enthalpy change ($\Delta H°$) The change in enthalpy for a process when all reactants and products are in their standard states. (6.8)

standard enthalpy of formation ($\Delta H°_f$) The change in enthalpy when 1 mol of a compound forms from its constituent elements in their standard states. (6.8)

standard entropy change (ΔS_{rxn}) The change in entropy for a process when all reactants and products are in their standard states. (17.6)

standard entropy change for a reaction ($\Delta S°_{rxn}$) The change in entropy for a process in which all reactants and products are in their standard states. (17.6)

Standard Hydrogen Electrode (SHE) The half-cell consisting of an inert platinum electrode immersed in 1 M HCl with hydrogen gas at 1 atm bubbling through the solution; used as the standard of a cell potential of zero. (18.4)

standard molar entropy ($S°$) A measure of the energy dispersed into one mole of a substance at a particular temperature. (17.6)

standard state For a gas the standard state is the pure gas at a pressure of exactly 1 atm; for a liquid or solid the standard state is the pure substance in its most stable form at a pressure of 1 atm and the temperature of interest (often taken to be 25 °C); for a substance in solution the standard state is a concentration of exactly 1 M. (6.8)

standard temperature and pressure (STP) The conditions of $T = 0$ °C (273 K) and $P = 1$ atm; used primarily in reference to a gas. (5.5)

starch A polysaccharide that consists of glucose units bonded together by α-glycosidic linkages; the main energy storage medium for plants. (21.3)

state A classification of the form of matter as a solid, liquid, or gas. (1.3)

state function A function whose value depends only on the state of the system, not on how the system got to that state. (6.2)

stereoisomers Molecules in which the atoms are bonded in the same order, but have a different spatial arrangement. (20.3, 24.4)

steroid A lipid composed of four fused hydrocarbon rings. (21.2)

stock solution A highly concentrated form of a solution used in laboratories to make less concentrated solutions via dilution. (4.4)

stoichiometery The numerical relationships between amounts of reactants and products in a balanced chemical equation. (4.2)

strong acid An acid that completely ionizes in solution. (4.5, 15.4)

strong base A base that completely dissociates in solution. (15.7)

strong electrolyte A substance that completely dissociates into ions when dissolved in water. (4.5)

strong force Of the four fundamental forces of physics, the one that is the strongest but acts over the shortest distance; the strong force is responsible for holding the protons and neutrons together in the nucleus of an atom. (19.4)

strong-field complex A complex ion in which the crystal field splitting is large. (24.5)

structural formula A molecular formula that shows how the atoms in a molecule are connected or bonded to each other. (3.3, 20.3)

structural isomers   Molecules with the same molecular formula but different structures. (20.3, 24.4)

sublevel (subshell)   Those orbitals in the same principle level with the same value of n and l. (7.5)

sublimation   The phase transition from solid to gas. (11.6)

substitutional alloy   An alloy in which one metal atom substitutes for another in the crystal structure. (23.4)

substrate   The reactant molecule of a biochemical reaction that binds to an enzyme at the active site. (13.7)

supersaturated solution   An unstable solution in which more than the equilibrium amount of solute is dissolved. (12.4)

surface tension   The energy required to increase the surface area of a liquid by a unit amount; responsible for the tendency of liquids to minimize their surface area, giving rise to a membrane-like surface. (11.4)

surroundings   In thermodynamics, everything in the universe which exists outside the system under investigation. (6.1)

system   In thermodynamics, the portion of the universe which is singled out for investigation. (6.1)

systematic error   Error that tends towards being consistently either too high or too low. (1.7)

systematic name   An official name for a compound, based of well-established rules, that can be determined by examining its chemical structure. (3.5)

temperature   A measure of the average kinetic energy of the atoms or molecules that compose a sample of matter. (1.6)

termolecular   An elementary step of a reaction in which three particles collide and go on to form products. (13.6)

tertiary structure   The large-scale bends and folds produced by interactions between the R groups of amino acids that are separated by large distances in the linear sequence of a protein chain. (21.5)

tetrahedral geometry   The molecular geometry of five atoms with 109.5° bond angles. (10.2)

tetrahedral hole   A space that exists directly above the center point of three closest packed metal atoms in one plane, and a fourth metal located directly above the center point in the adjacent plane in a crystal lattice. (23.4)

theoretical yield   The greatest possible amount of product that can be made in a chemical reaction based on the amount of limiting reactant. (4.3)

theory   A proposed explanation for observations and laws based on well-established and tested hypotheses, that presents a model of the way nature works and predicts behavior beyond the observations and laws on which it was based. (1.2)

thermal energy   A type of kinetic energy associated with the temperature of an object, arising from the motion of individual atoms or molecules in the object; see also *heat*. (1.5, 6.1)

thermal equilibrium   The point at which there is no additional net transfer of heat between a system and its surroundings. (6.3)

thermochemistry   The study of the relationship between chemistry and energy. (6.1)

thermodynamics   The general study of energy and its interconversions. (6.2)

third law of thermodynamics   The law stating that the entropy of a perfect crystal at absolute zero (0 K) is zero. (17.6)

titration   A laboratory procedure in which a substance in a solution of known concentration is reacted with another substance in a solution of unknown concentration in order to determine the unknown concentration; see also *acid–base titration*. (4.8)

transition elements (transition metals)   Those elements found in the d block of the periodic table whose properties tend to be less predictable based simply on their position in the table. (2.7)

transmutation   The transformation of one element into another as a result of nuclear reactions. (19.10)

triglyceride   Triesters composed of glycerol with three fatty acids attached. (21.2)

trigonal bipyramidal   The molecular geometry of six atoms with 120° bond angles between the three equatorial electron groups and 90° bond angles between the two axial electron groups and the trigonal plane. (10.2)

trigonal planar geometry   The molecular geometry of four atoms with 120° bond angles in a plane. (10.2)

trigonal pyramidal   The molecular geometry of a molecule with tetrahedral electron geometry and one lone pair. (10.3)

triple bond   The bond that forms when three electron pairs are shared between two atoms. (9.5)

triple point   The unique set of conditions at which all three phases of a substance are equally stable and in equilibrium. (11.8)

triprotic acid   An acid that contains three ionizable protons. (15.4)

T-shaped   The molecular geometry of a molecule with trigonal bipyramidal electron geometry and two lone pairs in axial positions. (10.3)

two-phase region   The region between the two phases in a metal alloy phase diagram, where the amount of each phase depends upon the composition of the alloy. (23.4)

Tyndall effect   The scattering of light by a colloidal dispersion. (12.8)

ultraviolet (UV) radiation   Electromagnetic radiation with slightly smaller wavelengths than visible light. (7.2)

unimolecular   Describes a reaction that involves only one particle that goes on to form products. (13.6)

unit cell   The smallest divisible unit of a crystal that, when repeated in three dimensions, reproduces the entire crystal lattice. (11.11)

units   Standard quantities used to specify measurements. (1.6)

unsaturated fat   A triglyceride with one or more double bonds in the hydrocarbon chain; unsaturated fats tend to be liquid at room temperature. (21.2)

unsaturated hydrocarbon   A hydrocarbon that includes one or more double or triple bonds. (20.5)

unsaturated solution   A solution containing less than the equilibrium amount of solute; any added solute will dissolve until equilibrium is reached. (12.4)

valence bond theory   An advanced model of chemical bonding in which electrons reside in quantum-mechanical orbitals localized on individual atoms that are a hybridized blend of standard atomic orbitals; chemical bonds result from an overlap of these orbitals. (10.6)

valence electrons   Those electrons that are important in chemical bonding. For main-group elements, the valence electrons are those in the outermost principal energy level . (8.4)

valence shell electron pair repulsion (VSEPR) theory   A theory that allows prediction of the shapes of molecules based on the idea that electrons—either as lone pairs or as bonding pairs—repel one another. (10.2)

van der Waals equation   The extrapolation of the ideal gas law that considers the effects of intermolecular forces and particle volume in a nonideal gas: $P + a\left(\dfrac{n}{V}\right)^2 \times (V - nb) = nRT$ (5.9)

van der Waals radius (nonbonding atomic radius)   Defined as one-half the distance between the centers of adjacent, nonbonding atoms in a crystal. (8.6)

van't Hoff factor (i)   The ratio of moles of particles in a solution to moles of formula units dissolved. (12.7)

vapor pressure   The partial pressure of a vapor in dynamic equilibrium with its liquid. (5.6, 11.5)

vaporization   The phase transition from liquid to gas. (11.5)

viscosity   A measure of the resistance of a liquid to flow. (11.4)

visible light   Those frequencies of electromagnetic radiation that can be detected by the human eye. (7.2)

volatile   Tending to vaporize easily. (1.3, 11.5)

voltaic (galvanic) cell   An electrochemical cell which produces electrical current from a spontaneous chemical reaction. (18.3)

volume (V)   A measure of space. Any unit of length, when cubed (raised to the third power), becomes a unit of volume. (1.6)

washing soda   The hydrated crystal of sodium carbonate, $Na_2CO_3 \, 10 \, H_2O$. (22.5)

wave function (ψ)   A mathematical function that describes the wavelike nature of the electron. (7.5)

wavelength (λ)   The distance between adjacent crests of a wave. (7.2)

weak acid   An acid that does not completely ionize in water. (4.5, 15.4)

weak base   A base that only partially ionizes in water. (15.7)

weak electrolyte   A substance that does not completely ionize in water and only weakly conducts electricity in solution. (4.5)

weak-field complex   A complex ion in which the crystal field splitting is small. (24.5)

white phosphorus   An unstable allotrope of phosphorus consisting of P_4 molecules in a tetrahedral shape, with the phosphorus atoms at the corners of the tetrahedron. (22.6)

work (w)   The result of a force acting through a distance. (1.5, 6.1)

X-rays   Electromagnetic radiation with wavelengths slightly longer than those of that gamma rays; used to image bones and internal organs. (7.2)

X-ray diffraction   A powerful laboratory technique that allows for the determination of the arrangement of atoms in a crystal and the measuring of the distance between them. (11.10)

Credits

Index

A

A.O. Smith Company, 1033
Absolute scale (Kelvin scale), 17
Absolute zero, 195
  of entropy, 793
Absorption spectra, 294-95
Accuracy, 27-28
Acetaldehyde (ethanal), 120, 931, 933
Acetate, 95
Acetic acid, 118, 120, 153, 161, 664, 665, 671, 719,
    720, 934, 935
  acid ionization constant for, 672
  reaction between salicylic acid and, 936
Acetone, 120, 478, 523, 524, 931, 933
  boiling point of, 479
  heat of fusion of, 488
  heat of vaporization of, 479
Acetonitrile, 588-89
Acetylene, 118, 433, 918, 1002
  common uses of, 119
  representations of, 88, 119
Acetylide ion, 1002
Acetylsalicylic acid, 456, 677, 936
Acid(s), 152-53. See also Acid-base chemistry;
    Acid-base titration
  binary, 99
  dilution of, 146
  diprotic, 161, 670, 671, 698
    anions in weak, 701-2
    titration with strong base, 743-44
  ionization of, 152-53
  metals dissolved in, 98, 833
  naming, 98-99
  polyprotic, 161
  rusting promoted by, 855
  strong, 152-53
  weak, 153
Acid anhydride, 936
Acid-base chemistry, 160-65, 662-715. See also
    Buffers; pH
  acid-base properties of ions, 690-96
    anions as weak bases, 691-94
    cations as weak acids, 695-96
  acid ionization constant (K_a), 671-72
  acid rain, 706-7
    effects of, 707
    fossil fuels and, 706
    legislation on, 707
  acid strength, 669-71, 702-4
    strong acid, 669-70
    weak acids, 669, 670-72
  addition to buffer, 724-28
  Arrhenius model of, 161, 666-67
  autoionization of water, 672-74
  base solutions, 687-90
    hydroxide ion concentration and pH of,
      688-90
    strong bases, 687
    weak bases, 687-88
  Brønsted-Lowry model of, 667-68, 702, 704-5
  definitions of, 666-69

equations, 163
  net ionic, 162
  heartburn and, 663-64
  Lewis model of, 704-6
  molecular structure, 702-4
    binary acids, 703
    oxyacids, 703-4
  nature of, 664-66
  neutralization reactions, 936
  pOH and other p scales, 676-77
  polyprotic acids, 698-702
    acid ionization constants for, 698, 699
    concentration of anions for weak diprotic
      acid solution, 701-2
    dissociation of, 702-3
    ionization of, 698
    pH of, 698-701
  salt solutions as acidic, basic, or neutral, 696-98
Acid-base titration, 163-65, 733-47
  endpoint of, 744, 745
  indicators, 744-47
  of polyprotic acid, 743-44
  of strong acid with strong base, 734-37
    equivalence point, 734
    overall pH curve, 736
  titration curve or pH curve, 734
  of weak acid with strong base, 737-43
    equivalence point, 737
    overall pH curve, 741
Acid dissociation constant. See Acid ionization
    constant(s) (K_a)
Acidic solution, 673
Acid ionization constant(s) (K_a), 671-72, 693
  for polyprotic acids, 698, 699
  for weak acids, 678
Acidity of blood, 718
Acidosis, 718
Acid rain, 100, 136, 225, 226, 268, 706-7, 1008
  effects of, 707
  fossil fuels and, 706
  legislation on, 707
  sulfur dioxide and, 1016
Acid rebound, 691
Acid reflux, 663
Actinides (inner transition elements), 330
  electron configurations of, 332
Activated carbon (activated charcoal), 1000
Activated complex (transition state), 588-89
Activation energy (activation barrier), 587,
    588-89, 590-92
  Arrhenius plots of, 590-92
  catalysis and, 599
  enzymes and, 602-3
Active metals, reaction of alcohols with, 930
Active site, 363, 406, 602, 604
Actual yield, 138
Acute radiation damage, 892
Addition polymer, 939, 940
Addition reactions
  of aldehydes and ketones, 934
  of alkenes and alkynes, 923-24

Adduct, 705
Adenine, 473, 474, 975
Adhesive forces, 476-77
Adipic acid, 940
Aerosol, 557
Aerosol cans, 198-99
Air
  composition of dry, 204
  fractionation of, 1011
Air pollution, 225-26, 268
Alanine, 966, 968
Alaska, 247
Alchemists, 890
Alcohol(s), 119, 928-31
  boiling point of, 479
  elimination (dehydration) reactions of, 929
  functional group of, 927, 928
  general formula for, 119, 120
  heat of fusion of, 488
  heat of vaporization of, 479
  naming, 929
  reactions, 929-31
Alcoholic beverages, ethanol in, 929
Aldehydes, 931-34
  functional group of, 928
  general formula for, 120
  naming, 932
  reactions of, 933-34
Aldohexose, 962
Aldose, 962
Aldrin, 538
Aliphatic hydrocarbons, 906. See also Alkanes;
    Alkenes; Alkynes
Aliquot, 575
Alizarin, 747
Alkali metals. See Group 1A metals
    (alkali metals)
Alkaline batteries, 843
Alkaline earth metals. See Group 2A metals
    (alkaline earth metals)
Alkaloids, 666
Alkanes, 118, 906, 912-16
  boiling points of, 467
  geometric (cis-trans) isomerism, 923-24
  n-alkanes, 912-13
  naming, 914-16
  reactions of, 922-23
  viscosity of, 475
Alkenes, 118, 906, 917-21
  geometric (cis-trans) isomerism in, 920-21
  hydrogenation of double bonds within, 601-2
  naming, 919-20
  reactions of, 923-24
Alkyl groups, 914
Alkynes, 118, 906, 917-21
  geometric (cis-trans) isomerism in, 920-21
  naming, 919-20
  reactions of, 923
Allotropes, 795, 996
  phosphorus, 1005-6
  sulfur, 1014

Alloys, 1028, 1034-39
　copper, 1043, 1044
　defined, 1034
　interstitial, 1034, 1038-39
　with limited solubility, 1035-37
　nickel, 1044
　steel, 1041, 1042
　substitutional, 1034-35
　zinc, 1044
Alpha (α) decay, 867, 868-69, 871
Alpha (α) particles, 54, 55, 868
Altitude, 241
Aluminosilicates, 993
Aluminum, 58, 89
　charge of, 94
　density of, 20
　Hall process to obtain, 1032
　ionization energies of, 346, 347
　specific heat of, 247
Aluminum acetate, 697
Aluminum-based antacids, 691
Aluminum hydroxide, 763
Aluminum ion, 706
Aluminum nitrate, 696
Aluminum oxide crystals, 1051
Amines, 688, 938
　functional group of, 928
　reactions of, 938
Amino acids, 402, 456, 604, 664, 956, 966-70
　as building blocks of proteins, 966-68
　common, 968
　essential, 970
　genetic code identifying, 977-78
　intramolecular acid-base reaction in, 967
　peptide bonding between, 969-70
　protein synthesis and, 980
L-Amino acids, 967
Ammonia, 161, 649, 666, 667, 687, 688, 1005, 1006
　buffer containing, 728
　Haber-Bosch process for making, 1006
　Lewis acid-base model and, 705
　molecular geometry of, 410-11
　molecular representation of, 88
　nitrogen orbitals in, 428
　nitrogen-to-hydrogen mass ratio of, 49
　pH of, 675
　reaction between boron trifluoride and, 705
　reaction between silver ion and, 1057
　in water, Henry's law constants for, 533
Ammonia ligand, 1058
Ammonium, 95
　in gas-evolution reactions, 166
Ammonium bromide, 696
Ammonium chloride, 728
Ammonium dichromate, 1042
Ammonium ion, 758
Ammonium nitrate, 95, 526, 1008
Ammonium nitrite, 697
Ammonium salts, 938
Amorphous solid, 8, 464
Ampere (A), 825, 852
Amphiboles, 994, 995
Amphogel, 691
Amphotericity, 672-73
Amphoteric substances, 667
　metal hydroxides, solubility of, 761-63
Amplitude, 283-84
Amylopectin, 965
Amylose, 965

Anemia, sickle-cell, 972
Angle of reflection, 496-97
Angular momentum, orbital, 446, 447
Angular momentum quantum number, 301, 302, 306
Aniline, 925
Anion(s), 60, 94
　as conjugate base of acid, 691-92
　electrolysis of mixture of cations and, 849
　Lewis structure of, 367
　periodic table and, 66
　radii of, 342
　in salts, 696-97
　as weak bases, 691-94
　in weak diprotic acid solutions, 701-2
Anionic ligands, naming, 1060, 1061
Anode, 51, 824
　in batteries, 843, 844, 845, 846
　in electrolytic cell, 848
　in voltaic cell, 824-25
Anodic regions, 854
Antacids, 160, 664, 677, 691
Antarctica, ozone hole over, 103, 227, 600-601
Anthracene, 926
Anthracite, 1000
Antibodies, radioactively tagged, 866
Antibonding orbital, 440-41
　σ_{2p}^{*} 444
Antifluorite structure, 506
Antifreeze, 549, 717-18
　in frogs, 550-51
Antiparticle, 870
Apatite, 1005
Appalachian red spruce, 100
Appendicitis, diagnosing, 865-66
Apples, pH of, 675
Aqueous (aq), 98
Aqueous reactions, enthalpy of reaction measured for, 257
Aqueous solution(s), 144, 150-54, 521. See also
　　Acid-base chemistry; Buffers; Solubility
　　equilibria; Solution stoichiometry
　electrolysis in, 849-52
　　aqueous sodium chloride and overvoltage, 851
　electrolyte and nonelectrolyte, 151-53
　of ethylene glycol, 549
　heats of hydration and, 528-29
　hydroxide and hydronium ions in, 673
　of ionic compounds, 472
　solubility of ionic compounds, 154-55
Arc-melting, 1040
Arginine, 968
Argon, 58, 63, 356, 522
　in air, 204
　ionization energies of, 347
　molar volume under pressure, 222
　properties of, 355
　van der Waals constants for, 223
Aristotle, 47
Aromatic hydrocarbons, 906, 924-27
　naming, 925-27
　reactions of aromatic compounds, 927
Arrhenius, Svante, 161, 587, 666
Arrhenius acids and bases, 161, 666-67
Arrhenius equation, 587-94
　activation energy (activation barrier), 587, 588-89, 590-92
　Arrhenius plots, 590-92

　collision model of, 592-93
　exponential factor, 587, 589, 590
　frequency factor (pre-exponential factor), 587, 588, 589, 590-92
　rate constant and, 587
Arrhenius plots, 590-92
Arsenic, 63
Arsenic pentafluoride, 390, 434
Artificial sweeteners, 405-6
Asbestos, 994, 995
Ascorbic acid, 699
Asimov, Isaac, 319
Asparagine, 968
Aspartame (Nutrasweet), 405, 406
Aspartic acid, 967, 968
Aspirin, 456, 677, 936
Atmosphere, 225-27
　air pollution, 225-26, 268
　Earth's early, 1011
　hydrocarbons released into, 388-89
　nitrogen in, 1005
　oxygen in, 1011
　ozone depletion, 226-27
　ozone in, 1013
　structure of, 225
Atmosphere (atm), 189
Atmospheric pressure, 185-87
Atom(s), 3-5, 6, 44-81. See also Periodic table;
　　Quantum-mechanical model of atom
　diamagnetic, 339
　early ideas about, 47
　electron, discovery of, 51-53
　　cathode rays and, 51-52
　　Millikan's oil drop experiment, 53
　elements and, 47
　humans and, 51
　imaging of, 45-47
　interaction energy of, 423-24
　interactions among, 364-65
　modern atomic theory, 47-51
　　Dalton and, 51
　　law of conservation of mass and, 47-48
　　law of definite proportions and, 48-49, 50
　　law of multiple proportions and, 49-50
　molar mass, 69-75
　moving individual, 45-47
　nuclear theory of, 55
　paramagnetic, 339
　plum-pudding model of, 54, 55
　properties of matter and, 2, 3-4
　radioactive, 866
　size of. See Atomic radius/radii
　structure of, 54-55
　subatomic particles, 56-60
Atomic bomb, 884
Atomic elements, 89
Atomic machines, 46
Atomic mass, 67-69, 101
　calculation of, 68
　mass spectrometry and, 68-69
Atomic mass unit (amu), 56
Atomic number (Z), 57, 867
　atomic radius vs., 335
　beta decay and, 869
　electron capture and, 871
　first ionization energy vs., 344
　instability of all atomic nuclei beyond 83 (bismuth), 875
　positron emission and, 870

Atomic orbitals, 301-3
  atomic radius and, 335
  degenerate, 323, 324, 326
  electron configuration and, 321
  energy ordering of, for multielectron atoms, 325
  hybridized. *See* Hybridization
  periodic table and, 330-31
  shapes of, 306-12
Atomic radius/radii, 334, 335-38
  of alkali metals, 352
  atomic number vs., 335
  effective nuclear charge and, 336-37
  electronegativity and, 378
  of halogens, 353
  of main-group elements, 991
  of noble gases, 355
  of nonmetals, 991
  period trends in, 334-38
  transition elements and, 337-38
  of transition metals, 1054-55
Atomic solids, 504, 506-7
Atomic spectroscopy
  Bohr model and, 292-95
  explanation of, 303-6
Atomic theory, 6, 7
Atomos, 47
Atto prefix, 19
Attractive strong force among nucleons, 872
Aufbau principle, 326
Auto industry, products made by powder metallurgy in, 1033
Autoionization of water, 672-74, 677
Average rate of reaction, 572-73
Avogadro, Amedeo, 70, 197
Avogadro's law, 197, 198, 217
  kinetic molecular theory and, 215
Avogadro's number, 70, 75
Axial positions, 409
Azimuthal quantum number, 301, 302

B

Bacon, Francis, 47
Bacterial infection, ulcers from, 677
Baking powder, 1004
Baking soda (sodium bicarbonate), 91, 374, 666, 690-91, 1004
  reaction between hydrochloric acid and, 165-66
Balancing equations, 114-17
  oxidation-reduction equations, 820-23
  procedure for, 115-16
Ball-and-stick models, 87-88
Band gap, 508-9
Band theory, 508-9, 1028
Bar codes, 294
Barium
  charge of, 94
  emission spectrum of, 292, 293
  flame tests for, 295
Barium chloride hexahydrate, 96
Barium fluoride, 747
Barium hydroxide, 161
Barium sulfate, 374, 747
Barometer, 188
Base(s), 160. *See also* **Acid-base chemistry;**
  Acid-base titration
  Arrhenius definition of, 161, 666-67
  complementary, 975, 976, 978-79
  in nucleic acids, 975

organic, 473-74
  purine, 975, 976
  pyrimidine, 975, 976
Base ionization constant (K_b), 687, 693
Base-pairing in DNA, 976
Basic solution, 673
Battery(ies), 842-46
  dry-cell, 842-43
  energy loss in, 804
  lead-acid storage, 843
  rechargeable, 776, 843-45
Bauxite, 1032
Bayer process, 1032
Becquerel, Antoine-Henri, 54, 866-67
Beers, pH of, 675
Bent geometry, 411, 414, 417
Benzaldehyde, 932
Benzene, 451
  disubstituted, 926
  freezing point depression and boiling point elevation constants for, 550
  molecular representation of, 88
  monosubstituted, 925
  ring structure of, 924-25
  substitution reactions of, 927
Benzoic acid, 672
Berkelium, 58
Beryllium, 58
  charge of, 94
  effective nuclear charge for, 337
  electron configuration for, 326
  incomplete octets of, 407
  ionization energy of, 346
  Lewis structure of, 366
  MO diagram for, 326, 443
Beryllium aluminum silicate crystals, 1051
Beta (β) decay, 867, 869-70, 871
Beta (β) particles, 54, 869
β-Pleated sheet, 971, 973
Bicarbonate ion, 664, 688, 690, 718
Bicarbonates, 95
  in gas-evolution reactions, 166
Bidentate ligands, 1057, 1059
Bimolecular elementary step, 594, 595
Binary acids, 99, 703
Binary compounds, 94
Binary phase diagram, 1035
  chromium and vanadium, 1035, 1036
  copper and nickel, 1035
Binding energy per nucleon, 888-89
Binnig, Gerd, 45-46
Bioamplification, 538
Biochemistry, 954-87
  carbohydrates, 961-65
    complex, 964-65
    simple, 961-64
    structure of, 961
  defined, 956
  diabetes and the synthesis of human insulin, 955-56
  lipids, 956-61
    fats and oils, 958-60
    fatty acids, 956-57
  proteins, 956, 966-81
    amino acids as building blocks of, 956, 966-68
    nucleic acids as blueprints for, 974-78
    peptide bonding between amino acids, 969-70

structure of, 970-74
  synthesis of, 978-81
Biological effectiveness factor (RBE), 893
Biological systems
  entropy and, 788
  nonspontaneous reactions coupled with highly spontaneous ones in, 802
Biomolecules, 1074-76
Bisulfate (hydrogen sulfate), 95
Bisulfite (hydrogen sulfite), 95
  in gas-evolution reactions, 166
Bituminous coal, 1000
Black holes, 42
Black phosphorus, 1006
Bleached hair, 173
Blood, human
  acidity of, 718
  buffer effectiveness in, 732
  pH of, 675, 718, 732
Blood pressure, 191
Body-centered cubic unit cell, 498, 499-500
Bohr, Neils, 280, 281, 293-94
Bohr model, 301, 308
  emission spectra and, 293
Boiling, 11
Boiling point(s)
  of *n*-alkanes, 467
  defined, 482
  dipole moment and, 468
  of group 4A and 6A compounds, 471
  of halogens, 353
  of hydrides, 494
  of molecular compounds, 376
  of noble gases, 355, 466, 467
  normal, 482
  temperature dependence of, 482-83
Boiling point elevation, 549-50, 551-52, 554
Boltzmann, Ludwig, 780
Boltzmann constant, 780
Bomb calorimeter, 251-52
Bond(s), 84-86, 362-459
  AIDS drugs and models of, 363-64
  covalent. *See* Covalent bond(s)
  double. *See* Double bond(s)
  electronegativity difference and, 379-80
  electron sea model of, 366, 397, 506, 508, 1028
  formation of, 365
  ionic. *See* Ionic bond(s)
  Lewis theory, 362, 364, 366-403, 430
    bond polarity and, 377, 379-81
    bond types under, 367-69, 430
    of covalent bonding, 374-76
    electronegativity and, 377-78
    formal charge and, 386-87
    of ionic bonding, 367-72
    of molecular compounds, 382-83
    octet rule exceptions, 387-91
    of polyatomic ions, 382, 383-84
    resonance and, 384-85
    valence electrons represented with dots, 366-67
  metallic, 365, 366, 506
  in metals, 397
  molecular orbital theory, 362, 404, 426, 438-51, 508
    linear combination of atomic orbitals (LCAO), 439-43, 449
    period two homonuclear diatomic molecules, 443-48

Bond(s), *(continued)*
   polyatomic molecules, 450-51
   second-period heteronuclear diatomic
      molecules, 448-50
   trial mathematical functions in, 439
peptide, 604, 969, 980
   between amino acids, 969-70
pi (π) 430
polar, 418-19, 703
rotation about, 430-31
sigma (σ), 430
single, 87, 374-75, 393
strength of, 703
triple. *See* Triple bond(s)
valence bond theory, 362, 404, 423-38
   hybridization of atomic orbitals, 425-38
   summarizing main concepts of, 424
valence shell electron pair repulsion (VSEPR)
      theory, 404, 406-22, 1019
   bent geometry, 411, 414, 417
   linear geometry, 407, 408, 413, 414, 415, 417
   lone pairs effect, 410-14
   molecular shape and polarity, 418-22
   octahedral geometry, 413, 415, 417, 435
   predicting molecular geometries with,
      415-18
   seesaw geometry, 412, 415, 417
   square planar geometry, 413, 415, 417
   square pyramidal geometry, 413, 415
   summary of, 414
   tetrahedral geometry, 408, 411, 414, 417,
      426-28
   trigonal bipyramidal geometry, 409-10, 413,
      415, 417, 434
   trigonal planar geometry, 407, 408, 414,
      417, 428
   trigonal pyramidal geometry, 410, 411,
      414, 417
   T-shaped geometry, 412, 415
writing hybridization and bonding schemes,
      435-39
Bond energy(ies), 424
   average, 392-95
   bond type and, 393
   defined, 392
   enthalpy changes of reaction estimated from,
      391-95
Bonding
   of carbon, 904-5
      ability to form double and triple bonds,
         904-5
      tendency to form four covalent bonds, 904
   in coordination compounds, 1066-73
      color of complex ions and crystal field
         strength, 1068-70
      crystal field theory, 1067-73
      magnetic properties, 1071-72
      octahedral complexes, 1067-72
      tetrahedral and square planar complexes,
         1072-73
      valence bond theory, 1066-67
   ionization energies and, 348
   peptide bonding between amino acids, 969-70
Bonding atomic radius (covalent radius), 334
Bonding orbital, 440-41
   π_{2p}, 444, 445
   σ_{2p}, 444, 445
Bonding pair, 374-75
Bonding theories of metals, 1028

Bond length
   average, 395-96
   equilibrium, 424
Bond order, 441-42, 446
Bone density, 21
Bone scan, 894
Boranes, 998
Borax, 996
Born-Haber cycle, 369-71
Boron, 996-98
   boron-halogen compounds, 997
   boron-hydrogen compounds, 998
   boron-oxygen compounds, 997
   electron configuration for, 326
   elemental, 996-97
   incomplete octet formation by, 389-90
   ionization energy of, 346
   Lewis structure of, 366
   orbital diagram for, 326, 444-46
   sources of, 996
   uses for, 997
Boron nitride, 989
Boron nitride nanotubes, 989-90
Boron trifluoride, 705
Boyle, Robert, 47, 192
Boyle's law, 192-94, 198, 217
   kinetic molecular theory and, 215
Bragg's law, 497
Brand, Henning, 1005
Brass, 1044
Breathalyzer, fuel-cell, 846
Bridging hydrogens, 382
Bromcresol green, 747
Bromcresol purple, 747
Bromide, 94
Bromine, 63, 65
   properties of, 353, 1017
   reaction between fluorine and, 354
   reaction between potassium and, 593
Bromine pentafluoride, 413
Bromobenzene, 925
1-Bromo-2-chlorobenzene, 926
Bromphenol blue, 747
Bromthymol blue, 747
Brønsted-Lowry acids and bases, 667-68, 702,
      704-5
Bronze, 1043
Brownian motion, 557
Buckminsterfullerene, 1001
Buckyballs (fullerenes), 1001, 1002
Buffers, 718-33
   acid-base titration, 733-47
      endpoint of, 744, 745
      indicators, 744-47
      of polyprotic acid, 743-44
      of strong acid with strong base, 734-37
      titration curve or pH curve, 734
      of weak acid with strong base, 737-43
   action of, 725
   adding acid or base to, 724-28
   containing a base and its conjugate acid, 728-29
   effectiveness of, 729-33
      absolute concentrations of acid and
         conjugate base and, 730-31
      capacity, 732-33
      in human blood, 732
      range, 731-32
      relative amounts of acid and base and,
         729-30

   formation of, 719
   importance of, 716
   pH of, 720-28
      equilibrium calculation for, 724
      equilibrium calculation of changes in, 725-26
      Henderson-Hasselbalch equation for, 722-25
      stoichiometry calculation of changes in,
         724-25, 726
Burning, 12
Bush, George H.W., 227
Butanal, 932
Butane, 906, 907
   condensed structural formula for, 907-8
n-**Butane,** 912
   common uses of, 119
   molecular formula for, 119
   space-filling model of, 119
   structural formula for, 119
Butanoic acid, 934
1-Butanol, 928
2-Butanol, 933
Butanone, 931
2-Butanone, 933
1-Butene, 917
Butyl substituent, 914
1-Butyne, 918
Butyric acid, 957

C

Cadmium, in batteries, 844
Caffeine, 456, 487
Calamine (zinc silicate), 1044
Calcination, 1030, 1032
Calcite, 92
Calcium
   charge of, 94
   reaction between water and, 820
Calcium acetate, 696
Calcium-based antacids, 691
Calcium borate, 996
Calcium carbide, 1002
Calcium carbonate, 91, 92, 155, 749, 752
   solubility product constant for, 747
Calcium fluoride, 505, 506, 747, 750, 751, 753
   solubility product constant for, 747
Calcium hydroxide, 161, 687
   solubility product constant for, 747
Calcium nitrate, 696
Calcium oxide, 1002
Calcium phosphate, 1010
Calcium sulfate, 374
   solubility product constant for, 747
Calcium sulfate hemihydrate, 96
Calculations, significant figures in, 25-27
Californium, 891
calorie (cal), 239-40
Calorie (Cal), 240
Calorimetry, 251
   bomb, 251-52
   coffee-cup, 257-59
   constant-pressure, 257-59
   constant-volume, 251-53, 259
Cancer(s)
   cisplatin as anticancer agent, 1076
   from radiation, 354, 892
   radiotherapy for, 286, 895
   radiotracers to locate, 894
   skin, 286
   thyroid, 354-55

Capillary action, 476-77
Capric acid, 957
Carbaloy, 1043
Carbides, 1002-3
Carbohydrates, 961-65
  complex, 964-65
  simple, 961-64
  structure of, 961
Carbon, 58, 63, 999-1002. *See also* Hydrocarbons;
    Organic chemistry
  ability to form double and triple bonds, 904-5
  chemistry of life and, 67
  electron configuration for, 326
  elemental, 999
  hybridization in, 426-27
  Lewis structure of, 366
  molar entropies of allotropes of, 795
  orbital diagram for, 326, 446
  organic compounds, 117-18
  reactions of
    with oxygen, 242
    with sulfur, 171
    with water, 260
  tendency to catenate, 905
  tendency to form four covalent bonds, 904
  uniqueness of, 902, 904-5
α-Carbon, 966, 967
Carbon-14 dating, 879
Carbonate(s), 95, 707, 1004
  in gas-evolution reactions, 166
  solubilities of, 752
Carbonate ion, 688, 732
  solubility and, 155
Carbon black, 1000
Carbon dioxide, 4, 49-50, 1003
  atmospheric, 204, 268
  chemical formula for, 86
  formula mass of, 101
  from fossil fuel combustion, 133-35, 268
  as greenhouse gas, 131, 132
  molar mass of, 102
  molecular geometry of, 407, 419
  phase diagrams of, 493
  reaction between water and, 705-6, 1076
  solid (dry ice), 277, 487, 505, 1003
    sublimation of, 12
  solubility of, 1004
  supercritical, 487
  van der Waals constants for, 223
  from volcanoes, 132, 135
  in water, Henry's law constants for, 533
Carbonic acid, 664, 671, 705-6, 718, 732, 1004
  ionization constants for, 699
Carbonic anhydrase, 732, 1076
Carbonization, 1000
Carbon monoxide, 3-4, 22, 49-50, 226, 600, 1000,
    1003
  reaction between hydroxyl radical and, 388
  reaction between nitrogen dioxide and, 595-96
Carbon monoxide ligand, 1058
Carbon nanotubes, 989, 1001-2
Carbon oxides, 1003
Carbon tetrachloride, 524
  chemical formula for, 86
  freezing point depression and boiling point
    elevation constants for, 550
  van der Waals constants for, 223
Carbonyl chloride (phosgene), 402, 1003
Carbonyl group in aldehydes and ketones, 931

charge density plots of, 934
Carbonyl sulfide, 1003
Carboxylic acids, 665, 934-37
  functional group of, 928
  general formula for, 120
  naming, 935
  reactions of, 936-37
Carlsbad Caverns National Park, 752
Carnotite, 1029
Cars, hybrid, 240
Carvone, 933
Cassiterite, 1029
Catalysis, 599-604
  enzymes, 602-4
  homogeneous and heterogeneous, 600-602
Catalyst, 227, 599
Catalytic converter, 599-600
Catalytic destruction of ozone, 227, 599
Cataracts, 286
Catenation of carbon, 905
Cathode, 824
  in electrolytic cell, 848
  in voltaic cell, 824-25
Cathode rays, 51-52
Cathode ray tube, 51-52
Cathode reaction in batteries, 843, 844, 845, 846
Cathodic regions, 854
Cation(s), 60, 85
  as conjugate acids of weak bases, 695
  as counterions of strong bases, 695
  electrolysis of mixture of anions and, 849
  electron configuration and, 333, 339
  as Lewis acids, 706
  metal, 695
  periodic table and, 66
  radii of, 341
  in salts, 696-97
  as weak acids, 695-96
Cell diagram, 826-27
Cell potential (E_{cell}) or cell emf, 826
  concentration and, 837-42
Cells, taste, 406
Cellular fluids, 521
Cellulose, 964, 965
Celsius (°C) scale, 17, 18
Centipoise (cP), 475
Centi prefix, 19
Cesium, 6
  charge of, 94
  properties of, 352
Cesium chloride, 505
CFCs, 103, 226-27, 580, 601, 1013
Chadwick, James, 55
Chain reaction in fission of uranium-235, 884
Chalcopyrite, 1043
  extracting copper from, 1032
Charcoal, 8, 1000
  density of, 20
Charles, J.A.C., 195
Charles's law, 195-96, 198, 214
  kinetic molecular theory and, 215
Chelate, 1059
Chelating agents, 1059, 1073
Chemical analysis, coordination compounds used
    in, 1073
Chemical bonds. *See* Bond(s); Bonding
Chemical changes, 11-13. *See also* Reaction(s)
Chemical energy, 15, 238, 239
  transformation of, 239

Chemical equations. *See* Equation(s)
Chemical formula(s), 86-87
  composition of compounds from, 106-9
  conversion factors from, 106-9
  determining from experimental data, 109-14
    combustion analysis, 112-14
    for compounds, 110-12
  elemental composition and, 109
  for ionic compounds, 92
  from mass percent composition, 109
Chemical gradient, 319
Chemical kinetics. *See* Reaction rate(s)
Chemical potential, 778
  Gibbs free energy, 788-92
Chemical property, 11
Chemical reactions. *See* Reaction(s)
Chemical symbol, 57-58
Chemistry, defined, 5
Chernobyl nuclear accident (1986), 354, 886
Cherries, pH of, 675
Chiral, 1065
Chiral complex, 1065
Chiral environment, chemical behavior in, 911
Chiral molecule, 910
Chlorate, 95
Chlordane, 538
Chloride ion, 1058
Chloride minerals, 1029
Chlorides, 94
  insoluble, 758
Chlorine, 58, 65, 379
  catalytic ozone destruction and, 599, 600-601
  dipole moment of, 380
  electron affinity of, 348
  electron configuration of, 331, 342
  electron configuration of anion, 342
  elemental, 1019
  ionization energies of, 344, 347
  Lewis structure of, 375
  mass spectrum of, 69
  properties of, 84, 353, 1017
  reactions of
    with chloromethane, 922
    with ethene, 923
    with hydrogen, 168
    with ozone, 227
    with potassium, 367
    with sodium, 47-48, 168, 352, 368
  van der Waals constants for, 223
Chlorine dioxide, 1021
Chlorine oxides, 1021
Chlorite, 95
Chlorobenzene, 925
Chloroethane, 922, 923
Chloroethene, 939
1-Chloro-3-ethylbenzene, 926
Chlorofluorocarbons (CFCs), 103, 226-27, 580,
    601, 1013
Chloroform, 523
  freezing point depression and boiling point
    elevation constants for, 550
Chloromethane, 922
  reaction between chlorine and, 922
Chlorophyll, 1075
1-Chloropropane, 923
2-Chloropropane, 923
Chlorous acid, 672
Cholesterol, 961
Chromate, 95

Chromate compounds, 1041
Chromate ion, 1041
Chromite, 1040, 1041
Chromium, 63
    cations formed by, 95
    chromium-vanadium phase diagram, 1035, 1036
    colors of compounds of, 1040
    functions in human body, 1074
    nickel-chromium phase diagram, 1036-37
    oxidation states of, 1041
    sources, properties, and products of, 1040-42
Chromosomes, 978, 980
    in genetic structure, 977
Chymotrypsin in digestion, 604
Cinnabar, 1029
Cinnamaldehyde, 117, 932
Cis fats, 959
Cisplatin, 1076
Cis-trans (geometric) isomerism, 431, 921, 1062-64
    in alkenes, 920-21
Cities, air pollution and, 225, 226
Citric acid, 664, 665, 935
    ionization constants for, 699
Citrus fruits, 677
Classical physics, 299
Clausius, Rudolf, 775
Clausius-Clapeyron equation, 483-86
Claus process, 1014
Clean Air Act, 22, 100, 143, 226, 707
Closest-packed crystal structures, 502-4, 506
    cubic, 503-4
    hexagonal, 502-3
Closo-Boranes, 998
Clostridium botulinum, 1009
Clouds, polar stratospheric (PSCs), 601
Club soda, 521, 531
Coal, 268, 706, 1000
    composition of main types of, 1000
Cobalt
    cations formed by, 95
    sources, properties, and products of, 1042-43
Cobalt(II) chloride hexahydrate, 96
Cobaltite, 1042
Codon(s), 977, 980
    gene as sequence of, 978
    in genetic structure, 977
Coffee-cup calorimeter, 257
Cohesive forces, 476-77
Coke, 1000
    reaction between silicon oxide and, 1003
    reaction of calcium oxide with, 1002
Cold pack, chemical, 254
Collagen, 971
Colligative properties, 549
    of ionic solutions, 554-56
    medical solutions and, 555-56
Collision(s)
    elastic, 214
    inelastic, 214
    intermolecular forces and, 223
    mean free path, 220
Collision frequency, 592
Collision model, 592-93
Colloids (colloidal dispersion), 557-59
Color, 284
    absorption of light energy and, 1052, 1068-69
    complementary, 1069

of complex ions, crystal field strength and, 1068-70
    of gemstones, 1051-52
Coloring agents, 1074
Color wheel, 1069
Columbite, 1029
Combustion, 5, 114, 1011
    bomb calorimetry for, 251-52
    conservation of energy in, 775-76
    empirical formula from analysis of, 112-14
    of fossil fuel. See Fossil fuel combustion
    hydrocarbon, 922
    as redox reaction, 173-74
Common ion effect, 720
Common names, 93, 97
Complementary base pairing, 474
Complementary bases, 975, 976, 978-79
Complementary color, 1069
Complementary properties, 298, 301
Complete ionic equations, 159, 160
Complete protein, 970
Complex carbohydrates, 964-65
Complex ion(s), 759, 1056
    color of, crystal field strength and, 1068-70
    common geometries of, 1059-60
    coordination compound and, 1056, 1057
    naming, 1060
    valence bond model hybridization schemes in, 1067
Complex ion equilibria, 759-63
    effect on solubility, 761
    formation constant (K_f), 759
    solubility of amphoteric metal hydroxides, 761-63
Compound(s), 9, 10
    atomic-level view of, 89-92
    binary, 94
    classification of, 89
    composition of, 103-9
        from chemical formulas, 106-9
        mass percent determination of, 103-6
    coordination. See Coordination compound(s)
    formula mass for, 101-3
    inorganic, 117, 905
    insoluble, 154
    interhalogen, 354
    ionic. See Ionic compound(s)
    mixtures vs., 84
    molar mass of, 101-3
    mole concept for, 101-3
    molecular. See Molecular compound(s)
    molecular formulas for, 110-12
    organic. See Organic compound(s)
    properties of, 84
    Proust's observations on, 48
    representing, 86-88
        chemical formulas, 86-87
        molecular models, 87-88
    soluble, 154
    specific heat of, 247
    standard enthalpy of formation for, 262-63
    undergoing gas-evolution reactions, 166
    volatile organic (VOCs), 226
Compressibility of gases, 9
Concentrated solution, 144, 535, 542
Concentration(s), 144-45
    cell potential and, 837-42
    at equilibrium. See Equilibrium constant (K)
    from equilibrium constant, 636-45

given all but one of equilibrium concentrations of reactants and products, 636-37
    given initial concentrations or pressures, 637-41
    simplifying approximations, 641-45
    equilibrium constant in terms of, 622
    finding equilibrium constant from, 630-33
    of ideal gas, 627
    Le Châtelier's principle on change in, 646-48
    reaction rate and. See Rate law
    time and. See Integrated rate law
Concentration cells, 840-42
    Cu/Cu^{2+}, 840-41
    in human nerve cells, 842
Conceptual plan, 29, 30, 34
Condensation, 477-78, 480
    of amines, 938
    of carboxylic acids, 936
    entropy of surroundings increased by, 785-86
Condensation polymers, 940, 941
Condensed phase(s), 460, 465-74. See also Liquid(s); Solid(s)
    intermolecular forces and, 462
Conduction band, 508-9
Conductivity
    of copper, 1043
    of ionic compounds, 373
    of semiconductors, 509
Conductors, 508
    metals as, 1028, 1029
Cones, 432
Conformation of protein, 970-74
Coniine, 666
Conjugate acid-base pair, 668-69, 670-71
    cations in, 695
    strength of, 692
Conservation
    of energy, 14, 237, 238-39, 240, 775-76. See also First law of thermodynamics
    of mass, 5, 6, 47-48
Constant-pressure calorimetry, 257-59
Constant-volume calorimetry, 251-53, 259
Constructive interference, 287, 288, 440, 496
Consumer products, radiation exposure from, 893
Contact process, 1017
Contaminants, 495
Contractile and motile proteins, 966
Controlled disequilibrium, 624
Conversion factor(s), 29
    coefficients of equations as, 211
    density as, 32-33
    from formulas, 106-9
    mass percent composition as, 105-6
Coordinate covalent bond, 424, 1057, 1066-67
Coordination compound(s), 1050, 1052, 1056-76
    applications of, 1073-76
    bonding in, 1066-73
        color of complex ions and crystal field strength, 1068-70
        crystal field theory, 1067-73
        magnetic properties, 1071-72
        octahedral complexes, 1067-72
        tetrahedral and square planar complexes, 1072-73
        valence bond theory, 1066-67
    complex ion and, 1056, 1057
    defined, 1056

naming, 1060-61
structure and isomerization in, 1062-66
stereoisomerism, 1062-66
structural isomerism, 1062
Coordination isomers, 1062
Coordination numbers, 498, 499, 500, 1056, 1059
common geometries of complex ions and, 1059-60
Copernicus, Nicolaus, 47
Copolymers, 940
Copper, 63, 397
cations formed by, 95
crystal structure of, 1034
density of, 20
functions in human body, 1074
nickel and copper alloy, 1034, 1035
refinement of, 1032-33
sources, properties, and products of, 1043-44
specific heat of, 247
Copper electrolysis cell, 1033
Copper(II) sulfate pentahydrate, 96
Copper(II) sulfide, 747
Copper ion, 823-26
Copper plating, 852-53
Core electrons, 329, 337
Corrosion, 854-55
nickel's resistance to, 1044
preventing, 855
zinc as anticorrosion coating, 1044, 1045
Coulomb's law, 364, 465
Counterions, 695, 1056, 1057
Covalent bond(s), 84, 85-86, 365, 366, 397
carbon's tendency to form four, 904
coordinate, 424, 1057, 1066-67
directionality of, 376
double, 375
Lewis theory of, 367, 374-76
nonpolar, 379
polar, 377
shapes of atomic orbitals and, 306
single, 374-75
triple, 375
Covalent carbides, 1003
Covalent radius (bonding atomic radius), 334
o-Cresolphthalein, 747
Crick, Francis H.C., 83, 473, 497, 954, 955, 978
Critical mass, 884
Critical point, 486-87, 491, 492
Critical pressure, 486
Critical temperature, 486
Cryogenic liquids, 355
Crystal field, 1051-52
Crystal field splitting energy, 1068, 1069-70, 1071
Crystal field theory, 1052, 1067-73
basic principles of, 1067-68
color of complex ions and crystal field strength, 1068-70
magnetic properties of transition metal complex and, 1071-72
Crystalline lattice, 85, 496, 498
Crystalline solid(s), 8, 464, 495-509
band theory of, 508-9
fundamental types, 504-7
atomic solids, 504, 506-7
ionic solids, 504, 505
molecular solids, 504, 505
structures of, 495-97, 502-4
closest-packed, 502-4, 506

unit cells, 498-501
body-centered cubic, 498, 499-500
for closest-packed structures, 503-4
face-centered cubic, 498, 500, 503-4
for ionic solids, 505-6
simple cubic, 498-99, 502
X-ray crystallography of, 495-97
Crystallography, X-ray, 363, 495-97
Crystal structure of metal, 1033-34
Crystal violet, 747
Cubic closest packing, 503-4
Cubic measure, 19
Cubic unit cells, 498-500
body-centered, 498, 499-500
face-centered, 498, 500, 503-4
simple, 498-99, 502
Curie, Marie Sklodowska, 54, 58, 867, 890
Curie, Pierre, 867
Curie (Ci), 892
Curium, 58
Current, tunneling, 45-46
Cyanide, 95
Cyanide ion, 1058
Cycles per second (cycle/s), 284
Cyclohexane, 118, 905
Cyclooctasulfur, 1014
Cyclotron, 890, 891
Cysteine, 968
Cytochrome C, 1074
Cytosine, 473, 474, 975

D

Dalton, John, 6, 7, 45, 46, 47, 49, 58, 67
atomic theory of, 51
Dalton's law of partial pressures, 204
kinetic molecular theory and, 215
Data gathering, integrity in, 28
Daughter nuclide, 868
DDT, 538
Dead Sea Scrolls, 879-80
De Broglie, Louis, 280, 295
De Broglie relation, 297
De Broglie wavelength, 297
Debye (D), 380
n-Decane, 913
Decane, 907
Decanting, 10
Deci prefix, 19
Decomposition, standard heat of formation for, 264
Deep-freezing, 487
Deep-sea diving, partial pressures and, 206-8
Deep wells, 185-86
Definite proportions, law of, 48-49, 50, 84
Degenerate orbitals, 323, 324, 326
Dehydrating agent, sulfuric acid as, 1016
Dehydration reactions of alcohols, 929
Democritus, 47
Density(ies)
of alkali metals, 352
of bone, 21
calculating, 20-21
as conversion factor, 32-33
of gas, 201-3
of noble gases, 355
probability, 307, 308, 309
SI unit for, 20
Deoxyribonucleic acid. See DNA
Deoxyribose, 974

Deposition, 487
Derived unit, 19-21
Destructive interference, 287, 288, 440, 496
Detergents, 1010
Deterministic laws, 299-300
Deuterium-tritium fusion reaction, 889
Dextrorotatory isomer, 911
Diabetes, 955-56
Diagnostic medical procedures, radioactivity in, 893, 894-95
Diamagnetism, 339, 447
Diamond, 8, 506, 507, 999-1000
molar entropies of, 795
structure, 999
Diastolic blood pressure, 191
Diatomic molecule(s), 89
bond order of, 441
heteronuclear, 448-50
homonuclear, 443-48
Diazomethane, 402
Diborane, 382
Dichlorobenzene, 926
Dichlorodifluoromethane, 580
Dichloroethane, 923
1,2-Dichloroethane, 431, 920-21
physical properties of cis- and trans-, 921
1,2-Dichlorotetrafluoroethane, 580
Dichromate, 95
Dichromate compounds, 1041-42
Dichromate ion, 1041
Dieldrin, 538
Diet, nature's heat tax and, 776
Diethyl ether, 120, 524, 937
boiling point of, 479
Clausius-Clapeyron plot for, 484
freezing point depression and boiling point elevation constants for, 550
heat of fusion for, 488
heat of vaporization of, 479
vapor pressure of, 482
Differential rate law, 580-81. See also Integrated rate law
Diffraction, 287-88
Diffraction patterns, 496
Diffusion, 220
Digestion
chymotrypsin in, 604
of disaccharides by hydrolysis, 964
of fats and oils, 959
Dihydrogen phosphate, 95
Dihydrogen sulfide (hydrogen sulfide), 424-25, 1014, 1015
Dilute solution, 144, 535
Dimensional analysis, 28
Dimer, 940, 941
Dimethyl ether, 937
hydrogen bonding in, 470, 471
Dimethylglyoxime, 1073
Dinitrogen monoxide (nitrous oxide), 97, 1008
Dinitrogen trioxide, 1007
2,4-Dinitrophenol, 747
Diodes, 509
Dioxin, 538
Dipeptide, 969
Dipolar ion, 967
Dipole(s)
permanent, 468
temporary (instantaneous), 466
Dipole-dipole forces, 468-70, 472, 473, 523

Dipole moment (μ), 380-81, 386
  boiling point and, 468
  molecule polarity and, 419-21
  as vector quantities, 419
Diprotic acid(s), 161, 670, 671, 698
  anions in weak, 701-2
  titration with strong base, 743-44
Dirac, P.A.M., 280, 298
Dirty bomb, 354
Dirty dozen chemicals (persistent organic
      pollutants), 538
Disaccharide, 964
Diseases, water quality and, 495
Disequilibrium, controlled, 624
Dispersion forces, 466-67, 472, 473, 523
Disproportionation, 1009
Dissociation constant for water. See Ion product
      constant for water (K_w)
Dissociation of polyprotic acids, 702-3
Dissolution, 12
  entropy and, 780
  relative standard entropies and, 796
Distillation, 10
Disubstituted benzenes, 926
Disulfide linkages, 972, 973
Diving, 193
DNA, 497, 974
  base-pairing in, 976
  bases in, 975
  hydrogen bonding in, 473-74
  short strand of, 976
  structure of, 974
    discovery of, 978
  sugars in, 974
DNA polymerase, 979
DNA replication, 978-79
Döbereiner, Johann, 320
Dopants, 509
Doping, 509
d orbitals, 302, 306, 310, 311
  gemstone color and splitting of, 1051-52
Double bond(s), 87, 375
  bond energy of, 393
  carbon's ability to form, 904-5
  hydrogenation of, 601-2
  in Lewis theory, 430
  rotation about, 920, 921
  single bond vs., 920, 921
  sp^2 hybridization and, 428-31
  in structural formulas, 908
  in valence bond theory, 430
Double helix, 978-79
Double silicate chains, 994, 995
Dowager's hump, 21
Drug(s)
  acidic, 677
  coordination compounds, 1076
  ionic compounds as, 374
  knowledge of human genome and development
      of, 981
Dry-cell batteries, 842-43
Dry ice, 277, 487, 505, 1003
  sublimation of, 12
Ductility, 62, 349
  of metals, 397
Duet, 367
Dynamic equilibrium. See also
      Equilibrium/equilibria
  concept of, 619-21

defined, 619
population analogy for, 620-21
in solution, 529-30
vapor pressure of solution and, 540-41

E

E. coli, irradiation of foods to kill, 895
Ears, pressure imbalance and, 188, 189
Earth
  crust of
    major elements in, 992
    metals in, 1029
    silicates as most abundant matter in, 992
  metals in core of, 1029
  uranium/lead radiometric dating to estimate
      age of, 883
Earth metals, 1029
Ectotherms, 569-70
EDTA ligand, 1058-59
Effective nuclear charge (Z_{eff}), 324, 336-37
Effusion, 220-21
Egg-white lysozyme, 972, 978
Egg whites, pH of, 675
Einstein, Albert, 3, 58, 280, 289-91, 884, 885
  energy equation of, 887
Einsteinium, 58
Eka-aluminum, 321
Eka-silicon, 62, 321
Elastic collision, 214
Eldrin, 538
Electrical charge, 52
  properties of, 52
  of subatomic particles, 56
Electrical current, 823
  amperes (A), measuring electron flow, 825
  driving force for, 825-26
Electrical resistivity of metals, 1028, 1029
Electric field, 282
Electricity
  driving nonspontaneous chemical reactions
      with, 846-53
  generating
    with batteries, 842-46
    with fission, 884-86
    from spontaneous chemical reactions, 823-
      27
  heating home with natural gas vs., 776, 777
  power grid distributing, 819-20
Electrochemical cell, 823-27
  concentration cells, 840-41
  predicting spontaneous redox reactions and
      sketching, 832-33
  standard free energy change for, 835
  standard potential of, 829-31
  Zn/Cu^{2+}, under standard and nonstandard
      conditions, 838
Electrochemical cell notation, 826-27
Electrochemistry, 818-63
  balancing oxidation-reduction equations, 820-
      23
  batteries, 842-46
  cell potential, free energy, and the equilibrium
      constant, 834-37
  cell potential and concentration, 837-42
  corrosion, 854-55
  electrolysis, 846-53
    applications of, 847
    in electrometallurgy, 1032-33
    predicting products of, 849-52

stoichiometry of, 852-53
  of water, 847
  standard reduction potentials, 827-33
  voltaic (galvanic) cells, 823-27
Electrode
  inert platinum, 827
  sacrificial, 855
Electrolysis, 846-53
  applications of, 847
  in electrometallurgy, 1032-33
  predicting products of, 849-52
  stoichiometry of, 852-53
  of water, 847
Electrolyte(s), 151
  rusting promoted by, 855
  strong, 151, 153, 669
  weak, 153, 669
Electrolyte solutions, 151-53
Electrolytic cells, 823, 846-53
  for copper plating, 852-53
  for silver plating, 847
  solar-powered, 847
  voltaic versus, 848
Electromagnetic radiation, 282-83, 870.
      See also Light
  atomic spectroscopy and, 292-95
Electromagnetic spectrum, 285-87
Electrometallurgy, 1032-33
Electromotive force (emf), 826
Electron(s)
  bonding pair, 374-75
  charge of, 53, 56
  charge-to-mass ratio of, 52
  core, 329, 337
  discovery of, 51-53
    cathode rays and, 51-52
    Millikan's oil drop experiment, 53
  excitation of, 303-4
  ions and, 60
  Lewis acid-base model and, 704-5
  lone pair, 374-75
  mass and size of, 281-82
  observation of, 282
  orbitals for. See Orbital(s)
  outermost, 337
  photon release by, 304
  position of, 299
  positron as antiparticle of, 870
  shielding and penetration of, 324
  valence, 329, 330
    chemical properties and, 332-33
  velocity of, 299
  wave nature of, 295-300
    de Broglie wavelength and, 297
    indeterminacy and probability distribution
        maps, 299-300
    uncertainty principle and, 298-99
Electron affinity(ies) (EA), 348-49, 366
  electronegativity vs., 377
  of halogens, 353
Electron capture, 867, 870-71
Electron cloud, dispersion force and, 466
Electron configuration(s), 321-32
  of alkali metals, 352
  chemical properties and, 328-29
  electron spin and Pauli exclusion principle,
      322-23
  of halogens, 353
  inner, 327, 331

of inner transition elements, 332
for multielectron atoms, 326-28
of noble gases, 355
orbital blocks in periodic table, 330-31
outer, 331
sublevel energy splitting in multielectron atoms, 323-25
of transition metals, 332, 1052-54
valence electrons, 329
writing, from periodic table, 331-32
Electron diffraction experiment, 298
Electronegativity, 377-78
electron affinity vs., 377
of halogens, 1017
oxyacids and, 704
of transition metals, 1055
Electronegativity difference (ΔEN), 379-80
Electron geometry, 410
hybridization scheme from, 435-36
linear, 414, 436
octahedral, 415, 436
tetrahedral, 414, 436
trigonal bipyramidal, 415, 436
trigonal planar, 414, 436
Electron groups
defined, 406
five, 409
with lone pairs, 412-15
four, 408
with lone pairs, 410-11
hybridization scheme and, 435-36
repulsion between, 406
variation in, 411
six, 409
with lone pairs, 413
three, 407
two, 407
Electron pairs, nonbonding vs. bonding, 411
Electron sea model, 366, 397, 506, 508, 1028
Electron spin, 322-23
Electron symbol, 868
Electron transfer, ionic bonding and, 367-68
Electron volt (eV), 888
Electrostatic forces, 52
Element(s), 9, 10
absorption spectra of, 294-95
atomic, 89
atomic-level view of, 89-92
atomic mass of, 67-69
atoms and, 47
classification of, 89
electron configurations for, 326
electronegativities of, 378
emission spectra of, 292, 294
family (group) of, 63
of life, 67
main-group. *See* Main-group elements
molecular, 89-90
properties of, 84
periodic. *See* Periodic property(ies)
proton number as definitive of, 57
specific heat of, 247
standard enthalpy of formation for, 262
transition. *See* Transition metal(s)
transuranium, 891
Elementary steps, 594, 596
rate laws for, 594-95
Elimination reactions of alcohols, 929
Emeralds, color of, 1051-52

Emf (electromotive force), 826, 827
Emission spectra, 292
Bohr model and, 293
Empirical formula, 86, 87
from combustion analysis, 112-14
from experimental data, 109-10
molecular formula from, 110-12
Empirical formula molar mass, 111
Emulsion, 557
Enantiomers, 910, 1065
in chiral environment, 911
L-Enantiomers, 967
Endothermicity of vaporization, 478, 479
Endothermic processes
entropy of surroundings decreased by, 786, 787
spontaneous, 779-80
Endothermic reaction(s), 254-55, 650, 651, 790, 791
bond energies and, 394
Endpoint of titration, 744, 745
Energy(ies), 14-15. *See also* **Chemical energy; Kinetic energy; Potential energy; Thermal energy; Thermochemistry**
from combustion reactions, 173
conservation of, 14, 237, 238-39, 240, 775-76. *See also* First law of thermodynamics
conversion factors, 240
defined, 14, 238
fission to generate electricity, 884-86
greatest dispersal of
entropy change associated with change in state and, 784
state with highest entropy and, 782-83
interaction, of atoms, 423-24
internal (*E*), 241-46
change in, 242-44
ionization. *See* Ionization energy(ies) (IE)
lattice, 368-72
ion charge and, 371-72
ion size and, 371
loss in most energy transactions, 776
nature of, 237-39
nature's heat tax and, 775-76
nuclear fusion as sun's source of, 889-90
of photon, 289-90, 294
"places" for, in gaseous NO, 795
renewable, 270-71
rotational, 784, 795
total, 14
transfer of, 238, 239
transformation of, 238-39
translational, 784
units of, 239-40
velocity and, as complementary properties, 301
wavelength, amplitude, and, 283-84
Energy use, environment and, 267-71
Engine, pressure-volume work by, 249-50
English system of measurement, 15, 19
Enthalpy(ies) (*H*), 253-57, 779. *See also* **Heat(s)**
defined, 253
exothermic and endothermic processes, 254
of solution (ΔH_soln), 527, 528-29
Enthalpy change (ΔH), 253
effect on spontaneity, 790-92
exothermic and endothermic processes, 255
magnitude or absolute value of, 529
to quantify change in entropy for surroundings, 787
of reaction (ΔH_rxn), 255-67, 262, 391

bond energies to estimate, 391-95
measuring, 257-59
relationships involving, 259-61
from standard heats of formation, 262-67
stoichiometry involving, 255-57
total energy change vs., 253-54
Entropy(ies), 774, 779-88
absolute zero of, 793
biological systems and, 788
change in, 781
associated with change in state, 784-85
effect on spontaneity, 790-92
standard, for reaction (ΔS°_rxn), 793-97
definition of, 522, 780
direction of chemical and physical change determined by, 781-82
as measure of energy dispersal per unit temperature, 786
relative standard, 794-96
second law of thermodynamics and, 779-85
solutions and, 522, 541-42
as state function, 781
of surroundings, 785-88
quantifying, 787-88
temperature dependence of, 786
Environment
acid rain, 100, 706-7
air pollution, 225-26
energy use and, 267-71
free radicals and, 388-89
Lake Nyos carbon dioxide accumulation, 534
MTBE in gasoline, 143
ozone, Lewis structure of, 396
ozone depletion, 226-27
persistent organic pollutants (POPs), 538
renewable energy, 270-71
water pollution, 495
Environmental radon, 877
Enzymes, 570, 602-4, 966
EPA, 107, 226, 495, 538, 877
Epsom salts, 96
Equation(s), 28
for acid-base reactions, 163
for aqueous reactions, 159-60
coefficients as conversion factors, 211
complete ionic, 159, 160
equilibrium constant and changes in, 624-26
for gas-evolution reactions, 167
molecular, 159, 160
net ionic, 159, 160
nuclear, 868-71
for fission reaction, 883
for precipitation reactions, 158-59
problems involving, 34-35
thermochemical, 255-57
writing and balancing, 114-17
procedure for, 115-16
Equatorial positions, 409
Equilibrium, thermal, 246
Equilibrium constant (*K*), 616, 621-28. *See also* **Acid ionization constant(s) (*K*_a)**
chemical equations and, 624-26
defined, 618, 622
equilibrium concentrations from, 636-45
given all but one of equilibrium concentrations of reactants and products, 636-37
given initial concentrations or pressures, 637-41

Equilibrium constant (*K*), *(continued)*
    simplifying approximations, 641-45
    expressing, 622
    for fetal vs. adult hemoglobin, 618
    free energy and, 807-9
    from measured equilibrium concentrations, 630-33
    reaction quotient vs., 633-34
    for redox reaction, 836-37
    significance of, 623-24
    temperature and, 630, 809
    in terms of concentrations, 622
    in terms of pressure, 626-28
    units of, 628
Equilibrium/equilibria, 616-61. *See also* **Acid-base chemistry; Buffers; Complex ion equilibria; Dynamic equilibrium; Solubility equilibria**
    acid strength and, 669
    controlled disequilibrium, 624
    fetal hemoglobin and, 617-19
    heterogeneous, 629-30
    Le Châtelier's principle, 645-52
        concentration change, 646-48
        population analogy for, 645-46
        temperature change, 650-52
        volume (or pressure) change, 649-50
    life and, 624
    reaction quotient, 633-35
Equivalence point, 163-64, 733
Eriochrome Black T, 747
Error
    random, 27
    systematic, 28
Erythrosin B, 747
Esophageal sphincter, 664
Essential amino acids, 970
Ester(s), 934-37
    functional group of, 928
    general formula for, 120
    naming, 935
    reactions, 936-37
Ester linkages, 958
Estimation
    order of magnitude, 33-34
    in weighing, 23
β-Estradiol, 961
Ethanal (acetaldehyde), 120, 931, 933
Ethane, 904, 912
    dipole-dipole forces in, 468
Ethanoic acid. *See* **Acetic acid**
Ethanol, 120, 521, 523, 524, 718, 928, 929
    breathalyzer to measure, 846
    density of, 20
    freezing point depression and boiling point elevation constants for, 550
    hydrogen bonding in, 470-71
    oxidized to acetic acid, 930
    oxidized to ethanal, 933
    reaction between hydrobromic acid and, 929
    reaction between oxygen and, 174
    specific heat of, 247
    vapor pressure of, 482
Ethene, 118, 917, 938
    common uses of, 119
    molecular formula for, 119
    reactions of
        with chlorine gas, 923
        with hydrogen, 593, 601-2

with hydrogen chloride, 923
    space-filling model of, 119
    structural formula for, 119
Ethers, 937
    functional group of, 928
    general formula for, 120
    naming, 937
Ethyl alcohol. *See* **Ethanol**
Ethylamine, 120, 688, 938
Ethylbenzene, 925
Ethyl butanoate, 936
Ethylene. *See* **Ethene**
Ethylenediamine ligand, 1057, 1058, 1059
Ethylenediaminetetraacetate (EDTA) ligand, 1058-59
Ethylenediaminetetraacetate ion, 1073
Ethylene glycol, 549, 717-18
    vapor pressure of, 482
Ethylmethylamine, 938
Ethyl methyl ether, 937
Ethyl pentanoate, 935
Ethyl propanoate, 934
Ethyl substituent, 914
Ethyne. *See* **Acetylene**
Europium, 58
Evaporation, 478
    entropy and, 780
Exact numbers, 24-25
Exa prefix, 19
Exothermicity of condensation, 478
Exothermic process(es)
    entropy of surroundings increased by, 786, 787
    spontaneity and, 787
Exothermic reaction(s), 254-55, 650-51, 790-91
    bond energies and, 394
Expanded octets, 390-91, 434
Experiment, 5, 6
Exponential factor, 587, 589
Extensive property, 20
External arrangement (macrostate), 781-82
Extractive metallurgy, 1029

F

Face-centered cubic unit cell, 498, 500, 503-4
Fac-mer isomerism, 1064
Fahrenheit (°F) scale, 17
Falsifiability, 5
Family(ies), 120
    of elements, 63. *See also specific groups*
    of organic compounds, 120
Faraday, Michael, 818
Faraday's constant (F), 834, 835, 853
Fats, 958-60
Fatty acids, 956-57
    structure for, 956
FDA, 355, 895, 959, 960
Feldspar, 993
Femto prefix, 19
Fermi, Enrico, 883
Ferrochrome, 1041
Ferromagnetic materials, 1043
Ferromanganese, 1042
Fertilizers
    ammonium nitrate, 1008
    nitrogen-containing, 1006
    phosphoric acid and production of, 1010
Fetal hemoglobin (HbF), 617-19
Fetuses, mercury exposure and, 107
Feynman, Richard P., 51, 237

Fibrous proteins, 970-71, 973
Film-badge dosimeters, 874, 875
Filtration, 10
Fireworks, 295
First ionization energy (*E*), 343, 344-46
First law of thermodynamics, 240-46, 775-76
    internal energy (*E*), 241-46
First-order integrated rate law, 580-82, 587
First-order reaction, 575, 576, 587
First-order reaction half-life, 584-85
Fish, methylmercury in, 107
Fission, nuclear
    converting mass to energy, 887-89
        mass defect and, 887-89
        nuclear binding energy, 888-89
    discovery of, 883-86
        atomic bomb and, 884
        nuclear power to generate electricity, 884-86
Flame tests, 294, 295, 758
Flash freezing, 495
Flintstones, The (TV show), 194
Fluid(s)
    cellular, 521
    intravenous, 70
    supercritical, 486-87
Fluoride, 94
Fluoride ion, electron configuration of, 339
Fluorine, 60, 65
    electron configuration of, 327, 339
    elemental, 1018
    high reactivity of, 1018
    Lewis structure of, 366
    orbital diagram for, 327
    oxidation state for, 169
    properties of, 353
    reaction between bromine and, 354
    reaction with noble gases, 356
    selected properties of, 1017
Fluorine-18, 894
    PET scan using, 895
Fluorite (CaF_2) structure, 506
Fluorspar, 1018
Flux, 1031
Foam, 557
Fog, 557, 558
Food(s)
    acidic, 677
    caloric value of, 405
    irradiation of, 895
    preservatives in, 1009
    taste of, 406
Food industry, phosphoric acid and phosphates used in, 1010-11
f orbital, 302, 310, 311
Force(s)
    adhesive, 476-77
    cohesive, 476-77
    defined, 42
    dispersion, 466-67, 472, 473, 523
    electrostatic, 52
    intermolecular. *See* **Intermolecular force(s)**
    intramolecular, 376
    SI unit of, 186
Forests, acid rain damage to, 100
Formal charge, Lewis structures and, 386-87
Formaldehyde, 931, 932
    dipole-dipole forces in, 468
    molecular geometry of, 407
Formalin, 932

Formation constant (K_f), 759
Formic acid, 671, 737-43
Formic acid (methanoic acid), 935
    acid ionization constant for, 672
Formula mass, 101-3
Formulas. *See* Chemical formula(s)
Formula unit, 90, 101
Fossil fuel combustion, 131-32, 133-35
    energy from, 267-68
    environmental problems associated with, 268-69
        acid rain, 100, 136, 225, 226, 268, 706-7, 1008, 1016
        carbon dioxide emission, 133-35
        global warming, 268-69
    incomplete, 226
    ozone production and, 226
Fossils, radiocarbon dating of, 879
Fractionation of air, 1011
Fragrances, 903-4
Franklin, Rosalind, 497
Franklinite, 1044
Frasch process, 1013-14
Free energy
    of formation (ΔG_f°), 799-801
    Gibbs (G), 788-92
    theoretical limit of available, 803
    why it is "free," 803-4
Free energy change of reaction
    under equilibrium conditions, 805
    under nonstandard conditions (ΔG_{rxn}), 804-7
    standard (ΔG_{rxn}°), 797-809, 837
        calculating, 797-804
        equilibrium constant (K) and, 807-9
        standard cell potential and, 834-36
Free radical(s), 388-89, 402
    chlorine, 922-23
Freezer burn, 487
Freezing, 488, 495
    energetics of, 488
    of water, entropy of surroundings increased by, 785-86
Freezing point depression, 549-51, 554
Freon, 227
Frequency (ν), 284, 285
    threshold, 289, 291
Frequency factor (A), 587, 588, 589, 590-92
    Arrhenius plots of, 590-92
    collision model and, 592-93
Friction, 244
Frogs, antifreeze in, 550-51
Fructose, 603, 962, 963
    conversion of sucrose to glucose and, 574
    formation of glycosidic linkage with glucose, 964
    molecular formula for, 111
Fuel(s). *See also* Fossil fuel combustion
    hydrocarbons as, 906
    oxygenated, 143
Fuel cell, 270, 819-20, 845-46
Fuel-cell breathalyzer, 846
Fuel-cell power plants, 819-20
Fuel fragments in exhaust, 600
Fuller, R. Buckminster, 989, 990, 1001
Fullerenes (buckyballs), 1001, 1002
Fulton, Robert, 241
Functional groups, 119, 120, 927-28
    common, 928
Functionalized hydrocarbons, 119-20

Furan, 538
Furnaces, 237
Fusion, 487-88. *See also* Melting
    energetics of, 488
    nuclear, 889-90
Fusion curve, 491, 492, 493

G

Galactose, 963
Galena, 1029
Galileo Galilei, 47
Gallium, 321, 509
    ionization energy of, 346
Gallium arsenide, 6
Galvanic cells. *See* Voltaic (galvanic) cells
Galvanized nails, 855
Galvanizing steel, 1044-45
Gamma (γ) rays, 54, 285, 286, 867, 870, 871
Gangue, 1030, 1031
Garnet, 1052
Gas(es), 184-235
    in atmosphere, 225-27
        air pollution, 225-26
        ozone depletion, 226-27
    Avogadro's law, 197, 198, 215, 217
    Boyle's law, 192-94, 198, 215, 217
    Charles's law, 195-96, 198, 214, 215
    in chemical reactions, 211-13
    collecting, over water, 209-10
    compressibility of, 9
    density of, 201-3
    diffusion of, 220
    effusion of, 220-21
    entropy change associated with change in state of, 784
    greenhouse, 131-32, 268
    ideal. *See* Ideal gas(es)
    kinetic molecular theory of, 214-19
        ideal gas law and, 215-17
        postulates of, 214
        pressure and, 215
        simple gas laws and, 215
        temperature and molecular velocities, 217-19
    mass of, 9
    mean free path of, 220
    mixtures of, 204-6
    molecular comparison with other phases, 462-65
    natural, 114, 237, 268
        heating home with electricity vs., 776, 777
        reaction between oxygen and, 174
    noble. *See* Noble gas(es)
    partial pressures, 204-10
        deep-sea diving and, 206-8
        vapor pressure, 209-10
    physical properties of, 191
    pressure, 185-91
        atmospheric, 185-87
        blood, 191
        defined, 185-86, 188
        manometer to measure, 190
        particle density and, 188
        temperature and, 199
        total, 200
        units of, 188-90
        volume and, 192-94
    properties of, 463
    real, 221-24
        finite volume of gas particles and, 222, 224
        intermolecular forces and, 223, 224

    molar volumes of, 222
    van der Waals constants for, 223
    van der Waals equation for, 224
    relative standard entropies of, 794
    solubility in water, 531-34
    standard state for, 262
Gas chromatograph, 575
Gaseous matter, 9
Gaseous solution, 521
Gas-evolution reactions, 160, 165-67
Gasoline, 15
    combustion of, 249-50
    MTBE in, 143
Gastric juice, pH of, 675
Gastroesophageal reflux disease (GERD), 664
Gauge pressure, 200
Gay-Lussac's law, 199, 215
Geckos, 461-62
Geiger-Müller counter, 875
Gemstones, color of, 1051-52
Gene(s), 956, 978, 980
    expression of, 978
    in genetic structure, 977
Genentech, 956
General Motors, 270
Genetic code, 977-78
Genetic defects, radiation exposure and, 892
Genetic structure, 977
Geometric (cis-trans) isomerism, 431, 921, 1062-64
    in alkenes, 920-21
Geometry
    electron, 410
        hybridization scheme from, 435-36
        linear, 414, 436
        octahedral, 415, 436
        tetrahedral, 414, 436
        trigonal bipyramidal, 415, 436
        trigonal planar, 414, 436
    molecular
        bent, 411, 414, 417
        electron group repulsion and, 406
        linear, 407, 408, 413, 414, 415, 417
        lone pairs effect, 410-14
        octahedral, 413, 415, 417, 435
        polarity and, 418-22
        predicting, with VSEPR, 415-18
        seesaw, 412, 415, 417
        square planar, 413, 415, 417
        square pyramidal, 413, 415
        tetrahedral, 408, 411, 414, 417, 426-28
        trigonal bipyramidal, 409-10, 413, 415, 417, 434
        trigonal planar, 407, 408, 414, 417, 428
        trigonal pyramidal, 410, 411, 414, 417
        T-shaped, 412, 415
Germanium, 62, 321
German or silver brass, 1044
Gibbs free energy (G), 788-92
    change in (ΔG), 789
Gibbsite, 92
Giga prefix, 19
Given information, 30
Glass, 8, 507, 992
    boron used in manufacture of, 997
    density of, 20
    etching, 1019
    reaction between hydrofluoric acid and, 1018-19
    specific heat of, 247

Global warming, 131-32, 268
  fossil fuels and, 268-69
Globular proteins, 971, 973
Glucose, 603, 961, 962
  conversion of sucrose to, 574
  formation of glycosidic linkage with fructose, 964
  molecular representation of, 88
  oxidation of, 802
  rearrangement to form ring, 963
Glutamic acid, 968
Glutamine, 968
Glycerol, 958
Glycine, 402, 967, 968
  molecular geometry of, 417
Glycogen, 964, 965
Glycolic acid, 718
Glycolipids, 960
Glycosidic linkage, α and β, 964, 965
Gold, 63, 1029
  density of, 20
  electronegativity of, 1055
  leaching process to obtain, 1031
  specific heat of, 247
Gold foil experiment, 54-55
Gout, 772
Gradient, chemical, 319
Graham, Thomas, 220
Graham's law of effusion, 220-21
Granite, specific heat of, 247
Graphite, 506, 507, 999
  molar entropies of, 795
  structure, 999
Gravitational potential energy, 14
Gravitational pull, 16
Gray (Gy), 892
Greenhouse gas, 131-32, 268
Ground-level ozone, 226
Ground state, 321
Group 1A metals (alkali metals), 64, 66, 330, 758
  electron configurations of, 351
  ion formation by, 66
  properties of, 352
    periodic, 352-53
  reactivity of, 333
  as reducing agents, 172
Group 1 metals, 320
Group 2A metals (alkaline earth metals), 64, 65, 330
  reactivity of, 333
  as reducing agents, 172
Group 4A compounds, boiling points of, 471
Group 5A elements, oxidation state for, 169
Group 6A compounds
  acidity trends of hydrides, 703
  boiling points of, 471
Group 6A elements, oxidation state for, 169
Group 7A elements. See Halogens
Group 7A hydrides, acidity trends of, 703
Group 8A elements. See Noble gas(es)
Group (family) of elements, 63
Guanine, 473, 474, 975

H

H_3O^+. See Hydronium ion; Hydronium ion concentration
Haber, Fritz, 1006
Haber-Bosch process, 1006
Hahn, Otto, 883

Hair, bleached, 173
Half-cell, 824
Half-cell potential, 827
  measuring, with SHE, 828-29
  standard, 829-31
Half-life of reaction, 584-86, 587, 876
  first-order, 584-85
  second-order, 585-86
  zero-order, 586
Half-reaction method of balancing, 820-23
Halides
  hydrogen, 353-54
  metal, 353
Halite. See Sodium chloride
Hall process, 1032
Halogen-nitrogen single bonds, 396
Halogen oxides, 1020-21
Halogens, 65, 330, 1017-21
  boron-halogen compounds, 997
  compounds, 1019-21
  diatomic molecules formed by, 375
  electron affinities for, 348
  electron configurations of, 351
  electronegativity of, 1017
  oxidation states of, 169, 1017
  properties of, 353-54, 1017
  reaction between phosphorus and, 1009
  reactions of alkali metals with, 352
  reactivity of, 333
Halogen substitution reactions, 922-23
Hamiltonian operator, 301
Hard water, 155, 749
Hardystonite, 993
Harpoon mechanism, 593
Hausmannite, 1042
Hawaii, 247
Heartburn, 663-64
Heat(s), 17, 246-49. See also Enthalpy(ies) (H)
  absorbed by or lost from the solution (q_{soln}), 257
  from combustion reactions, 114, 174
  at constant volume, 251
  defined, 246
  energy transferred through, 238
  of fusion (ΔH_{fus}), 488
  of hydration ($\Delta H_{hydration}$), 528-29
  internal energy change and, 244-46
  as product or reactant in reaction, 650-52
  of reaction. See under Enthalpy change (ΔH)
  temperature vs., 246
  of vaporization (ΔH_{vap}), 479-80, 488
    Clausius-Clapeyron equation to determine, 485
Heat capacity (C), temperature changes and, 246-47
Heating curve, 483
  for water, 488-91
Heating with natural gas vs. electricity, 776, 777
Heat tax, 804
  nature's, 775-76
Heat transfer
  changes in entropy of surroundings and, 785-88
  second law of thermodynamics and, 783
Heavy metal poisoning, medical treatment of, 1073
Heisenberg, Werner, 280, 299
Heisenberg's uncertainty principle, 298-99
Heliox, 207
Helium, 58, 63, 89

electron configuration of, 323, 367
  emission spectrum of, 292, 293
  Lewis structure of, 367
  liquid, 356
  properties of, 355
  real gas behavior of, 224
  van der Waals constants for, 223
  in water, Henry's law constants for, 533
Helium ion, 336
α-Helix, 971, 972
Hematite, 1029
Heme, 1074
Hemlock, 666
Hemoglobin, 3-4, 558
  fetal (HbF), 617-19
  as globular protein, 971
  oxygen and carbon monoxide binding to, 4
  oxygen-carrying site on, 1050, 1074-75
  sickle-cell anemia and, 972
  subunits, 972, 973
Henderson-Hasselbalch equation, 722-25
Henry, William, 1027
Henry's law, gas solubility and, 533-34
Henry's law constant, 533
Heptachlor, 538
Heptane, 523
n-Heptane, 475, 912, 913
2-Heptanone, 933
Hertz (Hz), 284
Hess's law, 260-61
Heterogeneous catalysis, 600-602
Heterogeneous equilibria, 629-30
Heterogeneous mixture, 9, 10
Heteronuclear diatomic molecules, 448-50
Hexachlorobenzene, 538
Hexagonal closest packing, 502-3
Hexamethylenediamine, 940
Hexane, 523, 524, 907
n-Hexane, 475, 912, 913
3-Hexanone, 932
1-Hexene, 917
Hexose, 962, 963
1-Hexyne, 918
High-spin complexes, 1071, 1072
Hinshelwood, Cyril N., 568, 569
Histidine, 968
HIV-protease, 363-64
Hoffmann, Roald, 461, 477
Homogeneous catalysis, 600-602
Homogeneous mixtures, 9, 10. See also Solution(s)
Homonuclear diatomic molecules, 443-48
Hooke, Robert, 192
Hormones, 966
Hot-air balloon, 195
Huheey, James E., 663
Human Genome Project, 954, 980-81
Human immunodeficiency virus (HIV), 363
Humans. See also Blood, human; Medicine
  atoms and, 51
  elemental composition of, 67
  nerve cells of, 842
  transition metals in, 1074
Hund's rule, 326, 327, 1071
Hwang Woo Suk, 28
Hybrid cars, 240
Hybridization, 384-85, 425-38
  defined, 426
  sp, 432-33, 436
  sp^2, 428-31, 436

sp^3, 426-28, 436
sp^3d^2, and sp^3d, 434-35, 436
writing, 435-37
Hybridization and bonding schemes, 435-39
Hybrid orbitals, 426
Hydrated ionic compounds, 96
Hydrates, 96
Hydration
heat of ($\Delta H_{hydration}$), 528-29
waters of, 96
Hydrazine, 1006-7
Hydride(s)
acidity trends of, 703
boiling points of, 494
Hydriodic acid, 670
Hydrobromic acid, 161, 670
reaction between ethanol and, 929
Hydrocarbons, 118-20, 906-28
alkanes (saturated hydrocarbons), 118, 906,
912-16
boiling points of, 467
geometric (cis-trans) isomerism, 923-24
n-alkanes, 912-13
naming, 914-16
reactions of, 922-23
viscosity of, 475
alkenes and alkynes (unsaturated
hydrocarbons), 118, 906, 917-21
geometric (cis-trans) isomerism in alkenes,
920-21
hydrogenation of double bonds within, 601-2
naming, 919-20
reactions of, 923-24
aromatic, 906, 924-27
naming, 925-27
reactions of compounds, 927
functional groups, 927-28
functionalized, 119-20
names for, 118
polarity of, 422
reactions of, 922-24
released into the atmosphere, 388-89
stereoisomerism and optical isomerism of, 909-
12
structures, 906-9
uses of, 906, 913
viscosity of, 475
Hydrocarbon tails of fatty acids, 957
Hydrochloric acid, 146, 152-53, 160, 161, 418, 663,
664, 666, 667, 669, 670-71
mixed with weak acid, 684-85
reactions of
with metal sulfides, 1015
with methylamine, 938
with sodium bicarbonate, 165-66
with sodium hydroxide, 162
with zinc, 833
titration with sodium hydroxide, 734-37
Hydrocyanic acid, acid ionization constant for,
672
Hydroelectric power, 270
Hydrofluoric acid, 161, 664, 670, 671, 692, 1018-
19
acid ionization constant for, 672
reaction between glass and, 1018-19
Hydrogen, 83-84
bond order for, 441
boron-hydrogen compounds, 998
bridging, 382

electron configuration for, 321-22, 323
emission spectrum of, 292
as fuel, 108
interaction energy of two atoms of, 423-24
Lewis structure of, 374
nonmetallic behavior of, 351
oxidation state for, 169
properties of, 83
reactions of
with chlorine, 168
with ethene, 593, 601-2
with iodine, 578, 619
with iodine monochloride, 594
with nitrogen, 651
with nitrogen monoxide, 578, 597
as reducing agent, 172
Schrödinger equation for, 301-3
transitions in, 304-5
weighted linear sum of molecular orbitals for,
439-41
Hydro-Gen3 (fuel cell car), 270
Hydrogenation
of alkenes and alkynes, 923
of double bonds, 601-2
of fats, 959-60
Hydrogen azide, 1007
Hydrogen bonding, 470-73, 494, 523, 973
in DNA, 473-74
Hydrogen carbonate, 95, 688
Hydrogen chloride, 666
reaction between ethene and, 923
reaction between propene and, 923
Hydrogen fluoride, 470
dipole moment of, 380
MO diagram for, 449
polar bonding in, 377-78
Hydrogen gas
reaction between propene and, 923
van der Waals constants for, 223
Hydrogen halides, 353-54
Hydrogen ions, 98
Hydrogen-oxygen fuel cell, 819, 820, 845-46
Hydrogen peroxide, 4-5, 86, 173, 376
Hydrogen phosphate, 95
Hydrogen sulfate (bisulfate), 95
Hydrogen sulfide (dihydrogen sulfide), 424-25,
1014, 1015
Hydrogen sulfite (bisulfite), 95
in gas-evolution reactions, 166
Hydroiodic acid, 161
Hydrolysis, 964
of disaccharides, 964
Hydrometallurgy, 1031
Hydronium ion, 157, 161, 376
Hydronium ion concentration, 667, 673
of mixture of acids, 684-86
strong-weak mixture, 684-85
weak-weak mixture, 686
of strong acids, 677-78
of weak acids, 678-82, 684
Hydrophilic molecule, 960
Hydrophobic interactions, 973
Hydrophobic molecule, 960
Hydroxide ion (OH^-) concentration, 667, 673
of weak bases, 688-90
Hydroxides, 95
base-insoluble, 758
solubilities of, 752
Hydroxyl group, 928

Hydroxyl radical (atmospheric vacuum cleaner),
388
Hyperbaric oxygen therapy, 1012
Hyperosmotics, 565
Hyperosmotic solutions, 555-56
Hypertension, 191
Hypochlorite, 95, 96
Hypochlorite ion, 91
Hypochlorous acid, acid ionization constant for,
672
Hyposmotic solutions, 555-56
Hypothesis, 5
Hypoxia, 206

I

Ice, 505
density of, 20
melting of, 779
ICE table, 631, 637
Icosahedron, 996
Ideal gas(es)
concentration of, 627
solution of, 522
Ideal gas constant, 198
Ideal gas law, 198-204, 627
breakdown of, 221
corrected for intermolecular forces, 223
corrected for volume of gas particles, 222
density of gas, 201-3
kinetic molecular theory and, 215-17
molar mass of gas, 203-4
molar volume at standard temperature and
pressure, 201
partial pressure computed from, 204-5
simple gas laws and, 198
Ideal solution, 545-46
Igneous rocks, uranium/lead dating of, 881-83
Ilmenite, 1039-40
Inches of mercury (in Hg), 189
Incomplete octets, 389-90
Indeterminacy, 299-300
Indicators, 164, 733, 744-47
Indinavir, 363
Inelastic collision, 214
Inert platinum electrode, 827
Infection, radiotracers to locate, 894
Infrared (IR) radiation, 287
Infrared waves, 285
Initial rates, method of, 577, 578
Inner electron configuration, 327, 331
Inner transition elements (actinides), 330
electron configurations of, 332
Inorganic chemistry, 988
Inorganic compounds, 117, 905
Inosilicates (pyroxenes), 994, 995
Insoluble compounds, 154
Instantaneous dipole (temporary dipole), 466
Instantaneous rate of reaction, 573
Insulated nanowires, 989-90
Insulators, 508
Insulin, 955
as globular protein, 971
synthesis of human, 955-56
Insulin receptors, 970
Integrated rate law, 580-87, 877-79
first-order, 580-82, 587
half-life of reaction, 584-86, 587
second-order, 583-84, 587
zero-order, 584, 587

Integrity in data gathering, 28
Intensive properties, 20, 247
Interaction energy of atoms, 423-24
Interference, 287-88, 496
    constructive, 287, 288, 440, 496
    destructive, 287, 288, 440, 496
Interference pattern, 288, 295-96, 298, 496
Interhalogen compounds (interhalides), 354,
    1019-20
Intermediates, reaction, 594, 596, 597
Intermolecular force(s), 376, 461-77
    bonding forces vs., 465-66
    capillary action, 476-77
    condensed phases and, 462, 465-74
    dipole-dipole, 468-70, 472, 473, 523
    dispersion, 466-67, 472, 473, 523
    geckos and, 461-62
    hydrogen bonding, 470-72, 473, 494, 523
        in DNA, 473-74
    ion-dipole, 472-73, 523, 528-29
    molecular solids and, 505
    real gases and, 223, 224
    solutions and, 522-26
    surface tension and, 474-75
    temperature and, 465
    viscosity, 475-76
Internal energy (E), 241-46
Internal energy change (ΔE), 242-44
    for chemical reactions (ΔE_{rxn}), 251-53
    enthalpy change vs., 253-54
International Bureau of Weights and Measures at
    Sèvres, France, 16
International System of Units (SI), 15. *See also* SI
    unit(s)
International Union of Pure and Applied
    Chemistry (IUPAC) nomenclature
    system, 914
Internuclear distance, 423-24
Interstitial alloys, 1034, 1038-39
Intramolecular forces, 376
Intravenous fluids, 70
Intravenous solutions, 556
Iodide, 94
Iodine, 63, 65
    phase diagrams of, 493
    properties of, 353, 1017
    reaction between hydrogen and, 578, 619
Iodine-131, 894
Iodine monochloride, 594
Ion(s), 60, 338-48. *See also* Anion(s); Cation(s)
    acetylide, 1002
    acid-base properties of, 690-96
        anions as weak bases, 691-94
        cations as weak acids, 695-96
    complex, 759
    electrolysis of mixture of, 849
    electron configurations of, 339-40
    formation of, with predictable charges, 333
    ionic radii, 341-43
    ionization energy, 343-48
    isoelectronic series of, 342
    magnetic properties of, 339-40
    periodic table and, 65-66
    polyatomic, 90-92, 95
        Lewis structure for, 382, 383-84
    spectator, 159
Ion channels, 319-20
Ion-dipole forces, 472-73, 523, 528-29
Ionic bond(s), 84, 85, 90, 365, 366, 377, 397

dipole moment of, 380
directionality of, 376
electron transfer and, 367-68
lattice energy and, 368-72
Lewis theory of, 367-72
model for, 372-73
Ionic carbides, 1002
Ionic compound(s), 85, 90-92, 101
    aqueous solutions of, 472
    colligative properties of solutions, 554-56
    dissolution of, 151
    as drugs, 374
    formulas for, 92-93
    hydrated, 96
    identifying, 93
    lattice energy of, 368-71
    melting of, 372
    naming, 93-96
        containing metal that forms more than one
            kind of cation, 94-95
        containing metal that forms only one type of
            cation, 94
        containing polyatomic ions, 95-96
    solubility of, 154-55, 716, 747-53
        common ion effect on, 751-52
        pH and, 752-53
        solubility product constant (K_{sp}), 747-49,
            753
    types of, 93
Ionic equations
    complete, 159, 160
    net, 159, 160
Ionic solids, 504, 505-6
Ionic solutes, 544-45, 554
    vapor pressure and, 544-45
Ionization
    of acids, 669-70
    of polyprotic acids, 698
    of strong base, 687
    of weak acid, 683-84
    of weak base, 687
Ionization energy(ies) (IE), 343-48, 366
    of alkali metals, 352
    bonding and, 348
    defined, 343
    first, 343, 344-46
    of noble gases, 355, 356
    second and successive, 343, 347-48
    of transition metals, 1055
Ionizing power, 869
    of alpha radiation, 869
    of beta radiation, 870
    of gamma rays, 870
Ionizing radiation, 286
Ionone, 933
Ion pairing, 554
Ion product constant for water (K_w), 673, 693
Ion pumps, 319
Iron, 89, 506
    cations formed by, 95
    charge of, 93
    corrosion of (rusting), 854-55
    density of, 20
    functions in human body, 1074
    MO diagram for, 446
    specific heat of, 247
Iron-59, 894
Iron blue, 1074
Iron(II) carbonate, 750

solubility product constant for, 747
Iron(II) disulfide (iron pyrite), 1014
Iron(II) hydroxide, solubility product constant
    for, 747
Iron(II) sulfate, 95
Iron(II) sulfide, solubility product constant for,
    747
Iron(III) chloride, 696
Iron(III) fluoride, 697
Iron sponge, 1033
Irradiation of foods, 895
Irreversible reactions, 803-4
Isobutane, 118, 905, 906, 907
Isobutyl substituent, 914
Isoleucine, 968
Isomerism
    cis-trans, 431, 921, 1062-64
        in alkenes, 920-21
    in coordination compounds, 1062-66
        stereoisomerism, 1062-66
        structural, 1062
    optical, 909-10, 1062, 1065-66
        carbohydrates and, 962-63
        defined, 909
        in hydrocarbons, 909-12
Isomerization, 432
Isomers, 431
    coordination, 1062
    geometric, 1062-64
    optical, 1062, 1065-66
    structural, 907, 962-63
Isopropyl alcohol, 119, 120, 927, 929
    heat of fusion for, 488
    oxidized to 2-propanone, 933
    properties of, 39
Isopropyl substituent, 914
Isosmotic (isotonic) solutions, 556
Isotones, 80
Isotope(s), 58-59, 343
    mass of, 69
    natural abundance of, 58
    notation for symbolizing, 867
    radioactive, as radiotracers, 894
    relative abundance of, 69
IUPAC nomenclature system, 914

J

Joliot-Curie, Irène and Frédéric, 890
Joule, James, 239
Joule (J), 239, 240
J-tube, 192

K

Kekulé, Friedrich August, 924
Kelvin (K), 17
Kelvin scale (absolute scale), 17
Kepler, Johannes, 47
Keratin, 971, 972
Kernite, 996
Ketohexose, 962
Ketones, 931-34, 962
    functional group of, 928
    general formula for, 120
    naming, 932
    reactions, 933-34
Ketose, 962
Kevlar, 941
Kilocalorie (kcal), 240
Kilogram (kg), 16

Kilojoules (kJ), 239
Kilo prefix, 19
Kilowatt-hour (kWh), 240
Kinetic energy, 14, 238
    defined, 239
    of ejected electron, 291
    temperature and, 214, 217-18
    transformation of, 239
Kinetic molecular theory, 214-19
    ideal gas law and, 215-17
    postulates of, 214
    pressure and, 215
    simple gas laws and, 215
    temperature and molecular velocities, 217-19
Kinetics, chemical. *See* Reaction rate(s)
Knowledge, scientific approach to, 5-7
Krypton, 63, 334
    properties of, 355
    reaction between fluorine and, 356
    van der Waals constants for, 223
Kuhn, Thomas S., 7, 405
Kwolek, Stephanie, 941

L

Lactic acid, 732, 935
Lakes, acid rain and, 707
Lanthanide contraction, 1054
Lanthanides, 330
Lattice energy(ies), 368-72
    ion charge and, 371-72
    ion size and, 371
Lattice point, 498
Laughing gas, 97
Lavoisier, Antoine, 5, 47, 1011
Lavoisier, Marie, 5
Law(s)
    of conservation of energy, 14, 237, 238-39, 240,
        775-76
    of conservation of mass, 5, 6, 47-48
    of definite proportions, 48-49, 50, 84
    deterministic, 299-300
    of mass action, 622, 625, 627, 804
    of multiple proportions, 49-50
    scientific, 5
Leaching, 1031
Lead, 63
    density of, 20
    specific heat of, 247
Lead-acid storage batteries, 843
Lead(II) bromide, solubility product constant for,
    747
Lead(II) chloride, solubility product constant for,
    747
Lead(II) nitrate, reaction between potassium
        iodide and, 155-56
Lead(II) sulfate, solubility product constant for, 747
Lead(II) sulfide
    roasting, 1031
    solubility product constant for, 747
Le Châtelier, Henri, 617
Le Châtelier's principle, 481, 645-52, 684, 838
    concentration change, 646-48
    free energy changes and, 807
    population analogy for, 645-46
    temperature change, 650-52
    volume (or pressure) change, 649-50
Lemons, pH of, 675
Length, SI unit of, 16, 19
Leucine, 968

Leucippus, 47
Lever rule, 1037
Levorotatory isomer, 911
Lewis, G.N., 364, 704
Lewis acid-base adduct, metal-ligand complex as,
        1057
Lewis acids and bases, 704-6
    boron trihalides as strong Lewis acids, 997
Lewis electron-dot structures (Lewis structures),
        364
Lewis theory, 362, 364, 366-403, 820
    bond polarity and, 377, 379-81
    bond types under, 367-69
    of covalent bonding, 374-76
    double bonds in, 430
    electronegativity and, 377-78
    formal charge and, 386-87
    of ionic bonding, 367-72
    of molecular compounds, 382-83
    octet rule exceptions, 387-91
    of polyatomic ions, 382, 383-84
    resonance and, 384-85
    valence bond theory and, 436
    valence electrons represented with dots, 366-67
Libby, Willard, 879
Life
    as controlled disequilibrium, 624
    effects of radiation on, 891-94
    elements of, 67
Life of Pi (Martel), 519
Ligand(s), 1056
    bidentate, 1057, 1059
    capable of linkage isomerization, 1062, 1063
    common, 1058
        names and formulas of, 1060, 1061
    coordination isomerization and, 1062
    geometric isomerism and, 1062-65
    monodentate, 1057
    naming, 1060
    nitro, 1062
    polydentate, 1058, 1059
    strong-field, 1070, 1071
    used in chemical analysis, 1073
    weak-field, 1070, 1071
Light, 282-92
    absorption of, by elements, 294-95
    diffraction, 287-88
    electromagnetic spectrum, 285-87
    interference, 287-88
    packet of. *See* Photon(s)
    particle nature of, 288-92
    rotation of polarized, 911
    visible, 284, 285, 286
    wave nature of, 282-85
    wave-particle duality of, 282, 292
Lightning, 56
Lignite, 1000
Like dissolves like, 524
Limes, pH of, 675
Limestone, 92, 100, 707, 749
Limiting reactant, 137-43
Linear accelerator, 890-91
Linear combination of atomic orbitals (LCAO),
        439-43, 449
Linear geometry
    of complex ions, 1059
    electron, 414, 436
    molecular, 407, 408, 413, 414, 415, 417
Line notation, 826-27

Linkage isomers, 1062, 1063
Linoleic acid, 957
Linolenic acid, 957
Lipid bilayer, 960
Lipids, 956-61
    fats and oils, 958-60
    fatty acids, 956-57
Liquid(s), 460
    cryogenic, 355
    entropy change associated with change in state
        of, 784
    equilibria involving, 629-30
    molecular comparison with other phases, 462-
        65
    nonvolatile, 478
    properties of, 463-64
    relative standard entropies of, 794
    standard state for, 262
    volatile, 478
Liquid helium, 356
Liquid matter, 9
Liquid solution, 521
Liter (L), 19
Lithium, 60, 65
    charge of, 94
    effective nuclear charge for, 336
    electron configuration for, 326, 339
    energy levels of molecular orbitals in, 508
    flame tests for, 295
    Lewis structure for, 366
    orbital diagram for, 326, 443
    properties of, 352
Lithium bromide, 529
Lithium carbonate, 374
Lithium fluoride, dipole moment of, 380
Lithium hydroxide, 161, 687
Lithium ion, electron configuration of, 324, 339
Lithium ion battery, 844-45
Lithium sulfide, reaction between sulfuric acid
        and, 165
Litmus paper, 666
Lizards, 569-70
Logarithm, 674
London force. *See* Dispersion forces
Lone pairs, 374-75
    electron groups with, 410-15
        five, 412-15
        four, 410-11
    hybridization and, 428
    in weak bases, 688
Los Angeles County, carbon monoxide
        concentrations in, 22
Low-spin complexes, 1071, 1072
Lukens, Isaiah, 241
Lysine, 967, 968
Lysozyme, egg-white, 972, 978

M

Maalox, 691
Machines
    atomic, 46
    perpetual motion, 240, 241, 776
Macromolecules, 956
Macrostate (external arrangement), 781-82
Magic numbers, 874
Magnesium
    charge of, 94
    electron configuration of, 347
    ionization energies of, 347

Magnesium-based antacids, 691
Magnesium carbonate, 155, 749
  solubility product constant for, 747
Magnesium hydroxide, 374, 750, 752
  solubility product constant for, 747
Magnesium sulfate, 374
Magnesium sulfate heptahydrate, 96
Magnetic field, 282
Magnetic quantum number, 301, 302
Magnetism
  cobalt and, 1043
  unpaired electrons and, 446
Main-group elements, 63, 64, 65, 66, 988, 990-91
  atomic radii of, 991
    trends in, 334
  effective nuclear charge for, 337
  electron affinities for, 348-49
  ion formation by, 65, 66
  ionization energy of, 344
  valence electrons for, 329, 330
Malachite, 1043
Malic acid, 665
Malleability, 62, 349
  of metals, 397, 1028
Manganese
  functions in human body, 1074
  sources, properties, and products of, 1042
Manganese oxides, 1042
Manhattan Project, 884
Manometer, 190
Maps, probability distribution, 299-300
Markovnikov's rule, 923
Mars, 132
  water on, 494
Mars Climate Orbiter, 15
Marshall, Barry J., 677
Martel, Yann, 519
Mass(es)
  atomic, 67-69, 101
    calculation of, 68
    mass spectrometry and, 68-69
  conservation of, 5, 6, 47-48
  converted to energy by nuclear fission, 887-89
  converting between moles and, 71-74
  of gas, 9
  of isotope, 69
  molar. *See* Molar mass
  SI unit of, 16, 19, 239
Mass action, law of, 622, 625, 627, 804
Mass defect, 887-89
Mass number (*A*), 58-59, 867
  binding energy per nucleon as function of, 889
Mass percent composition, 103-6
  chemical formula from, 109
  as conversion factor, 105-6
Mass spectrometry, 68-69, 575
Matter
  classification of, 7-10
    by composition, 9-10
    by state, 7-9
  defined, 7
  phases of. *See* Gas(es); Liquid(s); Solid(s)
  physical and chemical changes in, 11-13
  properties of, 2
  states of, 7-9, 11
    entropy change associated with change in, 784-85
Maximum contaminant levels (MCLs), 495, 538
Maxwell, James Clerk, 185

Mean free path, 220
Measurement, 15-28
  reliability of, 22-28
    exact numbers, 24-25
    precision and accuracy, 23, 27-28
    significant figures, 23-27
  units of, 15-21. *See also* SI unit(s)
    derived units, 19-21
    English system, 15, 19
    metric system, 15
    prefix multipliers, 19
Mechanical potential energy, 778
Medicine
  antacids, 691
  blood pressure, 191
  bone density, 21
  buffer effectiveness in human blood, 732
  chelating agents used in, 1073
  chymotrypsin in digestion, 604
  colligative properties and, 555-56
  coordination compounds used in, 1076
  definition of life, 624
  elements of life, 67
  ionic compounds as drugs, 374
  methylmercury in fish, 107
  nuclear, 865-66
  oxygen used in, 1012
  potassium iodide in radiation emergencies, 354-55
  radiation exposure from diagnostic medical procedures, 893
  radioactivity in, 866, 893, 894-95
  radiotherapy in, 286, 895
  ulcers, 677
Megaelectron volt (MeV), 888
Mega prefix, 19
Meitner, Lise, 883
Melanin, 173
Melting, 487-88. *See also* Fusion
  of ice, 779
  of ionic compounds, 372
Melting point(s), 487-88
  of alkali metals, 352
  of fatty acids, 957
  of halogens, 353
  of ionic compounds, 372-73
  of molecular compounds, 376
Mendeleev, Dmitri, 60-62, 318, 320-21, 328
Mercury, 58
  absorption spectrum of, 295
  cations formed by, 95
  density of, 20
  emission spectrum of, 292, 295
  in fish, 107
  meniscus of, 476-77
Mesopause, 225
Mesosphere, 225
Messenger rNA (mRNA), 979-80
Metal(s), 62, 1026-49. *See also* Alloys; Group 1A metals (alkali metals); Group 2A metals (alkaline earth metals)
  bonding atomic radii for, 334
  bonding in, 397
  bonding theories of, 1028
  closest-packed crystal structures in, 506
  dissolved in acids, 98, 833
  ductility of, 397
  earth, 1029
  electrolysis to plate metals onto other, 847

  electron affinities for, 348
  forming more than one kind of cation, 94-95
  forming only one type of cation, 94
  general properties and natural distribution of, 1028-29
  group 1, 320
  ionization energies of, 366
  malleability of, 397
  mineral sources for, 1029
  name in anionic complex, 1061
  noble, 1029
  oxidation (corrosion) of, 854-55
  properties of, 62
  reaction between nonmetals and, 85, 168-69. *See also* Oxidation-reduction reaction(s)
  reaction of alcohols with active, 930
  structures of, 1033-34
  transition. *See* Transition metal(s)
Metal chlorides, lattice energies of, 371
Metal halides, 353
Metal hydride, in batteries, 844
Metal hydroxides, 687
  amphoteric, 761-63
Metallic atomic solids, 504, 506
Metallic bond(s), 365, 366, 506
  electron sea model of, 366
Metallic carbides, 1003
Metallic character, periodic trends in, 349-51
Metalloid(s), 62, 63
  covalent carbides composed of carbon and, 1003
  properties of, 63
Metallurgy, 1026, 1029-33
  defined, 1028
  electrometallurgy, 1032-33
  extractive, 1029
  hydrometallurgy, 1031
  powder, 1033
  pyrometallurgy, 1030-31
  separation, 1030
Metal nitrates, 1009
Metal nitrides, 1005
Metal oxides, electrolysis of, 847
Metal sulfides, 1014, 1015-16
  common, 1016
  reaction between hydrochloric acid and, 1015
Meter (m), 16
Methanal. *See* Formaldehyde
Methane, 120, 904, 912, 928
  combustion of, 115, 389
    standard heat of reaction for, 264-65
  common uses of, 119
  molecular formula for, 87, 119
  molecular geometry for, 408, 426-27
  space-filling model for, 117, 119
  standard enthalpy of formation for, 262
  structural formula for, 117, 119
  van der Waals constants for, 223
Methanoic acid (formic acid), 672, 935
Methanol, 119, 120, 524, 927, 928, 929
  reaction between sodium and, 930
Methionine, 968
Method of successive approximations, 642
Methy butanoate, 934
Methyl acetate, 120
Methylamine, 688
  reaction between hydrochloric acid and, 938
Methylammonium chloride, 938
Methyl butanoate, 936

3-Methylhexane, optical isomers of, 910
Methyl isonitrile, 588-89
Methylmercury in fish, 107
Methyl propanoate, 935
Methyl red, 746, 747
Methyl substituent, 914
Methyl tertiary butyl ether (MTBE), 143
Metric system, 15
Meyer, Julius Lothar, 320
Mica, 995
Micelles, 558-59
Michaelis-Menten equation, 987
Micro prefix, 19
Microscope, scanning tunneling (STM), 45-46
Microwaves, 285, 287
Milk, 557
Milk of magnesia, 160, 663, 664, 677, 691
    pH of, 675
Milligrams solute/per liter, 536
Millikan, Robert, 53
Millikan's oil drop experiment, 53
Milliliter (mL), 19
Millimeter of mercury (mmHg), 188, 189
Milli prefix, 19
Minerals, 1029
    chloride, 1029
    extracting metal from, 1030-33
    oxide, 1029
    separating from gangue, 1030
Mirex, 538
Miscibility, 469
Miscible solid solutions, 1034-35
Mixing
    spontaneous, 520
    tendency toward, 541-42
Mixture(s), 9
    of cations or anions, electrolysis of, 849
    compounds vs., 84
    defined, 10
    of gas, 204-6
    homogeneous. *See* Solution(s)
    separating, 10
Models, theories as, 7
Moissanite, 1003
Molality (*m*), 535, 536, 540
Molar heat capacity. *See* Specific heat capacity
    (CS)
Molarity (M), 144-46, 149, 535, 540
    in calculations, 145-46
Molar mass, 69-75, 101-3, 794-95
    defined, 71
    dispersion force and, 466-67
    empirical formula, 111
    of gas, 203-4
    variation of velocity distribution with, 218
    viscosity and, 475
Molar solubility, 748-50
Molar volume
    at standard temperature and pressure, 201
    stoichiometry and, 212-13
Mole(s), 70, 75
    for compounds, 101-3
    converting between mass and, 71-74
    converting between number of atoms and
        number of, 70-71
Molecular complexity, relative standard entropies
        and, 795-96
Molecular compound(s), 86, 89, 90-92, 96-100
    formulas for, 96

identifying, 97
    Lewis structures for, 382-83
    melting and boiling points of, 376
    naming, 97-100
        acids, 98-99
        binary acids, 99
        oxyacids, 99-100
Molecular elements, 89-90
Molecular equation, 159, 160
Molecular formula, 86, 87
Molecular geometry
    bent, 411, 414, 417
    electron group repulsion and, 406
    linear, 407, 408, 413, 414, 415, 417
    lone pairs effect, 410-14
    octahedral, 413, 415, 417, 435
    polarity and, 418-22
    predicting, with VSEPR, 415-18
    seesaw, 412, 415, 417
    square planar, 413, 415, 417
    square pyramidal, 413, 415
    tetrahedral, 408, 411, 414, 417, 426-28
    trigonal bipyramidal, 409-10, 413, 415,
        417, 434
    trigonal planar, 407, 408, 414, 417, 428
    trigonal pyramidal, 410, 411, 414, 417
    T-shaped, 412, 415
Molecularity, 594
Molecular mass. *See* Formula mass
Molecular models, 87-88
    size of molecules and, 103
Molecular orbitals
    antibonding, 440-41, 444
    bonding, 440-41, 444, 445
    nonbonding, 449
Molecular orbital theory, 362, 404, 426, 438-51,
        508
    linear combination of atomic orbitals (LCAO),
        439-43, 449
    period two homonuclear diatomic molecules,
        443-48
    polyatomic molecules, 450-51
    second-period heteronuclear diatomic
        molecules, 448-50
    trial mathematical functions in, 439
Molecular solids, 504, 505-6
Molecular structure, acid strength and, 702-4
    binary acids, 703
    oxyacids, 703-4
Molecular velocities, temperature and, 217-19
Molecular weight. *See* Formula mass
Molecule(s), 3-5
    diatomic. *See* Diatomic molecule(s)
    formula mass, 101-3
    as Lewis acids, 705-6
    mass spectrum of, 69
    nonpolar, 381
    organic, 904
    polar, 381, 468-69
    polyatomic, 89
        molecular orbital theory applied to, 450-51
    properties of matter and, 2, 3-4
    shapes of, 406. *See also* Molecular geometry
        dispersion force and, 467
    temperature and motion of, 17
Mole fraction (χ_{solute}), 205, 535, 539-40
Mole percent (mol %), 535, 539-40
Molybdenite, 1029
Molybdenum, functions in human body, 1074

Mond process, 1044
Monel (alloy), 1044
Monochloropentafluoroethane, 580
Monodentate ligands, 1057
Monomeric proteins, 973
Monomers, 938, 939, 940
Monoprotic acids, 670, 672, 683
Monosaccharides, 961-63, 964
Monosubstituted benzenes, 925
Monoun, Lake, 534
Monounsaturated fat, 959
Monounsaturated fatty acids, 957
Moseley, Henry, 321
Motile proteins, 966
Motion
    Brownian, 557
    Newton's laws of, 299
Motor oil, 478
    viscosity of, 476
mRNA, 979-80
MTBE in gasoline, 143
Multimeric proteins, 973
Multiple proportions, law of, 49-50
Multiwalled nanotubes (MWNT), 1001
Mylanta, 160, 664, 691
Myristic acid, 956-57

N

Names. *See* Nomenclature
Nanomachines, 1002
Nano prefix, 19
Nanotechnology, 42, 46
Nanotubes, 1001-2
    boron nitride, 989-90
    carbon, 989, 1001-2
Nanowires, insulated, 989-90
Naphthalene, 926
National Air Quality and Emissions Trends
        Report, 226
National Institutes of Health (NIH), 980
Natural abundance of isotopes, 58
Natural gas, 114, 237, 268
    heating home with electricity vs., 776, 777
    reaction between oxygen and, 174
Natural radiation, exposure to, 893
Nature of Chemical Bond, The (Pauling), 377-78
Nature's heat tax, 775-76
Negative charge, 824
Neon, 63, 522
    electron configuration for, 327
    emission spectrum of, 294
    isotopes of, 58-59
    Lewis structure of, 366
    MO diagram for, 446
    orbital diagram for, 327
    properties of, 355
    van der Waals constants for, 223
Neopentane, 467
Nernst equation, 838, 841
Nerve cells, 319
    concentration cells in human, 842
Nerve signal transmission, 319-20
Nesosilicates (orthosilicates), 993, 995
Net ionic equations, 159, 160
Network covalent atomic solids, 504, 506-7
Neutralization reactions of carboxylic acids,
        936
Neutral ligands, naming, 1060, 1061
Neutral solution, 673

Neutron(s), 55, 872
  actual number of, 874
  charge of, 56
  mass of, 56
  number of, 867
  N/Z ratio and, 873-74
Neutron stars, 55, 80
Neutron symbol, 868
Newlands, John, 320
Newton, Isaac, 47
Newton's laws of motion, 299
Newtons (N), 186
Nickel, 506
  nickel-chromium phase diagram, 1036-37
  nickel-copper alloy, 1034, 1035
  sources, properties, and products of, 1044
Nickel-cadmium (NiCad) battery, 844
Nickel-metal hydride (NiMH) battery, 844
Nido-Boranes, 998
NIH, 980
Niobium, 1029
Nitrate, 95
Nitrate ion, solubility and, 154-55
Nitrates, 1008-9
Nitric acid, 100, 136, 161, 664, 665, 670, 706, 1008
  reduction half-reaction oxidizing metals, 833
Nitric oxide (nitrogen monoxide), 388, 600, 1005, 1007-8
  electron density map of, 449
  MO diagram for, 448-49
  reaction between hydrogen and, 578, 597
Nitride, 94
Nitrite, 95, 1009
Nitrogen, 1004-9
  in air, 204
  compounds, 1006-9
  electron affinity of, 348
  electron configuration for, 327
  elemental, 1005
  ionization energy of, 346
  Lewis structure of, 366
  MO diagram for, 446
  orbital diagram for, 327
  reaction between hydrogen and, 651
  sources for, 1005
  in water, Henry's law constants for, 533
Nitrogen dioxide, 97, 1007
  reaction between carbon monoxide and, 595-96
Nitrogen gas, van der Waals constants for, 223
Nitrogen-halogen single bonds, 396
Nitrogen hydrides, 1006-7
Nitrogen monoxide. *See* Nitric oxide (nitrogen monoxide)
Nitrogen narcosis (rapture of the deep), 207
Nitrogen oxides, 226, 706, 1007-8
Nitrogen tetroxide, 646-48
Nitro ligand, 1062
m-Nitrophenol, 747
Nitrosamines, 1009
Nitrosyl chloride, 593
Nitrous acid, acid ionization constant for, 672
Nitrous oxide (dinitrogen monoxide), 97, 1008
Noble gas(es), 63, 64, 330, 333
  boiling points of, 466, 467
  electron configurations of, 351
  ionization energy and, 344
  properties of, 355-56
  standard entropies of, 794-95
Noble metals, 1029

Nomenclature
  for acids, 98
  common names, 93
  for hydrocarbons, 118
  for ionic compounds, 93-96
    containing metal that forms more than one kind of cation, 94-95
    containing metal that forms only one type of cation, 94
    containing polyatomic ions, 95-96
  IUPAC system, 914
  for molecular compounds, 97-100
    acids, 98-99
    binary acids, 99
    oxyacids, 99-100
  for neutral ligands, 1060, 1061
  systematic names, 93
n-Nonane, 475, 913
Nonbonding atomic radius (van der Waals radius), 334
Nonbonding atomic solids, 504, 506
Nonbonding orbitals, 449
Nonelectrolyte solutions, 152-53
Nonmetal(s), 62-63, 349, 988-1025. *See also* Boron; Carbon; Main-group elements; Nitrogen; Oxygen; Phosphorus; Sulfur
  atomic size, 991
  bonding atomic radii for, 334
  bonds, types of, 991
  carbides, 1002-3
  carbonates, 1004
  carbon oxides, 1003
  electron affinities of, 366
  halogens. *See* Halogens
  insulated nanowires, 989-90
  ion formation by, 65, 66
  ionization energies of, 366
  main-group elements. *See* Main-group elements
  nitrogen. *See* Nitrogen
  oxidation states for, 169
  properties of, 62-63
  reaction between metal and, 85, 168-69. *See also* Oxidation-reduction reaction(s)
  reactions of alkali metals with, 352
  silicates, 992-96
Nonpolar covalent bond, 379
Nonpolar molecules, 381
Nonpolar solvents, 524
Nonspontaneous process, 778, 786
  increase in Gibbs free energy and, 789
  made spontaneous, 802
Nonvolatile liquids, 478
Normal boiling point, 482
Normal science, 7
n-type semiconductor, 509
Nuclear binding energy, 888-89
Nuclear charge, 324
Nuclear chemistry. *See also* Radioactivity
  effects of radiation on life, 891-94
  fission, 883-86
    atomic bomb and, 884
    converting mass to energy, 887-89
    mass defect and, 887-89
    nuclear binding energy, 888-89
    nuclear power, to generate electricity, 884-86
  nuclear fusion, 889-90
  nuclear transmutation and transuranium elements, 890-91

Nuclear energy industry
  boron used in, 997
  interhalide used in, 1020
Nuclear equation(s), 868-71
  for fission reaction, 883
Nuclear fission, 883-89
  converting mass to energy in, 887-89
    mass defect and, 887-89
    nuclear binding energy and, 888-89
  discovery of, 883-86
Nuclear fusion, 889-90
Nuclear medicine, 865-66
Nuclear power, 884-86
Nuclear power plant, 885-86
Nuclear stability, 872-74
  magic numbers and, 874
  ratio of neutrons to protons (N/Z) and, 873
Nuclear theory of atom, 55
Nuclear transmutation, 890-91
Nuclear waste disposal, 886
Nuclear weapons, 884
Nucleic acids
  basic structure of, 974-76
  as blueprints for proteins, 974-78
  genetic code and, 977-78
  mass spectrum of, 69
Nucleon(s)
  attractive strong force among, 872
  binding energy per, 888-89
  number of stable nuclides with even and odd numbers of, 874
Nucleotide(s), 473, 974-75
  fake, viral drug therapies using, 987
  in genetic structure, 977
Nucleus, 55
  predicting the mode of decay for, 872-74
  strong force binding, 872
Nuclide(s), 867
  binding energy per nucleon for, 888-89
  daughter, 868
  half-lives of selected, 876
Nutrasweet (aspartame), 405, 406
Nylon 6,6#, 940, 941
Nyos, Lake, carbon dioxide accumulation in, 534
N/Z ratio, 873

O

Observation, 5
Octahedral complexes
  cis-trans isomerism in, 1063
  d orbital energy changes for, 1067-72
  fac-mer isomerism in, 1064
  high-spin, 1071
  low-spin, 1071
  optical isomerism in, 1066
Octahedral geometry
  of complex ions, 1059
  electron, 415, 436
  molecular, 413, 415, 417, 435
Octahedral hole, 1038-39
Octane, combustion of, 133-34, 143
n-Octane, 475, 912, 913
Octet(s), 366
  expanded, 434
Octet rule, 367, 376, 387-91
  expanded, 390-91
  incomplete, 389-90
  odd-electron species, 388-89
Odd-electron species, 388-89

Odors, 903-4, 932-33. *See also* Aromatic hydrocarbons
OH⁻. *See* Hydroxide ion (OH⁻) concentration
-OH functional group, 119
Oil(s), 958-60
  motor, 476, 478
  vanadium contamination in, 1027-28
Oil drop experiment, Millikan's, 53, 56
Oleic acid, 957
Oleum, 1017
Oligopeptides, 969
Olivines, 993
On the Revolution of the Heavenly Orbs (Copernicus), 47
Opal, 557
Opaque, metals as, 1028
Oppenheimer, J.R., 884
Optical isomerism, 909-10, 1062, 1065-66
  carbohydrates and, 962-63
  defined, 909
  in hydrocarbons, 909-12
Orbital(s)
  atomic, 301-3
    atomic radius and, 335
    degenerate, 323, 324, 326
    electron configuration and, 321
    energy ordering of, for multielectron atoms, 325
    hybridized. *See* Hybridization
    periodic table and, 330-31
    shapes of, 306-12
  molecular
    antibonding, 440-41, 444
    bonding, 440-41, 444, 445
    nonbonding, 449
Orbital angular momentum, 446, 447
Orbital diagram, 322, 323
Orbital overlap. *See* Valence bond theory
Order of magnitude estimations, 33-34
Ores, 1029
  chromium, 1041
  cobalt, 1042
  copper, 1043
  processing metal-containing, 1029-33
  zinc, 1044
Organic bases, 473-74
Organic chemistry, 67, 902-53. *See also* Hydrocarbons
  alcohols, 928-31
  aldehydes, 931-34
  amines, 938
  carbon and, uniqueness of, 902, 904-5
  carboxylic acids, 934-37
  esters, 934-37
  ethers, 937
  ketones, 931-34
  polymers, 938-41
Organic compound(s), 117-20, 905
  carbon and, 117-18
  decomposition of, 117
  families of, 120
  hydrocarbons. *See* Hydrocarbons
  properties of, 117
Organic molecules, 904
Orientation factor, 592-93
Orthosilicates (nesosilicates), 993, 995
Osmosis, 552-54
  defined, 552
Osmosis cell, 552-53

Osmotic pressure, 552-53, 554
Osteoporosis, 21
Ostwald process, 1008
Outer electron configuration, 331
Outermost electrons, 337
Overvoltage, electrolysis of aqueous sodium chloride and, 851
Oxalate ion, 1057, 1058
Oxalic acid, ionization constants for, 699
Oxidation, definition of, 168, 820
Oxidation-reduction reaction(s), 167-74
  of alcohols, 930, 933
  in aqueous solutions, balancing, 820-23
    acidic solution, 821-22
    basic solution, 822-23
  in batteries, 842-46
  combustion reactions, 173-74
  corrosion as undesirable, 854-55
  fuel cells based on, 820
  identifying, 171-72
    through changes in oxidation states, 820
  with oxygen, 167
  without oxygen, 168
  with partial electron transfer, 168, 169
  periodic trends and the direction of spontaneity for, 836
  predicting the spontaneous direction of, 831-33
  spontaneous, generating electricity from, 823-27
Oxidation state(s) (oxidation number(s)), 169-71, 820
  fractional, 170
  to identify redox reactions, 171
  rules for assigning, 169
  of transition metals, 1056
Oxide(s), 94, 1012
  halogen, 1020-21
  nitrogen, 226, 706, 1007-8
  phosphorus, 1010
  types of, 1012
Oxide minerals, 1029
Oxidizing agents, 172, 353, 1008, 1013, 1018, 1021, 1041-42
  nonmetals as, 991
  positive reduction half-cell potentials of, 831
Oxyacids (oxoacids), 703-4
  naming, 99-100
Oxyanions, 95
Oxygen, 58, 83-84, 1011-13
  in air, 204
  boron-oxygen compounds, 997
  electron configuration for, 327
  elemental, 1011-12
  emission spectrum of, 294
  ionization energy of, 346
  Lewis structure of, 366, 374
  liquid, 447
  orbital diagram for, 327, 446
  oxidation state for, 169
  as oxidizing agent, 172
  paramagnetism of, 446
  partial pressure limits of, 207
  production of, 1011-12
  properties of, 83
  reactions of
    with carbon, 242
    with ethanol, 174
    with hemoglobin, 617-19
    with natural gas, 174

    with sodium, 168
    with white phosphorus, 1010
  redox reactions with, 167
  redox reactions without, 168
  silicon and, in silicate tetrahedron, 992
  uses for, 1012
  van der Waals constants for, 223
  in water, Henry's law constants for, 533
Oxygenated fuel, 143
Oxygen toxicity, 207
Oxygen transport, vanadium compounds used for, 1028
Ozone, 226, 1012-13
  from fuel fragments in exhaust, 600
  Lewis structure of, 396, 450-51
  molecular orbital model of, 451
  photodecomposition of, 388
  properties of, 39
  reaction between chlorine and, 227
  use of, 1013
  valence bond model of, 450-51
Ozone layer, 580, 1013
  catalytic destruction of, 227, 599, 600-601
  depletion of, 226-27
  hole over Antarctica, 103, 227, 600-601

P

Packet of light. *See* Photon(s)
Packing efficiency, 499, 500, 502
Palmitic acid, 277, 957
Pancreas, 604
Parabolic troughs, 270
Paramagnetic atom or ion, 339
Paramagnetism, 446
Parent nuclide, 868
Partially hydrogenated vegetable oil, 923
Partial pressures, 204-10, 626-27
  Dalton's law of, 204, 215
  deep-sea diving and, 206-8
  vapor pressure, 209-10
Particle nature of light, 288-92
Parts by mass, 535, 536-37, 540
Parts by volume, 536-37, 540
Parts per billion by mass (ppb), 535, 536
Parts per million by mass (ppm), 535, 536
Pascal (Pa), 189
Patchouli alcohol, 903, 904
Patina, 1044
Pauli, Wolfgang, 323
Pauli exclusion principle, 322-23, 326
Pauling, Linus, 131, 377-78
Peaches, pH of, 675
Penetrating power, 869
  of alpha radiation, 869
  of gamma rays, 870
Penetration, 324-25
Pentaamminenitritocobalt(III), 1062, 1063
Pentaamminenitrocobalt(III), 1062, 1063
Pentagonal bipyramidal geometry, 1020
Pentanal, 932
Pentane, 469, 523, 907
n-Pentane, 467, 475, 912, 913
  common uses of, 119
  critical point transition for, 486
  dynamic equilibrium in, 481
  molecular formula for, 119
  space-filling model of, 119
  structural formula for, 119
Pentanoic acid, 935

2-Pentanone, 932

1-Pentene, 917

1-Pentyne, 918

Peptide bond(s), 604, 969, 980
  between amino acids, 969-70

Percent by mass (%), 535, 536

Percent ionic character, 380, 381

Percent ionization of weak acid, 683-84

Percent mass to volume, 556

Percent yield, 138-43

Perchlorate, 95, 96

Perchloric acid, 161, 670

Peridot, 1052

Periodic law, 60-62, 321

Periodic property(ies), 318-61
  of alkali metals (group 1A), 351, 352-53
  defined, 320
  electron affinities, 348-49
  electron configurations and, 321-32, 339-40
    electron spin and the Pauli exclusion
        principle, 322-23
    element's properties and, 328-29
    for multielectron atoms, 326-28
    orbital blocks in periodic table, 330-31
    sublevel energy splitting in multielectron
        atoms, 323-25
    valence electrons, 329
    writing, from periodic table, 331-32
  of halogens (group 7A), 351, 353-54
  of ions, 338-48
    electron configurations of, 339-40
    ionic radii, 341-43
    ionization energy, 343-48
    magnetic properties, 339-40
  metallic character, 349-51
  nerve signal transmission and, 319-20
  of noble gases (group 8A), 351, 355-56
  periodic trends in size of atoms, 335-38
    effective nuclear charge and, 336-37
    transition elements and, 337-38

Periodic table, 60-67, 318, 991. See also specific
        elements; specific families or groups
  atomic mass, 67-69
  development of, 320-21
  groups 3A-7A, 990
  ions and, 65-66
  metalloids, 62, 63
  metals, 62
  modern, 62
  noble gases, 63, 64
  nonmetals, 62-63
  orbital blocks in, 330-31
  organization of, 57
  quantum-mechanical theory and, 318, 321-28
  transition elements or transition metals, 63, 64,
        66
  writing electron configuration from, 331-32

Permanent dipoles, 468

Permanganate, 95

Permanganate ion, 1042

Peroxide, 95, 1012

Perpetual motion machine, 240, 241, 776

Persistent organic pollutants (POPs), 538

Perturbation theory, 423

PET, 895

Peta prefix, 19

Petroleum, 268

pH, 674-86. See also Acid-base chemistry; Buffers
  of blood, 718, 732

of buffers, 720-28
  equilibrium calculation for, 724
  equilibrium calculation of changes in, 725-
      26
  Henderson-Hasselbalch equation for, 722-25
  stoichiometry calculation of changes in, 724-
      25, 726
of mixture of acids, 684-86
pH curve, 734. See also Acid-base titration
pH meter, 744
pH scale, 662
of polyprotic acids, 698-701
of salt solutions, 696-98
solubility and, 752-53
of solution with anion acting as weak base, 693-
      94
of solution with conjugate acid of weak base,
      695
of strong acid solutions, 677-78
of weak acid solutions, 678-82, 684
of weak bases, 688-90

Phase changes. See Phase transition(s)

Phase diagram(s), 491-93
  binary, 1035, 1036
  chromium and nickel, 1036-37
  guidelines for interpreting, 1035
  major features of, 491-92
  navigation within, 492-93
  of other substances, 493
  for water, 491-93

Phases of matter, molecular comparison of, 462-
        65. See also Gas(es); Liquid(s); Solid(s)

Phase transition(s), 464-65. See also
        Condensation; Vaporization
  critical point, 486-87
  deposition, 487
  entropy and, 784
  freezing, 488, 495
  melting or fusion, 487-88
  sublimation, 12, 487, 576

Phenol, 925
  acid ionization constant for, 672

Phenolphthalein, 744-45, 747

Phenol red, 747

4-Phenyl-1-hexene, 925

Phenylalanine, 967, 968

Phenyl group, 925

3-Phenylheptane, 925

Phosgene (carbonyl chloride), 402, 1003

Phosphate(s), 95, 1010
  in food industry, 1011
  insoluble, 758

Phosphate links, in nucleic acids, 975-76

Phosphatidylcholine, 960

Phosphide, 94

Phosphine, 1009

Phospholipids, 960

Phosphorescence, 866

Phosphoric acid, 664, 671, 1010
  ionization constants for, 699

Phosphorus, 89, 1004
  black, 1006
  compounds, 1009-11
  elemental, 1005-6
  ionization energies of, 347
  red, 1005-6
  silicon doped with, 509
  white, 1005, 1009

Phosphorus-30, 870

Phosphorus-32, 894

Phosphorus halides, 1009

Phosphorus oxides, 1010

Phosphorus oxychloride, 1009

Phosphorus oxyhalides, 1009

Phosphorus pentachloride, 409

Photochemical smog, 226

Photoelectric effect, 288-92

Photon(s), 289-91
  electron relaxation and release of, 304
  energy of, 294

Photosynthesis, 135, 1003
  atmospheric oxygen from, 1011
  chlorophyll and, 1075

Phyllosilicates, 995

Physical changes, 11-13

Physical property, 11

Physics, classical, 299

π_{2p} bonding orbital, 444, 445

Pi (π) bond, 430

Pickling fluids, 677

Pico prefix, 19

pK_a scale, 677

Planck, Max, 280, 289

Planck's constant, 289

Plane-polarized light, 911

Plastic products, 938-39

Platinum
  density of, 20
  inert electrode of, 827

Plato, 5, 47

Plum-pudding model, 54, 55

Plums, pH of, 675

p-n junctions, 509

pOH scale, 676-77

Poise (P), 475

Polar bonds, 418-19

Polar covalent bond, 377

Polarimetry, 574, 575

Polarity
  bond, 379-81, 703
  molecular shape and, 418-22

Polarized light, rotation of, 911

Polar molecules, 381, 468-69

Polar solvents, 524

Polar stratospheric clouds (PSCs), 227, 601

Pollutant(s), 100
  persistent organic (POPs), 538
  sulfuric acid as, 1016

Pollution, 3
  air, 225-26, 268
  catalytic converters and, 600
  water, 495

Polonium, 58, 867

Polyatomic ion(s), 90-91, 95
  Lewis structures for, 382, 383-84
  solubility and, 155

Polyatomic molecules, 89

Polychlorinated biphenyls (PCBs), 538

Polycyclic aromatic hydrocarbons, 926

Polydentate ligands, 1058, 1059

Polyethylene, 939, 940

Polyethylene terephthalate, 940

Polymer(s), 938-41
  addition, 939, 940
  coiled, 476
  of commercial importance, 940
  condensation, 940, 941

Polypeptides, 969

Polypropylene, 940
Polyprotic acids, 161, 671
  acid ionization constants for, 698, 699
  concentration of anions for weak diprotic acid
    solution, 701-2
  dissociation of, 702-3
  ionization of, 698
  pH of, 698-701
Polysaccharides, 964-65
Polystyrene, 940
Polyunsaturated fatty acids, 957
Polyurethane, 940
Polyvinyl chloride (PVC), 939, 940
Popper, Karl, 362, 363
p orbitals, 302, 306, 308-9, 310
  2p, 308, 310, 323-24
Porphyrin, 1050, 1074-75
Position
  in classical mechanics, 299
  velocity and, as complementary properties, 301
Positive charge, 824
Positron, 870
Positron emission, 867, 870, 871
Positron emission tomography (PET), 895
Potassium
  charge of, 94
  flame tests for, 295
  properties of, 352
  reaction between bromine and, 593
  reaction between chlorine and, 367
Potassium bromide, 696
Potassium chloride (sylvite), 92, 94, 1029
Potassium hydroxide, 161, 529, 666, 687
  reaction between sulfuric acid and, 162
Potassium iodide, 374, 754
  in radiation emergencies, 354-55
  reaction between lead(II) nitrate and, 155-56
  reaction between sodium chloride and, 156-57
Potassium ions, nerve signal transmission and,
    319-20
Potassium nitrate, 531
Potassium nitrite, 696
Potassium permanganate, 374
Potential difference, 825-26, 834
  SI unit of, 825
Potential energy, 14, 15, 238
  of charged particles, 364-65
  Coulomb's law and, 465
  exothermic chemical reaction and, 255
  solution formation and, 522
  stability of covalent bond and, 85-86
  transformation of, 239
Potential energy per unit charge, difference of,
    834
Pounds per square inch (psi), 189
Powder metallurgy, 1033
Power grid, 819-20
Power plants, fuel-cell, 819-20
Precipitate, 155
Precipitation, 155-59, 753-56
  reaction quotient and, 753-54
  selective, 755-56
    qualitative analysis by, 756-59
  solubility and, 156-58
  writing equations for, 158-59
Precision, 23, 27-28
Pre-exponential factor. See Frequency factor (A)
Prefixes
  for base names of alkane chains, 914

hydrate, 96, 97
  in naming molecular compounds, 97
Prefix multipliers, 19
Pressure(s), 185-91. See also Gas(es)
  atmospheric, 185-87
  blood, 191
  calculating, 216-17
  critical, 486
  defined, 185-86, 188
  dynamic equilibrium and, 481
  equilibrium constant in terms of, 626-28
  gas, 186
  gas solubility in water and, 532-34
  gauge, 200
  kinetic molecular theory and, 215
  Le Châtelier's principle on change in, 649-50
  manometer to measure, 190
  osmotic, 552-53, 554
  partial, 204-10, 626-27
    Dalton's law of, 204, 215
    deep-sea diving and, 206-8
    vapor pressure, 209-10
  particle density and, 188
  phase changes and, 464
  reaction and, 213
  reaction rates and, 574-75
  SI unit of, 189
  temperature and, 199
  total, 200
  units of, 188-90
  vapor, 209-10
  volume and, 192-94
Pressure-volume work, 249-50
Priestley, Joseph, 1011
Primary structure of protein, 971, 972, 974
Primary valence, 1056
Principal level (principal shell), 302
Principal quantum numbers, 301, 344
Principles, 5
Probability density, 307, 309
  radial distribution function vs., 308
Probability distribution maps, 299-300
  for electron states, 301
Problem solving, 28-35
  general strategy for, 30-31
  involving equations, 28, 34-35
  order of magnitude estimations, 33-34
  unit conversion problems, 28-29, 31-33
    units raised to a power, 31-32
Products, 114, 131
Proline, 968
Propanal, 931
Propane, 118, 905, 912, 923
  burning of, 12
  common uses of, 119
  liquid, 464
  molecular formula for, 119
  space-filling model of, 119
  structural formula for, 119
Propanoic acid, 935
  reaction between sodium hydroxide and, 936
2-Propanol. See Isopropyl alcohol
Propanone. See Acetone
Propene, 917
  reaction between hydrogen chloride and, 923
  reaction between hydrogen gas and, 923
  structural formula of, 908
Property
  extensive, 20

intensive, 20
Propyl substituent, 914
Propyne, 918
  structural formula of, 908
Protactinium-234, 874
Protease inhibitors, 364
Protective proteins, 966
Protein(s), 363, 956, 966-81. See also Amino acids
  active site of, 406
  classes of, 966
  complete, 970
  digestion of, 604
  fibrous, 970-71, 973
  functions, 966
  globular, 971, 973
  mass spectrum of, 69
  monomeric, 973
  multimeric, 973
  nucleic acids as blueprints for, 974-78
  structure of, 970-74
  synthesis of, 978-81
Proton(s), 42, 55, 56
  actual number of, 874
  Brønsted-Lowry definition of acids and bases
    and, 667-68, 702, 704-5
  charge of, 56
  ionizable, 670, 671
  mass of, 56
  number of, as definitive of element, 57
  N/Z ratio and, 873-74
  repulsive electrostatic force among, 872
Proton symbol, 868
Proust, Joseph, 48
Pseudogout, 772
p-type semiconductor, 509
Pump(s)
  ion, 319
  water, 186-87
Pure compounds, standard enthalpy of formation
    for, 262
Pure elements, standard enthalpy of formation
    for, 262
Pure substances, 9, 10
Purine bases, 975, 976
PVC, 939, 940
Pyrene, 926, 927
Pyrex, specific heat of, 247
Pyridine, 688
Pyrimidine bases, 975, 976
Pyrolusite, 1042
Pyrometallurgy, 1030-31
Pyrosilicates (sorosilicates), 993, 995
Pyroxenes (inosilicates), 994, 995

Q

Quadratic equations, 638
Qualitative chemical analysis, 295, 756-59
  acid-insoluble sulfides, 758
  alkali metals and NH_4^+, 758
  base-insoluble sulfides and hydroxides, 758
  general scheme for, 757-59
  insoluble chlorides, 758
  insoluble phosphates, 758
Quantitative analysis, 295, 756
Quantum-mechanical model of atom, 280-317
  atomic spectroscopy
    Bohr model and, 292-95
    explanation of, 303-6
    explanatory power of, 332-33

Quantum-mechanical model of atom, *(continued)*
  light, 282-92
    diffraction, 287-88
    electromagnetic spectrum, 285-87
    interference, 287-88
    particle nature of, 288-92
    visible, 284, 285, 286
    wave nature of, 282-85
    wave-particle duality of, 282, 292
  periodic table and, 321-28
  Schrödinger equation for hydrogen atom, 301-3
  shapes of atomic orbitals, 306-12
    d orbitals, 306, 310, 311
    f orbitals, 310, 311
    p orbitals, 306, 308-9, 310
    s orbitals, 306, 307-8, 309
  wave nature of electron, 295-300
    de Broglie wavelength and, 297
    indeterminacy and probability distribution maps, 299-300
    uncertainty principle and, 298-99
Quantum mechanics, 62
  periodic table and, 318
Quantum numbers, 301-3
  Pauli exclusion principle and, 323
Quartz, 507
  structure of, 992
Quaternary structure of protein, 971, 973, 974

R

Racemic mixture, 911
Radial distribution function, 308, 309, 323-24
Radiation
  in cancer treatment, 286
  electromagnetic, 282-83. *See also* Light
  of hydrogen energy, 304-5
  ionizing, 286
Radiation emergencies, potassium iodide in, 354-55
Radiation exposure
  effects of, 891-93
  measuring, 892-94
  by source, in U.S., 893
Radicals. *See* **Free radical(s)**
Radioactive atoms, 866
Radioactive decay, kinetics of, 875-83
Radioactive decay series, 874
Radioactivity, 864. *See also* **Nuclear chemistry**
  defined, 866
  detecting, 874-75
  discovery of, 54, 866-67
  kinetics of radioactive decay and radiometric dating, 875-83
    integrated rate law, 877-79
    radiocarbon dating, 879-81
    uranium/lead dating, 881-83
  in medicine, 866, 893, 894-95
  other applications, 895
  types of, 867-72
    alpha (α) decay, 867, 868-69, 871
    beta (β) decay, 867, 869-70, 871
    electron capture, 867, 870-71
    gamma (γ) ray emission, 867, 870, 871
    positron emission, 867, 870, 871
    predicting, 872-74
Radiometric dating, 879-83
  radiocarbon dating, 879-81
  uranium/lead dating, 881-83
Radiotherapy in medicine, 895

Radiotracer, 894
Radio waves, 285, 287
Radium, 867, 1029
Radium-228, nuclear equation for beta decay of, 869
Radon, 877
Radon-220, decay of, 876
Rain, acid. *See* Acid rain
Rainwater, pH of, 675
Rana sylvatica (wood frogs), 550-51
Random coils, 973
Random error, 27
Raoult, François-Marie, 519
Raoult's law, 543, 545-48
  deviations from, 546-48
  ideal solution and, 545-46
Rapture of the deep (nitrogen narcosis), 207
Rate constant(s) (k), 575, 576
  for second-order reactions, 578
  temperature dependence of, 587
  for zero-order reaction, 578
Rate-determining steps, 595-96
Rate law, 575-79, 596
  containing intermediates, 596, 597
  determining order of reaction, 577-78
  differential, 580-81
  for elementary steps, 594-95
  first-order reaction, 575, 576, 587
  integrated, 580-87, 877-79
    first-order, 580-82, 587
    half-life of reaction, 584-86, 587
    second-order, 583-84, 587
    zero-order, 584, 587
  overall, 595-96
  reaction order for multiple reactants, 578-79
  second-order reaction, 575, 576-78, 587
  zero-order reaction, 575, 576, 577-78, 587
RBE, 893
Reactant(s), 114, 131
  in excess, 138
  limiting, 137-43
  reaction order for multiple, 578-79
Reaction(s), 114. *See also* **Equilibrium/equilibria;**
    specific kinds of reactions
  acid-base, 160-65
    equations for, 163
    titrations, 163-65
  calculating standard changes in entropy, 793-97
  combustion. *See* Combustion
  direction of, 633-35
  endothermic, 254-55, 650, 651, 790, 791
    bond energies and, 394
  enthalpy change for (ΔH_{rxn}), 255-59, 391
    measuring, 257-59
    relationships involving, 259-61
  enthalpy(ies) of (ΔH_{rxn}), 255-67
    constant-pressure calorimetry to measure, 257-59
    relationships involving, 259-61
    from standard heats of formation, 262-67
    stoichiometry involving, 255-57
  exothermic, 254-55, 650-51, 790-91
    bond energies and, 394
  gas-evolution, 160, 165-67
  half-life of, 584-86, 587
    first-order, 584-85
    second-order, 585-86
    zero-order, 586
  heat evolved in, at constant pressure, 253-57

  internal energy change for (ΔE_{rxn}), 242-43, 251-53
  pressure and, 213
  rates of. *See* Reaction rate(s)
  reversible, 619, 803
  side, 143
  spontaneity of, 778
  standard enthalpy change for, 264-67
Reaction intermediates, 594, 596, 597
Reaction mechanisms, 594-99
  defined, 594
  with fast initial step, 596-98
  rate-determining steps and overall reaction rate laws, 595-96
  rate laws for elementary steps, 594-95
Reaction order (n), 575, 576
  determining, 577-78
  for multiple reactants, 578-79
Reaction quotient (Q), 633-35
  precipitation and, 753-54
Reaction rate(s), 568-615, 619
  average, 572-73
  catalysis, 599-604
    enzymes, 602-4
    homogeneous and heterogeneous, 600-602
  instantaneous, 573
  integrated rate law, 580-87
    first-order, 580-82, 587
    half-life of reaction, 584-86, 587
    second-order, 583-84, 587
    zero-order, 584, 587
  measuring, 574-75
  of radioactive decay and radiometric dating, 875-79
  rate law, 575-79, 596
    containing intermediates, 596, 597
    determining the order of reaction, 577-78
    differential, 580-81
    first-order reaction, 575, 576, 587
    reaction order for multiple reactants, 578-79
    second-order reaction, 575, 576-78, 587
    zero-order reaction, 575, 576, 577-78, 587
  reaction mechanisms and, 594-99
  with respect to product, 571, 572
  with respect to reactant, 571, 572
  with respect to time, 570-71
  temperature effect on (Arrhenius equation), 587-94
    activation energy (activation barrier), 587, 588-89, 590-92
    Arrhenius plots, 590-92
    collision model of, 592-93
    exponential factor, 587, 589
    frequency factor (pre-exponential factor), 587, 588, 589, 590-92
    rate constant and, 587
  thermodynamics and, 778
Reaction stoichiometry, 133-43
  actual yield, 138
  gases, 211-13
  limiting reactant, 137-43
  mass-to-mass conversions, 134-37
  mole-to-mole conversions, 133-34
  percent yield, 138-43
  reactant in excess, 138
  theoretical yield, 137-43
Reactive elements, halogens as, 1017, 1018
Real gas(es), 221-24
  finite volume of gas particles and, 222, 224

intermolecular forces and, 223, 224
molar volumes of, 222
van der Waals constants for, 223
van der Waals equation for, 224
Rechargeable batteries, 776, 843-45
Recrystallization, 531
Redheffer, Charles, 241
Redheffer's perpetual motion machine, 241
Redox reactions. *See* Oxidation-reduction reaction(s)
Red phosphorus, 1005-6
Reducing agents, 172
negative reduction half-cell potentials of, 831
Reduction
of aldehydes and ketones, 933
definition of, 168, 820
Reductionism, 51
Refining, 1029
of copper, 1032-33
Reflection, angle of, 496-97
Relative solubility, 750
Relative standard entropies, 794-96
Reliability of measurement, 22-28
exact numbers, 24-25
precision and accuracy, 27-28
significant figures, 23-27
Rem (roentgen equivalent man), 893
Renewable energy, 270-71
Repulsive electrostatic force among protons, 872
Resonance, Lewis structures and, 384-85
Resonance hybrid, 384-85
Resonance structures, 384, 924-25
benzene as hybrid of two, 925
Respiration, water-carbon dioxide reaction in, 1076
Resting potential, 842
Retina, 432
Retinal isomers, 432
Reversible reaction, 619, 803
R group (side chains) of amino acids, 966, 967
Rhodochrosite, 1042
Ribonucleic acid (RNA), 974, 975
Ribose, 974
Ribosomes, 979-80
Ripening agent, ethene as, 917
RNA, 974, 975
Roasting, 1031
Rock candy, 531
Rocks, uranium/lead dating of, 881-83
Rock salt structure, 505
Rods, 432
Roentgen equivalent man (rem), 893
Rohrer, Heinrich, 45-46
Roosevelt, Franklin, 884, 885
Root mean square velocity, 217-20
Rotational energy, 784, 795
Rotation of polarized light, 911
Ru-92, 871
Rubbing alcohol. *See* Isopropyl alcohol
Rubidium
charge of, 94
properties of, 352
Rubies, color of, 1051-52
Rusting, 11
of iron, 854-55
Rutherford, Ernest, 54-55, 58, 867, 890
Rutherfordium, 58
Rutile, 1029, 1039
Rydberg, Johannes, 293

Rydberg constant, 293
Rydberg equation, 301

S

Saccharin, 405-6
Sacramento, 247
Sacrificial electrode, 855
SAE scale, 476
Safe Drinking Water Act (SDWA), 495, 565
Salicylic acid, reaction between ethanoic acid and, 936
Salmonella, irradiation of foods to kill, 895
Salt, 117, 505
Salt(s), 690-91
from acid-base reactions, 162
as de-icer, 548
density of, 20
electrolysis of molten, 849
solutions as acidic, basic, or neutral, 696-98
table. *See* Sodium chloride
Salt bridges, 825, 973
Salt water, 150-52
Sand, specific heat of, 247
San Francisco, 247
Sanger, Frederick, 956
Saturated fat, 958-60
Saturated fatty acid, 957
Saturated hydrocarbons. *See* Alkanes
Saturated solution, 530, 753
Scandium, charge of, 94
Scanning tunneling microscope (STM), 45-46
Schrödinger, Erwin, 280
Schrödinger equation, 301
for hydrogen atom, 301-3
for molecules, 439
for multielectron atoms, 322, 323
Science, 7
Scientific approach to knowledge, 5-7
Scientific law, 5
Scientific method, 6, 214
Scientific notation, 19
Scientific revolution, 7, 47
Scientific theory, 217
Scintillation counter, 875
SCN^- ligand, chemical analysis with, 1073
Screening (shielding), 324-25
effective nuclear charge and, 336-37
types of, 337
Scuba diving, 193, 207
Seawater, 519-20, 521, 755
Secondary structure of protein, 971, 972-73, 974
Secondary valence, 1056
Second ionization energy (IE_2), 343, 347-48
Second law of thermodynamics, 776
defined, 783
entropy and, 779-85
Second-order integrated rate law, 583-84, 587
Second-order reaction, 575, 576-78, 587
Second-order reaction half-life, 585-86
Second (s), 16
Seesaw geometry, 412, 415, 417
Selective precipitation, 755-56
qualitative analysis by, 756-59
Selenium, 58
Semiconductor(s), 63, 508-9
conductivity of, 509
n-type, 509
p-type, 509
Semipermeable membrane, 552

Sequel (fuel cell SUV), 270
Serine, 967, 968, 977
Setae, 461
Shallow wells, 185, 186-87
SHE half-cell, 828-29
-SH groups, 173
Shielding (screening), 324-25
effective nuclear charge and, 336
types of, 337
Shroud of Turin, radiocarbon dating and, 881
Sickle-cell anemia, 972
Side reactions, 143
σ^*_{2p} antibonding orbital, 444
σ_{2p} bonding orbital, 444, 445
Sigma σ bond, 430
Significant figures, 23-27
in calculations, 25-27
Silica, 507, 992
Silicates, 507, 992-96
aluminosilicates, 993
quartz and glass, 992
types of silicate structures, 993-96
Silicon, 63
doping of, 509
electron configuration of, 329
ionization energies of, 347
Silicon carbide, 1003
Silicon dioxide, 506
Silicon oxide, reaction between coke and, 1003
Silver
charge of, 94
magnetic properties of, 339
specific heat of, 247
Silver bromide, solubility product constant for, 747
Silver chloride, 154, 748
solubility product constant for, 747
Silver chromate, solubility product constant for, 747
Silver iodide, solubility product constant for, 747
Silver ions, 759
reaction between ammonia and, 1057
Silver nitrate, 154, 374, 754
Silver plating, 847
Simple carbohydrates, 961-64
Simple cubic structure, 502
Simple cubic unit cell, 498-99
Single bond, 87
bond energy of, 393
covalent, 374-75
double bond vs., 920, 921
Single-nucleotide polymorphisms (SNPs), 981
Single-walled nanotubes (SWNT), 1001
SI unit(s)
base units, 16
of density, 20
derived units, 19
of energy, 239
of force, 186
of length, 16, 19
of mass, 16, 19, 239
prefix multipliers, 19
of pressure, 189
of speed, 19
of temperature, 17-18
of time, 16
Skin, wrinkling of, 286
Skin cancer, 286
Skunk, smell of, 904

SLAC, 890, 891
Slag, 1031
Smell, sense of, 903-4
Smelting, 1031
Smithsonite, 1044
Smog, photochemical, 226
Smoke, 557
Snowflake, 496
SNPs, 981
Soap, 422, 558, 559
Socrates, 666
Soda ash, 92
Sodium, 58
    charge of, 93, 94
    electron affinity of, 348
    electron configuration of, 341, 347
    electron sea model for, 397
    emission spectrum of, 294
    flame tests for, 295
    ionization energies of, 344, 347
    properties of, 84, 352
    reactions of
        with chlorine, 47-48, 168, 352, 368
        with methanol, 930
        with oxygen, 168
        with sulfur, 367-68
        with water, 930
    second ionization energy of, 343
Sodium acetate, 530, 719, 754
Sodium bicarbonate (baking soda), 91, 374, 666,
        690-91, 1004
    reaction between hydrochloric acid and, 165-66
Sodium borates, 996
Sodium carbide, reaction between water and,
        1002
Sodium carbonate, 155, 666
Sodium chloride, 84, 92, 168, 352, 696, 1029
    Born-Haber cycle for production of, 369-71
    chemical formula for, 86
    density of, 20
    electrical conductivity of, 373
    electrolysis of aqueous, overvoltage and, 851
    electrolysis of molten, 849
    formation of, 85
    formula unit for, 90
    lattice energy of, 368, 369-70
    melting of, 372
    mixed with water, 472
    reaction between potassium iodide and, 156-57
    in seawater, 520
    solubility in water, 521
    solute and solvent interactions in solution, 151
    unit cell for, 505
    in water, 472, 526, 529-30, 544
Sodium fluoride, 374, 696, 751
Sodium hydroxide, 161, 526, 666, 687, 930
    pH of, 675
    reactions of
        with hydrochloric acid, 162
        with propanoic acid, 936
    titrations of
        with formic acid, 737-43
        with hydrochloric acid, 734-37
        with sulfurous acid, 743-44
Sodium hypochlorite, 90
Sodium iodide solution, electrolysis of, 850
Sodium ion(s)
    electron configuration of, 341
    nerve signal transmission and, 319-20

solubility and, 154
Sodium methoxide, 930
Sodium nitrite, 91, 95, 1009
Sodium oxide, 168
    formula mass of, 101
Sodium phosphate compounds, 1010
Soft drinks, pH of, 675
Solar power, 270
Solar-powered electrolytic cell, 847
Solid(s), 460. See also Crystalline solid(s)
    amorphous, 464
    crystalline, 464
    entropy change associated with change in state
        of, 784
    equilibria involving, 629-30
    molecular comparison with other phases, 462-
        65
    properties of, 463-64
    relative standard entropies of, 794
    solubility of, temperature dependence of, 531
    standard state for, 262
    vapor pressure of, 487
Solid aerosol, 557
Solid emulsion, 557
Solid matter, 7-8
Solid solution, 521
Solubility
    alloys with limited, 1035-37
    of amphoteric metal hydroxides, 761-63
    complex ion equilibria and, 761
    defined, 521
    of gases in water, 531-34
    molar, 748-50
    precipitation reactions and, 156-58
    relative, 750
    of solids, temperature dependence of, 531
Solubility equilibria, 154-55, 716, 747-53
    common ion effect on, 751-52
    pH and, 752-53
    solubility product constant (K_{sp}), 747-49, 753
        molar solubility and, 748-50
        relative solubility and, 750
Solubility product constant (K_{sp}), 747-49, 753
    molar solubility and, 748-50
    relative solubility and, 750
Solubility rules, 154-55, 747
Soluble compounds, 154
Solute(s), 144, 520
    intermolecular forces acting on, 522-23, 546
    ionic, 544-45, 554
        vapor pressure and, 544-45
    van't Hoff factors for, 554
Solute-solute interactions, 150, 522-23
Solute-solvent interactions, 545, 546
Solution(s), 518-67. See also Acid-base chemistry;
        Aqueous solution(s)
    acidic, 673
    aqueous, 144, 150-54
        electrolyte and nonelectrolyte, 151-53
        solubility of ionic compounds, 154-55
    basic, 673
    boiling point elevation, 549-50, 551-52, 554
    colligative properties of, 549, 554-56
    colloids, 557-59
    components of, 520
    concentrated, 144, 535, 542
    concentration of, 535-40
        molality, 535, 536, 540
        molarity, 535, 540

mole fraction and mole percent, 535, 539-40
parts by mass and parts by volume, 535, 536-
        37, 540
defined, 144, 520
dilute, 144, 535
dilution of, 146-48
energetics of formation, 526-29
    enthalpy of solution (ΔH_{soln}), 527, 528-29
entropy and, 522, 541-42
equilibrium processes in, 529-30
examples of, 518
freezing point depression, 549-51, 554
gaseous, 521
hyperosmotic, 555-56
hyposmotic, 555-56
ideal, 545-46
intermolecular forces in, 522-26
intravenous, 556
isosmotic (isotonic), 556
liquid, 521
miscible solid, 1034-35
molarity of, 144-46, 149
neutral, 673
nonelectrolyte, 152-53
nonideal, 546-48
osmosis, 552-54
saturated, 530, 753
seawater, 519-20
solid, 521
stock, 146
supersaturated, 530, 754
thirsty, 519, 521, 541-42
transition metal ions in. See Complex ion
        equilibria
unsaturated, 530, 753
vapor pressure of, 540-49
    ideal and nonideal solutions, 545-48
    ionic solutes and, 544-45, 554
    nonvolatile solute and, 548-49
    Raoult's law, 543, 545-48
Solution concentration, 144-45
Solution stoichiometry, 149-67
    acid-base reactions, 160-65
    aqueous solutions and solubility, 150-55
    gas-evolution reactions, 160, 165-67
    precipitation reactions, 155-59
    representing aqueous reactions, 159-60
Solvent(s), 144, 520, 521
    intermolecular forces acting on, 522-23, 546
    laboratory, 524
    nonpolar, 524
    polar, 524
Solvent-solute interactions, 150, 522-23
Solvent-solvent interactions, 522-23, 546
Soot, 1000
s orbitals, 302, 306, 307-8, 309
    1s, 307-8
    2s, 309, 323-24
    3s, 309
Sorosilicates (pyrosilicates), 993, 995
Sound, speed of, 283
sp hybridization, 432-33, 436
sp^2, hybridization, 428-31, 436
sp^3, hybridization, 426-28, 436
sp^3d^2, and sp^3d, hybridization, 434-35, 436
Space-filling molecular models, 87-88
Space Shuttle, 475
Spatulae, 461
Specific heat capacity (C_S), 247

Spectator ions, 159
Spectrochemical series, 1070
Spectrometer, 574
Spectroscopy, 574, 575
Speed, SI unit for, 19
Sphalerite, 1029, 1044
Sphygmomanometer, 191
Spin-pairing, 424-25
Spin quantum number, 323
Spin up and spin down, 322, 323
Spontaneity
    change in Gibbs free energy as criterion for, 789-90
    effect of change in entropy (ΔS), change in enthalpy (ΔH), and temperature on, 790-92
    in oxidation-reduction reactions, 823-27, 836
Spontaneous process(es), 777-78
    decrease in Gibbs free energy and, 789
    endothermic, 779-80
    entropy of universe as criterion for spontaneity, 783
    exothermic processes as, 787
    mixing, 520
    nonspontaneous process made, 802
    in voltaic (or galvanic) cells, 823-27
Square planar complex(es)
    cis-trans isomerism in, 1063
    d orbital energy changes for, 1072-73
    low-spin, 1072
Square planar geometry
    of complex ions, 1059, 1060
    molecular, 413, 415, 417
Square pyramidal geometry, molecular, 413, 415
Stability, valley (or island) of, 873
Stainless steels, 1041
Stalactites, 752
Stalagmites, 752
Standard cell potential ($E°_{cell}$) or standard emf, 826, 827
    relationship between equilibrium constant (K) for redox reaction and, 836-37
    relationship between $\Delta G°$ and, 834-36
Standard change in free energy ($\Delta G°$)
    free energy (ΔG) and, 838
    for reaction ($\Delta G°_{rxn}$) 797-809, 837
        calculating, 797-804
        equilibrium constant (K) and, 807-9
        free energy change of reaction under nonstandard conditions and, 804-7
        standard cell potential and, 834-36
Standard enthalpy change ($\Delta H°$) 262
    for reaction ($\Delta H°_{rxn}$), 264-67
Standard enthalpy of formation ($\Delta H°_f$) 262-67
Standard entropy change ($S°_{rxn}$)
    calculating, 796-97
    for reaction ($\Delta S°_{rxn}$) 793-97
Standard hydrogen electrode (SHE) half-cell, 828-29
Standard molar entropies (S°), 793-97
Standard molar free energies of formation, 800
Standard reduction potentials, 827-33
Standard state, 262
Standard temperature and pressure (STP), molar volume at, 201
Stanford Linear Accelerator (SLAC), 890, 891
Starch, 964, 965
Stars, neutron, 55, 80
State function, 241-42

entropy as, 781
States of matter, 7-9, 11
    entropy change associated with change in, 784-85
Stationary states, 293-94, 296
Steam burn, 478
Stearic acid, 957
Steel
    galvanizing, 1044-45
    stainless, 1041
Steel alloys, 1041, 1042
Steel production
    cobalt and, 1043
    oxygen used in, 1012
    phosphoric acid used in, 1010
Stepwise reaction, determining standard change in free energy for, 801-2
Stereoisomerism, 1062-66
    geometric (or cis-trans) isomerism, 920-21, 1062-65
    optical isomerism, 909-10, 962-63, 1062, 1065-66
    in hydrocarbons, 909-12
Stereoisomers, 1062
    defined, 909
Stern-Gerlach experiment, 322
Steroids, 961
Stock solutions, 146
Stoichiometry, 130, 133-37
    defined, 133
    of electrolysis, 852-53
    involving enthalpy change, 255-57
    molar volume and, 212-13
    oxidation-reduction reactions, 167-74
        combustion reactions, 173-74
        identifying, 171-72
        oxidation states (oxidation number), 169-71
        with oxygen, 167
        without oxygen, 168
        with partial electron transfer, 168, 169
    reaction, 133-43
        actual yield, 138
        gases, 211-13
        limiting reactant, 137-43
        mass-to-mass conversions, 134-37
        mole-to-mole conversions, 133-34
        percent yield, 138-43
        reactant in excess, 138
        theoretical yield, 137-43
    solution, 149-67
        acid-base reactions, 160-65
        aqueous solutions and solubility, 150-55
        gas-evolution reactions, 160, 165-67
        precipitation reactions, 155-59
        representing aqueous reactions, 159-60
Storage proteins, 966
Strassmann, Fritz, 883
Stratopause, 225
Stratosphere, 225
Stratospheric ozone, 226
Strong acid(s), 152-53, 669-70
    hydronium ion sources in, 677-78
    pH of, mixed with weak acid, 684-85
    titration with strong base, 734-37
        equivalence point, 734
        overall pH curve, 736
Strong base(s), 687
    cations as counterions of, 695
    hydroxide ion concentration and pH of, 688-89
    titration of diprotic acid with, 743-44

titration with strong acid, 734-37
    equivalence point, 734
    overall pH curve, 736
titration with weak acid, 737-43
    equivalence point, 737
    overall pH curve, 741
Strong electrolytes, 151, 153, 669
Strong-field complexes, 1068
Strong-field ligands, 1070, 1071
Strong force, 872-73
Strontium, 63
    charge of, 94
Strontium hydroxide, 687
Structural formula, 86, 87, 907
    of hydrocarbons, 907-9
Structural isomers, 907, 962-63, 1062
Structural proteins, 966
Structure of Scientific Revolutions, The (Kuhn), 7
Styrene, 925
Subatomic particles, 56-60. See also Electron(s); Neutron(s); Proton(s)
Sublevels (subshells), 302
    energy splitting in multielectron atoms, 323-25
Sublimation, 12, 487
    as zero-order reaction, 576
Sublimation curve, 491, 492
Substance(s)
    distillation of, 10
    filtration of, 10
    pure, 9, 10
    in solution, standard state for, 262
Substituents, 914
Substitutional alloys, 1034-35
Substitution reactions
    alcohol, 929
    of alkanes, 922
    of aromatic compounds, 927
Substrate, 602-3
Successive approximations, method of, 642
Sucrase, 603
Sucrose, 405, 964
    catalytic breakup of, 603
    density of, 20
    hydrolysis of, 574, 575
    sulfuric acid and dehydration of, 1016
Sugar(s), 117
    density of, 20
    dissolution of, 12
    in nucleic acids, 974-75
Sugar water, 150-52
Sulfate, 95
Sulfide(s), 94, 1029
    acid-insoluble, 758
    base-insoluble, 758
    in gas-evolution reactions, 166
    solubilities of, 752
Sulfites, in gas-evolution reactions, 166
Sulfonic acid groups, 173
Sulfur, 58, 63, 1013-17
    elemental, 1013-14
    ionization energies of, 347
    production, sources of, 1013-14
    reactions of
        with carbon, 171
        with sodium, 367-68
Sulfur dioxide, 707, 1016
Sulfur fluoride
    electron geometry of, 412
    molecular geometry of, 412

Sulfur hexafluoride, 390, 409, 435
Sulfuric acid, 100, 136, 161, 390-91, 664, 665, 670,
    671, 706, 1016-17
    industrial "scrubbers" to remove, 1016
    ionization constants for, 699
    reactions of
        with calcium phosphate, 1010
        with fluorspar, 1018
        with lithium sulfide, 165
        with potassium hydroxide, 162
Sulfurous acid, 100, 671, 698
    ionization constants for, 699
    titration with sodium hydroxide, 743-44
Sulfur oxides, 225, 226, 706
Sun, 270
    power of, 889-90
Sunburns, 286
Suntans, 286
Supercritical fluid, 486-87
Superoxide, 1012
Supersaturated solution, 530, 754
Surface tension, 474-75
Surroundings, 239. See also System-surroundings
        energy exchange; Thermochemistry
    energy flow in, 242-44
    entropy of, 785-88
Suspensions, 691
Sweating, 478
SWNT, 1001
Sylvite (potassium chloride), 92, 94, 1029
System(s), 239
    energy flow in, 242-44
    internal energy change of, 244-45
    state of, 241-42
Systematic error, 28
Systematic names, 93
System-surroundings energy exchange. See also
        Thermochemistry
    heat, 246-49
    pressure-volume work, 249-50
Systolic blood pressure, 191

T

T1r3 protein, 406
Talc, 995
Tantalite, 1029
Tantalum, 1029
Tastant, 406
Taste cells, 406
Taste of food, 406
Taste receptors, 406
Technetium-99m, 894
Tellurium, 321
Temperature(s)
    absolute zero, 195
    boiling point and, 482-83
    critical, 486
    defined, 246
    effect on spontaneity, 790-92
    entropy of surroundings and, 786
    equilibrium constant and, 630, 809
    gas solubility in water and, 531-32
    global, 132
    heat capacity and, 246-47
    heat vs., 246
    intermolecular forces and, 465
    kinetic energy and, 214, 217-18
    Le Châtelier's principle on change in, 650-52
    molecular velocities and, 217-19

phase changes and, 464
pressure and, 199
reaction rate and. See Arrhenius equation
scale conversions, 18
SI unit of, 17-18
solubility of solids and, 531
vapor pressure and, 482, 483
viscosity and, 475
volume and, 195-96
water's moderating effect on, 494
Temporary dipole (instantaneous dipole), 466
10W-40 oil, 476
Tera prefix, 19
Termolecular steps, 595
Terrorist strikes, 354
tert-Butyl substituent, 914
Tertiary structure of protein, 971, 973, 974
Testosterone, 961
Tetracene, 927
Tetrahedral complex(es)
    d orbital energy changes for, 1072
    high-spin, 1072
    optical isomerism in, 1066
Tetrahedral geometry, 87
    electron, 414, 436
    molecular, 408, 411, 414, 417, 426-28
Tetrahedral geometry of complex ions, 1059-60
Tetrahedral hole, 505, 1038, 1039
Tetrahedron, 408
    silicate (SiO4), 992
Tetrapeptide, 969
Tetraphosphorus decaoxide, 1010
Tetraphosphorus hexaoxide, 1010
Thallium-201, 894
Theoretical yield, 137-43
Theories
    testing, 6
    as true models, 7
Therapeutic agents, coordination compounds,
        1076
Therapeutic techniques, use of radioactivity in,
        894
Thermal conductivity of metals, 1028, 1029
Thermal energy, 14, 17, 238, 239, 246, 255, 462,
        568. See also Heat(s)
    dispersal of, 522
    distribution of, 589
    vaporization and, 477-80
Thermal equilibrium, 246
Thermochemical equations, 255-57
Thermochemistry, 236-79. See also Energy(ies)
    defined, 237
    enthalpy. See Enthalpy(ies) (H)
    first law of thermodynamics, 240-46, 775-76
        internal energy (E), 241-46
    heat, 246-49
        defined, 246
        temperature vs., 246
    internal energy change for chemical reactions
        (ΔE_{rxn}), 251-53
    pressure-volume work, 249-50
Thermodynamics. See also Equilibrium/equilibria
    defined, 240
    first law of, 240-46, 775-76
        internal energy (E), 241-46
    goal of predicting spontaneity, 777-78
    kinetics and, 778
    reversible reaction in, 803
    second law of, 776

defined, 783
entropy and, 779-85
third law of, 793-97
Thermosphere, 225
Thiocyanate ion, 1058
Thiols, 173
Third ionization energy (IE₃), 343
Third law of thermodynamics, 793-97
Thirsty solution, 519, 521, 541-42
Thomson, J.J., 51-52, 53, 54
Thorium-232, decay of, 876
Threonine, 968, 977
Threshold frequency, 289, 291
Thymine, 473, 474, 975
Thymol blue, 747
Thymolphthalein, 747
Thyroid cancers, 354-55
Thyroid gland, radiotracer used for, 894
Thyroxine, 355
Time
    concentration and. See Integrated rate law
    SI unit of, 16
Tin, 58
    alloy of copper and (bronze), 1043, 1044
    cations formed by, 95
Titanium
    density of, 20
    sources, properties, and products of, 1039-40
Titanium dioxide, 1040
Titanium oxide, 1040
Titration, 575. See also Acid-base titration
Titration curve, 734
Tokamak fusion reactor, 890
Toluene, 524, 925
Torr, 188, 189
Torricelli, Evangelista, 188
Total pressure, 200
Toxaphene, 538
Trajectory, classical vs. quantum concepts of, 299-
        300
Trans-cis isomers. See Cis-trans (geometric)
        isomerism
Trans fats, 959-60
Transition(s), 293-94, 303
    in hydrogen atom, 304-5
Transition metal(s), 63, 64, 93, 330, 1029, 1050-81
    atomic radii and, 337-38
    coordination compounds, 1056-76
        applications of, 1073-76
        bonding in, 1066-73
        naming, 1060-61
        structure and isomerization in, 1062-66
    crystal structures for 3d, 1034
    functions in human body, 1074
    in host crystals, colors of, 1051-52, 1068
    inner, 330
    ion formation by, 66
    properties of, 1052-56
        atomic size, 1054-55
        electron configurations, 332, 339, 1052-54
        electronegativity, 1055
        ionization energy, 1055
        oxidation states, 1056
    sources, properties, and products of some of 3d,
        1039-45
        chromium, 1040-42
        cobalt, 1042-43
        copper nickel, 1043-44
        manganese, 1042

nickel, 1044
  titanium, 1039-40
  zinc, 1044-45
  valence electrons for, 329
Transition metal ions. *See also* Complex ion equilibria
  electron configuration of cations, 339
  magnetic properties of, 339
Transition state (activated complex), 588-89
Translational energy, 784
Translational motion, energy in form of, 795
Transmutation, nuclear, 890-91
Transport proteins, 966
Transuranium elements, 891
Tremolite, 994
Trichlorofluoromethane, 580
1,1,2-Trichloro-1,2,2-trifluoroethane, 580
Triglycerides, 958-60
Trigonal bipyramidal geometry, 1020
  electron, 415, 436
  molecular, 409-10, 413, 415, 417, 434
Trigonal planar geometry
  electron, 414, 436
  molecular, 407, 408, 414, 417, 428
Trigonal pyramidal geometry, 410, 411, 414, 417
Trihalides, 997
Trimethylamine, 938
Triolein, 959
Tripeptide, 969
Triple bond(s), 375
  bond energy of, 393
  carbon's ability to form, 904-5
  sp hybridization and, 432-33
  in structural formulas, 908
Triple point, 491, 492
Triprotic acids, 671, 698
Tristearin, 958
Trona, 92
Tropopause, 225
Troposphere, 225
Tryptophan, 968
T-shaped geometry, 412, 415
Tums, 664, 677, 691
Tungsten carbide, 1003
Tunneling current, 45-46
Turquoise, 1052
Two-phase region, 1036-37
  lever rule and, 1037
Tyndall effect, 558
Tyrosine, 968

U

Ulcers, 677
Ultraviolet (UV) radiation, 226-27, 285, 286, 396
Uncertainty principle, 298-99
Unimolecular elementary step, 594, 595
Unit cells, 498-501
  body-centered cubic, 498, 499-500
  for closest-packed structures, 503-4
  face-centered cubic, 498, 500, 503-4
  for ionic solids, 505-6
  simple cubic, 498-99, 502
Unit conversion problems, 28-29, 31-33
  units raised to a power, 31-32
United States, energy consumption in, 267, 706
U.S. Department of Agriculture (USDA), 895
U.S. Department of Energy (DOE), 134, 267, 980
U.S. Environmental Protection Agency (EPA), 107, 226, 495, 538, 877

U.S. Food and Drug Administration (FDA), 355, 895, 959, 960
U.S. Food and Drug Administration (FDA) action level, 107
Units of measurement, 15-21. *See also* SI unit(s)
  derived units, 19-21
  English system, 15, 19
  metric system, 15
  prefix multipliers, 19
Unsaturated fat, 958-60
Unsaturated fatty acids, 957
Unsaturated hydrocarbons, 917-21, 923. *See also* Alkenes; Alkynes
Unsaturated solutions, 530, 753
Unsaturation, effect of, 957
Uracil, 975
Uranic rays, 867
Uranium, 58, 1029
Uranium-235, 883
  energy produced per mole of, 887
  nuclear-powered electricity generation using, 885-86
  self-amplifying chain reaction in fission of, 884
Uranium-238
  nuclear equation for alpha decay of, 868
  radioactive decay series, 874
Uranium fuel rods, 885
Uranium/lead dating, 881-83
Urea, 905

V

Valence
  primary, 1056
  secondary, 1056
Valence band, 508-9
Valence bond theory, 362, 404, 423-38
  bonding in coordination compounds and, 1066-67
  double bonds in, 430
  hybridization of atomic orbitals, 425-38
    in carbon, 426-27
    sp, 432-33, 436
    sp^2, 428-31, 436
    sp^3, 426-28, 436
    sp^3d^2 and sp^3d 434-35, 436
    writing, 435-37
  Lewis theory and, 430, 436
  summarizing main concepts of, 424
Valence electrons, 329, 330
  chemical properties and, 332-33
Valence shell electron pair repulsion (VSEPR) theory, 404, 406-22, 1019
  bent geometry, 411, 414, 417
  linear geometry, 407, 408, 413, 414, 415, 417
  lone pairs effect, 410-14
  molecular shape and polarity, 418-22
  octahedral geometry, 413, 415, 417, 435
  predicting molecular geometries with, 415-18
  seesaw geometry, 412, 415, 417
  square planar geometry, 413, 415, 417
  square pyramidal geometry, 413, 415
  summary of, 414
  tetrahedral geometry, 408, 411, 414, 417, 426-28
  trigonal bipyramidal geometry, 409-10, 413, 415, 417, 434
  trigonal planar geometry, 407, 408, 414, 417, 428
  trigonal pyramidal geometry, 410, 411, 414, 417
  T-shaped geometry, 412, 415

Valine, 968
Valley (or island) of stability, 873
Vanadinite, 1029
Vanadium, 1027-28
  chromium-vanadium phase diagram, 1035, 1036
  electron configuration of, 339
  mineral sources for, 1029
Vanadium ion, electron configuration of, 339
Van der Waals, Johannes, 222, 223
Van der Waals constants, 223
Van der Waals equation, 224
Van der Waals radius (nonbonding atomic radius), 334
Vanillin, 932
Van't Hoff factor (i), 554
Vaporization, 13, 477-87
  Clausius-Clapeyron equation and, 483-86
  critical point, 486-87
  energetics of, 478-79
  heat of (ΔH_{vap}), 479-80
  process of, 477-78
  vapor pressure and dynamic equilibrium, 480-83
Vaporization curve, 491, 492
Vapor pressure, 209-10
  defined, 480
  ionic solutes and, 544-45
  of solid, 487
  of solutions, 540-49
    ideal and nonideal solutions, 545-48
    ionic solutes and, 544-45, 554
    nonvolatile solute and, 548-49
    Raoult's law, 543, 545-48
  temperature and, 483
  temperature dependence of, 482-83
Variational method, 439
Vector addition, 420
Vector quantities, 419
Vegetable shortening, 959
Vegetarian diet, nature's heat tax and, 776
Velocity(ies)
  in classical mechanics, 299
  of electron, 299
  energy and, as complementary property, 301
  molecular, 217-19
  position and, as complementary property, 301
  root mean square, 217-20
Venus, 132
Vibrational motion, energy in form of, 795
Vinegar, 665
Viscosity, 475
Visible light, 284, 285, 286
Vision, 432
Vital force, 905
Vitalism, 905
Vitamins, 457
Volatile organic compounds (VOCs), 226
Volatility, 10, 478
Volcanoes, carbon dioxide emitted by, 132, 135
Voltaic (galvanic) cells, 823-27
  batteries as, 842-46
  concentration cells, 840-41
  electrolytic cells vs., 848
Volt (V), 825
Volume
  gas amount and, 197
  Le Châtelier's principle on change in, 649-50
  molar, 201

Volume *(continued)*
    stoichiometry and, 212-13
  pressure and, 192-94
  SI unit for, 19
  temperature and, 195-96
VSEPR. *See* Valence shell electron pair repulsion
    (VSEPR) theory

W

Warren, J. Robin, 677
Washing soda, 1004
Water, 4-5, 83-84, 494-95
  amphotericity of, 672-73
  Arrhenius definition of acids and bases and,
    667
  atomization of, 1033
  autoionization of, 672-74, 677
  boiling point of, 479
    normal, 482
  charge distribution in, 150
  chemical formula for, 86
  collecting gases over, 209-10
  decomposition of, 48, 270
  density of, 20
  electrolysis of, 847
    in aqueous solutions, 849-52
  electron geometry of, 411
  empirical formula for, 109
  from fossil fuel combustion, 268
  free energy versus pressure for, 805
  freezing of, entropy of surroundings increased
    by, 785-86
  freezing point depression and boiling point
    elevation constants for, 550
  hard, 155, 749
  heat capacity of, 248
  heating curve for, 488-91
  heat of fusion for, 488
  heat of vaporization of, 479
  hexane mixed with, 523
  hydrogen bonding in, 471, 494
  Lewis structure of, 374, 375-76
  as ligand, 1058
  on Mars, 494
  meniscus of, 476-77
  molecular geometry of, 411, 419
  phase diagram for, 491-93
  phases of, 462
  polarity of, 150, 419, 422, 469
  properties of, 83-84, 494-95
  reactions of
    with alkali metals, 352-53
    with calcium, 820
    with carbon, 260
    with carbon dioxide, 705-6, 1076
    with sodium, 930
    with sodium carbide, 1002
  real gas behavior of, 224
  in seawater, 520

sodium chloride in, 472, 526, 529-30, 544
solubility of gases in, 531-34
solubility of sodium chloride in, 521
as solvent, 524
specific heat of, 247
thermal energy distributions for, 477
van der Waals constants for, 223
vapor pressure of, 482
viscosity of, 475
from wells, 185-87
Water pollution, 495
Waters of hydration, 96
Water vapor, condensation of, 478
Watson, James, 473, 497, 954, 978
Watt (W), 239
Wave, electromagnetic, 283-84
Wave function, 301
Wavelength (λ), 283-84, 285
  de Broglie, 297
  frequency and, 284
Wave nature
  of electron, 295-300
    de Broglie wavelength and, 297
    indeterminacy and probability distribution
      maps, 299-300
    uncertainty principle and, 298-99
  of light, 282-85
Wave-particle duality of light, 282, 292
Weak acid(s), 153, 669, 670-72
  acid ionization constants for, 678
  in buffer solution, 718-19
  cations as, 695-96
  hydronium ion sources in, 678-82, 684
  percent ionization of, 683-84
  pH of, mixed with weak acid, 686
  titration with strong base, 737-43
    equivalence point, 737
    overall pH curve, 741
Weak base(s), 687-88
  anions as, 691-94
  in buffer solution, 718-19
  hydroxide ion concentration and pH of,
    689-90
Weak electrolytes, 153, 669
Weak-field complexes, 1068
Weak-field ligands, 1070, 1071
Weather, 188
Weighing, estimation in, 23
Weight, 16
Wells, water from, 185-87
Werner, Alfred, 1056, 1060
Western bristlecone pine trees, calibrating
    radiocarbon dating with age of, 880
Wet chemistry, 756
Whipped cream, 557
White light spectrum, 284, 292, 293
White phosphorus, 1005
  disproportionation of, 1009
  reaction between oxygen and, 1010

Wilhelmy, Ludwig, 570, 574, 575
Wilkins, Maurice, 497
Willemite, 993
Wind, 188
Wind power, 270, 271
Wines, pH of, 675
Winkler, Clemens, 62
Witt, Otto N., 717
Wˆhler, Friedrich, 902, 903, 905
Wood alcohol. *See* Methanol
Wood frogs (*Rana sylvatica*), 550-51
Work
  defined, 14, 238, 249
  internal energy change and, 244-46
  pressure-volume, 249-50

X

Xenon, 63, 506
  non-ideal behavior of, 223
  properties of, 355
  reaction between fluorine and, 356
  van der Waals constants for, 223
Xenon difluoride, 413
x is small approximation, 641-42, 679-82, 702, 724
X-ray crystallography, 363, 496
  of crystalline solids, 495-97
X-rays, 285, 286, 866

Y

Yield of reactions, 137-43
  actual, 138
  percent, 138-43
  theoretical, 137-43

Z

Zeolites, 1005
Zero entropy, 793
Zero-order integrated rate law, 584, 587
Zero-order reaction, 575, 576, 577-78, 587
Zero-order reaction half-life, 586
Zinc, 339, 506
  alloy of copper and (brass), 1044
  charge of, 94
  in dry-cell batteries, 842-43
  functions in human body, 1074
  galvanized nails coated with thin layer of, 855
  reaction between hydrochloric acid and, 833
  sources, properties, and products of, 1044-45
  in spontaneous redox reaction with copper
    ions, 823-26
  standard half-cell potential for oxidation of, 828
Zinc blende structure, 505
Zinc ion, magnetic properties of, 339-40
Zinc oxide, 374
  smelting of, 1031
Zinc phosphate, 1045
Zinc silicate (calamine), 1044
Zinc sulfide, 505
Zwitterion, 967

Conversion Factors and Relationships

Length

SI unit: meter (m)

$1\,m = 1.0936\,yd$
$1\,cm = 0.39370\,in$
$1\,in = 2.54\,cm\,(exactly)$
$1\,km = 0.62137\,mi$
$1\,mi = 5280\,ft$
$\quad\;\; = 1.6093\,km$
$1\,\text{Å} = 10^{-10}\,m$

Temperature

SI unit: kelvin (K)

$0\,K = -273.15\,°C$
$\quad\;\; = -459.67\,°F$
$K = °C + 273.15$
$°C = \dfrac{(°F - 32)}{1.8}$
$°F = 1.8\,(°C) + 32$

Energy (derived)

SI unit: joule (J)

$1\,J = 1\,kg \cdot m^2/s^2$
$\quad\; = 0.23901\,cal$
$\quad\; = 1\,C \cdot V$
$\quad\; = 9.4781 \times 10^{-4}\,Btu$
$1\,cal = 4.184\,J$
$1\,eV = 1.6022 \times 10^{-19}\,J$

Pressure (derived)

SI unit: pascal (Pa)

$1\,Pa = 1\,N/m^2$
$\quad\;\; = 1\,kg/(m \cdot s^2)$
$1\,atm = 101,325\,Pa$
$\quad\;\;\; = 760\,torr$
$\quad\;\;\; = 14.70\,lb/in^2$
$1\,bar = 10^5\,Pa$
$1\,torr = 1\,mmHg$

Volume (derived)

SI unit: cubic meter (m^3)

$1\,L = 10^{-3}\,m^3$
$\quad\; = 1\,dm^3$
$\quad\; = 10^3\,cm^3$
$\quad\; = 1.0567\,qt$
$1\,gal = 4\,qt$
$\quad\;\;\; = 3.7854\,L$
$1\,cm^3 = 1\,mL$
$1\,in^3 = 16.39\,cm^3$
$1\,qt = 32\,fluid\,oz$

Mass

SI unit: kilogram (kg)

$1\,kg = 2.2046\,lb$
$1\,lb = 453.59\,g$
$\quad\; = 16\,oz$
$1\,amu = 1.66053873 \times 10^{-27}\,kg$
$1\,ton = 2000\,lb$
$\quad\;\; = 907.185\,kg$
$1\,metric\,ton = 1000\,kg$
$\quad\qquad\quad = 2204.6\,lb$

Geometric Relationships

π	$= 3.14159\ldots$
Circumference of a circle	$= 2\pi r$
Area of a circle	$= \pi r^2$
Surface area of a sphere	$= 4\pi r^2$
Volume of a sphere	$= \dfrac{4}{3}\pi r^3$
Volume of a cylinder	$= \pi r^2 h$

Fundamental Constants

Atomic mass unit	1 amu 1 g	$= 1.66053873 \times 10^{-27}\,kg$ $= 6.02214199 \times 10^{23}\,amu$
Avogadro's number	N_A	$= 6.02214199 \times 10^{23}/mol$
Bohr radius	a_0	$= 5.29177211 \times 10^{-11}\,m$
Boltzmann's constant	k	$= 1.38065052 \times 10^{-23}\,J/K$
Electron charge	e	$= 1.60217653 \times 10^{-19}\,C$
Faraday's constant	F	$= 9.64853383 \times 10^4\,C/mol$
Gas constant	R	$= 0.08205821\,(L \cdot atm/(mol \cdot K)$ $= 8.31447215\,J/(mol \cdot K)$
Mass of an electron	m_e	$= 5.48579909 \times 10^{-4}\,amu$ $= 9.10938262 \times 10^{-31}\,kg$
Mass of a neutron	m_n	$= 1.00866492\,amu$ $= 1.67492728 \times 10^{-27}\,kg$
Mass of a proton	m_p	$= 1.00727647\,amu$ $= 1.67262171 \times 10^{-27}\,kg$
Planck's constant	h	$= 6.62606931 \times 10^{-34}\,J \cdot s$
Speed of light in vacuum	c	$= 2.99792458 \times 10^8\,m/s\,(exactly)$

SI Unit Prefixes

a	f	p	n	μ	m	c	d	k	M	G	T	P	E
atto	femto	pico	nano	micro	milli	centi	deci	kilo	mega	giga	tera	peta	exa
10^{-18}	10^{-15}	10^{-12}	10^{-9}	10^{-6}	10^{-3}	10^{-2}	10^{-1}	10^{3}	10^{6}	10^{9}	10^{12}	10^{15}	10^{18}